S0-FKS-634

CRYPTOCOCCUS

FROM HUMAN PATHOGEN TO MODEL YEAST

CRYPTOCOCCUS

FROM HUMAN PATHOGEN TO MODEL YEAST

EDITED BY

Joseph Heitman
Duke University Medical Center
Durham, NC 27710

Thomas R. Kozel
University of Nevada School of Medicine
Reno, NV 89557-0320

Kyung J. Kwon-Chung
National Institute of Allergy and Infectious Diseases
Bethesda, MD 20892

John R. Perfect
Duke University Medical Center
Durham, NC 27710

Arturo Casadevall
Albert Einstein College of Medicine
Bronx, NY 10461

WASHINGTON, DC

Cover photo credits:

Top left: Basidium (blue) and chains of infectious basidiospores (purple) produced during sexual reproduction of *Cryptococcus neoformans*, visualized by scanning electron microscopy (pseudocolored). Photo by Chaoyang Xue, Kasey Carroll, and Joseph Heitman.

Top right: *C. neoformans* yeast cells stained with an anticapsular antibody (blue) and fluorescein as a cell wall marker (green), visualized by indirect immunofluorescence microscopy. Photo by Lynda Pierini and Tamara Doering.

Lower left: *Cryptococcus gattii* NIH444 (gold) undergoing phagocytosis by a neutrophil (pink); pseudocolored scanning electron micrograph. Photo by Deborah Springer and Vishnu Chaturvedi.

Lower right: *C. neoformans* budding yeast cell stained with anticapsular monoclonal antibody 339 (red) and antibody to the human C3 component of complement (green), visualized by immunofluorescence microscopy. Photo by Marcellene Gates-Hollingsworth and Thomas Kozel.

ASM Press
American Society for Microbiology
1752 N Street, N.W.
Washington, DC 20036-2904

Library of Congress Cataloging-in-Publication Data

Cryptococcus : from human pathogen to model yeast / edited by Joseph Heitman ... [et al.].
p. ; cm.
Includes bibliographical references and index.
ISBN 978-1-55581-501-1
1. Cryptococcus. I. Heitman, Joseph. II. American Society for Microbiology.
[DNLM: 1. Cryptococcus—pathogenicity. 2. Cryptococcus—physiology.
3. Cryptococcosis—microbiology. QW 180.5.D38 C954 2011]
QK625.C76C79 2011
579.5′9—dc22 2010019350

10 9 8 7 6 5 4 3 2 1

Address editorial correspondence to: ASM Press, 1752 N St., N.W., Washington, DC 20036-2904, U.S.A.

Send orders to: ASM Press, P.O. Box 605, Herndon, VA 20172, U.S.A.
Phone: 800-546-2416; 703-661-1593
Fax: 703-661-1501
Email: Books@asmusa.org
Online: estore.asm.org

Contents

Contributors

BARBARA D. ALEXANDER
Departments of Medicine and Pathology, Duke University Medical Center, Durham, NC 27710

J. ANDREW ALSPAUGH
Department of Medicine and Department of Molecular Genetics and Microbiology, Duke University School of Medicine, Durham, NC 27710

YONG-SUN BAHN
Department of Biotechnology, Center for Fungal Pathogenesis, College of Life Science and Biotechnology, Yonsei University, Seoul 120-749, Republic of Korea

UMA BANERJEE
Department of Microbiology, All India Institute of Medical Sciences, New Delhi, India

KAREN BARTLETT
School of Environmental Health, University of British Columbia, 372-2206 East Mall, Vancouver, BC, Canada V6T 1Z3

TIHANA BICANIC
Centre for Infection, St. George's University of London, Cranmer Terrace, London SW17 ORE, United Kingdom

TEUN BOEKHOUT
CBS Fungal Diversity Center, Uppsalalaan 8, 3584 CT Utrecht, The Netherlands

MARY E. BRANDT
Mycotic Diseases Branch, Centers for Disease Control and Prevention, Atlanta, GA 30333

EDMOND BYRNES
Duke University Medical Center, 312 CARL Building, Box 3546, Research Drive, Durham, NC 27710

LEONA CAMPBELL
School of Molecular Bioscience, Building G08, University of Sydney, NSW 2006, Australia

PAUL J. CANFIELD
Faculty of Veterinary Science, The University of Sydney, Sydney, NSW Australia 2006

DEE A. CARTER
School of Molecular Bioscience, Building G08, University of Sydney, NSW 2006, Australia

ARTURO CASADEVALL
Department of Microbiology and Immunology and Department of Medicine, Albert Einstein College of Medicine, 1300 Morris Park Ave., Bronx, NY 10461

ELIZABETH CASTAÑEDA
Emerita Investigator, Instituto Nacional de Salud, Bogotá, Colombia

SUDHA CHATURVEDI
Mycology Laboratory, Wadsworth Center, New York State Department of Health, and Department of Biomedical Sciences, University at Albany School of Public Health, Albany, NY 12208

VISHNU CHATURVEDI
Mycology Laboratory, Wadsworth Center, New York State Department of Health, and Department of Biomedical Sciences, University at Albany School of Public Health, Albany, NY 12208

JANGHAN CHEN
Shanghai Changzheng Hospital, Second Military Medical University, Shanghai, PR China

SHARON C.-A. CHEN
Centre for Infectious Diseases and Microbiology, Westmead Hospital and the University of Sydney, Westmead, NSW 2145 Australia

YUCHONG CHEN
Shanghai Changzheng Hospital, Second Military Medical University, Shanghai, PR China

TOM M. CHILLER
Mycotic Diseases Branch, Centers for Disease Control and Prevention, Atlanta, GA 30333

CARA J. CHRISMAN
Albert Einstein College of Medicine, Department of Microbiology and Immunology, Bronx, NY 10461

KARL V. CLEMONS
California Institute for Medical Research, San Jose, CA 95128; Div. of Infectious Diseases, Santa Clara Valley Medical Center, San Jose, CA 95128; and Div. of Infectious Diseases and Geographic Medicine, Stanford University, Stanford, CA 94305

MASSIMO COGLIATI
Laboratory of Medical Mycology, Department of Public Health – Microbiology – Virology, Università degli Studi di Milano, Via Pascal 36, 20133 Milan, Italy

JEFFREY J. COLEMAN
Division of Infectious Diseases, Harvard Medical School, Massachusetts General Hospital, Boston, MA 02114

GARY M. COX
Department of Medicine Mycology Research Unit, Duke University, Durham, NC 27710

EKATERINA DADACHOVA
Departments of Nuclear Medicine and Microbiology & Immunology, Albert Einstein College of Medicine, 1300 Morris Park Ave., Bronx, NY 10461

MAURIZIO DEL POETA
Department of Biochemistry and Molecular Biology, Department of Microbiology and Immunology, and Division of Infectious Diseases, Medical University of South Carolina, Charleston, SC 29425

MARA R. DIAZ
Rosenstiel School of Marine and Atmospheric Science, University of Miami, Miami, FL 33149

JULIANNE T. DJORDJEVIC
Centre for Infectious Diseases and Microbiology, ICPMR and Westmead Millennium Institute, University of Sydney at Westmead Hospital, Westmead, 2145 NSW, Australia

TAMARA L. DOERING
Department of Molecular Microbiology, Washington University School of Medicine, St. Louis, MO 63110

FRANÇOISE DROMER
Institut Pasteur, Molecular Mycology Unit, CNRS URA3012, 25, rue du Dr. Roux, 75724 Paris cedex 15, France

COLLEEN DUNCAN
Colorado State University Veterinary Diagnostic Laboratory, 300 West Drake Ave., Fort Collins, CO 80523

JACK W. FELL
Rosenstiel School of Marine and Atmospheric Science, University of Miami, Key Biscayne, FL 33149

KEISHA FINDLEY
Department of Molecular Genetics and Microbiology, Duke University Medical Center, Durham, NC 27710

MATTHEW FISHER
Imperial College, London, United Kingdom

DEBORAH S. FOX
Department of Pediatrics and Department of Microbiology, Immunology and Parasitology, The Research Institute for Children and LSU Health Sciences Center, Children's Hospital, New Orleans, LA 70118

BETTINA C. FRIES
Departments of Medicine, Microbiology and Immunology, Albert Einstein College of Medicine, Bronx, NY 10804

MURRAY FYFE
Office of the Medical Health Officer, Vancouver Island Health Authority, 430 - 1900 Richmond Ave., Victoria, BC, Canada V8R 4R2

ELENI GALANIS
British Columbia Centre for Disease Control and School of Population and Public Health, University of British Columbia, 655 W. 12th Ave., Vancouver, BC, Canada V5Z 4R4

SCARLETT GEUNES-BOYER
Department of Cell Biology, 438 Nanaline Duke Bldg., Box 3709, Duke University Medical Center, Durham, NC 27710

MAHMOUD A. GHANNOUM
Center for Medical Mycology, Case Western Reserve University, Cleveland, OH 44106

NICOLE M. GILBERT
Department of Biochemistry and Molecular Biology, Saint Louis University, St. Louis, MO 63104

FELIX GILGADO
Molecular Mycology Research Laboratory, Centre for Infectious Diseases and Microbiology, The University of Sydney, Sydney Medical School - Westmead at Westmead Hospital, Westmead Millennium Institute, Westmead, NSW, 2145, Australia

NELESH P. GOVENDER
Mycology Reference Unit, National Institute for Communicable Diseases, a division of the National Health Laboratory Service, and Faculty of Health Sciences, University of the Witwatersrand, Johannesburg, South Africa

FERRY HAGEN
CBS-KNAW Fungal Diversity Centre, Uppsalalaan 8, NL-3584CT Utrecht, The Netherlands

THOMAS S. HARRISON
Centre for Infection, St. George's University of London, Cranmer Terrace, London SW17 ORE, United Kingdom

JOSEPH HEITMAN
Department of Molecular Genetics and Microbiology, Duke University Medical Center, Durham, NC 27710

LINDA HOANG
British Columbia Centre for Disease Control, Dept. of Pathology and Laboratory Medicine, University of British Columbia, Vancouver, BC, Canada V5Z 4R4

YEN-PING HSUEH
Division of Biology, California Institute of Technology, Pasadena, CA 91125

GARY B. HUFFNAGLE
Internal Medicine - Pulmonary Division, 6301 MSRB III, Box 5642, 1150 W. Medical Center Drive, University of Michigan Medical Center, Ann Arbor, MI 48109-5642

SHAUNNA M. HUSTON
The Snyder Institute for Infection, Inflammation and Immunity and Department of Internal Medicine, University of Calgary, Calgary, Alberta, Canada T2N 4N1

ALEXANDER IDNURM
Division of Cell Biology and Biophysics, School of Biological Sciences, University of Missouri-Kansas City, 5100 Rockhill Rd., Kansas City, MO 64110

KOICHI IZUMIKAWA
Department of Molecular Microbiology and Immunology, Nagasaki University Graduate School of Biomedical Sciences, Nagasaki, Japan

GUILHEM JANBON
Unité des Aspergillus, Institut Pasteur, Paris, France

JOSEPH N. JARVIS
Centre for Infection, St. George's University of London, Cranmer Terrace, London SW17 ORE, United Kingdom

SIMON A. JOHNSTON
School of Biosciences, University of Birmingham, Birmingham B15 2TT, United Kingdom

SARAH KIDD
Department of Medicine, Monash University, Central Clinical School, Alfred Hospital, Melbourne, VIC 3004, Australia

SHIGERU KOHNO
Department of Molecular Microbiology and Immunology, Nagasaki University Graduate School of Biomedical Sciences, Nagasaki, Japan

THOMAS R. KOZEL
Department of Microbiology and Immunology, University of Nevada School of Medicine, Reno, NV 89557

MARK B. KROCKENBERGER
Faculty of Veterinary Science, McMaster Building, B14, The University of Sydney, Sydney, NSW 2006, Australia

JAMES W. KRONSTAD
The Michael Smith Laboratories, University of British Columbia, Vancouver, British Columbia, Canada V6T 1Z4

KYUNG J. KWON-CHUNG
Molecular Microbiology Section, Laboratory of Clinical Infectious Diseases, National Institute of Allergy and Infectious Diseases, National Institutes of Health, Bethesda, MD 20892

MARCIA S. LAZÉRA
Mycology Laboratory, Instituto de Pesquisa Clínica Evandro Chagas-Fiocruz, Rio de Janeiro, RJ, Brazil

STUART M. LEVITZ
Department of Medicine, University of Massachusetts Medical School, Worcester, MA 01605

XIAORONG LIN
Department of Biology, Texas A&M University, College Station, TX 77843

ANASTASIA P. LITVINTSEVA
Department of Molecular Genetics and Microbiology, Duke University Medical Center, Durham, NC 27710

SHAWN R. LOCKHART
Mycotic Diseases Branch, Centers for Disease Control and Prevention, Atlanta, GA 30333

JENNIFER K. LODGE
Department of Molecular Microbiology, Box 8230, Washington University School of Medicine, 660 South Euclid Ave., St. Louis, MO 63110

BRENDAN J. LOFTUS
Conway Institute, School of Medicine and Medical Science, University College Dublin, Belfield, Dublin 4, Ireland

LAURA MacDOUGALL
British Columbia Centre for Disease Control, Vancouver, BC, Canada V5Z 4R4

SUNNY MAK
British Columbia Centre for Disease Control, Vancouver, BC, Canada V5Z 4R4

RICHARD MALIK
Centre for Veterinary Education, Conference Centre, B22, The University of Sydney, Sydney, NSW Australia 2006

WEERAWAT MANOSUTHI
Bamrasnaradura Infectious Diseases Institute, Ministry of Public Health, Nonthaburi, Thailand

KIEREN A. MARR
Johns Hopkins University School of Medicine, Baltimore, MD 21205

ROBIN C. MAY
School of Biosciences, University of Birmingham, Birmingham B15 2TT, United Kingdom

TRAVIS McQUISTON
Departments of Biochemistry and Molecular Biology, Medical University of South Carolina, Charleston, SC 29425

BANU METIN
Department of Molecular Genetics and Microbiology, Duke University Medical Center, Durham, NC 27710

WIELAND MEYER
Molecular Mycology Research Laboratory, Centre for Infectious Diseases and Microbiology, The University of Sydney, Sydney Medical School - Westmead at Westmead Hospital, Westmead Millennium Institute, Westmead, NSW, 2145, Australia

KATHLEEN J. MIGLIA
Department of Molecular Genetics and Microbiology, Duke University Medical Center, Durham, NC 27710

THOMAS G. MITCHELL
Department of Molecular Genetics and Microbiology, Duke University Medical Center, Durham, NC 2771

CHRISTOPHER H. MODY
The Snyder Institute for Infection, Inflammation and Immunity, Department of Internal Medicine, and Department of Microbiology and Infectious Diseases, University of Calgary, Calgary, Alberta, Canada T2N 4N1

MUHAMMAD MORSHED
British Columbia Centre for Disease Control, Dept. of Pathology and Laboratory Medicine, University of British Columbia, Vancouver, BC, Canada V5Z 4R4

FRITZ A. MÜHLSCHLEGEL
School of Biosciences, University of Kent, Canterbury, Kent CT2 7NJ, and East Kent Hospitals University NHS Foundation Trust, Clinical Microbiology Service, William Harvey Hospital, Ashford, Kent TN24 0LZ, United Kingdom

ELEFTHERIOS MYLONAKIS
Division of Infectious Diseases, Harvard Medical School, Massachusetts General Hospital, Boston, MA 02114

POPCHAI NGAMSKULRUNGROJ
Molecular Mycology Research Laboratory, Centre for Infectious Diseases and Microbiology, The University of Sydney, Sydney Medical School - Westmead at Westmead Hospital, Westmead Millennium Institute, Westmead, NSW, 2145, Australia

M. HONG NGUYEN
University of Pittsburgh, Pittsburgh, PA 15261

CONNIE B. NICHOLS
Department of Medicine and Department of Molecular Genetics and Microbiology, Duke University School of Medicine, Durham, NC 27710

KIRSTEN NIELSEN
Department of Microbiology, Medical School, University of Minnesota, Minneapolis, MN 55455

JOSHUA D. NOSANCHUK
Department of Medicine and Department of Microbiology and Immunology, Albert Einstein College of Medicine, Jack and Pearl Resnick Campus, 1300 Morris Park Ave., Ullmann Building, Room 107, Bronx, NY 10461

CAROLYN R. O'BRIEN
Faculty of Veterinary Science, University of Melbourne, Werribee, Victoria, Australia 3030

PETER G. PAPPAS
Division of Infectious Diseases, University of Alabama at Birmingham, 1900 University Blvd., THT 229, Birmingham, AL 35294-0006

BENJAMIN J. PARK
Mycotic Diseases Branch, Centers for Disease Control and Prevention, Atlanta, GA 30333

JOHN R. PERFECT
Department of Medicine, Division of Infectious Diseases, Duke University Mycology Research Unit (DUMRU), 0557 Hospital South, Box 3353, Duke University Medical Center, Durham, NC 27710

MICHAEL A. PFALLER
Departments of Pathology and Epidemiology, University of Iowa College of Medicine and College of Public Health, 200 Hawkins Dr., Iowa City, IA 52242-1009

PETER PHILLIPS
Infectious Diseases Unit, St. Paul's Hospital, Vancouver, BC, Canada V6Z 1YC

LIISE-ANNE PIROFSKI
Division of Infectious Diseases, Albert Einstein College of Medicine, 1300 Morris Park Ave., Bronx, NY 10461

MARCIO L. RODRIGUES
Laboratório de Estudos Integrados em Bioquímica Microbiana, Universidade Federal do Rio de Janeiro, Instituto de Microbiologia, Avenida Carlos Chagas Filho, 373. Cidade Universitária CCS, Bloco I, Rio de Janeiro - RJ, 21941-902, Brazil

MARIANELA RODRIGUEZ-CARRES
Department of Biology, Duke University, Durham, NC 27710

JULIAN C. RUTHERFORD
Institute for Cell and Molecular Biosciences (ICaMB), Medical School, Newcastle University, Catherine Cookson Building, Framlington Place, Newcastle upon Tyne, NE2 4HH, United Kingdom

NATHAN SAUL
Faculty of Veterinary Science, Building B14, University of Sydney, Sydney, NSW 2006, Australia

WEI-CHIANG SHEN
Department of Plant Pathology and Microbiology, National Taiwan University, No. 1 Roosevelt Road, Sec 4, 10617 Taipei, Taiwan

NINA SINGH
VA Pittsburgh Healthcare System and University of Pittsburgh, Pittsburgh, PA 15240

TANIA C. SORRELL
Centre for Infectious Diseases and Microbiology, Sydney Medical School-Western & Westmead Millennium Institute Level 3 ICPMR, University of Sydney at Westmead Hospital, Westmead 2145 NSW, Australia

CHARLES A. SPECHT
Department of Medicine, University of Massachusetts, Worcester, MA 01605

DAVID A. STEVENS
California Institute for Medical Research, San Jose, CA 95128; Div. of Infectious Diseases, Santa Clara Valley Medical Center, San Jose, CA 95128; and Div. of Infectious Diseases and Geographic Medicine, Stanford University, Stanford, CA 94305

SOMNUEK SUNGKANUPARPH
Faculty of Medicine, Ramathibodi Hospital, Mahidol University, Bangkok, Thailand

LUCIANA TRILLES
Mycology Laboratory, Instituto de Pesquisa Clínica Evandro Chagas (IPEC), Fundação Oswaldo Cruz (FIOCRUZ), Av. Brasil, 4365, 21040-900 Rio de Janeiro, Brazil

DAVID TROFA
Department of Medicine and Department of Microbiology and Immunology, Albert Einstein College of Medicine, Jack and Pearl Resnick Campus, 1300 Morris Park Ave., Ullmann Building, Room 107, Bronx, NY 10461

MARIA ANNA VIVIANI
Laboratory of Medical Mycology, Department of Public Health – Microbiology – Virology, Università degli Studi di Milano, Via Pascal 36, 20133 Milan, Italy

KERSTIN VOELZ
School of Biosciences, University of Birmingham, Birmingham B15 2TT, United Kingdom

PING WANG
The Research Institute for Children, New Orleans, LA 70118

BODO WANKE
Mycology Laboratory, Instituto de Pesquisa Clínica Evandro Chagas-Fiocruz, Rio de Janeiro, RJ, Brazil

SARAH WEST
Divisions of General Internal Medicine and Infectious Diseases, Oregon Health and Science University, Portland, OR 97239

BRIAN L. WICKES
Department of Microbiology, University of Texas Health Center, University of Texas, San Antonio, TX 78284-7758

KRISTI L. WILLIAMS
Department of Cell Biology, 438 Nanaline Duke Bldg., Box 3709, Duke University Medical Center, Durham, NC 27710

PETER R. WILLIAMSON
Laboratory of Clinical Infectious Diseases, National Institutes of Allergy and Infectious Diseases, National Institutes of Health, Bethesda, MD 20892

FLOYD L. WORMLEY, JR.
The University of Texas at San Antonio, Department of Biology, One UTSA Circle, San Antonio, TX 78249

KAREN L. WOZNIAK
Department of Biology, University of Texas at San Antonio, San Antonio, TX 78249

JO RAE WRIGHT
Department of Cell Biology, 438 Nanaline Duke Bldg., Box 3709, Duke University Medical Center, Durham, NC 27710

JIANPING XU
Department of Biology and Institute of Infectious Disease Research, Michael G. DeGroote School of Medicine, McMaster University, 1280 Main St. West, Hamilton, Ontario, L8S 4K1, Canada

CHAOYANG XUE
Public Health Research Institute Center, University of Medicine and Dentistry of New Jersey, Newark, NJ 07103

OSCAR ZARAGOZA
Servicio de Micología, Centro Nacional de Microbiología, Instituto de Salud Carlos III, Carretera Majadahonda-Pozuelo, Km2, Majadahonda 28220, Madrid, Spain

LI-PING ZHU
Department of Infectious Diseases, Huashan Hospital, Fudan University, Shanghai, PR China

Foreword

Cryptococcus neoformans is an important opportunistic fungal pathogen in T-cell-immunosuppressed patients, especially those with AIDS or solid organ transplants and others receiving prolonged immunosuppressive therapy. For example, the global burden of HIV-associated cryptococcosis approaches 1 million cases annually. This unique yeast also causes disease in immunocompetent hosts, but much less frequently. Although cryptococcosis is recognized most often in selected geographic areas, e.g., the United States, Western Europe, Australia, and sub-Saharan Africa, this disease is seen throughout the world.

In this multiauthored book, the five editors planned a comprehensive approach into the understanding of *Cryptococcus* at many different levels. Over a decade ago, the original book on *Cryptococcus neoformans* written by two of the present editors (Casadevall and Perfect) set the stage for how an entire book could be devoted to this fungal pathogen. In this new volume, the plan was to allow the entire expert field of cryptococcal investigators to examine in detail the life cycle, pathophysiology (from immunology to virulence factors), molecular biology, diagnosis, clinical futures, and management of this encapsulated yeast. Cryptococcosis in this book covers both the major pathogenic species, *Cryptococcus neoformans* and *Cryptococcus gattii*. The attention to detail and the comprehensive nature of the discussion within each chapter make this book a new standard for written work on a single pathogenic fungal species. There are several outstanding comprehensive medical mycology books in print, but for a single fungal pathogen there are few books that approach the completeness and reference standard of this treatise.

Cryptococcosis remains a deadly disease, with many victims and an enlarging pool of risk patients. Clearly, we need more insights into the pathobiology of this encapsulated yeast to help us treat and prevent its impact on the human condition. This book provides a landmark to acknowledge that this yeast and its study have become a model system for the study and understanding of fungal pathogenesis.

Any molecular biologist, microbiologist, infectious disease epidemiologist, or infectious disease clinician with an interest in mycology, especially cryptococcosis, will find this clearly written, highly organized, well-referenced, and all-inclusive book to be a valuable and up-to-date resource. This latest book devoted entirely to *Cryptococcus*, edited by Heitman, Kozel, Kwon-Chung, Perfect, and Casadevall, is the premier reference on the biology and pathogenesis of this fascinating and deadly yeast in our increasing population of immunocompromised hosts.

WILLIAM E. DISMUKES, MD
Professor Emeritus of Medicine
University of Alabama at Birmingham
School of Medicine

Preface

Cryptococcosis is a dynamic fungal infection that continues to evolve in the second century after its initial recognition as a human pathogen in the 1890s. From its first modest clinical appearance in a case report in 1894–1895, *Cryptococcus* has advanced as a human pathogen to the point where it causes infection in approximately 1 million individuals per year, with over 600,000 attributable annual mortalities caused by this pathogenic yeast, resulting in approximately one-third of all AIDS-associated deaths. From isolated cases in the clinical practice of immunocompromised hosts to major disease outbreaks in animals, from humans with HIV infection in sub-Saharan Africa to a recent geographically based outbreak in Vancouver, Canada, and the northwestern United States, this fungus has grown prominent in the clinical landscape of 21st-century medicine.

For over 50 years, the ecological path and the biological and clinical features of this encapsulated yeast have been intensely characterized. With the advent over the past two decades of a molecular biology infrastructure, allowing the study of *Cryptococcus* to expand into genome-wide investigations, the detailed understanding of cryptococcal disease mechanisms has been amplified. Control of the yeast through molecular manipulations, coupled with the years of understanding its immunological properties, has resulted in the development of a pathogenic yeast model for study. Clearly, the mysteries of the sugar-encapsulated yeasts are being stripped away.

This book's format of multiple authors and chapters provides a comprehensive understanding of what it signifies for *Cryptococcus neoformans* and *Cryptococcus gattii* to exist and to produce disease. This is the first multiauthored book on *C. neoformans*, and that in itself reflects the progress that has been made in recent years, as a vast amount of information has been accumulated by many experts. The goal is to provide the research investigator, clinician, biologist, and mycologist a single source of information and a description of the principles that define this fungal pathogen at the onset of the second decade in the 21st century.

In proximity to publication of this book, the Infectious Diseases Society of America published *Clinical Practice Guidelines for the Management of Cryptococcal Disease* in 2010. In these guidelines, management of cryptococcal meningoencephalitis is divided into three risk groups: (i) HIV-infected individuals, (ii) organ transplant recipients, and (iii) non-HIV-infected and non-transplant hosts. There are specific recommendations for unique risk populations such as children; pregnant women; individuals in resource-limited environments; patients with infection in sites other than brain, such as the lungs; and patients with *C. gattii* infections. These guidelines emphasize recommendations for the specific management of complications including increased intracranial pressure, immune reconstitution inflammatory syndrome, drug resistance, and cryptococcomas. We recommend that for these clinical management issues, the Guidelines site (idsaglobalhealth.org/WorkArea/linkit.aspx?LinkIdentifier=id&ItemID=15977) should be visited.

It is our sincere hope that this book frames our knowledge base of what *Cryptococcus* is, what it has done and can do, what we need to know, and where to look for it. Cryptococcosis is a model infectious disease with a story that will continue to unfold now and in the future.

In closing, it is our pleasure to have served as authors and coeditors for this project. Our goal in this endeavor was not only to highlight advances and progress in the field, but also to serve to bring together the diverse members of our community to give voice to the myriad perspectives on this unique pathogenic yeast. We thank all of our colleagues who are represented here as coauthors for their contributions to the success of these aspirations. We wish to thank our families for their forbearance and patience during the gestation of this project. We also want to graciously and explicitly acknowledge our editors at ASM Press, Gregory Payne and Ellie Tupper, without whose support, encouragement, and tireless efforts this project would not have been realized. We hope that the ultimate success of our efforts will lie in stimulating the field to advance the understanding, diagnosis, treatment, and prevention of this pathogen to such a degree that a future edition of this volume becomes unnecessary.

JOSEPH HEITMAN, Duke University
THOMAS R. KOZEL, University of Nevada, Reno
KYUNG J. KWON-CHUNG, National Institute of Allergy and Infectious Diseases
JOHN R. PERFECT, Duke University
ARTURO CASADEVALL, Albert Einstein College of Medicine

GENERAL PRINCIPLES

I

Cryptococcus: From Human Pathogen to Model Yeast
Edited by J. Heitman et al.

1

Systematics of the Genus *Cryptococcus* and Its Type Species *C. neoformans*

KYUNG J. KWON-CHUNG, TEUN BOEKHOUT, BRIAN L. WICKES, AND JACK W. FELL

DISCOVERY OF CRYPTOCOCCOSIS AND ITS ETIOLOGIC AGENTS

Historical Review of Cryptococcosis and Discovery of *Cryptococcus neoformans*

According to Freeman, cryptococcosis was first described by Zenker in 1861 (45). However, the validity of Zenker's case was questionable since there was no evidence of a microbial culture. Thus, credit for the first description of cryptococcosis goes to Busse and Buschke. In 1894, Otto Busse (Fig. 1), a pathologist who was working at the Institute of Pathology at Greifswald University, in Germany, observed round-to-oval "corpuscles" within and outside the giant cells in a surgical specimen obtained from a sarcoma-like lesion of the tibia from a 31-year-old woman (20). Busse isolated a culture from this lesion, and its pathogenicity was confirmed by reinoculation into the patient's skin. The patient eventually died due to disseminated infection (21). Buschke, a physician who took care of the patient, cultured the same pathogen from skin eruptions of the patient and thought that the etiologic agent was a coccidium (19). In the same year of Busse's first observation, Francesco Sanfelice (Fig. 2) isolated an encapsulated yeast from peach juice in Italy and named it *Saccharomyces neoformans*, thereby establishing the taxonomic priority of the species epithet "*neoformans*" for the fungus (103). In the meantime, Busse called the fungus *Saccharomyces* and the disease saccharomycosis hominis (21). In 1895 Sanfelice recognized the similarity between *S. neoformans* and Busse's fungus. He also isolated a yeast similar to *S. neoformans* from the lymph node of an ox and confirmed the pathogenic nature of the fungus (104). In 1901 Vuillemin transferred both Busse's and Sanfelice's fungi to the genus *Cryptococcus* as *C. hominis* and *C. neoformans*, respectively, based on their inability to ferment carbon sources and lack of ascospore formation, which characterize the genus *Saccharomyces* (124).

In 1902, Frothingham recovered a pathogenic yeast from a pulmonary lesion in a horse in Massachusetts that was similar to the fungus isolated by Busse and Buschke (46). This finding, together with Sanfelice's isolation of the yeast from an ox, established that Busse and Buschke's fungus causes infection in both humans and animals. Cryptococcal meningitis was first recognized in 1914 by Verse from an antemortem diagnosis of the disease in a woman (120). Two years later, Stoddard and Cutler recorded two new meningitis cases and named the causative fungus *Torula histolytica* (111). They misinterpreted the fungal capsule as evidence of histolytic action of the fungus in host tissue. The name torulosis, which was derived from the misinterpretation, persisted in the literature until the 1950s.

In 1935, Benham conducted a comprehensive study with numerous yeasts, which included 22 strains of pathogenic and nonpathogenic cryptococci isolated from humans. These 22 strains included those isolated by Busse, Curtis, and others that had been designated under the three genera *Cryptococcus*, *Saccharomyces*, and *Torula*. Based on morphological, pathological, and serological observations, she concluded that all human isolates were probably one species, *C. hominis*, and comprised two varieties (7). In 1949 Evans discovered serological differences between strains isolated from cryptococcosis patients and identified three serotypes, A, B, and C (36). The fourth serotype, D, was identified by Vogel nearly 20 years later in an isolate (123) and was confirmed in more isolates by Wilson et al. (128). In 1950 Benham proposed that the name torulosis and torula meningitis be replaced by cryptococcosis and that *Cryptococcus neoformans* become a *nomen conservandum* (a taxon name formally accepted under the International Code of Botanical Nomenclature as the correct name). Her work also resolved the confusion surrounding the term "blastomycosis," which until 1950 had been used for both cryptococcosis and North American blastomycosis (8).

Kyung J. Kwon-Chung, Molecular Microbiology Section, Laboratory of Clinical Infectious Diseases, National Institute of Allergy and Infectious Diseases, National Institutes of Health, Bethesda, MD 20892. **Teun Boekhout,** CBS Fungal Diversity Center, Uppsalalaan 8, 3584 CT Utrecht, The Netherlands. **Brian L. Wickes,** Department of Microbiology, University of Texas Health Center, University of Texas, San Antonio, TX 78284-7758. **Jack W. Fell,** Rosenstiel School of Marine and Atmospheric Science, University of Miami, Key Biscayne, FL 33149.

FIGURE 1 Otto Busse (1867–1922).

FIGURE 2 Francesco Sanfelice (1861–1945).

FIGURE 3 Ferdinand Curtis (1858–1937).

Discovery of *Cryptococcus gattii*, the Second Etiologic Agent of Cryptococcosis

A year after the discovery of cryptococcosis by Busse and Buschke, Ferdinand Curtis (Fig. 3), a professor of pathology at the Faculty of Medicine of Lille, in France, described a "vegetal parasite" that belonged to a yeast species to be the etiologic agent of inguino-crural tumors and a lumbar ulcerated abscess in a young, otherwise healthy French man. The clinical observation was reported in 1895, and a detailed description was published in 1896 (27). Curtis noticed elongated, oval, or bacilliform yeasts in addition to round cells and the absence of fermentation in his strain. He named the fungus *Saccharomyces subcutaneous tumefaciens*, considering it to be distinct from *S. neoformans* of Sanfelice and *S. hominis* of Busse based on morphological differences and its affinity for subcutaneous tissue in experimental animals (27). Benham rejected the species status of *S. subcutaneous tumefaciens* and proposed it to be a variety of *C. hominis* (*C. hominis* var. *tumefaciens*) (7), an earlier name of *C. neoformans*.

In 1970 Gatti and Eeckels isolated a *C. neoformans* strain from a 7-year-old leukemic patient in Zaire (currently the Republic of Congo) who suffered from meningoencephalitis (47). Vanbreuseghem and Takashio reported that the isolate produced elongated cells in vivo as well as in culture but was otherwise the same as the typical isolate of *C. neoformans*. To accommodate this atypical isolate, they described a new variety and named it *C. neoformans* var. *gattii* (117). However, Kwon-Chung et al. determined that *C. neoformans* var. *neoformans* and *C. neoformans* var. *gattii* differed considerably in morphological, biochemical, and serological characteristics and that the isolate from Zaire was identical to the type strain of *Cryptococcus bacillisporus*, the anamorph of *Filobasidiella bacillispora* (72). These findings indicated that *C. neoformans* var. *gattii* was a synonym of *C. bacillisporus*. In 2002 the name *C. bacillisporus* was replaced with *C. gattii* on the basis that the epithet "*gattii*" has been more widely used worldwide than "*bacillisporus*."

In 2001 molecular analysis of CBS 1622, which would have been an ex-syntype of the Curtis strain, indicated that this strain was identical to *Cryptococcus gattii* (13). These findings indicated that cryptococcosis caused by *C. gattii* was first described by Ferdinand Curtis one year after Busse and Bushke's discovery of cryptococcosis. Since there is no information on the patient's travel record, it is not known if it represents an autochthonous case of *C. gattii* infection in France.

ESTABLISHMENT OF THE GENERA *CRYPTOCOCCUS* AND *FILOBASIDIELLA* AND THEIR SYSTEMATIC POSITION IN THE KINGDOM FUNGI

Establishment of the Genus *Cryptococcus*

The genus *Cryptococcus* was first established by Kützing in 1833 to accommodate the new organism *Cryptococcus mollis*, which was isolated from a dirty window (64). The genus was characterized as "Globuli mucosi hyaline non colorati microscopici in stratum indeterminatum mucosum facile secedens sine ordine aggergati." The generic description was accompanied by a specimen dried onto a small square of glass. Since then, the generic name has been subjected to a complex history of amendments. In 1901 Vuillemin considered that the organisms isolated by Busse and Sanfelice in 1894, one from a necrotic lesion of a patient in Germany and the other from peach juice in Italy, which were classified in the genus *Saccharomyces*, should belong to the genus *Cryptococcus*. He transferred both organisms and amended the description of the genus *Cryptococcus* to include only the asexual pathogenic yeasts, thus excluding all other species except for *C. neoformans* (124). In 1935 Benham broadened the definition of the genus *Cryptococcus* to include both pathogenic and nonpathogenic, fermenting and nonfermenting yeasts that did not produce hyphae (7). In 1952 Lodder and Kreger-van Rij accepted the generic definition amended by Vuillemin but expanded it to include only nonfermenting encapsulated yeasts that produce starch (82). Lodder and Kreger-van Rij designated *C. neoformans* as the type species of the genus. Phaff and Spencer abandoned starch formation as a diagnostic criterion and delimited the genus to nonfermenting yeasts that can utilize *myo*-inositol and either produce pseudohyphae or true hyphae or produce no hyphae at all (96).

In 1979, however, Von Arx and Weijman reexamined Kützing's type specimen and found that it contained at least three different organisms including algae, mold, and yeast (123a). They proposed, therefore, that the genus *Cryptococcus* is a *nomen dubium* (doubtful name). Subsequently, Rodrigues de Miranda and his coworker examined Kützing's type material using electron as well as light microscopy and detected basidiomycetous yeast cells. They concluded that *C. mollis* was a legitimate type of the genus (100, 101). Because it was not clear that the organism Kützing referred to as *C. mollis* was a yeast and because the nomenclatural stability of *Cryptococcus* was extremely important, Fell et al. proposed conserving the genus name as *Cryptococcus* Vuillemin instead of *Cryptococcus* Kützing, designating *C. neoformans* as the type species (38). At present, the genus includes both pathogenic and nonpathogenic yeasts and comprises 70 species (41). Some of these species produce a teleomorphic state belonging to genera such as *Filobasidium*, *Filobasidiella*, *Cystofilobasidium*, and *Kwoniella* in the Tremellomycetes of Basidiomycota (12).

Establishment of *Filobasidiella*, the Teleomorphic Genus of Both *C. neoformans* and *C. gattii*

Prior to the discovery of the genus *Filobasidiella*, the teleomorph of *C. neoformans* and *C. gattii*, yeast taxonomists had speculated that *C. neoformans* belonged to the Basidiomycota for the following reasons. The G+C ratio of *Cryptococcus* DNA was higher than that of the ascomycetous fungi and was within the range of fungi belonging to the Basidiomycota (112). *C. neoformans* produced starch and assimilated *myo*-inositol, which are common features of basidiomycetous yeasts rather than ascomycetous yeasts (95). More importantly, Shadomy observed hyphae with clamp-like structures in a strain of *C. neoformans* (NIH12) isolated from a clinical sample (106). The clamps were not fused to hyphae, and in retrospect, these hyphae were likely the result of monokaryotic fruiting.

Four years after Shadomy's observation, the teleomorph of *C. neoformans* described as *Filobasidiella neoformans* was found by matings between two compatible strains, *MAT*α and *MAT***a**, of serotype D as well as between compatible strains of serotypes A and D (68). Analysis of the progeny showed that *C. neoformans* was a heterothallic fungus with a bipolar mating system (67). The second species of *Filobasidiella*, *F. bacillispora*, was found to be produced by matings between compatible strains of *C. gattii*, serotypes B and C (69). The existence of two distinct teleomorphs among the agents of cryptococcosis indicated that the differences between *C. neoformans* and *C. gattii* were not limited to their antigenic properties. Subsequent studies have shown that the two species are distinct in their ecological niche (34, 35), epidemiology (71), biochemistry (32, 72, 76, 77, 88), and genomic arrangement (53, 127). The genus *Filobasidiella* currently contains five accepted species: *F. neoformans*, *F. bacillispora*, *F. depauperata*, *F. lutea* (65), and "*F. amylolenta*" (K. Findley and J. Heitman, unpublished data).

Filobasidiella is one of the most unusual genera in the phylum Basidiomycota. First, it is the only basidiomycetous genus that contains human pathogens that produce life-threatening infection of the central nervous system in immunocompromised as well as immunocompetent patients (65). Second, *Filobasidiella* is the only genus in the phylum Basidiomycota that produces long chains of basidiospores on the apex of the holobasidia by repetitive basipetal budding (68) (Fig. 4). Third, the genus contains two groups of species, one with and one without ontogenetic yeast stages. *F. neoformans*, *F. bacillispora*, and "*F. amylolenta*" produce a haploid yeast stage, while *F. depauperata* lacks a yeast stage (75). The presence or absence of a yeast stage in *F. lutea* is unclear since the life cycle of the species has not been characterized (99) and a pure culture is not available for analysis of its life cycle.

Interestingly, the organization of 5S rRNA is different between the species with and without an ontogenetic yeast stage. While the 5S rRNA gene is embedded in the nontranscribed spacer regions of the rDNA repeat unit in

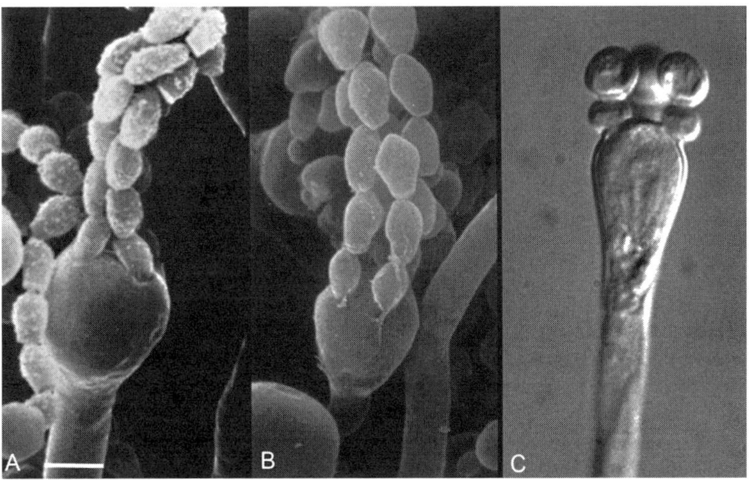

FIGURE 4 Basidial morphology of (A) *Filobasidiella neoformans*, (B) *F. depauperata*, and (C) *F. lutea*. (A and B courtesy of R. A. Samson and Antonie van Leeuwenhoek.) Bar = 5 μm.

the species with an ontogenic yeast stage, as is the case with other basidiomycetous species (25, 121), the 5S rRNA gene is dispersed throughout the genome in *F. depauperata* (which lacks an ontogenic yeast stage), as in some species belonging to the Ascomycota (74). The other unique aspect of the genus is that the members of *Filobasidiella* are either pathogenic or parasitic for a wide variety of hosts. The two species with yeast stages, *F. neoformans* and *F. bacillispora*, are broad-spectrum pathogens that can infect protozoan species and a wide variety of vertebrate (78, 87, 109, 110) and invertebrate species (83). The third species of the genus with an ontogenic yeast stage, "*F. amylolenta*," has been isolated from insect frass in South Africa (118). Whether "*F. amylolenta*" is an insect parasite or is associated with organisms that parasitize insects is yet to be determined. The remaining two species that lack ontogenic yeast stages are mycoparasites with a narrow host range. *F. depauperata* is a parasite of the entomogenous ascomycetous fungus *Verticillium lecanii* (48), while *F. lutea* is a parasite of the basidiomycete *Hypochnicium vellereum* (99). *Filobasidiella xianghuijun*, a species described in the genus by Zang (130), was isolated from *Tremella fuciformis* collected in China. This species, however, was not accepted in the genus *Filobasidiella* due to the basidial characteristics that better fit the description of the genus *Chionosphaera* (65).

Phylogenetic Position of the Genera *Cryptococcus* and *Filobasidiella*

Our knowledge of the biodiversity and taxonomy of the anamorphic basidiomycetous yeast genus *Cryptococcus* has changed considerably in recent years. New cryptococcal species are being continuously discovered (10, 41), and some traditionally recognized species, such as *Cryptococcus albidus* and *Cryptococcus laurentii*, were found to represent species aggregates (10, 42, 114). Various phylogenetic studies based mainly on sequence analysis of parts of the ribosomal DNA (rDNA) locus have contributed to the recognition of an extended species concept based on dissimilarities among rDNA sequences (37, 105). Through use of multigene analysis, this rDNA-based species concept may eventually evolve into a phylogenetic species concept (15). Furthermore, it has become clear that the genus *Cryptococcus* as currently characterized (an anamorphic basidiomycetous budding yeast that utilizes D-glucuronate and usually *myo*-inositol, usually synthesizes extracellular starch-like compounds, and has xylose present in the cell wall) (41) is polyphyletic. The species belong to five major lineages within the Tremellomycetes (subphylum Agaricomycotina, Basidiomycota), namely the orders Tremellales, Filobasidiales, Trichosporonales, Cystofilobasidiales, and the *Holtermannia* lineage (41) (Fig. 5). In addition to these five lineages, some species, e.g., *Cryptococcus marinus*, occur as single species on long branches in the rDNA phylogenetic trees. It is possible that these branches may emerge as well-supported separate lineages that represent higher taxa (e.g., at the family or ordinal levels) when more species are added (37, 41, 105). It is important to note that the support values for many of the internal lineages in rDNA-based trees are not high. Additional sampling of species and analysis of more DNA loci are needed to improve the phylogenetic resolution of the trees to assess the proper phylogenetic relationships among the species concerned.

The genus *Cryptococcus* presently contains 70 species that are only known in their anamorph state (41) (Fig. 5). However, this number is increasing rapidly due to the recognized importance of DNA sequence analysis in understanding fungal biodiversity. Several well-known *Cryptococcus* species such as *C. neoformans* and *C. gattii* are classified in the teleomorph state, *F. neoformans* and *F. bacillisporus*, respectively (65, 68, 69). Similarly, the teleomorphic state of *Cryptococcus uniguttulattus* is classified as *Filobasidium uniguttulatum* (66, 70). Due to taxonomic confusion about the status of the type species of the genus *Cryptococcus*, as well as the widely recognized importance of *C. neoformans* and *C. gattii* as human pathogens, the genus was neotypified by *C. neoformans* (38, 49). This revision implies that if one uses a modern phylogenetic genus concept, only those anamorphic species that belong to the *Filobasidiella* clade should be classified as *Cryptococcus* species, e.g., *C. neoformans*, *C. gattii*, and *C. amylolentus*. However, this concept has not yet been implemented, mainly because it would result in a significant number of taxonomic changes. In the fifth edition of *The Yeasts: a Taxonomic Study*, the genus is presented in its current polyphyletic format (12, 41), and we will adhere to this practice in this chapter.

Species of the polyphyletic anamorph genus *Cryptococcus* belong to the subphylum Agaricomycotina of the Basidiomycota (51). Within this subphylum the species of the genus cluster in the orders Tremellales, Filobasidiales, Trichosporonales, Cystofilobasidiales, and the *Holtermannia* lineage (41) (Fig. 5). The orders Tremellales and Filobasidiales contain the majority of the species, whereas 10 species belong to Trichosporonales, four species belong to the *Holtermannia* lineage, and one to the Cystofilobasidiales. Species that have been reported from clinical sources are known in all lineages, except for the *Holtermannia* lineage (Fig. 5). This organization clearly demonstrates that pathogenicity has evolved many times in this group of yeasts. Among the Tremellales, species classified as *Cryptococcus* occur intermingled with species of diverse genera such as *Bullera*, *Tremella*, *Papiliotrema*, *Auriculibuller*, *Trimorphomyces*, *Kwoniella*, *Fibulobasidium*, and *Bulleribasidium* (12). It is important to note that both of the anamorphic genera, *Cryptococcus* and *Bullera*, are largely polyphyletic, whereas most of the teleomorph genera, except *Tremella*, are monophyletic. Despite this taxonomically confusing situation, it is interesting to note that species of *Bullera* that occur intermingled among cryptococcoid yeasts are not clinically important.

The genus *Filobasidiella*, the teleomorph states of *C. neoformans* and *C. gattii*, is a small monophyletic genus containing five species: *F. neoformans*, *F. bacillispora*, *F. depauperata*, *F. lutea*, and "*F. amylolenta*" (Findley and Heitman, unpublished data). Small-subunit rDNA analysis of a number of fungi demonstrated that *F. depauperata* was closely related to *F. neoformans* (75). Based on a phylogenetic study of the internal transcribed spacer (ITS) and small-subunit rDNA sequence data, the mycoparasitic *F. lutea* was found to be a member of *Filobasidiella* (107). However, the phylogenetic position of *F. xianghuijun* remains to be determined when a culture becomes available (130). Based on morphological similarities between the basidia of *Filobasidium* and *Filobasidiella*, both genera have been previously classified in the Filobasidiales (4, 11, 57). It is noteworthy that Olive classified *Filobasidium floriforme*, the type species of the genus *Filobasidium*, in the family Filobasidiaceae of Ustilaginales (91). In the fourth edition of *The Yeasts: a Taxonomic Study* (11), the family Filobasidiaceae, which included *Filobasidium*, *Filobasidiella*, *Mrakia*, *Xanthophyllomyces*, and probably *Cystofilobasidium*, was placed in the Filobasidiales mainly because of the presence of aseptate clavate basidia that occur in all these genera (11). Molecular phylogenetic studies based on the sequence of the D1/D2 domains of the LSU (Large Subunit)

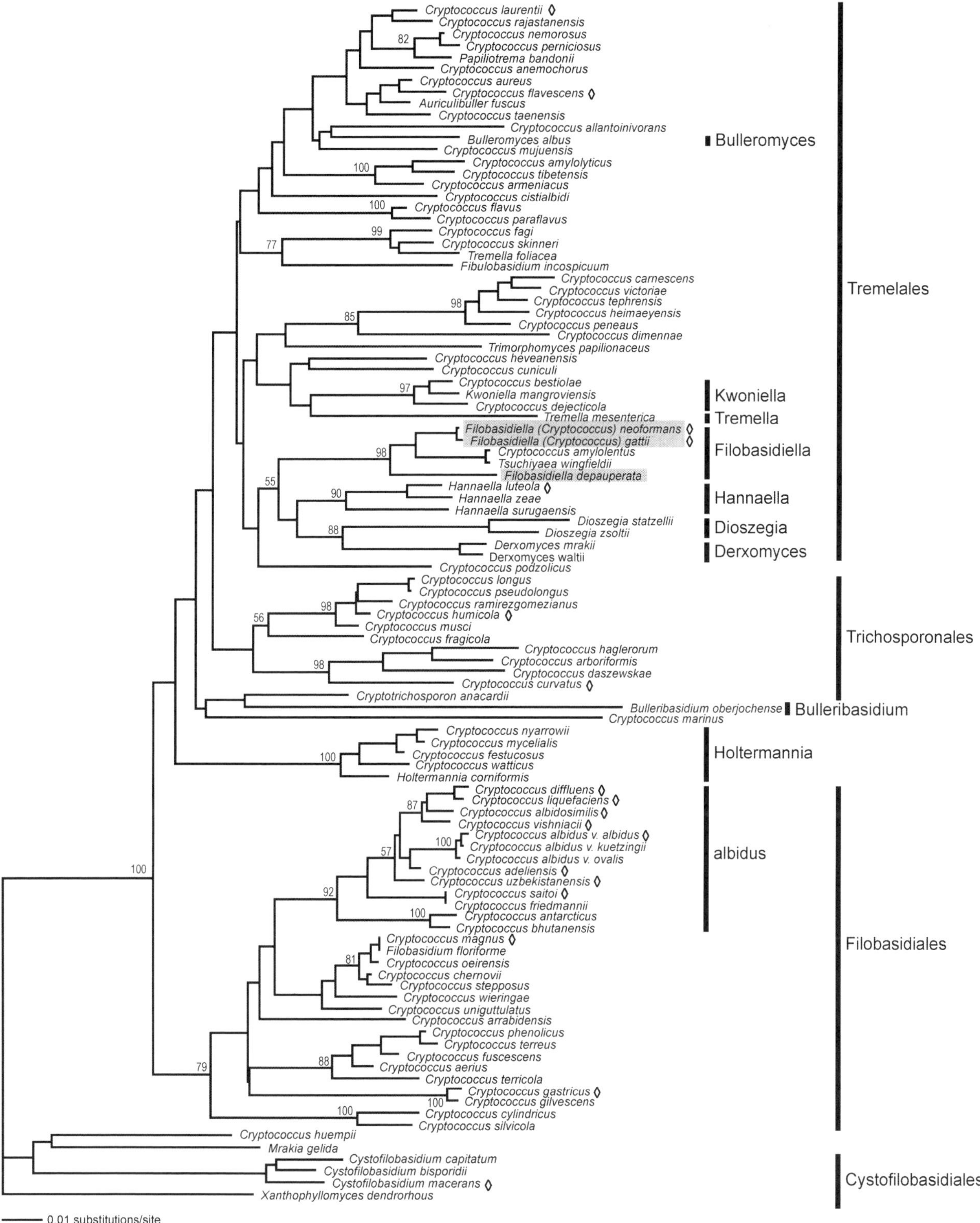

FIGURE 5 Phylogenetic tree of representative members of the Tremellomycetes to demonstrate the diversity of the genus *Cryptococcus* within the four orders of the class. Neighbor joining analysis of the D1/D2 region of the LSU rDNA. Numbers at the branch nodes represent bootstrap percentages (>50%) from 1,000 full heuristic replicates (PAUP 4.0b10). Diamond symbol indicates species that have been isolated from clinical sources. The genus *Filobasidiella* is shaded for emphasis.

rRNA gene and the ITS 1 and 2 regions of the rDNA, however, demonstrated that species of *Filobasidium* cluster in the Filobasidiales, while *Filobasidiella* species form a monophyletic lineage in the Tremellales (12, 37, 50, 105).

Within the Tremellales, the *Filobasidiella* clade appears to have a weakly supported sister group relationship with the *luteolus* clade (i.e., *Bullera coprosmaensis*, *Bullera kunmingensis*, *Bullera derxii*, *Bullera sinensis*, *Bullera oryzae*, *Cryptococcus luteolus*, *Cryptococcus zeae*, and *Cryptococcus surugaensis*), which recently has been reclassified in the genus *Hannaella* (125). Together, these two clades form a poorly supported sister group with the genus *Dioszegia* and the *mrakii* clade of *Bullera* (Fig. 5). The latter has been reclassified as the genus *Derxomyces* (125). In the study of Scorzetti et al. (105), in which both the D1/D2 domains of the LSU rDNA and the ITS 1 and 2 regions were separately studied, it was noted that the *Filobasidiella* clade occupied different positions in the respective trees. In the D1/D2 tree the *Filobasidiella* lineage was found to be separate from all other lineages, whereas in the ITS-based tree a sister clade comprised *Bullera dendrophila*, *Cryptococcus heveanensis*, and *Cryptococcus* species CBS 8507, which is now classified in a teleomorphic yeast genus as *Kwoniella mangroviensis* (108).

A recent phylogenetic study based on the analyses of several genes (i.e., *RPB1*, *RPB2*, *EF1*α, mitochondrial small-subunit rDNA, nuclear LSU rDNA, and the ITS regions) confirmed the monophyly of the *Filobasidiella* clade with high support values (39): *Cryptococcus amylolentus* and *Tsuchiyaea wingfieldii* formed a clade (100% support), but these two species are considered to be conspecific in the fifth edition of *The Yeasts: a Taxonomic Study* based on the D1/D2 domains of the LSU rDNA and the ITS 1 and 2 regions of the rDNA sequence (41). These two species (or single species?) occupied a sister group relationship with *C. neoformans* and *C. gattii*. *F. depauperata* occupied a basal position to these species.

It is interesting to note that a sexual state typical of the genus *Filobasidiella* (slender clavate basidia with four chains of basidiospores) was recently discovered in *C. amylolentus* (Findley and Heitman, unpublished data), thus providing a synapomorphic character state for the monophyletic *Filobasidiella* lineage. In this study the *Kwoniella* clade occupied a well-supported sister group relationship to the *Filobasidiella* lineage. The *Kwoniella* clade comprises *B. dendrophila*, *K. mangroviensis*, and three anamorphic species, *Cryptococcus bestiolae*, *Cryptococcus dejecticola*, and *C. heveanensis* (39). Importantly, only representatives of *C. neoformans* and *C. bacillispora* grew well at 37°C, produced capsules, and formed melanin. Two species, *C. heveanensis* and *Cryptococcus humicola*, which are not closely related to the *Filobasidiella* species, showed a faint coloration on melanin-inducing media, indicating that they may have potential to produce melanin. Interestingly, even more distantly related species, namely *Cryptococcus podzolicus* and *Cryptotrichosporon anacardii*, produce melanin (90, 93). Laccase activity, which is responsible for melanin formation, has also been demonstrated in *C. albidus*, *C. laurentii*, and *Cryptococcus curvatus* (54), all of which have been reported from clinical cases, albeit rarely (41). Based on a *Galleria mellonella* virulence model, *C. gattii* and the two varieties of *C. neoformans* showed the highest virulence, and two other species in the *Filobasidiella* clade, namely *C. amylolentus* (including *T. wingfieldii*) and *F. depauperata*, were avirulent, whereas some species from the *Kwoniella* clade showed intermediate levels of virulence (39).

In conclusion, it is apparent that our knowledge of phylogenetic relationships among these fungi is in its infancy. By applying a multigene approach (56) or a phylogenomics approach (see, e.g., 33, 40, 63, 85), we can make considerable progress to improve the phylogenetic understanding of these fungi, which would be an important basis of the evolutionary origin of virulence and its determinant traits.

NOMENCLATURE OF CRYPTOCOCCOSIS AGENTS

Until Benham's critical studies in 1935 (7) and 1950 (8), the nomenclature of *C. neoformans* was in disarray. According to Rodrigues de Miranda, the name *C. neoformans* had 40 synonyms assigned under seven different genera (100). The most commonly known synonyms were *S. neoformans* (104), *C. hominis* (124), *Torula neoformans* (126), *T. histolytica* (111), and *Debaryomyces hominis* (116). Benham showed that all the pathogenic isolates assigned under *Torula*, *Cryptococcus*, *Saccharomyces*, and *Debaryomyces* were of the same species and proposed to conserve the name *C. neoformans*.

After the discovery of the teleomorphs of *C. neoformans*, *F. neoformans*, and *F. bacillispora*, the nomenclature of the fungus underwent further changes. The anamorphic state of *F. bacillispora*, which contains strains of serotypes B and C, was elevated to a separate species as *C. bacillisporus* based on antigenic, biochemical, and epidemiological differences compared to *C. neoformans* (72). Since then, the name has undergone changes from *C. bacillisporus* to *C. neoformans* var. *gattii* to *C. gattii* (73). The basis of these nomenclatural changes has already been described above in "Discovery of *Cryptococcus gattii*, the Second Etiologic Agent of Cryptococcosis." The nomenclature of *C. neoformans* also has undergone revision: *C. neoformans* to *C. neoformans* var. *neoformans* for the strains of serotype D and *C. neoformans* var. *grubii* for the strains of serotype A (43). The classification of the strains of serotype A and serotype D into two different varieties was based first on the diversity in the *URA5* gene sequence, which showed a difference in about seven nucleotides between the two groups, in addition to distinct DNA fingerprinting patterns. Subsequent analysis of various gene sequences also showed a significant divergence between strains of serotypes A and D. However, division of the species *C. neoformans* into two varieties, based on serotype, left a few unsettling problems. First, serotype is based on the antigenic properties of the strain, which are not always stable, while the varietal status is based on DNA sequence (43). Second, the serotype and genotype do not always concur, and third, some diploid strains have been serotyped as AD, A, or D depending on the laboratory in which the serotyping was performed (76a). Finally, stable AD serotype strains can be identified as neither *C. neoformans* var. *grubii* nor as *C. neoformans* var. *neoformans*. Rather than serotyping, molecular strain typing based on the PCR patterns of microsatellite DNA, such as M13, or the sequence of rDNA intergenic spacer (IGS) regions, and multilocus sequence typing yielded much more reproducible results on the genetic diversity between the two groups (15, 30, 86).

GENETIC DIVERSITY OF CRYPTOCOCCOSIS AGENTS

Many molecular genotyping methods have been used to distinguish groups within the *C. neoformans–C. gattii* species complex. Six haploid genotypic groups have been found based on molecular fingerprinting methods, such as PCR fingerprinting, restriction fragment length polymorphism, randomly amplified polymorphic DNA, and amplified fragment length polymorphism (AFLP) analyses (13, 34, 62, 79, 86, 102), as well as sequence analysis of coding and noncoding regions (9, 14, 23, 26, 30). Note that we will focus this chapter on genetic diversity as assessed by sequence analysis of DNA (or rRNA), as the various DNA fingerprint approaches that are applied to type isolates of *Cryptococcus* will be discussed elsewhere (see chapter 24, this volume).

Many sequence typing studies used a single locus (9, 23, 26, 30), and the studies that sequenced multiple loci used either *C. neoformans* isolates (81) or *C. gattii* isolates (44, 61). Three studies used *C. neoformans* as well as *C. gattii* isolates (15, 113, 129). ITS, IGS, orotidine monophosphate pyrophosphorylase (*URA5*), mitochondrial cytochrome *b* (*cytb*), mitochondrial large ribosomal subunit RNA (MtLrRNA), *PRP8* intein, mating-pheromone (*MF*α), topoisomerase (*TOP1*), laccase (*LAC*), and capsule gene (*CAP59*) sequences demonstrated that *C. gattii*, *C. neoformans* var. *grubii*, and *C. neoformans* var. *neoformans* formed distinct clusters (9, 23, 26, 30, 58,113, 129).

In a recent study, Bovers et al. (15) characterized 117 isolates from clinical and environmental origins representing all known haploid molecular genotypes of *C. neoformans* and *C. gattii*. Six nuclear loci, including two rDNA regions, namely ITS 1 and 2, 5.8S rDNA and the intergenic spacer 1 (IGS1) region, the laccase gene (*CNLAC1*), the largest and second-largest subunits of RNA polymerase II (*RPB1* and *RPB2*), and translation elongation factor 1α (*TEF1*), revealed six monophyletic lineages that correspond to previously recognized molecular genotypes (13, 30, 86). Concatenated sequences of *C. neoformans* and *C. gattii* were found to be 84–86% similar. Sequence similarity of the six concatenated loci was 91–92% between the two *C. neoformans* varieties and 95–96% between the *C. gattii* genotypes. These similarities are in agreement with those observed in other molecular studies (2, 23, 26, 30, 59). The two varieties of *C. neoformans* share 85–90% nucleotide identity at the genomic level (59), and preliminary data estimate the level of nucleotide divergence between the genotypes of *C. gattii* VGI and VGII between 85 and 95% and those between *C. gattii* and *C. neoformans* between 80 and 95% (J. W. Kronstad, unpublished data).

C. neoformans

The phylogenetic structure within *C. neoformans* revealed 4 haplotypes for ITS, 11 for *CNLAC1*, 12 for *RPB1*, 13 for *RPB2*, 16 for IGS1, and 18 for *TEF1* (15). All isolates belonged to either var. *grubii* or var. *neoformans* with strong support values (var. *grubii* had bootstrap values of 99% for maximum parsimony [MP] and 100% for neighbor joining [NJ], maximum likelihood [ML], and Bayesian analyses, and var. *neoformans* had 100% for NJ, MP, ML, and Bayesian analyses).

The six concatenated loci supported the *C. neoformans* var. *neoformans* clade containing AFLP2/VNIV (var. *neoformans* molecular type IV) isolates and included the type strain of *Torula nasalis* (CBS 882) and the VNIV reference strain (WM629). Phylogenetic analysis revealed the presence of three clusters within this clade, which correspond to previously described AFLP genotypes 1, 1A, and 1B (5, 13). The reference strain of genotype VNI (WM148) clustered within AFLP1/VNI together with the type strains of *C. neoformans* var. *grubii* (H99 = CBS 8710 = CBS 10515) and *Candida psicrophylicus* (CBS996), the latter of which is considered a synonym of the former. AFLP1B isolates clustered together with the reference strain of VNII (WM626). All AFLP1A isolates fell into the cluster that contained VNB (var. *neoformans* Botswana molecular type) isolates that originated from Botswana (81; see also chapter 8).

C. gattii

Based on molecular fingerprinting methods, *C. gattii* formed four genotypes (34, 79, 86, 102), which were supported by sequence analysis (9, 15, 23, 25, 30, 44, 61). Phylogenetic analysis of *C. gattii* ($n = 51$) revealed 9 haplotypes for ITS, 11 for *RPB1*, 11 for *RPB2*, 16 for *CNLAC1*, 16 for *TEF1*, and 20 for IGS1 (15). The AFLP4/VGI cluster received bootstrap support values of 59% for MP, 64% for ML, 77% for NJ, and 93% for Bayesian analyses. The AFLP5/VGIII, AFLP6/VGII, and AFLP7/VGIII groups received strong support. The AFLP4/VGI and AFLP5/VGIII clusters formed a sister group relationship (bootstrap support values of 78% for NJ, 88% for ML, 99% for MP, and 100% for Bayesian analyses), and AFLP7/VGIV clustered basal to this AFLP4/VGI–AFLP5/VGIII clade, whereas the AFLP6/VGII cluster occurred basal to all other *C. gattii* genotypic groups. The same topology has been found in analyses of the IGS1+5S+IGS2 (30) and *TEF1* region (44). Phylogenetic analyses of *PRP8* intein and *URA5* sequences supported a topology where AFLP7/VGIV clustered basal to the AFLP4/VGI and AFLP5/VGIII sister groups (23, 61), and AFLP4/VGI and AFLP5/VGIII formed sister groups after analysis of *GPD1* data (44).

C. gattii topologies that are in conflict with the above-described topology have been published. For instance, *CNLAC1* analysis indicated that AFLP4/VGI clustered basal to the AFLP6/VGII and AFLP7/VGIV sister groups (15, 44, 61). The topology obtained by analysis of the *FTR1* gene showed that AFLP4/VGI clustered basal to the AFLP5/VGIII and AFLP7/VGIV sister groups (61), and IGS1 analysis showed AFLP6/VGII as a basal lineage to the AFLP4/VGI and AFLP5/VGIII sister groups (15). Analysis of the complete IGS1+5S+IGS2 region (30) resulted in a topology identical to that obtained with the six concatenated loci as described above. An analysis of concatenated data sets consisting of four (61) and six (15) loci, respectively, resulted in the same topology. These latter studies have used different loci, and therefore, we consider the topology obtained with the concatenated data sets as the most accurate one to date. Thus, in brief, AFLP4/VGI and AFLP5/VGIII are the two *C. gattii* groups that are related to the AFLP7/VGIV sister clade, and the AFLP6/VGII clade appears to be the most basal group in the *C. gattii* cluster.

In the past, several species have been described that are currently considered synonyms under *C. gattii* (14, 65). The type or otherwise authentic material of some of these synonymous species has been preserved in culture collections, and hence, they could be studied. The type strains of *Torulopsis neoformans* var. *sheppei* (CBS 919), *C. neoformans* var. *shanghaiensis* (CBS 7229), the syntype (one of several strains cited by an author when originally proposing a name but where no holotype was selected) of *Cryptococcus hondurianus* (CBS 883), and the reference strain of VGI (WM179) clustered with AFLP4/VGI isolates. The type strain of *C. gattii* (RV20186 = CBS 6289 = CBS 8273) clustered in the AFLP4A/VGI subcluster of AFLP4/VGI. The type strain of *C. bacillisporus* (CBS 6955), a current synonym of *C. gattii* (14), clustered with AFLP5/VGIII isolates. The reference strain of VGIII (WM161) belonged to the AFLP5B subcluster.

Several studies indicated that both serotype B and C isolates may be present in a single *C. gattii* genotype (see, e.g., 86), but other studies challenged this observation. Bovers et al. (15) observed that isolates of both serotype B and serotype C were present in only the AFLP5/VGIII and the AFLP7/VGIV lineages, and reinvestigation indicated that AFLP5/VGIII isolates reported to be serotype B are in fact strains of serotype C (F. Hagen, T. Boekhout, and K. Tintelnot, unpublished data). Importantly, some of the *C. gattii* genotypic groups have interesting attributes. For instance, AFLP4/VGI is the *C. gattii* parental genotype present in all currently known *C. neoformans–C. gattii* hybrids (16, 17); AFLP6/VGII is the *C. gattii* genotype responsible for the outbreak of cryptococcosis on Vancouver Island (62) that expanded to mainland Canada and the United States (6, 24); and AFLP7/VGIV is associated with infections of HIV-positive patients in Africa (15, 80; F. Hagen and T. Boekhout, unpublished data).

Hybrids

Several diploid or aneuploid hybrids occur in the *C. neoformans–C. gattii* complex. The best-studied hybrids are the serotype AD hybrids, which are formed after conjugation of yeast cells from *C. neoformans* var. *neoformans* and *C. neoformans* var. *grubii*. The clinical relevance of these AD hybrids is clearly demonstrated by their incidence in some parts of the world. In a European study covering 311 cryptococcal isolates collected during 30 months, 30% of the isolates belonged to this hybrid. A high incidence of serotype AD hybrid strains was found in Portugal (50%), Greece (48%), and Spain (45%) (122). These AD hybrids will be treated in a separate chapter (see chapter 25) and are not further addressed here. Recently, some *C. neoformans–C. gattii* hybrids were described that originated from clinical material (16, 17). In a retrospective study of Dutch *C. neoformans* isolates, three anomalous isolates were observed that did not match any of the known AFLP genotypes. One isolate was obtained in 1977 from a 23-year-old male who was treated for an apparent brain tumor and died during treatment. It is unknown whether this patient was immunocompromised, but it is not likely that he suffered from an HIV infection, because the patient was treated before the onset of the AIDS epidemic. The other two isolates originated from a 35-year-old male patient in February and May 2001, respectively. Further study of these three isolates using flow cytometry; light microscopy; canavanine-glycine-bromothymol blue medium; serology; AFLP; cloning and sequencing of ITS, IGS, *LAC*, and *RPB2*; Luminex genotype analysis; and mating- and serotype-specific PCRs revealed that the isolates are hybrids between *C. neoformans* (serotype D, VNIV/AFLP genotype 2) and *C. gattii* (serotype B, VGI/AFLP genotype 4) that were designated as BD hybrids (16).

Another *C. neoformans–C. gattii* hybrid was found during reinvestigation of an atypical serotype B isolate of *C. gattii* from Canada. The isolate was obtained from a 31-year-old Canadian AIDS patient who traveled to Mexico approximately 15 months before cryptococcosis was diagnosed. Unfortunately, the patient died despite extensive treatment. The isolate was found to be monokaryotic, diploid or aneuploid, and the reinvestigation demonstrated the serotype as AB. AFLP and extensive sequence analysis showed that the isolate contained alleles of both *C. neoformans* var. *grubii* (VNI/AFLP1) and *C. gattii* (VGI/AFLP4). Therefore, it was concluded that the isolate represented another hybrid between *C. neoformans*, namely *C. neoformans* var. *grubii* (serotype A-VNI/AFLP1) and *C. gattii* (serotype B-VGI/AFLP4) and was designated as an AB hybrid (17).

EMERGING PATHOGENIC SPECIES OF *CRYPTOCOCCUS*

In addition to *Cryptococcus neoformans* (both varieties *neoformans* and *grubii*) and *C. gattii*, which are important human and animal pathogens, there are approximately 15 other members of this genus, referred to as non-*neoformans*, which have appeared as human clinical isolates with varying frequency (Fig. 5 and 6). These are *C. adeliensis*, *C. albidosimilis*, *C. albidus*, *C. curvatus*, *C. diffluens*, *C. flavescens*, *C. gastricus*, *C. humicola*, *C. laurentii*, *C. liquefaciens*, *C. luteolus* (recently reclassified as *Hannaella luteola*), *C. macerans* (= *Cystofilobasidium macerans*), *C. magnus*, *C. saitoi*, *C. uzbekistanensis*, and *C. vishniacii* (41). Moreover, some teleomorph species

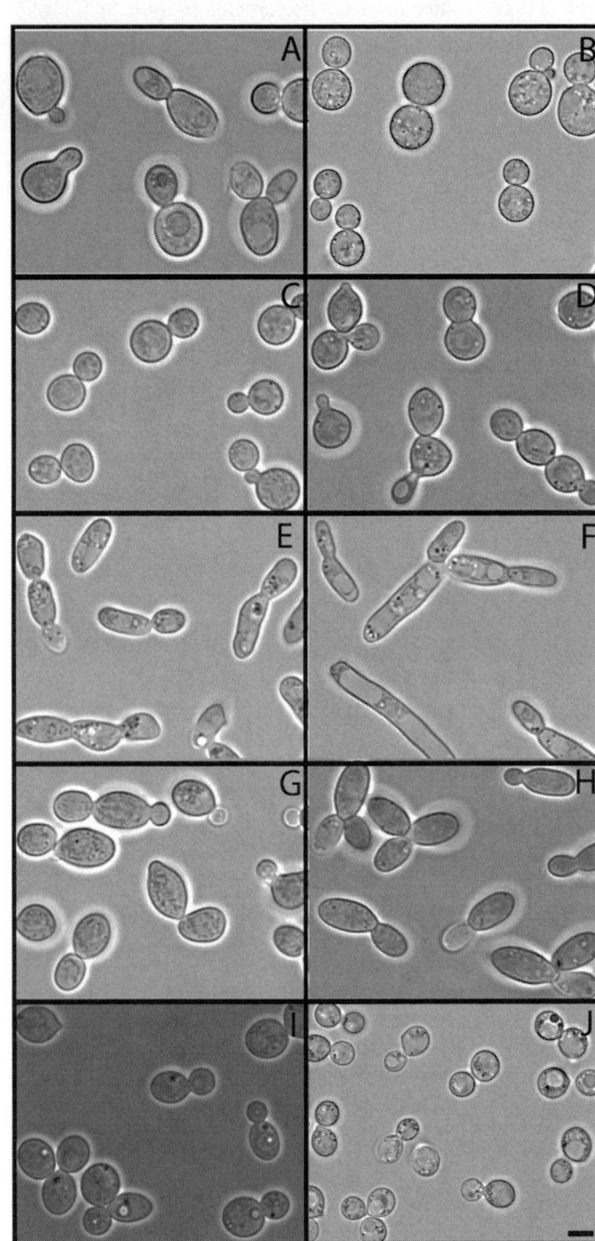

FIGURE 6 Yeast morphology of diverse non-*neoformans* species of *Cryptococcus* that may occur in clinical samples. Cells were grown in yeast nitrogen base with 2% glucose for 48 h at 25°C. (A) *C. laurentii* CBS 139; (B) *C. magnus* CBS 140; (C) *C. albidus* CBS 142; (D) *C. aerius* CBS 155; (E) *C. curvatus* CBS 570; (F) *C. humicola* CBS 571; (G) *C. luteolus* CBS 943; (H) *C. macerans* CBS 2206; (I) *C. gastricus* CBS 2288; (J) *C. gattii* CBS 6955. Bar = 5 μm.

(e.g., *F. uniguttulatum, Cystofilobasidium macerans*) with a cryptococcal anamorph (viz., *C. uniguttulatus* and *C. macerans*, respectively) are known from clinical sources as well. Infections from these species are usually acquired from the environment, with a variety of sources serving as potential reservoirs. These sources include birds (excreta), food (cheese, fruit), soil, plants (trees), and water (41, 62). Most or all of these species produce the hallmark virulence factors, capsule and melanin, which are associated with *C. neoformans* (54, 92); however, laccase activity (required for melanin production) was found to be lower in the non-*neoformans* species studied (54).

C. laurentii and *C. albidus*, which are recovered at a high enough frequency to be considered opportunistic pathogens of humans, comprise about 80% of the non-*neoformans* cases (60). The remaining species, which have been rarely reported, likely only cause infections in extreme cases. However, it is reasonable to expect continued recovery of these rare isolates due to the increasing application of aggressive treatment regimens that lead to more severe and/or prolonged immunosuppression and due to increasingly sensitive identification methods. Nevertheless, the other non-*neoformans* species, such as *C. adeliensis, C. albidosimilis, C. curvatus, C. humicola, C. luteolus, C. macerans* (*Cystofilobasidium macerans*), and *C. uniguttulatus* (*Filobasidium uniguttulatum*), can be considered nonpathogenic, with their isolation attributed to noninvasive colonization or extreme host debilitation in cases of invasive infections. It is interesting to note that *C. adeliensis*, which occurs in cold habitats such as on decaying algae in Antarctica or in glacier water or melt water (22, 29), is also present in immunocompromised human hosts, bird droppings, and sheep feces (98, 115, 119). In spite of the rarity of *C. laurentii* and *C. albidus* in the clinical laboratory, patients with decreased cell-mediated immunity can be at risk for fatal infections, which should be considered when clinical specimens yield encapsulated non-*neoformans*, basidiomycetous yeasts. We must keep in mind, however, that the recent taxonomic revision of these species has demonstrated that *C. albidus* and *C. laurentii* represent species complexes (42, 114), and consequently, the older reports on the occurrence of these species in clinical sources should be interpreted with caution.

The accurate identification of species is currently a pressing issue in medical mycology and extends to a variety of fungi (3), as these organisms can display significant clinical characteristics (e.g., different drug susceptibilities) that affect treatment choices. In the case of *C. albidus*, approximately 70–100 clinical isolates have been reported in the literature, with most strains recovered from debilitated patients and from diverse sites in the body including lungs, blood, cerebrospinal fluid, and skin (18, 28, 31, 52, 60, 84, 89, 94, 97). Most of these isolates are not maintained in one of the main public microbial repositories, and therefore, their identity remains doubtful, as most were identified by nonmolecular methods. *C. albidus* sensu lato was found to contain 12 species, of which *C. aerius, C. albidus* sensu stricto, *C. diffluens, C. liquefaciens, C. saitoi*, and *C. uzbekistanensis* have been reported from clinical materials (41). Therefore, it is unknown whether or not one of these new species constitutes the majority of reported *C. albidus* isolates from clinical specimens. Unfortunately, the conclusive identification of *C. albidus* with conventional biochemical and physiological assays, which are routinely used in a clinical laboratory, can be difficult or even impossible. Serological assays may also not be sensitive enough to identify *C. albidus*. Antigenic analysis of *C. albidus* showed that some isolates are indistinguishable from *C. neoformans* serotype A (55).

CONCLUSIONS

During the first 60 years since the first diagnosis of cryptococcosis, nomenclature of the disease and the taxonomic classification of its etiology were chaotic and have undergone numerous revisions. By 1950 *C. neoformans* was accepted as the sole etiologic agent of cryptococcosis. With the discovery of the two sexual states in *C. neoformans* with typical basidiomycetous features, the etiologic agent of cryptococcosis was formally placed in the phylum Basidiomycota in the mid 1970s. Molecular analysis of rDNA confirmed the agent of cryptococcosis to be a complex species. Currently, two etiologic agents, *C. neoformans* and *C. gattii*, are firmly established in the order Tremellales within Basidiomycota. Phylogenetic studies based on DNA sequence revealed both species, *C. neoformans* and *C. gattii*, to be highly complex, and genetic analysis between the intraspecific groups suggested that they are reproductively isolated. These two species appear to have evolved into six genetically distinct subspecies. Proper taxonomic treatment for these cryptic species will be forthcoming.

Cryptococcus species other than the *C. neoformans*–*C. gattii* complex, such as *C. albidus* and *C. laurentii*, are increasingly isolated from various patients at high risk for infectious diseases. DNA sequence-based identification has shown that the species traditionally identified as *C. albidus* and *C. laurentii* are also polyphyletic, containing more than one species. Consequently the identification of clinically important species of *Cryptococcus*, whether rare or not, has potential to be confusing until the taxonomic and nomenclatural issues are stabilized. Issues of proper fungal nomenclature continue to compound the identification of even the common species and varieties of *Cryptococcus* (*neoformans* and *gattii*) since these species have teleomorphs. However, current mycological instruction to medical students rarely insists on using the teleomorphic name when describing the causative agent of cryptococcosis. Therefore, continued investigation of this genus using molecular, genetic, and other methods to characterize these fungi needs to proceed and should go hand in hand with classical mycological input.

K. J. Kwon-Chung was supported by funds from the intramural program of the National Institute of Allergy and Infectious Diseases, National Institutes of Health, and B. L. Wickes is supported by Grant No. PR054228 from the U.S. Army Medical Research and Material Command, Office of Congressionally Directed Medical Research Programs. J. W. Fell was supported, in part, by the National Science Foundation Biotic Surveys and Inventories Grant No. DEB 0206521. We thank Bart Theelen for his help in preparation of the phylogenetic tree and Ferry Hagen for helpful discussions.

REFERENCES

1. [Reference deleted.]
2. **Aulakh, H. S., S. E. Straus, and K. J. Kwon-Chung.** 1981. Genetic relatedness of *Filobasidiella neoformans* (*Cryptococcus neoformans*) and *Filobasidiella bacillispora* (*Cryptococcus bacillisporus*). *Int. J. Syst. Bacteriol.* **31:**97–103.
3. **Balajee, S. A., A. M. Borman, M. E. Brandt, J. Cano, M. Cuenca-Estrella, E. Dannaoui, J. Guarro, G. Haase, C. C. Kibbler, W. Meyer, K. O'Donnell, C. A. Petti, J. L. Rodriguez-Tudela, D. Sutton, A. Velegraki, and B. L. Wickes.** 2008. Sequence-based identification of *Aspergillus*, *Fusarium* and *Mucorales* in the clinical mycology laboratory: where are we and where should we go from here? *J. Clin. Microbiol.* doi:10.1128/JCM.01685-08.

4. **Bandoni, R. J.** 1995. Dimorphic heterobasidiomycetes: taxonomy and parasitism. *Stud. Mycol.* **38:**13–27.
5. **Barreto de Oliviera, M. T., T. Boekhout, B. Theelen, F. Hagen, F. A. Baroni, M. S. Lazera, K. B. Lengler, J. Heitman, I. N. Revera, and C. R. Paula.** 2004. *Cryptococcus neoformans* shows a remarkable genotypic diversity in Brazil. *J. Clin. Microbiol.* **42:**1356–1359.
6. **Bartlett, K. H., S. E. Kidd, and J. W. Kronstad.** 2008. The emergence of *Cryptococcus gattii* in British Columbia and the Pacific Northwest. *Curr. Infect. Dis. Rep.* **10:**58–65.
7. **Benham, R. W.** 1935. Cryptococci: their identification by morphology and serology. *J. Infect. Dis.* **57:**255–274.
8. **Benham, R. W.** 1950. Cryptococcosis and blastomycosis. *Ann. NY Acad. Sci.* **50:**1299–1314.
9. **Biswas, S. K., L. Wang, K. Yokoyama, and K. Nishimura.** 2003. Molecular analysis of *Cryptococcus neoformans* mitochondrial cytochrome b gene sequences. *J. Clin. Microbiol.* **41:**5572–5576.
10. **Boekhout, T.** 2005. Biodiversity: gut feeling for yeasts. *Nature* **434:**449–451.
11. **Boekhout, T., R. J. Bandoni, J. W. Fell, and K. J. Kwon-Chung.** 1998. Discussion of teleomorphic and anamorphic genera of heterobasidiomycetous yeasts, p. 609–625. *In* C. P. Kurtzman and J. W. Fell (ed.), *The Yeasts: a Taxonomic Study*, 4th ed. Elsevier, Amsterdam, The Netherlands.
12. **Boekhout, T., A. Fonseca, J. P. Sampaio, R. J. Bandoni, J. W. Fell, and K. J. Kwon-Chung.** 2010. Discussion of teleomorphic and anamorphic basidiomycetous yeasts, p. 1339–1372. *In* C. P. Kurtzman, J. W. Fell, and T. Boekhout (ed.), *The Yeasts: a Taxonomic Study*, 5th ed. Elsevier, Amsterdam, The Netherlands.
13. **Boekhout, T., B. Theelen, M. R. Diaz, J. W. Fell, W. C. Hop, E. C. Abeln, F. Dromer, and W. Meyer.** 2001. Hybrid genotypes in the pathogenic yeast *Cryptococcus neoformans*. *Microbiology* **147:**891–907.
14. **Bovers, M., F. Hagen, and T. Boekhout.** 2008. Diversity of the *Cryptococcus neoformans-Cryptococcus gattii* species complex. *Rev. Iberoam. Micol.* **25:**S4–12.
15. **Bovers, M., F. Hagen, E. E. Kuramae, and T. Boekhout.** 2008. Six monophyletic lineages identified within *Cryptococcus neoformans* and *Cryptococcus gattii* by multi-locus sequence typing. *Fungal Genet. Biol.* **45:**400–421.
16. **Bovers, M., F. Hagen, E. E. Kuramae, M. R. Diaz, L. Spanjaard, F. Dromer, H. L. Hoogveld, and T. Boekhout.** 2006. Unique hybrids between the fungal pathogens *Cryptococcus neoformans* and *Cryptococcus gattii*. *FEMS Yeast Res.* **6:**599–607.
17. **Bovers, M., F. Hagen, E. E. Kuramae, H. L. Hoogveld, F. Dromer, G. St-Germain, and T. Boekhout.** 2008. AIDS patient death caused by novel *Cryptococcus neoformans* x *C. gattii* hybrid. *Emerg. Infect. Dis.* **14:**1105–1108.
18. **Burnik, C., N. D. Altintas, G. Ozkaya, T. Serter, Z. T. Selcuk, P. Firat, S. Arikan, M. Cuenca-Estrella, and A. Topeli.** 2007. Acute respiratory distress syndrome due to *Cryptococcus albidus* pneumonia: case report and review of the literature. *Med. Mycol.* **45:**469–473.
19. **Buschke, A.** 1895. Über eine durch coccidien hervorgerufene Krankheit des Menschen. *Dtsch. Med. Wochenschr.* **21:**14.
20. **Busse, O.** 1894. Über parasitäre Zelleinschlüsse und ihre Züchtung. *Zentralbl. Bakteriol.* **16:**175–180.
21. **Busse, O.** 1895. Über Saccharomycosis hominis. *Virchows Arch. A.* **140:**23–46.
22. **Butinar, L., I. Spencer-Martins, and N. Gunde-Cimerman.** 2007. Yeasts in high Arctic glaciers: the discovery of a new habitat for eukaryotic microorganisms. *Antonie Van Leeuwenhoek* **91:**277–289.
23. **Butler, M. I., and R. T. Poulter.** 2005. The PRP8 inteins are a source of phylogenetic and epidemiological information. *Fungal Genet. Biol.* **42:**452–463.
24. **Byrnes, E. J., III, R. J. Bildfell, P. L. Dearing, B. A. Valentine, and J. Heitman.** 2009. *Cryptococcus gattii* with bimorphic colony types in a dog in western Oregon: additional evidence for expansion of the Vancouver Island outbreak. *J. Vet. Diagn. Invest.* **21:**133–136.
25. **Cassidy, J. R., D. Moor, B. C. Lu, and P. J. Pukkila.** 1984. Unusual organization and lack of recombination in the ribosomal RNA genes of *Coprinus cinereus*. *Curr. Genet.* **8:**607–613.
26. **Chaturvedi, V., J. Fan, B. Stein, M. J. Behr, W. A. Samsonoff, B. L. Wickes, and S. Chaturvedi.** 2002. Molecular genetic analyses of mating pheromones reveal intervariety mating or hybridization in *Cryptococcus neoformans*. *Infect. Immun.* **70:**5225–5235.
27. **Curtis, F.** 1896. Contribution a l'etude de la saccharomycose humaine. *Ann. Inst. Pasteur* **10:**449–468.
28. **de Castro, L. E., O. A. Sarraf, J. M. Lally, H. P. Sandoval, K. D. Solomon, and D. T. Vroman.** 2005. *Cryptococcus albidus* keratitis after corneal transplantation. *Cornea* **24:**882–883.
29. **de Garcia, V., S. Brizzio, D. Libkind, P. Buzzini, and M. van Broock.** 2007. Biodiversity of cold-adapted yeasts from glacial meltwater rivers in Patagonia, Argentina. *FEMS Microbiol. Ecol.* **59:**331–341.
30. **Diaz, M. R., T. Boekhout, T. Kiesling, and J. W. Fell.** 2005. Comparative analysis of the intergenic spacer regions and population structure of the species complex of the pathogenic yeast *Cryptococcus neoformans*. *FEMS Yeast Res.* **12:**1129–1140.
31. **Drancourt, M., P. Berger, C. Terrada, B. Bodaghi, J. Conrath, D. Raoult, and P. LeHoang.** 2008. High prevalence of fastidious bacteria in 1520 cases of uveitis of unknown etiology. *Medicine* (Baltimore) **87:**167–176.
32. **Dufait, R., R. Velho, and C. De Vroey.** 1987. Rapid identification of the two varieties of *Cryptococcus neoformans* by D-proline assimilation. *Mykosen* **30:**483.
33. **Dutilh, B. E., V. van Noort, R. T. van der Heijden, T. Boekhout, B. Snel, and M. A. Huynen.** 2007. Assessment of phylogenomic and orthology approaches for phylogenetic inference. *Bioinformatics* **23:**815–824.
34. **Ellis, M., M. Richardson, and B. de Pauw.** 2000. Epidemiology. *Hosp. Med.* **61:**605–609.
35. **Emmons, C. W.** 1951. Isolation of *Cryptococcus neoformans* from soil. *J. Bacteriol.* **62:**685–690.
36. **Evans, E. E.** 1949. An immunologic comparison of twelve strains of *Cryptococcus neoformans* (*Torula histolitica*). *Proc. Soc. Exp. Biol. Med.* **71:**644–646.
37. **Fell, J. W., T. Boekhout, A. Fonseca, G. Scorzetti, and A. Statzell-Tallman.** 2000. Biodiversity and systematics of basidiopmycetous yeasts as determined by large subunit rDNA D1/D2/domain sequence analysis. *Int. J. Syst. Bacteriol.* **50:**1352–1371.
38. **Fell, J. W., C. P. Kurtzman, and K. J. Kwon-Chung.** 1989. Proposal to conserve *Cryptococcus*. *Taxon* **38:** 151–152.
39. **Findley, K., M. Rodriguez-Carres, B. Metin, J. Kroiss, A. Fonseca, R. Vilgalys, and J. Heitman.** 2009. Phylogeny and phenotypic characterization of pathogenic *Cryptococcus* species and closely related saprobic taxa in the Tremellales. *Eukaryot Cell.* **8:**353–361.
40. **Fitzpatrick, D. A., M. E. Logue, J. E. Stajich, and G. Butler.** 2006. A fungal phylogeny based on 42 complete genomes derived from supertree and combined gene analysis. *BMC Evol. Biol.* **6:**99.
41. **Fonseca, A., J. W. Fell, and T. Boekhout.** 2010. *Cryptococcus* Vuillemin, p. 1665–1740. *In* C. P. Kurtzman, J. W. Fell, and T. Boekhout (ed.), *The Yeasts: a Taxonomic Study*, 5th ed. Elsevier, Amsterdam, The Netherlands.
42. **Fonseca, A., G. Scorzetti, and J. W. Fell.** 2000. Diversity in the yeast *Cryptococcus albidus* and related species as

revealed by ribosomal DNA sequence analysis. *Can. J. Microbiol.* **46:**7–27.

43. **Franzot, S. P., I. F. Salkin, and A. Casadevall.** 1999. *Cryptococcus neoformans* var. *grubii*: separate varietal status for *Cryptococcus neoformans* serotype A isolates. *J. Clin. Microbiol.* **37:**838–840.
44. **Fraser, J. A., S. S. Giles, E. C. Wenink, S. G. Geunes-Boyer, J. R. Wright, S. Diezmann, A. Allen, J. E. Stajich, F. S. Dietrich, J. R. Perfect, and J. Heitman.** 2005. Same-sex mating and the origin of the Vancouver Island *Cryptococcus gattii* outbreak. *Nature* **437:**1360–1364.
45. **Freeman, W. J.** 1931. *Torula* infection of the central nervous system. *J. Psychol. Neurol.* **43:**236.
46. **Frothingham, L. A.** 1902. A tumor-like lesion in the lung of a horse caused by a blastomyces (torula). *J. Med Res.* **3:**31–43.
47. **Gatti, E., and R. Eeckels.** 1970. An atypical strain of *Cryptococcus neoformans* (San Felice) Vuillemin 1984. Part 1. Description of the disease and of the strain. *Ann. Soc. Belg. Med. Trop.* **50:**689–694.
48. **Ginns, J., and D. W. Malloch.** 2003. *Filobasidiella depauperata* (Tremellales): Haustorial branches and parasitism of *Verticillium lecanii*. *Mycol. Progr.* **2:**137–140.
49. **Greuter, W., F. R. Barrie, H. M. Burdet, W. G. Chaloner, V. Demoulin, D. L. Hawksworth, P. M. Jorgensen, D. H. Nicolson, P. C. Silva, P. Trehane, and J. McNeil (ed).** 1994. *International Code of Botanical Nomenclature*. Koeltz Scientific Books, Konigstein, Germany.
50. **Gueho, E., L. Improvisi, R. Christen, and G. S. de Hoog.** 1993. Phylogenetic relationships of *Cryptococcus neoformans* and some related basidiomycetous yeasts determined from partial large subunit rRNA sequences. *Antonie Van Leeuwenhoek* **63:**175–189.
51. **Hibbett, D. S., M. Binder, J. F. Bischoff, M. Blackwell, P. F. Cannon, O. E. Eriksson, S. Huhndorf, T. James, P. M. Kirk, R. Lucking, H. T. Lumbsch, F. Lutzoni, P. B. Matheny, D. J. McLaughlin, M. J. Powell, S. Redhead, C. L. Schoch, J. W. Spatafora, J. A. Stalpers, R. Vilgalys, M. C. Aime, A. Aptroot, R. Bauer, D. Begerow, G. L. Benny, L. A. Castlebury, P. W. Crous, Y. C. Dai, W. Gams, D. M. Geiser, G. W. Griffith, C. Gueidan, D. L. Hawksworth, G. Hestmark, K. Hosaka, R. A. Humber, K. D. Hyde, J. E. Ironside, U. Koljalg, C. P. Kurtzman, K. H. Larsson, R. Lichtwardt, J. Longcore, J. Miadlikowska, A. Miller, J. M. Moncalvo, S. Mozley-Standridge, F. Oberwinkler, E. Parmasto, V. Reeb, J. D. Rogers, C. Roux, L. Ryvarden, J. P. Sampaio, A. Schussler, J. Sugiyama, R. G. Thorn, L. Tibell, W. A. Untereiner, C. Walker, Z. Wang, A. Weir, M. Weiss, M. M. White, K. Winka, Y. J. Yao, and N. Zhang.** 2007. A higher-level phylogenetic classification of the Fungi. *Mycol. Res.* **111:**509–547.
52. **Hoang, J. K., and J. Burruss.** 2007. Localized cutaneous *Cryptococcus albidus* infection in a 14-year-old boy on etanercept therapy. *Pediatr. Dermatol.* **24:**285–288.
53. **Hu, G., I. Liu, A. Sham, J. E. Stajich, F. S. Dietrich, and J. W. Kronstad.** 2008. Comparative hybridization reveals extensive genome variation in the AIDS-associated pathogen *Cryptococcus neoformans*. *Genome Biol.* **9:**R41.
54. **Ikeda, R., T. Sugita, E. S. Jacobson, and T. Shinoda.** 2002. Laccase and melanization in clinically important *Cryptococcus* species other than *Cryptococcus neoformans*. *J. Clin. Microbiol.* **40:**1214–1218.
55. **Ikeda, R., T. Sugita, and T. Shinoda.** 2002. Serologic relationships of *Cryptococcus* spp.: distribution of antigenic factors in *Cryptococcus* and intraspecies diversity. *J. Clin. Microbiol.* **38:**4021–4025.
56. **James, T. Y., F. Kauff, C. L. Schoch, P. B. Matheny, V. Hofstetter, C. J. Cox, G. Celio, C. Gueidan, E. Fraker, J. Miadlikowska, H. T. Lumbsch, A. Rauhut, V. Reeb, A. E. Arnold, A. Amtoft, J. E. Stajich, K. Hosalka, G. H. Sung, D. Johnson, B. O'Rourke, M. Crockett, M. Binder, J. M. Curtis, J. C. Slot, Z. Wang, A. W. Wilson, A. Schüßler, J. E. Longcore, K. O'Donnell, S. Mozley-Standridge, D. Porter, P. M. Letcher, M. J. Powell, J. W. Taylor, M. M. White, G. W. Griffith, D. R. Davies, R. A. Humber, J. B. Morton, J. Sugiyama, A. Y. Rossman, J. D. Rogers, D. H. Pfister, D. Hewitt, K. Hansen, S. Hambleton, R. A. Shoemaker, J. Kohlmeyer, B. Volkmann-Kohlmeyer, R. A. Spotts, M. Serdani, P. W. Crous, K. W. Hughes, K. Matsuura, E. Langer, G. Langer, W. A. Untereiner, R. Lücking, B. Büdel, D. M. Geiser, A. Aptroot, P. Diederich, I. Schmitt, M. Schultz, R. Yahr, D. S. Hibbett, F. Lutzoni, D. J. McLaughlin, J. W. Spatafora, and R. Vilgalys.** 2006. Reconstructing the early evolution of fungi using a six-gene phylogeny. *Nature* **443:**758–761.
57. **Julich, W.** 1983. Parasitic heterobasidiomycetes on other fungi. *Int. J. Mycol. Lichen* **1:**189–203.
58. **Katsu, M., S. Kidd, A. Ando, M. L. Moretti-Branchini, Y. Mikami, K. Nishimura, and W. Meyer.** 2004. The internal transcribed spacers and 5.8S rRNA gene show extensive diversity among isolates of the *Cryptococcus neoformans* species complex. *FEMS Yeast Res.* **4:**377–388.
59. **Kavanaugh, L. A., J. A. Fraser, and F. S. Dietrich.** 2006. Recent evolution of the human pathogen *Cryptococcus neoformans* by intervarietal transfer of a 14-gene fragment. *Mol. Biol. Evol.* **23:**1879–1890.
60. **Khawcharoenporn, T., A. Apisarnthanarak, and L. M. Mundy.** 2007. Non-*neoformans* cryptococcal infections: a systematic review. *Infection* **35:**51–58.
61. **Kidd, S. E., H. Guo, K. H. Bartlett, J. Xu, and J. W. Kronstad.** 2005. Comparative gene genealogies indicate that two clonal lineages of *Cryptococcus gattii* in British Columbia resemble strains from other geographical areas. *Eukaryot Cell* **4:**1629–1638.
62. **Kidd, S. E., F. Hagen, R. L. Tscharke, M. Huynh, K. H. Bartlett, M. Fyfe, L. Macdougall, T. Boekhout, K. J. Kwon-Chung, and W. Meyer.** 2004. A rare genotype of *Cryptococcus gattii* caused the cryptococcosis outbreak on Vancouver Island (British Columbia, Canada). *Proc. Natl. Acad. Sci. USA* **101:**17258–17263.
63. **Kuramae, E. E., V. Robert, B. Snel, M. Weiss, and T. Boekhout.** 2006. Phylogenomics reveal a robust fungal tree of life. *FEMS Yeast Res.* **6:**1213–1220.
64. **Kützing, F.** 1833. Systematische Zusammenstellung der neiden Algen-gattungen und Arten. *Linneae* **8:**365.
65. **Kwon-Chung, K. J.** 2010. *Filobasidiella* Kwon-Chung, p. 1443–1455. *In* C. P. Kurtzman, J. W. Fell, and T. Boekhout (ed.), *The Yeasts: a Taxonomic Study*, 5th ed. Elsevier, Amsterdam, The Netherlands.
66. **Kwon-Chung, K. J.** 2010. *Filobasidium* Olive, p. 1457–1465. *In* C. P. Kurtzman, J. W. Fell, and T. Boekhout (ed.), *The Yeasts: a Taxonomic Study*, 5th ed. Elsevier, Amsterdam, The Netherlands.
67. **Kwon-Chung, K. J.** 1976. Morphogenesis of *Filobasidiella neoformans*, the sexual state of *Cryptococcus neoformans*. *Mycologia* **68:**821–833.
68. **Kwon-Chung, K. J.** 1975. A new genus, *Filobasidiella*, the perfect state of *Cryptococcus neoformans*. *Mycologia* **67:**1197–1200.
69. **Kwon-Chung, K. J.** 1976. A new species of *Filobasidiella*, the sexual state of *Cryptococcus neoformans* B and C serotypes. *Mycologia* **68:**943–946.
70. **Kwon-Chung, K. J.** 1977. Perfect state of *Cryptococcus uniguttulatus*. *Int. J. Syst. Bacteriol.* **27:**293–299.
71. **Kwon-Chung, K. J., and J. E. Bennett.** 1984. Epidemiologic differences between the two varieties of *Cryptococcus neoformans*. *Am. J. Epidemiol.* **120:**123–130.

72. **Kwon-Chung, K. J., J. E. Bennett, and T. S. Theodore.** 1978. *Cryptococcus bacillisporus* sp. nov.: serotype B-C of *Cryptococcus neoformans*. *Int. J. Syst. Bacteriol.* **28:**616–620.
73. **Kwon-Chung, K. J., T. Boekhout, J. W. Fell, and M. Diaz.** 2002. Proposal to conserve the name *Cryptococcus gattii* against *C. hondurianus* and *C. bacillisporus* (*Basidiomycota, Hymenomycetes, Tremellomycetidae*). *Taxon* **51:**804–806.
74. **Kwon-Chung, K. J., and Y. C. Chang.** 1994. Gene arrangement and sequence of the 5S rRNA in *Filobasidiella neoformans* (*Cryptococcus neoformans*) as a phylogenetic indicator. *Int. J. Syst. Bacteriol.* **44:**209–213.
75. **Kwon-Chung, K. J., Y. C. Chang, R. Bauer, E. C. Swann, J. W. Taylor, and R. Goel.** 1995. The characteristics that differentiate *Filobasidiella depauperata* from *Filobasidiella neoformans*. *Stud. Mycol.* **38:**67–79.
76. **Kwon-Chung, K. J., I. Polacheck, and J. E. Bennett.** 1982. Improved diagnostic medium for separation of *Cryptococcus neoformans* var. *neoformans* (serotypes A and D) and *Cryptococcus neoformans* var. *gattii* (serotypes B and C). *J. Clin. Microbiol.* **15:**535–537.
76a. **Kwon-Chung, K. J., and A. Varma.** 2006. Do major species concepts support one, two or more species within *Cryptococcus neoformans*? *FEMS Yeast Res.* **6:**574–587.
77. **Kwon-Chung, K. J., B. L. Wickes, J. L. Booth, H. S. Vishniac, and J. E. Bennett.** 1987. Urease inhibition by EDTA in the two varieties of *Cryptococcus neoformans*. *Infect. Immun.* **55:**1751–1754.
78. **Kwon-Chung, K. J., and J. E. Bennett.** 1992. *Medical Mycology*, p. 397–446. Lea & Febiger, Philadelphia, PA.
79. **Latouche, G. N., M. Huynh, T. C. Sorrell, and W. Meyer.** 2003. PCR-restriction fragment length polymorphism analysis of the phospholipase B (PLB1) gene for subtyping of *Cryptococcus neoformans* isolates. *Appl. Environ. Microbiol.* **69:**2080–2086.
80. **Litvintseva, A. P., R. Thakur, L. B. Reller, and T. G. Mitchell.** 2005. Prevalence of clinical isolates of *Cryptococcus gattii* serotype C among patients with AIDS in sub-Saharan Africa. *J. Infect. Dis.* **192:**888–892.
81. **Litvintseva, A. P., R. Thakur, R. Vilgalys, and T. G. Mitchell.** 2006. Multilocus sequence typing reveals three genetic subpopulations of *Cryptococcus neoformans* var. *grubii* (serotype A), including a unique population in Botswana. *Genetics* **172:**2223–2238.
82. **Lodder, J., and N. J. W. Kreger-Van Rij.** 1952. *The Yeasts: a Taxonomic Study*. North Holland, Amsterdam, The Netherlands.
83. **London, R., B. S. Orozco, and E. Mylonakis.** 2006. The pursuit of cryptococcal pathogenesis: heterologous hosts and the study of cryptococcal host-pathogen interactions. *FEMS Yeast Res.* **6:**567–573.
84. **Manzano-Gayosso, P., F. Hernandez-Hernandez, L. J. Mendez-Tovar, Y. Palacios-Morales, E. Cordova-Martinez, E. Bazan-Mora, and R. Lopez-Martinez.** 2008. Onychomycosis incidence in type 2 diabetes mellitus patients. *Mycopathologia* **166:**41–45.
85. **Marcet-Houben, M., and T. Gabaldon.** 2009. The tree versus the forest: the fungal tree of life and the topological diversity within the yeast phylome. *PLoS ONE* **4:**e4357.
86. **Meyer, W., A. Castaneda, S. Jackson, M. Huynh, and E. Castaneda.** 2003. Molecular typing of IberoAmerican *Cryptococcus neoformans* isolates. *Emerg. Infect. Dis.* **9:**189–195.
87. **Miller, W. G., A. A. Padhye, W. van Bonn, E. Jensen, M. E. Brandt, and S. H. Ridgway.** 2002. Cryptococcosis in a bottlenose dolphin (*Tursiops truncatus*) caused by *Cryptococcus neoformans* var. *gattii*. *J. Clin. Microbiol.* **40:**721–724.
88. **Min, K. H., and K. J. Kwon-Chung.** 1986. The biochemical basis for the distinction between the two *Cryptococcus neoformans* varieties with CGB medium. *Zentralbl. Bakteriol. Mikrobiol. Hyg.* A **261:**471–480.
89. **Narayan, S., K. Batta, P. Colloby, and C. Y. Tan.** 2000. Cutaneous cryptococcus infection due to *C. albidus* associated with Sezary syndrome. *Br. J. Dermatol.* **143:**632–634.
90. **Okoli, I., C. A. Oyeka, K. J. Kwon-Chung, B. Theelen, V. Robert, J. Z. Groenewald, D. C. McFadden, A. Casadevall, and T. Boekhout.** 2007. *Cryptotrichosporon anacardii* gen. nov., sp. nov., a new trichosporonoid capsulate basidiomycetous yeast from Nigeria that is able to form melanin on niger seed agar. *FEMS Yeast Res.***7:**339–350.
91. **Olive, L. S.** 1968. An unusual heterobasidiomycete with *Tilletia*-like basidia. *J. Elisha Mitchell Sc. Soc.* **84:**261–266.
92. **Perfect, J. R., and A. Casadevall.** 2002. Cryptococcosis. *Infect. Dis. Clin. North Am.* **16:**837–874.
93. **Petter, R., B. S. Kang, T. Boekhout, B. J. Davis, and K. J. Kwon-Chung.** 2001. A survey of heterobasidiomycetous yeasts for the presence of the genes homologous to virulence factors of *Filobasidiella neoformans*, *CNLAC1* and *CAP59*. *Microbiology* **147:**2029–2036.
94. **Pfaller, M. A., D. J. Diekema, D. L. Gibbs, V. A. Newell, H. Bijie, D. Dzierzanowska, N. N. Klimko, V. Letscher-Bru, M. Lisalova, K. Muehlethaler, C. Rennison, and M. Zaidi.** 2009. Results from the ARTEMIS DISK Global Antifungal Surveillance Study, 1997 to 2007: 10.5-year analysis of susceptibilities of noncandidal yeast species to fluconazole and voriconazole determined by CLSI standardized disk diffusion testing. *J. Clin. Microbiol.* **47:**117–123.
95. **Phaff, H. J., and J. W. Fell.** 1971. *Cryptococcus* Kützing emend. Phaff et Spencer, p. 1089–1145. *In* L. Lodder (ed.), *The Yeasts: a Taxonomic Study*, 2nd ed. Elsevier-North Holland, Amsterdam, The Netherlands.
96. **Phaff, H. J., and J. F. T. Spencer.** 1969. Improved parameters in the separation of species in the genera *Rhodotorula* and *Cryptococcus*, p. 59–67. *In* Proceedings of the Second International Symposium on Yeasts, Bratislava, 1966.
97. **Ramchandren, R., and D. E. Gladstone.** 2004. *Cryptococcus albidus* infection in a patient undergoing autologous progenitor cell transplant. *Transplantation* **77:**956.
98. **Rimek, D., G. Haase, A. Luck, J. Casper, and A. Podbielski.** 2004. First report of a case of meningitis caused by *Cryptococcus adeliensis* in a patient with acute myeloid leukemia. *J. Clin. Microbiol.* **42:**481–483.
99. **Roberts, P.** 1997. New Heterobasidiomycetes from Great Britain. *Mycotaxon* **63:**195–216.
100. **Rodrigues de Miranda, L.** 1984. *Cryptococcus* Kützing emend. Phaff et Spencer, p. 845–872. *In* N. J. W. Kreger-van Rij (ed.), *The Yeasts: a Taxonomic Study*, 3rd ed. Elsevier, Amsterdam, The Netherlands.
101. **Rodrigues de Miranda, L. and W. H. Batenburn-van der Vegte.** 1981. *Cryptococcus mollis* Kützing, type species of the genus *Cryptococcus*: investigation of the type material. *Antonie Van Leeuwenhoek* **47:**65–72.
102. **Ruma, P., S. C. Chen, T. C. Sorrell, and A. G. Brownlee.** 1996. Characterization of *Cryptococcus neoformans* by random DNA amplification. *Lett. Appl. Microbiol.* **23:**312–316.
103. **Sanfelice, F.** 1894. Contributo alla morfologia e biologia dei blastomiceti che si sviluppano nei succhidi alcuni frutti. *Ann. Igien* **4:**463–495.
104. **Sanfelice, F.** 1895 Über einen neuen Pathogen Blastomyceten welcher innerhalb der Gewebe unter Bildung, Kalkartig aussehender massendegeneriert. *Zentralbl. Bakteriol. Parasit. Infekt. Hyg.* **18:**521–526.
105. **Scorzetti, G., J. W. Fell, A. Fonseca, and A. Statzell-Tallman.** 2002. Systematics of basidiomycetous yeasts: a comparison of large subunit D1/D2 and internal transcribed spacer rDNA regions. *FEMS Yeast Res.* **2:**495–517.
106. **Shadomy, H. J.** 1970. Clamp connections in two strains of *Cryptococcus neoformans*. *Spectrum Monogr. Ser.* **1:**67–72.

107. **Sivakumaran, S., P. Bridge, and P. Roberts.** 2002. Genetic relatedness among *Filobasidiella* species. *Mycopathologia* **156:**157–162.
108. **Statzell-Tallman, A., C. Belloch, and J. W. Fell.** 2008. *Kwoniella mangroviensis* gen. nov., sp.nov. (Tremellales, Basidiomycota), a teleomorphic yeast from mangrove habitats in the Florida Everglades and Bahamas. *FEMS Yeast Res.* **8:**103–113.
109. **Steenbergen, J. N., J. D. Nosanchuk, S. D. Malliaris, and A. Casadevall.** 2003. *Cryptococcus neoformans* virulence is enhanced after growth in the genetically malleable host *Dictyostelium discoideum. Infect. Immun.* **71:**4862–4872.
110. **Steenbergen, J. N., H. A. Shuman, and A. Casadevall.** 2001. *Cryptococcus neoformans* interactions with amoebae suggest an explanation for its virulence and intracellular pathogenic strategy in macrophages. *Proc. Natl. Acad. Sci. USA* **98:**15245–15250.
111. **Stoddard, J. L., and E. C. Cutler.** 1916. *Torula* infection in man. *Rockefeller Inst. Med. Res. Monogr.* **6:**1–98.
112. **Storck, R., C. J. Alexopoulos, and H. J. Phaff.** 1969. Nucleotide composition of deoxyribonucleic acid of some species of *Cryptococcus*, *Rhodotorula* and *Sporobolomyces. J. Bacteriol.* **98:**1069–1072.
113. **Sugita, T., M. Takashima, R. Ikeda, T. Nakase, and T. Shinoda.** 2001. Intraspecies diversity of *Cryptococcus albidus* isolated from humans as revealed by sequences of the internal transcribed spacer regions. *Microbiol. Immunol.* **45:**291–297.
114. **Takashima, M., T. Sugita, T. Shinoda, and T. Nakase.** 2003. Three new combinations from the *Cryptococcus laurentii* complex: *Cryptococcus aureus*, *Cryptococcus carnescens* and *Cryptococcus peneaus*. *Int. J. Syst. Evol. Microbiol.* **53:**1187–1194.
115. **Tintelnot, K., and H. Losert.** 2005. Isolation of *Cryptococcus adeliensis* from clinical samples and the environment in Germany. *J. Clin. Microbiol.* **43:**1007.
116. **Todd, R. L., and W. W. Herrmann.** 1936. The life cycle of the organism causing yeast meningitis. *J. Bacteriol.* **32:**89–97.
117. **Vanbreuseghem, R., and M. Takashio.** 1970. An atypical strain of *Cryptococcus neoformans* (Sanfelice) Vuillemin, 1894, II. *Cryptococcus neoformans* var. *gattii* var.nov. *Ann. Soc. Belg. Med. Trop.* **50:**695–702.
118. **Van der Walt, J. P., D. B. Scott, and W. C. van der Klift.** 1972. Six new *Candida* species from South African insect sources. *Mycopathol. Mycol. Appl.* **47:**221–236.
119. **Velazquez, E., M. del Villar, I. Grondona, E. Monte, and T. Gonzalez-Villa.** 2006. Ultrastructural and chemotaxonomic analysis of a xylanolytic strain of *Cryptococcus adeliensis* isolated from sheep droppings in Spain. *Arch. Microbiol.* **186:**195–202.
120. **Verse, M.** 1914. Über einen Full von general isofierter Blastomycose beim Menschen. *Dtsch. Pathol. Ges.* **17:**275–278.
121. **Vilgalys, R., and D. Gonzales.** 1990. Organization of ribosomal DNA in the basidiomycete *Thanatephorus praticola. Curr. Genet.* **18:**277–280.
122. **Viviani, M. A., M. Cogliati, M. C. Esposto, K. Lemmer, K. Tintelnot, M. F. Colom Valiente, D. Swinne, A. Velegraki, and R. Velho.** 2006. Molecular analysis of 311 *Cryptococcus neoformans* isolates from a 30-month ECMM survey of cryptococcosis in Europe. *FEMS Yeast Res.* **6:**614–619.
123. **Vogel, R. A.** 1966. The indirect fluorescent antibody test for the detection of antibody in human cryptococcal disease. *J. Infect. Dis.* **116:**573–580.
123a. **von Arx, J. A., and A. C. M. Weijman**. 1979. Conidiation and carbohydrate composition in some *Candida* and *Torulopsis* species. *Antonie Van Leeuwenhoek* **45:**547–555.
124. **Vuillemin, J. P.** 1901. Les blastomycetes pathogenes. *Rev. Gen. Sci.* **12:**732–751.
125. **Wang, Q. M., and F. Y. Bai.** 2008. Molecular phylogeny of basidiomycetous yeasts in the *Cryptococcus luteolus* lineage (Tremellales) based on nuclear rRNA and mitochondrial cytochrome b gene sequence analyses: proposal of *Derxomyces* gen. nov. and *Hannaella* gen. nov., and description of eight novel *Derxomyces* species. *FEMS Yeast Res.* **8:**799–814.
126. **Weis, J. D.** 1902. Four pathogenic torulae (Blastomycetes). *J. Med. Res.* **7:**280–311.
127. **Wickes, B. L., T. D. Moore, and K. J. Kwon-Chung.** 1994. Comparison of the electrophoretic karyotypes and chromosomal location of ten genes in the two varieties of *Cryptococcus neoformans. Microbiology* **140**(Pt. 3):543–550.
128. **Wilson, D. E., J. E. Bennett, and J. W. Bailey.** 1968. Serologic grouping of *Cryptococcus neoformans*. *Proc. Soc. Exp. Biol. Med.* **127:**820–823.
129. **Xu, J., R. Vilgalys, and T. G. Mitchell.** 2000. Multiple gene genealogies reveal recent dispersion and hybridization in the human pathogenic fungus *Cryptococcus neoformans*. *Mol. Ecol.* **9:**1471–1481.
130. **Zang, M.** 1999. A new taxon, *Filobasidiella xianghuijun*. *Edible Fung. China* **18:**43–44.

Cryptococcus: From Human Pathogen to Model Yeast
Edited by J. Heitman et al.

2

The History of *Cryptococcus* and Cryptococcosis

JOHN R. PERFECT AND ARTURO CASADEVALL

THE NAMING OF A FUNGAL PATHOGEN

This chapter will take into account over a century of clinical experience to examine the encapsulated yeast *Cryptococcus*, to identify the highlights of its maturation in identifying, understanding, diagnosing, and managing this major emerging fungal pathogen. We have previously provided a detailed description of the first century of clinical cryptococcosis with a series of comprehensive historical facts, and we encourage readers to review in detail this early history of cryptococcosis (11). In this historical review, we will attempt to concisely show where the pathogenic yeast has been and where it is going in clinical medicine and basic research.

The genus *Cryptococcus*, from the Greek word *kryptus*, meaning "hidden," was first created by Kutzing in 1833 for a group of yeasts that lacked the ability to produce endospores (44). This general categorization would eventually plague the early nomenclature of this poorly defined group of yeasts. It was in 1894–1895 that the yeast received its first clinical description of disease and the nomenclatural battle for identifying the yeast began. Just before the start of the 20th century, the medical history of this sugar-coated yeast began its journey as a clinically described human pathogen with fatal consequences. In 1894, the German pathologist Busse observed the fungus in a sarcoma-like tibial lesion in a 31-year-old woman. Similarly, a German surgeon, Buschke, reviewed this case and described the tibial lesion along with lymphadenopathy and secondary skin lesions (8–10). It was reported that the disease had spread to the knee joint, and eventually the patient died with multiple lesions in the lungs, spleen, kidney, bones, and skin. It is ironic that this initial fatal disseminated case of cryptococcosis did not focus on the central nervous system, but the case report clearly chronicled the yeast's ability to invade multiple organs in the human host. This case vividly set the stage for the appreciation that this yeast can disseminate throughout the body and cause lethal disease. At the same time as the first clinical description of cryptococcosis, the Italian scientist Sanfelice, at the Hygiene Institute of the University of Cagliari (Sardinia), performed the first isolation of this yeast from an environmental source (78). This discovery of being able to grow the yeast from peach juice was particularly important from a nomenclatural standpoint, and Sanfelice further showed the yeast could produce disease as cancer-like tumors formed in animals at sites of direct yeast inoculations. In 1895, he named the yeast *Saccharomyces neoformans*. The name of the yeast went through a series of creative name changes, which we have previously detailed (11).

Benham improved the mycologic and taxonomic classification of cryptococcus in 1935 when she made a careful classification of 22 strains of cryptococci, in which the original cultures of Busse, Buschke, and Curtis were found to be similar, and formally placed them under a single genus, *Cryptococcus* (7). In addition, Benham showed that strains could be divided into multiple serotypes. Finally, in 1952, after a definitive taxonomic study of yeasts by Lodder and Kreger-van Rij, the fungal pathogen received its present name of *Cryptococcus neoformans* (55).

Despite this progress in identification and taxonomy, which helped stabilize the name of the yeast, understanding of its life cycle was not complete, and further studies were needed to appreciate fully the ancestry of this yeast. Thus, over a decade after its formal name acceptance, Shadomy and coworkers made the challenging observation that some strains of *C. neoformans* could produce hyphae with clamp connections (58). It was this morphologic observation that ushered in ground-breaking studies of the life cycle of this yeast by Kwon-Chung et al. in 1975–1976 (45). A series of elegant studies clearly demonstrated that this yeast has a bipolar mating system. Yeast of different mating types produced filamentous forms with clamp connections and formed basidium and basidiospores when placed in close physical proximity. These results made clear that this pathogenic yeast has a perfect or sexual stage, and the teleomorph was called *Filobasidiella neoformans*. It was also observed that some of these yeast strains produced different-shaped basidiospores, and with the prior knowledge from Benham that there were strains with four different serotypes (A–D), Kwon-Chung identified another teleomorph, *Filobasidiella bacillispora*, since in the heterothallic crosses these strains produced bacilliform-shaped basidiospores. Serotype A and D isolates formed *F. neoformans*, while serotype B and C isolates produced a teleomorph stage identified as *F. bacillispora*.

John R. Perfect, Duke University Medical Center, Durham, NC 27705. **Arturo Casadevall,** Albert Einstein College of Medicine, Bronx, NY 10461.

It then took another 25 to 30 years to propose formally that these two yeasts with differing sexual structures represented two separate species. Kwon-Chung et al. separated these major cryptococcal pathogenic yeast forms into two species (*Cryptococcus neoformans* and *Cryptococcus gattii*) in 2003 (46). While the first clinical isolate of Busse-Buschke is *C. neoformans*, in retrospect it appears that the isolate from the second reported clinical case of cryptococcosis, a myxomatous tumor-like infection of the hip in 1896 (14), was similar to an isolate from a meningitis case in Africa imported by Gatti in 1970 and named serotype B.C. by Vanbrueuseghem and Takashio (86). Therefore, it is likely that both cryptococcal species were identified to cause human disease within a similar time frame. However, it is also apparent from years of clinical experience that, both environmentally and clinically, *C. neoformans* is the most common pathogenic species.

The two species designations for the pathogenic cryptococcal complex continue to be challenged. Investigations and discussions around nomenclature persist on whether the present serotype A- and D-designated isolates should be separated into distinct species from their varietal status: *C. neoformans* var. *grubii* (serotype A) and *C. neoformans* var. *neoformans* (serotype D). There is a series of subtle molecular, physical, and clinical differences between these varieties, which were formally named varieties in 1999 (25). Molecular evolutionary studies demonstrated that these two varieties have been genetically separated from each other for approximately 19 million years (93). However, at present the serotype D and A strains are considered to be varieties: *C. neoformans* var. *neoformans* and *C. neoformans* var. *grubii*, respectively. With the AIDS pandemic, *C. neoformans* var. *grubii* is by a large margin the most common clinical variety of pathogenic cryptococcal species to cause disease. Its worldwide case numbers of clinical disease (71) even rival the disease incidence of the common fungal pathogen *Candida* if mucosal candidiasis is excluded.

YEASTS, SPORES, AND HYPHAE

Further recent insights into the complex morphologic life cycle of this basidiomycete pathogen may eventually tell us more about it and its disease. In 1998, Wickes et al. demonstrated that cryptococcus could demonstrate morphological flexibility in that under certain environmental conditions the yeasts produce hyphae and spores without a mating reaction; in other words, some cryptococcal strains were found to have the ability to haploid fruit, and thus a new part of the life cycle of cryptococcus was identified (92). The significance of haploid fruiting with its spore production in pathobiology and/or epidemiology for this fungus needs further study. For example, since the identification of soil and bird guano as environmental sources for *C. neoformans* (21, 22) and trees such as eucalyptus for *C. gattii* (20), the infectious propagule (yeast versus spore) in nature remains uncertain, and clearly this fungus has a complex morphology that can adapt to environmental conditions. It can even complete its sexual cycle on a plant (94). Furthermore, Lin et al. have shown that cryptococcus can perform same-sex mating (50), and recently the consequences of this same-sex ability for recombination in nature have been demonstrated in that *C. gattii* isolates appear to have used this flexibility in DNA exchange for the creation of a hypervirulent strain in the recent Vancouver and Northwest U.S. outbreak of infections (26).

Finally, even the yeast form of cryptococcus has shown pleomorphic morphology issues. Historically, several investigators appreciated the sectoring of yeast colonies (smooth, rough, mucoid, etc.) (25, 53). These observations have now been studied at a molecular and phenotypic level, and the examination includes the impact of morphological changes in the yeast on the disease (27). The ability of cryptococcus to morphologically transition is part of what makes it pathogenic. From its ability as a yeast to enhance its polysaccharide capsule in response to certain signals or within the host environment (ambient CO_2 concentration, iron limitation, or exposure to serum), to its hyphal structures with basidium and basidiospores for recombination, to its switching colony morphology, its impressive structural flexibility is critically important to this complex pathogen and the disease it produces.

CLINICAL DISEASE

There are several critical clinical landmarks in the history of this encapsulated yeast over the last century. After both *C. neoformans* and *C. gattii* were described as soft tissue infections in 1894 and 1896, respectively, Frothingham observed this yeast in the lung of a horse in 1902 (28), which established the fungus as a pulmonary pathogen and demonstrated a wider host range ability to cause disease within mammals. The description of cryptococcus as a central nervous system (CNS) pathogen was elegantly done by von Hansemann when he reported a postmortem case of meningoencephalitis with gelatinous cysts containing yeasts in the brain (31). However, it was approximately a decade later that Verse, Stoddard, and Cutler presented cases of cryptococcal meningoencephalitis diagnosed antemortem (29, 90). Consequently, the ability of this encapsulated yeast to invade the CNS and produce disease was reported in the early decades of the 1900s, but it required many years of clinical experience to appreciate the real magnitude of its CNS propensity for disease. In many of these early studies, the initial observation was based on histological analysis of affected tissues, and this led to the misconception that confused the field for several years. The early investigators postulated that the yeast produced a local reaction that seemed to lyse cells and tissue surrounding the yeast, and this incorrect histopathological interpretation even led to naming this yeast, in which this histological finding was induced in tissue and body fluids, *Torula histolytica* (91). However, this histopathological phenomenon was eventually shown to be caused by the surrounding capsule occupying space in tissue and appearing as a translucent zone around the yeast cell, a phenomenon that was further accentuated by enlargement of the capsule structure in vivo.

Littman and Zimmerman elegantly described the tracing of the early history of cryptococcosis and its knowledge base for the understanding of the pathogen during the first half century of its clinical existence in a monograph published in 1956 (53). At the time of this monograph there were approximately 300 cases of reported cryptococcosis worldwide to help provide the clinical characteristics and outcome of this invasive fungal infection. In 1998, we further chronicled and updated the recent detailed history of cryptococcosis over the first century of its pathobiological understandings and provided a framework for discussion issues around this fungus (11).

In particular, several critical events in clinical cryptococcosis occurred in the second half of the 20th century. First was the development and use of a serum cryptococcal polysaccharide antigen test for diagnosis. This assay can arguably be considered one of the most successful microbial antigen detection tests in clinical diagnostic use today. The use of latex agglutination or enzyme-linked immunosorbent assay testing to detect polysaccharide antigen shed from the yeast into the host's blood or cerebrospinal fluid (CSF) had simple beginnings in 1951 when Neill initially described the idea (66) and later in 1968 when Kauffman and Blumer brought it into clinical reality (37). Importantly, the cryptococcal polysaccharide antigen test was taken from laboratory studies to be commercialized, and in our opinion it now represents the best serological test for diagnosis of any invasive mycosis. It has sensitivity and specificity profiles for diagnosis of over 95%, and it carries both diagnostic and prognostic features. A future goal of this diagnostic tool would be to provide a cheap, simple antigen test for undeveloped countries since it could provide substantial improvement in preemptive management strategies in AIDS patients with a risk for cryptococcosis.

Second, after discovery of amphotericin B (AMB), it became apparent that a uniformly fatal CNS infection with cryptococcus could be successfully treated with this polyene. This discovery was made by a single case successfully managed by Appelbaum and Shtokalko and not through a randomized clinical trial (3). The case involved the successful management of cryptococcal meningitis in a 46-year-old female diabetic laboratory technician with approximately 3 g of AMB treatment in 1956.

Third, a landmark study by Bennett et al. in 1979 introduced the rigorous methodology of comparative, randomized trials into antifungal therapy and particularly cryptococcosis. This study validated the efficacy of combination antifungal therapy for cryptococcosis (7). The study represented a major starting point not just for cryptococcosis but also for other invasive fungal infections. It was the first to use a prospective, randomized, and multicenter design for fungal disease that is so frequently used today for our clinical trials to obtain robust results and conclusions. This study initially provided much of the infrastructure for understanding the management of cryptococcal meningitis. It was also the platform study for the formation of a large collaborative National Institute of Allergy and Infectious Diseases Mycoses Study Group that has been together for over 30 years. This collaborative group under an independent banner continues to run clinical trials for invasive mycoses today.

Fourth, it was clear as modern medicine progressed in the treatment of cancers and transplantation of organs that the collateral complications of these conditions and their management would be immunosuppression and fungal infections. The term "awakening giant" for cryptococcosis was first coined by Kauffman and Blumer in 1977 (38), as these clinical mycologists clearly saw the trend of increasing numbers of cases associated with the rising numbers of higher-risk groups due to the immunosuppressive state. However, no one in the mid-20th century could have predicted the magnitude of what has happened to the giant in the last quarter of a century. With the first series of cases of cryptococcosis linked to AIDS in 1983–1984 (81), a spark lit a fire under the giant (cryptococcosis), which now rages at a million cases per year (71).

GENETICS, MOLECULAR BIOLOGY, AND PATHOGENESIS

As the 20th century came to an end and with it came an increasing need to provide clinical care for a rapidly enlarging immunocompromised population, this morphologically flexible fungus with its ability to produce disease in a vast number of patients and with known virulence phenotypes was poised for further investigations. Between 1990 and 2000 was the beginning era of genetics (36) and molecular biology for pathobiological discovery in cryptococcus. During this time, there was much research into the infrastructure development and the discovery of molecular pathogenic principles with guidance from the model fungal pathogen *Saccharomyces cerevisiae*. The scientific stage was set for the entry of cryptococcus into the field of molecular pathogenesis and a substantial leap in our understanding of the basic pathobiology of this yeast. Typically, in the focused early investigative studies, specific fungal strains were used to develop a platform for genetic and molecular studies, and in cryptococcosis research there was no exception to this design. Several primary cryptococcal strains were used for study. First, the serotype D strains 3501 and 3502 were used (19, 38) to study the life cycle of *C. neoformans*, and these well-characterized strains were further backcrossed to create congenic strains named JEC 20 and JEC 21. The history of these strains and their creation is well chronicled in a careful review dedicated to their origins (32).

Second, a serotype A strain of cryptococcus named H99 was developed for molecular pathogenic studies. H99 was first isolated from the CSF of a 27-year-old Caucasian male who had been treated for Hodgkin's disease and lived in the Piedmont area of North Carolina. His initial lumbar puncture on 14 February 1978 was India ink positive with normal CSF protein, and the CSF glucose was 30 mg/dl with simultaneous serum glucose of 157 mg/dl. CSF culture grew an encapsulated yeast. His opening pressure was 270 mm of CSF, and he had 58 white blood cells/ml of CSF with a predominance of lymphocytes. The patient was treated with AMB at 0.3 mg/kg/day and flucytosine 150 mg/kg/day for 6 weeks, and at 1-year follow-up he was considered cured with this treatment. The strain was labeled H99 at Duke and used as the type culture for *C. neoformans* var. *grubii* (25) and disseminated to the research community for further study.

H99's value has been that (i) it represents the major clinical serotype, (ii) it is a standard clinical human isolate from CSF whose pathogenicity is now well described in animal models, (iii) its ability to be transformed for gene transfer is established, and (iv) its entire genome has been sequenced, a process that was catalyzed by the leadership of Jenny Lodge, who organized the community for an effort that produced one of the earliest complete fungal genomes. It is a mating-type alpha strain and has been genetically manipulated to create congenic pairs (KN99 "alpha" and "a" strains) for use in genetic and pathology studies (68). It is these strains (3501, 3502, JEC 20, JEC 21, H99, KN99 alpha and a) and their mutants that form the basis for most of the molecular studies performed today in *C. neoformans*. Recently, molecular studies conducted with several *C. gattii* strains have started to appear in the literature to widen our appreciation of both the differences and similarities between the two cryptococcal species at the functional molecular level (41, 67).

A critical point in the development of the molecular pathogenesis field for cryptococcosis has been the development of a usable transformation system. The first reported transformation system in 1992 used complementation of uracil auxotrophs with the *URA3* gene during electroporation DNA delivery. While the efficiency of DNA transformation of the serotype D strain was acceptable for molecular use, it produced primarily ectopic integrative events or extrachromosomal transformants and therefore was very cumbersome for making site-directed stable mutants. In 1993, using the serotype A (H99) strain and the delivery of the *ADE2* gene into adenine auxotrophs with biolistic DNA delivery, researchers measured the rate of homologous integrative events at around 5% (85). This technical advancement allowed the routine use of the transformation system to efficiently produce site-specific gene knockouts or mutants, and now the field has exploded with the creation of site-specific mutants for impact on pathogenesis. Over 150 knockout mutants have been studied for pathogenesis impact on disease in a variety of animal models (72). These models of disease range from the use of mammals such as mice, rats, and rabbits (43, 74) to invertebrates such as worms and grubs (46, 64). It is clear that with the present molecular technical tools, the pathogenic understanding of the "sugar-coated killer yeast" is being stripped away one gene at a time. In 2005, with the complete sequencing of the serotype D strain, JEC 21 (56), the infrastructure is now in place for genome-wide studies from large mutant libraries (54) to whole-organism microarray transcriptional studies (42).

The history of studies of the pathogenesis of cryptococcosis is rich with discoveries involving this complex infection. Space limitations do not allow us to record all these insights in this review, and many of these findings will be articulated very well in other chapters of this book. However, several landmark studies deserve some mention. First, early studies in anatomical pathology clearly showed the granulomatous host responses to this yeast. Furthermore, in 1954 pathologist Roger Baker and colleagues, while performing autopsies on healthy trauma victims, made a major histological finding by identifying the pulmonary–lymph node complex of cryptococcosis (4). The proposed scenario of primary infection, dormancy, and reactivation during immunosuppression that has dominated our current understanding of the pathogenesis of cryptococcosis was born from this report. Second, from 1994 to 1996, seminal papers were published concerning the identification of specific genes linked to the three major virulence phenotypes: capsule (12), melanin (77), and high-temperature growth (69). These three manuscripts used site-directed gene disruption and then gene reconstitution for strain creations to satisfy Falkow's postulates for virulence gene identification by the association of a gene, phenotype, and animal model outcomes. These manuscripts illustrate the paradigm for basic molecular pathogenesis studies today and illustrate the powerful proof of concept of these molecular types for investigation.

Third, the ability to pinpoint host responses against this encapsulated yeast has been shown multiple times (48, 63). The many immunological findings are too numerous to mention, but the use of immune-deficient animals and hybridoma technology to begin to dissect cell-mediated and humoral immunity, respectively, has provided critical tools in furthering our understanding of host defenses. The early immunosuppressive studies on the use of cortisone in animal models of cryptococcosis by Gadebush and Gikas represent the elegant experimental platform for this fungus that can now also use selective gene knockout mice. Furthermore, multiple arms of immunity, from innate, cell-mediated to humoral immune systems, are clearly important to control and resolve this infection. In fact, when humoral immunity was demonstrated to have a definitive role in the effective host response against cryptococcus (62), it changed the view of cell-mediated immunity as being the only protective host factor; this eventually led to a monoclonal antibody (MAb) being considered as a treatment strategy in humans for cryptococcosis (47).

Simultaneously with these efforts, Kozel and collaborators carried out elegant studies on the interaction of complement with cryptococcal cells (41). The investigations into the basic importance of cell-mediated immunity translated into the first randomized study of the use of the immune modulator gamma interferon as an adjunctive therapy in cryptococcal meningoencephalitis (70). Although this study did not give a definitive answer to gamma interferon's value, its importance in 2004 as the first randomized study in the treatment of invasive mycoses with immune modulators should be recognized as future studies improve on its design.

In the areas of pathogenesis, diagnosis, and treatment of cryptococcosis, there are myriad events to report. Therefore, in Table 1, we have attempted to collate and highlight studies that we believe were pivotal in our understanding of this yeast and the disease it produces. Of course, there are many deserving events, and with limited space we may miss many worthy works that should be listed. We apologize for these omissions, but our attempt is to list landmark studies in cryptococcosis, and we believe this table is a good starting spot.

THE VIEW OF CRYPTOCOCCOSIS "BIG AND SMALL"

Mitchell and Perfect summarized the history of cryptococcosis over the first century of its known existence as a lethal pathogen in 1995 (60), and in Table 2 we have attempted to list the major comprehensive treatises on *Cryptococcus* spp. and cryptococcosis. While each comprehensive review attempts to summarize the information to date, we note that the rapidity of developments in clinical practice may generate new problems as old ones remain unsolved. For instance, currently we are witnessing an ongoing outbreak of *C. gattii* in Vancouver and the Pacific Northwest United States produced by a recombinant strain with increased virulence, and a persistent large epidemic of cryptococcosis in sub-Saharan Africa associated with AIDS goes on without an end in sight (Table 3). Nevertheless, there is much left to do to understand the epidemiology of this invasive mycosis. The first large outbreak of cryptococcosis was associated with cows and mastitis in the 1950s, probably related to direct inoculation with contaminated fomites (40, 76). Now, in the first decade of the new millennium, we are facing two large outbreaks, with one localized to a specific geographical area and another linked to a worldwide epidemic of a retroviral infection; cryptococcosis is clearly a public health problem.

However, most cases of cryptococcosis today are not linked to a specific known environmental exposure but occur one case at a time without the advantages of other cases occurring at the same time and place, and person-to-person transmission does not generally occur. On the other hand, the epidemiology of cryptococcosis continues to evolve as the fungus adapts to new hosts and the risk groups enlarge. If it is not an increasing HIV population or transplant recipient group, the rise of potent immune-modulating MAbs such as anti-tumor necrosis factor or anti-CD52 will surely bring us more cryptococcosis as their use matures (64, 65). Clinicians must include cryptococcosis in their differential diagnosis at all times in immunocompromised hosts and particularly those with chronic pneumonias or meningitis.

In 1989, 64 investigators met in Jerusalem at the First International Conference on *Cryptococcus* and Cryptococcosis. The highlights were (i) early results with fluconazole

TABLE 1 Landmarks in the clinical history of cryptococcosis

Date	Event	Importance	Reference(s)
1895	First clinical case description of *C. neoformans*	*Cryptococcus* causes human disease, disseminates in body, and causes fatal infection	8–10
1896	Clinical disease caused by *C. gattii*	Disease caused by more than one strain	14
1902	Horse infection	Host range of infection includes mammals besides humans	28
1906–1916	Description of CNS cryptococcosis	The recognition of the clinical features of cryptococcosis in the CNS	84, 90
1935	First mycological classification of cryptococcosis	The creation of mycological and taxonomic structure of the genus along with appreciation that there were four serotypes	6
1951–1955	Isolation of cryptococcus from soil and bird guano	The ability to find yeast in the environmental traffic of humans	21, 22
1952	Prolonged, untreated cryptococcosis of CNS	Appreciation that this yeast can persist in body for a prolonged period of time	5
1955	First case of cryptococcal meningitis successfully treated with AMB	The beginning of the era of successful treatment of invasive mycoses and, specifically, fatal CNS cryptococcosis	3
1955	Autopsy studies show lung–lymph node complex	Supports the dormancy theory of cryptococcosis	4
1960–1965	Relationship of pigeons to *C. neoformans* clarified	Littman carried out several seminal studies establishing that pigeon guano is a fertile site for the growth of *C. neoformans* in cities	51, 52
1962	Development of bird seed agar	Staib developed a new agar type that facilitates identification of *C. neoformans* in clinical and environmental samples	75
1968	Creation of cryptococcal polysaccharide antigen test	One of the best diagnostic and prognostic fungal tests available to clinicians today	37
1975	Identification of the perfect state	Major understanding of life cycle with implications for genetics, evolution, and transmission	45
1978	Recovery of H99 from a patient in North Carolina	This strain became the prototypical *C. neoformans* var. *grubii* strain and by the early 21st century was widely used in experimental studies	74
1979	Randomized, prospective trial comparing AMB and AMB + 5-flucytosine for cryptococcal meningoencephalitis	A new standard for antifungal therapy; successful introduction of combination therapy for invasive mycoses	7
1983–1984	Link established between AIDS and cryptococcosis; although AIDS was first clinically described in 1981, by 1984 clearly AIDS-opportunistic infections included cryptococcosis	HIV infection remains the single biggest risk factor for appearance of cryptococcal disease in the world today	81
1987	Generation of the first MAbs to *C. neoformans* capsular polysaccharide	The availability of MAbs to *C. neoformans* allowed new approaches to the study of capsular antigenic structure and to analyze the contribution of humoral immunity to host defense. Dromer's pioneering work describing the first protective MAb catalyzed numerous studies in the 1990s.	16
1990	*Cryptococcus* found to be associated with eucalyptus trees	The identification of trees as an environmental source for this fungus	20
1990–1999	Monoclonal anticapsular antibody found to have positive impact on disease	Beginning in 1991, several laboratories employed MAbs to dissect the role of humoral immunity in host defense. This work established a new paradigm whereby antibody-mediated immunity can be useful against intracellular pathogens.	24, 61, 79
1990–1999	Cell-mediated immunity established to have critical role in host defense	Although granuloma formation had long been established as the effective tissue reaction, several laboratories conclusively associated cell-mediated immunity with host defense in the 1990s	33–35, 39, 49, 89

(Continued on next page)

TABLE 1 *(Continued)*

Date	Event	Importance	Reference(s)
1990–1999	Mechanisms of polysaccharide-mediated immunosuppression identified	Numerous laboratories showed that capsular polysaccharide has protean immunomodulatory effects	88
1992–1993	Development of a transformation system	Opened molecular studies of cryptococcal virulence composite	18, 85
1993	A capsular-conjugated vaccine was protective in mice	Principle of protective fungal vaccine is theoretically possible	15
1994–1997	Creation of capsule, melanin, and high-temperature growth mutants	Use of site-directed mutants to study the major virulence phenotypes	12, 15, 69, 77
1994	Standard polysaccharide structural motifs assigned	After more than two decades of investigation by several laboratories Cherniak and Sunstrum proposed that six mannose triads are responsible for polysaccharide structure	13
1997	Integration of induction, consolidation, and suppressive phases of treatment for meningoencephalitis	Allowed a strategy for initial aggressive fungicidal approach with prolonged therapy for residual disease	87
1998	Identification of haploid fruiting	Morphologic and life cycle features with epidemiologic implications	92
1998	Description of phenotypic switching in *C. neoformans*	The capacity of *C. neoformans* for morphological change is associated with changes in virulence and persistence	29
2000	*C. neoformans* established as a facultative intracellular pathogen	Although *C. neoformans* was known since the 1960s to replicate inside phagocytes in vitro, the recognition that this occurred in vivo ushered in a new phase of pathogenesis studies	23
2001	Virulence hypothesized to emerge from protista interactions	Provided an explanation for the ready-made virulence of *C. neoformans* in soils and led to the exploration of several alternative nonmammalian host systems	82
2002	Outbreak of *C. gattii* infection in humans and animals identified	An ecological and environmental shift in habitat and infections in apparently normal hosts	83
2004	Gamma interferon used as adjunctive therapy in meningitis	First randomized trial with the use of immune modulators for fungal treatment	70
2005	Genome of *C. neoformans* sequenced	*C. neoformans* studies can go genome wide	56
2005	Classification of two pathogenic species	*C. neoformans* and *C. gattii* considered evolved into separate species for taxonomy purposes	46
2005	Clinical description of IRIS with HAART[a]	IRIS is a major clinical condition in AIDS, organ transplant recipients, and normal hosts	57, 80
2005	*C. gattii* outbreak strains found to represent recombinants with improved fitness	Example of same-sex mating and creation in nature producing hyper-virulent strains and clinical outbreak	26
2006	Nonlytic exocytosis of *C. neoformans* from phagocytic cells	*C. neoformans* provided a new paradigm in host-microbe interactions by demonstrating nonlytic exit from infected cells	2, 59
2010	Infectious Disease Society of America (IDSA) cryptococcal treatment guidelines	Updated comprehensive version of the previous 2000 IDSA guidelines	73

[a]IRIS, immune reconstitution inflammatory syndrome; HAART, highly active antiretroviral therapy.

treatment of cryptococcosis, (ii) identification of *C. gattii* and its association with eucalyptus trees, and (iii) some very preliminary studies on molecular biology with this yeast. In 2008 (almost 20 years later), the Seventh International Conference on *Cryptococcus* and Cryptococcosis met in Nagasaki, Japan, with hundreds in attendance. The basic science understanding of this yeast was so advanced that it had now become a model fungal pathogen for studying general fungal pathogenesis, and careful documentation of two raging epidemics with *C. neoformans* and *C. gattii* was presented. Clearly, the clinical and basic science infrastructures are attempting to provide the fertile ground to plant the seeds of knowledge to help us effectively coexist with this human pathogen.

THE ROAD AHEAD

The history of cryptococcosis has been dynamic and continues to evolve. From its humble beginnings as single case reports to its model system status in molecular studies for fungal pathogenesis, it has become a sophisticated complex organism for study and management. The cryptococcal field now comprises a large and diverse set of international investigators studying a wide variety of problems ranging from

TABLE 2 Monographs, reviews, and books with critical descriptions of *Cryptococcus* through history

Date	Authors	Title	Focus	Reference
1916	Stoddard and Cutler	*Torula Infection in Man*	Description of CNS cryptococcosis	84
1956	Littman and Zimmerman	*Cryptococcosis*	History and clinical description prior to effective treatment	53
1971	Al-Doory	A bibliography of cryptococcosis	Listing of published papers on cryptococcosis	1
1995	Mitchell and Perfect	Cryptococcosis in the era of AIDS 100 years after the discovery of *Cryptococcus neoformans*	Cryptococcosis with an emphasis on AIDS era	60
1997	Drouhet	Milestones in the history of cryptococcus and cryptococcosis	Historical view of cryptococcus	17
1998	Casadevall and Perfect	*Cryptococcus neoformans*	Over 500-page book devoted in depth to cryptococcosis	11

population structure to ecology to pathogenesis. *Cryptococcus* spp. emerged as major human pathogens in the 20th century largely as a consequence of a rising number of immunosuppressed individuals, linked to medical progress and the HIV epidemic. Given the ongoing health-related issues that challenge humanity, it is likely that the medical importance of *Cryptococcus* spp. will continue to rise. Furthermore, given the hints that even subclinical cryptococcal infection may be a cofactor in other chronic diseases such as asthma (30), it is possible that the spectrum of diseases associated with this genus will dramatically increase in the years ahead. The history of cryptococcosis is integrally linked to the rise of modern medicine and its science. The final chapters of this story of cryptococcosis are not yet told, but the infrastructure, in the sense of a mature and dynamic field, is now in place to write them.

In 1900, the mathematician David Hilbert posed a set of 23 unsolved problems in mathematics and by stating them in a list was able to galvanize the field to focus on their solution. Borrowing a page from Hilbert, we thought we would list some of the greatest unsolved problems with *Cryptococcus* spp. in the hope that stating them will catalyze research into their solution. Although this list may differ from a mathematical problem list in that for many items there is already considerable information, we sought to identify problems that remain unsolved in the sense that a definitive accounting is absent. Our list is as follows:

- The molecular underpinning for capsule assembly, structure, and architecture
- The cellular form responsible for human infection
- The mechanisms for extrapulmonary dissemination and meningeal invasion
- The mechanism for nonlytic exocytosis
- The architecture of the cell wall
- The microbial and host factors responsible for latency
- The microbial basis for persistence in tissue despite antifungal therapy
- Host factors that predispose to disease in the absence of immune suppression
- The role of fungal export mechanisms in pathogenesis
- The role of sexual reproduction in generating diversity and modulating virulence
- The relationship between phenotypic variation and virulence
- The identification of natural nonvertebrate hosts in the environment
- The potential of the human-pathogenic *Cryptococcus* spp. for plant virulence
- B and T cell epitopes associated with protective immunity
- Deep mapping of the global yeast population structure
- The contribution of fungus-induced immunosuppression to pathogenesis
- The role of immunological memory in host defense
- The interplay between innate and adoptive humoral immunity and fungal components
- The complete mapping of signal transduction pathways and their effects on virulence determinants
- The mechanisms for daughter cell transport and separation through the polysaccharide capsule
- Better diagnosis and management of immune reconstitution inflammatory syndrome (IRIS) in cryptococcosis
- Therapeutic challenges: (i) delivery of cryptococcosis care in undeveloped countries; (ii) development of safe, rapidly fungicidal drugs; (iii) consistent management of increased intracranial pressure
- Host genetic susceptibility to cryptococcosis

TABLE 3 Outbreaks of cryptococcal disease

Description	Epidemiology	Pathogenesis	Outcome	Reference(s)
Bovine mastitis	Several outbreaks with hundreds of animals involved	Likely direct inoculation related to contaminated equipment	Stopped with infection control practices	40, 76
Vancouver and NW United States: humans and animals	Hundreds with disease	Ecologic change with recombinant strain and improved fungal fitness	Ongoing; clinicians aware of disease in area	83
Worldwide with HIV pandemic	Millions with disease	Severe immunosuppression and reactivation of disease	Continues as HIV pandemic remains	71

REFERENCES

1. **Al-Doory, Y.** 1971. A bibliography of cryptococcosis. *Mycopathol. Mycol. Appl.* **45:**2–60.
2. **Alvarez, M., and A. Casadevall.** 2006. Phagosome extrusion and host-cell survival after *Cryptococcus neoformans* phagocytosis by macrophages. *Curr. Biol.* **16:**2161–2165.
3. **Appelbaum, E., and S. Shitokalko.** 1956. Cryptococcus meningitis arrested with amphotericin B. *Ann. Intern. Med.* **47:**346–351.
4. **Baker, R. D., and R. K. Haugen.** 1955. Tissue changes and tissue diagnosis of cryptococcosis: a study of 26 cases. *Am. J. Pathol.* **25:**14.
5. **Beeson, P. B.** 1952. Cryptococcic meningitis of nearly sixteen years' duration. *AMA Arch. Intern. Med.* **89:** 797–801.
6. **Benham, R. W.** 1935. Cryptococci, their identification by morphology and serology. *J. Infect. Dis.* **57:**255–274.
7. **Bennett, J. E., W. Dismukes, R. J. Duma, G. Medoff, M. A. Sande, H. Gallis, J. Leonard, B. T. Fields, M. Bradshaw, H. Haywood, Z. A. McGee, T. R. Cate, C. G. Cobbs, J. F. Warner, and D. W. Alling.** 1979. A comparison of amphotericin B alone and combined with flucytosine in the treatment of cryptococcal meningitis. *N. Engl. J. Med.* **301:**126–131.
8. **Buschke, A.** 1895. Ueber eine durch coccidien hemorgerufene krankheit des menschen. *Dtsch. Med. Wochenschr.* **21:**14.
9. **Busse, O.** 1895. Ueber Saccharomycosis hominis. *Virchows. Arch. F. Path. Anat.* **140:**23–46.
10. **Busse, O.** 1894. Ueber parasitare zelleinschulusse and ihre zuchtung. *Int. J. Med. Microbiol.* **16:**175–180.
11. **Casadevall, A., and J. Perfect.** 1998. *Cryptococcus neoformans.* ASM Press, Washington, DC.
12. **Chang, Y. C., and K. J. Kwon-Chung.** 1994. Complementation of a capsule-deficiency mutation of *Cryptococcus neoformans* restores its virulence. *Mol. Cell. Biol.* **14:** 4912–4919.
13. **Cherniak, R., and J. B. Sundstrom.** 1994. Polysaccharide antigens of the capsule of *Cryptococcus neoformans. Infect. Immun.* **62:**1507–1512.
14. **Curtis, F.** 1896. Contribution a l'etude de la saccharomycose humaine. *Ann Inst. Pasteur* **10:**449–468.
15. **Devi, S. J., R. Scheerson, W. Egan, T. J. Ulrich, D. Bryla, J. B. Robbins, and J. E. Bennett.** 1991. *Cryptococcus neoformans* serotype A glucuronoxylomannan protein conjugate vaccines: synthesis, characterization, and immunogenicity. *Infect. Immun.* **59:**3700–3707.
16. **Dromer, F., J. Charreire, A. Contrepois, C. Carbon, and P. Yeni.** 1987. Protection of mice against experimental cryptococcosis by anti-*Cryptococcus neoformans* monoclonal antibody. *Infect. Immun.* **55:**749–752.
17. **Drouhet, E.** 1997. Milestones in the history of cryptococcus and cryptococcosis. *J. Mycol. Med.* **7:**10–17.
18. **Edman, J. C., and K. J. Kwon-Chung.** 1990. Isolation of the *URA5* gene from *Cryptococcus neoformans* var. *neoformans* and its use as a selective marker for transformation. *Mol. Cell. Biol.* **10:**4538–4544.
19. **Edwards, J. E., Jr., J. Morrison, D. K. Henderson, and J. Z. Montgomerie.** 1980. Combined effect of amphotericin B and rifampin on Candida species. *Antimicrob. Agents Chemother.* **17:**484–487.
20. **Ellis, D. H., and T. J. Pfeiffer.** 1990. Natural habitat of *Cryptococcus neoformans* var *gattii. J. Clin. Microbiol.* **28:**1642–1644.
21. **Emmons, C. W.** 1955. Saprophytic sources of *Cryptococcus neoformans* associated with the pigeon. *Am. J. Hyg.* **62:**227–232.
22. **Emmons, C. W.** 1951. Isolation of *Cryptococcus neoformans* from soil. *J. Bacteriol.* **62:**685–690.
23. **Feldmesser, M., Y. Kress, P. Novikoff, and A. Casadevall.** 2000. Cryptococcus neoformans is a facultative intracellular pathogen in murine pulmonary infection. *Infect. Immun.* **68:**4225–4237.
24. **Fleuridor, R., Z. Zhong, and L.-A. Pirofski.** 1998. A human IgM monoclonal antibody prolongs survival of mice with lethal cryptococcosis. *J. Infect. Dis.* **178:** 1213–1216.
25. **Franzot, S. P., I. F. Salkin, and A. Casadevall.** 1999. *Cryptococcus neoformans* var. *grubii*: separate variety status for *Cryptococcus neoformans* serotype A isolates. *J. Clin. Microbiol.* **37:**838–840.
26. **Fraser, J. A., S. S. Giles, E. C. Wenink, S. G. Geunes-Boyer, J. R. Wright, S. Diezmann, A. Allen, J. E. Stajich, F. S. Dietrich, J. R. Perfect, and J. Heitman.** 2005. Same-sex mating and the origin of the Vancouver Island *Cryptococcus gattii* outbreak. *Nature* **437:**1360–1364.
27. **Fries, B. C., D. L. Goldman, and A. Casadevall.** 2002. Phenotypic switching in *Cryptococcus neoformans. Microbiol. Infect. Dis.* **4:**1345–1352.
28. **Frothingham, L. A.** 1902. A tumor-like lesion in the lung of a horse caused by a blastomyces (torula). *J. Med. Res.* **3:**31–43.
29. **Goldman, D. L., B. C. Fries, S. P. Franzot, L. Montella, and A. Casadevall.** 1998. Phenotypic switching in the human pathogenic fungus, *Cryptococcus neoformans*, is associated with changes in virulence and pulmonary inflammatory response in rodents. *Proc. Natl. Acad. Sci. USA* **95:**14967–14972.
30. **Goldman, D. L., and G. B. Huffnagle.** 2009. Potential contribution of fungal infection and colonization to the development of allergy. *Med. Mycol.* **47**(5)**:**445–456.
31. **Hansemann, D. V.** 1905. Uber eine bisher nicht beobachtete Gehirner Krankung durch Hefen. *Verh. Dtsch. Ges. Pathol.* **9:**21–24.
32. **Heitman, J., B. Allen, J. A. Alspaugh, and K. J. Kwon-Chung.** 1999. On the origins of congesic MAT alpha and MAT a strains of the pathogenic yeast *Cryptococcus neoformans. Fungal Genet. Biol.* **28:**1–5.
33. **Hidore, M. R., N. Nabavi, F. Sonnleitner, and J. W. Murphy.** 2010. Murine natural killer cells are fungicidal to *Cryptococcus neoformans. Infect. Immun.* **59**(5)**:**1747–1754.
34. **Hill, J. O.** 1992. CD4+ T cells cause multinucleated giant cells to form around *Cryptococcus neoformans* and confine the yeast within the primary site of infection in the respiratory tract. *J. Exp. Med.* **175:**1685–1695.
35. **Huffnagle, G. B., J. L. Yates, and M. F. Lipscomb.** 1991. Immunity to a pulmonary *Cryptococcus neoformans* infection requires both CD4+ and CD8+ T cells. *J. Exp. Med.* **173:**793–800.
36. **Hull, C. M., and J. Heitman.** 2002. Genetics of *Cryptococcus neoformans. Annu. Rev. Genet.* **36:**557–615.
37. **Kauffman, L., and S. Blumer.** 1968. Value and interpretation of serological tests for the diagnosis of cryptococcosis. *Appl. Microbiol.* **16:**1907–1912.
38. **Kauffman, L., and S. Blumer.** 2009. Cryptococcosis: the awakening giant. Proceedings of the Fourth International Conference on the mycoses. PAHO Scientific publication no. 356, p. 176–182.
39. **Kawakami, K., M. H. Qureshi, T. Zhang, H. Okamura, M. Kurimoto, and A. Saito.** 1997. IL-18 protects mice against pulmonary and disseminated infection with *Cryptococcus neoformans* by inducing IFN-gamma production. *J. Immunol.* **159:**5528–5534.
40. **Klein, E.** 1901. Pathogenic microbes in milk. *J. Hygiene* **1:**78–95.
41. **Kozel, T. R., and G. S. T. Pfrommer.** 1986. Activation of the complement system by *Cryptococcus neoformans* leads to binding of iC3b to the yeast. *Infect. Immun.* **52:**1–5.

42. **Kraus, P. R., M. J. Boily, S. S. Giles, J. E. Stajich, A. Allen, G. M. Cox, F. S. Dietrich, J. R. Perfect, and J. Heitman.** 2004. Identification of *Cryptococcus neoformans* temperature-regulated genes with a genomic-DNA microarray. *Eukaryot. Cell* **3:**1249–1260.
43. **Kuhn, L. R.** 1939. Growth and viability of *Cryptococcus hominis* at mouse and rabbit body temperatures. *Proc. Soc. Exp. Biol. Med.* **4:**573–574.
44. **Kutzing, F.** 1833. Algarum Aquae dulcis germaniae decas III. New York Botanical Garden, New York, NY.
45. **Kwon-Chung, K. J.** 1975. A new genus, *Filobasidiella*, the perfect state of *Cryptococcus neoformans*. *Mycologia* **67:**1197–1200.
46. **Kwon-Chung, K. J., T. Boekhout, J. W. Fell, and M. Diaz.** 2002. Proposal to conserve the name *Cryptococcus gattii* against *C. hondurianus* and *C. bacillisporus* (*Basidiomycota, Hymenomycetes, Tremellomycetidae*). *Taxon* **51:** 804–806.
47. **Larsen, R. A., P. G. Pappas, J. Perfect, J. A. Aberg, A. Casadevall, G. A. Cloud, R. James, S. Filler, and W. E. Dismukes.** 2005. Phase I evaluation of the safety and pharmacokinetics of murine-derived anticryptococcal antibody 18B7 in subjects with treated cryptococcal meningitis. *Antimicrob. Agents Chemother.* **49:**952–958.
48. **Levitz, S. M.** 1992. Overview of host defenses in fungal infections. *Clin. Infect. Dis.* **14:**S37–S42.
49. **Levitz, S. M., and M. P. Dupont.** 1993. Phenotypic and functional characterization of human lymphocytes activated by interleukin-2 to directly inhibit growth of *Cryptococcus neoformans* in vitro. *J. Clin. Invest.* **91:**1490–1498.
50. **Lin, X., C. M. Hull, and J. Heitman.** 2005. Sexual reproduction between partners of the same mating type in *Cryptococcus neoformans*. *Nature* **434:**1017–1021.
51. **Littman, M. L., and R. Borok.** 1968. Relation of the pigeon to cryptococcosis: natural carrier state, heat resistance and survival of *Cryptococcus neoformans*. *Mycopathologia* **35:**329–345.
52. **Littman, M. L., R. Borok, and T. J. Dalton.** 1965. Experimental avian cryptococcosis. *Am. J. Epidemiol.* **82:**197–207.
53. **Littman, M. L., and L. E. Zimmerman.** 1956. *Cryptococcosis*, p. 121–146. Grune & Stratton, New York, NY.
54. **Liu, O. W., C. D. Chun, E. D. Chow, C. Chen, and H. D. Madhani.** 2008. Systematic genetic analysis of virulence in the human fungal pathogen *Cryptococcus neoformans*. *Cell* **135:**174–188.
55. **Lodder, J., and N. J. W. Kreger-van Rij.** 2009. *The Yeasts: a Taxonomic Study*. Interscience, New York, NY.
56. **Loftus, B. J., E. Fung, P. Roncaglia, D. Rowley, P. Amedeo, D. Bruno, J. Vamathevan, M. Miranda, I. J. Anderson, J. A. Fraser, J. E. Allen, I. E. Bosdet, M. R. Brent, R. Chiu, T. L. Doering, M. J. Donlin, C. A. D'Souza, D. S. Fox, V. Grinberg, J. Fu, M. Fukushima, B. J. Haas, J. C. Huang, G. Janbon, S. J. Jones, H. L. Koo, M. I. Krzywinski, J. K. Kwon-Chung, K. B. Lengeler, R. Maiti, M. A. Marra, R. E. Marra, C. A. Mathewson, T. G. Mitchell, M. Pertea, F. R. Riggs, S. L. Salzberg, J. E. Schein, A. Shvartsbeyn, H. Shin, M. Shumway, C. A. Specht, B. B. Suh, A. Tenney, T. R. Utterback, B. L. Wickes, J. R. Wortman, N. H. Wye, J. W. Kronstad, J. K. Lodge, J. Heitman, R. W. Davis, C. M. Fraser, and R. W. Hyman.** 2005. The genome of the basidiomycetous yeast and human pathogen *Cryptococcus neoformans*. *Science* **307:** 1321–1324.
57. **Lortholary, O., A. Fontanet, N. Memain, A. Martin, K. Sitbon, and F. Dromer.** 2005. Incidence and risk factors of immune reconstitution inflammatory syndrome complicating HIV-associated cryptococcosis in France. *AIDS* **19:**1043–1049.
58. **Lurie, H., and H. J. Shadomy.** 1971. Morphological variations of a hypha-forming strain of *Cryptococcus neoformans* (Coward strain) in tissues of mice. *Sabouraudia* **9:**10–14.
59. **Ma, H., J. E. Croudace, D. A. Lammas, and R. C. May.** 2006. Expulsion of live pathogenic yeast by macrophages. *Curr. Biol.* **16:**2156–2160.
60. **Mitchell, T. G., and J. R. Perfect.** 1995. Cryptococcosis in the era of AIDS 100 years after the discovery of *Cryptococcus neoformans*. *Clin. Microbiol. Rev.* **8:**515–548.
61. **Mukherjee, J., M. D. Scharff, and A. Casadevall.** 1992. Protective murine monoclonal antibodies to *Cryptococcus neoformans*. *Infect. Immun.* **60:**4534–4541.
62. **Mukherjee, S., S. Lee, J. Mukherjee, M. D. Scharff, and A. Casadevall.** 1994. Monoclonal antibodies to *Cryptococcus neoformans* capsular polysaccharide modify the course of intravenous infection in mice. *Infect. Immun.* 62**:**1079–1088.
63. **Murphy, J. W.** 1992. Cryptococcal immunity and immunostimulation. *Adv. Exp. Med. Biol.* **319:**225–230.
64. **Mylonakis, E., F. M. Ausubel, J. R. Perfect, J. Heitman, and S. B. Calderwood.** 2002. Killing of *Caenorhabditis elegans* by *Cryptococcus neoformans* as a model of yeast pathogenesis. *Proc. Natl. Acad. Sci. USA* **99:**15675–15680.
65. **Nath, D. S., R. Kandaswamy, R. Gruessner, D. E. Sutherland, D. L. Dunn, and A. Humar.** 2005. Fungal infections in transplant recipients receiving alemtuzumab. *Transplant. Proc.* **37:**934–936.
66. **Neill, J. M.** 1951. Serologically reactive material in spinal fluid, blood, and urine from a human case of cryptococcosis (torulosis). *Proc. Soc. Exp. Biol. Med.* **77:**775–778.
67. **Ngamskulrungroj, P., U. Himmelreich, J. A. Breger, C. Wilson, M. Chayakulkeeree, M. B. Krockenberger, R. Malik, H. M. Daniel, D. Toffaletti, J. T. Djordjevic, E. Mylonakis, W. Meyer, and J. R. Perfect.** 2009. The trehalose synthesis pathway is an integral part of the virulence composite for *Cryptococcus gattii*. *Infect. Immun.* **77:**4584–4596.
68. **Nielsen, K., G. M. Cox, P. Wang, D. L. Toffaletti, J. R. Perfect, and J. Heitman.** 2003. Sexual cycle of *Cryptococcus neoformans* var. *grubii* and virulence of congenic a and alpha isolates. *Infect. Immun.* **71:**4831–4841.
69. **Odom, A., J. R. Perfect, D. Toffaletti, and J. Heitman.** 1996. Calcineurin A is a temperature-regulated virulence gene for *Cryptococcus neoformans*. *EMBO J.* **16:**2576-2589.
70. **Pappas, P. G., B. Bustamante, E. Ticona, R. J. Hamill, P. C. Johnson, A. Reboli, J. Aberg, R. Hasbun, and H. H. Hsu.** 2004. Recombinant interferon-gamma 1b as adjunctive therapy for AIDS-related acute cryptococcal meningitis. *J. Infect. Dis.* **189:**2185–2191.
71. **Park, B. J., K. A. Wannemuehler, B. J. Marston, N. Govender, P. G. Pappas, and T. M. Chiller.** 2009. Estimation of the current global burden of cryptococcal meningitis among persons living with HIV/AIDS. *AIDS* **23:**525–530.
72. **Perfect, J. R.** 2009. *Cryptococcus neoformans*. *Microbiol. Infect. Dis.* **45:**395–404.
73. **Perfect, J. R., W. E. Dismukes, F. Dromer, D. L. Goldman, J. R. Graybill, R. J. Hamill, T. S. Harrison, R. A. Larsen, O. Lortholary, M. H. Nguyen, P. G. Pappas, W. G. Powderly, N. Singh, J. D. Sobel, and T. C. Sorrell.** 2010. Clinical practice guidelines for the management of cryptococcal disease: 2010 update by the Infectious Diseases Society of America. *Clin. Infect. Dis.* **50:**291–322.
74. **Perfect, J. R., S. D. R. Lang, and D. T. Durack.** 1980. Chronic cryptococcal meningitis: a new experimental model in rabbits. *Am. J. Pathol.* **101:**177–194.
75. **Polacheck, I.** 1991. The discovery of melanin production in *Cryptococcus neoformans* and its impact on diagnosis and the study of virulence. *Zentralbl. Bakteriol. Hyg. A* **276:**120–123.

76. **Pounden, W. D., J. M. Amberson, and R. F. Jaeger.** 1952. A severe mastitis problem associated with *Cryptococcus neoformans* in a large dairy herd. *Am. J. Vet. Res.* **13:**121–128.
77. **Salas, S. D., J. E. Bennett, K. J. Kwon-Chung, J. R. Perfect, and P. R. Williamson.** 1996. Effect of the laccase gene, CNLAC1, on virulence of *Cryptococcus neoformans*. *J. Exp. Med.* **184:**377–386.
78. **Sanfelice, F.** 1894. Contributo alla morfologia e biolgia dei blastomiceti che si sviluppano nei succhi di alcuni frutti. *Ann. Igiene* **4:**463–495.
79. **Sanford, J. E., D. M. Lupan, A. M. Schlageter, and T. R. Kozel.** 1990. Passive immunization against *Cryptococcus neoformans* with an isotype-switch family of monoclonal antibodies reactive with cryptococcal polysaccharide. *Infect. Immun.* **58:**1919–1923.
80. **Shelburne, S. A., III, J. Darcourt, A. C. White, Jr., S. B. Greenberg, R. J. Hamill, R. L. Atmar, and F. Visnegarwala.** 2005. The role of immune reconstitution inflammatory syndrome in AIDS-related *Cryptococcus neoformans* disease in the era of highly active antiretroviral therapy. *Clin. Infect. Dis.* **40:**1049–1052.
81. **Snider, W. D., D. M. Simpson, S. Nielsen, J. W. M. Gold, C. E. Metroka, and J. B. Posner.** 1983. Neurological complications of acquired immune deficiency syndrome: analysis of 50 patients. *Ann. Neurol.* **14:**403–418.
82. **Steenbergen, J. N., H. A. Shuman, and A. Casadevall.** 2001. *Cryptococcus neoformans* interactions with amoebae suggest an explanation for its virulence and intracellular pathogenic strategy in macrophages. *Proc. Natl. Acad. Sci. USA* **98:**15245–15250.
83. **Stephen, C., S. Lester, W. Black, M. Fyfe, and S. Raverty.** 2002. Multispecies outbreak cryptococcosis on Southern Vancouver Island, British Columbia. *Can. J. Vet. Res.* **43:**792–794.
84. **Stoddard, J. L., and E. C. Cutler.** 1916. *Torula Infection in Man.* Rockefeller Institute for Medical Research, Monograph no. 6, p. 1–98.
85. **Toffaletti, D. L., T. H. Rude, S. A. Johnston, D. T. Durack, and J. R. Perfect.** 1993. Gene transfer in *Cryptococcus neoformans* by use of biolistic delivery of DNA. *J. Bacteriol.* **175:**1405–1411.
86. **Vanbreuseghem, R., and M. Takashio.** 1970. An atypical strain of *Cryptococcus neoformans* (Sanfelice) Vuillemin. Part II. *Cryptococcus neoformans* var. *gatti Nov. Ann. Soc. Belg. Med. Trop.* **50:**695–702.
87. **van der Horst, C. M., M. S. Saag, G. A. Cloud, R. J. Hamill, J. R. Graybill, J. D. Sobel, P. C. Johnson, C. U. Tuazon, T. Kerkering, B. L. Moskovitz, W. G. Powderly, and W. E. Dismukes.** 1997. Treatment of cryptococcal meningitis associated with the acquired immunodeficiency syndrome. National Institute of Allergy and Infectious Diseases Mycoses Study Group and AIDS Clinical Trials Group. *N. Engl. J. Med.* **337:**15–21.
88. **Vecchiarelli, A.** 2010. Fungal capsular polysaccharide and T-cell suppression: the hidden nature of poor immunogenicity. *Crit. Rev. Immunol.* **27**(6)**:**547–557.
89. **Vecchiarelli, A., C. Monari, F. Baldelli, D. Pietrella, C. Retini, C. Tascini, D. Francisci, and F. Bistoni.** 1995. Beneficial effect of recombinant human granulocyte colony-stimulating factor on fungicidal activity of polymorphonuclear leukocytes from patients with AIDS. *J. Infect. Dis.* **171:**1448–1454.
90. **Verse, M.** 1914. Uber einen Fall von generalisierter Blastomykose beim menschen. *Verh. Dtsch. Pathol. Ges.* **17:** 275–278.
91. **Vuillemin, P.** 1901. Les blastomycetes pathogenes. *Rev. Gen. Sci. Pures Appl.* **12:**732–751.
92. **Wickes, B. L., M. E. Mayorga, U. Edman, and J. C. Edman.** 1996. Dimorphism and haploid fruiting in *Cryptococcus neoformans*: association with the alpha-mating type. *Proc. Natl. Acad. Sci. USA* **93:**7327–7331.
93. **Xu, J., R. Vilgalys, and T. G. Mitchell.** 2000. Multiple gene genealogies reveal recent dispersion and hybridization in the human pathogenic fungus *Cryptococcus neoformans*. *Mol. Ecol.* **9:**1471–1481.
94. **Xue, C., Y. Tada, X. Dong, and J. Heitman.** 2007. The human fungal pathogen Cryptococcus can complete its sexual cycle during a pathogenic association with plants. *Cell Host. Microbe* **1:**263–273.

Cryptococcus: From Human Pathogen to Model Yeast
Edited by J. Heitman et al.

3

Biosynthesis and Genetics of the *Cryptococcus* Capsule

GUILHEM JANBON AND TAMARA L. DOERING

The capsule of *Cryptococcus neoformans* has been noted as its distinguishing feature since the first description of this microbe at the end of the 19th century. Pioneer investigators in this field suggested that this polysaccharide structure was the major virulence factor of the fungus and posed fundamental questions: What is the role of the capsule in pathogenicity? What are its components? How is the capsule synthesized? By the time of the publication of the first edition of this book in 1998 (20), some progress had been made on the first two questions, but little was known about capsule synthesis. In one breakthrough, four genes necessary for the presence of a visible capsule had been identified, but their functions were unknown, leaving capsule polysaccharide biosynthesis as a "black box." As will be reviewed below, the past 10 years have been highly productive in this area of *C. neoformans* research. Although numerous questions remain, each advance in our knowledge has revealed the capsule as an ever more fascinating subject.

EARLY OBSERVATIONS OF THE CRYPTOCOCCAL CAPSULE

In 1894, Sanfelice first reported the isolation of *C. neoformans* and showed that this yeast, isolated from fermented fruit juice, induced disease when inoculated into animals (132). He also isolated *C. neoformans* from a bovine ganglion and noted a capsule surrounding the cells (133). The same year Busse independently isolated *C. neoformans* from a gelatinous abscess in a patient who died a few days later (19). Like Sanfelice, he observed that the yeast was encapsulated in vivo but not in vitro (19). The following year, Curtis isolated *C. neoformans* from a young patient who died of meningitis (40). He described the presence of a capsule after in vitro growth on a gelatin-based medium (Fig. 1) and reported that the capsule could be stained with Ziehl's fuchsin, although the coloration was difficult to fix (40).

Guilhem Janbon, Unité des Aspergillus, Institut Pasteur, Paris, France. **Tamara L. Doering,** Department of Molecular Microbiology, Washington University School of Medicine, St. Louis, MO 63110.

In the mid-1930s, Benham first suggested that the cryptococcal capsule was important in virulence (3), although several decades elapsed before publication of experiments designed to analyze its role in the pathophysiology of infection. In 1950, Drouhet, Secretain, and Aubert hypothesized that the capsule was important in phagocytosis and virulence (46), and in the ensuing years they demonstrated the role of capsule polysaccharide in the regulation of leukocyte migration (47). Subsequent studies, reviewed elsewhere in this volume, demonstrated that the capsule is antiphagocytic (46, 90, 91) and is required for replication in macrophages (140) and that capsule polysaccharides may alter antigen presentation, affect cytokine production, inhibit leukocyte migration into infected sites, and induce complement depletion in the host (4, 16, 146). More generally, the capsule represents the surface of interaction between the host and the pathogen, and shed polysaccharides extend the impact of this structure. Any change in composition, conformation, or regulation of these polysaccharides may modify these interactions and potentially alter fungal virulence.

CAPSULE COMPOSITION AND STRUCTURE

Until the late 1940s, little was known about cryptococcal capsule composition or structure. In 1947, Mager and Aschner reported the presence of shed polysaccharide related to the capsule (106); the same year Kligman published the first method for isolating capsule polysaccharides by ethanol precipitation (86). A few years later, two independent teams used paper chromatography to analyze capsule composition (46, 53). Both groups reported mannose, xylose, glucuronic acid, and galactose as the components of *C. neoformans* capsule polysaccharide from serotypes A, B, and C; the same monosaccharides were later found in the capsule of serotype D (9). The first three serotypes were defined in 1949 based on reactivity of cryptococcal strains with rabbit sera (51) (see chapter 4). At that time it had been hypothesized that capsular polysaccharide structure determined *C. neoformans* antigenicity and therefore serotype (51, 52), but confirmation of this suggestion was not obtained until studies of capsule mutant strains in the 1970s and 1980s (29, 76, 90).

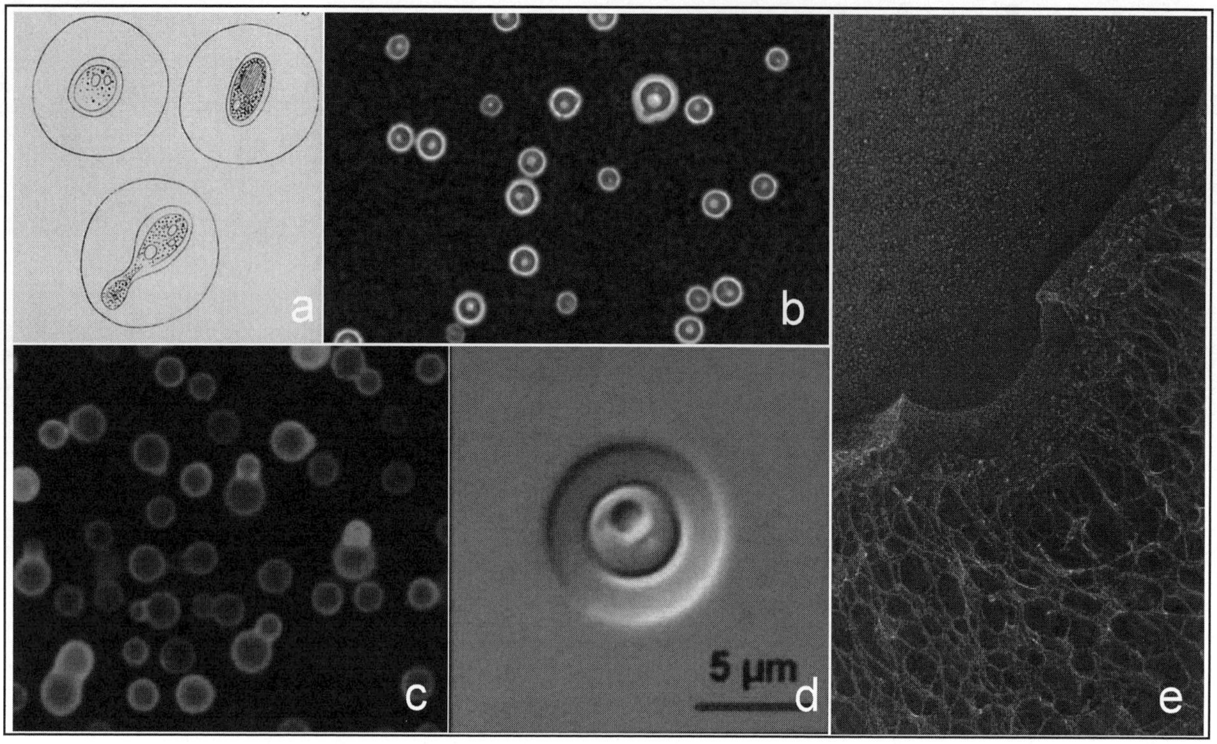

FIGURE 1 The *C. neoformans* capsule. (a) The capsule as drawn by Curtis in 1896 (40). (b) India ink negative staining demonstrates the capsule as a zone of exclusion surrounding the cell. (c) Immunofluorescence microscopy of cryptococci after labeling with an anti-GXM monoclonal antibody. (d) Capsule quelling reaction that occurs when *C. neoformans* is mixed with anti-GXM monoclonal antibody as visualized by differential interference contrast microscopy. Reprinted, with permission, from reference 92. (e) Quick-freeze deep-etch image of the edge of a budding cell. The plasma membrane (upper left) is separated by the cell wall from the fibrous capsule meshwork (extending toward lower right). Both wall and capsule surround both the parent cell and the emerging bud. Reprinted, with permission, from the cover image associated with reference 125. Cryptococcal cell bodies are typically 4 to 6 µm in diameter; the capsule can range from undetectable to ~30 µm in radius.

GXM

In 1958, differential reactivity with antipneumococcal sera was used to resolve two populations of *C. neoformans* capsule polysaccharides. One of the resulting fractions was enriched in galactose, suggesting the presence of at least two distinct polysaccharides in the capsule (123). Bhattacharjee and colleagues, the first to try to solve the structure of the *C. neoformans* capsule polysaccharides, showed in the late 1970s that the dominant polysaccharide of serotypes A, B, C, and D lacked galactose (8–12). The molecular weight of this polysaccharide, termed glucuronoxylomannan (GXM) for its components (see "Polymer Nomenclature: What's in a Name?" below), is on the order of 10^6 Da (62, 108).

GXM structure was studied intensively during the final 2 decades of the past century (13, 31). Nuclear magnetic resonance (NMR) analysis of GXM from numerous strains of different serotypes determined that this polymer consists of a linear α-1,3-linked mannose backbone bearing β-xylose, β-1,2-glucuronic acid, and 6-O-acetyl moieties. The xylose residues can be 1,2- or 1,4-linked. Originally, GXMs from serotypes A and D were grouped as containing only β-1,2-xylose, while B and C were substituted with both β-1,2-xylose and β-1,4-xylose; the overall xylose:mannose:glucuronic acid molar ratios determined for serotypes D, A, B, and C were 1:3:1, 2:3:1, 3:3:1, and 4:3:1, respectively (13). Subsequent study has shown this model to be an oversimplification of GXM structures, which are far less homogeneous than initially believed. Careful analysis of over 100 isolates identified six repeating mannosyl triads, each substituted with a single β-1,2-glucuronic acid residue and characterized by different linkage patterns of xylose (Fig. 2) (33). In the currently accepted model, the GXM of each cryptococcal strain is composed of one or more repeated motifs (33), with recent data suggesting that heterogeneity of repeats may also occur at the level of individual GXM molecules (109).

Beyond the basic carbohydrate structure of GXM, this polymer is significantly O-acetylated on mannose, with the mole percent of O-acetyl groups in one report measured as 8 to 12%, 6.5 to 9.5%, 3%, and 9% for serotypes A, B, C, and D, respectively (142). Despite the abundance of this modification, which may vary with growth conditions (81), structural studies have not identified the position of the O-acetyl residues on the GXM triads (Fig. 2). The only data concerning the position of O-acetylation derive from analysis of a serotype D mutant strain lacking O-acetyl groups, *cas1Δ* (81). Comparison of two-dimensional NMR spectra from this strain and wild-type cells suggests that only two mannose positions of the M1 triad can be 6-O-acetylated: the unbranched

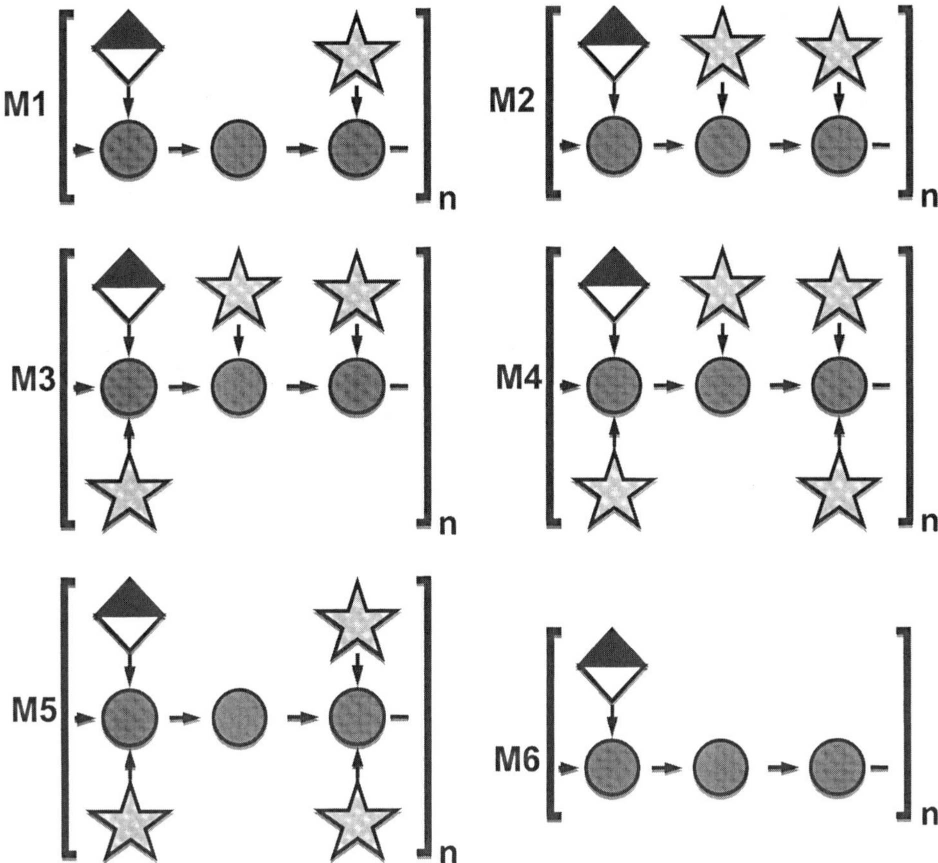

FIGURE 2 The six structural reporter groups of GXM defined by Cherniak et al. (33). Mannose residues (shaded circles) are α-1,3-linked; xylose (stars) and glucuronic acid residues (half-filled diamonds) shown above the mannose backbone are β-1,2-linked; xylose shown below the backbone is β-1,4-linked. All sugars are in pyranose form, and mannose acetylation is not shown. Shapes follow the recommendations of the *Essentials of Glycobiology* (http://www.ncbi.nlm.nih.gov/bookshelf/br.fcgi?book=glyco2). Modified, with permission, from reference 45.

mannose and the mannose substituted with glucuronic acid (81). Studies performed in a serotype A background similarly indicate that the mannose bearing xylose in the M2 triad is never O-acetylated (G. Janbon, unpublished data).

It is important to note that chemical analyses of GXM structure have all been performed on material purified from culture supernatant fluid, even though the relationship between this material and capsule material bound to cells has not yet been defined. Although a common working hypothesis is that these populations are mechanistically related, physical differences have been noted in populations of GXM depending on whether they are collected from culture supernatants or actively released from cells (30, 57). As equally significant differences in GXM are induced by the mode of polysaccharide isolation or release (30, 57), further work will be required to resolve this question, but data interpretation should consider the source of polysaccharide.

GalXM (GXMGal)

The first evidence that the polysaccharide material shed from C. *neoformans* consisted of several molecular populations came from differential reactivity with antisera (123). Methods based on polymer charge, such as differential precipitation with cetyltrimethlylammonium bromide (119), allowed preparative separation of these populations for more detailed investigation (110). Consistent with earlier work, the less acidic polysaccharide was found to be smaller and to include galactose.

In 1982, Cherniak and colleagues reported the first detailed study of the galactose-rich capsule polysaccharide (29). This polymer was found to be 275,000 Da and to contain galactose, mannose, and xylose along with minor amounts of glucuronic acid and acetyl moieties; it was termed galactoxylomannan (GalXM). A few years later these investigators reported that concanavalin A could be used to further separate this GalXM fraction into two pools: one that tightly bound the lectin and had the characteristics of mannoproteins and a second, containing xylose, mannose, and galactose, that was considered to be purified GalXM (141). A decade later, James and Cherniak assessed GalXM isolated from strains of serotypes A, C, and D, characterizing them as a group of "closely related complex polysaccharides" (80). They noted that the bulk of the GalXM bound to anion exchange resin and used this fraction for further analysis. These analyses confirmed the content of mannose, xylose, and galactose (some in the form of galactofuranose), along with some glucose postulated to be of separate origin. Methylation analysis suggested a complex branched structure, in which

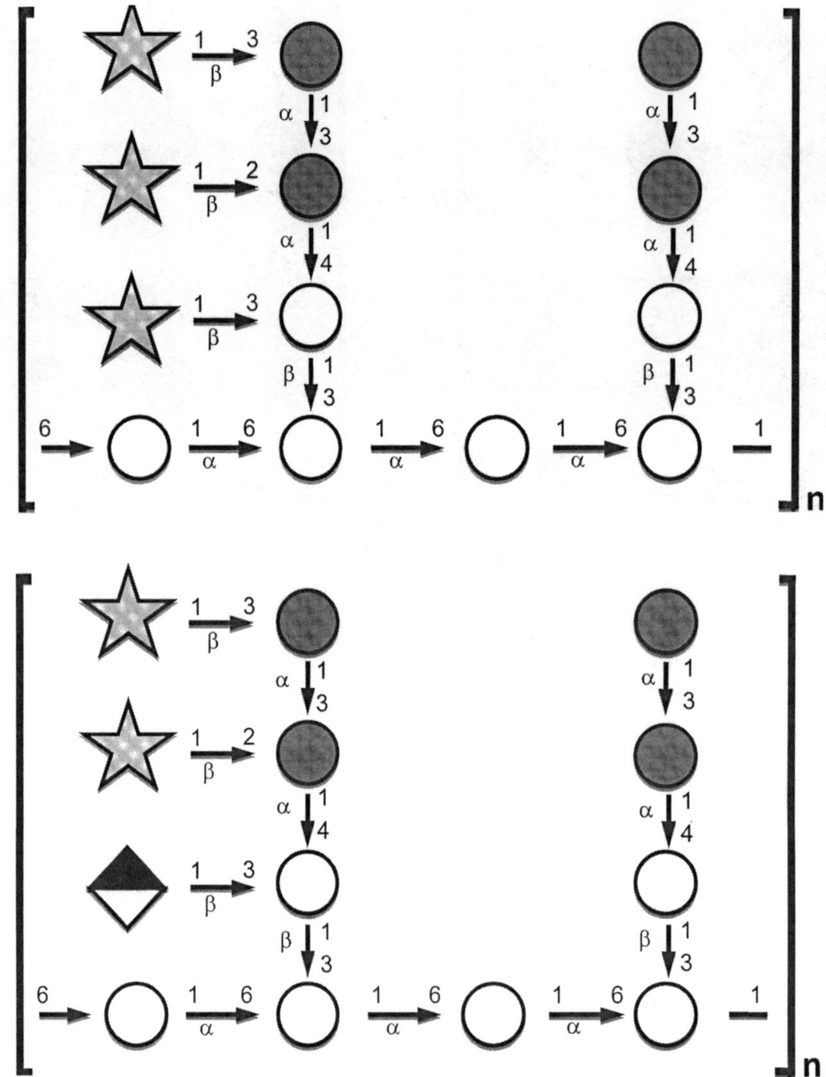

FIGURE 3 The GalXM (GXMGal) structure. (Top) The structure of GalXM proposed by Vaishnav et al. (144). (Bottom) The upper structure as revised by Heiss et al. (73). The side branches that occur on every second galactose of the polymer backbone may be substituted on all three residues, on only the more distal two residues, on only the proximal two residues, or not at all (144). The two extreme cases are shown. Symbols are as in Fig. 2, with galactose shown as open circles and linkages as indicated. Modified, with permission, from reference 45.

the xylose and galactofuranose residues were terminal, but the complexity of the GalXM structure precluded full structure determination by this method.

To address the technical challenge of determining GalXM structure, Vaishnav and coworkers isolated this polymer from a serotype D acapsular strain, Cap67 (144). This eliminated potential contamination from the more abundant GXM and allowed detailed analysis. GalXM was determined to be an α-1,6-linked polymer of galactose, substituted at alternate residues with β-1,3-linked galactose (Fig. 3, top). The latter galactose is extended by an α-1,4-linked dimer of mannose-α1,3-mannose, completing the trimeric side chain. Up to three of the side chain moieties may be substituted with xylose, as shown in Fig. 3. This variable substitution creates an interesting parallel to the variable xylose addition that has been demonstrated for GXM (33), although analyses of this polymer from additional strains of multiple serotypes will be needed to generalize this observation. The positions of galactofuranose modification and acetylation have not been determined.

The NMR spectra used to determine GalXM structure are complicated, due to the branched structure and heterogeneous side chains of this polymer. Substantial simplification of these data was achieved recently by analyzing cells defective in xylose addition (*cxt1*Δ; see below). This study (73) revealed that the GalXM from Cap67 cells contains glucuronic acid, a result different from the original report of the polymer structure. This glucuronic acid is linked to the side chain galactose in a position originally deduced, but not directly demonstrated, to be occupied by xylose (see Fig. 3, bottom). This correction reconciles the polymer structure with its binding to anion exchange resin, a feature that was previously unexplained, and with reported compositional analyses that indicated the presence of glucuronic acid (29, 37).

Polymer Nomenclature: What's in a Name? (*Romeo and Juliet,* II, 2)

Biological nomenclature strives to strike a balance between accuracy, historical continuity, and ease of use. The International Union of Pure and Applied Chemistry was formed in 1919 and soon afterward began to systematize chemical nomenclature based on structural data, information that was lacking for the C. *neoformans* capsule for decades after its discovery. Early indications of capsule polysaccharide structure came from studies in the 1960s that suggested that the backbone of the dominant polysaccharide was composed of mannose (14, 112). Based on this information and compositional analyses, the major polymer was termed glucuronoxylomannan, abbreviated as G_AXM (13) or more commonly as GXM (28). This name adheres to the International Union of Pure and Applied Chemistry guidelines for carbohydrate nomenclature, which suggest that heteropolysaccharides be named for the principal chain component if that is a homopolymer (in this case a mannan), with the other substituents indicated by prefixes in alphabetical order (http://www.chem.qmul.ac.uk/iupac/2carb/). GXM is an appropriate term for all chemotypes of this polymer (Fig. 2), as the name does not specify the stoichiometry of the components. Notably, this name does not indicate the O-acetylation of GXM mannose residues, despite the antigenic importance of this modification.

Studies of the second capsule polymer indicated that its major components were galactose, mannose, and xylose; by analogy to the GXM nomenclature it was termed galactoxylomannan (GalXM). Despite subsequent determination that this structure is actually a galactan (144), the original nomenclature designating it a mannan was retained, presumably for historical reasons. The recent correction to the GalXM structure (73), however, necessitates a reexamination of this nomenclature. Continuing to term this polysaccharide (Fig. 3, bottom) galactoxylomannan (GalXM) is problematic because (i) it is a galactan, not a mannan; (ii) it contains stoichiometric glucuronic acid, which should be indicated in the name; and (iii) GalXM designates a structure (Fig. 3, top) that has not been shown to occur in nature. How should we proceed to rename what may be a family of polymers, as is the case with GXM? One possibility is to use the term GXMGal as an abbreviation for glucuronoxylomannogalactan. This name would maintain parallels to glucuronoxylomannan (GXM) while acknowledging unique features of this galactose-based polymer (45).

Other Capsule Components

Although most of the capsule is polysaccharide, other components have been, or could be, considered as elements of the capsule. Classically, the capsule is described as composed of GXM, GalXM, and mannoproteins (31, 44), based on the original fractionation of shed polysaccharide material (141). Mannoproteins act as powerful immunological modulators (118), as exemplified by MP-98's induction of large quantities of interleukin-2 by CD4+ T cells (96). This mannoprotein is highly homologous to chitin deacetylases (see chapter 6) and contains consensus motifs for glycosylphosphatidylinositol anchorage, both features that suggest surface localization (96). In a recent proteomic study, Eigenheer and colleagues identified 29 extracellular proteins from intact cells and cell walls of an acapsular mutant (*cap59*Δ) (49). All of these proteins contain putative N-terminal signal sequences, and most include predicted glycosylphosphatidylinositol anchor motifs (49); they are likely to be mannoproteins. Such polypeptides may be harvested with the shed polysaccharide that is most often used for capsule analysis, but the criteria for considering them as part of the capsule have not been defined.

Intriguing recent studies have reported hyaluronic acid at the surface of C. *neoformans* cells (26, 83). Although hyaluronic acid may be important for host-fungal interactions (84), how this polymer relates to the capsule is not clear. The presence of lipids as part of the capsule has not been addressed, although several lines of evidence link lipid biosynthesis and virulence (103, 104). Notably, several enzymes required for GXM and GalXM biosynthesis are also important for synthesis of cryptococcal glycosinositolphosphoryl ceramides (21, 70, 72).

Finally, cell wall polymers are involved in capsule association with the cell, although they are not considered part of the capsule itself. Reese and colleagues showed that cell wall α-glucan is required for the association of GXM with the cryptococcal surface (124, 125), and Rodrigues and colleagues recently described chitin-like structures associated with GXM and budding in C. *neoformans* (130). These topics will be addressed in chapter 6.

REGULATION OF CAPSULE FORMATION

Capsule Size

Busse and Sanfelice's earliest observations of the capsules of cells grown in vivo and in vitro suggested that the environment influences capsule synthesis (19, 133), a hypothesis confirmed by the demonstration that capsule size increases dramatically during mammalian infection (5, 102, 128). Capsule size also depends on the site of infection: Rivera and colleagues have shown that the C. *neoformans* cells in lung tissue have much larger capsules than those infecting the brain (128).

The observed variation in capsule size has stimulated systematic in vitro studies to understand the factors involved in this process (Fig. 4), particularly because in vivo observations imply that it is relevant to pathogenesis. In 1958, Littman demonstrated the dependence of capsule size on constituents of culture medium such as vitamins, amino acids, and carbon sources (97). Some years later Granger and colleagues showed that CO_2 induces capsule growth (68); capsule size also increases when cells are incubated with mammalian serum (152). The concentration of iron in growth medium dramatically affects capsule size, such that low-iron medium has been extensively used in the literature to generate large capsules (145). In contrast, other growth parameters have a less marked effect. For example, basic pH positively influences capsule growth but is not sufficient to induce this process, as increasing the pH of complete medium does not result in increased capsule size (48, 55, 68, 152). Osmotic pressure may influence capsule biosynthesis, as high concentrations of glucose, glycerol, or NaCl have been reported to repress capsule size (48, 55, 79). Divalent ion concentration has also been shown to influence capsule size, leading to the hypothesis that size may be influenced by GXM molecule aggregation (120). Extensive data have been published on signal transduction pathways that influence capsule biosynthesis in C. *neoformans* (see chapters 12, 14, and 17 in this book), but how in vitro and in vivo signals affect the size, secretion, aggregative properties, and/or

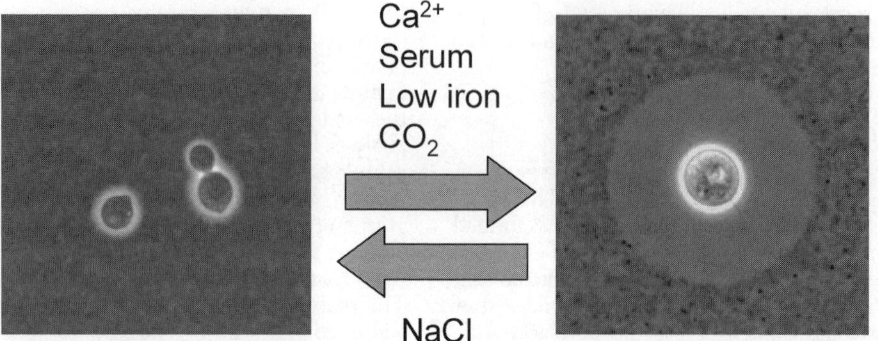

FIGURE 4 Stimuli that lead to an alteration of the capsule size. Additional factors have been studied for their ability to modulate capsule but have a less marked effect. See text for details.

structures of capsule polymers remains to be determined. A recent clue as to one mechanism involved originated with the observation that cells from late stationary phase cultures have larger capsules than cells from repeatedly diluted cultures (150). Interestingly, this capsule enlargement is associated with an apparent size increase in individual polysaccharide molecules shed from those cells, manifested as reduced electromobility (150). This study was complemented by physical analysis that demonstrated that larger GXM molecules are released from cells with induced capsules than from cells with small capsules (58). Together, these reports support a model in which changes in capsule size are mediated, at least in part, at the level of individual GXM molecules (58, 150). These changes in GXM are likely to occur intracellularly because they are observed even in *ags1*Δ mutants (124) that cannot display GXM on the cell surface (150).

Capsule Structure

The capsule shows tremendous variability in multiple biological contexts. At the highest level, the structure of capsule polysaccharides varies between serotypes, as demonstrated using anti-GXM antibodies (Fig. 5), and can differ between distinct isolates of the same serotype (33, 142, 143). Even when dealing with a single strain, capsule structure in both laboratory strains and patient isolates can change over time (36, 56, 63). By comparing GXM structures in sequential isolates from patients with recurrent cryptococcosis, Cherniak and colleagues identified several isolates that underwent a change in GXM structure during the course of infection (32). Additional studies of single strains have demonstrated alterations of capsule structure associated with prolonged in vitro growth and/or experimental cryptococcosis (36, 63). Some variations of capsule

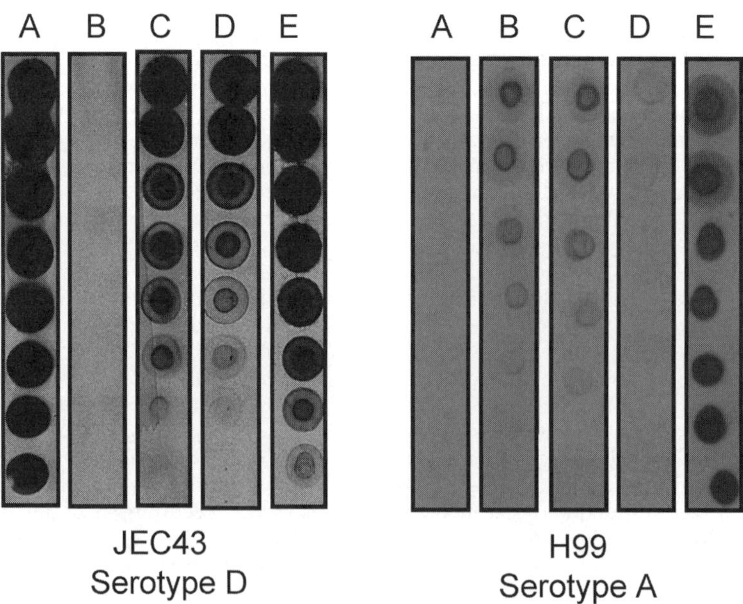

FIGURE 5 Reactivity of different GXM-specific monoclonal antibodies with two strains representative of serotypes A and D. Serial twofold dilutions were spotted (starting with 3×10^4 cells on the first spot at the top of each lane) on a nitrocellulose membrane and probed with a panel of anti-GXM monoclonal antibodies (designated A through E). Reprinted, with permission, from reference 113.

structure are associated with altered colony morphology and have been termed "switching" by analogy with a process described in *Candida albicans* (59, 67, 134). Importantly, this switching phenomenon has been shown to occur during infection and to correlate with altered virulence (60, 61). Variation of capsule structures between cells in a clonal population has been also reported (36, 63, 109, 138). This variation even extends to the level of single cells: microscopic examination and studies of capsule antigen arrangement show that the density of the capsule, and perhaps its structure, varies according to distance from the cell wall and growth conditions (65, 108, 122), testifying to the dynamic nature of this material.

Finally, although no in vitro studies have been published on this subject, the changes in capsule antigenicity observed in vivo (27, 35, 140) suggest that environmental conditions can influence capsule structure in addition to size.

BIOSYNTHETIC PATHWAYS OF CAPSULE POLYSACCHARIDES

Early observations of variation in C. *neoformans* capsule size became the basis for genetic studies initiated by Bulmer and colleagues in the late 1960s. These investigators performed UV mutagenesis of C. *neoformans* and isolated seven mutants that appeared acapsular by negative staining with India ink and were avirulent in a mouse model of infection. Reversion of the mutations restored the capsular phenotype and strain virulence (17). Although this work was not completely rigorous from a genetic standpoint, these pioneering experiments strongly suggested genetic regulation of capsule formation and a link between the integrity of the mutated genes and cryptococcal virulence.

Fifteen years after the original mutagenesis studies, Jacobson and colleagues isolated additional capsule mutant strains and characterized genetic complementation groups (76, 136). This work, combined with a growing understanding of capsule structure, suggested that a large cohort of genes was involved in capsule biosynthesis. Nonetheless, the first "capsule gene" was only cloned in the early 1990s. In a groundbreaking study, Chang and Kwon-Chung used functional complementation of the Jacobson mutant strains to clone the first gene involved in capsule biosynthesis (22). This gene, termed *CAP59*, encodes a novel protein that is essential for GXM biosynthesis and for the virulence of C. *neoformans*. Importantly, this demonstrated that the capsule was a virulence factor as defined by molecular Koch's postulates (54). Using a similar strategy, these researchers successively cloned three other genes: *CAP60*, *CAP64*, and *CAP10* (23–25). As with *CAP59*, these genes are required for GXM biosynthesis and virulence, but sequence homology provides no hints as to function. Although a role in secretion has been proposed for Cap59p (64), the specific functions of these four Cap proteins in GXM biosynthesis are still unknown. Surprisingly, genes encoding Cap-homologous proteins occur in the genomes of organisms devoid of capsule (15), suggesting that such proteins may also be involved in polysaccharide biosynthetic processes distinct from capsule synthesis (82). Multiple *CAP*-homologous genes (Table 1) are present in the C. *neoformans* genomes (100), some of

TABLE 1 The Cap proteins[a]

Protein	Mutant	Protein function	Comments	Reference(s)
Cap10p	A and D	Unknown	Necessary for GXM biosynthesis	25, 114
Cxt1p	A and D	β-1,2-Xylosyltransferase	Important for GXM, GalXM, and GIPC xylosylation	87, 88, 115
Cxt2p	A	β-1,2-Xylosyltransferase		88, 115
Cap1p	A	Unknown		115
Cap2p	A	Unknown		115
Cap4p	A	Unknown		115
Cap59p	A and D	Unknown	Necessary for GXM biosynthesis	22, 114
Cap60p	A and D	Unknown	Necessary for GXM biosynthesis	24, 114
Cmt1p	A and D	α-1,3-Mannosyltransferase	Copurifies with Cas31p	114, 134
Cap6p	A	Unknown		
Cap64p	A and D	Unknown	Necessary for GXM biosynthesis	23, 114
Cas3p	A and D	Unknown	Regulates GXM O-acetylation and xylosylation pattern	113, 115
Cas31p	A and D	Unknown	Regulates GXM O-acetylation and xylosylation pattern; copurifies with Cmt1p	113, 115
Cas32p	A and D	Unknown	Regulates GXM xylosylation pattern	113, 115
Cas33p	A and D	Unknown	Regulates GXM xylosylation pattern	113, 115
Cas34p	A and D	Unknown	Regulates GXM xylosylation pattern	113, 115
Cas35p	A and D	Unknown	Regulates GXM xylosylation pattern and capsule size	113, 115

[a] Three families of Cap proteins have been identified on the basis of sequence homology. Blank lines separate groups of homologs; the original Cap mutants identified by Chang and Kwon-Chung (see text) are indicated in bold. GIPC, glycosinositolphosphoryl ceramide.

them encoding proteins that are directly involved in glycan synthesis or interact with such proteins (88, 135) (see below). This supports the hypothesis that Cap proteins actively participate in capsule polysaccharide synthesis, perhaps as members of the extensive glycosyltransferase families that are characteristic of fungi (87, 95, 105, 111). At the end of the 20th century, only the four *CAP* genes had been implicated in capsule biosynthesis. Since then, a powerful combination of biochemistry, genetics, and genomics has allowed the identification of more than 60 genes potentially involved in this process, although for most of them a specific function has yet to be defined (15, 82, 99, 116). Below we review our current understanding of these genes and their products.

The Synthesis of Capsule Precursors

Polysaccharide synthesis starts with the generation of precursor molecules, the most common being the activated sugars discovered by Leloir (94). Based on the known composition, capsule synthesis would require GDP-mannose, UDP-xylose, UDP-glucuronic acid, and UDP-galactose. A number of proteins involved in the synthesis of these precursors (Fig. 6) have been identified and studied in *C. neoformans*, as will be reviewed below. This work has been facilitated by the fact that many of these enzymes are highly conserved in terms of sequence, as would be expected from the participation of activated sugar precursors in multiple synthetic pathways across biology. In 2001, the gene *MAN1*, encoding phosphomannose isomerase, was isolated from a cryptococcal genomic library using a probe amplified by degenerate PCR (148). Consistent with the central role of mannose in the capsule, gene deletion experiments showed that this enzyme was required for the synthesis of a visible capsule and for virulence (148). In 2001 and 2002, two laboratories independently used distinct strategies to identify the *UXS1* gene encoding UDP-xylose synthase. This enzyme forms UDP-xylose from UDP-glucuronic acid (1, 113), and its activity had previously been detected in *C. neoformans* (77). The biochemical work of Bar-Peled and colleagues on Uxs1p identified a new family of enzymes that are broadly conserved in eukaryotes (1, 71). Across the Atlantic, the generation of a *UXS1* deletion in *C. neoformans* serotype D by Moyrand and colleagues showed that it was required for the presence of xylose residues in the capsule and for virulence (113).

Another step required for capsule construction is the synthesis of UDP-glucuronic acid from UDP-glucose, catalyzed by UDP-glucose dehydrogenase (78). A cryptococcal gene encoding this protein, named *UGD1*, was identified by sequence homology and demonstrated to have the expected activity (69, 115). Biochemical studies showed that, in contrast to what has been reported in other eukaryotes, the cryptococcal protein is a dimer and is membrane associated (2, 69). *UGD1* deletion yields cells that are acapsular and temperature sensitive, suggesting a central role of glucuronic acid in the biology and pathogenicity of *C. neoformans* (69, 115). The absence of UDP-glucuronic acid in *ugd1*Δ mutant cells suggests that alternative pathways for synthesis of this precursor, such as described in plants, do not occur in *Cryptococcus* (18).

UDP-galactose is synthesized via an evolutionarily conserved process called the Leloir pathway (75). A key enzyme of this pathway is a UDP-glucose epimerase that in human cells regulates the equilibrium between UDP-glucose and UDP-galactose (41). In *C. neoformans* two enzymes, Uge1p and Uge2p, catalyze this interconversion (116, 117). Gene deletion experiments showed that *UGE1* is necessary for GalXM biosynthesis, growth at 37°C, and colonization of the brain by *C. neoformans* (116). The presence of galactofuranose residues in GalXM (see above) suggests that the corresponding sugar precursor (UDP-galactofuranose) is required for capsule synthesis. The UDP-galactopyranose mutase that synthesizes this precursor has been identified in *C. neoformans* (7), but deletion of the corresponding gene (*GLF1*) does not yield detectable

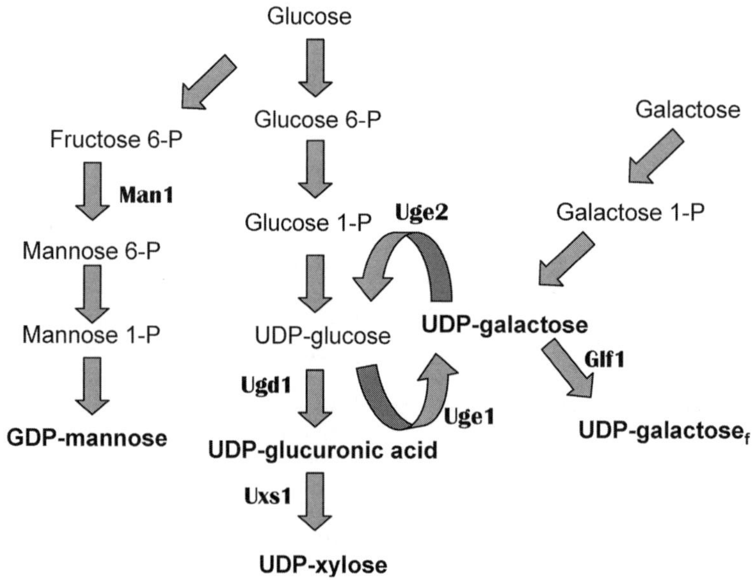

FIGURE 6 Nucleotide sugar metabolism in *C. neoformans*. A simplified reaction scheme, highlighting synthetic steps discussed in the text. Plain text indicates biosynthetic intermediates; bold text on arrows indicates enzyme names; other bold text indicates nucleotide sugars.

changes in cryptococcal morphology (H. Liu and T. L. Doering, unpublished data). This result is not surprising given the low abundance of this modification in the capsule polysaccharide (144).

Transport of Precursors to the Site of Polysaccharide Synthesis

Nucleotide sugar precursors are generally synthesized in the cytosol (127) and are either consumed in that compartment or transported to appropriate sites for glycan polymerization. In eukaryotic cells most glycosylation reactions occur in the lumen of the endoplasmic reticulum and the Golgi apparatus. Nucleotide sugars must therefore be translocated across the membranes of these organelles to become available for use by glycosyltransferases (74, 105, 137). Such translocation is mediated by specific transporters operating via an antiport mechanism that exchanges each nucleotide sugar for the corresponding nucleoside monophosphate (6, 74). Although the site(s) of capsule polysaccharide polymerization is not known, several studies suggest that it occurs at least partly within organelles (see below), so nucleotide sugar transporters must be considered as potential participants in capsule synthesis.

Analysis of the *C. neoformans* genome reveals 10 genes that potentially encode nucleotide sugar transporters (Janbon, unpublished data), but a limited amount has been published concerning their functions. GDP-mannose transport activity has been experimentally demonstrated for Gmt1 and Gmt2 (38), which share substantial amino acid sequence homology with the *Saccharomyces cerevisiae* GDP-mannose transporter Vrg4p (42). Deletion of *GMT1* yields cells with small capsules and a defect in capsule induction, consistent with a role in supplying GDP-mannose for capsule synthesis. Recently, *UGT1* was identified by homology with known UDP-galactose transporter genes (121, 139) and shown to be required for GalXM biosynthesis and virulence (116). The expected biochemical activity of the corresponding protein has also been demonstrated (A. Ashikov, R. Gerardy-Schahn, and T. L. Doering, unpublished data). Further work will be needed to identify the functions and roles in cryptococcal biology of the remaining candidate transporters, some of which may act in capsule regulation (Janbon, unpublished data).

The Polymerization of Capsule Polysaccharides

The complex and varied structures of the cryptococcal capsular polysaccharides suggest that an array of glycosyltransferases will be required for their synthesis. Together, GXM and GalXM exhibit nine distinct carbohydrate linkages, which are likely to require an equal number of independent glycosyltransferases. This may be an underestimate, as fungi frequently employ families of functionally redundant glycosyltransferases, and distinct enzymes may be needed for polymer initiation and elongation, a question that has not been addressed.

Although the genomes of several strains of *C. neoformans* have been sequenced (100), lack of precedent and the heterogeneity of known glycosyltransferases have prevented the sequence-based identification of glycosyltransferases potentially involved in capsule biosynthesis (39, 87, 93). Biochemical approaches have thus been the most productive way to investigate these enzymes. As early as 1990, White and colleagues detected xylose and glucuronic acid transferase activities in crude *C. neoformans* protein preparations, but these were not purified (147). Only recently has the first glycosyltransferase definitively involved in capsule biosynthesis been identified. Klutts and colleagues used an in vitro assay to purify a large protein with β-1,2-xylosyltransferase activity from *C. neoformans* membrane protein preparations (88). They then demonstrated that deletion of the corresponding gene, *CXT1*, in a serotype D background altered the structures of both GXM and GalXM and affected the ability of the corresponding strain to grow in the lungs (89). The same gene, previously named *CAP3*, was also deleted in a serotype A background as part of a systematic gene deletion study; the resulting mutant showed a similar defect in lung colonization (116). Interestingly, Cxt1p belongs to the same protein family as Cap10p, one of the *CAP* mutants described above (25), and defines a new family of glycosyltransferases (88). One of the other homologs also encodes an active xylosyltransferase (Cxt2p) that may similarly act in synthesis of capsule polysaccharides (J. S. Klutts and T. L. Doering, unpublished data).

A biochemical approach has been taken to identify mannosyltransferases involved in capsule biosynthesis, leading to the purification of a membrane protein with α-1,3-mannosyltransferase activity appropriate for a role in synthesis of either capsule polysaccharide (43). However, the deletion of the corresponding gene (*CMT1*) in a serotype D background had no apparent effect on capsule size or morphology (135) and deletion of the same gene in the serotype A background did not produce any in vitro or in vivo phenotype (116). It may be that this protein causes a defect in capsule structure that does not alter virulence or microscopic appearance, that it acts in another cellular pathway, or that other proteins can compensate for its loss; further studies will be needed to resolve these possibilities. Interestingly, the Cmt1p sequence is similar to that of Cap59p, a protein necessary for GXM biosynthesis (22). Cmt1p also copurified with Cas31p, a protein that is probably involved in O-acetylation and/or xylosylation of the capsule (114, 135) (see below). This suggests the existence of protein complexes involved in capsule synthesis, perhaps similar to those demonstrated for the synthesis of N- and O-linked polysaccharides in eukaryotic cells (66, 85).

A large-scale program of systematic gene deletion in *C. neoformans* (99) has recently identified two other genes required for the synthesis of normal GXM molecules, *PBX1* and *PBX2* (98). Strains in which one or both of these genes were deleted produced GXM bearing terminal glucose, although the specific structure was not determined. An interesting suggestion is that these parallel β-helix-containing proteins might edit incorrectly synthesized GXM molecules (98).

Capsule Polysaccharide O-Acetylation

Studies of chemically modified capsule polysaccharides and differentially O-acetylated natural variants of these molecules have implicated capsule O-acetylation in various pathophysiological parameters of cryptococcal infection. Despite this, nothing was known about the pathway of capsule O-acetylation until the past few years (31, 151). In 2001, the first gene involved in O-acetylation of the capsule was isolated by functional complementation of a mutant with a corresponding defect in capsule structure (81). This gene, *CAS1*, encodes a highly conserved membrane protein of unknown function and is required for GXM O-acetylation (81). Although the protein structure of Cas1p suggests it catalyzes O-acetyl transfer (81), definitive proof of its function has not yet been obtained. *cas1*Δ mutants have been used to study the role of O-acetylation in the binding

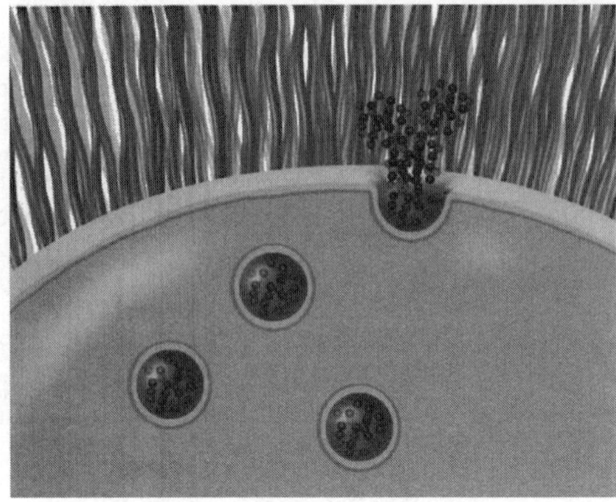

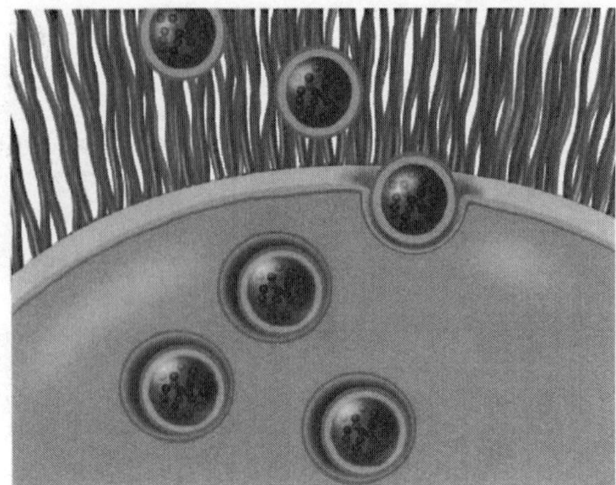

FIGURE 7 Cryptococcal vesicles. Two vesicle types observed in *C. neoformans*. (Left) Vesicles of the classical secretory pathway fusing with the plasma membrane to release their cargo extracellularly. (Right) Vesicles formed within intracellular membrane-bound structures are released intact at the cell surface. Modified, with permission, from reference 45.

of anti-GXM antibodies (92, 107) and to demonstrate the importance of capsule structure in parameters relevant to cryptococcal pathophysiology (50, 92) and in virulence (81). Importantly, comparison of NMR profiles of wild-type and mutant GXM suggests preferential O-acetylation of unbranched mannose over mannose bearing glucuronic acid; xylosylated mannose is generally devoid of O-acetyl residues (81).

Beyond *CAS1*, two additional genes involved in capsule O-acetylation have been cloned. *CAS3* was isolated by functional complementation of a capsule structure mutant (114). Surprisingly, Cas3p belongs to a protein family containing Cap64 (one of the original *cap* mutants [23]) (see above) and five other proteins (see Table 1). Analysis of GXM structures from single and double mutant strains revealed that Cas3p and its closest homolog (Cas31p) are both needed for normal O-acetylation of GXM (114). Furthermore, all tested mutations of genes in this family yield alterations in GXM xylosylation (114), but the biochemical functions of the corresponding proteins are not known.

Finally, three putative proteins with homology to the bacterial O-acetyltransferase Nodl (101) are encoded by *C. neoformans* genes (*CAS9*, *CAS91*, and *CAS92*), but single and multiple mutations in these genes did not cause changes in capsule phenotype (116) (Janbon, unpublished data); these proteins may act in other cellular processes.

Capsule Polysaccharide Intermediates and Secretion

The past few years have seen steady progress in understanding the machinery of capsule synthesis. However, simply identifying these components will not be sufficient to define this process. We must also understand how they work in concert to produce capsule polysaccharides. One interesting open question concerns the intermediates in polymer biosynthesis. For example, are individual trimer repeats of GXM or segments of GalXM first synthesized and then linked, as with synthesis of bacterial peptidoglycan? Or is an extended polymer synthesized and then sequentially modified, as with the conversion of chitin to chitosan (see chapter 6) or the synthesis of mammalian heparan sulfate? Current data on capsule polymer structures are potentially consistent with either scenario, and numerous intermediate models could also be hypothesized; this promises to be an exciting area of future research.

Another central question is the site, or sites, of capsule synthetic reactions. A variety of studies over several decades have addressed this question using ultrastructural approaches sometimes combined with immunodetection, but these have led to varying conclusions (reviewed in 149). One fundamental distinction in eukaryotic glycan formation is whether synthetic reactions occur in the lumen of secretory pathway organelles, as is the case with fungal mannan, or at the cell surface, as with some cell wall glucan and chitin. Materials synthesized in the secretory pathway are packaged into vesicles that fuse with the plasma membrane to release their contents outside of the cell (Fig. 7, left). In support of a lumenal location for capsule synthesis, a conditional mutant generated in both serotypes A and D that is defective in vesicle targeting to the plasma membrane accumulates post-Golgi vesicles containing GXM (149). The conclusion from this study is that GXM is made within the classical secretory pathway, consistent with the requirement for nucleotide sugar transporters to achieve normal capsule synthesis (see above). Other recent studies have suggested that GXM is packaged in vesicles that exit the cell and reach the extracellular compartment intact (129) (Fig. 7, right). These vesicles are hypothesized to be formed intracellularly as multivesicular bodies (131), suggesting that they export capsule materials from a compartment distinct from the secretory pathway. Future studies, including direct localization of proteins involved in capsule synthesis, should help identify the sources of these vesicular cargos. They should also address the nature of the vesicle contents, a critical point because the antibody reactivity used to identify GXM in all of these studies does not define the structure or size of the material detected. Such investigations will help elucidate the mechanism and intermediates of capsule synthesis, allowing the next generation of models for this process to be developed and tested.

FURTHER QUESTIONS

Although the past decade has seen tremendous progress in our understanding of cryptococcal capsule synthesis, multiple aspects of this process remain mysterious. Numerous genes critical for capsule synthesis have been identified, but details of their function remain obscure. Many interactions, both genetic and physical, between components of the capsule synthetic machinery have yet to be explored. How capsule polymer synthesis is initiated and whether nonglycan components are involved in this process is not known. Where the capsule is synthesized is a critical question that may have a complex answer. How capsule polysaccharides are exported remains controversial, as does the related question of how the capsule expands. Finally, the regulation of capsule structure is another key question whose answer may link capsule gene expression with observed structures and serospecificity. There is clearly much work to do to understand the sequence of events that generates a mature capsule and enable its response to environmental stimuli. This should certainly occupy researchers in this field until the next edition of this book and beyond.

REFERENCES

1. **Bar-Peled, M., C. L. Griffith, and T. L. Doering.** 2001. Functional cloning and characterization of a UDP-glucuronic acid decarboxylase: the pathogenic fungus *Cryptococcus neoformans* elucidates UDP-xylose synthesis. *Proc. Natl. Acad. Sci. USA* **98:**12003–12008.
2. **Bar-Peled, M., C. L. Griffith, J. J. Ory, and T. L. Doering.** 2004. Biosynthesis of UDP-GlcA, a key metabolite for capsular polysaccharide synthesis in the pathogenic fungus *Cryptococcus neoformans*. *Biochem. J.* **381:**131–136.
3. **Benham, R. W.** 1935. Cryptococci: their identification by morphology and by serology. *J. Infect. Dis.* **57:**255–274.
4. **Bennett, J. E., and H. F. Hasenclever.** 1965. *Cryptococcus neoformans* polysaccharide: studies on serologic properties and role in infection. *J. Immunol.* **94:**916–920.
5. **Bergman, F.** 1965. Studies on capsule synthesis of *Cryptococcus neoformans*. *Sabouraudia* **4:**23–31.
6. **Berninsone, P. M., and C. B. Hirschberg.** 2000. Nucleotide sugar transporters of the Golgi apparatus. *Curr. Opin. Struct. Biol.* **10:**542–547.
7. **Beverley, S. M., K. L. Owens, M. Showalter, C. L. Griffith, T. L. Doering, V. C. Jones, and M. R. McNeil.** 2005. Eukaryotic UDP-galactopyranose mutase (*GLF* gene) in microbial and metazoal pathogens. *Eukaryot. Cell* **4:**1147–1154.
8. **Bhattacharjee, A. K., K. J. Kwon-Chung, and C. P. Glaudemans.** 1978. On the structure of the capsular polysaccharide from *Cryptococcus neoformans* serotype C. *Immunochemistry* **15:**673–679.
9. **Bhattacharjee, A. K., K. J. Kwon-Chung, and C. P. Glaudemans.** 1979. The structure of the capsular polysaccharide from *Cryptococcus neoformans* serotype D. *Carbohydr. Res.* **73:**183–192.
10. **Bhattacharjee, A. K., K. J. Kwon-Chung, and C. P. Glaudemans.** 1979. On the structure of the capsular polysaccharide from *Cryptococcus neoformans* serotype C. II. *Mol. Immunol.* **16:**531–532.
11. **Bhattacharjee, A. K., K. J. Kwon-Chung, and C. P. Glaudemans.** 1980. Structural studies on the major capsular polysaccharide from *Cryptococcus bacillisporus* serotype B. *Carbohydr. Res.* **82:**103–111.
12. **Bhattacharjee, A. K., K. J. Kwon-Chung, and C. P. J. Glaudemans.** 1981. Capsulated polysaccharides from a parent strain and a possible, mutant strain of *Cryptococcus neoformans* serotype A. *Carbohydr. Res.* **95:**237–248.
13. **Bhattacharjee, A. K., J. E. Bennett, and C. P. J. Glaudemans.** 1984. Capsular polysaccharides of *Cryptococcus neoformans*. *Rev. Infect. Dis.* **6:**619–624.
14. **Blandamer, A., and I. Danishefsky.** 1966. Investigations on the structure of the capsular polysaccharide from *Cryptococcus neoformans* type B. *Biochim. Biophys. Acta* **117:** 305–313.
15. **Bose, I., A. J. Reese, J. J. Ory, G. Janbon, and T. L. Doering.** 2003. A yeast under cover: the capsule of *Cryptococcus neoformans*. *Eukaryot. Cell.* **2:**655–663.
16. **Buchanan, K. L., and J. W. Murphy.** 1998. What makes *Cryptococcus neoformans* a pathogen? *Emerg. Infect. Dis.* **4:** 71–83.
17. **Bulmer, G. S., M. D. Sans, and C. M. Gunn.** 1967. *Cryptococcus neoformans*. I. Nonencapsulated mutants. *J. Bacteriol.* **94:**1475–1479.
18. **Bülter, T., and L. Elling.** 1999. Enzymatic synthesis of nucleotide sugars. *Glycoconj. J.* **16:**147–159.
19. **Busse, O.** 1894. Ueber parasitärezellen schlüsse und thre zuchtung. *Centralbl. Bakt. Parasit.* **16:**175–180.
20. **Casadevall, A., and J. R. Perfect.** 1998. *Cryptococcus neoformans*. ASM Press, Washington, DC.
21. **Castle, S. A., E. A. Owuor, S. H. Thompson, M. R. Garnsey, J. S. Klutts, T. L. Doering, and S. B. Levery.** 2008. β1,2-Xylosyltransferase Cxt1p is solely responsible for xylose incorporation into *Cryptococcus neoformans* glycosphingolipids. *Eukaryot. Cell* **7:**1611–1615.
22. **Chang, Y. C., and K. J. Kwon-Chung.** 1994. Complementation of a capsule-deficient mutation of *Cryptococcus neoformans* restores its virulence. *Mol. Cell. Biol.* **14:**4912–4919.
23. **Chang, Y. C., L. A. Penoyer, and K. J. Kwon-Chung.** 1996. The second capsule gene of *Cryptococcus neoformans*, *CAP64*, is essential for virulence. *Infect. Immun.* **64:**1977–1983.
24. **Chang, Y. C., and K. J. Kwon-Chung.** 1998. Isolation of the third capsule-associated gene *CAP60*, required for virulence in *Cryptococcus neoformans*. *Infect. Immun.* **66:** 2230–2236.
25. **Chang, Y. C., and K. J. Kwon-Chung.** 1999. Isolation, characterization, and localization of a capsule-associated gene, *CAP10*, of *Cryptococcus neoformans*. *J. Bacteriol.* **181:** 5636–5643.
26. **Chang, Y. C., A. Jong, S. Huang, P. Zerfas, and K. J. Kwon-Chung.** 2006. *CPS1*, a homolog of the *Streptococcus pneumoniae* type 3 polysaccharide synthase gene, is important for the pathobiology of *Cryptococcus neoformans*. *Infect. Immun.* **74:**3930–3938.
27. **Charlier, C., F. Chretien, M. Baudrimont, E. Mordelet, O. Lortholary, and F. Dromer.** 2005. Capsule structure changes associated with *Cryptococcus neoformans* crossing of the blood-brain barrier. *Am. J. Pathol.* **166:**421–432.
28. **Cherniak, R., E. Reiss, M. E. Slodki, R. D. Plattner, and S. O. Blumer.** 1980. Structure and antigenic activity of the capsular polysaccharide of *Cryptococcus neoformans* serotype A. *Mol. Immunol.* **17:**1025–1032.
29. **Cherniak, R., E. Reiss, and S. H. Turner.** 1982. A galactomannan antigen of *Cryptococcus neoformans* serotype A. *Carbohydr. Res.* **103:**239–250.
30. **Cherniak, R., L. C. Morris, B. C. Anderson, and S. A. Meyer.** 1991. Facilitated isolation, purification and analysis of glucuronoxylomannan of *Cryptococcus neoformans*. *Infect. Immun.* **59:**59–64.
31. **Cherniak, R., and J. B. Sundstrom.** 1994. Polysaccharide antigens of the capsule of *Cryptococcus neoformans*. *Infect. Immun.* **62:**1507–1512.
32. **Cherniak, R., L. C. Morris, T. Belay, E. D. Spitzer, and A. Casadevall.** 1995. Variation in the structure of glucuronoxylomannan in isolates from patients with recurrent cryptococcal meningitis. *Infect. Immun.* **63:**1899–1905.
33. **Cherniak, R., H. Valafar, L. C. Morris, and F. Valafar.** 1998. *Cryptococcus neoformans* chemotyping by quantitative

analysis ^{1}H nuclear magnetic resonance spectra of glucuronoxylomannans with a computer-simulated artificial neural network. *Clin. Diagn. Lab. Immunol.* **5:**146–159.
34. [Reference deleted.]
35. **Chretien, F., O. Lortholary, I. Kansau, S. Neuville, F. Gray, and F. Dromer.** 2002. Pathogenesis of cerebral *Cryptococcus neoformans* infection after fungemia. *J. Infect. Dis.* **186:**522–530.
36. **Cleare, W., R. Cherniak, and A. Casadevall.** 1999. In vitro and in vivo stability of *Cryptococcus neoformans* glucuronoxylomannan epitope that elicits protective antibodies. *Infect. Immun.* **67:**3096–3107.
37. **Corradini, C., G. Canali, A. Cavazza, D. Delfino, and G. Teti.** 1998. Compositional analysis of the major capsular polysaccharides of *Cryptococcus neoformans* by high performance anion-exchange chromatography with pulsed amperometric detection (HPAEC-PAD). *J. L. Chromatogr. Related Technol.* **21:**941–951.
38. **Cottrell, T. R., C. L. Griffith, H. Liu, A. A. Nenninger, and T. L. Doering.** 2007. The pathogenic fungus *Cryptococcus neoformans* expresses two functional GDP-mannose transporters with distinct expression patterns and roles in capsule synthesis. *Eukaryot. Cell* **6:**776–785.
39. **Coutinho, P. M., E. Deleury, G. J. Davis, and B. Henrissat.** 2003. An evolving hierarchical family classification for glycosyltransferases. *J. Mol. Evol.* **328:**307–317.
40. **Curtis, F.** 1896. Contribution à l'étude de la saccharomycose humaine. *Ann. Inst. Pasteur* **15:**449–468.
41. **Daude, N., T. K. Gallaher, M. Zeschnigk, A. Starzinski-Powitz, K. G. Petry, I. S. Haworth, and J. K. Reichardt.** 1995. Molecular cloning, characterization, and mapping of a full-length cDNA encoding human UDP-galactose 4'-epimerase. *Biochem. Mol. Med.* **56:**1–7.
42. **Dean, N., Y. B. Zhang, and J. B. Poster.** 1997. The *VRG4* gene is required for GDP-mannose transport into the lumen of the Golgi in the yeast, *Saccharomyces cerevisiae*. *J. Biol. Chem.* **272:**31908–31914.
43. **Doering, T. L.** 1999. A unique α-1,3 mannosyltransferase of the pathogenic fungus *Cryptococcus neoformans*. *J. Bacteriol.* **181:**5482–5488.
44. **Doering, T. L.** 2000. How does *Cryptococcus* get its coat? *Trends Microbiol.* **8:**547–553.
45. **Doering, T. L.** 2009. How sweet it is! Capsule formation and cell wall biogenesis in *Cryptococcus neoformans*. *Ann. Rev. Microbiol.* **63:**223–247.
46. **Drouhet, E., G. Secretain, and J. P. Aubert.** 1950. Polyoside capsulaire d'un champignon pathogene *Torulopsis neoformans* relation avec la virulence. *Ann. Inst. Pasteur* **79:**891–900.
47. **Drouhet, E., and G. Secretain.** 1951. Inhibition de la migration leucocytaire *in vitro* par un polyoside capsulaire de *Torulopsis* (*Cryptococcus*) *neoformans*. *Ann. Inst. Pasteur* **81:**674–676.
48. **Dykstra, M. A., L. Friedman, and J. W. Murphy.** 1977. Capsule size of *Cryptococcus neoformans*: control and relationship to virulence. *Infect. Immun.* **16:**129–135.
49. **Eigenheer, R. A., Y. Jin Lee, E. Blumwald, B. S. Phinney, and A. Gelli.** 2007. Extracellular glycosylphosphatidylinositol-anchored mannoproteins and proteases of *Cryptococcus neoformans*. *FEMS Yeast Res.* **7:**499–510.
50. **Ellerbroek, P. M., D. J. Lefeber, R. van Veghel, J. Scharringa, E. Brouwer, G. J. Gerwig, G. Janbon, A. I. Hoepelman, and F. E. Coenjaerts.** 2004. O-acetylation of cryptococcal capsular glucuronoxylomannan is essential for interference with neutrophil migration. *J. Immunol.* **173:**7513–7520.
51. **Evans, E. E.** 1949. An immunologic comparison of twelve strains of *Cryptococcus neoformans* (*Torula histoytica*). *Proc. Soc. Exp. Biol. Med.* **71:**644–646.
52. **Evans, E. E.** 1950. The antigenic composition of *Cryptococcus neoformans*. *J. Immunol.* **64:**423–430.
53. **Evans, E. E., and J. W. Mehl.** 1950. A qualitative analysis of capsular polysaccharides from *Cryptococcus neoformans* by filter paper chromatography. *Science* **114:**10–11.
54. **Falkow, S.** 1988. Molecular Koch's postulates applied to microbial pathogenicity. *Rev. Infect. Dis.* **10:**5274–5276.
55. **Farhi, F., G. S. Bulmer, and J. R. Tacker.** 1970. *Cryptococcus neoformans*. IV. The not-so-encapsulated yeast. *Infect. Immun.* **1:**526–531.
56. **Franzot, S. P., J. Mukherjee, R. Cherniak, L. Chen, J. S. Hamdan, and A. Casadevall.** 1998. Microevolution of a standard strain of *Cryptococcus neoformans* resulting in differences in virulence and other phenotypes. *Infect. Immun.* **66:**89–97.
57. **Frases, S., L. Nimrichter, N. B. Viana, A. Nakouzi, and A. Casadevall.** 2008. *Cryptococcus neoformans* capsular polysaccharide and exopolysaccharide fractions manifest physical, chemical, and antigenic differences. *Eukaryot. Cell.* **7:**319–327.
58. **Frases, S., B. Pontes, L. Nimrichter, N. B. Viana, M. L. Rodrigues, and A. Casadevall.** 2009. Capsule of *Cryptococcus neoformans* grows by enlargement of polysaccharide molecules. *Proc. Natl. Acad. Sci. USA* **106:**1228–1233.
59. **Fries, B. C., D. L. Goldman, R. Cherniak, R. Ju, and A. Casadevall.** 1999. Phenotypic switching in *Cryptococcus neoformans* results in changes in cellular morphology and glucuronoxylomannan structure. *Infect. Immun.* **67:**6076–6083.
60. **Fries, B. C., C. P. Taborda, E. Serfass, and A. Casadevall.** 2001. Phenotypic switching of *Cryptococcus neoformans* occurs in vivo and influences the outcome of infection. *J. Clin. Invest.* **108:**1639–1648.
61. **Fries, B. C., S. C. Lee, R. Kennan, W. Zhao, A. Casadevall, and D. L. Goldman.** 2005. Phenotypic switching of *Cryptococcus neoformans* can produce variants that elicit increased intracranial pressure in a rat model of cryptococcal meningoencephalitis. *Infect. Immun.* **73:**1779–1787.
62. **Gadebusch, H. H., P. A. Ward, and E. P. Frenkel.** 1964. Natural host resistance to infection with *Cryptococcus neoformans*. III. The effect of cryptococcal polysacharide upon physiology of the reticuloendothelial system of laboratory animals. *J. Infect. Dis.* **114:**95–106.
63. **Garcia-Hermoso, D., F. Dromer, and G. Janbon.** 2004. *Cryptococcus neoformans* capsule structure evolution in vitro and during murine infection. *Infect. Immun.* **72:**3359–3365.
64. **García-Rivera, J., Y. C. Chang, K. J. Kwon-Chung, and A. Casadevall.** 2004. *Cryptococcus neoformans CAP59* (or Cap59p) is involved in the extracellular trafficking of capsular glucuronoxylomannan. *Eukaryot. Cell* **3:**385–392.
65. **Gates, M. A., P. Thorkildson, and T. R. Kozel.** 2004. Molecular architecture of the *Cryptococcus neoformans* capsule. *Mol. Microbiol.* **52:**13–24.
66. **Girrbach, V., and S. Strahl.** 2003. Members of the evolutionary conserved PMT family of protein O-mannosyltransferases from distinct protein complexes among themselves. *J. Biol. Chem.* **278:**12554–12562.
67. **Goldman, D. L., B. C. Fries, S. P. Franzot, L. Montella, and A. Casadevall.** 1998. Phenotypic switching in the human pathogenic fungus *Cryptococcus neoformans* is associated with change in virulence and pulmonary inflammatory response in rodents. *Proc. Natl. Acad. Sci. USA* **95:**14967–14972.
68. **Granger, D. L., J. R. Perfect, and D. T. Durack.** 1985. Virulence of *Cryptococcus neoformans*. Regulation of capsule synthesis by carbon dioxide. *J. Clin. Invest.* **76:**508–516.
69. **Griffith, C. L., J. S. Klutts, L. Zhang, S. B. Levery, and T. L. Doering.** 2004. UDP-glucose dehydrogenase plays

multiple roles in the biology of the pathogenic fungus *Cryptococcus neoformans*. *J. Biol. Chem.* **279:**51669–51676.

70. **Gutierrez, A. L., L. Farage, M. N. Melo, R. S. Mohana-Borges, Y. Guerardel, B. Coddeville, J. M. Wieruszeski, L. Mendonça-Previato, and J. O. Previato.** 2007. Characterization of glycoinositolphosphoryl ceramide structure mutant strains of *Cryptococcus neoformans*. *Glycobiology* **17:**1C–11C.
71. **Harper, A. D., and M. Bar-Peled.** 2002. Biosynthesis of UDP-xylose. Cloning and characterization of a novel *Arabidopsis* gene family, *UXS*, encoding soluble and putative membrane-bound UDP-glucuronic acid decarboxylase isoforms. *Plant Physiology* **130:**2188–2198.
72. **Heise, N., A. L. Gutierrez, K. A. Mattos, C. Jones, R. Wait, J. O. Previato, and L. Mendonça-Previato.** 2002. Molecular analysis of a novel family of complex glycoinositolphosphoryl ceramides from *Cryptococcus neoformans*: structural differences between encapsulated and acapsular yeast forms. *Glycobiology* **12:**409–420.
73. **Heiss, C., J. S. Klutts, Z. Wang, T. L. Doering, and P. Azadi.** 2009. The structure of *Cryptococcus neoformans* galactoxylomannan contains beta-D-glucuronic acid. *Carbohydr. Res.* **344:**915–920.
74. **Hirschberg, C. B., P. W. Robbins, and C. Abeijon.** 1998. Transporters of nucleotide sugars, ATP, and nucleotide sulfate in the endoplasmic reticulum and Golgi apparatus. *Annu. Rev. Biochem.* **67:**49–69.
75. **Holden, H. M., I. Rayment, and J. B. Thoden.** 2003. Structure and function of enzymes of the Leloir pathway for galactose metabolism. *J. Biol. Chem.* **278:**43885–43888.
76. **Jacobson, E. S., D. J. Ayers, A. C. Harrel, and C. C. Nicholas.** 1982. Genetic and phenotypic characterization of capsule mutants of *Cryptococcus neoformans*. *J. Bacteriol.* **150:**1292–1296.
77. **Jacobson, E. S., and W. R. Payne.** 1982. UDP-glucuronate decarboxylase and synthesis of capsular polysaccharide in *Cryptococcus neoformans*. *J. Bacteriol.* **152:**932–934.
78. **Jacobson, E. S.** 1987. Cryptococcal UDP-glucose dehydrogenase: enzymatic control of capsular biosynthesis. *J. Med. Vet. Mycol.* **25:**131–135.
79. **Jacobson, E. S., M. J. Tingler, and P. L. Quynn.** 1989. Effect of hypersolutes upon the polysaccharide capsule in *Cryptococcus neoformans*. *Mycoses* **32:**14–23.
80. **James, P. G., and R. Cherniak.** 1992. Galactoxylomannans of *Cryptococcus neoformans*. *Infect. Immun.* **60:**1084–1088.
81. **Janbon, G., U. Himmelreich, F. Moyrand, L. Improvisi, and F. Dromer.** 2001. Cas1p is a membrane protein necessary for the O-acetylation of the *Cryptococcus neoformans* capsular polysaccharide. *Mol. Microbiol.* **42:**453–467.
82. **Janbon, G.** 2004. *Cryptococcus neoformans* capsule biosynthesis and regulation. *FEMS Yeast Res.* **48:**765–771.
83. **Jong, A., C. H. Wu, H. M. Chen, F. Luo, K. J. Kwon-Chung, Y. C. Chang, C. W. Lamunyon, A. Plaas, and S. H. Huang.** 2007. Identification and characterization of *CPS1* as a hyaluronic acid synthase contributing to the pathogenesis of *Cryptococcus neoformans* infection. *Eukaryot. Cell.* **6:**1486–1496.
84. **Jong, A., C. H. Wu, G. M. Shackleford, K. J. Kwon-Chung, Y. C. Chang, H. M. Chen, Y. Ouyang, and S. H. Huang.** 2008. Involvement of human CD44 during *Cryptococcus neoformans* infection of brain microvascular endothelial cells. *Cell. Microbiol.* **10:**1313–1326.
85. **Jungmann, J., and S. Munro.** 1998. Multiprotein complexes in the cis Golgi of *Saccharomyces cerevisiae* with a-1,6-mannosyltransferase activity. *EMBO J.* **17:**423–434.
86. **Kligman, A. M.** 1947. Studies of the capsular substance of *Torula histolytica* and the immunologic properties of Torula cells. *J. Immunol.* **57:**395–401.
87. **Klutts, J. S., A. Yoneda, M. C. Reilly, I. Bose, and T. L. Doering.** 2006. Glycosyltransferases and their products: cryptococcal variations on fungal themes. *FEMS Yeast Res.* **6:**499–512.
88. **Klutts, J. S., S. B. Levery, and T. L. Doering.** 2007. A beta-1,2-xylosyltransferase from *Cryptococcus neoformans* defines a new family of glycosyltransferases. *J. Biol. Chem.* **282:**17890–17899.
89. **Klutts, J. S., and T. L. Doering.** 2008. Cryptococcal xylosyltransferase 1 (Cxt1p) from *Cryptococcus neoformans* plays a direct role in the synthesis of capsule polysaccharides. *J. Biol. Chem.* **283:**14327–14334.
90. **Kozel, T. R., and J. Cazin, Jr.** 1971. Nonencapsulated variant of *Cryptococcus neoformans*. *Infect. Immun.* **3:** 287–294.
91. **Kozel, T. R.** 1995. Virulence factors of *Cryptococcus neoformans*. *Trends Microbiol.* **3:**295–299.
92. **Kozel, T. R., S. M. Levitz, F. Dromer, M. A. Gates, P. Thorkildson, and G. Janbon.** 2003. Antigenic and biological characteristics of mutant strains of *Cryptococcus neoformans* lacking capsular O-acetylation or xylosyl side chains. *Infect. Immun.* **71:**2868–2875.
93. **Lairson, L. L., B. Henrissat, G. J. Davies, and S. G. Withers.** 2008. Glycosyltransferases: structures, functions, and mechanisms. *Annu. Rev. Biochem.* **77:**521–555.
94. **Leloir, L. F.** 1970. Two decades of research on the biosynthesis of saccharides. *Science* **172:**1299–1303.
95. **Lengeler, K. B., D. Tielker, and J. F. Ernst.** 2008. Protein-O-mannosyltransferases in virulence and development. *Cell. Mol. Life Sci.* **65:**528–544.
96. **Levitz, S. M., S. Nong, M. K. Mansour, C. Huang, and C. A. Specht.** 2001. Molecular characterization of a mannoprotein with homology to chitin deacetylases that stimulates T-cell responses to *Cryptococcus neoformans*. *Proc. Natl. Acad. Sci. USA* **98:**10422–10427.
97. **Littman, M. L.** 1958. Capsule synthesis by *Cryptococcus neoformans*. *Trans. N.Y. Acad. Sci.* **20:**623–648.
98. **Liu, O. W., M. J. Kelly, E. D. Chow, and H. D. Madhani.** 2007. Parallel beta-helix proteins required for accurate capsule polysaccharide synthesis and virulence in the yeast *Cryptococcus neoformans*. *Eukaryot. Cell* **6:** 630–640.
99. **Liu, O. W., C. D. Chun, E. D. Chow, C. Chen, H. D. Madhani, and S. M. Noble.** 2008. Systematic genetic analysis of virulence in the human fungal pathogen *Cryptococcus neoformans*. *Cell* **135:**174–188.
100. **Loftus, B., E. Fung, P. Roncaglia, D. Rowley, P. Amedeo, D. Bruno, J. Vamathevan, M. Miranda, I. Anderson, J. A. Fraser, J. Allen, I. Bosdet, M. R. Brent, R. Chiu, T. L. Doering, M. J. Donlin, C. A. D'Souza, D. S. Fox, V. Grinberg, J. Fu, M. Fukushima, B. Haas, J. C. Huang, G. Janbon, S. Jones, M. I. Krzywinski, K. J. Kwon-Chung, K. B. Lengeler, R. Maiti, M. Marra, R. E. Marra, C. Mathewson, T. G. Mitchell, M. Pertea, F. Riggs, S. L. Salzberg, J. Schein, A. Shvartsbeyn, H. Shin, C. Specht, B. Suh, A. Tenney, T. Utterback, B. L. Wickes, N. Wye, J. W. Kronstad, J. K. Lodge, J. Heitman, R. W. Davis, C. M. Fraser, and R. W. Hyman.** 2005. The genome and transcriptome of *Cryptococcus neoformans*, a basidiomycetous fungal pathogen of humans. *Science* **307:**1321–1324.
101. **López-Lara, I. M., D. Kafetzopoulos, H. P. Spaink, and J. E. Thomas-Oates.** 2001. Rhizobial NodL O-acetyl transferase and NodS N-methyl transferase functionally interfere in production of modified Nod factors. *J. Bacteriol.* **183:**3408–3416.
102. **Love, G. L., G. D. Boyd, and D. L. Greer.** 1985. Large *Cryptococcus neoformans* from brain abscess. *J. Clin. Microbiol.* **22:**1068–1070.

103. **Luberto, C., D. L. Toffaletti, E. A. Wills, S. C. Tucker, A. Casadevall, J. R. Perfect, Y. A. Hannun, and M. Del Poeta.** 2001. Roles for inositol-phosphoryl ceramide synthase 1 (*IPC1*) in pathogenesis of *C. neoformans*. *Genes Dev.* **12:**201–212.
104. **Luberto, C., B. Martinez-Marino, D. Taraskiewicz, B. Bolanos, P. Chitano, D. Toffaletti, G. Cox, J. Perfect, Y. Hannun, E. Balish, and M. Del Poeta.** 2003. Identification of App1 as a regulator of phagocytosis and virulence of *Cryptococcus neoformans*. *J. Clin. Invest.* **112:** 1080–1094.
105. **Lussier, M., A. M. Sdicu, and H. Bussey.** 1999. The *KTR* and *MNN1* mannosyltransferase families of *Saccharomyces cerevisiae*. *Biochim. Biophys. Acta* **1426:**323–334.
106. **Mager, J., and M. Aschner.** 1947. Biological studies on capsulated yeasts. *J. Bacteriol.* **53:**283–295.
107. **McFadden, D. C., and A. Casadevall.** 2004. Unexpected diversity in the fine specificity of monoclonal antibodies that use the same V region gene to glucuronoxylomannan of *Cryptococcus neoformans*. *J. Immunol.* **172:**3670–3677.
108. **McFadden, D. C., M. De Jesus, and A. Casadevall.** 2006. The physical properties of the capsular polysaccharides from *Cryptococcus neoformans* suggest features for capsule construction. *J. Biol. Chem.* **281:**1868–1875.
109. **McFadden, D. C., B. C. Fries, F. Wang, and A. Casadevall.** 2007. Capsule structural heterogeneity and antigenic variation in *Cryptococcus neoformans*. *Eukaryot. Cell* **6:**1464–1473.
110. **Merrifield, E. H., and A. M. Stephen.** 1980. Structural investigations of two capsular polysaccharides from *Cryptococcus neoformans*. *Carbohydr. Res.* **86:**69–76.
111. **Mille, C., P. Bobrowicz, P. A. Trinel, H. Li, E. Maes, Y. Guerardel, C. Fradin, M. Martínez-Esparza, R. C. Davidson, G. Janbon, D. Poulain, and S. Wildt.** 2008. Identification of a new family of genes involved in β1,2-mannosylation of glycans in *Pichia pastoris* and *Candida albicans*. *J. Biol. Chem.* **283:**9724–9736.
112. **Miyazaki, T.** 1961. Studies on fungal polysaccharides. III. Chemical structure of the capsular polysaccharide from *Cryptococcus neoformans*. *Chem. Pharm. Bull.* **9:**829– 833.
113. **Moyrand, F., B. Klaproth, U. Himmelreich, F. Dromer, and G. Janbon.** 2002. Isolation and characterization of capsule structure mutant strains of *Cryptococcus neoformans*. *Mol. Microbiol.* **45:**837–849.
114. **Moyrand, F., Y. C. Chang, U. Himmelreich, K. J. Kwon-Chung, and G. Janbon.** 2004. Cas3p belongs to a seven member family of capsule structure designer proteins. *Eukaryot. Cell* **3:**1513–1524.
115. **Moyrand, F., and G. Janbon.** 2004. *UGD1* encoding the *Cryptococcus neoformans* UDP-glucose dehydrogenase is essential for growth at 37°C and for capsule biosynthesis. *Eukaryot. Cell* **3:**1601–1608.
116. **Moyrand, F., T. Fontaine, and G. Janbon.** 2007. Systematic capsule gene disruption reveals the central role of galactose metabolism on *Cryptococcus neoformans* virulence. *Mol. Microbiol.* **64:**771–781.
117. **Moyrand, F., I. Lafontaine, T. Fontaine, and G. Janbon.** 2008. *UGE1* and *UGE2* regulate the UDP-glucose/ UDP-galactose equilibrium in *Cryptococcus neoformans*. *Eukaryot. Cell* **7:**2069–2077.
118. **Murphy, J. W.** 1998. Protective cell-mediated immunity against *Cryptococcus neoformans*. *Res. Immunol.* **149:** 373–386.
119. **Nimmich, W.** 1968. Isolation and qualitative component analysis of *Klebsiella* K-antigens. *Z. Med. Mikrobiol. Immunol.* **154:**117–131.
120. **Nimrichter, L., S. Frases, L. P. Cinelli, N. B. Viana, A. Nakouzi, L. R. Travassos, A. Casadevall, and M. L. Rodrigues.** 2007. Self-aggregation of *Cryptococcus neoformans* capsular glucuronoxylomannan is dependent on divalent cations. *Eukaryot. Cell* **6:**1400–1410.
121. **Norambuena, L., L. Marchant, P. Berninsone, C. B. Hirschberg, H. Silva, and A. Orellana.** 2002. Transport of UDP-galactose in plants. Identification and functional characterization of AtUTr1, an *Arabidopsis thaliana* UDP-galactose/UDP-glucose transporter. *J. Biol. Chem.* **277:**32923–32929.
122. **Pierini, L. M., and T. L. Doering.** 2001. Spacial and temporal sequence of capsule construction in *Cryptococcus neoformans*. *Mol. Microbiol.* **41:**105–115.
123. **Rebers, P. A., S. A. Barker, M. Heidelberger, Z. Dische, and E. E. Evans.** 1958. Precipitation of the specific polysaccharide of *Cryptococcus neoformans* A by types II and XIV antipneumococcal sera. *J. Am. Chem. Soc.* **80:** 1135–1137.
124. **Reese, A. J., and T. L. Doering.** 2003. Cell wall α-1,3-glucan is required to anchor the *Cryptococcus neoformans* capsule. *Mol. Microbiol.* **50:**1401–1409.
125. **Reese, A. J., A. Yoneda, J. A. Breger, A. Beauvais, H. Liu, C. L. Griffith, I. Bose, M. J. Kim, C. Skau, S. Yang, J. A. Sefko, M. Osumi, J. P. Latge, E. Mylonakis, and T. L. Doering.** 2007. Loss of cell wall alpha(1-3) glucan affects *Cryptococcus neoformans* from ultrastructure to virulence. *Mol. Microbiol.* **63:**1385–1398.
126. [Reference deleted.]
127. **Reiter, W. D., and G. F. Vanzin.** 2001. Molecular genetics of nucleotide sugar interconversion pathways in plants. *Plant Mol. Biol.* **47:**95–113.
128. **Rivera, J., M. Feldmesser, M. Cammer, and A. Casadevall.** 1998. Organ-dependent variation of capsule thickness in *Cryptococcus neoformans* during experimental murine infection. *Infect. Immun.* **66:**5027–5030.
129. **Rodrigues, M. L., L. Nimrichter, D. L. Oliveira, S. Frases, K. Miranda, O. Zaragoza, M. Alvarez, A. Nakouzi, M. Feldmesser, and A. Casadevall.** 2007. Vesicular polysaccharide export in *Cryptococcus neoformans* is a eukaryotic solution to the problem of fungal trans-cell wall transport. *Eukaryot. Cell* **6:**48–59.
130. **Rodrigues, M. L., M. Alvarez, F. L. Fonseca, and A. Casadevall.** 2008. Binding of the wheat germ lectin to *Cryptococcus neoformans* suggests an association of chitinlike structures with yeast budding and capsular glucuronoxylomannan. *Eukaryot. Cell* **7:**602–609.
131. **Rodrigues, M. L., E. S. Nakayasu, D. L. Oliveira, L. Nimrichter, J. D. Nosanchuk, I. C. Almeida, and A. Casadevall.** 2008. Extracellular vesicles produced by *Cryptococcus neoformans* contain protein components associated with virulence. *Eukaryot. Cell* **7:**58–67.
132. **Sanfelice, F.** 1894. Contributo alla morphologia e biologica dei blastomiceti che sisciluppano nei succhi di alcuni frutti. *Ann. Ist. Ig. R. Univ. Roma* **4:**463–469.
133. **Sanfelice, F.** 1895. Ueber einen neuen pathogenen Blastomyceten, welcher innerhalb der Gewebe unter Bildung kalkartig aussehender Massen degeneriert. *Zentralbl. Bakt. Parasit.* **18:**521–526.
134. **Soll, D. R.** 1992. High-frequency switching in *Candida albicans*. *Clin. Microbiol. Rev.* **5:**183–203.
135. **Sommer, U., H. Liu, and T. L. Doering.** 2003. An α-1,3-mannosyltransferase of *Cryptococcus neoformans*. *J Biol Chem* **278:**47724–47730.
136. **Still, C. N., and E. S. Jacobson.** 1983. Recombinational mapping of capsule mutations in *Cryptococcus neoformans*. *J. Bacteriol.* **156:**460–462.
137. **Strahl-Bolsinger, S., M. Gentzsch, and W. Tanner.** 1999. Protein O-mannosylation. *Biochim. Biophys. Acta* **1426:**297–307.
138. **Sumner, E. R., and S. V. Avery.** 2002. Phenotypic heterogeneity: differential stress resistance among individual

cells of the yeast *Saccharomyces cerevisiae*. *Microbiology* **148:**345–351.
139. **Sun-Wada, G. H., S. Yoshioka, N. Ishida, and M. Kawakita.** 1998. Functional expression of the human UDP-galactose transporters in the yeast *Saccharomyces cerevisiae*. *J. Biochem.* **123:**912–917.
140. **Tucker, S. C., and A. Casadevall.** 2002. Replication of *Cryptococcus neoformans* in macrophages is accompanied by phagosomal permeabilization and accumulation of vesicles containing polysaccharide in the cytoplasm. *Proc. Natl. Acad. Sci. USA* **99:**3165–3170.
141. **Turner, S. H., R. Cherniak, and E. Reiss.** 1984. Fractionation and characterization of galactoxylomannan from *Cryptococcus neoformans*. *Carbohydr. Res.* **125:**343–349.
142. **Turner, S. H., and R. Cherniak.** 1991. Multiplicity in the structure of the glucuronoxylomannan of *Cryptococcus neoformans*, p. 123–142. In J. P. Latgé and D. Boucias (ed.), *Fungal Cell Wall and Immune Response*. NATO ASI Series, Heidelberg, Germany.
143. **Turner, S. H., R. Cherniak, E. Reiss, and K. J. Kwon-Chung.** 1992. Structural variability in the glucuronoxylomannan of *Cryptococcus neoformans* serotype A isolates determined by ^{13}C-NMR spectroscopy. *Carbohydr. Res.* **233:**205–218.
144. **Vaishnav, V. V., B. E. Bacon, M. O'Neill, and R. Cherniak.** 1998. Structural characterization of the galactoxylomannan of *Cryptococcus neoformans* Cap67. *Carbohydr. Res.* **306:**315–330.
145. **Vartivarian, S. E., E. J. Anaissie, R. E. Cowart, H. A. Sprigg, M. J. Tingler, and E. S. Jacobson.** 1993. Regulation of cryptococcal capsular polysaccharide by iron. *J. Infect. Dis.* **167:**186–190.
146. **Vecchiarelli, A.** 2000. Immunoregulation by capsular components of *Cryptococcus neoformans*. *Med. Mycol.* **38:** 407–417.
147. **White, C. W., R. Cherniak, and E. S. Jacobson.** 1990. Side group addition by xylosyltransferase and glucuronyltransferase in the biosynthesis of capsular polysaccharide in *Cryptococcus neoformans*. *J. Med. Vet. Mycol.* **28:**289–301.
148. **Wills, E. A., I. S. Roberts, M. Del Poeta, J. Rivera, A. Casadevall, G. M. Cox, and J. R. Perfect.** 2001. Identification and characterization of the *Cryptococcus neoformans* phosphoisomannose isomerase-encoding gene, *MAN1*, and its impact on pathogenicity. *Mol. Microbiol.* **40:**610–620.
149. **Yoneda, A., and T. L. Doering.** 2006. A eukaryotic capsular polysaccharide is synthesized intracellularly and secreted via exocytosis. *Mol. Biol. Cell* **17:**5131–5140.
150. **Yoneda, A., and T. L. Doering.** 2008. Regulation of *Cryptococcus neoformans* capsule size is mediated at the polymer level. *Eukaryot. Cell* **7:**546–459.
151. **Young, B. J., and T. Kozel.** 1993. Effects of strain variation, serotype, and structural modification on kinetics for activation and binding of C3 to *Cryptococcus neoformans*. *Infect. Immun.* **61:**2966–2972.
152. **Zaragoza, O., B. C. Fries, and A. Casadevall.** 2003. Induction of capsule growth in *Cryptococcus neoformans* by mammalian serum and CO_2. *Infect. Immun.* **71:**6155–6164.

Cryptococcus: From Human Pathogen to Model Yeast
Edited by J. Heitman et al.

4

The Architecture and Antigenic Composition of the Polysaccharide Capsule

MARCIO L. RODRIGUES, ARTURO CASADEVALL, AND OSCAR ZARAGOZA

The most distinctive morphological characteristic of *Cryptococcus neoformans* and *Cryptococcus gattii* when visualized in India ink preparations is the presence of a polysaccharide capsule around the yeast cell body. This structure has been the central focus of many studies for several decades, given its critical role in promoting virulence. The mechanism by which the capsule contributes to virulence includes interference with immune response and protection of the fungal cell from immune mechanisms. These immunological effects include apoptosis induction, inhibition of antibody production, complement depletion, inhibition of leukocyte migration, interference with phagocytosis, reduced antigen presentation, and in vitro invasive growth (23, 25, 43, 50, 57, 58, 60, 87). On the other hand, the capsule also provides physical protection against host microbicidal mechanisms by quenching cytotoxic free radicals (88). For many of these effects, the architecture of the capsule is critical. This chapter will focus on chemical, physical, antigenic, architectural, and dynamical properties of the capsule, and we will examine how changes in these parameters can influence the interaction with the host and the virulence of the yeast. Furthermore, we will pay special attention to other important processes involved in the physical organization of the capsule, such as the polysaccharide transport mechanisms described and the anchoring of the polysaccharide fibers to the cell wall.

CHEMICAL AND PHYSICAL PROPERTIES OF THE CAPSULE

The capsule is composed mainly of polysaccharide. Classical studies identified two different types of polysaccharides: glucuronoxylomannan (GXM) and galactoxylomannan (GalXM) (6, 9, 15, 18, 19, 78). The former consists of a chain of mannose residues with substitutions of glucuronic acid and xylose. Mannosyl residues can also be 6-O-acetylated and substituted with xylosyl units in β-(1,2) or β-(1,4) linkages depending on the serotype (15, 19, 54, 55). GalXM is formed by a chain of galactose residues with substitutions of xylose and mannose. Consequently, GalXM is more properly referred to as a galactan, but for this chapter we will use the standard terminology, with the caveat that a change in nomenclature may occur in the future. In addition to polysaccharide, capsular preparations often contain mannoproteins, but their role, if any, in capsule architecture is unknown (59, 69). Acetylation of the mannose residues is an important process that confers antigenic properties to the capsule (20). Although the capsule is composed mainly of polysaccharide, it is highly hydrated since the polysaccharides are highly hydrophilic. It has been estimated that water accounts for 99% of the total volume and weight of the capsule (51).

The polysaccharides that form the capsule can be found not only attached to the cell forming the capsule, but also soluble in the supernatant, and are known as exopolysaccharides (14). Historically, exopolysaccharides have provided an abundant source of material for physical and chemical studies. The mass and molar proportion of these molecules in exopolysaccharide preparations is different. GXM is believed to account for 90–95% of the total polysaccharide composition, while GalXM is approximately 5–10% (18). A key aspect to consider when thinking about the structure of these polysaccharides is the method used to prepare the samples for analysis. Most studies of capsular polysaccharides have focused on the GXM structure since it is the most abundant in the capsule. GXM has been normally purified from supernatants of overgrown cultures (more than 2 weeks) after precipitation with the cationic detergent cetyltrimethylammonium bromide (CTAB) (16, 20). This approach was used to determine that GXM is a large polysaccharide, with an average molecular weight of 10^6 to 5×10^6 Da, depending on the strain and sample analyzed (20, 54, 76). However, this size is an average since the size distribution in the same sample is very heterogeneous, as demonstrated by dynamic light scattering and by electrophoretic separation (54, 55). More recently, new methods

Marcio L. Rodrigues, Laboratório de Estudos Integrados em Bioquímica Microbiana, Universidade Federal do Rio de Janeiro, Instituto de Microbiologia, Rio de Janeiro - RJ, 21941-902, Brasil. **Arturo Casadevall,** Departments of Microbiology & Immunology and Medicine, Albert Einstein College of Medicine, Bronx, NY 10461. **Oscar Zaragoza,** Servicio de Micología, Centro Nacional de Microbiología. Instituto de Salud Carlos III, Carretera Majadahonda-Pozuelo, Km2, Majadahonda 28220, Madrid, Spain.

to purify exopolysaccharides have been developed (61), based on ultrafiltration of the culture supernatant and consequent formation of highly viscous films containing almost pure polysaccharide. In addition, methods (dimethylsulfoxide extraction and γ-radiation) have been explored to efficiently release the capsular polysaccharide attached to the cell. Although there is always the concern that the isolation method could alter polysaccharide structure, the availability of these new methods has allowed the study of structural properties, and the comparison of polysaccharides recovered by the various methods has provided new insights into the capsule structure.

GXM fractions recovered from culture supernatants differ significantly from capsular material in terms of glycosyl composition, molecular mass, diameter, charge, viscosity, spectral properties, and reactivity with monoclonal antibodies (MAbs) (32). A comparison of supernatant-derived fractions of purified GXM revealed that the native (obtained by ultrafiltration) polysaccharide was more reactive with MAbs than preparations obtained by detergent precipitation (61). In addition, light scattering analysis revealed that the average molecular mass of GXM obtained by filtration and cation-dependent aggregation was ninefold smaller than that obtained by CTAB precipitation (17, 23). This result revealed that the structure of polysaccharides recovered from culture supernatants varied with the purification method and suggested that CTAB-derived preparations may preferentially isolate larger molecules and/or irreversibly aggregate molecules.

The other component of the capsule, GalXM, has significantly smaller molecular mass, around 10^5 Da (20, 54). Although GalXM constitutes only about 5–8% of the total polysaccharide composition, a calculation of the molar ratios revealed that GalXM was the major component of the exopolysaccharides in terms of numbers of molecules (54). Meanwhile, whereas GXM appears to be uniformly distributed throughout the capsule, the exact location of GalXM remains unknown. Initial reports suggested that GalXM was a cell wall-associated capsular component (78). However, recent findings indicate that GalXM is distributed in discrete pockets that are associated with vesicles present within the capsule, implying that this polysaccharide is primarily synthesized for export and is not necessarily a capsular structural component (24). In fact, GalXM has strong immunomodulatory properties (65), which could be more potent than the ones exhibited by GXM.

Electron microscopy (transmission and scanning) has been a useful tool to examine the capsule structure (22, 52, 66, 71). Electron microscopy revealed that the capsule is organized as a fibrillar network that is attached to the cell wall. However, these results must be evaluated with caution since electron microscopy necessitates exhaustive dehydration of the sample, and this can produce artifacts from collapse of polysaccharides into thick fibers. Nevertheless, electron microscopy clearly establishes that there are radial density differences in the capsule such that inner regions have higher density than the outer regions (see Fig. 1) (52, 66). These differences are more marked in cells with a large capsule, being more difficult to observe or even not noticeable when the capsule is smaller. Based on scanning electron microscopy, three structural regions with different densities and architecture have been recently defined (see Fig. 1) (33). Differences in the distribution of capsular polysaccharides have been demonstrated using other approaches, including labeling with fluorescent compounds and antibodies and estimates of the density by direct measurement of polysaccharide concentration in different capsular regions (38, 52). For cells with small capsules, the density of the polysaccharide is lower (Fig. 1D). In addition, the physical orientation of the fibers seems to be different according to capsule region. As shown in Fig. 1, the outer fibers (R3) seem to be tangentially oriented compared to the fibers found close to the cell wall, which seem to be parallel to the cell. This inner orientation is clearly visible when the outer layers of the capsule are released by γ-radiation treatment (see Fig. 1C).

Recently, this difference in the spatial organization of the capsule has also been observed by measuring different topo-optical reactions in the different regions of the capsule (35). Scanning transmission electron microscopy revealed two types of fibrils in GXM preparations: the so-called structure I, composed of long fibers entangled among themselves and with an average width of 4 nm, and structure II, formed by smaller fibers extending in multiple directions (54). Calculations of the mass and size of the polysaccharide fibers and GXM molecules indicated that the capsule is formed by multiple polysaccharide fibers formed by self-association of GXM (54). The physical organization of the capsule in fibers that differ in density could have an important effect on the outcome of the interaction with the host, since it could influence the recognition of cell wall components by mammalian receptors involved in phagocytosis or transcytosis processes.

The chemical composition of the capsule is responsible for several physical characteristics. For instance, the high content of glucuronic acid confers a negative charge, as shown by the very low zeta potential value of encapsulated cells (about −30 to −40 mV) (63). In contrast, the zeta potential of nonencapsulated cells is much higher and closer to 0 mV and similar to other yeasts. This property could have profound consequences during interaction with the host, since it could affect binding to mammalian cells due to charge repulsion or attraction.

DYNAMIC PROPERTIES OF THE CAPSULE: CHANGES IN ANTIGENIC PROPERTIES AND EFFECT ON THE HOST

Although the capsule visible by India ink seems to be very homogeneous, multiple studies have demonstrated that the capsule is highly dynamic, able to change its structure and size in different situations. In this section, we will review the antigenic properties of the capsule and the structural properties that could have important consequences during infection. Many different MAbs that specifically recognize the *C. neoformans* capsule have been obtained (10, 11, 26, 29, 67, 75, 79). The study of these antibodies has contributed to the identification of multiple structural features of the capsule, which we will discuss in the following section.

In Vitro Antigenic Properties and Variations in Capsule Structure

The capsule composition can vary greatly between strains. This was confirmed in early studies analyzing the GXM composition of 106 isolates (20). The authors observed six structural motifs in the GXM molecule, and their distribution in 106 isolates allowed the identification of six structure reporter groups (see Fig. 2A–F) and eight chemotypes (Chem1–8), suggesting that each strain has a unique capsular structure. Despite the identification of eight chemotypes, the classical classification of *C. neoformans* varieties based on immunoreactivity with polyclonal rabbit sera yielded five serotypes, A, B, C, D, and A/D, which meant that strains from the eight chemotypes shared antigenic properties. Molar ratios of xylose, mannose, and glucuronic acid residues in GXM are different depending on the serotype (19). Initial studies demonstrated variations in acetylation and xylosyl substitutions between the main serotypes (20). A seventh GXM repeating unit was recently described (61), which had been previously found only in polysaccharide fractions from a hypocapsular mutant (2). More recently, copolymerization of different GXM repeating units in one polysaccharide have also been identified (55). In some *C. neoformans* strains, GXM is composed of a single repeating unit, whereas in other strains the polysaccharide contains multiple units. Furthermore, strains that utilize the same set of repeating units often differ from one another in the ratio of those units within the GXM molecule. These structural variations result in slight structural differences that translate into antigenic differences (55). Concerning GalXM (see basic structure in Fig. 2G), the molar ratios of galactose, xylose, and mannose are different in different serotypes (as happens with GXM), suggesting that GalXM is in fact a group of complex, closely related polysaccharides (41, 78).

Although there are clear variations in the capsule associated with the serotype, differences in capsule structure have also been reported between strains of the same serotype, as shown by chemical analysis and ^{13}C nuclear magnetic resonance spectroscopy (77). The development of MAbs against the major component of the capsule, GXM, provided powerful tools to study the variation in the chemical and antigenic composition of the capsule at multiple levels, including inter- and intraserotype and also at the intrastrain level. A set of MAbs differing in their affinity and specificity for the same GXM molecules has provided evidence consistent with the notion that GXM structure is highly heterogeneous at the molecular level and that this structural heterogeneity translates into the expression of multiple epitopes (10, 11, 26, 29, 67, 75, 79). The binding of MAbs to the capsule is variable, but two binding patterns, annular and punctuate, have been identified (64). This situation has been extensively described in the case of immunoglobulin M MAbs to the capsule. A correlation has been established between the fluorescence binding pattern and the protective efficacy of the MAbs, with the annular and punctuate patterns correlating with protective and nonprotective effects, respectively (64). These data were supported by the fact that the binding of antibodies produced during infection, which are associated with a nonprotective effect, yield a punctuate binding pattern (85). However, in some cases, the binding pattern of the MAb depends on the serotype, being annular in serotype A strains and punctuate in

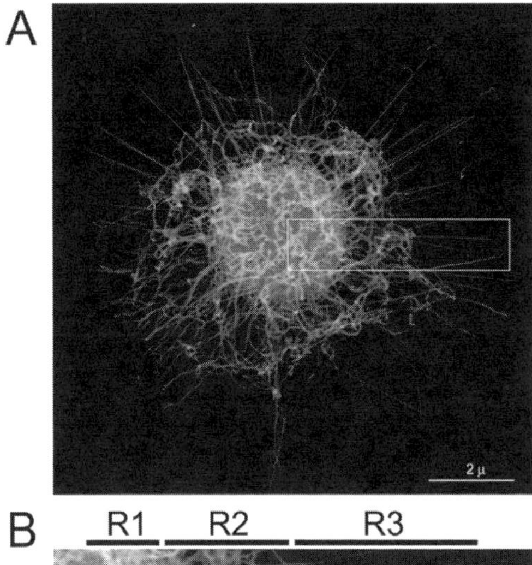

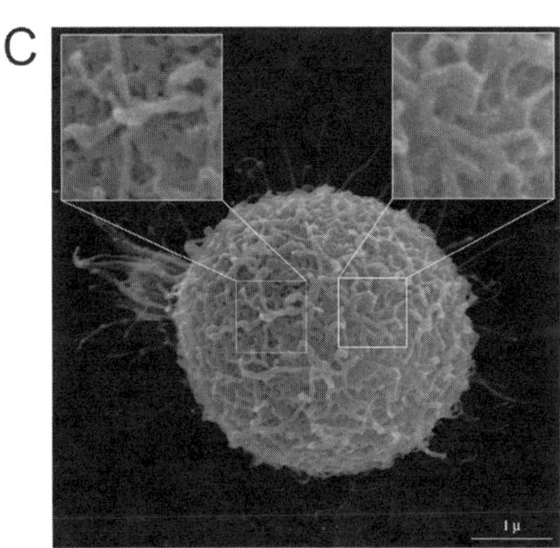

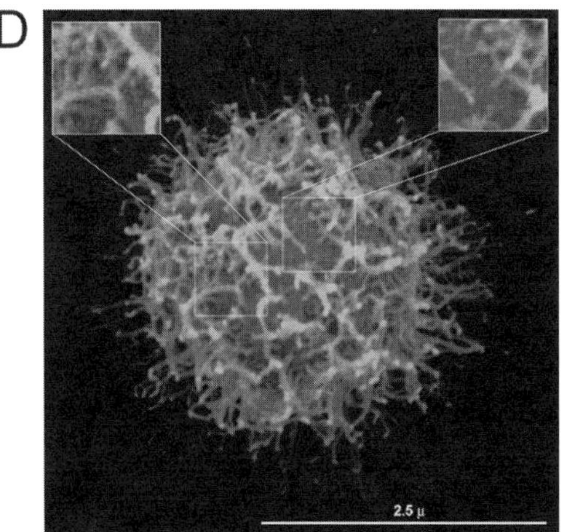

FIGURE 1 Capsule features shown by scanning electron microscopy. (A) Cells with enlarged capsule, where differences in polysaccharide density are clearly observed. (B) Detailed magnification of the three defined capsular regions based on the differential density. The region observed corresponds to the inset highlighted in panel A. (C) Detail of the inner region of the capsule. After growth in capsule-inducing medium, the outer regions were removed by γ-irradiation treatment, making visible the high-polysaccharide-density region close to the cell wall. (D) Cell grown in non-capsule-inducing conditions. Note how in this cell, the density of polysaccharide in the regions close to the cell wall is much lower than in cells with enlarged capsules (panels A, B, and C).

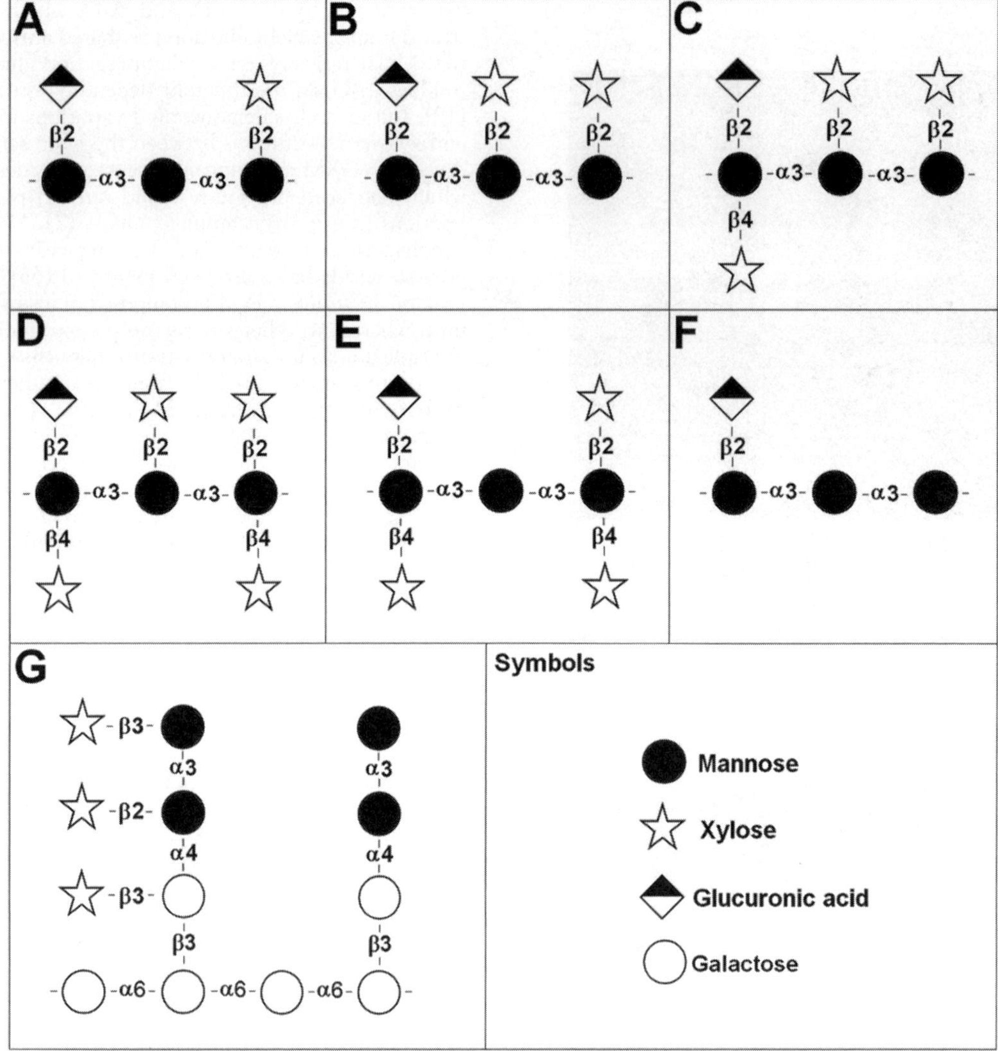

FIGURE 2 Basic composition of GXM and GalXM. (A–F) Structure of the six basic motifs described in GXM. (G) GalXM structure.

serotype D, supporting the notion that the capsule structure is highly heterogeneous, dependent on the serotype.

The heterogeneity of the capsular antigenic properties is also observed at the intrastrain level. The proportion of cells in the same culture that are able to bind MAbs can vary depending on growth conditions and in vitro microevolution (36, 55). Moreover, there is large variation in the intensity of the binding of the MAb to the cell, which suggests that even in the same cell population, there is a variation in the number of epitopes present in each cell. These findings indicate that the antigenic structure of the capsule in the same population is heterogeneous, which must reflect multiple structural features. In agreement with this idea is the finding that during budding, MAbs bind differently to the mother cell and to the bud (91), which suggests that capsule structure is also influenced by age-related phenomena. In fact, aging was proven to influence capsule structure, thereby producing changes in polysaccharide density (33, 52).

Another example of this heterogeneity is provided by the analysis of the capsules of isogenic strains that have undergone morphological switching. Two morphologies have been described in *C. neoformans*, smooth and mucoid, both derived from the same genetic background. Analysis of the capsule from these variants has shown that morphological switching is associated with changes in the basic properties of the capsule (34, 55).

Variation in the basic components of the capsule, such as acetylation and xylose composition, can influence capsule structure. The antigenic properties of the capsule are affected by acetylation (5, 17, 29, 53), so the degree of acetyl substitutions could have a profound effect during the interaction with the host. In fact, mutants defective in acetylation were hypervirulent (42). In contrast, mutants unable to incorporate xylose in the capsule were avirulent (42), and they accumulated complement protein 3 on the capsule at a higher rate than the isogenic wild-type strain (44). Xylose content can also affect the binding of some MAbs to the capsule, but its effect is not as dramatic as the one produced by acetylation (44). Both components also influenced the kinetics of GXM clearance from the host. These findings confirm that changes in the capsule structure have a profound effect on interaction with the host.

Antigenic Variations During In Vivo Infection

The heterogeneity in capsule structure has the potential to translate into antigenic and biological variation, a property that could have important consequences during interaction with the host. The notion that changes in capsule structure can influence the outcome of interaction with the host gained support from the observation that addition of purified capsular polysaccharide isolated from different serotype A strains to acapsular mutants had different effects on phagocytosis (74). Consequently, changes in capsule structure could have a profound effect on the recognition of *C. neoformans* by phagocytic cells and, by extension, to the development of different immune responses, which can contribute to evasion of the immune response. Structural and antigenic changes in the capsule during interaction with the host have been described. During the course of experimental infection, the *C. neoformans* capsule undergoes significant structural changes as demonstrated by the accumulation of certain MAb epitopes (36). In the same study, the capsule antigenic structure was found to differ depending on the organ infected, the fungal cell population being more homogeneous in the lungs and spleen than in the brain. These findings highlight the importance of the environment in the structural and antigenic heterogeneity of the capsule. Capsular changes have been correlated with the crossing of fungal cells through the blood-brain barrier (13). Furthermore, that study also showed that *C. neoformans* cells expressed epitopes for a MAb that were not expressed in vitro, indicating the expression of new polysaccharides in tissue.

DYNAMIC PROPERTIES OF THE CAPSULE: CHANGES IN CAPSULE SIZE AND CONSEQUENCES IN THE HOST

A striking characteristic feature of the *C. neoformans* capsule is its ability to vary in size depending on the strain and environmental conditions (45, 56, 90). Apart from interstrain differences, yeast cells also manifest variations in capsule size within a strain depending on the environmental conditions (see review in reference 86). This morphological change seems to play an important role during infection, since capsule enlargement occurs during the first hours of infection. Capsule enlargement can be reproduced in vitro by placing the fungal cells in several media that induce capsule enlargement. The phenomenon of capsule enlargement in vitro was first reported in the 1950s by Littman (47), who described several minimal media in which capsule enlargement was induced. Subsequently, other conditions were described including iron limitation, a high CO_2 concentration, and mammalian serum (39, 80, 89). More recently, media with a low amount of nutrients and slightly basic pH were also able to induce capsule growth (86). Factors that inhibit capsule enlargement include an increase in osmotic pressure and sugar concentration (28).

The mechanisms by which these conditions translate into changes in capsule size remain unknown. Curiously, the factors that induce capsule growth resemble the physiological conditions found during mammalian infection. In addition, in the media where capsule enlargement occurs, there is a slower growth rate, which suggests a link between capsule growth and the metabolic conditions in the cell. The importance of the environment has been clearly established using mutants defective in signal transduction, such as cyclic AMP, Ras1, and mitogen-activated protein kinase (Hog1, Ste12α) pathways (3, 12, 27, 84). Curiously, the involvement of these pathways is different depending on the inducing signal (89), indicating that multiple pathways participate in the process. In addition, the final size of the capsule after enlargement seems to be highly regulated, since it reaches a limit that is not passed when the cells are transferred repeatedly to fresh inducing medium (91). Interestingly, this limit seems to be determined by the size of the cell body, since a strong positive correlation between capsule size and cell body size is consistently found (91).

Capsule enlargement could conceivably occur by expansion of the preexisting polysaccharide fibers and/or by addition of new polysaccharide molecules. Recent studies have demonstrated that capsule enlargement occurs by addition of new polysaccharide to the capsule (51), which produces a significant increase in the density of the capsule, especially in regions close to the cell wall (52). Binding of the polysaccharide molecules to the preexisting ones appears to occur by self-association (54). Interestingly, when the capsule enlarges, the new polysaccharide added seems to be different from the older, preexisting material, as shown by their different chemical and physical properties (32). Yoneda and Doering found that during capsule growth, the size distribution of the polysaccharide molecules was different according to gel electrophoresis migration, an observation that suggested changes in mass and/or charge as the new types of molecules are formed (83). Measurements of capsular polysaccharide before and after capsule growth showed that the diameter of polysaccharide molecules increased during capsular enlargement (33). Other investigators have studied the antigenic composition and the physical properties of the capsule after induction of its size (52). In addition to changes in capsule density, capsule enlargement results in a heterogeneous distribution of epitopes and in differences in the zeta potential between different regions of the capsule (52). These findings support the idea that capsule enlargement is achieved by accumulation of new polysaccharide molecules that have different physical and antigenic properties than preexisting molecules. The differences in structure according to the spatial region of the capsule could also have important consequences during the interaction, because they could affect the localization of antibodies and complement proteins, in addition to the recognition by phagocytic cells.

Several groups have studied the mechanisms by which capsule enlargement is achieved. The general approach of these studies involves labeling the capsule prior to its enlargement, so the localization of the signal after induction makes it possible to trace the way in which the new polysaccharide is accumulated. The first study used a MAb to label the capsule. Using this strategy, the authors found that MAb labeling was displaced to the capsular edge, suggesting that the polysaccharide is added to the inner region of the capsule, displacing the old polysaccharide to the outer regions (66). However, the fact that the MAb does not covalently bind to the polysaccharide raised the issue of whether the location of the MAb remained fixed during capsule growth. To overcome this issue, another study used complement proteins as geographical markers (91). Unlike antibody, complement protein 3 binds to the capsule by a covalent thioester bond, and consequently, its location can be expected to remain fixed during capsular enlargement.

In contrast to the result obtained with antibody, the use of complement protein 3 as a geographic marker revealed that the complement proteins remained in the inner regions, a finding that was interpreted as suggesting that capsule growth occurred by apical addition of new molecules. This interpretation was also supported by the radioactive incorporation studies that showed preferential incorporation in the outer regions of the capsule during capsular enlargement. The model of apical growth is also consistent with data from other authors that have shown that, in vivo, the appearance of new epitopes for some MAbs occurs at the edge of the capsule (13). The current view posits that capsule enlargement involves the addition of new molecules with larger molecular weights that are then intercalated with preexisting fibers by self-association (see Fig. 3), possibly involving divalent cation bridges, in a process that results in a significant increase in the density of the capsule, especially in the inner regions and apical extension of the capsule.

An interesting feature related to capsule growth and its architecture is the fact that the capsule undergoes rearrangements during budding. Although the presence of a dense net of polysaccharide fibers should in principle pose a problem for the separation of the bud from the mother cell, the process of budding bypasses this hurdle by rearrangements of capsule in the budding area (91). These rearrangements result in the formation of a pore where the bud will move through to cross the capsule of the mother cell. The molecular basis for these capsular rearrangements and changes is not known, but chitin-like structures are supposedly involved in this process (70). These findings indicate that the capsule is a very dynamic structure able to change and rearrange its structure in many different situations. In this context, it is reasonable to believe that the cell must also have mechanisms for capsule degradation. However, the few available data show that when cells with enlarged capsules are transferred to capsule-noninducing medium, the capsule is not degraded, but the new buds formed have a small capsule (91). This suggests that the cells do not have a mechanism to degrade the existing polysaccharide. This subject, however, has not been addressed in detail and requires further study.

Capsular enlargement can profoundly affect the interaction of *C. neoformans* with the host. Capsule enlargement is an early morphological response during infection (31). The role of capsule growth in this stage of the infection is not known, but it has been shown that capsule enlargement interferes with complement-mediated phagocytosis (90). This inhibitory effect has been observed in other infection models, such as phagocytosis by amoebas (88). Interestingly, capsule growth occurs not only when *C. neoformans* is found in the tissues, but also during intracellular pathogenesis in phagocytic cells (see supplemental movie 1 in reference 49). Capsular enlargement within macrophages would have the effect of increasing the volume of phagosomes, which would dilute microbicidal substances. In this sense, capsule growth confers resistance to stress factors produced by macrophages to induce microbial killing, such as free radicals and antimicrobial peptides (88). All these findings suggest that capsule growth is an important process to avoid engulfment by macrophages and killing in the phagolysosome, processes that will in turn allow the survival of the fungal cells in the host and contribute to virulence.

A few isolates unable to induce capsule growth have been identified. Although the affected genes have not been identified, these mutants showed reduced virulence and were efficiently cleared from the host (7, 8, 39), which confirms that this process participates in the virulence of the yeast. Several groups have investigated if there is a correlation between capsule size in vitro and virulence. Since capsule synthesis in vivo is induced, it would be expected that the size of the capsule in vitro should not predict the virulence of the organism. In fact, several studies have confirmed this idea (21, 48). One study found that mice infected with heavily encapsulated cells in vitro died earlier than mice infected with the same strain grown in conditions where capsule size was significantly smaller, although there was no difference in their mortality rate once the mice started to die (28). However, in this work, unlike with heavily capsulated cells,

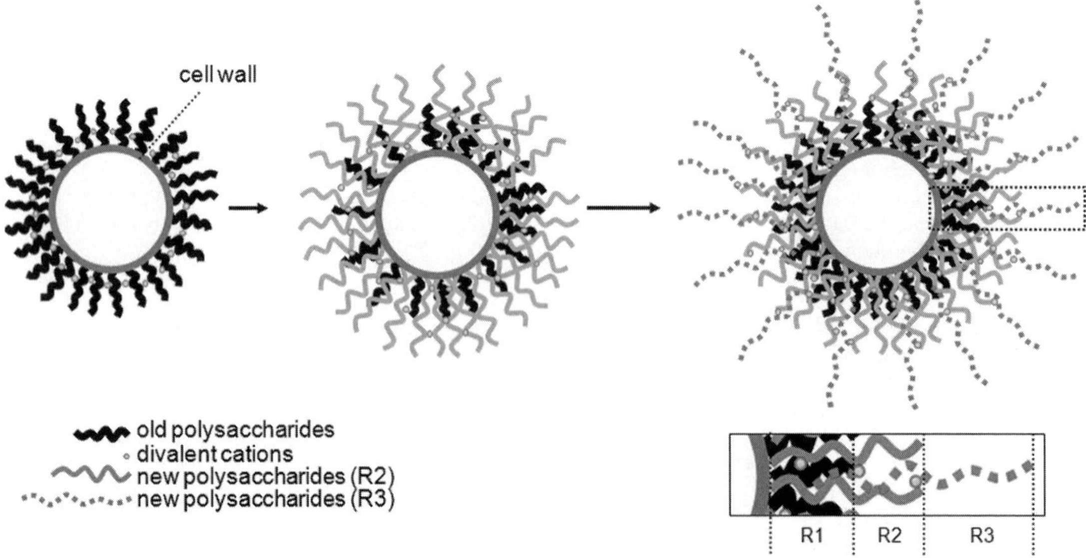

FIGURE 3 Proposed model for capsule growth. (Left) Cells with small capsules. After capsule enlargement, new fibers with larger molecular weight are intercalated between the old fibers, yielding a high-density region in the inner part of the capsule (middle). Finally, the capsule grows by apical addition of longer fibers (right), which yields an intermediate-density region (R2) and a low-density region in the outer layer (R3). (Bottom right) Magnification showing the three density regions.

cells with small capsule were obtained by exposure to high osmolarity. This could result in stressed cells that in the host would be compromised compared to the cells with larger capsule, delaying death of the host.

POLYSACCHARIDE TRANSPORT ACROSS THE CELL WALL

Polysaccharide biosynthesis in microorganisms has been studied in detail in several microbial species, due to the potential of microbial cell walls and capsules to play important roles as targets of antimicrobial drugs and vaccines. In bacteria, the cell wall structural polymer peptidoglycan is polymerized on the outer side of the plasma membrane (4). Bacterial capsular polysaccharides are synthesized by plasma membrane glycosyltransferases whose active sites face the extracytoplasmic environment (33). In fungi, chitin and glucan synthases, the enzymes responsible for the synthesis of the structural cell wall polysaccharides chitin and glucan, have the plasma membrane as their final cellular sites (5, 10), supporting the idea that polysaccharide polymerization occurs at extracytoplasmic sites. In summary, extracellular polymerization of microbial polysaccharides appears to be a common property of organisms of different kingdoms.

In contrast to most microbial polysaccharides, the cryptococcal GXM is apparently synthesized in cytoplasmic sites (12, 22, 34), at least in part. For *C. neoformans*, the capsular polysaccharide is also released to the extracellular milieu, implying that GXM traffic in *Cryptococcus* spp. includes mechanisms of trans-cell wall transport. The possible existence of trans-cell wall mechanisms of polysaccharide export was first hinted at in early studies by Takeo and coworkers, who demonstrated that *C. neoformans* secretes vesicles outside the cell membrane by freeze-etching electron microscopy (74a, 74b). Although no molecular evidence was provided at that time, the authors proposed that "capsular material is synthesized in and released via the vesicles." Almost 20 years after the report by Takeo and colleagues, immunogold electron microscopy experiments revealed clumped labeling of the cryptococcal cell wall that was suggestive of vesicular transport of GXM (31). A subsequent study showed that antibodies to GXM recognized cytoplasmic and cell wall clusters consistent with vesicular transport (37). Altogether, these results provided strong evidence that GXM is synthesized in the cytoplasm and exported to the exterior of the cell in secretory vesicles (Fig. 4).

Secretory post-Golgi vesicles involved in protein traffic to the extracellular space have been studied in detail in the model yeast *Saccharomyces cerevisiae* (reviewed in reference 72). Secretion in yeast requires the coordinated action of different proteins, including the small GTPase Sec4p. In *S. cerevisiae*, *sec4* mutants accumulate post-Golgi vesicles under restrictive conditions (81). Based on these studies, Yoneda and Doering (82) generated a *C. neoformans* strain defective in the production of Sav1p, a homolog of the *S. cerevisiae* small GTPase Sec4p. As demonstrated for *S. cerevisiae* cells lacking Sec4p expression (81), the *C. neoformans sav1* mutant also showed accumulation of post-Golgi

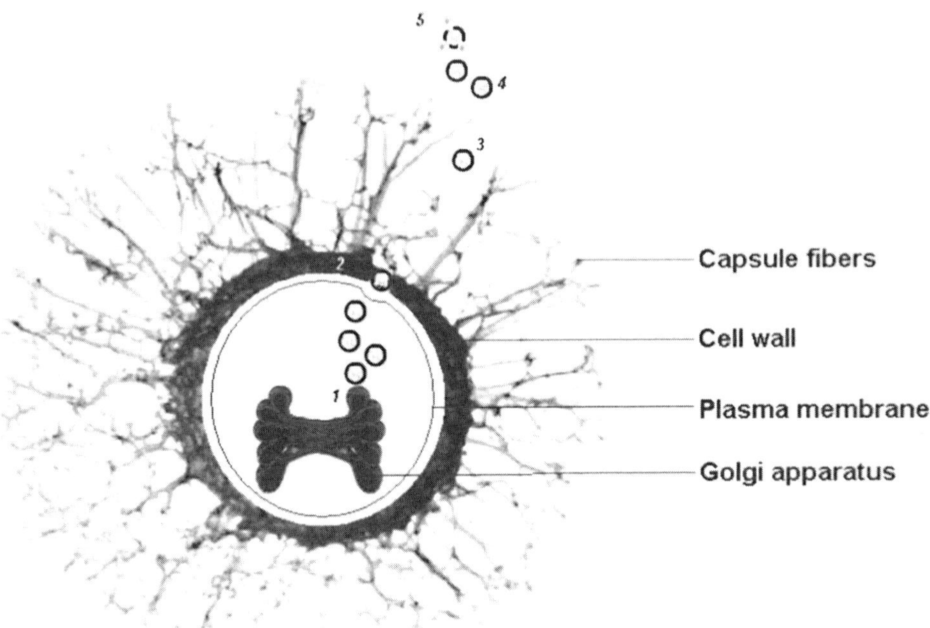

FIGURE 4 Model proposed for polysaccharide export in *C. neoformans*. According to the current literature, GXM is packaged in post-Golgi vesicles (1) that are targeted to the cell surface. The vesicles move outside the plasma membrane, then cross the cell wall (2) and the existing capsular network (3) by still unknown mechanisms to be released into the extracellular space (4), where they are supposedly lysed (5) for polysaccharide incorporation into the growing capsule.

vesicles under restrictive conditions. More precisely, the *sav1* mutant of *C. neoformans* had defective protein secretion and accumulated exocytic vesicles at the septum and the bud during cell division. Strikingly, these vesicles were strongly recognized by an antibody to GXM (82), reinforcing the supposition that GXM is packaged into secretory vesicles required for its transport to extracellular sites.

Given that capsules of *C. neoformans* can enlarge by apical growth (91), one must envision a process that requires transport of capsular components to the extracellular space followed by incorporation into the external layers of the growing capsule. Therefore, extracellular release of intracellularly synthesized GXM would be an essential step required for capsule assembly. Consequently, the studies showing the existence of GXM-containing secretory vesicles in *C. neoformans* (12, 22, 34, 74a, 74b) and the necessity of extracellular release of the polysaccharide for apical growth of the capsule (91) led to the hypothesis that extracellular vesicles loaded with capsular components would exist in cryptococcal culture supernatants. If true, this hypothesis would suggest the existence of polysaccharide-containing vesicles in association with the cell wall, a step that would be required for extracellular release of the secretory compartments. Using procedures previously designed for the purification of secreted vesicles in mammalian systems, lipid fractions that reacted with antibodies to GXM were isolated from *C. neoformans* supernatants. Morphological analysis of these fractions by transmission electron microscopy revealed the presence of round vesicles, in the range of 20 to 400 nm, containing bilayered membranes (24, 25). Importantly, some of the isolated vesicles were recognized by an antibody to GXM in immunogold assays (25), supporting the notion that these structures have a role in polysaccharide trans-cell wall secretion. These results suggest that the mechanisms by which intracellularly synthesized GXM reaches the extracellular space apparently involve polysaccharide packaging into secretory vesicles that move across the cell wall and are released extracellularly.

Vesicular secretion of GXM is putatively relevant for capsule assembly, since induction of capsule expression is accompanied by an enhanced detection of GXM in vesicular fractions (25). Acapsular cells were able to extract GXM from vesicular fractions and incorporate the polysaccharide into its cell surface, indicating that the fungus has the metabolic apparatus to extract GXM from membrane vesicles. The mechanisms potentially involved in GXM release from vesicles and incorporation into the growing capsule are still not clear. According to Siafakas and coworkers (73a), the cell wall-associated enzyme phospholipase B could be involved in the release of the molecules transported across the cell wall inside membrane-bound vesicles.

The supposition that vesicle secretion is related to capsule assembly suggested that other capsular components might also require vesicular transport to reach the extracellular space. In this regard, a recent study revealed discrete GalXM-containing lipid bodies in the surface of *C. neoformans* (24), a finding consistent with the notion that this polysaccharide is also delivered in vesicles. In fact, chromatographic analysis of vesicle preparations of acapsular cells revealed the presence of galactose, xylose, and mannose, the components that make up GalXM (24). As also pointed out by the authors, these results do not conclusively establish that GalXM is secreted in vesicles but are consistent with the notion that secretion of this capsular polysaccharide is vesicle associated.

A key unsolved question is the mechanism by which polysaccharide-containing vesicles cross the cell wall to reach the extracellular space. Transmission electron microscopy has shown that the cell walls of *C. neoformans* and *Histoplasma capsulatum* contain numerous vesicle-like structures (1, 24, 25). Atomic force microscopy has shown that fungal cell walls contain pores of dimensions that would allow vesicle passage (9, 11). Considering that the cell wall is a compact although malleable structure that is likely to suffer rearrangements (62), vesicular passage through pores could represent a solution to the problem of trans-cell wall transport of secretory vesicles. Another possibility is that trans-cell wall vesicle secretion could also involve vesicle-mediated cell wall remodeling. For example, *H. capsulatum* extracellular vesicles contain enzymes regulating synthesis and hydrolysis of cell wall components (1). In yeast cells, cell wall remodeling for vesicle passage could be facilitated during budding, when the cell wall is thinner in the bud than in nondividing areas of the cell surface (46). Finally, the fungal cell wall contains enzymes that regulate ATP metabolism (14) and motor proteins (30). Therefore, vesicle passage through the cell wall could require the energy-demanding activity of motor proteins, which also contribute to the cytoplasmic movement of transport vesicles (73).

POLYSACCHARIDE AND CELL SURFACE CONNECTIONS IN *C. NEOFORMANS*

Anchoring of the capsule to the cell wall is crucial for the pathogenicity of *C. neoformans*, since capsular components are required for virulence. The interactions between the components responsible for attaching the capsule to the cell wall remain poorly understood, but there is increasing evidence that polysaccharide-surface connections are complex and involve multiple components (19, 20, 27). α-1,3-Glucan was the first cell wall component associated with capsule anchoring (68). Cryptococcal cells with disrupted α-1,3-glucan synthase genes had normal levels of synthesis of capsular material but lacked a visible capsule (20). In addition, glucanase-treated acapsular cells bound to GXM in a defective manner (68). Although glucans and GXM could be directly connected, cell wall glucans can also anchor other polysaccharides and proteins at the fungal cell wall (62). Since loss of glucans can disturb cell wall assembly, the hypothesis that glucan-bound cell wall components interact with the capsule cannot be discarded. In fact, mutant cells lacking α-1,3-glucan synthesis have modified cell walls (20), strengthening the notion that the lack of cell wall glucans could lead to loss of other structural components in this compartment.

N-Acetylglucosamine or its deacetylated derivative play key roles in the assembly of the *C. neoformans* cell wall. For example, mutations in the genes responsible for the expression of chitin synthase 3 or of the biosynthetic regulator Csr2p lost the ability to retain the virulence-related pigment melanin in the cell wall (2, 3). These mutants are also defective in the synthesis of chitosan, which has also been demonstrated to regulate retention of cell wall melanin (2). Treatment of *C. neoformans* acapsular mutants with chitinase, the enzyme responsible for chitin hydrolysis, interfered with the incorporation of capsular components into the cell wall (70). Chitinase-treated acapsular mutants of *C. neoformans* still bound the soluble polysaccharide, but they did so in a defective manner that resulted in the formation of a loose polysaccharide coat at the cell surface (70). In this context, it is likely that chitin-related structures also play a role in capsule anchoring to the cell wall of *C. neoformans*. In fact, the existence of chitooligomers forming singular round or hook-like connections between the cell wall and the capsule of

C. neoformans was recently described (70). The connections of GXM to chitooligomers were punctual and clearly limited to certain areas of the cell surface, especially those involved in yeast budding, which could suggest that chitin-related structures are involved in capsule modeling during cell division. These observations suggested that, in addition to its expected role in the synthesis of cell wall components, chitin metabolism in cryptococci could be linked to capsule anchoring.

POLYSACCHARIDE-POLYSACCHARIDE INTERACTIONS

Anchoring of capsular components to the cell wall is crucial for capsule assembly. However, additional interpolysaccharide interactions are expected to occur at the capsular microenvironment. Although capsular enlargement is related to lengthening of polysaccharide fibers (33) and regulation of capsule size is mediated at the level of individual polysaccharide molecules (83), scanning electron microscopy images strongly suggest that capsular components are capable of self-aggregation (7, 16).

The aggregative properties of GXM were apparent from the observation that increasing the polysaccharide concentration in cryptococcal supernatants by concentration in ultrafiltration cells resulted in the formation of highly viscous films (61). Serological approaches, nuclear magnetic resonance, and chromatographic methods combined with mass spectrometry revealed that the films contained concentrated GXM in its native form (61). Analyses of polysaccharide solutions in an optical tweezers model for viscosity determination strongly suggested that polysaccharide aggregation in vitro was a consequence of the formation of metal bridges connecting negatively charged units of glucuronic acid in different GXM fibers. Consistent with this finding, divalent metal-mediated GXM aggregation also regulated capsule enlargement in *C. neoformans* (61). These results are in agreement with the early observation that growth of *C. neoformans* in high concentrations of the monovalent ion Na^+ results in capsule reduction (40), which probably results in unilateral neutralization of glucuronic acid residues and inhibition of GXM aggregation.

Apparently, polysaccharide aggregation within the capsular microenvironment may also involve GalXM. The addition of purified GalXM to GXM solutions resulted in a marked increase in the polysaccharide molecular mass (32), suggesting that GalXM binds to GXM or that it promotes its aggregation. In fact, analysis of monosaccharide composition in polysaccharide aggregates revealed that GXM is the major constituent of the viscous films obtained by filtration of supernatants, but galactose, the major component of GalXM, was constantly detected in those fractions (32). In addition, immunofluorescence studies showed that GalXM was deposited near the capsule edge and was abundant on the nascent capsules of budding daughter cells (24). Taken together, these results suggest that GalXM and GXM might somehow interact in the capsule of *C. neoformans*. Therefore, the current literature indicates that at least four types of polysaccharide-polysaccharide interaction occur at the cryptococcal capsule-cell wall interface, including GXM-GXM, GXM-GalXM, GXM-glucans, and GXM-chitin (8, 19, 20, 23, 27).

In addition to its potential relevance for the capsular assembly in *Cryptococcus* spp., the aggregative properties of GXM exposed new structural aspects of the major cryptococcal capsular component. GXM has been classically purified by crude precipitation of polysaccharides in fungal cultures with ethanol, followed by separation from neutral polysaccharides by selective precipitation with CTAB (16). CTAB is then supposedly removed from GXM fractions by extensive dialysis against saline followed by distilled water. However, a general concern about the purification of GXM by precipitation with CTAB is the uncertainty of total detergent removal. The presence of CTAB in GXM preparations could conceivably influence many of the biological effects attributed to the polysaccharide, including relevant host responses, since biological tests could be affected by the detergent itself or by modified polysaccharide-detergent complexes. In fact, it was reported that exposure of GXM to relatively small concentrations of CTAB resulted in detectable alterations in polysaccharide structure (32).

Considering the recently described new aspects of GXM structure and their relationship with polysaccharide functions, many questions related to its impact on fungal physiology and pathogenesis remain unanswered. The use of ultrafiltration systems allowed the generation of detergent-free fractions of purified GXM, isolating the polysaccharide from a pool of other exocellular components secreted by cryptococci by forming a film. Although such fractions could be considered as native GXM, they clearly differ from cell wall-derived fraction in biophysical properties (32). Since the biological properties of GXM must reflect its structure, the finding that the structure can vary depending on different parameters of isolation raises questions about what could be considered as native GXM. Of note, virtually all the immunological observations made were based on detergent preparations of the polysaccharide. Therefore, the structural divergence of GXM fractions suggests that there may be yet-to-be discovered properties that may only become apparent when polysaccharide fractions representing native structures are used. This hypothesis still needs to be explored, especially because the studies cited above included a restricted number of *C. neoformans* strains that may not fully represent the *Cryptococcus* complex of pathogenic species.

REFERENCES

1. **Albuquerque, P. C., E. S. Nakayasu, M. L. Rodrigues, S. Frases, A. Casadevall, R. M. Zancope-Oliveira, I. C. Almeida, and J. D. Nosanchuk.** 2008. Vesicular transport in *Histoplasma capsulatum*: an effective mechanism for trans-cell wall transfer of proteins and lipids in ascomycetes. *Cell. Microbiol.* **10:**1695–1710.
2. **Bacon, B. E., R. Cherniak, K. J. Kwon-Chung, and E. S. Jacobson.** 1996. Structure of the O-deacetylated glucuronoxylomannan from *Cryptococcus neoformans* Cap70 as determined by 2D NMR spectroscopy. *Carbohydr. Res.* **283:** 95–110.
3. **Bahn, Y. S., K. Kojima, G. M. Cox, and J. Heitman.** 2005. Specialization of the HOG pathway and its impact on differentiation and virulence of *Cryptococcus neoformans*. *Mol. Biol. Cell.* **16:**2285–2300.
4. **Barreteau, H., A. Kovac, A. Boniface, M. Sova, S. Gobec, and D. Blanot.** 2008. Cytoplasmic steps of peptidoglycan biosynthesis. *FEMS Microbiol. Rev.* **32:**168–207.
5. **Belay, T., and R. Cherniak.** 1995. Determination of antigen binding specificities of *Cryptococcus neoformans* factor sera by enzyme-linked immunosorbent assay. *Infect. Immun.* **63:**1810–1819.
6. **Bhattacharjee, A. K., J. E. Bennett, and C. P. Glaudemans.** 1984. Capsular polysaccharides of *Cryptococcus neoformans*. *Rev. Infect. Dis.* **6:**619–624.

7. **Blackstock, R., K. L. Buchanan, A. M. Adesina, and J. W. Murphy.** 1999. Differential regulation of immune responses by highly and weakly virulent *Cryptococcus neoformans* isolates. *Infect. Immun.* **67:**3601–3609.
8. **Blackstock, R., and J. W. Murphy.** 1997. Secretion of the C3 component of complement by peritoneal cells cultured with encapsulated *Cryptococcus neoformans. Infect. Immun.* **65:**4114–4121.
9. **Bose, I., A. J. Reese, J. J. Ory, G. Janbon, and T. L. Doering.** 2003. A yeast under cover: the capsule of *Cryptococcus neoformans. Eukaryot. Cell* **2:**655–663.
10. **Casadevall, A., M. DeShaw, M. Fan, F. Dromer, T. R. Kozel, and L. A. Pirofski.** 1994. Molecular and idiotypic analysis of antibodies to *Cryptococcus neoformans* glucuronoxylomannan. *Infect. Immun.* **62:**3864–3872.
11. **Casadevall, A., J. Mukherjee, S. J. Devi, R. Schneerson, J. B. Robbins, and M. D. Scharff.** 1992. Antibodies elicited by a *Cryptococcus neoformans*-tetanus toxoid conjugate vaccine have the same specificity as those elicited in infection. *J. Infect. Dis.* **165:**1086–1093.
12. **Chang, Y. C., B. L. Wickes, G. F. Miller, L. A. Penoyer, and K. J. Kwon-Chung.** 2000. *Cryptococcus neoformans* STE12alpha regulates virulence but is not essential for mating. *J. Exp. Med.* **191:**871–882.
13. **Charlier, C., F. Chretien, M. Baudrimont, E. Mordelet, O. Lortholary, and F. Dromer.** 2005. Capsule structure changes associated with *Cryptococcus neoformans* crossing of the blood-brain barrier. *Am. J. Pathol.* **166:**421–432.
14. **Cherniak, R.** 1988. Soluble polysaccharides of *Cryptococcus neoformans. Curr. Top. Med. Mycol.* **2:**40–54.
15. **Cherniak, R., R. G. Jones, and E. Reiss.** 1988. Structure determination of *Cryptococcus neoformans* serotype A-variant glucuronoxylomannan by 13C-n.m.r. spectroscopy. *Carbohydr. Res.* **172:**113–138.
16. **Cherniak, R., L. C. Morris, B. C. Anderson, and S. A. Meyer.** 1991. Facilitated isolation, purification, and analysis of glucuronoxylomannan of *Cryptococcus neoformans. Infect. Immun.* **59:**59–64.
17. **Cherniak, R., E. Reiss, M. E. Slodki, R. D. Plattner, and S. O. Blumer.** 1980. Structure and antigenic activity of the capsular polysaccharide of *Cryptococcus neoformans* serotype A. *Mol. Immunol.* **17:**1025–1032.
18. **Cherniak, R., E. Reiss, and S. Turner.** 1982. A galactoxylomannan antigen of *Cryptococcus neoformans* serotype A. *Carbohydr. Res.* **103:**239–250.
19. **Cherniak, R., and J. B. Sundstrom.** 1994. Polysaccharide antigens of the capsule of *Cryptococcus neoformans. Infect. Immun.* **62:**1507–1512.
20. **Cherniak, R., H. Valafar, L. C. Morris, and F. Valafar.** 1998. *Cryptococcus neoformans* chemotyping by quantitative analysis of 1H nuclear magnetic resonance spectra of glucuronoxylomannans with a computer-simulated artificial neural network. *Clin. Diagn. Lab. Immunol.* **5:**146–159.
21. **Clancy, C. J., M. H. Nguyen, R. Alandoerffer, S. Cheng, K. Iczkowski, M. Richardson, and J. R. Graybill.** 2006. *Cryptococcus neoformans* var. *grubii* isolates recovered from persons with AIDS demonstrate a wide range of virulence during murine meningoencephalitis that correlates with the expression of certain virulence factors. *Microbiology* **152:**2247–2255.
22. **Cleare, W., and A. Casadevall.** 1999. Scanning electron microscopy of encapsulated and non-encapsulated *Cryptococcus neoformans* and the effect of glucose on capsular polysaccharide release. *Med. Mycol.* **37:**235–243.
23. **Collins, H. L., and G. J. Bancroft.** 1991. Encapsulation of *Cryptococcus neoformans* impairs antigen-specific T-cell responses. *Infect. Immun.* **59:**3883–3888.
24. **De Jesus, M., A. M. Nicola, M. L. Rodrigues, G. Janbon, and A. Casadevall.** 2008. Capsular localization of the *Cryptococcus neoformans* polysaccharide component galactoxylomannan. *Eukaryot. Cell* **8:**96–103.
25. **Dong, Z. M., and J. W. Murphy.** 1995. Effects of the two varieties of *Cryptococcus neoformans* cells and culture filtrate antigens on neutrophil locomotion. *Infect. Immun.* **63:**2632–2644.
26. **Dromer, F., J. Salamero, A. Contrepois, C. Carbon, and P. Yeni.** 1987. Production, characterization, and antibody specificity of a mouse monoclonal antibody reactive with *Cryptococcus neoformans* capsular polysaccharide. *Infect. Immun.* **55:**742–748.
27. **D'Souza, C. A., and J. Heitman.** 2001. Conserved cAMP signaling cascades regulate fungal development and virulence. *FEMS Microbiol. Rev.* **25:**349–364.
28. **Dykstra, M. A., L. Friedman, and J. W. Murphy.** 1977. Capsule size of *Cryptococcus neoformans*: control and relationship to virulence. *Infect. Immun.* **16:**129–135.
29. **Eckert, T. F., and T. R. Kozel.** 1987. Production and characterization of monoclonal antibodies specific for *Cryptococcus neoformans* capsular polysaccharide. *Infect. Immun.* **55:**1895–1899.
30. **Esnault, K., B. el Moudni, J. P. Bouchara, D. Chabasse, and G. Tronchin.** 1999. Association of a myosin immunoanalogue with cell envelopes of *Aspergillus fumigatus* conidia and its participation in swelling and germination. *Infect. Immun.* **67:**1238–1244.
31. **Feldmesser, M., Y. Kress, and A. Casadevall.** 2001. Dynamic changes in the morphology of *Cryptococcus neoformans* during murine pulmonary infection. *Microbiology* **147:** 2355–2365.
32. **Frases, S., L. Nimrichter, N. B. Viana, A. Nakouzi, and A. Casadevall.** 2008. *Cryptococcus neoformans* capsular polysaccharide and exopolysaccharide fractions manifest physical, chemical, and antigenic differences. *Eukaryot. Cell* **7:**319–327.
33. **Frases, S., B. Pontes, L. Nimrichter, N. B. Viana, M. L. Rodrigues, and A. Casadevall.** 2009. Capsule of *Cryptococcus neoformans* grows by enlargement of polysaccharide molecules. *Proc. Natl. Acad. Sci. USA* **106:** 1228–1233.
34. **Fries, B. C., D. L. Goldman, R. Cherniak, R. Ju, and A. Casadevall.** 1999. Phenotypic switching in *Cryptococcus neoformans* results in changes in cellular morphology and glucuronoxylomannan structure. *Infect. Immun.* **67:** 6076–6083.
35. **Gahrs, W., Z. Tigyi, L. Emody, and J. Makovitzky.** 2009. Polarization optical analysis of the surface structures of various fungi. *Acta Histochem.* **111:**308–315.
36. **Garcia-Hermoso, D., F. Dromer, and G. Janbon.** 2004. *Cryptococcus neoformans* capsule structure evolution in vitro and during murine infection. *Infect. Immun.* **72:**3359–3365.
37. **Garcia-Rivera, J., Y. C. Chang, K. J. Kwon-Chung, and A. Casadevall.** 2004. *Cryptococcus neoformans* CAP59 (or Cap59p) is involved in the extracellular trafficking of capsular glucuronoxylomannan. *Eukaryot. Cell* **3:**385–392.
38. **Gates, M. A., P. Thorkildson, and T. R. Kozel.** 2004. Molecular architecture of the *Cryptococcus neoformans* capsule. *Mol. Microbiol.* **52:**13–24.
39. **Granger, D. L., J. R. Perfect, and D. T. Durack.** 1985. Virulence of *Cryptococcus neoformans*. Regulation of capsule synthesis by carbon dioxide. *J. Clin. Invest.* **76:**508–516.
40. **Jacobson, E. S., M. J. Tingler, and P. L. Quynn.** 1989. Effect of hypertonic solutes upon the polysaccharide capsule in *Cryptococcus neoformans. Mycoses* **32:**14–23.
41. **James, P. G., and R. Cherniak.** 1992. Galactoxylomannans of *Cryptococcus neoformans. Infect. Immun.* **60:**1084–1088.

42. **Janbon, G., U. Himmelreich, F. Moyrand, L. Improvisi, and F. Dromer.** 2001. Cas1p is a membrane protein necessary for the O-acetylation of the *Cryptococcus neoformans* capsular polysaccharide. *Mol. Microbiol.* **42:**453–467.
43. **Kozel, T. R., and E. C. Gotschlich.** 1982. The capsule of *Cryptococcus neoformans* passively inhibits phagocytosis of the yeast by macrophages. *J. Immunol.* **129:**1675–1680.
44. **Kozel, T. R., S. M. Levitz, F. Dromer, M. A. Gates, P. Thorkildson, and G. Janbon.** 2003. Antigenic and biological characteristics of mutant strains of *Cryptococcus neoformans* lacking capsular O acetylation or xylosyl side chains. *Infect. Immun.* **71:**2868–2875.
45. **Kozel, T. R., G. S. Pfrommer, A. S. Guerlain, B. A. Highison, and G. J. Highison.** 1988. Strain variation in phagocytosis of *Cryptococcus neoformans*: dissociation of susceptibility to phagocytosis from activation and binding of opsonic fragments of C3. *Infect. Immun.* **56:**2794–2800.
46. **Linnemans, W. A., P. Boer, and P. F. Elbers.** 1977. Localization of acid phosphatase in *Saccharomyces cerevisiae*: a clue to cell wall formation. *J. Bacteriol.* **131:**638–644.
47. **Littman, M. L.** 1958. Capsule synthesis by *Cryptococcus neoformans*. *Trans. NY Acad. Sci.* **20:**623–648.
48. **Littman, M. L., and E. Tsubura.** 1959. Effect of degree of encapsulation upon virulence of *Cryptococcus neoformans*. *Proc. Soc. Exp. Biol. Med.* **101:**773–777.
49. **Ma, H., J. E. Croudace, D. A. Lammas, and R. C. May.** 2006. Expulsion of live pathogenic yeast by macrophages. *Curr. Biol.* **16:**2156–2160.
50. **Macher, A. M., J. E. Bennett, J. E. Gadek, and M. M. Frank.** 1978. Complement depletion in cryptococcal sepsis. *J. Immunol.* **120:**1686–1690.
51. **Maxson, M. E., E. Cook, A. Casadevall, and O. Zaragoza.** 2007. The volume and hydration of the *Cryptococcus neoformans* polysaccharide capsule. *Fungal Genet. Biol.* **44:** 180–186.
52. **Maxson, M. E., E. Dadachova, A. Casadevall, and O. Zaragoza.** 2007. Radial mass density, charge, and epitope distribution in the *Cryptococcus neoformans* capsule. *Eukaryot. Cell* **6:**95–109.
53. **McFadden, D. C., and A. Casadevall.** 2004. Unexpected diversity in the fine specificity of monoclonal antibodies that use the same V region gene to glucuronoxylomannan of *Cryptococcus neoformans*. *J. Immunol.* **172:**3670–3677.
54. **McFadden, D. C., M. De Jesus, and A. Casadevall.** 2006. The physical properties of the capsular polysaccharides from *Cryptococcus neoformans* suggest features for capsule construction. *J. Biol. Chem.* **281:**1868–1875.
55. **McFadden, D. C., B. C. Fries, F. Wang, and A. Casadevall.** 2007. Capsule structural heterogeneity and antigenic variation in *Cryptococcus neoformans*. *Eukaryot. Cell* **6:**1464–1473.
56. **Mitchell, T. G., and L. Friedman.** 1972. In vitro phagocytosis and intracellular fate of variously encapsulated strains of *Cryptococcus neoformans*. *Infect. Immun.* **5:**491–498.
57. **Monari, C., F. Paganelli, F. Bistoni, T. R. Kozel, and A. Vecchiarelli.** 2008. Capsular polysaccharide induction of apoptosis by intrinsic and extrinsic mechanisms. *Cell Microbiol.* **10:**2129–2137.
58. **Monari, C., E. Pericolini, G. Bistoni, A. Casadevall, T. R. Kozel, and A. Vecchiarelli.** 2005. *Cryptococcus neoformans* capsular glucuronoxylomannan induces expression of fas ligand in macrophages. *J. Immunol.* **174:** 3461–3468.
59. **Murphy, J. W.** 1988. Influence of cryptococcal antigens on cell-mediated immunity. *Rev. Infect. Dis.* **10**(Suppl 2): S432–S435.
60. **Murphy, J. W., and G. C. Cozad.** 1972. Immunological unresponsiveness induced by cryptococcal capsular polysaccharide assayed by the hemolytic plaque technique. *Infect. Immun.* **5:**896–901.
61. **Nimrichter, L., S. Frases, L. P. Cinelli, N. B. Viana, A. Nakouzi, L. R. Travassos, A. Casadevall, and M. L. Rodrigues.** 2007. Self-aggregation of *Cryptococcus neoformans* capsular glucuronoxylomannan is dependent on divalent cations. *Eukaryot. Cell* **6:**1400–1410.
62. **Nimrichter, L., M. L. Rodrigues, E. G. Rodrigues, and L. R. Travassos.** 2005. The multitude of targets for the immune system and drug therapy in the fungal cell wall. *Microbes Infect.* **7:**789–798.
63. **Nosanchuk, J. D., and A. Casadevall.** 1997. Cellular charge of *Cryptococcus neoformans*: contributions from the capsular polysaccharide, melanin, and monoclonal antibody binding. *Infect. Immun.* **65:**1836–1841.
64. **Nussbaum, G., W. Cleare, A. Casadevall, M. D. Scharff, and P. Valadon.** 1997. Epitope location in the *Cryptococcus neoformans* capsule is a determinant of antibody efficacy. *J. Exp. Med.* **185:**685–694.
65. **Pericolini, E., E. Cenci, C. Monari, M. De Jesus, F. Bistoni, A. Casadevall, and A. Vecchiarelli.** 2006. *Cryptococcus neoformans* capsular polysaccharide component galactoxylomannan induces apoptosis of human T-cells through activation of caspase-8. *Cell. Microbiol.* **8:**267–275.
66. **Pierini, L. M., and T. L. Doering.** 2001. Spatial and temporal sequence of capsule construction in *Cryptococcus neoformans*. *Mol. Microbiol.* **41:**105–115.
67. **Pirofski, L., R. Lui, M. DeShaw, A. B. Kressel, and Z. Zhong.** 1995. Analysis of human monoclonal antibodies elicited by vaccination with a *Cryptococcus neoformans* glucuronoxylomannan capsular polysaccharide vaccine. *Infect. Immun.* **63:**3005–3014.
68. **Reese, A. J., and T. L. Doering.** 2003. Cell wall alpha-1,3-glucan is required to anchor the *Cryptococcus neoformans* capsule. *Mol. Microbiol.* **50:**1401–1409.
69. **Reiss, E., M. Huppert, and R. Cherniak.** 1985. Characterization of protein and mannan polysaccharide antigens of yeasts, moulds, and actinomycetes. *Curr. Top. Med. Mycol.* **1:**172–207.
70. **Rodrigues, M. L., M. Alvarez, F. L. Fonseca, and A. Casadevall.** 2008. Binding of the wheat germ lectin to *Cryptococcus neoformans* suggests an association of chitinlike structures with yeast budding and capsular glucuronoxylomannan. *Eukaryot. Cell* **7:**602–609.
71. **Sakaguchi, N.** 1993. Ultrastructural study of hepatic granulomas induced by *Cryptococcus neoformans* by quick-freezing and deep-etching method. *Virchows Arch. B Cell Pathol. Incl. Mol. Pathol.* **64:**57–66.
72. **Schekman, R. W.** 1994. Regulation of membrane traffic in the secretory pathway. *Harvey Lect.* **90:**41–57.
73. **Schliwa, M., and G. Woehlke.** 2003. Molecular motors. *Nature* **422:**759–765.
73a. **Siafakas, A. R., T. C. Sorrell, L. C. Wright, C. Wilson, M. Larsen, R. Boadle, P. R. Williamson, and J. T. Djordjevic.** 2007. Cell wall-linked cryptococcal phospholipase B1 is a source of secreted enzyme and a determinant of cell wall integrity. *J. Biol. Chem.* **282:** 37508–37514.
74. **Small, J. M., and T. G. Mitchell.** 1989. Strain variation in antiphagocytic activity of capsular polysaccharides from *Cryptococcus neoformans* serotype A. *Infect. Immun.* **57:** 3751–3756.
74a. **Takeo, K., I. Uesaka, K. Uehira, and M. Nishiura.** 1973. Fine structure of *Cryptococcus neoformans* grown in vitro as observed by freeze-etching. *J. Bacteriol.* **113:**1442–1448.
74b. **Takeo, K., I. Uesaka, K. Uehira, and M. Nishiura.** 1973. Fine structure of *Cryptococcus neoformans* grown in vivo as observed by freeze-etching. *J. Bacteriol.* **113:**1449–1454.

75. **Todaro-Luck, F., E. Reiss, R. Cherniak, and L. Kaufman.** 1989. Characterization of *Cryptococcus neoformans* capsular glucuronoxylomannan polysaccharide with monoclonal antibodies. *Infect. Immun.* **57:**3882–3887.
76. **Turner, S. H., and R. Cherniak.** 1991. Glucuronoxylomannan of *Cryptococcus neoformans* serotype B: structural analysis by gas-liquid chromatography-mass spectrometry and 13C-nuclear magnetic resonance spectroscopy. *Carbohydr. Res.* **211:**103–116.
77. **Turner, S. H., R. Cherniak, E. Reiss, and K. J. Kwon-Chung.** 1992. Structural variability in the glucuronoxylomannan of *Cryptococcus neoformans* serotype A isolates determined by 13C NMR spectroscopy. *Carbohydr. Res.* **233:**205–218.
78. **Vaishnav, V. V., B. E. Bacon, M. O'Neill, and R. Cherniak.** 1998. Structural characterization of the galactoxylomannan of *Cryptococcus neoformans* Cap67. *Carbohydr. Res.* **306:**315–330.
79. **van de Moer, A., S. L. Salhi, R. Cherniak, B. Pau, M. L. Garrigues, and J. M. Bastide.** 1990. An anti-*Cryptococcus neoformans* monoclonal antibody directed against galactoxylomannan. *Res. Immunol.* **141:**33–42.
80. **Vartivarian, S. E., E. J. Anaissie, R. E. Cowart, H. A. Sprigg, M. J. Tingler, and E. S. Jacobson.** 1993. Regulation of cryptococcal capsular polysaccharide by iron. *J. Infect. Dis.* **167:**186–190.
81. **Walworth, N. C., B. Goud, A. K. Kabcenell, and P. J. Novick.** 1989. Mutational analysis of SEC4 suggests a cyclical mechanism for the regulation of vesicular traffic. *EMBO J.* **8:**1685–1693.
82. **Yoneda, A., and T. L. Doering.** 2006. A eukaryotic capsular polysaccharide is synthesized intracellularly and secreted via exocytosis. *Mol. Biol. Cell* **17:**5131–5140.
83. **Yoneda, A., and T. L. Doering.** 2008. Regulation of *Cryptococcus neoformans* capsule size is mediated at the polymer level. *Eukaryot. Cell* **7:**546–549.
84. **Yue, C., L. M. Cavallo, J. A. Alspaugh, P. Wang, G. M. Cox, J. R. Perfect, and J. Heitman.** 1999. The STE12alpha homolog is required for haploid filamentation but largely dispensable for mating and virulence in *Cryptococcus neoformans*. *Genetics* **153:**1601–1615.
85. **Zaragoza, O., and A. Casadevall.** 2004. Antibodies produced in response to *Cryptococcus neoformans* pulmonary infection in mice have characteristics of non-protective antibodies. *Infect. Immun.* **72:**4271–4274.
86. **Zaragoza, O., and A. Casadevall.** 2004. Experimental modulation of capsule size in *Cryptococcus neoformans*. *Biol. Proced. Online* **6:**10–15.
87. **Zaragoza, O., M. Cuenca-Estrella, J. Regadera, and J. L. Rodriguez Tudela.** 2008. The capsule of the fungal pathogen *Cryptococcus neoformans* paradoxically inhibits invasive growth. *Open Mycol. J.* **2:**29–39.
88. **Zaragoza, O., C. J. Chrisman, M. V. Castelli, S. Frases, M. Cuenca-Estrella, J. L. Rodriguez-Tudela, and A. Casadevall.** 2008. Capsule enlargement in *Cryptococcus neoformans* confers resistance to oxidative stress suggesting a mechanism for intracellular survival. *Cell. Microbiol.* **10:**2043–2057.
89. **Zaragoza, O., B. C. Fries, and A. Casadevall.** 2003. Induction of capsule growth in *Cryptococcus neoformans* by mammalian serum and CO(2). *Infect. Immun.* **71:**6155–6164.
90. **Zaragoza, O., C. P. Taborda, and A. Casadevall.** 2003. The efficacy of complement-mediated phagocytosis of *Cryptococcus neoformans* is dependent on the location of C3 in the polysaccharide capsule and involves both direct and indirect C3-mediated interactions. *Euro. J. Immnunol.* **33:**1957–1967.
91. **Zaragoza, O., A. Telzak, R. A. Bryan, E. Dadachova, and A. Casadevall.** 2006. The polysaccharide capsule of the pathogenic fungus *Cryptococcus neoformans* enlarges by distal growth and is rearranged during budding. *Mol. Microbiol.* **59:**67–83.

Cryptococcus: From Human Pathogen to Model Yeast
Edited by J. Heitman et al.
©2011 ASM Press, Washington, DC

5

Melanin: Structure, Function, and Biosynthesis in *Cryptococcus*

DAVID TROFA, ARTURO CASADEVALL, AND JOSHUA D. NOSANCHUK

One of the defining characteristics of *Cryptococcus neoformans* is its ability to synthesize a dark cell wall-associated pigment when grown in media containing phenolic compounds. This dark pigment is a melanin, a negatively charged, high-molecular-mass, hydrophobic, amorphous, and insoluble compound that is difficult to characterize and study. Melanin pigments are produced by organisms within all biological kingdoms and are associated with a wide variety of biological phenomena, including microbial pathogenesis (12, 136), malignant melanoma (40, 103), Parkinson's disease (22, 27, 140), traumatic anterior-chamber uveitis (7, 8, 60), and vitiligo (68, 135). The structure of melanins remains unsolved because current biochemical and biophysical techniques are unable to provide a reliable chemical formula for this complex polymer (37, 136). At present, the chemical definition of melanins is vague. A substance is considered to be a melanin if it is black or brown, insoluble in aqueous or organic fluids, resistant to concentrated acid, and susceptible to bleaching by oxidizing agents (12).

Melanin production by *C. neoformans* was first described by Staib, who in 1962 documented brown cryptococcal colonies on agar containing *Guizotia abyssinica* seed extracts (115). Melanization remains a distinctive feature used for *C. neoformans* diagnostic microbiology (62, 112, 118). Compared to other members of the *Cryptococcus* genus, pigment production is quicker, more pronounced, and can be elicited by compounds not active in other *Cryptococcus* spp. (17). The synthesis of melanin in *C. neoformans* necessitates exogenous dihydroxyphenolic or polyaminobenzene compounds and molecular oxygen, and it is regulated by a variety of factors including iron, copper, and glucose levels (2, 48, 97, 111, 142, 144). *C. neoformans* lacks tyrosinase and is therefore incapable of generating endogenous dihydroxyphenols. Cryptococcal melanin synthesis is suggested to follow a scheme adapted after the classical tyrosinase pathway for the production of melanin in mammals (93) with minor modifications (98). Through the action of a phenoloxidase, specifically a laccase, that follows the Mason-Roper model as modified by Ito (46), exogenous substrates are oxidized to their corresponding quinine and undergo a spontaneous rearrangement through sequential polymerization, followed by auto-oxidation to form melanochrome and, finally, melanin (Fig. 1) (121). Melanin synthesis in *C. neoformans* from 3,4-dihydroxyphenylalanine (L-dopa) was validated by electron spin resonance spectroscopy criteria (129).

C. neoformans melanin "ghosts" have permitted structural studies of melanin composition. Melanin ghosts are generated from melanized cryptococcal cells by detergent treatment followed by hydrolysis with hot concentrated HCl acid (130). Figure 2 depicts a melanin ghost and identifies the laminate structure of melanin in *C. neoformans* yeast cells. Melanin ghosts from cells grown in L-dopa are negatively charged (85). In heavily melanized cells, the pigment can account for approximately 15% of the dry mass of the cells (130), although the amount of melanin per *C. neoformans* cell grown in L-dopa can vary as much as eightfold between strains (85, 130). Cell wall deposition of melanin can be visualized by electron microscopy in melanized cells (129). Increased incubation time of cells grown with L-dopa results in progressively melanized cells (130).

There are two major types of melanin, eumelanin and pheomelanin, and many biological melanins are a combination of these two pigments. Eumelanin is a dark-brown to black pigment composed of 5,6-dihydroxyindole (DHI) and 5,6-dihydroxyindole-2-carboxylic acid (DHICA) monomers with a nitrogen composition ranging between 6 and 9%. Pheomelanin is reddish-brown, comprised of benzothiazine monomers with 8–11% nitrogen and 9–12% sulfur (47, 125). Elemental studies performed on cryptococcal melanin have found a sulfur composition of 0.4% and a C/N ratio of 7.5:1 (20). High-performance liquid chromatography on cryptococcal melanin supported these findings (20). While these data do not quantify eumelanin and pheomelanin levels, they do show that DHICA subunits predominate in cryptococcal melanin compared to 1,3-thiazole-4,5 dicarboxylic acid (TDCA) at a ratio of 47.7:1.

There is an extensive body of evidence from several laboratories that establishes a role for melanization in virulence,

David Trofa, Arturo Casadevall, and Joshua D. Nosanchuk, Departments of Medicine and Microbiology & Immunology, Albert Einstein College of Medicine, Jack and Pearl Resnick Campus, Bronx, NY 10461.

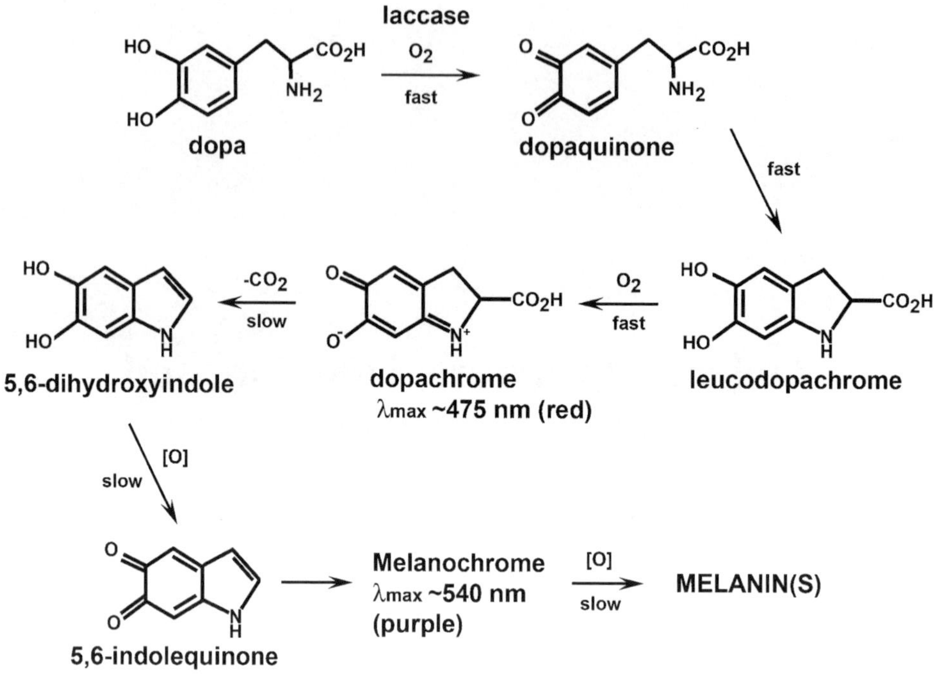

dopa → (laccase, O_2, fast) → dopaquinone → (fast) → leucodopachrome → (O_2, fast) → dopachrome (λ_{max} ~475 nm (red)) → ($-CO_2$, slow) → 5,6-dihydroxyindole → ([O], slow) → 5,6-indolequinone → Melanochrome (λ_{max} ~540 nm (purple)) → ([O], slow) → MELANIN(S)

FIGURE 1 Mason-Roper model for melanogenesis of *C. neoformans* beginning with L-dopa as the phenolic substrate.

providing a fascinating example of how a microorganism can utilize a ubiquitous pigment to undermine host defense mechanisms. First, melanin-deficient mutant strains generated by UV-irradiation are avirulent in murine models of cryptococcal infection (64, 102). Reversion to virulence was associated with recovery of melanin production (64). Microevolution within a laboratory strain of *C. neoformans* resulted in the generation of isolates with attenuated virulence displaying reduced melanin production as well as small polysaccharide capsules and slower growth rates (31). Furthermore, disruption of the laccase gene *LAC1* significantly attenuates cryptococcal virulence (100, 111). Despite the strong association of melanin with virulence, we note occasional reports of nonmelanotic strains recovered from patients (75). In those strains, other virulence factors may contribute to generating a virulent phenotype (77). In one strain the contribution of melanin to the overall virulence phenotype was approximately 10.5% (77). Although this appears small as an overall percentage, melanin ranks only second to capsule in overall importance.

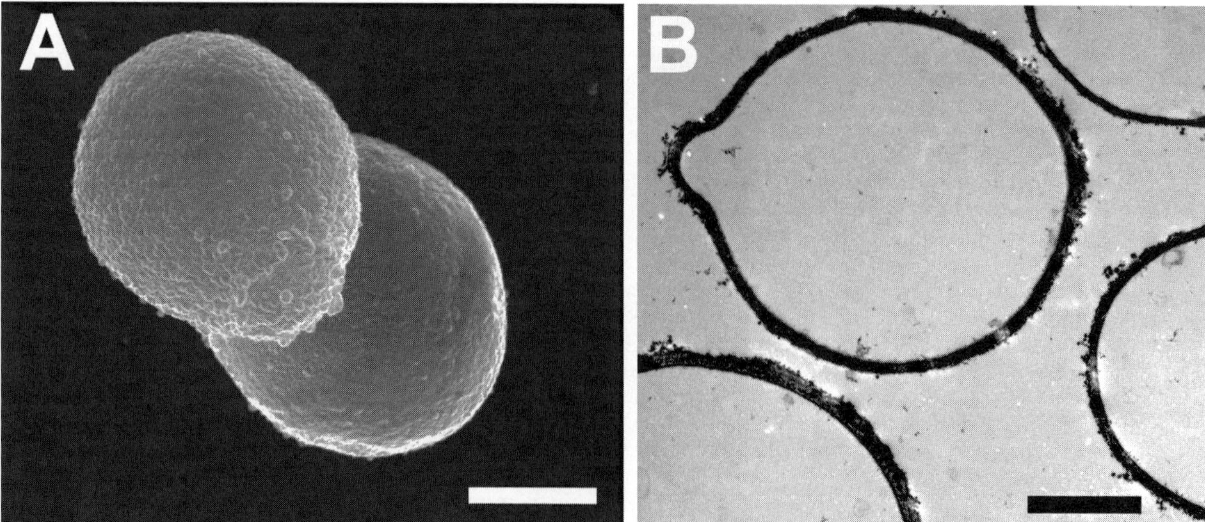

FIGURE 2 (A) Scanning and (B) transmission electron micrographs of melanin ghosts. Scale bars, 1 μm. Transmission electron micrograph courtesy of H. Eisenman.

Various studies suggest that melanin contributes to virulence by reducing the susceptibility of fungal cells to host defense mechanisms and interrupting normal immune responses (64, 102, 111). Although melanins are immunogenic, capable of inducing a strong T-cell-independent antibody response in mice (89) and eliciting anti-inflammatory responses (4, 81), they can influence normal activities of the immune system (89, 91, 107). For instance, cryptococcal melanin alters the levels of inflammatory cytokines during experimental animal infections (41, 78, 81), and melanization is associated with decreased rates of phagocytosis and killing by macrophages (129).

LACCASE

The enzyme responsible for melanin synthesis was characterized as a laccase based upon its substrate specificity (137). Cryptococcal laccase is a glycosylated copper-containing protein of 624 amino acids and 14 introns (137) that may exist as a dimeric species in cell extracts (44). The majority of *C. neoformans* melanin is produced by Lac1, although a second laccase enzyme, Lac2, adjacent to *LAC1* in the genome exists (80, 100, 142, 144). Mutation of the *LAC1* gene prevents melanin production, while deletion of *LAC2* does not (100, 126). *LAC1* contains a single open reading frame, a TATA and a CAAT box at positions –539 and –503, respectively, and a canonical Sp1-like enhancer site at position –1727 (141). As a member of the multicopper oxidase family, laccase contains an essential copper-binding site that when altered by a single amino acid base through site-directed mutagenesis abrogates laccase activity without preventing transcription. Typical of fungal laccases, the amino acid sequence of *LAC1* has a great deal of sequence diversity, with homology only in the copper-binding regions. *LAC1* mutants have delayed dissemination from the lungs but grow normally in lung tissue (92).

The oxidative pathway employed by *C. neoformans* is biochemically distinct from the polyketide pathway used by other melanotic fungal pathogens to produce pigment, such as *Wangiella dermatitidis* (98, 136). The mammalian pathway can utilize only tyrosine and dihydroxyphenylalanine as substrates for melanogenesis, while the *C. neoformans* phenoloxidase is a less specific enzyme capable of oxidizing a wide variety of substrates. Even among the *Cryptococcus* species, the *C. neoformans* laccase has a broad spectrum of specificity and generates pigments from many structurally diverse compounds including diphenols and indole compounds (18, 25, 66, 97, 99, 127). Among the diphenols that serve as substrates for pigment production, the *ortho*-diphenols (hydroxyl groups at the 2,3 or 3,4 position of the phenyl ring) result in cellular melanin deposition (18), while *para*-diphenols (OH- groups at the 1,4 or 2,5 position of the phenyl ring) result in pigments that diffuse into the culture medium (18). Monophenols (tyrosine, phenol) are poor substrates for phenoloxidase (137). Many structurally diverse indole compounds are substrates for melanin production (66), and the color of the pigment produced is dependent upon the substrate used. L-Dopa produces a black pigment, while other substrates result in green-black, brown, or orange-red pigments (18). Cryptococcal melanin derived from homogentisic acid, a substrate many bacterial species use to synthesize melanin, is brown (32), and esculin (6-β-D-glucose-dihydroxycoumarin) can also serve as a substrate for laccase, through the oxidation of the 6,7-dihydroxycoumarin component to a melanin-like pigment (25).

Several studies have addressed the question of the cellular location of laccase. Early work suggested that the majority of phenoloxidase activity was membrane bound (97), a theory supported by later evidence demonstrating that the laccase enzyme attached to the cell wall through a disulfide or thioester bond (142). However, other investigations found the most enzymatic activity in soluble fractions (44). The discrepancy in results may result from enzyme purification protocol differences (44), but it appears that the enzyme can be detected in both membrane and soluble fractions. Immunogold studies with antibody to laccase have shown that it is found in both the cell wall and cytoplasm (35). This topic was further clarified by Missall et al., who showed that the two *C. neoformans* laccases are differentially localized (80): Lac1 to the cell wall, while Lac2 is cytoplasmic. Functionally, Lac1 is situated where it is more likely to encounter substrates in the medium and where deposition of melanin into the cell wall would not require further transport (80). The same group found that in the absence of Lac1, Lac2 can be localized to the cell wall as well as the cytoplasm, perhaps due to differential expression of the enzyme or the production of a new protein product to fill the void in the cell wall.

Laccase activity can be measured through the oxidation of L-epinephrine to melanin (101), and *LAC1* is regulated by glucose concentration, temperature, the available nitrogen source, iron ion concentration, and copper concentration. The reaction of *C. neoformans* laccase with L-dopa results in consumption of oxygen (112). Glucose starvation and lower temperatures (25°C versus 37°C) increase the amount of enzyme (44, 97). The reduction in laccase activity at 37°C may appear unusual for a virulence factor of mammals, but the reduced activity at this temperature appears to translate into minor biological effects, as no differences in melanization could be detected by electron spin resonance spectroscopy at 30°C and 37°C (130). *C. neoformans* laccase produced at 25°C and 37°C exhibits differences in activity, electrophoretic mobility, and concanavalin A binding (43, 44, 50). The ability of *C. neoformans* to produce pigment also depends upon the nitrogen source provided for growth. Amino acids like glutamine, asparagine, and glycine are strong inducers of pigment synthesis (18). Biochemical differences have been noted in the activity of the laccase from two varieties of *C. neoformans* by the ability of glutamine and $(NH_4)_2SO_4$ to influence enzyme activity (93). Glutamine suppresses laccase activity in all isolates except those from serotype B, while $(NH_4)_2SO_4$ suppresses laccase activity of serotype A isolates (93).

LAC1 transcription is responsive to iron and copper (48, 97, 111, 143), and both ions are important for melanization, as demonstrated through insertion mutagenesis of the genes *CCC2* and *ATX1* (a copper transporter and copper chaperone, respectively) (126). Additionally, copper has the ability to suppress the melanin-deficient phenotype in certain mutants, suggesting a role in a signaling pathway with a copper-containing or transport enzyme that aids in melanin regulation (122). This is notable, as there seems to be discordant regulation of certain *C. neoformans* virulence determinants. For instance, stimuli that suppress capsular polysaccharide formation, such as iron, increase laccase activity (48).

There is evidence that laccase may serve a dual role, producing melanin and oxidizing ferrous iron during infection. Iron acquisition in *C. neoformans* is mediated through enzymatic processes in the form of high- and low-affinity iron uptake systems mediated by cell surface reductases (51) and nonenzymatic reduction of ferric iron via 3-hydroxyanthranilic acid and melanin (94). Melanin is assumed to play a role in ferric iron reduction, as nonmelanized C. *neoformans* cells produce ferrous iron at a rate of 0.1 nmol per 10^6 cells/h versus 2.1 for melanized cells (57, 94). Ferrous iron may also enhance C. *neoformans* pathogenicity by binding to melanin and increasing its extracellular redox buffering capacity, thus providing the pathogen better protection against oxidative killing (52). In murine infections, laccase expression can be detected early (24 h) during infection but steadily decreases as the fungal burden rises (35). Laccase release from the cell wall in vivo has been demonstrated, suggesting a possible role for laccase as an antioxidant or iron scavenger (35). Laccase possesses strong ferrous iron oxidase activity in the absence of substrates that may protect cells from hydroxyl radicals generated from host macrophages and neutrophils while also aiding in iron transport (72). Interestingly, Cir1, a central regulator of the C. *neoformans* iron regulon, targets *LAC1* and *LAC2* with expression patterns similar to *FTR1* and *FTR3*, genes of the reductive iron transport pathway (58). Furthermore, *cir1* mutants cannot appropriately sense external iron levels and are avirulent as the production of major virulence factors is impaired, including the formation of melanin in the cell wall (58). Furthermore, *LAC1* is upregulated in *cir1* mutants, and laccase activity is constitutively expressed in these mutants regardless of the glucose concentration of growth media. Taken together, increased laccase expression at the onset of infection fits well with laccase's role in iron acquisition and protection from host hydroxyl radical attack (58).

The uptake of iron in many fungi and bacteria is, in part, mediated by exported siderophores that bind ferric iron and are subsequently imported into the cell. Although siderophores have not been identified in C. *neoformans* (55), the transcript for a putative siderophore transporter (*SIT1*) has recently been identified in C. *neoformans* by serial analysis of gene expression (119), providing evidence that the pathogen may utilize siderophores from other microbes as a mechanism of obtaining iron. Interestingly, the ability of serotype D mutants deficient in SIT1 (B3501A background) to produce melanin on agar medium containing DOPA is enhanced, and laccase activity is slightly higher compared to wild type (119). This has led to the hypothesis that the loss of *SIT1* either alters metal homeostasis in the pathogen and/or changes laccase expression or localization, influencing copper loading on the copper-requiring laccase. Additional tests were conducted with wild-type cells incubated with copper-enhanced laccase activity in wild type, matching the level of activity in the *SIT1* disruptant.

MICROSTRUCTURE OF CELL WALL-ASSOCIATED MELANIN

The hypothesis that melanin is a cross-linked polymer of phenol and indole subunits presents numerous questions concerning the basic biological mechanisms of C. *neoformans*, including nutrient transportation across the melanin layer and budding through melanin (26). For example, the mechanism by which this polymer is synthesized and attached to the cell wall remains largely undefined.

A recently proposed structural model of C. *neoformans* melanin (Fig. 3) has implications for cell division, cell wall remodeling, and antifungal drug discovery (26). Various microscopy techniques were used to develop the melanin model, including atomic force and scanning electron microscopy to examine the surface structure of melanin, transmission electron microscopy (TEM) to study cross sections of melanin, and nuclear magnetic resonance (NMR) cryoporometry to investigate the porosity of melanin. The high-resolution microscopy techniques revealed

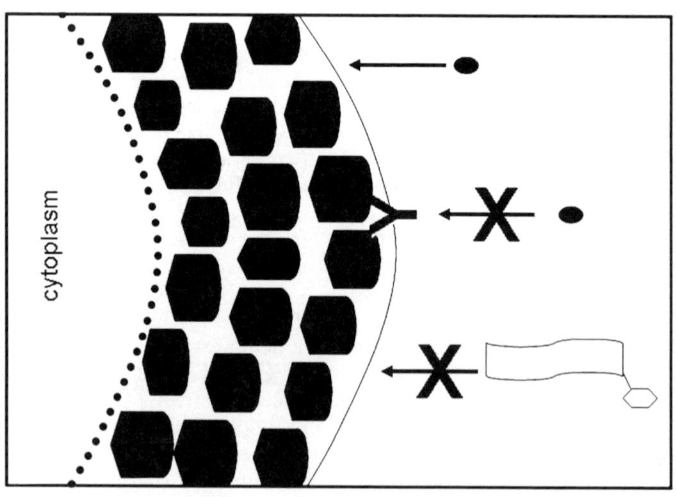

FIGURE 3 Cross-sectional depiction of the microstructure of a melaninzed C. *neoformans* cell wall. Multiple layers of melanin granules form into a dense layer in the cell wall. The packing of the granules allows for the acquisition of nutrients, such as sugars and amino acids. However, the melanin-binding antibody can block the passage of nutrients. The packing further inhibits the passage of certain large antifungal compounds, such as polyenes and pneumocandins. (Adapted from reference 26.)

that melanin ghosts are formed by discrete, spherical granules with roughly similar dimensions, approximately 40–130 nm in diameter. This finding is consistent with previous reports describing melanin assembly by smaller melanin particles from other biological sources (reviewed in 26). TEM analysis revealed that melanin ghosts are arranged in a concentric pattern of two to five layers. Interestingly, melanin ghosts isolated from budding cells have less melanin, although a thicker melanin layer was observed as the bud increased in size relative to the mother cell. These data are consistent with the observation that melanin accumulation increases with age, presumably by the addition of more layers of melanin particles. TEM also showed that during budding, there is a break in the melanin wall that allows budding to occur. Melanin formation occurs de novo in the bud rather than deriving melanin from the mother cell (86), and budding cells similarly generate their own capsule (96).

The data obtained by the TEM study of melanin pose an interesting question about how melanin granules may be held together within the *C. neoformans* wall (26), for which there are two plausible hypotheses. First, the granules may be simply cross-linked together. A more complex possibility involves a lattice or scaffold within the *C. neoformans* cell wall comprised of melanin and/or proteins and polysaccharides keeping the melanin in place. As melanin layers in ghosts are similar to cell wall layers (1, 117), and multilaminate structures are a common feature of fungal cell walls, there is a good possibility that structures exist in the cell wall that direct the deposition of melanin into layers. Furthermore, formation of a scaffold would have implications for *C. neoformans* growth and budding, as melanin remodeling could be accomplished through scaffold degradation by specific processes and enzymes. As discussed below in "Genetic Studies," chitin may play a role in scaffolding for melanin formation (5, 126). Future gene disruption studies looking for less structured melanin formation within the cell wall may discover the pathway and enzymes used for melanin deposition and formation into layers of granules (26).

Nutrient acquisition necessitates pores. NMR cryoporometry revealed pores of 1–4 nm in diameter in melanin ghosts, the porosity of which decreased with increasing culture age (26). A small number of larger, 30-nm-diameter pores were also identified. Additionally, incubation with melanin-binding antibody reduced the pore size. Thus, it can be inferred that the antibody binds melanin granules and physically blocks smaller pores, but since the antibodies had no detectable effect on larger pores, they must be inaccessible to the antibody. From this result, a proposed model of melanin structure and function can be depicted where irregularly shaped granules are arranged in layers. The small spaces between granules make up the smaller pores, while the internal spaces between layers create the larger pores, allowing melanin to act as a sieve. Studies with different-molecular-mass dextrans support this model (53). As a result of its highly charged and aromatic surface, melanin may also bind and sequester various molecules (85, 91). In summary, the porosity of melanin is a property of its structure, and the fungus does not appear to require specialized pore structures to attain nutrients.

BIOLOGICAL PHENOMENA

In Vivo Melanization

C. neoformans produces melanin during experimental rodent infection and in human disease (90, 91, 108). Melanization is presumably facilitated by the presence of various phenolic compounds that can serve as substrates for the *C. neoformans* laccase. Also, the high concentrations of norepinephrine, 3,4-dihydroxyphenylacetic acid, homovanillic acid, 5-hydroxyindoleacetic acid, serotonin, and dopamine in brain tissue may provide an explanation for the neurotropism of the pathogen (90, 97). Indeed, areas of the brain rich in catecholamines, such as the basal ganglia, are among the most frequently invaded by *C. neoformans* (71). Notably, the *C. neoformans* laccase gene (*LAC1*) transcript has been detected in cells in the subarachnoid space by reverse transcription-PCR during experimental cryptococcal meningitis infections in rabbits (111).

Melanin Contribution to *C. neoformans* Survival in the Environment

In the environment, melanin is thought to protect *C. neoformans* from various stresses including enzymatic degradation (105), radiation (UV, solar, gamma) (104, 131), and heavy metals (silver nitrate) (34) while providing thermotolerance (104, 139) and structural integrity to withstand osmotic challenges (13). Melanization is also associated with increased survival after ingestion from predatory microorganisms such as the nematode *Caenorhabditis elegans* (82, 116) and the amoeba *Acanthamoeba castellani* (82, 116), both of which produce hydrolytic enzymes to aid in the digestion of microbes. These findings are consistent with in vitro studies showing that melanization affords *C. neoformans* protection against hydrolytic enzyme digestion, whereas nonmelanized *C. neoformans* cells are highly susceptible to degradation by hydrolytic enzymes (11, 83, 105, 110). While the exact mechanism of protection has not been determined (105), melanin can bind certain hydrolytic enzymes as well as their substrates to prevent their degradation (9, 10, 63). Thus, given that melanin accumulates in the cryptococcal cell wall, it is possible that melanin protects cell wall components by directly inactivating hydrolytic enzymes and/or binding with the substrates and shielding the yeast from the enzymes.

Various ecological niches have been hypothesized for *C. neoformans* based upon its ability to synthesize melanin and survive in hostile environments (105, 109). For example, melanin facilitates *C. neoformans* survival at extremes of temperature, humidity, and acidity. Furthermore, melanin protects the yeast against various types of radiation (20, 131). One of the organism's possible ecological niches is in decaying trees, and melanin production may be related to *C. neoformans*' ability to break down polyphenolic compounds (61, 70). In this regard, phenoloxidases have been found in lignolytic mushrooms, and laccases are used commercially for degrading wood pulp lignin. There is also evidence that pigeon guano is an important ecological niche for *C. neoformans*, supporting growth and reproduction (84), and likely represents a common source of human infection (19).

Melanized fungal species are found in environments exposed to high constant radiation levels, such as the damaged reactor at Chernobyl (79). This finding, in accordance with the discovery of bacterial melanin-producing species in nuclear reactor pool water (114) and laboratory observation, suggests a role for melanin in radioprotection. Recent studies have attributed the radioprotective effects of *C. neoformans* melanin to be a function of its chemical composition, stable radical presence, and spatial arrangement (20). Accordingly, melanized *C. neoformans* cells are more resistant to external gamma radiation at levels ranging from 0 to 0.22 kGy than nonmelanized cells. In comparison, 0.005 kGy is lethal to humans. The same study found that C. *neoformans* ghosts were able to scatter X-ray radiation, resulting in a physical shielding effect.

Melanin Contribution to Pathogenesis

The broad subject of melanin's contribution to microbial virulence and clinical resistance to antimicrobial compounds was recently reviewed (87). Melanin is an important virulence factor that protects the fungus from the host immune response (14, 36, 138) and functions as an immunomodulator (78). Melanin increases C. *neoformans* resistance to phagocytosis in vitro and in vivo (76, 129). The inhibition of phagocytosis in melanized cells may, in part, be a result of the additional negative charge imparted to the cryptococcal cellular wall by melanin (6, 85, 129). Melanization was shown to increase the cellular negative charge as measured by zeta potential by 3–33% in nine encapsulated strains and by 86% in an acapsular strain (85, 87). Melanization also protects C. *neoformans* from macrophage killing (129) and permits survival in alveolar macrophages by providing resistance to oxidative stress and facilitating extrapulmonary dissemination to the brain (36).

Melanin is a powerful antioxidant with a broad capacity to protect fungal cells against oxygen- and nitrogen-derived oxidants produced by host effector cells (49, 54, 56, 129, 133, 136). The antioxidant capabilities of melanin are similar to those of superoxide dismutase, although melanin has greater chemical reactivity and could, at least in theory, protect against a wider variety of oxidants (54). In vitro studies have investigated the antioxidative capacity of melanin. Phagocytosis assays incorporating an oxidative system show killing of non-melanin-producing mutants but not wild-type strains (99). Genetic studies have confirmed a link between oxygen-sensitive mutants and their inability to produce melanin (49), while cellular investigations have supported the hypothesis that melanin protects C. *neoformans* from both oxygen- and nitrogen-derived oxidative damage from host cells (129, 134). Melanins are scavengers of free radicals (113) and have electron transfer properties (33). Electron spin resonance spectroscopy illustrated the capacity of C. *neoformans*-derived melanin for transferring electrons from free radical species generated in solution (130).

There is substantial evidence that melanin contributes to C. *neoformans* pathogenicity as an immunomodulatory compound by interfering with immune responses. In an experimental murine pulmonary infection model studying the early stages of infection, infection with melanized cells resulted in a greater fungal burden and was associated with higher levels of interleukin-4 and monocyte chemoattractant protein 1 (MCP-1), as well as a greater number of infiltrating leukocytes, than infection with nonmelanized cells (78). The same study also found that infection with laccase-positive organisms elicited higher MCP-1 levels and more infiltrating leukocytes than did mutants lacking a functioning laccase enzyme and that melanization conferred protection against host phagocytosis in vivo (78). Earlier investigations have also noted changes in the host cytokine/chemokine response attributed to C. *neoformans* melanization (41, 78). Furthermore, injection of C. *neoformans* melanin particles into mice promotes granuloma formation and activates complement (106). Melanin may also play a role in host immune response to C. *neoformans*, as strains with varying degrees of melanin induce different amounts of tumor necrosis factor alpha and differentially stimulate T-cell lymphoproliferation (41).

The complement system plays a key role in the resistance to cryptococcal infections, particularly with the deposition of complement fragment C3 onto the polysaccharide capsule, which results in optimal phagocytosis and fungal clearance (21, 22). In in vitro assays, C. *neoformans* melanin particles are bound as early as 1 minute after incubation with ^{125}I-labeled complement C3, with maximum binding occurring after 12 minutes (106). By blocking the classical complement pathway, the kinetics and total deposition of C3 onto melanin particles is not affected, illustrating that melanins activate complement through the alternative pathway. An in vivo immunofluorescence assay involving an intratracheal murine injection of cryptococcal melanin particles also demonstrates deposition of C3 onto the ghosts. However, although melanins have the ability to activate complement, they are present in the cryptococcal cell wall (129) and therefore are unlikely to be significantly exposed to the extracellular milieu due to the surrounding polysaccharide capsule (106). As such, it is improbable that melanin interacts with components of the complement system during C. *neoformans* infection unless the capsule is small or significantly perturbed.

Melanin and Antifungals

The charge, size, and amorphous shape of melanin facilitates its binding to diverse compounds (59, 69), including antifungals. Melanin adsorbs microbicidal peptides, which may prevent certain host-toxic compounds from reaching their final targets in melanized C. *neoformans* cells (23). Melanized C. *neoformans* cells are less susceptible to the antifungals amphotericin B and caspofungin than nonmelanized cells. Incubation of amphotericin and caspofungin with melanin prior to exposure with the fungus significantly reduces their antifungal activities against C. *neoformans*, while incubation of itraconazole, fluconazole, and flucytosine had no effect on antifungal effectiveness (45, 123, 124, 132). Interestingly, a 16-fold increase in MIC is attributed to the adsorption of amphotericin B with melanin particles derived from C. *neoformans*, and the increase in MIC correlates with the amount of melanin particles added to the antifungal drug solution before incubation with C. *neoformans* (45). Time-kill assays have also shown that adding melanin particles to amphotericin B or caspofungin significantly reduces their toxicities against C. *neoformans* (123). For instance, 79% and 71% of yeast cells survived exposure to caspofungin and two times the MIC of amphotericin B preincubated with synthetic melanin, respectively. On the other hand, C. *neoformans*' survival rate decreased to 11% and 8% for untreated caspofungin and amphotericin B, respectively. Melanized C. *neoformans* biofilms are also protected from amphotericin B and caspofungin (76). In the case of biofilms exposed to 32 μg/ml of amphotericin B, 40% of melanized C. *neoformans* cells were metabolically active, as opposed to 20% of nonpigmented cells. Additionally, the elemental composition of melanin was altered after incubation with amphotericin B and caspofungin, but not with flucytosine or the azoles (123, 124). From these data, it can be inferred that cell wall melanin binds caspofungin and amphotericin B and impairs them from reaching their targets. Other studies have confirmed these results (45).

As described above, the granular arrangement of cell wall-associated melanins in C. *neoformans* (26) helps explain why certain drugs are prevented from diffusing into the cell body. First, the distribution of the polymer affords a large surface area for it to bind certain types of compounds.

Furthermore, the tight spaces between melanin granules may prevent or impede the entry of large drugs such as amphotericin B (molecular mass, 924 g/mol) and caspofungin (molecular mass, 1,093.5 g/mol) into *C. neoformans* cells (87). This effect may be especially important for amphotericin B, as the deoxycholate form of the drug may form large aggregates in solution (67), and delivery of liposomal amphotericin B may be further impaired in this setting.

As described by Nosanchuk and Casadevall (87), the finding that melanin can bind to amphotericin B and caspofungin, in combination with the recent analysis of *C. neoformans* melanin microstructure, provides a possible explanation for the difficulty of these drugs to treat infections. It is known that *C. neoformans* melanizes in vivo (90), that melanin production increases with time after infection (29, 91), and that amphotericin B often fails to eradicate *C. neoformans* from patients (95, 145). Because melanized cells show resistance to amphotericin B in vitro, while nonmelanized cells are susceptible, the efficacy of amphotericin B may be largely due to its activity against nonmelanized or poorly melanized *C. neoformans* buds or by its immunomodulatory effects (87). Like amphotericin B, caspofungin is active against nonmelanized cells in vitro but ineffective in an experimental infection model with *C. neoformans*. Interestingly, acapsular strains of *C. neoformans* are also resistant to caspofungin, meaning the large polysaccharide capsule does not prevent the drug from engaging the cell. Thus, the resistance to caspofungin in vivo could in part be attributed to melanization (30).

Targeting Melanin

Melanin-binding drugs may be useful as therapeutics against *C. neoformans* infection (134). In fact, administration of melanin-binding monoclonal antibodies (MAbs) prolongs the survival of mice in a lethal intravenous infection model (107). In this study, the fungal burdens of collected brains and lungs of infected mice were taken 7 days after infection, and mice administered MAbs to melanin had significantly lower fungal burdens than control mice administered irrelevant immunoglobulin. The same study found that in vitro growth rates of melanized *C. neoformans* cells decreased when incubated with MAbs to melanin. Melanin-binding MAbs must diffuse through the polysaccharide capsule before binding at the cell surface where melanin is localized (91, 108). Since MAbs to melanin were not found to induce phagocytosis in vitro by the murine macrophage-like cell line J774.16 and/or to cause agglutination, MAbs can be assumed to interfere with cell growth and replication by altering the properties of the pigment within the fungal cell wall (107).

A second example of the biological impact of targeting *C. neoformans* melanin is the use of glyphosate during experimental cryptococcosis (88). Glyphosate is a systemic herbicide that interferes with amino acid synthesis via the shikimate acid pathway. Administration of glyphosate to mice infected with *C. neoformans* significantly delayed cryptococcal melanization and prolonged survival.

GENETIC STUDIES

As briefly described above, early studies that linked melanin to *C. neoformans* virulence used classical genetic techniques to generate mutants through nonspecific mutagenesis to identify yeast cells deficient in melanin production (64, 65, 99, 102). Albino mutants (Mel^-) were subsequently found to be less virulent, and interestingly, Mel^--infected mice had high rates of reversion to melanin production, which restored virulence (102). Although such investigations indicated a potential role for melanin in *C. neoformans* pathogenesis, they did not implicate a mechanism for yeast protection.

Recently, extensive genetic investigations of *C. neoformans* have illustrated the abundance of genes that participate in effective melanogenesis. Screening of a genetic library of 1,201 signature-tagged, targeted gene deletion *C. neoformans* mutants identified 38 mutants that exhibited decreased melanization when grown on media containing L-dopa, and 33 novel genes were identified that regulate melanization, supporting the hypothesis that melanin synthesis involves a number of novel pathways (73). These pathways include metal ion homeostasis, cell wall maintenance, and the mitogen-activated protein kinase (MAPK) cascade. In addition to the link between metal ion homeostasis and melanin production discussed in the "Laccase" section above, genetic studies have identified two genes involved in maintaining the proper ion balance and metal homeostasis that are important for melanogenesis: *VPH1* and *CLC1*, genes involved in the vesicular movement of protons and chloride ions, respectively (28, 42, 143).

Interference of the genes involved in cell wall integrity, such as those that regulate chitin synthesis, can result in severe cellular alterations (5, 126). Chitin, a linear polymer of β-(1,4)-linked *N*-acetylglucosamine, is an essential component of the fungal cell wall that provides strength and integrity. Chitosan, the deacylated form of chitin, is generated enzymatically by chitin deacetylases, and *C. neoformans* possesses four chitin deacetylase genes (5). Mutants lacking chitosan exhibit the "leaky melanin" phenotype that is attributed to strains in which pigment is found in the medium supernatant where the yeasts have lost the ability to either attach the laccase enzyme within the cell wall or retain the melanin or a melanin-like pigment within the cell wall (5, 126). These data suggest that chitosan may serve as a scaffolding protein within the cell wall, which is necessary for the attachment of Lac1 and Lac2 and/or melanin.

Numerous *C. neoformans* genes and proteins involved, directly or indirectly, in diverse pathways can impact melanin regulation and synthesis. For instance, pleiotropic effects are observed when proteins involved in the MAPK phosphorylation cascades are deleted. The PKC1-MAPK pathway mediates cell wall integrity in response to various environmental responses, and various signaling molecules of the cascade, such as *STE12* (15, 16) and cyclophilin A (128), are involved in laccase regulation. The IPC1-DAG-PKC1 pathway plays an important role in the localization and anchoring of laccase to the cell wall as well as in the maintenance of cell wall integrity (38). The sphingolipid enzyme inositol-phosphoryl ceramide synthase 1 (IPC1) regulates melanin production through the generation of diacylglycerol (DAG) that activates protein kinase C1 (PKC1) (39). Deletion of the putative DAG binding domain on PKC1 decreases *C. neoformans* laccase activity by up to 65% in addition to other adverse effects, such as the production of a smaller capsule than in the wild-type strain and defects in cell wall integrity (38). Blocking the IPC1-DAG-PKC1 pathway does not completely prevent melanin formation, providing evidence that other pathways are working in tandem to produce pigment. There is also evidence that *C. neoformans* melanin production is also controlled by the Gα protein-cyclic AMP protein kinase (2, 3, 24).

CONCLUSION

Melanin plays many vital roles contributing to the survival and pathogenesis of *C. neoformans*, and laccase is required for the dissemination of yeast cells from the lung to the brain. However, details about the process of melanization are lacking. Despite advances in the cryptococcal field and the study of melanin in recent years, there is still great interest in elucidating the pigment production pathway to better understand fungal pathogenesis and, potentially, to develop antifungal drugs to protect and prevent cryptococcal disease. There are numerous outstanding questions at the fore of melanin research. For instance, how does a melanized yeast cell grow and replicate in the face of a rigid, dense, and tenacious polymer? No specific melanin-degrading enzymes have been described, yet such enzymes must exist given the abundance of this material in nature. The existence of enzymes in fungi is supported by such facts as that *Aspergillus fumigatus* can utilize environmental melanins as a sole carbon source for growth, indicating that the fungus is capable of degrading melanin (74). *C. neoformans* melanin granules appear to be anchored in the cell wall in an orderly pattern where they form concentric layers that contribute to structural strength while being permeable and flexible enough to allow replication. At this time, there is no information on how this is accomplished, except for high-resolution microscopy showing layers upon layers of relatively uniform melanin granules (26, 119). Although significant progress has been made to identify the basic units of cryptococcal melanin, we remain in the dark about how this pigment is made into granules. Recently, a new approach to this problem based on metabolic incorporation of labeled precursors combined with solid-state NMR has shown that melanin ghosts from *C. neoformans* contain polysaccharide and aliphatic compounds (120).

REFERENCES

1. **al-Doory, Y.** 1971. The ultrastructure of *Cryptococcus neoformans*. *Sabouraudia* **9:**115–118.
2. **Alspaugh, J. A., J. R. Perfect, and J. Heitman.** 1997. *Cryptococcus neoformans* mating and virulence are regulated by the G-protein alpha subunit GPA1 and cAMP. *Genes Dev.* **11:**3206–3217.
3. **Alspaugh, J. A., R. Pukkila-Worley, T. Harashima, L. M. Cavallo, D. Funnell, G. M. Cox, J. R. Perfect, J. W. Kronstad, and J. Heitman.** 2002. Adenylyl cyclase functions downstream of the Galpha protein Gpa1 and controls mating and pathogenicity of *Cryptococcus neoformans*. *Eukaryot. Cell* **1:**75–84.
4. **Avramidis, N., A. Kourounakis, L. Hadjipetrou, and V. Senchuk.** 1998. Anti-inflammatory and immunomodulating properties of grape melanin. Inhibitory effects on paw edema and adjuvant induced disease. *Arzneimittelforschung* **48:**764–771.
5. **Banks, I. R., C. A. Specht, M. J. Donlin, K. J. Gerik, S. M. Levitz, and J. K. Lodge.** 2005. A chitin synthase and its regulator protein are critical for chitosan production and growth of the fungal pathogen Cryptococcus neoformans. *Eukaryot. Cell* **4:**1902–1912.
6. **Blasi, E., R. Barluzzi, R. Mazzolla, B. Tancini, S. Saleppico, M. Puliti, L. Pitzurra, and F. Bistoni.** 1995. Role of nitric oxide and melanogenesis in the accomplishment of anticryptococcal activity by the BV-2 microglial cell line. *J. Neuroimmunol.* **58:**111–116.
7. **Broekhuyse, R. M., E. D. Kuhlman, and H. J. Winkens.** 1992. Experimental autoimmune anterior uveitis (EAAU). II. Dose-dependent induction and adoptive transfer using a melanin-bound antigen of the retinal pigment epithelium. *Exp. Eye Res.* **55:**401–411.
8. **Broekhuyse, R. M., E. D. Kuhlmann, and R. J. Winkens.** 1993. Experimental autoimmune anterior uveitis (EAAU): induction by melanin antigen and suppression by various treatments. *Pigment Cell Res.* **6:**1–6.
9. **Bull, A. T.** 1970. Inhibition of polysaccharases by melanin: enzyme inhibition in relation to mycolysis. *Arch. Biochem. Biophys.* **137:**345–356.
10. **Bull, A. T.** 1970. Kinetics of cellulase inactivation by melanin. *Enzymologia.* **39:**333–347.
11. **Bunting, L. A., J. B. Neilson, and G. S. Bulmer.** 1979. *Cryptococcus neoformans*: gastronomic delight of a soil ameba. *Sabouraudia.* **17:**225–232.
12. **Butler, M. J., and A. W. Day.** 1998. Fungal melanins: a review. *Can.. J. Microbiol.* **44:**1115–1136.
13. **Casadevall, A., and J. R. Perfect.** 1998. *Cryptococcus neoformans*. ASM Press, Washington, DC.
14. **Casadevall, A., A. L. Rosas, and J. D. Nosanchuk.** 2000. Melanin and virulence in *Cryptococcus neoformans*. *Curr. Opin. Microbiol.* **3:**354–358.
15. **Chang, Y. C., L. A. Penoyer, and K. J. Kwon-Chung.** 2001. The second STE12 homologue of *Cryptococcus neoformans* is MATa-specific and plays an important role in virulence. *Proc. Natl. Acad. Sci. USA* **98:**3258–3263.
16. **Chang, Y. C., B. L. Wickes, G. F. Miller, L. A. Penoyer, and K. J. Kwon-Chung.** 2000. *Cryptococcus neoformans* STE12alpha regulates virulence but is not essential for mating. *J. Exp. Med.* **191:**871–882.
17. **Chaskes, S., and R. L. Tyndall.** 1978. Pigment production by *Cryptococcus neoformans* and other *Cryptococcus* species from aminophenols and diaminobenzenes. *J. Clin. Microbiol.* **7:**146–152.
18. **Chaskes, S., and R. L. Tyndall.** 1975. Pigment production by *Cryptococcus neoformans* from para- and ortho-diphenols: effect of the nitrogen source. *J. Clin. Microbiol.* **1:**509–514.
19. **Currie, B. P., L. F. Freundlich, and A. Casadevall.** 1994. Restriction fragment length polymorphism analysis of *Cryptococcus neoformans* isolates from environmental (pigeon excreta) and clinical sources in New York City. *J. Clin. Microbiol.* **32:**1188–1192.
20. **Dadachova, E., R. A. Bryan, R. C. Howell, A. D. Schweitzer, P. Aisen, J. D. Nosanchuk, and A. Casadevall.** 2008. The radioprotective properties of fungal melanin are a function of its chemical composition, stable radical presence and spatial arrangement. *Pigment Cell Melanoma Res.* **21:**192–199.
21. **Diamond, R. D., J. E. May, M. A. Kane, M. M. Frank, and J. E. Bennett.** 1974. The role of the classical and alternate complement pathways in host defenses against *Cryptococcus neoformans* infection. *J. Immunol.* **112:**2260–2270.
22. **d'Ischia, M., and G. Prota.** 1997. Biosynthesis, structure, and function of neuromelanin and its relation to Parkinson's disease: a critical update. *Pigment Cell Res.* **10:**370–376.
23. **Doering, T. L., J. D. Nosanchuk, W. K. Roberts, and A. Casadevall.** 1999. Melanin as a potential cryptococcal defence against microbicidal proteins. *Med. Mycol.* **37:**175–181.
24. **D'Souza, C. A., J. A. Alspaugh, C. Yue, T. Harashima, G. M. Cox, J. R. Perfect, and J. Heitman.** 2001. Cyclic AMP-dependent protein kinase controls virulence of the fungal pathogen *Cryptococcus neoformans*. *Mol. Cell. Biol.* **21:**3179–3191.
25. **Edberg, S. C., S. J. Chaskes, E. Alture-Werber, and J. M. Singer.** 1980. Esculin-based medium for isolation and

identification of *Cryptococcus neoformans*. *J. Clin. Microbiol.* **12:**332–335.

26. **Eisenman, H. C., J. D. Nosanchuk, J. B. Webber, R. J. Emerson, T. A. Camesano, and A. Casadevall.** 2005. Microstructure of cell wall-associated melanin in the human pathogenic fungus *Cryptococcus neoformans*. *Biochemistry* **44:**3683–3693.
27. **Enochs, W. S., T. Sarna, L. Zecca, P. A. Riley, and H. M. Swartz.** 1994. The roles of neuromelanin, binding of metal ions, and oxidative cytotoxicity in the pathogenesis of Parkinson's disease: a hypothesis. *J. Neural Transm. Park. Dis. Dement. Sect.* **7:**83–100.
28. **Erickson, T., L. Liu, A. Gueyikian, X. Zhu, J. Gibbons, and P. R. Williamson.** 2001. Multiple virulence factors of *Cryptococcus neoformans* are dependent on VPH1. *Mol. Microbiol.* **42:**1121–1131.
29. **Feldmesser, M., Y. Kress, and A. Casadevall.** 2001. Dynamic changes in the morphology of *Cryptococcus neoformans* during murine pulmonary infection. *Microbiology* **147:**2355–2365.
30. **Feldmesser, M., Y. Kress, A. Mednick, and A. Casadevall.** 2000. The effect of the echinocandin analogue caspofungin on cell wall glucan synthesis by *Cryptococcus neoformans*. *J. Infect. Dis.* **182:**1791–1795.
31. **Franzot, S. P., J. Mukherjee, R. Cherniak, L. C. Chen, J. S. Hamdan, and A. Casadevall.** 1998. Microevolution of a standard strain of *Cryptococcus neoformans* resulting in differences in virulence and other phenotypes. *Infect. Immun.* **66:**89–97.
32. **Frases, S., A. Salazar, E. Dadachova, and A. Casadevall.** 2007. Cryptococcus neoformans can utilize the bacterial melanin precursor homogentisic acid for fungal melanogenesis. *Appl. Environ. Microbiol.* **73:**615–621.
33. **Gan, E. V., H. F. Haberman, and I. A. Menon.** 1976. Electron transfer properties of melanin. *Arch. Biochem. Biophys.* **173:**666–672.
34. **Garcia-Rivera, J., and A. Casadevall.** 2001. Melanization of *Cryptococcus neoformans* reduces its susceptibility to the antimicrobial effects of silver nitrate. *Med. Mycol.* **39:**353–357.
35. **Garcia-Rivera, J., S. C. Tucker, M. Feldmesser, P. R. Williamson, and A. Casadevall.** 2005. Laccase expression in murine pulmonary *Cryptococcus neoformans* infection. *Infect. Immun.* **73:**3124–3127.
36. **Gomez, B. L., and J. D. Nosanchuk.** 2003. Melanin and fungi. *Curr. Opin. Infect. Dis.* **16:**91–96.
37. **Henson, J. M., M. J. Butler, and A. W. Day.** 1999. The dark side of the mycelium: melanins of phytopathogenic fungi. *Annu. Rev. Phytopathol.* **37:**447–471.
38. **Heung, L. J., A. E. Kaiser, C. Luberto, and M. Del Poeta.** 2005. The role and mechanism of diacylglycerol-protein kinase C1 signaling in melanogenesis by *Cryptococcus neoformans*. *J. Biol. Chem.* **280:**28547–28555.
39. **Heung, L. J., C. Luberto, A. Plowden, Y. A. Hannun, and M. Del Poeta.** 2004. The sphingolipid pathway regulates Pkc1 through the formation of diacylglycerol in *Cryptococcus neoformans*. *J. Biol. Chem.* **279:**21144–21153.
40. **Hill, H. Z.** 1991. Melanins in the photobiology of skin cancer and the radiobiology of melanomas, p. 31–53. *In* S. H. Wilson (ed.), *Cancer Biology and Biosynthesis*. Telford Press, Caldwell, NJ.
41. **Huffnagle, G. B., G. H. Chen, J. L. Curtis, R. A. McDonald, R. M. Strieter, and G. B. Toews.** 1995. Down-regulation of the afferent phase of T cell-mediated pulmonary inflammation and immunity by a high melanin-producing strain of *Cryptococcus neoformans*. *J. Immunol.* **155:**3507–3516.
42. **Idnurm, A., J. L. Reedy, J. C. Nussbaum, and J. Heitman.** 2004. *Cryptococcus neoformans* virulence gene discovery through insertional mutagenesis. *Eukaryot. Cell* **3:**420–429.
43. **Ikeda, R., and E. S. Jacobson.** 1992. Heterogeneity of phenol oxidases in *Cryptococcus neoformans*. *Infect. Immun.* **60:**3552–3555.
44. **Ikeda, R., T. Shinoda, T. Morita, and E. S. Jacobson.** 1993. Characterization of a phenol oxidase from *Cryptococcus neoformans* var. *neoformans*. *Microbiol. Immunol.* **37:**759–764.
45. **Ikeda, R., T. Sugita, E. S. Jacobson, and T. Shinoda.** 2003. Effects of melanin upon susceptibility of *Cryptococcus* to antifungals. *Microbiol. Immunol.* **47:**271–277.
46. **Ito, S.** 1993. Biochemistry and physiology of melanin, pp. 33–59. *In* N. Levine (ed.), *Pigmentation and Pigmentary Disorders*. CRC Press, Boca Raton, FL.
47. **Ito, S., and K. Fujita.** 1985. Microanalysis of eumelanin and pheomelanin in hair and melanomas by chemical degradation and liquid chromatography. *Anal. Biochem.* **144:**527–536.
48. **Jacobson, E. S., and G. M. Compton.** 1996. Discordant regulation of phenoloxidase and capsular polysaccharide in *Cryptococcus neoformans*. *J. Med. Vet. Mycol.* **34:** 289–291.
49. **Jacobson, E. S., and H. S. Emery.** 1991. Catecholamine uptake, melanization, and oxygen toxicity in *Cryptococcus neoformans*. *J. Bacteriol.* **173:**401–403.
50. **Jacobson, E. S., and H. S. Emery.** 1991. Temperature regulation of the cryptococcal phenoloxidase. *J. Med. Vet. Mycol.* **29:**121–124.
51. **Jacobson, E. S., A. P. Goodner, and K. J. Nyhus.** 1998. Ferrous iron uptake in *Cryptococcus neoformans*. *Infect. Immun.* **66:**4169–4175.
52. **Jacobson, E. S., and J. D. Hong.** 1997. Redox buffering by melanin and Fe(II) in *Cryptococcus neoformans*. *J. Bacteriol.* **179:**5340–5346.
53. **Jacobson, E. S., and R. Ikeda.** 2005. Effect of melanization upon porosity of the cryptococcal cell wall. *Med. Mycol.* **43:**327–333.
54. **Jacobson, E. S., N. D. Jenkins, and J. M. Todd.** 1994. Relationship between superoxide dismutase and melanin in a pathogenic fungus. *Infect. Immun.* **62:**4085–4086.
55. **Jacobson, E. S., and M. J. Petro.** 1987. Extracellular iron chelation in *Cryptococcus neoformans*. *J. Med. Vet. Mycol.* **25:**415–418.
56. **Jacobson, E. S., and S. B. Tinnell.** 1993. Antioxidant function of fungal melanin. *J. Bacteriol.* **175:**7102–7104.
57. **Jung, W. H., and J. W. Kronstad.** 2008. Iron and fungal pathogenesis: a case study with *Cryptococcus neoformans*. *Cell. Microbiol.* **10:**277–284.
58. **Jung, W. H., A. Sham, R. White, and J. W. Kronstad.** 2006. Iron regulation of the major virulence factors in the AIDS-associated pathogen *Cryptococcus neoformans*. *PLoS Biol.* **4:**e410.
59. **Kaliszan, R., A. Kaliszan, and I. W. Wainer.** 1993. Prediction of drug binding to melanin using a melanin-based high-performance liquid chromatographic stationary phase and chemometric analysis of the chromatographic data. *J. Chromatogr.* **615:**281–288.
60. **Kaya, M., D. P. Edward, H. Tessler, and R. L. Hendricks.** 1992. Augmentation of intraoccular inflammation by melanin. *Invest. Ophthamol. Vis. Sci.* **33:**522–531.
61. **Kojima, Y., Y. Tsukuda, Y. Kawai, A. Tsukamoto, J. Sugiura, M. Sakaino, and Y. Kita.** 1990. Cloning, sequence analysis, and expression of ligninolytic phenoloxidase genes of the white-rot basidiomycete *Coriolus hirsutus*. *J. Biol. Chem.* **265:**15224–15230.
62. **Korth, H., and G. Pulverer.** 1971. Pigment formation for differentiating *Cryptococcus neoformans* from *Candida albicans*. *Appl. Microbiol.* **21:**541–542.
63. **Kuo, M. J., and M. Alexander.** 1967. Inhibition of the lysis of fungi by melanins. *J. Bacteriol.* **94:**624–629.

64. **Kwon-Chung, K. J., I. Polacheck, and T. J. Popkin.** 1982. Melanin-lacking mutants of *Cryptococcus neoformans* and their virulence for mice. *J. Bacteriol.* **150:**1414–1421.
65. **Kwon-Chung, K. J., and J. C. Rhodes.** 1986. Encapsulation and melanin formation as indicators of virulence in *Cryptococcus neoformans*. *Infect. Immun.* **51:**218–223.
66. **Kwon-Chung, K. J., W. K. Tom, and J. L. Costa.** 1983. Utilization of indole compounds by *Cryptococcus neoformans* to produce a melanin-like pigment. *J. Clin. Microbiol.* **18:** 1419–1421.
67. **Lamy-Freund, M. T., S. Schreier, R. M. Peitzsch, and W. F. Reed.** 1991. Characterization and time dependence of amphotericin B: deoxycholate aggregation by quasielastic light scattering. *J. Pharm. Sci.* **80:**262–266.
68. **Langhof, H., M. Feuerstein, and G. Schabinski.** 1965. Melaninantikorperbildung bei Vitiligo. *Hautarzt* **16:**209–212.
69. **Larsson, B. S.** 1993. Interaction between chemicals and melanin. *Pigment Cell Res.* **6:**127–133.
70. **Lazera, M. S., F. D. Pires, L. Camillo-Coura, M. M. Nishikawa, C. C. Bezerra, L. Trilles, and B. Wanke.** 1996. Natural habitat of *Cryptococcus neoformans* var. *neoformans* in decaying wood forming hollows in living trees. *J. Med. Vet. Mycol.* **34:**127–131.
71. **Lee, S. C., D. W. Dickson, and A. Casadevall.** 1996. Pathology of cryptococcal meningoencephalitis: analysis of 27 patients with pathogenic implications. *Hum. Pathol.* **27:**839–847.
72. **Liu, L., R. P. Tewari, and P. R. Williamson.** 1999. Laccase protects *Cryptococcus neoformans* from antifungal activity of alveolar macrophages. *Infect. Immun.* **67:**6034–6039.
73. **Liu, O. W., C. D. Chun, E. D. Chow, C. Chen, H. D. Madhani, and S. M. Noble.** 2008. Systematic genetic analysis of virulence in the human fungal pathogen *Cryptococcus neoformans*. *Cell* **135:**174–188.
74. **Luther, J. P., and H. Lipke.** 1980. Degradation of melanin by *Aspergillus fumigatus*. *Appl. Environ. Microbiol.* **40:**145–155.
75. **Mandal, P., U. Banerjee, A. Casadevall, and J. D. Nosanchuk.** 2005. Dual infections with pigmented and albino strains of *Cryptococcus neoformans* in patients with or without human immunodeficiency virus infection in India. *J. Clin. Microbiol.* **43:**4766–4772.
76. **Martinez, L. R., and A. Casadevall.** 2006. Susceptibility of *Cryptococcus neoformans* biofilms to antifungal agents in vitro. *Antimicrob. Agents Chemother.* **50:**1021–1033.
77. **McClelland, E. E., P. Bernhardt, and A. Casadevall.** 2006. Estimating the relative contributions of virulence factors for pathogenic microbes. *Infect. Immun.* **74:**1500–1504.
78. **Mednick, A. J., J. D. Nosanchuk, and A. Casadevall.** 2005. Melanization of *Cryptococcus neoformans* affects lung inflammatory responses during cryptococcal infection. *Infect. Immun.* **73:**2012–2019.
79. **Mironenko, N. V., I. A. Alekhina, N. N. Zhdanova, and S. A. Bulat.** 2000. Intraspecific variation in gamma-radiation resistance and genomic structure in the filamentous fungus *Alternaria alternata*: a case study of strains inhabiting Chernobyl reactor no. 4. *Ecotoxicol. Environ. Safety* **45:**177–187.
80. **Missall, T. A., J. M. Moran, J. A. Corbett, and J. K. Lodge.** 2005. Distinct stress responses of two functional laccases in *Cryptococcus neoformans* are revealed in the absence of the thiol-specific antioxidant Tsa1. *Eukaryot. Cell* **4:**202–208.
81. **Mohagheghpour, N., N. Waleh, S. J. Garger, L. Dousman, L. K. Grill, and D. Tuse.** 2000. Synthetic melanin suppresses production of proinflammatory cytokines. *Cell. Immunol.* **199:**25–36.
82. **Mylonakis, E., F. M. Ausubel, J. R. Perfect, J. Heitman, and S. B. Calderwood.** 2002. Killing of *Caenorhabditis elegans* by *Cryptococcus neoformans* as a model of yeast pathogenesis. *Proc. Natl. Acad. Sci. USA* **99:**15675–15680.
83. **Neilson, J. B., M. H. Ivey, and G. S. Bulmer.** 1978. *Cryptococcus neoformans*: pseudohyphal forms surviving culture with *Acanthamoeba polyphaga*. *Infect. Immun.* **20:**262–266.
84. **Nielsen, K., A. L. De Obaldia, and J. Heitman.** 2007. *Cryptococcus neoformans* mates on pigeon guano: implications for the realized ecological niche and globalization. *Eukaryot. Cell* **6:**949–959.
85. **Nosanchuk, J., and A. Casadevall.** 1997. Cellular charge of *Cryptococcus neoformans*: contributions from the capsular polysaccharide, melanin, and monoclonal antibody binding. *Infect. Immun.* **65:**1836–1841.
86. **Nosanchuk, J. D., and A. Casadevall.** 2003. Budding of melanized *Cryptococcus neoformans* in the presence or absence of L-dopa. *Microbiology* **149:**1945–1951.
87. **Nosanchuk, J. D., and A. Casadevall.** 2006. Impact of melanin on microbial virulence and clinical resistance to antimicrobial compounds. *Antimicrob. Agents Chemother.* **50:**3519–3528.
88. **Nosanchuk, J. D., R. Ovalle, and A. Casadevall.** 2001. Glyphosate inhibits melanization of *Cryptococcus neoformans* and prolongs survival of mice after systemic infection. *J. Infect. Dis.* **183:**1093–1099.
89. **Nosanchuk, J. D., A. L. Rosas, and A. Casadevall.** 1998. The antibody response to fungal melanin in mice. *J. Immunol.* **160:**6026–6031.
90. **Nosanchuk, J. D., A. L. Rosas, S. C. Lee, and A. Casadevall.** 2000. Melanisation of *Cryptococcus neoformans* in human brain tissue. *Lancet* **355:**2049–2050.
91. **Nosanchuk, J. D., P. Valadon, M. Feldmesser, and A. Casadevall.** 1999. Melanization of *Cryptococcus neoformans* in murine infection. *Mol. Cell. Biol.* **19:**745–750.
92. **Noverr, M. C., P. R. Williamson, R. S. Fajardo, and G. B. Huffnagle.** 2004. CNLAC1 is required for extrapulmonary dissemination of *Cryptococcus neoformans* but not pulmonary persistence. *Infect. Immun.* **72:** 1693–1699.
93. **Nurudeen, T. A., and D. G. Ahearn.** 1979. Regulation of melanin production by *Cryptococcus neoformans*. *J. Clin. Microbiol.* **10:**724–729.
94. **Nyhus, K. J., A. T. Wilborn, and E. S. Jacobson.** 1997. Ferric iron reduction by *Cryptococcus neoformans*. *Infect. Immun.* **65:**434–438.
95. **Perfect, J. R., K. A. Marr, T. J. Walsh, R. N. Greenberg, B. DuPont, J. de la Torre-Cisneros, G. Just-Nubling, H. T. Schlamm, I. Lutsar, A. Espinel-Ingroff, and E. Johnson.** 2003. Voriconazole treatment for less-common, emerging, or refractory fungal infections. *Clin. Infect. Dis.* **36:**1122–1131.
96. **Pierini, L. M., and T. L. Doering.** 2001. Spatial and temporal sequence of capsule construction in *Cryptococcus neoformans*. *Mol. Microbiol.* **41:**105–115.
97. **Polacheck, I., V. J. Hearing, and K. J. Kwon-Chung.** 1982. Biochemical studies of phenoloxidase and utilization of catecholamines in *Cryptococcus neoformans*. *J. Bacteriol.* **150:**1212–1220.
98. **Polacheck, I., and K. J. Kwon-Chung.** 1988. Melanogenesis in *Cryptococcus neoformans*. *J. Gen. Microbiol.* **134:**1034–1041.
99. **Polacheck, I., Y. Platt, and J. Aronovitch.** 1990. Catecholamines and virulence of *Cryptococcus neoformans*. *Infect. Immun.* **58:**2919–2922.
100. **Pukkila-Worley, R., Q. D. Gerrald, P. R. Kraus, M. J. Boily, M. J. Davis, S. S. Giles, G. M. Cox, J. Heitman, and J. A. Alspaugh.** 2005. Transcriptional network of multiple capsule and melanin genes governed by the

Cryptococcus neoformans cyclic AMP cascade. *Eukaryot. Cell* **4:**190–201.

101. **Rhodes, J. C., and K. J. Kwon-Chung.** 1985. Production and regeneration of protoplasts from *Cryptococcus*. *Sabouraudia* **23:**77–80.
102. **Rhodes, J. C., I. Polacheck, and K. J. Kwon-Chung.** 1982. Phenoloxidase activity and virulence in isogenic strains of *Cryptococcus neoformans*. *Infect. Immun.* **36:** 1175–1184.
103. **Riley, P. A.** 1991. Melanogenesis: a realistic target for antimelanoma therapy? *Eur. J. Cancer* **27:**1172–1177.
104. **Rosas, A. L., and A. Casadevall.** 1997. Melanization affects susceptibility of *Cryptococcus neoformans* to heat and cold. *FEMS Microbiol. Lett.* **153:**265–272.
105. **Rosas, A. L., and A. Casadevall.** 2001. Melanization decreases the susceptibility of *Cryptococcus neoformans* to enzymatic degradation. *Mycopathologia* **151:**53–56.
106. **Rosas, A. L., R. S. MacGill, J. D. Nosanchuk, T. R. Kozel, and A. Casadevall.** 2002. Activation of the alternative complement pathway by fungal melanins. *Clin. Diagn. Lab. Immunol.* **9:**144–148.
107. **Rosas, A. L., J. D. Nosanchuk, and A. Casadevall.** 2001. Passive immunization with melanin-binding monoclonal antibodies prolongs survival of mice with lethal *Cryptococcus neoformans* infection. *Infect. Immun.* **69:**3410–3412.
108. **Rosas, A. L., J. D. Nosanchuk, M. Feldmesser, G. M. Cox, H. C. McDade, and A. Casadevall.** 2000. Synthesis of polymerized melanin by *Cryptococcus neoformans* in infected rodents. *Infect. Immun.* **68:**2845–2853.
109. **Ruiz, A., R. A. Fromtling, and G. S. Bulmer.** 1981. Distribution of *Cryptococcus neoformans* in a natural site. *Infect. Immun.* **31:**560–563.
110. **Ruiz, A., J. B. Neilson, and G. S. Bulmer.** 1982. Control of *Cryptococcus neoformans* in nature by biotic factors. *Sabouraudia* **20:**21–29.
111. **Salas, S. D., J. E. Bennett, K. J. Kwon-Chung, J. R. Perfect, and P. R. Williamson.** 1996. Effect of the laccase gene CNLAC1, on virulence of *Cryptococcus neoformans*. *J. Exp. Med.* **184:**377–386.
112. **Shaw, C. E., and L. Kapica.** 1972. Production of diagnostic pigment by phenoloxidase activity of *Cryptococcus neoformans*. *Appl. Microbiol.* **24:**824–830.
113. **Sichel, G., C. Corsaro, M. Scalia, A. J. Di Bilio, and R. P. Bonomo.** 1991. In vitro scavenger activity of some flavonoids and melanins against O2-(.). *Free Radic. Biol. Med.* **11:**1–8.
114. **Sinilova, N. G., Z. G. Pershina, A. P. Duplitseva, and I. B. Pavlova.** 1969. A radio-resistant pigmented bacterial culture isolated from atomic reactor water. *Zh. Mikrobiol. Epidemiol. Immunobiol.* **46:**94–99. (In Russian.)
115. **Staib, F.** 1962. *Cryptococcus neoformans* and *Guizotia abyssicnica* (syn. G. *Oleifera* D.C.). *Z. Hyg.* **148:**466–475.
116. **Steenbergen, J. N., H. A. Shuman, and A. Casadevall.** 2001. *Cryptococcus neoformans* interactions with amoebae suggest an explanation for its virulence and intracellular pathogenic strategy in macrophages. *Proc. Natl. Acad. Sci. USA* **98:**15245–15250.
117. **Stoetzner, H., and C. Kemmer.** 1971. The morphology of *Cryptococcus neoformans* in human cryptococcosis. A light-, phase-contrast and electron-microscopic study. *Mycopathol. Mycol. Appl.* **45:**327–335.
118. **Strachan, A. A., R. J. Yu, and F. Blank.** 1971. Pigment production of *Cryptococcus neoformans* grown with extracts of *Guizotia abyssinica*. *Appl. Microbiol.* **22:**478–479.
119. **Tangen, K. L., W. H. Jung, A. P. Sham, T. Lian, and J. W. Kronstad.** 2007. The iron- and cAMP-regulated gene SIT1 influences ferrioxamine B utilization, melanization and cell wall structure in *Cryptococcus neoformans*. *Microbiology* **153:**29–41.
120. **Tian, S., J. Garcia-Rivera, B. Yan, A. Casadevall, and R. E. Stark.** 2003. Unlocking the molecular structure of fungal melanin using 13C biosynthetic labeling and solid-state NMR. *Biochemistry* **42:**8105–8109.
121. **Torres-Guererro, H., and J. C. Edman.** 1994. Melanin-deficient mutants of *Cryptococcus neoformans*. *J. Med. Vet. Mycol.* **32:**303–313.
122. **Valentine, J. S., and E. B. Gralla.** 1997. Delivering copper inside yeast and human cells. *Science* **278:**817–818.
123. **Van Duin, D., A. Casadevall, and J. D. Nosanchuk.** 2002. Melanization of *Cryptococcus neoformans* and *Histoplasma capsulatum* reduces their susceptibility to amphotericin B and caspofungin. *Antimicrob. Agents Chemother.* **46:**3394–3400.
124. **Van Duin, D., W. Cleare, O. Zaragoza, A. Casadevall, and J. D. Nosanchuk.** 2004. Effects of voriconazole on *Cryptococcus neoformans*. *Antimicrob. Agents Chemother.* **48:**2014–2020.
125. **Wakamatsu, K., and S. Ito.** 2002. Advanced chemical methods in melanin determination. *Pigment Cell Res.* **15:**174–183.
126. **Walton, F. J., A. Idnurm, and J. Heitman.** 2005. Novel gene functions required for melanization of the human pathogen *Cryptococcus neoformans*. *Mol. Microbiol.* **57:** 1381–1396.
127. **Wang, H. S., R. T. Zeimis, and G. D. Roberts.** 1977. Evaluation of a caffeic acid-ferric citrate test for rapid identification of *Cryptococcus neoformans*. *J. Clin. Microbiol.* **6:**445–449.
128. **Wang, P., M. E. Cardenas, G. M. Cox, J. R. Perfect, and J. Heitman.** 2001. Two cyclophilin A homologs with shared and distinct functions important for growth and virulence of *Cryptococcus neoformans*. *EMBO Rep.* **2:**511–518.
129. **Wang, Y., P. Aisen, and A. Casadevall.** 1995. *Cryptococcus neoformans* melanin and virulence: mechanism of action. *Infect. Immun.* **63:**3131–3136.
130. **Wang, Y., P. Aisen, and A. Casadevall.** 1996. Melanin, melanin "ghosts," and melanin composition in *Cryptococcus neoformans*. *Infect. Immun.* **64:**2420–2424.
131. **Wang, Y., and A. Casadevall.** 1994. Decreased susceptibility of melanized *Cryptococus neoformans* to UV light. *Appl. Environ. Microbiol.* **60:**3864-3866.
132. **Wang, Y., and A. Casadevall.** 1994. Growth of *Cryptococcus neoformans* in presence of L-dopa decreases its susceptibility to amphotericin B. *Antimicrob. Agents Chemother.* **38:**2648–2650.
133. **Wang, Y., and A. Casadevall.** 1994. Susceptibility of melanized and nonmelanized *Cryptococcus neoformans* to nitrogen- and oxygen-derived oxidants. *Infect. Immun.* **62:**3004–3007.
134. **Wang, Y., and A. Casadevall.** 1996. Susceptibility of melanized and nonmelanized *Cryptococcus neoformans* to the melanin-binding compounds trifluoperazine and chloroquine. *Antimicrob. Agents Chemother.* **40:**541–545.
135. **Wassermann, H. P., and J. J. Van Der Walt.** 1973. Antibodies against melanin: the significance of negative results. *S. Afr. Med. J.* **47:**7–9.
136. **Wheeler, M. H., and A. A. Bell.** 1988. Melanins and their importance in pathogenic fungi. *Curr. Top. Med. Mycol.* **2:**338–387.
137. **Williamson, P. R.** 1994. Biochemical and molecular characterization of the diphenol oxidase of *Cryptococcus neoformans*: identification as a laccase. *J. Bacteriol.* **176:** 656–664.
138. **Williamson, P. R.** 1997. Laccase and melanin in the pathogenesis of *Cryptococcus neoformans*. *Front. Biosci.* **2:**e99–e107.
139. **Yang, Z., R. C. Pascon, A. Alspaugh, G. M. Cox, and J. H. McCusker.** 2002. Molecular and genetic analysis

of the *Cryptococcus neoformans* MET3 gene and a met3 mutant. *Microbiology* **148:**2617–2625.
140. **Youdim, M. B., D. Ben-Sachar, and P. Riederer.** 1994. The enigma of neuromelanin in Parkinson's disease substantia nigra. *J. Neural Transm. Suppl.* **43:**113–122.
141. **Zhang, S., A. Varma, and P. R. Williamson.** 1999. The yeast *Cryptococcus neoformans* uses 'mammalian' enhancer sites in the regulation of the virulence gene, CNLAC1. *Gene* **227:**231–240.
142. **Zhu, X., J. Gibbons, J. Garcia-Rivera, A. Casadevall, and P. R. Williamson.** 2001. Laccase of *Cryptococcus neoformans* is a cell wall-associated virulence factor. *Infect. Immun.* **69:**5589–5596.
143. **Zhu, X., J. Gibbons, S. Zhang, and P. R. Williamson.** 2003. Copper-mediated reversal of defective laccase in a Deltavph1 avirulent mutant of *Cryptococcus neoformans*. *Mol. Microbiol.* **47:**1007–1014.
144. **Zhu, X., and P. R. Williamson.** 2004. Role of laccase in the biology and virulence of *Cryptococcus neoformans*. *FEMS Yeast Res.* **5:**1–10.
145. **Zuger, A., E. Louie, R. S. Holzman, M. S. Simberkoff, and J. J. Rahal.** 1986. Cryptococcal disease in patients with the acquired immunodeficiency syndrome: diagnostic features and outcome of treatment. *Ann. Intern. Med.* **104:**234–240.

Cryptococcus: From Human Pathogen to Model Yeast
Edited by J. Heitman et al.

6

The Cell Wall of *Cryptococcus*

NICOLE M. GILBERT, JENNIFER K. LODGE, AND CHARLES A. SPECHT

The fungal cell wall is an essential organelle, vital for maintaining cell integrity against various chemical, physical, and biological stressors. This complex structure, immediately adjacent to the plasma membrane, serves several biological functions. Its cohesive structure, linked by covalent and hydrogen bonding or other types of interactions, provides great strength and stability. In this way, the cell wall prevents bursting that would be caused by internal turgor pressure and protects against external mechanical injury. The cell wall is the primary determinant of cell morphogenesis, constantly being remodeled to allow for growth, cell division, and transitioning from yeast to branching hyphal forms to structures of sporulation. Since it is present at the cell surface, the cell wall interacts with the environment. In the case of pathogenic fungi, host defense systems are often directed against cell wall components. The cell wall has long been recognized as an ideal target for antifungal therapies because it is essential for fungal cell viability, and many of its components are absent from the mammalian host.

Fungal cell walls are composed primarily of polysaccharides and proteins, principally mannoprotein, along with other components, including lipids, pigments, and inorganic salts. Polysaccharides account for about 80–90% of the cell wall, and although the basic framework is generally conserved, there is substantial diversity across the fungal kingdom (12). In common are polymers of glucose (α- and β-glucan), mannose (N- and O-linked mannan attached to protein), and *N*-acetylglucosamine (chitin). Synthesis and maintenance of the cell wall involves a large number of biosynthetic and signaling pathways. *Saccharomyces cerevisiae* utilizes more than 60 genes for synthesis and remodeling of its polysaccharides and the mannan side chains of cell wall proteins (37). Biochemical, genomic, and proteomic analyses of model fungal systems such as *S. cerevisiae*, *Candida albicans*, *Schizosaccharomyces pombe*, *Histoplasma capsulatum*, and *Aspergillus fumigatus* have greatly advanced the understanding of the fungal cell wall and provided the groundwork for studies of *Cryptococcus*. This chapter covers work on the cell wall of *Cryptococcus*, drawing on studies from the above-mentioned fungi for introduction and comparison.

Nicole M. Gilbert, Department of Biochemistry and Molecular Biology, Saint Louis University, St. Louis, MO 63104. **Jennifer K. Lodge,** Department of Molecular Microbiology, Washington University, St. Louis, MO 63110. **Charles A. Specht,** Department of Medicine, University of Massachusetts, Worcester, MA 01605.

STRUCTURE AND SYNTHESIS

Understanding the cell wall structure of a fungal species has historically been initiated by taking a biochemical inventory of its components. Table 1 lists the cell wall composition of some medically important fungi. Purified cell walls are insoluble, but through a series of chemical and enzymatic extractions, the various proteins and polysaccharides can be sequentially made soluble and analyzed. It became evident from studies that portions of the cell wall were resistant to mild alkaline hydrolysis and remained a mixture of components. This led to the perception of the cell wall as a covalently cross-linked meshwork of glucans and protein. Studies of the *Saccharomyces* cell wall have elucidated how chitin, β-1,3-glucan, β-1,6-glucan, and glycosylphosphatidylinositol (GPI)-anchored protein are linked to one another (64, 65, 68). Some of the enzymes that form the cross-links have also been recently identified (18). The pursuit of a comprehensive understanding of cell wall structure and dynamics of *S. cerevisiae* has directly benefited studies of other fungi, as many aspects of cell wall biosynthesis are conserved.

Observation of the *Cryptococcus neoformans* cell wall by electron microscopy reveals the presence of an electron-dense inner layer and an electron-lucid outer layer (Fig. 1). Biochemical analysis of the cell wall of *Cryptococcus* is complicated by the presence of the exopolysaccharide capsule, associated with the outer layer of the cell wall (Fig. 1). The highly abundant capsular glucuronoxylomannan and galactoxylomannan polymers copurify with the cell wall and interfere with standard polysaccharide analysis. To circumvent this problem, acapsular strains, such as *cap67*Δ and *cap59*Δ, have been used for defining cell wall composition (57, 104). These studies have revealed that the cryptococcal cell wall contains α-1,3-glucan, β-1,3– and β-1,6–linked glucans, and chitin. The caveat to these studies is that since the defects in the acapsular mutants are not well defined, it is not clearly established that the acapsular wall is a good model for the wild-type *C. neoformans* wall. Additionally, *C. neoformans* contains significant amounts of chitosan, the deacetylated derivative of chitin, in its vegetative cell wall (9).

TABLE 1 Major cell wall polysaccharides of select fungal organisms

Organism	Component (reference)		
	β-Glucan	α-Glucan	Chitin [Chitosan]
Cryptococcus neoformans	78–80%[a] β-1,6; 10–12%[a] β-1,3 side chains (57)	97%[b] α-1,3; 3%[b] α-1,4 (57)	9.8%[c] (57) [>65%[d] (9)]
Saccharomyces cerevisiae	50%[c] β-1,3; 10%[c] β-1,6 (71)	None (71)	1–3%[c] (71)
Candida albicans	40%[d] β-1,3; 20%[d] β-1,6 (61)	None (61)	1–2%[c] (61)
Schizosaccharomyces pombe	55%[c] β-1,3; 6%[c] β-1,6 (53)	28%[c] α-1,3 (53)	0.5%[c] (53)
Aspergillus fumigatus (mycelium)	30%[e] β-1,3 (83)	42%[f] α-1,3 (83)	13%[e] [4%[e]] (83)
Aspergillus fumigatus (conidia)	38%[e] β-1,3; 5%[f] β-1,3 (83)	14%[f] α-1,3 (83)	1.7%[e]; 0.3%[e] [3.9%[e]; 0.2%[f]] (83)

[a] Percentage of water-soluble cell wall fraction.
[b] Percentage of water-insoluble cell wall fraction.
[c] Percentage of total cell wall mass.
[d] Percentage of cell wall chitin.
[e] Percentage of alkali-insoluble cell wall fraction.
[f] Percentage of alkali-soluble cell wall fraction.

Monosaccharide analysis determined that glucose and *N*-acetylglucosamine/glucosamine are the two primary sugars, comprising 86% and 7.3% of purified cell walls, respectively (57). The cryptococcal cell wall has also been shown to contain nonpolysaccharide components including mannoproteins, lipids, and the pigment melanin.

Most cell wall polysaccharides are synthesized de novo by a synthase protein in the plasma membrane. These enzymes are highly conserved among fungi, sharing domains for catalyzing the exchange of sugar monomers from nucleotide sugar donors found in the cytoplasm. As a nascent chain lengthens, it is extruded through a pore in the plasma membrane formed by the synthase for final assembly with other cell wall components. Homology to proteins well characterized in *S. cerevisiae* and other fungi has been the key approach for identification of genes responsible for the synthesis of the major cell wall polysaccharides in *Cryptococcus*, discussed individually below.

α-1,3-Glucan

The cell walls of several *Cryptococcus* species, both pathogenic and nonpathogenic, have been shown to contain α-1,3-glucan (5, 57, 104). Immunoelectron microscopy studies have localized α-1,3-glucan primarily to the outer cell wall (104). While termed α-1,3-glucan for its predominant linkage joining glucoses, interspersed in the polymer, about 3% of the linkages are α-1,4 (57). A proposed mechanism for its stepwise synthesis utilizes a short primer of α-1,4–linked glucose residues (49). When a stretch of α-1,3-glucan has been added, two polymers are linked with the α-1,4 primer of one polymer embedded in the middle. Studies of *S. pombe* identified the protein, Ags1p (alpha-glucan synthase), required for α-1,3-glucan synthesis (53). *A. fumigatus* has three such enzymes, Ags1p, Ags2p, and Ags3p, while *H. capsulatum* encodes only a single Ags1 (13, 78, 102). A single *AGS1* gene was first identified in the *C. neoformans* H99 serotype A genome by protein sequence homology to *S. pombe* and *A. fumigatus* Ags1p (103). Each Ags1 protein is about 2,500 amino acids with conserved domains: a proposed cytoplasmic synthase domain, a transmembrane, pore-forming domain for extruding the polymer through the plasma membrane, and an extracellular transglycosylase domain for linking polymers. Recent studies of *S. pombe* Ags1p have demonstrated synthesis of only α-1,4 oligosaccharides (the primer) and have suggested that an unidentified protein catalyzes the α-1,3–linked glucose cross-links (130). It is therefore inferred that *Cryptococcus* may also have an unidentified ortholog.

AGS1 of *Cryptococcus* has been disrupted by RNA interference in serotype D JEC43 or deletion of part of the transglycosylase domain in serotype D JEC21 and serotype A H99 strains. Loss of *AGS1* resulted in slow growth and increased SDS sensitivity at 30°C and temperature sensitivity at 37°C. Biochemical analysis of *ags1Δ* cell walls revealed complete loss of α-1,3-glucan, as expected. Also, a portion of β-1,3-glucan was shifted from the alkali-soluble to the alkali-insoluble cell wall fraction, and chitin/chitosan content was nearly doubled. These shifts reflect a compensation

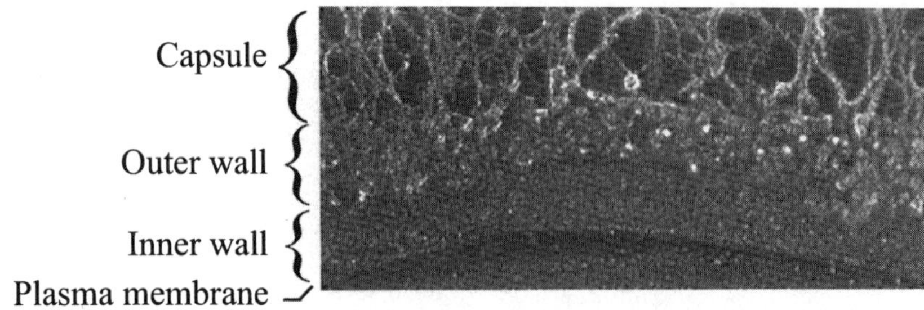

FIGURE 1 Quick-freeze deep-etch scanning electron micrograph of the cell wall of *Cryptococcus* illustrating the attachment of capsule and the two layers of the cell wall. Micrograph kindly provided by Tamara Doering and John Heuser.

mechanism for the stress of losing α-1,3-glucan (104). The *ags1*Δ strain was unable to grow in a mouse model of infection, likely because of its temperature sensitivity, but caused death when used to infect the nematode *Caenorhabditis elegans* at 25°C (104).

Interestingly, cells disrupted in *AGS1* displayed no visible capsule polysaccharide on their surface. An elegant set of capsule-transfer assays demonstrated that both *AGS-i* (RNAi strain) and *ags1*Δ failed to bind exogenous capsule material from wild-type cells. However, the mutants synthesized capsular components, which were released into the culture medium and could be bound to acceptor acapsular strains (103, 104). These results indicate that cells lacking *AGS1* and α-1,3-glucan do not contain the determinants necessary to assemble a capsule on the cell surface. The localization of α-1,3-glucan to the outer cell wall supports the possibility that this polymer participates directly in the binding of capsular polysaccharides, but the exact nature of the cell wall–capsule linkage is yet to be determined.

β-1,3-Glucan

In *S. cerevisiae* and other ascomycetous fungi, β-1,3-glucan is a major structural component and is the most abundant β-glucan, comprising between 30 and 60% of the total cell wall mass (45, 62). This percentage is significantly lower in *C. neoformans* (57). A β-1,3-glucan polymer from *S. cerevisiae* is estimated to be 1,500 glucose residues. β-1,6 side branches are spaced approximately every 30 glucoses and link the linear sections of β-1,3 (76). Triple helices are proposed to form from the linear regions, adding strength to the fibers and water insolubility.

The mechanism of β-1,3-glucan synthesis and the function and structure of β-1,3-glucan synthase β-1,3-GS have been well characterized in *S. cerevisiae* and found to be conserved in *C. albicans* and *Aspergillus* (38, 59, 82). β-1,3-GS is a heterodimeric complex composed of either Fks1p (215 kDa) or Fks2p (217 kDa), which are 88% identical, integral membrane proteins, and one small GTP-binding protein, Rho1p (39, 40, 79, 80, 100). The Fks1 proteins contain the catalytic activity, while Rho1p serves a regulatory function, with many roles during the cell cycle (63, 88, 99, 135). The *S. cerevisiae* genome encodes a third Fks homolog (Fks3p) with approximately 55% identity, whose catalytic activity has not been confirmed. *FKS1* and *FKS2* are expressed during vegetative growth, with *FKS1* expression being predominant when glucose is the carbon source. *FKS2* repression by glucose is reversible by addition of calcium to the medium and dependent on calcineurin for this induction (39, 80). Likewise, β-1,3-GS activity is sensitive to calcineurin inhibitor FK506 in proportion to the cell's dependence on *FKS2* expression. Fks2p is also the dominant β-1,3-GS activity during mating and sporulation, induced in a calcineurin-dependent manner by mating pheromone and starvation, which activate these processes. Deletion of either *FKS1* or *FKS2* yields cells diminished in β-1,3-glucan, and only an *fks1*Δ*fks2*Δ strain is not viable. Other fungal species have one or more *FKS* genes. For example, *C. albicans* has three homologs, *FKS1/GSC1*, *GSL1* (*FKS3* homolog), and *GSL2*, while *A. fumigatus* has a single *FKS1* homolog (39, 82).

The cell wall has been recognized as an ideal target for development of antifungal therapies. Inhibition of β-1,3-glucan synthesis by the aculeacin, papulocandin, and especially, echinocandin classes of drugs has proven very successful (52, 137). These broad-spectrum antifungal β-1,3-glucan inhibitors are effective against *Candida* and *Aspergillus* spp. and other less common pathogenic molds (10, 32, 94). However, *C. neoformans* is much less susceptible to β-1,3-glucan inhibitors (10, 55, 67). Several hypotheses have been proposed in the literature to explain this lack of sensitivity. One possibility is that since *C. neoformans* contains less β-1,3-glucan than the susceptible ascomycetes, blocking its synthesis is not sufficiently damaging. Thompson et al. (125) found that *C. neoformans* has a single *FKS1* gene, and its deletion causes loss of viability. Unless the essential function of *CnFKS1* is other than that of β-1,3-glucan synthesis, which is unlikely since there is no precedent for such bifunctional characteristics in this class of enzymes, this evidence indicates that β-1,3-glucan synthesis is required for cell viability in *C. neoformans* (125).

A second possibility was that the cryptococcal β-1,3-GS enzymes are significantly diverged from other fungal enzymes, rendering them insensitive. However, *CnFKS1* is highly homologous to *S. cerevisiae*, *C. albicans*, and *Aspergillus* spp. (125). Treatment of serotype A and D strains of *C. neoformans* with the echinocandin caspofungin resulted in quantitatively decreased immunogold staining for both β-1,3-glucan and β-1,6-glucan, indicating Fks1 inhibition (44). Maligie and Selitrennikoff (75) developed an in vitro assay for β-1,3-glucan synthesis in *C. neoformans* and discovered that this activity is very sensitive to both caspofungin and cilofungin (K_i, 0.17 ± 0.02 and 22 ± 5.7 μM, respectively). However, growth of *C. neoformans* in concentrations of caspofungin that are fungicidal to *Candida* spp. produced only a one-third decrease in β-1,3-glucan (44). Furthermore, echinocandin treatment of *C. albicans* resulted in changes not directly related to the cell wall, such as organelle damage, vacuolation, and mitochondrial swelling (22). Treatment of *Pneumocystis carinii* caused altered export of surface glycoproteins by trophozoites (11). These data suggest that the antifungal effect against other fungi may result from other mechanisms additional to blocking β-1,3-glucan synthesis (44).

A final hypothesis yet to be fully explored is that β-1,3-glucan inhibitors fail to access the cryptococcal β-1,3-GS complex, possibly due to the exopolysaccharide capsule or the presence of cell surface transporters that pump the compounds out of the cell (21, 125). It is unlikely that the presence of capsule alone fully accounts for resistance because acapsular strains of *C. neoformans* have MIC values similar to that of their respective parental strain (C. P. Selitrennikoff, personal communication). Furthermore, deletion of 23 putative pump-encoding genes individually in *C. neoformans* serotype A H99 did not alter sensitivity to caspofungin (W. Lam, J. A. Lodge, and C. P. Selitrennikoff, unpublished).

While echinochandins are not effective against *Cryptococcus* at clinically accepted concentrations, synergistic growth inhibition with calcineurin inhibitors may be achieved (31). Unlike *S. cerevisiae* Fks2p, cited above, *CnFKS1* transcription is increased by inactivation of calcineurin (66). Because the calcineurin pathway has diverged from that of *S. cerevisiae* to one mediating a cell wall stress response, further studies are needed to understand how upregulation of *CnFKS1* increases sensitivity to echinochandins.

β-1,6-Glucan

β-1,6-Glucan is the most abundant β-glucan in the cryptococcal cell wall (57), in contrast to many ascomycetous fungi, which have significantly higher levels of β-1,3-glucan (14, 53, 61, 71). Although a minor component in *S. cerevisiae*, β-1,6-glucan plays a critical cross-linking function, having been found to be covalently linked to all the other cell wall polysaccharides and to the majority of cell wall proteins (65). The mechanism of β-1,6-glucan synthesis is poorly understood compared to that of other cell wall polysaccharides, mainly because no synthase enzyme has been definitively identified in any species. Large-scale genetic studies in *S. cerevisiae* identified several genes whose disruption resulted in decreased levels of cell wall β-1,6-glucan (116). These include *KRE5*, *KRE6*, *KRE1*, *KRE9*, *KRE11*, *KNH1*, *SKN1*, *ROT1*, *CWH41*, and *CNE1*. *C. neoformans* encodes seven genes homologous to *S. cerevisiae KRE5*, *KRE6* (five homologs), and *SKN1* (46a). There are no apparent cryptococcal homologs of *KRE1*, *KRE9*, *KRE11*, or *KNH1*, indicating that the mechanism of β-1,6-glucan synthesis may be distinct in *C. neoformans* or that the protein sequences have significantly diverged (46a). The biochemical functions of these genes have not been determined, but they likely affect β-1,6-glucan levels indirectly, either via the synthesis of a precursor or the processing of the actual synthase protein(s). Several have a predicted localization in the secretory pathway, leading to one hypothesis that β-1,6-glucan synthesis occurs in a stepwise manner beginning within the endoplasmic reticulum (116).

Deletion of *KRE5* or *KRE6* together with *SKN1* in *C. neoformans* resulted in virtually complete loss of cell wall β-1,6-glucan (46a). Loss of any of the remaining *KRE6* homologs alone had no detectable effect on cell wall β-1,6-glucan levels. The loss of β-1,6-glucan from *kre5*Δ and *kre6*Δ*skn1*Δ compromised cell integrity, as evidenced by their aberrant morphology (enlarged and multibudded), slow growth, and increased sensitivity to high temperature and various chemical stressors (46a). These mutants were also avirulent in a mouse inhalation model of infection, likely due to their inability to survive at host temperature. They also had enlarged capsules with altered architecture (46a).

In *S. cerevisiae* and *C. albicans*, the majority of cell wall proteins are attached to β-1,6-glucan via a remnant of their GPI anchor (29). Homology searches of the *Cryptococcus* genome uncovered 55 predicted GPI-anchored proteins (70). Sixteen were identified by mass spectrometry analysis as extracellular proteins of the acapsular (*cap59*Δ) strain of *C. neoformans* (42), and 15 of those 16 were also among the predicted GPI-anchored proteins described by Levitz and Specht (70). Among these proteins were three chitin deacetylases (CDAs), responsible for the enzymatic conversion of chitin to chitosan, which are discussed in detail below. Although not identified in this analysis, the established cryptococcal virulence factor phospholipase B1 (Plb1), discussed in the next section, has also been shown to contain a GPI-anchor and to be concentrated in the cell wall (37). Digestion of wild-type cryptococcal cells with β-1,3-glucanase released Plb1 containing β-1,6-glucan, indicating covalent attachment between Plb1 and β-1,6-glucan (118). In further support of this idea, the *kre5*Δ and *kre6*Δ*skn1*Δ strains lacking β-1,6-glucan had decreased cell wall localization and increased secretion of Plb1 (46a).

Chitin/Chitosan

Chitin, a homopolymer of β-1,4-linked *N*-acetylglucosamine (GlcNAc), is one of the most abundant polymers in nature, found in the shells of crustaceans and the cell walls of virtually all fungi. Chitin chains arrange into microfibrils via H^+ bonding, therefore providing excellent strength and rigidity to the cell (81). Chitin may be covalently linked to β-1,3- or β-1,6-glucans (64, 65). Chitin comprises 1–2% of the dry weight of yeast, such as *S. cerevisiae*, and up to 40% of some filamentous fungi, such as *Mucor rouxii* (12, 60).

Chitin polymerization from cytoplasmic pools of UDP-GlcNAc occurs via the action of the multiple transmembrane protein chitin synthase (CHS). Chitin chains extrude across the plasma membrane as they are synthesized and become integrated into the cell wall. There are six classes of CHS (I, II, III, IV, V, VI) based on the protein sequence of the catalytic domain. These classes are divided into two families. The first family (classes I–III) are ~900-amino-acid proteins with the catalytic domain near the N-terminus, while the second (classes IV–VI) are ~1,200 amino acids with a C-terminal catalytic domain (15, 85, 123). Various CHS enzymes are important for virulence in important animal and plant pathogens, including *Botrytis cinerea* (122), *C. albicans* (86), *Exophiala* (*Wangiella*) *dermatitidis* (72), *Fusarium oxysporum* (74), and *Ustilago maydis* (46). The mechanism of chitin synthesis has been well characterized for *S. cerevisiae*. Chs1p repairs the daughter cell wall after division (19, 20). Chs2p synthesizes the primary septum during cytokinesis (117). Chs3p generates 90% of the vegetative cell wall chitin, the chitin ring found at the bud neck and in the spore wall (16, 117, 128). Chs3p requires the regulator protein Skt5p for enzymatic activity and localization to the bud neck and a homologous regulator, Shc1p, for spore wall synthesis (33, 91, 126).

C. neoformans produces three to six times the amount of chitin as *S. cerevisiae* during vegetative growth, and levels increase with culture density (9). This is in contrast to the virtually constant levels of chitin observed in *S. cerevisiae* and *C. albicans* (9). *C. neoformans* encodes eight CHSs in various classes: Chs1 and Chs3 (class IV), Chs2 and Chs7 (class II), Chs4 and Chs5 (class V), and Chs6 and Chs8 (class I and II) (9). The enzymes in classes I–III and IV–VI have seven and six predicted transmembrane domains, respectively, and the catalytic domain of each is predicted to face the cytoplasm (5). The presence of so many *CHS* genes is shared among fungal species that form elaborate structures during growth, such as *Aspergillus*, *U. maydis*, and *Phycomyces blakesleeanus*. It is possible that different *C. neoformans CHS* genes are involved in various steps of mating. Interestingly, Chs5 contains a myosin motor domain (9), which has been shown in other fungi to mediate interaction with the actin cytoskeleton (124). *C. neoformans* also has three homologs of Skt5p, named chitin synthase regulators CSR1, CSR2, and CSR3 (9).

Chitosan, the deacetlyated form of chitin, is generally less abundant in nature than chitin but is found in the cell wall of several fungal species. *S. cerevisiae* and other ascomycetes only contain chitosan in the ascospore wall (27). Therefore, initial studies of chitosan production were performed using *M. rouxii*, which generates significant levels of cell wall chitosan (4, 92, 133). Chitosan is generated from chitin through enzymatic conversion of *N*-acetylglucosamine to glucosamine by CDAs (3, 58). Cryptococcus cells in the yeast phase of growth deacetylate >65% of their chitin (9).

C. neoformans encodes three CDAs and produces chitosan during vegetative growth that increases with culture density (6, 9). All three CDAs have an N-terminal signal sequence, internal catalytic domain, and C-terminal ω-site, for the attachment of a GPI anchor. Such an organization suggests that these proteins are secreted to the surface, where they are anchored in either the plasma membrane or cell wall. In support of this prediction, all three CDAs were found in the cryptococcal extracellular proteome (42).

Each of the genes associated with chitin and chitosan production in *C. neoformans* have been deleted, and their effect on cells in culture were evaluated (6, 9). Deletion of *CHS4* or *CHS5* resulted in a 50% decrease in chitin at 48 h of growth, and *chs5* was slightly temperature sensitive (9). Interestingly, *chs3*Δ, *csr2*Δ, and the *cda1*Δ*cda2*Δ*cda3*Δ triple mutant completely lacked chitosan and had a two- to three-fold increase in chitin by 48 h of growth (6, 9). These mutants also had several shared in vitro phenotypes, including temperature sensitivity; increased sensitivity to Congo red, caffeine, and SDS; and altered morphology, with cells being enlarged and having difficulty budding (6, 9). Another interesting observation was that when grown in the presence of L-DOPA, *chs3*Δ, *csr2*Δ and *cda1*Δ*cda2*Δ*cda3*Δ strains had darkened culture supernatants (6, 9). This phenotype, termed "leaky melanin," could be caused by a reduced ability to retain melanin in the cell wall, due to the loss of chitosan. Alternatively, laccase, the enzyme responsible for melanin production in *C. neoformans*, may be mislocalized in these strains. A final possibility is that the pigment in the media is not melanin and may be produced by a chemical, as opposed to an enzymatic, reaction. None of the remaining single deletion strains displayed phenotypes.

The *chs3*Δ, *csr2*Δ, and *cda1*Δ*cda2*Δ*cda3*Δ strains also displayed enlarged capsules compared to wild-type cells when grown under capsule-inducing conditions (6). A possible connection between the capsule and chitin/chitosan has been observed in *C. neoformans*. Wild-type cryptococcal cells have been shown to bind wheat-germ agglutinin, which is specific for sialic acids and β-1,4-linked GlcNAc oligomers (2, 98) at distinct polarized sites on the cell surface (106). Confocal microscopy revealed that this wheat-germ agglutinin–reactive structure formed a hook-like projection extending from the cell wall into the capsule. This structure appeared to be firmly associated with the cell wall, and evidence indicated that it is made up of β-1,4-linked GlcNAc and not sialic acid. This evidence suggests an association of chitin-like structures, either chitin or chitosan, with the exopolysaccharide capsule of *C. neoformans*. The mechanism of this interaction and the biological importance of this structure have not been determined. One hypothesis could be that this wheat-germ agglutinin–reactive structure is important for anchoring the capsule to the cell surface, due to its tight association with the cell wall (106).

The data from Banks et al. (9) and Baker et al. (6) indicate that Chs3 is the most important CHS for vegetative growth in *C. neoformans*. The shared phenotypes of *chs3*Δ and *csr2*Δ suggest that these proteins are complexed, with Csr2 regulating Chs3 in a similar relationship as Chs3p and Skt5p in *S. cerevisiae*. The lack of chitosan in both mutants leads to a model in which chitosan is produced specifically from the chitin generated by Chs3/Csr2. These proteins may complex with one or more CDAs, allowing deacetylation to occur before chitin microfibril formation.

Other Components

The pigment melanin is an established and well-studied cryptococcal virulence determinant associated with the cell wall. Strains of *C. neoformans* that fail to produce melanin have decreased virulence and a lower rate of phagocytosis by macrophages. Melanin is immunogenic, and its presence in the cell wall elicits production of antibodies that inhibit growth. Melanized cells have increased resistance to antifungals (56). Melanin is not only important in the context of infection, but has important functions in the environment as well. It is critical for protection against environmental stresses, including toxins, extreme temperature, and UV light. It protects against oxidants by functioning as a redox buffer (104).

Melanin is produced by the action of laccase. The *C. neoformans* genome contains two paralogs, *LAC1* and *LAC2*, of which the first expresses a cell wall–associated enzyme that has the dominant activity under glucose starvation conditions (described below) (134, 139). The structure of *C. neoformans* melanin is thought to be cross-linked polymers of phenol and indole. It is deposited in the cell wall to form a highly stable network that may be cross-linked to cell wall polysaccharides or proteins (138). Melanin is resistant to harsh chemical treatments that remove cytoplasm, lipids, and proteins, resulting in the production of melanin "ghosts," which retain the shape of the cell. These melanin ghosts have been extensively analyzed by scanning and transmission electron microscopy, atomic force microscopy, and nuclear magnetic resonance (43). These investigations revealed that melanin is deposited in the cell wall in concentric layers of densely packed granules ranging from 40–>100 nm and containing 10- to 300-Å-diameter pores (43). The melanin layer appears to thicken with age while the porosity decreases (43, 56).

Visualization of acapsular and encapsulated cryptococcal cells by transmission electron microscopy revealed the presence membrane-bound vesicles within the cell wall (109). These vesicles contain various types of cargo, including the capsule component glucuronoxylomannan, proteins, pigments (melanin), and glucosylceramide. It has been hypothesized that melanin is synthesized within these vesicles that are then trapped in the wall, producing the spherical melanin granules (109). Similar vesicles were also detected at the capsular edge and purified from culture supernatants, indicating that they are actively secreted from *C. neoformans* cells. RNAi knockdown of *SEC6*, a component of a protein complex known as the "exocyst," in *C. neoformans* completely abolished vesicle secretion and resulted in decreased virulence in a mouse model of infection (93). Extracellular vesicles have also been observed in *H. capsulatum* (1), suggesting that secretion of vesicles is a mechanism for fungal trans–cell wall transport and secretion of proteins and lipids. Several major questions remain to be answered regarding these vesicles. Do the vesicles fuse with the plasma membrane and then reform? If these vesicles traverse the wall, how are the cell wall polysaccharides remodeled to allow this transfer? And, finally, how are the contents of these vesicles eventually released at the appropriate time and location?

Sialic acids, which are derivatives of neuraminic acids, have been found on the cell surface and characterized in several fungal species including *Paracoccidioides brasiliensis*, *P. carinii*, *C. albicans*, *and Aspergillus* spp. (36, 120, 121, 131). *C. neoformans* contains two cell surface sialic acids: *N*-acetylneuraminic acid (Neu5Ac) and the 9-*O*-acetylated derivative (Neu5,9,Ac_2). Interestingly, *C. neoformans* is the only pathogenic fungi shown to express Neu5,9,Ac_2 (110). Two sialylglycoproteins, 38 and 67 kDa, and no sialylglycolipids have been detected in *C. neoformans* (50, 107). Sialic acid units protect cryptococcal cells from phagocytosis and therefore could be important for initial defense against the host innate immune response.

Cross-Linking and Rearrangement

Although once viewed as a rigid, static structure, the current view of the cell wall is that of a dynamic organelle, constantly rearranging to account for growth, division, and development of higher-order cellular structures. Certain cell wall components can be up- or downregulated in response to various external conditions. As described above, the fungal cell wall contains several different components, and much is understood regarding their structures and mechanisms of synthesis. However, the understanding of how these components are rearranged and linked together is far less advanced.

Cell wall rearrangement is essential for budding during vegetative growth and for the formation of complex mating structures, such as hyphae. In *S. cerevisiae* and other fungi, chitinases are essential for bud formation and are involved in sexual reproduction (41). *Cryptococcus* encodes four chitinase genes and a single *N*-acetylhexosaminidase (*HEX1*). Surprisingly, deletion of all four chitinase genes or three in combination with *HEX1* did not adversely affect asexual (budding) growth, although it did disrupt sexual growth (6a). These data indicate that the mechanism of cell wall remodeling during budding in *C. neoformans* is very different from other fungi.

Some studies have defined various linkages, such as that between β-1,6-glucan and GPI-anchored cell wall proteins or β-1,3-glucan and chitin (64, 65). However, the enzymes responsible for creating these linkages and their mechanisms of action are, for the most part, unidentified. Certain proteins, such as Gas1p (97, 101) and Bgl2p (84), have been proposed to function in cell wall organization, but definitive studies are still needed. Future investigations in this area promise to dramatically increase our understanding of cell wall assembly and dynamics.

Recent studies have identified the first, and only, genes directly implicated in cross-linking cell wall components in *S. cerevisiae* (17, 18, 111). These genes, named *CRH1* and *CRH2* for their Congo red hypersensitivity, share significant sequence similarity to bacterial endoglucanases and plant endotransglycosidases (111). Both contain an N-terminal signal sequence, C-terminal GPI-anchor attachment motif, and Ser/Thr-rich region (111), all characteristics of cell wall proteins. A green fluorescent protein–Crh1p fusion protein localized to chitin-rich regions of the cell surface during the cell cycle (11), much like Chs3p (115). The cell wall of a *crh1*Δ*crh2*Δ strain had a nearly twofold increase of glucan in the alkali-soluble fraction. These data suggested a decrease in the cross-linking of glucan to chitin, since this linkage is the primary determinant of glucan solubility in alkali (51, 64). Using a novel method for analyzing chitin cross-links, Cabib et al. (17) found that chitin linked to β-1,6-glucan is decreased in the *crh1*Δ and *crh2*Δ single mutants and not present in *crh1*Δ*crh2*Δ. Growth of wild-type cells at 38°C resulted in increased chitin cross-linking to β-1,6-glucan at the cell cortex, and both Crh1p and Crh2p were enriched in these areas. Based on these data, the authors propose a model in which Crh1p and Crh2p transport chitin chains from Chs3p and cross-link them specifically to β-1,6-glucan (17). *Cryptococcus* encodes one Crh/Crr homolog, which has not been characterized.

CELL WALL PROTEINS

Fungal cell walls contain protein either covalently or noncovalently attached to the polysaccharide matrix; some are temporally associated and then released into the surrounding medium. Cell wall proteins are collectively important for cell integrity, providing strength and osmotic support by controlling the cell's porosity (34, 35). They function in the remodeling of polysaccharides, as well as being involved in cellular adhesion, mating interactions, and biofilm formation (29). The covalently linked cell wall proteins can be divided into two main classes: those that are attached to β-1,3-glucan through a mild alkali-sensitive linkage (ASL-CWP) and those with a GPI anchor that can be transferred through a portion of the GPI to β-1,6-glucan (GPI-CWP). These glycan-linked proteins can bind other proteins through disulfide bridges. Noncovalently bound protein may interact with cell wall glycans through a carbohydrate-binding domain.

Cell wall proteins are typically identified by mass spectrometry using different chemical or enzymatic treatments of cell walls for their release. For *S. cerevisiae* 18 GPI-CWPs and 7 ASL-CWPs have been identified; for *C. albicans* 13 GPI-CWPs and 2 ASL-CWPs have been identified (29). Algorithms that predict the addition of GPI anchors to protein suggest that >60 proteins may be extracellular, anchored in either the plasma membrane or the cell wall of *S. cerevisiae*, and >100 proteins for *C. albicans* (29). As these proteins are identified, it has become clear that fungal species encode a unique complement of cell wall proteins. For example, there are no homologs of *S. cerevisiae* PIR (proteins with internal repeats) proteins, which are the best characterized ASL-CWPs encoded by *Cryptococcus*, and several of the predicted cryptococcal GPI-anchored proteins are so far unique to *Cryptococcus* (70). Our understanding of the evolution of fungal cell wall proteins will advance with fungal genome sequencing.

These proteins are commonly referred to as mannoproteins for bearing branched N-linked glycan structures containing as many as 200 mannose residues and short O-linked chains of up to five mannoses for *S. cerevisiae* and *C. albicans* (8, 48). Other fungi have O-glycans of about five monosaccharides, but with diverse structures incorporating galactose and glucose in addition to mannose (48). The protein portion of mannoprotein accounts for approximately 3 to 4% of the cell wall dry weight of yeast and the mannan, ranging from 15% for *S. pombe* to 30% for *S. cerevisiae* and *C. albicans* (30). Mannoproteins in the cell wall of *S. cerevisiae* and *C. albicans* form an electron-dense fibrillar layer visible by electron microscopy on the outer surface of the cell wall. In contrast, mannoproteins in *C. neoformans* are found primarily in the inner cell wall, as evidenced by electron microscopy and immunogold labeling (129). Furthermore, nonencapsulated cryptococcal cells are not agglutinated by concanavalin A, a lectin that binds terminal mannose residues, or antimannoprotein antibodies (129), indicating that mannoproteins are not present on the outer surface of the cell.

Mannoproteins of *C. neoformans* have been implicated in infection and modulation of the host immune response. Antibodies against *C. neoformans* mannoproteins have been recovered from patients of cryptococcal infection, indicating that they are immunogenic (105). Mannoproteins stimulate the production of cytokines and proliferation of lymphocytes (54, 69). They also can induce human and murine macrophages to produce tumor-necrosis factor alpha and interleukin-12 (23, 95, 96). Injection of a preparation of cryptococcal mannoproteins elicits a delayed-type hypersensitivity response in mice (87). Preparations of cryptococcal mannoproteins have been studied as potential vaccines to protect against cryptococcal infection (77).

The structures of the N- and O-linked glycans of *Cryptococcus* have not been determined. The glycosyltransferases used in their synthesis, however, are conserved among fungi, and their presence predicts some aspects of their final structure. Cryptococcal N-linked glycans are predicted to not have the outer chain mannan that would yield the 200 mannoses per N-glycan observed for *S. cerevisiae* and *C. albicans*. *Cryptococcus* lacks the Mnn9 family of mannosyltransferases. O-glycan synthesis uses mannosyltransferases of the Pmt family to add the first mannose to Ser/Thr residues on protein. *Cryptococcus* encodes three Pmts (90) and one α1,2-mannosyl transferase for elongation (48); the complete length of the O-glycan cannot be predicted. Disruption of the attachment of these O-glycan side chains by deletion of protein O-mannosyltransferase 4 (Pmt4) resulted in abnormal morphology, increased sensitivity to cell wall perturbing agents, and alterations in the cell wall proteome (90). The O-glycan side chains also appear to be important for induction of an optimal immune response, as O-deglycosylation decreased the protective efficacy of mannoprotein immunization (77).

Identification and Characterization of Cell Wall Proteins

There has been some debate in the literature as to whether the cryptococcal mannoproteins are covalently associated with the cell wall polysaccharides. Chemical analysis has found purified cell walls to be free of mannose, indicating that mannoproteins are not covalently bound to the cell wall matrix but are instead a nonstructural cell wall component (57, 129). However, in a separate study, mannan was detected in the alkali-insoluble cell wall fraction of purified cryptococcal cell walls (104). This discrepancy may be due to differences in the methods of cell wall preparation and solubilization of polysaccharides for analysis. Whether covalently or noncovalently linked, the cryptococcal cell wall clearly contains several proteins important for the biology and virulence of *C. neoformans* (Table 2).

Eigenheer et al. (42) performed the first proteomic analysis of *C. neoformans* secreted and cell surface proteins using an acapsular strain. Proteins were released from cryptococcal cells or purified cell walls by trypsin or β-glucanase digestion and purified and identified by liquid chromatography–mass spectrometry. Twenty-nine proteins were identified that contained the typical N-terminal signal sequence (Table 2). Seventeen of these also contained a GPI-anchor attachment motif. This suggested that *C. neoformans* is capable of targeting GPI proteins to the cell wall. Release by β-glucanase digestion indicated that these proteins were covalently attached to the polysaccharide matrix. However, digestion could also serve to loosen the cell wall structure, thereby facilitating release of noncovalently associated proteins.

Among the proteins identified were seven proteases. Proteases may play a role in host infection by facilitating host tissue penetration from the alveolar space to the lung parenchyma (47). They have also been shown to cleave the basement membrane proteins collagen, laminin, and fibronectin and host complement factors and immunoglobulins (108, 136). Four polysaccharide deacetylases, including the three CDAs described above, were also identified. Their presence in the cell wall has implications regarding the possible mechanisms of chitosan production.

Phospholipase B1

A few cryptococcal proteins that were not detected in the extracellular proteome analysis described above have also been localized to the cell wall. One such protein is the virulence factor Plb1 (24, 26). Plb1 can hydrolyze phospholipids present in the mammalian plasma membrane and pulmonary surfactant (24–26, 113). It is required for invasion of host lung tissue and dissemination of infection (114). Plb1 has been found in the cryptococcal plasma membrane, where it is concentrated in lipid rafts and the cell wall and secreted from the cell surface (37, 118, 119). The Plb1 protein sequence contains a C-terminal GPI-anchor attachment motif, which is required for its association with both the plasma membrane and cell wall (28, 37). Plb1 transport to the plasma membrane and cell wall also requires N-glycosylation (127). Digestion of purified cell walls with β-1,3-glucanase released Plb1 that contained β-1,6-glucan, indicating that there is a covalent association between Plb1 and β-1,6-glucan (118). In further support of this idea, the levels of Plb1 are decreased in the cell wall and increased in the secreted fraction of two *C. neoformans* mutants that have decreased cell wall β-1,6-glucan (46a). A *plb1*Δ strain was sensitive to SDS and Congo red, and heat stress increased the levels of Plb1 in wild-type cryptococcal cell walls (118). These data suggest that cell-wall-localized Plb1 plays an important role in the maintenance of cell integrity.

Laccase

Another important protein in the *C. neoformans* cell wall is laccase 1 (Lac1), one of the two enzymes responsible for the production of the melanin pigment (described above). Laccase also prevents the formation of toxic iron-dependent Fenton products, which normally would accumulate in the phagolysosome of macrophages, suggesting a possible role for laccase in the survival of *C. neoformans* within macrophages (73). Laccase is clearly important for virulence. A *lac1*Δ strain had attenuated virulence in an intravenous mouse model of infection (112). *LAC1* is required for dissemination in an intratracheal mouse model, but not for pulmonary persistence (89).

Peripheral localization of laccase would facilitate participation in immune modulation and access to the exogenous catecholamine substrates required for melanin synthesis (132, 139). Laccase has been localized to the cell wall of *C. neoformans* in vitro by immunoelectron microscopy and utilization of a green fluorescent protein–tagged expression construct (139). This cell wall localization was observed in cryptococcal cells in alveolar macrophages, lungs, and brains from intravenously infected mice (132). Laccase activity was also detected in the secreted vesicles incubated with L-DOPA (109), supporting the idea that melanin production occurs within vesicles in the cell wall. The presence of laccase in the cell wall is a unique feature of *C. neoformans*, as most other fungal laccases are completely secreted (7). An extended C-terminal region of *CNLAC1* absent from other fungi appears to facilitate cell wall targeting, as a Lac1 C-terminal truncation mutant was localized to cytosolic vesicles and not the cell wall (132). The precise mechanism by which this unique sequence promotes cell wall association and how this corresponds to potential association with vesicles is unknown.

CONCLUSIONS AND PERSPECTIVES

Pioneering studies in model organisms such as *S. cerevisiae* set the foundation for cell wall exploration of other medically and environmentally important fungi. Further research has revealed that there is great variety in the cell wall across the fungal kingdom, with each species having its own unique features that are distinct from *Saccharomyces*. Such complexity demands in-depth studies of the cell walls of other

TABLE 2 Extracellular proteins of *Cryptococcus neoformans*

Protein Name	Protein Accession #	Function	GPI-anchor	Reference
Cda1	AAW44209	Chitin deacetylase	+	70
Cda2/MP98	AAW43254	Chitin deacetylase	+	70
Cda3/MP84	AAW42893	Chitin deacetylase	+	70
Celc	CAD31110	Cellulase	–	14b
Cig	CAC78984	Glycoprotein involved in iron metabolism	–	70a
Dha1	AAG01395	Unknown	–	75a
Fpd1/D25	CAD10036	Deacetylase	–	14a
Gas1	AAW47197	Putative 1,3-β-glucanosyltransferase	+	14b
Glo1	AAW43726	Glyoxal oxidase	+	70
Glo2	AAW44259	Glyoxal oxidase	+	70
Glo3	AAW41343	Glyoxal oxidase	+	70
Lac1	ABI58272	Laccase	–	139
MP88	AAL87197	Unknown		54a
MP115	CAI79615	Carboxylesterase 1	–	14b
MP-A	AAW46870	Unknown	+	70
MP-B	AAW45184	Unknown	+	70
MP-C	EAL19195	Unknown	+	70
MP-D	AAW46410	Phosphatidylethanolamine binding protein	+	70
MP-E	AAW44733	Unknown	+	70
MP-G	AAW44978	Unknown	+	70
MP-H	AAW43326	Unknown	+	70
MP-I	AAW40819	Unknown	+	70
MP-J	AAW41559	Unknown	+	70
Plb1	AAW46882	Phosholipase B	+	37
Sod	XP_570285	Superoxide dismutase	–	49a
Unnamed	AAW43607	α-Amylase	+	70
Unnamed	AAW41268	α-Amylase	+	70
Unnamed	XP_777848	Aspartic protease	–	42
Unnamed	AAW43946	Aspartic protease	+	70
Unnamed	AAW44235	Aspartic protease	+	70
Unnamed	AAW45503	Aspartic protease	+	70
Unnamed	AAW45937	Aspartic protease	+	70
Unnamed	AAW40834	β-Endoglucanase	+	70
Unnamed	AAW45003	β-Endoglucanase	+	70
Unnamed	AAW46063	β-Endoglucanase	+	70
Unnamed	AAW46065	β-Endoglucanase	+	70
Unnamed	XP_776557	Carboxylesterase 2	–	42
Unnamed	AAW46493	CFEM	+	70
Unnamed	XP_776749	Endo-1,3-β-glucanase	–	42
Unnamed	AAW43707	Fasciclin	+	70
Unnamed	AAW46674	Glucuronyl hydrolase	+	70
Unnamed	AAW41079	GPI-glucanosyltransferase	+	70
Unnamed	XP_776261	Lipoprotein	–	42
Unnamed	XP_773186	Metalloprotease	–	42
Unnamed	AAW40860	Metallo-peptidase	+	70
Unnamed	XP_772274	MP88-like		54a
Unnamed	XP_775708	PG-PI transfer protein	+	42
Unnamed	XP_776649	Serine peptidase 1	–	42
Unnamed	XP_774571	Serine peptidatse 2	–	42
Unnamed	XP_778132	Serine peptidase 1 (subtilase family)	–	42
Unnamed	XP_773011	Serine peptidase 2 (subtilase family)	–	42
Unnamed	AAW44587	Similar to alginate lyase	+	70
Unnamed	XP_774109	Trehalase	–	42
Unnamed	XP_772373	Unknown	+	14b
Unnamed	XP_774464	Unknown	+	42

organisms. This is especially urgent for pathogenic fungi, as these advances in understanding have the potential to lead to new therapeutic targets.

Cryptococcus is an excellent organism for these types of studies because it is genetically manipulable, and many biochemical and molecular biology tools have been developed. Based upon the data presented in this chapter, we propose a working model for the cell wall of *Cryptococcus* depicting the components known to be present (see Color Plate 1). Studies of *C. neoformans* have revealed that its cell wall is very different from and more complex than other well-characterized fungal cell walls. At the composition level, such differences include the presence of abundant α-glucan and chitosan polymers, both absent from *Saccharomyces* vegetative cells. β-Glucans in *Cryptococcus* are predominately linked via β-1,6 bonds, with minor levels of β-1,3, which is opposite of that in *S. cerevisiae*. *C. neoformans* also lacks obvious homologs to the PIR class of proteins that are readily found in the cell walls of many fungal species, although some noncanonical proteins are present in the cryptococcal cell wall. Beyond the basic composition of the cell wall, other distinct features include the ability to deposit melanin, the association with capsular polysaccharides, and the apparent lack of requirement of chitinases for budding.

Not only is the *C. neoformans* cell wall complex, but data have revealed its dynamic nature. Mutants of *Cryptococcus* lacking β-1,6-glucan, and therefore having decreased retention of cell wall proteins, appear to have shifts in chitosan localization. Yet despite these changes, the cells remain viable. *C. neoformans* cells can lose α-1,3-glucan, a major component, with minimal effect on viability. Disruption of chitosan production results in increased chitin levels and affects pigments and mating. Cells can also survive when secretion of vesicles across the cell wall is abolished.

The fungal cell wall remains the most attractive target for the next generation of antifungal drugs. Decades of research have advanced the field of cell wall biology to our current understanding of the dynamic and complex nature of this essential organelle, but many more questions remain to be answered. Studies into the cell wall of fungi have an exciting future, and many important discoveries are predicted in the years to come.

We thank John Heuser and Tamara Doering for providing scanning electron microscopy images.

REFERENCES

1. **Albuquerque, P. C., E. S. Nakayasu, M. L. Rodrigues, S. Frases, A. Casadevall, R. M. Zancope-Oliveira, I. C. Almeida, and J. D. Nosanchuk.** 2008. Vesicular transport in *Histoplasma capsulatum*: an effective mechanism for trans-cell wall transfer of proteins and lipids in ascomycetes. *Cell. Microbiol.* **10:**1695–1710.
2. **Allen, A. K., A. Neuberger, and N. Sharon.** 1973. The purification, composition and specificity of wheat-germ agglutinin. *Biochem. J.* **131:**155–162.
3. **Araki, Y., and E. Ito.** 1975. A pathway of chitosan formation in *Mucor rouxii*. Enzymatic deacetylation of chitin. *Eur. J. Biochem.* **55:**71–78.
4. **Arcidiacono, S., and D. L. Kaplan.** 1992. Molecular weight distribution of chitosan isolated from *Mucor rouxii* under different culture and processing conditions. *Biotechnol. Bioeng.* **39:**281–286.
5. **Bacon, J. S., D. Jones, V. C. Farmer, and D. M. Webley.** 1968. The occurrence of alpha(1-3)glucan in *Cryptococcus*, *Schizosaccharomyces* and *Polyporus* species, and its hydrolysis by a *Streptomyces* culture filtrate lysing cell walls of *Cryptococcus*. *Biochim. Biophys. Acta* **158:**313–315.
6. **Baker, L. G., C. A. Specht, M. J. Donlin, and J. K. Lodge.** 2007. Chitosan, the deacetylated form of chitin, is necessary for cell wall integrity in *Cryptococcus neoformans*. *Eukaryot. Cell* **6:**855–867.

6a. **Baker, L. G., C. A. Specht, and J. K. Lodge.** 2009. Chitinases are essential for sexual development but not vegetative growth in *Cryptococcus neoformans*. *Eukaryot. Cell.* **8:**1692–1705.

7. **Baldrian, P.** 2006. Fungal laccases: occurrence and properties. *FEMS Microbiol. Rev.* **30:**215–242.
8. **Ballou, L., L. M. Hernandez, E. Alvarado, and C. E. Ballou.** 1990. Revision of the oligosaccharide structures of yeast carboxypeptidase Y. *Proc. Natl. Acad. Sci. USA* **87:**3368–3372.
9. **Banks, I. R., C. A. Specht, M. J. Donlin, K. J. Gerik, S. M. Levitz, and J. K. Lodge.** 2005. A chitin synthase and its regulator protein are critical for chitosan production and growth of the fungal pathogen *Cryptococcus neoformans*. *Eukaryot. Cell* **4:**1902–1912.
10. **Bartizal, K., C. J. Gill, G. K. Abruzzo, A. M. Flattery, L. Kong, P. M. Scott, J. G. Smith, C. E. Leighton, A. Bouffard, J. F. Dropinski, and J. Balkovec.** 1997. In vitro preclinical evaluation studies with the echinocandin antifungal MK-0991 (L-743,872). *Antimicrob. Agents Chemother.* **41:**2326–2332.
11. **Bartlett, M. S., W. L.Current, M. P. Goheen, C. J. Boylan, C. H. Lee, M. M. Shaw, S. F. Queener, and J. W. Smith.** 1996. Semisynthetic echinocandins affect cell wall deposition of *Pneumocystis carinii* in vitro and in vivo. *Antimicrob. Agents Chemother.* **40:**1811–1816.
12. **Bartnicki-Garcia, S., and E. Lippman.** 1969. Fungal morphogenesis: cell wall construction in *Mucor rouxii*. *Science* **165:**302–304.
13. **Beauvais, A., D. Maubon, S. Park, W. Morelle, M. Tanguy, M. Huerre, D. S. Perlin, and J. P. Latgé.** 2005. Two alpha(1-3) glucan synthases with different functions in *Aspergillus fumigatus*. *Appl. Environ. Microbiol.* **71:**1531–1538.
14. **Bernard, M., and L. P. Latgé.** 2001. *Aspergillus fumigatus* cell wall: composition and biosynthesis. *Med. Mycol.* **39**(Suppl)**:**19–17.

14a. **Biondo, C., C. Beninati, D. Delfino, M. Oggioni, G. Mancuso, A. Midiri, M. Bombaci, G. Tomaselli, and G. Teti.** 2002. Identification and cloning of a cryptococcal deacetylase that produces protective immune responses. *Infect. Immun.* **70:**2383–2391.

14b. **Biondo, C., G. Mancuso, A. Midiri, M. Bombaci, L. Messina, C. Beninati, and G. Teti.** 2006. Identification of major proteins secreted by *Cryptococcus neoformans*. *FEMS Yeast Res.* **6:**645–651.

15. **Bowen, A. R., J. L. Chen-Wu, M. Momany, R. Young, P. J. Szaniszlo, and P. W. Robbins.** 1992. Classification of fungal chitin synthases. *Proc. Natl. Acad. Sci. USA* **89:**519–523.
16. **Bulawa, C. E.** 1992. CSD2, CSD3, and CSD4, genes required for chitin synthesis in *Saccharomyces cerevisiae*: the CSD2 gene product is related to chitin synthases and to developmentally regulated proteins in *Rhizobium* species and *Xenopus laevis*. *Mol. Cell. Biol.* **12:**1764–1776.
17. **Cabib, E., N. Blanco, C. Grau, J. M. Rodríguez-Peña, and J. Arroyo.** 2007. Crh1p and Crh2p are required for the cross-linking of chitin to beta(1-6)glucan in the *Saccharomyces cerevisiae* cell wall. *Mol. Microbiol.* **63:**921–935.
18. **Cabib, E., V. Farkas, O. Kosík, N. Blanco, J. Arroyo, and P. McPhie.** 2008. Assembly of the yeast cell wall. Crh1p and Crh2p act as transglycosylases in vivo and in vitro. *J. Biol. Chem.* **283:**29859–29872.

19. **Cabib, E., A. Sburlati, B. Bowers, and S. J Silverman.** 1989. Chitin synthase 1, an auxiliary enzyme for chitin synthesis in *Saccharomyces cerevisiae*. *J. Cell Biol.* **108:** 1665–1672.
20. **Cabib, E., S. J. Silverman, and J. A. Shaw.** 1992. Chitinase and chitin synthase 1: counterbalancing activities in cell separation of *Saccharomyces cerevisiae*. *J. Gen. Microbiol.* **138:**97–102.
21. **Cannon, R. D., E. Lamping, A. R. Holmes, K. Niimi, P. V. Baret, M. V. Keniya, K. Tanabe, M. Niimi, A. Goffeau, and B. C. Monk.** 2009. Efflux-mediated antifungal drug resistance. *Clin. Microbiol. Rev.* **22:**291–321.
22. **Cassone, A., R. E. Mason, and D. Kerridge.** 1981. Lysis of growing yeast-form cells of *Candida albicans* by echinocandin: a cytological study. *Sabouraudia* **19:**97–110.
23. **Chaka, W., A. F. Verheul, V. V. Vaishnav, R. Cherniak, J. Scharringa, J. Verhoef, H. Snippe, and A. I. Hoepelman.** 1997. Induction of TNF-alpha in human peripheral blood mononuclear cells by the mannoprotein of *Cryptococcus neoformans* involves human mannose binding protein. *J. Immunol.* **159:**2979–2985.
24. **Chen, S. C., M. Muller, J. Z. Zhou, L. C. Wright, and T. C. Sorrell.** 1997. Phospholipase activity in *Cryptococcus neoformans*: a new virulence factor? *J. Infect. Dis.* **175:** 414–420.
25. **Chen, S. C., L. C. Wright, J. C. Golding, and T. C. Sorrell.** 2000. Purification and characterization of secretory phospholipase B, lysophospholipase and lysophospholipase/transacylase from a virulent strain of the pathogenic fungus *Cryptococcus neoformans*. *Biochem. J.* **347**(Pt 2): 431–439.
26. **Chen, S. C., L. C. Wright, R. T. Santangelo, M. Muller, V. R. Moran, P. W. Kuchel, and T. C. Sorrell.** 1997. Identification of extracellular phospholipase B, lysophospholipase, and acyltransferase produced by *Cryptococcus neoformans*. *Infect. Immun.* **65:**405–411.
27. **Christodoulidou, A., V. Bouriotis, and G. Thireos.** 1996. Two sporulation-specific chitin deacetylase-encoding genes are required for the ascospore wall rigidity of *Saccharomyces cerevisiae*. *J. Biol. Chem.* **271:**31420–31425.
28. **Cox, G. M., H. C. McDade, S. C. Chen, S. C. Tucker, M. Gottfredsson, L. C. Wright, T. C. Sorrell, S. D. Leidich, A. Casadevall, M. A. Ghannoum, and J. R. Perfect.** 2001. Extracellular phospholipase activity is a virulence factor for *Cryptococcus neoformans*. *Mol. Microbiol.* **39:** 166–175.
29. **De Groot, P. W. J., A. F. Ram, and F. M. Klis.** 2005. Features and functions of covalently linked proteins in fungal cell walls. *Fungal Genet. Biol.* **42:**657–675.
30. **De Groot, P. W. J., Q. Y. Yin, M. Weig, G. J. Sosinska, F. M. Klis, and C. G. de Koster.** 2007. Mass spectrometric identification of covalently bound cell wall proteins from the fission yeast *Schizosaccharomyces pombe*. *Yeast* **24:** 267–278.
31. **Del Poeta, M., M. C. Cruz, M. E. Cardenas, J. R. Perfect, and J. Heitman.** 2000. Synergistic antifungal activities of bafilomycin A(1), fluconazole, and the pneumocandin MK-0991/caspofungin acetate (L-743,873) with calcineurin inhibitors FK506 and L-685,818 against *Cryptococcus neoformans*. *Antimicrob. Agents Chemother.* **44:**739–746.
32. **Del Poeta, M., W. A. Schell, and J. R. Perfect.** 1997. In vitro antifungal activity of pneumocandin L-743,872 against a variety of clinically important molds. *Antimicrob. Agents Chemother.* **41:**1835–1836.
33. **DeMarini, D. J., A. E. Adams, H. Fares, C. De Virgilio, G. Valle, J. S. Chuang, and J. R. Pringle.** 1997. A septin-based hierarchy of proteins required for localized deposition of chitin in the *Saccharomyces cerevisiae* cell wall. *J. Cell Biol.* **139:**75–93.
34. **De Nobel, J. G., F. M. Klis, T. Munnik, J. Priem, and H. van den Ende.** 1990. An assay of relative cell wall porosity in *Saccharomyces cerevisiae, Kluyveromyces lactis* and *Schizosaccharomyces pombe*. *Yeast* **6:**483–490.
35. **De Nobel, J. G., F. M. Klis, J. Priem, T. Munnik, and H. van den Ende.** 1990. The glucanase-soluble mannoproteins limit cell wall porosity in *Saccharomyces cerevisiae*. *Yeast* **6:**491–499.
36. **De Stefano, J. A., M. T. Cushion, V. Puvanesarajah, and P. D. Walzer.** 1990. Analysis of *Pneumocystis carinii* cyst wall. II. Sugar composition. *J. Protozool.* **37:**436–441.
37. **Djordjevic, J. T., M. Del Poeta, T. C. Sorrell, K. M. Turner, and L. C. Wright.** 2005. Secretion of cryptococcal phospholipase B1 (PLB1) is regulated by a glycosylphosphatidylinositol (GPI) anchor. *Biochem. J.* **389**(Pt 3):803–812.
38. **Douglas, C. M., J. A. D'Ippolito, G. J. Shei, M. Meinz, J. Onishi, J. A. Marrinan, W. Li, G. K. Abruzzo, A. Flattery, K. Bartizal, A. Mitchell, and M. B. Kurtz.** 1997. Identification of the FKS1 gene of *Candida albicans* as the essential target of 1,3-beta-D-glucan synthase inhibitors. *Antimicrob. Agents Chemother.* **41:**2471–2479.
39. **Douglas, C. M., F. Foor, J. A. Marrinan, N. Morin, J. B. Nielsen, A. M. Dahl, P. Mazur, W. Baginsky, W. Li, and M. el-Sherbeini.** 1994. The *Saccharomyces cerevisiae* FKS1 (ETG1) gene encodes an integral membrane protein which is a subunit of 1,3-beta-D-glucan synthase. *Proc. Natl. Acad. Sci. USA* **91:**12907–12911.
40. **Drgonová, J., T. Drgon, K. Tanaka, R. Kollár, G. C. Chen, R. A. Ford, C. S. Chan, Y. Takai, and E. Cabib.** 1996. Rho1p, a yeast protein at the interface between cell polarization and morphogenesis. *Science* **272:**277–279.
41. **Duo-Chuan, L.** 2006. Review of fungal chitinases. *Mycopathologia* **161:**345–360.
42. **Eigenheer, R. A., Y. Jin Lee, E. Blumwald, B. S. Phinney, and A. Gelli.** 2007. Extracellular glycosylphosphatidylinositol-anchored mannoproteins and proteases of *Cryptococcus neoformans*. *FEMS Yeast Res.* **7:**499–510.
43. **Eisenman, H. C., J. D. Nosanchuk, J. B. W. Webber, R. J. Emerson, T. A. Camesano, and A. Casadevall.** 2005. Microstructure of cell wall-associated melanin in the human pathogenic fungus *Cryptococcus neoformans*. *Biochemistry* **44:**3683–3693.
44. **Feldmesser, M., Y. Kress, A. Mednick, and A. Casadevall.** 2000. The effect of the echinocandin analogue caspofungin on cell wall glucan synthesis by *Cryptococcus neoformans*. *J. Infect. Dis.* **182:**1791–1795.
45. **Fleet, G. H.** 1985. Composition and structure of yeast cell walls. *Curr. Top. Med. Mycol.* **1:**24–56.
46. **Garcerá-Teruel, A., B. Xoconostle-Cázares, R. Rosas-Quijano, L. Ortiz, C. León-Ramírez, C. A. Specht, R. Sentandreu, and J. Ruiz-Herrera.** 2004. Loss of virulence in *Ustilago maydis* by Umchs6 gene disruption. *Res. Microbiol.* **155:**87–97.
46a. **Gilbert, N. M., M. J. Donlin, K. J. Gerik, C. A. Specht, J. T. Djordjevic, C. F. Wilson, T. C. Sorrell, and J. K. Lodge.** 2010. *KRE* genes are required for beta-1,6-glucan synthesis, maintenance of capsule architecture and cell wall protein anchoring in *Cryptococcus neoformans*. *Mol. Microb.* **76:**517–534.
47. **Goldman, D., S. C. Lee, and A. Casadevall.** 1994. Pathogenesis of pulmonary *Cryptococcus neoformans* infection in the rat. *Infect. Immun.* **62:**4755–4761.
48. **Goto, M.** 2007. Protein O-glycosylation in fungi: diverse structures and multiple functions. *Biosci. Biotechnol. Biochem.* **71:**1415–1427.
49. **Grün, C. H., F. Hochstenbach, B. M. Humbel, A. J. Verkleij, J. H. Sietsma, F. M. Klis, J. P. Kamerling, and J. F. G. Vliegenthart.** 2005. The structure of cell wall alpha-glucan from fission yeast. *Glycobiology* **15:**245–257.

49a. **Hamilton, A. J., and M. D. Holdom.** 1997. Biochemical comparison of the Cu,Zn superoxide dismutases of *Cryptococcus neoformans* var. *neoformans* and *Cryptococcus neoformans* var. *gattii*. *Infect. Immun.* **65:**488–494.

50. **Hamilton, A. J., L. Jeavons, P. Hobby, and R. J. Hay.** 1992. A 34- to 38-kilodalton *Cryptococcus neoformans* glycoprotein produced as an exoantigen bearing a glycosylated species-specific epitope. *Infect. Immun.* **60:**143–149.

51. **Hartland, R. P., C. A. Vermeulen, F. M. Klis, J. H. Sietsma, and J. G. Wessels.** 1994. The linkage of (1-3)-beta-glucan to chitin during cell wall assembly in *Saccharomyces cerevisiae*. *Yeast* **10:**1591–1599.

52. **Hector, R. F.** 1993. Compounds active against cell walls of medically important fungi. *Clin. Microbiol. Rev.* **6:**1–21.

53. **Hochstenbach, F., F. M. Klis, H. van den Ende, E. van Donselaar, P. J. Peters, and R. D. Klausner.** 1998. Identification of a putative alpha-glucan synthase essential for cell wall construction and morphogenesis in fission yeast. *Proc. Natl. Acad. Sci. USA* **95:**9161–9166.

54. **Hoy, J. F., J. W. Murphy, and G. G. Miller.** 1989. T cell response to soluble cryptococcal antigens after recovery from cryptococcal infection. *J. Infect. Dis.* **159:**116–119.

54a. **Huang, C., S. H. Nong, M. K. Mansour, C. A. Specht, and S. M. Levitz.** 2002. Purification and characterization of a second immunoreactive mannoprotein from *Cryptococcus neoformans* that stimulates T-Cell responses. *Infect. Immun.* **70:**5485–5493.

55. **Iwata, K., Y. Yamamoto, H. Yamaguchi, and T. Hiratani.** 1982. In vitro studies of aculeacin A, a new antifungal antibiotic. *J. Antibiot.* (Tokyo) **35:**203–209.

56. **Jacobson, E. S., and R. Ikeda.** 2005. Effect of melanization upon porosity of the cryptococcal cell wall. *Med. Mycol.* **43:**327–333.

57. **James, P. G., R. Cherniak, R. G. Jones, C. A. Stortz, and E. Reiss.** 1990. Cell-wall glucans of *Cryptococcus neoformans* Cap 67. *Carbohydr. Res.* **198:**23–38.

58. **Kafetzopoulos, D., A. Martinou, and V. Bouriotis.** 1993. Bioconversion of chitin to chitosan: purification and characterization of chitin deacetylase from *Mucor rouxii*. *Proc. Natl. Acad. Sci. USA* **90:**2564–2568.

59. **Kelly, R., E. Register, M. J. Hsu, M. Kurtz, and J. Nielsen.** 1996. Isolation of a gene involved in 1,3-beta-glucan synthesis in *Aspergillus nidulans* and purification of the corresponding protein. *J. Bacteriol.* **178:**4381–4391.

60. **Klis, F. M.** 1994. Review: cell wall assembly in yeast. *Yeast* **10:**851–869.

61. **Klis, F. M., P. de Groot, and K. Hellingwerf.** 2001. Molecular organization of the cell wall of *Candida albicans*. *Med. Mycol.* **39**(Suppl)**:**11–18.

62. **Klis, F. M., P. Mol, K. Hellingwerf, and S. Brul.** 2002. Dynamics of cell wall structure in *Saccharomyces cerevisiae*. *FEMS Microbiol. Rev.* **26:**239–256.

63. **Kohno, H., K. Tanaka, A. Mino, M. Umikawa, H. Imamura, T. Fujiwara, Y. Fujita, K. Hotta, H. Qadota, T. Watanabe, Y. Ohya, and Y. Takai.** 1996. Bni1p implicated in cytoskeletal control is a putative target of Rho1p small GTP binding protein in *Saccharomyces cerevisiae*. *EMBO J.* **15:**6060–6068.

64. **Kollár, R., E. Petráková, G. Ashwell, P. W. Robbins, and E. Cabib.** 1995. Architecture of the yeast cell wall. The linkage between chitin and beta(1→3)-glucan. *J. Biol. Chem.* **270:**1170–1178.

65. **Kollár, R., B. B. Reinhold, E. Petráková, H. J. Yeh, G. Ashwell, J. Drgonová, J. C. Kapteyn, F. M. Klis, and E. Cabib.** 1997. Architecture of the yeast cell wall. Beta (1→6)-glucan interconnects mannoprotein, beta(1→) 3-glucan, and chitin. *J. Biol. Chem.* **272:**17762–17775.

66. **Kraus, P. R., D. S. Fox, G. M. Cox, and J. Heitman.** 2003. The *Cryptococcus neoformans* MAP kinase Mpk1 regulates cell integrity in response to antifungal drugs and loss of calcineurin function. *Mol. Microbiol.* **48:**1377–1387.

67. **Krishnarao, T. V., and J. N. Galgiani.** 1997. Comparison of the in vitro activities of the echinocandin LY303366, the pneumocandin MK-0991, and fluconazole against *Candida* species and *Cryptococcus neoformans*. *Antimicrob. Agents Chemother.* **41:**1957–1960.

68. **Lesage, G., and H. Bussey.** 2006. Cell wall assembly in *Saccharomyces cerevisiae*. *Microbiol. Mol. Biol. Rev.* **70:**317–343.

69. **Levitz, S. M., and E. A. North.** 1997. Lymphoproliferation and cytokine profiles in human peripheral blood mononuclear cells stimulated by *Cryptococcus neoformans*. *J. Med. Vet. Mycol.* **35:**229–236.

70. **Levitz, S. M., and C. A. Specht.** 2006. The molecular basis for the immunogenicity of *Cryptococcus neoformans* mannoproteins. *FEMS Yeast Res.* **6:**513–524.

70a. **Lian, T., M. I. Simmer, C. A. D'Souza, B. R. Steen, S. D. Zuyderduyn, S. J. Jones, M. A. Marra, and J. W. Kronstad.** 2005. Iron-regulated transcription and capsule formation in the fungal pathogen *Cryptococcus neoformans*. *Mol. Microbiol.* **55:**1452–1472.

71. **Lipke, P. N., and R. Ovalle.** 1998. Cell wall architecture in yeast: new structure and new challenges. *J. Bacteriol.* **180:**3735–3740.

72. **Liu, H., S. Kauffman, J. M. Becker, and P. J. Szaniszlo.** 2004. *Wangiella (Exophiala) dermatitidis* WdChs5p, a class V chitin synthase, is essential for sustained cell growth at temperature of infection. *Eukaryot. Cell* **3:**40–51.

73. **Liu, L., R. P. Tewari, and P. R. Williamson.** 1999. Laccase protects *Cryptococcus neoformans* from antifungal activity of alveolar macrophages. *Infect. Immun.* **67:**6034–6039.

74. **Madrid, M. P., A. Di Pietro, and M. I. G. Roncero.** 2003. Class V chitin synthase determines pathogenesis in the vascular wilt fungus *Fusarium oxysporum* and mediates resistance to plant defence compounds. *Mol. Microbiol.* **47:**257–266.

75. **Maligie, M. A., and C. P. Selitrennikoff.** 2005. *Cryptococcus neoformans* resistance to echinocandins: (1,3)beta-glucan synthase activity is sensitive to echinocandins. *Antimicrob. Agents Chemother.* **49:**2851–2856.

75a. **Mandel, M. A., G. G. Grace, K. I. Orsborn, F. Schafer, J. W. Murphy, M. J. Orbach, and J. N. Galgiani.** 2000. The *Cryptococcus neoformans* gene *DHA1* encodes an antigen that elicits a delayed-type hypersensitivity reaction in immune mice. *Infect. Immun.* **68:**6196–6201.

76. **Manners, D. J., A. J. Masson, and J. C. Patterson.** 1973. The structure of a beta-(1→3)-D-glucan from yeast cell walls. *Biochem. J.* **135:**19–30.

77. **Mansour, M. K., L. E. Yauch, J. B. Rottman, and S. M. Levitz.** 2004. Protective efficacy of antigenic fractions in mouse models of cryptococcosis. *Infect. Immun.* **72:**1746–1754.

78. **Maubon, D., S. Park, M. Tanguy, M. Huerre, C. Schmitt, M. C. Prévost, D. S. Perlin, J. P. Latgé, and A. Beauvais.** 2006. AGS3, an alpha(1-3)glucan synthase gene family member of *Aspergillus fumigatus*, modulates mycelium growth in the lung of experimentally infected mice. *Fungal Genet. Biol.* **43:**366–375.

79. **Mazur, P., and W. Baginsky.** 1996. In vitro activity of 1,3-beta-D-glucan synthase requires the GTP-binding protein Rho1. *J. Biol. Chem.* **271:**14604–14609.

80. **Mazur, P., N. Morin, W. Baginsky, M. el-Sherbeini, J. A. Clemas, J. B. Nielsen, and F. Foor.** 1995. Differential expression and function of two homologous subunits of yeast 1,3-beta-D-glucan synthase. *Mol. Cell. Biol.* **15:**5671–5681.

81. **Minke, R., and J. Blackwell.** 1978. The structure of alpha-chitin. *J. Mol. Biol.* 120:167–181.

82. **Mio, T., M. Adachi-Shimizu, Y. Tachibana, H. Tabuchi, S. B. Inoue, T. Yabe, T. Yamada-Okabe, M. Arisawa, T. Watanabe, and H. Yamada-Okabe.** 1997. Cloning of the *Candida albicans* homolog of *Saccharomyces cerevisiae* GSC1/FKS1 and its involvement in beta-1,3-glucan synthesis. *J. Bacteriol.* **179:**4096–4105.
83. **Mouyna, I., and T. Fontaine.** 2009. Cell wall of *Aspergillus fumigatus*: a dynamic structure, p. 169–183. *In* J. P. Latgé and W. J. Steinbach (ed.), *Aspergillus fumigatus and Aspergillosis*. ASM Press, Washington, DC.
84. **Mrsa, V., F. Klebl, and W. Tanner.** 1993. Purification and characterization of the *Saccharomyces cerevisiae* BGL2 gene product, a cell wall endo-beta-1,3-glucanase. *J. Bacteriol.* **175:**2102–2106.
85. **Munro, C. A., and N. A. Gow.** 2001. Chitin synthesis in human pathogenic fungi. *Med. Mycol.* **39**(Suppl)**:** 141–153.
86. **Munro, C. A., K. Winter, A. Buchan, K. Henry, J. M. Becker, A. J. Brown, C. E. Bulawa, and N. A. Gow.** 2001. Chs1 of *Candida albicans* is an essential chitin synthase required for synthesis of the septum and for cell integrity. *Mol. Microbiol.* **39:**1414–1426.
87. **Murphy, J. W., R. L. Mosley, R. Cherniak, G. H. Reyes, T. R. Kozel, and E. Reiss.** 1988. Serological, electrophoretic, and biological properties of *Cryptococcus neoformans* antigens. *Infect. Immun.* **56:**424–431.
88. **Nonaka, H., K. Tanaka, H. Hirano, T. Fujiwara, H. Kohno, M. Umikawa, A. Mino, and Y. Takai.** 1995. A downstream target of RHO1 small GTP-binding protein is PKC1, a homolog of protein kinase C, which leads to activation of the MAP kinase cascade in *Saccharomyces cerevisiae*. *EMBO J.* **14:**5931–5938.
89. **Noverr, M. C., P. R. Williamson, R. S. Fajardo, and G. B. Huffnagle.** 2004. CNLAC1 is required for extrapulmonary dissemination of *Cryptococcus neoformans* but not pulmonary persistence. *Infect. Immun.* **72:**1693–1699.
90. **Olson, G. M., D. S. Fox, P. Wang, J. A. Alspaugh, and K. L. Buchanan.** 2007. Role of protein O-mannosyltransferase Pmt4 in the morphogenesis and virulence of *Cryptococcus neoformans*. *Eukaryot. Cell* **6:**222–234.
91. **Ono, N., T. Yabe, M. Sudoh, T. Nakajima, T. Yamada-Okabe, M. Arisawa, and H. Yamada-Okabe.** 2000. The yeast Chs4 protein stimulates the trypsin-sensitive activity of chitin synthase 3 through an apparent protein-protein interaction. *Microbiology* **146**(Pt 2)**:**385–391.
92. **Orlowski, M.** 1991. Mucor dimorphism. *Microbiol. Rev.* **55:**234–258.
93. **Panepinto, J., K. Komperda, S. Frases, Y. D. Park, J. T. Djordjevic, A. Casadevall, and P. R. Williamson.** 2009. Sec6-dependent sorting of fungal extracellular exosomes and laccase of *Cryptococcus neoformans*. *Mol. Microbiol.* **71:**1165–1176.
94. **Pfaller, M. A., S. A. Messer, and S. Coffman.** 1997. In vitro susceptibilities of clinical yeast isolates to a new echinocandin derivative, LY303366, and other antifungal agents. *Antimicrob. Agents Chemother.* **41:**763–766.
95. **Pietrella, D., R. Cherniak, C. Strappini, S. Perito, P. Mosci, F. Bistoni, and A. Vecchiarelli.** 2001. Role of mannoprotein in induction and regulation of immunity to *Cryptococcus neoformans*. *Infect. Immun.* **69:**2808–2814.
96. **Pitzurra, L., R. Cherniak, M. Giammarioli, S. Perito, F. Bistoni, and A. Vecchiarelli.** 2000. Early induction of interleukin-12 by human monocytes exposed to *Cryptococcus neoformans* mannoproteins. *Infect. Immun.* **68:**558–563.
97. **Popolo, L., and M. Vai.** 1999. The Gas1 glycoprotein, a putative wall polymer cross-linker. *Biochim. Biophys. Acta* **1426:**385–400.
98. **Privat, J. P., F. Delmotte, G. Mialonier, P. Bouchard, and M. Monsigny.** 1974. Fluorescence studies of saccharide binding to wheat-germ agglutinin (lectin). *Eur. J. Biochem.* **47:**5–14.
99. **Qadota, H., I. Ishii, A. Fujiyama, Y. Ohya, and Y. Anraku.** 1992. RHO gene products, putative small GTP-binding proteins, are important for activation of the CAL1/CDC43 gene product, a protein geranylgeranyltransferase in *Saccharomyces cerevisiae*. *Yeast* **8:**735–741.
100. **Qadota, H., C. P. Python, S. B. Inoue, M. Arisawa, Y. Anraku, Y. Zheng, T. Watanabe, D. E. Levin, and Y. Ohya.** 1996. Identification of yeast Rho1p GTPase as a regulatory subunit of 1,3-beta-glucan synthase. *Science* **272:**279–281.
101. **Ram, A. F., J. C. Kapteyn, R. C. Montijn, L. H. Caro, J. E. Douwes, W. Baginsky, P. Mazur, H. van den Ende, and F. M. Klis.** 1998. Loss of the plasma membrane-bound protein Gas1p in *Saccharomyces cerevisiae* results in the release of beta1,3-glucan into the medium and induces a compensation mechanism to ensure cell wall integrity. *J. Bacteriol.* **180:**1418–1424.
102. **Rappleye, C. A., J. T. Engle, and W. E. Goldman.** 2004. RNA interference in *Histoplasma capsulatum* demonstrates a role for alpha-(1,3)-glucan in virulence. *Mol. Microbiol.* **53:**153–165.
103. **Reese, A. J., and T. L Doering.** 2003. Cell wall alpha-1,3-glucan is required to anchor the *Cryptococcus neoformans* capsule. *Mol. Microbiol.* **50:**1401–1409.
104. **Reese, A. J., A. Yoneda, J. A. Breger, A. Beauvais, H. Liu, C. L. Griffith, I. Bose, M. Kim, C. Skau, S. Yang, J. A. Sefko, M. Osumi, J. Latge, E. Mylonakis, and T. L. Doering.** 2007. Loss of cell wall alpha(1-3) glucan affects *Cryptococcus neoformans* from ultrastructure to virulence. *Mol. Microbiol.* **63:**1385–1398.
105. **Reiss, E., R. Cherniak, R. Eby, and L. Kaufman.** 1984. Enzyme immunoassay detection of IgM to galactoxylomannan of *Cryptococcus neoformans*. *Diagn. Immunol.* **2:**109–115.
106. **Rodrigues, M. L., M. Alvarez, F. L Fonseca, and A. Casadevall.** 2008. Binding of the wheat germ lectin to *Cryptococcus neoformans* suggests an association of chitinlike structures with yeast budding and capsular glucuronoxylomannan. *Eukaryot. Cell* **7:**602–609.
107. **Rodrigues, M. L., A. S. Dobroff, J. N. Couceiro, C. S. Alviano, R. Schauer, and L. R. Travassos.** 2002. Sialylglycoconjugates and sialyltransferase activity in the fungus *Cryptococcus neoformans*. *Glycoconj. J.* **19:**165–173.
108. **Rodrigues, M. L., F. C. G. dos Reis, R. Puccia, L. R. Travassos, and C. S. Alviano.** 2003. Cleavage of human fibronectin and other basement membrane-associated proteins by a *Cryptococcus neoformans* serine proteinase. *Microb. Pathog.* **34:**65–71.
109. **Rodrigues, M. L., L. Nimrichter, D. L. Oliveira, S. Frases, K. Miranda, O. Zaragoza, M. Alvarez, A. Nakouzi, M. Feldmesser, and A. Casadevall.** 2007. Vesicular polysaccharide export in *Cryptococcus neoformans* is a eukaryotic solution to the problem of fungal trans-cell wall transport. *Eukaryot. Cell* **6:**48–59.
110. **Rodrigues, M. L., S. Rozental, J. N. Couceiro, J. Angluster, C. S. Alviano, and L.R. Travassos.** 1997. Identification of N-acetylneuraminic acid and its 9-O-acetylated derivative on the cell surface of *Cryptococcus neoformans*: influence on fungal phagocytosis. *Infect. Immun.* **65:**4937–4942.
111. **Rodríguez-Peña, J. M., V. J. Cid, J. Arroyo, and C. Nombela.** 2000. A novel family of cell wall-related proteins regulated differently during the yeast life cycle. *Mol. Cell. Biol.* **20:**3245–3255.
112. **Salas, S. D., J. E. Bennett, K. J. Kwon-Chung, J. R. Perfect, and P. R. Williamson.** 1996. Effect of the lac-

case gene CNLAC1, on virulence of *Cryptococcus neoformans*. *J. Exp. Med.* **184:**377–386.

113. **Santangelo, R. T., M. H. Nouri-Sorkhabi, T. C. Sorrell, M. Cagney, S. C. Chen, P. W. Kuchel, and L. C. Wright.** 1999. Biochemical and functional characterisation of secreted phospholipase activities from *Cryptococcus neoformans* in their naturally occurring state. *J. Med. Microbiol.* **48**(8)**:**731–740.
114. **Santangelo, R., H. Zoellner, T. Sorrell, C. Wilson, C. Donald, J. Djordjevic, Y. Shounan, and L. Wright.** 2004. Role of extracellular phospholipases and mononuclear phagocytes in dissemination of cryptococcosis in a murine model. *Infect. Immun.* **72:**2229–2239.
115. **Santos, B., and M. Snyder.** 1997. Targeting of chitin synthase 3 to polarized growth sites in yeast requires Chs5p and Myo2p. *J. Cell Biol.* **136:**95–110.
116. **Shahinian, S., and H. Bussey.** 2000. beta-1,6-Glucan synthesis in *Saccharomyces cerevisiae*. *Mol. Microbiol.* **35:** 477–489.
117. **Shaw, J. A., P. C. Mol, B. Bowers, S. J. Silverman, M. H. Valdivieso, A. Durán, and E. Cabib.** 1991. The function of chitin synthases 2 and 3 in the *Saccharomyces cerevisiae* cell cycle. *J. Cell Biol.* **114:**111–123.
118. **Siafakas, A. R., T. C. Sorrell, L. C. Wright, C. Wilson, M. Larsen, R. Boadle, P. R. Williamson, and J. T. Djordjevic.** 2007. Cell wall-linked cryptococcal phospholipase B1 is a source of secreted enzyme and a determinant of cell wall integrity. *J. Biol. Chem.* **282:**37508–37514.
119. **Siafakas, A. R., L. C. Wright, T. C. Sorrell, and J. T. Djordjevic.** 2006. Lipid rafts in *Cryptococcus neoformans* concentrate the virulence determinants phospholipase B1 and Cu/Zn superoxide dismutase. *Eukaryot. Cell* **5:**488–498.
120. **Soares, R. M., C. S. Alviano, J. Angluster, and L. R. Travassos.** 1993. Identification of sialic acids on the cell surface of hyphae and yeast forms of the human pathogen *Paracoccidioides brasiliensis*. *FEMS Microbiol. Lett.* **108:** 31–34.
121. **Soares, R. M., R. M. de A. Soares, D. S. Alviano, J. Angluster, C. S. Alviano, and L. R. Travassos.** 2000. Identification of sialic acids on the cell surface of *Candida albicans*. *Biochim. Biophys. Acta.* **1474:**262–268.
122. **Soulié, M., A. Piffeteau, M. Choquer, M. Boccara, and A. Vidal-Cros.** 2003. Disruption of *Botrytis cinerea* class I chitin synthase gene Bcchs1 results in cell wall weakening and reduced virulence. *Fungal Genet. Biol.* **40:**38–46.
123. **Specht, C. A., Y. Liu, P. W. Robbins, C. E. Bulawa, N. Iartchouk, K. R. Winter, P. J. Riggle, J. C. Rhodes, C. L. Dodge, D. W. Culp, and P. T. Borgia.** 1996. The chsD and chsE genes of *Aspergillus nidulans* and their roles in chitin synthesis. *Fungal Genet. Biol.* **20:**153–167.
124. **Takeshita, N., A. Ohta, and H. Horiuchi.** 2005. CsmA, a class V chitin synthase with a myosin motor-like domain, is localized through direct interaction with the actin cytoskeleton in *Aspergillus nidulans*. *Mol. Biol. Cell.* **16:**1961–1970.
125. **Thompson, J. R., C. M. Douglas, W. Li, C. K. Jue, B. Pramanik, X. Yuan, T. H. Rude, D. L. Toffaletti, J. R. Perfect, and M. Kurtz.** 1999. A glucan synthase FKS1 homolog in *Cryptococcus neoformans* is single copy and encodes an essential function. *J. Bacteriol.* **181:** 444–453.
126. **Trilla, J. A., T. Cos, A. Duran, and C. Roncero.** 1997. Characterization of CHS4 (CAL2), a gene of *Saccharomyces cerevisiae* involved in chitin biosynthesis and allelic to SKT5 and CSD4. *Yeast* **13:**795–807.
127. **Turner, K. M., L. C. Wright, T. C. Sorrell, and J. T. Djordjevic.** 2006. N-linked glycosylation sites affect secretion of cryptococcal phospholipase B1, irrespective of glycosylphosphatidylinositol anchoring. *Biochim. Biophys. Acta* **1760:**1569–1579.
128. **Valdivieso, M. H., P. C. Mol, J. A. Shaw, E. Cabib, and A. Durán.** 1991. CAL1, a gene required for activity of chitin synthase 3 in *Saccharomyces cerevisiae*. *J. Cell Biol.* **114:**101–109.
129. **Vartivarian, S. E., G. H. Reyes, E. S. Jacobson, P. G. James, R. Cherniak, V. R. Mumaw, and M. J Tingler.** 1989. Localization of mannoprotein in *Cryptococcus neoformans*. *J. Bacteriol.* **171:**6850–6852.
130. **Vos, A., N. Dekker, B. Distel, J. A. M. Leunissen, and F. Hochstenbach.** 2007. Role of the synthase domain of Ags1p in cell wall alpha-glucan biosynthesis in fission yeast. *J. Biol. Chem.* **282:**18969–18979.
131. **Wasylnka, J. A., M. I. Simmer, and M. M. Moore.** 2001. Differences in sialic acid density in pathogenic and non-pathogenic *Aspergillus* species. *Microbiology* **147**(Pt 4)**:**869–877.
132. **Waterman, S. R., M. Hacham, J. Panepint, G. Hu, S. Shin, and P. R. Williamson.** 2007. Cell wall targeting of laccase of *Cryptococcus neoformans* during infection of mice. *Infect. Immun.* **75:**714–722.
133. **White, S. A., P. R. Farina, and I. Fulton.** 1979. Production and isolation of chitosan from *Mucor rouxii*. *Appl. Environ. Microbiol.* **38:**323–328.
134. **Williamson, P. R.** 1994. Biochemical and molecular characterization of the diphenol oxidase of *Cryptococcus neoformans*: identification as a laccase. *J. Bacteriol.* **176:**656–664.
135. **Yamochi, W., K. Tanaka, H. Nonaka, A. Maeda, T. Musha, and Y. Takai.** 1994. Growth site localization of Rho1 small GTP-binding protein and its involvement in bud formation in *Saccharomyces cerevisiae*. *J. Cell Biol.* **125:**1077–1093.
136. **Yoo Ji, J., Y. S. Lee, C. Song, and B. S. Kim.** 2004. Purification and characterization of a 43-kilodalton extracellular serine proteinase from *Cryptococcus neoformans*. *J. Clin. Microbiol.* **42:**722–726.
137. **Zaas, A. K.** 2008. Echinocandins: a wealth of choice—how clinically different are they? *Curr. Opin. Infect. Dis.* **21:**426–432.
138. **Zhong, J., S. Frases, H. Wang, A. Casadevall, and R. E. Stark.** 2008. Following fungal melanin biosynthesis with solid-state NMR: biopolymer molecular structures and possible connections to cell-wall polysaccharides. *Biochemistry* **47:**4701–4710.
139. **Zhu, X., J. Gibbons, J. Garcia-Rivera, A. Casadevall, and P. R. Williamson.** 2001. Laccase of *Cryptococcus neoformans* is a cell wall-associated virulence factor. *Infect. Immun.* **69:**5589–5596.

Cryptococcus: From Human Pathogen to Model Yeast
Edited by J. Heitman et al.

7

Sexual Reproduction of *Cryptococcus*

YEN-PING HSUEH, XIAORONG LIN, KYUNG J. KWON-CHUNG, AND JOSEPH HEITMAN

Cryptococcus neoformans and *Cryptococcus gattii* have a laboratory-defined sexual cycle that has been known for more than 3 decades (76, 77). Both species have a bipolar mating system orchestrated by an unusual mating type (*MAT*) locus, and mating involves fusion between cells of opposite mating type (**a** and α), resulting in a dimorphic transition from a budding yeast form to a dikaryotic filamentous hypha that ultimately produces basidia and basidiospores, a suspected infectious propagule (Fig. 1) (59). Yet the finding that the vast majority of clinical and environmental isolates are predominantly one of the two mating types (α) raised central questions about how sexual reproduction could occur, and diversity be maintained, in largely unisexual populations. Furthermore, the discovery that the α mating type can be linked to virulence has further focused interest on the molecular nature of the *MAT* locus and how, when, and where sexual reproduction might occur (84). The discovery that monokaryotic fruiting under laboratory conditions represents a novel type of sexual reproduction involving only one of the two mating types, most commonly α, revealed that same-sex or unisexual reproduction could profoundly influence the population structure (Fig. 1) (96). In fact, recent population genetics studies provide robust evidence that both **a**-α opposite-sex mating and α-α unisexual mating occur in nature in both *C. neoformans* and *C. gattii*, with the potential to influence the evolutionary trajectory and the production of infectious spores (17, 21, 22, 51). We review here the discovery of both opposite-sex and unisexual reproduction and illustrate how these pathways are molecularly controlled and the central cell biology questions that remain to be addressed. The structure, function, and evolution of the *MAT* locus are the purview of chapter 11 of this volume.

Yen-Ping Hsueh, Division of Biology, California Institute of Technology, Pasadena, CA 91125. **Joseph Heitman,** Department of Molecular Genetics and Microbiology, Duke University Medical Center, Durham, NC 27710. **Xiaorong Lin,** Department of Biology, Texas A & M University, College Station, TX 77843-3258. **Kyung J. Kwon-Chung,** Molecular Microbiology Section, Laboratory of Clinical Infectious Diseases, National Institute of Allergy and Infectious Diseases, National Institutes of Health, Bethesda, MD 20892.

DISCOVERY OF THE SEXUAL CYCLE OF *C. NEOFORMANS*

In 1966, Shadomy and Utz (137) reported hyphal formation in a strain of *C. neoformans* (Coward strain = NIH12) that was isolated from a patient with cryptococcal cellulitis and osteomyelitis who had sarcoidosis as an underlying condition. They initially hypothesized that the strain was a hyphae-forming mutant of *C. neoformans* since one of the diagnostic criteria at that time for the genus *Cryptococcus* was the absence of hyphae. Subsequently, Shadomy observed clamp connections, a characteristic of basidiomycetous fungi, in the hyphae formed by two strains of *C. neoformans*, including the NIH12 strain. It was not discernible whether the clamp connections shown in the micrograph were fused to the hyphal wall, and the details on the clamp connections were lacking in the article (136). On the basis of hyphae with clamp connections observed in the two strains and the fact that all basidiomycetous yeasts belonged to heterobasidiomycetes, Shadomy proposed that *C. neoformans* be classified in the genus *Leucosporidium* of the Ustilaginales (136). The genus *Leucosporidium* was established by Fell et al. in 1969 to accommodate teleomorphs of several white to cream-colored yeasts that had been classified in the genus *Candida* such as *C. scottii*, *C. gelida*, and *C. frigida* (39). All species of *Leucosporidium* form thick-walled teliospores that germinate to produce promycelia (39). The hyphae produced in the *C. neoformans* strains failed to develop any structures that could be determined as teliospores or promycelia. The only common feature between species of *Leucosporidium* and the hypha-forming *C. neoformans* strains, besides the presence of clamp connections, was that both fungi formed white to cream-colored mucoid colonies composed of starch-producing yeast cells.

Because the clamp connection is a hallmark of the Basidiomycota and only rare strains of *C. neoformans* produced hyphae on conventional agar media, Kwon-Chung hypothesized that *C. neoformans* may be a heterothallic basidiomycete that requires mating with sexually compatible strains to complete the sexual cycle. To determine whether sexually compatible mating type strains exist, the NIH12 strain (serotype D) was crossed with 20 randomly chosen strains of serotype D, mostly of environmental origin. An NIH12

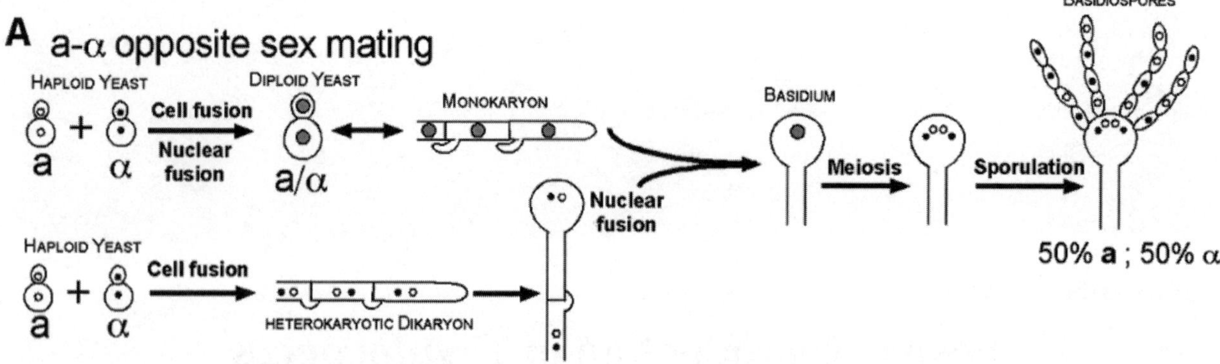

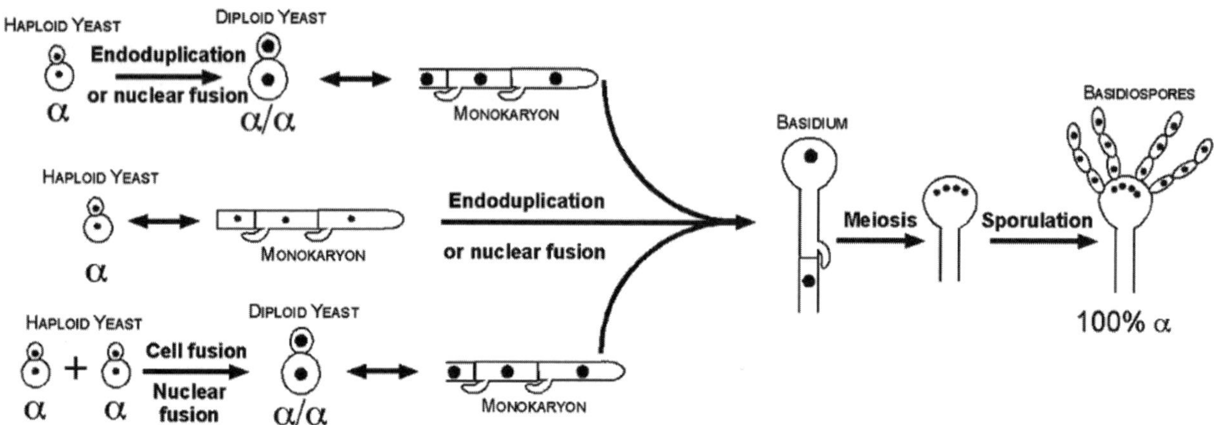

FIGURE 1 Sexual cycle of C. *neoformans*. (A) **a**-α opposite-sex mating. **a** and α haploid yeast cells secrete peptide pheromones that trigger **a**-α cell-cell fusion under nutrient-limiting conditions. After fusion, cells switch to filamentous growth, and the two nuclei congress but do not fuse in the resulting dikaryotic hyphae. Clamp cells are formed to ensure that two nuclei are faithfully segregated during hyphal growth. Blastospores (yeast-like cells) can bud from the hyphae and divide mitotically. Some hyphal cells can enlarge and form chlamydospores (94). In the basidium, two nuclei (**a** and α) fuse and undergo meiosis to produce four chains of basidiospores. In cases where cell fusion is immediately followed by nuclear fusion between **a** and α haploid yeast cells, heterozygous diploid **a**/α yeast cells are created. These diploid cells produce monokaryotic hyphae with unfused clamp cells. Eventually meiosis occurs in the basidium, and basidiospores with **a** and α mating types are generated. (B) α-α same-sex mating (fruiting). During same-sex mating, cells of one mating type become diploid either by endoduplication or by cell-cell and nuclear fusion. The diploid monokaryotic hyphae form rudimentary clamp connections that do not fuse to the preceding cell. Blastospores and chlamydospores are also produced during fruiting. In the basidium, meiosis occurs and haploid basidiospores are produced in four chains. In an alternative model, haploid α cells produce monokaryotic hyphae with haploid nuclei, and diploidization occurs in the basidium, followed by meiosis to produce spores with only the α mating type.

derivative that had been maintained by passage and no longer produced hyphae on malt extract agar was used in the search for mating partners. Of the 20 strains tested, NIH430 and NIH433, which were both recovered from Danish pigeon droppings, mated with NIH12 and produced a hitherto unknown teleomorph. The remaining 17 strains also successfully mated with NIH433 and NIH430, suggesting a significantly skewed ratio of opposite mating types. Based on the mating results with these 20 strains, *Filobasidiella neoformans* was described as the sexual state of C. *neoformans* (76).

The feature of the genus *Filobasidiella* that distinguishes it from other members of the Basidiomycota is formation of four long basipetal chains of small basidiospores (1.8–2.5 μm in diameter) on the apex of flask-shaped slender holobasidia (76, 78). The chains of basidiospores are produced by repetitious budding from four loci on the apex of basidia, and each basidiospore is a product of postmeiotic mitosis (59a, 78, 79).

Shortly before the discovery of F. *neoformans*, Olive established the family Filobasidiaceae to accommodate a newly discovered heterothallic fungus, *Filobasidium floriforme*, which produces a *Cryptococcus* anamorph (116). Unlike mating of *C. neoformans*, *F. floriforme* produced only one set of six to eight large basidiospores (6–9 by 10–16.5 μm) terminally on the apex of the basidia in a petal-like whirl, giving the basidium a flower-like appearance in an apical view. Although the manner of basidiospore formation in *C. neoformans* is distinct from species of the genus *Filobasidium*, *C. neoformans* was first considered as a member of the Filobasidiacea on the basis of their similarity in the morphology of holobasidia, septal ultrastructure, and their asexual properties (78, 85). The phylogenetic trees constructed on the basis of ribosomal DNA sequence, however, indicated that the genus *Filobasidiella* clusters with members of Tremellaceae rather than Filobasidiacea in the order Tremellales (see chapter 1).

The NIH12 strain was designated *MAT*α, while the strains NIH430 and NIH433 were designated *MAT***a**. Genetic analysis of the progeny obtained from a cross between B-3501 (*MAT*α) and B-3502 (*MAT***a**), the F_1 strains of NIH12 × NIH433, indicated that *C. neoformans* is a species with bipolar heterothallism (78, 79). In 1976, Erke (37) reported homothallism in strain NIH12 along with two other hyphae forming strains of *C. neoformans* based on their ability to form basidia and basidiospores without apparent mating. Basidiospore formation in these so-called homothallic strains was subsequently defined as "monokaryotic or haploid fruiting" (163) and later found to represent a novel form of unisexual reproduction (96).

DISCOVERY OF THE SEXUAL CYCLE OF *C. GATTII*

Soon after the discovery of *C. neoformans* sexual reproduction, it became clear that the typical *F. neoformans* teleomorph was formed in the cross between *MAT***a** and *MAT*α strains of either serotype A or D but not between strains of serotype A or D (*C. neoformans*) with strains of serotype B or C (*C. gattii*). Crossing sexually compatible clinical strains of *C. gattii*, serotype B and C, resulted in the discovery of the second teleomorph of *Filobasidiella*, *Filobasidiella bacillispora*, in 1976 (77). The environmental strains of *C. gattii* became available nearly 15 years after the discovery of *F. bacillispora* (36), and fertile environmental strains were also found to produce a typical *F. bacillispora* state upon crossing with *MAT***a** or *MAT*α tester strains of clinical origin (22, 43). Bacillary-shaped, smooth basidiospores produced by the strains of *C. gattii* are readily distinguishable from the finely rough globose to oval-shaped basidiospores of *C. neoformans* (78, 128). In certain crosses between strains of *C. neoformans* and *C. gattii*, viable basidiospores of mixed morphology were produced (82). However, no evidence of recombination was found among the viable progeny since they were either diploid/aneuploid or parental types of serotype D, supporting the evidence that the strains of serotype AD and BC are two different biological species (86). There have been reports on the isolation of serotype AB or DB hybrid strains (15, 16) from clinical or environmental sources. Interestingly, similar diploid hybrid strains can be readily generated in the laboratory by mixing cells of *C. gattii* with cells of a *C. neoformans* strain belonging to the opposite mating type (86).

SEXUAL REPRODUCTION IS ORCHESTRATED BY THE MATING-TYPE LOCUS (*MAT*)

Fungal sexual reproduction is genetically regulated by the mating-type locus (*MAT*), a specialized region of the genome that is idiomorphic or allelic between different sexes (mating types). The molecular structure of *MAT* was first determined in the budding yeast *Saccharomyces cerevisiae*. In *S. cerevisiae*, there are two different *MAT* alleles, **a** and α, encoding transcriptional activators and repressors that govern the expression of cell-type-specific genes (50). In α cells, the α domain transcription factor α1 is responsible for activating the expression of α-specific genes (αsg), and the homeodomain transcription α2 functions to repress the expression of **a**-specific genes (asg). In **a** cells, asgs are expressed independent of the homeodomain transcription factor **a**1, and so far, no known haploid-specific function has been ascribed to **a**1. Both **a** and α cells are capable of mating to form **a**/α diploids, the third cell type that undergoes meiosis under appropriate conditions (49). In diploid cells, the homeodomain proteins **a**1 and α2 form a heterodimer that functions to repress expression of haploid-specific genes and to promote expression of genes governing meiosis and sporulation (49).

In addition to species in the Ascomycete lineage, the *MAT* loci have been identified and characterized in the Basidiomycota and, more recently, in the Zygomycota, a basal group in the fungal kingdom (20, 42, 54, 62, 88). The gene content of *MAT* varies between species; however, a general theme is that transcription factors that have conserved homeodomain, α-domain, or High-Mobility Group (HMG) domains are encoded by the *MAT* locus and serve as master regulators to orchestrate sexual reproduction. As mentioned above, a more complex tetrapolar mating system has evolved in the phylum Basidiomycota; the corn smut fungus *Ustilago maydis* is one such example. *U. maydis* has two unlinked *MAT* loci, the *a* and the *b* loci. The *a* locus is biallelic (*a1* and *a2*) and encodes pheromones and pheromone receptors. The pheromone receptor encoded by the *a1* allele can only sense pheromone secreted from *a2* but not *a1* cells and vice versa and thus is only activated via a paracrine signaling loop. The *b* locus encodes two divergently transcribed homeodomain protein genes, *bE* and *bW*, and at least 19 alleles exist in nature (65). bE and bW of different alleles (for example, bE1 and bW2) form an active heterodimeric transcription factor complex, but proteins encoded by the same allele (for example, bE1 and bW1) do not dimerize. To initiate a mating response, cells must first differ at the *a* locus to enable pheromone signaling, which leads to mate recognition and cell-cell fusion. After cell fusion, an active heterodimeric complex composed of different allelic bE and bW is required to regulate the expression of a set of ~350 genes (65). If cells lack a functional bE/bW heterodimer, dikaryotic hypha development and meiosis fail to occur after cell-cell fusion (11, 66, 134). Thus, to be mating compatible, cells must differ at both the *a* and *b* loci.

C. neoformans has a bipolar mating system involving two mating types, **a** and α. The *MAT* structure of *C. neoformans* is much more complex compared to other well-studied fungal *MAT* loci such as those of *S. cerevisiae* and *U. maydis* (see chapter 11 for a detailed discussion of the evolution of *MAT* in *C. neoformans*). Among the ~20 genes encoded by the *MAT* locus, the homeodomain proteins Sxi1α and Sxi2**a** are key regulators that establish sexual identity of the cells (57, 58). The *MAT* locus was first identified to encompass an ~50-kb region in the genome (67); however, deleting this defined *MAT*α region in diploid cells did not eliminate **a**/α cell identity (58). This observation suggested that additional components outside the previously defined *MAT* locus were able to confer sexual identity and prompted the discovery of the homeodomain transcription factor Sxi1α (58). The corresponding homeodomain protein in the *MAT***a** allele was discovered later, and all lines of evidence support models in which the two homeodomain transcription factors are required and sufficient to establish **a**/α cell identity (57, 58). One key open question is determining the elements that function downstream and upstream of this homeodomain heterocomplex that lead to the heterokaryon development, meiosis, and sporulation after cell-cell fusion.

The targets of the Sxi1α/Sxi2**a** complex remain largely unknown. Employing a candidate gene approach, a recent study identified *CLP1* (clampless1) as a direct target of Sxi1α/Sxi2**a** (34). The *CLP1* gene was first identified from the model mushroom *Coprinopsis cinerea*; mutation of this gene caused a clampless phenotype in the self-compatible AmutBmut strain, and *CLP1* was shown to be required and sufficient to induce the A (homeodomain locus)–regulated pathway (63). The *CLP1* homolog in *U. maydis* was also shown to be a *b* (homeodomain locus)–regulated gene and required for proliferation of dikaryotic filaments in the host (132). In *C. neoformans*, disruption of *CLP1* showed that Clp1 is required for dikaryotic filament formation (34). In cell-cell fusion assays, the *clp1* mutants failed to produce colonies that could grow on selective medium. However, microscopic examination revealed that the *clp1* mutants were able to fuse with the mating partner, but the fusion products were then arrested without further hyphal development (34). A requirement for Clp1 in postfusion growth was also observed in *U. maydis*, suggesting a conserved function for Clp1 in the Basidiomycota (132). Gel-shift assays demonstrated that Sxi2**a** exhibits strong binding affinity to the *CLP1* promoter, implicating *CLP1* as a direct target of the Sxi1α/Sxi2**a** complex. The discovery of *CLP1* provides a foundation to assess the potential genome-wide targets of Sxi1α/Sxi2**a**.

ENVIRONMENTAL SIGNALS THAT TRIGGER SEXUAL REPRODUCTION OF *C. NEOFORMANS*

C. neoformans is predominantly haploid in nature, and sexual reproduction (mating and meiosis) occurs in response to nutritional limitation or specific signals. This is in stark contrast to *S. cerevisiae*, which is typically diploid in nature. In *S. cerevisiae*, mating occurs between haploid isolates in the presence of abundant nutrients, whereas meiosis is triggered by nitrogen limitation and the presence of acetate. Similar to the model fission yeast *Schizosaccharomyces pombe*, mating and meiosis of *C. neoformans* occur sequentially in response to limitation or specific nutrient cues. In addition, other environmental cues such as light and temperature also affect mating of *C. neoformans*. Those factors that have been connected to the sexual reproduction of *C. neoformans* are summarized in Fig. 2.

After the discovery of the *C. neoformans* sexual cycle, efforts were made to define the conditions that induce mating. One of the first important findings was the discovery that mating occurs efficiently on solid agar medium containing V8 juice, which enables strains to be crossed routinely in the laboratory (84). V8 is derived from the juice of eight vegetables and is readily available commercially. Nitrogen-limiting medium such as SLAD (super-low-ammonium dextrose) [50 μM $(NH_4)_2SO_4$] and filament agar (no added nitrogen source) also support mating but not as efficiently as V8 juice agar. Addition of 2% glucose, 38 μM $(NH_4)_2SO_4$, or both does not prevent mating on V8 medium, and thus, certain active compounds promoting mating are present in V8 medium. Earlier studies suggested that the active compounds are water-soluble, heat-stable, and have a molecular weight less than 10 kDa (59).

In addition to V8 juice, carrot and tomato juice also support mating. Furthermore, several *Cryptococcus* species can be isolated from trees and fermenting fruits, suggesting that there might be potential plant-fungus interactions that contribute to sexual reproduction of *C. neoformans* and *C. gattii* in nature. To explore this hypothesis, *Arabidopsis thaliana* and *Eucalyptus camaldulensis* seedlings were used as models to study potential fungus-plant interactions. In these models, Xue et al. discovered that mating can occur on the plant surface, and *myo*-inositol and the plant hormone indoleacetic acid (IAA) are two major plant-derived compounds that were found to stimulate mating (167). On Murashige and Skoog (MS) medium that is widely used in plant tissue culture, robust mating of *C. neoformans* was observed, generating abundant long chains of basidiospores on the edges of colonies. Even for the standard serotype A reference strains

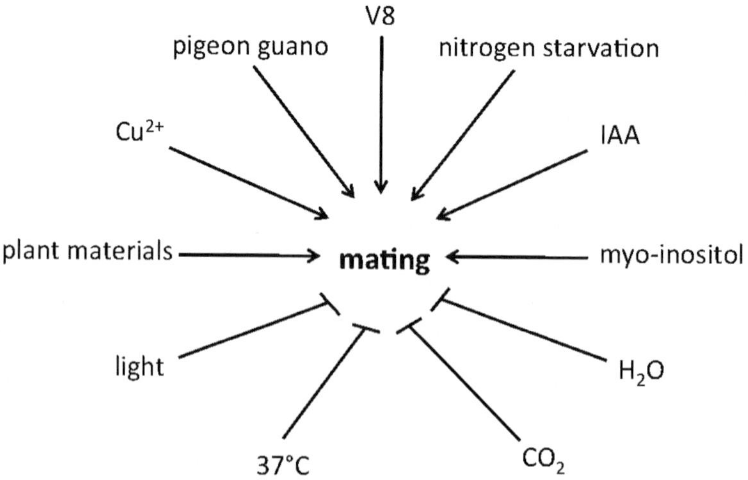

FIGURE 2 Factors that affect mating in *C. neoformans*. Plant materials, copper ions, pigeon guano, nitrogen starvation, V8 juice, IAA, and *myo*-inositol are compounds and conditions that promote sexual development of *C. neoformans*. Light, high temperature (37°C), high CO_2, and water, on the other hand, are known to inhibit mating.

H99 and KN99**a** that produce a limited amount of spores and filaments on V8 mating medium, coculturing the two strains on MS medium results in extremely abundant sporulation (167). MS is a defined medium, and by omission and readdition studies, *myo*-inositol (100 mg/liter) was determined to be the compound in MS medium that induces mating of *C. neoformans*. A previous study reported that V8 juice contains 16.675 mg/liter *myo*-inositol, and MS medium containing this level of inositol supports robust mating. Therefore, *myo*-inositol is a factor in V8 juice that stimulates mating of *Cryptococcus* (27, 167). In addition to *myo*-inositol, IAA was also identified to be another plant-derived compound that enhances mating at a low concentration (<100 μM) and acts synergistically with *myo*-inositol in vitro. This is interesting because it has been shown previously that IAA induces invasive growth in *S. cerevisiae* (120). These lines of evidence all suggest that fungus-plant interactions in nature might play an important role in the sexual cycles of *C. neoformans* and *C. gattii*.

To further identify the mating-inducing factor in V8 mating medium, Kent et al. carried out complete composition analysis of V8 juice by a fractionation approach (68). Interestingly, they found that V8 juice medium contains ~0.5 mM inorganic nitrogen and thus is not limiting for nitrogen, in accord with previous studies. Furthermore, although there is no single factor that accounts for the mating-inducing capacity of the medium, copper ions (Cu^{2+}) were found to play a key role in inducing the expression of pheromone genes and to support mating (68). This is congruent with the previous finding that mutations in genes regulating copper homeostasis (i.e., *CCC2*) caused mating defects, and copper restored mating in some melanin-deficient sterile mutants (150, 156). Moreover, copper has previously been shown to enhance filamentation during fruiting, and the copper-regulated transcription factor Mac1 is one of several quantitative trait loci (QTLs) that were shown to have a significant impact on hyphal growth (95). Copper is also used to assist cell fusion in mating between isolates of the same mating type.

As mentioned above, mating also occurs under nutrient-limiting conditions such as nitrogen starvation. Nitrogen-limiting medium with 50 μm $(NH_4)_2SO_4$ (SLAD), or no added nitrogen source (filament agar), all support sexual development of *C. neoformans*. In addition, low-nitrogen conditions also trigger invasive growth of wild-type haploid cells (130). The ability to sense nitrogen is mediated by Amt2, a high-affinity ammonium transporter. Mutant cells that lack this sensor are unable to undergo the sexual cycle under low-ammonium conditions while retaining the ability to mate on MS medium that is not limited for nitrogen sources (130).

Apart from nutritional cues, other environmental signals also influence sexual reproduction. Among these, temperature is a key factor. High temperature inhibits the dimorphic transition from yeast to hypha, and thus, mating and fruiting do not occur at 37°C. Even in strains that are hyperfilamentous (i.e., *gpa3* mutant, XL280, Cpr2 overexpression strains), high temperature effectively blocks the dimorphic transition. Diploid **a**/α strains are also thermally dimorphic (142). This is strikingly similar to the thermo-dimorphism observed in the dimorphic fungal pathogens such as *Histoplasma capsulatum* and *Coccidioides* spp. that grow as yeast at 37°C but switch to a filamentous state at 25°C. In *H. capsulatum*, three regulatory elements, Ryp1, 2, and 3, have been shown to regulate the dimorphic transition at 37°C, but the pathways upstream of these regulators remain to be defined (109, 161). How *C. neoformans* senses temperature is not well understood. Signaling components including calcineurin and Ras1 have been associated with the ability to grow at high temperatures (2, 74, 114), but how these pathways contribute to temperature-dependent morphogenesis remains to be elicidated (29). Interestingly, a recent study of *C. albicans* demonstrated that the chaperone Hsp90 governs yeast to hypha development via repressing the Ras-protein kinase A (PKA) pathway in a temperature-dependent fashion (138).

Light is a critical environmental cue that a majority of organisms can sense and respond to. In the fungal kingdom, light regulates several physiological and developmental processes including circadian rhythm, pigmentation, and sexual and asexual reproduction, whereas some species are blind. The mechanisms of light sensing are best understood in the model fungus *Neurospora crassa*, in which the white collar proteins (WC-1 and WC-2) form a central complex that regulates light responses (99). The light response of *C. neoformans* was revealed when investigators found that wrapping mating plates in foil enhances mating, while exposing mating plates under constant white light inhibits mating. At the molecular level, the conserved white collar genes *BWC1* and *BWC2* were identified to encode the photoreceptors that regulate light responses (61, 100, 168). Interestingly, the *bwc1* and *bwc2* mutants exhibit attenuated virulence, suggesting that in addition to mating, light-sensing might also be critical for the virulence of *C. neoformans* (10, 61). Also of note, the inhibition of mating by light can be overcome by coculture with plants (167), indicating that plant-derived signals can override Bwc1/Bwc2-mediated inhibition.

Why should mating be inhibited by light? The Bwc1/Bwc2 proteins contain an LOV (light, oxygen, or voltage) domain that is known to bind a flavin chromophore and thereby serve to sense blue light. Sunlight also contains abundant UV irradiation; the flavin chromophore can also absorb in the UV range, and *bwc1* and *bwc2* mutants are sensitive to UV light (61). We hypothesize that the genome may be more vulnerable to UV damage during the meiotic phase of sexual reproduction, and limiting mating to darkness may avoid this genomic threat. As discussed above, high temperature is a key factor that blocks mating and might also be correlated to light, because temperature is higher during the daytime. It will be interesting to investigate whether the temperature- and light-sensing pathways utilize overlapping components.

Water and CO_2 are two other ubiquitous signals that also affect mating of *C. neoformans*. Mating of *C. neoformans* occurs on solid surfaces and has never been observed in liquid cultures. Standard V8 mating medium contains 4% agar (regular medium contains 2% agar), and it is observed that mating is more robust under more desiccated conditions (no condensation on plate lids). How water is sensed is not known, but it does not appear to involve the aquaporin homolog Aqp1 (X. Lin and J. Heitman, unpublished results). CO_2 plays multiple roles in mating. At a high CO_2 concentration (5%), cell-cell fusion during the early stage of mating is decreased to <1% when compared to that under normal ambient conditions (~0.036%) (6). This is likely a result of inhibition of pheromone gene induction early in the mating process under high CO_2 conditions (6). Interestingly, CO_2 is required for basidiospore formation, which is a downstream event of the sexual development cascade. Inside fungal cells, CO_2 is hydrated to bicarbonate via the carbonic anhydrase Can2; abnormal basidia and very few basidiospores were observed in the bilateral cross of the *can2* mutant, suggesting that a proper level of bicarbonate is important for the final stage of sexual development (6). It was also demonstrated that adenylyl cyclase is one of the downstream targets of bicarbonate, which also links CO_2 sensing to the cyclic AMP (cAMP) signaling pathway (72, 105). These results illustrate that dissolved gases also affect mating in *C. neoformans* (10).

C. neoformans is commonly isolated from avian excreta, and early studies showed that medium containing pigeon guano supports growth and sexual development of *C. neoformans* serotype D isolates (144, 145). In a more recent study, Nielsen et al. further characterized the growth and development of both *C. neoformans* var. *grubii* and *C. gattii* on pigeon guano medium. Interestingly, they found that although both *C. neoformans* (including var. *grubii* and var. *neoformans*) and *C. gattii* grow on pigeon guano medium, only *C. neoformans* mated robustly when cells of both mating types were cocultured (113). More filaments, basidia, and basidiospores were observed in crosses cultured on pigeon guano medium when compared to that cultured on standard V8 mating medium. On the contrary, pigeon guano medium does not support mating of *C. gattii*, which may reflect the fact that *C. gattii* is typically isolated from trees but not from avian excreta (113). These findings demonstrate that pigeon guano is a realized ecological niche for *C. neoformans* (113), and *C. gattii* might respond to different environmental signals to complete the sexual cycle in nature.

SIGNALING PATHWAYS THAT REGULATE SEXUAL DEVELOPMENT IN *C. NEOFORMANS*

Conserved signaling pathways are known to regulate development and physiological responses in fungi (89). In *C. neoformans*, several of these conserved pathways have been shown to regulate sexual reproduction (Fig. 3). Among these, the pheromone-activated mitogen-activated protein kinase (MAPK) signaling cascade and the nutrient-sensing cAMP pathway are among the best-understood signaling pathways in eukaryotes.

The pheromone-signaling pathway is conserved in both the Ascomycota and Basidiomycota. For example, homologs of the most upstream component in this pathway, the pheromone receptors, have been identified and characterized in many species (5, 13, 19, 23, 26, 45, 53, 64, 69, 75, 107, 118, 119, 127, 135, 166, 169). In *C. neoformans*, the *MAT*-encoded pheromone receptor genes *STE3***a** (*CPR***a**) and *STE3α* (*CPRα*) are both required for pheromone sensing and sexual development (23, 26), and expression of phero-

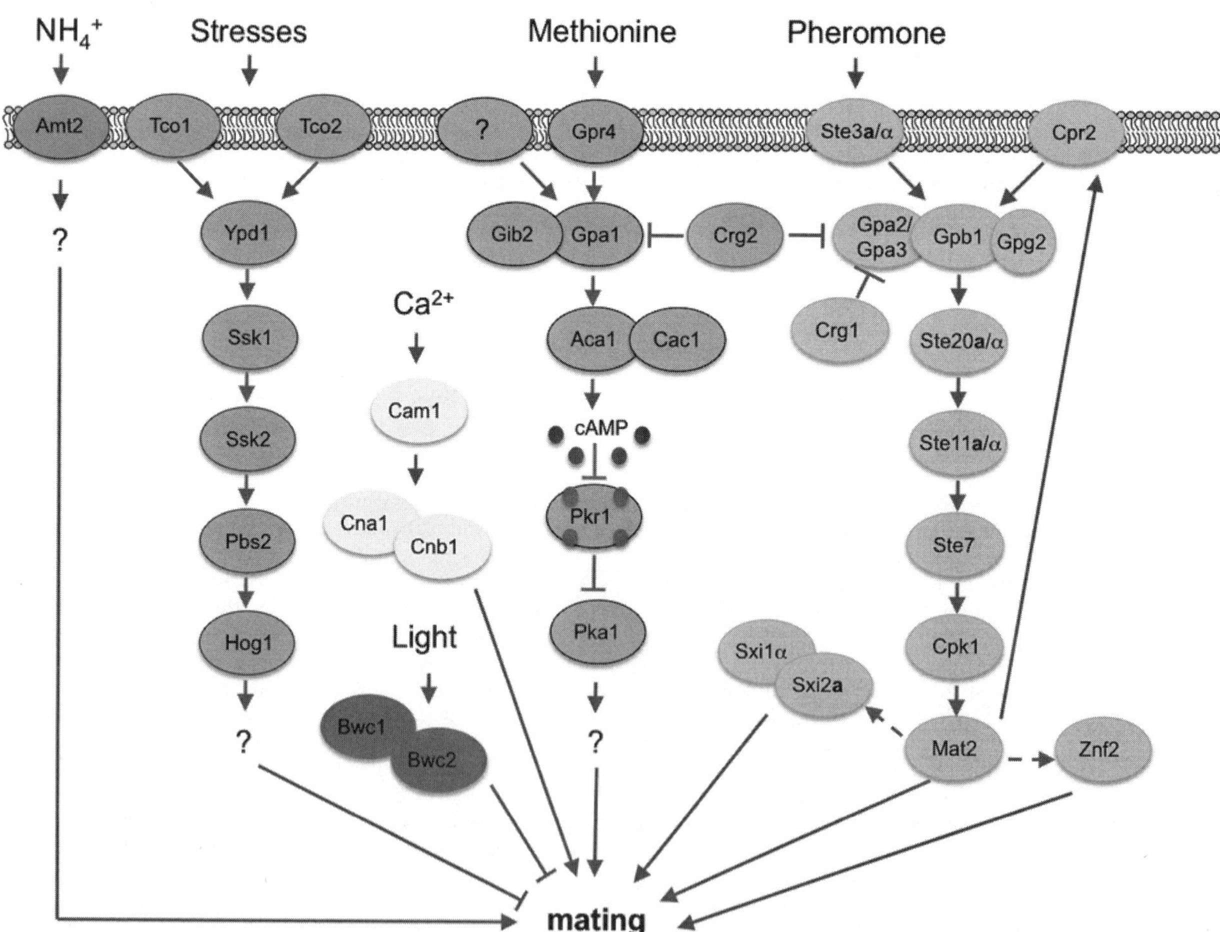

FIGURE 3 Signaling cascades that contribute to sexual development of *C. neoformans*. The pheromone response pathway, cAMP-PKA pathway, Ca^{2+}-calcineurin pathway, and the ammonium sensor Amt2 all positively regulate sexual development, while the light- and stress-sensing pathways negatively regulate this process (see text for details).

mones and pheromone receptors specifies the α- and **a**-haploid cell types (145a). The expression of the pheromone receptors is induced during mating on V8 mating medium and is barely detectable under nutrient-rich conditions such as YPD (yeast extract, peptone, dextrose) medium. Dramatic induction was observed at early mating time points (4 to 10 h), whereas later expression was reduced to the basal level (23). Mutants that lack the pheromone receptors exhibit severe mating defects and are impaired for pheromone sensing (26). Interestingly, the *ste3***a** mutant has been reported to be less virulent in a murine virulence model, suggesting that it might play additional roles in regulating virulence inside the host (23).

The pheromone receptors are coupled to heterotrimeric G proteins that transduce signals into the cell upon receptor activation. Three G protein α subunits, Gpa1, Gpa2, and Gpa3, are encoded in the *C. neoformans* genome. Gpa1 was first characterized to play a role in the nutrient-sensing cAMP pathway, while the functions of the two other Gα subunits have been recently elucidated (3, 4, 33, 55, 92). On the other hand, one Gβ subunit (Gpb1) and a Gβ-like subunit (Gib2) are encoded by the genome, and Gpb1 is required for pheromone responsiveness and mating (117, 159). The *gpb1* mutant exhibits a strong unilateral mating defect and is insensitive to pheromone (159). Thus, analogous to the function of the Ste4/Ste18 (Gβγ) complex in *S. cerevisiae*, Gpb1 plays a positive role in the pheromone signaling pathway. More recently, two putative Gγ subunits, Gpg1 and Gpg2, were identified (117). Biochemical analyses demonstrated that the Gγ subunits interact with the Gβ subunit Gpb1, and deletion of *GPG2* caused a unilateral sterile phenotype that resembled the *gpb1* mutant, suggesting that the Gpb1-Gpg2 heterodimer has a more prominent role in activating the pheromone-signaling pathway (55, 92, 117).

Gpa2 and Gpa3 were found to function as the Gα subunits that participate in the pheromone-signaling pathway. Genetic analyses showed that neither the *gpa2* nor the *gpa3* single mutant was sterile, even in bilateral crosses, suggesting that the mutants could still sense pheromone. These observations led to the hypothesis that Gpa2 and Gpa3 might have a redundant role in the pheromone-signaling pathway. In accord with this model, *gpa2 gpa3* double mutants are sterile in bilateral crosses, and cell-cell fusion assays demonstrated that in the *gpa2 gpa3* double mutant bilateral crosses, although the pheromone genes were highly expressed, cells failed to fuse with the mating partner (55, 92). Northern blot analyses showed that the two Gα subunits are differentially expressed; *GPA2* is induced during mating, while *GPA3* is constantly expressed on V8 medium even in the absence of cells of the opposite mating type (55). *gpa3* mutants are constitutively active for pheromone signaling as a result of activation of the liberated Gβγ complex. The pheromone genes are highly expressed in both the serotype A and D *gpa3* mutant monocultures, and the serotype D *gpa3* mutants are hyper self-filamentous (55).

Interestingly, in addition to sequestering the Gβγ complex, studies have shown that the GTP-bound Gpa2 and Gpa3 could exert additional signaling activities. Overexpression of the dominant active $GPA2^{Q203L}$ allele increased pheromone signaling, while overexpression of the dominant active $GPA3^{Q206L}$ allele decreased pheromone signaling (55). These lines of evidence support the idea that both G proteins have been recruited into pheromone signaling and share both overlapping and divergent roles in the pheromone response pathway (55, 92). Gpa3 functions to suppress filamentous growth in the absence of a mating partner, while Gpa2 promotes mating. Compared to the *S. cerevisiae* pheromone signaling paradigm, in which only one G protein α subunit (Gpa1) is involved, the addition of extra Gα subunits increases signaling complexity. Similar regulatory circuits could also function in other fungal species.

Other critical components in the pathway are the regulator of G-protein signaling (RGS) proteins that stimulate GTP hydrolysis on the Gα subunit in the active state (GTP-bound form) and function as negative regulators to enable cell desensitization to pheromone activation (32, 149). Three proteins containing the RGS domain have been identified in *C. neoformans* (Crg1, Crg2, and Crg3), and evidence has shown that Crg1 and Crg2 play a role in the pheromone signaling pathway (139, 157, 165). Deletion of *CRG1* dramatically enhances pheromone responsiveness; serotype A *crg1* mutants produce abundant conjugation tubes in confrontation assays (**a** *crg1* confronted with α *crg1*), which is not observed in wild-type cells (112, 157). In addition, mating is more robust in *crg1* mutants in both *C. neoformans* and *C. gattii* backgrounds, and therefore, the *crg1* mutation can be introduced into less fertile isolates to enable sexual crosses (43, 112). Mutation of the second RGS gene *CRG2* increases pheromone gene expression, and the mutant cells produced longer hyphae during mating (55). The expression of *CRG1* and *CRG2* is also differentially regulated; *CRG1* is induced during mating, by nutrient deprivation and pheromone, while *CRG2* is constitutively expressed. Interestingly, parallel studies showed that Crg2 also negatively regulates the Gpa1-cAMP signaling pathway, establishing Crg2 as a multiregulatory RGS (139, 165).

A MAPK cascade functions downstream of the pheromone receptors and the G proteins and exhibits both canonical and novel features in that several elements are encoded by the *MAT* locus as divergent alleles (30). Upon pheromone exposure, the liberated Gβγ subunit activates Ste20**a**/α, a signal-transducing kinase of the p21-activated kinase family that triggers the three-tiered phosphorelay on Ste11**a**/α (MAPKKK), Ste7 (MAPKK), and Cpk1 (MAPK), leading to the activation of the pheromone response (30, 110, 158). The MAPK signaling cascade is conserved with that in *S. cerevisiae*, but a key distinction is that Ste20**a**/α and Ste11**a**/α are encoded by the *MAT* locus and are cell-type specific (30, 59, 89). Deletion of Ste20**a**/α, Ste11**a**/α, Ste7, or Cpk1 causes a prominent mating defect, indicating that these components are required for sexual development (30, 110, 158). Cell-cell fusion during early mating is blocked in the *ste11***a**/α, *ste7*, and *cpk1* mutants, but not in the *ste20***a**/α mutants, suggesting that the mating defects of the *ste20***a**/α mutants are due to a defect in postfusion hyphal growth (30, 110). The MAPK signaling cascade is also required for monokaryotic fruiting, suggesting that both mating and fruiting may be regulated by the same signaling pathway (30). Although virulence has been linked with mating type, and the pheromoneless and pheromone receptor mutants are moderately attenuated for virulence, the Cpk1 signaling cascade is dispensable for virulence (23, 30, 84, 140). These lines of evidence demonstrate that the Cpk1 MAPK signaling cascade is mainly dedicated to regulate sexual morphogenesis of *C. neoformans*. Because the *MAT* locus is one of several QTLs that contribute to promote hyphal elongation (95), it will be interesting to further dissect whether the mating-type-specific *STE20***a**/α, *STE11***a**/α, or *STE12***a**/α alleles are among the major contributors to this activity of the *MAT* locus.

After the core components of the pheromone-signaling pathway were identified and characterized, it was a surprise to discover another *STE3* homolog unlinked to the *MAT* locus (Y. C. Chang and K. J. Kwon-Chung, ASM General meeting 2001, abstr. F-55). This gene, *CPR2* (*Cryptococcus* pheromone receptor 2), was found to be induced modestly when cells were exposed to pheromones of the opposite mating type and massively induced post–cell-cell fusion during mating (56). Deletion of *CPR2* decreased cell-cell fusion efficiency, and the fusion products of the *cpr2* mutants were heterogeneous in size. Although *cpr2* mutants are able to produce mating hyphae and basidiospores comparable to wild type, microscopic examination showed that a significant proportion of the mating hyphae were abnormal. Multiple budding at the clamp cells and thinner, irregular hyphae that share features with the "haustorial hyphae" described in the classic literature (80) were observed, suggesting that Cpr2 has a role in maintaining the dikaryon. Interestingly, overexpression of Cpr2 dramatically enhances fruiting; this self-filamentous phenotype is even more prominent when Cpr2 is overexpressed in the pheromone receptor mutant *ste3*α background (56). When heterologously expressed in *S. cerevisiae*, Cpr2 constitutively activates the pheromone-signaling pathway in the absence of any apparent ligand (56). When overexpressed in *C. neoformans*, Cpr2 also induces pheromone responses such as the induction of pheromone gene expression, and cells become self-filamentous. These activities require a leucine residue in transmembrane domain 6, which substitutes a proline residue that is conserved in 90% of G protein-coupled receptors (GPCRs) (56). Substitution of the proline in transmembrane domain 6 has been shown to result in constitutive activity in fungal pheromone receptors (73, 87, 146). Epistasis analysis showed that the G protein β subunit Gpb1, MAPKK Ste7, and the downstream transcription factor Mat2 are all required for Cpr2-elicited signaling (56), supporting models in which Cpr2 activates the same signaling pathway as Ste3α. Physical interaction analyses also showed that Cpr2 and Ste3α interact with the same G protein α subunits (56). Therefore, when both the Cpr2 and Ste3 receptors are expressed, they compete for G protein binding and signaling.

Each clamp cell represents a cell-cell fusion event that involves pheromone signaling, and the unusual clamp structures and mating hyphae observed in *cpr2* mutant crosses suggest a role for Cpr2 in this process. One possible model is that Cpr2 may function to shield the clamp cells from pheromone secreted by nearby yeasts or hyphae to ensure fusion fidelity. The morphology of the unusual hyphae produced by *cpr2* mutants resembles the features of haustorial hyphae that have been described and reported in many basidiomycetes, including *C. neoformans* (80). The function of this type of hypha is unknown in *C. neoformans*, but in other mycoparasitic basidiomycetes, they function to invade and parasitize another fungal species (12). It is also plausible that pheromones or other molecules secreted by other species can function as inverse agonists for Cpr2. Mutants with altered clamp structures have also been observed in other basidiomycetes such as *C. cinerea* and *Schizophyllum commune*. For example, expressing the constitutively active Cdc42 or deleting the Ras GTPase-activating protein Gap1, which results in accumulation of Ras-GTP, both lead to formation of defective clamp cells (133, 160).

Recent studies of *S. cerevisiae* revealed additional roles for the pheromone receptors in morphogenesis. For example, interactions between the activated Ste2 and Ste3 have been shown to play a role in later cell-fusion stages of mating (141). An intact receptor-heterotrimeric G protein module is essential for pheromone-induced chemotropism (147). In addition, when the **a**-factor receptor Ste3 is inappropriately expressed in **a** cells, pheromone signaling is inhibited, and cells become resistant to pheromone treatment, a process termed receptor inhibition (28, 52). This inhibition is mediated by suppressing Gβγ signaling and requires the **a**-specific gene *ASG7* (70, 71, 129). *asg7* mutants exhibit wild-type mating efficiency; however, the zygotes formed after cell-cell fusion exhibit abnormal morphology, which resembles shmoos produced by haploid cells responding to pheromones (129). This suggests that in diploid cells, Ste3 and Asg7 may function coordinately to inhibit pheromone signaling in postmating zygotes and to promote the transition to vegetative growth (129). These lines of evidence all indicate that the pheromone receptors play multiple roles during sexual reproduction, and the involvement of additional pheromone receptors in development might be a common theme in many basidiomycetes that express mating-type-independent pheromone receptors (1, 64, 102, 111, 127).

The well-studied nutrient-sensing cAMP-PKA signaling pathway is known to regulate not only mating, but also the production of melanin and capsule, two critical virulence factors (59, 60, 89). Mutations in the major components of this pathway, including the upstream Gα subunit Gpa1, a cognate receptor Gpr4, adenylyl cyclase Cac1, cAMP-dependent PKA (Pka1/2), and the adenylyl cyclase-associated protein Aca1 all cause prominent mating defects (3, 4, 7, 33, 164). Quantitative cell-cell fusion assays showed that the fusion efficiency in many mutants was decreased (7, 164). Surprisingly, the Gpa1-interacting GPCR Gpr4 does not sense glucose. Instead, it senses the amino acid methionine, which when added to medium, stimulates filamentation (164). Furthermore, a recent two-hybrid screen identified one Gβ-like protein Gib2 as an atypical Gβ in the Gpa1-mediated cAMP pathway, and Gib2 serves as a multifunctional protein that is required for cell growth in serotype D strains (117).

The fact that both the pheromone-signaling and cAMP pathways orchestrate sexual development suggests the existence of coincidence signaling between the two pathways. The recent observation that at least one component, the RGS protein Crg2, is involved in the regulation of both signaling pathways further supports this hypothesis (139, 165). Expression of genes regulated by the MAPK cascade, such as *MF*α, are also coordinately regulated by nutrient sensing via the Gpa1-cAMP pathway.

Several other signaling cascades have also been shown to play a role in mating; however, further investigation is needed to understand how these pathways regulate sexual development at the molecular level. For example, the stress-activated Hog1 MAPK was shown to antagonize both pheromone production and sexual development. Deletion of the *HOG1* gene enhanced filamentation and pheromone production in the serotype A background (8). Furthermore, mating and confrontation assays demonstrated that the upstream response regulators Ssk1 and Skn7 also negatively regulate sexual development (9). In contrast, it was surprising to find that one of the upstream two-component histidine kinase sensors, Tco1, is required for hyphal growth in mating, although it remains unclear whether this phenotype is directly linked to the regulation of the Hog1 MAPK pathway (9). The Ca^{2+}-calcineurin pathway that is essential for growth at 37°C also has a role in mating. Cells lacking the catalytic (*CNA1*) or the regulatory subunit (*CNB1*) of calcineurin are impaired in mating (29). The calcineurin mutants were able to undergo cell-cell fusion during mating, but the postfusion cells were unable to initiate hyphal growth, indicating that calcineurin has a critical role in hyphal growth of the dikaryon (29). This phenotype is reminiscent of *clp1* mutant cells and illustrates a key difference from yeast in that specialized components are necessary to ensure survival and maintenance of the filamentous dikaryons.

UNISEXUAL MATING IN *CRYPTOCOCCUS*

As discussed earlier, the canonical mating process of this heterothallic basidiomycetous fungus involves haploid cells of α and **a** mating type (76) and produces an equal proportion of **a** and α basidiospores in four chains (Fig. 1) (59, 76, 78). However, the predominance of the α mating type (>99%) in the *Cryptococcus* population represents a paradox as to how sexual reproduction might occur in this essentially unisexual population (81). Before traditional **a**-α mating was discovered, hyphal formation was documented in some *MAT*α strains (e.g., the Coward or NIH12 strain) in the absence of an **a** mating partner. This was the earliest evidence that haploid *C. neoformans* strains could undergo a transition from yeast to filamentous growth and sporulation through a process later called monokaryotic fruiting (haploid fruiting) (37, 38, 96, 137, 151, 163). Because only cells of a single mating type are involved (commonly α) and only spores of that particular mating type are produced during this process, monokaryotic fruiting was considered to be strictly haploid, mitotic, and asexual.

However, it was puzzling why this fungus would produce elaborate fruiting structures similar to those produced during sexual **a**-α matings simply as another means of mitotic propagation. Yeast growth by budding likely represents a more energy-efficient mechanism for mitotic propagation. In addition, *Cryptococcus* yeast cells are very resilient to dehydration and other stresses and are suitable as both infectious propagules and long-term survival structures in the natural environment (18). On the other hand, filamentous growth and sporulation could yield survival benefits in response to certain conditions.

When ploidy doubling and reduction was observed during the fruiting process, it led to the hypothesis that fruiting is a sexual process instead of a mitotic event (96). The hypothesis is also consistent with observations that both the morphological differentiation and the environmental cues required for the monokaryotic fruiting process resemble those for **a**-α mating. For example, the unfused monokaryotic hyphae produced during fruiting are remarkably similar to the hyphae produced by **a**/α diploid cells during self-filamentation. High levels of recombination are observed in the progeny produced by monokaryotic fruiting, on par with that observed among progeny produced by **a**-α mating, further supporting the hypothesis that fruiting is a modified form of sexual reproduction. Furthermore, highly conserved key meiotic regulators such as Dmc1 (a recombinase) and Spo11 (a key enzyme that initiates meiotic recombination by introducing double strand breaks) are both critical for generation and viability of spores produced by fruiting (96). The *spo11* and *dmc1* mutants exhibit altered sporulation patterns. Two short chains of spores, instead of four, were often observed on the basidia of these mutants. In addition, the basidiospore germination frequency decreased from ~27% in wild type to 1.5% in the *dmc1* mutant.

All lines of evidence argue that monokaryotic fruiting represents a modified form of sexual reproduction that occurs between strains of the same mating type. During the fruiting process, the nuclear content doubles (to diploid α/α) either through cell fusion between two haploid α cells or endoduplication. Subsequent nuclear reduction occurs through meiosis during sporulation. It appears that there are several stages where diploidization could occur, including before initiation of filamentation, during filamentation, or in the basidia immediately preceding sporulation, and the timing of duplication appears to be dependent on the strain background (96).

The earlier observations that fruiting is only observed in α isolates and a later description of fruiting in some **a** isolates has raised questions regarding the role of mating type α in dimorphic differentiation (151, 163). Recently, fruiting ability was reinvestigated as a quantitative trait instead of a simple Mendelian trait (95). By applying QTL mapping to an inbred population, it was demonstrated that the *MAT*α allele enhances hyphal growth during monokaryotic fruiting as a QTL. *MAT* was one of five QTLs identified in a scan of 25% of the genome and was found to be the most significant QTL influencing hyphal growth, and the α allele promoted fruiting to a greater extent than **a**. Thus, the *MAT* locus works in concert with other genomic loci to govern fruiting. Moreover, *MAT* genes are therefore subject to different selection forces during α-α and **a**-α mating, which may influence their evolutionary trajectory. These observations evoke an explanation for the phenomenon that monokaryotic fruiting is most commonly observed in α isolates, and this has resolved the conflicting conclusions regarding the role of mating type α in dimorphic yeast to hyphal growth (151, 163).

The discovery of same-sex mating in *C. neoformans* resulted in a paradigm shift in considering how genetic diversity is generated in a unisexual population and the evolutionary role of a unisexual reproductive mode. This concept was applied to the interpretation of the origin of the cryptococcosis outbreak on Vancouver Island: mating between two different α strains may have given rise to more virulent strains occupying a new environmental niche and causing the Vancouver outbreak (41). Several population genetic studies have revealed sexual recombination in natural populations of both *C. neoformans* serotype A and *C. gattii* where only the α mating type exists (17, 41, 131). For example, studies of 31 serotype A isolates that infected dogs or cats revealed genetic recombination among these α isolates, 4 of which were diploid and exhibited a self-filamentous phenotype (17). These lines of evidence suggest that same-sex mating is likely to occur in nature. The discovery of αADα hybrids that arose by fusion between two α cells of serotype A and D, and the recent finding that natural unisexual mating generates diploid *Cryptococcus* isolates including intravarietal allodiploid hybrids produced by fusion of two genetically distinct α cells, indicate that same-sex mating indeed occurs in nature (97, 98). These studies indicate the significant impact of this unisexual reproduction mode on the current population structure of all of the major lineages in the *C. neoformans*/*C. gattii* species complex.

Multiple pathways have been shown to be involved in unisexual mating. The pheromone-sensing pathway is certainly important, as deletion of the components including Ste20α (a p21-activated kinase), mating pheromone MFα, pheromone receptor Ste3α, the downstream MAPK pathway (Ste11α, Ste7, Cpk1), and Mat2 (a transcription factor) lead to filamentation defects during both traditional **a**-α mating and α-α fruiting (40, 96a, 110, 140, 162). In contrast, activation of the pheromone response pathway in the absence of the opposite mating-type partner enhances fruiting. For example, overexpression of *CPR2*, a constitutively active GPCR that also activates pheromone signaling pathway, or the pheromone genes, results in much more robust α-α unisexual mating compared to wild-type cells (56, 140). The role of Ste12 in same-sex mating is less clear, as it appears to be dependent on the genetic background (24, 25, 126, 162, 170, 171). The cell identity protein complex Sxi1α and Sxi2**a** can also regulate the same-sex mating process. For example, transgenic **a** strains with the α cell identity gene *SXI1α*, or transgenic α strains with the **a** cell identity gene *SXI2***a**, mimic the diploid or dikaryotic state and enable more robust same-sex mating in response to temperature and environmental cues (57, 58). However, deletion of the *SXI1α* gene does not impair same-sex mating of the α isolates tested, indicating that *SXI1α* is not an essential component of the regulatory elements for this process, in contrast to its essential role in traditional **a**-α matings. In addition to all of the genetic components discussed above, ploidy appears to influence same-sex mating, and diploid isolates are in general enhanced in fruiting filamentation. Genes that are involved in the general processes of morphogenesis (e.g., *ZNF2* [96a]), genome duplication, and meiosis are thus likely to be critical for same-sex mating (93).

Although monokaryotic fruiting has only been demonstrated to be a meiotic process in *C. neoformans*, it might be a fairly common process in other basidiomycetes. For example, *Filobasidiella depauperata*, a species closely related to *C. neoformans*, lacks a yeast-like form and grows exclusively filamentously. The hyphae are monokaryotic and lack clamp connections (83). These monokaryotic hyphae derived from single spore culture eventually lead to production of four chains of basidiospores through meiosis as evidenced by the appearance of synaptonemal complexes (44, 83). The homothallic life cycle of *F. depauperata* could be similar to the same-sex mating observed in *C. neoformans*. In some basidiomycetes, it has been known for a long time that monokaryons can form fruiting bodies under certain conditions such as nutritional depletion (35, 46, 90, 104, 106, 143, 152, 154, 155, 163). Mutations both at and unlinked to the mating-type locus can also enable fruiting, including *su-A* (31), *fis*^{c} (103, 115, 152, 153), *pcc1* (108), *Amut*, and *Bmut* (148). All these lines of evidence suggest that genes that are essential in establishing and maintaining compatible heterokaryotic status may not be essential for meiosis and fruiting morphogenesis (91). Whether the homothallic fruiting of these other fungi is a sexual process involving karyogamy and meiosis needs to be investigated further. The full significance and impact of homokaryotic fruiting is not yet clear, and future studies will be necessary to address the molecular mechanisms of this process and its evolutionary impact.

Regardless of being homothallic or heterothallic, sexual development typically involves a cell fusion event and then a nuclear fusion event that creates a diploid state, which is followed by meiosis to generate recombinant progeny (121–125). What distinguishes heterothallic from homothallic species is the initiation event for the sexual cycle. Heterothallic fungi require a compatible partner for mating, whereas homothallic fungi do not and are able to self-mate. Therefore, the ability to undergo meiosis and generate recombinant progeny through both means is likely important for evolutionary adaptation (96, 98, 101).

SPORES AS INFECTIOUS PROPAGULES

Mating and fruiting serve as the major routes that lead to the production of basidiospores, which have long been hypothesized as the infectious propagule of *C. neoformans*. Compared to yeast cells, the size of spores (~1–2 μm) is significantly smaller, which allows them to enter alveoli more readily following inhalation by the host (47). Although spores have been long suspected as the infectious propagules, direct evidence linking spores to infection remains to be established.

Despite the evidence that spores could be critical in the *C. neoformans* infectious cycle, our knowledge about spores is extremely limited, and one major reason is that it is relatively difficult to obtain enough pure spores for molecular analyses. To overcome this difficulty, recently a gradient-based centrifugation method was developed to efficiently purify spores from serotype D crosses, providing a new approach to study the roles of spores during pathogenesis (14).

Morphological analyses observed by both scanning electron microscopy and transmission electron microscopy revealed that serotype D spores have a textured surface, a polar structure, and a large, uneven surface coat. It was hypothesized that the polar structure, with a "stalk-like" structure at one end and a narrower tip at the other, may be important for spore-chain formation. Scanning electron microscopy imaging also revealed that spores from serotype A isolates (var. *grubii*) exhibited a structure similar to serotype D spores but are slightly larger in size (2~3 μm in A vs. 2~2.5 μm in D), whereas spores of *C. gattii* exhibited distinctly elongated morphology (3~5 μm) (154a). In addition, spores produced by monokaryotic fruiting in the serotype D background were structurally similar to spores produced by mating but were modestly longer (2–3 μm in length) (154a).

Fluorescence staining with lectin antibodies demonstrated that *Datura stramonium* lectin and concanavalin A antibodies exhibited spore-specific staining, suggesting that spores and yeasts have different polysaccharide composition at the surface (14). Furthermore, when spores were challenged by various stresses such as high temperature, desiccation, oxidative stress, and general chemical challenge, they were shown to be more resistant compared to yeast cells (14). Interestingly, *C. neoformans* spores are not more resistant to UV irradiation, indicating that UV irradiation is highly adverse to this organism, and this might be one of the reasons why mating is inhibited under sunlight (14).

After basidiospores are produced by either **a**-α opposite-sex or α-α same-sex mating, they can be dispersed and germinate into yeast cells to complete the life cycle. The spore chains are directly exposed to the environment without any outer protecting structure, and it has been thought that the basidiospores are passively discharged by natural physical forces such as wind or rain. Interestingly, in the laboratory, vigorous shaking of mating plates does not release spores; however, a strong airstream mimicking the wind in nature efficiently triggers spore dispersal (154a). Furthermore, in the laboratory, spores can readily germinate on water-agarose medium with no added nutrients, indicating that spores contain enough nutrients to support germination and initial growth (154a). This result is surprising, as spores do not germinate while they are still in chains on mating plates, suggesting that other nonnutritional signals, such as contact sensing, might be involved in spore germination.

If spores are the infectious propagule, they must be virulent to cause disease. Indeed, when mice were infected with a spore-enriched inocula (500 CFU contain ~95% spores), they caused 100% lethality, although the mortality was modestly delayed compared to yeast cells (154a). These findings provide direct evidence that spores can serve as infectious propagules, and their ability to readily disseminate may be a key feature of the virulence cycle.

CONCLUSIONS

In the past two decades, knowledge of how sex is orchestrated in *C. neoformans* and how this might affect virulence and the infectious cycle has advanced tremendously. Nonetheless, the answers to several interesting biological questions regarding the development of the dikaryotic hyphae and same-sex mating remain open. For example, how are nuclear migration and fusion controlled? In the dikaryon, the two nuclei are often closely associated with each other (congressed), but they do not fuse. What signals are required for the juxtaposition of the two nuclei in a large hyphal compartment? The two nuclei fuse and undergo meiosis in the basidium. What triggers these events? Similarly, in same-sex mating, meiosis of the diploid nuclei also takes place in the basidium. How is this event regulated, and what signals are required for the development of basidia? It is also observed that in some strains that undergo monokaryotic fruiting, diploidization of the nuclear content may only occur late in the basidia. How is the timing of diploidization controlled? Finally, in contrast to opposite-sex mating in which meiosis is governed by the Sxi1α/Sxi2**a** homeodomain protein complex, how is meiosis achieved in same-sex mating when this central regulator is absent (Sxi2**a** is absent and Sxi1α is not essential)?

It is also clear that although *C. neoformans* and *C. gattii* have retained the entire machinery for sexual reproduction, they limit their access to this reproduction mode, possibly because of costs associated with sex. Recombination and random assortment are beneficial for the organism to achieve a diversified population, but on the other hand, they risk disrupting favorable gene combinations that might contribute to the success of a pathogen (48). Future studies of the two different modes of sexual reproduction, **a**-α opposite-sex mating and α-α same-sex mating, to understand the underlying molecular mechanisms, the environmental and laboratory conditions that induce these developmental processes, and how these might affect the infectious cycle will certainly provide additional insights into how disease outbreaks might occur and equip us with new tools for antifungal therapy and disease prevention. Moreover, the transitions that occur in the fungal kingdom between heterothallic outbreeding and homothallic inbreeding modes promise to reveal general features by which sexual reproduction enhances fitness and enables evolutionary success throughout biology.

REFERENCES

1. **Aimi, T., R. Yoshida, M. Ishikawa, D. Bao, and Y. Kitamoto.** 2005. Identification and linkage mapping of the genes for the putative homeodomain protein (*hox1*) and the putative pheromone receptor protein homologue (*rcb1*) in a bipolar basidiomycete, *Pholiota nameko*. *Curr. Genet.* **48:** 184–194.
2. **Alspaugh, J. A., L. M. Cavallo, J. R. Perfect, and J. Heitman.** 2000. *RAS1* regulates filamentation, mating and growth at high temperature of *Cryptococcus neoformans*. *Mol. Microbiol.* **36:**352–365.
3. **Alspaugh, J. A., J. R. Perfect, and J. Heitman.** 1997. *Cryptococcus neoformans* mating and virulence are regulated by the G-protein alpha subunit GPA1 and cAMP. *Genes Dev.* **11:**3206–3217.
4. **Alspaugh, J. A., R. Pukkila-Worley, T. Harashima, L. M. Cavallo, D. Funnell, G. M. Cox, J. R. Perfect, J. W. Kronstad, and J. Heitman.** 2002. Adenylyl cyclase functions downstream of the Galpha protein Gpa1 and controls mating and pathogenicity of *Cryptococcus neoformans*. *Eukaryot. Cell* **1:**75–84.
5. **Anderson, C. M., D. A. Willits, P. J. Kosted, E. J. Ford, A. D. Martinez-Espinoza, and J. E. Sherwood.** 1999. Molecular analysis of the pheromone and pheromone receptor genes of *Ustilago hordei*. *Gene* **240:**89–97.
6. **Bahn, Y. S., G. M. Cox, J. R. Perfect, and J. Heitman.** 2005. Carbonic anhydrase and CO_2 sensing during *Cryptococcus neoformans* growth, differentiation, and virulence. *Curr. Biol.* **15:**2013–2020.
7. **Bahn, Y. S., J. K. Hicks, S. S. Giles, G. M. Cox, and J. Heitman.** 2004. Adenylyl cyclase-associated protein Aca1 regulates virulence and differentiation of *Cryptococcus neoformans* via the cyclic AMP-protein kinase A cascade. *Eukaryot. Cell* **3:**1476–1491.
8. **Bahn, Y. S., K. Kojima, G. M. Cox, and J. Heitman.** 2005. Specialization of the HOG pathway and its impact on differentiation and virulence of *Cryptococcus neoformans*. *Mol. Biol. Cell* **16:**2285–2300.
9. **Bahn, Y. S., K. Kojima, G. M. Cox, and J. Heitman.** 2006. A unique fungal two-component system regulates stress responses, drug sensitivity, sexual development, and virulence of *Cryptococcus neoformans*. *Mol. Biol. Cell* **17:**3122–3135.
10. **Bahn, Y. S., C. Xue, A. Idnurm, J. C. Rutherford, J. Heitman, and M. E. Cardenas.** 2007. Sensing the environment: lessons from fungi. *Nat. Rev. Microbiol.* **5:**57–69.
11. **Banuett, F.** 1995. Genetics of *Ustilago maydis*, a fungal pathogen that induces tumors in maize. *Annu. Rev. Genet.* **29:**179–208.
12. **Bauer, R., and F. Oberwinkler.** 2004. Cellular basidiomycete-fungus interactions, p. 267–279. *In* A. Varma, L. Abbott, D. Werner, and R. Hampp (ed.), *Plant Surface Microbiology*. Springer-Verlag, Heidelberg, Germany.
13. **Bolker, M., M. Urban, and R. Kahmann.** 1992. The a mating type locus of *U. maydis* specifies cell signaling components. *Cell* **68:**441–450.
14. **Botts, M. R., S. S. Giles, M. A. Gates, T. R. Kozel, and C. M. Hull.** 2009. Isolation and characterization of *Cryptococcus neoformans* spores reveal a critical role for capsule biosynthesis genes in spore biogenesis. *Eukaryot. Cell* **8:**595–605.
15. **Bovers, M., F. Hagen, E. E. Kuramae, M. R. Diaz, L. Spanjaard, F. Dromer, H. L. Hoogveld, and T. Boekhout.** 2006. Unique hybrids between the fungal pathogens *Cryptococcus neoformans* and *Cryptococcus gattii*. *FEMS Yeast Res.* **6:**599–607.
16. **Bovers, M., F. Hagen, E. E. Kuramae, H. L. Hoogveld, F. Dromer, G. St-Germain, and T. Boekhout.** 2008. AIDS patient death caused by novel *Cryptococcus neoformans* × *C. gattii* hybrid. *Emerg. Infect. Dis.* **14:** 1105–1108.
17. **Bui, T., X. Lin, R. Malik, J. Heitman, and D. Carter.** 2008. Isolates of *Cryptococcus neoformans* from infected animals reveal genetic exchange in unisexual, α mating type populations. *Eukaryot. Cell* **7:**1771–1780.
18. **Bulmer, G. S.** 1990. Twenty-five years with *Cryptococcus neoformans*. *Mycopathologia* **109:**111–122.
19. **Burkholder, A. C., and L. H. Hartwell.** 1985. The yeast alpha-factor receptor: structural properties deduced from the sequence of the *STE2* gene. *Nucleic Acids Res.* **13:** 8463–8475.
20. **Butler, G.** 2007. The evolution of *MAT*: the ascomycetes, p. 3–18. *In* J. Heitman, J. W. Kronstad, J. W. Taylor, and L. A. Casselton (ed.), *Sex in Fungi*. ASM Press, Washington, DC.
21. **Campbell, L. T., B. J. Currie, M. Krockenberger, R. Malik, W. Meyer, J. Heitman, and D. Carter.**

2005. Clonality and recombination in genetically differentiated subgroups of *Cryptococcus gattii*. *Eukaryot. Cell* **4:**1403–1409.

22. **Campbell, L. T., J. A. Fraser, C. B. Nichols, F. S. Dietrich, D. Carter, and J. Heitman.** 2005. Clinical and environmental isolates of *Cryptococcus gattii* from Australia that retain sexual fecundity. *Eukaryot. Cell* **4:**1410–1419.
23. **Chang, Y. C., G. F. Miller, and K. J. Kwon-Chung.** 2003. Importance of a developmentally regulated pheromone receptor of *Cryptococcus neoformans* for virulence. *Infect. Immun.* **71:**4953–4960.
24. **Chang, Y. C., L. A. Penoyer, and K. J. Kwon-Chung.** 2001. The second STE12 homologue of *Cryptococcus neoformans* is MATa-specific and plays an important role in virulence. *Proc. Natl. Acad. Sci. USA* **98:**3258–3263.
25. **Chang, Y. C., B. L. Wickes, G. F. Miller, L. A. Penoyer, and K. J. Kwon-Chung.** 2000. *Cryptococcus neoformans STE12α* regulates virulence but is not essential for mating. *J. Exp. Med.* **191:**871–882.
26. **Chung, S., M. Karos, Y. C. Chang, J. Lukszo, B. L. Wickes, and K. J. Kwon-Chung.** 2002. Molecular analysis of *CPRalpha*, a *MATalpha*-specific pheromone receptor gene of *Cryptococcus neoformans*. *Eukaryot. Cell* **1:**432–439.
27. **Clements, R. S., Jr., and B. Darnell.** 1980. Myo-inositol content of common foods: development of a high-myo-inositol diet. *Am. J. Clin. Nutr.* **33:**1954–1967.
28. **Couve, A., and J. P. Hirsch.** 1996. Loss of sustained Fus3p kinase activity and the G1 arrest response in cells expressing an inappropriate pheromone receptor. *Mol. Cell. Biol.* **16:**4478–4485.
29. **Cruz, M. C., D. S. Fox, and J. Heitman.** 2001. Calcineurin is required for hyphal elongation during mating and haploid fruiting in *Cryptococcus neoformans*. *EMBO J.* **20:**1020–1032.
30. **Davidson, R. C., C. B. Nichols, G. M. Cox, J. R. Perfect, and J. Heitman.** 2003. A MAP kinase cascade composed of cell type specific and non-specific elements controls mating and differentiation of the fungal pathogen *Cryptococcus neoformans*. *Mol. Microbiol.* **49:**469–485.
31. **Day, P. R.** 1963. Mutations of the A mating type factor in *Coprinus lagopus*. *Genet. Res. Camb.* **4:**55–64.
32. **Dohlman, H. G., J. Song, D. Ma, W. E. Courchesne, and J. Thorner.** 1996. Sst2, a negative regulator of pheromone signaling in the yeast *Saccharomyces cerevisiae*: expression, localization, and genetic interaction and physical association with Gpa1 (the G-protein alpha subunit). *Mol. Cell. Biol.* **16:**5194–5209.
33. **D'Souza, C. A., J. A. Alspaugh, C. Yue, T. Harashima, G. M. Cox, J. R. Perfect, and J. Heitman.** 2001. Cyclic AMP-dependent protein kinase controls virulence of the fungal pathogen *Cryptococcus neoformans*. *Mol. Cell. Biol.* **21:**3179–3191.
34. **Ekena, J. L., B. C. Stanton, J. A. Schiebe-Owens, and C. M. Hull.** 2008. Sexual development in *Cryptococcus neoformans* requires *CLP1*, a target of the homeodomain transcription factors Sxi1alpha and Sxi2**a**. *Eukaryot. Cell* **7:**49–57.
35. **Elliott, T. J.** 1985. Developmental genetics: from spore to sporephore, p. 451–465. *In* D. Moore, L. A. Casselton, D. A. Wood, and J. C. Frankland (ed.), *Developmental Biology of Higher Fungi*. Cambridge University Press, Cambridge, UK.
36. **Ellis, D. H., and T. J. Pfeiffer.** 1990. Natural habitat of *Cryptococcus neoformans* var. *gattii*. *J. Clin. Microbiol.* **28:** 1642–1644.
37. **Erke, K. H.** 1976. Light microscopy of basidia, basidiospores, and nuclei in spores and hyphae of *Filobasidiella neoformans* (*Cryptococcus neoformans*). *J. Bacteriol.* **128:** 445–455.
38. **Erke, K. H., and J. D. Schneidau, Jr.** 1973. Relationship of some *Cryptococcus neoformans* hypha-forming strains to standard strains and to other species of yeasts as determined by deoxyribonucleic acid base ratios and homologies. *Infect. Immun.* **7:**941–948.
39. **Fell, J. W., A. C. Statzell, I. L. Hunter, and H. J. Phaff.** 1969. *Leucosporidium* gen. n., the heterobasidiomycetous stage of several yeasts of the genus *Candida*. *Antonie Van Leeuwenhoek* **35:**433–462.
40. **Fraser, J. A., S. Diezmann, R. L. Subaran, A. Allen, K. B. Lengeler, F. S. Dietrich, and J. Heitman.** 2004. Convergent evolution of chromosomal sex-determining regions in the animal and fungal kingdoms. *PLoS Biol.* **2:**e384.
41. **Fraser, J. A., S. S. Giles, E. C. Wenink, S. G. Geunes-Boyer, J. R. Wright, S. Diezmann, A. Allen, J. E. Stajich, F. S. Dietrich, J. R. Perfect, and J. Heitman.** 2005. Same-sex mating and the origin of the Vancouver Island *Cryptococcus gattii* outbreak. *Nature* **437:**1360–1364.
42. **Fraser, J. A., Y. P. Hsueh, K. M. Findley, and J. Heitman.** 2007. Evolution of the mating-type locus: the basidiomycetes, p. 19–34. *In* J. Heitman, J. W. Kronstad, J. W. Taylor, and L. A. Casselton (ed.), *Sex in Fungi*. ASM Press, Washington, DC.
43. **Fraser, J. A., R. L. Subaran, C. B. Nichols, and J. Heitman.** 2003. Recapitulation of the sexual cycle of the primary fungal pathogen *Cryptococcus neoformans* var. *gattii*: implications for an outbreak on Vancouver Island, Canada. *Eukaryot. Cell* **2:**1036–1045.
44. **Gueho, E., L. Improvisi, R. Christen, and G. S. de Hoog.** 1993. Phylogenetic relationships of *Cryptococcus neoformans* and some related basidiomycetous yeasts determined from partial large subunit rRNA sequences. *Antonie Van Leeuwenhoek* **63:**175–189.
45. **Hagen, D. C., G. McCaffrey, and G. F. Sprague, Jr.** 1986. Evidence the yeast *STE3* gene encodes a receptor for the peptide pheromone a factor: gene sequence and implications for the structure of the presumed receptor. *Proc. Natl. Acad. Sci. USA* **83:**1418–1422.
46. **Hanna, W. F.** 1928. Sexual stability in monosporous mycelia of *Coprinus lagopus*. *Ann. Bot.* **42:**379–388.
47. **Hatch, T. F.** 1961. Distribution and deposition of inhaled particles in respiratory tract. *Bacteriol. Rev.* **25:**237–240.
48. **Heitman, J.** 2006. Sexual reproduction and the evolution of microbial pathogens. *Curr. Biol.* **16:**R711–725.
49. **Herskowitz, I.** 1989. A regulatory hierarchy for cell specialization in yeast. *Nature* **342:**749–757.
50. **Herskowitz, I., and Y. Oshima.** 1981. Control of cell type in *Saccharomyces cerevisiae*: mating type and mating-type interconversion, p. 181–209. *In* J. N. Strathern, E. W. Jones, and J. R. Broach (ed.), *The Molecular Biology of the Yeast Saccharomyces: Life Cycle and Inheritance*. Cold Spring Harbor Laboratory Press, Cold Spring Harbor, NY.
51. **Hiremath, S. S., A. Chowdhary, T. Kowshik, H. S. Randhawa, S. Sun, and J. Xu.** 2008. Long-distance dispersal and recombination in environmental populations of *Cryptococcus neoformans* var. *grubii* from India. *Microbiology* **154:**1513–1524.
52. **Hirsch, J. P., and F. R. Cross.** 1993. The pheromone receptors inhibit the pheromone response pathway in *Saccharomyces cerevisiae* by a process that is independent of their associated G alpha protein. *Genetics* **135:**943–953.
53. **Hoff, B., S. Poggeler, and U. Kuck.** 2008. Eighty years after its discovery, Fleming's *Penicillium* strain discloses the secret of its sex. *Eukaryot. Cell* **7:**465–470.
54. **Hsueh, Y. P., and J. Heitman.** 2008. Orchestration of sexual reproduction and virulence by the fungal mating-type locus. *Curr. Opin. Microbiol.* **11:**517–524.
55. **Hsueh, Y. P., C. Xue, and J. Heitman.** 2007. G protein signaling governing cell fate decisions involves opposing

Galpha subunits in *Cryptococcus neoformans*. *Mol. Biol. Cell* **18:**3237–3249.

56. **Hsueh, Y. P., C. Xue, and J. Heitman.** 2009. A constitutively active GPCR governs morphogenic transitions in *Cryptococcus neoformans*. *EMBO J.* **28:**1220–1233.
57. **Hull, C. M., M. J. Boily, and J. Heitman.** 2005. Sex-specific homeodomain proteins Sxi1alpha and Sxi2a coordinately regulate sexual development in *Cryptococcus neoformans*. *Eukaryot. Cell* **4:**526–535.
58. **Hull, C. M., R. C. Davidson, and J. Heitman.** 2002. Cell identity and sexual development in *Cryptococcus neoformans* are controlled by the mating-type-specific homeodomain protein Sxi1alpha. *Genes Dev.* **16:**3046–3060.
59. **Hull, C. M., and J. Heitman.** 2002. Genetics of *Cryptococcus neoformans*. *Annu. Rev. Genet.* **36:**557–615.

59a. **Idnurm, A.** 2010. A tetrad analysis of the basidiomycete fungus *Cryptococcus neoformans*. *Genetics* **185:**153–163.

60. **Idnurm, A., Y. S. Bahn, K. Nielsen, X. Lin, J. A. Fraser, and J. Heitman.** 2005. Deciphering the model pathogenic fungus *Cryptococcus neoformans*. *Nat. Rev. Microbiol.* **3:**753–764.
61. **Idnurm, A., and J. Heitman.** 2005. Light controls growth and development via a conserved pathway in the fungal kingdom. *PLoS Biol.* **3:**e95.
62. **Idnurm, A., F. J. Walton, A. Floyd, and J. Heitman.** 2008. Identification of the *sex* genes in an early diverged fungus. *Nature* **451:**193–196.
63. **Inada, K., Y. Morimoto, T. Arima, Y. Murata, and T. Kamada.** 2001. The *clp1* gene of the mushroom *Coprinus cinereus* is essential for A-regulated sexual development. *Genetics* **157:**133–140.
64. **James, T. Y., P. Srivilai, U. Kues, and R. Vilgalys.** 2006. Evolution of the bipolar mating system of the mushroom *Coprinellus disseminatus* from its tetrapolar ancestors involves loss of mating-type-specific pheromone receptor function. *Genetics* **172:**1877–1891.
65. **Kahmann, R., and J. Schirrawski.** 2007. Mating in the smut fungi: from *a* to *b* to the downstream cascades, p. 377–387. *In* J. Heitman, J. W. Kronstad, J. W. Taylor, and L. A. Casselton (ed.), *Sex in Fungi*. ASM Press, Washington, DC.
66. **Kamper, J., M. Reichmann, T. Romeis, M. Bolker, and R. Kahmann.** 1995. Multiallelic recognition: nonself-dependent dimerization of the bE and bW homeodomain proteins in *Ustilago maydis*. *Cell* **81:**73–83.
67. **Karos, M., Y. C. Chang, C. M. McClelland, D. L. Clarke, J. Fu, B. L. Wickes, and K. J. Kwon-Chung.** 2000. Mapping of the *Cryptococcus neoformans* *MAT*α locus: presence of mating type-specific mitogen-activated protein kinase cascade homologs. *J. Bacteriol.* **182:**6222–6227.
68. **Kent, C. R., P. Ortiz-Bermudez, S. S. Giles, and C. M. Hull.** 2008. Formulation of a defined V8 medium for induction of sexual development of *Cryptococcus neoformans*. *Appl. Environ. Microbiol.* **74:**6248–6253.
69. **Kim, H., and K. A. Borkovich.** 2006. Pheromones are essential for male fertility and sufficient to direct chemotropic polarized growth of trichogynes during mating in *Neurospora crassa*. *Eukaryot. Cell* **5:**544–554.
70. **Kim, J., E. Bortz, H. Zhong, T. Leeuw, E. Leberer, A. K. Vershon, and J. P. Hirsch.** 2000. Localization and signaling of G(beta) subunit Ste4p are controlled by a-factor receptor and the a-specific protein Asg7p. *Mol. Cell. Biol.* **20:**8826–8835.
71. **Kim, J., A. Couve, and J. P. Hirsch.** 1999. Receptor inhibition of pheromone signaling is mediated by the Ste4p Gbeta subunit. *Mol. Cell. Biol.* **19:**441–449.
72. **Klengel, T., W. J. Liang, J. Chaloupka, C. Ruoff, K. Schroppel, J. R. Naglik, S. E. Eckert, E. G. Mogensen, K. Haynes, M. F. Tuite, L. R. Levin, J. Buck, and F. A. Muhlschlegel.** 2005. Fungal adenylyl cyclase integrates CO_2 sensing with cAMP signaling and virulence. *Curr. Biol.* **15:**2021–2026.
73. **Konopka, J. B., S. M. Margarit, and P. Dube.** 1996. Mutation of Pro-258 in transmembrane domain 6 constitutively activates the G protein-coupled alpha-factor receptor. *Proc. Natl. Acad. Sci. USA* **93:**6764–6769.
74. **Kozubowski, L., S. C. Lee, and J. Heitman.** 2009. Signalling pathways in the pathogenesis of *Cryptococcus*. *Cell Microbiol.* **11:**370–380.
75. **Kulkarni, R. D., M. R. Thon, H. Pan, and R. A. Dean.** 2005. Novel G-protein-coupled receptor-like proteins in the plant pathogenic fungus *Magnaporthe grisea*. *Genome Biol.* **6:**R24.
76. **Kwon-Chung, K. J.** 1975. A new genus, *Filobasidiella*, the perfect state of *Cryptococcus neoformans*. *Mycologia* **67:**1197–1200.
77. **Kwon-Chung, K. J.** 1976. A new species of *Filobasidiella*, the sexual state of *Cryptococcus neoformans* B and C serotypes. *Mycologia* **68:**943–946.
78. **Kwon-Chung, K. J.** 1976. Morphogenesis of *Filobasidiella neoformans*, the sexual state of *Cryptococcus neoformans*. *Mycologia* **68:**821–833.
79. **Kwon-Chung, K. J.** 1980. Nuclear genotypes of spore chains in *Filobasidiella neoformans* (*Cryptococcus neoformans*). *Mycologia* **72:**418–422.
80. **Kwon-Chung, K. J.** 1998. *Filobasidiella* Kwon-Chung, p. 656–662. *In* C. P. Kurtzman and J. W. Fell (ed.), *The Yeasts: a Taxonomic Study*, 4th ed. Elsevier Science, Amsterdam, The Netherlands.
81. **Kwon-Chung, K. J., and J. E. Bennett.** 1978. Distribution of alpha and **a** mating types of *Cryptococcus neoformans* among natural and clinical isolates. *Am. J. Epidemiol.* **108:**337–340.
82. **Kwon-Chung, K. J., J. E. Bennett, and J. C. Rhodes.** 1982. Taxonomic studies on *Filobasidiella* species and their anamorphs. *Antonie Van Leeuwenhoek* **48:**25–38.
83. **Kwon-Chung, K. J., Y. C. Chang, R. Bauer, E. C. Swann, J. W. Taylor, and R. Goel.** 1995. The characteristics that differentiate *Filobasidiella depauperata* from *Filobasidiella neoformans*. *Stud. Mycol.* **38:**67–79.
84. **Kwon-Chung, K. J., J. C. Edman, and B. L. Wickes.** 1992. Genetic association of mating types and virulence in *Cryptococcus neoformans*. *Infect. Immun.* **60:**602–605.
85. **Kwon-Chung, K. J., and T. J. Popkin.** 1976. Ultrastructure of septal complex in *Filobasidiella neoformans* (*Cryptococcus neoformans*). *J. Bacteriol.* **126:**524–528.
86. **Kwon-Chung, K. J., and A. Varma.** 2006. Do major species concepts support one, two or more species within *Cryptococcus neoformans*? *FEMS Yeast Res.* **6:**574–587.
87. **Ladds, G., K. Davis, A. Das, and J. Davey.** 2005. A constitutively active GPCR retains its G protein specificity and the ability to form dimers. *Mol. Microbiol.* **55:**482–497.
88. **Lee, S. C., N. Corradi, E. J. Byrnes, 3rd, S. Torres-Martinez, F. S. Dietrich, P. J. Keeling, and J. Heitman.** 2008. Microsporidia evolved from ancestral sexual fungi. *Curr. Biol.* **18:**1675–1679.
89. **Lengeler, K. B., R. C. Davidson, C. D'Souza, T. Harashima, W. C. Shen, P. Wang, X. Pan, M. Waugh, and J. Heitman.** 2000. Signal transduction cascades regulating fungal development and virulence. *Microbiol. Mol. Biol. Rev.* **64:**746–785.
90. **Leonard, T. J., and S. Dick.** 1968. Chemical induction of haploid fruiting bodies in *Schizophyllum commune*. *Proc. Natl. Acad. Sci. USA* **59:**745–751.
91. **Leslie, J. F., and T. J. Leonard.** 1984. Nuclear control of monokaryotic fruiting in *Schizophyllum commune*. *Mycologia* **76:**760–763.

92. **Li, L., G. Shen, Z. G. Zhang, Y. L. Wang, J. K. Thompson, and P. Wang.** 2007. Canonical heterotrimeric G proteins regulating mating and virulence of *Cryptococcus neoformans*. *Mol. Biol. Cell* **18:**4201–4209.

93. **Lin, X.** 2009. *Cryptococcus neoformans*: morphogenesis, infection, and evolution. *Infect. Genet. Evol.* **9:**401–416.

94. **Lin, X., and J. Heitman.** 2005. Chlamydospore formation during hyphal growth in *Cryptococcus neoformans*. *Eukaryot. Cell* **4:**1746–1754.

95. **Lin, X., J. C. Huang, T. G. Mitchell, and J. Heitman.** 2006. Virulence attributes and hyphal growth of *C. neoformans* are quantitative traits and the *MATalpha* allele enhances filamentation. *PLoS Genet.* **2:**e187.

96. **Lin, X., C. M. Hull, and J. Heitman.** 2005. Sexual reproduction between partners of the same mating type in *Cryptococcus neoformans*. *Nature* **434:**1017–1021.

96a. **Lin, X., J. C. Jackson, M. Feretzaki, C. Xue, and J. Heitman.** 2010. Transcription factors Mat2 and Znf2 operate cellular circuits orchestrating opposite- and same-sex mating in *Cryptococcus neoformans*. *PLoS Genet.* **6:**e1000953.

97. **Lin, X., A. Litvintseva, K. Nielsen, S. Patel, Z. Kapadia, A. Floyd, T. G. Mitchell, and J. Heitman.** 2007. αADα hybrids of *Cryptococcus neoformans*: evidence of same sex mating in nature and hybrid fitness. *PLoS Genet.* **3:**1975–1990.

98. **Lin, X., S. Patel, A. P. Litvintseva, A. Floyd, T. G. Mitchell, and J. Heitman.** 2009. Diploids in the *Cryptococcus neoformans* serotype A population homozygous for the alpha mating type originate via unisexual mating. *PLoS Pathog.* **5:**e1000283.

99. **Liu, Y., Q. He, and P. Cheng.** 2003. Photoreception in *Neurospora*: a tale of two White Collar proteins. *Cell. Mol. Life Sci.* **60:**2131–2138.

100. **Lu, Y. K., K. H. Sun, and W. C. Shen.** 2005. Blue light negatively regulates the sexual filamentation via the Cwc1 and Cwc2 proteins in *Cryptococcus neoformans*. *Mol. Microbiol.* **56:**480–491.

101. **Marra, R. E., J. C. Huang, E. Fung, K. Nielsen, J. Heitman, R. Vilgalys, and T. G. Mitchell.** 2004. A genetic linkage map of *Cryptococcus neoformans* variety *neoformans* serotype D (*Filobasidiella neoformans*). *Genetics* **167:**619–631.

102. **Martin, F., A. Aerts, D. Ahren, A. Brun, E. G. Danchin, F. Duchaussoy, J. Gibon, A. Kohler, E. Lindquist, V. Pereda, A. Salamov, H. J. Shapiro, J. Wuyts, D. Blaudez, M. Buee, P. Brokstein, B. Canback, D. Cohen, P. E. Courty, P. M. Coutinho, C. Delaruelle, J. C. Detter, A. Deveau, S. DiFazio, S. Duplessis, L. Fraissinet-Tachet, E. Lucic, P. Frey-Klett, C. Fourrey, I. Feussner, G. Gay, J. Grimwood, P. J. Hoegger, P. Jain, S. Kilaru, J. Labbe, Y. C. Lin, V. Legue, F. Le Tacon, R. Marmeisse, D. Melayah, B. Montanini, M. Muratet, U. Nehls, H. Niculita-Hirzel, M. P. Oudot-Le Secq, M. Peter, H. Quesneville, B. Rajashekar, M. Reich, N. Rouhier, J. Schmutz, T. Yin, M. Chalot, B. Henrissat, U. Kues, S. Lucas, Y. Van de Peer, G. K. Podila, A. Polle, P. J. Pukkila, P. M. Richardson, P. Rouze, I. R. Sanders, J. E. Stajich, A. Tunlid, G. Tuskan, and I. V. Grigoriev.** 2008. The genome of *Laccaria bicolor* provides insights into mycorrhizal symbiosis. *Nature* **452:**88–92.

103. **Miyake, H., K. Tanaka, and T. Ishikawa.** 1980. Basidiospore formation in monokaryotic fruiting bodies of a mutant strain of *Coprinus macrorhizus*. *Arch. Microbiol.* **126:**207–211.

104. **Mizushina, Y., L. Hanashima, T. Yamaguchi, M. Takemura, F. Sugawara, M. Saneyoshi, A. Matsukage, S. Yoshida, and K. Sakaguchi.** 1998. A mushroom fruiting body-inducing substance inhibits activities of replicative DNA polymerases. *Biochem. Biophys. Res. Commun.* **249:**17–22.

105. **Mogensen, E. G., G. Janbon, J. Chaloupka, C. Steegborn, M. S. Fu, F. Moyrand, T. Klengel, D. S. Pearson, M. A. Geeves, J. Buck, L. R. Levin, and F. A. Muhlschlegel.** 2006. *Cryptococcus neoformans* senses CO_2 through the carbonic anhydrase Can2 and the adenylyl cyclase Cac1. *Eukaryot. Cell* **5:**103–111.

106. **Moore, D.** 1998. *Fungal Morphogenesis*. Cambridge University Press, Cambridge, United Kingdom.

107. **Muller, H., C. Hennequin, J. Gallaud, B. Dujon, and C. Fairhead.** 2008. The asexual yeast *Candida glabrata* maintains distinct a and alpha haploid mating types. *Eukaryot. Cell* **7:**848–858.

108. **Murata, Y., M. Fujii, M. E. Zolan, and T. Kamada.** 1998. Molecular analysis of *pcc1*, a gene that leads to A-regulated sexual morphogenesis in *Coprinus cinereus*. *Genetics* **149:**1753–1761.

109. **Nguyen, V. Q., and A. Sil.** 2008. Temperature-induced switch to the pathogenic yeast form of *Histoplasma capsulatum* requires Ryp1, a conserved transcriptional regulator. *Proc. Natl. Acad. Sci. USA* **105:**4880–4885.

110. **Nichols, C. B., J. A. Fraser, and J. Heitman.** 2004. PAK kinases Ste20 and Pak1 govern cell polarity at different stages of mating in *Cryptococcus neoformans*. *Mol. Biol. Cell* **15:**4476–4489.

111. **Niculita-Hirzel, H., J. Labbe, A. Kohler, F. le Tacon, F. Martin, I. R. Sanders, and U. Kues.** 2008. Gene organization of the mating type regions in the ectomycorrhizal fungus *Laccaria bicolor* reveals distinct evolution between the two mating type loci. *New Phytol.* **180:**329–342.

112. **Nielsen, K., G. M. Cox, P. Wang, D. L. Toffaletti, J. R. Perfect, and J. Heitman.** 2003. Sexual cycle of *Cryptococcus neoformans* var. *grubii* and virulence of congenic **a** and alpha isolates. *Infect. Immun.* **71:**4831–4841.

113. **Nielsen, K., A. L. De Obaldia, and J. Heitman.** 2007. *Cryptococcus neoformans* mates on pigeon guano: implications for the realized ecological niche and globalization. *Eukaryot. Cell* **6:**949–959.

114. **Odom, A., S. Muir, E. Lim, D. L. Toffaletti, J. Perfect, and J. Heitman.** 1997. Calcineurin is required for virulence of *Cryptococcus neoformans*. *EMBO J.* **16:**2576–2589.

115. **Oishi, K., I. Uno, and T. Ishikawa.** 1982. Timing of DNA replication during the meiotic process in monokaryotic basidiocarps of *Coprinus macrorhizus*. *Arch. Microbiol.* **132:**372–374.

116. **Olive, L. S.** 1968. An unusual heterobasidiomycete with tilletia-like basidia. *J. Elisha Mitchell Sci. Soc.* **84:**261–266.

117. **Palmer, D. A., J. K. Thompson, L. Li, A. Prat, and P. Wang.** 2006. Gib2, a novel Gbeta-like/RACK1 homolog, functions as a Gbeta subunit in cAMP signaling and is essential in *Cryptococcus neoformans*. *J. Biol. Chem.* **281:**32596–32605.

118. **Paoletti, M., C. Rydholm, E. U. Schwier, M. J. Anderson, G. Szakacs, F. Lutzoni, J. P. Debeaupuis, J. P. Latge, D. W. Denning, and P. S. Dyer.** 2005. Evidence for sexuality in the opportunistic fungal pathogen *Aspergillus fumigatus*. *Curr. Biol.* **15:**1242–1248.

119. **Poggeler, S., and U. Kuck.** 2001. Identification of transcriptionally expressed pheromone receptor genes in filamentous ascomycetes. *Gene* **280:**9–17.

120. **Prusty, R., P. Grisafi, and G. R. Fink.** 2004. The plant hormone indoleacetic acid induces invasive growth in *Saccharomyces cerevisiae*. *Proc. Natl. Acad. Sci. USA* **101:**4153–4157.

121. **Raju, N. B.** 1978. Meiosis nuclear behaviour and ascospore formation in five homothallic species of *Neurospora*. *Can. J. Bot.* **56:**754–763.

122. **Raju, N. B.** 1980. Meiosis and ascospore genesis in *Neurospora*. *Eur. J. Cell Biol.* **23:**208–223.
123. **Raju, N. B.** 1992. Functional heterothallism resulting from homokaryotic conidia and ascospores in *Neurospora tetrasperma*. *Mycol. Res.* **96:**103–116.
124. **Raju, N. B.** 1992. Genetic control of the sexual cycle in *Neurospora*. *Mycol. Res.* **96:**241–262.
125. **Raju, N. B., and D. D. Perkins.** 1994. Diverse programs of ascus development in pseudohomothallic species of *Neurospora*, *Gelasinospora*, and *Podospora*. *Dev. Genet.* **15:**104–118.
126. **Ren, P., D. J. Springer, M. J. Behr, W. A. Samsonoff, S. Chaturvedi, and V. Chaturvedi.** 2006. Transcription factor STE12alpha has distinct roles in morphogenesis, virulence, and ecological fitness of the primary pathogenic yeast *Cryptococcus gattii*. *Eukaryot. Cell* **5:**1065–1080.
127. **Riquelme, M., M. P. Challen, L. A. Casselton, and A. J. Brown.** 2005. The origin of multiple B mating specificities in *Coprinus cinereus*. *Genetics* **170:**1105–1119.
128. **Rogers, A. L., K. J. Kwon-Chung, and S. L. Flegler.** 1980. A scanning electron microscope comparison of basidial structures in *Filobasidiella neoformans* and *Filobasidiella bacillispora*. *Sabouraudia* **18:**85–89.
129. **Roth, A. F., B. Nelson, C. Boone, and N. G. Davis.** 2000. Asg7p-Ste3p inhibition of pheromone signaling: regulation of the zygotic transition to vegetative growth. *Mol. Cell. Biol.* **20:**8815–8825.
130. **Rutherford, J. C., X. Lin, K. Nielsen, and J. Heitman.** 2008. Amt2 permease is required to induce ammonium-responsive invasive growth and mating in *Cryptococcus neoformans*. *Eukaryot. Cell* **7:**237–246.
131. **Saul, N., M. Krockenberger, and D. Carter.** 2008. Evidence of recombination in mixed-mating-type and alpha-only populations of *Cryptococcus gattii* sourced from single eucalyptus tree hollows. *Eukaryot. Cell* **7:**727–734.
132. **Scherer, M., K. Heimel, V. Starke, and J. Kamper.** 2006. The Clp1 protein is required for clamp formation and pathogenic development of *Ustilago maydis*. *Plant Cell* **18:**2388–2401.
133. **Schubert, D., M. Raudaskoski, N. Knabe, and E. Kothe.** 2006. Ras GTPase-activating protein Gap1 of the homobasidiomycete *Schizophyllum commune* regulates hyphal growth orientation and sexual development. *Eukaryot. Cell* **5:**683–695.
134. **Schulz, B., F. Banuett, M. Dahl, R. Schlesinger, W. Schafer, T. Martin, I. Herskowitz, and R. Kahmann.** 1990. The *b* alleles of *U. maydis*, whose combinations program pathogenic development, code for polypeptides containing a homeodomain-related motif. *Cell* **60:**295–306.
135. **Seo, J. A., K. H. Han, and J. H. Yu.** 2004. The *gprA* and *gprB* genes encode putative G protein-coupled receptors required for self-fertilization in *Aspergillus nidulans*. *Mol. Microbiol.* **53:**1611–1623.
136. **Shadomy, H. J.** 1970. Clamp connections in two strains of *Cryptococcus neoformans*. Spectrum Monograph Series **1:**67–72. Georgia State University, Atlanta, GA.
137. **Shadomy, H. J., and J. P. Utz.** 1966. Preliminary studies on a hypha-forming mutant of *Cryptococcus neoformans*. *Mycologia* **58:**383–390.
138. **Shapiro, R. S., P. Uppuluri, A. K. Zaas, C. Collins, H. Senn, J. R. Perfect, J. Heitman, and L. E. Cowen.** 2009. Hsp90 orchestrates temperature-dependent *Candida albicans* morphogenesis via Ras1-PKA signaling. *Curr. Biol.* **19:**621–629.
139. **Shen, G., Y. L. Wang, A. Whittington, L. Li, and P. Wang.** 2008. The RGS protein Crg2 regulates pheromone and cyclic AMP signaling in *Cryptococcus neoformans*. *Eukaryot. Cell* **7:**1540–1548.
140. **Shen, W. C., R. C. Davidson, G. M. Cox, and J. Heitman.** 2002. Pheromones stimulate mating and differentiation via paracrine and autocrine signaling in *Cryptococcus neoformans*. *Eukaryot. Cell* **1:**366–377.
141. **Shi, C., S. Kaminskyj, S. Caldwell, and M. C. Loewen.** 2007. A role for a complex between activated G protein-coupled receptors in yeast cellular mating. *Proc. Natl. Acad. Sci. USA* **104:**5395–5400.
142. **Sia, R. A., K. B. Lengeler, and J. Heitman.** 2000. Diploid strains of the pathogenic basidiomycete *Cryptococcus neoformans* are thermally dimorphic. *Fungal Genet. Biol.* **29:**153–163.
143. **Stahl, U., and K. Esser.** 1976. Genetics of fruit body production in higher basidiomycetes. I. Monokaryotic fruiting and its correlation with dikaryotic fruiting in *Polyporus ciliatus*. *Mol. Gen. Genet.* **148:**183–197.
144. **Staib, F.** 1981. The perfect state of *Cryptococcus neoformans*, *Filobasidiella neoformans*, on pigeon manure filtrate agar. *Zentralbl. Bakteriol.* A **248:**575–578.
145. **Staib, F., and A. Blisse.** 1982. Bird manure filtrate agar for the formation of the perfect state of *Cryptococcus neoformans*, *Filobasidiella neoformans*. A comparative study of the agars prepared from pigeon and canary manure. *Zentralbl. Bakteriol. Mikrobiol. Hyg.* A **251:**554–562.
145a. **Stanton, B. C., S. S. Giles, M. W. Staudt, E. K. Kruzel, and C. M. Hull.** 2010. Allelic exchange of pheromones and their receptors reprograms sexual identity in *Cryptococcus neoformans*. *PLos Genet.* **6:**e1000860.
146. **Stefan, C. J., M. C. Overton, and K. J. Blumer.** 1998. Mechanisms governing the activation and trafficking of yeast G protein-coupled receptors. *Mol. Biol. Cell* **9:**885–899.
147. **Strickfaden, S. C., and P. M. Pryciak.** 2008. Distinct roles for two Galpha-Gbeta interfaces in cell polarity control by a yeast heterotrimeric G protein. *Mol. Biol. Cell* **19:**181–197.
148. **Swamy, S., I. Uno, and T. Ishikawa.** 1984. Morphogenetic effects of mutations at the A and B incompatibility factors of *Coprinus cinereus*. *J. Gen. Microbiol.* **130:**3219–3224.
149. **Tesmer, J. J., D. M. Berman, A. G. Gilman, and S. R. Sprang.** 1997. Structure of RGS4 bound to AlF4-activated G(i alpha1): stabilization of the transition state for GTP hydrolysis. *Cell* **89:**251–261.
150. **Torres-Guererro, H., and J. C. Edman.** 1994. Melanin-deficient mutants of *Cryptococcus neoformans*. *J. Med. Vet. Mycol.* **32:**303–313.
151. **Tscharke, R. L., M. Lazera, Y. C. Chang, B. L. Wickes, and K. J. Kwon-Chung.** 2003. Haploid fruiting in *Cryptococcus neoformans* is not mating type alpha-specific. *Fungal Genet. Biol.* **39:**230–237.
152. **Uno, I., and T. Ishikawa.** 1971. Chemical and genetical control of induction of monokaryotic fruiting bodies in *Coprinus macrorhizus*. *Mol. Gen. Genet.* **113:**229–239.
153. **Uno, I., and T. Ishikawa.** 1973. Purification and identification of the fruiting-inducing substances in *Coprinus macrorhizus*. *J. Bacteriol.* **113:**1240–1248.
154. **Urayama, T.** 1969. Stimulative effect of extracts from fruit bodies of *Agaricus bisporus* and some other hymenomycetes on primordia formation in *Marasmius* sp. *Trans. Mycol. Soc. Jpn.* **10:**73–78.
154a. **Velagapudi, R., Y. P. Hsueh, S. Geunes-Boyer, J. R. Wright, and J. Heitman.** 2009. Spores as infectious propagules of *Cryptococcus neoformans*. *Infect. Immun.* **77:**4345–4355.
155. **Verrinder-Gibbins, A. M., and B. C. Lu.** 1984. Induction of normal fruiting on originally monokaryotic cultures of *Coprinus cinereus*. *Trans. Br. Mycol. Soc.* **82:**331–335.

156. **Walton, F. J., A. Idnurm, and J. Heitman.** 2005. Novel gene functions required for melanization of the human pathogen *Cryptococcus neoformans*. *Mol. Microbiol.* **57:** 1381–1396.
157. **Wang, P., J. Cutler, J. King, and D. Palmer.** 2004. Mutation of the regulator of G protein signaling Crg1 increases virulence in *Cryptococcus neoformans*. *Eukaryot. Cell* **3:**1028–1035.
158. **Wang, P., C. B. Nichols, K. B. Lengeler, M. E. Cardenas, G. M. Cox, J. R. Perfect, and J. Heitman.** 2002. Mating-type-specific and nonspecific PAK kinases play shared and divergent roles in *Cryptococcus neoformans*. *Eukaryot. Cell* **1:**257–272.
159. **Wang, P., J. R. Perfect, and J. Heitman.** 2000. The G-protein beta subunit Gpb1 is required for mating and haploid fruiting in *Cryptococcus neoformans*. *Mol. Cell. Biol.* **20:**352–362.
160. **Weber, M., V. Salo, M. Uuskallio, and M. Raudaskoski.** 2005. Ectopic expression of a constitutively active Cdc42 small GTPase alters the morphology of haploid and dikaryotic hyphae in the filamentous homobasidiomycete *Schizophyllum commune*. *Fungal Genet. Biol.* **42:** 624–637.
161. **Webster, R. H., and A. Sil.** 2008. Conserved factors Ryp2 and Ryp3 control cell morphology and infectious spore formation in the fungal pathogen *Histoplasma capsulatum*. *Proc. Natl. Acad. Sci. USA* **105:**14573–14578.
162. **Wickes, B. L., U. Edman, and J. C. Edman.** 1997. The *Cryptococcus neoformans STE12α* gene: a putative *Saccharomyces cerevisiae STE12* homologue that is mating type specific. *Mol. Microbiol.* **26:**951–960.
163. **Wickes, B. L., M. E. Mayorga, U. Edman, and J. C. Edman.** 1996. Dimorphism and haploid fruiting in *Cryptococcus neoformans*: association with the alpha-mating type. *Proc. Natl. Acad. Sci. USA* **93:**7327–7331.
164. **Xue, C., Y. S. Bahn, G. M. Cox, and J. Heitman.** 2006. G protein-coupled receptor Gpr4 senses amino acids and activates the cAMP-PKA pathway in *Cryptococcus neoformans*. *Mol. Biol. Cell* **17:**667–679.
165. **Xue, C., Y. P. Hsueh, L. Chen, and J. Heitman.** 2008. The RGS protein Crg2 regulates both pheromone and cAMP signalling in *Cryptococcus neoformans*. *Mol. Microbiol.* **70:**379–395.
166. **Xue, C., Y. P. Hsueh, and J. Heitman.** 2008. Magnificent seven: roles of G protein-coupled receptors in extracellular sensing in fungi. *FEMS Microbiol. Rev.* **32:** 1010–1032.
167. **Xue, C., Y. Tada, X. Dong, and J. Heitman.** 2007. The human fungal pathogen *Cryptococcus* can complete its sexual cycle during a pathogenic association with plants. *Cell Host Microbe* **1:**263–273.
168. **Yeh, Y. L., Y. S. Lin, B. J. Su, and W. C. Shen.** 2009. A screening for suppressor mutants reveals components involved in the blue light-inhibited sexual filamentation in *Cryptococcus neoformans*. *Fungal Genet. Biol.* **46:**42–54.
169. **Yi, S., N. Sahni, K. J. Daniels, C. Pujol, T. Srikantha, and D. R. Soll.** 2008. The same receptor, G protein, and mitogen-activated protein kinase pathway activate different downstream regulators in the alternative white and opaque pheromone responses of *Candida albicans*. *Mol. Biol. Cell* **19:**957–970.
170. **Young, L. Y., M. C. Lorenz, and J. Heitman.** 2000. A STE12 homolog is required for mating but dispensable for filamentation in *Candida lusitaniae*. *Genetics* **155:**17–29.
171. **Yue, C., L. M. Cavallo, J. A. Alspaugh, P. Wang, G. M. Cox, J. R. Perfect, and J. Heitman.** 1999. The STE12α homolog is required for haploid filamentation but largely dispensable for mating and virulence in *Cryptococcus neoformans*. *Genetics* **153:**1601–1615.

Cryptococcus: From Human Pathogen to Model Yeast
Edited by J. Heitman et al.

8

Population Structure and Ecology of *Cryptococcus neoformans* and *Cryptococcus gattii*

ANASTASIA P. LITVINTSEVA, JIANPING XU, AND THOMAS G. MITCHELL

Cryptococcus neoformans and *Cryptococcus gattii* comprise the pathogenic *Cryptococcus* species complex (47, 75, 76). In recent years, there has been significant progress in determining the population biology and molecular epidemiology of these important mammalian pathogens. The other pathogenic species of *Cryptococcus* (e.g., *C. albidus*, *C. laurentii*, etc.) will not be discussed here because they rarely cause infections, and they are phylogenetically quite divergent from *C. neoformans* and *C. gattii*. Rather, this chapter will focus on the population structure and ecology of *C. neoformans* and *C. gattii*.

This chapter summarizes current knowledge of the population structure of *C. neoformans* and *C. gattii* and discusses the associations of the genetically isolated subpopulations with their ecological niches. "Population structure" refers to (i) the genetic diversity among individuals comprising the population, (ii) the operative mode(s) of reproduction and genetic exchange, and (iii) the formation of subdivisions, which may be driven by geographical, temporal, ecological, or genetic factors that distribute the individuals into separate groups. The relationships among component subpopulations often affect or reflect the gene flow among them and their evolution. Subpopulations may be geographically isolated, or they may occupy the same environment or adjacent but nonoverlapping niches. Ecology has a profound influence on population structure. The availability of nutrients and other beneficial growth conditions usually determines the size and genetic history of the population. Thus, population subdivisions are often created by adaptation to different ecological niches.

METHODS OF MOLECULAR GENOTYPING

Fungal population genetics would not be possible without molecular genotyping tools, which can differentiate individual members of the population. Numerous methods have been developed to genotype individuals. Applications to cryptococcal isolates are reviewed in chapter 24 as well as several recent reviews (105–107, 150, 152). The most apt and reliable methods to genotype *Cryptococcus* use markers that detect polymorphisms among microsatellites and similar repetitive DNA elements, restriction fragment length polymorphisms (RFLP), amplified fragment length polymorphisms (AFLP), and direct DNA sequencing, such as multilocus sequence typing (MLST). In addition to genotyping isolates to analyze their population structure, the ability to genotype clinical strains serves many other purposes. For example, genotypes are frequently utilized to establish the accurate diagnosis of atypical isolates, to resolve taxonomic controversies, to investigate the molecular epidemiology of an outbreak, and to identify relevant phenotype-genotype associations (107).

Most studies of the population genetics of the pathogenic species of *Cryptococcus* use reference strains to recognize the major clades or subpopulations, which are designated as VNI through VNIV and VNB for *C. neoformans* and VGI through VGIV for *C. gattii*. These subgroups were identified by a standard method, termed PCR fingerprinting, which was developed by Wieland Meyer and his colleagues (101–103). PCR fingerprinting uses the single M13 primer to amplify genomic DNA, and the electrophoresed amplicons are compared to reference strains. This method has been refined by the addition of a PCR-RFLP step. A portion of the *URA5* gene is amplified, and the amplicon is digested with two endonucleases, Sau96I and HhaI (101). This simple, relatively fast and inexpensive protocol has been implemented in numerous laboratories around the world. Since the PCR fingerprinting patterns may vary in different laboratories, the inclusion of reference strains representing each of the major subpopulations is required as controls, and these reference strains are readily available to researchers.

Other typing systems have also been developed to recognize the subpopulations of the *C. neoformans*/*C. gattii* complex. Teun Boekhout and associates devised a robust AFLP method, and their major AFLP clades correspond to those defined by the VN-VG nomenclature (7, 8, 70, 141). The molecular types associated with these systems are summarized in Table 1. AFLP has the advantage of generating numerous polymorphic markers for analysis. However, as with

Anastasia P. Litvintseva and Thomas G. Mitchell, Department of Molecular Genetics and Microbiology, Duke University Medical Center, Durham, NC 27710. **Jianping Xu,** Department of Biology, McMaster University, Hamilton, Ontario, Canada L8S 4K1.

TABLE 1 Nomenclature of the subpopulations of *C. neoformans* and *C. gattii*

Species	Serotype (variety)	PCR fingerprint	AFLP genotype	MSLT diversity[a]
C. neoformans	A (var. *grubii*)	VNI	1	+++
		VNII	1B	+
		VNB	1A	++++
	AD	VNIII	3	++
	D (var. *neoformans*)	VNIV	2	++
C. gattii	B	VGI	4	+
		VGII	6	++
	C or B[b]	VGIII	5	++
	C	VGIV	7	+

[a] Diversity was estimated from several MLST analyses.
[b] The great majority of isolates of serotype B have the molecular type VGI or VGII, but rare isolates may be VGIII. In addition, a large collection of *C. gattii* strains has been reported to include strains of serotype B and C with each of the four VG molecular types (W. Meyer, personal communication).
References: 7, 8, 10, 11, 19, 66, 100, 101.

the VN-VG typing, AFLP data are laboratory specific, even when the same protocols are followed, and therefore, reference strains are essential (107). Another disadvantage is that the genomic locations and distribution of AFLP loci are usually unknown. Other genotyping methods identify the major subpopulations using markers generated by PCR-RFLP (46), microsatellites (59, 160), or selected ribosomal DNA sequences (36, 66).

In contrast, MLST markers reflect specific DNA sequence polymorphisms, are absolute, and can be reproduced in any laboratory. Consequently, MLST is the preferred method of genotyping *C. neoformans* and *C. gattii* (11, 20, 30, 46, 92, 111, 127). The most recent, universal and unambiguous set of MLST markers was developed by The *Cryptococcus* Species Genotyping Working Group (http://www.isham.org/Groups.html), organized under the auspices of the International Society for Human and Animal Mycology. The participants established a minimum set of consensus MLST markers that are sufficient to assign new isolates to the major subpopulations (100). The most current MLST markers for isolates of *C. neoformans* var. *grubii* can be found at http://cneoformans.mlst.net/.

Table 1 shows the current designation of the nine major subpopulations of the *C. neoformans*/*C. gattii* complex and their relationships to the conventional serotypes. Unfortunately, a standardized set of commercial serotyping reagents is no longer available. This system used specific antisera to identify the four immunodominant epitopes associated with the major capsular polysaccharide, glucuronoxylomannan. This method of serotyping served for years to separate *C. neoformans* (serotypes A, D, and AD) from *C. gattii* (serotypes B and C). Currently, the only facile, if imperfect, nonmolecular methods to distinguish these species exploit physiological differences between them (74). More importantly, the subpopulations (or cryptic species) are best identified using DNA-based methods.

METHODS OF POPULATION GENETICS

Computational methods of population structure rely on statistical analyses of the frequencies of genotypes and alleles among the isolates that comprise the population under investigation. Analyses of the population structure presented in this chapter are concerned with (i) genetic diversity, (ii) mode of reproduction (clonality versus recombination), and (iii) population subdivision. However, other relevant questions about the population biology of *C. neoformans* and *C. gattii* can also be addressed, including the effects of natural selection, adaptation of specific genotypes to different ecological conditions, and migration and gene flow among populations. A more extensive discourse on the methodology of population genetics is beyond the scope of this chapter, but excellent texts are available to learn more about analyzing population structure (17, 60, 61).

Genetic diversity in the populations can be estimated by calculating standard diversity indices, such as (i) gene diversity (h), which is the probability that two randomly selected haplotypes (alleles) in the sample are different (109); (ii) nucleotide diversity (π), which is the probability that two randomly selected homologous nucleotides are different (110); and (iii) pairwise difference (d), which is the mean number of base-pair differences between all pairs of haplotypes in the sample. Genetic diversity in the population can also be estimated by calculating the index of genotypic diversity by dividing the number of different genotypes in the population sample by the number of isolates in the sample. Since the degree of genetic diversity of a population is a reflection of its phylogenetic age, comparing the indices of diversity among populations can provide valuable insight about their histories. Older populations tend to be more genetically diverse, emergent younger populations are significantly less variable, and ancestral populations exhibit the greatest levels of genetic variation.

The mode of reproduction is assessed by estimating the amount of clonality and recombination in the populations. A recombining population is expected to have (i) an equal distribution of mating types, and (ii) the loci should display random associations or linkage equilibrium. Conversely, a population that reproduces clonally is expected to have nonrandom association among physically unlinked loci. In a clonal population, loci on different chromosomes will segregate as if they were physically linked. For *C. neoformans* and *C. gattii*, the degree of linkage equilibrium among the loci in the populations has been estimated by calculating the index of association (I_A) (98), or $\bar{r}_d$, an estimate of multilocus linkage equilibrium (18). I_A values significantly different from zero suggest linkage disequilibrium among the loci and indicate that recombination in the population is rare or absent. The amount of recombination in a population can also be assessed by phylogenetic methods. In a clonal population, the genealogies of multiple genes will be congruent, indicating that all parts of the genome have the same phylogenetic history. Conversely, in a recombining population, different genes often have different phylogenies. Statistically significant conflict among the genealogies of different genes provides strong evidence for recent or past recombination in the population (138).

Compound populations are often subdivided into smaller genetically isolated subpopulations, which may be associated with different geographic locations or ecological niches. Several methods can be used to detect evidence of subdivision within a population. Wright's fixation index (F_{ST}) estimates the rate of genetic exchange between the populations (64, 147). F_{ST} has a theoretical minimum value of 0 (indicating free genetic exchange with no isolation) and a theoretical maximum of 1 (indicating complete genetic isolation). However, F_{ST} rarely reaches the maximum of 1, and an F_{ST} value above 0.15 denotes considerable isolation between populations.

For multilocus data, differentiation between subpopulations can also be detected by ordination methods, such as principle coordinate, principle component, or nonmetric multidimensional scaling analyses (97). These methods assign multilocus genotypes to subpopulations based on the number of shared alleles.

Methods of phylogenetic analysis can be used to detect subpopulations that have been isolated for a considerable time frame. Limited gene flow among subpopulations over time allows the accumulation of specific mutations in each subpopulation, and eventually, a sufficient number of mutations are accrued to distinguish these subpopulations phylogenetically. Subpopulations that are supported by robust phylogenies and strong statistical significance are considered cryptic species (137, 138).

METHODS OF ECOLOGICAL SAMPLING

Although *C. neoformans* and *C. gattii* are notable for their pathogenicity, these species normally occupy a saprobic existence. Therefore, to understand the population structure of these pathogens, it is necessary to examine environmental samples. To establish the habitats of cryptococcal isolates in nature requires extensive field work. Investigators must collect the appropriate terrestrial, aquatic, and airborne samples. In the laboratory, these environmental samples are processed and examined for the presence of *Cryptococcus*. The first environmental isolations were accomplished by Chester W. Emmons, who collected samples of soil and feces from avian roosting areas and used a mouse-inoculation method to recover *C. neoformans* (40, 41). These pioneering studies also presaged the recognition of significant phenotypic and genetic variation among isolates of *C. neoformans* (105). For example, Hasenclever and Emmons compared the murine virulence of natural isolates of *C. neoformans* and noted thousandfold differences in the 50% lethal doses of isolates from the same fecal sample (62).

The methods to select and identify pathogenic *Cryptococcus* in environmental samples are facilitated by the unique ability of these yeasts to produce laccase (diphenoloxidase), an enzyme that oxidizes diphenolic substrates to form long polymers of melanin. This helpful phenotype was first discovered by Friedrich Staib, who observed that the growth of *C. neoformans* on agar supplemented with seeds of the Ethiopian herb *Guizotia abyssinica* (niger seed) stimulated the production of melanin, which is associated with the cell walls and imparts a brownish, nondiffusible pigment to the colonies (129). Today, the preferred selective medium for the isolation of *C. neoformans* from nonsterile samples continues to be Staib's medium, an aqueous infusion of ground niger seeds with the addition of glucose, biphenyl, and antibiotics. In addition to niger seed, caffeic acid, catecholamines, and other diphenolic compounds also serve as substrates for laccase. Very few other fungi produce a laccase with this specificity. Furthermore, as described in other chapters, laccase is one of three properties, along with the polysaccharide capsule and ability to grow at 37°C, that are almost always required for cryptococcal virulence (108). The melanin that results from laccase protects the yeast cells from several potent host defenses (115, 161), as well as environmental competitors and predators. As described in the following section on environmental niches, the medium of Friedrich Staib has become a crucial tool for the isolation of *Cryptococcus* from a wide variety of natural sites.

C. NEOFORMANS

Ecology

The two genetically isolated varieties of *C. neoformans*, *C. neoformans* var. *grubii* (serotype A) and *C. neoformans* var. *neoformans* (serotype D), share 95–96% DNA sequence identity (67). In addition, diploid AD hybrids are commonly isolated from environmental samples and patient specimens (73, 86, 114, 143). Although the ecology of *C. neoformans* has been studied extensively, as detailed in chapter 18 of this volume, its ecology has not been completely defined. Two vastly different habitats have been established as natural reservoirs for *C. neoformans*: the excreta of feral pigeons (*Columba livia*) and decayed wood. There have been numerous, global isolations of strains of serotype A, D, and AD hybrids from the decayed wood of many species of trees (44, 55, 68, 78, 86, 120–122) and the feces of pigeons and other birds (2, 27, 29, 41, 42, 45, 50, 65, 72, 86, 96, 117, 123, 126, 128).

The association of *C. neoformans* with decayed wood and vegetation is not surprising, and *C. gattii* is commonly isolated from similar environments. Closely related basidiomycetous yeasts within the order Tremellales are lignicolous and commonly associated with wood (47). Current evidence suggests that decayed wood is a primary ecological niche for *C. gattii*, and depending upon the geographical location, a variety of indigenous trees may harbor this species. For example, *C. gattii* has been perennially isolated from eucalyptus trees in Australia, almond trees in Colombia, and fir trees in southwestern Canada.

In contrast, *C. neoformans* has a well-documented affinity for pigeon feces (2, 27, 29, 41, 42, 45, 50, 65, 72, 86, 96, 117, 123, 126, 128). Much less frequently, it has been isolated from habitats associated with the avian excreta of other birds (1, 21, 93, 128). Considering the vast environmental supply and global distribution of excreta from vertebrate animals, why is avian feces, and especially pigeon feces, particularly attractive to *C. neoformans*? Why does *C. gattii* not prefer pigeon habitats?

Species of *Cryptococcus* are not fermentative, and in saprobic settings, they are inhibited by UV light and temperatures above 44°C. Beginning with the earliest isolations from avian habitats, analyses of pigeon feces determined that desiccated pigeon excrement provided an excellent natural substrate for the growth and enrichment of *C. neoformans*. In this milieu, the yeast cells are able to catabolize the high concentrations of urea, catecholamines, and other nitrogenous compounds (48). They also produce laccase and become melanized in avian manure (116), which provides some protection against UV radiation, temperature extremes, and oxidative compounds. The melanin also chelates silver and perhaps other toxic heavy metals (53), and it protects against degradative enzymes (124). In addition, avian feces may deter competing or deleterious microbes. For example, certain species of *Candida* and *Pichia* produce killer toxins that are active against *C. neoformans*, and these killer strains, which are most active at pH 4 to 6, have been isolated from the feces of pigeons and canaries (34); however, *C. neoformans* may prefer more acidic conditions (146). Avian feces may prevent the growth of potential predators, such as *Acanthamoeba* (132). Birds are rarely infected with *Cryptococcus*, probably because the average avian body temperature of 42.5°C is too high (85). However, the beaks, crops, claws, and feathers of pigeons have been shown to serve as vectors for the dissemination of *C. neoformans* (135), and the yeast cells can survive passage through their digestive tracts (85). In laboratory studies, pigeon manure medium supports the growth and mating of *C. neoformans* (113, 130).

This environmental niche is a highly plausible source of human cryptococcosis. In sites where pigeon manure has been shown to harbor *C. neoformans*, air sampling has detected aerosols of the yeast cells or basidiospores or a mixture of both. Yeast cells isolated from pigeon manure tend to have minimally sized capsules (6, 125, 131).

The advances in genotyping described in this chapter will continue to improve, and with additional markers, MLST will become more discriminating. In the near future, it will be feasible and affordable to perform comprehensive microarray typing or whole-genome sequencing on environmental isolates. These advances will provide the means to address the following questions: Is there any evidence of ecological specialization among strains from pigeon feces and wood? Can the same strains or genotypes occupy both ecological niches? Are some isolates better able to propagate on feces, and others more suited to arboreal growth? Assuming that ecological adaption is not an epigenetic phenomenon, these genomic methods may also identify the genes and pathways that govern the preference for one niche over another. Although the answers to these questions are largely unknown, there has been substantial progress in addressing these issues. In the following sections, we will discuss advances in our understanding of the intertwining effects of ecology and genetics on the lifestyle of *C. neoformans*.

Population Structure of *C. neoformans* var. *grubii* (Serotype A)

C. neoformans var. *grubii* is responsible for more than 90% of all cases of cryptococcosis worldwide. In the environment, strains of serotype A are commonly isolated from pigeon excreta and decayed wood (91, 120, 141), and its population structure and molecular epidemiology have been extensively studied. Earlier reports on the molecular epidemiology of serotype A utilized methods of RFLP (50), multilocus enzyme electrophoresis (MLEE) (14), and PCR fingerprinting (15, 103, 104), and they detected low genetic diversity and clonality in the populations. These results were later supported by the first phylogeographic study of *C. neoformans* by Xu et al. (153), who analyzed the gene genealogies of three nuclear loci and one mitochondrial locus in 14 serotype A isolates and detected no evidence of phylogeographic structure in the global population sample. Similar results were obtained in a study by Boekhout et al. (8), who used AFLP genotyping and analyzed genetic diversity among 96 clinical and environmental isolates of serotype A from 13 countries and detected no evidence of geographic structure. Similarly, Litvintseva et al. (86) used AFLP genotyping and analyzed the genetic diversity of clinical and environmental isolates of serotype A from the United States and reported an extremely low level of genetic variability, as only 12 genotypes were detected among 826 isolates of serotype A. In addition, all earlier studies demonstrated that both clinical and environmental populations of serotype A are dominated by isolates with a single mating type allele, *MAT*α (8, 86, 145, 157).

AFLP and DNA fingerprinting were used to recognize the existence and global dispersion of the VNI and VNII subpopulations (8, 101). VNI strains are common in clinical and environmental populations around the world. This subpopulation is characterized by very low genetic diversity and dominated by isolates with the *MAT*α mating type. Conversely, VNII isolates are less common, they are primarily isolated from clinical samples, and only a small number of environmental strains of VNII have been described (92). However, all known VNII strains have the *MAT*α mating type. All of these population studies agreed that strains with the VNI and VNII molecular type are globally distributed and lack any apparent geographic structure (8, 92, 101, 111).

This understanding of the population structure of *C. neoformans* var. *grubii* changed in 2003, when an unusual set of clinical isolates of serotype A was discovered in Botswana (88). This population was highly genetically variable and contained an unprecedented proportion of isolates with the *MAT***a** mating type, which were capable of mating and recombining in the laboratory. Isolates with the *MAT***a** mating type were previously thought to be extinct, but they were first reported to exist in 2000 and 2001 (80, 144). In a subsequent investigation, an MLST genotyping method was implemented using 12 unlinked nuclear loci that are located on nine chromosomes. This MLST protocol was applied to 102 isolates of serotype A from 15 countries (92). The results confirmed that the unusual Botswanan population was extraordinarily diverse and geographically restricted to southern Africa (92). This analysis determined that the global population of serotype A consists of three genetically isolated subpopulations, designated VNI, VNII, and VNB; strains of VNI and VNII were globally dispersed, whereas VNB strains were endemic to Africa (92). Figure 1 depicts these clades of serotype A.

The existence of these three genetically isolated clades of serotype A was independently substantiated in a comprehensive analysis by Bovers et al. (11). They used a different collection of strains and six MLST loci to conduct an independent phylogenetic analysis of 117 haploid strains of *C. neoformans* and *C. gattii* (11). Unlike the original MLST study (92), this analysis did not include a substantial number of VNB strains and therefore did not identify the VNB group as the third major molecular type of serotype A. Nevertheless, the combined gene genealogy included four strains that clustered with previously described VNB strains from Botswana (11). However, only one of these strains was isolated from Africa; the other three were obtained from South America (11), which suggested that VNB strains are not entirely restricted to southern Africa. Similar results were obtained in another MLST study by Ngamskulrungroj et al. (111), who also detected a small number of VNB strains from South America. These independent MLST studies suggest that the genetic diversity among non-African VNB strains is significantly lower than that of VNB isolates from Africa; however, these studies only included a few South American VNB strains (11, 111), whereas 32 VNB genotypes were detected in Africa (92). In addition, no isolates with the *MAT***a** mating type have been discovered among the non-African isolates of VNB. Overall, the current data suggest that although a small number of VNB strains may exist outside Africa, most VNB isolates are geographically confined to sub-Saharan Africa.

A more recent analysis of a large number of environmental isolates from South Africa and Botswana demonstrated

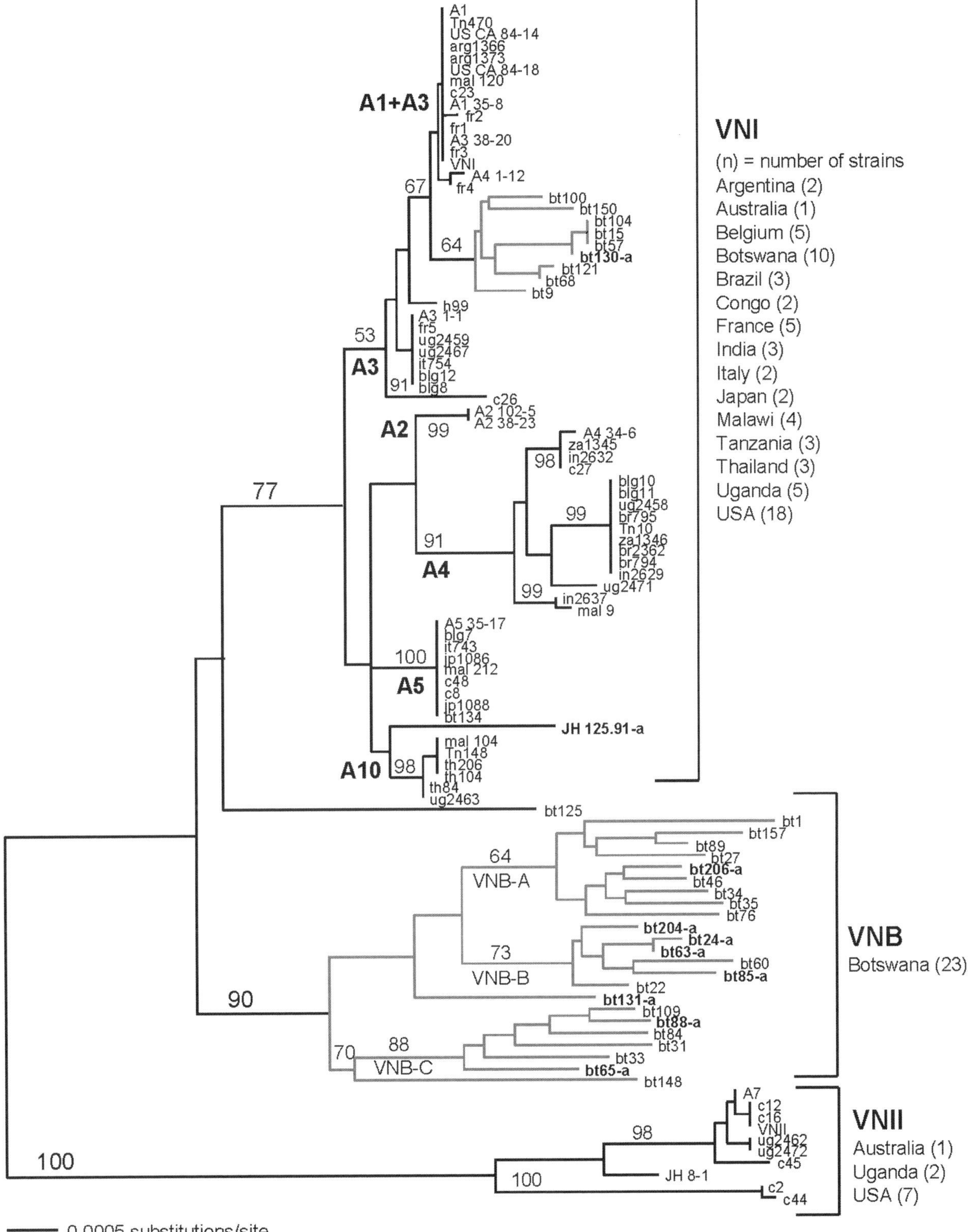

FIGURE 1 Genetic relationships of MLST genotypes among 98 isolates of *C. neoformans* serotype A visualized by the neighbor-joining dendrogram. Numbers on each branch indicate the bootstrap values >50%, based on 500 replications. Vertical lines represent strains with identical genotypes. The clades labeled A1 to A5 and A10 correspond to AFLP genotypes (88). VNB-A and VNB-B are two groups within the VNB subpopulation from Botswana. Isolates with the *MAT***a** mating type are shown in bold and designated with "**a**." From Litvintseva et al. (92) and used with permission.

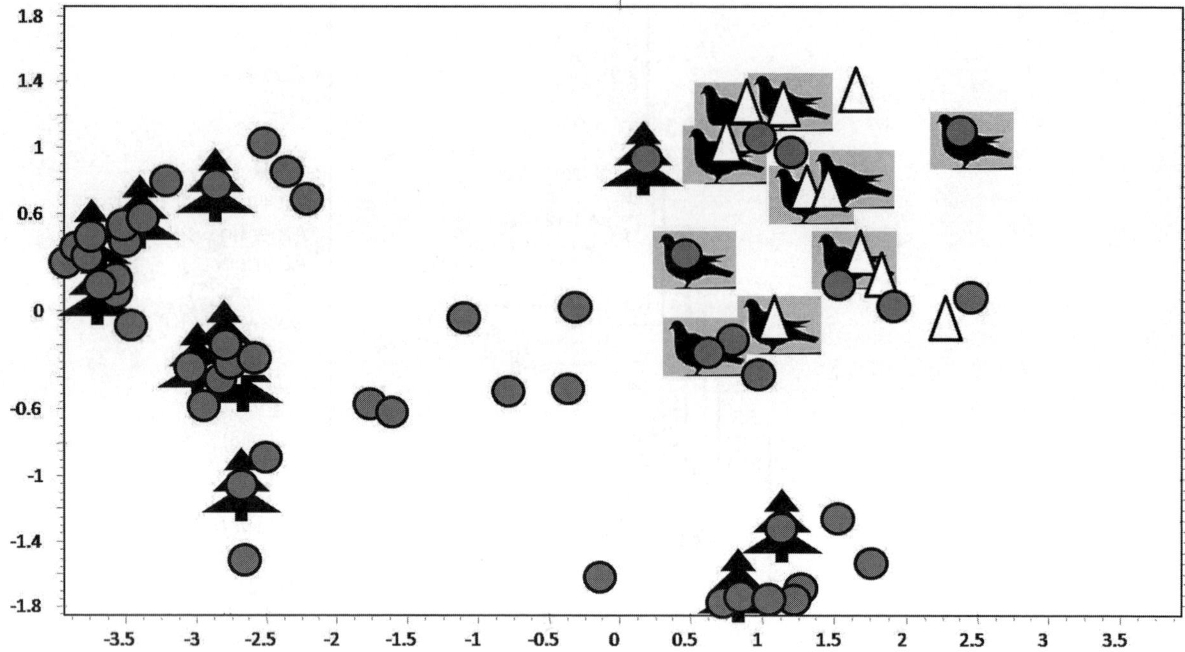

FIGURE 2 Genetic relationships among 58 MLST genotypes of *C. neoformans* serotype A visualized by principal component analysis. Each circle or triangle represents a genotype with a unique eight-digit allelic profile. Gray circles denote genotypes of strains that are endemic to Africa, and open triangles denote genotypes of cosmopolitan strains. Genotypes that have been isolated from African trees are marked with tree symbols, and genotypes that have been isolated from pigeon feces are marked with pigeon symbols. Genotypes without a tree or pigeon symbol are clinical strains. Note that all global strains are closely related and associated with pigeon habitats. Conversely, there is significant diversity among genotypes that are unique to Africa, and many of these African genotypes are associated with African trees (90, 92).

that southern Africa harbors two ecologically and geographically isolated subpopulations of serotype A, which include (i) the indigenous population of VNB isolates, which has an ecological niche in rural native trees, and (ii) an introduced population of VNI isolates, which are restricted to urban areas and associated with pigeon feces (90). Significantly, strains of both the native and introduced lineages were isolated from patients (90). The arboreal population consisted of isolates that were endemic to Africa, and the pigeon coprophilic population consisted of the cosmopolitan VNI isolates (Fig. 2). Although arboreal strains were isolated from a number of indigenous or introduced trees in southern Africa, the highest number of isolates was obtained from wood hollows of the native African mopane tree, *Colophospermum mopane*, which is endemic to southern Africa (90).

Recombination in *C. neoformans* var. *grubii* (Serotype A)

Mating and recombination are extant among isolates of the African VNB subpopulation of serotype A. One-fourth of the clinical VNB isolates from Botswanan patients were shown to possess the rare *MAT***a** mating type allele, and linkage equilibrium was detected between AFLP loci, implying a history of sexual reproduction (88). Subsequent studies of environmental isolates in Africa revealed (i) fertile isolates with the *MAT***a** mating type, (ii) incongruence among genealogies constructed from the DNA sequences of multiple genes, and (iii) evidence of genetic recombination within the loci by analysis of minimal ancestral recombination graphs (88, 90, 94).

Among the global subpopulations of VNI and VNII isolates, the extent of mating and recombination remains to be determined. These isolates exhibit relatively little genetic variation (30, 92), and isolates with the *MAT***a** mating type are extremely rare (80, 112, 144, 145). Both observations suggest that sexual reproduction and recombination in these subpopulations is limited (14, 91). However, these isolates possess a functional and expressed mating type locus, which provides the potential for mating to occur in nature (80, 112).

Indeed, several studies have determined that the population structure of global isolates of serotype A cannot be explained wholly by clonal reproduction. For example, a sample of clinical isolates was genotyped with protein markers using MLEE (15), and subsequent calculations of the I_A and r_d values revealed linkage equilibrium among the loci, an indication of recent or past recombination (18). Similarly, linkage equilibrium was observed among loci in a highly monomorphic environmental population of serotype A from the United States (86). Other groups have reported evidence of recombination among (i) veterinary isolates from Australia with the *MAT*α mating type (16), (ii) environmental isolates of serotype A strains from trees in India (63), and (iii) global samples of VNI, VNII, and VNIII (serotype D) (111). In addition to evidence of linkage equilibrium, vestigial or current recombination is supported by the incongruence of the genealogies of several genic loci in serotype A (63, 90) and AD hybrids (151). Thus, despite the dearth of *MAT***a** strains, sexual reproduction among global isolates is a distinct possibility.

Other, nonexclusive mechanisms have been proposed to explain the apparent linkage equilibrium in the global population of VNI. The capacity for same-sex mating (81) may result in occasional recombinational events among isolates with the same mating type (16, 23), and analyses of diploid hybrid strains possessing two different *MAT*α alleles have confirmed that same-sex mating occurs in nature (82, 84). Another possibility is that sexual recombination is geographically restricted to Africa (90), but contemporary populations of VNI and VNII may have originated in Africa, and their linkage equilibrium is evidence of ancient, ancestral recombination. Both mechanisms may have contributed to the current population structure of the VNI group.

Population Structure of *C. neoformans* var. *neoformans* (Serotype D)

Since population studies of *C. neoformans* include relatively few isolates of serotype D (i.e., VNIV), there is limited information on their population structure. An initial exploration of genetic diversity in *C. neoformans* used 10 MLEE markers and identified five genotypes among six geographically distant strains of serotype D (13). Seven of these loci showed no variation, and there was no evidence of recombination at the population level among alleles at the three polymorphic loci (13). Subsequent studies using different genetic markers and strains from different sources found no evidence of recombination within serotype D (11, 14). The lack of evidence of recombination among the samples of serotype D is not unexpected because most, if not all, of the analyzed strains had the same mating type, *MAT*α (11, 157).

Although clonality has appeared to be dominant, analyses of AD hybrid strains discerned evidence of recombination in populations of serotype D as well as serotype A. In contrast to the predominance of the *MAT*α mating type in clinical and environmental isolates of both serotypes A and D (157), the ratios of mating type alleles in AD hybrid strains (see below) are generally balanced because most hybrids possess both mating type alleles. During the hybridization process, some acquired *MAT***a** from serotype D and *MAT*α from serotype A, and others received *MAT***a** from serotype A and *MAT*α from serotype D (79, 157). In a sequence analysis of two polymorphic nuclear genes, each locus of 14 AD hybrids contained distinct alleles; one allele was similar to the A strains, and the other to the D strains. However, within both A and D allelic groups, there was significant incongruence between the gene genealogies, denoting recombination in natural populations of both serotypes A and D (151). Indeed, the A and D alleles from the two loci also indicated multiple hybridization events. Since the hybrids were all of recent origin, the results suggested that mating and recombination must have occurred in recent history in natural serotype D (and A) populations (151). Direct evidence of recombination among serotype D isolates was provided by a study of 58 environmental strains of serotype D from North Carolina that were analyzed by AFLP markers, and five genotypes were identified. Although the genotypic diversity was limited and there was overrepresentation of each genotype, both indications of clonality, in the clone-corrected sample, the associations of alleles (i.e., AFLP bands) were not significantly different from random, which is consistent with genesis by sexual reproduction (86).

Because relatively few samples of serotype D strains are available for analysis, relationships among geographic populations of this serotype have not been thoroughly investigated. However, the accretion of evidence is consistent with widespread dispersion of serotype D (VNIV) strains. For example, strains from distant geographical areas often share identical genotypes (9, 13, 153). The potential means of dispersal include wind currents, birds such as pigeons (3, 86), and anthropogenic activities such as the transportation of goods and human travel. More rigorous analyses of large serotype D samples from diverse areas are required to quantify the extent of gene flow among geographic populations in nature.

Despite their global distribution, it has long been recognized that proportionately more clinical cases of serotype D occur in Europe (37, 73, 140).

AD Hybrids

Serotype AD strains are produced by hybridization of an isolate of serotype A and an isolate of serotype D. Most isolates of serotypes A and D are haploid, but AD strains are diploid or aneuploid, as they contain two sets of chromosomes and often two mating type alleles, *MAT***a** and *MAT*α, one from each serotype (32, 79, 149). As mentioned above, AD hybrid strains are regularly isolated from clinical and environmental samples (73, 86, 114, 143).

Genotyping studies have elucidated the population structure and evolution of each component genome. For example, a comparison of *LAC1* and *URA5* gene sequences of 14 clinical AD isolates identified abundant polymorphisms and two haplotypes corresponding to the A and D genomes. The gene genealogies were consistent with the hypothesis that multiple recent hybridization events between strains of serotypes A and D are responsible for the current distribution of serotype AD strains (149, 151). These and other data confirm that serotypes A and D represent monophyletic lineages.

As we previously noted, the global population of serotype A is dominated by isolates with the *MAT*α mating type (Aα); however, some AD strains were discovered to possess the extremely rare serotype A *MAT***a** allele (A**a**) (79, 86, 157). In a study to determine the origin of these A**a** strains, AFLP genotyping and partial sequences of three genes were used to compare 45 representative strains of serotype A with 12 AD hybrid strains from China, Italy, Kuwait, and the United States (87). Six of the AD hybrids possessed the Aα allele and clustered phylogenetically with VNI strains of serotype A. The other six AD hybrids had the A**a** genotype and clustered with the VNB clade, indicating that the serotype A contribution to these AD hybrids from Asia, Europe, and North America was likely acquired from members of the VNB subpopulation in Africa (87). In addition, these AD hybrids demonstrated enhanced fitness or hybrid vigor, such as greater resistance to UV irradiation than haploid VNB strains (87). Other studies have shown that AD hybrids have the ability to undergo same-sex mating (82). Laboratory-constructed AD hybrids exhibited comparatively more chromosomal rearrangements (134) but comparable murine virulence when they possessed both mating type alleles (83). These cumulative results support the hypotheses that (i) A**a**Dα hybrid strains originated in southern Africa from a cross between strains of serotypes A and D and (ii) this fusion produced hybrid strains with increased fitness, enabling the serotype A *MAT***a** genome, which is apparently geographically restricted, to survive, emigrate, and propagate throughout the world.

Although most naturally occurring AD hybrid strains are incapable of mating in the laboratory, some are self-fertile. That is, when stimulated by growth on mating medium in the absence of a mating partner, they produce hyphae, basidia, and basidiospores, although few of the basidiospores are able to germinate (33, 79, 136). Nevertheless, this process provides another potential mechanism for genetic exchange and evolution of *C. neoformans*.

In addition to these studies, which were focused on nuclear genetic markers, recombination in the mitochondrial genome was reported in 2004 by Toffaletti et al., who demonstrated nonparental segregation of mitochondrial markers in the progeny of a cross between strains of serotype A and serotype D (139). Most of the AD hybrids in this study inherited mitochondria from only one parent, but a few inherited mitochondria from each parent and possessed recombinant mitochondria (139).

Out-of-Africa Origin of Serotype A

Earlier studies of clinical isolates from Botswana and the recent examination of the environmental and clinical samples from South Africa and Botswana indicated that the population of serotype A in southern Africa is more genetically variable than the global population (88, 90, 92). The first MLST study involved 102 global isolates and detected 32 VNB, 18 VNI, and 7 VNII genotypes (92). The second study used eight of the original MLST loci and detected an even greater level of genetic diversity in the African population (90): 49 MLST genotypes were identified in the African population, and only 9 and 5 genotypes were detected among global VNI and VNII populations, respectively.

This unique southern African population is characterized by extraordinary genetic diversity and an isolated environmental niche in the mopane tree. To test the hypothesis that this sequestered population of serotype A strains is the progenitor of the species, haplotype network analysis was performed for each MLST locus, and the ecological and geographical origins of each haplotype were mapped on haplotype networks. As shown in Fig. 3, the results indicated that numerous haplotypes (VNB strains) from mopane tree isolates (black circles) occupy both ancestral (internal) and derived (distal) positions, which supports an ancient association between these strains and mopane trees. Conversely, there is a scarcity of the haplotypes of VNI strains (gray circles), which represent the global population. They occupy distal positions in the networks and are almost always associated with pigeons. Most tellingly, the ancestral haplotypes for all eight loci are found among the mopane tree strains.

These findings support the "out-of-Africa" hypothesis for the evolution of serotype A. According to this model, (i) the ancestral population of serotype A is endemic to the southern region of sub-Saharan Africa; (ii) this ancestral population underwent a range expansion, and a small number of strains established a new ecological niche in the excreta of pigeons (*C. livia*), where they proliferated clonally and became genetically isolated; and (iii) in recent centuries, as pigeons were introduced and dispersed globally, they functioned as vectors to disseminate these strains of serotype A.

Although the data indicate that southern Africa represents the ancestral origin of serotype A, the number of expansion events and the extent of emigration from Africa have not been determined. For example, the VNB subpopulation was thought to be endemic to southern Africa, but recent reports describe the isolation of two VNB strains from Colombia and Brazil, respectively (11, 111). The Colombian isolate was clinical, and the travel history of the patient is unknown, but the Brazilian isolate (Hamden C3-1) was recovered from pigeon droppings, which clearly suggests the possibility of a reservoir of VNB strains in South America.

There are several explanations for the presence of VNB strains in South America. As noted above, isolates of serotype A with the *MAT***a** mating type are extremely rare outside Africa, but half of the global AD hybrid strains carry this rare A**a** allele (87). They originated from a cross with the African strains and emerged from Africa as AD hybrids. Some of these hybrids may have undergone a reduction to haploidy, giving rise to VNB genotypes in South America, or the VNB isolates from South America may represent unrecognized AD hybrids.

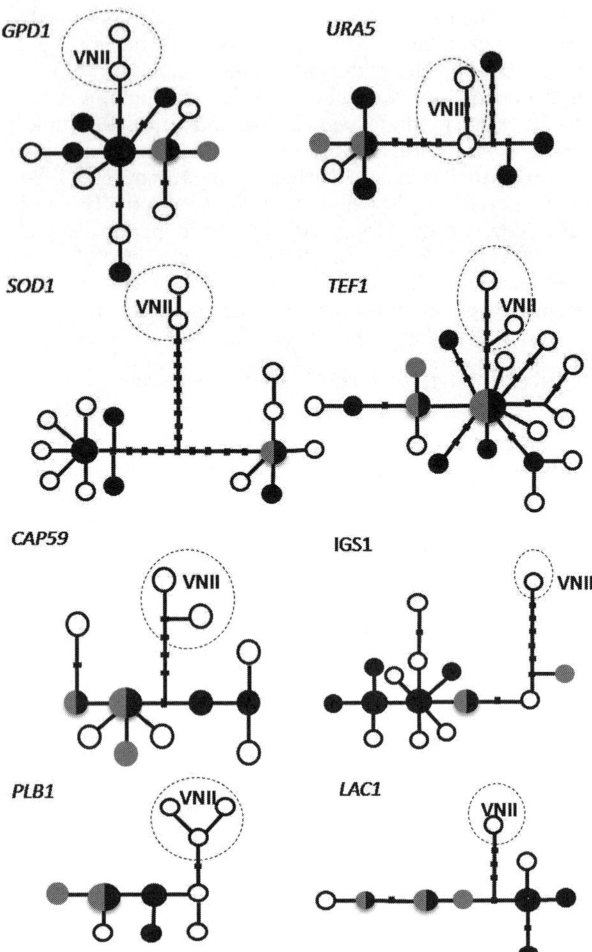

FIGURE 3 Haplotype networks of eight MLST loci of *C. neoformans* serotype A. The strain haplotypes (alleles) that are unique to Africa and associated with African trees are shown in black circles. Haplotypes of cosmopolitan strains are associated with pigeon feces and shown in gray circles. Circles that are half black and half gray indicate haplotypes of strains that were isolated from both African trees and pigeon feces. Haplotypes from the VNII group are used as an outgroup and labeled. The ancestral haplotypes are internal, and derived haplotypes occupy tip or distal positions. Small dashes on lines connecting haplotypes represent mutational steps separating the alleles. Recombinant haplotypes have been excluded. For each individual gene, the ancestral haplotype is always associated with an African tree. The analyses indicate significantly greater genetic diversity among haplotypes from Africa compared to the strains with cosmopolitan haplotypes (90, 92).

Another explanation for the phylogeny of serotype A is possible. There may have been a historical, global expansion of many genotypes from Africa, leading to the geographical dispersion of both VNI and VNB strains from Africa. The detection of VNB genotypes in South America and AD hybrids is consistent with such an event. Subsequently, genetic drift and selection may have eliminated VNB strains from most regions of the world, although haploid survivors exist in Africa and South America. Chapter 20 of this volume provides additional information regarding the diversity of African isolates.

C. GATTII

Molecular studies using PCR fingerprinting (101), scoring AFLP markers (8, 91), and MLST (11, 51, 111) identified four genetically isolated groups (molecular types or subpopulations) within *C. gattii*, designated VGI, VGII, VGIII, and VGIV, which may represent cryptic species (11, 51, 111).

Ecology

Environmental sampling suggests that strains from all four molecular types occupy similar ecological niches and are associated with decayed wood, plant debris, and soil. In particular, eucalyptus trees represent the most common and well-described source of *C. gattii*. However, it is unclear whether the higher incidence of isolation of *C. gattii* from eucalyptus trees is caused by a "true" association with this tree or whether this apparent association is the result of biased sampling. The initial environmental isolations of *C. gattii* on three continents, Australia, Asia, and North America, implicated *Eucalyptus camaldulensis* and other species of *Eucalyptus* (28, 39, 118). Review of the literature confirms that eucalyptus trees have been sampled more frequently than other species (44, 54, 122). Since the discovery of *C. gattii* in the hollows of *E. camaldulensis* in Australia (38, 119), the association of *C. gattii* with *Eucalyptus* species has been well documented in Australia (58, 142), South America (4, 27, 54, 72, 101, 122), North America (118), and Asia (28).

Most environmental as well as clinical isolates of *C. gattii* have been strains of serotype B, which are VGI or VGII (see Table 1). Strains of molecular types VGI, VGII, and VGIII (serotype C) have all been isolated from eucalyptus trees (11). Other trees have also been shown to harbor *C. gattii* (5, 22, 44, 49, 54–56, 63, 68, 77, 120, 121, 141). Multiple lines of evidence suggest that *C. gattii* does not have a strong preference for a specific tree and frequently becomes associated with native vegetation. For example, *E. camaldulensis* is native to Australia, and several reports have confirmed that Australian eucalyptus trees are heavily colonized by *C. gattii* (58, 142). In India, *C. gattii* has been isolated from native Indian trees, such as *Azadirachta indica*, *Mimusops elengi*, *Tamarindus indica*, *Mangifera indica*, *Pithecolobium dulce*, and *Syzygium cumini* (55, 120). In South America, *C. gattii* is frequently isolated from the naturalized eucalyptus and almond trees (*Terminalia catappa*) (22); however, it is also common in native trees, such as *Moquilea tomentosa*, *Licania tomentosa*, and *Ficus soatensis* (54, 77, 141). In North America, strains causing the outbreak on Vancouver Island are associated with Douglas fir (*Pseudotsuga menziesii*), alder (*Alnus rubra*), and other trees that are native to the Pacific Northwest (68, 70). Chapter 18 in this volume provides a current list of the geographic distribution and arboreal isolations of *C. gattii*.

C. gattii has also been isolated from soil, water, and air (68). Because *C. gattii* is not restricted to particular host species and can rapidly colonize native tree populations, the introduction of nonnative trees provides a plausible explanation for the global distribution and widespread clonality among global populations. *C. gattii* may have been introduced to Asia and the Western Hemisphere with eucalyptus and almond trees. The population structure of *C. gattii* is consistent with this hypothesis.

Population Structure

Because strict genetic isolation between different molecular types of *C. gattii* has been well established (11, 51, 69), the population structures of each molecular type (VGI, VGII, VGIII, and VGIV) should be analyzed separately. Unfortunately, most of the phylogeographic studies of *C. gattii* used combined data from different molecular types, which do not provide sufficient resolution for assessing diversity within each group (7, 11, 44). The most comprehensive phylogeographic analysis of *C. gattii* was reported by Kidd et al., who addressed the genetic diversity and global population structure of the VGII molecular group (69). In this study, the authors analyzed gene genealogies of four loci in the VGI and VGII molecular types and detected no evidence of geographic isolation among populations from different geographic areas, as isolates with identical genotypes were detected on different continents. In addition, they observed no correlation between sequence divergence among the isolates and the geographic distance among the populations, suggesting that the global spread of VGII occurred relatively recently (69). These data are supported by the results of Fraser et al. (51), who analyzed the genetic diversity of *C. gattii* populations by using multilocus genetic analyses and found identical isolates of VGI, VGII, and VGIV molecular types in geographically isolated locations. Similarly, Bovers et al. (11) detected no evidence of phylogeographic structure within these four major subpopulations of *C. gattii*.

However, all studies agree that the genotypes associated with the ongoing outbreak on Vancouver Island are genetic variants of VGII, designated VGIIa and VGIIb, which have spread to mainland British Columbia (11, 19, 35, 51, 69). Strains of VGIIa appear to be geographically restricted to the Pacific Northwest (see chapter 23 in this volume). In 2008 and 2009, it was clear that the outbreak had expanded to the U.S. states of Washington and Oregon with additional human and veterinary cases (35). Furthermore, a novel MLST genotype, VGIIc, was discovered among the animal and clinical isolates from Oregon (19). Overall, current data on the phylogeographic structure of *C. gattii* are consistent with the hypothesis of recent global dispersal of VGII (69). However, comprehensive geographic sampling and more sophisticated genetic analyses, including methods of coalescent analysis (26, 71), will improve understanding of the global population structure and origins of *C. gattii*.

Recombination

Whether natural isolates of *C. gattii* can undergo sexual reproduction and recombination in the environment remains an open subject. Although the genome of *C. gattii* contains an intact, functional mating type locus (52), and some isolates are capable of mating and recombination in the laboratory (25, 52), most natural populations are dominated by isolates with the *MAT*α mating type, and many isolates are unable to produce basidia and basidiospores in the laboratory (25, 52, 58). More specifically, most known isolates of VGIII and VGIV (serotype C) possess the *MAT*α mating type allele (11, 44, 69, 91, 111). Many natural populations of the VGII molecular type are also dominated by isolates with the *MAT*α mating type allele. An Australian population of VGII (25) and the isolates from Vancouver Island (51, 52, 70, 133) contain only *MAT*α isolates. However, a number of fertile VGII isolates with the *MAT***a** mating type were discovered in Colombia (43, 44), but to our knowledge, their genetic relationship to other VGII strains, especially those causing the ongoing outbreak in Canada and the northwestern United States, has not yet been established.

Dee Carter and her collaborators have documented the only population of C. *gattii* that maintains equal proportions of isolates with the *MAT***a** and *MAT*α mating types, which is found in certain geographic locations in southern Australia, where 50% of the environmental VGI strains possess the rare *MAT***a** allele (57); however, whether or not this population undergoes sexual reproduction remains an open question. The earlier studies detected very little evidence of genetic exchange between *MAT*α and *MAT***a** isolates in this Australian population (57). These investigators subsequently performed AFLP analysis of 30 strains of *C. gattii* isolated from tree hollows of *E. camaldulensis* in southern Australia and demonstrated that strains with *MAT*α and *MAT***a** mating types formed two distinct, well-supported clades on the unweighted-pair group method using average linkages dendrogram (58). There was significant linkage disequilibrium among loci in this population, which was consistent with a predominantly clonal mode of reproduction (58). However, later reports from the same group suggested an alternative explanation for observed clonality in this population. The apparent linkage disequilibrium among the loci in these Australian isolates may be an artifact of sampling several highly subdivided populations residing in different tree hollows, and their physical separation likely precluded contact, mating, and genetic exchange among the isolates (24, 127). Conversely, strains with the *MAT*α and *MAT***a** mating types were isolated from the same tree hollow. They were very closely related, and linkage equilibrium was detected among their AFLP loci (24). The authors hypothesized that mating and recombination occur among strains within the same tree hollow, but there are barriers to this interaction among strains from different tree hollows. Another study detected evidence of recombination in subpopulations consisting of isolates with only a single *MAT*α mating type from the same tree hollow (127), which supports the concept that same-sex as well as conventional mating may occur in these isolated subpopulations (51).

The above studies have focused on nuclear genes. Unambiguous evidence of recombination and hybridization was identified in the mitochondrial genome of *C. gattii*. Bovers et al. reconstructed a phylogeny of the mitochondria of 51 *C. gattii* strains using the DNA sequences of two mitochondrial loci, *MtLrRNA* and *ATP6* (12). This analysis generated monophyletic lineages for each of two subpopulations, VGII and VGIII; however, mitochondria of the VGI group evinced five haplotypes (12). Many VGI isolates contained fragments of mitochondrial DNA (mtDNA) from the VGII group. In addition, half of the analyzed VGIV strains contained the VGI allele at the *MtLrRNA* locus, indicating the possibility of recombination between isolates of *C. gattii* from different molecular types. These results were confirmed by Xu et al. (154), who analyzed sequences of five mtDNA genic loci (*ND2*, *ND4*, *ND5*, *ATP6*, and *COX1/ATP8*) of 50 isolates belonging to the two subpopulations, VGI and VGII. The analyses revealed greater sequence diversity among the strains of VGI than the sample of VGII, which is consistent with the observed diversity among the nuclear genes of VGI and VGII. The combined analyses of all five gene fragments indicated significant divergence between VGI and VGII. However, the five individual genealogies showed different relationships among the isolates, consistent with recent hybridization and mitochondrial gene transfer between the two lineages (154). The population genetic analyses of these multilocus data revealed evidence for predominantly clonal mitochondrial reproduction within both lineages. The data also provided clear signatures of recombination among mitochondrial genes within the VGII lineage.

This evidence for mtDNA recombination in *C. gattii* was unexpected because earlier studies showed that sexual crosses between VNI and VNIV strains of *C. neoformans* produced a pattern of exclusively uniparental mitochondrial inheritance (148, 148, 155, 159). Several hypotheses were proposed to explain the observed hybridization and recombination in the mitochondrial genomes of these two lineages, including biparental mtDNA inheritance in *C. gattii*, same-sex mating, and/or exposure to stresses such as high temperature and UV irradiation during mating (154). Same-sex mating and exposure to high temperature and UV irradiation during mating were previously determined to promote biparental mtDNA inheritance and mtDNA recombination in *C. neoformans* (156, 158).

Unlike most eukaryotes, the option of biparental mtDNA inheritance and recombination offers species of *Cryptococcus* the opportunity for mitochondrial genetic exchange and a potentially rapid response to the stress of adverse environmental conditions. In support of this supposition, many of the Vancouver Island outbreak strains of VGII were shown to be "hypervirulent" in mice and to possess comparatively upregulated mitochondrial virulence genes that promoted survival within macrophages (95). These mitochondrial changes may have resulted from recombinational events and precipitated the outbreak.

CONCLUSIONS

As this review has documented, strains of *C. neoformans* and *C. gattii* are globally distributed in a range of overlapping environmental habitats. Considering that most of the studies comparing the virulence phenotypes of natural isolates exhibit wide variation among clinical and environmental isolates (31, 62, 89, 99, 105), it is not surprising that their genomes, as proven by genotyping and phylogenetics, are similarly diverse. The implications of this ongoing research are profound because the data have clearly shown that cryptococcosis has multiple etiologies.

REFERENCES

1. **Abegg, M. A., F. L. Cella, J. Faganello, P. Valente, A. Schrank, and M. H. Vainstein.** 2006. *Cryptococcus neoformans* and *Cryptococcus gattii* isolated from the excreta of *Psittaciformes* in a Southern Brazilian zoological garden. *Mycopathologia* **161:**83–91.
2. **Abraham, M., M. S. Matthews, and T. J. John.** 1997. Environmental isolation of *Cryptococcus neoformans* var. *neoformans* from Vellore. *Indian J. Med. Res.* **106:**458–459.
3. **Ansheng, L., K. Nishimura, H. Taguchi, R. Tanaka, S. Wu, and M. Miyaji.** 1993. The isolation of *Cryptococcus neoformans* from pigeon droppings and serotyping of naturally and clinically sourced isolates in China. *Mycopathologia* **124:**1–5.
4. **Arguero Licea, B., D. Garza Garza, V. Flores Urbieta, and R. A. Cervantes Olivares.** 1999. Isolation and characterization of *Cryptococcus neoformans* var. *gattii* from samples of *Eucalyptus camaldulensis* in Mexico city. *Rev. Iberoam. Micol.* **16:**40–42. (In Spanish.)
5. **Baltazar, L. D., and M. A. Ribeiro.** 2008. First isolation of *Cryptococcus gattii* from the environment in the State of

Espirito Santo. *Rev. Soc. Bras. Med. Trop.* **41:**449–453. (In Portuguese.)

6. **Baroni, F. D., C. R. Paula, E. G. da Silva, F. C. Viani, I. N. G. Rivera, M. T. Barreto de Oliveira, and W. Gambale.** 2006. *Cryptococcus neoformans* strains isolated from church towers in Rio de Janeiro City, RJ, Brazil. *Rev. Inst. Med. Trop. São Paulo* **48:**71–75.
7. **Barreto de Oliveira, M. T., T. Boekhout, B. Theelen, F. Hagen, F. D. Baroni, M. S. Lazéra, K. B. Lengeler, J. Heitman, I. N. G. Rivera, and C. R. Paula.** 2004. *Cryptococcus neoformans* shows a remarkable genotypic diversity in Brazil. *J. Clin. Microbiol.* **42:**1356–1359.
8. **Boekhout, T., B. Theelen, M. R. Diaz, J. W. Fell, W. C. J. Hop, E. C. Abeln, F. Dromer, and W. Meyer.** 2001. Hybrid genotypes in the pathogenic yeast *Cryptococcus neoformans*. *Microbiology* **147:**891–907.
9. **Boekhout, T., A. van Belkum, A. C. A. P. Leenders, H. A. Verbrugh, P. Mukamurangwa, D. Swinne, and W. A. Scheffers.** 1997. Molecular typing of *Cryptococcus neoformans*: taxonomic and epidemiological aspects. *Int. J. Syst. Bacteriol.* **47:**432–442.
10. **Bovers, M., F. Hagen, and T. Boekhout.** 2008. Diversity of the *Cryptococcus neoformans-Cryptococcus gattii* species complex. *Rev. Iberoam. Micol.* **25:**S4–S12.
11. **Bovers, M., F. Hagen, E. E. Kuramae, and T. Boekhout.** 2008. Six monophyletic lineages identified within *Cryptococcus neoformans* and *Cryptococcus gattii* by multi-locus sequence typing. *Fungal Genet. Biol.* **45:**400–421.
12. **Bovers, M., F. Hagen, E. E. Kuramae, and T. Boekhout.** 2009. Promiscuous mitochondria in *Cryptococcus gattii*. *FEMS Yeast Res.* **9:**489–503.
13. **Brandt, M. E., S. L. Bragg, and R. W. Pinner.** 1993. Multilocus enzyme typing of *Cryptococcus neoformans*. *J. Clin. Microbiol.* **31:**2819–2823.
14. **Brandt, M. E., L. C. Hutwagner, L. A. Klug, W. S. Baughman, D. Rimland, E. A. Graviss, R. J. Hamill, C. Thomas, P. G. Pappas, A. L. Reingold, R. W. Pinner, and Cryptococcal Disease Active Surveillance Group.** 1996. Molecular subtype distribution of *Cryptococcus neoformans* in four areas of the United States. *J. Clin. Microbiol.* **34:**912–917.
15. **Brandt, M. E., L. C. Hutwagner, R. J. Kuykendall, R. W. Pinner, and Cryptococcal Disease Active Surveillance Group.** 1995. Comparison of multilocus enzyme electrophoresis and random amplified polymorphic DNA analysis for molecular subtyping of *Cryptococcus neoformans*. *J. Clin. Microbiol.* **33:**1890–1895.
16. **Bui, T., X. Lin, R. Malik, J. Heitman, and D. A. Carter.** 2008. Isolates of *Cryptococcus neoformans* from infected animals reveal genetic exchange in unisexual, α mating type populations. *Eukaryot. Cell* **7:**1771–1780.
17. **Burnett, J.** 2003. *Fungal Populations and Species*. Oxford University Press, Oxford, UK.
18. **Burt, A., V. Koufopanou, and J. W. Taylor.** 2000. Population genetics of human pathogenic fungi, p. 229–244. *In* R. C. A. Thompson (ed.), *Molecular Epidemiology of Infectious Diseases*, 1st ed. Arnold, London, UK.
19. **Byrnes, E. J., III, R. Bildfell, S. A. Frank, T. G. Mitchell, K. A. Marr, and J. Heitman.** 2009. Molecular evidence that the range of the Vancouver Island outbreak of *Cryptococcus gattii* infection has expanded into the Pacific Northwest in the United States. *J. Infect. Dis.* **199:**1081–1086.
20. **Byrnes, E. J., III, W. Li, Y. Lewit, J. R. Perfect, D. A. Carter, G. M. Cox, and J. Heitman.** 2009. First reported case of *Cryptococcus gattii* in the Southeastern USA: implications for travel-associated acquisition of an emerging pathogen. *PLoS ONE* **4:**e5851.
21. **Cafarchia, C., D. Romito, R. Iatta, A. Camarda, M. T. Montagna, and D. Otranto.** 2006. Role of birds of prey as carriers and spreaders of *Cryptococcus neoformans* and other zoonotic yeasts. *Med. Mycol.* **44:**485–492.
22. **Callejas, A., N. Ordóñez, M. C. Rodríguez, and E. Castañeda.** 1998. First isolation of *Cryptococcus neoformans* var. *gattii*, serotype C, from the environment in Colombia. *Med. Mycol.* **36:**341–344.
23. **Campbell, L. T., and D. A. Carter.** 2006. Looking for sex in the fungal pathogens *Cryptococcus neoformans* and *Cryptococcus gattii*. *FEMS Yeast Res.* **6:**588–598.
24. **Campbell, L. T., B. J. Currie, M. B. Krockenberger, R. Malik, W. Meyer, J. Heitman, and D. A. Carter.** 2005. Clonality and recombination in genetically differentiated subgroups of *Cryptococcus gattii*. *Eukaryot. Cell* **4:**1403–1409.
25. **Campbell, L. T., J. A. Fraser, C. B. Nichols, F. S. Dietrich, D. A. Carter, and J. Heitman.** 2005. Clinical and environmental isolates of *Cryptococcus gattii* from Australia that retain sexual fecundity. *Eukaryot. Cell* **4:**1410–1419.
26. **Carbone, I., and L. M. Kohn.** 2001. A microbial population-species interface: nested cladistic and coalescent inference with multilocus data. *Mol. Ecol.* **10:**947–964.
27. **Casali, A. K., L. Goulart, L. K. Rosa e Silva, Â. M. Ribeiro, A. A. Amaral, S. H. Alves, A. Schrank, W. Meyer, and M. H. Vainstein.** 2003. Molecular typing of clinical and environmental *Cryptococcus neoformans* isolates in the Brazilian state Rio Grande do Sul. *FEMS Yeast Res.* **3:**405–415.
28. **Chakrabarti, A., M. Jatana, P. Kumar, L. Chatha, A. Kaushal, and A. A. Padhye.** 1997. Isolation of *Cryptococcus neoformans* var. *gattii* from *Eucalyptus camaldulensis* in India. *J. Clin. Microbiol.* **35:**3340–3342.
29. **Chee, H. Y., and K. B. Lee.** 2005. Isolation of *Cryptococcus neoformans* var. *grubii* (serotype A) from pigeon droppings in Seoul, Korea. *J. Microbiol.* **43:**469–472.
30. **Chen, J., A. Varma, M. R. Diaz, A. P. Litvintseva, K. K. Wollenberg, and K. J. Kwon-Chung.** 2008. *Cryptococcus neoformans* strains and infection in apparently immunocompetent patients, China. *Emerg. Infect. Dis.* **14:**755–762.
31. **Clancy, C. J., M. H. Nguyen, R. Alandoerffer, S. Cheng, K. A. Iczkowski, M. Richardson, and J. R. Graybill.** 2006. *Cryptococcus neoformans* var. *grubii* isolates recovered from persons with AIDS demonstrate a wide range of virulence during murine meningoencephalitis that correlates with the expression of certain virulence factors. *Microbiology* **152:**2247–2255.
32. **Cogliati, M., M. C. Esposto, D. L. Clarke, B. L. Wickes, and M. A. Viviani.** 2001. Origin of *Cryptococcus neoformans* var. *neoformans* diploid strains. *J. Clin. Microbiol.* **39:**3889–3894.
33. **Cogliati, M., M. C. Esposto, A. M. Tortorano, and M. A. Viviani.** 2006. *Cryptococcus neoformans* population includes hybrid strains homozygous at mating-type locus. *FEMS Yeast Res.* **6:**608–613.
34. **Criseo, G., M. Gallo, and A. Pernice.** 1999. Killer activity at different pHs against *Cryptococcus neoformans* var. *neoformans* serotype A by environmental yeast isolates. *Mycoses* **42:**601–608.
35. **Datta, K., K. H. Bartlett, R. Baer, E. J. Byrnes, III, E. Galanis, J. Heitman, L. M. N. Hoang, M. J. Leslie, L. MacDougall, S. S. Magill, M. G. Morshed, and K. A. Marr.** 2009. Spread of *Cryptococcus gattii* into Pacific Northwest region of the United States. *Emerg. Infect. Dis.* **15:**1185–1191.
36. **Diaz, M. R., and J. W. Fell.** 2005. Use of a suspension array for rapid identification of the varieties and genotypes of the *Cryptococcus neoformans* species complex. *J. Clin. Microbiol.* **43:**3662–3672.
37. **Dromer, F., S. Mathoulin, B. Dupont, L. Letenneur, O. Ronin, and French Cryptococcosis Study Group.** 1996.

Individual and environmental factors associated with infection due to *Cryptococcus neoformans* serotype D. *Clin. Infect. Dis.* **23:**91–96.
38. **Ellis, D. H., and T. J. Pfeiffer.** 1990. Natural habitat of *Cryptococcus neoformans* var. *gattii*. *J. Clin. Microbiol.* **28:**1642–1644.
39. **Ellis, D. H., and T. J. Pfeiffer.** 1992. The ecology of *Cryptococcus neoformans*. *Eur. J. Epidemiol.* **8:**321–325.
40. **Emmons, C. W.** 1951. Isolation of *Cryptococcus neoformans* from soil. *J. Bacteriol.* **62:**685–690.
41. **Emmons, C. W.** 1955. Saprophytic sources of *Cryptococcus neoformans* associated with the pigeon (*Columba livia*). *Am. J. Hyg.* **62:**227–232.
42. **Emmons, C. W.** 1960. Prevalence of *Cryptococcus neoformans* in pigeon habitats. *Public Health Rep.* **75:**362–364.
43. **Escandón, P., P. Ngamskulrungroj, W. Meyer, and E. Castañeda.** 2007. In vitro mating of Colombian isolates of the *Cryptococcus neoformans* species complex. *Biomédica* **27:**308–314. (In Spanish.)
44. **Escandón, P., A. Sánchez, M. Martinez, W. Meyer, and E. Castañeda.** 2006. Molecular epidemiology of clinical and environmental isolates of the *Cryptococcus neoformans* species complex reveals a high genetic diversity and the presence of the molecular type VGII mating type a in Colombia. *FEMS Yeast Res.* **6:**625–635.
45. **Eswaran, V., H. Harpending, and A. R. Rogers.** 2005. Genomics refutes an exclusively African origin of humans. *J. Hum. Evol.* **49:**1–18.
46. **Feng, X., Z. Yao, D. Ren, W. Liao, and J. Wu.** 2008. Genotype and mating type analysis of *Cryptococcus neoformans* and *Cryptococcus gattii* isolates from China that mainly originated from non-HIV-infected patients. *FEMS Yeast Res.* **8:**930–938.
47. **Findley, K., M. Rodriguez-Carres, B. Metin, J. Kroiss, Á. Fonseca, R. J. Vilgalys, and J. Heitman.** 2009. Phylogeny and phenotypic characterization of pathogenic *Cryptococcus* species and closely related saprobic taxa in the *Tremellales*. *Eukaryot. Cell* **8:**353–361.
48. **Fiskin, A. M., M. C. Zalles, and R. G. Garrison.** 1990. Electron cytochemical studies of *Cryptococcus neoformans* grown on uric acid and related sources of nitrogen. *J. Med. Vet. Mycol.* **28:**197–207.
49. **Fortes, S. T., M. S. Lazéra, M. M. Nishikawa, R. C. L. Macedo, and B. Wanke.** 2001. First isolation of *Cryptococcus neoformans* var. *gattii* from a native jungle tree in the Brazilian Amazon rainforest. *Mycoses* **44:**137–140.
50. **Franzot, S. P., J. S. Hamdan, B. P. Currie, and A. Casadevall.** 1997. Molecular epidemiology of *Cryptococcus neoformans* in Brazil and the United States: evidence for both local genetic differences and a global clonal population structure. *J. Clin. Microbiol.* **35:**2243–2251.
51. **Fraser, J. A., S. S. Giles, E. C. Wenink, S. G. Geunes-Boyer, J. R. Wright, S. Diezmann, A. Allen, J. E. Stajich, F. S. Dietrich, J. R. Perfect, and J. Heitman.** 2005. Same-sex mating and the origin of the Vancouver Island *Cryptococcus gattii* outbreak. *Nature* **437:**1360–1364.
52. **Fraser, J. A., R. L. Subaran, C. B. Nichols, and J. Heitman.** 2003. Recapitulation of the sexual cycle of the primary fungal pathogen *Cryptococcus neoformans* var. *gattii*: implications for an outbreak on Vancouver Island, Canada. *Eukaryot. Cell* **2:**1036–1045.
53. **García-Rivera, J., and A. Casadevall.** 2001. Melanization of *Cryptococcus neoformans* reduces its susceptibility to the antimicrobial effects of silver nitrate. *Med. Mycol.* **39:**353–357.
54. **Granados, D. P., and E. Castañeda.** 2006. Influence of climatic conditions on the isolation of members of the *Cryptococcus neoformans* species complex from trees in Colombia from 1992–2004. *FEMS Yeast Res.* **6:**636–644.
55. **Grover, N., S. R. Nawange, J. Naidu, S. M. Singh, and A. Sharma.** 2007. Ecological niche of *Cryptococcus neoformans* var. *grubii* and *Cryptococcus gattii* in decaying wood of trunk hollows of living trees in Jabalpur City of Central India. *Mycopathologia* **164:**159–170.
56. **Gugnani, H. C., T. G. Mitchell, A. P. Litvintseva, K. B. Lengeler, J. Heitman, A. Kumar, S. Basu, and A. Paliwal-Johsi.** 2005. Isolation of *Cryptococcus gattii* and *Cryptococcus neoformans* var. *grubii* from the flowers and bark of *Eucalyptus* trees in India. *Med. Mycol.* **43:**565–569.
57. **Halliday, C. L., T. Bui, M. B. Krockenberger, R. Malik, D. H. Ellis, and D. A. Carter.** 1999. Presence of α and a mating types in environmental and clinical collections of *Cryptococcus neoformans* var. *gattii* strains from Australia. *J. Clin. Microbiol.* **37:**2920–2926.
58. **Halliday, C. L., and D. A. Carter.** 2003. Clonal reproduction and limited dispersal in an environmental population of *Cryptococcus neoformans* var. *gattii* isolates from Australia. *J. Clin. Microbiol.* **41:**703–711.
59. **Hanafy, A. M., S. Kaocharoen, A. Jover-Botella, M. Katsu, S. Iida, T. Kogure, T. Gonoi, Y. Mikami, and W. Meyer.** 2008. Multilocus microsatellite typing for *Cryptococcus neoformans* var. *grubii*. *Med. Mycol.* **46:**685–696.
60. **Hartl, D. L.** 2000. *A Primer of Population Genetics*. Sinauer Associates, Sunderland, MA.
61. **Hartl, D. L., and A. G. Clark.** 2007. *Principles of Population Genetics*. Sinauer Associates, Sunderland, MA.
62. **Hasenclever, H. F., and C. W. Emmons.** 1963. The prevalence and mouse virulence of *Cryptococcus neoformans* strains isolated from urban areas. *Am. J. Hyg.* **78:**227–231.
63. **Hiremath, S. S., A. Chowdhary, T. Kowshik, H. S. Randhawa, S. Sun, and J. Xu.** 2008. Long-distance dispersal and recombination in environmental populations of *Cryptococcus neoformans* var. *grubii* from India. *Microbiology* **154:**1513–1524.
64. **Holsinger, K. E., and B. S. Weir.** 2009. Genetics in geographically structured populations: defining, estimating and interpreting F_{ST}. *Nat. Rev. Genet.* **10:**639–650.
65. **Kao, C. J., and J. Schwarz.** 1957. The isolation of *Cryptococcus neoformans* from pigeon nests; with remarks on the identification of virulent cryptococci. *Am. J. Clin. Pathol.* **27:**652–663.
66. **Katsu, M., S. E. Kidd, A. Ando, M. L. Moretti-Branchini, Y. Mikami, K. Nishimura, and W. Meyer.** 2004. The internal transcribed spacers and 5.8S rRNA gene show extensive diversity among isolates of the *Cryptococcus neoformans* species complex. *FEMS Yeast Res.* **4:**377–388.
67. **Kavanaugh, L. A., J. A. Fraser, and F. S. Dietrich.** 2006. Recent evolution of the human pathogen *Cryptococcus neoformans* by inter-varietal transfer of a 14 gene fragment. *Mol. Biol. Evol.* **23:**1879–1890.
68. **Kidd, S. E., Y. Chow, S. Mak, P. J. Bach, H. Chen, A. O. Hingston, J. W. Kronstad, and K. H. Bartlett.** 2007. Characterization of environmental sources of the human and animal pathogen, *Cryptococcus gattii*, in British Columbia, Canada, and the Pacific Northwest of the United States. *Appl. Environ. Microbiol.* **73:**1433–1443.
69. **Kidd, S. E., H. Guo, K. H. Bartlett, J. Xu, and J. W. Kronstad.** 2005. Comparative gene genealogies indicate that two clonal lineages of *Cryptococcus gattii* in British Columbia resemble strains from other geographical areas. *Eukaryot. Cell* **4:**1629–1638.
70. **Kidd, S. E., F. Hagen, R. L. Tscharke, M. Huynh, K. H. Bartlett, M. Fyfe, L. MacDougall, T. Boekhout, K. J. Kwon-Chung, and W. Meyer.** 2004. A rare genotype of *Cryptococcus gattii* caused the cryptococcosis outbreak on Vancouver Island (British Columbia, Canada). *Proc. Natl. Acad. Sci. USA* **101:**17258–17263.

71. **Knowles, L. L., and W. P. Maddison.** 2002. Statistical phylogeography. *Mol. Ecol.* **11:**2623–2635.
72. **Kobayashi, C. C., L. K. D. Souza, O. D. Fernandes, S. C. de Brito, A. C. Silva, E. D. de Sousa, and M. D. Silva.** 2005. Characterization of *Cryptococcus neoformans* isolated from urban environmental sources in Goiânia, Goiás State, Brazil. *Rev. Inst. Med. Trop. São Paulo* **47:**203–207.
73. **Kwon-Chung, K. J., and J. E. Bennett.** 1984. Epidemiologic differences between the two varieties of *Cryptococcus neoformans*. *Am. J. Epidemiol.* **120:**123–130.
74. **Kwon-Chung, K. J., and J. E. Bennett.** 1992. *Medical Mycology*, p. 397–446. Lea & Febiger, Philadelphia, PA.
75. **Kwon-Chung, K. J., T. Boekhout, J. W. Fell, and M. Diaz.** 2004. Proposal to conserve the name *Cryptococcus gattii* against *C. hondurianus* and *C. bacillisporus* (*Basidiomycota, Hymenomycetes, Tremellomycetidae*). *Taxon* **51:**804–806.
76. **Kwon-Chung, K. J., and A. Varma.** 2006. Do major species concepts support one, two or more species within *Cryptococcus neoformans*? *FEMS Yeast Res.* **6:**574–587.
77. **Lazéra, M. S., M. A. S. Cavalcanti, L. Trilles, M. M. Nishikawa, and B. Wanke.** 1998. *Cryptococcus neoformans* var. *gattii*: evidence for a natural habitat related to decaying wood in a pottery tree hollow. *Med. Mycol.* **36:**119–122.
78. **Lazéra, M. S., M. A. Salmito Cavalcanti, A. T. Londero, L. Trilles, M. M. Nishikawa, and B. Wanke.** 2000. Possible primary ecological niche of *Cryptococcus neoformans*. *Med. Mycol.* **38:**379–383.
79. **Lengeler, K. B., G. M. Cox, and J. Heitman.** 2001. Serotype AD strains of *Cryptococcus neoformans* are diploid or aneuploid and are heterozygous at the mating-type locus. *Infect. Immun.* **69:**115–122.
80. **Lengeler, K. B., P. Wang, G. M. Cox, J. R. Perfect, and J. Heitman.** 2000. Identification of the *MATa* mating-type locus of *Cryptococcus neoformans* reveals a serotype A *MATa* strain thought to have been extinct. *Proc. Natl. Acad. Sci. USA* **97:**14455–14460.
81. **Lin, X., C. M. Hull, and J. Heitman.** 2005. Sexual reproduction between partners of the same mating type in *Cryptococcus neoformans*. *Nature* **434:**1017–1021.
82. **Lin, X., A. P. Litvintseva, K. Nielsen, S. Patel, A. Floyd, T. G. Mitchell, and J. Heitman.** 2007. αADα hybrids of *Cryptococcus neoformans*: evidence of same-sex mating in nature and hybrid fitness. *PLoS Genet.* **3:**e186.
83. **Lin, X., K. Nielsen, S. Patel, and J. Heitman.** 2008. Impact of mating type, serotype, and ploidy on the virulence of *Cryptococcus neoformans*. *Infect. Immun.* **76:**2923–2938.
84. **Lin, X., S. Patel, A. P. Litvintseva, A. Floyd, R. Hicks, T. G. Mitchell, and J. Heitman.** 2009. Diploids in the *Cryptococcus neoformans* serotype A population homozygous for the α mating type originate via unisexual mating. *PLoS Pathog.* **5:**e1000283.
85. **Littman, M. L., and R. Borok.** 1968. Relation of the pigeon to cryptococcosis: natural carrier state, heat resistance and survival of *Cryptococcus neoformans*. *Mycopathol. Mycol. Appl.* **35:**329–345.
86. **Litvintseva, A. P., L. Kestenbaum, R. J. Vilgalys, and T. G. Mitchell.** 2005. Comparative analysis of environmental and clinical populations of *Cryptococcus neoformans*. *J. Clin. Microbiol.* **43:**556–564.
87. **Litvintseva, A. P., X. Lin, I. Templeton, J. Heitman, and T. G. Mitchell.** 2007. Many globally isolated AD hybrid strains of *Cryptococcus neoformans* originated in Africa. *PLoS Pathog.* **3:**e114.
88. **Litvintseva, A. P., R. E. Marra, K. Nielsen, J. Heitman, R. J. Vilgalys, and T. G. Mitchell.** 2003. Evidence of sexual recombination among *Cryptococcus neoformans* serotype A isolates in sub-Saharan Africa. *Eukaryot. Cell* **2:**1162–1168.
89. **Litvintseva, A. P., and T. G. Mitchell.** 2009. Most environmental isolates of *Cryptococcus neoformans* var. *grubii* are not lethal for mice. *Infect. Immun.* **77:**3188–3195.
90. **Litvintseva, A. P., T. G. Mitchell, and N. Govender.** 2009. Out-of-Africa origin of *Cryptococcus neoformans* var. *grubii* (serotype A). Abstract EP-04-2, p. 205. 17th Congress, International Society for Human and Animal Mycology, Tokyo, Japan, 25–29 May 2009.
91. **Litvintseva, A. P., R. Thakur, L. B. Reller, and T. G. Mitchell.** 2005. Prevalence of clinical isolates of *Cryptococcus gattii* serotype C among patients with AIDS in sub-Saharan Africa. *J. Infect. Dis.* **192:**888–892.
92. **Litvintseva, A. P., R. Thakur, R. J. Vilgalys, and T. G. Mitchell.** 2006. Multilocus sequence typing reveals three genetic subpopulations of *Cryptococcus neoformans* var. *grubii* (serotype A), including a unique population in Botswana. *Genetics* **172:**2223–2238.
93. **Lugarini, C., C. S. Goebel, L. A. Condas, M. D. Muro, M. R. de Farias, F. M. Ferreira, and M. H. Vainstein.** 2008. *Cryptococcus neoformans* isolated from passerine and psittacine bird excreta in the state of Parana, Brazil. *Mycopathologia* **166:**61–69.
94. **Lyngso, R. B., Y. S. Song, and J. Hein.** 2005. Minimum recombination histories by branch and bound, p. 239–250. *In* R. Casadio and G. Myers (eds), *Algorithms in Bioinformatics: Fifth International Workshop*. Springer, Berlin, Germany.
95. **Ma, H., F. Hagen, D. J. Stekel, S. A. Johnston, E. Sionov, R. Falk, I. Polacheck, T. Boekhout, and R. C. May.** 2009. The fatal fungal outbreak on Vancouver Island is characterized by enhanced intracellular parasitism driven by mitochondrial regulation. *Proc. Natl. Acad. Sci. USA* **106:**12980–12985.
96. **Mahmoud, Y. A. G.** 1999. First environmental isolation of *Cryptococcus neoformans* var. *neoformans* and var. *gatti* from the Gharbia Governorate, Egypt. *Mycopathologia* **148:**83–86.
97. **Manly, B. F. J.** 2004. *Multivariate Statistical Methods: A Primer*. Chapman & Hall/CRC, Boca Raton, FL.
98. **Maynard Smith, J. M., N. H. Smith, M. O'Rourke, and B. G. Spratt.** 1993. How clonal are bacteria? *Proc. Natl. Acad. Sci. USA* **90:**4384–4388.
99. **McFadden, D. C., B. C. Fries, F. Wang, and A. Casadevall.** 2007. Capsule structural heterogeneity and antigenic variation in *Cryptococcus neoformans*. *Eukaryot. Cell* **6:**1464–1473.
100. **Meyer, W., D. M. Aanensen, T. Boekhout, M. Cogliati, M. R. Diaz, M. C. Esposto, M. C. Fisher, F. Gilgado, F. Hagen, S. Kaocharoen, A. P. Litvintseva, T. G. Mitchell, S. P. Simwami, L. Trilles, M. A. Viviani, and K. J. Kwon-Chung.** 2009. Consensus multi-locus sequence typing scheme for *Cryptococcus neoformans* and *Cryptococcus gattii*. *Med. Mycol.* **47:**561–570.
101. **Meyer, W., A. Castañeda, S. Jackson, M. Huynh, and E. Castañeda.** 2003. Molecular typing of IberoAmerican *Cryptococcus neoformans* isolates. *Emerg. Infect. Dis.* **9:**189–195.
102. **Meyer, W., K. Marszewska, M. Amirmostofian, R. P. Igreja, C. Hardtke, K. Methling, M. A. Viviani, A. Chindamporn, S. Sukroongreung, M. A. John, D. H. Ellis, and T. C. Sorrell.** 1999. Molecular typing of global isolates of *Cryptococcus neoformans* var. *neoformans* by polymerase chain reaction fingerprinting and randomly amplified polymorphic DNA: a pilot study to standardize techniques on which to base a detailed epidemiological survey. *Electrophoresis* **20:**1790–1799.
103. **Meyer, W., and T. G. Mitchell.** 1995. PCR fingerprinting in fungi using single primers specific to minisatellites

and simple repetitive DNA sequences: strain variation in *Cryptococcus neoformans*. *Electrophoresis* **16:**1648–1656.

104. **Meyer, W., T. G. Mitchell, E. Z. Freedman, and R. J. Vilgalys.** 1993. Hybridization probes for conventional DNA fingerprinting used as single primers in the polymerase chain reaction to distinguish strains of *Cryptococcus neoformans*. *J. Clin. Microbiol.* **31:**2274–2280.
105. **Mitchell, T. G.** 2008. Genomic approaches to investigate the pathogenicity of *Cryptococcus neoformans*. *Curr. Fungal Infect. Rep.* **2:**172–179.
106. **Mitchell, T. G.** 2010. Population genetics of pathogenic fungi in humans and other animals, p. 139–158. *In* J. Xu (ed.), *Microbial Population Genetics*. Horizon Scientific Press, Hethersett, UK.
107. **Mitchell, T. G., and A. P. Litvintseva.** 2010. Typing species of *Cryptococcus* and epidemiology of cryptococcosis, p. 167–190. *In* H. R. Ashbee and E. M. Bignell (ed.), *Pathogenic Yeasts*. Springer-Verlag, Berlin, Germany.
108. **Mitchell, T. G., and J. R. Perfect.** 1995. Cryptococcosis in the era of AIDS: 100 years after the discovery of *Cryptococcus neoformans*. *Clin. Microbiol. Rev.* **8:**515–548.
109. **Nei, M.** 1973. Analysis of gene diversity in subdivided populations. *Proc. Natl. Acad. Sci. USA* **70:**3321–3323.
110. **Nei, M., and W. H. Li.** 1979. Mathematical model for studying genetic variation in terms of restriction endonucleases. *Proc. Natl. Acad. Sci. USA* **76:**5269–5273.
111. **Ngamskulrungroj, P., F. Gilgado, J. Faganello, A. P. Litvintseva, A. L. Leal, K. M. Tsui, T. G. Mitchell, M. H. Vainstein, and W. Meyer.** 2009. Genetic diversity of the *Cryptococcus* species complex suggests that *Cryptococcus gattii* deserves to have varieties. *PLoS ONE* **4:**e5862.
112. **Nielsen, K., G. M. Cox, P. Wang, D. L. Toffaletti, J. R. Perfect, and J. Heitman.** 2003. The sexual cycle of *Cryptococcus neoformans* variety *grubii* and virulence of congenic **a** and α isolates. *Infect. Immun.* **71:**4831–4841.
113. **Nielsen, K., A. L. De Obaldia, and J. Heitman.** 2007. *Cryptococcus neoformans* mates on pigeon guano: implications for the realized ecological niche and globalization. *Eukaryot. Cell* **6:**949–959.
114. **Nishikawa, M. M., M. S. Lazéra, G. G. Barbosa, L. Trilles, B. R. Balassiano, R. C. Macedo, C. C. F. Bezerra, M. A. Perez, P. Cardarelli, and B. Wanke.** 2003. Serotyping of 467 *Cryptococcus neoformans* isolates from clinical and environmental sources in Brazil: analysis of host and regional patterns. *J. Clin. Microbiol.* **41:**73–77.
115. **Nosanchuk, J. D., and A. Casadevall.** 2003. The contribution of melanin to microbial pathogenesis. *Cell. Microbiol.* **5:**203–223.
116. **Nosanchuk, J. D., J. Rudolph, A. L. Rosas, and A. Casadevall.** 1999. Evidence that *Cryptococcus neoformans* is melanized in pigeon excreta: implications for pathogenesis. *Infect. Immun.* **67:**5477–5479.
117. **Pal, M.** 1997. First report of isolation of *Cryptococcus neoformans* var. *neoformans* from avian excreta in Kathmandu, Nepal. *Rev. Iberoam. Micol.* **14:**181–183.
118. **Pfeiffer, T. J., and D. H. Ellis.** 1991. Environmental isolation of *Cryptococcus neoformans gattii* from California. *J. Infect. Dis.* **163:**929–930.
119. **Pfeiffer, T. J., and D. H. Ellis.** 1992. Environmental isolation of *Cryptococcus neoformans* var. *gattii* from *Eucalyptus terreticornis*. *J. Med. Vet. Mycol.* **30:**407–408.
120. **Randhawa, H. S., T. Kowshik, A. Chowdhary, S. K. Preeti, Z. U. Khan, S. Sun, and J. Xu.** 2008. The expanding host tree species spectrum of *Cryptococcus gattii* and *Cryptococcus neoformans* and their isolations from surrounding soil in India. *Med. Mycol.* **46:**823–833.
121. **Randhawa, H. S., T. Kowshik, and Z. U. Khan.** 2003. Decayed wood of *Syzygium cumini* and *Ficus religiosa* living trees in Delhi/New Delhi metropolitan area as natural habitat of *Cryptococcus neoformans*. *Med. Mycol.* **41:**199–209.
122. **Refojo, N., D. E. Perrotta, M. Brudny, R. Abrantes, A. I. Hevia, and G. Davel.** 2009. Isolation of *Cryptococcus neoformans* and *Cryptococcus gattii* from trunk hollows of living trees in Buenos Aires City, Argentina. *Med. Mycol.* **47:**177–184.
123. **Ribeiro, M. A., and P. Ngamskulrungroj.** 2008. Molecular characterization of environmental *Cryptococcus neoformans* isolated in Vitória, ES, Brazil. *Rev. Inst. Med. Trop. São Paulo* **50:**315–320.
124. **Rosas, A. L., and A. Casadevall.** 2001. Melanization decreases the susceptibility of *Cryptococcus neoformans* to enzymatic degradation. *Mycopathologia* **151:**53–56.
125. **Ruiz, A., D. Velez, and R. A. Fromtling.** 1989. Isolation of saprophytic *Cryptococcus neoformans* from Puerto Rico: distribution and variety. *Mycopathologia* **106:**167–170.
126. **Saraçli, M. A., S. T. Yildiran, K. Sener, A. Gönlüm, L. Doganci, S. M. Keller, and B. L. Wickes.** 2006. Genotyping of Turkish environmental *Cryptococcus neoformans* var. *neoformans* isolates by pulsed field gel electrophoresis and mating type. *Mycoses* **49:**124–129.
127. **Saul, N., M. B. Krockenberger, and D. A. Carter.** 2008. Evidence of recombination in mixed mating type and α-only populations of *Cryptococcus gattii* sourced from single *Eucalyptus* tree hollows. *Eukaryot. Cell* **7:**727–734.
128. **Soogarun, S., V. Wiwanitkit, A. Palasuwan, P. Pradniwat, J. Suwansaksri, T. Lertlum, and T. Maungkote.** 2006. Detection of *Cryptococcus neoformans* in bird excreta. *Southeast Asian J. Trop. Med. Public Health* **37:**768–770.
129. **Staib, F.** 1962. *Cryptococcus neoformans* und *Guizotia abyssinica* (Syn. *G. oleifera*) Farbreaktion für *Cr. neoformans*. *Zbl. Hyg.* **148:**466–475.
130. **Staib, F.** 1981. The perfect state of *Cryptococcus neoformans*, *Filobasidiella neoformans*, on pigeon manure filtrate agar. *Zbl. Bakt. A* **248:**575–578.
131. **Staib, F.** 1985. Sampling and isolation of *Cryptococcus neoformans* from indoor air with the aid of the Reuter Centrifugal Sampler (RCS) and *Guizotia abyssinica* creatinine agar. A contribution to the mycological-epidemiological control of *Cr. neoformans* in the fecal matter of caged birds. *Zbl. Bakt. Mikrobiol. Hyg. B* **180:**567–575.
132. **Steenbergen, J. N., H. A. Shuman, and A. Casadevall.** 2001. *Cryptococcus neoformans* interactions with amoebae suggest an explanation for its virulence and intracellular pathogenic strategy in macrophages. *Proc. Natl. Acad. Sci. USA* **98:**15245–15250.
133. **Stephen, C., S. J. Lester, W. Black, M. Fyfe, and S. Raverty.** 2002. Multispecies outbreak of cryptococcosis on southern Vancouver Island, British Columbia. *Can. Vet. J.* **43:**792–794.
134. **Sun, S., and J. Xu.** 2009. Chromosomal rearrangements between serotype A and D strains in *Cryptococcus neoformans*. *PLoS ONE* **4:**e5524.
135. **Swinne-Desgain, D.** 1975. *Cryptococcus neoformans* of saprophytic origin. *Sabouraudia* **13:**303–308.
136. **Tanaka, R., K. Nishimura, and M. Miyaji.** 1999. Ploidy of serotype AD strains of *Cryptococcus neoformans*. *Jpn. J. Med. Mycol.* **40:**31–34.
137. **Taylor, J. W., D. M. Geiser, A. Burt, and V. Koufopanou.** 1999. The evolutionary biology and population genetics underlying fungal strain typing. *Clin. Microbiol. Rev.* **12:**126–146.
138. **Taylor, J. W., D. J. Jacobson, and M. C. Fisher.** 1999. The evolution of asexual fungi: reproduction, speciation and classification. *Annu. Rev. Phytopathol.* **37:**197–246.
139. **Toffaletti, D. L., K. Nielsen, F. S. Dietrich, J. Heitman, and J. R. Perfect.** 2004. *Cryptococcus neoformans*

mitochondrial genomes from serotype A and D strains do not influence virulence. *Curr. Genet.* **46:**193–204.

140. **Tortorano, A. M., M. A. Viviani, A. L. Rigoni, M. Cogliati, A. Roverselli, and A. Pagano.** 1997. Prevalence of serotype D in *Cryptococcus neoformans* isolates from HIV positive and HIV negative patients in Italy. *Mycoses* **40:**297–302.
141. **Trilles, L., M. S. Lazéra, B. Wanke, B. Theelen, and T. Boekhout.** 2003. Genetic characterization of environmental isolates of the *Cryptococcus neoformans* species complex from Brazil. *Med. Mycol.* **41:**383–390.
142. **Vilcins, I., M. B. Krockenberger, H. Agus, and D. A. Carter.** 2002. Environmental sampling for *Cryptococcus neoformans* var. *gattii* from the Blue Mountains National Park, Sydney, Australia. *Med. Mycol.* **40:**53–60.
143. **Viviani, M. A., M. Cogliati, M. C. Esposto, K. Lemmer, K. Tintelnot, M. F. Valiente, D. Swinne, A. Velegraki, and R. Velho.** 2006. Molecular analysis of 311 *Cryptococcus neoformans* isolates from a 30-month ECMM survey of cryptococcosis in Europe. *FEMS Yeast Res.* **6:**614–619.
144. **Viviani, M. A., M. C. Esposto, M. Cogliati, M. T. Montagna, and B. L. Wickes.** 2001. Isolation of a *Cryptococcus neoformans* serotype A *MATa* strain from the Italian environment. *Med. Mycol.* **39:**383–386.
145. **Viviani, M. A., R. Nikolova, M. C. Esposto, G. Prinz, and M. Cogliati.** 2003. First European case of serotype A *MATa Cryptococcus neoformans* infection. *Emerg. Infect. Dis.* **9:**1179–1180.
146. **Walter, J. E., and R. B. Yee.** 1968. Factors that determine the growth of *Cryptococcus neoformans* in avian excreta. *Am. J. Epidemiol.* **88:**445–450.
147. **Wright, S.** 1978. *Variability within and among Natural Populations. Evolution and the Genetics of Populations.* University of Chicago Press, Chicago.
148. **Xu, J., R. Y. Ali, D. A. Gregory, D. Amick, S. E. Lambert, H. J. Yoell, R. J. Vilgalys, and T. G. Mitchell.** 2000. Uniparental mitochondrial transmission in sexual crosses in *Cryptococcus neoformans. Curr. Microbiol.* **40:**269–273.
149. **Xu, J., G. Luo, R. J. Vilgalys, M. E. Brandt, and T. G. Mitchell.** 2002. Multiple origins of hybrid strains of *Cryptococcus neoformans* with serotype AD. *Microbiology* **148:**203–212.
150. **Xu, J., and T. G. Mitchell.** 2002. Strain variation and clonality in *Candida* spp. and *Cryptococcus neoformans*, p. 739–749. *In* R. A. Calderone and R. L. Cihlar (ed.), *Fungal Pathogenesis: Principles and Clinical Applications.* Marcel Dekker, New York, NY.
151. **Xu, J., and T. G. Mitchell.** 2003. Comparative gene genealogical analyses of strains of serotype AD identify recombination in populations of serotypes A and D in the human pathogenic yeast *Cryptococcus neoformans. Microbiology* **149:**2147–2154.
152. **Xu, J., and T. G. Mitchell.** 2003. Population genetic analyses of medically important fungi, p. 703–722. *In* D. H. Howard (ed.), *Pathogenic Fungi in Humans and Animals.* 2nd ed. Marcel Dekker, New York, NY.
153. **Xu, J., R. J. Vilgalys, and T. G. Mitchell.** 2000. Multiple gene genealogies reveal recent dispersion and hybridization in the human pathogenic fungus *Cryptococcus neoformans. Mol. Ecol.* **9:**1471–1481.
154. **Xu, J., Z. Yan, and H. Guo.** 2009. Divergence, hybridization, and recombination in the mitochondrial genome of the human pathogenic yeast *Cryptococcus gattii. Mol. Ecol.* **18:**2628–2642.
155. **Yan, Z., C. M. Hull, J. Heitman, S. Sun, and J. Xu.** 2004. *SXI1α* controls uniparental mitochondrial inheritance in *Cryptococcus neoformans. Curr. Biol.* **14:**R743–R744.
156. **Yan, Z., C. M. Hull, S. Sun, J. Heitman, and J. Xu.** 2007. The mating type-specific homeodomain genes *SXI1α* and *SXI2***a** coordinately control uniparental mitochondrial inheritance in *Cryptococcus neoformans. Curr. Genet.* **51:**187–195.
157. **Yan, Z., X. Li, and J. Xu.** 2002. Geographic distribution of mating type alleles of *Cryptococcus neoformans* in four areas of the United States. *J. Clin. Microbiol.* **40:**965–972.
158. **Yan, Z., S. Sun, M. Shahid, and J. Xu.** 2007. Environment factors can influence mitochondrial inheritance in the fungus *Cryptococcus neoformans. Fungal Genet. Biol.* **44:**315–322.
159. **Yan, Z., and J. Xu.** 2003. Mitochondria are inherited from the *MATa* parent in crosses of the basidiomycete fungus *Cryptococcus neoformans. Genetics* **163:**1315–1325.
160. **Zhu, J., Y. Kang, J. Uno, H. Taguchi, Y. Liu, M. Ohata, R. Tanaka, M. L. Moretti, and Y. Mikami.** 2009. Comparison of genotypes between environmental and clinical isolates of *Cryptococcus neoformans* var. *grubii* based on microsatellite patterns. *Mycopathologia* **169:**47–55.
161. **Zhu, X., and P. R. Williamson.** 2004. Role of laccase in the biology and virulence of *Cryptococcus neoformans. FEMS Yeast Res.* **5:**1–10.

GENETICS AND GENOMICS

Cryptococcus: From Human Pathogen to Model Yeast
Edited by J. Heitman et al.

9

The *Cryptococcus* Genomes: Tools for Comparative Genomics and Expression Analysis

JAMES W. KRONSTAD, BRENDAN J. LOFTUS, AND JENNIFER K. LODGE

The establishment of an international genome consortium resulted in the development of genomic resources to support the analysis of the biology and pathogenesis of *Cryptococcus neoformans* and *Cryptococcus gattii*. These resources include genome sequences for two related strains of *C. neoformans* var. *neoformans*, one strain of *C. neoformans* var. *grubii*, and two strains of *C. gattii* representing the VGI and VGIIa molecular subtypes. The genomes have been exploited for comparative studies by sequence analysis and comparative genome hybridization with tiling microarrays. Importantly, the genome sequences enabled the development and use of spotted oligonucleotide microarrays and the application of serial analysis of gene expression for transcriptome studies, as well as high-throughput methods to construct sets of deletion mutants. The work to date provides only a small glimpse of the potential of genomic studies to accelerate molecular studies of cryptococcal pathogenesis. In this chapter, we describe the genomic resources available for *C. neoformans* and *C. gattii* and discuss selected examples of the application of the resources to address questions relevant to virulence.

FIVE CRYPTOCOCCAL GENOME SEQUENCES

A project to sequence the genome of *C. neoformans* was formulated at a 1999 meeting in St. Louis organized by J. Lodge (21). The meeting resulted in the establishment of an international genome consortium as well as collaborations with genome sequencing centers to prepare funding applications. The justification for funding was based on the presence of a large, highly collaborative research community that would rapidly make use of the sequence information. In addition, sequencing was justified by the fact that *C. neoformans* represents an excellent model for the examination of fungal pathogenesis because it has haploid and diploid phases, classical and molecular tools are available for genetic analysis, several selectable markers support DNA transformation, and a gene disruption system is in place to make targeted mutations (21). Additional meetings of the international genome consortium were held in St. Louis in 2001 and Vancouver in 2003, and genomics workshops have become a feature of the International Conferences on *Cryptococcus* and Cryptococcosis held in 2002 (Adelaide), 2005 (Boston), and 2008 (Nagasaki). These meetings have helped focus community efforts to sequence additional genomes from serotype A and B strains, to produce a deletion collection of all nonessential genes, to establish microarray resources, and to develop plans for a curated genome database. Table 1 summarizes the current status of the sequence projects for *C. neoformans* and *C. gattii*.

Serotype D Genomes

The initial efforts of the genome consortium resulted in sequences for two related serotype D strains of *C. neoformans*, JEC21 and B3501A (40). The Institute for Genomic Research sequenced the genome of JEC21, and the Stanford Genome Technology Center sequenced the B3501A genome. The JEC21 genome was sequenced because this strain and the congenic mating type **a** isolate JEC20 had been developed as useful experimental strains for genetic analyses (21, 22). JEC21 and B3501A are highly related inbred strains of the α mating type, and this is reflected in the similarity at the genomic level, with these strains showing approximately 99.5% nucleotide identity over their entire genomes. However, this level of identity is split into large blocks of 100% identical sequences coupled with regions that are approximately 99% identical at the nucleotide level, most likely reflecting the recombination events that occurred during the generation of the strains. Indeed, by far the most dramatic differences between the strains are a small chromosomal translocation and an exact ~60-kb duplication that are present in the JEC21 genome but not in B3501A (18). Although the B3501A genome sequence is not assembled into completed chromosomes, it is possible to tell that the gene sets for each organism are almost identical, with genes for a Ras guanosine triphosphatase–activating protein and two proteins of unknown function being specific to B3501A. Genes for four proteins of unknown function appear to be specific to JEC21.

James W. Kronstad, The Michael Smith Laboratories, University of British Columbia, Vancouver, British Columbia, Canada V6T 1Z4. **Brendan J. Loftus,** Conway Institute, School of Medicine and Medical Science, University College Dublin, Belfield, Dublin 4, Ireland. **Jennifer K. Lodge,** Washington University School of Medicine, Dept. of Molecular Microbiology, St. Louis, MO 63110.

TABLE 1 C. *neoformans* and C. *gattii* strains with sequenced genomes

Strain	Variety and serotype	Subtype (PCR)	Coverage	Sequencing center	Websites for sequence access
JEC21	*neoformans*, D	VNIII	~12.5 X	J. Craig Venter Institute (TIGR)	http://www.tigr.org/tdb/e2k1/cna1/
B3501A	*neoformans*, D	VNIII	~10X	Stanford Genome Technology Center	http://www-sequence.stanford.edu/group/C.neoformans/index.html
H99	*grubii*, A	VNI	~11X	Broad Institute, Duke University	www.broad.mit.edu/annotation/genome/cryptococcus_neoformans/Home.html http://cneo.genetics.duke.edu/
R265	*gattii*, B	VGIIa	~5X	Broad Institute	www.broad.mit.edu/annotation/genome/cryptococcus_neoformans_b/Home.html
WM276	*gattii*, B	VGI	~6X	Canada's Michael Smith Genome Sciences Centre	http://www.bcgsc.bc.ca/cgi-bin/crypto_data/blast_wm276.pl

Note: The BAC physical maps and BAC end sequences for each strain were obtained by Canada's Michael Smith Genome Sciences Centre. The University of Oklahoma's Advanced Center for Genome Technology has also collected expressed sequence tag sequences for some of the strains (http://www.genome.ou.edu/cneo.html).

These minor differences, on top of the additional exact sequence copies of the 22 genes on the 60-kb duplication in the JEC21 genome, delineate the strains. The vast majority (99.7%) of the genes in common to the strains share >98% identity at the nucleotide level. These striking similarities at the genic level are intriguing given that the phenotypes of JEC21 and B3501 differ markedly, with B3501A being more thermotolerant and more virulent in animal models than JEC21. This suggests that factors other than the presence and absence of individual genes define these differences.

The gene structures, as determined by extensive sequencing of full-length cDNAs for JEC21 and B3501A, were also unexpected (40). In contrast to gene structure in ascomycete fungi, the C. *neoformans* genes showed a remarkable number of introns per gene as well as numerous examples of alternative splicing and antisense transcripts. Although with recent publications showing increased transcriptional complexity across a broad range of organisms, this might not now draw attention, these findings were remarkable at the time and may in time prove a fruitful source for exploring the role of all aspects of the transcriptome in virulence. In this context, a promising technology that looks set to transform transcriptomics is RNA.seq, where instead of using microarrays, mRNA, ribosome-depleted RNA, or small RNA is fragmented and sequenced directly for alignment to the genome (65). This requires no a priori knowledge of the genome in order to construct the transcriptional complement of an organism. As the use of this technology becomes pervasive, a new era of qualitatively and quantitatively improved transcript data will provide a greater appreciation for the role of transcriptional complexity and noncoding RNA in the biology of C. *neoformans* and other organisms.

Genome Sequencing for Serotype A Strains

The genome of the serotype A strain H99 of C. *neoformans* has been sequenced at Duke University and the Broad Institute as part of the Fungal Genome Initiative (Table 1). This highly virulent strain is often used to investigate the genetic basis of virulence. The genome sequence has greatly facilitated this analysis by allowing the identification of candidate genes, by supporting insertional mutagenesis studies, and by enabling transcriptome and proteome studies. The genome is also serving as a template for resequencing variants of strain H99 with altered virulence and mating (J. Heitman, personal communication). Importantly, the sequence also supported the construction of sets of deletion mutants for ~1,700 genes from strain H99 (39; J. Lodge, personal communication; http://genome.slu.edu/delete/index.html); these deletion sets are available from the American Type Culture Collection and from the Fungal Genetic Stock Center.

Serotype B Genomes

The genomes of two C. *gattii* strains, R265 and WM276, have also been sequenced, and this effort was motivated in part by the outbreak of infections caused by this species in British Columbia (see chapter 23, this volume) (5). The R265 strain has the VGIIa molecular subtype of C. *gattii*, and this subtype has caused the majority of infections in British Columbia; WM276 represents the VGI subtype that is the most common worldwide (42). The R265 genome was sequenced by the Broad Institute, and the WM276 genome was sequenced by Canada's Michael Smith Genome Sciences Centre. The latter group also provided physical maps for the three C. *neoformans* genomes and both C. *gattii* genomes. These maps were prepared by fingerprinting bacterial artificial chromosome (BAC) libraries and the end sequences of the BAC clones were used to support assembly of the genomes (54).

Genetic Maps

In addition to the physical maps described above, genetic maps have also been constructed to characterize cryptococcal genomes, facilitate genetic studies, and assist in assembly of the genome sequences (17, 41, 58). Marra et al. (41) constructed a map by crossing the *MAT*α strain B3501 with the *MAT***a** strain B3502 and analyzing 94 progeny for segregation of 301 polymorphic markers. These included 228 restriction site polymorphisms, 63 minisatellite markers, two indels, and eight *MAT*-associated markers. The mapping resulted in an average marker density of 5.4 centimorgan and 20 linkage groups. Importantly, selected markers were hybridized to electrophoretically separated chromosomes to correlate the linkage groups with chromosomes. This analysis identified 14 chromosomes with sizes from 0.8 to 2.3 Mb and also provided evidence for at least one reciprocal translocation. Sun and Xu (58) analyzed a hybrid population of 163 progeny from a cross between a serotype A strain (CDC15) and a serotype D strain (JEC20) using 114 codominant PCR–restriction fragment length polymorphism markers and one direct PCR marker. Marker analysis revealed that

162 of the progeny had at least one heterozygous marker, suggesting that they were diploid, and all of the progeny showed uniparental inheritance of the mitochondria from the MAT**a** parent JEC20 (66). Linkage analysis of the markers revealed independent assortment on different linkage groups, and the map length from this hybrid cross was smaller than the map from the cross of serotype D strains (41). This result suggests that recombination is suppressed in the cross between the serotype A and D strains.

The maps have proven useful in several genetic studies. For example, Lin et al. (37) used quantitative trait loci mapping to discover that hyphal length during monokaryotic fruiting is a quantitative trait. In addition, the virulence traits of melanin production and growth at elevated temperature are also quantitative and shared a common locus with hyphal growth. The Mac1 transcription factor was encoded at this locus and influences filamentation, melanization, and growth at high temperatures. In another study, Bahn et al. (2) employed the mapping progeny from the cross between strains B3501 and B3502 to attribute differences in the Hog1 signaling pathway to the *SSK2* gene. The two strains differ in their Hog1-dependent sensitivity to the antifungal drug fludioxonil, and segregation analysis mapped the trait responsible for the difference, thus leading to the identification of polymorphisms in *SSK2*.

Genomic Analysis of Intron Structure

The complexity of the gene structures revealed by the various *Cryptococcus* genomes, in particular the high number of introns per gene relative to other fungal species, have been further investigated through various analyses of the selection pressures on the number and types of introns (26, 53, 55). These studies conclude that rather than intron gain being a feature of cryptococcal genes, mRNA-mediated intron loss appears to be a defining theme. The rate of loss appears to be much slower than for other fungal lineages. In this, the cryptococcal species bear a resemblance to other eukaryotic lineages that have retained much of their ancestral intron complement and where first introns are significantly longer than subsequent introns within genes. It appears that in *C. neoformans*, natural selection is gradually lengthening short introns and shortening longer introns toward a modal size, which presumably provides an increased level of evolutionary fitness. The current picture, however, may be refined in light of future sequencing of related genomes.

GENOME COMPARISON BY SEQUENCE ANALYSIS AND COMPARATIVE GENOME HYBRIDIZATION

The availability of genome sequences opens up opportunities for detailed comparisons of variability between species, varieties, molecular subtypes within varieties, and variants arising spontaneously or as a result of mutagenesis. The study conducted by Kavanaugh et al. (31) illustrates the types of comparative analyses that are possible with sequenced genomes, in this case the genomes of the strains JEC21 and H99 (representing var. *neoformans* and *grubii*, respectively). These varieties diverged from a common ancestor ~18.5 million years ago (67), thus providing an opportunity to examine divergence and introgression during the likely transition toward speciation. The genomes have 85–90% nucleotide sequence identity overall, and synteny is highly conserved. However, Kavanaugh et al. (31) found two regions that had higher than expected sequence identity between the two genomes. One region of ~8 kb carries four genes and is found in a subtelomeric region; the sequence identity is ~95% for this region. A second region contains 14 genes and has a length of ~40 kb showing 98.5% identity. This region, termed the identity island, likely arose from a nonreciprocal sequence transfer from var. *grubii* to var. *neoformans* that is estimated to have occurred ~2 million years ago. The majority of clinical and environmental isolates of var. *neoformans* from a global collection have the ~40-kb sequence. However, exceptions include two strains of var. *neoformans* that lack the sequence, indicating that the introgression happened after the expansion of the variety.

The hypothesized mechanism for the introgression of the ~40-kb identity island is that it occurred as a result of hybrid formation between strains of the two varieties. Hybrid strains of the AD serotype are found in nature, and these strains may sequentially lose chromosomes to transition from a diploid strain to aneuploidy or eventual restoration of haploidy (23, 33). It is possible that the sequence exchange for the identity island occurred in a hybrid intermediate and that repetitive elements associated with the region may have participated in the introgression. A role for repetitive elements is supported by the finding that the non–long terminal repeat retrotransposable element Cnl1 flanks the identity island in both varieties. In general, the discovery of the identity island indicates that introgression events can contribute to genetic exchange and speciation in *C. neoformans*.

Sun and Xu (59) recently made use of the sequenced genomes of strains JEC21 and H99 to perform an extensive analysis of the chromosomal rearrangements between the genomes of var. *neoformans* (serotype D) and *grubii* (serotype A). They also included an analysis of the genomes of serotype AD hybrid strains. The availability of the sequenced genomes provided opportunities to categorize the types of rearrangements and to estimate the effects of chromosomal rearrangements on recombination frequencies. The detailed view of the rearrangements also served as a solid platform to survey natural isolates of the three serotypes (A, D, and AD) for polymorphism associated with the rearrangements.

The analysis by Sun and Xu (59) revealed that the genomes are mostly syntenic but that 32 unambiguous chromosome rearrangements are present including five translocations, nine simple inversions, and 18 complex rearrangements. The large translocations included an intrachromosomal rearrangement on chromosome 3 of JEC21 and four interchromosomal rearrangements on chromosomes 3, 8, and 11 (ranging in size from 212 to 868 kb). Fraser et al. (18) previously described the interchromosomal translocations that occurred between chromosomes 8 and 12. The 18 complex rearrangements between H99 and JEC21 varied in size from 13 to 166 kb, and the majority of these (13) were found near proposed centromeric regions (40). The nine simple inversions were located on six chromosomes, with the smallest (3 kb) containing three genes and the largest (~394 kb) carrying 151 genes. Transposable elements, which may contribute to genome rearrangements, were found at the boundaries of only one of the translocated regions and in high density near 16 of the 18 complex rearrangements; in contrast, only five of the nine simple inversion regions were found to have transposable elements nearby in the JEC21 and/or H99 genomes. The availability of genetic linkage maps (41, 58) provided an opportunity to calculate the recombination frequencies between markers associated with the rearrangements. In general, the overall recombination frequencies surrounding simple inversions and complex rearrangements were lower than the frequencies in the syntenic chromosomal regions.

Finally, Sun and Xu (59) examined the distribution of 13 chromosomal rearrangements in a collection of 64 natural isolates of serotypes A, D, and AD strains. For one translocation juncture, all of the natural isolates had the same chromosomal type as JEC21, suggesting that this translocation happened only in the evolution of the H99 genome. The analysis of the nine simple inversions and the three complex rearrangements indicated that the chromosomal types carrying these differences were fixed within the two varieties. The 21 serotype AD strains showed different levels of heterozygosity among the rearranged regions, and the data indicated that the loss of chromosomes or chromosomal segments was likely not random. That is, some biases were observed. For example, all the chromosomal types for four simple inversion regions on chromosome 1 were biased toward the serotype A type; the strong linkage disequilibrium for these regions among serotype AD strains raised the possibility of strong epistatic interactions among loci along chromosome 1. This finding is consistent with the similar bias toward the serotype A version of chromosome 1 in serotype AD strains observed by Hu et al. (23) (described below).

The genomes of strains JEC21 and H99 have also been employed in comparative genome hybridization (CGH) experiments to examine the extent of variation in isolates of different mating types, different ploidies, and different molecular subtypes within varieties (23). The CGH experiments were performed with tiling microarrays from NimbleGen, Inc. that consisted of unique 45- to 85-bp oligonucleotides spaced at an average interval of 44 bp on one strand of each chromosome. As an initial calibration experiment to examine sequence divergence and deletions, the arrays were used to compare the log2 ratio of hybridization on the arrays with the percent sequence identity for 20 genes in the ~100-kb *MAT***a** and *MAT*α mating type loci. These calibrations revealed that log2 ratios between –3.77 and 0.49 corresponded to sequence identities from 75 to 100% (23).

After calibration with the MAT sequences, the JEC21 array was used to examine the serotype D genomes from the strains NIH12 and NIH433 that were the progenitors for the series of backcrosses that eventually yielded JEC21 (22). The CGH analysis revealed, as expected, that approximately 50% of the genetic content of JEC21 was derived from each of the progenitor genomes. Interestingly, the log2 ratios across each chromosome were sufficiently divergent (on the order of 2%) upon hybridization of NIH12 and NIH433 to the genome of JEC21 that the sites of recombination from the backcrossing experiment could be discerned (Fig. 1). This result indicated that CGH was sufficiently sensitive to detect sequence divergence within the same variety. It is also consistent with the regions of different nucleotide identity observed between the genomes of strains B3501A and JEC21 (described above) because B3501A was an intermediate strain in the backcrossing experiment (22).

CGH analysis with the genome of the serotype A strain H99 (var. *grubii*) was also performed to examine isolates representing the three molecular subtypes that have been identified by PCR fingerprinting, amplified fragment length polymorphism analysis, and multilocus sequence typing studies (38). Genomic DNA from strains representing the VNI, VNII, and VNB types was hybridized with the array based on the genome of H99, a VNI strain. Extensive variation was observed for all of the strains, with the VNII and VNB genomes being more variable than those of the VNI strains. The different levels of variability (including divergent sequence, deletions, and amplifications) were consistent with the classification of the strains into different subtypes. In addition to the extensive variation, the CGH experiments also revealed that two strains, CBS7779 (VNI) and WM626 (VNII), were disomic for chromosome 13. This finding was confirmed by quantitative real time PCR for both strains and by integrating a marker on chromosome 13 in strain WM626 to distinguish each copy of the chromosome in transformants. Strains CBS7779 and WM626 are clinical isolates, and the CGH results raise the possibility that disomy may occur in strains of *C. neoformans* during the infection of patients. Certainly, karyotype instability has been documented during infection (19). Chromosome copy number variation has not been previously reported for *C. neoformans*, but recent work indicates that disomy occurs in *C. neoformans* and *C. gattii* strains displaying heteroresistance to fluconazole (A. Varma, Y. C. Chang, G. Hu, J. Kronstad, and K. J. Kwon-Chung, presented at the 25th Fungal Genetics Conference at Asilomar, CA, 17 to 22 March 2009; H. Lee, E. Sionov, Y. C. Chang, and K. J. Kwon-Chung, presented at the 25th Fungal Genetics Conference at Asilomar, CA, 17 to 22 March 2009).

The opportunities presented by the CGH approach with the JEC21 (serotype D) and H99 (serotype A) genomes were extended to characterize the genomes of three strains with the hybrid AD capsular serotype (strains CDC228, CDC304, and KW5) (23). As noted above, these hybrid strains are thought to arise from mating interactions between strains of different serotypes, and they are generally diploid or aneuploid (33). The application of CGH revealed that the hybrid strains did not retain a full set of chromosomes from each parent because all three strains only had the serotype A version of chromosome 1; this finding suggests that the serotype D version had been lost. Furthermore, the serotype A versions of chromosomes 6 and 7 were present in strain KW5, and chromosome 8 in this strain was from the serotype D genome. Similarly, the serotype D version of chromosome 5 was retained in strain CDC304. PCR–restriction fragment length polymorphism analysis was used to confirm these results and to examine 16 additional strains. The serotype A version of chromosome 1 was preferentially retained in 11 of the strains, and the other five strains carried the serotype A and D versions of the chromosome. Although this analysis needs to be extended to additional strains, it is possible that the serotype A version of chromosome 1 may confer a selective advantage on hybrid strains during the presumed transition back to haploidy.

TRANSCRIPTOME ANALYSIS

The genome sequences for the *C. neoformans* and *C. gattii* strains provide an opportunity to characterize the transcriptomes of the pathogens under a variety of growth conditions in vitro and in vivo. Although several studies employed subtractive hybridization methods and differential display to identify genes with specific patterns of regulation, we will focus in this chapter on experiments with microarrays and serial analysis of gene expression (SAGE) (64) methods because of space limitations. As described in the following sections, whole-transcriptome studies have focused on several virulence-related conditions including the response to different temperatures, stress and hypoxia, expression during infection, and specific defects in signaling pathways.

Temperature

The SAGE approach was initially used to examine transcript levels in cells of a serotype A strain (H99) and a serotype D strain (B3501A) grown at 25°C and 37°C (56).

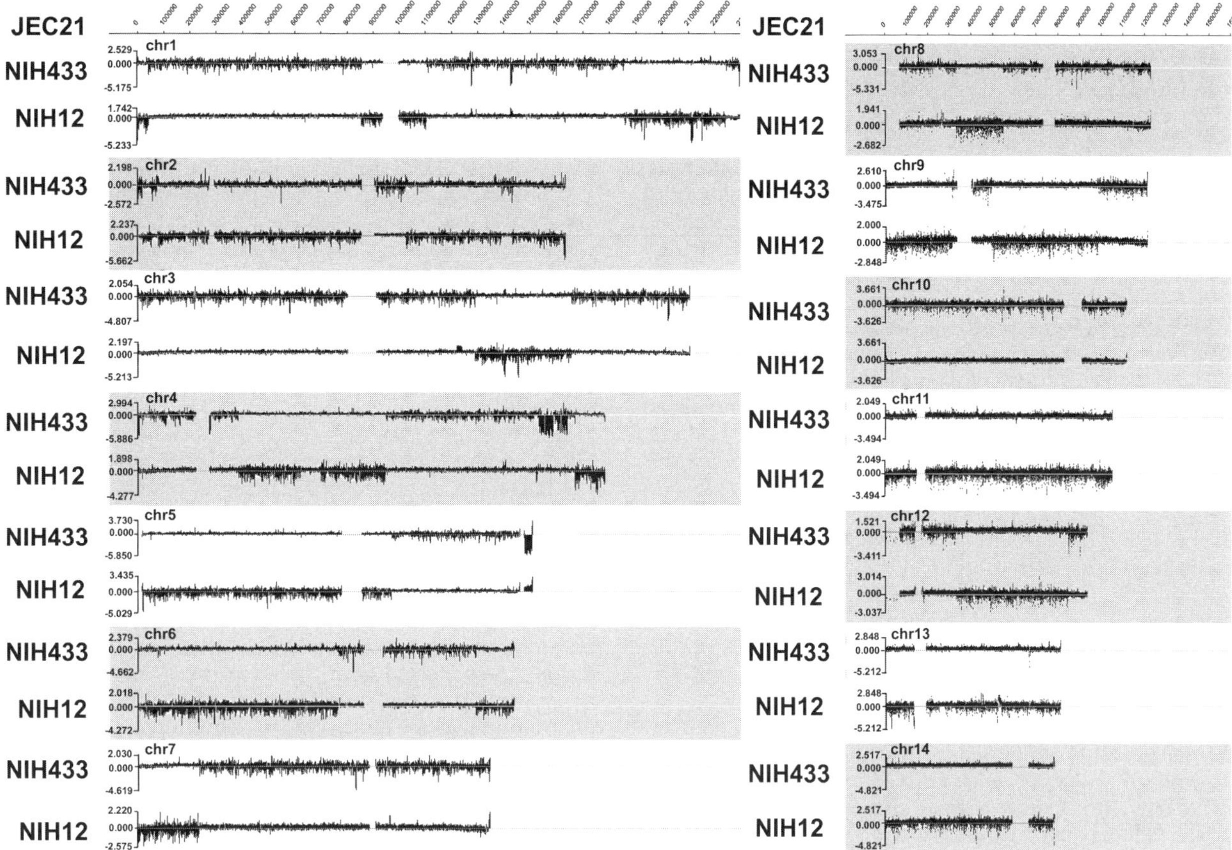

FIGURE 1 Hybridization of the genomes of the progenitor strains NIH12 (*MAT*α) and NIH433 (*MAT*a) to the tiling array of the reference strain JEC21. Regions with higher variability in log2 ratios in the NIH12 and NIH433 genomes are more divergent from the JEC21 sequence; regions with log2 ratios close to zero have greater similarity. A reciprocal pattern of similar and divergent segments is found upon hybridization of genomes of NIH12 and NIH433 to the JEC21 array. The scale of chromosome coordinates for the JEC21 genome is indicated at the top of the figure, and gaps in the chromosomes represent putative centromeric regions (40). The borders of segments are likely sites of recombination events that occurred during the cross between strains NIH12 and NIH433 and the subsequent backcrossing to obtain JEC21 (22). Reproduced from Hu et al. (23).

SAGE involves preparing libraries of short sequence tags (e.g., 14 bases) that represent the mRNAs present in cells grown under different conditions (64). A count of each tag frequency by large-scale sequencing provides an indication of the abundance of the corresponding transcript. For strain B3501A, 13,615 distinct tag sequences were identified, and 4.9% showed a differential expression level between the two temperatures. For comparison, 12,056 tag sequences were found for strain H99, and 12.5% were differentially expressed. Matching tags to genes revealed that the most abundant transcripts encoded functions known to also be highly expressed in other organisms. These included transcripts for ribosomal proteins, glyceraldehyde phosphate dehydrogenase, translation elongation factor, stress-related functions such as superoxide dismutase, and various metabolic enzymes. These results help to validate the SAGE data and identify promoters of genes encoding abundant transcripts that have potential utility in overexpression studies.

The data for strain B3501A were analyzed in detail to identify patterns of differential gene expression for 25°C and 37°C. The data revealed that the tags for the histones H1, H3, and H4 were more abundant at 25°C, suggesting that changes in chromatin may occur as a function of growth temperature. Similarly, elevated tags at 25°C were also found for genes encoding enzymes in sterol and lipid metabolism. The analysis of tags with higher levels at 37°C suggested that this condition results in elevated expression of heat shock proteins, translation machinery components, mitochrondrial proteins, and stress proteins such as superoxide dismutase. Taken together, this initial study of the influence of temperature on gene expression provided insights into specific patterns of transcriptome changes and, importantly, revealed that strains of different serotypes respond with distinct patterns. Therefore, caution is needed in interpreting transcriptional profiling results because they may be serotype- or even strain-specific.

Kraus et al. (32) also evaluated the influence of temperature on gene expression with a shotgun genomic DNA microarray containing 6,274 elements. The elements included 130 PCR-amplified cDNAs from the *C. neoformans* strains JEC21 and H99 and PCR-amplified DNA fragments from 6,144 shotgun genomic clones of strain H99. This array was estimated to provide 0.5× coverage of the genome. The influence of temperature was evaluated by two methods. First, cultures of the serotype A strain H99 were shifted from 25°C to 37°C and sampled at multiple time points. Second, steady state transcript levels were compared for cultures in the logarithmic phase of growth at 25°C or 37°C. From these experiments, a total of 239 elements showed differential expression. The genes induced at the higher temperature generally encoded functions related to stress and resistance to reactive oxygen species. Some of the genes with lower transcript levels at 37°C encoded functions for amino acid and pyrimidine biosynthesis, as well as the flavohemoglobin denitrosylase (Fhb1) that mediates resistance to nitrosative stress. One of the transcripts induced at 37°C encoded the candidate transcription factor Mga2. This gene was deleted, and the mutant had slow growth at 25°C and 30°C and an even more pronounced growth defect at 37°C. Based on results in *Saccharomyces cerevisiae*, it was predicted that Mga2 might regulate genes involved in fatty acid metabolism and membrane remodeling. Consistent with this prediction, the *mga2* mutant of *C. neoformans* was hypersensitive to the sterol synthesis inhibitors fluconazole and fenpropimorph. Furthermore, the shotgun microarray was used in additional transcriptome profiling experiments to identify targets of Mga2, and these included genes for fatty acid biosynthesis. Overall, this study serves to focus attention on fatty acid and sterol metabolism as important aspects of the response to temperature and other stresses.

Stress

Missall et al. (44) employed the JEC21 oligonucleotide array to compare transcript levels in H99 cells with and without treatment with acidified sodium nitrite to induce nitrosative stress. A JEC21 serotype D array of 7,737 70-mer oligonucleotides was employed to analyze transcriptional changes in strain H99. This study explored the utility of using an array from one serotype to examine transcripts in a different serotype. Inconsistent hybridization showed on 982 spots, which could have resulted from low expression, poor signal due to mismatch, or uneven spotting of the oligonucleotide on the array. More recently, oligonucleotides have been added to the original array design for those genes where mismatches interfered with hybridization between serotypes (http://genome.wustl.edu/services/microarray/cryptococcus_neoformans). Differential signals showed on 1,714 spots, and 421 genes showed a nitrosative stress-induced change in transcript level of twofold or greater. Of these, 205 were down between 2- and 27-fold, and 216 were up between 2- and 51-fold. The genes that were upregulated in response to nitrosative stress included 20 that encoded enzymes for the biosynthesis of the amino acids arginine, histidine, leucine, lysine, methionine, and serine. The subsequent examination of two methionine auxotrophic mutants did not reveal a change in sensitivity to nitric oxide, suggesting that the influence on gene expression may be indirect. Other induced genes encoded functions for respiration and carbohydrate metabolism, cell wall maintenance and biogenesis, transport, and the stress response. The latter category included heat shock proteins as well as Fhb1 and thioredoxin reductase. Overall, this study, which also included a parallel proteomics analysis (see below), provided a genome-wide view of the responses to nitrosative stress and identified specific functions such as glutathione reductase with relevance to virulence.

Chow et al. (10) employed the draft genome sequence of strain H99 to prepare a 70-mer oligonucleotide microarray and to study the transcriptional responses to heat shock, nitric oxide, and elevated temperature. The response to a shift from 30°C to 42°C included elevated expression of genes for heat shock proteins, the thioredoxin system, and a subunit of RNA polymerase II. Reduced expression was seen for genes encoding components of the ergosterol biosynthetic pathway and for the gene encoding Fhb1. The analysis of growth at 30°C versus 37°C revealed seven genes with higher expression at 37°C and 53 genes with lower expression. Dipropylenetriamine-NO treatment was also used to provoke nitric oxide stress with the identification of 24 genes that showed a transcriptional response. These genes included *FHB1*, *CAT3* (catalase), and *AOX1* (alternative oxidase) along with genes for other oxidoreductases. In addition, a gene for a protein with a hemerythrin cation binding motif was induced, as was a gene for mannitol synthesis and a novel four-gene family of unknown function.

Hypoxia

Chang et al. (9) characterized the response of the serotype D strain B3501A to a low-oxygen environment and identified the *SRE1* gene as encoding a homologue of the mammalian sterol regulatory element binding protein. Microarray analysis with the oligonucleotide array for the JEC21 genome revealed that Sre1 influences the expression of genes for ergosterol biosynthesis, iron and copper uptake, transporters, and various other functions under low-oxygen conditions. As expected from these results, an *sre1* mutant is hypersensitive to azole antifungal drugs and does not grow on iron-limited medium. The transcriptional profiling revealed that 100 genes showed higher expression in wild-type cells compared with the *sre1* mutant when cultures were incubated for 3 h in 1% oxygen. Under these conditions, another 414 genes showed higher transcript levels in the mutant cells; some of these genes encoded functions related to stress including catalase genes, glutathione *S*-transferases, and chaperones.

In a parallel study, Chun et al. (11) also examined genes involved in the response to hypoxia in *C. neoformans*, this time in the serotype A strain H99. They identified a histidine kinase, Tco1, and characterized the role of Sre1 and hypoxia in the regulation of gene expression and virulence. Chun et al. (11) employed the same H99 oligonucleotide array as was used for the study of nitrosative stress (10) and found that the transcript levels for 347 genes responded to hypoxia in the wild-type strain. Genes encoding functions related to stress and carbohydrate metabolism were found to be elevated. Transcripts for sterol, heme, and fatty acid metabolism were also upregulated in response to the low-oxygen environment. Downregulated genes encoded functions for protein synthesis, vesicle trafficking, and cell wall and capsule synthesis. Of the 347 genes, 54 were influenced by Sre1 under hypoxia including genes for ergosterol biosynthesis. Interestingly, loss of Tco1 did not influence expression of the set of 347 genes, indicating that the kinase may not influence the response to hypoxia at the transcriptional level.

Gene Expression during Phagocytosis and Infection of Host Tissue

Fan et al. (16) employed the shotgun genomic array for strain H99 to examine gene expression during murine macrophage infection. Two time points were employed: 2 h and 24 h after phagocytosis and fungal cells incubated in the same medium without macrophages. A total of 525 genes were analyzed further based on statistical evaluation of the result for all 6,274 elements on the array. These genes included 157 that were downregulated in the internalized cells compared with 123 upregulated genes. The group of upregulated genes included a set of predicted membrane transporters for a variety of substrates including hexoses, amino acids, iron, ammonium, nicotinic acid, and phosphate. Transcripts for peroxisomal fatty acid transporters were also elevated, as was the expression of genes for autophagy, peroxisome function, and lipid metabolism. Phagocytosis also appears to provoke a stress response in *C. neoformans* because genes encoding oxidative stress functions such as Fhb1 were upregulated. In addition, components of the cyclic AMP (cAMP) signal transduction pathway and known virulence factors were more highly expressed upon phagocytosis. Remarkably, approximately half (11) of the genes clustered at the mating type locus were activated for expression upon phagocytosis. It was hypothesized that this may reflect the responsiveness of the genes to nutrient limitation that may occur in the intracellular environment of the macrophage. Many of the genes with reduced expression upon phagocytosis encoded functions related to translation such as initiation and elongation factors and rRNA processing proteins. The view that emerges from the transcriptional analysis is one in which the phagocytosed fungal cells experience nutrient limitations and a stressful environment that is distinct from temperature stress alone. In addition, the cAMP signaling pathway appears to play an important role in coordinating the expression of capsule and melanin synthesis with the response to nutrient deprivation.

The transcriptional response has also been characterized for *C. neoformans* cells collected from the cerebral spinal fluid of infected rabbits (57). Specifically, fungal cells were collected from 12 rabbits at 5, 7, and 9 days after infection, and the pooled cells (8.3×10^7) were used to isolate RNA for the generation of a SAGE library. Subsequent sequencing of the library revealed that 16,207 tag sequences and gene matches were obtained for 164 of the 304 most abundant tags. The analysis of these genes was performed to gain insights into the functions that are most highly expressed during infection of the central nervous system. In this regard, the most abundant category contained ribosomal protein genes and other functions related to translation. Additional abundant tags represented genes for protein degradation, small-molecule transport, and signaling. A large number of tags also matched genes encoding functions for energy production, suggesting active growth in the host. Additional functions were similar to those observed following phagocytosis, and these included abundantly expressed genes related to the stress response (e.g., heat shock proteins), carbohydrate and amino acid metabolism, and transport (e.g., phosphate, iron, hexoses). Overall, this study provides insights into the environmental conditions that *C. neoformans* encounters in cerebral spinal fluid. In particular, the transporters emerge as common functions that yield insights into the nutrients available to the fungus in vivo.

Hu et al. (24) carried out an additional SAGE analysis of *C. neoformans* gene expression during infection. In this case, fungal cells of strain H99 were recovered from the lungs of infected mice at 8 and 24 h after intranasal inoculation and used for the preparation of SAGE libraries. The availability of SAGE data for cells grown at different temperatures (56) and iron levels (see below), and recovered from cerebrospinal fluid, allowed comparisons with the libraries from the lung infections. These comparisons focused attention on genes that were highly expressed under specific conditions. In particular, it was noted that genes encoding functions in central carbon metabolism were highly expressed in cells from the lungs. These genes encoded functions for the production and utilization of acetyl-coenzyme A (acetyl-CoA), for the glyoxylate cycle, and for gluconeogenesis. As was found with gene expression upon phagocytosis (16), genes for lipid metabolism showed elevated expression in the cells from the pulmonary infection. For example, the transcript for a triacylglycerol lipase (Cgl1) was elevated in both data sets. These results indicate that lipid and other alternative carbon sources are likely to be important during growth in host tissue. In this light, it is clear that acetate is not an important carbon source in the host because deletion of the *ACS1* gene encoding acetyl-CoA synthase resulted in a mutant that could not use acetate as a carbon source but could still cause disease in mice.

Genes encoding transport functions were also upregulated during lung infection, and this category also overlapped with the observed expression changes seen upon phagocytosis. Candidate transporters were identified for hexose, trehalose, amino acids, metals (copper and iron), acetate, and phosphate. One candidate hexose transporter was encoded by the most abundant transcript upon growth in the lung environment. However, abundance does not always indicate importance because deletion of this gene did not attenuate virulence in a mouse model, although the mutant did show early production of melanin. In contrast, the transcript for the iron permease gene *CFT1* was elevated during lung infection, and deletion of this gene did attenuate virulence (28). Other categories of upregulated genes included stress functions and known virulence factors. Again, there were similarities with the expression patterns seen after phagocytosis.

In general, the patterns of regulation for genes involved in carbon metabolism and transport suggested that the lung environment might be limited for glucose. In fact, follow-up quantitative real-time PCR measurements of transcript levels for several of the genes revealed that they were regulated by glucose levels in vitro. Examples of these genes include those encoding the hexose, copper, and iron transporters; the acetyl-CoA synthase Acs1; and the triacylglycerol lipase Cgl1. The pattern of gene expression was reminiscent of the regulation by the Snf1 protein kinase in *S. cerevisiae* (68). This kinase regulates alternative carbon source utilization as well as the response to stress. An *snf1* mutant was therefore constructed for *C. neoformans* and tested for glucose regulation of target genes such as *ACS1*. Surprisingly, Snf1 did not influence the response to glucose, but it did regulate the expression of the *LAC1* gene encoding the laccase for melanin production, and the *SNF1* gene was required for virulence in mice.

Iron Regulation of Gene Expression

Iron is an important factor in the virulence of *C. neoformans* due to nutritional competition during infection and because iron starvation induces capsule formation (63). In addition, iron overload exacerbates many infectious diseases including cryptococcosis (4). The pioneering work of Eric Jacobson and his colleagues established the genetic and physiological framework for iron acquisition by *C. neoformans* (27, 47, 48). Recent transcriptional profiling studies built on this framework to further define an iron regulatory network and iron acquisition functions that may be important during growth in mammalian hosts (28–30, 36, 60). Initially, SAGE was used to evaluate the response of *C. neoformans* to iron deprivation (36), and this study identified general patterns of gene expression as well as specific iron-responsive functions. Specific iron-regulated genes were identified, and these encoded iron acquisition functions such as the iron permease Cft1, which is highly expressed during infection, and a predicted mannoprotein described as a cytokine-inducing glycoprotein (Cig1) (6, 7). As mentioned above, Cft1 is important for virulence and is required for high-affinity, reductive uptake of iron from inorganic sources and transferrin (28). Cig1 is interesting because the transcript was the most abundant message detected in cells upon iron starvation, and it was 10-fold higher than in cells grown in iron-replete medium. A mutant defective in Cig1 showed poor growth in low-iron medium, and capsule formation in the mutant was no longer suppressed by growth in iron-replete medium. These observations suggest that Cig1 may function as a sensor of iron levels in the environment.

The characterization of iron-regulated genes in *C. neoformans* and analysis of the genome sequence led to identification of the key iron-related transcription factor, Cir1 (*Cryptococcus* iron regulator) (29). Cir1 was initially identified based on sequence similarity to GATA-type zinc finger transcription factors that mediate iron regulation in other fungi, such as Urbs1 in *Ustilago maydis* (29). Urbs1 is a negative regulator that represses the transcription of iron acquisition functions under iron-replete conditions. Cir1 also participates in negative regulation because *cir1* mutants are derepressed for cell-surface reductase activity. Remarkably, *cir1* mutants also have phenotypic changes in all of the major known virulence factors of *C. neoformans*, and these include decreased capsule formation, poor growth at 37°C, and elevated melanin production. These results extend the connection between iron and virulence beyond just the regulation of capsule size, and they also suggest positive regulation by Cir1 because of the loss of capsule in the mutants.

Microarray experiments with the arrays for the JEC21 genome were performed to compare the transcriptomes of the *cir1* mutant and the wild-type strain under low- and high-iron conditions. The surprising result was that loss of Cir1 eliminated practically all of the transcriptional response to iron. That is, 483 genes were downregulated and 250 were upregulated in low-iron versus high-iron media for the wild-type strain, but no differentially expressed genes were identified in the *cir1* mutant in response to iron levels. Gene ontology categories were defined for the genes regulated by Cir1, and the top groupings were for iron ion transport and siderophore transport. As predicted from the phenotypes of *cir1* mutants, some of the genes were negatively regulated by Cir1, and others were positively regulated. Examples of the negatively regulated genes included those encoding the high-affinity iron-uptake system (e.g., iron permease, ferric reductase, etc.) as well as the laccases Lac1 and Lac2 for melanin production. The positively regulated genes included siderophore transporters such as the *SIT1* gene. Deletion of this gene revealed that it is required for iron acquisition from the siderophore ferroxamine but that it is not required for virulence (60). Overall, the microarray results revealed that Cir1 may function as both a transcriptional repressor and an activator and indicated that protein is a key component of an iron regulatory network that links virulence factor elaboration to iron acquisition.

Transcriptional Analysis of cAMP Signaling

Pukkila-Worley et al. (50) employed two types of microarrays to identify genes regulated by the Gpa1 protein (Gα subunit) of the cAMP signaling pathway. Loss of Gpa1 function is known to influence mating, capsule formation, melanin production, and virulence in *C. neoformans* (1). The first array contained PCR-amplified cDNAs from 111 known *C. neoformans* genes. A comparison of RNAs isolated from the *gpa1* mutant and the wild-type strain upon glucose starvation identified 15 genes with transcript levels increased by 2.5-fold or more in the wild-type strain. Glucose starvation was employed because this condition results in melanin production. The regulated genes include nine that are thought to function in capsule synthesis or assembly. These included the *CAS1*, *CAS2*, *CAP10*, *CAP59*, and *CAP64* genes. The second type of array was the whole-genome shotgun array developed by Kraus et al. (32). Experiments with this array identified six genomic fragments that showed fivefold or higher transcript levels in the wild-type strain compared with the *gpa1* mutant. One of these fragments carried the *LAC1* gene, and two others carried the *LAC2* gene, a previously undefined homolog of *LAC1*. The discovery of *LAC2* with the shotgun microarray analysis prompted a detailed analysis of the role of this gene in melanin production and virulence. In general, Lac2 appears to have a redundant function with Lac1; its basal transcription level is lower, and loss of Lac2 causes only a slight delay in melanin formation.

One of the genes identified by Pukkila-Worley et al. (50) as being Gpa1 regulated encoded a homolog of the Nrg1 family of transcription factors. Cramer et al. (12) characterized the role of the *C. neoformans NRG1* gene by examining the phenotypes of a disruption mutant and by employing a microarray to identify downstream genes. Loss of Nrg1 resulted in production of wild-type levels of melanin, a reduction in capsule size, and attenuated virulence in a mouse model. The *nrg1* mutants also displayed a defect in cell separation and a decreased ability to mate. Transcriptome analysis to compare gene expression in the *nrg1* mutant and the wild-type strain identified 71 genes that were induced or repressed at least 2.5-fold in the *nrg1* mutant. Of these genes, 21 did not have assignable functions, and most of the remaining group encoded functions for chitin biosynthesis, carbohydrate metabolism, sugar transport, and the response to oxidative stress. Taken together, this analysis demonstrates the utility of microarray analysis as one strategy for mutant characterization and, when combined with the analysis of Gpa1 (1, 50), nicely illustrates the application of transcriptome profiling to the dissection of a signaling pathway.

The SAGE technique has also been employed to compare patterns of gene expression in the serotype A strain H99 and mutants defective in the regulatory and catalytic subunits of protein kinase A (PKA) after growth in low-iron medium (24). Part of the motivation for the analysis was the observation that mutation of the *pka1* gene for the PKA catalytic subunit resulted in a small capsule; in contrast, loss of the *pkr1* for the regulatory subunit causes an enlarged capsule (14). The SAGE analysis revealed that PKA influences the transcript abundance of genes encoding functions for cell wall synthesis, transport (including iron uptake), the tricarboxylic acid cycle, and glycolysis. In addition, differential expression was observed for genes encoding ribosomal proteins, stress and chaperone functions, virulence functions, secretory pathway components, and phospholipid biosynthetic enzymes. These patterns of gene expression suggested that the *pka1* and *pkr1* mutants might have different phenotypes under specific conditions. For example, it was found that the *pka1* mutant was more resistant to a 50°C heat shock than the wild-type or *pkr1* strains. The expression data for secretion components also indicated that inhibitors of this pathway (e.g., brefeldin A) would block capsule elaboration on the cell surface, and this turned out to be the case. This analysis was extended to include lithium as an agent to interfere with inositol metabolism because genes for myoinositol-1-phosphate synthase, inositol-1- (or 4)-monophosphatase, and a myoinositol transporter were regulated by PKA. A connection between inositol and phospholipid metabolism was established by the observation that treatment with lithium blocked capsule formation in both the wild-type and *pkr1* strains. Glycerol is also known to influence phospholipid metabolism, and it caused a similar suppression of capsule formation.

A particularly interesting category of genes that were regulated by PKA encoded cell wall and extracellular proteins. One of these, Ova1, has similarity to phosphatidylethanolamine binding proteins, and the corresponding transcript was elevated in the *pka1* mutant. Deletion of the *OVA1* gene resulted in cells with an enlarged capsule, and loss of Ova1 in a *pka1* mutant background also enhanced capsule size. These observations indicate that *OVA1* is a downstream target that is negatively regulated by PKA and that contributes to the PKA influence on capsule size. A similar connection was also found for Ova1, PKA, and melanin production. Overall, Ova1 may be a regulatory protein that is targeted by PKA as part of a general regulation of secretion (24).

PROTEOMICS

Direct analysis of protein expression levels is highly complementary to the analysis of mRNA expression levels, since there is not a direct correlation between mRNA abundance and protein abundance (20). Some genes may be regulated by transcriptional control, but others may be regulated by translational control, by protein degradation, or by regulation of activities by posttranslational modifications. Proteomics also has the advantage of being able to identify the components of a specific cellular compartment.

Protein identification can be accomplished by several methods, but peptide mass fingerprinting using matrix-assisted laser desorption ionization (MALDI) or internal peptide sequencing using liquid chromatography tandem mass spectrometry (LC-MS/MS) have both been used for *C. neoformans* proteins. Both methods require a high-quality database of predicted proteins, and this essential resource has been provided through the genome projects with high-coverage, accurate genome sequences as well as extensive bioinformatic and experimental data on gene coding sequences (40, 62; Table 1). Peptide mass fingerprinting compared the experimentally determined molecular mass of peptides generated by digesting the protein in a spot with a protease-like trypsin that produces specific cleavages to a database of predicted peptide masses.

Techniques such as two-dimensional (2D) gel electrophoresis followed by identification of specific spots by peptide mass fingerprinting or internal peptide sequencing have been used to determine which proteins are regulated under specific conditions and can be used to identify proteins that are modified in response to a signal. Although useful, this technique is limiting since it is confined to analysis of proteins that can be separated on a 2D protein gel, generally soluble proteins with pIs between 4 and 10 that are mid-range in molecular weight. Therefore, it is not a comprehensive analysis of proteomic changes. In *C. neoformans*, the protein compositions of specific compartments have been identified by proteinase digestion of all of the proteins in the compartment, followed by peptide sequencing. Newer methods hold promise for analyzing differential expression of proteins but have not yet been applied to *Cryptococcus*. The isotope-coded affinity tag method covalently attaches tags to specific amino acids (e.g., cysteines). Two different tags are used, so that two different protein lysates can be labeled. The proteins are cleaved into peptides by a protease, and the labeled peptides are purified. The two samples are mixed together and separated by LC. Peptides present in a different ratio than most of the other peptides are then chosen for sequencing and identification.

Proteomics has been used extensively in other fungal systems to identify cell wall proteins and responses to drugs and other stresses. To date, several studies have been published using proteomics in *C. neoformans*. For example, a proteomic comparison of *C. neoformans* strain H99 grown at 25°C versus 37°C using 2D gel analysis revealed that two proteins were highly upregulated at 37°C (43). Both of these proteins were identified as thiol peroxidases, proteins that have been shown to be important for resistance to oxidative stress in other systems. The deletion of both genes, as well as a third thiol peroxidase that was identified through a genomic search, demonstrated that only one of these, Tsa1, had a role in protection against exogenous oxidative stresses and in virulence. The *tsa1*Δ strain also was sensitive to nitrosative stress. In addition, 2D gels of the lysates from the deletion strains demonstrated conclusively that the correct identification of the protein spots had been made: the spots that had been identified as Tsa1 and Tsa3 were absent in the lysates from the appropriate deletion strain.

The effects on the proteome of var. *grubii* of other stress conditions including nitrosative (44) and oxidative (S. Brown and J. Lodge, submitted for publication) stress have been examined. To determine the appropriate concentration of the stressor, time and concentration effects on growth were determined for acidified nitrite, which produces nitric oxide or H_2O_2. 2D gel analysis of the response to acidified nitrate at two relevant inhibitory concentrations revealed over 30 protein spots that changed in intensity. The results from the proteomic analysis were compared to microarray and/or real-time PCR analysis of mRNA levels, and in some cases, there was not complete concordance of protein and RNA expression changes, in that some protein levels appeared to increase, but the mRNA decreased, or the protein or mRNA change was modest, but the other was very high. The proteins that were increased in response to NO were in two major classes: stress response and metabolism. Spot patterns of three of them, Tsa1, aconitase, and transaldolase, suggested changes in protein modifications in response to NO, as well as changes in overall abundance. Surprisingly, the proteomic response to peroxide stress is minimal, with only a few proteomic changes detected, even at several time points and concentrations.

There is substantial interest in the extracellular proteome of fungal pathogens, since these proteins are often immunodominant, can be critical for fungal adherence, and can modify host components. Traditional methods of protein purification have been used to separate extracellular proteins and sequence and clone the corresponding genes. Proteins with chitin/polysaccharide deacetylase domains (MP98, d25), superoxide dismutase, glucosyltransferases, as well as proteins of unknown function have been identified (6, 7, 25, 35).

Mannoproteins are a major component of fungal cell walls and can trigger the mannose receptor on antigen-presenting cells (see chapter 6 for more detail). These proteins have a secretion signal sequence to export the protein, Ser/Thr-rich domains that are the acceptors of the O-linked mannosyl moieties, and often a glycosylphosphatidylinositol (GPI) anchor that is cleaved to produce a residue that can be linked to β-1,6-glucan in the cell wall. Levitz and Specht (34) searched the *C. neoformans* genome and identified over 40 putative mannoproteins with the appropriate signal sequence, Ser/Thr-rich region, and site for a GPI anchor. The majority of the predicted proteins were annotated as cell wall or capsule modifying, cellulose degrading, proteases, or proteins of unknown function.

Eigenheer et al. (15) used an acapsular strain to determine the extracellular proteome of C. *neoformans* var. *neoformans*, since few proteins were found using encapsulated cells. Cells and cell wall fractions were treated with peptide: *N*-glycosidase F to remove N-glycosylation, since this deglycosylation step increases the number of proteins identified (61) and then delipidated. Proteins were enzymatically released from cells or purified cell walls by trypsin or β-glucanase digestion, and the resulting peptides were identified by internal sequencing using LC-MS/MS. Of the 30 proteins that were identified, 29 contained the typical N-terminal signal sequence, and of these, 17 also contained a GPI-anchor attachment motif. Of the 30 empirically identified proteins, 14 had been predicted to be mannoproteins in the analysis by Levitz and Specht (34) described above, validating their predictions. Eleven of the 30 were identified following release by β-glucanase, suggesting that these may be attached to the glucans in the cell wall. Similar to the analysis by Levitz and Specht (34), the major functional groups of proteins identified through direct experimentation were also proteases and carbohydrate modification or catabolism enzymes. The proteases may be involved in release of GPI-anchored proteins from the cell wall.

Among the proteins identified were seven proteinases. Secreted fungal proteinases are thought to have functions in proteolysis of mating pheromone, release of GPI-linked proteins from the cell wall, and proteolysis of host proteins. C. *neoformans* proteinases have been implicated in cleavage of host proteins such as fibronectin but have not been extensively characterized for their role in virulence. Four polysaccharide deacetylases were also confirmed as cell wall associated, three of which have been demonstrated to have chitin deacetylase activity to produce chitosan, the deacetylated form of chitin in *Cryptococcus* (3).

Extracellular vesicles containing the capsular material, glucuronoxylomannan, have been found in the supernatants of *Cryptococcus* (52). The mechanism of how these vesicles are secreted is unknown, although an exocyst component, Sec6, seems to be required for secretion of the vesicles, but not for secretion of other extracellular components such as capsule and the secreted protein phospholipase (49). Purification of the extracelluar vesicles followed by trypsin digest and LC-MS/MS identified 78 proteins (51). Most of these proteins were not classical extracellular proteins with a secretion signal, and many are normally considered to be cytoplasmic. These include proteins involved in carbohydrate, lipid, and amino acid metabolism, ribosome, and stress resistance. There have been proteomic analyses of cell wall proteins of several fungi that do not include stringent washing, and these analyses have also identified what were traditionally considered to be cytoplasmic proteins (8, 46). The combination of these cell wall and vesicle analyses suggests that there may be a mechanism for transporting nontraditional or "moonlighting" proteins to the extracellular milieu, although the role that they play in that compartment is undefined.

Probing the total extracellular vesicular proteins with human serum from infected patients suggested that several of the proteins were reactive. In other systems, enolase, a moonlighting protein with cytoplasmic functions in glycolysis, has also been shown to be a major antigen of fungal infection (13, 45). In C. *neoformans*, enolase was also identified as a major immunodominant protein (69). In this study, immune sera from mice inoculated with an acapsular or a heat-killed wild-type C. *neoformans* strain and subsequently infected with wild type were compared for their ability to recognize C. *neoformans* proteins. 2D protein gels of total lysates were transferred to membranes and probed with the different antisera, and proteins were identified from the spots that were recognized by the mouse serum inoculated with live cryptococcal cells. The major spots were identified as enolase, transaldolase, and Hsp70, once again highlighting the potential for cytoplasmic proteins to be extracellular. No mannoproteins were found in this analysis, potentially because there was not an enzymatic release of proteins from the cell wall matrix or because the glycosylation interferes with the ability of the proteins to be resolved on 2D gel electrophoresis.

SUMMARY

Genome sequencing projects have now been completed for two related strains of C. *neoformans* var. *neoformans*, one strain of C. *neoformans* var. *grubii*, and two strains of C. *gattii* representing the VGI and VGIIa molecular subtypes. To date, only the genomes of two related serotype D strains of var. *neoformans* have been described in a publication, but it is anticipated that the other genomes will be formally described in the near future. As outlined in this chapter, these sequencing efforts have allowed the genomes to be exploited for comparative studies by sequence analysis and comparative genome hybridization with tiling microarrays. An extension of this analysis is the rapid sequencing of the genomes of additional strains and variants using next-generation sequencing technology. The genome sequences have enabled a series of transcriptome analyses with oligonucleotide microarrays and serial analysis of gene expression. These studies are becoming standard in the analysis of complex traits (e.g., growth temperature) and the impact of mutations and drug treatments. Finally, the genomes have facilitated proteomic studies and the application of high-throughput methods with the ultimate goal of collecting deletion mutations in all of the nonessential genes. There is a clear need for a central, curated database for the cryptococcal genome sequences and related resources such as mutant phenotypes, transcriptome and proteome data, protein interaction data, literature, and community information. This resource is critically needed to move cryptococcal genomics to the next level of sophistication.

REFERENCES

1. **Alspaugh, J. A., J. R. Perfect, and J. Heitman.** 1997. *Cryptococcus neoformans* mating and virulence are regulated by the G-protein alpha subunit GPA1 and cAMP. *Genes Dev.* **11:**3206–3217.
2. **Bahn, Y. S., S. Geunes-Boyer, and J. Heitman.** 2007. Ssk2 mitogen-activated protein kinase kinase kinase governs divergent patterns of the stress-activated Hog1 signaling pathway in *Cryptococcus neoformans*. *Eukaryot. Cell* **6:**2278–2289.
3. **Baker, L. G., C. A. Specht, M. J. Donlin, and J. K. Lodge.** 2007. Chitosan, the deacetylated form of chitin, is necessary for cell wall integrity in *Cryptococcus neoformans*. *Eukaryot. Cell* **6:**855–867.
4. **Barluzzi, R., S. Saleppico, A. Nocentini, J. R. Boelaert, R. Neglia, F. Bistoni, and E. Blasi.** 2002. Iron overload exacerbates experimental meningoencephalitis by *Cryptococcus neoformans*. *J. Neuroimmunol.* **132:**140–146.
5. **Bartlett, K. H., S. E. Kidd, and J. W. Kronstad.** 2007. The emergence of *Cryptococcus gattii* in British Columbia and the Pacific Northwest. *Curr. Fungal Infect. Rep.* **1:**108–115.

6. **Biondo, C., C. Beninati, D. Delfino, M. Oggioni, G. Mancuso, A. Midiri, M. Bombaci, and G. Teti.** 2002. Identification and cloning of a cryptococcal deacetylase that produces protective immune responses. *Infect. Immun.* **70:**2383–2391.
7. **Biondo, C., G. Mancuso, A. Midiri, M. Bombaci, L. Messina, C. Beninati, and G. Teti.** 2006. Identification of major proteins secreted by *Cryptococcus neoformans.* *FEMS Yeast Res.* **6:**645–651.
8. **Chaffin, W. L.** 2008. *Candida albicans* cell wall proteins. *Microbiol. Mol. Biol. Rev.* **72:**495–544.
9. **Chang, Y. C., C. M. Bien, H. Lee, P. J. Espenshade, and K. J. Kwon-Chung.** 2007. Sre1p, a regulator of oxygen sensing and sterol homeostasis, is required for virulence in *Cryptococcus neoformans*. *Mol. Microbiol.* **64:**614–629.
10. **Chow, E. D., O. W. Lui, S. O'Brien, and H. D. Madhani.** 2007. Exploration of whole-genome responses of the human AIDS-associated yeast pathogen *Cryptococcus neoformans* var *grubii*: nitric oxide stress and body temperature. *Curr. Genet.* **52:**137–148.
11. **Chun, C. D., O. W. Liu, and H. D. Madhani.** 2007. A link between virulence and homeostatic responses to hypoxia during infection by the human fungal pathogen *Cryptococcus neoformans*. *PLoS Pathog.* **3:**e22.
12. **Cramer, K. L., Q. D. Gerrald, C. B. Nichols, M. S. Price, and J. A. Alspaugh.** 2006. Transcription factor Nrg1 mediates capsule formation, stress response, and pathogenesis in *Cryptococcus neoformans*. *Eukaryot. Cell* **5:**1147–1156.
13. **Denikus, N., F. Orfaniotou, G. Wulf, P. F. Lehmann, M. Monod, and U. Reichard.** 2005. Fungal antigens expressed during invasive aspergillosis. *Infect. Immun.* **73:**4704–4713.
14. **D'Souza, C. A., J. A. Alspaugh, C. Yue, T. Harashima, G. M. Cox, J. R. Perfect, and J. Heitman.** 2001. Cyclic AMP-dependent protein kinase controls virulence of the fungal pathogen *Cryptococcus neoformans*. *Mol. Cell Biol.* **21:**3179–3191.
15. **Eigenheer, R. A., Y. Jin Lee, E. Blumwald, B. S. Phinney, and A. Gelli.** 2007. Extracellular glycosylphosphatidylinositol-anchored mannoproteins and proteases of *Cryptococcus neoformans*. *FEMS Yeast Res.* **7:**499–510.
16. **Fan, W., P. R. Kraus, M. J. Boily, and J. Heitman.** 2005. *Cryptococcus neoformans* gene expression during murine macrophage infection. *Eukaryot. Cell* **4:**1420–1433.
17. **Forche, A., J. Xu, R. Vilgalys, and T. G. Mitchell.** 2000. Development and characterization of a genetic linkage map of *Cryptococcus neoformans* var. *neoformans* using amplified fragment length polymorphisms and other markers. *Fungal Genet. Biol.* **31:**189–203.
18. **Fraser, J. A., J. C. Huang, R. Pukkila-Worley, J. A. Alspaugh, T. G. Mitchell, and J. Heitman.** 2005. Chromosomal translocation and segmental duplication in *Cryptococcus neoformans*. *Eukaryot. Cell* **4:**401–406.
19. **Fries, B. C., F. Chen, B. P. Currie, and A. Casadevall.** 1996. Karyotype instability in *Cryptococcus neoformans* infection. *J. Clin. Microbiol.* **34:**1531–1534.
20. **Gygi, S. P., Y. Rochon, B. R. Franza, and R. Aebersold.** 1999. Correlation between protein and mRNA abundance in yeast. *Mol. Cell Biol.* **19:**1720–30.
21. **Heitman, J., A. Casadevall, J. K. Lodge, and J. R. Perfect.** 1999. The *Cryptococcus neoformans* genome sequencing project. *Mycopathologia* **148:**1–7.
22. **Heitman, J., B. Allen, J. A. Alspaugh, and K. J. Kwon-Chung.** 1999. On the origins of congenic MATalpha and MATa strains of the pathogenic yeast *Cryptococcus neoformans*. *Fungal Genet. Biol.* **28:**1–5.
23. **Hu, G., I. Liu, A. Sham, J. E. Stajich, F. S. Dietrich, and J. W. Kronstad.** 2008. Comparative hybridization reveals extensive genome variation in the AIDS-associated pathogen *Cryptococcus neoformans*. *Genome Biol.* **9:**R41.
24. **Hu, G., B. R. Steen, T. Lian, A. P. Sham, N. Tam, K. L. Tangen, and J. W. Kronstad.** 2007. Transcriptional regulation by protein kinase A in *Cryptococcus neoformans*. *PLoS Path.* **3:**e42.
25. **Huang, C., S. H. Nong, M. K. Mansour, C. A. Specht, and S. M. Levitz.** 2002. Purification and characterization of a second immunoreactive mannoprotein from *Cryptococcus neoformans* that stimulates T-cell responses. *Infect. Immun.* **70:**5485–5493
26. **Hughes, S. S., C. O. Buckley, and D. E. Neafsey.** 2008. Complex selection on intron size in *Cryptococcus neoformans*. *Mol. Biol. Evol.* **25:**247–253.
27. **Jacobson, E. S., A. P. Goodner, and K. J. Nyhus.** 1998. Ferrous iron uptake in *Cryptococcus neoformans*. *Infect. Immun.* **66:**4169–4175.
28. **Jung, W. H., A. P. Sham, T. S. Lian, A. Singh, D. Kosman, and J. W. Kronstad.** 2008. Iron source preference and regulation of iron uptake in the AIDS-associated pathogen *Cryptococcus neoformans*. *PLoS Pathog.* **4:**e45.
29. **Jung, W. H., A. Sham, R. White, and J. W. Kronstad.** 2006. Iron regulation of the major virulence factors in the AIDS-associated pathogen *Cryptococcus neoformans*. *PLoS Biol.* **4:**e410.
30. **Jung, W. H., and J. W. Kronstad.** 2007. Iron and fungal pathogenesis: a case study with *Cryptococcus neoformans*. *Cell. Microbiol.* **10:**277–284.
31. **Kavanaugh, L. A., J. A. Fraser, and F. S. Dietrich.** 2006. Recent evolution of the human pathogen *Cryptococcus neoformans* by intervarietal transfer of a 14-gene fragment. *Mol. Biol. Evol.* **23:**1879–1890.
32. **Kraus, P. R., M. J. Boily, S. S. Giles, J. E. Stajich, A. Allen, G. M. Cox, F. S. Dietrich, J. R. Perfect, and J. Heitman.** 2004. Identification of *Cryptococcus neoformans* temperature-regulated genes with a genomic-DNA microarray. *Eukaryot. Cell* **3:**1249–1260.
33. **Lengeler, K. B., G. M. Cox, and J. Heitman.** 2001. Serotype AD strains of *Cryptococcus neoformans* are diploid or aneuploid and are heterozygous at the mating-type locus. *Infect. Immun.* **69:**115–122.
34. **Levitz, S. M., and C. A. Specht.** 2006. The molecular basis for the immunogenicity of *Cryptococcus neoformans* mannoproteins. *FEMS Yeast Res.* **6:**513–524.
35. **Levitz, S. M., S. Nongn, M. K. Mansour, C. Huang, and C. A. Specht.** 2001. Molecular characterization of a mannoprotein with homology to chitin deacetylases that stimulates T cell responses to *Cryptococcus neoformans*. *Proc. Natl. Acad. Sci. USA* **98:**10422–10427.
36. **Lian, T., M. I. Simmer, C. A. D'Souza, B. R. Steen, S. D. Zuyderduyn, S. J. Jones, M. A. Marra, and J. W. Kronstad.** 2005. Iron-regulated transcription and capsule formation in the fungal pathogen *Cryptococcus neoformans*. *Mol. Microbiol.* **55:**1452–1472.
37. **Lin, X., J. C. Huang, T. G. Mitchell, and J. Heitman.** 2006. Virulence attributes and hyphal growth of *C. neoformans* are quantitative traits and the MATalpha allele enhances filamentation. *PLoS Genet.* **2:**e187.
38. **Litvintseva, A. P., R. Thakur, R. Vilgalys, and T. G. Mitchell.** 2006. Multilocus sequence typing reveals three genetic subpopulations of *Cryptococcus neoformans* var. *grubii* (serotype A), including a unique population in Botswana. *Genetics* **172:**2223–2238.
39. **Liu, O. W., C. D. Chun, E. D. Chow, C. Chen, H. D. Madhani, and S. M. Noble.** 2008. Systematic genetic analysis of virulence in the human fungal pathogen *Cryptococcus neoformans*. *Cell* **135:**174–188.
40. **Loftus, B. J., E. Fung, P. Roncaglia, D. Rowley, P. Amedeo, D. Bruno, J. Vamathevan, M. Miranda, I. J. Anderson,**

J. A. Fraser, J. E. Allen, I. E. Bosdet, M. R. Brent, R. Chiu, T. L. Doering, M. J. Donlin, C. A. D'Souza, D. S. Fox, V. Grinberg, J. Fu, M. Fukushima, B. J. Haas, J. C. Huang, G. Janbon, S. J. Jones, H. L. Koo, M. I. Krzywinski, K. J. Kwon-Chung, K. B. Lengeler, R. Maiti, M. A. Marra, R. E. Marra, C. A. Mathewson, T. G. Mitchell, M. Pertea, F. R. Riggs, S. L. Salzberg, J. E. Schein, A. Shvartsbeyn, H. Shin, M. Shumway, C. A. Specht, B. B. Suh, A. Tenney, T. R. Utterback, B. L. Wickes, J. R. Wortman, N. H. Wye, J. W. Kronstad, J. K. Lodge, J. Heitman, R. W. Davis, C. M. Fraser, and R. W. Hyman. 2005. The genome of the basidiomycetous yeast and human pathogen *Cryptococcus neoformans*. *Science* **307:**1321–1324.

41. **Marra, R. E., J. C. Huang, E. Fung, K. Nielsen, J. Heitman, R. Vilgalys, and T. G. Mitchell.** 2004. A genetic linkage map of *Cryptococcus neoformans* variety *neoformans* serotype D (*Filobasidiella neoformans*). *Genetics* **167:**619–631.
42. **Meyer, W., A. Castaneda, S. Jackson, M. Huynh, E. Castaneda, and the IberoAmerican Cryptococcal Study Group.** 2003. Molecular typing of IberoAmerican *Cryptococcus neoformans* isolates. *Emerg. Infect. Dis.* **9:**189–195.
43. **Missall, T. A., M. E. Pusateri, and J. K. Lodge.** 2004. Thiol peroxidase is critical for virulence and resistance to nitric oxide and peroxide in the fungal pathogen, *Cryptococcus neoformans*. *Mol. Microbiol.* **51:**1447–1458.
44. **Missall, T. A., M. E. Pusateri, M. J. Donlin, K. T. Chambers, J. A. Corbett, and J. K. Lodge.** 2006. Post-translational, translational, and transcriptional responses to nitric oxide stress in *Cryptococcus neoformans*: implications for virulence. *Eukaryot. Cell* **5:**518–529.
45. **Montagnoli, C., S. Sandini, A. Bacci, L. Romani, and R. La Valle.** 2004. Immunogenicity and protective effect of recombinant enolase of *Candida albicans* in a murine model of systemic candidiasis. *Med. Mycol.* **42:**319–324.
46. **Nombela, C., C. Gil, and W. L. Chaffin.** 2006. Non-conventional protein secretion in yeast. *Trends Microbiol.* **14:**15–21.
47. **Nyhus, K. J., A. T. Wilborn, and E. S. Jacobson.** 1997. Ferric iron reduction by *Cryptococcus neoformans*. *Infect. Immun.* **65:**434–438.
48. **Nyhus, K. J., and E. S. Jacobson.** 1999. Genetic and physiologic characterization of ferric/cupric reductase constitutive mutants of *Cryptococcus neoformans*. *Infect. Immun.* **5:**2357–2365.
49. **Panepinto, J., K. Komperda, S. Frases, Y. D. Park, J. T. Djordjevic, A. Casadevall, and P. R. Williamson.** 2009. Sec6-dependent sorting of fungal extracellular exosomes and laccase of *Cryptococcus neoformans*. *Mol. Microbiol.* **71:**1165–1176.
50. **Pukkila-Worley, R., Q. D. Gerrald, P. R. Kraus, M. J. Boily, M. J. Davis, S. S. Giles, G. M. Cox, J. Heitman, and J. A. Alspaugh.** 2005. Transcriptional network of multiple capsule and melanin genes governed by the *Cryptococcus neoformans* cyclic AMP cascade. *Eukaryot. Cell* **4:**190–201.
51. **Rodrigues, M. L., E. S. Nakayasu, D. L. Oliveira, L. Nimrichter, J. D. Nosanchuk, I. C. Almeida, and A. Casadevall.** 2008. Extracellular vesicles produced by *Cryptococcus neoformans* contain protein components associated with virulence. *Eukaryot. Cell* **7:**58–67.
52. **Rodrigues, M. L., L. Nimrichter, D. L. Oliverira, S. Frases, K. Miranda, O. Zaragoza, M. Alvarez, A. Nakouzi, M. Feldmesser, and A. Casadevall.** 2007. Vesicular polysaccharide export in *Cryptococcus neoformans* is a eukaryotic solution to the problem of fungal trans-cell wall transport. *Eukaryot. Cell* **6:**48–59.
53. **Roy, S. W., D. Penny, and D. E. Neafsey.** 2007. Evolutionary conservation of UTR intron boundaries in Cryptococcus. *Mol. Biol. Evol.* **24:**1140–1148.
54. **Schein, J., K. Tangen, R. Chiu, H. Shin, K. Lengeler, K. MacDonald, I. Bosdet, J. Heitman, S. J. M. Jones, M. Marra, and J. Kronstad.** 2002. Physical maps for sequence analysis of the genomes of serotype A and D strains of the fungal pathogen *Cryptococcus neoformans*. *Genome Res.* **12:**1445–1453.
55. **Sharpton, T. J., D. E. Neafsey, J. E. Galagan, and J. W. Taylor.** 2008. Mechanisms of intron gain and loss in *Cryptococcus*. *Genome Biol.* **9:**R24.
56. **Steen, B. R., T. Lian, S. Zuyderduyn, W. K. MacDonald, M. Marra, S. J. Jones, and J. W. Kronstad.** 2002. Temperature-regulated transcription in the pathogenic fungus *Cryptococcus neoformans*. *Genome Res.* **12:**1386–1400.
57. **Steen, B. R., S. Zuyderduyn, D. L. Toffaletti, M. Marra, S. J. Jones, J. R. Perfect, and J. Kronstad.** 2003. *Cryptococcus neoformans* gene expression during experimental cryptococcal meningitis. *Eukaryot. Cell* **2:**1336–1349.
58. **Sun, S., and J. Xu.** 2007. Genetic analyses of a hybrid cross between serotypes A and D strains of the human pathogenic fungus *Cryptococcus neoformans*. *Genetics* **177:**1475–1486.
59. **Sun, S., and J. Xu.** 2009. Chromosomal rearrangements between serotype A and D strains in *Cryptococcus neoformans*. *PLoS ONE* **4:**e5524.
60. **Tangen, K. T., W. H. Jung, A. Sham, T. S. Lian, and J. W. Kronstad.** 2007. The iron and cAMP regulated gene SIT1 influences siderophore utilization, melanization and cell wall structure in *Cryptococcus neoformans*. *Microbiology* **153:**29–41.
61. **Tarentino, A. L., C. M. Gómez, and T. H. Plummer, Jr.** 1985. Deglycosylation of asparagine-linked glycans by peptide:N-glycosidase F. *Biochemistry* **24:**4665–4671.
62. **Tenney, A. E., R. H. Brown, C. Vaske, J. K. Lodge, T. L. Doering, and M. R. Brent.** 2004. Gene prediction and verification in a compact genome with numerous small introns. *Genome Res.* **14:**2330–2335.
63. **Vartivarian, S. E., E. J. Anaissie, R. E. Cowart, H. A. Sprigg, M. J. Tingler, and E. S. Jacobson.** 1993. Regulation of cryptococcal capsular polysaccharide by iron. *J. Infect. Dis.* **167:**186–190.
64. **Velculescu, V. E., L. Zhang, B. Vogelstein, and K. W. Kinzler.** 1995. Serial analysis of gene expression. *Science* **270:**484–487.
65. **Wang, Z., M. Gerstein, and M. Snyder.** 2009. RNA-Seq: a revolutionary tool for transcriptomics. *Nat. Rev. Genet.* **10:**57–63.
66. **Xu, J., R. Y. Ali, D. A. Gregory, D. Amick, S. E. Lambert, H. J. Yoell, R. J. Vilgalys, and T. G. Mitchell.** 2000. Uniparental mitochondrial transmission in sexual crosses in *Cryptococcus neoformans*. *Curr. Microbiol.* **40:**269–273.
67. **Xu, J., R. Vilgalys, and T. G. Mitchell.** 2000. Multiple gene genealogies reveal recent dispersion and hybridization in the human pathogenic fungus *Cryptococcus neoformans*. *Mol. Ecol.* **9:**1471–1481.
68. **Young, E. T., K. M. Dombek, C. Tachibana, and T. Ideker.** 2003. Multiple pathways are co-regulated by the protein kinase Snf1 and the transcription factors Adr1 and Cat8. *J. Biol. Chem.* **278:**26146–26158.
69. **Young, M., S. Macias, D. Thomas, and F. L. Wormley, Jr.** 2009. A proteomic-based approach for the identification of immunodominant *Cryptococcus neoformans* proteins. *Proteomics* **9:**2578–2588.

Cryptococcus: From Human Pathogen to Model Yeast
Edited by J. Heitman et al.

10

Genetic and Genomic Approaches to *Cryptococcus* Environmental and Host Responses

ALEXANDER IDNURM AND PETER R. WILLIAMSON

Cryptococcus neoformans has been studied for over a century, often with the emphasis on gaining insight into how this pathogen causes disease (4, 40). Experimental and technological developments in the past decade, since the publication of the first book on *Cryptococcus* and cryptococcosis (4), have led to rapid advances in understanding gene function in this fungus. One major development was the completion of genome sequences for multiple *C. neoformans* strains. Particularly effective techniques have been insertional mutagenesis that produces genome-wide mutations tagged with known DNA fragments, as well as microarray analysis that allows genome-wide comparisons of transcription. The latter technique can be used to compare transcription of wild-type strains grown under different conditions or to compare mutant versus wild-type strains to elucidate regulatory targets, and this approach is described in detail in chapter 9 of this volume. The benefit of developing these tools is that *C. neoformans* represents an excellent experimental system to study gene function in fungi and is among the best pathogenic eukaryotes in which to study the genetic basis of virulence.

Despite the long battle with HIV and the development of highly active antiretroviral therapy (HAART) in the 1990s, cryptococcosis still remains a problem, especially in developing countries, with an estimate that more than 600,000 people die each year from this disease (53). While infection with *Cryptococcus* species is often considered an AIDS-defining illness, the outbreak of *C. gattii* on Vancouver Island underscores the broad potential of the *C. neoformans* species complex to evade host immune defenses and cause disease. Thus, as a major pathogen with an excellent tool set, it is unsurprising that in the past decade more than 400 papers have been reported in PubMed that include the words "gene" and "*Cryptococcus*" in either the title or abstract. Nearly 1,500 mutants in specific genes have been generated. Thus, it is not possible to describe in detail in this chapter the functions of the majority of genes that have been investigated to date. Here the emphasis is on experimental techniques and current resources, using specific examples to illustrate points with a summary of the major findings about the biology of this fungus as revealed by these approaches. Caveats, limitations, and resources that are lacking are also discussed, in the anticipation that these will be developed in the near future.

Alexander Idnurm, Division of Cell Biology and Biophysics, School of Biological Sciences, University of Missouri-Kansas City, Kansas City, MO 64110. **Peter R. Williamson,** Laboratory of Clinical Infectious Diseases, National Institutes of Allergy and Infectious Diseases, National Institutes of Health, Bethesda, MD 20892.

THE FUNCTION AND IMPACT OF THE COMPLETE GENOME SEQUENCES

A motivating factor behind research on *C. neoformans* was the impetus to develop strategies to control and treat this fungus, which is still responsible for a death toll greater than half a million annually (53). In 1999, a group within the research community discussed the possibility that *C. neoformans* could be sequenced, at a time when only a single eukaryotic genome sequence had been completed (17). In 2005, the first report of the genome sequence of two closely related serotype D strains was published (35; see chapter 9). At 19 Mb and with an estimated 6,500 genes, the genome is physically about 50% larger than that of *Saccharomyces cerevisiae* but contains a similar number of genes. Sequencing is completed and in the annotation phase for another three *C. neoformans* strains, those being serotype A H99 and serotypes B R265 and WM276. Thus, *C. neoformans* researchers are in an unusual position among the mycology research community in having five complete genome sequences representing isolates that differ in evolutionary divergence and pathogenic capability. These sequences were generated by Sanger dideoxy chemistry. Alternative sequencing techniques will generate even greater genome resources to provide information about the diversity of isolates at the DNA level.

For some reading this book, it may seem a frustratingly slow process to have attempted research in an era before complete genomes were available. The impact of genome sequencing projects on molecular biology research will probably appear in future years to be of the same magnitude as the development of PCR in the late 1980s. However, genome sequencing by itself is inherently a descriptive science: it is rare that major insights about an organism emerge from defining the arrangement of all the nucleotides in a genome. It is therefore not surprising that the genome sequence of C. *neoformans* did not lead immediately to new insights into the pathogenic ability of this fungus (35). The acquisition of structural sequence information alone does not necessarily lead to conclusions about gene function or fungal pathogenesis. As a second example of the limitations of genome sequences, the data for the genome sequences of the other three *Cryptococcus* strains have been available since 2005. Nevertheless, in this time little information has emerged to explain, for example, the clinical prevalence of serotype A over the other serotypes or why C. *gattii* can cause disease in apparently immunocompetent individuals.

The impact of the genome sequencing projects comes from their benefits in understanding gene function in C. *neoformans* through subsequent experimental approaches. This is via both forward and reverse genetics and whole-genome expression profiling. The types of studies described in more detail below that generated 1,200 targeted gene deletion strains (34) or 30,000 transfer DNA (T-DNA) insertion lines (22) would not have been feasible or practical without an available genome sequence. The completed genomic sequence enables genes to be identified based on sequence similarity with those of known function, and specifically mutated by homologous recombination. With forward genetic approaches, insertional mutants enable the rapid identification of flanking regions and therefore gene identification. Availability of the complete genome also saves the time and labor of cloning and sequencing DNA each time a gene is characterized.

One widespread scientific benefit of the C. *neoformans* genome project developed from the challenge of predicting genes in fungal genomes through bioinformatics approaches. Because of the divergence of C. *neoformans* from the better known (at a DNA level) ascomycetes and the relatively high density of 5.3 introns per gene, an expressed sequence tag project was undertaken to define the gene content. The fungus was grown under 14 different conditions, RNA was extracted, corresponding cDNAs were cloned, and the ends were sequenced from 23,000 clones. This enabled 80% of the genes reported in the genome project to be assigned to a transcribed gene, greatly aiding genome annotation and confidence about gene content based on predictions (60). A genomic DNA shotgun approach together with sequencing cDNA libraries is now almost standard practice for producing a de novo genome sequence for a fungus species.

There are several simple developments that need to be made to maximize the use of C. *neoformans* genomes. The first is the publication of the genomes of the serotype A and C. *gattii* strains and a finalized nomenclature for each gene in each strain sequenced, preferably as similar to other designations as possible. Second—and this is more an issue to be addressed by databases such as NCBI, EMBL, and DDBJ—there needs to be a way of reannotating the original submissions when new information about gene function emerges. For a gene deposited in one of these databases, it is currently difficult for nonprimary authors or submitters of data to reannotate accessions with functional information that is subsequently gained from mutant or other studies. Alternatively, a community-organized Web-based resource like the *Saccharomyces* Genome Database or *Candida* Genome Database should be established to track gene functions as they are defined, although this would require a continuous source of funding for the curators of the database.

REVERSE GENETICS

Reverse genetics is an approach to identifying a gene either in a sequence or through cloning based on a previously characterized homolog, and defining the function of this gene by making a specific disruption strain and analyzing its phenotype. For C. *neoformans* the original approach was to identify genes via complementation of mutant organisms (6, 12) or PCR with degenerate oligonucleotide primers designed from protein sequences (66), followed by gene disruption (57). With the presence of genome sequences, it is easier to find homologous genes through bioinformatic approaches and then disrupt them. Targeted disruption has become an acceptable technique to assess gene function and is particularly aided by the generation of deletion alleles through overlap PCR, although subcloning, split-marker, and yeast in vivo recombination approaches have also been used successfully to create chimeric DNA constructs used for homologous replacement. As illustrated in Fig. 1, overlap PCR requires four gene-specific primers and two common drug marker primers. The original overlap method for C. *neoformans* (10) recommended that two additional 36-nucleotide (nt) primers are required specifically for each overlap, but usually 18-nt common primers are just as efficient (15). Additionally, primers on either side and within the transformation marker for screening transformants are invaluable to confirm that a correct gene replacement has occurred (Fig. 1).

There are many examples in the literature of the reverse genetics approach for C. *neoformans*, with much focus on genes that may be required for virulence. One illustration of the impact on the genome sequence and research progress on this fungus is to compare a review published in 1998 (25), in which six genes were reported to be associated with virulence (based on the phenotypes of gene disruption strains), and a review published in 2006 (54), in which the list had grown sevenfold to 43 genes. This has been further dramatically increased since then, especially by Liu et al.; in this recent study 1,200 knockout strains were generated and analyzed, resulting in 70 new genes being identified as required for the fungus to grow in the mouse lung (34).

Incorporating signature tags in strains has enabled multiple strains to be tested for virulence in mouse models. A signature tag is a foreign piece of DNA (currently 40 bp for C. *neoformans* studies) placed adjacent to the drug marker used in gene disruption. The tag can be used to distinguish between different strains, including strain abundance in mixes by either dot blot or quantitative PCR. The assay for virulence is a competition of multiple strains within the lung (or other organ), and their loss from the population is measured by a decrease in tag abundance used to indicate reduced virulence. As developed in the laboratory of Jennifer Lodge (44), the signature tag plasmids (nourseothricin acetyltransferase and hygromycin phosphotransferase) for C. *neoformans* have been distributed widely and are increasingly being used. For example, in mouse models signature-tagged strains have been used to analyze the role of different genes required for capsule biosynthesis in virulence (41) in one forward genetic screen discussed later (21) and in the large-scale analysis by Liu et al. (34).

The analysis of Liu et al. represents the first approach of a global analysis of virulence factors in C. *neoformans*. The authors generated 1,201 strains in which the wild-type copy of the gene was replaced with the nourseothricin acetyltransferase gene coupled to a unique 40-bp signature tag. Eighty-five strains had defects in melanin, capsule, or high-temperature growth, providing new insights into the genetic

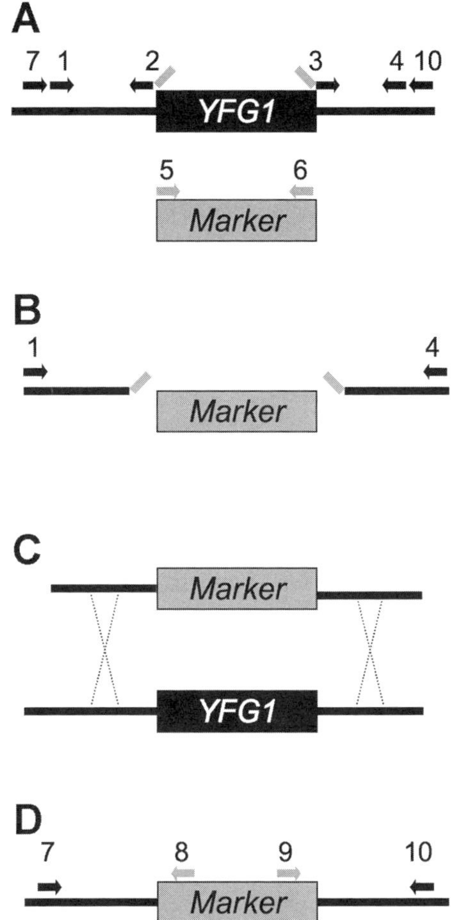

FIGURE 1 Overlap PCR for generating gene disruption constructs. The overlap PCR process first generates three products (A): a 5′ (primers 1 and 2) and 3′ (primers 3 and 4) region of 1.0–1.5 kb flanking the gene to be mutated (*your favorite gene 1* [*YFG1*]) and one for the marker (primers 5 and 6) used to select for transformed strains. All primers are standard 18 nt in length, with the exception of numbers 2 and 3, which are 36 nt long and composed of a chimera of the *YFG1* gene and the selectable marker. (B) The three pieces are mixed in equimolar amounts and amplified a second time using primers 1 and 4. (C) This product is transformed into *C. neoformans* using a biolistic apparatus, where the construct in many cases undergoes a double homologous recombination event at the *YFG1* locus. (D) The specific targeting events are screened using primers (7 and 8; 9 and 10) that only amplify when the marker is inserted, and gene disruption and the presence of only one copy of the construct in the genome is subsequently confirmed by Southern blot analysis. Primers 7 and 10 are distal to 1 and 4. Primers 5, 6, 8, and 9 can be a common set used for multiple gene disruption events.

basis of these traits. Pools of 48 mutant strains were inoculated into the lungs of mice, and the abundance of each strain relative to the level in the inocula, as measured by a quantitative real-time PCR of the signature tags, was used to derive a signature tag mutant score. Seventy new genes play a role in growth in the murine lung, and 40 of these with altered growth in the mouse lung were affected in genes that had no known virulence trait change. Many genes were of unknown function, while others were homologs of genes required for transcription, chromatin remodeling, DNA repair, and signal transduction. This number (i.e., 70) of genes that control virulence was derived from the phenotypes of at least two independent knockout strains. Furthermore, an additional 100 strains appear to change in frequency in the lung beyond what is likely a highly rigorous signature tag score of 2.5 as established in the study: these genes were not independently checked with second mutant strains, but many will likely be confirmed to regulate virulence in subsequent experiments.

One strain, with a mutation in a gene named *GAT201*, exhibited a major decrease in growth in the mouse lung as measured through a decrease in signature tag signal and was further characterized in detail (34). The predicted Gat201 protein contains a putative GATA-type transcription factor domain. A microarray comparison of transcript abundance between wild-type and *gat201* strains revealed over 500 Gat201-dependent genes, including many other transcription factors, secreted proteins, and cell wall glycosyl hydrolyases and transferases, and thus this gene functions as a global regulator of transcription. A major in vitro phenotype for the *gat201* mutants is a reduction in capsule biosynthesis. The *gat201* mutants are phagocytosed by macrophages with high efficiency, to an extent not seen in the *cap10*, *cap60*, or *cap64* mutants that have abolished capsule biosynthesis. Double *cap gat201* mutants exhibit no capsule but, surprisingly, show the same high rate of phagocytosis as the *gat201* single strain, implicating this global regulator as controlling phagocytosis independently of capsule formation.

The basis for this phenotype is unclear. The identification of a novel major determinant of virulence without a clear mechanism for this effect illustrates the power of a systematic deletion approach coupled with virulence studies. At present, approximately one-fourth to one-fifth of the genes in *C. neoformans* have been deleted and analyzed by this approach, so there are still a large number of insights to be made about the fungus. In addition, the mouse lung assay cannot distinguish strains that exhibit dissemination defects, such as *lac1* or *ure1* mutants (34, 48, 49). "Bottlenecking" events, that is, situations where only a small subset of the total population of cells are selected to reproduce while spreading throughout the body, make it challenging to study the genetic basis of this trait. However, it is of high clinical relevance in the development of the most severe meningitis symptoms of cryptococcosis.

FORWARD GENETICS

Forward genetics is an approach whereby an organism is mutated at random positions throughout the genome, strains with a mutant phenotype of interest are isolated, and the gene affected is then identified by any number of techniques. A selected mutant may be complemented with an expression library to identify the gene or the gene identified as facilitated by insertional mutagenesis with a piece of known DNA that can be used to isolate the DNA regions adjacent to the insertion. The availability of the genome sequence has dramatically streamlined approaches for forward genetics in *C. neoformans* and other fungi, particularly the plant pathogens in which large-scale collections have been generated and mutant strains screened for altered pathogenicity on plants (23, 39). One example to illustrate this point is to compare the informational content generated in the same laboratory by the discovery of the *CAP59* gene by transforming a gene library into the *cap59* mutant strain in 1994 (6) with the isolation of 35 mutants and corresponding genes required for growth under hypoxic conditions as mimicked by cobalt chloride, identified as *Agrobacterium* T-DNA insertions, in 2008 (22). Examples of larger-scale screens (set arbitrarily at >10,000 isolates analyzed; the Clark and Carbon formula [7] estimates 95% confidence of hitting any average-sized gene by screening just under 30,000 isolates) include a series on melanin discussed below, morphology (63), suppressor screens of light sensitivity of mating (69), and growth under hypoxic conditions (22). An example of the process of an insertional mutant screen and identifying the mutated *AGS1* gene is illustrated in Fig. 2.

A major hindrance to forward genetic screens that use virulence as a phenotype is the difficulty in adapting animal models for high-throughput analysis. Even using a small number of animals per strain, testing thousands of insertional mutants in mice is unfeasible for most facilities. Two approaches have been explored to facilitate forward screens in animals. In the first, "alternative" hosts have been used (42), with the most success obtained thus far in the nematode *Caenorhabditis elegans*. Two mutants with altered fitness in *C. elegans* led to the identification of the *KIN1* and *ROM2* genes. Importantly, both genes were found to be required for maximal virulence in murine inhalation models (43, 59). Other systems that have been tried are discussed in chapter 19 on host models for the fungus. A second approach is to use signature tags and screen pools of strains in animals. In the first study in *C. neoformans*, attenuated strains were identified, but no new virulence genes emerged, while a number of technical issues were identified (44). The signature-tagged plasmids were made available to others in the research community. In a second study, these tags and newly created ones were incorporated into plasmids that enabled their delivery as T-DNA molecules from *Agrobacterium tumefaciens*. Attenuated strains were identified by a reduction in signature tag signal in strains recovered from mouse lungs, including one strain with a T-DNA insertion within the *ENA1* gene. *ENA1* was then shown to be required for virulence in conventional virulence assays that employ one strain per set of mice (21). This represents one

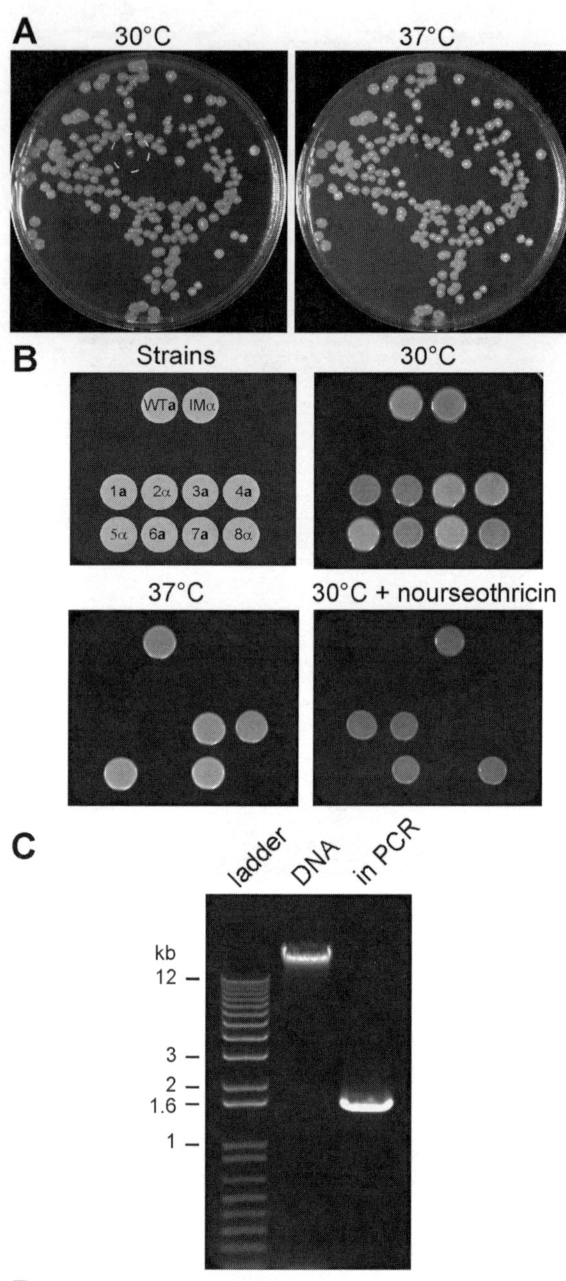

FIGURE 2 Example of a forward genetic screen, using T-DNA insertions delivered by *Agrobacterium* and temperature sensitivity as the phenotype. (A) *C. neoformans* strains are replicated on yeast extract peptone dextrose medium and incubated overnight at 30°C or 37°C. One strain (circled) shows impaired growth at 37°C. (B) The insertional mutant strain (IMα) is crossed to the wild-type strain of opposite mating type (WT**a**), and progeny are isolated. Here a subset of eight progeny from a genetic analysis show no growth at 37°C if they are also nourseothricin resistant, or show wild-type growth at 37°C and nourseothricin sensitivity, indicating linkage between the T-DNA insertion event and the temperature-sensitive phenotype. Mating type (**a** or α adjacent to the progeny number) can be used as an independent marker to demonstrate genetic recombination in the population of the progeny. (C) Genomic DNA is extracted, digested with restriction enzyme, and self-ligated, and the regions flanking the T-DNA insertion are obtained by inverse PCR using the ligation as a template. (D) The inverse PCR product is sequenced, defining the junctions between T-DNA (gray) and *C. neoformans* DNA. (E) Sequence comparison against the genome database reveals the nature of the gene affected. In this case the T-DNA insertion is 321 bp upstream of the start codon of a gene previously characterized with a role in growth at 37°C [*AGS1*, which encodes an α(1–3)glucan synthase] (55, 56).

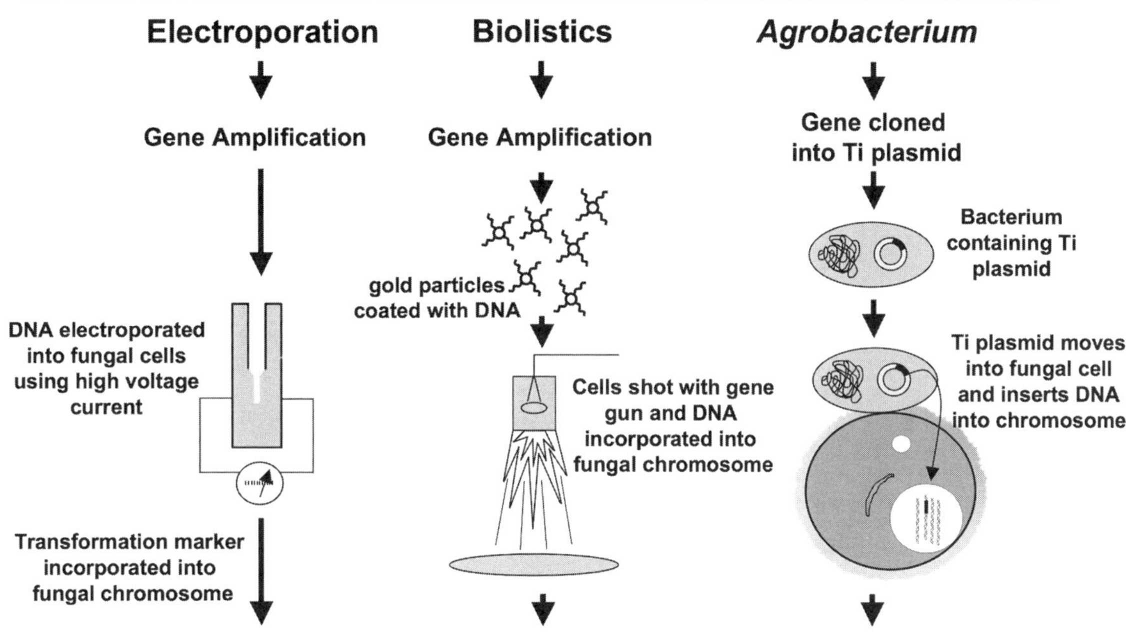

Transformation Marker
(URA5, Nat^R, Hyg^R, Neo^R)
Electroporation
Biolistics
Agrobacterium
Gene Amplification
Gene Amplification
Gene cloned into Ti plasmid
Bacterium containing Ti plasmid
gold particles coated with DNA
DNA electroporated into fungal cells using high voltage current
Ti plasmid moves into fungal cell and inserts DNA into chromosome
Cells shot with gene gun and DNA incorporated into fungal chromosome
Transformation marker incorporated into fungal chromosome
Transformed cells identified using selectable marker and screened for the phenotype of interest

FIGURE 3 Overview of three alternative methods to create insertional mutants of *C. neoformans* for use in forward genetic screens. Tagged DNA constructs containing the transformation markers such as the cryptococcal *URA5* gene or the synthetic antibiotic-resistant cassettes Nat^r, Hyg^r, and Neo^r, conveying resistance to nourseothricin, hygromycin, and neomycin, respectively, are inserted within the cryptococcal genome using at least three methods: electroporation, biolistics, and *Agrobacterium*. Electroporation uses a high-voltage pulse delivered within a metal-coated cuvette. Biolistic transformation uses gold beads coated with transforming DNA that is delivered by a helium gas pulse directly on a thin layer of fungal cells inoculated on agar plates. *Agrobacterium* methods use the transforming bacterium *A. tumefaciens* to incorporate a specialized Ti plasmid containing the selected cryptococcal transformation marker and tagged elements into the cryptococcal DNA chromosome. Transformants are then isolated using the selective media appropriate for the transformation marker.

of the first discoveries of a gene required for virulence in pathogenic fungi from a forward genetic screen directly in a mammalian host; however, the *ENA1* gene was also identified in the systemic deletion study described above (34).

Another approach to gain insight into the pathogenesis of *C. neoformans* has been to screen insertional mutants using a surrogate marker for virulence, such as the virulence factor laccase (57). Laccase, encoded by the *LAC1* gene, is a well-known virulence factor in *C. neoformans* that utilizes host cell catecholamines to produce a number of immunomodulatory molecules (see chapter 5). The screen exploits the visible melanin pigments produced by the enzyme in vitro, and laccase mutants are thus easily identified as white colonies when grown on catecholamine-containing media. The screen has been surprisingly successful in that all candidate mutants also show reduced virulence, usually much greater than mutants of the laccase structural gene itself. This is indicative of a central role of laccase in cryptococcal virulence, not only in regard to its enzymatic activity as an oxidoreductase, but also by virtue of its central placement within virulence-related expression networks, the so-called virulome (29). This latter phenomenon may be important to explain laccase's historical role as a phenotypic marker of pathogenic strains of *C. neoformans* in environmental surveys and in clinical laboratories.

Insertional mutagenesis can be achieved by a number of methods, including biolistic, electroporation, or *Agrobacterium* delivery of the DNA molecule used as the mutagen (Fig. 3). Each technique has advantages and disadvantages. For instance, electroporation and biolistic transformation generate a large number of unstable transformants, ranging from about 70% for electroporation to 85% for biolistic delivery (20, 61; P. R. Williamson, unpublished). However, genomic insertions achieved using electroporation are typically single site and strongly associated with the observed phenotype. *Agrobacterium*-mediated T-DNA mutagenesis was adapted from a method first utilized in plants (11) and uses the α-proteobacterium *A. tumefaciens*, which inserts a cryptococcal transformation marker from its Ti plasmid to create genomic insertions (20, 37). This method has the

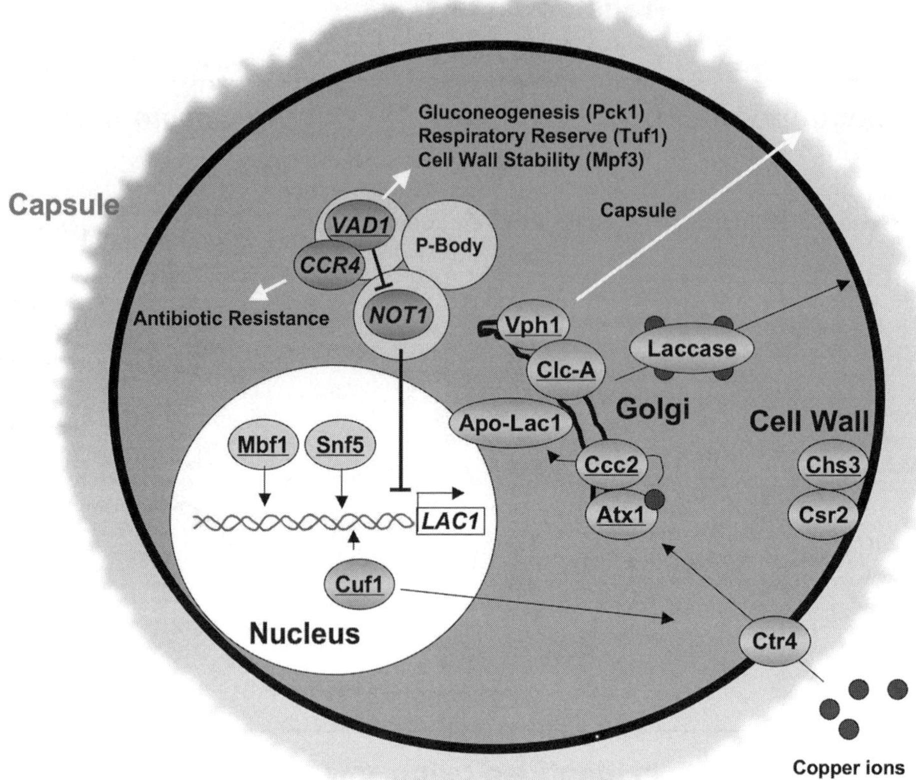

FIGURE 4 Genes required for melanin biosynthesis in serotype A *C. neoformans*. Those that have emerged from forward genetic screens are underlined. Laccase is the central enzyme for catalyzing the polymerization of phenolic substrates into melanin. Laccase is a copper-dependent enzyme; copper metabolism is therefore required for melanization. Other genes play regulatory roles for laccase directly or through copper metaliation. Laccase localizes to the cell wall, and thus mutations affecting cell wall integrity may have altered melanin profiles, such as chitin synthase (Chs3) and its regulator (Csr2). Additional genes that are known to regulate melanization (34), but through unknown mechanisms, have been omitted.

advantage of a higher rate of mutational frequency and rates of transformation, but half to one-third of mutant strains exhibit phenotypes not linked to the DNA insertion event. However, linked genomic insertions derived from any transformation method can readily be identified by performing genetic crosses to congenic wild-type strains, as long as the strains are fertile and it is not a screen that affects mating. For example, note the perfect correlation between nourseothricin resistance/sensitivity with temperature sensitivity in the progeny of the cross in Fig. 2B.

Insertional events by any of these methods are probably not truly random, and bias may be strain and method dependent, so that it is difficult to apply mathematical formulas to predict completeness of mutagenesis. In the case of *Agrobacterium* transformation of serotype A strain H99, there is clear bias for T-DNA inserting into some regions of the genome, with the best example being the promoter of *LAC1* in which five T-DNA insertions have been isolated, with none found in the coding region for the gene (21, 64). Insertion into such "recombinant hot spots" is likely technique dependent, as use of electroporation for transformation has led to less than 2% of *LAC1* mutants with insertions in the promoter (Williamson, unpublished). However, electroporation in *C. neoformans* most likely also shows bias in targeting, as no laccase mutants were identified from over 50,000 transformants screened in serotype D, whereas in serotype A the rate is close to 1 in 500. Nonetheless, good representation of mutagenesis has been shown by these methods, as many previously uncharacterized genes have been identified by these techniques. Furthermore, there has been redundant overlap in the discovery of some genes, like that encoding a vesicular chloride channel that was isolated from insertional mutants generated by both electroporation and *Agrobacterium* (20, 72). Clearly, in designing mutant screens, differences in technique and strain may greatly affect the gene discovery process.

As shown in Fig. 4, laccase-associated genes have been identified by insertional mutagenesis methods, and all of these genes, when tested, have proven to be significant virulence determinants. These include genes involved in transcriptional control of laccase such as the copper-dependent regulator Cuf1 (65) as well as the transcriptional coactivator Mbf1 and the chromatin-remodeling enzyme Snf5 (64). Cuf1 was also identified by a Mendelian genetic analysis of quantitative trait loci in serotype D *C. neoformans* (30). Forward genetic screens can then be combined with whole-genome expression studies to identify new virulence determinants expressed in parallel with laccase. The hypothesis is that expression of virulence is a multigenetic trait, and therefore investigation of coregulated genes may identify novel virulence attributes. For example, examination of copper-dependent Cuf1 regulatory pathways identified a high-affinity copper transporter, *CTR4*, which was found to facilitate copper acquisition by the fungus during brain infection (65). A requirement for Cuf1-mediated *CTR4* induction identified copper acquisition by the fungus and sequestration by the host as an important host-pathogen interaction during brain infections. Acquisition of copper may be particularly important for neurotropic pathogens such as *C. neoformans*, as an increased requirement for copper for cytochrome-mediated respiration in the relatively hypoxic environment of the brain mandates an aggressive strategy for copper acquisition. Interestingly, the hypoxic regulator Sre1 was also found to regulate *CTR4* during hypoxic conditions by microarray methods (5), suggesting an intersection between hypoxic signaling and copper-related pathogenesis in brain infections.

New collections of *C. neoformans* isolates with robust clinical information have now allowed genetic studies to be applied to "real-life" clinical problems. To this end, the first multicenter collection of *C. neoformans* isolates with detailed clinical information, including patient outcome, was obtained by Nina Singh at the University of Pittsburgh, PA (24). Studying clinically derived strains is of particular importance in light of the recent observation that most environmental isolates are nonpathogenic, at least in a mouse model (32). The acquisition of such collections now allows molecular studies in genomic diversity and pathogen virulence to be combined in clinical outcomes studies. For example, from within this collection a cohort of matched primary cryptococcal isolates from solid-organ transplant patients was analyzed for *CTR4* transcription under uniform conditions and was found to exhibit over a 100-fold diversity of *CTR4* transcript levels between isolates. Furthermore, *CTR4* expression levels were found to be independently predictive of dissemination of cryptococcal isolates to the brains of these patients. Identification of genetic relationships, such as that between *CTR4* expression and clinical outcome, may thus allow development of clinical genetic markers to differentiate high-risk patients from low-risk patients with cryptococcal disease. This could be particularly important in organ transplant patients, where current recommendations are to reduce immune suppression in all individuals exhibiting cryptococcal disease, but this results in a life-threatening immune reconstitution syndrome in a significant subpopulation of these patients (58). Additional tools such as genetic markers could thus help differentiate those patients that could benefit from more aggressive therapeutic interventions or reductions in immunosuppressive medication from those that will do well with standard therapy. Application of genetic methods to stratify patients for individualized care is an example of so-called "personalized medicine" and has been proposed to be a priority for research by the National Institutes of Health in the United States (70).

Genes involved in copper metalation of laccase have also been identified from laccase-deficient mutants, further implicating a role for copper acquisition in virulence, and have included those encoding the vacuolar proton pump, Vph1 (13), and the chloride channel Clc-A (20, 72). These genes were also found to have roles in additional virulence processes besides laccase metalation including the synthesis of the extensive polysaccharide capsule of the fungus, an important antiphagocytic virulence factor (1). In addition, homologs of the copper transporter Ccc2 and the copper chaperone Atx1 were also identified by a mutant screen using *Agrobacterium*-mediated insertional mutagenesis (64), establishing the importance of copper homeostasis in effective laccase expression and cryptococcal virulence.

Further analysis of melanin-deficient insertional mutants has identified genes involved in the regulation of mRNA stability, designated as transcription degradative factors (TDFs), including a virulence-associated DEAD box protein, *VAD1* (51). Again, the combination of forward genomics with expression methodologies (see chapter 9) identified additional virulence-related regulatory networks controlled by this TDF. For example, investigation of the *VAD1* transcriptome by differential display (DD) analysis, in turn, identified a diverse array of virulence-related processes involved in cell wall integrity (Mpf3), mitochondrial function (Tuf1), and gluconeogenesis (Pck1). DD, like serial analysis of gene expression (SAGE; as covered in chapter 9), is an example of an "open method" in that it does not rely on a previously identified set of annotated genes such as in a microarray. While more tedious than methods such as microarray, analysis of the *VAD1*-related transcriptome by DD identified the presence of retained introns in *vad1*Δ strains (Williamson, unpublished observation), suggesting differential splicing as a mechanism of *VAD1*-related virulence gene regulation. Analysis of related TDFs such as *CCR4* demonstrated a role for mRNA stability in antibiotic resistance and binding of the immune effector molecule mannose binding protein (52). The complexity of the phenotypes of the *vad1*Δ and *ccr4*Δ mutants demonstrate an intersecting role for post-transcriptional regulation of gene expression not only in virulence factor production (laccase), but also in the ability of *C. neoformans* to adapt to and thrive within the host environment (using the gene products Mpk1, Tuf1, and Pck1). Further exploration of the laccase-associated transcriptome will likely identify additional novel virulence-related cell processes.

DOWNSTREAM SUPPRESSOR AND TWO-HYBRID SCREENS TO ELUCIDATE PROTEIN INTERACTIONS AND PATHWAYS

While ascomycete model yeasts such as *S. cerevisiae* and *Schizosaccharomyces pombe* have been fruitful sources of candidate virulence genes using comparative and reverse genetics, evolutionary divergence in regulatory and signal transduction pathways has required, in many instances, whole-genome methods to identify downstream regulatory targets or interacting partners involved in virulence. For example, a whole-genome expression library developed by Fox et al. (14) was used to identify a novel downstream suppressor of calcineurin, *CTS1*. Cts1 was found to control septation and restored growth at 37°C but not mating of calcineurin mutant strains, providing evidence that calcineurin regulates growth at elevated temperatures and regulates mating via different targets (9, 14).

Protein-protein interactions are important in a number of regulatory events involved in cryptococcal virulence and have been detected in *C. neoformans* by using a variety of methods. Single gene interactions have been detected by coimmunoprecipitation studies of the nuclear heat shock transcription factor, Hsf1, with a coactivating heat shock homolog, Ssa1, during laccase induction (71). Yeast two-hybrid analysis has been used to assess the interactions between specific pairs of proteins, such as the Sxi1α-Sxi2a homeodomain proteins required for mating (18), the Bwc1-Bwc2 light-sensing complex (19), the RAM (regulation of Ace2p activity and cellular morphogenesis) morphology pathway (63), and Crg2 with Gα subunits (68), including more recently the split ubiquitin technique that can assess interactions between membrane-associated proteins (67). Discovery screens of protein-protein interactions have adopted yeast two-hybrid approaches using a *C. neoformans* cDNA library. Such a method was used to identify the Cbp1 protein that interacts with the calcineurin A subunit (16). The interaction was subsequently confirmed using coimmunoprecipitation and by largely overlapping phenotypes exhibited between *cbp1* and *cna1* mutant strains. In a second example, a yeast two-hybrid screen was used to identify a RACK1 homolog, Gib2, that was found using the previously identified Gα subunit Gpa1 as bait (50). A role for Gib2 in Gpa1 signaling was then confirmed by studies showing that overexpression of Gib2 suppressed defects of *gpa1* mutation in both melanization and capsule formation. In a similar fashion, a yeast two-hybrid approach was used to identify novel interactions between a calcium channel Cch1 and a previously identified translation elongation factor, EF3 (3). Coimmunoprecipitation and mutational studies confirmed the interaction and novel function of the translation elongation factor in calcium channel signaling (33).

MENDELIAN GENETIC ANALYSIS

A major experimental advantage of *C. neoformans*, uncommon among the human pathogenic fungi, is the established complete sexual cycle that enables classical Mendelian crosses and genetics. This can be used to infer that a single gene causes a phenotype, to demonstrate a linkage between insertional mutation events and phenotype (Fig. 2B), or to generate strains bearing multiple mutations. Furthermore, congenic strains, that is, strains in which the DNA is identical for both parents except for the mating type locus (*MAT*), are available for serotype A and serotype D. These types of strains prevent problems of genetic background interactions in subsequent segregation analysis.

A congenic set of *MAT***a** and *MAT*α serotype D strains was developed in the early 1990s (27). They became one of the most widely used strain sets (e.g., JEC21 and the precursor parent B-3501 were the strains used in the genome sequencing project). The difference in virulence between the congenic *MAT***a** and *MAT*α strains JEC20 and JEC21 also heavily influenced the field to explore the link between mating type and virulence (27). This observation was reproduced independently by backcrossing the same strains, but appears to be strain dependent because for an independent set of congenic serotype D strains there is no difference in virulence between the *MAT***a** and *MAT*α parents (47). Worldwide, serotype A predominates as the environmental as well as the clinical form of *C. neoformans*. A genetic system for this serotype appeared to be elusive, until 2000 when a strain of the **a** mating type was identified (28). Nevertheless, even today, with the exception of a small population in Botswana and sporadic isolation in Europe (31, 62), no *MAT***a** strain has been found in many areas, including the United States. The development of congenic pairs in this mating type has thus facilitated genetic studies using serotype A strains (46). However, the mating efficiency of many serotype A strains is lower compared to serotype D strains, and for researchers wishing to perform rapid genetic analysis or study the sexual process in *C. neoformans*, serotype D offers many advantages.

One caveat when using the serotype A congenic set is that while the **a** and α strains exhibit identical virulence when inoculated individually in mice, in competition assays there may be a fitness decrease for the **a** parent (45). There is other evidence for a "bottlenecking" event for strains entering the brain, independent of their mating type (44). While this **a** versus α observation has yet to be reproduced, experiments that assess brain colonization in pooled strain populations derived from genetic crosses (e.g., if marked with signature tags) should take this into consideration in the experimental design and choice of strain mating type.

Map-based cloning of genes in which a mutant is produced and the gene is identified by mapping is not a fast approach for gene discovery in fungi, including *C. neoformans*. Thus, the mutation or trait would have to be exceptionally important or complex (i.e., multigenic) to warrant this approach. However, genetic mapping has proven invaluable for gene discovery in a number of cases. On completion of the genome sequences for the two serotype D strains, a complete genetic map was generated covering the genome as much as was possible since half of the genomes of the two parental strains used (B-3501 and B-3502) are identical (36). The genetic markers developed for this map were subsequently used in a quantitative trait locus analysis to understand the differences in efficiency in mating between the parents B-3501 and B-3502 (30). Two major findings were that the *MAT* locus allele conferred filament length differences and that a regulator of copper homeostasis (named *MAC1*) also regulated filamentation as well as melanization. This same gene (named *CUF1*) was also shown independently to regulate melanization (65). A second example of mapping gene function is by Bahn et al., who identified the *SSK2* gene as regulating Hog1 phosphorylation based on the observation that the two parents (B-3501 and B-3502) used in the mapping population differ in phosphorylation of the Hog1 kinase (2). This was significant because of a natural level of variation in resistance to multiple stresses seen in *C. neoformans* isolates and is at least in part attributed to a polymorphic form of the Ssk2 protein that functions upstream of Hog1. These two studies clearly demonstrate that mapping can be used to identify a single gene as well as multiple genes affecting a trait.

Mating in *C. gattii* has yet to be developed into an efficient system compared to *C. neoformans*. In part this may be due to the smaller number of researchers, reduced mating ability compared to the other serotypes, and a concern based on a hypothesis that spores are the infectious particle (8) and that *C. gattii* tends to infect people without a known underlying immunodeficiency, thus potentially placing laboratory workers at risk. Nevertheless, conditions are known that induce mating (15, 26), and the development of a robust genetic system is certainly a valuable direction to pursue in understanding the evolution of gene functions in the *C. neoformans*/*C. gattii* species complex.

FUTURE DIRECTIONS

The past decade has witnessed *C. neoformans* emerge into an unprecedented position among the pathogenic fungi: there are five completed genomes (compared to one for most other pathogens), targeted and random gene disruptions are easy techniques, one-fifth of the genes have been deleted and arrayed as a deletion strain set, and Mendelian genetics is a facile system in most serotypes. As a member of the phylum Basidiomycota, *C. neoformans* offers insights into the evolutionary trajectory of genes in the fungal kingdom and is a key species for comparative genomics between other basidiomycetous human pathogens such as *Malassezia* spp. and the plant-pathogenic rusts and smuts. In addition, combining genomics methods and virulence studies allows probing of the role of genetic diversity in real-life fungal pathogens that may provide novel methods to control and prevent cryptococcosis. Application of genomic methods to isolates from clinical infections, for example, may help to identify genes whose expression is predictive of clinical outcomes of human infections and may also serve to focus basic science studies on virulence attributes most relevant to clinical problems in medicine.

There is still a need to develop specific tools to achieve better understanding of the responses of *C. neoformans* to its hosts and environment. It is challenging to identify point mutations within genes, and a system for *C. gattii* mating with congenic pairs is yet to be developed, as is a set of robust plasmids for fluorescent reporters. Greater coordination is needed between research groups. One practical solution is to deposit plasmids and newly developed useful strains as part of a "Crypto Tool Kit" at the Fungal Genetics Stock Center (University of Missouri-Kansas City) (38). This facility distributes these resources at minimal cost around the world and already houses gene deletion collections for *C. neoformans* as well as large numbers of other fungi. Completion of the deletion set would generate an outstanding resource. At present *C. neoformans* offers disadvantages for researchers coming in from outside fields. Gene disruption requires a biolistic apparatus for transformation, micromanipulation of basidiospores requires a specialized microscope, and BSL-2 or equivalent compliant procedures are required for working on *Cryptococcus* species. The expense of these items and the administrative burden hamper the curiosity-driven researcher and may reduce the recruitment of outsiders as can occur in more economical species (e.g., *S. cerevisiae* or the basidiomycetous yeast *Ustilago maydis*). However, access to a complete gene deletion set and the fact that many yeast laboratories are equipped with microscopes with micromanipulators alleviates some of these issues. A final consideration is to continue the active recruitment of new investigators from a range of disciplines to take up the challenge of combating an important fungal disease through understanding the biology of this fungus.

REFERENCES

1. **Baba, T.** 1988. Electron microscopic cytochemical analysis of hepatic granuloma induced by *Cryptococcus neoformans*. *Mycopathologia* **104:**37–46.
2. **Bahn, Y.-S., S. Geunes-Boyer, and J. Heitman.** 2007. Ssk2 mitogen-activated protein kinase kinase kinase governs divergent patterns of the stress-activated Hog1 signaling pathway in *Cryptococcus neoformans*. *Eukaryot. Cell* **6:**2278–2289.
3. **Blakely, G., J. Hekman, K. Chakraburtty, and P. R. Williamson.** 2001. Evolutionary divergence of an elongation factor 3 from *Cryptococcus neoformans*. *J. Bacteriol.* **183:**2241–2248.
4. **Casadevall, A., and J. Perfect.** 1998. *Cryptococcus neoformans*. ASM Press, Washington, DC.
5. **Chang, Y. C., C. M. Bien, H. Lee, P. J. Espenshade, and K. J. Kwon-Chung.** 2007. Sre1p, a regulator of oxygen sensing and sterol homeostasis, is required for virulence in *Cryptococcus neoformans*. *Mol. Microbiol.* **64:**614–629.
6. **Chang, Y. C., and K. J. Kwon-Chung.** 1994. Complementation of a capsule-deficient mutation of *Cryptococcus neoformans* restores its virulence. *Mol. Cell. Biol.* **14:**4912–4919.
7. **Clarke, L., and J. Carbon.** 1976. A colony bank containing synthetic Col El hybrid plasmids representative of the entire *E. coli* genome. *Cell* **9:**91–99.
8. **Cohen, J., J. R. Perfect, and D. T. Durack.** 1982. Cryptococcosis and the basidiospore. *Lancet* **1:**1301.
9. **Cruz, M. C., D. S. Fox, and J. Heitman.** 2001. Calcineurin is required for hyphal elongation during mating and haploid fruiting in *Cryptococcus neoformans*. *EMBO J.* **20:**1020–1032.
10. **Davidson, R. C., J. R. Blankenship, P. R. Kraus, M. de Jesus Berrios, C. M. Hull, C. D'Souza, P. Wang, and J. Heitman.** 2002. A PCR-based strategy to generate integrative targeting alleles with large regions of homology. *Microbiology* **148:**2607–2615.
11. **Dhaese, P., H. De Greve, H. Decraemer, J. Schell, and M. Van Montagu.** 1979. Rapid mapping of transposon insertion and deletion mutations in the large Ti-plasmids of *Agrobacterium tumefaciens*. *Nucleic Acids Res.* **7:**1837–1849.
12. **Edman, J. C., and K. J. Kwon-Chung.** 1990. Isolation of the *URA5* gene from *Cryptococcus neoformans* var. *neoformans* and its use as a selective marker for transformation. *Mol. Cell. Biol.* **10:**4538–4544.
13. **Erickson, T., L. Liu, A. Gueyikian, X. Zhu, J. Gibbons, and P. R. Williamson.** 2001. Multiple virulence factors of *Cryptococcus neoformans* are dependent on *VPH1*. *Mol. Microbiol.* **42:**1121–1131.
14. **Fox, D. S., G. M. Cox, and J. Heitman.** 2003. Phospholipid-binding protein Cts1 controls septation and functions coordinately with calcineurin in *Cryptococcus neoformans*. *Eukaryot. Cell* **2:**1025–1035.
15. **Fraser, J. A., R. L. Subaran, C. B. Nichols, and J. Heitman.** 2003. Recapitulation of the sexual cycle of the primary fungal pathogen *Cryptococcus neoformans* var. *gattii*: implications for an outbreak on Vancouver Island, Canada. *Eukaryot. Cell* **2:**1036–1045.
16. **Görlach, J., D. S. Fox, N. S. Cutler, G. M. Cox, J. R. Perfect, and J. Heitman.** 2000. Identification and characterization of a highly conserved calcineurin binding protein, CBP1/calcipressin, in *Cryptococcus neoformans*. *EMBO J.* **19:**3618–3629.
17. **Heitman, J., A. Casadevall, J. K. Lodge, and J. R. Perfect.** 1999. The *Cryptococcus neoformans* genome sequencing project. *Mycopathologia* **148:**1–7.
18. **Hull, C. M., M.-J. Boily, and J. Heitman.** 2005. Sex-specific homeodomain proteins Sxi1a and Sxi2a coordinately regulate sexual development in *Cryptococcus neoformans*. *Eukaryot. Cell* **4:**526–535.
19. **Idnurm, A., and J. Heitman.** 2005. Light controls growth and development via a conserved pathway in the fungal kingdom. *PLoS Biol.* **3:**615–626.
20. **Idnurm, A., J. L. Reedy, J. C. Nussbaum, and J. Heitman.** 2004. *Cryptococcus neoformans* virulence gene discovery through insertional mutagenesis. *Eukaryot Cell* **3:**420–429.
21. **Idnurm, A., F. J. Walton, A. Floyd, J. L. Reedy, and J. Heitman.** 2009. Identification of *ENA1* as a virulence gene of the human pathogenic fungus *Cryptococcus neoformans* through signature-tagged insertional mutagenesis. *Eukaryot. Cell* **8:**315–326.
22. **Ingavale, S. S., Y. C. Chang, H. Lee, C. M. McClelland, M. L. Leong, and K. J. Kwon-Chung.** 2008. Importance of mitochondria in survival of *Cryptococcus neoformans* under low oxygen conditions and tolerance to cobalt chloride. *PLoS Pathog.* **4:**e1000155.
23. **Jeon, J., S.-Y. Park, M.-H. Chi, J. Choi, J. Park, H.-S. Rho, S. Kim, J. Goh, S. Yoo, J. Choi, J.-Y. Park, M. Yi, S. Yang, M.-J. Kwon, S.-S. Han, B. R. Kim, C. H. Khang, B. Park, S.-E. Lim, K. Jung, S. Kong, M. Karunakaran, H.-S. Oh, H. Kim, S. Kim, J. Park, S. Kang, W.-B. Choi, S. Kang, and Y.-H. Lee.** 2007. Genome-wide functional analysis of pathogenicity genes in the rice blast fungus. *Nat. Genet.* **39:**561–565.
24. **Kontoyiannis, D. P., R. E. Lewis, B. D. Alexander, O. Lortholary, F. Dromer, K. L. Gupta, G. T. John, R. del**

Busto, G. B. Klintmalm, J. Somani, G. M. Lyon, K. Pursell, V. Stosor, P. Munoz, A. P. Limaye, A. C. Kalil, T. L. Pruett, J. Garcia-Diaz, A. Humar, S. Houston, A. A. House, D. Wray, S. Orloff, L. A. Dowdy, R. A. Fisher, J. Heitman, N. D. Albert, M. M. Wagener, and N. Singh.** 2008. Calcineurin inhibitor agents interact synergistically with antifungal agents in vitro against *Cryptococcus neoformans* isolates: correlation with outcome in solid organ transplant recipients with cryptococcosis. *Antimicrob. Agents Chemother.* **52:**735–738.

25. **Kwon-Chung, K.** 1998. Gene disruption to evaluate the role of fungal candidate virulence genes. *Curr. Opin. Microbiol.* **1:**381–389.
26. **Kwon-Chung, K. J.** 1976. A new species of Filobasidiella, the sexual state of *Cryptococcus neoformans* B and C serotypes. *Mycologia* **68:**943–946.
27. **Kwon-Chung, K. J., J. C. Edman, and B. L. Wickes.** 1992. Genetic association of mating types and virulence in *Cryptococcus neoformans*. *Infect. Immun.* **60:**602–605.
28. **Lengeler, K. B., P. Wang, G. M. Cox, J. R. Perfect, and J. Heitman.** 2000. Identification of the MATa mating-type locus of *Cryptococcus neoformans* reveals a serotype A MATa strain thought to have been extinct. *Proc. Natl. Acad. Sci. USA* **97:**14455–14460.
29. **Liautard, J.-P., V. Jubier-Maurin, R.-A. Boigegrain, and S. Köhler.** 2006. Antimicrobials: targeting virulence genes necessary for intracellular multiplication. *Trends Microbiol.* **14:**109–113.
30. **Lin, X., J. C. Huang, T. G. Mitchell, and J. Heitman.** 2006. Virulence attributes and hyphal growth of *C. neoformans* are quantitative traits and the *MAT*α allele enhances filamentation. *PLoS Genet.* **2:**e187.
31. **Litvintseva, A. P., R. E. Marra, K. Nielsen, J. Heitman, R. Vilgalys, and T. G. Mitchell.** 2003. Evidence of sexual recombination among *Cryptococcus neoformans* serotype A isolates in sub-Saharan Africa. *Eukaryot. Cell* **2:**1162–1168.
32. **Litvintseva, A. P., and T. G. Mitchell.** 2009. Most environmental isolates of *Cryptococcus neoformans* var. *grubii* (serotype A) are not lethal for mice. *Infect. Immun.* **77:**3188–3195.
33. **Liu, M., and A. Gelli.** 2008. Elongation factor 3, EF3, associates with the calcium channel Cch1 and targets Cch1 to the plasma membrane in *Cryptococcus neoformans*. *Eukaryot. Cell* **7:**1118–1126.
34. **Liu, O. W., C. D. Chun, E. D. Chow, C. Chen, H. D. Madhani, and S. M. Noble.** 2008. Systematic genetic analysis of virulence in the human fungal pathogen *Cryptococcus neoformans*. *Cell* **135:**174–188.
35. **Loftus, B. J., E. Fung, P. Roncaglia, D. Rowley, P. Amedeo, D. Bruno, J. Vamathevan, M. Miranda, I. J. Anderson, J. A. Fraser, J. E. Allen, I. E. Bosdet, M. R. Brent, R. Chiu, T. L. Doering, M. J. Donlin, C. A. D'Souza, D. S. Fox, V. Grinberg, J. Fu, M. Fukushima, B. J. Haas, J. C. Huang, G. Janbon, S. J. Jones, H. L. Koo, M. I. Krzywinski, J. K. Kwon-Chung, K. B. Lengeler, R. Maiti, M. A. Marra, R. E. Marra, C. A. Mathewson, T. G. Mitchell, M. Pertea, F. R. Riggs, S. L. Salzberg, J. E. Schein, A. Shvartsbeyn, H. Shin, M. Shumway, C. A. Specht, B. B. Suh, A. Tenney, T. R. Utterback, B. L. Wickes, J. R. Wortman, N. H. Wye, J. W. Kronstad, J. K. Lodge, J. Heitman, R. W. Davis, C. M. Fraser, and R. W. Hyman.** 2005. The genome of the basidiomycetous yeast and human pathogen *Cryptococcus neoformans*. *Science* **307:**1321–1324.
36. **Marra, R. E., J. C. Huang, E. Fung, K. Nielsen, J. Heitman, R. Vilgalys, and T. G. Mitchell.** 2004. A genetic linkage map of *Cryptococcus neoformans* variety *neoformans* serotype D (*Filobasidiella neoformans*). *Genetics* **167:**619–631.
37. **McClelland, C. M., Y. C. Chang, and K. J. Kwon-Chung.** 2005. High frequency transformation of *Cryptococcus neoformans* and *Cryptococcus gattii* by *Agrobacterium tumefaciens*. *Fungal Genet. Biol.* **42:**904–913.
38. **McCluskey, K.** 2003. The Fungal Genetics Stock Center: from molds to molecules. *Adv. Appl. Microbiol.* **52:**245–262.
39. **Michielse, C. B., R. van Wijk, L. Reijnen, B. J. Cornelissen, and M. Rep.** 2009. Insight into the molecular requirements for pathogenicity of *Fusarium oxysporum* f. sp. *lycopersici* through large-scale insertional mutagenesis. *Genome Biol.* **10:**R4.
40. **Mitchell, T. G., and J. R. Perfect.** 1995. Cryptococcosis in the era of AIDS: 100 years after the discovery of *Cryptococcus neoformans*. *Clin. Microbiol. Rev.* **8:**515–548.
41. **Moyrand, F., T. Fontaine, and G. Janbon.** 2007. Systematic capsule gene disruption reveals the central role of galactose metabolism on *Cryptococcus neoformans* virulence. *Mol. Microbiol.* **64:**771–781.
42. **Mylonakis, E., A. Casadevall, and F. M. Ausubel.** 2007. Exploiting amoeboid and non-vertebrate animal model systems to study the virulence of human pathogenic fungi. *PLoS Pathog.* **3:**e101.
43. **Mylonakis, E., A. Idnurm, R. Moreno, J. El Khoury, J. B. Rottman, F. M. Ausubel, J. Heitman, and S. B. Calderwood.** 2004. *Cryptococcus neoformans* Kin1 protein kinase homologue, identified through a *Caenorhabditis elegans* screen, promotes virulence in mammals. *Mol. Microbiol.* **54:**407–419.
44. **Nelson, R. T., J. Hua, B. Pryor, and J. K. Lodge.** 2001. Identification of virulence mutants of the fungal pathogen *Cryptococcus neoformans* using signature-tagged mutagenesis. *Genetics* **157:**935–947.
45. **Nielsen, K., G. M. Cox, A. P. Litvintseva, E. Mylonakis, S. D. Malliaris, D. K. Benjamin, Jr., S. S. Giles, T. G. Mitchell, A. Casadevall, J. R. Perfect, and J. Heitman.** 2005. *Cryptococcus neoformans* α strains preferentially disseminate to the central nervous system during coinfection. *Infect. Immun.* **73:**4922–4933.
46. **Nielsen, K., G. M. Cox, P. Wang, D. L. Toffaletti, J. R. Perfect, and J. Heitman.** 2003. Sexual cycle of *Cryptococcus neoformans* var. *grubii* and virulence of congenic **a** and α isolates. *Infect. Immun.* **71:**4831–4841.
47. **Nielsen, K., R. E. Marra, F. Hagen, T. Boekhout, T. G. Mitchell, G. M. Cox, and J. Heitman.** 2005. Interaction between genetic background and the mating-type locus in *Cryptococcus neoformans* virulence potential. *Genetics* **171:**975–983.
48. **Noverr, M. C., P. R. Williamson, R. S. Fajardo, and G. B. Huffnagle.** 2004. *CNLAC1* is required for extrapulmonary dissemination of *Cryptococcus neoformans* but not pulmonary persistence. *Infect. Immun.* **72:**1693–1699.
49. **Olszewski, M. A., M. C. Noverr, G. H. Chen, G. B. Toews, G. M. Cox, J. R. Perfect, and G. B. Huffnagle.** 2004. Urease expression by *Cryptococcus neoformans* promotes microvascular sequestration, thereby enhancing central nervous system invasion. *Am. J. Pathol.* **164:**1761–1771.
50. **Palmer, D. A., J. K. Thompson, L. Li, A. Prat, and P. Wang.** 2006. Gib2, a novel Gβ-like/RACK1 homolog, functions as a Gβ subunit in cAMP signaling and is essential in *Cryptococcus neoformans*. *J. Biol. Chem.* **281:**32596–32605.
51. **Panepinto, J., L. Liu, J. Ramos, X. Zhu, T. Valyi-Nagy, S. Eksi, J. Fu, H. A. Jaffe, B. Wickes, and P. Williamson.** 2005. The DEAD-box RNA helicase Vad1 regulates multiple virulence-associated genes in *Cryptococcus neoformans*. *J. Clin. Invest.* **115:**632–641.
52. **Panepinto, J. C., K. W. Komperda, M. Hacham, S. Shin, X. Liu, and P. R. Williamson.** 2007. Binding of serum

mannan binding lectin to a cell integrity-defective *Cryptococcus neoformans ccr4Δ* mutant. *Infect. Immun.* **75:** 4769–4779.
53. **Park, B. J., K. A. Wannemuehler, B. J. Marston, N. Govender, P. G. Pappas, and T. M. Chiller.** 2009. Estimation of the current global burden of cryptococcal meningitis among persons living with HIV/AIDS. *AIDS* **23:**525–530.
54. **Perfect, J. R.** 2005. *Cryptococcus neoformans*: a sugar-coated killer with designer genes. *FEMS Immunol. Med. Microbiol.* **45:**395–404.
55. **Reese, A. J., and T. L. Doering.** 2003. Cell wall α-1,3-glucan is required to anchor the *Cryptococcus neoformans* capsule. *Mol. Microbiol.* **50:**1401–1409.
56. **Reese, A. J., A. Yoneda, J. A. Breger, A. Beauvais, H. Liu, C. L. Griffith, I. Bose, M.-J. Kim, C. Skau, S. Yang, J. A. Sefko, M. Osumi, J.-P. Latge, E. Mylonakis, and T. L. Doering.** 2007. Loss of cell wall α(1-3) glucan affects *Cryptococcus neoformans* from ultrastructure to virulence. *Mol. Microbiol.* **63:**1385–1398.
57. **Salas, S. D., J. E. Bennett, K. J. Kwon-Chung, J. R. Perfect, and P. R. Williamson.** 1996. Effect of the laccase gene *CNLAC1*, on virulence of *Cryptococcus neoformans*. *J. Exp. Med.* **184:**377–386.
58. **Singh, N., O. Lortholary, B. D. Alexander, K. L. Gupta, G. T. John, K. Pursell, P. Munoz, G. B. Klintmalm, V. Stosor, R. del Busto, A. P. Limaye, J. Somani, M. Lyon, S. Houston, A. A. House, T. L. Pruett, S. Orloff, A. Humar, L. Dowdy, J. Garcia-Diaz, A. C. Kalil, R. A. Fisher, and S. Husain.** 2005. An immune reconstitution syndrome-like illness associated with *Cryptococcus neoformans* infection in organ transplant recipients. *Clin. Infect. Dis.* **40:**1756–1761.
59. **Tang, R. J., J. Breger, A. Idnurm, K. J. Gerik, J. K. Lodge, J. Heitman, S. B. Calderwood, and E. Mylonakis.** 2005. *Cryptococcus neoformans* gene involved in mammalian pathogenesis identified by a *Caenorhabditis elegans* progeny-based approach. *Infect. Immun.* **73:**8219–8225.
60. **Tenney, A. E., R. H. Brown, C. Vaske, J. K. Lodge, T. L. Doering, and M. R. Brent.** 2004. Gene prediction and verification in a compact genome with numerous small introns. *Genome Res.* **14:**2330–2335.
61. **Toffaletti, D. L., T. H. Rude, S. A. Johnston, D. T. Durack, and J. R. Perfect.** 1993. Gene transfer in *Cryptococcus neoformans* by use of biolistic delivery of DNA. *J. Bacteriol.* **175:**1405–1411.
62. **Viviani, M. A., M. C. Esposto, M. Cogliati, M. T. Montagna, and B. L. Wickes.** 2001. Isolation of a *Cryptococcus neoformans* serotype A *MAT*a strain from the Italian environment. *Med. Mycol.* **39:**383–386.
63. **Walton, F. J., J. Heitman, and A. Idnurm.** 2006. Conserved elements of the RAM signaling pathway establish cell polarity in the basidiomycete *Cryptococcus neoformans* in a divergent fashion from other fungi. *Mol. Biol. Cell* **17:**3768–3780.
64. **Walton, F. J., A. Idnurm, and J. Heitman.** 2005. Novel gene functions required for melanization of the human pathogen *Cryptococcus neoformans*. *Mol. Microbiol.* **57:**1381–1396.
65. **Waterman, S. R., M. Hacham, G. Hu, X. Zhu, Y. D. Park, S. Shin, J. Panepinto, T. Valyi-Nagy, C. Beam, S. Husain, N. Singh, and P. R. Williamson.** 2007. Role of a CUF1-CTR4 copper regulatory axis in the virulence of *Cryptococcus neoformans*. *J. Clin. Invest.* **117:**794–802.
66. **Williamson, P. R.** 1994. Biochemical and molecular characterization of the diphenol oxidase of *Cryptococcus neoformans*: identification as a laccase. *J. Bacteriol.* **176:** 656–664.
67. **Xue, C., Y.-S. Bahn, G. M. Cox, and J. Heitman.** 2006. G protein-coupled receptor Gpr4 senses amino acids and activates the cAMP-PKA pathway in *Cryptococcus neoformans*. *Mol. Biol. Cell* **17:**667–679.
68. **Xue, C., Y. P. Hsueh, L. Chen, and J. Heitman.** 2008. The RGS protein Crg2 regulates both pheromone and cAMP signalling in *Cryptococcus neoformans*. *Mol. Microbiol.* **70:**379–395.
69. **Yeh, Y. L., Y. S. Lin, B. J. Su, and W. C. Shen.** 2009. A screening for suppressor mutants reveals components involved in the blue light-inhibited sexual filamentation in *Cryptococcus neoformans*. *Fungal Genet. Biol.* **46:**42–54.
70. **Zerhouni, E.** 2005. Translational and clinical science: time for a new vision. *N. Engl. J. Med.* **353:**1621–1623.
71. **Zhang, S., M. Hacham, J. Panepinto, G. Hu, S. Shin, X. Zhu, and P. R. Williamson.** 2006. The Hsp70 member, Ssa1 acts as a DNA-binding transcriptional co-activator in *Cryptococcus neoformans*. *Mol. Microbiol.* **62:**1090–1101.
72. **Zhu, X., and P. R. Williamson.** 2003. A CLC-type chloride channel gene is required for laccase activity and virulence in *Cryptococcus neoformans*. *Mol. Microbiol.* **50:**1271–1281.

Cryptococcus: From Human Pathogen to Model Yeast
Edited by J. Heitman et al.

11

The Mating-Type Locus of *Cryptococcus*: Evolution of Gene Clusters Governing Sex Determination and Sexual Reproduction from the Phylogenomic Perspective

YEN-PING HSUEH, BANU METIN, KEISHA FINDLEY, MARIANELA RODRIGUEZ-CARRES, AND JOSEPH HEITMAN

Sex is favored by evolution and often directly linked to the fitness of an organism. In many species, such as humans, sex is obligatory; offspring are only produced via sexual reproduction. In microbes, although cells can propagate via asexual reproduction, sex remains critical for their fitness and serves to recombine parental genomes as well as purge deleterious mutations. Mechanisms of sex determination are myriad and diverse among different organisms. In vertebrates, sex can be genetically regulated (genetic sex determination) or environmentally regulated (environmental sex determination). For example, in mammals and birds, sex is determined by a pair of sex chromosomes XX/XY or ZZ/ZW, while in many reptiles sex is determined by temperature (7, 49, 66, 68). Some species of annelids, echinoderms, crustaceans, mollusks, and fish can even undergo sex changes during their life cycles (53). It was thought that genetic sex determination and environmental sex determination were completely distinct pathways, but recent studies have shown that in some reptile species, elements of both pathways are both employed in sex determination (62). Sex is also well studied in a number of model organisms such as *Drosophila* and *Caenorhabditis elegans*, where a dosage-dependent mechanism orchestrates sex determination (61). A similar pattern of dosage-dependent sex determination involving the Z-linked gene *DMRT1* in chickens was also recently reported, supporting the Z dosage hypothesis for avian sex determination (67).

Sexual reproduction in fungi is governed by a specialized region of the genome known as the mating-type (*MAT*) locus that orchestrates cell type identity, cell-cell fusion, and the fate of the zygote. There are two patterns of sex in heterothallic fungi: bipolar and tetrapolar. In bipolar species, sex is governed by a single *MAT* locus that is usually biallelic, while in tetrapolar species, two unlinked *MAT* loci that are often multiallelic coordinately orchestrate the sexual reproduction (6, 17, 20, 38, 52). Despite the central importance of this genomic region to fungal biology and species identity and success, the *MAT* locus is remarkably plastic, and the genes resident therein differ considerably between fungal phyla and among species within phyla. A general principle is that *MAT* loci encode DNA binding transcription factors (with homeodomain, alpha domain, or high-mobility group [HMG] domains), and in the basidiomycetes, *MAT* loci also encode pheromones and pheromone receptors. But beyond this, there is an array of different organizations, sizes, and gene content that can differ considerably, even between closely related species.

The *Cryptococcus* *MAT* locus has emerged as an exemplary model to elucidate general principles governing the assembly and function of gene clusters. Moreover, comparative genomic studies of the evolutionary trajectory of *MAT* in the pathogenic *Cryptococcus* species complex mirror evolutionary steps that have been hypothesized to have given rise to the more complex sex chromosomes in mammals, insects, fish, and plants (14–16). These then represent examples of convergent evolution to specialized genomic regions governing sexual reproduction throughout the eukaryotic tree of life. Recent studies further reveal tremendous plasticity in both *MAT* loci in fungi and sex chromosomes and their functions in metazoans.

THE *MAT* LOCUS OF *C. NEOFORMANS* AND *C. GATTII*

Cryptococcus neoformans and *Cryptococcus gattii* are basidiomycetous yeast pathogens that cause pneumonia and meningitis in humans and animals. Both species have a bipolar mating system involving **a** and α cells. One unique feature of the *C. neoformans* serotype A and *C. gattii* populations is that

Yen-Ping Hsueh, Division of Biology, California Institute of Technology, Pasadena, CA 91125. **Banu Metin, Keisha Findley, and Joseph Heitman,** Department of Molecular Genetics and Microbiology, Duke University Medical Center, Durham, NC 27710. **Marianela Rodriguez-Carres,** Department of Biology, Duke University, Durham, NC 27710.

the vast majority (>99.9% in serotype A) of clinical and environmental isolates are mating type α; mating-type **a** strains are extremely rare, except in a population isolated from the sub-Saharan region in Africa, in which **a** cells make up ~10% of the population (48). The drastic distortion between the prevalence of **a** and α cells among clinical and environmental isolates led to the hypothesis that *MAT*α cells may be more fit than *MAT***a** cells under some conditions. Virulence studies have been conducted comparing congenic **a** and α strains. In both the serotype D JEC20/JEC21 and NIH433**a**/NIH433α backgrounds, α cells are more virulent than **a** cells (42, 57). However, in the serotype D B-3501 background, the congenic strains (B-3501**a**/B-3501α) are equally virulent (57). No difference in virulence was observed between the congenic serotype A strains H99 (*MAT*α) and KN99**a**, but when two strains were coinfected in mice, α cells exhibited a greater ability to cross the blood-brain barrier to enter the central nervous system (55, 56). In addition, the ability to undergo monokaryotic fruiting has been shown to be linked to α cells (73).

These interesting links between mating type, virulence, and development focused interest on the molecular nature of the *MAT* locus. Moore and Edman identified an ~35- to 45-kb region specific to mating type α by differential cloning, indicating that *MAT* has expanded in *C. neoformans* compared to other known fungi (51). To identify genes within this locus that are involved in mating, **a** cells were transformed with various fragments derived from subcloning, and a 2.1-kb region was found to induce conjugation tube formation in **a** but not α cells. Sequence analysis revealed that one open reading frame shared similarity with other fungal pheromone precursors (51). This gene, *MF*α*1*, became the first gene identified in the *MAT* locus. Further characterization revealed that genes involved in the mitogen-activated protein kinase (MAPK) signaling cascades including *STE11*, *STE20*, and *STE12* are embedded in an ~50-kb *MAT* region (35, 46, 70, 72). Two additional pheromone genes, *MF*α*2* and *3*, the pheromone receptor *CPR*α (*STE3*α), and a *MAT*α-specific myosin gene were also found to be encoded by the *MAT* locus, indicating that *MAT* had expanded to at least ~50 kb in *C. neoformans* (35).

Surprisingly, eliminating the entire ~50-kb *MAT*α allele in an **a**/α diploid strain did not abolish **a**/α cell identity; the **a**/Δ cells were still self-fertile and produced viable **a** basidiospores (spores with the 50-kb deletion were inviable due to essential genes present in the deleted region) (28). This finding indicated that the previously identified ~50-kb *MAT*α allele was not necessary for sexual identity, and additional components outside of this region must confer identity in **a**/α cells (28). The previously identified ~50-kb *MAT*α allele lacks obvious DNA-binding homeodomain proteins that are commonly present in the fungal *MAT* locus (35). These DNA-binding proteins are transcription factors that govern the expression of sex-related genes and are central regulators of sex determination (20). By a candidate gene approach, Hull et al. identified a *MAT*α-specific homeodomain protein, Sxi1α, that is essential for the identity of **a**/α cells (28).

The discovery that *SXI1*α is located ~55 kb 5′ of the previously identified *MAT* locus region suggested that either *SXI1*α represented a second *MAT* locus or the size of *MAT* was much larger than anticipated. To determine the complete molecular structure of *MAT*, Lengeler et al. cloned and sequenced the *MAT* locus from the serotype D strains JEC20 (**a**) and JEC21 (α) and the serotype A strains H99 (α) and 125.91 (**a**) (45). Based on the nucleotide sequence and synteny analyses, the size of *MAT* spans ~105 to ~130 kb, indicating that the *MAT* locus of *C. neoformans* has dramatically expanded compared to most other known fungal *MAT* loci, which typically range from <1 to a few kilobases in length (6, 19, 30). The entire *MAT* locus is inherited as a single unit in a genetic cross, indicating that recombination is suppressed across the locus, which is a key feature of sex-determining regions that prevents the generation of self-fertile and sterile progeny (45). Furthermore, *MAT* was stably inherited in a series of 10 backcrosses, as no difference in Southern pattern was observed between the original parents and progeny derived from 10 backcrosses, also supporting the idea that recombination is suppressed within this region (45).

What is the gene content of this expanded *MAT* locus? Sequence analyses revealed that ~25 genes are encoded in this locus, comprising ~50% of *MAT* (Color Plate 2). Multiple transposon remnants and repetitive sequences are also present in both the **a** and α alleles (13, 45). Among the genes that are encoded in *MAT*, the homeodomain proteins and the pheromone/receptor (P/R) genes are the expected central regulators for sex identity that are universally present in tetrapolar basidiomycetes such as *Ustilago maydis*. In tetrapolar fungi, one of the two unlinked *MAT* loci encodes homeodomain proteins and the other encodes pheromones and pheromone receptors (17). In addition to these components that are critical for sex determination, several genes that may have a role in mating have also been incorporated into the locus. For example, *STE20***a**/α, *STE11***a**/α, and *STE12***a**/α are involved in pheromone signaling (see chapter 7), and *SPO14*, *RUM1*, and *BSP1* may participate in meiosis or sporulation (13, 45). *CID1* is a protein that is required for the S-M checkpoint control in *Schizosaccharomyces pombe* (71).

Based on gene disruption studies, five essential genes (*MYO2*, *PRT1*, *RPL39*, *RPL22*, and *RPO41*) are present in the *C. neoformans MAT* locus, which is a distinct feature compared to other fungal *MAT* alleles (13, 45). These genes are involved in actin-based transport of cargos, protein synthesis, and mitochondrial transcription. The presence of essential genes in the *C. neoformans MAT* locus is a unique and interesting feature. It is obvious that they have been incorporated into the locus during *MAT* expansion. One possible explanation is that these essential genes influence the evolution of the locus by constraining rearrangements to only those that do not result in loss of one or more essential genes. They may also serve to prevent the loss of the entire *MAT* locus, which could occur in a population with limited sexual reproduction. Finally, the function of several genes in the *MAT* locus is still unclear, including several genes specific to the basidiomycete lineage (*BSP2*, *BSP3*), and it will be of interest to examine whether they also have functions related to sexual reproduction (Table 1).

The gene content between different serotypes and mating types is highly similar, suggesting that they diverged from a common ancestor. However, the gene order has been subject to extensive rearrangement, in contrast to sequences outside of *MAT*, where synteny is maintained over longer distances between different mating types, varieties, and sibling species (13, 45). One exception is that three genes (*IKS1*, *BSP3*, and *NCM1*) are present in the flanking region in the serotype D lineage instead of being an integral component of the *MAT* locus, which is likely caused by gene conversion that fixed the α allele in both mating types, resulted in the loss of the ancestral **a** allele, and as a consequence, evicted these genes from *MAT* (13).

In addition to coding sequences, large numbers of transposon-related sequences, including two complete copies of

TABLE 1 Gene content of *MAT* locus and linked regions in *C. neoformans*/*C.gattii* and the closely related species

Species	Homeodomain genes	Pheromone genes	Pheromone receptors	Essential genes	Genes involved in mating, meiosis, or sporulation	Genes of unknown function
C. neoformans and *C. gattii* (**a** allele)	*SXI2*	MF**a**1-4	*STE3***a**	*MYO2, PRT1, RPL39, RPL22, RPO41*	*STE20, STE11, STE12, SPO14, RUM1, BSP1*	*IKS1, LPD1, BSP3, ZNF1, PRT1, ETF1, CAP1, GEF1, CID1, BSP2, FAO1, NOG2*
T. wingfieldii	*SXI1, SXI2*[a]	MF**a**1-3	*STE3***a**	All	All except *STE11*	All except *IKS1* and *FAO1*
C. amylolentus	*SXI1, SXI2*[a]	MF**a**1-2	*STE3***a**	All except *RPL22*	*STE20, STE12*	All except *IKS1* and *FAO1*
C. heveanensis	*SXI1, SXI2*[a]	MF**a**1	*STE3***a**	All	All except *RUM1*	All except *NOG2*

[a] The two HD genes are both present in the same *MAT* allele in these species.

Cnirt3, are present in the *MAT* locus, as well as repetitive sequences such as simple sequence repeats. The positions of these transposon remnants and repetitive sequences are highly variable between different alleles, suggesting that these elements were acquired independently during different time points in evolution and likely have contributed to structural rearrangement of the *MAT* locus (14, 45).

One interesting question is how this unusual *MAT* structure evolved as a large gene cluster. To gain further insight into the evolutionary trajectory of the *MAT* locus, Fraser et al. cloned and sequenced *MAT* alleles from two *C. gattii* strains, WM276 (α) and E566 (**a**) (13). *C. gattii* and *C. neoformans* are closely related species that diverged ~40 million years ago (39). Therefore, *C. gattii* represents a unique vantage point to study the evolutionary trajectory of *MAT*. Cloning and sequencing *MAT* alleles from *C. gattii* revealed that *MAT* in *C. gattii* exhibits a similar architecture in terms of size and gene content (13, 14). Nonetheless, the order of genes is also drastically rearranged, further supporting the idea that the **a** and α alleles diverged from a common ancestor and have been subject to extensive rearrangements including inversions, gene conversion, and nucleotide substitutions resulting in recombination suppression between the two alleles.

Phylogenetic analyses of the *MAT* genes in the three *Cryptococcus* lineages demonstrated that the genes of the *MAT* locus could be classified into four different strata (13). The first category is the "ancient class," including genes such as the pheromones and pheromone receptors and genes that are involved in the MAPK signaling pathway. They cluster into a mating-type-specific pattern. The second and third categories, termed "intermediate I" and "intermediate II," are genes that are also clustered into a mating-type-specific pattern, but the alleles of the two mating types share a higher percentage of similarity; thus, the phylogenetic pattern between the two alleles is not as divergent as the ancient class. The last category is the "recent class," which consists of genes that display a species-specific phylogenetic pattern, and the sequence between the **a** and α alleles shares more than 90% nucleotide identity (13). The distinct patterns of the phylogenetic trees of the *MAT* genes reflect their evolutionary origins and suggest that these genes were acquired gradually at different time points of evolution (13). These analyses led to the proposal of an evolutionary model for the *Cryptococcus MAT* locus (Color Plate 3) (13). Fraser et al. proposed that this unusually large *MAT* cluster evolved from an ancestral tetrapolar system, in which one locus encodes homeodomain protein transcription factors and the other encodes pheromones and pheromone receptors (13). The two ancestral loci expanded by a series of gene acquisitions, and a chromosomal translocation took place and fused the two clusters into one contiguous *MAT* allele in one mating type. At this transitional stage, one of the alleles has completed the transition from two unlinked *MAT* loci to one single *MAT* locus, while the other still exhibits two unlinked *MAT* loci. We termed this transitional mating state "tripolar," which further collapsed into the bipolar state when the second allele underwent fusion of the two *MAT* loci via interallelic recombination (Color Plate 3) (13).

Comparative genomics is a fruitful approach to studying the evolution of the *Cryptococcus MAT* gene cluster and led to a proposed evolutionary model (13). Therefore, we set out to extend this analysis to the structure of *MAT* in the closely related species among the monophyletic *Filobasidiella* clade, including *Cryptococcus amylolentus*, *Tsuchiyaea wingfieldii*, and *Filobasidiella depauperata*, as well as the sensu lato species *Cryptococcus heveanensis* that lies in a sister clade to the pathogenic *Cryptococcus* species complex (12).

EVOLUTION OF *MAT* IN THE *FILOBASIDIELLA* SPECIES COMPLEX

In the mid-1970s the discovery of sex in *C. neoformans* and later in *C. gattii* (see chapter 7) resulted in the taxonomic classification of these fungi into their own teleomorphic genus, *Filobasidiella*. This classification is based on both biochemical characteristics and morphological features of their sexual structures. Recently, phylogenetic analyses employing a multilocus sequence typing approach revealed that two yeast species with an unknown sexual cycle (*C. amylolentus* and *T. wingfieldii*) and a homothallic strictly filamentous fungus (*F. depauperata*) are members of the monophyletic *Filobasidiella* clade composed of the human-pathogenic species *C. neoformans* var. *neoformans*, *C. neoformans* var. *grubii*, and *C. gattii* (12). *F. depauperata* had previously been included in the *Filobasidiella* genus based on shared morphological and molecular features with pathogenic *Cryptococcus* species (40, 41, 50, 63a). Today the genus *Filobasidiella* is a well-defined monophyletic clade composed of five to nine species: *C. neoformans* var. *neoformans* (serotype D), *C. neoformans* var. *grubii* (serotype A), *C. gattii* (including four cryptic species: VGI, VGII, VGIII, and VGIV), *T. wingfieldii*, *C. amylolentus*, and *F. depauperata* (12).

The sexual cycle of *C. neoformans* and *C. gattii* was discovered decades ago, but sex in the other three species is relatively poorly understood. Only one strain of *T. wingfieldii* is available worldwide. This strain does not produce any sexual structures without a mating partner under the conditions tested thus far, suggesting a heterothallic sexual cycle; however, because of the lack of additional isolates, crossing isolates of different mating types is not yet possible. On the other hand, it has recently been demonstrated that the two *C. amylolentus* strains of opposite mating types are fertile when crossed, producing basidia and spores when cocultured, indicating a heterothallic sexual cycle (K. Findley and J. Heitman, unpublished). Based on light and electron microscopic observations, *F. depauperata* appears to be homothallic (41, 63a). The mating structures produced by the mating-competent species in the *Filobasidiella* clade are very similar; terminal hyphae differentiate into aseptated basidia (the site of meiosis), and subsequently four basipetal chains of basidiospores are formed.

T. wingfieldii

T. wingfieldii CBS7118 is a saprobic yeast-like species isolated from the frass of beetles in South Africa (5). It reproduces by budding and does not produce hyphae. Phenotypically, *T. wingfieldii* does not display the virulence attributes associated with pathogenic *Cryptococcus* species. It is unable to produce melanin or capsule and grows at room temperature (22–25°C) but not higher temperatures. Additionally, in the *Galleria mellonella* (greater wax moth) virulence model, *T. wingfieldii* is avirulent when compared to the pathogenic *C. neoformans* var. *grubii* strain H99 (12). Sexual reproduction has never been observed in this species, and only one strain is available, limiting the ability to perform mating assays.

Although *T. wingfieldii* currently lacks a known sexual cycle, a genomic region related to the *MAT* locus has been identified. Fosmid libraries were constructed and probed with genes present in the *MAT* locus of *C. neoformans*. Positive clones were sequenced, and the assembled *MAT* sequences were further analyzed. Two candidate *MAT* loci, one containing the pheromone/receptor (P/R) and the other containing the homeodomain proteins (HD), were identified. The P/R locus spans ~70 kb and the HD locus spans ~56 kb (Color Plate 2). Sequence analyses revealed that over 20 genes are encoded in *MAT* (Findley and Heitman, unpublished). The overall gene content of the two loci is quite similar to that of the *C. neoformans* and *C. gattii* *MAT* locus (Color Plate 2; Table 1), with the exception of the absence of *STE11* in the P/R locus of *T. wingfieldii* (Findley and Heitman, unpublished).

An interesting feature of *MAT* in *T. wingfieldii* that provides insight into the ancestral form of the HD locus is the presence of orthologs of both the *SXI1*α and *SXI2***a** genes (Color Plate 2). The homeodomain transcription factors *SXI1*α and *SXI2***a** are adjacent and likely divergently transcribed, resembling the organization of the homeodomain genes *bE* and *bW* in *U. maydis* (34). This is in contrast to *C. neoformans* and *C. gattii*, in which only one HD gene is present in each *MAT* allele: *SXI1*α is specific to the α allele and *SXI2***a** to the **a** allele. In the *T. wingfieldii* P/R locus, a pheromone receptor gene, *STE3*, and three mating pheromone genes are present. Annotating the amino acid sequence of the predicted mature mating pheromones of *T. wingfieldii* shows that they are highly homologous to the MF**a** gene product of *C. neoformans* (97% identity). The five most recently acquired genes (*LPD1*, *RPO41*, *BSP2*, *CID1*, and *GEF1*) are all present in the P/R locus, suggesting that the ancestral form of the P/R locus already included these five genes or they were linked.

To determine whether the HD and P/R loci are physically unlinked in the genome, we performed pulsed-field gel electrophoresis, and approximately 8–10 chromosomes (ranging in size from 800 kb to 2.2 Mb) were resolved. Probing the chromoblot with probes that are specific to the HD or P/R locus revealed that the two subloci are located on different chromosomes and therefore physically unlinked (Findley and Heitman, unpublished). However, the lack of additional *T. wingfieldii* strains makes it difficult to address experimentally whether the two loci are both required for sexual identity.

C. amylolentus

C. amylolentus strains CBS6039 and CBS6273 are saprobic yeast-like species isolated from the frass of scolytid beetles in South Africa. They reproduce by budding and produce hyphae or pseudohyphae on yeast peptone dextrose agar (5). Similar to *T. wingfieldii*, *C. amylolentus* is unable to survive at 37°C but grows at room temperature (22–25°C) and at 30°C. Furthermore, capsule and melanin production are also absent, and the fungus is avirulent in the greater wax moth (*G. mellonella*) model (12). Although previous studies indicated that *C. amylolentus* is asexual, we discovered that the two strains (CBS6039 and CBS6273) produce sexual structures when cocultured on V8 solid medium (Findley and Heitman, unpublished). In the span of a week (in the dark at room temperature), basidia terminating in four individual spore chains and filaments with fused clamp connections are produced. The sexual state of *C. amylolentus* has been named *Filobasidiella amylolenta*. Thus, elucidating the structure of the *C. amylolentus* *MAT* locus will not only provide information on *MAT* locus evolution in the *Filobasidiella* clade, but will also be critical to understanding its sexual cycle.

The *MAT* locus of *C. amylolentus* has been cloned and sequenced employing approaches similar to those described above. Fosmid libraries were generated and probed with genes present in the *MAT* locus of *C. neoformans*. Positive clones were sequenced, assembled, and further analyzed. The final assembly is still in progress but the size of the HD and P/R loci appears to be similar to those of *T. wingfieldii*. From their close phylogenetic relationship, the gene order of *MAT* in *C. amylolentus* and *T. wingfieldii* could be conserved, given that the *MAT* locus presents remarkable synteny between the closely related VGI and VGII cryptic species of *C. gattii* (13, 14).

The overall structure of the *MAT* locus is related between *C. amylolentus* and *T. wingfieldii*. *STE11* is also missing in the *MAT* assembly and present elsewhere in the genome of *C. amylolentus*. Orthologs of both *SXI1* and *SXI2* are also present in the HD locus, linked, and likely divergently transcribed. The mating pheromone genes and the pheromone receptor gene *STE3* are located in the P/R locus along with the five most recently acquired genes. The P/R locus and HD locus in *C. amylolentus* are also unlinked, based on a *C. amylolentus* chromoblot analysis. Furthermore, it will be of interest to know which genes in each locus are required for sexual identity and reproduction. Taken together, the analysis of the *MAT* assemblies for these two sibling species suggests that both *SXI1* and *SXI2* were present in the ancestral HD locus, and the five most recently acquired genes (*LPD1*, *RPO41*, *BSP2*, *CID1*, and *GEF1*) were already linked to the P/R locus. Also, in both species the HD and P/R loci are unlinked, similar to tetrapolar species or bipolar species such as *Coprinellus disseminatus* in which the P/R locus is either constitutively active or no longer linked to sex determination (32). Thus, the organization of *MAT* in these sibling species mirrors key aspects of the proposed intermediates in the evolution of *MAT* in the pathogenic *Cryptococcus* species.

F. depauperata

F. depauperata strains CBS7841 and CBS7855 are strictly filamentous fungi isolated from a dead spider in Canada and a dead *Carpocapsa* caterpillar in the Czech Republic (CBS Fungal Biodiversity Center). Although *F. depauperata* appears to be associated with insects, it does not appear to be pathogenic in the greater wax moth (*G. mellonella*) model (12). *F. depauperata* grows exclusively filamentously and forms sexual structures regardless of environmental cues, such as light, media, and pH (63a). *F. depauperata* has aseptated basidia, four basipetal spore chains, monokaryotic hyphae lacking clamp connections, and monokaryotic basidiospores (41, 63a, 65). Furthermore, transmission electron microscopy of the basidium of *F. depauperata* revealed the presence of synaptonemal complexes, suggesting that the process involves meiosis and is sexual (41). Interestingly, these morphological features are reminiscent of a modified form of the sexual cycle involving same-sex mating (monokaryotic fruiting) in *C. neoformans* (47, 73). Thus, *F. depauperata* appears to be continuously engaged in sexual reproduction and, as such, might be the first fungal species that is obligately sexual.

To gain insight into the *MAT* structure of *F. depauperata*, we set out to clone and sequence the *MAT* locus. Three fosmid clones containing genes homologous to those linked to the P/R locus in *C. neoformans* were identified, sequenced, and assembled. Phylogenetic analyses of these genes suggest that two of the genes found in this region, *MYO2* and *STE20*, appear to be clustered with the *MAT***a** allele of *C. neoformans*. Furthermore, synteny analyses between *C. neoformans* and *F. depauperata* showed that some genes linked to the *MAT* locus in *C. neoformans* are present in different chromosomes in *F. depauperata*. In contrast, synteny comparisons of three additional fosmids that correspond to different chromosomal locations in *C. neoformans* displayed conserved gene order in *F. depauperata* (63a). Therefore, chromosomal rearrangements appear to be a major force driving speciation and sexual divergence in these closely related pathogenic and saprobic species. Although the molecular nature of the homothallic sexual cycle of *F. depauperata* remains to be elucidated, possible models for the homothallic life cycle of *F. depauperata* are discussed below.

Model 1: The Presence of a Single *MAT* Allele Governs the Homothallic Sexual Cycle

The P/R genes are part of the gene cluster that includes several genes (*STE20*, *PRT1*, *ZNF1*, *RPL39*, and *STE11*) in the genomes of the heterothallic relatives, *C. neoformans* and *C. gattii*, and the more distant relatives *C. heveanensis* and *Tremella mesenterica* (49a), which indicates that this gene cluster arrangement is ancestral to divergence of the *Filobasidiella* lineage. However, in *F. depauperata* the homologous genomic region that contains these genes does not include the P/R genes (63a). Parsimony-based phylogenetic analyses showed that at least two of these genes, *STE20* and *MYO2*, in *F. depauperata* are closer to the homologous genes in the *MAT***a** allele of *C. neoformans*, *STE20***a** and *MYO20***a**, rather than to their counterparts in the *MAT*α allele (63a). Given that monokaryotic fruiting occurs in *C. neoformans* in the absence of compatible P/R and HD genes, it is possible that the *F. depauperata* homothallic life cycle occurs in the absence of compatible *MAT* alleles.

Model 2: The Presence of Compatible Mating-Type Genes in the Genome of *F. depauperata* Governs Homothallism

Most of the genes linked to the *MAT* locus of *C. neoformans* and *C. gattii* in *F. depauperata* appear to be species-specific, except for the *STE20* and the *MYO2* genes, which appear closer to the *C. neoformans* and *C. gattii* *STE20***a** and *MYO2***a** alleles. Although phylogenetic analyses of the *STE11* gene from *F. depauperata* showed that this gene follows the expected species phylogeny, DNA identity plots suggest that the *STE11* gene shares a higher identity to the *STE11*α allele of *C. neoformans* (63a). Thus, circumstantial evidence supports the hypothesis that the homothallic sexual cycle of *F. depauperata* could be the result of compatible mating-type loci (*MAT***a** and *MAT*α) present in one nucleus. If this is the case, *F. depauperata*–compatible alleles might be fused or linked in one locus, similar to the majority of examples of homothallic euascomycetes. Alternatively, compatible mating alleles might be found in different genomic locations, such as in the homothallic fungi *Aspergillus nidulans* and *Neosartorya fischeri* (18, 60, 64). Interestingly, haploid laboratory strains of *C. neoformans* engineered to contain both compatible HD genes become self-fertile (23, 27, 29, 31) and form monokaryotic hyphae with unfused clamp connections (23), similar to monokaryotic fruiting in *C. neoformans* and *F. depauperata*. Given that paired, divergently oriented, self-incompatible homeodomain genes are extant in *C. amylolentus*, *T. wingfieldii*, and *C. heveanensis*, it is possible that two linked self-compatible *SXI1*α and *SXI2***a** alleles may drive the homothallic life cycle of *F. depauperata*.

Model 3: A Modified Signaling Cascade Enables Homothallic Sexual Reproduction

Synteny comparisons between regions linked to the *MAT* locus of *C. neoformans* and homologous regions in *F. depauperata* suggest that an intrachromosomal translocation event between the genes linked to the *MAT* locus and genes located in the subtelomeric regions of the same chromosome, chromosome 4 in *C. neoformans*, gave rise to the extant genomic architecture in *F. depauperata*. Interestingly, one of these translocated regions involves the *STE11* gene. This gene is usually linked to the P/R locus in the majority of closely related fungi. Furthermore, the *STE11* gene functions in the MAPK cascade that regulates filamentation and mating in fungi (44). Mutations and changes in the regulation of the *STE11* gene have been shown to interfere with filamentation during heterothallic mating and during monokaryotic fruiting in *C. neoformans* (9, 11). Thus, the translocation of the *STE11* gene in *F. depauperata* may have led to modifications in the signaling cascade, resulting in morphological changes, such as constitutive hyphal growth in *F. depauperata*. In this case, the requirement for *MAT* genes might be bypassed completely, and at least some *MAT* genes may have even been subsequently lost from the genome. Recent studies of *C. neoformans* also showed that a pheromone receptor homolog, Cpr2, is a constitutively active G protein-coupled receptor (GPCR) that activates the pheromone-signaling pathway without any apparent ligand.

When *CPR2* is overexpressed, it dramatically enhances hyphal growth and same-sex mating, which is also a form of homothallic sexual reproduction (26). Thus, it is possible that such a constitutively active GPCR may function in a similar fashion to support homothallism in *F. depauperata*. Continuous sexual reproduction in *F. depauperata*, regardless of its molecular structure and *MAT* organization, could be the result of epistatic changes in the cascade that controls filamentation.

C. heveanensis

C. heveanensis is a nonpathogenic yeast that lies in a sister clade to the pathogenic *Cryptococcus* species complex (12). Therefore, it is closely related to but distinct from the *Cryptococcus* pathogenic species complex, providing a vantage point from which to study the evolution of MAT. Until recently, there was only a single strain of *C. heveanensis* (type strain CBS 569) and two isolates that represent cryptic sibling species. Therefore, the sexual cycle was a complete mystery. Isolation of new strains of *C. heveanensis* by Nakase et al. enabled us to test the sexual cycle by pairwise incubation of their isolates (54). Coincubation of compatible *C. heveanensis* isolates on V8 mating medium resulted in dikaryotic hyphae with closed clamp connections and the production of basidia and basidiospores (49a). The basidia with cross-shaped septa are quite different from the nonseptate basidial structure of *C. neoformans* and are more similar to the closely related species *Kwoniella mangroviensis*, whose sexual cycle was recently identified (69). Because of this similarity in the sexual structure, the sexual state of *C. heveanensis* was named *Kwoniella heveanensis* (49a).

The *MAT* locus of *C. heveanensis* was cloned and sequenced by probing a fosmid library with genes specific to the *C. neoformans MAT* locus. Positive fosmids were sequenced and assembled, revealing two large genomic regions, one corresponding to the HD locus (~80 kb) and the other corresponding to the P/R locus (~180 kb) (49a).

The HD locus of *C. heveanensis* contains two divergently transcribed homeodomain genes, *SXI1* and *SXI2*, with an arrangement similar to the genes encoding bE and bW in *U. maydis* and HD1 and HD2 in *Coprinopsis cinerea*. This structure again supports the hypothesis that one of two paired homeodomain genes was recently lost in both the **a** and α alleles of *C. neoformans* and *C. gattii*. Thus, the ancestral form of the HD locus contained two divergently transcribed homeodomain genes, as observed in *T. wingfieldii*, *C. amylolentus*, and *C. heveanensis* (Color Plate 2). Genes that are either part of (*SPO14*, *CAP1*, and *RPL22*) or flanking (*FAO1*, *FCY1*, and *UAP1*) the *C. neoformans MAT* locus were also found linked to the HD genes (Color Plate 2). However, sequence analysis in multiple *C. heveanensis* strains demonstrated that these genes are likely to be *MAT*-flanking genes, and the *MAT*-specific region is limited to the *SXI1* and *SXI2* genes (49a).

Interestingly, the HD genes *SXI1* and *SXI2* are multiallelic in *C. heveanensis*. Southern blot analysis of *SXI1* and *SXI2* in different *C. heveanensis* strains showed a highly variable pattern. An ~12-kb region spanning the *SXI1* and *SXI2* genes was sequenced from seven isolates, revealing six different alleles (49a). This analysis suggests that additional alleles likely remain to be discovered.

The P/R locus and flanking regions of *C. heveanensis* contain the pheromone gene *MF***a***1/2*; the pheromone receptor gene *STE3*; the pheromone-sensing pathway genes *STE11*, *STE12*, and *STE20*; and genes that are part of the *MAT* locus in *C. neoformans* and *C. gattii* (*LPD1*, *GEF1*, *CID1*, *BSP1*, *ETF1*, *RPO41*, *BSP2*, *ZNF1*, *PRT1*, *NCP1*, *MYO2*, *IKS1*, *RPL39*, and *BSP3*). Scattered around these are (i) genes that are not located in the *MAT* locus but located elsewhere in the genome of *C. neoformans* or *C. gattii* and (ii) genes that do not have apparent *C. neoformans* or *C. gattii* homologs. Sequence analysis showed that the *LPD1*, *STE11*, *ZNF1*, and *IKS1* genes are highly similar among different *C. heveanensis* isolates (99–100% sequence identity), indicating that these genes lie outside the *MAT*-specific P/R locus region.

Analysis of the pheromone genes of seven *C. heveanensis* strains revealed two different alleles, indicating at least a biallelic pheromone receptor locus with the *MAT*-specific region containing at least *STE3*, *STE12*, *MF***a***1/2*, and *CNG04540*.

The finding of two apparently unlinked genomic loci indicates a tetrapolar *MAT* organization for *C. heveanensis*. This is further supported by the fact that each of these loci contains a *MAT*-specific region showing sequence divergence among different isolates surrounded by regions with at least 99% sequence identity. Additionally, the organization of a multiallelic homeodomain locus and a biallelic P/R locus is similar to the closely related tetrapolar species *T. mesenterica*, as well as to the smut fungus *U. maydis* (4, 74). The multiallelic nature of the homeodomain locus is also a characteristic of the tetrapolar mushroom fungi such as *C. cinerea* and *Schizophyllum commune* (8).

In summary, comparative genomic analysis of the structure of MAT loci from *Cryptococcus*-related species has important implications for the hypothesized evolutionary model. First, the organization of *MAT* in two unlinked loci in *T. wingfieldii*, *C. amylolentus*, and *C. heveanensis* provides evidence that the ancestral state of the *C. neoformans MAT* locus was tetrapolar. Second, the *MAT* genes present in these species have implications for the gene content of the ancestral homeodomain and P/R loci in the model. Third, the presence of two divergently transcribed homeodomain genes, similar to other tetrapolar species, indicates a recent loss of one or the other homeodomain gene in the pathogenic *Cryptococcus* species. Fourth, the P/R locus boundaries of *C. heveanensis* have expanded to include not only the pheromone and pheromone receptor genes but also at least one of the pheromone response pathway genes, but not all of the *C. neoformans* and *C. gattii MAT* locus gene homologs, which illuminates the serial expansions that created evolutionary strata in the *MAT* locus genes of the pathogenic *Cryptococcus* species. Taken together, the *MAT* locus of *C. heveanensis* represents an intermediate stage in the hypothesized *MAT* evolutionary model where the mating system is tetrapolar, and mating-related genes have been acquired into two different loci but not yet all captured into the *MAT*-specific region.

EVOLUTION OF BIPOLAR AND TETRAPOLAR MATING SYSTEMS

The Basidiomycota includes species with both bi- and tetrapolar mating systems, and the tetrapolar mating system is thought to have evolved within this phylum. Three major groups, the Ustilaginomycotina (smut), the Pucciniomycotina (rust), and the Agaricomycotina (mushroom), constitute the phylum of Basidiomycota, and the pathogenic *Cryptococcus* species reside in the Agaricomycotina together with mushrooms and jelly fungi (21, 22). The mating system is best understood in the smut and mushroom lineages, and relatively less is known about the molecular structure of the

TABLE 2 Mating systems of basidiomycete fungi

Species	Mating system	Host	No. of alleles	Inbreeding (%)	Outcrossing (%)
C. neoformans	Bipolar-biallelic	Human	2	50	<50
C. amylolentus	Tetrapolar?	Saprobe/insect frass	2	25?	50
F. depauperata	Homothallic	Saprobe/dead spider	?	100	0
C. heveanensis	Tetrapolar-bi/ multiallelic	Saprobe/sheet rubber	2 (A locus) >6 (*B* locus)	25	50
U. maydis	Tetrapolar-bi/ multiallelic	Corn	2 (*a* locus) >25 (*b* locus)	25	50
U. hordei	Bipolar-biallelic	Corn	2	50	50
C. cinerea	Tetrapolar-multi/ multiallelic	Saprotrophic	>90 for both A and *B* loci	25	99.5
M. globosa[a]	Bipolar?	Human	Unknown	–	–

[a]Sexual cycle remains to be identified. Pictures of meiotic spores are shown in the table except for *M. globosa*.

mating-type locus in rust fungi (10). Within the mushroom fungi lineage, the predominant mating system (~65% of the species) is tetrapolar (63). Nonetheless, conversions between bi- and tetrapolar mating systems in closely related species are common and have been observed in different lineages; genomic analysis of the *MAT* locus structure supports the idea that these transitions in different lineages have occurred independently (2, 23, 24, 32, 43, 75). In a single meiotic event, bipolar species generate progeny of two different mating types, while tetrapolar species give rise to progeny with four different mating types. Therefore, 50% of the progeny are interfertile in a bipolar cross, but only 25% of the progeny are interfertile in a tetrapolar cross, suggesting that inbreeding is more restricted in a tetrapolar mating system (Table 2).

A model has been proposed to highlight the key evolutionary steps that facilitated the transition from tetrapolarity to bipolarity in the pathogenic *Cryptococcus* species (Color Plate 3). To provide experimental evidence to support this hypothesis, we genetically engineered *C. neoformans* strains to relocate the homeodomain gene to a genomic location unlinked to *MAT* to recapitulate the ancestral tetrapolar mating system (23). In these strains, the homeodomain gene *SXI1*α or *SXI2***a** is physically unlinked to the pheromone and the receptor genes; therefore, the HD and the P/R loci segregate independently in a genetic cross. The modified "tetrapolar" *C. neoformans* strains can complete the sexual cycle and produce fertile, viable spores. We also experimentally tested the transitional tripolar state by crossing two strains in which one harbors a contiguous *MAT* locus while the other harbors unlinked HD and P/R loci. Progeny derived from this tripolar cross have decreased fertility; 50% of the progeny are sterile, and the chance for inbreeding between the progeny is only 12.5% (23). Thus, the transitional tripolar state is under strong selection pressure, which could be disadvantageous and directly or indirectly facilitate the transition from a tripolar to a bipolar system (23).

Another factor that might have contributed to the evolutionary trajectory of the *MAT* gene cluster involves the *MAT*-linked recombination hotspots. These GC-rich recombination hotspots increase the meiotic recombination frequency up to ~50-fold, and it is hypothesized that they may have driven several key events such as gene acquisition and translocation during the early steps of *MAT* locus evolution (25). Furthermore, a prominent GC peak was also observed within the *MAT* locus, between the *BSP2* and *RPO41* genes, which are highly similar between the **a** and α alleles. This observation led to the hypothesis that the GC-rich region inside the *MAT* locus might have triggered gene conversion events inside the *MAT* locus (25).

Transitions between bi- and tetrapolarity are commonly observed in closely related lineages in basidiomycetes (2). For example, in the smut fungi lineage, *U. maydis* is tetrapolar, while *Ustilago hordei* is bipolar; the model mushroom *C. cinerea* is tetrapolar, while a closely related species *C. disseminatus* is bipolar. In the 1960s, John Raper proposed mechanisms that could lead to this transition (63). These hypotheses are now supported by the molecular structure of the *MAT* locus. For example, the *C. neoformans*, *U. hordei*, and *Malassezia globosa MAT* structural architectures argue that chromosomal translocations have independently occurred in each lineage, linking the two sex-determining regions and converting a tetrapolar mating system to a bipolar one (3, 43, 75). On the other hand, the *MAT* locus of *C. disseminatus* favors the model that the P/R locus has become self-activated and only the HD locus is responsible for sexual identity (32). Mutants that render cells self-activated have also been isolated in the laboratory (58, 59). This provides evidence that not only did the transition from tetrapolarity to bipolarity occur multiple times independently during evolution, but it occurred via at least two different mechanisms.

EVOLUTION OF MULTIALLELISM

Coincident with the emergence of the tetrapolar mating system is the evolution of the multiallelic nature of the sex-determining genes. Most tetrapolar species described so far are multiallelic in at least one of the HD and P/R loci (4, 8, 33, 37). As discussed above, tetrapolar mating systems restrict inbreeding; furthermore, as mating is coordinately controlled by two loci, it is possible that the potential mates are only compatible at one of the loci. For example, cells can be compatible at the P/R locus but not at the HD locus. In species in which cell-cell fusion during the early process of mating is dependent on the P/R recognition, this situation will lead to successful fusion, but the fused zygotes will fail to undergo meiosis, as they lack a pair of compatible homeodomain proteins. Such events are deleterious for the population, and we hypothesize that multiallelism arose to circumvent this issue and maximize the chance of compatibility after cell-cell fusion.

The number of mating-type alleles existing in the population determines the level of outcrossing for a given species. For example, the tetrapolar mushrooms *C. cinerea* and *S. commune* are both estimated to have a large number of alleles at both the HD and P/R loci (36, 63). As a result, the chance of outcrossing for both species approximates 99% (Table 2). On the other hand, there are also multiallelic bipolar species: *C. disseminatus* and *Pholiota nameko* are both bipolar and multiallelic at the HD locus (1, 32, 76). *C. disseminatus* is estimated to have 123 alleles, and at least 6 alleles have been discovered for *P. nameko* (32, 76). Therefore, the probability of outcrossing is also high in these bipolar species (Table 2).

Multiallelism is absent in the pathogenic *Cryptococcus* species complex, so it was surprising to find that the closely related species *C. heveanensis* exhibits six different alleles among seven characterized strains (49a). It will be of interest to explore whether there is a relationship between pathogenicity and the loss of multiallelism at the *MAT* locus.

CONCLUSION

In this chapter we have considered the structure and evolution of the *Cryptococcus* mating-type locus from a phylogenomic perspective. This analysis began with an emphasis on the pathogenic species cluster including *C. neoformans* var. *grubii* (VNI/II/B), *C. neoformans* var. *neoformans* (VNIV), and *C. gattii* (VGI, VGII, VGIII, and VGIV). This analysis resulted in an evolutionary model in which the extant *MAT* and bipolar mating-type system evolved from an ancestral tetrapolar mating system via a series of evolutionary steps that include gene acquisition to two large sex-related gene clusters, gene fusion to yield a tripolar intermediate state, collapse to a true bipolar system with only two alleles, and then ongoing gene conversion events that have punctuated the evolution of *MAT*.

More recently, the scope of this comparative phylogenomic analysis has been extended to include also related but more diverged species, including *C. amylolentus*, *C. heveanensis*, *F. depauperata*, and *T. mesenterica*. This analysis reveals extant species that mirror the hypothesized evolutionary intermediates, including examples in which two large sex-related gene clusters have been assembled, but not yet fused, and examples in which one of these two clusters is still multiallelic, like other fungi with tetrapolar mating systems. This analysis has also revealed a pair of divergently transcribed genes encoding homeodomain proteins resident in *MAT* of these species.

This analysis reveals tremendous plasticity in the evolution of *MAT* and illustrates transitions from a multiallelic tetrapolar ancestral state to a biallelic bipolar extant species complex. Moreover, the fact that closely aligned species have retained the paired divergent homeodomain organization, whereas the pathogenic species complex *MAT* alleles retain only one or the other homeodomain gene resident at *MAT*, indicates that one or the other HD gene was lost recently. These transitions all impose a restriction on outcrossing and promote inbreeding, perhaps concomitant with the emergence of this species complex as successful pathogens of animals, including humans.

REFERENCES

1. **Aimi, T., R. Yoshida, M. Ishikawa, D. Bao, and Y. Kitamoto.** 2005. Identification and linkage mapping of the genes for the putative homeodomain protein (*hox1*) and the putative pheromone receptor protein homologue (*rcb1*) in a bipolar basidiomycete, *Pholiota nameko*. *Curr. Genet.* **48:**184–194.
2. **Bakkeren, G., J. Kamper, and J. Schirawski.** 2008. Sex in smut fungi: structure, function and evolution of mating-type complexes. *Fungal Genet. Biol.* **45**(Suppl 1)**:**S15–S21.
3. **Bakkeren, G., and J. W. Kronstad.** 1994. Linkage of mating-type loci distinguishes bipolar from tetrapolar mating in basidiomycetous smut fungi. *Proc. Natl. Acad. Sci. USA* **91:**7085–7089.
4. **Banuett, F.** 2007. History of the mating types in *Ustilago maydis*, p. 351–375. *In* J. Heitman, J. W. Kronstad, J. W. Taylor, and L. A. Casselton (ed.), *Sex in Fungi: Molecular Determination and Evolutionary Implications*. ASM Press, Washington, DC.
5. **Barnett, J. A., R. W. Payne, and D. Yarrow.** 1990. *Yeasts: Characteristics and Identification*, 2nd ed. Cambridge University Press, Cambridge, UK.
6. **Butler, G.** 2007. The evolution of *MAT*: the ascomycetes, p. 3–18. *In* J. Heitman, J. W. Kronstad, J. W. Taylor, and L. A. Casselton (ed.), *Sex in Fungi: Molecular Determination and Evolutionary Implications*. ASM Press, Washington, DC.
7. **Capel, B.** 1998. Sex in the 90s: SRY and the switch to the male pathway. *Annu. Rev. Physiol.* **60:**497–523.
8. **Casselton, L. A., and N. S. Olesnicky.** 1998. Molecular genetics of mating recognition in basidiomycete fungi. *Microbiol. Mol. Biol. Rev.* **62:**55–70.

9. **Chung, S., M. Karos, Y. C. Chang, J. Lukszo, B. L. Wickes, and K. J. Kwon-Chung.** 2002. Molecular analysis of *CPRalpha*, a *MATalpha*-specific pheromone receptor gene of *Cryptococcus neoformans*. *Eukaryot. Cell* **1:**432–439.
10. **Coelho, M. A., A. Rosa, N. Rodrigues, A. Fonseca, and P. Goncalves.** 2008. Identification of mating type genes in the bipolar basidiomycetous yeast *Rhodosporidium toruloides*: first insight into the *MAT* locus structure of the Sporidiobolales. *Eukaryot. Cell* **7:**1053–1061.
11. **Davidson, R. C., C. B. Nichols, G. M. Cox, J. R. Perfect, and J. Heitman.** 2003. A MAP kinase cascade composed of cell type specific and non-specific elements controls mating and differentiation of the fungal pathogen *Cryptococcus neoformans*. *Mol. Microbiol.* **49:**469–485.
12. **Findley, K., M. Rodriguez-Carres, B. Metin, J. Kroiss, A. Fonseca, R. Vilgalys, and J. Heitman.** 2009. Phylogeny and phenotypic characterization of pathogenic *Cryptococcus* species and closely related saprobic taxa in the *Tremellales*. *Eukaryot. Cell* **8:**353–361.
13. **Fraser, J. A., S. Diezmann, R. L. Subaran, A. Allen, K. B. Lengeler, F. S. Dietrich, and J. Heitman.** 2004. Convergent evolution of chromosomal sex-determining regions in the animal and fungal kingdoms. *PLoS Biol.* **2:**e384.
14. **Fraser, J. A., S. S. Giles, E. C. Wenink, S. G. Geunes-Boyer, J. R. Wright, S. Diezmann, A. Allen, J. E. Stajich, F. S. Dietrich, J. R. Perfect, and J. Heitman.** 2005. Same-sex mating and the origin of the Vancouver Island *Cryptococcus gattii* outbreak. *Nature* **437:**1360–1364.
15. **Fraser, J. A., and J. Heitman.** 2004. Evolution of fungal sex chromosomes. *Mol. Microbiol.* **51:**299–306.
16. **Fraser, J. A., and J. Heitman.** 2005. Chromosomal sex-determining regions in animals, plants and fungi. *Curr. Opin. Genet. Dev.* **15:**645–651.
17. **Fraser, J. A., Y. P. Hsueh, K. M. Findley, and J. Heitman.** 2007. Evolution of the mating-type locus: the basidiomycetes, p. 19–34. *In* J. Heitman, J. W. Kronstad, J. W. Taylor, and L. A. Casselton (ed.), *Sex in Fungi: Molecular Determination and Evolutionary Implications*. ASM Press, Washington, DC.
18. **Galagan, J. E., S. E. Calvo, C. Cuomo, L. J. Ma, J. R. Wortman, S. Batzoglou, S. I. Lee, M. Baştürkmen, C. C. Spevak, J. Clutterbuck, V. Kapitonov, J. Jurka, C. Scazzocchio, M. Farman, J. Butler, S. Purcell, S. Harris, G. H. Braus, O. Draht, S. Busch, C. D'Enfert, C. Bouchier, G. H. Goldman, D. Bell-Pedersen, S. Griffiths-Jones, J. H. Doonan, J. Yu, K. Vienken, A. Pain, M. Freitag, E. U. Selker, D. B. Archer, M. A. Peñalva, B. R. Oakley, M. Momany, T. Tanaka, T. Kumagai, K. Asai, M. Machida, W. C. Nierman, D. W. Denning, M. Caddick, M. Hynes, M. Paoletti, R. Fischer, B. Miller, P. Dyer, M. S. Sachs, S. A. Osmani, and B. W. Birren.** 2005. Sequencing of *Aspergillus nidulans* and comparative analysis with *A. fumigatus* and *A. oryzae*. *Science* **438:**1105–1115.
19. **Herskowitz, I., and Y. Oshima.** 1981. Control of cell type in *Saccharomyces cerevisiae*: mating type and mating-type interconversion, p. 181–209. *In* J. N. Strathern (ed.), *The Molecular and Cellular Biology of the Yeast Saccharomyces: Life Cycle and Inheritance*. Cold Spring Harbor Laboratory, Cold Spring Harbor, NY.
20. **Herskowitz, I., J. Rine, and J. Strathern.** 1992. Mating-type determination and mating-type interconversion in *Saccharomyces cerevisiae*, p. 583–656. *In* J. N. Strathern, Elizabeth W. Jones, and James R. Broach (ed.), *The Molecular and Cellular Biology of the Yeast Saccharomyces*. Cold Spring Harbor Laboratory Press, Cold Spring Harbor, NY.
21. **Hibbett, D. S.** 2006. A phylogenetic overview of the *Agaricomycotina*. *Mycologia* **98:**917–925.
22. **Hibbett, D. S., M. Binder, J. F. Bischoff, M. Blackwell, P. F. Cannon, O. E. Eriksson, S. Huhndorf, T. James, P. M. Kirk, R. Lucking, H. Thorsten Lumbsch, F. Lutzoni, P. B. Matheny, D. J. McLaughlin, M. J. Powell, S. Redhead, C. L. Schoch, J. W. Spatafora, J. A. Stalpers, R. Vilgalys, M. C. Aime, A. Aptroot, R. Bauer, D. Begerow, G. L. Benny, L. A. Castlebury, P. W. Crous, Y. C. Dai, W. Gams, D. M. Geiser, G. W. Griffith, C. Gueidan, D. L. Hawksworth, G. Hestmark, K. Hosaka, R. A. Humber, K. D. Hyde, J. E. Ironside, U. Koljalg, C. P. Kurtzman, K. H. Larsson, R. Lichtwardt, J. Longcore, J. Miadlikowska, A. Miller, J. M. Moncalvo, S. Mozley-Standridge, F. Oberwinkler, E. Parmasto, V. Reeb, J. D. Rogers, C. Roux, L. Ryvarden, J. P. Sampaio, A. Schussler, J. Sugiyama, R. G. Thorn, L. Tibell, W. A. Untereiner, C. Walker, Z. Wang, A. Weir, M. Weiss, M. M. White, K. Winka, Y. J. Yao, and N. Zhang.** 2007. A higher-level phylogenetic classification of the Fungi. *Mycol. Res.* **111:**509–547.
23. **Hsueh, Y. P., J. A. Fraser, and J. Heitman.** 2008. Transitions in sexuality: recapitulation of an ancestral tri- and tetrapolar mating system in *Cryptococcus neoformans*. *Eukaryot. Cell* **7:**1847–1855.
24. **Hsueh, Y. P., and J. Heitman.** 2008. Orchestration of sexual reproduction and virulence by the fungal mating-type locus. *Curr. Opin. Microbiol.* **11:**517–524.
25. **Hsueh, Y. P., A. Idnurm, and J. Heitman.** 2006. Recombination hotspots flank the *Cryptococcus* mating-type locus: implications for the evolution of a fungal sex chromosome. *PLoS Genet.* **2:**e184.
26. **Hsueh, Y. P., C. Xue, and J. Heitman.** 2009. A constitutively active GPCR governs morphogenic transitions in *Cryptococcus neoformans*. *EMBO J.* **28:**1220–1233.
27. **Hull, C. M., M.-J. Boily, and J. Heitman.** 2005. Sex-specific homeodomain proteins Sxi1α and Sxi2a coordinately regulate sexual development in *Cryptococcus neoformans*. *Eukaryot Cell.* **4:**526–535.
28. **Hull, C. M., R. C. Davidson, and J. Heitman.** 2002. Cell identity and sexual development in *Cryptococcus neoformans* are controlled by the mating-type-specific homeodomain protein Sxi1alpha. *Genes Dev.* **16:**3046–3060.
29. **Hull, C. M., and J. Heitman.** 2002. Genetics of *Cryptococcus neoformans*. *Annu. Rev. Genet.* **36:**557–615.
30. **Hull, C. M., and A. D. Johnson.** 1999. Identification of a mating type-like locus in the asexual pathogenic yeast *Candida albicans*. *Science* **285:**1271–1275.
31. **Idnurm, A., and J. Heitman.** 2005. Light controls growth and development via a conserved pathway in the fungal kingdom. *PLoS Biol.* **3:**e95.
32. **James, T. Y., P. Srivilai, U. Kues, and R. Vilgalys.** 2006. Evolution of the bipolar mating system of the mushroom *Coprinellus disseminatus* from its tetrapolar ancestors involves loss of mating-type-specific pheromone receptor function. *Genetics* **172:**1877–1891.
33. **Kahmann, R., and J. Schirrawski.** 2007. Mating in the smut fungi: from *a* to *b* to the downstream cascades, p. 377–387. *In* J. Heitman, J. W. Kronstad, J. W. Taylor, and L. A. Casselton (ed.), *Sex in Fungi: Molecular Determination and Evolutionary Implications*. ASM Press, Washington, DC.
34. **Kamper, J., M. Reichmann, T. Romeis, M. Bolker, and R. Kahmann.** 1995. Multiallelic recognition: nonself-dependent dimerization of the bE and bW homeodomain proteins in *Ustilago maydis*. *Cell* **81:**73–83.
35. **Karos, M., Y. C. Chang, C. M. McClelland, D. L. Clarke, J. Fu, B. L. Wickes, and K. J. Kwon-Chung.** 2000. Mapping of the *Cryptococcus neoformans MAT*alpha locus: presence of mating type-specific mitogen-activated protein kinase cascade homologs. *J. Bacteriol.* **182:**6222–6227.

36. **Kimura, K.** 1952. Studies on the sex of *Coprinus macrorhizus* Rea f. *microsporus* Hongo. I. Introductory experiments. *Biol. J. Okayama Univ.* **1:**72–79.
37. **Koltin, Y., J. R. Raper, and G. Simchen.** 1967. The genetic structure of the incompatibility factors of *Schizophyllum commune*: the B factor. *Proc. Natl. Acad. Sci. USA* **57:**55–62.
38. **Kronstad, J. W., and C. Staben.** 1997. Mating type in filamentous fungi. *Annu. Rev. Genet.* **31:**245–276.
39. **Kwon-Chung, K. J., T. Boekhout, J. W. Fell, and M. Diaz.** 2002. Proposal to conserve the name *Cryptococcus gattii* against *C. hondurianus* and *C. bacillisporus* (*Basidiomycota, Hymenomycetes, Tremellomycetidae*). *Taxon* **51:**804–806.
40. **Kwon-Chung, K. J., Y. C. Chang, and L. Penoyer.** 1997. Species of the genus *Filobasidiella* differ in the organization of their 5S rRNA genes. *Mycologia* **89:**244–249.
41. **Kwon-Chung, K. J., Y. C. Chang, R. Bauer, E. C. Swann, J. W. Taylor, and R. Goel.** 1995. The characteristics that differentiate *Filobasidiella depauperata* from *Filobasidiella neoformans*. *Stud. Mycol.* **38:**67–79.
42. **Kwon-Chung, K. J., J. C. Edman, and B. L. Wickes.** 1992. Genetic association of mating types and virulence in *Cryptococcus neoformans*. *Infect. Immun.* **60:**602–605.
43. **Lee, N., G. Bakkeren, K. Wong, J. E. Sherwood, and J. W. Kronstad.** 1999. The mating-type and pathogenicity locus of the fungus *Ustilago hordei* spans a 500-kb region. *Proc. Natl. Acad. Sci. USA* **96:**15026–15031.
44. **Lengeler, K. B., R. C. Davidson, C. D'Souza, T. Harashima, W. C. Shen, P. Wang, X. Pan, M. Waugh, and J. Heitman.** 2000. Signal transduction cascades regulating fungal development and virulence. *Microbiol. Mol. Biol. Rev.* **64:**746–785.
45. **Lengeler, K. B., D. S. Fox, J. A. Fraser, A. Allen, K. Forrester, F. S. Dietrich, and J. Heitman.** 2002. Mating-type locus of *Cryptococcus neoformans*: a step in the evolution of sex chromosomes. *Eukaryot. Cell* **1:**704–718.
46. **Lengeler, K. B., P. Wang, G. M. Cox, J. R. Perfect, and J. Heitman.** 2000. Identification of the *MAT*a mating-type locus of *Cryptococcus neoformans* reveals a serotype A *MAT*a strain thought to have been extinct. *Proc. Natl. Acad. Sci. USA* **97:**14455–14460.
47. **Lin, X., C. M. Hull, and J. Heitman.** 2005. Sexual reproduction between partners of the same mating type in *Cryptococcus neoformans*. *Nature* **434:**1017–1021.
48. **Litvintseva, A. P., R. E. Marra, K. Nielsen, J. Heitman, R. Vilgalys, and T. G. Mitchell.** 2003. Evidence of sexual recombination among *Cryptococcus neoformans* serotype A isolates in sub-Saharan Africa. *Eukaryot. Cell* **2:**1162–1168.
49. **Marshall Graves, J. A.** 2008. Weird animal genomes and the evolution of vertebrate sex and sex chromosomes. *Annu. Rev. Genet.* **42:**565–586.
49a. **Metin, B., K. Findley, and J. Heitman.** 2010. The mating type locus (*MAT*) and sexual reproduction of *Cryptococcus heveanensis*: insights into the evolution of sex and sex-determining chromosomal regions in fungi. *PLoS Genet.* **6:**e1000961.
50. **Mitchell, T. G., T. J. White, and J. W. Taylor.** 1992. Comparison of 5.8S ribosomal DNA sequences among the basidiomycetous yeast genera *Cystofilobasidium, Filobasidium* and *Filobasidiella*. *J. Med. Vet. Mycol.* **30:**207–218.
51. **Moore, T. D., and J. C. Edman.** 1993. The alpha-mating type locus of *Cryptococcus neoformans* contains a peptide pheromone gene. *Mol. Cell. Biol.* **13:**1962–1970.
52. **Morrow, C. A., and J. A. Fraser.** 2009. Sexual reproduction and dimorphism in the pathogenic basidiomycetes. *FEMS Yeast Res.* **9:**161–177.
53. **Munday, P. L., P. M. Buston, and R. R. Warner.** 2006. Diversity and flexibility of sex-change strategies in animals. *Trends Ecol. Evol.* **21:**89–95.
54. **Nakase, T., S. Jindamorakot, S. Am-In, W. Potacharoen, and M. Tanticharoen.** 2006. Yeast biodiversity in tropical forests of Asia, p. 441–460. *In* G. P. A. Rosa (ed.), *The Yeast Handbook: Biodiveristy and Ecophysiology of Yeasts*. Springer, Berlin, Germany.
55. **Nielsen, K., G. M. Cox, A. P. Litvintseva, E. Mylonakis, S. D. Malliaris, D. K. Benjamin, Jr., S. S. Giles, T. G. Mitchell, A. Casadevall, J. R. Perfect, and J. Heitman.** 2005. *Cryptococcus neoformans* alpha strains preferentially disseminate to the central nervous system during coinfection. *Infect. Immun.* **73:**4922–4933.
56. **Nielsen, K., G. M. Cox, P. Wang, D. L. Toffaletti, J. R. Perfect, and J. Heitman.** 2003. Sexual cycle of *Cryptococcus neoformans* var. *grubii* and virulence of congenic **a** and alpha isolates. *Infect. Immun.* **71:**4831–4841.
57. **Nielsen, K., R. E. Marra, F. Hagen, T. Boekhout, T. G. Mitchell, G. M. Cox, and J. Heitman.** 2005. Interaction between genetic background and the mating-type locus in *Cryptococcus neoformans* virulence potential. *Genetics* **171:**975–983.
58. **Olesnicky, N. S., A. J. Brown, S. J. Dowell, and L. A. Casselton.** 1999. A constitutively active G-protein-coupled receptor causes mating self-compatibility in the mushroom *Coprinus*. *EMBO J.* **18:**2756–2763.
59. **Olesnicky, N. S., A. J. Brown, Y. Honda, S. L. Dyos, S. J. Dowell, and L. A. Casselton.** 2000. Self-compatible *B* mutants in *Coprinus* with altered pheromone-receptor specificities. *Genetics* **156:**1025–1033.
60. **Paoletti, M., F. A. Seymour, M. J. Alcocer, N. Kaur, A. M. Calvo, D. B. Archer, and P. S. Dyer.** 2007. Mating type and the genetic basis of self-fertility in the model fungus *Aspergillus nidulans*. *Curr. Biol.* **17:**1384–1389.
61. **Parkhurst, S. M., and P. M. Meneely.** 1994. Sex determination and dosage compensation: lessons from flies and worms. *Science* **264:**924–932.
62. **Quinn, A. E., A. Georges, S. D. Sarre, F. Guarino, T. Ezaz, and J. A. Graves.** 2007. Temperature sex reversal implies sex gene dosage in a reptile. *Science* **316:**411.
63. **Raper, J.** 1966. *Genetics of Sexuality in Higher Fungi*. The Ronald Press, New York, NY.
63a. **Rodriguez-Carres, M., K. Findley, S. Sun, F. S. Dietrich, and J. Heitman.** 2010. Morphological and genomic characterization of *Filobasidiella depauperata*: a homothallic sibling species of the pathogenic *Cryptococcus* species complex. *PLoS One* **5:**e9620.
64. **Rydholm, C., P. S. Dyer, and F. Lutzoni.** 2007. DNA sequence characterization and molecular evolution of *MAT1* and *MAT2* mating-type loci of the self-compatible ascomycete mold *Neosartorya fischeri*. *Eukaryot. Cell* **6:**868–874.
65. **Samson, R. A., J. A. Stalpers, and A. C. Weijman.** 1983. On the taxonomy of the entomogenous fungus *Filobasidiella arachnophila*. *Antonie Van Leeuwenhoek* **49:**447–456.
66. **Sarre, S. D., A. Georges, and A. Quinn.** 2004. The ends of a continuum: genetic and temperature-dependent sex determination in reptiles. *Bioessays* **26:**639–645.
67. **Smith, C. A., K. N. Roeszler, T. Ohnesorg, D. M. Cummins, P. G. Farlie, T. J. Doran, and A. H. Sinclair.** 2009. The avian Z-linked gene *DMRT1* is required for male sex determination in the chicken. *Nature* **461:**267–271.
68. **Smith, C. A., and A. H. Sinclair.** 2004. Sex determination: insights from the chicken. *Bioessays* **26:**120–132.
69. **Statzell-Tallman, A., C. Belloch, and J. W. Fell.** 2008. *Kwoniella mangroviensis* gen. nov., sp.nov. (*Tremellales, Basidiomycota*), a teleomorphic yeast from mangrove habitats in the Florida Everglades and Bahamas. *FEMS Yeast Res.* **8:**103–113.
70. **Wang, P., C. B. Nichols, K. B. Lengeler, M. E. Cardenas, G. M. Cox, J. R. Perfect, and J. Heitman.** 2002.

Mating-type-specific and nonspecific PAK kinases play shared and divergent roles in *Cryptococcus neoformans*. *Eukaryot. Cell* **1:**257–272.

71. **Wang, S. W., T. Toda, R. MacCallum, A. L. Harris, and C. Norbury.** 2000. Cid1, a fission yeast protein required for S-M checkpoint control when DNA polymerase delta or epsilon is inactivated. *Mol. Cell. Biol.* **20:**3234–3244.
72. **Wickes, B. L., U. Edman, and J. C. Edman.** 1997. The *Cryptococcus neoformans STE12alpha* gene: a putative *Saccharomyces cerevisiae STE12* homologue that is mating type specific. *Mol. Microbiol.* **26:**951–960.
73. **Wickes, B. L., M. E. Mayorga, U. Edman, and J. C. Edman.** 1996. Dimorphism and haploid fruiting in *Cryptococcus neoformans*: association with the alpha-mating type. *Proc. Natl. Acad. Sci. USA* **93:**7327–7331.
74. **Wong, G. J., K. Wells, and R. J. Bandoni.** 1985. Interfertility and comparative morphological studies of *Tremella mesenterica*. *Mycologia* **77:**36–49.
75. **Xu, J., C. W. Saunders, P. Hu, R. A. Grant, T. Boekhout, E. E. Kuramae, J. W. Kronstad, Y. M. Deangelis, N. L. Reeder, K. R. Johnstone, M. Leland, A. M. Fieno, W. M. Begley, Y. Sun, M. P. Lacey, T. Chaudhary, T. Keough, L. Chu, R. Sears, B. Yuan, and T. L. Dawson, Jr.** 2007. Dandruff-associated *Malassezia* genomes reveal convergent and divergent virulence traits shared with plant and human fungal pathogens. *Proc. Natl. Acad. Sci. USA* **104:**18730–18735.
76. **Yi, R., T. Tachikawa, M. Ishikawa, H. Mukaiyama, D. Bao, and T. Aimi.** 2009. Genomic structure of the A mating-type locus in a bipolar basidiomycete, *Pholiota nameko*. *Mycol. Res.* **113:**240–248.

SIGNALING AND VIRULENCE

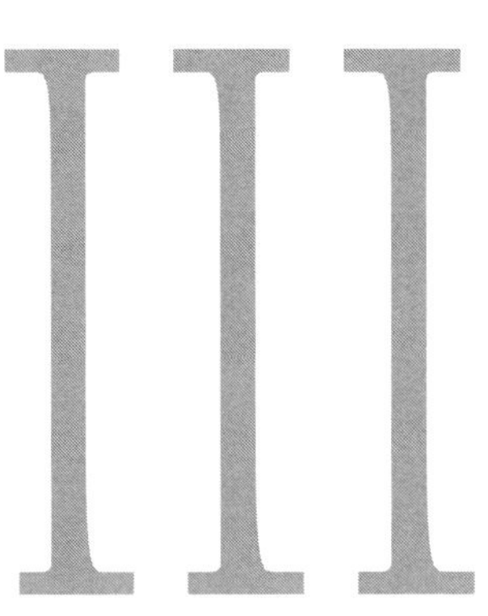

Cryptococcus: From Human Pathogen to Model Yeast
Edited by J. Heitman et al.

12

G-Protein Signaling Pathways: Regulating Morphogenesis and Virulence of *Cryptococcus*

J. ANDREW ALSPAUGH, CONNIE B. NICHOLS, CHAOYANG XUE, WEI-CHIANG SHEN, AND PING WANG

HETEROTRIMERIC AND MONOMERIC G PROTEINS

As pathogenic microorganisms infect their hosts, they must rapidly adapt to new environmental conditions and protect themselves from the hosts' immune system. Many of the core signaling molecules responsible for survival in the host act in highly conserved signal transduction pathways. However, these conserved proteins are used in novel ways by microbial pathogens to interpret host-specific signals and to result in an adaptive cellular response.

Guanine nucleotide-binding proteins (G proteins) regulate the signal transduction cascades that allow pathogenic fungi to survive within their hosts. These proteins act as molecular switches by cycling between an active (GTP-bound) form and an inactive (GDP-bound) form. G proteins may exist either as monomeric signaling molecules, such as Ras proteins, or as multimeric signaling complexes. Heterotrimeric G proteins, consisting of alpha, beta, and gamma subunits, offer a well-established mechanism by which cells precisely control the cellular response to extracellular signals (Fig. 1).

In addition to revealing their relevance for microbial pathogenesis, studying these processes in microbial pathogens also offers important insight into developing models of G-protein signaling in small eukaryotes. Similar to *Cryptococcus neoformans*, the yeasts *Saccharomyces cerevisiae* and *Schizosaccharomyces pombe* regulate morphological and stress responses through Gα proteins. Both of these fungi possess two Gα protein genes, as well as single genes encoding classical Gβ and Gγ proteins. In these divergent yeasts, each Gα protein is specialized to control either a pheromone response/mitogen-activated protein (MAP) kinase pathway or a glucose response/cyclic AMP (cAMP) pathway (61, 74). The ligand(s) for the cAMP/stress response pathways are still unclear, but they likely include host-specific signals in the case of microbial pathogens. Additionally, some investigators have proposed new paradigms involving monomeric or unconventional Gα protein signaling in small eukaryotes (38, 47). Therefore, G-protein signaling in microorganisms will likely be quite distinct from the models developed in mammalian systems. Moreover, studying cAMP signaling in classical model organisms as well as in microbial pathogens offers a new perspective for how conserved signaling proteins are used in the pathogenesis of human diseases.

This chapter will address the roles of heterotrimeric and monomeric G proteins in the growth and development of *C. neoformans*. These proteins control the activation of specific signal transduction pathways mediating important cellular processes such as stress response, mating, and morphogenesis. Several of these pathways are also required for survival within the infected host.

J. Andrew Alspaugh and Connie B. Nichols, Departments of Medicine and Molecular Genetics & Microbiology, Duke University School of Medicine, Durham, NC 27710. **Chaoyang Xue,** Public Health Research Institute Center, University of Medicine and Dentistry of New Jersey, Newark, NJ 07103. **Wei-Chiang Shen,** Department of Plant Pathology and Microbiology, National Taiwan University, Taipei, Taiwan 10617. **Ping Wang,** The Research Institute for Children, New Orleans, LA 70118.

THE Gα PROTEIN (Gpa1)/cAMP SIGNALING PATHWAY

Introduction

The central motifs of the cAMP signal transduction pathway are highly conserved in eukaryotes, allowing diverse cell types to respond to different environmental stresses. However, the role of this pathway in microbial virulence was not predicted in studies involving nonpathogenic model organisms. Many investigators have demonstrated that Gα protein/cAMP pathways are required for pathogenesis in several other microorganisms. For example, adenylyl cyclase, the enzyme that makes cAMP within the cell, is required for hyphal formation and pathogenicity of the human fungal pathogen *Candida albicans* (79). Additionally, fungal pathogens of plants, such as the maize pathogen *Ustilago maydis* and the rice blast fungus *Magnaporthe grisea*, use cAMP pathways to regulate morphogenesis and virulence (21, 36). In the mould pathogen *Aspergillus fumigatus*, the cAMP pathway controls the production of secondary metabolites,

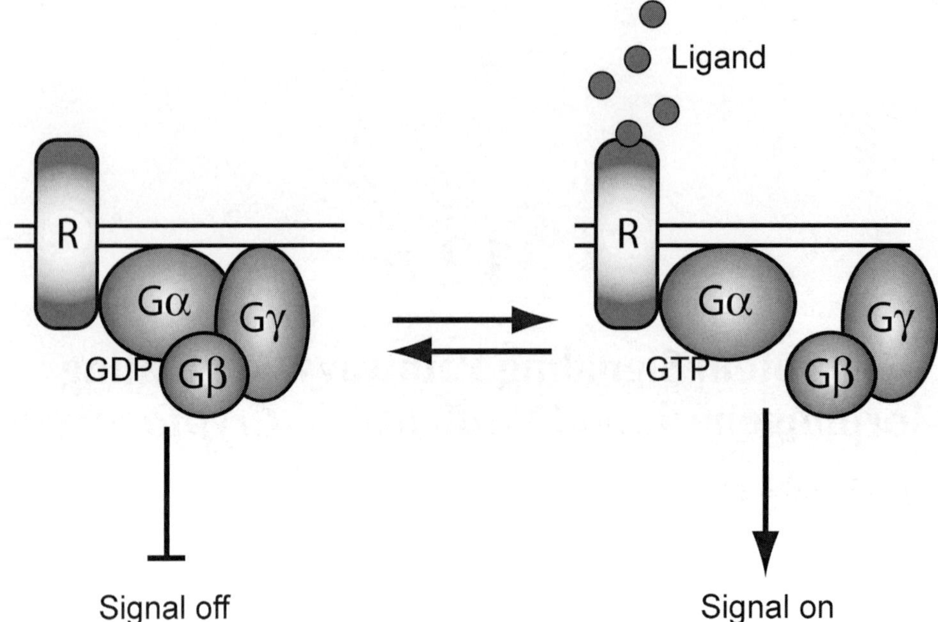

FIGURE 1 Model of heterotrimeric G-protein signaling. Upon activation of the upstream GPCR (R), the Gα protein exchanges GDP for GTP, resulting in a dissociation of the α, β, and γ subunits and subsequent activation of the downstream signal transduction pathways.

and these compounds are likely related to the pathogenesis of *Aspergillus* infections (59).

The Gpa1 Gα Subunit

The *C. neoformans* Gpa1 Gα protein controls the expression of the two most important inducible phenotypes required for this organism's pathogenesis: melanin and capsule (3). *C. neoformans gpa1* mutant strains demonstrate markedly attenuated capsule production in response to capsule-inducing conditions in vitro. Additionally, *gpa1* mutants have a significant defect in melanin production (Fig. 2). The induction of melanin and capsule is required for pathogenesis of *C. neoformans*, and *gpa1* mutant strains are highly attenuated for virulence in animal models of cryptococcosis (3).

Genetic and chemical epistasis experiments defined the downstream signaling pathways activated by the Gpa1 Gα protein. In contrast to wild type, the *gpa1* mutant fails to increase intracellular cAMP levels in response to glucose stimulation (4). Also, the addition of exogenous cAMP fully restores capsule and melanin to the *gpa1* mutant strains (3). These observations confirmed that the *C. neoformans* Gpa1 Gα protein acts upstream of a cAMP signaling cascade to regulate virulence trait expression and pathogenesis (3, 77).

Gpa1 Activates Conserved Elements of a cAMP Pathway

Adenylyl Cyclase

The intracellular messenger cAMP is generated by the enzyme adenylyl cyclase. The single adenylyl cyclase gene in *C. neoformans* (*CAC1*) is not essential for growth (4). Furthermore, adenylyl cyclase (*cac1*) mutants are phenotypically similar to *gpa1* mutant strains, with diminished capsule, melanin, and mating (4). Also, overexpression of a wild-type *CAC1* gene in a *gpa1* mutant background suppresses the mutant defects in melanin and capsule formation (4). Taken together, these data provide further evidence that the *C. neoformans* Gα protein Gpa1 activates the adenylyl cyclase protein Cac1 to produce cAMP.

PKA

Protein kinase A (PKA), or cAMP-dependent protein kinase, exists as a tetrameric protein composed of dimeric regulatory subunits and two monomeric catalytic subunits. cAMP causes activation of this enzyme by inducing the release of the regulatory subunits from the catalytic subunits (Fig. 2). As predicted from this model, mutation of the gene encoding the Pka1 catalytic subunit results in identical defects in capsule, melanin, mating, and virulence as the *gpa1* and *cac1* mutants. However, the addition of exogenous cAMP has no effect on the *pka1* mutant phenotypes, consistent with the predicted signaling role of Pk1 downstream of cAMP (33).

Mutation of the *PKR1* gene, encoding the regulatory subunit of cAMP-dependent protein kinase, would be predicted to constitutively activate the cAMP signaling pathway (Fig. 2). Accordingly, the *pkr1* mutation fully suppresses the *gpa1* mutant capsule and melanin defects (Fig. 2). Moreover, *pkr1* mutant strains make more capsule than isogenic wild-type strains. The effect of the *pkr1* mutation was tested in several animal models of cryptococcosis; in each case, *pkr1* mutant strains were hypervirulent compared to wild type (33, 67). These studies indicate that the core elements of the highly conserved cAMP signaling pathway are important for *C. neoformans* virulence factor induction and for pathogenesis.

Gβ Protein Function in the cAMP Pathway (Gib2)

As described above, heterotrimeric G proteins consist of three subunits: α, β, and γ proteins (35). Upon ligand stimulation of a specific G-protein coupled receptor (GPCR), the activated GTP-Gα disassociates from the βγ heterodimer to stimulate downstream effector molecules. βγ remains tightly bound and reassociates with the GDP-bound Gα, forming a stable heterotrimeric protein complex upon signal termination or

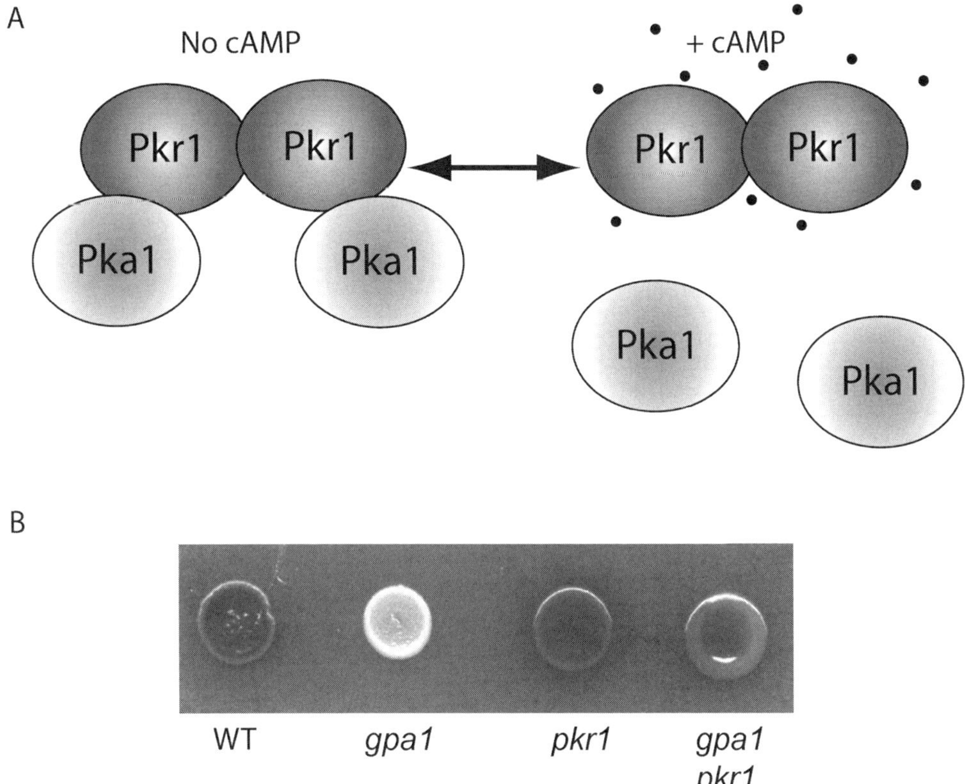

FIGURE 2 Pka1 and Pkr1 are two components of the *C. neoformans* PKA holoenzyme. (A) In the presence of cAMP, the dimeric regulatory subunits of PKA (Pkr1) dissociate from the catalytic subunits (Pka1), allowing downstream transmission of the cAMP signal. (B) The wild-type (WT), *gpa1* mutant, *pkr1* mutant, and *gpa1 pkr1* double mutant strains were assessed for melanin production by incubation on Niger seed medium for 5 days. A *pkr1* mutation constitutively activates downstream effectors of this pathway, restoring melanin to the *gpa1* mutant (33).

desensitization (Fig. 1). The βγ heterodimer, in addition, can also activate certain downstream target molecules (23, 24, 83, 84).

In *C. neoformans*, which encodes three Gα proteins, the Gα Gpa1-mediated cAMP signaling pathway exhibits remarkable conservation in both pathway composition and functional mechanism with those that are operational in mammalian and other eukaryotic organisms (3–5, 33, 93). However, in a fashion that is similar to many fungi, including the budding yeast *S. cerevisiae* and fission yeast *S. pombe*, but distinct from other eukaryotes in which multiple α and β subunits were found, only a single conventional Gβ has been identified that functionally couples with one of the multiple α subunits in either mating or cAMP signaling (26, 54, 55, 89). Since G proteins function within a heterotrimeric complex in which there also exists great specificity between α and β(γ) subunits, the lack of additional β subunits has prompted speculation that certain fungal Gαs might function as a monomeric G protein without the presence of a dimeric βγ protein complex or that additional proteins could serve as a functional Gβ subunit by coupling with Gα to regulate its activation cycle, aiding in its interaction with specific GPCRs (10, 38, 42, 75).

There is no definitive evidence supporting the hypothesis that fungal Gαs can function as a monomeric G protein without coupling to βγ. However, recent studies have gathered evidence indicating that proteins capable of exhibiting a β-transducin-like structure, containing either seven WD-repeat (tryptophan-aspartate repeat) or Kelch repeat motifs, can mimic the Gβ function by stabilizing Gα and/or assisting its functions (10, 38, 75, 98). The *S. cerevisiae* Kelch repeat proteins Gpb1/Krh2 and Gpb2/Krh1 were found to functionally couple with Gpa2 in a cAMP signaling pathway that senses nutrients to regulate pseudohyphal differentiation (10, 38, 39, 76). More recently, a β-like/receptor for activated C protein kinase 1 (RACK1) protein homolog, Asc1, was found to exhibit guanine nucleotide dissociation inhibitor activity toward Gpa2, a conserved function of Gβ subunits (98).

In *C. neoformans*, no conventional Gβ was found that couples to Gpa1 and functions in the Gpa1-cAMP signaling pathway. Remarkably, in a yeast two-hybrid screen using Gpa1 as bait, Palmer and colleagues identified a Gβ-like/RACK1 protein homolog, Gib2 (for Gpa1 interacting beta), which binds to Gpa1 (75). Gib2 also exhibited in vitro and in vivo binding activity with the Gγ subunits Gpg1 and Gpg2, suggesting that Gib2 could be part of a heterotrimeric protein complex with Gpa1 and Gpg1 or Gpg2 (75). In addition, overexpression of Gib2 in a *C. neoformans* var. *neoformans gpa1* mutant resulted in partial restoration of melanin formation and full restoration of capsule production, demonstrating that Gib2 also positively regulates cAMP signaling (75). Additional studies may demonstrate whether Gib2 exhibits the guanine nucleotide dissociation inhibitor activity of classical Gβ proteins as well as the mechanism by which Gib2 promotes downstream activation of the cAMP signaling pathway.

The identification of *S. cerevisiae* Gpb1, Gpb2, and Asc1, as well as *C. neoformans* Gib2, all add to a repertoire of proteins that function as bona fide Gβs in G-protein signaling. This new paradigm highlights the evolutionary divergence of fungi from their mammalian counterparts through adaptation to a limited number of canonical β-subunits existing in the fungal genome.

Modulators of the Gpa1/cAMP Pathway

The activity of G proteins is finely regulated by a specific set of modulators. G-protein pathways are typically activated by ligand binding with a membrane-associated GPCR, promoting the Gα protein switch from an inactive GDP-bound form to an active GTP-bound stage (Fig. 1). While Gα proteins contain an intrinsic GTPase activity, regulator of G-protein signaling (RGS) proteins accelerate the hydrolysis of GTP, negatively regulating pathway activation. Other proteins also influence G-protein/cAMP signaling, such as the phosphodiesterases and adenylyl cyclase–associated proteins.

RGS Proteins

Four putative proteins (*Cryptococcus* regulator of G-protein signaling [Crg]1 through Crg4) containing RGS domains have been identified from the completed genome sequence of *C. neoformans*, and two have clear roles in G-protein signaling regulation and fungal development (44, 71, 81, 87, 94). Crg1 and Crg2 both regulate the *C. neoformans* pheromone response pathway, but the functions of Crg3 and Crg4 are still unclear. Crg1 is the first reported RGS protein in *Cryptococcus* and is an important negative regulator of pheromone-responsive mating signaling (71, 87). Crg1 may also be linked to virulence by negatively regulating melanin production, a key virulence-associated phenotype (87). Crg2 has also been found to regulate the pheromone response pathway by associating with Gpa2 and Gpa3, two Gα proteins controlling mating pathway activation in concert with Crg1 (44).

Besides controlling the activity of the *C. neoformans* mating pathway, Crg2 also plays a role in the Gpa1-cAMP signaling pathway. Crg2 interacted with Gpa1 in the yeast two-hybrid and in vitro binding assays. Additionally, *crg2* mutants displayed increased cAMP production, suggesting that Crg2 negatively regulates the cAMP pathway by interacting with Gpa1. Although *crg2* mutations and the Gpa1 dominant active allele $GPA1^{Q284L}$ both showed enhanced melanin production under normally repressive conditions, virulence of the *crg2* mutant was attenuated in a murine model, suggesting Crg2 may have additional roles to modulate fungal virulence. Alternatively, overproduction of intracellular cAMP may also impair virulence through mechanisms such as stress tolerance modulation. Consistent with this hypothesis, strains expressing the dominant active $GPA1^{Q284L}$ allele also display attenuated virulence (94).

A fourth protein containing an RGS domain (Crg4/Liv8) has been recently reported. The *crg4/liv8* mutant strain was sensitive to acidified nitrite (NO), sodium dodecyl sulfate, and KCN, and the mutant also exhibited attenuated liver infectivity (60). A role for Crg4/Liv8 in G-protein regulation has not been established. Studies of RGS proteins in *Cryptococcus* have revealed that the regulation of G-protein signaling is more intricate and potentially more complicated than in *S. cerevisiae*. This also illustrates the importance of comparative studies of G-protein signaling and regulation in *Cryptococcus* and other fungal pathogens such as *Candida* species to allow a better understanding of the signaling mechanisms affecting fungal growth and development.

Phosphodiesterase (Pde1/2)

Detailed studies of the Gpa1-activated PKA-cAMP signaling have also revealed additional regulatory mechanisms for this pathway. Phosphodiesterases regulate the intracellular cAMP level to modulate Gpa1 signaling. In *S. cerevisiae*, two phosphodiesterases (Pde1 and Pde2) have been identified. The high-affinity enzyme Pde2 regulates the basal cAMP level and is important in determining stress tolerance. The low-affinity enzyme Pde1 exhibits a significant role in regulating cAMP levels induced by glucose and does not affect the basal levels of cAMP (63). Similarly, two phosphodiesterase protein homologs (Pde1 and Pde2) have been found in *C. neoformans*. While the function of Pde2 in cAMP-level regulation is still unclear, genetic studies revealed that Pde1 moderately regulates cAMP level, which is sufficient to rescue the melanin defect of *gpa1* mutants. The activity of Pde1 is controlled by Pka1, and *pde1 pka1* double mutants produce cAMP levels significantly higher than those of *pka1* single mutants (~15-fold). This finding suggests that, besides the Pka1-dependent regulation of Pde1, there is an additional negative feedback loop to tightly control intracellular cAMP production, perhaps through regulation of Cac1 adenylyl cyclase activity (40).

Adenylyl Cyclase–Associated Protein (Aca1)

An adenylyl cyclase (Cac1)–associated protein (Aca1) has been identified to control Cac1 function in parallel with Gpa1 (5). The presence of a Cac1 regulator in addition to Gpa1 was suggested by the more severe defects in capsule and melanin production in the *cac1* mutant strain compared to the *gpa1* mutant. Aca1 was demonstrated to interact with the C terminus of Cac1, modulating the effects of this protein on mating, capsule, and melanin. *gpa1 aca1* double mutants displayed melanin and capsule phenotypes similar to those of *cac1* mutants, suggesting that both Aca1 and Gpa1 are activators of the Cac1/cAMP signaling pathway (5). These studies further define the conserved nature of cAMP signaling in this fungus and emphasize the important role of this pathway in cryptococcal development and pathogenesis.

Upstream Activation of the Gpa1/cAMP Pathway (GPCRs)

The activity of G proteins is controlled by upstream signaling molecules, including GPCRs. The GPCR family of proteins is a large group of plasma membrane proteins that contain seven transmembrane spans. Their common function is to sense extracellular signals and to activate intracellular pathways by association with G-protein α and/or βγ subunits. Mammalian GPCRs are the subject of extensive research interest because of their importance in drug discovery. The most extensively studied GPCRs in fungi are the pheromone receptors, such as the Ste2 and Ste3 proteins of *S. cerevisiae*, which sense pheromone molecules and activate the pheromone-responsive mating pathways (14, 37).

In *Cryptococcus*, only Ste3 homologs have been identified, even though fungal pheromone response pathways are relatively well conserved. There is no Ste2 homolog in *Cryptococcus* or other basidiomycetes, suggesting that pheromone/pheromone receptor recognition mechanisms differ from *S. cerevisiae*. In *S. cerevisiae*, Ste2 and Ste3 recognize the peptide pheromone α factor and the lipid-modified peptide **a** factor, respectively, while all pheromones in basidiomycetes are lipid modified and recognized by the Ste3-like pheromone receptors (15, 50, 95). Moreover, the pheromone receptors in *Cryptococcus* are integrated within the mating-type (*MAT*) locus and are therefore mating-type specific.

In *Cryptococcus*, there are at least 24 putative GPCRs based on a genome-wide analysis of potential seven-transmembrane proteins (93). Two pheromone receptors, Ste3α (Cprα) and Ste3**a** (Cpra), were identified in *MAT*α and *MAT***a** strains, respectively, and shown to sense **a**/α pheromone molecules and activate the Gpa2 and Gpa3 Gα proteins (6, 22).

Besides pheromone receptors, GPCRs have also been identified as nutrient sensors in fungi. A well-characterized example is the sugar sensor Gpr1 in *S. cerevisiae*. Yeasts developed multiple ways to sense and transport fermentable sugars like glucose, the primary carbon source for cells. Gpr1 is an unusually large GPCR that contains 961 amino acids with a long third cytoplasmic loop and C-terminal tail. Gpr1 senses glucose and sucrose and activates the Gα protein Gpa2 to control the cAMP signaling, which regulates pseudohyphal differentiation (49, 56, 62, 80, 96). The sequence of Gpr1 is conserved in other ascomycetes but not in basidiomycetes. By direct sequence comparison, there is no apparent Gpr1 ortholog in *Cryptococcus*, even though glucose is also utilized as an important carbon source for this fungus and induces the cAMP level.

The GPCR protein Gpr4 was identified in the *Cryptococcus* genome, exhibiting structural features similar to *S. cerevisiae* Gpr1, although the two proteins do not share much sequence homology. Gpr4 contains 841 amino acids with a long third cytoplasmic loop and C-terminal tail. Interestingly, Gpr4 interacted in vitro with Gpa1, and it appeared to function upstream of the Gpa1-cAMP pathway. Gpr4 is also important for capsule and mating, but not melanin, and the defects of capsule and mating in *gpr4* mutants are somewhat weaker compared to those in *gpa1* mutants, indicating that Gpr4 may not be the only upstream activator for Gpa1. Interestingly, Gpr4 was not glucose specific since the *gpr4* mutants still responded to glucose stimulation. Further studies have revealed that Gpr4 might instead function as an amino acid sensor. Amino acids like methionine can trigger the internalization of the Gpr4-DsRed fusion protein, and additional methionine induces mating in the wild-type strain but not in the *gpr4* mutants (93). Interestingly, cAMP production was enhanced in the wild-type cells with the addition of both glucose and methionine, indicating that cAMP signaling may be activated by multiple extracellular nutrient signals. How *Cryptococcus* cells sense glucose still remains to be understood. It is possible that other unidentified GPCRs may sense glucose and activate Gpa1, or *C. neoformans* may use a different mechanism in sensing glucose and activating the cAMP cascade. A Gpr1 homolog in *C. albicans* has also been identified to sense methionine and is important for morphological transitions (64), suggesting that amino acid sensing by GPCRs is shared in other pathogenic fungi. However, the detailed reasons that fungi use methionine to activate the cAMP pathway remain to be elucidated.

Downstream Targets of cAMP and PKA

To further define the mechanism by which the Gpa1/cAMP pathway regulates capsule and melanin, gene microarray studies evaluated differences in the transcriptomes of the wild-type and *gpa1* mutant strains. Via the use of partial genome microarrays, genes were identified whose transcription was at least 2.5-fold greater in the wild type than in the *gpa1* mutant (78). Nine of these genes had previously been defined as functioning in the capsule biosynthesis pathway (the *CAS* and *CAP* genes), and two of these genes encode laccase proteins involved in melanin synthesis (78).

Among the other genes suggested by these microarrays to have altered transcription in the *gpa1* mutant versus the wild-type strain, several were identified with significant homology to DNA-binding proteins and transcription factors. One of these is homologous to the *S. cerevisiae NRG1* gene, encoding a protein that inhibits the transcription of glucose-regulated genes in budding yeast (11). The *C. albicans NRG1* ortholog represses the transcription of several genes required for the yeast-hyphal transition, and this gene is required for *Candida* virulence (12).

Consistent with its hypothesized role in cAMP signaling, the *C. neoformans nrg1* mutant has a capsule defect, and this phenotype is not suppressed by exogenous cAMP. Interestingly, the *C. neoformans* Nrg1 protein contains a potential consensus sequence for PKA phosphorylation, and mutation of this sequence negatively affects Nrg1 function. Therefore the Gpa1/cAMP pathway likely controls Nrg1 activity both by regulating the transcription of the *NRG1* gene and by direct PKA phosphorylation of the Nrg1 protein (27).

Although Nrg1 acts as a cAMP-regulated transcription factor involved in capsule expression, the current data suggest that this protein is not the only transcription factor by which the cAMP pathway controls virulence factor expression. First, several of the capsule and melanin genes whose transcription is controlled by cAMP do not appear to be targets of Nrg1. Additionally, the altered capsule phenotype of the *nrg1* mutant strain is less severe than the acapsular phenotype of strains with defective cAMP signaling. Lastly, melanin is not significantly altered in the *nrg1* mutant (27). Therefore, other transcriptional regulators likely act with Nrg1 to mediate the downstream effects of cAMP on *C. neoformans* growth, development, and pathogenesis.

Summary

The cAMP signal transduction pathway plays a central role in the stress response and virulence of microbial pathogens. Upon entering the infected host, the human fungal pathogen *C. neoformans* uses this pathway to induce the expression of phenotypes required for survival in the host: capsule and melanin (Fig. 3). The cAMP pathway is also required for full mating competence. Using conserved and unique effectors, this pathway directs the induction of these disparate cellular processes, each playing central roles in microbial development or pathogenesis.

THE Gpa2/Gpa3 PHEROMONE RESPONSE PATHWAY

Introduction

Sexual reproduction is an important process for generating genetic diversity in eukaryotes. In fungi, sexual development involves fundamental events such as plasmogamy, karyogamy, meiosis, and production of sexual spores. The small, secreted, peptide pheromone molecules play critical roles at initial mating steps in higher fungi such as ascomycetes and basidiomycetes (7). Perception of pheromones by mating partners induces the pheromone response pathway to tune cellular physiology for mating. The paradigm of the pheromone response pathway in fungi has been established mostly based on the studies of the budding yeast *S. cerevisiae* (9, 31, 32). Conserved molecules for pheromone signaling have been identified, and the basic framework for their regulation has been generated. Studies of other fungi have

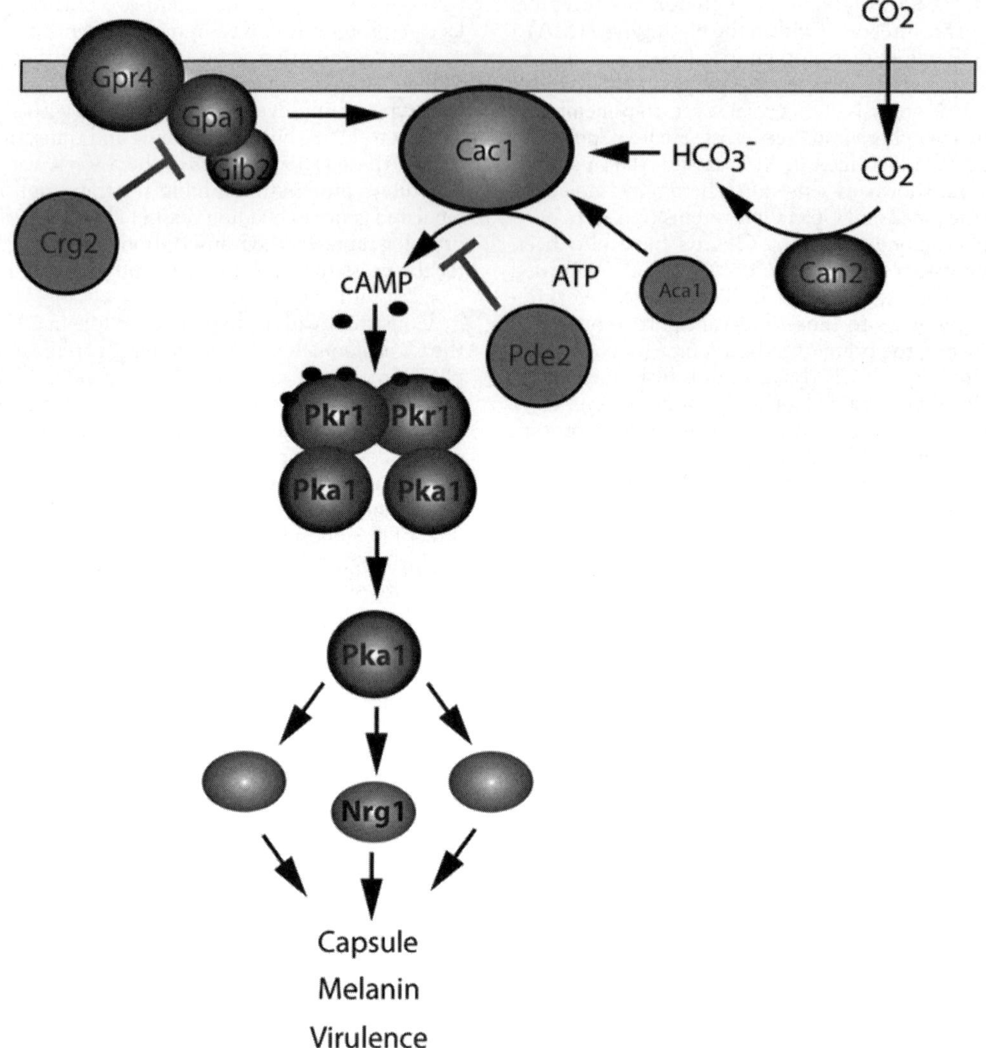

FIGURE 3 Model of the *C. neoformans* Gpa1/cAMP signaling pathway (48).

revealed certain levels of conservation; however, unique compositions and regulations and additional roles of the pheromone response pathway in different fungal systems have also been found, suggesting in-depth studies are required for each fungal organism.

C. neoformans is a heterothallic basidiomycete with a bipolar mating system with two distinct mating types, *MAT*α and *MAT***a** (2, 51). Two important observations have motivated research efforts to isolate the *MAT* loci and characterize the sexual development in *C. neoformans*. First, there is a unique ratio of *MAT*α to *MAT***a** isolates observed in the clinic and environment, with *MAT*α predominating. Second, animal modeling of cryptococcal infection with *MAT*α and *MAT***a** strains suggest that *MAT*α strains are intrinsically more virulent (52, 53, 72).

By difference cloning, a *MAT*α-specific fragment was identified encoding a pheromone gene (66). Further characterization revealed that *C. neoformans MAT*α and *MAT***a** are unusually large for mating-type loci, each including over 20 genes (34, 57). In addition to pheromones, pheromone receptors, and homeodomain-containing transcriptional regulators, there are also other components of these loci, including members of conserved MAP kinase cascades and genes with no known mating functions (34, 57).

Multiple copies of the mating-type-specific pheromone genes are present in *C. neoformans MAT*α and *MAT***a** loci (28, 65, 82). The peptide pheromone molecules produced by the *C. neoformans MAT***a** and *MAT*α strains are both lipophilic, as reported in other basidiomycetous fungi. Production of the pheromones is dramatically induced by nutrient deprivation and the presence of mating partners (28, 82). Pheromone is required for normal fusion and mating differentiation (82). Secretion of mature pheromones is achieved via a multidrug-resistance-related transporter, Ste6 (43). Binding of extracellular pheromone molecules to the surface pheromone receptors triggers an intracellular signaling relay.

C. neoformans pheromone receptors are GPCRs encoded by the *CPR*α*/STE3*α and *CPR***a***/STE3***a** genes (17, 22). Strains containing pheromone receptor gene mutations fail to induce pheromone genes during mating, dramatically decreasing mating efficiency. Interestingly, the *C. neoformans CPR***a** gene also has been reported to be important for survival and growth of the organism in host tissue (17). In response to pheromone binding, activated pheromone receptor causes guanine nucleotide exchange within the Gα subunit and dissociation of the G-protein complex to activate the downstream pathways. Components of the pheromone-responsive heterotrimeric G-protein complex and their regulators have been identified in *C. neoformans*.

Interestingly, an additional pheromone receptor located outside of the *MAT* locus has been identified in *C. neoformans*. Cpr2 (*Cryptococcus* pheromone receptor 2) is a homolog of Ste3 but is not mating-type specific, yet its expression is induced during mating. Studies revealed that Cpr2 is a constitutively active GPCR capable of activating pheromone response in the absence of a ligand in a heterologous expression system (45). Its activity requires the presence of an important amino acid residue, L222, and a single point mutation, L222P, is sufficient to abolish its activity (45). *cpr2* mutants were fertile, albeit with reduced cell fusion efficiency, while *CPR2* overproduction stimulated unisexual reproduction. Moreover, Cpr2 also interacted with Gpa2 and Gpa3 to regulate mating signaling by competing with Ste3α/**a** for G-protein binding. Similar GPCRs have also been identified in other basidiomycetes, such as the mushroom *Coprinopsis cinerea*, suggesting that Cpr2-like GPCRs may be important for certain fungi, allowing additional flexibility in signaling involved in morphogenesis and mating recognition.

Gα Proteins and Interaction with Pheromone Receptors

C. neoformans contains three genes encoding Gα subunits. As described above, the Gpa1 Gα protein acts upstream of the cAMP-PKA pathway. In contrast, Gpa2 and Gpa3 interact with pheromone receptors and regulate mating via the Cpk1 MAP kinase pathway (3, 44, 58). To study the roles of *GPA2* and *GPA3* in mating, both genes were deleted singly or in combination in both mating types of *C. neoformans* serotype A and D strains. The single *gpa2* or *gpa3* mutants exhibited no significant mating alterations compared to the sterile phenotype of the bilateral crosses in which both mating partners have a *gpa2 gpa3* double mutation. These data suggest that both of these Gα subunits are functionally redundant and collectively required for normal mating in *C. neoformans* (44, 58).

Gpa2 and Gpa3 were shown to play both positive and negative roles in pheromone production when placed in confrontation with *crg1* mutants and assessed for conjugation tube production (58). The negative function of *GPA3* is also supported by the enhanced self-filamentation phenotype observed in the serotype D *gpa3* mutant. During growth on V8 or filament agar medium, the *gpa3* mutant produces prolific hyphae along the periphery of the colony. This self-hyperfilamenting phenotype does not require Gpa2 but is dependent on the Gβ subunit Gpb1 (44). The hyperfilamentous phenotype of the *gpa3* mutant is a result of constitutive activation of the pheromone response pathway. The expression of pheromone genes is dramatically induced in the *gpa3* mutant; additionally, the *GPB1* transcript level in the *gpa3* cells is also 1.3- to 1.8-fold higher than in the wild-type cells. These results suggest that Gpa2 and Gpa3 both play roles in pheromone responses, with Gpa3 playing primarily a negative regulatory role in pheromone response (44, 58).

Studying gene expression under mating conditions reveals that *GPA2* and *GPA3* are differentially regulated during mating in response to different extracellular cues. *GPA3* is induced when *MAT***a** or *MAT*α wild-type cells are incubated alone on V8 mating medium, and further induction is not observed when cells are exposed to the opposite mating type. In contrast, the expression pattern of *GPA2* is similar to those of pheromone genes. The transcript level of *GPA2* remains low when incubated alone, but its expression is strongly induced by the presence of the opposite mating partner. *GPA2* is also induced in the *gpa3* mutant background, suggesting that the regulation of *GPA2* is subjected to the activation of the pheromone response pathway (44).

To further determine the intrinsic signaling roles of Gpa2 and Gpa3 in mating, GTPase-deficient, dominant active versions of these Gα subunits were created, and their signaling effects on mating were observed (44). Strains containing the dominant active *GPA3*^{Q206L} allele show responses similar to the wild-type strain when incubated alone on V8 mating medium, and no filamentation or pheromone induction is observed. Additionally, the *GPA3*^{Q206L} strain fails to respond to a wild-type mating partner: pheromone gene induction is abolished, and reduced mating filamentation is observed compared to crosses involving congenic wild-type strains. In contrast, when incubated alone on V8 mating medium, strains containing the *GPA2*^{Q203L} allele are self-filamentous and show high pheromone gene induction in a reporter assay. In the presence of a wild-type mating partner, this strain shows even more robust filamentation and pheromone gene induction. These responses are blocked by the *gpb1* and *ste7* mutations, suggesting that activation of the pheromone response by Gpa2 requires a functional Gβ subunit, Gpb1, and the MAP kinase cascade (44).

In addition to the genetic evidence, physical interactions between Gpa2/Gpa3 and other pheromone signaling components were also demonstrated. Using the split-ubiquitin and yeast two-hybrid systems, the pheromone receptor Ste3α was shown to interact with both Gα subunits, especially Gpa2 (44, 58). Furthermore, yeast two-hybrid and in vitro binding assays demonstrated that Gpa2, but not Gpa3, also interacts with the Gβ subunit Gpb1, suggesting the negative effect of Gpa3 on mating may act through a Gpb1-independent mechanism. However, the observation that the *gpa3*Δ mutant phenotype depends on Gpb1 indicates that an interaction between Gpa3 and Gpb1 is still possible (44, 58).

Gβ Protein Gpb1

With the unique findings that both Gα proteins, Gpa2 and Gpa3, are involved in pheromone responses and mating in *C. neoformans*, the role of the sole Gβ protein, Gpb1, remains remarkably conserved with that of the *S. cerevisiae* Gβ protein Ste4. Overexpression of Gpb1 activates the conserved MAP kinase pathway, resulting in the production of conjugation tubes in the absence of pheromone stimulation (89). Consistent with this finding, the *gpb1* mutant strain in which the *GPB1* gene was disrupted was unable to mate with strains of the opposite mating type (89). Further genetic and biochemical studies had shown that Gpb1 couples with Gpa2 as well as with the Gγ proteins Gpg1 or Gpg2 in both in vivo and in vitro conditions, indicating that Gpa2, Gpb1, and Gpg1/Gpg2 could constitute a classic heterotrimeric G-protein complex to mediate pheromone-activated responses and mating (58).

Because of a close association suspected between mating, in particular the MATα mating type, and virulence in *C. neoformans* (53), the *gpb1* mutant strain was subjected to a virulence study by use of a rabbit model of cryptococcosis. Interestingly, the *gpb1* mutant strain was as virulent as the wild-type control strain, which effectively uncouples mating from fungal virulence (89).

Gpg1 and Gpg2

The Gγ subunit functions by binding to Gβ, tethering the dimeric βγ protein complex to the plasma membrane through its prenylated carboxyl terminal site (41). A total of 12 Gγ subunits have been found in the human and mouse genomes that could cross-bind to multiple Gβ subunits, in which binding of various combinations is suspected to confer signal specificity or intensity (46). A single γ subunit has been found in various yeast and filamentous fungal models. In contrast, two genes encoding Gγ subunits (*GPG1* and *GPG2*) are present in the *C. neoformans* genome (44, 58).

No convincing data have yet demonstrated a direct role for either of these two Gγ proteins in the *C. neoformans* cAMP pathway. Hsueh et al. reported that the Gγ protein Gpg2 is primarily involved in the mating process since the *gpg2* mutant strain is defective in pheromone responses and mating (44). Additionally, Li et al. described that Gpg1 exhibits a clear but less prominent role in mating (58). Mating is attenuated in a *gpg1* mutant, whereas the *gpg1 gpg2* double mutant is absolutely sterile, even in crosses with wild-type strains (58). The functional distinction between Gpg1 and Gpg2 may suggest that Gpg2 exhibits stronger in situ binding affinity to Gpb1, the Gβ protein involved in regulating the *C. neoformans* pheromone response. Since both *C. neoformans* Gγ proteins (Gpg1 and Gpg2) bind to both Gβ proteins (Gpb1 and Gib2) in vitro and in vivo, Gpg1 and Gpg2 may also have unexplored functions in addition to mating, such as an involvement in the Gpa1/cAMP pathway.

Regulators of the Gpa2 and Gpa3 Signaling Pathways

In *S. cerevisiae* an RGS protein is known to control pheromone responses by regulating G-protein activity and interacting with the pheromone receptor (8). In *C. neoformans*, three RGS proteins are present in the genome, and Crg1 and Crg2 have been shown to negatively regulate pheromone responses. Mutation of *CRG1* or *CRG2* caused elevation of the pheromone transcript. Crg1 and Crg2 interacted with the pheromone receptor Ste3α via a DEP (Dishevelled, EGL-10, Pleckstrin) domain and C-terminal transmembrane helices, respectively (44). Crg1 also interacted with Gpa2 and Gpa3 as shown in yeast two-hybrid and protein pull-down assays (58). Mutation of the *CRG1* gene results in enhanced pheromone and mating responses (87). The mating hypersensitive phenotype requires the presence of the Gpa2 or Gpa3 protein (44). A biochemical study demonstrated that Crg1 can accelerate the GTPase activity of Gpa2 (58). Therefore, one of the mechanisms by which the pheromone response may be downregulated by Crg1 is through direct physical contact with Gpa2 to accelerate GTP hydrolysis, resulting in conversion from the active GTP-bound form to the inactive GDP-bound form.

Mutation of *CRG2* also enhances dikaryotic filamentation but reduces cell fusion efficiency and sporulation in mating (94). Crg2 interacted with all three *C. neoformans* Gα proteins, depending on assay conditions. In addition to controlling mating via Gpa2 and Gpa3, Crg2 is also involved in Gpa1-cAMP signaling by negatively regulating cAMP production in response to glucose, suggesting broader roles and regulatory mechanisms of Crg2 in *C. neoformans* (81, 94).

Downstream Targets of the Pheromone Response Pathway

Upon activation of the G-protein complex, a series of kinases are sequentially involved in transducing the signal to downstream transcriptional regulators. Both cell-type-specific and cell-type-nonspecific components are uniquely present in *C. neoformans* to execute their regulatory functions. First, two Ste20/p21-activated kinase (PAK) kinases, Pak1 and Ste20, are found in *C. neoformans*, and mating-type-specific alleles *STE20*α and *STE20***a**, respectively, are present in α and **a** cells (88). Bilateral crosses involving the *ste20*α/*ste20***a** or *pak1* mutant strains showed defects in sexual filamentation. Epistasis analyses indicated that Ste20α functions downstream of the Gβ subunit Gpb1 and upstream of the MAP kinase cascade in the mating pathway (88). Further studies revealed that PAK kinases are critical for establishing polarity during morphogenesis, and these proteins physically interact with the small GTPase Cdc42. Pak1 is required for cell fusion during mating, whereas Ste20 homologs are required to maintain polarity in the heterokaryotic mating filaments (69).

Components of the MAP kinase cascade regulating mating in *C. neoformans* include MAP kinase (MAPK) Cpk1, MAP kinase kinase (MAPKK) Ste7, and MAP kinase kinase kinase (MAPKKK) Ste11 (25, 29). Cpk1 and Ste7 are homologs of *S. cerevisiae* Fus3/Kss1 and Ste7, respectively, and both proteins are necessary for mating and involved in mating partner cell fusion in *C. neoformans* (29). Mating-type-specific alleles of *STE11*α and *STE11***a** are also present in *MAT*α and *MAT***a** cells. Deletion of *STE11*α causes a sterile phenotype (25). These studies confirmed that a conserved MAP kinase cascade plays critical roles during the mating process but that it is uniquely constructed by the cell-type-specific and cell-type-nonspecific components for its regulation in *C. neoformans*. Downstream of the Cpk1 MAP kinase cascade, the conserved transcription factor Ste12 was described by identifying genes that induce hyphae when expressed from a high copy vector (92). *STE12* is also mating-type specific and contains the conserved homeodomain and C_2H_2 zinc finger motifs. *C. neoformans ste12*α and *ste12***a** mutants are fertile but with reduced mating efficiency, suggesting these homologs contribute to mating and that additional factor(s) downstream of Cpk1 might play more dominant roles in mating. Interestingly, *STE12*α is required for monokaryotic fruiting, and both *STE12* alleles regulate virulence in *C. neoformans* serotype D strains (18, 19, 97), and site-specific mutations of Ste12 further indicated that the homeodomain is important for mating and monokaryotic fruiting (20). Interestingly, Ste12 is required for full virulence expression in serotype D strains, whereas it is dispensable for virulence in a serotype A strain (18, 19, 97).

Summary

The *C. neoformans* pheromone response pathway directs important mating and morphogenic events. Although the mating response itself is not required for infection of the host, individual elements of the mating response pathway are involved in sporulation, environmental dispersal and survival, and mammalian pathogenesis. The G-protein signaling pathways that control the mating response share common features with similar pathways in well-described model systems. However, unique features of this pathway include the large size and complexity of the mating-type loci, the relative abundance of mating-type-specific genes, and the application of this pathway to a bipolar mating system in a basidiomycete.

MONOMERIC G PROTEIN SIGNALING: *C. NEOFORMANS* Ras PROTEINS

Introduction

In addition to heterotrimeric G proteins, *C. neoformans* contains several genes that encode monomeric G proteins, including Ras, Rho, and Rac homologs. Two Ras proteins have been identified in *C. neoformans*; however, only one

(Ras1) appears to be relevant for pathogenesis. The *RAS2* gene is expressed at very low levels, and deletion of *RAS2* has no discernable phenotype (90). In contrast, deletion of the *C. neoformans RAS1* gene results in defects in mating, morphogenesis, and stress survival (1).

In *S. cerevisiae*, Ras proteins function upstream of cAMP signal transduction pathways (85). However, Ras proteins play a very minor role in *C. neoformans* cAMP signaling. For example, *ras1* mutants do not exhibit defects in melanin or capsule production, two phenotypes controlled by cAMP activation. Although mating defects occur in *ras1* mutants as well as in mutants in the cAMP pathway, the qualitative and quantitative nature of these sterile phenotypes is quite distinct. In contrast to the cAMP pathway mutants, the *ras1* mutants have a much more severe defect in mating, and this phenotype is only minimally suppressed by exogenous cAMP (1). The cAMP-independent aspects of Ras signaling in *C. neoformans* are therefore quite distinct from models of Ras protein function developed in *S. cerevisiae*. However, similar to developing models in other fungal systems including *S. pombe*, *C. neoformans* Ras proteins regulate two distinct signal transduction pathways to control two equally distinct cellular processes: mating and morphogenesis (16, 68, 70, 73).

Ras Proteins and *C. neoformans* Mating

Early investigations demonstrated that *C. neoformans* Ras1 signaling is required for mating and pheromone response. *ras1* mutants displayed profound defects in cell fusion and subsequent mating events. It was subsequently determined that Ras1 was required for both pheromone production and response. Confrontation assays, which assess the morphogenic response to pheromone, indicated that the *ras1* mutant neither produced abundant pheromone nor responded to pheromone produced by wild-type strains (1, 91).

Transcriptional analysis revealed that the induction of pheromone-responsive genes, such as *GPB1* and *MF*α*1*, was dependent on the presence of intact Ras1 signaling. Moreover, overexpression of either *GPB1* or *MF*α*1* restored mating in the *ras1* deletion strain (91). These data indicate that Ras1 functions during the fusion step of mating, inducing genes involved in pheromone production and response. However, the signaling proteins directly upstream and downstream of Ras1 in this pathway are currently unknown. Although genetic epistasis data places *RAS1* upstream of *GPB1*, it is still unclear whether Ras1 is directly involved in activating the Gpb1 protein, whether it activates the cascade leading to the transcriptional induction of *GPB1*, or whether it induces both processes in a positive feedback loop.

Ras Proteins and *C. neoformans* Morphogenesis

In addition to its mating defects, the *ras1* mutant strain also exhibits temperature-sensitive alterations in growth and morphology. At 37°C, *ras1* mutants arrest as large unbudded cells (1). Other mild cell stresses, such as salt and hypoxic stress, cause a similar cell arrest in the *ras1* mutant strain. In wild-type cells, such nonlethal cell stresses cause a transient depolarization of the actin cytoskeleton, followed by the rapid resumption of the cell budding cycle. In contrast, *ras1* mutant strains lack the ability to recover from this type of stress, and the actin cytoskeleton remains depolarized (70). Because of this temperature-sensitive growth defect, *ras1* mutants are unable to cause disease in animal model systems (1). These data suggest that Ras1 functions in a "morphogenesis pathway" that is distinct from the mating/pheromone response pathway. Recent investigations have demonstrated that this pathway includes several conserved proteins involved in morphogenesis in divergent species, such as Cdc24, Cdc42, and PAK kinases (69, 70).

Cdc24 is a conserved guanine nucleotide exchange factor that activates Cdc42, a major determinant of cell polarity. In *C. neoformans*, Cdc24 is a direct effector of Ras1. *cdc24* mutant strains have a morphogenesis defect identical to *ras1* mutants, but the *cdc24* mutant does not exhibit any mating defects. Cdc24 interacts with wild-type Ras1, as well as a dominant active form of Ras1, in the yeast two-hybrid assay. As expected from a downstream effector, Cdc24 is unable to interact with dominant negative Ras1 proteins. Overexpression of *CDC24* suppresses the *ras1* morphology defect but does not suppress the mating defect of the *ras1* mutant strain (70). Together these data support a model in which Cdc24 acts downstream of Ras1 to regulate morphogenesis signals, but Cdc24 does not control the Ras1 effects on pheromone response and mating.

In contrast to other fungi, *C. neoformans* contains two paralogous copies of the *CDC42* gene (*CDC42* and *CDC420*). Overexpression of either of these two genes suppresses the *ras1* and *cdc24* mutant growth defects at high temperature, suggesting that both Cdc42-like proteins are downstream effectors of the Ras1 morphogenesis pathway (70). In addition to Cdc42, *C. neoformans* also contains Rac proteins, which demonstrate striking homology to Cdc42 and regulate morphogenesis in many species (30). *S. cerevisiae* and *S. pombe* each contain a single Cdc42 protein and no Rac homologs. The *C. neoformans RAC1* gene was initially isolated as a high-copy suppressor of the *ras1* mutant temperature-sensitive growth defect. However, subsequent analysis of strains deleted for *rac1* demonstrates that the primary role of Rac1 is likely related to hyphal morphogenesis (86).

PAK kinases often function as direct effectors for Cdc42 and Rac proteins. The mating-specific PAK kinase protein Ste20 appears to function downstream of Ras1 in the morphogenesis pathway. Overexpression of *STE20* suppresses the morphogenesis defects of both *ras1* and *cdc24* mutant strains (70). In contrast to *CDC24*, *STE20* overexpression also restores mating to the *ras1* mutant strain, suggesting that there is cross-talk between the morphogenesis and mating arms of Ras1 signaling. The *ste20* mutants also exhibit a mating defect that, unlike the *ras1* mutant mating defect, occurs postfusion. In the *ste20* mutant strain, fusion occurs normally, but the mating filaments are unable to maintain proper tip polarity when *STE20* is deleted from both mating partners (69).

Ras Signaling Specificity and Lipid Modification

The C termini of all Ras proteins are posttranslationally modified by the addition of lipid groups to cysteine residues. These lipid modifications direct Ras proteins to distinct subcellular locations. All Ras proteins are cleaved at the C terminus and farnesylated. This modification is irreversible and absolutely required for Ras protein function (13). In addition, some Ras proteins, including *C. neoformans* Ras1, are palmitoylated on cysteine residues located immediately upstream of the farnesylated cysteine. In addition to being a reversible modification, the cycling between palmitoylated and nonpalmitoylated forms is important for certain Ras functions.

Distinct from many other well-characterized palmitoylated Ras proteins, *C. neoformans* Ras1 has two putative palmitoylation sites, C203 and C204. Recently, this double palmitoylation motif was found to be conserved in a subset of Ras proteins from filamentous fungi. In addition, both cysteine residues were documented to be true targets of palmitoylation in vitro, and the presence of either cysteine supports both palmitoylation and complete Ras1 protein function (68).

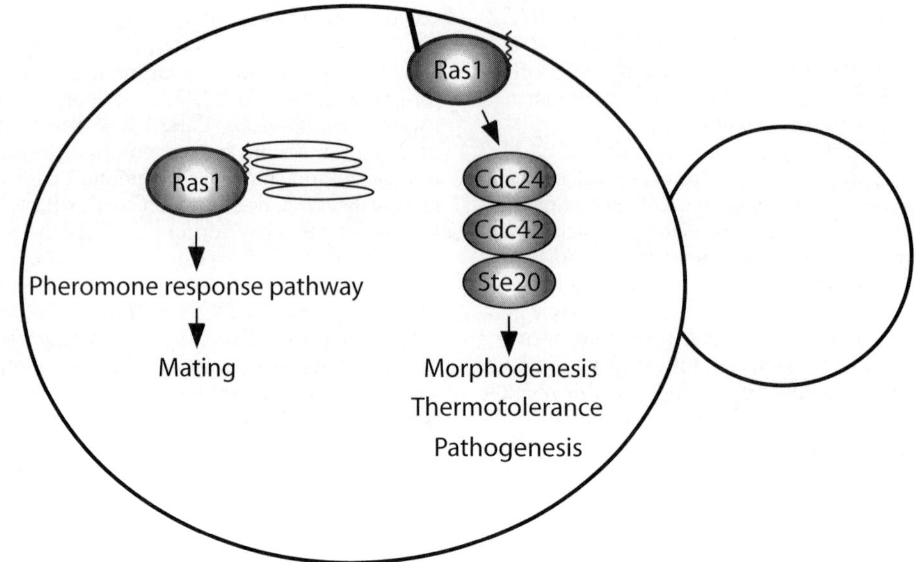

FIGURE 4 Model of the *C. neoformans* Ras1 pathway. The downstream signaling specificity of the Ras1 protein is largely determined by its subcellular localization. Ras1 farnesylation (thin wavy line) is required for all protein functions, and this posttranslational modification is sufficient to direct the protein to the endomembranes and support mating. In contrast, Ras1 palmitoylation (thick straight line) is required to localize Ras1 to the plasma membrane, and this modification is required to activate the Ras1 morphogenesis pathway.

Palmitoylation specifies the differential membrane localization of Ras proteins. After farnesylation, Ras proteins are palmitoylated at the endoplasmic reticulum (ER) by protein acyl transferases. Once palmitoylated, Ras proteins travel from the ER to the plasma membrane through the secretory system. Depalmitoylation is mediated by acylprotein thioesterases, resulting in recycling of the Ras protein from the plasma membrane. In *S. pombe*, the regulation of Ras palmitoylation has been demonstrated to mediate pathway specificity: plasma membrane-localized Ras1 mediates mating, while ER-localized Ras1 mediates morphogenesis (16). Recent data indicate that palmitoylation of *C. neoformans* Ras1 also mediates its pathway specificity. In contrast to *S. pombe*, *C. neoformans* Ras1 protein palmitoylation and plasma membrane localization are important for activating the thermotolerance/morphogenesis pathway, and they are not required for mating (68). Therefore, controlling differential subcellular localization is a conserved mechanism by which fungi direct the signaling specificity of Ras proteins (Fig. 4). However, different species have "rewired" these conserved Ras pathways to control the specific cellular outputs required for their survival.

Summary

Microorganisms use conserved signal transduction pathways to control fundamental cellular processes such as morphogenesis, mating, and response to cell stress. Microbial pathogens such as *C. neoformans* have adapted these pathways to regulate specific phenotypes that are required for survival in the infected host. Many of these pathways use either monomeric or heterotrimeric G proteins to internalize extracellular signals. Activation of these conserved signaling pathways allows *C. neoformans* to develop a coordinated cellular response to the host, ultimately leading to survival within this harsh environment and pathogenesis.

Research in the Wang lab is supported in part by National Institutes of Health grant AI054958. The Alspaugh laboratory is supported by NIH grants AI063242 and AI050128.

REFERENCES

1. **Alspaugh, J. A., L. M. Cavallo, J. R. Perfect, and J. Heitman.** 2000. RAS1 regulates filamentation, mating and growth at high temperature of *Cryptococcus neoformans*. *Mol. Microbiol.* **36:**352–365.
2. **Alspaugh, J. A., R. C. Davidson, and J. Heitman.** 2000. Morphogenesis of *Cryptococcus neoformans*, p. 217–238. *In* J. F. Ernst, and A. Schmidt (ed.), *Dimorphism in Human Pathogenic and Apathogenic Yeasts*, Contributions to Microbiology, vol. 5. Karger, Basel, Switzerland.
3. **Alspaugh, J. A., J. R. Perfect, and J. Heitman.** 1997. *Cryptococcus neoformans* mating and virulence are regulated by the G-protein alpha subunit GPA1 and cAMP. *Genes Dev.* **11:**3206–3217.
4. **Alspaugh, J. A., R. Pukkila-Worley, T. Harashima, L. M. Cavallo, D. Funnell, G. M. Cox, J. R. Perfect, J. W. Kronstad, and J. Heitman.** 2002. Adenylyl cyclase functions downstream of the G-alpha protein Gpa1 and controls mating and pathogenicity of *Cryptococcus neoformans*. *Eukaryot. Cell* **1:**75–84.
5. **Bahn, Y. S., J. K. Hicks, S. S. Giles, G. M. Cox, and J. Heitman.** 2004. Adenylyl cyclase-associated protein Aca1 regulates virulence and differentiation of *Cryptococcus neoformans* via the cyclic AMP-protein kinase A cascade. *Eukaryot. Cell* **3:**1476–1491.
6. **Bahn, Y. S., C. Xue, A. Idnurm, J. C. Rutherford, J. Heitman, and M. E. Cardenas.** 2007. Sensing the environment: lessons from fungi. *Nat. Rev.* **5:**57–69.
7. **Bakkeren, G., J. Kamper, and J. Schirawski.** 2008. Sex in smut fungi: structure, function and evolution of mating-type complexes. *Fungal Genet. Biol.* **45**(Suppl. 1)**:**S15–S21.

8. **Ballon, D. R., P. L. Flanary, D. P. Gladue, J. B. Konopka, H. G. Dohlman, and J. Thorner.** 2006. DEP-domain-mediated regulation of GPCR signaling responses. *Cell* **126:**1079–1093.
9. **Bardwell, L.** 2004. A walk-through of the yeast mating pheromone response pathway. *Peptides* **25:**1465–1476.
10. **Batlle, M., A. Lu, D. A. Green, Y. Xue, and J. P. Hirsch.** 2003. Krh1p and Krh2p act downstream of the Gpa2p G(alpha) subunit to negatively regulate haploid invasive growth. *J. Cell Sci.* **116:**701–710.
11. **Berkey, C. D., V. K. Vyas, and M. Carlson.** 2004. Nrg1 and Nrg2 transcriptional repressors are differently regulated in response to carbon source. *Eukaryot. Cell* **3:**311–317.
12. **Braun, B. R., D. Kadosh, and A. D. Johnson.** 2001. NRG1, a repressor of filamentous growth in *C. albicans*, is down-regulated during filament induction. *EMBO J.* **20:**4753–4761.
13. **Broach, J. R., and R. J. Deschenes.** 1990. The functions of *RAS* genes in *Saccharomyces cerevisiae*. *Adv. Cancer Res.* **54:**79–138.
14. **Burkholder, A. C., and L. H. Hartwell.** 1985. The yeast alpha-factor receptor: structural properties deduced from the sequence of the STE2 gene. *Nucleic Acids Res.* **13:**8463–8475.
15. **Casselton, L. A.** 2002. Mate recognition in fungi. *Heredity* **88:**142–147.
16. **Chang, E. C., and M. R. Philips.** 2006. Spatial segregation of Ras signaling: new evidence from fission yeast. *Cell Cycle* **5:**1936–1939.
17. **Chang, Y. C., G. F. Miller, and K. J. Kwon-Chung.** 2003. Importance of a developmentally regulated pheromone receptor of *Cryptococcus neoformans* for virulence. *Infect. Immun.* **71:**4953–4960.
18. **Chang, Y. C., L. A. Penoyer, and K. J. Kwon-Chung.** 2001. The second *STE12* homologue of *Cryptococcus neoformans* is MATa-specific and plays an important role in virulence. *Proc. Natl. Acad. Sci. USA* **98:**3258–3263.
19. **Chang, Y. C., B. L. Wickes, G. F. Miller, L. A. Penoyer, and K. J. Kwon-Chung.** 2000. *Cryptococcus neoformans* Ste12alpha regulates virulence but is not essential for mating. *J. Exp. Med.* **191:**871–882.
20. **Chang, Y. C., L. C. Wright, R. L. Tscharke, T. C. Sorrell, C. F. Wilson, and K. J. Kwon-Chung.** 2004. Regulatory roles for the homeodomain and C2H2 zinc finger regions of *Cryptococcus neoformans* Ste12alphap. *Mol. Microbiol.* **53:**1385–1396.
21. **Choi, W., and R. A. Dean.** 1997. The adenylate cyclase gene MAC1 of *Magnaporthe grisea* controls appressorium formation and other aspects of growth and development. *Plant Cell* **9:**1973–1983.
22. **Chung, S., M. Karos, Y. C. Chang, J. Lukszo, B. L. Wickes, and K. J. Kwon-Chung.** 2002. Molecular analysis of *CPRalpha*, a MATalpha-specific pheromone receptor gene of *Cryptococcus neoformans*. *Eukaryot. Cell* **1:**432–439.
23. **Clapham, D. E., and E. J. Neer.** 1993. New roles for G-protein beta gamma-dimers in transmembrane signalling. *Nature* **365:**403–406.
24. **Clapham, D. E., and E. J. Neer.** 1997. G protein beta gamma subunits. *Annu. Rev. Pharmacol. Toxicol.* **37:**167–203.
25. **Clarke, D. L., G. L. Woodlee, C. M. McClelland, T. S. Seymour, and B. L. Wickes.** 2001. The *Cryptococcus neoformans* STE11alpha gene is similar to other fungal mitogen-activated protein kinase kinase kinase (MAPKKK) genes but is mating type specific. *Mol. Microbiol.* **40:**200–213.
26. **Cole, G. M., and S. I. Reed.** 1992. Pheromone-induced phosphorylation of a G protein β subunit in *S. cerevisiae* is associated with an adaptive response to mating pheromone. *Cell* **64:**703–716.
27. **Cramer, K. L., Q. D. Gerrald, C. B. Nichols, M. S. Price, and J. A. Alspaugh.** 2006. The transcription factor Nrg1 mediates capsule, stress response, and pathogenesis in *Cryptococcus neoformans*. *Eukaryot. Cell* **5:**1147–1168.
28. **Davidson, R. C., T. D. Moore, A. R. Odom, and J. Heitman.** 2000. Characterization of the MFalpha pheromone of the human fungal pathogen *Cryptococcus neoformans*. *Mol. Microbiol.* **38:**1017–1026.
29. **Davidson, R. C., C. B. Nichols, G. M. Cox, J. R. Perfect, and J. Heitman.** 2003. A MAP kinase cascade composed of cell type specific and non-specific elements controls mating and differentiation of the fungal pathogen *Cryptococcus neoformans*. *Mol. Microbiol.* **49:**469–485.
30. **Didsbury, J., R. F. Weber, G. M. Bokoch, T. Evans, and R. Snyderman.** 1989. rac, a novel ras-related family of proteins that are botulinum toxin substrates. *J. Biol. Chem.* **264:**16378–16382.
31. **Dohlman, H. G.** 2002. G proteins and pheromone signaling. *Annu. Rev. Physiol.* **64:**129–152.
32. **Dohlman, H. G., and J. W. Thorner.** 2001. Regulation of G protein-initiated signal transduction in yeast: paradigms and principles. *Annu. Rev. Biochem.* **70:**703–754.
33. **D'Souza, C. A., J. A. Alspaugh, C. Yue, T. Harashima, G. M. Cox, J. R. Perfect, and J. Heitman.** 2001. Cyclic AMP-dependent protein kinase controls virulence of the fungal pathogen *Cryptococcus neoformans*. *Mol. Cell. Biol.* **21:**3179–3191.
34. **Fraser, J. A., S. Diezmann, R. L. Subaran, A. Allen, K. B. Lengeler, F. S. Dietrich, and J. Heitman.** 2004. Convergent evolution of chromosomal sex-determining regions in the animal and fungal kingdoms. *PLoS Biol.* **2:**e384.
35. **Gilman, A. G.** 1987. G-proteins: transducers of receptor-generated signals. *Annu. Rev. Biochem.* **56:**615–649.
36. **Gold, S. E., S. M. Brogdon, M. E. Mayorga, and J. W. Kronstad.** 1997. The *Ustilago maydis* regulatory subunit of a cAMP-dependent protein kinase is required for gall formation in maize. *Plant Cell* **9:**1585–1594.
37. **Hagen, D. C., G. McCaffrey, and G. F. Sprague, Jr.** 1986. Evidence the yeast STE3 gene encodes a receptor for the peptide pheromone a-factor: gene sequence and implications for the structure of the presumed receptor. *Proc. Natl. Acad. Sci. USA* **83:**1418–1422.
38. **Harashima, T., and J. Heitman.** 2002. The Galpha protein Gpa2 controls yeast differentiation by interacting with kelch repeat proteins that mimic Gbeta subunits. *Mol. Cell* **10:**163–173.
39. **Harashima, T., and J. Heitman.** 2005. Galpha subunit Gpa2 recruits kelch repeat subunits that inhibit receptor-G protein coupling during cAMP-induced dimorphic transitions in *Saccharomyces cerevisiae*. *Mol. Biol. Cell* **16:**4557–4571.
40. **Hicks, J. K., Y. S. Bahn, and J. Heitman.** 2005. Pde1 phosphodiesterase modulates cyclic AMP levels through a protein kinase A-mediated negative feedback loop in *Cryptococcus neoformans*. *Eukaryot. Cell* **4:**1971–1981.
41. **Higgins, J. B., and P. J. Casey.** 1996. The role of prenylation in G-protein assembly and function. *Cell. Signal.* **8:**433–437.
42. **Hoffman, C. S.** 2007. Propping up our knowledge of G protein signaling pathways: diverse functions of putative noncanonical Gbeta subunits in fungi. *Sci. STKE* **2007:**pe3.
43. **Hsueh, Y. P., and W. C. Shen.** 2005. A homolog of Ste6, the **a**-factor transporter in *Saccharomyces cerevisiae*, is required for mating but not for monokaryotic fruiting in *Cryptococcus neoformans*. *Eukaryot. Cell* **4:**147–155.
44. **Hsueh, Y. P., C. Xue, and J. Heitman.** 2007. G protein signaling governing cell fate decisions involves opposing Galpha subunits in *Cryptococcus neoformans*. *Mol. Biol. Cell* **18:**3237–3249.

45. **Hsueh, Y. P., C. Xue, and J. Heitman.** 2009. A constitutively active GPCR governs morphogenic transitions in *Cryptococcus neoformans*. *EMBO J.* **28:**1220–1233.
46. **Hurowitz, E. H., J. M. Melnyk, Y. J. Chen, H. Kouros-Mehr, M. I. Simon, and H. Shizuya.** 2000. Genomic characterization of the human heterotrimeric G protein alpha, beta, and gamma subunit genes. *DNA Res.* **7:**111–120.
47. **Ivey, F. D., and C. S. Hoffman.** 2002. Pseudostructural inhibitors of G protein signaling during development. *Dev. Cell* **3:**154–155.
48. **Kozubowski, L., S. C. Lee, and J. Heitman.** 2009. Signalling pathways in the pathogenesis of *Cryptococcus*. *Cell. Microbiol.* **11:**370–380.
49. **Kraakman, L., K. Lemaire, P. Ma, A. W. Teunissen, M. C. Donaton, P. Van Dijck, J. Winderickx, J. H. de Winde, and J. M. Thevelein.** 1999. A *Saccharomyces cerevisiae* G-protein coupled receptor, Gpr1, is specifically required for glucose activation of the cAMP pathway during the transition to growth on glucose. *Mol. Microbiol.* **32:**1002–1012.
50. **Kronstad, J. W., and C. Staben.** 1997. Mating type in filamentous fungi. *Annu. Rev. Genet.* **31:**245–276.
51. **Kwon-Chung, K. J.** 1976. Morphogenesis of *Filobasidiella neoformans*, the sexual state of *Cryptococcus neoformans*. *Mycologia* **68:**821–833.
52. **Kwon-Chung, K. J., and J. E. Bennett.** 1978. Distribution of α and **a** mating types of *Cryptococcus neoformans* among natural and clinical isolates. *Am. J. Epidemiol.* **108:**337–340.
53. **Kwon-Chung, K. J., J. C. Edman, and B. L. Wickes.** 1992. Genetic association of mating types and virulence in *Cryptococcus neoformans*. *Infect. Immun.* **60:**602–605.
54. **Landry, S., and C. S. Hoffman.** 2001. The git5 Gβ and git11 Gγ form an atypical Gβγ dimer acting in the fission yeast glucose/cAMP pathway. *Genetics* **157:**1159–1168.
55. **Landry, S., M. T. Pettit, E. Apolinario, and C. S. Hoffman.** 2000. The fission yeast *git5* gene encodes a Gβ subunit required for glucose-triggered adenylate cyclase activation. *Genetics* **154:**1463–1471.
56. **Lemaire, K., S. Van de Velde, P. Van Dijck, and J. M. Thevelein.** 2004. Glucose and sucrose act as agonist and mannose as antagonist ligands of the G protein-coupled receptor Gpr1 in the yeast *Saccharomyces cerevisiae*. *Mol. Cell* **16:**293–299.
57. **Lengeler, K. B., D. S. Fox, J. A. Fraser, A. Allen, K. Forrester, F. S. Dietrich, and J. Heitman.** 2002. Mating-type locus of *Cryptococcus neoformans*: a step in the evolution of sex chromosomes. *Eukaryot. Cell* **1:**704–718.
58. **Li, L., G. Shen, Z. G. Zhang, Y. L. Wang, J. K. Thompson, and P. Wang.** 2007. Canonical heterotrimeric G proteins regulating mating and virulence of *Cryptococcus neoformans*. *Mol. Biol. Cell* **18:**4201–4209.
59. **Liebmann, B., M. Muller, A. Braun, and A. A. Brakhage.** 2004. The cyclic AMP-dependent protein kinase A network regulates development and virulence in *Aspergillus fumigatus*. *Infect. Immun.* **72:**5193–5203.
60. **Liu, O. W., C. D. Chun, E. D. Chow, C. Chen, H. D. Madhani, and S. M. Noble.** 2008. Systematic genetic analysis of virulence in the human fungal pathogen *Cryptococcus neoformans*. *Cell* **135:**174–188.
61. **Lorenz, M. C., and J. Heitman.** 1997. Yeast pseudohyphal growth is regulated by GPA2, a G protein α homolog. *EMBO J.* **16:**7008–7018.
62. **Lorenz, M. C., X. Pan, T. Harashima, M. E. Cardenas, Y. Xue, J. P. Hirsch, and J. Heitman.** 2000. The G protein-coupled receptor Gpr1 is a nutrient sensor that regulates pseudohyphal differentiation in *Saccharomyces cerevisiae*. *Genetics* **154:**609–622.
63. **Ma, P., S. Wera, P. Van Dijck, and J. M. Thevelein.** 1999. The *PDE1*-encoded low-affinity phosphodiesterase in the yeast *Saccharomyces cerevisiae* has a specific function in controlling agonist-induced cAMP signaling. *Mol. Biol. Cell* **10:**91–104.
64. **Maidan, M. M., L. De Rop, J. Serneels, S. Exler, S. Rupp, H. Tournu, J. M. Thevelein, and P. Van Dijck.** 2005. The G protein-coupled receptor Gpr1 and the Galpha protein Gpa2 act through the cAMP-protein kinase A pathway to induce morphogenesis in *Candida albicans*. *Mol. Biol. Cell* **16:**1971–1986.
65. **McClelland, C. M., J. Fu, G. L. Woodlee, T. S. Seymour, and B. L. Wickes.** 2002. Isolation and characterization of the *Cryptococcus neoformans* MATa pheromone gene. *Genetics* **160:**935–947.
66. **Moore, T. D., and J. C. Edman.** 1993. The alpha-mating type locus of *Cryptococcus neoformans* contains a peptide pheromone gene. *Mol. Cell. Biol.* **13:**1962–1970.
67. **Mylonakis, E., F. M. Ausubel, J. R. Perfect, J. Heitman, and S. B. Calderwood.** 2002. Killing of *Caenorhabditis elegans* by *Cryptococcus neoformans* as a model of yeast pathogenesis. *Proc. Natl. Acad. Sci. USA* **99:**15675–15680.
68. **Nichols, C. B., J. Ferreyra, E. R. Ballou, and J. A. Alspaugh.** 2009. Subcellular localization directs signaling specificity of the *Cryptococcus neoformans* Ras1 protein. *Eukaryot. Cell* **8:**181–189.
69. **Nichols, C. B., J. A. Fraser, and J. Heitman.** 2004. PAK kinases Ste20 and Pak1 govern cell polarity at different stages of mating in *Cryptococcus neoformans*. *Mol. Biol. Cell* **15:**4476–4489.
70. **Nichols, C. B., Z. Perfect, and J. A. Alspaugh.** 2007. A Ras1-Cdc24 signal transduction pathway mediates thermotolerance in the fungal pathogen *Cryptococcus neoformans*. *Mol. Microbiol.* **63:**1118–1130.
71. **Nielsen, K., G. M. Cox, P. Wang, D. L. Toffaletti, J. R. Perfect, and J. Heitman.** 2003. Sexual cycle of *Cryptococcus neoformans* var. *grubii* and virulence of congenic **a** and alpha isolates. *Infect. Immun.* **71:**4831–4841.
72. **Nielsen, K., R. E. Marra, F. Hagen, T. Boekhout, T. G. Mitchell, G. M. Cox, and J. Heitman.** 2005. Interaction between genetic background and the mating-type locus in *Cryptococcus neoformans* virulence potential. *Genetics* **171:**975–983.
73. **Nielsen, O., J. Davey, and R. Egel.** 1992. The *ras1* function of *Schizosaccharomyces pombe* mediates pheromone-induced transcprition. *EMBO J.* **11:**1391–1395.
74. **Nocero, M., T. Isshiki, M. Yamamoto, and C. S. Hoffman.** 1994. Glucose repression of *fbp1* transcription in *Schizosaccharomyces pombe* is partially regulated by adenylate cyclase activation by a G protein α subunit encoded by *gpa2* (*git8*). *Genetics* **138:**39–45.
75. **Palmer, D. A., J. K. Thompson, L. Li, A. Prat, and P. Wang.** 2006. Gib2, a novel Gbeta-like/RACK1 homolog, functions as a Gbeta subunit in cAMP signaling and is essential in *Cryptococcus neoformans*. *J. Biol. Chem.* **281:**32596–32605.
76. **Peeters, T., W. Louwet, R. Gelade, D. Nauwelaers, J. M. Thevelein, and M. Versele.** 2006. Kelch-repeat proteins interacting with the Galpha protein Gpa2 bypass adenylate cyclase for direct regulation of protein kinase A in yeast. *Proc. Natl. Acad. Sci. USA* **103:**13034–13039.
77. **Pukkila-Worley, R., and J. A. Alspaugh.** 2004. Cyclic AMP signaling in *Cryptococcus neoformans*. *FEMS Yeast Res.* **4:**361–367.
78. **Pukkila-Worley, R., Q. D. Gerrald, P. R. Kraus, M. J. Boily, M. J. Davis, S. S. Giles, G. M. Cox, J. Heitman, and J. A. Alspaugh.** 2005. Transcriptional network of multiple capsule and melanin genes governed by the *Cryptococcus neoformans* cyclic AMP cascade. *Eukaryot. Cell* **4:**190–201.
79. **Rocha, C. R., K. Schroppel, D. Harcus, A. Marcil, D. Dignard, B. N. Taylor, D. Y. Thomas, M. Whiteway, and**

E. Leberer. 2001. Signaling through adenylyl cyclase is essential for hyphal growth and virulence in the pathogenic fungus *Candida albicans*. *Mol. Biol. Cell* **12:**3631–3643.

80. **Rolland, F., J. Winderickx, and J. M. Thevelein.** 2002. Glucose-sensing and -signalling mechanisms in yeast. *FEMS Yeast Res.* **2:**183–201.
81. **Shen, G., Y. L. Wang, A. Whittington, L. Li, and P. Wang.** 2008. The RGS protein Crg2 regulates pheromone and cyclic AMP signaling in *Cryptococcus neoformans*. *Eukaryot. Cell* **7:**1540–1548.
82. **Shen, W. C., R. C. Davidson, G. M. Cox, and J. Heitman.** 2002. Pheromones stimulate mating and differentiation via paracrine and autocrine signaling in *Cryptococcus neoformans*. *Eukaryot. Cell* **1:**366–377.
83. **Smrcka, A. V.** 2008. G protein betagamma subunits: central mediators of G protein-coupled receptor signaling. *Cell. Mol. Life Sci.* **65:**2191–2214.
84. **Sternweis, P. C.** 1994. The active role of βγ in signal transduction. *Curr. Opin. Cell. Biol.* **6:**198–203.
85. **Toda, T., I. Uno, T. Ishikawa, S. Powers, T. Kataoka, D. Broek, S. Cameron, J. Broach, K. Matsumoto, and M. Wigler.** 1985. In yeast, *RAS* proteins are controlling elements of adenylate cyclase. *Cell* **40:**27–36.
86. **Vallim, M. A., C. B. Nichols, L. Fernandes, K. L. Cramer, and J. A. Alspaugh.** 2005. A Rac homolog functions downstream of Ras1 to control hyphal differentiation and high-temperature growth in the pathogenic fungus *Cryptococcus neoformans*. *Eukaryot. Cell* **4:**1066–1078.
87. **Wang, P., J. Cutler, J. King, and D. Palmer.** 2004. Mutation of the regulator of G protein signaling Crg1 increases virulence in *Cryptococcus neoformans*. *Eukaryot. Cell* **3:**1028–1035.
88. **Wang, P., C. B. Nichols, K. B. Lengeler, M. E. Cardenas, G. M. Cox, J. R. Perfect, and J. Heitman.** 2002. Mating-type-specific and nonspecific PAK kinases play shared and divergent roles in *Cryptococcus neoformans*. *Eukaryot. Cell* **1:**257–272.
89. **Wang, P., J. R. Perfect, and J. Heitman.** 2000. The G-protein beta subunit GPB1 is required for mating and haploid fruiting in *Cryptococcus neoformans*. *Mol. Cell. Biol.* **20:**352–362.
90. **Waugh, M. S., C. B. Nichols, C. M. DeCesare, G. M. Cox, J. Heitman, and J. A. Alspaugh.** 2002. Ras1 and Ras2 contribute shared and unique roles in physiology and virulence of *Cryptococcus neoformans*. *Microbiology* **148:**191–201.
91. **Waugh, M. S., M. A. Vallim, J. Heitman, and J. A. Alspaugh.** 2003. Ras1 controls pheromone expression and response during mating in *Cryptococcus neoformans*. *Fungal Genet. Biol.* **38:**110–121.
92. **Wickes, B. L., U. Edman, and J. C. Edman.** 1997. The *Cryptococcus neoformans STE12alpha* gene: a putative *Saccharomyces cerevisiae STE12* homologue that is mating type specific. *Mol. Microbiol.* **26:**951–960.
93. **Xue, C., Y. S. Bahn, G. M. Cox, and J. Heitman.** 2006. G protein-coupled receptor Gpr4 senses amino acids and activates the cAMP-PKA pathway in *Cryptococcus neoformans*. *Mol. Biol. Cell* **17:**667–679.
94. **Xue, C., Y. P. Hsueh, L. Chen, and J. Heitman.** 2008. The RGS protein Crg2 regulates both pheromone and cAMP signalling in *Cryptococcus neoformans*. *Mol. Microbiol.* **70:**379–395.
95. **Xue, C., Y. P. Hsueh, and J. Heitman.** 2008. Magnificent seven: roles of G protein-coupled receptors in extracellular sensing in fungi. *FEMS Microbiol. Rev.* **32:** 1010–1032.
96. **Xue, Y., M. Batlle, and J. P. Hirsch.** 1998. *GPR1* encodes a putative G protein-coupled receptor that associates with the Gpa2p Gα subunit and functions in a Ras-independent pathway. *EMBO J.* **17:**1996–2007.
97. **Yue, C., L. M. Cavallo, J. A. Alspaugh, P. Wang, G. M. Cox, J. R. Perfect, and J. Heitman.** 1999. The *STE12alpha* homolog is required for haploid filamentation but largely dispensable for mating and virulence in *Cryptococcus neoformans*. *Genetics* **153:**1601–1615.
98. **Zeller, C. E., S. C. Parnell, and H. G. Dohlman.** 2007. The RACK1 ortholog Asc1 functions as a G-protein beta subunit coupled to glucose responsiveness in yeast. *J. Biol. Chem.* **282:**25168–25176.

Cryptococcus: From Human Pathogen to Model Yeast
Edited by J. Heitman et al.

13

A Role for Mating in Cryptococcal Virulence

KIRSTEN NIELSEN AND KYUNG J. KWON-CHUNG

Cryptococcus neoformans and *Cryptococcus gattii*, the two etiologic agents of cryptococcosis, belong to a group of about 25% of the Basidiomycota that have heterothallic bipolar mating systems (1). Strains of both species occur as either mating type (MAT) *MAT***a** or *MAT*α idiomorphs, and they propagate in a haploid state even though a small fraction of the population exist as diploids or aneuploids (41). Upon contact of *MAT***a** with *MAT*α cells on appropriate substrata, the two mating-type cells fuse (Fig. 1) and undergo the sexual life cycle to produce meiotic products termed basidiospores. These spores can be readily isolated and subjected to classical genetic analysis, and have recently been used in virulence studies. Although laboratory analysis of progeny resulting from mating between *MAT***a** and *MAT*α strains show Mendelian inheritance of the *MAT* loci, occurrence of the two mating types in nature as well as in clinical sources is severely skewed toward the *MAT*α type. Furthermore, genetic analysis has linked the mating type and genes associated with mating, such as pheromone receptor genes, to virulence in *C. neoformans*. In light of the importance of mating type in epidemiology, ecology, and pathogenicity, the mating system in *C. neoformans* has received considerable attention since it was first discovered three decades ago. Consequently, significant progress has been made in our understanding of the molecular and genetic network that controls mating in *C. neoformans*. While it is clear that the agents of cryptococcosis do mate, the effect that mating has on pathogenesis and virulence still remains unclear. In this chapter, we discuss the effect of mating type, pheromone signaling, spore production, and ploidy on cryptococcal virulence.

MATING TYPE AND VIRULENCE

Cryptococcus may have the most unusual mating-type locus of all the human-pathogenic fungi. The majority of basidiomycetous fungi have tetrapolar mating systems with two unlinked pheromone and sex-determining loci in which strains must differ at both loci for mating to succeed. In *Cryptococcus*, two traditional mating loci appear to have fused to generate one very large mating-type (*MAT*) locus that contains not only the sex-determining homeodomain transcription factors, pheromones, and pheromone receptors, but also genes from many other functional categories including several essential genes (23, 33, 34, 48). Due to the large size and complexity of the *MAT* locus, simple gene exchange experiments may not be sufficient to elucidate the role of this large genomic region with regard to virulence. Instead, epidemiological studies and congenic strains have been used to determine the role of this genomic region in virulence.

Epidemiological studies show that the vast majority of clinical isolates are *MAT*α type. In *C. neoformans* var. *neoformans*, *MAT*α strains outnumber *MAT***a** strains by a ratio of almost 45:1 in environmental isolates and by almost 30:1 in clinical isolates (44). Thus, *MAT***a** strains only account for less than 2% of the var. *neoformans* population. This predominance of *MAT*α strains is even more dramatic in *C. neoformans* var. *grubii*, where only 17 out of several thousand strains have been identified to date as *MAT***a** strains (49, 56, 63, 79, 80). Interestingly, many of these *MAT***a** strains belong to the unique VNB (var. *neoformans* Botswana molecular type) subpopulation, where *MAT***a** strains account for 25% of the population (55, 56). In *C. gattii*, the distribution of the two mating types among clinical isolates has also indicated a predominance of the *MAT*α type. Of the 25 clinical *C. gattii* strains that had been isolated in the United States prior to the advent of the AIDS epidemic, 21 were *MAT*α and 4 were *MAT***a**, while all isolates obtained from the recent cryptococcosis outbreak on Vancouver Island are *MAT*α strains (24, 25, 39, 42).

In *C. neoformans* var. *neoformans*, a comparison of the levels of virulence in heterogeneous populations revealed that *MAT*α strains are more virulent at lower inoculum levels than *MAT***a** strains (44). While *MAT***a** strains of var. *grubii* are exceptionally rare and have only recently been identified, a similar trend holds true for this variety. Barchiesi et al. observed that nonisogenic strains of var. *grubii* *MAT***a** are less virulent than *MAT*α strains (2). These data suggest that global populations of *MAT*α strains may be more virulent than *MAT***a** strains.

To directly examine the role of mating type in virulence, congenic strains in both varieties *neoformans* and *grubii* have been generated (43, 59, 60). The var. *neoformans* congenic

Kirsten Nielsen, Department of Microbiology, Medical School, University of Minnesota, Minneapolis, MN 55455. **Kyung J. Kwon-Chung,** Molecular Microbiology Section, Laboratory of Clinical Infectious Diseases, National Institute of Allergy and Infectious Diseases, NIH, Bethesda, MD 20892.

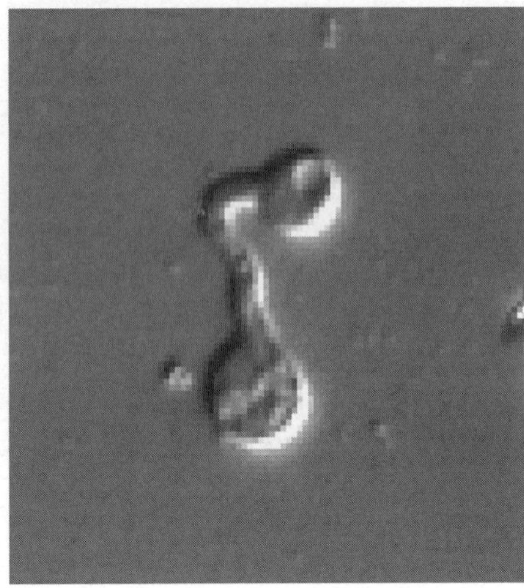

FIGURE 1 Conjugation between a large *MAT***a** cell (lower) and a smaller *MAT*α cell on V8 juice agar media at 30°C.

strains were generated by crossing a highly virulent clinical isolate, NIH12 (*MAT*α), with a less virulent environmental isolate, NIH433 (*MAT***a**). A progeny from this cross, B-3501 (*MAT*α), was then backcrossed into the *MAT***a** background (B-3502) to generate the congenic strains JEC20 (*MAT***a**) and JEC21 (*MAT*α) (30, 43). The *MAT*α strain JEC21 was more virulent than the *MAT***a** strain in the murine systemic model of cryptococcosis as well as in the heterologous host *Caenorhabditis elegans* system (43, 57). Additional congenic strains were also generated in the NIH433 (*MAT***a**) and B-3501 (*MAT*α) genetic backgrounds (62). Similar to the JEC20/21 congenic strains, the congenic strains in the NIH433 genetic background showed that the *MAT*α strain was more virulent than the *MAT***a** strain. However, the B-3501 congenic strains showed no difference in virulence between the two mating types. The same result was obtained with a B-3501 congenic pair constructed using the JEC20 and B-3501 backgrounds (K. J. Kwon-Chung, unpublished data). Congenic strains have also been generated in the var. *grubii* H99 genetic background (60). In this case the two parental mating types were clinical isolates. The strain H99 is a clinical isolate from North Carolina and is the var. *grubii* genome reference strain (76). The *MAT***a** type parental strain, 125.91, is a clinical isolate from Dar Es Salaam, Tanzania (49). Interestingly, the two parental mating-type donors, H99 (*MAT*α) and 125.91 (*MAT***a**), were unable to directly mate. Instead, the 125.91 strain had to first be mated with another clinical *MAT*α strain, 8-1 (60). One of the progeny from this cross was able to mate with an H99 *crg1* mutant strain, and subsequent *MAT***a** progeny were backcrossed into the wild-type H99 *MAT*α genetic background. The resulting congenic strains KN99**a** and KN99α had equivalent virulence in murine inhalation and tail vein injection models, in the rabbit meningitis model, as well as in the heterologous host *C. elegans* and amoebae systems (59, 60).

The role of mating type in pathogenicity of *C. neoformans* var. *grubii* has been further characterized by examining the colonization of various organs by the KN99**a**/α congenic strains in animal models (60, 64). When infected individually, both mating types accumulate equivalently in all organs examined at early, intermediate, and late stages of the infection (60). However, when the two mating types were coinfected into the same animal simultaneously, the *MAT*α strains preferentially accumulated in the central nervous system (59).

Taken together, the congenic strain studies in both varieties *neoformans* and *grubii* suggest that virulence potential of mating type can differ with the genetic background of the strain. These results suggest that genes in the *MAT* locus may interact with other genes in the genome to affect virulence. Since *MAT*α and *MAT***a** congenic strains have not been generated yet in *C. gattii*, the role of mating type has not been directly studied in this species. However, given that *MAT*α strains are also predominant in this species, results similar to *C. neoformans* might be expected. Interestingly, in all cases where a difference in virulence between the *MAT*α and *MAT***a** strains was observed, it was always the *MAT*α mating type that was more virulent. These results suggest the prevalence of *MAT*α strains among clinical as well as environmental isolates is in part due to distinct differences between the *MAT*α and *MAT***a** strains and that characterization of these differences will likely provide insight into how *Cryptococcus* responds to various environmental stresses and causes disease in humans.

PHEROMONE SIGNALING AND VIRULENCE

The signaling pathway involved in mating is well characterized in *Cryptococcus*, and many genes involved in this pathway appear to have a role in pathogenesis. Pheromone sensing enables *Cryptococcus* to identify the presence of cells of the opposite mating type and to initiate a signaling cascade that ultimately leads to mating and possibly quorum sensing (45, 46, 69). Pheromone signaling is initiated when a pheromone binds to its cognate receptor, which leads to activation of heterotrimeric G proteins and subsequent mitogen-activated protein (MAP) kinase signaling cascade activation (Fig. 2). The *STE3* pheromone receptor genes (also referred to as *CPR1*) and many of the pheromone signaling pathway MAP kinase genes are embedded in the mating-type locus and thus have two alleles (48). Additionally, the sex-determining homeodomain protein genes (*SXI1*α and *SXI2***a**) also reside in the *MAT* locus and are distant paralogs (33). Many of the genes present in the pheromone signaling pathway have been shown to be important for mating, and some are upregulated in response to pheromone.

Studies examining virulence of single gene deletions in the pheromone response pathway indicate that many of these genes are involved in virulence (Fig. 2). Interestingly, most of these studies used infections with only one mating type, where pheromone sensing and signaling should be minimal. Most early mutant gene analysis studies used var. *neoformans* strains. Strains lacking pheromone production (*mf*α*1*Δ*mf*α*2*Δ*mf*α*3*Δ triple mutant) or the *STE3***a** pheromone receptor (*ste3***a**Δ, also referred to as *cpr1***a**Δ) had attenuated virulence (11, 69). No virulence defect was observed in mutant strains lacking *GPB1* (not in the locus but examined in *MAT*α strains), *STE20*α, *STE7*α, or *CPK1*α (17, 83, 84). Conflicting results were seen with *STE11*α mutants that could be due to subtle differences in strain background or differences in the mutations themselves (16, 17).

The *Cryptococcus STE12* gene encodes a protein similar to the *Saccharomyces cerevisiae* Ste12p transcriptional activator. In *S. cerevisiae*, Ste12p is a member of the MAP kinase signal transduction pathway involved in mating, and disruption of this gene abolishes mating (22). *STE12* homologs in other ascomycetes are all associated with mating and reproduction. Unlike deletion of *STE12* in *S. cerevisiae*, deletion of

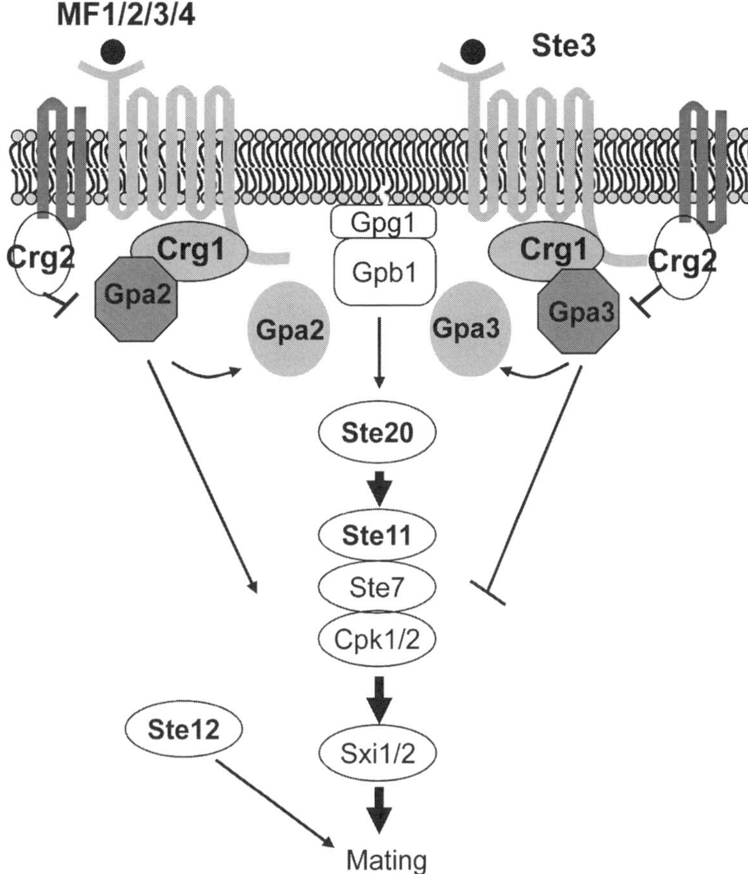

FIGURE 2 *Cryptococcus* pheromone signaling pathway. Gene deletions or disruptions of components indicated in bold resulted in altered virulence during individual infections in at least one of the *C. neoformans* varieties.

either *STE12α* or *STE12***a** in var. *neoformans* did not cause sterility but instead resulted in reduced mating frequency (12–14). These results suggest that Ste12 does not function directly in the pheromone signaling cascade but may be involved further downstream. Further characterization of *ste12α* mutants revealed that mutation of the homeodomain DNA binding region of the protein increased virulence, whereas mutations in the C_2H_2 zinc finger domain decreased virulence (14). These data suggest that Ste12 may have multiple functions in var. *neoformans*.

The effect of mutation in the G proteins and their regulators (regulators of G-protein signaling [RGS]) on virulence have been characterized in var. *grubii*. Two G-protein α subunits, Gpa2 and Gpa3, and two RGS proteins, Crg1 and Crg2, regulate the pheromone response pathway in *C. neoformans* (32, 51, 68). Mutation of either Gpa2 or Gpa3 individually had no effect on virulence (51). However, *gpa2Δ gpa3Δ* double mutants did show modest virulence defects in severely immunocompromised SCID mice (51). The effect of *CRG1* and *CRG2* mutations is quite interesting. RGS proteins are GTPase-activating proteins for the Gα subunit of the G-protein complex and function as GTPase accelerating proteins to increase hydrolysis of GTP bound to the Gα subunits to return Gα to an inactive state. Deletion of *CRG1* had variable virulence in independent mutants, while deletion of *CRG2* attenuated virulence (82, 85). These results are yet another indication of the complexity of the pheromone response pathway in *Cryptococcus* and the role it may play in virulence.

Single gene deletions in var. *grubii* strains have revealed differences between the two varieties of *C. neoformans* in the role of the pheromone response pathway in virulence. In contrast to var. *neoformans*, mutation of the *STE3***a** gene in var. *grubii* abolishes mating but does not affect pathogenicity in the murine inhalation model of cryptococcosis (64). In var. *neoformans*, mutation of the *STE20α* gene had no effect on virulence, but it attenuated virulence in var. *grubii* (83). Inverse results were observed for *STE12α*, where mutation in var. *grubii* had no effect on virulence, while virulence was attenuated in var. *neoformans* (13, 87). The *STE12α* gene has also been mutated in *C. gattii*, where it was shown to result in severely reduced virulence (66). These results suggest differences in the role of pheromone sensing or in the signaling cascade between the *C. neoformans* varieties that could ultimately affect their virulence and explain the predominance of var. *grubii* in clinical as well as environmental isolates.

The single gene deletion studies have shown that blocking pheromone signaling decreases virulence in individual infections when pheromone levels are expected to be low. Yet preliminary evidence suggests that *MFα* pheromone may be expressed in vivo in the absence of a mating partner (18). Coinfection studies with both *MAT***a** and *MATα* strains simultaneously have been used to further investigate the

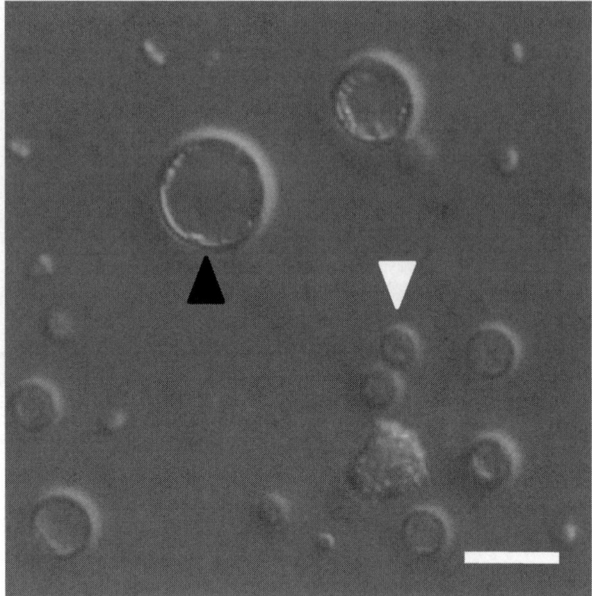

FIGURE 3 *MAT***a** cells overproduce enlarged titan cells in response to pheromone signaling. Bronchoalveolar lavage of coinfected mice reveals enlarged *MAT***a** titan cells (black arrow) compared to smaller *MAT*α normal cells (white arrow). Bar = 20 μm.

role of mating and/or pheromone signaling in virulence (59, 64). While coinfection does not appear to affect the virulence of var. *grubii MAT*α cells, *MAT***a** cells have decreased trafficking to the central nervous system. This reduced central nervous system penetration is caused by pheromone signaling as *ste3***a** mutants, which lack the pheromone receptor and so are unable to sense the presence of *MAT*α cells, accumulate in the brain (Okagaki et al., submitted). The increased pheromone signaling observed during coinfection resulted in a morphological change by the *MAT***a** cells, which may affect their ability to interact with host phagocytes (Fig. 3) (64).

Taken together, these data suggest that the pheromone response pathway, or elements of this pathway, may play a role in virulence. However, it is still unclear whether this pathway plays a direct role in virulence or whether mutations in this signaling pathway lead to global alterations in gene expression that indirectly affect virulence.

SEX AND VIRULENCE

Sex increases genetic variation within a population and allows populations to respond to selective pressures (9, 10, 50). While sexual recombination disrupts accumulation of deleterious mutations by reducing linkage, it can also disrupt the accumulation of beneficial mutations. Because virulence is thought to be a complex polygenic trait, too much sexual reproduction may be detrimental to pathogenesis. In *C. neoformans* and *C. gattii*, a nearly unisexual population limits its sexual reproduction to rare encounters between *MAT***a** and *MAT*α cells (63). Additionally, while many basidiomycetous fungi have a tetrapolar mating system that promotes outbreeding, the human-pathogenic *Cryptococcus* species have developed a bipolar mating system that promotes inbreeding (25, 31). Thus, the *Cryptococcus* species that are pathogenic to humans appear to combat the possible negative attributes of recombination on virulence by limiting their sexual reproduction while still maintaining a sexual state in order to allow adaptation to a changing environment (63).

Monokaryotic fruiting/same-sex mating, either within a single strain or between different strains of the same mating type, also significantly promotes inbreeding. Yet monokaryotic fruiting allows for sexual recombination under conditions that would normally only support clonal propagation. *Cryptococcus* may employ monokaryotic fruiting in order to retain polygenic traits, such as those required for virulence or environmental predation, while allowing airborne dispersal and limited sexual recombination to withstand changing environments. This hypothesis suggests that sexual recombination via monokaryotic fruiting could promote virulence. Recent evidence suggests that the novel hypervirulent genotype responsible for the *C. gattii* outbreak in British Columbia and the Pacific Northwest United States could be due to migration of an α strain from Australia or South America with subsequent adaptation to the new environment via monokaryotic fruiting (24, 39, 40).

Thus, both the bipolar mating system as well as the unique sexual monokaryotic fruiting observed in *Cryptococcus* may sustain, and possibly promote, the virulence of this organism.

SPORES AND VIRULENCE

What direct role, if any, sexual reproduction plays in cryptococcosis remains unclear. Yet sexual reproduction, either by traditional **a**/α mating or by monokaryotic fruiting, leads to the production of spores. Spores are the infectious propagules for many human-pathogenic fungi including dermatophytes, *Penicillium marneffei*, *Histoplasma capsulatum*, *Blastomyces dermatitidis*, *Coccidioides immitis*, and *Aspergillus* species. The infectious particles produced by *Cryptococcus* in nature still remain unknown, but recent evidence suggests both spores and desiccated yeast cells are highly infectious (26, 77).

Based on clinical data, the initial site of infection appears to be the lungs (or nasal cavities in animals). Efficient mucociliary airway clearance is observed with particles >5 μm, suggesting that only smaller infectious propagules will be readily deposited deep in the lung alveolar spaces (29). Air-sampling studies indicate that cryptococcal particles in the environment can be as small as 1.5 μm (38, 58, 67). Recent studies have shown that *C. neoformans* and *C. gattii* spores generated by both classical mating and monokaryotic fruiting are approximately 1–1.5 μm in diameter and 2–5 μm in length (4, 77). Typical yeasts grown in the laboratory are 3–10 μm in diameter, suggesting these cells are too large for efficient alveolar penetration. However, smaller desiccated yeast cells have been observed both in the laboratory and in the environment (<3 μm in diameter) (20, 58, 65, 67, 77). *C. neoformans* fruiting bodies are produced on aerial hyphae with exposed spore chains that allow efficient aerosolization and subsequent dispersal. Recent studies by Velagapudi et al. have shown that *C. neoformans* var. *grubii* spores can become easily aerosolized and disperse more rapidly than yeast cells (77).

The role of spores in the pathogenesis of *C. neoformans* continues to remain elusive. Early studies examining the pathogenesis of var. *neoformans* strains found that spores were more infectious than yeast cells grown in the laboratory (70, 88). More recent studies with var. *grubii* strains show no difference in mouse mortality after infection with equivalent numbers of spores or hydrated yeast cells; all treatments cause 100% mortality (26, 77). These observations demonstrate that both spores and yeast cells are infectious propagules.

However, at low inoculum levels, a modest delay in the time required to produce a lethal infection is observed with spore inocula compared to yeasts (77). This difference in pathogenesis between spores and yeast cells could be due to differential interactions with the host innate immune system. While yeast cells require opsonization for phagocytosis by host macrophages, spores are readily phagocytosed in the absence of an opsonin (26, 77). Studies by Giles et al. have shown that interaction between β-(1,3)-glucan on the fungal spore surface and the Dectin-1 receptor along with CD11b (a component of the leukocyte adhesion receptor CR3) are important for spore phagocytosis (26).

C. neoformans yeast cells possess a highly redundant antioxidant defense system that provides protection against reactive oxygen-nitrogen species generated upon macrophage stimulation (8, 27, 28). Interestingly, spores are readily killed by activated macrophages via production of reactive oxygen-nitrogen species (26). These data suggest that the survival of intracellular *C. neoformans* spores is dependent on their ability to germinate into yeast cells prior to activation of the host phagocyte. Spore germination in alveolar macrophages, and subsequent yeast cell escape into the extracellular space without damage to the host cell, has been observed (26). Moreover, it has been shown that intracellular yeast replicates faster than extracellular yeast, suggesting that alveolar macrophages serve as a protected niche and nutrient source for *C. neoformans* (19, 26). Thus, the initial state of cryptococcal infection could involve inhalation of aerosolized spores (either from mating or monokaryotic fruiting), phagocytosis, and subsequent germination to produce intracellular yeast cells. Spores are generated via **a**-α mating on pigeon guano and in association with plants, suggesting these substances can be reservoirs from which aerosolized spores could originate in nature (61, 86). However, given that both *MAT***a** and *MAT*α strains have not been isolated from the same pigeon guano and only rarely from plant samples, the environmental source for spores still remains unknown.

PLOIDY AND VIRULENCE

Hybridization between *C. neoformans* var. *grubii* (serotype A) and var. *neoformans* (serotype D) to generate diploid or aneuploid AD strains was first shown in the early 1990s and accounts for ~20% of clinical isolates in Europe (2, 3, 7, 21, 35, 36, 47, 71–74, 78, 81). Hybrids between *C. neoformans* and *C. gattii* have also recently been identified (5, 6). These hybrids emerge when the two mating strains are of either a different variety or a different species. Cell-cell fusion proceeds normally, but meiosis is impaired due to 10 to 15% nucleotide polymorphism between the divergent strains (37). Despite impaired meiosis, some hybrid cells propagate mitotically as uninucleate diploid/aneuploid cells after nuclear fusion events. These hybrids usually contain both mating-type alleles, implying their formation is a consequence of classical mating. Some hybrid strains contain only a single mating type, suggesting they arose through mating between two different strains of the same mating type (52). In addition to hybrid strains, diploid strains of var. *grubii* and *neoformans* have also been identified with either one or both mating types (53, 54).

Approximately 8% of *C. neoformans* strains are diploid (54, 55). Because of the abundance of clinical and environmental AD hybrid and var. *grubii* diploid strains, the importance of ploidy in virulence potential has recently been examined. Studies comparing naturally occurring diploid AD hybrid strains and haploid strains reported variable results depending on the strains used for analysis (2, 15, 47, 75). To address this concern, hybrid and diploid strains in varieties *grubii* and *neoformans* were developed in the laboratory to differentiate between ploidy and differences in genetic background, variety, and mating type (53). In these studies, diploid AD hybrid strains with any mating-type combination (αAD**a**, αADα, **a**ADα, **a**AD**a**) were significantly more virulent than the var. *neoformans* (serotype D) parental strain but were attenuated compared to the var. *grubii* (serotype A) parental strain. This modest attenuation is likely due to ploidy, as var. *grubii MAT*α diploid strains (αAAα) also had attenuated virulence compared to the parental strains (53). Interestingly, **a**AD**a** isolates were severely attenuated compared to the var. *grubii* (serotype A) parental strain even though both **a**AA**a** and **a**DD**a** diploid strains had equivalent virulence to the parental strains. These data suggest that genetic background may be more important than ploidy in determining virulence potential. Thus, ploidy changes due to mating and same-sex monokaryotic fruiting may not have a direct effect on virulence. Instead, these strains may increase the mutational capacity of cells to allow faster adaptation to changing environments, either for survival in nature or during infection of a host.

CONCLUDING REMARKS

While many studies have addressed the relationship between sex and virulence in *Cryptococcus* over the 30 years since the **a**-α sexual cycle was first defined in this organism, many questions still remain unanswered. Predominant among these questions is whether mating, same-sex α-α monokaryotic fruiting, or signaling through the pheromone response pathway plays a direct role in pathogenesis. Or is sex in cryptococcal pathogenesis simply a method to generate genetic diversity in the population and a means by which to generate infectious propagules?

REFERENCES

1. **Alexopoulos, C. J., C. W. Mims, and M. Blackwell.** 1996. *Introductory Mycology*. John Wiley & Sons, Inc., New York, NY.
2. **Barchiesi, F., M. Cogliati, M. C. Esposto, E. Spreghini, A. M. Schimizzi, B. L. Wickes, G. Scalise, and M. A. Viviani.** 2005. Comparative analysis of pathogenicity of *Cryptococcus neoformans* serotypes A, D and AD in murine cryptococcosis. *J. Infect.* **51:**10–16.
3. **Boekhout, T., and A. van Belkum.** 1997. Variability of karyotypes and RAPD types in genetically related strains of *Cryptococcus neoformans*. *Curr. Genet.* **32:**203–208.
4. **Botts, M. R., S. S. Giles, M. A. Gates, T. R. Kozel, and C. M. Hull.** 2009. Isolation and characterization of *Cryptococcus neoformans* spores reveal a critical role for capsule biosynthesis genes in spore biogenesis. *Eukaryot. Cell* **8:**595–605.
5. **Bovers, M., F. Hagen, E. E. Kuramae, M. R. Diaz, L. Spanjaard, F. Dromer, H. L. Hoogveld, and T. Boekhout.** 2006. Unique hybrids between the fungal pathogens *Cryptococcus neoformans* and *Cryptococcus gattii*. *FEMS Yeast Res.* **6:**599–607.
6. **Bovers, M., F. Hagen, E. E. Kuramae, H. L. Hoogveld, F. Dromer, G. St-Germain, and T. Boekhout.** 2008. AIDS patient death caused by novel *Cryptococcus neoformans* x *C. gattii* hybrid. *Emerg. Infect. Dis.* **14:**1105–1108.
7. **Brandt, M. E., S. L. Bragg, and R. W. Pinner.** 1993. Multilocus enzyme typing of *Cryptococcus neoformans*. *J. Clin. Microbiol.* **31:**2819–2823.
8. **Brown, S. M., L. T. Campbell, and J. K. Lodge.** 2007. *Cryptococcus neoformans*, a fungus under stress. *Curr. Opin. Microbiol.* **10:**320–325.

9. **Burt, A.** 2000. Perspective: sex, recombination, and the efficacy of selection: was Weismann right? *Evolution* **54:**337–351.
10. **Butlin, R.** 2002. Evolution of sex: the costs and benefits of sex: new insights from old asexual lineages. *Nat. Rev. Genet.* **3:**311–317.
11. **Chang, Y. C., G. F. Miller, and K. J. Kwon-Chung.** 2003. Importance of a developmentally regulated pheromone receptor of *Cryptococcus neoformans* for virulence. *Infect. Immun.* **71:**4953–4960.
12. **Chang, Y. C., L. A. Penoyer, and K. J. Kwon-Chung.** 2001. The second *STE12* homologue of *Cryptococcus neoformans* is *MAT***a**-specific and plays an important role in virulence. *Proc. Natl. Acad. Sci. USA* **98:**3258–3263.
13. **Chang, Y. C., B. L. Wickes, G. F. Miller, L. A. Penoyer, and K. J. Kwon-Chung.** 2000. *Cryptococcus neoformans STE12α* regulates virulence but is not essential for mating. *J. Exp. Med.* **191:**871–882.
14. **Chang, Y. C., L. C. Wright, R. L. Tscharke, T. C. Sorrell, C. F. Wilson, and K. J. Kwon-Chung.** 2004. Regulatory roles for the homeodomain and C2H2 zinc finger regions of *Cryptococcus neoformans* Ste12αp. *Mol. Microbiol.* **53:**1385–1396.
15. **Chaturvedi, V., J. Fan, B. Stein, M. J. Behr, W. A. Samsonoff, B. L. Wickes, and S. Chaturvedi.** 2002. Molecular genetic analyses of mating pheromones reveal intervariety mating or hybridization in *Cryptococcus neoformans. Infect. Immun.* **70:**5225–5235.
16. **Clarke, D. L., G. L. Woodlee, C. M. McClelland, T. S. Seymour, and B. L. Wickes.** 2001. The *Cryptococcus neoformans STE11α* gene is similar to other fungal mitogen-activated protein kinase kinase kinase (MAPKKK) genes but is mating type specific. *Mol. Microbiol.* **40:**200–213.
17. **Davidson, R. C., C. B. Nichols, G. M. Cox, J. R. Perfect, and J. Heitman.** 2003. A MAP kinase cascade composed of cell type specific and non-specific elements controls mating and differentiation of the fungal pathogen *Cryptococcus neoformans. Mol. Microbiol.* **49:**469–485.
18. **del Poeta, M., D. L. Toffaletti, T. H. Rude, S. D. Sparks, J. Heitman, and J. R. Perfect.** 1999. *Cryptococcus neoformans* differential gene expression detected in vitro and in vivo with green fluorescent protein. *Infect. Immun.* **67:**1812–1820.
19. **Diamond, R. D., and J. E. Bennett.** 1973. Growth of *Cryptococcus neoformans* within human macrophages *in vitro. Infect. Immun.* **7:**231–236.
20. **Ellis, D. H., and T. J. Pfeiffer.** 1990. Ecology, life cycle, and infectious propagule of *Cryptococcus neoformans. Lancet* **336:**923–925.
21. **Esposto, M. C., M. Cogliati, A. M. Tortorano, and M. A. Viviani.** 2009. Electrophoretic karyotyping of *Cryptococcus neoformans* AD-hybrid strains. *Mycoses* **52:**16–23.
22. **Fields, S., and I. Herskowitz.** 1987. Regulation by the yeast mating-type locus of *STE12*, a gene required for cell-type-specific expression. *Mol. Cell. Biol.* **7:**3818–3821.
23. **Fraser, J. A., S. Diezmann, R. L. Subaran, A. Allen, K. B. Lengeler, F. S. Dietrich, and J. Heitman.** 2004. Convergent evolution of chromosomal sex-determining regions in the animal and fungal kingdoms. *PLoS Biol.* **2:**e384.
24. **Fraser, J. A., S. S. Giles, E. C. Wenink, S. G. Geunes-Boyer, J. R. Wright, S. Diezmann, J. R. Wright, J. R. Perfect, and J. Heitman.** 2005. Same-sex mating and the origin of the Vancouver Island *Cryptococcus gattii* outbreak. *Nature* **437:**1360–1364.
25. **Fraser, J. A., R. L. Subaran, C. B. Nichols, and J. Heitman.** 2003. Recapitulation of the sexual cycle of the primary fungal pathogen *Cryptococcus neoformans* var. *gattii*: implications for an outbreak on Vancouver Island, Canada. *Eukaryot. Cell* **2:**1036–1045.
26. **Giles, S. S., T. R. Dagenais, M. R. Botts, N. P. Keller, and C. M. Hull.** 2009. Elucidating the pathogenesis of spores from the human fungal pathogen *Cryptococcus neoformans. Infect. Immun.* **77:**3491–3500.
27. **Giles, S. S., J. R. Perfect, and G. M. Cox.** 2005. Cytochrome c peroxidase contributes to the antioxidant defense of *Cryptococcus neoformans. Fungal Genet. Biol.* **42:**20–29.
28. **Giles, S. S., J. E. Stajich, C. Nichols, Q. D. Gerrald, J. A. Alspaugh, F. Dietrich, and J. R. Perfect.** 2006. The *Cryptococcus neoformans* catalase gene family and its role in antioxidant defense. *Eukaryot. Cell* **5:**1447–1459.
29. **Hatch, T. F.** 1961. Distribution and deposition of inhaled particles in respiratory tract. *Bacteriol. Rev.* **25:**237–240.
30. **Heitman, J., B. Allen, J. A. Alspaugh, and K. J. Kwon-Chung.** 1999. On the origins of congenic *MATα* and *MAT***a** strains of the pathogenic yeast *Cryptococcus neoformans. Fungal Genet. Biol.* **28:**1–5.
31. **Hsueh, Y. P., J. A. Fraser, and J. Heitman.** 2008. Transitions in sexuality: recapitulation of an ancestral tri- and tetrapolar mating system in *Cryptococcus neoformans. Eukaryot. Cell* **7:**1847–1855.
32. **Hsueh, Y. P., C. Xue, and J. Heitman.** 2007. G protein signaling governing cell fate decisions involves opposing Gα subunits in *Cryptococcus neoformans. Mol. Biol. Cell* **18:**3237–3249.
33. **Hull, C. M., M. J. Boily, and J. Heitman.** 2005. Sex-specific homeodomain proteins Sxi1α and Sxi2a coordinately regulate sexual development in *Cryptococcus neoformans. Eukaryot. Cell* **4:**526–535.
34. **Hull, C. M., R. C. Davidson, and J. Heitman.** 2002. Cell identity and sexual development in *Cryptococcus neoformans* are controlled by the mating-type-specific homeodomain protein Sxi1α. *Genes Dev.* **16:**3046–3060.
35. **Ikeda, R., A. Nishikawa, T. Shinoda, and Y. Fukazawa.** 1985. Chemical characterization of capsular polysaccharide from *Cryptococcus neoformans* serotype A-D. *Microbiol. Immunol.* **29:**981–991.
36. **Kabasawa, K., H. Itagaki, R. Ikeda, T. Shinoda, K. Kagaya, and Y. Fukazawa.** 1991. Evaluation of a new method for identification of *Cryptococcus neoformans* which uses serologic tests aided by selected biological tests. *J. Clin. Microbiol.* **29:**2873–2876.
37. **Kavanaugh, L. A., J. A. Fraser, and F. S. Dietrich.** 2006. Recent evolution of the human pathogen *Cryptococcus neoformans* by intervarietal transfer of a 14-gene fragment. *Mol. Biol. Evol.* **23:**1879–1890.
38. **Kidd, S. E., Y. Chow, S. Mak, P. J. Bach, H. Chen, A. O. Hingston, J. W. Kronstad, and K. H. Bartlett.** 2007. Characterization of environmental sources of the human and animal pathogen *Cryptococcus gattii* in British Columbia, Canada, and the Pacific Northwest of the United States. *Appl. Environ. Microbiol.* **73:**1433–1443.
39. **Kidd, S. E., H. Guo, K. H. Bartlett, J. Xu, and J. W. Kronstad.** 2005. Comparative gene genealogies indicate that two clonal lineages of *Cryptococcus gattii* in British Columbia resemble strains from other geographical areas. *Eukaryot. Cell* **4:**1629–1638.
40. **Kidd, S. E., F. Hagen, R. L. Tscharke, M. Huynh, K. H. Bartlett, M. Fyfe, L. Macdougall, T. Boekhout, K. J. Kwon-Chung, and W. Meyer.** 2004. A rare genotype of *Cryptococcus gattii* caused the cryptococcosis outbreak on Vancouver Island (British Columbia, Canada). *Proc. Natl. Acad. Sci. USA* **101:**17258–17263.
41. **Kwon-Chung, K. J.** 1975. A new genus, *Filobasidiella*, the perfect state of *Cryptococcus neoformans. Mycologia* **67:**1197–1200.
42. **Kwon-Chung, K. J., and J. E. Bennett.** 1978. Distribution of **a** and α mating types of *Cryptococcus neoformans* among natural and clinical isolates. *Am. J. Epidemiol.* **108:**337–340.

43. **Kwon-Chung, K. J., J. C. Edman, and B. L. Wickes.** 1992. Genetic association of mating types and virulence in *Cryptococcus neoformans*. *Infect. Immun.* **60:**602–605.
44. **Kwon-Chung, K. J., and W. B. Hill.** 1981. Sexuality and pathogenicity of *Filobasidiella neoformans* (*Cryptococcus neoformans*), p. 243–250. *In* R. Vanbreuseghem and C. DeVroy (ed.), *Sexuality and Pathogenicity of Fungi*. Masson, Paris, France.
45. **Lee, H., Y. C. Chang, and K. J. Kwon-Chung.** 2005. *TUP1* disruption reveals biological differences between *MAT***a** and *MAT*α strains of *Cryptococcus neoformans*. *Mol. Microbiol.* **55:**1222–1232.
46. **Lee, H., Y. C. Chang, G. Nardone, and K. J. Kwon-Chung.** 2007. *TUP1* disruption in *Cryptococcus neoformans* uncovers a peptide-mediated density-dependent growth phenomenon that mimics quorum sensing. *Mol. Microbiol.* **64:**591–601.
47. **Lengeler, K. B., G. M. Cox, and J. Heitman.** 2001. Serotype AD strains of *Cryptococcus neoformans* are diploid or aneuploid and are heterozygous at the mating-type locus. *Infect. Immun.* **69:**115–122.
48. **Lengeler, K. B., D. S. Fox, J. A. Fraser, A. Allen, K. Forrester, F. S. Dietrich, and J. Heitman.** 2002. Mating-type locus of *Cryptococcus neoformans*: a step in the evolution of sex chromosomes. *Eukaryot. Cell* **1:**704–718.
49. **Lengeler, K. B., P. Wang, G. M. Cox, J. R. Perfect, and J. Heitman.** 2000. Identification of the MATa mating-type locus of *Cryptococcus neoformans* reveals a serotype A MATa strain thought to have been extinct. *Proc. Natl. Acad. Sci. USA* **97:**14455–14460.
50. **Lenski, R. E.** 2001. Genetics and evolution. Come fly, and leave the baggage behind. *Science* **294:**533–534.
51. **Li, L., G. Shen, Z. G. Zhang, Y. L. Wang, J. K. Thompson, and P. Wang.** 2007. Canonical heterotrimeric G proteins regulating mating and virulence of *Cryptococcus neoformans*. *Mol. Biol. Cell* **18:**4201–4209.
52. **Lin, X., A. P. Litvintseva, K. Nielsen, S. Patel, A. Floyd, T. G. Mitchell, and J. Heitman.** 2007. αADα hybrids of *Cryptococcus neoformans*: evidence of same-sex mating in nature and hybrid fitness. *PLoS Genet.* **3:**1975–1990.
53. **Lin, X., K. Nielsen, S. Patel, and J. Heitman.** 2008. Impact of mating type, serotype, and ploidy on the virulence of *Cryptococcus neoformans*. *Infect. Immun.* **76:**2923–2938.
54. **Lin, X., S. Patel, A. P. Litvintseva, A. Floyd, T. G. Mitchell, and J. Heitman.** 2009. Diploids in the *Cryptococcus neoformans* serotype A population homozygous for the α mating type originate via unisexual mating. *PLoS Pathog.* **5:**e1000283.
55. **Litvintseva, A. P., L. Kestenbaum, R. Vilgalys, and T. G. Mitchell.** 2005. Comparative analysis of environmental and clinical populations of *Cryptococcus neoformans*. *J. Clin. Microbiol.* **43:**556–564.
56. **Litvintseva, A. P., R. E. Marra, K. Nielsen, J. Heitman, R. Vilgalys, and T. G. Mitchell.** 2003. Evidence of sexual recombination among *Cryptococcus neoformans* serotype A isolates in sub-Saharan Africa. *Eukaryot. Cell* **2:**1162–1168.
57. **Mylonakis, E., F. M. Ausubel, J. R. Perfect, J. Heitman, and S. B. Calderwood.** 2002. Killing of *Caenorhabditis elegans* by *Cryptococcus neoformans* as a model of yeast pathogenesis. *Proc. Natl. Acad. Sci. USA* **99:**15675–15680.
58. **Neilson, J. B., R. A. Fromtling, and G. S. Bulmer.** 1977. *Cryptococcus neoformans*: size range of infectious particles from aerosolized soil. *Infect. Immun.* **17:**634–638.
59. **Nielsen, K., G. M. Cox, A. P. Litvintseva, E. Mylonakis, S. D. Malliaris, D. K. Benjamin, Jr., S. S. Giles, T. G. Mitchell, A. Casadevall, J. R. Perfect, and J. Heitman.** 2005. *Cryptococcus neoformans* α strains preferentially disseminate to the central nervous system during coinfection. *Infect. Immun.* **73:**4922–4933.
60. **Nielsen, K., G. M. Cox, P. Wang, D. L. Toffaletti, J. R. Perfect, and J. Heitman.** 2003. Sexual cycle of *Cryptococcus neoformans* var. *grubii* and virulence of congenic **a** and α isolates. *Infect. Immun.* **71:**4831–4841.
61. **Nielsen, K., A. L. De Obaldia, and J. Heitman.** 2007. *Cryptococcus neoformans* mates on pigeon guano: implications for the realized ecological niche and globalization. *Eukaryot. Cell* **6:**949–959.
62. **Nielsen, K., R. E. Marra, F. Hagen, T. Boekhout, T. G. Mitchell, G. M. Cox, and J. Heitman.** 2005. Interaction between genetic background and the mating-type locus in *Cryptococcus neoformans* virulence potential. *Genetics* **171:**975–983.
63. **Nielsen, K., and J. Heitman.** 2007. Sex and virulence of human fungal pathogens. *Adv. Genet.* **57:**143–173.
64. **Okagaki, L. H., A. K. Strain, J. N. Nielsen, C. Charlier, N. J. Baltes, F. Chrétien, J. Heitman, F. Dromer, and K. Nielsen.** 2010. Cryptococcal cell morphology affects host cell interactions and pathogenicity. *PLoS Pathog.* **6:**e1000953.
65. **Powell, K. E., B. A. Dahl, R. J. Weeks, and F. E. Tosh.** 1972. Airborne *Cryptococcus neoformans*: particles from pigeon excreta compatible with alveolar deposition. *J. Infect. Dis.* **125:**412–415.
66. **Ren, P., D. J. Springer, M. J. Behr, W. A. Samsonoff, S. Chaturvedi, and V. Chaturvedi.** 2006. Transcription factor *STE12*alpha has distinct roles in morphogenesis, virulence, and ecological fitness of the primary pathogenic yeast *Cryptococcus gattii*. *Eukaryot. Cell* **5:**1065–1080.
67. **Ruiz, A., and G. S. Bulmer.** 1981. Particle size of airborne *Cryptococcus neoformans* in a tower. *Appl. Environ. Microbiol.* **41:**1225–1229.
68. **Shen, G., Y. L. Wang, A. Whittington, L. Li, and P. Wang.** 2008. The RGS protein Crg2 regulates pheromone and cyclic AMP signaling in *Cryptococcus neoformans*. *Eukaryot. Cell* **7:**1540–1548.
69. **Shen, W. C., R. C. Davidson, G. M. Cox, and J. Heitman.** 2002. Pheromones stimulate mating and differentiation via paracrine and autocrine signaling in *Cryptococcus neoformans*. *Eukaryot. Cell* **1:**366–377.
70. **Sukroongreung, S., K. Kitiniyom, C. Nilakul, and S. Tantimavanich.** 1998. Pathogenicity of basidiospores of *Filobasidiella neoformans* var. *neoformans*. *Med. Mycol.* **36:**419–424.
71. **Takeo, K., R. Tanaka, H. Taguchi, and K. Nishimura.** 1993. Analysis of ploidy and sexual characteristics of natural isolates of *Cryptococcus neoformans*. *Can. J. Microbiol.* **39:**958–963.
72. **Tanaka, R., K. Nishimura, and M. Miyaji.** 1999. Ploidy of serotype AD strains of *Cryptococcus neoformans*. *Nippon Ishinkin Gakkai Zasshi* **40:**31–34.
73. **Tanaka, R., H. Taguchi, K. Takeo, M. Miyaji, and K. Nishimura.** 1996. Determination of ploidy in *Cryptococcus neoformans* by flow cytometry. *J. Med. Vet. Mycol.* **34:**299–301.
74. **Tintelnot, K., K. Lemmer, H. Losert, G. Schar, and A. Polak.** 2004. Follow-up of epidemiological data of cryptococcosis in Austria, Germany and Switzerland with special focus on the characterization of clinical isolates. *Mycoses* **47:**455–464.
75. **Toffaletti, D. L., K. Nielsen, F. Dietrich, J. Heitman, and J. R. Perfect.** 2004. *Cryptococcus neoformans* mitochondrial genomes from serotype A and D strains do not influence virulence. *Curr. Genet.* **46:**193–204.
76. **Toffaletti, D. L., T. H. Rude, S. A. Johnston, D. T. Durack, and J. R. Perfect.** 1993. Gene transfer in *Cryptococcus neoformans* by use of biolistic delivery of DNA. *J. Bacteriol.* **175:**1405–1411.
77. **Velagapudi, R., Y.-P. Hsueh, S. Geunes-Boyer, J. R. Wright, and J. Heitman.** 2009. Spores as infectious

propagules of *Cryptococcus neoformans*. *Infect. Immun.* **77:**4345–4355.

78. **Viviani, M. A., M. Cogliati, M. C. Esposto, K. Lemmer, K. Tintelnot, M. F. Colom Valiente, D. Swinne, A. Velegraki, and R. Velho.** 2006. Molecular analysis of 311 *Cryptococcus neoformans* isolates from a 30-month ECMM survey of cryptococcosis in Europe. *FEMS Yeast Res.* **6:**614–619.
79. **Viviani, M. A., M. C. Esposto, M. Cogliati, M. T. Montagna, and B. L. Wickes.** 2001. Isolation of a *Cryptococcus neoformans* serotype A MATa strain from the Italian environment. *Med. Mycol.* **39:**383–386.
80. **Viviani, M. A., R. Nikolova, M. C. Esposto, G. Prinz, and M. Cogliati.** 2003. First European case of serotype A MATa *Cryptococcus neoformans* infection. *Emerg. Infect. Dis.* **9:**1179–1180.
81. **Viviani, M. A., H. Wen, A. Roverselli, R. Caldarelli-Stefano, M. Cogliati, P. Ferrante, and A. M. Tortorano.** 1997. Identification by polymerase chain reaction fingerprinting of *Cryptococcus neoformans* serotype AD. *J. Med. Vet. Mycol.* **35:**355–360.
82. **Wang, P., J. Cutler, J. King, and D. Palmer.** 2004. Mutation of the regulator of G protein signaling Crg1 increases virulence in *Cryptococcus neoformans*. *Eukaryot. Cell* **3:**1028–1035.
83. **Wang, P., C. B. Nichols, K. B. Lengeler, M. E. Cardenas, G. M. Cox, J. R. Perfect, and J. Heitman.** 2002. Mating-type-specific and nonspecific PAK kinases play shared and divergent roles in *Cryptococcus neoformans*. *Eukaryot. Cell* **1:**257–272.
84. **Wang, P., J. R. Perfect, and J. Heitman.** 2000. The G-protein beta subunit GPB1 is required for mating and haploid fruiting in *Cryptococcus neoformans*. *Mol. Cell. Biol.* **20:**352–362.
85. **Xue, C., Y. P. Hsueh, L. Chen, and J. Heitman.** 2008. The RGS protein Crg2 regulates both pheromone and cAMP signalling in *Cryptococcus neoformans*. *Mol. Microbiol.* **70:**379–395.
86. **Xue, C., Y. Tada, X. Dong, and J. Heitman.** 2007. The human fungal pathogen *Cryptococcus* can complete its sexual cycle during a pathogenic association with plants. *Cell Host Microbe* **1:**263–273.
87. **Yue, C., L. M. Cavallo, J. A. Alspaugh, P. Wang, G. M. Cox, J. R. Perfect, and J. Heitman.** 1999. The *STE12*α homolog is required for haploid filamentation but largely dispensable for mating and virulence in *Cryptococcus neoformans*. *Genetics* **153:**1601–1615.
88. **Zimmer, B. L., H. O. Hempel, and N. L. Goodman.** 1984. Pathogenicity of the basidiospores of *Filobasidiella neoformans*. *Mycopathologia* **85:**149–153.

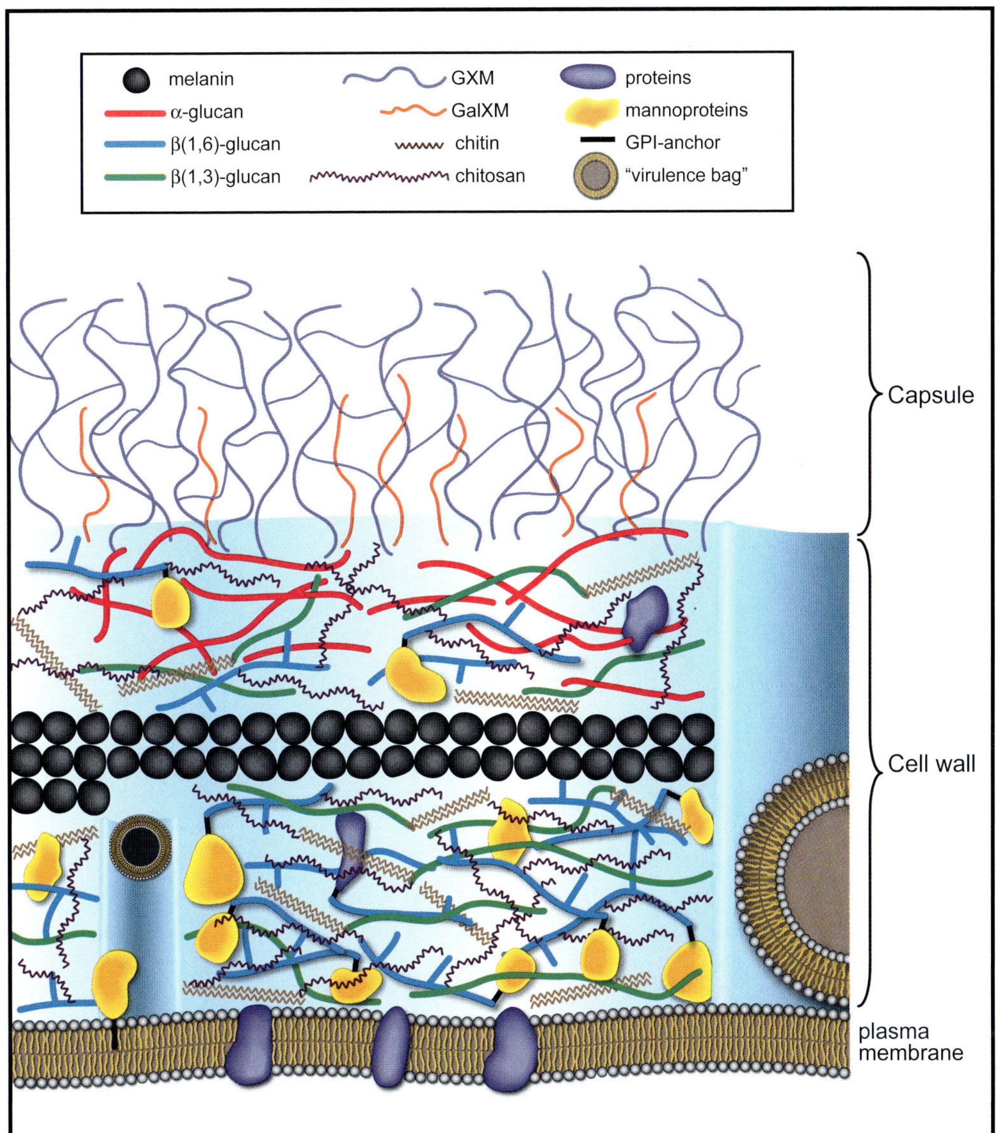

COLOR PLATE 1 (chapter 6) Working model of the cell wall of *Cryptococcus*. This schematic depicts the components known to be present, shown in the key. However, several aspects of this model remain to be investigated. The distribution of individual components within the wall and the exact linkages between them are largely undetermined. Chitin is shown linked to β-1,3-glucan based upon data in *S. cerevisiae*, but this linkage has not been examined directly in *Cryptococcus*. Also, the model illustrates chitosan as linked to chitin or β-1,6-glucan, but this remains to be determined biochemically. The distribution of chitin and chitosan is still under investigation. Chitosan in a wild-type cell is normally stained by Eosin Y but can be inaccessible to the dye in the absence of β-1,6-glucan, suggesting that it is normally localized to the outside of the wall. We show α-1,3-glucan in the outer wall since electron microscopy immunogold labeling suggests that it is largely found in the periphery, and it is known to be an essential component for capsule binding. The β-glucans are shown distributed throughout since there is little evidence to the contrary. GPI-CWPs are shown linked to β-1,6-glucan, but the mechanism of cell wall association of noncanonical cell wall proteins (those lacking a Ser/Thr-rich region and a GPI anchor, depicted in purple) is unknown. We have depicted vesicles traversing the wall, or being deposited in the wall, based on recent findings. There may be vesicles of different sizes and cargos that are important for delivery of cell wall, capsule, and other secreted material, and the mechanism by which secreted vesicles traverse the cell wall matrix is yet to be determined. Concentric layers of melanin granules have been observed within the cryptococcal cell wall, but it is unclear whether the layers are present in the inner or outer wall or are present throughout.

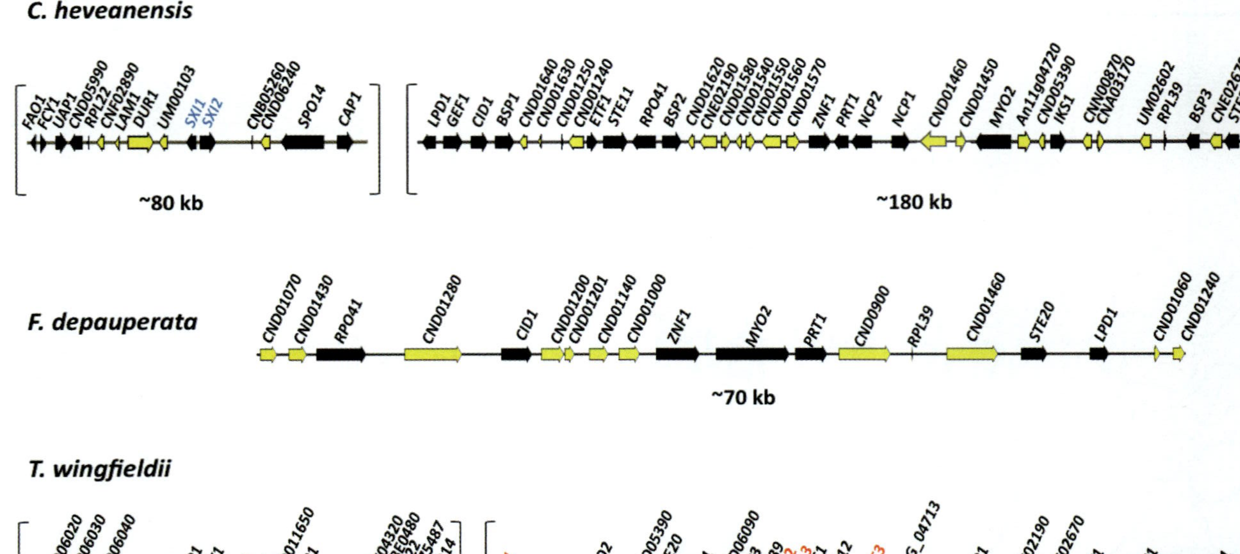

COLOR PLATE 2 (chapter 11) *MAT* locus structure of *C. neoformans* and closely related species. Homeodomain genes are highlighted in blue and pheromone and receptor genes are highlighted in red. Genes in black are present in the *MAT* locus of the pathogenic *Cryptococcus* species, while genes in yellow are additional genes identified in the other closely related species. The size of each locus is indicated below.

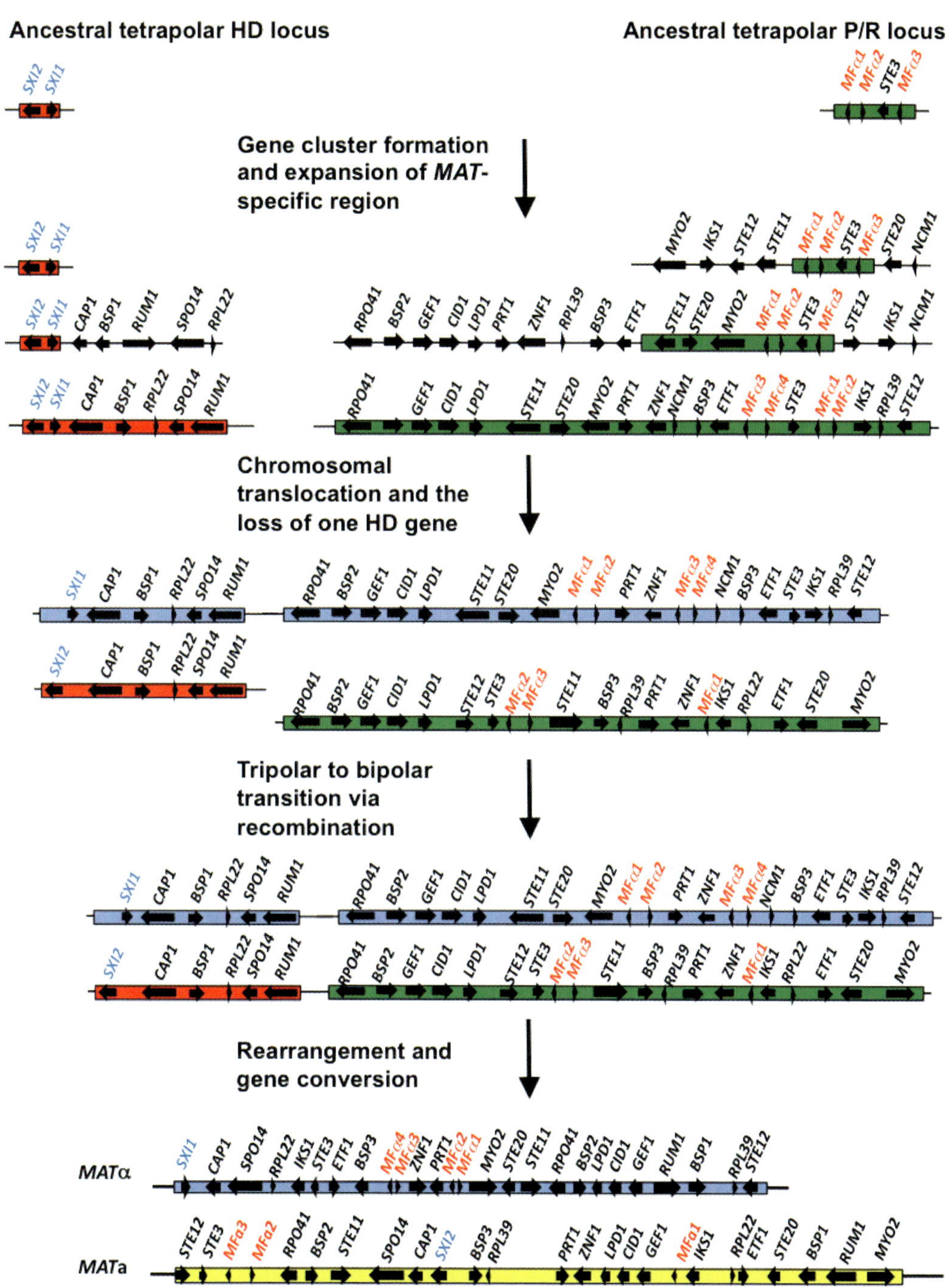

COLOR PLATE 3 (chapter 11) A model of the evolution of the bipolar *MAT* locus in the pathogenic *Cryptococcus* species. We proposed that the *C. neoformans MAT* locus descended from an ancestral tetrapolar system with two unlinked HD and P/R loci. These loci expanded via gene acquisition, forming two unlinked gene clusters, and additional genes were gradually incorporated into the *MAT*-specific region at both the HD and P/R loci. Chromosomal translocation then occurred to fuse the two unlinked loci to one contiguous allele, resulting in the intermediate tripolar state. The tripolar system then further collapsed into a bipolar system via recombination between the two *MAT* alleles. Lastly, ongoing gene rearrangements, gene inversions, and gene conversions have contributed to shape the gene order and gene content of the *MAT* locus.

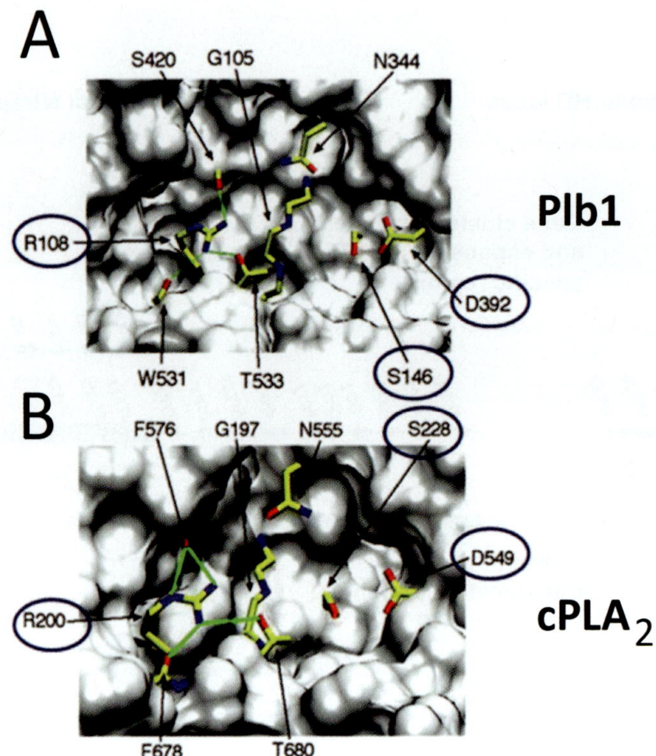

COLOR PLATE 4 (chapter 17) View of the active site cleft of Plb1/human cPLA$_2$ showing conserved residues. The catalytic residues in both proteins are circled. Cap domain residues 415 to 424 for cPLA$_2$, which partially occlude the active site in the X-ray structure (PDB: 1CYJ), have been removed for clarity. Labeled residues are shown in stick form and colored: carbon yellow, nitrogen blue, and oxygen red. Other residues are depicted as a surface representation using a probe radius of 1.4 Å. Residues 577 and 578 have been rendered as transparent surface to reveal the backbone carbonyl oxygen of F576. Hydrogen bonds between the side chain of R200 and the backbone carbonyl oxygen of F576 and F678, as well as the side chain of T680, are depicted as green lines. The backbone nitrogen atoms of G197 (labeled) and G198 (visible beneath T680) form a putative oxyanion hole, with S228 and D549 forming the catalytic dyad. R200 is thought to bind the phosphate moiety of the lipid head group and may also stabilize the geometry of the oxyanion hole. Published with permission from the American Chemical Society; license # 2171070482164.

COLOR PLATE 5 (chapter 23) Map of (A) reported cases, (B) environmental sampling, and (C) forecasted ecological niche of *C. gattii* on Vancouver Island and the British Columbia mainland. (A) Human and animal cases (1999 to 2007) are mapped by place of residence at a geographic scale that does not identify individuals. Cases on the British Columbia mainland with travel to Vancouver Island or other *C. gattii* endemic areas prior to onset of illness are excluded from the map. (B) Environmental sampling (2001 to 2007) for *C. gattii* and the biogeoclimatic zones of British Columbia. The Coastal Douglas Fir and Coastal Western Hemlock (very dry) zones are of particular interest for *C. gattii* epidemiology in British Columbia. (C) Forecasted ecological niche model for *C. gattii* in British Columbia. Geographic areas with the optimal environmental conditions to support *C. gattii* in British Columbia are illustrated in red and orange.

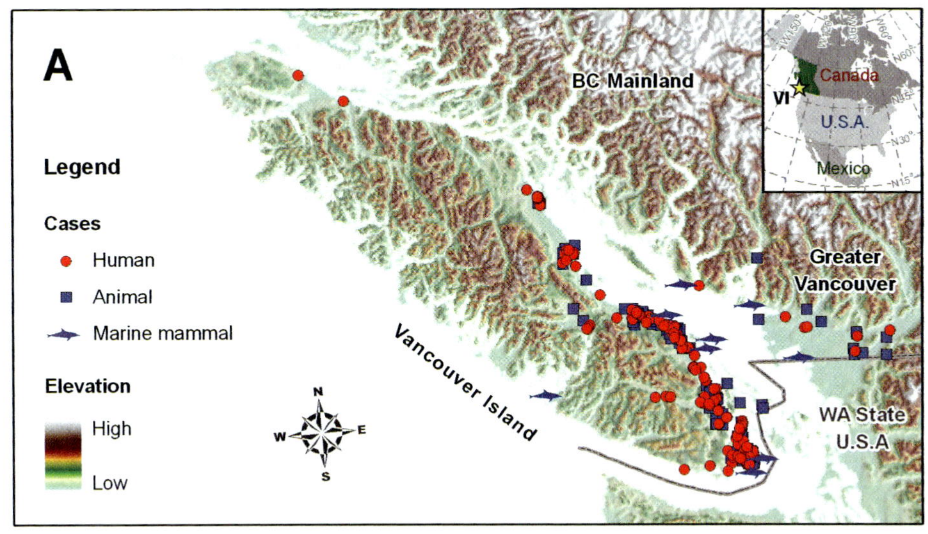
A
BC Mainland
Canada
VI
U.S.A.
Mexico
Legend
Cases
Human
Animal
Marine mammal
Elevation
High
Low
Greater Vancouver
Vancouver Island
WA State U.S.A
N
E
S
W

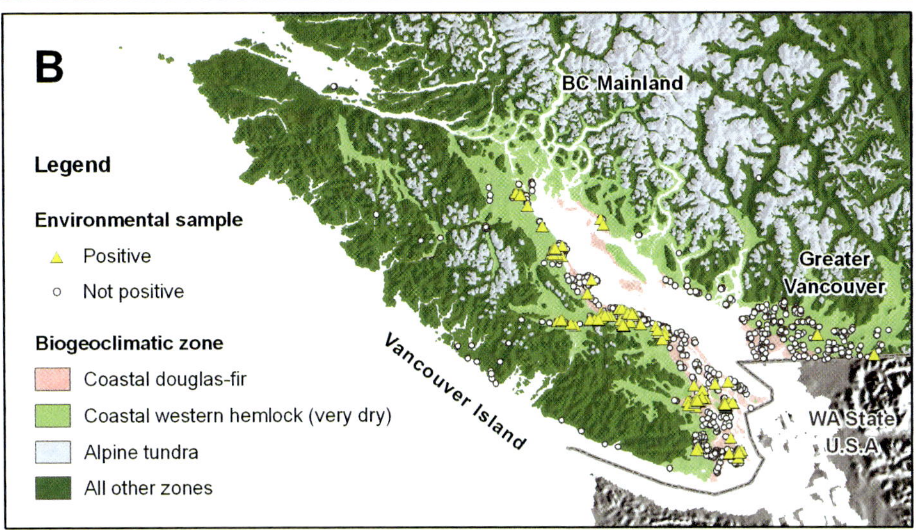
B
BC Mainland
Legend
Environmental sample
Positive
Not positive
Biogeoclimatic zone
Coastal douglas-fir
Coastal western hemlock (very dry)
Alpine tundra
All other zones
Greater Vancouver
Vancouver Island
WA State U.S.A

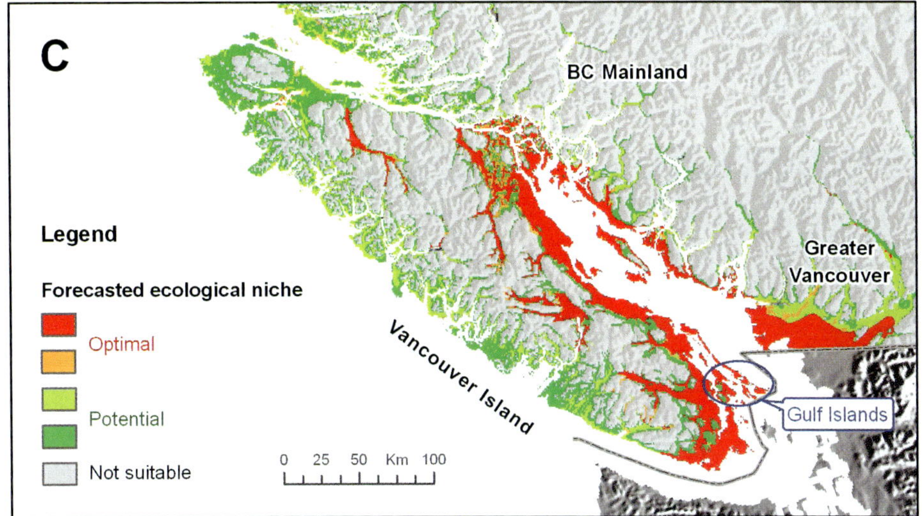
C
BC Mainland
Legend
Forecasted ecological niche
Optimal
Potential
Not suitable
Greater Vancouver
Vancouver Island
Gulf Islands
0 25 50 Km 100

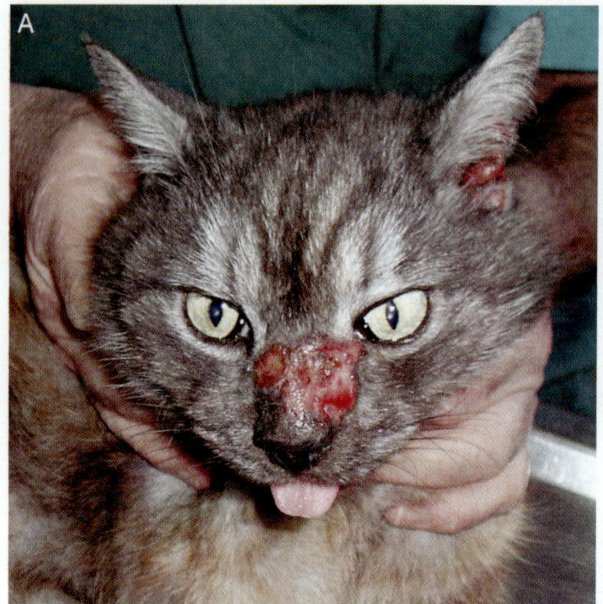

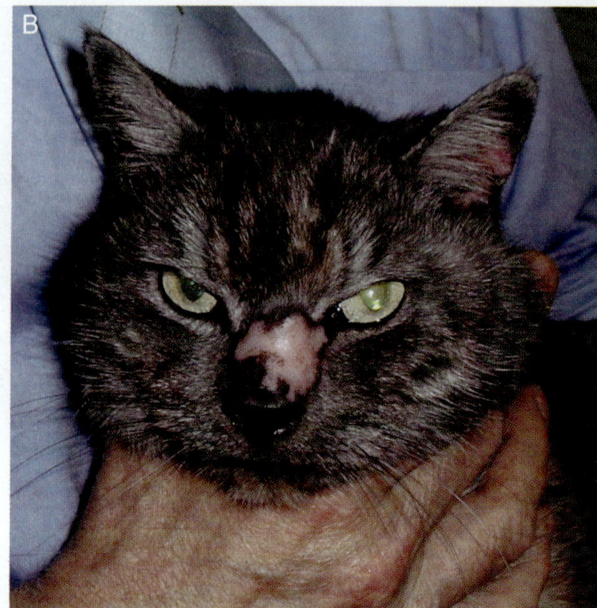

COLOR PLATE 6 (chapter 36) (A) Cat with cryptococcal rhinosinusitis with invasion of the tissues over the nasal bones. This case was erroneously given corticosteroids, which caused the disease to suddenly progress. Note the secondary lesion on the skin of the pinna (ear). (B) The cat responded very favorably to long-term monotherapy with fluconazole, although some scarring with depigmentation of the nasal bridge region is evident. Photographs courtesy of Phillip Druce.

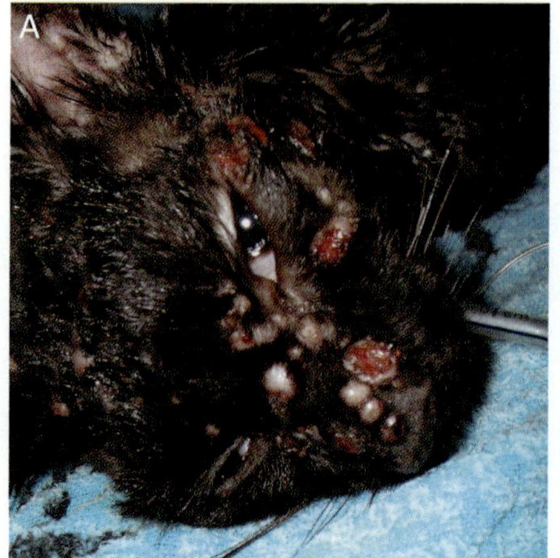

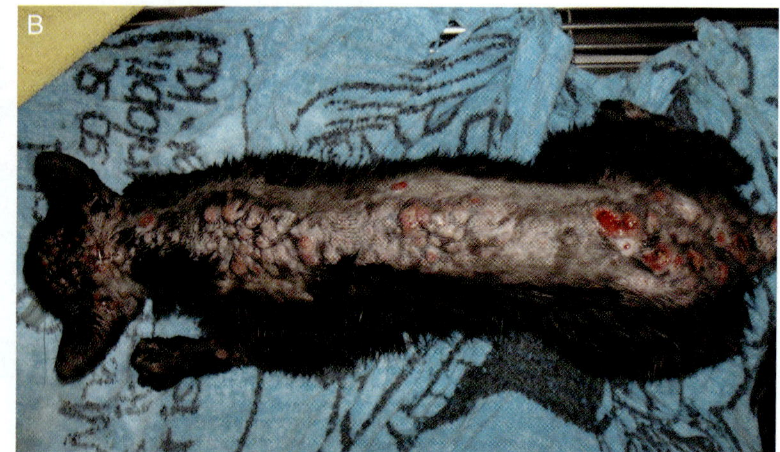

COLOR PLATE 7 (chapter 36) Multifocal skin lesions on (A) the head and (B) the trunk of an FIV-positive cat with disseminated cutaneous cryptococcosis. The primary site of infection was the nasal cavity. Many of the skin lesions have ulcerated.

COLOR PLATE 8 (chapter 36) Invasive deforming cryptococcal rhinitis due to C. *gattii* (VGIIb). There is deformity of the right-hand side of the nasal planum (arrow), and erosion of a blood vessel has resulted in severe epistaxis (bleeding from the nose). This koala subsequently responded to a long course of subcutaneous amphotericin B, plus consolidation therapy with fluconazole.

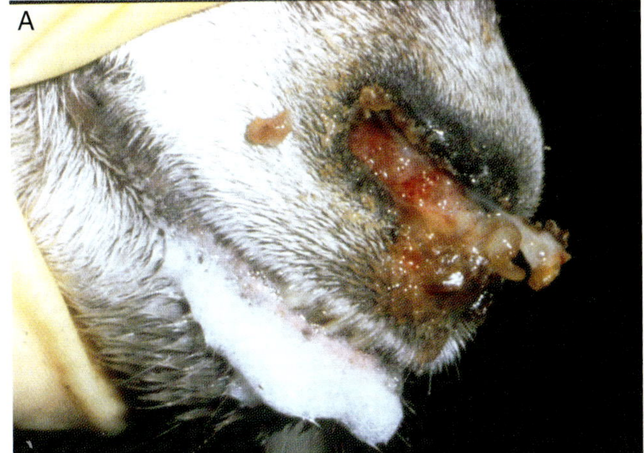

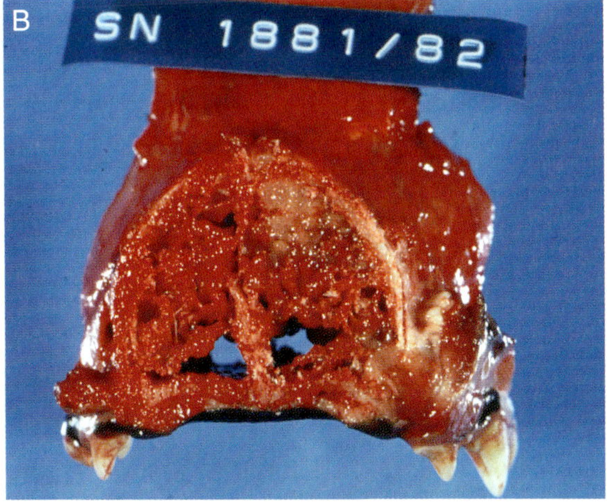

COLOR PLATE 9 (chapter 36) Invasive and proliferative cryptococcal rhinitis in a goat from southwestern Western Australia. We suspect this was a C. *gattii* VGIIb infection. (A) Photograph of the animal's head. (B) A cross-section through the nasal cavity of a similarly affected greyhound.

Cryptococcus: From Human Pathogen to Model Yeast
Edited by J. Heitman et al.
©2011 ASM Press, Washington, DC

14

Sensing Extracellular Signals in *Cryptococcus neoformans*

ALEXANDER IDNURM, YONG-SUN BAHN, WEI-CHIANG SHEN, JULIAN C. RUTHERFORD, AND FRITZ A. MÜHLSCHLEGEL

The extracellular environment is a complex assembly of physical and biological factors profoundly affecting every microorganism. Among the variables are gases, water and ion concentrations, and carbon, nitrogen, and other nutrient sources, as well as metabolites produced by adjacent cells and other organisms. *Cryptococcus neoformans* is a globally distributed fungus found most commonly associated with soil, bird guano, and trees. The organism is exposed to dramatic environmental changes during infection of a mammalian host, and it must be able to sense these changes and adapt for successful colonization and proliferation. This flexibility to adapt to different conditions implies a strong capacity to sense and respond to extracellular signals.

A thorough understanding of how *C. neoformans* senses its environment is critical in light of the fact that many of these signals are known to regulate the virulence of this fungus. There have been many recent advances in understanding the sensory process that *C. neoformans* uses to adapt successfully to its changing environment. Sensors are required to transmit information about the environment that subsequently activates appropriate response pathways. Indeed, *C. neoformans* has developed a delicate sensing machinery to perceive diverse external stimuli and intricately adjusts its physiology for proper responses. Sensing nutrient deprivation (starvation), pheromones, and a small pheromone-like molecule are described in other chapters. In the following chapter we focus on the environmental signals that differ most dramatically between the outside and inside of the human host, since perturbation of a sensory mechanism may serve as an attractive lead for future drug development processes.

Alexander Idnurm, Division of Cell Biology and Biophysics, School of Biological Sciences, University of Missouri-Kansas City, 5100 Rockhill Rd., Kansas City, MO 64110. **Yong-Sun Bahn**, Department of Biotechnology, Center for Fungal Pathogenesis, College of Life Science and Biotechnology, Yonsei University, Seoul 120-749, Republic of Korea. **Wei-Chiang Shen**, Department of Plant Pathology and Microbiology, National Taiwan University, No. 1 Roosevelt Road, Sec 4, 10617 Taipei, Taiwan. **Julian C. Rutherford**, Institute for Cell and Molecular Biosciences (ICaMB), Medical School, Newcastle University, Catherine Cookson Bldg., Framlington Place, Newcastle upon Tyne, NE2 4HH, UK. **Fritz A. Mühlschlegel**, School of Biosciences, University of Kent, Canterbury, Kent, CT2 7NJ, UK and East Kent Hospitals University NHS Foundation Trust, Clinical Microbiology Service, William Harvey Hospital, Ashford, Kent, TN24 0LZ, UK.

SENSING STRESS

"Stress" can be defined in many ways: for fungi, it can be considered to be external forces that exert physiological constraints on the organism. Two signaling pathways are primarily involved in sensing, response, and adaptation of *C. neoformans* to a wide variety of environmental stresses. One is the high-osmolarity glycerol response (HOG) pathway in association with the two-component-like phosphorelay system, and the other is the protein kinase C (PKC)/Mpk1 mitogen-activated protein kinase (MAPK) pathway (see reviews in references 6 and 15). The HOG pathway not only is involved in defense against a range of environmental stimuli, including osmotic shock, high temperature, UV irradiation, oxidative damages, and antifungal drug and toxic metabolite treatment, but also modulates diverse cellular functions of *C. neoformans*, such as production of two major virulence factors, melanin and capsule, and sexual differentiation (4, 7, 8). The PKC/Mpk1 MAPK pathway plays key roles in both sensing stress and maintaining cell wall integrity under stressed conditions (15). Besides these two major stress-sensing signaling pathways, the Ca^{2+}/calcineurin-signaling pathway also plays a significant role in responding to external cues, such as high-temperature shock and cell wall stress (59, 76).

The *Cryptococcus* HOG pathway shares common features observed in HOG pathways of other fungi and yet exhibits unique features not observed in other fungi. In common with other fungi, the *Cryptococcus* HOG pathway consists of two signaling modules, the two-component-like phosphorelay system and the Hog1 MAPK modules. The phosphorelay system is composed of hybrid sensor kinases harboring a histidine kinase and response regulator (RR) domains in a single polypeptide, a histidine-containing phosphotransfer protein (HPt), and RRs. The Hog1 MAPK module comprises the Ssk2-like MAPK kinase kinase (MAPKKK), the

Pbs2-like MAPK kinase (MAPKK), and the Hog1 MAPK. A total of seven hybrid sensory histidine kinases (Tco1 to Tco7 [two-component-like proteins]), a single HPt (Ypd1), and two RRs (Ssk1 and Skn7) have been identified in *C. neoformans*, all of which are evolutionarily conserved in other fungi, although the numbers of hybrid sensor kinases vary between species. The general regulatory mechanism of the HOG pathway in fungi is as follows. A hybrid sensor kinase senses external signals, autophosphorylates a His residue in the histidine kinase domain, and transfers its phosphate group to an Asp residue in the RR domain. The phosphate is subsequently transferred to a His residue in the HPt protein, which is relayed to the Asp residue in one of two RRs. The phosphorylated RRs are either activated or inactivated and appear to interact with the Ssk2-like MAPKKKs. The interaction between RRs and Ssk2 triggers the autophosphorylation of the Ssk2-like MAPKKK for its activation. Activated Ssk2-like MAPKKK in turn activates the Pbs2-like MAPKK, which dually phosphorylates Thr and Tyr residues of the Hog1 MAPK. Phosphorylated Hog1 MAPK dimerizes and translocates into the nucleus to activate downstream target genes, such as glycerol biosynthesis genes, in response to osmotic shock (6, 45, 46) (Fig. 1).

Several unique features have been uncovered in the *Cryptococcus* HOG pathway. Notably, a number of *C. neoformans* clinical and environmental strains contain constitutively phosphorylated Hog1 under unstressed, normal conditions (8, 57). Interestingly, the constitutive phosphorylation level of Hog1 appears to be correlated with levels of stress resistance (the more Hog1 is phosphorylated, the higher the stress resistance) and virulence (4, 7, 8, 57). Furthermore, *C. neoformans* strains containing constitutively phosphorylated Hog1 (e.g., strain H99) exhibit other stress-independent phenotypes, such as increased melanin and capsule production and enhanced sexual differentiation, possibly by cross-talking with other signaling pathways such as cyclic AMP (cAMP) signaling and pheromone-responsive MAPK pathways, whereas *C. neoformans* strains with no constitutive Hog phosphorylation (e.g., strain JEC21) do not (4, 7, 8). In response to stresses, such as osmotic shock or treatment with fludioxonil (the phenylpyrrole class of antifungal drugs) or methylglyoxal (a toxic metabolite), Hog1 in strain H99 is rapidly dephosphorylated, whereas Hog1 in strain JEC21 is phosphorylated, similar to Hog1 in other fungi (8). The constitutive and induced phosphorylation of Hog1 is solely dependent on Pbs2 MAPKK, the deletion of which generates phenotypes highly comparable to those of the *hog1*Δ mutant (8). Unlike *Saccharomyces cerevisiae* and *Schizosaccharomyces pombe*, which encode multiple MAPKKKs, *C. neoformans* contains a single MAPKKK, named Ssk2, upstream of Pbs2 MAPKK (7). Interestingly, it has been found that Ssk2 interfaces between the MAPK module and the phosphorelay

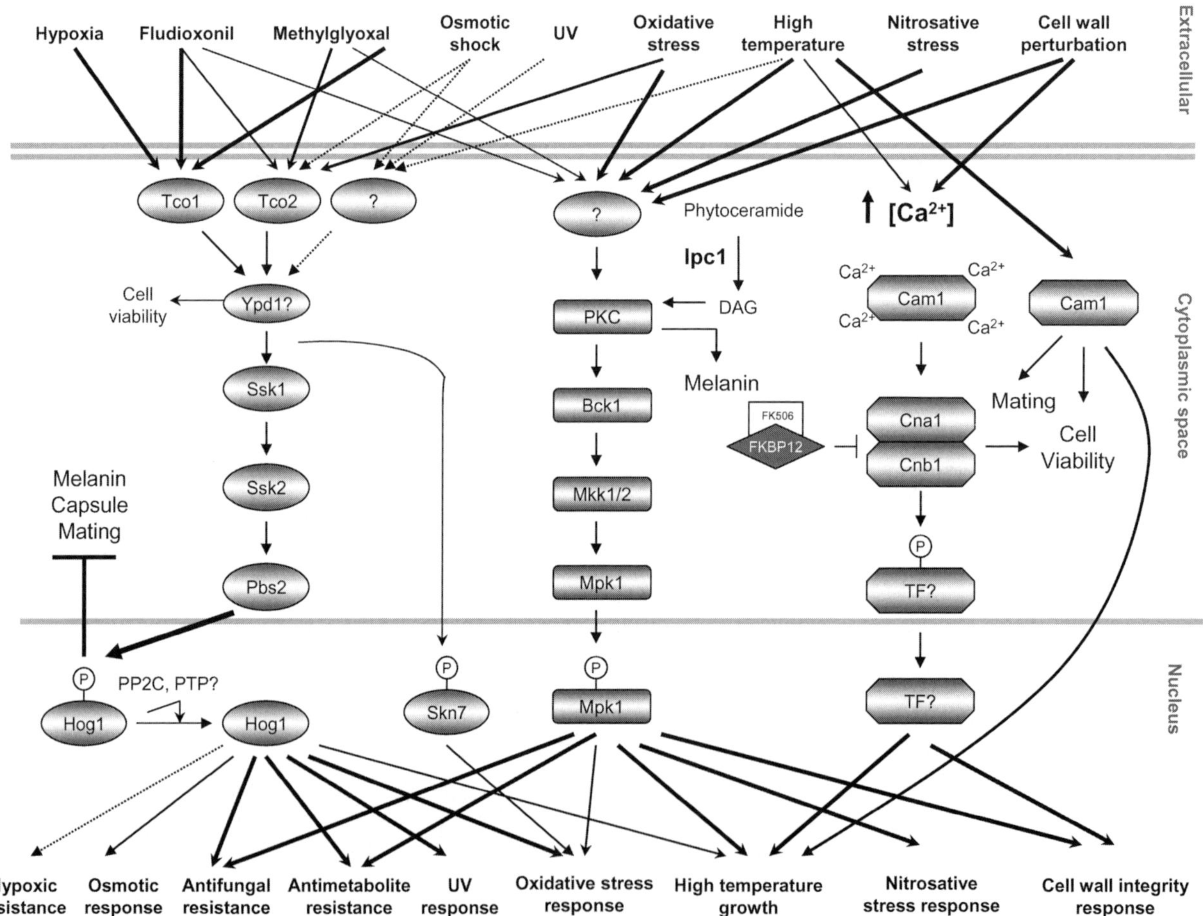

FIGURE 1 Stress-sensing pathways of *C. neoformans*. Extracellular stresses are sensed and the signal transmitted via signaling pathways. At present only a handful of regulators with overlapping functions are known that enter, or are predicted to enter, the nucleus to alter transcription.

system to play a critical role in phosphorylation of Hog1. Different alleles of Ssk2 underlie whether or not strains under unstressed conditions contain constitutively phosphorylated Hog1 (7).

Of the two RRs, Ssk1 and Skn7, Ssk1 is the major upstream signaling regulator of the Ssk2-Pbs2-Hog1 MAPK pathway since the *ssk1*Δ mutant phenotypes are highly comparable to those of the *ssk2*Δ, *pbs2*Δ, and *hog1*Δ mutants (4). Skn7, which contains a DNA-binding domain in addition to the RR domain, is also involved in responses to high-salt and oxidative stresses, melanin production, and virulence (4, 100). Although it has not been demonstrated that the two RRs interact with the single phosphotransfer protein Ypd1, Ypd1 appears to be essential for viability in *C. neoformans* (Y.-S. Bahn et al., unpublished data). Among the seven Tco proteins, Tco1 and Tco2 play redundant and distinctive roles in regulating the phosphorelay system. Both Tco1 and Tco2 play shared roles in conferring sensitivity to the antifungal agent fludioxonil (4). However, Tco2, but not Tco1, is partly involved in protecting cells from oxidative damages and methylglyoxal treatment (4). In contrast, Tco1 is involved in sexual differentiation and melanin production in a positive and negative manner, respectively (4). Importantly, Tco1 is also implicated in sensing hypoxia, and disruption of *TCO1* attenuates virulence of *C. neoformans* (discussed later in this chapter) (21).

The PKC/Mpk1 MAPK pathway also plays a vital role in defending *C. neoformans* against various forms of stresses, including osmotic shock, high temperature, and oxidative and nitrosative damages (Fig. 1). Besides its role in stress responses, the PKC/Mpk1 pathway is involved in capsule and melanin production as well as maintenance of cell wall integrity. The central components of the pathway in fungi include the small GTP-binding protein Rho1, PKC, Bck1 MAPKKK, Mkk1 (also known as Mkk2) MAPKK, and Mpk1 (also known as Slt2) MAPK (62). In *C. neoformans* the Mpk1 MAPK is activated by phosphorylation in response to cell wall–damaging agents (e.g., calcofluor white), oxidative stress (e.g., diamide and H_2O_2), and nitrosative stress (e.g., $NaNO_2$) (34). Activation of Mpk1 in response to stresses is completely dependent on Pkc1 (34). Furthermore, the *mpk1*Δ, *mkk1*Δ, and *bck1*Δ mutants show hypersensitivity to fludioxonil, methylglyoxal, oxidative stress, and high temperature in an equivalent manner (7), supporting the model that Bck1, Mkk1/Mkk2, and Mpk1 function in a linear signaling pathway. Deletion of the *PKC1* gene generates stress-related phenotypes comparable to those of the *mpk1*Δ mutant, including hypersensitivity to cell wall inhibitors, oxidative and nitrosative stresses, and high temperature (34). However, unlike the *bck1*Δ, *mkk1*Δ, and *mpk1*Δ mutants, the *pkc1*Δ mutant is nonviable under normal growth conditions, but can be rescued by osmotic stabilization (e.g., 1 M sorbitol) (34), indicating that Pkc1 regulates multiple signaling pathways other than the Mpk1 MAPK pathway. Supporting this interpretation, the *pkc1*Δ mutant shows enhanced capsule production and reduced melanin production (34, 44), which are phenotypes that are not observed in the *mpk1*Δ, *mkk1*Δ, and *bck1*Δ mutants. Of particular relevance for altered melanin production in *pkc1*Δ mutants is that previous studies had shown that inositol-phosphorylceramide synthase-1 (Ipc1) regulates melanin biosynthesis (65), and the basis for this observation is that Pkc1 is activated by diacylglycerol produced by Ipc1 (43, 44).

The Ca^{2+}/calcineurin pathway is also involved in a variety of stress responses, including growth at high temperature and membrane and cell wall stress (33, 59). In response to certain external stimuli, intracellular Ca^{2+} concentrations are transiently increased, and subsequently calmodulin, a calcium-binding protein, is activated (Fig. 1). Ca^{2+}-bound calmodulin in turn activates the Ser/Thr-specific phosphatase calcineurin, which is a heterodimer composed of Cna1 (a catalytic A subunit) and Cnb1 (a Ca^{2+}-binding regulatory B subunit). Calcineurin is dispensable for viability at low temperature but required for growth at body temperature (37°C) (32, 76). Calcineurin promotes cell wall integrity in concert with the PKC/Mpk1 MAPK pathway (58). Perturbation of the Ca^{2+}/calcineurin pathway leads to induction of the *FKS1* gene, encoding the sole β-1,3-glucan synthase found in *C. neoformans*, in an Mpk1-dependent manner, indicating that the two signaling pathways compensate for each other to maintain cell wall integrity (58). Calmodulin mediates multiple signaling pathways since it is essential for viability of *C. neoformans* regardless of growth temperature. Analysis of a *C. neoformans* strain in which calmodulin expression is compromised indicates that calmodulin is involved in cellular morphogenesis and growth at high temperature in Ca^{2+} and calcineurin-independent and -dependent manners (60). Normal calcineurin function is blocked by the pharmacological agents cyclosporine A and FK506 (tacrolimus) via the cyclophilin A and FKBP12 proteins, respectively, which are targeted to the hydrophobic interface between the catalytic A and regulatory B subunits. Therefore, FK506 and its derivatives are considered to be antifungal agents by themselves or in combination with other antifungal agents, such as fluconazole, that show synergistic inhibitory effects on cell viability (75, 78).

SENSING LIGHT

Light is one of the most important energy sources for life on Earth. Light is also a ubiquitous signal that may provide clues about location in the environment and the time of day. Light can also act as a stress or a deleterious form of radiation, particularly in the UV wavelengths. Photoresponses are widely observed in many organisms, including the fungi (80). The molecular mechanisms of light perception and responses in diverse filamentous fungi have been the subjects of intensive studies since 1996, when the first photoreceptor gene was cloned from *Neurospora crassa* (9, 22). In vitro, *C. neoformans* grows vegetatively as unpigmented, budding yeast cells, and its responses to light are less pronounced than in some of the filamentous fungal species. A photoresponse in *C. neoformans* was initially observed during the study of sexual differentiation (88): light was found to inhibit the production of dikaryotic filaments associated with the mating process among different serotype strains (51, 64) (Fig. 2). As with the majority of fungi, *C. neoformans* is capable of sensing blue wavelengths in the visible light spectrum (51, 64). Bioinformative analysis reveals three candidate photoreceptors in the genome, encoding an opsin (*OPS1*), a phytochrome (*PHY1/TCO3*), and a white collar 1 homolog (*BWC1/CWC1*). From the analysis of mutant strains, it is only the homolog of *N. crassa wc-1* that functions to modulate the currently known light responses of *C. neoformans*. However, the full response of *C. neoformans* to light remains to be elucidated.

The white collar 1 gene is known either as *BWC1* or *CWC1* (basidiomycete or *Cryptococcus* white collar 1) (51, 64). The phylum designation reflects a difference in the domain arrangement compared to the homologs from ascomycetes and zygomycetes, as most basidiomycete homologs do not contain a zinc finger domain. For simplicity, the name

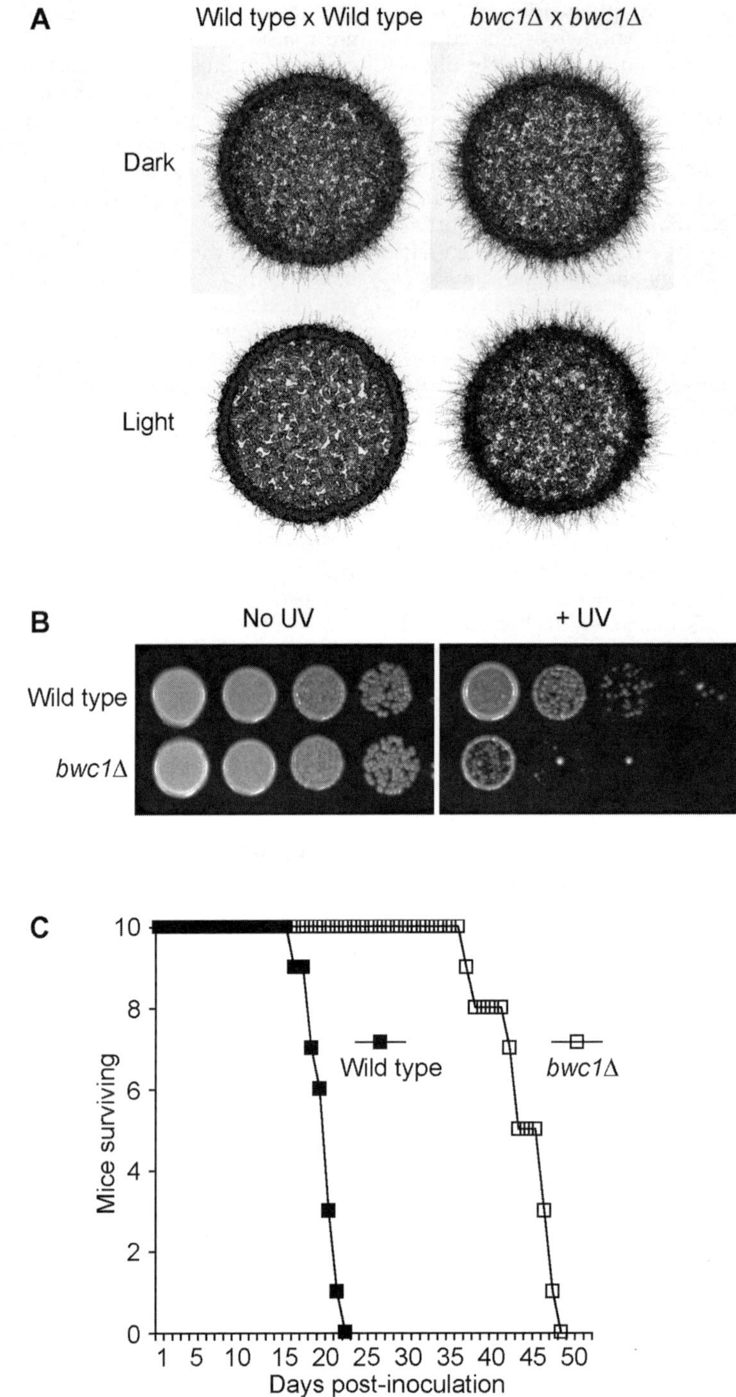

FIGURE 2 Light sensing is mediated by homologs of white collar 1 and 2 to control at least three responses in *C. neoformans*. Deletion of either the *BWC1* or *BWC2* gene renders *C. neoformans* insensitive to the inhibition of mating by light, increases UV sensitivity, and reduces virulence. (A) Strains of both mating types were cocultured in the light or dark for 2 days at room temperature on Murashige-Skoog medium. (B) Tenfold serial dilutions of wild-type and *bwc1* mutant strains were made on yeast extract peptone dextrose medium, and one set was irradiated with UV (100 J/m^2). Plates were incubated for 2 days at 30°C. (C) Ten mice were infected with *C. neoformans* strains, and survival was monitored over time. Data from Idnurm and Heitman (51).

WC1 (rather than *BWC1/CWC1*) will be used in this chapter. The *WC1* gene regulates sexual development (mating and monokaryotic fruiting), UV sensitivity, and virulence as determined through the phenotypic analysis of gene deletion strains in both serotype A and D isolates (51, 64). Wild-type *C. neoformans* mating and monokaryotic fruiting are inhibited by light, with blue wavelengths being those sensed, while green, red, and far-red have no effect (51, 64). Notably, mutation of the *WC1* gene abolishes this inhibition, and this light sensitivity phenotype is dosage-dependent since partial photoresponses are seen in the unilateral mutant crosses. Light negatively regulates at least two stages of sexual reproduction, that of cell fusion and hyphal development subsequent to cell fusion. Deletion of the *WC1* gene also renders the strains more sensitive to UV irradiation. Due to a lack of photolyase in the *C. neoformans* genome, Wc1 is also proposed to mediate UV resistance in response to UV light, but the mechanism is unlikely to be enzymatic. More intriguingly, Wc1 contributes to the full virulence of *C. neoformans* (Fig. 2). The interior of a host can be considered a dark or red environment, and the absence of light or specific wavelengths may be sensed by the infecting organism. The *wc1* mutant strains exhibit a decrease in virulence to a degree that infected mice survive about twice as long compared to those infected with the wild-type isolate in a nasal inhalation assay (51). An independent study searching for components required to infect an animal host also identified and confirmed a role for Wc1 (and Wc2; see below) in virulence (63).

Blue light sensing in at least three major lineages of the fungi is mediated by white collar homologs (for the most recent reviews see references 22, 41, and 80). They were first identified and named in the ascomycete fungus *N. crassa*, and their descriptive name derives from the fact that the *N. crassa* mutants no longer produce the orange carotenoid pigmentation in hyphae in response to light, giving rise to colonies with white edges (collars) in response to light. The white collar 1 proteins invariably comprise a light-oxygen-voltage (LOV) domain that binds a flavin chromophore that acts as the photoreceptor part of the protein. Then two PAS (Per-Arnt-Sim) domains that interact with other proteins, and in most fungi, but not *C. neoformans* and most other basidiomycetes, also have a zinc finger DNA binding domain such that they can act as transcription factors. In addition to white collar 1, all characterized light-sensing fungi also have an interacting component known as white collar 2. This protein features a PAS domain and a zinc finger domain.

In *C. neoformans*, the homologous gene to *N. crassa wc-2*, *BWC2/CWC2* (hereafter *WC2*), was identified through both bioinformatic reverse genetic and forward genetic approaches. Of particular note was the use of two independent *Agrobacterium* insertional mutagenesis studies that revealed mutant strains that no longer exhibited mating inhibition when exposed to light. The white collar 2 homolog emerged from both screens (51, 102). The phenotypes of the *wc2*Δ mutant are identical to those of the *wc1*Δ mutant in terms of mating, UV sensitivity, and virulence, and *wc1 wc2* double mutants show phenotypes identical to the single mutants with respect to mating and UV sensitivity. Both proteins physically interact in a yeast two-hybrid assay (51). These data indicate that the two proteins function cooperatively in a complex.

Based on what is known about *N. crassa* WC-1 and WC-2, it is hypothesized that the corresponding *C. neoformans* proteins function similarly. Light is perceived via a flavin molecule within the LOV domain of Wc1. A structural change occurs that is transmitted via the PAS-PAS domain interaction between the two proteins. Transcription of downstream target genes, whose promoters are physically bound to Wc2 via its zinc finger domain, is then altered. However, at present the above model remains largely hypothetical. Deletion of the LOV and PAS domains from Wc1 and the PAS and zinc finger domains of Wc2 demonstrates that these regions are required for protein function (102).

WC2 transcript abundance is regulated by light and requires the presence of *WC1*, while *WC1* is expressed at a constant level (51). Interestingly, overexpression of *WC1* or *WC2* leads to an enhanced repression of mating, but only upon light irradiation and as long as a wild-type copy of the other gene is present (64, 102). This strongly suggests that posttranscriptional modification, possibly phosphorylation, by light is required for the proper functions of both proteins. How this is controlled is currently unclear, but such posttranscriptional modifications may be required to tune the photosensory machinery to exhibit inhibitory responses to light.

One question that remains unanswered is why *C. neoformans* would prefer to mate in the dark. At present it is only possible to speculate that undergoing sexual reproduction in darkness provides some evolutionary benefit to the fungus. It could be that during mating DNA repair enzymes are downregulated, and as such, harmful UV irradiation would be detrimental to the fungus if mating is initiated or progresses in the light. On the other hand, the timing from cell fusion to basidiospore production in the laboratory lasts more than a single day, so at some stage during mating the fungus is exposed to light.

To understand downstream components in light signaling, one approach has been forward genetics. From a large-scale screen of more than 10,000 transfer DNA (T-DNA) insertional mutants, 134 strains were isolated that showed restored filamentation in light (102). Seventeen T-DNA flanks were identified. Insertions were found in the *WC2* gene, and the link between T-DNA insertion event and mutant phenotype was confirmed by genetic crosses or complementation analysis in the *SSN8* homolog (associated with the monokaryotic fruiting phenotype), a histone methyltransferase, cyclic dependent kinase, and a predicted pyridoxal kinase required for bud-site selection in *S. cerevisiae*. The *SSN8* gene was characterized in further detail. It encodes a predicted component of RNA polymerase II and thus may regulate a number of different transcriptional processes in the cell, perhaps directly affecting the photosensory genes or their targets (102).

There are several future paths toward understanding the light responses of *C. neoformans*. First, the functions of the opsin and phytochrome in light sensing, if any, are currently unknown. Second, the conformational effects of light on the white collar complex are also unknown in *C. neoformans* or in detail from any fungal system since the proteins are notoriously difficult to purify in abundance for structural and other studies. Third, the genes that are regulated by light remain to be fully elucidated, although this is being addressed through transcript profiling microarray and genetic screens, particularly since these genes should control *C. neoformans*' ability to proliferate in the wild and cause disease in humans.

SENSING AMMONIA AND OTHER NITROGEN SOURCES

Nitrogen assimilation is fundamental for all life, and fungi can utilize a wide range of nitrogen sources. Generally, nitrogen metabolism has not been extensively characterized in *C. neoformans*. However, studies of other model fungi have identified concepts relating to nitrogen utilization that may be conserved in *C. neoformans*. Nitrogen sources can be divided between those that are readily assimilated, such as the preferred nitrogen sources ammonium or glutamine, and others that are only acquired from the environment in the absence of a preferred nitrogen source, such as proline. Specific catabolic pathways and permeases are required for the assimilation of individual nitrogen sources, and these are regulated at the transcriptional level (reviewed in reference 99). Activation of these pathways occurs when cellular nitrogen levels become low and, in many cases, requires the additional presence of a specific nitrogen source as an inducer. This mechanism of control is known as nitrogen catabolite, or metabolite, repression. *C. neoformans* contains a homolog of the Gat1/AreA/Nit2 family of transcription factors that mediate nitrogen catabolite repression in fungi, although the role of this homolog is yet to be established (99). Furthermore, a diagnostic tool for *C. neoformans* is its inability to utilize nitrate as a sole nitrogen source, and its genome does not contain the nitrate and nitrite reductase genes that are commonly used model genes for understanding catabolite repression in ascomycete species (82, 90). Clearly, a full understanding of nitrogen metabolism is required so that the influence of nitrogen utilization by *C. neoformans* during infection can be established. Two areas of nitrogen metabolism that have been studied illustrate the significance of this area in the life cycle of *C. neoformans*. These relate to nitrogen source availability in pigeon guano and the role of ammonium sensing during invasive growth and mating.

C. neoformans survives well in avian excreta, which is a selective medium for this yeast in the environment (1). Pigeon guano contains creatinine and uric acid, both of which support *C. neoformans* growth under laboratory conditions (79, 93). In one study, the distribution of *C. neoformans* within a particular environmental site correlated with the level of uric acid and creatinine in the pigeon excreta from which the fungus was recovered (49). Creatinine deiminase catalyzes the hydrolysis of creatinine to provide ammonium, and in *C. neoformans* expression of this enzyme is induced by the presence of creatinine and specifically repressed by ammonium (79). This is consistent with the gene encoding creatinine deiminase being regulated via nitrogen catabolite repression. Growth of *C. neoformans* on uric acid or xanthine, a precursor of uric acid, results in the formation of microbodies within the cytoplasm that contain the catabolic enzymes that are required to assimilate these nitrogen sources (31). Microbody induction in response to uric acid or xanthine in otherwise minimal media is also consistent with regulation via nitrogen catabolite repression.

Urea, another product of purine degradation, has been linked to the ability of *C. neoformans* to cause disease (24). Urea is converted to ammonium and carbamate by the urease enzyme, which in *C. neoformans* is encoded by the *URE1* gene. A *C. neoformans ure1* mutant was significantly less virulent than a wild-type strain when analyzed using both the murine inhalation and intravenous models of cryptococcosis (24). A detailed analysis of the localization of wild type and the urease-lacking mutant in vivo concluded that urease facilitates the sequestration of the yeast within small capillaries and therefore promotes the aggregation of fungal cells in organs with closed capillary beds (77). Urease does not appear to be required for survival in the lungs, the blood, or the central nervous system or for dissemination to the spleen (77). The suggestion, therefore, is that urease-mediated sequestration within capillaries enables *C. neoformans* to take hold and accumulate within the small vessels of the brain. Cryptococcal cells are then able to gain access to the brain, and yeast proliferation takes place. However, the mechanistic link between urease activity and microvascular sequestration of *C. neoformans* is not known. Indeed, there have been reports of urease-defective strains isolated from patients, which complicates interpretation of the work carried out using the murine models of infection (10, 83). One such strain was used to identify, by complementation, the *URE2* gene that restored growth of the strain on urea medium, but its role in urease activity is not clear (98). No doubt this area will be the focus of future research, as animals including humans do not contain urease, and urease is also required for the ability of *Helicobacter pylori* to cause gastric ulcers (28), making this enzyme an interesting potential therapeutic target.

Nitrogen availability can influence fungal development, and *C. neoformans* mating and invasive growth are initiated in response to growth on low levels of ammonium. Amt1 and Amt2, which are low- and high-affinity permeases, respectively, mediate ammonium uptake, and a mutant lacking both permeases grows poorly on low-ammonium medium (84). Amt2 is also required for the induction of mating and invasive growth in response to ammonium availability (84). Studies of other model pathogenic fungi such as *Candida albicans* have identified homologs of Amt2 that are involved in the ammonium-responsive induction of pseudohyphal growth (14, 91). Various studies support the hypothesis that certain ammonium transporters have roles in both ammonium transport and the sensing of ammonium availability. The role of Amt2 in mating and invasive growth suggests that it may also have similar functions.

The mechanism of this transporter-mediated sensing is not clear, although current evidence supports the hypothesis that the permease interacts directly with a signal transduction pathway. In this model the permease acts like a receptor, which responds to the binding and/or transport of ammonium through the protein. Analysis of bacterial homologs of the fungal ammonium transporters has identified certain classes that are fused to domains with a known sensor and/or regulatory function, supporting the model that ammonium transport is linked to sensing (97). Such a mechanism would therefore be distinct from the permease-altering intracellular levels of ammonium, with a consequent change in intracellular metabolite levels that are then sensed by some other internal mechanism. The potential mechanistic similarities between filamentous growth in other yeasts and ammonium-induced mating in *C. neoformans* is supported by the finding that rapamycin inhibits both processes (25). Rapamycin inhibits the Tor protein kinases that control cellular growth in response to nutrient levels and that in *S. cerevisiae* regulate the cellular localization of transcription factors involved in nitrogen catabolite repression (11, 13, 18). Of considerable interest will be the extent to which the Amt2 ammonium signal pathway interacts with other pathways that control mating and the role of the Tor pathway generally in the regulation of nitrogen metabolism in *C. neoformans*.

SENSING CO_2

CO_2 is a ubiquitous, linear chemical compound consisting of one carbon atom covalently bonded to two oxygen atoms. It has a molecular mass (M) of 44 g/mol and is found as a colorless, odorless trace gas at ambient temperature and pressure. CO_2 plays an important role in our ecosystem by comprising about 0.0365% of the earth's atmosphere (42, 81).

It is a greenhouse gas, whose levels in the atmosphere are rising due to increasing combustion of fossil fuels combined with excessive deforestation. In fact, CO_2 levels in the atmosphere have increased significantly since the industrial revolution, contributing to a global greenhouse effect responsible for the gradual warming of the earth (42, 81).

CO_2 is also an essential molecule for plants and photosynthetic prokaryotes such as cyanobacteria that, in the process of carbon assimilation, fix this molecule to generate three-carbon sugar phosphates in the Calvin cycle (16). Furthermore, CO_2 is produced by aerobically respiring organisms as a by-product of cellular metabolism. Here the breakdown of sugars and fatty acids leads to the production of CO_2, which in higher animals circulates in the body before being excreted (39). As a consequence, substantially increased levels (5%) of CO_2 can be found in the bloodstream and tissues when compared to the surrounding atmosphere. In blood the vast majority (up to 80%) of CO_2 is carried in the form of bicarbonate ions (HCO_3^-) generated by the erythrocyte enzyme carbonic anhydrase (CA) (39).

CO_2 also serves as an important signaling molecule, and consequently many organisms have developed sensing mechanisms for its detection. For example, increased levels of CO_2 are sensed by female mosquitoes to determine the location of prey, whereas a decrease in CO_2 affects gamete formation (exflagellation) in malaria parasites (19, 29). Thus, CO_2 sensing is an integral component in the spread of malaria. *Bacillus anthracis* detects elevated CO_2 levels in its environment, which induce the synthesis of the antiphagocytic polysaccharide capsule (66). In fact, capsule operon transcription is more than 20-fold increased at CO_2 levels above 5%, and the synthesis of the *B. anthracis* tripartite toxin, which is encoded by the *lef*, *cya*, and *pagA* genes situated on the pXO1 plasmid, is enhanced about 10-fold (89). In *Streptococcus pyogenes* elevated CO_2 levels increase the production of the antiphagocytic M protein, which is controlled by the regulatory Mga protein, a transcription factor with similarity to *B. anthracis* AtxA (17). In addition, elevated CO_2 concentrations influence the expression of fungal virulence determinants promoting the switch between yeast and filamentous growth forms in the fungal pathogen *C. albicans* and stabilize the opaque mating-competent cell type (48, 56). This reversible morphological transition is considered an important virulence attribute.

During its life cycle *C. neoformans* is exposed to dramatic changes of CO_2 levels in the surrounding atmosphere. While in its natural habitat, it is exposed to the low ambient CO_2 concentrations of the normal atmosphere, whereas upon inhalation of cells and subsequent infection it encounters the relatively high concentrations (5%) of CO_2 found in the body. These very different conditions profoundly influence the physiology of this fungus (Fig. 3). One example is capsule biosynthesis, which is regulated by host factors. Serum, pH and, importantly, the body's elevated CO_2 all promote capsule biosynthesis (38, 103). Another example is the mating pheromone production pathway, which is critically important for basidiospore generation. *C. neoformans* readily mates at 24°C and atmospheric CO_2 levels. In contrast, high concentrations of CO_2 block sexual reproduction in wild-type strains by repressing normal pheromone production in the presence of the opposite mating partner (3).

Molecular characterization of the CO_2-sensing system in *C. neoformans* not only provides novel insight into microbial pathogenesis, but has already revealed a number of compounds that interfere with this pathway (53, 54, 87). The CO_2-sensing system of *C. neoformans* is currently composed of two enzymes, adenylyl cyclase (AC) and CA, that play major roles in this fungus's physiology and metabolism (5) (Fig. 3). Indeed, *C. neoformans* capsule biosynthesis is mediated by the second messenger cAMP synthesized by the AC Cac1p (2). The critical role of this enzyme is further emphasized by the fact that *C. neoformans cac1Δ* mutants are avirulent (2). Elements of the Gpa1-cAMP signaling cascade pathway have been recently investigated for their involvement in fungal CO_2 sensing. Thus far, no direct links have been reported between heterotrimeric G proteins and CO_2 chemosensing, narrowing the focus to Cac1p. Recent reports show that the enzymatic activity of purified, recombinant *C. neoformans* Cac1p was stimulated in a dose-dependent manner by physiological concentrations of bicarbonate (25 mM) as found in blood (56, 72). This identifies Cac1p as a major sensor for ambient CO_2. Although HCO_3^- seems to be the AC-activating form of CO_2 in the cells, recent reports clearly show that a subset of ACs is also activated directly by molecular carbon dioxide (96) (Fig. 4). In fact, both classical transmembrane ACs and the more recently identified soluble ACs found in mammals along with other fungal and bacterial ACs are equally responsive to elevated concentrations of bicarbonate and/or CO_2, suggesting that AC-mediated CO_2 sensing is evolutionarily conserved (55, 94). This remarkable observation is further supported by the analysis of high-resolution crystal structures of bacterial soluble AC-like proteins that undergo conformational changes in response to bicarbonate, thus increasing the production of cAMP (55, 94).

CAs form a large family of metalloenzymes catalyzing the conversion of CO_2 to bicarbonate ions (HCO_3^-) in a two-step process. However, this reaction can occur freely when the levels of CO_2 are sufficiently high (95). The CA family is subdivided into five evolutionarily distinct classes, and all known mammalian isoforms belong to the α class. The β class comprises a diverse set of CAs found in plants, algae, bacteria, and archaea, and the γ, δ, and ζ classes currently consist of a single or a few microbial enzymes (95).

C. neoformans possesses two CA genes designated *CAN1* and *CAN2* (3, 72). *CAN2*, which is expressed at high levels, is the major CA isoform in *C. neoformans*. Interestingly, *can2Δ* mutants are only able to grow in atmospheres enriched with carbon dioxide (5%), and this phenotype was annotated as high-CO_2 requiring (HCR) (3, 5). In contrast, *CAN1* shows low expression levels, and strains lacking this gene grow in the presence of low or high concentrations of CO_2. Therefore, *CAN2*, but not *CAN1*, is essential for growth in environments where CO_2 is limiting. However, the fact that heterologous overexpression of *CAN1* is able to complement partly the HCR phenotypes of the *S. cerevisiae* CA mutant (*nce103Δ*) indicates that Can1 also has a CA activity. The inability of the *can2Δ* mutant to grow in low CO_2 conditions can be rescued by the addition of intermediates of fatty acid biosynthesis or purine and pyrimidine biosynthesis to the growth medium (3). Palmitate, and to a lesser degree myristate, rescue the growth of the *can2Δ* strain, albeit not as effectively as supplementing the atmosphere with CO_2 (3). These findings suggest that the growth defect of the *can2Δ* mutant is partially attributable to insufficient bicarbonate being provided for the synthesis of fatty acids during intermediate metabolism.

Recently, Can2p was fully characterized biochemically (87). These studies demonstrated that Can2p has significant CA activity and is assigned to the β-CAs. Furthermore,

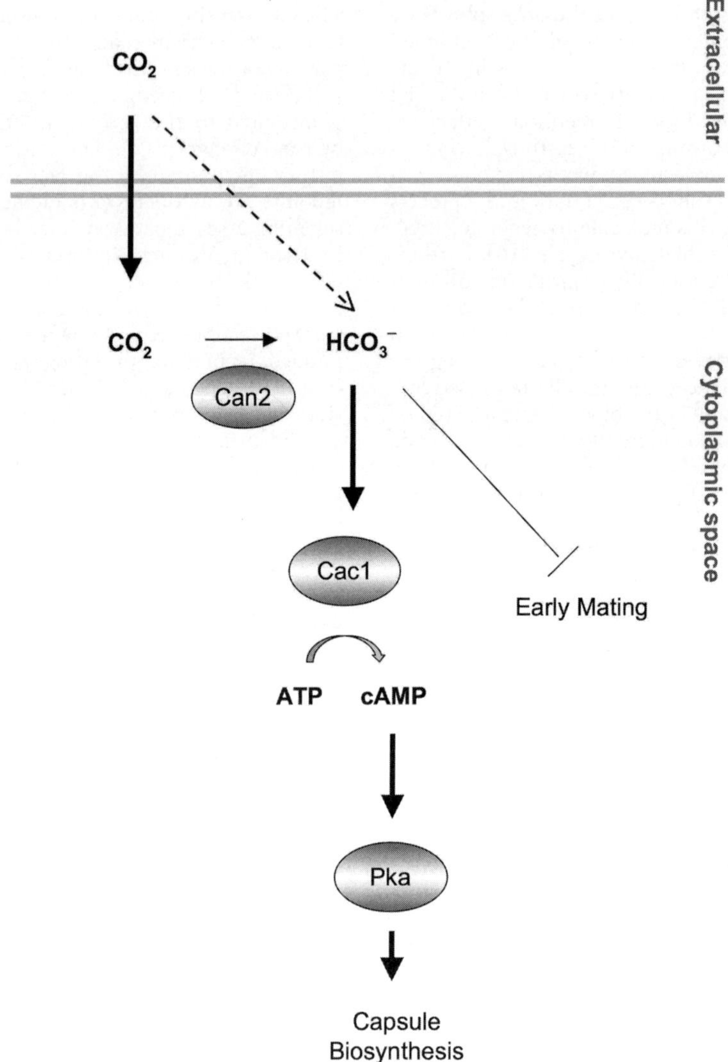

FIGURE 3 Carbon dioxide sensing in *C. neoformans*. CO_2 affects both capsule biosynthesis and mating. Carbon dioxide diffuses into the cell and is either spontaneously hydrated (elevated environmental concentrations of CO_2) to bicarbonate or alternatively enzymatically hydrated (low environmental concentrations of CO_2) by the action of the CA Can2. Bicarbonate stimulates the AC Cac1.

purified Can2p is inhibited by a range of CA inhibitors, some of which show good selectivity over the human isoforms (53, 54, 87). The Can2p crystal structure has been solved, revealing that this protein carries a unique N-terminal extension that can interact with the active site entrance of its own dimer or a thus far unidentified partner molecule (87), which could be AC.

CO_2 sensing via Can2p is not only essential for saprophytic growth of *C. neoformans* but may also play a role in cutaneous cryptococcosis, where this fungus must sustain growth when exposed to the normal atmosphere. In fact, the skin is the third most common organ for appearance of *C. neoformans* infection, and lesions occur in 10% of patients (67). By contrast, mutations in either *can1Δ* or *can2Δ* do not interfere with *C. neoformans* capsule biosynthesis upon exposure to elevated CO_2 or impact on deep-seated infections in virulence models (3). There are two potential explanations for these interesting results. First, since spontaneous conversion of CO_2 to bicarbonate occurs in high-CO_2 environments (including the host), sufficient amounts of bicarbonate could be generated to induce the capsule. Alternatively, Can1p and Can2p are not required for capsule biosynthesis and therefore virulence, favoring the possibility that it is molecular CO_2 and not bicarbonate directly impacting on capsule biosynthesis. Notably, incubation of *C. neoformans* in atmospheres of high CO_2 inhibits mating (3). In contrast, *can2Δ* mutants showed an increase in mating filaments compared to wild-type strains, suggesting a role of CA in mating. Although *can2Δ* mutants produce mating filaments, they are deficient in sporulation. Increased cAMP signaling is required for sporulation in *S. cerevisiae*, and it is thus possible that increased cAMP levels similarly promote sporulation in *C. neoformans* (3). One possibility is that Can2p generates sufficient intracellular bicarbonate to directly activate AC, which is essential for spore formation.

The current model of the CO_2-sensing system in *C. neoformans* suggests that under limiting concentrations this molecule diffuses into the cell and is subsequently hydrated to

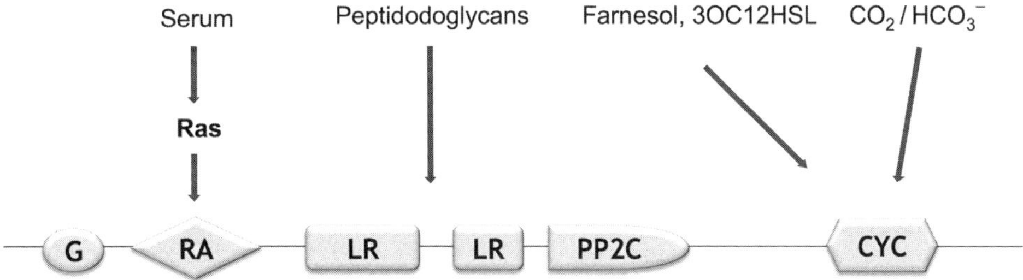

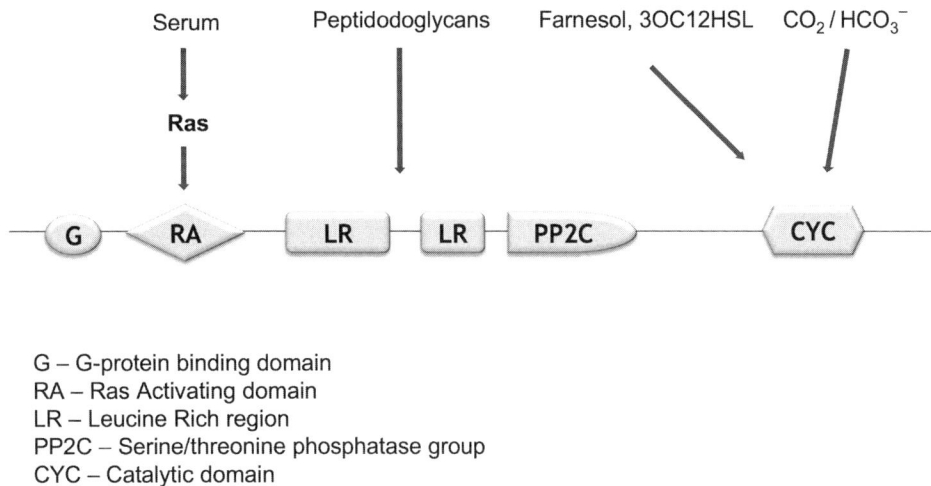

G – G-protein binding domain
RA – Ras Activating domain
LR – Leucine Rich region
PP2C – Serine/threonine phosphatase group
CYC – Catalytic domain

FIGURE 4 The fungal AC is a multidomain sensor. Serum, CO_2, and the quorum sensing molecules farnesol and 3-oxy-C12-homoserine lactone (3OC12HSL) influence AC activity in the fungal pathogen *C. albicans*.

HCO_3^- and fixed inside the cell by the CA Can2p. Bicarbonate would then serve as a substrate in a number of carboxylation reactions during the intermediate metabolism required for saprophytic growth. HCO_3^- would also stimulate the AC Cac1p, activating the cAMP signaling pathway that controls important virulence determinants including capsule biosynthesis and impacts on mating. In contrast, when this fungus is exposed to high levels of CO_2, Can2p activity is dispensable, as sufficient amounts of HCO_3^- are generated spontaneously to promote metabolism and stimulate cyclase activity.

ENVIRONMENTAL SENSING IN OTHER FUNGI: THE ROLE OF ACs

ACs also serve as environmental sensors in other fungi including the fungal pathogen of humans, *C. albicans*. In fact, *C. albicans* Cyr1p is thought to be able to directly sense and respond to changes in the environmental CO_2 concentrations (56). In *C. albicans* elevated CO_2 promotes a change in morphology, from yeast to filamentous, which is an important virulence determinant of this fungus. Importantly, it has also been shown that Cyr1p binds bacterial peptidoglycan-derived molecules found in the serum of the mammalian host (101). In contrast to CO_2/bicarbonate, which is thought to bind directly to Cyr1p's catalytic domain, the peptidoglycans were shown to bind a leucine-rich repeat domain located within the center of the *C. albicans* AC molecule (101). Thus Cyr1p is a multidomain environmental sensor, and serum-mediated activation of Cyr1p depends on Ras1p and occurs through a Ras1p-activating domain located in the amino terminus of the molecule (30). In addition to a role in regulating yeast-hyphal morphology, CO_2 regulates the white-opaque phenotypic switch that is required for mating competence (48). Opaque cells switch rapidly back to white cells in vitro at 37°C, providing a puzzle as to how mating could occur in animals. However, recent experiments show that high CO_2 levels promote the white-opaque switch and stabilize the opaque form at 37°C. This response is mediated at least at atmospheric and low (1%) CO_2 concentrations by AC and Ras1 because switching is reduced in strains with these genes mutated (48).

An additional link between *C. albicans* morphology and CO_2 is the observation that arginine and urea induce *C. albicans* filamentation in a density-dependent manner (35). The fungal arginase (Car1p) converts arginine to urea, which in turn is degraded by urea amidolyase (Dur1, Dur2p) to produce CO_2. When constructing a urea amidolyase deletion mutant, Ghosh et al. found that the latter was unable to filament subsequent to phagocytosis by macrophages and was compromised to form germ tubes in the presence of arginine or urea but not elevated CO_2 (35). These studies suggest a link between arginine metabolism, CO_2/bicarbonate activation of AC, and a mechanism for escaping attack by the host's defense.

Quorum sensing involves the regulation of genes in a cell-density-dependent manner. Quorum sensing contributes to the pathogenicity of some bacteria through the regulation of important virulence factors (92). Hornby and colleagues identified *C. albicans* as a eukaryotic organism with quorum sensing properties by showing that inhibition of *C. albicans* filamentation at high cell densities was an effect of the sesquiterpene quorum sensing molecule, farnesol (47). Recently, Davis-Hanna and colleagues have shown that addition of exogenous cAMP into farnesol-containing media restored *C. albicans* filamentation (26), suggesting that inhibition of cAMP signaling is a major target of farnesol. Furthermore, they showed that a common analog of the *Pseudomonas aeruginosa* quorum sensing molecule 3-oxy-C12-homoserine lactone (3OC12HSL) displayed similar effects on *C. albicans* filamentation and the cAMP signaling cascade as farnesol (26). Therefore, it is possible that activity of the fungal AC Cyr1 is also controlled by microbial quorum sensing molecules like farnesol and 3OC12HSL (40).

Asexual spore development in *N. crassa* is subject to circadian regulation (85). Circadian-regulated spore formation, also termed conidial banding, is suppressed by elevated CO_2 (86). Belden and colleagues found that CO_2 suppression of conidial banding is bypassed in a strain containing a dominant *ras-1* allele (12). These findings suggest cross-talk between CO_2 sensing, cAMP signaling, and the RAS/MAPK pathway in controlling circadian outputs in *N. crassa*.

SENSING O_2 AND HYPOXIA

In a situation similar to the difference in CO_2 concentrations between the environment and inside the human host, the oxygen levels encountered by *C. neoformans* vary dramatically, from 21% in the air to almost anoxic conditions in the human host, depending on the tissue. *C. neoformans* must be able to adapt to these different conditions, particularly since it is unable to grow anaerobically.

Two overlapping studies identified a set of conserved proteins involved in growth under hypoxic conditions (20, 21). Both partially originate from the discovery of a similar system found in *S. pombe*, in which an endoplasmic reticulum–associated putative sensor, Scp1 (SREBP cleavage-activating protein 1), triggers cleavage of an inactive transcription factor, Sre1 (sterol regulatory element-binding protein 1), to an active form (50). Sre1 is then transported to the nucleus, where it activates gene transcription. In research by Chang et al. (20), the homologs of Sre1 and Scp1 were identified, the genes mutated, and their phenotypes analyzed. In addition, the transcriptional response to low-oxygen conditions was assessed by microarray analysis. This study was performed in the serotype D background. In the studies by Chun et al. (21), the authors screened 1,200 deletion mutants for impaired growth under low oxygen, revealing four genes: *SCP1*, *SRE1*, *STP1*, and *TCO1*. Transcript profiles in response to hypoxic conditions were also analyzed. This study was performed in the serotype A background. Thus, both studies were conducted in independent laboratories with different strains, yet reached similar conclusions about the role of Sre1 and Scp1 in responding to low oxygen, providing robust evidence for the function of these genes. In both studies, deletion of either *SRE1* or *SCP1* reduced the ability of *C. neoformans* to grow under low oxygen concentrations (<3%). Knockout mutants have reduced virulence and are affected in growth in the brain and lungs of the infected animals. Western blot analysis demonstrated that Sre1 is cleaved in response to low oxygen. The transcript profile analyses showed that Sre1 regulates 50 to 100 genes, with key ones being required for sterol biosynthesis and metal ion homeostasis. Mutation of *SRE1* or *SCP1* causes changes in sterol composition, thus leading to increased sensitivity of *C. neoformans* to the azole antifungal drugs. Host hypoxia may explain in part the surprising efficiency of azoles against fungi at drug concentrations that are lower than those that are effective in vitro (21).

Analysis of deletion mutants identified two genes in addition to *SRE1* and *SCP1* that were required for full growth under hypoxic conditions (21). Stp1 is predicted to be the activating protease for Sre1. The other gene encodes Tco1, which is a histidine kinase that is required for full virulence and involved in activating the Hog1 pathway described in the first part of this chapter (4). How this gene is involved in this sensory process is currently unclear, although the lack of any specific transcriptional response suggests a posttranslational role. The double *tco1 sre1* mutant is more sensitive to low oxygen than each of the single mutants, providing evidence that at least two separate pathways exist for responding to low oxygen concentrations. This is further confirmed by the microarray studies, since not all hypoxia-induced genes are regulated by Sre1 (20, 21). For example, the heme biosynthesis enzyme *HEM13* is induced by hypoxia but independently of the presence of Sre1, and in *C. neoformans* this may implicate other sensors of low oxygen such as the homologs or equivalent of the *S. cerevisiae* Hap1 or Rox1 transcriptional regulators that are oxygen-dependent (20).

To further elucidate the mechanisms of oxygen sensing in *C. neoformans*, the response of the fungus to cobalt chloride ($CoCl_2$) has been recently examined (52, 61). $CoCl_2$ is believed to mimic hypoxia, although the mechanism of action is unknown. One hypothesis is that its action may in part be through inhibition of iron-dependent enzymes. In *C. neoformans*, there is a greater response to $CoCl_2$ than to hypoxia as assessed by transcription profiling, and furthermore the *SRE1* or *SCP1* mutants are highly sensitive to this metal (61). A multicopy suppressor screen in the *sre1* mutant strain showed that this sensitivity could be alleviated by overexpression of the *ERG25* gene and that the Sre1 protein directly binds to the *ERG25* promoter (61). The response to $CoCl_2$ was further studied by a large-scale screen of 30,000 T-DNA insertion mutants for those with impaired growth in the presence of $CoCl_2$ (52). Thirty-seven strains were identified and the mutated genes were identified, including multiple T-DNA insertions into *STE1* and *SCP1*, further confirming that these are the key players in responding to low-oxygen conditions. Nine other genes are implicated in mitochondrial function, and many of the insertional mutants exhibit elevated levels of reactive oxygen species (ROS), leading to a hypothesis that hypoxia induces production of ROS.

In addition to an oxygen requirement for respiratory functions, *C. neoformans* must also be able to evade ROS produced as part of the host defense system. While genes such as flavohemoglobin (*FHB1*) and superoxide dismutases (*SOD1* and *SOD2*) are required for full virulence (23, 27, 36, 73, 74), others such as the four catalases or two glutathione peroxidases are not (37, 68). Furthermore, the thioredoxins, thioredoxin reductases, and thiol peroxidases are essential for viability or indeed normal growth (69–71). Thus, detoxification of ROS is key for successful infection, and even growth, but which ROS types and genes are important remains to be fully established. At present it is unclear whether there is a specific sensing mechanism involved in ROS detection or whether the production of inactivating enzymes is mediated through general stress response pathways. It is of interest to note that mutation of *SRE1* and *SCP1* leads to an increased sensitivity to ROS, which led Ingavale et al. to propose that there is a link in oxygen sensing between hypoxia, mitochondrial function, and ROS generation (52). However, this would be an additional sensory mechanism from the current model in other fungi, which is that oxygen is sensed indirectly via the sterol biosynthesis pathway, which requires molecular oxygen.

In summary, at present at least two independent pathways have been identified as required for growth at low concentrations of oxygen, and a set of mutant strains has been identified through screening on $CoCl_2$ that may yield insights into additional mechanisms. One major question in this area is what *C. neoformans* actually senses under low oxygen concentrations. One hypothesis is that low oxygen inhibits the oxygen-dependent sterol biosynthesis pathway, and that Scp1 senses changes in sterol composition in the endoplasmic reticulum membrane to trigger the cascade of events leading to Sre1 activation based on models in other fungal species.

CONCLUSIONS

The preceding sections on stress, light, nitrogen, carbon dioxide, and oxygen describe the abilities of *C. neoformans* to sense and adapt to a subset of environmental conditions it encounters. The mechanisms for sensing these extracellular signals are emerging. Currently, most research has, from an understandably reductionist perspective, examined single signals and pathways at a time. The transition from environment to host comprises multiple sensory inputs, so it will be of interest in the future to gain an understanding of cross-talk and interactions between these pathways.

REFERENCES

1. **Abou-Gabal, M., and M. Atia.** 1978. Study of the role of pigeons in the dissemination of *Cryptococcus neoformans* in nature. *Sabouraudia* **16:**63–68.
2. **Alspaugh, J. A., R. Pukkila-Worley, T. Harashima, L. M. Cavallo, D. Funnell, G. M. Cox, J. R. Perfect, J. W. Kronstad, and J. Heitman.** 2002. Adenylyl cyclase functions downstream of the Gα protein Gpa1 and controls mating and pathogenicity of *Cryptococcus neoformans*. *Eukaryot. Cell* **1:**75–84.
3. **Bahn, Y.-S., G. M. Cox, J. R. Perfect, and J. Heitman.** 2005. Carbonic anhydrase and CO_2 sensing during *Cryptococcus neoformans* growth, differentiation, and virulence. *Curr. Biol.* **15:**2013–2020.
4. **Bahn, Y.-S., K. Kojima, G. M. Cox, and J. Heitman.** 2006. A unique fungal two-component system regulates stress responses, drug sensitivity, sexual development, and virulence of *Cryptococcus neoformans*. *Mol. Biol. Cell* **17:**3122–3135.
5. **Bahn, Y.-S., and F. A. Mühlschlegel.** 2006. CO_2 sensing in fungi and beyond. *Curr. Opin. Microbiol.* **9:**572–578.
6. **Bahn, Y. S.** 2008. Master and Commander in fungal pathogens: the two-component system and the HOG signaling pathway. *Eukaryot. Cell* **7:**2017–2036.
7. **Bahn, Y. S., S. Geunes-Boyer, and J. Heitman.** 2007. Ssk2 mitogen-activated protein kinase kinase kinase governs divergent patterns of the stress-activated Hog1 signaling pathway in *Cryptococcus neoformans*. *Eukaryot. Cell* **6:**2278–2289.
8. **Bahn, Y. S., K. Kojima, G. M. Cox, and J. Heitman.** 2005. Specialization of the HOG pathway and its impact on differentiation and virulence of *Cryptococcus neoformans*. *Mol. Biol. Cell* **16:**2285–2300.
9. **Ballario, P., P. Vittorioso, A. Magrelli, C. Talora, A. Cabibbo, and G. Macino.** 1996. White collar-1, a central regulator of blue light responses in *Neurospora*, is a zinc finger protein. *EMBO J.* **15:**1650–1657.
10. **Bava, A. J., R. Negroni, and M. Bianchi.** 1993. Cryptococcosis produced by a urease negative strain of *Cryptococcus neoformans*. *J. Med. Vet. Mycol.* **31:**87–89.
11. **Beck, T., and M. N. Hall.** 1999. The TOR signalling pathway controls nuclear localization of nutrient-regulated transcription factors. *Nature* **402:**689–692.
12. **Belden, W. J., L. F. Larrondo, A. C. Froehlich, M. Shi, C. H. Chen, J. J. Loros, and J. C. Dunlap.** 2007. The *band* mutation in *Neurospora crassa* is a dominant allele of *ras-1* implicating RAS signaling in circadian output. *Genes Dev.* **21:**1494–1505.
13. **Bertram, P. G., J. H. Choi, J. Carvalho, W. Ai, C. Zeng, T. F. Chan, and X. F. Zheng.** 2000. Tripartite regulation of Gln3p by TOR, Ure2p, and phosphatases. *J. Biol. Chem.* **275:**35727–35733.
14. **Biswas, K., and J. Morschhäuser.** 2005. The Mep2p ammonium permease controls nitrogen starvation-induced filamentous growth in *Candida albicans*. *Mol. Microbiol.* **56:**649–669.
15. **Brown, S. M., L. T. Campbell, and J. K. Lodge.** 2007. *Cryptococcus neoformans*, a fungus under stress. *Curr. Opin. Microbiol.* **10:**320–325.
16. **Bryant, D. A., and N.-U. Frigaard.** 2006. Prokaryotic photosynthesis and phototrophy illuminated. *Trends Microbiol.* **14:**488–496.
17. **Caparon, M. G., R. T. Geist, J. Perez-Casal, and J. R. Scott.** 1992. Environmental regulation of virulence in group A streptococci: transcription of the gene encoding M protein is stimulated by carbon dioxide. *J. Bacteriol.* **174:**5693–5701.
18. **Cardenas, M. E., N. S. Cutler, M. C. Lorenz, C. J. Di Como, and J. Heitman.** 1999. The TOR signaling cascade regulates gene expression in response to nutrients. *Genes Dev.* **13:**3271–3279.
19. **Carter, R., and M. M. Nijhout.** 1977. Control of gamete formation (exflagellation) in malaria parasites. *Science* **195:**407–409.
20. **Chang, Y. C., C. M. Bien, H. Lee, P. J. Espenshade, and K. J. Kwon-Chung.** 2007. Sre1p, a regulator of oxygen sensing and sterol homeostasis, is required for virulence in *Cryptococcus neoformans*. *Mol. Microbiol.* **64:**614–629.
21. **Chun, C. D., O. W. Liu, and H. D. Madhani.** 2007. A link between virulence and homeostatic responses to hypoxia during infection by the human fungal pathogen *Cryptococcus neoformans*. *PLoS Pathog.* **3:**e22.
22. **Corrochano, L. M.** 2007. Fungal photoreceptors: sensory molecules for fungal development and behaviour. *Photochem. Photobiol. Sci.* **6:**725–736.
23. **Cox, G. M., T. S. Harrison, H. C. McDade, C. P. Taborda, G. Heinrich, A. Casadevall, and J. R. Perfect.** 2003. Superoxide dismutase influences the virulence of *Cryptococcus neoformans* by affecting growth within macrophages. *Infect. Immun.* **71:**173–180.
24. **Cox, G. M., J. Mukherjee, G. T. Cole, A. Casadevall, and J. R. Perfect.** 2000. Urease as a virulence factor in experimental cryptococcosis. *Infect. Immun.* **68:**443–448.
25. **Cutler, N. S., X. Pan, J. Heitman, and M. E. Cardenas.** 2001. The TOR signal transduction cascade controls cellular differentiation in response to nutrients. *Mol. Biol. Cell* **12:**4103–4113.
26. **Davis-Hanna, A., A. E. Piispanen, L. I. Stateva, and D. A. Hogan.** 2008. Farnesol and dodecanol effects on the *Candida albicans* Ras1-cAMP signalling pathway and the regulation of morphogenesis. *Mol. Microbiol.* **67:**47–62.
27. **de Jesús-Berríos, M., L. Liu, J. C. Nussbaum, G. M. Cox, J. S. Stamler, and J. Heitman.** 2003. Enzymes that counteract nitrosative stress promote fungal virulence. *Curr. Biol.* **13:**1963–1968.
28. **Eaton, K. A., C. L. Brooks, D. R. Morgan, and S. Krakowka.** 1991. Essential role of urease in pathogenesis of gastritis induced by *Helicobacter pylori* in gnotobiotic piglets. *Infect. Immun.* **59:**2470–2475.
29. **Enserink, M.** 2002. What mosquitoes want: secrets of host attraction. *Science* **298:**90–92.
30. **Fang, H. M., and Y. Wang.** 2006. RA domain-mediated interaction of Cdc35 with Ras1 is essential for increasing cellular cAMP level for *Candida albicans* hyphal development. *Mol. Microbiol.* **61:**484–496.
31. **Fiskin, A. M., M. C. Zalles, and R. G. Garrison.** 1990. Electron cytochemical studies of *Cryptococcus neoformans* grown on uric acid and related sources of nitrogen. *J. Med. Vet. Mycol.* **28:**197–207.
32. **Fox, D. S., M. C. Cruz, R. A. Sia, H. Ke, G. M. Cox, M. E. Cardenas, and J. Heitman.** 2001. Calcineurin regulatory subunit is essential for virulence and mediates interactions with FKBP12-FK506 in *Cryptococcus neoformans*. *Mol. Microbiol.* **39:**835–849.
33. **Fox, D. S., and J. Heitman.** 2002. Good fungi gone bad: the corruption of calcineurin. *Bioessays* **24:**894–903.
34. **Gerik, K. J., S. R. Bhimireddy, J. S. Ryerse, C. A. Specht, and J. K. Lodge.** 2008. *PKC1* is essential for protection against both oxidative and nitrosative stresses, cell integrity, and normal manifestation of virulence factors in the pathogenic fungus *Cryptococcus neoformans*. *Eukaryot. Cell* **7:**1685–1698.
35. **Ghosh, S., D. H. Navarathna, D. D. Roberts, J. T. Cooper, A. L. Atkin, T. M. Petro, and K. W. Nickerson.**

2009. Arginine induced germ tube formation in *Candida albicans* is essential for escape from murine macrophage cell line RAW264.7. *Infect. Immun.* **77:**1596–1605.

36. **Giles, S. S., I. Batinic-Haberle, J. R. Perfect, and G. M. Cox.** 2005. *Cryptococcus neoformans* mitochondrial superoxide dismutase: an essential link between antioxidant function and high-temperature growth. *Eukaryot. Cell* **4:**46–54.
37. **Giles, S. S., J. E. Stajich, C. Nichols, Q. D. Gerrald, J. A. Alspaugh, F. Dietrich, and J. R. Perfect.** 2006. The *Cryptococcus neoformans* catalase gene family and its role in antioxidant defense. *Eukaryot. Cell* **5:**1447–1459.
38. **Granger, D. L., J. R. Perfect, and D. T. Durack.** 1985. Virulence of *Cryptococcus neoformans*. Regulation of capsule synthesis by carbon dioxide. *J. Clin. Invest.* **76:**508–516.
39. **Guyton, A. C., and J. E. Hall.** 2000. *Textbook of Medical Physiology*, 10th ed. W.B. Saunders, Philadelphia, PA.
40. **Hall, R. A., F. Cottier, and F. A. Mühlschlegel.** 2009. Molecular networks in the fungal pathogen *Candida albicans*. *Adv. Appl. Microbiol.* **67:**191–212.
41. **Herrera-Estrella, A., and B. A. Horwitz.** 2007. Looking through the eyes of fungi: molecular genetics of photoreception. *Mol. Microbiol.* **64:**5–15.
42. **Hetherington, A. M., and J. A. Raven.** 2005. The biology of carbon dioxide. *Curr. Biol.* **15:**R406–R410.
43. **Heung, L. J., A. E. Kaiser, C. Luberto, and M. Del Poeta.** 2005. The role and mechanism of diacylglycerol-protein kinase C1 signaling in melanogenesis by *Cryptococcus neoformans*. *J. Biol. Chem.* **280:**28547–28555.
44. **Heung, L. J., C. Luberto, A. Plowden, Y. A. Hannun, and M. Del Poeta.** 2004. The sphingolipid pathway regulates Pkc1 through the formation of diacylglycerol in *Cryptococcus neoformans*. *J. Biol. Chem.* **279:**21144–21153.
45. **Hohmann, S.** 2002. Osmotic stress signaling and osmoadaptation in yeasts. *Microbiol. Mol. Biol. Rev.* **66:**300–372.
46. **Hohmann, S., M. Krantz, and B. Nordlander.** 2007. Yeast osmoregulation. *Methods Enzymol.* **428:**29–45.
47. **Hornby, J. M., E. C. Jensen, A. D. Lisec, J. J. Tasto, B. Jahnke, R. Shoemaker, P. Dussault, and K. W. Nickerson.** 2001. Quorum sensing in the dimorphic fungus *Candida albicans* is mediated by farnesol. *Appl. Environ. Microbiol.* **67:**2982–2992.
48. **Huang, G., T. Srikantha, N. Sahni, S. Yi, and D. R. Soll.** 2009. CO_2 regulates white-to-opaque switching in *Candida albicans*. *Curr. Biol.* **19:**330–334.
49. **Hubálek, Z.** 1975. Distribution of *Cryptococcus neoformans* in a pigeon habitat. *Folia Parasitol.* **22:**73–79.
50. **Hughes, A. L., B. L. Todd, and P. J. Espenshade.** 2005. SREBP pathway responds to sterols and functions as an oxygen sensor in fission yeast. *Cell* **120:**831–842.
51. **Idnurm, A., and J. Heitman.** 2005. Light controls growth and development via a conserved pathway in the fungal kingdom. *PLoS Biol.* **3:**615–626.
52. **Ingavale, S. S., Y. C. Chang, H. Lee, C. M. McClelland, M. L. Leong, and K. J. Kwon-Chung.** 2008. Importance of mitochondria in survival of *Cryptococcus neoformans* under low oxygen conditions and tolerance to cobalt chloride. *PLoS Pathog.* **4:**e1000155.
53. **Innocenti, A., R. A. Hall, C. Schlicker, F. A. Mühlschlegel, C. Steegborn, and C. T. Supuran.** 2009. Carbonic anhydrase inhibitors. Inhibition of the β-class enzymes from the fungal pathogens *Candida albicans* and *Cryptococcus neoformans* with aliphatic and aromatic carboxylates. *Bioorg. Med. Chem.* **17:**2654–2657.
54. **Innocenti, A., F. A. Mühlschlegel, R. A. Hall, C. Steegborn, A. Scozzafava, and C. T. Supuran.** 2008. Carbonic anhydrase inhibitors: inhibition of the β-class enzymes from the fungal pathogens *Candida albicans* and *Cryptococcus neoformans* with simple anions. *Bioorg. Med. Chem. Lett.* **18:**5066–5070.
55. **Kamenetsky, M., S. Middelhaufe, E. M. Bank, L. R. Levin, J. Buck, and C. Steegborn.** 2006. Molecular details of cAMP generation in mammalian cells: a tale of two systems. *J. Mol. Biol.* **362:**623–639.
56. **Klengel, T., W.-J. Liang, J. Chaloupka, C. Ruoff, K. Schröppel, J. R. Naglik, S. E. Eckert, E. G. Mogensen, K. Haynes, M. F. Tuite, L. R. Levin, J. Buck, and F. A. Mühlschlegel.** 2005. Fungal adenylyl cyclase integrates CO_2 sensing with cAMP signaling and virulence. *Curr. Biol.* **15:**2021–2026.
57. **Kojima, K., Y. S. Bahn, and J. Heitman.** 2006. Calcineurin, Mpk1 and Hog1 MAPK pathways independently control fludioxonil antifungal sensitivity in *Cryptococcus neoformans*. *Microbiology* **152:**591–604.
58. **Kraus, P. R., D. S. Fox, G. M. Cox, and J. Heitman.** 2003. The *Cryptococcus neoformans* MAP kinase Mpk1 regulates cell integrity in response to antifungal drugs and loss of calcineurin function. *Mol. Microbiol.* **48:**1377–1387.
59. **Kraus, P. R., and J. Heitman.** 2003. Coping with stress: calmodulin and calcineurin in model and pathogenic fungi. *Biochem. Biophys. Res. Commun.* **311:**1151–1157.
60. **Kraus, P. R., C. B. Nichols, and J. Heitman.** 2005. Calcium- and calcineurin-independent roles for calmodulin in *Cryptococcus neoformans* morphogenesis and high-temperature growth. *Eukaryot. Cell* **4:**1079–1087.
61. **Lee, H., C. M. Bien, A. L. Hughes, P. J. Espenshade, K. J. Kwon-Chung, and Y. C. Chang.** 2007. Cobalt chloride, a hypoxia-mimicking agent, targets sterol synthesis in the pathogenic fungus *Cryptococcus neoformans*. *Mol. Microbiol.* **65:**1018–1033.
62. **Levin, D. E.** 2005. Cell wall integrity signaling in *Saccharomyces cerevisiae*. *Microbiol. Mol. Biol. Rev.* **69:**262–291.
63. **Liu, O. W., C. D. Chun, E. D. Chow, C. Chen, H. D. Madhani, and S. M. Noble.** 2008. Systematic genetic analysis of virulence in the human fungal pathogen *Cryptococcus neoformans*. *Cell* **135:**174–188.
64. **Lu, Y.-K., K.-H. Sun, and W.-C. Shen.** 2005. Blue light negatively regulates the sexual filamentation via the Cwc1 and Cwc2 proteins in *Cryptococcus neoformans*. *Mol. Microbiol.* **56:**280–291.
65. **Luberto, C., D. L. Toffaletti, E. A. Wills, S. C. Tucker, A. Casadevall, J. R. Perfect, Y. A. Hannun, and M. Del Poeta.** 2001. Roles for inositol-phosphoryl ceramide synthase 1 (*IPC1*) in pathogenesis of *C. neoformans*. *Genes Dev.* **15:**201–212.
66. **Makino, S., C. Sasakawa, I. Uchida, N. Terakado, and M. Yoshikawa.** 1988. Cloning and CO_2-dependent expression of the genetic region for encapsulation from *Bacillus anthracis*. *Mol. Microbiol.* **2:**371–376.
67. **Mandell, G. L., J. E. Bennett, and R. Dolin.** 2000. *Principles and Practice of Infectious Diseases*. Churchill Livingstone, New York, NY.
68. **Missall, T. A., J. F. Cherry-Harris, and J. K. Lodge.** 2005. Two glutathione peroxidases in the fungal pathogen *Cryptococcus neoformans* are expressed in the presence of specific substrates. *Microbiology* **151:**2573–2581.
69. **Missall, T. A., and J. K. Lodge.** 2005. Function of the thioredoxin proteins in *Cryptococcus neoformans* during stress or virulence and regulation by putative transcriptional modulators. *Mol. Microbiol.* **57:**847–858.
70. **Missall, T. A., and J. K. Lodge.** 2005. Thioredoxin reductase is essential for viability in the fungal pathogen *Cryptococcus neoformans*. *Eukaryot. Cell* **4:**487–489.
71. **Missall, T. A., M. E. Pusateri, and J. K. Lodge.** 2004. Thiol peroxidase is critical for virulence and resistance to

nitric oxide and peroxide in the fungal pathogen, *Cryptococcus neoformans*. *Mol. Microbiol.* **51**:1447–1458.

72. **Mogensen, E. G., G. Janbon, J. Chaloupka, C. Steegborn, M. S. Fu, F. Moyrand, T. Klengel, D. S. Pearson, M. A. Geeves, J. Buck, L. R. Levin, and F. A. Mühlschlegel.** 2006. *Cryptococcus neoformans* senses CO_2 through the carbonic anhydrase Can2 and the adenylyl cyclase Cac1. *Eukaryot. Cell* **5**:103–111.
73. **Narasipura, S. D., J. G. Ault, M. J. Behr, V. Chaturvedi, and S. Chaturvedi.** 2003. Characterization of Cu,Zn superoxide dismutase (*SOD1*) gene knock-out mutant of *Cryptococcus neoformans* var. *gattii*: role in biology and virulence. *Mol. Microbiol.* **47**:1681–1694.
74. **Narasipura, S. D., V. Chaturvedi, and S. Chaturvedi.** 2005. Characterization of *Cryptococcus neoformans* variety *gattii SOD2* reveals distinct roles of the two superoxide dismutases in fungal biology and virulence. *Mol. Microbiol.* **55**:1782–1800.
75. **Odom, A., M. Del Poeta, J. Perfect, and J. Heitman.** 1997. The immunosuppressant FK506 and its nonimmunosuppressive analog L-685,818 are toxic to *Cryptococcus neoformans* by inhibition of a common target protein. *Antimicrob. Agents Chemother.* **41**:156–161.
76. **Odom, A., S. Muir, E. Lim, D. L. Toffaletti, J. Perfect, and J. Heitman.** 1997. Calcineurin is required for virulence of *Cryptococcus neoformans*. *EMBO J.* **16**:2576–2589.
77. **Olszewski, M. A., M. C. Noverr, G.-H. Chen, G. B. Toews, G. M. Cox, J. R. Perfect, and G. B. Huffnagle.** 2004. Urease expression by *Cryptococcus neoformans* promotes microvascular sequestration, thereby enhancing central nervous system invasion. *Am. J. Pathol.* **164**:1761–1771.
78. **Onyewu, C., J. R. Blankenship, M. Del Poeta, and J. Heitman.** 2003. Ergosterol biosynthesis inhibitors become fungicidal when combined with calcineurin inhibitors against *Candida albicans*, *Candida glabrata*, and *Candida krusei*. *Antimicrob. Agents Chemother.* **47**:956–964.
79. **Polacheck, I., and K. J. Kwon-Chung.** 1980. Creatinine metabolism in *Cryptococcus neoformans* and *Cryptococcus bacillisporus*. *J. Bacteriol.* **142**:15–20.
80. **Purschwitz, J., S. Müller, C. Kastner, and R. Fischer.** 2006. Seeing the rainbow: light sensing in fungi. *Curr. Opin. Microbiol.* **9**:566–571.
81. **Reilly, J. M., and K. R. Richards.** 1993. Climate change damage and the trace gas index issue. *Environ. Resource Econ.* **3**:41–61.
82. **Rhodes, J. C., and G. D. Roberts.** 1975. Comparison of four methods for determining nitrate utilization by cryptococci. *J. Clin. Microbiol.* **1**:9–10.
83. **Ruane, P. J., L. J. Walker, and W. L. George.** 1988. Disseminated infection caused by urease-negative *Cryptococcus neoformans*. *J. Clin. Microbiol.* **26**:2224–2225.
84. **Rutherford, J. C., X. Lin, K. Nielsen, and J. Heitman.** 2008. Amt2 permease is required to induce ammonium-responsive invasive growth and mating in *Cryptococcus neoformans*. *Eukaryot. Cell* **7**:237–246.
85. **Sargent, M. L., W. R. Briggs, and D. O. Woodward.** 1966. Circadian nature of a rhythm expressed by an invertaseless strain of *Neurospora crassa*. *Plant Physiol.* **41**:1343–1349.
86. **Sargent, M. L., and S. H. Kaltenborn.** 1972. Effects of medium composition and carbon dioxide on circadian conidiation in Neurospora. *Plant Physiol.* **50**:171–175.
87. **Schlicker, C., R. A. Hall, D. Vullo, S. Middelhaufe, M. Gertz, C. T. Supuran, F. A. Mühlschlegel, and C. Steegborn.** 2009. Structure and inhibition of the CO_2-sensing carbonic anhydrase Can2 from the pathogenic fungus *Cryptococcus neoformans*. *J. Mol. Biol.* **385**:1207–1220.
88. **Shen, W.-C., R. C. Davidson, G. M. Cox, and J. Heitman.** 2002. Pheromones stimulate mating and differentiation via paracrine and autocrine signaling in *Cryptococcus neoformans*. *Eukaryot. Cell* **1**:366–377.
89. **Sirard, J.-C., M. Mock, and A. Fouet.** 1994. The three *Bacillus anthracis* toxin genes are coordinately regulated by bicarbonate and temperature. *J. Bacteriol.* **176**:5188–5192.
90. **Slot, J. C., and D. S. Hibbett.** 2007. Horizontal transfer of a nitrate assimilation gene cluster and ecological transitions in fungi: a phylogenetic study. *PLoS ONE* **2**:e1097.
91. **Smith, D. G., M. D. Garcia-Pedrajas, S. E. Gold, and M. H. Perlin.** 2003. Isolation and characterization from pathogenic fungi of genes encoding ammonium permeases and their roles in dimorphism. *Mol. Microbiol.* **50**:259–275.
92. **Smith, R. S., and B. H. Iglewski.** 2003. *P. aeruginosa* quorum-sensing systems and virulence. *Curr. Opin. Microbiol.* **6**:56–60.
93. **Staib, F., S. K. Mishra, T. Able, and A. Blisse.** 1976. Growth of *Cryptococcus neoformans* on uric acid agar. *Zentralbl. Bakteriol. Orig. A* **236**:374–385.
94. **Steegborn, C., T. N. Litvin, L. R. Levin, J. Buck, and H. Wu.** 2005. Bicarbonate activation of adenylyl cyclase via promotion of catalytic active site closure and metal recruitment. *Nat. Struct. Mol. Biol.* **12**:32–37.
95. **Supuran, C. T.** 2008. Carbonic anhydrases: novel therapeutic applications for inhibitors and activators. *Nat. Rev. Drug Discov.* **7**:168–181.
96. **Townsend, P. D., P. M. Holliday, S. Fenyk, K. C. Hess, M. A. Gray, D. R. Hodgson, and M. J. Cann.** 2009. Stimulation of mammalian G-protein-responsive adenylyl cyclases by carbon dioxide. *J. Biol. Chem.* **284**:784–791.
97. **Tremblay, P.-L., and P. C. Hallenbeck.** 2009. Of blood, brains and bacteria, the Amt/Rh transporter family: emerging role of Amt as a unique microbial sensor. *Mol. Microbiol.* **71**:12–22.
98. **Varma, A., S. Wu, N. Guo, W. Liao, G. Lu, A. Li, Y. Hu, G. Bulmer, and K. J. Kwon-Chung.** 2006. Identification of a novel gene, *URE2*, that functionally complements a urease-negative clinical strain of *Cryptococcus neoformans*. *Microbiology* **152**:3723–3731.
99. **Wong, K. H., M. J. Hynes, and M. A. Davis.** 2008. Recent advances in nitrogen regulation: a comparison between *Saccharomyces cerevisiae* and filamentous fungi. *Eukaryot. Cell* **7**:917–925.
100. **Wormley, F. L., Jr., G. Heinrich, J. L. Miller, J. R. Perfect, and G. M. Cox.** 2005. Identification and characterization of an *SKN7* homologue in *Cryptococcus neoformans*. *Infect. Immun.* **73**:5022–5030.
101. **Xu, X.-L., R. T. H. Lee, H.-M. Fang, Y.-M. Wang, R. Li, H. Zou, Y. Zhu, and Y. Wang.** 2008. Bacterial peptidoglycan triggers *Candida albicans* hyphal growth by directly activating the adenylyl cyclase Cyr1p. *Cell Host Microbe* **4**:28–39.
102. **Yeh, Y.-L., Y.-S. Lin, B.-J. Su, and W.-C. Shen.** 2009. A screening for suppressor mutants reveals components involved in the blue light-inhibited sexual filamentation in *Cryptococcus neoformans*. *Fungal Genet. Biol.* **46**:42–54.
103. **Zaragoza, O., and A. Casadevall.** 2004. Experimental modulation of capsule size in *Cryptococcus neoformans*. *Biol. Proced. Online* **6**:10–15.

Cryptococcus: From Human Pathogen to Model Yeast
Edited by J. Heitman et al.

15

Virulence Mechanisms of *Cryptococcus gattii*: Convergence and Divergence

SUDHA CHATURVEDI AND VISHNU CHATURVEDI

Cryptococcus gattii has emerged as a significant global health problem nearly 4 decades after its first report from Africa (154). Although it is responsible for a wide spectrum of infections in humans and animals, our understanding of its epidemiology and ecology is still elementary and in a state of flux. A noticeable increase in the incidence of *C. gattii* infections in apparently healthy individuals in the Pacific Northwest in the United States, close on the heels of a notable outbreak on Vancouver Island, Canada, has renewed scientific interest in this primary pathogenic yeast (24, 70, 80, 143). Prior to the outbreak in Canada, it was widely believed that *C. gattii* infections were restricted to healthy individuals in tropical and subtropical countries, as the fungal ecological niche was confined to *Eucalyptus* trees (16, 86, 141). Most of these concepts were based upon early reports from Australia, where almost all *C. gattii* infections were seen in non–HIV/AIDS patients. This earlier paradigm might have also led to less attention being paid to *C. gattii* vis-à-vis *Cryptococcus neoformans*, which was the sole focus of scientific interest for a considerable period due to its impact on AIDS patients in the developed world. This picture has slowly but surely changed with the reported isolations of *C. gattii* from more than 50 species of angiosperms and gymnosperms globally (143). These environmental reports were matched in many instances with case reports of *C. gattii* cryptococcosis from temperate and tropical countries, although the exact incidence rates still remain to be determined (143). Perhaps the delays in clinical recognition of the true incidence of *C. gattii* infections were either due to a lack of awareness among clinicians or unavailability of specialized laboratory reagents or the result of a combination of both factors.

Experimental studies aimed at explaining the unique ecological niche of *C. neoformans* in pigeon manure and its virulence for patients with impaired cell-mediated immunity were greatly facilitated by early applications of classical genetics, molecular genetics, and immunological tools (16, 68, 86). A great body of literature has emerged to explain the specific roles of capsule, mating, melanin, and thermal tolerance in *C. neoformans* pathogenesis and exquisite regulatory mechanisms that control the development of infection from the perspective of the host and the pathogen. In a similar vein, experimental studies directed toward understanding *C. gattii* structure, function, and regulation have picked up momentum in the past few years, and interesting glimpses of its virulence are now beginning to emerge. This chapter will describe what we have learned so far as well as many outstanding questions regarding the virulence of *C. gattii*. A special attempt will be made to document available evidence for convergence and/or divergence of virulence mechanisms in the closely related pathogen, *C. neoformans*.

Sudha Chaturvedi and Vishnu Chaturvedi, Mycology Laboratory, Wadsworth Center, New York State Department of Health, and Department of Biomedical Sciences, University at Albany School of Public Health, Albany, NY 12208.

RATIONALE FOR STUDY OF *C. GATTII* VIRULENCE MECHANISMS

Emerging clinical and epidemiological data suggest that *C. gattii* cryptococcosis follows a distinct and unique pattern, which is dissimilar to the disease caused by *C. neoformans*. A limited number of experimental studies also suggest that *C. gattii* virulence properties are somewhat similar to those of *C. neoformans*, but noticeable differences exist in their virulence mechanisms. Furthermore, there are too many outstanding questions regarding *C. gattii* cryptococcosis that cannot be resolved by extrapolation from our understanding of *C. neoformans* virulence mechanisms. For example, *C. gattii* isolates obtained from most parts of the world belong to serotype B, while *C. gattii* serotype C isolates are more geographically restricted. However, the reasons for and significance of this disparity in the global distribution of the two serotypes are not clear. More recent molecular analyses indicated that the majority of the clinical, veterinary, and environmental isolates from Canada belonged to the *MAT*α, VGII/AFLP6 genotype, and their hypervirulence trait could be behind this inexplicable outbreak. However, this assertion must be balanced with the fact that a similar strain of *C. gattii*, NIH444 (serotype B, *MAT*α, ATCC 32609), was isolated in 1970 from sputum of a patient from Seattle, WA (85).

We do not know what conditions cause emergence or reemergence of highly virulent strains. More importantly, why C. *gattii* is found mostly in association with trees and how it infects apparently healthy humans and animals remains a mystery. Compared to cases of cryptococcosis caused by C. *neoformans*, those caused by C. *gattii* exhibit more pulmonary and cerebral nodules (cryptococcomas), increased neurological morbidity, and slower response to antifungal treatments (31, 58, 136, 142). The host immune status could largely account for distinct immunological and cellular responses seen against these two closely related species. However, the fact that C. *gattii* can frequently infect healthy humans by itself indicates that it possesses a certain virulence repertoire, which is unique to this species. Alternatively, C. *gattii* possesses a virulence repertoire similar to that of C. *neoformans*, but the levels of gene expression or alteration in posttranslational modifications could account for differences in clinical manifestations of the two species.

DISTINGUISHING FEATURES OF *C. GATTII*

C. *gattii* and C. *neoformans* are closely related pathogenic species, but there are distinguishing features in their biological, ecological, genetic, and disease-associated attributes (16, 26, 68, 69, 83–86, 89, 134, 141). Some key features that distinguish C. *gattii* from C. *neoformans* are summarized in Table 1. These distinguishing features reveal major differences in shape, antigenicity, and assimilation abilities that might underpin differences in the virulence mechanisms of C. *gattii* and C. *neoformans*. Furthermore, these two pathogens are believed to have diverged from a common ancestor approximately 38.5 million years ago as determined from phylogenetic analyses of a number of genes (163). The recent formal recognition of two close but separate species was based upon reconsideration of morphological, physiological, and genetic features and other distinctions comprehensively documented in the literature over 30 years (53, 88, 89).

TABLE 1 Comparison of features that distinguish C. *gattii* from C. *neoformans*

Feature	C. *gattii*	C. *neoformans*[a]
Yeast	Spherical or elongate	Spherical or ellipsoidal (resembling shape of the earth)
Basidiospore	Reniform (kidney or bean-shaped)	Spherical
Thermal tolerance[b]	37°C +W	37°C
Capsular serotype	B, C, BC	A, D, AD
Assimilation	Nitrate+, D-proline+, D-tryptophan+	Nitrate−, D-proline−, D-tryptophan−
CDB agar[c]	Blue/blue-green	Yellow-orange
CGB agar	Blue/blue-green	Yellow-orange
GCP agar	Bright red	Orange/gold
Chromosome no.	~13–15	~12–14
Whole genome size	~18–20 Mb	~18–19 Mb
Mitochondrial genome size	34.7 Kb	24–32 Kb
Genotype/AFLP	VGI–VGIV	VNI–VNVI

[a] C. *neoformans* includes var. *grubii* and var. *neoformans*.

[b] +W growth of C. *gattii* reported to be not as robust as C. *neoformans*, while some C. *gattii* strains (mostly environmental?) are unable to grow at 37°C.

[c] CDB agar, creatinine dextrose bromothymol blue; CGB agar, canavanine glycine bromothymol blue; GCP agar, glycine cycloheximide phenol red.

TOOLBOX TO STUDY *C. GATTII* VIRULENCE

Fungal Genetic Manipulation

C. *gattii* has a bipolar, heterothallic mating system that was used for the discovery of its teleomorphic state, *Filobasidiella bacillispora*, around the same time the perfect stage of C. *neoformans* was named *Filobasidiella neoformans* (84, 85). C. *neoformans* *MAT*α and *MAT***a** strains including congenic pairs have been extensively used in genetic analyses to delineate the role of capsule, melanin, and mating loci in virulence (73, 90, 91, 118). However, similar analyses of C. *gattii* virulence traits by meiotic segregation are still lacking, especially since a congenic pair of mating strains is unavailable. One particular area where this tool could prove immediately useful is to examine the significance of the occurrence of disproportionate numbers of *MAT*α over *MAT***a** strains in the clinical and environmental samples (80, 87, 135).

Unlike the situation for classical genetic analyses, there has been considerable progress in C. *gattii* gene manipulations and construction of targeted knockout mutants. A number of high-efficiency DNA transformation methods have been described based upon both auxotrophic and dominant selection markers (52, 63, 114, 115, 117). A notable technical advance in homologous reconstitution of mutant strains is the use of a gene of interest and marker gene integration to their native loci with the restoration of the gene phenotype to the wild-type level (114, 116). The overlap PCR strategy for the construction of a disruption cassette without cloning was also successfully applied to gene disruption in C. *gattii* (43, 63, 116). Gene disruption by insertional mutagenesis based on in vitro transposition was described for C. *gattii* (65); this approach combined with genomic resources might facilitate high-throughput genetic analysis. Likewise, an insertional mutagenesis approach using *Agrobacterium tumefaciens*–mediated transformation is another technology now available for gene disruption in C. *gattii*, and it can be used for a large-scale genetic analysis of the molecular determinants of virulence in both *Cryptococcus* species (103).

The studies of genetic manipulations of C. *gattii* benefited greatly from the experience gained from similar work done with C. *neoformans*, but the reagents were not necessarily interchangeable between these two pathogens, which underscored their close but distinct biological identities. As described later, the availability of gene manipulation systems has accelerated characterization of C. *gattii* gene functions and evaluations of their relevance in virulence.

Animal Models of *C. gattii* Cryptococcosis

A systematic study of host-pathogen interactions is critically dependent upon the availability of genetic tools for manipulation of the pathogens and tractable animal models to investigate the host factors presumed to be critical in the infectious processes. A variety of nonprimate models are now available for the study of C. *gattii* pathogenesis. Inbred immunocompetent mice and mice with inborn or drug-induced immunodeficiency have been used for a considerable period to model various facets of host reactions to fungal inhalation or injection and subsequent development of pulmonary, cerebral, and systemic cryptococcosis (15, 39, 149). These models have allowed further delineation of various virulence factors, comparisons of the virulence of various strains, key roles played by a variety of genes in infection, and variable host immune responses in these interactions. Mice remain by far the preferred model for tests of *C. gattii* virulence, but a number of other invertebrate models have been shown to be efficacious and complementary to murine testing (4, 100, 111, 112). These newly available models are especially useful for high-throughput screens that require exceedingly large numbers of animals for examination of conditional fungal mutants that are deficient for growth in physiological conditions, for screening of promising antifungal compounds, and for proof-of-principle studies that require more than one animal model system.

Special mention must be made of the work being done to use the model plant system *Arabidopsis thaliana* for *C. gattii* virulence studies (143a, 165). This work builds upon an earlier report of survival of *C. gattii* for 100 days in inoculated almond tree (*Terminalia catappa*) seedlings (67). Not only has *A. thaliana* proven very useful for dissection of bacterial pathogenesis, but it is uniquely appealing for *C. gattii* virulence studies, as it could also be used to model the ecological niche of this fungus and to examine the role of natural habitats in fungal pathogenesis.

PRIMARY DETERMINANTS OF *C. GATTII* VIRULENCE

Capsule

The overwhelming evidence of the role of capsule in virulence is based upon the fact that clinical strains are almost always encapsulated, capsule size is relatively larger when strains are tested by mimicking physiological conditions, and mutants lacking capsule are avirulent. In the early 1960s, stable antigenic differences in the polysaccharide capsule of *C. neoformans* strains were revealed, which allowed production of serotyping reagents and classification of *C. neoformans* strains as serotypes A, D, and AD and *C. gattii* strains as serotypes B and C (69). Extensive investigations have been carried out on the structure, function, synthesis, and secretion of capsular polysaccharide in *C. neoformans* (37, 131, 152, 166). Although its capsule has yet to receive similar experimental attention, *C. gattii* strains vary greatly in their degree of encapsulation, and those with the most exuberant capsule production have mucoid appearance in the laboratory.

A carbohydrate-rich outer capsule composed primarily of glucuronoxylomannan (GXM) with smaller proportions of galactoxylomannan (GalXM) and mannoprotein is the major virulence factor of *C. neoformans* (16, 34, 36, 44, 86). The GXMs of *C. neoformans* and *C. gattii* differ in the amount of xylose substitutions on the mannose backbone and in the extent of O acetylation (35, 119, 167). The role of two other minor capsular components, GalGM and mannoprotein, is far from clear. Interestingly, other *Cryptococcus* species (*Cryptococcus albidus* and *Cryptococcus laurentii*) produce large capsule, but they are not noticeably pathogenic (42, 81). This suggests that the capsule is necessary but not sufficient for *Cryptococcus* species to cause disease and that its role in infection is enhanced and/or supplemented by other virulence factors. It has been suggested that the degree of capsule O acetylation might contribute to virulence (7), and support for this tenet came from a study where deacetylation of GXM significantly reduced neutrophil recruitments both in vitro and in vivo (48). The absence of the acetyl group in the capsule of *C. albidus* and *C. laurentii* could be evidence of their low pathogenic potential (132).

GXM has been shown to be an important virulence factor of *C. neoformans*; it mediates many deleterious effects on the immune system, including alteration of cytokine regulation (128, 139), apoptosis (38), and interference with leukocyte migration (45, 46). Interestingly, the culture filtrates of *C. gattii* inhibit chemotaxis and chemokinesis responses of human neutrophils, while culture filtrates of *C. neoformans* stimulate these responses (45, 46). Since the capsule is secreted heavily in the supernatant and GXM is the predominant component, the structural complexity of this molecule might be crucial in neutrophil locomotion leading to differential pathogenic response. Capsule also serves as a site for activation and deposition of complement component C3, which promotes phagocytosis of *C. neoformans* (82). Kinetic studies have shown that C3 binds less efficiently to *C. gattii* compared to *C. neoformans*, suggesting that *C. gattii* enhances virulence by preferential evasion of immune recognition (167) (Table 2). The association of encapsulation with resistance to phagocytosis has been recognized for many years, but the role of encapsulation in intracellular survival has been recognized only recently, and similar pathogenic strategies are used by *C. neoformans* and *C. gattii* (3). Thus, this would be evidence of convergence of virulence mechanisms in these two pathogens.

Capsule mutants of *C. neoformans* (*CAP10*, *CAP59*, *CAP60*, and *CAP64*) secrete GalXM in the supernatant, but since these mutants are not pathogenic, one can predict that GalXM may not be directly involved in the *C. neoformans* virulence mechanism (57). Similarly, four mannoproteins including MP84, MP88, MP98, and MP115 have been identified in *C. neoformans* (11, 66, 96). Precise roles of these proteins in the virulence of *C. neoformans* are not clear except for MP98, a chitin deacetylase encoded by the *CDA2* gene. Deletion of *CDA2* resulted in a thinner capsule and leaky melanin phenotypes, with possible adverse effects on virulence (5). To the best of our knowledge, nothing is known about these minor capsular components in *C. gattii* pathogenesis. Compared with an encapsulated strain, an acapsular isogenic mutant strain of *C. neoformans* was found to be avirulent in the mouse model of cryptococcosis (18–22). Similarly, a *cap59* null mutant strain of *C. gattii* was also avirulent in the mouse model of cryptococcosis (143a; P. Ren, V. Chaturvedi, and S. Chaturvedi, unpublished data). These investigations strongly suggest that the capsule is a crucial virulence factor for both *C. neoformans* and *C. gattii*, but the available evidence does not suffice to explain to what extent each capsular component might be critical in the virulence of these two closely related species. A number of other putative virulence factors of *C. gattii* affect capsule size and expression of polysaccharides and by extension *C. gattii* virulence: trehalose, transcription factor Ste12α, and Gα protein cyclic AMP (cAMP)–protein kinase A (PKA) subunits (63, 117, 127). Future studies are likely to yield interesting clues about the roles of capsular components and other genes in the structure and function of this important cell envelope.

Melanin

Melanin, a brown to black pigment and a product of multistep oxidation of diphenolic compounds through catalytic activities of laccase or diphenol oxidase, is a known virulence factor of *C. neoformans* (16, 72, 133, 169). There are several mechanisms by which melanin contributes to *C. neoformans* virulence, the two most prominent being scavenging activities against reactive oxygen and nitrogen-derived radicals (74, 122, 155). *C. neoformans* has two laccase genes in its genome, *LAC1* and *LAC2* (157, 158). The enzyme Lac1p is the primary contributor to melanin production, while Lac2p

TABLE 2 Notable structural-functional differences between *C. gattii* and *C. neoformans*[a]

Feature	Structure-function	Reference(s)
Capsular polysaccharide	GXM inhibits chemotaxis and chemokine response of neutrophils Variability in xylose substitution and extent of O acetylation on GXM Inefficient binding of complement component C3 GXM inhibits T-cell proliferation	7–10, 45, 46, 152, 166, 168
MAT locus, MFα, MFa	Rearrangements in order and orientation of genes; mating-specific genes with higher nonsynonymous mutation rates Two copies of MFα gene versus three or four copies in *C. neoformans* Significant nucleotide differences (30%)	27, 78, 126
Transcription factor *STE12α*	Essential for melanin/laccase; not essential for same-sex mating (haploid fruiting)	23, 127, 169
Cu, Zn-superoxide dismutase (*SOD1*)	Critical for the expression of other virulence factors 20–29% difference in deduced aa with 6% difference in nonsynonymous aa substitution Larger nonreduced molecular mass (145 kDa); highly sensitive to copper chelator (DDC) and sodium azide Questionable release in the extracellular environment	25, 40, 61, 114
Mn-superoxide dismutase (*SOD2*)	Essential for growth at 37°C but not at 30°C Release in the extracellular environment (Chaturvedi and Hauer, personal communication)	54, 115
Phospholipase (*PLB1*)	Larger native mass (275 kDa) with 17 possible N glycosylation and 4 possible O glycosylation sites Plb1 enzyme has LPL/PLB ratio of 6.6, and LPTA/LPL ratio of 0.3–0.4; preferable substrate DPPC; strongly inhibited by $FeCl_3$ and Triton X-100 Enzyme shown to degrade mammalian membrane lipids and lung surfactants	30, 41, 141, 163
Laccase (*LAC1*)	20–23% difference in deduced aa	71
Trehalose-6-phosphate synthase (*TSP1*)	Required for growth at 37°C on nonfermentable sugar and critical for cell wall integrity and expression of virulence factors, melanin, and capsule	117
Gα protein cAMP-PKA (*PKA1B, PKA2B*)	Both catalytic subunits (Pka1 and Pka2) retain mating and virulence factor production, with the exception of Pka1 having lost its ability to regulate melanin	63

[a] *C. neoformans* includes both var. *grubii* and var. *neoformans*. aa, amino acids.

plays a minor role (106). *LAC1* of *C. neoformans* appears to be quite diverged from *LAC1* of *C. gattii* (113, 146, 147), and whether this divergence has any impact on melanin production and virulence of *C. gattii* remains to be investigated. There is compelling evidence to indicate that increased laccase production was directly linked to several outbreaks associated with *C. gattii* strains in Spain, where immunocompetent goats died of lung and disseminated cryptococcosis (2). Reduced virulence of *C. gattii sod1* and *ste12α* mutants was also associated with defects in melanization and reduced laccase enzyme activity (114, 127). It is reasonable to suggest that synthesis, regulation, and functions of *C. gattii* laccases are an important area for future investigations for a full understanding of virulence mechanisms.

Thermal Tolerance

All human-pathogenic fungi display the ability to grow at physiological temperature (37°C), which distinguishes these pathogens from the myriad of nonpathogenic fungi in nature that generally prefer lower temperatures. Surprisingly, *C. gattii* strains have been reported to be not as robust in their growth at 37°C as were *C. neoformans* strains, at least initially (61, 86). It has been found that optimum growth for some strains is at 35°C instead, and that they exhibit mycelial phenotype in tissues instead of yeast cells (129). It is relevant to recall that the thermal dimorphic phenotype is a characteristic of primary pathogenic fungi such as *Blastomyces dermatitidis*, *Histoplasma capsulatum*, *Paracoccidioides brasiliensis*, and *Sporothrix schenckii* (86, 129). An interesting explanation of the apparent slow growth rate of *C. gattii* at 37°C suggested that it was more typical of an environmental strain since more robust growth was seen in clinical strains (33). However, temperature sensitivity is not exclusive to *C. gattii*, as a *C. neoformans* strain with inability to grow at 37°C but with good growth at 35°C was reported from a case of feline granuloma (6).

It is relevant to recall that there is variable tolerance to higher temperature in two *C. neoformans* varieties (*C. neoformans* var. *grubii* and *C. neoformans* var. *neoformans*), and a switch from lower to higher temperature causes important changes in the transcriptional regulation of a great many genes (83, 102, 144). Such detailed transcriptional mapping in response to higher temperature has not yet been performed for *C. gattii*. However, many gene knockout mutants of *C. gattii* have revealed that the ability to grow at 37°C is a complex trait dependent upon the coordinated expression of a number of genes and that this property has a positive correlation with virulence (40, 63, 114, 115, 117, 127). Thus, the discovered linkage between 37°C growth and virulence appears to be common between *C. gattii and C. neoformans* and provides evidence of convergent evolution.

Superoxide Dismutase

Some of the earliest and most comprehensive elucidations of virulence mechanisms of *C. gattii* and *C. neoformans* have come from studies of superoxide dismutases; almost simultaneous work in two laboratories has provided opportunities for side-by-side comparison of roles of these antioxidants. *C. gattii* and *C. neoformans* each have one copy of *SOD1*, encoding cytosolic copper-zinc superoxide dismutase (Sod1p), and one copy of *SOD2*, encoding mitochondrial manganese superoxide dismutase (Sod2p) (40, 54, 114, 115). Initial biochemical, molecular, and structural comparisons of Sod1p between *C. gattii* and *C. neoformans* strongly suggested that this antioxidant has divergent roles in these two closely related species (25, 61) (Table 2). These initial suggestions were verified by construction of targeted gene deletion strains, which suggested a critical role for Sod1p. Indeed, Sod1p has a more profound effect on the biology of *C. gattii* than that of *C. neoformans*, as the *sod1* mutant of *C. gattii* showed vacuolar fragmentation as well as marked reduction in the expression of several known virulence factors including melanin, laccase, phospholipase, urease, and capsule. These results indicated that a reduced environment in the cytosol might be particularly critical for the stability as well as posttranslational modification of proteins or transport mechanism(s), particularly Golgi bodies that might also be fragmented and result in defective secretion. The absence of this phenotype in the *sod1* mutant of *C. neoformans* indicated either a minor role for Sod1p or the presence of other compensatory antioxidant mechanisms.

Interestingly, Sod2p was found to be primarily responsible for scavenging of intracellularly produced O_2^-, and its antioxidant function was linked to high temperature and stationary-phase growth both in *C. gattii* and *C. neoformans*. Careful analysis indicated that *SOD2*-induced temperature sensitivity was more severe in *C. neoformans* than in *C. gattii*. The *C. neoformans sod2* mutant exhibited a temperature-sensitive phenotype at a much lower temperature of 30°C, and this could be suppressed by the addition of ascorbic acid (10 mM) at 30°C but not at 37°C. On the contrary, the *C. gattii sod2* mutant showed temperature sensitivity only at a higher temperature of 37°C, which could also be suppressed by addition of ascorbic acid (10 mM). The results of these studies suggest that the *C. gattii* and *C. neoformans* antioxidant systems are composed of several functionally overlapping and compensatory components that provide protection against exogenous and endogenous oxidative stresses.

The nonclassical (without the use of endoplasmic reticulum [ER]-Golgi) route of secretion has been implicated in the release of this protein in *Mycobacterium tuberculosis* (SodA) and in mammalian cells (Sod1p) (62, 109). Indeed, researchers from Australia recently demonstrated that cytosolic Sod1p of *C. neoformans* is localized in membrane lipid rafts, suggesting a possible secretion mechanism through a process of membrane blebbing, a pathway earlier described for another nonclassical secretory protein, galectin, in mammalian cells (104, 140). Interestingly, in *C. gattii* Sod2p but not Sod1p is secreted as an active protein in the supernatant as analyzed by proteomics and activity gel assays (S. Chaturvedi and C. Hauer, personal communication). Recently, it has been proposed that O_2^- serves a key function in maintaining the alkaline pH (8.0 to 8.5) in the phagosome for lytic enzymes to function optimally (125). These results underscore the important role for O_2^- in the generation of other reactive oxygen species (phagocyte's oxidative machinery) and for optimal functioning of lytic enzymes (phagocyte's nonoxidative machinery) (125). The extracellular localization of Sod1p in *C. neoformans* and Sod2p in *C. gattii* might be a mechanism to remove phagocyte-generated O_2^-, thereby foiling both oxidative and nonoxidative machineries of the phagocytes. The dichotomy in the extracellular localization of Sod1p versus Sod2p in *C. neoformans* and *C. gattii* provides another unique difference between these two closely related species.

Additionally, cytochrome *c* peroxidase (Ccp1), alternative oxidase (Aox1), thiol peroxidase (Tsa1), and glutathione peroxidase (Gpx1, Gpx2) have been shown to contribute to protection against oxidative stress in *C. neoformans* (1, 55, 106, 107). The precise roles of many of these antioxidants are not clear in *C. gattii*, but one can assume that they might also be critical for protection against oxidative stress similar to the role seen in *C. neoformans*.

Trehalose

Trehalose, a disaccharide with stress protectant and reserve carbohydrate functions, has been investigated in depth in *C. gattii* and *C. neoformans* (117, 120). Initial studies of *C. neoformans* found that trehalose-6-phosphate (T6P) synthase, encoded by the *TPS1* (trehalose-6-phosphate synthase) gene, and T6P phosphatase, encoded by the *TPS2* gene, have distinct and overlapping functions in thermotolerance, glycolysis, and virulence. A subsequent comparative study of *C. gattii* revealed evidence of functional convergence and divergence. The two pathogens shared TPS functions in thermal tolerance and virulence. More importantly, *C. gattii tps1* and *tps2* mutant strains also exhibited functional divergence from respective *C. neoformans* mutants in that the absence of TPS1 adversely affected the expression levels of important virulence determinants, capsule and melanin, as well as mating, cell wall integrity, and protein secretion. The same study found that these profound roles were only linked to TPS, since a neutral trehalase (NTH1) gene knockout mutant had no change in phenotypes or virulence from the parent strains. Overall, these studies affirm a trend reported for other prominent *C. gattii* antioxidants, Sod1p and Sod2p.

Phenotype Switching

Changes in colony morphology between smooth and mucoid or rough phenotypes is known as an important pathogenic attribute for host adaptation and microevolution. Phenotypic switching has been extensively studied in *C. neoformans* and *C. gattii* (59, 75, 76). Both pathogens exhibit this phenomenon in vitro and in vivo, and it is associated with change in capsular polysaccharide and virulence. In a significant divergence, *C. gattii* phenotype switches in vivo are reversible, and specific phenotypes are associated with lungs or the central nervous system tissue compartment, which underscores the evidence of evolution of a sophisticated mechanism for host survival and organ tropism in *C. gattii*, unlike those seen in *C. neoformans*. Further examination of this phenomenon by differential expression techniques is likely to provide important insights into how *C. gattii* cells differ in their in vivo colonization from *C. neoformans* strains.

Phospholipases

Phospholipase is a known virulence factor of *C. neoformans* revealed by biochemical analyses and gene knockout experiments (32, 41); the Plb1p enzyme has been extensively characterized in *C. gattii* and *C. neoformans*, and although the *PLB1* gene shows considerable similarity (at least 85% nucleotide identity) between these two closely related species, some important differences were seen in Plb1 proteins (93) (Table 2). Characterization of Plb1p from *C. gattii* showed that it differed from that of *C. neoformans* in its larger native mass (275 kDa), high Plb1p activity relative to lysophospholipase and lysophosphotransacylase, and preference for saturated lipid substrates such as dipalmitoyl phosphatidylcholine. These findings are of interest, since human infection with *C. gattii* typically presents with large lung lesions, or cryptococcomas, and dipalmitoyl phosphatidylcholine is abundant in the surfactant lining within the small air spaces (air sacs) of the lung (162). Also, the highly glycosylated nature of the *C. gattii* Plb1p with 17 possible N glycosylation sites and 4 possible O glycosylation sites (162) is suggestive of the stable and immunogenic nature of this protein. Clearly, targeted knockout strains of *C. gattii* are needed to obtain further evidence of the role of this enzyme in fungal virulence.

ACCESSORY DETERMINANTS OF VIRULENCE

Mating Locus

An intriguing observation about cryptococcosis is that in many geographical areas, the clinical and environmental isolates of *C. gattii* are *MAT*α, and this has been well documented for *C. neoformans* (52, 60, 87). A number of studies suggested that *C. neoformans MAT*α strains are more virulent than *MAT***a** strains (90, 92). It was also observed that in coinfection experiments with both mating types, *MAT*α strains preferentially disseminate to the central nervous system (118). The predominance of the α mating type was linked to haploid or same-sex mating in *C. neoformans* in the laboratory (97, 156). This possibility was supported by a large-scale genealogical analysis conducted on various *C. gattii* isolates collected worldwide (51). Consistent with the notion that sex or mating is critical for virulence, it is feasible that these *C. gattii* isolates are propagating through sexual or same-sex mating, but the conditions required for these processes to occur are yet to be discovered. Mating locus is critical in fungal virulence, but how genes in this locus govern overall virulence of *Cryptococcus* species is far from clear. An interesting example of mating pheromone α (MFα) gene duplication has been documented for *C. gattii* and *C. neoformans*. Two copies of *C. gattii MF*α genes, three *C. neoformans* var. *neoformans MF*α copies, and four *C. neoformans* var. *grubii MF*α copies have been reported (27, 95, 126). Remarkably, pheromone genes are the only open reading frames duplicated in this locus; all other genes are present in single copies in *Cryptococcus* species. Additionally, the rearrangements in order and orientation of *MAT*α locus genes in *C. gattii* vis-à-vis *C. neoformans* and mating-specific genes with higher nonsynonymous mutation rates are all intriguing observations, but their relative influence on virulence awaits careful investigations (126, 137).

STE12α

Transcription factor Ste12 is found only in fungi and controls mating and response to environmental signals. Targeted gene deletions in *STE12*α revealed its critical role for *C. gattii* virulence and ecological fitness, as *ste12*α mutants were markedly attenuated in virulence in a murine model of meningoencephalitis, pigmented poorly on wood extract agar, and showed poor survival and multiplication in wood blocks (127). A comparative analysis of *ste12*α mutants in *C. gattii* and *C. neoformans* (23, 127, 168) showed that only regulation of mating was a common feature of this conserved transcription factor (23). Moreover, a loss of virulence in *C. gattii* was not linked to a defect in the capsule or haploid fruiting as observed in *C. neoformans* (Table 2). These results suggest that not only functional genes but also regulatory genes could yield important clues as to how shared virulence mechanisms show subtle adaptations for presumably different virulence mechanisms of *C. gattii* and *C. neoformans*.

PKA

A comprehensive body of literature now exists on the regulatory elements that control *C. neoformans* developmental pathways and virulence. Although similar intense studies have not yet been reported for *C. gattii*, there is no doubt that this field of investigation offers many opportunities to delineate intricate control of a host of cellular processes. It was discovered earlier that the Gα protein cAMP-PKA signaling pathway regulates mating, capsule, and melanin in *C. neoformans* (47). A comparative investigation of PKA subunits in *C. gattii* showed subtle differences in roles for two subunits (63). Unlike *C. neoformans*, *C. gattii* mating and capsule were not affected when either *PKA1B* or *PKA2B* were deleted. However, deletion of both of these genes caused *C. gattii* to show defects in mating and capsule similar to the *C. neoformans pka1A* mutant strain. The authors termed this "evolutionary reconfiguration of a signaling cascade," which is consistent with the theme of convergence and divergence in functions of various genes and genetic process in these two closely related pathogens (63).

Comparative Genomics

Whole-genome analyses of *C. gattii* strains that vary in important virulence properties would allow global assessment of virulence-specific genes. A similar comparison to genomes of *C. neoformans* would provide the extent of conserved and divergent genes in these closely related pathogens. Currently, two strains of *C. gattii* (R265 and W276), one strain of *C. neoformans* var. *grubii* (H99), and two strains of *C. neoformans* var. *neoformans* (B-3501, JEC21) are part of ongoing or completed genome sequencing projects at the Broad Institute, Boston, MA, and Michael Smith Genome Sciences Center, Vancouver, BC, Canada. Interestingly, the genome size of the Vancouver Island outbreak strain of *C. gattii* R265 is the largest (approximately 20 Mb), followed by more closely related genome sizes of an Australian environmental *C. gattii* W276 (~18.37 Mb), *C. neoformans* var. *neoformans* JEC21 (~19 Mb) and B3501 (~18.52 Mb), and *C. neoformans* var. *grubii* H99 (~18.9 Mb). The large genome size of R265 could be interpreted as suggestive of gene acquisitions via horizontal gene transfer, leading to its new ecological niche colonization as well as enhanced virulence. Notably, preliminary evidence for intervarietal gene transfer has been documented in *C. neoformans* (77).

As of December 2009, a whole-genome sequence has been published for only *C. neoformans* var. *neoformans* JEC21 (98). The completion of other genome projects could eventually lead to the availability of valuable tools to compare differential expression of critical genes under various conditions. So far, cDNA library subtraction, serial analysis of gene expression, and partial genome-DNA microarray have been used to identify genes that are preferentially expressed at mammalian body temperature and by implication in *C. neoformans* virulence (1, 83, 144, 145). Along these lines, a recent microarray assessment of *C. gattii* strains revealed that genes either encoded by mitochondria or related to mitochondrial functions are overexpressed in virulent *C. gattii* strains (99). Future whole-genome investigations along these lines could surely reveal invaluable insights into *C. gattii* virulence mechanisms.

Extracellular Secretory Products

C. gattii elaborates several extracellular products, not all of which have been well characterized. The extracellular products identified in the culture supernatant of *C. gattii* include polysaccharide capsule, laccase, proteinase, and phospholipase (17, 161, 162), which have been characterized as potential virulence factors in *C. neoformans*. Additionally, several other factors have been identified in the culture supernatant of *C. gattii* including those involved in oxidative stress (catalase A, heat shock protein, mannitol 1-phosphate dehydrogenase, Sod2p) and those with unknown functions (glyoxal peroxidase, glucose 6-phosphate isomerase, polysaccharide deacetylase, adenosylhomocysteinase, aspartate aminotransferase, and cellulase) (S. Chaturvedi and C. Hauer, unpublished observations). However, secretome analysis of *C. gattii* does not provide any striking difference from what has already been reported in *C. neoformans*. In *C. neoformans*, the recent understanding is that the proteins destined for secretion (transport) are packaged in heterogeneous extracellular vesicles, which pass through the cell wall and are released in the extracellular environment (131). The proteomic analysis of these vesicles identified some 76 proteins of classical (with ER signal) and nonclassical (without ER signal) origin, and some of these proteins have already been characterized as important determinants of virulence as well as a strong inducer of immune response and oxidative stress (16, 130). We believe a similar transport system or one with subtle variation might exist in *C. gattii*.

Apart from secreted proteins as well as polysaccharide capsule, other metabolites released by *C. gattii* and *C. neoformans* might also be critical in the overall scheme of pathogenesis. Support for these observations comes from several in vivo and in vitro studies indicating that neutrophil infiltrates are more prominent in the lungs of mice infected with *C. neoformans* than with *C. gattii* (39, 50), that culture supernatants from *C. neoformans* but not *C. gattii* contain neutrophil chemotactic and chemokinetic activity, and that the culture supernatants from *C. gattii* inhibit neutrophil chemotaxis (45). Another study indicated that intraperitoneal injection of a very high concentration of *C. gattii*–derived supernatants reduced the extent of inflammation in a mouse model of staphylococcal arthritis (105). Magnetic resonance spectroscopy showed that the nonenzymatic extracellular metabolites identified in *C. neoformans* with the potential to influence the function of mammalian phagocytes were ethanol, acetic acid, mannitol, glycerol, glucitol, erythritol, γ-aminobutyric acid, choline and ethanolamine derivatives, nucleosides, and amino acids (14). Similarly, acetic acid, mannitol, glycerophosphorylcholine, ethanol, and small amounts of γ-aminobutyric acid were identified by magnetic resonance spectroscopy in cryptococcomas harvested from the brains of experimental rats infected with *C. gattii* (64). Mannitol production has been associated with brain edema (meningoencephalitis) and protection of yeast against oxidative, heat, and osmotic stress (28, 29, 159). Comparative analysis of released metabolites on human neutrophil functions indicated that in general, metabolites released from *C. gattii* are potentially less proinflammatory than those metabolized and released from *C. neoformans*. These results suggest that the cryptococcal products in the microenvironment of *C. neoformans* cause neutrophil necrosis and recruitment of host inflammatory and immune cells with subsequent containment of cryptococcal growth. On the other hand, inhibition of neutrophil function by products of *C. gattii* at the site of infection results in survival of extracellular organisms and local multiplication to form cryptococcomas (160). The precise role of each metabolite in fungal pathogenesis awaits more in-depth investigation.

Ecological Fitness

It is reasonable to suggest that the virulence of *C. gattii* for the mammalian host is a consequence of adaptation to plants and plant-based substrates. This rationale is supported by recent reports that many bacteria, especially enteric pathogens, acquire ecological fitness by growth on plants (13). Indeed, the model plant *A. thaliana* has proven to be a tractable host for the study of bacterial pathogenesis to unravel virulence factors that are shared between plants and animals (123, 153). To date, *C. gattii* has been documented from the decayed hollows of 50 species of angiosperm and gymnosperm trees (143). It has been deduced that *C. gattii* is long established among fungal flora of native trees in many parts of the world (94, 124, 143). The general composition of the decayed hollow includes limiting nutrients, water, and microbial communities (101, 138). It is reasonable to extrapolate that *C. gattii* utilizes the hollow's microenvironment to survive, and by defending itself against a number of predators found in this niche, it sustains its virulence attributes. Some experimental studies also provide initial evidence for *C. gattii*–plant association. Huérfano et al. (67) demonstrated that *C. gattii* can survive and be recovered from experimentally infected almond seedlings 100 days postinoculation, thereby demonstrating its ability to colonize and thrive in live plants. Xue et al. (164) reported that congenic pairs of *C. gattii* can colonize, cause leaf chlorosis, and produce basidiospores on young seedlings of *A. thaliana* and *Eucalyptus camaldulensis* plants in the laboratory, suggesting a possibility that infectious propagules are likely to be produced on plants in nature. We have recently reported that *C. gattii* produces unique extracellular fibrils on *A. thaliana* and plant-derived substrates and that these fibrils are associated with enhanced virulence (Springer et al., submitted).

Earlier studies suggested a correlation between the distribution of human cryptococcal disease and that of *Eucalyptus* trees in Australia (49, 121). *Eucalyptus* bark is rich in dihydroxyphenylalanine, a substrate metabolized by *Cryptococcus* species to produce melanin; this has been hypothesized to be a reason for their environmental survival (12). Interestingly, our recent report indicates that *C. gattii* can grow profusely in the laboratory on wood and wood extracts from a variety of tree species and produce melanin-like pigment (24, 127; Springer et al., submitted). It is likely that these associations might serve to enhance overall virulence traits in *C. gattii*. Limited investigations of *C. gattii*–plant or plant-based substrate interactions show increased production of melanin or melanin-like pigments, suggesting that there is a link between ecological fitness and virulence (24, 127). This also provides a possible explanation of why outbreak strain R265, with a genotype identical to that of NIH 444, is relatively more virulent in a murine intranasal infection model (51). In summary, these results suggest that more careful investigations are needed to understand the virulence attributes of *C. gattii*, and one obvious place to look is its natural habitats.

Antifungal Susceptibility Profiles

A more virulent pathogen could be distinguished from its less virulent sibling species by the outcome data from clinical cases and by correlation of this information with antifungal susceptibility patterns of their respective clinical isolates. On this score, the available data for *C. gattii*/*C. neoformans* cryptococcosis are very limited and suggest contradictory findings. Earlier, Australian investigators reported that patients with cryptococcal central nervous system disease due to *C. gattii* have worse outcomes than patients with non–*C. gattii* disease (108). More recently, a population surveillance study from South Africa found no difference in the outcome in patients with *C. gattii* or non–*C. gattii* cryptococcal meningitis (110). Among laboratory studies, a number of investigators have reported that *C. gattii* isolates are relatively less susceptible to antifungal drugs than *C. neoformans* isolates (110, 149). The higher MIC values are exhibited by both clinical and environmental *C. gattii* isolates and various genotypes (70, 79). However, other investigators have found no evidence in antifungal susceptibility patterns of *C. gattii* and *C. neoformans* (56, 148). Thus, we do not know for certain if *C. gattii* or *C. neoformans* infection outcomes are dependent upon their clinical response to antifungal drugs and if such differences highlight any cryptic difference in the virulence mechanism(s) of these pathogens.

SUMMARY AND FUTURE PROSPECTS

The studies summarized in this chapter should convince the reader about the availability of good experimental tools and a number of valuable clues for carrying out further virulence studies in *C. gattii*. The emerging themes in this area are suggestive of both convergence as well as divergence of virulence mechanisms from *C. neoformans*. Moving forward, it will be prudent to do some side-by-side comparative studies with *C. neoformans*, but it would be more productive to focus on the unique aspects of *C. gattii* pathogenesis. It will be critical to choose the strains that will serve as workhorses for laboratory studies, considering that there is now evidence of great heterogeneity among clinical and environmental populations of *C. gattii*. The benefits of whole-genome sequencing could be soon realized, especially with the availability of tools for genome-wide analyses and comparative genomics. Fundamental questions such as the connection between the *C. gattii* ecological niche on trees and its virulence could now be systematically investigated using excellent model systems. Features that underlie host predilections and immune responses are beginning to be understood, but much remains to be done. Clinical manifestations of cryptococcosis due to *C. gattii* and *C. neoformans* need to be further delineated to better understand the course of disease caused by these two distinct entities. Ultimately, a better understanding of *C. gattii* virulence mechanisms would lead to more effective therapeutic options and the possible development of a *C. gattii* vaccine strategy, especially for areas of high incidence and for veterinary use.

Investigations in our laboratories were supported by funds from the National Institutes of Health (AI-41968, AI-48462, AI-05887701, and AI-053732), Pfizer Inc., and Clinical Laboratory Reference Systems of the Wadsworth Center. We were privileged to work with many collaborators, trainees, and students in the past decade (Andrew J. Hamilton, Birgit Stein née Rodeghier, Brian L. Wicks, the late Charles Lowry, Deborah J. Springer, Guan Zhu, Jinjiang Fan, Madhu Dyavaiah, Manoj Iyer, Melissa J. Behr, Ping Ren, Srinivas D. Narasipura, Soumitra K. Saha, and William A. Samsonoff).

REFERENCES

1. **Akhter, S., H. C. McDade, J. M. Gorlach, G. Heinrich, G. M. Cox, and J. R. Perfect.** 2003. Role of alternative oxidase gene in pathogenesis of *Cryptococcus neoformans*. *Infect. Immun.* **71:**5794–5802.
2. **Alvarado-Ramírez, E., J. M. Torres-Rodríguez, M. Sellart, and V. Vidotto.** 2008. Laccase activity in *Cryptococcus gattii* strains isolated from goats. *Rev. Iberoam. Micol.* **25:**150–153.
3. **Alvarez, M., C. Saylor, and A. Casadevall.** 2008. Antibody action after phagocytosis promotes *Cryptococcus neoformans* and *Cryptococcus gattii* macrophage exocytosis with biofilm-like microcolony formation. *Cell Microbiol.* **10:**1622–1633.
4. **Apidianakis, Y., L. G. Rahme, J. Heitman, F. M. Ausubel, S. B. Calderwood, and E. Mylonakis.** 2004. Challenge of *Drosophila melanogaster* with *Cryptococcus neoformans* and role of the innate immune response. *Eukaryot. Cell* **3:**413–419.
5. **Baker, L. G., C. A. Specht, M. J. Donlin, and J. K. Lodge.** 2007. Chitosan, the deacetylated form of chitin, is necessary for cell wall integrity in *Cryptococcus neoformans*. *Eukaryot. Cell* **6:**855–867.
6. **Bemis, D. A., D. J. Krahwinkel, L. A. Bowman, P. Mondon, and K. J. Kwon-Chung.** 2000. Temperature-sensitive strain of *Cryptococcus neoformans* producing hyphal elements in a feline nasal granuloma. *J. Clin. Microbiol.* **38:**926–928.
7. **Bhattacharjee, A. K., J. E. Bennett, and C. P. Glaudemans.** 1984. Capsular polysaccharides of *Cryptococcus neoformans*. *Rev. Infect. Dis.* **6:**619–624.
8. **Bhattacharjee, A. K., K. J. Kwon-Chung, and C. P. Glaudemans.** 1992. The major capsular polysaccharide of *Cryptococcus neoformans* serotype B. *Carbohydr. Res.* **233:**271–272.
9. **Bhattacharjee, A. K., K. J. Kwon-Chung, and C. P. Glaudemans.** 1978. On the structure of the capsular polysaccharide from *Cryptococcus neoformans* serotype C. *Immunochemistry* **15:**673–679.
10. **Bhattacharjee, A. K., K. J. Kwon-Chung, and C. P. Glaudemans.** 1979. On the structure of the capsular polysaccharide from *Cryptococcus neoformans* serotype C-II. *Mol. Immunol.* **16:**531–532.
11. **Biondo, C., L. Messina, M. Bombaci, G. Mancuso, A. Midiri, C. Beninati, V. Cusumano, E. Gerace, S. Papasergi, and G. Teti.** 2005. Characterization of two novel cryptococcal mannoproteins recognized by immune sera. *Infect. Immun.* **73:**7348–7355.
12. **Bland, D. E., and A. F. Logan.** 1969. Lignification in eucalyptus: metabolism of 3-methoxy-4-hydroxyphenylalanine. *Phytochemistry* **8:**575–578.
13. **Brandl, M. T.** 2006. Fitness of human enteric pathogens on plants and implications for food safety. *Annu. Rev. Phytopathol.* **44:**367–392.
14. **Bubb, W. A., L. C. Wright, M. Cagney, R. T. Santangelo, T. C. Sorrell, and P. W. Kuchel.** 1999. Heteronuclear NMR studies of metabolites produced by *Cryptococcus neoformans* in culture media: identification of possible virulence factors. *Magn. Reson. Med.* **42:**442–453.
15. **Capilla, J., C. M. Maffei, K. V. Clemons, R. A. Sobel, and D. A. Stevens.** 2006. Experimental systemic infection with *Cryptococcus neoformans* var. *grubii* and *Cryptococcus gattii* in normal and immunodeficient mice. *Med. Mycol.* **44:**601–610.
16. **Casadevall, A., and J. R. Perfect.** 1998. *Cryptococcus neoformans*. ASM Press, Washington, DC.
17. **Chan, M. Y., and S. T. Tay.** 2009. Enzymatic characterisation of clinical isolates of *Cryptococcus neoformans*, *Cryptococcus gattii* and other environmental *Cryptococcus* spp. *Mycoses* **53:**26–31.
18. **Chang, Y. C., and K. J. Kwon-Chung.** 1994. Complementation of a capsule-deficient mutation of *Cryptococcus neoformans* restores its virulence. *Mol. Cell. Biol.* **14:**4912–4919.
19. **Chang, Y. C., and K. J. Kwon-Chung.** 1998. Isolation of the third capsule-associated gene, *CAP60*, required for virulence in *Cryptococcus neoformans*. *Infect. Immun.* **66:**2230–2236.
20. **Chang, Y. C., and K. J. Kwon-Chung.** 1999. Isolation, characterization, and localization of a capsule-associated gene, *CAP10*, of *Cryptococcus neoformans*. *J. Bacteriol.* **181:**5636–5643.
21. **Chang, Y. C., L. A. Penoyer, and K. J. Kwon-Chung.** 1996. The second capsule gene of *Cryptococcus neoformans*, *CAP64*, is essential for virulence. *Infect. Immun.* **64:**1977–1983.

22. **Chang, Y. C., B. L. Wickes, and K. J. Kwon-Chung.** 1995. Further analysis of the *CAP59* locus of *Cryptococcus neoformans*: structure defined by forced expression and description of a new ribosomal protein-encoding gene. *Gene* **167:**179–183.
23. **Chang, Y. C., B. L. Wickes, G. F. Miller, L. A. Penoyer, and K. J. Kwon-Chung.** 2000. *Cryptococcus neoformans STE12*alpha regulates virulence but is not essential for mating. *J. Exp. Med.* **191:**871–882.
24. **Chaturvedi, S., M. Dyavaiah, R. A. Larsen, and V. Chaturvedi.** 2005. *Cryptococcus gattii* in AIDS patients, southern California. *Emerg. Infect. Dis.* **11:**1686–1692.
25. **Chaturvedi, S., A. J. Hamilton, P. Hobby, G. Zhu, C. V. Lowry, and V. Chaturvedi.** 2001. Molecular cloning, phylogenetic analysis and three-dimensional modeling of Cu,Zn superoxide dismutase (*CnSOD1*) from three varieties of *Cryptococcus neoformans*. *Gene* **268:**41–51.
26. **Chaturvedi, S., B. Rodeghier, J. Fan, C. M. McClelland, B. L. Wickes, and V. Chaturvedi.** 2000. Direct PCR of *Cryptococcus neoformans MAT*α and *MAT*a pheromones to determine mating types, ploidy, and variety: a tool for epidemiological and molecular pathogenesis studies. *J. Clin. Microbiol.* **38:**2007–2009.
27. **Chaturvedi, V., J. Fan, B. Stein, M. J. Behr, W. A. Samsonoff, B. L. Wickes, and S. Chaturvedi.** 2002. Molecular genetic analyses of mating pheromones reveal intervariety mating or hybridization in *Cryptococcus neoformans*. *Infect. Immun.* **70:**5225–5235.
28. **Chaturvedi, V., T. Flynn, W. G. Niehaus, and B. Wong.** 1996. Stress tolerance and pathogenic potential of a mannitol mutant of *Cryptococcus neoformans*. *Microbiology* **142:**937–943.
29. **Chaturvedi, V., B. Wong, and S. L. A. Newman.** 1996. Oxidative killing of *Cryptococcus neoformans* by human neutrophils. Evidence that fungal mannitol protects by scavenging reactive oxygen intermediates. *J. Immunol.* **156:**3836–3840.
30. **Chen, L. C., L. A. Pirofski, and A. Casadevall.** 1997. Extracellular proteins of *Cryptococcus neoformans* and host antibody response. *Infect. Immun.* **65:**2599–2605.
31. **Chen, S., T. Sorrell, G. Nimmo, B. Speed, B. Currie, D. Ellis, D. Marriott, T. Pfeiffer, D. Parr, and K. Byth.** 2000. Epidemiology and host- and variety-dependent characteristics of infection due to *Cryptococcus neoformans* in Australia and New Zealand. Australasian Cryptococcal Study Group. *Clin. Infect. Dis.* **31:**499–508.
32. **Chen, S. C., M. Muller, J. Z. Zhou, L. C. Wright, and T. C. Sorrell.** 1997. Phospholipase activity in *Cryptococcus neoformans*: a new virulence factor? *J. Infect. Dis.* **175:**414–420.
33. **Cheng, P. Y., A. Sham, and J. W. Kronstad.** 2009. *Cryptococcus gattii* isolates from the British Columbia cryptococcosis outbreak induce less protective inflammation in a murine model of infection than *Cryptococcus neoformans*. *Infect. Immun.* **77:**4284–4294.
34. **Cherniak, R., R. G. Jones, and E. Reiss.** 1988. Structure determination of *Cryptococcus neoformans* serotype A-variant glucuronoxylomannan by 13C-n.m.r. spectroscopy. *Carbohydr. Res.* **172:**113–138.
35. **Cherniak, R., L. C. Morris, and S. A. Meyer.** 1992. Glucuronoxylomannan of *Cryptococcus neoformans* serotype C: structural analysis by gas-liquid chromatography-mass spectrometry and 13C-nuclear magnetic resonance spectroscopy. *Carbohydr. Res.* **225:**331–337.
36. **Cherniak, R., E. B. O'Neill, and S. Sheng.** 1998. Assimilation of xylose, mannose, and mannitol for synthesis of glucuronoxylomannan of *Cryptococcus neoformans* determined by 13C nuclear magnetic resonance spectroscopy. *Infect. Immun.* **66:**2996–2998.
37. **Cherniak, R., E. Reiss, M. E. Slodki, R. D. Plattner, and S. O. Blumer.** 1980. Structure and antigenic activity of the capsular polysaccharide of *Cryptococcus neoformans* serotype A. *Mol. Immunol.* **17:**1025–1032.
38. **Chiapello, L. S., M. P. Aoki, H. R. Rubinstein, and D. T. Masih.** 2003. Apoptosis induction by glucuronoxylomannan of *Cryptococcus neoformans*. *Med. Mycol.* **41:**347–353.
39. **Church, D., and R. Washburn.** 1992. Comparison of virulence among serotypes of *Cryptococcus neoformans* in a murine inhalation model, abstr. no. B-328, p. 80. Abstr. 92nd Gen Meet. Am. Soc. Microbiol. American Society for Microbiology, Washington, DC.
40. **Cox, G. M., T. S. Harrison, H. C. McDade, C. P. Taborda, G. Heinrich, A. Casadevall, and J. R. Perfect.** 2003. Superoxide dismutase influences the virulence of *Cryptococcus neoformans* by affecting growth within macrophages. *Infect. Immun.* **71:**173–180.
41. **Cox, G. M., H. C. McDade, S. C. Chen, S. C. Tucker, M. Gottfredsson, L. C. Wright, T. C. Sorrell, S. D. Leidich, A. Casadevall, M. A. Ghannoum, and J. R. Perfect.** 2001. Extracellular phospholipase activity is a virulence factor for *Cryptococcus neoformans*. *Mol. Microbiol.* **39:**166–175.
42. **David, M., M. Gabriel, and M. Kopecka.** 2007. Cytoskeletal structures, ultrastructural characteristics and the capsule of the basidiomycetous yeast *Cryptococcus laurentii*. *Antonie Van Leeuwenhoek* **92:**29–36.
43. **Davidson, R. C., M. C. Cruz, R. A. Sia, B. Allen, J. A. Alspaugh, and J. Heitman.** 2000. Gene disruption by biolistic transformation in serotype D strains of *Cryptococcus neoformans*. *Fungal Genet. Biol.* **29:**38–48.
44. **Doering, T. L.** 2000. How does *Cryptococcus* get its coat? *Trends Microbiol.* **8:**547–553.
45. **Dong, Z. M., and J. W. Murphy.** 1995. Effects of the two varieties of *Cryptococcus neoformans* cells and culture filtrate antigens on neutrophil locomotion. *Infect. Immun.* **63:**2632–2644.
46. **Dong, Z. M., and J. W. Murphy.** 1995. Intravascular cryptococcal culture filtrate (CneF) and its major component, glucuronoxylomannan, are potent inhibitors of leukocyte accumulation. *Infect. Immun.* **63:**770–778.
47. **D'Souza, C. A., J. A. Alspaugh, C. Yue, T. Harashima, G. M. Cox, J. R. Perfect, and J. Heitman.** 2001. Cyclic AMP-dependent protein kinase controls virulence of the fungal pathogen *Cryptococcus neoformans*. *Mol. Cell. Biol.* **21:**3179–3191.
48. **Ellerbroek, P. M., D. J. Lefeber, R. van Veghel, J. Scharringa, E. Brouwer, G. J. Gerwig, G. Janbon, A. I. Hoepelman, and F. E. Coenjaerts.** 2004. O-acetylation of cryptococcal capsular glucuronoxylomannan is essential for interference with neutrophil migration. *J. Immunol.* **173:**7513–7520.
49. **Ellis, D., and T. Pfeiffer.** 1990. Ecology, life cycle, and infectious propagule of *Cryptococcus neoformans*. *Lancet* **336:**923–925.
50. **Feldmesser, M., Y. Kress, P. Novikoff, and A. Casadevall.** 2000. *Cryptococcus neoformans* is a facultative intracellular pathogen in murine pulmonary infection. *Infect. Immun.* **68:**4225–4237.
51. **Fraser, J. A., S. S. Giles, E. C. Wenink, S. G. Geunes-Boyer, J. R. Wright, S. Diezmann, A. Allen, J. E. Stajich, F. S. Dietrich, J. R. Perfect, and J. Heitman.** 2005. Same-sex mating and the origin of the Vancouver Island *Cryptococcus gattii* outbreak. *Nature* **437:**1360–1364.
52. **Fraser, J. A., R. L. Subaran, C. B. Nichols, and J. Heitman.** 2003. Recapitulation of the sexual cycle of the primary fungal pathogen *Cryptococcus neoformans* var. *gattii*: implications for an outbreak on Vancouver Island, Canada. *Eukaryot. Cell* **2:**1036–1045.

53. **Gams, W.** 2005. Proposal to conserve or reject: report of the committee for fungi: 12. *Taxon* **54:**520–522.
54. **Giles, S. S., I. Batinic-Haberle, J. R. Perfect, and G. M. Cox.** 2005. *Cryptococcus neoformans* mitochondrial superoxide dismutase: an essential link between antioxidant function and high-temperature growth. *Eukaryot. Cell* **4:**46–54.
55. **Giles, S. S., J. R. Perfect, and G. M. Cox.** 2005. Cytochrome c peroxidase contributes to the antioxidant defense of *Cryptococcus neoformans*. *Fungal Genet. Biol.* **42:**20–29.
56. **Gomez-Lopez, A., O. Zaragoza, M. Dos Anjos Martins, M. C. Melhem, J. L. Rodriguez-Tudela, and M. Cuenca-Estrella.** 2008. *In vitro* susceptibility of *Cryptococcus gattii* clinical isolates. *Clin. Microbiol. Infect.* **14:**727–730.
57. **Grijpstra, J., G. J. Gerwig, H. Wosten, J. P. Kamerling, and H. de Cock.** 2009. Production of extracellular polysaccharides by CAP mutants of *Cryptococcus neoformans*. *Eukaryot. Cell* **8:**1165–1173.
58. **Grosse, P., K. Tintelnot, O. Söllner, and B. Schmitz.** 2001. Encephalomyelitis due to *Cryptococcus neoformans* var *gattii* presenting as spinal tumour: case report and review of the literature. *J. Neurol. Neurosurg. Psychiatry* **70:**113–116.
59. **Guerrero, A., N. Jain, D. L. Goldman, and B. C. Fries.** 2006. Phenotypic switching in *Cryptococcus neoformans*. *Microbiology* **152:**3–9.
60. **Halliday, C. L., T. Bui, M. Krockenberger, R. Malik, D. H. Ellis, and D. A. Carter.** 1999. Presence of alpha and a mating types in environmental and clinical collections of *Cryptococcus neoformans* var. *gattii* strains from Australia. *J. Clin. Microbiol.* **37:**2920–2926.
61. **Hamilton, A. J., and M. D. Holdom.** 1997. Biochemical comparison of the Cu,Zn superoxide dismutases of *Cryptococcus neoformans* var. *neoformans* and *Cryptococcus neoformans* var. *gattii*. *Infect. Immun.* **65:**488–494.
62. **Harth, G., and M. A. Horwitz.** 1999. Export of recombinant *Mycobacterium tuberculosis* superoxide dismutase is dependent upon both information in the protein and mycobacterial export machinery. A model for studying export of leaderless proteins by pathogenic mycobacteria. *J. Biol. Chem.* **274:**4281–4292.
63. **Hicks, J. K., and J. Heitman.** 2007. Divergence of protein kinase A catalytic subunits in *Cryptococcus neoformans* and *Cryptococcus gattii* illustrates evolutionary reconfiguration of a signaling cascade. *Eukaryot. Cell* **6:**413–420.
64. **Himmelreich, U., C. Allen, S. Dowd, R. Malik, B. P. Shehan, C. Mountford, and T. C. Sorrell.** 2003. Identification of metabolites of importance in the pathogenesis of pulmonary cryptococcoma using nuclear magnetic resonance spectroscopy. *Microbes Infect.* **5:**285–290.
65. **Hu, G., and J. W. Kronstad.** 2006. Gene disruption in *Cryptococcus neoformans* and *Cryptococcus gattii* by *in vitro* transposition. *Curr. Genet.* **49:**341–350.
66. **Huang, C., S. H. Nong, M. K. Mansour, C. A. Specht, and S. M. Levitz.** 2002. Purification and characterization of a second immunoreactive mannoprotein from *Cryptococcus neoformans* that stimulates T-cell responses. *Infect. Immun.* **70:**5485–5493.
67. **Huérfano, S., A. Castañeda, and E. Castañeda.** 2001. Experimental infection of almond trees seedlings (*Terminalia catappa*) with an environmental isolate of *Cryptococcus neoformans* var. *gattii*, serotype C. *Rev. Iberoam. Micol.* **18:**131–132.
68. **Idnurm, A., Y. S. Bahn, K. Nielsen, X. Lin, J. A. Fraser, and J. Heitman.** 2005. Deciphering the model pathogenic fungus *Cryptococcus neoformans*. *Nat. Rev. Microbiol.* **3:**753–764.
69. **Ikeda, R., T. Shinoda, Y. Fukazawa, and L. Kaufman.** 1982. Antigenic characterization of *Cryptococcus neoformans* serotypes and its application to serotyping of clinical isolates. *J. Clin. Microbiol.* **16:**22–29.
70. **Iqbal, N., E. E. Debess, R. Wohrle, B. Sun, R. J. Nett, A. M. Ahlquist, T. Chiller, and S. R. Lockhart for the *Cryptococcus gattii* Public Health Working Group.** 2010. Correlation of genotype and *in vitro* susceptibilities of *Cryptococcus gattii* from the Pacific Northwest of the United States. *J. Clin. Microbiol.* **48:**539–544.
71. **Ito-Kuwa, S., K. Nakamura, B. Valderrama, S. Aoki, V. Vidotto, and T. Osafune.** 2008. Diversity of laccase among *Cryptococcus neoformans* serotypes. *Microbiol. Immunol.* **52:**492–498.
72. **Jacobson, E. S.** 2000. Pathogenic roles for fungal melanins. *Clin. Microbiol. Rev.* **13:**708–717.
73. **Jacobson, E. S., D. J. Ayers, A. C. Harrell, and C. C. Nicholas.** 1982. Genetic and phenotypic characterization of capsule mutants of *Cryptococcus neoformans*. *J. Bacteriol.* **150:**1292–1296.
74. **Jacobson, E. S., and S. B. Tinnell.** 1993. Antioxidant function of fungal melanin. *J. Bacteriol.* **175:**7102–7104.
75. **Jain, N., and B. C. Fries.** 2008. Phenotypic switching of *Cryptococcus neoformans* and *Cryptococcus gattii*. *Mycopathologia* **166:**181–188.
76. **Jain, N., L. Li, D. C. McFadden, U. Banarjee, X. Wang, E. Cook, and B. C. Fries.** 2006. Phenotypic switching in a *Cryptococcus neoformans* variety *gattii* strain is associated with changes in virulence and promotes dissemination to the central nervous system. *Infect. Immun.* **74:**896–903.
77. **Kavanaugh, L. A., J. A. Fraser, and F. S. Dietrich.** 2006. Recent evolution of the human pathogen *Cryptococcus neoformans* by intervarietal transfer of a 14-gene fragment. *Mol. Biol. Evol.* **23:**1879–1890.
78. **Keller, S. M., M. A. Viviani, M. C. Esposto, M. Cogliati, and B. L. Wickes.** 2003. Molecular and genetic characterization of a serotype A *MAT*a *Cryptococcus neoformans* isolate. *Microbiology* **149:**131–142.
79. **Khan, Z. U., H. S. Randhawa, T. Kowshik, A. Chowdhary, and R. Chandy.** 2007. Antifungal susceptibility of *Cryptococcus neoformans* and *Cryptococcus gattii* isolates from decayed wood of trunk hollows of *Ficus religiosa* and *Syzygium cumini* trees in north-western India. *J. Antimicrob. Chemother.* **60:**312–316.
80. **Kidd, S. E., F. Hagen, R. L. Tscharke, M. Huynh, K. H. Bartlett, M. Fyfe, L. Macdougall, T. Boekhout, K. J. Kwon-Chung, and W. Meyer.** 2004. A rare genotype of *Cryptococcus gattii* caused the cryptococcosis outbreak on Vancouver Island (British Columbia, Canada). *Proc. Natl. Acad. Sci. USA* **101:**17258–17263.
81. **Kovalenko, O. H., A. O. Barkalova, and S. S. Nahorrna.** 2005. Antiphytoviral activity of capsule substances of *Cryptococcus albidus* yeast. *Mikrobiol. Z.* **67:**81–84. (In Ukranian.)
82. **Kozel, T. R.** 1993. Opsonization and phagocytosis of *Cryptococcus neoformans*. *Arch. Med. Res.* **24:**211–218.
83. **Kraus, P. R., M.-J. Boily, S. S. Giles, J. E. Stajich, A. Allen, G. M. Cox, F. S. Dietrich, J. R. Perfect, and J. Heitman.** 2004. Identification of *Cryptococcus neoformans* temperature-regulated genes with a genomic-DNA microarray. *Eukaryotic Cell.* **3:**1249–1260.
84. **Kwon-Chung, K. J.** 1975. A new genus, *Filobasidiella*, the perfect state of *Cryptococcus neoformans*. *Mycologia* **67:**1197–1200.
85. **Kwon-Chung, K. J.** 1976. A new species of *Filobasidiella*, the sexual state of *Cryptococcus neoformans* B and C serotypes. *Mycologia* **68:**943–946.
86. **Kwon-Chung, K. J., and J. E. Bennett.** 1992. *Medical Mycology*. Lea & Febiger, Philadelphia, PA.
87. **Kwon-Chung, K. J., and J. E. Bennett.** 1978. Distribution of alpha and alpha mating types of *Cryptococcus neoformans*

among natural and clinical isolates. *Am. J. Epidemiol.* **108:**337–340.

88. **Kwon-Chung, K. J., J. E. Bennett, and T. S. Theodore.** 1978. *Cryptococcus bacillisporus* sp. nov.: serotype B-C of *Cryptococcus neoformans. Int. J. Syst. Bacteriol.* **28:**616–620.
89. **Kwon-Chung, K. J., T. Boekhout, J. W. Fell, and M. Diaz.** 2002. (1557) Proposal to conserve the name *Cryptococcus gattii* against *C. hondurianus* and *C. bacillisporus* (*Basidiomycota, Hymenomycetes, Tremellomycetidae*). *Taxon* **51:**804–806.
90. **Kwon-Chung, K. J., J. C. Edman, and B. L. Wickes.** 1992. Genetic association of mating types and virulence in *Cryptococcus neoformans. Infect. Immun.* **60:**602–605.
91. **Kwon-Chung, K. J., I. Polacheck, and T. J. Popkin.** 1982. Melanin-lacking mutants of *Cryptococcus neoformans* and their virulence for mice. *J. Bacteriol.* **150:**1414–1421.
92. **Kwon-Chung, K. J., B. L. Wickes, L. Stockman, G. D. Roberts, D. Ellis, and D. H. Howard.** 1992. Virulence, serotype, and molecular characteristics of environmental strains of *Cryptococcus neoformans* var. *gattii. Infect. Immun.* **60:**1869–1874.
93. **Latouche, G. N., T. C. Sorrell, and W. Meyer.** 2002. Isolation and characterization of the phospholipase B gene of *Cryptococcus neoformans* var. *gattii. FEMS Yeast Res.* **2:**551–561.
94. **Lazera, M. S., M. A. Cavalcanti, L. Trilles, M. M. Nishikawa, and B. Wanke.** 1998. *Cryptococcus neoformans* var. *gattii*: evidence for a natural habitat related to decaying wood in a pottery tree hollow. *Med. Mycol.* **36:**119–122.
95. **Lengeler, K. B., D. S. Fox, J. A. Fraser, A. Allen, K. Forrester, F. S. Dietrich, and J. Heitman.** 2002. Mating-type locus of *Cryptococcus neoformans*: a step in the evolution of sex chromosomes. *Eukaryot. Cell* **1:**704–718.
96. **Levitz, S. M., S. Nong, M. K. Mansour, C. Huang, and C. A. Specht.** 2001. Molecular characterization of a mannoprotein with homology to chitin deacetylases that stimulates T cell responses to *Cryptococcus neoformans. Proc. Natl. Acad. Sci. USA* **98:**10422–10427.
97. **Lin, X., C. M. Hull, and J. Heitman.** 2005. Sexual reproduction between partners of the same mating type in *Cryptococcus neoformans. Nature* **434:**1017–1021.
98. **Loftus, B. J., E. Fung, P. Roncaglia, D. Rowley, P. Amedeo, D. Bruno, J. Vamathevan, M. Miranda, I. J. Anderson, J. A. Fraser, J. E. Allen, I. E. Bosdet, M. R. Brent, R. Chiu, T. L. Doering, M. J. Donlin, C. A. D'Souza, D. S. Fox, V. Grinberg, J. Fu, M. Fukushima, B. J. Haas, J. C. Huang, G. Janbon, S. J. Jones, H. L. Koo, M. I. Krzywinski, J. K. Kwon-Chung, K. B. Lengeler, R. Maiti, M. A. Marra, R. E. Marra, C. A. Mathewson, T. G. Mitchell, M. Pertea, F. R. Riggs, S. L. Salzberg, J. E. Schein, A. Shvartsbeyn, H. Shin, M. Shumway, C. A. Specht, B. B. Suh, A. Tenney, T. R. Utterback, B. L. Wickes, J. R. Wortman, N. H. Wye, J. W. Kronstad, J. K. Lodge, J. Heitman, R. W. Davis, C. M. Fraser, and R. W. Hyman.** 2005. The genome of the basidiomycetous yeast and human pathogen *Cryptococcus neoformans. Science* **307:**1321–1324.
99. **Ma, H., F. Hagen, D. J. Stekel, S. A. Johnston, E. Sionov, R. Falk, I. Polacheck, T. Boekhout, and R. C. May.** 2009. The fatal fungal outbreak on Vancouver Island is characterized by enhanced intracellular parasitism driven by mitochondrial regulation. *Proc. Natl. Acad. Sci. USA* **106:**12980–12985.
100. **Malliaris, S. D., J. N. Steenbergen, and A. Casadevall.** 2004. *Cryptococcus neoformans* var. *gattii* can exploit *Acanthamoeba castellanii* for growth. *Med. Mycol.* **42:**149–158.
101. **Manion, P. D.** 1991. *Tree Disease Concepts*, 2nd ed. Prentice Hall Career & Technology, Englewood Cliffs, NJ.
102. **Martinez, L. R., J. Garcia-Rivera, and A. Casadevall.** 2001. *Cryptococcus neoformans* var. *neoformans* (Serotype D) strains are more susceptible to heat than *C. neoformans* var. *grubii* (Serotype A) strains. *J. Clin. Microbiol.* **39:**3365–3367.
103. **McClelland, C. A., Y. C. Chang, and K. J. Kwon-Chung.** 2005. High frequency transformation of *Cryptococcus neoformans* and *Cryptococcus gattii* by *Agrobacterium tumefaciens. Fungal Genet. Biol.* **42:**904–913.
104. **Mehul, B., and R. C. Hughes.** 1997. Plasma membrane targetting, vesicular budding and release of galectin 3 from the cytoplasm of mammalian cells during secretion. *J. Cell Sci.* **110:**1169–1178.
105. **Mirshafiey, A., F. Mehrabian, A. Razavi, M. R. Shidfar, and S. Namaki.** 2000. Novel therapeutic approach by culture filtrate of *Cryptococcus neoformans* var. *gattii* (CneF) in experimental immune complex glomerulonephritis. *Gen. Pharmacol.* **34:**311–319.
106. **Missall, T. A., J. M. Moran, J. A. Corbett, and J. K. Lodge.** 2005. Distinct stress responses of two functional laccases in *Cryptococcus neoformans* are revealed in the absence of the thiol-specific antioxidant Tsa1. *Eukaryot. Cell* **4:**202–208.
107. **Missall, T. A., M. E. Pusateri, and J. K. Lodge.** 2004. Thiol peroxidase is critical for virulence and resistance to nitric oxide and peroxide in the fungal pathogen, *Cryptococcus neoformans. Mol. Microbiol.* **51:**1447–1458.
108. **Mitchell, D. H., T. C. Sorrell, A. M. Allworth, C. H. Heath, A. R. McGregor, K. Papanaoum, M. J. Richards, and T. Gottlieb.** 1995. Cryptococcal disease of the CNS in immunocompetent hosts: influence of cryptococcal variety on clinical manifestations and outcome. *Clin. Infect. Dis.* **20:**611–616.
109. **Mondola, P., T. Annella, M. Santillo, and F. Santangelo.** 1996. Evidence for secretion of cytosolic CuZn superoxide dismutase by Hep G2 and human fibroblast. *Int. J. Biochem. Cell Biol.* **28:**677–681.
110. **Morgan, J., K. M. McCarthy, S. Gould, K. Fan, B. Arthington-Skaggs, N. Iqbal, K. Stamey, R. A. Hajjeh, and M. E. Brandt.** 2006. *Cryptococcus gattii* infection: characteristics and epidemiology of cases identified in a South African province with high HIV seroprevalence, 2002-2004. *Clin. Infect. Dis.* **43:**1077–1080.
111. **Mylonakis, E., F. M. Ausubel, J. R. Perfect, J. Heitman, and S. B. Calderwood.** 2002. Killing of *Caenorhabditis elegans* by *Cryptococcus neoformans* as a model of yeast pathogenesis. *Proc. Natl. Acad. Sci. USA* **99:**15675–15680.
112. **Mylonakis, E., R. Moreno, J. B. El Khoury, A. Idnurm, J. Heitman, S. B. Calderwood, F. M. Ausubel, and A. Diener.** 2005. *Galleria mellonella* as a model system to study *Cryptococcus neoformans* pathogenesis. *Infect. Immun.* **73:**3842–3850.
113. **Nakamura, Y.** 2001. Molecular analyses of the serotype of *Cryptococcus neoformans. Nippon Ishinkin Gakkai Zasshi* **42:**69–74.
114. **Narasipura, S. D., J. G. Ault, M. J. Behr, V. Chaturvedi, and S. Chaturvedi.** 2003. Characterization of Cu,Zn superoxide dismutase (*SOD1*) gene knock-out mutant of *Cryptococcus neoformans* var. *gattii:* role in biology and virulence. *Mol. Microbiol.* **47:**1681–1694.
115. **Narasipura, S. D., V. Chaturvedi, and S. Chaturvedi.** 2005. Characterization of *Cryptococcus neoformans* variety *gattii SOD2* reveals distinct roles of the two superoxide dismutases in fungal biology and virulence. *Mol. Microbiol.* **55:**1782–1800.
116. **Narasipura, S. D., P. Ren, M. Dyavaiah, I. Auger, V. Chaturvedi, and S. Chaturvedi.** 2006. An efficient method for homologous gene reconstitution in *Cryptococcus gattii* using *URA5* auxotrophic marker. *Mycopathologia* **162:**401–409.

117. **Ngamskulrungroj, P., U. Himmelreich, J. A. Breger, C. Wilson, M. Chayakulkeeree, M. B. Krockenberger, R. Malik, H.-M. Daniel, D. Toffaletti, J. T. Djordjevic, E. Mylonakis, W. Meyer, and J. R. Perfect.** 2009. The trehalose synthesis pathway is an integral part of the virulence composite for *Cryptococcus gattii*. *Infect. Immun.* **77:**4584–4596.

118. **Nielsen, K., G. M. Cox, A. P. Litvintseva, E. Mylonakis, S. D. Malliaris, D. K. Benjamin, Jr., S. S. Giles, T. G. Mitchell, A. Casadevall, J. R. Perfect, and J. Heitman.** 2005. *Cryptococcus neoformans* a strains preferentially disseminate to the central nervous system during coinfection. *Infect. Immun.* **73:**4922–4933.

119. **Otteson, E. W., W. H. Welch, and T. R. Kozel.** 1994. Protein-polysaccharide interactions. A monoclonal antibody specific for the capsular polysaccharide of *Cryptococcus neoformans*. *J. Biol. Chem.* **269:**1858–1864.

120. **Petzold, E. W., U. Himmelreich, E. Mylonakis, T. Rude, D. Toffaletti, G. M. Cox, J. L. Miller, and J. R. Perfect.** 2006. Characterization and regulation of the trehalose synthesis pathway and its importance in the pathogenicity of *Cryptococcus neoformans*. *Infect. Immun.* **74:**5877–5887.

121. **Pfeiffer, T. J., and D. H. Ellis.** 1992. Environmental isolation of *Cryptococcus neoformans* var. *gattii* from *Eucalyptus tereticornis*. *J. Med. Vet. Mycol.* **30:**407–408.

122. **Polacheck, I., Y. Platt, and J. Aronovitch.** 1990. Catecholamines and virulence of *Cryptococcus neoformans*. *Infect. Immun.* **58:**2919–2922.

123. **Rahme, L. G., F. M. Ausubel, H. Cao, E. Drenkard, B. C. Goumnerov, G. W. Lau, S. Mahajan-Miklos, J. Plotnikova, M. W. Tan, J. Tsongalis, C. L. Walendziewicz, and R. G. Tompkins.** 2000. Plants and animals share functionally common bacterial virulence factors. *Proc. Natl. Acad. Sci. USA* **97:**8815–8821.

124. **Randhawa, H. S., T. Kowshik, A. Chowdhary, K. Preeti Sinha, Z. U. Khan, S. Sun, and J. Xu.** 2008. The expanding host tree species spectrum of *Cryptococcus gattii* and *Cryptococcus neoformans* and their isolations from surrounding soil in India. *Med. Mycol.* **46:**823–833.

125. **Reeves, E. P., H. Lu, H. L. Jacobs, C. G. Messina, S. Bolsover, G. Gabella, E. O. Potma, A. Warley, J. Roes, and A. W. Segal.** 2002. Killing activity of neutrophils is mediated through activation of proteases by K+ flux. *Nature* **416:**291–297.

126. **Ren, P., P. Roncaglia, D. J. Springer, J. J. Fan, and V. Chaturvedi.** 2005. Genomic organization and expression of 23 new genes from MAT alpha locus of *Cryptococcus neoformans* var. *gattii*. *Biochem. Biophys. Res. Commun.* **326:**233–241.

127. **Ren, P., D. J. Springer, M. J. Behr, W. A. Samsonoff, S. Chaturvedi, and V. Chaturvedi.** 2006. Transcription factor STE12alpha has distinct roles in morphogenesis, virulence, and ecological fitness of the primary pathogenic yeast *Cryptococcus gattii*. *Eukaryot. Cell* **5:**1065–1080.

128. **Retini, C., A. Vecchiarelli, C. Monari, C. Tascini, F. Bistoni, and T. R. Kozel.** 1996. Capsular polysaccharide of *Cryptococcus neoformans* induces proinflammatory cytokine release by human neutrophils. *Infect. Immun.* **64:**2897–2903.

129. **Rippon, J. W.** 1988. *Medical Mycology: The Pathogenic Fungi and the Pathogenic Actinomycetes*, 3rd ed. W.B. Saunders, Philadelphia, PA.

130. **Rodrigues, M. L., E. S. Nakayasu, D. L. Oliveira, L. Nimrichter, J. D. Nosanchuk, I. C. Almeida, and A. Casadevall.** 2008. Extracellular vesicles produced by *Cryptococcus neoformans* contain protein components associated with virulence. *Eukaryot. Cell* **7:**58–67.

131. **Rodrigues, M. L., L. Nimrichter, D. L. Oliveira, S. Frases, K. Miranda, O. Zaragoza, M. Alvarez, A. Nakouzi, M. Feldmesser, and A. Casadevall.** 2007. Vesicular polysaccharide export in *Cryptococcus neoformans* is a eukaryotic solution to the problem of fungal trans-cell wall transport. *Eukaryot. Cell* **6:**48–59.

132. **Ross, A., and I. E. Taylor.** 1981. Extracellular glycoprotein from virulent and avirulent *Cryptococcus* species. *Infect. Immun.* **31:**911–918.

133. **Salas, S. D., J. E. Bennett, K. J. Kwon-Chung, J. R. Perfect, and P. R. Williamson.** 1996. Effect of the laccase gene *CNLAC1*, on virulence of *Cryptococcus neoformans*. *J. Exp. Med.* **184:**377–386.

134. **Salkin, I. F., and N. J. Hurd.** 1982. New medium for differentiation of *Cryptococcus neoformans* serotype pairs. *J. Clin. Microbiol.* **15:**169–171.

135. **Saul, N., M. Krockenberger, and D. Carter.** 2008. Evidence of recombination in mixed-mating-type and alpha-only populations of *Cryptococcus gattii* sourced from single eucalyptus tree hollows. *Eukaryot. Cell* **7:**727–734.

136. **Seaton, R. A., S. Naraqi, J. P. Wembri, and D. A. Warrell.** 1997. Cell-mediated immunity in HIV seronegative patients recovered from *Cryptococcus neoformans* var. *gattii* meningitis. *J. Med. Vet. Mycol.* **35:**7–11.

137. **Shen, W.-C., R. C. Davidson, G. M. Cox, and J. Heitman.** 2002. Pheromones stimulate mating and differentiation via paracrine and autocrine signaling in *Cryptococcus neoformans*. *Eukaryot. Cell* **1:**366–377.

138. **Shigo, A. L.** 1984. Compartmentalization: a conceptual framework for understanding how trees grow and defend themselves. *Annu. Rev. Phytopathol.* **22:**189–214.

139. **Shoham, S., C. Huang, J. M. Chen, D. T. Golenbock, and S. M. Levitz.** 2001. Toll-like receptor 4 mediates intracellular signaling without TNF-alpha release in response to *Cryptococcus neoformans* polysaccharide capsule. *J. Immunol.* **166:**4620–4626.

140. **Siafakas, A. R., L. C. Wright, T. C. Sorrell, and J. T. Djordjevic.** 2006. Lipid rafts in *Cryptococcus neoformans* concentrate the virulence determinants phospholipase B1 and Cu/Zn superoxide dismutase. *Eukaryot. Cell* **5:**488–498.

141. **Sorrell, T. C.** 2001. *Cryptococcus neoformans* variety *gattii*. *Med. Mycol.* **39:**155–168.

142. **Speed, B., and D. A. Dunt.** 1995. Clinical and host differences between infections with two varieties of *Cryptococcus neoformans*. *Clin. Infect. Dis.* **21:**28–34.

143. **Springer, D. J., and V. Chaturvedi.** 2010. Projecting global occurrence of *Cryptococcus gattii*. *Emerg. Infect. Dis.* **16:**14–20.

143a. **Springer, D. J., P. Ren, R. Raina, Y. Dong, M. J. Behr, B. F. McEwen, S. Bowser, W. A. Samsonoff, S. Chaturvedi, and V. Chaturvedi.** 2010. Extracellular fibrils of pathogenic yeast *Cryptococcus gattii* are important for ecological niche, murine virulence, and human neutrophil interactions. *PLoS One* **5:**e10978.

144. **Steen, B. R., T. Lian, S. Zuyderduyn, W. K. MacDonald, M. Marra, S. J. M. Jones, and J. W. Kronstad.** 2002. Temperature-regulated transcription in the pathogenic fungus *Cryptococcus neoformans*. *Genome Res.* **12:**1386–1400.

145. **Steen, B. R., S. Zuyderduyn, D. L. Toffaletti, M. Marra, S. J. Jones, J. R. Perfect, and J. Kronstad.** 2003. *Cryptococcus neoformans* gene expression during experimental cryptococcal meningitis. *Eukaryot. Cell* **2:**1336–1349.

146. **Sugita, T., R. Ikeda, and T. Shinoda.** 2001. Diversity among strains of *Cryptococcus neoformans* var. *gattii* as revealed by a sequence analysis of multiple genes and a chemotype analysis of capsular polysaccharide. *Microbiol. Immunol.* **45:**757–768.

147. **Tanaka, E., S. Ito-Kuwa, K. Nakamura, S. Aoki, V. Vidotto, and M. Ito.** 2005. Comparisons of the laccase gene

among serotypes and melanin-deficient variants of *Cryptococcus neoformans*. *Microbiol. Immunol.* **49:**209–217.

148. **Tay, S. T., T. Tanty Haryanty, K. P. Ng, M. Y. Rohani, and H. Hamimah.** 2006. *In vitro* susceptibilities of Malaysian clinical isolates of *Cryptococcus neoformans* var. *grubii* and *Cryptococcus gattii* to five antifungal drugs. *Mycoses* **49:**324–330.

149. **Torres-Rodríguez, J. M., Y. Morera, T. Baró, J. M. Corominas, and E. Castañeda.** 2003. Pathogenicity of *Cryptococcus neoformans* var. *gattii* in an immunocompetent mouse model. *Med. Mycol.* **41:**59–63.

150. **Trilles, L., B. Fernàndez-Torres, S. Lazéra Mdos, B. Wanke, and J. Guarro.** 2004. In vitro antifungal susceptibility of *Cryptococcus gattii*. *J. Clin. Microbiol.* **42:**4815–4817.

151. **Turner, S. H., R. Cherniak, and E. Reiss.** 1984. Fractionation and characterization of galactoxylomannan from *Cryptococcus neoformans*. *Carbohydr. Res.* **125:** 343–349.

152. **Vaishnav, V. V., B. E. Bacon, M. O'Neill, and R. Cherniak.** 1998. Structural characterization of the galactoxylomannan of *Cryptococcus neoformans* Cap67. *Carbohydr. Res.* **306:**315–330.

153. **van Baarlen, P., A. van Belkum, and B. P. Thomma.** 2007. Disease induction by human microbial pathogens in plant-model systems: potential, problems and prospects. *Drug Discov. Today* **12:**167–173.

154. **Vanbreuseghem, R., and M. Takashio.** 1970. An atypical strain of *Cryptococcus neoformans* (San Felice) Vuillemin 1894. II. *Cryptococcus neoformans* var. *gattii* var. nov. *Ann. Soc. Belg. Med. Trop.* **50:**695–702.

155. **Wang, Y., and A. Casadevall.** 1994. Susceptibility of melanized and nonmelanized *Cryptococcus neoformans* to nitrogen- and oxygen-derived oxidants. *Infect. Immun.* **62:**3004–3007.

156. **Wickes, B. L., M. E. Mayorga, U. Edman, and J. C. Edman.** 1996. Dimorphism and haploid fruiting in *Cryptococcus neoformans*: association with the α-mating type. *Proc. Natl. Acad. Sci. USA* **93:**7327–7331.

157. **Williamson, P. R.** 1994. Biochemical and molecular characterization of the diphenol oxidase of *Cryptococcus neoformans*: identification as a laccase. *J. Bacteriol.* **176:**656–664.

158. **Williamson, P. R., K. Wakamatsu, and S. Ito.** 1998. Melanin biosynthesis in *Cryptococcus neoformans*. *J. Bacteriol.* **180:**1570–1572.

159. **Wong, B., J. R. Perfect, S. Beggs, and K. A. Wright.** 1990. Production of the hexitol D-mannitol by *Cryptococcus neoformans* in vitro and in rabbits with experimental meningitis. *Infect. Immun.* **58:**1664–1670.

160. **Wright, L., W. Bubb, J. Davidson, R. Santangelo, M. Krockenberger, U. Himmelreich, and T. Sorrell.** 2002. Metabolites released by *Cryptococcus neoformans* var. *neoformans* and var. *gattii* differentially affect human neutrophil function. *Microbes Infect.* **4:**1427–1438.

161. **Wright, L. C., S. C. Chen, C. F. Wilson, M. F. Simpanya, R. Blackstock, G. M. Cox, J. W. Murphy, and T. C. Sorrell.** 2002. Strain-dependent effects of environmental signals on the production of extracellular phospholipase by *Cryptococcus neoformans*. *FEMS Microbiol. Lett.* **209:**175–181.

162. **Wright, L. C., J. Payne, R. T. Santangelo, M. F. Simpanya, S. C. A. Chen, F. Widmer, and T. C. Sorrell.** 2004. Cryptococcal phospholipases: a novel lysophospholipase discovered in the pathogenic fungus *Cryptococcus gattii*. *Biochem. J.* **384:**377–384.

163. **Xu, J., R. Vilgalys, and T. G. Mitchell.** 2000. Multiple gene genealogies reveal recent dispersion and hybridization in the human pathogenic fungus *Cryptococcus neoformans*. *Mol. Ecol.* **9:**1471–1481.

164. **Xue, C., Y. Tada, X. Dong, and J. Heitman.** 2007. The human fungal pathogen *Cryptococcus* can complete its sexual cycle during a pathogenic association with plants. *Cell Host Microbe.* **1:**263–273.

165. **Yauch, L. E., J. S. Lam, and S. M. Levitz.** 2006. Direct inhibition of T-cell responses by the *Cryptococcus* capsular polysaccharide glucuronoxylomannan. *PLoS Pathog.* **2:**e120.

166. **Yoneda, A., and T. L. Doering.** 2006. A eukaryotic capsular polysaccharide is synthesized intracellularly and secreted via exocytosis. *Mol. Biol. Cell* **17:**5131–5140.

167. **Young, B. J., and T. R. Kozel.** 1993. Effects of strain variation, serotype, and structural modification on kinetics for activation and binding of C3 to *Cryptococcus neoformans*. *Infect. Immun.* **61:**2966–2972.

168. **Yue, C., L. M. Cavallo, J. A. Alspaugh, P. Wang, G. M. Cox, J. R. Perfect, and J. Heitman.** 1999. The *STE12*alpha homolog is required for haploid filamentation but largely dispensable for mating and virulence in *Cryptococcus neoformans*. *Genetics* **153:**1601–1615.

169. **Zhu, X., and P. R. Williamson.** 2004. Role of laccase in the biology and virulence of *Cryptococcus neoformans*. *FEMS Yeast Res.* **5:**1–10.

Cryptococcus: From Human Pathogen to Model Yeast
Edited by J. Heitman et al.

16

Drug Resistance in *Cryptococcus*: Epidemiology and Molecular Mechanisms

M. A. PFALLER, J. K. LODGE, AND M. A. GHANNOUM

Cryptococcosis is a systemic mycosis caused by the encapsulated, basidiomycetous yeast-like fungi *Cryptococcus neoformans* and *Cryptococcus gattii* (23). More than 30 species are included in the genus *Cryptococcus*; however, infections due to species other than *C. neoformans* and *C. gattii* are rare and will not be addressed in this chapter on antifungal resistance.

C. neoformans is the most common species of *Cryptococcus* affecting patients with AIDS and other immunocompromising conditions (23, 35, 61, 78, 88), whereas *C. gattii* infections occur mainly in immunocompetent hosts in endemic regions throughout the world (24, 25, 45, 60, 92). The emergence of *C. gattii* infections in immunocompetent humans and animals in the Pacific Northwest region of North America is nothing short of spectacular (9). The incidence in this new endemic area has reached 35 cases per million population annually, markedly higher than rates reported in other endemic areas such as Australia (1.2 cases per million per year) (58, 91, 92).

Until the first half of the 20th century cryptococcosis was rarely reported. Prior to the AIDS outbreak, cryptococcal infection was diagnosed in less than 1,000 patients per year in the United States (42, 43). Indeed, among persons without AIDS the incidence of cryptococcosis in the United States has been shown to be 0.2 to 0.8 infections per 100,000 population per year and has remained unchanged for well over a decade (43, 61).

A dramatic increase in the incidence of cryptococcosis was observed with the advent of the AIDS pandemic, and subsequently HIV infection has been associated with more than 80% of cryptococcosis cases worldwide (34, 35, 43, 61). During the 1990s, the prevalence of cryptococcosis progressively declined in developed countries, first as a result of the widespread use of fluconazole (77, 87) and later due to successful treatment of HIV infection with the use of highly active antiretroviral therapy (HAART) (35, 61).

M. A. Pfaller, Departments of Pathology and Epidemiology, University of Iowa College of Medicine and College of Public Health, Iowa City, IA 52242-1009. **J. K. Lodge,** Department of Molecular Microbiology, Washington University of School of Medicine, St. Louis, MO 63110. **M. A. Ghannoum,** Center for Medical Mycology, Case Western Reserve University, Cleveland, OH 44106.

Despite this progress, cryptococcosis continues to carry a significant morbidity and mortality: the annual case-fatality ratio in a recent U.S. population-based survey was 11% among HIV-infected persons and 21% among HIV-uninfected individuals and did not change significantly over the 8-year study period (61). In industrialized countries cryptococcosis continues to occur in those with undiagnosed HIV infection and in socioeconomically disadvantaged HIV-infected people without access to HAART or other HIV-supportive care (35, 61). Lack of access to HAART and antifungal therapy (especially amphotericin B and flucytosine) is a major problem in resource-limited regions such as Africa and Southeast Asia, where both AIDS and cryptococcosis are rampant (11–13, 30, 35, 44, 56, 60, 65). Together with tuberculosis, cryptococcosis represents the main killers of HIV-infected persons in sub-Saharan Africa.

Untreated cryptococcal meningitis carries a 100% mortality rate (65). The introduction of amphotericin B deoxycholate in the 1950s improved the prognosis of this disease, and the cure rate for cryptococcal meningitis rose to over 50% (68). With the currently recommended antifungal treatment regimens (amphotericin B with or without flucytosine followed by fluconazole maintenance therapy) coupled with HAART, the prognosis for patients diagnosed with AIDS-associated central nervous system (CNS) cryptococcosis has improved dramatically (57). However, the acute mortality remains in the range of 6 to 15% in industrialized settings (18, 57, 61, 80, 82) and is 25% to greater than 50% in resource-limited regions such as Africa (12, 13, 44, 56, 60).

Prior to HAART, high rates of fungal persistence and frequent disease relapse contributed to a growing concern among clinicians regarding the potential for the emergence of antifungal resistance in *C. neoformans* (10, 17, 68, 71). These concerns remain today, especially among individuals with limited access to HAART, antifungal therapy, and other supportive measures, and are especially acute in the developing world (12, 13, 44, 56).

In considering drug resistance in cryptococcosis, Perfect and Cox (68) identified several factors that may result in clinical resistance to antifungal therapy. Clinical resistance in *C. neoformans* may occur because of one or more of the following: (i) faulty host defenses, (ii) site of infection (CNS versus prostate), (iii) intrinsic virulence of the organism, (iv) ineffective drug pharmacokinetics and dosages, (v) poor compliance with either HAART or antifungal regimens, and (vi) development of primary or secondary resistance of the infecting organism to the specific antifungal drug prescribed (68). Regarding the latter factor, the minimum inhibitory concentrations (MICs) of antifungal agents for *C. neoformans* have been shown both in animal models (98) and in clinical cases of relapse to be predictive of treatment failures for polyenes (51), flucytosine (16), and azoles (7, 10, 15, 66, 67, 99, 100). In this chapter we will devote our attention to describing the epidemiology of antifungal drug resistance as identified by in vitro susceptibility testing and studies of the mechanisms of resistance in *C. neoformans* and *C. gattii*.

IN VITRO ANTIFUNGAL SUSCEPTIBILITY TESTING

Antifungal susceptibility testing in vitro is assuming an increasing role in antifungal drug selection, as an aid in drug discovery and development studies, and as a means of tracking the emergence of antifungal resistance in epidemiological studies (3, 4, 68, 71, 73, 79). The Clinical and Laboratory Standards Institute (CLSI) Subcommittee for Antifungal Testing has developed standardized broth microdilution (BMD) (28, 29) and disk diffusion (26, 27) methods for in vitro susceptibility testing of *Candida* spp. and *C. neoformans* (36). Although there were some early concerns that the CLSI method was not optimal for testing *C. neoformans* (39, 104), the CLSI BMD and disk diffusion methods have now been applied worldwide to study the emergence of resistance in serial isolates of *C. neoformans* (Table 1), to provide a comparison of the activity of new and established agents against *C. neoformans* and *C. gattii* (Table 2), to identify geographic and temporal trends in antifungal resistance (Tables 3–5), and to identify clinical and laboratory strains that can be used to study mechanisms of antifungal resistance (Table 6).

These methods are reproducible and accurate and provide clinically useful information that is comparable to that of antibacterial testing (4, 79). There is substantial support from controlled in vivo animal model experiments that the in vitro antifungal drug susceptibility studies with *C. neoformans* can predict the outcome of treatment (84, 98) as well as numerous reports and case series of elevated MICs of polyenes, flucytosine, and azoles associated with clinical treatment failures in cryptococcosis (Tables 1, 4, and 6).

MIC and zone diameter interpretive breakpoints have not been developed for any antifungal agent and *C. neoformans* to date. It is clear, however, that patients infected with *C. neoformans* for which the fluconazole MICs are ≤8 μg/ml respond better to treatment with fluconazole than those infected with strains for which MICs are ≥32 μg/ml (4). Taking into consideration the MIC distribution profiles for the various antifungal agents, the pharmacology of the antifungal drugs, and studies of resistance mechanisms and clinical outcomes, it is reasonable to adapt the breakpoints developed by CLSI for *Candida* spp. for use in this discussion of antifungal resistance in *C. neoformans* and *C. gattii*. The CLSI breakpoints are as follows: amphotericin B susceptible (S) ≤1 μg/ml, resistant (R) ≥2 μg/ml; flucytosine S ≤4 μg/ml, intermediate (I) 8 to 16 μg/ml, R ≥32 μg/ml; fluconazole S ≤8 μg/ml (zone diameter ≥19 mm), susceptible dose dependent (SDD) 16 to 32 μg/ml (15 to 18 mm), R ≥64 μg/ml

TABLE 1 In vitro susceptibility changes to fluconazole in serial isolates of *C. neoformans*

Study	No. of patients	Change in MIC[a]	Magnitude of change			
			None	Twofold	Fourfold	≥Eightfold
Casadevall et al. (21)	5	None	3			
		Increase				
		Decrease		1		1
Paugam et al. (66)	1	None				
		Increase				1
		Decrease				
Pfaller et al. (70)	17	None	10			
		Increase		4		
		Decrease		3		
Klepser and Pfaller (53)	14	None	8			
		Increase		3		
		Decrease		3		
Davey et al. (33)	6	None				
		Increase				6
		Decrease				
Friese et al. (38)	1	None				
		Increase				1
		Decrease				
Brandt et al. (17)	71	None	33			
		Increase		25	4	3
		Decrease			5	1
Yildiran et al. (107)	20	None	9			
		Increase			6	
		Decrease			5	

[a] Comparison of initial versus final isolate fluconazole MICs.

TABLE 2 In vitro susceptibility of *C. neoformans* and *C. gattii* to systemically active antifungal agents determined by CLSI M27-A3 BMD methods[a]

Organism	Antifungal agent	No. tested	MIC (μg/ml)[b]			% by category[c]	
			Range	50%	90%	S	R
C. neoformans	Amphotericin B	1,811	0.25–2	1	1	99	1
	Flucytosine	1,811	0.06–>64	8	16	44	1
	Fluconazole	1,811	0.12–>64	4	16	83	1
	Itraconazole	1,615	0.03–8	0.25	4	26	20
	Posaconazole	1,646	0.015–2	0.12	0.5	99	0
	Voriconazole	1,800	0.007–4	0.06	0.25	99	1
C. gattii	Amphotericin B	42	0.12–1	0.25	0.25	100	0
	Flucytosine	42	0.06–16	2	8	NA[d]	0
	Fluconazole	42	0.5–32	2	8	NA	0
	Posaconazole	42	<0.015–0.25	0.03	0.12	100	0
	Voriconazole	42	<0.015–0.5	0.12	0.25	100	0

[a] Data compiled from Pfaller et al. (71, 72) and Thompson et al. (94).
[b] 50% and 90%, MIC encompassing 50% and 90% of isolates tested, respectively.
[c] Susceptibility category: S, susceptible; R, resistant. Interpretative breakpoints as described in CLSI document M27-S3 (29): amphotericin B, S ≤1 μg/ml, R ≥2 μg/ml; flucytosine, S ≤4 μg/ml, R ≥32 μg/ml; fluconazole, S ≤8 μg/ml, R ≥64 μg/ml; voriconazole S ≤1 μg/ml; R ≥4 μg/ml; posaconazole, same as voriconazole.
[d] NA, data not available.

(≤14 mm); itraconazole S ≤0.12 μg/ml, SDD 0.25 to 0.5 μg/ml, R ≥1 μg/ml; voriconazole S ≤1 μg/ml (≥17 mm), SDD 2 μg/ml (14 to 16 mm), R ≥4 μg/ml (≤13 mm). Since there are no interpretive breakpoints yet established for posaconazole, we have elected to use the voriconazole MIC breakpoints for purposes of discussion here.

ACTIVITY OF SYSTEMICALLY ACTIVE ANTIFUNGAL AGENTS AGAINST *C. NEOFORMANS* AND *C. GATTII*

Clearly, the driving force behind the concerns about emerging antifungal resistance in *C. neoformans* has been the extensive use of fluconazole as an agent in chronic antifungal maintenance therapy and primary antifungal prophylaxis in AIDS patients (10, 76, 77). Although the emergence of clinically significant resistance to fluconazole has been reported in individual cases of recurrent cryptococcal infection (4, 7, 10, 15, 66, 97, 98), data from larger series of recurrent cryptococcosis suggest that the development of antifungal resistance is not a major factor in recurrent cryptococcal disease (Table 1). In the large population-based surveillance study of cryptococcosis conducted by the Centers for Disease Control and Prevention in the United States from 1992 through 1998, 71 patients with recurrent cryptococcosis were identified (172 serial isolates), of which only 7 (9.9%) showed a fourfold or greater increase in fluconazole MIC over the course of study (17) (Table 1). Similar findings were reported by Casadevall et al. (21), Pfaller et al. (70), and Klepser and Pfaller (53) (Table 1), whereas Yildiran et al. (107) found six patients out of 20 (30%) with relapsing infections where the fluconazole MIC of the infecting isolate increased by fourfold, and Davey et al. (33) reported six patients for which the isolates showed an eightfold or greater increase in fluconazole MIC over the course of their infections (Table 1).

Despite these findings, large surveys in which incident isolates of *C. neoformans* and *C. gattii* were tested by the CLSI BMD method against the available systemically active antifungal agents look quite positive in regard to low levels of antifungal resistance (71, 72, 94) (Table 2). With the exception of itraconazole and flucytosine, high levels of susceptibility

TABLE 3 Fluconazole susceptibility of *C. neoformans* over time and by geographic region: ARTEMIS DISK Global Antifungal Surveillance Study, 2001–2007[a]

Region	No. of study sites[c]	Susceptibility[b]								
		2001–2004			2005–2007			2001–2007		
		N	S	R	N	S	R	N	S	R
Africa-Middle East	11	635	70	12	234	70	15	869	70	12
Latin America	16	308	86	15	258	77	14	595	78	14
Asia-Pacific	28	321	83	10	209	74	12	530	79	11
Europe	66	213	89	6	257	83	7	470	86	7
North America	13	221	82	9	26	92	0	247	84	8

[a] Data compiled from Pfaller et al. (73).
[b] Fluconazole disk diffusion testing was performed in accordance with CLSI document M44-A (26). The interpretive breakpoints were as follows: susceptible, (S), ≥19 mm; resistant (R), ≤14 mm.
[c] Number of institutions contributing data or isolates to the study.

TABLE 4 Temporal and geographic variation in the in vitro susceptibilities of *C. neoformans* isolates to systemically active antifungal agents

Location (ref.)	Time period	Antifungal agent	No. of isolates	MIC (μg/ml)[a]			% by category[b]	
				Range	50%	90%	S	R
United States (17)	1992–1994	Amphotericin B	368	0.25–1	1	1	100	0
		Flucytosine	368	0.5–≥128	8	16	NA[c]	1.6
		Fluconazole	253	≤8–256	8	16	71	2.4
		Itraconazole	253	0.006–2	0.5	1	4.7	12.6
	1996–1998	Amphotericin B	364	0.25–2	1	1	99.4	0.4
		Flucytosine	364	0.5–≥128	8	16	NA	2.2
		Fluconazole	269	≤8–256	8	16	89	0.7
		Itraconazole	269	0.03–2	0.25	0.5	19	4.8
United States (53)	1987–1994	Amphotericin B	98	0.25–1	1	1	100	0
		Flucytosine	98	0.5–≥128	8	8	NA	NA
		Fluconazole	98	0.5–8	4	8	100	0
		Itraconazole	98	0.03–0.5	0.25	0.25	NA	NA
United States (107)	1990–1999	Fluconazole	213	≤0.12–≥64	2	8	93	0.9
		Voriconazole	213	≤0.12–32	≤0.12	≤0.12	93	NA
		Posaconazole	213	≤0.015–≥8	≤0.015	0.06	97	NA
United Kingdom (33)	1971–1985	Fluconazole	36	1–≥64	2	32	NA	NA
		Itraconazole	36	≤0.03–1	0.25	0.5	NA	NA
	1986–1989	Fluconazole	41	≤0.25–≥64	8	32	NA	4.9
		Itraconazole	41	≤0.03–4	0.06	0.5	NA	NA
	1994–1996	Fluconazole	143	≤0.25–≥64	8	32	NA	5.6
		Itraconazole	143	≤0.03–4	0.12	1	NA	NA
Africa (22)	1994–2000	Amphotericin B	52	<0.002–0.09	0.006	0.012	100	0
		Fluconazole	52	0.03–≥256	8	32	50	5.8
		Voriconazole	52	0.004	0.12	0.25	100	0
Africa (6)	Pre-2000	Amphotericin B	16	1–2	2	2	31	69
		Flucytosine	16	1–16	4	8	69	0
		Fluconazole	16	4–32	8	16	87	0
		Itraconazole	16	0.03–0.12	0.03	0.12	100	0
Africa (14)	2003–2004	Amphotericin B	80	0.12–1	0.5	0.5	100	0
		Flucytosine	80	0.5–64	64	64	1.2	51
		Fluconazole	80	1–64	16	64	24	11
		Itraconazole	80	0.06–8	0.12	0.5	38	6.3
Africa (12)	2003–2005	Fluconazole	20	≤8–≥256	≥256	≥256	20	70
Africa (33)	1996	Fluconazole	43	2–≥64	8	32	NA	NA
		Itraconazole	43	0.06–2	0.25	1	NA	NA
Africa (70)	Pre-1998	Fluconazole	164	1–16	8	8	94	0
		Itraconazole	164	0.06–1	0.25	0.5	NA	NA
		Voriconazole	164	0.015–0.5	0.12	0.25	100	0
		Flucytosine	164	4–>128	8	16	NA	1
Egypt (1)	NA	Amphotericin B	29	0.25–1	0.5	1	100	0
		Flucytosine	29	0.5–8	2	8	NA	0
		Fluconazole	29	0.5–16	2	4	NA	0
		Itraconazole	29	0.03–0.25	0.12	0.25	NA	0
Cambodia (85)	2000–2001	Amphotericin B	134	0.002–1	0.19	0.5	100	0
		Fluconazole	134	0.5–>256	4	12	NA	2.5
	2001–2002	Amphotericin B	268	0.002–1	0.09	0.5	100	0
		Fluconazole	268	0.5–>256	12	96	NA	14
Cambodia (22)	1998–2001	Amphotericin B	110	<0.02–0.12	0.016	0.094	100	0
		Fluconazole	110	0.064–>256	16	32	22	6.4
		Voriconazole	110	0.012–>32	0.19	0.38	99.1	0.9
Malaysia (93)	1980–1990	Amphotericin B	17	0.002–0.38	0.25	0.38	100	0
		Flucytosine	17	0.003–>32	0.016	>32	NA	11.8
		Fluconazole	17	0.5–>256	2	16	NA	5.9
		Itraconazole	17	0.002–1.5	0.09	0.5	NA	5.9

(Continued on next page)

TABLE 4 *(Continued)*

Location (ref.)	Time period	Antifungal agent	No. of isolates	MIC (μg/ml)[a]			% by category[b]	
				Range	50%	90%	S	R
	2002–2004	Amphotericin B	31	0.12–0.75	0.25	0.38	100	0
		Flucytosine	31	0.006–>32	0.02	0.19	NA	6.5
		Fluconazole	31	0.5–48	3	8	NA	3.2
		Itraconazole	31	0.008–0.75	0.19	0.5	NA	0
Taiwan (25)	1982–1997	Amphotericin B	59	0.12–1	0.5	1	100	0
		Flucytosine	59	0.12–16	4	16	54	0
		Fluconazole	59	0.12–32	4	8	NA	0
Taiwan (47)	2003	Amphotericin B	70	0.12–2	0.5	1	97.1	2.9
		Flucytosine	70	0.12–8	1	8	NA	0
		Fluconazole	70	0.12–16	2	8	96	0
		Itraconazole	70	0.03–0.25	0.12	0.25	NA	0
		Voriconazole	70	0.12–0.5	0.12	0.12	100	0
Thailand (6)	Pre-2000	Amphotericin B	29	0.5–2	1	2	72	28
		Flucytosine	29	2–8	4	8	72	0
		Fluconazole	29	4–16	8	16	83	0
		Itraconazole	29	0.03–0.12	0.06	0.06	100	0
India (32)	1997–2000	Fluconazole	44	≤8–32	4	16	84	0
		Itraconazole	44	≤0.12–1	0.03	0.12	93	0
India (20)	2005–2007	Amphotericin B	35	0.06–1	0.5	1	100	0
		Fluconazole	35	0.25–16	4	8	91	0
Cuba (48)	NA	Amphotericin B	117	0.03–1	0.25	0.25	100	0
		Flucytosine	117	0.06–8	4	8	NA	0
		Fluconazole	117	0.25–8	2	4	100	0
		Itraconazole	117	0.016–1	0.03	0.25	NA	NA
Spain (69)	1995–2004	Amphotericin B	317	0.03–2	0.25	1	94.7	5.3
		Flucytosine	317	0.12–128	4	16	54	4.7
		Fluconazole	317	0.12–32	4	16	53.4	0
		Itraconazole	317	0.015–2	0.25	1	NA	15.8
		Voriconazole	317	0.015–4	0.12	0.5	99.1	0.9
France (31)	1997–2001	Amphotericin B	94	0.03–2	0.25	1	95.7	4.3
		Flucytosine	94	0.12–64	2	8	89.4	3.2
		Fluconazole	94	0.12–4	1	2	100	0
Latin America (19)	NA	Amphotericin B	89	NA	0.12	0.5	100	0
		Flucytosine	89	NA	4	4	90	1
		Fluconazole	89	NA	4	8	99	NA
		Itraconazole	89	NA	0.06	0.12	NA	NA
Latin America (5)	1996–2000	Amphotericin B	69	0.06–5	0.12	0.25	100	0
		Fluconazole	69	0.12–16	2	8	NA	0
		Itraconazole	69	0.03–0.25	0.12	0.25	NA	0

[a] 50% and 90%, MIC encompassing 50% and 90% of isolates tested, respectively.
[b] Category: S, susceptible; R, resistant; interpretive breakpoints are those described in CLSI document M27-S3 (29).
[c] NA, data not available.

(83 to 100%) are seen for amphotericin B, fluconazole, posaconazole, and voriconazole, with only 0 to 1% resistance. Although flucytosine exhibits only modest activity when tested alone against cryptococci, it is always used in combination with amphotericin B (and sometimes fluconazole), and synergy may be reliably achieved with the combination irrespective of the level of flucytosine activity when tested singly (86). Despite these comforting findings, it is important to take into account both the time period during which the organisms were obtained as well as the geographic origin of the isolates (73). Clearly, incident isolates are less likely to appear resistant than isolates from recurrent infections. In addition, isolates from some geographic locations appear to show improvement in susceptibility over time to both fluconazole and flucytosine (17, 71, 73). This may reflect in part a decrease in overall drug pressure concomitant with a decrease in cryptococcosis among HIV-infected individuals receiving HAART (71), whereas more recent isolates from resource-limited countries, where fluconazole monotherapy is the only course of treatment, may exhibit declining susceptibility to this important antifungal agent (12, 14, 44, 85) (Tables 3 and 4).

TABLE 5 Temporal and geographic variation in the in vitro susceptibility of *C. gattii* to systemically active antifungal agents

Location (ref.)	Time period	Antifungal agent	No. of isolates	MIC (μg/ml)[a]			% by category[b]	
				Range	50%	90%	S	R
Africa (64)	2002–2004	Fluconazole	41	0.51–8	1	4	100	0
		Itraconazole	41	0.03–0.5	0.12	0.25	NA	0
		Voriconazole	41	0.015–0.12	0.03	0.06	100	0
Brazil (97)	NA[c]	Amphotericin B	57	0.25–2	0.59[d]	1	96.5	3.5
		Flucytosine	57	0.5–>64	6.16[d]	16	NA	NA
		Fluconazole	57	1–64	9.54[d]	32	NA	NA
		Itraconazole	57	0.03–0.5	0.28[d]	0.5	NA	0
		Voriconazole	57	<0.03–1	0.15[d]	0.25	100	0
Brazil (41)	NA	Amphotericin B	23	0.03–0.25	0.12	0.12	100	0
		Flucytosine	23	0.25–32	1	16	NA	4.3
		Fluconazole	23	4–>64	16	25.6	NA	NA
		Itraconazole	23	0.12–2	0.5	1	NA	NA
		Posaconazole	23	0.03–0.5	0.25	0.5	100	0
		Voriconazole	23	0.03–1	0.5	1	100	0
Latin America (63)	NA	Fluconazole	48	0.5–32	4	16	85	0
		Voriconazole	48	0.008–0.12	0.06	0.06	100	0
Latin America (19)	Pre-2001	Amphotericin B	11	NA	0.25	0.5	100	0
		Flucytosine	11	NA	2	8	90	9
		Fluconazole	11	NA	8	16	73	NA
		Itraconazole	11	NA	0.12	0.25	NA	NA
Malaysia (93)	2002–2004	Amphotericin B	18	0.02–0.38	0.25	0.38	100	0
		Flucytosine	18	0.004–>32	0.016	0.19	NA	5.6
		Fluconazole	18	0.75–64	4	16	77.8	5.6
		Itraconazole	18	0.002–0.5	0.09	0.38	NA	0
Australia (24)	1994–1997	Fluconazole	18	0.25–64	4	64	50	11
Various[e] (94)	NA	Amphotericin B	42	0.12–1	0.25	0.25	100	0
		Flucytosine	42	0.06–16	2	8	NA	0
		Fluconazole	42	0.5–32	2	8	NA	0
		Posaconazole	42	<0.015–0.25	0.03	0.12	100	0
		Voriconazole	42	<0.015–0.5	0.12	0.25	100	0

[a] 50% and 90%, MIC encompassing 50% and 90% of isolates tested, respectively.
[b] Interpretive breakpoints are those described in CLSI document M27-S3 (29); S, susceptible; R, resistant.
[c] NA, data not available.
[d] Values are geometric mean MICs.
[e] Includes isolates from the United States, Australia, France, Denmark, Italy, Thailand, and Africa.

EPIDEMIOLOGY OF ANTIFUNGAL RESISTANCE IN *C. NEOFORMANS* AND *C. GATTII*

One of the important by-products of the standardization process for antifungal susceptibility testing has been the ability to conduct surveillance of resistance to antifungal agents among cryptococci by uniform methods (14, 17, 20, 22, 32, 33, 47, 72, 73, 85, 107). Meaningful large-scale surveys of antifungal susceptibility and resistance conducted over time would not be possible without a standardized BMD or disk diffusion method for performing the in vitro studies. Studies of trends in resistance among *C. neoformans* and *C. gattii* to commonly used antifungal agents such as fluconazole (14, 17, 73) and comparative analyses of licensed and newly introduced agents against these organisms (22, 47, 48, 71–73, 107) have provided large amounts of useful data and allow one to form a more accurate picture of the evolving epidemiology of antifungal resistance among the pathogenic cryptococci.

Geographic and Temporal Trends in Antifungal Resistance among Clinical Isolates of *C. neoformans*

Certainly the greatest effort in defining trends in resistance among *C. neoformans* isolates has been devoted to fluconazole (68). A recent global (134 study sites in 39 countries) antifungal surveillance study, covering the years 2001 through 2007 and encompassing 2,711 isolates of *C. neoformans* (73), found that resistance to fluconazole was less prominent in Europe and North America compared to that seen in the Asia-Pacific, Africa–Middle East, and Latin American regions (Table 3). Notably, a decrease in susceptibility to fluconazole was observed in both the Asia-Pacific and Latin American regions over the course of the study, whereas susceptibility improved in North America.

These data are corroborated by other studies conducted throughout the world as shown in Table 4. Studies of clinical isolates of *C. neoformans* by Brandt et al. (17), Davey et al. (33), Klepser and Pfaller (53), Pfaller et al. (71), and

TABLE 6 Mechanisms involved in the development of resistance to antifungal agents in *C. neoformans*

Reference	Antifungal agent	Mechanism of resistance	Comments
Block et al. (16)	Flucytosine	Defect in cytosine permease/ deaminase; mutation in uridine-5 monophosphate pyrophosphorylase (UMP) or uracil phosphoribosyl transferase	Antifungal resistance correlated with failure of flucytosine monotherapy Syngeristic interaction with amphotericin B related to the mechanisms of flucytosine resistance: defect in permease/ deaminase (86)
Kim et al. (52)	Polyenes	Metabolic block in sterol Δ^8-Δ^7 isomerization step	High-level resistance to nystatin and pimaricin; marginal to no resistance to amphotericin B
Kelly et al. (51)	Amphotericin B	Defective sterol Δ^8-Δ^7 isomerase; depletion of ergosterol	Amphotericin B resistance (MIC = 4 μg/ml) emerged in a clinical isolate following fluconazole and amphotericin B therapy
Joseph-Horne et al. (50)	Amphotericin B and fluconazole	Decreased intracellular content of fluconazole; possible multidrug efflux pump	Laboratory selection of antifungal-resistant mutants; two mutants isolated that were cross-resistant to fluconazole and amphotericin B
Lamb et al. (55)	Fluconazole	Decreased sensitivity of P450-mediated sterol 14α-demethylase to inhibition by fluconazole; slight increase in P450 and decrease in intracellular fluconazole content	Four clinical isolates from AIDS patients who failed fluconazole therapy; cross-resistance to ketoconazole and itraconazole
Thornewell et al. (96)	Multidrug resistance (MDR)	CNEMDR1 efflux pump; CNEMDR2 efflux pump	In vitro detection of the genes *CneMDR1* and *CneMDR2* encoding for proteins with structural characteristics of ATP binding cassette (ABC) proteins. No evidence presented for a role in antifungal resistance
Venkateswarlu et al. (99)	Fluconazole and amphotericin B	Reduced ergosterol content, defect in sterol Δ^8-Δ^7 isomerase (amphotericin B); reduced sensitivity of P450 14α-demethylase to inhibition by fluconazole (low-level resistance); reduced cellular content of fluconazole (high-level resistance)	11 clinical isolates with varying patterns of susceptibility to amphotericin B and fluconazole. Data suggest that efflux most important in high-level (MIC >64 μg/ml) fluconazole resistance
Rodero et al. (81)	Fluconazole	G484S amino acid substitution in P450 sterol 14α-demthylase	Isogenic series of five isolates from an AIDS patient on long-term fluconazole therapy. Fluconazole MIC increased from 2 μg/ml to 32 μg/ml. Mutation results in decreased fluconazole binding. No cross-resistance to itraconazole or voriconazole detected
Posteraro et al. (74)	Fluconazole	ABC transporter CNAFR1; *CnAFR1* overexpression important in determining increased azole resistance	In vitro selection of fluconazole-resistant strain (MIC = 64 μg/ml). Identification of cDNA overexpressed; sequence analysis identified ABC transporter encoding gene, *C. neoformans* Antifungal Resistance 1 (*CnAFR1*). Knockout mutant susceptible to fluconazole; reintroduction of gene restored resistance phenotype. Cross-resistance to ketoconazole and itraconazole
Sanguinetti et al. (84)	Fluconazole	Overexpression of ABC transporter–encoding gene, *AFR1* (antifungal resistance 1; formerly known as *CnAFR1*)	Upregulation of *AFR1* in murine model of cryptococcosis essential for in vivo fluconazole resistance and also enhances the virulence of *C. neoformans*

Yildiran et al. (107) provide antifungal susceptibility data generated by CLSI reference MIC methods (28, 29), indicating that in vitro resistance to fluconazole (as well as to amphotericin B and flucytosine) remains uncommon and has not increased with time in the United States and the United Kingdom (Table 4). Analysis of trends in susceptibility to fluconazole from 1992 through 1998 in the context of a population-based study conducted in the United States (17) shows that the susceptibility profile has actually improved since the introduction of HAART (Table 4). Comparing the percentage of U.S. isolates susceptible (MIC ≤8 μg/ml) to fluconazole from the time periods 1992 to 1994 and 1996 to 1998 shows an improvement from 71 to 89% susceptible. This trend continues to the present day, with 92 to 96% of North American isolates susceptible to fluconazole (71, 73) (Table 3). In contrast, reports from Cambodia (22, 85), Africa (11–14), and Spain (69) indicate that recent isolates from those areas exhibit decreased susceptibility to fluconazole and other azoles (Table 4). In one report from Africa (12), 70% of isolates from patients with a clinical relapse following treatment of cryptococcal meningitis with fluconazole as the first-line therapy exhibited resistance (MIC ≥64 μg/ml) to this antifungal (overall 75% had decreased susceptibility [MIC ≥32 μg/ml]) (Table 4).

Factors underlying these therapeutic failures and the emerging resistance include (44) (i) profound immunodeficiency, (ii) an increased CNS microbiological burden (median 340,000 yeast cells/mm^3), (iii) absence of fungicidal activity of fluconazole (11), and (iv) likely inadequate CNS levels of fluconazole due to unappreciated drug-drug interactions (44% of patients with drug-resistant isolates received rifampin along with fluconazole). These data underscore the fact that fluconazole monotherapy is unreliable for the treatment of cryptococcal meningitis in AIDS and demonstrates how continued host exposure to fluconazole coupled with multiple recurrences of meningitis and a high organism burden provides the "perfect" setting for the emergence of high-level fluconazole resistance.

The various studies demonstrate that amphotericin B, as well as the newer triazoles (posaconazole and voriconazole), retain potent activity against *C. neoformans*, with more than 99% of isolates showing MICs of ≤1 μg/ml, irrespective of the geographic origin of the isolates (Table 4). The issue of cross-resistance between fluconazole and the newer triazoles is not completely clear; however, 80% of fluconazole-resistant strains of *C. neoformans* remain susceptible to voriconazole (73). Notably, all cryptococci are intrinsically resistant to the echinocandin class of antifungal agents (2, 8).

Resistance to flucytosine remains uncommon in most geographic regions with the exception of certain parts of Africa (51%) (14) and Malaysia (11.8%) (92). Monotherapy with this agent reliably results in the emergence of secondary resistance (16). Hospenthal and Bennett (46) reviewed the clinical experience of flucytosine monotherapy in the 1960s and 1970s and found a 57% failure rate in treatment of cryptococcosis.

Geographic and Temporal Trends in Antifungal Resistance among Clinical Isolates of *C. gattii*

Although less well studied than *C. neoformans*, several recent studies have provided antifungal susceptibility data on clinical isolates of *C. gattii* (Table 5). Studies from the United States, South Africa, Latin America, Malaysia, and Australia all document low levels of resistance and overall susceptibilities comparable to those of *C. neoformans* for amphotericin B, flucytosine, fluconazole, itraconazole, voriconazole, and posaconazole (Table 5). A few studies (41, 63, 97) show minor differences in the MICs for *C. gattii* when compared to *C. neoformans*: *C. gattii* isolates were more susceptible to amphotericin B and flucytosine and less susceptible to azoles than were *C. neoformans* isolates. Presently, resistance to antifungal agents does not seem to pose a problem with *C. gattii*. Notably, even recent isolates from Africa remain completely susceptible to fluconazole (64), in marked contrast to African isolates of *C. neoformans* from the same time period (11, 14).

MECHANISMS OF RESISTANCE TO ANTIFUNGAL AGENTS IN *C. NEOFORMANS*

Although most of the effort to elucidate the mechanisms of antifungal resistance has been expended in studies of *Candida* (83, 103), the results of these studies have directed similar work devoted to resistance mechanisms in *C. neoformans* (68) (Table 6).

Many parallels exist between the well-studied antibacterial resistance mechanisms and antifungal resistance mechanisms (40); however, there are no data to suggest that destruction or modification of antifungal agents is an important component of antifungal resistance. Likewise, it does not appear that fungi can employ the genetic exchange mechanisms that allow rapid transmission of antimicrobial resistance in bacteria (40). On the other hand, it is apparent that multidrug efflux pumps, target alterations, and reduced access to targets are important mechanisms of resistance to antifungal agents, just as they are important in antibacterial resistance (16, 50–52, 55, 74, 81, 83, 96, 103) (Table 6). In contrast to the rapid emergence of high-level antimicrobial resistance that occurs among bacteria, antifungal resistance develops more slowly and involves the emergence of intrinsically resistant species (*C. neoformans* is intrinsically resistant to the echinocandins) or a stepwise alteration of cellular structures or functions that results in resistance to an agent to which there has been prior exposure (40, 68).

Flucytosine

Despite reports in the older literature of a high frequency of primary resistance to flucytosine among *C. neoformans* (16), more recent studies using validated, standardized test methods indicate that primary resistance is actually uncommon among clinical isolates of *C. neoformans* (17, 71) (Tables 2, 4, and 5). Secondary resistance to flucytosine, on the other hand, is well documented to occur in *C. neoformans* during monotherapy with this agent (16, 46).

Resistance to flucytosine may develop from decreased uptake (loss of permease activity) or loss of enzymatic activity required for the conversion of flucytosine to 5-fluorouracil (cytosine deaminase) and 5-fluorouridylic acid (FUMP pyrophosphorylase) (16, 101) (Table 6). Of these possible mechanisms, the most important appear to be the loss of cytosine deaminase activity or the loss of FUMP pyrophosphorylase activity. Notably, Schwarz et al. (86) have shown that among isolates of *C. neoformans* with flucytosine resistance due to defects in either the permease or the deaminase activity, synergy with amphotericin B may still be achieved. Uracil phosphoribocil transferase, another enzyme in the pyrimidine salvage pathway, is also important in the formation of FUMP, and loss of its activity is sufficient to confer resistance to flucytosine (101, 102).

Polyenes

Resistance to amphotericin B remains uncommon among *C. neoformans* and *C. gattii* despite extensive utilization for 50 years. Primary resistance to amphotericin B has not been reported, and secondary resistance appears rare (25, 51) (Tables 2, 4, and 5).

Our understanding of the mechanism of resistance to amphotericin B in *C. neoformans* stems largely from studies of laboratory-derived mutants (50, 52) and from characterization of rare serial isolates from patients failing amphotericin B therapy (51, 99) (Table 6). The mechanism of amphotericin B resistance appears to be from a quantitative or qualitative alteration in the sterol content of cells (51, 52, 99). Because ergosterol is the primary sterol target for amphotericin B in the fungal cell membrane, resistant cells with altered sterol content should bind smaller amounts of amphotericin B than do susceptible cells (40, 68). Accordingly, mutants of *C. neoformans* resistant to amphotericin B have been shown to have reduced ergosterol content and replacement of polyene binding sterols by intermediates that bind polyenes less well (51, 52, 99). Studies of sterol profiles of both laboratory mutants and clinical isolates indicate a block in the sterol Δ^8-Δ^7 isomerization step with depletion of ergosterol (Table 6). Although Joseph-Horne et al. (50) isolated two laboratory mutants of *C. neoformans* that were cross-resistant to amphotericin B and fluconazole, neither isolate had reduced ergosterol content, and both displayed a multidrug efflux pump phenotype. Multidrug efflux mediated amphotericin B and fluconazole cross-resistance has yet to be reported in clinical isolates. Isolating such strains is highly unlikely since amphotericin B is not taken in by the fungal cell and exerts its activity through physiochemical interaction with sterols in the outer cell envelope, particularly the cell membrane.

Azoles

As noted previously, primary resistance to fluconazole and other azoles is uncommon among isolates of *C. neoformans* (Table 2), although secondary resistance has been described among individuals with AIDS and relapsing cryptococcal meningitis (4, 7, 10, 15, 66, 99, 100).

Resistance to azoles among yeasts can be from a modification in the quality or quantity of the target enzyme (sterol 14α-demethylase), reduced access of the drug to the target enzyme, or some combination of these mechanisms (55, 74, 81, 83, 96, 103) (Table 6). In the first instance, point mutations in the gene (*ERG11*) encoding for the target enzyme, 14α-demethylase, leads to an altered target with decreased affinity for azoles (55, 81, 99). Overexpression of *ERG11* (with or without a mutation) results in production of high concentrations of the target enzyme, creating the need for higher intracellular azole concentrations to inhibit all enzyme molecules in the cell (55). The latter mechanism appears to be less important than an altered target in producing azole resistance in *C. neoformans* (55).

The second major mechanism involves active efflux of azole antifungal agents out of the cell through the action of two multidrug efflux transporters: the major facilitators (encoded by *MDR* genes) and those of the ATP binding cassette (ABC) superfamily (encoded by the antifungal resistance 1 [*AFR1*] gene) (74, 83, 84, 96). Although Thornewell et al. (96) characterized genes encoding two MDR-type efflux pumps (*CneMDR1* and *CneMDR2*), *CneMDR1* or *CneMDR2* expression has yet to be linked with azole resistance in *C. neoformans* clinical isolates or in laboratory-derived resistant mutants.

In contrast, overexpression of the *AFR1* (previously called *CnAFR1*) gene, which encodes the ABC transporter AFR1, leads to azole resistance that can be demonstrated in vitro and in vivo (74, 84). Posteraro et al. (74) demonstrated that knockout mutants without *AFR1* were hypersusceptible to fluconazole and also showed that reintroduction of the gene restored the resistant phenotype. Sanguinetti et al. (84) subsequently showed that upregulation of *AFR1* in a murine model of cryptococcosis was essential for in vivo fluconazole resistance and also enhanced the virulence of *C. neoformans* and improved the survival of the fungus in macrophages.

The development of fluconazole resistance in *C. neoformans* has been shown to be a dynamic and heterogeneous mutational process (105). It is estimated that approximately 5% of clinical isolates exhibited a so-called fluconazole-heteroresistant phenotype (62, 106) in which each isolate produces cultures with heterogeneous compositions in which most of the cells are susceptible, but cells highly resistant (MIC ≥64 μg/ml) to fluconazole are recovered at a variable frequency. Recent studies have suggested that heteroresistance is intrinsic to most strains, and the higher MIC clones are more virulent (89). Heteroresistant *C. neoformans* isolates grow on solid media at four- to eightfold higher fluconazole concentrations than their BMD MIC would suggest (105). It is notable that *AFR1* appears to be associated with this phenotype (89). For example, under selective azole pressure some *C. neoformans* colonies will become aneuploid and have increased copy numbers of chromosomes. This leads to increased copies of *AFR1*, if its chromosome is duplicated, and thus higher-level fluconazole resistance in the colony (90). These changes in chromosomal numbers are unstable without selection pressure, and these azole resistance changes are dynamic. In the original description of fluconazole heteroresistance by Mondon et al. (62), the emergence of a highly resistant clone of *C. neoformans* was demonstrated in a patient with continued exposure to fluconazole and multiple recurrences of meningitis. The clone persisted as long as fluconazole was present. This same clinical scenario exists in AIDS patients with cryptococcal meningitis who do not have access to, or are resistant to, HAART and who do not have the benefit of the fungicidal combination of amphotericin B and flucytosine (44). Heteroresistance may be a primary resistance mechanism in vivo that can be missed if the strain reverts back to the susceptible phenotype under no drug selection in vitro.

Echinocandins

The echinocandin class of antifungals is relatively ineffective against *Cryptococcus* sp. (2, 8, 54), but the mechanism of this resistance is still unknown. In other fungi, the echinocandins are known to inhibit β-1,3-glucan synthase. The evidence suggests that *C. neoformans* has the appropriate target for the echinocandins. There is β-1,3-glucan present in the cryptococcal cell wall (49), and there is a single, essential *FKS1* gene (95), suggesting that β-1,3-glucan is necessary for a viable cell wall. In vitro enzymatic studies showed that the *C. neoformans* glucan synthase activity is sensitive to caspofungin (59). Cryptococci treated with caspofungin at noninhibitory concentrations show reduced levels of β-1,3-glucan (37), leading to the hypothesis that there may be other targets of echinocandins present in other fungi. However, the lower limit of β-1,3-glucan that is necessary for robust growth in *C. neoformans* has not been established, and it is possible that *C. neoformans* may be better able to compensate for reduction in the β-1,3-glucan component of its cell wall compared to echinocandin-susceptible organisms.

SUMMARY AND CONCLUSIONS

The epidemiology of antifungal resistance in the pathogenic cryptococci is a dynamic process that differs considerably according to the status of health care, especially that devoted to HIV and AIDS, in the region of interest. In industrialized settings the cumulative effect of HAART and optimal antifungal regimens that include initial treatment of cryptococcal meningitis with amphotericin B and flucytosine followed by fluconazole maintenance and prophylactic therapy, result in an improved antifungal susceptibility profile of *C. neoformans* with increased susceptibility to fluconazole and flucytosine and sustained activity of amphotericin B and the new triazoles (Tables 2–5). The situation in many developing countries and regions is quite different, where the lack of HIV-supportive care leads to delayed diagnosis of HIV infection and poor access to HAART, with resultant profound immunocompromise and frequent infection with *C. neoformans*. Patients suffering from cryptococcal meningitis generally receive medical attention late in the course of the infection and thus have a very high organism burden. Primary antifungal therapy in these settings is limited to the administration of fluconazole, often at inadequate doses for treating CNS disease. Concomitant administration of rifampin, as prophylaxis for mycobacterial infection, without fluconazole dose adjustment virtually ensures that CNS levels of fluconazole will be suboptimal to treat the infection. Such conditions are ideal for the generation of antifungal resistance, as is now evidenced by numerous case reports, case series, and the findings of broad surveillance efforts (Tables 1–5).

This situation will continue unless developing countries can obtain adequate resources and the necessary clinical expertise to manage the many complications associated with advanced AIDS (11, 44, 56, 75). Although access to HAART and fluconazole is improving in many developing regions, additional strategies for the treatment of cryptococcal infections are needed, including access to amphotericin B and flucytosine for primary therapy, modification of dosing strategies for fluconazole, and improved ability to recognize and manage the immune reconstitution syndrome (IRIS) (11, 44, 56).

Improved understanding of the mechanisms of antifungal resistance in cryptococci will ultimately help in the battle against emerging resistance. Given the demonstrated importance of efflux pumps in the resistance to fluconazole and other azoles (Table 6), further investigation of the use of efflux pump inhibitors is a highly promising area.

REFERENCES

1. **Abdel-Salam, H. A.** 2005. In vitro susceptibility of *Cryptococcus neoformans* clinical isolates from Egypt to seven antifungal drugs. *Mycoses* **48:**327–332.
2. **Abruzzo, G. K., A. M. Flattery, C. J. Gill, L. Kong, J. G. Smith, V. B. Pikounis, J. M. Balkovec, A. F. Boufard, J. F. Dropinski, H. Rosen, H. Kropp, and K. Bartizal.** 1997. Evaluation of the echinocandin antifungal MK-0991 (L-743,872): efficacies in mouse models of disseminated aspergillosis, candidiasis and cryptococcosis. *Antimicrob. Agents Chemother.* **41:**2333–2336.
3. **Alexander, B. D., and M. A. Pfaller.** 2006. Contemporary tools for the diagnosis and management of invasive mycoses. *Clin. Infect. Dis.* **43:**515–527.
4. **Aller, A. I., E. Martin-Manuelos, F. Lozano, J. Gomez-Mateos, L. Steele-Moore, W. J. Holloway, M. J. Gutierrez, F. J. Recio, and A. Espinel-Ingroff.** 2000. Correlation of fluconazole MICs with clinical outcome in cryptococcal infection. *Antimicrob. Agents Chemother.* **44:**1544–1548.
5. **Alves, S. H., L. T. Oliveira, J. M. Costa, I. Lubeck, A. K. Casali, and M. H. Vainstein.** 2001. In vitro susceptibility to antifungal agents of clinical and environmental *Cryptococcus neoformans* isolated in Southern Brazil. *Rev. Inst. Med. Trop. S. Paulo* **43:**267–270.
6. **Archibald, L. K., M. J. Tuohy, D. A. Wilson, O. Nwanyanwu, P. N. Kazembe, S. Tansuphasawadikul, B. Eampokalup, A. Chaovavanich, L. B. Reller, W. R. Jarvis, G. S. Hall, and G. W. Procop.** 2004. Antifungal susceptibilities of *Cryptococcus neoformans*. *Emerg. Infect. Dis.* **10:**143–145.
7. **Armengou, A., C. Porear, J. Mascaro, and F. Garcia-Bragado.** 1996. Possible development of resistance to fluconazole during suppressive therapy for AIDS-associated cryptococcal meningitis. *Clin. Infect. Dis.* **23:**1337–1338.
8. **Bartizal, K., C. Gill, G. Abruzzo, A. M. Flattery, L. Kong, P. M. Scott, J. G. Smith, C. E. Leighton, A. Bouffard, J. F. Dropinsky, and J. Balkovec.** 1997. In vitro preclinical evaluation studies with the echinocandin antifungal MK-0991 (L-743,872). *Antimicrob. Agents Chemother.* **41:**2326–2332.
9. **Bartlett, K. H., S. E. Kidd, and J. W. Kronstad.** 2007. The emergence of *Cryptococcus gattii* in British Columbia and the Pacific Northwest. *Curr. Fungal Infect. Rep.* **1:**108–115.
10. **Berg, J., C. J. Clancy, and M. H. Nguyen.** 1998. The hidden danger of primary fluconazole prophylaxis for patients with AIDS. *Clin. Infect. Dis.* **26:**186–187.
11. **Bicanic, T., R. Wood, L. G. Bekker, M. Darder, G. Meintjes, and T. S. Harrison.** 2005. Antiretroviral roll-out, antifungal roll-back: access to treatment for cryptococcal meningitis. *Lancet Infect. Dis.* **5:**530–531.
12. **Bicanic, T., T. Harrison, A. Niepieklo, N. Dyakopu, and G. Meintjes.** 2006. Symptomatic relapse of HIV-associated cryptococcal meningitis after initial fluconazole monotherapy: the role of fluconazole resistance and immune reconstitution. *Clin. Infect. Dis.* **43:**1069–1073.
13. **Bicanic, T., G. Meintjes, R. Wood, M. Hayes, K. Rebe, L. G. Bekker, and T. Harrison.** 2007. Fungal burden, early fungicidal activity, and outcome in cryptococcal meningitis in antiretroviral-naïve or antiretroviral-experienced patients treated with amphotericin B or fluconazole. *Clin. Infect. Dis.* **45:**76–80.
14. **Bii, C. C., K. Makimura, S. Abe, H. Taguchi, O. M. Mugasia, G. Revathi, N. C. Wamu, and S. Kamiya.** 2006. Antifungal drug susceptibility of *Cryptococcus neoformans* from clinical sources in Nairobi, Kenya. *Mycoses* **50:**25–30.
15. **Birley, H. D., E. M. Johnson, P. McDonald, C. Parry, P. B. Carey, and D. W. Warnock.** 1995. Azole drug resistance as a cause of clinical relapse in AIDS patients with cryptococcal meningitis. *Int. J. STD AIDS* **6:**353–355.
16. **Block, E. R., A. E. Jennings, and J. E. Bennett.** 1973. 5-Fluorocytosine resistance in *Cryptococcus neoformans*. *Antimicrob. Agents Chemother.* **3:**649–656.
17. **Brandt, M. E., M. A. Pfaller, R. A. Hajjeh, R. J. Hamill, P. G. Pappas, A. L. Reingold, D. Rimland, and D. W. Warnock for the Cryptococcal Disease Active Surveillance Group.** 2001. Trends in antifungal drug susceptibility of *Cryptococcus neoformans* isolates in the United States: 1992 to 1994 and 1996 to 1998. *Antimicrob. Agents Chemother.* **45:**3065–3069.
18. **Brouwer, A. E., A. Rajanuwong, W. Chierakul, G. E. Griffin, R. A. Larsen, N. J. White, and T. S. Harrison.** 2004. Combination antifungal therapies for HIV-associated cryptococcal meningitis: a randomized trial. *Lancet* **363:**1764–1767.
19. **Calvo, B. M., A. L. Colombo, O. Fischman, A. Santiago, L. Thompson, M. Lazera, F. Telles, K. Fukushima, K. Nichimura, R. Tanaku, M. Myiajy, and M. L. Moretti-Branchini.** 2001. Antifungal susceptibilities, varieties, and

electrophoretic karyotypes of clinical isolates of *Cryptococcus neoformans* from Brazil, Chile, and Venezuela. *J. Clin. Microbiol.* **39:**2348–2350.

20. **Capoor, M. R., P. Mandal, M. Deb, P. Aggarwal, and U. Banerjee.** 2008. Current scenario of cryptococcosis and antifungal susceptibility pattern in India: a cause for reappraisal. *Mycoses* **51:**258–265.
21. **Casadevall, A., E. D. Spitzer, D. Webb, and M. G. Rinaldi.** 1993. Susceptibilities of serial *Cryptococcus neoformans* isolates from patients with recurrent cryptococcal meningitis to amphotericin B and fluconazole. *Antimicrob. Agents Chemother.* **37:**1383–1386.
22. **Chandenier, J., K. D. Adou-Bryn, C. Douchet, B. Sar, M. Kombila, D. Swinne, M. Therizol-Ferly, Y. Buisson, and B. Richard-Lenoble.** 2004. In vitro activity of amphotericin B, fluconazole and voriconazole against 162 *Cryptococcus neoformans* isolates from Africa and Cambodia. *Eur. J. Clin. Microbiol. Infect. Dis.* **23:**506–508.
23. **Chayakulkeeree, M., and J. R. Perfect.** 2006. Cryptococcosis. *Infect. Dis. Clin. N. Am.* **20:**507–544.
24. **Chen, S., T. Sorrell, G. Nimmo, B. Speed, B. Currie, D. Ellis, D. Marriott, T. Pfeiffer, D. Parr, K. Byth, and the Australasian Cryptococcal Study Group.** 2000. Epidemiology and host- and variety-dependent characteristics of infection due to *Cryptococcus neoformans* in Australia and New Zealand. *Clin. Infect. Dis.* **31:**499–508.
25. **Chen, Y. C., S. C. Chang, C. C. Shih, C. C. Hung, K. T. Luh, Y. S. Pan, and W. C. Hsieh.** 2000. Clinical features and in vitro susceptibilities of two varieties of *Cryptococcus neoformans* in Taiwan. *Diagn. Microbiol. Infect. Dis.* **36:**175–183.
26. **Clinical and Laboratory Standards Institute.** 2004. Method for antifungal disk diffusion susceptibility testing of yeasts; approved standard M44-A. Clinical and Laboratory Standards Institute, Wayne, PA.
27. **Clinical and Laboratory Standards Institute.** 2008. Zone diameter interpretive standards, corresponding minimal inhibitory concentration (MIC) interpretive breakpoints, and quality control limits for antifungal disk diffusion susceptibility testing of yeasts; informational supplement, 2nd ed. M44-S2. Clinical and Laboratory Standards Institute, Wayne, PA.
28. **Clinical and Laboratory Standards Institute.** 2008. Reference method for broth dilution antifungal susceptibility testing of yeasts; approved standard, 3rd ed. M27-A3. Clinical and Laboratory Standards Institute, Wayne, PA.
29. **Clinical and Laboratory Standards Institute.** 2008. Reference method for broth dilution antifungal susceptibility testing of yeasts; informational supplement, 3rd ed. M27-S3. Clinical and Laboratory Standards Institute, Wayne, PA.
30. **Corbett, E. L., G. J. Churchyard, S. Charalambos, B. Samb, V. Moloi, T. C. Clayton, A. D. Grant, J. Murray, R. J. Hayes, and K. M. DeCock.** 2002. Morbidity and mortality in South African gold miners: impact of untreated disease due to human immunodeficiency virus. *Clin. Infect. Dis.* **34:**1251–1258.
31. **Dannaoui, E., M. Abdul, M. Arpin, A. Michel-Nguyen, M. A. Piens, A. Favel, O. Lortholary, F. Dromer, and the French Cryptococcosis Study Group.** 2006. Results obtained with various antifungal susceptibility testing methods do not predict early clinical outcome in patients with cryptococcosis. *Antimicrob. Agents Chemother.* **50:** 2464–2470.
32. **Datta, K., N. Jain, S. Sethi, A. Rattan, A. Casadevall, and U. Banerjee.** 2003. Fluconazole and itraconazole susceptibility of clinical isolates of *Cryptococcus neoformans* at a tertiary care center in India: a need for care. *J. Antimicrob. Chemother.* **52:**683–686.
33. **Davey, K. G., E. M. Johnson, A. D. Holmes, A. Szekely, and D. W. Warnock.** 1998. In-vitro susceptibility of *Cryptococcus neoformans* isolates to fluconazole and itraconazole. *J. Antimicrob. Chemother.* **42:**217–220.
34. **Dromer, F., S. Mathoulin, B. Dupont, A. Laporte, and the French Cryptococcosis Study Group.** 1996. Epidemiology of cryptococcosis in France: 9-year survey (1985–1993). *Clin. Infect. Dis.* **23:**82–90.
35. **Dromer, F., S. Mathoulin-Pelissier, A. Fontanet, O. Renin, B. Dupont, O. Lortholary, on behalf of the French Cryptococcosis Study Group.** 2004. Epidemiology of HIV-associated cryptococcosis in France (1985–2001): comparison of the pre- and post-HAART eras. *AIDS* **18:**555–562.
36. **Espinel-Ingroff, A. V., and M. A. Pfaller.** 2007. Susceptibility test methods: yeasts and filamentous fungi, p. 1880–1893. In P. R. Murray, E. J. Baron, J. H. Jorgensen, M. L. Landry, and M. A. Pfaller (ed.), *Manual of Clinical Microbiology*, 9th ed. ASM Press, Washington, DC.
37. **Feldmesser, M., Y. Kress, A. Mednick, and A. Casadevall.** 2000. The effect of the echinocandin analogue caspofungin on cell wall glucan synthesis of *Cryptococcus neoformans*. *J. Infect. Dis.* **182:**1791–1795.
38. **Friese, G., T. Discher, R. Fussle, A. Schmalreck, and J. Lohmeger.** 2001. Development of azole resistance during fluconazole maintenance therapy for AIDS-associated cryptococcal disease. *AIDS* **15:**2344–2345.
39. **Ghannoum, M. A., A. S. Ibrahim, Y. Fu, M. C. Shafia, J. E. Edwards, Jr., and R. S. Criddle.** 1992. Susceptibility testing of *Cryptococcus neoformans*: a microdilution technique. *J. Clin. Microbiol.* **30:**2881–2886.
40. **Ghannoum, M. A., and L. B. Rice.** 1999. Antifungal agents: mode of action, mechanisms of resistance, and correlation of these mechanisms with bacterial resistance. *Clin. Microbiol. Rev.* **12:**501–517.
41. **Gomez-Lopez, A., O. Zaragoza, M. Dos Anjos Martins, M. C. Melhem, J. L. Rodriguez-Tudela, and M. Cuenca-Estrella.** 2008. In vitro susceptibility of *Cryptococcus gattii* clinical isolates. *Clin. Microbiol. Infect.* **14:**727–730.
42. **Graybill, J. R., J. Sobel, M. Saag, C. van der Horst, W. Powderly, G. Cloud, L. Riser, R. Hamill, W. Dismukes, and the NIAID Mycoses Study Group and AIDS Cooperative Treatment Groups.** 2000. Diagnosis and management of increased intracranial pressure in patients with AIDS and cryptococcal meningitis. *Clin. Infect. Dis.* **30:**47–54.
43. **Hajjeh, R. A., L. A. Conn, D. S. Stephens, W. Baughman, R. Hamill, E. Graviss, P. G. Pappas, C. Thomas, A. Reingold, G. Rothrock, L. C. Hutwagner, A. Schuchat, M. E. Brandt, R. W. Pinner, and the Cryptococcal Active Surveillance Group.** 1999. Cryptococcosis: population-based multistate active surveillance and risk factors in human immunodeficiency virus-infected persons. *J. Infect. Dis.* **179:**449–454.
44. **Hamill, R. J.** 2006. Free fluconazole for cryptococcal meningitis: too little of a good thing? *Clin. Infect. Dis.* **43:**1074–1076.
45. **Hoang, L. M. N., J. A. Maguire, P. Doyle, M. Fyfe, and D. L. Roscoe.** 2004. *Cryptococcus neoformans* infections at Vancouver Hospital and Health Science Centre (1997–2002): epidemiology, microbiology and histopathology. *J. Med. Microbiol.* **53:**935–940.
46. **Hospenthal, D. R., and J. E. Bennett.** 1998. Flucytosine monotherapy for cryptococcosis. *Clin. Infect. Dis.* **27:**260–264.
47. **Hsueh, P. R., Y. J. Lau, Y. C. Chuang, J. H. Wan, W. K. Huong, J. M. Shyr, J. J. Yan, K. W. Yu, J. J. Wu, W. C. Ko, Y. C. Yang, Y. C. Liu, L. J. Teng, C. Y. Liu, and K. T. Luh.** 2005. Antifungal susceptibilities of clinical isolates of *Candida* species, *Cryptococcus neoformans* and

Aspergillus species from Taiwan: surveillance of multicenter antimicrobial resistance in Taiwan program data from 2003. *Antimicrob. Agents Chemother.* **49:**512–517.

48. **Illnait-Zaragozi, M. T., G. F. Martinez, I. Curfs-Breuker, C. M. Fernandez, T. Boekhout, and J. F. Meis.** 2008. In vitro activity of the new azole isavuconazole (BAL4815) compared with six other antifungal agents against 162 *Cryptococcus neoformans* isolates from Cuba. *Antimicrob. Agents Chemother.* **52:**1580–1582.
49. **James, P., R. Cherniak, R. G. Jones, C. A. Stortz, and E. Reiss.** 1990. Cell-wall glucans of *Cryptococcus neoformans* CAP67. *Carbohydr. Res.* **198:**23–38.
50. **Joseph-Horne, T., D. Holloman, R. S. T. Loeffler, and S. L. Kelly.** 1995. Cross-resistance to polyene and azole drugs in *Cryptococcus neoformans. Antimicrob. Agents Chemother.* **39:**1526–1529.
51. **Kelly, S. L., D. C. Lamb, M. Taylor, A. J. Corran, B. C. Baldwin, and W. G. Powderly.** 1994. Resistance to amphotericin B associated with defective sterol $\Delta^{8\rightarrow7}$ isomerase in a *Cryptococcus neoformans* strain from an AIDS patient. *FEMS Microbiol. Lett.* **122:**39–42.
52. **Kim, S. J., K. J. Kwon-Chung, G. W. A. Milne, W. B. Hill, and G. Patterson.** 1975. Relationship between polyene resistance and sterol compositions in *Cryptococcus neoformans. Antimicrob. Agents Chemother.* **7:**99–106.
53. **Klepser, M. E., and M. A. Pfaller.** 1998. Variation in electrophoretic karyotype and antifungal susceptibility of clinical isolates of *Cryptococcus neoformans* at a university-affiliated teaching hospital from 1987 to 1994. *J. Clin. Microbiol.* **36:**3653–3656.
54. **Krishnarao, T., and J. Galgiani.** 1997. Comparison of the in vitro activities of the echinocandin LY303366, the pneumocandin MK-0991 and fluconazole against *Candida* species and *Cryptococcus neoformans. Antimicrob Agents Chemother.* **41:**1957–1960
55. **Lamb, D. C., A. Corran, B. C. Baldwin, J. Kwon-Chung, and S. L. Kelly.** 1995. Resistant P45051A1 activity in azole antifungal tolerant *Cryptococcus neoformans* from AIDS patients. *FEBS Lett.* **368:**326–330.
56. **Lortholary, O.** 2007. Management of cryptococcal meningitis in AIDS: the need for specific studies in developing countries. *Clin. Infect. Dis.* **45:**81–83.
57. **Lortholary, O., G. Poizat, V. Zeller, S. Neuville, A. Boibieux, M. Alvarez, P. Dellamonica, F. Botterel, F. Dromer, and G. Chene.** A 2006. Long-term outcome of AIDS-associated cryptococcosis in the era of combination antiretroviral therapy. *AIDS* **20:**2183–2191.
58. **MacDougall, L., S. E. Kidd, E. Galanis, S. Mak, M. J. Leslie, P. R. Cieslak, J. W. Kronstad, M. G. Morshed, and K. H. Bartlett.** 2007. Spread of *Cryptococcus gattii* in British Columbia, Canada, and detection in the Pacific Northwest, USA. *Emerg. Infect. Dis.* **13:**42–50.
59. **Maligie, M. A., and C. P. Selitrennikoff.** 2005. *Cryptococcus neoformans* resistance to echinocandins: (1, 3)beta-glucan synthase activity is sensitive to echinocandins. *Antimicrob Agents Chemother.* **49:**2851–2856.
60. **McCarthy, K. M., J. Morgan, K. A. Wannemuehler, S. A. Mirza, S. M. Gould, N. Mhlongo, P. Moeng, B. R. Maloba, H. H. Crewe-Brown, M. E. Brandt, and R. A. Hajjeh for the Gauteng Cryptococcal Surveillance Initiative Group.** 2006. Population-based surveillance for cryptococcosis in an antiretroviral-naïve South African province with a high HIV seroprevalence. *AIDS* **20:**2199–2206.
61. **Mirza, S. A., M. Phelan, D. Rimland, E. Graviss, R. Hamill, M. E. Brandt, T. Gardner, M. Sattah, G. Ponce de Leon, W. Baughman, and R. A. Hajjeh.** 2003. The changing epidemiology of cryptococcosis: an update from population-based active surveillance in 2 large metropolitan areas, 1992–2000. *Clin. Infect. Dis.* **36:**789–794.
62. **Mondon, P., R. Petter, G. Amalfitano, R. Luzzati, E. Concia, I. Polacheck, and K. J. Kwon-Chung.** 1999. Heteroresistance to fluconazole and voriconazole in *Cryptococcus neoformans. Antimicrob. Agents Chemother.* **43:**1856–1861.
63. **Morera-Lopez, Y., J. M. Torres-Rodriguez, T. Jimenez-Cabello, and T. Baro-Tomas.** 2005. *Cryptococcus gattii*: in vitro susceptibility to the new antifungal albaconazole versus fluconazole and voriconazole. *Med. Mycol.* **43:**505–510.
64. **Morgan, J., K. M. McCarthy, S. Gould, K. Fan, B. Arthington-Skaggs, N. Iqbal, K. Stamey, R. A. Hajjeh, and M. E. Brandt for the Gauteng Cryptococcal Surveillance Initiative Group.** 2006. *Cryptococcus gattii* infection: characteristics and epidemiology of cases identified in a South African province with a high HIV seroprevalence, 2002-2004. *Clin. Infect. Dis.* **43:**1077–1080.
65. **Mwaba, P., J. Mwansa, C. Chintu, J. Pobee, M. Scarborough, S. Portsmouth, and A. Zumia.** 2001. Clinical presentation, natural history, and cumulative death rates of 230 adults with primary cryptococcal meningitis in Zambian AIDS patients treated under local conditions. *Postgrad. Med.* **77:**769–773.
66. **Paugam, A., J. Dupouy-Camet, P. Blanche, J. P. Gangneux, C. Tourte-Schaefer, and D. Sicard.** 1994. Increased fluconazole resistance of *Cryptococcus neoformans* isolated from a patient with AIDS and recurrent meningitis. *Clin. Infect. Dis.* **19:**975–976.
67. **Peetermans, W., H. Bobbaers, J. Verhaegen, and J. Vanderkette.** 1993. Fluconazole-resistant *Cryptococcus neoformans* var *gattii* in an AIDS patient. *Acta Clin. Belgica* **48:**405–409.
68. **Perfect, J. R., and G. M. Cox.** 1999. Drug resistance in *Cryptococcus neoformans. Drug Resist. Updates* **2:**259–269.
69. **Perkins, A., A. Gomez-Lopez, E. Mellado, J. L. Rodriquez-Tudella, and M. Cuenca-Estrella.** 2005. Rates of antifungal resistance among Spanish clinical isolates of *Cryptococcus neoformans* var. *neoformans. J. Antimicrob. Chemother.* **56:**1144–1147.
70. **Pfaller, M., J. Zhang, S. Messer, M. Tumberland, E. Mbidde, C. Jessup, and M. Ghannoum.** 1998. Molecular epidemiology and antifungal susceptibility of *Cryptococcus neoformans* isolates from Ugandan AIDS patients. *Diagn. Microbiol. Infect. Dis.* **32:**191–199.
71. **Pfaller, M. A., S. A. Messer, L. Boyken, C. Rice, S. Tendolkar, R. J. Hollis, G. V. Doern, and D. J. Diekema.** 2005. Global trends in the antifungal susceptibility of *Cryptococcus neoformans* (1990 to 2004). *J. Clin. Microbiol.* **43:**2163–2167.
72. **Pfaller, M. A., L. Boyken, R. J. Hollis, S. A. Messer, S. Tendolkar, and D. J. Diekema.** 2005. In vitro susceptibilities of clinical isolates of *Candida* species, *Cryptococcus neoformans*, and *Aspergillus* species to itraconazole: global survey of 9,359 isolates tested by Clinical and Laboratory Standards Institutes broth microdilution methods. *J. Clin. Mircobiol.* **43:**3801–3810.
73. **Pfaller, M. A., D. J. Diekema, D. L. Gibbs, V. A. Newell, H. Bijie, D. Dzierzanowska, N. N. Klimko, V. Letscher-Bru, M. Lisalova, K. Muchlethaler, C. Rennison, M. Zaida, and the Global Antifungal Surveillance Group.** 2009. Results from the ARTEMIS DISK Global Antifungal Surveillance Study, 1997 to 2007: 10.5-year analysis of susceptibilities of noncandidal yeast species to fluconazole and voriconazole determined by CLSI standardized disk diffusion testing. *J. Clin. Microbiol.* **47:**117–123.
74. **Posteraro, B., M. Sanguinetti, D. Sanglard, M. La Sorda, S. Boccia, L. Romano, G. Morace, and G. Fadda.** 2003. Identification and characterization of a *Cryptococcus neoformans* ATP binding cassette (ABC) transporter-encoding gene, *CnAFR1*, involved in resistance to fluconazole. *Mol. Microbiol.* **47:**357–371.

75. **Powderly, W. G.** 2008. Dosing amphotericin B in cryptococcal meningitis. *Clin. Infect. Dis.* **47:**131–132.
76. **Powderly, W. G., M. S. Saag, G. A. Cloud, P. Robinson, R. D. Meyer, J. M. Jacobson, J. R. Graybill, A. M. Sugar, V. J. McAuliffe, S. E. Follansbee, and W. Dismukes.** 1992. A controlled trial of fluconazole or amphotericin B to prevent relapse of cryptococcal meningitis in patients with the acquired immunodeficiency syndrome. *N. Engl. J. Med.* **326:**793–798.
77. **Powderly, W. G., D. M. Finkelstein, J. Feinberg, P. Frame, W. He, C. Van der Horst, S. L. Koletar, M. E. Eyster, J. Carey, H. Waskin, T. M. Hooton, N. Hyslop, S. A. Spector, and S.A. Bozzette, for The NIAID AIDS Clinical Trials Group.** 1995. A randomized trial comparing fluconazole with clotrimazole troches for the prevention of fungal infections in patients with advanced human immunodeficiency virus infection. *N. Engl. J. Med.* **332:**700–705.
78. **Pukkila-Worley, R., and E. Mylonakis.** 2008. Epidemiology and management of cryptococcal meningitis: developments and challenges. *Expert Opin. Pharmacother.* **9:**551–560.
79. **Rex, J. H., and M. A. Pfaller.** 2002. Has antifungal susceptibility testing come of age? *Clin. Infect. Dis.* **35:**983–989.
80. **Robinson, P. A., M. Bauer, M. A. E. Leal, S. G. Evans, P. D. Holtom, D. M. Diamond, J. M. Leedom, and R. A. Larson.** 1999. Early mycological treatment failure in AIDS-associated cryptococcal meningitis. *Clin. Infect. Dis.* **28:**82–92.
81. **Rodero, L., E. Mellado, A. C. Rodriquez, A. Salve, L. Guelfond, P. Cahn, M. Cuenca-Estrella, G. Davel, and J. L. Rodriguez-Tudela.** 2003. G484S amino acid substitution in lanosterol 14- α demethylase (*ERG11*) is related to fluconazole resistance in a recurrent *Cryptococcus neoformans* clinical isolate. *Antimicrob. Agents Chemother.* **47:**3653–3656.
82. **Saag, M. S., W. G. Powderly, G. A. Cloud, P. Robinson, M. H. Grieco, P. K. Sharkey, S. E. Thompson, A. M. Sugar, C. U. Tuazon, J. F. Fisher, N. Hyslop, J. M. Jacobson, R. Hafner, W. E. Dismukes, and the NIAID Mycoses Study Group and the AIDS Clinical Trials Group.** 1992. Comparison of amphotericin B with fluconazole in the treatment of acute AIDS-associated cryptococcal meningitis. *N. Engl. J. Med.* **326:**83–89.
83. **Sanglard, D., and F. C. Odds.** 2002. Resistance of *Candida* species to antifungal agents: molecular mechanisms and clinical consequences. *Lancet Infect. Dis.* **2:**73–85.
84. **Sanguinetti, M., B. Posteraro, M. LaSorda, R. Torelli, B. Fiori, R. Santangelo, G. Delogu, and G. Fadda.** 2006. Role of *AFR1*, an ABC transporter encoding gene, in the in vivo response to fluconazole and virulence of *Cryptococcus neoformans*. *Infect. Immun.* **74:**1352–1359.
85. **Sar, B., D. Monchy, M. Vann, C. Keo, J. L. Sarthou, and Y. Buisson.** 2004. Increasing in vitro resistance to fluconazole in *Cryptococcus neoformans* Cambodian isolates: April 2000 to March 2002. *J. Antimicrob. Chemother.* **54:**563–565.
86. **Schwarz, P., G. Janbon, F. Dromer, O. Lortholary, and E. Dannaoui.** 2007. Combination of amphotericin B with flucytosine is active in vitro against flucytosine-resistant isolates of *Cryptococcus neoformans*. *Antimicrob. Agents Chemother.* **51:**383–385.
87. **Singh, N., M. J. Barnish, S. Berman, B. S. Bender, M. M. Wagener, M. G. Rinaldi, and V. L. Yu.** 1996. Low-dose fluconazole as primary prophylaxis for cryptococcal infection in AIDS patients with CD4 cell counts of ≤100/mm^3 : demonstration of efficacy in a prospective, multicenter trial. *Clin. Infect. Dis.* **23:**1282–1286.
88. **Singh, N., F. Dromer, J. R. Perfect, and O. Lortholary.** 2008. Cryptococcosis in solid organ transplant recipients: current state of the science. *Clin. Infect. Dis.* **47:**1321–1327.
89. **Sionov, E., Y. C. Chang, H. M. Garraffo, and K. J. Kwon-Chung.** 2009. Heteroresistance to fluconazole in *Cryptococcus neoformans* is intrinsic and associated with virulence. *Antimicrob. Agents Chemother.* **53:**2804–2815.
90. **Sionov, E., H. S. Lee, Y. C. Chang, J. E. Bennett, and J. Kwon-Chung.** 2008. Molecular and biological evidence that the strain H99 does not represent serotype A *C. neoformans*. 7th International Conference on *Cryptococcus* and Cryptococcosis, Nagasaki, Japan. Abstr.
91. **Sorrell, T. C.** 2001. *Cryptococcus neoformans* variety *gattii*. *Med. Mycol.* **39:**155–168.
92. **Speed, B., and D. Dunt.** 1995. Clinical and host differences between infections with the two varieties of *Cryptococcus neoformans*. *Clin. Infect. Dis.* **21:**28–34.
93. **Tay, S. T., T. T. Haryanty, K. P. Ng, M. Y. Rohani, and H. Hamimah.** 2006. In vitro susceptibilities of Malaysian clinical isolates of *Cryptococcus neoformans* var. *grubii* and *Cryptococcus gattii* to five antifungal drugs. *Mycoses* **49:**324–330.
94. **Thompson, G .R., III, N. P. Wiederhold, A. W. Fothergill, A. C. Vallor, B. L. Wickes, and T. F. Patterson.** 2009. Antifungal susceptibilities among different serotypes of *Cryptococcus gattii* and *Cryptococcus neoformans*. *Antimicrob. Agents Chemother.* **53:**309–311.
95. **Thompson, J. R., C. M. Douglas, W. Li, C. K. Jue, B. Pramanik, X. Yuan, T. H. Rude, D. L. Toffaletti, J. R. Perfect, and M. Kurtz.** 1999. A glucan synthase *FKS1* homolog in *Cryptococcus neoformans* is single copy and encodes an essential function. *J. Bacteriol.* **181:**444–453.
96. **Thornewell, S. J., R. B. Peery, and P. L. Skatrud.** 1997. Cloning and characterization of *CneMDR1*: a *Cryptococcus neoformans* gene encoding a protein related to multidrug resistance proteins. *Gene* **201:**21–29.
97. **Trilles, L., B. Fernandez-Torres, M. dos Santos Lazera, B. Wanke, and J. Guarro.** 2004. In vitro antifungal susceptibility of *Cryptococcus gattii*. *J. Clin. Microbiol.* **42:**4815–4817.
98. **Velez, J. D., R. Allendoerfer, M. Luther, M. G. Rinaldi, and J. R. Graybill.** 1993. Correlation of in vitro azole susceptibility with in vivo response in a murine model of cryptococcal meningitis. *J. Infect. Dis.* **168:**508–510.
99. **Venkateswarlu, K., M. Taylor, N. J. Manning, M. G. Rinaldi, and S. L. Kelly.** 1997. Fluconazole tolerance in clinical isolates of *Cryptococcus neoformans*. *Antimicrob. Agents Chemother.* **41:**748–751.
100. **Viard, J. P., C. Hennequin, N. Fortineau, N. Pertuiset, C. Rothschild, and H. Zylberberg.** 1995. Fulminant cryptococcal infections in HIV-infected patients on oral fluconazole. *Lancet* **346:**118.
101. **Whelan, W. L.** 1987. The genetic basis of resistance to 5-fluorocytosine in *Candida* species and *Cryptococcus neoformans*. *Crit. Rev. Microbiol.* **15:**45–56.
102. **Whelan, W. L., and D. Kerridge.** 1984. Decreased activity of UMP pyrophosphorylase associated with resistance to 5-fluorocytosine in *Candida albicans*. *Antimicrob. Agents Chemother.* **26:**570–574.
103. **White, T. C., K. A. Marr, and R. A. Bowden.** 1998. Clinical, cellular, and molecular factors that contributed to antifungal resistance. *Clin. Microbiol. Rev.* **19:**382–402.
104. **Witt, M. D., R. J. Lewis, R. A. Larsen, E. N. Milefchik, M. A. Leal, R. H. Haubrick, J. A. Richie, J. E. Edwards, Jr., and M. A. Ghannoum.** 1996. Identification of patients with acute AIDS-associated cryptococcal meningitis who can be effectively treated with fluconazole: the role of antifungal susceptibility testing. *Clin. Infect. Dis.* **22:**322–328.

105. **Xu, J., C. Onyewu, H. J. Yoell, R. Y. Ali, R. J. Vilgalys, and T. G. Mitchell.** 2001. Dynamic and heterogeneous mutations to fluconazole resistance in *Cryptococcus neoformans*. *Antimicrob. Agents Chemother.* **45:**420–427.
106. **Yamazumi, T., M. A. Pfaller, S. A. Messer, A. K. Houston, L. Boyken, R. J. Hollis, I. Furuta, and R. N. Jones.** 2003. Characterization of heteroresistance to fluconazole among clinical isolates of *Cryptococcus neoformans*. *J. Clin. Microbiol.* **41:**267–272.
107. **Yildiran, S. T., A. W. Fothergill, D. A. Sutton, and M. G. Rinaldi.** 2002. In vitro susceptibilities of cerebrospinal fluid isolates of *Cryptococcus neoformans* collected during a ten-year period against fluconazole, voriconazole, and posaconazole (SCH56592). *Mycoses* **45:**378–383.

Cryptococcus: From Human Pathogen to Model Yeast
Edited by J. Heitman et al.
©2011 ASM Press, Washington, DC

17

Signaling Cascades and Enzymes as *Cryptococcus* Virulence Factors

DEBORAH S. FOX, JULIANNE T. DJORDJEVIC, AND TANIA C. SORRELL

Signal transduction cascades are utilized by all organisms to convey signals perceived at the cell surface to effectors within the cell. These enzymatic signaling cascades are important in the pathogenesis of many infections, including cryptococcosis. Our understanding of the links between environmental events and virulence in *Cryptococcus neoformans* has dramatically increased within the past decade and includes mechanisms involved in fungal survival within the host, vesicle export to the cell surface, secretion, cell wall synthesis, and antifungal drug interactions. This chapter summarizes current knowledge of the significance and functional interactions involved in the cell wall integrity, phospholipase, and calcineurin signaling pathways for the establishment of *C. neoformans* virulence.

PHOSPHOLIPASES AND VIRULENCE

Phospholipases are categorized into five classes (A_1, A_2, B, C, and D) depending on the site of hydrolysis of the phospholipid ester linkage (Fig. 1) (174). They are produced and secreted by a wide range of pathogenic microorganisms, including fungi, as part of their virulence repertoire (61). Two phospholipase classes, phospholipase B (PLB) and phospholipase C (PLC), have been implicated in cryptococcal virulence. Cryptococcal Plb1 is a product of the *PLB1* gene and is secreted into the extracellular environment, where it hydrolyzes host phospholipids, particularly phosphatidylcholine (PC). It is a multifunctional enzyme with PLB, lysophospholipase (LPL), and LPL-transacylase (LPTA) activities. PLB removes fatty acyl chains from the *sn-1* and *sn-2* positions on the glycerol backbone of its phospholipid substrate, LPL hydrolyzes the remaining fatty acyl chain on a lysophospholipid (a phospholipid containing only one fatty acyl chain at either the *sn-1* or the *sn-2* position), and LPTA catalyzes the transfer of a free fatty acyl chain onto the unoccupied alcohol in a lysophospholipid (174). It has been speculated that PLB activity is in fact a combination of phospholipase A_2 (PLA_2) and LPL activities (174). As their name implies, membrane lysophospholipids are potentially cellulolytic and are short-lived in biological membranes, perhaps due to the activity of LPLs. The fungal Plc enzymes referred to in this review preferentially hydrolyze phosphatidylinositol (PI)-based substrates within the cryptococcal cell and affect multiple cellular functions, including the secretion of Plb1.

Extracellular Plb and Its Potential Role in Virulence

Extracellular phospholipase activity in *C. neoformans* was first demonstrated independently by Chen et al. (26) and Vidotto et al. (177), as a zone of precipitation due to substrate hydrolysis, surrounding colonies growing on egg-yolk agar medium, a rich source of phospholipid. In vitro, strains of *C. neoformans* serotype A secreted more phospholipase activity than those of serotypes B and C (*Cryptococcus gattii*) (24) and nonpathogenic fungi such as *Saccharomyces cerevisiae* (41). Chen et al. (26) reported that the radial zone of precipitation around colonies on egg-yolk agar, and hence Plb secretion, correlated with virulence in mice as assessed by cryptococcal organ burden. Since the egg-yolk agar assay does not distinguish accurately between classes of phospholipase or between phospholipases and triglycerol lipases, this group also developed a radiometric, thin-layer chromatographic separation enzyme assay and used ^1H and ^{31}P nuclear magnetic resonance to further define phospholipase activity in cryptococcal secretions (24). Both methods revealed that glycerophosphocholine (PC minus its two fatty acid chains) was the sole degradation product of the reaction between PC and supernatants of stationary phase cultures of *C. neoformans*, indicative of PLB activity (or a combination of PLA and LPL activity (Fig. 1). No lysoPC or triglyceride degradation products were detected, and there was no evidence of extracellular PLA, PLC, or PLD activities. LPL and LPTA activities were present in cryptococcal secretions, since radiolabeled free fatty acids were released from labeled lysoPC, with PC formed from labeled lysoPC. Triglycerol lipase is a virulence determinant of the pathogenic yeast

Deborah S. Fox, Departments of Pediatrics and Microbiology, Immunology and Parasitology, The Research Institute for Children and LSU Health Sciences Center, Children's Hospital, New Orleans, LA 70118. **Julianne Djordjevic and Tania C. Sorrell,** Centre for Infectious Diseases and Microbiology, ICPMR and Westmead Millennium Institute, University of Sydney at Westmead Hospital, Westmead, 2145 NSW, Australia.

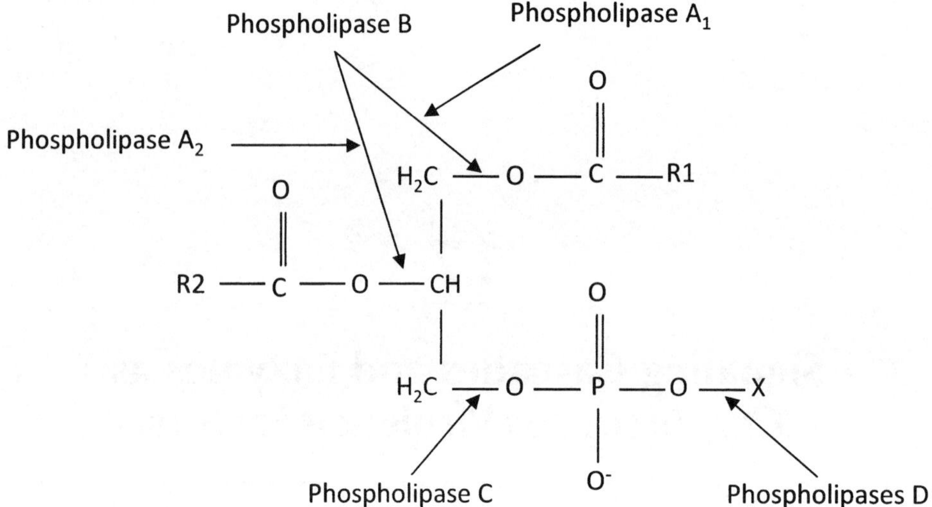

FIGURE 1 Site of attack of phospholipases on a phospholipid. R1 and R2 are fatty acid chains; X is a head group. Phospholipase A_1 hydrolyzes the fatty acyl ester bond at the *sn-1* position of the phospholipid to form a 2-acyl phospholipid (lysophospholipid) and a free fatty acid. Phospholipase A_2 cleaves the fatty acyl ester bond at the *sn-2* position of this molecule, resulting in the formation of a lysophospholipid. Phospholipase B (PLB) catalyzes the simultaneous hydrolysis of fatty acids from both the *sn-1* and *sn-2* position of the phospholipid. Phospholipase C cleaves the phosphodiester bond in the phospholipid backbone to form 1,2-diacylglycerol and a phosphorylated head group. Phospholipase D removes the head group from the phospholipid, forming phosphatidic acid. Plb1 has the strongest preference for PC, where the head group (X) is choline, and shows weaker preference for PI, phosphatidylethanolamine, and phosphatidylserine, where the head groups are inositol, ethanolamine, and serine, respectively. Adapted from reference 177.

Candida parapsilosis (57), and many pathogenic fungi produce lipases belonging to this superfamily (76). However, no matching sequences are present in the *C. neoformans* genomic database, suggesting that triglycerol lipases are absent in this pathogen.

Secreted Plb Activity Is Attributable to One Enzyme (Plb1)

Chen et al. (25) provided the first complete description of the purification and properties of a cryptococcal Plb enzyme, which was later called Plb1 when it was identified as the product of the *PLB1* gene. They found Plb1 to be an acidic, highly glycosylated protein with pI values of 5.5 and 3.5 and a molecular weight (MW) of 70 to 90 kDa on sodium dodecyl sulfate-polyacrylamide gel electrophoresis. Subsequently, it was found that the Plb1 MW could be as high as 125 kDa due to extensive asparagine N-linked glycosylation, which is responsible for at least 30% of the MW of Plb1 and essential for its activity (25, 154, 173). The native Plb1 MW, as determined by gel filtration chromatography, was 160 to 180 kDa, suggesting that the protein may exist in a dimeric form. The purified protein displayed PLB, LPL, and LPTA activities in a ratio similar to that in cryptococcal secretions (0.01:1.6:1 for PLB:LPL:LPTA, respectively) and with similar acidic pH optima of 4.0 for PLB and 4.0 to 5.0 for LPL and LPTA. All three enzymatic activities were present in the absence of an energy source. The apparent V_{max} values for the PLB and LPL activities were 12.3 and 870 μmol/min per mg of protein, respectively, and the corresponding K_m values were approximately 185.3 and 92.2 μM. Enzyme activity was unaffected by the addition of most cations, but was inhibited by Fe^{3+}, and LPL and LPTA activities were decreased by 50% in the presence of 0.1% (vol/vol) Triton X-100. Palmitoylcarnitine (0.5 mM), an LPTA inhibitor that prevents *Candida albicans* adherence in vitro (134), competitively inhibited PLB activity by 97%, and LPL and LPTA activities by 35%. Although all phospholipids, except phosphatidic acid, were degraded by PLB, dipalmitoyl phosphatidylcholine (DPPC) and dioleoyl PC were the preferred substrates, followed by phosphatidylethanolamine, phosphatidylserine, and PI. Short protein sequences, including one at the N terminus and five internal peptides, were identified by protein sequencing and later determined to be encoded by a cryptococcal gene (*PLB1*), identified and cloned by Cox et al. (30).

Using degenerate primers, Cox and coworkers used a PCR product amplified from a genomic library to probe a Southern blot of a H99 karyotype gel and identified a 4.7-kbp SacI restriction fragment containing a full-length gene of 2,218 bp. The product of this gene, designated *PLB1*, had high homology to the products of other *PLB* homologues from *C. albicans* and *S. cerevisiae* (37% and 36%, respectively) and was predicted to encode a protein containing 617 amino acids with an estimated MW of 65 kDa (30). However, sequencing of the actual *PLB1* cDNA revealed a gene product of only 614 amino acids, due to splicing out of 9 bp at a less common splice acceptor site between nucleotide positions 774 and 782 (CCGGAAAAG) of the cDNA (J. Djordjevic and M. Del Poeta, unpublished observation). The expected Plb1 MW of 65 kDa corresponds to that of the purified, deglycosylated enzyme as reported by Chen et al. (25). *PLB1* in fact encodes 17 potential N-linked glycosylation sites and contains a hydrophobic stretch of amino acids at the N terminus characteristic of a secretory signal leader peptide. Selective disruption of *PLB1* abolished PLB, LPL, and LPTA activities in cryptococcal supernatants, and activity was restored following *PLB1* reconstitution within the *PLB1* deletion mutant, confirming that the gene encodes the secreted Plb1 enzyme purified by Chen et al. (25).

Plb1: Tissue Invasion, Immune Response Evasion, and Fungal Dissemination

Plb enzymes are of particular interest in disease pathogenesis, as they are the only proven cryptococcal "invasins," are essential for virulence of *C. neoformans* and *C. albicans*, and are produced by several pathogenic fungi (30, 99, 115, 122, 146, 153, 168). By comparing organ burdens of *C. neoformans* in mice infected with wild-type (H99), *plb1*Δ, and *plb1*Δ^{rec}, it was shown that *PLB1* is essential for the initiation of interstitial pulmonary infections and for dissemination from the lung, by way of the blood and lymphatics, to the brain (30, 122, 146), but not for transfer across the blood-brain barrier (146). There is evidence that cryptococcal Plb1 is produced during human infection, as specific antibodies are present in sera from patients with cryptococcosis (147). DPPC and phosphatidylglycerol, which are abundant in lung surfactant (the fluid lining pulmonary alveoli), and phospholipids such as dioleoyl PC, which is present in the outer leaflet of mammalian cell membranes, are preferred substrates of Plb1. Hydrolysis of these phospholipids by cryptococcal Plb1 would plausibly result in penetration of host tissue and establishment of cryptococcal infection. Evidence that DPPC is a substrate for cryptococcal Plb1 in vivo comes from an experimental rat model, where larger amounts of glycerophosphorylcholine, the product of PLB hydrolysis, are found in lung, which is rich in DPPC, than in brain, which is not (70, 71). Furthermore, partially purified Plb1 hydrolyzes lung surfactant (148) and facilitates adherence of cryptococci to lung epithelial cells (58).

Plb1 and Survival in Macrophages

C. neoformans is a facultative intracellular pathogen that survives and replicates within the intracellular phagolysosomal compartment of macrophages. Several outcomes of this interaction, all of which potentially drive the establishment, progression, and dissemination of infection, have been described. These include macrophage destruction (48, 171), and/or expulsion into the extracellular environment, or direct transfer to neighboring host cells without macrophage destruction (108). Macrophages also appear to provide a vehicle for dissemination of cryptococci to the brain since intravenous inoculation of cryptococci within autologous macrophages increases brain fungal burden (20, 146). The *PLB1* gene is essential for optimal uptake of cryptococci by, and replication within, macrophages (30) and for dissemination of cryptococcal infection to the brain (30, 122, 146). Specifically, Cox et al. (30) reported that budding of intracellular *plb1*Δ in the macrophage-like cell line J774, was reduced compared with H99, but that phagocytosis was not affected. In contrast, phagocytosis of strain H99 (7.3% ± 1.2%) was significantly greater than *plb1*Δ (4.3% ± 0.8%) in THP1 cells (182). Using the murine alveolar macrophage cell line MHS-1, which displays many of the properties of primary alveolar macrophages (145), and which cannot exert fungicidal activity unless activated, Noverr et al. (122) showed that Plb1 is essential for the optimal survival of *C. neoformans*. Cryptococcal Plb1 may also assist in combating the host immune system by establishing conditions that favor a chronic and disseminated infection in the host and/or provide access to a source of nutrients, as discussed below.

Plb1 and Eicosanoid Production

Eicosanoids of fungal origin may promote yeast survival by suppressing the host immune response (122). Using the *plb1*Δ mutant, Noverr et al. found that Plb1 is essential for the synthesis of bioactive anti-inflammatory eicosanoids by *C. neoformans*, both in vitro and in vivo (122, 123). Though cryptococci lack arachidonic acid (AA), which is the precursor of eicosanoids in mammalian cells, arachidonate is readily taken up by cryptococci in vitro (122, 182). In a series of elegant experiments, Wright et al. showed that cryptococci undergo transcellular metabolism of macrophage-derived arachidonate during phagocytosis and that the incorporation of this arachidonate into cryptococcal lipids is Plb1-dependent (182). In vitro, Noverr et al. (122) also showed that cryptococcal Plb1 was essential for the synthesis of eicosanoids from diarachidonyl PC. However, though palmitoylarachidyl PC is the physiological phospholipid in macrophages, neither the wild type nor the *plb1*Δ mutant could release arachidonate from the *sn-2* position (182). This suggests that free AA generated by the host, presumably following activation of mammalian cytosolic phospholipase A_2 enzyme (cPLA_2) by the membrane perturbation induced by phagocytosis, is the substrate for Plb1-dependent eicosanoid production in *C. neoformans* (182). The Th1/Th2 inflammatory profile generated during experimental murine cryptococcosis, as defined by the type of cytokine, chemokine, and acute neutrophil response in the lung, favors a Th2 response and chronic and disseminated infection in mice infected with wild-type H99. Conversely, infection with *plb1*Δ elicits a protective cell-mediated immune response, with a shift in the Th1/Th2 profile toward a Th1-type response and clearance of pulmonary infection (122).

Plb1 and Fungal Nutrition

During infection, hydrolysis of host phospholipids by cryptococcal Plb1 may provide a source of energy. Both phospholipids and the fatty acids derived from them can be used as sole carbon sources by *C. neoformans* (122, 182). This would confer a survival advantage on the fungus within low-glucose environments of the host, such as in the phagolysosome and within cryptococcomas, as Plb1 activity is not energy-dependent (182). Consistent with this hypothesis, the wild-type strain H99, *plb1*Δ, and the *PLB1* reconstituted *plb1*Δ strain (*plb1*Δ^{rec}) can all grow intracellularly, although growth of *plb1*Δ is compromised (122, 182). In further support of this, Wright et al. (182) demonstrated that metabolism of the physiologically relevant lipids, DPPC and dioleoyl PC, the uptake and esterification of the fatty acids (arachidonic, palmitic, and oleic acid) and cellulolytic lysoPC under stress conditions, and the detoxification (by acylation) of cellulolytic lysoPC administered as the sole carbon source are each Plb1-dependent.

Plb1 and Fungal Membrane Homeostasis, Exosome Disruption, and Cell Wall Integrity

The location of cryptococcal Plb1 in lipid raft membranes in the outer leaflet of the cell membrane bilayer, in close proximity to its phospholipid substrates, suggests that it plays a role in membrane homeostasis, and hence growth, in *C. neoformans*. However, the growth rate of *plb1*Δ under nutrient-rich conditions in vitro is comparable to that of the H99 wild type at either 30 or 37°C. This observation suggests either that other phospholipases are also involved in membrane turnover or that membrane Plb1 is inactive. When attached to raft membranes by its glycosylphosphatidylinositol (GPI) anchor, PLB enzyme activity is in fact suppressed (155), suggesting that sequestration of Plb1 in lipid rafts may be a mechanism to prevent fungal membrane damage. Thus, presumably any membrane homeostatic effect of Plb1 only occurs after its release from the GPI anchor. Although lipid rafts generally facilitate signal activation, there is precedent for suppression of enzyme activity within the raft environment, as has been reported for a metalloproteinase (89). It is not known whether raft-associated Plb1 has a role in cell signaling, though it is plausible, as hydrolysis of the Plb1 GPI anchor releases the secondary intracellular messenger diacylglycerol (DAG). A role in signal transduction has been proposed for GPI-anchored proteins in mammalian cells (166).

Recently, *SEC6*-dependent exosomes were identified as the primary means of transport of laccase, urease, and soluble glucuronoxylomannan (GXM) polysaccharide across the cryptococcal cell wall barrier to the exterior of the cell (16, 128, 138, 139). The mechanism of release of these virulence determinants from exosomes is unknown, but cell wall lipases and phospholipases such as Plb1 are likely to be involved. Furthermore, Plb1 contributes to maintenance of cell wall integrity, as *plb1*Δ exhibits compromised growth relative to wild type, in the presence of cell wall-perturbing agents (154).

Plb1 Molecular Modeling: What It Reveals about Function

As determined by analysis of the *PLB1* gene, the predicted Plb1 protein contained catalytic motifs common to mammalian phospholipases, namely the GX<u>S</u>G(G/S) lipase motif, which is found in all lipolytic enzymes, the putative phospholipase catalytic motif, SGGGX<u>R</u>A(M/L), and the subtilisin protease aspartate motif, IXVV<u>D</u>SGLXXXN. Residues analogous to Ser146, Asp392, and Arg108, underlined in the above sequences, are involved in catalysis of the human $cPLA_2$ enzyme (131, 152) and are conserved in $cPLA_2$ from six mammalian species and in nearly all fungal Plb enzymes, except in *Schizosaccharomyces pombe*. Jones et al. determined that Plb1 contains a single active site by performing site-directed mutagenesis of each of the three putative catalytic residues and expressing the respective cDNAs in *S. cerevisiae* (38, 82). The mutant Plb1s were deficient in all three enzymatic activities, as expected for an enzyme with a single active site. In addition, trafficking of Plb1 was defective in all three mutants. In two (D392 and R108), Plb1 failed to reach the cell membrane, and in the third (S146A), Plb1 was transported to the membrane but did not reach the cell wall and was not secreted. Presumably, substitution of these residues results in abnormal protein conformation and/or an inability to acquire N-linked glycosylation, a prerequisite for secretion.

The three-dimensional structure of Plb1 has not been solved. To better understand the structure of the Plb1 active site and its catalytic mechanism, Jones et al. (82) developed a homology model based on the X-ray crystal structure of the mammalian $cPLA_2$ catalytic domain and the significant degree of secondary structural similarity between Plb1 and the $cPLA_2$ enzyme. Comparison of the Plb1 molecular model and the $cPLA_2$ crystal structure revealed a hydrophobic binding pocket, which accommodated all three catalytic residues in identical positions, forming a single active site. S146/S228 and D392/D549 were similarly distanced in both models, suggesting the formation of a catalytic dyad. In silico substrate docking studies with $cPLA_2$ confirmed that the catalytic mechanism of Plb1 and $cPLA_2$ is similar. We also determined the mechanism that allows different substrates to be accommodated by both fungal and human enzymes and why various inhibitors differentially affect the three Plb1 enzymatic activities (59). In both structures, R108/R200 is sufficiently distanced from the S/D catalytic dyad to act as a stabilizer for the phospholipid head group and to allow cleavage/reacylation of the acyl chain of the phospholipid/lysophospholipid substrate, respectively, by the catalytic dyad. With respect to binding of the two phospholipid acyl chains, it was confirmed that cleavage of the *sn-2* ester bond can only occur within one of two separate hydrophobic binding tracts (the upper tract) that accommodate the two lipid acyl chains. For cleavage of the second palmitoyl group to occur in the case of Plb1, the lysoPC would have to flip vertically to place the palmitoyl group in the upper binding tract (Color Plate 4).

Sequence hypervariability was present in the Plb1 and $cPLA_2$ upper binding tracts where cleavage occurs at the *sn-2* position of a phospholipid. $cPLA_2$ has a preference for AA in the *sn-2* position and removes only this fatty acid chain (36), whereas Plb1 prefers either palmitoyl or oleoyl fatty acid chains in both the *sn-1* and *sn-2* positions and can remove both via the PLB activity but cannot remove AA from the *sn-2* position (25, 182). Binding residues specific for AA and palmitic acid were identified in the $cPLA_2$ and Plb1 structures, respectively.

Our previous studies have demonstrated that several agents, a Plb1 peptide antibody, carnitine, palmitoylcarnitine, *N*-ethylmaleimide, 5,5′-dithiobis (2-nitrobenzoic acid), dioctadecyldimethylammonium bromide, 1,12-bis(tributylphosphonium)dodecane dibromide, Triton X-100, and alexidine dihydrochloride, differentially inhibit either the LPL/LPTA or the PLB activities of secreted cryptococcal Plb1 (25, 58, 59, 148). We initially concluded that Plb1 contains two active sites. However, the finding from the docking study that each lipase has one substrate-binding site composed of two separate binding tracts for the *sn-1* and *sn-2* acyl chains of the ligand provides an alternative explanation for the differential inhibition of LPL/LPTA and PLB activities of secreted Plb1 by various acyl chain–containing agents. The different inhibitors may in fact exhibit a preference for interaction with one or the other of the binding tracts. For example, an inhibitor that affects PLB, but not LPL, activity may sit only within the lower binding tract where it would not sterically hinder the binding of a lysophospholipid, as the single acyl chain could still be accommodated within the upper binding tract and undergo catalytic cleavage. Alternatively, the initial point of contact of the substrate with the enzyme may not be confined to the enzyme active site and may be different for phospholipids and lysophospholipids.

One difference between Plb1 and $cPLA_2$ revealed by the homology modeling is the absence of the central "lid" within the cap domain of Plb1 which, in $cPLA_2$, occludes the active site and is thought to be involved in interfacial activation. However, since Plb1 also displays interfacial activation in the absence of the lid, an alternative, and more likely, explanation is that this region is involved in substrate specificity. This is consistent with the hypervariability in this region among $cPLA_2$ isoforms and between $cPLA_2$ and Plb1. Hence the lid region may function to recognize the appropriate lipid substrate in a membrane.

How Is Plb1 Secreted?

For Plb1 to function as a virulence determinant, it must first be exported to the external milieu. The *PLB1* gene contains both a secretory signal leader motif at the N terminus and a hydrophobic motif at the C terminus, which conforms to that of a GPI anchor attachment site. The presence of both motifs indicates that Plb1 is exported from the endoplasmic reticulum (ER)/Golgi apparatus to the cell surface as a GPI-anchored protein. Blocking acquisition of the GPI anchor in the ER, by treating cryptococci with the GPI anchor biosynthetic inhibitor YW3548, prevents transport of Plb1 from the ER/Golgi apparatus to the cell surface and its localization in lipid raft membranes (154). It is known from studies in *S. cerevisiae* that the transport of proteins, including GPI-anchored proteins, to the cell periphery occurs via COP II (coat protein complex II)-secretory vesicles that originate in the ER (39, 117, 167). The protein cargo is housed inside the vesicle lumen. GPI-anchored proteins are physically anchored via the GPI to the inner membrane of the vesicle bilayer. Following fusion of the secretory vesicles with the plasma membrane, their internal cargo is expelled into the periplasmic space, whereas GPI-anchored proteins remain anchored at the cell surface within the lipid raft environment, which contains lipids derived from the secretory vesicle (117). The active part of the protein protrudes into the periplasmic space. The clustering of Plb1 and another virulence determinant, superoxide dismutase, within lipid rafts implicates these membrane domains in fungal pathogenesis (155).

Not only does the GPI anchor regulate secretion of Plb1, but it may also signal secretory vesicle sorting, as has been proposed for other GPI-anchored proteins in *S. cerevisiae* (117). Truncation of the Plb1 protein proximal to the GPI anchor attachment site by deletion of the predicted GPI anchor consensus motif resulted in Plb1 hypersecretion in an *S. cerevisiae* expression system (38). However, deletion of the GPI anchor consensus motif resulted in a greater than 50% reduction in transport of the protein through the secretory pathway, suggesting that the truncated protein is exported from the ER/Golgi apparatus via an alternative, and less efficient, trafficking route. Since treatment of wild-type H99 cells with the GPI anchor biosynthetic inhibitor, YW3548, halted export of Plb1, it is probable that, as in *S. cerevisiae*, the GPI anchor represents a signal for vesicle sorting (117). In the presence of YW3548, incomplete synthesis of GPI anchors would prevent their addition to proteins temporarily anchored in ER membranes via the GPI anchor consensus motif, resulting in sequestration of the protein in the ER and eventual degradation. As truncated Plb1 lacks this temporary ER membrane anchor, it is most likely seen by the secretion machinery as a soluble protein and packaged into secretory vesicles with other soluble secretory proteins.

Following transport of Plb1 to the fungal cell membrane, some is relocated to the cell wall and/or secreted to the external milieu. The mechanism of release of Plb1 from the cell membrane anchor is unknown. The GPI anchor is a potential substrate for enzymes such as (G)PI-specific phospholipase C (PI-PLC). Alternatively, secretion of Plb1 could follow cleavage of the GPI anchor by a (G)PI-PLD (33, 97) and/or a number of different proteases (116, 129) and/or an α-mannosidase (56, 90) (Fig. 2). Like many other GPI-anchored proteins, Plb1 attaches to β-1,6-glucans in the outer layer of the cell wall and contributes to cell wall integrity (154). These β-1,6-glucans are present on Plb1 secreted into the culture supernatant, indicating that the cell wall is a source of secreted enzyme, and implicating β-glucanases in the secretion of virulence determinants, in addition to their traditional role in cell wall remodeling (Fig. 2). Finally, physiological temperature stress is associated with accumulation of Plb1 protein in the cell wall, as well as secretion of an enzyme with higher specific activity, facilitating both maintenance of cell wall integrity and invasion of host tissue (154).

Both N-linked glycosylation and acquisition of the GPI anchor are required for transport of Plb1 from the ER/Golgi apparatus to the fungal cell periphery, as the traffic of recombinant cryptococcal Plb1 through the secretory pathway of *S. cerevisiae* was compromised when the glycosylation sites, Asn56, Asn430, and Asn550, were changed to Ala by site-directed mutagenesis (173). Two lines of evidence indicate that O-linked sugars are not involved in secretion of Plb1. First, in contrast to many other secreted and cell wall-associated mannoproteins of *C. neoformans* (102), Plb1 lacks O-linked glycosylation sites. Second, Plb secretion from O-linked mannosylation-deficient *C. albicans* mutants is not affected (29).

The class of secretory vesicle that transports GPI-anchored proteins, including Plb1, to the fungal cell periphery is not known. *SAV1*/Sec4 and *SEC6*-dependent secretory vesicles transport capsular polysaccharide and capsular polysaccharide, laccase, and urease, respectively (128, 184). Plb1 secretion was little affected in a cryptococcal mutant strain in which *SEC6* mRNA was inhibited using RNAi, suggesting that Plb1 is primarily transported in another class of vesicle. We recently showed that Plb1 secretion, but not the secretion of insoluble capsular polysaccharide, is dependent on the presence of an intact *PLC1* gene. *PLC1* encodes a PI-PLC (21) and is discussed in more detail below. Plb1 protein was synthesized, but accumulated in a membrane-enriched fraction. The role of PI-PLC in secretion of Plb1 is not yet understood.

In summary, secretion of Plb1 by *C. neoformans*, as defined by the release of cell-associated enzyme into the external milieu, appears to be regulated by the acquisition of a GPI anchor at the beginning of the pathway (at the ER/Golgi apparatus), allowing sorting of Plb1 into secretory vesicles, and at the end of the pathway (at the cell periphery) by hydrolysis of the GPI anchor. Interestingly, not all Plb enzymes secreted by pathogenic fungi contain a GPI anchor. In *C. albicans*, Plb3, 4, and 5, but not Plb1 and 2, are GPI anchored, and Plb1 and 5 are proven virulence determinants (115, 168). In *Aspergillus fumigatus*, Plb1 and 3, but not Plb2, are GPI anchored (153). It is not clear why some Plb enzymes involved in virulence are GPI anchored and some are not. It is likely that the GPI-anchored virulence determinants evolved an additional role in cell wall integrity through the acquisition of a GPI anchor attachment motif (154).

Other Plb-Like Phospholipases Produced by *C. neoformans*

In *C. neoformans*, two enzymes with activity restricted to LPL and LPTA only have been identified. *CnLYSO1* contains two GXSXG lipase catalytic motifs and was characterized in serotype D (27). *CnLPL1* contains one GXSXG lipase motif and was characterized initially in *C. gattii* (181). Though *CnLYSO1* lacks a secretory signal leader peptide, it appears to affect the secretion of Plb1. The *CnLPL1* gene sequence contains a putative secretory signal leader peptide and GPI anchor. Its function is unknown. Its catalytic properties differ from those of Plb1, and disruption of the *CnLPL1* homologue in *C. neoformans* serotype A did not alter growth in vitro or virulence in a mouse model (J. Djordjevic and T. Sorrell, unpublished observation). An enzyme with phospholipase A_1 activity was also recently identified in *C. neoformans* serotype A. It may function in conjunction with the LPLs mentioned above to remove lysophospholipids that

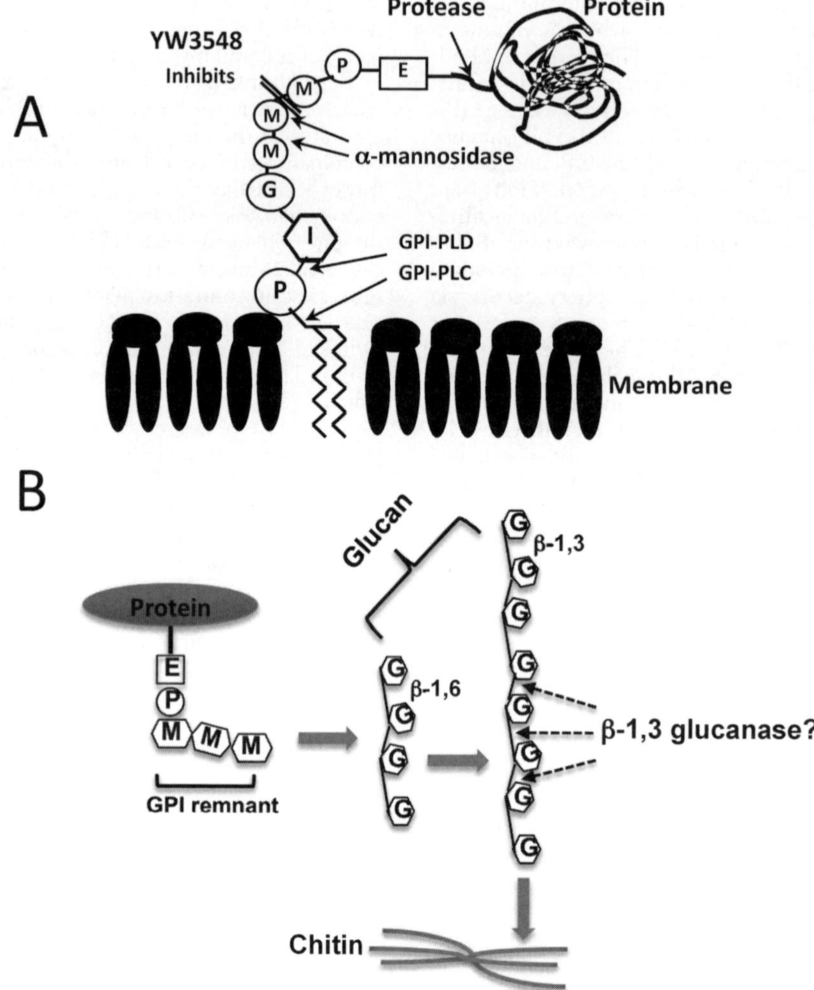

FIGURE 2 Possible surface release mechanisms for GPI-anchored proteins. (A) Model of a GPI-anchored protein showing potential sites of GPI anchor hydrolysis at the membrane. E, ethanolamine; P, phosphate; M, mannose; G, glucosamine; I, inositol. Arrows indicate potential cleavage sites (protein release mechanisms). Mannose residues from top to bottom are α-1,2-, 1,6-, and 1,4-linked. The PI moiety is attached to a diacyl group embedded in the membrane. The site of YW3548 inhibition of GPI anchor biosynthesis (prior to GPI anchor addition to the protein) is also shown. (B) Organization of the fungal cell wall. Potential sites of β-1,3-glucanase cleavage, releasing GPI-anchored proteins linked to β-1,6-glucan polymer, are indicated (in this case the GPI-anchored protein was released from the membrane by α-mannosidase). Modified from reference 164, with permission from Landes Bioscience.

are toxic to C. *neoformans* (182). These three enzymes may provide functional redundancy by maintaining cryptococcal growth in the absence of Plb1.

PI-PLC/Plc

Unlike Plb enzymes, which are found mainly in fungi, PI-PLC/Plc enzymes are ubiquitous in nature (86). They differ from Plb in that their preferred substrates are PI phospholipids and/or their phosphorylated derivatives [PI(4)P and PI(4,5)P/PIP_2], rather than PC. Plc enzymes hydrolyze the membrane-proximal phosphoester bond of the phosphoryl head group (Fig. 1) and have a role in signal transduction. Plc is involved in bacterial virulence (61) and has recently been recognized as a fungal virulence determinant (21). Only one of the six Plc isoforms described in higher eukaryotes is present in fungal genomes (124). Fungal Plc enzymes are most similar to the Plc-δ isoform of higher eukaryotes and may represent the archetypal isoform. Fungal Plcs lack some of the membrane-binding and regulatory domains present in their mammalian counterparts (Fig. 3), suggesting that their mode of regulation is unique and that they have potential for use as antifungal drug targets.

Plc-δ Functional Domains and Catalysis

Eukaryotic Plc-δ enzymes, including some of the fungal enzymes, are known to contain catalytic X and Y domains. Other fungal Plc enzymes are more similar to prokaryotic Plcs in that they have only an X domain. However, it is the X domain that contains the catalytic residues, while regions in the Y domain determine the distinct substrate preferences (66).

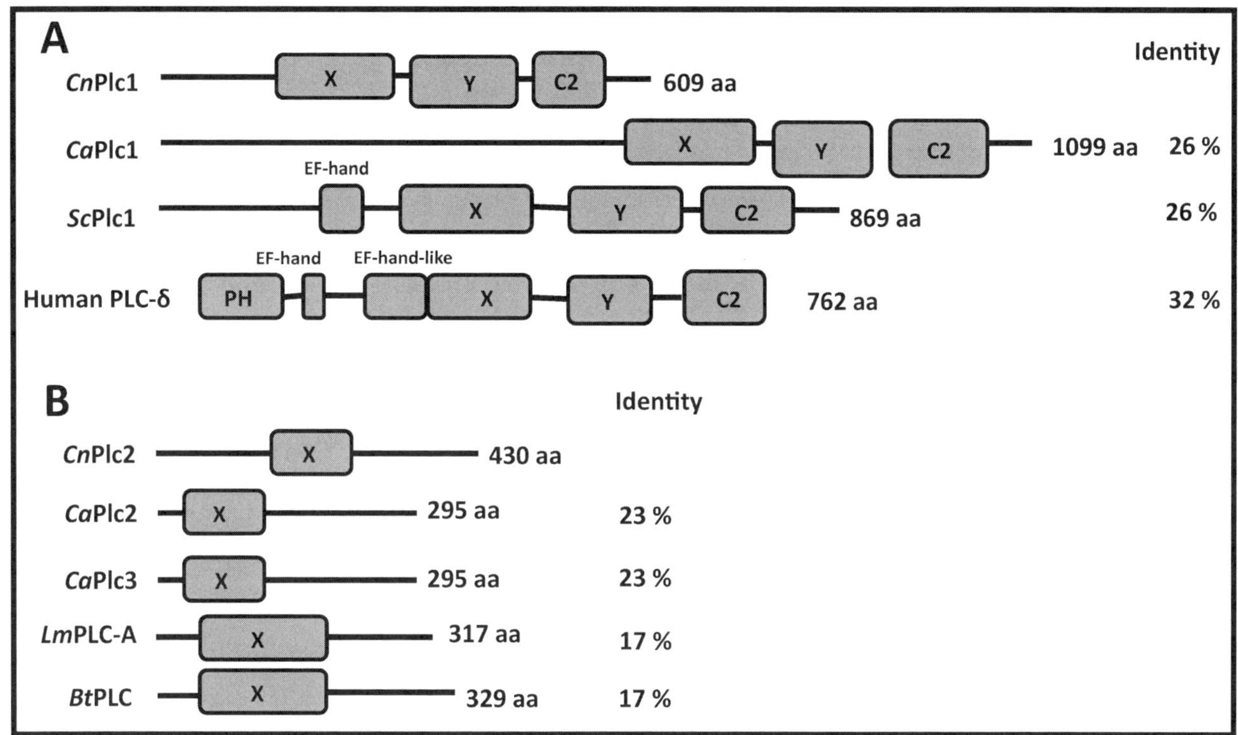

FIGURE 3 Comparison of the predicted Plc/PI-PLC protein motifs. (A) Plc1 proteins in all three yeast species and in the human PLC-δ4 isoform contain an X and Y catalytic domain and a C2 "calcium-binding" domain. *S. cerevisiae* is the only yeast species with an EF-hand domain. (B) Plc2 proteins from *C. neoformans* and *C. albicans*, the Plc3 protein from *C. albicans*, and Plcs from *L. monocytogenes* and *Bacillus thuringiensis* all contain only an X catalytic domain. Published with permission from Blackwell Publishing Ltd. (21).

Most eukaryotic Plc-δ isoforms contain three additional domains that are absent in both Plcs from prokaryotes and eukaryotic Plcs bearing only an X domain. These domains are the C2, pleckstrin homology, and EF-hand (EF) domains. The C2 domain is a calcium-binding motif that engages Plc with the membrane and orients the catalytic domain with the phospholipid membrane substrates (66, 124). The pleckstrin homology domain also has an affinity for PIP_2 substrate in the membrane (130). The EF-hand motif can also bind calcium and contributes to activity by an undefined mechanism (118).

Eukaryotic Plcs with an X and Y domain preferentially catalyze the breakdown of phosphorylated PI derivatives [PI(4)P and PI(4,5)P/PIP_2] rather than PI, in a Ca^{2+}-dependent manner, generating two intracellular secondary messengers, 1,2-DAG and inositol 1,4,5-trisphosphate (IP_3), respectively (121, 137, 156, 185). DAG is a potent activator of the Ca^{2+}-dependent enzyme protein kinase C (Pkc) (120). In both *S. cerevisiae* and *C. neoformans*, Pkc is an integral part of the mitogen-activated protein kinase (Mpk1) pathway that restores cell wall integrity following cell wall injury (4, 23, 77) and regulates fungal virulence (144). Triggers for this pathway include high temperature and cell wall-perturbing agents (96). IP_3 either releases Ca^{2+} from intracellular stores, enabling modulation of Ca^{2+}- and calmodulin-regulated pathways (7, 8), or is further converted by several kinases and phosphatases to a variety of inositol phosphates, some of which are involved in cell signaling. Bacterial Plcs and GPI-Plcs from certain trypanosomes preferentially hydrolyze PI and GPI substrates releasing DAG (66).

Cryptococcal PI-PLC (Plc1): a Global Regulator of Virulence

We recently demonstrated that PI-PLC1 (Plc1) regulates cryptococcal virulence, acting in part through interactions with the Pkc/Mpk1 cell wall integrity pathway. Specifically, Plc1 was essential for multiple virulence phenotypes, including growth at 37°C, cell separation following cytokinesis, production of laccase-derived extracellular melanin, secretion of Plb1, cell wall integrity, and virulence that was independent of the 37°C growth defect (21). As in *C. albicans* and some strains of *S. cerevisiae* (2, 50, 98), Plc1 was also essential for phenotypes relevant to cellular viability in vitro, including normal cell morphology and tolerance of osmotic stress.

Two putative Plc-encoding genes (*PLC1* and *PLC2*, GenBank accession numbers EU196232 and EU19623, respectively) are present in the *C. neoformans* var. *grubii* genomic database. *CnPLC1* encodes a putative 609-amino-acid protein with X/Y catalytic and C2 domains, similar to Plc1 of *C. albicans* and *S. cerevisiae* (98, 183), and shares a 26% amino acid sequence identity with these fungal homologues. *CnPLC2* encodes a putative 430-amino-acid protein, contains only an X catalytic domain, lacks a C2 domain, and shares a 23% amino acid sequence identity to Plc2/Plc3 from *C. albicans*. Compared to CnPlc1, CnPlc2 has a higher degree of sequence homology with the Plc (PLC-A protein) from

Listeria monocytogenes (Fig. 3). Conservation of the topology, and parts of the X domain active site in prokaryotic and eukaryotic Plcs, suggests that divergent evolution from a common ancestral protein has taken place (66).

It was demonstrated that *CnPLC1* encodes a functional Plc1 protein with a potential role in signaling, since hydrolysis of both PIP_2 and, to a lesser extent, PI was reduced in a deletion mutant, *plc1*Δ (21). Similarly, a semipurified recombinant CnPlc1 enzyme preferentially hydrolyzed PIP_2 in a Ca^{2+}-dependent manner (J. Djordjevic, unpublished observation). Deletion of *PLC2* did not affect substrate hydrolysis, indicating that if *PLC2* does encode an active enzyme, it cannot functionally compensate for the absence of *PLC1*. Moreover, unlike *PLC1*, *PLC2* is dispensable for growth, expression of virulence traits, and virulence per se (21). Finally, treatment of wild-type H99 with the commercially available Plc inhibitor U73122 (≥2.5 μM) inhibited growth at 30°C and 37°C, with growth inhibition being most prominent at 37°C, thus mimicking the temperature-sensitive phenotype displayed by *plc1*Δ.

Plc1 and Plb1 Secretion

PLC1 is essential for optimal secretory function since Plb1 was absent from cryptococcal culture supernatants of *plc1*Δ and accumulated in a membrane-enriched fraction. The mode of Plc1-regulated Plb1 secretion is unknown. Plc1 could directly hydrolyze the Plb1 GPI anchor, either in the Golgi apparatus or at the plasma membrane, but is unlikely to have direct access to Plb1 within the secretory pathway, as it lacks a secretory signal leader peptide. Plc2 neither affects Plb1 secretion nor has a secretory signal leader peptide. Two pieces of evidence emphasize that the direct hydrolysis of the Plb1 GPI anchor by Plc1 cannot be ruled out until further studies are performed. First, in contrast to Plcs from higher eukaryotes, Plcs from the parasite *Trypanosoma brucei* preferentially hydrolyze the GPI anchor of variant surface glycoprotein or GPI biosynthetic intermediates, in addition to PI, but not the phosphorylated intermediates (12), despite their localization to the peripheral cytoplasmic face of intracellular vesicles (15). Second, metabolic labeling studies performed in *S. cerevisiae* implicated a Plc enzyme and a secondary-acting protease (both unidentified) in hydrolysis of the GPI anchor of certain proteins in the plasma membrane, resulting in their subsequent localization in the cell wall (116). ScPlc1, the only Plc1 in *S. cerevisiae*, like CnPlc1, lacks a secretory signal leader peptide.

Plc1 and Signaling

High-Temperature Growth and Cell Wall Integrity

In contrast to *C. albicans*, *plc1*Δ in *C. neoformans* was viable under all growth conditions tested, but like *Scplc1*Δ and a conditional *CaPLC1* mutant, could not adapt to growth at 37°C (50, 98). The cell walls of *plc1*Δ were also defective, as indicated by clumping, and most distinctly by slower growth in the presence of cell wall-perturbing agents. Kraus et al. (96) demonstrated that the calcineurin and Pkc/Mpk1 signaling pathways coordinately regulate responses to high-temperature stress in *C. neoformans*, with cell wall integrity directly regulated by the Pkc/Mpk1 pathway (60). The observations that Plc1 is essential for high-temperature growth, cell wall integrity, and hydrolysis of PI/PIP_2 to produce DAG, a known activator of cryptococcal Pkc (68, 69), while Mpk1 is not activated in *plc1*Δ following cell wall injury (21), are together consistent with Plc1-induced signaling through the Pkc/Mpk1 pathway (21). Comparative transcriptome analysis of wild type and *plc1*Δ did not reveal differential expression of *MPK1*, suggesting that if Mpk1 signaling is involved, the cell wall defect in *plc1*Δ is due to a lack of phosphorylation and activation of Mpk1 rather than merely downregulation of transcription and translation (Djordjevic, unpublished observation). High-temperature growth could also be compromised in *plc1*Δ if Plc1 is linked to the calcineurin pathway via the production of IP_3, which in turn releases Ca^{2+} from intracellular stores causing activation of calmodulin (Cam1). The potential link between Plc1 and both the Pkc/Mpk1 and calcineurin pathways was further supported by analysis of the transcriptome of *plc1*Δ, where, relative to the wild type, mRNA levels of *FKS1*, which encodes the essential β-1,3-glucan synthase catalytic subunit, were increased (Djordjevic, unpublished observation). An increase in *FKS1* expression was also observed following a loss of calcineurin function, and this increase was dependent on the presence of Mpk1 (Fig. 4) (96).

The functional relationship between calcineurin and the C2 domain-containing and phospholipid-binding protein Cts1 also underscores the linkages between Plc, cell wall integrity, and calcineurin signaling pathways (51). In *C. neoformans*, overexpression of the *CTS1* gene restored growth of a calcineurin mutant at elevated (37°C) temperature, and expression of the *CTS1* gene was found to be necessary for growth at 37°C, hyphal elongation, cell separation following cytokinesis, and virulence in a murine model of cryptococcosis (51). Analysis of the importance of the calcium-binding C2 domain within the N terminus of Cts1 revealed that Cts1 function was dependent on the presence of this domain and that the C2 domain conferred phospholipid-binding capacity specifically for PI(4)P and PI(5)P interactions, suggesting potential functional interactions with Plc-dependent signaling (Fig. 4) (51).

Secretion

In eukaryotes, Plcs modulate membrane lipid levels and cell signaling processes essential for intracellular trafficking and secretion. Evidence suggests that Plc1 may regulate secretion of Plb1, and possibly other cell wall proteins, by means of the cyclic AMP (cAMP)/Pka pathway. Comparative transcriptomic studies performed by Hu et al. (75) and Chayakulkeeree et al. (22) implicate the cAMP/Pka pathway in secretion, as both studies identified differentially expressed genes encoding secretion pathway components, enzymes involved in glycosylation and/or cell wall biosynthesis and phospholipid homeostasis, in strains expressing defective protein kinase catalytic (Pka) or regulatory (Pkr) subunits and in *plc1*Δ, respectively. Finally, Plb1 secretion could be mediated by IP_3 and Ca^{2+} production, as has been suggested in a recent review (95). Further studies are required to confirm this. Determination of the location of a Plb1 secretion block in *plc1*Δ by immune and/or fluorescence microscopy will shed more light on the role of Plc1 in cryptococcal secretory processes.

Melanin Production

Laccase-induced melanin production in *C. neoformans* has been linked to activation of Pkc1 by DAG (68), which is independent of the activation of Mpk1 (60), and to the cAMP/Pka pathway, as the two laccase-encoding genes, *LAC1* and *LAC2*, are downregulated in a *GPA1* (encodes a G protein α subunit in the cAMP/Pka pathway) deletion mutant (134), leading to loss of melanin and capsule production (1). Chayakulkeeree et al. (22) demonstrated that the absence

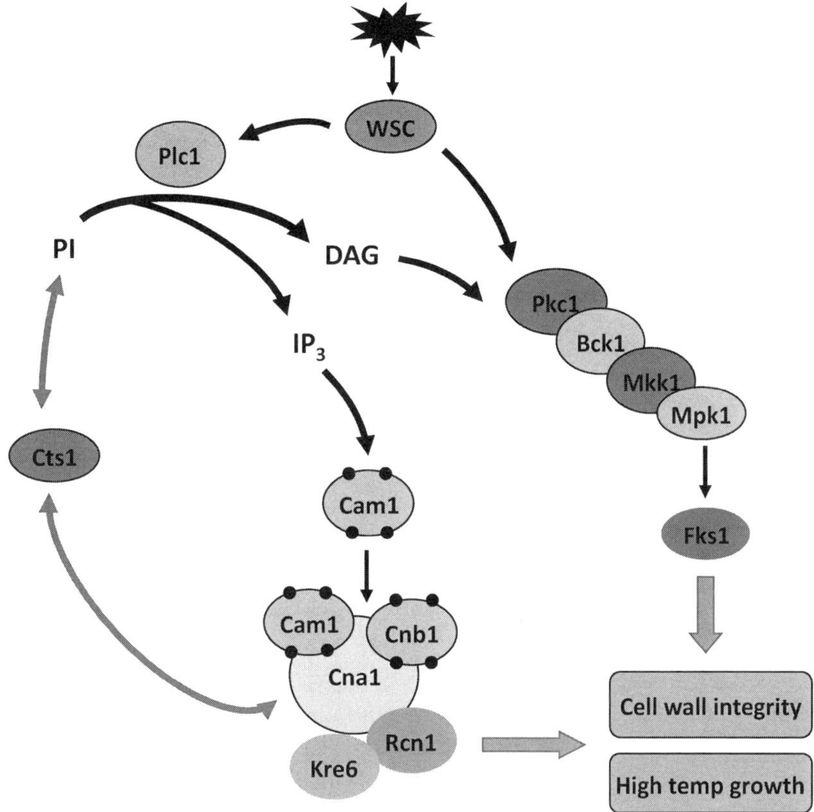

FIGURE 4 Model depicting the functional linkages of the Pkc1/Mpk1, phospholipase, and calcineurin signaling pathways in the regulation of cell wall integrity in *C. neoformans*. Phospholipase C (Plc1) signaling mediates cell wall integrity and high-temperature growth through activation of the Pkc1/Mpk1 and calcineurin pathways. The phospholipid-binding protein Cts1 may direct communication between Plc1 and calcineurin signaling pathways. The calcineurin regulator and effector Rcn1 may facilitate some aspects of calcineurin activity by modulation of calcineurin substrate specificity, while Kre6 may serve as a possible calcineurin substrate and regulator of β-glucan synthesis. The putative cell wall/membrane sensor of cell wall integrity is depicted as WSC. PI, phosphatidylinositol; DAG, 1,2-diacylglycerol; IP_3, inositol 1,4,5-triphosphate. Adapted from references 21, 99.

of CnPlc1 in *plc1*Δ abolished melanin production by suppressing *LAC1* mRNA transcription but did not affect capsule production, arguing against Plc1 regulating melanin production through the cAMP/Pka pathway. On the other hand, evidence for a link between Plc1 and the cAMP/Pka pathway has been demonstrated in *S. cerevisiae*, even though this organism does not make melanin.

First, the results of a recent comparative genome-wide expression analysis study between *S. cerevisiae* wild type and its *plc1*Δ were consistent with a model in which Plc1 acts together with the membrane receptor, Gpr1, and associated Gα protein, Gpa2, in a pathway converging on Pka (35). Second, Ansari et al. (2) demonstrated that Gpr1 physically interacts with ScPlc1 and with the Gα protein Gpa2. CnPlc1 may also play an analogous role to ScPlc1 in nutrient sensing and cell cycle regulation via the cAMP/Pka pathway (49) and by transducing glucose signals (170). Finally, PI3 kinase (PI3K), which synthesizes phosphatidylinositol 3-phosphate (PIP_3) from PI, also regulates melanin production by a mechanism that may be related to the cellular traffic of laccase to the cell wall and extracellular environment. Both melanin production and autophagy, a trafficking process involving fusion of cytoplasmic vesicles and defective organelles with vacuoles as part of a recycling process during stress and nutrient starvation, are decreased within a PI3K deletion mutant after phagocytosis by macrophages, suggesting that PI3K signaling is a regulatory mechanism for survival in stress conditions (74). In *S. cerevisiae*, PI3K (the class III Vps34) forms a membrane-associated signaling complex to regulate protein sorting of hydrolases to the vacuole lumen and to trigger autophagy in nutrient-limited conditions (11, 67, 84, 160). Vps34 is activated by the pheromone-activated GTP-bound Gα subunit (Gpa1) at the endosomal membrane (157). Whether melanin production is regulated by PI3K-mediated transport of laccase to the cell surface by means of vacuoles remains to be determined. Plc1 competes with PI3K for PI substrate and could therefore potentially regulate laccase-induced melanin production (and autophagy) by means of PI3K signaling. However, the link between the two signaling pathways, and laccase trafficking to the cell surface, is impossible to study in *plc1*Δ strains, as Plc1 protein is essential for transcription of *LAC1* mRNA.

In conclusion, Plc1 may exert its effect on high-temperature growth, secretion, and virulence through many signal transduction pathways, including the cell wall integrity (Pkc/Mpk1) and calcineurin signaling pathways, and these pathways may have overlapping functions (Fig. 4). Further biochemical and molecular analyses of *plc1*Δ, and of the various signaling mutants that have been constructed throughout the cryptococcal research community, will shed more light on the potentially complex links between Plc and the various signal transduction cascades.

Are Fungal Phospholipases Potential Targets for Antifungal Drug Development?

Since Plb is a virulence determinant of both *C. neoformans* and *C. albicans* and is produced by other pathogenic fungi including *A. fumigatus* (9), Plb itself, or components of its secretory pathway, are potential targets for antifungal drug development. The feasibility of selective inhibition of Plb was established by demonstrating that compounds with structural similarities to phospholipids selectively inhibited cryptococcal Plb compared with mammalian PLA_2 and exhibited activity against strains of *C. neoformans* and *C. albicans* (59). Subsequent structure-function studies on a series of structurally related bis- (quaternary ammonium) alkane compounds further validated Plb as a potential antifungal target, since inhibition of Plb1 correlated with antifungal activity (119). Inhibition of Plb may also be one of several targets inhibited by similar classes of compounds. An example of this is the phosphocholine group of compounds, one member of which, miltefosine, is currently marketed for the treatment of the parasitic disease leishmaniasis. Miltefosine has broad-spectrum antifungal activity against several fungal pathogens, including *Scedosporium prolificans*, that are resistant to most available drugs (179). Oral administration of miltefosine prevented or delayed the establishment of pulmonary and cerebral infection in a mouse model of systemic cryptococcosis (179) and contributed to a therapeutic response in a child with severe and progressive scedosporiosis (88). Miltefosine inhibits cryptococcal Plb1, but only at concentrations more than six times the minimal inhibitory concentration, as determined by broth microdilution, suggesting that phospholipase inhibition is not the primary mode of action. Given its physicochemical properties and chemical similarity to the membrane phospholipid PC, miltefosine would be expected to interact with subcellular structures and enzymes associated with membranes. Multiple sites of action have been demonstrated or proposed in studies in *Leishmania* (31). Plc has not yet been investigated directly as a potential antifungal drug target. However, given our increasing understanding of Plc1 and Plb1 and their secretory pathways in *C. neoformans*, the identification of additional antifungal drug targets will stimulate further research into antifungal drug development.

CALCINEURIN: A CENTRAL REGULATOR OF *C. NEOFORMANS* VIRULENCE

The utilization of signal transduction cascades has been shown to be a common mechanism for many pathogenic fungi to promote disease by allowing the organisms to readily adapt to rapidly changing environmental conditions within the host. The primary components of signal transduction cascades are protein kinases and phosphatases, which act on protein targets via a mechanism of reversible phosphorylation. Calcineurin, a Ca^{2+}-calmodulin-activated serine/threonine protein phosphatase, is composed of two subunits, a catalytic A subunit and a regulatory B subunit, that form a functional heterodimer that is necessary for the pathogenesis of *C. neoformans* and other medically important fungi (reviewed in 54, 95, 162, 165). Conserved in all eukaryotes, from yeast to human, calcineurin regulates many physiological processes necessary for life, including gene expression, morphogenesis, cell wall biosynthesis, and ion homeostasis. Calcineurin is the target of the immunosuppressive drugs FK506 and cyclosporin A, which, when bound to intracellular receptors known as immunophilins, act to specifically inhibit the phosphatase function of the heterodimer (104).

In *C. neoformans*, calcineurin is intimately involved in cellular processes necessary for pathogenesis, as it is essential for growth at 37°C and for virulence in animal models of cryptococcosis (52, 125). In addition to its indispensable role in growth at 37°C, calcineurin is also required for morphogenic events that control hyphal elongation in *C. neoformans*, a process vital to the development of infectious spores (32). In addition to the well-described roles for the catalytic and regulatory subunits of calcineurin in virulence, additional studies have demonstrated the importance of calcineurin in antifungal drug resistance, cell wall biogenesis, and ion homeostasis, facilitating the identification of several candidate effectors of calcineurin that regulate growth, virulence, and hyphal elongation: the highly conserved calcineurin-binding protein Rcn1 (Cbp1), the phospholipid-binding protein Cts1, the β-1,3-glucan synthase subunit Fks1, the plasma membrane calcium channel Cch1, the ER calcium pump Eca1, and the proton transporter Ena1 (46, 51, 53, 62, 78, 96, 105).

The Importance of Calcineurin in Cell Wall Biogenesis

The fungal cell wall is a unique structure that provides mechanical strength and acts as a physical barrier to protect the cell from damage. In addition to its structural role, the cell wall is also an important mediator of events necessary for cell-to-cell recognition, growth, and morphogenesis (reviewed in 91, 157). In *C. neoformans*, the cell wall is also important for pathogenesis, as the polysaccharide capsule and the antifungal protectant melanin are cell wall-associated virulence factors (44, 79, 136, 175, 178). To permit dynamic processes such as growth and morphogenesis, which require rapid changes in cell wall composition and structure, the machinery controlling the synthesis and assembly of the cell wall must be responsive to both internal and external stimuli. The cell wall is composed of complexes of proteins and polysaccharides, which vary in composition and structure, depending on growth conditions (18, 19, 85, 100, 107). The polysaccharide components of the cell wall, including β-1,3-glucan, β-1,6-glucan, α-glucan, and chitin, provide the framework for the structure of the cell wall, with the β-1,6-glucan component acting as the anchor for the assembly of the other components contributing to cell wall structure (14, 65, 91–94, 151, 158).

The integral membrane proteins Fks1 and Fks2 of *S. cerevisiae* are differentially expressed catalytic subunits of the β-1,3-glucan synthase complex (41, 80, 111, 186), while the GTP-binding protein Rho1 has been shown to function as an activator of β-1,3-glucan synthesis and interacts with the catalytic subunits Fks1/Fks2 to promote β-1,3-glucan synthesis (3, 111). Other Fks-associated proteins, including the H^+-ATPase homologue Pma1 and homologues of the ABC glucan transporter, have been identified by enrichment for glucan synthase components by product entrapment in *S. cerevisiae*, *C. albicans*, *A. fumigatus*, *Aspergillus nidulans*, and *Neurospora crassa* (6, 55, 80, 87, 149). Additional studies have shown that the cell wall sensor protein Wsc1/Slg1, containing a glucan-binding WSC domain (28, 132), is involved in the activation of Rho1 and the reorganization of the actin cytoskeleton in response to stress and functions as a suppressor of the *fks1-1154* mutation to restore β-1,3-glucan synthesis in *S. cerevisiae* (64, 150, 176). As recent data suggest that additional components of the β-1,3-glucan synthase complex, as well as proteins involved in trafficking and actin polarization, may exist that act to facilitate the interaction between the Rho1 regulatory subunit and the putative catalytic subunits Fks1 and Fks2 and permit the proper localization of the β-1,3-glucan synthase complex (45, 72, 143), the identification of other subunits of the β-1,3-glucan synthase complex, as well as the characterization of the role of these proteins in the regulation of β-1,3-glucan synthesis, would dramatically increase our understanding of fungal cell wall synthesis.

In the budding yeast *S. cerevisiae*, the serine/threonine protein phosphatase calcineurin acts as a positive regulator of the expression of *FKS2*, which encodes a component of the β-1,3-glucan synthase complex necessary for cell wall integrity, and it is also required for viability in the absence of functional Fks1, the alternatively expressed β-1,3-glucan synthase component (41, 112, 186). In addition, calcineurin may regulate β-1,6-glucan synthesis, as calcineurin is found in a complex with the phosphoprotein Kre6, a putative subunit of the β-1,6-glucan synthase complex that also interacts with the actin patch assembly proteins Las17 and Sla1 (10, 72, 103, 114, 140–142). Interestingly, Kre6 is also implicated in β-1,3-glucan synthesis, as mutation of *KRE6* results in reduced levels of both β-1,3- and β-1,6-glucan (140).

In contrast to *S. cerevisiae*, the abrogation of calcineurin function through deletion or inhibition in *C. neoformans* results in a compensatory induction of *FKS1* expression by the cell wall integrity Pkc/Mpk1 pathway, demonstrating a central, though divergent, role for calcineurin in the regulation of cell wall synthesis and integrity in fungi (96). In support of this observation, ultrastructural and microscopic analysis of *C. neoformans* by both transmission electron and light microscopy revealed that, compared to their wild-type counterparts, calcineurin mutants have grossly abnormal cellular morphologies, with apparent defects in the composition and structure of the cell wall (D. S. Fox, unpublished observation). Analysis of the carbohydrate composition of the cell walls of wild-type and calcineurin-deficient (*cnb1*Δ) strains grown under non-capsular-inducing conditions revealed that the wild-type strain contained only modest amounts of cell wall β-1,3-glucan, with β-1,6-glucan comprising the majority of the alkali-insoluble glucan component (Fox, unpublished observation). In contrast, the cell wall glucan composition of the *cnb1*Δ mutant strain was dramatically reduced for all three glucan forms, with the most obvious reduction in β-1,6-glucan, where the *cnb1*Δ mutant contained only 6% of the amount present within the cell wall of the wild-type control (Fox, unpublished observation). However, analyses of the chitin and protein content of purified cell walls from calcineurin mutants revealed no significant differences from the wild-type strain (17, 43, 81).

As β-1,6-glucan levels were most affected in the calcineurin mutants, and β-1,6-glucan functions as the scaffold for the interconnection of all cell wall components (94), the results suggest a role for calcineurin in the regulation of β-1,6-glucan synthesis in *C. neoformans* (Fig. 2). As mentioned above, a recent approach to identify protein complexes in *S. cerevisiae* found calcineurin in a complex with Kre6, a proposed component of the β-1,6-glucan synthase machinery. The *KRE6* and *SKN1* genes are differentially expressed and encode functionally redundant membrane proteins required for β-1,6-glucan synthesis in both *S. cerevisiae* and *C. albicans* (113, 114, 140, 141). Therefore, to examine the functional link between calcineurin and Kre6 in *C. neoformans*, we identified the *C. neoformans* Kre6 homolog (locus CND00550) by analysis of the annotated *C. neoformans* JEC21 and B-3501A genome sequence entries within the NCBI database with the *S. cerevisiae* Kre6 amino acid sequence (106). Due to the presence of five additional predicted *KRE6/SKN1* gene family members in the *C. neoformans* genome, we designed allele-specific primers to selectively amplify the CND00550 gene and cDNA for cloning and expression analysis (Fox, unpublished observation). CND00550 was chosen for analysis in this study due to its overall identity to Kre6 from *S. cerevisiae* compared to other family members (33% identity, 45% homology).

To exclude the possibility that calcineurin directs β-1,6-glucan synthesis in *C. neoformans* by regulating *KRE6* gene expression, we performed reverse-transcriptase PCR with two sets of *KRE6* allele-specific primers to amplify *KRE6* from total RNA isolated from wild-type (*CNB1*), calcineurin-deficient (*cnb1*Δ), and FK506-resistant (51) (*CNB1-1*) cells. Our results indicated that *KRE6* gene expression in *C. neoformans* was not calcineurin dependent and suggested that calcineurin may direct β-1,6-glucan synthesis via posttranslational regulation of the Kre6 phosphorylation state (Fox, unpublished observation). To examine this hypothesis, the predicted cytosolic domain of Kre6 was cloned into the DNA-binding domain yeast two-hybrid plasmid, pGBKT7, and coexpressed with the activation-domain yeast two-hybrid plasmid, pGADT7, containing the full-length Cna1 subunit (residues 1 to 642), the C-terminal truncation mutant Cna1-Cam-AID (residues 1 to 400), or the Cna1 C-terminal AID and tail region (residues 501 to 642), in yeast two-hybrid reporter strains containing the wild-type *S. cerevisiae CNB1* allele or the *cnb1::ADE2* mutation (Fig. 5). Although full-length Cna1 displayed demonstrable binding to the Kre6 cytosolic domain (Kre6-pS), the C-terminal Cna1 (501 to 642) bound weakly, while the Cna1-Cam-AID truncation was no longer able to bind (Fig. 5) (Fox, unpublished observation). Taken together, our results suggest that (i) calcineurin associates with Kre6 via the catalytic subunit, Cna1, and (ii) the calmodulin-binding domain of Cna1 is important for Kre6 association. Thus, the interaction of calcineurin with Kre6 is consistent with a model in which calmodulin-activated calcineurin directs β-1,6-glucan synthesis by modulation of the Kre6 phosphorylation state. Taken together, these findings lend further support to a model in which calcineurin plays a vital role in the regulation of cell wall synthesis and integrity in *C. neoformans*.

Antifungal Drug Tolerance Is Mediated by Calcineurin Signaling

The echinocandin class of antifungal agents, which include caspofungin, micafungin, and anidulafungin, are potent inhibitors of cell wall synthesis in many pathogenic fungi but have very low potency against *C. neoformans* (13, 63, 73, 83). The proposed catalytic subunits of β-1,3-glucan synthesis, the Fks proteins, have been implicated as the target of these agents, as specific mutations in *FKS* genes confer resistance to this class of inhibitors, although additional echinocandin targets may exist in *C. albicans* and *S. cerevisiae* (5, 40, 42, 45, 101, 110, 126, 135). However, speculation exists as to the exact nature of the resistance of *C. neoformans* to the echinocandins. Several possibilities have been proposed, including (i) the inaccessibility of the cell to echinocandins due to the presence of the polysaccharide capsule, (ii) the lack of β-1,3-glucan in the cell wall, (iii) Fks1-mediated

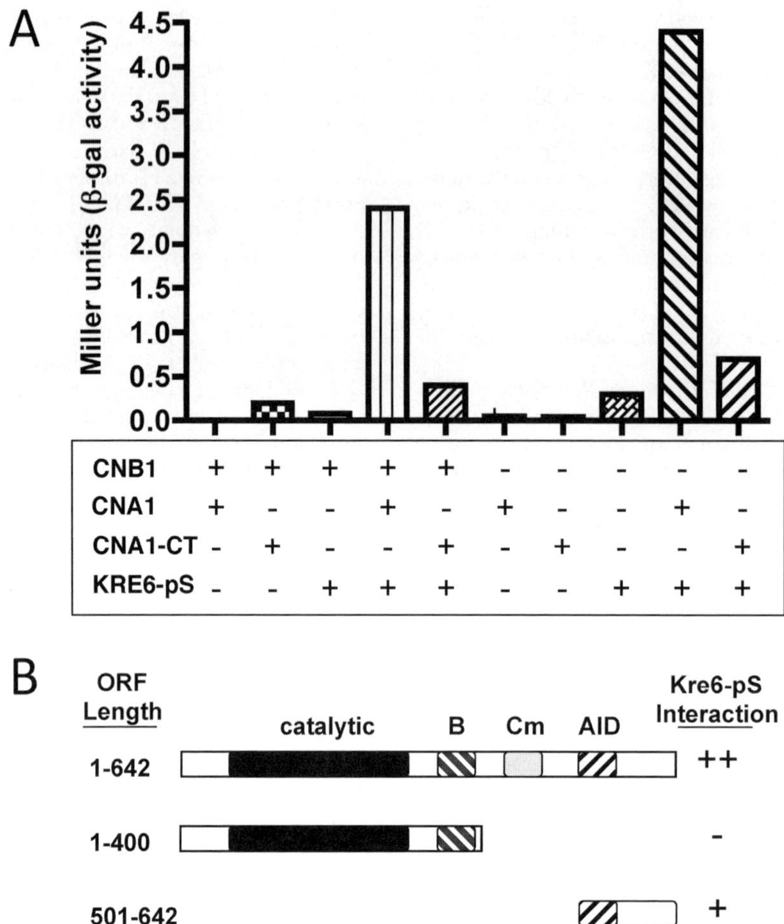

FIGURE 5 Calcineurin catalytic subunit Cna1 interacts with the predicted cytosolic domain of Kre6. (A) Isogenic two-hybrid reporter strains expressing calcineurin B (Y190) or lacking calcineurin B (SMY7-7) were cotransformed with plasmids expressing the Gal4 DNA-binding domain alone or fused to calcineurin A: full length (CNA1) or the C-terminal tail (CNA1-CT) and the Gal4 activation domain alone or fused to the Kre6 cytoplasmic domain (KRE6-pS) (alleles listed on *x* axis). Interaction values are shown in Miller units. (B) Two-hybrid analysis to map the Kre6 binding region within calcineurin A. Open reading frame (ORF) length is shown in amino acid residues. The shaded boxes indicate the catalytic, calcineurin B-binding (B), calmodulin-binding (Cm), and autoinhibitory (AID) domains of calcineurin A.

resistance, (iv) transport-related reduced susceptibility (127), and (v) lack of efficacy due to a fundamental difference in the manner in which Fks1, the integral membrane component of the β-1,3-glucan synthase complex, is regulated in *C. neoformans*. However, unpublished data have shown that capsule synthesis does not contribute to echinocandin resistance (J. R. Perfect, personal communication), and analysis of the *C. neoformans* cell wall indicates that β-1,3-glucan linkages are present (81). Additionally, it has been determined that *FKS1* is essential and does not harbor any mutations that have been shown to confer resistance in other fungi (40, 169).

Analysis of cell wall composition in *C. neoformans* has provided tantalizing clues that suggest the lack of susceptibility of this organism to inhibitors of β-1,3-glucan synthesis is reflected in a substantially reduced β-1,3-glucan content in cell walls of *C. neoformans* as compared to other fungi (47, 81). This apparent decrease in the β-1,3-glucan content is compensated by a concomitant increase in the abundance of other components of the cell wall, including chitin, α-glucan, and β-1,6-glucan, suggesting that the regulation of cell wall biogenesis and assembly is quite different in *C. neoformans* and may result in a lack of efficacy for many currently available antifungal drugs that target β-1,3-glucan synthesis (81).

Additional discoveries suggest a potential role for the Fks proteins in the synthesis of both the β-1,3- and β-1,6-glucans, as *fks* mutants in *S. cerevisiae* have reduced levels of both (37). The ultrastructural examination of glucan synthesis in *C. neoformans* revealed that both β-1,3- and β-1,6-glucan linkages are significantly decreased following exposure to the echinocandin caspofungin, suggesting that, in addition to the inhibition of β-1,3-glucan synthesis, the echinocandins may negatively influence β-1,6-glucan synthesis as well (47). The decrease in β-glucan linkages in the cell wall of *C. neoformans* in response to echinocandin exposure demonstrates that the target or targets of this antifungal do exist in this pathogen, and implicate Fks1 as the target of the echinocandins in *C. neoformans*. However, these studies do not address the role of Fks1 in β-glucan synthesis in *C. neoformans* or whether Fks1 serves as the target of the echinocandins, although recent findings by Maligie and Selitrennikoff demonstrate that the inherent echinocandin resistance in *C. neoformans* is independent of Fks1 (109).

Previous studies have shown that (i) the β-1,3-glucan synthase inhibitor caspofungin exhibits synergistic antifungal activity against *C. neoformans* when used in combination with the calcineurin inhibitor FK506 (34), (ii) loss of calcineurin function activates expression of *FKS1*, a proposed subunit of the β-1,3-glucan synthase complex, through a compensatory mechanism (96), and (iii) calcineurin-deficient strains have reduced β-glucan levels (Fox, unpublished observation), demonstrating a role for calcineurin in the regulation of cell wall synthesis and integrity in this fungal pathogen. Recent studies regarding the importance of calcineurin and the cell wall integrity response for the attenuation of echinocandin activity in *C. albicans* and *A. fumigatus* lend support to these findings (161, 163, 164, 180). Therefore, the observation that mutation or loss of function of the conserved phosphatase calcineurin confers echinocandin sensitivity in *C. neoformans* suggests that a mechanism exists to enhance the susceptibility of this fungus to the echinocandin class of antifungals through the targeting of specific calcineurin effectors and/or components of parallel enzymatic processes, including those of the cell wall integrity and phospholipase signaling pathways.

REFERENCES

1. **Alspaugh, J. A., J. R. Perfect, and J. Heitman.** 1997. *Cryptococcus neoformans* mating and virulence are regulated by the G-protein alpha subunit GPA1 and cAMP. *Genes Dev.* **11:**3206–3217.
2. **Ansari, K., S. Martin, M. Farkasovsky, I. M. Ehbrecht, and H. Kuntzel.** 1999. Phospholipase C binds to the receptor-like GPR1 protein and controls pseudohyphal differentiation in *Saccharomyces cerevisiae*. *J. Biol. Chem.* **274:**30052–30058.
3. **Arellano, M., A. Duran, and P. Perez.** 1996. Rho 1 GTPase activates the (1-3)beta-D-glucan synthase and is involved in *Schizosaccharomyces pombe* morphogenesis. *EMBO J.* **15:**4584–4591.
4. **Bahn, Y. S., C. Xue, A. Idnurm, J. C. Rutherford, J. Heitman, and M. E. Cardenas.** 2007. Sensing the environment: lessons from fungi. *Nat. Rev. Microbiol.* **5:**57–69.
5. **Balashov, S. V., S. Park, and D. S. Perlin.** 2006. Assessing resistance to the echinocandin antifungal drug caspofungin in *Candida albicans* by profiling mutations in FKS1. *Antimicrob. Agents Chemother.* **50:**2058–2063.
6. **Beauvais, A., J. M. Bruneau, P. C. Mol, M. J. Buitrago, R. Legrand, and J. P. Latge.** 2001. Glucan synthase complex of *Aspergillus fumigatus*. *J. Bacteriol.* **183:**2273–2279.
7. **Berridge, M. J.** 1993. Cell signalling. A tale of two messengers. *Nature* **365:**388–389.
8. **Berridge, M. J.** 1993. Inositol trisphosphate and calcium signalling. *Nature* **361:**315–325.
9. **Birch, M., G. Robson, D. Law, and D. W. Denning.** 1996. Evidence of multiple extracellular phospholipase activities of *Aspergillus fumigatus*. *Infect. Immun.* **64:**751–755.
10. **Bowen, S., and A. E. Wheals.** 2004. Incorporation of Sed1p into the cell wall of *Saccharomyces cerevisiae* involves KRE6. *FEMS Yeast Res.* **4:**731–735.
11. **Burda, P., S. M. Padilla, S. Sarkar, and S. D. Emr.** 2002. Retromer function in endosome-to-Golgi retrograde transport is regulated by the yeast Vps34 PtdIns 3-kinase. *J. Cell Sci.* **115:**3889–3900.
12. **Butikofer, P., M. Boschung, U. Brodbeck, and A. K. Menon.** 1996. Phosphatidylinositol hydrolysis by *Trypanosoma brucei* glycosylphosphatidylinositol phospholipase C. *J. Biol. Chem.* **271:**15533–15541.
13. **Cabello, M. A., G. Platas, J. Collado, M. T. Diez, I. Martin, F. Vicente, M. Meinz, J. C. Onishi, C. Douglas, J. Thompson, M. B. Kurtz, R. E. Schwartz, G. F. Bills, R. A. Giacobbe, G. K. Abruzzo, A. M. Flattery, L. Kong, and F. Pelaez.** 2001. Arundifungin, a novel antifungal compound produced by fungi: biological activity and taxonomy of the producing organisms. *Int. Microbiol.* **4:**93–102.
14. **Cabib, E., T. Drgon, J. Drgonova, R. A. Ford, and R. Kollar.** 1997. The yeast cell wall, a dynamic structure engaged in growth and morphogenesis. *Biochem. Soc. Trans.* **25:**200–204.
15. **Cardoso De Almeida, M. L., M. Geuskens, and E. Pays.** 1999. Cell lysis induces redistribution of the GPI-anchored variant surface glycoprotein on both faces of the plasma membrane of *Trypanosoma brucei*. *J. Cell Sci.* **112**(Pt 23)**:**4461–4473.
16. **Casadevall, A., J. D. Nosanchuk, P. Williamson, and M. L. Rodrigues.** 2009. Vesicular transport across the fungal cell wall. *Trends Microbiol.* **17:**158–162.
17. **Casadevall, A., and J. R. Perfect.** 1998. *Cryptococcus neoformans*. ASM Press, Washington, DC.
18. **Chaffin, W. L., J. L. Lopez-Ribot, M. Casanova, D. Gozalbo, and J. P. Martinez.** 1998. Cell wall and secreted proteins of *Candida albicans*: identification, function, and expression. *Microbiol. Mol. Biol. Rev.* **62:**130–180.
19. **Chaffin, W. L., and D. M. Stocco.** 1983. Cell wall proteins of *Candida albicans*. *Can. J. Microbiol.* **29:**1438–1444.
20. **Charlier, C., K. Nielsen, S. Daou, M. Brigitte, F. Chretien, and F. Dromer.** 2009. Evidence of a role for monocytes in dissemination and brain invasion by *Cryptococcus neoformans*. *Infect. Immun.* **77:**120–127.
21. **Chayakulkeeree, M., T. C. Sorrell, A. R. Siafakas, C. F. Wilson, N. Pantarat, K. J. Gerik, R. Boadle, and J. T. Djordjevic.** 2008. Role and mechanism of phosphatidylinositol-specific phospholipase C in survival and virulence of *Cryptococcus neoformans*. *Mol. Microbiol.* **69:**809–826.
22. **Chayakulkeeree, M., C. Wilson, T. Sorrell, and J. Djordjevic.** 2008. Delineating phospholipase C1 (PLC1)-related signalling pathways in *Cryptococcus neoformans*. Abstract ID 1002, 77th International Conference on Cryptococcus and Cryptococcosis, Nagasaki, Japan.
23. **Chen, R. E., and J. Thorner.** 2007. Function and regulation in MAPK signaling pathways: lessons learned from the yeast *Saccharomyces cerevisiae*. *Biochim. Biophys. Acta* **1773:**1311–1340.
24. **Chen, S. C., M. Muller, J. Z. Zhou, L. C. Wright, and T. C. Sorrell.** 1997. Phospholipase activity in *Cryptococcus neoformans*: a new virulence factor? *J. Infect. Dis.* **175:**414–420.
25. **Chen, S. C., L. C. Wright, J. C. Golding, and T. C. Sorrell.** 2000. Purification and characterization of secretory phospholipase B, lysophospholipase and lysophospholipase/transacylase from a virulent strain of the pathogenic fungus *Cryptococcus neoformans*. *Biochem. J.* **347:**431–439.
26. **Chen, S. C., L. C. Wright, R. T. Santangelo, M. Muller, V. R. Moran, P. W. Kuchel, and T. C. Sorrell.** 1997. Identification of extracellular phospholipase B, lysophospholipase, and acyltransferase produced by *Cryptococcus neoformans*. *Infect. Immun.* **65:**405–411.
27. **Coe, J. G., C. F. Wilson, T. C. Sorrell, N. G. Latouche, and L. C. Wright.** 2003. Cloning of CnLYSO1, a novel extracellular lysophospholipase of the pathogenic fungus *Cryptococcus neoformans*. *Gene* **316:**67–78.
28. **Cohen-Kupiec, R., K. E. Broglie, D. Friesem, R. M. Broglie, and I. Chet.** 1999. Molecular characterization of a

novel beta-1,3-exoglucanase related to mycoparasitism of *Trichoderma harzianum*. *Gene* **226:**147–154.

29. **Corbucci, C., E. Cenci, F. Skrzypek, E. Gabrielli, P. Mosci, J. F. Ernst, F. Bistoni, and A. Vecchiarelli.** 2007. Immune response to *Candida albicans* is preserved despite defect in O-mannosylation of secretory proteins. *Med. Mycol.* **45:**709–719.
30. **Cox, G. M., H. C. McDade, S. C. Chen, S. C. Tucker, M. Gottfredsson, L. C. Wright, T. C. Sorrell, S. D. Leidich, A. Casadevall, M. A. Ghannoum, and J. R. Perfect.** 2001. Extracellular phospholipase activity is a virulence factor for *Cryptococcus neoformans*. *Mol. Microbiol.* **39:**166–175.
31. **Croft, S. L., and J. Engel.** 2006. Miltefosine: discovery of the antileishmanial activity of phospholipid derivatives. *Trans. R. Soc. Trop. Med. Hyg.* **100**(Suppl 1)**:**S4–S8.
32. **Cruz, M. C., D. S. Fox, and J. Heitman.** 2001. Calcineurin is required for hyphal elongation during mating and haploid fruiting in *Cryptococcus neoformans*. *EMBO J.* **20:**1020–1032.
33. **de Almeida, M. L., and N. Heise.** 1993. Proteins anchored via glycosylphosphatidylinositol and solubilizing phospholipases in *Trypanosoma cruzi*. *Biol. Res.* **26:**285–312.
34. **Del Poeta, M., M. C. Cruz, M. E. Cardenas, J. R. Perfect, and J. Heitman.** 2000. Synergistic antifungal activities of bafilomycin A(1), fluconazole, and the pneumocandin MK-0991/caspofungin acetate (L-743,873) with calcineurin inhibitors FK506 and L-685,818 against *Cryptococcus neoformans*. *Antimicrob. Agents Chemother.* **44:**739–746.
35. **Demczuk, A., N. Guha, P. H. Nguyen, P. Desai, J. Chang, K. Guzinska, J. Rollins, C. C. Ghosh, L. Goodwin, and A. Vancura.** 2008. *Saccharomyces cerevisiae* phospholipase C regulates transcription of Msn2p-dependent stress-responsive genes. *Eukaryot. Cell* **7:**967–979.
36. **Dessen, A.** 2000. Phospholipase A(2) enzymes: structural diversity in lipid messenger metabolism. *Structure* **8:**R15–R22.
37. **Dijkgraaf, G. J., M. Abe, Y. Ohya, and H. Bussey.** 2002. Mutations in Fks1p affect the cell wall content of beta-1,3- and beta-1,6-glucan in *Saccharomyces cerevisiae*. *Yeast* **19:**671–690.
38. **Djordjevic, J. T., M. Del Poeta, T. C. Sorrell, K. M. Turner, and L. C. Wright.** 2005. Secretion of cryptococcal phospholipase B1 (PLB1) is regulated by a glycosylphosphatidylinositol (GPI) anchor. *Biochem. J.* **389:**803–812.
39. **Doering, T. L., and R. Schekman.** 1996. GPI anchor attachment is required for Gas1p transport from the endoplasmic reticulum in COP II vesicles. *EMBO J.* **15:**182–191.
40. **Douglas, C. M., J. A. D'Ippolito, G. J. Shei, M. Meinz, J. Onishi, J. A. Marrinan, W. Li, G. K. Abruzzo, A. Flattery, K. Bartizal, A. Mitchell, and M. B. Kurtz.** 1997. Identification of the FKS1 gene of *Candida albicans* as the essential target of 1,3-beta-D-glucan synthase inhibitors. *Antimicrob. Agents Chemother.* **41:**2471–2479.
41. **Douglas, C. M., F. Foor, J. A. Marrinan, N. Morin, J. B. Nielsen, A. M. Dahl, P. Mazur, W. Baginsky, W. Li, M. el-Sherbeini, et al.** 1994. The *Saccharomyces cerevisiae* FKS1 (ETG1) gene encodes an integral membrane protein which is a subunit of 1,3-beta-D-glucan synthase. *Proc. Natl. Acad. Sci. USA* **91:**12907–12911.
42. **Douglas, C. M., J. A. Marrinan, W. Li, and M. B. Kurtz.** 1994. A *Saccharomyces cerevisiae* mutant with echinocandin-resistant 1,3-beta-D-glucan synthase. *J. Bacteriol.* **176:**5686–5696.
43. **Dubois, M., K. A. Gilles, J. K. Hamilton, P. A. Rebers, and F. Smith.** 1956. Colorimetric method for determination of sugars and related substances. *Anal. Chem.* **28:**350–356.
44. **Eisenman, H. C., J. D. Nosanchuk, J. B. Webber, R. J. Emerson, T. A. Camesano, and A. Casadevall.** 2005. Microstructure of cell wall-associated melanin in the human pathogenic fungus *Cryptococcus neoformans*. *Biochemistry* **44:**3683–3693.
45. **el-Sherbeini, M., and J. A. Clemas.** 1995. Cloning and characterization of GNS1: a *Saccharomyces cerevisiae* gene involved in synthesis of 1,3-beta-glucan in vitro. *J. Bacteriol.* **177:**3227–3234.
46. **Fan, W., A. Idnurm, J. Breger, E. Mylonakis, and J. Heitman.** 2007. Eca1, a sarcoplasmic/endoplasmic reticulum Ca2+-ATPase, is involved in stress tolerance and virulence in *Cryptococcus neoformans*. *Infect. Immun.* **75:**3394–3405.
47. **Feldmesser, M., Y. Kress, A. Mednick, and A. Casadevall.** 2000. The effect of the echinocandin analogue caspofungin on cell wall glucan synthesis by *Cryptococcus neoformans*. *J. Infect. Dis.* **182:**1791–1795.
48. **Feldmesser, M., Y. Kress, P. Novikoff, and A. Casadevall.** 2000. *Cryptococcus neoformans* is a facultative intracellular pathogen in murine pulmonary infection. *Infect. Immun.* **68:**4225–4237.
49. **Flick, J. S., and J. Thorner.** 1998. An essential function of a phosphoinositide-specific phospholipase C is relieved by inhibition of a cyclin-dependent protein kinase in the yeast *Saccharomyces cerevisiae*. *Genetics* **148:**33–47.
50. **Flick, J. S., and J. Thorner.** 1993. Genetic and biochemical characterization of a phosphatidylinositol-specific phospholipase C in *Saccharomyces cerevisiae*. *Mol. Cell. Biol.* **13:**5861–5876.
51. **Fox, D. S., G. M. Cox, and J. Heitman.** 2003. Phospholipid-binding protein Cts1 controls septation and functions coordinately with calcineurin in *Cryptococcus neoformans*. *Eukaryot. Cell* **2:**1025–1035.
52. **Fox, D. S., M. C. Cruz, R. A. Sia, H. Ke, G. M. Cox, M. E. Cardenas, and J. Heitman.** 2001. Calcineurin regulatory subunit is essential for virulence and mediates interactions with FKBP12-FK506 in *Cryptococcus neoformans*. *Mol. Microbiol.* **39:**835–849.
53. **Fox, D. S., and J. Heitman.** 2005. Calcineurin-binding protein Cbp1 directs the specificity of calcineurin-dependent hyphal elongation during mating in *Cryptococcus neoformans*. *Eukaryot. Cell* **4:**1526–1538.
54. **Fox, D. S., and J. Heitman.** 2002. Good fungi gone bad: the corruption of calcineurin. *Bioessays* **24:**894–903.
55. **Frost, D., K. Brandt, C. Estill, and R. Goldman.** 1997. Partial purification of (1,3)-beta-glucan synthase from *Candida albicans*. *FEMS Microbiol. Lett.* **146:**255–261.
56. **Fujii, T., H. Shimoi, and Y. Iimura.** 1999. Structure of the glucan-binding sugar chain of Tip1p, a cell wall protein of *Saccharomyces cerevisiae*. *Biochim. Biophys. Acta* **1427:**133–144.
57. **Gacser, A., D. Trofa, W. Schafer, and J. D. Nosanchuk.** 2007. Targeted gene deletion in *Candida parapsilosis* demonstrates the role of secreted lipase in virulence. *J. Clin. Invest.* **117:**3049–3058.
58. **Ganendren, R., E. Carter, T. Sorrell, F. Widmer, and L. Wright.** 2006. Phospholipase B activity enhances adhesion of *Cryptococcus neoformans* to a human lung epithelial cell line. *Microbes Infect.* **8:**1006–1015.
59. **Ganendren, R., F. Widmer, V. Singhal, C. Wilson, T. Sorrell, and L. Wright.** 2004. In vitro antifungal activities of inhibitors of phospholipases from the fungal pathogen *Cryptococcus neoformans*. *Antimicrob. Agents Chemother.* **48:**1561–1569.
60. **Gerik, K. J., M. J. Donlin, C. E. Soto, A. M. Banks, I. R. Banks, M. A. Maligie, C. P. Selitrennikoff, and J. K. Lodge.** 2005. Cell wall integrity is dependent on the PKC1 signal transduction pathway in *Cryptococcus neoformans*. *Mol. Microbiol.* **58:**393–408.
61. **Ghannoum, M. A.** 2000. Potential role of phospholipases in virulence and fungal pathogenesis. *Clin. Microbiol. Rev.* **13:**122–143.

62. **Gorlach, J., D. S. Fox, N. S. Cutler, G. M. Cox, J. R. Perfect, and J. Heitman.** 2000. Identification and characterization of a highly conserved calcineurin binding protein, CBP1/calcipressin, in *Cryptococcus neoformans*. *EMBO J.* **19:**3618–3629.
63. **Groll, A. H., and T. J. Walsh.** 2001. Caspofungin: pharmacology, safety and therapeutic potential in superficial and invasive fungal infections. *Expert Opin. Investig. Drugs* **10:**1545–1558.
64. **Gualtieri, T., E. Ragni, L. Mizzi, U. Fascio, and L. Popolo.** 2004. The cell wall sensor Wsc1p is involved in reorganization of actin cytoskeleton in response to hypo-osmotic shock in *Saccharomyces cerevisiae*. *Yeast* **21:**1107–1120.
65. **Hartland, R. P., C. A. Vermeulen, F. M. Klis, J. H. Sietsma, and J. G. Wessels.** 1994. The linkage of (1-3)-beta-glucan to chitin during cell wall assembly in *Saccharomyces cerevisiae*. *Yeast* **10:**1591–1599.
66. **Heinz, D. W., L. O. Essen, and R. L. Williams.** 1998. Structural and mechanistic comparison of prokaryotic and eukaryotic phosphoinositide-specific phospholipases C. *J. Mol. Biol.* **275:**635–650.
67. **Herman, P. K., and S. D. Emr.** 1990. Characterization of VPS34, a gene required for vacuolar protein sorting and vacuole segregation in *Saccharomyces cerevisiae*. *Mol. Cell. Biol.* **10:**6742–6754.
68. **Heung, L. J., A. E. Kaiser, C. Luberto, and M. Del Poeta.** 2005. The role and mechanism of diacylglycerol-protein kinase C1 signaling in melanogenesis by *Cryptococcus neoformans*. *J. Biol. Chem.* **280:**28547–28555.
69. **Heung, L. J., C. Luberto, A. Plowden, Y. A. Hannun, and M. Del Poeta.** 2004. The sphingolipid pathway regulates Pkc1 through the formation of diacylglycerol in *Cryptococcus neoformans*. *J. Biol. Chem.* **279:**21144–21153.
70. **Himmelreich, U., C. Allen, S. Dowd, R. Malik, B. P. Shehan, C. Mountford, and T. C. Sorrell.** 2003. Identification of metabolites of importance in the pathogenesis of pulmonary cryptococcoma using nuclear magnetic resonance spectroscopy. *Microbes Infect.* **5:**285–290.
71. **Himmelreich, U., T. E. Dzendrowskyj, C. Allen, S. Dowd, R. Malik, B. P. Shehan, P. Russell, C. E. Mountford, and T. C. Sorrell.** 2001. Cryptococcomas distinguished from gliomas with MR spectroscopy: an experimental rat and cell culture study. *Radiology* **220:**122–128.
72. **Ho, Y., A. Gruhler, A. Heilbut, G. D. Bader, L. Moore, S. L. Adams, A. Millar, P. Taylor, K. Bennett, K. Boutilier, L. Yang, C. Wolting, I. Donaldson, S. Schandorff, J. Shewnarane, M. Vo, J. Taggart, M. Goudreault, B. Muskat, C. Alfarano, D. Dewar, Z. Lin, K. Michalickova, A. R. Willems, H. Sassi, P. A. Nielsen, K. J. Rasmussen, J. R. Andersen, L. E. Johansen, L. H. Hansen, H. Jespersen, A. Podtelejnikov, E. Nielsen, J. Crawford, V. Poulsen, B. D. Sorensen, J. Matthiesen, R. C. Hendrickson, F. Gleeson, T. Pawson, M. F. Moran, D. Durocher, M. Mann, C. W. Hogue, D. Figeys, and M. Tyers.** 2002. Systematic identification of protein complexes in *Saccharomyces cerevisiae* by mass spectrometry. *Nature* **415:** 180–183.
73. **Hoang, A.** 2001. Caspofungin acetate: an antifungal agent. *Am. J. Health Syst. Pharm.* **58:**1206–1214.
74. **Hu, G., M. Hacham, S. R. Waterman, J. Panepinto, S. Shin, X. Liu, J. Gibbons, T. Valyi-Nagy, K. Obara, H. A. Jaffe, Y. Ohsumi, and P. R. Williamson.** 2008. PI3K signaling of autophagy is required for starvation tolerance and virulence of *Cryptococcus neoformans*. *J. Clin. Invest.* **118:**1186–1197.
75. **Hu, G., B. R. Steen, T. Lian, A. P. Sham, N. Tam, K. L. Tangen, and J. W. Kronstad.** 2007. Transcriptional regulation by protein kinase A in *Cryptococcus neoformans*. *PLoS Pathog.* **3:**e42.
76. **Hube, B., F. Stehr, M. Bossenz, A. Mazur, M. Kretschmar, and W. Schafer.** 2000. Secreted lipases of *Candida albicans*: cloning, characterisation and expression analysis of a new gene family with at least ten members. *Arch. Microbiol.* **174:**362–374.
77. **Idnurm, A., Y. S. Bahn, K. Nielsen, X. Lin, J. A. Fraser, and J. Heitman.** 2005. Deciphering the model pathogenic fungus *Cryptococcus neoformans*. *Nat. Rev. Microbiol.* **3:**753–764.
78. **Idnurm, A., F. J. Walton, A. Floyd, J. L. Reedy, and J. Heitman.** 2009. Identification of ENA1 as a virulence gene of the human pathogenic fungus *Cryptococcus neoformans* through signature-tagged insertional mutagenesis. *Eukaryot. Cell* **8:**315–326.
79. **Ikeda, R., T. Sugita, E. S. Jacobson, and T. Shinoda.** 2003. Effects of melanin upon susceptibility of *Cryptococcus* to antifungals. *Microbiol. Immunol.* **47:**271–277.
80. **Inoue, S. B., N. Takewaki, T. Takasuka, T. Mio, M. Adachi, Y. Fujii, C. Miyamoto, M. Arisawa, Y. Furuichi, and T. Watanabe.** 1995. Characterization and gene cloning of 1,3-beta-D-glucan synthase from *Saccharomyces cerevisiae*. *Eur. J. Biochem.* **231:**845–854.
81. **James, P. G., R. Cherniak, R. G. Jones, C. A. Stortz, and E. Reiss.** 1990. Cell-wall glucans of *Cryptococcus neoformans* Cap 67. *Carbohydr. Res.* **198:**23–38.
82. **Jones, P. M., K. M. Turner, J. T. Djordjevic, T. C. Sorrell, L. C. Wright, and A. M. George.** 2007. Role of conserved active site residues in catalysis by phospholipase B1 from *Cryptococcus neoformans*. *Biochemistry* **46:**10024–10032.
83. **Kahn, J. N., M. J. Hsu, F. Racine, R. Giacobbe, and M. Motyl.** 2006. Caspofungin susceptibility in aspergillus and non-aspergillus molds: inhibition of glucan synthase and reduction of beta-D-1,3 glucan levels in culture. *Antimicrob. Agents Chemother.* **50:**2214–2216.
84. **Kametaka, S., T. Okano, M. Ohsumi, and Y. Ohsumi.** 1998. Apg14p and Apg6/Vps30p form a protein complex essential for autophagy in the yeast, *Saccharomyces cerevisiae*. *J. Biol. Chem.* **273:**22284–22291.
85. **Kapteyn, J. C., R. C. Montijn, G. J. Dijkgraaf, and F. M. Klis.** 1994. Identification of beta-1,6-glucosylated cell wall proteins in yeast and hyphal forms of *Candida albicans*. *Eur. J. Cell. Biol.* **65:**402–407.
86. **Katan, M.** 1998. Families of phosphoinositide-specific phospholipase C: structure and function. *Biochim. Biophys. Acta.* **1436:**5–17.
86. **Kelly, R., E. Register, M. J. Hsu, M. Kurtz, and J. Nielsen.** 1996. Isolation of a gene involved in 1,3-beta-glucan synthesis in *Aspergillus nidulans* and purification of the corresponding protein. *J. Bacteriol.* **178:**4381–4391.
88. **Kesson, A. M., M. C. Bellemore, T. J. O'Mara, D. H. Ellis, and T. C. Sorrell.** 2009. *Scedosporium prolificans* osteomyelitis in an immunocompetent child treated with a novel agent, hexadecylphospocholine (miltefosine), in combination with terbinafine and voriconazole: a case report. *Clin. Infect. Dis.* **48:**1257–1261.
89. **Kim, H. N., and H. S. Chung.** 2008. Caveolin-1 inhibits membrane-type 1 matrix metalloproteinase activity. *BMB Rep.* **41:**858–862.
90. **Kitagaki, H., H. Wu, H. Shimoi, and K. Ito.** 2002. Two homologous genes, DCW1 (YKL046c) and DFG5, are essential for cell growth and encode glycosylphosphatidylinositol (GPI)-anchored membrane proteins required for cell wall biogenesis in *Saccharomyces cerevisiae*. *Mol. Microbiol.* **46:**1011–1022.
91. **Klis, F., P. Mol, K. Hellingwerf, and S. Brul.** 2002. Dynamics of cell wall structure in *Saccharomyces cerevisiae*. *FEMS Microbiol. Rev.* **26:**239.
92. **Klis, F. M., P. de Groot, and K. Hellingwerf.** 2001. Molecular organization of the cell wall of *Candida albicans*. *Med. Mycol.* **39:**1–8.

93. **Kollar, R., E. Petrakova, G. Ashwell, P. W. Robbins, and E. Cabib.** 1995. Architecture of the yeast cell wall. The linkage between chitin and beta(1–>3)-glucan. *J. Biol. Chem.* **270:**1170–1178.
94. **Kollar, R., B. B. Reinhold, E. Petrakova, H. J. Yeh, G. Ashwell, J. Drgonova, J. C. Kapteyn, F. M. Klis, and E. Cabib.** 1997. Architecture of the yeast cell wall. Beta(1–>6)-glucan interconnects mannoprotein, beta(1–>)3-glucan, and chitin. *J. Biol. Chem.* **272:**17762–17775.
95. **Kozubowski, L., S. C. Lee, and J. Heitman.** 2009. Signalling pathways in the pathogenesis of *Cryptococcus*. *Cell. Microbiol.* **11:**370–380.
96. **Kraus, P. R., D. S. Fox, G. M. Cox, and J. Heitman.** 2003. The *Cryptococcus neoformans* MAP kinase Mpk1 regulates cell integrity in response to antifungal drugs and loss of calcineurin function. *Mol. Microbiol.* **48:** 1377–1387.
97. **Kristiansen, S., and E. A. Richter.** 2002. GLUT4-containing vesicles are released from membranes by phospholipase D cleavage of a GPI anchor. *Am. J. Physiol. Endocrinol. Metab.* **283:**E374–E382.
98. **Kunze, D., I. Melzer, D. Bennett, D. Sanglard, D. MacCallum, J. Norskau, D. C. Coleman, F. C. Odds, W. Schafer, and B. Hube.** 2005. Functional analysis of the phospholipase C gene CaPLC1 and two unusual phospholipase C genes, CaPLC2 and CaPLC3, of *Candida albicans*. *Microbiology* **151:**3381–3394.
99. **Leidich, S. D., A. S. Ibrahim, Y. Fu, A. Koul, C. Jessup, J. Vitullo, W. Fonzi, F. Mirbod, S. Nakashima, Y. Nozawa, and M. A. Ghannoum.** 1998. Cloning and disruption of caPLB1, a phospholipase B gene involved in the pathogenicity of *Candida albicans*. *J. Biol. Chem.* **273:**26078–26086.
100. **Lesage, G., and H. Bussey.** 2006. Cell wall assembly in *Saccharomyces cerevisiae*. *Microbiol. Mol. Biol. Rev.* **70:**317–343.
101. **Lesage, G., A. M. Sdicu, P. Menard, J. Shapiro, S. Hussein, and H. Bussey.** 2004. Analysis of beta-1,3-glucan assembly in *Saccharomyces cerevisiae* using a synthetic interaction network and altered sensitivity to caspofungin. *Genetics* **167:**35–49.
102. **Levitz, S. M., and C. A. Specht.** 2006. The molecular basis for the immunogenicity of *Cryptococcus neoformans* mannoproteins. *FEMS Yeast Res.* **6:**513–524.
103. **Li, H., N. Page, and H. Bussey.** 2002. Actin patch assembly proteins Las17p and Sla1p restrict cell wall growth to daughter cells and interact with cis-Golgi protein Kre6p. *Yeast* **19:**1097–1112.
104. **Liu, J., J. D. Farmer, W. S. Lane, J. Friedman, I. Weissman, and S. L. Schreiber.** 1991. Calcineurin is a common target of cyclophilin-cyclosporin A and FKBP-FK506 complexes. *Cell* **66:**807–815.
105. **Liu, M., P. Du, G. Heinrich, G. M. Cox, and A. Gelli.** 2006. Cch1 mediates calcium entry in *Cryptococcus neoformans* and is essential in low-calcium environments. *Eukaryot. Cell* **5:**1788–1796.
106. **Loftus, B. J., E. Fung, P. Roncaglia, D. Rowley, P. Amedeo, D. Bruno, J. Vamathevan, M. Miranda, I. J. Anderson, J. A. Fraser, J. E. Allen, I. E. Bosdet, M. R. Brent, R. Chiu, T. L. Doering, M. J. Donlin, C. A. D'Souza, D. S. Fox, V. Grinberg, J. Fu, M. Fukushima, B. J. Haas, J. C. Huang, G. Janbon, S. J. Jones, H. L. Koo, M. I. Krzywinski, J. K. Kwon-Chung, K. B. Lengeler, R. Maiti, M. A. Marra, R. E. Marra, C. A. Mathewson, T. G. Mitchell, M. Pertea, F. R. Riggs, S. L. Salzberg, J. E. Schein, A. Shvartsbeyn, H. Shin, M. Shumway, C. A. Specht, B. B. Suh, A. Tenney, T. R. Utterback, B. L. Wickes, J. R. Wortman, N. H. Wye, J. W. Kronstad, J. K. Lodge, J. Heitman, R. W. Davis, C. M. Fraser, and R. W. Hyman.** 2005. The genome of the basidiomycetous yeast and human pathogen *Cryptococcus neoformans*. *Science* **307:**1321–1324.
107. **Lopez-Ribot, J. L., D. A. Cortlandt, D. C. Straus, K. J. Morrow, and W. L. Chaffin.** 1995. Complex interaction between different proteinaceous components within the cell-wall structure of *Candida albicans*. *Mycopathologia* **132:**87–93.
108. **Ma, H., J. E. Croudace, D. A. Lammas, and R. C. May.** 2006. Expulsion of live pathogenic yeast by macrophages. *Curr. Biol.* **16:**2156–2160.
109. **Maligie, M. A., and C. P. Selitrennikoff.** 2005. *Cryptococcus neoformans* resistance to echinocandins: (1,3)beta-glucan synthase activity is sensitive to echinocandins. *Antimicrob. Agents Chemother.* **49:**2851–2856.
110. **Markovich, S., A. Yekutiel, I. Shalit, Y. Shadkchan, and N. Osherov.** 2004. Genomic approach to identification of mutations affecting caspofungin susceptibility in *Saccharomyces cerevisiae*. *Antimicrob. Agents Chemother.* **48:**3871–876.
111. **Mazur, P., and W. Baginsky.** 1996. In vitro activity of 1,3-beta-D-glucan synthase requires the GTP-binding protein Rho1. *J. Biol. Chem.* **271:**14604–14609.
112. **Mazur, P., N. Morin, W. Baginsky, M. el-Sherbeini, J. A. Clemas, J. B. Nielsen, and F. Foor.** 1995. Differential expression and function of two homologous subunits of yeast 1,3-beta-D-glucan synthase. *Mol. Cell. Biol.* **15:**5671–5681.
113. **Mio, T., T. Yamada-Okabe, T. Yabe, T. Nakajima, M. Arisawa, and H. Yamada-Okabe.** 1997. Isolation of the *Candida albicans* homologs of *Saccharomyces cerevisiae* KRE6 and SKN1: expression and physiological function. *J. Bacteriol.* **179:**2363–2372.
114. **Montijn, R. C., E. Vink, W. H. Muller, A. J. Verkleij, H. Van Den Ende, B. Henrissat, and F. M. Klis.** 1999. Localization of synthesis of beta1,6-glucan in *Saccharomyces cerevisiae*. *J. Bacteriol.* **181:**7414–7420.
115. **Mukherjee, P. K., K. R. Seshan, S. D. Leidich, J. Chandra, G. T. Cole, and M. A. Ghannoum.** 2001. Reintroduction of the PLB1 gene into *Candida albicans* restores virulence in vivo. *Microbiology* **147:**2585–2597.
116. **Muller, G., E. Gross, S. Wied, and W. Bandlow.** 1996. Glucose-induced sequential processing of a glycosylphosphatidylinositol-anchored ectoprotein in *Saccharomyces cerevisiae*. *Mol. Cell. Biol.* **16:**442–456.
117. **Muniz, M., and H. Riezman.** 2000. Intracellular transport of GPI-anchored proteins. *EMBO J.* **19:**10–15.
118. **Nakashima, S., Y. Banno, T. Watanabe, Y. Nakamura, T. Mizutani, H. Sakai, Y. Zhao, Y. Sugimoto, and Y. Nozawa.** 1995. Deletion and site-directed mutagenesis of EF-hand domain of phospholipase C-delta 1: effects on its activity. *Biochem. Biophys. Res. Commun.* **211:**365–369.
119. **Ng, C. K., D. Obando, F. Widmer, L. C. Wright, T. C. Sorrell, and K. A. Jolliffe.** 2006. Correlation of antifungal activity with fungal phospholipase inhibition using a series of bisquaternary ammonium salts. *J. Med. Chem.* **49:**811–816.
120. **Nishizuka, Y.** 1992. Intracellular signaling by hydrolysis of phospholipids and activation of protein kinase C. *Science* **258:**607–614.
121. **Noh, D. Y., S. H. Shin, and S. G. Rhee.** 1995. Phosphoinositide-specific phospholipase C and mitogenic signaling. *Biochim. Biophys. Acta* **1242:**99–113.
122. **Noverr, M. C., G. M. Cox, J. R. Perfect, and G. B. Huffnagle.** 2003. Role of PLB1 in pulmonary inflammation and cryptococcal eicosanoid production. *Infect. Immun.* **71:**1538–1547.
123. **Noverr, M. C., G. B. Toews, and G. B. Huffnagle.** 2002. Production of prostaglandins and leukotrienes by pathogenic fungi. *Infect. Immun.* **70:**400–402.

124. **Ochocka, A. M., and T. Pawelczyk.** 2003. Isozymes delta of phosphoinositide-specific phospholipase C and their role in signal transduction in the cell. *Acta. Biochim. Pol.* **50:**1097–1110.
125. **Odom, A., S. Muir, E. Lim, D. L. Toffaletti, J. Perfect, and J. Heitman.** 1997. Calcineurin is required for virulence of *Cryptococcus neoformans*. *EMBO J.* **16:**2576–2589.
126. **Osherov, N., G. S. May, N. D. Albert, and D. P. Kontoyiannis.** 2002. Overexpression of Sbe2p, a Golgi protein, results in resistance to caspofungin in *Saccharomyces cerevisiae*. *Antimicrob. Agents Chemother.* **46:**2462–2469.
127. **Paderu, P., S. Park, and D. S. Perlin.** 2004. Caspofungin uptake is mediated by a high-affinity transporter in *Candida albicans*. *Antimicrob. Agents Chemother.* **48:**3845–3849.
128. **Panepinto, J., K. Komperda, S. Frases, Y. D. Park, J. T. Djordjevic, A. Casadevall, and P. R. Williamson.** 2009. Sec6-dependent sorting of fungal extracellular exosomes and laccase of *Cryptococcus neoformans*. *Mol. Microbiol.* **71:**1165–1176.
129. **Parkin, E. T., N. T. Watt, A. J. Turner, and N. M. Hooper.** 2004. Dual mechanisms for shedding of the cellular prion protein. *J. Biol. Chem.* **279:**11170–11178.
130. **Paterson, H. F., J. W. Savopoulos, O. Perisic, R. Cheung, M. V. Ellis, R. L. Williams, and M. Katan.** 1995. Phospholipase C delta 1 requires a pleckstrin homology domain for interaction with the plasma membrane. *Biochem. J.* **312**(Pt 3)**:**661–666.
131. **Pickard, R. T., X. G. Chiou, B. A. Strifler, M. R. DeFelippis, P. A. Hyslop, A. L. Tebbe, Y. K. Yee, L. J. Reynolds, E. A. Dennis, R. M. Kramer, and J. D. Sharp.** 1996. Identification of essential residues for the catalytic function of 85-kDa cytosolic phospholipase A2. Probing the role of histidine, aspartic acid, cysteine, and arginine. *J. Biol. Chem.* **271:**19225–19231.
132. **Ponting, C. P., K. Hofmann, and P. Bork.** 1999. A latrophilin/CL-1-like GPS domain in polycystin-1. *Curr. Biol.* **9:**R585–R588.
133. **Prakobphol, A., H. Leffler, C. I. Hoover, and S. J. Fisher.** 1997. Palmitoyl carnitine, a lysophospholipase-transacylase inhibitor, prevents *Candida* adherence in vitro. *FEMS Microbiol. Lett.* **151:**89–94.
134. **Pukkila-Worley, R., Q. D. Gerrald, P. R. Kraus, M. J. Boily, M. J. Davis, S. S. Giles, G. M. Cox, J. Heitman, and J. A. Alspaugh.** 2005. Transcriptional network of multiple capsule and melanin genes governed by the *Cryptococcus neoformans* cyclic AMP cascade. *Eukaryot. Cell* **4:**190–201.
135. **Radding, J. A., S. A. Heidler, and W. W. Turner.** 1998. Photoaffinity analog of the semisynthetic echinocandin LY303366: identification of echinocandin targets in *Candida albicans*. *Antimicrob. Agents Chemother.* **42:**1187–1194.
136. **Reese, A. J., and T. L. Doering.** 2003. Cell wall alpha-1,3-glucan is required to anchor the *Cryptococcus neoformans* capsule. *Mol. Microbiol.* **50:**1401–1409.
137. **Rhee, S. G.** 2001. Regulation of phosphoinositide-specific phospholipase C. *Annu. Rev. Biochem.* **70:**281–312.
138. **Rodrigues, M. L., E. S. Nakayasu, D. L. Oliveira, L. Nimrichter, J. D. Nosanchuk, I. C. Almeida, and A. Casadevall.** 2008. Extracellular vesicles produced by *Cryptococcus neoformans* contain protein components associated with virulence. *Eukaryot. Cell* **7:**58–67.
139. **Rodrigues, M. L., L. Nimrichter, D. L. Oliveira, S. Frases, K. Miranda, O. Zaragoza, M. Alvarez, A. Nakouzi, M. Feldmesser, and A. Casadevall.** 2007. Vesicular polysaccharide export in *Cryptococcus neoformans* is a eukaryotic solution to the problem of fungal trans-cell wall transport. *Eukaryot. Cell* **6:**48–59.
140. **Roemer, T., and H. Bussey.** 1991. Yeast beta-glucan synthesis: KRE6 encodes a predicted type II membrane protein required for glucan synthesis in vivo and for glucan synthase activity in vitro. *Proc. Natl. Acad. Sci. USA* **88:**11295–11299.
141. **Roemer, T., S. Delaney, and H. Bussey.** 1993. SKN1 and KRE6 define a pair of functional homologs encoding putative membrane proteins involved in beta-glucan synthesis. *Mol. Cell. Biol.* **13:**4039–4048.
142. **Roemer, T., G. Paravicini, M. A. Payton, and H. Bussey.** 1994. Characterization of the yeast (1–>6)-beta-glucan biosynthetic components, Kre6p and Skn1p, and genetic interactions between the PKC1 pathway and extracellular matrix assembly. *J Cell. Biol.* **127:**567–579.
143. **Roh, D. H., B. Bowers, H. Riezman, and E. Cabib.** 2002. Rho1p mutations specific for regulation of beta(1–>3)glucan synthesis and the order of assembly of the yeast cell wall. *Mol. Microbiol.* **44:**1167–1183.
144. **Roman, E., D. M. Arana, C. Nombela, R. Alonso-Monge, and J. Pla.** 2007. MAP kinase pathways as regulators of fungal virulence. *Trends Microbiol.* **15:**181–190.
145. **Sankaran, K., and H. B. Herscowitz.** 1995. Phenotypic and functional heterogeneity of the murine alveolar macrophage-derived cell line MH-S. *J. Leukoc. Biol.* **57:**562–568.
146. **Santangelo, R., H. Zoellner, T. Sorrell, C. Wilson, C. Donald, J. Djordjevic, Y. Shounan, and L. Wright.** 2004. Role of extracellular phospholipases and mononuclear phagocytes in dissemination of cryptococcosis in a murine model. *Infect. Immun.* **72:**2229–2239.
147. **Santangelo, R. T., S. C. Chen, T. C. Sorrell, and L. C. Wright.** 2005. Detection of antibodies to phospholipase B in patients infected with *Cryptococcus neoformans* by enzyme-linked immunosorbent assay (ELISA). *Med. Mycol.* **43:**335–341.
148. **Santangelo, R. T., M. H. Nouri-Sorkhabi, T. C. Sorrell, M. Cagney, S. C. Chen, P. W. Kuchel, and L. C. Wright.** 1999. Biochemical and functional characterisation of secreted phospholipase activities from *Cryptococcus neoformans* in their naturally occurring state. *J. Med. Microbiol.* **48:**731–740.
149. **Schimoler-O'Rourke, R., S. Renault, W. Mo, and C. P. Selitrennikoff.** 2003. *Neurospora crassa* FKS protein binds to the (1,3)beta-glucan synthase substrate, UDP-glucose. *Curr. Microbiol.* **46:**408–412.
150. **Sekiya-Kawasaki, M., M. Abe, A. Saka, D. Watanabe, K. Kono, M. Minemura-Asakawa, S. Ishihara, T. Watanabe, and Y. Ohya.** 2002. Dissection of upstream regulatory components of the Rho1p effector, 1,3-beta-glucan synthase, in *Saccharomyces cerevisiae*. *Genetics* **162:**663–676.
151. **Shahinian, S., and H. Bussey.** 2000. beta-1,6-Glucan synthesis in *Saccharomyces cerevisiae*. *Mol. Microbiol.* **35:**477–489.
152. **Sharp, J. D., D. L. White, X. G. Chiou, T. Goodson, G. C. Gamboa, D. McClure, S. Burgett, J. Hoskins, P. L. Skatrud, J. R. Sportsman, et al.** 1991. Molecular cloning and expression of human Ca(2+)-sensitive cytosolic phospholipase A2. *J. Biol. Chem.* **266:**14850–14853.
153. **Shen, D. K., A. D. Noodeh, A. Kazemi, R. Grillot, G. Robson, and J. F. Brugere.** 2004. Characterisation and expression of phospholipases B from the opportunistic fungus *Aspergillus fumigatus*. *FEMS Microbiol. Lett.* **239:**87–93.
154. **Siafakas, A. R., T. C. Sorrell, L. C. Wright, C. Wilson, M. Larsen, R. Boadle, P. R. Williamson, and J. T. Djordjevic.** 2007. Cell wall-linked cryptococcal phospholipase B1 is a source of secreted enzyme and a determinant of cell wall integrity. *J. Biol. Chem.* **282:**37508–37514.
155. **Siafakas, A. R., L. C. Wright, T. C. Sorrell, and J. T. Djordjevic.** 2006. Lipid rafts in *Cryptococcus neoformans* concentrate the virulence determinants phospholipase B1 and Cu/Zn superoxide dismutase. *Eukaryot. Cell* **5:**488–498.

156. **Singer, W. D., H. A. Brown, and P. C. Sternweis.** 1997. Regulation of eukaryotic phosphatidylinositol-specific phospholipase C and phospholipase D. *Annu. Rev. Biochem.* **66:**475–509.
157. **Slessareva, J. E., S. M. Routt, B. Temple, V. A. Bankaitis, and H. G. Dohlman.** 2006. Activation of the phosphatidylinositol 3-kinase Vps34 by a G protein alpha subunit at the endosome. *Cell* **126:**191–203.
158. **Smits, G. J., J. C. Kapteyn, H. van den Ende, and F. M. Klis.** 1999. Cell wall dynamics in yeast. *Curr. Opin. Microbiol.* **2:**348–352.
159. **Sorrell, T., and S. C. Chen.** 2009. Fungal-derived immune modulating molecules. *Adv. Exp. Med. Biol.* **666:**108–120.
160. **Stack, J. H., P. K. Herman, P. V. Schu, and S. D. Emr.** 1993. A membrane-associated complex containing the Vps15 protein kinase and the Vps34 PI 3-kinase is essential for protein sorting to the yeast lysosome-like vacuole. *EMBO J.* **12:**2195–2204.
161. **Steinbach, W. J., R. A. Cramer, B. Z. Perfect, C. Henn, K. Nielsen, J. Heitman, and J. R. Perfect.** 2007. Calcineurin inhibition or mutation enhances cell wall inhibitors against *Aspergillus fumigatus*. *Antimicrob. Agents Chemother.* **5:**418–430.
162. **Steinbach, W. J., J. L. Reedy, R. A. Cramer, Jr., J. R. Perfect, and J. Heitman.** 2007. Harnessing calcineurin as a novel anti-infective agent against invasive fungal infections. *Nat. Rev. Microbiol.* **5:**418–430.
163. **Steinbach, W. J., W. A. Schell, J. R. Blankenship, C. Onyewu, J. Heitman, and J. R. Perfect.** 2004. In vitro interactions between antifungals and immunosuppressants against *Aspergillus fumigatus*. *Antimicrob. Agents Chemother.* **48:**1664–1669.
164. **Stevens, D. A., M. Espiritu, and R. Parmar.** 2004. Paradoxical effect of caspofungin: reduced activity against *Candida albicans* at high drug concentrations. *Antimicrob. Agents Chemother.* **48:**3407–3411.
165. **Stie, J., and D. Fox.** 2008. Calcineurin regulation in fungi and beyond. *Eukaryot. Cell* **7:**177–186.
166. **Stulnig, T. M., M. Berger, T. Sigmund, H. Stockinger, V. Horejsi, and W. Waldhausl.** 1997. Signal transduction via glycosyl phosphatidylinositol-anchored proteins in T cells is inhibited by lowering cellular cholesterol. *J. Biol. Chem.* **272:**19242–19247.
167. **Sutterlin, C., T. L. Doering, F. Schimmoller, S. Schroder, and H. Riezman.** 1997. Specific requirements for the ER to Golgi transport of GPI-anchored proteins in yeast. *J. Cell. Sci.* **110**(Pt 21)**:**2703–2714.
169. **Theiss, S., G. Ishdorj, A. Brenot, M. Kretschmar, C. Y. Lan, T. Nichterlein, J. Hacker, S. Nigam, N. Agabian, and G. A. Kohler.** 2006. Inactivation of the phospholipase B gene PLB5 in wild-type *Candida albicans* reduces cell-associated phospholipase A2 activity and attenuates virulence. *Int. J. Med. Microbiol.* **296:**405–420.
169. **Thompson, J. R., C. M. Douglas, W. Li, C. K. Jue, B. Pramanik, X. Yuan, T. H. Rude, D. L. Toffaletti, J. R. Perfect, and M. Kurtz.** 1999. A glucan synthase FKS1 homolog in *Cryptococcus neoformans* is single copy and encodes an essential function. *J. Bacteriol.* **181:**444–453.
170. **Tisi, R., S. Baldassa, F. Belotti, and E. Martegani.** 2002. Phospholipase C is required for glucose-induced calcium influx in budding yeast. *FEBS Lett.* **520:**133–138.
171. **Tucker, S. C., and A. Casadevall.** 2002. Replication of *Cryptococcus neoformans* in macrophages is accompanied by phagosomal permeabilization and accumulation of vesicles containing polysaccharide in the cytoplasm. *Proc. Natl. Acad. Sci. USA* **99:**3165–3170.
172. **Turner, K. M.** 2006. Identification of residues critical for the function of the fungal virulence factor phospholipase B1 from *Cryptococcus neoformans*. Doctoral thesis.
173. **Turner, K. M., L. C. Wright, T. C. Sorrell, and J. T. Djordjevic.** 2006. N-linked glycosylation sites affect secretion of cryptococcal phospholipase B1, irrespective of glycosylphosphatidylinositol anchoring. *Biochim. Biophys. Acta* **1760:**1569–1579.
174. **van den Bosch, J. S., A. G. Smals, I. M. Valk, and P. W. Kloppenborg.** 1982. Lack of difference in growth stimulating effect between weekly single and multiple human chorionic gonadotrophin administration in boys with delayed puberty. *Clin. Endocrinol.* **16:**1–9.
175. **van Duin, D., A. Casadevall, and J. D. Nosanchuk.** 2002. Melanization of *Cryptococcus neoformans* and *Histoplasma capsulatum* reduces their susceptibilities to amphotericin B and caspofungin. *Antimicrob. Agents Chemother.* **46:**3394–3400.
176. **Verna, J., A. Lodder, K. Lee, A. Vagts, and R. Ballester.** 1997. A family of genes required for maintenance of cell wall integrity and for the stress response in *Saccharomyces cerevisiae*. *Proc. Natl. Acad. Sci. USA* **94:**13804–13809.
177. **Vidotto, V., A. Sinicco, D. Di Fraia, S. Cardaropoli, S. Aoki, and S. Ito-Kuwa.** 1996. Phospholipase activity in *Cryptococcus neoformans*. *Mycopathologia* **136:**119–123.
178. **Wang, Y., P. Aisen, and A. Casadevall.** 1995. *Cryptococcus neoformans* melanin and virulence: mechanism of action. *Infect. Immun.* **63:**3131–3136.
179. **Widmer, F., L. C. Wright, D. Obando, R. Handke, R. Ganendren, D. H. Ellis, and T. C. Sorrell.** 2006. Hexadecylphosphocholine (miltefosine) has broad-spectrum fungicidal activity and is efficacious in a mouse model of cryptococcosis. *Antimicrob. Agents Chemother.* **50:**414–421.
180. **Wiederhold, N. P., D. P. Kontoyiannis, R. A. Prince, and R. E. Lewis.** 2005. Attenuation of the activity of caspofungin at high concentrations against candida albicans: possible role of cell wall integrity and calcineurin pathways. *Antimicrob. Agents Chemother.* **49:**5146–5148.
181. **Wright, L. C., J. Payne, R. T. Santangelo, M. F. Simpanya, S. C. Chen, F. Widmer, and T. C. Sorrell.** 2004. Cryptococcal phospholipases: a novel lysophospholipase discovered in the pathogenic fungus *Cryptococcus gattii*. *Biochem. J.* **384:**377–384.
182. **Wright, L. C., R. M. Santangelo, R. Ganendren, J. Payne, J. T. Djordjevic, and T. C. Sorrell.** 2007. Cryptococcal lipid metabolism: phospholipase B1 is implicated in transcellular metabolism of macrophage-derived lipids. *Eukaryot. Cell* **6:**37–47.
183. **Yoko-o, T., Y. Matsui, H. Yagisawa, H. Nojima, I. Uno, and A. Toh-e.** 1993. The putative phosphoinositide-specific phospholipase C gene, PLC1, of the yeast *Saccharomyces cerevisiae* is important for cell growth. *Proc. Natl. Acad. Sci. USA* **90:**1804–1808.
184. **Yoneda, A., and T. L. Doering.** 2006. A eukaryotic capsular polysaccharide is synthesized intracellularly and secreted via exocytosis. *Mol. Biol. Cell.* **17:**5131–5140.
185. **York, J. D., S. Guo, A. R. Odom, B. D. Spiegelberg, and L. E. Stolz.** 2001. An expanded view of inositol signaling. *Adv. Enzyme Regul.* **41:**57–71.
186. **Zhao, C., U. S. Jung, P. Garrett-Engele, T. Roe, M. S. Cyert, and D. E. Levin.** 1998. Temperature-induced expression of yeast FKS2 is under the dual control of protein kinase C and calcineurin. *Mol. Biol. Cell* **18:**1013–1022.

ENVIRONMENTAL INTERACTIONS AND POPULATION GENETICS

Cryptococcus: From Human Pathogen to Model Yeast
Edited by J. Heitman et al.

18

Environmental Niches for *Cryptococcus neoformans* and *Cryptococcus gattii*

THOMAS G. MITCHELL, ELIZABETH CASTAÑEDA, KIRSTEN NIELSEN, BODO WANKE, AND MARCIA S. LAZÉRA

This chapter addresses the most fundamental and crucial problem regarding cryptococcosis: how do people acquire this infection? Before we review this issue, it will be helpful to frame the context and justify formulating, investigating, and resolving this question. The overwhelming majority of cases of cryptococcosis are caused by either of two sibling species, *Cryptococcus neoformans* or *Cryptococcus gattii*, which were discovered over 100 and 40 years ago, respectively (21, 22, 49, 84, 216, 260). During this span of time, the natural history of cryptococcosis has been elucidated (139, 142). Although these yeasts have evolved the ability to infect mammals, they are accidental pathogens.

Species of *Cryptococcus* reside in the environment, and during their saprobic growth, the yeast cells (blastoconidia) or possibly basidiospores may become airborne, inhaled by humans or other vertebrates, and establish a pulmonary infection, which will elicit innate and immune responses in the host (Fig. 1). This process may lead to (i) resolution, (ii) sequestration of the yeast cells in a latent or dormant state with the potential for reactivation in the future, or (iii) active disease with the propensity to disseminate to the central nervous system, as well as other organs. One or more of these outcomes may ensue. They are not mutually exclusive and may occur sequentially or simultaneously. The progression to cryptococcal disease may be subacute or chronic with single- or multiorgan involvement. Manifold variations of these events have been described, and the cumulative evidence indicates that nearly every systemic cryptococcal infection terminates with the individual host (170).

The spectrum of vertebrate hosts is expansive. Most human cases of cryptococcosis occur in people who are immunocompromised, but immunocompetency does not ensure protection from disease. The disease occurs more often in men, and in most locales, pediatric cases are unaccountably rare (1, 185, 244). Veterinary reports have confirmed that both species of *Cryptococcus* may infect other mammals, including domestic pets (dogs, cats), livestock (alpacas), and wild animals (koalas, squirrels) (57, 124, 157, 184, 245), as well as birds (159, 199) and aquatic mammals (porpoises) (243). No vertebrate animal host has emerged as a significant, much less requisite, component of the cryptococcal life cycle, and the rare transmission of cryptococcosis among humans or animals is an anomaly (129, 267). Therefore, because infection with *Cryptococcus* is invariably acquired by exposure to yeasts (or perhaps basidiospores) from an exogenous source, the environmental reservoirs of *C. neoformans* and *C. gattii* command paramount interest.

Comparisons with the habitats of related fungi have not been instructive. The genus *Cryptococcus* encompasses close to 100 species and varieties, but very few of these taxa are capable of causing infection. *Cryptococcus* species are unified by the lack of fermentation, the elaboration of a polysaccharide capsule, and the secretion of urease. Overall, species of this diverse genus occupy an enormous range of habitats, including seawater (125, 176, 250), arctic climates (19, 23, 53), pH extremes (79, 212), and thermal springs (251). *Cryptococcus* is an anamorphic genus, and many species have teleomorphic or sexual states represented by the genus *Filobasidiella*. Species of both genera are exogenous and are polyphyletic members of the class Tremellomycetes of the phylum Basidiomycota (69, 73, 98, 221). Similar to their tremellomycetous relatives, numerous *Cryptococcus* species are associated with lignaceous environments, such as vegetation, trees, flowering plants, humus, bark, and rotting wood (214). Others are found in soil, excreta, and insects (11, 63). They grow vegetatively as budding yeast cells, but as noted, many species, including *C. neoformans* and *C. gattii*, have a teleomorphic state and are capable of sexual reproduction, which leads to the production of small, potentially infectious basidiospores (127). Consequently, an environmental niche that is conducive for mating may be crucial for infectivity.

In addition to *C. neoformans* and *C. gattii*, rare infections with several other species of *Cryptococcus* have been reported, including *C. albidus*, *C. laurentii*, *C. adeliensis*, *C. humicolus*,

Thomas G. Mitchell, Department of Molecular Genetics and Microbiology, Duke University Medical Center, Durham, NC 27710. **Elizabeth Castañeda,** Emerita Investigator, Instituto Nacional de Salud, Bogotá, Colombia. **Kirsten Nielsen,** Department of Microbiology, Medical School, University of Minnesota, Minneapolis, MN 55455. **Bodo Wanke and Marcia S. Lazéra,** Mycology Laboratory, Instituto de Pesquisa Clínica Evandro Chagas-Fiocruz, Rio de Janeiro, RJ, Brazil.

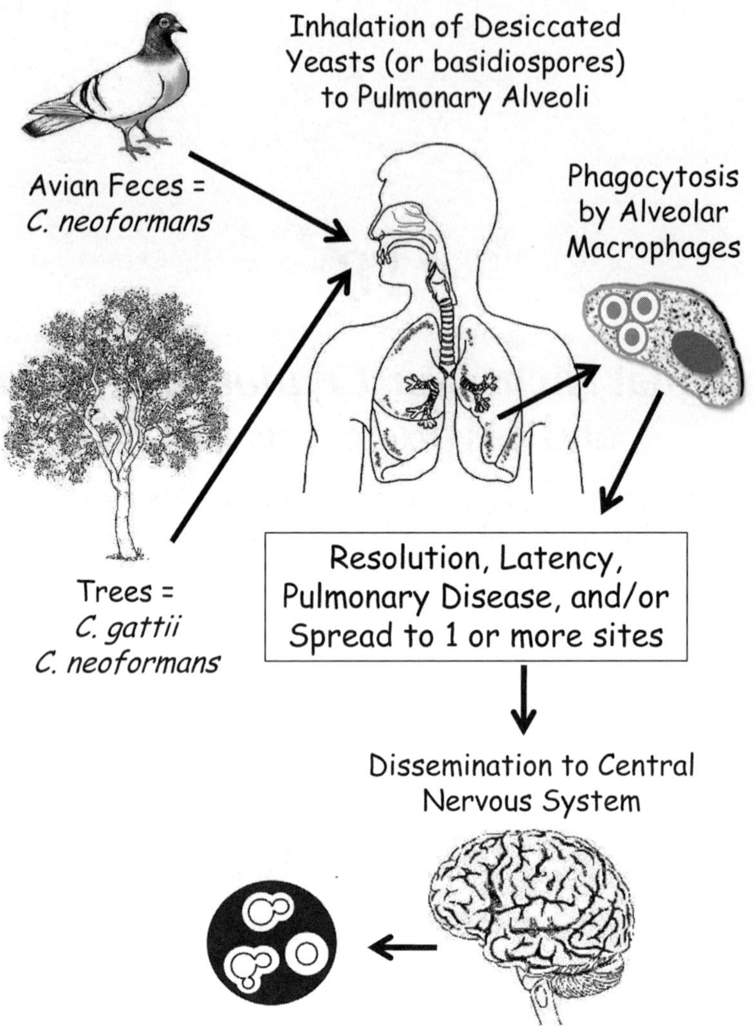

FIGURE 1 Crucial importance of the environmental source of *Cryptococcus* in the natural history of cryptococcosis.

C. *uniguttulatus*, C. *luteolus*, and C. *curvatus* (113). This chapter will focus on the environmental habitats of the dominant human-pathogenic species, C. *neoformans* and C. *gattii*, and their varieties and subgroups. However, this knowledge is incomplete, and the definitive ecological reservoirs, as well as the molecular and phenotypic boundaries of each species or clade, continue to be investigated.

The plethora of clinical manifestations of cryptococcosis can include lesions in almost any organ of the body, and there is a broad range of individual responses to treatment. This protean nature of cryptococcal disease is usually attributed to some impairment of the defense mechanisms of the host. However, a significant minority of patients with C. *neoformans* have no discernible immunocompromising condition or underlying disease (42, 58, 75, 155, 215, 225). An equally tenable hypothesis to explain the extensive variation in cryptococcal pathogenesis imputes the infecting strain of *Cryptococcus*. Certain pathobiological properties are significantly correlated with virulence, such as the abilities to grow at 37°C, develop a capsule, and produce laccase (204). However, strains of C. *neoformans* possessing these virulence factors have been shown in experimental infections and cellular assays to vary by orders of magnitude in their pathogenicity (43, 96, 109, 145, 167, 238). To determine the extent to which the infecting isolate of *Cryptococcus* affects the host-fungal dynamic, it is necessary to characterize the pathogen as well as the patient. Thus, it has become increasingly enlightening and useful to analyze and compare the phenotypes and genotypes of clinical and environmental isolates.

METHODS OF ENVIRONMENTAL ISOLATION OF *C. NEOFORMANS*

Since the 1950s, pathogenic species of *Cryptococcus* have been successfully isolated from soil and avian habitats (4, 61, 62, 110, 141). These early investigations were directed by eminent medical mycological pioneers, Chester W. Emmons, Libero Ajello, and of course, Maxwell L. Littman, who conducted the most comprehensive early studies of C. *neoformans* and cryptococcosis (139). The first environmental isolations were accomplished by collecting samples of soil or avian feces, mixing and diluting the samples in saline, and inoculating small quantities into experimental animals (e.g., mice, guinea pigs, hamsters) by intraperitoneal, intravenous, or intracerebral injection (61). Broad-spectrum antibacterial antibiotics were often used to prevent acute bacterial infections. The animals were monitored for illness and subsequently necropsied. Target organs (e.g., lungs, spleen, brain, liver, peritoneal exudates) were cultured to recover *C. neoformans*, as well as other pathogens. This general protocol proved successful, and by the mid-1960s, the association of *C. neoformans* with pigeons and other avian reservoirs was well documented (61, 62, 78, 110, 141, 162, 187, 226). Furthermore, the murine method of isolation confirmed that the connection between *C. neoformans* and pigeon habitats occurred on a global scale, as positive isolations were reported from North America (94, 141, 162), South America (226), Europe (14, 187), and Australia (78).

The environmental isolation of pathogenic *Cryptococcus* species was supremely enhanced when Friedrich Staib discovered that cultivating *C. neoformans* on medium supplemented with niger seeds (*Guizotia abyssinica*) induces the formation of melanin within the cell walls, and this phenotype produces colonies with a brown to black color (234). Formation of this pigment is catalyzed by laccase, which oxidizes a variety of diphenolic substrates to generate long polymers of melanin. After a few days' growth on an appropriate substrate, colonies acquire an easily recognizable chocolaty, occasionally mucoid appearance. Staib's agar medium is also used in clinical laboratories to confirm the identification of *C. neoformans* and *C. gattii*, especially in nonsterile specimens (e.g., skin, urine), as these are the only positively reacting pathogenic fungi (127).

As noted above, laccase is an important virulence factor for these species of *Cryptococcus*, and its substrates include catecholamines. Melanization of the cell walls protects the yeasts from host defenses (182, 275). The cells have been shown to produce melanin under natural conditions, such as growth on pigeon feces (83, 179, 183, 209), and melanin protects the cells from adverse environmental conditions, such as metallic ions, degradative enzymes, elevated temperatures, and UV light (83, 137, 209). Melanized cryptococcal yeasts are also more resistant to killing by phagocytic environmental predators, such as amoebae and slime molds (15, 240, 241). The combined antiphagocytic properties of melanin and the capsule may explain the ability to infect mammals (241).

For environmental studies, the variety of samples that have been analyzed include solid substances, such as soil, sand, rock, plants and vegetation (e.g., bark, tree hollows, flowers, wood chips, humus), scrapings or swabs of similar materials and other fomites (e.g., shoe soles, tires, roads and buildings, dirt and dust on floors, houses, vehicles); semisolid and liquid samples (fresh, sea, or brackish water; mud; tree sap; swamp debris); animal droppings and habitats (e.g., insects, nests, roosts, and lairs); and air (using exposed plates or sophisticated air samplers). The methods of collecting, processing, and isolating *Cryptococcus* from all of these sample types have been detailed in multiple reports (27, 63, 86, 114, 115, 133, 143, 157, 195, 236). Procedures for sampling live animals (e.g., skin, fur, respiratory and gastrointestinal tracts, mucosal surfaces) are similar to those used for patients (45, 57, 208).

After the environmental samples are collected, they should be processed as soon as possible for optimal recovery of *Cryptococcus*. Current formulations of Staib's medium supplement the niger seed extract with glucose, agar, antibacterial antibiotics (e.g., chloramphenicol and gentamicin), and biphenyl (115, 143, 236). Solid and semisolid samples are vigorously mixed in sterile water or saline, particulates are allowed to settle, and the clarified upper portion is diluted, streak-plated on Staib's agar, incubated at 30°C, and observed daily for the appearance of brownish colonies. Because the number and growth rate of saprobic molds varies greatly, multiple dilutions are often plated. Potential cryptococcal colonies are subcultured, purified, and identified (see below). Aqueous and other liquid samples are plated directly. Air-sampling instruments are usually loaded with Staib's agar medium.

Because negative results are rarely reported, it is impossible to know how many potential niches have been sampled in vain. From accounts in the literature and personal experience, the percentage of positive environmental samples usually varies from zero to approximately 10% and occasionally higher. Even when permanently colonized sites are resampled, the recovery is typically less than 50%.

METHODS OF ENVIRONMENTAL ISOLATION OF *C. GATTII*

David H. Ellis was the first to isolate *C. gattii* from *Eucalyptus camaldulensis*. This prescient discovery ignited the global search for *Cryptococcus* in arboreal reservoirs (59), leading eventually to the realization that both *C. neoformans* and *C. gattii* may inhabit many species of trees. The procedures for isolating *C. gattii* from any environmental sample are the same as those described above for *C. neoformans*. However, the crucial identification of *C. gattii* was enabled by the development of the differential medium, canavanine-glycine-bromothymol blue agar, on which only *C. gattii* imparts a blue color to the medium (127).

METHODS TO IDENTIFY AND CHARACTERIZE ISOLATES OF *C. NEOFORMANS* AND *C. GATTII*

Beyond the identification of *C. neoformans* and *C. gattii*, the genetic characterization of these species is required to discern the source of infection, as well as any implications pertaining to the clinical manifestations, management, and prognosis of cryptococcosis. Because species of *Cryptococcus* are eukaryotes, they possess large genomes, and their determinants of virulence are polygenic and complex. For instance, *C. neoformans* has a genome size of ≈20.2 Mb, 14 chromosomes, and an estimated 6,500 genes (151, 160). Since the advent of this millennium, the rapid accrual of new data has confirmed that clinical and environmental isolates of *Cryptococcus* manifest considerable phenotypic and genetic variation. This diversity may reflect historical and/or current evolutionary processes, as well as multiple mechanisms for the exchange of genes and chromosomal rearrangements. For example, the remarkable plasticity of *Cryptococcus* has been demonstrated by comparative genomics (102), by the ongoing outbreak of unique strains in the U.S. Pacific Northwest (24), by the discovery of novel mechanisms to exchange genetic information (138), and by analyses of the population structure of African isolates (148). Several chapters discuss these processes in detail.

For continued progress, future studies of the diversity, ecology, and evolution of these pathogenic yeasts depend on the vigilance of physicians, clinical microbiologists, and public health officers, who must exercise the initiative to retain and transmit clinical and veterinary isolates to laboratories investigating the molecular genetics and pathogenicity of *Cryptococcus* (50, 51, 164).

Since there are manifold differences among strains and well-defined subpopulations, it is minimally informative to identify only the species or variety of *Cryptococcus*. To begin to investigate the epidemiology and ecology of novel isolates of *Cryptococcus*, it is necessary to employ robust methods of genotyping. Fortunately, rigorous and innovative methods have been developed to compare strains.

New clinical or environmental isolates can be rapidly identified as *C. neoformans* or *C. gattii* by key phenotypes. They are encapsulated, budding yeasts, and similar to other species of *Cryptococcus*, they produce urease, but only the pathogens grow at 37°C and produce laccase, developing brown colonies on solid medium with a diphenolic substrate, such as Staib's niger seed agar (73, 189, 213, 268). Unfortunately, commercial serotyping reagents are no longer available to identify the two species. In the past, latex particles coated with specific antisera were used in a slide agglutination test to identify the four immunodominant epitopes of the major capsular polysaccharide, glucuronoxylomannan. This method of serotyping was employed for many years to differentiate *C. neoformans* (serotype A, D, or AD) and *C. gattii* (serotype B or C) (32, 127, 170). Currently, the only facile, if imperfect, nonmolecular methods to distinguish these species exploit physiological differences between them, such as growth on canavanine-glycine-bromothymol blue agar (127). The species and subpopulations (or cryptic species) of *C. neoformans* and *C. gattii* are most reliably identified using DNA-based methods.

Several methods can be used to determine the "molecular type" (clade, subgroup, or subpopulation) of an isolate, as well as to characterize the genotypes of individual strains. Figure 2 presents the molecular nomenclature for the nine well-established subgroups of *C. neoformans* and *C. gattii*: VNI–VNIV, VNB, and VGI–VGIV. Detailed descriptions of the molecular markers and methods of genotyping these species as well as characterizing their population genetics are provided in chapters 8 and 24 of this volume, and several reviews (167–169, 269, 270). The most popular method to identify eight of the nine subpopulations is PCR fingerprinting, which is quick and cost-effective. Wieland Meyer and colleagues standardized this protocol, which in brief, entails the following steps. The genomic DNA of an isolate is extracted and amplified with a single M13 primer, the PCR products are electrophoresed, and band sizes are compared with those of reference strains of each molecular type (166). In another PCR method, the *URA5* gene is amplified and digested with two endonucleases, and the electrophoretic bands are compared with reference strains (165). Reference strains of each subgroup are required because the band sizes can vary with different thermocyclers, reagents, and PCR conditions. Detection of amplified fragment length polymorphisms (AFLPs) and other methods can be used to achieve similar results (16, 17, 44, 70, 134). However, AFLP patterns are also laboratory-specific, and reference strains are needed to identify each molecular type.

Although more costly, multilocus sequence typing (MLST) is preferable because the polymorphic genotyping markers are specific DNA sequences and are reproducible in any laboratory. MLST obviates the need for reference strains. A consensus set of primers to identify several unlinked MLST polymorphisms has been established to identify the major subpopulations of *C. neoformans* and *C. gattii* (18, 26, 40, 71, 148, 164, 178, 219). The *Cryptococcus* MLST website permits the global community of researchers to access and augment the database (http://cneoformans.mlst.net/). Because VNB strains (Fig. 2) are exceptionally diverse, they are seldom recognized as a distinct subgroup by PCR fingerprinting but are usually typed as isolates of the VNI or VNII subgroup (178). However, MLST markers for *C. neoformans* will identify the diverse VNB clade, which is largely restricted to Africa (17, 18, 148). MLST markers have been used to analyze the population genetics of environmental isolates, and in future studies, it will be increasingly useful to genotype multiple isolates from each positive sample. Results to date suggest that multiple isolates from avian feces are more diverse, perhaps due to increased exposure and opportunities

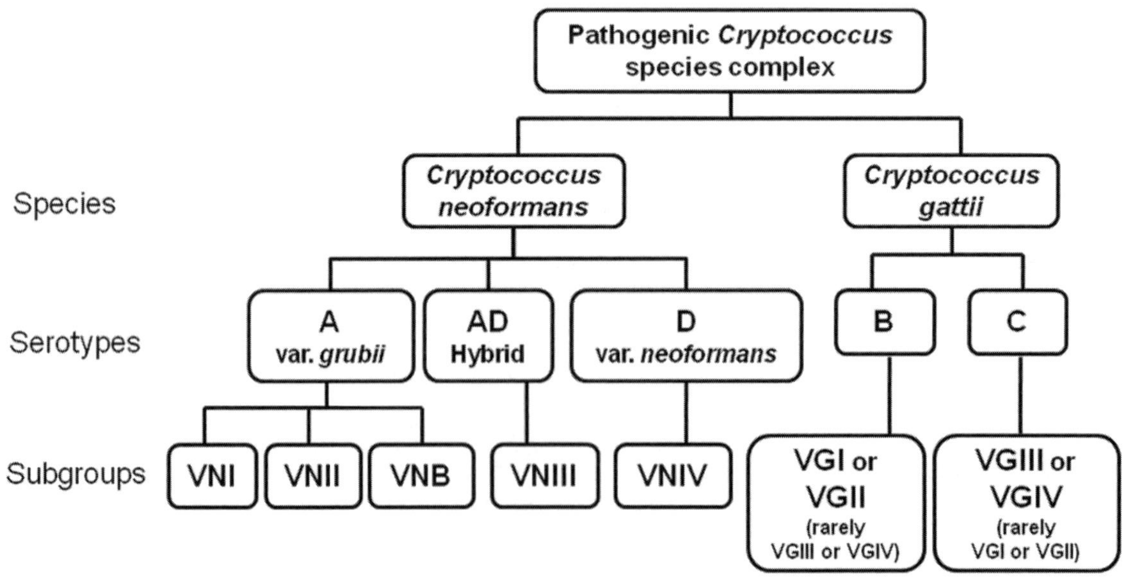

FIGURE 2 Schematic presentation of the relationships and nomenclature of *C. neoformans* and *C. gattii*.

for colonization, than arboreal isolates, which tend to be highly clonal (29).

When reviewing the century of literature on *Cryptococcus* and cryptococcosis, readers must be aware that the serotypes were not recognized before 1949 (67, 68, 266). *C. gattii*, which was only identified in 1970 (260), was considered a variety of *C. neoformans* until acquiring species status in 2004 (128). Although there are currently at least nine well-defined subpopulations, if not cryptic species, of *C. neoformans* and *C. gattii*, the older scientific publications use the nomenclature of their times. Consequently, the early ecological studies did not differentiate these species, and the tools for genotyping were not available.

GEOGRAPHICAL DISTRIBUTION AND EPIDEMIOLOGY OF *C. NEOFORMANS* AND *C. GATTII*

C. neoformans

The vast majority of AIDS-associated infections with *C. neoformans* are caused by the cosmopolitan population of VNI strains, which can be readily isolated from the environment (17, 36, 40, 118, 136, 143, 148, 161, 166, 232, 253, 254). Strains of VNII have apparently also infected people on every continent, but they are much less frequent (17, 20, 36, 136, 143, 148, 161, 166, 253, 257), and there have been very few isolations of VNII from the environment. To date, all the clinical cases with VNB strains have occurred in sub-Saharan Africa or Brazil (148, 178). Strains of VNIV (serotype D) and VNIII (AD hybrids) have been isolated globally from the environment and patients. However, infections with VNIII and VNIV are significantly concentrated in Europe (17, 71, 143, 257, 265).

Despite the strong association with HIV/AIDS and other immunocompromising, underlying conditions, there are numerous reports of *C. neoformans* infections in immunocompetent persons. In these ostensibly "normal" hosts, pulmonary infections are more common, but dissemination also occurs (3, 7, 40, 175, 185).

C. gattii

In contrast to *C. neoformans*, global epidemiological studies have consistently shown that the majority of infections with *C. gattii* occur in immunocompetent persons. However, the traditional serotypic designation as B or C is not useful because genotyping methods are more discriminating (Fig. 2). In geographical regions where significant numbers of clinical isolates have been genotyped, the relative prevalence of each subgroup has tended to vary, and more data are needed to determine whether the currently observed patterns are stable or evolving, as is the current outbreak of VGII infections in Canada (British Columbia) and the western United States (Washington and Oregon) (50). In South America, all four subgroups have been reported, but more cases are attributable to VGII and VGIII (56, 257). The prevalence of HIV/AIDS is highest in Africa, but the incidence of *C. gattii* is low, and most reported infections were caused by VGIV and occurred in AIDS patients (111, 147, 174). The incidence in Europe and Asia is also comparatively low and usually due to VGI or VGII (10, 71, 126, 254, 262). Most clinical cases in Australia are caused by VGI, followed in frequency by VGII, but regional differences have been noted (29, 41, 230). A large, multiyear review of veterinary cases in Australia found that VGII predominated in Western Australia, but *C. gattii* was less common and was usually due to VGI on the eastern coast (163).

ENVIRONMENTAL SOURCES OF *C. NEOFORMANS*

As in patients, both varieties of *C. neoformans*, *C. neoformans* var. *grubii* and *C. neoformans* var. *neoformans*, are globally distributed in the environment, but serotype A is ubiquitous, and serotype D is more prevalent in Europe. Serotype A strains are responsible worldwide for most cases of cryptococcosis and more than 95% of cases in patients with AIDS (32, 170). Although *C. neoformans* is a ubiquitous species, it is less prevalent in regions such as northern Russia (259), Canada (222), and Scandinavia (121, 256) compared to tropical and temperate regions.

Pigeons and Other Birds

Avian habitats are a significant natural reservoir for *C. neoformans*, especially the roosting areas claimed by pigeons, where the yeast is enriched within the aged avian excreta (207). While birds may inadvertently spread the yeast cells, they are rarely infected because the avian body temperature ranges from 41.5 to 43°C, which is too high to support the growth of *Cryptococcus* (62, 140). Nevertheless, at least 26 avian cases of cryptococcosis due to both species have been documented, but the temperature tolerance of these strains was not reported (159). However, most of these avian infections were limited to cutaneous sites and the upper respiratory tract (159). In many tropical and semitropical climates, *C. neoformans* can be readily isolated from avian roosting areas that are shielded from direct sunlight and UV light, and in this milieu, the yeasts secrete urease and apparently thrive on urea and other nitrogenous substrates. They also produce airborne yeast cells that are small enough to reach the alveoli if inhaled (190, 249). Both *Cryptococcus* species grow well on pigeon feces in the laboratory and become melanized (179, 183). *C. neoformans* can also produce potentially infectious basidiospores under these conditions (179).

The prevalence of *C. neoformans* in avian environments tends to correspond with the endemic areas for cryptococcosis. Indeed, the association of *C. neoformans* with avian habitats has long been considered a major source of human cryptococcosis, and many surveys have noted the proximity of patients to pigeons and other birds (33, 48, 54, 82, 93, 154, 188, 272). More convincingly, isolates of *C. neoformans* with the same genotype have been recovered from patients and pigeon feces (48, 76, 143). Overall, the evidence supporting a significant convergence of pigeons, *C. neoformans*, and cryptococcal disease in humans is substantial but largely circumstantial. Since all three species share overlapping global habituation, proof of this pathway of causation requires rigorous genotyping and comparisons of clinical and avian isolates. Despite the popular consensus that pigeons pose a threat of cryptococcosis, reported exceptions to this theory indicate that *C. neoformans* can be acquired from other sources, such as soil or household dust (4, 130–132, 152, 246, 247). Since the advent of genotyping, there have been few reports of the pathogenicity of environmental isolates, and results have not been conclusive. For example, three separate studies of isolates from different locales reported that 100% of environmental isolates were virulent for mice when administered intracerebrally (48), 50% were lethal when injected intravenously (105), and only 9% were lethal when mice were challenged intranasally (145). The latter report demonstrated that pigeon isolates were less virulent for mice than human isolates with the same AFLP and MLST genotypes (145). Although this observation may be an artifact of the experimental infections or a limitation of the genotyping methods, it could also indicate that C. *neoformans* has other significant reservoirs in nature.

Trees

In recent years, there have been a growing number of reported isolations of C. *neoformans* from a variety of trees in various parts of the world, including Argentina (201), Colombia (28, 87), Brazil (122, 181, 202), India (89, 90, 193), South Africa (146), and Thailand (233). Thus, C. *neoformans* can be found in arboreal as well as avian environs. Tables 1 and 2 list many of the reports documenting these niches.

Other Sites

There are numerous reports of isolating C. *neoformans* from a variety of other sites, including bat caves, rural farms and farm animals, fruits, and insects (11, 32, 63, 153, 171, 248). However, none of these sources has proven to be a perennial niche. Perhaps most relevant are the scattered reports of capturing airborne cells of C. *neoformans* (190, 210, 211).

ENVIRONMENTAL SOURCES OF *C. GATTII*

C. *gattii* differs from C. *neoformans* in geographical distribution, serotypes, ecology, clinical manifestations, incidence in immunocompetent hosts, several in vitro phenotypes, and incidence in compromised patients. The prevalence of C. *gattii* is more limited in patients and the environment. In the environment, C. *gattii* has long been associated with trees, tree hollows, and decayed wood, and this species has been isolated from *E. camaldulensis* and other varieties of *Eucalyptus* as well as other species of trees in Australia and elsewhere (59, 60, 229). Although *Eucalyptus* trees have been sampled more than other trees, it is now clear that C. *gattii* populates native and introduced trees wherever it becomes established, and the tree species that harbor C. *gattii* differ in Argentina, Australia, Brazil, Canada, Colombia, India, and the United States.

The roster of tree species from which C. *gattii* has been isolated continues to increase. As indicated in Table 3, C. *gattii* has been isolated most often in countries where it has been more sedulously sought. Certain tree species (denoted by asterisks in Table 3) are recurrently or permanently colonized. These signature trees vary in each country, but there is little apparent correlation with the subgroup(s) of C. *gattii*. *Eucalyptus* trees appear to be associated with VGI more than other subpopulations of serotype B, but they are a major habitat only in Australia (230). In addition, eucalypts from Australia have been widely exported, and large, well-established populations exist in Brazil, China, Colombia, India, South Africa, southern Europe, and the Pacific coastal United States, but they are not the foremost habitat for serotype B outside Australia. Furthermore, despite extensive sampling, the environmental niche of serotype C beyond almond trees in Colombia has not been elucidated (28).

The affinity of C. *gattii* for different trees in different parts of the world does not mirror the census of frequently colonized trees. Other factors undoubtedly contribute to the colonization of particular trees. As described below, several investigators have carefully recorded geographic, climatic, and telluric data associated with the environmental prevalence of C. *neoformans* or C. *gattii*. Trends have been noted with respect to factors that may correlate with the cryptococcal yield in specific niches (e.g., elevation, temperature, precipitation, humidity, pH, salt concentrations), but unique or essential parameters have not been identified.

Historically, C. *gattii* was thought to be restricted to tropical areas of Australia, Asia, Africa, and the Americas, until it emerged this millennium in southwestern Canada and the northwestern United States (24, 50, 51, 55, 77, 116, 117). The predominant clinical genotypes (VGIIa and VGIIb) have been isolated from regional trees (Douglas fir, alder, red cedar, arbutus, Garry oak, etc.), soil, air, fresh and sea water, and fomites (shoes, auto tires) (114, 115). For a complete description of this ongoing outbreak, refer to chapter 23 of this volume.

As with clinical isolates, the geographical distribution of the subpopulations of C. *gattii* varies considerably, and more genotyping surveys are needed to confirm and extend the current data. Overall, VGI and VGII are far more common than VGIII and VGIV. In Australia, VGI isolates of serotype B are most prevalent in environmental samples (230), but in South America, VGII isolates of serotype B predominate (66, 87, 149, 181, 257). Nearly all environmental isolates of C. *gattii* have been VGI, VGII, or VGIV. As noted above, the ecology of serotype C was a mystery until isolates were recovered from almond trees in Colombia (28).

In addition to trees, C. *gattii* has been isolated on occasion from other exogenous sources. Strains of serotype B (VGI, *MAT*α) have been cultured from psittacine birds in Porto Alegre and São Paulo, Brazil (2, 199). Serotype B was also recovered from bat feces in Rio de Janeiro (133). In British Columbia, VGII strains were obtained from fomites and air and water samples (114, 115).

LONGITUDINAL STUDIES

For obvious logistical reasons, the most sustained environmental studies have been conducted in circumscribed regions. Several researchers have sampled specific sites temporally and spatially. Many have included a variety of telluric, arboreal, aquatic, and aerial niches. Several of these focused reports have generated significant detail and insights, and four, which were conducted in Brazil, Colombia, India, and the United States, are highlighted below. A common theme is that species and subgroups of *Cryptococcus* may colonize a particular niche transiently, intermittently, or perennially. The more robust the genotyping, the more accurately the environmental niche can be clarified. Three important series of reports are not included here because they are covered in detail in other chapters:

- Chapter 20 presents the African, arboreal association of the unique and extraordinarily diverse VNB strains from which the global population of C. *neoformans* may have evolved.
- Chapter 22 explains the fascinating implications of the enduring relationship between C. *gattii* and *Eucalyptus* trees in Australia.
- Chapter 23 describes the ecologic range, habitats, and potential modes of dispersion of VGIIa, VGIIb, and VGIIc in the province of British Columbia (Canada) and the states of Washington and Oregon (United States).

Brazil

Since 1990, Marcia S. Lazéra and her colleagues have conducted extensive, pioneering studies in several geographically diverse areas of Brazil. Initially, they used the murine inoculation method to isolate C. *neoformans* from several urban habitats in Rio de Janeiro associated with pigeons, other birds, and bats (133). During 1990 to 1992, they isolated C. *neoformans* from 25% of 32 pigeon nesting areas.

TABLE 1 Isolations of *C. neoformans* var. *grubii* (serotype A), *C. neoformans* var. *neoformans* (serotype D), and AD hybrids from excreta and habitats of *Columba livia* and other birds

Continent	Country	*C. livia* and other birds (references)	Serotype, genotype, mating type
North America	Mexico	*C. livia* (35, 153)	
	United States	*C. livia* (48, 94, 211)	
		C. livia (143)	A, VNI, VNII D AD
South America	Brazil	*C. livia* (11, 31, 33, 54, 72, 76, 101, 122, 172, 206, 226, 227)	A A, VNI, *MAT*α A, *MAT*α
		Psittacine birds (2, 46)	A, VNI, *MAT*α
		Passerine and psittacine birds (154)	A, VNI, *MAT*α
	Colombia	*C. livia* (66, 86, 191)	A A, VNI, *MAT*α
		Polyborus plancus (27)	
Asia	China	*C. livia* (135)	A D AD
	India	*C. livia* (92)	
	Japan	Chicken (123)	
		C. livia (112, 177, 252, 272)	A D AD
	Korea	*C. livia* (39)	A
	Malaysia	*C. livia* (253)	A, VNI
	Nepal	*C. livia* (186)	A, *MAT*α
	Thailand	*C. livia* (9, 228, 233, 255)	A
		Streptopelia chinensis (233)	A
		Streptopelia decaocto (233)	A
		Geopelia striata (233)	A
	Turkey	*C. livia* (218, 273, 274)	A, *MAT*α A, *MAT***a** AD, *MAT*α/**a**
Europe	Belgium	Canary (249)	
		C. livia (249)	
	Czech Republic	*C. livia* (97, 103)	
	Denmark	*C. livia* (242)	
	France	*C. livia* (82)	D
	Germany	*Ara militaris* (239)	
		C. livia (120, 220)	A D
		C. livia and other birds (237, 261)	A
		Pet birds (119, 120)	A
		Podargus strigoides (239)	
		Probosciger aterrimus (239)	
		Psittacus erithacus (239)	
	Italy	Canary (47)	
		C. livia (80, 171, 263)	A D
	Spain	*C. livia* (81, 173, 208)	A
	United Kingdom	*C. livia* (192)	

(Continued on next page)

TABLE 1 *(Continued)*

Continent	Country	*C. livia* and other birds (references)	Serotype, genotype, mating type
Africa	Botswana	*C. livia* (146)	A, VNI
	Burundi	*C. livia* (247)	
	Egypt	*C. livia*, sparrow (158, 200)	
	Ethiopia	*C. livia* (5)	
	Nigeria	*C. livia* (91)	
		Dendrocygna viduata (107)	A
		Bubo africanus cinerascene (107)	A
		Pavo cristatus (107)	A
		B. africanus cinerascene (107)	D
	South Africa	*C. livia* (146)	A, VNI

Expanding the known roster of cryptococcal niches, they were the first to recover *C. neoformans* from decayed wood in the hollow of a Java plum tree (*Syzygium jambolana*) that sheltered bats and birds, and they provided the first reports of the isolation of *C. neoformans* from a bat habitat, as well as *C. gattii* from bat excreta in the attic of an abandoned house, which was located far from any *Eucalyptus* trees (133).

They proceeded to confirm these findings and to explore the range and duration of established Brazilian reservoirs. From 1994 to 1996, they sampled 31 hollows of seven tree species in urban areas of Rio de Janeiro. *C. neoformans* was recovered from 62 samples of decayed wood from the hollows of five pink shower trees (*Cassia grandis*), two fig trees (*Ficus microcarpa*), and one November shower tree (*Senna multijuga*) (131). This high yield of *C. neoformans* was obtained from decayed wood scraped from the inner surfaces of the hollows of viable trees, but other arboreal sites, such as the bark on the trunks, were negative. One year later, these same hollows remained positive, establishing their perennial colonization with *C. neoformans*, and the negative trees did not become colonized (131). These studies led Lazéra et al. to propose that decaying wood in the hollows of many species of local trees constitutes a newly recognized natural habitat of *C. neoformans* (131).

This group then shifted their focus from Rio de Janeiro to the northeastern city of Teresina, capital of the state of Piauí, 2,700 km (ca. 1,680 miles) north of Rio. The majority of patients diagnosed with cryptococcal meningitis in Teresina resided in Piauí or the adjacent state of Maranhão. Most of these cases occurred in immunocompetent persons and were caused by *C. gattii*. Furthermore, a large percentage (25 to 30%) of the patients were children or adolescents. From 1993 to 1997, these researchers collected serial samples of decayed wood from the inner walls of tree hollows of four species of living trees in an old downtown square: pottery trees (*Moquilea tomentosa*), pink shower trees (*C. grandis*), mango trees (*Mangifera indica*), and fig trees (*Ficus* species). Except for the mango trees, the other trees yielded both species of *Cryptococcus* (130, 132, 203). This investigation also found the first incidences of *C. neoformans* and *C. gattii* occupying the same trees, a pink shower, a fig, and pottery trees, and in the pink shower tree, they were isolated from the same hollow. The trees were colonized for periods of 19 to 36 months or more, and the census of *C. neoformans* ranged from 150 to 21,700 CFU/g sample (132). These results forecasted recurring themes of other longitudinal studies, such as the colonization of decayed wood in the hollows of indigenous trees by either species of *Cryptococcus* and a similarity in the regional prevalence of environmental and clinical isolates.

In another investigation in Rio de Janeiro, these researchers sought to determine the risk of cryptococcosis among patients with HIV/AIDS who were likely exposed to environmental sources of *Cryptococcus*. They collected 824 samples of household dust, outdoor soil, and avian excreta from 154 houses, and 20 (13%) of the houses were positive for *C. neoformans* (188). They compared the percentages of positive houses that were occupied by healthy individuals, persons with HIV/AIDS but without cryptococcosis, and those with HIV/AIDS and cryptococcosis. The data supported the conclusion that patients with HIV/AIDS had an increased risk of cryptococcosis if they resided in homes that were proximal to avian excreta contaminated with *C. neoformans* (188).

In another revealing study, they traveled to the northwestern Brazilian state of Roraima and sampled trees on an undeveloped island in the Amazonian rainforest. From this site, which was uninhabited by humans and contained no introduced plants, they isolated *C. gattii* from a native jungle tree, *Guettarda acreana* (74). This finding highlights the cumulative global evidence that both species of *Cryptococcus* are able to colonize a variety of trees under a broad range of geographical and meteorological conditions (Tables 2 and 3). The regional relevance of this report is underscored by the description of 43 cases of cryptococcal meningitis (due to strains of VNI, VGI, and VGII) among people living in the eastern Amazonian state of Pará (217).

Colombia

Extensive multiyear studies in Colombia have provided a wealth of data regarding the distribution and ecology of *C. neoformans* and *C. gattii*. From 1997 through 2005, infections due to all four conventional serotypes were documented in Colombia. During this period, 788 clinical isolates were analyzed; 95.9% were serotype A, 3.3% were serotype B, 0.5% were serotype C, and 0.3% were serotype D (149). As typical everywhere, the vast majority of cases were caused by serotype A, but second most prevalent were infections caused by isolates of serotype B. Although 76 clinical laboratories throughout Colombia participated in this study, most patients infected with serotype B (11/18) were HIV-negative, and they resided in the northeastern city of Cúcuta (149, 150).

TABLE 2 Arboreal isolations of *C. neoformans* var. *grubii* (serotype A), *C. neoformans* var. *neoformans* (serotype D), and AD hybrids from bark, tree hollows, or decaying wood

Continent	Country	Trees[a] (references)	Serotype, genotype, mating type
North America	United States	*Eucalyptus camaldulensis* (261)	A
		Quercus alba (143)	D, VNIV
South America	Argentina	*Cedrus deodara* (201)	A, VNI
		Cupressus sempervirens (201)	A, VNI
		Eucalyptus species (201)	A, VNI
		Phoenix species (201)	A
		Tipuana tipu (201)	A, VNI
	Brazil	*Anadenanthera peregrina* (202)	A, MATα
		Caesalpinia peltophoroides (202)	A, MATα
		**Cassia grandis* (122, 132, 258)	A A, VNI, MATα
		E. camaldulensis (33, 122, 172, 181, 205)	A D, VNIV, MATα
		Eucalyptus tereticornis (33, 205)	D, VNIV, MATα
		Eucalyptus species (33, 122, 172, 205)	A A, MATα, MAT**a** D, VNIV, MATα
		Guettarda acreana (131, 132, 181, 258)	A
		**Ficus microcarpa* (122)	A
		Licania (Moquilea) tomentosa (132, 258)	D, VNIV, MAT**a**
		**Moquilea tomentosa* (131, 132, 181, 258)	A
		Myroxylon peruiferum (131, 132, 181, 258)	A
		Senna multijuga (131, 132, 181, 258)	A
		Senna siamea (46)	A, VNI, MATα
		Syzygium cumini (*S. jambolanum*) (131–133, 181, 258)	A D, MATα, MAT**a**
		Theobroma cacao (131, 132, 181, 258)	A, VNI, MATα
	Colombia	*Acacia decurrens* (66, 86)	A, VNI, MATα
		E. camaldulensis (86)	A
		Eucalyptus species (66, 86, 87, 191)	A, VNI, MATα
		Ficus tequendamae (66, 87)	A, VNI, MATα
		Laurus species (66, 86)	A, VNI, MATα
		Licania (Moquilea) tomentosa (66, 86, 165)	A, VNI, MATα
		Pinus species (66, 86)	A, VNI, MATα
		Prosopis juliflora (66, 86, 165)	A, VNI, MATα
		Prunus dulcis (66, 86)	A, VNI, MATα
		Tabebuia guayacan (66, 86)	A, VNI, MATα
		Terminalia catappa (66, 165)	A, VNI, MATα
Asia	India	*Acacia nilotica* (193)	A, MATα
		Aegle marmelos (193)	A, MATα
		Alstonia scholaris (193)	A, MATα
		Azadirachta indica (89, 193)	A, MATα
		Butea monosperma (197)	A
		Cassia fistula (89, 193)	A
		Dalbergia sissoo (193)	A, MATα
		E. camaldulensis (90)	A
		Eucalyptus globulus (85)	A
		Eucalyptus species (193, 197)	A, MATα

(*Continued on next page*)

TABLE 2 *(Continued)*

Continent	Country	Trees[a] (references)	Serotype, genotype, mating type
		Ficus religiosa (193–195)	A, MATα
		Mangifera indica (89, 193)	A, MATα
		Manilkara hexandra (193)	A, MATα
		Mimusops elengi (193)	A, MATα
		Polyalthia longifolia (193)	A, MATα
		**Syzygium cumini* (89, 193–196)	A, MATα
		Tamarindus indica (89, 197)	A
		Terminalia arjuna (89)	A
	Thailand	*Eucalyptus deglupta* (233)	A
Europe	Italy	*E. camaldulensis* (30)	
Africa	Ethiopia	*Eucalyptus* species (5)	
	Botswana	*Adansonia digitata* (146)	A, VNB
		Colophospermum mopane (146)	A, VNI A, VNB
	Democratic Republic of the Congo	*Entandophragma* (245)	
	South Africa	*Eucalyptus* species (146)	A, VNB

[a]*, Perennial colonization of these trees in certain locations.

Following the discovery in the 1990s that isolates of serotype B (*C. gattii*) in Australia were associated with *Eucalyptus* trees, David Ellis and his colleagues hypothesized that cases of cryptococcosis caused by serotype B corresponded with the global distribution of some *Eucalyptus* species (59, 60, 231). To test this hypothesis, Elizabeth Castañeda and her colleagues organized a large investigation in Cúcuta. Over a period of 11 months, they collected 1,579 samples of detritus, seeds, bark, leaves, and flowers from four indigenous arboreal genera, *Eucalyptus*, *Moquilia* (oití), *Delonix* (acacia), and *Terminalia* (almonds) (28). They obtained 186 isolates of *Cryptococcus* species: 40 isolates of *C. neoformans* serotype A, 121 *C. laurentii*, and 25 *C. albidus*. No isolates of serotype B were recovered, but 2 of 68 almond trees (*Terminalia catappa*) yielded the first environmental isolates of serotype C (28). The almond trees were repeatedly positive over a period of 19 months (28, 34), and all these cryptococcal isolates had the same molecular type by PCR fingerprinting (VGIII) (165) and AFLP genotyping (type 5) (18). To explore the interaction between serotype C and almond trees, the stems of 30 seedlings were inoculated with yeast cells and then sampled at different times (104). These cryptococcal "infections" caused no visible lesions or microscopic phytopathology, but the yeasts could be cultured from the plants for at least 1 year after inoculation (64, 104). Furthermore, the cryptococcal cells were observed to spread from the stems to the soil and back to the plants with no apparent damage to the young almond trees, indicating the affinity of serotype C for almond trees and the ready establishment of a stable, potentially long-term balance between these species (64).

In pursuit of the environmental niche of serotype B, another survey was conducted in a forest in the province of Cundinamarca, where strains of serotype B were successfully isolated from 46 *Eucalyptus* samples (65). The yield was higher during the rainy season and more prevalent from bark than tree hollows or soil at the base of the trees. All of these serotype B isolates displayed robust growth at 37°C, produced capsules and melanin, and possessed the *MAT***a** mating-type allele. The virulence of these strains was evaluated by challenging BALB/c mice intravenously with 5×10^5 CFU; however, none of the infected mice succumbed during the 70-day period of observation, although viable yeasts could be cultured from the lungs, brains, and spleens of sacrificed mice (65). As echoed below, the nonlethality of environmental isolates of *C. neoformans* has also been observed (145).

In a large follow-up study, pigeon feces and *Eucalyptus* and almond trees were sampled in 26 Colombian municipalities, and nearly half the sites yielded positive cultures (191). This project confirmed that strains of serotype C reside on almond trees; in addition, isolates of serotypes A were recovered from pigeon feces, and strains of serotype B were obtained from eucalypts. The researchers quantified the cryptococcal colony-forming units per sample, but the census varied considerably and inconsistently. However, the recovery from all sample types was greater when the ambient temperatures were cooler (191).

This same research group also investigated whether environmental factors such as pH, humidity, temperature, and sunlight affected the prevalence of *C. neoformans* in avian feces and vegetation in urban areas around Bogotá (86). Over several months, they collected 480 samples of bark, tree-base soil, hollows, and flowers of 32 trees, as well as 89 avian fecal samples. The arboreal samples from nine tree species yielded 97 isolates of serotype B and one serotype A. The yeasts were more frequently recovered from bark than soil or hollows and from trees with hollows or rotting wood than from trees with nesting birds. The yeasts were more numerous in the rainy season and more likely to be cultured from microhabitats where the temperature and humidity

TABLE 3 Arboreal isolations of *C. gattii* (serotype B or C) from bark, tree hollows, or decaying wood

Continent	Country	Trees[a] (references)	Serotype, genotype, mating type
North America	Canada	*Acer* species (114, 115, 157)	B, VGII
		Alnus rubra (114, 115, 157)	B, VGII
		Arbutus menziesii (114, 115, 157)	B, VGII
		Picea species (114, 115, 157)	B, VGII
		Pinus species (114, 115, 157)	B, VGII
		Prunus emarginata (114, 115, 157)	B, VGII
		Pseudotsuga menziesii var. *menziesii* (114, 115, 157)	B, VGII
		Quercus garryana (114, 115, 157)	B, VGII
		Thuja plicata (114, 115, 157)	B, VGII
	Mexico	*Eucalyptus camaldulensis* (6)	B, VGIIa
	United States	*E. camaldulensis* (230)	B, VGII
		Eucalyptus citriodora (230)	VGI
		Eucalyptus tereticornis (230)	VGI
		Eucalyptus species (230)	VGIII
		Multiple species (157)	
South America	Argentina	*Acacia visca* (201)	B, VGI
		Cedrus deodara (201)	B, VGI
		Eucalyptus species (201)	B, VGI
		Phoenix species (201)	B, VGI
		Tipuana tipu (201)	B, VGI
		Ulmus campestrus (201)	B, VGI
	Brazil	**Cassia grandis* (130, 132, 181, 258)	B, VGII, *MAT*α, *MAT***a**
		E. camaldulensis (172, 181)	B
		Eucalyptus species (172)	B
		**Ficus microcarpa* (130, 132, 181, 258)	B, VGII, *MAT*α, *MAT***a**
		Guettarda acreana (74)	B
		**Moquilea tomentosa* (130, 132, 181, 258)	B, *MAT*α, *MAT***a**
		Senna siamea (46)	B, VGII, *MAT*α
	Colombia	*Acacia decurrens* (66, 86, 87)	B, VGII, *MAT***a**
		Coussapoa species (86, 87)	B
		Croton bogotanus (66, 86, 87)	B, VGII, *MAT***a**
		Croton funckianus (66, 86, 87)	B, VGII, *MAT***a**
		Cupressus lusitanica (66, 86, 87)	B, VGII, *MAT***a**
		E. camaldulensis (86, 191)	B C
		Eucalyptus globulus (66, 86)	B, VGIV, *MAT***a**
		Eucalyptus species (65, 66, 86, 87, 191)	B, VGIV, *MAT*α B, VGII, *MAT***a** C, VGIV, *MAT*α
		Ficus soatensis (66, 86, 87)	B, VGII, *MAT***a**
		Laurus species (66, 87)	B, VGII, *MAT*α
		Licania tomentosa (66,87)	B, VGII, *MAT*α
		Pinus radiata (86, 87)	B, VGII, *MAT***a**
		Prunus dulcis (66)	B, VGII, *MAT***a** C, VGIII, *MAT*α
		**Terminalia catappa* (28, 66, 87, 165, 191)	C, VGI, *MAT*α C, VGIII, *MAT*α C, VGIV, *MAT*α

(Continued on next page)

TABLE 3 *(Continued)*

Continent	Country	Trees[a] (references)	Serotype, genotype, mating type
Asia	India	*Acacia nilotica* (193)	B
		Azadirachta indica (89, 193)	B
		Cassia fistula (193)	B
		Cassia marginata (85)	B
		E. camaldulensis (38)	B
		Eucalyptus tereticornis (90)	B
		Eucalyptus species (193)	B
		Mangifera indica (89, 193)	B
		Manilkara hexandra (193)	B
		Mimusops elengi (89, 193)	B
		Pithecolobium dulce (89)	B
		Polyalthia longifolia (193)	B
		Tamarindus indica (89, 193)	B
		**Syzygium cumini* (85, 89, 194–196)	B
Australia		*E. camaldulensis* (95,230)	B, VGI
		E. tereticornis (230)	B, VGI
		Syncarpia glomulifera (264)	
Africa	Egypt	*E. camaldulensis* (158)	

[a]*, Perennial colonization of these trees in certain locations.

slightly exceeded that of the surrounding environment (86). The positive trees were *Ficus soatensis*, *Ficus tequendamae*, *Croton bogotanus*, *Croton funckianus*, *Coussapoa* species, *Pinus radiata*, *Cupressus lusitanica*, *Acacia decurrens*, and three *Eucalyptus* species. Six avian fecal samples were positive for isolates of serotype A, and the average concentration of 1,525 CFU/g excreta was significantly higher than the yeasts in the arboreal sites, which averaged 168 CFU/g (86). These counts were not obviously affected by the location of the site or exposure to sunlight. Similarly, the water content and pH of the samples varied widely and did not correlate with the recovery of *Cryptococcus* (86).

These extensive field studies by Dr. Castañeda and her collaborators resulted in the accumulation of much information about the distribution of the three indigenous serotypes in Colombia, their prevalence in different environmental niches, and the effects of geographical and climatic factors on their ecology (28, 34, 65, 86, 87, 191). The overall rank order prevalence of serotypes in the environment reflected their clinical incidence. Serotype A was most common, followed by B, and then C; to date, serotype D and AD hybrids have not been found in the environment in Colombia (66). Molecular studies determined that most Colombian isolates of serotypes A, B, and C belonged to subgroups VNI, VGII, and VGIII, respectively (66). More specifically, 269 clinical and environmental isolates of serotype A were VNI and three were VNII; 118 isolates of serotype B were VGII and 1 was VGI; and of 34 isolates of serotype C, 1 each was VGI and VGII, 22 were VGIII, and 10 were VGIV (66). All isolates of serotype A had the *MAT*α mating-type allele, and of the serotype B isolates, 37 were *MAT*α and 116 were *MAT***a** (66). It is intriguing that most of the 96% of VGII isolates from Colombia were *MAT***a**, and most of the expanding outbreak isolates from British Columbia are *MAT*α strains of VGII (117). Could Colombian strains of VGII have been transported 6,700 km (4,200 miles) north along the eastern Pacific coast to Vancouver Island? Why are the mating types isolated? Population and phylogenetic analyses of these VGII populations would determine how closely they are related (see chapter 23).

Taking advantage of their sustained ecological research, Granados and Castañeda conducted a retrospective analysis of arboreal sample sites in four Colombian cities over a period of 12 years (1992 to 2004) (87). These data were amassed from 97 collection dates and 8,220 arboreal samples, of which 2.63%, representing 35 or more tree species, were positive for serotypes A ($n = 36$), B ($n = 119$), or C ($n = 60$). All but one of the serotype C isolates were obtained from almond trees (*T. catappa*). The arboreal prevalence of *Cryptococcus* in Bogotá and Cúcuta was more than six times greater than in the more western cities of Medellín and Cali; this difference was not attributable to the tree sampling but was consistent with the differential rates of cryptococcosis in these cities (87). Other factors, such as temperature, elevation, humidity, and precipitation, were not clearly associated with these differences. However, the arboreal prevalence of serotype B (VGII) was greater under conditions of low temperature and high humidity; conversely, serotypes A (VNI) and C (VGIII and VGIV) seemed to prefer higher temperatures and lower humidity (87). Two provocative findings to emerge from these studies are the significant association of VGIII and VGIV with almond trees and the preference of VGII for temperate climates, which is consistent with the establishment of VGII in British Columbia, the U.S. Pacific Northwest, and Argentina (Buenos Aires) (87).

India

As elsewhere, *C. neoformans* is the dominant clinical species in India, and since the 1960s, mycologists in India have confirmed the long-standing association of *C. neoformans* with pigeon feces (92, 223, 224). Most of these reports are listed in Table 1. In addition, the colonization of trees by *C. neoformans* (Table 2), as well as *C. gattii* (Table 3), has been firmly established in India. The most extensive and sustained research has been directed by the esteemed medical mycologist Harbans S. Randhawa. Unlike in Australia, *C. gattii* was not isolated from over 700 samples of *Eucalyptus* trees in northern India (197). However, Randhawa et al. discovered a significant niche in the hollows of *Syzygium cumini* trees (common names include jambul, jamun, and Java plum), which yielded the highest arboreal prevalence of *Cryptococcus* in India (195). In the initial report, 84% (26/31) of sampled *S. cumini* trees were positive, and they harbored isolates of serotype A and serotype B (195). This finding was followed with an expanded survey of *S. cumini* trees in four northwestern Indian cities. The recovery of *C. neoformans* was most prevalent among the trees sampled in Amritsar (54%), and *C. gattii* was highest in the trees in Delhi (89%) (196). Over a period of 5 years, repeated sampling of debris and air in the hollows of seven *S. cumini* trees in Delhi determined that they were perennially colonized with strains of both serotypes A and B (196).

These results led to more extensive sampling of decayed wood in hollows and soil around the bases of 14 species of trees in the south (Tamil Nadu) and north (Uttar Pradesh, Haryana, and Union Territory) (193). Overall, 20% of 311 samples yielded isolates of serotype A (51 trees), B (24 trees), or both (11 trees) (193). All the serotype A and B isolates had the *MAT*α mating type and were VNI and VGI, respectively. The species of trees and *Cryptococcus* were not correlated. In addition, many samples of basal soil as well as hollows were positive, indicating a telluric reservoir for *Cryptococcus*. Eight tree species were resampled for periods of up to 2.5 years, and perennial colonization was also demonstrated for the mast tree (*Polyalthia longifolia*), bullet wood (*Mimusops elengi*), and khirni (*Manilkara hexandra*) (193).

In a fruitful collaboration between the laboratories of H. S. Randhawa and Jianping Xu, 78 arboreal isolates of *C. neoformans* from nine tree species in five distinct regions of northwest India were genotyped using five MLST markers (99). Analysis of the population structure provided solid evidence of both clonality and recombination, even though all the isolates were genetically similar VNI strains with the *MAT*α mating-type allele (99). This discovery of recombination supported their hypothesis that strains of *C. neoformans* can undergo sexual reproduction on the decaying wood of several tree species in India. Their analysis of environmental isolates also affirmed the observation that compatible mating types of both *Cryptococcus* species are able to complete the sexual cycle and produce fertile basidiospores when grown in the laboratory on plants (*Eucalyptus* and *Arabidopsis*) (271).

United States

There have been numerous reports of the isolation of *C. neoformans* from a variety of environmental sites in the United States, and they have revealed that most isolates are serotype A and, lagging far behind in prevalence, serotypes D and AD. However, very few rigorous genotyping analyses have been conducted on environmental isolates, and fewer studies have compared clinical and environmental isolates that were collected in spatial and temporal proximity. Investigations of this type could be helpful in determining the source of human infections. In 1994, Brian P. Currie and colleagues genotyped clinical and environmental isolates of *C. neoformans* collected from pigeon feces near the residences of patients in New York City (48). Although these 25 isolates were quite diverse, a few with the same genotype were found in both patients and pigeon excreta, but 10 patient isolates were not among the environmental strains. All eight environmental strains were virulent for intracerebrally infected mice (48).

Ten years later, Anastasia P. Litvintseva and associates conducted a similar but larger comparison of clinical and environmental genotypes (143). They collected environmental isolates of *C. neoformans* from avian excreta at 24 locations in the eastern state of North Carolina, genotyped each isolate by using AFLP markers, analyzed the genetic structure of this population, and compared these strains with clinical isolates from North Carolina. Among 762 environmental isolates from North Carolina, 82.8% were VNI, 2.5% were VNII, 7.6% were VNIV, and 7.1% were VNIII, and no isolates of *C. gattii* were recovered. These environmental populations of VNI and VNII strains consisted of only 10 and 2 distinct genotypes, respectively. Analyses of their population structures determined that there was significant clonality, which was supported by the overrepresentation of a few genotypes and low genotypic diversity (143). Neither repeated sampling of multiple sites in the eastern coastal, central, and western mountains of North Carolina nor repeated sampling in the central area over a period of 15 months detected any notable geographical or temporal differences in the distribution of the genotypes in pigeon sites (143). Similar genotyping of clinical isolates from 42 local patients revealed a higher proportion of VNII strains (25 VNI, 15 VNII, and 1 each VNIII and VNIV) than the surrounding environment, where only 2.5% were VNII. Considering the proportion of each molecular type among these patient isolates, more environmental isolations of VNII might be expected, but the elusiveness of VNII in the environment has been recognized globally, and a similar disparity between clinical and environmental VNII isolations was reported in Brazil (257). Overall, the clinical isolates were more diverse than the environmental isolates. Six genotypes associated with the environmental strains were also isolated from patients, but one genotype that was highly prevalent in the environment was not detected in patients (143).

An additional 136 clinical isolates from other regions of the United States (Alabama, Georgia, Texas, and California) were also found to include strains with the same six genotypes present in both the environmental and clinical isolates from North Carolina. However, two other genotypes were common in both patient cohorts, but they were not isolated from the environment. Thus, most but not all of the genotypes of the environmental strains were found in patients across the United States, but the clinical samples included genotypes that were not isolated from this large environmental sampling (143).

These studies are generally consistent with other reports that clinical and environmental isolates of serotype A are often indistinguishable, leading to the conclusion that patients acquired their infections from the exogenous source of the environmental isolates with the same molecular genotype. If true, their phenotypes should be similar as well, including experimental virulence. Before Staib's medium was developed, mice were used to select and recover virulent strains from environmental samples. Most subsequent assessments of the virulence of environmental isolates employed intravenous challenges and obtained mixed results. Following her investigation of the genotypes of clinical and environmental isolates, Litvintseva developed a more robust set of MLST markers (148). To evaluate the hypothesis of a causative relationship between environmental and clinical strains with identical genotypes, she compared the murine virulence of several clinical and pigeon fecal isolates of VNI with the same AFLP and MLST genotypes (145). To simulate the natural route of infection, BALB/c mice were challenged intranasally with comparable inocula. Seven of 10 clinical strains were lethal at median times of 19 to 40 days, but only 1 of 11 environmental strains caused disease within the 60-day period of postinfection observation (145). The nonlethal strains were present in low numbers in the lungs and brains of the mice that survived, but passaging these environmental strains up to three times in mice did not significantly increase their virulence (145).

These results revealed a significant disparity between clinical isolates, which were lethal, and genotypically identical environmental isolates, which were not lethal. Thus, most of these VNI isolates from pigeon feces were nonlethal for mice, and AFLP/MLST genotypes of VNI strains did not correlate with murine virulence. These results bring into question the consensus that environmental isolates are virulent for mice. They also invoke other nonexclusive possibilities. For example, the course of cryptococcosis in mice may not necessarily correlate with cryptococcosis in humans, and although pigeon excreta harbor strains of VNI, human cases of cryptococcal disease may be acquired by exposure to other significant niche(s).

EVOLVING ENVIRONMENTAL NICHES

Specific environmental conditions are necessary for a species to exist in a particular physical environment, obtain energy and nutrients, reproduce, and evade predators. Consequently, the abundance and distribution of a species in its environment are determined by the available resources and physical conditions. In ecology, these requirements govern the niche for a species or population in an ecosystem (88). The full range of environmental conditions (biological and physical) under which an organism can exist define its fundamental niche. However, as a result of pressure from interactions with other organisms, as well as changes in the environment, species are usually forced to occupy a more restrictive niche. This constrained habitat is termed the "realized niche" and represents the environment in which a species becomes most highly adapted (106). Identification of the true realized niche for most organisms is challenging due to the complexity of the global ecosystem.

C. neoformans and *C. gattii* have different geographic distributions, which suggests the species have different realized niches. As summarized in Table 4, pigeon feces is a common reservoir for VNI and is postulated to play a central role in the dispersion of this species and its transmission from the environment to humans (179). Strains of VNI can readily be isolated from pigeon feces, and they are able to grow and mate on media containing pigeon feces (179, 235). As cited in this chapter, the rare environmental strains of VNII have included isolations from pigeon and arboreal habitats, but their numbers have been too low to define the ecological niche of this molecular type. Serotype D (VNIV) has also been isolated from pigeon feces in Europe and North America but is less prevalent in South America (13, 82). In addition, *C. neoformans* has established arboreal niches in Asia, Africa, and South America (Table 2). *C. gattii* is rarely isolated from pigeon feces but is associated with various tree species in Australia, Asia, North America, and South America (Table 3).

These different environmental niches have led to the hypothesis that *C. neoformans* is ubiquitous in the environment due to dissemination by pigeons following migratory and trade routes and that *C. gattii* is restricted to tropical and subtropical regions because it is not associated with pigeons (180).

Population studies of environmental isolates of both species have provided evidence for recombination as well as clonality (29, 99, 144). However, it has not been determined whether, or to what extent, either species reproduces sexually in nature. In the laboratory, both *C. neoformans* and *C. gattii* can mate on plant-based media (271), but pigeon feces inhibits the sexual reproduction of *C. gattii* (179).

As noted previously, before 1999, *C. gattii* seemed to be restricted to *Eucalyptus* trees in the tropics and subtropics (59, 60, 229). Ecological data have been used to track the expanding outbreak of *C. gattii* that started on Vancouver Island and has spread to mainland British Columbia (Canada) and contiguous regions in the northwestern United States (100, 157). Using a combination of environmental sampling, genotyping, geographic information system mapping, and growth analyses in the laboratory, recent studies suggest the emergence of *C. gattii* in temperate environments is likely due to an expansion or alteration of the ecological niche by a subset of the population that appears to have established a realized niche in acidic soil with low moisture and organic substrates in the coastal Douglas fir and western hemlock biogeoclimatic zones (50, 115).

In addition to the long-recognized association of serotype B with *Eucalyptus* trees in Australia, several new realized environmental niches have been discovered since 2000: serotype C with almond trees in Colombia, subpopulations

TABLE 4 Summary ecology of molecular types of *C. neoformans* and *C. gattii*

Subgroup	Most common niches	Distribution (more prevalent regions)
C. neoformans		
VNI	Pigeons, other birds, trees, soil	Global
VNII	Unknown	Global (Australia, Africa, North America)
VNB	Trees (mopane)	Southern Africa (Botswana, South Africa)
VNIII	Pigeons, soil, other birds	Global (Europe)
VNIV	Pigeons, soil, other birds	Global (Europe)
C. gattii		
VGI	Trees (eucalypt)	Global (Australia)
VGII	Trees (eucalypt)	North America (SE Canada, NW United States), South America (Colombia)
VGIII	Trees (almond)	South America
VGIV	Trees	North America (Africa?)

VNI and VGI with Java plum trees in India, subtype VGIIb with soil in specific biogeoclimatic zones of British Columbia, and VNB strains with mopane trees in South Africa and Botswana (described in chapter 20).

The public health impact of each newly discovered environmental niche must be evaluated. Verification of the niche as a source of human infection requires a powerful method(s) of determining the molecular fingerprint of each isolate and then comparing environmental isolates with clinical strains. This process is exemplified by the superb collaborations that have served to subtype VGII strains responsible for the expanding Vancouver Island isolates and trace them in patients, veterinary infections, and the environment (24, 25, 50, 114, 115, 117). Most cases of *C. gattii* associated with the Vancouver Island outbreak have occurred in immunocompetent persons, and the incubation period has been estimated in months rather than years (156). Thus, it has been possible to attribute new infections to recent exposure to environmental isolates.

With *C. neoformans*, there is mounting evidence that latent infection may precede disease for an extended period and may even be acquired in childhood (see Fig. 1). Asymptomatic infections have been detected by autopsy studies (8), serological data (52, 108), and isolation of the yeasts from respiratory specimens (3, 198). Therefore, the lag between acquisition of the infection and manifestation of disease in humans may be separated by a period of extensive time and travel. This dynamic confounds efforts to implicate an environmental niche as the source of cryptococcosis in vicinal patients. Perhaps a more logical strategy to investigate the clinical relevance of local environmental strains would be to compare their genotypes with those of infected nonhuman animals. Veterinary cases are plentiful and include domestic pets, especially dogs and cats, wild animals, farm animals, and aquatic mammals (12, 37, 163, 184). Unfortunately, most cases are not diagnosed unless the animals are expensive (e.g., alpacas), captives in zoos, or beloved house pets. Since cryptococcosis is not a zoonotic infection, veterinary cases are not reportable to most public health agencies. Even when the diagnosis is proven by positive culture, veterinary isolates are rarely retained or genotyped.

Clearly, there are no definitive answers to the question posed in the first sentence of this chapter: how do people acquire this infection? However, several realized niches of *C. neoformans* and *C. gattii* have been established, and in these regions, there is often a correlation between the molecular types of the environmental and clinical isolates, but not always (143, 261). This conundrum will be resolved by more rigorous genotyping to identify individual strains and by genotyping more clinical and environmental strains to detect identical clones in both samples. This effort will require the engagement of mycologists, clinical microbiologists, physicians, and veterinarians. The goal is worthy because it will determine not only the source of cryptococcal disease but the genotypes of the most pathogenic strains.

We greatly appreciate the following research support: U.S. Public Health Service NIH grants K22 AI070152 (K.N.) and 5R01 AI25783 (T.G.M.); Colciencias, Colombia, grants 21-01-173-95 and 21-04-1182 (E.C.); and the support of the Instituto Nacional de Salud, Bogotá, Colombia (E.C.).

REFERENCES

1. **Abadi, J., S. Nachman, A. B. Kressel, and L.-A. Pirofski.** 1999. Cryptococcosis in children with AIDS. *Clin. Infect. Dis.* **28:**309–313.
2. **Abegg, M. A., F. L. Cella, J. Faganello, P. Valente, A. Schrank, and M. H. Vainstein.** 2006. *Cryptococcus neoformans* and *Cryptococcus gattii* isolated from the excreta of Psittaciformes in a southern Brazilian zoological garden. *Mycopathologia* **161:**83–91.
3. **Aberg, J. A., L. M. Mundy, and W. G. Powderly.** 1999. Pulmonary cryptococcosis in patients without HIV infection. *Chest* **115:**734–740.
4. **Ajello, L.** 1958. Occurrence of *Cryptococcus neoformans* in soils. *Am. J. Hyg.* **67:**72–77.
5. **Amanuel, Y. W., L. Jemaneh, and D. Abate.** 2001. Isolation and characterization of *Cryptococcus neoformans* from environmental sources in Ethiopia. *Ethiop. J. Health Dev.* **15:**45–49.
6. **Arguero Licea, B., D. Garza Garza, V. Flores Urbieta, and R. A. Cervantes Olivares.** 1999. Isolation and characterization of *Cryptococcus neoformans* var. *gattii* from samples of *Eucalyptus camaldulensis* in Mexico city. *Rev. Iberoam. Micol.* **16:**40–42. (In Spanish.)
7. **Baddley, J. W., J. R. Perfect, R. A. Oster, R. A. Larsen, G. A. Pankey, H. Henderson, D. W. Haas, C. A. Kauffman, R. Patel, A. K. Zaas, and P. G. Pappas.** 2008. Pulmonary cryptococcosis in patients without HIV infection: factors associated with disseminated disease. *Eur. J. Clin. Microbiol. Infect. Dis.* **27:**937–943.
8. **Baker, R. D., and R. K. Haugen.** 1955. Tissue changes and tissue diagnosis in cryptococcosis. A study of twenty-six cases. *Am. J. Clin. Pathol.* **25:**14–24.
9. **Balankura, P.** 1974. Isolation of *Cryptococcus neoformans* from soil contaminated with pigeon droppings in Bangkok. *J. Med. Assoc. Thai.* **57:**158–159.
10. **Baró, T., J. M. Torres-Rodríguez, M. H. De Mendoza, Y. Morera, and C. Alía.** 1998. First identification of autochthonous *Cryptococcus neoformans* var. *gattii* isolated from goats with predominantly severe pulmonary disease in Spain. *J. Clin. Microbiol.* **36:**458–461.
11. **Baroni, F. A., C. R. Paula, E. G. Silva, F. C. Viani, I. N. G. Rivera, M. T. Oliveira, and W. Gambale.** 2006. *Cryptococcus neoformans* strains isolated from church towers in Rio de Janeiro City, RJ, Brazil. *Rev. Inst. Med. Trop. São Paulo* **48:**71–75.
12. **Bartlett, K. H., S. E. Kidd, and J. W. Kronstad.** 2007. The emergence of *Cryptococcus gattii* in British Columbia and the Pacific Northwest. *Curr. Infect. Dis. Rep.* **1:**108–115.
13. **Bennett, J. E., K. J. Kwon-Chung, and D. H. Howard.** 1977. Epidemiologic differences among serotypes of *Cryptococcus neoformans*. *Am. J. Epidemiol.* **105:**582–586.
14. **Bergman, F.** 1963. Occurrence of *Cryptococcus neoformans* in Sweden. *Acta Med. Scand.* **174:**651–655.
15. **Bliska, J. B., and A. Casadevall.** 2009. Intracellular pathogenic bacteria and fungi: a case of convergent evolution? *Nat. Rev. Microbiol.* **7:**165–171.
16. **Bovers, M., M. R. Diaz, F. Hagen, L. Spanjaard, B. Duim, C. E. Visser, H. L. Hoogveld, J. Scharringa, A. I. M. Hoepelman, J. W. Fell, and T. Boekhout.** 2007. Identification of genotypically diverse *Cryptococcus neoformans* and *Cryptococcus gattii* isolates by Luminex xMAP™ technology. *J. Clin. Microbiol.* **45:**1874–1883.
17. **Bovers, M., F. Hagen, and T. Boekhout.** 2008. Diversity of the *Cryptococcus neoformans-Cryptococcus gattii* species complex. *Rev. Iberoam. Micol.* **25:**S4–S12.
18. **Bovers, M., F. Hagen, E. E. Kuramae, and T. Boekhout.** 2008. Six monophyletic lineages identified within *Cryptococcus neoformans* and *Cryptococcus gattii* by multi-locus sequence typing. *Fungal Genet. Biol.* **45:**400–421.
19. **Brizzio, S., B. Turchetti, V. de Garcia, D. Libkind, P. Buzzini, and M. R. van Broock.** 2007. Extracellular enzymatic activities of basidiomycetous yeasts isolated from glacial and subglacial waters of northwest Patagonia (Argentina). *Can. J. Microbiol.* **53:**519–525.

20. **Bui, T., X. Lin, R. Malik, J. Heitman, and D. A. Carter.** 2008. Isolates of *Cryptococcus neoformans* from infected animals reveal genetic exchange in unisexual, alpha mating type populations. *Eukaryot. Cell* **7:**1771–1780.
21. **Buschke, A.** 1895. Über eine durch Coccidien Hervergerufene Krankheit des Menschen. *Dtsch. Med. Wochenschr.* **21:**14.
22. **Busse, O.** 1894. Über parasitäre Zelleinschlüsse und ihre Züchtung. *Zbl. Bakt.* **16:**175–180.
23. **Butinar, L., I. Spencer-Martins, and N. Gunde-Cimerman.** 2007. Yeasts in high Arctic glaciers: the discovery of a new habitat for eukaryotic microorganisms. *Antonie Van Leeuwenhoek* **91:**277–289.
24. **Byrnes, E. J., III, R. Bildfell, S. A. Frank, T. G. Mitchell, K. A. Marr, and J. Heitman.** 2009. Molecular evidence that the range of the Vancouver Island outbreak of *Cryptococcus gattii* infection has expanded into the Pacific Northwest in the United States. *J. Infect. Dis.* **199:**1081–1086.
25. **Byrnes, E. J., III, R. J. Bildfell, P. L. Dearing, B. A. Valentine, and J. Heitman.** 2009. *Cryptococcus gattii* with bimorphic colony types in a dog in western Oregon: additional evidence for expansion of the Vancouver Island outbreak. *J. Vet. Diagn. Invest.* **21:**133–136.
26. **Byrnes, E. J., III, W. Li, Y. Lewit, J. R. Perfect, D. A. Carter, G. M. Cox, and J. Heitman.** 2009. First reported case of *Cryptococcus gattii* in the Southeastern USA: implications for travel-associated acquisition of an emerging pathogen. *PLoS ONE* **4:**e5851.
27. **Caicedo, L. D., M. I. Alvarez, M. L. Delgado, and A. Cárdenas.** 1999. *Cryptococcus neoformans* in bird excreta in the city zoo of Cali, Colombia. *Mycopathologia* **147:**121–124.
28. **Callejas, A., N. Ordóñez, M. C. Rodríguez, and E. Castañeda.** 1998. First isolation of *Cryptococcus neoformans* var. *gattii*, serotype C, from the environment in Colombia. *Med. Mycol.* **36:**341–344.
29. **Campbell, L. T., B. J. Currie, M. B. Krockenberger, R. Malik, W. Meyer, J. Heitman, and D. A. Carter.** 2005. Clonality and recombination in genetically differentiated subgroups of *Cryptococcus gattii*. *Eukaryot. Cell* **4:**1403–1409.
30. **Campisi, E., F. Mancianti, G. Pini, E. Faggi, and G. Gargani.** 2003. Investigation in central Italy of the possible association between *Cryptococcus neoformans* var. *gattii* and *Eucalyptus camaldulensis*. *Eur. J. Epidemiol.* **18:**357–362.
31. **Carvalho, V. G., M. S. Terceti, A. L. T. Dias, C. R. Paula, J. P. Lyon, A. M. Siqueira, and M. C. Franco.** 2007. Serotype and mating type characterization of *Cryptococcus neoformans* by multiplex PCR. *Rev. Inst. Med. Trop. São Paulo* **49:**207–210.
32. **Casadevall, A., and J. R. Perfect.** 1999. *Cryptococcus neoformans*. ASM Press, Washington, DC.
33. **Casali, A. K., L. Goulart, L. K. Rosa e Silva, Â. M. Ribeiro, A. A. Amaral, S. H. Alves, A. Schrank, W. Meyer, and M. H. Vainstein.** 2003. Molecular typing of clinical and environmental *Cryptococcus neoformans* isolates in the Brazilian state Rio Grande do Sul. *FEMS Yeast Res.* **3:**405–415.
34. **Castañeda, A., S. Huérfano, M. C. Rodriguez, and E. Castañeda.** 2001. Recuperación de *Cryptococcus neoformans* var. *gattii* serotipo C a partir de detritos de almendros. *Biomédica* **21:**70–74.
35. **Castañón-Olivares, L. R., and R. López Martínez.** 1994. Isolation of *Cryptococcus neoformans* from pigeon (*Columba livia*) droppings in Mexico City. *Mycoses* **37:**325–327.
36. **Castañón-Olivares, L. R., K. Martínez Martínez, R. M. Bermúdez Cruz, M. A. Martínez Rivera, W. Meyer, R. A. Arreguín Espinosa, R. López Martínez, and G. M. Ruiz Palacios y Santos.** 2009. Genotyping of Mexican *Cryptococcus neoformans* and *C. gattii* isolates by PCR-fingerprinting. *Med. Mycol.* **47:**713–721.
37. **Castellá, G., M. L. Abarca, and F. J. Cabañes.** 2008. Cryptococcosis and domestic animals. *Rev. Iberoam. Micol.* **25:**S19–S24. (In Spanish.)
38. **Chakrabarti, A., M. Jatana, P. Kumar, L. Chatha, A. Kaushal, and A. A. Padhye.** 1997. Isolation of *Cryptococcus neoformans* var. *gattii* from *Eucalyptus camaldulensis* in India. *J. Clin. Microbiol.* **35:**3340–3342.
39. **Chee, H. Y., and K. B. Lee.** 2005. Isolation of *Cryptococcus neoformans* var. *grubii* (serotype A) from pigeon droppings in Seoul, Korea. *J. Microbiol.* **43:**469–472.
40. **Chen, J., A. Varma, M. R. Diaz, A. P. Litvintseva, K. K. Wollenberg, and K. J. Kwon-Chung.** 2008. *Cryptococcus neoformans* strains and infection in apparently immunocompetent patients, China. *Emerg. Infect. Dis.* **14:**755–762.
41. **Chen, S. C. A., B. J. Currie, H. M. Campbell, D. A. Fisher, T. J. Pfeiffer, D. H. Ellis, and T. C. Sorrell.** 1997. *Cryptococcus neoformans* var. *gattii* infection in northern Australia: existence of an environmental source other than known host eucalypts. *Trans. R. Soc. Trop. Med. Hyg.* **91:**547–550.
42. **Choe, Y. H., H. Moon, S. J. Park, S. R. Kim, H. J. Han, K. S. Lee, and Y. C. Lee.** 2009. Pulmonary cryptococcosis in asymptomatic immunocompetent hosts. *Scand. J. Infect. Dis.* **41:**602–607.
43. **Clancy, C. J., M. H. Nguyen, R. Alandoerffer, S. Cheng, K. A. Iczkowski, M. Richardson, and J. R. Graybill.** 2006. *Cryptococcus neoformans* var. *grubii* isolates recovered from persons with AIDS demonstrate a wide range of virulence during murine meningoencephalitis that correlates with the expression of certain virulence factors. *Microbiology* **152:**2247–2255.
44. **Cogliati, M., M. Allaria, A. M. Tortorano, and M. A. Viviani.** 2000. Genotyping *Cryptococcus neoformans* var. *neoformans* with specific primers designed from PCR-fingerprinting bands sequenced using a modified PCR-based strategy. *Med. Mycol.* **38:**97–103.
45. **Connolly, J. H., M. B. Krockenberger, R. Malik, P. J. Canfield, D. I. Wigney, and D. B. Muir.** 1999. Asymptomatic carriage of *Cryptococcus neoformans* in the nasal cavity of the koala (*Phascolarctos cinereus*). *Med. Mycol.* **37:**331–338.
46. **Costa, S. P. S. E., M. S. Lazéra, W. R. A. Santos, B. P. Morales, C. C. F. Bezerra, M. M. Nishikawa, G. G. Barbosa, L. Trilles, J. L. M. Nascimento, and B. Wanke.** 2009. First isolation of *Cryptococcus gattii* molecular type VGII and *Cryptococcus neoformans* molecular type VNI from environmental sources in the city of Belém, Pará, Brazil. *Mem. Inst. Oswaldo Cruz* **104:**662–664.
47. **Criseo, G., M. S. Bolignano, F. De Leo, and F. Staib.** 1995. Evidence of canary droppings as an important reservoir of *Cryptococcus neoformans*. *Zbl. Bakt.* **282:**244–254.
48. **Currie, B. P., L. F. Freundlich, and A. Casadevall.** 1994. Restriction fragment length polymorphism analysis of *Cryptococcus neoformans* isolates from environmental (pigeon excreta) and clinical sources in New York City. *J. Clin. Microbiol.* **32:**1188–1192.
49. **Curtis, F.** 1896. Contribution à l'étude de la saccharomycose humaine. *Ann. Inst. Pasteur* **10:**449–468.
50. **Datta, K., K. H. Bartlett, R. Baer, E. J. Byrnes, III, E. Galanis, J. Heitman, L. M. N. Hoang, M. J. Leslie, L. MacDougall, S. S. Magill, M. G. Morshed, and K. A. Marr.** 2009. Spread of *Cryptococcus gattii* into Pacific Northwest region of the United States. *Emerg. Infect. Dis.* **15:**1185–1191.
51. **Datta, K., K. H. Bartlett, and K. A. Marr.** 2009. *Cryptococcus gattii*: emergence in western North America: exploitation of a novel ecological niche. *Interdiscip. Perspect. Infect. Dis.* **2009:**176532.

52. **Davis, J., W. Y. Zheng, A. Glatman-Freedman, J. A. Ng, M. R. Pagcatipunan, H. Lessin, A. Casadevall, and D. L. Goldman.** 2007. Serologic evidence for regional differences in pediatric cryptococcal infection. *Pediatr. Infect. Dis. J.* **26:**549–551.
53. **de Garcia, V., S. Brizzio, D. Libkind, P. Buzzini, and M. R. van Broock.** 2007. Biodiversity of cold-adapted yeasts from glacial meltwater rivers in Patagonia, Argentina. *FEMS Microbiol. Ecol.* **59:**331–341.
54. **Delgado, A. C. N., H. Taguchi, Y. Mikami, M. Myiajy, M. C. B. Villares, and M. L. Moretti.** 2005. Human cryptococcosis: relationship of environmental and clinical strains of *Cryptococcus neoformans* var. *neoformans* from urban and rural areas. *Mycopathologia* **159:**7–11.
55. **Dixit, A., S. F. Carroll, and S. T. Qureshi.** 2009. *Cryptococcus gattii*: an emerging cause of fungal disease in North America. *Interdiscip. Perspect. Infect. Dis.* **2009:**840452.
56. **dos Santos, W. R., W. Meyer, B. Wanke, S. P. Costa, L. Trilles, J. L. M. Nascimento, R. Medeiros, B. P. Morales, C. C. F. Bezerra, R. C. de Macêdo, S. O. Ferreira, G. G. Barbosa, M. A. Perez, M. M. Nishikawa, and M. S. Lazéra.** 2008. Primary endemic *Cryptococcosis gattii* by molecular type VGII in the state of Pará, Brazil. *Mem. Inst. Oswaldo Cruz* **103:**813–818.
57. **Duncan, C. G., H. Schwantje, C. Stephen, J. Campbell, and K. H. Bartlett.** 2006. *Cryptococcus gattii* in wildlife of Vancouver Island, British Columbia, Canada. *J. Wildl. Dis.* **42:**175–178.
58. **Ecevit, I. Z., C. J. Clancy, I. M. Schmalfuss, and M. H. Nguyen.** 2006. The poor prognosis of central nervous system cryptococcosis among nonimmunosuppressed patients: a call for better disease recognition and evaluation of adjuncts to antifungal therapy. *Clin. Infect. Dis.* **42:**1443–1447.
59. **Ellis, D. H., and T. J. Pfeiffer.** 1990. Natural habitat of *Cryptococcus neoformans* var. *gattii*. *J. Clin. Microbiol.* **28:**1642–1644.
60. **Ellis, D. H., and T. J. Pfeiffer.** 1992. The ecology of *Cryptococcus neoformans*. *Eur. J. Epidemiol.* **8:**321–325.
61. **Emmons, C. W.** 1951. Isolation of *Cryptococcus neoformans* from soil. *J. Bacteriol.* **62:**685–690.
62. **Emmons, C. W.** 1955. Saprophytic sources of *Cryptococcus neoformans* associated with the pigeon (*Columba livia*). *Am. J. Hyg.* **62:**227–232.
63. **Ergin, C., M. Ilkit, and O. Kaftanoglu.** 2004. Detection of *Cryptococcus neoformans* var. *grubii* in honeybee (*Apis mellifera*) colonies. *Mycoses* **47:**431–434.
64. **Escandón, P., S. Huérfano, and E. Castañeda.** 2002. Experimental inoculation of *Terminalia catappa* seedlings with an environmental isolate of *Cryptococcus neoformans* var. *gattii* serotype C. *Biomédica* **22:**524–528. (In Spanish.)
65. **Escandón, P., E. Quintero, D. P. Granados, S. Huérfano, A. Ruiz, and E. Castañeda.** 2005. Isolation of *Cryptococcus gattii* serotype B from detritus of *Eucalyptus* trees in Colombia. *Biomédica* **25:**390–397. (In Spanish.)
66. **Escandón, P., A. Sánchez, M. Martinez, W. Meyer, and E. Castañeda.** 2006. Molecular epidemiology of clinical and environmental isolates of the *Cryptococcus neoformans* species complex reveals a high genetic diversity and the presence of the molecular type VGII mating type **a** in Colombia. *FEMS Yeast Res.* **6:**625–635.
67. **Evans, E. E.** 1950. The antigenic composition of *Cryptococcus neoformans*. I. A serologic classification by means of the capsular agglutinations. *J. Immunol.* **64:**423–430.
68. **Evans, E. E., and J. F. Kessel.** 1951. The antigenic compostion of *Cryptococcus neoformans*. II. Serologic studies with the capsular polysaccharide. *J. Immunol.* **67:**109–114.
69. **Fell, J. W., T. Boekhout, Á. Fonseca, G. Scorzetti, and A. Statzell-Tallman.** 2000. Biodiversity and systematics of basidiomycetous yeasts as determined by large-subunit rDNA D1/D2 domain sequence analysis. *Int. J. Syst. Evol. Microbiol.* **50:**1351–1371.
70. **Feng, X., Z. Yao, D. Ren, and W. Liao.** 2008. Simultaneous identification of molecular and mating types within the *Cryptococcus* species complex by PCR-RFLP analysis. *J. Med. Microbiol.* **57:**1481–1490.
71. **Feng, X., Z. Yao, D. Ren, W. Liao, and J. Wu.** 2008. Genotype and mating type analysis of *Cryptococcus neoformans* and *Cryptococcus gattii* isolates from China that mainly originated from non-HIV-infected patients. *FEMS Yeast Res.* **8:**930–938.
72. **Filiú, W. F., B. Wanke, S. M. Agüena, V. O. Vilela, R. C. L. Macedo, and M. S. Lazéra.** 2002. Avian habitats as sources of *Cryptococcus neoformans* in the city of Campo Grande, Mato Grosso do Sul, Brazil. *Rev. Soc. Bras. Med. Trop.* **35:**591–595. (In Portuguese.)
73. **Findley, K., M. Rodriguez-Carres, B. Metin, J. Kroiss, Á. Fonseca, R. J. Vilgalys, and J. Heitman.** 2009. Phylogeny and phenotypic characterization of pathogenic *Cryptococcus* species and closely related saprobic taxa in the Tremellales. *Eukaryot. Cell* **8:**353–361.
74. **Fortes, S. T., M. S. Lazéra, M. M. Nishikawa, R. C. L. Macedo, and B. Wanke.** 2001. First isolation of *Cryptococcus neoformans* var. *gattii* from a native jungle tree in the Brazilian Amazon rainforest. *Mycoses* **44:**137–140.
75. **Fox, D. L., and N. L. Müller.** 2005. Pulmonary cryptococcosis in immunocompetent patients: CT findings in 12 patients. *Am. J. Roentgenol.* **185:**622–626.
76. **Franzot, S. P., J. S. Hamdan, B. P. Currie, and A. Casadevall.** 1997. Molecular epidemiology of *Cryptococcus neoformans* in Brazil and the United States: evidence for both local genetic differences and a global clonal population structure. *J. Clin. Microbiol.* **35:**2243–2251.
77. **Fraser, J. A., S. S. Giles, E. C. Wenink, S. G. Geunes-Boyer, J. R. Wright, S. Diezmann, A. Allen, J. E. Stajich, F. S. Dietrich, J. R. Perfect, and J. Heitman.** 2005. Same-sex mating and the origin of the Vancouver Island *Cryptococcus gattii* outbreak. *Nature* **437:**1360–1364.
78. **Frey, D., and E. B. Durie.** 1964. The isolation of *Cryptococcus neoformans* (*Torula histolytica*) from soil in New Guinea and pigeon droppings in Sydney, New South Wales. *Med. J. Aust.* **1:**947–949.
79. **Gadanho, M., and J.-P. Sampaio.** 2009. *Cryptococcus ibericus* sp. nov., *Cryptococcus aciditolerans* sp. nov. and *Cryptococcus metallitolerans* sp. nov., a new ecoclade of anamorphic basidiomycetous yeast species from an extreme environment associated with acid rock drainage in São Domingos pyrite mine, Portugal. *Int. J. Syst. Evol. Microbiol.* **59:**2375–2379.
80. **Gallo, M. G., P. Cabeli, and V. Vidotto.** 1989. Presence of pathogenic yeasts in the feces of the semi-domesticated pigeon (*Columba livia*, Gmelin 1789, urban type) from the city of Turin. *Parassitologia* **31:**207–212. (In Italian.)
81. **García-Bermejo, M. J., J. Antón, C. A. Ferrer, I. Meseguer, J. L. Abad, and M. F. Colom.** 2001. Chromosome length polymorphism in *Cryptococcus neoformans* clinical and environmental isolates. *Rev. Iberoam. Micol.* **18:**174–179.
82. **García-Hermoso, D., S. Mathoulin-Pélissier, B. Couprie, O. Ronin, B. Dupont, and F. Dromer.** 1997. DNA typing suggests pigeon droppings as a source of pathogenic *Cryptococcus neoformans* serotype D. *J. Clin. Microbiol.* **35:**2683–2685.
83. **García-Rivera, J., and A. Casadevall.** 2001. Melanization of *Cryptococcus neoformans* reduces its susceptibility to the antimicrobial effects of silver nitrate. *Med. Mycol.* **39:**353–357.
84. **Gatti, F., and R. Eeckels.** 1970. An atypical strain of *Cryptococcus neoformans* (San Felice) Vuillemin 1894. Part I.

Description of the disease and of the strain. *Ann. Soc. Belges Med. Trop. Parasitol. Mycol.* **50:**689–693.

85. **Girish Kumar, C. P., D. Prabu, H. Mitani, Y. Mikami, and T. Menon.** 2010. Environmental isolation of *Cryptococcus neoformans* and *Cryptococcus gattii* from living trees in Guindy National Park, Chennai, South India. *Mycoses* **53:**262–264.
86. **Granados, D. P., and E. Castañeda.** 2005. Isolation and characterization of *Cryptococcus neoformans* varieties recovered from natural sources in Bogota, Colombia, and study of ecological conditions in the area. *Microb. Ecol.* **49:**282–290.
87. **Granados, D. P., and E. Castañeda.** 2006. Influence of climatic conditions on the isolation of members of the *Cryptococcus neoformans* species complex from trees in Colombia from 1992–2004. *FEMS Yeast Res.* **6:**636–644.
88. **Grinnell, J.** 1917. The niche-relationships of the California thrasher. *Auk* **34:**427–433.
89. **Grover, N., S. R. Nawange, J. Naidu, S. M. Singh, and A. Sharma.** 2007. Ecological niche of *Cryptococcus neoformans* var. *grubii* and *Cryptococcus gattii* in decaying wood of trunk hollows of living trees in Jabalpur City of Central India. *Mycopathologia* **164:**159–170.
90. **Gugnani, H. C., T. G. Mitchell, A. P. Litvintseva, K. B. Lengeler, J. Heitman, A. Kumar, S. Basu, and A. Paliwal-Johsi.** 2005. Isolation of *Cryptococcus gattii* and *Cryptococcus neoformans* var. *grubii* from the flowers and bark of *Eucalyptus* trees in India. *Med. Mycol.* **43:**565–569.
91. **Gugnani, H. C., and A. N. U. Njoku-Obi.** 1973. Occurrence of *Cryptococcus neoformans* in pigeon excreta in Enugu (Nigeria). *West Afr. Med. J. Niger. Med. Dent. Pract.* **22:**121–122.
92. **Gugnani, H. C., R. S. Sandhu, and S. K. Shome.** 1976. Prevalence of *Cryptococcus neoformans* in avian excreta in India. *Mykosen* **19:**183–187.
93. **Haag-Wackernagel, D., and H. Moch.** 2004. Health hazards posed by feral pigeons. *J. Infect.* **48:**307–313.
94. **Halde, C. J., and M. A. Fraher.** 1966. *Cryptococcus neoformans* in pigeon feces in San Francisco. *Calif. Med.* **104:**188–190.
95. **Halliday, C. L., and D. A. Carter.** 2003. Clonal reproduction and limited dispersal in an environmental population of *Cryptococcus neoformans* var. *gattii* isolates from Australia. *J. Clin. Microbiol.* **41:**703–711.
96. **Hasenclever, H. F., and C. W. Emmons.** 1963. The prevalence and mouse virulence of *Cryptococcus neoformans* strains isolated from urban areas. *Am. J. Hyg.* **78:**227–231.
97. **Hejtmankova, N., and M. Cisecky.** 1984. Isolation, cultivation and identification of *Cryptococcus neoformans* from pigeons. *Acta Univ. Palacki. Olomuc. Fac. Med.* **107:**43–49.
98. **Hibbett, D. S., M. Binder, J. F. Bischoff, M. Blackwell, P. F. Cannon, O. E. Eriksson, S. Huhndorf, T. Y. James, P. M. Kirk, R. Lücking, H. Thorsten Lumbsch, F. M. Lutzoni, P. B. Matheny, D. J. McLaughlin, M. J. Powell, S. Redhead, C. L. Schoch, J. W. Spatafora, J. A. Stalpers, R. J. Vilgalys, M. C. Aime, A. Aptroot, R. Bauer, D. Begerow, G. L. Benny, L. A. Castlebury, P. W. Crous, Y.-C. Dai, W. Gams, D. M. Geiser, G. W. Griffith, C. Gueidan, D. L. Hawksworth, G. Hestmark, K. Hosaka, R. A. Humber, K. D. Hyde, J. E. Ironside, U. Kõljalg, C. P. Kurtzman, K.-H. Larsson, R. Lichtwardt, J. E. Longcore, J. Miadlikowska, A. Miller, J.-M. Moncalvo, S. E. Mozley-Standridge, F. Oberwinkler, E. Parmasto, V. Reeb, J. D. Rogers, C. Roux, L. Ryvarden, J.-P. Sampaio, A. Schüssler, J. Sugiyama, R. G. Thorn, L. Tibell, W. A. Untereiner, C. Walker, Z. Wang, A. Weir, M. Weiss, M. M. White, K. Winka, Y.-J. Yao, and N. Zhang.** 2007. A higher-level phylogenetic classification of the Fungi. *Mycol. Res.* **111:**509–547.
99. **Hiremath, S. S., A. Chowdhary, T. Kowshik, H. S. Randhawa, S. Sun, and J. Xu.** 2008. Long-distance dispersal and recombination in environmental populations of *Cryptococcus neoformans* var. *grubii* from India. *Microbiol.* **154:**1513–1524.
100. **Hoang, L. M. N., J. A. Maguire, P. Doyle, M. Fyfe, and D. L. Roscoe.** 2004. *Cryptococcus neoformans* infections at Vancouver Hospital and Health Sciences Centre (1997–2002): epidemiology, microbiology and histopathology. *J. Med. Microbiol.* **53:**935–940.
101. **Horta, J. A., C. C. Staats, A. K. Casali, Â. M. Ribeiro, I. S. Schrank, A. Schrank, and M. H. Vainstein.** 2002. Epidemiological aspects of clinical and environmental *Cryptococcus neoformans* isolates in the Brazilian state Rio Grande do Sul. *Med. Mycol.* **40:**565–571.
102. **Hu, G., I. Liu, A. P. Sham, J. E. Stajich, F. S. Dietrich, and J. W. Kronstad.** 2008. Comparative hybridization reveals extensive genome variation in the AIDS-associated pathogen *Cryptococcus neoformans*. *Genome Biol.* **9:**R41.
103. **Hubálek, Z.** 1975. Distribution of *Cryptococcus neoformans* in a pigeon habitat. *Folia Parasitol.* **22:**73–79.
104. **Huérfano, S., A. Castañeda, and E. Castañeda.** 2001. Experimental infection of almond trees seedlings (*Terminalia catappa*) with an environmental isolate of *Cryptococcus neoformans* var. *gattii*, serotype C. *Rev. Iberoam. Micol.* **18:**131–132.
105. **Huérfano, S., M. C. Cepero, and E. Castañeda.** 2003. Phenotype characterization of environmental *Cryptococcus neoformans* isolates. *Biomédica* **23:**328–340. (In Spanish.)
106. **Hutchinson, G. E.** 1957. Concluding remarks. *Cold Spring Harb. Symp. Quant. Bio.* **22:**415–427.
107. **Irokanulo, E. A. O., A. A. Makinde, C. O. Akueshi, and M. Ekwonu.** 1997. *Cryptococcus neoformans* var. *neoformans* isolated from droppings of captive birds in Nigeria. *J. Wildl. Dis.* **33:**343–345.
108. **Jarvis, J. N., S. D. Lawn, M. Vogt, N. Bangani, R. Wood, and T. S. Harrison.** 2009. Screening for cryptococcal antigenemia in patients accessing an antiretroviral treatment program in South Africa. *Clin. Infect. Dis.* **48:**856–862.
109. **Kagaya, K., T. Yamada, Y. Miyakawa, Y. Fukazawa, and S. Saito.** 1985. Characterization of pathogenic constituents of *Cryptococcus neoformans*. *Microbiol. Immunol.* **29:**517–532.
110. **Kao, C. J., and J. Schwarz.** 1957. The isolation of *Cryptococcus neoformans* from pigeon nests; with remarks on the identification of virulent cryptococci. *Am. J. Clin. Pathol.* **27:**652–663.
111. **Karstaedt, A. S., H. H. Crewe-Brown, and F. Dromer.** 2002. Cryptococcal meningitis caused by *Cryptococcus neoformans* var. *gattii*, serotype C, in AIDS patients in Soweto, South Africa. *Med. Mycol.* **40:**7–11.
112. **Katsu, M., A. Ando, R. Ikeda, Y. Mikami, and K. Nishimura.** 2003. Immunomagnetic isolation of *Cryptococcus neoformans* by beads coated with anti-*Cryptococcus* serum. *Jpn. J. Med. Mycol.* **44:**139–144.
113. **Khawcharoenporn, T., A. Apisarnthanarak, and L. M. Mundy.** 2007. Non-*neoformans* cryptococcal infections: a systematic review. *Infection* **35:**51–58.
114. **Kidd, S. E., P. J. Bach, A. O. Hingston, S. Mak, Y. Chow, L. MacDougall, J. W. Kronstad, and K. H. Bartlett.** 2007. *Cryptococcus gattii* dispersal mechanisms, British Columbia, Canada. *Emerg. Infect. Dis.* **13:**51–57.
115. **Kidd, S. E., Y. Chow, S. Mak, P. J. Bach, H. Chen, A. O. Hingston, J. W. Kronstad, and K. H. Bartlett.** 2007. Characterization of environmental sources of the human and animal pathogen, *Cryptococcus gattii*, in British Columbia, Canada, and the Pacific Northwest of the United States. *Appl. Environ. Microbiol.* **73:**1433–1443.
116. **Kidd, S. E., H. Guo, K. H. Bartlett, J. Xu, and J. W. Kronstad.** 2005. Comparative gene genealogies indicate that two clonal lineages of *Cryptococcus gattii* in British

Columbia resemble strains from other geographical areas. *Eukaryot. Cell* **4:**1629–1638.

117. **Kidd, S. E., F. Hagen, R. L. Tscharke, M. Huynh, K. H. Bartlett, M. Fyfe, L. MacDougall, T. Boekhout, K. J. Kwon-Chung, and W. Meyer.** 2004. A rare genotype of *Cryptococcus gattii* caused the cryptococcosis outbreak on Vancouver Island (British Columbia, Canada). *Proc. Natl. Acad. Sci. USA* **101:**17258–17263.
118. **Kidd, S. E., C. A. Hitchcock, K. Maszewska, P. Balakrishnan, C. Hunt, P. Pajendran, A. R. Usha, S. P. Thyagarajan, T. C. Sorrell, and W. Meyer.** 2001. Molecular epidemiology of clinical and environmental isolates of *Cryptococcus neoformans* from Tamil Nadu, India. *Austral. Mycol.* **20:**105–114.
119. **Kielstein, P., and H. Hotzel.** 1998. Distribution, serovar affiliation and epidemiologic behavior of *Cryptococcus neoformans* isolates from ornamental bird breeds. *Dtsch. Tierarztl. Wochenschr.* **105:**349–353. (In German.)
120. **Kielstein, P., H. Hotzel, and A. Schmalreck.** 2002. Which are the conditions for *Cryptococcus neoformans* var. *neoformans*-strains from avian excrements as a cause for human infections? *Mycoses* **45**(Suppl.3)**:**61–64. (In German.)
121. **Knudsen, J. D., L. Jensen, T. L. Sorensen, T. Jensen, H. Kjersem, J. Stenderup, and C. Pedersen.** 1997. Cryptococcosis in Denmark: an analysis of 28 cases in 1988–1993. *Scand. J. Infect. Dis.* **29:**51–55.
122. **Kobayashi, C. C., L. K. Souza, O. F. Fernandes, S. C. Brito, A. C. Silva, E. D. Sousa, and M. R. Silva.** 2005. Characterization of *Cryptococcus neoformans* isolated from urban environmental sources in Goiânia, Goiás State, Brazil. *Rev. Inst. Med. Trop. São Paulo* **47:**203–207.
123. **Kohno, S.** 2003. Clinical and mycological features of cryptococcosis. *Jpn. J. Med. Mycol.* **44:**159–162. (In Japanese.)
124. **Krockenberger, M. B., P. J. Canfield, and R. Malik.** 2003. *Cryptococcus neoformans* var. *gattii* in the koala (*Phascolarctos cinereus*): a review of 43 cases of cryptococcosis. *Med. Mycol.* **41:**225–234.
125. **Kutty, S. N., and R. Philip.** 2008. Marine yeasts: a review. *Yeast* **25:**465–483.
126. **Kwon-Chung, K. J., and J. E. Bennett.** 1984. High prevalence of *Cryptococcus neoformans* var. *gattii* in tropical and subtropical regions. *Zbl. Bakt. Hyg. A* **257:**213–218.
127. **Kwon-Chung, K. J., and J. E. Bennett.** 1992. Cryptococcosis, p. 397–446. *In Medical Mycology*. Lea & Febiger, Philadelphia, PA.
128. **Kwon-Chung, K. J., T. Boekhout, J. W. Fell, and M. Diaz.** 2004. Proposal to conserve the name *Cryptococcus gattii* against *C. hondurianus* and *C. bacillisporus* (*Basidiomycota, Hymenomycetes, Tremellomycetidae*). *Taxon* **51:**804–806.
129. **Lagrou, K., J. Van Eldere, S. Keuleers, F. Hagen, R. Merckx, J. Verhaegen, W. E. Peetermans, and T. Boekhout.** 2005. Zoonotic transmission of *Cryptococcus neoformans* from a magpie to an immunocompetent patient. *J. Intern. Med.* **257:**385–388.
130. **Lazéra, M. S., M. A. S. Cavalcanti, L. Trilles, M. M. Nishikawa, and B. Wanke.** 1998. *Cryptococcus neoformans* var. *gattii*: evidence for a natural habitat related to decaying wood in a pottery tree hollow. *Med. Mycol.* **36:**119–122.
131. **Lazéra, M. S., F. D. A. Pires, L. Camillo-Coura, M. M. Nishikawa, C. C. F. Bezerra, L. Trilles, and B. Wanke.** 1996. Natural habitat of *Cryptococcus neoformans* var. *neoformans* in decaying wood forming hollows in living trees. *J. Med. Vet. Mycol.* **34:**127–131.
132. **Lazéra, M. S., M. A. Salmito Cavalcanti, A. T. Londero, L. Trilles, M. M. Nishikawa, and B. Wanke.** 2000. Possible primary ecological niche of *Cryptococcus neoformans*. *Med. Mycol.* **38:**379–383.
133. **Lazéra, M. S., B. Wanke, and M. M. Nishikawa.** 1993. Isolation of both varieties of *Cryptococcus neoformans* from saprophytic sources in the city of Rio de Janeiro, Brazil. *J. Med. Vet. Mycol.* **31:**449–454.
134. **Lemmer, K., D. Naumann, B. Raddatz, and K. Tintelnot.** 2004. Molecular typing of *Cryptococcus neoformans* by PCR fingerprinting, in comparison with serotyping and Fourier transform infrared-spectroscopy-based phenotyping. *Med. Mycol.* **42:**135–147.
135. **Li, A., K. Nishimura, H. Taguchi, R. Tanaka, S. Wu, and M. Miyaji.** 1993. The isolation of *Cryptococcus neoformans* from pigeon droppings and serotyping of naturally and clinically sourced isolates in China. *Mycopathologia* **124:**1–5.
136. **Liaw, S. J., H. C. Wu, and P. R. Hsueh.** 2010. Microbiological characteristics of clinical isolates of *Cryptococcus neoformans* in Taiwan: serotypes, mating types, molecular types, virulence factors, and antifungal susceptibility. *Clin. Microbiol. Infect.* **16:**696–703.
137. **Lin, X., J. C. Huang, T. G. Mitchell, and J. Heitman.** 2006. Virulence attributes and hyphal growth of *C. neoformans* are quantitative traits and the *MATa* allele enhances filamentation. *PLoS Genet.* **2:**e187.
138. **Lin, X., C. M. Hull, and J. Heitman.** 2005. Sexual reproduction between partners of the same mating type in *Cryptococcus neoformans*. *Nature* **434:**1017–1021.
139. **Littman, M. L.** 1959. Cryptococcosis (torulosis). Current concepts and therapy. *Am. J. Med.* **27:**976–998.
140. **Littman, M. L., and R. Borok.** 1968. Relation of the pigeon to cryptococcosis: natural carrier state, heat resistance and survival of *Cryptococcus neoformans*. *Mycopathol. Mycol. Appl.* **35:**329–345.
141. **Littman, M. L., and S. S. Schneierson.** 1959. *Cryptococcus neoformans* in pigeon excreta in New York City. *Am. J. Hyg.* **69:**49–59.
142. **Littman, M. L., and J. E. Walter.** 1968. Cryptococcosis: current status. *Am. J. Med.* **45:**922–932.
143. **Litvintseva, A. P., L. Kestenbaum, R. J. Vilgalys, and T. G. Mitchell.** 2005. Comparative analysis of environmental and clinical populations of *Cryptococcus neoformans*. *J. Clin. Microbiol.* **43:**556–564.
144. **Litvintseva, A. P., R. E. Marra, K. Nielsen, J. Heitman, R. J. Vilgalys, and T. G. Mitchell.** 2003. Evidence of sexual recombination among *Cryptococcus neoformans* serotype A isolates in sub-Saharan Africa. *Eukaryot. Cell* **2:**1162–1168.
145. **Litvintseva, A. P., and T. G. Mitchell.** 2009. Most environmental isolates of *Cryptococcus neoformans* var. *grubii* are not lethal for mice. *Infect. Immun.* **77:**3188–3195.
146. **Litvintseva, A. P., T. G. Mitchell, and N. Govender.** 2009. Out-of-Africa origin of *Cryptococcus neoformans* var. *grubii* (serotype A). 17th Congress, International Society for Human and Animal Mycology, Abstract EP-04-2, p. 205.
147. **Litvintseva, A. P., R. Thakur, L. B. Reller, and T. G. Mitchell.** 2005. Prevalence of clinical isolates of *Cryptococcus gattii* serotype C among patients with AIDS in sub-Saharan Africa. *J. Infect. Dis.* **192:**888–892.
148. **Litvintseva, A. P., R. Thakur, R. J. Vilgalys, and T. G. Mitchell.** 2006. Multilocus sequence typing reveals three genetic subpopulations of *Cryptococcus neoformans* var. *grubii* (serotype A), including a unique population in Botswana. *Genetics* **172:**2223–2238.
149. **Lizarazo, J., M. Linares, C. de Bedout, A. Restrepo, C. I. Agudelo, and E. Castañeda.** 2007. Results of nine years of the clinical and epidemiological survey on cryptococcosis in Colombia, 1997–2005. *Biomédica* **27:**94–109. (In Spanish.)
150. **Lizarazo, J., M. Mendoza, D. Palacios, A. Vallejo, A. Bustamante, E. Ojeda, A. Restrepo, and E. Castañeda.** 2000.

Cryptococcosis caused by *Cryptococcus neoformans* variety *gattii*. *Acta Med. Colombiana* **25:**171–178.

151. **Loftus, B. J., E. Fung, P. Roncaglia, D. Rowley, P. Amedeo, D. Bruno, J. Vamathevan, M. Miranda, I. J. Anderson, J. A. Fraser, J. E. Allen, I. E. Bosdet, M. R. Brent, R. Chiu, T. L. Doering, M. J. Donlin, C. A. D'Souza, D. S. Fox, V. Grinberg, J. Fu, M. Fukushima, B. J. Haas, J. C. Huang, G. Janbon, S. J. M. Jones, H. L. Koo, M. I. Krzywinshi, K. J. Kwon-Chung, K. B. Lengeler, R. Maiti, M. A. Marra, R. E. Marra, C. A. Mathewson, T. G. Mitchell, M. Pertea, F. R. Riggs, S. L. Salzberg, J. E. Schein, A. Shvartsbeyn, H. Shin, M. Shumway, C. A. Specht, B. B. Suh, A. E. Tenney, T. R. Utterback, B. L. Wickes, J. R. Wortman, N. H. Wye, J. W. Kronstad, J. K. Lodge, J. Heitman, R. W. Davis, C. M. Fraser, and R. W. Hyman.** 2005. The genome of the basidiomycetous yeast and human pathogen *Cryptococcus neoformans*. *Science* **307:**1321–1324.
152. **Lo Passo, C., I. Pernice, M. Gallo, C. Barbara, F. T. Luck, G. Criseo, and A. Pernice.** 1997. Genetic relatedness and diversity of *Cryptococcus neoformans* strains in the Maltese Islands. *J. Clin. Microbiol.* **35:**751–755.
153. **López Martínez, R., and L. R. Castañón-Olivares.** 1995. Isolation of *Cryptococcus neoformans* var. *neoformans* from bird droppings, fruits and vegetables in Mexico City. *Mycopathologia* **129:**25–28.
154. **Lugarini, C., C. S. Goebel, L. A. Condas, M. D. Muro, M. R. de Farias, F. M. Ferreira, and M. H. Vainstein.** 2008. *Cryptococcus neoformans* isolated from passerine and psittacine bird excreta in the state of Parana, Brazil. *Mycopathologia* **166:**61–69.
155. **Lui, G., N. Lee, M. Ip, K. W. Choi, Y. K. Tso, E. Lam, S. Chau, R. Lai, and C. S. Cockram.** 2006. Cryptococcosis in apparently immunocompetent patients. *Q. J. Med.* **99:**143–151.
156. **MacDougall, L., and M. Fyfe.** 2006. Emergence of *Cryptococcus gattii* in a novel environment provides clues to its incubation period. *J. Clin. Microbiol.* **44:**1851–1852.
157. **MacDougall, L., S. E. Kidd, E. Galanis, S. Mak, M. J. Leslie, P. R. Cieslak, J. W. Kronstad, M. G. Morshed, and K. H. Bartlett.** 2007. Spread of *Cryptococcus gattii* in British Columbia, Canada, and detection in the Pacific Northwest, USA. *Emerg. Infect. Dis.* **13:**42–50.
158. **Mahmoud, Y. A. G.** 1999. First environmental isolation of *Cryptococcus neoformans* var. *neoformans* and var. *gatti* from the Gharbia Governorate, Egypt. *Mycopathologia* **148:**83–86.
159. **Malik, R., M. B. Krockenberger, G. Cross, R. Doneley, D. N. Madill, D. Black, P. McWhirter, A. Rozenwax, K. Rose, M. Alley, D. Forshaw, I. Russell-Brown, A. C. Johnstone, P. Martin, C. R. O'Brien, and D. N. Love.** 2003. Avian cryptococcosis. *Med. Mycol.* **41:**115–124.
160. **Marra, R. E., J. C. Huang, E. Fung, K. Nielsen, J. Heitman, R. J. Vilgalys, and T. G. Mitchell.** 2004. A genetic linkage map of *Cryptococcus neoformans* variety *neoformans* serotype D (*Filobasidiella neoformans*). *Genetics* **167:**619–631.
161. **Matsumoto, M. T., A. M. F. Almeida, L. C. Baeza, M. S. C. Melhem, and M. J. S. Mendes-Giannini.** 2007. Genotyping, serotyping and determination of mating-type of *Cryptococcus neoformans* clinical isolates from São Paulo State, Brazil. *Rev. Inst. Med. Trop. São Paulo* **49:**41–47.
162. **McDonough, E. S., A. L. Lewis, and L. A. Penn.** 1966. Relationship of *Cryptococcus neoformans* to pigeons in Milwaukee, Wisconsin. *Public Health Rep.* **81:**1119–1123.
163. **McGill, S., R. Malik, N. Saul, S. Beetson, C. Secombe, I. Robertson, and P. Irwin.** 2009. Cryptococcosis in domestic animals in Western Australia: a retrospective study from 1995–2006. *Med. Mycol.* **47:**625–639.
164. **Meyer, W., D. M. Aanensen, T. Boekhout, M. Cogliati, M. R. Diaz, M. C. Esposto, M. C. Fisher, F. Gilgado, F. Hagen, S. Kaocharoen, A. P. Litvintseva, T. G. Mitchell, S. P. Simwami, L. Trilles, M. A. Viviani, and K. J. Kwon-Chung.** 2009. Consensus multi-locus sequence typing scheme for *Cryptococcus neoformans* and *Cryptococcus gattii*. *Med. Mycol.* **47:**561–570.
165. **Meyer, W., A. Castañeda, S. Jackson, M. Huynh, E. Castañeda, and the IberoAmerican Cryptococcal Study Group.** 2003. Molecular typing of IberoAmerican *Cryptococcus neoformans* isolates. *Emerg. Infect. Dis.* **9:**189–195.
166. **Meyer, W., K. Marszewska, M. Amirmostofian, R. P. Igreja, C. Hardtke, K. Methling, M. A. Viviani, A. Chindamporn, S. Sukroongreung, M. A. John, D. H. Ellis, and T. C. Sorrell.** 1999. Molecular typing of global isolates of *Cryptococcus neoformans* var. *neoformans* by polymerase chain reaction fingerprinting and randomly amplified polymorphic DNA–a pilot study to standardize techniques on which to base a detailed epidemiological survey. *Electrophoresis* **20:**1790–1799.
167. **Mitchell, T. G.** 2008. Genomic approaches to investigate the pathogenicity of *Cryptococcus neoformans*. *Curr. Fungal Infect. Rep.* **2:**172–179.
168. **Mitchell, T. G.** 2010. Population genetics of pathogenic fungi in humans and other animals, p. 139–158. *In* J. Xu (ed.), *Microbial Population Genetics*. Horizon Scientific Press, Hethersett, UK.
169. **Mitchell, T. G., and A. P. Litvintseva.** 2010. Typing species of *Cryptococcus* and epidemiology of cryptococcosis, p. 167–190. *In* H. R. Ashbee and E. M. Bignell (ed.), *Pathogenic Yeasts*. Springer-Verlag, Berlin, Germany.
170. **Mitchell, T. G., and J. R. Perfect.** 1995. Cryptococcosis in the era of AIDS: 100 years after the discovery of *Cryptococcus neoformans*. *Clin. Microbiol. Rev.* **8:**515–548.
171. **Montagna, M. T., M. P. Santacroce, G. Caggiano, D. Tato, and L. Ajello.** 2003. Cavernicolous habitats harbouring *Cryptococcus neoformans*: results of a speleological survey in Apulia, Italy, 1999–2000. *Med. Mycol.* **41:**451–455.
172. **Montenegro, H., and C. R. Paula.** 2000. Environmental isolation of *Cryptococcus neoformans* var. *gattii* and *C. neoformans* var. *neoformans* in the city of São Paulo, Brazil. *Med. Mycol.* **38:**385–390.
173. **Morera-López, Y., J. M. Torres-Rodríguez, T. Jiménez-Cabello, T. Baró-Tomás, C. Alia-Aponte, and M. S. Lazéra.** 2005. DNA fingerprinting pattern and susceptibility to antifungal drugs in *Cryptococcus neoformans* variety *grubii* isolates from Barcelona city and rural environmental samples. *Mycopathologia* **160:**9–14.
174. **Morgan, J., K. M. McCarthy, S. M. Gould, K. Fan, B. A. Arthington-Skaggs, N. J. Iqbal, K. Stamey, R. A. Hajjeh, and M. E. Brandt.** 2006. *Cryptococcus gattii* infection: characteristics and epidemiology of cases identified in a South African province with high HIV seroprevalence, 2002–2004. *Clin. Infect. Dis.* **43:**1077–1080.
175. **Nadrous, H. F., V. S. Antonios, C. L. Terrell, and J. H. Ryu.** 2003. Pulmonary cryptococcosis in nonimmunocompromised patients. *Chest* **124:**2143–2147.
176. **Nagahama, T., M. Hamamoto, T. Nakase, Y. Takaki, and K. Horikoshi.** 2003. *Cryptococcus surugaensis* sp. nov., a novel yeast species from sediment collected on the deep-sea floor of Suruga Bay. *Int. J. Syst. Evol. Microbiol.* **53:**2095–2098.
177. **Nakamura, Y., R. Kano, H. Sato, S. Watanabe, H. Takahashi, and A. Hasegawa.** 1998. Isolates of *Cryptococcus neoformans* serotype A and D developed on canavanine-glycine-bromthymol blue medium. *Mycoses* **41:**35–40.
178. **Ngamskulrungroj, P., F. Gilgado, J. Faganello, A. P. Litvintseva, A. L. Leal, K. M. Tsui, T. G. Mitchell,**

M. H. Vainstein, and W. Meyer. 2009. Genetic diversity of the *Cryptococcus* species complex suggests that *Cryptococcus gattii* deserves to have varieties. *PLoS ONE* **4:**e5862.
179. **Nielsen, K., A. L. De Obaldia, and J. Heitman.** 2007. *Cryptococcus neoformans* mates on pigeon guano: implications for the realized ecological niche and globalization. *Eukaryot. Cell* **6:**949–959.
180. **Nielsen, K., and J. Heitman.** 2007. Sex and virulence of human pathogenic fungi. *Adv. Genet.* **57:**143–173.
181. **Nishikawa, M. M., M. S. Lazéra, G. G. Barbosa, L. Trilles, B. R. Balassiano, R. C. Macedo, C. C. F. Bezerra, M. A. Perez, P. Cardarelli, and B. Wanke.** 2003. Serotyping of 467 *Cryptococcus neoformans* isolates from clinical and environmental sources in Brazil: analysis of host and regional patterns. *J. Clin. Microbiol.* **41:**73–77.
182. **Nosanchuk, J. D., and A. Casadevall.** 2003. The contribution of melanin to microbial pathogenesis. *Cell. Microbiol.* **5:**203–223.
183. **Nosanchuk, J. D., J. Rudolph, A. L. Rosas, and A. Casadevall.** 1999. Evidence that *Cryptococcus neoformans* is melanized in pigeon excreta: implications for pathogenesis. *Infect. Immun.* **67:**5477–5479.
184. **O'Brien, C. R., M. B. Krockenberger, D. I. Wigney, P. Martin, and R. Malik.** 2004. Retrospective study of feline and canine cryptococcosis in Australia from 1981 to 2001: 195 cases. *Med. Mycol.* **42:**449–460.
185. **Othman, N., N. A. Abdullah, and Z. A. Wahab.** 2004. Cryptococcal meningitis in an immunocompetent child: a case report and literature review. *Southeast.Asian J. Trop. Med. Public Health* **35:**930–934.
186. **Pal, M.** 1997. First report of isolation of *Cryptococcus neoformans* var. *neoformans* from avian excreta in Kathmandu, Nepal. *Rev. Iberoam. Micol.* **14:**181–183.
187. **Partridge, B. M., and H. I. Winner.** 1965. *Cryptococcus neoformans* in bird droppings in London. *Lancet* **1:**1060–1061.
188. **Passoni, L. F., B. Wanke, M. M. Nishikawa, and M. S. Lazéra.** 1998. *Cryptococcus neoformans* isolated from human dwellings in Rio de Janeiro, Brazil: an analysis of the domestic environment of AIDS patients with and without cryptococcosis. *Med. Mycol.* **36:**305–311.
189. **Polacheck, I., and K. J. Kwon-Chung.** 1988. Melanogenesis in *Cryptococcus neoformans*. *J. Gen. Microbiol.* **134:**1037–1041.
190. **Powell, K. E., B. A. Dahl, R. J. Weeks, and F. E. Tosh.** 1972. Airborne *Cryptococcus neoformans*: particles from pigeon excreta compatible with alveolar deposition. *J. Infect. Dis.* **125:**412–415.
191. **Quintero, E., E. Castañeda, and A. Ruiz.** 2005. Environmental distribution of *Cryptococcus neoformans* in the department of Cundinamarca, Colombia. *Rev. Iberoam. Micol.* **22:**93–98. (In Spanish.)
192. **Randhawa, H. S., Y. M. Clayton, and R. W. Riddell.** 1965. Isolation of *Cryptococcus neoformans* from pigeon habitats in London. *Nature* **208:**801.
193. **Randhawa, H. S., T. Kowshik, A. Chowdhary, S. K. Preeti, Z. U. Khan, S. Sun, and J. Xu.** 2008. The expanding host tree species spectrum of *Cryptococcus gattii* and *Cryptococcus neoformans* and their isolations from surrounding soil in India. *Med. Mycol.* **46:**823–833.
194. **Randhawa, H. S., T. Kowshik, and Z. U. Khan.** 2003. Decayed wood of *Syzygium cumini* and *Ficus religiosa* living trees in Delhi/New Delhi metropolitan area as natural habitat of *Cryptococcus neoformans*. *Med. Mycol.* **41:**199–209.
195. **Randhawa, H. S., T. Kowshik, and Z. U. Khan.** 2005. Efficacy of swabbing versus a conventional technique for isolation of *Cryptococcus neoformans* from decayed wood in tree trunk hollows. *Med. Mycol.* **43:**67–71.
196. **Randhawa, H. S., T. Kowshik, S. K. Preeti, A. Chowdhary, Z. U. Khan, Z. Yan, J. Xu, and A. Kumar.** 2006. Distribution of *Cryptococcus gattii* and *Cryptococcus neoformans* in decayed trunk wood of *Syzygium cumini* trees in north-western India. *Med. Mycol.* **44:**623–630.
197. **Randhawa, H. S., A. Y. Mussa, and Z. U. Khan.** 2001. Decaying wood in tree trunk hollows as a natural substrate for *Cryptococcus neoformans* and other yeast-like fungi of clinical interest. *Mycopathologia* **151:**63–69.
198. **Randhawa, H. S., and D. K. Paliwal.** 1979. Survey of *Cryptococcus neoformans* in the respiratory tract of patients with bronchopulmonary disorders and in the air. *Sabouraudia* **17:**399–404.
199. **Raso, T. F., K. Werther, E. T. Miranda, and M. J. S. Mendes-Giannini.** 2004. Cryptococcosis outbreak in psittacine birds in Brazil. *Med. Mycol.* **42:**355–362.
200. **Refai, M., M. Taha, S. A. Selim, F. Elshabourii, and H. H. Yousseff.** 1983. Isolation of *Cryptococcus neoformans*, *Candida albicans* and other yeasts from pigeon droppings in Egypt. *Sabouraudia* **21:**163–165.
201. **Refojo, N., D. E. Perrotta, M. Brudny, R. Abrantes, A. I. Hevia, and G. Davel.** 2009. Isolation of *Cryptococcus neoformans* and *Cryptococcus gattii* from trunk hollows of living trees in Buenos Aires City, Argentina. *Med. Mycol.* **47:**177–184.
202. **Reimão, J. Q., E. D. Drummond, M. S. Terceti, J. P. Lyon, M. C. Franco, and A. M. de Siqueira.** 2007. Isolation of *Cryptococcus neoformans* from hollows of living trees in the city of Alfenas, MG, Brazil. *Mycoses* **50:**261–264.
203. **Restrepo, A., D. J. Baumgardner, E. Bagagli, C. R. Cooper, Jr., M. R. McGinnis, M. S. Lazéra, F. H. Barbosa, S. M. Bosco, Z. P. Camargo, K. I. R. Coelho, S. T. Fortes, M. Franco, M. R. Montenegro, A. Sano, and B. Wanke.** 2000. Clues to the presence of pathogenic fungi in certain environments. *Med. Mycol.* **38**(Suppl.1)**:**67–77.
204. **Rhodes, J. C., I. Polacheck, and K. J. Kwon-Chung.** 1982. Phenoloxidase activity and virulence in isogenic strains of *Cryptococcus neoformans*. *Infect. Immun.* **36:**1175–1184.
205. **Ribeiro, Â. M., L. K. Rosa e Silva, I. S. Schrank, A. Schrank, W. Meyer, and M. H. Vainstein.** 2006. Isolation of *Cryptococcus neoformans* var. *neoformans* serotype D from eucalypts in South Brazil. *Med. Mycol.* **44:**707–713.
206. **Ribeiro, M. A., and P. Ngamskulrungroj.** 2008. Molecular characterization of environmental *Cryptococcus neoformans* isolated in Vitória, ES, Brazil. *Rev. Inst. Med. Trop. São Paulo* **50:**315–320.
207. **Rosario, I., B. Acosta, and M. F. Colom.** 2008. Pigeons and other birds as a reservoir for *Cryptococcus* spp. *Rev. Iberoam. Micol.* **25:**S13–S18. (In Spanish.)
208. **Rosario, I., M. Hermoso de Mendoza, S. Déniz, G. Soro, I. Álamo, and B. Acosta.** 2005. Isolation of *Cryptococcus* species including *C. neoformans* from cloaca of pigeons. *Mycoses* **48:**421–424.
209. **Rosas, A. L., and A. Casadevall.** 2001. Melanization decreases the susceptibility of *Cryptococcus neoformans* to enzymatic degradation. *Mycopathologia* **151:**53–56.
210. **Ruiz, A., and G. S. Bulmer.** 1981. Particle size of airborne *Cryptococcus neoformans* in a tower. *Appl. Environ. Microbiol.* **41:**1225–1229.
211. **Ruiz, A., R. A. Fromtling, and G. S. Bulmer.** 1981. Distribution of *Cryptococcus neoformans* in a natural site. *Infect. Immun.* **31:**560–563.
212. **Russo, G., D. Libkind, J.-P. Sampaio, and M. R. van Broock.** 2008. Yeast diversity in the acidic Rio Agrio-Lake Caviahue volcanic environment (Patagonia, Argentina). *FEMS Microbiol. Ecol.* **65:**415–424.

213. **Salas, S. D., J. E. Bennett, K. J. Kwon-Chung, J. R. Perfect, and P. R. Williamson.** 1996. Effect of the laccase gene *CNLAC1*, on virulence of *Cryptococcus neoformans*. *J. Exp. Med.* **184:**377–386.
214. **Sampaio, A., J.-P. Sampaio, and C. Leão.** 2007. Dynamics of yeast populations recovered from decaying leaves in a nonpolluted stream: a 2-year study on the effects of leaf litter type and decomposition time. *FEMS Yeast Res.* **7:**595–603.
215. **Sanchetee, P.** 1998. Cryptococcal meningitis in immunocompetent patients. *J. Assoc. Physicians India* **46:**617–619.
216. **Sanfelice, F.** 1894. Contributo alla morfologia e biologia dei blastomiceti che si sviluppano nei succhi di alcuni frutti. *Ann. Igien.* **4:**463–495.
217. **Santos, W. R. A., W. Meyer, B. Wanke, S. P. S. E. Costa, L. Trilles, J. L. M. Nascimento, R. Medeiros, B. P. Morales, C. C. F. Bezerra, R. C. L. Macedo, S. O. Ferreira, G. G. Barbosa, M. A. Perez, M. M. Nishikawa, and M. S. Lazéra.** 2008. Primary endemic *Cryptococcosis gattii* by molecular type VGII in the state of Pará, Brazil. *Mem. Inst. Oswaldo Cruz* **103:**813–818.
218. **Saraçli, M. A., S. T. Yildiran, K. Sener, A. Gönlüm, L. Doganci, S. M. Keller, and B. L. Wickes.** 2006. Genotyping of Turkish environmental *Cryptococcus neoformans* var. *neoformans* isolates by pulsed field gel electrophoresis and mating type. *Mycoses* **49:**124–129.
219. **Saul, N., M. B. Krockenberger, and D. A. Carter.** 2008. Evidence of recombination in mixed mating type and α-only populations of *Cryptococcus gattii* sourced from single *Eucalyptus* tree hollows. *Eukaryot. Cell* **7:**727–734.
220. **Schonborn, C., B. Schutze, and H. Pohler.** 1969. Budding fungi in feces of zoo birds, wild indigenous birds and pigeons gone wild. (Studies of the occurrence of *Cryptococcus neoformans* in birds). *Mykosen* **12:**471–490. (In German.)
221. **Scorzetti, G., J. W. Fell, Á. Fonseca, and A. Statzell-Tallman.** 2002. Systematics of basidiomycetous yeasts: a comparison of large subunit D1/D2 and internal transcribed spacer rDNA regions. *FEMS Yeast Res.* **2:**495–517.
222. **Sekhon, A. S., S. N. Bannerje., B. M. Mielke, H. Idikio, G. Wood, and J. M. S. Dixon.** 1990. Current status of cryptococcosis in Canada. *Mycoses* **33:**73–80.
223. **Sethi, K. K., and H. S. Randhawa.** 1968. Survival of *Cryptococcus neoformans* in the gastrointestinal tract of pigeons following ingestion of the organism. *J. Infect. Dis.* **118:**135–138.
224. **Sethi, K. K., H. S. Randhawa, S. Abraham, S. K. Mishra, and V. N. Damodaran.** 1966. Occurrence of *Cryptococcus neoformans* in pigeon excreta in Delhi. *Indian J. Chest Dis.* **8:**207–210.
225. **Shih, C. C., Y.-C. Chen, S.-C. Chang, K.-T. Luh, and W.-C. Hsieh.** 2000. Cryptococcal meningitis in non-HIV-infected patients. *Q. J. Med.* **93:**245–251.
226. **Silva, M. E., and L. A. Paula.** 1963. Isolation of *Cryptococcus neoformans* from excrement and nests of pigeons (*Columba livia*) in Salvador, Bahia (Brazil). *Rev. Inst. Med. Trop. São Paulo* **5:**9–11. (In Portuguese.)
227. **Soares, M. C., C. R. Paula, A. L. T. Dias, M. M. Caseiro, and S. O. Costa.** 2005. Environmental strains of *Cryptococcus neoformans* variety *grubii* in the city of Santos, SP, Brazil. *Rev. Inst. Med. Trop. São Paulo* **47:**31–36.
228. **Soogarun, S., V. Wiwanitkit, A. Palasuwan, P. Pradniwat, J. Suwansaksri, T. Lertlum, and T. Maungkote.** 2006. Detection of *Cryptococcus neoformans* in bird excreta. *Southeast Asian J. Trop. Med. Public Health* **37:**768–770.
229. **Sorrell, T. C.** 2001. *Cryptococcus neoformans* variety *gattii*. *Med. Mycol.* **39:**155–168.
230. **Sorrell, T. C., S. C. A. Chen, P. Ruma, W. Meyer, T. J. Pfeiffer, D. H. Ellis, and A. G. Brownlee.** 1996. Concordance of clinical and environmental isolates of *Cryptococcus neoformans* var. *gattii* by random amplification of polymorphic DNA analysis and PCR fingerprinting. *J. Clin. Microbiol.* **34:**1253–1260.
231. **Sorrell, T. C., and D. H. Ellis.** 1997. Ecology of *Cryptococcus neoformans*. *Rev. Iberoam. Micol.* **14:**42–43.
232. **Souza, L. K., A. H. Souza, Jr., C. R. Costa, J. Faganello, M. H. Vainstein, A. L. Chagas, A. C. Souza, and M. R. Silva.** 2009. Molecular typing and antifungal susceptibility of clinical and environmental *Cryptococcus neoformans* species complex isolates in Goiânia, Brazil. *Mycoses* **53:**62–67.
233. **Sriburee, P., S. Khayhan, C. Khamwan, S. Panjaisee, and P. Tharavichitkul.** 2004. Serotype and PCR-fingerprints of clinical and environmental isolates of *Cryptococcus neoformans* in Chiang Mai, Thailand. *Mycopathologia* **158:**25–31.
234. **Staib, F.** 1962. *Cryptococcus neoformans* und *Guizotia abyssinica* (Syn. G. *oleifera*) Farbreaktion für *Cr. neoformans*. *Zbl. Hyg.* **148:**466–475.
235. **Staib, F.** 1981. The perfect state of *Cryptococcus neoformans, Filobasidiella neoformans*, on pigeon manure filtrate agar. *Zbl. Bakt. A* **248:**575–578.
236. **Staib, F.** 1985. Sampling and isolation of *Cryptococcus neoformans* from indoor air with the aid of the Reuter Centrifugal Sampler (RCS) and *Guizotia abyssinica* creatinine agar. A contribution to the mycological-epidemiological control of *Cr. neoformans* in the fecal matter of caged birds. *Zbl. Bakt. Mikrobiol. Hyg. B* **180:**567–575.
237. **Staib, F., and M. Heissenhuber.** 1989. *Cryptococcus neoformans* in bird droppings: a hygienic-epidemiological challenge. *AIDS Forsch.* **12:**649–655.
238. **Staib, F., and S. K. Mishra.** 1975. Contributions to the strain-specific virulence of *Cryptococcus neoformans*. Animal experiments with two *C. neoformans*-strains isolated from bird manure. Preliminary report. *Zbl. Bakt. Orig. A* **230:**81–85. (In German.)
239. **Staib, F., and J. Schulz-Dieterich.** 1984. *Cryptococcus neoformans* in fecal matter of birds kept in cages: control of *Cr. neoformans* habitats. *Zbl. Bakt. Mikrobiol. Hyg. B* **179:**179–186.
240. **Steenbergen, J. N., J. D. Nosanchuk, S. D. Malliaris, and A. Casadevall.** 2003. *Cryptococcus neoformans* virulence is enhanced after growth in the genetically malleable host *Dictyostelium discoideum*. *Infect. Immun.* **71:**4862–4872.
241. **Steenbergen, J. N., H. A. Shuman, and A. Casadevall.** 2001. *Cryptococcus neoformans* interactions with amoebae suggest an explanation for its virulence and intracellular pathogenic strategy in macrophages. *Proc. Natl. Acad. Sci. USA* **98:**15245–15250.
242. **Stenderup, J., K. Flensted, C. Jorgensen, A. H. Sorensen, N. C. Hansen, and H. C. Siersted.** 1989. Occurrence of the yeast, *Cryptococcus (Cr) neoformans*, in pigeon droppings. *Ugeskr. Laeger* **151:**2974–2975. (In Danish.)
243. **Stephen, C., S. J. Lester, W. Black, M. Fyfe, and S. Raverty.** 2002. Multispecies outbreak of cryptococcosis on southern Vancouver Island, British Columbia. *Can. Vet. J.* **43:**792–794.
244. **Sweeney, D. A., M. T. Caserta, D. N. Korones, A. Casadevall, and D. L. Goldman.** 2003. A ten-year-old boy with a pulmonary nodule secondary to *Cryptococcus neoformans*: case report and review of the literature. *Pediatr. Infect. Dis. J.* **22:**1089–1093.
245. **Swinne, D.** 1988. Ecology of *Cryptococcus neoformans* and epidemiology and cryptococcosis in the old world, p. 113–119. *In* J. M. Torres-Rodríguez (ed.), *Proceedings of the X Congress of the International Society for Human and Animal Mycology*. Prous Scientific, Barcelona, Spain.

246. **Swinne, D., M. Deppner, R. Laroche, J. J. Floch, and P. Kadende.** 1989. Isolation of *Cryptococcus neoformans* from houses of AIDS-associated cryptococcosis patients in Bujumbura (Burundi). *AIDS* **3:**389–390.
247. **Swinne, D., M. Deppner, S. Maniratunga, R. Laroche, J. J. Floch, and P. Kadende.** 1991. AIDS-associated cryptococcosis in Bujumbura, Burundi: an epidemiological study. *J. Med. Vet. Mycol.* **29:**25–30.
248. **Swinne, D., K. D. Kayembe, and M. Niyimi.** 1986. Isolation of saprophytic *Cryptococcus neoformans* var. *neoformans* in Kinshasa, Zaire. *Ann. Soc. Belge Med. Trop.* **66:**57–61.
249. **Swinne-Desgain, D.** 1975. *Cryptococcus neoformans* of saprophytic origin. *Sabouraudia* **13:**303–308.
250. **Takashima, M., and T. Nakase.** 1999. Molecular phylogeny of the genus *Cryptococcus* and related species based on the sequences of 18S rDNA and internal transcribed spacer regions. *Microbiol. Cult. Coll.* **15:**35–47.
251. **Takashima, M., T. Sugita, Y. Toriumi, and T. Nakase.** 2009. *Cryptococcus tepidarius* sp. nov., a thermotolerant yeast species isolated from a stream from a hot-spring area in Japan. *Int. J. Syst. Evol. Microbiol.* **59:**181–185.
252. **Tanaka, R., K. Nishimura, and M. Miyaji.** 1999. Ploidy of serotype AD strains of *Cryptococcus neoformans*. *Jpn. J. Med. Mycol.* **40:**31–34.
253. **Tay, S. T., H. C. Lim, T. H. Tajuddin, M. Y. Rohani, H. Hamimah, and K. L. Thong.** 2006. Determination of molecular types and genetic heterogeneity of *Cryptococcus neoformans* and *C. gattii* in Malaysia. *Med. Mycol.* **44:**617–622.
254. **Tay, S. T., M. Y. Rohani, T. S. Soo Hoo, and H. Hamimah.** 2009. Epidemiology of cryptococcosis in Malaysia. *Mycoses* **52:**1–6.
255. **Taylor, R. L., and C. Duangmani.** 1968. Occurrence of *Cryptococcus neoformans* in Thailand. *Am. J. Epidemiol.* **87:**318–322.
256. **Torfoss, D., and P. Sandven.** 2005. Invasive fungal infections at The Norwegian Radium Hospital 1998–2003. *Scand. J. Infect. Dis.* **37:**585–589.
257. **Trilles, L., M. S. Lazéra, B. Wanke, R. V. Oliveira, G. G. Barbosa, M. M. Nishikawa, B. P. Morales, and W. Meyer.** 2008. Regional pattern of the molecular types of *Cryptococcus neoformans* and *Cryptococcus gattii* in Brazil. *Mem. Inst. Oswaldo Cruz* **103:**455–462.
258. **Trilles, L., M. S. Lazéra, B. Wanke, B. Theelen, and T. Boekhout.** 2003. Genetic characterization of environmental isolates of the *Cryptococcus neoformans* species complex from Brazil. *Med. Mycol.* **41:**383–390.
259. **Tsinzerling, V. A., D. V. Komarova, M. V. Vasil'eva, and V. E. Karev.** 2003. Pathological anatomy of HIV infection according to data in Saint-Petersburg. *Arkh. Patol.* **65:**42–45. (In Russian.)
260. **Vanbreuseghem, R., and M. Takashio.** 1970. An atypical strain of *Cryptococcus neoformans* (San Felice) Vuillemin 1894. Part II. *Cryptococcus neoformans* var. *gatti* var. nov. *Ann. Soc. Belge Med. Trop.* **50:**695–702.
261. **Varma, A., D. Swinne, F. Staib, J. E. Bennett, and K. J. Kwon-Chung.** 1995. Diversity of DNA fingerprints in *Cryptococcus neoformans*. *J. Clin. Microbiol.* **33:**1807–1814.
262. **Velegraki, A., V. G. Kiosses, H. Pitsouni, D. Toukas, V. D. Daniilidis, and N. J. Legakis.** 2001. First report of *Cryptococcus neoformans* var. *gattii* serotype B from Greece. *Med. Mycol.* **39:**419–422.
263. **Vidotto, V., and M. G. Gallo.** 1985. Study on the presence of yeasts in the feces of the rock pigeon (*Columba livia* Gmelin 1789) from rural areas. *Parassitologia* **27:**313–320. (In Italian.)
264. **Vilcins, I., M. B. Krockenberger, H. Agus, and D. A. Carter.** 2002. Environmental sampling for *Cryptococcus neoformans* var. *gattii* from the Blue Mountains National Park, Sydney, Australia. *Med. Mycol.* **40:**53–60.
265. **Viviani, M. A., M. Cogliati, M. C. Esposto, K. Lemmer, K. Tintelnot, M. F. Valiente, D. Swinne, A. Velegraki, and R. Velho.** 2006. Molecular analysis of 311 *Cryptococcus neoformans* isolates from a 30-month ECMM survey of cryptococcosis in Europe. *FEMS Yeast Res.* **6:**614–619.
266. **Vogel, R. A.** 1966. The indirect fluorescent antibody test for the detection of antibody in human cryptococcal disease. *J. Infect. Dis.* **116:**573–580.
267. **Wang, C.-Y., H.-D. Wu, and P.-R. Hsueh.** 2005. Nosocomial transmission of cryptococcosis. *N. Engl. J. Med.* **352:**1271–1272.
268. **Williamson, P. R.** 1994. Biochemical and molecular characterization of the diphenol oxidase of *Cryptococcus neoformans*: identification as a laccase. *J. Bacteriol.* **176:**656–664.
269. **Xu, J., and T. G. Mitchell.** 2002. Strain variation and clonality in *Candida* spp. and *Cryptococcus neoformans*, p. 739–749. *In* R. A. Calderone and R. L. Cihlar (ed.), *Fungal Pathogenesis: Principles and Clinical Applications*. Marcel Dekker, New York, NY.
270. **Xu, J., and T. G. Mitchell.** 2003. Population genetic analyses of medically important fungi, p. 703–722. *In* D. H. Howard (ed.), *Pathogenic Fungi in Humans and Animals*, 2nd ed. Marcel Dekker, New York, NY.
271. **Xue, C., Y. Tada, X. Dong, and J. Heitman.** 2007. The human fungal pathogen *Cryptococcus* can complete its sexual cycle during a pathogenic association with plants. *Cell Host Microbe* **1:**263–273.
272. **Yamamoto, Y., S. Kohno, H. Koga, H. Kakeya, K. Tomono, M. Kaku, T. Yamazaki, M. Arisawa, and K. Hara.** 1995. Random amplified polymorphic DNA analysis of clinically and environmentally isolated *Cryptococcus neoformans* in Nagasaki. *J. Clin. Microbiol.* **33:**3328–3332.
273. **Yildiran, S. T., M. A. Saraçli, A. Gönlüm, and H. Gün.** 1998. Isolation of *Cryptococcus neoformans* var. *neoformans* from pigeon droppings collected throughout Turkey. *Med. Mycol.* **36:**391–394.
274. **Yilmaz, A., G. Goral, S. Helvaci, K. Kilicturgay, F. Gokirmak, O. Tore, and S. Gedikoglu.** 1989. Distribution of *Cryptococcus neoformans* in pigeon feces. *Mikrobiyol. Bul.* **23:**121–126. (In Turkish.)
275. **Zhu, X., and P. R. Williamson.** 2004. Role of laccase in the biology and virulence of *Cryptococcus neoformans*. *FEMS Yeast Res.* **5:**1–10.

Cryptococcus: From Human Pathogen to Model Yeast
Edited by J. Heitman et al.

19

Cryptococcus neoformans: Nonvertebrate Hosts and the Emergence of Virulence

JEFFREY J. COLEMAN, CARA J. CHRISMAN, ARTURO CASADEVALL, AND ELEFTHERIOS MYLONAKIS

A variety of nonvertebrate hosts has been used to investigate virulence traits and therapies for several bacterial and fungal pathogens. These hosts include nonvertebrate animals and protozoan species and are henceforth referred to as "heterologous hosts" to distinguish them from mammalian hosts. Heterologous hosts have had two major uses in the cryptococcal field. First, heterologous hosts such as protozoa have provided many insights into the emergence of virulence for organisms acquired from the environment, such as *Cryptococcus neoformans*. Second, heterologous hosts have emerged as major experimental systems for the screening of virulence characteristics since they are less costly, are amenable to high-throughput protocols, and are exempt from vertebrate animal study regulations. A central theme that has emerged from these studies is that the ability of *C. neoformans* to cause disease in both mammalian and alternative hosts strongly suggests that certain virulence traits and certain corresponding host responses are conserved evolutionarily. This chapter details the use of heterologous hosts for the study of *C. neoformans*.

ACANTHAMOEBA CASTELLANII AND OTHER AMOEBAE

Amoebae are found in practically every habitat of the biosphere, including air currents, deserts, and hot springs (31, 32, 35). Amoebae are natural predatory cells that feed on bacteria, fungi, and other microorganisms. Amoebae and *Cryptococcus* spp. share ecologic sites, and it is reasonable to posit that they interact and that this interaction can exert selective pressures on each entity. In fact, amoebae have been proposed to be predators of *C. neoformans* and thus, involved in biological control of this fungus in the environment (33, 34). Amoebae and cryptococci have been shown to interact readily under laboratory conditions (Fig. 1), where the outcome of the interaction is dependent on the identity of both the protozoan and fungal strains. In this regard, it is noteworthy that the commonly used amoeba *A. castellanii* was first isolated in the 1930s as a contaminant of a cryptococcal culture (6).

The first studies of amoeba-cryptococcus interactions date to the 1950s when Castellani showed that *A. castellanii* ingested and destroyed a strain of *C. neoformans* (7). There the story rested for almost two decades until Bulmer and his collaborators revisited the problem in a remarkable series of experiments. While doing studies of cryptococcal pathogenesis, Bulmer and colleagues noted that attempts to recover yeast cells from infected mice were sometimes marred by contamination with amoebae. They studied this phenomenon and showed that *Acanthamoeba polyphaga* was a rapacious predator of yeast *C. neoformans* cells (25). An interesting finding in these studies was the observation that surviving colonies of *C. neoformans* were composed of pseudohyphal cells, suggesting that the protozoan either ignored or could not phagocytose the morphologic variants. Given that phenotypic switching of *C. neoformans* can result in pseudohyphal variants (14), this observation linked fungal morphological transitions with resistance against amoeboid predators.

In recent years the major impetus for studying cryptococcus-amoeba interactions was the emergent consensus that the fungus was a facultative intracellular pathogen with a distinctive and unusual intracellular survival strategy (12). The fact that *C. neoformans* could be ingested and survive inside phagocytic cells from many mammalian hosts raised the question of why a soil organism would have a sophisticated intracellular virulence strategy in the absence of any apparent need for animal infection for its survival or life cycle (reviewed in reference 4). The need to explain the origin of *C. neoformans*' intracellular virulence led to a renewed interest in environmental hosts, and Steenbergen et al. showed that the cryptococcal intracellular pathogenic strategy in *A. castellanii* was remarkably similar to that observed in mammalian macrophages (39). A striking finding was the correlation between known mammalian virulence factors such as the capsule, melanin, and phospholipase and their necessity for survival in amoebae.

Jeffrey J. Coleman and Eleftherios Mylonakis, Harvard Medical School, Massachusetts General Hospital, Division of Infectious Diseases, Boston, MA 02114. **Cara J. Chrisman and Arturo Casadevall,** Albert Einstein College of Medicine, Department of Microbiology and Immunology, Bronx, NY 10461.

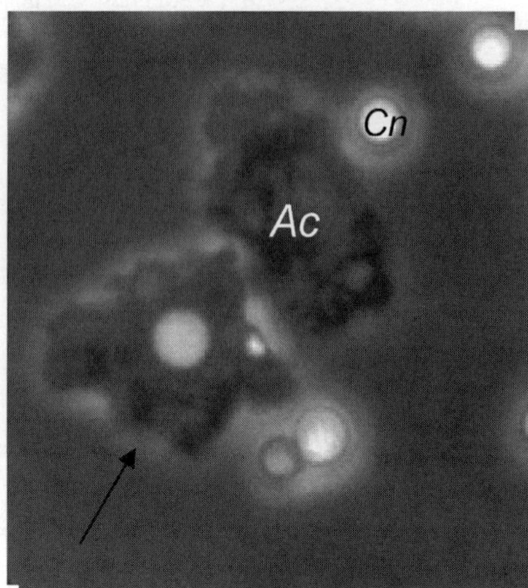

FIGURE 1 Interaction of *A. castellanii* and *C. neoformans* in liquid culture. The two organisms are shown after 2 h of incubation in phosphate-buffered saline. The *C. neoformans* (Cn) strain 24067 ingested by the *A. castellanii* (Ac) is indicated by the arrow. Image was taken at 40×.

Studies with amoebae have also provided a potential explanation for the phenomenon of capsular enlargement that follows introduction of *C. neoformans* into mammalian hosts. Recently, the concordance of capsule enlargement as a protective mechanism for both amoebae and macrophages was shown in vitro (42). When presented with *A. castellanii* cells in culture, *C. neoformans* reacted with an increase in the size of the capsule; often the overall capsule volume increased up to 68% when compared to the preincubation capsule volume. This increase in size can also be shown to decrease the ability of a phagocyte to ingest the cell, as *C. neoformans* with enlarged capsules is phagocytosed at a lower rate by both macrophages and *A. castellanii* (9, 42). This may indicate a further link between the virulence factors observed in interactions with both macrophages and *A. castellanii* and may further explain how interactions with *A. castellanii* in the environment enabled *C. neoformans* to develop a variety of highly specialized virulence characteristics.

Although many virulence traits have shown some concordance in their requirement for both mammalian and protozoan virulence, there are some discrepancies. For example, the *C. neoformans* α mating locus is important for mammalian virulence but does not appear to be involved in the interaction with *A. castellanii* (26). Furthermore, the interaction of *A. castellanii* with cryptococcal species can vary with the fungal species and strain. *C. neoformans* cells are readily ingested, while *C. gattii* cells are resistant to phagocytosis, a finding that was attributed to differences in the polysaccharide structure (19). Given that there are thousands of amoeba species in the environment, and that cryptococcal species are likely to interact with many types of phagocytic predators, it is prudent to be cautious when drawing inferences and when making conclusions from confrontational studies between single amoebic and cryptococcal species. Nevertheless, amoebae have proven to be a very useful system for drawing insights into the origin of cryptococcal virulence and remain the most useful single-celled phagocytic heterologous host to study cryptococcal biology.

One of the great advantages of amoebae as heterologous hosts for studying *C. neoformans* is that they are free-living organisms with which one can study the outcome of single-cell interactions in highly controlled conditions. The majority of recent studies have been done with *A. castellanii*, a workhorse of amoeba studies because it has been adapted for growth in axenic media. This obviates the need for providing other sources of food, such as live bacteria, which could complicate the interpretation of the results through their effects on either host. However, the gain in simplicity from having to dispense with live bacteria as food is tempered by the fact that laboratory strains of *A. castellanii* may have lost some of their fitness as predators. This laboratory adaptation may explain the observation that *C. neoformans* killed amoebae in recent studies (39), while in earlier reports the amoebae killed the fungus (7, 25). Another advantage of *A. castellanii* is that these amoebae are tolerant of higher than ambient temperatures, and thus, it is possible to carry out studies at mammalian temperatures.

The general protocol for amoeba experiments involves working with *A. castellanii* grown in liquid media in tissue culture flasks at 28°C. The preferred medium is peptone-yeast extract-glucose (ATCC medium 354), although studies can be conducted using other media such as AC buffer (a minimal medium) and phosphate-buffered saline. Being rather hardy, *A. castellanii* can be washed, spun down, counted on a hemacytometer, and used in experiments in much the same way as macrophages. The cells are generally spun at 1,200 rpm (350 × *g*) for 10 min at room temperature and washed with phosphate-buffered saline. Much like macrophages, *A. castellanii* is usually used when confluent on the bottom of the flask, but because they are not adherent in the same way, care must be taken in experiments to avoid washing off the cells. Many of the assays used when working with macrophages can be modified for use with *A. castellanii*, including killing assays, trypan blue staining, and phagocytosis assays.

DICTYOSTELIUM DISCOIDEUM

D. discoideum is the second protozoal host that has been used to study *C. neoformans* interactions with potential amoeboid predators (38). *D. discoideum* is the prototype of social amoebae and represents an organism that exists at the interface of unicellular and multicellular organisms since the individual ameboid cells can congregate into multicellular stalk structures. The interaction of ameboid cells of *D. discoideum* with cryptococci results in phagocytosis and has an outcome similar to that observed with *A. castellanii* (38). Passage of a hypovirulent *C. neoformans* strain through sequential cultures of *D. discoideum* resulted in a significant increase in virulence as evidenced by regaining the ability to kill mice (38). Acapsular *C. neoformans* cells were not able to grow in wild-type *D. discoideum* cells, but when *D. discoideum* mutants defective in either myosin VII or endosomal vesicle fusion were used, the acapsular yeasts were pathogenic (38). This experiment showed that yeast cell mutants harboring mutations that abrogated virulence retained pathogenic potential that could be expressed in a slime host cell debilitated by a mutation in a phenomenon that echoed the microbial opportunism observed with immunodeficient animal hosts. The major advantage of the *D. discoideum* system is the availability of mutant collections and many genetic tools that could be used to study the traits that affect host susceptibility and resistance. Such tools are not available for *A. castellanii*. However, in contrast to

A. castellanii, *D. discoideum* cells are smaller and therefore less effective at ingesting encapsulated organisms. Slime mold cultures require live bacteria as a food source. This requirement for live bacteria in plates led to the discovery that certain gram-negative bacteria, such as *Klebsiella aerogenes*, used to maintain *D. discoideum*, made precursors for cryptococcal melanization in the form of dopamine (13).

CAENORHABDITIS ELEGANS

The free-living soil nematode *C. elegans* has been used extensively as a host for *C. neoformans*. Several reasons account for the appeal of the nematode for host-pathogen interaction studies, as follows. (i) The worm is small (~1 mm), facilitating studies in microtiter plate wells. (ii) *C. elegans* is hermaphroditic, self-fertilizes, and has a short reproductive cycle producing many progeny, which allows a large, genetically identical population to be produced quickly. (iii) The cell lineage of the nematode has been mapped and is the same in all adult worms. (iv) Several molecular techniques have been devised or adapted to *C. elegans*, including RNA interference, as well as the availability of the annotated genome sequence. Furthermore, *C. elegans* is a suitable host to evaluate virulence of temperature-sensitive mutants of *C. neoformans*, since the worm normally lives at ambient temperatures.

When *C. elegans* is fed *C. neoformans*, the yeast accumulates inside the gastrointestinal tract of the nematode (Fig. 2), ultimately causing death. This pathogenicity is observed for all serotypes of *C. neoformans* tested (21). Pathogenicity in *C. elegans* mimics mouse studies, as worms fed nonpathogenic *Cryptococcus kuetzingii* and *Cryptococcus laurentii* isolates lived as long as worms fed the control food source (21). Additionally, *C. neoformans* mutants of pathogenicity genes displayed the same trends (either hypervirulence or reduced virulence) in the *C. elegans*–*C. neoformans* interaction as previously described in murine or rabbit models (21).

Two assays have been developed using *C. elegans* as a host to study *C. neoformans*. The first assay, termed the "killing assay," was initially utilized to identify *C. neoformans* mutants reduced in virulence (23). In the assay the normal nematode food source is replaced with the *C. neoformans* isolate to be studied, which is then ingested, resulting in death. Briefly, worms are grown and maintained on a lawn of *Escherichia coli* on nematode growth medium. Freshly grown isolates of *C. neoformans* to be tested are inoculated as lawns to nematode growth medium plates and grown 24 h at 30°C. After the lawn of the yeast has developed, nematodes at the L4 developmental stage are then placed on the *C. neoformans* lawn, and survival is monitored daily by removing dead worms. The number of *C. elegans* worms to include in the assay can be varied depending of the number of *C. neoformans* isolates to be assayed, as a large preliminary killing screen can be accomplished with 25 nematodes per isolate and increased to 120 to 150 nematodes per isolate in refined assays to calculate statistically different survival times (23).

Another assay has been developed based on the observation of smaller nematode brood size when infected with pathogenic *C. neoformans* (40). This screen is set up in a similar fashion as the killing assay, but fewer worms are added to the lawns of *C. neoformans* to be tested, and the number of progeny produced by the nematodes is monitored over the next several days (40). This assay simplifies the screening process when compared to the killing assay, as the survival of the worms does not need to be recorded every day and the dead worms do not need to be removed from the assayed petri plates.

As previously mentioned, *C. neoformans* accumulates in the gastrointestinal tract of *C. elegans*, despite the presence

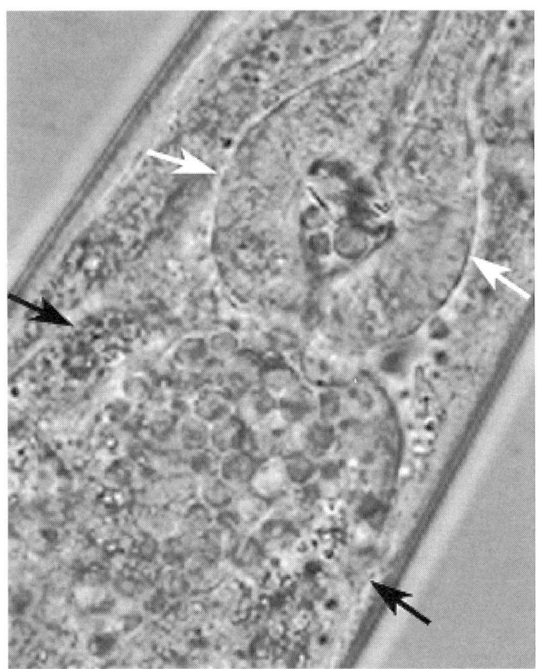

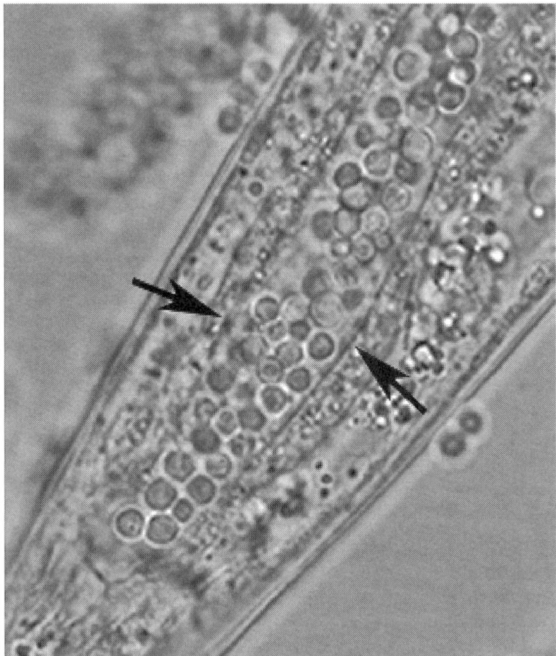

FIGURE 2 Accumulation of *C. neoformans* in the gastrointestional tract of *C. elegans*. Intact yeast cells are visible in the gastrointestional tract (black arrows) after feeding for 36 h on *C. neoformans* strain KN99α. The white arrows indicate the pharyngeal grinder of the digestive tract of *C. elegans*. Photos reprinted from reference 22.

of the pharyngeal grinder which functions as a physical obstacle that macerates ingested organisms (21). Although it is frequently used as a model to test fungal virulence, little is known about the innate immune response of *C. elegans* to fungal infection. The *C. elegans* genome encodes several homologs for components of a Toll pathway that is involved in immune response to *C. neoformans* in *Drosophila melanogaster*. Toll-interleukin 1 receptor has been identified, which controls the antimicrobial peptide NLP 31, which has antifungal activity on a number of fungi (10). The receptor itself, tol-1, is involved in nematode development (29) and appears to play a role in *C. neoformans* recognition and avoidance in the worm (E. Mylonakis, unpublished data). Recently, two *C. elegans* β-glucan–binding scavenger receptors termed CED-1 and C03F11.3 were found to mediate host defense against *C. neoformans*, as the receptors were involved in antimicrobial peptide production in the worm and were necessary for nematode survival following challenge with the yeast (20). Analysis of the *C. elegans* antimicrobial peptides suggests that they are rapidly evolving in response to selective pressures from pathogens (30). Interestingly, male *C. elegans* worms have a longer life span when placed on *C. neoformans* lawns as compared to their hermaphroditic counterparts (41), although the small number of male worms in a given population should not cause a significant difference when assaying a large number of worms.

D. MELANOGASTER

The fruit fly, *D. melanogaster*, has also been developed as a host to study *C. neoformans* virulence. Like *C. elegans*, *C. neoformans* is pathogenic to flies that have ingested the yeast (1). However, wild-type flies are able to overcome a systemic infection when 400 CFU of *C. neoformans* is injected into the hemolymph within 4 days postinoculation (1). Multiple mechanisms of inoculation have been developed for *D. melanogaster*, providing local and systemic infection assays. Microbial pathogens unable to infect through the exoskeleton of the fly can be inoculated by puncturing the dorsal thorax with a needle dipped in a culture of *C. neoformans*. Alternatively, the abdomen can be inoculated in a similar fashion, allowing a greater inoculation volume, but trauma develops more frequently when compared to the thorax inoculation. Microinjection can also be used for precise inoculation volumes into the body cavity. Another inoculation method used for *C. neoformans* to *D. melanogaster* is by spiking the food of the fly. It is important to note that when inoculating the food source, *C. neoformans* must be grown on a limited nutrient source such as one-third yeast-peptone-dextrose agar, as growth on a rich nutrient medium results in low virulence. The inoculated *D. melanogaster* flies are then placed at 25°C, and survival is monitored.

Several aspects of the *D. melanogaster* immune system have been investigated, and two immune signaling pathways have been identified. A Toll receptor has been shown to play an integral part in fungal resistance (18). This receptor interacts with a cytokine-like protein, Spätzle, which is released from fat-body cells, causing the threonine-serine kinase Pelle to phosphorylate Cactus. This ultimately activates transcription factors that control antifungal peptides (18). Of the 20 best-characterized antimicrobials in *D. melanogaster*, 8 appear to be most effective against fungi (17). Wild-type *D. melanogaster* flies are able to overcome systemic infection of *C. neoformans*, but mutants of proteins involved in the Toll signaling cascade are susceptible to systemic infection by *C. neoformans* (1). Interestingly, the Toll cascade is not involved in immune response to ingested *C. neoformans*, suggesting the fly does not have this immune response in the digestive system or *C. neoformans* is able to block the Toll signaling pathway when ingested (1). Another immune signaling pathway in *D. melanogaster* mediated through Imd appears to play little, if any, role in *C. neoformans* resistance, as *imd* mutant flies had similar survival rates as wild-type flies, and there was little difference in survival of *spätzle* mutants versus the double mutant of *spätzle* and *imd* (1).

GALLERIA MELLONELLA

The lepidopteran model organism *G. mellonella* has been used as a host to study *C. neoformans* virulence and antifungal therapies (24). The final instar larvae of the greater wax moth are inoculated with the *C. neoformans* isolates of interest, usually at the last left proleg. If antifungal agents are also to be used in the assay, they may be injected into one of the remaining uninoculated prolegs. Following *C. neoformans* inoculation, the larvae are placed in a 37°C incubator and survival is monitored daily.

The *G. mellonella* larvae utilize a hemolymph system in which *C. neoformans* proliferates within the hemocoel. Six types of hemocytes have been identified within this system (prohemocytes, coagulocytes, spherulocytes, oenocytoids, plasmatocytes, and granulocytes) (16). In addition to phagocytosis of the hemocytes, *C. neoformans* also overcomes encapsulation by the hemocytes and melanization of the larvae. Also of note, *G. mellonella* hemocytes produce reactive oxygen species in a similar fashion as human neutrophils (2).

COMPARISON OF THE METAZOAN MODEL HOSTS

There are distinct advantages and disadvantages to each of the animal heterologous hosts (Table 1). *C. elegans* is an attractive model host to screen large quantities of *C. neoformans* mutants, as many progeny can be obtained quickly, there is minimal cost, and the screens can be accomplished in microtiter plates. *C. elegans* has also been used to identify evolutionarily conserved immune response pathways (20). However, nematodes cannot survive at human physiological temperature, requiring the screen to be conducted at a suboptimal temperature.

D. melanogaster is an appealing model host because of immune system similarities to humans. Unfortunately, *C. neoformans* is unable to establish systemic infections in wild-type flies; studies of systemic infection in the fly require the use of a *Spätzle* mutant, possibly complicating further genetic approaches. Like *C. elegans*, *D. melanogaster* is also unable to survive at 37°C.

The larvae of *G. mellonella* have many alluring characteristics as a *C. neoformans* host. A known quantity of yeast cells can be injected into the arthropod, followed by antifungal treatment, if desired. *G. mellonella* is also the only host that can survive for an extended period of time at 37°C. Unlike *C. elegans* and *D. melanogaster*, the genome is not sequenced, and few molecular techniques have been developed for use with *G. mellonella*, making it difficult to study the host immune response, but it is possible (2).

EVOLUTION OF PATHOGENICITY

Genes associated with cryptococcal pathogenesis in mammalian infection assays also display a similar trend in heter-

TABLE 1 Comparison of heterologous model hosts for *C. neoformans*[a]

Parameter	*A. castellanii*	*D. discoideum*	*C. elegans*	*D. melanogaster*	*G. mellonella*
Precise inoculum	+	+	–	++	+++
Evaluation of antifungal compounds	Unknown	Unknown	+++	+	++
Sequenced genome	–	+++	+++	+++	–
Molecular biology techniques	++	+++	+++	++	+
Microarrays	–	+	+++	+++	–
Similarity to human systems	–	–	+	++	++
Short reproductive cycle	+++	+++	+++	++	N/A
Survival at 37°C	+	–	–	–	++
Overall cost	+++	+++	+++	++	+++

[a]The number of plus signs, ranging from + to +++, indicates how suitable the host is concerning the specified characteristic, while a minus sign (–) indicates the host system is not amenable for the characteristic. N/A, not applicable, as *G. mellonella* is usually obtained from a vendor.

ologous hosts (summarized in Table 2). The capsule of *C. neoformans* is an important virulence factor for the fungus. Synthesis of the capsule is regulated by the cyclic AMP–dependent protein kinase A, which is composed of the catalytic (PKA) and regulatory (PKR) subunits. Mutants of Pka1 or Gpa1 (the α subunit) are reduced in virulence in *C. elegans*, similar to the observed virulence in mice studies (21). However, acapsular mutants of *C. neoformans* also killed *C. elegans* despite being unable to accumulate and survive ingestion, suggesting the capsule was not required for virulence in the nematode, but rather was necessary for survival in the nematode intestine (21). As mentioned previously, acapsular *C. neoformans* mutants were also phagocytosed more frequently by *A. castellanii* when compared to encapsulated *C. neoformans* yeast cells, demonstrating that the capsule provides protection against phagocytosis (39). Acapsular mutants also displayed attenuated virulence in *D. discoideum* and *G. mellonella* infection assays, although they remain virulent in *D. melanogaster* infection assays (1, 24, 38). Additionally, the hyaluronic acid synthase Cps1 produces hyaluronic acid that is localized to the capsular layer of the outer cell wall (15). Hyaluronic acid is involved in the adhesion process of endothelial cells with *C. neoformans*, and *cps1* mutants exhibit attenuated virulence in a *C. elegans* killing assay (15). The conservation of the *C. neoformans* capsule as a virulence trait clearly illustrates that virulence factors may transcend several evolutionarily diverged hosts.

C. neoformans transporters have been identified as virulence factors in *C. elegans*, including two members of the major facilitator superfamily and a sarcoplasmic/endoplasmic reticulum Ca^{2+}-ATPase (SERCA)-type calcium pump (11, 23). The SERCA pump Eca1 is probably involved in maintaining endoplasmic reticulum function and therefore contributing to stress tolerance (11), and although it may not directly function as a virulence trait per se, mutants of *eca1* are reduced in virulence on a number of heterologous hosts and are reduced in ability to proliferate in murine macrophages (11).

TABLE 2 *C. neoformans* virulence factors involved in pathogenesis on heterologous model hosts

Pathogenic protein	Description	Heterologous hosts	Reference
Cap59	Capsule	*C. elegans*	21
		G. mellonella	24
Cap67	Capsule	*A. castellanii*	39
		D. discoideum	38
Plb1	Phospholipase B	*A. castellanii*	39
Plc1	Phospholipase C	*C. elegans*	8
Gpa1	G protein α subunit	*C. elegans*	21
Pka1	Cyclic AMP–dependent protein kinase catalytic subunit	*C. elegans*	21
		D. melanogaster	1
Ras1	GTPase involved in signaling	*C. elegans*	21
Lac1	Diphenoloxidase	*C. elegans*	21
Ade2	Phosphoribosylaminoimidazole	*C. elegans*	21
Kin1	Serine/threonine kinase	*C. elegans*	23
Rom2	Guanylnucleotide exchange factor	*C. elegans*	40
Eca1	Sarcoplasmic/endoplasmic reticulum Ca^{2+}-ATPase (SERCA)–type Ca^{2+} pump	*A. castellanii*	11
		C. elegans	11
		G. mellonella	11
Cps1	Hyaluronic acid synthase	*C. elegans*	15
Tps1	Trehalose-6-phosphate synthase	*C. elegans*	28

A central question in the biology of *C. neoformans* is why a soil-dwelling organism with no obligate requirement for animal colonization for survival or completion of its life cycle is pathogenic to so many animals including mammals, birds, reptiles, and insects. The complexity of this question is enhanced by the realization that the virulence phenotype is extremely complex and involves a large panoply of microbial virulence traits. The observation that the interactions of *C. neoformans* with amoebae have a striking resemblance to those with mammalian macrophages suggests selection pressures in the environment were also applicable to defense against animal phagocytic cells (5, 36, 37, 39). According to this conceptual formulation, the presence of well-defined virulence factors in *C. neoformans*, such as the capsule and melanin synthesis, were selected not for mammalian virulence per se, but instead for survival in the environment. Hence, experiments with heterologous hosts, and amoebae in particular, have provided important insights that have led to the concept of acquisition of pathogenicity through stochastic processes and the concept of accidental virulence (4). According to this view, interactions between microbes, such as *C. neoformans*, and their predators have selected and honed mechanisms of intracellular survival and other "pathogenicity" traits, which when combined with other characteristics, such as thermotolerance and adhesins, have facilitated the evolution of the *C. neoformans*' trait set to survive in animal hosts (3, 4).

This stochastic view also suggests an explanation as to why *C. neoformans* is the major pathogenic species in the *Cryptococcus* genus, despite the fact that several species share virulence factors (27). Since the *Cryptococcus* genus includes a large set of organisms, all of which are also presumably under ameboid predator selection in soils and are able to survive in their environments, the avirulence of most species represents the absence of a complete trait set for animal hosts (4). Perhaps running a gauntlet through a series of diverse heterologous hosts would enable selection and endow these nonpathogenic *Cryptococcus* spp. with the ability to infect mammalian hosts.

CONCLUSIONS

C. neoformans virulence factors on alternative hosts and mammalian models have distinct overlap, suggesting pathogenicity may have evolved by passage through potential hosts present in the surrounding everyday environment. This provides researchers with a unique opportunity, and a number of heterologous hosts have been identified, each having its distinct advantages and disadvantages, and selecting the most appropriate for experimentation is important. These *C. neoformans* hosts are ethical, low-cost alternatives for experimentation. We encourage researchers to utilize heterologous hosts as potential models for studies in host immune response and fungal pathogen studies.

REFERENCES

1. **Apidianakis, Y., L. G. Rahme, J. Heitman, F. M. Ausubel, S. B. Calderwood, and E. Mylonakis.** 2004. Challenge of *Drosophila melanogaster* with *Cryptococcus neoformans* and role of the innate immune response. *Eukaryot. Cell* **3:**413–419.
2. **Bergin, D., E. P. Reeves, J. Renwick, F. B. Wientjes, and K. Kavanagh.** 2005. Superoxide production in *Galleria mellonella* hemocytes: identification of proteins homologous to the NADPH oxidase complex of human neutrophils. *Infect. Immun.* **73:**4161–4170.
3. **Casadevall, A.** 2007. The cards of virulence and the global virulome. *Microbe* **1:**359–364.
4. **Casadevall, A., and L. A. Pirofski.** 2007. Accidental virulence, cryptic pathogenesis, martians, lost hosts, and the pathogenicity of environmental microbes. *Eukaryot. Cell* **6:**2169–2174.
5. **Casadevall, A., J. N. Steenbergen, and J. D. Nosanchuk.** 2003. "Ready made" virulence and "dual use" virulence factors in pathogenic environmental fungi:- the *Cryptococcus neoformans* paradigm. *Curr. Opin. Microbiol.* **6:**332–337.
6. **Castellani, A.** 1931. An amoeba growing in cultures of a yeast. *J. Trop. Med. Hyg.* **33:**188–191.
7. **Castellani, A.** 1955. Phagocytic and destructive action of *Hartmanella castellanii* (*Amoeba castellanii*) on pathogenic encapsulated yeast-like fungus *Torulopsis neoformans* (*Cryptococcus neoformans*). *Ann. Inst. Pasteur* **89:**1–7.
8. **Chayakulkeeree, M., T. C. Sorrell, A. R. Siafakas, C. F. Wilson, N. Pantarat, K. J. Gerik, R. Boadle, and J. T. Djordjevic.** 2008. Role and mechanism of phosphatidylinositol-specific phospholipase C in survival and virulence of *Cryptococcus neoformans*. *Mol. Microbiol.* **69:**809–826.
9. **Chrisman, C. J., and A. Casadevall.** 2005. The interaction of *Cryptococcus neoformans* with *Acanthamoeba castellanii* results in capsular enlargement, abstr. F-085. Abstr. 105th Gen. Meet. Am. Soc. Microbiol. American Society for Microbiology, Washington, DC.
10. **Couillault, C., N. Pujol, J. Reboul, L. Sabatier, J. F. Guichou, Y. Kohara, and J. J. Ewbank.** 2004. TLR-independent control of innate immunity in *Caenorhabditis elegans* by the TIR domain adaptor protein TIR-1, an ortholog of human SARM. *Nat. Immunol.* **5:**488–494.
11. **Fan, W., A. Idnurm, J. Breger, E. Mylonakis, and J. Heitman.** 2007. Eca1, a sarcoplasmic/endoplasmic reticulum Ca2+-ATPase, is involved in stress tolerance and virulence in *Cryptococcus neoformans*. *Infect. Immun.* **75:**3394–3405.
12. **Feldmesser, M., Y. Kress, P. Novikoff, and A. Casadevall.** 2000. *Cryptococcus neoformans* is a facultative intracellular pathogen in murine pulmonary infection. *Infect. Immun.* **68:**4225–4237.
13. **Frases, S., S. Chaskes, E. Dadachova, and A. Casadevall.** 2006. Induction by *Klebsiella aerogenes* of a melanin-like pigment in *Cryptococcus neoformans*. *Appl. Environ. Microbiol.* **72:**1542–1550.
14. **Fries, B. C., D. L. Goldman, R. Cherniak, R. Ju, and A. Casadevall.** 1999. Phenotypic switching in *Cryptococcus neoformans* results in changes in cellular morphology and glucuronoxylomannan structure. *Infect. Immun.* **67:**6076–6083.
15. **Jong, A., C.-H. Wu, H.-M. Chen, F. Luo, K. J. Kwon-Chung, Y. C. Chang, C. W. LaMunyon, A. Plaas, and S.-H. Huang.** 2007. Identification and characterization of CPS1 as a hyaluronic acid synthase contributing to the pathogenesis of *Cryptococcus neoformans* infection. *Eukaryot. Cell* **6:**1486–1496.
16. **Kavanagh, K., and E. P. Reeves.** 2004. Exploiting the potential of insects for in vivo pathogenicity testing of microbial pathogens. *FEMS Microbiol. Rev.* **28:**101–112.
17. **Lemaitre, B., and J. Hoffmann.** 2007. The host defense of *Drosophila melanogaster*. *Annu. Rev. Immunol.* **25:**697–743.
18. **Lemaitre, B., E. Nicolas, L. Michaut, J.-M. Reichhart, and J. A. Hoffmann.** 1996. The dorsoventral regulatory gene cassette spätzle/Toll/cactus controls the potent antifungal response in *Drosophila* adults. *Cell* **86:**973–983.
19. **Malliaris, S. D., J. N. Steenbergen, and A. Casadevall.** 2004. *Cryptococcus neoformans* var. *gattii* can exploit *Acanthamoeba castellanii* for growth. *Med. Mycol.* **42:**149–158.
20. **Means, T. K., E. Mylonakis, E. Tampakakis, L. Puckett, C. R. Stewart, R. Pukkila-Worley, S. E. Hickman, K. J. Moore, S. B. Calderwood, N. Hacohen, A. D. Luster, and J. El-Khoury.** 2009. Evolutionarily conserved recognition

and innate immunity to fungal pathogens by the scavenger receptors SCARF1 and CD36. *J. Exp. Med.* **206:**637–653.

21. **Mylonakis, E., F. M. Ausubel, J. R. Perfect, J. Heitman, and S. B. Calderwood.** 2002. Killing of *Caenorhabditis elegans* by *Cryptococcus neoformans* as a model of yeast pathogenesis. *Proc. Natl. Acad. Sci. USA* **99:**15675–15680.
22. **Mylonakis, E., A. Casadevall, and F. M. Ausubel.** 2007. Exploiting amoeboid and non-vertebrate animal model systems to study the virulence of human pathogenic fungi *PLoS Pathog.* **3:**e101.
23. **Mylonakis, E., A. Idnurm, R. Moreno, J. El Khoury, J. B. Rottman, F. M. Ausubel, J. Heitman, and S. B. Calderwood.** 2004. *Cryptococcus neoformans* Kin1 protein kinase homologue, identified through a *Caenorhabditis elegans* screen, promotes virulence in mammals. *Mol. Microbiol.* **54:**407–419.
24. **Mylonakis, E., R. Moreno, J. B. El Khoury, A. Idnurm, J. Heitman, S. B. Calderwood, F. M. Ausubel, and A. Diener.** 2005. *Galleria mellonella* as a model system to study *Cryptococcus neoformans* pathogenesis. *Infect. Immun.* **73:**3842–3850.
25. **Neilson, J. B., M. H. Ivey, and G. S. Bulmer.** 1978. *Cryptococcus neoformans*: pseudohyphal forms surviving culture with *Acanthamoeba polyphaga. Infect. Immun.* **20:**262–266.
26. **Nielsen, K., G. M. Cox, A. P. Litvintseva, E. Mylonakis, S. D. Malliaris, D. K. Benjamin, Jr., S. S. Giles, T. G. Mitchell, A. Casadevall, J. R. Perfect, and J. Heitman.** 2005. *Cryptococcus neoformans* α strains preferentially disseminate to the central nervous system during coinfection. *Infect. Immun.* **73:**4922–4933.
27. **Petter, R., B. S. Kang, T. Boekhout, B. J. Davis, and K. J. Kwon-Chung.** 2001. A survey of heterobasidiomycetous yeasts for the presence of the genes homologous to virulence factors of *Filobasidiella neoformans*, CNLAC1 and CAP59. *Microbiology* **147:**2029–2036.
28. **Petzold, E. W., U. Himmelreich, E. Mylonakis, T. Rude, D. Toffaletti, G. M. Cox, J. L. Miller, and J. R. Perfect.** 2006. Characterization and regulation of the trehalose synthesis pathway and its importance in the pathogenicity of *Cryptococcus neoformans. Infect. Immun.* **74:**5877–5887.
29. **Pujol, N., E. M. Link, L. X. Liu, C. L. Kurz, G. Alloing, M. W. Tan, K. P. Ray, R. Solari, C. D. Johnson, and J. J. Ewbank.** 2001. A reverse genetic analysis of components of the Toll signaling pathway in *Caenorhabditis elegans. Curr. Biol.* **11:**809–821.
30. **Pujol, N., O. Zugasti, D. Wong, C. Couillault, C. L. Kurz, H. Schulenburg, and J. J. Ewbank.** 2008. Anti-fungal innate immunity in *C. elegans* is enhanced by evolutionary diversification of antimicrobial peptides. *PLoS Pathog.* **4:**e1000105.
31. **Robinson, B. S., S. S. Bamforth, and P. J. Dobson.** 2002. Density and diversity of protozoa in some arid Australian soils. *J. Eukaryot. Microbiol.* **49:**449–453.
32. **Rodriguez-Zaragoza, S., F. Rivera, P. Bonilla, E. Ramirez, E. Gallegos, A. Calderon, R. Ortiz, and D. Hernandez.** 1993. Amoebological study of the atmosphere of San Luis Potosi, SLP, Mexico. *J. Expo. Anal. Environ. Epidemiol.* **3**(Suppl. 1)**:**229–241.
33. **Ruiz, A., R. A. Fromtling, and G. S. Bulmer.** 1981. Distribution of *Cryptococcus neoformans* in a natural site. *Infect. Immun.* **31:**560–563.
34. **Ruiz, A., J. B. Neilson, and G. S. Bulmer.** 1982. Control of *Cryptococcus neoformans* in nature by biotic factors. *Sabouraudia* **20:**21–29.
35. **Sheehan, K. B., J. A. Fagg, M. J. Ferris, and J. M. Henson.** 2003. PCR detection and analysis of the free-living amoeba *Naegleria* in hot springs in Yellowstone and Grand Teton National Parks. *Appl. Environ. Microbiol.* **69:**5914–5918.
36. **Steenbergen, J. N.** 2003. The evolution of fungal virulence, Ph.D. thesis. Albert Einstein College of Medicine, Bronx, NY.
37. **Steenbergen, J. N., and A. Casadevall.** 2003. The origin and maintenance of virulence for the human pathogenic fungus *Cryptococcus neoformans. Microbes Infect.* **5:**667–675.
38. **Steenbergen, J. N., J. D. Nosanchuk, S. D. Malliaris, and A. Casadevall.** 2003. *Cryptococcus neoformans* virulence is enhanced after growth in the genetically malleable host *Dictyostelium discoideum. Infect. Immun.* **71:**4862–4872.
39. **Steenbergen, J. N., H. A. Shuman, and A. Casadevall.** 2001. *Cryptococcus neoformans* interactions with amoebae suggest an explanation for its virulence and intracellular pathogenic strategy in macrophages. *Proc. Natl. Acad. Sci. USA* **98:**15245–15250.
40. **Tang, R. J., J. Breger, A. Idnurm, K. J. Gerik, J. K. Lodge, J. Heitman, S. B. Calderwood, and E. Mylonakis.** 2005. *Cryptococcus neoformans* gene involved in mammalian pathogenesis identified by a *Caenorhabditis elegans* progeny-based approach. *Infect. Immun.* **73:**8219–8225.
41. **van den Berg, M. C. W., J. Z. Woerlee, H. Ma, and R. C. May.** 2006. Sex-dependent resistance to the pathogenic fungus *Cryptococcus neoformans. Genetics* **173:**677–683.
42. **Zaragoza, O., C. J. Chrisman, M. V. Castelli, S. Frases, M. Cuenca-Estrella, J. L. Rodríguez-Tudela, and A. Casadevall.** 2008. Capsule enlargement in *Cryptococcus neoformans* confers resistance to oxidative stress suggesting a mechanism for intracellular survival. *Cell. Microbiol.* **10:**2043–2057.

Cryptococcus: From Human Pathogen to Model Yeast
Edited by J. Heitman et al.

20

Cryptococcosis in Africa

NELESH P. GOVENDER, THOMAS G. MITCHELL,
ANASTASIA P. LITVINTSEVA, AND KATHLEEN J. MIGLIA

The world's highest incidence of coinfection with cryptococcosis and HIV/AIDS occurs in sub-Saharan Africa. According to the UNAIDS/WHO 2008 Status Report on the global AIDS epidemic, sub-Saharan Africans account for two-thirds of all people living with HIV and 75% of the deaths from AIDS (140). Advances in treating patients with AIDS have extended lives, but new infections outnumber deaths, and the overall number of people living with HIV has persistently increased.

Throughout the history of the HIV epidemic, coinfections of HIV/AIDS and cryptococcosis have continued to be disproportionately concentrated in sub-Saharan Africa. For over a century, cryptococcosis has been recognized as a global infection of immunocompromised and immunocompetent patients, but the incidence exploded with the HIV epidemic (112). In parallel with the evolution of HIV, some of the earliest cases of AIDS-associated cryptococcal disease may have occurred in the 1960s. During this period, cases of cryptococcosis among young people were reported in the Democratic Republic of Congo (formerly Zaire), possibly linked to early, undiagnosed HIV infection (113). With the recognized start of the HIV pandemic in the early 1980s, there was a rapid rise in the number of cases reported in sub-Saharan Africa (21). Also striking was the simultaneous increase in reported cases among African immigrants with features of AIDS diagnosed in European countries (29, 37, 38, 50, 77, 143).

As the HIV pandemic steadily progressed in Africa during the 1980s and 1990s, *Cryptococcus neoformans* became the most common reported cause of laboratory-confirmed meningitis among adults in several sub-Saharan countries (Fig. 1). These increases in incidence paralleled the rise of cryptococcosis among HIV-infected persons in developed countries (48).

However, the incidence of cryptococcosis in developed nations peaked in the mid-1980s and has further declined in the past decade since the widespread administration of highly active antiretroviral therapy (HAART).

Nelesh P. Govender, Mycology Reference Unit, National Institute for Communicable Diseases, a division of the National Health Laboratory Service, Johannesburg, South Africa. **Thomas G. Mitchell, Anastasia P. Litvintseva, and Kathleen J. Miglia,** Department of Molecular Genetics and Microbiology, Duke University Medical Center, Durham, NC 27710.

Developing countries have been less fortunate. A recent study of the global burden of cryptococcal meningitis among patients with HIV/AIDS analyzed data from a variety of sources (120). Comparing incidence reports from several geographical regions, this report estimated that in 2006, there were 957,900 (range: 0.37 million to 1.54 million) global cases of cryptococcal meningitis. The highest number of estimated cases—three-fourths of the total—occurred in sub-Saharan Africa (ca. 720,000; range: 0.14 million to 1.3 million) (120). To estimate the mortality, the authors assumed a 10-week case-fatality rate of 9% for coinfected persons living in developed regions and 55% in primarily less-developed regions but excluding sub-Saharan Africa, where the estimate was 70% (120). In a comparison of the overall causes of death in sub-Saharan Africa, cryptococcal meningitis ranked fourth, after malaria, diarrheal diseases, and childhood-cluster diseases, and exceeded deaths from tuberculosis (120). Consistent with data from HIV-infected cohorts and natural history studies, where 13 to 44% of HIV/AIDS deaths in sub-Saharan Africa were attributable to cryptococcosis, this study estimated that 504,000 deaths were associated with cryptococcosis (range: 100,200 to 907,200). Many of these data, especially in less-developed countries, were reported before the availability of HAART. However, the introduction of HAART in southern Africa has failed to diminish both the incidence and early mortality rates due to cryptococcal meningitis (54, 65, 93).

There are several potential, nonexclusive explanations for the inordinate incidence and severity of cryptococcosis in sub-Saharan Africa. For cultural, financial, and public health reasons, the diagnosis is frequently not propitious, and treatment is less accessible. The region suffers from severe shortages of diagnostic microbiology laboratories and treatment options. Even when health care is available, many patients are reluctant to be tested for HIV infection because a positive diagnosis often leads to social stigmatization and familial ostracism. Some patients are averse to diagnostic lumbar punctures; others are noncompliant with their medications.

Other possible explanations for the disparity between the incidence of cryptococcal meningitis in sub-Saharan Africa and the rest of the world have not been fully explored. Are AIDS patients in Africa more susceptible or more exposed to *Cryptococcus* than AIDS patients in other parts of the world?

Are strains of *Cryptococcus* in Africa more virulent? This region is the global hotbed of cryptococcal disease and diversity among the species of *Cryptococcus*. Isolates of the dominant etiology, *C. neoformans* var. *grubii*, from patients in Botswana and South Africa are more diverse than isolates from other continents. They are partitioned into genetic subpopulations and capable of sexual reproduction and recombination (88).

To address these issues and highlight areas requiring further investigation, this chapter will review the epidemiology, diagnosis, clinical manifestations, treatment, and prognosis of cryptococcosis in Africa and the ecology and population genetics of African isolates of *Cryptococcus*.

CRYPTOCOCCOSIS IN AFRICA

History and Scope

Until recent years, the prevalence and epidemiology of cryptococcosis in sub-Saharan African nations were poorly documented (63, 120). Reports from northern Africa have been minimal (45, 140). Among African patients with AIDS, cryptococcosis is often the sentinel opportunistic infection and the leading cause of meningitis (21, 60). Most reports indicate an alarming incidence of cryptococcal meningitis in sub-Saharan Africa, but the geographic distribution of reported cases appears to vary. Many of these discrepancies are undoubtedly attributable to shortages of resources for surveillance and clinical studies. In general, the prevalence of cryptococcosis tends to increase in the more southerly nations and in parallel with the prevalence of HIV/AIDS. For example, although cryptococcosis has been documented in patients with AIDS in the West African nation of Ghana (5), a prospective study of 1,570 cerebrospinal fluid (CSF) specimens from patients in Kumasi, Ghana, detected 314 cases of meningitis, which were either aseptic or due to bacterial infections (47%), but no cases of cryptococcosis were detected; however, in this study, only 28 subjects were known to be HIV-infected (49). Based on clinical studies, Fig. 1 and 2

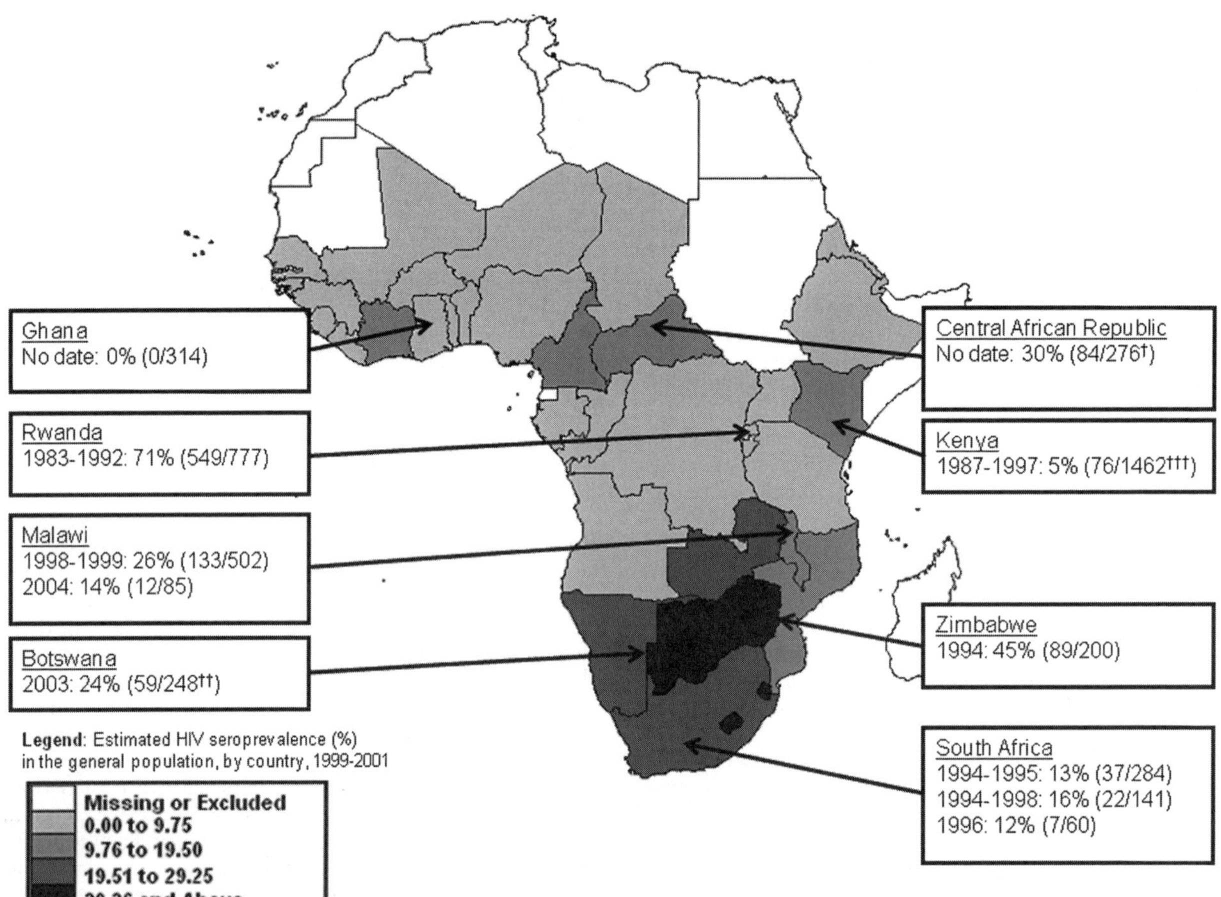

FIGURE 1 Summary of studies conducted in sub-Saharan Africa between 1983 and 2003 that reported the percentage of laboratory-confirmed meningitis caused by *Cryptococcus* species (7, 8, 18, 21, 33, 49, 52, 60, 118, 130, 133, 141). Data are overlaid on a choropleth map of Africa, which illustrates estimated HIV seroprevalence in the general population, by country, 1999 to 2001. For each study, boxes denote the country, date(s) of study, and percentage of people with confirmed cryptococcal meningitis. Most fractions denote the number with cryptococcal meningitis divided by the number with laboratory-confirmed meningitis (i.e., any abnormal CSF parameters with or without detection of a pathogen). Exceptions: Denominator included: †, people with clinically suspected meningitis; ††, people with elevated leukocytes in the CSF; or †††, people whose CSF specimens were examined with India ink.

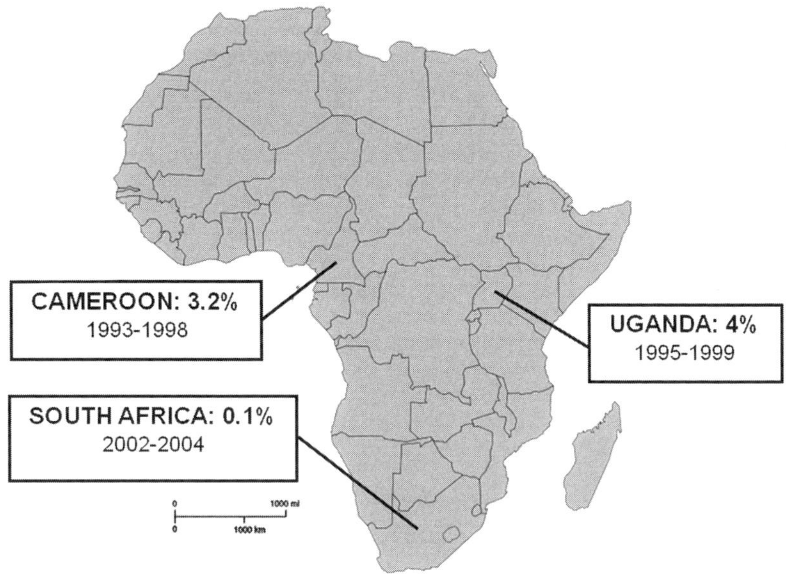

FIGURE 2 Published incidence rates for cryptococcosis from studies conducted in antiretroviral treatment-naive, HIV-infected, adult populations in sub-Saharan Africa (47, 97, 99).

depict the prevalence and incidence of reported cases of cryptococcal meningitis.

A retrospective analysis of 1,144 CSF specimens in Dar es Salaam, Tanzania, detected microbial pathogens in 55 pediatric (<15 years of age) and 167 adult CSF specimens, and among these 222 positive specimens, *C. neoformans* was isolated from 22% of the pediatric and 98% of the adult specimens (95). Additional CSF studies reported positive rates for *C. neoformans* of 5% in Kenya (118), 6 to 12% in Zaire (41, 121), 9% in the Congo (28), 19% in Rwanda (21), 22% in Zimbabwe (60, 62), 7% in Ethiopia (151), 13% in South Africa (8), and 67% in Malawi (150). In addition to infecting African adults, *C. neoformans* also occurs in African children (58, 95, 134). In contrast, pediatric cases of cryptococcal meningitis, even among AIDS patients, are rare in other parts of the world (29, 112).

Among individual studies, the mortality due to cryptococcal meningitis in adult patients with AIDS has been as high as 100%, as reported for 230 patients who were seen from 1998 to 1999 at the University Teaching Hospital in Lusaka, Zambia (116). Approximately half the patients were treated with fluconazole and half were untreated, and the median survival times were 19 and 10 days, respectively (116). Reported mortality rates for similar patients with HIV/AIDS and cryptococcal meningitis were 17% in Uganda (47), 39% in Zimbabwe (60, 62), 42% in Burkina Faso (107), 83% in Equatorial Guinea (148), 87% in Côte d'Ivoire (17), and 87% in Burundi (78).

Current Burden of HIV-Associated Cryptococcosis and HAART

Prior to the availability of HAART, the incidence rates of cryptococcosis among HIV-infected persons in developed countries were estimated to be 17 to 20 cases per 1,000 person-years (36, 127). The incidence declined rapidly and substantially in most developed countries during the mid- to late-1990s, initially reflecting the widespread use of fluconazole (59) and, subsequently, the introduction of HAART (48, 72, 108). For example, the incidence in the United States in 1998 had dropped to 0.1 cases per 1,000 person-years (72). In the post-HAART era, cases of cryptococcosis have been detected in North America and Europe among people who do not consider themselves at risk for HIV infection or who have not accessed HAART due to socioeconomic barriers (43, 108).

With the limited availability of effective therapies to mitigate the impact of cryptococcosis, even into the early 21st century, sub-Saharan Africa remains the region currently most affected by this disease, with an estimated 504,000 new cases (range: 100,800 to 907,200) occurring per year among persons with HIV/AIDS (120). This estimate of the burden of disease was based on incidence data reported in studies of HIV-infected and antiretroviral-naive populations in South Africa (0.1%), Cameroon (3.2%), and Uganda (4%) (see Fig. 2) (47, 97, 99). Two other studies indicated a similar range of incidence rates in Côte D'Ivoire (5.5%) and South Africa (2.2%) (40, 44, 63).

Although antiretroviral treatment programs have been rapidly implemented, recent laboratory-based surveillance from South Africa has indicated no reduction in the number of new cases of cryptococcosis (55). This trend is reflected in data from 2003 through 2008 at a single urban hospital in Cape Town (65). Similar findings were reported in Malawi, where more than 2,000 new, India ink–positive cases were detected per year from 2005 through 2007, despite the rapid increase during this time period in the number of antiretroviral treatment sites from 60 to 109 (93). This lack of any dramatic effect may be partially explained by the evolution and maturation of the HIV pandemic in southern Africa, as well as the continuing delay before patients are diagnosed and treated. Despite the rapid expansion of treatment programs, the number of patients with advanced HIV disease (i.e., $CD4^+$ T-cell count <100 cells/μl) who have not accessed HAART has either stabilized or increased. Patients who fall into this so-called treatment gap remain at high risk for opportunistic infections such as cryptococcosis (65). Hence, cryptococcosis is likely to remain a substantial problem in sub-Saharan Africa for many more years.

Epidemiology and Clinical Manifestations

In the pre-HAART era in sub-Saharan Africa, cryptococcosis was often a sentinel opportunistic infection among HIV-infected adults, heralding the diagnosis of AIDS in more than 88% of cryptococcal cases (62, 116). This trend has continued in the early post-HAART era (GERMS-SA and N. P. Govender, unpublished data). At the time of diagnosis, most patients continue to have a very low median $CD4^+$ T cell count (99). A large survey evaluated 2,753 primary cases of cryptococcosis among HIV-infected South Africans living in 2002 to 2004 (99). The median age of the patients was 34 years (range: 1 month to 74 years), the sex ratio was equal, and 97% presented with meningitis (99). The admitting signs and symptoms of patients enrolled into this population-based study included headache (78%), neck stiffness (69%), fever (55%), nausea and vomiting (41%), altered mental status (31%), seizures (9%), coma (3%), and sixth cranial nerve palsy (1%) (99).

In contrast to the developed world, where respiratory cryptococcosis and nonmeningeal dissemination is more commonly diagnosed, pulmonary disease remains underdiagnosed in resource-limited, African settings (99). There are several reasons. First, the index of suspicion among clinicians is low, and appropriate specimens are infrequently submitted to the laboratory. Second, coinfection with other respiratory pathogens tends to mask cryptococcal pneumonia. Cryptococcal pneumonia, as a clinical entity, may resemble the presentation of *Pneumocystis jirovecii* pneumonia or pulmonary tuberculosis, and coinfection with *P. jirovecii* and *Mycobacterium tuberculosis* has also been described (152). Third, diagnostic laboratories may not differentiate *Cryptococcus* species in mixed respiratory tract cultures, which often include yeasts such as *Candida albicans* (67). A recent autopsy study, which was conducted on a population of South African miners, reported a 7% (589/8,421) incidence rate of cryptococcal pneumonia (152). Of note, 16% of the miners with cryptococcal pneumonia had a concomitant respiratory infection, most commonly *P. jirovecii* pneumonia and mycobacterial infection. Meningitis was diagnosed premortem in approximately half the miners with autopsy-diagnosed, cryptococcal pneumonia, but before death, cryptococcal pneumonia was diagnosed in only 3% of the cases (152).

Despite the large number of children with AIDS in sub-Saharan Africa, pediatric cryptococcosis remains uncommon compared to other opportunistic infections (58). While the rarity of childhood cases of cryptococcosis is a global phenomenon (1), pediatric cryptococcosis is relatively more commonly described in sub-Saharan Africa. The largest series of pediatric cases of cryptococcosis was recently described among South African children (101), and the rates of meningitis were similar to those for *Haemophilus influenzae* in the postvaccination era (100). As described below, a subset of isolates of *C. neoformans* from this cohort of pediatric cases was genotyped and found to be exceptionally diverse (106).

A very small proportion of patients are diagnosed with cryptococcosis in sub-Saharan Africa with risk factors other than HIV disease (99). Infections with *Cryptococcus gattii* also cause a small proportion (<1%) of diagnosed cases of cryptococcosis (54), and they are clinically indistinguishable from disease caused by *C. neoformans* in HIV-infected South Africans (114).

Diagnosis

The initial diagnosis of cryptococcal meningitis in HAART-naive, HIV-infected patients is relatively straightforward and involves microscopy and culture of clinical specimens, as well as serology. At many diagnostic laboratories, specimens of CSF obtained from HIV-infected adult patients are routinely processed to detect *Cryptococcus* species. Classically described CSF parameters include a low glucose concentration, normal-to-high protein concentration, and high lymphocyte count, although a poor inflammatory response is common (116). The usually high fungal burden at the time of diagnosis improves the sensitivity of detecting encapsulated yeasts in the CSF by India ink staining up to 98% (10, 99). The specificity of the test is also improved by experienced laboratory personnel who diagnose cryptococcal meningitis frequently. Culturing of *Cryptococcus* is performed at laboratories with adequate facilities, and in most instances, growth on nonselective media is rapid (within 72 h). Identification of the colonies is simple and specific, viz., a positive test for urease on a urea-containing medium and/or the development of brown-pigmented colonies on Staib's niger seed agar (77). Facilities for histopathological diagnoses are usually restricted to larger, academic facilities (99). The diagnosis is less specific but improved by the observation of the large cryptococcal polysaccharide capsule with the mucicarmine stain.

The cryptococcal antigen test, which has excellent sensitivity and specificity, is not consistently available, partly due to the relatively high cost of the test at present. In South Africa, a recently conducted public sector survey of diagnostic-laboratory practices indicated that only 70% of laboratories offered the cryptococcal antigen test on site (N. P. Govender, unpublished data). Very few laboratories have the capacity to perform antifungal drug susceptibility testing. Fortunately, drug resistance has rarely been observed in isolates from either primary or recurrent episodes (30, 53, 123).

Treatment

Based on the results of several randomized clinical trials, the Infectious Diseases Society of America recommended in 2000 that HIV-infected persons with cryptococcal meningitis should be initially treated with a combination of the conventional formulation of amphotericin B deoxycholate (0.7 to 1 mg/kg/day) and flucytosine (100 mg/kg/day) for 2 weeks, followed by fluconazole (or itraconazole) consolidation-phase therapy for 8 to 10 weeks (126). The addition of flucytosine was justified in the induction-phase regimen because it was independently shown to accelerate the rate of sterilization of CSF to 2 weeks posttreatment (144) and to reduce the risk of relapse (125). Although the amphotericin B–based induction phase was fungicidal and rapidly effective, a switch to the azole-based consolidation regimen was recommended to reduce toxicity. It was recognized that following the induction and consolidation phases, HIV-infected patients were less likely to completely eradicate the cryptococcal infection, and hence the primary objective was to control the infection by resolving the clinical manifestations of disease (126). The subsequent administration of a low-dose azole-based maintenance therapy was intended to prevent relapses (126). A recent evaluation of patients from South Africa, Uganda, and Thailand confirmed that the rate of reduction and clearance of cryptococcal census from the CSF is an effective method to evaluate the efficacy of therapy (13).

Despite difficulties related to the administration of amphotericin B in resource-limited settings (i.e., the requirement for intravenous delivery and monitoring for toxicity and, hence, the need for a prolonged hospital admission), an increasing proportion of South African patients have been treated with amphotericin B in recent years (Govender, unpublished data). Amphotericin B is also the standard of care at tertiary centers in other African countries (19). In the absence of flucytosine and based on South African data (15), the southern African management guidelines recommend higher-dose amphotericin B (1 mg/kg/day) (98).

Through the auspices of a philanthropic partnership program, fluconazole remains widely available at no cost in many sub-Saharan African countries. Fluconazole is easy to administer and relatively nontoxic, even at high doses. It is used extensively in areas where resources prohibit other options. However, low-dose fluconazole monotherapy was associated with very unfavorable outcomes in African studies (96, 129), partly because fluconazole is a fungistatic agent and associated with slower rates of fungal clearance during the first 2 weeks (12). A recent fluconazole, dose-escalation study from Uganda randomized 30 newly diagnosed patients with cryptococcal meningitis to receive either 800 mg or 1,200 mg oral fluconazole per day (89). The rate of cryptococcal clearance in the CSF of these patients was measured by serial, quantitative cultures, and clearance was significantly greater in the group receiving 1,200 mg fluconazole. The regimen was also well tolerated. Unfortunately, mortality rates at 2 weeks (30%) and 10 weeks (54%) remained high, and there was no statistically significant difference between the dosage groups, although the study was not designed to detect such a difference (89).

Flucytosine is currently not widely available in sub-Saharan Africa, despite being recommended as a first-line agent (14). Flucytosine in combination with amphotericin B was favorably assessed in a clinical trial in South Africa (14, 15). Treatment with flucytosine is also very costly (70). Oral flucytosine, which exhibits similar pharmacokinetics as the intravenous formulation in HIV-infected patients (25), may be an attractive option in combination with higher doses of fluconazole in areas where prolonged intravenous therapy is difficult. Further advocacy effort is required to reduce the cost of flucytosine and improve the availability of this off-patent drug.

Maintenance therapy with fluconazole is usually prescribed for those patients who survive their initial episode of disease (54). However, the vast majority of South African patients who are prescribed maintenance therapy do not return for subsequent prescriptions within 1 year of diagnosis (39). This is most likely due to the very high early mortality associated with cryptococcosis. Adjunctive therapies, such as interferon gamma, and newer drugs, such as voriconazole, are generally unavailable in most of Africa. The in vitro susceptibilities of African isolates of *C. neoformans* to fluconazole, amphotericin B, and voriconazole are comparable to those of other regions (30, 122, 123).

Raised intracranial pressure is poorly managed in patients with cryptococcal meningitis in sub-Saharan Africa (see below), and inappropriate management of pressure has been associated with poorer outcomes in a developed-country setting (56). In a recent study including patients from Thailand and South Africa, two-thirds of the patients had substantially raised CSF opening pressures (9). The elevated baseline pressures were mitigated by frequent lumbar punctures over the course of treatment, and this practice was associated with comparable mortality rates in the groups with and without raised pressures (9). However, this procedure has not been implemented in routine clinical settings because of the limited availability of manometers and the reluctance of patients to allow repeated lumbar punctures.

HAART and IRIS

Although HAART has improved the long-term outcome of patients who survive the initial episode to access treatment, HAART may paradoxically precipitate a relapse of the current disease or unmask undiagnosed cryptococcal disease through the immune reconstitution inflammatory syndrome (IRIS). The proportion of patients diagnosed with cryptococcal disease who subsequently developed IRIS on follow-up range from 8% (10/120) in a French cohort (90) to 34% (20/59) (including both "unmasking" and "paradoxical" forms) in a North American cohort (132). In a cohort of South African patients treated with fluconazole at the onset of cryptococcosis, the incidence of cryptococcal IRIS was midway between these estimates: 18 cases per 100 person-years-observations (PYO), 6 cases per PYO, and 0 cases per PYO at 1, 2, and 3 months, respectively (79). In another South African cohort that was treated with amphotericin B and prospectively followed for 1 year, the incidence of cryptococcal IRIS was 17% (11/65) (11). Patients receiving low-dose fluconazole induction-phase treatment have a higher fungal burden at the time of HAART initiation and hence a theoretically increased chance of developing IRIS. Two-thirds of the described cases of cryptococcal IRIS in the first South African cohort were the paradoxical form (79). All cases of paradoxical IRIS developed within a month of initiation of HAART in the second cohort (11). All 11 patients had a recurrence of headache, while two patients had focal neurological signs, and no patients were detected with unusual extraneural syndromes. There was also no significant difference in survival between those patients who did or did not develop IRIS (11).

The diagnosis of cryptococcal disease in HAART-treated patients with paradoxical IRIS is slightly more difficult. A Cape Town study that compared CSF parameters from initial and IRIS-related episodes found no significant differences in total leukocyte counts, protein, glucose, or opening pressures, but there were increased neutrophils and decreased cryptococcal antigen titers in CSF from IRIS-related episodes (11). Depending on the definition of IRIS, CSF cultures are usually negative, but India ink examinations are positive in some patients, and patients usually have positive antigen tests (11).

Mortality

During a landmark clinical trial in North America, conducted during the pre-HAART era, where patients were randomized to receive induction-phase treatment with amphotericin B and flucytosine, fewer than 10% of the patients died at 8 weeks, despite having advanced HIV disease at baseline (median $CD4^+$ T cell count of 18 and 20 cells/μl for the two randomized groups) (144). In France, 18% of the patients died at 3 months postdiagnosis with the availability of HAART, which was similar to the 3-month case-fatality rates for patients treated during the pre-HAART era (21%) (91). Although introduction of HAART did not affect early mortality, it led to substantial improvements in long-term outcome for patients who survived the episode and continued treatment (91).

In sub-Saharan Africa, survival rates of patients in the absence of antifungal drugs and HAART have been dismal (62). As noted above, fluconazole became more widely available through a philanthropic partnership program initiated in 2000 in South Africa and expanded to include several other African countries (149). However, use of low-dose fluconazole monotherapy for induction-phase and subsequent therapy had little effect on survival rates (116, 129). Similarly, access to amphotericin B–based regimens and HAART has not improved early mortality. Ten-week mortality was 37% in a clinical trial involving patients in Cape Town with access to HAART who were treated with amphotericin B (12). In a Ugandan study, high early mortality rates were documented in the pre-HAART era among patients treated with amphotericin B (51% at 14 days) and persisted in the era of HAART (20% at 14 days) (70). In this study, the improvement in early mortality was possibly due to improved fluid management of patients receiving amphotericin B; indeed, mortality rates improved even before patients had access to HAART (70).

Several factors may contribute to this high early mortality, which occurs before patients have an opportunity to receive HAART. First, most patients are not diagnosed until they have advanced cryptococcal disease (with a high fungal burden). Second, management of raised intracranial pressure is not optimal, despite the proven survival benefit of therapeutic lumbar punctures (119, 126). In some instances, manometers are not available to measure the pressure (15). Furthermore, Kambugu et al. report that a very low proportion of Ugandan patients consent to therapeutic spinal taps, even in the presence of documented elevated pressure. They suggest that there is a strong cultural bias against lumbar punctures, which is related to the inconsistent use of a local anesthetic during the procedure and high mortality associated with cryptococcal meningitis (70). Third, advanced immunosuppression leads to a poor inflammatory response in the CSF, which may diminish control of the infection and worsen the outcome (70). Reassuringly, data from sub-Saharan Africa suggest that patients who survive to access HAART may have long-term survival rates comparable to those of patients from developed countries (12, 70).

Prevention and Screening

For HIV-infected patients, especially those with low CD4$^+$ T cell counts, the best protection against infection with *Cryptococcus* is timely access to HAART and rapid restoration of the immune system. This strategy is preferable to the administration of prophylactic fluconazole (or itraconazole), for which a clear survival benefit has not been demonstrated (31). Guidelines for management of patients in southern Africa have emphasized access to HAART as a means of prevention (98). However, despite a substantial improvement in the usage of HAART in sub-Saharan African countries, many people still lack access to treatment (140). Even if they do receive HAART, patients with advanced HIV infection remain at substantial risk for cryptococcosis until immune reconstitution occurs.

Prior to HAART availability, primary prophylaxis with fluconazole or itraconazole was shown to be a useful strategy for prevention of disease among persons with CD4$^+$ T cell counts of <100 cells/μl in Thailand, a high-incidence, developing-world setting (32, 35). To date, there are no published data on the efficacy of this strategy in sub-Saharan Africa, although this may still be under investigation in East Africa (66). Another preventive strategy is to test HAART-naive patients for cryptococcal antigenemia prior to the administration of antiretroviral treatment. A South African study retrospectively tested stored, pre-HAART plasma samples of HIV patients with CD4$^+$ T-cell counts of <100 cells/μl. The study found that 13% (42/336) of the subjects had a positive serum cryptococcal antigen test, and 29% (6/21) of those without a history of cryptococcosis developed disease during the first year of HAART (68). It was estimated that 52 patients with a CD4$^+$ T-cell count of <100 cells/μl and no history of cryptococcosis needed to be screened to identify and prevent one case of disease. The high negative predictive value of this test was also striking: none of the patients who had a negative test for antigenemia developed cryptococcosis during follow-up. However, the latter finding is not consistent in all such studies (83). A similar investigation in Uganda found a prevalence of cryptococcosis of 15% (48/311) among HAART-naive patients with CD4$^+$ T cell counts of <100 cells/μl (102). Of the 29 who were promptly treated with fluconazole, none developed disease, whereas 18 of 19 untreated patients developed "unmasking" cryptococcal IRIS. The estimated cost in 2009 to detect one patient with asymptomatic antigenemia was $108, and the cost to prevent one death was $300. This strategy may be justified in high-incidence African settings where the unmasking form of cryptococcal IRIS occurs frequently (69).

Challenges in Sub-Saharan Africa

The formidable challenges of cryptococcosis in sub-Saharan Africa involve significant problems with diagnosis, management, and prevention. First, there are substantial barriers to accessing health care, especially in rural settings. As a result, many patients remain unaware of their HIV status until they develop an indicator opportunistic infection such as cryptococcosis. With the onset of cryptococcal disease, patients who cannot access health care may die. Second, even if patients are able to access health care, assessment may be delayed, and patients are often diagnosed with advanced cryptococcal disease. This situation has a direct impact on early mortality. Third, the cryptococcal antigen test, which is an excellent diagnostic tool, is not widely available, partly due to cost. Fourth, not all effective antifungal drugs are available for treatment. Fifth, although HAART is the best way to protect against cryptococcal infection, treatment coverage is still inadequate and unlikely to affect either the burden of disease in the near future or the early mortality rates associated with cryptococcosis in sub-Saharan Africa.

CHARACTERIZATION OF *CRYPTOCOCCUS* SPECIES

From a clinical specimen, a typical isolate can be identified as a member of the *C. neoformans*–*C. gattii* complex by its growth at 37°C, production of a capsule, and demonstration of laccase by the development of melanized cell walls when cultured on media with an appropriate diphenolic substrate. At this point, most clinical laboratories proceed no further because the diagnosis of cryptococcosis has been established. Regardless of the species or subpopulation of *Cryptococcus*, the management of the patient will be the same. The frequent variation in clinical responses to treatment of cryptococcosis is usually attributed to the immune status of the host or to the occasional drug-resistant isolate. Alternatively, there are compelling scientific and practical reasons to identify the species and molecular type of both clinical and environmental isolates. The evidence is mounting that pathogenic yeasts identified as *Cryptococcus* by routine phenotypic tests are likely to include unrecognized species or varieties with different life cycles, different levels of pathogenicity, and different ecological niches. This phenomenon is not unique to *Cryptococcus*. Until the HIV/AIDS epidemic, *Candida dubliniensis* was mistakenly identified as *C. albicans* because both species share the same phenotypes that are routinely used to identify yeasts in the clinical laboratory (135). Because all isolates of *C. neoformans* or *C. gattii* are not the

TABLE 1 Nomenclature of the subpopulations of *C. neoformans* and *C. gattii*

Species	Serotype (variety)	Molecular type[a]	MSLT diversity[b]	Prevalence in Africa[c]
C. neoformans	A (var. *grubii*)	VNI	+++	++++
		VNII	+	+
		VNB	++++	+++
	AD	VNIII	++	+
	D (var. *neoformans*)	VNIV	++	+
C. gattii	B (or C)	VGI	+	+
		VGII	++	+
	C (or B)	VGIII	++	+
		VGIV	+	++

[a] Most, but not all, isolates of serotypes B and C are VGI or VGII and VGIII or VGIV, respectively.
[b] Diversity was estimated from several MLST analyses (6, 20, 23, 24, 26, 74, 103, 104).
[c] Relative prevalence among clinical and environmental isolates based on current data.

same, epidemiological and ecological studies require more precise strain characterization.

Identification of the species of *Cryptococcus* by phenotypic methods is problematic. It has long been recognized that the capsular serotypes of *C. neoformans* are either A, D, or AD, and those of *C. gattii* are B or C. Unfortunately, this classical method of species identification is obsolete because commercial serotyping kits are no longer available. The species can usually be identified by growth on canavanine-glycine-bromothymol blue agar (77), which distinguishes *C. gattii* from *C. neoformans*, but this method is imperfect, and erroneous results are not uncommon (80).

DNA-based approaches provide the most accurate and robust methods to identify each species, as well as determine the subpopulations. The first step to characterize a new wild-type strain is to determine its molecular type. Several methods have been developed to identify the major subpopulations of *C. neoformans* and *C. gattii*, as well as individual strains. For details about the molecular markers and methods of genotyping and investigating the population genetics of *Cryptococcus* consult chapter 8, chapter 24, or any of several reviews (109–111, 153, 154).

PCR fingerprinting is the most widely adopted method to identify eight of the nine established subpopulations of the *C. neoformans–C. gattii* species complex, VNI to VNIV and VGI to VGIV, respectively. This schema is summarized in Table 1. PCR fingerprinting is relatively rapid and inexpensive. Genomic DNA of the test isolate is extracted and PCR-amplified using a single primer based on a sequence of the M13 phage; the amplicon is then electrophoresed, and the band sizes are compared with those of reference strains of each molecular type (105). An additional step involves amplification of the *URA5* gene, digestion with two endonucleases, and comparison of the electrophoretic bands with reference strains (104). Reference strains representing the major molecular types are required because the banding patterns can vary in different laboratories.

In contrast, multilocus sequence typing (MLST) is more reliable because the markers represent specific DNA sequence polymorphisms. The MLST genotype of any strain can be reproduced in any laboratory, and reference strains are not required. Consensus gene sequence markers have been established to identify the major subpopulations of *C. neoformans* and *C. gattii* (24, 27, 34, 46, 88, 103, 117, 128). In addition, the MLST markers for *C. neoformans* will identify the highly diverse VNB clade, which is largely restricted to Africa. Table 1 indicates the current designation of the nine major subpopulations of the *C. neoformans–C. gattii* complex, their relationships to the conventional serotypes, and their relative diversity and prevalence in Africa.

ECOLOGY AND GEOGRAPHIC DISTRIBUTION OF *CRYPTOCOCCUS* SPECIES IN AFRICA

The environmental reservoirs of *C. neoformans* and *C. gattii* determine who will be exposed and at risk for infection. Infections with species of *Cryptococcus* have been investigated for more than a century (112), and with the exception of a few unusual cases, there is no evidence of routine transmission of cryptococcosis among humans or other animals. (The rare exceptions involve primary cutaneous cases, contaminated organ transplants, or nosocomial transmission [29, 147].) Thus, species of *Cryptococcus* are exogenous pathogens, and primary infections are almost invariably acquired by the inhalation of airborne yeast cells or basidiospores that were aerosolized from an environmental source.

In sub-Saharan Africa, cases of cryptococcosis due to *C. neoformans* far exceed the number of infections caused by *C. gattii*. As detailed in chapter 18, subpopulations of *C. neoformans* can be isolated from avian feces, a variety of trees, as well as soil and other sites throughout the world. The high number of infections with *C. neoformans* in sub-Saharan Africa indicates that this species is widespread in this region, which must provide a rich environment for the propagation and dispersion of cryptococcal yeast cells, basidiospores, or both. To elucidate the natural history of cryptococcosis in Africa, and perhaps develop strategies for preventative public health measures, it would be propitious to determine the source of human and veterinary infections.

In retrospect, the reports of fulminant cases of cryptococcosis in young Central Africans in the 1950s may have been early signs of the prevalence of *Cryptococcus* and harbingers of the HIV pandemic (113). Subsequently, a comparison in France of the genotypes of clinical isolates of serotype A from HIV-infected African émigrés and French natives revealed that the Africans' isolates differed genetically from European strains, which provided compelling evidence that the African patients were likely infected with *Cryptococcus* before they emigrated (50).

C. neoformans

In the 1970s and 1980s, *C. neoformans* was isolated from the excrement of pigeons (*Columba livia*) in Nigeria (57), Egypt (124), and Zaire (Congo) (138). In Burundi, Swinne et al. isolated *C. neoformans* from pigeon feces, soil, domestic dust, and wood samples from the homes of 7 of 20 patients with AIDS and cryptococcosis (136). The authors warned of the potential for reinfection of patients living in proximity with *C. neoformans*. These isolates were subsequently determined to be serotype A (145). In a follow-up study in Bujumbura, an additional 800 environmental samples were obtained from soil, wood, pigeon excreta, and house dust in and around the homes of patients with AIDS, and 54% of the patients' residences, as well as 20% of random control houses, were colonized by *C. neoformans* (137). The practice of raising pigeons was a likely source, as five of six dovecotes harbored *C. neoformans*, and the isolations from house dust suggested that *C. neoformans* was abundant in the air (137). Infrequent reports have since documented the presence of serotypes A and D in the feces of captive birds at the Wildlife Park and Zoo in Jos, Nigeria (64), as well pigeon excreta in Egypt (92) and Ethiopia (2). Figure 3 maps the subpopulations of clinical isolates.

C. gattii

Although *C. gattii* is much less common in the environment and among patients in Africa, the first isolate of *C. gattii* was obtained from a child in the Democratic Republic of the Congo (Zaire) (51, 142). Kwon-Chung and Bennett identified six strains of serotype B that were isolated before 1970 from patients in central Africa, namely, Cameroon, the Democratic Republic of the Congo, Kenya, and Côte d'Ivoire (75, 76). For many years, *C. gattii* was considered to be almost exclusively a primary pathogen of immunocompetent persons. Indeed, the vast majority of reported cases have occurred in tropical, equatorial, and Southern Hemispheric regions of the world, where *C. gattii* is prevalent in the environment, and infections are more common in previously healthy individuals (29). Consistent with this pattern, the currently expanding outbreak of VGII strains of serotype B in British Columbia involves mostly immunocompetent individuals (see chapter 23). Since the 1990s, more cases of cryptococcal meningitis due to *C. gattii* in HIV-infected persons have been reported from Africa, including Botswana (87), the Democratic Republic of the Congo (71, 81, 115), Kenya (16, 81), Malawi (87), Rwanda (21, 22), South Africa (73, 114), and Zimbabwe (62, 81). Several isolates of *C. gattii* from sub-Saharan Africa have been genotyped, and as shown in Fig. 3, most were serotype C isolates of VGIV, and they exhibited little diversity (73, 81, 87). The VGIV molecular type is quite rare elsewhere in the world (87). As mentioned in the previous section, the clinical course of *C. gattii* meningitis in HIV-infected patients is similar to the presentation of *C. neoformans*, with no significantly greater resistance to antifungal drugs (114).

In the environment, *C. gattii* was isolated from *Eucalyptus camaldulensis* trees in Egypt (92). In 2007, multiple strains of VGI were isolated from tree hollows in the North West and Mpumalanga provinces of South Africa (A. P. Litvintseva et al., unpublished data).

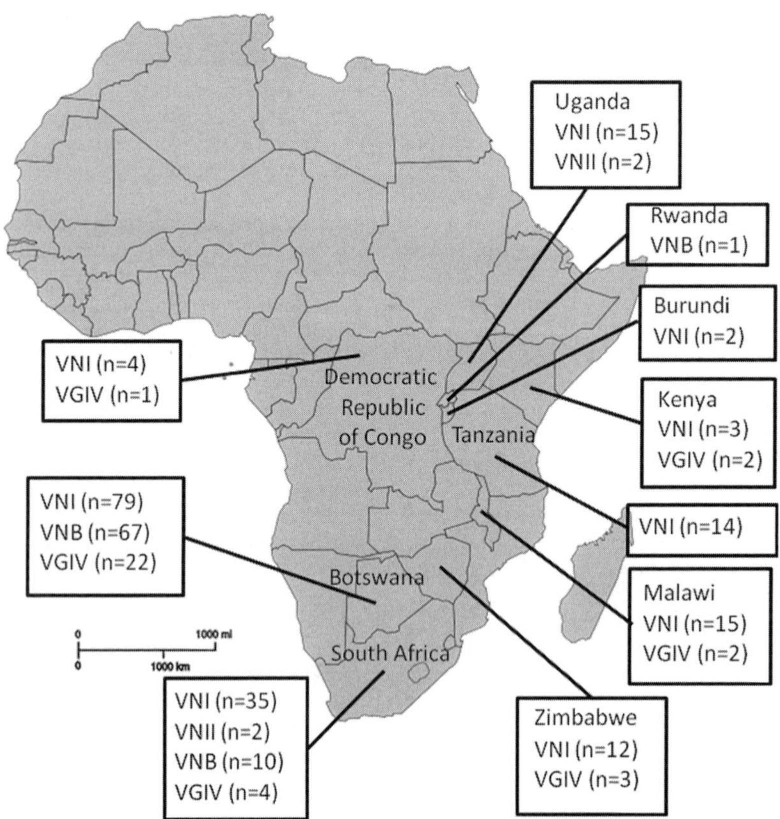

FIGURE 3 Distribution of molecular types in Africa as determined by MLST, amplified fragment length polymorphisms, and PCR fingerprinting (20, 24, 84–86, 104, 117).

Tables 2 and 3 list most of the reports of environmental and clinical isolates of species, varieties, serotypes, and subpopulations of *Cryptococcus*.

DIVERSITY OF AFRICAN STRAINS OF *C. NEOFORMANS*

In 2003, an atypical population of serotype A was isolated from patients with AIDS in Botswana. This subpopulation was termed "VNB" and was characterized by extraordinary genetic diversity and the inclusion of a high proportion of isolates with the normally rare *MAT***a** mating-type allele and evidence of recombination (85). Subsequently, a rigorous MLST method was developed to genotype a global sampling of isolates of serotype A (88). Comparative analyses of the MLST genotypes of strains from southern Africa and 13 other countries, representing North and South America, Europe, and Asia, indicated that the unique VNB population is geographically restricted to southern Africa.

To investigate the origin of the VNB subpopulation, 264 environmental isolates of serotype A were obtained from a variety of substrates in southern Africa (86). Most of the strains were recovered from arboreal samples (67%) and avian habitats (29%), but the largest number of isolates was obtained from wood hollows of a native African tree, *Colophospermum mopane* (mopane tree). Almost a third of the sampled mopane trees (9 of 31) were colonized by VNB strains. The correlation of VNB genotypes and this ecological niche was striking and significant (86). Thus, strains with genotypes that are restricted to Africa were associated with trees, especially the mopane, but African isolates with global genotypes (viz., VNI), which have been isolated from five continents, were associated with pigeon excreta (Fig. 4). (The results of this geophylogenetic analysis of serotype A are discussed in more depth in chapter 8.)

These data confirmed earlier observations and substantiated the consensus that *C. neoformans* is widespread in the African environment. In addition, the analyses suggested that southern Africa harbors two genetically and ecologically isolated subpopulations: (i) an indigenous population associated with endemic African trees and (ii) an introduced population residing in avian, mostly pigeon, habitats (Fig. 4). Phylogenetic studies determined that these exceptionally diverse African populations represent the ancestral population of serotype A (86).

Whether or not the mopane tree provides the main ecological reservoir for VNB in southern Africa requires further investigation. However, the observed high frequency of colonization of mopane trees may have implications for public health. The mopane tree is endemic to sub-Saharan Africa, where it has significantly affected the economy and culture (61, 131). Mopane timber is frequently used for firewood and construction. According to one report, the walls and roofs of traditional huts in the eastern part of Limpopo Province,

TABLE 2 Environmental isolations of *Cryptococcus* species from Africa

Country	Ecological niche	Variety, serotype, genotype	Reference(s)
Botswana	Trees: mopane trees (*Colophospermum mopane*)	A, VNI, VNB	86
	Trees: baobab trees (*Adansonia digitata*)	A, VNB	86
	Avian habitats: pigeons (*C. livia*)	A, VNI	86
Burundi	Avian habitats: pigeons (*C. livia*)	var. *grubii*	137
	House dust	var. *grubii*	136, 137, 145
Congo (Zaire)	House dust	var. *grubii*	139
	Soil	var. *grubii*	138, 145
Egypt	Trees: *Eucalyptus* trees (*E. camaldulensis*)	B or C	92
	Avian habitats: pigeons (*C. livia*)	var. *grubii*	92, 124
	Avian habitats: sparrows (*Passer* species)	var. *grubii*	92
Ethiopia	Avian habitats: pigeons (*C. livia*)	var. *grubii*	2
Nigeria	Avian habitats: pigeons (*C. livia*)	var. *grubii*	57
	Avian habitats: white face duck (*Dendrocygna viduata*)	A	64
	Avian habitats: spotted eagle owl (*Bubo africanus cinerascene*)	A and D	64
	Avian habitats: peacock (*Pavo cristatus*)	A	64
Rwanda	House dust	var. *grubii*	139
South Africa	Trees: *Eucalyptus* species	A, VNB	86
	Avian habitats: pigeons (*C. livia*)	A, VNI	86
	Tree hollows	B, VGI	A. P. Litvintseva, unpublished data

TABLE 3 Clinical isolates of *Cryptococcus* species from Africa

Country	Region	No. of isolates	Species, serotype, subgroup	Reference(s)
Botswana		139	A, VNI, VNB, C, VGIV	85, 87, 88
		7	A, VNI, VNB	86
Burundi		2	C. *neoformans*	145
			A, VNI	81
Congo (Zaire)		1	C. *gattii*, VGIV	81
Egypt		3	A	45
Kenya		3	A, VNI	81
		2	C. *gattii*, VGIV	81
Malawi		17	A, VNI C, VGIV	4, 87, 88
Rwanda		1	A, VNB	24, 42
South Africa	Johannesburg and Durban	48	A, VNI	105
	Johannesburg and Durban	9	A, VNII	105
	Johannesburg	4	C, VGIV	73
	Limpopo Province	42	A, VNI, VNII, VNB AD, VNIII	86
	All provinces	82	A, VNI, VNII, VNB, VNIII	106
Tanzania		14	A, VNI	4, 82, 88
Uganda		1 17	A, VNI A, VNI, VNII	81, 88
Zimbabwe		12	A, VNI	81
		3	C, VGIV	81

South Africa, are constructed almost entirely of debarked mopane poles, and mopane bark, wood, leaves, and seeds are used extensively in traditional medicine (94, 146). Mopane trees also serve as a substrate for the cultivation of edible caterpillars of the *Gonimbrasia belina* moth (mopane worm), which are used in local cuisine in Botswana, South Africa, Angola, and Namibia. Considering the high degree of colonization of mopane trees by C. *neoformans* and the high census of HIV-infected persons in these countries who may be potentially exposed to contaminated mopane products, the public health implications of using mopane trees for construction, medicine, and cultivation of mopane worms require careful investigation.

Pediatric Isolates

As described above, the global occurrence of cryptococcosis in children is inexplicably rare. Most of the reported pediatric cases have occurred in children aged 6 to 12 years with HIV/AIDS. Even though sub-Saharan Africa has the highest prevalence of children with HIV/AIDS, the number of children who are coinfected with HIV and *Cryptococcus* is disproportionately low compared to adults. However, the absolute number of pediatric coinfections in this region is comparatively high. The well-established medical and public health resources in South Africa provided an opportunity to examine perhaps the largest cohort of pediatric cases and elucidate the population genetics of isolates of *Cryptococcus* from children (106).

Cases of cryptococcosis that occur in all provinces of South Africa are routinely reported to the National Institute for Communicable Diseases (NICD), which is a division of the National Health Laboratory Service. The NICD manages the South African infectious diseases surveillance program, GERMS-SA, which records patient demographic data and receives clinical isolates for confirmation of their identification and storage in a repository. From January 2005 through December 2006, 10,991 cases of cryptococcosis were reported. Of the 9,952 cases for which the age was reported, 199 (2%) were less than 15 years old. Viable isolates were available from the CSF and blood specimens of 82 cases. The gender ratio was approximately 1, and the patients ranged in age from 2 days to 14 years. For patients whose HIV status was recorded, 96% were HIV-infected. The vast majority of these pediatric isolates (>98%) were C. *neoformans*, and they were genotyped using 11 unlinked MLST markers. They included representatives of four major molecular types, VNI, VNII, VNB, and VNIII, although most were VNI strains, and no VNIV (serotype D) isolates were found. To determine whether pediatric isolates differed from adult isolates in the distribution of molecular types or mating types, they were compared with a similar number of isolates from adult patients that were collected during the same 2-year period. As shown in Table 4, the pediatric and adult samples were comparable with regard to the proportion of each molecular type, mating-type allele, and ploidy (106). All of these features, including the relatively high percentage of diploids and *MAT***a** isolates, confirm the recognized diversity of C. *neoformans* in sub-Saharan Africa, and the dearth of serotype D is also consistent with previous reports (84, 85, 88).

The MLST analysis of these pediatric isolates revealed extensive genetic diversity, as 24 different genotypes were identified, and 10 (42%) were unique to South Africa. The estimated phylogeny of these genotypes is depicted in Fig. 5.

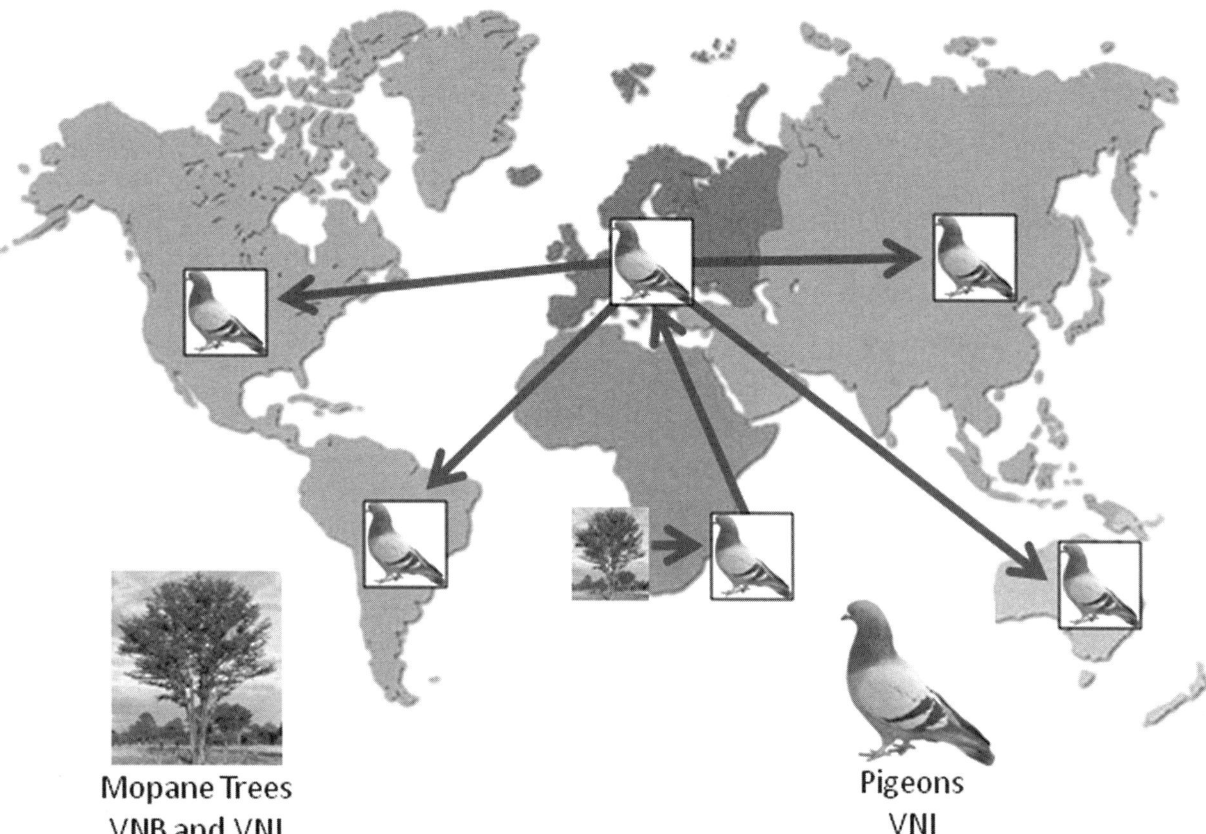

FIGURE 4 Strains of *C. neoformans* endemic to sub-Saharan Africa possess ancestral haplotypes of all MLST loci that have been analyzed, and these strains are associated with native African trees, including the mopane and baobab. This model suggests that a small number of strains emerged from Africa, established a new ecological niche in the excreta of *Columba livia* (rock pigeon), and then disseminated around the world.

TABLE 4 Comparison of clinical isolates of *C. neoformans* from adult and pediatric patients with cryptococcosis in South Africa[a]

Feature	Adult isolates ($n = 86$)	Pediatric isolates ($n = 82$)	% Adult isolates per type (n)	% Pediatric isolates per type (n)
Serotype, molecular type, %				
A, VNI	79.1	81.7	50 (68)	50 (67)
A, VNII	7.0	7.3	50 (6)	50 (6)
A, VNB	10.5	9.8	53 (9)	47 (8)
AD, VNIII	3.5	1.2	75 (3)	25 (1)
D, VNIV	0	0	0	0
MAT type allele, %				
*MAT*α	95.3	93.9	52 (82)	48 (77)
*MAT***a**	4.7	6.1	44 (4)	56 (5)
Ploidy, %				
n	87.2	91.5	51 (78)	49 (75)
$2n$[b]	12.8	8.5	61 (11)	39 (7)

[a] Isolates and clinical data were obtained by GERMS-SA, NICD, in 2005 and 2006 (106). Among adults (≥17 years of age) and children (≤16 years of age) for whom HIV status was known, 97% (73/75) and 96% (43/45), respectively, were HIV-infected. One isolate of *C. gattii* was obtained from a child, but that isolate is not included in this comparison.

[b] Diploids included serotype A diploids (AA) and hybrids (AD).

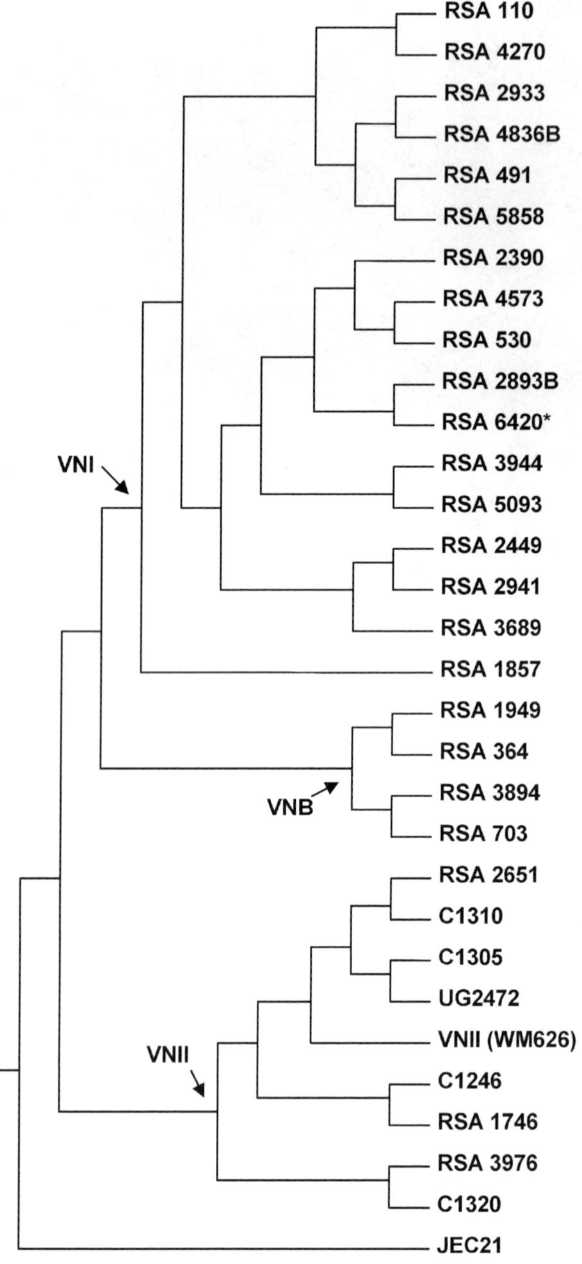

FIGURE 5 Pediatric isolates from the Republic of South Africa (RSA) were analyzed using Bayesian concordance analysis (3). The nuclear DNA sequence data were obtained from 10 unlinked loci: *CAP10, GPD1, IGS1, ISCA, LAC1, MPD1, PLB1, SOD1, TEF1*, and *TOP1*, which total 6,259 base pairs, consisting of exons, introns, and an intragenic spacer region. Each strain in the Bayesian tree has a different sequence. The branch lengths for each genotype are presented as equal in length for aesthetic purposes only. The subpopulations were identified as follows: VNII by inclusion of the reference strain (WM626) and five other global VNII strains; VNI by RSA6420, which has the same genotype as the VNI reference strain (WM148); and VNB genotypes were identified from a separate Bayesian phylogeny of each locus that included several VNB reference strains (bt34, bt63, bt85, bt88, bt89, and bt131) (88). The outgroup is the sequenced VNIV strain of serotype D (JEC21).

Seventeen genotypes were members of the VNI subpopulation, four were VNB, and three were VNII. Ten genotypes were specific for individual isolates. Conversely, approximately half of the 75 haploid isolates possessed one of three dominant genotypes, which included 19, 9, and 7 isolates and are represented in Fig. 5 by VNI strains RSA110, RSA1857, and RSA2893B, respectively. Further studies will determine whether strains with these genotypes are more virulent and/or more prevalent in the environment, and if so, why. There was no correlation between genotype and the geographical location or province in which the patients were treated. However, isolates with the most prevalent genotype were significantly associated with male patients (106). Consistent with their diversity, analyses of the population structure and phylogenetics of this pediatric sample produced evidence for both clonality and recombination.

CONCLUSIONS

Sub-Saharan Africa is the global epicenter of cryptococcosis and the HIV/AIDS pandemic. Both pathogens, *C. neoformans* and HIV, evolved in Africa, albeit at different rates. But their coexistence in this region has created the "perfect storm" that has produced the world's highest incidence of both infections and caused the greatest comortality. This chapter has provided a review of the clinical features of cryptococcosis with emphasis on particularly African aspects of the epidemiology, diagnosis, treatment, and prognosis of cryptococcal meningitis. The chapter also described the remarkable ecology, genetic diversity, and uniqueness of African strains of *Cryptococcus*.

The evidence is compelling that *C. neoformans* is continuing to evolve in the crucible of sub-Saharan Africa. In spite of the human tragedy wrought by cryptococcosis on this continent, there is the opportunity to scrutinize the interactions of *C. neoformans* with its primary host and the environment. It is possible to discover how and where people become infected with *Cryptococcus*. It is possible to determine whether specific genotypes impart clinically predictive information about the prognosis of cryptococcal disease. It is possible to elucidate the evolutionary relationships among African strains of *Cryptococcus* and their global ancestors.

We greatly appreciate research support from U.S. Public Health Service NIH grant 5R01 AI25783 and a grant from the Duke University Center for AIDS Research. Cryptococcal surveillance activities, undertaken by the National Institute for Communicable Diseases, South Africa, were partially supported through cooperative agreements with the U.S. Centers for Disease Control and Prevention.

REFERENCES

1. **Abadi, J., S. Nachman, A. B. Kressel, and L.-A. Pirofski.** 1999. Cryptococcosis in children with AIDS. *Clin. Infect. Dis.* **28:**309–313.
2. **Amanuel, Y. W., L. Jemaneh, and D. Abate.** 2001. Isolation and characterization of *Cryptococcus neoformans* from environmental sources in Ethiopia. *Ethiop. J. Health Dev.* **15:**45–49.
3. **Ané, C., B. Larget, D. A. Baum, S. D. Smith, and A. Rokas.** 2007. Bayesian estimation of concordance among gene trees. *Mol. Biol. Evol.* **24:**412–426.
4. **Archibald, L. K., M. O. den Dulk, K. J. Pallangyo, and L. B. Reller.** 1998. Fatal *Mycobacterium tuberculosis* bloodstream infections in febrile hospitalized adults in Dar es Salaam, Tanzania. *Clin. Infect. Dis.* **26:**290–296.
5. **Ayisi, N. K., E. K. Wiredu, T. Sata, C. Nyadedzor, V. K. Tsiagbe, M. Newman, C. N. Cofie, and K. Taneguchi.**

1997. T-lymphocytopaenia, opportunistic infections and pathological findings in Ghanaian AIDS patients and their sexual partners. *East Afr. Med. J.* **74:**784–791.

6. **Barreto de Oliveira, M. T., T. Boekhout, B. Theelen, F. Hagen, F. A. Baroni, M. S. Lazéra, K. B. Lengeler, J. Heitman, I. N. G. Rivera, and C. R. Paula.** 2004. *Cryptococcus neoformans* shows a remarkable genotypic diversity in Brazil. *J. Clin. Microbiol.* **42:**1356–1359.
7. **Bekondi, C., C. Bernede, N. Passone, P. Minssart, C. Kamalo, D. Mbolidi, and Y. Germani.** 2006. Primary and opportunistic pathogens associated with meningitis in adults in Bangui, Central African Republic, in relation to human immunodeficiency virus serostatus. *Int. J. Infect. Dis.* **10:**387–395.
8. **Bergemann, A., and A. S. Karstaedt.** 1996. The spectrum of meningitis in a population with high prevalence of HIV disease. *Q. J. Med.* **89:**499–504.
9. **Bicanic, T., A. E. Brouwer, G. A. Meintjes, K. Rebe, D. Limmathurotsakul, W. Chierakul, P. Teparrakkul, A. Loyse, N. J. White, R. Wood, S. Jaffar, and T. S. Harrison.** 2009. Relationship of cerebrospinal fluid pressure, fungal burden and outcome in patients with cryptococcal meningitis undergoing serial lumbar punctures. *AIDS* **23:**701–706.
10. **Bicanic, T., T. S. Harrison, A. Niepieklo, N. Dyakopu, and G. A. Meintjes.** 2006. Symptomatic relapse of HIV-associated cryptococcal meningitis after initial fluconazole monotherapy: the role of fluconazole resistance and immune reconstitution. *Clin. Infect. Dis.* **43:**1069–1073.
11. **Bicanic, T., G. A. Meintjes, K. Rebe, A. Williams, A. Loyse, R. Wood, M. Hayes, S. Jaffar, and T. S. Harrison.** 2009. Immune reconstitution inflammatory syndrome in HIV-associated cryptococcal meningitis: a prospective study. *J. Acquir. Immune Defic. Syndr.* **51:**130–134.
12. **Bicanic, T., G. A. Meintjes, R. Wood, M. Hayes, K. Rebe, L. G. Bekker, and T. S. Harrison.** 2007. Fungal burden, early fungicidal activity, and outcome in cryptococcal meningitis in antiretroviral-naive or antiretroviral-experienced patients treated with amphotericin B or fluconazole. *Clin. Infect. Dis.* **45:**76–80.
13. **Bicanic, T., C. Muzoora, A. E. Brouwer, G. A. Meintjes, N. Longley, K. Taseera, K. Rebe, A. Loyse, J. N. Jarvis, L. G. Bekker, R. Wood, D. Limmathurotsakul, W. Chierakul, K. Stepniewska, N. J. White, S. Jaffar, and T. S. Harrison.** 2009. Independent association between rate of clearance of infection and clinical outcome of HIV-associated cryptococcal meningitis: analysis of a combined cohort of 262 patients. *Clin. Infect. Dis.* **49:**702–709.
14. **Bicanic, T., R. Wood, L. G. Bekker, M. Darder, G. A. Meintjes, and T. S. Harrison.** 2005. Antiretroviral roll-out, antifungal roll-back: access to treatment for cryptococcal meningitis. *Lancet Infect. Dis.* **5:**530–531.
15. **Bicanic, T., R. Wood, G. A. Meintjes, K. Rebe, A. E. Brouwer, A. Loyse, L. G. Bekker, S. Jaffar, and T. S. Harrison.** 2008. High-dose amphotericin B with flucytosine for the treatment of cryptococcal meningitis in HIV-infected patients: a randomized trial. *Clin. Infect. Dis.* **47:**123–130.
16. **Bii, C. C., K. Makimura, S. Abe, H. Taguchi, O. M. Mugasia, G. Revathi, N. C. Wamae, and S. Kamiya.** 2007. Antifungal drug susceptibility of *Cryptococcus neoformans* from clinical sources in Nairobi, Kenya. *Mycoses* **50:**25–30.
17. **Bissagnene, E., J. Ouhon, O. Kra, and A. Kadio.** 1994. Current aspects of neuromeningitic cryptococcosis in Abidjan. *Med. Mal. Inf.* **24:**580–585.
18. **Bisson, G. P., J. Lukes, R. Thakur, I. Mtoni, and R. R. MacGregor.** 2008. *Cryptococcus* and lymphocytic meningitis in Botswana. *S. Afr. Med. J.* **98:**724–725.
19. **Bisson, G. P., R. Nthobatsong, R. Thakur, G. Lesetedi, K. Vinekar, P. Tebas, J. E. Bennett, S. Gluckman, T. Gaolathe, and R. R. MacGregor.** 2008. The use of HAART is associated with decreased risk of death during initial treatment of cryptococcal meningitis in adults in Botswana. *J. Acquir. Immune Defic. Syndr.* **49:**227–229.
20. **Boekhout, T., B. Theelen, M. R. Diaz, J. W. Fell, W. C. J. Hop, E. C. Abeln, F. Dromer, and W. Meyer.** 2001. Hybrid genotypes in the pathogenic yeast *Cryptococcus neoformans. Microbiology* **147:**891–907.
21. **Bogaerts, J., D. Rouvroy, H. Taelman, A. Kagame, M. A. Aziz, D. Swinne, and J. Verhaegen.** 1999. AIDS-associated cryptococcal meningitis in Rwanda (1983–1992): epidemiologic and diagnostic features. *J. Infect.* **39:**32–37.
22. **Bogaerts, J., H. Taelman, J. Batungwanayo, P. Van de Perre, and D. Swinne.** 1993. Two cases of HIV-associated cryptococcosis due to the variety *gattii* in Rwanda. *Trans. R. Soc. Trop. Med. Hyg.* **87:**63–64.
23. **Bovers, M., F. Hagen, and T. Boekhout.** 2008. Diversity of the *Cryptococcus neoformans-Cryptococcus gattii* species complex. *Rev. Iberoam. Micol.* **25:**S4–S12.
24. **Bovers, M., F. Hagen, E. E. Kuramae, and T. Boekhout.** 2008. Six monophyletic lineages identified within *Cryptococcus neoformans* and *Cryptococcus gattii* by multi-locus sequence typing. *Fungal Genet. Biol.* **45:**400–421.
25. **Brouwer, A. E., H. J. M. van Kan, E. M. Johnson, A. Rajanuwong, P. Teparrukkul, V. Wuthiekanun, W. Chierakul, N. P. Day, and T. S. Harrison.** 2007. Oral versus intravenous flucytosine in patients with human immunodeficiency virus-associated cryptococcal meningitis. *Antimicrob. Agents Chemother.* **51:**1038–1042.
26. **Byrnes, E. J., III, R. Bildfell, S. A. Frank, T. G. Mitchell, K. A. Marr, and J. Heitman.** 2009. Molecular evidence that the range of the Vancouver Island outbreak of *Cryptococcus gattii* infection has expanded into the Pacific Northwest in the United States. *J. Infect. Dis.* **199:**1081–1086.
27. **Byrnes, E. J., III, W. Li, Y. Lewit, J. R. Perfect, D. A. Carter, G. M. Cox, and J. Heitman.** 2009. First reported case of *Cryptococcus gattii* in the Southeastern USA: implications for travel-associated acquisition of an emerging pathogen. *PLoS ONE* **4:**e5851.
28. **Carme, B., P. M'Pele, A. Mbitsi, A. M. Kissila, G. M. Aya, G. Mouanga-Yidika, J. Mboussa, and A. Itoua-Ngaporo.** 1988. Opportunistic parasitic diseases and mycoses in AIDS. Their frequencies in Brazzaville (Congo). *Bull. Soc. Pathol. Exot. Filiales.* **81:**311–316. (In French.)
29. **Casadevall, A., and J. R. Perfect.** 1999. *Cryptococcus neoformans*. ASM Press, Washington, DC.
30. **Chandenier, J., K. D. Adou-Bryn, C. Douchet, B. Sar, M. Kombila, D. Swinne, M. Therizol-Ferly, Y. Buisson, and D. Richard-Lenoble.** 2004. In vitro activity of amphotericin B, fluconazole and voriconazole against 162 *Cryptococcus neoformans* isolates from Africa and Cambodia. *Eur. J. Clin. Microbiol. Infect. Dis.* **23:**506–508.
31. **Chang, L. W., W. T. Phipps, G. E. Kennedy, and G. W. Rutherford.** 2005. Antifungal interventions for the primary prevention of cryptococcal disease in adults with HIV. *Cochrane Database Syst. Rev.* **Jul 20:**CD004773.
32. **Chariyalertsak, S., K. Supparatpinyo, T. Sirisanthana, and K. E. Nelson.** 2002. A controlled trial of itraconazole as primary prophylaxis for systemic fungal infections in patients with advanced human immunodeficiency virus infection in Thailand. *Clin. Infect. Dis.* **34:**277–284.
33. **Checkley, A. M., Y. Njalale, M. Scarborough, and E. E. Zjilstra.** 2008. Sensitivity and specificity of an index for the diagnosis of TB meningitis in patients in an urban teaching hospital in Malawi. *Trop. Med. Int. Health* **13:**1042–1046.
34. **Chen, J., A. Varma, M. R. Diaz, A. P. Litvintseva, K. K. Wollenberg, and K. J. Kwon-Chung.** 2008. *Cryptococcus*

neoformans strains and infection in apparently immunocompetent patients, China. *Emerg. Infect. Dis.* **14:**755–762.

35. **Chetchotisakd, P., S. Sungkanuparph, B. Thinkhamrop, P. Mootsikapun, and P. Boonyaprawit.** 2004. A multicentre, randomized, double-blind, placebo-controlled trial of primary cryptococcal meningitis prophylaxis in HIV-infected patients with severe immune deficiency. *HIV Med.* **5:**140–143.
36. **Chuck, S. L., and M. A. Sande.** 1989. Infections with *Cryptococcus neoformans* in the acquired immunodeficiency syndrome. *N. Engl. J. Med.* **321:**794–799.
37. **Clumeck, N., F. Mascart-Lemone, J. De Maubeuge, D. Brenez, and L. Marcelis.** 1983. Acquired immune deficiency syndrome in Black Africans. *Lancet* **1:**642.
38. **Clumeck, N., J. Sonnet, H. Taelman, F. Mascart-Lemone, M. De Bruyere, P. Vandeperre, J. Dasnoy, L. Marcelis, M. Lamy, and C. Jonas.** 1984. Acquired immunodeficiency syndrome in African patients. *N. Engl. J. Med.* **310:**492–497.
39. **Collett, G., and A. Parrish.** 2007. Fluconazole donation and outcomes assessment in cryptococcal meningitis. *S. Afr. Med. J.* **97:**175–176.
40. **Corbett, E. L., G. J. Churchyard, S. Charalambous, B. Samb, V. Moloi, T. C. Clayton, A. D. Grant, J. Murray, R. J. Hayes, and K. M. De Cock.** 2002. Morbidity and mortality in South African gold miners: impact of untreated disease due to human immunodeficiency virus. *Clin. Infect. Dis.* **34:**1251–1258.
41. **Desmet, P., K. D. Kayembe, and C. de Vroey.** 1989. The value of cryptococcal serum antigen screening among HIV-positive/AIDS patients in Kinshasa, Zaire. *AIDS* **3:**77–78.
42. **Diaz, M. R., T. Boekhout, B. Theelen, and J. W. Fell.** 2000. Molecular sequence analyses of the intergenic spacer (IGS) associated with rDNA of the two varieties of the pathogenic yeast, *Cryptococcus neoformans*. *Syst. Appl. Microbiol.* **23:**535–545.
43. **Dromer, F., S. Mathoulin-Pélissier, A. Fontanet, O. Ronin, B. Dupont, and O. Lortholary.** 2004. Epidemiology of HIV-associated cryptococcosis in France (1985–2001): comparison of the pre- and post-HAART eras. *AIDS* **18:**555–562.
44. **Eholie, S. P., L. N'gbocho, E. Bissagnene, M. Coulibaly, E. Ehui, O. Kra, A. Assoumou, E. Aoussi, and A. Kadio.** 1997. Profound mycoses in AIDS in Abidjan (Cote d'Ivoire). *Bull. Soc. Pathol. Exot.* **90:**307–311. (In French.)
45. **Elias, M. L., A. K. Soliman, F. J. Mahoney, A. Z. Karam El-Din, R. A. El-Kebbi, T. F. Ismail, M. M. Wasfy, A. M. Mansour, Y. A. Sultan, G. Pimentel, and K. C. Earhart.** 2009. Isolation of *Cryptococcus*, *Candida*, *Aspergillus*, *Rhodotorula* and nocardia from meningitis patients in Egypt. *J. Egypt. Public Health Assoc.* **84:**169–181.
46. **Feng, X., Z. Yao, D. Ren, W. Liao, and J. Wu.** 2008. Genotype and mating type analysis of *Cryptococcus neoformans* and *Cryptococcus gattii* isolates from China that mainly originated from non-HIV-infected patients. *FEMS Yeast Res.* **8:**930–938.
47. **French, N., K. Gray, C. Watera, J. Nakiyingi, E. Lugada, M. Moore, D. Lalloo, J. A. Whitworth, and C. F. Gilks.** 2002. Cryptococcal infection in a cohort of HIV-1-infected Ugandan adults. *AIDS* **16:**1031–1038.
48. **Friedman, G. D., F. W. Jeffrey, N. V. Udaltsova, and L. B. Hurley.** 2005. Cryptococcosis: the 1981–2000 epidemic. *Mycoses* **48:**122–125.
49. **Frimpong, E. H., and R. A. Lartey.** 1998. Study of the aetiologic agents of meningitis in Kumasi, Ghana, with special reference to *Cryptococcus neoformans*. *East Afr. Med. J.* **75:**516–519.
50. **García-Hermoso, D., G. Janbon, and F. Dromer.** 1999. Epidemiological evidence for dormant *Cryptococcus neoformans* infection. *J. Clin. Microbiol.* **37:**3204–3209.
51. **Gatti, F., and R. Eeckels.** 1970. An atypical strain of *Cryptococcus neoformans* (San Felice) Vuillemin 1894. Part I. Description of the disease and of the strain. *Ann. Soc. Belge Med. Trop.* **50:**689–695.
52. **Gordon, S. B., A. L. Walsh, M. Chaponda, M. A. Gordon, D. Soko, M. Mbwvinji, M. E. Molyneux, and R. C. Read.** 2000. Bacterial meningitis in Malawian adults: pneumococcal disease is common, severe, and seasonal. *Clin. Infect. Dis.* **31:**53–57.
53. **Govender, N. P.** 2009. Trends in antifungal drug susceptibility of *Cryptococcus* species in South Africa, 2002-2008, p. 201. Abstract 17th Congress of the International Society for Human and Animal Mycology.
54. **Govender, N. P., and C. Cohen. (ed.).** 2008. Group for enteric, respiratory and meningeal disease surveillance in South Africa. GERMS-South Africa Annual Report 2008. National Institute for Communicable Diseases, Johannesburg, South Africa.
55. **Govender, N. P., C. Cohen, S. Meiring, V. Quan, J. Patel, H. Dawood, A. S. Karstaedt, Y. M. Coovadia, A. Hoosen, O. Perovic, and K. M. McCarthy.** 2008. Surveillance for cryptococcosis in South Africa, 2005–2006, p. 101. Abstract 7th International Conference on *Cryptococcus* & Cryptococcosis.
56. **Graybill, J. R., J. D. Sobel, M. S. Saag, C. M. van der Horst, W. G. Powderly, G. A. Cloud, L. Riser, R. J. Hamill, and W. E. Dismukes.** 2000. Diagnosis and management of increased intracranial pressure in patients with AIDS and cryptococcal meningitis. *Clin. Infect. Dis.* **30:**47–54.
57. **Gugnani, H. C., and A. N. U. Njoku-Obi.** 1973. Occurrence of *Cryptococcus neoformans* in pigeon excreta in Enugu (Nigeria). *West Afr. Med. J. Niger. Med. Dent. Pract.* **22:**121–122.
58. **Gumbo, T., G. Kadzirange, J. Mielke, I. T. Gangaidzo, and J. G. Hakim.** 2002. *Cryptococcus neoformans* meningoencephalitis in African children with acquired immunodeficiency syndrome. *Pediatr. Infect. Dis. J.* **21:**54–56.
59. **Hajjeh, R. A., L. A. Conn, D. A. Stephens, W. S. Baughman, R. J. Hamill, E. A. Graviss, P. G. Pappas, C. Thomas, A. L. Reingold, G. A. Rothrock, L. C. Hutwagner, A. Schuchat, M. E. Brandt, R. W. Pinner, and Cryptococcal Disease Active Surveillance Group.** 1999. Cryptococcosis: population-based multistate active surveillance and risk factors in human immunodeficiency virus-infected persons. *J. Infect. Dis.* **179:**449–454.
60. **Hakim, J. G., I. T. Gangaidzo, R. S. Heyderman, J. Mielke, E. Mushangi, A. Taziwa, V. J. Robertson, P. Musvaire, and P. R. Mason.** 2000. Impact of HIV infection on meningitis in Harare, Zimbabwe: a prospective study of 406 predominantly adult patients. *AIDS* **14:**1401–1407.
61. **Hempson, G. P., E. C. February, and G. A. Verboom.** 2007. Determinants of savanna vegetation structure: insights form *Colophospermum mopane*. *Austral Ecol.* **32:**429–435.
62. **Heyderman, R. S., I. T. Gangaidzo, J. G. Hakim, J. Mielke, A. Taziwa, P. Musvaire, V. J. Robertson, and P. R. Mason.** 1998. Cryptococcal meningitis in human immunodeficiency virus-infected patients in Harare, Zimbabwe. *Clin. Infect. Dis.* **26:**284–289.
63. **Holmes, C. B., E. Losina, R. P. Walensky, Y. Yazdanpanah, and K. A. Freedberg.** 2003. Review of human immunodeficiency virus type 1-related opportunistic infections in sub-Saharan Africa. *Clin. Infect. Dis.* **36:**652–662.
64. **Irokanulo, E. A. O., A. A. Makinde, C. O. Akueshi, and M. Ekwonu.** 1997. *Cryptococcus neoformans* var. *neoformans* isolated from droppings of captive birds in Nigeria. *J. Wildl. Dis.* **33:**343–345.
65. **Jarvis, J. N., A. Boulle, A. Loyse, T. Bicanic, K. Rebe, A. Williams, T. S. Harrison, and G. A. Meintjes.** 2009.

High ongoing burden of cryptococcal disease in Africa despite antiretroviral roll out. *AIDS* **23:**1182–1183.

66. **Jarvis, J. N., and T. S. Harrison.** 2007. HIV-associated cryptococcal meningitis. *AIDS* **21:**2119–2129.
67. **Jarvis, J. N., and T. S. Harrison.** 2008. Pulmonary cryptococcosis. *Semin. Respir. Crit. Care Med.* **29:**141–150.
68. **Jarvis, J. N., S. D. Lawn, M. Vogt, N. Bangani, R. Wood, and T. S. Harrison.** 2009. Screening for cryptococcal antigenemia in patients accessing an antiretroviral treatment program in South Africa. *Clin. Infect. Dis.* **48:**856–862.
69. **Jarvis, J. N., S. D. Lawn, R. Wood, and T. S. Harrison.** 2009. Reducing mortality associated with opportunistic infections among patients with advanced HIV infection in sub-Saharan Africa. *Clin. Infect. Dis.* **49:**812–813.
70. **Kambugu, A., D. B. Meya, J. Rhein, M. O'Brien, E. N. Janoff, A. R. Ronald, M. R. Kamya, H. Mayanja-Kizza, M. A. Sande, P. R. Bohjanen, and D. R. Boulware.** 2008. Outcomes of cryptococcal meningitis in Uganda before and after the availability of highly active antiretroviral therapy. *Clin. Infect. Dis.* **46:**1694–1701.
71. **Kapend'a, K., K. Komichelo, D. Swinne, and J. Vandepitte.** 1987. Meningitis due to *Cryptococcus neoformans* biovar *gattii* in a Zairean AIDS patient. *Eur. J. Clin. Microbiol.* **6:**320–321.
72. **Kaplan, J. E., D. Hanson, M. S. Dworkin, T. Frederick, J. Bertolli, M. L. Lindegren, S. Holmberg, and J. L. Jones.** 2000. Epidemiology of human immunodeficiency virus-associated opportunistic infections in the United States in the era of highly active antiretroviral therapy. *Clin. Infect. Dis.* **30**(Suppl. 1)**:**S5–S14.
73. **Karstaedt, A. S., H. H. Crewe-Brown, and F. Dromer.** 2002. Cryptococcal meningitis caused by *Cryptococcus neoformans* var. *gattii*, serotype C, in AIDS patients in Soweto, South Africa. *Med. Mycol.* **40:**7–11.
74. **Katsu, M., S. E. Kidd, A. Ando, M. L. Moretti-Branchini, Y. Mikami, K. Nishimura, and W. Meyer.** 2004. The internal transcribed spacers and 5.8S rRNA gene show extensive diversity among isolates of the *Cryptococcus neoformans* species complex. *FEMS Yeast Res.* **4:**377–388.
75. **Kwon-Chung, K. J., and J. E. Bennett.** 1984. Epidemiologic differences between the two varieties of *Cryptococcus neoformans*. *Am. J. Epidemiol.* **120:**123–130.
76. **Kwon-Chung, K. J., and J. E. Bennett.** 1984. High prevalence of *Cryptococcus neoformans* var. *gattii* in tropical and subtropical regions. *Zbl. Bakt. Hyg.* A **257:**213–218.
77. **Kwon-Chung, K. J., and J. E. Bennett.** 1992. *Medical Mycology*, p. 397–446. Lea & Febiger, Philadelphia, PA.
78. **Laroche, R., M. Deppner, J. J. Floch, P. Kadende, J. Goasguen, J. F. Sauniere, and B. Dupont.** 1990. Cryptococcosis in Bujumbura, Burundi. Apropos of 80 observed cases in 42 months. *Bull. Soc. Pathol. Exot.* **83:**159–169. (In French.)
79. **Lawn, S. D., L. G. Bekker, L. Myer, C. Orrell, and R. Wood.** 2005. Cryptococcocal immune reconstitution disease: a major cause of early mortality in a South African antiretroviral programme. *AIDS* **19:**2050–2052.
80. **Leal, A. L., J. Faganello, M. C. Bassanesi, and M. H. Vainstein.** 2008. *Cryptococcus* species identification by multiplex PCR. *Med. Mycol.* **46:**377–383.
81. **Lemmer, K., D. Naumann, B. Raddatz, and K. Tintelnot.** 2004. Molecular typing of *Cryptococcus neoformans* by PCR fingerprinting, in comparison with serotyping and Fourier transform infrared-spectroscopy-based phenotyping. *Med. Mycol.* **42:**135–147.
82. **Lengeler, K. B., P. Wang, G. M. Cox, J. R. Perfect, and J. Heitman.** 2000. Identification of the *MATa* mating-type locus of *Cryptococcus neoformans* reveals a serotype A *MATa* strain thought to have been extinct. *Proc. Natl. Acad. Sci. USA* **97:**14455–14460.
83. **Liechty, C. A., P. Solberg, W. Were, J. P. Ekwaru, R. L. Ransom, P. J. Weidle, R. Downing, A. Coutinho, and J. Mermin.** 2007. Asymptomatic serum cryptococcal antigenemia and early mortality during antiretroviral therapy in rural Uganda. *Trop. Med. Int. Health* **12:**929–935.
84. **Litvintseva, A. P., X. Lin, I. Templeton, J. Heitman, and T. G. Mitchell.** 2007. Many globally isolated AD hybrid strains of *Cryptococcus neoformans* originated in Africa. *PLoS Pathog.* **3:**e114.
85. **Litvintseva, A. P., R. E. Marra, K. Nielsen, J. Heitman, R. J. Vilgalys, and T. G. Mitchell.** 2003. Evidence of sexual recombination among *Cryptococcus neoformans* serotype A isolates in sub-Saharan Africa. *Eukaryot. Cell* **2:**1162–1168.
86. **Litvintseva, A. P., T. G. Mitchell, and N. P. Govender.** 2009. Out-of-Africa origin of *Cryptococcus neoformans* var. *grubii* (serotype A). 17th Congress, International Society for Human and Animal Mycology, Abstract EP-04-2, 205.
87. **Litvintseva, A. P., R. Thakur, L. B. Reller, and T. G. Mitchell.** 2005. Prevalence of clinical isolates of *Cryptococcus gattii* serotype C among patients with AIDS in sub-Saharan Africa. *J. Infect. Dis.* **192:**888–892.
88. **Litvintseva, A. P., R. Thakur, R. J. Vilgalys, and T. G. Mitchell.** 2006. Multilocus sequence typing reveals three genetic subpopulations of *Cryptococcus neoformans* var. *grubii* (serotype A), including a unique population in Botswana. *Genetics* **172:**2223–2238.
89. **Longley, N., C. Muzoora, K. Taseera, J. Mwesigye, J. Rwebembera, A. Chakera, E. Wall, I. Andia, S. Jaffar, and T. S. Harrison.** 2008. Dose response effect of high-dose fluconazole for HIV-associated cryptococcal meningitis in southwestern Uganda. *Clin. Infect. Dis.* **47:**1556–1561.
90. **Lortholary, O., A. Fontanet, N. Memain, A. Martin, K. Sitbon, and F. Dromer.** 2005. Incidence and risk factors of immune reconstitution inflammatory syndrome complicating HIV-associated cryptococcosis in France. *AIDS* **19:**1043–1049.
91. **Lortholary, O., G. Poizat, V. Zeller, S. Neuville, A. Boibieux, M. Alvarez, P. Dellamonica, F. Botterel, F. Dromer, and G. Chene.** 2006. Long-term outcome of AIDS-associated cryptococcosis in the era of combination antiretroviral therapy. *AIDS* **20:**2183–2191.
92. **Mahmoud, Y. A. G.** 1999. First environmental isolation of *Cryptococcus neoformans* var. *neoformans* and var. *gatti* from the Gharbia Governorate, Egypt. *Mycopathologia* **148:**83–86.
93. **Makombe, S. D., A. Nkhata, E. J. Schouten, K. Kamoto, and A. D. Harries.** 2009. Burden of cryptococcal meningitis in Malawi. *Trop. Doct.* **39:**32–34.
94. **Mashabane, L. G. W., D. C. J. Wessels, and M. J. Potgieter.** 2001. The utilization of *Colophospermum mopane* by the Vatsonga in the Gazankulu region (eastern Northern Province, South Africa). *S. Afr. J. Bot.* **67:**199–205.
95. **Matee, M. I. N., and R. Matre.** 2001. Pathogenic isolates in meningitis patients in Dar Es Salaam, Tanzania. *East Afr. Med. J.* **78:**458–460.
96. **Mayanja-Kizza, H., K. Oishi, S. Mitarai, H. Yamashita, K. Nalongo, K. Watanabe, T. Izumi, J. Ococi, K. Augustine, R. Mugerwa, T. Nagatake, and K. Matsumoto.** 1998. Combination therapy with fluconazole and flucytosine for cryptococcal meningitis in Ugandan patients with AIDS. *Clin. Infect. Dis.* **26:**1362–1366.
97. **Mbanya, D. N., R. Zebaze, E. M. Minkoulou, F. Binam, S. Koulla, and A. Obounou.** 2002. Clinical and epidemiologic trends in HIV/AIDS patients in a hospital setting of Yaounde, Cameroon: a 6-year perspective. *Int. J. Infect. Dis.* **6:**134–138.
98. **McCarthy, K. M., G. A. Meintjes, B. A. Arthington-Skaggs, T. Bicanic, T. M. Chiller, M. P. Cotton, N. P. Govender, T. S. Harrison, A. S. Karstaedt, G. Maartens,**

E. Varavia, F. Venter, and H. F. Vismer. 2007. Guidelines for the diagnosis, management and prevention of cryptococcal meningitis and disseminated cryptococcosis in HIV-infected patients. *S. Afr. J. HIV Med.* **Spring:**25–35.

99. **McCarthy, K. M., J. Morgan, K. A. Wannemuehler, S. A. Mirza, S. M. Gould, N. Mhlongo, P. Moeng, B. R. Maloba, H. H. Crewe-Brown, M. E. Brandt, and R. A. Hajjeh.** 2006. Population-based surveillance for cryptococcosis in an antiretroviral-naive South African province with a high HIV seroprevalence. *AIDS* **20:**2199–2206.
100. **Meiring, S.** 2009. Bacterial and fungal meningitis amongst children <5 years old, South Africa, 2007. Federation of Infectious Diseases, Societies of Southern Africa, Congress 2009. Abstract.
101. **Meiring, S., N. P. Govender, and V. Quan.** 2008. Paediatric cyptococcosis in South Africa, 2005-2007: a comparison with adult cryptococcosis, Abstract WEPE0127. XVII International AIDS Conference, Mexico City, Mexico.
102. **Meya, D. B., B. Castelnuovo, A. Kambugu, Y. C. Manabe, P. R. Bohjanen, M. R. Kamya, and D. R. Boulware.** 2009. Cost effectiveness of serum cryptococcal antigen (CRAG) screening to prevent death in HIV-infected persons with CD4$^+$<100/μL in sub-Saharan Africa, Abstract TUAD104. 5th Conference on HIV Pathogenesis, Treatment and Prevention. Economic Evaluation and Financing of HIV Interventions and Programmes.
103. **Meyer, W., D. M. Aanensen, T. Boekhout, M. Cogliati, M. R. Diaz, M. C. Esposto, M. C. Fisher, F. Gilgado, F. Hagen, S. Kaocharoen, A. P. Litvintseva, T. G. Mitchell, S. P. Simwami, L. Trilles, M. A. Viviani, and K. J. Kwon-Chung.** 2009. Consensus multi-locus sequence typing scheme for *Cryptococcus neoformans* and *Cryptococcus gattii*. *Med. Mycol.* **47:**561–570.
104. **Meyer, W., A. Castañeda, S. Jackson, M. Huynh, and E. Castañeda.** 2003. Molecular typing of IberoAmerican *Cryptococcus neoformans* isolates. *Emerg. Infect. Dis.* **9:**189–195.
105. **Meyer, W., K. Marszewska, M. Amirmostofian, R. P. Igreja, C. Hardtke, K. Methling, M. A. Viviani, A. Chindamporn, S. Sukroongreung, M. A. John, D. H. Ellis, and T. C. Sorrell.** 1999. Molecular typing of global isolates of *Cryptococcus neoformans* var. *neoformans* by polymerase chain reaction fingerprinting and randomly amplified polymorphic DNA: a pilot study to standardize techniques on which to base a detailed epidemiological survey. *Electrophoresis* **20:**1790–1799.
106. **Miglia, K. J., S, Meiring, N. P. Govender, J. Rossouw, and T. G. Mitchell.** 2008. Molecular analysis of 82 strains of *Cryptococcus* species obtained from South African pediatric patients during 2005 and 2006, Abstract 10105. 7th International Conference on Cryptococcus and Cryptococcosis, Nagasaki, Japan.
107. **Millogo, A., G. A. Ki-Zerbo, J. B. Andonaba, D. Lankoande, A. Sawadogo, I. Yameogo, and A. B. Sawadogo.** 2004. Cryptococcal meningitis in HIV-infected patients at Bobo-Dioulasso hospital (Burkina Faso). *Bull. Soc. Pathol. Exot.* **97:**119–121. (In French.)
108. **Mirza, S. A., M. A. Phelan, D. Rimland, E. A. Graviss, R. J. Hamill, M. E. Brandt, T. Gardner, M. Sattah, G. P. de Leon, W. S. Baughman, and R. A. Hajjeh.** 2003. The changing epidemiology of cryptococcosis: an update from population-based active surveillance in 2 large metropolitan areas, 1992–2000. *Clin. Infect. Dis.* **36:**789–794.
109. **Mitchell, T. G.** 2008. Genomic approaches to investigate the pathogenicity of *Cryptococcus neoformans*. *Curr. Fungal Infect. Rep.* **2:**172–179.
110. **Mitchell, T. G.** 2010. Population genetics of pathogenic fungi in humans and other animals, p. 139–158. *In* J. Xu (ed.), *Microbial Population Genetics*. Horizon Scientific Press, Hethersett, UK.
111. **Mitchell, T. G., and A. P. Litvintseva.** 2010. Typing species of *Cryptococcus* and epidemiology of cryptococcosis, p. 167–190. *In* H. R. Ashbee and E. M. Bignell (ed.), *Pathogenic Yeasts*. Springer-Verlag, Berlin, Germany.
112. **Mitchell, T. G., and J. R. Perfect.** 1995. Cryptococcosis in the era of AIDS—100 years after the discovery of *Cryptococcus neoformans*. *Clin. Microbiol. Rev.* **8:**515–548.
113. **Molez, J. F.** 1998. The historical question of acquired immunodeficiency syndrome in the 1960s in the Congo River basin area in relation to cryptococcal meningitis. *Am. J. Trop. Med. Hyg.* **58:**273–276.
114. **Morgan, J., K. M. McCarthy, S. M. Gould, K. Fan, B. A. Arthington-Skaggs, N. J. Iqbal, K. Stamey, R. A. Hajjeh, and M. E. Brandt.** 2006. *Cryptococcus gattii* infection: characteristics and epidemiology of cases identified in a South African province with high HIV seroprevalence, 2002-2004. *Clin. Infect. Dis.* **43:**1077–1080.
115. **Muyembe, T. J. J., K. D. Mupapa, L. Nganda, D. Ngwala-Bikindu, T. Kuezina, I. Kela-We, and J. Vandepitte.** 1992. Cryptococcosis caused by *Cryptococcus neoformans* var. *gattii*. A case associated with acquired immunodeficiency syndrome (AIDS) in Kinshasa, Zaire. *Med. Trop.* (Mars.) **52:**435–438. (In French.)
116. **Mwaba, P., J. Mwansa, C. Chintu, J. Pobee, M. Scarborough, S. Portsmouth, and A. Zumla.** 2001. Clinical presentation, natural history, and cumulative death rates of 230 adults with primary cryptococcal meningitis in Zambian AIDS patients treated under local conditions. *Postgrad. Med. J.* **77:**769–773.
117. **Ngamskulrungroj, P., F. Gilgado, J. Faganello, A. P. Litvintseva, A. L. Leal, K. M. Tsui, T. G. Mitchell, M. H. Vainstein, and W. Meyer.** 2009. Genetic diversity of the *Cryptococcus* species complex suggests that *Cryptococcus gattii* deserves to have varieties. *PLoS ONE* **4:**e5862.
118. **Odhiambo, F. A., E. M. Murage, W. Ngare, and J. O. Ndinya-Achola.** 1997. Detection rate of *Cryptococcus neoformans* in cerebrospinal fluid specimens at Kenyatta National Hospital, Nairobi. *East Afr. Med. J.* **74:**576–578.
119. **Pappas, P. G.** 2005. Managing cryptococcal meningitis is about handling the pressure. *Clin. Infect. Dis.* **40:**480–482.
120. **Park, B. J., K. A. Wannemuehler, B. J. Marston, N. P. Govender, P. G. Pappas, and T. M. Chiller.** 2009. Estimation of the current global burden of cryptococcal meningitis among persons living with HIV/AIDS. *AIDS* **23:**525–530.
121. **Perriens, J. H., M. Mussa, M. K. Luabeya, K. Kayembe, B. Kapita, C. Brown, P. Piot, and R. Janssen.** 1992. Neurological complications of HIV-1-seropositive internal-medicine inpatients in Kinshasa, Zaire. *J. Acquir. Immune Defic. Syndr.* **5:**333–340.
122. **Pfaller, M. A., S. A. Messer, L. Boyken, C. Rice, S. Tendolkar, R. J. Hollis, G. V. Doern, and D. J. Diekema.** 2005. Global trends in the antifungal susceptibility of *Cryptococcus neoformans* (1990 to 2004). *J. Clin. Microbiol.* **43:**2163–2167.
123. **Pfaller, M. A., J. Zhang, S. A. Messer, M. E. Brandt, R. A. Hajjeh, C. J. Jessup, M. Tumberland, E. K. Mbidde, and M. A. Ghannoum.** 1999. In vitro activities of voriconazole, fluconazole, and itraconazole against 566 clinical isolates of *Cryptococcus neoformans* from the United States and Africa. *Antimicrob. Agents Chemother.* **43:**169–171.
124. **Refai, M., M. Taha, S. A. Selim, F. Elshabourii, and H. H. Yousseff.** 1983. Isolation of *Cryptococcus neoformans*, *Candida albicans* and other yeasts from pigeon droppings in Egypt. *Sabouraudia* **21:**163–165.
125. **Saag, M. S., G. A. Cloud, J. R. Graybill, J. D. Sobel, C. U. Tuazon, P. C. Johnson, W. J. Fessel, B. L. Mos-**

kovitz, B. Wiesinger, D. Cosmatos, L. Riser, C. Thomas, R. Hafner, and W. E. Dismukes. 1999. A comparison of itraconazole versus fluconazole as maintenance therapy for AIDS-associated cryptococcal meningitis. National Institute of Allergy and Infectious Diseases Mycoses Study Group. *Clin. Infect. Dis.* **28:**291–296.

126. **Saag, M. S., J. R. Graybill, R. A. Larsen, P. G. Pappas, J. R. Perfect, W. G. Powderly, J. D. Sobel, and W. E. Dismukes.** 2000. Practice guidelines for the management of cryptococcal disease. *Clin. Infect. Dis.* **30:**710–718.

127. **Sacktor, N., R. H. Lyles, R. Skolasky, C. Kleeberger, O. A. Selnes, E. N. Miller, J. T. Becker, B. Cohen, and J. C. McArthur.** 2001. HIV-associated neurologic disease incidence changes: multicenter AIDS Cohort Study, 1990-1998. *Neurology* **56:**257–260.

128. **Saul, N., M. B. Krockenberger, and D. A. Carter.** 2008. Evidence of recombination in mixed mating type and α-only populations of *Cryptococcus gattii* sourced from single *Eucalyptus* tree hollows. *Eukaryot. Cell* **7:**727–734.

129. **Schaars, C. F., G. A. Meintjes, C. Morroni, F. A. Post, and G. Maartens.** 2006. Outcome of AIDS-associated cryptococcal meningitis initially treated with 200 mg/day or 400 mg/day of fluconazole. *BMC Infect. Dis.* **6:**118.

130. **Schutte, C. M., C. H. van der Meyden, and D. S. Magazi.** 2000. The impact of HIV on meningitis as seen at a South African Academic Hospital (1994 to 1998). *Infection* **28:**3–7.

131. **Sebego, R. J., W. Arnberg, B. Lunden, and S. Ringrose.** 2008. Mapping of *Colophospermum mopane* using Landsat™ in eastern Botswana. *S. Afr. Geogr. J.* **90:**41–53.

132. **Shelburne, S. A., III, F. Visnegarwala, J. Darcourt, E. A. Graviss, T. P. Giordano, A. C. White, Jr., and R. J. Hamill.** 2005. Incidence and risk factors for immune reconstitution inflammatory syndrome during highly active antiretroviral therapy. *AIDS* **19:**399–406.

133. **Silber, E., P. Sonnenberg, K. C. Ho, H. J. Koornhof, S. Eintracht, L. Morris, and D. Saffer.** 1999. Meningitis in a community with a high prevalence of tuberculosis and HIV infection. *J. Neurol. Sci.* **162:**20–26.

134. **Subramanyam, V. R., E. Mtitimila, C. A. Hart, and R. L. Broadhead.** 1997. Cryptococcal meningitis in African children. *Ann. Trop. Paediatr.* **17:**165–167.

135. **Sullivan, D. J., and D. C. Coleman.** 1998. *Candida dubliniensis*: characteristics and identification. *J. Clin. Microbiol.* **36:**329–334.

136. **Swinne, D., M. Deppner, R. Laroche, J. J. Floch, and P. Kadende.** 1989. Isolation of *Cryptococcus neoformans* from houses of AIDS-associated cryptococcosis patients in Bujumbura (Burundi). *AIDS* **3:**389–390.

137. **Swinne, D., M. Deppner, S. Maniratunga, R. Laroche, J. J. Floch, and P. Kadende.** 1991. AIDS-associated cryptococcosis in Bujumbura, Burundi: an epidemiological study. *J. Med. Vet. Mycol.* **29:**25–30.

138. **Swinne, D., K. D. Kayembe, and M. Niyimi.** 1986. Isolation of saprophytic *Cryptococcus neoformans* var. *neoformans* in Kinshasa, Zaire. *Ann. Soc. Belge Med. Trop.* **66:**57–61.

139. **Swinne, D., H. Taelman, J. Batungwanayo, A. Bigirankana, and J. Bogaerts.** 1994. Ecology of *Cryptococcus neoformans* in central Africa. *Med. Trop.* (Mars.) **54:**53–55. (In French.)

140. **UNAIDS08.** 2008. 2008 Report on the Global AIDS Epidemic. XVII International AIDS Conference, Mexico City. 1-362. World Health Organization, Geneva, Switzerland.

141. **U.S. Census Bureau.** 2004. The AIDS pandemic in the 21st century. International Population Reports, WP/02-2. US Government Printing Office, Washington, DC.

142. **Vanbreuseghem, R., and M. Takashio.** 1970. An atypical strain of *Cryptococcus neoformans* (San Felice) Vuillemin 1894. Part II. *Cryptococcus neoformans* var. *gatti* var. nov. *Ann. Soc. Belge Med. Trop.* **50:**695–702.

143. **Vandepitte, J., R. Verwilghen, and P. Zachee.** 1983. AIDS and cryptococcosis (Zaire, 1977). *Lancet* **1:**925–926.

144. **van der Horst, C. M., M. S. Saag, G. A. Cloud, R. J. Hamill, J. R. Graybill, J. D. Sobel, P. C. Johnson, C. U. Tuazon, T. M. Kerkering, B. L. Moskovitz, W. G. Powderly, and W. E. Dismukes.** 1997. Treatment of cryptococcal meningitis associated with the acquired immunodeficiency syndrome. National Institute of Allergy and Infectious Diseases Mycoses Study Group and AIDS Clinical Trials Group. *N. Engl. J. Med.* **337:**15–21.

145. **Varma, A., D. Swinne, F. Staib, J. E. Bennett, and K. J. Kwon-Chung.** 1995. Diversity of DNA fingerprints in *Cryptococcus neoformans*. *J. Clin. Microbiol.* **33:**1807–1814.

146. **Venter, F., and J.-A. Venter.** 1996. *Making the Most of Indigenous Trees.* Briza Publications, Pretoria, South Africa.

147. **Wang, C.-Y., H.-D. Wu, and P.-R. Hsueh.** 2005. Nosocomial transmission of cryptococcosis. *N. Engl. J. Med.* **352:**1271–1272.

148. **Wang, W., and A. R. Carm.** 2001. Clinical manifestations of AIDS with cryptococcal meningitis. *Chin. Med. J.* **114:**841–843.

149. **Wertheimer, A. I., T. M. Santella, and H. J. Lauver.** 2004. Successful public/private donation programs: a review of the Diflucan Partnership Program in South Africa. *J. Int. Assoc. Physicians AIDS Care* (Chic. Ill.) **3:**74–75.

150. **Wilson, L. K., A. Phiri, D. Soko, M. Mbwvinji, A. L. Walsh, and M. E. Molyneux.** 2003. Surveillance of invasive infections in children and adults admitted to QECH, Blantyre, 1996–2002. *Malawi Med. J.* **15:**52–55.

151. **Woldemanuel, Y., and T. Haile.** 2001. Cryptococcosis in patients from Tikur Anbessa Hospital, Addis Ababa, Ethiopia. *Ethiop. Med. J.* **39:**185–192.

152. **Wong, M. L., P. Back, G. Candy, G. Nelson, and J. Murray.** 2007. Cryptococcal pneumonia in African miners at autopsy. *Int. J. Tuberc. Lung Dis.* **11:**528–533.

153. **Xu, J., and T. G. Mitchell.** 2002. Strain variation and clonality in *Candida* spp. and *Cryptococcus neoformans*, p. 739–749. *In* R. A. Calderone and R. L. Cihlar (ed.), *Fungal Pathogenesis: Principles and Clinical Applications.* Marcel Dekker, New York, NY.

154. **Xu, J., and T. G. Mitchell.** 2003. Population genetic analyses of medically important fungi, p. 703–722. *In* D. H. Howard (ed.), *Pathogenic Fungi in Humans and Animals*, 2nd ed. Marcel Dekker, New York, NY.

Cryptococcus: From Human Pathogen to Model Yeast
Edited by J. Heitman et al.
©2011 ASM Press, Washington, DC

21

Cryptococcosis in Asia

JIANGPING XU, WEERAWAT MANOSUTHI, UMA BANERJEE, LI-PING ZHU, JANGHAN CHEN, SHIGERU KOHNO, KOICHI IZUMIKAWA, YUCHONG CHEN, SOMNUEK SUNGKANUPARPH, THOMAS S. HARRISON, AND MATTHEW FISHER

Asia is the world's largest and most populous continent and includes the two most populous countries in the world: China with over 1.3 billion and India with over 1.1 billion people. Asia is extremely diverse in its geography, climate, and culture. However, the majority of the people in Asia reside in the tropical and subtropical regions, a climate very conducive to fungal growth and spread. However, relatively little is known about the diversity and distribution of fungi in Asia, including *Cryptococcus neoformans* and *Cryptococcus gattii*. Similarly, the impact of environmental fungi on human health and how such impact differs between Asia and other parts of the world remains largely unexplored.

For several reasons, a better understanding of cryptococcal and fungal infection in general will become increasingly important in both Asia and other parts of the world. First, the number of immunocompromised patients due to HIV infection has increased dramatically in the past decade, now estimated to be between 3 million and 6 million in Asia as a whole (54). These patients are highly susceptible to a variety of opportunistic fungal infections. Second, diabetes mellitus, another condition predisposing to cryptococcal infection, affects over an astonishing 30 million people in India alone, and this patient population is still increasing rapidly (54). Third, a variety of other medical practices and procedures such as solid organ and bone marrow transplantations, immunosuppressive chemotherapy, intravenous drug use, and the application of broad-spectrum antibiotics are becoming increasingly common in Asia as a whole (10).

To minimize the potential impact of environmental fungi on human health and to develop effective strategies for the control and prevention of fungal infections, we need a comprehensive understanding of the organisms. Over the past decade, progress has been made in the ecology, epidemiology, and population genetics of several prominent fungal pathogens, including *Candida albicans*, *Aspergillus fumigatus*, and *C. neoformans*. However, most of the progress has come from developed countries, mainly those in Europe and North America. In comparison, relatively little is known about human fungal pathogens in Asia.

Fortunately, one area where progress has been made in the recent past is the ecology, epidemiology, and population genetics of *C. neoformans* and *C. gattii* in several regions in Asia. Recent work, discussed below, has identified both unique and common features for Asian populations of these two closely related species. In addition, the chapter includes sections on aspects of the clinical epidemiology and features of cryptococcal infection particular to Asia and management practices and experience from Asia, especially where this differs in some respect from other regions.

ECOLOGY AND MOLECULAR EPIDEMIOLOGY

Environmental Niches

Several research groups have investigated the potential environmental niches of *C. neoformans* and *C. gattii* in India. The main results are summarized in Table 1. Most of the surveys have focused on trees, initially prompted by the works of Ellis and Pfeiffer (18, 19) that showed an association between *C. gattii* and *Eucalyptus* species such as *Eucalyptus camaldulensis* (river red gum), *Eucalyptus tereticornis* (forest red gum), *Eucalyptus blakelyi* (Blakely's red gum), *Eucalyptus gomphocephala* (tuart), and *Eucalyptus rudis* (flooded gum) in Australia. To examine whether *Eucalyptus* trees were a natural reservoir for the *C. neoformans* species complex in India, Chakrabarti et al. (9) conducted an extensive sampling from around *Eucalyptus* trees in the state of Punjab in northwest

Jiangping Xu, Department of Biology and Institute of Infectious Disease Research, Michael G. DeGroote School of Medicine, McMaster University, Hamilton, Ontario, Canada L8S 4K1. **Weerawat Manosuthi,** Bamrasnaradura Infectious Diseases Institute, Ministry of Public Health, Nonthaburi, Thailand. **Uma Banerjee,** Department of Microbiology, All India Institute of Medical Sciences, New Delhi, India. **Li-Ping Zhu,** Department of Infectious Diseases, Huashan Hospital, Fudan University, Shanghai, PR China. **Janghan Chen and Yuchong Chen,** Shanghai Changzheng Hospital, Second Military Medical University, Shanghai, PR China. **Shigeru Kohno and Koichi Izumikawa,** Department of Molecular Microbiology and Immunology, Nagasaki University Graduate School of Biomedical Sciences, Nagasaki, Japan. **Somnuek Sungkanuparph,** Faculty of Medicine, Ramathibodi Hospital, Mahidol University, Bangkok, Thailand. **Thomas S. Harrison,** Centre for Infection, St George's University of London, London SW17 ORE, United Kingdom. **Matthew Fisher,** Imperial College, London, U.K.

TABLE 1 Frequency of isolation of *C. neoformans* and *C. gattii* from decaying wood in tree trunk hollows and bird droppings in various locations in India

Geographic location	Host	Sample type	Frequency (%) of *C. neoformans*	Frequency (%) of *C. gattii*	Reference
Jabalpur, central India	*Mangifera indica*	Debris from hollow	13.3	2.2	22
	Tamarindus indica	Debris from hollow	4.4	15.5	
	Terminalia arjuna	Debris from hollow	25	0	
	Pithecolobium dulce	Debris from hollow	0	12.5	
	Syzygium cumini	Bark	4	2	
	Cassia fistula	Debris from hollow	4.5	0	
	Butea monosperma	Bark	0	0	
		Debris from hollow	0	0	
		Flower	0	0	
Amritsar, Panjab	*S. cumini*	Debris from hollow	54	12.5	58
Meerut Canttonment	*S. cumini*	Debris from hollow	0	9	
Bulandshahr	*S. cumini*	Debris from hollow	9	27	
Delhi	*S. cumini*	Debris from hollow	44	89	
	Polyalthia longifolia	Debris from hollow	14.6	4.9	59
	Azadirachta indica	Debris from hollow	60	35	
	C. fistula	Debris from hollow	47.4	5.2	
	Acacia nilotica	Debris from hollow	14.3	0	
	Alstonia scholaris	Debris from hollow	5.2	0	
	Eucalyptus spp.	Debris from hollow	0	5.2	
	Ficus religiosa	Debris from hollow	6.7	0	
	Mimusops elengi	Debris from hollow	15.4	30.8	
	Dalbergia sissoo	Debris from hollow	7.7	0	
	Manilkara hexandra	Debris from hollow	50	33.3	
Amrouli, Haryana	*M. indica*	Debris from hollow	14.3	0	
	A. indica	Debris from hollow	80	0	
Hathrus, Uttar Pradesh	*A. indica*	Debris from hollow	40	20	
Bulandshahar	*A. indica*	Debris from hollow	40	40	
Tivuvannamalai, Tamil Nadu	*T. indica*	Debris from hollow	0	10	
Chandigarh, UT	*S. cumini*	Debris from hollow	14.2	0	
Madras	Pigeon	Fresh droppings	14.1	0	21
		Old droppings	13.2	0	
	Fowl	Fresh droppings	12.7	0	
		Old droppings	8.3	0	
	Crow	Fresh droppings	15.1	0	
		Old droppings	0	0	
	Rose-ringed parakeets	Fresh/old droppings	29.6	0	
	Turkey	Old droppings	25.6	0	
	Common myna	Old droppings	16.7	0	
	House sparrow	Old droppings	0	0	
	Cattle egret	Old droppings	40	0	
	Little egret	Old droppings	0	0	
	Pond heron	Old droppings	20	0	
	Cormorant	Old droppings	0	0	
	Common babbler	Old droppings	5.3	0	

India. Australian eucalypts were first introduced into Punjab in 1860, and *Eucalyptus citriodora*, *Eucalyptus globulus*, *E. tereticornis*, and *E. camaldulensis* are commonly found there. A total of 695 samples were collected and analyzed over a 2-year period from around *E. tereticornis*, *E. citriodora*, and *E. camaldulensis*. These samples included 236 air samples from under eucalypt trees, 118 samples from flowers, 35 from fruits, 50 from leaves, 61 from bark, 90 from debris collected under the trees, 98 from soil underneath trees, and 8 from the combs of wild bees' nests. Among these samples, only those from flowers from two trees of *E. camaldulensis* in the Chak Sarkar forest and one from the village of Periana near the Ferozepur area were positive for *C. gattii*, yielding a total of 5 isolates from among 31 samples from the *Eucalyptus* flowers in these two regions.

The sporadic presence of *C. gattii* from among eucalypts prompted other researchers to examine other potential ecological niches, specifically in decaying wood trunks of eucalypts and a variety of other trees. For example, Randhawa et al. (56) examined the distributions of *C. neoformans* and other yeast-like fungi in decaying tree trunks, bark, and other plant materials. They found that 3 of 45 (6.6%) decaying wood samples were positive for *C. neoformans* and 1 of 390 *Eucalyptus* bark samples was positive. The low isolation rate was consistent with that found by Chakrabarti et al. (9). Gugnani et al. conducted a similar study and showed similarly sporadic presence of *C. neoformans* and *C. gattii* in eucalypts (23). They analyzed 233 samples from *Eucalyptus* trees (120 flowers, 81 fragments of bark, and 32 leaves) and identified that two samples of flowers of *E. tereticornis* contained *C. gattii*, while *C. neoformans* var. *grubii* was recovered twice from the bark of *E. camaldulensis*.

While eucalypts were identified as a potentially minor environmental reservoir for *C. neoformans* and *C. gattii* in India, other plant species emerged to be of greater ecological importance. For example, in the study by Randhawa et al., positive cultures of *C. neoformans* and *C. gattii* were obtained from the tree trunk hollows of *Butea monosperma* and *Tamarindus indica* (56). Of special interest was the high isolation rate of *C. gattii* and *C. neoformans* var. *grubii* from decayed wood inside trunk hollows of *Syzygium cumini* and of *C. neoformans* var. *grubii* from *Ficus religiosa* trees in the Delhi–New Delhi metropolitan area (57). Specifically, 21% of the investigated *S. cumini* trees were positive for *C. neoformans* or *C. gattii*. *C. neoformans* var. *grubii* was also isolated from 3 of 17 *F. religiosa* trees, with 2 from decayed wood and 1 from bark. The colonizations seemed long-term, as repeated sampling from the same tree trunks were often successful for *C. gattii* and *C. neoformans* var. *grubii*. In addition, the population densities in some of the trees were extremely high, up to 6×10^5 CFU per g of decayed wood for *C. gattii* and 8×10^4 CFU/g for *C. neoformans* var. *grubii*. Because no eucalypt trees were near the positive *S. cumini* and *F. religiosa* trees and the densities of *C. neoformans* and *C. gattii* in these trees were comparable to or exceeded those found previously in *Eucalyptus* trees, the authors concluded that *C. gattii* and *C. neoformans* were not specific to woody or other debris of particular tree species, but instead were more generalized.

This conclusion was subsequently strengthened in two recent studies. Grover et al. repeatedly isolated *C. neoformans* var. *grubii* and *C. gattii* from decaying wood of trunk hollows in a variety of trees growing in Jabalpur City in central India (Table 1) (22). Specifically, they found *C. gattii* from decayed wood inside trunk hollows of *T. indica*, *Mangifera indica*, *Pithecolobium dulce*, and *S. cumini* and from bark of *S. cumini*. *C. neoformans* var. *grubii* was isolated from decaying wood debris of *T. indica*, *M. indica*, *Terminalia arjuna*, *S. cumini*, and *Cassia fistula* and from two samples from the bark of *S. cumini*. While the same tree species can be hosts for both fungal pathogens, these researchers did not observe co-occurrence of the two fungi in the same hollow. Long-term colonization was confirmed by repeated isolation of the same species up to 820 days. However, the genotypes of these isolates have not been examined. Another recent survey by Randhawa et al. found *C. gattii* and *C. neoformans* in decayed wood inside trunk hollows of many species from across broad geographic areas in northwest India: Amritsar (Panjab), Meerut Canttonment and Bulandshahr (Uttar Pradesh), and Delhi (59). An astonishing 89% of *S. cumini* trees in Delhi were found colonized by *C. gattii*. In contrast, *C. neoformans* had the highest prevalence (54%) in Amritsar. Furthermore, 44% of the *S. cumini* trees in Delhi, 9% in Bulandshahr, and 8% in Amritsar were concomitantly colonized by both *C. gattii* and *C. neoformans*. These observations suggest opportunities exist for mating to occur between strains of *C. neoformans* and *C. gattii*.

Similar to earlier findings (Table 1) (58), a long-term surveillance over 4.8 to 5.2 years of seven selected *S. cumini* trees in Delhi revealed perennial colonization by both *C. neoformans* and *C. gattii*. Taken together, the results from this and previous surveys suggest that in northwestern India, while many tree species can be hosts for *C. neoformans* and *C. gattii*, decayed wood in trunk hollows of *S. cumini* trees might be their primary environmental niche (58). At present, the contribution of *S. cumini* to human cryptococcal infection remains unknown. Future surveys should examine the relationship between the incidences of cryptococcosis and the density of and accessibility to *S. cumini* trees in targeted geographic areas.

Aside from surveys of trees, Gokulshankar et al. (21) examined avian guano, an important reservoir of *C. neoformans* identified in other parts of the world, for evidence of *C. neoformans* colonization. Among the total of 887 pellets of different avifauna in Madras, 106 (12%) yielded positive cultures of *C. neoformans* (Table 1). Of special note were the isolations of this yeast from crow droppings. Crows are ubiquitous in India, and both mating type **a** (MAT**a**) and MATα strains were isolated from crow droppings. In addition, serotype D strains were more prevalent than strains of serotype A, with 74% of the isolates belonging to serotype D and 26% to serotype A. No isolates of *C. gattii* were found. This pattern is very different from those observed for samples from trees where serotype B (as well as serotype A) was commonly isolated but no serotype D has been found (Table 1). The study adds to prior reports of the isolation of *C. neoformans* from excreta of pigeons (64), munia birds (50), and canaries (51) in India.

Aside from *C. neoformans* and *C. gattii*, other *Cryptococcus* species have also been identified from tree bark in India, including *Cryptococcus albidus*, *Cryptococcus laurentii*, *Cryptococcus saitoi*, *Cryptococcus friedmannii*, and *Cryptococcus albidosimilis* (7, 56). The clinical importance of these species in India remains to be defined.

Fewer published ecological studies have examined the prevalence of *Cryptococcus* in the environment in other regions of Asia. However, what information there is suggests that both *C. gattii* and *C. neoformans* occur widely in the environment. *C. neoformans* isolates have been regularly found infecting pigeon droppings (1, 68, 71, 76). In China, 40 of 52 *C. neoformans* strains isolated from pigeon droppings were serotype A and the remainder AD, the two serotypes being found together in some samples (1). In Malaysia, all isolates from bird guano were found to be *C. neoformans* VNI (71), while clinical isolates in the same study were 95.5% VNI and 4.5% VNII. Similarly, *C. neoformans* was recovered from pigeon droppings in Bangkok, showing that this is also a putative source of human infection (68). Chariyalertsak et al. examined the seasonal incidence of 793 cases of *C. neoformans* infection in northern Thailand and compared these against a similar number (550) of cases of *Penicillium marneffei*. While infections of *P. marneffei* were more frequent in the rainy season, cases of cryptococcosis showed no temporal variation, suggesting that rainfall is not an important factor in driving rates of infection (12).

Characterization of Clinical Isolates in Asia

Studies from India

Clinical isolates from India are dominated by serotype A strains, with a small percentage of serotypes B, D, and AD, a pattern more similar to the serotypes found in tree samples than in avian guano in India. For example, Padhye et al. examined 18 clinical strains of the cryptococcal species complex from India and found 15 belonged to *C. neoformans* (13 were serotype A and 2 were serotype AD), and 3 were *C. gattii* (serotype B) (49). In a retrospective study, Banerjee et al. analyzed the serotypes of *Cryptococcus* isolates from 36 patients from northern India (5). The majority of the isolates were serotype A (87%), and 11% were serotype B. One unusual isolate was not typable with the Iatron factor sera. Interestingly, one immunocompetent patient was colonized by isolates of serotypes A and B simultaneously from two different body sites, lung and scalp abscess, respectively. This was the first reported case in which an individual was infected with two serotypes at the same time. Interestingly, the same institution also reported isolation of both serotypes A and D with different melanizing characteristics from cerebrospinal fluid (CSF) from a single patient (42).

Several studies have reported the molecular genotypic analyses of isolates of *C. neoformans* from India. Through PCR-fingerprinting by two primers, the M13 core sequence and a simple sequence repeat, $(GACA)_4$, Meyer et al. surveyed the genotypes of a global collection of 356 isolates of *C. neoformans* (47). Their analyses identified four major genotype groups (VNI and VNII, serotype A; VNIII, serotype AD; and VNIV, serotype D). The majority (78%) of isolates belonged to VNI, compared with 18% VNII, 1% VNIII, and 3% VNIV. Among the 22 clinical isolates from Vellore, India, analyzed in this study, 19 (86.7%) belonged to VNI, while 3 (13.3%) belonged to VNII, similar to the overall global pattern. A recent report from Bangalore confirmed the predominanace of serotype A VNI isolates among a collection of strains from 139 patients, nearly all of whom were HIV-seropositive. In addition, four *C. gattii* serotype C VGIV isolates were also identified, for the first time in India (17).

Jain et al. reported the genotypic characterizations of 57 clinical isolates from several regions in India (26). They used a variety of molecular typing techniques, including PCR fingerprinting using the M13 core sequence, karyotyping using pulsed-field gel electrophoresis, restriction fragment length polymorphisms with the *C. neoformans* transposon 1, and *URA5* DNA sequence analysis. Among the 51 isolates of *C. neoformans* var. *grubii*, 3 were MAT**a** and 48 were MATα. The lone *C. neoformans* var. *neoformans* isolate was MATα, while the five *C. gattii* isolates included one MAT**a** and four MATα strains. Interestingly, their sequence analyses of the *URA5* gene did not separate the *C. neoformans* var. *neoformans* and *C. neoformans* var. *grubii* into distinct genotype groups. Rather, the *C. neoformans* var. *neoformans* cluster contained sequences from strains of *C. neoformans* var. *grubii*, suggesting hybridization between the two varieties. Furthermore, the molecular analyses of 18 additional sequential isolates from 14 patients revealed microevolution of the infecting strains, and in 1 patient simultaneous coinfection with two distinct *C. neoformans* strains (26).

Hiremath et al. (25) recently investigated the modes of reproduction and potential for gene flow among populations of *C. neoformans*. This study analyzed samples of *C. neoformans* var. *grubii* colonizing decaying wood in tree hollows of nine tree species in five geographical locations (Delhi, Bulandshahar, Hathras, Amritsar, and Amrouli) in northwestern India. Multilocus sequence typing (MLST) was conducted using five gene fragments for each of 78 isolates. Different from the clinical samples reported by Jain et al. (26) and Meyer et al. (47), all 78 isolates belonged to MATα and the VNI genotype group. Population genetic analyses identified no evidence for significant differentiation among populations belonging to either different geographical areas or different host tree species. However, despite the lack of MAT**a** strains in their survey, there was unambiguous evidence for recombination in the environmental populations of *C. neoformans* var. *grubii*. The authors proposed that either same-sex mating between MATα strains or opposite-sex mating between MAT**a** and MATα strains could have generated the recombinant genotypes. These results support the hypothesis of long-distance dispersal and recombination in environmental populations of *C. neoformans* var. *grubii* in India.

Studies from Thailand and China

In Thailand, prior to the AIDS epidemic, *C. gattii* was the most common cause of cryptococcosis and accounted for over half of all isolates that were recovered from human patients (36, 69). However, *C. neoformans* has become by far the most common species in Thailand since the arrival of the AIDS pandemic. Serotyping studies since the onset of AIDS have shown that 93 to 96% of isolates are *C. neoformans* serotype A, with 3 to 4% *C. gattii* serotype B (55, 69). Similarly, in China, *C. neoformans*, MATα, serotype A, and genotype VNI accounted for 120/129 isolates collected from both immunocompetent and immunocompromised patients (15). The VNI strains all seemed to have an identical subtype based on fingerprinting with the M13 core sequence as PCR primer. All six strains subjected to MLST clustered with strains belonged to the M5 type reported by Litvintseva et al. (38). The remaining nine isolates were *C. gattii*, belonging to serotype B, MATα, and genotype group VGI (15). Most of these *C. gattii* isolates were from provinces south of Shanghai, and while seven of nine were from apparently immunocompetent patients, two were from AIDS patients, as has also been reported occasionally in India (4, 5). Other serotypes and genotypes have only occasionally been reported, including a few serotypes D, AD, and C strains from Thailand (69) and VNIII (serotype AD), VNIV (serotype D), and VGII (serotype B) genotypes from China (20).

While valuable for local-scale studies on patterns of genetic diversity, PCR fingerprinting data using the M13 core sequence and a simple sequence repeat, $(GACA)_4$, as primers generates data that are inappropriate for most population genetic analyses. In certain cases, the fingerprinting patterns may not be reproducible between laboratories (48). As detailed elsewhere in this volume, modern methods of genotyping *C. neoformans* rely on using MLST to discriminate between isolates by sequencing a set of internationally agreed upon regions of the genome. The MLST scheme for *C. neoformans* and *C. gattii* has been recently established by an ISHAM working group led by June Kwon-Chung. This working group has chosen seven loci (detailed in reference 48) for typing *Cryptococcus* species globally. The ISHAM MLST scheme has been used to genotype a panel of 186 isolates of *C. neoformans* from different regions of Thailand (S. Simwami, T. Boukhout, and M. Fisher, unpublished data). These data can be accessed at http://cneoformans.mlst.net/. From these isolates, 12 unique multilocus sequence types (STs) have been recovered from Thailand. These STs were all MATα and all belonged to the VNI group (except for one isolate that was VNII), reflecting the findings of Hiremath et al. (25) from India and confirming that VNI is the most globally prevalent lineage of *C. neoformans*.

However, integrating the Thailand STs into the 43 STs sequenced from other Asian countries (India) and continents (North America and Africa) showed that 10 of the 12 STs from Thailand are unique, and only 2 (ST4 and ST6) are found elsewhere in the world (Africa). Further, the majority of isolates found in Thailand are composed of two closely related sequence types, ST44 (37% of Thai isolates) and ST45 (42% of Thai isolates). These data show that the *C. neoformans* population within Thailand may be relatively isolated from those in other parts of the world and that they are highly clonal. In fact, it is doubtful that the two STs shared between Africa and Thailand are examples of natural long-distance dispersal. Instead, they may well represent the intercontinental movements of patients who acquired their infections in one region before moving to another. If this is the case, then the population of *C. neoformans* that occurs in Thailand forms a unique, genetically isolated population when compared against VNI isolates from elsewhere. The current MLST data set is biased to clinical isolates of *C. neoformans*, with only 7.7% of isolates being recovered from environmental sources, such as pigeon droppings and trees. There is a clear need to ascertain a greater number of STs from isolates that have been recovered from a wider range of environmental substrates in order to assess the distribution of STs that are recovered from patients as well as the environment. Such studies will not only indicate the principal environmental sources of infection, but will also determine whether infectious isolates represent a subset of the genotypes that are found in the environment. These molecular epidemiological data will be important in ascertaining the risk of infection that particular environments and ecotypes pose to susceptible humans.

CLINICAL EPIDEMIOLOGY

The number of cryptococcosis cases reported across Asia has been increasing over recent decades (3, 80). In China, investigators at Shanghai Changzheng Hospital conducted a literature review and identified 2,196 documented cases of cryptococcosis from 1985 to 2007 (Y. Chen, unpublished). In common with other reports from China (78, 80), HIV infection was identified in a relatively small proportion (10% of cases). Underlying diseases included tuberculosis (12%), liver disease (12%), systemic lupus erythematosus (9%), diabetes (6%), kidney disease (6%), and cancer (5%). Other miscellaneous diseases were found in 25% of the cases. In the remaining 15% of the cases, no underlying conditions were identified (15; Chen, unpublished). Of the 120 *C. neoformans* isolates analyzed for their genotypic characteristics, 70% were from apparently healthy hosts, 22.5% from patients with risk factors other than HIV, and 7.5% from HIV-infected patients (15).

A high percentage of patients without recognized risk factors has also been reported recently from a series of 154 patients with cryptococcal meningitis without documented HIV infection (141 had a negative HIV serology and/or a normal CD4:CD8 cell ratio) seen at a single hospital in Shanghai between 1997 and 2007 (80). The number of both apparently healthy hosts and patients with predisposing conditions steadily increased up to 2005. There was a predominance of men (male:female ratio, 1.6:1) as seen in many studies globally. Of these patients, 67% were apparently healthy, while common predisposing conditions were corticosteroid use, autoimmune disease, liver and kidney disease, diabetes, and cancer (Table 2) (80). Nine patients (12% of those tested) had idiopathic CD4+ T cell lymphopenia.

The completeness of HIV serotesting is uncertain in some of these series in the earlier years (15). Despite this caveat, the balance of evidence suggests a higher proportion of apparently healthy hosts among non-HIV-infected patients with cryptococcal infection in China than most other parts of the world. Data from American, French, and Thai series

TABLE 2 Demographic features and underlying diseases of 154 non-HIV patients with cryptococcal meningitis from Huashan Hospital, Shanghai[a]

Items	Number	Percentage
Sex		
Male	94	61.0
Age		
<30	39	25.3
30~59	104	67.5
≥60	12	7.8
Time to diagnosis		
<1 week	3	2.0
1 week–4 months	132	85.7
>4 months	19	12.3
Definite cases	149	96.8
CSF culture (+)[b]	76	78.4
India ink smear (+)[c]	131	88.5
Histopathology (+)[d]	5	3.3
Predisposing factors	51	33.1
Corticosteroids	21	13.6
Autoimmune diseases	17	11.0
Liver cirrhosis	15	9.7
Diabetes mellitus	14	9.1
Immunosuppression	13	8.4
Chronic kidney diseases	11	7.1
Splenectomy	2	1.3
Solid malignancy	1	0.7
Hematologic malignancy	1	0.7
Kidney transplantation	1	0.7
Idiopathic CD4+ lymphocytopenia[e]	9	12.5
Extraneural involvement		
Pulmonary cryptococcosis	14	9.1
Cryptococcemia[f]	4	33.3

[a] The total number of cases examined was 154 unless otherwise specified. Data from reference 80.
[b] In 97 patients.
[c] In 148 patients.
[d] In five patients.
[e] CD4+ lymphocyte count tests were taken in 72 patients.
[f] Blood cultures were performed in 12 patients before starting antifungal treatment.

all show a relatively low proportion of healthy hosts, ranging from 17 to ~35%, with the majority of cryptococcal patients having some underlying condition (31, 52). The two-thirds of patients with apparently normal immunity seen in recent series from China is consistent with several earlier studies that focused on cryptococcal meningitis in Chinese populations. Studies from Hong Kong (46 cases, 1995 to 2005 [41]) and Taiwan (94 cases, 1977 to 1996 [66], and 71 cases, 1986 to 1997 [40]) found that normal hosts accounted for 44%, 55%, and 61% of the patients, respectively. A study in the 1970s from Singapore, a country where Chinese is the predominant ethnic group, reported that 96% (24/25) of the patients were evidently immunocompetent (73). Similarly, in an early study of cryptococcal meningitis from Malaysia in the 1970s, in which 20 out of the 30 patients were Chinese, only two patients were immunosuppressed, on corticosteroid therapy (61). Together, these observations suggest that in the Chinese population, cryptococcal patients are predominantly immunocompetent and raise the possibility that the Chinese population may be more susceptible than other ethnic groups to developing clinical cryptococcal disease (15, 80). Further studies, including broader surveillance measures and more detailed medical history information from hosts, are needed to fully address this issue.

Also, in Japan, most cryptococcal infection is probably not HIV-associated, although epidemiological data are limited. Two retrospective reports of the visceral mycoses in autopsied cases indicated that the frequency of cryptococcosis was less than 0.4% from 1969 to 2001, and cryptococcosis ranked as the third most common deep-seated mycosis (35, 77). Among patients with leukemia and the myelodysplastic disorder syndrome, the frequency of cryptococcosis was 0.9 to 2.8% from 1989 to 2001, and *Cryptococcus* ranked as the fourth most common fungal agent (35). The most common sites of *Cryptococcus* infection were lung (53.8%), followed by liver (14.9%), brain and meninges (10.6%), and spleen (10.6%) (35).

Regarding HIV-associated cryptococcosis, the first AIDS case was reported in China in 1985. Since then, only one paper has reported the prevalence of HIV-associated cryptococcosis, among HIV-infected in-patients at a hospital in Shanghai between 2004 and 2006. Of 85 patients, 6 (7%) had cryptococcal meningitis (65). More studies are needed.

Data on the prevalence of cryptococcosis in HIV patients in Japan are also very limited. The registered adult HIV prevalence is low (1,126 per 100,000 as of 2008). The national surveillance funded by the Japanese Ministry of Health, Labor, and Welfare found only 4 to 13 HIV patients with cryptococcosis per year among 400 medical institutes between 1995 and 2007 (79). The accumulative prevalence of cryptococcosis in Japanese HIV patients was estimated as 2.7%, and cryptococcosis was the 10th most common cause of AIDS-related opportunistic infections between 1995 and 2007. Although the prevalence of cryptococcosis was lower than other opportunistic infections such as pneumocystis pneumonia, cytomegalovirus infection, and candidiasis, the mortality was remarkably higher and reached around 30%.

In contrast to the reports to date from China and Japan, HIV-associated cryptococcal disease constitutes a large proportion of cases in Thailand and India, where HIV seroprevalence is higher. In a study from a Bangkok hospital, between 1996 and 2005, 149 of 179 (83%) cases were HIV-associated (29). Of the non-HIV-associated cases at the same hospital, 65% had identified underlying conditions such as immunosuppressive drug treatment, systemic lupus erythematosus, malignancies, and diabetes mellitus (31).

In India, a series of studies from the All India Institute of Medical Sciences, New Delhi, have tracked the effect of the HIV pandemic on cryptococcal disease (6). The percentage of cryptococcal cases associated with HIV infection increased from 20% of 39 cases between 1992 and 1996, to 37% of 61 cases between 1996 and 2000, and to 49% of 81 cases between 2000 and 2004. Patients with no apparent predisposing factor constituted approximately half of HIV-seronegative cases in 1992 to 1996 and over half of HIV-seronegative cases in 1996 to 2000 (6). Nonmeningeal presentations (especially undifferentiated fever) were common in both HIV- and non-HIV-associated disease. In a large cohort of HIV-infected patients in Chennai, cryptococcosis accounted for 5% of all AIDS-defining opportunistic infections (34). In Pune, cryptococcal meningitis was seen in 3% of 655 hospitalized HIV-infected patients and was the cause of 8% of all deaths (67).

CLINICAL MANIFESTATIONS

The series of cryptococcal infections from China provides an opportunity to investigate cryptococcal meningitis in non-HIV patients, and particularly, in apparently healthy hosts. In the study of Zhu et al., cryptococcal patients with apparently normal immune status were overall about 10 years younger than immunocompromised ones (Table 3) (80). The duration from onset of symptoms to diagnosis of cryptococcal meningitis in immunocompetent patients was longer than their predisposed counterparts. In immunocompetent patients, the CSF white cell count tended to be higher, and seizures, hydrocephalus, and shunt procedures were more frequent. Immunocompromised patients were more frequently found to have high fever and parenchymal lesions in cranial magnetic resonance imaging. Some of the same findings, e.g., longer time to presentation and a more marked meningeal inflammatory response in apparently immunocompetent hosts and more disseminated infection (fungemia and higher serum antigenemia) in immunocompromised hosts, have also been reported in series from Hong Kong and Taiwan (Table 3) (41, 66). Thus, despite the lack of serious underlying conditions, delayed diagnosis and a more intense inflammatory response in immunocompetent patients may lead to severe illness. Indeed, in the series of Zhu et al., there was no difference between immunocompetent and immunocompromised hosts in treatment response or 1-year survival (80). The all-cause mortality at 1 year was 29%. In a multivariate analysis, a prolonged time from onset of symptoms to diagnosis (>4 months), initial antifungal therapy without amphotericin B, advanced age (≥60 years), and coma were all independently associated with mortality. Other series of non-HIV-associated cryptococcal infections from Hong Kong and Taiwan have reported altered mental status, high CSF antigen titers, and underlying malignancy as poor prognostic factors (40, 41, 66).

A large series of non-HIV-associated pulmonary cryptococcosis has been reported from Nagasaki prefecture in the west of Japan (32). Of the 127 cases, 57 were in apparently immunocompetent and 70 in immunocompromised patients. In immunocompetent patients, cough was the most common symptom, followed by chest pain and fever. However, 36 cases (63%) were asymptomatic. The asymptomatic cases were identified by abnormal findings on chest radiograph examination and diagnosed by culture and pathology. Medical checkups that include chest radiograph and blood tests are commonly undertaken in Japan within companies, local government, and communities, which may

TABLE 3 Comparison between non-HIV-associated cryptococcal meningitis in predisposed hosts and healthy hosts in three series in Chinese populations

Parameter[a]	Zhu et al. (80) *n*/Total/median (%/range)			Shih et al. (66) *n*/Total/median (%/range)			Lui et al. (41) Mean ± SD, *n*/total (%)	
	Predisposed	Normal hosts	*P*	Predisposed	Normal hosts	*P*	Predisposed	Normal hosts
Number	51	103		30	64		8	16
Age (yr)	48 (14–67)	35 (9–75)	0.0001					
Time to diagnosis	30 (1–124)	40 (6–2,890)	0.006	14 (1–900)	29 (1–300)	0.015	15 ± 7	34 ± 8
Fever >39°C	21/51 (41.2)	17/103 (16.5)	0.001					
Seizure	9/51 (17.7)	35/103 (34.0)	0.035					
CSF WBC	63 (0–756)	100 (0–1,030)	0.06	20 (0–306)	77 (0–1,098)	0.004	54 ± 37	108 ± 25
CSF cryptococcal antigen	1,280 (10–>1,280)	1,280 (1–>1,280)	0.8	512 (8–16,348)	192 (0–16,348)	NS	643 ± 576	810 ± 355
Parenchymal lesions (MRI)	16/18 (88.9)	29/47 (61.7)	0.03					
Hydrocephalus	2/27 (7.4)	14/67 (20.9)	0.1	4/17 (23)	23/47 (49)	0.07	1/8 (14)	5/16 (31)
Surgical procedure	3/51 (5.9)	24/103 (23.3)	0.007					
Fungemia				7/30 (23)	6/64 (9)	0.07	2/8 (25)	0/16 (0)
1-year all-cause mortality	13/49 (26.5)	28/94 (29.8)	0.7					

[a]WBC, white blood cell; MRI, magnetic resonance imaging.

help explain the number of asymptomatic cryptococcosis cases identified. Either single or multiple pulmonary nodules larger than 1 cm in diameter were the most common findings on chest radiological examination. Four cases (11%) of pulmonary cryptococcosis among the apparently immunocompetent population also had meningoencephalitis proved by CSF examination. Almost all cases, including asymptomatic cases, were treated with azoles, and there were no associated deaths even in those with meningoencephalitis. Of the 70 cases of pulmonary cryptococcosis with underlying diseases other than HIV infection, diabetes (30%) was the most common underlying disease, followed by hematological (23%), collagen (21%), and renal (17%) diseases. Thirty cases (43%) were related to corticosteroid treatment. Seven cases (10%) developed meningoencephalitis, and 12 patients (17%) died (32).

Series of cases from India and Thailand do not suggest that the presentation of HIV-associated cryptococcal meningitis in Asia is different from the rest of the world (6, 29, 33, 75). The contrasting presentations of cryptococcosis in HIV- and non-HIV-infected patients have been compared in Thailand. Central nervous system (CNS) involvement was more common among HIV-infected patients (92% versus 21%), whereas pulmonary involvement was more common among non-HIV-infected patients (35% versus 3%) (29). In CNS infection, CSF India ink preparation was positive more often (80% versus 50%), and CSF antigen titers were higher in HIV-infected patients compared with non-HIV-infected patients (29, 31). For pulmonary cryptococcosis, clinical manifestations varied from severe pneumonitis to an asymptomatic lung mass, the latter being more common in non-HIV-infected patients (29, 31).

Reports from Thailand (29, 30, 70) and India (3) have also documented less common sites of infection, including bone and joint, urinary tract, gastrointestinal tract, peritoneum, and scalp. From India, Banerjee and colleagues have provided a comprehensive review of case reports and series from the first published case in India in 1941 (60) up to 2001 (Table 2 in reference 3) and a more recent update of HIV-associated cryptococcal infection (6). This literature informs several important clinical issues, including the co-occurrence of cryptococcosis and tuberculosis. There are many reports of patients with both cryptococcosis and pulmonary tuberculosis (3, 6), including many where cryptococcosis occurs in patients with healed or existing tuberculosis. In the series from the All India Institute of Medical Sciences in New Delhi, of the laboratory-confirmed cases of cryptococcosis, concomitant pulmonary tuberculosis was detected in 2.5 to 6.5% of patients. In HIV-infected cohorts, up to one-third of cryptococcosis patients were diagnosed with tuberculosis (37). Because tuberculosis is endemic in India, it is difficult to determine whether these reports reflect a level of coinfection greater than one might expect. However, it is tempting to speculate that tuberculosis may predispose hosts to subsequent cryptococcosis, rather than coinfection reflecting an underlying predisposition to both infections. The Indian reports are consistent with emerging data from HIV-infected cohorts in Africa suggesting that tuberculosis is a risk factor, independent of CD4 count, for subsequent cryptococcal disease (J. Jarvis et al., personal communication). If such an association could also be demonstrated in HIV-uninfected patients in India, it would have implications for the mechanism of the interaction. It is worth noting that in the Chinese series of non-HIV-associated cryptococcosis, 12% and 9% of patients were reported to have had prior tuberculosis (Chen, unpublished; 40).

On a practical note, it is worth remembering that in India, as in other resource-limited settings, because cryptococcosis has few distinctive clinical or radiological features and mycological diagnostic facilities are restricted to the major metropolitan centers, patients are often first diagnosed clinically as having tuberculosis and given antitubercular treatment, even when chronic meningitis develops (3, 6). Only in the absence of clinical response are other possibilities, including cryptococcosis, considered.

CLINICAL MANAGEMENT: EVIDENCE AND PRACTICE FROM ASIA

Management issues and experience in Asia are discussed below. From Thailand, in particular, a number of studies on the prevention and management of HIV-associated cryptococcal infection have influenced practice, not only in that country but around the world.

Primary Prophylaxis

Thailand is one country that has adopted a policy of primary prophylaxis against cryptococcosis for HIV-infected patients with low CD4 cell counts. A recent Cochrane review identified five randomized, controlled trials of antifungal therapy for primary prevention of cryptococcal disease in a total of 1,316 HIV-infected patients who mostly had CD4 cell counts less than 150 cells/μl (11). Overall, the incidence of cryptococcosis was significantly lower in patients receiving either fluconazole or itraconazole; however, there was no significant effect on the mortality rate. In a randomized, controlled trial in Thailand, itraconazole, compared with placebo, was effective in reducing invasive fungal infection, mostly cryptococcosis and penicilliosis, from 17.5 to 1.5%, especially among patients with CD4 cell counts less than 100 cells/μl. Prophylaxis, however, was not found to be associated with a survival benefit (14). A second study in Thailand showed lower mortality rates in advanced, immunosuppressed HIV patients who received oral fluconazole, 400 mg once weekly, for primary prophylaxis, although the study was small and most causes of death in the placebo group were not apparently due to fungal infections (16). As primary prophylaxis, oral dosing of fluconazole at 200 mg once daily versus 400 mg weekly was no different in terms of invasive fungal infections, but the rate of oral candidiasis was higher in the group taking daily fluconazole (24).

Guidelines from the United States do not recommend fluconazole for primary prophylaxis against cryptococcal disease because of the low frequency of cryptococcosis there, the lack of a survival benefit, and the relative cost effectiveness (62). Given the higher incidence of cryptococcal meningitis among AIDS patients in Thailand (13), the studies discussed above from Thailand providing supportive evidence, and the low cost of generic fluconazole in Thailand, *Thai Treatment Guidelines for HIV-Infected Patients* (2007, Thai version), which reflects the consensus opinion among experts from the Department of Disease Control, the Thai Ministry of Public Health, and the Thai AIDS Society, recommended oral fluconazole, 400 mg once weekly, for adult AIDS patients with (i) CD4 cell counts less than 100 cells/μl, (ii) no symptoms of active cryptococcal infection, and (iii) negative results on serum cryptococcal antigen testing (if available). As penicilliosis is endemic in the northern region of Thailand and is a frequent opportunistic infection there, itraconazole rather than fluconazole is used for primary fungal prophylaxis in that region. Additionally, Thai guidelines recommend discontinuing primary prophylaxis after the CD4 cell count has been greater than 100 cells/μl for at least 3 months.

One concern regarding the use of fluconazole for primary prophylaxis is whether long-term use will lead to the emergence of less susceptible fungal strains causing significant infections. Some evidence has linked fluconazole use for prophylaxis with an increase in fluconazole-resistant *Candida* isolates (2), although a study in Thais who did or did not receive fluconazole prophylaxis and subsequently developed cryptococcal meningitis found that the minimal inhibitory concentrations for fluconazole against the *C. neoformans* isolates did not differ significantly between these patient groups (43).

Additionally, the supportive data were generated, and the primary prophylaxis policy adopted, prior to the ready availability of antiretroviral therapy (ART), leaving open the question of whether such a policy is still justified if patients can be started promptly on ART. The supportive data were also generated in the absence of cryptococcal antigen screening. If antigen screening is available, whether prophylaxis is justified in antigen-negative patients, who have a low incidence of clinical infection if ART is started promptly (27, 28), is unclear. On the other hand, if prophylaxis is undertaken without screening, it is also unclear whether low-dose fluconazole is the optimal management for the approximately 10% of patients with low CD4 counts who are antigen positive.

Antifungal Therapy and Secondary Prophylaxis

In the study of Zhu et al., most non-HIV-infected patients received amphotericin B–based initial therapy (80). The average dosage used was 0.5 mg/kg per day. The best outcomes were seen in combination with flucytosine, the combination being given for a median duration of 13 weeks. However, the relative efficacy of this lower dosage of amphotericin B (<0.7 mg/kg per day) plus flucytosine for a long duration for initial therapy in non-HIV patients remains to be explored in future controlled studies. Patients given amphotericin B for initial therapy achieved higher response rates at weeks 2 and 10 than those receiving initial fluconazole therapy. A unique aspect of practice in this and other series from China is the continued use of intrathecal amphotericin B therapy in some patients. This form of therapy is no longer used in most countries because of concerns over drug-induced arachnoiditis and bacterial infection of the Ommaya reservoirs often used for administration. At Shanghai Changzheng Hospital, between 1986 and 2001, Yao and colleagues treated 40 non-HIV-infected patients with CNS cryptococcosis with combined intrathecal and intravenous amphotericin B, with or without flucytosine, followed by azole consolidation (78). Intravenous amphotericin B was given at 0.5 mg/kg/day, and intrathecal amphotericin B was diluted in water for injection and mixed with 4 to 5 ml of auto-CSF and 1 to 2 mg of dexamethasone and infused slowly via lumbar injection, increased from 0.1 to 1 mg, two or three times per week. This regimen was continued until CSF culture was negative, which reportedly took a mean of 8 weeks. Reversible retention of urine and paralysis of sacrococcygeal skin and/or lower extremities was reported in eight patients, but there was a high cure rate (78).

Flucytosine is currently unavailable in Thailand, and Thai treatment guidelines recommend amphotericin B alone during the 2-week induction phase of treatment. The immediately implementable alternative, across most of Asia, for induction therapy is amphotericin B plus fluconazole. Pappas and colleagues found a trend toward superiority in a phase II study conducted in Thailand and the United States using amphotericin B combined with fluconazole at 800 mg/day compared with amphotericin B alone or amphotericin B plus fluconazole at 400 mg/day (53). A phase III study including an arm treated with amphotericin B plus fluconazole at 800 mg/day is under way in Vietnam (trial ISRCTN95123928, www.controlled-trials.com).

The phase II study also provided further evidence supporting a dose-response effect with fluconazole, as suggested by studies of oral antifungal regimens in Africa (39). Patients in the amphotericin B plus fluconazole 800 mg/day arm had higher serum and CSF levels of fluconazole, and increased fluconazole area under the curve (AUC) was associated with more favorable 6-week and later clinical and composite end points (46). The findings support the use of higher doses of fluconazole in any areas where intravenous treatment with amphotericin B is not yet feasible. Significant QT prolongation was not found more frequently in patients treated with amphotericin B plus fluconazole 800 mg/day versus amphotericin B alone (45).

A randomized trial of discontinuation of secondary fluconazole prophylaxis was conducted in HIV-infected Thai patients who (i) were successfully treated for acute cryptococcal meningitis, (ii) responded to ART with undetectable plasma HIV viral loads for at least 3 months, and (iii) had greater than a 100-cells/μl increase in their CD4 cell counts. These patients were randomized to continue or discontinue secondary cryptococcal prophylaxis, and 48 weeks after randomization, there was no difference in the rate of cryptococcal relapse between the two groups (74). Thus, Thai guidelines recommend discontinuing secondary prophylaxis against cryptococcosis in patients who meet the inclusion criteria of this study.

Management of Increased Intracranial Pressure

A study from Thailand showed that the use of temporary external lumbar drainage can be a reasonably safe and effective management strategy for intractable elevated CSF pressure not responding to serial lumbar puncture in HIV-infected patients with cryptococcal meningitis (44), even in relatively resource-limited settings. Possible complications include secondary bacterial infections, nerve root irritation, and brain herniation or subdural hematoma from overaggressive CSF drainage (8, 44, 63, 72). The risks can be minimized, however, by strict adherence to aseptic technique and appropriate counseling of the patient and training of the involved medical and nursing staff.

REFERENCES

1. **Ansheng, L., K. Nishimura, H. Taguchi, R. Tanaksa, W. Shaoxi, and M. Miyaji.** 1993. The isolation of *Cryptococcus neoformans* from pigeon droppings and serotyping of naturally and clinically sourced isolates in China. *Mycopathologia* **124:**1–5.
2. **Apisarnthanarak, A., and L. M. Mundy.** 2008. The impact of primary prophylaxis for cryptococcosis on fluconazole resistance in *Candida* species. *J. AIDS.* **47:**644–645.
3. **Banerjee, U., K. Datta, T. Majumdar, and K. Gupta.** 2001. Cryptococcosis in India: the awakening of a giant? *Med. Mycol.* **39:**51–67.
4. **Banerjee, U., K. Datta, M. Diwedi, and S. Sethi.** 2001. Cryptococcosis due to var *gattii*: a short review and Indian clinical scenario. *Nat. J. Infect. Dis.* **2:**32–36.
5. **Banerjee, U., K. Datta, and A. Casadevall.** 2004. Serotype distribution of *Cryptococcus neoformans* in patients in a tertiary care center in India. *Med. Mycol.* **42:**181–186.
6. **Banerjee, U.** 2005. Progress in diagnosis of opportunistic infections in HIV/AIDS. *Indian J. Med. Res.* **121:**395–406.
7. **Bhadra, B., R. S. Rao, P. K. Singh, P. K. Sarkar, S. Shivaji.** 2008. Yeasts and yeast-like fungi associated with tree bark: diversity and identification of yeasts producing extracellular endoxylanases. *Curr. Microbiol.* **56:**489–594.
8. **Bloch, J., and L. Regli.** 2003. Brain stem and cerebellar dysfunction after lumbar spinal fluid drainage: case report. *J. Neurol. Neurosurg. Psychiatry* **74:**992–924.
9. **Chakrabarti, A., M. Jatana, P. Kumar, L. Chatha, A. Kaushal, and A. A. Padhye.** 1997. Isolation of *Cryptococcus neoformans* var. *gattii* from *Eucalyptus camaldulensis* in India. *J. Clin. Microbiol.* **35:**3340–3342.
10. **Chakrabarti, A., S. S. Chatterjee, and M. R. Shivaprakash.** 2008. Overview of opportunistic fungal infections in India. *Jpn. J. Med. Mycol.* **49:**165–172.
11. **Chang, L. W., W. T. Phipps, G. E. Kennedy, and G. W. Rutherford.** 2005. Antifungal interventions for the primary prevention of cryptococcal disease in adults with HIV. *Cochrane Database Syst. Rev.* **Jul 20**(3):CD004773.
12. **Chariyalertsak, S., T. Sirisanthana, K. Supparatpinyo, and K. E. Nelson.** 1996. Seasonal variation of disseminated *Penicillium marneffei* infections in northern Thailand: a clue to the reservoir? *J. Infect. Dis.* **173:**1490–1493.
13. **Chariyalertsak, S., T. Sirisanthana, O. Saengwonloey, and K. E. Nelson.** 2001. Clinical presentation and risk behaviors of patients with acquired immunodeficiency syndrome in Thailand, 1994–1998: regional variation and temporal trends. *Clin. Infect. Dis.* **32:**955–962.
14. **Chariyalertsak, S., K. Supparatpinyo, T. Sirisanthana, and K. E. Nelson.** 2002. A controlled trial of itraconazole as primary prophylaxis for systemic fungal infections in patients with advanced human immunodeficiency virus infection in Thailand. *Clin. Infect. Dis.* **34:**277–284.
15. **Chen, J., A. Varma, M. R. Diaz, A. P. Litvintseva, K. K. Wollenberg, and K. J. Kwon-Chung.** 2008. *Cryptococcus neoformans* strains and infection in apparently immunocompetent patients, China. *Emerg. Infect. Dis.* **14:**755–762.
16. **Chetchotisakd, P., S. Sungkanuparph, B. Thinkhamrop, P. Mootsikapun, and P. Boonyaprawit.** 2004. A multicentre, randomized, double-blind, placebo-controlled trial of primary cryptococcal meningitis prophylaxis in HIV-infected patients with severe immune deficiency. *HIV Med.* **5:**140–143.
17. **Cogliati, A. N., A. Chandrashekar, A. Prigitano, M. C. Esposto, B. Petrini, A. Chandramuki, and M. A. Viviani.** 2008. Clinical isolates from an Indian Hospital: an unexpected detection of a serotype C *Cryptococcus gattii* population. Abstract of 7th International Conference on Cryptococcus and Cryptococcosis, Nagasaki, Japan 11–14, September 2008, ICCC Abstract SY-05-04, p 52.
18. **Ellis, D. H., and T. J. Pfeiffer.** 1990. Natural habitat of *Cryptococcus neoformans* var. *gattii*. *J. Clin. Microbiol.* **28:**1642–1644.
19. **Ellis, D. H., and T. J. Pfeiffer.** 1990. Ecology, life cycle, and infectious propagule of *Cryptococcus neoformans*. *Lancet* **336:**923–925.
20. **Feng, X., Z. Yao, D. Ren, W. Liao, and J. Wu.** 2008. Genotype and mating type analysis of *Cryptococcus neoformans* and *Cryptococcus gattii* isolates from China that mainly originated form non-HIV-infected patients. *FEMS Yeast Res.* **8:**930–938.
21. **Gokulshankar, S., S. Ranganathan, M. S. Ranjith, and A. J. Ranjithsingh.** 2004. Prevalence, serotypes and mating patterns of *Cryptococcus neoformans* in the pellets of different avifauna in Madras, India. *Mycoses* **47:**310–314.
22. **Grover, N., S. R. Nawange, J. Naidu, S. M. Singh, and A. Sharma.** 2007. Ecological niche of *Cryptococcus neoformans* var. *grubii* and *Cryptococcus gattii* in decaying wood of trunk hollows of living trees in Jabalpur City of Central India. *Mycopathologia* **164:**159–170.

23. **Gugnani, H. C., T. G. Mitchell, A. P. Litvintseva, K. B. Lengeler, J. Heitman, A. Kumar, S. Basu, and A. Paliwal-Joshi.** 2005. Isolation of *Cryptococcus gattii* and *Cryptococcus neoformans* var. *grubii* from the flowers and bark of *Eucalyptus* trees in India. *Med. Mycol.* **43:**565–569.
24. **Havlir, D. V., M. P. Dube, J. A. McCutchan, D. N. Forthal, C. A. Kemper, M. W. Dunne, D. M. Parenti, P. N. Kumar, A. C. White, Jr., M. D. Witt, S. D. Nightingale, K. A. Sepkowitz, R. R. MacGregor, S. H. Cheeseman, F. J. Torriani, M. T . Zelasky, F. R. Sattler, and S. A. Bozzette.** 1998. Prophylaxis with weekly versus daily fluconazole for fungal infections in patients with AIDS. *Clin. Infect. Dis.* **27:**1369–1375.
25. **Hiremath, S. S., A. Chowdhary, T. Kowshik, H. S. Randhawa, S. Sun, and J. Xu.** 2008. Long-distance dispersal and recombination in environmental populations of *Cryptococcus neoformans* var. *grubii* from India. *Microbiology* **154:**1513–1524.
26. **Jain, N., B. L. Wickes, S. M. Keller, J. Fu, A. Casadevall, P. Jain, M. A. Ragan, U. Banerjee, and B. C. Fries.** 2005. Molecular epidemiology of clinical *Cryptococcus neoformans* strains from India. *J. Clin. Microbiol.* **43:**5733–5742.
27. **Jarvis, J. N., S. D. Lawn, M. Vogt, N. Bangani, R. Wood, and T. S. Harrison.** 2009. Screening for cryptococcal antigenaemia in patients accessing an antiretroviral treatment program in South Africa. *Clin. Infect. Dis.* **48:**856–862.
28. **Jarvis, J. N., S. D. Lawn, M. Vogt, N. Bangani, R. Wood, and T. S. Harrison.** 2009. Reducing mortality associated with opportunistic infections among patients with advanced HIV infection in sub-Saharan Africa. Reply to Di-Nubile. *Clin. Infect. Dis.* **49:**812–813.
29. **Jongwutiwes, U., S. Sungkanuparph, and S. Kiertiburanakul.** 2008. Comparison of clinical features and survival between cryptococcosis in human immunodeficiency virus (HIV)-positive and HIV-negative patients. *Jpn. J. Infect. Dis.* **61:**111–115.
30. **Kiertiburanakul, S., S. Sungkanuparph, B. Buabut, and R. Pracharktam.** 2004. Cryptococcuria as a manifestation of disseminated cryptococcosis and isolated urinary tract infection. *Jpn. J. Infect. Dis.* **57:**203–205.
31. **Kiertiburanakul, S., S. Wirojtananugoon, R. Pracharktam, and S. Sungkanuparph.** 2006. Cryptococcosis in human immunodeficiency virus-negative patients. *Int. J. Infect. Dis.* **10:**72–78.
32. **Kohno, S.** 2003. Clinical and mycological features of cryptococcosis. *Nippon Ishinkin Gakkai Zasshi* **44:**159–162. (In Japanese.)
33. **Kumar, S., A. Wanchu, A. Chakrabarti, A. Sharma, P. Bambery, and S. Singh.** 2008. Cryptococcal meningitis in HIV infected. Experience from a North Indian tertiary center. *Neurol. India* **56:**444–449.
34. **Kumarasamy, N., S. Solomon, T. P. Flanigan, R. Hemalatha, S. P. Thyagarajan, and K. H. Mayer.** 2003. Natural history of human immunodeficiency virus disease in southern India. *Clin Infect. Dis.* **36:**79–85.
35. **Kume, H., T. Yamazaki, M. Abe, H. Tanuma, M. Okudaira, and I. Okayasu.** 2006. Epidemiology of visceral mycoses in patients with leukemia and MDS: analysis of the data in annual of pathological autopsy cases in Japan in 1989, 1993, 1997 and 2001. *Nippon Ishinkin Gakkai Zasshi* **47:**15–24. (In Japanese.)
36. **Kwon-Chung, K. J., and J. E. Bennett.** 1984. Epidemiologic differences between the two varieties of *Cryptococcus neoformans*. *Am. J. Epidemiol.* **120:**123–130.
37. **Lakshmi, V., T. Sudha, V. D. Teja, and P. Umabala.** 2007. Prevalence of central nervous system cryptococcosis in human immunodeficiency virus reactive hospitalized patients. *Indian J. Med. Microbiol.* **25:**146–149.
38. **Litvintseva, A. P., R. Thakur, R. Vilgalys, and T. G. Mitchell.** 2006. Multilocus sequence typing reveals three genetic subpopulations of *Cryptocococcus neoformans* var. *grubii* (serotype A), including a unique population in Botswana. *Genetics* **172:**2223–2238.
39. **Longley. N., C. Muzoora, K. Taseera, J. Mwesigye, J. Rwebembera, A. Chakera, E. Wall, I. Andia, S. Jaffar, and T. S. Harrison.** 2008. Dose-response effect of high dose fluconazole for HIV-associated cryptococcal meningitis in Southwest Uganda. *Clin. Infect. Dis.* **47:**1556–1561.
40. **Lu, C. H., W. N. Chang, H. W. Chang, and Y. C. Chuang.** 1999. The prognositic factors of cryptococcal meningitis in HIV-negative patients. *J. Hosp. Infect.* **42:**313–320.
41. **Lui, G., N. Lee, M. Ip, K. W. Choi, Y. K. Tso, E. Lam, S. Chau, R. Lai, and C. S. Cockram.** 2006. Cryptococcosis in apparently immunocompetent patients. *Q. J. Med.* **99:**143–151.
42. **Mandal, P., U. Banerjee, A. Casadevall, and J. D. Nosanchuk.** 2005. Dual infections with pigmented and albino strains of *Cryptococcus neoformans* in patients with and without human immunodeficiency virus infection in India. *J. Clin. Microbiol.* **43:**4766–4772.
43. **Manosuthi, W., S. Sungkanuparph, S. Thongyen, N. Chumpathat, B. Eampokalap, U. Thawornwan, and S. Foongladda.** 2006. Antifungal susceptibilities of *Cryptococcus neoformans* cerebrospinal fluid isolates and clinical outcomes of cryptococcal meningitis in HIV-infected patients with/without fluconazole prophylaxis. *J. Med. Assoc. Thai.* **89:**795–802.
44. **Manosuthi, W., S. Sungkanuparph, S. Chottanapund, S. Tansuphaswadikul, S. Chimsuntorn, P. Limpanadusadee, and P. G. Pappas.** 2008. Temporary external lumbar drainage for reducing elevated intracranial pressure in HIV-infected patients with cryptococcal meningitis. *Int. J. STD AIDS* **19:**268–271.
45. **Manosuthi, W., S. Sungkanuparph, T. Anekthananon, K. Supparatpinyo, T. Nolen, L. Zimmer, P. Pappas, R. Larsen, S. Filler, and P. Chetchotisakd.** 2009. Effect of high-dose fluconazole (FLU) on QT interval in patients with HIV-associated cryptococcal meningitis. *Int. J. Antimicrob. Agents* **34:**494–496.
46. **Manosuthi, W., P. Chetchotisakd, T. L. Nolen, D. Wallace, S. Sungkanuparph, T. Anekthananon, K. Supparatpinyo, P. G. Pappas, R. A. Larsen, S. G. Filler, and D. Andes.** 2010. Monitoring and impact of fluconazole serum and cerebrospinal fluid concentration in HIV-associated cryptococcal meningitis infected patients. *HIV Med.* **11:**276–281.
47. **Meyer, W., K. Marszewska, M. Amirmostofian, R. P. Igreja, C. Hardtke, K. Methling, M. A. Viviani, A. Chindamporn, S. Sukroongreung, M. A. John, D. H. Ellis, and T. C. Sorrell.** 1999. Molecular typing of global isolates of *Cryptococcus neoformans* var. *neoformans* by polymerase chain reaction fingerprinting and randomly amplified polymorphic DNA: a pilot study to standardize techniques on which to base a detailed epidemiological survey. *Electrophoresis* **20:**1790–1799.
48. **Meyer, W., D. M. Aanensen, T. Boekhout, M. Cogliati, M. R. Diaz, M. C. Esposto, M. Fisher, F. Gilgado, F. Hagen, S. Kaocharoen, A. P. Litvintseva, T. G. Mitchell, S. P. Simwami, L. Trilles, M. A. Viviani, and J. Kwon-Chung.** 2009. Consensus multi-locus sequence typing scheme for *Cryptococcus neoformans* and *Cryptococcus gattii*. *Med. Mycol.* **47:**561–570.
49. **Padhye, A. A., A. Chakrabarti, J. Chander, and L. Kaufman.** 1993. *Cryptococcus neoformans* var. *gattii* in India. *J. Med. Vet. Mycol.* **31:**165–168.
50. **Pal, M.** 1989. *Cryptococcus neoformans* var *neoformans* and munia birds. *Mycoses* **32:**250–252.

51. **Pal, M.** 1995. Natural occurrence of *Cryptococcus neoformans* var *neoformans* in wooden canary cage. *Rev Iberoam. Micol.* **12:**93–94.
52. **Pappas, P. G., J. R. Perfect, G. A. Cloud, R. A. Larsen, G. A. Pankey, D. J. Lancaster, H. Henderson, C. A. Kauffman, D. W. Haas, M. Saccente, R. J. Hamill, M. S. Holloway, R. M. Warren, and W. E. Dismukes.** 2001. Cryptococcosis in human immunodeficiency virus-negative patients in the era of effective azole therapy. *Clin. Infect. Dis.* **33:**690–699.
53. **Pappas, P. G., P. Chetchotisakd, R. A. Larsen, W. Manosuthi, M. I. Morris, T. Anekthananon, S. Sungkanuparph, K. Supparatpinyo, T. L. Nolen, L. O. Zimmer, A. S. Kendrick, P. Johnson, J. D. Sobel, and S. G. Filler.** 2009. A Phase II randomized trial of amphotericin B alone or combined with fluconazole in the treatment of HIV-associated cryptococcal meningitis. *Clin. Infect. Dis.* **48:**1775–1783.
54. **Park, K.** 2007. Epidemiology of chronic non-communicable diseases and conditions. *In Park's Textbook of Preventive and Social Medicine*, 19th ed. Banarasidas Bhanot, Jabalpur, India.
55. **Poonwan, N., Y. Mikami, S. Poosuwan, J. Boon-Long, N. Mekha, M. Kusum, K. Yazawa, R. Tanaka, K. Nishimura, and K. Konyama.** 1997. Serotyping of *Cryptococcus neoformans* strains isolated from clinical specimens in Thailand and their susceptibility to various antifungal agents. *Eur. J. Epidemiol.* **13:**335–340.
56. **Randhawa, H. S., A. Y. Mussa, and Z. U. Khan.** 2001. Decaying wood in tree trunk hollows as a natural substrate for *Cryptococcus neoformans* and other yeast-like fungi of clinical interest. *Mycopathologia* **151:**63–69.
57. **Randhawa, H. S., T. Kowshik, and Z. U. Khan.** 2003. Decayed wood of *Syzygium cumini* and *Ficus religiosa* living trees in Delhi/New Delhi metropolitan area as natural habitat of *Cryptococcus neoformans*. *Med. Mycol.* **41:**199–209.
58. **Randhawa, H. S., T. Kowshik, K. Preeti Sinha, A. Chowdhary, Z. U. Khan, Z. Yan, J. Xu, and A. Kumar.** 2006. Distribution of *Cryptococcus gattii* and *Cryptococcus neoformans* in decayed trunk wood of *Syzygium cumini* trees in northwestern India. *Med. Mycol.* **44:**623–630.
59. **Randhawa, H. S., T. Kowshik, A. Chowdhary, K. Preeti Sinha, Z. U. Khan, S. Sun, and J. Xu.** 2008. The expanding host tree species spectrum of *Cryptococcus gattii* and *Cryptococcus neoformans* and their isolations from surrounding soil in India. *Med Mycol.* **46:**823–833.
60. **Reeves, D. L., E. M. Butt, and R. W. Hammack.** 1941. Torula infection of lung and central nervous system: reports of six cases with three autopsies. *Arch. Intern. Med.* **68:**57–79.
61. **Richardson, P. M., A. Mohandas, and N. Arumugasamy.** 1976. Cerebral cryptococcosis in Malaysia. *J. Neurol. Neurosurg. Psychiatry* **39:**330–337.
62. **Saag, M. S., R. J. Graybill, R. A. Larsen, P. G. Pappas, J. R. Perfect, W. G. Powderly, J. D. Sobel, and W. E. Dismukes.** 2000. Practice guidelines for the management of cryptococcal disease. Infectious Diseases Society of America. *Clin. Infect. Dis.* **30:**710–718.
63. **Schade, R. P., J. Schinkel, L. G. Visser, J. M. Van Dijk, J. H. Voormolen, and E. J. Kuijper.** 2005. Bacterial meningitis caused by the use of ventricular or lumbar cerebrospinal fluid catheters. *J. Neurosurg.* **102:**229–234.
64. **Sethi, K. K., H. S. Randhawa, S. Abraham, S. K. Misra, and V. N. Damodaran.** 1966. Occurrence of *Cryptococcus neoformans* in pigeon excreta in Delhi. *Indian J. Chest Dis.* **8:**207–210.
65. **Shen, Y. Z., T. K. Qi, J. X. Ma, X. Y. Jiang, J. R. Wang, Q. N. Xu, Q. Huang, X. N. Liu, H. Q. Sun, and H. Z. Lu.** 2007. Invasive fungal infections among inpatients with acquired immune deficiency syndrome at a Chinese university hospital. *Mycoses* **50:**475–480.
66. **Shih, C. C., Y. C. Chen, S. C. Chang, K. T. Luh, and W. C. Hsieh.** 2000. Cryptococcal meningitis in non-HIV-infected patients. *Q. J. Med.* **93:**245–251.
67. **Sobhani, R., A. Basavaraj, A. Gupta, A. S. Bhave, D. B. Kadam, S. A. Sangle, H. B. Prasad, J. Choi, J. Josephs, K. A. Gebo, S. N. Morde, R. C. Bollinger, Jr., and A. L. Kakrani.** 2007. Mortality & clinical characteristics of hospitalized adult patients with HIV in Pune, India. *Indian J. Med. Res.* **126:**116–121.
68. **Soogarun, S., V. Wiwanitkit, A. Palasuwan, P. Pradniwat, J. Suwansaksri, T. Lertlum, and T. Maungkote.** 2006. Detection of *Cryptococcus neoformans* in bird excreta. *Southeast Asian J. Trop. Med. Public Health* **37:**768–770.
69. **Sukroongreung, S., C. Nilakul, O. Ruangsomboon, W. Chuakul, and B. Eampokalap.** 1996. Serotypes of *Cryptococcus neoformans* isolated from patients prior to and during the AIDS era in Thailand. *Mycopathologia* **135:**75–78.
70. **Sungkanuparph, S., A. Vibhagool, and R. Pracharktam.** 2002. Spontaneous cryptococcal peritonitis in cirrhotic patients. *J. Postgrad. Med.* **48:**201–202.
71. **Tay, S. T., H. C. Lim, T. H. Tajuddin, M. Y. Rohani, H. Hamimah, and K. L. Thong.** 2006. Determination of molecular types and genetic heterogeneity of *Cryptococcus neoformans* and *C. gattii* in Malaysia. *Med. Mycol.* **44:**617–22.
72. **Thompson, H. J.** 2000. Managing patients with lumbar drainage devices. *Crit. Care Nurse* **20:**59–68.
73. **Tjia, T. L., Y. K. Yeow, and C. B. Tan.** 1985. Cryptococcal meningtis. *J. Neurol. Neurosurg. Psychiatry* **48:**853–858.
74. **Vibhagool, A., S. Sungkanuparph, P. Mootsikapun, P. Chetchotisakd, S. Tansuphaswaswadikul, C. Bowonwatanuwong, and A. Ingsathit.** 2003. Discontinuation of secondary prophylaxis for cryptococcal meningitis in human immunodeficiency virus-infected patients treated with highly active antiretroviral therapy: a prospective, multicenter, randomized study. *Clin. Infect. Dis.* **36:**1329–1331.
75. **Wadhwa, A., R. Kaur, S. K. Agarwal, S. Jain, and P. Bhalla.** 2007. AIDS- related opportunistic mycoses seen in a tertiary care hospital in North India. *J. Med. Microbiol.* **56:**1101–1106.
76. **Yamamoto, Y., S. Kohno, T. Noda, H. Kakeya, K. Yanagihara, H. Ohno, K. Ogawa, S. Kawamura, T. Ohtsubo, K. Tomono, et al.** 1995. Isolation of *Cryptococcus neoformans* from environments (pigeon excreta) in Nagasaki. *Kansenshogaku Zasshi* **69:**642–645. (In Japanese.)
77. **Yamazaki, T., H. Kume, S. Murase, E. Yamashita, and M. Arisawa.** 1999. Epidemiology of visceral mycoses: analysis of data in annual of the pathological autopsy cases in Japan. *J. Clin. Microbiol.* **37:**1732–1728.
78. **Yao, Z., W. Liao, and R. Chen.** 2005. Management of cryptococcosis in non-HIV-related patients. *Med. Mycol.* **43:**245–251.
79. **Yasuoka, A.** 2009. Trend of AIDS defining illness in Japan. The investigation in foothold institutes in Japan research ID: H18-AIDS-ippan-008. Nagasaki University Infection Control and Education Center, Nagasaki, Japan.
80. **Zhu, L. P., J. Q. Wu, B. Xu, X. T. Ou, Q. Q. Zhang, and X. H. Weng.** 2010. Cryptococcal meningitis in non-HIV-infected patients in a Chinese tertiary care hospital, 1997–2007. *Med. Mycol.* **48:**570–579.

Cryptococcus: From Human Pathogen to Model Yeast
Edited by J. Heitman et al.

22

Sexual Reproduction of *Cryptococcus gattii*: a Population Genetics Perspective

DEE CARTER, LEONA CAMPBELL, NATHAN SAUL, AND MARK KROCKENBERGER

Cryptococcus gattii is a sexual fungus with two well-characterized mating types, α and **a** (see chapter 7, this volume). Sex and recombination might therefore be expected to be a normal part of the *C. gattii* life cycle. Population genetic studies have revealed a far more complex and enigmatic relationship with sex, however. The recognition that *C. gattii* divides into distinct molecular genotypes with varying levels of fertility and recombination, the recognition that populations can be genetically differentiated at fine scales but connected over longer distances, and the apparent ability to undergo unisexual α-α unions have contributed to our understanding of this environmental yeast species. Here, we review the current knowledge of sexual reproduction of *C. gattii* from a population genetics perspective.

WHAT SEX AND RECOMBINATION CAN TELL US ABOUT *C. GATTII*

Sexual reproduction plays a vital role in the evolution and maintenance of eukaryotic species. In the fungi, sex is not only important for generating recombinant organisms and thereby increasing the potential to evolve and adapt; it is also often the means by which resistant propagules are produced for survival and for long-range dispersal and colonization. Studies of the population structure of *Cryptococcus* began in the mid 1990s, when population genetics was first being applied to understand the life histories of fungal pathogens. Initial population studies of the sibling species *Cryptococcus neoformans* had indicated that, while there was considerable genetic diversity in this species, it was globally clonal (3, 5, 16). In addition, *C. neoformans* isolates were overwhelmingly of the α mating type. Together, this indicated that although a sexual species, *C. neoformans* did not appear to be sexually recombining.

Would this also be the case in *C. gattii*? *C. gattii* offered the advantage to population analyses that it could be easily cultivated from its environmental niche, providing the ability to collect samples from a very defined space: for example, from a collection of trees from a small area, a single host tree, or even a single tree hollow. By assessing whether or not sex was occurring, we aimed to address the following questions: (i) Does *C. gattii* complete its life cycle in association with the *Eucalyptus* tree host, thereby indicating that this is the true ecological niche for *C. gattii* in the environment? (ii) Does *C. gattii* have the ability to produce recombinant progeny with a greater potential to evolve virulence and drug resistance mechanisms? (iii) Is the production of sexual basidiospores, which are hypothesized to be the infectious propagule, likely? And (iv) Would sex contribute to long-distance dispersal of *C. gattii*, and could we find evidence of this in geographically distinct regions?

POPULATION STUDIES REVEAL CRYPTIC SPECIATION IN *C. GATTII*

Before tackling the question of recombination, it is very important to ensure that the population under study consists of a single species. Many population genetic studies of fungi have found significant differences among groups within what was traditionally considered to be one species (19, 23, 25, 39). Generally these "cryptic species" cannot be distinguished readily by morphology, although subsequent studies often reveal differences in their physiology, biochemistry, host range, or virulence characteristics (15, 27). Speciation may arise via allopatry, where populations have become geographically isolated by physical barriers. Fungal populations have also been found to contain sympatric species, which occupy the same geographic niche but are genetically differentiated. As cryptic species diverge, they become fixed for different alleles at polymorphic loci (Fig. 1). In genetic studies, cryptic species are usually detected as exclusive groups of organisms occupying strongly supported branches on phylogenetic trees that have been derived from different, independent loci.

Early molecular studies of *C. gattii* revealed distinct differences that divided isolates into four groups, designated molecular types VGI–VGIV (38, 42) or AFLP4–7 (2) (where VGI = AFLP4, VGII = AFLP6, VGIII = AFLP5, and

Dee Carter and Leona Campbell, School of Molecular Bioscience, University of Sydney, Sydney, NSW 2006, Australia. **Nathan Saul and Mark Krockenberger,** Faculty of Veterinary Science, University of Sydney, Sydney, NSW 2006, Australia.

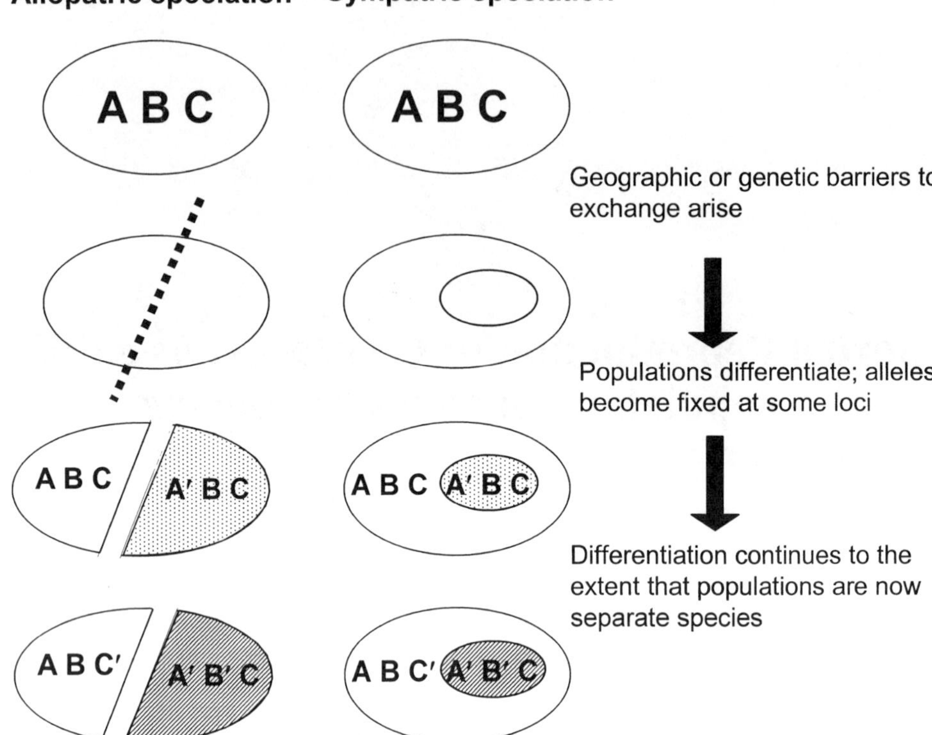

FIGURE 1 Mechanisms of allopatric and sympatric speciation. Geographic (allopatric speciation) or genetic (sympatric speciation) barriers arise within a population that limit genetic exchange. Over time the two populations differentiate and become fixed for different alleles at independent loci. ABC, three independent loci; A′B′C′, alternative alleles at each locus.

VGIV = AFLP7). Multilocus sequence typing (MLST) and phyogenetic analysis subsequently found each of these groups to be exclusive and monophyletic (4, 17). Under the Genealogical Concordance Phylogenetic Species Concept, which defines a species as "a basal, exclusive group of organisms all of whose genes coalesce more recently with each other than with those of any organism outside the group, and that contains no exclusive group within it," Bovers et al. (4) argued that the four major genotypes of *C. gattii* represented distinct taxa that might be cryptic species. Phylogenetic analysis of six concatenated genes indicated that VGI and VGIII were sister taxa, that VGIV was basal to these, and that VGII was the ancestral group (4) (Fig. 2). So far, researchers have stopped at giving the groups species status, as it is possible to mate isolates from different genotypes and to produce basidia and basidiospores. The viability of these spores has not yet been determined, however, and this information is needed to determine if these are in fact distinct biological species (4).

MATING-TYPE BIAS IN *C. GATTII* POPULATIONS

Mating in *Cryptococcus* generally involves the union of single cells of α and **a** mating types, which undergo fusion, plasmogamy, and karyogomy, followed by meiosis, to produce recombinant basidiospores (chapter 7). Theoretically, this will result in equal numbers of viable spores of the two different mating types, which has been verified in laboratory crosses of *C. neoformans* and is assumed to also occur in *C. gattii* (18, 26). Populations in which sexual **a**-α reproduction occurs to an appreciable extent should therefore contain approximately equal numbers of the two mating types. Assessing the relative numbers of α and **a** cells is therefore an important aspect of determining if populations are likely to have undergone sexual recombination.

The initial studies of mating type in *C. gattii* found 84% of clinical isolates to be of the α mating type (34). This was a global collection and included strains from Africa, Europe, and Asia. Mating type was assessed by coculture with tester strains, and the majority of strains (~90%) were fertile, producing basidia and basidiospores. Later studies have focused on more confined regions, have included environmental and veterinary as well as clinical isolates, and have largely assessed mating by the amplification of α- and **a**-specific genes from the mating type locus (10, 12, 14, 17, 21, 22, 24, 33, 36, 37, 40, 43). Surprisingly, these studies have found that populations from most areas are extremely biased toward one or the other mating type (Fig. 3). Mating type α predominates in the majority of populations, but in two populations, one from a single source in Balranald, Australia, and the other from Colombia, >90% of isolates are of the **a** mating type. Only a single environmental population from Renmark, Australia, has been found to date that contains α and **a** cells in the approximately 50:50 ratio expected from sexual recombination (21). There is no relationship between the bias in mating type and the molecular type of the population.

Studies of the sibling species *C. neoformans* have found mating-type bias to an even greater extent; indeed, in *C. neoformans* var. *grubii*, which causes the majority of cryptococcal infections worldwide, the **a** mating type was so rare

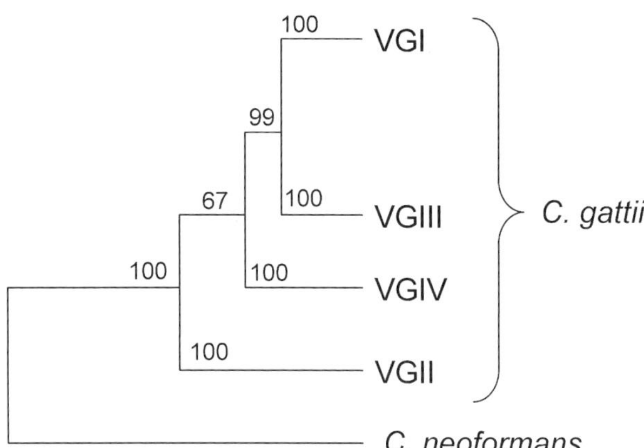

FIGURE 2 Phylogenetic relationships of the different *C. gattii* molecular genotypes that may represent separate cryptic species. Tree based on 459 parsimony informative sites from six independent MLST markers. Adapted from reference 3.

that until recently it was thought to be extinct (28, 32, 45). In an attempt to explain this bias, Wickes et al. (46) examined strains of α and **a** mating type for their ability to undergo monokaryotic or haploid fruiting, where under starvation conditions a single strain produces basidia and basidiospores. Initial studies indicated that this was indeed restricted to isolates of α mating type. However, subsequent analyses found some strains of **a** mating type that could undergo haploid fruiting (44). Quantitative trait locus analysis showed that fruiting is a quantitative trait and that the α allele of MAT promotes fruiting to a greater extent than the **a** allele (29).

More recent studies have indicated that, rather than being a strictly asexual process, haploid fruiting may be the reduction of diploid cells produced via unisexual mating (30). Both genetic and population genetic studies have found evidence of α-α mating in certain *C. neoformans* populations (6, 31). Extensive sequence analysis of the mating-type region of the *C. gattii* strain responsible for the current outbreak of cryptococcosis on Vancouver Island suggested α-α mating might also occur in this species (17). Thus, populations that show an extreme bias in mating types might be just as capable of undergoing sex, with the associated production of recombinant progeny and potentially infectious spores, as bisexual populations.

POPULATION STUDIES OF *C. GATTII*: THE DEVELOPMENT AND ANALYSIS OF MOLECULAR MARKERS

Unlike many basidiomycetes, which make visible fruiting bodies such as mushrooms, puffballs, and brackets, the sexual structures of *Cryptococcus* species are microscopic and cannot be detected in the environment. Matings can be done in the laboratory, but this tells us little about whether sex occurs at an appreciable frequency in nature. Indirect methods that assess whether populations have a history of genetic exchange are therefore required. In these methods molecular markers, usually based on DNA polymorphism, are developed to allow multilocus genotypes to be assigned to all members of a population. Phylogenetic and population genetic approaches can then be used to assess whether alleles at the different loci have been exchanged among members of the population. There are numerous different methods that have been developed to assess recombination in microorganisms, and a complete review of these is outside the scope of this article (for excellent reviews see references 7 and 47). Instead, we present here the methods that have been used to assess recombination in *C. gattii*.

Methods to Analyze Recombination: Tree Length (T_L) and the Index of Association (I_A)

Clonal lineages of organisms that do not recombine with one another effectively behave as separate species and therefore fit into well-resolved phylogenetic trees, whereas recombining populations do not have a clear path of descent among their members (Fig. 4A). The length of the phylogenetic tree (T_L) required to accommodate all evolutionary steps in the population can therefore be used to assess the extent of recombination; asexual populations have a significantly shorter T_L than recombining populations. There are a number of other hallmarks of populations that are strictly asexual:

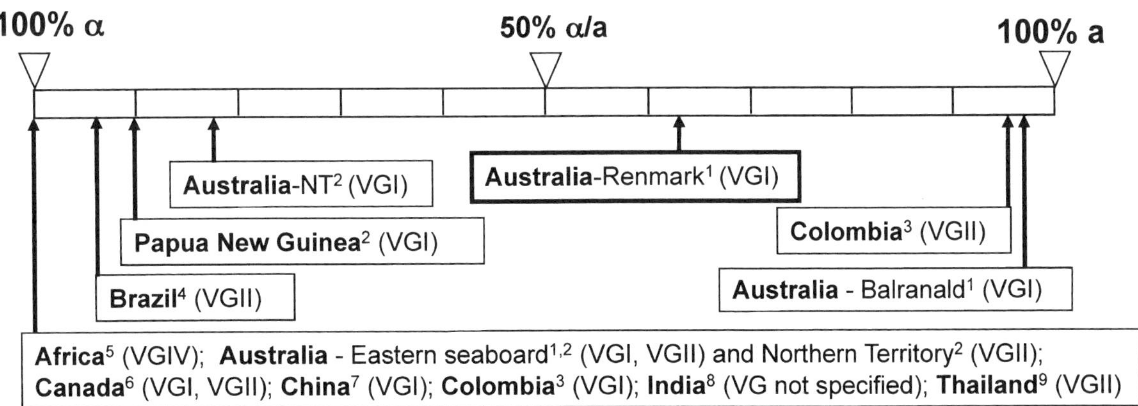

FIGURE 3 Frequency of mating types α and **a** in different *C. gattii* populations. Most populations show an extreme bias of one mating type over the other. Renmark, Australia, is the only region found to date with close to the 50:50 ratio of α:**a** expected for sexual reproduction. (Note: only populations with ≥10 isolates have been included in this survey.) References: 1 (21), 2 (9), 3 (13), 4 (43), 5 (33), 6 (24), 7 (14), 8 (37), 9 (36).

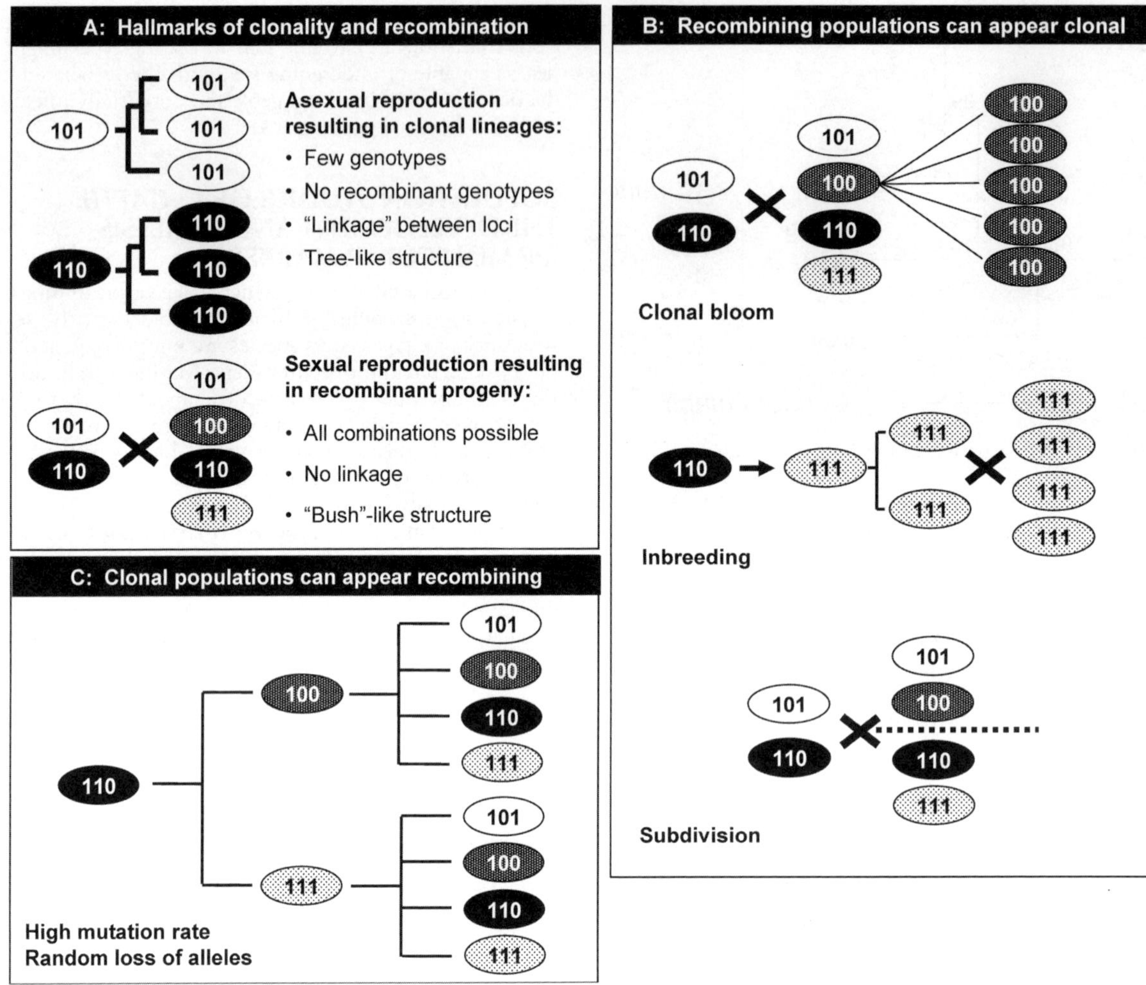

FIGURE 4 Multilocus molecular genotypes can indicate whether a population is clonal or recombining. (A) Hallmarks of clonal and recombining populations: clonal populations inherit their genotypes intact from their parents so that there are few genotypes present, there are no recombinant genotypes, loci appear to be genetically linked, and the population fits into a well-resolved phylogenetic tree as the individual taxa effectively behave like different species. In contrast, a sexual population can contain all possible combinations of genotypes, loci are not linked, and the population will not fit into a well-resolved phylogenetic tree, instead forming a bush-like structure. Complications: (B) Recombining populations can appear clonal if (i) they contain an overwhelming number of asexually derived isolates, as with a clonal bloom; (ii) there is inbreeding, which prevents genetic reassortment from being detected; or (iii) they are genetically subdivided, as loci fixed on either side of the division will appear linked. (C) Clonal populations can appear recombining if molecular markers have a high mutation rate leading to homoplasy or if there is a random loss of alleles.

they generally contain relatively few genotypes; there is an overrepresentation of some genotypes and an absence of recombinant genotypes; and although loci are not physically linked, they appear to be genetically linked, as they are not being reassorted among different genotypes. The extent of genetic linkage can be measured as an index of association (I_A) (35). In recombining populations, all genotype combinations are possible, as the alleles at each locus assort randomly and $I_A \sim 0$.

There are complications to these analyses, however, particularly in sexually reproducing microbial populations that can undergo massive population expansion asexually or can inbreed. In both cases recombination will be masked and the populations will appear to be asexual (Fig. 4B). If a population consists of two or more cryptic species, as outlined above, this can cause problems, as loci that are fixed in each population will appear to be linked, and clonality might be identified erroneously (Fig. 4B). Finally, if the genetic markers chosen are subject to high rates of mutation, or if there are experimental artifacts that cause alleles to be lost randomly, the loci can appear to randomly associate; in this case an asexual population can appear to be recombining (Fig. 4C). Care must therefore be taken in both the choice of population and the choice of marker. Population sampling needs to be sufficiently broad to reduce the chance of inbreeding or oversampling asexually derived clones, but not so broad that allopatric speciation is likely; markers need to be sufficiently discriminatory to differentiate among members of the population, but not subject to very high mutation rates and homoplasy. A very useful first step in analyzing a population is to fit the data to a phylogenetic tree and to look for distinct structures, such as groups on strongly supported branches that might be differentiated subpopulations, or clusters of identical genotypes that could indicate a clonal bloom. The analysis can then be targeted to specific subpopulations to identify recombination within them, and identical clones can be pruned from the population to remove their influence on population structure. Finally, the analysis should ideally be repeated on two or more separate occasions to eliminate experimental artifacts that might cause alleles to be lost or gained at random.

Once suitable multilocus genotypes have been developed and analyzed for linkage and phylogeny, statistical methods are used to determine whether or not the population is significantly different to one in which markers are assorted randomly. To do this, recombining populations are generated by taking the multilocus data for the observed population and randomizing alleles at each locus among the population members—effectively allowing the population to undergo virtual cyber sex. This can be done hundreds of times to produce a large number of recombining data sets that are based on the original multilocus genotypes. T_L and I_A are assessed for each of these to provide a range of values that might be obtained for recombining populations. The T_L and I_A values for the observed population are then compared to those of the derived, recombining population; if they are significantly different, one can reject the null hypothesis of random recombination; if they are not significantly different, it is concluded that recombination has occurred.

Methods to Analyze Recombination: Character Compatibility among Locus Pairs

The T_L and I_A tests are statistical and provide an assessment of recombination in a population or subpopulation as a whole but can overlook low levels of genetic exchange among isolates. A simple method for assessing if isolate genotypes have recombined is to look for "character compatibility" among loci. Here, informative loci are compared in pairs, and one looks for the presence of all four possible pairs of alleles (e.g., 1,1; 1,0; 0,1; 0,0). Such locus pairs are deemed incompatible (as their alleles are not genetically linked), and in the absence of parallel mutations, this can be considered evidence of recombination. Character compatibility is a useful test when characters are too limited to assess recombination by T_L or I_A, as it can detect a low level of recombination even when the overall signal is strongly clonal. It can also provide a graphical picture of recombining groups of isolates and can show whether any individuals or groups fall outside these groups.

ANALYSIS OF RECOMBINATION IN AN ENVIRONMENTAL *C. GATTII* POPULATION

Clonality and Short-Range Differentiation among Isolates from Different Host Trees

The first analysis of recombination in *C. gattii* was done following a survey of environmental *C. gattii* populations from the southeast of Australia, which found the first (and so far only; see above) population in which both mating types occurred in statistically equal frequencies, in Renmark, South Australia (21). The Renmark sampling site consists of a group of mature *Eucalyptus camaldulensis* trees along Ral Ral Creek, an anastomosing branch of the Murray River in the Riverland Biosphere Reserve, near the border with New South Wales. This area is predominantly open mallee bushland and Murray River floodplains dominated by *E. camaldulensis*, and the surrounding region is significantly degraded by overgrazing and drought, with little vegetation away from the river and creek banks. Most trees at this site have large hollows that are colonized by *C. gattii*, and in some cases the two *C. gattii* mating types have been found together on single trees or even in single tree hollows. Thirty isolates from 13 trees within ~5 km along the river were cultured, and their molecular type (all VGI) and mating type (16 α and 14 **a**) were determined (22).

Preliminary analysis revealed that the level of genetic diversity was low, which was consistent with other studies of VGI isolates from across Australia (38). Amplified fragment length polymorphisms (AFLPs), which are highly discriminatory molecular markers, were therefore selected to establish multilocus genotypes. Thirty-eight different AFLP loci were scored for presence (1) or absence (0), and each isolate was found to have a unique multilocus genotype. However, both the T_L and the I_A results were significantly different from those produced by 1,000 artificially randomized data sets, and both caused the null hypothesis of random recombination to be rejected with high significance ($P < 0.001$). We concluded that, despite having the two mating types in approximately equal amounts, the Renmark population was predominantly clonal and that sexual recombination was not occurring (22). Supporting this conclusion, it was not possible to induce matings in the laboratory between Renmark isolates of different mating types or between them and established mating tester strains. Finally, pulsed-field gel electrophoresis revealed considerable diversity among the karyotypes of the Renmark isolates, which would not be expected in a population undergoing regular meiotic recombination and segregation (20).

Further analyses of the data indicated that the story might be more complex, however. First, when all isolates were fitted to a phylogenetic tree, isolates of α and **a** mating type were interspersed to some extent, which would not be expected in an asexual population, as different mating types should be on separate evolutionary paths. Second, when the AFLP data were reduced to their principal components and used to produce a canonical variant analysis plot, isolates were found to segregate strongly according to their tree of origin, indicating a high level of differentiation according to geographic site. It was therefore possible that this population, although geographically confined, was genetically subdivided, which might cause an erroneous conclusion of clonality as noted above. Analysis on a finer scale might therefore be required.

Recombination in Individual Tree Hollows

The Renmark site was revisited in 2006, this time concentrating on two individual tree hollows that had been found to harbor both *C. gattii* mating types. A third population from a single hollow present on a tree in Mundarlo, New South Wales, was also examined (40). Mundarlo is in the Gundagai region and is located approximately 600 km to the east of Renmark on a major tributary of the Murray River. This site is an area where only α mating-type cells have ever been recovered; the aim was to assess whether α-α mating might also occur within the very limited scale of a single tree hollow. AFLP markers were again used to generate multilocus genotypes. Phylogenetic (T_L) and population genetic (I_A) analyses were used to assess whether recombination was occurring among genotypes.

Figure 5 shows a representative phylogram produced from isolates obtained from a single hollow present on one of the trees in Renmark. Phylograms from the other two populations

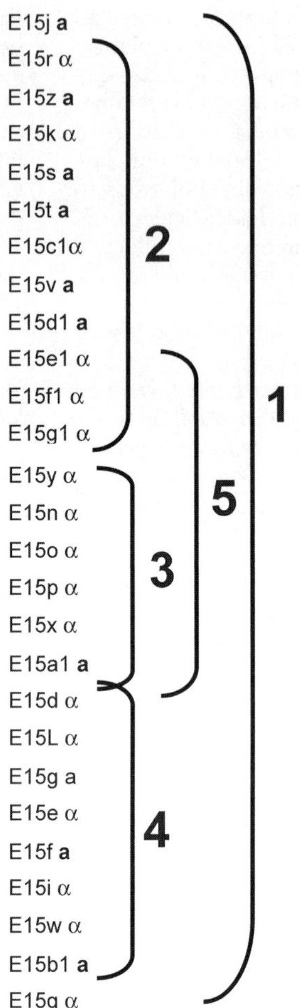

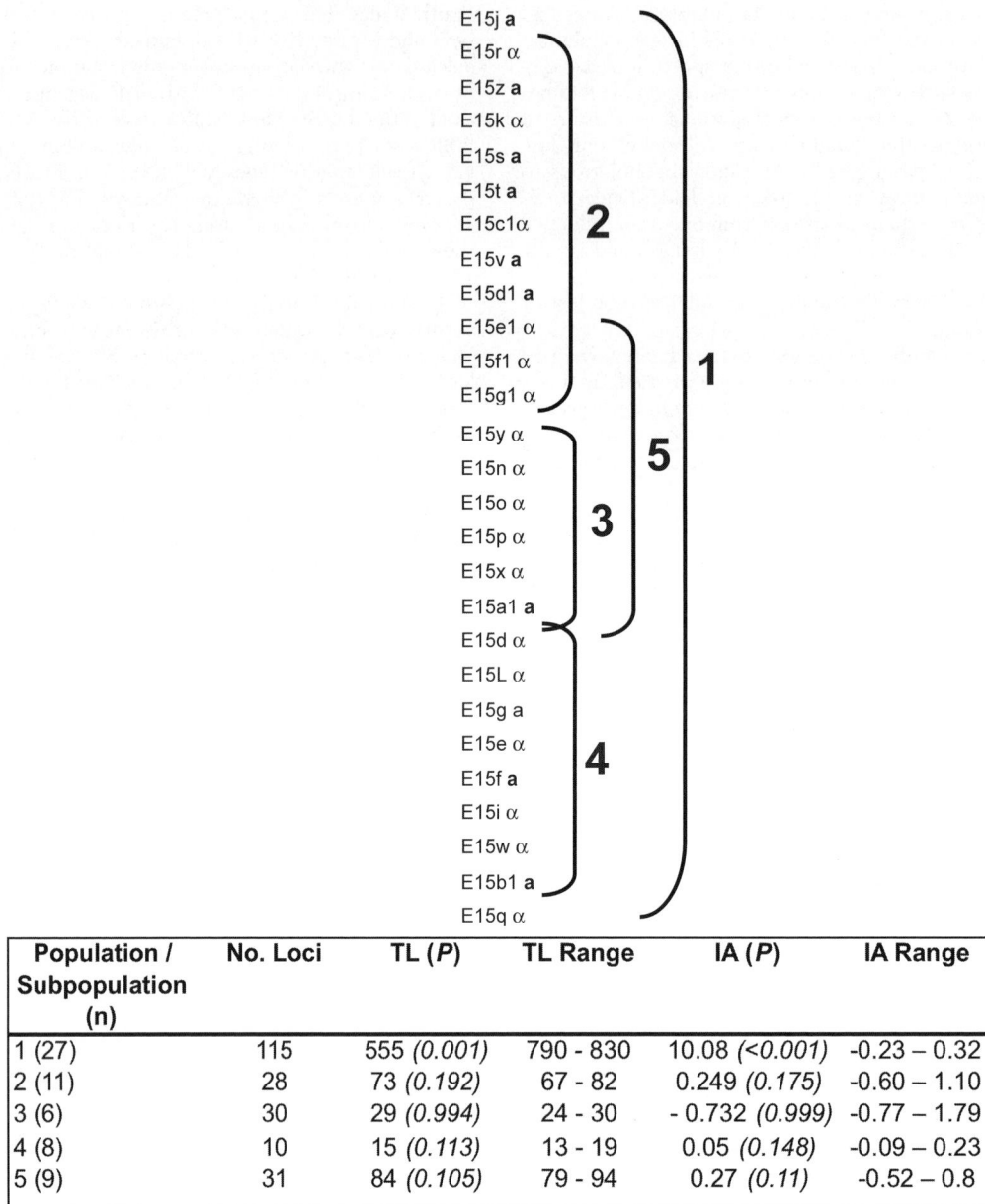

Population / Subpopulation (n)	No. Loci	TL (*P*)	TL Range	IA (*P*)	IA Range
1 (27)	115	555 *(0.001)*	790 - 830	10.08 *(<0.001)*	-0.23 – 0.32
2 (11)	28	73 *(0.192)*	67 - 82	0.249 *(0.175)*	-0.60 – 1.10
3 (6)	30	29 *(0.994)*	24 - 30	- 0.732 *(0.999)*	-0.77 – 1.79
4 (8)	10	15 *(0.113)*	13 - 19	0.05 *(0.148)*	-0.09 – 0.23
5 (9)	31	84 *(0.105)*	79 - 94	0.27 *(0.11)*	-0.52 – 0.8

FIGURE 5 Recombination in a C. *gattii* population derived from a single hollow on tree #15 in Renmark. When the entire population derived from one tree hollow was analyzed (population #1), both T_L and I_A were outside the range of the derived recombining populations, and the population appeared clonal, despite containing closely related and interspersed α and **a** isolates. By focusing on particular groups identified on the phylogram, it was possible to identify four recombining subpopulations (#2 to 5). Figure adapted from reference 40 with permission.

were similar overall in structure. On first examination, populations from the tree hollows were significantly different from their associated artificially recombined data sets, indicating that recombination was not occurring (population 1 in Fig. 5). However, it was noted that isolates on the phylograms grouped into some distinct clusters and that in the Renmark populations these clusters contained very closely related isolates of both α and **a** mating type. The analysis was then confined to each group that contained five or more isolates. Using this approach, it was possible to identify small, genetically differentiated subpopulations within each Renmark population, each consisting of α and **a** cells, and each with clear evidence of genetic reassortment indicating recombination. Furthermore, recombining subpopulations were also evident in the Gundagai tree hollow that contained α mating-type cells only. It was concluded that both α-**a** and α-α mating were occurring in *E. camaldulensis* tree hollows, but that fine-scale partitioning of isolates meant that even an individual hollow consisted of a number of genetically differentiated subpopulations (40).

Pairwise locus compatibility of MLST alleles for *C. gattii* isolates from Renmark

Isolate	*SXI1* α	*SXI2* **a**	*IGS*	*TEF1*	*GPD1*	*LAC1*	*CAP10*	*PLB1*	*MPD1*
E566	-	**1**	3	4	**5**	5	4	5	3
E567	-	**1**	3	4	**5**	5	4	5	3
E572	-	**1**	3	4	**5**	5	4	5	3
E555	-	**1**	3	4	**5**	5	4	5	3
E312	-	**1**	3	4	**5**	5	4	5	3
E287	-	**1**	3	4	**5**	5	4	5	3
E283	-	**1**	3	4	**5**	5	4	5	3
E276	-	**1**	3	4	**5**	5	4	5	3
E554	-	**1**	3	4	**14**	5	4	5	3
E316	**7**	-	3	4	**5**	5	4	5	3
E275	**7**	-	3	4	**5**	5	4	5	3
E286	**7**	-	3	4	**14**	5	4	5	3
E280	**7**	-	3	4	**14**	5	4	5	3
E307	**7**	-	3	4	**14**	5	4	5	3
E569	**7**	-	3	4	**14**	5	4	5	3
E310	**7**	-	3	4	**14**	5	4	5	3
E306	**7**	-	3	4	**14**	5	4	5	3
E549	**7**	-	3	4	**14**	5	4	5	3
E278	**7**	-	3	4	**14**	5	4	5	3
E296	**7**	-	**3**	**4**	**14**	5	4	5	3

Four combinations of alleles at loci *SXI1α*/*SXI2***a** + *GPD1*: *SXI1α* **-7** + *GPD1* **-5**;
SXI1α **-7** + *GPD1* **-14**;
*SXI2***a** **-1** + *GPD1* **-5**;
*SXI2***a** **-1** + *GPD1* **-14**

FIGURE 6 Pairwise compatibility of MLST alleles for a subset of *C. gattii* isolates from Renmark. The Renmark population is very restricted in MLST diversity, with only two polymorphic loci identified: *SXI1α*/*SXI2***a** and *GPD1*. These loci segregate independently to give four possible combinations, as would be expected in a recombining population.

Further evidence of sexual recombination in the Renmark population was evident in data produced by Fraser et al. (17), who analyzed a subset of Renmark isolates by MLST. Overall, genetic diversity among isolates was extremely limited, such that in over 5,740 and 6,925 base pairs sequenced from each of 9 **a** and 11 α strains, respectively, there were only two polymorphic loci: *SXI1α*/*SXI2***a** and *GDP1*. However, segregation of these alleles indicated that they had assorted independently, producing four possible combinations (*SXI1α*-**7** + *GPD1*-**5**; S *SXI1α*-**7** + *GPD1*-**14**; *SXI2***a**-**1** + *GPD1*-**5**; *SXI2***a**-**1** + *GPD1*-**14**) that would be expected to occur via sexual recombination (Fig. 6).

ANALYSIS OF RECOMBINATION IN CLINICAL POPULATIONS: ISOLATES FROM INFECTED HUMANS AND ANIMALS LIVING IN AUSTRALIA AND PAPUA NEW GUINEA

The population genetics of clinical collections is generally more complicated than the study of environmental populations as humans travel and may acquire an infection far from where they eventually present with clinical disease and an isolate is obtained. However, understanding clinical populations is vital for assessing how the human and animal populations interact with environmental pathogens and the likely risk of acquiring disease. We are fortunate in having access to three very defined clinical populations of *C. gattii* isolates: (i) from infected people living in the Northern Territory, an area with a particularly high level of *C. gattii* cryptococcosis, particularly that caused by the VGII genotype; (ii) from infected people in Papua New Guinea, where infection levels are also high and the VGI genotype predominates; and (iii) from infected domestic animals in the Sydney region. For cultural and socioeconomic reasons, the two human groups are relatively unlikely to have traveled far from their home regions, and domestic animals are likewise unlikely to travel long distances and probably acquired their infections close to their place of domicile. We were interested in targeting these clinical populations, first, because disease must be acquired by the inhalation of an infectious propagule, which we hypothesize is most likely to be a basidiospore, and second, because the clinical populations allowed us to sample *C. gattii* genotypes in addition to VGI.

AFLP markers were developed from ~30 isolates from each region (Northern Territory, human clinical = NT; Papua New Guinea, human clinical = PNG; Sydney veterinary = V), and the resulting data were used to produce a phylogram (9). From this, it was immediately apparent that the total population was subdivided according to molecular genotype (Fig. 7). Furthermore, it was clear that the VGI and VGII groups were quite different in terms of relatedness and molecular diversity: all VGI isolates partitioned strongly according to place of origin, and the populations from PNG and Sydney consisted of isolates that were very closely related on short branches, resembling clonal blooms. In contrast, the VGII isolates from Sydney grouped together in a cluster that also contained isolates from NT, and the VGII populations were considerably more diverse.

Each of the subgroups evident on the tree was analyzed for recombination using T_L and I_A tests as above. The results from this indicated that all VGI groups had a strongly clonal structure, whereas two VGII groups had evidence of

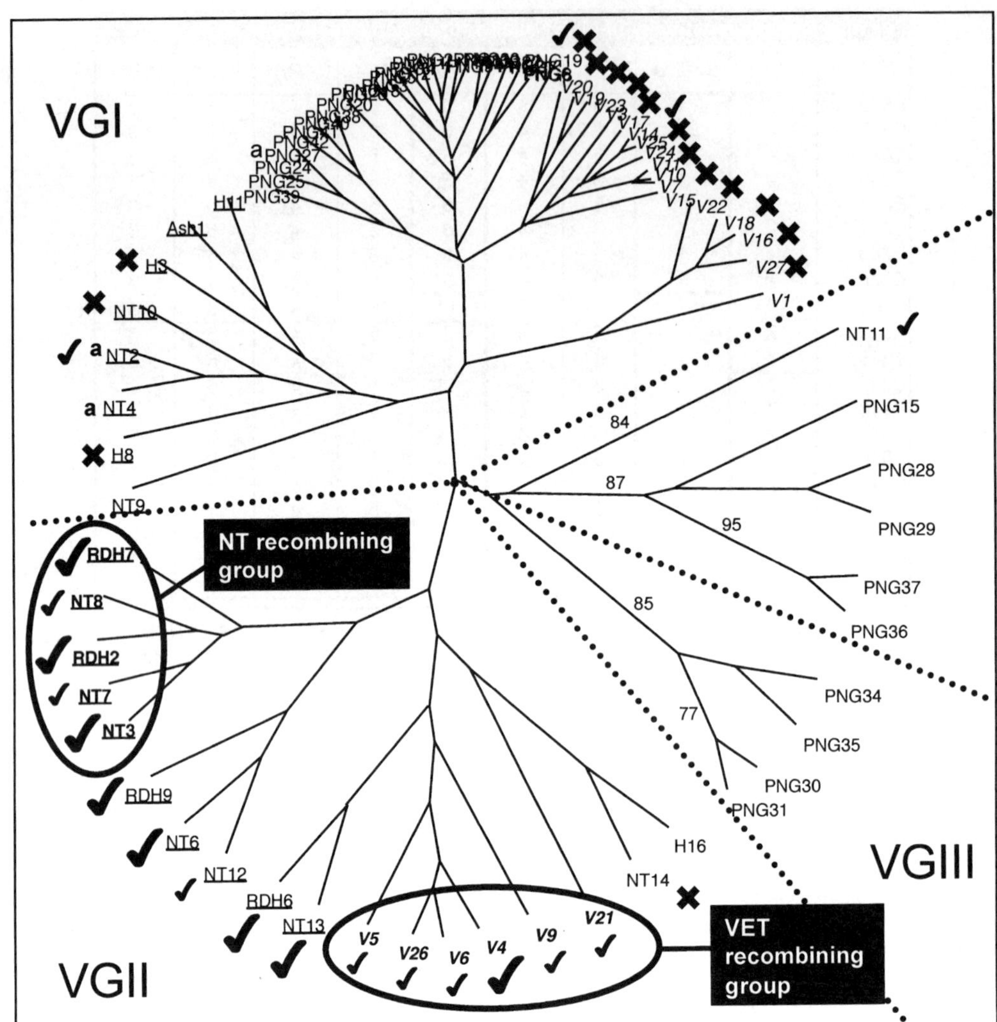

FIGURE 7 Phylogram of relatedness of clinical *C. gattii* isolates derived from AFLP data. Isolates were obtained from (i) infected people living in Papua New Guinea (prefixed by PNG); (ii) infected people living in the Northern Territory, Australia (underlined); and (iii) infected animals living in the Sydney region (in italics). Isolates are largely divided into the VG groups, and VGI isolates are grouped according to their origins. Two recombining subpopulations were identified in the VGII group. Mating tests found fertility to segregate with molecular genotype: almost all VGI isolates were infertile in laboratory crosses (indicated by an x), whereas the majority of VGII isolates were fertile (indicated by a check mark), with many robustly fertile (indicated by a large check mark). Figure adapted from reference 8 with permission.

recombination: the population of veterinary isolates from Sydney and one group of isolates from the Northern Territory. When some of the isolates were analyzed for fertility, a remarkable correlation with genotype was found. Almost all VGI isolates were unable to mate or mated only weakly (to produce filaments but not basidiospores), while almost all VGII isolates were fertile, and many were robustly so, producing copious basidia and basidiospores (Fig. 7) (10).

These results for the clinical VGII populations were particularly interesting, as it had recently been recognized that isolates of VGII genotype were causing the outbreak of cryptococcosis on Vancouver Island, and in this setting there appeared to be a high level of aerosolized propagules of 1 to 2 μm, which is a size consistent with basidiospores (24). Analysis of the Vancouver Island isolates so far indicates that the outbreak is the result of the clonal expansion of a distinct hypervirulent strain (known as the "major" or VGIIa genotype), as diversity appears to be very limited and all strains are mating type α. Fraser et al. (17) undertook an extensive analysis of the mating-type region and proposed that the hypervirulent VGII isolate causing the outbreak had been produced following α-α mating. Ongoing α-α mating between genetically similar isolates in this region, with the production of spores, is consistent with the observed low level of genetic diversity, absence of **a** cells, and presence of aerosolized propagules, which characterize the Vancouver Island *C. gattii* population.

A subsequent analysis of the two Australian recombining VGII groups using MLST data produced by Fraser et al. (17) indicated that these were distinctly different at almost all of the sequenced loci (11). The NT recombining group consists of a unique genotype that has not been seen in any isolates outside this region. In contrast, the VET recombining group is indistinguishable by MLST from isolates from NT, the Caribbean, and some isolates from Vancouver Island (known as the "minor" or VGIIb genotype) that are distinct from the VGIIa major outbreak strain. These two genotypes occur sympatrically in the Northern Territory, and a recent survey of *C. gattii* in this region found they can even occur together on a single tree (41). To date, the only evidence of recombination between these groups is the presence of an uncommon *SXI1α* allele (allele 20) in a single isolate from each group (11).

RECOMBINATION ON GLOBAL SCALES IN MOLECULAR GENOTYPES VGI TO VGIV

The results above indicated significant differences between the structures of VGI and VGII populations in Australia. VGI populations appeared to be restricted in diversity and genetically differentiated, with recombination occurring on very geographically confined scales, while VGII populations were genetically connected with evidence of recombination in two separate clinical populations. Would these patterns also be seen in global populations of these genotypes? And what of the other genotypes, VGIII and VGIV? To address this we have looked for evidence of character compatibility in the extensive MLST data produced by Fraser and colleagues (17). Each pair of polymorphic MLST loci was examined for the presence of alleles in the four possible combinations. Some loci were not sufficiently polymorphic to allow this analysis; others had too many alleles such that each was present in only a few isolates. Nonetheless, it was possible to identify independent assortment at a number of locus pairs. The results are presented in Fig. 8. Independently assorting alleles are shown in bold and boxed, and the locus pairs where these occurred are indicated below or beside the allele charts.

VGI: GLOBALLY CLONAL, LOCALLY RECOMBINING

The global VGI population consists of isolates from Australia, Asia, India, the United States, Canada, France, Brazil, and Mexico (Fig. 8A). Most isolates had a unique MLST genotype, but one group of 20 isolates from four regions, including Renmark, shared a genotype with the VGI type strain WM276. The pairwise test of compatibility indicated independent assortment of two loci: *SXI1α/2***a** and *GPD1*. The four combinations of alleles at these loci were not significantly different from the ratios expected under Hardy-Weinberg equilibrium (Fisher's exact test, $P = 0.11$). These are the same two loci found to segregate independently in the Renmark population (Fig. 6), but what was notable from the global analysis was that two of the four combinations occurred only in Renmark isolates, indicating that recombination at these loci has most likely occurred only in Renmark. This suggests two possible explanations: either sexual recombination is highly unusual in VGI and is restricted to Renmark, or recombination can only be detected when the analysis includes a sufficient number of isolates from a single, geographically confined region. Overall, we conclude that VGI appears to be globally clonal with little evidence of genetic exchange among regions. Results for the Renmark population (and the Gundagai population from a single tree hollow; see above) suggest there is the potential for recombination on a local scale, but more isolates are required to determine if recombining subpopulations are common.

VGII: GLOBALLY RECOMBINING WITH GENETICALLY DIFFERENTIATED SUBPOPULATIONS

The pattern of pairwise compatibility among global VGII isolates (Fig. 8B) presents a striking contrast to that seen in VGI. Here the majority of isolates possess loci that have undergone independent reassortment with other global isolates. The regions involved include Australia, Canada, Brazil, the Caribbean, Aruba, France, Uruguay, Senegal, and French Guiana. However, there are two groups of isolates that do not have evidence of allele exchange with other global isolates: the group represented by the Vancouver Island outbreak strain R265, which is restricted to Vancouver Island and neighboring areas in the United States, and the NT clinical group, which is the unique recombining group identified in the Northern Territory (see Fig. 7) (9). From these data, it appears that VGII is a panmictic genotype with evidence of genetic exchange across large distances, consistent with spore dispersal, but also with the potential to spawn genetically differentiated and isolated subpopulations.

VGIII: GENETICALLY DIVERSE AND GLOBALLY RECOMBINING

VGIII presents a similar pattern to VGII with evidence of allele exchange among most global isolates (Fig. 8C). There is no evidence of any genetically differentiated subpopulations of VGIII isolates, however, and overall there is a high level of genetic diversity in this molecular genotype. Data from other studies suggest that VGIII isolates, like VGII, are fertile and the *C. gattii* mating tester strains are in fact all VGIII isolates (18). The compatibility analysis suggests a high level of genetic connectivity, which, as with VGII, is consistent with the production and dispersal of aerosolized spores.

VGIV: RECOMBINATION ACROSS AFRICAN REGIONS

The VGIV genotype is largely restricted to Africa, and all isolates examined by Fraser et al. (17) were from the African continent (Fig. 8D). Two loci (*SXI1α* and *IGS*) are in Hardy-Weinberg equilibrium in a group of isolates from Botswana, South Africa, and Malawi ($P = 0.58$). A clonal group of 14 clinical isolates from Botswana and Malawi was outside this recombining group, suggesting that, similar to VGII, VGIV is characterized by clonal propagation in some populations as well as long-range genetic exchange.

SEX AND RECOMBINATION IN *C. GATTII*: CONCLUSIONS AND REMAINING CONUNDRUMS

With the recognition of the different molecular genotypes within *C. gattii* and their evolutionary relationships, the use of independent genetic markers and analysis methods, and the incorporation of biological tests of fertility and spore production, we are starting to see patterns emerging in *C. gattii* and its relationship to sex. VGI is probably the most enigmatic of the *C. gattii* genotypes in this regard, where we see evidence of recombination only on extremely limited geographic scales and between isolates that are extremely similar

A: Pairwise locus compatibility of MLST alleles for global isolates of VGI

Name	*SXI1 α*	*SXI2* **a**	*IGS*	*TEF1*	*GPD1*	*LAC1*	*CAP10*	*PLB1*	*MPD1*	Strain origin (n)
B5751	1	-	7	8	8	6	8	8	7	Clinical
Env 71	2	-	3	14	5	5	4	5	3	Aus ENV
NT-10	3	-	3	4	5	11	4	5	3	NT CLIN
V24/571_134	4	-	3	4	5	12	4	5	3	Sydney, VET
571_094	4	-	3	4	5	5	4	5	3	Sydney, VET
327/99	5	-	3	4	5	5	4	5	3	Sydney, VET
MMRL2648	6	-	3	4	15	5	4	5	10	Aus ENV
MYA867	7	-	3	4	15	5	4	5	10	San Diego VET
ICB94	4	-	3	4	5	5	4	5	15	Brazil CLIN
ICB181	4	-	3	4	5	5	4	5	3	Brazil ENV, PNG ENV (3)
S25L	7	-	3	4	5	5	4	5	3	Asia CLIN
C2077	-	2	12	4	5	5	4	5	3	South Australia ENV
RV20186	8	-	13	13	11	13	9	13	4	CLIN
B4492	8	-	13	4	11	5	9	13	12	Aus CLIN
R794	8	-	13	4	11	19	9	15	13	Vancouver Island CLIN
VPCI 87	30	-	3	4		13	9	13	13	India ENV
2005/215	29	-	24	4	11	13	9	13	4	France CLIN
94/943-8	7	-	3	4	15	5	13	5	10	Mexico CLIN (2)
V23/571_123	**7**	-	3	18	**5**	5	4	5	3	Sydney, VET
V13	**7**	-	20	4	**5**	5	4	5	3	Sydney, VET
V15/571_103	**7**	-	12	4	**5**	5	4	5	3	Sydney, VET (5)
WM276	**7**	-	3	4	**5**	5	4	5	3	**Renmark ENV (3)**, Sydney ENV =WM276 (1), Aus CLIN (2), Sydney VET (1), Asia CLIN (2) CLIN (9) France CLIN, Mexico CLIN = 20 isolates in total
E286	**7**	-	3	4	**14**	5	4	5	3	**Renmark ENV (11)**
E566	-	**1**	3	4	**5**	5	4	5	3	**Renmark ENV (8),** NT ENV (2) = 10 isolates in total
E554	-	**1**	3	4	**14**	5	4	5	3	**Renmark ENV (1)**

Four combinations of alleles at loci *SXI1α/SXI2***a** + *GPD1*: *SXI1α* -**7** + *GPD1* -**5**; *SXI1α* -**7** + *GPD1* -**14**; *SXI2***a** -**1** + *GPD1* -**5**; *SXI2***a** -**1** + *GPD1* -**14**

B: Pairwise locus compatibility of MLST alleles for global isolates of VGII

Name	*SX1 α*	*SXI2* ***a***	*IGS*	*TEF1*	*GPD1*	*LAC1*	*CAP10*	*PLB1*	*MPD1*	Strain origin (n)
R265	18	-	4	7	1	4	1	1	5	Vancouver Is CLIN (9), Vancouver Is ENV (10), Vancouver Is VET (3), San Francisco ENV (1), Seattle CLIN (1) **= Vancouver Is outbreak strain / VGIIa**
ICB107	18	-	4	7	1	24	1	1	5	Brazil CLIN
R272	19	-	10	**5**	**6**	**4**	**1**	**2**	5	Vancouver Is CLIN (1), Vancouver Is ENV (1), NT ENV (4), NT CLIN (3), Caribbean Islands CLIN (1), **Sydney VET (6)** **= Vancouver Is non-outbreak strain VGIIb**
NT-12	20	-	10	**5**	**6**	**4**	**1**	**2**	5	NT CLIN
2004/335	23	-	22	**7**	**6**	**4**	**1**	**2**	5	French Guiana CLIN (3), Caribbean Is CLIN (1)
CBS1930	-	5	25	**7**	**6**	**4**	**1**	18	5	Aruba VET
ICB184	23	-	15	**19**	**6**	**4**	**1**	**2**	5	Brazil ENV
2003/125	19	-	26	**7**	25	**21**	**1**	9	5	France CLIN
98/1037-2	28	-	26	**7**	25	**21**	**1**	9	5	France CLIN (2)
CBS8684	27	-	26	**5**	6	**21**	1	22	5	Uruguay ENVT
ICB182	23	-	21	20	6	21	1	2	5	Brazil CLIN
97/170	23	-	28	**5**	**21**	**21**	1	**2**	5	French Guiana CLIN
93/980	23	-	27	**5**	22	**21**	1	19	5	France CLIN
ICB183	23	-	30	**7**	**21**	**4**	14	18	5	Brazil ENV
ICB97	28	-	25	**7**	24	**4**	**10**	24	5	Brazil CLIN
99/901-1	26	-	21	**19**	**21**	**4**	**10**	**16**	5	France CLIN (5), Senegal CLIN (1)
ICB179	28	-	16	**7**	6	**4**	**10**	**16**	5	Brazil ENV (2)
2001/571	25	-	29	**7**	6	**4**	**10**	14	5	France CLIN
WM178	22	-	16	5	17	16	1	14	5	Sydney CLIN
WA 861	23	-	15	**7**	16	**4**	**10**	9	5	Western Australia VET
RDH-2	20	-	6	6	2	7	5	1	5	**NT CLIN (1)**
RDH-7	24	-	6	6	2	7	5	1	5	Sydney, CLIN (1), **NT CLIN (4)**

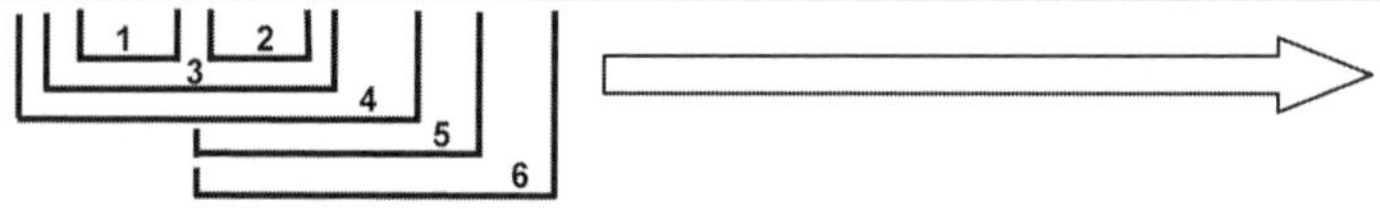

Locus pairs with 4 allele combinations:

1. *TEF1* -**5** + *GPD1* -**6**
   TEF1 -**5** + *GPD1* -**21**
   TEF1 -**7** + *GPD1* -**6**
   TEF1 -**7** + *GPD1* -**21**
2. *GPD1* -**6** + *LAC1* -**4**
   GPD1 -**6** + *LAC1* -**21**
   GPD1 -**21** + *LAC1* -**4**
   GPD1 -**21** + *LAC1* -**21**
3. *TEF1* -**5** + *LAC1* -**4**
   TEF1 -**5** + *LAC1* -**21**
   TEF1 -**7** + *LAC1* -**4**
   TEF1 -**7** + *LAC1* -**21**
4. *TEF1* -**7** + *CAP10* -**1**
   TEF1 -**7** + *CAP10* -**10**
   TEF1 -**19** + *CAP10* -**1**
   TEF1 -**19** + *CAP10* -**10**
5. *GPD1* -**6** + *CAP10* -**10**
   GPD1 -**6** + *CAP10* -**1**
   GPD1 -**21** + *CAP10* -**10**
   GPD1 -**21** + *CAP10* -**1**
6. *GPD1* -**6** + *PLB1* -**2**
   GPD1 -**6** + *PLB1* -**16**
   GPD1 -**21** + *PLB1* -**2**
   GPD1 -**21** + *PLB1* -**16**

FIGURE 8 Pairwise compatibility of MLST alleles for global *C. gattii* isolates belonging to molecular genotypes VGI to VGIV. (A) For VGI, only locus pair *SXI1α/SXI2*a and *GPD1* possessed four possible combinations of alleles, two of which were present in isolates from Renmark only, suggesting that recombination is largely confined to Renmark. (B) In contrast, in the VGII population there are six incompatible locus pairs, and most global isolates appear to participate in recombination. Two groups are exceptions: one group of isolates that have the genotype of the Vancouver Island outbreak strain (VGIIa) and the small group of distinct isolates identified as a recombining population in the Northern Territory. (C) Recombination at two or more loci is apparent in most isolates in the global VGIII population. (D) The VGIV population consists of a recombining group from Botswana, South Africa, and Malawi. A large number of genetically identical isolates from Botswana and Malawi are outside the recombining group. CLIN, clinical isolate; ENV, environmental isolate; VET, veterinary isolate. Only one representative isolate name is given for groups of two or more isolates; for details of isolates see reference 17.

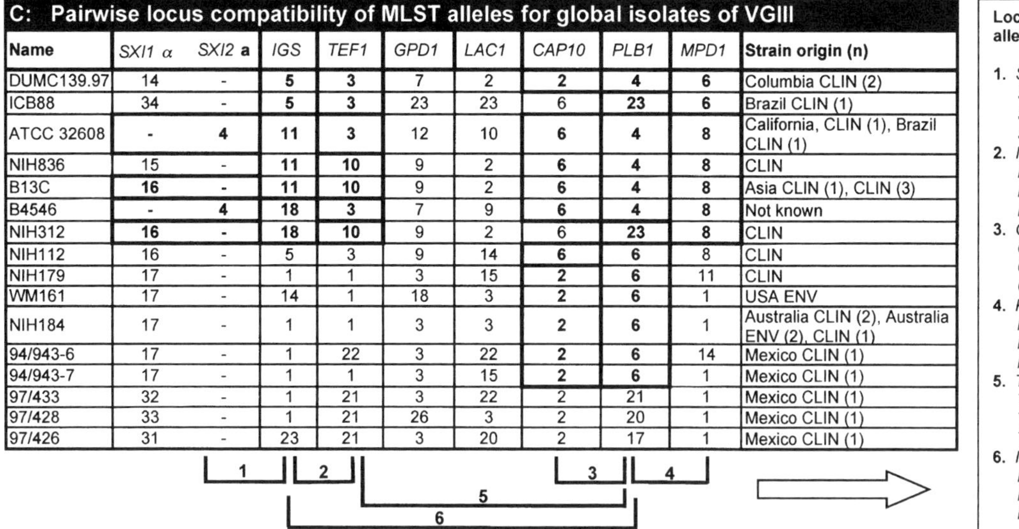

C: Pairwise locus compatibility of MLST alleles for global isolates of VGIII

Name	*SXI1* α	*SXI2* a	*IGS*	*TEF1*	*GPD1*	*LAC1*	*CAP10*	*PLB1*	*MPD1*	Strain origin (n)
DUMC139.97	14	-	**5**	**3**	7	2	**2**	**4**	**6**	Columbia CLIN (2)
ICB88	34	-	**5**	**3**	23	23	6	**23**	**6**	Brazil CLIN (1)
ATCC 32608	**-**	**4**	**11**	**3**	12	10	**6**	**4**	**8**	California, CLIN (1), Brazil CLIN (1)
NIH836	15	-	**11**	**10**	9	2	**6**	**4**	**8**	CLIN
B13C	**16**	**-**	**11**	**10**	9	2	**6**	**4**	**8**	Asia CLIN (1), CLIN (3)
B4546	-	**4**	**18**	**3**	7	9	**6**	**4**	**8**	Not known
NIH312	**16**	**-**	**18**	**10**	9	2	6	**23**	**8**	CLIN
NIH112	16	-	5	3	9	14	**6**	**6**	8	CLIN
NIH179	17	-	1	1	3	15	**2**	**6**	11	CLIN
WM161	17	-	14	1	18	3	**2**	**6**	1	USA ENV
NIH184	17	-	1	1	3	3	**2**	**6**	1	Australia CLIN (2), Australia ENV (2), CLIN (1)
94/943-6	17	-	1	22	3	22	**2**	**6**	14	Mexico CLIN (1)
94/943-7	17	-	1	1	3	15	**2**	**6**	1	Mexico CLIN (1)
97/433	32	-	1	21	3	22	2	21	1	Mexico CLIN (1)
97/428	33	-	1	21	26	3	2	20	1	Mexico CLIN (1)
97/426	31	-	23	21	3	20	2	17	1	Mexico CLIN (1)

Locus pairs with 4 allele combinations:

1. *SXI1α*–**16** + IGS-**11**; *SXI1α*–**16** + IGS-**18**; *SXI2a*–**4** + IGS-**11**; *SXI2a*–**4** + IGS-**18**
2. *IGS*-**18** + *TEF1*-**3**; *IGS*-**18** + *TEF1*-**10**; *IGS*-**11** + *TEF1*-**3**; *IGS*-**11** + *TEF1*-**10**
3. *CAP10*-**2** + *PLB1*-**4**; *CAP10*-**2** + *PLB1*-**6**; *CAP10*-**6** + *PLB1*-**4**; *CAP10*-**6** + *PLB1*-**6**
4. *PLB1*-**4** + *MPD1*-**6**; *PLB1*-**4** + *MPD1*-**8**; *PLB1*-**23** + *MPD1*-**6**; *PLB1*-**23** + *MPD1*-**8**
5. *TEF1*-**3** + *PLB1*-**4**; *TEF1*-**3** + *PLB1*-**23**; *TEF1*-**10** + *PLB1*-**4**; *TEF1*-**10** + *PLB1*-**23**
6. *IGS*-**5** + *PLB1*-**4**; *IGS*-**5** + *PLB1*-**23**; *IGS*-**18** + *PLB1*-**4**; *IGS*-**18** + *PLB1*-**23**

D: Pairwise locus compatibility of MLST alleles for global isolates of VGIV

Name	*SXI1* α	*IGS*	*TEF1*	*GPD1*	*LAC1*	*CAP10*	*PLB1*	*MPD1*	Strain origin (n)
M391	9	2	11	4	1	3	7	2	Botswana CLIN (13), Malawi CLIN (1)
MMRL2924	9	2	2	4	1	3	10	2	Botswana CLIN
MMRL2980	13	17	12	10	17	7	12	2	Botswana CLIN
MMRL2933	10	2	9	4	1	3	10	2	Botswana CLIN
MMRL2879	**11**	**8**	15	10	1	3	3	2	Botswana CLIN
MMRL3013	**12**	**2**	2	19	1	11	3	2	Botswana CLIN
MMRL2872	**12**	**2**	2	4	1	3	3	2	Botswana CLIN
MMRL2651	**12**	**8**	2	4	1	7	3	2	Botswana CLIN
WM779	**12**	**8**	17	10	18	3	3	2	South Africa VET
B5748	**12**	**8**	2	13	1	7	3	2	CLIN (2)
M250	**11**	**2**	2	10	17	15	3	2	Malawi CLIN

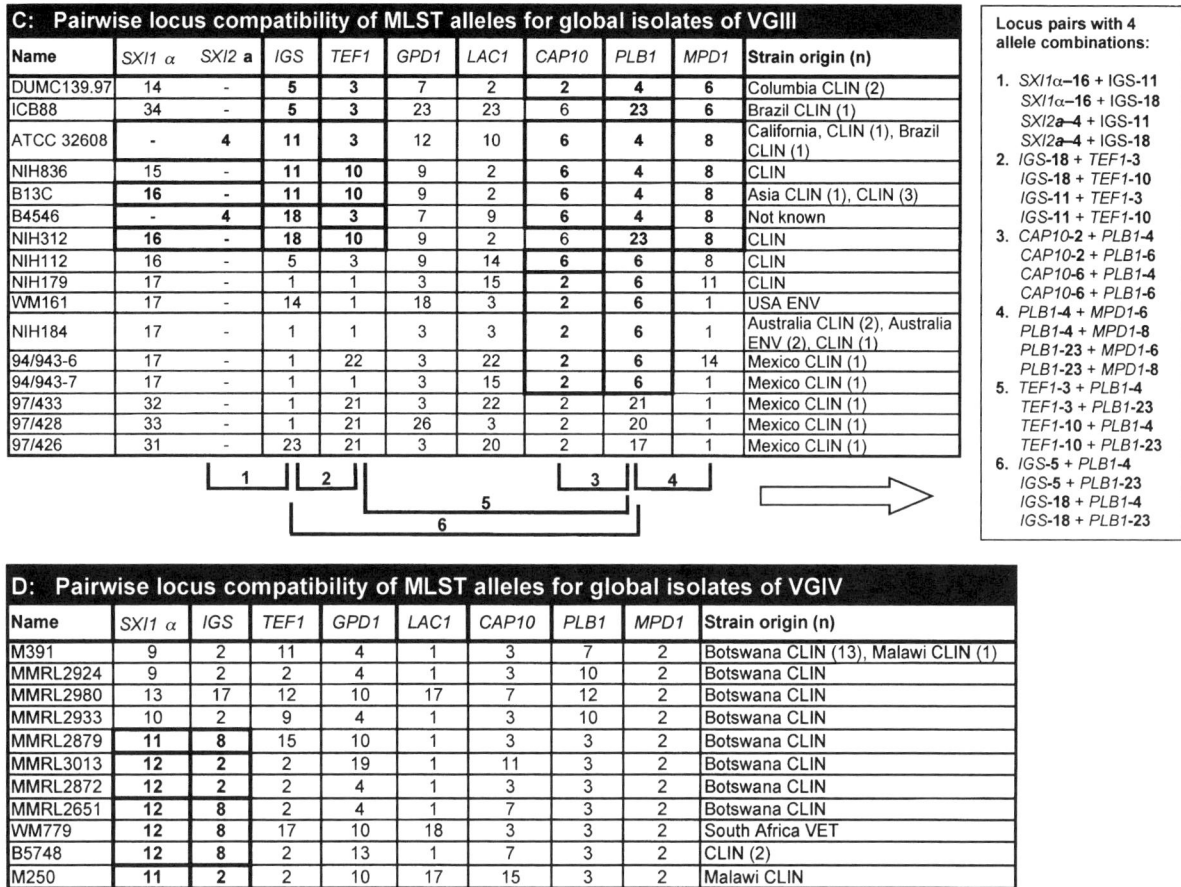

FIGURE 8 (*Continued*)

in genetic makeup, such that they are effectively inbreeding. Although both mating types occur in some environmental samples, mating studies have found that most isolates are only weakly fertile or are infertile, and we do not know the mechanism involved that mediates genetic exchange among VGI isolates, or whether spores are being produced. Using specially modified media with a high inositol concentration, it has recently been possible to induce mating between some Renmark VGI isolates and highly mating-competent tester strains, suggesting the Renmark isolates remain mating competent (E. J. Byrnes and J. Heitman, unpublished). However, Renmark isolates of the α and **a** mating types, and other VGI strains tested, remain stubbornly recalcitrant toward mating with one another in the laboratory. The fact that VGI populations are genetically differentiated to the extent that isolates are even associated with their host tree is further evidence that airborne basidiospores are not a frequent event. However, VGI cryptococcosis, although sporadic, occurs to a significant level in humans and animals and must involve the inhalation of a small, infectious propagule. More studies are required to assess medium- to long-distance dispersal by VGI strains, as this has obvious implications for the acquisition of disease.

We have mounting evidence that sex and recombination occur in the other *C. gattii* VG groups. VGII isolates are fertile in laboratory crosses, and in both geographically confined, clinical populations from NT and Sydney, and among global isolates, we have clear evidence of genetic reassortment indicating recombination. All VGII populations studied to date are heavily biased for one or the other mating type, and the most probable scenario is that mating occurs between isolates of the same sex. VGII is responsible for a high incidence of cryptococcosis in the Northern Territory as well as being the cause of the ongoing outbreak on Vancouver Island, and it is tempting to speculate that the ability to mate and produce abundant, easily aerosolized propagules is at least partly responsible for the increased levels of disease in these regions. We do not yet know the ecological niche for VGII, VGIII, or VGIV and how this relates to sex and spore production. Understanding the ecological relationship of *C. gattii* with its environment is the next important goal for determining the infective potential of this environmental yeast pathogen.

REFERENCES

1. **Bartlett, K. H., L. MacDougall, S. Mak, C. Duncan, S. Kidd, and M. Fyfe.** 2004. Abstr. 16th Biometeorol. Aerobiol. Meet. abstr. 5.5. http://ams.confex.com/ams/pdfpapers/80027.pdf.
2. **Boekhout, T., B. Theelen, M. Diaz, J. W. Fell, W. C. J. Hop, E. C. A. Abeln, F. Dromer, and W. Meyer.** 2001. Hybrid genotypes in the pathogenic yeast *Cryptococcus neoformans*. *Microbiology* **147:**891–907.
3. **Boekhout, T., A. van Belkum, A. C. Leenders, H. A. Verbrugh, P. Mukamurangwa, D. Swinne, and W. A. Scheffers.** 1997. Molecular typing of *Cryptococcus neoformans*: taxonomic

and epidemiological aspects. *Int. J. Syst. Evol. Bacteriol.* **47:**432–442.
4. **Bovers, M., F. Hagen, E. E. Kuramae, and T. Boekhout.** 2008. Six monophyletic lineages identified within *Cryptococcus neoformans* and *Cryptococcus gattii* by multi-locus sequence typing. *Fungal Genet. Biol.* **45:**400–421.
5. **Brandt, M. E., L. C. Hutwagner, L. A. Klug, W. S. Baughman, D. Rimland, E. A. Graviss, R. J. Hamill, C. Thomas, P. G. Pappas, A. L. Reingold, and R. W. Pinner.** 1996. Molecular subtype distribution of *Cryptococcus neoformans* in four areas of the United States. Cryptococcal Disease Active Surveillance Group. *J. Clin. Microbiol.* **34:**912–917.
6. **Bui, T., X. Lin, R. Malik, J. Heitman, and D. Carter.** 2008. Isolates of *Cryptococcus neoformans* from infected animals reveal genetic exchange in unisexual, alpha mating type populations. *Eukaryot. Cell* **7:**1771–1780.
7. **Burnett, J.** 2003. *Fungal Populations and Species.* Oxford University Press, Oxford, UK.
8. **Campbell, L. T., and D. A. Carter.** 2006. Looking for sex in the fungal pathogens *Cryptococcus neoformans* and *Cryptococcus gattii. FEMS Yeast Res.* **6:**588–598.
9. **Campbell, L. T., B. J. Currie, M. Krockenberger, R. Malik, W. Meyer, J. Heitman, and D. Carter.** 2005. Clonality and recombination in genetically differentiated subgroups of *Cryptococcus gattii. Eukaryot. Cell* **4:**1403–1409.
10. **Campbell, L. T., J. A. Fraser, C. B. Nichols, F. S. Dietrich, D. Carter, and J. Heitman.** 2005. Clinical and environmental isolates of *Cryptococcus gattii* from Australia that retain sexual fecundity. *Eukaryot. Cell* **4:**1410–1419.
11. **Carter, D., N. Saul, L. Campbell, T. Bui, and M. Krockenberger.** 2007. Sex in natural populations of *Cryptococcus gattii*, p. 477 –487. *In* J. Heitman, J. Kronstad, J. W. Taylor, and L. A. Casselton (ed.), *Sex in Fungi.* ASM Press, Washington, DC.
12. **Escandón, P., P. Ngamskulrungroj, M. Meyer, and E. Castañeda.** 2007. In vitro mating of Colombian isolates of the *Cryptococcus neoformans* species complex. *Biomédica* **27:**308–314.
13. **Escandón, P., A. Sanchez, M. Martinez, W. Meyer, and E. Castañeda.** 2006. Molecular epidemiology of clinical and environmental isolates of the *Cryptococcus neoformans* species complex reveals a high genetic diversity and the presence of the molecular type VGII mating type a in Colombia. *FEMS Yeast Res.* **6:**625–635.
14. **Feng, X., Z. Yao, D. Ren, W. Liao, and J. Wu.** 2008. Genotype and mating type analysis of *Cryptococcus neoformans* and *Cryptococcus gattii* isolates from China that mainly originated from non-HIV-infected patients. *FEMS Yeast Res.* **8:**930–938.
15. **Fisher, M. C., G. L. Koenig, T. J. White, and J. W. Taylor.** 2002. Molecular and phenotypic description of *Coccidioides posadasii* sp. nov., previously recognized as the non-California population of *Coccidioides immitis. Mycologia* **94:**73–84.
16. **Franzot, S. P., J. S. Hamdan, B. P. Currie, and A. Casadevall.** 1997. Molecular epidemiology of *Cryptococcus neoformans* in Brazil and the United States: evidence for both local genetic differences and a global clonal population structure. *J. Clin. Microbiol.* **35:**2243–2251.
17. **Fraser, J. A., S. S. Giles, E. C. Wenink, S. G. Geunes-Boyer, J. R. Wright, S. Diezmann, A. Allen, J. E. Stajich, F. S. Dietrich, J. R. Perfect, and J. Heitman.** 2005. Same-sex mating and the origin of the Vancouver Island *Cryptococcus gattii* outbreak. *Nature* **437:**1360–1364.
18. **Fraser, J. A., R. L. Subaran, C. B. Nichols, and J. Heitman.** 2003. Recapitulation of the sexual cycle of the primary fungal pathogen *Cryptococcus neoformans* var. *gattii*: implications for an outbreak on Vancouver Island, Canada. *Eukaryot. Cell* **2:**1036–1045.
19. **Geiser, D. M., J. I. Pitt, and J. W. Taylor.** 1998. Cryptic speciation and recombination in the aflatoxin-producing fungus *Aspergillus flavus. Proc. Natl. Acad. Sci. USA* **95:**388–393.
20. **Halliday, C. L.** 2000. PhD thesis, University of Sydney, Sydney, Australia.
21. **Halliday, C. L., T. Bui, M. Krockenberger, R. Malik, D. H. Ellis, and D. A. Carter.** 1999. Presence of alpha and **a** mating types in environmental and clinical collections of *Cryptococcus neoformans* var. *gattii* strains from Australia. *J. Clin. Microbiol.* **37:**2920–2926.
22. **Halliday, C. L., and D. A. Carter.** 2003. Clonal reproduction and limited dispersal in an environmental population of *Cryptococcus neoformans* var. *gattii* isolates from Australia. *J. Clin. Microbiol.* **41:**703–711.
23. **Kasuga, T., T. J. White, G. Koenig, J. McEwen, A. Restrepo, E. Castañeda, C. Da Silva Lacaz, E. M. Heins-Vaccari, R. S. De Freitas, R. M. Zancope-Oliveira, Z. Qin, R. Negroni, D. A. Carter, Y. Mikami, M. Tamura, M. L. Taylor, G. F. Miller, N. Poonwan, and J. W. Taylor.** 2003. Phylogeography of the fungal pathogen *Histoplasma capsulatum. Mol. Ecol.* **12:**3383–3401.
24. **Kidd, S. E., F. Hagen, R. L. Tscharke, M. Huynh, K. H. Bartlett, M. Fyfe, L. MacDougall, T. Boekhout, K. J. Kwon-Chung, and W. Meyer.** 2004. A rare genotype of *Cryptococcus gattii* caused the cryptococcosis outbreak on Vancouver Island (British Columbia, Canada). *Proc. Natl. Acad. Sci. USA* **101:**17258–17263.
25. **Koufopanou, V., A. Burt, T. Szaro, and J. W. Taylor.** 2001. Gene genealogies, cryptic species, and molecular evolution in the human pathogen *Coccidioides immitis* and relatives (Ascomycota, Onygenales). *Mol. Biol. Evol.* **18:**1246–1258.
26. **Kwon-Chung, K. J.** 1976. Morphogenesis of *Filobasidiella neoformans*, the sexual state of *Cryptococcus neoformans. Mycologia* **68:**821–833.
27. **Le Gac, M., M. E. Hood, E. Fournier, and T. Giraud.** 2007. Phylogenetic evidence of host-specific cryptic species in the anther smut fungus. *Evolution* **61:**15–26.
28. **Lengeler, K. B., P. Wang, G. M. Cox, J. R. Perfect, and J. Heitman.** 2000. Identification of the MATa mating-type locus of *Cryptococcus neoformans* reveals a serotype A MATa strain thought to have been extinct. *Proc. Natl. Acad. Sci. USA* **97:**14455–14460.
29. **Lin, X., J. C. Huang, T. G. Mitchell, and J. Heitman.** 2006. Virulence attributes and hyphal growth of *Cryptococcus neoformans* are quantitative traits and the MATalpha allele enhances filamentation. *PLoS Genet.* **2:**e187.
30. **Lin, X., C. M. Hull, and J. Heitman.** 2005. Sexual reproduction between partners of the same mating type in *Cryptococcus neoformans. Nature* **434:**1017–1021.
31. **Lin, X., A. P. Litvintseva, K. Nielsen, S. Patel, A. Floyd, T. G. Mitchell, and J. Heitman.** 2007. alphaADalpha hybrids of *Cryptococcus neoformans:* evidence of same-sex mating in nature and hybrid fitness. *PLoS Genet.* **3:**1975–1990.
32. **Litvintseva, A. P., R. E. Marra, K. Nielsen, J. Heitman, R. Vilgalys, and T. G. Mitchell.** 2003. Evidence of sexual recombination among *Cryptococcus neoformans* serotype A isolates in sub-Saharan Africa. *Eukaryot. Cell* **2:**1162–1168.
33. **Litvintseva, A. P., R. Thakur, L. B. Reller, and T. G. Mitchell.** 2005. Prevalence of clinical isolates of *Cryptococcus gattii* serotype C among patients with AIDS in sub-Saharan Africa. *J. Infect. Dis.* **192:**888–892.
34. **Madrenys, N., C. De Vroey, C. Raes-Wuytack, and J. M. Torres-Rodriguez.** 1993. Identification of the

perfect state of *Cryptococcus neoformans* from 195 clinical isolates including 84 from AIDS patients. *Mycopathologia* **123:**65–68.

35. **Maynard-Smith, J., N. H. Smith, M. O'Rourke, and B. G. Spratt.** 1993. How clonal are bacteria? *Proc. Natl. Acad. Sci. USA* **90:**4384–4388.
36. **Ngamskulrungroj, P., T. C. Sorrell, A. Chindamporn, A. Chaiprasert, N. Poonwan, and W. Meyer.** 2008. Association between fertility and molecular sub-type of global isolates of *Cryptococcus gattii* molecular type VGII. *Med. Mycol.* **46:**665–673.
37. **Randhawa, H. S., T. Kowshik, A. Chowdhary, K. Preeti Sinha, Z. U. Khan, S. Sun, and J. Xu.** 2008. The expanding host tree species spectrum of *Cryptococcus gattii* and *Cryptococcus neoformans* and their isolations from surrounding soil in India. *Med. Mycol.* **46:**823–833.
38. **Ruma, P., S. C. Chen, T. C. Sorrell, and A. G. Brownlee.** 1996. Characterization of *Cryptococcus neoformans* by random DNA amplification. *Lett. Appl. Microbiol.* **23:**312–316.
39. **Sato, H., and N. Murakami.** 2008. Reproductive isolation among cryptic species in the ectomycorrhizal genus *Strobilomyces*: population-level CAPS marker-based genetic analysis. *Mol. Phylogen. Evol.* **48:**326–334.
40. **Saul, N., M. Krockenberger, and D. Carter.** 2008. Evidence of recombination in mixed mating type and α-only populations of *Cryptococcus gattii* sourced from single *Eucalyptus* tree hollows. *Eukaryot. Cell* **7:**727–734.
41. **Saul, N. T.** 2009. PhD thesis. University of Sydney, Sydney, Australia.
42. **Sorrell, T. C., S. C. Chen, P. Ruma, W. Meyer, T. J. Pfeiffer, D. H. Ellis, and A. G. Brownlee.** 1996. Concordance of clinical and environmental isolates of *Cryptococcus neoformans* var. *gattii* by random amplification of polymorphic DNA analysis and PCR fingerprinting. *J. Clin. Microbiol.* **34:**1253–1260.
43. **Trilles, L., M. Lazera, B. Wanke, B. Theelen, and T. Boekhout.** 2003. Genetic characterization of environmental isolates of the *Cryptococcus neoformans* species complex from Brazil. *Med. Mycol.* **41:**383–390.
44. **Tscharke, R. L., M. Lazera, Y. C. Chang, B. L. Wickes, and K. J. Kwon-Chung.** 2003. Haploid fruiting in *Cryptococcus neoformans* is not mating type alpha-specific. *Fungal Genet. Biol.* **39:**230–237.
45. **Viviani, M. A., M. C. Esposto, M. Cogliati, M. T. Montagna, and B. L. Wickes.** 2001. Isolation of a *Cryptococcus neoformans* serotype A MAT**a** strain from the Italian environment. *Med. Mycol.* **39:**383–386.
46. **Wickes, B. L., M. E. Mayorga, U. Edman, and J. C. Edman.** 1996. Dimorphism and haploid fruiting in *Cryptococcus neoformans*: association with the alpha-mating type. *Proc. Natl. Acad. Sci. USA* **93:**7327–7331.
47. **Xu, J.** 2005. Fundamentals of fungal molecular population genetic analyses, p. 87–116. *In* J. Xu (ed.), *Evolutionary Genetics of Fungi*. Horizon Bioscience, Norfolk, UK.

Cryptococcus: From Human Pathogen to Model Yeast
Edited by J. Heitman et al.

23

The Emergence of *Cryptococcus gattii* Infections on Vancouver Island and Expansion in the Pacific Northwest

KAREN BARTLETT, EDMOND BYRNES, COLLEEN DUNCAN, MURRAY FYFE, ELENI GALANIS, JOSEPH HEITMAN, LINDA HOANG, SARAH KIDD, LAURA MACDOUGALL, SUNNY MAK, KIEREN MARR, MUHAMMAD MORSHED, SARAH WEST, AND JAMES KRONSTAD

A dramatic emergence of human and animal cases of cryptococcosis caused by *Cryptococcus gattii* has occurred over the past decade on Vancouver Island in British Columbia. These cases stand in contrast to the sporadic and AIDS-associated cases of cryptococcosis caused by *Cryptococcus neoformans* that are well described in the medical and veterinary literature (9). The emergence of cryptococcosis caused by *C. gattii* on Vancouver Island and the more recent appearance of the disease on the British Columbia mainland and in the greater Pacific Northwest provides a unique opportunity to combine clinical and epidemiological studies with a broader exploration of the environmental aspects of an outbreak. In addition, cryptococcal infections in the Pacific Northwest have dramatically illustrated the propensity of *C. gattii* to infect immunocompetent people and animals, in contrast to the opportunistic nature of *C. neoformans*–associated disease. The emergence of *C. gattii* has also been instructive in terms of developing and deploying public health responses that include the need to create multidisciplinary teams to identify and evaluate risks to the community from a pathogen adapting to a previously unexploited niche (1, 14; K. H. Bartlett, L. MacDougall, S. Mak, C. Duncan, S. Kidd, and M. Fyfe, presented at the 16th Conference on Biometeorology & Aerobiology, Vancouver, British Columbia, Canada, 23 to 27 August 2004). Additionally, the appearance of *C. gattii* as a "multispecies-specific" pathogen has created unusual challenges in public health communications involving the public and human and veterinary health care professionals, unlike recent experiences with other emerging pathogens such as coronavirus and West Nile virus (51).

Cryptococcosis in the Pacific Northwest arose against a backdrop of surveys that indicated a connection between *C. gattii* and tropical and subtropical climates (39). In addition, pioneering environmental sampling in Australia revealed an association of *C. gattii* with native tree species including eucalypts (21, 22, 24, 25). The congruence of environmental and human or animal isolates of *C. gattii* cases was well described by the Australian Cryptococcus Working Group (52–55). However, nowhere in the world has there been an example of an outbreak of *C. gattii*–associated cryptococcal disease in animals and humans with the concurrent evidence of environmental colonization by *C. gattii* (30, 34–37, 40, 56). This chapter outlines, we hope with a sense of discovery, issues that were quickly explored, starting with the realization in 2001 that Vancouver Island was a hot spot for a pathogen not previously described as endemic and not restricted to tropical and subtropical climates. The picture of the outbreak that has developed has implications for global travel, climate change, land use patterns, and environmental colonization. Importantly, cryptococcosis caused by *C. gattii* now serves as an excellent illustration of the impact of pathogen spread into a clement ecological niche, in this case, one

Karen Bartlett, School of Environmental Health, University of British Columbia, Vancouver, BC, Canada V6T 1Z3. **Edmond Byrnes and Joseph Heitman,** Duke University Medical Center, Durham, NC 27710. **Colleen Duncan,** Colorado State University Veterinary Diagnostic Laboratory, Fort Collins, CO 80523. **Murray Fyfe,** Office of the Medical Health Officer, Vancouver Island Health Authority, Victoria, BC, Canada V8R 4R2. **Eleni Galanis,** British Columbia Centre for Disease Control and School of Population and Public Health, University of British Columbia, Vancouver, BC, Canada V5Z 4R4. **Linda Hoang and Muhammad Morshed,** British Columbia Centre for Disease Control, Dept. of Pathology and Laboratory Medicine, University of British Columbia, Vancouver, BC, Canada V5Z 4R4. **Sarah Kidd,** Department of Medicine, Monash University, Central Clinical School, Alfred Hospital, Melbourne VIC 3004, Australia. **Laura MacDougall and Sunny Mak,** British Columbia Centre for Disease Control, Vancouver, BC, Canada V5Z 4R4. **Kieren Marr,** Johns Hopkins University School of Medicine, Baltimore, MD 21205. **Sarah West,** Divisions of General Internal Medicine and Infectious Diseases, Oregon Health and Science University, Portland, OR 97239. **James Kronstad,** The Michael Smith Laboratories, University of British Columbia, Vancouver, BC, Canada V6T 1Z4.

that happened to be in a major population center of western Canada and the Pacific Northwest (1, 14).

VANCOUVER ISLAND: A TEMPERATE LOCATION FOR THE EMERGENCE OF *C. GATTII*

Vancouver Island has a land area of 32,000 km^2 and is the largest island on the Pacific Coast of North America (Color Plate 5). The island spans four biogeoclimatic zones and has an overall temperate climate, with the east coast having less precipitation than the west. In undeveloped areas, vast tracts of forest dominate the landscape, providing habitat for native mammalian wildlife including insectivores, bats (chiroptera), rabbits (lagomorphs), rodents, carnivores, and horses and other ungulates. Marine mammals (pinnipedia and cetacean) are found in waters offshore. The human population is 752,000, with approximately half inhabiting the largest population center, Greater Victoria, on the southern tip of the island. Much of the remaining population is distributed in communities along the southeast coast that are connected by a major highway running north-south. Displaying a similar geographic distribution, companion animals and animals of agricultural significance are largely found along the southeast coast of Vancouver Island. Ferry service between Vancouver Island and the British Columbia mainland transports 17 million passengers annually, providing an important social and economic link for residents and visitors. Primary economic activities include forestry, tourism, fishing, and government services.

VETERINARY ASPECTS OF *C. GATTII* INFECTION

Historically, cryptococcosis was a condition uncommonly encountered by veterinarians practicing in British Columbia. Diagnostic laboratories confirmed the disease infrequently in the service area and only sporadically on Vancouver Island (19). However, a marked increase in cases was noted by veterinarians and diagnosticians in 2001, and *C. gattii*, previously unreported in animals in Canada, was identified as the causative organism. The significance of this finding prompted further investigation into the emergence and triggered public health and veterinary responses to the first recorded outbreak of cryptococcosis in British Columbia.

The first published report on *C. gattii* in animals from Vancouver Island focused on the spectrum of domestic and wild animals, geographic distribution, and diagnostic modalities (56); this information led to a larger study of clinical disease and associated risk factors, asymptomatic infection, and exposure. A series of presumed or confirmed *C. gattii* cases was compiled through record reviews of veterinary laboratories; this information summarized the general pattern of host, spatial, and temporal distribution of clinical disease in 1999 to 2003 (19). During this time period, there was an increase in the annual number of animal cases, although no seasonality was observed. Overall, there were 50% more feline than canine cases, and disease appeared more commonly in middle-aged cats and younger dogs. There was no sex predilection for either species.

Symptoms and clinicopathological changes in animals on Vancouver Island were consistent with disease reported elsewhere (40). The most common primary system involved was respiratory, followed by the central nervous system (CNS), in both cats and dogs. There was, however, a higher percentage of CNS disease in dogs relative to cats, and cats were much more likely to have subcutaneous or dermal masses relative to dogs. The presence of neurological symptoms was identified as a statistically significant predictor of mortality; those animals exhibiting CNS symptoms were over four times more likely to die than those never showing neural signs.

Dogs and cats were the most common veterinary species diagnosed with cryptococcosis, but disease was also reported in both Dall's and harbor porpoises (*Phocoenoides dalli*, *Phocoena phocoena*), llamas (*Lama glama*), three avian species (parrot [*Eclectus roratus*], lesser sulphur-crested cockatoo [*Cacatua sulphurea*], and a cockatoo of unknown species), domestic ferrets (*Mustela putorius furo*), and a horse (*Equus caballus*). It is possible that the diagnosis of animals and their inclusion in this case series may have been limited by owner willingness to seek veterinary assistance. Cases were distributed along the southeast coast of the island (Color Plate 5) (56).

Risk factors for clinical disease were identified through a case-control study (18). Variables associated with illness in animals include residing within 10 km of commercial logging or soil disruption, travel on Vancouver Island in the previous year, increased percentage of time spent outside a 10-km radius of the home, increased animal activity level, owners hiking or visiting botanical gardens, and association with other cases of cryptococcosis. These findings suggested that individual animal risk is increased when *C. gattii* is redistributed through large-scale environmental disturbance or when the animal has increased opportunities for exposure through travel or activity level (22).

The pattern of asymptomatic infection and nasal colonization was similar to that of clinical disease. Serum samples and materials for fungal culture were collected from dogs, cats, horses, and terrestrial mammal species residing within the region where clinical cases had been diagnosed (18, 20; C. Duncan, unpublished data). Nasal colonization was identified in squirrels, horses, dogs, and cats. Most of the animals sampled had no signs of systemic infection as determined by postmortem examination or cryptococcal antigen testing. This suggested environmental exposure with nasal colonization without overt cryptococcal disease. Asymptomatic infection, defined as the presence of cryptococcal antigen in the bloodstream in the absence of clinical symptoms, was identified in a small number of dogs and cats. Fourteen months of follow-up testing of asymptomatic animals revealed that animals could progress to clinical disease, remain subclinically infected, or clear the organism (21).

Animal cryptococcosis due to *C. gattii* is a nonregulated disease in Canada. Reports of new animal cases have not been actively sought since 2004, but veterinarians and laboratories have communicated cases that emerged in new areas, occurred in new species, or had a unique presentation. This information provided direction to researchers investigating the ecology of the organism and allowed the veterinary and medical communities to make comparisons. For example, it was found that there were almost 75% more animal cases than human cases between 1999 and 2003 even though it was hypothesized that animal cases are more likely to go undiagnosed or unreported when compared to humans. Animal cryptococcosis cases were identified on Vancouver Island prior to 1999, suggesting that the organism may have emerged in the region prior to its identification as a causative agent for human disease. In 2003, animal cases also predated new, non–travel-related human cases of cryptococcosis on the British Columbia mainland. This information suggests that animals, by virtue of increased case counts and, potentially, earlier onset of disease, may serve as good sentinels for human cryptococcosis (45).

ENVIRONMENTAL STUDIES AND CHARACTERIZATION OF *C. GATTII* ISOLATES

Molecular typing of *C. gattii* environmental isolates from Vancouver Island using PCR fingerprinting and/or restriction fragment length polymorphism methodologies revealed that the majority of isolates belonged to the VGII molecular type (where two subtypes have been observed among the Vancouver Island population, designated VGIIa and VGIIb), and a small number belonged to the VGI molecular type (34, 36). The VGIIa, VGIIb, and VGI types exist in proportions that are consistent with their involvement with human *C. gattii* infection in the region; for example, a survey of first-attempt samples (i.e., excluding replicates) indicated frequencies of 84%, 12%, and 4%, respectively (36). The prevalence of VGIIa could to be due to adaptation through sexual recombination, possibly occurring through α-α mating (27), allowing the fungus to thrive in a climate that was, until the turn of the millennium, considered an unlikely habitat for *C. gattii* (39).

In most sampled areas of Vancouver Island, VGIIa and VGIIb were codistributed and coisolated, suggesting concomitant introduction, concomitant dispersal, or synergistic colonization. Interestingly, at one specific sampling site, VGIIa, VGIIb, VGI, and a *C. neoformans* serotype AD hybrid strain were all routinely isolated (and often coisolated) over a 6-year period (36). At this sampling site, the genotype frequencies appear to be less skewed toward VGIIa, and the population dynamics could be very complex (S. Kidd and K. Bartlett, unpublished data). The heavy and sustained colonization with such a diverse *C. gattii* population at this particular site is an important finding in light of speculation about the possible points of introduction of *C. gattii* into British Columbia.

Comparisons of VGII isolates from British Columbia to those in other areas of the world indicated a number of similarities, suggestive of potential historical origins of these strains. Isolates with multilocus sequence typing (MLST) profiles identical to the British Columbia VGIIa genotype have been identified among an archived human clinical isolate from Seattle, Washington (NIH444, ca. 1970), and an environmental isolate from San Francisco, California (CBS7750, ca. 1992); a clinical isolate from Brazil (ICB107) that is related to the VGIIa genotype but differs at one of 30 MLST markers has also been identified (27, 37). Similarly, isolates with MLST profiles matching that of the Vancouver Island VGIIb genotype at 30 of 30 MLST loci have been identified among a fertile, interbreeding population of Australian isolates, and a related isolate has also been reported from a Thai collection (27, 37, 38). Recent characterization of clinical and environmental isolates indicate that the VGII genotype is prevalent in South America, although MLST analysis has not yet been performed to examine subtypes (17, 26, 58). Several theories have been put forward as to how and when the VGIIa and VGIIb genotypes became established in British Columbia, including dispersal via migratory birds and sea mammals, imported eucalypts, and ocean currents (27, 50).

More recently, VGIIa isolates have been identified among human and veterinary cases as well as environmental sources on mainland British Columbia, other British Columbia islands, and the states of Oregon and Washington (4, 45, 60; K. Bartlett, unpublished data; see below). Dispersal of *C. gattii* is consistent with the recovery of the fungus from air and water sources (fresh- and seawater) on Vancouver Island (4, 35, 36). Systematic sampling strategies revealed high concentrations of *C. gattii* in areas of human activity and recreation, while no *C. gattii* was recovered in isolated areas with little or no human activity, indicating the potential of dispersal and spread through human action. Mechanisms for human-mediated dispersal include forestry, landscaping, and construction activities leading to soil disturbances and aerosolization of *C. gattii* propagules and mechanical transfer on footwear and vehicles (35). Passive transfer of *C. gattii* on animals is also likely.

MLST studies also indicated that the VGIIa and VGIIb subtypes from British Columbia each represent a genotypically homogeneous group, while VGI appears to be more heterogeneous (27, 37). Although the VGIIa genotype is theorized to have originated through sexual recombination (27), there is no evidence for ongoing recombination events (e.g., between VGIIa and VGIIb), given that there have to date been no reported examples of isolates possessing genetic material from prospective parent strains from different lineages. However, there may be mating occurring within these lineages that is undetectable by MLST analyses (e.g., VGIIa mating with other VGIIa isolates). A diploid isolate has been reported in which the VGIIa genome appears to have been duplicated in an identical fashion, based on 30-marker MLST analysis, which may represent an intermediate in same-sex mating (27).

Approximately 8% of culture-confirmed human infections and a number of animal infections have been caused by isolates belonging to the VGI molecular type (34, 37; S. Kidd, unpublished data). The significance of the small number of these VGI infections remains unclear, despite the identification of environmental VGI isolates from a single sampling site on Vancouver Island as well as on an adjacent Gulf Island. The VGI isolates from this region appear to possess comparatively more genetic diversity than VGII isolates, as determined by MLST studies, and no strong epidemiological link has been made between human or animal cases and potential environmental sources of infection (34). For example, a VGI isolate obtained from a wild porpoise found dead on the shore of Vancouver Island shared an MLST genotype with soil and tree isolates from Vancouver Island, and an isolate obtained from the wheel well of a Vancouver Island–based motor vehicle shared another VGI MLST genotype with isolates from several human cases (S. Kidd and K. Bartlett, unpublished data). These links are somewhat tenuous given the potential travel histories of the vehicle, porpoise, and humans. Indeed, it is possible that some of the reported VGI infections were acquired through travel to other regions: as discussed below, a VGI infection of a man from North Carolina was recently reported, with a travel history to California but not British Columbia (4). However, sampling attempts in North Carolina and California have so far failed to isolate VGI strains (6). Further characterization of *C. gattii* isolates from North America and from global collections will be required to clarify the genetic diversity, population structure, and virulence of VGI isolates in a broader context. This characterization will help to clarify differences among the at least 20 sequence types that have been identified so far by MLST analysis (4).

EPIDEMIOLOGICAL AND CLINICAL ASPECTS OF *C. GATTII* ON VANCOUVER ISLAND AND THE MAINLAND OF BRITISH COLUMBIA

Emergence of *C. gattii* on Vancouver Island

Following the recent period of emergence of C. *gattii* as an important source of morbidity and mortality for Vancouver Island residents and visitors, the area is now recognized as having one of the highest incidence rates of C. *gattii* infections globally (12, 30, 31). While it cannot be determined when C. *gattii* initially established an environmental presence on Vancouver Island, there is little doubt that cryptococcal illnesses caused by the fungus first appeared in the late 1990s and have continued to occur since that time. As described in the following paragraphs, the evidence for this is fourfold.

First, in 2001 a veterinary laboratory in British Columbia identified an increase in cryptococcal infections in companion animals from Vancouver Island (56; see above). This laboratory was the major supplier of diagnostic services for veterinarians in British Columbia. Over the course of approximately 1 year, this laboratory used standard histopathology methods to identify *Cryptococcus* as the etiologic agent responsible for the illness or death of 12 cats and dogs and one ferret that resided on Vancouver Island. This was a considerable increase over the number of cryptococcal infections historically identified by this laboratory, without changes in methods used to identify pathogens. During this time frame, there was no concurrent increase in cryptococcal infections among mainland British Columbia companion animals. These findings were reported to public health officials (R. Lewis, M. Fyfe, and S. Lester, personal communication).

Second, laboratory physicians on Vancouver Island recognized an increasing number of cryptococcal infections occurring among humans. This observation was completely independent of the discovery by veterinarians. The increase appeared to start in 1999, and 27 cases of cryptococcosis were identified by the end of 2001. Prior to 1999, two or fewer cases per year were diagnosed on Vancouver Island. The patients since 1999 had presented mostly with pulmonary nodules and, in some cases, CNS infections (30). Diagnosis was made through standard histopathology, India ink staining of cerebrospinal fluid, and, where possible, fungal culture. The patients ranged in age from 20 to 76 years and, unexpectedly, none had HIV infection or other evidence of immunocompromising conditions. Laboratory physicians reported this finding to public health officials on Vancouver Island in 2001.

Third, a review of anonymous hospitalization data requested from the British Columbia Ministry of Health showed an increase in cryptococcal disease in HIV-negative individuals on Vancouver Island from 1995 to 2004 (30). The analysis of hospital discharges was restricted to HIV-negative individuals, given the limitation that international classification of disease codes do not distinguish between C. *neoformans* and C. *gattii* infection. This review identified a marked increase in cryptococcosis coded on hospital discharges beginning in 1999 on Vancouver Island (Color Plate 5A). A similar increase in cases was not seen at British Columbia mainland hospitals at that time. In fact, the major hospital providing care to patients with AIDS in Vancouver had seen a reduced number of cryptococcal infections over the preceding 5 years, coincident with institution of highly active antiretroviral therapy in the mid-1990s (M. Fyfe, unpublished data).

Finally, a review of all available clinical isolates from historic human cryptococcal cases in British Columbia completed by the Vancouver Island Health Authority and the British Columbia Centre for Disease Control (BCCDC) showed increased cases caused by C. *gattii*. Thirty-six stored isolates were available from 1995 to 2001. Since veterinary cases were diagnosed through histopathology but not culture, isolates from animals were not available. Serotyping of the clinical isolates was completed using CryptoCheck® (Iatron, Japan). Three of the 36 isolates were C. *gattii* (serotype B), with the remainder being serotype A or D (30). The serotype B isolates were all from cases that resided on Vancouver Island, and the earliest serotype B strain was from a case that presented in January 1999. Together, this information provided evidence that C. *gattii* began to cause illness in humans and animals on Vancouver Island around 1999.

Surveillance and Laboratory Diagnosis of *C. gattii* in British Columbia

Because of the unusual finding of clinical isolates that were serotype B in the face of an increase in the occurrence of cryptococcosis on Vancouver Island, public health officials initiated surveillance for the pathogen. The goals of surveillance were (i) to determine the extent of morbidity and mortality associated with C. *gattii* infections on Vancouver Island, (ii) to determine the geographic extent of infections, and (iii) to identify risk factors and environmental exposures that may be associated with C. *gattii* infection. Beginning in 2003, cryptococcal infections were made reportable in British Columbia, and physicians and laboratories were required to report cases that they identified to the medical health officer in their area. It was recognized that, for a proportion of cases, diagnosis of cryptococcosis would not be made through culture and isolates would not be available to determine serotype or genotype. A case definition therefore included those individuals diagnosed through histopathology, antigen latex agglutination, and microscopy who were not infected with HIV. Minor changes were made to the surveillance case definition over time as new knowledge became available (Table 1). It was felt that, in the absence of confirmed serotype information, cryptococcosis cases in patients with AIDS were more likely to be caused by C. *neoformans* than C. *gattii*; recognizing that some cases of C. *gattii* infection have now been diagnosed in people with HIV infection as a primary risk, the data generated may represent an underestimate of disease burden.

TABLE 1 *Cryptococcus gattii* case definitions in British Columbia as of December 2008

Case	Definition
Confirmed	British Columbia resident AND Culture-confirmed C. *gattii* infection
Probable	British Columbia resident AND HIV negative AND Lives in or visited a local or international area of endemicity AND Laboratory evidence of infection: • Positive latex agglutination test (1:8) AND symptoms compatible with C. *gattii* infection OR • Positive histopathology OR • Positive microscopy OR • Culture-confirmed *Cryptococcus* without information on species

Patients were interviewed with a standardized, enhanced surveillance questionnaire. Information collected included demographic variables, medical history, smoking status, clinical symptoms, occupation, travel history, recreational activities, gardening and landscaping activities, construction activities, and exposure to botanical gardens, zoos, aquariums, agriculture, animals, compost, bark mulch, various tree species, and wooded areas. Patients were asked to consider activities in the 3 months prior to onset of symptoms. Over the ensuing years, as new knowledge was gained on risk factors and the incubation period, the questionnaire was shortened and the time frame for exposure was expanded to 12 months prior to onset. Until 2006, physicians caring for patients were also interviewed using a standard questionnaire. Information collected included clinical presentation, underlying medical conditions, radiological and laboratory findings, and treatment. Overall, these procedures resulted in a more thorough surveillance of *C. gattii* cases in British Columbia. In addition to enhanced surveillance methods, the BCCDC implemented a standard set of tests to confirm the diagnosis of *C. gattii*. The test menu is sample dependent (Table 2) and consists of microscopy, cryptococcal capsular antigen test, culture, PCR, and other molecular testing (http://www.mycology.adelaide.edu.au/Fungal_Descriptions/Yeasts/Cryptococcus/C_gattii.html). Although the species or varieties cannot be distinguished by microscopy, the use of routine fungal culture media and capsular antigen testing, coupled with a specialized medium such as canavanine glycine-bromothymol blue agar provides preliminary identification of *C. gattii* (see below).

Cryptococcal cells are generally visualized microscopically with both calcofluor white (fluorescence) and India ink stains, and the cells may be as large as 20 μm in diameter if the capsule is included (http://www.mycology.adelaide.edu.au/Fungal_Descriptions/Yeasts/Cryptococcus/C_gattii.html). Detection of cryptococcal capsular antigen uses latex particles coated with polyclonal cryptococcal capsular antibodies from a commercial kit (Immuno-Mycologics, Inc., Oklahoma). This highly sensitive (93 to 100%) test is primarily done on serum (57). However, when performed on cerebrospinal fluid, it is the method of choice for establishing a diagnosis of cryptococcal meningitis.

A number of solid culture media, biochemical tests, and stains can confirm the diagnosis of *Cryptococcus* to the genus level. Isolates are generally cultured on canavanine-glycine-bromothymol blue agar for the presumptive identification of *C. gattii* (32). *C gattii* turns the agar from yellow to deep blue in 2 to 5 days, based on its ability to grow in the presence of L-canavanine and to assimilate glycine as a sole carbon source. *C. neoformans* is sensitive to canavanine and cannot use glycine as its sole carbon source, resulting in no growth or coloration on this selective indicator medium.

The BCCDC uses genotyping for definitive characterization of cryptococcal isolates. This has replaced serology now that commercial antisera are no longer available. As described earlier, PCR fingerprinting and/or restriction fragment length polymorphism methodologies separate *C. gattii* into the VGI, VGII, VGIII, or VGIV molecular types. MLST, an unambiguous procedure for characterizing isolates based on the partial sequence of six genes (*CAP1*, *FTR1*, *IGS1*, *LAC1*, *PLB1*, and *URA5*), is also conducted at the BCCDC.

Currently, there are no published guidelines with interpretative breakpoints for susceptibility testing of *C. gattii* against antifungal agents. *C. gattii* isolates from British Columbia have been tested for susceptibility to six antifungal drugs (amphotericin B, 5-fluorocytosine, itraconazole, fluconazole, voriconazole, and caspofungin) by the microbroth dilution method (13). Preliminary data suggest that *C. gattii* isolates from British Columbia are susceptible to most of the antifungal drugs tested except caspofungin (L. Hoang and M. Morshed, unpublished data).

TABLE 2 *Cryptococcus gattii* diagnostic tests used in BCCDC laboratories

Preferred samples	Tests
Serum	*Cryptococcus* capsular antigen test
Cerebrospinal fluid	Microscopy-calcofluor white stain or India ink stain *Cryptococcus* capsular antigen test Culture
Respiratory samples	Microscopy-calcofluor white stain or India ink stain Culture
Tissue specimens	Microscopy-calcofluor white stain or India ink stain Culture
Culture isolate	Genotyping (PCR, restriction fragment length polymorphism, MLST, sequencing)

Surveillance Findings in British Columbia

Between 1999 and 2007, 218 cases of *C. gattii* infection were reported in British Columbia (5.8/million British Columbia population per year) (31; E. Galanis, L. MacDougall, M. Li, C. Marra, R. Mitchell, Y. Peng, M. Morshed, and L. Hoang, presented at the 7th International *Cryptoccocus* and Cryptococcosis Conference, Nagasaki, Japan, 11 to 14 September 2008). There was an annual average of 24.2 cases and a steady increase from 6 cases reported in 1999 to 38 in 2006 (Fig. 1). There was no seasonality in month of onset. Close to three-quarters (73.9%) of the cases lived on Vancouver Island, where the average annual incidence rate was 25.1/million population between 1999 and 2008. The number of cases reported per year reached a plateau on Vancouver Island in 2002, whereas it has increased on the mainland of British Columbia since 2005. Just over half (55.5%) of the cases were male (31). The mean age was 58.7 years, with a range of 2 to 92 years. Only four cases were less than 19 years old. The incidence rate increased with age, with the highest age-specific rate seen in the 70- to 79-year-old group.

Based on information provided through patient and physician interviews, upon diagnosis, 167 (76.6%) cases presented with pulmonary signs or symptoms, 17 (7.8%) with CNS findings, 22 (10.1%) with both, and 1 each with a combination of skin and respiratory, skin and CNS, and sepsis and respiratory presentations (31). Among those who presented with respiratory complaints exclusively, the most common symptoms reported were cough, shortness of breath, and chest pain. Among those who presented with CNS findings exclusively, the most common symptoms reported were headache and neck stiffness. Some patients presented with systemic complaints of night sweats, weight loss, and anorexia. Fifteen cases (6.9%) were asymptomatic. Among cases with available information, 98 patients (60.9%) were hospitalized. Among the 118 cases with abnormal chest radiographs, 89 (75.4%) had single or multiple pulmonary nodules. Among the 124 culture-confirmed cases (56.9%), the most common genotype was VGIIa, and this genotype was responsible for

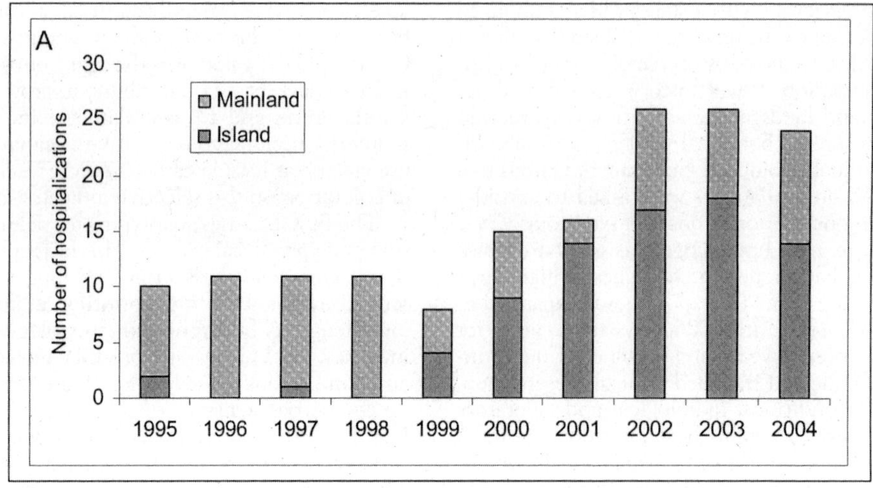

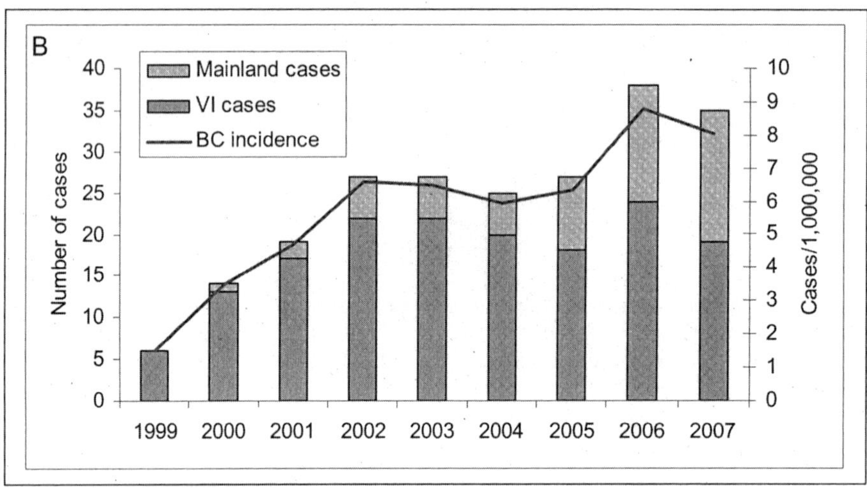

FIGURE 1 (A) Hospitalizations for cryptococcal infection in HIV-negative residents of British Columbia, 1995 to 2004. Reproduced with permission from reference 30. (B) Number of cases and incidence rate of *C. gattii* infection by place of residence, British Columbia, 1999 to 2007.

107 (86.3%) of the cases (31). Eight people were infected with VGI strains and nine with VGIIb. All isolates typed were serotype B (30; S. Mithani, personal communication). Based on data from the British Columbia Vital Statistics Agency, 19 patients died from or with *C. gattii* infection, for a case-fatality ratio of 8.7% (31). The *C. gattii*–specific mortality rate was 2.3/million on Vancouver Island and 0.5/million in all of British Columbia.

Place of Exposure

Between 1999 and 2004, the individuals affected by *C. gattii* had either lived within the Coastal Douglas Fir or xeric Coastal Western Hemlock biogeoclimatic zones on Vancouver Island or had traveled there recently (Color Plate 5). Environmental investigations during this period isolated the organism from trees native to this zone, while provincial sampling in off-island environments was negative (34; Bartlett et al., 16th Conference on Biometeorology & Aerobiology, Vancouver, British Columbia, 23 to 27 August 2004). Travel has played an important role in the understanding of the emergence of *C. gattii* disease. Travel histories of *C. gattii*–infected patients have been compared with travel patterns of the general public to validate and refine risk areas on Vancouver Island (10). A significantly greater proportion of visits to cities in central Vancouver Island was reported for patients than for Tourism British Columbia visitors, which concurs with *C. gattii* risk based on place of residence. When analysis was restricted to earlier (1999 to 2002) and later (2003 to 2006) time periods, no significant differences in results were found, suggesting that minimal spatial progression of risk areas has occurred over time.

Ecological Niche Modeling

Ecological niche modeling (also known as species distribution modeling) was employed to identify geographical areas in British Columbia with suitable environmental conditions to support the permanent colonization of *C. gattii* in the environment (47). The model identified "optimal" ecological niche areas of *C. gattii* along the central and southeastern coast of Vancouver Island, the Gulf Islands, and the southwestern coast of the British Columbia mainland (Color Plate 5C). These areas are characterized by low-lying elevations (below 770 m and averaging 100 m above sea level), above-freezing daily January average temperatures, and presence within the Coastal Douglas Fir and Coastal Western Hemlock biogeoclimatic zones. The association of daily January average temperatures above 0°C with the forecasted ecological niche areas of *C. gattii* and the general absence (or transiently positive nature) of *C. gattii* in geographic areas where the daily January average temperatures are below 0°C suggests that *C. gattii* is not able to survive in areas that commonly experience freezing temperatures.

Incubation Period

Previously, the incubation period remained largely ill defined, relying on a limited number of case reports in individuals returning from different areas of endemicity (2, 3, 59). Using British Columbia travelers to the endemic zone, an incubation period for *C. gattii* could be calculated by taking the difference between symptom onset and travel to Vancouver Island. Culture-confirmed off-island cases reported symptoms 2 to 11 months after exposure to Vancouver Island (median 6 to 7 months) (44). A subsequent case report of an international traveler to Vancouver Island suggested a shorter incubation period of 6 weeks, widening the accepted range to between 6 weeks and 11 months (41). Considerable variation in the length of incubation period, even among individuals exposed to the same environment, may reflect differences in host susceptibility, exposure dose, or variation in recognition of symptom onset.

Spread of the *C. gattii* Range to the British Columbia and U.S. Mainland

An understanding of the length of the incubation period is critical to the ability to track the spread of *C. gattii*. Given the above findings, detection of disease in people with no travel to areas of endemicity within a year of symptom onset may signal the emergence of the fungus in a new environment. Until late 2004, all human cases of *C. gattii* infection reported to the BCCDC were among those living on or traveling to Vancouver Island in the year before symptoms appeared. In December 2004, the first evidence of disease in humans without exposure to Vancouver Island or other known *C. gattii*–endemic areas was detected (45). As mentioned earlier, several culture-confirmed human cases have now been reported on the British Columbia mainland, and in Washington and Oregon, with molecular profiles identical to (British Columbia, Washington) or slightly different from (Oregon) Vancouver Island isolates by MLST (4, 45, 60; see below). From 2004 to 2007, seven *C. gattii* infections were believed to be acquired on the British Columbia mainland, and multiple cases in Washington and Oregon have been documented without travel to Vancouver Island (14, 15, 31). The emergence of disease in human populations in the Pacific Northwest is supported by the parallel emergence of *C. gattii* in animal populations on the British Columbia mainland and in Oregon and Washington and the identification of positive environmental samples in these areas (4, 45). While the number of cases reported per year in British Columbia has increased on the mainland since 2005, very few are thought to have been acquired on the mainland since they had also traveled to Vancouver Island. Rates of acquisition may be truly lower on the mainland, given significantly lower concentrations of the fungus in positive air samples in this environment compared with Vancouver Island or the Gulf Islands (45). To date, no point source (e.g., tree, soil) of the VGIIa isolates detected in air on the British Columbia mainland has been identified despite repeated sampling. As such, these results may suggest transient rather than permanent colonization of these areas.

Demographic, Medical, and Environmental Risk Factors

Apart from geographic exposures, a number of risk factors for cryptococcal infection have been identified (9). Most studies have not considered risk factors for *C. gattii* specifically, have often been conducted among specific subpopulations (e.g., HIV-positive patients), and have not compared the occurrence of risk factors in infected individuals to population controls. In a sex-matched case-control study of early *C. gattii* cases in British Columbia, cases were more likely than controls to report having ever been diagnosed by a physician with pneumonia (matched odds ratio [MOR], 2.71; 95% confidence interval [CI], 1.05 to 6.98) or other lung diseases (MOR, 3.21; 95% CI, 1.08 to 9.52), but not asthma. Cases were also more likely to have taken systemic corticosteroids (MOR, 8.11; 95% CI, 1.74 to 37.8) (L. MacDougall et al., manuscript submitted). Environmental exposures, such as chopping wood, pruning trees, and branch cleanup, were significantly less common among cases than controls, suggesting a protective aspect to these activities. *C. gattii* cases reported in British Columbia from 1999 to 2007 were compared with the general British Columbia population using existing data from population-based surveys and studies. Cases of *C. gattii* infection were more likely than the general population to be 50 years of age or older, be smokers ($P < 0.001$), be infected with HIV ($P < 0.001$), and/or have a history of cancer ($P < 0.001$) (MacDougall et al., manuscript submitted).

Cost and Quality of Life

A telephone survey of 35 patients diagnosed from 2000 to 2002 was conducted in 2004 to assess the costs and quality of life associated with *C. gattii* infection (Galanis et al., 7th International *Cryptococcus* and Cryptococcosis Conference, Nagasaki, Japan, 11 to 14 September 2008). Costs were estimated from the third-party payer (British Columbia Ministry of Health) perspective (including costs for physician visits, emergency department visits, days in hospital, and antifungal treatment) and the patient perspective (including loss of income and other patient-reported costs such as medications and transportation; www.health.gov.bc.ca). Assuming all cases received fluconazole at 400 mg/day for the duration reported or for 12 months and/or amphotericin B at 100 mg/day for the duration reported, total costs (including third-party payer and personal costs) ranged from $1,000 to $215,000 (Canadian) per patient. The mean quality-of-life (SF-6D) score was 0.73. This is significantly lower than the score for the general population at 0.83 ($P < 0.0001$) and for individuals with asthma (0.79) but higher than for those with rheumatoid arthritis (0.63) (48, 49).

Public Health Response

Based on the ubiquitous distribution of the pathogen in the Vancouver Island environment, it was recognized that *C. gattii* infections could not be prevented. The initial and continuing focus of the public health response was therefore to raise awareness among physicians and the public about the pathogen, clinical presentations, diagnostics, and treatment options to ensure early diagnosis and therapy. All physicians on Vancouver Island were notified through a newsletter sent in August 2001. Subsequent physician newsletters on Vancouver Island and throughout British Columbia provided updated information. Information was also transferred through educational rounds and conferences and through publications in local and peer-reviewed journals (10, 16, 30, 33–37, 44, 45, 56). Public notification initially took place through a media release in August 2001, followed by another in June 2005 on the spread of C. *gattii* to the British Columbia mainland, repeated media stories since that time, and information posted on Regional Health Authority websites. In addition, posters with information about C. *gattii* were posted at a provincial park in central Vancouver Island in the summer of 2002. This park was thought to be the site where a number of initial cases acquired their infections. The signs were removed when it became apparent that the fungus was more broadly distributed in the environment, primarily over 264 km along the eastern coast.

OREGON AND WASHINGTON: EPIDEMIOLOGY, EXPANSION, MOLECULAR CHARACTERIZATION, AND CLINICAL CHARACTERIZATION

Expansion of the *C. gattii* Vancouver Island Outbreak into the United States

Expansion of the C. *gattii* outbreak from Vancouver Island into the United States was first documented in humans in February 2006, at which time a confirmed case caused by the outbreak VGIIa/major genotype was identified in a 74-year-old male patient with leukemia from the Puget Sound region, Washington (60). The patient resided on Orcas in the San Juan Islands and had not traveled to Vancouver Island or elsewhere outside of the United States (60). The isolate was recovered from a 13-mm lingular nodule that was resected and cultured. This sentinel case of the Vancouver Island VGIIa genotype in the United States is geographically consistent with the continuing emergence from Vancouver Island onto the British Columbia mainland, including the identification of VGIIa in animal cases in Washington (45), and the first harbinger of expansion to the Pacific Northwest. This isolate has been extensively analyzed with MLST at 28 unlinked loci, showing it to be identical at all MLST loci with the VGIIa type strain R265 (60). While no contemporary human cases of the VGIIa outbreak genotype were reported prior to 2006 in the United States, two confirmed cases of C. *gattii* VGII molecular type were reported in Oregon in 2005. Each of the corresponding isolates had a distinct VGII genotype that diverged from the VGIIa genotype at one or four MLST loci, and thus their relationship to the Vancouver Island outbreak VGIIa/VGIIb genotypes, if any, is unknown (45).

In both of these cases, the patients had no travel history to Vancouver Island or other known endemic regions. Although neither case isolate was identical to the outbreak genotypes, one of the cases represents the index case for VGIIc (A6MR38), an emerging genotype thus far recovered only in Oregon, and the other (isolate KB11632) represents a unique genotype not seen in any other case to date (Table 3; K. Bartlett and S. Kidd, unpublished data). Based on retrospective chart reviews in two major Washington hospitals that revealed no C. *gattii* cases from 1999 though 2004, it appears that the sentinel Orcas Island case and possibly the Oregon cases were among the earliest cases of an emergent outbreak expansion rather than an underlying, and previously undiscovered, disease burden in the region. However, lack of case reporting and routine species identification in the United States clouds our understanding of the history of expansion.

TABLE 3 *Cryptococcus gattii* clinical isolates in Washington and Oregon: 2005 to 2008

Isolate	Host	Residence	Molecular type
T67707	Human	Washington	VGIIa/major
W15209	Human	Washington	VGIIa/major
EJB4	Human	Washington	VGIIa/major
EJB5	Human	Washington	VGIIa/major
EJB6	Human	Washington	VGIIa/major
EJB7	Human	Washington	VGIIa/major
EJB8	Human	Washington	VGIIa/major
EJB9	Human	Washington	VGIIa/major
EJB13	Human	Washington	VGIIa/major
KB11632	Human	Oregon	VGIIa/major[a]
EJB3	Human	Oregon	VGIIa/major
EJB19	Human	Oregon	VGIIa/major
MMC08-1042	Human	Oregon	VGIIa/major
EJB10	Human	Oregon	VGIIb/minor
A6MR38	Human	Oregon	VGIIc
EJB12	Human	Oregon	VGIIc
EJB18	Human	Oregon	VGIIc
EJB14	Cat	Oregon	VGIIc
EJB15	Alpaca	Oregon	VGIIc
EJB16	Alpaca	Oregon	VGIIa/major
EJB17	Dog	Oregon	VGIIa/major
3700(1)	Porpoise	Washington	VGIIa/major
3700(2)	Porpoise	Washington	VGIIa/major
3635	Porpoise	Washington	VGIIa/major
3059	Porpoise	Washington	VGIIa/major
EJB21	Porpoise	Oregon	VGIIa/major
EJB22	Dog	Oregon	VGIIa/major
MMC08-896	Dog	Oregon	VGIIb/minor
EJB11	Human	Washington	VGIII
MMC08-897	Cat	Oregon	VGIII

[a] The MLST profile of isolate KB11632 differs from the VGIIa major genotype at one of six loci (45).

Shortly after the initial expansion cases in the United States were reported, dozens of additional C. *gattii* illnesses in otherwise healthy and immunocompromised patients and animals were reported between 2006 and 2008 (4). Retrospective and prospective identification of human cases by informal communication has identified 21 human cases and 12 veterinary cases caused by microbiologically confirmed C. *gattii* as of February 2009 in Washington and Oregon (Table 3). Two-thirds of the cases occurred in males and one-third in females, with an age range of patients between 15 and 87 years old. The veterinary cases involve a range of domestic, wild, and agrarian animals, including harbor porpoises, alpacas, cats, and dogs (4, 5). The confirmed cases in the United States illustrate a clear expansion southward covering northern Washington to southern Oregon. This spread illustrates an increased risk for infection in both humans and animals, as well as an expansion of the native endemic zone for this emerging pathogenic fungus. Based on travel history, most of these cases were not likely to be the result of exposure on Vancouver Island or in Canada. As a consequence, *C. gattii* infections are increasing as an emerging infectious disease in the Pacific Northwest. Despite only two environmental isolates of VGIIa in northern Washington, this genotype has clearly expanded temporally and spatially from Vancouver Island to the mainland of British Columbia to the San Juan Islands to the Puget Sound in Washington, and is now generally distributed across Washington and into Oregon. As the endemic region expands, it is likely that infections will increase in number and geographic area and possibly expand to the neighboring states of Idaho, Montana, California, and Nevada.

In efforts to monitor this emergence in the United States and Canada, the scientific community has established the multidisciplinary *Cryptococcus gattii* Working Group of the Pacific Northwest (14). This group includes experts in the fields of medicine (infectious diseases), public health, epidemiology, environmental hygiene, veterinary medicine, and basic research, with representation from both the United States and Canada. To increase reporting and tracking of this emerging health threat, the U.S. Centers for Disease Control and Prevention in Atlanta, Georgia, has formed a *C. gattii* study group (S. Lockhart, personal communication). The enhancement of a multidisciplinary awareness of *C. gattii* infections in the region will encourage prompt and accurate diagnosis of patients and animals with compatible clinical symptoms (14). In addition, research efforts will help characterize the epidemiology and ecology of the organism as it possibly adapts to temperate climates, and basic research will enable a broader understanding of population structure and virulence of collected human, veterinary, and environmental isolates.

Molecular Characterization of *C. gattii* Outbreak Isolates in the United States

As described earlier, isolates of *C. gattii* are phenotypically and molecularly differentiated from the closely related species *C. neoformans* by several laboratory tests. For high-resolution MLST analysis, up to 30 unlinked loci can be examined (4, 27, 43, 46). MLST has the power to demonstrate that isolates are not clonally related when one or more loci differ. When all MLST loci analyzed are identical between two isolates, it indicates that they are closely related and may be clonal or differ at genomic regions not analyzed. This discriminatory power increases with analysis of more MLST loci, and in the future should also be extended to noncoding and more variable regions of the genome.

The data generated by MLST from different laboratories are directly comparable, and the approach can also establish mating type (α or **a**). To determine the mating type of each isolate, PCR and sequence analysis are used to detect mating-type-specific alleles of the *SXI* genes. Based on sequence analysis of the sex-specific *SXI1*α mating-type gene, as well as the absence of the *SXI2***a** gene, all 30 isolates in Washington and Oregon thus far analyzed are exclusively α, consistent with similar results from Vancouver Island and mainland British Columbia (4, 27, 29, 37).

Based on MLST analysis of the 30 isolates collected from humans and animals in the U.S. Pacific Northwest to date, only four molecular subtypes have been identified (4). Twenty-eight of the isolates are of the VGII type, which is the primary molecular type responsible for the emergence. Two others belong to the VGIII molecular type and are discussed further below. In Washington, 13 of the 14 isolates appear identical to the VGIIa genotype; the exception is a VGIII case in the southeastern region of the state (Table 3). With the exception of one clinical case from the Seattle metropolitan region, all VGII cases in Washington were located in the coastal Puget Sound region, in close proximity to Vancouver Island. The characterization of isolates from Washington correlates well with the epidemiologic and population data from Vancouver Island, which show that the majority of cases in the region were caused by the VGIIa genotype (31, 34, 37).

While isolates from Washington appear to be clonal or very closely related, the isolates from Oregon appear to be more diverse, although they are predominantly VGII (15/16). The 15 VGII isolates from Oregon group into three subtypes: VGIIa ($n = 8$), VGIIb ($n = 2$), and VGIIc ($n = 5$) (4). The VGIIb/minor genotype initially found on Vancouver Island is also found in both Australia and Thailand (7, 8, 37), while the VGIIc novel genotype is distinct from all other global VGII isolates analyzed thus far, although it shares some MLST alleles with both the VGIIa and VGIIb genotypes. The VGIIb/minor outbreak genotype has not yet been identified in Washington, indicating that there may be two independent foci (Vancouver Island and Oregon), a sampling bias that has not allowed isolation in Washington due to the low incidence, or that distribution occurred directly between Vancouver Island and Oregon. While the VGIIc group has only recently been categorized as a cohort of five cases, the index case of this genotype appears to be a clinical case from Oregon in 2005 that was initially reported as a divergent VGIIb/minor isolate (4, 45). The isolate from the other 2005 Oregon case appeared to be a variant from the VGIIa genotype (45). The VGIIc isolates are noteworthy because they have not yet been found on Vancouver Island, mainland British Columbia, Washington, or globally, but this novel genotype has been recovered from both patients and animals in Oregon.

Of the five VGIIc isolates, the MLST analysis was extended to cover a total of 15 unlinked loci, and all five isolates were identical across all loci (4). Phylogenetic analysis using concatenated sequences from the MLST profiles indicates that VGIIc may be more closely related to the VGIIa than to the VGIIb genotype. Previous studies show that the VGIIa genotype is highly virulent, while the VGIIb genotype is attenuated in murine virulence (27). Virulence of the VGIIc genotype is presently unknown, and infection experiments in mice and heterologous hosts are being conducted.

Based on genomic sequence analysis, the VGIIc genotype is unique despite sharing alleles with global VGII strains, including the VGIIa and VGIIb genotypes. The discovery of VGIIc, coupled with the recent cases of cryptococcal disease in humans and animals, marks the expansion of the outbreak and the emergence of a new VGII genotype in this region. The origins of the VGIIc genotype are unclear. It may have originated from outside the region via transfer from South America, Australia, or elsewhere; it may be the result of mating and recombination between strains in the Pacific Northwest; or it may be the result of mitotic expansion and mutation accumulation. Increased sampling both globally and in the Pacific Northwest may shed light on the origin of the novel VGIIc genotype.

The two isolates with the VGIII molecular type are distinct from VGII and therefore not directly related to the outbreak caused by VGIIa/VGIIb genotype isolates. One is from a patient in Washington, and the other from a cat in Oregon. The VGIII molecular type has yet to be identified in the environment on Vancouver Island or in Washington or Oregon, indicating that these cases could be the result of recent or past travel history and possible latency. The VGIII infection in the cat could be indicative of a local reservoir not yet identified. The two VGIII isolates have identical MLST profiles at eight loci with each other and with human and veterinary isolates from Australia, providing possible insight into their potential geographic origins (4, 27). The human isolate was recovered in southeastern Washington, distant from the outbreak expansion zone, and the veterinary isolate was collected from a cat in the vicinity of Eugene, Oregon. Many VGIII isolates harbor the uncommon C capsular serotype, and more commonly infect immunosuppressed patients, such as patients with AIDS (11, 27, 42).

While these VGIII cases are not part of the Vancouver Island outbreak expansion, their discovery suggests an increased regional incidence of C. *gattii* and highlights the need to molecularly classify clinical isolates of *Cryptococcus*. The evidence of VGIII in Washington and Oregon, together with evidence that ~5% of C. *gattii* infections in British Columbia are caused by VGI isolates, shows that in addition to the VGII outbreak, the VGI and VGIII types of C. *gattii* may also be colonizing the Pacific Northwest, but further sampling will be necessary to determine the environmental source of these infections. We note that one additional veterinary case of VGIII was found in the Canadian series in an exotic animal (woolly tapir) that had traveled extensively through various zoos in the United States prior to coming to British Columbia (S. Kidd, K. Bartlett, and S. Raverty, unpublished data), and again, this supports the hypothesis that at least in some cases the incidence of VGIII is attributable to acquisition in other geographic locations in which this molecular type is endemic, such as southern California (11).

While the reported incidence in Washington and Oregon appears to be increasing in both humans and animals, environmental sources of C. *gattii* remain elusive in these regions. In September of 2004, 397 diverse environmental samples were collected from the San Juan Islands, Washington, with no positive results for C. *gattii* (28). In 2005, sampling was conducted in Washington and Oregon in regions near the residences of human and animal cases; out of 274 samples, only 2 positive samples were identified, and both originated from northern Washington at a site within 10 km of the Canadian border (45). Of the two positive samples (from the same park in Washington), both were reported to belong to the VGIIa genotype based on six-marker MLST analysis (45). In November 2007 and February 2008, additional environmental sampling was conducted in Oregon. During this period a total of 191 samples were collected from a broad range of areas, including regions where human and veterinary cases were likely acquired, with no positive samples found (E. Byrnes, T. Mitchell, and J. Heitman, unpublished data). Sampling during October 2007 in San Francisco also yielded no C. *gattii* from 70 samples collected from red and blue gum *Eucalyptus* trees (E. Byrnes and J. Heitman, unpublished data). While accumulating cases in this region indicate an emergence, increased environmental sampling, with a particular emphasis in Oregon (where no isolates have been obtained thus far), remains a high priority.

Clinical Features of *C. gattii* Infections in the Pacific Northwest of the United States

Retrospective identification of cases, and ongoing prospective surveillance, has been initiated by investigators in Oregon, and has allowed for description of clinical features in human cases identified in Washington and Oregon. In this study, approved by the Oregon Health and Sciences University Institutional Review Board, and preliminarily presented in abstract form (S. K. West, E. J. Byrnes, III, S. Mostad, R. Thompson, R. Barnes, C. Beiser, J. Heitman, and K. A. Marr, presented at the 48th Annual ICAAC/IDSA 46th Annual Meeting, Washington, DC, 25 to 28 October 2008), cases were identified, and case report forms were completed to detail characteristics of hosts and progression of disease. Table 4 summarizes host and infection characteristics among the 21 Oregon cases identified to date. There are a number of salient features that are worthy of note. First, few people had a history of travel to the endemic regions in British Columbia, supporting the expansion of the epidemic. Only approximately half of the cases recognized have occurred in people who have some degree of immunosuppression, and only one case has been recognized among patients known to have HIV infection. Presenting findings included the lung in the majority of cases, with CNS involvement documented in approximately half. Outcomes, in general, were poor, with delayed diagnosis, prolonged hospitalization, and prolonged receipt of antifungal therapy being the norm.

TABLE 4 Summary data for cryptococcosis cases caused by *Cryptococcus gattii* in Oregon

Variable	No. of cases ($n = 21$)
Age (range)	51 (15–87) years
Gender	
Male	14 (67%)
Female	7 (33%)
Travel to British Columbia	3 (14%)
Travel to Vancouver Island	0
Underlying disease of host	
Cancer	4 (19%)
Organ transplant	3 (14%)
Connective tissue disease	3 (14%)
Chronic lung disease	1 (5%)
HIV/AIDS	1 (5%)
None identified	9 (43%)
Presenting signs or symptoms	
Headache	8 (38%)
Cough	7 (33%)
Shortness of breath	5 (24%)
Pleurisy	4 (19%)
Altered mental status	6 (28%)
Asymptomatic pulmonary nodule	3 (14%)
Pulmonary findings ($n = 15$)	
Cryptococcoma	7 (47%)
Nodular pneumonia	2 (13%)
Consolidation	5 (24%)
CNS finding ($n = 10$)	
Meningoencephalitis	9 (90%)
Increased intracranial pressure	5 (50%)
Drain required	4 (40%)
Cryptococcomas	1 (10%)
Parameningeal focus	2 (20%)
Outcomes	
Died	6 (28%)
Delay in diagnosis >1 week[a]	13 (65%)
Hospitalization >11 days[b]	15 (83%)
Prolonged therapy with antifungal other than fluconazole[c]	13 (87%)

[a] Of 20 patients with diagnosis premortem.
[b] Of 18 patients who required hospitalization.
[c] Of 15 patients who survived.

Clinical features of infection with *C. gattii*, with comparison to *C. neoformans*, are discussed in more depth elsewhere in this book (see chapter 44).

One potentially important finding in this recent study is that many of the *C. gattii* isolates recovered in this case series exhibited relatively high MICs to azoles used commonly to treat this infection. Of 11 isolates recovered in culture and tested for sensitivities in vitro, 5 had fluconazole MICs of >4 μg/ml. Whether this relative lack of in vitro susceptibility is associated with poor clinical outcomes and relative sensitivity to other drugs awaits further study; however, one can conclude that the historical treatment recommendations of providing fluconazole only for "pulmonary cryptococcosis" in non-HIV-infected hosts may not be adequate for known infection with these variants of *C. gattii*. Given the results of preliminary drug sensitivity tests for the British Columbia isolates, it is clear that a broader comparison of all of the Pacific Northwest isolates is needed.

SUMMARY

The emergence of cryptococcosis in the Pacific Northwest is unprecedented because of the large number of cases in individuals who do not have underlying HIV/AIDS. As outlined in this chapter, several troubling or worrisome aspects of the outbreak are worthy of further investigation. For example, a substantial number of cases occurred in companion pets and in more unusual animals such as porpoises and cockatoos. These animals play an important role as sentinels of environmental exposure, and further sampling of wild species may shed light on mechanisms of local and long-range spread of *C. gattii*. The fact that the majority (~90%) of isolates recovered from patients, animals, and the environment are represented by a single genotype, VGIIa, highlights the importance of what may be a predominantly clonal, hypervirulent isolate to the outbreak and its expansion. Given that a clinical isolate from Seattle in the 1970s appears genotypically identical to the current outbreak isolates, the organism may have been present in the Pacific Northwest for almost 30 years before the outbreak began. How it emerged to cause the outbreak may have involved dispersal, climate change, or sexual reproduction. Although the majority of the outbreak isolates are VGIIa, the presence of multiple other genotypes of *C. gattii* in environmental, clinical, and veterinary samples at a lower proportion is also likely to be important and may provide insights into how the species was introduced into the Pacific Northwest and possibly also from where.

There is clear interest in understanding the worldwide distribution of all of the *C. gattii* genotypes. Furthermore, the recent discovery of two VGIII clinical and veterinary isolates in Washington and Oregon reinforces the need for studies to examine molecular types to ascertain which cases are directly attributable to the VGII isolates causing the outbreak and raises the questions of whether this rare molecular type may have a local environmental reservoir and, if so, its relationship with host exposure. Along these lines, there is a pressing need to understand potential underlying causes of human susceptibility given likely widespread exposure, especially on Vancouver Island. The ongoing, detailed investigation of the outbreak in the Pacific Northwest should also help develop an understanding of climatic and environmental factors that may identify other human and animal populations that are at risk. Finally, the focused attention on *C. gattii* resulting from the outbreak should lead to an appreciation of the primary pathogenic nature of *C. gattii* and comparisons with *C. neoformans* in terms of interactions with host immune systems.

The authors thank the following individuals and groups for their contributions to the information summarized in this chapter: Sultana Mithani, Min-Kuang Lee, Marc Romney, Marty Pearce, Mike Starr, Pamela Kibsey, Louise Stein, Ron Lewis, Tania Sorrell, David Ellis, Sally Lester, William Black, Stephen Raverty, the environmental health officers in British Columbia, the state public health officials in Oregon and Washington, as well as the physicians and medical microbiologists in British Columbia who contributed to the collection of isolates and case information.

REFERENCES

1. **Bartlett, K. H., S. E. Kidd, and J. W. Kronstad.** 2007. The emergence of *Cryptococcus gattii* in British Columbia and the Pacific Northwest. *Curr. Fungal Infect. Rep.* **1:**108–115.
2. **Bodasing, N., R. A. Seaton, G. S. Shankland, and D. Kennedy.** 2004. *Cryptococcus neoformans* var. *gattii* meningitis in an HIV-positive patient: first observation in the United Kingdom. *J. Infect.* **49:**253–255.
3. **Bottone, E. J., P. A. Kirschner, and I. F. Salkin.** 1986. Isolation of highly encapsulated *Cryptococcus neoformans* serotype B from a patient in New York City. *J. Clin. Microbiol.* **23:**186–188.
4. **Byrnes, E. J., 3rd, R. J. Bildfell, P. L. Dearing, B. A. Valentine, and J. Heitman.** 2009. *Cryptococcus gattii* with bimorphic colony types in a dog in western Oregon: additional evidence for expansion of the Vancouver Island outbreak. *J. Vet. Diagn. Invest.* **21:**133–136.
5. **Byrnes, E. J., 3rd·, R. Bildfell, S. A. Frank, T. G. Mitchell, K. A. Marr, and J. Heitman.** 2009. Molecular evidence that the Vancouver Island *Cryptococcus gattii* outbreak has expanded into the United States Pacific Northwest. *J. Infect. Dis.* **199:**1081–1086.
6. **Byrnes, E. J., 3rd, W. Li, Y. Lewit, J. R. Perfect, D. A. Carter, G. M. Cox, and J. Heitman.** 2009. First reported clinical case of *Cryptococcus gattii* in the Southeastern USA: implications for travel-associated acquisition of an emerging fungal pathogen. *PLoS One* **4:**e5851.
7. **Campbell, L. T., B. J. Currie, M. Krockenberger, R. Malik, W. Meyer, J. Heitman, and D. Carter.** 2005. Clonality and recombination in genetically differentiated subgroups of *Cryptococcus gattii*. *Eukaryot. Cell.* **4:**1403–1409.
8. **Campbell, L. T., J. A. Fraser, C. B. Nichols, F. S. Dietrich, D. Carter, and J. Heitman.** 2005. Clinical and environmental isolates of *Cryptococcus gattii* from Australia that retain sexual fecundity. *Eukaryot. Cell.* **4:**1410–1419.
9. **Casadevall, A., and J. R. Perfect.** 1998. *Cryptococcus neoformans*, p. 357–363. ASM Press, Washington, DC.
10. **Chambers, C., L. MacDougall, M. Li, and E. Galanis.** 2008. Tourism and specific risk areas for *Cryptococcus gattii*, Vancouver Island, Canada. *Emerg. Infect. Dis.* **14:**1781–1783.
11. **Chaturvedi, S., M. Dyavaiah, R. A. Larsen, and V. Chaturvedi.** 2005. *Cryptococcus gattii* in AIDS patients, southern California. *Emerg. Infect. Dis.* **11:**1686–1692.
12. **Chen, S., T. Sorrell, G. Nimmo, B. Speed, B. Currie, D. Ellis, D. Marriott, T. Pfeiffer, D. Parr, and K. Byth.** 2000. Epidemiology and host- and variety-dependent characteristics of infection due to *Cryptococcus neoformans* in Australia and New Zealand. Australasian Cryptococcal Study Group. *Clin Infect Dis.* **31:**499–508.
13. **Clinical and Laboratory Standards Institute.** 2002. *Reference Method for Broth Dilution Antifungal Susceptibility Testing of Filamentous Fungi; Approved Standard.* NCCLS document M38-A. NCCLS, Wayne, PA.
14. **Datta, K., K. Bartlett, R. Baer, E. Byrnes, E. Galanis, J. Heitman, L. Hoang, M. J. Leslie, L. MacDougall,**

S. S. Magill, M. G. Morshed, K. A. Marr, and the *Cryptococcus gattii* Working Group of the Pacific Northwest. 2009. Spread of *Cryptococcus gattii* into Pacific Northwest region of the United States. *Emerg Infect Dis.* **15:**1185–1191.

15. **Datta, K., K. H. Bartlett, and K. A. Marr.** 2009. *Cryptococcus gattii* emergence in Western North America: exploitation of a novel ecological niche. *Interdiscip. Perspect. Infect. Dis.* Article ID 176532. doi: 10.1155/2009/176532.
16. **Dewar, G. J., and K. J. Kelly.** 2008. *Cryptococcus gattii*: an emerging cause of pulmonary nodules. *Can. Respir. J.* **15:**153–157.
17. **dos Santos, W. R., W. Meyer, B. Wanke, S. P. Costa, L. Trilles, J. L. Nascimento, R. Medeiros, B. P. Morales, C. C. Bezerra, R. C. de Macêdo, S. O. Ferreira, G. G. Barbosa, M. A. Perez, M. M. Nishikawa, and M. S. Lazéra.** 2008. Primary endemic *Cryptococcosis gattii* by molecular type VGII in the state of Pará, Brazil. *Mem Inst Oswaldo Cruz.* **103:**813–818.
18. **Duncan, C., H. Schwantje, C. Stephen, J. Campbell, and K. Bartlett.** 2006. *Cryptococcus gattii* in wildlife of Vancouver Island, British Columbia, Canada. *J. Wildl. Dis.* **42:**175–178.
19. **Duncan, C., C. Stephen, and J. Campbell.** 2006. Clinical characteristics and predictors of mortality for *Cryptococcus gattii* infection in dogs and cats of southwestern British Columbia. *Can. Vet. J.* **47:**993–998.
20. **Duncan, C., C. Stephen, S. Lester, and K. H. Bartlett.** 2005. Sub-clinical infection and asymptomatic carriage of *Cryptococcus gattii* in dogs and cats during an outbreak of cryptococcosis. *Med. Mycol.* **43:**511–516.
21. **Duncan, C., C. Stephen, S. Lester, and K. H. Bartlett.** 2005. Follow-up study of dogs and cats with asymptomatic *Cryptococcus gattii* infection or nasal colonization. *Med. Mycol.* **43:**663–666.
22. **Duncan, C. G., C. Stephen, and J. Campbell.** 2006. Evaluation of risk factors for *Cryptococcus gattii* infection in dogs and cats. *J. Am. Vet. Med. Assoc.* **228:**377–382.
23. **Ellis, D., and T. Pfeiffer.** 1992. The ecology of *Cryptococcus neoformans*. *Eur J. Epidemiol.* **8:**321–325.
24. **Ellis, D. H., and T. J. Pfeiffer.** 1990. Ecology, life cycle and infectious propagule of *Cryptococcus neoformans*. *Lancet* **336:**923–925.
25. **Ellis, D. H., and T. J. Pfeiffer.** 1990. Natural habitat of *Cryptococcus neoformans* var. *gattii*. *J. Clin. Microbiol.* **28:**1642–1644.
26. **Escandón, P., A. Sánchez, M. Martínez, W. Meyer, and E. Castañeda.** 2006. Molecular epidemiology of clinical and environmental isolates of the *Cryptococcus neoformans* species complex reveals a high genetic diversity and the presence of the molecular type VGII mating type a in Colombia. *FEMS Yeast Res.* **6:**625–635.
27. **Fraser, J. A., S. S. Giles, E. C. Wenink, S. G. Geunes-Boyer, J. R. Wright, S. Diezmann, A. Allen, J. E. Stajich, F. S. Dietrich, J. R. Perfect, and J. Heitman.** 2005. Same-sex mating and the origin of the Vancouver Island *Cryptococcus gattii* outbreak. *Nature* **437:**1360–1364.
28. **Fraser, J. A., S. M. Lim, S. Diezmann, E. C. Wenink, C. G. Arndt, G. M. Cox, F. S. Dietrich, and J. Heitman.** 2006. Yeast diversity sampling on the San Juan Islands reveals no evidence for the spread of the Vancouver Island *Cryptococcus gattii* outbreak to this locale. *FEMS Yeast Res.* **6:**620–624.
29. **Fraser, J. A., R. L. Subaran, C. B. Nichols, and J. Heitman.** 2003. Recapitulation of the sexual cycle of the primary fungal pathogen *Cryptococcus neoformans* var. *gattii*: implications for an outbreak on Vancouver Island, Canada. *Eukaryot. Cell* **2:**1036–1045.
30. **Fyfe, M., L. MacDougall, M. Romney, M. Starr, M. Pearce, S. Mak, S. Mithani, and P. Kibsey.** 2008. *Cryptococcus gattii* infections on Vancouver Island, British Columbia, Canada: emergence of a tropical fungus in a temperate environment. *Can. Commun. Dis. Rep.* **34:**1–12.
31. **Galanis, E., and L. MacDougall.** 2010. Epidemiology of *Cryptococcus gattii*, British Columbia, Canada, 1999–2007. *Emerg. Infect. Dis.* **16:**251–257.
32. **Harrington, B., and G. Hageage.** 2003. Calcolfluor white: a review of its uses and applications in clinical mycology and parasitology. *Lab. Med.* **34:**361–367.
33. **Hoang, L. M. N., J. A. Maguire, P. Doyle, M. Fyfe, and D. L. Roscoe.** 2004. *Cryptococcus neoformans* infections at Vancouver Hospital and Health Sciences Centre (1997–2002): epidemiology, microbiology and histopathology. *J. Med. Microbiol.* **53:**935–940.
34. **Kidd, S., F. Hagen, M. Tscharke, M. Huynh, K. H. Bartlett, M. Fyfe, L. MacDougall, T. Boekhout, K. J. Kwon-Chung, and W. Meyer.** 2004. A rare genotype of *Cryptococcus gattii* caused the cryptococcosis outbreak on Vancouver Island (British Columbia, Canada). *Proc. Natl. Acad. Sci. USA* **101:**17258–17263.
35. **Kidd, S. E., P. J. Bach, A. O. Hingston, S. Mak, Y. Chow, L. MacDougall, J. W. Kronstad, and K. H. Bartlett.** 2007. *Cryptococcus gattii* dispersal mechanisms, British Columbia, Canada. *Emerg. Infect. Dis.* **13:**51–57.
36. **Kidd, S. E., Y. Chow, S. Mak, P. J. Bach, H. Chen, A. O. Hingston, J. W. Kronstad, and K. H. Bartlett.** 2007. Characterization of environmental sources of *Cryptococcus gattii* in British Columbia, Canada, and the Pacific Northwest. *Appl. Environ. Microbiol.* **73:**1433–1443.
37. **Kidd, S. E., H. Guo, K. H. Bartlett, J. W. Kronstad, and J. Xu.** 2005. Comparative gene genealogies indicate that two clonal lineages of *Cryptococcus gattii* in British Columbia resemble strains from other geographical areas. *Eukaryot. Cell* **4:**1629–1638.
38. **Kluger, E. K., H. K. Karaoglu, M. B. Krockenberger, P. K. Della Torre, W. Meyer, and R. Malik.** 2006. Recrudescent cryptococcosis, caused by *Cryptococcus gattii* (molecular type VGII), over a 13-year period in a Birman cat. *Med. Mycol.* **44:**561–566.
39. **Kwon-Chung, K. J., and J. E. Bennett.** 1984. Epidemiologic differences between the two varieties of *Cryptococcus neoformans*. *Am. J. Epidemiol.* **120:**123–130.
40. **Lester, S. J., N. J. Kowalewich, K. H. Bartlett, M. B. Krockenberger, T. M. Fairfax, and R. Malik.** 2004. Clinicopathologic features of an unusual outbreak of cryptococcosis in dogs, cats, ferrets, and a bird: 38 cases (January to July 2003). *J. Am. Vet. Med. Assoc.* **225:**1716–1722.
41. **Lindberg, J., F. Hagen, A. Laursen, J. Stenderup, and T. Boekhout.** 2007. *Cryptococcus gattii* risk for tourists visiting Vancouver Island, Canada. *Emerg. Infect. Dis.* **13:**178–179.
42. **Litvintseva, A. P., R. Thakur, L. B. Reller, and T. G. Mitchell.** 2005. Prevalence of clinical isolates of *Cryptococcus gattii* serotype C among patients with AIDS in sub-Saharan Africa. *J. Infect. Dis.* **192:**888–892.
43. **Litvintseva, A. P., R. Thakur, R. Vilgalys, and T. G. Mitchell.** 2006. Multilocus sequence typing reveals three genetic subpopulations of *Cryptococcus neoformans* var. *grubii* (serotype A), including a unique population in Botswana. *Genetics* **172:**2223–2238.
44. **MacDougall, L., and M. Fyfe.** 2006. Emergence of *Cryptococcus gattii* in a novel environment provides clues to its incubation period. *J. Clin. Microbiol.* **44:**1851–1852.
45. **MacDougall, L., S. E. Kidd, E. Galanis, S. Mak, M. J. Leslie, P. R. Cieslak, J. W. Kronstad, M. G. Morshed, and K. H. Bartlett.** 2007. Spread of *Cryptococcus gattii* in British Columbia, Canada, and detection in the Pacific Northwest, USA. *Emerg. Infect. Dis.* **13:**42–50.

46. **Maiden, M. C., J. A. Bygraves, E. Feil, G. Morelli, J. E. Russell, R. Urwin, Q. Zhang, J. Zhou, K. Zurth, D. A. Caugant, I. M. Feavers, M. Achtman, and B. G. Spratt.** 1998. Multilocus sequence typing: a portable approach to the identification of clones within populations of pathogenic microorganisms. *Proc. Natl. Acad. Sci. USA* **95:**3140–3145.
47. **Mak, S.** 2007. Ecological niche modeling of *Cryptococcus gattii* in British Columbia, Canada. MSc thesis. Department of Geography, University of British Columbia, Vancouver, Canada.
48. **Marra, C. A., J. M. Esdaile, D. Guh, J. A. Kopee, J. E. Brazier, B. E. Koehler, A. Chalmers, and A. H. Anis.** 2004. A comparison of four indirect methods of assessing utility values in rheumatoid arthritis. *Med. Care* **42:**1125–1131.
49. **McTaggart-Cowan, H. M., C. A. Marra, Y. Yang, J. E. Brazier, J. A. Kopec, J. M. FitzGerald, A. H. Anis, and L. D. Lynd.** 2008. The validity of generic and condition-specific preference-based instruments: the ability to discriminate asthma control status. *Qual. Life Res.* **17:**453–462.
50. **Ngamskulrungroj, P., T. C. Sorrell, A. Chindamporn, A. Chaiprasert, N. Poonwan, and W. Meyer.** 2008. Association between fertility and molecular sub-type of global isolates of *Cryptococcus gattii* molecular type VGII. *Med. Mycol.* **46:**665–673.
51. **Nicol, A.-M., C. Hurrell, W. McDowall, K. Bartlett, and N. Elmieh.** 2008. Communicating the risks of a new, emerging pathogen: the case of *Cryptococcus gattii*. *J. Risk Anal.* **28:**373–386.
52. **Pfeiffer, T., and D. Ellis.** 1991. Environmental isolation of *Cryptococcus neoformans* var. *gattii* from California. *J. Infect. Dis.* **163:**929–930.
53. **Pfeiffer, T. J., and D. H. Ellis.** 1992. Environmental isolation of *Cryptococcus neoformans* var. *gattii* from *Eucalyptus tereticornis*. *J. Med. Vet. Mycol.* **30:**407–408.
54. **Sorrell, T. C., A. G. Brownlee, P. Ruma, R. Malik, T. J. Pfeiffer, and D. H. Ellis.** 1996. Natural environmental sources of *Cryptococcus neoformans* var. *gattii*. *J. Clin. Microbiol.* **34:**1261–1263.
55. **Sorrell, T. C., S. C. Chen, P. Ruma, W. Meyer, T. J. Pfeiffer, D. H. Ellis, and A. G. Brownlee.** 1996. Concordance of clinical and environmental isolates of *Cryptococcus neoformans* var. *gattii* by random amplification of polymorphic DNA analysis and PCR fingerprinting. *J. Clin. Microbiol.* **34:**1253–1260.
56. **Stephen, C., S. Lester, W. Black, M. Fyfe, and S. Raverty.** 2002. Multispecies outbreak of cryptococcosis on southern Vancouver Island, British Columbia. *Can. Vet. J.* **43:**792–794.
57. **Tanner, D. C., M. P. Weinstein, B. Fedorciw, K. L. Joho, J. J. Thorpe, and L. Reller.** 1994. Comparison of commercial kits for detection of cryptococcal antigen. *J. Clin. Microbiol.* **32:**1680–1684.
58. **Trilles, L., M. S. Lazéra, B. Wanke, R. V. Oliveira, G. G. Barbosa, M. M. Nishikawa, B. P. Morales, and W. Meyer.** 2008. Regional pattern of the molecular types of *Cryptococcus neoformans* and *Cryptococcus gattii* in Brazil. *Mem. Inst. Oswaldo Cruz.* **103:**455–462.
59. **Tsunemi, T., T. Kamata, Y. Fumimura, M. Watanabe, M. Yamawaki, Y. Saito, T. Kanda, K. Ohashi, N. Suegara, S. Murayama, K. Makimura, H. Yamaguchi, and H. Mizusawa.** 2001. Immunohistochemical diagnosis of *Cryptococcus neoformans* var. *gattii* infection in chronic meningoencephalitis: the first case in Japan. *Int. Med.* **40:**1241–1244.
60. **Upton, A., J. A. Fraser, S. E. Kidd, C. Bretz, J. Heitman, K. H. Bartlett, S. Raverty, and K. A. Marr.** 2007. First contemporary case of human infection with *Cryptococcus gattii* in Puget Sound: evidence for spread of the Vancouver Island outbreak. *J. Clin. Microbiol.* **45:**3086–3088.

Cryptococcus: From Human Pathogen to Model Yeast
Edited by J. Heitman et al.

24

Molecular Typing of the *Cryptococcus neoformans*/ *Cryptococcus gattii* Species Complex

WIELAND MEYER, FELIX GILGADO, POPCHAI NGAMSKULRUNGROJ, LUCIANA TRILLES, FERRY HAGEN, ELIZABETH CASTAÑEDA, AND TEUN BOEKHOUT

Classification of strains within the *Cryptococcus neoformans*/*Cryptococcus gattii* species complex is still a matter of debate. Initially, the etiologic agent of cryptococcosis had been considered as a homogeneous anamorphic species, *C. neoformans*. In 1949 the existence of four serotypes, based on the antigenic properties of the polysaccharide capsule, revealed phenotypic diversity within the species (41). Subsequently this species was divided into two varieties: *C. neoformans* (Sanfelice) Vuillemin var. *neoformans* (108) and *C. neoformans* (Sanfelice) Vuillemin var. *gattii* Vanbreuseghem et Takashio (127). Based on morphological and biochemical differences, the anamorphic varieties have also been classified as two different species: *C. neoformans* and *Cryptococcus bacillisporus* (74) (Fig. 1). Mating experiments conducted by K. J. Kwon-Chung in the mid 1970s revealed two morphologically distinct teleomorphs: *Filobasidiella neoformans* Kwon-Chung (71), formed by strains of serotypes A and D, and *Filobasidiella bacillispora* Kwon-Chung (72), produced by strains of serotypes B and C. Currently, the etiological agent of cryptococcosis is classified into two closely related species (75, 76): *C. neoformans*, with two varieties (*C. neoformans* var. *grubii* [serotype A] [47] and *C. neoformans* var. *neoformans* [serotype D] [70], as well as an AD hybrid), and *C. gattii* (serotypes B and C) (75) (Fig. 1). Ensuing studies revealed numerous differences between the species and varieties with regard to their geographical distribution, ecological niches, epidemiology, pathobiology, and clinical and molecular characteristics. Intraspecies genetic diversity has been revealed as more genotyping methods have been applied to the two species and four serotypes. In addition, interspecies hybrids between serotypes AB and BD have been found (10, 13). As a result, the number of scientifically valid species within the *C. neoformans*/*C. gattii* species complex has become a controversial issue due to differing opinions among taxonomists as to the appropriate definition of a species.

Due to the importance of the *C. neoformans*/*C. gattii* species complex as human fungal pathogens, several research groups are currently focusing on the molecular determination of the number of genetically diverged subgroups within each species. A large number of molecular typing techniques has been applied over the years to discriminate between individual isolates that had been indistinguishable using conventional techniques (98) and to obtain further insights into the epidemiology and population structure of this species complex. The techniques employed to type strains and investigate the population genetic structure of the human/animal-pathogenic cryptococcal species include multilocus enzyme electrophoresis (MLEE) (4, 15); electrophoretic karyotyping (7, 104, 105, 137); Southern blot hybridization with DNA probes to repetitive DNA sequences (DNA fingerprinting) (32, 117, 129, 130); random amplification of polymorphic DNA (RAPD) (7, 8, 24, 93, 107, 115); PCR fingerprinting (20, 27, 38, 91–93, 96, 115, 133, 135); amplified fragment length polymorphism (AFLP) (6, 54, 85–88); restriction fragment length polymorphism (RFLP) of protein coding genes such as *PLB1* and *URA5* (78, 96, 131); sequencing of the internal transcribed spacer (ITS) region of the rRNA gene cluster including the 5.8S gene (64), the intergenic spacer (IGS) region (33, 34, 119), and more recently multigene sequence analysis (11, 102, 140); multilocus sequence typing (MLST) (45, 86, 88, 95); and multilocus microsatellite typing (MLMT) (56, 62, 63).

Despite this large number of typing tools applied to the *C. neoformans*/*C. gattii* species complex, our knowledge of the population structure is still very patchy, which is mainly due to the lack of a consensus between the results obtained by the different methods and the different nomenclatures used in the various typing studies. This chapter aims to summarize the diverse typing techniques applied to the *C. neoformans*/*C. gattii* species complex, to correlate the obtained results, and to describe the global distribution of the major genotypes.

Wieland Meyer, Felix Gilgado, and Popchai Ngamskulrungroj, Molecular Mycology Research Laboratory, Centre for Infectious Diseases and Microbiology, University of Sydney, Sydney Medical School - Westmead at Westmead Hospital, Westmead Millennium Institute, Westmead, NSW, 2145, Australia. **Luciana Trilles,** Mycology Laboratory, Instituto de Pesquisa Clínica Evandro Chagas (IPEC), Fundação Oswaldo Cruz (FIOCRUZ), 21040-900 Rio de Janeiro, Brazil. **Ferry Hagen and Teun Boekhout,** CBS-KNAW Fungal Diversity Centre, NL-3584CT Utrecht, The Netherlands. **Elizabeth Castañeda,** Emerita Investigator, Instituto Nacional de Salud, Bogotá, Colombia.

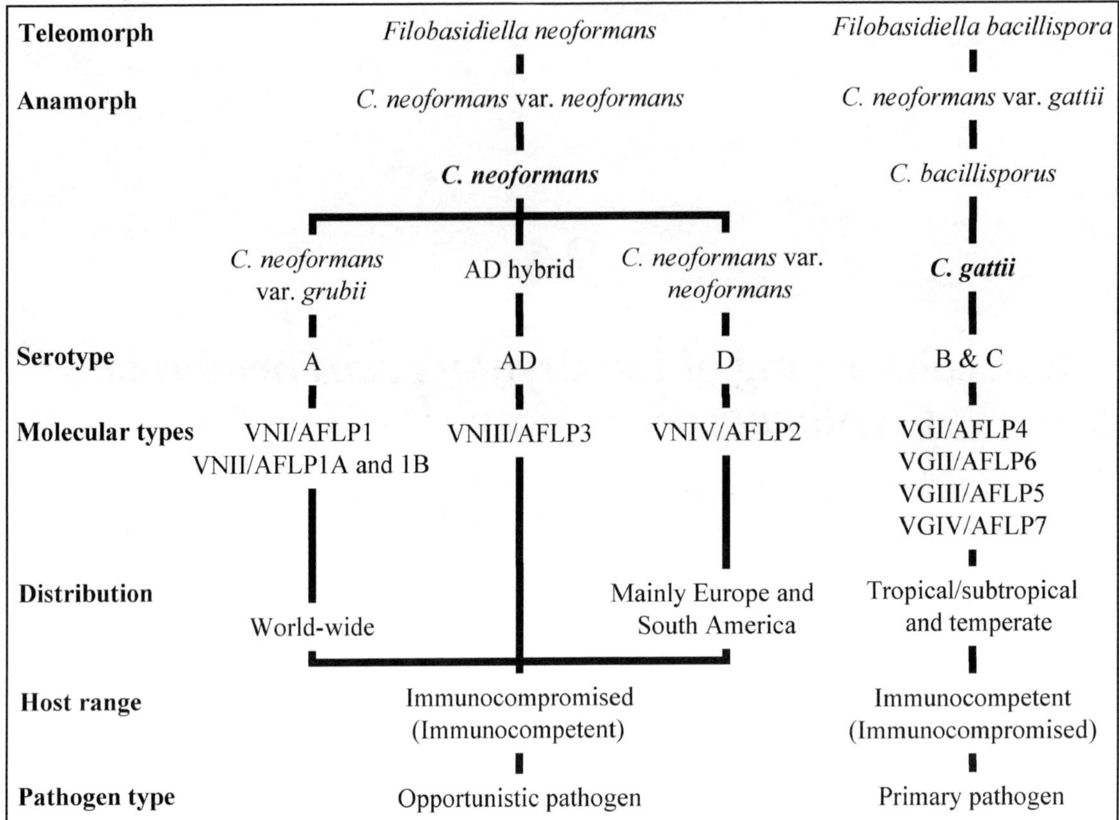

FIGURE 1 The taxonomic, molecular, epidemiological, and clinical correlation within the *C. neoformans* and *C. gattii* species complex. (Names given in bold are currently accepted taxonomic species names.)

TYPING METHODS

MLEE Analysis

MLEE is a technique that evaluates the polymorphisms of selected housekeeping enzymes. In this method, the enzymes of interest are extracted from the fungal cells, electrophoretically separated on a gel, and visualized by specific enzyme-staining procedures. Bands representing the presence of a particular enzyme are compared according to their different gel mobility (114). Most of these enzymes display a limited number of isotypes that are present mainly as a result of homologous amino acid substitutions, which are transferred from parental strains, thus enabling the establishment of phylogenetic relatedness by comparing adequate numbers of selected enzymes (53). If a panel of carefully selected enzymes is used, the information derived will enable the detection of microevolution within strains (114). Despite the ability of MLEE to differentiate between strains, this method is not adequately discriminatory for use as a single tool for strain typing. Nucleotide substitutions do not necessarily result in alteration in the amino acid composition, and subsequent changes in the electrophoretic profile may not be visualized. As a consequence, alleles, which are homologous in different individuals, may represent different gene alleles (53, 114).

Brandt et al. (15) applied MLEE to isolates representing each of the known varieties and serotypes of the *C. neoformans*/*C. gattii* species complex. MLEE was able to distinguish between varietal and species level. Variation in the mobility of the enzymes 6-phosphogluconate dehydrogenase, malate dehydrogenase, phosphoglucose isomerase, and phosphoglyceromutase differentiated *C. neoformans* var. *grubii* and *C. neoformans* var. *neoformans*. The enzymes malate dehydrogenase, alcohol dehydrogenase, phosphoglyceromutase, and glutamate dehydrogenase could separate *C. gattii* from *C. neoformans* (15). Four and five subtypes were distinguished within serotype A and serotype D, respectively. Among *C. neoformans* isolates, the hybrid serotype AD was distinguished by the mobility of 6-phosphogluconate dehydrogenase. A greater diversity was seen among *C. gattii* isolates, as all studied isolates showed a specific MLEE profile (15).

MLEE was also used to determine the relationships between genotype and serotype, and genotype and strain origin (i.e., clinical or environmental). Bertout et al. (4) investigated 107 environmental and clinical strains from various geographical areas. Of the 14 studied enzymes, 12 were polymorphic, containing between three and six alleles in each locus. These enzymes revealed 48 different MLEE types. Moreover, they observed a strong relationship between MLEE type and strain origin. However, factorial correspondence analysis clustered the isolates of the *C. neoformans*/*C. gattii* species complex together with *Cryptococcus laurentii* isolates. No separation between *C. neoformans* and *C. gattii* isolates was achieved. The obtained MLEE profiles showed no link between MLEE type and serotype (4).

Electrophoretic Karyotyping

Electrophoretic karyotyping differentiates between fungal strains by detecting the different numbers and sizes of chromosomes of each strain investigated after separating the whole chromosomes using pulsed-field gel electrophoresis (PFGE) (111). Karyotype variation in serial clinical isolates from individual patients and in clinical and environmental strains, before and after a murine passage, has been reported (48), indicating that chromosomal rearrangements can occur during infection, which may limit the applicability of electrophoretic karyotyping in epidemiological studies.

Electrophoretic karyotyping was for the first time applied to the *C. neoformans/C. gattii* species complex to study the genetic diversity between seven cryptococcal strains representing all four serotypes (104). Unique electrophoretic karyotype patterns of each strain were shown, suggesting that this technique could potentially be used for strain typing. Applying electrophoretic karyotyping to 40 clinical and environmental *C. neoformans* var. *grubii* and *C. neoformans* var. *neoformans* strains revealed karyotype variation in 90% of the studied strains (105). Fries et al. (48) found among the 32 clinical isolates 10 distinct electrophoretic karyotypes. Moreover, a subsequent study reported that electrophoretic karyotyping is able to detect genetic divergence by finding changes in the electrophoretic karyotype after genetic mutation or heterothallic mating (5). In addition, this study showed that the karyotypes obtained from *C. neoformans* and *C. gattii* differed substantially, and for both species extensive chromosomal length polymorphisms were observed, thus rendering PFGE a useful epidemiological tool. Subsequently, the clustering of isolates according to their electrophoretic karyotype has been correlated to RAPD profiles (7).

Electrophoretic karyotyping showed differences between the varieties and species within the *C. neoformans/C. gattii* species complex. Between 8 and 12 chromosomes were separated in *C. neoformans* var. *grubii* and *C. neoformans* var. *neoformans* isolates, with their sizes ranging from approximately 770 kb to 2.2 Mb, while in *C. gattii* isolates, between 11 and 13 chromosomes, with sizes ranging between 400 kb and 2.7 Mb, were found (5, 7, 104, 105). Based upon total chromosome sizes, the genome sizes for *C. neoformans* var. *grubii* and *C. neoformans* var. *neoformans* were estimated to range between 15 and 27 Mb, while that of *C. gattii* ranges between 12 and 18 Mb (5, 18), confirming the ability of electrophoretic karyotyping to separate the two species, *C. neoformans* and *C. gattii*, from each other, but also emphasizing that no correlation between the obtained karyotypes and the variety status could be detected.

DNA Fingerprinting

DNA fingerprinting is based on the recognition of hypervariable repetitive DNA sequences, such as simple sequence repeats or tandem repetitive elements, via hybridization of specific DNA probes, which are complementary to those repetitive sequences, with endonuclease-digested genomic DNA fixed on a nucleic acid-affinity membrane (39).

Genotypic differences between isolates of the *C. neoformans/C. gattii* species complex have been identified with this technique. DNA fingerprinting, using the CNRE-1 probe, recognized multiple profiles in clinical cryptococcal isolates (117). The differences in the detected DNA fingerprints are thought to be the result of genetic drift, chromosomal rearrangements, or sexual recombination. The CNRE-1 probe, when applied to clinical and environmental (i.e., pigeon excreta) cryptococcal isolates, showed a link between both sources (32). The CNRE-1 probe was used in combination with PFGE and RAPD analysis to show that persistent disease or reinfection is caused by the same strain in most cases (16). In an additional study by the same group, clinical and environmental isolates from two cities in Brazil (Belo Horizonte and Rio de Janeiro) were compared with isolates from New York City using electrophoretic karyotyping, DNA fingerprinting with the CNRE-1 probe, and *URA5* RFLP analysis. In this study, they found that the different typing methods revealed different degrees of variation in different geographic populations. The Brazilian isolates were found to be less variable than the others, and some of them were genotypically highly related to isolates from New York City, indicating either global dispersion of cryptococcal strains or a possible lack of resolution of the technique used. The obtained profiles indicated a possible link between clinical isolates and those from pigeon excreta (46). The obtained variation in CNRE-1 DNA fingerprinting profiles between serotype A and D isolates, in connection with other data, led to the proposal that serotype A isolates represent a separate variety within *C. neoformans* (47).

In another study, a radiolabeled probe of a 7-kb linear plasmid, UT-4p, revealed 21 different fingerprint patterns among 26 *C. neoformans* and *C. gattii* isolates (129). This probe could separate isolates to the variety and species level. The obtained DNA fingerprinting patterns were reproducible regardless of repeated subculturing, growth on different media, or murine passage of the investigated isolates (129). A subsequent study investigating clinical serotype D isolates, using the UT-4p probe, found a correlation between the obtained DNA fingerprinting patterns, specifically those with a DNA fingerprint of group I and patients in high-risk groups, such as drug addicts and homosexuals, or with a certain clinical manifestation, specifically with cryptococcal pneumonia (36). In addition, a possible linkage between isolates with a DNA fingerprint of group V (i.e., serotype D) and patients with pulmonary cryptococcosis was suggested (36). The UT-4p probe applied to investigate 156 strains from both cryptococcal species obtained from seven countries demonstrated 9 and 12 distinct DNA fingerprinting patterns for *C. neoformans* and *C. gattii*, respectively. None of the obtained profiles was specific for the isolates from AIDS patients. Certain profiles were specific for the isolates from a single country, but others were found to occur in several countries. Isolates from AIDS patients in Burundi and Zaire had identical profiles but different from those isolated from their respective houses. A high variation was found among 15 *C. gattii* isolates from Australia, and some environmental isolates had DNA fingerprinting patterns identical to those of clinical isolates (130). Environmental and clinical *C. neoformans* var. *neoformans* (i.e., serotype D) isolates from France were investigated using the UT-4p probe, and pigeon droppings were identified as a potential source of infection (50).

In an attempt to standardize the methods for strain typing, 17 environmental and 97 clinical *C. neoformans* var. *grubii* isolates were studied with two DNA fingerprinting probes: the pCNTel-1-labeled probe CENTEL and the CNRE-1 probe. The CENTEL probe revealed a higher degree of variation than the CNRE-1 probe. Cutoff points for strain relatedness were established as follows: a Dice coefficient (representing the similarity between strains) below 0.250 or above 0.449 was used for the CENTEL probe, and a Dice coefficient of 0.049 or above 0.149 for the CNRE-1 probe corresponded to strains that are either genetically linked or unrelated (49).

DNA fingerprinting with oligonucleotide probes specific to minisatellite DNA repeats $(GT)_8$, $(GTG)_5$, $(GATA)_4$, $(GACA)_4$, and $(GGAT)_4$ was used together with RAPD analysis to investigate serial isolates from five HIV-positive patients. Both techniques were found to have comparable discriminatory abilities and resulted in unique profiles for each patient. Isolates obtained in a single day from the same patient were indistinguishable from each other. It was shown that an infection can be caused by multiple strains and that recurrent infections may occur as a result of reinfection with a new strain (57).

RAPD Analysis

The technique of RAPD analysis is based on the amplification of random DNA fragments using several arbitrary, short primers (138). Since the primers bind randomly along the DNA template, no prior knowledge of the targeted DNA sequence is required. The detected polymorphisms can be used as genetic markers to construct genetic maps (138). RAPD analysis has been widely used, as it has the advantage of being highly discriminatory, rapid, and simple to use. Widespread adaptation, however, has been limited by the absence of standardized methodologies, inadequate reproducibility, and the requirement for strict quality control.

RAPD analysis was first applied by Crampin et al. (31) to investigate 12 *C. neoformans* isolates, revealing a high degree of variation, with a unique profile for each isolate investigated. Haynes et al. (57) used RAPD and DNA fingerprinting to show that multiple cryptococcal strains could be involved in a single infection. Clinical and environmental isolates from the southern Japanese prefecture of Nagasaki were divided into six major RAPD profiles, and corresponding profiles between clinical and environmental isolates have been shown (141). When RAPD analysis was applied to *C. gattii* isolates, it grouped all studied isolates into three major molecular types, namely VGI, VGII, and VGIII, with all of them corresponding to both serotypes of *C. gattii*, viz., B and C (107, 115, 116). When 62 clinical, 29 veterinary, and 45 environmental isolates were studied, it was revealed that the VGI molecular type was the most prevalent (115, 116). In the same studies, RAPD analysis identified a link between clinical and environmental isolates by revealing identical RAPD patterns between isolates from both sources (115, 116). Particular combinations of arbitrary primers CN1, MYC1, 5SOR, FPK1-07, FPK1-08, FPK1-13, and FPK1-20 enabled identification of variants within each of these molecular types. The molecular type VGI could be categorized into two subtypes by the primer pairs FPK1-08/FPK1-20 and FPK1-05/FPK1-13, whereas isolates of the molecular type VGII could be separated into two subtypes by amplification with the primer pairs 5SOR/CN1, MYC1/5SOR, and FPK1-05/FPK1-07 (116).

In the study conducted by Ruma et al. (107), the combination of the primer pair 5SOR/CN1 differentiated RAPD profiles obtained from either serotype B or C isolates. A common RAPD profile was identified among *C. neoformans* var. *grubii* isolates from AIDS patients (23, 24), whereas isolates from immunocompetent patients differed. In a subsequent RAPD study, the molecular type VGII was identified for the first time from seven *C. gattii* isolates from the Northern Territory of Australia. Until then all previously studied isolates obtained from *Eucalyptus* spp. in Australia belonged to molecular type VGI (25).

RAPD analysis in combination with PFGE was also used in a study conducted on 83 sequential cryptococcal clinical isolates obtained from 38 AIDS patients from São Paulo, Brazil. This study showed that the RAPD profiles and the PFGE profiles were highly related between all isolates and identified either microevolution of a strain during treatment or coinfection with multiple strains within a single patient (2).

Boekhout et al. (7) studied 91 strains of both varieties of *C. neoformans*, *C. neoformans* var. *grubii* and *C. neoformans* var. *neoformans*, using ERIC1 and ERIC2 primers, and recognized 16 different genotypes. In addition, they observed 8 different genotypes among 30 studied *C. gattii* strains. Their RAPD data suggested a geographical differentiation among *C. gattii*.

PCR Fingerprinting

PCR fingerprinting is based on the amplification of DNA sequences, which are flanked by hypervariable repeat units. The technique uses single primers specific to microsatellite or minisatellite DNA repeats, which were originally designed as hybridization probes used in DNA fingerprinting applied to other fungi (91). The primers used in PCR fingerprinting include the minisatellite-specific core sequence of the wild-type phage M13 (5′-GAGGGTGGCGGTTCT-3′) and the microsatellite-specific primers $(GTG)_5$ and $(GACA)_4$ (91). Following PCR amplification, the amplified interrepeat DNA sequences are separated by agarose gel electrophoresis, and the resulting banding patterns are analyzed using commercially available software programs such as BioloMICS (BioAware, Hannut, Belgium) or Bionumerics/GelCompar (Applied Maths, Kortrijk, Belgium). PCR fingerprinting produces species- and strain-specific multilocus profiles. However, despite the highly discriminatory ability of this approach, interlaboratory variation may occur due to varying experimental conditions and the use of different laboratory reagents. Consequently, PCR fingerprinting profiles may not be reproducible when undertaken in different laboratories, thus rendering this technique unsuitable for multi-institutional typing studies.

Oligonucleotides specific to highly repetitive DNA were first used in 1993 to study 42 cryptococcal strains (91). The DNA polymorphisms detected were able to separate individual strains and produced specific bands that separated *C. neoformans* var. *neoformans* (now *C. neoformans*) and *C. neoformans* var. *gattii* (now *C. gattii*). In addition, the primers M13, $(GTG)_5$, or $(GACA)_4$ were able to distinguish between isolates of serotype A, D, and B/C. In 1994, it was shown that the same oligonucleotides used as DNA hybridization probes in DNA fingerprinting or PCR primers in PCR fingerprinting revealed the same degree of polymorphism, allowing replacement of the methodological complex traditional DNA fingerprinting with the fast and easy-to-perform PCR fingerprinting (99).

In a follow-up study the same group investigated 27 clinical isolates from HIV-positive patients. No specific patterns corresponding to the HIV status were observed, as all isolates had their own specific PCR fingerprinting profile. The isolates clustered according to their serotype: A, D, and B/C (92). The same study demonstrated that the obtained PCR fingerprinting patterns were highly stable, even after repeated subculturing and reextraction of the genomic DNA over two and a half years. When PCR fingerprinting was applied to characterize two human cases of cryptococcosis in Germany, it was possible to show that both strains belonged to two different serotypes, A and D. In addition, a clear differentiation between *C. neoformans* and *C. gattii* strains was possible, and no correlation between potential environmental sources of the infection could be shown (59).

In a first attempt to standardize the PCR fingerprinting technique, 356 globally obtained cryptococcal strains were investigated using PCR fingerprinting and RAPD analysis, and the results obtained in two different laboratories were compared. Both laboratories grouped all isolates in the same subgroups, with both techniques resulting in four different major molecular types: VNI and VNII corresponded to *C. neoformans* var. *grubii*, serotype A, VNIII contained all AD hybrid isolates, and VNIV corresponded to *C. neoformans* var. *neoformans*, serotype D. A high degree of heterogeneity was observed among isolates from the United States when compared to those originating outside the United States (93). A global distribution of the major cryptococcal genotypes was reported for the first time in 2000. This study identified four major molecular types for *C. neoformans* (VNI, VNII, VNIII, and VNIV), as previously reported, and three major molecular types for *C. gattii* (VGI, VGII, and VGIII) (Fig. 1, Fig. 2A, Fig. 3, Table 1), with VNI and VGI being the most dominant molecular types worldwide (Fig. 3) (38; W. Meyer et al., unpublished data). The molecular epidemiology of clinical and environmental *C. neoformans* isolates from Tamil Nadu, India, was investigated by PCR fingerprinting and RAPD analysis and revealed that all isolates belonged to three major genotypes, with the majority belonging to VNI (68). PCR fingerprinting with the primers M13 and $(GACA)_4$ identified two strains isolated from insect frass obtained from the same *Eucalyptus tereticornis* tree hollow as belonging to two different molecular types, VGI and VGII, thus showing that both genotypes can apparently coexist in the same ecological niche (69).

In the late 1990s a collaborative network between researchers from IberoAmerican countries, including Argentina, Brazil, Chile, Colombia, Mexico, Peru, Venezuela, Guatemala, and Spain, was established to study the molecular epidemiology of the agents of cryptococcosis in Latin America and Spain. PCR fingerprinting using the primer M13 and *URA5* RFLP analysis grouped all 340 studied isolates into eight major molecular types (Table 1) and confirmed that the majority of infections were caused by the molecular type VNI (Fig. 3) (96). In contrast to earlier findings, this study showed that *C. gattii* isolates belonging to the molecular type VGII are more common in South America compared to the rest of the world, where VGI is predominant. The overall results showed similar banding patterns to be present in isolates obtained from a single country or even from different countries, thus suggesting partial clonal spread of this pathogen in South America. The majority of South American isolates showed a higher variability if compared to those from other parts of the world, which were less variable (46). The majority of VNI isolates were obtained from HIV-positive patients, confirming previous findings that most cryptococcosis cases in immunocompromised patients are caused by *C. neoformans* var. *grubii* strains. Most of the cryptococcosis cases in immunocompetent patients were caused by *C. gattii* strains (96).

PCR fingerprinting using the primers M13 and $(GACA)_4$ was used to study 105 clinical and 19 environmental (i.e., from pigeon and *Eucalyptus* spp.) isolates from the southern Brazilian state of Rio Grande do Sul. The majority of the clinical and environmental isolates belonged to *C. neoformans* var. *grubii*, molecular type VNI, including all isolates obtained from pigeon droppings, whereas the isolates obtained from *Eucalyptus* trees were all VGIII (20).

Another Brazilian study, using PCR fingerprinting with the primers M13, $(GTG)_5$, and $(GACA)_4$, investigating 60 serial cryptococcal isolates obtained from 19 HIV-positive patients over periods ranging from 18 to 461 days, showed that the majority of infections were caused by a single isolate. Persistent cryptococcal infections are caused by a relapse due to the same isolate, rather than reinfection with a different isolate. However, in exceptional cases, patients may be infected with more than one strain (61). Dual infections have also been reported in HIV-positive and HIV-negative patients from India based on M13 and $(GACA)_4$ PCR fingerprinting (90).

PCR fingerprinting using the primer $(GACA)_4$ was applied to 110 cryptococcal isolates obtained mainly from Germany and Africa as well as additional globally collected reference strains. This study divided *C. neoformans* strains into three major genotypes and introduced a novel genotype nomenclature: VNA1 to VNA3 for *C. neoformans* var. *grubii* isolates, VND1 and VND2 for *C. neoformans* var. *neoformans* isolates, and VNAD1 to VNAD3 for AD hybrid isolates. The small number of *C. gattii* isolates included in this study could be divided into the four major genotypes that were previously reported, namely, VGI, VGII, VGIII, and VGIV (80).

After the onset of the cryptococcosis outbreak on Vancouver Island (British Columbia, Canada) in 1999, PCR fingerprinting and AFLP analysis (see below) were applied to investigate its epidemiology and to identify its environmental source. All outbreak-related and environmental isolates from Vancouver Island were *C. gattii* and belonged to what until then was thought to be the rather rare molecular type VGII/AFLP6.

Two clonal subtypes have been identified to cause this ongoing outbreak, with the majority of clinical, veterinary, and environmental isolates belonging to VGIIa/AFLP6A, and a small group belonging to VGIIb/AFLP6B (67). A large number of environmental isolates with genotypes identical to either the major or minor outbreak genotype indicated the presence of the infective agents in the environment as the source of the infection. It is still unclear if the outbreak strains originated on Vancouver Island as the result of same-sex mating (45) or whether the genotypes were imported to Vancouver Island from other parts of the world (66, 94). The emergence of this usually tropical pathogen on Vancouver Island highlighted the potential of *C. gattii* to spread into new environmental niches (67).

The molecular epidemiology of 178 clinical and 247 environmental cryptococcal isolates from Colombia via PCR fingerprinting with the primers M13, $(GTG)_5$, and $(GACA)_4$ confirmed that the majority of cryptococcal infections are caused by *C. neoformans* var. *grubii*. This study also showed that the environmental isolates were mainly *C. gattii* VGII, and the majority were mating type **a**. Similar profiles between clinical and environmental isolates suggest that the patients may have acquired the infection from the environment (40).

A retrospective study of 443 Brazilian cryptococcal isolates determined the distribution of the major cryptococcal molecular types within Brazil. A high degree of genetic heterogeneity was observed, and all isolates were grouped into the eight major molecular types, with the most common type being VNI (64%), followed by VGII (21%), VNII (5%), VGIII (4%), VGI and VNIV (3% each), and VNIII (<1%) (125). As in the IberoAmerican study (96), VGII was the most common molecular type for *C. gattii* isolates (125), reemphasising a different distribution of the molecular types in South America if compared to the rest of the world, where VGI is the most common molecular type (38). An interesting finding of this study was that primary cryptococcosis caused by the molecular type VGII prevails in immunocompetent hosts in the north and northeast of Brazil, disclosing an endemic regional pattern for this specific molecular type in the north of Brazil (125).

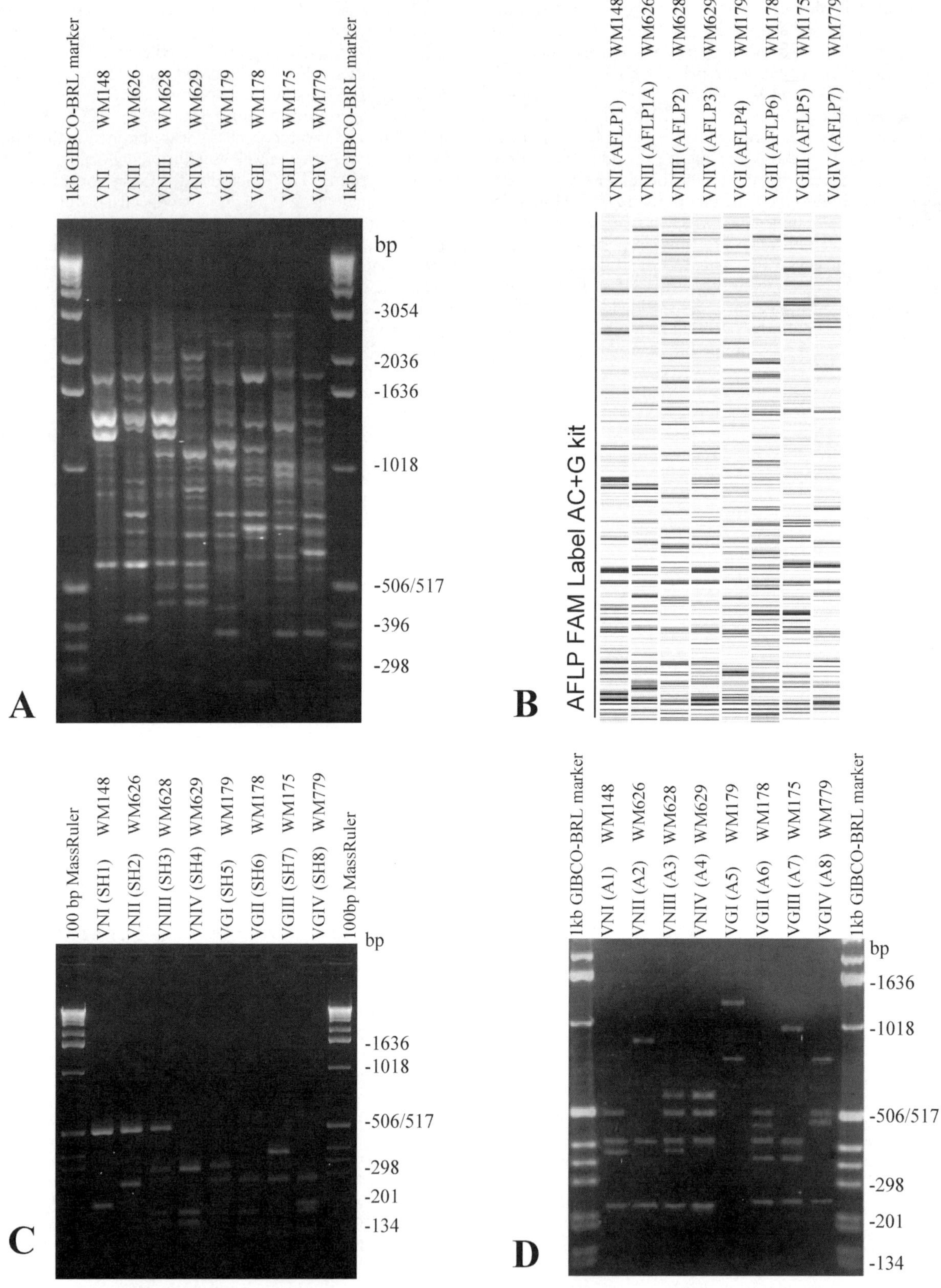

FIGURE 2 Assessment of the major molecular types within the *C. neoformans*/*C. gattii* species complex. (A) PCR fingerprints generated with the primer M13. (B) AFLP profiles generated with the FAM label AC+G kit. (C) *URA5* gene RFLP profiles identified via double digestion with Sau96I and HhaI. (D) *PLB1* gene RFLP profiles identified via digestion with AvaI obtained from the reference strains of each major genotype.

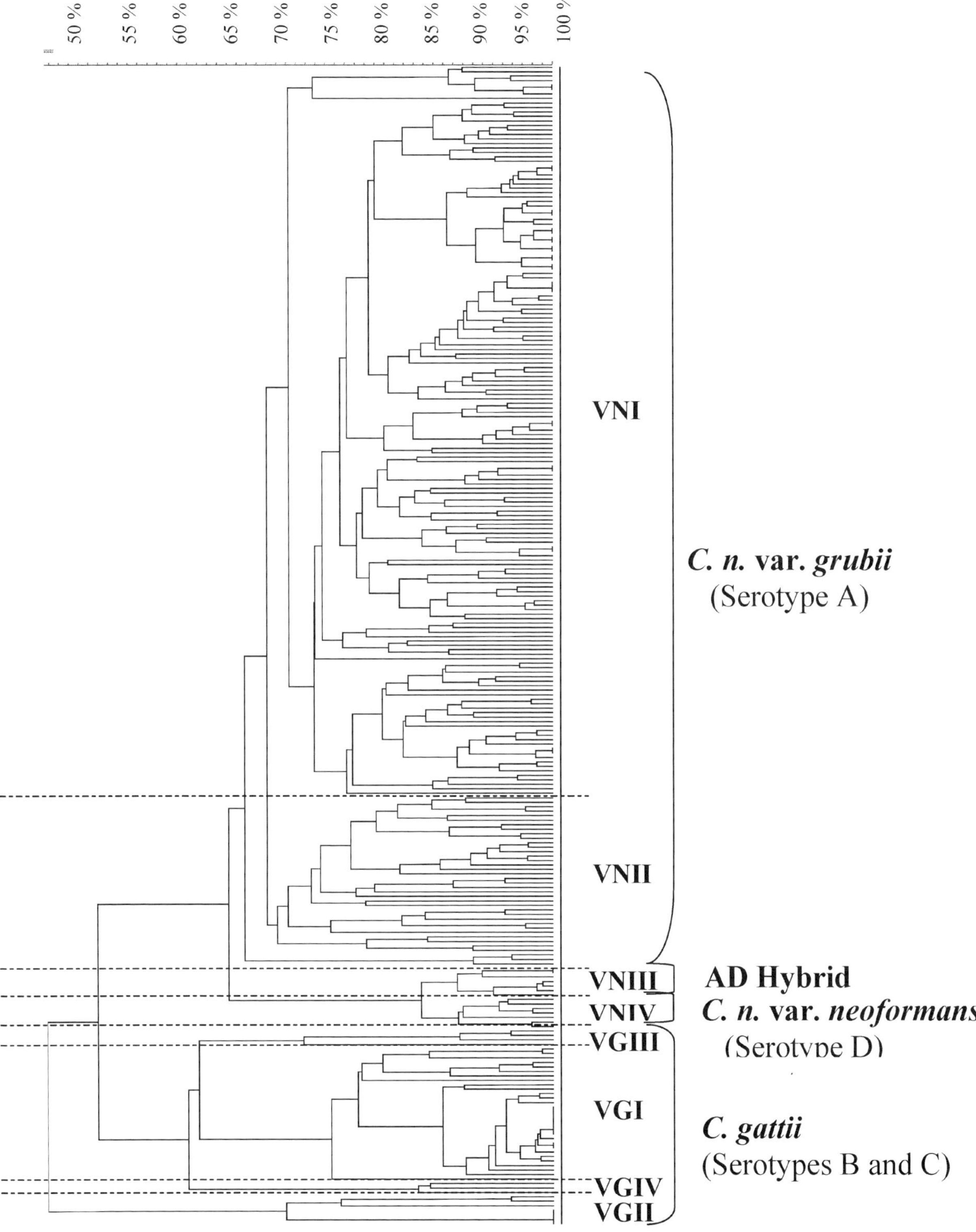

FIGURE 3 Phenogram of the combined PCR-fingerprinting data obtained with the primer M13 and $(GACA)_4$ from a selection of global cryptococcal isolates. All isolates fall into eight major molecular types, which correspond to *C. neoformans* var. *grubii*, serotype A, with two molecular types, VNI and VNII; *C. neoformans* var. *neoformans*, serotype D, with the molecular type VNIV; and *C. gattii*, serotypes B and C, with the molecular types VGI, VGII, VGIII, and VGIV. In addition to the three major clusters, we can see the intermediate molecular type VNIII, representing the AD hybrids (96; Meyer et al., unpublished data).

TABLE 1 Concordance of different molecular typing methods used for *C. neoformans* and *C. gattii*

Species/variety/ hybrid	Serotype	PCR fingerprinting molecular type (44, 93, 96)	PCR-fingerprinting molecular type (135)	AFLP genotype (6)	AFLP genotype (88)	*URA5* RFLP type (96)	*PLB1* RFLP type (78)	ITS genotype (64)	IGS genotype (33, 34)
C. neoformans var. *grubii*	A	VNI	VN6 (VN5)	AFLP1	VNI	VNI	A1	ITS1	1A/1B
	A	VNII		AFLP1A/ AFLP1B	VNB	VNII		ITS1	1A
	A	VNII	VN7	AFLP1A/ AFLP1B	VNII	VNII	A2	ITS1	1C
AD hybrid	AD	VNIII	VN3/VN4	AFLP3		VNIII	A3	ITS1/ITS2	2C
C. neoformans var. *neoformans*	D	VNIV	VN1 (VN2)	AFLP2		VNIV	A4	ITS2	2A/2B/2C
C. gattii	B/C	VGI		AFLP4A/ AFLP4B		VGI	A5	ITS3/ITS7	4
	B/C	VGII		AFLP6		VGII	A6	ITS4	3
	B/C	VGIII		AFLP5A/ AFLP5B/ AFLP5C		VGIII	A7	ITS5	5
	B/C	VGIV		AFLP7		VGIV	A8	ITS6	6

In a population genetic analysis of 120 *C. neoformans* and 9 *C. gattii* clinical strains from 16 provinces of mainland China, the majority of isolates were *C. neoformans* var. *grubii*, molecular type VNI, with 71% of those isolates being isolated from HIV-negative patients and only 8.5% from HIV-positive patients, which is in contrast to the situation in the rest of the world. Interestingly, the 120 *C. neoformans* isolates had identical PCR fingerprints when amplified with the primer M13 (22). Another study using PCR fingerprinting with the primer M13 identified 45 VNI, 2 VNIII, 1 VNIV, 9 VGI, and 1 VGIV isolate from 57 clinical and 1 veterinary isolate from China (43, 44).

One of the latest studies from Latin America has been conducted in Mexico. M13 PCR fingerprinting was used to investigate 72 Mexican clinical cryptococcal isolates (21). All eight major molecular types of the *C. neoformans/C. gattii* species complex were found to be present. Again, the majority of the infections in HIV-positive patients were caused by *C. neoformans* var. *grubii*, molecular type VNI (55 isolates). Five isolates were VNII, three were VNIII, one was VNIV, and two isolates of each molecular type of *C. gattii* (VGI, VGII, VGIII, and VGIV) were identified.

A slightly modified PCR fingerprinting technique with the microsatellite-specific primer $(GACA)_4$ was used to characterize *C. neoformans* isolates from Italy. Six genotype groups were distinguished initially as follows: VN1 to VN6, with VN1 corresponding to *C. neoformans* var. *neoformans* (serotype D), VN3 and VN4 corresponding to serotype AD hybrids, and VN6 corresponding to *C. neoformans* var. *grubii* (serotype A) (Table 1) (135). In a follow-up study, the same group identified specific bands corresponding to each of the four major molecular types: VN1, VN3, VN4, and VN6. They developed specific primers to enable a rapid identification of those four major groups (27) and characterized the amplified bands using the *C. neoformans* var. *grubii* and *C. neoformans* var. *neoformans* genomes (30). The same group confirmed the first mating type **a** strain from *C. neoformans* var. *grubii*, serotype A, by $(GACA)_4$ PCR fingerprinting (101, 134). Potential hybrid strains within *C. neoformans* were confirmed by PCR fingerprinting with the primer $(GACA)_4$ (29). PCR fingerprinting using the primer $(GACA)_4$ was applied to 311 *C. neoformans* strains obtained in a European Confederation of Medical Mycology prospective survey of cryptococcosis in Europe conducted from 1997 to 1999. Of all isolates studied, 51% were genotype VN6 *C. neoformans* var. *grubii*, serotype A, followed by 30% genotype VN1 isolates, *C. neoformans* var. *neoformans*, serotype D, and 19% of the isolates were AD hybrid strains exhibiting genotypes VN3 and VN4. This study revealed a high prevalence of serotype AD hybrid strains in Europe, mainly in Italy, Spain, and Portugal. Only six isolates were *C. gattii* for which the molecular type was not determined (133).

AFLP Analysis

AFLP is a universal genotyping method that targets the entire genome of the organism under study (136, 110). Information about the genome studied is not needed a priori, and only small amounts of DNA are required. The method is based on combined digestion with a frequent and a rare cutting restriction endonuclease, such as MseI and EcoRI, that has been applied in the AFLP analysis of the *C. neoformans/C. gattii* species complex (6) and ligation of an adaptor that creates PCR specificity at the restriction sites. Subsequent rounds of PCR use primers that may be more or less specific depending on the number of selective nucleotides added. Usually, the fluorescently labeled fragments are processed by a capillary sequencer, and the resulting banding patterns are further analyzed using standard software packages, such as Bionumerics (Applied Maths, Kortrijk, Belgium), Biolomics (BioAware, Hannut, Belgium), or Genescan (Applied Biosystems, Palo Alto, CA). To enhance reproducibility between runs and between different laboratories, an internal standard is present in every lane. Despite these precautions, it is sometimes difficult to standardize between different runs, and therefore it may be recommended to run important samples in a single run or to analyze the raw data of the different runs together so that the software enhancements (e.g., background correction) are identical for all samples. An additional drawback of AFLP is that variation of banding patterns may be artificial as a result of incomplete digestion. As a consequence, performing amplification in duplicate is necessary in critical studies.

In the case of the *C. neoformans/C. gattii* species complex, AFLP analysis has been applied in a number of typing studies. The first of them was an analysis of strains that were collected globally and that covered the genetic diversity of the species complex known at that time as much as was possible (6). The main result of this study was the distinction between *C. neoformans* (genotypes AFLP1, 2, and 3) and *C. gattii* (genotypes AFLP4-6) (Table 1) and the presence of significant genotypic diversity in each species. In *C. neoformans*, both varieties (*C. neoformans* var. *grubii* [serotype A = AFLP1/VNI] and *C. neoformans* var. *neoformans* [serotype D = AFLP2/VNIV]) could be discerned (Fig. 2B and Fig. 4A). Interestingly, the banding patterns from the serotype AD hybrids (genotype AFLP3/VNIII) were clearly different from those of the serotype A (AFLP1/VNI) and serotype D (AFLP2/VNIV) strains (Fig. 1, Fig. 2B, Fig. 4B). Next to the AD hybrids, another cluster of putative hybrids (AFLP1A/VNII) was recognized, but without understanding their genetic background (Fig. 4A). In later studies, three subtypes could be discerned within *C. neoformans* var. *grubii* that were designated AFLP1/VNI, AFLP1A, and AFLP1B (Fig. 4A). In subsequent studies, these genotypes of *C. neoformans* var. *grubii* were supported by multigene sequence analysis (11). AFLP1/VNI isolates corresponded to *C. neoformans* var. *grubii* (serotype A), showing a worldwide distribution. The strains of AFLP1A came from Rwanda, Portugal, and Brazil, whereas those belonging to AFLP1B originated from Brazil and Zaire (3, 6).

Based on sequence analysis of multiple genes, the genotype AFLP1A was found to belong to the VNB cluster (11), and more specifically to the VNB-A subclade, which is known to occur in Botswana (86, 88). In a multigene phylogenetic analysis, nine isolates were found to cluster inconsistently among the various gene genealogies, and alleles of both AFLP1/VNI and AFLP1A/VNB were found to be present. From this study, it was concluded that mating between these two genotypic groups may occur (88). Within *C. gattii* three AFLP genotypes could be discerned at that time (genotypes AFLP4, 5, and 6) (6). Serotype B strains belonged to all these three AFLP genotypes, but based on the set of strains analyzed, serotype C occurred only in AFLP5. Somewhat later, a fourth AFLP genotype was found to occur in *C. gattii* (AFLP7) (10) (Fig. 1, Fig. 2B, Fig. 4A).

A further important observation was that almost all isolates studied showed a unique AFLP pattern, thus rendering AFLP a potentially useful epidemiological tool. One has to keep in mind, however, that such fingerprinting methods suffer from a lack of recognition of the genetic nature of the similarly sized DNA fragments, which in theory may represent different parts of the genome. Moreover, the level of technical reproducibility is never 100%. AFLP genotyping, however, has been successfully applied in several epidemiological studies. For instance, in determining the cause of the ongoing and expanding outbreak of *C. gattii* on Vancouver Island (British Columbia, Canada), both PCR fingerprinting and AFLP analyses pointed toward the same culprit, namely strains of genotype AFLP6/VGII (67). Importantly, among the genotype AFLP6 isolates involved in this outbreak, two subtypes could be discerned, namely, a major genotype, AFLP6A/VGIIa, and a minor genotype called AFLP6B/VGIIb. Besides these two subgenotypes, isolates belonging to genotype AFLP4 were also observed to occur. However, these were later considered not to play a role in the actual outbreak (67). The amount of genetic diversity as assessed by AFLP of strains of both subgenotypes AFLP6A/VGIIa and AFLP6B/VGIIb was found to be limited, namely, 84.6% and 94.9% similarity after analysis using unweighted-pair group method using average linkages and single linkage, respectively (67). In this work a third subgenotype, AFLP6C, was initially discerned, but this could not be supported in later studies. AFLP analysis and multigene sequence analysis helped researchers understand tourist-related cases of *C. gattii* infections of visitors to Vancouver Island that were infected by the major genotype AFLP6A (52, 81, 83; F. Hagen, K. Tintelnot, and T. Boekhout, unpublished observation).

In a study of the genetic diversity of representatives of the *C. neoformans/C. gattii* species complex in Brazil, AFLP analysis revealed extensive genetic diversity among isolates collected throughout the country (3). A subset of environmental isolates originating from a single tree, however, showed high similarity values (>95%) (124). This latter work also demonstrated the occurrence of strains of both *C. neoformans* genotype AFLP1/VNI and *C. gattii* genotype AFLP6/VGII on a single pink shower tree in Piauí, northeast of Brazil. AFLP analysis also contributed to our understanding of a zoonotic transmission of *C. neoformans* var. *grubii* (serotype A, AFLP1/VNI) from a magpie to a 44-year-old immunocompetent woman who suffered from epilepsy (77). The genetic similarity between the genotype AFLP1/VNI isolates from the bird's excreta and the patient was 99.2%, thus rendering the conclusion of a direct transmission from the bird's environment to the patient. This was also supported by analysis of electrophoretic karyotypes of the isolates, which were found to be identical (77) and that are known to grossly differ between nonclonally related isolates of *C. neoformans* and *C. gattii* (7).

As AFLP analysis targets the entire genome, the method has been instrumental in the detection of various hybrids that occur within the *C. neoformans/C. gattii* species complex (Fig. 4B, C, and D), among them the so-called serotype AD hybrids (6). The AFLP patterns of the hybrid isolates were found to be a mixture of the respective AFLP patterns of each of the parental genotypes, viz., AFLP1/VNI and AFLP2/VNIV (Fig. 4B). In an epidemiological study of Dutch cryptococcal isolates, three isolates obtained from two HIV-negative patients isolated in 1977 and 2001, respectively, were found to be anomalous. A close inspection of their AFLP patterns, subsequently followed by fluorescence-activated cell sorter analysis of the nuclear DNA content, nuclear staining, and serology and cloning of the laccase (*LAC1*) gene, the ITS, the IGS, and the second-largest subunit of RNA polymerase II (*RPB2*), resulted in the discovery of the first interspecific hybrids between *C. neoformans* genotype AFLP2 (serotype D) and *C. gattii* genotype AFLP4/VGI (serotype B) (Fig. 4D). The hybrid strains (CBS10488 to 10490) were *MAT***a** serotype D (AFLP2/VNIV)/*MAT*α serotype B (AFLP4/VGI), and when investigated serologically, they were found to be serotype BD. Reinvestigation of an isolate that was obtained from a 31-year-old HIV-infected male that was described as the first Canadian isolate of *C. gattii* (118) revealed that this isolate also represented a *C. neoformans* × *C. gattii* hybrid (Fig. 4D). Analyses similar to those described above

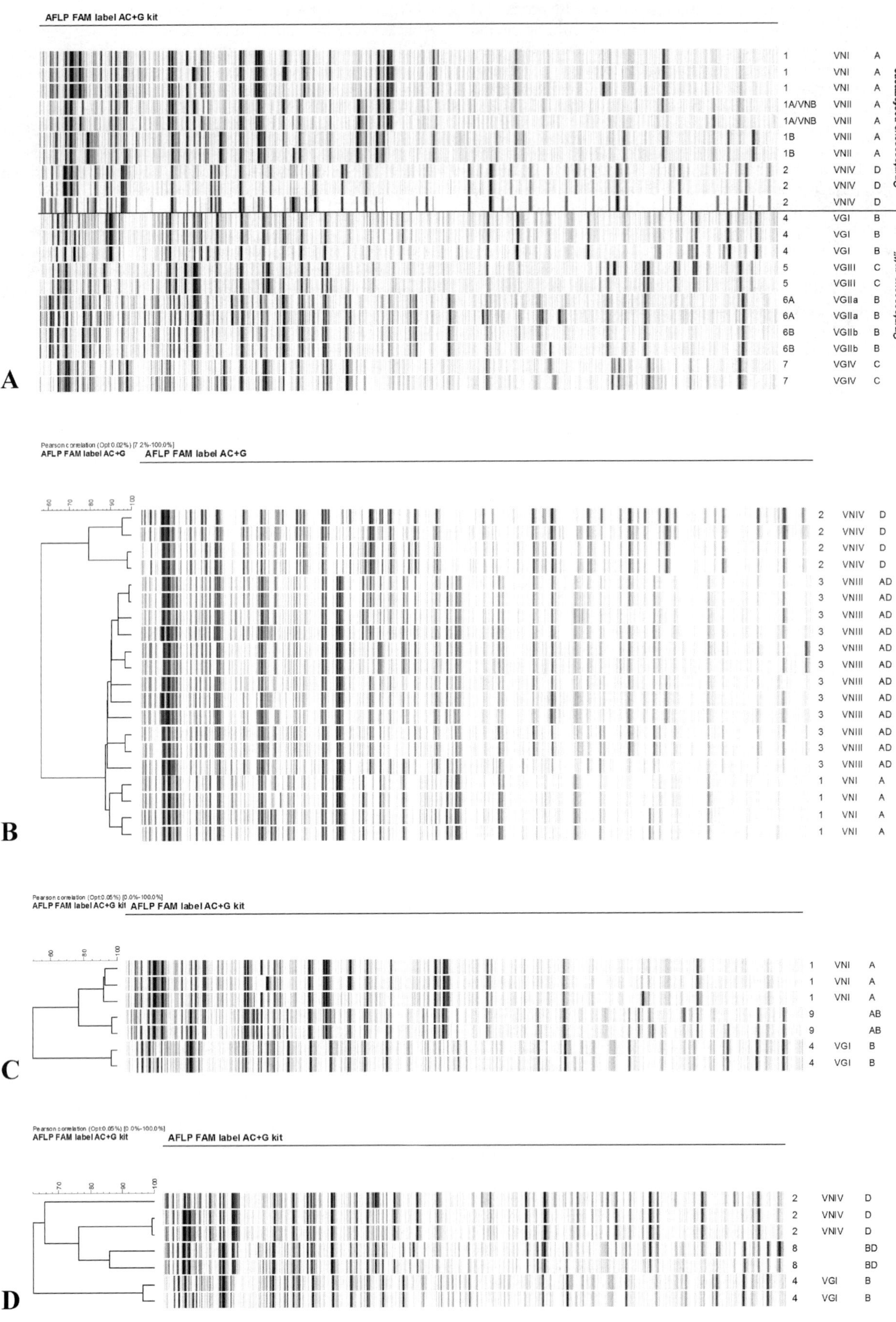

AFLP FAM label AC+G kit
1 VNI A
1 VNI A
1 VNI A
1A/VNB VNII A
1A/VNB VNII A
1B VNII A
1B VNII A
2 VNIV D
2 VNIV D
2 VNIV D
Cryptococcus neoformans
4 VGI B
4 VGI B
4 VGI B
5 VGIII C
5 VGIII C
6A VGIIa B
6A VGIIa B
6B VGIIb B
6B VGIIb B
7 VGIV C
7 VGIV C
Cryptococcus gattii
A
AFLP FAM label AC+G
2 VNIV D
2 VNIV D
2 VNIV D
2 VNIV D
3 VNIII AD
1 VNI A
B
AFLP FAM label AC+G kit
1 VNI A
9 AB
4 VGI B
C
AFLP FAM label AC+G kit
2 VNIV D
8 BD
4 VGI B
D

revealed that this isolate represented a hybrid between *C. neoformans* (AFLP1/VNI) serotype A and *C. gattii* (AFLP4/VGI) serotype B (Fig. 4C). Hence, the serotype of isolate CBS10496 was found to be serotype AB by agglutination tests (14). It is important to note that all these *C. gattii* × *C. neoformans* hybrids originated from clinical materials. Further detailed investigations of the genotypes, mating types, and serotypes undoubtedly will recover further anomalous strains that may represent further interspecies hybrids. This is already the case after an investigation of isolates from Germany that yielded some new *C. neoformans* × *C. gattii* hybrids (Hagen et al., unpublished observation).

AFLP has also been used as a method to infer reproduction strategies within the *C. neoformans/C. gattii* species complex. In an Australian study genetic diversity of environmental *C. gattii* isolates was assessed, and no genetic recombination could be detected, thus indicating the presence of a clonal mode of reproduction (55). Another study of *C. gattii* VGI revealed the presence of clonal propagation, whereas isolates of VGII from Australia and Papua New Guinea showed recombination (19). AFLP markers revealed recombination in VGI isolates from three *Eucalyptus* hollows in which both MAT**a** × MATα and MATα × MATα isolates were involved (109).

RFLP Analysis

RFLP analysis detects the variation present in DNA fragments after digestion by restriction enzymes. The RFLP patterns result from the presence of a restriction enzyme cleavage site at one place in the genome in one individual and the absence of that specific site in another individual (53). RFLP is relatively easy to use and is highly reproducible. Nevertheless, this technique is limited in its use to studies in which analysis of related isolates are considered, since it only allows identification at the species level.

Initially, several restriction enzymes (e.g., SalI, XbaI, and EcoRI) were used to digest the mitochondrial DNA of 20 cryptococcal isolates. The obtained RFLP patterns showed extensive heterogeneity among the isolates regardless of their serotype or species status. In addition, unique bands specific to either *C. neoformans* or *C. gattii* were observed (128). Subsequently, RFLP analysis of the repetitive multicopy cryptococcal rRNA gene cluster (rDNA) was performed. The obtained RFLP patterns differentiated *C. neoformans* var. *neoformans* (= *C. neoformans*) and *C. neoformans* var. *gattii* (= *C. gattii*), but they were found to be too conserved to efficiently determine the genetic diversity within each of the two species (42, 132).

The cryptococcal *URA5* gene encoding the orotidine monophosphate pyrophosphorylase has been used as a target for RFLP analysis, and this showed a link between environmental (pigeon excreta) and clinical isolates in New York City (32). In a follow-up study the same group compared RFLP patterns of the *URA5* gene of isolates from two Brazilian cities (i.e., Belo Horizonte and Rio de Janeiro) and New York City. This study suggested that long-distance dispersal of cryptococcal isolates occurred (46). The endonuclease AluI has been used to generate RFLP profiles that separated isolates of *C. neoformans* var. *grubii* and *C. neoformans* var. *neoformans* from *C. gattii*. Digestion by the endonuclease MspI revealed three distinct RFLP patterns corresponding to serotypes A, AD, and B/C/D (131). Double digestion with the endonucleases HhaI and Sau96I was able to differentiate all eight major molecular types of the *C. neoformans/C. gattii* species complex (Fig. 2C, Table 1) defined previously by PCR fingerprinting (96).

RFLP analysis of the *PLB1* gene, encoding phospholipase B, with the endonuclease AvaI showed the same results as the *URA5* RFLP analysis and separated all eight molecular types (Fig. 2D, Table 1). When the *PLB1* gene was digested with the endonuclease HindIII, all five serotypes (A, D, AD, B, and C) could be distinguished (78).

Recently, RFLP analysis of two genes (*CAP1* [combination of RFLP patterns obtained after individual digestions with the restriction enzymes AvaI, HgiCI, EcoRI, and PstI] and *GEF1* [combination of RFLP patterns obtained after individual digestions with the restriction enzymes EcoT14I and HapII]) enabled the simultaneous identification of each of the eight major molecular types (VNI to VNIV and VGI to VGIV) and the two mating types (α and **a**) by producing 16 RFLP profiles for each gene, with the following combinations of RFLP patterns of *CAP1* (C) and *GEF1* (G) corresponding to C1/G1 = VNIα, C2/G2 = VNI**a**, C3/G3 = VNIIα, C1/G15 = VNBα, C15/G2 = VNB**a**, C4/G4 = VNIIIα/**a**, C5/G5 = VNIII**a**/α, C16/G16 = VNIIIα/α, C6/G6 = VNIVα, C7/G7 = VNIV**a**, C8/G8 = VGIα, C9/G9 = VGI**a**, C10/G10 = VGIIα, C11/G11 = VGII**a**, C12/G12 = VGIIIα, C13/G13 = VGIII**a**, and C14/G14 = VGIVα (43, 44).

Sequence Analysis of the ITS Regions

The ITS region, containing the ITS1, the 5.8S rDNA gene, and the ITS2 region, is becoming the major target for species identification and barcoding of fungi. Since the rDNA gene cluster is a multicopy locus, it is easily amplified with universal primers, and ITS strain typing is highly reproducible, as it is based on sequence comparisons.

Katsu et al. (64) evaluated the value of the ITS region as a typing method to distinguish between the members of the *C. neoformans/C. gattii* species complex. In this study, the ITS regions of 94 globally selected cryptococcal isolates were sequenced, and seven sequence types have been delineated via specific combinations of eight nucleotide differences located at positions 10, 11, 5, 19, and 108 in the ITS1 region, position 221 in the 5.8S rDNA gene, and positions 298 and 346 in the ITS2 region. The seven ITS sequence types correlate to the previously identified major molecular types as follows: ITS type 1 (ATACTAGC) = *C. neoformans* var. *grubii*, molecular types VNI and VNII and the serotype A allele of the AD hybrid, molecular type VNIII (VNIIIA); ITS type 2 (ATATAGGC) = *C. neoformans* var. *neoformans*, molecular type VNIV and the serotype D allele of the AD hybrid, molecular type VNIII (VNIIIB); ITS type 3 (GCGCTGGC) and ITS type 7 (ACGCTGGC) = *C. gattii*, molecular type VGI; ITS type 4 (ACACTGAC) = *C. gattii*, molecular type VGII; ITS type 5 (ACACTGGG) = *C. gattii*, molecular type VGIII; and ITS type 6 (ACACTGGC) = *C. gattii*,

FIGURE 4 AFLP analysis of cryptococcal strains. (A) AFLP patterns representing all AFLP types obtained within the *C. neoformans/C. gattii* species complex. (B) AFLP patterns (AFLP3/VNIII) showing that the AD hybrid strains are a combination of the patterns obtained for serotype A (AFLP1/VNI) and serotype D (AFLP2/VNIV). (C) AFLP patterns indicating AB hybrids. (D) AFLP patterns indicating BD hybrids.

molecular type VGIV (Table 1). Cloned sequences of the serotype AD strains revealed that those hybrid strains are biallelic diploid at the ITS1-5.8S-ITS2 locus carrying both the ITS type 1 (ATACTAGC) and the ITS type 2 (ATATAGGC) alleles (64).

Sequence Analysis of the IGS Regions

The IGS region, containing the IGSI region, the 5S rDNA gene, and the IGSII region, is the most rapidly evolving part of the rDNA gene cluster separating the 28S (large subunit) and the 18S (small subunit) rRNA genes. With its average length of 2 kb, it has a much higher degree of sequence variation than the ITS region. Like the ITS region, the IGS region is part of the multicopy rDNA gene cluster and is easily amplified. IGS genotyping is sequence-based, making it highly reproducible. IGS sequencing is a suitable molecular tool for population studies and cryptococcal strain differentiation (34, 35).

The first 800 bp of the IGSI region located between the 3′ end of the large subunit and the 5′ end of the 5S rRNA gene were initially used to type 105 globally collected clinical and environmental isolates of the *C. neoformans/C. gattii* species complex (35). A highly conserved region was observed at the 5′ end, which was followed by a variable region starting at position 145. Five major genotypes were recognized (Table 1), with IGS genotype 1 corresponding mainly to serotype A, *C. neoformans* var. *grubii*, strains and IGS genotype 2 corresponding mainly to serotype D, *C. neoformans* var. *neoformans*, strains. However, the serotypes did not always correspond to one or the other genotypes. An even greater diversity was found among *C. gattii* strains, with three IGS genotypes. IGS genotypes 3 and 4 represented serotype B isolates, and IGS genotype 5 represented serotype B and C isolates. The highest nucleotide diversity was found between IGS genotypes 1 and 2, which was due to nucleotide variations and/or deletions/insertions (indels) that occur scattered throughout the whole IGS1 region. A high number of tandem repeats was found in all investigated IGS1 regions. Isolates from AIDS patients grouped mainly in IGS genotypes 1 and 2. There was no apparent correlation between the IGS genotypes 1 and 2 and the geographic origin of the isolates, which was in contrast to *C. gattii*, where IGS genotypes 3 and 5 originated from the Americas and IGS genotype 4 contained isolates from Australia, Asia, and Africa (35).

In a follow-up study, the same research group applied IGS sequencing to 107 global clinical and environmental isolates (33). This time, the entire IGS region, including the IGSI region, the 5S rDNA gene, and the IGSII region, was analyzed. They found the highest genetic variation in the IGSI region due to nucleotide substitutions and the presence of long indels. In contrast, the IGSII region showed less variation, and the indels were not as extensive as those displayed in the IGSI region. Both IGS regions contained short tandem repeat units, which accounted for the length polymorphisms observed. Six major IGS lineages were identified. Two corresponded to the two varieties of *C. neoformans*, *C. neoformans* var. *grubii* (IGS genotypes 1a, 1b, and 1c) and *C. neoformans* var. *neoformans* (IGS genotypes 2a, 2b, and 2c), and the other four to *C. gattii* (IGS genotypes 3, 4a, 4b, 4c, 5, and 6) (Table 1). In addition, a number of subgenotypes occurred within each of those lineages. Sequence analysis of cloned alleles of hybrid strains revealed that all hybrid isolates studied were monoallelic at the IGS locus. The IGSI region showed extensive length polymorphisms up to 194 bp, whereas the IGSII region displayed fewer length polymorphisms, with a maximum of 48 bp. Overall, the IGS regions of *C. gattii* differed up to 30.1% and 28.4% from those of *C. neoformans* var. *grubii* and *C. neoformans* var. *neoformans*, respectively. Again, a lack of concordance between IGS genotypes and serotypes was noted. The apparent lack of geographic concordance with phylogeny suggests that the *C. neoformans/C. gattii* species complex has undergone recent global dispersal (33).

IGS sequence analysis is a powerful tool to delineate the two varieties of *C. neoformans* and separate the four major molecular types of *C. gattii*. Due to its high genetic diversity, it can in addition identify subgenotypes in each of the species, varieties, and major molecular types.

MLMT

MLMT (multilocus microsatellite typing) indexes repeat variation within microsatellite sequences. Microsatellites, also known as short tandem repeats or simple sequence repeats (SSRs), are comprised of tandem repeated loci of one to six base pairs (122). MLMT analysis is highly reproducible. Microsatellite variation can either be detected via polyacrylamide gel electrophoresis during which fragments of the same size are separated or by direct sequence of the actual repeats. The results can differ depending on the detection methods being used. While sequencing detects the actual repeat numbers, polyacrylamide gel electrophoresis can result in misleading data since the resulting fragments are a combination of the repeat number and its flanking regions, which may result in the same fragment being present as "different" repeats.

The abundance of cryptococcal microsatellites was first reported in 2005, when the genome of *C. neoformans* var. *neoformans* strain JEC21 was screened. Within 18.9 Mb, a total of 1,973 microsatellites were found with a high frequency of mono- and dinucleotide repeats (62).

In 2008, MLMT was used to study the epidemiology of 87 clinical and environmental *C. neoformans* var. *grubii* isolates from 12 different countries. All isolates were typed by studying 15 SSR loci identified from the *C. neoformans* var. *grubii* strain H99 genome (56). Only three of the SSR loci, namely, CNG1 ("TA" repeats; primers CNG1F 5′-CACTTATGGTCTCAGAGGTA-3′ and CNG1R 5′-AACTTGGCTCGCTGCATCGT-3′), CNG2 ("GA" repeats; primers CNG2F 5′-CCGGAGAATGAGATTGTCGT-3′ and CNG2R 5′-TTATCGACGGCCATAGCTTC-3′), and CNG3 ("CAT" repeats; primers CNG3F 5′-AGATACTACCCGCAAACGTC-3′ and CNG3R 5′-TCCCAGTCCTATTCCTCACT-3′), were found to be polymorphic (56). In addition, these three polymorphic SSR loci were species-specific and amplified only from *C. neoformans* var. *grubii*, *C. neoformans* var. *neoformans*, and the AD hybrid strains. Among the 87 *C. neoformans* var. *grubii* strains studied, 30 MLMT genotypes were found. The highest discriminatory power (D) (60) was obtained for the locus CNG2 (D = 0.8), followed by CNG1 (D = 0.698) and CNG3 (D = 0.633). The combined discriminatory power of all three SSR loci was 0.992. The highest microsatellite variation was found in strains originating in South America (n = 47, MLMT types 1–4, 6, 8, 9, 11–28, 30; MLMT 13 was most frequent [n = 8]), followed by Asia (n = 29, MLMT types 5, 7, 10, 16, 17, 29; MLMT 17 was most frequent [n = 19]), and no microsatellite variation (MLMT type 22) was found in Africa (n = 8) and Europe (n = 3). One genotype was present on all continents except Asia (56) (Fig. 5).

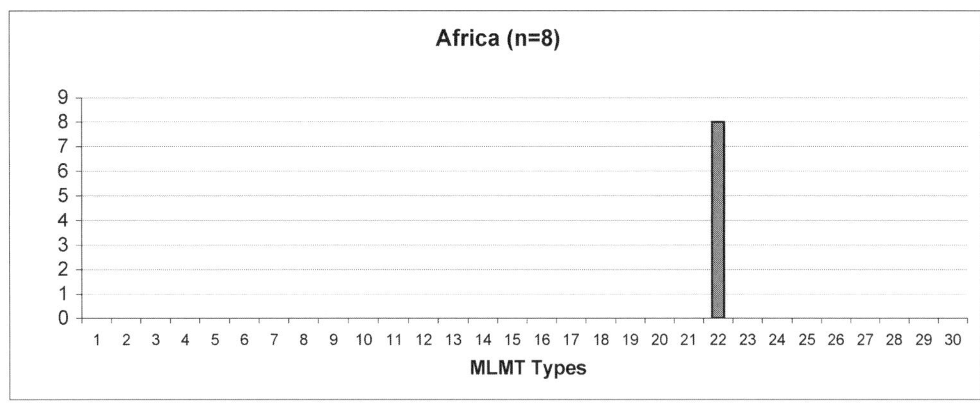

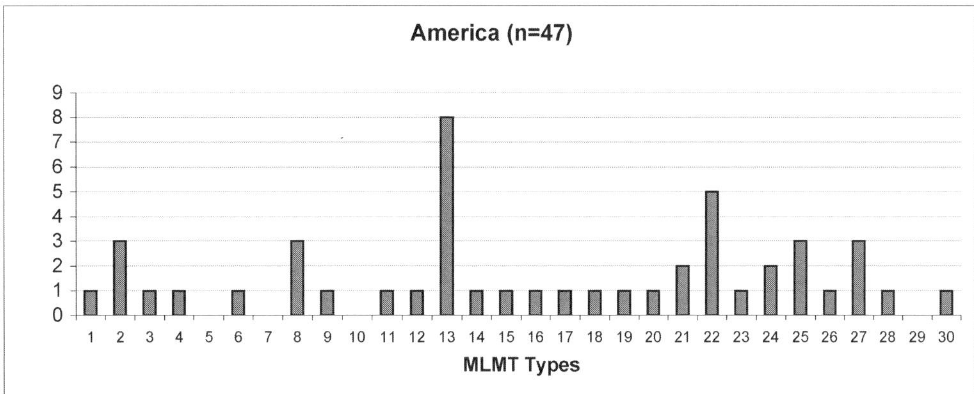

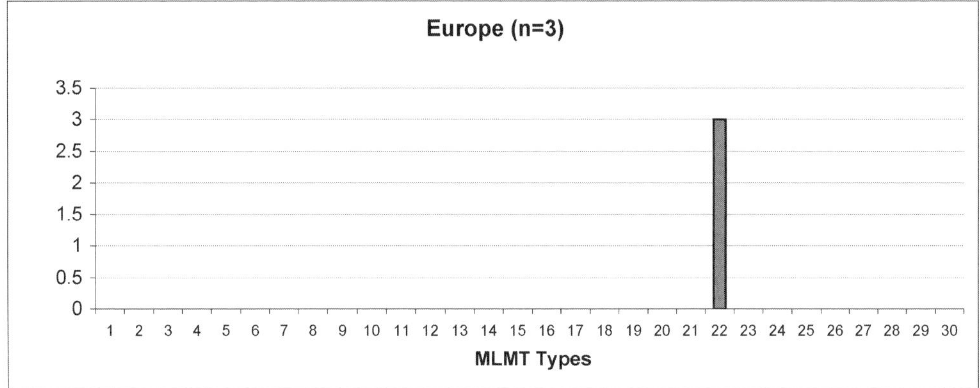

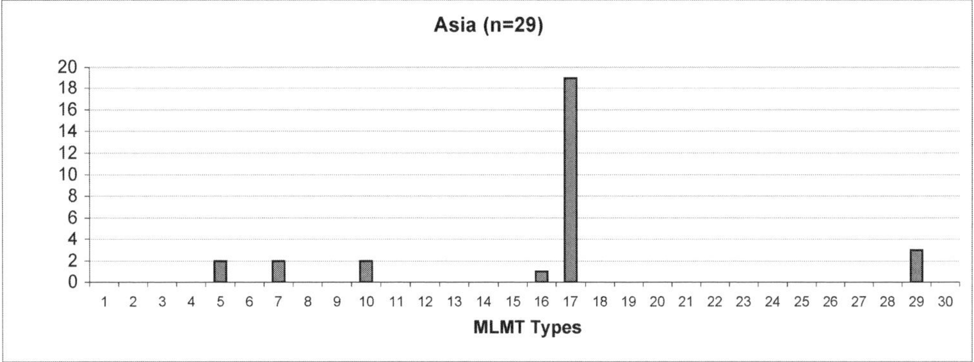

FIGURE 5 Geographical distribution of the 86 *C. neoformans* var. *grubii* strains studied with their MLMT type, made up of the combination of the three polymorphic microsatellite alleles. The number of strains given per MLMT locus is in parentheses, showing that the largest genetic variation is present in the Americas, especially in South and Latin America (reproduction from reference 56).

TABLE 2 Characterization of 11 microsatellite loci for *C. neoformans* var. *grubii*, the AD hybrid, and *C. neoformans* var. *neoformans* (63)

Locus	Repeat motif	Genome position (strain/chromosome #/start position)	Primer sequence (5′-3′)[a]	Size range (bp)
CNN1	$(GA)_9$	JEC21/10/567677	F:ATTCATGGGAAGATCGCTTGG R:GCGGCTCGGTCCTTACAATA	154–161
CNN2	$(TG)_{12}G(TG)_2$	B-3501/10/970135	F:AGAGACGGGCAAATCAGAGT R:GACTGGATTCTTGAACGGCT	145–173
CNN3	$(TCCTTT)_6$	JEC21/3/310108	F:TCGCTGTTGAGCGTGTTATC R:AGATGGATGGGAAAGGGAAG	161–174
CNN4	$(TTA)_{10}$	JEC21/2/1068035	F:TCTGTCTCCCTCGATATGTTGT R:TGATGTTTGAGGACAGCTTCTG	178–210
CNN5	$(AC)_{12}$	B-3501/7/591563	F:GATCATAATTGGCCTCGAT R:GCTGATGGGATGTAGGTTC	264–302
CNN6	$(CAA)_8$	JEC21/2/733919	F:AGAGCCCAAGGCAAAGAAGA R:GCCTTACCCATGTACTGAGC	440–467
CNN7	$(GT)_{12}$	B-3501/12/493289	F:CGGTGCTGGAAGGTCATA R:TTGAAATCCGGTCGCTAAA	147–190
CNN8	$(TATTT)_9$	JEC21/5/1766589	F:TATTGGATATGCTGGATCTGAC R:GGAGTATAAGGCTACCGTGTCT	148–174
CNN9	$(CCG)_8$	JEC21/8/1130826	F:TACTTTGCCCTCCACATACTGA R:CGCTTTCTTGCAGTCTGATGTC	186–198
CNN10	$(GTT)_9$	JEC21/6/432186	F:CCGACACCTTGGAGAATAAA R:TGAGGACGGGAGTCGAAAAG	190–211
CNN11	$(GGAT)_{13}$	B-3501/9/168199	F:GAGTTGATGCAGGATGCGTAAG R:CATCGGTAGCCTCGCCAGA	196–219

[a] F, forward; R, reverse.

The AD hybrid strains amplified two different alleles for the CNG1 locus, reflecting the diploid or aneuploid nature of these strains.

In a second study, 150 microsatellites identified from the *C. neoformans* var. *neoformans* genome sequencing projects of the strains B-3501 and JEC21 were screened for polymorphisms. Only 11 microsatellites (Table 2) were polymorphic for strains of *C. neoformans* var. *grubii*, *C. neoformans* var. *neoformans*, and the AD hybrid (63). The discriminatory power ranged from $D < 0.5$ (CNN4 locus) to $D = 0.93$ (CNN5 locus), and the combined discriminatory power of the seven most polymorphic loci was $D = 0.99$. D was generally higher for *C. neoformans* var. *neoformans*, molecular type VNIV if compared to the other molecular types, which may reflect a certain bias whereby loci are more polymorphic in the population from which they were developed because of preferential selection of long alleles. The two *C. neoformans* var. *grubii* molecular types VNI and VNII had similar numbers of alleles and nearly identical D values for all loci, indicating a similar level of genetic diversity in both groups. The same microsatellites could be amplified from *C. gattii* but showed little variation and very low D values for the four major molecular types (VGI to VGIV) of this species (63).

Since the variation detected in both MLMT studies is so far limited to *C. neoformans*, additional microsatellites need to be developed for *C. gattii*.

MLST

MLST (multilocus sequence typing), using partial sequence analysis of 7 to 10 housekeeping genes, has become the number-one typing approach for epidemiological investigations of microorganisms (123). For each locus studied, different genetic sequences present within a species are assigned as distinct alleles. The combination of the identified alleles at each of the loci defines the allelic profile or sequence type for each isolate. The data generated can be used to determine whether the fungal isolates are clonal or have undergone recombination. MLST is a sequence-based technique resulting in unambiguous, reproducible results that can be compared and made publicly available via electronic networking. However, the ability of the MLST analysis to give high strain discrimination relies on the choice of genes or loci selected in each study. Accurate sequence determination and comparison is crucial, as a single base difference denotes a different haplotype.

Two different MLST typing schemes have been introduced to type isolates of *C. neoformans* (88) and *C. gattii* (45), using 12 and 8 unlinked loci, respectively. The first study used 12 unlinked polymorphic loci, which are dispersed on 9 different chromosomes (*CAP10*, *CAP59*, *GPD1*, *LAC1*, *MPD1*, *MP88*, *PLB1*, *SOD1*, *TEF1α*, *TOP1*, *URE1*, and IGS1), to genotype 102 globally obtained serotype A strains (88). MLST analysis differentiated three major groups among the studied isolates, corresponding to VNI, VNII, and VNB (a Botswana-specific genotype closely related to VNI). In connection with this study, a central Web-based database was created at http://cneoformans.mlst.net/, allowing for an online determination of the alleles and sequence types of *C. neoformans* var. *grubii*, serotype A strains.

The second study, of 202 *C. gattii* strains, used seven unlinked polymorphic loci (*CAP10*, *GPD1*, IGS1, *LAC1*, *MPD1*, *PLB1*, *TEF1α*) and the two mating-type-locus-specific loci *SXI***a** and *SXIα*, which, however, cannot be amplified for all strains. These loci were further supplemented for a more detailed analysis of nine closely related strains by 22 additional gene loci (*HOG1*, *BWC1*, *CNB1*, *TOR1*, *CAC1*, *CRG1*, *URE1*, *FHB1*, *BWC2*, *CNA1*, *CBP1*, *TSA1*,

STE7, *FTR1*, *PAK1*, *CAP59*, *ICL1*, *GPA1*, *GPB1*, *RAS1*, *CCP1*, and *TRR1*) to investigate the origin of the Vancouver Island outbreak isolates (45). MLST differentiated all four major molecular types of *C. gattii* (VGI, VGII, VGIII, and VGIV), distinguished both Vancouver Island outbreak subgenotypes, VGIIa and VGIIb, and highlighted two possible origins (Australia or South America) for the outbreak strains. Subsequently, the same group described the first contemporary case of cryptococcosis in the United States caused by the major Vancouver Island outbreak genotype VGIIa using the same set of loci (126).

Another study used the four MLST loci *CAP1*, *FTR1*, *LAC1*, and *URA5* to investigate globally collected *C. gattii* strains and showed that strains genetically similar to those from the Vancouver island outbreak also occur in other parts of the world (66).

MLST, using six loci (ITS1/2, IGS1, *RPB1*, *RPB2*, *TEF1α*, and *LAC1*), was used to delineate six monophyletic lineages within the *C. neoformans/C. gattii* species complex and to show that the VNB genotype also occurs in South America (11). Each of the identified lineages corresponded to either the two varieties of *C. neoformans* (*C. neoformans* var. *grubii* and *C. neoformans* var. *neoformans*) or the four major molecular types (VGI, VGII, VGIII, and VGIV) within *C. gattii*.

Recently, an MLST study using five loci (ITS1/2, IGS1, *LAC1*, *RPB1*, *RPB2*, and *TEF1α*) was used to show recombination and long-distance dispersal in environmental populations of *C. neoformans* var. *grubii* from India (58).

To standardize genotyping within the *C. neoformans/C. gattii* species complex, a working group was formed under the umbrella of the International Society of Human and Animal Mycology (ISHAM) in 2007. A consensus set of genetic loci to be used in further typing studies was established by a statistical comparison of all loci using Simpson's index of diversity (113). This analysis revealed that a minimum of seven loci (*CAP59*, IGS1, *GPD1*, *LAC1*, *PLB1*, *MP88*, and *SOD1*, giving a Simpson's index of diversity of 0.9632 [88], and *GPD1*, IGS1, *TEF1*, *LAC1*, *MPD1*, *CAP10*, and *PLB1*, giving a Simpson's index of diversity of 0.9319 [45]) are required to differentiate between the sequence types of all studied strains. As a result, the following seven loci were selected as an international standard for MLST of strains of the *C. neoformans/C. gattii* species complex: six housekeeping genes, *CAP59*, *GPD1*, *LAC1*, *PLB1*, *SOD1*, and *URA5* (of which three genes code for cryptococcal virulence factors: the polysaccharide capsule [*CAP59*], melanin synthesis [*LAC1*], and cell invasion [*PLB1*]), and the IGS1 (Table 3) (95).

CORRELATION OF THE GENOTYPES AND THEIR DESIGNATED STANDARD STRAINS

Based on the comparison of the genotypes obtained for isolates included in some or all typing studies, it can be concluded that the different genotyping methods used by the different research groups lead to the recognition of concordant major genotypes for the agents of cryptococcosis (Table 1). From the relatively large array of different molecular typing techniques used for strain typing of isolates of the *C. neoformans/C. gattii* species complex, two main typing systems have been most commonly used: PCR fingerprinting using primers specific for microsatellite (M13) (91–93, 96) or minisatellite [$(GACA)_4$] DNA (27, 91, 92, 135) and AFLP analysis (6). Both typing schemes have over 2,000 isolates grouped into eight major molecular types. With some exceptions (33, 34), the molecular types of *C. neoformans* correlate well with the serotypes: *C. neoformans* var. *grubii*, serotype A, consists of molecular types VNI = VN6 = AFLP1 and VNII = VN7 = AFLP1A; the hybrid serotype AD comprises VNIII = VN3/VN4 = AFLP3; and *C. neoformans* var. *neoformans*, serotype D, corresponds to VNIV = VN1 = AFLP2. *C. gattii* consists of VGI = AFLP4, VGII = AFLP6, VGIII = AFLP5, and VGIV = AFLP7, which all correspond to both serotypes B or C (Fig. 1) (6, 96, 135). Based on these findings, the members of the ISHAM working group Genotyping of *Cryptococcus neoformans* and *C. gattii* agreed at their Torino meeting in 2007 to use the VNI to VNIV and VGI to VGIV nomenclature (96), since it correlates with the current concept of two species and represents the global population structure (95).

To enable a comparison of the results obtained in past and future studies, the ISHAM working group also designated a set of standard strains representing each of the eight major molecular types. This included the molecular type strains used in PCR fingerprinting or *URA5*-RFLP analysis (Fig. 2) (96) plus additional strains representing type cultures or strains, which are used in major cryptococcal genome projects (Table 4) (95). All standard strains are publicly available from either the CBS Fungal Biodiversity Centre (http://www.cbs.knaw.nl/), the American Type Culture Collection (http://www.atcc.org/), or the Fungal Genetic Stock Center (http://www.fgsc.net/). The corresponding collection numbers are listed in Table 4.

ENVIRONMENTAL ISOLATION OF THE MAJOR GENOTYPES

Most of the environmental isolates from IberoAmerican countries belonged to *C. neoformans* var. *grubii*, with 73.1% being VNI (*n* = 49) and 1.5% VNIII (*n* = 1) AD hybrids (96). The remaining 17 (25.3%) isolates were *C. gattii* with the molecular type VGI (*n* = 1), VGII (*n* = 3), VGIII (*n* = 12), and VGIV (*n* = 1). In general, the occurrence of the molecular types VGI to VGIV indicated that *C. gattii* is not restricted to tropical areas as described before, but that they have a much wider geographic distribution. The overall similarity between environmental *C. grubii* isolates from South America being as low as 50% indicates that South American isolates are more variable than those obtained in other studies. Interestingly, Chile and Spain had a large number of molecular type VNIII isolates (AD hybrids), 15.8% and 42.1%, respectively. In South America VNIV serotype D isolates were present in Chile (26.3%) only (96). Serotype D isolates have been described from Italy (133). The similarity in the molecular types obtained from Spanish and Chilean isolates provided further evidence that the cryptococcal strains present today in South America could have been introduced during the European colonization as suggested by Franzot et al. (46). They suggested that a major reservoir of *C. neoformans* in pigeon excreta may have originated in southern Europe and northern Africa followed by global dispersal due to human travel (46, 84).

Two recent studies carried out molecular typing of Colombian and Brazilian isolates. The first one studied the epidemiological relationship of clinical and environmental isolates of the *C. neoformans/C. gattii* species complex recovered from 1987 to 2004 in Colombia. Of the 247 environmental isolates, 44.2% were serotype A, 42.6% serotype B, and 13.2% serotype C. *URA5* RFLP analysis and PCR fingerprinting with the primers M13, $(GACA)_4$, and $(GTG)_5$ grouped all environmental serotype B isolates into the molecular type VGII, and 96.6% of serotype B isolates were mating type α. These results indicate close relationships

TABLE 3 Consensus MLST loci to be used for strain typing within the *C. neoformans/C. gattii* species complex (95)

Gene locus	Gene product	Chromosome location[a]	Primer name and sequence[b]	Amplification conditions	No. of bases analyzed (bp)[a]	Analyzed sequence fragment, start (5′) and end (3′) points[a]	Reference
CAP59	Capsular associated protein	1	**CAP59F** 5′-CTCTACGTCGAGCAAGTCAAG-3′ **CAP59R** 5′-TCCGCTGCACAAGTGATACCC-3′	94°C 3 min; 35 cycles: 94°C 30 s, 56°C 30 s, 72°C 1 min **Alternative conditions:** 30 cycles: 94°C 30 s, 64°C 30 s, 72°C 1 min **or:** 30 cycles: 95°C 3 min, 95°C 30 s, 54°C 30 s, 72°C 1 min	559	5′-ACGGTACGCGCCGGAGACAGAATG-3′	45
			Alternative primers: **CAP59LF** 5′-GTGAACAAGCTGCGGC-3′ **CAP59LR** 5′-GGATTCAGTGTGGTGGAAGA-3′	35 cycles: 94°C 30 s, 60°C 30 s, 72°C 1 min			95
GPD1	Glyceraldehyde-3-phosphate dehydrogenase	7	**GPD1F** 5′-CCACCGAACCCTTCTAGGATA-3′ **GPD1R** 5′-CTTCTTGGCACCTCCCTTGAG-3′	94°C 3 min; 35 cycles: 94°C 45 s, 63°C 1 min, 72°C 2 min **Alternative conditions:** 12 cycles; 62–56°C step-down 2°C every 2 cycles 95°C 3 min; 95°C 30 s, 62–56°C 30 s, 72°C 1 min; followed by 25 cycles: 95°C 30 s, 56°C 30 s, 72°C 1 min	543	5′-GGTTTCGGTACGGGACCCTGCCAA-3′	45
LAC1	Laccase	8	**LAC1F** 5′-AACATGTTCCCTGGGCCTGTG-3′ **LAC1R** 5′-ATGAGAATTGAATCGCCTTGT-3′	94°C 3 min; 30 cycles: 94°C 30 s, 58°C 30 s, 72°C 1 min **Alternative conditions:** 30 cycles: 95°C 30 s, 50°C 30 s, 72°C 1 min	469	5′-GTAAGTATCAGCTCAAGCTAAACA-3′	45
PLB1	Phospholipase	12	**PLB1F** 5′-CTTCAGGCGGAGAGAGGTTT-3′ **PLB1R** 5′-GATTTGGCGTTGGTTTCAGT-3′	94°C 3 min; 30 cycles: 94°C 45 s, 61°C 45 s, 72°C 1 min **Alternative conditions:** 12 cycles; 62–56°C step-down 2°C every 2 cycles 95°C 3 min; 95°C 30 s, 62–56°C 30 s, 72°C 1 min; followed by 25 cycles: 95°C 30 s, 56°C 30 s, 72°C 1 min	532	5′-TGTTACTTGGATTCTGGAACATCG-3′	88

SOD1	Cu, Zn superoxide dismutase	5	Primers for *C. neoformans:* **SOD1CNF** 5′-AAGCCTCTCATCCATATCTT-3′ **SOD1CNR** 5′-TTCAACCACGAATATGTA-3′ Primers for *C. gattii:* **SOD1CGF** 5′-GATCCTCACGCCATTACG-3′ **SOD1CGR** 5′-GAATGATGCGCTTAGTTGGA-3′	94°C 3 min; 35 cycles: 94°C 30 s, 52°C 30 s, 72°C 1.5 min	700	5′-**C**CACGTGCTC**G**C **A**CC**TGT**C**AA**TGC-3′	37
			Alternative primers for *C. neoformans*: **SOD1-f** 5′-TCTAATCGAAATGGTCAAGG-3′ **SOD1-r** 5′-CGCAGCTGTTCGTCTGGATA-3′	12 cycles; 62–56°C step-down 2°C every 2 cycles 95°C 3 min; 95°C 30 s, 62–56C° 30 s, 72°C 1 min; followed by 25 cycles: 95°C 30 s, 56°C 30 s, 72°C 1 min	535	5′-ATCGCTCACCGC TGCCCATTGTCA-3′	88
URA5	Orotidine monophosphate pyrophosphorylase	8	**URA5F** 5′-ATGTCCTCCCAAGCCCTCGAC-3′ **URA5R** 5′-TTAAGACCTCTGAACACCGTACTC-3′	94°C 3 min; 35 cycles: 94°C 45 s, 63°C 1 min, 72°C 2 min **Alternative conditions:** 30 cycles: 94°C 45 s, 63°C 1 min, 72°C 2 min (*C. neoformans*) 26 cycles: 94°C 30 s, 68°C 30 s, 72°C 30 s (*C. gattii*) **or:** 30 cycles: 95°C 3 min; 95°C 30 s, 63°C 30 s, 72°C 1min	601	5′-TTTTCGGCAACT CTTGGAA**A**GCTC-3′	96
IGS1	rRNA intergenic spacer	2	**IGSF** 5′-ATCCTTTGCAGACGACTTGA-3′ **IGSR** 5′-GTGATCAGTGCATTGCATGA-3′	94°C 3 min; 35 cycles: 94°C 30 s, 60°C 30 s, 72°C 1 min **Alternative conditions:** 30 cycles: 94°C 30 s, 56°C 30 s, 72°C 1 min	723	5′-TAAGCCCTTGTT A**A**AGAT**T**TA**TTG**-3′	88

[a] The sequences of the genome of strain H99 (*C. neoformans* var. *grubii*, VNI) at the Broad Institute (http://www.broad.mit.edu) were used as the master sequences. Nucleotide bases shown in bold denote nucleotide bases that could vary between the different molecular types.

[b] If not otherwise specified, primers listed will work for *C. neoformans* and *C. gattii*.

TABLE 4 Standard/reference strains for *C. neoformans* and *C. gattii* strain typing[a]

Species and variety	Molecular type[b]	CBS #	ATCC#	FGS#	Other numbers	MAT and serotype	Comments	References
C. neoformans								
C. neoformans var. *grubii*	VNI (93, 96) = AFLP1 (6) = VN6 (VN5) (135)	CBS 10085	ATCC MYA-4564	10415	WM 148; W10; Brown	αA	1989, Australia, NSW, Sydney, clinical, CSF[c], HIV–, isolated by Sharon Chen	24, 93
		CBS 8710	ATCC 48922	9487	DUMC 135.97; H99; NYSD 1649; CBS 10515; WM 04.15	αA	1978, U.S., NC, Durham, clinical, CSF, patient with Hodgkin's lymphoma, isolated by John Perfect/Wiley Schell, type culture of *C. neoformans* var. *grubii*, genome sequence strain	47
	VNII (93, 96) = AFLP1A (6) = VN7 (135)	CBS 10084	ATCC MYA-4565	10416	WM 626, W20; Cetin	αA	1993, Australia, NSW, Sydney, clinical, CSF, HIV–, isolated by Sharon Chen	24, 93
AD hybrid	VNIII (93, 96) = AFLP3 (6) = VN3 + VN4 (135)	CBS 10080	ATCC MYA-4566	10417	WM 628; 88B5400; Zapf	αA/aD	1988, Australia, VIC, Melbourne, clinical, CSF, HIV+, isolated by Bryan Speed	24, 93
		CBS 132	ATCC 32045	—	CCRC 20528; DBVPG 6010; IFO 0608; IGC 3957; NRRL Y-2534	αA/aD	1894, Italy, environmental, fermenting fruit juice, isolated by F. Sanfelice, type culture for *C. neoformans*	108
C. neoformans var. neoformans	VNIV (93, 96) = AFLP2 (6) = VN1 (VN2) (135)	CBS 10079	ATCC MYA-4567	10418	WM 629; B 87455, Borg, F 14	αD	1987, Australia, VIC, Melbourne, clinical, blood, HIV+, isolated by Bryan Speed	93
		CBS 6900	ATCC 34873	10423	B-3501; DBVPG 6228; CBS 7697	αD	1975, U.S., MD, Bethesda, NIH, crossing of NIH 12 × NIH 433, isolated by June Kwon-Chung	128
C. gattii	VGI (96) = AFLP4 (6)	CBS 10078	ATCC MYA-4560	10419	WM 179; Bryon; H33.1; MH56	αB	1993, Australia, NSW, Sydney, clinical, CSF, HIV–, isolated by Sharon Chen	24, 96
		CBS 6289	ATCC 32269	–	MUCL 30449, RV 20186; CBS 8273	aB	1966, Congo, Kinshasa, clinical, CSF, isolated by F. Gatti/R. Eeckels, type strain of *C. neoformans* var. *gattii*	51

	CBS 10510	–	–	WM 276; TCS –SC1	αB	1993, Australia, NSW, Mt Annan National Park, environmental, *Eucalyptus tereticornis* woody debris, isolated by Tania Sorrell/ Sharon Chen, genome sequence strain	96
VGII (96) = AFLP6 (6)	CBS 10082	ATCC MYA-4561	10420	WM 178; 49435; Colter; IFM 50894	αB	1991, Australia, NSW, Sydney, clinical, CSF, HIV–, isolated by Sharon Chen	96
	CBS 10514	–	–	CDC R265; WM 02.32	αB	2001, Canada, BC, Duncan, Vancouver Island, clinical, bronchial wash, isolated by British Columbia CDC, high virulent Vancouver Island outbreak strain, VGIIa, genome sequence strain	67
VGIII (96) = AFLP5 (6)	CBS 10081	ATCC MYA-4562	10421	WM 175; WM 161; E698; 689; TP 0689; D1.13H	αB	1992, U.S., California, San Diego, Blind Recreation Center/Park Boulevard UPAS street, environmental, *Eucalyptus* spp. woody debris, isolated by Tania Pfeiffer/David Ellis	96, 116
	CBS 6955	ATCC 32608	10424	DBVPG 6225; MUCL 30454; NIH 191; CBS 6916	aC	Before 1970, U.S., San Fernando, CA, clinical, CSF	74
VGIV (96) = AFLP7 (11)	CBS 10101	ATCC MYA-4563	10422	WM 779; King Cheetah; IFM 50896	αC	1994, South Africa, Johannesburg, veterinary, cheetah, isolated by Valarie Davis	9, 96

[a] See reference 95. CBS, CBS-KNAW Fungal Biodiversity Centre (http://www.cbs.knaw.nl); ATCC, American Type Culture Collection (http://www.atcc.org); FGS, Fungal Genetic Stock Center (http://www.fgsc.net).
[b] = indicates alternative designation(s); references given in parentheses.
[c] CSF, cerebrospinal fluid.

between environmental and clinical cryptococcal isolates in Colombia, suggesting that patients may have acquired the infection from the environment. The same study also revealed that the molecular type VGII is the most prevalent molecular type among *C. gattii* serotype B isolates. Six subtypes of VGII were identified (VGIIcoA and VGIIcoC to VGIIcoG), with VGIIcoA containing both environmental and clinical isolates (40).

In the second study from Brazil, all data from previously published cryptococcal strains and data obtained from new strains were reanalyzed in an effort to understand the regional distribution of the major molecular types using statistics and descriptive analysis (125). Of 320 *C. neoformans* isolates, 69 were environmental strains, and 37 out of the 123 *C. gattii* isolates were environmental isolates. PCR fingerprinting (M13, *URA5* RFLP, and AFLP analysis) identified VNI as the most common molecular type found, followed by VGII. *C. gattii* predominantly occurred in the north (odds ratio, 5.4; $P < 0.001$), and *C. neoformans* in the south (odds ratio, 2.6; $P < 0.001$). These Brazilian epidemiological data suggest a geographical north-south trend in the *C. gattii* population. In the northern Brazilian states (Amazonas, Bahia, Pernambuco, Piauí, and Roraima), *C. gattii* is endemic, but in the southern region (Mato Grosso do Sul, Minas Gerais, Paraná, Rio de Janeiro, Rio Grande do Sul, and São Paulo) infections due to *C. gattii* are only sporadically reported (125).

It is important to draw attention to the fact that *C. gattii* VGII is a common genotype in South America (96). Evidence shows that it is particularly well adapted to environmental biotopes associated with wood decay (79, 124). This eco-epidemiological characteristic of the VGII genotype is not a recent event but has been recognized for at least the past 20 years in Brazil and in other areas of Latin America (124, 125).

The unprecedented ongoing outbreak of *C. gattii*, mating type α isolates with high sexual fecundity in the temperate climate of Vancouver Island, British Columbia, Canada, was an exception to the ecological pattern recognized for *C. gattii* until 1999 (67) and reinforces the importance of continued surveillance of this primary pathogen. Numerous ecological and molecular studies have been carried out in order to determine the environmental source of the outbreak, the molecular and mating types of clinical and environmental isolates, and the extent of genetic diversity. PCR fingerprinting and AFLP analysis of 67 environmental *C. gattii* isolates (58 obtained from 25 native trees, 8 from air samples, and 1 from soil and tree debris) recovered from 732 environmental samples taken until June 2002 across 16 different sites in British Columbia indicated that they all belonged to the molecular type VGII/AFLP6, with 59 belonging to the subtype VGIIa/AFLP6A and 8 to VGIIb/AFLP6B (67). There was no genetic variation observed within the two different *C. gattii* lineages identified on Vancouver Island or the closely surrounding lower mainland of British Columbia. Moreover, the comparison of selected VGII/AFLP6 strains from different areas in the world with those from Vancouver Island revealed high genotypic similarities, thus suggesting that these genotypes have the potential to cause infection in temperate climates around the world (67).

A follow-up study compared the molecular variation among *C. gattii* isolates from British Columbia with isolates obtained from different areas of the world using a comparative gene genealogy approach based on four genes (*CAP1*, *FTR1*, *LAC1*, and *URA5*) (66). When only the VGII isolates were considered, the four genes had highly congruent genealogies ($P = 0.734$), with increased congruence observed when the four VGII *MAT***a** isolates were excluded from this analysis ($P = 0.810$). The results confirmed previous findings (67) that the *C. gattii* population in British Columbia is clonal, possesses a level of nucleotide sequence diversity that is equivalent to that of the global population, and shares identical MLST profiles with many isolates collected in other parts of the world (66). *C. gattii* isolates from Vancouver Island and the mainland of British Columbia do not appear to be exceptional in terms of their genotypes when compared to those from other areas of the world. These data suggest that sexual recombination had occurred on a global scale between *C. gattii* isolates of different VGII subgenotypes, which led to the *C. gattii* population in British Columbia, which is characterized by a predominantly clonal mode of reproduction.

In an attempt to investigate the spread of the two Vancouver Island outbreak genotypes across Vancouver Island, the mainland of British Columbia, and parts of the northwestern United States, environmental isolates were recovered from tree surfaces, soil, air, freshwater, and seawater. No seasonal prevalence was observed. For the first time it was established that *C. gattii* concentrations in air samples were significantly higher during the warm, dry summer months, although potentially infectious propagules (<3.3 μm in diameter) were present throughout the year. Some locations were designated colonization "hot spots" and were characterized for acidic soil, low moisture, and low organic carbon content (65). Isolates of VGIIa, VGIIb, and VGI were coisolated (65). Strains that are closely related, but not identical, to the Vancouver Island outbreak isolates have been identified in Oregon (17, 89). Based on MLST profiles, Vancouver Island outbreak VGIIa strains and other *C. gattii* isolates from North America are identical, as are the VGIIb strain and isolates from Australia and Southeast Asia, thus providing clues to the potential origins of the *C. gattii* isolates from British Columbia (45, 94, 97).

The environmental samples belonging to the *C. neoformans*/*C. gattii* species complex that have been collected worldwide are related to plant materials, different kinds of bird droppings, soil, dwellings, insects, air samples, fruit juice, animal cages, and bat guano (20, 46, 50, 65, 69, 98, 103, 121, 124, 125). Decomposing wood from tree hollows has been described as the natural source of both *C. neoformans* and *C. gattii* (69, 79). All four major molecular types of *C. gattii* (VGI, VGII, VGIII, and VGIV) and three molecular types of *C. neoformans* have been isolated from trees and plant materials in Argentina, Australia, Brazil, Canada, China, Colombia, India, Mexico, the United States, and Zaire, with VGII for *C. gattii* (45.2%) and VNI for *C. neoformans* (29.4%) being the most prevalent (6, 20, 40, 43, 44, 61, 64, 67–69, 96, 124, 125). Only the AD hybrid, VNIII, is an exception that has been isolated only from bird droppings (Europe and Japan) and fruit juice (98, 108). Unlike *C. gattii*, *C. neoformans* has been correlated with bird droppings in the environment. The molecular types VNI (86%), VNII (4%), VNIII (4%), and VNIV (5%) have been isolated from avian habitats and bird droppings in North and South America, Europe, Asia, and Australia (1, 6, 20, 40, 43, 44, 61, 64, 67–69, 96, 124, 125).

CURRENT KNOWLEDGE OF THE GLOBAL DISTRIBUTION OF THE MAJOR GENOTYPES

The worldwide distribution of the eight major molecular types within the *C. neoformans*/*C. gattii* species complex is based on the integrated analysis of 2,755 cryptococcal isolates investigated in several studies (6, 20–22, 40, 43, 61, 64, 67, 68, 93, 96, 124, 125, 133; W. Meyer, unpublished data) (Fig. 6A and Fig. 7). Among the 2,046 clinical, 68 veterinary, and 604 environmental isolates, the clear majority were *C. neoformans* isolates, with VNI (serotype A) being globally the major molecular type: 63% of clinical/veterinary and 41% of environmental origin (Fig. 6B and 6C). Infections due to any of the *C. gattii* major molecular types are substantially less common, totaling 20% compared to 80% caused by the four major molecular types of *C. neoformans*. Looking at only the environmental isolates, the two species seem to be equally present among the globally isolated strains: 48% *C. neoformans* and 52% *C. gattii* (Fig. 6C). A difference between clinical/veterinary and environmental isolates was noted: more of the clinical isolates were VGI (9%), and only 7% were VGII, compared to the environmental isolates, of which 35% were VGII and only 9% were VGI. This may be due to the extensive sampling efforts in connection with the Vancouver Island cryptococcosis outbreak that was caused by VGII isolates. A clear difference occurs in the distribution of *C. neoformans* and *C. gattii*, with a global prevalence of *C. neoformans* and a higher prevalence of *C. gattii* in the Americas and the Southern Hemisphere (Fig. 7). *C. gattii* was only rarely isolated from patients or the environment in Europe, including Russia and parts of Asia, especially China, Thailand, and Japan.

The majority of infections in immunocompetent and immunocompromised patients are caused by *C. neoformans*, molecular type VNI, serotype A. All major molecular types of *C. neoformans* caused consistently more infections in immunocompromised patients, and all major molecular types of *C. gattii* caused consistently more infections in immunocompetent patients (Fig. 8A). An interesting finding was observed in the distribution of the major molecular types

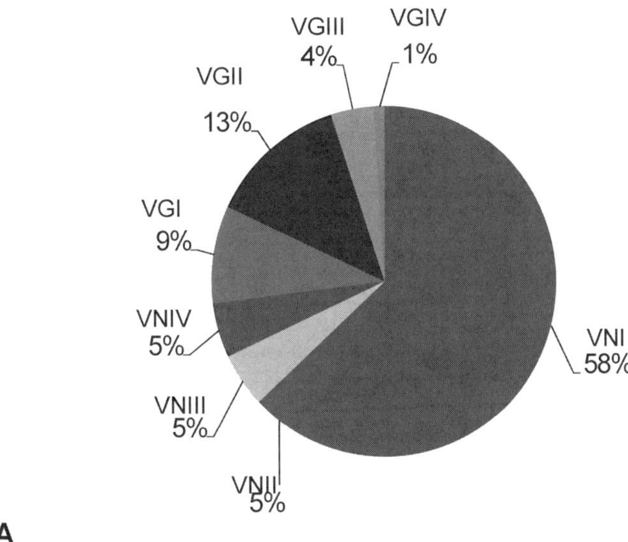

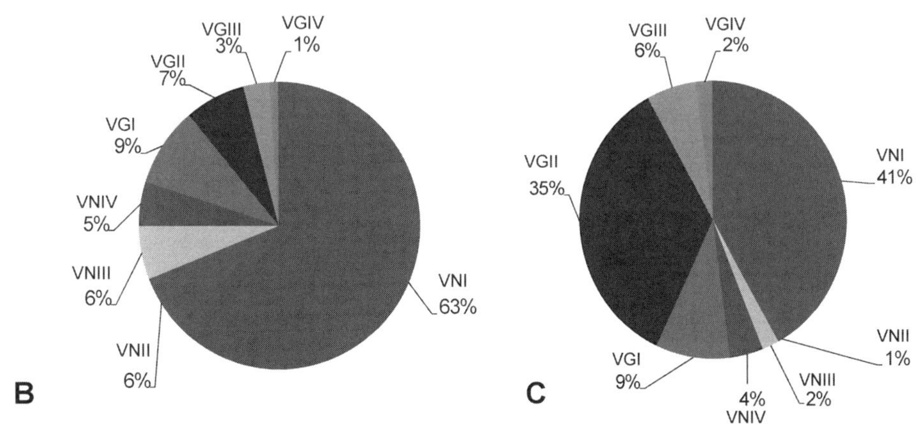

FIGURE 6 (A) Distribution of the eight major molecular types identified among 2,755 globally collected isolates (6, 20–22, 40, 43, 61, 64, 67, 68, 93, 96, 124, 125, 133). (B) Data for 2,046 clinical and 68 veterinary cryptococcal strains obtained from 48 countries. (C) Data for 604 environmental isolates obtained from 21 countries. Thirty-seven isolates were excluded from the analysis due to missing source information.

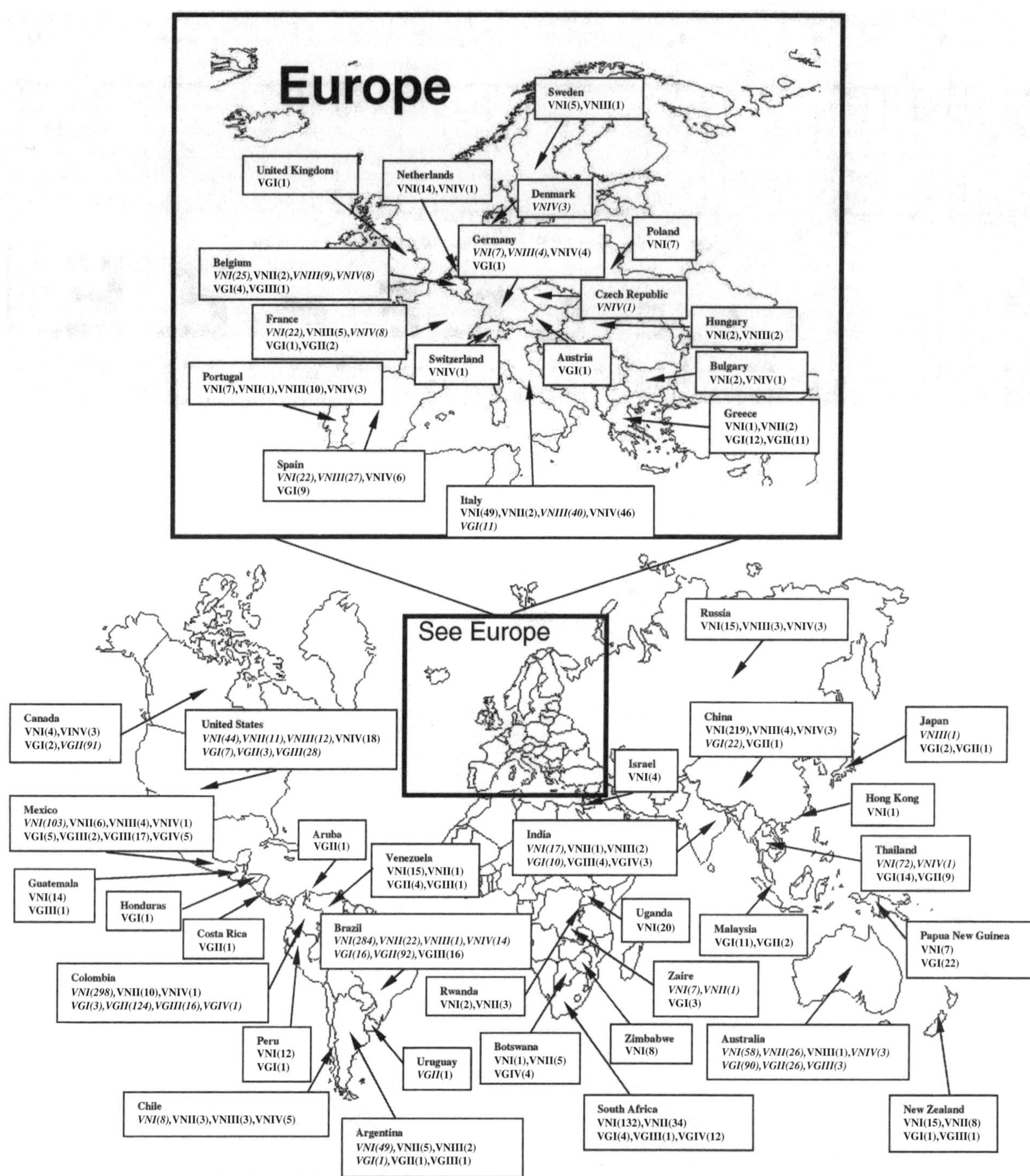

FIGURE 7 Global distribution of the eight major molecular types of the *C. neoformans*/*C. gattii* species complex. Numbers in italics indicate that clinical and environmental samples have been reported for this molecular type. Nonitalic numbers indicate that only clinical strains have been reported. Data have been combined from the following references: 6, 20–22, 40, 43, 61, 64, 67, 68, 93, 96, 124, 125, 133.

when the available data were analyzed with and without the Chinese isolates. Globally, the majority of the infections in immunocompetent patients were caused by isolates of *C. gattii* molecular types VGI and VGII (Fig. 8C). A strikingly different picture was obtained among the Chinese isolates, where the majority of infections in immunocompetent patients were caused by *C. neoformans* molecular type VNI (Fig. 8B) (22, 26, 43, 44, 142, 143). Similar findings have been reported from Korea (139). The reasons for this dramatic shift are unknown. This finding is especially

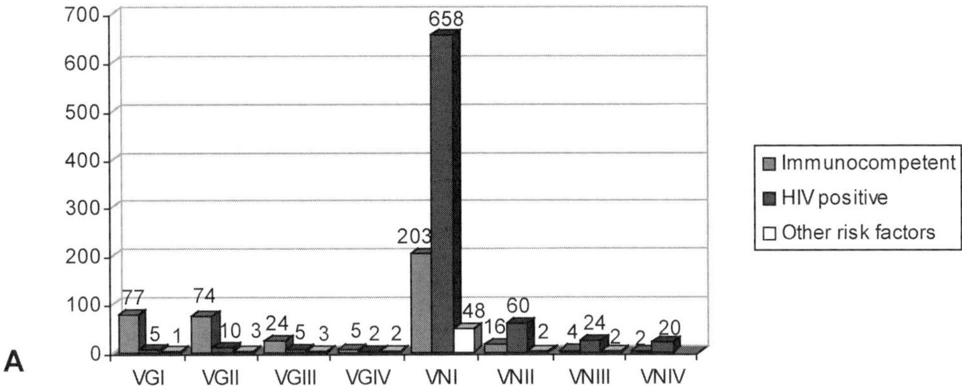

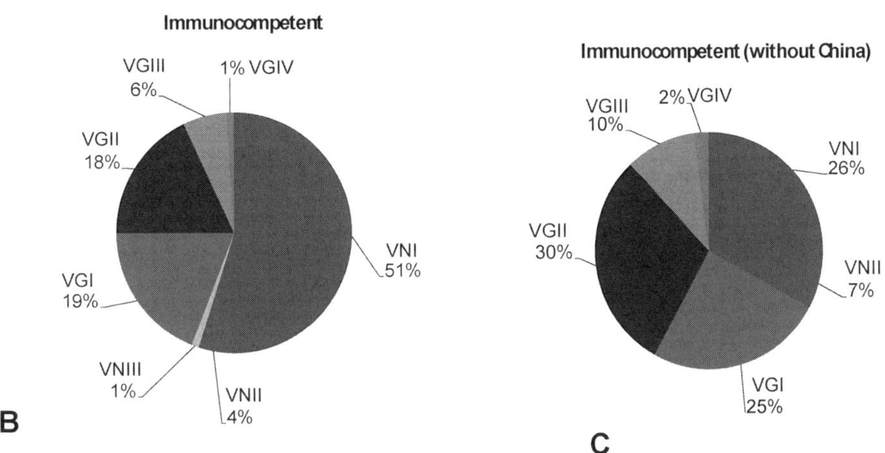

FIGURE 8 (A) Numbers of clinical isolates obtained from immunocompetent and immunocompromised patients (HIV-positive) and patients with other risk factors per major molecular type from a total of 1,250 clinical strains from a total of 2,046 strains investigated by PCR fingerprinting with the primer M13 and AFLP analysis for which the clinical data were available (6, 20–22, 40, 43, 61, 64, 67, 68, 93, 96, 124, 125, 133). Other risk factors are alcoholism, corticosteroid use, disorder T immunity, diabetes, leukemia, systemic lupus erythematosus, transplant, and tumor. (B) Distribution of the major molecular types including all clinical isolates obtained from immunocompetent patients. (C) Distribution of the major molecular types obtained from immunocompetent patients excluding the Chinese data (22, 26, 44).

intriguing since the opposite is the case in nearby countries such as India, Thailand, and Japan, where the majority of infections caused by *C. neoformans* molecular type VNI are found in immunocompromised patients (28, 58, 100, 120).

A shift in the geographical distribution was observed for molecular types VNIII (AD hybrids) and VNIV (serotype D), which have been previously described mainly from southern Europe including Italy and France (36, 50, 73, 133, 135). They occur also in relatively high numbers in Latin America, mainly Mexico, Colombia, Brazil, and Chile (Fig. 7) (96, 125).

During analysis of the environmental data, a dramatic shift in the distribution of the major molecular types was found. In South America the most common molecular types isolated in the environment are VNI (46%) and VGII (36%) (Fig. 9D). VGII is the most common (61%) in North America (Fig. 9C), which is probably due to the large Vancouver Island cryptococcosis outbreak in Canada caused by this genotype, followed by VNI (29%). In contrast, in Oceania, including Australia, the most commonly found molecular type is VGI (70%) (Fig. 9A), and in Europe the most common is VNI (40%) (Fig. 9B). No comments can be made for Africa and Asia, due to the small number of isolates, 5 and 13, respectively. The high number of VGII isolates found in the Americas (Fig. 7), in connection with the phylogenetic placement of VGII as an ancestral population within the *C. gattii* clade (Fig. 10), is in agreement with the original reporting of *C. gattii* being found in high numbers in South America (73), reinforcing the possibility of placing the origin of this species within South America (102), from where it dispersed globally. In addition, it also became clear that *C. gattii* has extended its ecological niche from being restricted to tropical and subtropical areas, as originally reported (73), to temperate regions, e.g., Argentina, Canada, and Greece (Fig. 7) (17, 28, 40, 67, 89, 131).

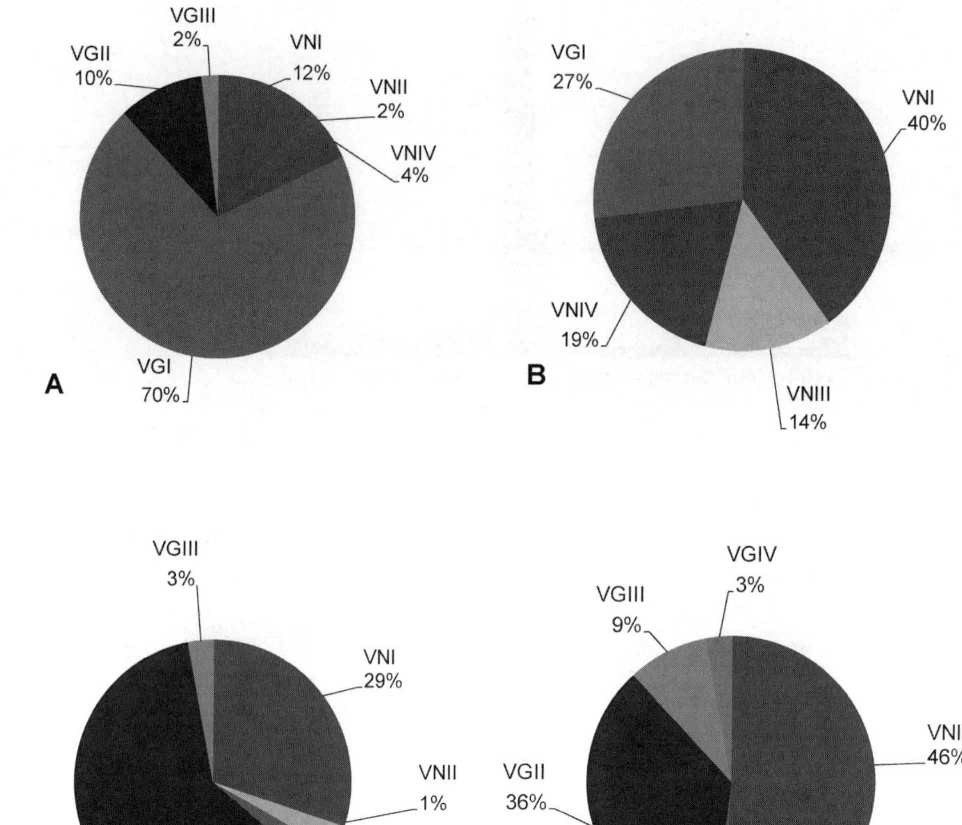

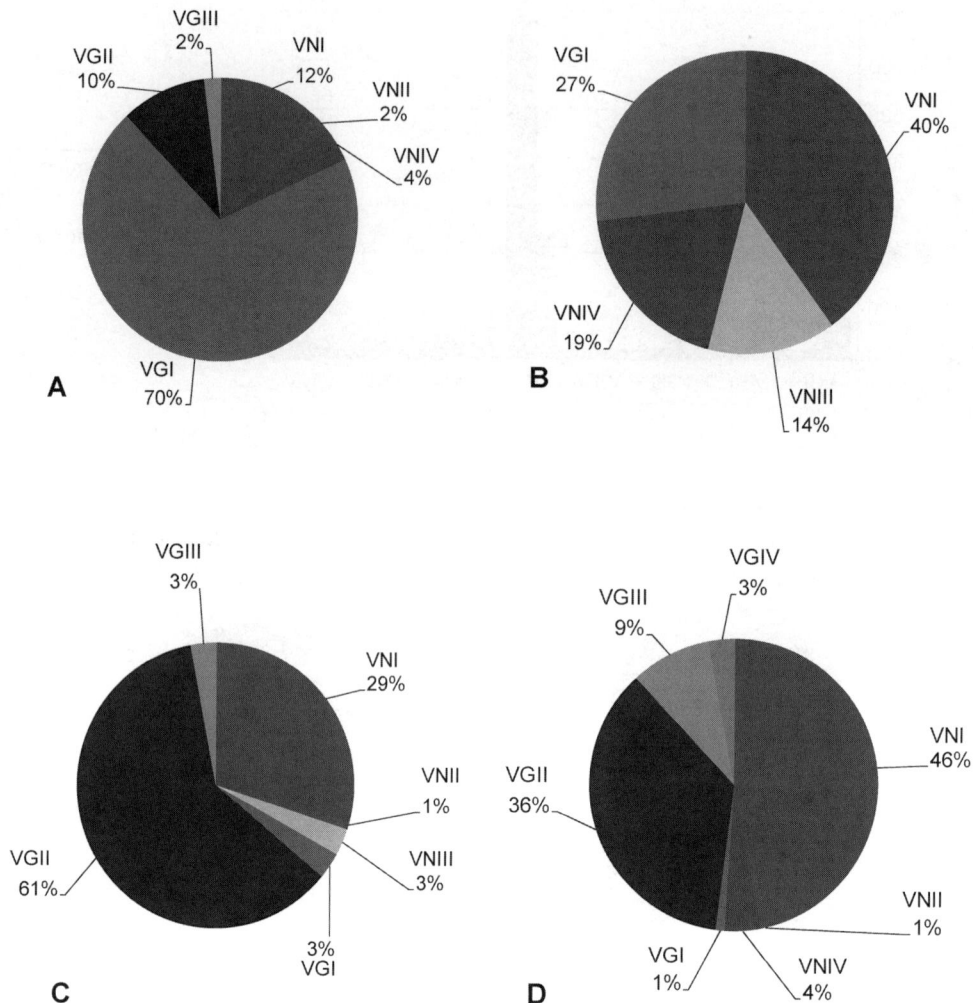

FIGURE 9 Distribution of the major molecular types among 586 environmental isolates, excluding the 5 African and 13 Asian environmental isolates (6, 20, 40, 44, 61, 64, 67–69, 96, 124, 125). (A) Data for Oceania ($n = 51$). (B) Data for Europe ($n = 37$). (C) Data for North America ($n = 111$). (D) Data for South America ($n = 387$), indicating a different prevalence between VGI and VGII in the Americas as compared to the rest of the world.

PHYLOGENETIC RELATIONSHIPS OF THE MAJOR MOLECULAR TYPES

The phylogenetic relationships between the major molecular types of the *C. neoformans*/*C. gattii* species complex have been investigated in two major studies (11, 102). The first study focused on the phylogenetic relationships of the AFLP types using six concatenated genetic loci (ITS1/2, IGS, *CNLAC1*, *RPB1*, *RPB2*, and *TEF1α*). The AFLP1/VNI and AFLP1A/VNII genotypes were found to have a sister group relationship, with genotype AFLP1B/VNII occurring as a basal lineage, and it was suggested that mating between isolates of genotype AFLP1A/VNII and AFLP1B/VNII may be possible (11), thus suggesting the presence of a panmictic entity (e.g., a species) comprising three subgenotypes that may represent infraspecific taxa. The currently recognized four AFLP types of *C. gattii* were found to be fully concordant with the results of a phylogenetic analysis using six concatenated nuclear loci (11). However, an analysis of two mitochondrial loci (MtLrRNA and *ATP6*) suggested recombination of mitochondrial loci, which was not detected in the six nuclear loci studied (12). In the nuclear gene analysis, AFLP6/VGII isolates occurred as a basal lineage to all other *C. gattii* genotypes, with AFLP7/VGIV clustering basal to the sister groups AFLP4/VGI and AFLP5/VGIII (11).

The second study focused on the phylogenetic relationships of the PCR fingerprinting types, using four concatenated genetic loci (*ACT1*, *IDE1*, *PLB1*, and *URA5*) (102). Separate or combined sequence analysis of all four loci revealed seven major clades, three within *C. neoformans* (VNI/VNB = AFLP1, VNII = AFLP1A/AFLP1B, and VNIV = AFLP2) and four within *C. gattii* (VGI = AFLP4, VGII = AFLP6, VGIII = AFLP5, and VGIV = AFLP7). The topology of the individual gene trees was identical for each of the clades of *C. neoformans*, but was found to be incongruent for the clades of *C. gattii*, indicating recent recombination events within *C. gattii* (102). The sequence variation among VGI, VGIII, and VGIV was in the same range as

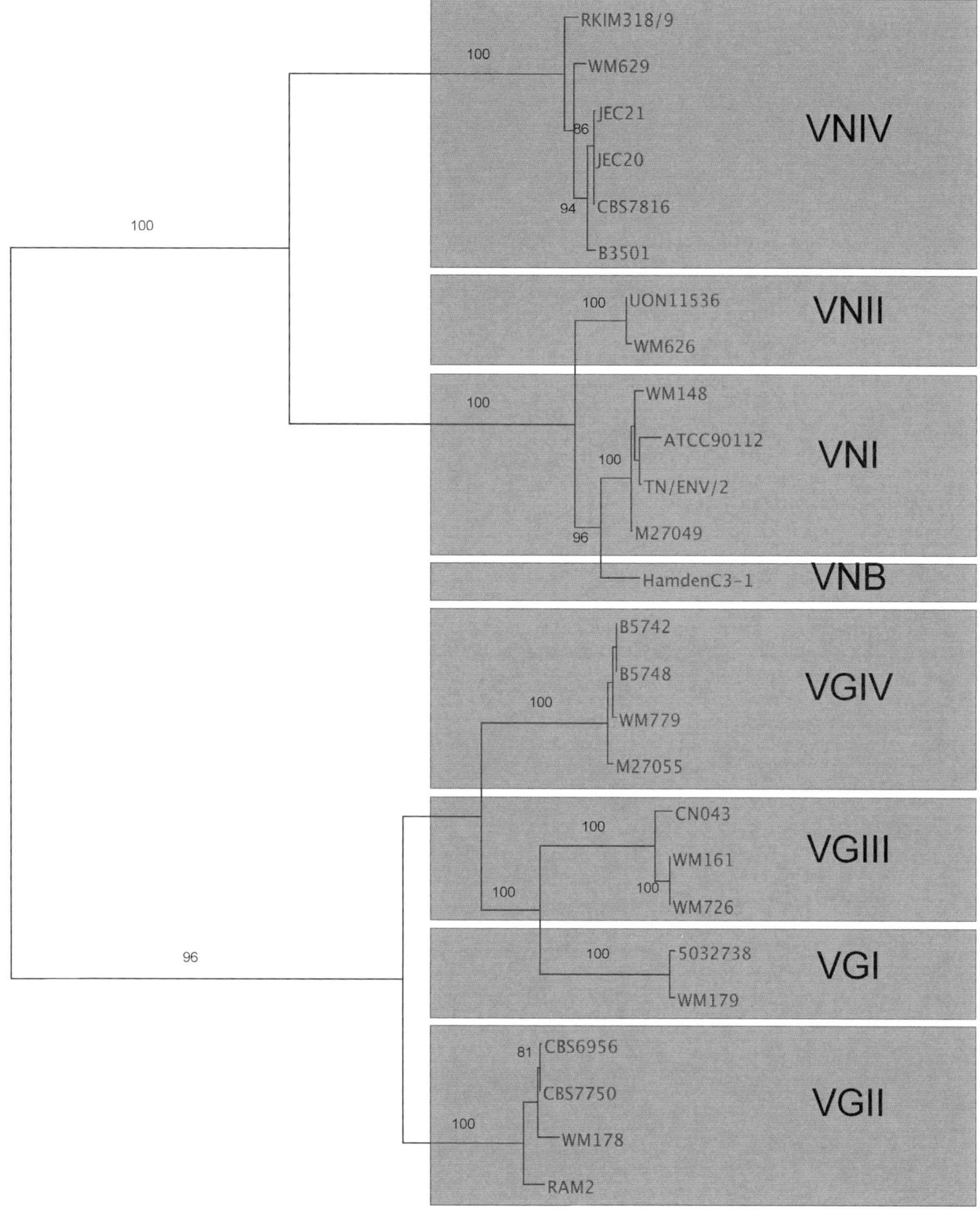

FIGURE 10 One of the two most parsimonious trees obtained from heuristic searches based on analysis produced from 10 combined genes (*ACT1*, *IDE1*, ITS1/2, IGS, *LAC1*, *PLB1*, *RPB1*, *RPB2*, *TEF1*α, and *URA5*). Bootstrap support values above 70% are indicated at the nodes. The tree was rooted at the midpoint.

that between VNI/VNB and VNII and VNIV, indicating that all seven haploid monophyletic lineages warrant at least varietal, if not species, status. This study also showed that the VGII = AFLP6 molecular type is basal to the *C. gattii* clade, as had been shown by Bovers et al. (11), thus supporting the proposal that VGII forms the ancestral group for all *C. gattii* clades (102). This may be relevant, as hypervirulent isolates of VGII/AFLP6 are involved in outbreaks on Vancouver Island (British Columbia, Canada) and in the Pacific Northwest (17, 89) as well as South America (97) including Colombia (40) and Brazil (106, 124, 125). Strong evidence of recombination in the global VGII population was found, providing the basis for the emergence of highly virulent strains (102).

A combined study performed for this chapter using sequences of all 10 loci (*ACT1*, *CNLAC1*, *IDE1*, ITS1/2, IGS, *PLB1*, *RPB1*, *RPB2*, *TEF1*α, and *URA5*) of strains included in both previous studies (11, 102) confirmed the three major monophyletic lineages within *C. neoformans*, with VNB being closely related and ancestral to VNI, and four major monophyletic lineages within *C. gattii* (Fig. 10). In addition, it also confirmed the basal position of VNIV for the *C. neoformans* clade and VGII for the *C. gattii* clade, reinforcing the ancestral role of both major molecular types for each of these two lineages (Fig. 10) (F. Gilgado and W. Meyer, unpublished data). This combined analysis supports the notion that each of those seven monophyletic lineages within the *C. neoformans*/*C. gattii* species complex deserves at least variety, if not species status.

The divergence times among the seven identified haploid monophyletic lineages within the *C. neoformans*/*C. gattii* species complex were estimated based on the phylogenetic study carried out by Ngamskulrungroj et al. (102) by applying a molecular clock hypothesis. Estimates of the time since divergence and hybridization were calculated assuming the consensus mutation rate of 2×10^{-9} per nucleotide per year for protein coding genes (82, 139). Maximum-likelihood estimations with and without a molecular clock demonstrated that the mutation rates of each gene did not differ significantly from molecular clock expectations ($P = 0.99$). Comparing the two species, the lineages within *C. gattii* evolved later (12.5 million years ago) than the major *C. neoformans* lineages, *C. neoformans* var. *grubii* and *C. neoformans* var. *neoformans* (24.5 million years ago), suggesting more recent speciation events (Fig. 11B). However, the clades of *C. gattii* diverged prior to the two monophyletic lineages within *C. neoformans* var. *grubii*, VNI and VNII, which seems to be the result of the most recent speciation event, which took place around 4.5 million years ago (Fig. 11B). The divergence time between each *C. gattii* molecular-type clade was estimated to be 11.7 million years ago for the split between VGIV and VGI/VGIII and 8.5 million years ago for the split between VGI and VGIII (102). The divergence between VGII and VGI/VGIII/VGIV, estimated to be 12.5 million years ago (102), is in a similar range as the previously obtained divergence between the two varieties of *C. neoformans*, *C. neoformans* var. *grubii* and *C. neoformans* var. *neoformans* (Fig. 11A) (139). The divergence times (102) are approximately in agreement with previous studies, which used a different set of genetic loci (139), but are closer to estimations based on a comparison of the available whole genomes, which suggested a split between the two species 80 (16 to 160) million years ago (Fig. 11C) (112).

CONCLUDING REMARKS

The past two decades have seen a large increase in molecular typing studies carried out to understand the genetic variation

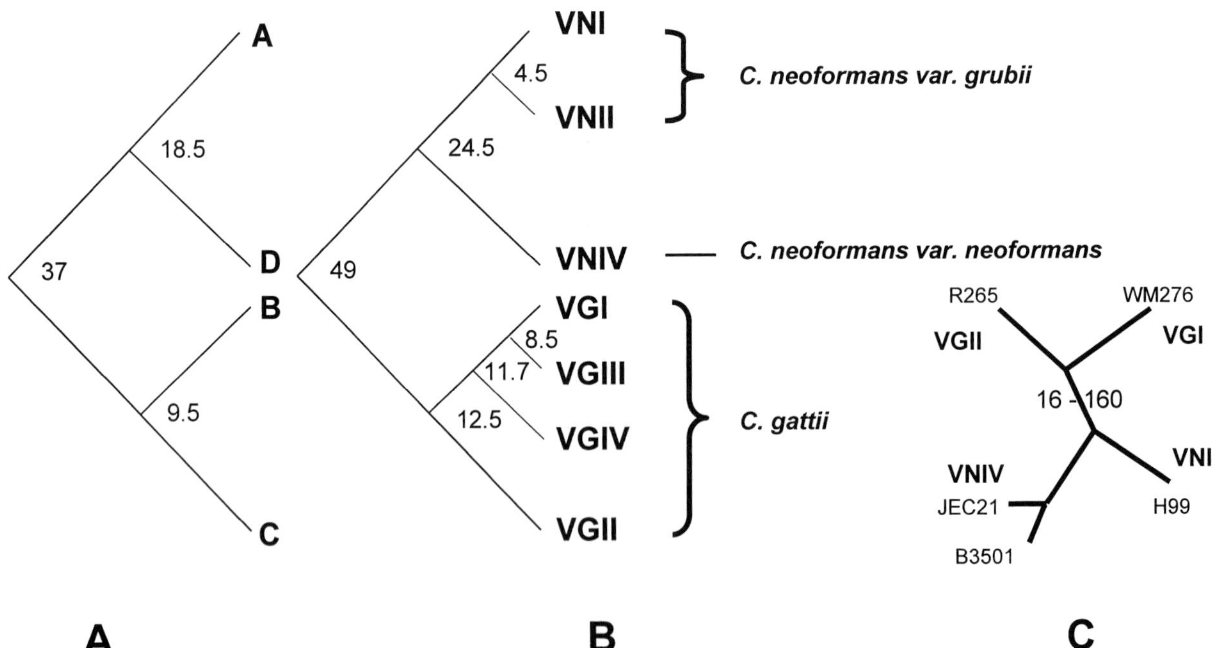

FIGURE 11 Molecular divergence between the major haploid molecular types of the *C. neoformans*/*C. gattii* species complex in million of years (A) based on the *LAC1* and *URA5* genes (139), (B) based on the *ACT1*, *IDE1*, *PLB1*, and *URA5* genes (102), and (C) based on the analysis of the whole genomes (112).

of cryptococcal strains, with two major typing methods emerging, namely, PCR fingerprinting (96) and AFLP analysis (6). This has led to the establishment of the *C. neoformans/C. gattii* species complex comprising at least seven monophyletic lineages (major molecular types: VNI/VNB, VNII, VNIV, VGI, VGII, VGIII, VGIV) (Fig. 10) and a better understanding of the distribution of those major molecular types (Fig. 7) (11, 102). In the case of this study it was found that *C. gattii* has a much larger global distribution than previously considered (73) and that this species is not restricted to tropical and subtropical regions (Fig. 7) (67, 96, 125). The molecular types VNIII (the AD hybrids) and VNIV (serotype D) are not only mainly found in southern Europe (36, 50, 133, 135), but are also present in the Americas (Fig. 7) (96, 125).

Due to the lack of interlaboratory reproducibility associated with the two typing techniques mentioned above, MLST (45, 88) and MLMT (56, 63) analysis have been developed. In a communal effort to standardize future genotyping efforts, an MLST typing scheme using seven unlinked genetic loci (*CAP59*, *GPD1*, IGS1, *LAC1*, *PLB1*, *SOD1*, and *URA5*) (Table 2) has recently been chosen as the universal typing method to be commonly applied for future studies. Furthermore, a set of standard reference strains (Table 4) has been selected, allowing a global population diversity study of the *C. neoformans/C. gattii* species complex to understand the origin of the species and their global spread (95).

REFERENCES

1. **Abegg, M. A., F. L. Cella, J. Faganello, P. Valente, A. Schrank, and M. H. Vainstein.** 2006. *Cryptococcus neoformans* and *Cryptococcus gattii* isolated from the excreta of *Psittaciformes* in a southern Brazilian zoological garden. *Mycopathologia* **161:**83–91.
2. **Almeida, A. M. F., M. T. Matsumoto, L. C. Baeza, R. B. de Oliveira e Silva, A. A. P. Kleiner, M. de Souza Carvalho Melhem, M. J. S. M. Giannini, and the Laboratory Group on Cryptococcosis.** 2006. Molecular typing and antifungal susceptibility of clinical sequential isolates of *Cryptococcus neoformans* from Sao Paulo State, Brazil. *FEMS Yeast Res.* **7:**152–164.
3. **Barreto de Oliveira, M. T., T. Boekhout, B. Theelen, F. Hagen, F. A. Baroni, M. S. Lazera, K. B. Lengeler, J. Heitman, I. N. Rivera, and C. R. Paula.** 2004. *Cryptococcus neoformans* shows a remarkable genotypic diversity in Brazil. *J. Clin. Microbiol.* **42:**1356–1359.
4. **Bertout, S., F. Renaud, D. Swinne, M. Mallie, and J. M. Bastide.** 1999. Genetic multilocus studies of different strains of *Cryptococcus neoformans*: taxonomy and genetic structure. *J. Clin. Microbiol.* **37:**715–720.
5. **Boekhout, T., M. Renting, W. A. Scheffers, and R. Bosboom.** 1993. The use of karyotyping in the systematics of yeasts. *Antonie Van Leeuwenhoek* **63:**157–163.
6. **Boekhout, T., B. Theelen, M. Diaz, J. W. Fell, W. C. Hop, E. C. Abeln, F. Dromer, and W. Meyer.** 2001. Hybrid genotypes in the pathogenic yeast *Cryptococcus neoformans*. *Microbiology* **147:**891–907.
7. **Boekhout, T., and A. van Belkum.** 1997. Variability of karyotypes and RAPD types in genetically related strains of *Cryptococcus neoformans*. *Curr. Genet.* **32:**203–208.
8. **Boekhout, T., A. van Belkum, A. C. Leenders, H. A. Verbrugh, P. Mukamurangwa, D. Swinne, and W. A. Scheffers.** 1997. Molecular typing of *Cryptococcus neoformans*: taxonomic and epidemiological aspects. *Int. J. Syst. Bacteriol.* **47:**432–442.
9. **Bolton, L. A., R. G. Lobetti, D. N. Evezard, J. A. Picard, J. W. Nesbit, J. van Heerden, and R. E. Burroughs.** 1990. Cryptococcosis in captive cheetah (*Acinonyx jubatus*): two cases. *J. S. Afr. Vet. Assoc.* **70:**35–39.
10. **Bovers, M., F. Hagen, and T. Boekhout.** 2008. Diversity of the *Cryptococcus neoformans - Cryptococcus gattii* species complex. *Rev. Iberoam. Micol.* **25:**S4–S12.
11. **Bovers, M., F. Hagen, E. E. Kuramae, and T. Boekhout.** 2008. Six monophyletic lineages identified within *Cryptococcus neoformans* and *Cryptococcus gattii* by multi-locus sequence typing. *Fungal Genet. Biol.* **45:**400–421.
12. **Bovers, M., F. Hagen, E. E. Kuramae, and T. Boekhout.** 2009. Promiscuous mitochondria in *Cryptococcus gattii*. *FEMS Yeast Res.* **9:**489–503.
13. **Bovers, M., F. Hagen, E. E. Kuramae, M. R. Diaz, L. Spanjaard, F. Dromer, H. L. Hoogveld, and T. Boekhout.** 2006. Unique hybrids between the fungal pathogens *Cryptococcus neoformans* and *Cryptococcus gattii*. *FEMS Yeast Res.* **6:**599–607.
14. **Bovers, M., F. Hagen, E. E. Kuramae, H. L. Hoogveld, F. Dromer, G. St-Germain, and T. Boekhout.** 2008c. AIDS patient death caused by novel *Cryptococcus neoformans x C. gattii* hybrid. *Emerg. Infect. Dis.* **14:**1105–1108.
15. **Brandt, M. E., S. L. Bragg, and R. W. Pinner.** 1993. Multilocus enzyme typing of *Cryptococcus neoformans*. *J. Clin. Microbiol.* **31:**2819–2823.
16. **Brandt, M. E., M. A. Pfaller, R. A. Hajjeh, E. A. Graviss, J. Rees, E. D. Spitzer, R. W. Pinner, L. W. Mayer, and the Cryptococcal Disease Active Surveillance Group.** 1996. Molecular subtypes and antifungal susceptibilities of serial *Cryptococcus neoformans* isolates in human immunodeficiency virus-associated cryptococcosis. *J. Infect. Dis.* **174:**812–820.
17. **Byrnes, E. J., 3rd, R. J. Bildfell, S. A. Frank, T. G. Mitchell, K. A. Marr, and J. Heitman.** 2009. Molecular evidence that the range of the Vancouver Island outbreak of *Cryptococcus gattii* infection has expanded into the Pacific Northwest in the United States. *J. Infect. Dis.* **199:** 1081–1086.
18. **Calvo, B. M., A. L. Colombo, O. Fischman, A. Santiago, L. Thompson, M. Lazera, F. Telles, K. Fukushima, K. Nishimura, R. Tanaka, M. Myiajy, and M. L. Moretti-Branchini.** 2001. Antifungal susceptibilities, varieties, and electrophoretic karyotypes of clinical isolates of *Cryptococcus neoformans* from Brazil, Chile, and Venezuela. *J. Clin. Microbiol.* **39:**2348–2350.
19. **Campbell, L. T., B. J. Currie, M. Krockenberger, R. Malik, W. Meyer, J. Heitman, and D. Carter.** 2005. Clonality and recombination in genetically differentiated subgroups of *Cryptococcus gattii*. *Eukaryot. Cell* **4:**1403–1409.
20. **Casali, A. K., L. Goulart, L. K. Silva, A. M. Ribeiro, A. A. Amaral, S. H. Alves, A. Schrank, W. Meyer, and M. H. Vainstein.** 2003. Molecular typing of clinical and environmental *Cryptococcus neoformans* isolates in the Brazilian State Rio Grande do Sul. *FEMS Yeast Res.* **3:**405–415.
21. **Castañón-Olivares, L. R., K. Martínez-Martínez, R. M. Bermúdez-Cruz, M. A. Martínez-Rivera, R. A. Arreguín-Espinosa, R. A. López-Martínez, W. Meyer, and G. M. Ruiz-Palacios y Santos.** 2009. Genotyping of Mexican *Cryptococcus neoformans* and *C. gattii* isolates by PCR-fingerprinting. *Med. Mycol.* **47:**713–721.
22. **Chen, J. A. Varma, M. R. Diaz, A. P. Litvintseva, K. K. Wollenberg, and K. J. Kwon-Chung.** 2008. *Cryptococcus neoformans* strains and infection in apparently immunocompetent patients, China. *Emerg. Infect. Dis.* **14:**755–762.
23. **Chen, S., T. Sorrell, G. Nimmo, B. Speed, B. Currie, D. Ellis, D. Marriott, T. Pfeiffer, D. Parr, K. Byth, and the Australasian Cryptococcal Study Group.** 2000. Epidemiology and host- and variety-dependent characteristics

of infection due to *Cryptococcus neoformans* in Australia and New Zealand. *Clin. Infect. Dis.* **31:**449–508.

24. **Chen, S. C. A., A. Brownlee, T. Sorrell, P. Ruma, D. H. Ellis, T. Pfeiffer, B. R. Speed, and G. Nimmo.** 1996. Identification by random amplification of polymorphic DNA (RAPD) of a common molecular type of *C. neoformans* var. *neoformans* in patients with AIDS. *J. Infect. Dis.* **173:**754–758.
25. **Chen, S. C. A., B. J. Currie, H. M. Campbell, D. A. Fisher, T. J. Pfeiffer, D. H. Ellis, and T. C. Sorrell.** 1997. *Cryptococcus neoformans* var. *gattii* infection in northern Australia: existence of an environmental source other than known host eucalypts. *Trans. R. Soc. Trop. Med. Hyg.* **91:**547–550.
26. **Chen, Y., F. Che, F. Wei, N. Xu, M. Yang, and J. Chen.** 2008. Cryptococcosis in China from 1985 to 2007. 7th International Conference on *Cryptococcus* & Cryptococcosis, Nagasaki, Japan, 11 to 14 September 2008, abstract #SY-01-01.
27. **Cogliati, M., M. Allaria, A. M. Tortorano, and M. A. Viviani.** 2000. Genotyping *Cryptococcus neoformans* var. *neoformans* with specific primers designed from PCR-fingerprinting bands sequenced using a modified PCR-based strategy. *Med. Mycol.* **38:**97–103.
28. **Cogliati, M., N. Chandrashekar, A. Prigitano, M. C. Esposto, B. Petrini, A. Chandramuki, and M. A. Viviani.** 2008. Clinical isolates from an Indian hospital: an unexpected detection of a serotype C *Cryptococcus gattii* population. 7th International Conference on *Cryptococcus* & Cryptococcosis, Nagasaki, Japan, 11 to 14 September 2008, abstract #SY-05-04.
29. **Cogliati, M., M. C. Esposto, A. M. Tortorano, and M. A. Viviani.** 2006. *Cryptococcus neoformans* population includes hybrid strains homozygous at mating-type locus. *FEMS Yeast Res.* **6:**608–613.
30. **Cogliati, M., M. C. Esposto, G. Liberi, A. M. Tortorano, and M. A. Viviani.** 2007. *Cryptococcus neoformans* typing by PCR fingerprinting using $(GACA)_4$ primer: a new light on the basis of *Cryptococcus neoformans* genome project data. *J. Clin. Microbiol.* **45:**3427–3430.
31. **Crampin, A. C., R. C. Matthews, D. Hall, and E. G. V. Evans.** 1993. PCR fingerprinting of *C. neoformans* by random amplification of polymorphic DNA. *J. Med. Vet. Mycol.* **31:**463–465.
32. **Currie, B. P., L. F. Freundlich, and A. Casadevall.** 1994. Restriction fragment length polymorphism analysis of *Cryptococcus neoformans* isolates from environmental (pigeon excreta) and clinical sources in New York City. *J. Clin. Microbiol.* **32:**1188–1192.
33. **Diaz, M. R., T. Boekhout, T. Kiesling, and J. W. Fell.** 2005. Comparative analysis of the intergenic spacer regions and population structure of the species complex of the pathogenic yeast *Cryptococcus neoformans*. *FEMS Yeast Res.* **5:**1129–1140.
34. **Diaz, M. R., T. Boekhout, B. Theelen, and J. W. Fell.** 2000. Molecular sequence analysis of the intergenic spacer (IGS) associated with rDNA of the two varieties of the pathogenic yeast, *Cryptococcus neoformans*. *Syst. Appl. Microbiol.* **23:**535–545.
35. **Diaz, M. R., and J. W. Fell.** 2000. Molecular analysis of the IGS & ITS regions of rDNA of the psychrophilic yeasts in the genus *Mrakia*. *Antonie van Leeuwenhoek* **77:**7–12.
36. **Dromer, F., A. Varma, O. Ronin, S. Mathoulin, and B. Dupont.** 1994. Molecular typing of *Cryptococcus neoformans* serotype D clinical isolates. *J. Clin. Microbiol.* **32:** 2364–2371.
37. **D'Souza, C. A., F. Hagen, T. Boekhout, G. M. Cox, and J. Heitman.** 2004. Investigation of the basis of virulence in serotype A strains of *Cryptococcus neoformans* from apparently immunocompetent individuals. *Curr. Genet.* **46:**92–102.
38. **Ellis, D., D. Marriott, R. A. Hajjeh, D. Warnock, W. Meyer, and R. Barton.** 2000. Epidemiology: surveillance of fungal infections. *Med. Mycol.* **38**(Suppl. 1)**:**173–182.
39. **Epplen, J., and T. Lubjuhn.** 1999. *DNA Profiling and DNA Fingerprinting*. Birkhauser, Heidelberg, Germany.
40. **Escandón, P., A. Sánchez, M. Martínez, W. Meyer, and E. Castañeda.** 2006. Molecular epidemiology of clinical and environmental isolates of the *Cryptococcus neoformans* species complex reveals a high genetic diversity and the presence of the molecular type VGII mating type a in Colombia. *FEMS Yeast Res.* **6:**625–635.
41. **Evans, E. E.** 1950. The antigenic composition of *Cryptococcus neoformans*. I. A serologic classification by means of the capsular and agglutination reactions. *J. Immunol.* **64:** 423–430.
42. **Fan, M., B. P. Currie, R. R. Gutell, M. A. Ragan, and A. Casadevall.** 1994. The 16S-like, 5.8S and 23S-like rRNAs of the two varieties of *Cryptococcus neoformans*: sequence, secondary structure, phylogenetic analysis and restriction fragment polymorphisms. *J. Med. Vet. Mycol.* **32:**163–180.
43. **Feng, X., Z. Yao, D. Ren, and W. Liao.** 2008. Simultaneous identification of molecular and mating types within the *Cryptococcus* species complex by PCR-RFLP analysis. *J. Med. Microbiol.* **57:**1481–1490.
44. **Feng, X., Z. Yao, D. Ren, W. Liao, and J. Wu.** 2008. Genotyping and mating type analyses of *Cryptococcus neoformans* and *Cryptococcus gattii* isolates from China that mainly originated from non-HIV-infected patients. *FEMS Yeast Res.* **8:**930–938.
45. **Fraser, J. A., S. S. Giles, E. C. Wenink, S. G. Geunes-Boyer, J. R. Wright, S. Diezmann, A. Allen, J. E. Stajich, F. S. Dietrich, J. R. Perfect, and J. Heitman.** 2005. Same-sex mating and the origin of the Vancouver Island *Cryptococcus gattii* outbreak. *Nature* **437:**1360–1364.
46. **Franzot, S. P., J. S. Hamdan, B. P. Currie, and A. Casadevall.** 1997. Molecular epidemiology of *Cryptococcus neoformans* in Brazil and the United States: evidence for both local genetic differences and a global clonal population structure. *J. Clin. Microbiol.* **35:**2243–2251.
47. **Franzot, S. P., I. F. Salkin, and A. Casadevall.** 1999. *Cryptococcus neoformans* var. *grubii*: separate varietal status for *Cryptococcus neoformans* serotype A isolates. *J. Clin. Microbiol.* **37:**838–840.
48. **Fries, B. C., F. Chen, B. P. Currie, A. Casadevall.** 1996. Karyotype instability in *Cryptococcus neoformans* infection. *J. Clin. Microbiol.* **34:**1531–1534.
49. **Garcia-Hermoso, D., F. Dromer, S. Mathoulin-Pelissier, and G. Janbon.** 2001. Are two *Cryptococcus neoformans* strains epidemiologically linked? *J. Clin. Microbiol.* **39:** 1402–1406.
50. **Garcia-Hermoso, D., S. Mathoulin-Pelissier, B. Couprie, O. Ronin, B. Dupont, and F. Dromer.** 1997. DNA typing suggests pigeon droppings as a source of pathogenic *Cryptococcus neoformans* serotype D. *J. Clin. Microbiol.* **35:**2683–2685.
51. **Gatti, F., and R. Eeckels.** 1970. An atypical strain of *Cryptococcus neoformans* (Sanfelice) Vuillemin. Part I. Description of the diseases and of the strain. *Ann. Soc. Belg. Med. Trop.* **50:**689–694.
52. **Georgi, A., M. Schneemann, K. Tintelnot, R. C. Calligaris-Maibach, S. Meyer, R. Weber, and P. P. Bosshard.** 2009. *Cryptococcus gattii* meningoencephalitis in an immunocompetent person 13 months after exposure. *Infection* **37:**370–373.
53. **Gil-Lamaignere, C, E. Roilides, J. Hacker, and F. M. Muller.** 2003. Molecular typing for fungi: a critical review

of the possibilities and limitations of currently and future methods. *Clin. Microbiol. Infect.* **9:**172–185.

54. **Halliday, C. L., T. Bui, M. Krockenberger, R. Malik, D. H. Ellis, and D. A. Carter.** 1999. Presence of alpha and a mating types in environmental and clinical collections of *Cryptococcus neoformans* var. *gattii* strains from Australia. *J. Clin. Microbiol.* **37:**2920–2926.
55. **Halliday, C. L., and D. A. Carter.** 2003. Clonal reproduction and limited dispersal in an environmental population of *Cryptococcus neoformans* var. *gattii* isolates from Australia. *J. Clin. Microbiol.* **41:**703–711.
56. **Hanafy, A., S. Kaocharoen, A. Jover-Botella, M. Katsu, S. Iida, T. Kogure, T. Gonoi, Y. Mikami, and W. Meyer.** 2008. Multilocus microsatellite typing for *C. neoformans* var. *grubii*. *Med. Mycol.* **46:**685–696.
57. **Haynes, K. A., D. J. Sullivan, D. C. Coleman, J. C. K. Clarke, R. Emilianus, C. Atkinson, and K. J. Cann.** 1995. Involvement of multiple *Cryptococcus neoformans* strains in a single episode of cryptococcosis and reinfection with novel strains in recurrent infection demonstrated by random amplification of polymorphic DNA and DNA fingerprinting. *J. Clin. Microbiol.* **33:**99–102.
58. **Hiremath, S. S., A. Chowdhary, T. Kowshik, A. H. Randhawa, S. Sun, and J. Xu.** 2008. Long-distance dispersal and recombination in environmental populations of *Cryptococcus neoformans* var. *grubii* from India. *Microbiology* **154:**1513–1524.
59. **Hotzel, H., P. Kielstein, R. Blaschke-Hellmessen, J. Wendisch, and W. Bär.** 1998. Phenotypic and genotypic differentiation of several human and avian isolates of *Cryptococcus neoformans*. *Mycoses* **41:**389–396.
60. **Hunter, P. R., and M. A. Gaston.** 1988. Numerical index of the discriminatory ability of typing systems: an application of Simpson's index of diversity. *J. Clin. Microbiol.* **26:** 2465–2466.
61. **Igreja, R. P., M. S. Lazera, B. Wanke, M. C. G. Galhardo, S. Kidd, and W. Meyer.** 2004. Molecular epidemiology of *Cryptococcus neoformans* isolates from AIDS patients of the Brazilian city, Rio de Janeiro. *Med. Mycol.* **42:**229–238.
62. **Karaoglu, H., C. M. Y. Lee, and W. Meyer.** 2004. Survey of simple sequence repeats in completed fungal genomes. *Mol. Biol. Evol.* **22:**639–649.
63. **Karaoglu, H., C. M. Y. Lee, D. Carter, and W. Meyer.** 2008. Development of polymorphic microsatellite markers for *Cryptococcus neoformans*. *Mol. Ecol. Resour.* **8:**1136–1138.
64. **Katsu, M., S. Kidd, A. Ando, M. L. Moretti-Branchini, Y. Mikami, K. Nishimura, and W. Meyer.** 2004. The internal transcribed spacers and 5.8S rRNA gene show extensive diversity among isolates of the *Cryptococcus neoformans* species complex. *FEMS Yeast Res.* **4:**377–388.
65. **Kidd, S. E., Y. Chow, S. Mak, P. J. Bach, H. Chen, A. O. Hingston, J. W. Kronstad, and K. H. Bartlett.** 2007. Characterization of environmental sources of the human and animal pathogen *Cryptococcus gattii* in British Columbia, Canada and the pacific Northwest of the United States. *App. Environ. Microbiol.* **73:**1433–1443.
66. **Kidd, S. E., H. Guo, K. H. Bartlett, J. Xu, and J. W. Kronstad.** 2005. Comparative gene genealogies indicate that two clonal lineages of *Cryptococcus gattii* in British Columbia resemble strains from other geographical areas. *Eukarot. Cell* **4:**1629–1638.
67. **Kidd, S. E., F. Hagen, R. L. Tscharke, M. Huynh, K. H. Bartlett, M. Fyfe, L. MacDougall, T. Boekhout, K. J. Kwon-Chung, and W. Meyer.** 2004. A rare genotype of *Cryptococcus gattii* caused the cryptococcosis outbreak on Vancouver Island (British Columbia, Canada). *Proc. Natl. Acad. Sci. USA* **101:**17258–17263.
68. **Kidd, S. E., C. Hitchcock, K. Maszewska, P. Balakrishnan, C. Hunt, P. Rajendran, A. R. Usha, S. P. Thyagarajan, T. C. Sorrell, and W. Meyer.** 2001. Molecular epidemiology of clinical and environmental isolates of *Cryptococcus neoformans* from Tamil Nadu, India. *Australas. Mycol.* **20:** 105–114.
69. **Kidd, S. E., T. C. Sorrell, and W. Meyer.** 2003. Isolation of two molecular types of *Cryptococcus neoformans* var. *gattii* from insect frass. *Med. Mycol.* **41:**171–176.
70. **Kwon-Chung, K. J.** 1975. A new genus, *Filobasidiella*, the perfect state of *Cryptococcus neoformans*. *Mycologia* **67:** 1197–1200.
71. **Kwon-Chung, K. J.** 1976. Morphogenesis of *Filobasidiella neoformans*, the sexual state of *Cryptococcus neoformans*. *Mycologia* **68:**821–833.
72. **Kwon-Chung, K. J.** 1976. A new species of *Filobasidiella*, the sexual state of *Cryptococcus neoformans* B and C serotypes. *Mycologia* **68:**943–946.
73. **Kwon-Chung, K. J., and J. E. Bennett.** 1984. Epidemiologic differences between the two varieties of *Cryptococcus neoformans*. *Am. J. Epidemiol.* **120:**123–130.
74. **Kwon-Chung, K. J., J. E. Bennett, and T. S. Theodore.** 1978. *Cryptococcus bacillisporus* sp. nov. serotype B-C of *Cryptococcus neoformans*. *Int. J. Syst. Bacteriol.* **28:**616–620.
75. **Kwon-Chung, K. J., T. Boekhout, J. W. Fell, and M. Diaz.** 2002. Proposal to conserve the name *Cryptococcus gattii* against *C. hondurianus* and *C. basillisporus* (Basidiomycota, Hymenomycetes, Tremellomycetidae). *Taxon* **51:**804–806.
76. **Kwon-Chung, K. J., and A. Varma.** 2006. Do major species concepts support one, two or more species within *Cryptococcus neoformans*? *FEMS Yeast Res.* **6:**574–587.
77. **Lagrou, K., J. van Eldere, S. Keuleers, F. Hagen, R. Merckx, J. Verhaegen, W. E. Peetermans, and T. Boekhout.** 2005. Zoonotic transmission of *Cryptococcus neoformans* from a magpie to an immunocompetent patient. *J. Intern. Med.* **257:**385–388.
78. **Latouche, G. N., M. Huynh, T. C. Sorrell, and W. Meyer.** 2003. PCR-restriction fragment length polymorphism analysis of the phospholipase B (*PLB1*) gene for subtyping of *Cryptococcus neoformans* isolates. *Appl. Environ. Microbiol.* **69:**2080–2086.
79. **Lazéra, M. S., M. A. S. Cavalcanti, A. T. Londero, L. Trilles, M. M. Nishikawa, and B. Wanke.** 2000 Possible primary ecological niche of *Cryptococcus neoformans*. *Med. Mycol.* **38:**379–383.
80. **Lemmer, K., D. Naumann, B. Raddatz, and K. Tintelnot.** 2004. Molecular typing of *Cryptococcus neoformans* by PCR fingerprinting, in comparison with serotyping and Fourier transform infrared-spectroscopy-based phenotyping. *Med. Mycol.* **42:**135–147.
81. **Levy, R., J. Pitout, P. Long, and M. J. Gill.** 2007. Late presentation of *Cryptococcus gattii* meningitis in a traveller to Vancouver Island: a case report. *Can. J. Infect. Dis. Med. Microbiol.* **18:**197–199.
82. **Li, W. H., M. Tanimura, and P. M. Sharp.** 1987. An evaluation of the molecular clock hypothesis using mammalian DNA sequences. *J. Mol. Evol.* **25:**330–342.
83. **Lindberg, J., F. Hagen, A. Laursen, J. Stenderup, and T. Boekhout.** 2007. *Cryptococcus gattii* risk for tourists visiting Vancouver Island, Canada. *Emerg. Infect. Dis.* **13:** 178–179.
84. **Litvintseva, A., J. Rosouw, R. Thakur, N. Govender, and T. G. Mitchell.** 2008. *Cryptococcus neoformans* var. *grubii* (serotype A) in Africa. 7th International Conference on *Cryptococcus* & Cryptococcosis, Nagasaki, Japan, 11 to 14 September 2008, abstract #SY-12-03.
85. **Litvintseva, A. P., L. Kestenbaum, R. Vilgalys, and T. G. Mitchell.** 2005. Comparative analysis of environmental and clinical populations of *Cryptococcus neoformans*. *J. Clin. Microbiol.* **43:**556–564.

86. **Litvintseva, A. P., X. Lin, I. Templeton, J. Heitman, and T. G. Mitchell.** 2007. Many globally isolated AD hybrid strains of *Cryptococcus neoformans* originated in Africa. *PLoS Pathog.* **3:**e114.
87. **Litvintseva, A. P., R. E. Marra, K. Nielsen, J. Heitman, R. Vilgalys, and T. G. Mitchell.** 2003. Evidence of sexual recombination among *Cryptococcus neoformans* serotype A isolates in sub-Saharan Africa. *Eukarot. Cell* **2:**1162–1168.
88. **Litvintseva, A. P., R. Thakur, R. Vilgalys, and T. G. Mitchell.** 2006. Multilocus sequence typing reveals three genetic subpopulations of *Cryptococcus neoformans* var. *grubii* (serotype A), including a unique population in Botswana. *Genetics* **172:**2223–2238.
89. **MacDougall, L., S. E. Kidd, E. Galanis, S. Mak, M. J. Leslie, P. R. Cieslak, J. W. Kronstad, M. G. Morshed, and K. H. Bartlett.** 2007. Spread of *Cryptococcus gattii* in British Columbia, Canada, and detection in the Pacific Northwest, USA. *Emerg. Infect. Dis.* **13:**42–50.
90. **Mandal, P., U. Banerjee, A. Casadevall, and J. D. Nosanchuk.** 2005. Dual infections with pigmented and albino strains of *Cryptococcus neoformans* in patients with and without human immunodeficiency virus infection in India. *J. Clin. Microbiol.* **43:**4766–4772.
91. **Meyer, W., T. G. Mitchell, E. Z. Freedman, and R. Vilgalys.** 1993. Hybridization probes for conventional DNA fingerprinting can be used as single primers in the PCR to distinguish strains of *Cryptococcus neoformans*. *J. Clin. Microbiol.* **30:**2274–2280.
92. **Meyer, W., and T. G. Mitchell.** 1995. PCR fingerprinting in fungi using single primers specific to minisatellites and simple repetitive DNA sequences: strain variation in *Cryptococcus neoformans*. *Electrophoresis* **16:**1648–1656.
93. **Meyer, W., K. Marszewska, M. Amirmostofian, R. P. Igreja, C. Hardtke, K. Methling, M. A. Viviani, A. Chindamporn, S. Sukroongreung, M. A. John, D. H. Ellis, and T. C. Sorrell.** 1999. Molecular typing of global isolates of *Cryptococcus neoformans* var. *neoformans* by polymerase chain reaction fingerprinting and randomly amplified polymorphic DNA: a pilot study to standardize techniques on which to base a detailed epidemiological survey. *Electrophoresis* **20:**1790–1799.
94. **Meyer, W.** 2007. The emergence of *Cryptococcus gattii* VGII as a super killer? *Microbiol. Austral.* **5:**70–71.
95. **Meyer, W., D. M. Aanensen, T. Boekhout, M. Cogliati, M. R. Diaz, M. E. Esposto, M. Fisher, F. Gilgado, F. Hagen, S. Kaocharoen, A. P. Litvintseva, T. G. Mitchell, S. P. Simwami, L. Trilles, M. A. Viviani, and J. Kwon-Chung.** 2009. Consensus multi-locus sequence typing scheme for *Cryptococcus neoformans* and *Cryptococcus gattii*. *Med. Mycol.* **47:**561–570.
96. **Meyer, W., A. Castaneda, S. Jackson, M. Huynh, E. Castaneda, and the Ibero-American Cryptococcal Study Group.** 2003. Molecular typing of IberoAmerican *Cryptococcus neoformans* isolates. *Emerg. Infect. Dis.* **9:**189–195.
97. **Meyer, W., F. Gilgado, S. Kaocharoen, L. Trilles, P. Ngamskulrungroj, P. Escandón, F. Hagen, T. Boekhout, A. Chindamporn, and M. S. Lazéra.** 2008. Global molecular epidemiology divides the *Cryptococcus* species complex into 8 major molecular types and reveals the possible origin of the Vancouver Island outbreak. 7th International Conference on *Cryptococcus* & Cryptococcosis, Nagasaki, Japan, 11 to 14 September 2008, abstract #SY-12-04.
98. **Mitchell, T. G., and J. R. Perfect.** 1995. Cryptococcosis in the era of AIDS: 100 years after the discovery of *Cryptococcus neoformans*. *Clin. Microbiol. Rev.* **8:**515–548.
99. **Mitchell, T. G., R. L. Sandin, B. H. Bowman, W. Meyer, and W. G. Merz.** 1994. Molecular mycology: probes and applications of PCR technology. *J. Med. Vet. Mycol.* **32**(Suppl. 1)**:**351–366.
100. **Miyazaki, Y.** 2008. Cryptococcosis in Japan. 7th International Conference on *Cryptococcus* & Cryptococcosis, Nagasaki, Japan, 11 to 14 September 2008, abstract #SY-01-04.
101. **Montagna, M. T.** 2002. A note on the isolation of *Cryptococcus neoformans* serotype A MATa strain from the Italian environment. *Med. Mycol.* **40:**593–595.
102. **Ngamskulrungroj, P., F. Gilgado, J. Faganello, A. P. Litvntseva, A. L. Leal, K. M. Tsui, T. G. Mitchell, M. H. Vainstein, and W. Meyer.** 2009. Genetic diversity of the *Cryptococcus* species complex suggest that *Cryptococcus gattii* deserves to have varieties. *PLoS ONE* **4**(6)**:**e5862. doi:10.1371/journal.pone.0005862.
103. **Passoni, L. F. C., B. Wanke, M. M. Nishikawa, and M. S. Lazéra.** 1998. *Cryptococcus neoformans* isolated from human dwellings in Rio de Janeiro, Brazil: an analysis of domestic environment of AIDS patients with and without cryptococcosis. *J. Med. Vet. Mycol.* **36:**305–311.
104. **Perfect, J. R., B. B. Magee, and P. T. Magee.** 1989. Separation of chromosomes of *Cryptococcus neoformans* by pulsed field gel electrophoresis. *Infect. Immun.* **57:** 2624–2627.
105. **Perfect, J. R., N. Ketabchi, G. M. Cox, C. W. Ingram, and C. L. Beiser.** 1993. Karyotyping of *Cryptococcus neoformans* as an epidemiological tool. *J. Clin. Microbiol.* **31:**3305–3309.
106. **Raso, T. F., K. Werther, E. T. Miranda, and M. J. Mendes-Giannini.** 2004. Cryptococcosis outbreak in psittacine birds in Brazil. *Med. Mycol.* **42:**355–362.
107. **Ruma, P., S. C. Chen, T. C. Sorrell, and A. G. Brownlee.** 1996. Characterization of *Cryptococcus neoformans* by random DNA amplification. *Lett. Appl. Microbiol.* **23:** 312–316.
108. **Sanfelice, F.** 1894. Contributo alla morfologia e biologia dei blastomiceti che si sviluppano nei succhi di alcuni frutti. *Ann. Igien.* **4:**463–495.
109. **Saul, N., M. Krockenberger, and D. Carter.** 2008. Evidence of recombination in mixed-mating-type and α-only populations of *Cryptococcus gattii* sourced from single *Eucalyptus* tree hollows. *Eukarot. Cell* **7:**727–734.
110. **Savelkoul, P. H., H. J. Aarts, J. de Haas, L. Dijkshoorn, B. Duim, M. Otsen, J. L. Rademaker, L. Schouls, and J. A. Lenstra.** 1999. Amplified-fragment length polymorphism analysis: the state of an art. *J. Clin. Microbiol.* **37:**3083–3091.
111. **Schwarz, D. C., and C. R. Cantor.** 1984. Separation of yeast chromosome-sized DNAs by pulsed field gradient gel electrophoresis. *Cell* **37:**67–75.
112. **Sharpton, T. J., D. E. Neafsey, J. E. Galagan, and J. W. Taylor.** 2008. Mechanisms of intron gain and loss in *Cryptococcus*. *Genome Biol.* **9:**R24.
113. **Simpson, E. H.** 1949. Measurement of diversity. *Nature* **163:**688.
114. **Soll, D. R.** 2000. The ins and outs of DNA fingerprinting the infectious fungi. *Clin. Microbiol. Rev.* **13:**332–370.
115. **Sorrell, T. C., S. C. Chen, P. Ruma, W. Meyer, T. J. Pfeiffer, D. H. Ellis, and A. G. Brownlee.** 1996. Concordance of clinical and environmental isolates of *Cryptococcus neoformans* var. *gattii* by random amplification of polymorphic DNA analysis and PCR fingerprinting. *J. Clin. Microbiol.* **34:**1253–1260.
116. **Sorrell, T. C., A. G. Brownlee, P. Ruma, R. Malik, T. J. Pfeiffer, and D. H. Ellis.** 1996. Natural environmental sources of *Cryptococcus neoformans* var. *gattii*. *J. Clin. Microbiol.* **34:**1261–1263.
117. **Spitzer, E. D., and S. G. Spitzer.** 1992. Use of a dispersed repetitive DNA element to distinguish clinical isolates of *Cryptococcus neoformans*. *J. Clin. Microbiol.* **30:**1094–1097.

118. **St-Germain, G., G. Noel, and K. J. Kwon-Chung.** 1988. Disseminated cryptococcosis due to *Cryptococcus neoformans* var. *gattii* in a Canadian patient with AIDS. *Eur. J. Clin. Microbiol. Infect. Dis.* **7:**587–588.
119. **Sugita, T., R. Ikeda, and T. Shinoda.** 2001. Diversity among strains of *Cryptococcus neoformans* var. *gattii* as revealed by a sequence analysis of multiple genes and a chemotype analysis of capsular polysaccharide. *Microbiol. Immunol.* **45:**757–768.
120. **Supparatpinyo, K.** 2008. Cryptococcosis in Thailand. 7th International Conference on *Cryptococcus* & Cryptococcosis, Nagasaki, Japan, 11 to 14 September 2008, abstract #SY-01-05.
121. **Swinne, D., M. Deppner, R. Laroche, J. J. Floch, and P. Kadende.** 1989. Isolation of *Cryptococcus neoformans* from houses of AIDS-associated cryptococcosis patients in Bujumbura (Burundi). *AIDS* **3:**389–390.
122. **Tautz, D., and M. Renz.** 1984. Simple sequences are ubiquitous repetitive components of eukaryotic genomes. *Nucleic Acids Res.* **12:**4127–4138.
123. **Taylor, J. W., and M. C. Fisher.** 2003. Fungal multilocus sequence typing: it's not just for bacteria. *Curr. Opin. Microbiol.* **6:**1–6.
124. **Trilles, L., M. S. Lazéra, B. Wanke, B. Theelen, and T. Boekhout.** 2003. Genetic characterization of environmental isolates of the *Cryptococcus neoformans* species complex from Brazil. *Med. Mycol.* **41:**383–390.
125. **Trilles, L., M. S. Lazéra, B. Wanke, R. V. Oliveira, G. G. Barbosa, M. M. Nishikawa, B. P. Morales, and W. Meyer.** 2008. Regional pattern of the molecular types of *Cryptococcus neoformans* and *Cryptococcus gattii* in Brazil. *Mem. Inst. Oswaldo Cruz* **103:**455–462.
126. **Upton, A., J. A. Fraser, S. E. Kidd, C. Bretz, K. H. Bartlett, J. Heitman, and K. A. Marr.** 2007. First contemporary case of human infection with *Cryptococus gattii* in Puget Sound: evidence of spread of the Vancouver Island outbreak. *J. Clin. Microbiol.* **45:**3086–3088.
127. **Vanbreuseghem, R., and M. Takashio.** 1970. An atypical strain of *Cryptococcus neoformans* (Sanfelice) Vuillemin. 1894, II. *Cryptococcus neoformans* var. *gattii* var. nov. *Ann. Soc. Belg. Med. Trop.* **50:**695–702.
128. **Varma, A., and K. J. Kwon-Chung.** 1989. Restriction fragment polymorphism in mitochondrial DNA of *Cryptococcus neoformans*. *J. Gen. Microbiol.* **135:**3353–3362.
129. **Varma, A., and K. J. Kwon-Chung.** 1992. DNA probe for strain typing of *Cryptococcus neoformans*. *J. Clin. Microbiol.* **30:**2960–2967.
130. **Varma, A., D. Swinne, F. Staib, J. E. Bennett, and K. J. Kwon-Chung.** 1995. Diversity of DNA fingerprinting in *Cryptococcus neoformans*. *J. Clin. Microbiol.* **33:** 1807–1814.
131. **Velegraki, A., V. G. Kiosses, A. Kansouzidou, S. Smilakou, A. Mitroussia-Ziouva, and N. J. Legakis.** 2001. Prospective use of RFLP analysis on amplified *Cryptococcus neoformans URA5* gene sequences for rapid identification of varieties and serotypes in clinical samples. *Med. Mycol.* **39:**409–417.
132. **Vilgalys, R., and M. Hester.** 1990. Rapid genetic identification and mapping of enzymatically amplified ribosomal DNA from several *Cryptococcus* species. *J. Bacteriol.* **172:**4238–4246.
133. **Viviani, M. A., M. Cogliati, M. C. Esposto, K. Lemmer, K. Tintelnot, M. F. C. Valiente, D. Swinne, A. Velegraki, R. Velho, and the European Confederation of Medical Mycology (ECMM) Cryptococcosis Working Group.** 2006. Molecular analyses of 311 *Cryptococcus neoformans* isolates from a 30-month ECMM survey of cryptococcosis in Europe. *FEMS Yeast Res.* **6:**614–619.
134. **Viviani, M. A., M. C. Esposto, M. Cogliati, M. T. Montagna, and B. L. Wickes.** 2001. Isolation of a *Cryptococcus neoformans* serotype A *MAT*a strain from the Italian environment. *Med. Mycol.* **39:**383–386.
135. **Viviani, M. A., H. Wen, A. Roverselli, R. Caldarelli-Stefano, M. Cogliati, P. Ferrante, A. M. Tortorano.** 1997. Identification by polymerase chain reaction fingerprinting of *Cryptococcus neoformans* serotype AD. *J. Med. Vet. Mycol.* **35:**355–360.
136. **Vos, P., R. Hogers, M. Bleeker, M. Reijans, T. van de Lee, M. Hornes, A. Frijters, J. Pot, J. Peleman, M. Kuiper, and M. Zabeau.** 1995. AFLP: a new technique for DNA fingerprinting. *Nucleic Acids Res.* **23:**4407–4414.
137. **Wickes, B. L., T. D. E. Moore, and K. J. Kwon-Chung.** 1994. Comparison of the electrophoretic karyotypes and chromosomal location of ten genes in the two varieties of *Cryptococcus neoformans*. *Microbiology* **140:**543–550.
138. **Williams, J. G. K., A. R. Kubelik, K. J. Livak, J. A. Raflski, S. V. Tingey.** 1990. DNA polymorphisms amplified by arbitrary primers are useful as genetic markers. *Nucleic Acids Res.* **18:**6532–6535.
139. **Woo, J. H.** 2008. Prognostic factors of cryptococcal meningitisin patients without HIV infection. 7th International Conference on *Cryptococcus* & Cryptococcosis, Nagasaki, Japan, 11 to 14 September 2008, abstract #SY-01-02.
140. **Xu, J., R. Vilgalys, and T. G. Mitchell.** 2000. Multiple gene genealogies reveal recent dispersion and hybridization in the human pathogenic fungus *Cryptococcus neoformans*. *Mol. Ecol.* **9:**1471–1481.
141. **Yamamoto, Y., S. Kohno, H. Koga, H. Kakeya, K. Tomono, M. Kaku, T. Yamazaki, M. Arisawa, and K. Hara.** 1995. Random amplified polymorphic DNA analysis of clinically and environmentally isolated *Cryptococcus neoformans* in Nagasaki. *J. Clin. Microbiol.* **33:**3328–3332.
142. **Yuanjie, Z., W. Hai, C. Jianhan, and G. Julin.** 2008. Mycological profile of HIV-negative patients with cryptococcal meningitis during treatment and after follow-up. 7th International Conference on *Cryptococcus* & Cryptococcosis, Nagasaki, Japan, 11 to 14 September 2008, abstract #SY-01-03.
143. **Zue, L. P., B. Xu, J. Q. Wu, X. T. Ou, Q. Q. Zhang, and X. H. Weng.** 2008. Cryptococcal meningitis in non-HIV patients: review of 11 years experience in a Chinese tertiary care hospital. 7th International Conference on *Cryptococcus* & Cryptococcosis, Nagasaki, Japan, 11 to 14 September 2008, abstract #SY-03-02.

Cryptococcus: From Human Pathogen to Model Yeast
Edited by J. Heitman et al.

25

Hybridization and Its Importance in the *Cryptococcus* Species Complex

MASSIMO COGLIATI, XIAORONG LIN, AND MARIA ANNA VIVIANI

THE DISCOVERY OF *CRYPTOCOCCUS* HYBRID STRAINS

Like many other fungal species, the two sibling *Cryptococcus* species, *C. neoformans* and *C. gattii*, are heterothallic and live in a haploid state for most of their life cycle. Earlier investigators considered the *Cryptococcus* species to be asexual haploid yeasts. A morphologically atypical *C. neoformans* isolate was first reported by Shadomy (67). This isolate was able to produce uninucleate hyphae with clamp connections and to produce spores apparently in the absence of a partner of the opposite mating type. This self-fertile isolate, renamed NIH12 by Kwon-Chung, was later used to study the basis of fertility in self-fertile isolates and their possible origin (39). The author showed that self-fertile strains were able to produce progeny including both self-fertile and self-sterile strains in an equal ratio and that self-fertility was lost during repeated subcultures. In addition, self-sterile progeny behaved as the α type since they mated only with *MAT***a** tester strains during crosses. A further important observation was that self-fertile isolates produced basidiospores of two different sizes; the larger ones were self-fertile, and the smaller ones were not able to germinate. These findings led the author to conclude that the most likely explanation for the cause of self-fertility in the isolates analyzed was that they might be diploid strains containing both **a** and α mating information.

Although the strain NIH12 was recently found to be a haploid α isolate that can undergo proficient monokaryotic fruiting and produce hyphae on its own (47, 48), the hypothesis that self-fertile strains could be **a**/α diploid was supported by an elegant experiment performed by White and Jacobson (84). They constructed one niacin auxotrophic strain in B3501 (*MAT*α) background and one pantothenate acid auxotrophic strain in B3502 (*MAT***a**) background, crossed the two auxotrophic strains, and then analyzed the progeny produced. The results showed that among the progeny, there were a few prototrophic strains, and all of these strains were self-fertile, suggesting that they contained information of both α and **a** parents. To examine the ploidy of these self-fertile progeny, their nucleus was stained, and the intensity of fluorescence was measured and compared with that of the parents. Self-fertile progeny were found to be uninucleate and contained twice as much DNA as the parental strains, indicating that they were diploid. Later, by a set of complementation tests including five genetic markers (*ilv1*, *cys1*, *cys2*, *cys3*, and *fcy1*), Whelan and Kwon-Chung (83) confirmed that the crossing of two auxotrophic parental strains (B3501α × B3502**a**) could generate some self-fertile prototrophic progeny and that the allelic pattern of the genetic markers was consistent with Mendelian segregation. In a more recent study, Sia et al. (69) crossed two congenic serotype D strains (JEC169 *MAT***a** *ade2 ura5 lys1* × JEC170 *MAT*α *ade2 lys2*) and obtained self-fertile diploid **a**/α strains. They also observed that these strains grow as diploid yeast cells at 37°C, whereas they filament and sporulate, producing recombinant haploid basidiospores at 24°C.

The introduction of fluorescence flow cytometry as a simple and reliable method to determine *Cryptococcus* DNA content (73) and the growing use of molecular techniques for *C. neoformans* typing facilitated later research on diploid strains. Viviani et al. (82) carried out a study of molecular typing of a large number of *C. neoformans* strains belonging to serotypes A, D, and, in particular, AD. The goal was to compare the serotyping method with the molecular typing method and to establish a possible correlation between the two. The authors showed that two specific genotypes were strictly correlated to serotypes A and D, respectively. In addition, a set of genotypes with combining patterns of serotypes A and D were correlated with serotype AD strains. These results were the preliminary evidence that serotype AD isolates might be diploid hybrids originated by fusing of serotype A and D isolates and have a combined genotype. Later, Tanaka et al. (72) investigated the ploidy of some serotype AD strains isolated mainly from the environment and from the progeny of one self-fertile isolate. As expected, all but one of the strains tested were diploid, further supporting the hybrid nature of serotype AD isolates. When

Massimo Cogliati and Maria Anna Viviani, Laboratory of Medical Mycology, Department of Public Health – Microbiology – Virology, Università degli Studi di Milano, 20133 Milano, Italy. **Xiaorong Lin,** Department of Biology, Texas A&M University, College Station, TX 77843.

Cogliati et al. (17) applied flow cytometry to determine the DNA content of the isolates with the genotype intermediate between serotypes A and D, all were diploid.

Furthermore, the sequence analysis of one fingerprint band common in all of the genotypes showed that the sequence of the intermediate genotypes presented points of heterozygosity at sites where the haploid serotype A and D sequences differed. This indicated that the isolates with the intermediate genotype likely contained both serotype A and D alleles. Identification of hybrids was also reported by the study of Boekhout et al. (6). This study identified six distinct clusters based on the genotypes of a large number of *C. neoformans* and *C. gattii* global isolates using amplified fragment length polymorphism (AFLP). At least two clusters of putative hybrids, one in the *C. neoformans* and the other in the *C. gattii* species group, were found in that study (6).

The full sequencing of *C. neoformans* **a** and α pheromone genes as well as other genes embedded in the mating-type locus supplied new tools to identify diploid strains and to elucidate their origin. Two different research groups took advantage of the sequence information of the mating-type alleles to study the origin of serotype AD strains. Lengeler et al. (44) analyzed 10 AD hybrid strains by flow cytometry and by the amplification of four serotype- and mating-type-specific alleles of the *STE20* gene located in the mating locus and three serotype-specific genes (*CNA1*, *GPA1*, and *CLA4*) outside the mating-type locus. They showed that all the AD hybrids contained about twice the DNA content (some were $2n$ and others $<2n$) compared to haploid control isolates, and these AD hybrids were heterozygous at most of the genetic loci analyzed. The mating-type locus of these strains harbors the *STE20*α allele from one serotype and the *STE20***a** allele from the other serotype in either of the two different combinations, αAD**a** and **a**ADα.

In the other study, Cogliati et al. (19) performed ploidy analysis of 45 putative AD hybrids identified by $(GACA)_4$ PCR fingerprinting and found that all putative hybrids were diploid. Sequencing the **a** and α pheromone genes from these hybrid isolates revealed the presence of two mating-type alleles at the *MAT* locus, consistent with the results obtained by Lengeler et al. (44). In addition, isolates presenting only the α or the **a** pheromone type were identified. The two studies both concluded that serotype AD isolates are diploid or aneuploid hybrids that originated by mating between serotype A and D strains. Yan et al. (90) determined the mating-type allelic pattern of 358 *C. neoformans* strains isolated in the United States by PCR of serotype- and mating-type-specific genes and reported that 5% of the strains were AD hybrids presenting different combinations of the allelic pattern, similar to the results previously described by Cogliati et al. (19) and Lengeler et al. (44).

As summarized in Table 1, *C. neoformans* AD hybrid strains contain about twice as much DNA as the haploid reference strains, and they present a genotype combining the serotype A and D genotypes. They can be self-fertile, but often they are self-sterile. They are often heterozygous at the mating-type locus, but some strains present only one mating-type allele (see "The Mating Type of Hybrid Strains" section for more details). Although serological typing techniques were traditionally used in identifying *Cryptococcus* strains, the results could be misleading, as some AD strains can express both A and D antigens, and others only express one of the two. A combination of genotype and ploidy analyses are able to discriminate AD hybrids from haploid serotype A or D isolates.

TABLE 1 Phenotypic and genotypic characteristics to distinguish hybrid from haploid strains of *C. neoformans*

Characteristic	True AD hybrid	Cryptic AD hybrid	True haploid
Fertility	Self-fertile	Self-sterile	Self-sterile
Serotype	AD	A or D	A or D
Genotype	Intermediate	Intermediate	Haplotype A or D
Ploidy	Diploid or aneuploid	Diploid or aneuploid	Haploid
Mating crossing	a/α	a or α	a or α
Mating-type PCR	aADα or αADa	aA, αA, aD or αD	aA, αA, aD or αD

IDENTIFICATION OF HYBRID STRAINS BY SEROLOGICAL AND GENOTYPING TECHNIQUES

Serological Typing of AD Hybrids

Serological reactions and a set of genotyping techniques have been used to determine the serotype of *Cryptococcus* strains. In the past, identification of hybrid strains was limited by the use of insensitive typing methods such as those based on serological reactions. The first serological classification system of *C. neoformans* reported by Evans (24, 25) distinguished only three serotypes, A, B, and C. Later, Wilson et al. (86) added a fourth serotype, D. Then, Ikeda et al. (33) established a serological agglutination test for *Cryptococcus* typing employing a set of eight polyclonal antisera from immunized rabbits. This test is able to distinguish four serotypes of *C. neoformans*: serotypes A, B, C, and D. In the Ikeda et al. study, 3 out of 62 Japanese isolates agglutinated with both serotype A- and D-specific antisera. Therefore, they concluded that the serological classification of *C. neoformans* might include a fifth serotype, AD (Fig. 1a). Subsequently, the same researchers reported that the *C. neoformans* type strain CBS132 belonged to serotype AD and further investigated the structure of its capsular polysaccharides by column chromatography (32). Their results indicated that the capsule polysaccharides of this serotype AD isolate contained both serotype A and D epitopes, which likely caused the cross-reaction of CBS132 to both serotype A- and D-specific antisera.

Cherniak et al. (16) compared the capsular glucuronoxylomannan (GXM) structure of isolates from patients with recurrent cryptococcosis and their reactivity with the antisera from Ikeda's group. The initial isolate and relapses from the same patient were shown to differ in capsular polysaccharide composition presenting a different proportion of serotype A- and D-specific GXMs. In addition, changes in capsule structure resulted in differences in the antisera reactions since isolates producing serotype A GXM as the major component reacted with antisera 7 (serotype A), isolates producing serotype D GXM as the major component reacted with antisera 8 (serotype D), whereas those producing an equal quantity of both the GXMs reacted with both antisera (serotype AD). Altogether these findings suggest that AD hybrid strains present two parallel pathways that produce two different types of GXM and that the yeast is able to modulate the combination of the two structures in response

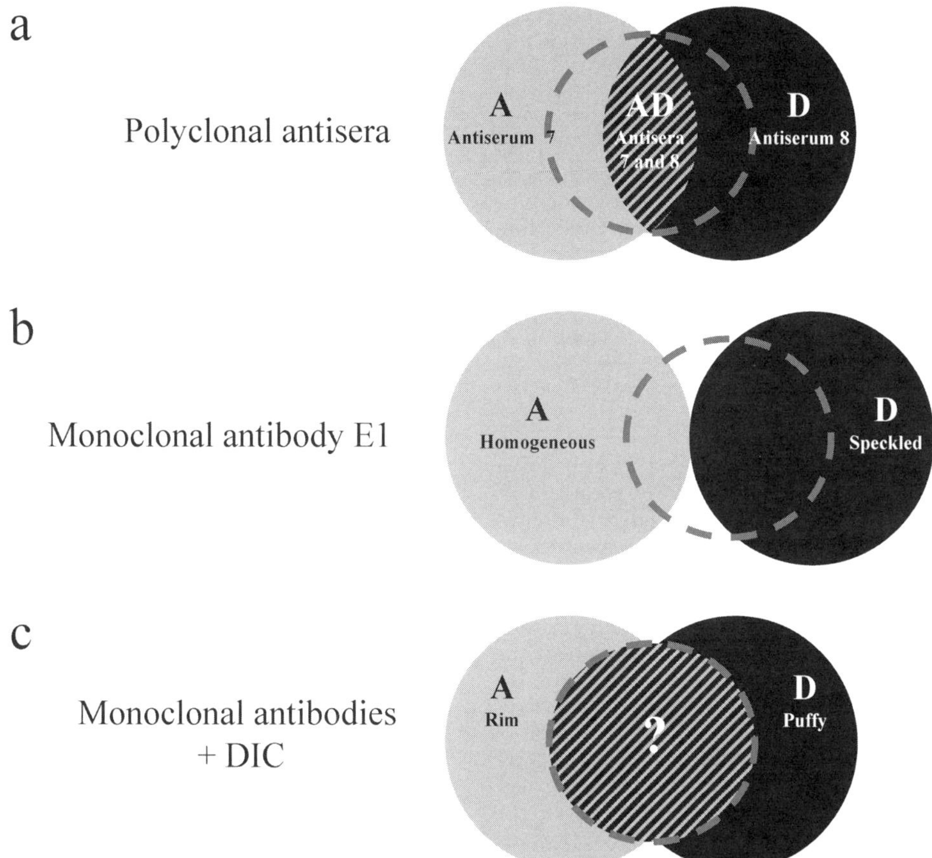

FIGURE 1 Identification of AD hybrid strains by (a) polyclonal antisera as described by Ikeda et al. (33), (b) E1 monoclonal antibody as described by Dromer et al. (22), and (c) monoclonal antibodies plus DIC as described by MacGill et al. (56). Gray circles represent the serotype A group, black circles the serotype D group, and dashed circles the AD hybrids group. Areas in black and gray stripes represent the portion of AD hybrid strains identified by the method. Black or gray areas included in the dashed circles represent the cryptic AD hybrids not identified by the method.

to environmental pressure. The detection of capsular polysaccharide antigens, therefore, is not a suitable method to identify AD hybrid strains because it is dependent on the relative quantity of the two different GXM types produced under a specific condition. Therefore, this method is not reliable to detect AD hybrids where serotype A- and D-specific GXMs are unequally expressed.

Dromer et al. (22) serotyped 156 *C. neoformans* and *C. gattii* isolates using both the monoclonal antibody E1 and the polyclonal antisera described by Ikeda et al. (33). The results showed that the E1 monoclonal antibody produced a homogeneous reaction with serotype A yeasts, a speckled reaction with serotype B and D isolates, and no reactions with serotype C strains. Combined with the canavanine-glycine-bromothymol blue agar test to distinguish *C. neoformans* from *C. gattii* species, this serological method was able to detect all of the four serotypes. The authors stressed that the method was also able to identify the serotype of isolates that were untypeable using polyclonal antisera. However, all serotype AD isolates were typed as either serotype A or D. The method could not detect AD hybrids, probably because the fluorescence pattern of the hybrids was indistinguishable from the speckled pattern of serotype D strains, and thus they were likely misidentified as serotype D (Fig. 1b). Therefore, with respect to AD hybrids, this combination typing method again failed to solve all the serotyping ambiguities.

An interesting study of MacGill et al. (56) revealed a new aspect of monoclonal antibodies' reaction with the *Cryptococcus* capsule. A set of monoclonal antibodies was tested against serotype A and D reference strains, and the capsular quellung-type reaction was analyzed by differential-interference contrast (DIC). The antibodies belonging to group II produced two different DIC images for serotypes A and D, respectively. Serotype A strains presented a rim pattern characterized by a sharp increase in the optical gradient at the capsular edge followed directly by an immediate decrease. In contrast, serotype D strains presented a puffy pattern characterized by an increase in the optical gradient at the capsular surface and the absence of the immediate decrease specific to the rim pattern. These experiments showed that although quantitative serological methods such as agglutination, immunofluorescence, and quantitative precipitation gave identical reactions for serotype A and D strains, DIC imaging of the capsular quellung reaction was a qualitative assessment of antibody-capsule interaction able to reflect the real differences between the capsule structure in the two serotypes. It would be interesting to investigate the quellung reaction in AD hybrid strains in order to evaluate if this method could be useful for their identification and characterization (Fig. 1c).

Genotyping of AD Hybrids

The limit of serology in typing AD hybrid strains was first pointed out by Viviani et al. (82), who investigated the reproducibility of serotyping compared to that of genotyping by $(GACA)_4$ PCR fingerprinting. A sample of serotype A, D, and AD isolates was subcultured every other day, and serotyping and genotyping were performed each week for 1 month. Serological tests showed that putative AD hybrids presented a variable serotype (A, D, or AD) depending on the time of testing. In contrast, the genotype was stable and showed a pattern combining the serotype A and D genotypes. The results showed that genotyping is a more reliable and informative method independent of the variable capsular structure. This and later studies (17, 19) confirmed that the intermediate genotypes VN3 and VN4 identified by $(GACA)_4$ PCR fingerprinting were correlated to diploid AD hybrid strains. Therefore, this genotyping method was suitable to distinguish *C. neoformans* AD hybrids from haploids. The typing method was then simplified by the setup of a multiplex PCR capable of simultaneously amplifying four PCR fingerprinting products recognized as markers of the different genotypes (18). The advantage was the simple interpretation of the results and the unequivocal identification of the genotypes on the basis of the combination of only four bands. Serotype A haploid strains displayed three products, and serotype D haploids displayed two products and hybrids containing all four products. However, the method could not subtype *C. gattii* isolates due to its high specificity to *C. neoformans* isolates.

To standardize a typing technique for *C. neoformans*, Meyer et al. (58) carried out a multicenter pilot study to evaluate two molecular methods, PCR fingerprinting and randomly amplified polymorphic DNA. They typed 356 global isolates, and both methods were shown to be able to distinguish the species in four genotypes: VNI, VNII, VNIII, and VNIV. Most importantly, the VNIII genotype included all of the serotype AD strains, making these methods useful for the identification of AD hybrids. In another study (57), 14 VNIII strains out of 340 IberoAmerican isolates of clinical, veterinary, or environmental origin were identified by M13 PCR fingerprinting. Most of these VNIII strains were polymorphic at the *URA5* locus based on restriction fragment length polymorphism (RFLP) analysis, confirming the heterozygous nature of these strains.

Boekhout et al. (6) applied AFLP to type 207 *C. neoformans* and *C. gattii* isolates and divided each species into three different clusters (AFLP1 to 6). Putative hybrid genotypes were found in both species, corresponding to genotype AFLP3 in *C. neoformans* and genotype AFLP6 in *C. gattii*. The AFLP6 genotype was later also recognized as the main genotype causing an unusual cryptococcosis outbreak on Vancouver Island, Canada (38), suggesting that a recent recombination event might have generated a highly virulent *C. gattii* genotype (28). More recently, four atypical AB and DB interspecies hybrids—three hybrids between the AFLP2 genotype (serotype D) and the AFLP4 genotype (serotype B) (8), and one hybrid between the AFLP1 genotype (serotype A) and the AFLP4 genotype (serotype B)—were identified using AFLP (9). AFLP has also been applied to identify serotype A intravarietal hybrids (51). In that study, close to 8% of global *Cryptococcus* isolates ($n = 500$) were identified as serotype A diploids based on ploidy analysis through a fluorescence activated cell sorter (FACS). Although the majority of the diploid serotype A isolates identified in that study belong to a single molecular type (VNI, VNII, or VNB), some diploid isolates were found to be hybrids likely derived from fusion between two serotype A isolates of different molecular types. However, because the AFLP used in that study did not have the power to distinguish the VNII and the VNB subgenotypes within serotype A, the definite genotype of these intravarietal diploids was established by multilocus sequence typing (MLST) analyses as discussed below.

MLST analysis has been applied as a strain genotyping method for *Cryptococcus*. Litvintseva et al. discovered the novel VNB subgenotype within serotype A using this technique (55). The VNB subgenotype is unique to Africa, which facilitated the subsequent discovery that many **a**ADα hybrids contain the Africa-specific A**a** parental genotype, and these hybrids likely originated from Africa and then were distributed globally (53). Lin et al. used the MLST technique to study the origins of serotype A diploid α strains (51). This study is one of the serial elegant studies of *Cryptococcus* unisexual mating (48, 49, 51). As discussed later in this chapter, the overwhelming majority of clinical and environmental isolates of *C. neoformans* are of the α mating type (>98 to 99.9%) (13, 40), and the authors have speculated, based on their previous findings that α-α mating occurs both under laboratory conditions and in nature (48, 49), that in addition to the conventional **a**-α mating, α-α mating might be a major means by which genetic diversity is generated in nature.

To study the impact of α-α mating on the population structure of the most virulent serotype A isolates, the authors surveyed ~500 global isolates and found that close to 8% of these isolates are diploid serotype A α strains (51). Characterization of these diploid strains by serotype- and mating-type-specific PCR, comparative genome hybridization, and FACS analysis indicate that all these serotype diploids contain two α alleles. Several of them, based on MLST analysis of loci in the mating-type locus and other genomic regions, contain not only two different alleles at other genomic regions, but also two different serotype A-specific α alleles in their mating-type locus. Thus, these diploids are intravarietal hybrids (Aα^1Aα^2) derived from mating between two genetically distinct serotype A-specific α isolates (for more discussion about unisexual mating see chapter 7).

Although the molecular typing methods described above are different and use different nomenclatures to classify *C. neoformans* and *C. gattii* genotypes, the results are consistent among these studies. All of the methods can discriminate the two species, recognize the same number of intraspecies clusters, and finally, they are able to identify AD hybrids and other less common hybrids. On this basis, the researchers involved in the molecular typing of the *C. neoformans* species complex have recently decided (56a) to adopt the reference genotype nomenclature (VNI, VNII, VNIII, VNIV, VGI, VGII, VGIII, VGIV) reported by Meyer et al. (57). The correspondence between previous nomenclatures with respect to the reference one is summarized in Table 2.

THE MATING TYPE OF HYBRID STRAINS

Methods based on mating reactions with reference strains and genotypes of the *MAT* locus have been used to characterize the mating type of *Cryptococcus* isolates. Strains to be tested can be crossed with reference strains of known mating type on mating-inductive media such as V8 medium. Positive mating reactions are characterized by dikaryotic hyphal

TABLE 2 Correspondence table of the genotype nomenclatures

Reference nomenclature	M13 PCR fingerprinting and *URA5* RFLP (57)	$(GACA)_4$ PCR fingerprinting (82)	AFLP (6)
VNI	VNI	VN6	AFLP1
VNII	VNII	VN6	AFLP1A
VNIII[a]	VNIII[a]	VN3–VN4[a]	AFLP3[a]
VNIV	VNIV	VN1	AFLP2
VGI	VGI	–	AFLP4
VGII	VGII	–	AFLP6
VGIII	VGIII	–	AFLP5
VGIV	VGIV	–	AFLP7

[a]AD-hybrid genotypes.

formation, fused clamps, and basidia. The mating reaction test offers a low-cost and easy way to identify mating type. However, the mating assay cannot identify the mating-type allelic pattern of self-fertile strains, nor can it be used to identify strains that are sterile under laboratory conditions.

On the contrary, genotyping methods, commonly based on the amplification of serotype- and mating-type-specific genes located within the *MAT* locus, bypass the problem related to strain sterility and are able to identify four different mating-type alleles in *C. neoformans*: Aα, Dα, A**a**, and D**a**, and α and **a** alleles from *C. gattii*. Most *C. neoformans* AD hybrids were shown to be heterozygous at the mating-type locus presenting two opposite mating-type alleles, αAD**a** or **a**ADα (19, 44). Because the α mating type predominates over **a** in *Cryptococcus* populations, and particularly A**a** isolates were thought to be extinct (45), the finding that some AD hybrid isolates contained the A**a** allele suggested that mating type A**a** haploid strains were likely present in the environment. Indeed, A**a** haploid strains were later identified in different parts of the world—Tanzania (45), Italy (80), Hungary (81), and Botswana (54)—confirming the previous hypothesis. Using MLST, Litvintseva et al. (53) genotyped A**a** haploid strains isolated in Botswana and compared their genotypes with those of A**a** haploid and **a**ADα hybrid strains from other geographical areas. The authors found that the A**a** genotypes of all the **a**ADα hybrids clustered with those of A**a** haploid isolates from Botswana, supporting the hypothesis that the A**a** allele of hybrid strains originated from an African progenitor.

Although the majority of hybrid strains contain both **a** and α information at their mating-type locus, some studies reported the presence of only a single mating-type allele in a small percentage of diploid hybrid isolates (19, 78, 79, 90). In order to understand the possible origin of these atypical strains, Cogliati et al. (20) investigated them further. The allelic pattern of five genes embedded in the mating-type locus (*STE20*, *STE3*, *STE11*, *STE12*, and *MF*) and one outside it (*URA5*) was determined for 14 strains. Only a single mating-type allele was detected in these loci in agreement with the previous results. In contrast, the *URA5* gene was heterozygous in most of the isolates, consistent with the hybrid nature of these strains established by PCR fingerprinting. These AD hybrids with only one mating-type allele could be derived from mating between serotype A and D isolates. To test this hypothesis, the same set of genes was analyzed among progeny from a cross between H99 (Aα) and JEC20 (D**a**). As expected, two diploid F1 isolates with a hybrid genotype presented only one mating-type allele. The observation that these strains showed recombinant electrophoretic karyotypes further confirmed their hybrid genotypes.

All these results showed that the atypical strains described here represented a diploid AD hybrid subpopulation. These results also suggested that they were probably strains with two identical alleles at the mating-type locus, and they were probably originated by the postmeiotic random fusion of two of the four recombinant nuclei during the mating process (Fig. 2a). Another interesting issue concerns the existence in nature of diploid *MAT*α isolates that, apparently, do not present an AD hybrid genotype but cluster with the haploid serotype A or D genotype (7, 19, 49, 51). Hull et al. (31) tried to create α/α diploid strains by crossing two haploid *MAT*α strains. They failed to obtain any fusion products. Interestingly, when they incubated the two α strains in the presence of one *MAT***a** helper strain, they isolated α/α diploids, and vice versa, **a**/**a** diploids were obtained when two *MAT***a** strains were incubated with a *MAT*α helper (Fig. 2b). In the study described above they identified the α determinant gene (*SXI1*α) and later (30) the **a** determinant one (*SXI2***a**), showing their key role in sexual reproduction.

Lin et al. analyzed the haploid fruiting phenomenon in *MAT*α strains where haploid α cells can produce hyphae and basidiospores in the absence of an opposite mating partner (48). Although some **a** isolates can also undergo fruiting and produce spores (77), the α mating-type locus is one of the significant quantitative trait loci to promote fruiting (47, 85). The authors hypothesized that the fruiting process represented a sexual reproduction mode involving nuclear diploidization, ploidy reduction, and meiotic basidiospore production (48). The genetic analysis of spores produced by a heterozygous α/α diploid showed that these basidiospores were recombinant and haploid. Furthermore, meiotic-specific proteins such as Dmc1 and Spo11 are required for normal sporulation during monokaryotic fruiting. This evidence indicates that a meiotic process occurred during fruiting. The authors further speculated that sexual reproduction between partners of the same mating type may occur in nature and be a *Cryptococcus* strategy to recombine in the absence of *MAT***a** strains (Fig. 2c). This hypothesis is supported by the recent identification of αADα environmental strains isolated in the United States (49, 52) and the finding that some natural diploid αAAα strains are derived from intravarietal matings between two genetically distinct Aα isolates based on MLST analyses (51) (see earlier discussion of MLST for more details).

Recent population genetic studies also provide evidence that same-sex mating occurs in nature. For example, phylogenetic analysis of the sibling species *C. gattii* has shown that same-sex mating between two different α strains may have given rise to a more virulent strain occupying a new environmental niche and causing the Vancouver Island outbreak (28). Other population genetic studies of *C. gattii* strains from Australian eucalyptus trees (66) and *C. neoformans* serotype A veterinary isolates in Sydney (11) also reveal evidence of recombination in a unisexual α population. Analyzing 78 *C. neoformans* serotype A strains isolated from tree hollows in northwestern India also finds unambiguous evidence for recombination in that α-only population (29). More evidence will likely emerge in the future to provide further indirect support for the occurrence of same-sex mating in natural populations.

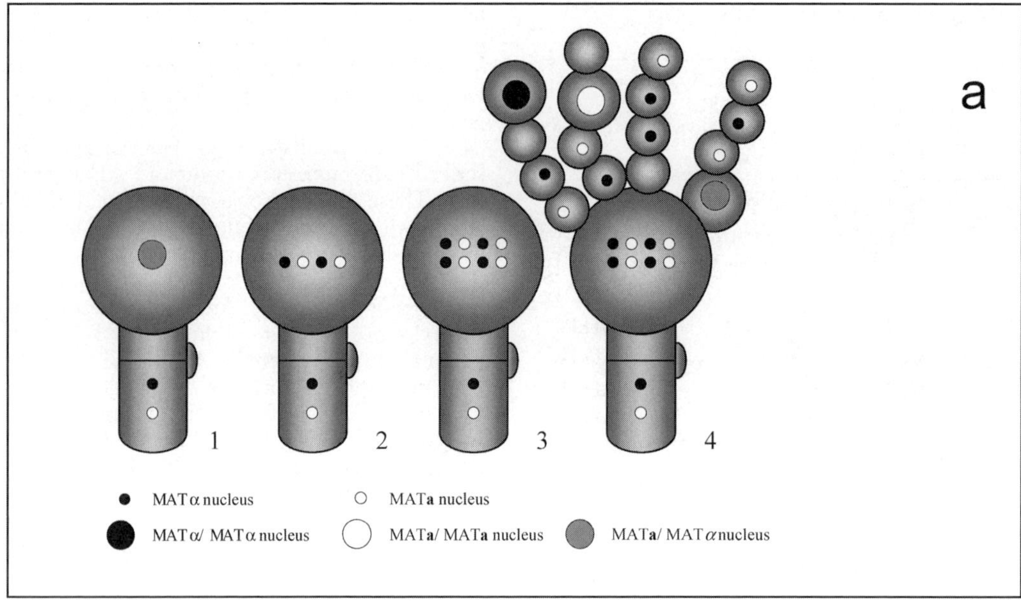

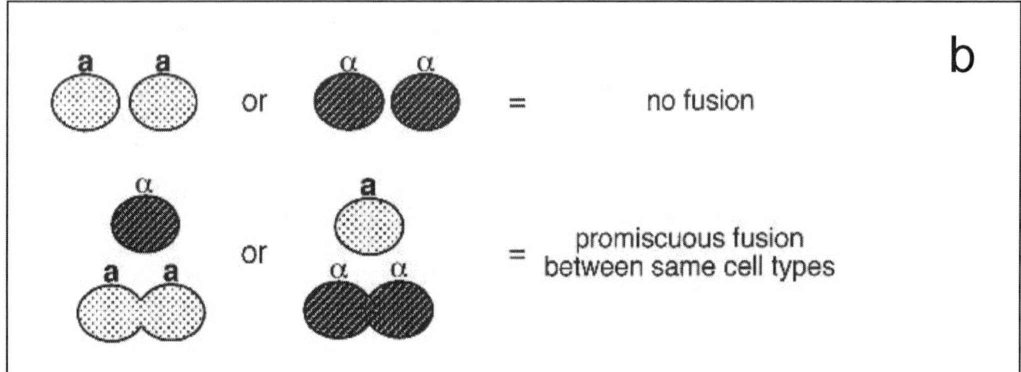

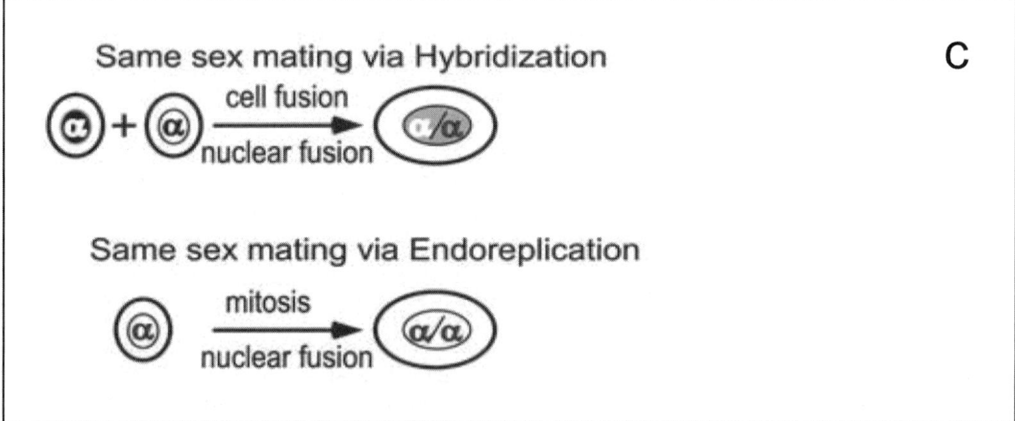

FIGURE 2 Hypothetical mechanisms yielding α/α and **a**/**a** homozygous isolates. (a) Postmeiotic fusion of two of the four recombinant nuclei. Image from Cogliati et al. (20). (b) Fusion of two strains with the same mating type by one helper strain of opposite mating type. Image from Hull et al. (31). (c) Same-sex fusion of two nonisogenic *MATα* strains or endoreplication of one *MATα* nucleus. Image from Lin et al. (51).

Although some laboratory strains are prolific in fruiting, the majority of clinical or environmental isolates are found to be incapable or inefficient in fruiting under laboratory conditions, and natural conditions that stimulate fruiting remain to be identified.

GEOGRAPHICAL DISTRIBUTION OF HYBRID STRAINS

Table 3 summarizes the prevalence of AD hybrids in different geographical regions based on studies using only serotyping techniques. It is clear that serotype AD strains have been consistently isolated from all the continents and are not endemic to any specific region. Table 4 summarizes the prevalence of AD hybrids based on genotyping and indicates a significantly increased percentage of hybrids compared to Table 3, especially in Europe. This discrepancy is largely due to the inability of the serological typing methods to identify all of the AD hybrid isolates. This comparison underscores the influence of the techniques on obtaining a reliable estimation of the hybridization phenomenon.

Table 4 also provides the mating-type allelic patterns of the hybrids, especially those from the United States and Europe. When the results obtained from these two geographical areas are compared, some differences in the allelic distribution are noted. In the United States the mating type **a**ADα seems to be more prevalent than the mating type αAD**a**. In Europe the two allelic patterns seem to be distributed more heterogeneously. In Spain, Portugal, and Germany there is a prevalence of the αAD**a** configuration, and in Italy the two alleles are equally distributed, whereas in Greece there is a prevalence of the **a**ADα configuration. The difference in the presence of αADα hybrid strains homozygous at the mating-type locus may be due to clonal expansion in the United States (49). A final remark on *C. neoformans* hybrid distribution is that their number is strictly related to the abundance of serotype D haploid strains in a specific region. This is not surprising given that serotype A is the predominant serotype worldwide, and serotype D isolates are relatively less common. Therefore, the greater the size of the serotype D population, the greater the possibility that hybridization events between serotype A and D isolates might occur. Only studies with a more relevant epidemiological interest are included in Table 4, and many other studies that have reported the molecular identification of AD hybrid strains (15, 23, 35, 43, 49, 88) are not reflected here.

TABLE 3 Geographical distribution of serotype AD strains from clinical and environmental sources

Geographical origin	No. of strains investigated	No. of serotype AD strains identified	Reference
Asia			
China	51	11 (21.5%)	46
India	15	2 (13.3%)	63
	18	2 (11.1%)	14
Japan	62	3 (4.8%)	33
Thailand	169	3 (1.8%)	71
	90	11.1%	34
Africa			
(Country not specified)	34	2 (5.9%)	5
Egypt	33	9 (27.3%)	1
North America			
Canada	78	5 (6.4%)	41
United States	288	11 (3.8%)	4
	321	13 (4.0%)	41
	251	15 (5.9%)	10
	40	1 (2.5%)	70
South America			
Argentina	44	2 (4.5%)	3
	129	3 (2.3%)	12
Brazil	31	1 (3.2%)	41
	467	6 (1.3%)	62
Venezuela	132	7 (5.3%)	64
Oceania			
Australia	17	2 (11.8%)	41
Europe			
(Country not specified)	44	1 (2.3%)	5
Belgium	14	8 (57.1%)	41
France	27	6 (22.2%)	41
Germany	21	3 (14.3%)	59
Italy	206	7 (3.4%)	75
Spain	154	13 (8.4%)	3
	43	9 (20.9%)	21
Switzerland	18	4 (22.2%)	41
The Netherlands	18	4 (22.2%)	41

VIRULENCE OF HYBRIDS

Lengeler et al. (44) first investigated the in vivo pathogenicity of a small number of *C. neoformans* hybrid isolates using a murine inhalation model of cryptococcosis. Four strains, two αAD**a** and two **a**ADα, were tested. The results showed that one αAD**a** isolate showed a level of virulence similar to that of the H99 reference strain, whereas the remaining isolates were moderately less virulent. In a different murine model (intravenous inoculum), Chaturvedi et al. (15) found that seven out of eight αAD**a** hybrids were as virulent as the highly pathogenic strain H99. In a later study, Barchiesi et al. (2) carried out a comparative analysis of the pathogenicity of 15 clinical isolates (not congenic) including haploid and hybrid strains belonging to six different combinations of serotype and mating type: Aα, A**a**, Dα, D**a**, αAD**a**, and **a**ADα. The authors showed that all the haploid and hybrid strains containing the Aα allele produced 100% mortality and an increased tissue burden compared to the H99 strain. Haploids and hybrids containing the Dα allele presented a variable moderate virulence. In contrast, D**a** and A**a** haploid strains were much less virulent. The authors concluded that the presence of Aα was associated with enhanced virulence in haploids and in AD hybrids.

However, several studies using haploid *Cryptococcus* strains have drawn different conclusions about the role of mating type in virulence. For example, congenic serotype A *MAT*α and one *MAT***a** strain in the H99 background that differ only in the mating-type locus show equivalent pathogenicity in murine and rabbit infection models (60). In several

TABLE 4 Geographical distribution of AD hybrid strains identified by DNA molecular typing

Geographical origin	No. of strains studied	No. of AD hybrids identified	Mating-type allelic pattern							Reference
			αADa	aADα	αADα	αAAα	αDDα	aAAa	aDDa	
Asia										
China	115	2 (1.7%)				1	1			26
Turkey	26	1 (3.8%)		1						65
North America										
United States	762	54 (7.0%)	4	9	41					52
	358	19 (5.3%)	2	14		1	1		1	90
Mexico	69	1 (1.5%)								57
South America										
Argentina	57	2 (3.5%)								57
Brazil	443	4 (0.9%)								76
Chile	19	3 (15.8%)								57
Europe										
Belgium	21	4 (19.0%)	1	1		1	1			79
Germany	114	13 (11.4%)								74
	44	8 (18.2%)	5	1		1	1			79
Greece	23	11 (47.8%)	2	7		2				79
Hungary	4	2 (50.0%)	1	1						79
Italy	129	39 (30.2%)	26	13						78
	123	37 (30.1%)	16	16		1	2	1	1	19
Portugal	20	10 (50.0%)	8	2						79
Spain	19	8 (42.1%)								57
	42	19 (45.2%)	17	2						79
Sweden	6	1 (16.7%)		1						79

other studies, natural serotype A isolates of the α mating type were shown to be more virulent than those of the **a** mating type, although the strains are not congenic in this example (2, 37, 45). Similarly, Kwon-Chung et al. (42) reported that a serotype D *MAT*α isolate was more virulent than the congenic *MAT***a** strain. To examine the potential influence of the genes outside the mating-type locus on *C. neoformans* virulence, Nielsen et al. (61) generated two pairs of congenic strains in two serotype D backgrounds, B3501 and NIH433. They showed that virulence was equivalent between the B3501 **a** and α congenic strains, but the α strain was more virulent than its **a** counterpart in the NIH433 genetic background. These experiments demonstrate that the genomic regions outside the mating-type locus may interact with the mating-type locus and contribute to the differences in virulence between **a** and α cells.

Because AD hybrids harbor the genomes of both serotypes A and D in a single cell and could have many different combinations at the mating-type locus, it is challenging to compare their virulence levels and find factors that contribute to their virulence potential. Recently, Lin et al. (50) investigated the influence of ploidy and mating type on virulence. They created a set of possible combinations of *C. neoformans* diploids (αAAα, αDDα, **a**DD**a**, **a**AA**a**, αADα, αAD**a**, **a**ADα, **a**AD**a**) in the backgrounds of serotype A and D reference strains H99 and JEC21, and tested the pathogenicity of these strains in murine tail vein injection and inhalation models. The analysis of the results revealed that ploidy had a minor effect on virulence, and the combination of mating types in AD hybrids (αAD**a**, **a**ADα, and αADα) did not influence the pathogenicity levels of the hybrid strains tested. The only exception was the **a**AD**a** hybrids. All three **a**AD**a** hybrids that were derived independently using different marked parental strains were significantly lower in virulence than the serotype A parental strains or other AD hybrids. These findings provide evidence of negative epistatic interactions between the A**a** and D**a** alleles of the mating-type locus. The molecular principles underlying the detrimental **a**A/D**a** interactions remain to be defined.

IMPLICATIONS OF HYBRIDIZATION ON THE EVOLUTION OF *CRYPTOCOCCUS* SPECIES COMPLEX

The genetic divergence between serotype A and D populations was first described by Franzot et al. (27), who proposed to separate the species *C. neoformans* into two varieties, var. *grubii* (serotype A) and var. *neoformans* (serotype D). The sequence analysis of the *URA5* gene (777 bp) revealed that sequences from serotype A and D strains differed for 42 nucleotides (5.4%), a sufficient distance to justify the varietal separation. Later, Xu et al. (89) performed a deeper analysis calculating the genetic divergence between the four serotypes comparing the sequences of *URA5* and *LAC1* genes. The authors estimated that *C. neoformans* separated from *C. gattii* species about 37 millions years ago, whereas serotype A diverged from the serotype D lineage about 18.5 million years ago. Five strains, likely hybrids, were inconsistently placed in the four genealogies, suggesting that the cause of this incongruence was the occurrence of hybridization among varieties and serotypes. The estimation of the time of hybridization showed that recombination occurred very recently (0 to 3.2 million years ago), indicating that hybridization is important during *Cryptococcus* evolution.

The same researchers (87) investigated 14 serotype AD isolates in order to calculate the time of hybridization on the basis of *LAC1* sequence divergence. Gene genealogical analysis supported the hypothesis of multiple origins of hybrid strains, suggesting that hybridization events occurred in different times probably coinciding with the continuing dispersion of *C. neoformans* in the environment that allows the contact between genetic divergent populations. A recent study (36) reports the finding of a 40-kb DNA region in a large number of serotype D isolates that represents an identity island common with the serotype A genome. It was hypothesized that this region might have originated with an introgression of var. *grubii* in the var. *neoformans* genome that likely occurred via an incomplete intervarietal sexual cycle, creating a hybrid intermediate where mobile elements common to both lineages promoted the DNA exchange.

Lin et al. (50) compared melanin production, cell size, high-temperature tolerance, and UV resistance between isogenic haploid and AD hybrid strains. They observed that AD hybrids presented an increased cell size, were more resistant to UV irradiation, and were more tolerant to high temperatures than the haploid parents. In contrast, melanization in hybrids is modestly decreased. These results were consistent with those of previous studies (49, 53) reporting the same characteristics for some αADα and **a**ADα natural isolates. In another study Shahid et al. (68) crossed two genetically diverged A and D parental strains that differed in their growth at different temperatures and on different media as well as in their susceptibility to antifungals. A total of 269 progeny were obtained, and their vegetative growth was determined in 40 environments that differed in nutrient, temperature, and fluconazole concentration. The results revealed that the vegetative fitness mean of hybrid progeny was intermediate between that of the parents, excluding heterosis or outbreeding depression. However, when the vegetative fitness of each of the progeny was compared with the parents, a small percentage of transgressive segregants were observed. Interestingly, the majority of these isolates presented a higher fitness level than the fitter parent, suggesting that hybridization contributes positively to environmental adaptation.

In conclusion, hybridization, in addition to a low rate of sexual reproduction, represents an important mechanism for *Cryptococcus* to introduce genetic variants in a population with a prevalent clonal expansion. Hybrids are often diploid or aneuploid, containing two different genomes that could interact in different ways depending on the genetic content of the particular hybrid strain. In addition, a diploid genome can protect cells against gene defects or damage, enhancing the survival of the species. Therefore, the high variability of the genome plays an important role in the environment adaptation and host interaction that allow *Cryptococcus* to colonize new areas and to protect itself against host defenses. Hybrids also represent a reservoir of rare alleles that might be lost over time due to selective pressures. An example might be the mating type **a** locus, which is important for sexual reproduction during **a**-α matings. The rarity of *MAT***a** isolates in *Cryptococcus* populations may have already significantly shifted its reproduction mode from a heterothallic bisexual style to a combination of bisexual and unisexual styles (11, 28, 48, 49, 51, 66). The identification of hybrids such as αADα and αAAα that originated from matings between partners of the same mating type provides unequivocal evidence of this shift (49, 51). The shift might have happened very recently and is still ongoing, as *Cryptococcus* species still prefer a heterothallic life cycle when opposite mating partners are available.

Finally, most of the hybrid strains contain both mating type **a** and α loci and are potentially self-fertile. In defined genetic crosses between serotype D (**a** × α) strains, self-fertile strains arise spontaneously. Kwon-Chung (39) reported that 2 out of 101 basidiospores analyzed were self-fertile **a**/α strains, whereas Sia et al. (69) reported 12.5% (3/24). When progeny from serotype A × D strains was investigated, the percentage of a/α strains increased up to 50% (20, 37). In contrast, self-fertile strains originating from an intravarietal cross produced a higher percentage (50 to 80%) of viable progeny (39, 44, 69) than AD hybrid self-fertile strains (5% viability) (44). In addition, the former self-fertile strains are able to generate recombinant haploid spores (69), whereas the latter are probably able to produce only recombinant diploids. Altogether, the data reflect the genetic distance between the two *C. neoformans* varieties. They also reflect the ability of *C. neoformans* to hybridize at high frequency, producing stable diploids or aneuploids. Although with lower efficiency, these self-fertile AD hybrid strains are able to produce recombinant progeny and create a sufficient genetic variability to colonize new areas or to meet the challenges of a changing environment. Further studies are needed to elucidate the fine mechanisms that control hybridization processes as well as the environmental conditions that induce hybridization.

REFERENCES

1. **Abdel-Salam, H. A.** 2003. Characterization of *Cryptococcus neoformans* var. *neoformans* serotype A and A/D in samples from Egypt. *Folia Microbiol.* **48:**261–268.
2. **Barchiesi, F., M. Cogliati, M. C. Esposto, E. Spreghini, M. Schimizzi, B. L. Wickes, G. Scalise, and M. A. Viviani.** 2005. Comparative analysis of pathogenicity of *Cryptococcus neoformans* serotypes A, D and AD in murine cryptococcosis. *J. Infect.* **51:**10–16.
3. **Barò, T., J. M. Torres-Rodriguez, Y. Morera, C. Alia, O. Lopez, and R. Mendez.** 1999. Serotyping of *Cryptococcus neoformans* isolates from clinical and environmental sources in Spain. *J. Clin. Microbiol.* **37:**1170–1172.
4. **Bennett, J. E., K. J. Kwon-Chung, and D. H. Howard.** 1977. Epidemiologic differences among serotypes of *Cryptococcus neoformans*. *Am. J. Epidemiol.* **105:**582–586.
5. **Bertout, S., F. Renaud, D. Swinne, M. Mallie, and J. M. Bastide.** 1999. Genetic multilocus studies of different strains of *Cryptococcus neoformans*: taxonomy and genetic structure. *J. Clin. Microbiol.* **37:**715–720.
6. **Boekhout, T., B. Theelen, M. Diaz, J. W. Fell, W. C. J. Hop, E. C. A. Abeln, F. Dromer, and W. Meyer.** 2001. Hybrid genotypes in the pathogenic yeast *Cryptococcus neoformans*. *Microbiology* **147:**891–907.
7. **Bovers, M., F. Hagen, E. E. Kuramae, and T. Boekhout.** 2008. Six monophyletic lineages identified within *Cryptococcus neoformans* and *Cryptococcus gattii* by multi-locus sequence typing. *Fungal Genet. Biol.* **45:**400–421.
8. **Bovers, M., F. Hagen, E. E. Kuramae, M. R. Diaz, L. Spanjaard, F. Dromer, H. L. Hoogveld, and T. Boekhout.** 2006. Unique hybrids between the fungal pathogens *Cryptococcus neoformans* and *Cryptococcus gattii*. *FEMS Yeast Res.* **6:**599–607.
9. **Bovers, M., F. Hagen, E. E. Kuramae, H. L. Hoogveld, F. Dromer, G. St-Germain, and T. Boekhout.** 2008. AIDS

patient death caused by novel *Cryptococcus neoformans* x *C. gattii* hybrid. *Emerg. Infect. Dis.* **14:**1105–1108.

10. **Brandt, M. E., L. C. Hutwagner, L. A. Klug, W. S. Baughman, D. Rimland, E. A. Graviss, R. J. Hamill, C. Thomas, P. G. Pappas, A. L. Reingold, R. W. Pinner, D. Stephens, M. Farley, M. Bardsley, B. Siegel, G. Jackson, C. Lao, J. Otte, C. Harvey, R. Gillespie, G. Rothrock, B. Pattni, and P. Daily.** 1996. Molecular subtype distribution of *Cryptococcus neoformans* in four areas of the United States. *J. Clin. Microbiol.* **34:**912–917.
11. **Bui, T., X. R. Lin, R. Malik, J. Heitman, and D. Carter.** 2008. Isolates of *Cryptococcus neoformans* from infected animals reveal genetic exchange in unisexual, alpha mating type populations. *Eukaryot. Cell* **7:**1771–1780.
12. **Canteros, C. E., M. Brudny, L. Rodero, D. Perrotta, and G. Davel.** 2002. Distribution of *Cryptococcus neoformans* serotypes associated with human infections in Argentina. *Rev. Argent. Microbiol.* **34:**213–218.
13. **Casadevall, A., and J. R. Perfect.** 1998. *Cryptococcus neoformans*. ASM Press, Washington, DC.
14. **Chander, J., R. K. Sapra, and P. Talwar.** 1994. Incidence of cryptococcosis in and around Chandigarh, India, during the period 1982–91. *Mycoses* **37:**23–26.
15. **Chaturvedi, V., J. J. Fan, B. Stein, M. J. Behr, W. A. Samsonoff, B. L. Wickes, and S. Chaturvedi.** 2002. Molecular genetic analyses of mating pheromones reveal intervariety mating or hybridization in *Cryptococcus neoformans*. *Infect. Immun.* **70:**5225–5235.
16. **Cherniak, R., L. C. Morris, T. Belay, E. D. Spitzer, and A. Casadevall.** 1995. Variation in the structure of glucuronoxylomannan in isolates from patients with recurrent cryptococcal meningitis. *Infect. Immun.* **63:**1899–1905.
17. **Cogliati, M., M. Allaria, G. Liberi, A. M. Tortorano, and M. A. Viviani.** 2000. Sequence analysis and ploidy determination of *Cryptococcus neoformans* var. *neoformans*. *J. Mycol. Med.* **10:**171–176.
18. **Cogliati, M., M. Allaria, A. M. Tortorano, and M. A. Viviani.** 2000. Genotyping *Cryptococcus neoformans* var. *neoformans* with specific primers designed from PCR-fingerprinting bands sequenced using a modified PCR-based strategy. *Med. Mycol.* **38:**97–103.
19. **Cogliati, M., M. C. Esposto, D. L. Clarke, B. L. Wickes, and M. A. Viviani.** 2001. Origin of *Cryptococcus neoformans* var. *neoformans* diploid strains. *J. Clin. Microbiol.* **39:** 3889–3894.
20. **Cogliati, M., M. C. Esposto, A. M. Tortorano, and M. A. Viviani.** 2006. *Cryptococcus neoformans* population includes hybrid strains homozygous at mating-type locus. *FEMS Yeast Res.* **6:**608–613.
21. **Colom, M. F., S. Frases, C. Ferrer, E. Martin-Mazuelos, M. Hermoso de Mendoza, J. M. Torres Rodriguez, and G. Quindos.** 2001. Epidemiological study of cryptococcosis in Spain: first results. *Rev. Iberoam. Micol.* **18:**99–104.
22. **Dromer, F., E. Gueho, O. Ronin, and B. Dupont.** 1993. Serotyping of *Cryptococcus neoformans* by using a monoclonal antibody specific for capsular polysaccharide. *J. Clin. Microbiol.* **31:**359–363.
23. **Enache-Angoulvant, A., J. Chandenier, F. Symoens, P. Lambe, J. Bolognini, C. Douchet, J. L. Poirot, and C. Hennequin.** 2007. Molecular identification of *Cryptococcus neoformans* serotypes. *J. Clin. Microbiol.* **45:**1261–1265.
24. **Evans, E. E.** 1950. The antigenic composition of *Cryptococcus neoformans*. 1. A serological classification by means of the capsular and agglutination reactions. *J. Immunol.* **64:**423–430.
25. **Evans, E. E.** 1951. The antigenic composition of *Cryptococcus neoformans*. 2. Serologic studies with the capsular polysaccharide. *J. Immunol.* **67:**109–114.
26. **Feng, X. B., Z. R. Yao, D. M. Ren, W. Q. Liao, and J. S. Wu.** 2008. Genotype and mating type analysis of *Cryptococcus neoformans* and *Cryptococcus gattii* isolates from China that mainly originated from non-HIV-infected patients. *FEMS Yeast Res.* **8:**930–938.
27. **Franzot, S. P., I. F. Salkin, and A. Casadevall.** 1999. *Cryptococcus neoformans* var. *grubii*: separate varietal status for *Cryptococcus neoformans* serotype A isolates. *J. Clin. Microbiol.* **37:**838–840.
28. **Fraser, J. A., S. S. Giles, E. C. Wenink, S. G. Geunes-Boyer, J. R. Wright, S. Diezmann, A. Allen, J. E. Stajich, F. S. Dietrich, J. R. Perfect, and J. Heitman.** 2005. Same-sex mating and the origin of the Vancouver Island *Cryptococcus gattii* outbreak. *Nature* **437:**1360–1364.
29. **Hiremath, S. S., A. Chowdhary, T. Kowshik, H. S. Randhawa, S. Sun, and J. Xu.** 2008. Long-distance dispersal and recombination in environmental populations of *Cryptococcus neoformans* var. *grubii* from India. *Microbiology* **154:**1513–1524.
30. **Hull, C. M., M. J. Boily, and J. Heitman.** 2005. Sex-specific homeodomain proteins Sxi1 alpha and Sxi2a coordinately regulate sexual development in *Cryptococcus neoformans*. *Eukaryot. Cell* **4:**526–535.
31. **Hull, C. M., R. C. Davidson, and J. Heitman.** 2002. Cell identity and sexual development in *Cryptococcus neoformans* are controlled by the mating-type-specific homeodomain protein Sxi1 alpha. *Genes Dev.* **16:**3046–3060.
32. **Ikeda, R., A. Nishikawa, T. Shinoda, and Y. Fukazawa.** 1985. Chemical characterization of capsular polysaccharide from *Cryptococcus neoformans* serotype AD. *Microbiol. Immunol.* **29:**981–991.
33. **Ikeda, R., T. Shinoda, Y. Fukazawa, and L. Kaufman.** 1982. Antigenic characterization of *Cryptococcus neoformans* serotypes and its application to serotyping of clinical isolates. *J. Clin. Microbiol.* **16:**22–29.
34. **Ito-Kuwa, S., K. Pienthaweechai, S. Aoki, V. Vidotto, M. Ohkusu, K. Hata, and K. Takeo.** 2000. Serotype, mating type and ploidy of *Cryptococcus neoformans* strains isolated from AIDS patients in Thailand. *J. Mycol. Med.* **10:**191–196.
35. **Katsu, M., S. Kidd, A. Ando, M. L. Moretti-Branchini, Y. Mikami, K. Nishimura, and W. Meyer.** 2004. The internal transcribed spacers and 5.8S rRNA gene show extensive diversity among isolates of the *Cryptococcus neoformans* species complex. *FEMS Yeast Res.* **4:**377–388.
36. **Kavanaugh, L. A., J. A. Fraser, and F. S. Dietrich.** 2006. Recent evolution of the human pathogen *Cryptococcus neoformans* by intervarietal transfer of a 14-gene fragment. *Mol. Biol. Evol.* **23:**1879–1890.
37. **Keller, S. M., M. A. Viviani, M. C. Esposto, M. Cogliati, and B. L. Wickes.** 2003. Molecular and genetic characterization of a serotype A MATa *Cryptococcus neoformans* isolate. *Microbiologym* **149:**131–142.
38. **Kidd, S. E., F. Hagen, R. L. Tscharke, M. Huynh, K. H. Bartlett, M. Fyfe, L. MacDougall, T. Boekhout, K. J. Kwon-Chung, and W. Meyer.** 2004. A rare genotype of *Cryptococcus gattii* caused the cryptococcosis outbreak on Vancouver Island (British Columbia, Canada). *Proc. Natl. Acad. Sci. USA* **101:**17258–17263.
39. **Kwon-Chung, K. J.** 1978. Heterothallism vs. self-fertile isolates of *Filobasidiella neoformans* (*Cryptococcus neoformans*). Presented at the IV International Conference on the Mycoses. The Black and White Yeasts, p. 204–213.
40. **Kwon-Chung, K. J., and J. E. Bennett.** 1978. Distribution of alpha and **a** mating types of *Cryptococcus neoformans* among natural and clinical isolates. *Am. J. Epidemiol.* **108:**337–340.
41. **Kwon-Chung, K. J., and J. E. Bennett.** 1984. Epidemiologic differences between the two varieties of *Cryptococcus neoformans*. *Am. J. Epidemiol.* **120:**123–130.

42. **Kwon-Chung, K. J., J. C. Edman, and B. L. Wickes.** 1992. Genetic association of mating types and virulence in *Cryptococcus neoformans*. *Infect. Immun.* **60:**602–605.
43. **Latouche, G. N., M. Huynh, T. C. Sorrell, and W. Meyer.** 2003. PCR-restriction fragment length polymorphism analysis of the phospholipase B (PLB1) gene for subtyping of *Cryptococcus neoformans* isolates. *Appl. Environ. Microbiol.* **69:**2080–2086.
44. **Lengeler, K. B., G. M. Cox, and J. Heitman.** 2001. Serotype AD strains of *Cryptococcus neoformans* are diploid or aneuploid and are heterozygous at the mating-type locus. *Infect. Immun.* **69:**115–122.
45. **Lengeler, K. B., P. Wang, G. M. Cox, J. R. Perfect, and J. Heitman.** 2000. Identification of the MATa mating-type locus of *Cryptococcus neoformans* reveals a serotype A MATa strain thought to have been extinct. *Proc. Natl. Acad. Sci. USA* **97:**14455–14460.
46. **Li, A. S., K. Nishimura, H. Taguchi, R. Tanaka, S. X. Wu, and M. Miyaji.** 1993. The isolation of *Cryptococcus neoformans* from pigeon droppings and serotyping of naturally and clinically sourced isolates in China. *Mycopathologia* **124:**1–5.
47. **Lin, X., J. C. Huang, T. G. Mitchell, and J. Heitman.** 2006. Virulence attributes and hyphal growth of *C. neoformans* are quantitative traits and the MATα allele enhances filamentation. *PLoS Genet.* **2:**e187.
48. **Lin, X., C. M. Hull, and J. Heitman.** 2005. Sexual reproduction between partners of the same mating type in *Cryptococcus neoformans*. *Nature* **434:**1017–1021.
49. **Lin, X., A. P. Litvintseva, K. Nielsen, S. Patel, A. Floyd, T. G. Mitchell, and J. Heitman.** 2007. alphaADalpha hybrids of *Cryptococcus neoformans*: evidence of same-sex mating in nature and hybrid fitness. *PLoS Genet.* **3:**e186.
50. **Lin, X., K. Nielsen, S. Patel, and J. Heitman.** 2008. Impact of mating type, serotype, and ploidy on the virulence of *Cryptococcus neoformans*. *Infect. Immun.* **76:**2923–2938.
51. **Lin, X., S. Patel, A. P. Litvintseva, A. Floyd, T. G. Mitchell, and J. Heitman.** 2009. Diploids in the *Cryptococcus neoformans* serotype A population homozygous for the alpha mating type originate via unisexual mating. *PLoS Pathog.* **5:**1–18.
52. **Litvintseva, A. P., L. Kestenbaum, R. Vilgalys, and T. G. Mitchell.** 2005. Comparative analysis of environmental and clinical populations of *Cryptococcus neoformans*. *J. Clin. Microbiol.* **43:**556–564.
53. **Litvintseva, A. P., X. Lin, I. Templeton, J. Heitman, and T. G. Mitchell.** 2007. Many globally isolated AD hybrid strains of *Cryptococcus neoformans* originated in Africa. *PLoS Pathog.* **3:**1109–1117.
54. **Litvintseva, A. P., R. E. Marra, K. Nielsen, J. Heitman, R. Vilgalys, and T. G. Mitchell.** 2003. Evidence of sexual recombination among *Cryptococcus neoformans* serotype A isolates in sub-Saharan Africa. *Eukaryot. Cell* **2:**1162–1168.
55. **Litvintseva, A. P., R. Thakur, R. Vilgalys, and T. G. Mitchell.** 2006. Multilocus sequence typing reveals three genetic subpopulations of *Cryptococcus neoformans* var. *grubii* (serotype A), including a unique population in Botswana. *Genetics* **172:**2223–2238.
56. **MacGill, T. C., R. S. MacGill, A. Casadevall, and T. R. Kozel.** 2000. Biological correlates of capsular (quellung) reactions of *Cryptococcus neoformans*. *J. Immunol.* **164:** 4835–4842.
56a. **Meyer, W., D. M. Aanensen, T. Boekhout, M. Cogliati, M. R. Diaz, M. C. Esposto, M. Fisher, F. Gilgado, F. Hagen, S. Kaocharoen, A. P. Litvintseva, T. Mitchell, S. P. Simwami, L. Trilles, M. A. Viviani, and J. K. Kwon Chung.** 2009. Consensus multi-locus sequence typing sheme for *Cryptococcus neoformans* and *Cryptococcus gattii*. Med. Mycol. **47:**561–570.
57. **Meyer, W., A. Castaneda, S. Jackson, M. Huynh, E. Castaneda, and IberoAmerican Cryptococcal Study Group.** 2003. Molecular typing of IberoAmerican *Cryptococcus neoformans* isolates. *Emerg. Infect. Dis.* **9:**189–195.
58. **Meyer, W., K. Marszewska, M. Amirmostofian, R. P. Igreja, C. Hardtke, K. Methling, M. A. Viviani, A. Chindamporn, S. Sukroongreung, M. A. John, D. H. Ellis, and T. C. Sorrell.** 1999. Molecular typing of global isolates of *Cryptococcus neoformans* var. *neoformans* by polymerase chain reaction fingerprinting and randomly amplified polymorphic DNA: a pilot study to standardize techniques on which to base a detailed epidemiological survey. *Electrophoresis* **20:**1790–1799.
59. **Mishra, S. K., F. Staib, U. Folkens, and R. A. Fromtling.** 1981. Serotypes of *Cryptococcus neoformans* strains isolated in Germany. *J. Clin. Microbiol.* **14:**106–107.
60. **Nielsen, K., G. M. Cox, P. Wang, D. L. Toffaletti, J. R. Perfect, and J. Heitman.** 2003. Sexual cycle of *Cryptococcus neoformans* var. *grubii* and virulence of congenic a and alpha isolates. *Infect. Immun.* **71:**4831–4841.
61. **Nielsen, K., R. E. Marra, F. Hagen, T. Boekhout, T. G. Mitchell, G. M. Cox, and J. Heitman.** 2005. Interaction between genetic background and the mating-type locus in *Cryptococcus neoformans* virulence potential. *Genetics* **171:** 975–983.
62. **Nishikawa, M. M., M. S. Lazera, G. G. Barbosa, L. Trilles, B. R. Balassiano, R. C. L. Macedo, C. C. F. Bezerra, M. A. Perez, P. Cardarelli, and B. Wanke.** 2003. Serotyping of 467 *Cryptococcus neoformans* isolates from clinical and environmental sources in Brazil: analysis of host and regional patterns. *J. Clin. Microbiol.* **41:**73–77.
63. **Padhye, A. A., A. Chakrabarti, J. Chander, and L. Kaufman.** 1993. *Cryptococcus neoformans* var. *gattii* in India. *J. Med. Vet. Mycol.* **31:**165–168.
64. **Perez, C., M. Dolande, M. Moya, A. Rosello, C. R. H. de Capriles, M. E. Landaeta, and S. Mata-Essayag.** 2008. *Cryptococcus neoformans*, *Cryptococcus gattii*: serotypes in Venezuela. *Mycopathologia* **166:**149–153.
65. **Saracli, M. A., S. T. Yildiran, K. Sener, A. Gonlum, L. Doganci, S. M. Keller, and B. L. Wickes.** 2006. Genotyping of Turkish environmental *Cryptococcus neoformans* var. *neoformans* isolates by pulsed field gel electrophoresis and mating type. *Mycoses* **49:**124–129.
66. **Saul, N., M. Krockenberger, and D. Carter.** 2008. Evidence of recombination in mixed-mating-type and alpha-only populations of *Cryptococcus gattii* sourced from single eucalyptus tree hollows. *Eukaryot. Cell* **7:**727–734.
67. **Shadomy, H. J.** 1970. Clamp-connection in two strains of *Cryptococcus neoformans*. Spectrum Monograph Series, Vol 1, Arts and Sci. p. 67–72, Georgia State University, Atlanta.
68. **Shahid, M., S. Han, H. Yoell, and J. P. Xu.** 2008. Fitness distribution and transgressive segregation across 40 environments in a hybrid progeny population of the human-pathogenic yeast *Cryptococcus neoformans*. *Genome* **51:**272–281.
69. **Sia, R. A., K. B. Lengeler, and J. Heitman.** 2000. Diploid strains of the pathogenic basidiomycete *Cryptococcus neoformans* are thermally dimorphic. *Fungal Genet. Biol.* **29:**153–163.
70. **Steenbergen, J. N., and A. Casadevall.** 2000. Prevalence of *Cryptococcus neoformans* var. *neoformans* (serotype D) and *Cryptococcus neoformans* var. *grubii* (serotype A) isolates in New York city. *J. Clin. Microbiol.* **38:**1974–1976.
71. **Sukroongreung, S., C. Nilakul, O. Ruangsomboon, W. Chuakul, and B. Eampokalap.** 1996. Serotypes of *Cryptococcus neoformans* isolated from patients prior to and during the AIDS era in Thailand. *Mycopathologia* **135:**75–78.

72. **Tanaka, R., K. Nishimura, and M. Miyaji.** 1999. Ploidy of serotype AD strains of *Cryptococcus neoformans*. *Jpn. J. Med. Mycol.* **40:**31–34.

73. **Tanaka, R., H. Taguchi, K. Takeo, M. Miyaji, and K. Nishimura.** 1996. Determination of ploidy in *Cryptococcus neoformans* by flow cytometry. *J. Med. Vet. Mycol.* **34:**299–301.

74. **Tintelnot, K., K. Lemmer, H. Losert, G. Schar, and A. Polak.** 2004. Follow-up of epidemiological data of cryptococcosis in Austria, Germany, and Switzerland with special focus on the characterization of clinical isolates. *Mycoses* **47:**455–464.

75. **Tortorano, A. M., M. A. Viviani, A. L. Rigoni, M. Cogliati, A. Roverselli, and A. Pagano.** 1997. Prevalence of serotype D in *Cryptococcus neoformans* isolates from HIV positive and HIV negative patients in Italy. *Mycoses* **40:**297–302.

76. **Trilles, L., M. D. Lazera, B. Wanke, R. V. Oliveira, G. G. Barbosa, M. M. Nishikawa, B. P. Morales, and W. Meyer.** 2008. Regional pattern of the molecular types of *Cryptococcus neoformans* and *Cryptococcus gattii* in Brazil. *Mem. Inst. Oswaldo Cruz* **103:**455–462.

77. **Tscharke, R.L., M. Lazera, Y. C. Chang, B. L. Wickes, and K. J. Kwon-Chung.** 2003. Haploid fruiting in *Cryptococcus neoformans* is not mating type alpha-specific. *Fungal Genet. Biol.* **39**: 230–237.

78. **Viviani, M. A., S. Antinori, M. Cogliati, M. C. Esposto, G. Pinsi, S. Casari, A. Bergamasco, M. L. De Santis, P. Ghirga, C. Bonaccorso, G. Jacchetti, F. Niero, A. Ammassari, G. Morace, C. U. Foppa, C. Ossi, M. T. Montagna, G. Angarano, C. Farina, F. Maggiolo, M. Moroni, S. Pierdomenico, G. Michelone, C. Cavanna, F. Viti, T. Carli, F. Barchiesi, C. Agrappi, M. Mena, M. Giola, G. Lombardi, M. Tinelli, A. Ceraminiello, M. Codeluppi, C. Casolari, S. Foresti, S. Bramati, A. Ruggieri, L. Caggese, A. Astolfi, S. Bonora, P. G. Scotton, R. Rigoli, G. Angioni, F. Meneghetti, G. Di Perri, and FIMUA Cryptococcosis Network.** 2002. European Confederation of Medical Mycology (ECMM) prospective survey of cryptococcosis: report from Italy. *Med. Mycol.* **40:**507–517.

79. **Viviani, M. A., M. Cogliati, M. C. Esposto, K. Lemmer, K. Tintelnot, M. F. C. Valiente, D. Swinne, A. Velegraki, R. Velho, and ECMM Cryptococcosis Network.** 2006. Molecular analysis of 311 *Cryptococcus neoformans* isolates from a 30-month ECMM survey of cryptococcosis in Europe. *FEMS Yeast Res.* **6:**614–619.

80. **Viviani, M. A., M. C. Esposto, M. Cogliati, M. T. Montagna, and B. L. Wickes.** 2001. Isolation of a *Cryptococcus neoformans* serotype A MATa strain from the Italian environment. *Med. Mycol.* **39:**383–386.

81. **Viviani, M. A., R. Nikolova, M. C. Esposto, G. Prinz, and M. Cogliati.** 2003. First European case of serotype A MATa *Cryptococcus neoformans* infection. *Emerg. Infect. Dis.* **9:**1179–1180.

82. **Viviani, M. A., H. Wen, A. Roverselli, R. Caldarelli-Stefano, M. Cogliati, P. Ferrante, and A. M. Tortorano.** 1997. Identification by polymerase chain reaction fingerprinting of *Cryptococcus neoformans* serotype AD. *J. Med. Vet. Mycol.* **35:**355–360.

83. **Whelan, W. L., and K. J. Kwon-Chung.** 1986. Genetic complementation in *Cyptococcus neoformans*. *J. Bacteriol.* **166:**924–929.

84. **White, C. W., and E. S. Jacobson.** 1985. Occurrence of diploid strains of *Cryptococcus neoformans*. *J. Bacteriol.* **161:** 1231–1232.

85. **Wickes, B. L., M. E. Mayorga, U. Edman, and J. C. Edman.** 1996. Dimorphism and haploid fruiting in *Cryptococcus neoformans*: association with the alpha-mating type. *Proc. Natl. Acad. Sci. USA* **93:**7327–7331.

86. **Wilson, D. E.** 1968. Serologic grouping of *Cryptococcus neoformans*. *Proc. Soc. Exp. Biol. Med.* **127:**820–823.

87. **Xu, J. P., G. Z. Luo, R. J. Vilgalys, M. E. Brandt, and T. G. Mitchell.** 2002. Multiple origins of hybrid strains of *Cryptococcus neoformans* with serotype AD. *Microbiology* **148:**203–212.

88. **Xu, J. P., and T. G. Mitchell.** 2003. Comparative gene genealogical analyses of strains of serotype AD identify recombination in populations of serotypes A and D in the human pathogenic yeast *Cryptococcus neoformans*. *Microbiology* **149:**2147–2154.

89. **Xu, J. P., R. Vilgalys, and T. G. Mitchell.** 2000. Multiple gene genealogies reveal recent dispersion and hybridization in the human pathogenic fungus *Cryptococcus neoformans*. *Mol. Ecol.* **9:**1471–1481.

90. **Yan, Z., X. G. Li, and J. P. Xu.** 2002. Geographic distribution of mating type alleles of *Cryptococcus neoformans* in four areas of the United States. *J. Clin. Microbiol.* **40:**965–972.

INTERACTIONS WITH THE IMMUNE SYSTEM

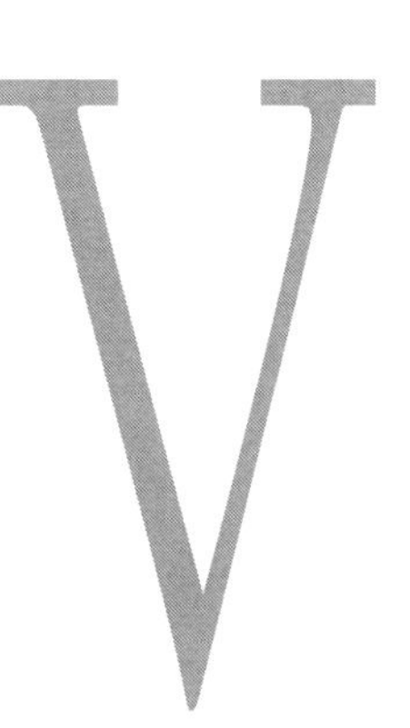

Cryptococcus: From Human Pathogen to Model Yeast
Edited by J. Heitman et al.

26

The Interaction of *Cryptococcus neoformans* with Host Macrophages and Neutrophils

TRAVIS MCQUISTON AND MAURIZIO DEL POETA

Upon inhalation of infectious propagules, *Cryptococcus neoformans* traverses through the host airways before lodging in the alveolar spaces of the lung. In the alveolar spaces, *C. neoformans* can survive and replicate in the extracellular environment or, following phagocytosis, within the phagolysosome of the resident alveolar macrophages (AMs). Thus, *C. neoformans* is considered a facultative intracellular pathogen.

To adapt to the host environment(s), *C. neoformans* responds to its new surroundings with an alteration of its gene expression, leading to phenotypic changes that enable the inhaled fungus to endure the lung environment. Perhaps the most compelling example of how *C. neoformans* cells adapt to the lung environment is the production of a polysaccharide capsule. By surrounding the cell wall or by being shed into the host extracellular or intracellular milieu, this factor protects *C. neoformans* from the host through a variety of mechanisms. In addition to the capsule, *C. neoformans* responds to the host environment by producing a variety of other factors. Studies have shown that factors such as antiphagocytic protein 1 (App1), mannoproteins, superoxide dismutases (SODs), laccases, phospholipase B1 (Plb1), phytoceramide, and glucosylceramide are deemed important to the survival of *C. neoformans* within the extracellular and/or intracellular compartments of the host. These fungal factors not only favor *C. neoformans* growth, but they specifically protect *C. neoformans* from a host immune response that should ultimately eradicate the fungus and thereby prevent infection and disease.

Since the AMs are the first component of the host immune response to confront *C. neoformans*, the outcome of this interaction is vital in determining if containment and clearance of the infection will occur or if dissemination of the pathogen will lead to life-threatening meningoencephalitis. This chapter will focus on the dynamic interaction of *C. neoformans* with AMs and how the consequential actions of both the host and fungal pathogen can affect the outcome of *C. neoformans* exposure. In addition, the role of neutrophils and their antifungal actions will be reviewed.

Travis McQuiston, Departments of Biochemistry and Molecular Biology, Medical University of South Carolina, Charleston, SC 29425. **Maurizio Del Poeta,** Departments of Biochemistry and Molecular Biology, Microbiology and Immunology, and Division of Infectious Diseases, Medical University of South Carolina, Charleston, SC 29425.

RECOGNITION AND BINDING OF *C. NEOFORMANS*

Recognition without Opsonins

As mentioned above, AMs are the first cells of the host innate immune system to confront *C. neoformans* cells in the lungs. Thus, phagocytosis is the initial effector action of these phagocytes and is dependent on the ability of the receptors on the surface of AMs to recognize and bind to fungal specific components or ligands localized on the surface of *C. neoformans*. This recognition can occur with or without opsonins. Non-opsonin-mediated interactions include the mannose and β-glucan receptors, which bind to their respective ligands found on the surface of *C. neoformans* (Fig. 1). These ligands are mainly localized to the *C. neoformans* cell wall, and thus, their involvement in phagocytosis has been demonstrated in conditions in which *C. neoformans* is not encapsulated (73). The presence of the polysaccharide capsule would shroud the molecular motifs located on the cell wall serving as ligands for these phagocyte receptors (74, 79). Epidemiological studies suggest that acapsular *C. neoformans* does not exist in nature because all environmental isolates of *C. neoformans* examined to date possess at least a thin capsule enveloping the fungal cell (110, 118). This would argue that phagocytosis through mannose and β-glucan receptors is not physiologically relevant unless poorly encapsulated *C. neoformans* cells still have the property to partially expose their cell wall ligands. Indeed, using various populations of primary macrophages and macrophage cell lines, it is proposed that the binding of mannose and β-glucan phagocyte receptors to encapsulated *C. neoformans* is minimal to nonexistent in vitro.

Besides, even if poorly encapsulated cells were inhaled, the low levels of iron and elevated bicarbonate concentrations within the alveolar spaces would induce the upregulation of capsule synthesis, thereby increasing capsule thickness and further diminishing the ability of AMs to recognize and

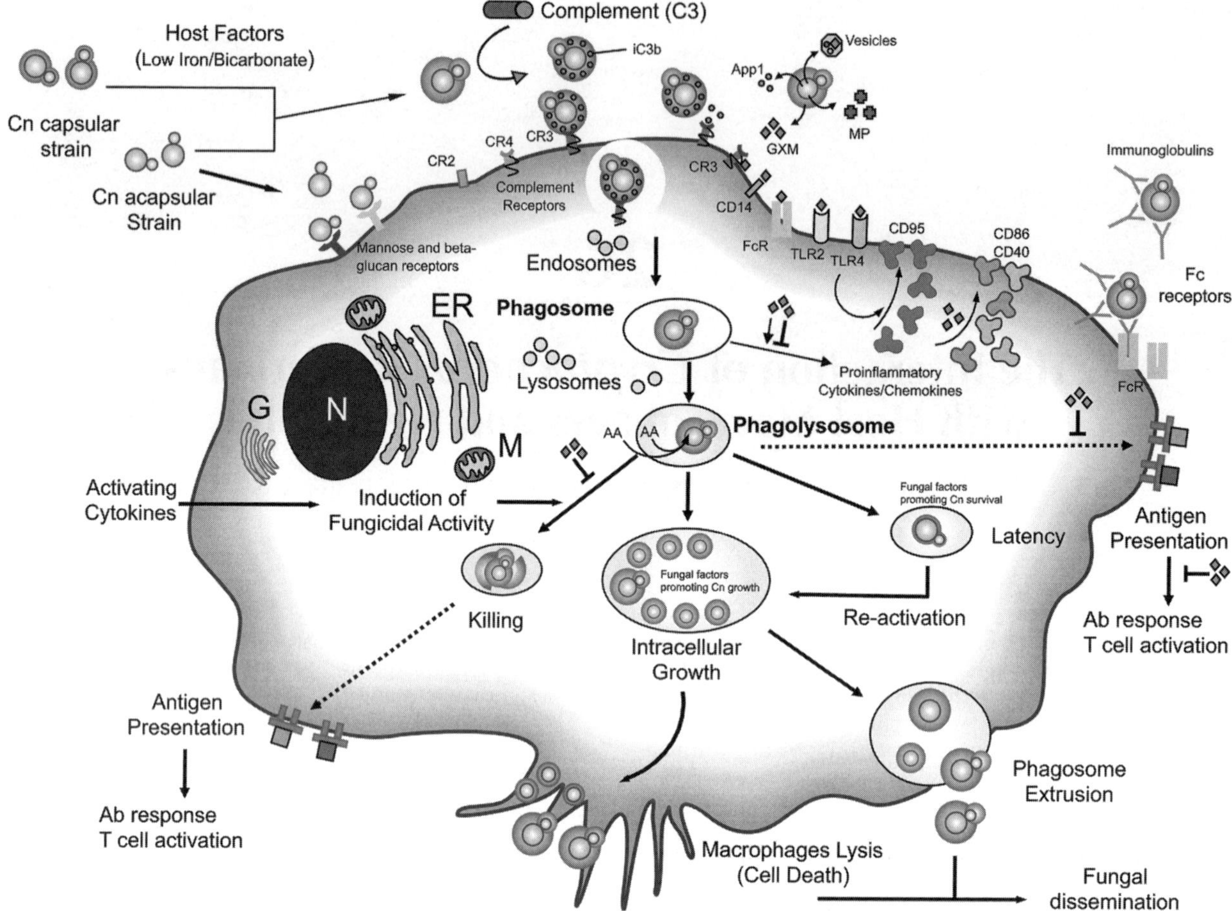

FIGURE 1 Interaction of *C. neoformans* with macrophages. *C. neoformans* is internalized by macrophages through a series of receptor-ligand interactions. Once internalized, *C. neoformans* can either be killed by the host, grow within the phagolysosome, or enter a dormant state. *C. neoformans* can move from the intracellular to the extracellular environment through macrophage lysis or by phagosome extrusion. Many *C. neoformans* and host factors have been identified to regulate these processes that can lead either to the containment of the infection or to the dissemination of *C. neoformans* cells and the development of cryptococcosis. CR, complement receptor; App1, antiphagocytic protein 1; GXM, glucuronoxylomannan; MP, mannoprotein; TLR, Toll-like receptor; AA, arachidonic acid; ER, endoplasmic reticulum; G, Golgi; N, nucleus; M, mitochondria.

bind mannose and β-glucan on the *C. neoformans* cell surface (54, 147, 153). Therefore, since the ligand-receptor interaction is impeded by the capsule, and the lung environment promotes an increase in capsule thickness, one could argue that these receptors would most likely not have a significant role in the recognition and binding of *C. neoformans* by macrophages. However, very recent studies showed that the β-glucan scavenger receptor family 1 (SCARF1) and CD36 are required for macrophage binding of the encapsulated *C. neoformans* (strain H99), which is important for cytokine production and for controlling *C. neoformans* infection in mice (93). Interestingly, their *Caenorhabditis elegans* orthologues, CED-1 and C03F11.3, are required for the induction of protective immune responses in worms against *C. neoformans*, suggesting an evolutionarily conserved pathway for these receptors (93). These results suggest that further studies are needed to fully elucidate the role of β-glucan receptors in fungal pathogenesis.

Several other receptors expressed by AMs recognize *C. neoformans* without the requirement for opsonization, but the activation of these receptors following binding to their respective ligands does not directly regulate phagocytosis. These receptors include CD14 and Toll-like receptors 2 (TLR2) and 4 (TLR4) (Fig. 1). All three receptors are expressed by macrophages and other phagocytes, such as neutrophils, which are the first proinflammatory cell recruited by AMs from the serum to the site of infection. They mostly bind to glucuronoxylomannan (GXM) and facilitate its uptake from serum and tissues (33, 56, 138, 167). In macrophages, GXM binding to TLR2 and TLR4, in conjunction with CD14, triggers NF-κB activation and its translocation to the nucleus (9, 81, 98). Since NF-κB is a key regulator of the transcription of many genes, the GXM-TLR interaction will evoke a different host response(s) that only very recently has become the focus of intense investigation.

Recognition with Opsonins

Recognition and binding of *C. neoformans* by phagocytes occurs with greater avidity when cells are opsonized. The alveolar spaces of the lung are coated with an epithelial lining fluid containing components essential to the induction of the host immune response, including complement and immunoglobulins (Ig). These opsonins can deposit onto microbial surfaces, thereby mediating binding to specific complement and Ig receptors expressed on the surface of phagocytes, and enhance microbial engulfment.

Activation of complement protects the host against *C. neoformans* because experimental hosts depleted of complement are more susceptible to cryptococcosis (30, 135). Complement opsonization occurs as a complex series of biochemical steps producing the active component 3b (C3b), which covalently associates with the pathogen surface (Fig. 1). C3b can be produced by the classical, lectin, or alternative pathway. However, the molecular events leading to the activation of each pathway differ significantly. The cascade of biochemical steps leading to complement opsonization is the focus of another chapter in this book and should be read for a detailed description of the molecular mechanisms leading to complement opsonization. However, in the context of AMs, the alternative pathway is an imperative facet during initial exposure to a pathogen in the lung environment since it does not require an adaptor molecule (Ab) or recognition motif for activation.

In the alveolar spaces, *C. neoformans* GXM activates the alternative complement cascade, leading to the deposition of C3b on the exterior surface and interior matrix of the capsule, which then is converted to iC3b (72, 75). Macrophages express three major complement receptors (CRs) termed CD35 (CR1), CD11b/CD18 (CR3), and CD11c/CD18 (CR4) that recognize iC3b fragments attached to the microbial surface (37). CR3 is the CR predominantly responsible for phagocytosis of microbes (76). CR3 possesses two subunits: CD11b and CD18. In addition to binding to iC3b, CD11b also binds to ICAM-1, fibrogen, and neutrophil inhibitory factor. The CD18 subunit binds to β-glucan but can also bind to GXM, and this recognition can be detrimental to the host because it can inhibit phagocytosis and leukocyte infiltration (34, 146).

Interestingly, the activation of the alternative pathway induced by the *C. neoformans* capsule differs significantly depending on the *C. neoformans* strain. For instance, the capsule of *C. neoformans* stimulates the alternative complement cascade to a greater degree than *Cryptococcus gattii* (161, 168), a *Cryptococcus* species that causes mainly pulmonary infections in immunocompetent patients (11, 17, 152). Recently, a more in-depth examination of the role of complement in *C. gattii* infection suggested that the alternative pathway of complement activation is not the only complement pathway contributing to protection against disease (95). Perhaps the differences in complement activation between these species are due to species-specific variations in capsule composition and structure (161, 168). Since complement opsonization is the major mechanism of cryptococcal recognition, capsule differences may play an important role in determining organ tropism and the type of clinical manifestation. Additional studies are clearly needed to determine whether the difference in complement activation may correlate with the different clinical manifestations of these species.

Antibodies are another major class of opsonins targeting microbes for host recognition and binding (79, 172) (Fig. 1). Macrophages express cell surface receptors recognizing the Fc portion of IgG antibodies, called FcγRs, which participate in many processes including phagocytosis of infectious organisms. Specifically, macrophages express four types of FcγRs: FcγRI, -II, -III, and -IV (24, 112). FcγRII is further classified as FcγRIIA and FcγRIIB isoforms. FcγRI, FcγRIIA, FcγRIII, and FcγRIV initiate cellular actions through cytoplasmic domains called immunoreceptor tyrosine-based activation motifs (51, 120, 155). In contrast, FcγRIIB has inhibitory functions because its immunoreceptor tyrosine-based inhibition motifs activate the SH2-containing inositol polyphosphate 5-phosphatase (121). FcγRI is a high-affinity receptor that binds monomeric IgG as well as immune complexes; FcγRII, III, and IV are low-affinity receptors that only bind immune complexes. The role of Fc receptors during *C. neoformans* infection is discussed in greater detail elsewhere in this book. But here we would like to emphasize that in addition to the variables illustrated above involving Fc and IgG isotype interaction, direct interaction between *C. neoformans* and AMs may add an additional layer of complexity. Intriguingly, it has been shown that the downstream effect of the interaction between anti-GXM antibody with its Fc receptor can be overturned by a direct interaction of GXM with the Fc receptor, suggesting that there is a complex interplay among *C. neoformans* antigen(s), antibody production, Fc receptors, and host immune response (100).

The importance of endogenous antibody production during *C. neoformans* infection as a mechanism of defense is uncertain, particularly as opsonins for phagocytosis following the initial exposure. In mice, the humoral immune response to pulmonary infection has been shown to require between 3 and 7 days to produce specific antibodies against *C. neoformans* (171). Furthermore, after 7 days, only IgG is found in the lungs following *C. neoformans* infection, and it is not opsonic, as it localizes within the capsule and is not accessible to the FcγRs of AMs. In addition, anti-GXM IgM, IgG, and IgA are ubiquitously present in the serum of infected mice and humans, but titers are low, when detectable, and are also usually not opsonic (13, 27, 28, 79). The most compelling evidence that humoral immunity is not pertinent to preventing cryptococcosis is that human subjects with agammaglobulinemia are not, in general, hypersusceptible to cryptococcal infection, as only a few cases have been documented (111). Therefore, although antibody-mediated phagocytosis may be a very efficient method for removing *C. neoformans* cells from the extracellular milieu, its overall role in protection needs further investigation.

Despite the inconclusive role of endogenous antibodies during cryptococcosis, passive administration of monoclonal antibodies (mAbs) against GXM has shown encouraging therapeutic effects (14, 25). However, the protective efficacy of mAb therapies against experimental cryptococcosis has varied greatly and depends on numerous factors including antibody isotype, epitope specificity, antibody dosage, the size of the infecting inoculum and its route of administration, and host immunomodulatory effects. Administration of mAbs against GXM can increase *C. neoformans* internalization and prolong survival in murine models of cryptococcosis via complement pathway-independent and/or -dependent mechanisms (35, 46, 107, 113, 128, 135, 146, 163, 164, 169, 170). Administration of IgG1, IgG2a, and IgG2b mAbs increased the internalization of *C. neoformans* cells by AMs and prolonged survival of C3-deficient mice, demonstrating that mAbs can serve as effective opsonins in vivo without involvement of the complement pathways (135). These selected mAbs also increase the survival of A/J mice, which have a deficiency in the chemoattractant C5

(C5a) that results in a decreased neutrophil infiltration at the site of the infection (122, 126, 163, 164).

Interestingly, mAbs can work synergistically with components comprising the complement-mediated opsonization and recognition pathway to increase the efficacy of the host immune response in addition to classical complement pathway activation. As previously stated, complement can deposit within the capsular matrix near the cell wall where, if the capsule is large, it is not accessible to the CRs expressed by AMs. However, iC3b localizes to the outer capsular surface in the presence of higher experimental concentrations (100 μg ml^{-1}) of anti-GXM antibodies through both activation of the classical complement pathway and steric hindrance that prevents complement penetration (173). In the absence of complement, mAbs can modulate a lectin recognition-like interaction of GXM and CR3 and CR4 leading to *C. neoformans* phagocytosis (146). These studies suggest that mAbs amplify the effectiveness of complement opsonization by augmenting its deposition on the surface of *C. neoformans* cells and, thus, an increase of complement-mediated phagocytosis.

Despite the positive acclaim that mAb administration can preclude *C. neoformans* infection, several studies have shown that administration of anti-GXM IgG3 mAbs does not offer protection from *C. neoformans* infection and can even exacerbate cryptococcosis in C5-deficient mice (106, 135). In contrast, IgG1 and IgG2 mAb isotypes targeting identical epitopes of GXM significantly prolonged survival of mice infected intravenously with *C. neoformans* (14, 25). These data suggest that passive immunity may enhance the host response against cryptococcosis, but the mechanisms of host defense should be further delineated. It is therefore important to fully define the Ig-FcR interaction for both protective and nonprotective antibodies, as specific receptor recognition and activation determines the actions of the host immune cell against *C. neoformans*.

In addition to preventing receptor recognition leading to phagocytosis by phagocytes, capsule components of *C. neoformans* inhibit the host response to infection, thereby enabling the fungus to elude other immune cells. For instance, GXM reduces neutrophil chemotaxis to the site of infection by decreasing L-selectin (CD62L) expression on the surface of human neutrophils and preventing the movement of neutrophils along the endothelium (34, 39–41).

But the interaction of extracellular GXM with phagocytes is more complex than anticipated. For instance, GXM interacts with many receptors, including CD14, CD18, TLR4, or FcγRII. Interaction of GXM with TLR4 stimulates the expression of other receptors, such as CD95, CD80, and CD40 (99–103, 159). Upon GXM internalization, its degradation is regulated differently depending on the type of phagocytic cell (103). In neutrophils, GXM is rapidly degraded, but in rat macrophages, GXM can persist for months (53). Thus, the role of GXM in the regulation of the host immune response through macrophages is clearly more complex than that occurring through neutrophils.

In addition to GXM, other *C. neoformans* factors are secreted, and they regulate different host immune responses (Fig. 1). For instance, *C. neoformans* mannoproteins desensitize neutrophils to tumor necrosis factor α by abolishing tumor necrosis factor α receptor expression (20, 34). *C. neoformans* can specifically inhibit complement-mediated phagocytosis through the production of App1, which binds to the CD11b subunit of CR3 and thereby impedes the binding and internalization of iC3b-coated *C. neoformans* cells (86, 91, 143, 151). *C. neoformans* can secrete vesicles containing a variety of proteins, some of which have unknown function (124, 125). These studies suggest that in addition to capsular materials, *C. neoformans* may have developed additional mechanisms to evade the recognition of phagocytic cells and to modulate the host response to its advantage.

THE INTERNALIZATION OF *C. NEOFORMANS*

The binding of *C. neoformans* to one of the phagocytic receptors described above can activate signaling cascades and initiate the process of phagocytosis. Unlike many pathogenic microbes, particularly obligate intracellular pathogens, *C. neoformans* does not possess a mechanism to actively penetrate host phagocytes. Therefore, internalization of inhaled cryptococcal cells is an active process completely dependent on host mechanisms induced by receptor activation.

If *C. neoformans* does not prevent the induction of phagocytosis, pseudopods extend around the fungal cell, engulfing it entirely and creating a phagocytic vacuole termed a phagosome. This newly formed vesicle undergoes drastic membrane remodeling through fusion with endomembrane compartments and orchestrated fission processes. The incorporation of these endosomes into the phagosome results in the integration of membrane proteins required for the maturation process and regulates the intracellular microenvironment while depositing a multitude of microbicidal agents (80, 150).

Some of the membrane proteins incorporated into the vesicle membrane of the maturing phagosome include the NADPH oxidase complex, SOD, myeloperoxidase (MPO), and vacuolar-type ATPase. MPO interacts with respiratory burst-derived hydrogen peroxide (created through the proceeding actions of the NADPH oxidase complex and SOD to oxidize chloride, producing the potent microbicidal hypochlorous acid) (70, 71). MPO also oxidizes bromide, iodine, and the pseudohalide ion thiocyanate to produce other antimicrobial agents. MPO has limited expression in macrophages, while it is abundantly found in neutrophils. This may be one of the reasons why macrophages and neutrophils differ in their killing efficiencies.

Vacuolar-type ATPase converts the energy of ATP hydrolysis to move protons across the phagosome membrane, resulting in the acidification of the phagosome and roping of the pH from slightly alkaline down to pH 4.5 to 5.5 (88, 89). In addition to being directly cytotoxic for many microbes, the acidification of the phagosome lumen is thought to regulate the maturation process and to be required for optimal activity of the microbicidal molecules populating the phagolysosome lumen. Interestingly, the degree of acidification can vary greatly between phagocytes, particularly between macrophages and neutrophils (130–132). Macrophage phagosomes display a profound and rapid acidification, decreasing from approximately pH 7.0 to between 4.5 and 5.5, whereas neutrophil phagosomes actually alkalinize initially, then undergo a modest acidification to pH 6.5 (57, 89). Data show that differences in proton delivery to the phagosome/phagolysosome lumen, the varying proportion of proton consumption by products of the NADPH oxidase, and increased electron leakage (via membrane permeability) into the cytosol are responsible for the discrepancy in luminal pH between macrophages and neutrophils (5–8, 66, 130–132). The roles of NADPH and MPO in resolving whether intracellular *C. neoformans* perish or subsist will be discussed in the next section of this chapter.

The phagosome maturation process is completed upon the fusion of lysosomes and the deposition of hydrolytic enzymes to create a phagolysosome. This process transforms the nascent phagosome enclosing the cryptococcal cell to a highly specialized antimicrobial vesicle termed the phagolysosome. In the case of most internalized microbes, the acidic environment and actions of the various microbicidal molecules contained within the phagolysosome generally lead to rapid killing and do not allow for colonization. To live parasitically, intracellular pathogens such as pathogenic *Mycobacterium* species, *Histoplasma capsulatum*, *Legionella pneumophila*, and *Toxoplasma gondii* have developed mechanisms preventing acidification of the phagolysosome in which they reside. In contrast, *C. neoformans* does not actively attempt to avoid the phagolysosome acidification and actually grows better in acidic compared to alkaline conditions (78). In fact, the acidification of the phagolysosome may be required for the virulence of intracellular *C. neoformans*, because alkalinizing the pH of the phagolysosome with chloroquine significantly improves the anticryptococcal activity of monocyte-derived macrophages (77). However, these studies examined the intracellular growth of *C. neoformans* in vitro using primary or macrophage-like cells. Although these systems provide excellent insight into the fungistatic activity of macrophages on internalized *C. neoformans*, the host environment is obviously different and, therefore, additional research is necessary to ascertain the rate of growth of intracellular *C. neoformans* specifically within the lung environment.

FATES OF INTERNALIZED *C. NEOFORMANS*

It is generally accepted that *C. neoformans* internalized by macrophages has one of three fates (although not necessarily mutually exclusive of each other): (i) *C. neoformans* can succumb to the antimicrobial actions of the phagocyte; (ii) *C. neoformans* can reside dormant within macrophages; and (iii) *C. neoformans* can replicate and escape cell killing (Fig. 1).

Killing of Intracellular *C. neoformans*

This killing mechanism is highly regulated by a helper T cell 1 (Th-1)–type response—hence, the high prevalence of cryptococcal infections in HIV-infected patients. Macrophages aid in the containment and killing of cryptococcal cells through granuloma formation involving a Th-1 response with cytokines such as interleukin (IL)-12, interferon γ, and tumor necrosis factor α (67, 68). A Th-1 response to cryptococcal infection has been demonstrated in clinical studies to be required for patient survival, while a predominant Th-2 response can exacerbate the infection (10, 15, 141). Therefore, macrophages can summon other immune cells in organizing an efficient host innate and adaptive immune response against *C. neoformans* through the induction of a series of chemokines and cytokines, such as IL-12, IL-18, monocyte chemoattractant protein 1, and macrophage inflammatory protein 1α (21, 45, 59, 158).

In addition to a concerted mechanism involving all arms of the host immune response, macrophages and neutrophils are capable of killing internalized microbes directly via oxygen-independent and oxygen-dependent mechanisms; however, research shows that macrophages differ greatly in their killing aptitude and response to cytokines. In fact, the mechanisms by which macrophages kill *C. neoformans* are still debatable. Fungicidal activity depends mostly on the type of macrophages used (e.g., human, mouse, rat) and whether the macrophages are primary cells or cell lines (26). In general, rats are a relatively resistant species, and their AMs efficiently inhibit yeast replication in vitro. In contrast, mice are highly susceptible, and the macrophages provide a permissive environment for *C. neoformans* replication. Humans appear to be highly susceptible to infection as evidenced by serology, and human macrophages are generally permissive for *C. neoformans* intracellular replication, at least in vitro.

Successful killing of internalized *C. neoformans* can enable AMs to serve as antigen-presenting cells (APCs) and regulators of T lymphocyte proliferation and activation (156, 157) (Fig. 1). Dead *C. neoformans* is processed, and *C. neoformans* mitogen is presented by APCs to naive T cells through HLA class II direct repeat molecules (145, 156, 158). In addition, B7-1 (CD80) or B7-2 (CD86) expressed by APCs interacts with CD28 or CTLA-4 on these T cells (157). Together with secretion of cytokines, these events activate T cells and induce proliferation, thereby mediating an antigen-specific T-cell-mediated immune response against *C. neoformans* (145, 156). The *C. neoformans* capsule limits the expression levels of B7-1 and B7-2, thereby decreasing antigen presentation by human AMs (157).

In contrast to discrepancies between the fungicidal activities of different types of macrophages, more agreement exists about neutrophils, as they readily kill *C. neoformans* in vitro (32, 96). Neutrophil killing is mediated by antimicrobial granules and a variety of reactive oxygen species. Cytosolic granules are trafficked to the *C. neoformans*–containing vesicle, where they fuse and release their contents. There are three specialized granule subsets involved in this concerted process, which fuse to the phagosome in the following order: the specific granules (also called secondary granules), the gelatinase granules (also named tertiary granules), and the azurophilic granules (also known as primary granules). The azurophilic granules contribute small antimicrobial peptides, α-defensins, seprocidins, and antibiotic proteases such as cathepsin G, proteinase 3, and neutrophil elastase. The azurophilic granules also contain MPO, and thus, they are also called "peroxidase-positive" granules. MPO is a neutrophil-specific enzyme that catalyzes the production of hypochlorous acid from hydrogen peroxide and chloride anion. Hypochlorous acid is an extremely potent microbicidal agent. The specific and gelatinase granules deliver antimicrobial proteins, such as lactoferrin, lipocalin, gelatinase, and metalloproteases (43). The specific and gelatinase granules do not contain MPO, and thus, they are also called "peroxidase-negative" granules. Both peroxidase-positive and -negative granules exert their antimicrobial effect mainly through the formation of multimeric transmembrane pores on microbial cells, leading to microbial cell lysis.

Although several different granules have been identified, their timing, sorting, targeting, and secretion are still poorly understood, especially during neutrophil maturation and activation. Studies have shown that undifferentiated human neutrophils (e.g., at the promyelocytic stage of maturation, such as the undifferentiated cell line HL-60) retain MPO and secrete lysozyme but not defensins. Interestingly, differentiated human HL-60 dramatically increases the transcription of defensins (60, 61), suggesting that production of defensin is regulated by neutrophil maturation. This is supported by clinical studies in which defensins are elevated in the plasma of patients with elevated or/and activated neutrophils (64, 114) and in patients with infections (64, 65, 115).

C. neoformans is particularly sensitive to defensin antifungal activity (1, 160), but how defensin is produced by neutrophils during *C. neoformans* infection is not known.

Also unknown is the mechanism by which defensin kills *C. neoformans*. Studies of the role and mechanisms by which defensins kill *C. neoformans* have been hampered by the fact that mouse neutrophils lack defensins (38). This could be the reason why this rodent species permits a more severe *C. neoformans* infection than humans and rats. But the era of genetic engineering has created a series of transgenic mice that express defensins. For instance, transgenic mice expressing human intestinal defensin exert protection against enteric salmonellosis (127), suggesting that an increase of the innate immunity can decrease the susceptibility of this rodent to infection(s).

C. neoformans killing by neutrophils is also mediated by the production of reactive oxygen species. Neutrophils from patients with chronic granulomatous disease lacking NADH oxidase have decreased bactericidal actions and completely lack fungicidal activity (29, 31, 130–132). Since MPO uses the products of NADPH to create a vast array of oxygen-derived microbicidal compounds, MPO is considered the enzyme ultimately responsible for the killing of *C. neoformans* by neutrophils (154). MPO deficiency results in a discernible decrease in neutrophil fungicidal activity in vitro (29, 31, 116). Furthermore, mice lacking MPO are highly susceptible to infection with *C. neoformans* var. *neoformans* (serotype D) strain ATCC 24067 (also known as 52D) (4). These studies suggest that neutrophils kill *C. neoformans* and are probably essential to containment and/or clearance of *C. neoformans* cells but perhaps not required for the initial confrontation with *C. neoformans*, which is a function performed by AMs. This hypothesis is supported by the fact that neutrophil depletion enhances host susceptibility to numerous intracellular pathogens, such as *Francisella tularensis*, *L. pneumophila*, *Mycobacterium* species, and *Candida albicans*, but not *C. neoformans* (48, 94, 117, 133, 148, 149). However, different studies have shown that neutropenia has no effect on the survival of wild-type BALB/c mice infected intravenously with *C. neoformans* strain 24067 (serotype D) and, actually, increased host survival when *C. neoformans* was administered intratracheally (94). Can neutrophils be detrimental for the host and exacerbate *C. neoformans* infection? Clearly, more studies are needed to fully understand their role during *C. neoformans* infection, and particular attention should be given to the mouse and *C. neoformans* strains that are used.

C. neoformans has developed mechanisms to survive within the phagolysosome. The enlargement of the polysaccharide capsule following phagocytosis provides resistance to microbicidal agents within the lumen of the phagolysosome, such as reactive oxygen species and antimicrobial peptides (174). *C. neoformans* also expresses superoxide dismutase 1 (SOD1) and 2 (SOD2) that can decrease the deleterious effects of superoxide by converting it to produce hydrogen peroxide and molecular oxygen (47). Considering their different biochemical requirements for activity and localization within the fungal cell, SOD1 and SOD2 have different roles in the defense against intracellular antimicrobial actions (22, 109). Although both enzymes provide protection against varying stresses, such as oxygen-derived free radicals, SOD1 protects against neutrophil killing, while SOD2 does not. SOD2 may be more important to the progression of cryptococcal infections, as a SOD2 deletion mutant created from *C. gattii* is avirulent, while deletion of SOD1 attenuates the virulence of *C. neoformans*. However, since deletion of SOD genes in *C. neoformans* species results in different phenotypic characteristics (22, 52, 108, 109), more studies are warranted to define their role in the *C. neoformans*-phagocyte interaction.

Another virulence factor protecting intracellular *C. neoformans* is melanin, which is produced by laccases (Lac1 and Lac2) (97, 165). In *C. neoformans*, Lac1 and Lac2 are cell wall-associated enzymes that oxidize exogenous catecholamines and iron to produce the pigment melanin. Laccases also possess iron oxidase activity, and they may protect internalized *C. neoformans* cells by impeding the phagocyte's ability to produce hydroxyl radicals through the oxidation of phagolysosomal iron and the generation of free radical scavenging molecules (82, 162, 175). This competition between host and intracellular pathogen for iron has been demonstrated to occur in macrophages with *Mycobacterium tuberculosis* (58). Interestingly, the sphingolipid-metabolizing enzyme inositol-phosphoryl ceramide synthase 1 (Ipc1) modulates laccase activity through its production of diacylglycerol, which binds and activates protein kinase C1 through its C1 domain. And Ipc1 has been found to be highly upregulated within the phagolysosome when compared to the extracellular environment (42, 62, 63, 87), suggesting that the mechanism of *C. neoformans* protection within the intracellular environment may be further controlled by the sphingolipid pathway.

In fact, in addition to Ipc1, other enzymes of the sphingolipid pathway have also been identified as essential regulators of *C. neoformans* growth in the intracellular environment. For instance, the inositol phosphosphingolipid-phospholipase C1 (Isc1), an enzyme that metabolizes fungal inositol sphingolipids into phytoceramide, is essential for intracellular growth, as it provides protection from acidic, oxidative, and nitrosative stresses of the phagolysosome microenvironment through a Pma1 (plasma membrane ATPase 1)-dependent mechanism(s) (49, 137). Very interestingly, Δ*isc1* cells are almost exclusively localized in the extracellular environment of the lung. Perhaps the Δ*isc1* strain is pressured in the extracellular environment, whose conditions (low iron and high CO_2) continuously stimulate capsule production. This phenotype is present in immunocompetent CBA/J mice and appears to be dependent on host immunity, because it is not observed in an immunodeficient mouse model Tgε26 (137). These studies suggest that *C. neoformans* may sense the immune condition of the host and respond accordingly to its benefit.

Whereas Isc1 is vital to *C. neoformans* survival within the host phagolysosome, glucosylceramide synthase (Gcs1) is essential for extracellular growth (69, 123). Gcs1 is involved in controlling cell-cycle progression and fungal growth in environments characterized by a neutral/alkaline pH and physiological concentrations of CO_2, such as the lung alveolar spaces and the bloodstream. The Δ*gcs1* mutant strain is contained within lung granulomas and therefore does not disseminate to other organs. Interestingly, infection by Δ*gcs1* is totally contained by immunocompetent CBA/J mice but not by the T- and NK-cell-immunodeficient Tgε26 mouse model (69, 123). Fascinatingly, depletion of AMs in Tgε26 significantly increased survival of mice infected with the Δ*gcs1* strain. These studies strongly suggest that under conditions of immune suppression the presence of AMs can be detrimental for the host and may facilitate *C. neoformans* dissemination (16, 69, 134, 137). Furthermore, these studies also present an intriguing role for enzymes of the *C. neoformans* sphingolipid pathway that appears to be differentiated in regulating both intra- and extracellular growth of *C. neoformans* but through different enzymes and sphingolipids.

Another *C. neoformans* factor that promotes the survival of *C. neoformans* at different junctions following host inhalation is the enzyme Plb1 (18, 23, 129, 166). Plb1 is a cell-membrane-associated enzyme attached by a glycosylphosphatidylinositol anchor (139, 140). Plb1 can be cleaved from the glycosylphosphatidylinositol anchor and secreted into the external environment (139). Plb1 can act as a phospholipase B, as a lysophospholipase, and as a lysophospholipase transacylase to cleave host lipids. Deletion of the gene encoding Plb1 produces a mutant strain (Δ*plb1*) with an intracellular growth defect that is attributed to the production of fungal eicosanoids that scavenge macrophage-derived arachidonic acid (166). The production of fungal eicosanoids may protect *C. neoformans* from the host response and, concomitantly, the incorporation of host arachidonic acid into *C. neoformans* metabolism may reduce the host cellular pool for critical macrophage functions.

Intracellular Dormancy/Latency of *C. neoformans*

A second possibility occurring upon internalization is that *C. neoformans* could reside within the phagolysosome indefinitely without causing tissue damage and disease (Fig. 1). This could occur upon the primary infection. These *C. neoformans* cells do not replicate, and they are controlled by host immune response, possibly within the confines of a fibrotic lung granuloma. It is widely accepted that this mechanism of *C. neoformans* survival is frequent in human hosts following initial exposure (85). Upon immunosuppression, these *C. neoformans* cells start to replicate in the lung and disseminate to other organs, especially the brain, causing disease. Studies have suggested that exposure to *C. neoformans* occurs during childhood with an asymptomatic infection (36, 50, 142). This model of *C. neoformans* pathogenesis in an immunocompromised individual fits soundly with the concept of the damage-response framework of microbial pathogenesis, as presented by Casadevall and Pirofski, where host damage serves as the essential factor in the host-pathogen interaction, facilitating disease by a normally nonpathogenic organism (12). However, cases of cryptococcosis in immune-competent individuals must be contemplated in different terms, as an immune deficiency is not required for *C. neoformans* to cause disease in normal subjects. Therefore, since both pathogen and host factors contribute to the outcome of infection and differences exist in the virulence potential among different *C. neoformans* strains or species, the physiopathology of *C. neoformans* infection is much more complex.

Intracellular Growth and Escape of *C. neoformans*

C. neoformans can evade the killing activity of the phagolysosome, replicate intracellularly, and ultimately, be released in the extracellular environment and readily infect other host cells (Fig. 1). Interestingly, release into the extracellular environment may occur through two different mechanisms: by lysis of the host macrophages or by expulsion without macrophage lysis (2, 3, 90). Over time, the replication of *C. neoformans* cells and the shedding of the polysaccharide capsule may alter the membrane of the phagolysosome, which may become leaky and discontinuous (44, 153). Eventually, the phagolysosome breaks down, followed by the lysis of the entire plasma membrane and the release of the *C. neoformans* back into the alveolar spaces. Since one macrophage will release many *C. neoformans* cells, this process may explain why in a given time during mouse lung infection, the majority of *C. neoformans* cells are mainly extracellular.

The *C. neoformans*-host phagocyte relationship became even more complex than had been assumed with the discovery that internalized cryptococcal cells can egress from a macrophage into the extracellular spaces without affecting the viability of the host cell (2, 3, 90). This phenomenon of *C. neoformans* egression from macrophages without cytotoxicity was observed by two separate groups independently and has been termed "phagosome extrusion" or "phagosome expulsion" (3, 90). This cellular exit strategy can occur in opsonin-free conditions or following opsonization with either complement or antibody. Interestingly, the state of activation did not affect the rate of the egression. Both groups also observed the increased size of the phagolysosome containing the internalized fungal cells prior to the *C. neoformans* escape back into the extracellular environment. Importantly, this unique event occurs in a variety of phagocytic cells including in vitro cell lines and human primary macrophages and appears to be mediated by an actin-dependent process (3). Additionally, it is possible that *C. neoformans* can spread from one macrophage to another through cell-to-cell transfer if macrophages are adjacent or attached to each other, thereby allowing the fungal cell to avoid contact with the extracellular environment (2). More studies are needed to establish their incidence and role during in vivo infection.

IDENTIFICATION OF NEW *C. NEOFORMANS* FACTORS IMPORTANT FOR LUNG INFECTION

Historically, the identification of new *C. neoformans* virulence factors has been difficult. The majority of research attempting to reveal novel *C. neoformans* virulence attributes mostly produces new mechanisms by which the three archetypal virulence factors (the ability to grow at human body temperature, the production of the polysaccharide capsule, and the production of melanin) modulate pathogenesis. However, recently, the availability of *C. neoformans* genome sequences with the progressive applications of classic molecular tools has enabled researchers to resolve traits essential for *C. neoformans* survival. For instance, microarray studies have identified new and validated previously discovered factors involved in the regulation of intracellular growth (42). For instance, the *C. neoformans* transcriptional gene profile shows that membrane transporters for essential nutrients and genes encoding for alternative carbon source and amino acid pathways were induced by *C. neoformans* following phagocytosis. This suggests that several nutrients available to internalized *C. neoformans* within the phagolysosome are clearly inadequate, and thus, *C. neoformans* overcomes this starvation response by upregulating the transport of various nutrients. Additional studies using gene deletion libraries have identified novel *C. neoformans* genes important to lung infectivity, defined as the ability to proliferate and survive within the host lung, suggesting that there are additional players in the *C. neoformans*–host lung interaction that need to be studied (83, 84, 104, 105, 119). Validation of the candidate gene by standard methods will reveal their role and mechanism in the regulation of *C. neoformans* pathogenesis.

CONCLUDING REMARKS

C. neoformans is a unique etiological agent because it is a free-living organism whose evolution to survive in its natural environmental niche has provided it with adaptive mechanisms to survive within a multitude of microenvironments within the mammalian host. Consequently, these survival mechanisms are the *C. neoformans* virulence factors and facilitate its ability to cause disease as both an extracellular and intracellular pathogen. Perhaps this unique ability to tolerate these two different host compartments might be attributed to the adaptation that *C. neoformans* has developed to survive against environmental predators, such as free-living amoebae, which show many similarities to mammalian phagocytes (55, 92, 144, 176).

It is also possible that the choice of the *C. neoformans* lifestyle (intra- or extracellular) is not randomly determined but is highly regulated by the conditions that *C. neoformans* finds in the host. For instance, in conditions of immunodeficiency it seems that the intracellular component has a major role in the development of cryptococcosis because (i) *C. neoformans* can replicate faster intra- versus extracellularly; (ii) *C. neoformans* can disseminate from the lung to other organs within macrophages; (iii) depletion of AMs in mice leads to a decrease of *C. neoformans* proliferation; and (iv) depletion of AMs in immunodeficient mice (Tgε26) increases mouse survival following infection with *C. neoformans* (19, 69, 86, 134, 136). Therefore, if upon phagocytosis, AMs cannot effectively kill *C. neoformans*, phagocytes would provide a protective environment and an opportunity for the fungus to produce disease. Under these conditions, the host would benefit if the fungal cells were forced into an extracellular lifestyle. Thus, the identification and characterization of pathogen or host factors that contribute to the intra- or extracellular growth mode will be useful not only because they contribute to a better understanding of host-pathogen adaptation but also because they can be exploited for the development of novel therapeutic strategies against cryptococcosis occurring in immunocompromised patients.

We thank all members of Del Poeta's laboratory for helpful discussions. Special thanks to Caroline Westwater for helpful comments and to Lisa M. Fennessy for helping with the figure. This work was supported in part by the Burroughs Wellcome Fund, in part by Grants AI56168, AI78439, and AI71142 (to M.D.P) from the National Institutes of Health, in part by RR17677 Project 2 (to M.D.P.) from the Centers of Biomedical Research Excellence Program of the National Center for Research Resources, and in part by NIH C06 RR015455 from the Extramural Research Facilities Program of the National Center for Research Resources. T.M. is supported by the Graduate Assistance in Areas of National Need (GAANN) training grant in Lipidology and New Technologies (to M.D.P.) from the United States Department of Education. Maurizio Del Poeta is a Burroughs Wellcome new investigator in pathogenesis of infectious diseases.

REFERENCES

1. **Alcouloumre, M. S., M. A. Ghannoum, A. S. Ibrahim, M. E. Selsted, and J. E. Edwards, Jr.** 1993. Fungicidal properties of defensin NP-1 and activity against *Cryptococcus neoformans* in vitro. *Antimicrob. Agents Chemother.* **37:**2628–2632.
2. **Alvarez, M., and A. Casadevall.** 2007. Cell-to-cell spread and massive vacuole formation after *Cryptococcus neoformans* infection of murine macrophages. *BMC Immunol.* **8:**16.
3. **Alvarez, M., and A. Casadevall.** 2006. Phagosome extrusion and host-cell survival after *Cryptococcus neoformans* phagocytosis by macrophages. *Curr. Biol.* **16:**2161–2165.
4. **Aratani, Y., F. Kura, H. Watanabe, H. Akagawa, Y. Takano, A. Ishida-Okawara, K. Suzuki, N. Maeda, and H. Koyama.** 2006. Contribution of the myeloperoxidase-dependent oxidative system to host defence against *Cryptococcus neoformans*. *J. Med. Microbiol.* **55:**1291–1299.
5. **Aratani, Y., F. Kura, H. Watanabe, H. Akagawa, Y. Takano, K. Suzuki, M. C. Dinauer, N. Maeda, and H. Koyama.** 2002. Critical role of myeloperoxidase and nicotinamide adenine dinucleotide phosphate-oxidase in high-burden systemic infection of mice with *Candida albicans*. *J. Infect. Dis.* **185:**1833–1837.
6. **Aratani, Y., F. Kura, H. Watanabe, H. Akagawa, Y. Takano, K. Suzuki, M. C. Dinauer, N. Maeda, and H. Koyama.** 2004. In vivo role of myeloperoxidase for the host defense. *Jpn. J. Infect. Dis.* **57:**S15.
7. **Aratani, Y., F. Kura, H. Watanabe, H. Akagawa, Y. Takano, K. Suzuki, M. C. Dinauer, N. Maeda, and H. Koyama.** 2002. Relative contributions of myeloperoxidase and NADPH-oxidase to the early host defense against pulmonary infections with *Candida albicans* and *Aspergillus fumigatus*. *Med. Mycol.* **40:**557–563.
8. **Aratani, Y., F. Kura, H. Watanabe, H. Akagawa, Y. Takano, K. Suzuki, N. Maeda, and H. Koyama.** 2000. Differential host susceptibility to pulmonary infections with bacteria and fungi in mice deficient in myeloperoxidase. *J. Infect. Dis.* **182:**1276–1279.
9. **Brightbill, H. D., D. H. Libraty, S. R. Krutzik, R. B. Yang, J. T. Belisle, J. R. Bleharski, M. Maitland, M. V. Norgard, S. E. Plevy, S. T. Smale, P. J. Brennan, B. R. Bloom, P. J. Godowski, and R. L. Modlin.** 1999. Host defense mechanisms triggered by microbial lipoproteins through toll-like receptors. *Science* **285:**732–736.
10. **Brouwer, A. E., A. A. Siddiqui, M. I. Kester, K. C. Sigaloff, A. Rajanuwong, S. Wannapasni, W. Chierakul, and T. S. Harrison.** 2007. Immune dysfunction in HIV-seronegative, *Cryptococcus gattii* meningitis. *J. Infect.* **54:**e165–e168.
11. **Capilla, J., C. M. Maffei, K. V. Clemons, R. A. Sobel, and D. A. Stevens.** 2006. Experimental systemic infection with *Cryptococcus neoformans* var. *grubii* and *Cryptococcus gattii* in normal and immunodeficient mice. *Med. Mycol.* **44:**601–610.
12. **Casadevall, A., and L. Pirofski.** 2001. Host-pathogen interactions: the attributes of virulence. *J. Infect. Dis.* **184:**337–344.
13. **Casadevall, A., and L. A. Pirofski.** 2001. Adjunctive immune therapy for fungal infections. *Clin. Infect. Dis.* **33:**1048–1056.
14. **Casadevall, A., and L. A. Pirofski.** 1999. Host-pathogen interactions: redefining the basic concepts of virulence and pathogenicity. *Infect. Immun.* **67:**3703–3713.
15. **Chaka, W., A. F. Verheul, V. V. Vaishnav, R. Cherniak, J. Scharringa, J. Verhoef, H. Snippe, and A. I. Hoepelman.** 1997. Induction of TNF-alpha in human peripheral blood mononuclear cells by the mannoprotein of *Cryptococcus neoformans* involves human mannose binding protein. *J. Immunol.* **159:**2979–2985.
16. **Charlier, C., K. Nielsen, S. Daou, M. Brigitte, F. Chretien, and F. Dromer.** 2009. Evidence of a role for monocytes in dissemination and brain invasion by *Cryptococcus neoformans*. *Infect. Immun.* **77:**120–127.
17. **Chayakulkeeree, M., and J. R. Perfect.** 2006. Cryptococcosis. *Infect. Dis. Clin. North Am.* **20:**507–544, v–vi.
18. **Chayakulkeeree, M., T. C. Sorrell, A. R. Siafakas, C. F. Wilson, N. Pantarat, K. J. Gerik, R. Boadle, and J. T. Djordjevic.** 2008. Role and mechanism of phosphatidylinositol-specific phospholipase C in survival and virulence of *Cryptococcus neoformans*. *Mol. Microbiol.* **69:**809–826.
19. **Chretien, F., O. Lortholary, I. Kansau, S. Neuville, F. Gray, and F. Dromer.** 2002. Pathogenesis of cerebral *Cryptococcus neoformans* infection after fungemia. *J. Infect. Dis.* **186:**522–530.
20. **Coenjaerts, F. E., A. M. Walenkamp, P. N. Mwinzi, J. Scharringa, H. A. Dekker, J. A. van Strijp, R. Cherniak, and A. I. Hoepelman.** 2001. Potent inhibition of neutrophil migration by cryptococcal mannoprotein-4-induced desensitization. *J. Immunol.* **167:**3988–3995.

21. **Collins, H. L., and G. J. Bancroft.** 1992. Cytokine enhancement of complement-dependent phagocytosis by macrophages: synergy of tumor necrosis factor-alpha and granulocyte-macrophage colony-stimulating factor for phagocytosis of *Cryptococcus neoformans*. *Eur. J. Immunol.* **22:**1447–1454.
22. **Cox, G. M., T. S. Harrison, H. C. McDade, C. P. Taborda, G. Heinrich, A. Casadevall, and J. R. Perfect.** 2003. Superoxide dismutase influences the virulence of *Cryptococcus neoformans* by affecting growth within macrophages. *Infect. Immun.* **71:**173–180.
23. **Cox, G. M., H. C. McDade, S. C. Chen, S. C. Tucker, M. Gottfredsson, L. C. Wright, T. C. Sorrell, S. D. Leidich, A. Casadevall, M. A. Ghannoum, and J. R. Perfect.** 2001. Extracellular phospholipase activity is a virulence factor for *Cryptococcus neoformans*. *Mol. Microbiol.* **39:** 166–175.
24. **Daeron, M.** 1997. Fc receptor biology. *Annu. Rev. Immunol.* **15:**203–234.
25. **Datta, K., and L. A. Pirofski.** 2006. Towards a vaccine for *Cryptococcus neoformans*: principles and caveats. *FEMS Yeast Res.* **6:**525–536.
26. **Del Poeta, M.** 2004. Role of phagocytosis in the virulence of *Cryptococcus neoformans*. *Eukaryot. Cell* **3:**1067–1075.
27. **Deshaw, M., and L. A. Pirofski.** 1995. Antibodies to the *Cryptococcus neoformans* capsular glucuronoxylomannan are ubiquitous in serum from HIV+ and HIV- individuals. *Clin. Exp. Immunol.* **99:**425–432.
28. **Diamond, R. D.** 1974. Antibody-dependent killing of *Cryptococcus neoformans* by human peripheral blood mononuclear cells. *Nature* **247:**148–150.
29. **Diamond, R. D., R. A. Clark, and C. C. Haudenschild.** 1980. Damage to *Candida albicans* hyphae and pseudohyphae by the myeloperoxidase system and oxidative products of neutrophil metabolism in vitro. *J. Clin. Invest.* **66:**908–917.
30. **Diamond, R. D., J. E. May, M. Kane, M. M. Frank, and J. E. Bennett.** 1973. The role of late complement components and the alternate complement pathway in experimental cryptococcosis. *Proc. Soc. Exp. Biol. Med.* **144:** 312–315.
31. **Diamond, R. D., F. Oppenheim, Y. Nakagawa, R. Krzesicki, and C. C. Haudenschild.** 1980. Properties of a product of *Candida albicans* hyphae and pseudohyphae that inhibits contact between the fungi and human neutrophils in vitro. *J. Immunol.* **125:**2797–2804.
32. **Diamond, R. D., R. K. Root, and J. E. Bennett.** 1972. Factors influencing killing of *Cryptococcus neoformans* by human leukocytes in vitro. *J. Infect. Dis.* **125:**367–376.
33. **Dong, Z. M., and J. W. Murphy.** 1997. Cryptococcal polysaccharides bind to CD18 on human neutrophils. *Infect. Immun.* **65:**557–563.
34. **Dong, Z. M., and J. W. Murphy.** 1996. Cryptococcal polysaccharides induce L-selectin shedding and tumor necrosis factor receptor loss from the surface of human neutrophils. *J. Clin. Invest.* **97:**689–698.
35. **Dromer, F., J. Charreire, A. Contrepois, C. Carbon, and P. Yeni.** 1987. Protection of mice against experimental cryptococcosis by anti-*Cryptococcus neoformans* monoclonal antibody. *Infect. Immun.* **55:**749–752.
36. **Dromer, F., O. Ronin, and B. Dupont.** 1992. Isolation of *Cryptococcus neoformans* var. *gattii* from an Asian patient in France: evidence for dormant infection in healthy subjects. *J. Med. Vet. Mycol.* **30:**395–397.
37. **Ehlers, M. R.** 2000. CR3: a general purpose adhesion-recognition receptor essential for innate immunity. *Microbes Infect.* **2:**289–294.
38. **Eisenhauer, P. B., and R. I. Lehrer.** 1992. Mouse neutrophils lack defensins. *Infect. Immun.* **60:**3446–3447.
39. **Ellerbroek, P. M., D. J. Lefeber, R. van Veghel, J. Scharringa, E. Brouwer, G. J. Gerwig, G. Janbon, A. I. Hoepelman, and F. E. Coenjaerts.** 2004. O-acetylation of cryptococcal capsular glucuronoxylomannan is essential for interference with neutrophil migration. *J. Immunol.* **173:**7513–75120.
40. **Ellerbroek, P. M., L. H. Ulfman, A. I. Hoepelman, and F. E. Coenjaerts.** 2004. Cryptococcal glucuronoxylomannan interferes with neutrophil rolling on the endothelium. *Cell Microbiol.* **6:**581–592.
41. **Ellerbroek, P. M., A. M. Walenkamp, A. I. Hoepelman, and F. E. Coenjaerts.** 2004. Effects of the capsular polysaccharides of *Cryptococcus neoformans* on phagocyte migration and inflammatory mediators. *Curr. Med. Chem.* **11:**253–266.
42. **Fan, W., P. R. Kraus, M. J. Boily, and J. Heitman.** 2005. *Cryptococcus neoformans* gene expression during murine macrophage infection. *Eukaryot. Cell* **4:**1420–1433.
43. **Faurschou, M., and N. Borregaard.** 2003. Neutrophil granules and secretory vesicles in inflammation. *Microbes Infect.* **5:**1317–1327.
44. **Feldmesser, M., Y. Kress, P. Novikoff, and A. Casadevall.** 2000. *Cryptococcus neoformans* is a facultative intracellular pathogen in murine pulmonary infection. *Infect. Immun.* **68:**4225–4237.
45. **Fidan, I., E. Yesilyurt, F. C. Gurelik, B. Erdal, and T. Imir.** 2008. Effects of recombinant interferon-gamma on cytokine secretion from monocyte-derived macrophages infected with *Salmonella typhi*. *Comp. Immunol. Microbiol. Infect. Dis.* **31:**467–475.
46. **Fleuridor, R., Z. Zhong, and L. Pirofski.** 1998. A human IgM monoclonal antibody prolongs survival of mice with lethal cryptococcosis. *J. Infect. Dis.* **178:**1213–1216.
47. **Fridovich, I.** 1995. Superoxide radical and superoxide dismutases. *Annu. Rev. Biochem.* **64:**97–112.
48. **Fulton, S. A., S. M. Reba, T. D. Martin, and W. H. Boom.** 2002. Neutrophil-mediated mycobacteriocidal immunity in the lung during *Mycobacterium bovis* BCG infection in C57BL/6 mice. *Infect. Immun.* **70:**5322–5327.
49. **Garcia, J., J. Shea, F. Alvarez-Vasquez, A. Qureshi, C. Luberto, E. O. Voit, and M. Del Poeta.** 2008. Mathematical modeling of pathogenicity of *Cryptococcus neoformans*. *Mol. Syst. Biol.* **4:**183.
50. **Garcia-Hermoso, D., G. Janbon, and F. Dromer.** 1999. Epidemiological evidence for dormant *Cryptococcus neoformans* infection. *J. Clin. Microbiol.* **37:**3204–3209.
51. **Gerber, J. S., and D. M. Mosser.** 2001. Stimulatory and inhibitory signals originating from the macrophage Fc-gamma receptors. *Microbes Infect.* **3:**131–139.
52. **Giles, S. S., I. Batinic-Haberle, J. R. Perfect, and G. M. Cox.** 2005. *Cryptococcus neoformans* mitochondrial superoxide dismutase: an essential link between antioxidant function and high-temperature growth. *Eukaryot. Cell* **4:**46–54.
53. **Goldman, D., S. C. Lee, and A. Casadevall.** 1994. Pathogenesis of pulmonary *Cryptococcus neoformans* infection in the rat. *Infect Immun.* **62:**4755–4761.
54. **Granger, D. L., J. R. Perfect, and D. T. Durack.** 1985. Virulence of *Cryptococcus neoformans*. Regulation of capsule synthesis by carbon dioxide. *J. Clin. Invest.* **76:**508–516.
55. **Greub, G., and D. Raoult.** 2004. Microorganisms resistant to free-living amoebae. *Clin. Microbiol. Rev.* **17:**413–433.
56. **Grinsell, M., L. C. Weinhold, J. E. Cutler, Y. Han, and T. R. Kozel.** 2001. In vivo clearance of glucuronoxylomannan, the major capsular polysaccharide of *Cryptococcus neoformans*: a critical role for tissue macrophages. *J. Infect. Dis.* **184:**479–487.
57. **Hackam, D. J., O. D. Rotstein, W. J. Zhang, N. Demaurex, M. Woodside, O. Tsai, and S. Grinstein.** 1997. Regulation of phagosomal acidification. Differential targeting of

Na+/H+ exchangers, Na+/K+-ATPases, and vacuolar-type H+-ATPases. *J. Biol. Chem.* **272:**29810–29820.

58. **Hancock, R. E., and G. Diamond.** 2000. The role of cationic antimicrobial peptides in innate host defences. *Trends Microbiol.* **8:**402–410.
59. **He, W., A. Casadevall, S. C. Lee, and D. L. Goldman.** 2003. Phagocytic activity and monocyte chemotactic protein expression by pulmonary macrophages in persistent pulmonary cryptococcosis. *Infect. Immun.* **71:**930–936.
60. **Herwig, S., Q. Su, and P. Tempst.** 1998. Drug-activated multiple pathways of defensin mRNA regulation in HL-60 cells are defined by reversed roles of participating protein kinases. *Leuk. Res.* **22:**913–925.
61. **Herwig, S., Q. Su, W. Zhang, Y. Ma, and P. Tempst.** 1996. Distinct temporal patterns of defensin mRNA regulation during drug-induced differentiation of human myeloid leukemia cells. *Blood* **87:**350–364.
62. **Heung, L. J., A. E. Kaiser, C. Luberto, and M. Del Poeta.** 2005. The role and mechanism of diacylglycerol-protein kinase C1 signaling in melanogenesis by *Cryptococcus neoformans*. *J. Biol. Chem.* **280:**28547–28555.
63. **Heung, L. J., C. Luberto, A. Plowden, Y. A. Hannun, and M. Del Poeta.** 2004. The sphingolipid pathway regulates Pkc1 through the formation of diacylglycerol in *Cryptococcus neoformans*. *J. Biol. Chem.* **279:**21144–21153.
64. **Ihi, T., M. Nakazato, H. Mukae, and S. Matsukura.** 1997. Elevated concentrations of human neutrophil peptides in plasma, blood, and body fluids from patients with infections. *Clin. Infect. Dis.* **25:**1134–1140.
65. **Isomoto, H., H. Mukae, H. Ishimoto, Y. Nishi, C. Y. Wen, A. Wada, K. Ohnita, T. Hirayama, M. Nakazato, and S. Kohno.** 2005. High concentrations of human beta-defensin 2 in gastric juice of patients with *Helicobacter pylori* infection. *World J. Gastroenterol.* **11:**4782–4787.
66. **Jankowski, A., C. C. Scott, and S. Grinstein.** 2002. Determinants of the phagosomal pH in neutrophils. *J. Biol. Chem.* **277:**6059–6066.
67. **Kawakami, K., Y. Koguchi, M. H. Qureshi, Y. Kinjo, S. Yara, A. Miyazato, M. Kurimoto, K. Takeda, S. Akira, and A. Saito.** 2000. Reduced host resistance and Th1 response to *Cryptococcus neoformans* in interleukin-18 deficient mice. *FEMS Microbiol. Lett.* **186:**121–126.
68. **Kawakami, K., Y. Koguchi, M. H. Qureshi, S. Yara, Y. Kinjo, K. Uezu, and A. Saito.** 2000. NK cells eliminate *Cryptococcus neoformans* by potentiating the fungicidal activity of macrophages rather than by directly killing them upon stimulation with IL-12 and IL-18. *Microbiol. Immunol.* **44:**1043–1050.
69. **Kechichian, T. B., J. Shea, and M. Del Poeta.** 2007. Depletion of alveolar macrophages decreases the dissemination of a glucosylceramide-deficient mutant of *Cryptococcus neoformans* in immunodeficient mice. *Infect. Immun.* **75:**4792–4798.
70. **Klebanoff, S. J.** 1999. Myeloperoxidase. *Proc. Assoc. Am. Physicians* **111:**383–389.
71. **Klebanoff, S. J.** 2005. Myeloperoxidase: friend and foe. *J. Leukoc. Biol.* **77:**598–625.
72. **Kozel, T. R.** 1996. Activation of the complement system by pathogenic fungi. *Clin. Microbiol. Rev.* **9:**34–46.
73. **Kozel, T. R.** 1977. Non-encapsulated variant of *Cryptococcus neoformans*. II. Surface receptors for cryptococcal polysaccharide and their role in inhibition of phagocytosis by polysaccharide. *Infect. Immun.* **16:**99–106.
74. **Kozel, T. R., and E. C. Gotschlich.** 1982. The capsule of *Cryptococcus neoformans* passively inhibits phagocytosis of the yeast by macrophages. *J. Immunol.* **129:**1675–1680.
75. **Kozel, T. R., and G. S. Pfrommer.** 1986. Activation of the complement system by *Cryptococcus neoformans* leads to binding of iC3b to the yeast. *Infect. Immun.* **52:**1–5.
76. **Kozel, T. R., A. Tabuni, B. J. Young, and S. M. Levitz.** 1996. Influence of opsonization conditions on C3 deposition and phagocyte binding of large- and small-capsule *Cryptococcus neoformans* cells. *Infect. Immun.* **64:**2336–2338.
77. **Levitz, S. M., T. S. Harrison, A. Tabuni, and X. Liu.** 1997. Chloroquine induces human mononuclear phagocytes to inhibit and kill *Cryptococcus neoformans* by a mechanism independent of iron deprivation. *J. Clin. Invest.* **100:**1640–1646.
78. **Levitz, S. M., S. H. Nong, K. F. Seetoo, T. S. Harrison, R. A. Speizer, and E. R. Simons.** 1999. *Cryptococcus neoformans* resides in an acidic phagolysosome of human macrophages. *Infect. Immun.* **67:**885–890.
79. **Levitz, S. M., and A. Tabuni.** 1991. Binding of *Cryptococcus neoformans* by human cultured macrophages. Requirements for multiple complement receptors and actin. *J. Clin. Invest.* **87:**528–535.
80. **Levy, O.** 2000. Antimicrobial proteins and peptides of blood: templates for novel antimicrobial agents. *Blood* **96:**2664–2672.
81. **Lien, E., T. J. Sellati, A. Yoshimura, T. H. Flo, G. Rawadi, R. W. Finberg, J. D. Carroll, T. Espevik, R. R. Ingalls, J. D. Radolf, and D. T. Golenbock.** 1999. Toll-like receptor 2 functions as a pattern recognition receptor for diverse bacterial products. *J. Biol. Chem.* **274:** 33419–33425.
82. **Liu, L., K. Wakamatsu, S. Ito, and P. R. Williamson.** 1999. Catecholamine oxidative products, but not melanin, are produced by *Cryptococcus neoformans* during neuropathogenesis in mice. *Infect. Immun.* **67:**108–112.
83. **Liu, O. W., C. D. Chun, E. D. Chow, C. Chen, H. D. Madhani, and S. M. Noble.** 2008. Systematic genetic analysis of virulence in the human fungal pathogen *Cryptococcus neoformans*. *Cell* **135:**174–188.
84. **Liu, X., G. Hu, J. Panepinto, and P. R. Williamson.** 2006. Role of a VPS41 homologue in starvation response, intracellular survival and virulence of *Cryptococcus neoformans*. *Mol. Microbiol.* **61:**1132–1146.
85. **Lortholary, O., H. Nunez, M. W. Brauner, and F. Dromer.** 2004. Pulmonary cryptococcosis. *Semin. Respir. Crit. Care. Med.* **25:**145–157.
86. **Luberto, C., B. Martinez-Marino, D. Taraskiewicz, B. Bolanos, P. Chitano, D. L. Toffaletti, G. M. Cox, J. R. Perfect, Y. A. Hannun, E. Balish, and M. Del Poeta.** 2003. Identification of App1 as a regulator of phagocytosis and virulence of *Cryptococcus neoformans*. *J. Clin. Invest.* **112:**1080–1094.
87. **Luberto, C., D. L. Toffaletti, E. A. Wills, S. C. Tucker, A. Casadevall, J. R. Perfect, Y. A. Hannun, and M. Del Poeta.** 2001. Roles for inositol-phosphoryl ceramide synthase 1 (IPC1) in pathogenesis of *C. neoformans*. *Genes Dev.* **15:**201–212.
88. **Lukacs, G. L., O. D. Rotstein, and S. Grinstein.** 1991. Determinants of the phagosomal pH in macrophages. In situ assessment of vacuolar H(+)-ATPase activity, counterion conductance, and H+ "leak." *J. Biol. Chem.* **266:** 24540–24548.
89. **Lukacs, G. L., O. D. Rotstein, and S. Grinstein.** 1990. Phagosomal acidification is mediated by a vacuolar-type H(+)-ATPase in murine macrophages. *J. Biol. Chem.* **265:** 21099–1107.
90. **Ma, H., J. E. Croudace, D. A. Lammas, and R. C. May.** 2006. Expulsion of live pathogenic yeast by macrophages. *Curr. Biol.* **16:**2156–2160.
91. **Mare, L., R. Iatta, M. T. Montagna, C. Luberto, and M. Del Poeta.** 2005. APP1 transcription is regulated by inositol-phosphorylceramide synthase 1-diacylglycerol pathway and is controlled by ATF2 transcription factor in *Cryptococcus neoformans*. *J. Biol. Chem.* **280:**36055–36064.

92. **Mayer, A. M., and R. C. Staples.** 2002. Laccase: new functions for an old enzyme. *Phytochemistry* **60:**551–565.
93. **Means, T. K., E. Mylonakis, E. Tampakakis, R. A. Colvin, E. Seung, L. Puckett, M. F. Tai, C. R. Stewart, R. Pukkila-Worley, S. E. Hickman, K. J. Moore, S. B. Calderwood, N. Hacohen, A. D. Luster, and J. El Khoury.** 2009. Evolutionarily conserved recognition and innate immunity to fungal pathogens by the scavenger receptors SCARF1 and CD36. *J. Exp. Med.* **206:**637–653.
94. **Mednick, A. J., M. Feldmesser, J. Rivera, and A. Casadevall.** 2003. Neutropenia alters lung cytokine production in mice and reduces their susceptibility to pulmonary cryptococcosis. *Eur. J. Immunol.* **33:**1744–1753.
95. **Mershon, K. L., A. Vasuthasawat, G. W. Lawson, S. L. Morrison, and D. O. Beenhouwer.** 2009. Role of complement in protection against *Cryptococcus gattii* infection. *Infect. Immun.* **77:**1061–1070.
96. **Miller, M. F., and T. G. Mitchell.** 1991. Killing of *Cryptococcus neoformans* strains by human neutrophils and monocytes. *Infect. Immun.* **59:**24–28.
97. **Missall, T. A., J. M. Moran, J. A. Corbett, and J. K. Lodge.** 2005. Distinct stress responses of two functional laccases in *Cryptococcus neoformans* are revealed in the absence of the thiol-specific antioxidant Tsa1. *Eukaryot. Cell* **4:**202–208.
98. **Modlin, R. L., H. D. Brightbill, and P. J. Godowski.** 1999. The toll of innate immunity on microbial pathogens. *N. Engl. J. Med.* **340:**1834–1835.
99. **Monari, C., F. Bistoni, A. Casadevall, E. Pericolini, D. Pietrella, T. R. Kozel, and A. Vecchiarelli.** 2005. Glucuronoxylomannan, a microbial compound, regulates expression of costimulatory molecules and production of cytokines in macrophages. *J. Infect. Dis.* **191:**127–137.
100. **Monari, C., T. R. Kozel, F. Paganelli, E. Pericolini, S. Perito, F. Bistoni, A. Casadevall, and A. Vecchiarelli.** 2006. Microbial immune suppression mediated by direct engagement of inhibitory Fc receptor. *J. Immunol.* **177:**6842–6851.
101. **Monari, C., F. Paganelli, F. Bistoni, T. R. Kozel, and A. Vecchiarelli.** 2008. Capsular polysaccharide induction of apoptosis by intrinsic and extrinsic mechanisms. *Cell. Microbiol.* **10:**2129–2137.
102. **Monari, C., E. Pericolini, G. Bistoni, A. Casadevall, T. R. Kozel, and A. Vecchiarelli.** 2005. *Cryptococcus neoformans* capsular glucuronoxylomannan induces expression of Fas ligand in macrophages. *J. Immunol.* **174:**3461–3468.
103. **Monari, C., C. Retini, A. Casadevall, D. Netski, F. Bistoni, T. R. Kozel, and A. Vecchiarelli.** 2003. Differences in outcome of the interaction between *Cryptococcus neoformans* glucuronoxylomannan and human monocytes and neutrophils. *Eur. J. Immunol.* **33:**1041–1051.
104. **Moyrand, F., T. Fontaine, and G. Janbon.** 2007. Systematic capsule gene disruption reveals the central role of galactose metabolism on *Cryptococcus neoformans* virulence. *Mol. Microbiol.* **64:**771–781.
105. **Moyrand, F., I. Lafontaine, T. Fontaine, and G. Janbon.** 2008. UGE1 and UGE2 regulate the UDP-glucose/UDP-galactose equilibrium in *Cryptococcus neoformans*. *Eukaryot. Cell* **7:**2069–2077.
106. **Mukherjee, J., G. Nussbaum, M. D. Scharff, and A. Casadevall.** 1995. Protective and nonprotective monoclonal antibodies to *Cryptococcus neoformans* originating from one B cell. *J. Exp. Med.* **181:**405–409.
107. **Mukherjee, J., M. D. Scharff, and A. Casadevall.** 1992. Protective murine monoclonal antibodies to *Cryptococcus neoformans*. *Infect. Immun.* **60:**4534–4541.
108. **Narasipura, S. D., J. G. Ault, M. J. Behr, V. Chaturvedi, and S. Chaturvedi.** 2003. Characterization of Cu,Zn superoxide dismutase (SOD1) gene knock-out mutant of *Cryptococcus neoformans* var. *gattii*: role in biology and virulence. *Mol. Microbiol.* **47:**1681–1694.
109. **Narasipura, S. D., V. Chaturvedi, and S. Chaturvedi.** 2005. Characterization of *Cryptococcus neoformans* variety *gattii* SOD2 reveals distinct roles of the two superoxide dismutases in fungal biology and virulence. *Mol. Microbiol.* **55:**1782–1800.
110. **Neilson, J. B., R. A. Fromtling, and G. S. Bulmer.** 1977. *Cryptococcus neoformans*: size range of infectious particles from aerosolized soil. *Infect. Immun.* **17:**634–638.
111. **Neto, R. J., M. C. Guimaraes, M. J. Moya, F. R. Oliveira, P. L. Louzada, Jr., and R. Martinez.** 2000. Hypogammaglobulinemia as risk factor for *Cryptococcus neoformans* infection: report of 2 cases. *Rev. Soc. Bras. Med. Trop.* **33:**603–608. (In Portuguese.)
112. **Nimmerjahn, F., P. Bruhns, K. Horiuchi, and J. V. Ravetch.** 2005. FcgammaRIV: a novel FcR with distinct IgG subclass specificity. *Immunity* **23:**41–51.
113. **Nussbaum, G., W. Cleare, A. Casadevall, M. D. Scharff, and P. Valadon.** 1997. Epitope location in the *Cryptococcus neoformans* capsule is a determinant of antibody efficacy. *J. Exp. Med.* **185:**685–694.
114. **Okazaki, T., Y. Ota, N. Yuki, A. Hayashida, A. Shoda, M. Nishikawa, K. Oshima, I. Fukasawa, H. Watanabe, and N. Inaba.** 2007. Plasma levels of alpha-defensins 1-3 are an indicator of neutrophil activation in pregnant and post-partum women. *J. Obstet. Gynaecol. Res.* **33:**645–650.
115. **Panyutich, A. V., E. A. Panyutich, V. A. Krapivin, E. A. Baturevich, and T. Ganz.** 1993. Plasma defensin concentrations are elevated in patients with septicemia or bacterial meningitis. *J. Lab. Clin. Med.* **122:**202–207.
116. **Parry, M. F., R. K. Root, J. A. Metcalf, K. K. Delaney, L. S. Kaplow, and W. J. Richar.** 1981. Myeloperoxidase deficiency: prevalence and clinical significance. *Ann. Intern. Med.* **95:**293–301.
117. **Pedrosa, J., B. M. Saunders, R. Appelberg, I. M. Orme, M. T. Silva, and A. M. Cooper.** 2000. Neutrophils play a protective nonphagocytic role in systemic *Mycobacterium tuberculosis* infection of mice. *Infect. Immun.* **68:**577–583.
118. **Powell, K. E., B. A. Dahl, R. J. Weeks, and F. E. Tosh.** 1972. Airborne *Cryptococcus neoformans*: particles from pigeon excreta compatible with alveolar deposition. *J. Infect. Dis.* **125:**412–415.
119. **Price, M. S., C. B. Nichols, and J. A. Alspaugh.** 2008. The *Cryptococcus neoformans* Rho-GDP dissociation inhibitor mediates intracellular survival and virulence. *Infect. Immun.* **76:**5729–5737.
120. **Ravetch, J. V., and S. Bolland.** 2001. IgG Fc receptors. *Annu. Rev. Immunol.* **19:**275–290.
121. **Ravetch, J. V., and L. L. Lanier.** 2000. Immune inhibitory receptors. *Science* **290:**84–89.
122. **Rhodes, J. C.** 1985. Contribution of complement component C5 to the pathogenesis of experimental murine cryptococcosis. *Sabouraudia* **23:**225–234.
123. **Rittershaus, P. C., T. B. Kechichian, J. C. Allegood, A. H. Merrill, Jr., M. Hennig, C. Luberto, and M. Del Poeta.** 2006. Glucosylceramide synthase is an essential regulator of pathogenicity of *Cryptococcus neoformans*. *J. Clin. Invest.* **116:**1651–1659.
124. **Rodrigues, M. L., E. S. Nakayasu, D. L. Oliveira, L. Nimrichter, J. D. Nosanchuk, I. C. Almeida, and A. Casadevall.** 2008. Extracellular vesicles produced by *Cryptococcus neoformans* contain protein components associated with virulence. *Eukaryot. Cell* **7:**58–67.
125. **Rodrigues, M. L., L. Nimrichter, D. L. Oliveira, S. Frases, K. Miranda, O. Zaragoza, M. Alvarez, A. Nakouzi,**

M. Feldmesser, and A. Casadevall. 2007. Vesicular polysaccharide export in *Cryptococcus neoformans* is a eukaryotic solution to the problem of fungal trans-cell wall transport. *Eukaryot. Cell* **6:**48–59.

126. **Saenz, H. L., and C. Dehio.** 2005. Signature-tagged mutagenesis: technical advances in a negative selection method for virulence gene identification. *Curr. Opin. Microbiol.* **8:**612–619.
127. **Salzman, N. H., D. Ghosh, K. M. Huttner, Y. Paterson, and C. L. Bevins.** 2003. Protection against enteric salmonellosis in transgenic mice expressing a human intestinal defensin. *Nature* **422:**522–526.
128. **Sanford, J. E., D. M. Lupan, A. M. Schlageter, and T. R. Kozel.** 1990. Passive immunization against *Cryptococcus neoformans* with an isotype-switch family of monoclonal antibodies reactive with cryptococcal polysaccharide. *Infect. Immun.* **58:**1919–1923.
129. **Santangelo, R., H. Zoellner, T. Sorrell, C. Wilson, C. Donald, J. Djordjevic, Y. Shounan, and L. Wright.** 2004. Role of extracellular phospholipases and mononuclear phagocytes in dissemination of cryptococcosis in a murine model. *Infect. Immun.* **72:**2229–2239.
130. **Segal, A. W.** 1981. The antimicrobial role of the neutrophil leukocyte. *J. Infect.* **3:**3–17.
131. **Segal, A. W., M. Geisow, R. Garcia, A. Harper, and R. Miller.** 1981. The respiratory burst of phagocytic cells is associated with a rise in vacuolar pH. *Nature* **290:** 406–409.
132. **Segal, A. W., A. Harper, R. Garcia, O. T. Jones, and A. R. Cross.** 1981. The nature and function of the microbicidal oxidase system of neutrophils. *Bull. Eur. Physiopathol. Respir.* **17**(Suppl)**:**187–191.
133. **Seiler, P., P. Aichele, B. Raupach, B. Odermatt, U. Steinhoff, and S. H. Kaufmann.** 2000. Rapid neutrophil response controls fast-replicating intracellular bacteria but not slow-replicating *Mycobacterium tuberculosis*. *J. Infect. Dis.* **181:**671–680.
134. **Shao, X., A. Mednick, M. Alvarez, N. van Rooijen, A. Casadevall, and D. L. Goldman.** 2005. An innate immune system cell is a major determinant of species-related susceptibility differences to fungal pneumonia. *J. Immunol.* **175:**3244–3251.
135. **Shapiro, S., D. O. Beenhouwer, M. Feldmesser, C. Taborda, M. C. Carroll, A. Casadevall, and M. D. Scharff.** 2002. Immunoglobulin G monoclonal antibodies to *Cryptococcus neoformans* protect mice deficient in complement component C3. *Infect. Immun.* **70:**2598–2604.
136. **Shea, J. M., J. L. Henry, and M. Del Poeta.** 2006. Lipid metabolism in *Cryptococcus neoformans*. *FEMS Yeast Res.* **6:**469–479.
137. **Shea, J. M., T. B. Kechichian, C. Luberto, and M. Del Poeta.** 2006. The cryptococcal enzyme inositol phosphosphingolipid-phospholipase C confers resistance to the antifungal effects of macrophages and promotes fungal dissemination to the central nervous system. *Infect. Immun.* **74:**5977–5988.
138. **Shoham, S., C. Huang, J. M. Chen, D. T. Golenbock, and S. M. Levitz.** 2001. Toll-like receptor 4 mediates intracellular signaling without TNF-alpha release in response to *Cryptococcus neoformans* polysaccharide capsule. *J. Immunol.* **166:**4620–4626.
139. **Siafakas, A. R., T. C. Sorrell, L. C. Wright, C. Wilson, M. Larsen, R. Boadle, P. R. Williamson, and J. T. Djordjevic.** 2007. Cell wall-linked cryptococcal phospholipase B1 is a source of secreted enzyme and a determinant of cell wall integrity. *J. Biol. Chem.* **282:**37508–37514.
140. **Siafakas, A. R., L. C. Wright, T. C. Sorrell, and J. T. Djordjevic.** 2006. Lipid rafts in *Cryptococcus neoformans* concentrate the virulence determinants phospholipase B1 and Cu/Zn superoxide dismutase. *Eukaryot. Cell* **5:** 488–498.
141. **Siddiqui, A. A., A. E. Brouwer, V. Wuthiekanun, S. Jaffar, R. Shattock, D. Irving, J. Sheldon, W. Chierakul, S. Peacock, N. Day, N. J. White, and T. S. Harrison.** 2005. IFN-gamma at the site of infection determines rate of clearance of infection in cryptococcal meningitis. *J. Immunol.* **174:**1746–1750.
142. **Spitzer, E. D., S. G. Spitzer, L. F. Freundlich, and A. Casadevall.** 1993. Persistence of initial infection in recurrent *Cryptococcus neoformans* meningitis. *Lancet* **341:**595–596.
143. **Stano, P., V. Williams, M. Villani, E. S. Cymbalyuk, A. Qureshi, Y. Huang, G. Morace, C. Luberto, S. Tomlinson, and M. Del Poeta.** 2009. App1: an antiphagocytic protein that binds to complement receptors 3 and 2. *J. Immunol.* **182:**84–91.
144. **Steenbergen, J. N., H. A. Shuman, and A. Casadevall.** 2001. *Cryptococcus neoformans* interactions with amoebae suggest an explanation for its virulence and intracellular pathogenic strategy in macrophages. *Proc. Natl. Acad. Sci. USA* **98:**15245–15250.
145. **Syme, R. M., J. C. Spurrell, L. L. Ma, F. H. Green, and C. H. Mody.** 2000. Phagocytosis and protein processing are required for presentation of *Cryptococcus neoformans* mitogen to T lymphocytes. *Infect. Immun.* **68:**6147–6153.
146. **Taborda, C. P., and A. Casadevall.** 2002. CR3 (CD11b/CD18) and CR4 (CD11c/CD18) are involved in complement-independent antibody-mediated phagocytosis of *Cryptococcus neoformans*. *Immunity* **16:**791–802.
147. **Takeo, K., I. Uesaka, K. Uehira, and M. Nishiura.** 1973. Fine structure of *Cryptococcus neoformans* grown in vivo as observed by freeze-etching. *J. Bacteriol.* **113:**1449–1454.
148. **Tateda, K., T. A. Moore, J. C. Deng, M. W. Newstead, X. Zeng, A. Matsukawa, M. S. Swanson, K. Yamaguchi, and T. J. Standiford.** 2001. Early recruitment of neutrophils determines subsequent T1/T2 host responses in a murine model of *Legionella pneumophila* pneumonia. *J. Immunol.* **166:**3355–3361.
149. **Tateda, K., T. A. Moore, M. W. Newstead, W. C. Tsai, X. Zeng, J. C. Deng, G. Chen, R. Reddy, K. Yamaguchi, and T. J. Standiford.** 2001. Chemokine-dependent neutrophil recruitment in a murine model of *Legionella* pneumonia: potential role of neutrophils as immunoregulatory cells. *Infect. Immun.* **69:**2017–2024.
150. **Tjelle, T. E., T. Lovdal, and T. Berg.** 2000. Phagosome dynamics and function. *Bioessays* **22:**255–263.
151. **Tommasino, N., M. Villani, A. Qureshi, J. Henry, C. Luberto, and M. Del Poeta.** 2008. Atf2 transcription factor binds to the APP1 promoter in *Cryptococcus neoformans*: stimulatory effect of diacylglycerol. *Eukaryot. Cell* **7:**294–301.
152. **Torda, A., R. K. Kumar, and P. D. Jones.** 2001. The pathology of human and murine pulmonary infection with *Cryptococcus neoformans* var. *gattii*. *Pathology* **33:**475–478.
153. **Tucker, S. C., and A. Casadevall.** 2002. Replication of *Cryptococcus neoformans* in macrophages is accompanied by phagosomal permeabilization and accumulation of vesicles containing polysaccharide in the cytoplasm. *Proc. Natl. Acad. Sci. USA* **99:**3165–170.
154. **Urban, C. F., S. Lourido, and A. Zychlinsky.** 2006. How do microbes evade neutrophil killing? *Cell. Microbiol.* **8:**1687–1696.
155. **Van den Herik-Oudijk, I. E., P. J. Capel, T. van der Bruggen, and J. G. Van de Winkel.** 1995. Identification of signaling motifs within human Fc gamma RIIa and Fc gamma RIIb isoforms. *Blood* **85:**2202–2211.
156. **Vecchiarelli, A., M. Dottorini, D. Pietrella, C. Monari, C. Retini, T. Todisco, and F. Bistoni.** 1994. Role of

human alveolar macrophages as antigen-presenting cells in *Cryptococcus neoformans* infection. *Am. J. Respir. Cell Mol. Biol.* **11:**130–137.

157. **Vecchiarelli, A., C. Monari, C. Retini, D. Pietrella, B. Palazzetti, L. Pitzurra, and A. Casadevall.** 1998. *Cryptococcus neoformans* differently regulates B7-1 (CD80) and B7-2 (CD86) expression on human monocytes. *Eur. J. Immunol.* **28:**114–121.
158. **Vecchiarelli, A., D. Pietrella, M. Dottorini, C. Monari, C. Retini, T. Todisco, and F. Bistoni.** 1994. Encapsulation of *Cryptococcus neoformans* regulates fungicidal activity and the antigen presentation process in human alveolar macrophages. *Clin. Exp. Immunol.* **98:**217–223.
159. **Villena, S. N., R. O. Pinheiro, C. S. Pinheiro, M. P. Nunes, C. M. Takiya, G. A. DosReis, J. O. Previato, L. Mendonca-Previato, and C. G. Freire-de-Lima.** 2008. Capsular polysaccharides galactoxylomannan and glucuronoxylomannan from *Cryptococcus neoformans* induce macrophage apoptosis mediated by Fas ligand. *Cell. Microbiol.* **10:**1274–1285.
160. **Wang, Y., Y. Jiang, T. Gong, X. Cui, W. Li, Y. Feng, B. Wang, Z. Jiang, and M. Li.** 2010. High-level expression and novel antifungal activity of mouse beta defensin-1 mature peptide in *Escherichia coli*. *Appl. Biochem. Biotechnol.* **160:**213–221.
161. **Washburn, R. G., B. J. Bryant-Varela, N. C. Julian, and J. E. Bennett.** 1991. Differences in *Cryptococcus neoformans* capsular polysaccharide structure influence assembly of alternative complement pathway C3 convertase on fungal surfaces. *Mol. Immunol.* **28:**465–470.
162. **Waterman, S. R., M. Hacham, J. Panepinto, G. Hu, S. Shin, and P. R. Williamson.** 2007. Cell wall targeting of laccase of *Cryptococcus neoformans* during infection of mice. *Infect. Immun.* **75:**714–722.
163. **Wetsel, R. A., D. T. Fleischer, and D. L. Haviland.** 1990. Deficiency of the murine fifth complement component (C5). A 2-base pair gene deletion in a 5'-exon. *J. Biol. Chem.* **265:**2435–2440.
164. **Wheat, W. H., R. Wetsel, A. Falus, B. F. Tack, and R. C. Strunk.** 1987. The fifth component of complement (C5) in the mouse. Analysis of the molecular basis for deficiency. *J. Exp. Med.* **165:**1442–1447.
165. **Williamson, P. R.** 1994. Biochemical and molecular characterization of the diphenol oxidase of *Cryptococcus neoformans*: identification as a laccase. *J. Bacteriol.* **176:** 656–664.
166. **Wright, L. C., R. M. Santangelo, R. Ganendren, J. Payne, J. T. Djordjevic, and T. C. Sorrell.** 2007. Cryptococcal lipid metabolism: phospholipase B1 is implicated in transcellular metabolism of macrophage-derived lipids. *Eukaryot. Cell* **6:**37–47.
167. **Yauch, L. E., M. K. Mansour, and S. M. Levitz.** 2005. Receptor-mediated clearance of *Cryptococcus neoformans* capsular polysaccharide in vivo. *Infect. Immun.* **73:** 8429–8432.
168. **Young, B. J., and T. R. Kozel.** 1993. Effects of strain variation, serotype, and structural modification on kinetics for activation and binding of C3 to *Cryptococcus neoformans*. *Infect. Immun.* **61:**2966–2972.
169. **Yuan, R., A. Casadevall, G. Spira, and M. D. Scharff.** 1995. Isotype switching from IgG3 to IgG1 converts a nonprotective murine antibody to *Cryptococcus neoformans* into a protective antibody. *J. Immunol.* **154:**1810–1816.
170. **Yuan, R. R., A. Casadevall, J. Oh, and M. D. Scharff.** 1997. T cells cooperate with passive antibody to modify *Cryptococcus neoformans* infection in mice. *Proc. Natl. Acad. Sci. USA* **94:**2483–2488.
171. **Zaragoza, O., and A. Casadevall.** 2004. Antibodies produced in response to *Cryptococcus neoformans* pulmonary infection in mice have characteristics of nonprotective antibodies. *Infect. Immun.* **72:**4271–4274.
172. **Zaragoza, O., and A. Casadevall.** 2004. Experimental modulation of capsule size in *Cryptococcus neoformans*. *Biol. Proced. Online* **6:**10–15.
173. **Zaragoza, O., and A. Casadevall.** 2006. Monoclonal antibodies can affect complement deposition on the capsule of the pathogenic fungus *Cryptococcus neoformans* by both classical pathway activation and steric hindrance. *Cell. Microbiol.* **8:**1862–1876.
174. **Zaragoza, O., C. J. Chrisman, M. V. Castelli, S. Frases, M. Cuenca-Estrella, J. L. Rodriguez-Tudela, and A. Casadevall.** 2008. Capsule enlargement in *Cryptococcus neoformans* confers resistance to oxidative stress suggesting a mechanism for intracellular survival. *Cell. Microbiol.* **10:**2043–2057.
175. **Zhu, X., J. Gibbons, S. Zhang, and P. R. Williamson.** 2003. Copper-mediated reversal of defective laccase in a Deltavph1 avirulent mutant of *Cryptococcus neoformans*. *Mol. Microbiol.* **47:**1007–1014.
176. **Zhu, X., and P. R. Williamson.** 2004. Role of laccase in the biology and virulence of *Cryptococcus neoformans*. *FEMS Yeast Res.* **5:**1–10.

Cryptococcus: From Human Pathogen to Model Yeast
Edited by J. Heitman et al.

27

T Cell and Dendritic Cell Immune Responses to *Cryptococcus*

KAREN L. WOZNIAK AND STUART M. LEVITZ

Cryptococcus neoformans is an opportunistic fungal pathogen that is typically acquired via inhalation of the organism. *C. neoformans* primarily infects individuals who have impaired T-cell function, particularly those with AIDS and lymphoid malignancies and recipients of immunosuppressive therapies (49, 61, 80, 84). The closely related fungus *Cryptococcus gattii* also has a propensity to cause disease in those with compromised T-cell-mediated immunity, although it commonly affects immunocompetent individuals too (86). *Cryptococcus* is surrounded by a large polysaccharide capsule, which is the organism's major virulence factor (60, 65, 73). As discussed elsewhere in this book, the capsule subverts both phagocytic and B-cell defenses. This forces the host to rely heavily on T-cell defenses.

In addition to clinical observations, data from animal models support the requirement of an adaptive T-cell-mediated immune response for protection against cryptococcosis (33, 37, 39, 40, 64). Also critical for development of protective responses are innate immune cells, particularly dendritic cells (DCs). DCs not only have direct antifungal activity, but also initiate adaptive responses and play a key role in determining the nature of the subsequent T-cell response (Fig. 1). This chapter reviews current knowledge regarding the role of DCs and T cells in the generation of protective immunity against *C. neoformans* infections.

DC RESPONSE TO *C. NEOFORMANS*

DCs were so named by Ralph Steinman in 1973 because their projections morphologically resembled the dendrites of neurons (88). DCs function as sentinels in the innate immune system. They are the most effective antigen-presenting cells (APCs) for inducing cell-mediated immune responses and are uniquely capable of activating naive T cells (18). DCs phagocytose pathogens, endocytose foreign antigens, process and present antigens to T cells, and are key mediators in the initiation of adaptive immune responses. DCs are uniquely capable of decoding fungal-associated information and translating it into different adaptive Th-type immune responses (10, 12, 22, 76). DCs are a heterogeneous population present throughout the body and are derived from both myeloid and lymphoid lineages. DC subpopulations include conventional (also known as myeloid) DCs, plasmacytoid (also known as lymphoid) DCs, and skin DCs such as dermal DCs and Langerhans DCs (1, 3, 16). These types of DCs are categorized by location, cell surface markers, cytokine secretion, and function (1). DCs are located particularly at portals of entry for many pathogens, such as mucosal tissues and skin, which allows them to quickly respond to invading pathogens (28). The respiratory tract, in particular, is composed of a dense network of DCs specialized for antigen uptake (79), which is important for responding to respiratory-acquired pathogens such as *C. neoformans*.

Resting or immature DCs are highly endocytic and have relatively low surface expression of major histocompatibility complex (MHC) class II and costimulatory molecules (CD40, CD80, and CD86). Upon recognition, they can rapidly phagocytose pathogens and their antigenic components. This causes a series of events to occur in DCs, resulting in DC maturation, which transforms them into highly efficient APCs. Stimuli for DC maturation include pattern-recognition receptor ligands (such as Toll-like receptor [TLR] ligands), proinflammatory cytokines, and T-cell ligands (such as CD40L).

DC maturation is evidenced through upregulation of costimulatory molecules and MHC II on the cell surface followed by migration to T-cell-rich areas of secondary lymphoid organs (92). Once DCs arrive in the lymphoid organs, they can present antigens to naive T cells and produce cytokines that direct the adaptive immune response (3, 94). Cytokines produced by DCs and lymphocytes instruct the $CD4^+$ T cell response by polarizing T-cell development or inducing tolerance (17, 92, 95). Thus, Th1-biased responses are promoted by interleukin-12 (IL-12) and interferon-γ (IFN-γ), Th2 responses by IL-4 (and the absence of IL-12), Treg responses by transforming growth factor β (and maybe IL-2), and Th17 responses by the combination of transforming growth factor β and IL-6 and propagated by IL-23 (although recent data downplay the role of IL-23 in mice) (48, 56, 95). This results in recruitment of other immune cells to the site

Karen L. Wozniak, Department of Biology, University of Texas at San Antonio, San Antonio, TX 78249. **Stuart M. Levitz,** Department of Medicine, University of Massachusetts Medical School, Worcester, MA 01605.

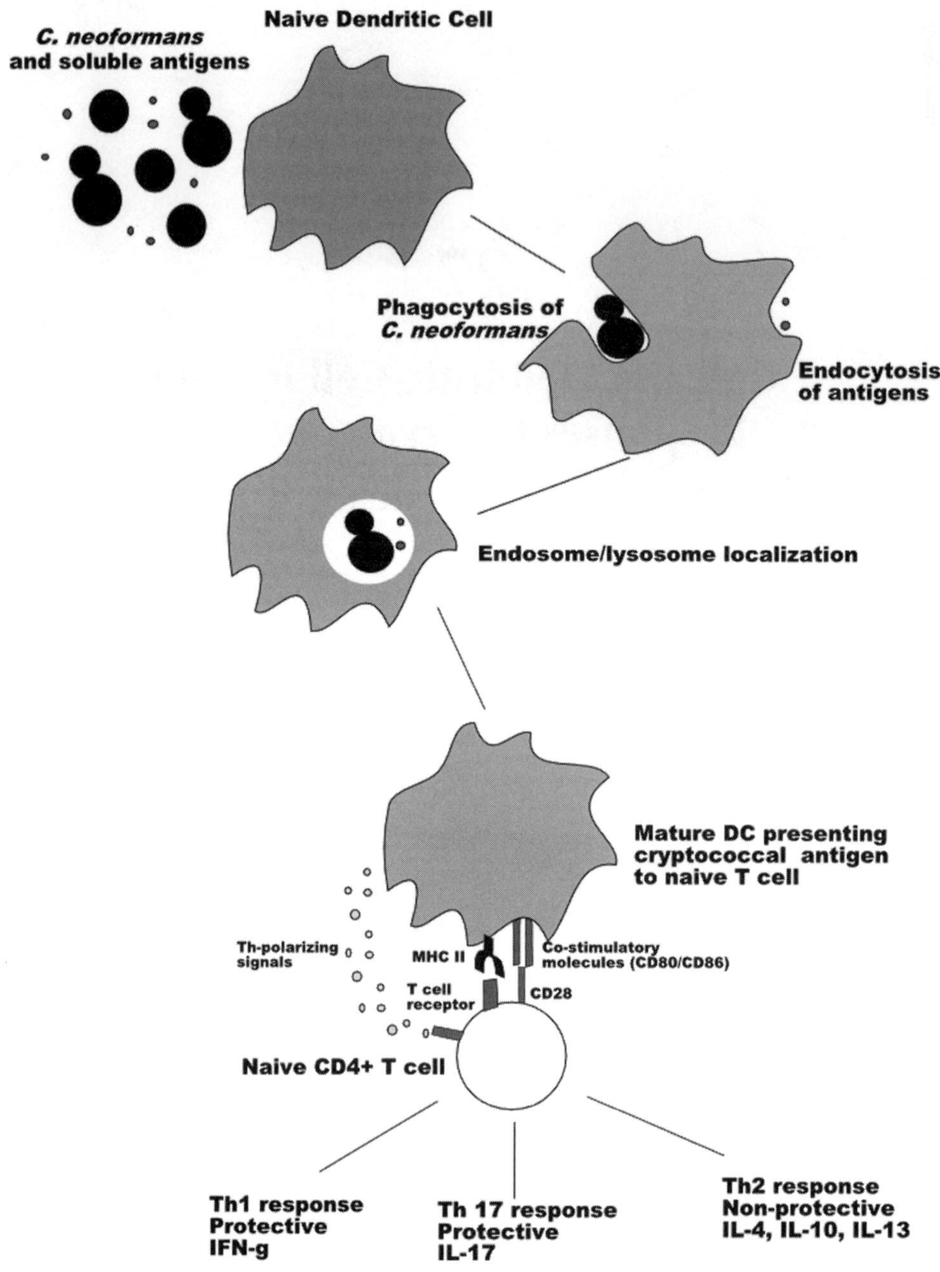

FIGURE 1 Overview of the link between innate recognition by DCs and the development of adaptive CD4+ T-cell immunity to *C. neoformans*. The immune response begins with DC phagocytosis of *C. neoformans*, followed by entry into the endolysosomal pathway. Antigen is then processed and presented to naive T cells in the presence of costimulatory molecules and polarizing cytokines. This leads to T-cell activation, Th skewing, and lymphoproliferation. Not shown in the figure are Treg cells, whose role in cryptococcal infections is poorly defined.

of infection or to the draining lymph nodes. There is a strong association between development of Th1-biased responses and protection against cryptococcosis (47, 67). The role of Treg and Th17 cells in cryptococcosis has not been well studied, although Th17 responses have been correlated with protection (46, 67).

DC Interaction with *C. neoformans* In Vitro

In vitro studies of DCs with *C. neoformans* have shown that DCs are involved in detection, binding, phagocytosis (Fig. 2), processing, antigen presentation, T-cell activation, and killing of the organism. Studies with an acapsular mutant of *C. neoformans* showed that DCs were the most efficient APCs for presenting cryptococcal antigen. Further, the DC surface receptors mannose receptor (MR) and Fcγ receptor II were required for presentation of the antigen as well as for uptake of the organism (89). In vitro studies of murine DCs, B cells, and macrophages with cryptococcal mannoproteins (MPs) demonstrated that only DCs could take up MP and effectively present it to MP-specific T cells. In addition, depletion

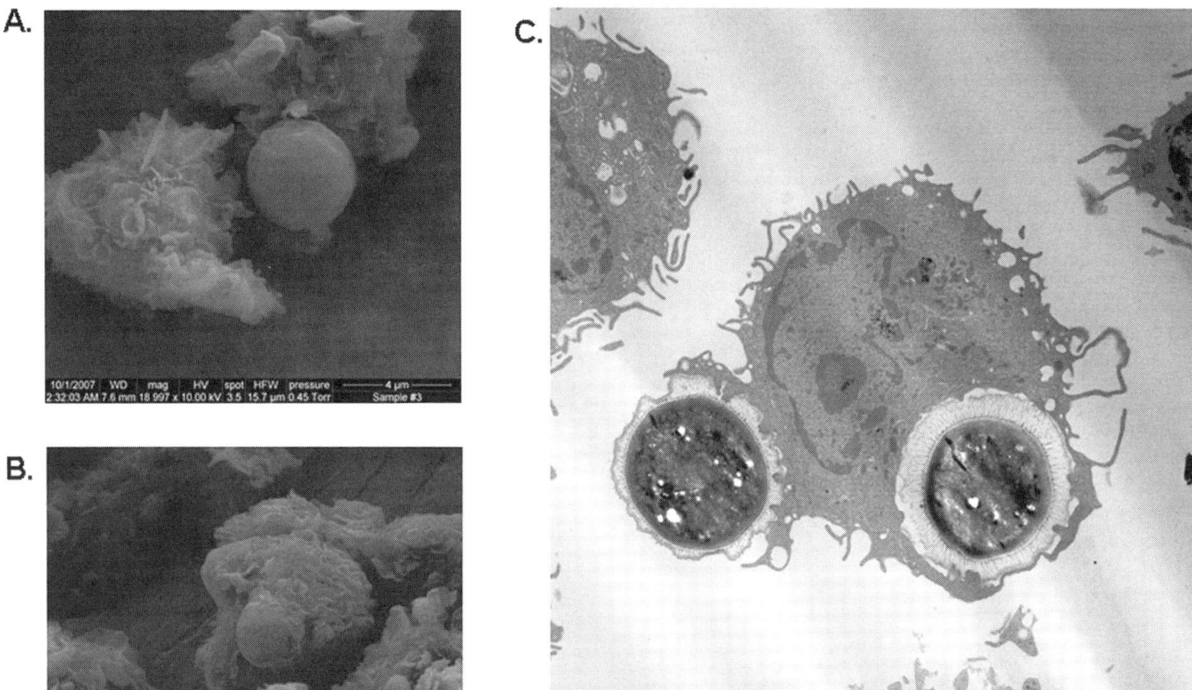

FIGURE 2 Electron microscopy of *C. neoformans* phagocytosis by murine bone-marrow-derived DCs. (A, B) Scanning electron microscopy of *C. neoformans* in the process of being phagocytosed by DCs. (C) Transmission electron microscopy of a DC that has completely phagocytosed one *C. neoformans* and has partially internalized a second yeast cell.

of DCs abrogated the T-cell response (57). These data suggest that DCs provide the crucial link between innate and adaptive immune responses to *C. neoformans*.

Similarly, another study observed that DCs loaded with cryptococcal MP can stimulate T cells and promote CD4 and CD8 T-cell proliferation in response to *C. neoformans*. These DCs also become mature and secrete cytokines such as IL-12 and tumor necrosis factor α (TNF-α) (75). Furthermore, DC phagocytosis of MP in the presence of the appropriate adjuvant induces production of Th1-type cytokines (19). This suggests that when stimulated with the appropriate cryptococcal antigen/adjuvant combination, DCs can induce protective Th1-type immune responses against *C. neoformans*. Additional studies revealed that the interaction of *C. neoformans* with DCs, but not macrophages, induced the production of IL-12 and IL-23, two cytokines associated with protection against cryptococcosis. This interaction resulted in maturation of DCs, evidenced by upregulation of surface MHC II and CD86 (82). These data suggest that DCs are critical for IL-12 and IL-23 production during cryptococcosis. Other studies have shown that DCs can also induce a protective Th1-type immune response when stimulated with *C. neoformans* DNA. DCs stimulated with DNA from *C. neoformans* release IL-12p40 and express CD40, a costimulatory molecule associated with DC maturation. Activation of DCs by cryptococcal DNA was almost completely abrogated in mice lacking TLR9 or MyD88 (70).

Additional in vitro work revealed that both human and murine DCs can phagocytose *C. neoformans* and that phagocytosis was dependent on opsonization with either antibody or complement (45). Also, both murine and human DCs are able to kill *C. neoformans*, by both oxidative and nonoxidative mechanisms (45). Independent of the method of opsonization, *C. neoformans* traffics into the endosomal/lysosomal pathway of DCs. Moreover, lysosomes isolated from DCs were able to kill *C. neoformans* (97).

DC Response to *C. neoformans* In Vivo

In vivo studies have provided additional information on the interactions of DCs with *C. neoformans* and the resultant T-cell responses. Studies using a crude protective antigen, *C. neoformans* culture filtrate (CneF), showed that DCs are important for the induction of a protective immune response in mice. Following subcutaneous immunization of mice with CneF, infiltration of Langerhans cells and myeloid DCs into the draining lymph nodes was induced (6). The protective immune responses correlated with accumulation of myeloid DCs in the draining lymph nodes, while nonprotective responses were associated with accumulation of lymphoid DCs (6). The accumulation of DCs in the lymph nodes of immunized mice was regulated by production of TNF-α (5). In a pulmonary infection model of *C. neoformans*, lung DCs were important in phagocytosis and antigen presentation of cryptococcal antigens to T cells (98).

DCs infiltrated into the lungs of infected mice in response to pulmonary infection. Uptake of *C. neoformans* by pulmonary DCs in mice resulted in DC maturation, evidenced by increased surface expression of MHC II, CD80, and CD86 (98). In addition, DCs isolated from infected lungs presented cryptococcal MP to MP-specific T cells and induced T-cell activation ex vivo (98). Recent data examining infection in CCR2-deficient mice further show the importance of DCs in the generation of protective immune responses. CCR2 is involved in trafficking of monocytes, and thus a deficiency in CCR2 impairs the infiltration of monocyte-derived DCs. Upon infection with *C. neoformans*, CCR2-deficient mice developed a nonprotective Th2-type immune response and persistent infection. DC recruitment was markedly impaired, bronchovascular infiltrates were diminished, and mice developed features of Th2-type responses, including bronchovascular collagen deposition and IL-4 production (72).

ACQUIRED IMMUNE RESPONSE TO *C. NEOFORMANS*

Evidence of T-Cell Importance in Human Cryptococcosis

As reviewed in greater detail elsewhere in this book, the clinical evidence unequivocally demonstrates that $CD4^+$ T-cell-mediated immunity is paramount to the control of cryptococcosis. The vast majority of patients with cryptococcosis have impaired T-cell function due to an underlying disease (particularly AIDS, lymphoma, and idiopathic $CD4^+$ lymphocytopenia) or receipt of immunosuppressive medications (particularly to prevent rejection of solid-organ transplants) (15, 23, 83). Restoring CD4 T-cell function by boosting CD4 counts with antiretroviral therapy or withdrawing immunosuppressive medication often results in an "immune response inflammatory syndrome," which can be life-threatening and often is paradoxically treated with corticosteroids (53, 83).

Peripheral blood mononuclear cells from HIV-infected donors have profoundly impaired proliferative and cytokine responses to cryptococcal antigens (29, 34). In addition, stimulation of HIV-infected cells in vitro with *C. neoformans* induces HIV replication, a factor that, if it occurs in vivo, could worsen the course of HIV infection (71, 74). In a study of AIDS patients with cryptococcal meningitis, cerebrospinal fluid concentrations of IFN-γ, a Th1-associated cytokine, were associated with survival and shown to be an independent factor predicting the rate of clearance of infection (81). In a small study, IFN-γ production in response to cryptococcal antigens peaked coincident with the occurrence of an immune response inflammatory syndrome following antiretroviral therapy (90).

Evidence of T-Cell Importance in Animal Models

In support of clinical findings, data from animal models have shown that an adaptive T-cell-mediated immune response is required for protection against cryptococcosis (33, 37, 39, 40, 64). Immunization studies have been useful in determining cells associated with protection against cryptococcosis. Immunization of mice with heat-killed *C. neoformans* conferred protection against challenge in wild-type mice but did not induce protection in nude mice (lacking T cells), demonstrating the importance of T cells in protection in the central nervous system (CNS) (9). A model of immunization of mice with CneF antigen resulted in influx of lymphocytes, including T cells (both $CD4^+$ and $CD8^+$), into the brain (11). In mouse models of infection, $CD4^+$ T cells have been shown to be critical for protection against pulmonary, intravenous, and CNS infections (32, 33, 37, 40, 64). In particular, $CD4^+$ T cells are credited with prevention of dissemination of disease to other organs (33).

While $CD4^+$ T cells are predominately associated with protection against *C. neoformans*, there is evidence that $CD8^+$ T cells also provide protection. Following pulmonary challenge with *C. neoformans*, mice deficient in either $CD4^+$ or $CD8^+$ T cells exhibited reduced leukocyte recruitment into the lungs (37). However, whereas $CD4^+$ deficiency did not affect recruitment of $CD8^+$ T cells, $CD8^+$ T-cell deficiency significantly reduced recruitment of $CD4^+$ T cells. In addition, $CD8^+$ T cell deficiency resulted in decreased levels of IFN-γ secretion, indicating that $CD8^+$ T cells play a role in both lymphocyte recruitment and IFN-γ production during cryptococcal infection (37).

Additional studies established that mice depleted of $CD8^+$ T cells had decreased pulmonary clearance of *C. neoformans* compared to wild-type mice. CD8 depletion abrogated delayed-type hypersensitivity (DTH) without affecting antigen recognition or lymphocyte proliferation in vitro, suggesting that $CD8^+$ T cells enhanced rather than suppressed host defenses against *C. neoformans* (62). $CD8^+$ T-cell-depleted mice infected with a more virulent strain of *C. neoformans* showed significantly reduced survival compared to wild-type mice, but $CD8^+$ T-cell depletion did not affect fungal burden (63). One caveat to the above studies is that some subsets of DCs also express CD8. Thus, both $CD8^+$ DCs and T cells could have been depleted by monoclonal antibody treatment.

The importance of T cells in cryptococcosis has been emphasized by infection models in mice lacking T cells, including nude mice (lacking T cells) (40, 77), severe combined immunodeficiency (SCID) mice (lacking T and B cells) (40), T-cell knockout mice (59), and mice depleted of CD4 and CD8 cells (37, 39). Each is more susceptible to *C. neoformans* infection than wild-type mice. Moreover, adoptive transfer of lung and lymph node T cells from immunocompetent mice enables SCID and nude mice to clear a pulmonary infection (40).

ROLE OF T CELLS AT SITES OF INFECTION

The lungs and brain are the most common sites of infection for *C. neoformans* and *C. gattii*. As the usual route of acquisition is inhalation of airborne organisms, this might account, in large measure, for the predilection for the lungs. In mice, the host's genetic background greatly impacts the pulmonary immune responses to cryptococcal infections (14). For example, the inbred mouse strain BALB/c clears *C. neoformans* infection in association with development of a Th1-skewed immune response. In contrast, C57BL/6 mice develop chronic pulmonary infections with a Th2-skewed immune response. In general, as with systemic models of cryptococcosis, conditions that skew toward Th1 responses are associated with improved outcomes, whereas those that induce Th2 bias are deleterious. Following pulmonary challenge of mice with *C. neoformans*, a $CD4^+$ T-cell-dependent inflammatory response appears that sequesters yeast within the pulmonary alveoli (31). A major component of this response is the enclosure of *C. neoformans* within multinucleated giant cells in granulomas.

Interestingly, a role for pulmonary cryptococcal infection in the pathogenesis of asthma has been postulated. In a rat model of disease, localized (but not disseminated) cryptococcal infection exacerbated allergic responses to the model antigen, ovalbumin (26). In a mouse model, allergic manifestations of pulmonary cryptococcosis required expression of the IL-4 receptor α-chain (which is part of the receptor for the Th2 cytokines IL-4 and IL-13) (66). In humans, a role for *C. neoformans* in the pathogenesis of asthma remains hypothetical. Circumstantial evidence favoring such a role includes serological studies demonstrating that exposure to *C. neoformans* is common in childhood in an inner city region where childhood asthma is high (20).

The basis for the neurotropism of *C. neoformans* is likely multifactorial. One area of interest is the T-cell response in the brain following cryptococcal infection. Following intravenous challenge of mice with *C. neoformans*, about 0.1% of the fungi localize to the brain (32). However, despite this low initial number, the mice eventually die of meningoencephalitis. CFUs increase more slowly in wild-type compared with SCID mice. As SCID mice lack both B cells and T cells, these data suggest that an acquired immune response controls infection. Further studies were performed that, taken together, established the critical importance of $CD4^+$ T cells. First, mice with preexisting anticryptococcal T-cell responses, either by virtue of immunization with cryptococcal antigens or prior pulmonary infection, had prolonged survival following brain inoculation (11, 32). Second, the protective effect of T cells could be abrogated by in vivo depletion of $CD4^+$ T cells, but not $CD8^+$ T cells (11, 32). Third, sensitized $CD4^+$ T cells could transfer partial protection to naive wild-type mice or SCID mice infected intracerebrally with *C. neoformans* (32, 93). These data suggest that vaccination strategies do not need to target CNS compartments because systemic T cells will traffic to the brain. Moreover, the findings have implications for understanding the neurological manifestations of the immune response inflammatory syndrome that often occurs following immune reconstitution in patients with cryptococcal meningitis (see above).

CYTOKINE RESPONSE TO CRYPTOCOCCOSIS

In pulmonary, systemic, and CNS infections, Th1-type cytokines are required for a protective cell-mediated immune response. Th1-type cytokines, including IL-12 and IFN-γ, as well as the proinflammatory cytokine TNF-α, have been associated with clearance of the organism. Th2-type cytokines, including IL-4, IL-5, IL-10, and IL-13, are associated with exacerbation of disease in animal models. As with Th1-type cytokines, the Th17-type cytokines, IL-23 and IL-17, have been associated with protective responses. Th17 cells were recognized relatively recently, and in the older literature, some effects attributed to Th1 cells may have actually been mediated by Th17 cells. This is because IL-12 and IL-23 are heterodimers that share a common p40 subunit, and some techniques that inhibit IL-12 affect IL-23 as well. Additionally, Th1 and Th17 cells (as well as $CD8^+$ T cells and NK cells) can produce IFN-γ. These caveats should be kept in mind when interpreting some of the studies cited below.

Production of Th1-type cytokines over Th2-type cytokines protects mice from infection, whereas infection is severe when Th2 cytokines predominate (42, 47). Mice depleted of Th1-type cytokines, such as IFN-γ and TNF-α, are more susceptible to cryptococcal infection (43) (38, 41). Upon infection with *C. neoformans*, mice deficient in IL-12 (IL-12p35 and IL-12p40 knockout) died significantly earlier than wild-type mice, had fewer granulomas in the lungs, and had increased production of IL-4, indicating the importance of IL-12 in inducing a protective response (21). Additional studies revealed that TNF-α was essential for development of a Th1-type immune response to *C. neoformans* (30). Mice lacking Th2 cytokines, including IL-4 and IL-10 knockout mice, had less severe infection than wild-type mice. When infected with a highly virulent isolate of *C. neoformans*, IL-10-deficient mice survived infection significantly longer than wild-type mice, suggesting that IL-10 is important in downregulating the protective immune response (8). Mice lacking IL-4 had prolonged survival and reduced fungal burden following *C. neoformans* infection compared to wild-type mice, suggesting that reduction in the Th2-type response contributes to protection (21). In a mouse model of infection, plasma levels of TNF-α and IL-10, but not IL-6, correlated with fungal burden in blood and spleen, but not brain. This result was similar to what was seen in HIV^+ patients with disseminated cryptococcal infection (54).

Immunization and immunotherapy studies have also demonstrated the importance of the Th1-type response in protection against cryptococcosis. Immunization with a melanin-deficient *C. neoformans* strain protected against challenge with a wild-type strain, and protection was associated with increased expression of IL-12, TNF-α, IL-1β, IFN-γ, and inducible nitric oxide synthase in CNS tissue (4). Additionally, immunotherapy with anti-CD40 and IL-2 prolonged the survival time of mice infected intracranially. However, this effect was lost in IFN-γ knockout mice, reinforcing the importance of IFN-γ for protection against cryptococcosis (100). Further studies examining a *C. neoformans* strain engineered to produce IFN-γ showed that infection with this organism resulted in higher Th1-type cytokine and chemokine expression, lower pulmonary fungal burden, and increased leukocyte recruitment compared to mice infected with the wild-type *C. neoformans* strain (96). Also, infection with the IFN-γ-producing strain led to complete protection against challenge by the wild-type *C. neoformans* strain.

Another cytokine involved in protection against cryptococcosis is IL-23, a heterodimeric cytokine composed of the p40 subunit (which is shared with IL-12) and a p19 subunit. In order to examine the relative contributions of IL-12 and IL-23 to anticryptococcal host defenses, mice lacking IL-12p35, IL-23p19, and IL-12/IL-23p40 were infected with *C. neoformans*. Mice deficient in p40 had a higher mortality than IL-12p35 knockout mice, and adding rIL-23 prolonged their survival. IL-23p19 knockout mice had a reduced survival time (although not as reduced as seen with the IL-12p35 knockout mice), delayed fungal clearance in the liver, and strongly impaired production of IL-17. Moreover, activation of microglia cells and expression of IL-1β, IL-6, and monocyte chemoattractant protein 1 in the brains of IL-12p19 knockout mice was impaired. The lack of IL-23 did not affect the Th1/Th2 balance. Although depletion of IL-23 resulted in a less pronounced phenotype compared with IL-12, these results suggest an independent role for Th17 cells in host defenses against cryptococcosis (46).

The importance of IL-13, a Th2 cytokine, in susceptibility to *C. neoformans* infection was investigated using IL-13 knockout mice and IL-13 transgenic mice. Following infection with *C. neoformans*, IL-13 transgenic mice had a higher mortality rate and fungal burden than wild-type mice. The IL-13 knockout mice were much more resistant to infection than wild-type mice. The increased susceptibility in the IL-13 transgenic mice was associated with Th2-type cytokine production, allergic inflammation, and reduced levels of IL-17. This indicated that IL-13 contributes the nonprotective Th2-type response during *C. neoformans* infection (67).

TABLE 1 Cytokines associated with protective and nonprotective responses to *C. neoformans* infection

Cytokine	Reference(s)
Protective	
IL-12	4, 21, 44
IFN-γ	4, 13, 43, 44, 96, 100
IL-23	46
IL-17	67
TNF-α	4, 5, 30, 38
IL-1β	4
Nonprotective	
IL-4	21
IL-10	8
IL-13	67

Not surprisingly, in addition to the presence of cytokines, cytokine signaling is important in host defense against cryptococcosis. Mice deficient in the IFN-γ receptor were more susceptible to pulmonary infection compared to wild-type mice, as demonstrated by increased pulmonary CFUs and increased dissemination from the lungs. IFN-γ receptor knockout mice also had increased pulmonary levels of the Th2-type cytokines, IL-4, IL-5, and IL-10, suggesting that IFN-γ signaling was required for development of protective Th1-type responses in the lung (13).

Other mediators have been shown to be important for cytokine production and Th skewing during experimental cryptococcosis. For example, mice lacking NADPH oxidase had increased levels of pulmonary Th1-type cytokines and improved pathogen containment within granulomas. The ratio of IFN-γ-producing T cells to IL-5-producing T cells was greatly augmented in the mice lacking NADPH oxidase. Also, the NADPH oxidase–deficient mice exhibited less dissemination of *C. neoformans* to the brain, suggesting that this enzyme plays a role in pathology and dissemination of the organism (85). As discussed above, *C. neoformans* infection in CCR2 knockout mice results in the development of a nonprotective Th2-type immune response and impaired DC recruitment (72). An earlier study demonstrated that CCR2 knockout mice infected with *C. neoformans* had higher CFU and greater dissemination to the spleen and brain compared to wild-type mice. Also, these mice had a reduction in macrophage and CD8+ T cell recruitment to the lungs, increased IL-4 and IL-5 in the lungs, decreased amounts of IFN-γ, and high levels of serum immunoglobulin E (91). Together, these studies suggest that CCR2 is required for the development of a Th1-type immune response and for recruitment of macrophages and DCs to the lungs following infection with *C. neoformans*. Table 1 summarizes the known protective and nonprotective cytokines associated with cryptococcosis.

C. NEOFORMANS ANTIGENS THAT STIMULATE T-CELL RESPONSES

The necessity of T cells for host defenses against cryptococcosis has prompted research into identifying immunoreactive cryptococcal antigens that could serve as vaccine candidates and as diagnostic reagents to measure T-cell responses in infected or at-risk patients (52). For the earliest studies, an antigen referred to as "cryptococcin" was prepared from a urea extract of *C. neoformans* (2). While cryptococcin elicited DTH reactions and lymphocyte transformation reactions in some patients with cryptococcosis, use of this antigen was limited by cross-reactivity in patients with other mycoses (2, 27). Studies with cryptococcin also provided the first clues that a subset of patients with cryptococcosis had abnormalities in T-cell-mediated immunity (27, 78). Other work with crude antigens established that a culture supernatant preparation that was formalin-fixed and dialyzed stimulated DTH responses in immunized mice (69). This preparation, designated CneF, partially protected mice from a lethal inoculum of *C. neoformans* (6). The vaccine needed to be administered with complete Freund's adjuvant in order to obtain protective responses, emphasizing the importance of both antigen and adjuvant.

The demonstration that unpurified antigens derived from *C. neoformans* could elicit T-cell responses stimulated investigators to search for the responsible component antigen(s). Fractionation of CneF and similar preparations revealed that it was the MP fraction, defined by the ability to adhere to a concanavalin A affinity column, that was predominantly responsible for the DTH responses in mice and the lymphoproliferative responses in humans recovered from cryptococcosis (35, 51, 68). However, partial protection against lethal fungal challenge was seen when mice were immunized with either the MP or the flow-through fraction that did not bind to the concanavalin A column (59). In this model, T cells were required for protection, while B cells were dispensable.

The above studies examined responses to complex mixtures of antigens. To determine individual antigens that stimulated CD4+ T cells, mice were immunized with *C. neoformans*, and CD4+ T cell hybridomas were generated by fusion of splenocytes with thymoma cells. Two immunoreactive MPs, designated MP98 and MP88, were identified on the basis of their ability to stimulate *C. neoformans*–reactive T-cell hybridomas (36, 50). Both MP98 and MP88 were purified to homogeneity, and the genes encoding the proteins were cloned and sequenced. MP88 and MP98 share structural features, including signal sequences, Ser/Thr-rich C-terminal regions (which likely serve as sites of extensive O glycosylation), and glycosylphosphatidylinositol anchor motifs (by which the protein can be attached to the membrane or cell wall). In addition, there are numerous potential sites for N-linked glycosylation. Over 50 additional *Cryptococcus* MPs were deduced by proteomic and genomic analyses (24, 52). Other studies have identified cryptococcal antigens that stimulate T-cell responses but do not have the signature S/T-rich sequences typical of MPs (7, 52, 55).

MPs are approximately 80% carbohydrate and the predominant carbohydrate is mannose (52). The extensive mannosylation of cryptococcal MP is the result of fundamental differences between how mammals and fungi glycosylate. N-linked and O-linked glycans on yeasts tend to be terminally mannosylated, whereas fully processed mammalian glycoproteins rarely have terminal mannose groups. N-linked glycans on yeast proteins can be quite long with extensive branching, while O-linked glycans tend to be linear chains of two to three mannoses (25, 52).

Mannosylation appears to be a critical component of the immunogenicity of MPs. Immature DCs are rich in MRs, including the MR CD206 and DC-SIGN (CD209). MRs avidly take up MPs by receptor-mediated endocytosis into MHC II+ compartments (57, 58). Blocking MRs or removing mannose from MPs profoundly inhibits antigen-specific CD4+ T cell responses (57, 58, 87). Nevertheless, MPs are relatively weak stimulators of DC responses in vitro and in vivo. While MRs are important for pathogen recognition, they also serve as endogenous receptors for secreted proteins and are critical for lymphocyte trafficking (19). Thus, it perhaps makes teleological sense that ligation of MRs would not elicit strong inflammatory responses. Interestingly though, stimulation of DCs with MPs plus a TLR ligand results in synergistic production of cytokines and chemokines as well as enhanced T-cell responses (19) (Fig. 3). The MR CD206 was key, as cooperative stimulation was not observed in DCs from CD206 knockout mice. As *C. neoformans* components are known to stimulate TLRs (70, 99), these data suggest that during the course of a cryptococcal infection, secreted MPs can combine with TLR ligands to stimulate potent T cell responses.

Development of a vaccine against cryptococcosis will be a formidable challenge. While many antigens have been identified that stimulate T cells, there does not appear to be a single antigen that is immunodominant in mice. There is added complexity in humans, as T-cell responses are dependent on histocompatibility antigens that are highly variable within the population. Vaccines that utilize multiple antigens are more likely to elicit protective responses, but this must be balanced against the likelihood of increased toxicity. In fact, nearly all successful vaccines that elicit protective T-cell responses in humans utilize live, attenuated organisms. This approach would not be realistic for a *Cryptococcus* vaccine because most of the target population is immunocompromised. However, an interesting proof of concept was established with the demonstration that mice challenged with live *C. neoformans* genetically engineered to express IFN-γ not only survived the infection, but were protected from a subsequent challenge with virulent *C. neoformans* (96). Mice that received the engineered strain had an influx of T lymphocytes and high levels of Th1-type and inflammatory cytokines and chemokines. Thus, a successful cryptococcal vaccine that stimulated T cell responses would appear to need the correct combination of antigens, adjuvants, and perhaps cytokines.

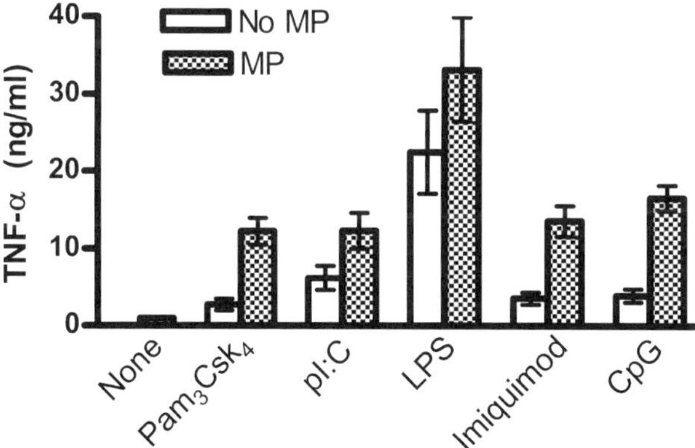

FIGURE 3 Cryptococcal MP synergizes with TLR ligands to enhance DC cytokine production. Murine bone-marrow-derived DCs were incubated for 24 h with 10 μg/ml MP; 10 μg/ml Pam_3CSK_4 (a synthetic bacterial lipoprotein), which activates TLR1/2; 10 μg/ml polyinosine-polycytidylic acid (pI:C), which activates TLR3; 1 μg/ml lipopolysaccharide (LPS), which activates TLR4; 10 μg/ml imiquimod, which activates TLR7/8; and 10 μg/ml CpG DNA, which activates TLR9. Supernatants were collected and analyzed for TNF-α by enzyme-linked immunosorbent assay. Data represent means ± standard error of four independent experiments, each of which was performed in singlicate. $P < 0.001$ comparing any TLR ligand alone with the TLR ligand plus MP by the two-tailed paired t-test. Figure adapted from Dan et al. (19).

CONCLUSIONS

DCs and T cells have been shown to play major roles in host defense against cryptococcosis in vitro and in vivo. While CD4+ T cells are of primary importance, CD8+ T cells also contribute to protection against cryptococcosis. It is presumed that the major function of T cells is to secrete cytokines and chemokines that recruit and activate phagocytes to inhibit and kill *C. neoformans*. Th1- and Th17-type cytokines are associated with protection against infection, while Th2-type responses are associated with exacerbation of disease. DCs have important roles in initiation of the immune response to *C. neoformans*. Antigen uptake can occur either via phagocytosis of whole organisms or via endocytosis/macropinocytosis of antigens secreted by the fungus. The immunodominant MPs are heavily mannosylated, which facilitates uptake by MRs. Antigens are then processed by DCs and presented to naive T cells. This induces T-cell proliferation, which, in turn, results in cytokine production and effector T-cell functions. DCs also have effector functions, as they can potently inhibit and kill phagocytosed *C. neoformans*. The challenge ahead will be to apply this knowledge to the development of vaccines to prevent cryptococcosis and immunotherapies that can be used as adjuvants to treat this often deadly mycosis.

REFERENCES

1. **Ardavin, C., G. M. del Hoyo, P. Martin, F. Anjuere, C. F. Arias, A. R. Marin, S. Ruiz, V. Parrillas, and H. Hernandez.** 2001. Origin and differentiation of dendritic cells. *Trends Immunol.* **22:**691–700.
2. **Atkinson, A. J., Jr., and J. E. Bennett.** 1968. Experience with a new skin test antigen prepared from *Cryptococcus neoformans*. *Am. Rev. Respir. Dis.* **97:**637–643.
3. **Banchereau, J., F. Briere, C. Caux, J. Davoust, S. Lebecque, Y. T. Liu, B. Pulendran, and K. Palucka.** 2000. Immunobiology of dendritic cells. *Annu. Rev. Immunol.* **18:**767–811.
4. **Barluzzi, R., A. Brozzetti, G. Mariucci, M. Tantucci, R. G. Neglia, F. Bistoni, and E. Blasi.** 2000. Establishment of protective immunity against cerebral cryptococcosis by means of an avirulent, non melanogenic *Cryptococcus neoformans* strain. *J. Neuroimmunol.* **109:**75–86.
5. **Bauman, S. K., G. B. Huffnagle, and J. W. Murphy.** 2003. Effects of tumor necrosis factor alpha on dendritic cell accumulation in lymph nodes draining the immunization site and the impact on the anticryptococcal cell-mediated immune response. *Infect. Immun.* **71:**68–74.
6. **Bauman, S. K., K. L. Nichols, and J. W. Murphy.** 2000. Dendritic cells in the induction of protective and nonprotective anticryptococcal cell-mediated immune responses. *J. Immunol.* **165:**158–167.
7. **Biondo, C., C. Beninati, D. Delfino, M. Oggioni, G. Mancuso, A. Midiri, M. Bombaci, G. Tomaselli, and G. Teti.** 2002. Identification and cloning of a cryptococcal deacetylase that produces protective immune responses. *Infect. Immun.* **70:**2383–2391.
8. **Blackstock, R., K. L. Buchanan, A. M. Adesina, and J. W. Murphy.** 1999. Differential regulation of immune responses by highly and weakly virulent *Cryptococcus neoformans* isolates. *Infect. Immun.* **67:**3601–3609.
9. **Blasi, E., R. Mazzolla, R. Barluzzi, P. Mosci, and F. Bistoni.** 1994. Anticryptococcal resistance in the mouse brain: beneficial effects of local administration of heat-inactivated yeast cells. *Infect. Immun.* **62:**3189–3196.

10. **Bozza, S., C. Montagnoli, R. Gaziano, G. Rossi, G. Nkwanyuo, S. Bellocchio, and L. Romani.** 2004. Dendritic cell-based vaccination against opportunistic fungi. *Vaccine* **22:**857–864.
11. **Buchanan, K. L., and H. A. Doyle.** 2000. Requirement for CD4+ T lymphocytes in host resistance against *Cryptococcus neoformans* in the central nervous system of immunized mice. *Infect. Immun* **68:**456–462.
12. **Buentke, E., and A. Scheynius.** 2003. Dendritic cells and fungi. *APMIS* **111:**789–796.
13. **Chen, G. H., R. A. McDonald, J. C. Wells, G. B. Huffnagle, N. W. Lukacs, and G. B. Toews.** 2005. The gamma interferon receptor is required for the protective pulmonary inflammatory response to *Cryptococcus neoformans*. *Infect. Immun.* **73:**1788–1796.
14. **Chen, G. H., D. A. McNamara, Y. Hernandez, G. B. Huffnagle, G. B. Toews, and M. A. Olszewski.** 2008. Inheritance of immune polarization patterns is linked to resistance versus susceptibility to *Cryptococcus neoformans* in a mouse model. *Infect. Immun.* **76:**2379–2391.
15. **Chuck, S. L., and M. A. Sande.** 1989. Infections with *Cryptococcus neoformans* in the acquired immunodeficiency syndrome. *N. Engl. J. Med.* **321:**794–799.
16. **Colonna, M., B. Pulendran, and A. Iwasaki.** 2006. Dendritic cells at the host-pathogen interface. *Nat. Immunol.* **7:**117–120.
17. **Corthay, A.** 2006. A three-cell model for activation of naive T helper cells. *Scand. J. Immunol.* **64:**93–96.
18. **Crowley, M., K. Inaba, and R. M. Steinman.** 1990. Dendritic cells are the principal cells in mouse spleen bearing immunogenic fragments of foreign proteins. *J. Exp. Med.* **172:**383–386.
19. **Dan, J. M., J. P. Wang, C. K. Lee, and S. M. Levitz.** 2008. Cooperative stimulation of dendritic cells by *Cryptococcus neoformans* mannoproteins and CpG oligodeoxynucleotides. *PLoS ONE* **3:**e2046.
20. **Davis, J., W. Y. Zheng, A. Glatman-Freedman, J. A. Ng, M. R. Pagcatipunan, H. Lessin, A. Casadevall, and D. L. Goldman.** 2007. Serologic evidence for regional differences in pediatric cryptococcal infection. *Pediatr. Infect. Dis. J.* **26:**549–551.
21. **Decken, K., G. Kohler, K. Palmer-Lehmann, A. Wunderlin, F. Mattner, J. Magram, M. K. Gately, and G. Alber.** 1998. Interleukin-12 is essential for a protective Th1 response in mice infected with *Cryptococcus neoformans*. *Infect. Immun.* **66:**4994–5000.
22. **d'Ostiani, C. F., G. Del Sero, A. Bacci, C. Montagnoli, A. Spreca, A. Mencacci, P. Ricciardi-Castagnoli, and L. Romani.** 2000. Dendritic cells discriminate between yeasts and hyphae of the fungus *Candida albicans*: implications for initiation of T helper cell immunity in vitro and in vivo. *J. Exp. Med.* **191:**1661–1673.
23. **Duncan, R. A., C. F. von Reyn, G. M. Alliegro, Z. Toossi, A. M. Sugar, and S. M. Levitz.** 1993. Idiopathic CD4+ T-lymphocytopenia: four patients with opportunistic infections and no evidence of HIV infection. *N. Engl. J. Med.* **328:**393–398.
24. **Eigenheer, R. A., Y. Jin Lee, E. Blumwald, B. S. Phinney, and A. Gelli.** 2007. Extracellular glycosylphosphatidylinositol-anchored mannoproteins and proteases of *Cryptococcus neoformans*. *FEMS Yeast Res.* **7:**499–510.
25. **Gemmill, T. R., and R. B. Trimble.** 1999. Overview of N- and O-linked oligosaccharide structures found in various yeast species. *Biochim. Biophys. Acta* **1426:**227–237.
26. **Goldman, D. L., J. Davis, F. Bommarito, X. Shao, and A. Casadevall.** 2006. Enhanced allergic inflammation and airway responsiveness in rats with chronic *Cryptococcus neoformans* infection: potential role for fungal pulmonary infection in the pathogenesis of asthma. *J. Infect. Dis.* **193:**1178–1186.
27. **Graybill, J. R., and R. H. Alford.** 1974. Cell-mediated immunity in cryptococcosis. *Cell. Immunol.* **14:**12–21.
28. **Guermonprez, P., J. Valladeau, L. Zitvogel, C. Thery, and S. Amigorena.** 2002. Antigen presentation and T cell stimulation by dendritic cells. *Annu. Rev. Immunol.* **20:**621–667.
29. **Harrison, T. S., and S. M. Levitz.** 1996. Role of IL-12 in peripheral blood mononuclear cell responses to fungi in persons with and without HIV infection. *J. Immunol.* **156:**4492–4497.
30. **Herring, A. C., J. Lee, R. A. McDonald, G. B. Toews, and G. B. Huffnagle.** 2002. Induction of interleukin-12 and gamma interferon requires tumor necrosis factor alpha for protective T1-cell-mediated immunity to pulmonary *Cryptococcus neoformans* infection. *Infect. Immun.* **70:**2959–2964.
31. **Hill, J. O.** 1992. CD4+ T cells cause multinucleated giant cells to form around *Cryptococcus neoformans* and confine the yeast within the primary site of infection in the respiratory tract. *J. Exp. Med.* **175:**1685–1695.
32. **Hill, J. O., and K. M. Aguirre.** 1994. CD4+ T cell-dependent acquired state of immunity that protects the brain against *Cryptococcus neoformans*. *J. Immunol.* **152:**2344–2350.
33. **Hill, J. O., and A. G. Harmsen.** 1991. Intrapulmonary growth and dissemination of an avirulent strain of *Cryptococcus neoformans* in mice depleted of CD4+ or CD8+ T-cells. *J. Exp. Med.* **173:**755–758.
34. **Hoy, J. F., D. E. Lewis, and G. G. Miller.** 1988. Functional versus phenotypic analysis of T cells in subjects seropositive for the human immunodeficiency virus: a prospective study of in vitro responses to *Cryptococcus neoformans*. *J. Infect. Dis.* **158:**1071–1078.
35. **Hoy, J. F., J. W. Murphy, and G. G. Miller.** 1989. T-cell response to soluble cryptococcal antigens after recovery from cryptococcal infection. *J. Infect. Dis.* **159:**116–119.
36. **Huang, C., S. H. Nong, M. K. Mansour, C. A. Specht, and S. M. Levitz.** 2002. Purification and characterization of a second immunoreactive mannoprotein from *Cryptococcus neoformans* that stimulates T-cell responses. *Infect. Immun.* **70:**5485–5493
37. **Huffnagle, G. B., M. F. Lipscomb, J. A. Lovchik, K. A. Hoag, and N. E. Street.** 1994. The role of CD4(+) and CD8(+) T-cells in the protective inflammatory response to a pulmonary cryptococcal infection. *J. Leukoc. Biol.* **55:**35–42.
38. **Huffnagle, G. B., G. B. Toews, M. D. Burdick, M. B. Boyd, K. S. McAllister, R. A. McDonald, S. L. Kunkel, and R. M. Strieter.** 1996. Afferent phase production of TNF-alpha is required for the development of protective T cell immunity to *Cryptococcus neoformans*. *J. Immunol.* **157:**4529–4536.
39. **Huffnagle, G. B., J. L. Yates, and M. F. Lipscomb.** 1991. Immunity to a pulmonary *Cryptococcus neoformans* infection requires both CD4+ and CD8+ T-cells. *J. Exp. Med.* **173:**793–800.
40. **Huffnagle, G. B., J. L. Yates, and M. F. Lipscomb.** 1991. T-cell-mediated immunity in the lung: a *Cryptococcus neoformans* pulmonary infection model using SCID and athymic nude-mice. *Infect. Immun.* **59:**1423–1433.
41. **Kawakami, K., X. Qifeng, M. Tohyama, M. H. Qureshi, and A. Saito.** 1996. Contribution of tumour necrosis factor-alpha (TNF-alpha) in host defence mechanism against *Cryptococcus neoformans*. *Clin. Exp. Immunol.* **106:**468–474.
42. **Kawakami, K., M. Tohyama, X. Qifeng, and A. Saito.** 1997. Expression of cytokines and inducible nitric oxide synthase mRNA in the lungs of mice infected with *Cryptococcus neoformans*: effects of interleukin-12. *Infect. Immun.* **65:**1307–1312.
43. **Kawakami, K., M. Tohyama, K. Teruya, N. Kudeken, Q. F. Xie, and A. Saito.** 1996. Contribution of interferon-gamma in protecting mice during pulmonary and disseminated

infection with *Cryptococcus neoformans. FEMS Immunol. Med. Microbiol.* **13:**123–130.

44. **Kawakami, K., M. Tohyama, Q. Xie, and A. Saito.** 1996. IL-12 protects mice against pulmonary and disseminated infection caused by *Cryptococcus neoformans. Clin. Exp. Immunol.* **104:**208–214.
45. **Kelly, R. M., J. M. Chen, L. E. Yauch, and S. M. Levitz.** 2005. Opsonic requirements for dendritic cell-mediated responses to *Cryptococcus neoformans. Infect. Immun.* **73:**592–598.
46. **Kleinschek, M. A., U. Muller, S. J. Brodie, W. Stenzel, G. Kohler, W. M. Blumenschein, R. K. Straubinger, T. McClanahan, R. A. Kastelein, and G. Alber.** 2006. IL-23 enhances the inflammatory cell response in *Cryptococcus neoformans* infection and induces a cytokine pattern distinct from IL-12. *J. Immunol.* **176:**1098–1106.
47. **Koguchi, Y., and K. Kawakami.** 2002. Cryptococcal infection and Th1-Th2 cytokine balance. *Int. Rev. Immunol.* **21:**423–438.
48. **Laurence, A., and J. J. O'Shea.** 2007. TH-17 differentiation: of mice and men. *Nat. Immunol.* **8:**903–905.
49. **Levitz, S. M.** 1991. The ecology of *Cryptococcus neoformans* and the epidemiology of cryptococcosis. *Rev. Infect. Dis.* **13:**1163–1169.
50. **Levitz, S. M., S. Nong, M. K. Mansour, C. Huang, and C. A. Specht.** 2001. Molecular characterization of a mannoprotein with homology to chitin deacetylases that stimulates T cell responses to *Cryptococcus neoformans. Proc. Natl. Acad. Sci. USA* **98:**10422–10427.
51. **Levitz, S. M., and E. A. North.** 1997. Lymphoproliferation and cytokines profiles in human peripheral blood mononuclear cells stimulated by *Cryptococcus neoformans. J. Med. Vet. Mycol.* **35:**229–236.
52. **Levitz, S. M., and C. A. Specht.** 2006. The molecular basis for the immunogenicity of *Cryptococcus neoformans* mannoproteins. *FEMS Yeast Res.* **6:**513–524.
53. **Lortholary, O., A. Fontanet, N. Memain, A. Martin, K. Sitbon, and F. Dromer.** 2005. Incidence and risk factors of immune reconstitution inflammatory syndrome complicating HIV-associated cryptococcosis in France. *AIDS* **19:**1043–1049.
54. **Lortholary, O., L. Improvisi, N. Rayhane, F. Gray, C. Fitting, J. M. Cavaillon, and F. Dromer.** 1999. Cytokine profiles of AIDS patients are similar to those of mice with disseminated *Cryptococcus neoformans* infection. *Infect. Immun.* **67:**6314–6320.
55. **Mandel, M. A., G. G. Grace, K. I. Orsborn, F. Schafer, J. W. Murphy, M. J. Orbach, and J. N. Galgiani.** 2000. The *Cryptococcus neoformans* gene DHA1 encodes an antigen that elicits a delayed-type hypersensitivity reaction in immune mice. *Infect. Immun.* **68:**6196–6201.
56. **Mangan, P. R., L. E. Harrington, D. B. O'Quinn, W. S. Helms, D. C. Bullard, C. O. Elson, R. D. Hatton, S. M. Wahl, T. R. Schoeb, and C. T. Weaver.** 2006. Transforming growth factor-[beta] induces development of the TH17 lineage. *Nature* **441:**231–234.
57. **Mansour, M. K., E. Latz, and S. M. Levitz.** 2006. *Cryptococcus neoformans* glycoantigens are captured by multiple lectin receptors and presented by dendritic cells. *J. Immunol.* **176:**3053–3061.
58. **Mansour, M. K., L. S. Schlesinger, and S. M. Levitz.** 2002. Optimal T cell responses to *Cryptococcus neoformans* mannoprotein are dependent on recognition of conjugated carbohydrates by mannose receptors. *J. Immunol.* **168:**2872–2879.
59. **Mansour, M. K., L. E. Yauch, J. B. Rottman, and S. M. Levitz.** 2004. Protective efficacy of antigenic fractions in mouse models of cryptococcosis. *Infect. Immun.* **72:**1746–1754.
60. **McFadden, D., O. Zaragoza, and A. Casadevall.** 2006. The capsular dynamics of *Cryptococcus neoformans. Trends Microbiol.* **14:**497–505.
61. **Mitchell, T. G., and J. R. Perfect.** 1995. Cryptococcosis in the era of AIDS: 100 years after the discovery of *Cryptococcus neoformans. Clin. Microbiol. Rev.* **8:**515–548.
62. **Mody, C. H., G. H. Chen, C. Jackson, J. L. Curtis, and G. B. Toews.** 1993. Depletion of murine CD8^{+} T cells in vivo decreases pulmonary clearance of a moderately virulent strain of *Cryptococcus neoformans. J. Lab. Clin. Med.* **121:**765–773.
63. **Mody, C. H., G. H. Chen, C. Jackson, J. L. Curtis, and G. B. Toews.** 1994. In vivo depletion of murine CD8 positive T cells impairs survival during infection with a highly virulent strain of *Cryptococcus neoformans. Mycopathologia* **125:**7–17.
64. **Mody, C. H., M. F. Lipscomb, N. E. Street, and G. B. Toews.** 1990. Depletion of CD4+ (L3T4+) lymphocytes in vivo impairs murine host defense to *Cryptococcus neoformans. J. Immunol.* **144:**1472–1477.
65. **Monari, C., F. Bistoni, and A. Vecchiarelli.** 2006. Glucuronoxylomannan exhibits potent immunosuppressive properties. *FEMS Yeast Res.* **6:**537–542.
66. **Muller, U., W. Stenzel, G. Kohler, T. Polte, M. Blessing, A. Mann, D. Piehler, F. Brombacher, and G. Alber.** 2008. A gene-dosage effect for interleukin-4 receptor α-chain expression has an impact on Th2-mediated allergic inflammation during bronchopulmonary mycosis. *J. Infect. Dis.* **198:**1714–1721.
67. **Muller, U., W. Stenzel, G. Kohler, C. Werner, T. Polte, G. Hansen, N. Schutze, R. K. Straubinger, M. Blessing, A. N. J. McKenzie, F. Brombacher, and G. Alber.** 2007. IL-13 induces disease-promoting type 2 cytokines, alternatively activated macrophages and allergic inflammation during pulmonary infection of mice with *Cryptococcus neoformans. J. Immunol.* **179:**5367–5377.
68. **Murphy, J. W.** 1988. Influence of cryptococcal antigens on cell-mediated immunity. *Rev. Infect. Dis.* **10:**S432–S435.
69. **Murphy, J. W., R. L. Mosley, R. Cherniak, G. H. Reyes, T. R. Kozel, and E. Reiss.** 1988. Serological, electrophoretic, and biological properties of *Cryptococcus neoformans* antigens. *Infect. Immun.* **56:**424–431.
70. **Nakamura, K., A. Miyazato, G. Xiao, M. Hatta, K. Inden, T. Aoyagi, K. Shiratori, K. Takeda, S. Akira, S. Saijo, Y. Iwakura, Y. Adachi, N. Ohno, K. Suzuki, J. Fujita, M. Kaku, and K. Kawakami.** 2008. Deoxynucleic acids from *Cryptococcus neoformans* activate myeloid dendritic cells via a TLR9-dependent pathway. *J. Immunol.* **180:**4067–4074.
71. **Orendi, J. M., H. S. Nottet, M. R. Visser, A. F. Verheul, H. Snippe, and J. Verhoef.** 1994. Enhancement of HIV-1 replication in peripheral blood mononuclear cells by *Cryptococcus neoformans* is monocyte-dependent but tumour necrosis factor-independent. *AIDS* **8:**423–429.
72. **Osterholzer, J. J., J. L. Curtis, T. Polak, T. Ames, G.-H. Chen, R. McDonald, G. B. Huffnagle, and G. B. Toews.** 2008. CCR2 mediates conventional dendritic cell recruitment and the formation of bronchovascular mononuclear cell infiltrates in the lungs of mice infected with *Cryptococcus neoformans. J. Immunol.* **181:**610–620.
73. **Perfect, J. R., and A. Casadevall.** 2002. Cryptococcosis. *Infect. Dis. Clin. N. Am.* **16:**837–874.
74. **Pettoello-Mantovani, M., A. Casadevall, T. R. Kollmann, A. Rubinstein, and H. Goldstein.** 1992. Enhancement of HIV-1 infection by the capsular polysaccharide of *Cryptococcus neoformans. Lancet* **339:**21–23.
75. **Pietrella, D., C. Corbucci, S. Perito, G. Bistoni, and A. Vecchiarelli.** 2005. Mannoproteins from *Cryptococcus neoformans* promote dendritic cell maturation and activation. *Infect. Immun.* **73:**820–827.

76. **Romani, L., C. Montagnoli, S. Bozza, K. Perruccio, A. Spreca, P. Allavena, S. Verbeek, R. A. Calderone, F. Bistoni, and P. Puccetti.** 2004. The exploitation of distinct recognition receptors in dendritic cells determines the full range of host immune relationships with *Candida albicans*. *Int. Immunol.* **16:**149–161.
77. **Salkowski, C. A., and E. Balish.** 1990. Pathogenesis of *Cryptococcus neoformans* in congenitally immunodeficient beige athymic mice. *Infect. Immun.* **58:**3300–3306.
78. **Schimpff, S. C., and J. E. Bennett.** 1975. Abnormalities in cell-mediated immunity in patients with *Cryptococcus neoformans* infection. *J. Allergy Clin. Immunol.* **55:**430–441.
79. **Schon-Hegrad, M. A., J. Oliver, P. G. McMenamin, and P. G. Holt.** 1991. Studies on the density, distribution, and surface phenotype of intraepithelial class-II major histocompatibility complex antigen (Ia)-bearing dendritic cells (DC) in the conducting airways. *J. Exp. Med.* **173:**1345–1356.
80. **Shoham, S., and S. M. Levitz.** 2005. The immune response to fungal infections. *Br. J. Haematol.* **129:**569–582.
81. **Siddiqui, A. A., A. E. Brouwer, V. Wuthiekanun, S. Jaffar, R. Shattock, D. Irving, J. Sheldon, W. Chierakul, S. Peacock, N. Day, N. J. White, and T. S. Harrison.** 2005. IFN-gamma at the site of infection determines rate of clearance of infection in cryptococcal meningitis. *J. Immunol.* **174:**1746–1750.
82. **Siegemund, S., and G. Alber.** 2008. *Cryptococcus neoformans* activates bone marrow-derived conventional dendritic cells rather than plasmacytoid dendritic cells and down-regulates macrophages. *FEMS Immunol. Med. Microbiol.* **52:**417–427.
83. **Singh, N., F. Dromer, J. R. Perfect, and O. Lortholary.** 2008. Cryptococcosis in solid organ transplant recipients: current state of the science. *Clin. Infect. Dis.* **47:**1321–1327.
84. **Singh, N., T. Gayowski, M. M. Wagener, and I. R. Marino.** 1997. Clinical spectrum of invasive cryptococcosis in liver transplant recipients receiving tacrolimus. *Clin. Transpl.* **11:**66–70.
85. **Snelgrove, R. J., L. Edwards, A. E. Williams, A. J. Rae, and T. Hussell.** 2006. In the absence of reactive oxygen species, T cells default to a Th1 phenotype and mediate protection against pulmonary *Cryptococcus neoformans* infection. *J. Immunol.* **177:**5509–5516.
86. **Sorrell, T. C.** 2001. *Cryptococcus neoformans* variety *gattii*. *Med. Mycol.* **39:**155–168.
87. **Specht, C. A., S. Nong, J. M. Dan, C. K. Lee, and S. M. Levitz.** 2007. Contribution of glycosylation to T cell responses stimulated by recombinant *Cryptococcus neoformans* mannoprotein. *J. Infect. Dis.* **196:**796–800.
88. **Steinman, R. M., and Z. A. Cohn.** 1973. Identification of a novel cell type in peripheral lymphoid organs of mice. I. Morphology, quantitation, tissue distribution. *J. Exp. Med.* **137:**1142–1162.
89. **Syme, R. M., J. C. L. Spurrell, E. K. Amankwah, F. H. Y. Green, and C. H. Mody.** 2002. Primary dendritic cells phagocytose *Cryptococcus neoformans* via mannose receptors and Fc gamma receptor II for presentation to T lymphocytes. *Infect. Immun.* **70:**5972–5981.
90. **Tan, D. B., Y. K. Yong, H. Y. Tan, A. Kamarulzaman, L. H. Tan, A. Lim, I. James, M. French, and P. Price.** 2008. Immunological profiles of immune restoration disease presenting as mycobacterial lymphadenitis and cryptococcal meningitis. *HIV Med.* **9:**307–316.
91. **Traynor, T. R., W. A. Kuziel, G. B. Toews, and G. B. Huffnagle.** 2000. CCR2 expression determines T1 versus T2 polarization during pulmonary *Cryptococcus neoformans* infection. *J. Immunol.* **164:**2021–2027.
92. **Trombetta, E. S., and I. Mellman.** 2005. Cell biology of antigen processing in vitro and in vivo. *Annu. Rev. Immunol.* **23:**975–1028.
93. **Uicker, W. C., J. P. McCracken, and K. L. Buchanan.** 2006. Role of $CD4^+$ T cells in a protective immune response against *Cryptococcus neoformans* in the central nervous system. *Med. Mycol.* **44:**1–11.
94. **Upham, J. W.** 2003. The role of dendritic cells in immune regulation and allergic airway inflammation. *Respirology* **8:**140–148.
95. **Weaver, C. T., R. D. Hatton, P. R. Mangan, and L. E. Harrington.** 2007. IL-17 family cytokines and the expanding diversity of effector T cell lineages. *Annu. Rev. Immunol.* **25:**821–852.
96. **Wormley, F. L., Jr., J. R. Perfect, C. Steele, and G. M. Cox.** 2007. Protection against cryptococcosis using a murine interferon-gamma producing *Cryptococcus neoformans* strain. *Infect. Immun.* **75:**1453–1462.
97. **Wozniak, K. L., and S. M. Levitz.** 2008. *Cryptococcus neoformans* enters the endolysosomal pathway of dendritic cells and is killed by lysosomal components. *Infect. Immun.* **76:**4764–4771.
98. **Wozniak, K. L., J. M. Vyas, and S. M. Levitz.** 2006. In vivo role of dendritic cells in a murine model of pulmonary cryptococcosis. *Infect. Immun.* **74:**3817–3824.
99. **Yauch, L. E., M. K. Mansour, S. Shoham, J. B. Rottman, and S. M. Levitz.** 2004. Involvement of CD14, toll-like receptors 2 and 4, and MyD88 in the host response to the fungal pathogen *Cryptococcus neoformans* in vivo. *Infect. Immun.* **72:**5373–5382.
100. **Zhou, Q., R. A. Gault, T. R. Kozel, and W. J. Murphy.** 2007. Protection from direct cerebral cryptococcus infection by interferon-{gamma}-dependent activation of microglial cells. *J. Immunol.* **178:**5753–5761.

Cryptococcus: From Human Pathogen to Model Yeast
Edited by J. Heitman et al.
©2011 ASM Press, Washington, DC

28

Acquired Humoral Immunity to *Cryptococcus neoformans*

LIISE-ANNE PIROFSKI AND ARTURO CASADEVALL

The importance of humoral immunity for resistance to natural infection with *Cryptococcus neoformans* remains uncertain, but a wealth of new knowledge gained since cryptococcal meningitis emerged as a complication of HIV-associated immunodeficiency in the 1980s shows that specific antibody (Ab) can be effective in host defense. Research initiated in the late 1980s has been the driving force behind a major expansion of our knowledge about Ab immunity (AI) to cryptococcal disease (cryptococcosis) and a sea change in our understanding of mechanisms of AI to *C. neoformans* and other microbes for which AI is not thought to be the predominant protective host defense mechanism. This chapter will highlight this information, placing emphasis on how new knowledge about AI to *C. neoformans* has advanced our understanding of pathogenesis, virulence, and mechanisms of Ab action. We begin by acknowledging the many investigators and their trainees who have contributed to our understanding of AI to *C. neoformans* as we apologize in advance for any information that might be omitted in this chapter, noting prior reviews on the subject (14, 17–20, 101).

AI AGAINST *C. NEOFORMANS*: EARLY INSIGHTS

Understanding the importance of AI in resistance to infectious diseases in general and microbial agents in particular has been informed by the use of available tools. Historically, the available tools were observational and experimental methods that made it possible to establish the importance of AI in resistance to infectious diseases by three main approaches: (i) by seeking an association between the presence or acquisition of Ab to the causative microbe and resistance to the disease, (ii) by demonstrating that individuals with Ab or B-cell disorders were susceptible to the disease, and (iii) by demonstrating that transfer of immune sera to a naive host was associated with protection against the disease or death due to challenge with the causative microbe (reviewed in reference 101). Use of the latter method in the late 19th and early 20th centuries established proof of principle that a specific Ab can protect against the host damage and/or death caused by a microbial agent. This landmark finding led to the awarding of the first Nobel Prize to Emil von Behring in 1901 and fueled the development of serum therapy (23) and, in parallel, the development of vaccines for active immunization. Serum therapy, which was the first form of antimicrobial therapy for infectious diseases, consisted of a specific Ab to the causative agent of the disease. This milestone in prevention of infectious diseases enabled previously unimaginable victories against tetanus, diphtheria, and ultimately viral diseases among populations with access to vaccines.

However, the advances against bacterial and viral causes of infectious diseases did not extend to fungi or the medical mycoses. This disparity became evident in last two decades of the 20th century when the prevalence of fungal diseases rose sharply in parallel with the almost simultaneous emergence of the HIV/AIDS pandemic and the advent of medical interventions and therapies that resulted in immune impairment. Efforts to develop Ab-based therapies against fungi were thwarted by an inability to demonstrate a clear-cut or consistent role for AI using previously established tools. Nonetheless, data suggesting Abs could alter the course of cryptococcal disease were obtained in a small clinical study in which adjunctive immune rabbit serum reduced the fungal burden in two of three patients with cryptococcal meningitis (51). However, the promise of this approach was counterbalanced by a body of negative data that found no role for AI in resistance to *C. neoformans*. This assessment stood until the application of new tools to the question of whether and how AI affects the course of infection with *C. neoformans*.

NATURAL AI TO *C. NEOFORMANS*

Historically, studies with polyclonal sera yielded inconsistent information on the effect of AI on the course of infection with *C. neoformans*. Similarly, studies of the natural Ab response failed to demonstrate a definite role for AI in resistance to cryptococcal disease. Nonetheless, serological surveys of susceptible and resistant populations produced important discoveries that advanced our understanding of why

Liise-anne Pirofski, Division of Infectious Diseases, Albert Einstein College of Medicine, Bronx, NY 10461. **Arturo Casadevall,** Department of Microbiology and Immunology, Albert Einstein College of Medicine, Bronx, NY 10461.

certain individuals develop cryptococcal disease, whereas others do not. In the 1990s, studies of the human serum Ab response to the C. *neoformans* polysaccharide glucuronoxylomannan (GXM) revealed three important features: (i) immunoglobulin M (IgM) and IgG to GXM were ubiquitous in human sera, being found in adults and children with and without HIV infection (3, 35, 44, 56, 102, 120); (ii) the human Ab response to GXM was restricted with respect to idiotype and isotype expression (33, 44); and (iii) individuals with an increased risk for cryptococcal disease, such as those with HIV infection and/or solid organ transplant recipients, had lower levels of IgM to GXM than resistant individuals, i.e., those who do not have a higher risk of disease (33, 44, 56, 57, 120). The observation that Ab to GXM is found in sera from adults and children from North America was complemented by the discovery that resistant and susceptible adults and children also have serum Ab to cryptococcal proteins (24, 49). The discovery that normal children have serum Ab to C. *neoformans* by the age of 2 years suggested that, as for other encapsulated microbes that infect the respiratory tract, infection with C. *neoformans* also occurs in childhood (49). This discovery also provided further evidence that human infection with C. *neoformans* results in a state of latency (46, 119).

The human serum IgG response to GXM is highly restricted to the IgG2 subtype (33, 56, 146). IgG2 is the predominant IgG subtype elicited by T-independent type 2 antigens (Ags), which include bacterial capsular polysaccharides (100), and GXM (122). Human Ab to GXM is also restricted with respect to variable (V) region gene use, expressing idiotypes that predominate in the Ab response to other capsular polysaccharides. These idiotypes are indicative of the use of V genes from the heavy chain (V_H) gene family V_{H3} (44, 100). Levels of IgG2 and V_{H3} are each reduced in the course of HIV infection, and differences between the specific IgG2 and V_{H3} responses of HIV-infected and HIV-uninfected individuals have been observed (11, 33, 44). In a retrospective study that evaluated banked sera prospectively, V_{H3} levels were lower among CD4 T-cell matched HIV-infected subjects who developed cryptococcal disease than those who did not (44) and among HIV-infected Ugandan patients without a history of pneumonia who developed cryptococcal disease (120). These studies suggested that cryptococcal disease is associated with a state of relative (V_{H3}) hypogammaglobulinemia. Hypogammaglobulinemia was associated with an increased risk for cryptococcal disease (54, 99, 138), but neither GXM-specific IgG2 nor V_{H3} levels were linked to immunity to HIV-associated cryptococcal disease.

Levels of IgG to GXM were higher in HIV-infected than HIV-uninfected individuals in studies using a direct GXM capture enzyme-linked immunosorbent assay (ELISA) (33, 44, 120), but lower in studies using different ELISA methods (35, 56). In contrast, IgM to GXM levels were higher in individuals who are resistant to cryptococcal disease than those who are susceptible, including those with HIV infection and solid organ transplant recipients (33, 56, 57, 120). However, a clear-cut association or causal relationship between lower levels of IgM to GXM and cryptococcal disease has not been established. The serum IgM response to polysaccharide Ags is derived from B cells that are part of the naturally occurring Ig repertoire. In humans, approximately 50% of this repertoire is composed of B cells that lack IgD and express IgM and the cell surface marker CD27 (IgD-IgM+CD27+) or memory IgM (126). Memory IgM was shown to have a role in resistance to pneumococcus (63) and the immune response to pneumococcal vaccine (116). Of relevance to the pathogenesis of cryptococcal disease, expression of memory IgM is reduced in HIV infection, X-linked hyper IgM, and common variable immunodeficiency, each of which is a condition that has been associated with an increased risk for cryptococcal disease (reviewed in reference 121). Recognition of this association has focused attention on the possible role of B cells in resistance to cryptococcal disease.

A recent study demonstrated that memory IgM B-cell expression (memory IgM) was lower in HIV-infected subjects with a history of, or who later developed, cryptococcal disease, than those who had no history of and/or did not subsequently develop cryptococcal disease (121). This finding suggests that memory IgM could enhance resistance to cryptococcal disease and holds promise as a surrogate marker for susceptibility to disease, although more work is needed to strengthen the association and establish a causal relationship between memory IgM expression and resistance to cryptococcal disease. Nonetheless, since memory IgM produces cytokines/chemokines that are important for host defense against C. *neoformans* (7, 48) and could be the source of Ab to GXM that promotes opsonic uptake of C. *neoformans* (147), memory IgM could enhance resistance to cryptococcal disease by a variety of mechanisms.

The role of B cells in resistance to cryptococcal disease in mouse models is complex. Mice rendered B-cell deficient by neonatal Ab depletion were not more susceptible to C. *neoformans* (78). However, XID (70) and B-cell-deficient (μMT knockout) mice were more susceptible to lethal challenge with C. *neoformans* (108). In addition, B cells were required for resistance to C. *neoformans* in reconstituted SCID mice (4) and for the efficacy of monoclonal Abs (MAbs) to GXM against lethal challenge with C. *neoformans* (108). In concert, the preceding observations and evidence that B1-derived mouse phagocytes phagocytose C. *neoformans* and have anticryptococcal activity (48) underscore the currently prevailing concept that AI to cryptococcal disease enhances cellular immune responses to cryptococcal disease (see below) (reviewed in references 17, 19, and 101). This novel concept emerged from the observation that Ab to GXM collaborates with underlying host cellular effector mechanisms to promote direct and indirect antifungal effects, which result in a beneficial outcome. The studies that led to the formulation of this paradigm span nearly two decades and have their roots in the use of new tools to study AI to cryptococcal disease.

ACQUIRED AI AGAINST *C. NEOFORMANS*

Studies with immune sera from experimental animals failed to reveal a consistent or predictable role for AI in resistance to cryptococcal disease (12, 52). Ultimately, new tools that could dissect individual components of the immune response were required to determine the ability of Ab to C. *neoformans* to mediate processes that enhance host defense against cryptococcal disease. One impediment in studies with polyclonal sera was that the development of effective immune sera was undermined by the poor immunogenicity of GXM, which is a T-independent type 2 Ag (122). The success of efforts to enhance the immunogenicity of GXM were inconsistent until the development of immunogens based on conjugate vaccines (34). The rationale for the development of GXM-protein conjugates was based on the success of other capsular polysaccharide-protein conjugate vaccines and their ability to elicit T-dependent Ab responses (110). Use of GXM-red blood cell complexes, the GXM of serotype A or D C. *neoformans*, or peptide mimotopes of GXM conjugated to tetanus toxoid (TT), diphtheria toxoid (DT), or keyhole limpet hemocyanin to induce Ab to GXM (8, 16, 34, 37, 39, 43, 68, 69, 74, 80) led to the production of a robust library of MAbs. Studies of the structure, biological activity in vitro, and efficacy of GXM-induced MAbs in vivo fueled a rapid expansion of knowledge about AI to *C. neoformans*. The horizons that were revealed by these studies informed countless studies of AI to other microbes, which in turn created an unprecedented body of new knowledge about mechanisms of AI to infectious diseases.

The driving force that led to the advancement of existing knowledge about AI to *C. neoformans* was the use of the hybridoma technology. This approach enabled the production of MAbs to GXM. MAbs recognize a single Ag determinant and are available in unlimited supply. The development of MAbs to GXM made it possible to study the effect of individual components of the Ab response on the course of experimental cryptococcal disease. Use of MAbs to GXM in lethal *C. neoformans* infection models in mice revealed that certain MAbs protected against death, whereas others had no effect on or enhanced the likelihood of death (38, 68, 81–84). The single most important discovery stemming from the use of MAbs in survival studies was that GXM is a complex moiety that expresses unique determinants that elicit protective, nonprotective, and disease-enhancing Ab(s). This breakthrough led to the realization that polyclonal sera are a mixture of different amounts of Abs with different specificities and different abilities to affect the course of cryptococcal disease, which in aggregate have an unpredictable effect on cryptococcal disease. Hence, the previously observed lack of and/or inconsistent efficacy of polyclonal sera against experimental cryptococcal disease was most likely in part due to insufficient protective Abs. On the other hand, the ability of certain MAbs to protect normal, but not immunodeficient, mice and vice versa (9, 37, 42, 107, 108, 141) (see below) revealed that Ab efficacy against *C. neoformans* is inextricably linked to and a function of the immune status of the host. The discovery of protective and nonprotective MAbs to GXM fueled studies designed to identify the characteristics of protective Abs, which in turn produced far-reaching, paradigm-shifting studies of mechanisms of Ab action and host–*C. neoformans* interaction.

STRUCTURE-FUNCTION RELATIONSHIPS OF Abs TO GXM

MAbs to GXM are highly restricted with respect to variable and constant (C)-region gene use (101). As mentioned above, T-independent type 2 Ags such as GXM induce restricted Ab responses. Mouse and human MAbs to GXM use restricted V- and C-region genes that are considered to be structural homologs. V-region gene use is largely limited to the use of a small number of V genes, predominantly members of the V_{H3} family in human Abs and the 7183 gene family in mouse Abs (reviewed in reference 101). These human and mouse gene families are members of the V-region clan III (59), which has been proposed to consist of genes that encode Ig receptors that have the capacity to bind similar structural determinants. Hence, clan III Abs could constitute a carbohydrate-reactive repertoire (100). Mouse MAbs to GXM have been further grouped into six groups, based on their V gene use (15). These groups feature shared molecular structure and serotype reactivity. Most GXM-elicited mouse MAbs are classified as class II Abs; these MAbs use V_{H7183}, $V_{\kappa5.1}$, J_{H2}, and $V_{\kappa1}$ and have an 11-amino acid CDR3 that confers a defined idiotype. Class II MAbs have been isolated from different mouse strains after immunization with GXM-protein conjugates, although it is composed of nonprotective, in addition to protective, MAbs (82, 95, 96). MAbs from five of the six groups have been shown to mediate protection against experimental cryptococcal disease in mice (see reference 15).

GXM specificity is a critical determinant of MAb efficacy. For mouse MAbs, heavy-chain V-region mutations in CDR1 and CDR2 discriminate protective and nonprotective IgM to GXM (89, 90). Mouse and human MAbs elicited by GXM-TT are somatically mutated (80, 102); however, human MAbs to GXM that were produced from GXM-DT-immunized human Ig transgenic mice use entirely germ line genes (68). Hence, the human germ line Ig repertoire has the capacity to produce protective MAbs by combinatorial mechanisms. In support of this, protective human IgM MAbs share a motif that includes germ line CDR1 and CDR2 residues that are expressed by 6 of 22 V_{H3} genes, including 4 of the 5 genes used by human MAbs to GXM (68). Human and mouse MAbs to GXM each use a very limited repertoire of light-chain V-region genes (15, 68, 80). To date, V_L gene use has not been correlated with MAb efficacy against cryptococcal disease, although mouse MAbs elicited by GXM-TT use a canonical V_L ($V_{\kappa5.1}$), and human MAbs elicited by GXM-DT use the same V_L (68, 80). Taken together, available information on the molecular response to GXM suggests that the ability to bind to any determinant on GXM (group reactivity) is conferred by V_L structure, whereas the ability to bind to a particular GXM determinant (fine specificity) is conferred by V_H- and/or C-region structure.

The C-region response to GXM is dominated by the use of IgG2 in human Abs and IgG3 in mouse Abs (101). There is a strong association between Ab responses to capsular polysaccharides and use of clan III (59) and/or class II genes (15), suggesting that the pairing of certain V_H genes with certain IgG subtypes results in an Ab structure that favors polysaccharide binding. A host benefit of Ab restriction has not been established, but it has been proposed that the propensity of human IgG2 to dimerize may promote preferential binding to polysaccharide Ags (140). Consistent with this hypothesis, IgG2 in normal sera from GXM-TT vaccine recipients promoted effector cell phagocytosis of *C. neoformans* (146) and human Ig transgenic mice expressing human IgG2, but not IgG4-produced Ab responses to a GXM mimotope that prolonged survival after lethal challenge with *C. neoformans* (69). Human serum IgG4 is often monovalent (1, 2). Possible advantages notwithstanding, there is ample evidence that restricted Ab responses could be detrimental in host defense. For example, mouse IgG3 to GXM is not protective against cryptococcal disease in normal mice (82, 142). In addition, human V_{H3} and IgG2 are depleted and dysregulated in the setting of HIV infection (reviewed in references 100 and 101). Hence, HIV-infected individuals could have a "hole" in their GXM reactive Ab repertoire that translates into an inability to mount a beneficial Ab response to GXM (100). There are serological data to support this hypothesis; sera from HIV-infected subjects that subsequently developed cryptococcal disease had lower levels of V_{H3} Abs than sera from those who did not develop disease (33, 44).

A major finding to emerge from structure-function studies with MAbs to GXM is that the heavy-chain C region is an important determinant of the specificity of IgG to GXM and, in turn, its efficacy against cryptococcal disease. For example, subclass switching an IgG3 to a V-region identical IgG1 converted a nonprotective IgG3 to a protective Ab (142). Although the initial interpretation of this result attributed the differences in IgG1 and IgG3 protective efficacy to C-region effects, recent observations suggest a more complex picture. Both the protective IgG1 and the nonprotective IgG3 were opsonic, implying that differences in efficacy were not a result of differences in the ability to promote *C. neoformans* ingestion by phagocytes. Instead, this phenomenon is most likely to involve a C-region-mediated alteration in MAb specificity in combination with a difference in the receptors engaged by IgG1 and IgG3.

Another layer of complexity in the phenomenon of C-region influences on specificity and Ab efficacy is that the efficacy of IgG1 and IgG3 differ depending on the mouse strain; IgG1 was protective in some strains and nonprotective in others (106). Hence, the functional efficacy of a MAb to GXM cannot be defined solely on Ig structural parameters but is also dependent on the immune status and genetic background of the host. Nonetheless, the C region was found to alter the specificity of Ab to a carbohydrate Ag expressed by group A streptococci (27, 28). Studies with switch variants of mouse MAbs to GXM expressing either mouse or human C regions revealed that the C region has a major independent influence on Ab specificity, and thermodynamic studies established proof of the principle that the C region influences the physical nature of MAb-GXM interaction, which in turn affects GXM specificity (76, 129–131). This discovery has far-reaching implications for the prospect of harnessing AI to *C. neoformans* for Ab-based therapeutics and vaccine development.

The influence of the MAb-GXM interaction on specificity and, ultimately, MAb efficacy was also demonstrated by the use of fluorescence and differential interference contrast microscopy to visualize MAb–*C. neoformans* interactions. These studies revealed distinct binding patterns by which protective and nonprotective MAbs and MAbs expressing different isotypes bound to the cryptococcal capsule (76, 90, 91). Annular or rim binding patterns were observed for MAbs, including V-region identical switch variants that protected against serotype D cryptococcal disease in mice, whereas punctate or puffy binding patterns were observed for nonprotective MAbs. However, MAbs that bind serotype D with a punctate pattern can bind serotype A with an annular pattern, and such MAbs are protective against serotype A (see references 18 and 96). Hence, Ab-GXM binding is highly complex, though annular binding appears to predict Ab efficacy. In concert with its influence on specificity, MAb binding to the *C. neoformans* capsule has additional effects, such as mediating changes in cellular charge (zeta potential), complement component 3 (C3) deposition, biofilm formation, and GXM secretion/release (see reference 18). Similar functions have been described for protective MAbs to *C. neoformans* melanin, which appear to be directly fungistatic against *C. neoformans* (92, 94, 112). Taken together, structure-function relationships for MAbs to GXM reflect V- and C-region collaboration that orchestrates binding to distinct determinants that translates into physical changes in capsular structure, which result in interactions with host mediators/receptors that can either confer protection against, accelerate lethality, or have no effect on the course of experimental cryptococcal disease.

BIOLOGICAL ACTIVITY AND FUNCTIONS OF Ab TO GXM

The recognition that GXM elicits protective and nonprotective Abs generated intense interest in how these outcomes are mediated. At the time this question was originally posed, hypotheses about how AI mediates host protection against a microbe were formulated based on the prevailing concept that Ab action was the result of one of the following mechanisms (reviewed in references 14, 17, and 19): (i) neutralization, (ii) complement fixation, (iii) enhancement of phagocytosis, and (iv) Ab-dependent cellular cytotoxicity. With the exception of neutralization, these mechanisms are distinguished by the way in which they depend on host factors; e.g., complement fixation requires an intact complement activation pathway(s), whereas phagocytosis and Ab-dependent cellular cytotoxicity require host phagocytes and effector cells, respectively. Ab action against *C. neoformans* does not include neutralization in the classical sense of viral or toxin neutralization. However, MAb-mediated changes in physical characteristics of the capsule (67, 90, 91, 125), reduction in GXM release (73), and GXM clearance (50, 53, 65) profoundly influence *C. neoformans* interaction with host receptors, the immune response, and levels of GXM. Since these functions can mitigate deleterious effects of GXM on the immune system, they can be viewed as a type of neutralization. Although they have not been a focus of studies of AI to cryptococcal disease, NK cells can exert antifungal activity (55). IgG1 to GXM failed to protect SCID mice (88, 141), but this could have been attributable to the absence of either T or B cells, both of which have been shown to be required for AI to cryptococcal disease (108, 141). Hence, it is likely that each classical mechanism of Ab action can mediate protection against cryptococcal disease. However, in addition to classical mechanisms, studies to unravel AI to *C. neoformans* revealed new mechanisms of Ab action (14).

Mechanisms of AI to *C. neoformans* can be divided into those that require components of cellular immunity to translate into a host benefit and those that do not (Table 1). However, it should be noted that this distinction is arbitrary, because ultimately, the host benefit of Ab action probably stems from enhancement of host immune function through effects that involve Fc receptors, complement activation, and the clearance of immunosuppressive polysaccharide (GXM).

Mechanisms of AI to *C. neoformans* That Depend on Cellular Immunity

The predominant mechanism of AI to *C. neoformans* that depends on cellular immunity is phagocytosis. Natural human IgG promotes phagocytosis of unencapsulated (62, 75), but not encapsulated (139), *C. neoformans*. The latter is probably due to insufficient specific opsonic Ab, because GXM-TT-elicited human immune sera promoted phagocytosis of *C. neoformans* (146). Mouse MAbs elicited by GXM-TT that are protective in vivo promote effector cell phagocytosis of *C. neoformans* in vitro, but nonprotective MAbs can also promote phagocytosis (90, 143). Macrophage phagocytosis of *C. neoformans* can occur in vivo in the absence of specific opsonins (41), but this does not translate into protection against cryptococcal disease in mice. Ab-dependent phagocytosis of *C. neoformans* in vitro is a strong correlate of protection against experimental cryptococcal disease in mice, with the caveat that the immune status of the host is a critical determinant of Ab efficacy (see below). The opsonic efficacy of mouse IgG induced by GXM immunization is a function of IgG subtype, whereby (for MAbs to GXM) IgG2a>IgG1>IgG1b (87, 114). For IgM MAbs, opsonic efficacy in vitro, in the absence of complement, is a function of specificity (25). Opsonic MAbs also enhance effector cell function and secretion/production of proinflammatory mediators and cytokines in vitro (77, 79, 134, 136, 137).

The major effect of MAb to GXM on effector cell function is achieved by enhancing expression of costimulatory

TABLE 1 Mechanisms of AI to cryptococcosis

Ab function	Mechanism of Ab action (reference[s])
Dependent on cellular immunity	Opsonization/phagocytosis[a,b,c] (18, 86, 87, 114, 146, 147)
	Ab-dependent cellular cytotoxicity[c] (88)
	Immunomodulation
	Induction of T-cell stimulation; cellular (monocyte) coreceptor expression[d] (135–137)
	Enhancement of Th1-type milieu[d] (133, 134)
	Effect on proinflammatory mediator release (increase or decrease)[d] (109, 118, 124)
	Requirement for B cells[d] (108)
	Reduced potential for damage due to inflammatory response[d] (107, 108, 124)
Independent of cellular immunity	Complement activation; C3 deposition on capsule[b,d,e] (47, 60, 61, 103, 144)
	Fungistasis[a] (94, 112)
	Enhancement of macrophage uptake via CRs[a,c] (91, 125, 144)
	Inhibition of GXM release during *C. neoformans* growth[a] (73)
	Inhibition of biofilm formation[a] (71)
	Clearance of GXM[a,d] (50, 53, 65)
Deleterious effects	Enhancement of lethal infection[b,c] (82, 97)
	Acute lethal toxicity[b,c,d] (66, 113)
	Prozone-like phenomenon[d] (123, 124)

[a] Direct Ab effect.
[b] Function determined by Ab isotype.
[c] Function determined by Ab specificity.
[d] Indirect Ab effect.
[e] Function determined by Ab isotype (IgM) and Ab specificity (IgG).

molecules, cytokine receptors, and cytokines that polarize the cellular response toward a Th1-like response (see references 134 and 136). Hence, in addition to promoting uptake of *C. neoformans* by phagocytes, MAbs to GXM stimulate effector cells to create an immune milieu that can augment antifungal activity and overcome the immunosuppressive and deleterious effects of GXM. Nonetheless, it should be noted that although Ab-dependent phagocytosis of *C. neoformans* can result in a reduction in fungal growth in vitro (86, 87, 147), Ab can also promote uptake of *C. neoformans* that results in intracellular replication, death of the host cell, and a unique extrusion event that leaves the host cell intact (5, 6, 41, 132). Hence, the role that Ab-mediated phagocytosis plays in inhibiting growth of *C. neoformans* remains to be determined. Since AI was not associated with an early reduction in the fungal burden in mice that were protected from death in a pulmonary model of infection (40), the beneficial effect of Ab is more likely to stem from the ability of Ab to enhance other aspects of host defense against *C. neoformans*.

There are ample data that link the in vitro opsonic activity of mouse MAbs to protection against cryptococcal disease in normal mice. However, for mouse-human chimeric MAbs to GXM expressing human IgG subtypes, opsonic activity did not predict Ab efficacy: IgG1 and IgG3 failed to protect BALB/c mice, whereas nonopsonic IgG2 and IgG4 were protective (10). A mechanism to explain the lack of efficacy of human IgG1/IgG3 against cryptococcal disease has not been established, but since a mouse-human chimeric IgG1 to GXM was protective in C5 (immune)-deficient A/J mice (145), the lack of efficacy of IgG1/IgG3 in immunologically normal mice could be associated with an overly exuberant inflammatory response. For mouse IgG1 to GXM, in vitro activity is not a sufficient correlate of Ab efficacy in the setting of immune deficiency. T cells, cytokines/other mediators, and/or cellular receptors have a profound influence on the ability of mouse IgG to GXM to mediate protection against experimental cryptococcal disease in mice (see references 14, 17, and 18). For example, an IgG1 MAb to GXM required $CD4^+$ T cells, interferon-γ, Th1 and Th2 cytokines (interleukin-12 [IL-12], IL-6, and IL-4), inducible nitric oxide synthase, and Fc receptors (143) to mediate protection against systemic and/or pulmonary challenge with *C. neoformans* (9, 107, 141). In contrast, an IgG3 that enhanced cryptococcal disease in normal and CD4-deficient mice was protective in CD8-deficient mice (141).

While these findings support the view that AI to cryptococcal disease depends on the presence or Ab-mediated induction of an inflammatory milieu that promotes phagocytosis and fungal killing, enhancement of the inflammatory response is not always associated with Ab efficacy. For example, Ab-mediated protection against pulmonary challenge was associated with either increased or decreased levels of interferon-γ, depending on the (genetic background of the) mouse strain (42, 107). Immune complex formation after administration of IgG1 to GXM to mice with existing *C. neoformans* infection precipitated massive macrophage inflammatory mediator release that was ameliorated by reducing the rate of immune complex formation and/or inhibiting (platelet activating) receptor-mediated chemokine release (66, 113). Hence, Ab efficacy against cryptococcal disease can be associated with a reduction or enhancement of the inflammatory response, depending on host immunity and the presence of immune mediators/receptors. Taken together, studies of mechanisms of AI that depend on cellular immunity reveal that the same MAb can be protective in one immune milieu but nonprotective or deleterious in another. These observations led to the paradigm-shifting discovery that AI to cryptococcal disease is a function of the immune status of the host, rather than an independent, invariant, or intrinsic feature of the Ab itself.

Mechanisms of AI to *C. neoformans* That Do Not Depend on Cellular Immunity

One layer of complexity for AI to *C. neoformans* that has linked Ab efficacy to cellular immunity is the immune status of the host. Another layer of complexity, which is independent of cellular immunity, is the dependence of AI on deposition of C3 and immune complex formation. Mouse IgG to GXM does not require C3 to mediate protection (115); however, C3 was required for the efficacy of human (45) and mouse IgM (115). There are important species differences in the location of C3 deposition on the *C. neoformans* capsule by nonimmune sera from mice, rats, and humans that correlate to the relative susceptibility of each species to cryptococcal disease (47, 103). Human sera deposit C3 close to the capsule edge, whereas mouse sera deposit C3 distal to the capsular surface (47), locations that are more likely to facilitate or mitigate against binding to C receptors (CRs), respectively. Protective mouse MAbs to GXM alter the amount of and manner in which C3 is deposited on the *C. neoformans* capsule in an isotype-dependent fashion; IgM inhibits C3 deposition by the alternative complement pathway, and protective IgG1 enhances C3 deposition by the classical complement pathway (67). The mechanism by which IgM-mediated C activation/C3 deposition enhances AI to cryptococcal disease has not been fully revealed.

The phenomenon of Ab-dependent C-mediated uptake of *C. neoformans* is well recognized (144, 147). However, Ab-dependent, C-independent uptake of *C. neoformans* has also been described (91, 125). Using different Ab reagents and *C. neoformans* strains, two groups described the ability of MAb $F(ab)_2$ and IgM to GXM to promote C-independent CR3/CR4 (CD18/CD11b/CD18/CD11c)-mediated uptake of *C. neoformans* (91, 125). The same process was found to enhance C-mediated MAb-dependent phagocytosis of *C. neoformans* (144). Uptake was shown to be mediated by Ab binding to GXM, which resulted in a change in capsular structure that enabled *C. neoformans* binding to macrophages CR3/CR4 via GXM and triggered uptake (125). Hence, the phenomenon of Ab-mediated C-independent enhancement of macrophage uptake of *C. neoformans* is a newly described, indirect mechanism of Ab action that stems from a direct Ab function, namely, Ab-induced changes in the structure of the *C. neoformans* capsule. Ab-mediated C-independent uptake by CRs establishes that a product of the adaptive immune system can promote phagocytosis through a receptor of the innate immune system.

Another newly described Ab function that is independent of cellular immunity is the ability of protective (IgM and IgG) MAb to GXM to inhibit GXM release during *C. neoformans* growth in culture (73). This direct Ab effect prevents biofilm formation by *C. neoformans* and serves as a surrogate marker for AI to cryptococcal disease in mice (71). At present, the ability of specific Abs to inhibit biofilm formation by other microbes is unknown. Nonetheless, the potential importance of the ability of AI to reduce microbial biofilm formation in vivo cannot be overstated and exemplifies the way in which studies of AI to *C. neoformans* can inform new directions in research in microbial pathogenesis and antimicrobial therapy.

One caveat to the ability of protective MAb to GXM to inhibit *C. neoformans* biofilm formation is that the same MAb that can enhance the activity of antifungal drugs against planktonic *C. neoformans* can antagonize the effect of these drugs against established *C. neoformans* biofilms (72). Protective MAbs to GXM synergize with antifungal drugs that are also able to induce alterations in *C. neoformans* charge and capsular structure, such as amphotericin B and fluconazole (36, 85, 93). The inability of MAb to GXM to enhance the activity of these antifungal drugs against *C. neoformans* biofilms underscores and reinforces the concept that Ab-induced change in capsular structure is a, if not the central, mechanism of MAb to GXM–mediated AI to cryptococcal disease. This mechanism could be precluded by the physical nature of *C. neoformans* biofilms. Another recent observation that highlights the complex relationship between Abs and *C. neoformans* is the fact that Ab-mediated immobilization of *C. neoformans* on polystyrene can itself promote biofilm formation (111). Although this phenomenon is unlikely to have a role in vivo, it illustrates that Ab binding can mediate growth-related changes.

A long-recognized mechanism that is central to AI to cryptococcal disease is Ab-mediated clearance of GXM (50, 53, 65). The ability of Ab to GXM to induce clearance of GXM from the circulation does not require elements of cellular immunity but can require Fc receptors and/or complement, depending on the Ab isotype (65). Fc-receptor-mediated clearance reflects interactions between immune complexes containing GXM and their ability to bind Fcγ receptors (143). An indirect mechanism by which protective MAbs to GXM affect the inflammatory response is through binding to effector cells, including Fc receptors (134). Stimulation of activating Fc-receptor-mediated release of inflammatory mediators by MAbs to GXM, which trigger phagocyte/effector cell antifungal effects, is a major mechanism by which specific Ab is thought to mediate protection against cryptococcal disease, particularly for cells in the central nervous system (79, 118).

However, it should be noted that protective MAbs can be nonprotective in the setting of an overly exuberant inflammatory response, such as that which was observed in the absence of B cells and inducible nitric oxide synthase (107, 108). Therefore, although the ability to clear GXM is a direct Ab function that does not depend on cellular immunity, the binding of Ab-GXM immune complexes to cellular receptors can affect the inflammatory response. This Ab function, which depends on cellular immunity, can be detrimental or beneficial depending on numerous variables including the relevant receptor and mouse strain (106). An Ab-GXM immune-complex-induced inflammatory response was implicated in an IgG-mediated prozone-like phenomenon that abrogated the ability of a protective Ab to mediate protection against cryptococcal disease (124). The discovery of a prozone-like phenomenon that can be mediated by either IgM (123) (including a human IgM [68]) or IgG (124) provided an explanation for the observation that the same MAb to GXM could be protective, nonprotective, or disease enhancing based on the ratio of the amount of MAb given to the challenge dose (inoculum). Although the mechanism by which prozone-like phenomena abrogate AI to cryptococcal disease has not been fully explained, the induction of an excessive inflammatory response and/or interference with underlying antifungal mechanisms have been linked to this phenomenon (see reference 18). Hence, untoward effects of certain direct mechanisms of Ab action that do not depend on cellular immunity, such as clearance of GXM and physical changes in the *C. neoformans* capsule, are likely to be the result of indirect mechanisms that depend on cellular immunity by engaging and/or activating cellular receptors that affect the immune response to *C. neoformans*.

AI TO CRYPTOCOCCAL DISEASE: THE FUTURE

The remarkable expansion of knowledge about AI to cryptococcal disease since the late 1980s has already translated into potential therapies to ameliorate human cryptococcal disease. A phase I clinical trial with a protective mouse MAb (18B7 [13]) was conducted in HIV-infected patients with a history of cryptococcal meningitis beginning in 2000 (64). Selection of this MAb was rationally based on its ability to bind *C. neoformans* serotypes A, D, and AD and to enhance human and mouse macrophage phagocytosis of *C. neoformans* (13) and the production of proinflammatory mediators with antifungal activity, including chemokines from human microglia (79, 118). The phase I trial of 18B7 provided proof of principle and held promise for further development of Ab-based therapy for *C. neoformans*; a reduction in serum GXM was observed after MAb administration to some subjects (64). The promise of 18B7 as a therapeutic MAb has been expanded with the application of a historically validated approach for treatment of malignancy that is novel in its application to treatment of an infectious disease, namely, radioimmunotherapy (30, 31). Radiolabeled 18B7 prolonged survival of mice when it was administered after lethal intravenous challenge (31). The mechanisms by which protection was achieved were multiple, including enhancement of macrophage-mediated killing of *C. neoformans*, apoptosis of *C. neoformans*, and immunomodulation (29). The efficacy of radioimmunotherapy against *C. neoformans* was a function of IgG subtype, with IgG1>IgG2b (29), with the caveat that the C region contributes to the specificity of MAb to GXM (76, 131). Hence, it is noteworthy that since radiolabeled MAb to GXM is used to target *C. neoformans*–infected cells, which are killed by radiation-induced effects, its ability to promote traditional effector functions is secondary, if not theoretically dispensable to its mechanism of action. This and other mechanisms by which radioimmunotherapy exerts antifungal activity are under active investigation.

The expansion of interest in Ab-based therapies for fungal diseases that stemmed from the demonstration that defined MAbs to many fungi could mediate protection against experimental disease (101) and the successful phase I trial of 18B7 for cryptococcal disease has led to the development of novel Ab reagents that could hold promise for clinical use. These reagents include nonspecific (22), in addition to specific, Abs. The identification of a MAb to β-1,3-glucan, originally produced in response to laminarin (2G8) (128), with broad reactivity with and efficacy against experimental disease with numerous fungi (21), including *C. neoformans* (105), has added new points to the roadmap for development of Ab-based therapies for fungal diseases. It is worth noting that although the MAb to laminarin is referred to herein and elsewhere as nonspecific, it binds β-1,3-glucan moieties that are found in the cell walls of many fungi. Hence, this and other so-called nonspecific Ab reagents, including a MAb to "killer" toxin that also has broad reactivity against numerous fungi (104), are in fact specific for determinants that are shared across many fungal species (22). While the hypothesis that Ags that are shared across fungal species can be harnessed as targets for Ab-based therapies might seem unconventional, promising preclinical results with the MAb 2G8 against different fungi, including *C. neoformans*, underscore the novelty and potential importance of cross-reactive Abs as therapeutic agents (22).

The emergence of cryptococcal disease in the setting of antiretroviral-therapy-induced immune reconstitution inflammatory syndrome in HIV-infected patients and in the setting of antifungal therapy and/or tapering of immunosuppression in solid organ transplant recipients (58, 117) has focused attention on cryptococcal disease as an inflammatory disease. This concept is in marked contrast to the late 1980s view that cryptococcal disease arose in the setting of profound T-cell immunodeficiency. While it has been recognized that MAb-induced inflammatory responses can be detrimental in experimental cryptococcal disease in mice, many beneficial mechanisms of Ab action against cryptococcal disease in normal and immunodeficient mouse models involve enhancement of the inflammatory response and polarization of the immune response to Th-1 (see Table 1). However, mechanisms of Ab action that induce inflammation would be contradicted in the treatment of immune reconstitution inflammatory syndrome–associated cryptococcal disease; instead, Ab reagents that reduce the inflammatory response and/or coadministration of immunomodulators would be required (Fig. 1). In theory, the development/engineering of

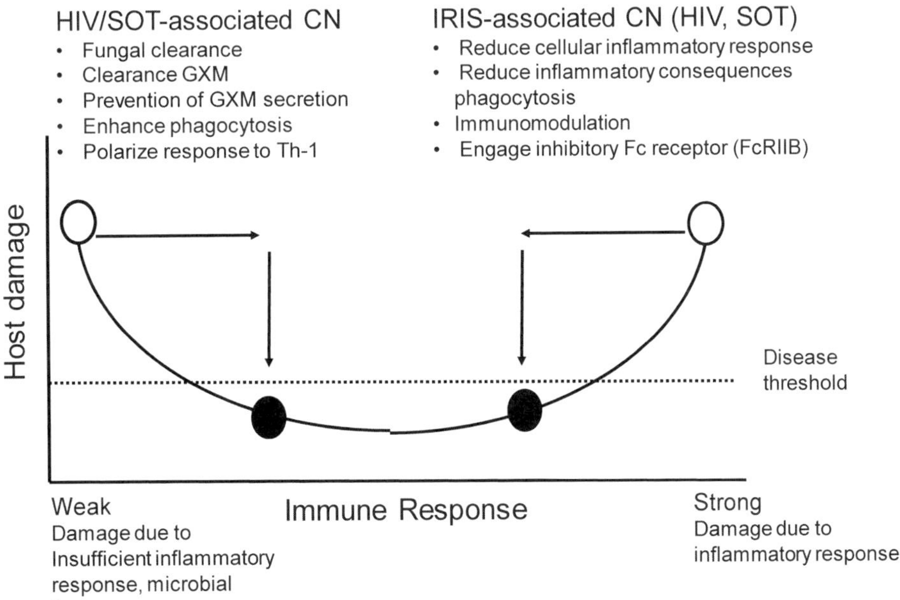

FIGURE 1 Damage-response curve depicting host damage (*y* axis) in the setting of HIV-associated cryptococcal disease as a function of the host inflammatory response (*x* axis). The open circles depict points on the curve at which host damage is maximal, and the black circles depict points on the curve at which host damage is below the disease threshold. The arrows depict the potential for the Ab-mediated processes noted under each disease heading to reduce host damage. Note that Ab functions associated with reduction of damage in the setting of a weak immune response involve microbial clearance, and functions associated with reduction of damage in the setting of a strong immune response involve a reduction in the inflammatory response. Abbreviations: HIV, human immunodeficiency virus; CN, *C. neoformans*; SOT, solid organ transplant; IRIS, immune reconstitution inflammatory syndrome; Th-1, T helper cell type 1.

Ab reagents with the ability to modulate the inflammatory response should be possible, since inhibitory (Fc) receptors are known to induce a reduction in the inflammatory response (26, 98), and MAbs that require these receptors to mediate protection against other encapsulated microbes have been identified (127).

The study of AI to *C. neoformans* has provided precedents that have informed and influenced our views on Ab immunity in general and the efficacy of Abs against other microbes. For example, it is now acknowledged that AI can be effective against intracellular pathogens (see reference 19). The concept that protective Ab against a microbe such as *C. neoformans* can be identified despite the inability to demonstrate a role for Ab in natural resistance has been extended to and validated by the isolation of protective MAbs to numerous pathogens for which AI was not previously thought to have a role, including *Listeria monocytogenes*, *Mycobacterium tuberculosis*, and *Histoplasma capsulatum*. Perhaps more importantly, studies of AI to *C. neoformans* have revealed new complexities in AI that were unsuspected two decades ago, which have in turn led to renewed efforts to understand the mechanisms of adaptive AI to infectious diseases and provided a roadmap for doing so.

REFERENCES

1. **Aalberse, R. C., and J. Schuurman.** 2002. IgG4 breaking the rules. *Immunology* **105:**9–19.
2. **Aalberse, R. C., J. Schuurman, and R. Van Ree.** 1999. The apparent monovalency of human IgG4 is due to bispecificity. *Int. Arch. Allergy Immunol.* **118:**187–189.
3. **Abadi, J., and L. Pirofski.** 1999. Antibodies reactive with the cryptococcal capsular polysaccharide glucuronoxylomannan are present in sera from children with and without HIV infection. *J. Infect. Dis.* **180:**915–919.
4. **Aguirre, K. M., and L. L. Johnson.** 1997. A role for B cells in resistance to *Cryptococcus neoformans* in mice. *Infect. Immun.* **65:**525–530.
5. **Alvarez, M., and A. Casadevall.** 2006. Phagosome extrusion and host-cell survival after *Cryptococcus neoformans* phagocytosis by macrophages. *Curr. Biol.* **16:**2161–2165.
6. **Alvarez, M., and A. Casadevall.** 2007. Cell-to-cell spread and massive vacuole formation after *Cryptococcus neoformans* infection of murine macrophages. *BMC Immunol.* **8:**16.
7. **Amu, S., A. Tarkowski, T. Dorner, M. Bokarewa, and M. Brisslert.** 2007. The human immunomodulatory CD25+ B cell population belongs to the memory B cell pool. *Scand. J. Immunol.* **66:**77–86.
8. **Beenhouwer, D. O., R. J. May, P. Valadon, and M. D. Scharff.** 2002. High affinity mimotope of the polysaccharide capsule of *Cryptococcus neoformans* identified from an evolutionary phage peptide library. *J. Immunol.* **169:**6992–6999.
9. **Beenhouwer, D. O., S. Shapiro, M. Feldmesser, A. Casadevall, and M. D. Scharff.** 2001. Both Th1 and Th2 cytokines affect the ability of monoclonal antibodies to protect mice against *Cryptococcus neoformans*. *Infect. Immun.* **69:**6445–6455.
10. **Beenhouwer, D. O., E. M. Yoo, C. W. Lai, M. A. Rocha, and S. L. Morrison.** 2007. Human immunoglobulin G2 (IgG2) and IgG4, but not IgG1 or IgG3, protect mice against *Cryptococcus neoformans* infection. *Infect. Immun.* **75:**1424–1435.
11. **Berberian, L., J. Shukla, R. Jefferis, and J. Braun.** 1994. Effects of HIV infection on V_H3 (D12 idiotope) B cells in vivo. *J. Acq. Immun. Def. Synd.* **7:**641–646.
12. **Casadevall, A.** 1995. Antibody immunity to invasive fungal infections. *Infect. Immun.* **63:**4211–4218.
13. **Casadevall, A., W. Cleare, M. Feldmesser, R. Glatman-Freedman, T. R. Kozel, N. Lendvai, J. Mukherjee, L. Pirofski, J. Rivera, A. L. Rosas, M. D. Scharff, P. Valadon, K. Westin, and Z. Zhong.** 1998. Characterization of a murine monoclonal antibody to *C. neoformans* polysaccharide which is a candidate for human therapeutic studies. *Antimicrob. Agents Chemother.* **42:**1437–1446.
14. **Casadevall, A., E. Dadachova, and L. A. Pirofski.** 2004. Passive antibody therapy for infectious diseases. *Nat. Rev. Microbiol.* **2:**695–703.
15. **Casadevall, A., M. DeShaw, M. Fan, F. Dromer, T. R. Kozel, and L. Pirofski.** 1994. Molecular and idiotypic analysis of antibodies to *Cryptococcus neoformans* glucuronoxylomannan. *Infect. Immun.* **62:**3864–3872.
16. **Casadevall, A., J. Mukherjee, S. J. Devi, R. Schneerson, J. B. Robbins, and M. D. Scharff.** 1992. Antibodies elicited by a *Cryptococcus neoformans*-tetanus toxoid conjugate vaccine have the same specificity as those elicited in infection. *J. Infect. Dis.* **165:**1086–1093.
17. **Casadevall, A., and L. Pirofski.** 2003. Antibody mediated regulation of cellular immunity and the inflammatory response. *Trends Immunol.* **24:**474–478.
18. **Casadevall, A., and L. Pirofski.** 2005. Insights into mechanisms of antibody-mediated immunity from studies with *Cryptococcus neoformans*. *Curr. Mol. Med.* **5:**421–433.
19. **Casadevall, A., and L. Pirofski.** 2006. A reappraisal of humoral immunity based on mechanisms of antibody-mediated protection against intracellular pathogens. *Adv. Immunol.* **91:**1–44.
20. **Casadevall, A., and L. A. Pirofski.** 2004. New concepts in antibody-mediated immunity. *Infect. Immun.* **72:**6191–6196.
21. **Casadevall, A., and L. A. Pirofski.** 2006. Polysaccharide-containing conjugate vaccines for fungal diseases. *Trends Mol. Med.* **12:**6–9.
22. **Casadevall, A., and L. A. Pirofski.** 2007. Antibody-mediated protection through cross-reactivity introduces a fungal heresy into immunological dogma. *Infect. Immun.* **75:**5074–5078.
23. **Casadevall, A., and M. D. Scharff.** 1994. Serum therapy revisited: animal models of infection and development of passive antibody therapy. *Antimicrob. Agents Chemother.* **38:**1695–1702.
24. **Chen, L. C., D. L. Goldman, T. L. Doering, L. Pirofski, and A. Casadevall.** 1999. Antibody response to *Cryptococcus neoformans* proteins in rodents and humans. *Infect. Immun.* **67:**2218–2224.
25. **Cleare, W., and A. Casadevall.** 1998. The different binding patterns of two immunoglobulin M monoclonal antibodies to *Cryptococcus neoformans* serotype A and D strains correlate with serotype classification and differences in functional assays. *Clin. Diagn. Lab. Immunol.* **5:**125–129.
26. **Clynes, R., J. S. Maizes, R. Guinamard, M. Ono, T. Takai, and J. V. Ravetch.** 1999. Modulation of immune complex-induced inflammation in vivo by the coordinate expression of activation and inhibitory Fc receptors. *J. Exp. Med.* **189:**179–185.
27. **Cooper, L. J., D. Robertson, R. Granzow, and N. S. Greenspan.** 1994. Variable domain-identical antibodies exhibit IgG subclass-related differences in affinity and kinetic constants as determined by surface plasmon resonance. *Mol. Immunol.* **31:**577–584.
28. **Cooper, L. J., J. C. Schimenti, D. D. Glass, and N. S. Greenspan.** 1991. H chain C domains influence the strength of binding of IgG for streptococcal group A carbohydrate. *J. Immunol.* **146:**2659–2663.
29. **Dadachova, E., R. A. Bryan, C. Apostolidis, A. Morgenstern, T. Zhang, T. Moadel, M. Torres, X. Huang, E. Revskaya, and A. Casadevall.** 2006. Interaction of radiolabeled antibodies with fungal cells and components of

the immune system in vitro and during radioimmunotherapy for experimental fungal infection. *J. Infect. Dis.* **193:**1427–1436.

30. **Dadachova, E., T. Burns, R. A. Bryan, C. Apostolidis, M. W. Brechbiel, J. D. Nosanchuk, A. Casadevall, and L. Pirofski.** 2004. Feasibility of radioimmunotherapy of experimental pneumococcal infection. *Antimicrob. Agents Chemother.* **48:**1624–1629.
31. **Dadachova, E., A. Nakouzi, R. A. Bryan, and A. Casadevall.** 2003. Ionizing radiation delivered by specific antibody is therapeutic against a fungal infection. *Proc. Natl. Acad. Sci. USA* **100:**10942–10947.
32. [Reference deleted.]
33. **DeShaw, M., and L. Pirofski.** 1995. Antibodies to *Cryptococcus neoformans* capsular polysaccharide glucuronoxylomannan are ubiquitous in the serum of HIV+ and HIV- individuals. *Clin. Exp. Immunol.* **99:**425–432.
34. **Devi, S. G. N., R. Schneerson, W. Egan, T. J. Ulrich, D. Bryla, J. B. Robbins, and J. E. Bennett.** 1991. *Cryptococcus neoformans* serotype A glucuronoxylomannan-protein conjugate vaccines: synthesis, characterization, and immunogenicity. *Infect. Immun.* **59:**3700–3707.
35. **Dromer, F., P. Aucouturier, J.-P. Clauvel, G. Saimot, and P. Yeni.** 1988. *Cryptococcus neoformans* antibody levels in patients with AIDS. *Scand. J. Infect. Dis.* **20:**283–285.
36. **Dromer, F., and J. Charreire.** 1991. Improved amphotericin B activity by a monoclonal anti-*Cryptococcus neoformans* antibody: study during murine cryptococcosis and mechanisms of action. *J. Infect. Dis.* **163:**1114–1120.
37. **Dromer, F., J. Charreire, A. Contrepois, C. Carbon, and P. Yeni.** 1987. Protection of mice against experimental cryptococcosis by an anti-*Cryptococcus neoformans* monoclonal antibody. *Infect. Immun.* **55:**749–752.
38. **Dromer, F., J. Salamero, A. Contrepois, C. Carbon, and P. Yeni.** 1987. Production, characterization, and antibody specificity of a mouse monoclonal antibody reactive with *Cryptococcus neoformans* capsular polysaccharide. *Infect. Immun.* **55:**742–748.
39. **Eckert, T. F., and T. R. Kozel.** 1987. Production and characterization of monoclonal antibodies specific for *Cryptococcus neoformans* capsular polysaccharide. *Infect. Immun.* **55:**1895–1899.
40. **Feldmesser, M., and A. Casadevall.** 1997. Effect of serum IgG1 to *Cryptococcus neoformans* glucuronoxylomannan on murine pulmonary infection. *J. Immunol.* **158:**790–799.
41. **Feldmesser, M., Y. Kress, P. Novikoff, and A. Casadevall.** 2000. *Cryptococcus neoformans* is a facultative intracellular pathogen in murine pulmonary infection. *Infect. Immun.* **68:**4225–4237.
42. **Feldmesser, M., A. Mednick, and A. Casadevall.** 2002. Antibody-mediated protection in murine *Cryptococcus neoformans* infection is associated with pleotrophic effects on cytokine and leukocyte responses. *Infect. Immun.* **70:**1571–1580.
43. **Fleuridor, R., A. Lees, and L. Pirofski.** 2001. A cryptococcal capsular polysaccharide mimotope prolongs the survival of mice with *Cryptococcus neoformans* infection. *J. Immunol.* **166:**1087–1096.
44. **Fleuridor, R., R. H. Lyles, and L. Pirofski.** 1999. Quantitative and qualitative differences in the serum antibody profiles of HIV-infected persons with and without *Cryptococcus neoformans* meningitis. *J. Infect. Dis.* **180:**1526–1536.
45. **Fleuridor, R., Z. Zhong, and L. Pirofski.** 1998. A human IgM monoclonal antibody prolongs survival of mice with lethal cryptococcosis. *J. Infect. Dis.* **178:**1213–1216.
46. **Garcia-Hermoso, D., G. Janbon, and F. Dromer.** 1999. Epidemiologic evidence for dormant *Cryptococcus neoformans* infection. *J. Clin. Microbiol.* **37:**3204–3209.
47. **Gates, M. A., and T. R. Kozel.** 2006. Differential localization of complement component 3 within the capsular matrix of *Cryptococcus neoformans*. *Infect. Immun.* **74:**3096–3106.
48. **Ghosn, E. E., M. Russo, and S. R. Almeida.** 2006. Nitric oxide-dependent killing of *Cryptococcus neoformans* by B-1-derived mononuclear phagocyte. *J. Leukoc. Biol.* **80:**36–44.
49. **Goldman, D. L., H. Khine, J. Abadi, D. L. Lindenberg, L. Pirofski, R. Niang, and A. Casadevall.** 2001. Serologic evidence for *Cryptococcus neoformans* infection in early childhood. *Pediatrics* **107:**e66.
50. **Goldman, D. L., S. C. Lee, and A. Casadevall.** 1995. Tissue localization of *Cryptococcus neoformans* glucuronoxylomannan in the presence and absence of specific antibody. *Infect. Immun.* **63:**3448–3453.
51. **Gordon, M. A., and A. Casadevall.** 1995. Serum therapy for cryptococcal meningitis. *Clin. Infect. Dis.* **21:**1477–1479.
52. **Gordon, M. A., and E. Lapa.** 1964. Serum protein enhancement of antibiotic therapy in cryptococcosis. *J. Infect. Dis.* **114:**373–377.
53. **Grinsell, M., L. C. Weinhold, J. E. Cutler, Y. Han, and T. R. Kozel.** 2001. In vivo clearance of glucuronoxylomannan, the major capsular polysaccharide of *Cryptococcus neoformans*: a critical role for tissue macrophages. *J. Infect. Dis.* **184:**479–487.
54. **Gupta, S., M. Ellis, T. Cesario, M. Ruhling, and B. Vayuvegula.** 1987. Disseminated cryptococcal infection in a patient with hypogammaglobulinemia and normal T cell functions. *Am. J. Med.* **82:**129–131.
55. **Hidore, M. R., N. Nabavi, F. Sonleitner, and J. W. Murphy.** 1991. Murine natural killer cells are fungicidal to *Cryptococcus neoformans*. *Infect. Immun.* **59:**1747–1754.
56. **Houpt, D. C., G. S. Pfrommer, B. J. Young, T. A. Larson, and T. R. Kozel.** 1994. Occurrences, immunoglobulin classes, and biological activities of antibodies in normal human serum that are reactive with *Cryptococcus neoformans* glucuronoxylomannan. *Infect. Immun.* **62:**2857–2864.
57. **Jalali, Z., L. Ng, N. Singh, and L. A. Pirofski.** 2006. Antibody response to *Cryptococcus neoformans* capsular polysaccharide glucuronoxylomannan in patients after solid-organ transplantation. *Clin. Vaccine Immunol.* **13:**740–746.
58. **Jenny-Avital, E. R., and M. Abadi.** 2002. Immune reconstitution cryptococcosis after initiation of successful highly active antiretroviral therapy. *Clin. Infect. Dis.* **35:**e128–e133.
59. **Kirkham, P. M., R. F. Mortari, J. A. Newton, and H. W. Schroeder.** 1992. Immunoglobulin V_H clan and family identity predicts variable domain structure and may influence antigen binding. *EMBO J.* **11:**603–609.
60. **Kozel, T. R., B. C. deJong, M. M. Grinsell, R. S. MacGill, and K. K. Wall.** 1998. Characterization of anticapsular monoclonal antibodies that regulate activation of the complement system by the *Cryptococcus neoformans* capsule. *Infect. Immun.* **66:**1538–1546.
61. **Kozel, T. R., and J. L. Follette.** 1981. Opsonization of encapsulated *Cryptococcus neoformans* by specific anticapsular antibody. *Infect. Immun.* **31:**978–984.
62. **Kozel, T. R., and T. G. McGaw.** 1979. Opsonization of *Cryptococcus neoformans* by human immunoglobulin G:role of immunoglobulin G in phagocytosis by macrophages. *Infect. Immun.* **18:**701–707.
63. **Kruetzmann, S., M. M. Rosado, H. Weber, U. Germing, O. Tournilhac, H. H. Peter, R. Berner, A. Peters, T. Boehm, A. Plebani, I. Quinti, and R. Carsetti.** 2003. Human immunoglobulin M memory B cells controlling *Streptococcus pneumoniae* infections are generated in the spleen. *J. Exp. Med.* **197:**939–945.
64. **Larsen, R. A., P. G. Pappas, J. Perfect, J. A. Aberg, A. Casadevall, G. A. Cloud, R. James, S. Filler, and W. E. Dismukes.** 2005. Phase I evaluation of the safety

and pharmacokinetics of murine-derived anticryptococcal antibody 18B7 in subjects with treated cryptococcal meningitis. *Antimicrob. Agents Chemother.* **49:**952–958.

65. **Lendvai, N., A. Casadevall, Z. Liang, D. L. Goldman, J. Mukherjee, and L. Zuckier.** 1998. Effect of immune mechanisms on the pharmacokinetics and organ distribution of cryptococcal polysaccharide. *J. Infect. Dis.* **177:**1647–1659.
66. **Lendvai, N., X. W. Qu, W. Hsueh, and A. Casadevall.** 2000. Mechanism for the isotype dependence of antibody-mediated toxicity in *Cryptococcus-neoformans*-infected mice. *J. Immunol.* **164:**4367–4374.
67. **MacGill, T. C., R. S. MacGill, A. Casadevall, and T. R. Kozel.** 2000. Biological correlates of capsular (quellung) reactions of *Cryptococcus neoformans*. *J. Immunol.* **164:**4835–4842.
68. **Maitta, R., K. Datta, Q. Chang, R. Luo, K. Subramaniam, B. Witover, and L. Pirofski.** 2004. Protective and nonprotective human IgM monoclonal antibodies to *Cryptococcus neoformans* glucuronoxylomannan manifest different specificity and gene use. *Infect. Immun.* **72:**4810–4818.
69. **Maitta, R. W., K. Datta, A. Lees, S. S. Belouski, and L. A. Pirofski.** 2004. Immunogenicity and efficacy of *Cryptococcus neoformans* capsular polysaccharide glucuronoxylomannan peptide mimotope-protein conjugates in human immunoglobulin transgenic mice. *Infect. Immun.* **72:**196–208.
70. **Marquis, G., S. Montplaisir, M. Pelletier, S. Mousseau, and P. Auger.** 1985. Genetic resistance to murine cryptococcosis: increased susceptibility in the CBA/N XID mutant strain of mice. *Infect. Immun.* **47:**282–287.
71. **Martinez, L. R., and A. Casadevall.** 2005. Specific antibody can prevent fungal biofilm formation and this effect correlates with protective efficacy. *Infect. Immun.* **73:**6350–6362.
72. **Martinez, L. R., E. Christaki, and A. Casadevall.** 2006. Specific antibody to *Cryptococcus neoformans* glucurunoxylomannan antagonizes antifungal drug action against cryptococcal biofilms in vitro. *J. Infect. Dis.* **194:**261–266.
73. **Martinez, L. R., D. Moussai, and A. Casadevall.** 2004. Antibody to *Cryptococcus neoformans* glucuronoxylomannan inhibits the release of capsular antigen. *Infect. Immun.* **72:**3674–3679.
74. **May, R. J., D. O. Beenhouwer, and M. D. Scharff.** 2003. Antibodies to keyhole limpet hemocyanin cross-react with an epitope on the polysaccharide capsule of *Cryptococcus neoformans* and other carbohydrates: implications for vaccine development. *J. Immunol.* **171:**4905–4912.
75. **McGaw, T. G., and T. R. Kozel.** 1979. Opsonization of *Cryptococcus neoformans* by human immunoglobulin G: masking of immunoglobulin G by cryptococcal polysaccharide. *Infect. Immun.* **25:**262–267.
76. **McLean, G. R., M. Torres, N. Elguezabal, A. Nakouzi, and A. Casadevall.** 2002. Isotype can affect the fine specificity of an antibody for a polysaccharide antigen. *J. Immunol.* **169:**1379–1386.
77. **Monari, C., A. Casadevall, C. Retini, F. Baldelli, F. Bistoni, and A. Vecchiarelli.** 1999. Antibody to capsular polysaccharide enhances the function of neutrophils from patients with AIDS against *Cryptococcus neoformans*. *AIDS* **13:**653–660.
78. **Monga, D. P., R. Kumar, L. N. Mohapatra, and A. N. Malaviya.** 1979. Experimental cryptococcosis in normal and B cell deficient mice. *Infect. Immun.* **26:**1–3.
79. **Mozaffarian, N., J. W. Berman, and A. Casadevall.** 1995. Immune complexes increase nitric oxide production by interferon-gamma-stimulated murine macrophage-like J774.16 cells. *J. Leukoc. Biol.* **57:**657–662.
80. **Mukherjee, J., A. Casadevall, and M. D. Scharff.** 1993. Molecular characterization of the antibody responses to *Cryptococcus neoformans* infection and glucuronoxylomannan-tetanus toxoid conjugate immunization. *J. Exp. Med.* **177:**1105–1116.
81. **Mukherjee, J., T. R. Kozel, and A. Casadevall.** 1998. Monoclonal antibodies reveal additional epitopes of serotype D *Cryptococcus neoformans* capsular glucuronoxylomannan that elicit protective antibodies. *J.Immunol.* **161:**3557–3567.
82. **Mukherjee, J., G. Nussbaum, M. D. Scharff, and A. Casadevall.** 1995. Protective and nonprotective monoclonal antibodies to *Cryptococcus neoformans* originating from one B cell. *J. Exp. Med.* **181:**405–409.
83. **Mukherjee, J., L. Pirofski, M. D. Scharff, and A. Casadevall.** 1993. Antibody-mediated protection in mice with lethal intracerebral *Cryptococcus neoformans* infection. *Proc. Natl. Acad. Sci. USA* **90:**3636–3640.
84. **Mukherjee, J., M. D. Scharff, and A. Casadevall.** 1992. Protective murine monoclonal antibodies to *Cryptococcus neoformans*. *Infect. Immun.* **60:**4534–4541.
85. **Mukherjee, J., L. S. Zuckier, M. D. Scharff, and A. Casadevall.** 1994. Therapeutic efficacy of monoclonal antibodies to *Cryptococcus neoformans* glucuronoxylomannan alone and in combination with amphotericin B. *Antimicrob. Agents Chemother.* **38:**580–587.
86. **Mukherjee, S., M. Feldmesser, and A. Casadevall.** 1996. J774 murine macrophage-like cell interactions with *Cryptococcus neoformans* in the presence and absence of opsonins. *J. Infect. Dis.* **173:**1222–1231.
87. **Mukherjee, S., S. C. Lee, and A. Casadevall.** 1995. Antibodies to *Cryptococcus neoformans* glucuronoxylomannan enhance antifungal activity of murine macrophages. *Infect. Immun.* **63:**573–579.
88. **Nabavi, N., and J. W. Murphy.** 1986. Antibody-dependent natural killer cell-mediated growth inhibition of *Cryptococcus neoformans*. *Infect. Immun.* **51:**556–562.
89. **Nakouzi, A., and A. Casadevall.** 2003. The function of conserved amino acids in or near the complementarity determining regions for related antibodies to *Cryptococcus neoformans* glucuronoxylomannan. *Mol. Immunol.* **40:**351–361.
90. **Nakouzi, A., P. Valadon, J. D. Nosanchuk, N. Green, and A. Casadevall.** 2001. Molecular basis for immunoglobulin M specificity to epitopes in *Cryptococcus neoformans* polysaccharide that elicit protective and non-protective antibodies. *Infect. Immun.* **69:**3398–3409.
91. **Netski, D., and T. R. Kozel.** 2002. Fc-dependent and Fc-independent opsonization of *Cryptococcus neoformans* by anticapsular monoclonal antibodies: importance of epitope specificity. *Infect. Immun.* **70:**2812–2819.
92. **Nosanchuk, J. D., and A. Casadevall.** 1997. Cellular charge of *Cryptococcus neoformans*: contributions from the capsular polysaccharide, melanin, and monoclonal antibody binding. *Infect. Immun.* **65:**1836–1841.
93. **Nosanchuk, J. D., W. Cleare, S. P. Franzot, and A. Casadevall.** 1999. Amphotericin B and fluconazole affect cellular charge, macrophage phagocytosis, and cellular morphology of *Cryptococcus neoformans* at subinhibitory concentrations. *Antimicrob. Agents Chemother.* **43:**233–239.
94. **Nosanchuk, J. D., A. L. Rosas, and A. Casadevall.** 1998. The antibody response to fungal melanin in mice. *J. Immunol.* **160:**6026–6031.
95. **Nussbaum, G., S. Anandasabapathy, J. Mukherjee, M. Fan, A. Casadevall, and M. D. Scharff.** 1999. Molecular and idiotypic analyses of the antibody response to *Cryptococcus neoformans* glucuronoxylomannan-protein conjugate vaccine in autoimmune and nonautoimmune mice. *Infect. Immun.* **67:**4469–4476.
96. **Nussbaum, G., W. Cleare, A. Casadevall, M. D. Scharff, and P. Valadon.** 1997. Epitope location in the *Cryptococcus neoformans* capsule is a determinant of antibody efficacy. *J. Exp. Med.* **185:**685–694.

97. **Nussbaum, G., R. Yuan, A. Casadevall, and M. D. Scharff.** 1996. Immunoglobulin G3 blocking antibodies to the fungal pathogen *Cryptococcus neoformans*. *J. Exp. Med.* **183:**1905–1909.
98. **Park-Min, K. H., N. V. Serbina, W. Yang, X. Ma, G. Krystal, B. G. Neel, S. L. Nutt, X. Hu, and L. B. Ivashkiv.** 2007. FcgammaRIII-dependent inhibition of interferon-gamma responses mediates suppressive effects of intravenous immune globulin. *Immunity* **26:**67–78.
99. **Pires Neto, R., Jr., M. C. Guimaraes, M. J. Moya, F. R. Oliveira, P. Louzada, Jr., and R. Martinez.** 2000. Hypogammaglobulinemia as predisposing factor for *Cryptococcus neoformans* infection: regarding two cases. *Rev. Soc. Bras. Med. Trop.* **33:**603–608. (In Portuguese.)
100. **Pirofski, L.** 2001. Polysaccharides, mimotopes and vaccines for encapsulated pathogens. *Trends Microbiol.* **9:**445–452.
101. **Pirofski, L., and A. Casadevall.** 2006. Acquired antibody mediated immunity to fungi, p. 487–503. *In*: J. Heitman, S. G. Filler, and A. P. Mitchell (ed.), *Molecular Principles of Fungal Pathogenesis*. ASM Press, Washington, DC.
102. **Pirofski, L., R. Lui, M. DeShaw, A. B. Kressel, and Z. Zhong.** 1995. Analysis of human monoclonal antibodies elicited by vaccination with a *Cryptococcus neoformans* glucuronoxylomannan capsular polysaccharide vaccine. *Infect. Immun.* **63:**3005–3014.
103. **Pirofski, L. A.** 2006. Of mice and men, revisited: new insights into an ancient molecule from studies of complement activation by *Cryptococcus neoformans*. *Infect. Immun.* **74:**3079–3084.
104. **Polonelli, L., S. Conti, M. Gerloni, W. Magliani, M. Castagnola, G. Morace, and C. Chezzi.** 1991. 'Antibiobodies': antibiotic-like anti-idiotypic antibodies. *J. Med. Vet. Mycol.* **29:**235–242.
105. **Rachini, A., D. Pietrella, P. Lupo, A. Torosantucci, P. Chiani, C. Bromuro, C. Proietti, F. Bistoni, A. Cassone, and A. Vecchiarelli.** 2007. An anti-beta-glucan monoclonal antibody inhibits growth and capsule formation of *Cryptococcus neoformans* in vitro and exerts therapeutic, anticryptococcal activity in vivo. *Infect. Immun.* **75:**5085–5094.
106. **Rivera, J., and A. Casadevall.** 2005. Mouse genetic background is a major determinant of isotype-related differences for antibody-mediated protective efficacy against *Cryptococcus neoformans*. *J. Immunol.* **174:**8017–8026.
107. **Rivera, J., J. Mukherjee, L. M. Weiss, and A. Casadevall.** 2002. Antibody efficacy in murine pulmonary *Cryptococcus neoformans* infection: a role for nitric oxide. *J. Immunol.* **168:**3419–3427.
108. **Rivera, J., O. Zaragoza, and A. Casadevall.** 2005. Antibody-mediated protection against *Cryptococcus neoformans* is dependent on B cells. *Infect. Immun.* **73:**1141–1150.
109. [Reference deleted.]
110. **Robbins, J. B., and R. Schneerson.** 1990. Polysaccharide-protein conjugates: a new generation of vaccines. *J. Infect. Dis.* **161:**821–832.
111. **Robertson, E. J., and A. Casadevall.** 2009. Antibody-mediated immobilization of *Cryptococcus neoformans* promotes biofilm formation. *Appl. Environ. Microbiol.* **75:**2528–2533.
112. **Rosas, A. L., J. D. Nosanchuk, and A. Casadevall.** 2001. Passive immunization with melanin-binding monoclonal antibodies prolongs survival of mice with lethal *Cryptococcus neoformans* infection. *Infect. Immun.* **69:**3410–3412.
113. **Savoy, A. L., D. M. Lupan, P. B. Manalo, J. C. Roberts, A. M. Schlageter, L. C. Weinhold, and T. R. Kozel.** 1997. Acute lethal toxicity following passive immunization for therapy of murine cryptococcosis. *Infect. Immun.* **65:**1800–1807.
114. **Schlageter, A. M., and T. R. Kozel.** 1990. Opsonization of *Cryptococcus neoformans* by a family of isotype-switch variant antibodies specific for the capsular polysaccharide. *Infect. Immun.* **58:**1914–1918.
115. **Shapiro, S., D. O. Beenhouwer, M. Feldmesser, C. Taborda, M. C. Carroll, A. Casadevall, and M. D. Scharff.** 2002. Immunoglobulin G monoclonal antibodies to *Cryptococcus neoformans* protect mice deficient in complement component C3. *Infect. Immun.* **70:**2598–2604.
116. **Shi, Y., T. Yamazaki, Y. Okubo, Y. Uehara, K. Sugane, and K. Agematsu.** 2005. Regulation of aged humoral immune defense against pneumococcal bacteria by IgM memory B cell. *J. Immunol.* **175:**3262–3267.
117. **Singh, N., O. Lortholary, B. D. Alexander, K. L. Gupta, G. T. John, K. Pursell, P. Munoz, G. B. Klintmalm, V. Stosor, R. del Busto, A. P. Limaye, J. Somani, M. Lyon, S. Houston, A. A. House, T. L. Pruett, S. Orloff, A. Humar, L. Dowdy, J. Garcia-Diaz, A. C. Kalil, R. A. Fisher, and S. Husain.** 2005. An immune reconstitution syndrome-like illness associated with *Cryptococcus neoformans* infection in organ transplant recipients. *Clin. Infect. Dis.* **40:**1756–1761.
118. **Song, X., S. Shapiro, D. L. Goldman, A. Casadevall, M. Scharff, and S. C. Lee.** 2002. Fcgamma receptor I- and III-mediated macrophage inflammatory protein 1alpha induction in primary human and murine microglia. *Infect. Immun.* **70:**5177–5184.
119. **Spitzer, E. D., S. G. Spitzer, L. F. Freundlich, and A. Casadevall.** 1993. Persistence of initial infection in recurrent *Cryptococcus neoformans* meningitis. *Lancet* **341:**595–596.
120. **Subramaniam, K., N. French, and L. A. Pirofski.** 2005. *Cryptococcus neoformans*-reactive and total immunoglobulin profiles of human immunodeficiency virus-infected and uninfected Ugandans. *Clin. Diagn. Lab. Immunol.* **12:**1168–1176.
121. **Subramaniam, K., B. Metzger, L. Hanau, A. Guh, S. Badri, and L. Pirofski.** 2009. Memory IgM B cell expression predicts HIV-associated *Cryptococcus neoformans* disease status. *J. Infect. Dis.* **200:**244–51.
122. **Sundstrom, J. B., and R. Cherniak.** 1992. The glucuronoxylomannan of *Cryptococcus neoformans* serotype A is a type 2 T-cell-independent antigen. *Infect. Immun.* **60:**4080–4087.
123. **Taborda, C., and A. Casadevall.** 2001. Immunoglobulin M efficacy against *Cryptococcus neoformans*: mechanism, dose dependence, and prozone-like effects in passive protection experiments. *J. Immunol.* **166:**2100–2107.
124. **Taborda, C., J. Rivera, O. Zaragoza, and A. Casadevall.** 2003. More is not necessarily better: prozone-like effects in passive immunization with IgG. *J. Immunol.* **170:**3621–3630.
125. **Taborda, C. P., and A. Casadevall.** 2002. CR3 (CD11b/CD18) and CR4 (CD11c/CD18) are involved in complement-independent antibody-mediated phagocytosis of *Cryptococcus neoformans*. *Immunity* **16:**791–802.
126. **Tangye, S. G., and K. L. Good.** 2007. Human IgM+ CD27+ B cells: memory B cells or "memory" B cells? *J. Immunol.* **179:**13–19.
127. **Tian, H., S. Weber, P. Thorkildson, T. R. Kozel, and L. Pirofski.** 2009. Efficacy of opsonic and non-opsonic serotype 3 pneumococcal capsular polysaccharide-specific antibodies against intransasal challenge with *Streptococcus pneumoniae* in mice. *Infect. Immun.* **77:**1502–1513.
128. **Torosantucci, A., C. Bromuro, P. Chiani, F. De Bernardis, F. Berti, C. Galli, F. Norelli, C. Bellucci, L. Polonelli, P. Costantino, R. Rappuoli, and A. Cassone.** 2005. A novel glyco-conjugate vaccine against fungal pathogens. *J. Exp. Med.* **202:**597–606.

129. **Torres, M., and A. Casadevall.** 2008. The immunoglobulin constant region contributes to affinity and specificity. *Trends Immunol.* **29:**91–97.
130. **Torres, M., N. Fernandez-Fuentes, A. Fiser, and A. Casadevall.** 2007. The immunoglobulin heavy chain constant region affects kinetic and thermodynamic parameters of antibody variable region interactions with antigen. *J. Biol. Chem.* **282:**13917–13927.
131. **Torres, M., R. May, M. D. Scharff, and A. Casadevall.** 2005. Variable-region identical antibodies differing in isotype demonstrate differences in fine specificity and idiotype. *J. Immunol.* **174:**2132–2142.
132. **Tucker, S. C., and A. Casadevall.** 2002. Replication of *Cryptococcus neoformans* in macrophages is accompanied by phagosomal permeabilization and accumulation of vesicles containing polysaccharide in the cytoplasm. *Proc. Natl. Acad. Sci. USA* **99:**3165–3170.
133. **Vecchiarelli, A.** 2000. Immunoregulation by capsular components of *Cryptococcus neoformans*. *Med. Mycol.* **38:**407–417.
134. **Vecchiarelli, A.** 2005. The cellular responses induced by the capsular polysaccharide of *Cryptococcus neoformans* differ depending on the presence or absence of specific protective antibodies. *Curr. Mol. Med.* **5:**413–420.
135. **Vecchiarelli, A., C. Monari, C. Retini, D. Pietrella, B. Palazzetti, L. Pitzurra, and A. Casadevall.** 1998. *Cryptococcus neoformans* differently regulates B7-1 (CD80) and B7-2 (CD86) expression on human monocytes. *Eur. J. Immunol.* **28:**114–121.
136. **Vecchiarelli, A., D. Pietrella, F. Bistoni, T. R. Kozel, and A. Casadevall.** 2002. Antibody to *Cryptococcus neoformans* capsular glucuronoxylomannan promotes expression of interleukin-12Rbeta2 subunit on human T cells in vitro through effects mediated by antigen-presenting cells. *Immunology* **106:**267–272.
137. **Vecchiarelli, A., C. Retini, C. Monari, and A. Casadevall.** 1998. Specific antibody to *Cryptococcus neoformans* alters human leukocyte cytokine synthesis and promotes T-cell proliferation. *Infect. Immun.* **66:**1244–1247.
138. **Wahab, J. A., M. J. Hanifah, and K. E. Choo.** 1995. Bruton's agammaglobulinemia in a child with cryptococcal empyema thoracis and periauricular pyogenic abscess. *Singapore Med. J.* **36:**686–689.
139. **Wilson, M. A., and T. R. Kozel.** 1992. Contribution of antibody in normal human serum to early deposition of C3 onto encapsulated and nonencapsulated *Cryptococcus neoformans*. *Infect. Immun.* **60:**754–761.
140. **Yoo, E. M., L. A. Wims, L. A. Chan, and S. L. Morrison.** 2003. Human IgG2 can form covalent dimers. *J. Immunol.* **170:**3134–3138.
141. **Yuan, R., A. Casadevall, and M. D. Scharff.** 1997. T cells cooperate with passive antibody to modify the course of *Cryptococcus neoformans* in mice. *Proc. Natl. Acad. Sci. USA* **94:**2483–2488.
142. **Yuan, R., A. Casadevall, G. Spira, and M. D. Scharff.** 1995. Isotype switching from IgG3 to IgG1 converts a nonprotective murine antibody to *Cryptococcus neoformans* to a protective antibody. *J. Immunol.* **154:**1810–1816.
143. **Yuan, R., R. Clynes, J. Oh, J. V. Ravetch, and M. D. Scharff.** 1998. Antibody-mediated modulation of *Cryptococcus neoformans* infection is dependent on distinct Fc receptor functions and IgG subclasses. *J. Exp. Med.* **187:**641–648.
144. **Zaragoza, O., C. P. Taborda, and A. Casadevall.** 2003. The efficacy of complement-mediated phagocytosis of *Cryptococcus neoformans* is dependent on the location of C3 in the polysaccharide capsule and involves both direct and indirect C3-mediated interactions. *Eur. J. Immunol.* **33:**1957–1967.
145. **Zebedee, S. L., R. K. Koduri, J. Mukherjee, S. Mukherjee, S. Lee, D. F. Sauer, M. D. Scharff, and A. Casadevall.** 1994. Mouse-human immunoglobulin G1 chimeric antibodies with activities against *Cryptococcus neoformans*. *Antimicrob. Agents Chemother.* **38:**1507–1514.
146. **Zhong, Z., and L. Pirofski.** 1996. Opsonization of *Cryptococcus neoformans* by human anti-glucuronoxylomannan antibodies. *Infect. Immun.* **64:**3446–3450.
147. **Zhong, Z., and L. Pirofski.** 1998. Antifungal activity of a human antiglucuronoxylomannan antibody. *Clin. Diag. Lab. Immunol.* **5:**58–64.

Cryptococcus: From Human Pathogen to Model Yeast
Edited by J. Heitman et al.

29

Interactions of Capsule with Antibody and Complement

THOMAS R. KOZEL

The capsule is the interface between *Cryptococcus neoformans* and its environment. In the environment, this interface prevents desiccation, serves as an adhesion in various environmental niches, and may prevent ingestion by predatory amebae. In vivo, the capsule is a barrier to phagocytosis. The interface between the cryptococcal capsule and its environment in vivo is dramatically altered by binding of anticapsular antibodies or proteins of the complement system to the capsular matrix. The primary constituent of the cryptococcal capsule is the high-molecular-weight polysaccharide glucuronoxylomannan (GXM). Additional components of the capsule include the polysaccharide galactoxylomannan and mannoprotein. The structure and synthesis of GXM, galactoxylomannan, and mannoprotein are described in chapter 3. GXM is the most prominent of the capsular constituents and is the primary target for anticapsular antibody and the most likely mediator of capsule-complement interactions.

The cryptococcal capsule is a molecular sieve that increases in porosity with distance from the cell wall (10). This variable porosity has consequences for interaction with serum proteins. The high density of the matrix at the capsule interior prevents penetration of large macromolecules, including antibody and complement proteins, to sites near the cell wall. In contrast, the polysaccharide matrix at the capsular edge that is the interface with phagocytes presents polysaccharide with a low density that exhibits considerable plasticity and permeability to macromolecules. As a consequence, macromolecules such as antibody or complement proteins freely penetrate the outer layers of the capsule and interact with the capsular matrix. The architecture of the cryptococcal capsule is described in chapter 4.

The goal of this chapter is to describe (i) variability in the interactions of GXM antibodies with the capsular matrix, (ii) the mechanisms for interaction of the capsule with proteins of the complement system, (iii) the nexus of the anticapsular antibody and complement system in capsule interactions, and (iv) the consequences of antibody and complement binding for modification of the capsular matrix.

Thomas R. Kozel, Department of Microbiology and Immunology, University of Nevada School of Medicine, Reno, NV 89557.

CAPSULE-ANTIBODY INTERACTIONS

Alteration in Capsule Architecture Produced by Binding of Anticapsular Antibody

Antibodies have long been known to bind to the capsule of *C. neoformans*. One of the first descriptions of anticapsular antibodies for *C. neoformans* was the report by Evans of the production of agglutinating antibodies in rabbits that could differentiate three serological types that he named serotypes A, B, and C (8).

The first evidence of an antibody-induced alteration in capsule architecture was reported by Neill et al. (39), who found that incubation of hyperimmune rabbit serum with cryptococci collected from culture medium or harvested from the peritoneal cavity of infected mice produced prominent capsules with a dark and sharply outlined border. This antibody-induced change in the optical appearance of the capsule occurred without any measurable change in capsule size. The authors attributed the antibody-induced increase in visibility of the capsule to an increase in the optical density at the perimeter of the capsule.

A major advance in the understanding of antibody-capsule interactions followed the generation of monoclonal antibodies (MAbs) to GXM (1, 5, 6). Initial studies using indirect immunofluorescence found that different MAbs showed distinct patterns for binding to the capsular matrix (38, 41). Some antibodies produced an annular fluorescence pattern at the capsule edge. Other antibodies bound throughout the capsule in a punctate pattern. These results suggested that epitopes recognized by GXM MAbs are not uniformly distributed throughout the capsule; epitopes recognized by annular MAbs are located at the capsular edge, and epitopes recognized by punctate MAbs are located at both the capsule edge and the capsule interior. Notably, MAbs that produced the annular fluorescence patterns were protective in a murine model of cryptococcosis, whereas MAbs producing the punctate pattern failed to protect. Human protective MAbs also manifest annular fluorescence patterns upon capsular binding (30). Antibody-induced structural changes in the fibrous capsular network have also been observed by scanning electron microscopy (35).

Further insight into the effects of antibody binding on the architecture of the cryptococcal capsule was provided when capsule reactions such as those observed by Neill et al. (39) were evaluated by differential interference contrast (DIC) microscopy (28). Incubation of encapsulated cryptococci with different GXM MAbs produced two distinct capsular reactions whose appearance depended on the specificity of the MAb and the serotype of the target cell. In the first pattern, termed "rim," the capsule appears transparent with a highly refractile outer edge. In the second pattern, termed "puffy," the capsule appears opaque and lacks a highly refractive outer rim. An advantage of DIC microscopy is the fact that the optical system provides insight to the physical basis for the capsular reaction. DIC systems detect optical gradients, preserve its sign, and convert the gradients into visible intensity gradients. The observed variation in light intensity reflects changes in optical gradient. The rim pattern is characterized by an initial increase in the optical gradient at the capsule edge, followed directly by a decrease in the gradient to background, giving the capsule interior a transparent appearance. This rim pattern identifies a rapid increase and rapid decrease in the refractive index at the capsular edge, with no change in the capsule interior. The most likely explanation for the refractile rim is accumulation of a large amount of antibody at the capsule edge.

In contrast to the rim pattern, the puffy pattern is characterized by an initial increase in the optical gradient at the capsule edge followed by a gradient of change throughout the capsule. The striking decrease in refractive index at the capsule edge that is characteristic of the rim pattern is not observed. The capsule reactions observed by DIC microscopy suggest that binding of rim and puffy MAbs have fundamentally different effects on the capsular structure.

The ability of a MAb to produce different capsule reactions when viewed by DIC microscopy has implications for the mechanisms by which antibodies interact with the cryptococcal capsule. There are two possible explanations for the abilities of antibodies to produce rim or puffy patterns. First, two binding patterns could be due to differential expression of epitopes recognized by the MAbs at different sites in the matrix. Binding of antibody to the capsule edge to produce the annular rim could be due to localization of epitopes only at the capsule perimeter; MAbs producing puffy patterns may bind to epitopes located throughout the capsule. Alternatively, production of a rim at the capsular edge could be due to cross-linking of the capsular surface by antibody such that additional antibody is unable to penetrate to the capsule interior. A failure to cross-link would allow for binding of antibodies throughout the capsule as occurs in the puffy-type reaction.

Insight into the molecular nature of rim- versus puffy-type reactions was provided by a study of Fab and $F(ab')_2$ fragments of MAbs that produced rim-type reactions (28). $F(ab')_2$ fragments of rim-type MAbs produced a rim capsule reaction; Fab fragments produced a puffy reaction. These results show that production of a rim-type reaction is a function of cross-linking of the capsular edge by antibody. The fact that Fab fragments of rim-type MAbs produced a puffy reaction demonstrated that the epitope recognized by the antibodies is distributed throughout the capsule.

The ability of antibodies to bind to different sites within the capsular matrix is influenced by the molecular sieving properties of the capsule (10). As noted above, the capsule is a sieve of variable porosity; the pore size is largest at the capsule periphery and becomes increasingly less porous as the cell wall is approached. The pore size is also influenced by culture conditions; capsules of cells grown in vitro under capsule induction conditions have considerable porosity, whereas capsules of cells harvested from infected tissue have smaller pores. This variable porosity has consequences for antibody binding. Intact immunoglobulin G (IgG) MAbs bind to the outer layers of the capsule but are unable to penetrate deep into the capsule interior. In contrast, Fab fragments of these same antibodies are able to penetrate deep into the capsule interior.

The impact of the sieving action of the capsule on the diffusion of antibody into the matrix has been modeled and experimentally tested (44). The model predicted rapid diffusion of antibody to all regions of the capsule where the pore size was greater than the Stokes diameter of the antibody. Maximum binding of antibody occurs at intermediate sites within the capsule where the GXM concentrations are high and the pore size is large enough to permit free diffusion of antibody to antigen. The depth of binding of antibodies of different isotypes is predictable on the basis of the Stokes diameter of each isotype.

Finally, binding of GXM MAbs to the capsule edge alters the permeability of the capsular matrix (55). Treatment of encapsulated cryptococci with a GXM MAb reduces the permeability of the capsule to penetration by dextrans and proteins of the complement cascade. It is possible that such antibody-induced changes in permeability could restrict availability of nutrients or prevent access to the cell wall by antimicrobial compounds.

Modes of Alteration in Capsular Architecture Correlate with the Ability of an Antibody to Protect

As with the case of MAbs that produced the annular or punctate patterns by immunofluorescence, the ability of a MAb to produce rim or puffy patterns when observed by DIC microscopy correlates with protection (28). MAbs that produce the rim pattern are protective; MAbs that produce the puffy pattern are not protective. A key question that remains is whether the ability of a MAb to remodel the capsule edge to produce the annular (immunofluorescence) or rim (DIC microscopy) patterns is causally related to protection. Alternatively, it is possible that production of the annular or rim patterns is a marker for protection without a causal link. Importantly, the ability of an antibody to produce a rim pattern by DIC microscopy is also linked to protection by MAbs specific for the capsular polypeptide of *Bacillus anthracis* (18).

Other Effects of Antibody on Capsule

A major consequence of binding of GXM antibodies to the cryptococcal capsule is opsonization for phagocytosis by macrophages and neutrophils. Enhanced phagocytosis of encapsulated cryptococci following opsonization with polyclonal and monoclonal anticapsular antibodies has been described by many investigators (4, 14, 25, 36, 37). Notably, MAbs that produce the rim-type capsule reaction are much more opsonic that MAbs that produce the puffy-type reaction (40), suggesting that the capsule rearrangement that occurs on binding of rim-type antibodies may facilitate Fc-dependent phagocytosis. Opsonization is dose dependent and functions through Fc receptors. However, there is a prozone-like effect in which decreased opsonic activity is observed at very high concentrations of opsonic antibody (29). In vivo studies of survival following passive immunization also found a prozone in which administration of high doses of antibody produced a reduction in protection against challenge (48, 49).

Opsonization by anticapsular antibodies and phagocytosis of encapsulated cryptococci is described in other chapters in this section.

Anticapsular antibodies also promote phagocytosis via Fc-independent mechanisms. Fc-independent opsonization takes two forms. First, $F(ab')_2$ fragments of GXM MAbs are synergistic with heat-labile factors in normal human serum for opsonization (40). Second, $F(ab')_2$ fragments or IgM can facilitate phagocytosis in a manner that is independent of serum factors (40, 47). Enhancement of serum-independent phagocytosis by $F(ab')_2$ is inhibited by soluble GXM, suggesting a direct interaction between capsular GXM and phagocyte receptors for GXM. Complement-independent opsonization by IgM is inhibited by antibodies to CR3 (CD11b/CD18) and CR4 (CD11c/CD18) and with macrophages from CD18-deficient mice, identifying a role for CR3 and CR4 in recognition of capsular GXM (47). The specific mechanism by which anticapsular antibodies facilitate the interaction of capsular GXM with CR3 and CR4 is not known; however, the opsonic action of $F(ab')_2$ fragments suggests a mechanism involving a change in capsular structure (40, 47). Possible mechanisms include an exposure of cryptic GXM motifs that could interact with macrophage receptors or an antibody-produced alteration in the fluidity of GXM at the capsular edge.

Binding of GXM MAbs to encapsulated cryptococci in culture significantly inhibits release of soluble GXM into the culture medium (31). Binding of MAbs also confers protection against gamma radiation-induced release of GXM from the cells (33). The extent to which such blockade of antigen release occurs in vivo is not known. However, given the numerous biological activities of GXM that might contribute to virulence (52), inhibition of GXM release could be one mechanism for the protective activity of GXM antibodies. Notably, a MAb that is protective and produced an annular ring by immunofluorescence was more effective in preventing GXM release than MAbs that were not protective and produced a punctate pattern.

CAPSULE-COMPLEMENT INTERACTIONS

The complement system is a group of proteins that have roles in both innate and antibody-dependent resistance to microbial infection. The complement system is activated in a cascade-like fashion. The consequences of activation of the complement cascade are threefold. First, activation of the complement system may lead to covalent binding of select complement proteins to the activating target cell or molecule. Binding of complement proteins produces a change in the cell surface and may lead to interactions between the bound complement protein and cellular receptors on phagocytes. Second, activation of the complement system leads to the generation of small-molecular-weight fragments of complement proteins that have potent biological activities that contribute to the inflammatory response to infection. Finally, activation may lead to formation of a membrane attack complex and lysis or death of the target cell. The thick cryptococcal cell wall appears to protect the yeast from the membrane attack complex.

Initiation of the complement system can occur by three pathways: the classical pathway, the lectin pathway, and the alternative pathway. The classical pathway depends on binding of antibody for initiation. The lectin pathway is activated when mannan-binding lectin (MBL) binds to carbohydrate structures found on a variety of microbial surfaces. The alternative pathway is a mechanism for initiation of the complement system that does not require the specific recognition provided by antibody or MBL. Depending on the circumstances, *C. neoformans* yeast cells can activate the complement system by any of these three pathways.

Role of the Complement System in Resistance to Cryptococcosis

Evidence of the importance of the complement system is provided in several reports of the course of cryptococcosis in normal versus complement-deficient mice. Mice deficient in C5 have decreased resistance to cryptococcosis (27, 45). C5 facilitates removal of cryptococci from the bloodstream within pulmonary vessels (27). A second approach to studying the role of the complement system in vivo is administration of cobra venom factor (CVF) to experimental animals. Treatment with CVF leads to a temporary depletion of C3 and C5 (51). CVF-treated experimental animals show a markedly increased susceptibility to disseminated cryptococcosis (2, 12). The availability of C3- and factor B–deficient mice has allowed for further evaluation of the role of the complement system in protection against infection. C3 plays critical roles in both the classical and alternative pathways, whereas factor B is a component of the alternative pathway. Mice deficient in either C3 or factor B show increased susceptibility to infection by *Cryptococcus gattii* (34). C3 appears to play a more important role than factor B in protection, suggesting that the alternative pathway of complement activation is not the only complement pathway contributing to protection.

Activation by the Alternative Pathway

Activation of the complement system occurs when encapsulated cryptococci are incubated in normal human serum. This activation occurs entirely via the action of the alternative pathway; normal serum lacks the amounts of anticapsular antibody needed for activation of the classical pathway. Activation of the alternative pathway leads to conversion of C3 to C3b, which in turn, binds to the cryptococcal cell.

The first evidence of activation of the complement system was provided by Goren and Warren, who noted that incubation of viable or killed cryptococci in fresh normal mouse, rabbit, guinea pig, or human sera led to deposition of globulins within the capsular matrix (11). Termed a psi reaction by the authors, the globulins were fragments of C3.

The importance of the alternative pathway in complement activation by encapsulated cryptococci was demonstrated by Diamond et al., who found that depletion of serum of the alternative pathway proteins properdin or factor B led to a marked reduction in the opsonic activity of the serum (3). Definitive evidence of C3 deposition via the alternative pathway was demonstrated when incubation of cryptococci in an alternative pathway reconstituted from purified alternative pathway protein factors B, D, H, I, and C3 and properdin led to activation and binding of C3 fragments to the capsule in a manner that is quantitatively and qualitatively identical to the pattern for activation and binding that occurs with normal human serum (22).

Examination of the cellular sites for C3 binding by immunofluorescence microscopy has provided considerable insight into the mechanism for activation of the alternative pathway (21). When cells are evaluated shortly after incubation in human serum (less than 2 min), no bound C3 is found. At 4 min of incubation, minute sites of bound C3 are observed. With increasing incubation time, the foci for bound C3 appear to expand with time until the entire capsule is filled with C3 fragments. Activation and binding of

C3 via the alternative pathway are asynchronous. That is, sites for initial binding of C3 occur at random locations and at random times. Those sites that are formed early in the process expand with time, likely reflecting amplification of the complement cascade. As the early sites enlarge with increased incubation time, new small sites are also initiated.

The number of bound C3 molecules is quite large: 10^7–10^8 C3 molecules per cell (17, 54). This is at least an order of magnitude greater than the number of C3 molecules binding to other fungi, e.g., zymosan particles (approximately 4×10^6 C3 fragments per particle [21]), *Candida albicans* blastoconidia (approximately 2×10^6 per cell [19]), conidia of *Aspergillus fumigatus* (2×10^6 [20]), or acapsular cryptococci (2×10^6 [21]). The large number of bound C3 molecules reflects the large volume for C3 binding that is presented by the three-dimensional capsule matrix. This three-dimensional volume contrasts with the more limited binding area presented by the two-dimensional surface on a cell that lacks a capsule.

The capsule itself is the site for C3 binding. The capsular location for C3 deposition has been demonstrated by both immunofluorescence and immune-electron microscopy. C3b binds to the capsule via ester linkages. The capsule is rich in hydroxyl groups that would be available for formation of ester bonds. Moreover, bound C3 is completely removed by treatment with hydroxylamine (16), a procedure that is very effective in stripping C3 bound to target cells by ester bonds (23, 24).

Once C3 is deposited in the capsule, it is rapidly degraded to iC3b (16, 43). C3b molecules bound to encapsulated cryptococci have a uniformly high susceptibility to conversion to iC3b by complement factors H and I (43). In contrast, a significant portion of C3b bound to nonencapsulated cryptococci is very resistant to conversion to iC3b by factors H and I. This rapid conversion of C3b to iC3b in the capsular matrix is a somewhat unexpected result because iC3b cannot support amplification of the alternative pathway. As noted above, activation of the complement cascade in the capsule is a nuclear event that is random and eventually expands to fill the capsule (21). It is likely that activation of the alternative pathway occurs as an expanding sphere from the nuclear initiation sites in which amplification occurs at the surface of the sphere, with rapid degradation of the bound C3b to iC3b occurring within and immediately beneath the surface of the sphere. The fact that most of the C3 is in the form of iC3b suggests that receptors with a high affinity for iC3b, i.e., CR3 and CR4, will be the primary receptors involved in phagocytosis of encapsulated cryptococci that are opsonized by incubation in normal serum. Thus, it is not surprising that CR3 and CR4 have been shown to play key roles in phagocytosis of the yeast (25).

The host species for serum can dramatically impact the site for C3 deposition in the cryptococcal capsule. Incubation of encapsulated cryptococci in human serum leads to binding of C3 at or very near the capsule edge (9). C3 deposition is further from the capsule edge with guinea pig or rat sera; in the case of mouse serum, there is C3 deposition deep within the capsule matrix, but there is little or no C3 at the capsule edge (9, 56). The differences in sites of binding of C3 from human or mouse sera are due to yet unknown differences in the behavior of C3 from the two species.

The density of the capsular matrix is another important variable in determining sites for C3 deposition (9, 10). For example, the capsule of yeast cells isolated from infected tissue have a high matrix density and deposit mouse C3 very near the capsular edge. In contrast, cryptococci grown in vitro under capsule induction conditions have a very low matrix density. Incubation of cryptococci with a low matrix density in mouse serum leads to deposition of C3 deep within the capsular matrix but not at the capsular edge.

There are at least two alternative explanations for binding of human and mouse C3 to different sites in the capsule matrix. First, human and mouse C3 could show different specificities for potential acceptors for binding of activated C3 that differ as a function of the position within the capsular matrix. For example, the studies of binding of GXM MAbs with different epitope specificities to sites within the capsular matrix suggest differences in epitope expression relative to the capsule edge (41). An alternative explanation is the possibility that human and mouse metastable C3b differ in their molecular lifetimes. With a decrease in potential acceptors for metastable C3b at the low capsule density at the capsule edge, metastable C3b might become inactivated before binding to the putative acceptor. If activated mouse C3b showed a shorter lifetime than activated human C3b, the mouse C3b might require a higher matrix density to allow for acceptor binding before inactivation occurs.

Activation via the Lectin Pathway

MBL is a C-type serum lectin that binds with high affinity to the simple carbohydrates *N*-acetylglucosamine and mannose that may be found on the surfaces of viruses, bacteria, and fungi. Mannose-associated serine proteases that are associated with MBL become activated when MBL binds to a microbial surface and activate the complement cascade.

MBL binds to acapsular cryptococci in a calcium-dependent manner (46). This binding enhances complement activation and uptake of acapsular cells by polymorphonuclear cells (50). In contrast to acapsular cryptococci, MBL shows little binding to encapsulated cryptococci (26, 46), and encapsulated cryptococci show little or no activation of the complement cascade via the lectin pathway (50). GXM, the predominant constituent of the cryptococcal capsule, has a backbone of mannose residues. The lack of binding of MBL to the capsular matrix indicates that the mannose residues in GXM are not exposed in a position needed for binding of MBL. Given the limited binding of MBL to encapsulated cryptococci, it is not surprising that no correlation has been found between MBL deficiency and cryptococcosis in nonimmunocompromised patients (7).

The failure of MBL to bind to encapsulated cryptococci further suggests that the molecular sieving activity of the capsule prevents access of this high-molecular-weight lectin to binding sites at the cell wall. Recent studies indicate that wild-type cell wall architecture is also a factor that inhibits binding of MBL to encapsulated cryptococci (42). MBL binds to the cell wall of encapsulated cryptococci when cell integrity is perturbed by mutation of *CCR4*. This result suggests that integrity of the cell wall architecture is sufficient to protect the cell from MBL binding.

Activation by the Classical Pathway

The lag in activation and binding of C3 to encapsulated cryptococci via the alternative pathway is greatly reduced if anticapsular antibody is present for activation of the classical pathway. However, the ability of an antibody to support C3 deposition via the classical pathway is dependent on the epitope specificity of the antibody (13). The importance of epitope specificity in complement activation has only been appreciated with the availability of MAbs that are reactive with GXM. Depending on the epitope specificity of the antibody, addition of an anti-GXM MAb to a mixture of normal serum and encapsulated cryptococci may (i) produce early C3 deposition via the classical pathway, (ii) have no effect on C3 deposition via the classical pathway, (iii) suppress C3 binding via the alternative pathway, or (iv) have no effect on the alternative pathway.

Using serotype A cells as the activating particle, MAbs that are reactive with an epitope that is shared by capsular serotypes A, B, C, and D produced very rapid C3 deposition on the cells. In contrast, MAbs reactive with only serotypes A and D failed to activate the classical pathway. The MAbs that activate the classical pathway are the same antibodies that produce the rim-type capsule reaction described in a preceding section; the MAbs that fail to activate the classical pathway produce puffy-type reactions on serotype A cells.

In a study of the effect of GXM MAbs on C3 deposition from mouse serum, Zaragoza and Casadevall found that incubation of encapsulated cryptococci in the presence of both GXM MAbs and normal mouse serum led to deposition of C3 at the capsular edge (55). This deposition at the capsular edge contrasts with deposition beneath the capsule surface that occurs via the alternative pathway with mouse serum. These results demonstrate that GXM MAbs alter both the kinetics and location for deposition of C3 that would normally occur via the alternative pathway in the absence of antibody.

Finally, early, antibody-mediated activation and binding of C3 to the capsular surface may occur at a price. As noted above, the cryptococcal capsule itself is a powerful activator of the complement system, leading to deposition of potentially opsonic fragments of C3 into the capsular matrix. This C3 deposition follows the relatively slow kinetics that are characteristic of the alternative pathway. C3 deposition in the presence of antibodies that produce the rim-type capsular reaction is much more rapid than the deposition that occurs via the alternative pathway alone (13, 28). However, the amount of C3 that is bound to the capsular matrix is much lower (approximately fivefold lower) than the amount that would be bound in the absence of the antibody. In contrast, MAbs that produce the puffy-type capsular reaction had no effect on C3 deposition via the alternative pathway. One explanation for the suppression of overall levels of C3 accumulation in the presence of rim-type antibodies is the likelihood that cross-linking of the capsular matrix by the antibodies prevents penetration of complement proteins to the capsule interior (15, 28). This explanation is supported by the fact that Fab fragments of MAbs that normally suppress C3 deposition via the alternative pathway bind to the capsule but do not have any effect on either the kinetics or the amount of C3 binding (15, 28).

Impact of C3 Deposition on the Capsular Matrix

Incubation of encapsulated cryptococci in normal human serum leads to deposition of large amounts of protein, i.e., C3 fragments, into the capsular matrix. Such protein alters the refractile properties of the capsule. For example, Diamond et al. found that incubation of cryptococci in fresh guinea pig or human serum led to a reaction similar to the quellung reaction (3). However, the extent of penetration of C3 into the capsule is limited by the molecular sieving action of the capsule (10). As a consequence, alterations in refractile properties of the capsule are limited to the outer region of the capsule that allows for penetration of complement proteins into the capsular matrix.

Acapsular Cryptococci: a Special Case

Unlike the slow, asynchronous binding of C3 that occurs when encapsulated cryptococci are incubated in normal serum, incubation of acapsular cryptococci in normal serum occurs rapidly and synchronously (21). This rapid deposition of C3 on acapsular cryptococci is due to activation of the classical complement pathway by antibodies to the cryptococcal cell wall that are found in normal human serum (53). Adsorption of human serum with acapsular cryptococci removes antibodies reactive with the cell wall. Incubation of acapsular cryptococci in adsorbed serum leads to deposition of C3 onto the cell wall, but C3 binding occurs via the slow kinetics characteristic of the alternative pathway. In contrast, encapsulated cryptococci are unable to adsorb the antibodies from normal serum that mediate the rapid, early deposition of C3 onto acapsular cryptococci. Such blockade of antigenic sites by the capsule may be due to molecular sieving to prevent penetration of antibodies to the capsule interior or to a direct blockade of sites at the cell wall itself.

SUMMARY

The cryptococcal capsule is composed largely of water that fills the loose polysaccharide capsular matrix (32). Binding of anticapsular antibody or proteins of the complement cascade within the capsule introduces a large amount of protein into the matrix. Binding of antibodies produces an additional effect. Some antibodies cross-link the matrix due to the multivalent nature of antibody binding. Such binding can be visualized by examination by DIC microscopy (rim pattern) or by indirect immunofluorescence (annular pattern). These antibodies typically activate the classical complement pathway, are highly opsonic, and tend to be protective in murine models of cryptococcosis. Other antibodies with different epitope specificities may fail to cross-link the matrix. Binding of these non-cross-linking antibodies is apparent by DIC microscopy as a puffy-type pattern or as a punctate pattern when viewed by immunofluorescence. These antibodies that fail to cross-link are typically nonprotective. Finally, the capsular matrix itself is a potent activator of the alternative complement pathway in the absence of anticapsular antibody. Activation of the alternative pathway by the cryptococcal capsule is delayed and begins as asynchronously appearing sites of nuclear C3 deposition that expand with incubation time to fill the capsule. This potent activation of the alternative pathway contrasts with the often reported inability of most microbial capsules to activate the complement system. The chemical or structural features of the cryptococcal capsule that make it such a powerful activator of the alternative pathway are not known.

REFERENCES

1. **Casadevall, A., J. Mukherjee, S. J. N. Devi, R. Schneerson, J. B. Robbins, and M. D. Scharff.** 1992. Antibodies elicited by a *Cryptococcus neoformans*-tetanus toxoid conjugate vaccine have the same specificity as those elicited in infection. *J. Infect. Dis.* **165:**1086–1093.
2. **Diamond, R. D., J. E. May, M. Kane, M. M. Frank, and J. E. Bennett.** 1973. The role of late complement components and the alternate complement pathway in experimental cryptococcosis. *Proc. Soc. Exp. Biol. Med.* **144:**312–315.
3. **Diamond, R. D., J. E. May, M. A. Kane, M. M. Frank, and J. E. Bennett.** 1974. The role of the classical and alternate complement pathways in host defenses against *Cryptococcus neoformans* infection. *J. Immunol.* **112:**2260–2270.
4. **Dromer, F., C. Perronne, J. Barge, J. L. Vilde, and P. Yeni.** 1989. Role of IgG and complement component C5 in the initial course of experimental cryptococcosis. *Clin. Exp. Immunol.* **78:**412–417.

5. **Dromer, F., J. Salamero, A. Contrepois, C. Carbon, and P. Yeni.** 1987. Production, characterization, and antibody specificity of a mouse monoclonal antibody reactive with *Cryptococcus neoformans* capsular polysaccharide. *Infect. Immun.* **55:**742–748.
6. **Eckert, T. F., and T. R. Kozel.** 1987. Production and characterization of monoclonal antibodies specific for *Cryptococcus neoformans* capsular polysaccharide. *Infect. Immun.* **55:**1895–1899.
7. **Eisen, D. P., M. M. Dean, M. V. O'Sullivan, S. Heatley, and R. M. Minchinton.** 2008. Mannose-binding lectin deficiency does not appear to predispose to cryptococcosis in non-immunocompromised patients. *Med. Mycol.* **46:**371–375.
8. **Evans, E. E.** 1950. The antigenic composition of *Cryptococcus neoformans*. I. A serologic classification by means of the capsular and agglutination reactions. *J. Immunol.* **64:**423–430.
9. **Gates, M. A., and T. R. Kozel.** 2006. Differential localization of complement component 3 within the capsular matrix of *Cryptococcus neoformans*. *Infect. Immun.* **74:**3096–3106.
10. **Gates, M. A., P. Thorkildson, and T. R. Kozel.** 2004. Molecular architecture of the *Cryptococcus neoformans* capsule. *Mol. Microbiol.* **52:**13–24.
11. **Goren, M. B., and J. Warren.** 1968. Immunofluorescence studies of reactions at the cryptococcal capsule. *J. Infect. Dis.* **118:**215–229.
12. **Graybill, J. R., and J. Ahrens.** 1981. Immunization and complement interaction in host defence against murine cryptococcosis. *J. Reticuloendothel. Soc.* **30:**347–357.
13. **Kozel, T. R., B. C. H. deJong, M. M. Grinsell, R. S. MacGill, and K. K. Wall.** 1998. Characterization of anticapsular monoclonal antibodies that regulate activation of the complement system by the *Cryptococcus neoformans* capsule. *Infect. Immun.* **66:**1538–1546.
14. **Kozel, T. R., and J. L. Follette.** 1981. Opsonization of encapsulated *Cryptococcus neoformans* by specific anticapsular antibody. *Infect. Immun.* **31:**978–984.
15. **Kozel, T. R., R. S. MacGill, and K. K. Wall.** 1998. Bivalency is required for anti-capsular monoclonal antibodies to optimally suppress activation of the alternative complement pathway by the *Cryptococcus neoformans* capsule. *Infect. Immun.* **66:**1547–1553.
16. **Kozel, T. R., and G. S. T. Pfrommer.** 1986. Activation of the complement system by *Cryptococcus neoformans* leads to binding of iC3b to the yeast. *Infect. Immun.* **52:**1–5.
17. **Kozel, T. R., G. S. T. Pfrommer, A. S. Guerlain, B. A. Highison, and G. J. Highison.** 1988. Strain variation in phagocytosis of *Cryptococcus neoformans*: dissociation of susceptibility to phagocytosis from activation and binding of opsonic fragments of C3. *Infect. Immun.* **56:**2794–2800.
18. **Kozel, T. R., P. Thorkildson, S. Brandt, W. H. Welch, J. A. Lovchik, D. P. AuCoin, J. Vilai, and C. R. Lyons.** 2007. Protective and immunochemical activities of monoclonal antibodies reactive with the *Bacillus anthracis* polypeptide capsule. *Infect. Immun.* **75:**152–163.
19. **Kozel, T. R., L. C. Weinhold, and D. M. Lupan.** 1996. Distinct characteristics of initiation of the classical and alternative complement pathways by *Candida albicans*. *Infect. Immun.* **64:**3360–3368.
20. **Kozel, T. R., M. A. Wilson, T. P. Farrell, and S. M. Levitz.** 1989. Activation of C3 and binding to *Aspergillus fumigatus* conidia and hyphae. *Infect. Immun.* **57:**3412–3417.
21. **Kozel, T. R., M. A. Wilson, and J. W. Murphy.** 1991. Early events in initiation of alternative complement pathway activation by the capsule of *Cryptococcus neoformans*. *Infect. Immun.* **59:**3101–3110.
22. **Kozel, T. R., M. A. Wilson, G. S. T. Pfrommer, and A. M. Schlageter.** 1989. Activation and binding of opsonic fragments of C3 on encapsulated *Cryptococcus neoformans* by using an alternative complement pathway reconstituted from six isolated proteins. *Infect. Immun.* **57:**1922–1927.
23. **Law, S. K., and R. P. Levine.** 1977. Interaction between the third complement protein and cell surface macromolecules. *Proc. Natl. Acad. Sci. USA* **74:**2701–2705.
24. **Law, S. K., N. A. Lichtenberg, and R. P. Levine.** 1979. Evidence for an ester linkage between the labile binding site of C3b and receptive surfaces. *J. Immunol.* **123:**1388–1394.
25. **Levitz, S. M., and A. Tabuni.** 1991. Binding of *Cryptococcus neoformans* by human cultured macrophages: requirements for multiple complement receptors and actin. *J. Clin. Invest.* **87:**528–535.
26. **Levitz, S. M., A. Tabuni, and C. Treseler.** 1993. Effect of mannose-binding protein on binding of *Cryptococcus neoformans* to human phagocytes. *Infect. Immun.* **61:**4891–4893.
27. **Lovchik, J. A., and M. F. Lipscomb.** 1993. Role of C5 and neutrophils in the pulmonary intravascular clearance of circulating *Cryptococcus neoformans*. *J. Respir. Cell Mol. Biol.* **9:**617–627.
28. **MacGill, T. C., R. S. MacGill, A. Casadevall, and T. R. Kozel.** 2000. Biological correlates of capsular (quellung) reactions of *Cryptococcus neoformans*. *J. Immunol.* **164:**4835–4842.
29. **Macura, N., T. Zhang, and A. Casadevall.** 2007. Dependence of macrophage phagocytic efficacy on antibody concentration. *Infect. Immun.* **75:**1904–1915.
30. **Maitta, R. W., K. Datta, Q. Chang, R. X. Luo, B. Witover, K. Subramaniam, and L. A. Pirofski.** 2004. Protective and nonprotective human immunoglobulin M monoclonal antibodies to *Cryptococcus neoformans* glucuronoxylomannan manifest different specificities and gene use profiles. *Infect. Immun.* **72:**4810–4818.
31. **Martinez, L. R., D. Moussai, and A. Casadevall.** 2004. Antibody to *Cryptococcus neoformans* glucuronoxylomannan inhibits the release of capsular antigen. *Infect. Immun.* **72:**3674–3679.
32. **Maxson, M. E., E. Cook, A. Casadevall, and O. Zaragoza.** 2007. The volume and hydration of the *Cryptococcus neoformans* polysaccharide capsule. *Fungal. Genet. Biol.* **44:**180–186.
33. **Maxson, M. E., E. Dadachova, A. Casadevall, and O. Zaragoza.** 2007. Radial mass density, charge, and epitope distribution in the *Cryptococcus neoformans* capsule. *Eukaryot. Cell* **6:**95–109.
34. **Mershon, K. L., A. Vasuthasawat, G. W. Lawson, S. L. Morrison, and D. O. Beenhouwer.** 2009. Role of complement in protection against *Cryptococcus gattii* infection. *Infect. Immun.* **77:**1061–1070.
35. **Mukherjee, J., W. Cleare, and A. Casadevall.** 1995. Monoclonal antibody mediated capsular reactions (quellung) in *Cryptococcus neoformans*. *J. Immunol. Methods* **184:**139–143.
36. **Mukherjee, J., T. R. Kozel, and A. Casadevall.** 1999. Monoclonal antibodies reveal additional epitopes of serotype D *Cryptococcus neoformans* capsular glucuronoxylomannan that elicit protective antibodies. *J. Immunol.* **161:**3557–3568.
37. **Mukherjee, J., S. C. Lee, and A. Casadevall.** 1995. Antibodies to *Cryptococcus neoformans* glucuronoxylomannan enhance antifungal activity of murine macrophages. *Infect. Immun.* **63:**573–579.
38. **Mukherjee, J., G. Nussbaum, M. D. Scharff, and A. Casadevall.** 1995. Protective and non-protective monoclonal antibodies to *Cryptococcus neoformans* originating from one B-cell. *J. Exp. Med.* **181:**405–409.
39. **Neill, J. M., C. G. Castillo, R. H. Smith, and C. E. Kapros.** 1949. Capsular reactions and soluble antigens of *Torula histolytica* and of *Sporotrichum schenckii*. *J. Exp. Med.* **89:**93–106.

40. **Netski, D., and T. R. Kozel.** 2002. Fc-dependent and Fc-independent opsonization of *Cryptococcus neoformans* by anticapsular monoclonal antibodies: importance of epitope specificity. *Infect. Immun.* **70:**2812–2819.
41. **Nussbaum, G., W. Cleare, A. Casadevall, M. D. Scharff, and P. Valadon.** 1997. Epitope location in the *Cryptococcus neoformans* capsule is a determinant of antibody efficacy. *J. Exp. Med.* **185:**685–695.
42. **Panepinto, J. C., K. W. Komperda, M. Hacham, S. Shin, X. Liu, and P. R. Williamson.** 2007. Binding of serum mannan binding lectin to a cell integrity-defective *Cryptococcus neoformans ccr4Δ* mutant. *Infect. Immun.* **75:**4769–4779.
43. **Pfrommer, G. S. T., S. M. Dickens, M. A. Wilson, B. J. Young, and T. R. Kozel.** 1993. Accelerated decay of C3b to iC3b when C3b is bound to the *Cryptococcus neoformans* capsule. *Infect. Immun.* **61:**4360–4366.
44. **Rakesh, V., A. D. Schweitzer, O. Zaragoza, R. Bryan, K. Wong, A. Datta, A. Casadevall, and E. Dadachova.** 2008. Finite-element model of interaction between fungal polysaccharide and monoclonal antibody in the capsule of *Cryptococcus neoformans*. *J. Phys. Chem. B* **112:**8514–8522.
45. **Rhodes, J. C.** 1985. Contribution of complement component C5 to the pathogenesis of experimental murine cryptococcosis. *Sabouraudia* **23:**225–234.
46. **Schelenz, S., R. Malhotra, R. B. Sim, U. Holmskov, and G. J. Bancroft.** 1995. Binding of host collectins to the pathogenic yeast *Cryptococcus neoformans*: human surfactant protein D acts as an agglutinin for acapsular yeast cells. *Infect. Immun.* **63:**3360–3366.
47. **Taborda, C. P., and A. Casadevall.** 2002. CR3 (CD11b/CD18) and CR4 (CD11c/CD18) are involved in complement-independent antibody-mediated phagocytosis of *Cryptococcus neoformans*. *Immunity* **16:**791–802.
48. **Taborda, C. P., and A. Casadevall.** 2001. Immunoglobulin M efficacy against *Cryptococcus neoformans*: mechanism, dose dependence, and prozone-like effects in passive protection experiments. *J. Immunol.* **166:**2100–2107.
49. **Taborda, C. P., J. Rivera, O. Zaragoza, and A. Casadevall.** 2003. More is not necessarily better: prozone-like effects in passive immunization with IgG. *J. Immunol.* **170:**3621–3630.
50. **van Asbeck, E. C., A. I. Hoepelman, J. Scharringa, B. L. Herpers, and J. Verhoef.** 2008. Mannose binding lectin plays a crucial role in innate immunity against yeast by enhanced complement activation and enhanced uptake of polymorphonuclear cells. *BMC Microbiol.* **8:**229.
51. **Van den Berg, C. W., P. C. Aerts, and H. van Dijk.** 1991. In vivo anti-complementary activities of the cobra venom factors from *Naja naja* and *Naja haje*. *J. Immunol. Methods* **136:**287–294.
52. **Vecchiarelli, A.** 2000. Immunoregulation by capsular components of *Cryptococcus neoformans*. *Med. Mycol.* **38:**407–417.
53. **Wilson, M. A., and T. R. Kozel.** 1992. Contribution of antibody in normal human serum to early deposition of C3 onto encapsulated and nonencapsulated *Cryptococcus neoformans*. *Infect. Immun.* **60:**754–761.
54. **Young, B. J., and T. R. Kozel.** 1993. Effects of strain variation, serotype and structural modification on the kinetics for activation and binding of C3 to *Cryptococcus neoformans*. *Infect. Immun.* **61:**2966–2972.
55. **Zaragoza, O., and A. Casadevall.** 2006. Monoclonal antibodies can affect complement deposition on the capsule of the pathogenic fungus *Cryptococcus neoformans* by both classical pathway activation and steric hindrance. *Cell Microbiol.* **8:**1862–1876.
56. **Zarogoza, O., C. P. Taborda, and A. Casadevall.** 2003. The efficacy of complement-mediated phagocytosis of *Cryptococcus neoformans* is dependent on the location of C3 in the polysaccharide capsule and involves both direct and indirect C3-mediated interactions. *Eur. J. Immunol.* **33:**1957–1967.

Cryptococcus: From Human Pathogen to Model Yeast
Edited by J. Heitman et al.

30

Cryptococcus Interactions with Innate Cytotoxic Lymphocytes

SHAUNNA M. HUSTON AND CHRISTOPHER H. MODY

Investigations of immunity to *Cryptococcus* have helped disassemble the paradigm that T cells, natural killer T (NKT) cells, and natural killer (NK) cells kill by recognizing combinations of self and either altered self or nonself. Moreover, studies of cryptococcal host defense have demonstrated innate T-cell responses to components of *Cryptococcus* in addition to their contribution to adaptive immunity. The traditional view is that the devastating potential of cytotoxic T cells, NK cells, and NKT cells is contained by highly specific interactions utilizing the exquisite sensitivity of the T-cell receptor (TCR) or the combination of activating and inhibitory NK receptors on the host cell. During this process, cytotoxic lymphocytes recognize and bind to their target. This results in stimulation of the lymphocyte through mechanisms that are ultimately responsible for killing the target. However, Murphy and McDaniel made a fascinating discovery in 1982, showing that NK cells could directly kill *Cryptococcus neoformans* (73), dismantling the paradigm that NK cells kill only tumor and microbe-infected cells. This work triggered a series of studies demonstrating that cytotoxic lymphocytes function in solitude to directly kill microbes. These studies have shown direct NK cell killing of a variety of pathogens (Table 1).

In addition to this cytotoxicity, these cells are also important in the production and secretion of cytokines such as gamma interferon (IFN-γ), which activates innate and adaptive immunity (reviewed in reference 8). Moreover, NKT cells and T cells also function as innate cytotoxic lymphocytes through the production of cytokines as well as the capacity to directly kill their targets (Table 1) (110). For example, production of IFN-γ by NKT cells in response to *Cryptococcus* results in enhanced elimination of the organism in murine models (44–46).

Conventional CD4+ and CD8+ T cells that express TCR and CD3 were previously thought to function only through the adaptive immune response in association with antigen specificity and major histocompatibility complex (MHC) restriction. However, more recently CD4+ and CD8+ T cells, cultured in vitro with *Cryptococcus*, exhibited direct killing of the microbe (Table 1) (54). The focus of this chapter is to portray our current understanding of the mechanisms involved in direct killing of *Cryptococcus* by innate cytotoxic lymphocytes. In addition, we have provided future insights into research in this area and speculated on potential clinical relevance.

NK CELLS AND HOST DEFENSE

NK cells are innate immune cells that recognize and directly kill altered host cells and pathogens. NK cells are large granular lymphocytes that do not express CD3 or TCR but do express CD16 (FcγRIII) and CD56 in humans or NK1.1 in mice (99; reviewed in references 16, 81, 83). To kill, NK cells must initially bind to the target cell (75). Binding results in NK-cell reorganization of the actin and microtubular cytoskeleton that is important for formation of the immunological synapse (NKIS) (13, 30). Although the receptors used in microbial killing have not yet been identified, formation of the NKIS results in increased localization of adhesion molecules, activation receptors, and signaling molecules. Adhesion molecules enhance the interactions between the NK cell and its target (18, 79) to allow for high concentrations of NK-cell effector molecules that kill more efficiently (68). Activation receptors move to the NKIS so that they are in close proximity to target ligands. Once the stimulatory threshold is reached, due to multiple receptor ligand interactions, the NK cell becomes activated (reviewed in reference 52). Localization of signaling molecules at the NKIS is also important to allow for initiation of signaling pathways involved in stimulation and release of effector molecules (Fig. 1).

It is important to note that during tumor killing NK cells utilize a unique balance of receptors. These receptors consist of various inhibitory and activation receptors (reviewed in reference 52). The balance upon stimulation of the combination of these receptors ultimately determines the fate of the signaling pathway. If the pendulum swings toward

Shaunna M. Huston, The Snyder Institute for Infection, Inflammation, and Immunity and Department of Internal Medicine, University of Calgary, Calgary, Alberta, Canada. **Christopher H. Mody,** The Snyder Institute for Infection, Inflammation, and Immunity and Departments of Internal Medicine, Microbiology, and Infectious Diseases, University of Calgary, Calgary, Alberta, Canada.

TABLE 1 Microorganisms directly or indirectly killed by innate cytotoxic lymphocytes

Innate effector lymphocyte	Microorganism	Reference
NK cells	*Mycobacterium tuberculosis*	9
	C. neoformans	73
	Paracoccidioides brasiliensis	40
	Coccidioides immitis	82
	Histoplasma capsulatum	98
	Leishmania major	53
NKT cells	*Streptococcus pneumoniae*	48
	Pseudomonas aeruginosa	77
	Borrelia burgdorferi	51
	L. major	39
	Salmonella typhimurium	6
	M. tuberculosis	4
	Schistosoma mansoni	24
	C. neoformans	45
T cells	*M. tuberculosis*	12
	Mycobacterium leprae	78
	S. mansoni	21
	Entamoeba histolytica	87
	Toxoplasma gondii	49
	Plasmodium falciparum	22
	P. aeruginosa	62
	C. neoformans	54
	C. albicans	5

recognition of self, stimulation of inhibitory receptors causes the NK cells to remain noncytotoxic (tolerant). On the other hand, exceeding the threshold of activation receptors results in triggering of the cytotoxic mechanisms responsible for killing the target. Therefore, the expression of both inhibitory and activation receptors allows NK cells to distinguish between self, altered self, and nonself. However, since there is no evidence that *Cryptococcus* possesses an inhibitory ligand, it is likely that NK responses are dominated by activating receptors. Moreover, while our past understanding of NK cell function comes mainly from observations made with tumor and virus-infected cells, a large contribution to our understanding of both the NK cells and host defense has been accomplished through studies of NK-cell interactions with *Cryptococcus*. This comes from the observation that killing of *Cryptococcus* by NK cells has similarities to NK-cell killing of tumor targets but also uses unique mechanisms to kill *Cryptococcus* compared to killing of tumor targets.

NK Cells and Host Defense against *Cryptococcus*

In 1982, Murphy and McDaniel pioneered the observation that murine splenic NK cells possess direct anticryptococcal activity (73). Splenic NK cells in these early studies were defined as nylon wool nonadherent, Thy1$^-$, surface Ig$^-$, Ia$^-$, asialo GM1-positive cells, with the functional capacity to kill YAC-1 tumor targets (73). It is important to note that NK cells were anticryptococcal without any prior stimulation (73). Thus, the stimulus provided by *Cryptococcus* itself is sufficient to activate the NK cells to kill. This observation was significant to both the *Cryptococcus* and NK-cell field. Previous to this, no studies had shown NK-cell killing of targets outside of the realm of virus-infected host cells or tumor targets, although many have followed since (Table 1).

Later, studies were employed to determine if NK cells were simply inhibiting the growth of *Cryptococcus* or if they were actually killing the organism. The evidence that NK cells kill *Cryptococcus* came from studies showing decreased 3-(4,5-dimethylthiazol-2-yl)-2,5-diphenyl tetrazolium bromide and fluoresein diacetate uptake, increased membrane permeability to propridium iodide, the ability to reduce the number of organisms below the initial inoculum, and limiting dilution killing assays (35; A. Islam, unpublished data).

Mouse models have been used extensively to determine the mechanisms involved in NK-cell killing of *Cryptococcus* both in vitro and in vivo. These early studies were facilitated

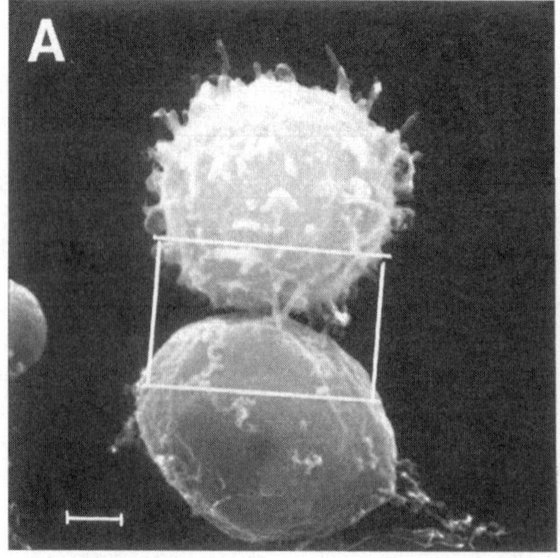

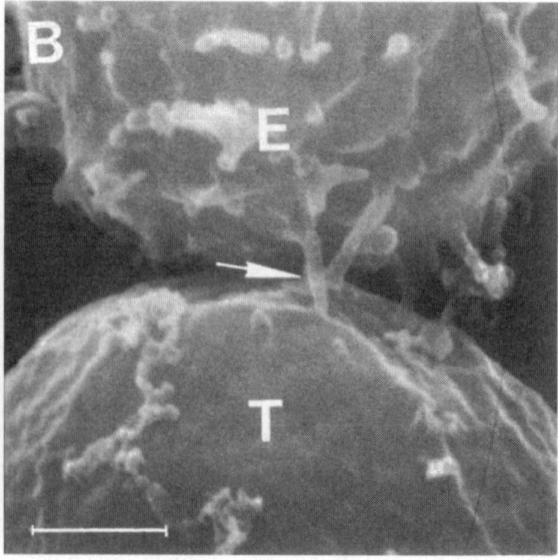

FIGURE 1 (A) Scanning electron micrograph of an NK-cell conjugate with *C. neoformans*. (B) Higher magnification of the area from panel A, showing appendages from the NK effector cell directed at the *C. neoformans* target. Image from Nabavi and Murphy (75). Reprinted from *Infection and Immunity* with permission of the publisher.

by previous observations that asialo GM1 was expressed on NK cells and that injection of mice with anti-asialo GM1 antibody abrogated NK-cell killing of tumor and virus-infected targets. To determine if NK cells play a significant role in elimination of *Cryptococcus* in vivo, investigators took advantage of this well-established model. Hidore and colleagues used cyclophosphamide to initially deplete mice of natural effector cells (32). Syngeneic, nylon wool nonadherent, splenic cells (NK cells) were then adoptively transferred into the depleted mice. The mice were injected intravenously with *Cryptococcus* in the presence or absence of anti-asialo GM1, which depletes NK cells. The results showed an increase in the number of *Cryptococcus* in the lungs, liver, and spleen in mice that had been injected with anti-asialo GM1 compared to controls, suggesting that NK cells were responsible for effective elimination of *Cryptococcus* in these organs (32). Further support for the importance of NK cells in vivo comes from the beige murine model and from the use of anti-NK1.1-depleting antibody. The beige mutation (*bg/bg*) results in suboptimal NK-cell function (84, 86), and studies showed an increase in *Cryptococcus* CFUs in the spleen and lungs of *bg/bg* mice compared to *bg/+* mice (34). Also, following the discovery of the NK1.1 surface molecule on murine NK cells, mice were injected with anti-NK1.1 to deplete the NK cells. Consistent with past results, there was an increase in lung CFUs when the mice were injected with the anti-NK1.1 compared to controls (58, 89). Together, the results of these studies implicate NK cells in optimal defense of *Cryptococcus*.

The response of NK cells to *Cryptococcus* has been extended to human studies. Human NK cells can be obtained from the blood, where they are present as a small percentage of the total lymphocyte population. Studies by Levitz and colleagues demonstrated that human primary NK cells are able to kill *Cryptococcus* in vitro (54). However, unlike murine NK cells, primary blood NK cells kill *Cryptococcus* even more efficiently in the presence of anticryptococcal antibody (69), indicating a role for antibody-dependent cellular cytotoxicity (ADCC). In another study, investigators chose to analyze primary NK cells from patients with cryptococcosis. NK cells from these patients showed limited capacity to kill *Cryptococcus* in vitro compared to healthy donors (29). Although these data do not exclude a defect produced by *C. neoformans*, they raise the intriguing possibility that a defect in NK cell function might contribute to development of infection. In addition, this study showed that the function of NK cells in killing of *Cryptococcus* is recovered in the presence of cytokines such as interleukin-2 (IL-2) (29). Therefore, growth inhibition of *Cryptococcus* in humans may require more stringent environmental conditions than their murine counterparts. It is also of interest that NK deficiencies have been cited as the etiology of other human mycoses (1, 27, 42, 50, 67).

Binding Is Required for NK-Cell Killing of *Cryptococcus*

Cryptococcus binds to NK cells, and binding is a prerequisite to NK-cell killing. This is similar to binding of tumor cells, which is a prerequisite to tumor killing (85). Initial studies, using transmission electron microscopy, showed intimate association of *Cryptococcus* with murine NK cells (Fig. 1) (75). In addition, binding with subsequent conjugate formation correlated with killing of *Cryptococcus* (75). Further evidence came from the human NK cell line, YT. In these experiments, YT cells were separated from *Cryptococcus* by a membrane, and the ability of YT cells to kill *Cryptococcus* was analyzed by CFUs (60). YT-cell killing was abrogated under these conditions (60). These data support the contention that binding was essential for NK-cell killing of *Cryptococcus*.

At least in part, the mechanisms involved in binding of *Cryptococcus* to NK cells requires formation of disulfide bonds and actin polymerization as assessed by experiments that employed scanning and transmission electron microscopy in conjunction with assays to assess NK-cell killing of *Cryptococcus* (71). Specifically, in the presence of the cytochalasin B that inhibits actin polymerization, conjugates did not form between *Cryptococcus* and NK cells, and NK cells were unable to kill *Cryptococcus* (71). These studies highlight the importance of actin in NK cell conjugate formation with *Cryptococcus* and killing of the organism.

While the receptor and ligand remain elusive, binding of *Cryptococcus* to NK cells was found to be magnesium-dependent (33). Murine splenic NK cell conjugates with *Cryptococcus* were observed by light microscopy, using alcian blue staining to identify *Cryptococcus* (33). Depletion of magnesium, but not calcium, resulted in decreased binding of *Cryptococcus* to the murine splenic NK cells (33). Using the same model, binding was also observed to be temperature independent (33). Although these are the minimal requirements for NK-cell binding to *Cryptococcus*, magnesium, calcium, and appropriate temperature (37°C) are required for NK-cell killing of *Cryptococcus* (33).

In the search for the receptor responsible for *Cryptococcus* conjugate formation, known receptors of NK cells have been considered. It is important to note that the binding receptor used by NK cells to form conjugates with *Cryptococcus* may be distinct from the receptor responsible for release of effector molecules that kill *Cryptococcus*. Nevertheless, an important receptor used by NK cells in binding to tumor targets is the heterodimer CD11a/CD18 beta 2 integrin, also known as lymphocyte-function-associated antigen-1 (LFA-1) (3, 11, 14, 36, 65, 66, 94). It is also of interest that CD18 is involved in other host-microbe interactions. Sewald and colleagues recently reported that beta 2 integrin (CD18) mediates entry of VacA, a pore-forming toxin of *Helicobacter pylori*, into human T cells (91). Attempts to explore the role of LFA-1 were motivated by observations that cryptococcal galactoxylomannan and glucuronoxylomannan (GXM) bind to CD18 on neutrophils (19), and macrophages from CD18-deficient mice showed reduced phagocytosis of *Cryptococcus* compared to wild-type mice (97). There is also evidence that CD18 may play a role in effective elimination of *Cryptococcus* in vivo. CD18-deficient mice injected with GXM show modest accumulation of GXM in the liver compared to wild-type mice (106). The sum of this evidence raised the possibility that *Cryptococcus* bound to NK cells via LFA-1 as observed with tumor targets. However, recent evidence indicates that LFA-1 is not the NK-cell receptor responsible for antibody-independent binding to *Cryptococcus* (41). Using LFA-1-sufficient and LFA-1-deficient YT cells, unlike tumor cells, binding of *Cryptococcus* to LFA-1-deficient YT cells was comparable to that of the LFA-1-sufficient cells. In addition, LFA-1-deficient YT cells killed *Cryptococcus* as well as LFA-1-sufficient YT cells (41). Therefore, it appears that unlike tumor-cell killing, LFA-1 is dispensable for binding and killing of *Cryptococcus*.

Studies have also examined the role of other potential receptors. Although GXM also binds to CD14, Toll-like receptor 2 (TLR2), and TLR4 (92, 107), there is no evidence that these molecules mediate NK-cell anticryptococcal activity, since CD14 is not expressed on NK cells, and anti-TLR2 and TLR4 blocking antibodies had no affect on the ability of NK cells to kill *C. neoformans* (41). We, and others, have demonstrated that dendritic cells recognize *C. neoformans* via interactions between lectin receptors and manosylated proteins found in the cryptococcal cell wall (61, 95). However, blocking lectin receptors by pretreating NK cells with soluble mannan failed to inhibit NK-cell anticryptococcal responses (41). Lastly, many independent studies have demonstrated that the immune system can respond to fungal pathogens such as *Candida albicans* and *Aspergillus fumigatus* through interaction of dectin-1 with fungal β-glucans (10, 26, 28, 80). However, dectin-1-deficient mice remain resistant to *C. neoformans* infection (76), and anticryptococcal activity persisted in NK cells despite being pretreated with soluble β-glucans (41); therefore, there is no evidence that recognition of *C. neoformans* β-glucans by dectin-1 underlies NK-cell anticryptococcal activity.

NK cells can also recognize tumor cells, virus-infected cells, or microbes by antibody-mediated opsonization. NK cells use Fc receptors to recognize tumor and virally infected target cells that are coated with antibody. The process of recognition and killing of the target in this manner is known as ADCC. Phenotypic alterations in the tumor or virus-infected cells allow for binding of antibodies to the cell surface. The Fc portion of these antibodies has affinity for the Fc receptors, including CD16 (also known as FcγRIII) found on subsets of NK cells. Ligation of CD16 stimulates an activation pathway that ultimately results in cytotoxicity, associated with killing of the target cell (103). With this in mind, killing of *Cryptococcus* by murine nylon wool nonadherent splenic cells (NK cells) was enhanced in the presence of *Cryptococcus*-specific antibody (74). In addition, human blood NK cells were observed to kill *Cryptococcus* with increasing efficiency in the presence of rabbit anticryptococcal antiserum (69), further demonstrating that NK cells recognize and kill *Cryptococcus* using ADCC, similar to tumor targets. However, the YT cell line that has been used in many studies does not express FcγRIII, and primary human NK cells kill in the absence of specific antibody (20), excluding ADCC as the sole mechanism of recognition.

NK-Cell Signaling During Host Defense to *Cryptococcus*

The signaling pathway involved in NK-cell recognition and killing of *Cryptococcus* requires both the p110 enzymatic and p85α regulatory subunits of phosphoinositide 3-kinase (PI3K), as well as the mitogen-activated kinase p42/44, extracellular signal-regulated kinases (ERK)1/2. Through the use of Western blot analysis, Akt phosphorylation was observed in the presence of *Cryptococcus*, which indicates PI3K activation. Inhibition of PI3K through the use of specific inhibitors and gene silencing resulted in decreased killing of *Cryptococcus* by NK cells (105). Downstream of PI3K, ERK1/2 was phosphorylated in the presence of *Cryptococcus*. Moreover, ERK1/2-specific inhibitors resulted in inhibition of NK-cell killing of *Cryptococcus* (105). These studies were conducted using both human primary blood NK cells and YT cells (105). While these studies establish a foothold into the signaling events that are required for NK-mediated killing of *Cryptococcus*, further studies are required to define the proximal and distal signaling events.

Calcium is also involved in NK-cell killing of *Cryptococcus*. As mentioned above, binding is a prerequisite to NK-cell killing of *Cryptococcus*. Although binding is magnesium dependent, killing of *Cryptococcus* requires magnesium, calcium, and 37°C (33). The significance of magnesium in culture may be limited to increasing binding affinity for conjugate formation. Calcium has been extensively studied during lymphocyte activation, as it is required for stimulation of specific signaling pathways associated with immunomodulatory tyrosine-based activation motif (ITAM)-bearing receptors (101). Most of the activation receptors for NK cells have, or associate with, membrane molecules that contain ITAMs in their cytoplasmic domains (52). Activation of these receptors results in increased cytoplasmic calcium flux; therefore, the requirement for calcium in NK-cell killing of *Cryptococcus* may be through signaling mechanisms (52). However, calcium may also be necessary for perforin polymerization that is required for NK-cell killing of *Cryptococcus* (discussed below). Overall, the minimum divalent cation requirement during NK-cell killing of *Cryptococcus* is consistent with NK-cell killing of tumor cells. In addition, in the future, it will be interesting to examine a potential role for intracellular calcium signaling in NK-cell killing of *Cryptococcus*.

NK-Cell Effector Molecules Targeting *Cryptococcus*

Perforin is the effector molecule required for NK-cell killing of *Cryptococcus*. Via exocytosis of lytic granules, NK cells secrete effector molecules that are involved in the killing of tumor targets. NK cells kill tumor cells by damaging the cell membrane with perforin, which facilitates the uptake of granzymes that stimulate apoptotic pathways within the target cells (93). Initial studies showed that strontium chloride–treated NK cells had decreased ability to kill *Cryptococcus* (60). Strontium chloride causes exocytosis of lytic granules, therefore depleting NK cells of specific effector molecules, including perforin. To function as an effector molecule, perforin requires calcium. Gene silencing using sRNAi knockdown, calcium chelators (EGTA), or inhibitors of the vacuolar ATPases that are required to maintain perforin in lytic granules (concanamycin A) were used to determine if perforin was the effector molecule used by NK cells in the killing of *Cryptococcus*. Under the conditions of perforin or calcium depletion or through inhibition of active perforin, respectively, there was a decrease in NK-cell killing of *Cryptococcus* (60).

In addition, flow cytometric analysis, using perforin-specific antibodies, indicates degranulation of intracellular perforin in the presence of *Cryptococcus* (63). These results indicate that perforin is required for NK-cell killing of *Cryptococcus*. It also appears that NK cells also have the capacity to re-arm through increased perforin production and that this results in enhanced killing of *Cryptococcus* (63). Remarkably, effector T cells use a completely different effector mechanism, granulysin, to kill *Cryptococcus* (59, 109). However, gene silencing of granulysin had no effect on killing of *Cryptococcus* by NK cells, suggesting that this alternate effector mechanism is not employed by NK cells to kill *Cryptococcus* (60). Despite convincing evidence of the role of perforin, the events that occur on the membrane of *C. neoformans* following exposure to perforin that leads to death of *Cryptococcus* have not yet been elucidated, nor have the contributions of other molecules been explored, such as granzymes that work in conjunction with perforin in the killing of tumor targets.

Reorientation of the NK-cell cytoskeleton is required for NK-cell killing of *Cryptococcus*. NK-cell conjugation with tumor cells results in temporal changes to the cytoskeleton. Although it has not been shown for microbial killing, cytoskeletal changes start with accumulation of actin and talin at the site of conjugation that is responsible for firm adhesion to tumor cells and for the delivery of cytotoxic effector molecules (102). Actin accumulation also stimulates the polarization of signaling molecules to the site of conjugate formation (102). Among other functions, signaling results in the formation of a microtubule organization center (MTOC) at the site of tumor contact (102). The function of the MTOC is to polarize cytolytic granules and the Golgi apparatus to the site of contact (102). Experiments using transmission electron and fluorescence microscopy show the presence of the Golgi apparatus and MTOC at the site of *Cryptococcus* contact (Fig. 2 and Fig. 3) (31; M. Timm-McCann and C. H. Mody, unpublished). In addition, the use of colchicine to block MTOC formation resulted in inhibition of NK-cell killing of *Cryptococcus* (31). These results confirm the significance of MTOC formation in the killing of *Cryptococcus* and suggest that the pathways involved in the killing of *Cryptococcus* are similar to those of tumor killing by NK cells.

Cytokines Produced by NK Cells Enhance the Immune Response to *Cryptococcus*

NK cells can receive signals from other innate cells in response to *Cryptococcus* as well as produce cytokines that stimulate other effector cells and shape adaptive immune responses. Host immune cells produce specific cytokines, in response to antigen- or pathogen-associated molecular patterns, resulting in either autocrine (*cis*) or paracrine (*trans*) stimulation of immune cells in the environment. NK-cell-associated cytokines are important for effective elimination of *Cryptococcus*. Nude mice, which are devoid of T cells, produce IFN-γ that is attributed to NK-cell secretion. Nude mice injected with anti-IFN-γ antibody during *Cryptococcus* inoculation showed decreased capacity to clear the fungus (88). These data stress the importance of NK-cell production of TH1 cytokines (IFN-γ) for effective elimination of *Cryptococcus*. Human primary NK cells also secrete IFN-γ that correlates with increased killing of *Cryptococcus* in vitro (56). In addition, IL-12 and IL-18, which are produced by macrophages or dendritic cells, stimulate NK cells to secrete IFN-γ, leading to elimination of *Cryptococcus*. IL-12-deficient mice produce less IFN-γ than wild-type mice and show decreased clearance of *Cryptococcus* (47). In addition, IL-12-deficient mice injected with neutralizing anti-IL-18 antibody, during *Cryptococcus* inoculation, result in decreased IFN-γ production and increased lung and brain CFUs (47).

By extension, NK cells from HIV patients also show increased in vitro killing of *Cryptococcus* in the presence of exogenous IL-12 (38). In gain-of-function experiments, injecting

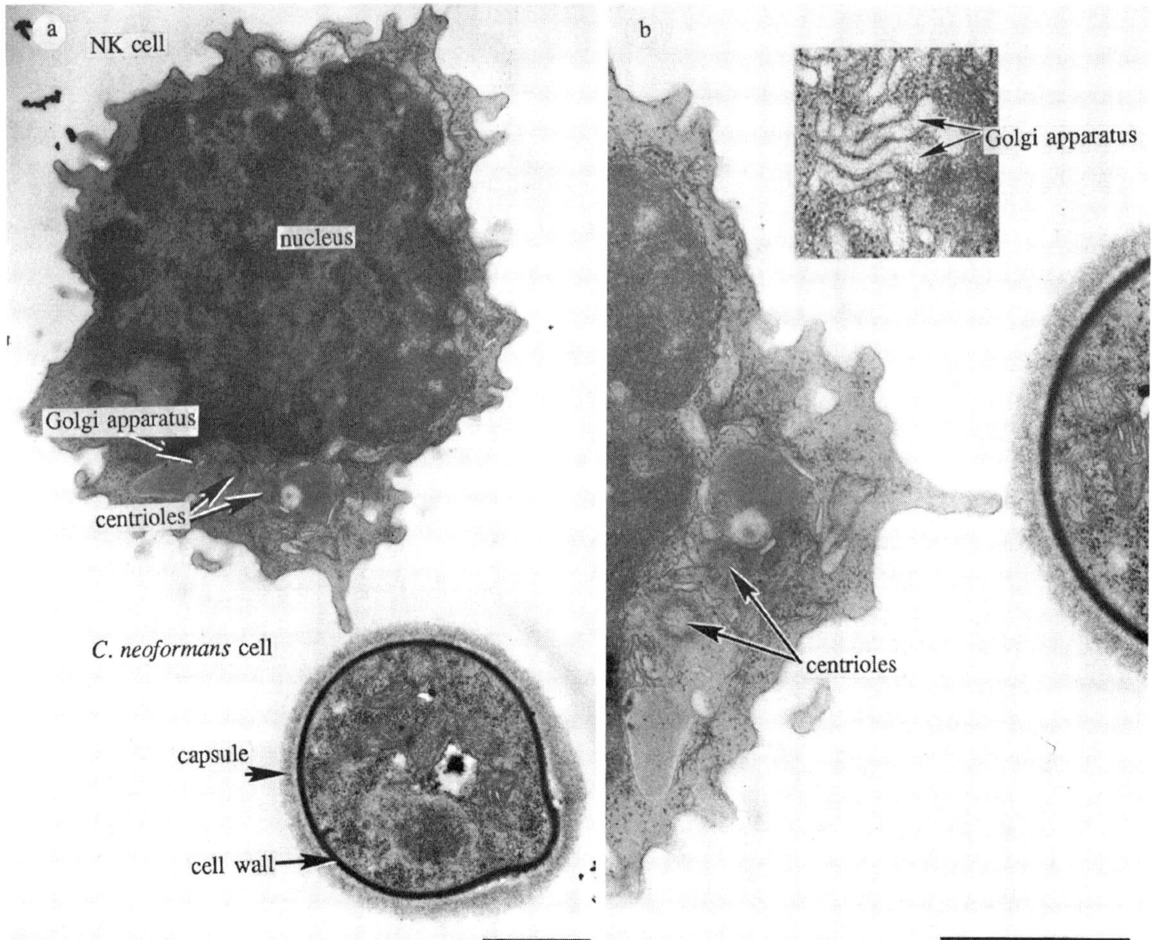

FIGURE 2 Percoll-fractionated splenic large granular lymphocyte (NK cell) Golgi apparatus and centrioles at the site of *Cryptococcus* contact. (a) Transmission electron microscopy of an NK cell conjugate with *C. neoformans*, identifying NK cell cytoskeletal changes. (b) Higher magnification including Golgi apparatus. Image from Hidore and associates (31). Reprinted from *Infection and Immunity* with permission of the publisher.

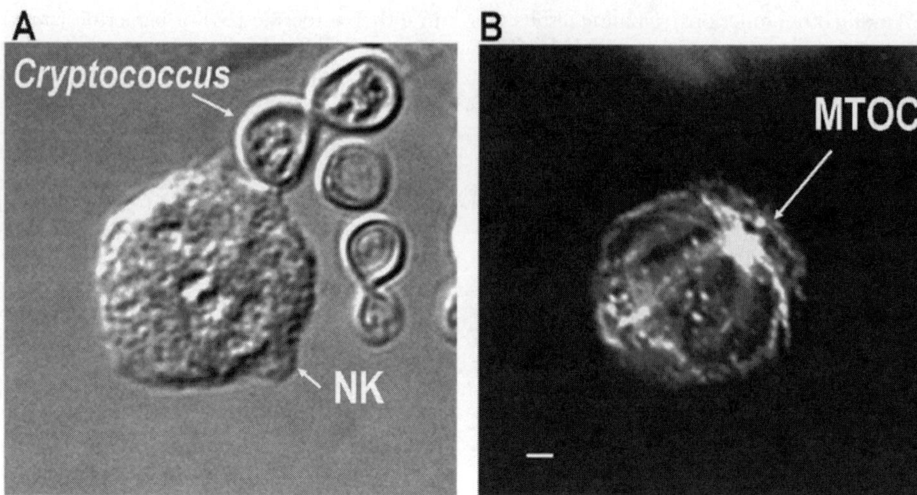

FIGURE 3 The NK cell MTOC forms near the site of contact with *Cryptococcus*. (A) NK cell in association with *Cryptococcus* as imaged by differential interference contrast imaging. (B) The same NK cell showing formation of the MTOC identified by fluorescence microscopy using antibody-specific labeling of tubulin in association with the site of contact of *Cryptococcus*. Bar represents 1 μm. Images acquired by Martina Timm-McCann.

IL-12 into mice inoculated with *Cryptococcus* significantly enhanced elimination of the infection. Mice injected with IL-12 showed decreased CFUs in the brain compared to those without IL-12 injections (15). This suggests that IL-12 is important in systemic *Cryptococcus* infections. Overall, these data suggest that the production of IL-12 and IL-18 by mononuclear cells resulted in increased IFN-γ production by NK cells. These studies point to the important role of these cells in cryptococcal host defense and the potential role of NK-associated cytokines as potential targets for drug therapy.

In summary, NK cells kill *Cryptococcus*. Killing requires magnesium-dependent conjugate formation. However, unlike tumor cells, the initial binding does not involve LFA-1 and does not appear to involve other receptors for fungi such as TLR2, TLR4, or lectin receptors (41). Binding of *Cryptococcus* to NK cells is enhanced with antibodies that lead to the killing of *Cryptococcus* in a manner that is similar to ADCC. However, the primary receptor has not yet been identified. Binding of *Cryptococcus*, along with stimulation of activation receptors, which may involve two distinct receptors, results in activation of signaling molecules. These signaling molecules include PI3K, ERK1/2, and calcium. Direct killing of *Cryptococcus* by NK cells is mediated by perforin. Perforin resides in lytic granules, and cytoskeletal rearrangement involving actin and MTOC localization to the NKIS may be the mechanism by which perforin granules move to the NKIS. In addition, cytokine (IL-12 and IL-18) production by other immune cells (such as macrophages) enhances NK-cell secretion of IFN-γ that shapes the T-cell subset repertoire and stimulates increased killing by macrophages and cytotoxic T lymphocytes.

NKT CELLS AND HOST DEFENSE

NKT cells are innate lymphocytes that shape the immune response and directly kill altered host cells and pathogens. NKT cells express both TCR and NK1.1 or Ly-49 (in mice). The TCR expressed by invariant (or type I) NKT cells consists of a conserved Vα chain and a semiconserved Vβ chain (7). The specificity of this TCR is limited to recognition of antigen in the context of CD1d, unlike the variable Vα and Vβ chains on T cells that recognize antigen in the context of MHC. Recently, studies implicated NKT cells in the innate and adaptive immune response to pathogens, including responses to *Cryptococcus* (Table 1) (25, 45). Pathogenic ligands that associate with NKT cells remain elusive. However, α-galactosylcerimide (α-Gal Cer), a glycolipid from a marine sponge, associates with CD1, and the association of α-Gal Cer with CD1 can activate both murine and human NKT cells (17). Most NKT studies use murine models injected with α-Gal Cer to enhance NKT function, in addition to inoculation of the pathogen. In this way, a change in the burden of the pathogen can be attributed to NKT cell activity. In response to altered tumor cells, virus-infected cells, or specific pathogens, α-Gal Cer-activated NKT cells respond by proliferating, increasing cell surface expression of specific molecules (including CD69 and CD40L), secreting cytokines (such as tumor necrosis factor α, IL-2, IFN-γ, and IL-4), and exhibiting cytotoxic activity (17, 64, 104). While many of these details need to be established in cryptococcal host defense, it is clear that NKT cells contribute to host defense against *Cryptococcus*.

NKT Cells and Host Defense against *Cryptococcus*

Current research indicates that NKT cells are involved in the killing of *Cryptococcus*, although we await the studies to determine whether this is an indirect or a direct effect. Decreased *Cryptococcus* lung and spleen CFUs were observed in mice treated with α-Gal Cer compared to untreated mice (45). In addition, an increase in IFN-γ production was seen on day 3 and continued to rise until it peaked on day 7 in the α-Gal Cer-treated animals. However, this effect was not observed in the NKT-knockout mice (45). Thus, using both gain-of-function and loss-of-function approaches, NKT cells have been implicated in the production of IFN-γ and optimal host defense against *Cryptococcus*.

Cytokines and Chemokines Produced by NKT Cells Enhance the Immune Response to *Cryptococcus*

Cytokines and chemokines are important mediators of immune cell function. Initial observations of mice that were injected with α-Gal Cer and infected with *Cryptococcus* showed increased production of the chemokine monocyte chemoattractant protein 1 (MCP-1) in the lung (44). Subsequent studies using MCP-1-deficient mice showed decreased numbers of NKT cells in the lungs of infected mice (44). These data indicate a role for MCP-1 in NKT-cell recruitment to the lungs of *Cryptococcus*-infected mice. In addition, similar to the observations of NK cells, NKT cells enhanced host defense against *Cryptococcus* in the presence of IL-12 and IFN-γ. Inoculation of IL-12-deficient mice with *Cryptococcus* and α-Gal Cer resulted in decreased IFN-γ production compared to inoculation of wild-type mice (46). These data suggest that the IFN-γ production by NKT cells is at least in part regulated by IL-12, similar to that observed for NK cells. Thus, similar to NK cells, cytokines are important in eliciting NKT cell responses to *Cryptococcus* and in direct killing of the organism. Overall, these studies indicate a role for NKT cells in the host immune response to *Cryptococcus* that results in decreased fungal burden.

INNATE T-CELL RESPONSES TO *CRYPTOCOCCUS*

T cells contribute to innate responses to *Cryptococcus* in a number of important ways. T cells can be triggered to undergo clonal expansion to a cryptococcal mitogen and release cytokines in addition to their ability to bind directly to *Cryptococcus* and kill the organism.

In the classic paradigm of adaptive immunity, antigen-presenting cells (APCs) bind, internalize, and process cryptococcal antigen before displaying the antigen in the context of MHC. A TCR that has undergone positive selection on class I or class II MHC recognizes the MHC-antigen complex, resulting in T-cell activation. Cytokines produced by the APCs or cells of the surrounding environment further shape the acquired immune response. In the case of *Cryptococcus*, these interactions result in a protective Th1 (IFN-γ-producing) or nonprotective Th2 (IL-5) immune response (37). However, in addition to this classical paradigm, T cells contribute to the innate response in important ways.

Mitogenic T-Cell Responses to *Cryptococcus*

Innate T-cell responses are well described in response to staphylococcal and streptococcal exotoxins (90). In the innate T-cell response to *Cryptococcus*, proliferation is APC and CD4+ T-cell dependent, but not MHC restricted (the stimulatory APC can be allogeneic to the responding CD4+ T cell), and the proportion of naive T cells responding is similar to the proportion of memory T cells. The lack of MHC restriction and high precursor frequency of responding T cells indicates that *Cryptococcus* possesses a T-cell mitogen (59, 70, 95, 96).

Binding of *Cryptococcus* to T Cells

T cells are also directly microbicidal to *Cryptococcus* (54, 55, 59, 72, 108, 109). Through a number of microscopic techniques including video microscopy, human CD3+ T cells were shown to form conjugates directly with live *Cryptococcus* (54, 72). Conjugates formed by T cells with *Cryptococcus* resulted in a broader area of binding than that observed for NK cells, which involves microvilli attachments (72). The exact mechanism of T-cell direct binding to *Cryptococcus* remains elusive. However, T cells treated with trypsin or bromelain that strips extracellular receptors show decreased direct binding to, and killing of, *Cryptococcus* (55).

Cytokines Alter the Direct T-Cell Response to *Cryptococcus*

Cytokines are required to prime T cells for direct killing of *Cryptococcus*. T-cell growth factors such as IL-2 and IL-15 are required for CD4+ and CD8+ T cells to become directly cytotoxic to *Cryptococcus* (59, 108, 109). Moreover, following binding of *Cryptococcus* directly to T cells, IFN-γ is transcribed and released from the T cells (56). This release of cytokines from T cells, due to direct interactions with microbial pathogens, was a novel finding in the field of host immunity. IFN-γ (produced either by this mechanism, during adaptive immunity, or by NK cells) results in enhanced direct cytotoxic activity by T lymphocytes. This suggests that the IFN-γ produced by T cells in response to *Cryptococcus* results in autocrine stimulation of T-cell cytotoxicity that kills *Cryptococcus*. However, unlike the observations with NK cells or MHC-restricted T cells, IL-12 did not enhance the innate ability of T cells to directly kill *Cryptococcus* (56).

T-Cell Effector Molecules Targeting *Cryptococcus*

All of the steps involved in direct killing of *Cryptococcus* by T cells have not been fully elucidated. However, advances have been made. While phagocytic innate immune cells kill their targets in part through the production of hydroxyl radicals, the role of hydroxyl radicals is less definite for T cells. Inhibition of hydroxyl radicals using catechin, but not diethyl urea or propyl gallate, resulted in decreased antifungal activity by T cells (57). In addition, cyclooxygenase inhibitors including piroxicam and indomethacin had no effect on direct T-cell killing of *Cryptococcus* (57). These data suggest that unlike phagocytic cells, the mechanism of T-cell direct killing of *Cryptococcus* is not mediated through a typical inflammatory response. More recently, it has been shown that cytotoxic CD8+ and CD4+ T-lymphocyte killing of *Cryptococcus* is dependent on granulysin (59, 108, 109), a member of the saposin-like protein family of molecules including amebapores, pulmonary surfactant protein-B, saposins, and NK-lysin (2). Granulysin interacts with lipids in the cell membrane and also activates lipid-degrading enzymes (100), leading to lesions in the cell membrane that result in death of the target cell (23, 43).

Overall, T cells form direct conjugates with *Cryptococcus* that result in IFN-γ production. IFN-γ, in an autocrine fashion, activates the T cells to directly kill *Cryptococcus* by a mechanism that involves release of granulysin. It has also been determined that mitogen-activated CD4+ T cells proliferate and are also able to license dendritic cells that in turn activate CD8+ T cells to become microbicidal (59, 70, 95, 96). These are novel findings that no doubt have increased our knowledge about host interactions with *Cryptococcus* that may also occur with other pathogens.

In summary, NK, NKT, and T cells participate in innate immunity through a number of mechanisms. In the presence of *Cryptococcus* these lymphocytes are triggered via non-MHC-restricted activation in addition to classical mechanisms of adaptive immunity, resulting in clonal expansion. It follows that these lymphocytes produce cytokines that activate other effector cells to kill *Cryptococcus*. Moreover, although there is much that remains to be elucidated, these cells can also bind and become directly cytotoxic to *Cryptococcus*.

REFERENCES

1. **Akiba, H., Y. Motoki, M. Satoh, K. Iwatsuki, and F. Kaneko.** 2001. Recalcitrant trichophytic granuloma associated with

NK-cell deficiency in a SLE patient treated with corticosteroid. *Eur. J. Dermatol.* **11:**58–62.

2. **Andreu, D., C. Carreno, C. Linde, H. G. Boman, and M. Andersson.** 1999. Identification of an anti-mycobacterial domain in NK-lysin and granulysin. *Biochem. J.* **344** (Pt 3)**:**845–849.
3. **Barber, D. F., and E. O. Long.** 2003. Coexpression of CD58 or CD48 with intercellular adhesion molecule 1 on target cells enhances adhesion of resting NK cells. *J. Immunol.* **170:**294–299.
4. **Behar, S. M., C. C. Dascher, M. J. Grusby, C. R. Wang, and M. B. Brenner.** 1999. Susceptibility of mice deficient in CD1D or TAP1 to infection with *Mycobacterium tuberculosis. J. Exp. Med.* **189:**1973–1980.
5. **Beno, D. W., and H. L. Mathews.** 1990. Growth inhibition of *Candida albicans* by interleukin-2-induced lymph node cells. *Cell. Immunol.* **128:**89–100.
6. **Berntman, E., J. Rolf, C. Johansson, P. Anderson, and S. L. Cardell.** 2005. The role of CD1d-restricted NK T lymphocytes in the immune response to oral infection with *Salmonella typhimurium. Eur. J. Immunol.* **35:**2100–2109.
7. **Berzins, S. P., A. P. Uldrich, D. G. Pellicci, F. McNab, Y. Hayakawa, M. J. Smyth, and D. I. Godfrey.** 2004. Parallels and distinctions between T and NKT cell development in the thymus. *Immunol. Cell Biol.* **82:**269–275.
8. **Billiau, A., and P. Matthys.** 2009. Interferon-gamma: a historical perspective. *Cytokine Growth Factor Rev.* **20:**97–113.
9. **Brill, K. J., Q. Li, R. Larkin, D. H. Canaday, D. R. Kaplan, W. H. Boom, and R. F. Silver.** 2001. Human natural killer cells mediate killing of intracellular *Mycobacterium tuberculosis* H37Rv via granule-independent mechanisms. *Infect. Immun.* **69:**1755–1765.
10. **Brown, G. D., and S. Gordon.** 2001. Immune recognition. A new receptor for beta-glucans. *Nature* **413:**36–37.
11. **Bryceson, Y. T., M. E. March, D. F. Barber, H. G. Ljunggren, and E. O. Long.** 2005. Cytolytic granule polarization and degranulation controlled by different receptors in resting NK cells. *J. Exp. Med.* **202:**1001–1012.
12. **Canaday, D. H., R. J. Wilkinson, Q. Li, C. V. Harding, R. F. Silver, and W. H. Boom.** 2001. CD4(+) and CD8(+) T cells kill intracellular *Mycobacterium tuberculosis* by a perforin and Fas/Fas ligand-independent mechanism. *J. Immunol.* **167:**2734–2742.
13. **Carpen, O., I. Virtanen, V. P. Lehto, and E. Saksela.** 1983. Polarization of NK cell cytoskeleton upon conjugation with sensitive target cells. *J. Immunol.* **131:**2695–2698.
14. **Chen, X., P. P. Trivedi, B. Ge, K. Krzewski, and J. L. Strominger.** 2007. Many NK cell receptors activate ERK2 and JNK1 to trigger microtubule organizing center and granule polarization and cytotoxicity. *Proc. Natl. Acad. Sci. USA* **104:**6329–6334.
15. **Clemons, K. V., E. Brummer, and D. A. Stevens.** 1994. Cytokine treatment of central nervous system infection: efficacy of interleukin-12 alone and synergy with conventional antifungal therapy in experimental cryptococcosis. *Antimicrob. Agents Chemother.* **38:**460–464.
16. **Cooper, M. A., T. A. Fehniger, and M. A. Caligiuri.** 2001. The biology of human natural killer-cell subsets. *Trends Immunol.* **22:**633–640.
17. **Crowe, N. Y., A. P. Uldrich, K. Kyparissoudis, K. J. Hammond, Y. Hayakawa, S. Sidobre, R. Keating, M. Kronenberg, M. J. Smyth, and D. I. Godfrey.** 2003. Glycolipid antigen drives rapid expansion and sustained cytokine production by NK T cells. *J. Immunol.* **171:** 4020–4027.
18. **Davis, D. M., I. Chiu, M. Fassett, G. B. Cohen, O. Mandelboim, and J. L. Strominger.** 1999. The human natural killer cell immune synapse. *Proc. Natl. Acad. Sci. USA* **96:**15062–15067.
19. **Dong, Z. M., and J. W. Murphy.** 1997. Cryptococcal polysaccharides bind to CD18 on human neutrophils. *Infect. Immun.* **65:**557–563.
20. **Drexler, H. G., and Y. Matsuo.** 2000. Malignant hematopoietic cell lines: in vitro models for the study of multiple myeloma and plasma cell leukemia. *Leukoc. Res.* **24:**681–703.
21. **Ellner, J. J., G. R. Olds, C. W. Lee, M. E. Kleinhenz, and K. L. Edmonds.** 1982. Destruction of the multicellular parasite *Schistosoma mansoni* by T lymphocytes. *J. Clin. Invest.* **70:**369–378.
22. **Elloso, M. M., H. C. van der Heyde, J. A. vande Waa, D. D. Manning, and W. P. Weidanz.** 1994. Inhibition of *Plasmodium falciparum* in vitro by human gamma delta T cells. *J. Immunol.* **153:**1187–1194.
23. **Ernst, W. A., S. Thoma-Uszynski, R. Teitelbaum, C. Ko, D. A. Hanson, C. Clayberger, A. M. Krensky, M. Leippe, B. R. Bloom, T. Ganz, and R. L. Modlin.** 2000. Granulysin, a T cell product, kills bacteria by altering membrane permeability. *J. Immunol.* **165:**7102–7108.
24. **Faveeuw, C., V. Angeli, J. Fontaine, C. Maliszewski, A. Capron, L. Van Kaer, M. Moser, M. Capron, and F. Trottein.** 2002. Antigen presentation by CD1d contributes to the amplification of Th2 responses to *Schistosoma mansoni* glycoconjugates in mice. *J. Immunol.* **169:**906–912.
25. **Gansert, J. L., V. Kiessler, M. Engele, F. Wittke, M. Rollinghoff, A. M. Krensky, S. A. Porcelli, R. L. Modlin, and S. Stenger.** 2003. Human NKT cells express granulysin and exhibit antimycobacterial activity. *J. Immunol.* **170:**3154–3161.
26. **Gantner, B. N., R. M. Simmons, and D. M. Underhill.** 2005. Dectin-1 mediates macrophage recognition of *Candida albicans* yeast but not filaments. *EMBO J.* **24:**1277–1286.
27. **Gazit, R., K. Hershko, A. Ingbar, M. Schlesinger, S. Israel, C. Brautbar, O. Mandelboim, and V. Leibovici.** 2008. Immunological assessment of familial tinea corporis. *J. Eur. Acad. Dermatol. Venereol.* **22:**871–874.
28. **Gersuk, G. M., D. M. Underhill, L. Zhu, and K. A. Marr.** 2006. Dectin-1 and TLRs permit macrophages to distinguish between different *Aspergillus fumigatus* cellular states. *J. Immunol.* **176:**3717–3724.
29. **Gonzalez-Amaro, R., J. F. Salazar-Gonzalez, L. Baranda, C. Abud-Mendoza, O. Martinez-Maza, M. Miramontes, and B. Moncada.** 1991. Natural killer cell-mediated cytotoxicity in cryptococcal meningitis. *Rev. Invest. Clin.* **43:**133–138.
30. **Graham, D. B., M. Cella, E. Giurisato, K. Fujikawa, A. V. Miletic, T. Kloeppel, K. Brim, T. Takai, A. S. Shaw, M. Colonna, and W. Swat.** 2006. Vav1 controls DAP10-mediated natural cytotoxicity by regulating actin and microtubule dynamics. *J. Immunol.* **177:**2349–2355.
31. **Hidore, M. R., T. W. Mislan, and J. W. Murphy.** 1991. Responses of murine natural killer cells to binding of the fungal target *Cryptococcus neoformans. Infect. Immun.* **59:**1489–1499.
32. **Hidore, M. R., and J. W. Murphy.** 1986. Correlation of natural killer cell activity and clearance of *Cryptococcus neoformans* from mice after adoptive transfer of splenic nylon wool-nonadherent cells. *Infect. Immun.* **51:**547–555.
33. **Hidore, M. R., and J. W. Murphy.** 1989. Murine natural killer cell interactions with a fungal target, *Cryptococcus neoformans. Infect. Immun.* **57:**1990–1997.
34. **Hidore, M. R., and J. W. Murphy.** 1986. Natural cellular resistance of beige mice against *Cryptococcus neoformans. J. Immunol.* **137:**3624–3631.
35. **Hidore, M. R., N. Nabavi, F. Sonleitner, and J. W. Murphy.** 1991. Murine natural killer cells are fungicidal to *Cryptococcus neoformans. Infect. Immun.* **59:**1747–1754.

36. **Hildreth, J. E., F. M. Gotch, P. D. Hildreth, and A. J. McMichael.** 1983. A human lymphocyte-associated antigen involved in cell-mediated lympholysis. *Eur. J. Immunol.* **13:**202–208.
37. **Hoag, K. A., N. E. Street, G. B. Huffnagle, and M. F. Lipscomb.** 1995. Early cytokine production in pulmonary *Cryptococcus neoformans* infections distinguishes susceptible and resistant mice. *Am. J. Respir. Cell Mol. Biol.* **13:**487–495.
38. **Horn, C. A., and R. G. Washburn.** 1995. Anticryptococcal activity of NK cell-enriched peripheral blood lymphocytes from human immunodeficiency virus-infected subjects: responses to interleukin-2, interferon-gamma, and interleukin-12. *J. Infect. Dis.* **172:**1023–1027.
39. **Ishikawa, H., H. Hisaeda, M. Taniguchi, T. Nakayama, T. Sakai, Y. Maekawa, Y. Nakano, M. Zhang, T. Zhang, M. Nishitani, M. Takashima, and K. Himeno.** 2000. CD4(+) v(alpha)14 NKT cells play a crucial role in an early stage of protective immunity against infection with *Leishmania major*. *Int. Immunol.* **12:**1267–1274.
40. **Jimenez, B. E., and J. W. Murphy.** 1984. In vitro effects of natural killer cells against *Paracoccidioides brasiliensis* yeast phase. *Infect. Immun.* **46:**552–558.
41. **Jones, G. J., J. C. Wiseman, K. J. Marr, S. Wei, J. Y. Djeu, and C. H. Mody.** 2009. In contrast to anti-tumor activity, YT cell and primary NK cell cytotoxicity for *Cryptococcus neoformans* bypasses LFA-1. *Int. Immunol.* **21:**423–32.
42. **Kahana, D. D., O. Cass, J. Jessurun, S. J. Schwarzenberg, H. Sharp, and K. Khan.** 2003. Sclerosing cholangitis associated with trichosporon infection and natural killer cell deficiency in an 8-year-old girl. *J. Pediatr. Gastroenterol. Nutr.* **37:**620–623.
43. **Kaspar, A. A., S. Okada, J. Kumar, F. R. Poulain, K. A. Drouvalakis, A. Kelekar, D. A. Hanson, R. M. Kluck, Y. Hitoshi, D. E. Johnson, C. J. Froelich, C. B. Thompson, D. D. Newmeyer, A. Anel, C. Clayberger, and A. M. Krensky.** 2001. A distinct pathway of cell-mediated apoptosis initiated by granulysin. *J. Immunol.* **167:**350–356.
44. **Kawakami, K., Y. Kinjo, K. Uezu, S. Yara, K. Miyagi, Y. Koguchi, T. Nakayama, M. Taniguchi, and A. Saito.** 2001. Monocyte chemoattractant protein-1-dependent increase of V alpha 14 NKT cells in lungs and their roles in Th1 response and host defense in cryptococcal infection. *J. Immunol.* **167:**6525–6532.
45. **Kawakami, K., Y. Kinjo, S. Yara, Y. Koguchi, K. Uezu, T. Nakayama, M. Taniguchi, and A. Saito.** 2001. Activation of Valpha14(+) natural killer T cells by alpha-galactosylceramide results in development of Th1 response and local host resistance in mice infected with *Cryptococcus neoformans*. *Infect. Immun.* **69:**213–220.
46. **Kawakami, K., Y. Kinjo, S. Yara, K. Uezu, Y. Koguchi, M. Tohyama, M. Azuma, K. Takeda, S. Akira, and A. Saito.** 2001. Enhanced gamma interferon production through activation of Valpha14(+) natural killer T cells by alpha-galactosylceramide in interleukin-18-deficient mice with systemic cryptococcosis. *Infect. Immun.* **69:**6643–6650.
47. **Kawakami, K., Y. Koguchi, M. H. Qureshi, S. Yara, Y. Kinjo, K. Uezu, and A. Saito.** 2000. NK cells eliminate *Cryptococcus neoformans* by potentiating the fungicidal activity of macrophages rather than by directly killing them upon stimulation with IL-12 and IL-18. *Microbiol. Immunol.* **44:**1043–1050.
48. **Kawakami, K., N. Yamamoto, Y. Kinjo, K. Miyagi, C. Nakasone, K. Uezu, T. Kinjo, T. Nakayama, M. Taniguchi, and A. Saito.** 2003. Critical role of Valpha14+ natural killer T cells in the innate phase of host protection against *Streptococcus pneumoniae* infection. *Eur. J. Immunol.* **33:**3322–3330.
49. **Khan, I. A., K. A. Smith, and L. H. Kasper.** 1988. Induction of antigen-specific parasiticidal cytotoxic T cell splenocytes by a major membrane protein (P30) of *Toxoplasma gondii*. *J. Immunol.* **141:**3600–3605.
50. **Krishnaraj, R., and A. Svanborg.** 1993. Low natural killer cell function in disseminated aspergillosis. *Scand. J. Infect. Dis.* **25:**537–541.
51. **Kumar, H., A. Belperron, S. W. Barthold, and L. K. Bockenstedt.** 2000. Cutting edge: CD1d deficiency impairs murine host defense against the spirochete, *Borrelia burgdorferi*. *J. Immunol.* **165:**4797–4801.
52. **Lanier, L. L.** 2008. Up on the tightrope: natural killer cell activation and inhibition. *Nat. Immunol.* **9:**495–502.
53. **Laskay, T., M. Rollinghoff, and W. Solbach.** 1993. Natural killer cells participate in the early defense against *Leishmania major* infection in mice. *Eur. J. Immunol.* **23:**2237–2241.
54. **Levitz, S. M., and M. P. Dupont.** 1993. Phenotypic and functional characterization of human lymphocytes activated by interleukin-2 to directly inhibit growth of *Cryptococcus neoformans* in vitro. *J. Clin. Invest.* **91:**1490–1498.
55. **Levitz, S. M., M. P. Dupont, and E. H. Smail.** 1994. Direct activity of human T lymphocytes and natural killer cells against *Cryptococcus neoformans*. *Infect. Immun.* **62:**194–202.
56. **Levitz, S. M., and E. A. North.** 1996. Gamma interferon gene expression and release in human lymphocytes directly activated by *Cryptococcus neoformans* and *Candida albicans*. *Infect. Immun.* **64:**1595–1599.
57. **Levitz, S. M., E. A. North, M. P. Dupont, and T. S. Harrison.** 1995. Mechanisms of inhibition of *Cryptococcus neoformans* by human lymphocytes. *Infect. Immun.* **63:**3550–3554.
58. **Lipscomb, M. F., T. Alvarellos, G. B. Toews, R. Tompkins, Z. Evans, G. Koo, and V. Kumar.** 1987. Role of natural killer cells in resistance to *Cryptococcus neoformans* infections in mice. *Am. J. Pathol.* **128:**354–361.
59. **Ma, L. L., J. C. Spurrell, J. F. Wang, G. G. Neely, S. Epelman, A. M. Krensky, and C. H. Mody.** 2002. CD8 T cell-mediated killing of *Cryptococcus neoformans* requires granulysin and is dependent on CD4 T cells and IL-15. *J. Immunol.* **169:**5787–5795.
60. **Ma, L. L., C. L. Wang, G. G. Neely, S. Epelman, A. M. Krensky, and C. H. Mody.** 2004. NK cells use perforin rather than granulysin for anticryptococcal activity. *J. Immunol.* **173:**3357–3365.
61. **Mansour, M. K., E. Latz, and S. M. Levitz.** 2006. *Cryptococcus neoformans* glycoantigens are captured by multiple lectin receptors and presented by dendritic cells. *J. Immunol.* **176:**3053–3061.
62. **Markham, R. B., J. Goellner, and G. B. Pier.** 1984. In vitro T cell-mediated killing of *Pseudomonas aeruginosa*. I. Evidence that a lymphokine mediates killing. *J. Immunol.* **133:**962–968.
63. **Marr, K. J., G. J. Jones, C. Zheng, S. M. Huston, M. Timm-McCann, A. Islam, B. M. Berenger, L. L. Ma, J. C. Wiseman, and C. H. Mody.** 2009. *Cryptococcus neoformans* directly stimulates perforin production and rearms NK cells for enhanced anticryptococcal microbicidal activity. *Infect. Immun.* **77:**2436–2446.
64. **Matsuda, J. L., O. V. Naidenko, L. Gapin, T. Nakayama, M. Taniguchi, C. R. Wang, Y. Koezuka, and M. Kronenberg.** 2000. Tracking the response of natural killer T cells to a glycolipid antigen using CD1d tetramers. *J. Exp. Med.* **192:**741–754.
65. **Matsumoto, G., M. P. Nghiem, N. Nozaki, R. Schmits, and J. M. Penninger.** 1998. Cooperation between CD44 and LFA-1/CD11a adhesion receptors in lymphokine-activated killer cell cytotoxicity. *J. Immunol.* **160:**5781–5789.

66. **Matsumoto, G., Y. Omi, U. Lee, T. Nishimura, J. Shindo, and J. M. Penninger.** 2000. Adhesion mediated by LFA-1 is required for efficient IL-12-induced NK and NKT cell cytotoxicity. *Eur. J. Immunol.* **30:**3723–3731.
67. **Matsuyama, W., S. Takenaga, K. Nakahara, M. Kawabata, Y. Iwakiri, K. Arimura, and M. Osame.** 1997. Idiopathic hypoparathyroidism with fungal seminal vesiculitis. *Intern. Med.* **36:**113–117.
68. **McCann, F. E., B. Vanherberghen, K. Eleme, L. M. Carlin, R. J. Newsam, D. Goulding, and D. M. Davis.** 2003. The size of the synaptic cleft and distinct distributions of filamentous actin, ezrin, CD43, and CD45 at activating and inhibitory human NK cell immune synapses. *J. Immunol.* **170:**2862–2870.
69. **Miller, M. F., T. G. Mitchell, W. J. Storkus, and J. R. Dawson.** 1990. Human natural killer cells do not inhibit growth of *Cryptococcus neoformans* in the absence of antibody. *Infect. Immun.* **58:**639–645.
70. **Mody, C. H., C. J. Wood, R. M. Syme, and J. C. Spurrell.** 1999. The cell wall and membrane of *Cryptococcus neoformans* possess a mitogen for human T lymphocytes. *Infect. Immun.* **67:**936–941.
71. **Murphy, J. W., M. R. Hidore, and N. Nabavi.** 1991. Binding interactions of murine natural killer cells with the fungal target *Cryptococcus neoformans. Infect. Immun.* **59:**1476–1488.
72. **Murphy, J. W., M. R. Hidore, and S. C. Wong.** 1993. Direct interactions of human lymphocytes with the yeast-like organism, *Cryptococcus neoformans. J. Clin. Invest.* **91:**1553–1566.
73. **Murphy, J. W., and D. O. McDaniel.** 1982. In vitro reactivity of natural killer (NK) cells against *Cryptococcus neoformans. J. Immunol.* **128:**1577–1583.
74. **Nabavi, N., and J. W. Murphy.** 1986. Antibody-dependent natural killer cell-mediated growth inhibition of *Cryptococcus neoformans. Infect. Immun.* **51:**556–562.
75. **Nabavi, N., and J. W. Murphy.** 1985. In vitro binding of natural killer cells to *Cryptococcus neoformans* targets. *Infect. Immun.* **50:**50–57.
76. **Nakamura, K., T. Kinjo, S. Saijo, A. Miyazato, Y. Adachi, N. Ohno, J. Fujita, M. Kaku, Y. Iwakura, and K. Kawakami.** 2007. Dectin-1 is not required for the host defense to *Cryptococcus neoformans. Microbiol. Immunol.* **51:**1115–1119.
77. **Nieuwenhuis, E. E., T. Matsumoto, M. Exley, R. A. Schleipman, J. Glickman, D. T. Bailey, N. Corazza, S. P. Colgan, A. B. Onderdonk, and R. S. Blumberg.** 2002. CD1d-dependent macrophage-mediated clearance of *Pseudomonas aeruginosa* from lung. *Nat. Med.* **8:**588–593.
78. **Ochoa, M. T., S. Stenger, P. A. Sieling, S. Thoma-Uszynski, S. Sabet, S. Cho, A. M. Krensky, M. Rollinghoff, E. Nunes Sarno, A. E. Burdick, T. H. Rea, and R. L. Modlin.** 2001. T-cell release of granulysin contributes to host defense in leprosy. *Nat. Med.* **7:**174–179.
79. **Orange, J. S., K. E. Harris, M. M. Andzelm, M. M. Valter, R. S. Geha, and J. L. Strominger.** 2003. The mature activating natural killer cell immunologic synapse is formed in distinct stages. *Proc. Natl. Acad. Sci. USA* **100:**14151–14156.
80. **Palma, A. S., T. Feizi, Y. Zhang, M. S. Stoll, A. M. Lawson, E. Diaz-Rodriguez, M. A. Campanero-Rhodes, J. Costa, S. Gordon, G. D. Brown, and W. Chai.** 2006. Ligands for the beta-glucan receptor, Dectin-1, assigned using "designer" microarrays of oligosaccharide probes (neoglycolipids) generated from glucan polysaccharides. *J. Biol. Chem.* **281:**5771–5779.
81. **Paul, W. E.** 2003. *Fundamental Immunology*, 5th ed. Lippincott Williams & Wilkins, Philadelphia, PA.
82. **Petkus, A. F., and L. L. Baum.** 1987. Natural killer cell inhibition of young spherules and endospores of *Coccidioides immitis. J. Immunol.* **139:**3107–3111.
83. **Robertson, M. J., and J. Ritz.** 1990. Biology and clinical relevance of human natural killer cells. *Blood* **76:**2421–2438.
84. **Roder, J. C.** 1979. The beige mutation in the mouse. I. A stem cell predetermined impairment in natural killer cell function. *J. Immunol.* **123:**2168–2173.
85. **Roder, J. C., R. Kiessling, P. Biberfeld, and B. Andersson.** 1978. Target-effector interaction in the natural killer (NK) cell system. II. The isolation of NK cells and studies on the mechanism of killing. *J. Immunol.* **121:**2509–2517.
86. **Roder, J. C., M. L. Lohmann-Matthes, W. Domzig, and H. Wigzell.** 1979. The beige mutation in the mouse. II. Selectivity of the natural killer (NK) cell defect. *J. Immunol.* **123:**2174–2181.
87. **Salata, R. A., J. G. Cox, and J. I. Ravdin.** 1987. The interaction of human T-lymphocytes and *Entamoeba histolytica*: killing of virulent amoebae by lectin-dependent lymphocytes. *Parasite Immunol.* **9:**249–261.
88. **Salkowski, C. A., and E. Balish.** 1991. A monoclonal antibody to gamma interferon blocks augmentation of natural killer cell activity induced during systemic cryptococcosis. *Infect. Immun.* **59:**486–493.
89. **Salkowski, C. A., and E. Balish.** 1991. Role of natural killer cells in resistance to systemic cryptococcosis. *J. Leukoc. Biol.* **50:**151–159.
90. **Schlievert, P. M., D. J. Schoettle, and D. W. Watson.** 1979. Nonspecific T-lymphocyte mitogenesis by pyrogenic exotoxins from group A streptococci and *Staphylococcus aureus. Infect. Immun.* **25:**1075–1077.
91. **Sewald, X., B. Gebert-Vogl, S. Prassl, I. Barwig, E. Weiss, M. Fabbri, R. Osicka, M. Schiemann, D. H. Busch, M. Semmrich, B. Holzmann, P. Sebo, and R. Haas.** 2008. Integrin subunit CD18 Is the T-lymphocyte receptor for the *Helicobacter pylori* vacuolating cytotoxin. *Cell Host Microbe* **3:**20–29.
92. **Shoham, S., C. Huang, J. M. Chen, D. T. Golenbock, and S. M. Levitz.** 2001. Toll-like receptor 4 mediates intracellular signaling without TNF-alpha release in response to *Cryptococcus neoformans* polysaccharide capsule. *J. Immunol.* **166:**4620–4626.
93. **Smyth, M. J., E. Cretney, J. M. Kelly, J. A. Westwood, S. E. Street, H. Yagita, K. Takeda, S. L. van Dommelen, M. A. Degli-Esposti, and Y. Hayakawa.** 2005. Activation of NK cell cytotoxicity. *Mol. Immunol.* **42:**501–510.
94. **Sugie, K., Y. Minami, T. Kawakami, and A. Uchida.** 1995. Stimulation of NK-like YT cells via leukocyte function-associated antigen (LFA)-1. Possible involvement of LFA-1-associated tyrosine kinase in signal transduction after recognition of NK target cells. *J. Immunol.* **154:**1691–1698.
95. **Syme, R. M., J. C. Spurrell, E. K. Amankwah, F. H. Green, and C. H. Mody.** 2002. Primary dendritic cells phagocytose *Cryptococcus neoformans* via mannose receptors and Fcgamma receptor II for presentation to T lymphocytes. *Infect. Immun.* **70:**5972–5981.
96. **Syme, R. M., J. C. Spurrell, L. L. Ma, F. H. Green, and C. H. Mody.** 2000. Phagocytosis and protein processing are required for presentation of *Cryptococcus neoformans* mitogen to T lymphocytes. *Infect. Immun.* **68:**6147–6153.
97. **Taborda, C. P., and A. Casadevall.** 2002. CR3 (CD11b/CD18) and CR4 (CD11c/CD18) are involved in complement-independent antibody-mediated phagocytosis of *Cryptococcus neoformans. Immunity* **16:**791–802.
98. **Tewari, R. P., and L. A. Von Behren.** 2000. Immune responses in histoplasmosis, a prototype of respiratory mycoses. *Indian J. Chest Dis. Allied Sci.* **42:**265–269.
99. **Timonen, T., C. W. Reynolds, J. R. Ortaldo, and R. B. Herberman.** 1982. Isolation of human and rat natural killer cells. *J. Immunol. Methods* **51:**269–277.

100. **Vaccaro, A. M., M. Tatti, F. Ciaffoni, R. Salvioli, A. Barca, and C. Scerch.** 1997. Effect of saposins A and C on the enzymatic hydrolysis of liposomal glucosylceramide. *J. Biol. Chem.* **272:**16862–16867.
101. **Vig, M., and J. P. Kinet.** 2009. Calcium signaling in immune cells. *Nat. Immunol.* **10:**21–27.
102. **Vyas, Y. M., H. Maniar, and B. Dupont.** 2002. Visualization of signaling pathways and cortical cytoskeleton in cytolytic and noncytolytic natural killer cell immune synapses. *Immunol. Rev.* **189:**161–178.
103. **Werfel, T., P. Uciechowski, P. A. Tetteroo, R. Kurrle, H. Deicher, and R. E. Schmidt.** 1989. Activation of cloned human natural killer cells via Fc gamma RIII. *J. Immunol.* **142:**1102–1106.
104. **Wilson, M. T., C. Johansson, D. Olivares-Villagomez, A. K. Singh, A. K. Stanic, C. R. Wang, S. Joyce, M. J. Wick, and L. Van Kaer.** 2003. The response of natural killer T cells to glycolipid antigens is characterized by surface receptor down-modulation and expansion. *Proc. Natl. Acad. Sci. USA* **100:**10913–10918.
105. **Wiseman, J. C., L. L. Ma, K. J. Marr, G. J. Jones, and C. H. Mody.** 2007. Perforin-dependent cryptococcal microbicidal activity in NK cells requires PI3K-dependent ERK1/2 signaling. *J. Immunol.* **178:** 6456–6464.
106. **Yauch, L. E., M. K. Mansour, and S. M. Levitz.** 2005. Receptor-mediated clearance of *Cryptococcus neoformans* capsular polysaccharide in vivo. *Infect. Immun.* **73:**8429–8432.
107. **Yauch, L. E., M. K. Mansour, S. Shoham, J. B. Rottman, and S. M. Levitz.** 2004. Involvement of CD14, toll-like receptors 2 and 4, and MyD88 in the host response to the fungal pathogen *Cryptococcus neoformans* in vivo. *Infect. Immun.* **72:**5373–5382.
108. **Zheng, C. F., G. J. Jones, M. Shi, J. C. Wiseman, K. J. Marr, B. M. Berenger, S. M. Huston, M. J. Gill, A. M. Krensky, P. Kubes, and C. H. Mody.** 2008. Late expression of granulysin by microbicidal CD4+ T cells requires PI3K- and STAT5-dependent expression of IL-2Rbeta that is defective in HIV-infected patients. *J. Immunol.* **180:**7221–7229.
109. **Zheng, C. F., L. L. Ma, G. J. Jones, M. J. Gill, A. M. Krensky, P. Kubes, and C. H. Mody.** 2007. Cytotoxic CD4+ T cells use granulysin to kill *Cryptococcus neoformans*, and activation of this pathway is defective in HIV patients. *Blood* **109:**2049–2057.
110. **Zlotnik, A., D. I. Godfrey, M. Fischer, and T. Suda.** 1992. Cytokine production by mature and immature CD4-CD8- T cells. Alpha beta-T cell receptor+ CD4-CD8- T cells produce IL-4. *J. Immunol.* **149:**1211–1215.

PATHOGENESIS OF CRYPTOCOCCOSIS

Cryptococcus: From Human Pathogen to Model Yeast
Edited by J. Heitman et al.

31

Cryptococcus neoformans: Latency and Disease

FRANÇOISE DROMER, ARTURO CASADEVALL, JOHN PERFECT, AND TANIA SORRELL

In the past decade considerable progress has been made in our understanding of the pathogenesis of cryptococcosis with regard to the relationship between infection and latency and disease and the temporal relationship between these two events. The major new tools that have allowed progress are the development of serological assays indicative of infection and molecular typing that discriminates among cryptococcal strains. These techniques together with several clinical studies have provided evidence that infection can progress directly to disease or be contained in a state of latency that can subsequently reactivate to cause disease.

Understanding the temporal relationship between *Cryptococcus neoformans* infection and latency and disease is essential for devising strategies to reduce the incidence of cryptococcosis through vaccines, prophylactic antifungal therapy, and therapies that medically improve or impair immunity. For example, if initial infection occurs in childhood and results in the state of latency, then a vaccine to prevent the establishment of infection would have to be administered early in life. On the other hand, knowledge of prior infection with the ability to reactivate could help in identifying "unmasking" immune reconstitution syndrome (IRIS) or approaching asymptomatic cryptococcal antigenemia in HIV-infected hosts. At this time, there is evidence of a complex temporal relationship between infection and disease in cryptococcosis, and this chapter will review the evidence of disease after acute infection and reactivation of latent infection, considering the newer epidemiological and serological studies and the older anatomical evidence.

Françoise Dromer, Institut Pasteur, Molecular Mycology Unit, CNRS URA3012, 75724 Paris cedex 15, France. **Arturo Casadevall,** Department of Microbiology & Immunology, Albert Einstein College of Medicine, Jack and Pearl Resnick Campus, Bronx, NY 10461. **John Perfect,** Department of Medicine, Division of Infectious Diseases, Duke University Mycology Research Unit (DUMRU), Durham, NC 27710. **Tania Sorrell,** Centre for Infectious Diseases and Microbiology & Westmead Millennium Institute, Western Clinical School, University of Sydney, Wentworthville NSW 2145, Australia.

CLINICAL AND EPIDEMIOLOGICAL EVIDENCE FOR DISEASE FOLLOWING ACUTE AND LATENT INFECTIONS

Evidence for Disease after Acute Infection

The association of *C. neoformans* with pigeon droppings was made in the 1950s (31, 64). The proximity of pigeons to human dwellings, especially in many urban environments, suggests the possibility that infection progressing rapidly to disease can originate from immediate domestic sources, especially for immunocompromised individuals. Circumstantial evidence for subclinical infection after exposure to pigeons comes from serological studies of pigeon fanciers in which higher rates of cryptococcal antibodies were found in these individuals compared to controls (35, 78, 111). In one study, cryptococcosis was more frequent among AIDS patients residing in dwellings from which *C. neoformans* var. *grubii* was isolated than among AIDS patients from whose domestic environments the fungus was not demonstrated by the methods used (odds ratio = 2.05) (83). The genetic relatedness between isolates recovered from pigeon excreta and from clinical specimens has been demonstrated for *C. neoformans* var. *grubii* and for var. *neoformans* (17, 36, 41). Currie et al. identified strains of *C. neoformans* var. *grubii* with indistinguishable genetic fingerprints in both clinical and environmental samples in New York City (17), as did Garcia-Hermoso et al. for var. *neoformans* (41). When studying isolates from the environment of patients, Varma and colleagues, using fingerprinting analysis, failed to demonstrate genetic relatedness between the strains (109). The lack of identity between environmental and clinical isolates in that study could be explained by the vast heterogeneity of the strains present in the environment and cannot rule out acute infection; it could also be explained by the multiple and remote sources of potential exposure if the concept of reactivation of latent infection holds true.

Several anecdotal cases report exposure to *C. neoformans*–contaminated dust prior to the onset of acute pulmonary infection and even meningitis. Cryptococcal meningitis was diagnosed in an HIV-infected patient who helped dismantle an aviary, and another patient had pigeons nesting in the ceiling above his desk (34). Nosanchuk and colleagues reported the case of a 72-year-old kidney transplant recipient who was diagnosed with cryptococcal meningitis. *C. neoformans* was recovered from the guano of a pet cockatoo (79). The clinical and the avian isolates were found to be indistinguishable by chemical and restriction fragment length polymorphism analysis. Recently, a case of cryptococcal pneumonia in a patient who was taking infliximab for rheumatoid arthritis was described (94). A temporally related exposure history raised the possibility that the patient acquired the infection from his pet cockatiel. Wang and colleagues reported a case of possible human-to-human transmission of *C. neoformans* between two elderly men hospitalized in beds next to each other (112). The isolates from both patients were found to be genetically identical using arbitrarily primed PCR with four primers, suggesting acute infection through nosocomial transmission. However, a common source of the infection of both patients with the same strain could not be ruled out. In another report, the diagnosis of cryptococcosis following the placement of a ventriculoperitoneal shunt for progressive hydrocephalus the same day by the same surgeon in two immunocompetent patients raised the question of a nosocomial source of infection (54). This was ruled out by a careful analysis of the isolates including karyotyping, leaving open the hypothesis of reactivation of a preexisting infection as the most likely explanation.

Skin lesions due to *C. neoformans* infection can also occur following acute exposure to contaminated debris, especially in rural areas or in individuals who engage in outdoor activities. It has also been reported following a peck by a pigeon (20) and an insect bite (77). Two reviews on primary cutaneous cryptococcosis were recently published (16, 77). In the analysis of skin lesions attributed to *C. neoformans* infections in France, patients with primary cryptococcosis ($n = 28$) differed significantly from those with secondary cutaneous cryptococcosis ($n = 80$) or other forms of the disease ($n = 1{,}866$) by reporting a higher proportion of outdoor activity, the lack of underlying disease, an older age, and an even proportion of men and women (77). Furthermore, patients with primary cutaneous cryptococcosis were significantly more often infected with var. *neoformans* (serotype D) isolates than by var. *grubii* (serotype A) (72). Given that serotype D isolates are more susceptible to heat than serotype A (72), the higher association of serotype D with cutaneous disease reinforces the idea that the skin lesions in these patients are the result of primary inoculation with contaminated soil rather than hematogenous dissemination.

It is not clear whether what some authors call "airway colonization," i.e., culture of *C. neoformans* from a respiratory tract specimen in the absence of symptoms or radiological pulmonary lesions (1, 86), is a real clinical entity or if it is the first manifestation of a recent infection that is not yet associated with clinical symptoms.

Despite the presence of multiple strains in the environment and the ubiquitous nature of the fungus, relapse of cryptococcosis is usually due to persistence of the original isolate (12, 36, 53, 69, 108). However, when the relapse and the original isolate are found to be indistinguishable by the genotyping method, several explanations can be found: (i) the initial isolate may not have been eradicated by the antifungal treatment, and the relapse and the original isolate are truly identical, thus ruling out acute infection for the relapse isolate; (ii) the methods used are not discriminatory enough, and the isolates appear to be the same but may not be identical, which prevents any interpretation of reactivation or acute infection; or (iii) the infecting genotype is common in the environment, and the patient has been reinfected by the same genotype by chance.

This last hypothesis finds its basis in the study published in 1997 by Franzot and colleagues, comparing isolates from Brazil and from New York City. The same genotypes were found in both locations, demonstrating that some genotypes are globally dispersed (36), which was also recently confirmed using multilocus sequence typing on strains isolated from different regions of India (51). A few anecdotal reports of reinfection (i.e., acute infection) with a different strain have been published (10, 49, 65, 109), making it possible that this is a much more common phenomenon than currently thought. Likewise, disease is supposed to be caused by a unique strain resuming from latency, the multiple strains that are likely to be inhaled being eradicated over time. Microevolutions have been reported during the course of infection (2, 11, 36, 39, 55), but rare cases of infection with multiple strains have also been published (2, 49, 53, 55, 58, 71, 102). These are caused by mixed serotypes and genotypes, and the occurrence of such cases is more consistent with acute infection, rather than reactivation, although this cannot be demonstrated easily. Of note, these mixed infections may not be as rare as previously thought, since up to 18% of cases were caused by multiple strains of *C. neoformans* in a recent study (19a).

Cryptococcal infection occurs in children as demonstrated by the existence of antibodies to *C. neoformans* in children. Goldman and collaborators conducted an analysis by immunoblotting the sera from 185 immunocompetent individuals ranging in age from 1 week to 21 years for antibodies to *C. neoformans* proteins (44). They showed that the proportion of children immunized against *C. neoformans* increased with age. Sera from children under the age of 2 years showed minimal reactivity, whereas 70% of samples from children over 5 years old reacted with cryptococcal antigens. The authors concluded that the low incidence of cryptococcosis in children is not the result of a lack of exposure to *C. neoformans*. Primary pulmonary cryptococcosis may be asymptomatic or confused with viral infections frequently observed during childhood. Of note, the reactivity of children's sera differed in three geographical regions and suggested regional differences in exposure (18). This would also partially explain the variations in the incidence of cryptococcosis around the world, even among HIV-infected patients, independent of the availability of highly active antiretroviral therapy (HAART).

Circumstantial evidence suggests that *C. gattii*–related disease is most commonly an outcome of primary infection with *C. gattii* rather than reactivation. *C. gattii* (then designated *C. neoformans* var. *gattii*) was recognized in 1970 (107), and its ecological niche was identified in Australia as the red gum group of eucalyptus trees (30, 85). It was noted that the distribution of human cases of cryptococcosis due to *C. gattii* in Australia generally corresponds to the distribution of host eucalypts (28, 29), in keeping with a later study that revealed a high incidence in Aboriginal people with a rural/semirural domicile (15). Furthermore, all eucalypt and virtually all clinical isolates were serotype B and of identical molecular type (VGI) when analyzed by randomly amplified polymorphic DNA and PCR fingerprinting (98). Confirmation of a direct link between an environmental source and a cluster of cases of human disease was also made in Arnhem Land in the Northern Territory of Australia (96) and in the recent outbreak of C *gattii*–related diseases on Vancouver Island and in the northwest United States. In all three locations a second molecular type, VGII, was isolated from both trees and infected individuals (101). The concept that cryptococcosis caused by *C. gattii* follows closely after primary infections is supported by veterinary reports of localized respiratory disease in cats, dogs, birds, and native koalas and by a localized outbreak in sheep in Western Australia (59, 70, 81, 97). Moreover, human cases present with pulmonary or concurrent pulmonary and neurological disease significantly more often in *C. gattii* infections than those due to *C. neoformans* (15). In British Columbia, 87% of cases presented with respiratory findings (10% with concurrent central nervous system findings), following an incubation period of 6 weeks to 11 months, as determined in infected travelers from outside the endemic area (63, 68) (see also chapter 23).

In contrast to the findings with *C. neoformans*, where serum antibodies to protein antigens were present in 70% of children over the age of 5 years in New York City, serological studies from Australia suggest that exposure to cryptococcosis is low in childhood and increases with age. Speed et al., using a class-specific enzyme immunoassay test that did not distinguish between *C. neoformans* and *C. gattii*, demonstrated specific immunoglobulin G antibodies to capsular antigens in only 4% of Australian children, compared with 69% of healthy adult blood donors (99). In Papua New Guinea, the seroprevalence of antibodies to noncapsular protein antigens of *C. gattii* in controls mirrored that of patients with cryptococcal meningitis, being highest in adults (males higher than females) and relatively low in children (92), consistent with the presumed frequency of exposure to environmental niches of *C. gattii*. Although immunoglobulin G antibodies can persist for at least 3 years following successful therapy of *C. gattii* disease in healthy hosts (100), and healthy hosts can have "high" antibody titers, it is not clear whether this represents persistent infection or repeated heavy exposure to environmental cryptococci.

These results suggest that many *C. gattii*–related diseases are acute responses to an initial exposure rather than reactivation of older latent primary infections. Clinical and serological studies distinguish poorly between persistence of infection, relapse, reactivation, and reinfection, and there are no long-term cohort studies in patients with cryptococcosis. Cryptococcal antigenemia was reported in 28 healthy koalas from different regions in Australia, two of which had small cryptococcal lesions in a paranasal sinus and lung, respectively, and one of which developed cryptococcal pneumonia 6 months later (59). Some animals became antigen negative ("recovered"), and though the fate of others with persisting low-level antigenemia was not determined, the authors speculated that small residual foci of infection may persist in the respiratory tract and constitute potential sites of reactivation (59). The high incidence of disease reported in the Vancouver Island outbreak may not be the norm, even though it appears from epidemiological studies that *C. neoformans* has a greater potential for reactivation of infection leading to disease than *C. gattii*. Specifically, high inocula from this outbreak might be extreme and push toward acute infection followed by disease and, since the outbreak strain could have been created from a recent recombination event (37), it might possess virulence features that favor acute infection. Notably, a higher proportion of veterinary cases of cryptococcosis in the Vancouver Island outbreak presented with neurological disease than has been observed in Australia, a potential indicator of genotype-dependent differences in virulence (61).

Evidence for Reactivation of Latent Infection

In contrast to the serological evidence from areas not affected by *C. gattii* that cryptococcal infection in children is frequent, pediatric cryptococcal disease due to both species of *Cryptococcus* is uncommon. Similarly, cryptococcal disease is also rare in immunocompetent individuals despite widespread evidence for prior infection, as reflected by serological studies showing that the presence of antibodies to *C. neoformans* is common in adults (19, 24, 50). Also, there are several cryptococcin skin testing studies that support frequent exposure to *C. neoformans* and/or cryptococcal antigens among laboratory workers and pigeon breeders that leads to a cell-mediated immune response with positive skin tests without development of clinical disease (6). In high-risk patients such as HIV-infected individuals, there is a well-described syndrome of "asymptomatic cryptococcal antigenemia" that in many patients represents incubating or reactivating disease (32, 62, 73). Nevertheless, disease is an important complication of cellular immune defects including HIV infection and solid organ transplantation. The exact risk factors for developing disease in solid organ transplant recipients are poorly understood, but serological studies in these transplant recipients do suggest that reactivation of latent infection was more likely than acute infection when cryptococcosis occurred early after surgery (89). For example, investigators analyzed antibodies to *C. neoformans* in serum samples obtained before and after surgery in solid organ transplant recipients who did and did not develop cryptococcosis after transplantation (89). Patients who developed cryptococcosis early after transplantation had more serum antibody reactivity than those who developed cryptococcosis later. In the former case, the preexisting antibodies suggested the existence of subclinical infection and reactivation following immunosuppression. In the latter case, the lack of prior immunization and the late occurrence of cryptococcosis may reflect the necessary time to be exposed and acquire new infection.

Perhaps the most persuasive evidence of the rarity of reactivation in cryptococcosis due to *C. gattii* is the rarity of disease in patients with immunosuppression due to HIV and other causes (15, 88). Only about 2% of HIV-associated cases are due to *C. gattii* in areas where this species is endemic (15, 75), and *C. gattii* similarly caused a minority of cases (~5%) in other classically immunocompromised patients compared with 44% in healthy hosts. Overall, 89% of infections due to *C. gattii* were in nonimmunocompromised hosts, compared with 20% of those caused by *C. neoformans* (15). These observations suggest that persisting infection and reactivation are rare, even though the occasional case of infection due to *C. gattii* in nonendemic areas, consistent with reactivation, has been reported (26). Furthermore, three episodes of cryptococcosis over 13 years, with isolates of *C. gattii* from the first and last infections identical by DNA fingerprinting and random amplification of polymorphic DNA analysis, were reported in a Birman cat (56). Although historically, cryptococcosis caused by *C. gattii* in nonendemic areas provided a powerful argument for latency and then reactivation, this argument is weakened given the ongoing outbreak on Vancouver Island, which shows that *C. gattii* is occupying new geographic areas (37).

On the other hand, analysis of clinical isolates of *C. neoformans* from patients diagnosed in France provided another hint that reactivation of latent infection is frequent. While var. *neoformans* (serotype D) accounts for approximately 26% of the infections in France, it is recovered in less than 10% of the patients born in Africa ($P < 0.001$) (French National Reference Center for Mycoses and Antifungals, unpublished data). Since isolates of var. *neoformans* are rarely identified in Africa, this epidemiological finding suggested that patients born in Africa and diagnosed with cryptococcosis in France were likely to have been infected by strains of var. *grubii* from Africa. This was demonstrated using restriction fragment length polymorphism analysis showing that isolates of var. *grubii* from patients born in Africa who developed disease in France clustered together, away from isolates of patients born in Europe (40). These data provided the first molecular evidence of reactivation of latent infection, and in fact one of the patients had not traveled back to Africa for the past 17 years before developing cryptococcosis.

Overall, clinical and epidemiological data, as well as analysis of strains, suggest that disease can follow both reactivation of latent infections and acute infection, especially in infections due to *C. neoformans*. Furthermore, some studies have provided tantalizing evidence consistent with the acquisition of new infections in the setting of ongoing or treated cryptococcosis and raise the possibility that new asymptomatic infections are common events in the course of life.

ANATOMICAL AND CELLULAR SITES FOR LATENCY

The best information about the anatomical sites of persistent infection and latency remains the classic autopsy studies carried out in the 1950s (8, 9, 48). Baker and collaborators reported an association between subpleural nodules and *C. neoformans* and proposed that the pathogenesis of cryptococcosis paralleled that of tuberculosis, although the frequency of finding this "primary complex" was much lower than in cases of tuberculosis (7, 90). According to this view, inhalation of infectious propagules resulted in an initial infection in the lung that elicited a granulomatous immune response and formed a subpleural nodule or primary lymph node complex. This early immune response contained the infection to the nodule and draining lymph nodes. With the passage of time, a normal healthy immune response would contain the infection to the nodule and/or lymph node or eradicate the yeasts. According to this hypothesis, eradication would represent cure of infection, while containment was indicated by persistence of infection, possibly in a latent form. The overwhelming majority of individuals with persistent/latent infection would remain asymptomatic as long as they retained an intact immune system. However, events that subsequently attack the immune system such as HIV infection and immunosuppressive therapies carry the potential for undermining the host immune containment, with subsequent reactivation of infection leading to the appearance of disease. From Baker's autopsy studies and proposed pathogenic scheme, we can infer at least two sites of persistence of infection: subpleural nodules and their draining lymph nodes. In recent decades, numerous studies have confirmed the fact that *C. neoformans* pulmonary infection is often associated with nodules that are often noted in radiographic studies (84, 103) and resemble lung cancer in appearance (87). In murine inhalational models acute pulmonary infection with *C. neoformans* is rapidly followed by the appearance of cryptococci both free and within macrophages in the lymphoid sinuses of hilar (draining) lymph nodes (80, 91). In men, the prostate could also represent a site for chronic latent infection (60).

It was the classic studies by Gadebusch and Gikas that elegantly framed the possibility of reactivating cryptococcal infections (38). With low aerosol inocula, they showed that disease could be reactivated with the use of corticosteroids in both guinea pigs and rats. Their studies found that reactivation could only occur in a defined window (4 weeks) after initial infection, and this idea of length of dormancy for a viable reactivation to disease remains uncertain in human disease, but the ability to return infection to disease with immunosuppression like in animals likely occurs in humans.

The best animal model available for the study of latency and reactivation is experimental rat pulmonary cryptococcosis (42). The rat is remarkably resistant to *C. neoformans* infection and is able to contain even large inocula in the lung through an effective granulomatous response. However, despite mounting an immune response that causes large reductions in fungal burden, rats often cannot fully eradicate the infection, and they contain these yeasts in subpleural granulomas (45). Administration of corticosteroid therapy to rats can reliably reactivate latent infection (45). Furthermore, despite the effectiveness of the rat pulmonary inflammatory response, cryptococcal infection alters the pulmonary inflammatory milieu such that the animals are then predisposed to allergic lung diseases (43), an observation that has suggested a link between cryptococcosis and urban asthma. Autopsy studies of rats that have successfully contained cryptococcal infection show persistence of infection in lung granulomas, thus providing an animal model correlate for the human autopsy studies cited above.

The most likely cellular sites for latency are macrophages, where *C. neoformans* can reside as a facultative intracellular pathogen (33). Containment of *C. neoformans* in tissue is invariably associated with a strong granulomatous response where one finds few yeasts, almost always inside macrophages and giant cells (reviewed in reference 13). Given the ability of cryptococci to survive inside macrophages, one might imagine that the intracellular residency could provide yeast cells with a protected environment for persistence despite strong granulomatous tissue responses. In rats, persistence of infection is invariably associated with intracellular yeasts (45). In this regard persistence may be facilitated by the recently described ability of *C. neoformans* to spread from cell to cell in macrophage cultures (3, 67).

CONSEQUENCES OF ACUTE AND LATENT INFECTIONS

Whereas in 1995 the U.S. Public Health Service and the Infectious Diseases Society of America recommended avoiding sites likely to be heavily contaminated with pigeon droppings (106), in the updated 2002 version of the recommendations for the prevention of opportunistic infections among HIV-infected persons, it now states that HIV-infected persons cannot completely avoid exposure to *C. neoformans* and that no evidence exists that exposure to pigeon droppings is associated with an increased risk for acquiring cryptococcosis (105). Although such recommendations reflect global epidemiological studies, the high prevalence of *C. neoformans* in urban environments, and serological evidence for widespread infection, we note that the absence of evidence does not rule out a connection. Given the reports of cryptococcosis after acute exposures to pigeons, caution is encouraged for immunocompromised patients being exposed to pigeon droppings. Therefore, one could recommend avoiding massive inhalation of pet avian or pigeon excreta such as cleaning a heavily contaminated environment (old buildings, bird cages, etc.) to all patients with cellular immune defects. The public health response to the Vancouver Island outbreak has been to raise awareness of *C. gattii* among physicians and the public and to provide education about cryptococcosis and its management (see chapter 23).

The risks associated with latency are probably higher and less controllable than those following exposure to a contaminated environment. As already mentioned, infections occur in more than 80% of the adult population based on serum reactivity against cryptococcal polysaccharide or protein antigen (14, 24). Dominiak and colleagues noted the existence of endogenous sources of infection in transplant recipients (23) and other immunosuppressed patients. Almost a third of the HIV-negative patients with cryptococcosis are under corticosteroid therapy for various underlying conditions (22, 25, 82). Since their start of clinical use, anti-CD52, anti-CD20, and anti–tumor necrosis factor alpha antibodies have been occasionally associated with invasive fungal infections including cryptococcosis (4, 46, 47, 52, 57, 76, 95, 104, 113). In all these patients, the risk of reactivating latent infections after prescription of immunosuppressive drugs should be considered. This implies extended workup with systematic lumbar puncture and serum cryptococcal antigen in cases with symptoms suggestive of disseminated fungal infection including unexplained fever and biopsy of new pulmonary lesions occurring on these immunosuppressive therapies.

A potential risk of latent infection concerns HIV-infected patients who are naive for antiretrovirals. Restoration of immunity after prescription of HAART can lead to inflammatory responses and clinical manifestations known as IRIS. This inflammation can be targeted at live or dead pathogens including fungi (21). Cases of this unmasking IRIS occur in HIV-infected patients who have been prescribed HAART, and shortly after, the diagnosis of latent opportunistic infections occurs when HIV infection is controlled (114). On the other hand, the risk factors for the development of "paradoxical" IRIS in the setting of recent cryptococcosis treatment have been assessed and provide a good basis for monitoring the patients (66, 93). In the cohort of HIV-infected patients with cryptococcal infections treated with HAART, Shelburne and colleagues (93) identified 18 patients with IRIS (30.5%) including 3 with IRIS unmasking latent infection. This clinical entity is probably misdiagnosed in routine clinical practice. In fact, if it manifests as enlarging lymph nodes or pulmonary infiltrates or fever, it may have minimal clinical consequences and improve without directed therapies. In central nervous system infections, IRIS requires more clinical attention.

Another potential consequence of latent infection is the association with chronic pulmonary diseases regardless of the prescription of corticosteroids. Colonization rather than disease has been evoked after isolation of *C. neoformans* from the respiratory tract of patients with chronic respiratory diseases such as asthma or chronic obstructive pulmonary diseases (27). One study from Japan suggested a link between allergic lung manifestations and immune reactions against *C. neoformans* (74). Allergic bronchopulmonary mycoses in humans are frequent in the setting of cystic fibrosis and are largely caused by molds including *Aspergillus* spp. The proposed pathogenesis is that inhalation of fungi leads to irritation of the airway epithelium and to secretion of chemokines and cytokines in reaction to fungal mediators and enzymes (5). In a normal host, these cytokine mediators include proinflammatory cytokines and lead to a Th1 response, while in an atopic host the fungal assault leads to the generation of anti-inflammatory and Th2 responses. In the rat model, Goldman and colleagues showed that localized pulmonary *C. neoformans* infection exacerbated allergic responses to respiratory challenges such as ovalbumin (43). The lung concentrations of Th2 cytokines increased, and those of Th1 cytokines decreased. They also observed a partial amelioration after fluconazole treatment. Experimental pulmonary cryptococcosis has been shown to induce chitinase in rats, and this enzyme has also been implicated in the pathogenesis of asthma (110). Altogether, these results suggest a potential role for latent pulmonary cryptococcosis in the pathogenesis of asthma. Whether these results can be extrapolated to humans remains to be determined. Analysis of serum reactivity against cryptococcal proteins in patients with asthma and other chronic pulmonary diseases compared to healthy individuals could provide useful information regarding *C. neoformans* involvement in the pathogenesis of these diseases and create a basis for antifungal drug prescriptions.

CONCLUSIONS

The majority of the epidemiological data support that humans are frequently exposed to environmental *C. neoformans* and *C. gattii*, and although outbreaks of disease are infrequent, infection is common worldwide with one or more cryptococcal strains. It appears from serological and skin test epidemiological studies, focused pathological studies of human tissue, animal models and intracellular macrophages, and the temporal relationship of immune dysregulation with appearance of disease that reactivation is a common consequence of infection with *C. neoformans*. For *C. gattii*, there is circumstantial evidence that disease following acute infection is the norm and that persistence of infection in a latent form with reactivation is a rarer event.

REFERENCES

1. **Aberg, J. A., L. M. Mundy, and W. G. Powderly.** 1999. Pulmonary cryptococcosis in patients without HIV infection. *Chest* **115:**734–740.
2. **Almeida, A. M., M. T. Matsumoto, L. C. Baeza, E. S. R. B. de Oliveira, A. A. Kleiner, S. Melhem Mde, and M. J. Mendes Giannini.** 2007. Molecular typing and antifungal susceptibility of clinical sequential isolates of *Cryptococcus neoformans* from Sao Paulo State, Brazil. *FEMS Yeast Res.* **7:**152–164.
3. **Alvarez, M., and A. Casadevall.** 2007. Cell-to-cell spread and massive vacuole formation after *Cryptococcus neoformans* infection of murine macrophages. *BMC Immunol.* **8:**16.
4. **Arend, S. M., E. J. Kuijper, C. F. Allaart, W. H. Muller, and J. T. Van Dissel.** 2004. Cavitating pneumonia after treatment with infliximab and prednisone. *Eur. J. Clin. Microbiol. Infect. Dis.* **23:**638–641.
5. **Arora, S., and G. B. Huffnagle.** 2005. Immune regulation during allergic bronchopulmonary mycosis: lessons taught by two fungi. *Immunol. Res.* **33:**53–68.
6. **Atkinson, A. J., Jr., and J. E. Bennett.** 1968. Experience with a new skin test antigen prepared from *Cryptococcus neoformans*. *Am. Rev. Respir. Dis.* **97:**637–643.
7. **Baker, R. D.** 1976. The primary pulmonary lymph node complex of cryptococcosis. *Am. J. Clin. Pathol.* **65:**83–92.
8. **Baker, R. D.** 1952. Resectable mycotic lesions and acutely fatal mycoses. *JAMA* **150:**1579–1581.
9. **Baker, R. D., and R. K. Haugen.** 1955. Tissue changes and tissue diagnosis in cryptococcosis; a study of 26 cases. *Am. J. Clin. Pathol.* **25:**14–24.
10. **Barchiesi, F., R. J. Hollis, S. A. Messer, G. Scalise, M. G. Rinaldi, and M. A. Pfaller.** 1995. Electrophoretic karyotype and in vitro antifungal susceptibility of *Cryptococcus neoformans* isolates from AIDS patients. *Diagn. Microbiol. Infect. Dis.* **23:**99–103.
11. **Blasi, E., A. Brozzetti, D. Francisci, R. Neglia, G. Cardinali, F. Bistoni, V. Vidotto, and F. Baldelli.** 2001. Evidence of microevolution in a clinical case of recurrent *Cryptococcus neoformans* meningoencephalitis. *Eur. J. Clin. Microbiol. Infect. Dis.* **20:**535–543.

12. **Brandt, M. E., L. C. Hutwagner, L. A. Klug, W. S. Baughman, D. Rimland, E. A. Graviss, R. J. Hamill, C. Thomas, P. G. Pappas, A. L. Reingold, and R. W. Pinner.** 1996. Molecular subtype distribution of *Cryptococcus neoformans* in four areas of the United States. Cryptococcal Disease Active Surveillance Group. *J. Clin. Microbiol.* **34:**912–917.
13. **Casadevall, A., and J. R. Perfect.** 1998. *Cryptococcus neoformans*. ASM Press, Washington, DC.
14. **Chen, L. C., D. L. Goldman, T. L. Doering, L. Pirofski, and A. Casadevall.** 1999. Antibody response to *Cryptococcus neoformans* proteins in rodents and humans. *Infect. Immun.* **67:**2218–2224.
15. **Chen, S., T. Sorrell, G. Nimmo, B. Speed, B. Currie, D. Ellis, D. Marriott, T. Pfeiffer, D. Parr, and K. Byth.** 2000. Epidemiology and host- and variety-dependent characteristics of infection due to *Cryptococcus neoformans* in Australia and New Zealand. Australasian Cryptococcal Study Group. *Clin. Infect. Dis.* **31:**499–508.
16. **Christianson, J. C., W. Engber, and D. Andes.** 2003. Primary cutaneous cryptococcosis in immunocompetent and immunocompromised hosts. *Med. Mycol.* **41:**177–188.
17. **Currie, B. P., L. F. Freundlich, and A. Casadevall.** 1994. Restriction fragment length polymorphism analysis of *Cryptococcus neoformans* isolates from environmental (pigeon excreta) and clinical sources in New York City. *J. Clin. Microbiol.* **32:**1188–1192.
18. **Davis, J., W. Y. Zheng, A. Glatman-Freedman, J. A. Ng, M. R. Pagcatipunan, H. Lessin, A. Casadevall, and D. L. Goldman.** 2007. Serologic evidence for regional differences in pediatric cryptococcal infection. *Pediatr. Infect. Dis. J.* **26:**549–551.
19. **DeShaw, M., and L. A. Pirofski.** 1995. Antibodies to the *Cryptococcus neoformans* capsular glucuronoxylomannan are ubiquitous in serum from HIV+ and HIV- individuals. *Clin. Exp. Immunol.* **99:**425–432.
19a. **Desnos-Ollivier, M., S. Patel, A. R. Spaulding, C. Charlier, D. Garcia-Hermoso, K. Nielsen, and F. Dromer.** 2010. Mixed infections and in vivo evolution in the human fungal pathogen *Cryptococcus neoformans*. *mBio* **1:**e00091-10. doi:10.1128/mBio.00091-10.
20. **Dev, D., G. S. Basran, D. Slater, P. Taylor, and M. Wood.** 1994. Immunodeficiency without HIV. Consider HIV negative immunodeficiency in cryptococcosis. *BMJ* **308:**1436.
21. **Dhasmana, D. J., K. Dheda, P. Ravn, R. J. Wilkinson, and G. Meintjes.** 2008. Immune reconstitution inflammatory syndrome in HIV-infected patients receiving antiretroviral therapy: pathogenesis, clinical manifestations and management. *Drugs* **68:**191–208.
22. **Diamond, R. D., and J. E. Bennett.** 1974. Prognostic factors in cryptococcal meningitis. A study in 111 cases. *Ann. Intern. Med.* **80:**176–181.
23. **Dominiak, A., B. Interewicz, E. Swoboda, and W. L. Olszewski.** 2006. Endogeneous sources of infection in transplant recipients. *Ann. Transplant.* **11:**30–37.
24. **Dromer, F., P. Aucouturier, J. P. Clauvel, G. Saimot, and P. Yeni.** 1988. *Cryptococcus neoformans* antibody levels in patients with AIDS. *Scand. J. Infect. Dis.* **20:**283–285.
25. **Dromer, F., S. Mathoulin, B. Dupont, and A. Laporte.** 1996. Epidemiology of cryptococcosis in France: a 9-year survey (1985–1993). French Cryptococcosis Study Group. *Clin. Infect. Dis.* **23:**82–90.
26. **Dromer, F., O. Ronin, and B. Dupont.** 1992. Isolation of *Cryptococcus neoformans* var. *gattii* from an Asian patient in France: evidence for dormant infection in healthy subjects. *J. Med. Vet. Mycol.* **30:**395–397.
27. **Duperval, R., P. E. Hermans, N. S. Brewer, and G. D. Roberts.** 1977. Cryptococcosis, with emphasis on the significance of isolation of *Cryptococcus neoformans* from the respiratory tract. *Chest* **72:**13–19.
28. **Ellis, D. H., and T. Pfeiffer.** 1994. Cryptococcosis and the ecology of *Cryptococcus neoformans*. *Jpn. J. Med. Mycol.* **35:**111–122.
29. **Ellis, D. H., and T. J. Pfeiffer.** 1990. Ecology, life cycle, and infectious propagule of *Cryptococcus neoformans*. *Lancet* **336:**923–925.
30. **Ellis, D. H., and T. J. Pfeiffer.** 1990. Natural habitat of *Cryptococcus neoformans* var. *gattii*. *J. Clin. Microbiol.* **28:**1642–1644.
31. **Emmons, C. W.** 1951. Isolation of *Cryptococcus neoformans* from soil. *J. Bacteriol.* **62:**685–690.
32. **Feldmesser, M., C. Harris, S. Reichberg, S. Khan, and A. Casadevall.** 1996. Serum cryptococcal antigen in patients with AIDS. *Clin. Infect. Dis.* **23:**827–830.
33. **Feldmesser, M., Y. Kress, P. Novikoff, and A. Casadevall.** 2000. *Cryptococcus neoformans* is a facultative intracellular pathogen in murine pulmonary infection. *Infect. Immun.* **68:**4225–4237.
34. **Fessel, W. J.** 1993. Cryptococcal meningitis after unusual exposures to birds. *N. Engl. J. Med.* **328:**1354–1355.
35. **Fink, J. N., J. J. Barboriak, and L. Kaufman.** 1968. Cryptococcal antibodies in pigeon breeders' disease. *J. Allergy* **41:**297–301.
36. **Franzot, S. P., J. S. Hamdan, B. P. Currie, and A. Casadevall.** 1997. Molecular epidemiology of *Cryptococcus neoformans* in Brazil and the United States: evidence for both local genetic differences and a global clonal population structure. *J. Clin. Microbiol.* **35:**2243–2251.
37. **Fraser, J. A., S. S. Giles, E. C. Wenink, S. G. Geunes-Boyer, J. R. Wright, S. Diezmann, A. Allen, J. E. Stajich, F. S. Dietrich, J. R. Perfect, and J. Heitman.** 2005. Same-sex mating and the origin of the Vancouver Island *Cryptococcus gattii* outbreak. *Nature* **437:**1360–1364.
38. **Gadebusch, H. H., and P. W. Gikas.** 1965. The effect of cortisone upon experimental pulmonary cryptococcosis. *Am. Rev. Respir. Dis.* **92:**64–74.
39. **Garcia-Hermoso, D., F. Dromer, and G. Janbon.** 2004. *Cryptococcus neoformans* capsule structure evolution in vitro and during murine infection. *Infect. Immun.* **72:**3359–3365.
40. **Garcia-Hermoso, D., G. Janbon, and F. Dromer.** 1999. Epidemiological evidence for dormant *Cryptococcus neoformans* infection. *J. Clin. Microbiol.* **37:**3204–3209.
41. **Garcia-Hermoso, D., S. Mathoulin-Pelissier, B. Couprie, O. Ronin, B. Dupont, and F. Dromer.** 1997. DNA typing suggests pigeon droppings as a source of pathogenic *Cryptococcus neoformans* serotype D. *J. Clin. Microbiol.* **35:**2683–2685.
42. **Goldman, D., S. C. Lee, and A. Casadevall.** 1994. Pathogenesis of pulmonary *Cryptococcus neoformans* infection in the rat. *Infect. Immun.* **62:**4755–4761.
43. **Goldman, D. L., J. Davis, F. Bommarito, X. Shao, and A. Casadevall.** 2006. Enhanced allergic inflammation and airway responsiveness in rats with chronic *Cryptococcus neoformans* infection: potential role for fungal pulmonary infection in the pathogenesis of asthma. *J. Infect. Dis.* **193:**1178–1186.
44. **Goldman, D. L., H. Khine, J. Abadi, D. J. Lindenberg, L. Pirofski, R. Niang, and A. Casadevall.** 2001. Serologic evidence for *Cryptococcus neoformans* infection in early childhood. *Pediatrics* **107:**E66.
45. **Goldman, D. L., S. C. Lee, A. J. Mednick, L. Montella, and A. Casadevall.** 2000. Persistent *Cryptococcus neoformans* pulmonary infection in the rat is associated with intracellular parasitism, decreased inducible nitric oxide synthase expression, and altered antibody responsiveness to cryptococcal polysaccharide. *Infect. Immun.* **68:**832–838.
46. **Hage, C. A., K. L. Wood, H. T. Winer-Muram, S. J. Wilson, G. Sarosi, and K. S. Knox.** 2003. Pulmonary

cryptococcosis after initiation of anti-tumor necrosis factor-alpha therapy. *Chest* **124:**2395–2397.

47. **Hamilton, C. D.** 2005. Immunosuppression related to collagen-vascular disease or its treatment. *Proc. Am. Thorac. Soc.* **2:**456–460.
48. **Haugen, R. K., and R. D. Baker.** 1954. The pulmonary lesions in cryptococcosis with special reference to subpleural nodules. *Am. J. Clin. Pathol.* **24:**1381–1390.
49. **Haynes, K. A., D. J. Sullivan, D. C. Coleman, J. C. Clarke, R. Emilianus, C. Atkinson, and K. J. Cann.** 1995. Involvement of multiple *Cryptococcus neoformans* strains in a single episode of cryptococcosis and reinfection with novel strains in recurrent infection demonstrated by random amplification of polymorphic DNA and DNA fingerprinting. *J. Clin. Microbiol.* **33:**99–102.
50. **Henderson, D. K., J. E. Bennett, and M. A. Huber.** 1982. Long-lasting, specific immunologic unresponsiveness associated with cryptococcal meningitis. *J. Clin. Invest.* **69:**1185–1190.
51. **Hiremath, S. S., A. Chowdhary, T. Kowshik, H. S. Randhawa, S. Sun, and J. Xu.** 2008. Long-distance dispersal and recombination in environmental populations of *Cryptococcus neoformans* var. *grubii* from India. *Microbiology* **154:**1513–1524.
52. **Horcajada, J. P., J. L. Pena, V. M. Martinez-Taboada, T. Pina, I. Belaustegui, M. E. Cano, D. Garcia-Palomo, and M. C. Farinas.** 2007. Invasive cryptococcosis and adalimumab treatment. *Emerg. Infect. Dis.* **13:**953–955.
53. **Igreja, R. P., M. S. Lazéra, B. Wanke, M. C. Galhardo, S. E. Kidd, and W. Meyer.** 2004. Molecular epidemiology of *Cryptococcus neoformans* isolates from AIDS patients of the Brazilian city, Rio de Janeiro. *Med. Mycol.* **42:**229–238.
54. **Ingram, C. W., H. B. Haywood, 3rd, V. M. Morris, R. L. Allen, and J. R. Perfect.** 1993. Cryptococcal ventricular-peritoneal shunt infection: clinical and epidemiological evaluation of two closely associated cases. *Infect. Control Hosp. Epidemiol.* **14:**719–722.
55. **Jain, N., B. L. Wickes, S. M. Keller, J. Fu, A. Casadevall, P. Jain, M. A. Ragan, U. Banerjee, and B. C. Fries.** 2005. Molecular epidemiology of clinical *Cryptococcus neoformans* strains from India. *J. Clin. Microbiol.* **43:**5733–5742.
56. **Kluger, E. K., H. K. Karaoglu, M. B. Krockenberger, P. K. Della Torre, W. Meyer, and R. Malik.** 2006. Recrudescent cryptococcosis, caused by *Cryptococcus gattii* (molecular type VGII), over a 13-year period in a Birman cat. *Med. Mycol.* **44:**561–566.
57. **Kozic, H., K. Riggs, F. Ringpfeil, and J. B. Lee.** 2008. Disseminated *Cryptococcus neoformans* after treatment with infliximab for rheumatoid arthritis. *J. Am. Acad. Dermatol.* **58:**S95–96.
58. **Krajden, S., R. C. Summerbell, J. Kane, I. F. Salkin, M. E. Kemna, M. G. Rinaldi, M. Fuksa, E. Spratt, C. Rodrigues, and J. Choe.** 1991. Normally saprobic cryptococci isolated from *Cryptococcus neoformans* infections. *J. Clin. Microbiol.* **29:**1883–1887.
59. **Krockenberger, M. B., P. J. Canfield, J. Barnes, L. Vogelnest, J. Connolly, C. Ley, and R. Malik.** 2002. *Cryptococcus neoformans* var. *gattii* in the koala (*Phascolarctos cinereus*): serological evidence for subclinical cryptococcosis. *Med. Mycol.* **40:**273–282.
60. **Larsen, R. A., S. Bozzette, J. A. McCutchan, J. Chiu, M. A. Leal, and D. D. Richman.** 1989. Persistent *Cryptococcus neoformans* infection of the prostate after successful treatment of meningitis. California Collaborative Treatment Group. *Ann. Intern. Med.* **111:**125–128.
61. **Lester, S. J., N. J. Kowalewich, K. H. Bartlett, M. B. Krockenberger, T. M. Fairfax, and R. Malik.** 2004. Clinicopathologic features of an unusual outbreak of cryptococcosis in dogs, cats, ferrets, and a bird: 38 cases (January to July 2003). *J. Am. Vet. Med. Assoc.* **225:**1716–1722.
62. **Liechty, C. A., P. Solberg, W. Were, J. P. Ekwaru, R. L. Ransom, P. J. Weidle, R. Downing, A. Coutinho, and J. Mermin.** 2007. Asymptomatic serum cryptococcal antigenemia and early mortality during antiretroviral therapy in rural Uganda. *Trop. Med. Int. Health* **12:**929–935.
63. **Lindberg, J., F. Hagen, A. Laursen, J. Stenderup, and T. Boekhout.** 2007. *Cryptococcus gattii* risk for tourists visiting Vancouver Island, Canada. *Emerg. Infect. Dis.* **13:**178–179.
64. **Littman, M. L., and S. S. Schneierson.** 1959. *Cryptococcus neoformans* in pigeon excreta in New York City. *Am. J. Hyg.* **69:**49–59.
65. **Litvintseva, A. P., L. Kestenbaum, R. Vilgalys, and T. G. Mitchell.** 2005. Comparative analysis of environmental and clinical populations of *Cryptococcus neoformans*. *J. Clin. Microbiol.* **43:**556–564.
66. **Lortholary, O., A. Fontanet, N. Memain, A. Martin, K. Sitbon, and F. Dromer.** 2005. Incidence and risk factors of immune reconstitution inflammatory syndrome complicating HIV-associated cryptococcosis in France. *AIDS* **19:**1043–1049.
67. **Ma, H., J. E. Croudace, D. A. Lammas, and R. C. May.** 2007. Direct cell-to-cell spread of a pathogenic yeast. *BMC Immunol.* **8:**15.
68. **MacDougall, L., and M. Fyfe.** 2006. Emergence of *Cryptococcus gattii* in a novel environment provides clues to its incubation period. *J. Clin. Microbiol.* **44:**1851–1852.
69. **Magee, J. T., C. Philpot, J. Yang, and I. K. Hosein.** 1994. Pyrolysis typing of isolates from a recurrence of systemic cryptococcosis. *J. Med. Microbiol.* **40:**165–169.
70. **Malik, R., M. B. Krockenberger, G. Cross, R. Doneley, D. N. Madill, D. Black, P. McWhirter, A. Rozenwax, K. Rose, M. Alley, D. Forshaw, I. Russell-Brown, A. C. Johnstone, P. Martin, C. R. O'Brien, and D. N. Love.** 2003. Avian cryptococcosis. *Med. Mycol.* **41:**115–124.
71. **Mandal, P., U. Banerjee, A. Casadevall, and J. D. Nosanchuk.** 2005. Dual infections with pigmented and albino strains of *Cryptococcus neoformans* in patients with or without human immunodeficiency virus infection in India. *J. Clin. Microbiol.* **43:**4766–4772.
72. **Martinez, L. R., J. Garcia-Rivera, and A. Casadevall.** 2001. *Cryptococcus neoformans* var. *neoformans* (serotype D) strains are more susceptible to heat than *C. neoformans* var. *grubii* (serotype A) strains. *J. Clin. Microbiol.* **39:** 3365–3367.
73. **Micol, R., O. Lortholary, B. Sar, D. Laureillard, C. Ngeth, J. P. Dousset, H. Chanroeun, L. Ferradini, P. J. Guerin, F. Dromer, and A. Fontanet.** 2007. Prevalence, determinants of positivity, and clinical utility of cryptococcal antigenemia in Cambodian HIV-infected patients. *J. Acquir. Immune Defic. Syndr.* **45:**555–559.
74. **Miyagawa, T., T. Ochi, and H. Takahashi.** 1978. Hypersensitivity pneumonitis with antibodies to *Cryptococcus neoformans*. *Clin. Allergy* **8:**501–509.
75. **Morgan, J., K. M. McCarthy, S. Gould, K. Fan, B. Arthington-Skaggs, N. Iqbal, K. Stamey, R. A. Hajjeh, M. E. Brandt, and Gauteng Cryptococcal Surveillance Initiative.** 2006. *Cryptococcus gattii* infection: characteristics and epidemiology of cases identified in a South African province with high HIV seroprevalence, 2002–2004. *Clin. Infect. Dis.* **43:**1077–1080.
76. **Munoz, P., M. Giannella, M. Valerio, T. Soria, F. Diaz, J. L. Longo, and E. Bouza.** 2007. Cryptococcal meningitis in a patient treated with infliximab. *Diagn. Microbiol. Infect. Dis.* **57:**443–446.
77. **Neuville, S., F. Dromer, O. Morin, B. Dupont, O. Ronin, and O. Lortholary.** 2003. Primary cutaneous

cryptococcosis: a distinct clinical entity. *Clin. Infect. Dis.* **36:**337–347.

78. **Newberry, W. M., Jr., J. E. Walter, J. W. Chandler, Jr., and F. E. Tosh.** 1967. Epidemiologic study of *Cryptococcus neoformans*. *Ann. Intern. Med.* **67:**724–732.
79. **Nosanchuk, J. D., S. Shoham, B. C. Fries, D. S. Shapiro, S. M. Levitz, and A. Casadevall.** 2000. Evidence of zoonotic transmission of *Cryptococcus neoformans* from a pet cockatoo to an immunocompromised patient. *Ann. Intern. Med.* **132:**205–208.
80. **Noverr, M. C., G. M. Cox, J. R. Perfect, and G. B. Huffnagle.** 2003. Role of PLB1 in pulmonary inflammation and cryptococcal eicosanoid production. *Infect. Immun.* **71:**1538–1547.
81. **O'Brien, C. R., M. B. Krockenberger, D. I. Wigney, P. Martin, and R. Malik.** 2004. Retrospective study of feline and canine cryptococcosis in Australia from 1981 to 2001: 195 cases. *Med. Mycol.* **42:**449–460.
82. **Pappas, P. G., J. R. Perfect, G. A. Cloud, R. A. Larsen, G. A. Pankey, D. J. Lancaster, H. Henderson, C. A. Kauffman, D. W. Haas, M. Saccente, R. J. Hamill, M. S. Holloway, R. M. Warren, and W. E. Dismukes.** 2001. Cryptococcosis in human immunodeficiency virus-negative patients in the era of effective azole therapy. *Clin. Infect. Dis.* **33:**690–699.
83. **Passoni, L. F., B. Wanke, M. M. Nishikawa, and M. S. Lazera.** 1998. *Cryptococcus neoformans* isolated from human dwellings in Rio de Janeiro, Brazil: an analysis of the domestic environment of AIDS patients with and without cryptococcosis. *Med. Mycol.* **36:**305–311.
84. **Paterson, D. L., N. Singh, T. Gayowski, and I. R. Marino.** 1998. Pulmonary nodules in liver transplant recipients. *Medicine* **77:**50–58.
85. **Pfeiffer, T. J., and D. H. Ellis.** 1992. Environmental isolation of *Cryptococcus neoformans* var. *gattii* from *Eucalyptus tereticornis*. *J. Med. Vet. Mycol.* **30:**407–408.
86. **Randhawa, H. S., and D. K. Paliwal.** 1979. Survey of *Cryptococcus neoformans* in the respiratory tract of patients with bronchopulmonary disorders and in the air. *Sabouraudia* **17:**399–404.
87. **Rolston, K. V., S. Rodriguez, N. Dholakia, E. Whimbey, and I. Raad.** 1997. Pulmonary infections mimicking cancer: a retrospective, three-year review. *Support Care Cancer* **5:**90–93.
88. **Rozenbaum, R., and A. J. Goncalves.** 1994. Clinical epidemiological study of 171 cases of cryptococcosis. *Clin. Infect. Dis.* **18:**369–380.
89. **Saha, D. C., D. L. Goldman, X. Shao, A. Casadevall, S. Husain, A. P. Limaye, M. Lyon, J. Somani, K. Pursell, T. L. Pruett, and N. Singh.** 2007. Serologic evidence for reactivation of cryptococcosis in solid-organ transplant recipients. *Clin. Vaccine Immunol.* **14:**1550–1554.
90. **Salyer, W. R., D. C. Salyer, and R. D. Baker.** 1974. Primary complex of *Cryptococcus* and pulmonary lymph nodes. *J. Infect. Dis.* **130:**74–77.
91. **Santangelo, R., H. Zoellner, T. Sorrell, C. Wilson, C. Donald, J. Djordjevic, Y. Shounan, and L. Wright.** 2004. Role of extracellular phospholipases and mononuclear phagocytes in dissemination of cryptococcosis in a murine model. *Infect. Immun.* **72:**2229–2239.
92. **Seaton, R. A., A. J. Hamilton, R. J. Hay, and D. A. Warrell.** 1996. Exposure to *Cryptococcus neoformans* var. *gattii*: a seroepidemiological study. *Trans. R. Soc. Trop. Med. Hyg.* **90:**508–512.
93. **Shelburne, S. A., 3rd, J. Darcourt, A. C. White, Jr., S. B. Greenberg, R. J. Hamill, R. L. Atmar, and F. Visnegarwala.** 2005. The role of immune reconstitution inflammatory syndrome in AIDS-related *Cryptococcus neoformans* disease in the era of highly active antiretroviral therapy. *Clin. Infect. Dis.* **40:**1049–1052.
94. **Shrestha, R. K., J. K. Stoller, G. Honari, G. W. Procop, and S. M. Gordon.** 2004. Pneumonia due to *Cryptococcus neoformans* in a patient receiving infliximab: possible zoonotic transmission from a pet cockatiel. *Respir. Care* **49:**606–608.
95. **Silveira, F. P., S. Husain, E. J. Kwak, P. K. Linden, A. Marcos, R. Shapiro, P. Fontes, J. W. Marsh, M. de Vera, K. Tom, N. Thai, H. P. Tan, A. Basu, K. Soltys, and D. L. Paterson.** 2007. Cryptococcosis in liver and kidney transplant recipients receiving antithymocyte globulin or alemtuzumab. *Transpl. Infect. Dis.* **9:**22–27.
96. **Sorrell, T. C.** 2001. *Cryptococcus neoformans* var. *gattii*. *Med. Mycol.* **39:**155–168.
97. **Sorrell, T. C., A. G. Brownlee, P. Ruma, R. Malik, T. J. Pfeiffer, and D. H. Ellis.** 1996. Natural environmental sources of *Cryptococcus neoformans* var. *gattii*. *J. Clin. Microbiol.* **34:**1261–1263.
98. **Sorrell, T. C., S. C. Chen, P. Ruma, W. Meyer, T. J. Pfeiffer, D. H. Ellis, and A. G. Brownlee.** 1996. Concordance of clinical and environmental isolates of *Cryptococcus neoformans* var. *gattii* by random amplification of polymorphic DNA analysis and PCR fingerprinting. *J. Clin. Microbiol.* **34:**1253–1260.
99. **Speed, B. R., and J. Kaldor.** 1997. Rarity of cryptococcal infection in children. *Pediatr. Infect. Dis. J.* **16:**536–537.
100. **Speed, B. R., J. Kaldor, B. Cairns, and M. Pegorer.** 1996. Serum antibody response to active infection with *Cryptococcus neoformans* and its varieties in immunocompetent subjects. *J. Med. Vet. Mycol.* **34:**187–193.
101. **Stephen, C., S. Lester, W. Black, M. Fyfe, and S. Raverty.** 2002. Multispecies outbreak of cryptococcosis on southern Vancouver Island, British Columbia. *Can. Vet. J.* **43:**792–794.
102. **Sullivan, D., K. Haynes, G. Moran, D. Shanley, and D. Coleman.** 1996. Persistence, replacement, and microevolution of *Cryptococcus neoformans* strains in recurrent meningitis in AIDS patients. *J. Clin. Microbiol.* **34:**1739–1744.
103. **Sweeney, D. A., M. T. Caserta, D. N. Korones, A. Casadevall, and D. L. Goldman.** 2003. A ten-year-old boy with a pulmonary nodule secondary to *Cryptococcus neoformans*: case report and review of the literature. *Pediatr. Infect. Dis. J.* **22:**1089–1093.
104. **True, D. G., M. Penmetcha, and S. J. Peckham.** 2002. Disseminated cryptococcal infection in rheumatoid arthritis treated with methotrexate and infliximab. *J. Rheumatol.* **29:**1561–1563.
105. **USPHS/IDSA Prevention of Opportunistic Infection Working Group.** 2002. Guidelines for preventing opportunistic infections among HIV-infected persons. *MMWR Recomm. Rep.* **51:**1–41.
106. **USPHS/IDSA Prevention of Opportunistic Infection Working Group.** 1995. USPHS/IDSA guidelines for the prevention of opportunistic infections in persons infected with the human immunodeficiency virus: disease-specific recommendations. *Clin. Infect. Dis.* **21**(Suppl. 1):S32–43.
107. **Vanbreuseghem R., and M. Takashio.** 1970. An atypical strain of *Cryptococcus neoformans* (San Felice) Vuillemin 1894. Part II. *Cryptococcus neoformans* var *gattii* var. nov. *Belg. Med. Trop.* **50:**695–702.
108. **Varma, A., and K. J. Kwon-Chung.** 1991. Rapid method to extract DNA from *Cryptococcus neoformans*. *J. Clin. Microbiol.* **29:**810–812.

109. **Varma, A., D. Swinne, F. Staib, J. E. Bennett, and K. J. Kwon-Chung.** 1995. Diversity of DNA fingerprints in *Cryptococcus neoformans*. *J. Clin. Microbiol.* **33:**1807–1814.
110. **Vicencio, A. G., S. Narain, Z. Du, W. Y. Zeng, J. Ritch, A. Casadevall, and D. L. Goldman.** 2008. Pulmonary cryptococcosis induces chitinase in the rat. *Respir. Res.* **9:**40.
111. **Walter, J. E., and R. W. Atchison.** 1966. Epidemiological and immunological studies of *Cryptococcus neoformans*. *J. Bacteriol.* **92:**82–87.
112. **Wang, C. Y., H. D. Wu, and P. R. Hsueh.** 2005. Nosocomial transmission of cryptococcosis. *N. Engl. J. Med.* **352:**1271–1272.
113. **Wilson, M. L., L. D. Sewell, and C. M. Mowad.** 2008. Primary cutaneous cryptococcosis during therapy with methotrexate and adalimumab. *J. Drugs. Dermatol.* **7:**53–54.
114. **Woods, M. L., 2nd, R. MacGinley, D. P. Eisen, and A. M. Allworth.** 1998. HIV combination therapy: partial immune restitution unmasking latent cryptococcal infection. *AIDS* **12:**1491–1494.

Cryptococcus: From Human Pathogen to Model Yeast
Edited by J. Heitman et al.

32

Intracellular Replication and Exit Strategies

KERSTIN VOELZ, SIMON A. JOHNSTON, AND ROBIN C. MAY

Pathogens have evolved a great variety of strategies to achieve survival within the host environment. Intracellular parasitism, as one of these strategies, has a significant impact on pathogenesis and important implications for the effectiveness of both the host's immune response and therapeutic reagents. This chapter discusses the ability of the pathogenic cryptococci to proliferate intracellularly and examines the different strategies cryptococci use to exit host cells. It will also address the importance of these two traits in pathogenesis and debate potential molecular mechanisms.

CRYPTOCOCCUS AS AN INTRACELLULAR PROLIFERATOR

In vitro, intracellular multiplication (Fig. 1) of *Cryptococcus neoformans* was first reported in isolated murine peritoneal macrophages (72) and in macrophages derived from human peripheral blood monocytes (24) and has since also been shown in human fetal microglia (58), in murine alveolar macrophages (105), and in human endothelial cells (18). Although an earlier study observed effective intracellular killing of *C. neoformans* by human leukocytes during the first 4 h of infection (103), a subsequent study demonstrated that *C. neoformans* first needs to adapt to the host cell environment before commencing replication (24). Thus, experiments over a shorter time frame might underestimate the intracellular proliferative potential.

In vivo, it remains difficult to demonstrate intracellular proliferation directly, relying instead on indirect evidence such as histological examinations, which provide data at fixed time points but do not allow direct observation of intracellular proliferation (30). However, observations made in rat (36) and mouse model organisms (28) and also from histopathological studies in patients (37) point toward intracellular replication of *Cryptococcus* in vivo. Feldmesser et al. (30) summarized the following observations as evidence for intracellular replication in vivo: (i) the higher number of intracellularly budding yeasts as compared with extracellularly budding yeasts; (ii) the fact that intracellular cryptococcal cells increase both in number and in the range of sizes they display (a sign of active budding); and (iii) the observation of intracellular budding yeast cells at all stages of budding.

Nowadays, *Cryptococcus* is generally accepted as a facultative intracellular pathogen that can replicate inside host cells while retaining the ability to grow in an extracellular niche within the host and in the environment (30). The intracellular location provides a dual benefit to *Cryptococcus*, both in avoiding extracellular host immune mechanisms, such as complement, and in reducing exposure to antifungal agents. Indeed, maintenance of *Cryptococcus* within phagocytic cells is important for progression of cryptococcosis (53). The establishment of different in vitro systems that utilize cell lines (58, 74) and the application of techniques such as live cell imaging (5, 68) have contributed to rapid advances in our understanding of the molecular mechanisms influencing *Cryptococcus*'s ability to proliferate intracellularly.

Factors Influencing Intracellular *Cryptococcus* Proliferation

Intracellular parasitism is associated with a continuous struggle between the pathogen and its host cell. Within host cells, *Cryptococcus* encounters a harsh environment of reactive oxygen and nitrogen species; oxygen, nutrient, and metal ion deprivation; and low pH and high temperatures. Therefore, the yeast expresses multiple virulence factors including a capsule, melanin, and a variety of secreted enzymes that can modify the host's defense mechanisms to achieve intracellular replication.

Part of the strategy by which phagocytic cells kill pathogens is the production of oxygen and nitrogen radicals (3, 25, 42, 64, 75, 108). The cryptococcal polysaccharide capsule is an important factor in protecting against oxidative stress by buffering reactive oxygen species (ROS). The capsule becomes enlarged upon interaction with phagocytic cells in vivo and in vitro (27) and is important for the yeast's ability to multiply intracellularly, since encapsulated, but not acapsular, yeast cells can replicate inside macrophages (28). Cryptococcal susceptibility to ROS is directly correlated with capsule size; ROS kill cells with larger capsules less efficiently than cells with smaller capsules (115). The effect can be mimicked in cells with a small capsule by adding soluble capsular polysaccharide to the medium, thus demonstrating that resistance is directly mediated by the presence of capsular material (115).

Kerstin Voelz, Simon A. Johnston, and Robin C. May, School of Biosciences, University of Birmingham, Birmingham B15 2TT, U.K.

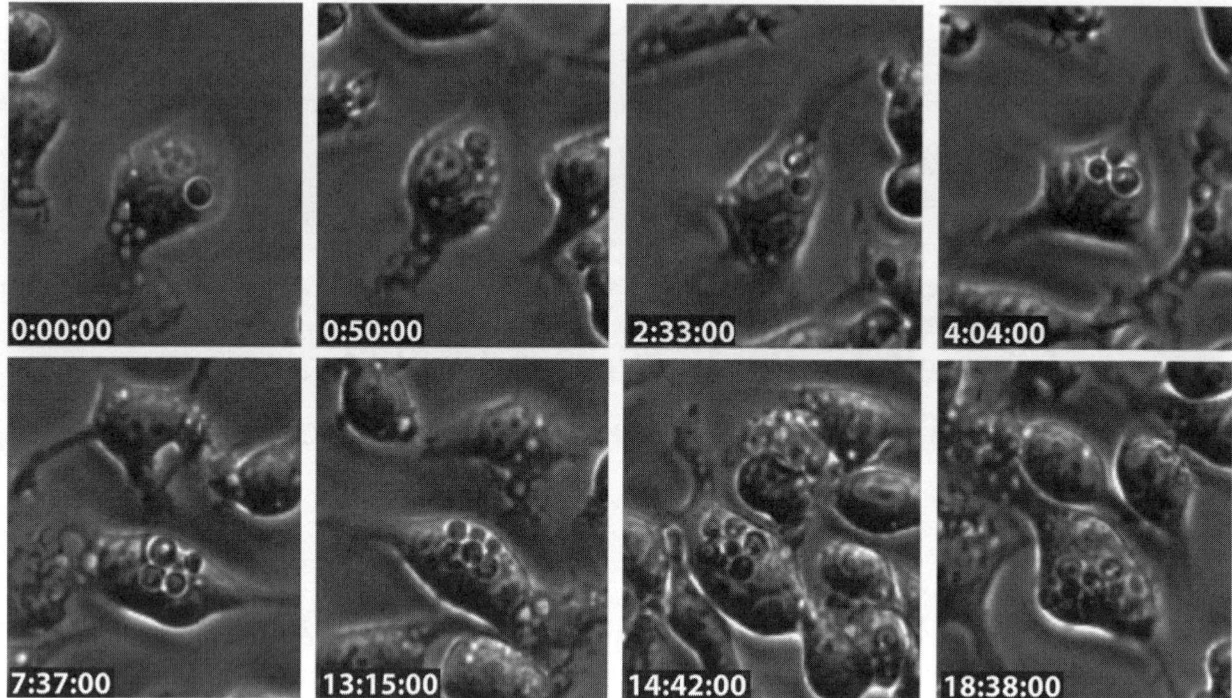

FIGURE 1 Intracellular proliferation of *C. gattii* within a J774 murine macrophage. The intracellular yeast cell starts budding after 50 min and continues replication until the macrophage is filled with yeast cells after 19 h. Times are shown in hours:minutes:seconds.

In fact, an earlier study showed that glucuronoxylomannan, the major component of the cryptococcal capsule, reduces cryptococcal killing and production of superoxide in primary human neutrophils (73). This effect may underlie the phenomenon of phenotypic switching, characterized by a switch in cryptococcal colony morphology between smooth and mucoid, with the latter showing increased survival in the murine macrophage cell line J774 (33, 52).

Furthermore, during intracellular growth a range of secreted enzymes are involved in detoxifying oxygen and nitrogen radicals. One example is the superoxide dismutase (SOD) that converts superoxide to hydrogen peroxide and oxygen (32). A *C. neoformans sod1* mutant strain exhibits significantly attenuated growth inside primary human and murine macrophages (19). Disruption of the alternative oxidase gene *aox1* also leads to significantly slower growth within murine alveolar macrophages (1). Similarly, FHB1, a flavohemoglobin denitrosylase, plays an important role in the response to nitrosative stress, and knockout of *fbh1* leads to reduced proliferation in the murine macrophage cell line MH-S (23). In addition, the urease URE1 (21) and the thiol peroxidase TSA1 (71) have been shown to be involved in counteracting oxidative and nitrosative stresses, although there are no direct data on the impact of these proteins on intracellular proliferation.

The pigment melanin, produced by the enzyme laccase, is thought to provide protection from both oxygen- and nitrogen-reactive species in *Cryptococcus* by virtue of its antioxidant properties (109, 110). The melanized *C. neoformans* strain 145 is more resistant to cell death caused by lymphocytes than the less melanized strain 52 (48). A model for the regulation of melanin production by the sphingolipid pathway in *Cryptococcus* proposes activation of protein kinase C1 (PKC1) via inositol-phosphoryl ceramide synthase (IPC1) through the regulation of diacylglycerol (DAG) and phytoceramide concentrations, resulting in subsequent PKC1-dependent laccase activation (41). An *ipc1* mutant shows a decrease in melanin production (66) and significant down-regulation of intracellular growth in J774 macrophages. IPC1 consumes phytoceramide to produce complex sphingolipids and DAG. Thus, IPC1 probably increases cryptococcal survival via the activation of melanin production (66). Interestingly, DAG has been implicated in activation of PKC1, an activator of laccase (41). The enzyme laccase is also involved in the synthesis of melanin by oxidizing catecholamine substrates (112). However, Liu et al. (62) reported a protective function of laccase independent of melanin synthesis. The *C. neoformans* strain 2E-TU, a laccase-deficient strain, is more sensitive toward killing by murine alveolar macrophages than the laccase-positive strain 2E-TUC. This laccase-mediated protection from killing by macrophages also occurs in media lacking suitable catecholamine substrates and thus does not require melanin production. Further investigation revealed an iron oxidase function of laccase that may maintain iron in an oxidized form, thereby inhibiting production of hydroxyl radicals by the host cell (62).

Within the intracellular environment, *Cryptococcus* is deprived of oxygen and nutrients. Under these conditions, autophagy appears to be one way in which *Cryptococcus* acquires additional nutrients upon starvation. A cryptococcal strain lacking *vps34*, a component of the type I kinase subcomplex that is required for autophagy, is rapidly killed by murine macrophages (55). In addition, no autophagic bodies were detected in the mutant cells (46). Strikingly, autophagy is normally a host response to control the replication of intracellular pathogens such as *Mycobacterium tuberculosis* (39) and *Legionella pneumophila* (7). *Cryptococcus* therefore represents a rare example of a pathogen, rather than its host, utilizing autophagy during intracellular growth.

There is growing evidence that cryptococci increase nutrient flow into the phagolysosome via membrane disruption. Phospholipases hydrolyze ester linkages in glycerophospholipids, which can result in destabilization of membranes, cell lysis, and release of second messengers (35, 88). Phagocytosed *C. neoformans* cells, negative for phospholipase B (PLB) expression, display a defect in the onset off budding. This leads to a lag in intracellular proliferation relative to wild-type cells (20). As phospholipases can cause membrane damage and are also involved in lipid metabolism (114), they may therefore be important for increasing nutrient availability to the pathogen.

Beside nutrient and oxygen starvation, intracellular cryptococci also experience low concentrations of metal ions. Many cryptococcal virulence enzymes such as laccase, SOD, catalase, and urease depend upon metal ions, and thus cation homeostasis is important for their functions. To inhibit cryptococcal replication, macrophages actively try to reduce the concentration of metal ions in the phagolysosome. The macrophage-expressed protein NRAMP1 (natural resistance-associated-macrophage protein 1), which is homologous to the membrane transporters SFM1 and SFM2 in *Saccharomyces cerevisiae*, has been suggested to be involved in transport of divalent cations, such as Mn^{2+}, Zn^{2+}, or Fe^{2+}, across the phagosome membrane (102). The *Nramp1*$^{-}$ bone-marrow-derived macrophage cell line 129.1 shows significantly reduced anticryptococcal activity in the first 6 h after infection (12). This might reflect an altered intracellular environment with better cation availability for cryptococcal metal-dependent enzymes and therefore higher enzymatic activity and better protection from oxidative and nitrosative stress. Notably, many studies of *Cryptococcus*–host cell interactions use macrophage cell lines (e.g., RAW) derived from BALB/c mice (10). However, it has been shown that these mice have defects in NRAMP1, and thus the intracellular proliferative potential might be overestimated in some cases.

Following uptake, the phagolysosome pH rapidly decreases to below pH 5.5 (100) to improve antimicrobial activity. However, *Cryptococcus* actively prefers an acidic environment, and phagosome alkalinization, by agents such as chloroquine, inhibits cryptococcal survival and intracellular proliferation (61). In addition, increased pH also reduces cation availability and might inhibit yeast enzyme functions. Nonetheless, *Cryptococcus* still needs to adapt to the low phagolysosomal pH. The enzyme inositol phosphosphingolipid-phospholipase C, an enzyme generating phytoceramide, seems to be important in providing protection from the acidic milieu. In *S. cerevisiae*, phytoceramide plays an important role in regulating PMA1 (31, 34), an ATPase that is involved in the regulation of intracellular pH (44, 92, 94) as well as oxidative (91) and nitrosative stress (116). In fact, intracellular survival in J774 macrophages of the *isc1* mutant strain was significantly reduced, and the strain was more sensitive to oxidative, nitric, and acidic stress (89). The cryptococcal VSP41 homolog has also been implicated in regulating metal ion concentrations by copper loading of the iron transporter FET3 in yeast (83). Disruption of *vps41* in *C. neoformans* results in abolishment of intracellular growth in J774 macrophages as well as reduced cryptococcal growth on iron- and copper-deficient media (63).

It is clear that the processes that allow the transition between extracellular and intracellular proliferation of *Cryptococcus* in host cells are controlled. However, very little is known about the signaling pathways regulating the expression of the virulence factors described above. The cryptococcal two-component stress regulator SKN7 has been shown to be involved in oxidative stress signaling in endothelial cells. Disruption of SKN7 results in significantly impaired growth in primary human umbilical-vein endothelial cells (18). In addition, the Rho-type GTPases CDC42 and RAC have been implicated in the regulation of high-temperature growth, mating, and cell morphology in *C. neoformans* (2, 77, 107, 111), while RDI1 (a Rho-GTPase dissociation inhibitor that regulates GTPase activity in fission yeast [56, 76]) is important for intracellular growth (82).

In addition to host cell and *Cryptococcus* factors, intracellular yeast proliferation is also regulated by immune-signaling molecules such as cytokines. Generally, the balance between Th1 and Th2 immune responses is important for the outcome of cryptococcosis. Resistance to cryptococcal infection is associated with a Th1 response and the consequent phagocyte activation (47, 90), whereas Th2-polarized host responses lead to inhibition of phagocyte activity and enhanced susceptibility to *C. neoformans* (57). In support of this, intracellular cryptococcal proliferation is significantly higher in macrophages activated by the Th2 cytokines interleukin-4 (IL-4) or IL-13 than in cells activated by Th1 cytokines, such as gamma interferon (IFN-γ) and tumor necrosis factor alpha (107a). Similarly, the number of intracellular cryptococci is increased in alveolar macrophages isolated from IFN-γ knockout mice (8). In contrast, IFN-γ treatment increases intracellular growth of cryptococci in human monocytes (59, 84), perhaps indicating a more complex regulation of Th1- and Th2-associated immunity in humans. To counteract Th1-related cytokine activation, *C. neoformans* expresses eicosanoids (79), for example, prostaglandins and leukotrienes, which are potent inhibitors of Th1-type immunity that shift the Th1-Th2 balance toward a Th2 immune response and thus more efficient intracellular proliferation (78).

In conclusion, the establishment of intracellular proliferation by *Cryptococcus* results from the activity of a great variety of virulence factors that coordinate to undermine the host's defense response. Although much progress has been made in identifying these virulence factors, the molecular mechanisms by which adaptation to the intracellular environment is mediated and how these factors are regulated remains to be elucidated.

Origin and Evolution of Cryptococcal Intracellular Proliferation

Interestingly, *Cryptococcus* does not seem to utilize manipulative strategies similar to those known to be used by other intracellular pathogens. In fact, the yeast seems to be well adapted to the harsh environment within the host cell. In contrast to pathogens such as *Listeria monocytogenes* and *Shigella flexneri*, *C. neoformans* resides in and does not escape from the phagosome (61, 90, 93). Moreover, *C. neoformans* does not inhibit phagosome-lysosome fusion in human macrophages (61), as has been shown for *L. pneumophila* (45), nor does the yeast interfere with phagosome maturation or acidification, as occurs during infections with *Histoplasma capsulatum* or *Mycobacterium avium* (61, 90, 99, 100). Instead, the yeast survives and replicates in the acidic phagolysosome and, in fact, any increase in phagosomal pH (e.g., by experimental addition of chloroquine or ammonium chloride) leads to reduced intracellular proliferation (60).

The question of how and why *Cryptococcus* has acquired and maintained virulence mechanisms that enable intracellular proliferation is still unanswered. However, the interactions between *C. neoformans* and environmentally widespread invertebrates and protists has led to the proposal that virulence toward mammalian cells might have arisen from protective mechanisms used by the yeast against environmental predators. Coincubation of the free-living soil amebae *Acanthamoeba castellanii* or *Dictyostelium discoideum* with *C. neoformans* results in intracellular budding and growth, eventually causing lysis of the host amebae (95, 96). Interestingly, mutations in known mammalian virulence factors significantly inhibited cryptococcal growth in this system. Interestingly, melanization protects acapsular, but not encapsulated, strains from ameboid killing (96), suggesting that melanization is not a critical virulence factor in this host. Notably, *D. discoideum* mutants with impaired phagocytosis or exocytosis (14, 106, 113) are both susceptible to the hypovirulent strain Cap67 (95). Thus, as in humans, the outcome of intracellular parasitism results from a complex interaction between host and pathogen factors. Intriguingly, repeated passage of *C. neoformans* through *D. discoideum* results in increased virulence in mice (95). This increase is correlated with phenotypic changes, such as increased capsule size and more rapid melanization (95), and is probably a microevolutionary process similar to that previously reported in HIV-infected patients with recurrent cryptococcal meningitis (101). Cryptococcal proliferation within amebae is remarkably similar to replication within mammalian phagocytes. Therefore, the selective pressure to survive and replicate within amebae may have driven evolution of adaptation mechanisms to mammalian hosts and may be critical for the maintenance of intracellular proliferation traits in this organism.

However, although these interactions offer explanations for mechanisms involved in virulence and in intracellular proliferation in the immunocompromised host, they do not explain the recent origin of certain hypervirulent *Cryptococcus gattii* strains (9, 43, 69, 97). Strikingly, data on the maximal intracellular proliferation rates (IPRs) of different *C. gattii* strains show that an enhanced ability to replicate within macrophages might have enabled certain *C. gattii* strains to cause symptomatic infections in otherwise healthy individuals. A comparison of *C. gattii* strains of different genotypes revealed an increase in IPRs in genotype VGIIa strains (AFLP6A), the hypervirulent genotype responsible for the Vancouver Island outbreak (54). Since there is also a clear correlation between IPR and virulence in mice, the exceptional virulence of the Vancouver Island *Cryptococcus* population seems to be a direct result of an enhanced ability to multiply intracellularly in mammalian macrophages (68a). Whole-genome microarray analysis has identified a large number of candidate genes that may influence proliferative capacity and revealed an unexpected role for mitochondrial genes in regulating cryptococcal hypervirulence in the Vancouver Island strains (68a). Previous work in the related species *C. neoformans* has demonstrated that mitochondrial genotype does not influence virulence (104). Thus, mitochondrial regulation of virulence must have evolved after *C. gattii* and *C. neoformans* diverged and may potentially explain the very different behaviors of these two pathogens.

Mitochondrial functions include oxidative respiration, resistance to oxidative stress (38), certain heme synthesis reactions (80), and sterol synthesis (86). *C. neoformans* mutants with impaired mitochondrial functions show reduced oxidative respiration and higher sensitivity to ROS under low-oxygen conditions (51). In this context, increased intracellular replication might be due to additional mitochondrial genes, polymorphisms, or greater mitochondrial plasticity that allow for more rapid adaptation to environmental changes such as available oxygen and oxidative stress. Future research will hopefully reveal how mitochondria influence cryptococcal intracellular proliferation at the molecular level.

CRYPTOCOCCAL ESCAPE FROM THE HOST CELL

To cause the disease state observed in cryptococcosis, cryptococci must eventually escape the phagocyte in which they initially replicate. Different escape routes influence both secondary disease states and the role that the adaptive and innate immune system will play in subsequent pathology. More generally, host cell escape may be important both for the multitissue passage of *Cryptococcus* through the body and for maintaining latency. At least three exit strategies for *Cryptococcus* are currently recognized: lysis, expulsion, and lateral transfer (Fig. 2). We address each of these in turn below.

Host Cell Lysis

Host cell lysis is the most common escape route for intracellular pathogens. Microorganisms are often enclosed in at least two membranes (an endocytic or phagocytic one, as well as the host cell plasma membrane) and must disrupt these to escape. Many virulence factors, such as pore-forming proteins, phospholipases, and proteases, are thought to function through lysing host cell membranes. The mechanisms of host cell membrane disruption are not well understood but are assumed to involve a combination of pore-forming and phospholipase proteins. To date, no pore-forming proteins that function to attack host membranes have been identified in the *Cryptococcus* genome, but PLB (*plb1*) is an obvious candidate cell-lysis factor. Although cryptococcal parasitism of macrophages does not obviously require escape from the phagosome, three experimental observations suggest that *Cryptococcus*-containing phagosomes may be permeabilized. First, phagosomal membranes containing *Cryptococcus* appear to have holes when viewed by transmission electron microscopy (29, 105). Second, fluorescent dextran that is phagocytosed along with cryptococci subsequently diffuses out of the phagosome into the cytoplasm (105). Finally, *Cryptococcus*-containing phagosomes cannot maintain an acidic pH (105). This permeabilization of the phagosome may be required to provide access to cytoplasmic nutrients to support proliferation, as the PLB1 mutant is growth-inhibited in macrophages (78). Nevertheless, the lysis of phagosomes or host cells by PLB has not been shown directly.

Mechanical burden alone, in the absence of permeabilizing factors, is not well described as a mechanism for host cell lysis for intracellular pathogens. However, the concurrent expansion of the cryptococcal capsule and rapid proliferation of many *Cryptococcus* strains within macrophages means that mechanical rupture of cell membranes may provide an alternative explanation for lysis (5, 105).

Expulsion

Many intracellular pathogens have mechanisms for controlled expulsion from host cells. These generally fall into two types: actin-based protrusion and membrane extrusion. Actin-based protrusion has been described for *L. monocytogenes*, *Rickettsia* spp., *S. flexneri*, and *Burkholderia pseudomallei*, all of which hijack the host actin cytoskeleton to form plasma membrane projections that are endocytosed by neighboring

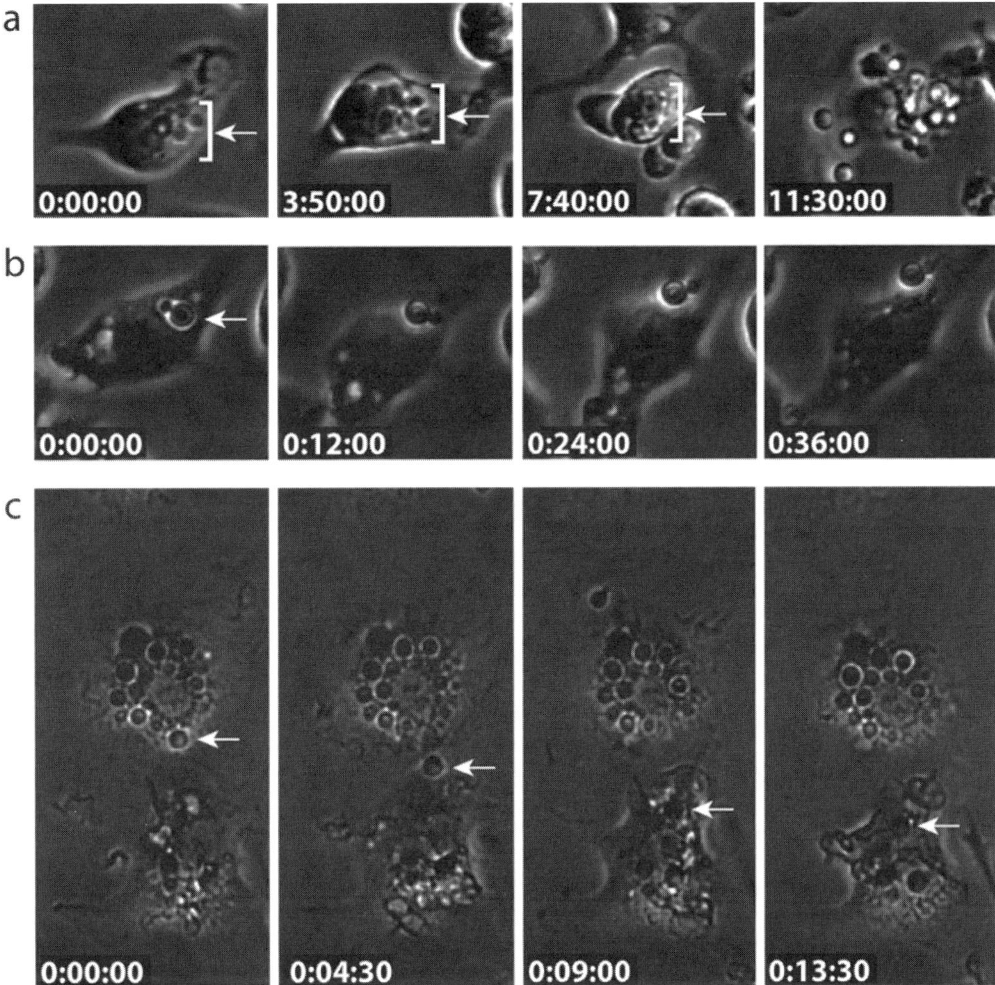

FIGURE 2 *Cryptococcus* has three strategies for escaping macrophages: (a) lysis, (b) expulsion, and (c) lateral transfer. Time lapse microscopy of J774 cells (a, b) and human primary macrophages (c). Arrows indicate intracellular cryptococci. Times are shown in hours:minutes:seconds.

cells (98). In contrast, cryptococcal expulsion occurs without any obvious reliance on the actin cytoskeleton (5, 68). Two other superficially similar exit strategies have been reported: the triple-membraned single-cell budding of *Orientia tsutsugamushi* (87) and the large-vacuole multicell extrusion of *Chlamydia* spp. (50).

Cryptococcal expulsion bears a striking resemblance to exocytosis, the process by which internal vesicles fuse with the plasma membrane and release their contents to the outside environment. The presentation of antigens also relies on the exocytic machinery, a process that is especially important in phagocytes. Exocytosis of large particles has been observed only once, though, in the social ameba *D. discoideum* (17).

Expulsion of *Cryptococcus* occurs in macrophage cell lines (5, 68), primary murine macrophages (5), and primary human macrophages (68). The rate of expulsion is increased 2.5-fold in primary human macrophages (68), suggesting that it may be a frequent event in infected patients. After expulsion of cryptococci, macrophages are still motile and able to divide (5, 68). Expelled cryptococci are also viable (5, 68) and can be rephagocytosed (6). Expulsion appears to be a process triggered by the yeast, since heat-killed cryptococci are not expelled (5, 68), a finding that contrasts with the expulsion of dead yeast fragments by *Dictyostelium* (17). Expulsion events are rare during the first 2 h of infection, but after this initial lag phase, events are randomly distributed over the period of observation (5, 68). The expulsion event itself takes approximately 60 seconds (5, 68), but detailed data on the kinetics of expulsion are not currently available. Expulsion occurs with both *C. neoformans* (in both varieties) and *C. gattii* (5, 68), but there is significant interstrain variation in the frequency of expulsion (H. Ma, K. Voelz, S. A. Johnston, and R. C. May, unpublished data).

A key unanswered question in the field concerns the role played by different phagocytic uptake routes on subsequent cryptococcal behavior. The molecular processes that underlie phagocytosis vary extensively with different receptors as well as with size and shape of the engulfed particle (15, 49). Although the route of phagocytic uptake is not predictive of cryptococcal expulsion (5, 68), frequency is reduced with complement opsonization (5, 68). Since the phagocytosis of *Cryptococcus*, even when heavily opsonized, will likely include multiple types of receptors, the behavior of each engulfed cell is likely to be influenced by a complex interplay of multiple signaling pathways.

Prior to expulsion or proliferation, phagosomes containing *Cryptococcus* mature normally and fuse with lysosomes (5, 68), and perturbation of this process modulates expulsion. Expulsion is reduced upon treatment with concanamycin A (5, 68), an inhibitor of the V-type ATPase that is responsible for acidification of endosomes and lysosomes that fuse with the phagosome (26). In contrast, chloroquine, a weak base that accumulates in acidic vesicles, buffering the pH, enhances the rate of expulsion (5, 68). Since chloroquine and concanamycin A should produce similar effects on the phagosome, the difference in modulating expulsion may be due to a secondary activity of chloroquine, for example, on the *Cryptococcus* vacuole (40). Chloroquine has been previously described to enhance anticryptococcal activity of bone marrow macrophages and is protective in murine infection (60). The discovery that chloroquine increases expulsion therefore raises the possibility that the enhancement of anticryptococcal activity may be due to increased expulsion of *Cryptococcus* into the extracellular milieu where it is susceptible to killing by complement.

Expulsion does not require activation of macrophages with particular cytokines or other agonists (5, 68). However, the rate of expulsion is significantly higher in macrophages activated by so-called type I cytokines (IFN-γ or tumor necrosis factor alpha) than in those activated by type II cytokines (e.g., IL-4 or IL-13) (107a).

Phagosomes and other cellular vesicles, even very small ones, are relatively flat at the scale of the fusion of membrane bilayers. Thus, for fusion to occur, a significant charge barrier must be overcome by a local and transient high membrane curvature. Phagosomes containing even a single cryptococcal cell are relatively large, and those in which substantial cryptococcal proliferation has occurred can represent a significant proportion of the host cell's volume. This and other energetic barriers make a purely mechanical explanation for cryptococcal expulsion highly unlikely. This is supported by experimental evidence that inert particles (heat-killed yeast or polystyrene beads) are never expelled by macrophages (5, 68). Normally, macrophages function in the lung to clear indigestible particles and then travel up and out of the lungs before being swallowed (13). Were macrophages to freely expel indigestible material, lung clearance would be severely impaired.

The molecular mechanism of cryptococcal expulsion is unknown, but since exocytosis is regulated by a large number of protein complexes that are responsible for membrane fusion and specificity, it is extremely unlikely that *Cryptococcus* provides all these components itself. Instead, one possibility is that cryptococcal expulsion involves the hijacking of host cell factors in order to harness the host cell's exocytic machinery for expulsion. An analogous mechanism occurs in other intracellular pathogens, such as the gram-positive bacterium *L. monocytogenes*, which uses the bacterial protein ActA to "hijack" the host Arp2/3 complex in order to stimulate the production of an actin tail for bacterial motility (70). Such hijacking factors, however, are yet to be identified in *Cryptococcus*.

Lateral Transfer

A number of intracellular bacteria can be transferred laterally between host cell types (11, 81, 98), but the mechanism of transmission of bacteria that spread from cell to cell is not well understood even when the mechanism of escape is. *L. monocytogenes* polymerizes a tail of host actin that propels the bacterium into the plasma membrane (22), whereupon the resulting membrane protrusion is engulfed by neighboring cells and the bacteria internalized. Internalization is an active process, since the force of the membrane protrusion alone is insufficient, but further details are unknown (85).

Similarly, *Cryptococcus* can be directly transferred from macrophage to macrophage in both macrophage cell lines and primary macrophages (4, 67). Lateral transfer shares the same basic requirements as expulsion: it does not occur with heat-killed *Cryptococcus* or latex beads (4) and is independent of both the phagocytic receptor type and the cryptococcal strain (67). Unlike expulsion, however, lateral transfer is dependent on the actin cytoskeleton since treatment with the actin-depolymerizing drug cytochalasin D completely inhibits lateral transfer. The relationship between expulsion and lateral transfer is unclear, but it is easy to imagine that lateral transfer may occur when expulsion occurs in close proximity to a suitable acceptor cell that can receive the ejected *Cryptococcus*. As cytochalasin D does not inhibit expulsion, it is therefore likely that it is the process of acceptance that is actin dependent, as with normal phagocytic processes. A second possibility is that a transient (actin-dependent) channel is opened between the two cells for transfer of *Cryptococcus*. Since macrophages form frequent and intimate contacts with each other and with other cell types (for example, in order to form Langerhans or foreign body giant cells or to activate T cells), it is likely that lateral transfer represents an important route for prolonged intracellular persistence in vivo.

CONCLUDING REMARKS

It is now clear that cryptococci are adept intracellular pathogens, with complex mechanisms that allow them to proliferate within, and eventually exit from, host phagocytes. However, a vital and unanswered question is how the different facets of intracellular behavior affect the progress of cryptococcal disease. There can be little doubt that intracellular proliferation ultimately benefits the pathogen, but the question of whether expulsion and/or lateral transfer slows or enhances disease progression remains unanswered. On the one hand, cryptococci expelled from circulating monocytes will likely be exposed to a far greater immune attack than those that remain intracellular. However, if many monocytes are infected and there has been substantial intracellular proliferation, then expulsion may lead to fungemia as the extracellular fungi overwhelm the circulation. In addition, the Trojan horse hypothesis suggests that *Cryptococcus* crosses the blood-brain barrier within parasitized macrophages (16, 65), in which case the subsequent expulsion of the pathogen within the brain tissue is likely to be critical. Finally, it is also possible that cryptococci may be transferred directly from macrophages to endothelial cells via lateral transfer, followed by polar expulsion of *Cryptococcus* into the central nervous system. In the years ahead, the development of more effective therapeutic regimes is likely to rely heavily on the elucidation of both the molecular mechanisms that underlie the complexity of host manipulation by *Cryptococcus* and the impact that these mechanisms have on the progress of disease in infected patients.

REFERENCES

1. **Akhter, S., H. C. McDade, J. M. Gorlach, G. Heinrich, G. M. Cox, and J. R. Perfect.** 2003. Role of alternative oxidase gene in pathogenesis of *Cryptococcus neoformans*. *Infect. Immun.* **71:**5794–5802.

2. **Alspaugh, J. A., L. M. Cavallo, J. R. Perfect, and J. Heitman.** 2000. RAS1 regulates filamentation, mating and growth at high temperature of *Cryptococcus neoformans*. *Mol. Microbiol.* **36:**352–365.
3. **Alspaugh, J. A., and D. L. Granger.** 1991. Inhibition of *Cryptococcus neoformans* replication by nitrogen oxides supports the role of these molecules as effectors of macrophage-mediated cytostasis. *Infect. Immun.* **59:**2291–2296.
4. **Alvarez, M., and A. Casadevall.** 2007. Cell-to-cell spread and massive vacuole formation after *Cryptococcus neoformans* infection of murine macrophages. *BMC Immunol.* **8:**16.
5. **Alvarez, M., and A. Casadevall.** 2006. Phagosome extrusion and host-cell survival after *Cryptococcus neoformans* phagocytosis by macrophages. *Curr. Biol.* **16:**2161–2165.
6. **Alvarez, M., C. Saylor, and A. Casadevall.** 2008. Antibody action after phagocytosis promotes *Cryptococcus neoformans* and *Cryptococcus gattii* macrophage exocytosis with biofilm-like microcolony formation. *Cell. Microbiol.* **10:**1622–1633.
7. **Amer, A. O., and M. S. Swanson.** 2005. Autophagy is an immediate macrophage response to *Legionella pneumophila*. *Cell. Microbiol.* **7:**765–778.
8. **Arora, S., Y. Hernandez, J. R. Erb-Downward, R. A. McDonald, G. B. Toews, and G. B. Huffnagle.** 2005. Role of IFN-gamma in regulating T2 immunity and the development of alternatively activated macrophages during allergic bronchopulmonary mycosis. *J. Immunol.* **174:**6346–6356.
9. **Bartlett, K. H., S. E. Kidd, and J. W. Kronstad.** 2008. The emergence of *Cryptococcus gattii* in British Columbia and the Pacific Northwest. *Curr. Infect. Dis. Rep.* **10:**58–65.
10. **Barton, C. H., S. H. Whitehead, and J. M. Blackwell.** 1995. Nramp transfection transfers Ity/Lsh/Bcg-related pleiotropic effects on macrophage activation: influence on oxidative burst and nitric oxide pathways. *Mol. Med.* **1:**267–279.
11. **Bernardini, M. L., J. Mounier, H. d'Hauteville, M. Coquis-Rondon, and P. J. Sansonetti.** 1989. Identification of icsA, a plasmid locus of *Shigella flexneri* that governs bacterial intra- and intercellular spread through interaction with F-actin. *Proc. Natl. Acad. Sci. USA* **86:**3867–3871.
12. **Blasi, E., B. Colombari, A. Mucci, A. Cossarizza, D. Radzioch, J. R. Boelaert, and R. Neglia.** 2001. Nramp1 gene affects selective early steps in macrophage-mediated anti-cryptococcal defense. *Med. Microbiol. Immunol.* **189:**209–216.
13. **Brain, J. D.** 1980. Macrophage damage in relation to the pathogenesis of lung diseases. *Environ. Health Perspect.* **35:**21–28.
14. **Brazill, D. T., D. R. Caprette, H. A. Myler, R. D. Hatton, R. R. Ammann, D. F. Lindsey, D. A. Brock, and R. H. Gomer.** 2000. A protein containing a serine-rich domain with vesicle fusing properties mediates cell cycle-dependent cytosolic pH regulation. *J. Biol. Chem.* **275:**19231–19240.
15. **Champion, J. A., and S. Mitragotri.** 2006. Role of target geometry in phagocytosis. *Proc. Natl. Acad. Sci. USA* **103:**4930–4934.
16. **Charlier, C., K. Nielsen, S. Daou, M. Brigitte, F. Chretien, and F. Dromer.** 2009. Evidence of a role for monocytes in dissemination and brain invasion by *Cryptococcus neoformans*. *Infect. Immun.* **77:**120–127.
17. **Clarke, M., J. Kohler, Q. Arana, T. Liu, J. Heuser, and G. Gerisch.** 2002. Dynamics of the vacuolar H(+)-ATPase in the contractile vacuole complex and the endosomal pathway of *Dictyostelium* cells. *J. Cell Sci.* **115:**2893–2905.
18. **Coenjaerts, F. E., A. I. Hoepelman, J. Scharringa, M. Aarts, P. M. Ellerbroek, L. Bevaart, J. A. Van Strijp, and G. Janbon.** 2006. The Skn7 response regulator of *Cryptococcus neoformans* is involved in oxidative stress signalling and augments intracellular survival in endothelium. *FEMS Yeast Res.* **6:**652–661.
19. **Cox, G. M., T. S. Harrison, H. C. McDade, C. P. Taborda, G. Heinrich, A. Casadevall, and J. R. Perfect.** 2003. Superoxide dismutase influences the virulence of *Cryptococcus neoformans* by affecting growth within macrophages. *Infect. Immun.* **71:**173–180.
20. **Cox, G. M., H. C. McDade, S. C. Chen, S. C. Tucker, M. Gottfredsson, L. C. Wright, T. C. Sorrell, S. D. Leidich, A. Casadevall, M. A. Ghannoum, and J. R. Perfect.** 2001. Extracellular phospholipase activity is a virulence factor for *Cryptococcus neoformans*. *Mol. Microbiol.* **39:**166–175.
21. **Cox, G. M., J. Mukherjee, G. T. Cole, A. Casadevall, and J. R. Perfect.** 2000. Urease as a virulence factor in experimental cryptococcosis. *Infect. Immun.* **68:**443–448.
22. **Dabiri, G. A., J. M. Sanger, D. A. Portnoy, and F. S. Southwick.** 1990. Listeria monocytogenes moves rapidly through the host-cell cytoplasm by inducing directional actin assembly. *Proc. Natl. Acad. Sci. USA* **87:**6068–6072.
23. **de Jesus-Berrios, M., L. Liu, J. C. Nussbaum, G. M. Cox, J. S. Stamler, and J. Heitman.** 2003. Enzymes that counteract nitrosative stress promote fungal virulence. *Curr. Biol.* **13:**1963–1968.
24. **Diamond, R. D., and J. E. Bennett.** 1973. Growth of *Cryptococcus neoformans* within human macrophages in vitro. *Infect. Immun.* **7:**231–236.
25. **Diamond, R. D., R. K. Root, and J. E. Bennett.** 1972. Factors influencing killing of *Cryptococcus neoformans* by human leukocytes in vitro. *J. Infect. Dis.* **125:**367–376.
26. **Drose, S., and K. Altendorf.** 1997. Bafilomycins and concanamycins as inhibitors of V-ATPases and P-ATPases. *J. Exp. Biol.* **200:**1–8.
27. **Feldmesser, M., Y. Kress, and A. Casadevall.** 2001. Dynamic changes in the morphology of *Cryptococcus neoformans* during murine pulmonary infection. *Microbiology* **147:**2355–2365.
28. **Feldmesser, M., Y. Kress, P. Novikoff, and A. Casadevall.** 2000. *Cryptococcus neoformans* is a facultative intracellular pathogen in murine pulmonary infection. *Infect. Immun.* **68:**4225–4237.
29. **Feldmesser, M., A. Mednick, and A. Casadevall.** 2002. Antibody-mediated protection in murine *Cryptococcus neoformans* infection is associated with pleotrophic effects on cytokine and leukocyte responses. *Infect. Immun.* **70:**1571–1580.
30. **Feldmesser, M., S. Tucker, and A. Casadevall.** 2001. Intracellular parasitism of macrophages by *Cryptococcus neoformans*. *Trends Microbiol.* **9:**273–278.
31. **Ferreira, T., A. B. Mason, and C. W. Slayman.** 2001. The yeast Pma1 proton pump: a model for understanding the biogenesis of plasma membrane proteins. *J. Biol. Chem.* **276:**29613–29616.
32. **Fridovich, I.** 1995. Superoxide radical and superoxide dismutases. *Annu. Rev. Biochem.* **64:**97–112.
33. **Fries, B. C., C. P. Taborda, E. Serfass, and A. Casadevall.** 2001. Phenotypic switching of *Cryptococcus neoformans* occurs in vivo and influences the outcome of infection. *J. Clin. Invest.* **108:**1639–1648.
34. **Gaigg, B., B. Timischl, L. Corbino, and R. Schneiter.** 2005. Synthesis of sphingolipids with very long chain fatty acids but not ergosterol is required for routing of newly synthesized plasma membrane ATPase to the cell surface of yeast. *J. Biol. Chem.* **280:**22515–22522.
35. **Ghannoum, M. A.** 2000. Potential role of phospholipases in virulence and fungal pathogenesis. *Clin. Microbiol. Rev.* **13:**122–143.
36. **Goldman, D. L., S. C. Lee, A. J. Mednick, L. Montella, and A. Casadevall.** 2000. Persistent *Cryptococcus neoformans* pulmonary infection in the rat is associated with intracellular parasitism, decreased inducible nitric oxide synthase

expression, and altered antibody responsiveness to cryptococcal polysaccharide. *Infect. Immun.* **68:**832–838.

37. **Granier, F., J. Kanitakis, C. Hermier, Y. Y. Zhu, and J. Thivolet.** 1987. Localized cutaneous cryptococcosis successfully treated with ketoconazole. *J. Am. Acad. Dermatol.* **16:**243–249.
38. **Grant, C. M., F. H. MacIver, and I. W. Dawes.** 1997. Mitochondrial function is required for resistance to oxidative stress in the yeast *Saccharomyces cerevisiae. FEBS Lett.* **410:**219–222.
39. **Gutierrez, M. G., S. S. Master, S. B. Singh, G. A. Taylor, M. I. Colombo, and V. Deretic.** 2004. Autophagy is a defense mechanism inhibiting BCG and *Mycobacterium tuberculosis* survival in infected macrophages. *Cell* **119:**753–766.
40. **Harrison, T. S., J. Chen, E. Simons, and S. M. Levitz.** 2002. Determination of the pH of the *Cryptococcus neoformans* vacuole. *Med. Mycol.* **40:**329–332.
41. **Heung, L. J., C. Luberto, A. Plowden, Y. A. Hannun, and M. Del Poeta.** 2004. The sphingolipid pathway regulates Pkc1 through the formation of diacylglycerol in *Cryptococcus neoformans. J. Biol. Chem.* **279:**21144–21153.
42. **Hibbs, J. B., Jr., R. R. Taintor, and Z. Vavrin.** 1987. Macrophage cytotoxicity: role for L-arginine deiminase and imino nitrogen oxidation to nitrite. *Science* **235:**473–476.
43. **Hoang, L. M., J. A. Maguire, P. Doyle, M. Fyfe, and D. L. Roscoe.** 2004. *Cryptococcus neoformans* infections at Vancouver Hospital and Health Sciences Centre (1997–2002): epidemiology, microbiology and histopathology. *J. Med. Microbiol.* **53:**935–940.
44. **Holyoak, C. D., M. Stratford, Z. McMullin, M. B. Cole, K. Crimmins, A. J. Brown, and P. J. Coote.** 1996. Activity of the plasma membrane H(+)-ATPase and optimal glycolytic flux are required for rapid adaptation and growth of *Saccharomyces cerevisiae* in the presence of the weak-acid preservative sorbic acid. *Appl. Environ. Microbiol.* **62:**3158–3164.
45. **Horwitz, M. A.** 1983. The Legionnaires' disease bacterium (*Legionella pneumophila*) inhibits phagosome-lysosome fusion in human monocytes. *J. Exp. Med.* **158:**2108–2126.
46. **Hu, G., M. Hacham, S. R. Waterman, J. Panepinto, S. Shin, X. Liu, J. Gibbons, T. Valyi-Nagy, K. Obara, H. A. Jaffe, Y. Ohsumi, and P. R. Williamson.** 2008. PI3K signaling of autophagy is required for starvation tolerance and virulence of *Cryptococcus neoformans. J. Clin. Invest.* **118:**1186–1197.
47. **Huffnagle, G. B.** 1996. Role of cytokines in T cell immunity to a pulmonary *Cryptococcus neoformans* infection. *Biol. Signals* **5:**215–222.
48. **Huffnagle, G. B., G. H. Chen, J. L. Curtis, R. A. McDonald, R. M. Strieter, and G. B. Toews.** 1995. Down-regulation of the afferent phase of T cell-mediated pulmonary inflammation and immunity by a high melanin-producing strain of *Cryptococcus neoformans. J. Immunol.* **155:**3507–3516.
49. **Huynh, K. K., J. G. Kay, J. L. Stow, and S. Grinstein.** 2007. Fusion, fission, and secretion during phagocytosis. *Physiology* **22:**366–372.
50. **Hybiske, K., and R. S. Stephens.** 2007. Mechanisms of host cell exit by the intracellular bacterium *Chlamydia. Proc. Natl. Acad. Sci. USA* **104:**11430–11435.
51. **Ingavale, S. S., Y. C. Chang, H. Lee, C. M. McClelland, M. L. Leong, and K. J. Kwon-Chung.** 2008. Importance of mitochondria in survival of *Cryptococcus neoformans* under low oxygen conditions and tolerance to cobalt chloride. *PLoS Pathog.* **4:**e1000155.
52. **Jain, N., L. Li, D. C. McFadden, U. Banarjee, X. Wang, E. Cook, and B. C. Fries.** 2006. Phenotypic switching in a *Cryptococcus neoformans* variety *gattii* strain is associated with changes in virulence and promotes dissemination to the central nervous system. *Infect. Immun.* **74:**896–903.
53. **Kechichian, T. B., J. Shea, and M. Del Poeta.** 2007. Depletion of alveolar macrophages decreases the dissemination of a glucosylceramide-deficient mutant of *Cryptococcus neoformans* in immunodeficient mice. *Infect. Immun.* **75:**4792–4798.
54. **Kidd, S. E., F. Hagen, R. L. Tscharke, M. Huynh, K. H. Bartlett, M. Fyfe, L. MacDougall, T. Boekhout, K. J. Kwon-Chung, and W. Meyer.** 2004. A rare genotype of *Cryptococcus gattii* caused the cryptococcosis outbreak on Vancouver Island (British Columbia, Canada). *Proc. Natl. Acad. Sci. USA* **101:**17258–17263.
55. **Kihara, A., T. Noda, N. Ishihara, and Y. Ohsumi.** 2001. Two distinct Vps34 phosphatidylinositol 3-kinase complexes function in autophagy and carboxypeptidase Y sorting in *Saccharomyces cerevisiae. J. Cell Biol.* **152:**519–530.
56. **Koch, G., K. Tanaka, T. Masuda, W. Yamochi, H. Nonaka, and Y. Takai.** 1997. Association of the Rho family small GTP-binding proteins with Rho GDP dissociation inhibitor (Rho GDI) in *Saccharomyces cerevisiae. Oncogene* **15:**417–422.
57. **Koguchi, Y., and K. Kawakami.** 2002. Cryptococcal infection and Th1-Th2 cytokine balance. *Int. Rev. Immunol.* **21:**423–438.
58. **Lee, S. C., Y. Kress, M. L. Zhao, D. W. Dickson, and A. Casadevall.** 1995. *Cryptococcus neoformans* survive and replicate in human microglia. *Lab. Invest.* **73:**871–879.
59. **Levitz, S. M., and T. P. Farrell.** 1990. Growth inhibition of *Cryptococcus neoformans* by cultured human monocytes: role of the capsule, opsonins, the culture surface, and cytokines. *Infect. Immun.* **58:**1201–1209.
60. **Levitz, S. M., T. S. Harrison, A. Tabuni, and X. Liu.** 1997. Chloroquine induces human mononuclear phagocytes to inhibit and kill *Cryptococcus neoformans* by a mechanism independent of iron deprivation. *J. Clin. Invest.* **100:**1640–1646.
61. **Levitz, S. M., S. H. Nong, K. F. Seetoo, T. S. Harrison, R. A. Speizer, and E. R. Simons.** 1999. *Cryptococcus neoformans* resides in an acidic phagolysosome of human macrophages. *Infect. Immun.* **67:**885–890.
62. **Liu, L., R. P. Tewari, and P. R. Williamson.** 1999. Laccase protects *Cryptococcus neoformans* from antifungal activity of alveolar macrophages. *Infect. Immun.* **67:**6034–6039.
63. **Liu, X., G. Hu, J. Panepinto, and P. R. Williamson.** 2006. Role of a VPS41 homologue in starvation response, intracellular survival and virulence of *Cryptococcus neoformans. Mol. Microbiol.* **61:**1132–1146.
64. **Lovchik, J. A., C. R. Lyons, and M. F. Lipscomb.** 1995. A role for gamma interferon-induced nitric oxide in pulmonary clearance of *Cryptococcus neoformans. Am. J. Respir. Cell. Mol. Biol.* **13:**116–124.
65. **Luberto, C., B. Martinez-Marino, D. Taraskiewicz, B. Bolanos, P. Chitano, D. L. Toffaletti, G. M. Cox, J. R. Perfect, Y. A. Hannun, E. Balish, and M. Del Poeta.** 2003. Identification of App1 as a regulator of phagocytosis and virulence of *Cryptococcus neoformans. J. Clin. Invest.* **112:**1080–1094.
66. **Luberto, C., D. L. Toffaletti, E. A. Wills, S. C. Tucker, A. Casadevall, J. R. Perfect, Y. A. Hannun, and M. Del Poeta.** 2001. Roles for inositol-phosphoryl ceramide synthase 1 (IPC1) in pathogenesis of *C. neoformans. Genes Dev.* **15:**201–212.
67. **Ma, H., J. E. Croudace, D. A. Lammas, and R. C. May.** 2007. Direct cell-to-cell spread of a pathogenic yeast. *BMC Immunol.* **8:**15.
68. **Ma, H., J. E. Croudace, D. A. Lammas, and R. C. May.** 2006. Expulsion of live pathogenic yeast by macrophages. *Curr. Biol.* **16:**2156–2160.

68a. **Ma, H., F. Hagen, D. J. Stekel, S. A. Johnston, E. Sionov, R. Falk, I. Polachek, T. Boekhout, and R. C. May.** 2009.

The fatal fungal outbreak on Vancouver Island is characterized by enhanced intracellular parasitism driven by mitochondrial regulation. *Proc. Natl. Acad. Sci. USA* **106:**12980–12985.

69. **MacDougall, L., S. E. Kidd, E. Galanis, S. Mak, M. J. Leslie, P. R. Cieslak, J. W. Kronstad, M. G. Morshed, and K. H. Bartlett.** 2007. Spread of *Cryptococcus gattii* in British Columbia, Canada, and detection in the Pacific Northwest, USA. *Emerg. Infect. Dis.* **13:**42–50.
70. **May, R. C., M. E. Hall, H. N. Higgs, T. D. Pollard, T. Chakraborty, J. Wehland, L. M. Machesky, and A. S. Sechi.** 1999. The Arp2/3 complex is essential for the actin-based motility of *Listeria monocytogenes*. *Curr. Biol.* **9:**759–762.
71. **Missall, T. A., M. E. Pusateri, and J. K. Lodge.** 2004. Thiol peroxidase is critical for virulence and resistance to nitric oxide and peroxide in the fungal pathogen, *Cryptococcus neoformans*. *Mol. Microbiol.* **51:**1447–1458.
72. **Mitchell, T. G., and L. Friedman.** 1972. In vitro phagocytosis and intracellular fate of variously encapsulated strains of *Cryptococcus neoformans*. *Infect. Immun.* **5:**491–498.
73. **Monari, C., C. Retini, A. Casadevall, D. Netski, F. Bistoni, T. R. Kozel, and A. Vecchiarelli.** 2003. Differences in outcome of the interaction between *Cryptococcus neoformans* glucuronoxylomannan and human monocytes and neutrophils. *Eur. J. Immunol.* **33:**1041–1051.
74. **Mukherjee, J., L. S. Zuckier, M. D. Scharff, and A. Casadevall.** 1994. Therapeutic efficacy of monoclonal antibodies to *Cryptococcus neoformans* glucuronoxylomannan alone and in combination with amphotericin B. *Antimicrob. Agents Chemother.* **38:**580–587.
75. **Murray, H. W., and D. M. Cartelli.** 1983. Killing of intracellular *Leishmania donovani* by human mononuclear phagocytes. Evidence for oxygen-dependent and -independent leishmanicidal activity. *J. Clin. Invest.* **72:**32–44.
76. **Nakano, K., T. Mutoh, R. Arai, and I. Mabuchi.** 2003. The small GTPase Rho4 is involved in controlling cell morphology and septation in fission yeast. *Genes Cells* **8:**357–370.
77. **Nichols, C. B., Z. H. Perfect, and J. A. Alspaugh.** 2007. A Ras1-Cdc24 signal transduction pathway mediates thermotolerance in the fungal pathogen *Cryptococcus neoformans*. *Mol. Microbiol.* **63:**1118–1130.
78. **Noverr, M. C., G. M. Cox, J. R. Perfect, and G. B. Huffnagle.** 2003. Role of PLB1 in pulmonary inflammation and cryptococcal eicosanoid production. *Infect. Immun.* **71:**1538–1547.
79. **Noverr, M. C., S. M. Phare, G. B. Toews, M. J. Coffey, and G. B. Huffnagle.** 2001. Pathogenic yeasts *Cryptococcus neoformans* and *Candida albicans* produce immunomodulatory prostaglandins. *Infect. Immun.* **69:**2957–2963.
80. **Oh-hama, T.** 1997. Evolutionary consideration on 5-aminolevulinate synthase in nature. *Orig. Life Evol. Biosph.* **27:**405–412.
81. **Portnoy, D. A., V. Auerbuch, and I. J. Glomski.** 2002. The cell biology of *Listeria monocytogenes* infection: the intersection of bacterial pathogenesis and cell-mediated immunity. *J. Cell Biol.* **158:**409–414.
82. **Price, M. S., C. B. Nichols, and J. A. Alspaugh.** 2008. The *Cryptococcus neoformans* Rho-GDP dissociation inhibitor mediates intracellular survival and virulence. *Infect. Immun.* **76:**5729–5737.
83. **Radisky, D. C., W. B. Snyder, S. D. Emr, and J. Kaplan.** 1997. Characterization of VPS41, a gene required for vacuolar trafficking and high-affinity iron transport in yeast. *Proc. Natl. Acad. Sci. USA* **94:**5662–5666.
84. **Reardon, C. C., S. J. Kim, R. P. Wagner, and H. Kornfeld.** 1996. Interferon-gamma reduces the capacity of human alveolar macrophages to inhibit growth of *Cryptococcus neoformans* in vitro. *Am. J. Respir. Cell. Mol. Biol.* **15:**711–715.
85. **Robbins, J. R., A. I. Barth, H. Marquis, E. L. de Hostos, W. J. Nelson, and J. A. Theriot.** 1999. *Listeria monocytogenes* exploits normal host cell processes to spread from cell to cell. *J. Cell. Biol.* **146:**1333–1350.
86. **Rossier, M. F.** 2006. T channels and steroid biosynthesis: in search of a link with mitochondria. *Cell Calcium* **40:**155–164.
87. **Schaechter, M., F. M. Bozeman, and J. E. Smadel.** 1957. Study on the growth of Rickettsiae. II. Morphologic observations of living Rickettsiae in tissue culture cells. *Virology* **3:**160–172.
88. **Schmiel, D. H., and V. L. Miller.** 1999. Bacterial phospholipases and pathogenesis. *Microbes Infect.* **1:**1103–1112.
89. **Shea, J. M., T. B. Kechichian, C. Luberto, and M. Del Poeta.** 2006. The cryptococcal enzyme inositol phosphosphingolipid-phospholipase C confers resistance to the antifungal effects of macrophages and promotes fungal dissemination to the central nervous system. *Infect. Immun.* **74:**5977–5988.
90. **Shoham, S., and S. M. Levitz.** 2005. The immune response to fungal infections. *Br. J. Haematol.* **129:**569–582.
91. **Sigler, K., and M. Hofer.** 1991. Activation of the plasma membrane H(+)-ATPase of *Saccharomyces cerevisiae* by addition of hydrogen peroxide. *Biochem. Int.* **23:**861–873.
92. **Soteropoulos, P., T. Vaz, R. Santangelo, P. Paderu, D. Y. Huang, M. J. Tamas, and D. S. Perlin.** 2000. Molecular characterization of the plasma membrane H(+)-ATPase, an antifungal target in *Cryptococcus neoformans*. *Antimicrob. Agents Chemother.* **44:**2349–2355.
93. **Southwick, F. S., and D. L. Purich.** 1996. Intracellular pathogenesis of listeriosis. *N. Engl. J. Med.* **334:**770–776.
94. **Stadler, N., M. Hofer, and K. Sigler.** 2001. Mechanisms of *Saccharomyces cerevisiae* PMA1 H+-ATPase inactivation by Fe2+, H2O2 and Fenton reagents. *Free Radic. Res.* **35:**643–653.
95. **Steenbergen, J. N., J. D. Nosanchuk, S. D. Malliaris, and A. Casadevall.** 2003. *Cryptococcus neoformans* virulence is enhanced after growth in the genetically malleable host *Dictyostelium discoideum*. *Infect. Immun.* **71:**4862–4872.
96. **Steenbergen, J. N., H. A. Shuman, and A. Casadevall.** 2001. *Cryptococcus neoformans* interactions with amoebae suggest an explanation for its virulence and intracellular pathogenic strategy in macrophages. *Proc. Natl. Acad. Sci. USA* **98:**15245–15250.
97. **Stephen, C., S. Lester, W. Black, M. Fyfe, and S. Raverty.** 2002. Multispecies outbreak of cryptococcosis on southern Vancouver Island, British Columbia. *Can. Vet. J.* **43:**792–794.
98. **Stevens, J. M., E. E. Galyov, and M. P. Stevens.** 2006. Actin-dependent movement of bacterial pathogens. *Nat. Rev. Microbiol.* **4:**91–101.
99. **Strasser, J. E., S. L. Newman, G. M. Ciraolo, R. E. Morris, M. L. Howell, and G. E. Dean.** 1999. Regulation of the macrophage vacuolar ATPase and phagosome-lysosome fusion by *Histoplasma capsulatum*. *J. Immunol.* **162:**6148–6154.
100. **Sturgill-Koszycki, S., P. H. Schlesinger, P. Chakraborty, P. L. Haddix, H. L. Collins, A. K. Fok, R. D. Allen, S. L. Gluck, J. Heuser, and D. G. Russell.** 1994. Lack of acidification in *Mycobacterium* phagosomes produced by exclusion of the vesicular proton-ATPase. *Science* **263:**678–681.
101. **Sullivan, D., K. Haynes, G. Moran, D. Shanley, and D. Coleman.** 1996. Persistence, replacement, and microevolution of *Cryptococcus neoformans* strains in

recurrent meningitis in AIDS patients. *J. Clin. Microbiol.* **34:**1739–1744.

102. **Supek, F., L. Supekova, H. Nelson, and N. Nelson.** 1996. A yeast manganese transporter related to the macrophage protein involved in conferring resistance to mycobacteria. *Proc. Natl. Acad. Sci. USA* **93:**5105–5110.

103. **Tacker, J. R., F. Farhi, and G. S. Bulmer.** 1972. Intracellular fate of *Cryptococcus neoformans. Infect. Immun.* **6:**162–167.

104. **Toffaletti, D. L., K. Nielsen, F. Dietrich, J. Heitman, and J. R. Perfect.** 2004. *Cryptococcus neoformans* mitochondrial genomes from serotype A and D strains do not influence virulence. *Curr. Genet.* **46:**193–204.

105. **Tucker, S. C., and A. Casadevall.** 2002. Replication of *Cryptococcus neoformans* in macrophages is accompanied by phagosomal permeabilization and accumulation of vesicles containing polysaccharide in the cytoplasm. *Proc. Natl. Acad. Sci. USA* **99:**3165–3170.

106. **Tuxworth, R. I., I. Weber, D. Wessels, G. C. Addicks, D. R. Soll, G. Gerisch, and M. A. Titus.** 2001. A role for myosin VII in dynamic cell adhesion. *Curr. Biol.* **11:**318–329.

107. **Vallim, M. A., C. B. Nichols, L. Fernandes, K. L. Cramer, and J. A. Alspaugh.** 2005. A Rac homolog functions downstream of Ras1 to control hyphal differentiation and high-temperature growth in the pathogenic fungus *Cryptococcus neoformans. Eukaryot. Cell* **4:**1066–1078.

107a. **Voelz, K., D. A. Lammas, and R. C. May.** 2009. Cytokine signaling regulates the outcome of intracellular macrophage parasitism by *Cryptococcus neoformans. Infect. Immun.* **77:**3450–3457.

108. **Walker, L., and D. B. Lowrie.** 1981. Killing of *Mycobacterium microti* by immunologically activated macrophages. *Nature* **293:**69–71.

109. **Wang, Y., P. Aisen, and A. Casadevall.** 1995. *Cryptococcus neoformans* melanin and virulence: mechanism of action. *Infect. Immun.* **63:**3131–3136.

110. **Wang, Y., and A. Casadevall.** 1994. Susceptibility of melanized and nonmelanized *Cryptococcus neoformans* to nitrogen- and oxygen-derived oxidants. *Infect. Immun.* **62:**3004–3007.

111. **Waugh, M. S., M. A. Vallim, J. Heitman, and J. A. Alspaugh.** 2003. Ras1 controls pheromone expression and response during mating in *Cryptococcus neoformans. Fungal Genet. Biol.* **38:**110–121.

112. **Williamson, P. R.** 1997. Laccase and melanin in the pathogenesis of *Cryptococcus neoformans. Front Biosci.* **2:**e99–e107.

113. **Wood, S. A., R. R. Ammann, D. A. Brock, L. Li, T. Spann, and R. H. Gomer.** 1996. RtoA links initial cell type choice to the cell cycle in *Dictyostelium. Development* **122:**3677–3685.

114. **Wright, L. C., R. M. Santangelo, R. Ganendren, J. Payne, J. T. Djordjevic, and T. C. Sorrell.** 2007. Cryptococcal lipid metabolism: phospholipase B1 is implicated in transcellular metabolism of macrophage-derived lipids. *Eukaryot. Cell* **6:**37–47.

115. **Zaragoza, O., C. J. Chrisman, M. V. Castelli, S. Frases, M. Cuenca-Estrella, J. L. Rodriguez-Tudela, and A. Casadevall.** 2008. Capsule enlargement in *Cryptococcus neoformans* confers resistance to oxidative stress suggesting a mechanism for intracellular survival. *Cell. Microbiol.* **10:**2043–2057.

116. **Zhao, L., F. Zhang, J. Guo, Y. Yang, B. Li, and L. Zhang.** 2004. Nitric oxide functions as a signal in salt resistance in the calluses from two ecotypes of reed. *Plant Physiol.* **134:**849–857.

Cryptococcus: From Human Pathogen to Model Yeast
Edited by J. Heitman et al.
©2011 ASM Press, Washington, DC

33

Pulmonary Innate and Adaptive Defenses against *Cryptococcus*

KRISTI L. WILLIAMS, FLOYD L. WORMLEY, JR., SCARLETT GEUNES-BOYER, JO RAE WRIGHT, AND GARY B. HUFFNAGLE

The most common route of entry of *Cryptococcus neoformans* and *Cryptococcus gattii* into the body is via inhalation, which allows these organisms to establish a primary infection in the lungs. Physical defense mechanisms such as upper airway turbulence and ciliary clearance function to prevent most inhaled organisms from reaching the alveoli. Those that do make it deep into the airways are controlled by the coordinated efforts of innate and adaptive host immune defenses. Clinical and experimental studies clearly substantiate the role of acquired cell-mediated immunity in the establishment of long-term protection against pulmonary cryptococcosis. However, the innate immune system is also centrally involved in both immediate antifungal immune responses as well as promoting adaptive immunity. Together, innate and adaptive immune mechanisms function in a coordinated stepwise process:

- Recognition/binding of the fungus (by pattern recognition receptors [PRRs] and opsonin receptors expressed on host cells)
- The release of inflammatory mediators (including cytokines, chemokines, eicosanoids, and reactive oxygen species)
- Phagosome-mediated antifungal immunity and the recruitment of new innate immune cells to fight infection (including neutrophils, monocytes, dendritic cells, and NK cells)
- Antigen presentation
- Polarization of adaptive immune responses
- Recruitment and activation of additional effector leukocytes
- Killing of cryptococci through effector-phase mechanisms
- Control/resolution of pulmonary inflammation by immunologic regulatory pathways

Kristi L. Williams, Scarlett Geunes-Boyer, and Jo Rae Wright, Department of Cell Biology, Duke University Medical Center, Durham, NC 27710. **Floyd L. Wormley, Jr.,** The University of Texas at San Antonio, Department of Biology, San Antonio, TX 78249. **Gary B. Huffnagle,** Internal Medicine - Pulmonary Division, University of Michigan Med Center, Ann Arbor, MI 48109-5642.

The resident phagocyte of the airspaces, the alveolar macrophage, together with surfactant proteins form the first line of innate defense against inhaled cryptococcal organisms. The second line of defense involves the recruitment of leukocytes into the lungs. Pulmonary dendritic cells, both resident and recruited, are able to internalize and carry processed cryptococcal antigens to the mediastinal nodes (the lung-associated lymph nodes), where as antigen-presenting cells they present cryptococcal antigens to T cells. The subsequent coordinated expression of a cascade of specific cytokines, chemokines, and eicosanoids by the innate immune response is critical for promoting the development of a protective Th1-type immune response in the lungs, which drives the clearance of the infection. In contrast, Th2-skewed immune responses result in a chronic infection (little is known about the role of Th17 and Th9 responses). More recent evidence is also now pointing to an important role for regulatory immune responses (both T-cell- and antibody-mediated) in limiting T-cell-mediated inflammation in the lungs, thereby controlling inflammatory tissue damage while promoting clearance of the infection.

RECOGNITION/BINDING OF *C. NEOFORMANS* IN THE LUNGS: INNATE HUMORAL MECHANISMS

Cryptococcus (or any microbe) is recognized and bound by host defense mechanisms through either (i) soluble (humoral) factors that form a bridge between the organism and specific receptors on host leukocytes or (ii) direct recognition of microbial structures by PRRs on host leukocytes. Adaptive humoral mediators include *C. neoformans*–specific antibodies, while innate humoral mediators include surfactant proteins and complement. Surfactant proteins are exclusively localized to the lung environment, while the role of complement in the airways is less clear. In addition, future studies may identify additional lung-specific innate humoral mechanisms because encapsulated cryptococci, which are more resistant to surfactant-mediated opsonization, are readily taken up by alveolar macrophages in the lungs.

SURFACTANT PROTEINS AND COMPLEMENT

Surfactant is secreted by type II cells in the alveolar space of the lung, and it is composed of a complex mixture of phospholipids and proteins, including surfactant proteins A and D (SP-A, SP-D), which play a role in innate host defense. SP-A and SP-D are relatively hydrophilic proteins, in contrast to other surfactant components, including the extremely hydrophobic proteins SP-B and SP-C. SP-A and SP-D have two distinct regions: the collagen-like region and the lectin-like domain. The structures of SP-A and SP-D are critical for their innate immune activity. The lectin-like domain, also known as the carbohydrate recognition domain, binds in a calcium-dependent manner to sugars that are commonly present on the surfaces of microorganisms, and consequently, SP-A and SP-D function as pattern recognition molecules.

As a result of this interaction, SP-A and SP-D enhance the clearance of many pathogens from the pulmonary space, playing both distinct and overlapping roles during innate immune responses, as well as regulating the progression of pulmonary inflammation. SP-A- and SP-D-deficient mice have been shown to be more susceptible to a variety of different microorganisms, including a number of fungi including *Candida albicans* (161) and *Aspergillus fumigatus* (115, 116). Interestingly, conflicting results have been obtained regarding the role of SP-D in murine *Pneumocystis* infection, as it has been shown to both exacerbate (164) and facilitate clearance of this fungus in a mouse model of pneumonia (6, 109).

To date, very little is known about the roles of surfactant proteins in *C. neoformans* infection. It has been proposed that both SP-A and SP-D bind to acapsular strains, resulting in aggregation of the yeast cells (150). Furthermore, SP-A has been shown to bind to both encapsulated and acapsular *C. neoformans* without enhancing uptake by macrophages (166). In that study, SP-A bound acapsular yeast cells threefold better than encapsulated yeast cells, raising the possibility that SP-A interacts with multiple binding partners on the surface of *C. neoformans* cells. Despite these findings, SP-A has been shown to be ineffective in modulating overall survival in a mouse model of *C. neoformans* infection (52), yet the question remains as to whether SP-A plays a more subtle role during early or very late stages of infection.

β-Glucans are powerful activators of innate immune responses, and the prominent role played by SP-D during *C. neoformans* infection appears to depend, at least in part, on its interaction with β-1,6-glucan present in the cell wall of *C. neoformans* cells. SP-D, but not SP-A, was found to bind *C. neoformans* isolated capsular components mannoprotein 1 (MP1) and glucuronoxylomannan (GXM) in a dose-dependent manner in vitro (160), and pustulan, an analog of β-1,6-glucan, has been shown to bind SP-D (2). Importantly, exogenous GXM inhibits SP-D-dependent aggregation of *C. neoformans* (160), suggesting that shed capsule components may affect SP-D functionality.

Recent data from our laboratory (J.R.W., S.G., K.L.W.) are beginning to support the hypothesis that binding of SP-D to *C. neoformans* modulates immune cell functions. For example, purified SP-D binds to *C. neoformans* cells and to isolated *C. neoformans* cell wall and capsular components. SP-D opsonization of hypocapsular *C. neoformans* in vitro enhances uptake of the yeast cells by murine macrophages, which results in impaired rather than enhanced killing compared to uptake of unopsonized cells (unpublished data). Intriguingly, SP-D appears to function as a risk factor for *C. neoformans* infection in a mouse inhalation model employing SP-D$^{-/-}$ and wild-type mice, as demonstrated by the delayed mean time to mortality among SP-D$^{-/-}$ compared to wild-type mice (unpublished data). The mechanism by which *C. neoformans* cells subvert the normal host-protection mechanisms of SP-D remains to be elucidated. However, *C. neoformans* can, in fact, replicate inside macrophages and grows at a faster rate intracellularly than it does extracellularly (40). In recent studies, *C. neoformans* was shown, under conditions including phagosomal maturation, to exit the macrophage through extrusion of the phagosome while both the yeast cell and the host cell remained viable, a process termed "vomocytosis" (3, 113). Thus, it is possible that SP-D serves to facilitate uptake of *C. neoformans* as a "Trojan horse" in which *C. neoformans* cells survive intracellularly and eventually disseminate from the lungs (18). Further investigation of the mechanism by which *C. neoformans* cells exploit SP-D in order to subvert innate immune responses may facilitate the design of novel therapeutic strategies that target the interaction between SP-D and this pathogenic fungus.

Complement plays an important role during innate immune responses to *C. neoformans* cells in extrapulmonary sites as a consequence of its ability to bind to encapsulated yeast cells and function as an opsonin. In the airways, the role of complement is less clear. Both C3- and C5a-deficient mice display greater susceptibility to *C. neoformans* infection (11). Interestingly, a secreted *C. neoformans*–specific protein, antiphagocytic protein 1 (App1), inhibits phagocytosis of *C. neoformans* cells by alveolar macrophages (112). This effect has been demonstrated to occur via binding of App1 to complement receptors 2 and 3 (CR2, CR3) (152), implicating the importance of App1 in both innate and adaptive immune responses, as CR2 is mainly present on the surface of B cells (41, 42, 143).

RECOGNITION/BINDING OF *C. NEOFORMANS* IN THE LUNGS: INNATE RECEPTORS

Recognition of microbe-associated molecular patterns by germ line-encoded PRRs on innate immune cells is essential for host defense against microorganisms (95, 154). The cell walls of fungi consist mainly of carbohydrates including mannose-based structures, β-glucans, and chitin. The innate immune system recognizes microbial pathogens including fungi via PRRs such as Toll-like receptors (TLRs), C-type lectin receptors including Dectin-1, and the mannose receptor (MR). PRR activation initiates an immediate response to the invading pathogen through the production of inflammatory mediators including cytokines, chemokines, antimicrobial peptides, and the production of reactive oxygen species.

TLRs have been implicated in airway inflammation, and recent evidence suggests the involvement of certain TLRs in mediating a protective *C. neoformans* response by promoting a Th1-type T-cell adaptive immune response. While TLRs are expressed on both hematopoietic and nonhematopoietic cell types, the initial innate immune response to *C. neoformans* pulmonary infection is mediated by resident innate immune cells including alveolar macrophages. In response to intranasal infection, MyD88 and TLR2 null mice have significantly reduced survival compared to wild-type mice, while CD14 null mice were only susceptible to intravenous infection, and TLR4 null mice were not significantly susceptible to either mode of infection (172). In addition to supporting a broad role for MyD88 in the host response to *C. neoformans* infection, these studies suggest that the route of infection may result in different outcomes. A separate group showed that MyD88 and TLR2, but not TLR4, were involved in host defense against *C. neoformans* infection (9). In this study, the investigators used macrophages isolated from TLR2 null, TLR4 mutant C3H/HeJ, and MyD88 null mice followed by infection with *C. neoformans* in vitro. Analysis of cytokine production suggested that TLR2 and MyD88, but not TLR4, are required for tumor necrosis factor alpha (TNF-α) production. Further, analysis of susceptibility in response to *C. neoformans* infection via intraperitoneal injection suggested that TLR2 and MyD88 null mice were more susceptible to infection than TLR4 null mice. Using a murine model of pulmonary infection and TLR2 and TLR4 null mice, Nakamura et al. suggested that TLR2 and TLR4 have little if any role in mediating host immunity to *C. neoformans* infection (135). It is interesting to note that in many of these studies, relatively few differences were observed in cytokine responses including TNF-α, interleukin-1β (IL-1β), IL-6, IL-12p40, and gamma interferon (IFN-γ). Taken together, these studies suggest that the common TLR signaling adaptor MyD88 plays a role in *C. neoformans* immunity, but a clear role for specific TLRs is still in question.

The C-type lectin receptor family of PRRs includes Dectin-1, the MR, and the dendritic cell ICAM3-grabbing nonintegrin (DC-SIGN/CD209), all of which recognize particular carbohydrate residues on the surface of *C. neoformans* via their extracellular carbohydrate recognition domain. The MR preferentially recognizes terminal mannose residues, while DC-SIGN recognizes internal mannose residues (13). Analysis of the major antigenic determinants of *C. neoformans* has suggested that mannoproteins (MPs), a group of glycoproteins with heavy carbohydrate decorations terminating in mannose residues, promote a full host immune response to infection (69, 103, 118, 119, 131, 132, 145). Our current understanding of the role of DC-SIGN in *C. neoformans* immunity is limited to a study investigating the ability of DC-SIGN to act as a potential receptor for MPs. Stable transfection of the K562 monocyte cell line with an expression vector for DC-SIGN demonstrated the ability of DC-SIGN to bind fluorescently labeled MP that could be competed by unlabeled mannan (117). These studies suggest that DC-SIGN can serve as a *C. neoformans* recognition receptor on dendritic cells. Future studies revealing the role of DC-SIGN using murine models of *C. neoformans* infection will likely yield important information regarding the in vivo role of the MR in host immunity.

The MR recognizes terminal mannose residues and has a broad expression pattern (120). In vitro evidence suggests the importance of the MR in recognition of MP by dendritic cells and the subsequent induction of the T-cell response (117). In a recent study by Levitz and colleagues, the in vivo role of the MR was examined in a model of *C. neoformans* pulmonary infection using MR null mice (28). MR null mice exhibited a statistically significant decrease in susceptibility based on survival curves and decreased fungal burden in the lung. The dendritic cells from wild-type and MR null mice exhibited similar abilities to phagocytose MP and similar levels of maturation markers on their cell surface. However, the $CD4^+$ T cells isolated from the lymph nodes of MR null mice 2 weeks postinfection had reduced proliferative capacity in response to MP in an in vitro antigen presentation assay. These data suggest that the MR is at least partially involved in mediating a full immune response to *C. neoformans* infection.

A recently identified C-type lectin receptor, known as Dectin-1, has been characterized as an important fungal pathogen receptor. Dectin-1 specifically recognizes β-1,3-glucan, which normally comprises approximately 50% of fungal cell walls. Dectin-1 receptor activation leads to activation of the proximal Syk tyrosine kinase, resulting in engagement of the caspase recruitment domain (CARD)-containing protein CARD9 (61). Recent data suggest that both TLR2 and Dectin-1 recognize a range of pathogenic fungi and work in concert to generate effective antifungal responses in phagosomal vesicles (136). However, the role of Dectin-1 in mediating the host immune response to *C. neoformans* is currently believed to be minimal, which may be due to steric hindrance in accessing β-1,3-glucan molecules on the surface of encapsulated cryptococci. Using mouse models of *C. neoformans* pulmonary and intravenous infection, responses of Dectin-1 null mice were comparable to wild-type mice in regard to susceptibility to infection, fungal burden, and cytokine production (134). These studies suggested that Dectin-1 was not required for a full host response to *C. neoformans* infection.

UPTAKE OF *C. NEOFORMANS* BY RESIDENT PULMONARY LEUKOCYTES

C. neoformans enters the pulmonary system in both a spore and yeast form and has the ability to remain dormant for many years within the phagosomal compartment before infection is activated (47–49). *C. neoformans* has also evolved different mechanisms to evade the destructive phagosomal pathway. Early research suggested that acapsular *C. neoformans* is easily ingested by macrophages and neutrophils, whereas encapsulated *C. neoformans* is resistant to phagocytosis (12, 97). A larger capsule inhibits phagocytosis of *C. neoformans* by macrophages and neutrophils (32, 34, 98, 176), dendritic cells (163), and endothelial cells (79). *C. neoformans* has developed a novel mechanism of intracellular survival by an encapsulated organism, whereby ingestion is followed by *C. neoformans*–induced damage to the host phagosomal membrane (157). This results in continuity with the cytoplasm and accumulation of polysaccharide-containing vesicles within the host cell. Subsequent phagosomes that develop from this process contain multiple *C. neoformans* cells per vesicle. App1, a secreted *C. neoformans*–specific protein, has been shown to inhibit phagocytosis of *C. neoformans* cells by alveolar macrophages and involves binding of this protein to complement receptors (112, 152). *C. gattii*, another species of *Cryptococcus*, also causes pulmonary cryptococcomas (22) that often require surgical removal (43). While the pathogenesis of these two cryptococcal species is remarkably different, the ability of *C. neoformans* and *C. gattii* to undergo antibody-mediated phagocytosis, followed by intracellular replication, host cell cytoplasmic accumulation, and phagosomal extrusion, are similar between the two strains (4).

Alveolar macrophages are able to phagocytize *C. neoformans*, but the fate of the ingested yeast depends on the host species source of the alveolar macrophages. Via intratracheal liposomal clodronate to deplete alveolar macrophages in rats and mice, the loss of alveolar macrophages was associated with increased fungal burden in rats and decreased fungal burden in mice (151). Alveolar macrophages in rats have significant anticryptococcal phagocytic activity, oxidative burst, lysozyme secretion, and ability to limit intracellular growth. Rat alveolar macrophages were more resistant to lysis due to intracellular cryptococcal infection. Activation of rat alveolar macrophages by IFN-γ or granulocyte-macrophage colony-stimulating factor enhances their killing activity (19, 125). In mice, alveolar macrophages may exacerbate *C. neoformans* infection and promote dissemination to the brain in conditions in which there is severe host immunodeficiency that prevents macrophage activation (92, 151).

Human alveolar macrophages are capable of growth-inhibiting C. *neoformans* in vitro, but this is diminished as the size of the capsule increases (14, 162, 165, 167). Thus, resident alveolar macrophages can be active contributors to host defense, or they can promote the pathogenesis of the infection, depending on the host species and immune status.

RECRUITMENT OF ANTIGEN-PRESENTING CELLS INTO THE LUNGS TO PROMOTE THE DEVELOPMENT OF ADAPTIVE IMMUNITY

Typically produced in abundance by alveolar macrophages after infection, the early production of TNF-α is required for the recruitment of CD11c^{+}/MHCII^{+} dendritic cells in the lung after infection and subsequent development of Th1-type cell-mediated immunity to C. *neoformans* infection (7, 64, 76). Dendritic cells are a large part of the host immune response to C. *neoformans* because of their unique ability to phagocytose C. *neoformans* and present cryptoccocal antigens important in initiating an adaptive immune response. C. *neoformans* will enter into endosomal and lysosomal pathways following dendritic cell phagocytosis, and lysosomal components can kill C. *neoformans* both in vivo and in vitro (169). The critical window for dendritic cell recruitment is early in the infection, as demonstrated by neutralization of TNF-α through repeated doses of anti-TNF-α during the first 2 weeks or a single injection of anti-TNF-α at the time of infection (76). Consistent with this observation, transient neutralization of TNF-α resulted in decreased expression of IL-12, monocyte chemotactic protein-1 (MCP-1)/CCL2, macrophage inflammatory protein (MIP)-1α/CCL3, and IFN-γ in the lungs of mice, leading to increased susceptibility (63, 64).

In a study identifying the different cell types recruited to the lungs after C. *neoformans* infection, it was observed that neutrophils were present in the lungs at day 7 postinfection in a TNF-α-dependent manner (63). The importance of Gr-1^{+} cells in host immunity to C. *neoformans* infection was examined by injection of an anti-Gr-1(Ly6C/Ly6G) antibody to deplete Gr-1^{+} populations (note that these and other studies originally identified Gr-1^{+} cells as solely neutrophils). Transient antibody-mediated deletion of GR-1^{+} cells suggested that this cell population is not required for the development of a protective antifungal immune response in the host. However, in another study, it was demonstrated that the absence of Gr-1^{+} cells in lung tissue during the initial stages of infection appeared to alter the inflammatory response in a manner that was subsequently beneficial to the host, suggesting that these cells may play an important function in modulating the development of the immune response. While both of these studies support the hypothesis that neutrophils are not required for clearance of C. *neoformans*, it is important to note that Gr-1(Ly6C/Ly6G) antibody depletion may also delete a subset of Ly6C-expressing monocytes (10, 50, 147).

Chemokines such as CCL2 and CCL3 are important during pulmonary C. *neoformans* infection since they promote leukocyte recruitment and a protective Th1-type immune response. Neutralization of CCL2 or CCL3 results in reduced recruitment of macrophages and neutrophils to the lungs of infected mice (74, 75). In contrast, neutralization of CCL2 but not CCL3 results in decreased recruitment of CD4^{+} and CD8^{+} T cells to the lungs of infected mice. In studies of C. *neoformans* pulmonary infection, treatment of wild-type mice with CCL2-neutralizing antibody or deletion of CCR2, the only known receptor for CCL2, results in similar decreases in macrophage and T-cell recruitment into the lungs of mice. A more detailed analysis of the differential role of CCL2 and CCR2 in the development of T-cell immunity to C. *neoformans* revealed that CCR2 is required for the development of Th1-cell-based immunity in the lymph nodes. In the absence of the chemokine CCL2, Th1 cells migrate to the lung-associated lymph nodes but do not traffic from the lymph nodes to the lungs, resulting in a nonprotective pulmonary Th2 response (155). CCL3 is a chemokine required for optimal leukocyte recruitment in response to cryptococcal antigens. While CCL3 null mice exhibit similar numbers of recruited leukocytes and macrophages to the lungs after infection, CCL3 null mice have higher numbers of eosinophils in the lungs after infection, suggesting the development of a Th2 immune response (138). In a later study the clear role for CCL3 in promoting a Th1 T-cell response was supported by the observation that CCL3 prevented the development of nonprotective eosinophilic pneumonia even in the absence of TNF-α, IFN-γ, and CCL2 (139).

The chemokine receptors CCR5 and CCR2, which are expressed on subtypes of innate immune cells, are important for a full host response to infection. CCR5 is expressed on many cell types including lymphocytes, macrophages, and granulocytes as well as nonhemopoietic cells (148). Studies using CCR5 gene deletion mice in a pulmonary C. *neoformans* infection model suggested that CCR5 was not important for leukocyte recruitment to the lungs but was required for leukocyte recruitment to the brain and for the elimination of cryptococcal polysaccharide from the brain (73). More important for pulmonary immune responses, numerous studies using gene deletion mice have identified the requirement for the chemokine receptor CCR2 in the pulmonary immune response to C. *neoformans* infection. CCR2 is expressed on monocytes, activated T cells, B cells, and NK cells (45). The chemokine CCL2 binds with high affinity to CCR2. Additional chemokines including CCL3, CCL7, and CCL12 are capable of binding CCR2 as well as other chemokine receptors. CCR2 null mice have severe defects in development of a protective immune response to C. *neoformans* (156). In an intratracheal model of C. *neoformans* infection, CCR2 null mice had increased fungal burdens and dissemination to the brain and dramatic reductions in macrophage CD8^{+} T-cell recruitment to the lungs after infection. CCR2-deficient mice developed a nonprotective Th2-type T-cell response characterized by a persistent infection. In a later study to determine the mechanism by which CCR2 null mice develop a nonprotective immune response, it was determined that recruitment of dendritic cells to the lung was markedly impaired in the absence of CCR2 (141).

Results from these studies suggested that CCR2 expression is required for recruitment of conventional dendritic cells to the lungs and for promoting the development of a protective Th1 response. In future studies, it would be interesting to determine if the dendritic cells observed in these mice developed from CCR2-expressing inflammatory monocytes recruited to the lungs after infection. Regardless of the developmental lineage of dendritic cells in the lung after C. *neoformans* infection, these studies clearly support the requirement of dendritic cells in the development of a protective Th1-type immune response.

EARLY INNATE SIGNALS THAT DRIVE THE DEVELOPMENT OF PROTECTIVE ADAPTIVE IMMUNITY

TNF-α production during the very early stages of the developing immune response to pulmonary C. *neoformans* infection is critical for the establishment of protective T-cell-mediated immunity (CMI). Neutralization of TNF-α using TNF-α-specific antiserum at the onset of experimental pulmonary *C. neoformans* infection was shown to inhibit CD4+ T-cell, macrophage, and polymorphonuclear leukocyte recruitment to the lungs by 64%, 98%, and 81%, respectively, and caused a two- to threefold increase in pulmonary fungal burden (76). A single dose of TNF-α antiserum was sufficient to prevent the development of *C. neoformans*–specific delayed-type hypersensitivity (DTH) reactivity and protective CMI up to 5 weeks postinfection. Interestingly, the negative effects of TNF-α ablation were not observed if the cytokine was neutralized beginning 2 weeks postinfection. Moreover, neutralization of TNF-α during the afferent phase of the *C. neoformans* immune response resulted in the inhibition of IFN-γ and IL-12 production by lung leukocytes, increased pulmonary eosinophilia, and a shift to a nonprotective Th2-type phenotype (63, 64). In support of this, mice treated with a TNF-α-expressing adenoviral vector and given an experimental pulmonary *C. neoformans* infection had significantly less fungal burden, increased pulmonary recruitment of classically activated macrophages and neutrophils, no development of pulmonary eosinophilia, and increased Th1-type cytokine production compared to mice given *C. neoformans* alone or *C. neoformans* with a control vector (121). Altogether, the studies suggest that the development of protective Th1-type CMI against pulmonary *C. neoformans* infection requires TNF-α production during the afferent phase of the immune response.

IL-12, similar to TNF-α, is critical during the initial immune response for the induction of protective Th1-type CMI responses to pulmonary *C. neoformans* infection. Mice with targeted disruptions of the genes encoding for IL-12p35 or IL-12p40 subunits succumbed to pulmonary *C. neoformans* infection significantly earlier than infected wild-type mice (29). Conversely, administration of recombinant IL-12 during pulmonary infection promoted the clearance of *C. neoformans* from the lungs and inhibited dissemination of cryptococci to the brain, resulting in reduced mortality (91). IL-12 was shown to help initiate protective Th1-type responses against pulmonary *C. neoformans* infection in resistant C.B-17 mice (67). The protective effect of IL-12 correlates with an induction of Th1-type polarized immune responses including increased IFN-γ, IL-18, and inducible nitric oxide synthase (iNOS) production and local production of mononuclear leukocyte chemokines such as MCP-1, MIP-1α, MIP-1β, regulated upon activation normal T-cell expressed and secreted (RANTES), and interferon-inducing protein-10 (64, 67, 86, 88, 89). However, the protection associated with IL-12 may predominantly be mediated by its induction of IFN-γ. Neutralization of endogenous IFN-γ production in mice with pulmonary cryptococcosis using anti-IFN-γ-specific monoclonal antibody (MAb) completely abrogated the positive effect of IL-12 on survival, pulmonary fungal burden, pulmonary inflammatory leukocyte infiltration, and local chemokine production (86).

IL-18 plays an important role in the development of protective immunity against pulmonary cryptococcosis. Pulmonary fungal clearance, serum IL-12 and IFN-γ levels, and cryptococcal-specific DTH responses are reduced in IL-18-deficient mice compared to infected wild-type controls (84). Administration of recombinant IL-18 almost completely restores the reduced responses in IL-18-deficient mice. IL-12 and IL-18 appear to act synergistically in protecting mice against experimental pulmonary cryptococcosis by enhancing local production of IFN-γ (85, 87, 88, 144). Interestingly, studies have shown that the protective effect of IL-18 against pulmonary cryptococcal infection was mediated not through inducing the development of Th1 cells, but by strengthening IL-12-induced Th1 cell development under normal conditions (105). Moreover, the protective effect of IL-18 in IL-12-deficient mice was predominantly mediated by the induction of IFN-γ production by NK cells. Thus, IL-18 may contribute to protection against pulmonary *C. neoformans* infections by optimizing the development of Th1-type immune responses induced by IL-12 and by activating IFN-γ production by NK cells.

A number of other cytokines have also been demonstrated to be involved in the host response to *C. neoformans* infection. The requirement for IFN-α/β receptor signaling and IFN-β cytokines in promoting a protective Th1-type immune response was observed when mice deleted for these genes developed Th2-polarized responses including IL-4, IL-13, and IL-10 secretion, as well as decreased expression of IFN-γ and iNOS (8). IL-23, a member of the IL-12 cytokine family, has recently been shown to complement the more dominant role of IL-12 in protection against *C. neoformans* by promoting increased IL-17, IL-1β, IL-6, and CCL2 inflammatory gene expression and increased survival after *C. neoformans* infection (94).

CMI AGAINST PULMONARY *C. NEOFORMANS* INFECTION

Both CD4+ and CD8+ T-cell subsets contribute to host defense against *C. neoformans*. Experimentally, congenitally athymic (nude) mice, which are deficient in CMI, are more susceptible to experimental *C. neoformans* infection than their heterozygous thymus-containing littermates (17, 60). In fact, *C. neoformans* infection in nude mice is characterized by the absence of granulomas, which are dependent on T-cell function for formation and are necessary for efficient containment of infection and resolution of disease. Moreover, passive immunization with sensitized T cells from immunocompetent donors into SCID mice that lack T and B cells resulted in pulmonary clearance of cryptococci following intratracheal inoculation (78). Pulmonary clearance was significantly better in SCID mice that received immune T cells from the lungs of wild-type mice compared to those that received immune T cells from the spleen, indicating that T cells activated in the lungs have better effector cell function against pulmonary cryptococcosis (78). Mice depleted of CD4+ or CD8+ T cell subsets by the administration of antibody are more susceptible to *C. neoformans* infection and have reduced inflammatory cell responses to the infection (72, 77, 122–124).

CD4+ and CD8+ T cells appear to work in synergy with each other to generate protective inflammatory responses against cryptococcal infections in the lung (77). CD4+ T cells recruit and activate inflammatory cells, whereas CD8+ T cells primarily function to either lyse unactivated phagocytes containing cryptococci or activate them through the production of IFN-γ. Depletion of CD4+ T cells results in enhanced dissemination from the lungs to the central nervous system and reduced survival following pulmonary infection (124). Depletion of CD8+ T cells also leads to reduced survival, but the effect on organ fungal burden appears to be dependent on the virulence of the cryptococcal strain (122, 123). Depletion of CD8+ T cells in mice following pulmonary infection led to impaired fungal clearance and DTH responses, whereas infection with a highly virulent strain led to reduced survival but did not alter the fungal burden.

$CD4^+$ T-cell help is not required for the generation of $CD8^+$ T-cell-mediated responses during pulmonary *C. neoformans* infection (108). Recruitment, activation, and differentiation of $CD8^+$ T cells into IFN-γ-producing effector cells can occur independently of $CD4^+$ T cells. Although $CD4^+$ and $CD8^+$ T cells were required for optimal reduction in pulmonary fungal burden, these studies suggested that $CD8^+$ T cells may play a protective role against pulmonary cryptococcosis in $CD4^+$ T-cell-deficient individuals. Studies showing that $CD4^+$ T cells are dispensable for vaccine-induced $CD8^+$ T-cell-mediated resistance against experimental *Blastomyces dermatitidis* and *Histoplasma capsulatum* pulmonary infections support this claim (170).

HUMORAL ADAPTIVE IMMUNITY AGAINST PULMONARY *C. NEOFORMANS* INFECTION

While CMI and macrophage activation make up the primary effector arm of adaptive immunity in controlling and clearing *C. neoformans* from the lungs, pulmonary infection does stimulate the production of anticryptococcal antibodies including antibodies against the capsular polysaccharide GXM (15). Anticryptococcal antibodies can enhance clearance or exacerbate disease in immunocompetent mice, depending on the isotype, magnitude, and epitope binding of the specific antibodies. For example, in rats pulmonary granuloma formation coincides with the appearance of opsonic serum antibody, suggesting a temporal relationship between humoral immunity and regulation of cell-mediated immunity/inflammation (55), and chronic infection in this model is associated with downregulation of both arms of adaptive immunity to *C. neoformans* (56). Pulmonary *C. neoformans* infection in mice elicits antibodies to both the capsular polysaccharide and protein antigens, including Hsp70 (81, 82). Although the mechanism of antibody action in vivo is not well characterized, one correlation has been made between the effectiveness of the antibody and the fluorescence pattern produced in vitro when the antibody is bound to the capsule. Protective anticapsular antibodies exhibit annular binding patterns by immunofluorescence, while nonprotective antibodies have punctate patterns (24, 126, 137, 173). Although both immunoglobulin M (IgM) and IgG were produced in response to infection, staining of yeast cells in alveolar spaces revealed only IgG, suggesting that only this isotype crossed into the alveolar space.

Overall, the role of the antibody response in immunity to *C. neoformans* pulmonary infection is not clear. In rats and humans, circumstantial evidence supports an association between this response and resistance to infection, which correlates with the fact that these species are relatively resistant to cryptococcosis. In contrast, the antibody response does not appear to promote clearance of the pulmonary cryptococcal infection in mice, which is a highly susceptible species. However, it is intriguing to note that between mouse strains, a genetic difference at the immunoglobulin heavy-chain locus between C.B-17 and BALB/c mice was associated with increased pulmonary clearance and a lack of a transient Th2 response (111).

POLARIZATION OF ADAPTIVE IMMUNE RESPONSES TO PULMONARY *C. NEOFORMANS* INFECTION

Experimental and clinical studies strongly suggest that CMI by Th1-type $CD4^+$ T cells is the protective host immune response to *C. neoformans* infection. Clinically, individuals with decreased CMI (i.e., individuals with HIV infection [23, 33, 36, 46, 96], lymphoproliferative disorders [27, 51, 93, 106], or undergoing extensive corticosteroid therapy [31, 35, 57]) have a high susceptibility to *C. neoformans* infections. Experimentally, mouse strains that are genetically more resistant to experimental pulmonary cryptococcosis, CBA/J, BALB/c, SJL/J, and 129/J mice, typically develop Th1-type $CD4^+$ T-cell-mediated immune responses. In comparison, C57BL/6 and C3H mice that are susceptible to pulmonary *C. neoformans* infection generate Th2-type $CD4^+$ T-cell-mediated immune responses. Th1-type $CD4^+$ T cells mediate protective anticryptococcal immune responses through the generation of Th1-type cytokine responses characterized by the production of IL-2, IL-12, IL-23, TNF-α, and IFN-γ. These cytokines, in turn, induce lymphocyte and phagocyte recruitment and activation of anticryptococcal DTH responses resulting in increased cryptococcal uptake and killing by effector phagocytes (1, 26, 44, 71, 99, 100, 125). Neutralization of TNF-α, IL-12, or IFN-γ in mice results in increased pulmonary fungal burden and prevents the induction of protective pulmonary Th1-type responses (29, 64, 86). Studies of mice and humans have shown some efficacy in using systemically administered recombinant Th1-type cytokines to stimulate anticryptococcal host responses and to enhance antifungal chemotherapy (25, 59, 80, 88, 91).

The development of protective immune responses in the lung against *C. neoformans* is dependent on the production of IFN-γ. Neutralization of IFN-γ in mice with anti-IFN-γ-specific antibody impaired the recruitment of inflammatory cells to the lung and promoted expansive fungal growth and pulmonary tissue destruction (90). In contrast, administration of recombinant IFN-γ to *C. neoformans*–infected mice resulted in increased survival and reduced pulmonary fungal burden. Pulmonary infection with a *C. neoformans* strain engineered to produce IFN-γ resulted in the development of a polarized Th1-type cell-mediated immune response and clearance of the fungal infection in mice (168). Mice lacking the IFN-γ receptor gene showed reduced survival and increased pulmonary fungal burden following pulmonary *C. neoformans* infection compared to infected wild-type mice (20). Interestingly, inflammatory cell recruitment to the lungs of mice lacking the IFN-γ receptor was not defective, but the increased number of cryptococci in the lungs suggested that macrophage antimicrobial function is impaired in the absence of IFN-γ signaling. Mice given a pulmonary infection with an IFN-γ-producing *C. neoformans* strain developed a polarized Th1-type cytokine response, increased neutrophil and T-cell recruitment, and resolved the acute infection (168). Moreover, these studies showed that mice immunized with the IFN-γ-producing strain were capable of completely resolving a subsequent pulmonary challenge with a highly virulent *C. neoformans* strain. These studies confirm the need for Th1-type CMI in the establishment of protective immunity against pulmonary *C. neoformans* infections. The chemokines MCP-1 and MIP-1α and chemokine receptor CCR2 are also important in promoting the development of protective anticryptococcal Th1-type cytokine responses, the prevention of eosinophilia, and the recruitment of phagocytes and $CD4^+$ T cells to the lungs in response to cryptococcal infection (75, 83, 139, 141, 155, 156).

Nonprotective Th2-type $CD4^+$ T-cell responses during experimental pulmonary cryptococcosis in mice are characterized by the production of IL-4, IL-5, IL-10, IL-13, high-serum IgE, and pulmonary eosinophilia (21, 62, 68, 70, 111). The development of a nonprotective host response to

C. neoformans infection can be characterized by pulmonary eosinophilia in certain mouse strains (70). CBA/J (highly resistant), BALB/c (moderately resistant), and C57BL/6 (susceptible) mice displayed very different patterns of eosinophil infiltrate in the lungs after *C. neoformans* infection. C57BL/6 mice developed chronic infection characterized by long-term eosinophil infiltrate into the lungs, resulting in a nonprotective immunopathology typically associated with a Th2-type immune response. The importance of these findings is underscored by the recent observation of a disseminated chronic cryptococcal infection in an apparently healthy individual with eosinophilia (171). IL-5 production by CD4$^+$ T cells appeared to be responsible for the pulmonary eosinophil infiltration and eosinophil YM1 crystal disposition (70). Decreased pulmonary Th2-type cytokine expression and the absence of pulmonary eosinophilia were observed in IL-4$^{(-/-)}$ and IL-10$^{(-/-)}$ mice compared to infected C57BL/6 parental control mice during pulmonary *C. neoformans* infection (62). Importantly, these studies demonstrated that although IL-4$^{(-/-)}$ and IL-10$^{(-/-)}$ mice produce fewer pulmonary Th2-type cytokines, increases in TNF-α and IL-12 production by lung leukocytes were observed in IL-10$^{(-/-)}$, but not IL-4$^{(-/-)}$, mice, suggesting that IL-10 plays the greater role directing Th2-type cytokine responses to *C. neoformans* in the lung. These data suggest that in both murine models and human infections, eosinophilia is a characteristic of chronic *C. neoformans* infection and is likely a contributing factor to the nonprotective Th2-type adaptive immune response.

Much less is known concerning the influence of the Th17 pathway (IL-23/IL-17) on the development of protective anti–*C. neoformans* immune responses in the lung. IL-23 is a member of the IL-12 cytokine family and is composed of a p40 subunit of IL-12 and a p19 subunit (140). Studies have shown that the IL-23/IL-17 developmental pathway promotes inflammation and susceptibility to *C. albicans* and *A. fumigatus* infections (174). In contrast, studies indicate that the IL-23/IL-17 pathway participates in host defense against *Pneumocystis carinii* infection (149). Although the role of IL-23 in pulmonary anticryptococcal immune responses has yet to be elucidated, IL-23 has been shown to enhance inflammatory responses and supplement a more dominant role of IL-12 following intraperitoneal and intravenous infection with *C. neoformans*. IL-23p19$^{(-/-)}$ mice were not observed to have an alteration in the Th1/Th2 type balance but significantly impaired IL-17 production, suggesting a potential role for IL-17. Thus, IL-17 may play an efficacious role in protective immune responses against pulmonary cryptococcosis. Significant amounts of IL-17 were produced in the lungs of BALB/c mice that resolved an experimental pulmonary infection with an IFN-γ-producing *C. neoformans* strain compared to wild-type infected mice that experienced a progressive infection (168).

While regulatory T-cell responses have yet to be investigated in depth during pulmonary *C. neoformans* infections, γ/δ T cells have been reported to accumulate in the lungs of *C. neoformans*–infected mice in a manner different from NK and NKT cells and play a down-modulatory role in the development of a Th1 response and host resistance against this organism (158).

EFFECTOR MECHANISMS OF ADAPTIVE IMMUNITY IN THE LUNGS

Phagocytosis and intracellular killing by "classically activated" macrophages is the predominant mechanism of anticryptococcal effector function driven by adaptive immunity, although others (described below) are also likely operational during the infection. As described earlier, *C. neoformans* can persist in the lungs as both an extracellular pathogen within the alveolar spaces and as an intracellular pathogen within phagocytes, so both CD4$^+$ and CD8$^+$ T cells are activated during pulmonary cryptococcal infection. Many cytokines produced during a protective response to the infection can contribute to the acquisition of effector function by recruited macrophages. However, of all the cytokines produced during the effector phase, IFN-γ is the most important because it is a potent activator of iNOS and subsequent induction of classically activated macrophages during pulmonary *C. neoformans* infections (5, 54, 110). The induction of protective cellular responses and fungal clearance is associated with increased iNOS expression following pulmonary *C. neoformans* infection in rats and mice (54, 56, 110). Blocking of either IFN-γ (using anti-IFN-γ-neutralizing antibodies) or NO synthesis (using N^G-monomethyl-L-arginine) impairs fungal clearance. However, pulmonary inflammatory cell recruitment remains intact, further suggesting that IFN-γ induction of iNOS is required for optimal inflammatory effector cell function. Altogether, these studies show that IFN-γ production during pulmonary cryptococcosis is associated with the development of protective Th1-type cell-mediated immune responses against infection.

The induction of alternative macrophage activation by Th2-type cytokines, namely IL-4 and IL-13, is also associated with progressive pulmonary cryptococcal infection (5, 30). Alternatively activated macrophages produce arginase, which decreases NO concentration levels by competing with iNOS for the substrate L-arginine (65). Alternatively activated macrophages also have increased macrophage MR, which increases phagocytosis but decreases intracellular killing and TNF-α production (5, 53, 58, 153). Pulmonary *C. neoformans* infection in C57BL/6 mice is partly characterized by enhanced alternative macrophage activation in association with nonprotective Th2-type cytokine responses and progressive pulmonary cryptococcal infection (5). Augmented Th2-type production and the generation of alternatively activated macrophages were also observed in C57BL/6 mice deficient in IFN-γ. Moreover, IL-13 contributed to the induction of alternative macrophage activation, allergic inflammation, and Th2-type cytokine responses during experimental pulmonary cryptococcosis in mice. Thus, the induction of macrophage activation by Th2 cytokines has a clear role in the development of resistance or susceptibility to pulmonary cryptococcosis (129).

Anticryptococcal antibody is likely to enhance both effector mechanisms and immunoregulation in the lungs, but there is often a dichotomy between the two. For example, delivery of an anti-GXM IgG1 MAb prior to intratracheal infection in A/J and C57BL/6 mice significantly enhanced survival without a concomitant decrease in pulmonary fungal burden (37, 38). The MAb-treated mice had a markedly different inflammatory response compared to isotype-control-treated mice, noted by a more localized inflammatory process. These MAbs are potent opsonins for *C. neoformans*, and the addition of this MAb to suspensions of macrophages and yeast cells could enhance killing of yeast cells by macrophages (127, 128). Despite the histological differences associated with the inflammatory response to *C. neoformans* in the presence and absence of exogenous IgG1 MAb, fluorescence-activated cell sorter analysis of lung homogenates revealed no significant differences in the number of leukocytes in MAb-treated and control mice (39). However, MAb-treated mice had more granulocytes at day 14, higher macrophage CD86 surface expression on day 28, higher amounts of IL-10 at day 7, and potentially lower amounts of IFN-γ and IL-4 at the same time point. Subsequent analysis of inflammatory responses in IgG1 anti-GXM MAb-treated iNOS-sufficient and -deficient C57BL/6 mice indicated that NO is involved in the regulation of cytokine expression in response to *C. neoformans* infection in the lungs and is necessary for antibody efficacy against this microbe in mice (146). Overall, it is likely that anti-GXM antibody can protect mice by reducing immune-mediated damage to the lung, which can promote a more effective inflammatory response that controls the infection and reduces mortality (16).

In addition to phagocytes, T cells may inhibit *C. neoformans* through a variety of mechanisms. Human CD4$^+$ and CD8$^+$ T cells have been shown to directly inhibit the replication of *C. neoformans* in vitro via a mechanism not associated with phagocytosis (101, 102, 104, 130). Subsequently, in vitro studies by Muth et al. demonstrated that T lymphocytes from immunized mice were able to kill *C. neoformans* via a mechanism that required direct cell contact (133). While the exact mechanisms by which CD4$^+$ T cells may directly inhibit *C. neoformans* are unknown, CD4$^+$ T cells are certainly the critical mediators of protective host responses against *C. neoformans* infection. Recent studies have suggested that CD4$^+$ T cells isolated from the peripheral blood of humans use granulysin, a saponin-like protein that interacts with cell membrane lipids and activates lipid-degrading enzymes (142, 159), to kill *C. neoformans* (175). The acquisition of CD4$^+$ cell effector function during pulmonary cryptococcal infection occurs in lung tissues and not the draining lymph nodes (107). Following pulmonary *C. neoformans* infection, CD4$^+$ T cells expand and proliferate in the lung-associated lymph nodes and then traffic to pulmonary tissues. Upon trafficking to the lungs, CD4$^+$ T cells lose their proliferative ability, upregulate surface activation markers, and gain the ability to produce IFN-γ.

Most of the effector cell function of CD4$^+$ T cells appears to be mediated through cytokine production. Cytokine production by Th1-type CD4$^+$ T cells induces phagocyte recruitment and activation resulting in increased cryptococcal uptake and killing (1, 26, 44, 70, 99, 100, 125). Neutralization of TNF-α, IFN-γ, or IL-12 in infected mice prevents the development of a protective Th1-type cytokine response and reduces pulmonary clearance (64, 67, 76). CD4$^+$ T cells are required for multinucleated giant cell formation around cryptococci within granulomas in the lung that surround and confine well-encapsulated yeast cells (66). Therefore, the generation of Th1-type CD4$^+$ T cells is essential for the long-term resolution of cryptococcal infections.

Studies have shown that CD8$^+$ T cells can mediate direct killing of *C. neoformans* by the granule exocytosis of granulysin (114). The granulysin-mediated antifungal effect was shown to be dependent on activation by IL-15. IL-15 is produced by accessory cells that are in turn activated by CD4$^+$ T cells. CD8$^+$ T cells may also be required for the optimal recruitment of CD4$^+$ T cells to the lungs during pulmonary cryptococcosis. The absence of CD8$^+$ T cells was shown to result in the generation of a predominant nonprotective Th2-type (IL-4, IL-5, and IL-10) CD4$^+$ T cell cytokine response to pulmonary *C. neoformans* infection (72). Effector CD8$^+$ T-cell function against pulmonary *C. neoformans* infection can occur in the absence of CD4$^+$ T cells. Depletion of CD8$^+$ T cells in CD4-deficient mice resulted in increases in pulmonary fungal burden, intracellular cryptococcal growth within macrophages, and the surface expression of activation markers. Altogether, CD8$^+$ T cells appear to have a key, if not absolute, role in promoting protective Th1-type CD4$^+$ T cell host immunity in response to *C. neoformans* infection.

SUMMARY

Pulmonary host defense against *C. neoformans* exposure is a coordinated process marked by a transition from innate responses to adaptive responses. The immune response to this organism is flexible and involves both cellular and humoral factors. However, environmental, genetic, or cryptococcal factors can exert pressures on this flexible program of host defense, leading to chronic and/or progressive infections, including disseminated disease. Experimental models have been critically useful in studying host defense against this organism. Variables such as infectious dose, route of infection, method of infection, animal species, animal strain, and *Cryptococcus* strain can all affect the outcome of the experiments. Thus, this poses both technical hurdles and tremendous opportunities for studies of the pathogenesis of this organism. The recent identification of and rise in *Cryptococcus*-associated inflammatory reconstitution disease cases in humans further illustrates the predictive power of these diverse animal models because chronic cryptococcal inflammatory disease in rodent models was already being investigated prior to the identification of the first cases of this human disease.

REFERENCES

1. **Aguirre, K., E. A. Havell, G. W. Gibson, and L. L. Johnson.** 1995. Role of tumor necrosis factor and gamma interferon in acquired resistance to *Cryptococcus neoformans* in the central nervous system of mice. *Infect. Immun.* **63:**1725–1731.
2. **Allen, M., D. Voelker, and R. Mason.** 2001. Interactions of surfactant proteins A and D with *Saccharomyces cerevisiae* and *Aspergillus fumigatus*. *Infect. Immun.* **69:**2037–2044.
3. **Alvarez, M., and A. Casadevall.** 2006. Phagosome extrusion and host-cell survival after *Cryptococcus neoformans* phagocytosis by macrophages. *Curr. Biol.* **16:**2161–2165.
4. **Alvarez, M., C. Saylor, and A. Casadevall.** 2008. Antibody action after phagocytosis promotes *Cryptococcus neoformans* and *Cryptococcus gattii* macrophage exocytosis with biofilm-like microcolony formation. *Cell. Microbiol.* **10:**1622–1633.
5. **Arora, S., Y. Hernandez, J. Erb-Downward, R. McDonald, G. Toews, and G. Huffnagle.** 2005. Role of IFN-gamma in regulating T2 immunity and the development of alternatively activated macrophages during allergic bronchopulmonary mycosis. *J. Immunol.* **174:**6346–6356.
6. **Atochina, E., A. Gow, J. Beck, A. Haczku, A. Inch, H. Kadire, Y. Tomer, C. Davis, A. Preston, F. Poulain, S. Hawgood, and M. Beers.** 2004. Delayed clearance of *Pneumocystis carinii* infection, increased inflammation, and altered nitric oxide metabolism in lungs of surfactant protein-D knockout mice. *J. Infect. Dis.* **189:**1528–1539.
7. **Bauman, S. K., G. B. Huffnagle, and J. W. Murphy.** 2003. Effects of tumor necrosis factor alpha on dendritic cell accumulation in lymph nodes draining the immunization site and the impact on the anticryptococcal cell-mediated immune response. *Infect. Immun.* **71:**68–74.
8. **Biondo, C., A. Midiri, M. Gambuzza, E. Gerace, M. Falduto, R. Galbo, A. Bellantoni, C. Beninati, G. Teti, T. Leanderson, and G. Mancuso.** 2008. IFN-alpha/beta signaling is required for polarization of cytokine responses toward a protective type 1 pattern during experimental cryptococcosis. *J. Immunol.* **181:**566–573.
9. **Biondo, C., A. Midiri, L. Messina, F. Tomasello, G. Garufi, M. R. Catania, M. Bombaci, C. Beninati, G. Teti, and G. Mancuso.** 2005. MyD88 and TLR2, but not TLR4,

are required for host defense against *Cryptococcus neoformans*. *Eur. J. Immunol.* **35:**870–878.

10. **Bronte, V., E. Apolloni, A. Cabrelle, R. Ronca, P. Serafini, P. Zamboni, N. P. Restifo, and P. Zanovello.** 2000. Identification of a CD11b(+)/Gr-1(+)/CD31(+) myeloid progenitor capable of activating or suppressing CD8(+) T cells. *Blood* **96:**3838–3846.
11. **Brummer, E., and D. Stevens.** 1994. Anticryptococcal activity of macrophages: role of mouse strain, C5, contact, phagocytosis, and L-arginine. *Cell Immunol.* **157:**1–10.
12. **Bulmer, G. S., and M. D. Sans.** 1967. *Cryptococcus neoformans*. II. Phagocytosis by human leukocytes. *J. Bacteriol.* **94:**1480–1483.
13. **Cambi, A., and C. G. Figdor.** 2003. Dual function of C-type lectin-like receptors in the immune system. *Curr. Opin. Cell Biol.* **15:**539–546.
14. **Cameron, M. L., D. L. Granger, J. B. Weinberg, W. J. Kozumbo, and H. S. Koren.** 1990. Human alveolar and peritoneal macrophages mediate fungistasis independently of L-arginine oxidation to nitrite or nitrate. *Am. Rev. Respir. Dis.* **142:**1313–1319.
15. **Casadevall, A., W. Cleare, M. Feldmesser, A. Glatman-Freedman, D. L. Goldman, T. R. Kozel, N. Lendvai, J. Mukherjee, L. A. Pirofski, J. Rivera, A. L. Rosas, M. D. Scharff, P. Valadon, K. Westin, and Z. Zhong.** 1998. Characterization of a murine monoclonal antibody to *Cryptococcus neoformans* polysaccharide that is a candidate for human therapeutic studies. *Antimicrob. Agents Chemother.* **42:**1437–1446.
16. **Casadevall, A., and L. A. Pirofski.** 2003. The damage-response framework of microbial pathogenesis. *Nat. Rev. Microbiol.* **1:**17–24.
17. **Cauley, L. K., and J. W. Murphy.** 1979. Response of congenitally athymic (nude) and phenotypically normal mice to *Cryptococcus neoformans* infection. *Infect. Immun.* **23:**644–651.
18. **Charlier, C., K. Nielsen, S. Daou, M. Brigitte, F. Chretien, and F. Dromer.** 2008. Evidence for a role of monocytes in dissemination and brain invasion by *Cryptococcus neoformans*. *Infect. Immun.* **77:**120–127.
19. **Chen, G. H., J. L. Curtis, C. H. Mody, P. J. Christensen, L. R. Armstrong, and G. B. Toews.** 1994. Effect of granulocyte-macrophage colony-stimulating factor on rat alveolar macrophage anticryptococcal activity in vitro. *J. Immunol.* **152:**724–734.
20. **Chen, G. H., R. A. McDonald, J. C. Wells, G. B. Huffnagle, N. W. Lukacs, and G. B. Toews.** 2005. The gamma interferon receptor is required for the protective pulmonary inflammatory response to *Cryptococcus neoformans*. *Infect. Immun.* **73:**1788–1796.
21. **Chen, G. H., D. A. McNamara, Y. Hernandez, G. B. Huffnagle, G. B. Toews, and M. A. Olszewski.** 2008. Inheritance of immune polarization patterns is linked to resistance versus susceptibility to *Cryptococcus neoformans* in a mouse model. *Infect. Immun.* **76:**2379–2391.
22. **Chen, Y. C., S. C. Chang, C. C. Shih, C. C. Hung, K. T. Luhd, Y. S. Pan, and W. C. Hsieh.** 2000. Clinical features and in vitro susceptibilities of two varieties of *Cryptococcus neoformans* in Taiwan. *Diagn. Microbiol. Infect. Dis.* **36:**175–183.
23. **Chuck, S. L., and M. A. Sande.** 1989. Infections with *Cryptococcus neoformans* in the acquired immunodeficiency syndrome. *N. Engl. J. Med.* **321:**794–799.
24. **Cleare, W., and A. Casadevall.** 1998. The different binding patterns of two immunoglobulin M monoclonal antibodies to *Cryptococcus neoformans* serotype A and D strains correlate with serotype classification and differences in functional assays. *Clin. Diagn. Lab. Immunol.* **5:**125–129.
25. **Clemons, K. V., E. Brummer, and D. A. Stevens.** 1994. Cytokine treatment of central nervous system infection: efficacy of interleukin-12 alone and synergy with conventional antifungal therapy in experimental cryptococcosis. *Antimicrob. Agents Chemother.* **38:**460–464.
26. **Collins, H. L., and G. J. Bancroft.** 1992. Cytokine enhancement of complement-dependent phagocytosis by macrophages: synergy of tumor necrosis factor-alpha and granulocyte-macrophage colony-stimulating factor for phagocytosis of *Cryptococcus neoformans*. *Eur. J. Immunol.* **22:**1447–1454.
27. **Collins, V. P., A. Gellhorn, and J. R. Trimble.** 1995. The coincidence of cryptococcosis and disease of the reticulo-endothelial and lymphatic systems. *Cancer* **4:**883–889.
28. **Dan, J. M., R. M. Kelly, C. K. Lee, and S. M. Levitz.** 2008. Role of the mannose receptor in a murine model of *Cryptococcus neoformans* infection. *Infect. Immun.* **76:**2362–2367.
29. **Decken, K., G. Kohler, K. Palmer-Lehmann, A. Wunderlin, F. Mattner, J. Magram, M. K. Gately, and G. Alber.** 1998. Interleukin-12 is essential for a protective Th1 response in mice infected with *Cryptococcus neoformans*. *Infect. Immun.* **66:**4994–5000.
30. **De Groot, P. W., A. F. Ram, and F. M. Klis.** 2005. Features and functions of covalently linked proteins in fungal cell walls. *Fungal Genet. Biol.* **42:**657–675.
31. **Diamond, R. D., and J. E. Bennett.** 1974. Prognostic factors in cryptococcal meningitis. *Ann. Intern. Med.* **80:**176–181.
32. **Diamond, R. D., R. K. Root, and J. E. Bennett.** 1972. Factors influencing killing of *Cryptococcus neoformans* by human leukocytes in vitro. *J. Infect. Dis.* **125:**367–376.
33. **Dismukes, W. E.** 1988. *Cryptococcal meningitis* in patients with AIDS. *J. Infect. Dis.* **157:**624–628.
34. **Dong, Z. M., and J. W. Murphy.** 1997. Cryptococcal polysaccharides bind to CD18 on human neutrophils. *Infect. Immun.* **65:**557–563.
35. **Duperval, R., P. E. Hermans, N. S. Brewer, and G. S. Roberts.** 1977. Cryptococcosis, with emphasis on the significance of isolation of *Cryptococcus neoformans* from the respiratory tract. *Chest* **72:**13–19.
36. **Eng, R. H., E. Bishburg, S. M. Smith, and R. Kapila.** 1986. Cryptococcal infections in patients with acquired immune deficiency syndrome. *Am. J. Med.* **81:**19–23.
37. **Feldmesser, M., and A. Casadevall.** 1997. Effect of serum IgG1 to *Cryptococcus neoformans* glucuronoxylomannan on murine pulmonary infection. *J. Immunol.* **158:**790–799.
38. **Feldmesser, M., and A. Casadevall.** 1998. Mechanism of action of antibody to capsular polysaccharide in *Cryptococcus neoformans* infection. *Front Biosci.* **3:**d136–d151.
39. **Feldmesser, M., A. Mednick, and A. Casadevall.** 2002. Antibody-mediated protection in murine *Cryptococcus neoformans* infection is associated with pleotrophic effects on cytokine and leukocyte responses. *Infect. Immun.* **70:**1571–1580.
40. **Feldmesser, M., S. Tucker, and A. Casadevall.** 2001. Intracellular parasitism of macrophages by *Cryptococcus neoformans*. *Trends Microbiol.* **9:**273–278.
41. **Fingeroth, J., M. Benedict, D. Levy, and J. Strominger.** 1989. Identification of murine complement receptor type 2. *Proc. Natl. Acad. Sci. USA* **86:**242–246.
42. **Fingeroth, J., M. Heath, and D. Ambrosino.** 1989. Proliferation of resting B cells is modulated by CR2 and CR1. *Immunol. Lett.* **21:**291–301.
43. **Fisher, D., J. Burrow, D. Lo, and B. Currie.** 1993. *Cryptococcus neoformans* in tropical northern Australia: predominantly variant *gattii* with good outcomes. *Aust. N. Z. J. Med.* **23:**678–682.
44. **Flesch, I. E., G. Schwamberger, and S. H. Kaufmann.** 1989. Fungicidal activity of IFN-gamma-activated macrophages

extracellular killing of *Cryptococcus neoformans*. *J. Immunol.* **142:**3219–3224.

45. **Frade, J. M., M. Mellado, G. del Real, J. C. Gutierrez-Ramos, P. Lind, and A. C. Martinez.** 1997. Characterization of the CCR2 chemokine receptor: functional CCR2 receptor expression in B cells. *J. Immunol.* **159:**5576–5584.
46. **Gal, A. A., M. N. Koss, J. Hawkins, S. Evans, and H. Einstein.** 1986. The pathology of pulmonary cryptococcal infections in the acquired immunodeficiency syndrome. *Arch. Pathol. Lab. Med.* **110:**502–507.
47. **Garcia-Hermoso, D., F. Dromer, and G. Janbon.** 2004. *Cryptococcus neoformans* capsule structure evolution in vitro and during murine infection. *Infect. Immun.* **72:**3359–3365.
48. **Garcia-Hermoso, D., F. Dromer, S. Mathoulin-Pelissier, and G. Janbon.** 2001. Are two *Cryptococcus neoformans* strains epidemiologically linked? *J. Clin. Microbiol.* **39:** 1402–1406.
49. **Garcia-Hermoso, D., G. Janbon, and F. Dromer.** 1999. Epidemiological evidence for dormant *Cryptococcus neoformans* infection. *J. Clin. Microbiol.* **37:**3204–3209.
50. **Geissmann, F., S. Jung, and D. R. Littman.** 2003. Blood monocytes consist of two principal subsets with distinct migratory properties. *Immunity* **19:**71–82.
51. **Gendel, B. R., M. Ende, and S. L. Norman.** 1950. Cryptococcosis: a review with special reference to apparent association with Hodgkin's disease. *Am. J. Med.* **9:**343–355.
52. **Giles, S., A. Zaas, M. Reidy, J. Perfect, and J. Wright.** 2007. *Cryptococcus neoformans* is resistant to surfactant protein A mediated host defense mechanisms. *PLoS ONE* **2:**e1370.
53. **Goerdt, S., O. Politz, K. Schledzewski, R. Birk, A. Gratchev, P. Guillot, N. Hakiy, C. D. Klemke, E. Dippel, V. Kodelja, and C. E. Orfanos.** 1999. Alternative versus classical activation of macrophages. *Pathobiology* **67:**222–226.
54. **Goldman, D., Y. Cho, M. Zhao, A. Casadevall, and S. C. Lee.** 1996. Expression of inducible nitric oxide synthase in rat pulmonary *Cryptococcus neoformans* granulomas. *Am. J. Pathol.* **148:**1275–1282.
55. **Goldman, D., S. C. Lee, and A. Casadevall.** 1994. Pathogenesis of pulmonary *Cryptococcus neoformans* infection in the rat. *Infect. Immun.* **62:**4755–4761.
56. **Goldman, D. L., S. C. Lee, A. J. Mednick, L. Montella, and A. Casadevall.** 2000. Persistent *Cryptococcus neoformans* pulmonary infection in the rat is associated with intracellular parasitism, decreased inducible nitric oxide synthase expression, and altered antibody responsiveness to cryptococcal polysaccharide. *Infect. Immun.* **68:**832–838.
57. **Goldstein, E., and O. N. Rambo.** 1962. Cryptococcal infection following steroid therapy. *Ann. Intern. Med.* **56:**114–120.
58. **Gordon, S.** 2003. Alternative activation of macrophages. *Nat. Rev. Immunol.* **3:**23–35.
59. **Graybill, J. R., R. Bocanegra, C. Lambros, and M. F. Luther.** 1997. Granulocyte colony stimulating factor therapy of experimental cryptococcal meningitis. *J. Med. Vet. Mycol.* **35:**243–247.
60. **Graybill, J. R., and D. J. Drutz.** 1978. Host defense in cryptococcosis. II. Cryptococcosis in the nude mouse. *Cell. Immunol.* **40:**263–274.
61. **Gross, O., A. Gewies, K. Finger, M. Schafer, T. Sparwasser, C. Peschel, I. Forster, and J. Ruland.** 2006. Card9 controls a non-TLR signalling pathway for innate anti-fungal immunity. *Nature* **442:**651–656.
62. **Hernandez, Y., S. Arora, J. R. Erb-Downward, R. A. McDonald, G. B. Toews, and G. B. Huffnagle.** 2005. Distinct roles for IL-4 and IL-10 in regulating T2 immunity during allergic bronchopulmonary mycosis. *J. Immunol.* **174:**1027–1036.
63. **Herring, A. C., N. R. Falkowski, G. H. Chen, R. A. McDonald, G. B. Toews, and G. B. Huffnagle.** 2005. Transient neutralization of tumor necrosis factor alpha can produce a chronic fungal infection in an immunocompetent host: potential role of immature dendritic cells. *Infect. Immun.* **73:**39–49.
64. **Herring, A. C., J. Lee, R. A. McDonald, G. B. Toews, and G. B. Huffnagle.** 2002. Induction of interleukin-12 and gamma interferon requires tumor necrosis factor alpha for protective T1-cell-mediated immunity to pulmonary *Cryptococcus neoformans* infection. *Infect. Immun.* **70:**2959–2964.
65. **Hesse, M., M. Modolell, A. C. La Flamme, M. Schito, J. M. Fuentes, A. W. Cheever, E. J. Pearce, and T. A. Wynn.** 2001. Differential regulation of nitric oxide synthase-2 and arginase-1 by type 1/type 2 cytokines in vivo: granulomatous pathology is shaped by the pattern of L-arginine metabolism. *J. Immunol.* **167:**6533–6544.
66. **Hill, J. O.** 1992. CD4+ T cells cause multinucleated giant cells to form around *Cryptococcus neoformans* and confine the yeast within the primary site of infection in the respiratory tract. *J. Exp. Med.* **175:**1685–1695.
67. **Hoag, K. A., M. F. Lipscomb, A. A. Izzo, and N. E. Street.** 1997. IL-12 and IFN-gamma are required for initiating the protective Th1 response to pulmonary cryptococcosis in resistant C B-17 mice. *Am. J. Respir. Cell. Mol. Biol.* **17:**733–739.
68. **Hoag, K. A., N. E. Street, G. B. Huffnagle, and M. F. Lipscomb.** 1995. Early cytokine production in pulmonary *Cryptococcus neoformans* infections distinguishes susceptible and resistant mice. *Am. J. Respir. Cell. Mol. Biol.* **13:**487–495.
69. **Huang, C., S. H. Nong, M. K. Mansour, C. A. Specht, and S. M. Levitz.** 2002. Purification and characterization of a second immunoreactive mannoprotein from *Cryptococcus neoformans* that stimulates T-cell responses. *Infect. Immun.* **70:**5485–5493.
70. **Huffnagle, G. B., M. B. Boyd, N. E. Street, and M. F. Lipscomb.** 1998. IL-5 is required for eosinophil recruitment, crystal deposition, and mononuclear cell recruitment during a pulmonary *Cryptococcus neoformans* infection in genetically susceptible mice (C57BL/6). *J. Immunol.* **160:**2393–2400.
71. **Huffnagle, G. B., and M. F. Lipscomb.** 1998. Cells and cytokines in pulmonary cryptococcosis. *Res. Immunol.* **149:**387–396.
72. **Huffnagle, G. B., M. F. Lipscomb, J. A. Lovchik, K. A. Hoag, and N. E. Street.** 1994. The role of CD4+ and CD8+ T cells in the protective inflammatory response to a pulmonary cryptococcal infection. *J. Leukoc. Biol.* **55:**35–42.
73. **Huffnagle, G. B., L. K. McNeil, R. A. McDonald, J. W. Murphy, G. B. Toews, N. Maeda, and W. A. Kuziel.** 1999. Cutting edge: role of C-C chemokine receptor 5 in organ-specific and innate immunity to *Cryptococcus neoformans*. *J. Immunol.* **163:**4642–4646.
74. **Huffnagle, G. B., R. M. Strieter, L. K. McNeil, R. A. McDonald, M. D. Burdick, S. L. Kunkel, and G. B. Toews.** 1997. Macrophage inflammatory protein-1alpha (MIP-1alpha) is required for the efferent phase of pulmonary cell-mediated immunity to a *Cryptococcus neoformans* infection. *J. Immunol.* **159:**318–327.
75. **Huffnagle, G. B., R. M. Strieter, T. J. Standiford, R. A. McDonald, M. D. Burdick, S. L. Kunkel, and G. B. Toews.** 1995. The role of monocyte chemotactic protein-1 (MCP-1) in the recruitment of monocytes and CD4+ T cells during a pulmonary *Cryptococcus neoformans* infection. *J. Immunol.* **155:**4790–4797.
76. **Huffnagle, G. B., G. B. Toews, M. D. Burdick, M. B. Boyd, K. S. McAllister, R. A. McDonald, S. L. Kunkel,**

and R. M. Strieter. 1996. Afferent phase production of TNF-alpha is required for the development of protective T cell immunity to *Cryptococcus neoformans*. *J. Immunol.* **157:**4529–4536.

77. **Huffnagle, G. B., J. L. Yates, and M. F. Lipscomb.** 1991. Immunity to a pulmonary *Cryptococcus neoformans* infection requires both CD4+ and CD8+ T cells. *J. Exp. Med.* **173:**793–800.
78. **Huffnagle, G. B., J. L. Yates, and M. F. Lipscomb.** 1991. T cell-mediated immunity in the lung: a *Cryptococcus neoformans* pulmonary infection model using SCID and athymic nude mice. *Infect. Immun.* **59:**1423–1433.
79. **Ibrahim, A. S., S. G. Filler, M. S. Alcouloumre, T. R. Kozel, J. E. Edwards, Jr., and M. A. Ghannoum.** 1995. Adherence to and damage of endothelial cells by *Cryptococcus neoformans* in vitro: role of the capsule. *Infect. Immun.* **63:**4368–4374.
80. **Joly, V., L. Saint-Julien, C. Carbon, and P. Yeni.** 1994. In vivo activity of interferon-gamma in combination with amphotericin B in the treatment of experimental cryptococcosis. *J. Infect. Dis.* **170:**1331–1334.
81. **Kakeya, H., H. Udono, N. Ikuno, Y. Yamamoto, K. Mitsutake, T. Miyazaki, K. Tomono, H. Koga, T. Tashiro, E. Nakayama, and S. Kohno.** 1997. A 77-kilodalton protein of *Cryptococcus neoformans*, a member of the heat shock protein 70 family, is a major antigen detected in the sera of mice with pulmonary cryptococcosis. *Infect. Immun.* **65:**1653–1658.
82. **Kakeya, H., H. Udono, S. Maesaki, E. Sasaki, S. Kawamura, M. A. Hossain, Y. Yamamoto, T. Sawai, M. Fukuda, K. Mitsutake, Y. Miyazaki, K. Tomono, T. Tashiro, E. Nakayama, and S. Kohno.** 1999. Heat shock protein 70 (hsp70) as a major target of the antibody response in patients with pulmonary cryptococcosis. *Clin. Exp. Immunol.* **115:**485–490.
83. **Kawakami, K., Y. Kinjo, K. Uezu, S. Yara, K. Miyagi, Y. Koguchi, T. Nakayama, M. Taniguchi, and A. Saito.** 2001. Monocyte chemoattractant protein-1-dependent increase of V alpha 14 NKT cells in lungs and their roles in Th1 response and host defense in cryptococcal infection. *J. Immunol.* **167:**6525–6532.
84. **Kawakami, K., Y. Koguchi, M. H. Qureshi, Y. Kinjo, S. Yara, A. Miyazato, M. Kurimoto, K. Takeda, S. Akira, and A. Saito.** 2000. Reduced host resistance and Th1 response to *Cryptococcus neoformans* in interleukin-18 deficient mice. *FEMS Microbiol. Lett.* **186:**121–126.
85. **Kawakami, K., Y. Koguchi, M. H. Qureshi, A. Miyazato, S. Yara, Y. Kinjo, Y. Iwakura, K. Takeda, S. Akira, M. Kurimoto, and A. Saito.** 2000. IL-18 contributes to host resistance against infection with *Cryptococcus neoformans* in mice with defective IL-12 synthesis through induction of IFN-gamma production by NK cells. *J. Immunol.* **165:**941–947.
86. **Kawakami, K., M. H. Qureshi, T. Zhang, Y. Koguchi, K. Shibuya, S. Naoe, and A. Saito.** 1999. Interferon-gamma (IFN-gamma)-dependent protection and synthesis of chemoattractants for mononuclear leucocytes caused by IL-12 in the lungs of mice infected with *Cryptococcus neoformans*. *Clin. Exp. Immunol.* **117:**113–122.
87. **Kawakami, K., M. H. Qureshi, T. Zhang, Y. Koguchi, S. Yara, K. Takeda, S. Akira, M. Kurimoto, and A. Saito.** 2000. Involvement of endogenously synthesized interleukin (IL)-18 in the protective effects of IL-12 against pulmonary infection with *Cryptococcus neoformans* in mice. *FEMS Immunol. Med. Microbiol.* **27:**191–200.
88. **Kawakami, K., M. H. Qureshi, T. Zhang, H. Okamura, M. Kurimoto, and A. Saito.** 1997. IL-18 protects mice against pulmonary and disseminated infection with *Cryptococcus neoformans* by inducing IFN-gamma production. *J. Immunol.* **159:**5528–5534.
89. **Kawakami, K., M. Tohyama, X. Qifeng, and A. Saito.** 1997. Expression of cytokines and inducible nitric oxide synthase mRNA in the lungs of mice infected with *Cryptococcus neoformans*: effects of interleukin-12. *Infect. Immun.* **65:**1307–1312.
90. **Kawakami, K., M. Tohyama, K. Teruya, N. Kudeken, Q. Xie, and A. Saito.** 1996. Contribution of interferon-gamma in protecting mice during pulmonary and disseminated infection with *Cryptococcus neoformans*. *FEMS Immunol. Med. Microbiol.* **13:**123–130.
91. **Kawakami, K., M. Tohyama, Q. Xie, and A. Saito.** 1996. IL-12 protects mice against pulmonary and disseminated infection caused by *Cryptococcus neoformans*. *Clin. Exp. Immunol.* **104:**208–214.
92. **Kechichian, T. B., J. Shea, and M. Del Poeta.** 2007. Depletion of alveolar macrophages decreases the dissemination of a glucosylceramide-deficient mutant of *Cryptococcus neoformans* in immunodeficient mice. *Infect. Immun.* **75:**4792–4798.
93. **Keye, J. D., and W. E. Magee.** 1956. Fungal diseases in a general hospital. *Am. J. Clin. Pathol.* **26** 1235–1253.
94. **Kleinschek, M. A., U. Muller, S. J. Brodie, W. Stenzel, G. Kohler, W. M. Blumenschein, R. K. Straubinger, T. McClanahan, R. A. Kastelein, and G. Alber.** 2006. IL-23 enhances the inflammatory cell response in *Cryptococcus neoformans* infection and induces a cytokine pattern distinct from IL-12. *J. Immunol.* **176:**1098–1106.
95. **Kopp, E., and R. Medzhitov.** 2003. Recognition of microbial infection by Toll-like receptors. *Curr. Opin. Immunol.* **15:**396–401.
96. **Kovacs, J. A., A. A. Kovacs, M. Polis, W. C. Wright, V. J. Gill, C. U. Tuazon, E. P. Gelmann, H. C. Lane, R. Longfield, and G. Overturf.** 1985. Cryptococcosis in the acquired immunodeficiency syndrome. *Ann. Intern. Med.* **103:**533–538.
97. **Kozel, T. R., and J. Cazin, Jr.** 1971. Nonencapsulated variant of *Cryptococcus neoformans* I. Virulence studies and characterization of soluble polysaccharide. *Infect. Immun.* **3:**287–294.
98. **Kozel, T. R., B. Highison, and C. J. Stratton.** 1984. Localization on encapsulated *Cryptococcus neoformans* of serum components opsonic for phagocytosis by macrophages and neutrophils. *Infect. Immun.* **43:**574–579.
99. **Levitz, S. M.** 1991. Activation of human peripheral blood mononuclear cells by interleukin-2 and granulocyte-macrophage colony-stimulating factor to inhibit *Cryptococcus neoformans*. *Infect. Immun.* **59:**3393–3397.
100. **Levitz, S. M., and D. J. DiBenedetto.** 1988. Differential stimulation of murine resident peritoneal cells by selectively opsonized encapsulated and acapsular *Cryptococcus neoformans*. *Infect. Immun.* **56:**2544–2551.
101. **Levitz, S. M., and M. P. Dupont.** 1993. Phenotypic and functional characterization of human lymphocytes activated by interleukin-2 to directly inhibit growth of *Cryptococcus neoformans* in vitro. *J. Clin. Invest.* **91:**1490–1498.
102. **Levitz, S. M., M. P. Dupont, and E. H. Smail.** 1994. Direct activity of human T lymphocytes and natural killer cells against *Cryptococcus neoformans*. *Infect. Immun.* **62:**194–202.
103. **Levitz, S. M., S. Nong, M. K. Mansour, C. Huang, and C. A. Specht.** 2001. Molecular characterization of a mannoprotein with homology to chitin deacetylases that stimulates T cell responses to *Cryptococcus neoformans*. *Proc. Natl. Acad. Sci. USA* **98:**10422–10427.
104. **Levitz, S. M., E. A. North, M. P. Dupont, and T. S. Harrison.** 1995. Mechanisms of inhibition of *Cryptococcus neoformans* by human lymphocytes. *Infect. Immun.* **63:**3550–3554.

105. **Levitz, S. M., and C. A. Specht.** 2006. The molecular basis for the immunogenicity of *Cryptococcus neoformans* mannoproteins. *FEMS Yeast Res.* **6:**513–524.
106. **Lewis, J. L., and S. Rabinovich.** 1972. The wide spectrum of cryptococcal infections. *Am. J. Med.* **53:**315–322.
107. **Lindell, D. M., T. A. Moore, R. A. McDonald, G. B. Toews, and G. B. Huffnagle.** 2006. Distinct compartmentalization of CD4+ T-cell effector function versus proliferative capacity during pulmonary cryptococcosis. *Am. J. Pathol.* **168:**847–855.
108. **Lindell, D. M., T. A. Moore, R. A. McDonald, G. B. Toews, and G. B. Huffnagle.** 2005. Generation of antifungal effector CD8+ T cells in the absence of CD4+ T cells during *Cryptococcus neoformans* infection. *J. Immunol.* **174:**7920–7928.
109. **Linke, M., A. Ashbaugh, J. Koch, R. Tanaka, and P. Walzer.** 2005. Surfactant protein A limits *Pneumocystis murina* infection in immunosuppressed C3H/HeN mice and modulates host response during infection. *Microbes Infect.* **7:**748–759.
110. **Lovchik, J. A., C. R. Lyons, and M. F. Lipscomb.** 1995. A role for gamma interferon-induced nitric oxide in pulmonary clearance of *Cryptococcus neoformans*. *Am. J. Respir. Cell. Mol. Biol.* **13:**116–124.
111. **Lovchik, J. A., J. A. Wilder, G. B. Huffnagle, R. Riblet, C. R. Lyons, and M. F. Lipscomb.** 1999. Ig heavy chain complex-linked genes influence the immune response in a murine cryptococcal infection. *J. Immunol.* **163:**3907–3913.
112. **Luberto, C., B. Martinez-Marino, D. Taraskiewicz, B. Bolanos, P. Chitano, D. L. Toffaletti, G. M. Cox, J. R. Perfect, Y. A. Hannun, E. Balish, and M. Del Poeta.** 2003. Identification of App1 as a regulator of phagocytosis and virulence of *Cryptococcus neoformans*. *J. Clin. Invest.* **112:**1080–1094.
113. **Ma, H., J. Croudace, D. Lammas, and R. May.** 2007. Direct cell-to-cell spread of a pathogenic yeast. *BMC Immunol.* **8:**15.
114. **Ma, L. L., J. C. Spurrell, J. F. Wang, G. G. Neely, S. Epelman, A. M. Krensky, and C. H. Mody.** 2002. CD8 T cell-mediated killing of *Cryptococcus neoformans* requires granulysin and is dependent on CD4 T cells and IL-15. *J. Immunol.* **169:**5787–5795.
115. **Madan, T., U. Kishore, M. Singh, P. Strong, H. Clark, E. Hussain, K. Reid, and P. Sarma.** 2001. Surfactant proteins A and D protect mice against pulmonary hypersensitivity induced by *Aspergillus fumigatus* antigens and allergens. *J. Clin. Invest.* **107:**467–475.
116. **Madan, T., U. Kishore, M. Singh, P. Strong, E. Hussain, K. Reid, and P. Sarma.** 2001. Protective role of lung surfactant protein D in a murine model of invasive pulmonary aspergillosis. *Infect. Immun.* **69:**2728–2731.
117. **Mansour, M. K., E. Latz, and S. M. Levitz.** 2006. *Cryptococcus neoformans* glycoantigens are captured by multiple lectin receptors and presented by dendritic cells. *J. Immunol.* **176:**3053–3061.
118. **Mansour, M. K., L. S. Schlesinger, and S. M. Levitz.** 2002. Optimal T cell responses to *Cryptococcus neoformans* mannoprotein are dependent on recognition of conjugated carbohydrates by mannose receptors. *J. Immunol.* **168:**2872–2879.
119. **Mansour, M. K., L. E. Yauch, J. B. Rottman, and S. M. Levitz.** 2004. Protective efficacy of antigenic fractions in mouse models of cryptococcosis. *Infect. Immun.* **72:**1746–1754.
120. **McGreal, E. P., J. L. Miller, and S. Gordon.** 2005. Ligand recognition by antigen-presenting cell C-type lectin receptors. *Curr. Opin. Immunol.* **17:**18–24.
121. **Milam, J. E., A. C. Herring-Palmer, R. Pandrangi, R. A. McDonald, G. B. Huffnagle, and G. B. Toews.** 2007. Modulation of the pulmonary type 2 T-cell response to *Cryptococcus neoformans* by intratracheal delivery of a tumor necrosis factor alpha-expressing adenoviral vector. *Infect. Immun.* **75:**4951–4958.
122. **Mody, C. H., G. H. Chen, C. Jackson, J. L. Curtis, and G. B. Toews.** 1993. Depletion of murine CD8+ T-cells in vivo decreases pulmonary clearance of a moderately virulent strain of *Cryptococcus neoformans*. *J. Lab. Clin. Med.* **121:**765–773.
123. **Mody, C. H., G. H. Chen, C. Jackson, J. L. Curtis, and G. B. Toews.** 1994. In vivo depletion of murine CD8 positive T cells impairs survival during infection with a highly virulent strain of *Cryptococcus neoformans*. *Mycopathologia* **125:**7–17.
124. **Mody, C. H., M. F. Lipscomb, N. E. Street, and G. B. Toews.** 1990. Depletion of CD4+ (L3T4+) lymphocytes in vivo impairs murine host defense to *Cryptococcus neoformans*. *J. Immunol.* **144:**1472–1477.
125. **Mody, C. H., C. L. Tyler, R. G. Sitrin, C. Jackson, and G. B. Toews.** 1991. Interferon-gamma activates rat alveolar macrophages for anticryptococcal activity. *Am. J. Respir. Cell. Mol. Biol.* **5:**19–26.
126. **Mukherjee, J., T. R. Kozel, and A. Casadevall.** 1998. Monoclonal antibodies reveal additional epitopes of serotype D *Cryptococcus neoformans* capsular glucuronoxylomannan that elicit protective antibodies. *J. Immunol.* **161:**3557–3568.
127. **Mukherjee, S., M. Feldmesser, and A. Casadevall.** 1996. J774 murine macrophage-like cell interactions with *Cryptococcus neoformans* in the presence and absence of opsonins. *J. Infect. Dis.* **173:**1222–1231.
128. **Mukherjee, S., S. C. Lee, and A. Casadevall.** 1995. Antibodies to *Cryptococcus neoformans* glucuronoxylomannan enhance antifungal activity of murine macrophages. *Infect. Immun.* **63:**573–579.
129. **Muller, U., W. Stenzel, G. Kohler, C. Werner, T. Polte, G. Hansen, N. Schutze, R. K. Straubinger, M. Blessing, A. N. McKenzie, F. Brombacher, and G. Alber.** 2007. IL-13 induces disease-promoting type 2 cytokines, alternatively activated macrophages and allergic inflammation during pulmonary infection of mice with *Cryptococcus neoformans*. *J. Immunol.* **179:**5367–5377.
130. **Murphy, J. W., M. R. Hidore, and S. C. Wong.** 1993. Direct interactions of human lymphocytes with the yeast-like organism, *Cryptococcus neoformans*. *J. Clin. Invest.* **91:**1553–1566.
131. **Murphy, J. W., R. L. Mosley, R. Cherniak, G. H. Reyes, T. R. Kozel, and E. Reiss.** 1988. Serological, electrophoretic, and biological properties of *Cryptococcus neoformans* antigens. *Infect. Immun.* **56:**424–431.
132. **Murphy, J. W., F. Schafer, A. Casadevall, and A. Adesina.** 1998. Antigen-induced protective and nonprotective cell-mediated immune components against *Cryptococcus neoformans*. *Infect. Immun.* **66:**2632–2639.
133. **Muth, S. M., and J. W. Murphy.** 1995. Direct anticryptococcal activity of lymphocytes from *Cryptococcus neoformans*-immunized mice. *Infect. Immun.* **63:**1637–1644.
134. **Nakamura, K., T. Kinjo, S. Saijo, A. Miyazato, Y. Adachi, N. Ohno, J. Fujita, M. Kaku, Y. Iwakura, and K. Kawakami.** 2007. Dectin-1 is not required for the host defense to *Cryptococcus neoformans*. *Microbiol. Immunol.* **51:**1115–1119.
135. **Nakamura, K., K. Miyagi, Y. Koguchi, Y. Kinjo, K. Uezu, T. Kinjo, M. Akamine, J. Fujita, I. Kawamura, M. Mitsuyama, Y. Adachi, N. Ohno, K. Takeda, S. Akira, A. Miyazato, M. Kaku, and K. Kawakami.** 2006. Limited contribution of Toll-like receptor 2 and 4 to the host

response to a fungal infectious pathogen, *Cryptococcus neoformans*. *FEMS Immunol. Med. Microbiol.* **47:**148–154.

136. **Netea, M. G., J. W. Van der Meer, and B. J. Kullberg.** 2006. Role of the dual interaction of fungal pathogens with pattern recognition receptors in the activation and modulation of host defence. *Clin. Microbiol. Infect.* **12:**404–409.
137. **Nussbaum, G., W. Cleare, A. Casadevall, M. D. Scharff, and P. Valadon.** 1997. Epitope location in the *Cryptococcus neoformans* capsule is a determinant of antibody efficacy. *J. Exp. Med.* **185:**685–694.
138. **Olszewski, M. A., G. B. Huffnagle, R. A. McDonald, D. M. Lindell, B. B. Moore, D. N. Cook, and G. B. Toews.** 2000. The role of macrophage inflammatory protein-1 alpha/CCL3 in regulation of T cell-mediated immunity to *Cryptococcus neoformans* infection. *J. Immunol.* **165:**6429–6436.
139. **Olszewski, M. A., G. B. Huffnagle, T. R. Traynor, R. A. McDonald, D. N. Cook, and G. B. Toews.** 2001. Regulatory effects of macrophage inflammatory protein 1alpha/CCL3 on the development of immunity to *Cryptococcus neoformans* depend on expression of early inflammatory cytokines. *Infect. Immun.* **69:**6256–6263.
140. **Oppmann, B., R. Lesley, B. Blom, J. C. Timans, Y. Xu, B. Hunte, F. Vega, N. Yu, J. Wang, K. Singh, F. Zonin, E. Vaisberg, T. Churakova, M. Liu, D. Gorman, J. Wagner, S. Zurawski, Y. Liu, J. S. Abrams, K. W. Moore, D. Rennick, R. de Waal-Malefyt, C. Hannum, J. F. Bazan, and R. A. Kastelein.** 2000. Novel p19 protein engages IL-12p40 to form a cytokine, IL-23, with biological activities similar as well as distinct from IL-12. *Immunity* **13:**715–725.
141. **Osterholzer, J. J., J. L. Curtis, T. Polak, T. Ames, G. H. Chen, R. McDonald, G. B. Huffnagle, and G. B. Toews.** 2008. CCR2 mediates conventional dendritic cell recruitment and the formation of bronchovascular mononuclear cell infiltrates in the lungs of mice infected with *Cryptococcus neoformans*. *J. Immunol.* **181:**610–620.
142. **Pena, S. V., and A. M. Krensky.** 1997. Granulysin, a new human cytolytic granule-associated protein with possible involvement in cell-mediated cytotoxicity. *Semin. Immunol.* **9:**117–125.
143. **Pillemer, S., G. Tsokos, S. Barbieri, J. Balow, and B. Golding.** 1990. The CR2 receptor (CD21) shows increased expression in the more differentiated cells of an antigen-specific B cell line. *Cell. Immunol.* **125:**386–395.
144. **Qureshi, M. H., T. Zhang, Y. Koguchi, K. Nakashima, H. Okamura, M. Kurimoto, and K. Kawakami.** 1999. Combined effects of IL-12 and IL-18 on the clinical course and local cytokine production in murine pulmonary infection with *Cryptococcus neoformans*. *Eur. J. Immunol.* **29:**643–649.
145. **Reeke, G. N., Jr., J. W. Becker, B. A. Cunningham, G. R. Gunther, J. L. Wang, and G. M. Edelman.** 1974. Relationships between the structure and activities of concanavalin A. *Ann. NY Acad. Sci.* **234:**369–382.
146. **Rivera, J., J. Mukherjee, L. M. Weiss, and A. Casadevall.** 2002. Antibody efficacy in murine pulmonary *Cryptococcus neoformans* infection: a role for nitric oxide. *J. Immunol.* **168:**3419–3427.
147. **Robben, P. M., M. LaRegina, W. A. Kuziel, and L. D. Sibley.** 2005. Recruitment of Gr-1+ monocytes is essential for control of acute toxoplasmosis. *J. Exp. Med.* **201:**1761–1769.
148. **Rottman, J. B., K. P. Ganley, K. Williams, L. Wu, C. R. Mackay, and D. J. Ringler.** 1997. Cellular localization of the chemokine receptor CCR5. Correlation to cellular targets of HIV-1 infection. *Am. J. Pathol.* **151:**1341–1351.
149. **Rudner, X. L., K. I. Happel, E. A. Young, and J. E. Shellito.** 2007. Interleukin-23 (IL-23)-IL-17 cytokine axis in murine *Pneumocystis carinii* infection. *Infect. Immun.* **75:**3055–3061.
150. **Schelenz, S., R. Malhotra, R. B. Sim, U. Holmskov, and G. J. Bancroft.** 1995. Binding of host collectins to the pathogenic yeast *Cryptococcus neoformans*: human surfactant protein D acts as an agglutinin for acapsular yeast cells. *Infect. Immun.* **63:**3360–3366.
151. **Shao, X., A. Mednick, M. Alvarez, N. van Rooijen, A. Casadevall, and D. L. Goldman.** 2005. An innate immune system cell is a major determinant of species-related susceptibility differences to fungal pneumonia. *J. Immunol.* **175:**3244–3251.
152. **Stano, P., V. Williams, M. Villani, E. Cymbalyuk, A. Qureshi, Y. Huang, G. Morace, C. Luberto, S. Tomlinson, and M. Del Poeta.** 2009. App1: an antiphagocytic protein that binds to complement receptors 3 and 2. *J. Immunol.* **182:**84–91.
153. **Stein, M., S. Keshav, N. Harris, and S. Gordon.** 1992. Interleukin 4 potently enhances murine macrophage mannose receptor activity: a marker of alternative immunologic macrophage activation. *J. Exp. Med.* **176:**287–292.
154. **Takeda, K., and S. Akira.** 2004. TLR signaling pathways. *Semin. Immunol.* **16:**3–9.
155. **Traynor, T. R., A. C. Herring, M. E. Dorf, W. A. Kuziel, G. B. Toews, and G. B. Huffnagle.** 2002. Differential roles of CC chemokine ligand 2/monocyte chemotactic protein-1 and CCR2 in the development of T1 immunity. *J. Immunol.* **168:**4659–4666.
156. **Traynor, T. R., W. A. Kuziel, G. B. Toews, and G. B. Huffnagle.** 2000. CCR2 expression determines T1 versus T2 polarization during pulmonary *Cryptococcus neoformans* infection. *J. Immunol.* **164:**2021–2027.
157. **Tucker, S. C., and A. Casadevall.** 2002. Replication of *Cryptococcus neoformans* in macrophages is accompanied by phagosomal permeabilization and accumulation of vesicles containing polysaccharide in the cytoplasm. *Proc. Natl. Acad. Sci. USA* **99:**3165–3170.
158. **Uezu, K., K. Kawakami, K. Miyagi, Y. Kinjo, T. Kinjo, H. Ishikawa, and A. Saito.** 2004. Accumulation of gammadelta T cells in the lungs and their regulatory roles in Th1 response and host defense against pulmonary infection with *Cryptococcus neoformans*. *J. Immunol.* **172:**7629–7634.
159. **Vaccaro, A. M., M. Tatti, F. Ciaffoni, R. Salvioli, A. Barca, and C. Scerch.** 1997. Effect of saposins A and C on the enzymatic hydrolysis of liposomal glucosylceramide. *J. Biol. Chem.* **272:**16862–16867.
160. **van de Wetering, J. K., F. E. Coenjaerts, A. B. Vaandrager, L. M. van Golde, and J. J. Batenburg.** 2004. Aggregation of *Cryptococcus neoformans* by surfactant protein D is inhibited by its capsular component glucuronoxylomannan. *Infect. Immun.* **72:**145–153.
161. **van Rozendaal, B., A. van Spriel, J. van De Winkel, and H. Haagsman.** 2000. Role of pulmonary surfactant protein D in innate defense against *Candida albicans*. *J. Infect. Dis.* **182:**917–922.
162. **Vecchiarelli, A., D. Pietrella, M. Dottorini, C. Monari, C. Retini, T. Todisco, and F. Bistoni.** 1994. Encapsulation of *Cryptococcus neoformans* regulates fungicidal activity and the antigen presentation process in human alveolar macrophages. *Clin. Exp. Immunol.* **98:**217–223.
163. **Vecchiarelli, A., D. Pietrella, P. Lupo, F. Bistoni, D. C. McFadden, and A. Casadevall.** 2003. The polysaccharide capsule of *Cryptococcus neoformans* interferes with human dendritic cell maturation and activation. *J. Leukoc. Biol.* **74:**370–378.
164. **Vuk-Pavlovic, Z., E. Mo, C. Icenhour, J. Standing, J. Fisher, and A. Limper.** 2006. Surfactant protein D

enhances *Pneumocystis* infection in immune-suppressed mice. *Am. J. Physiol. Lung Cell. Mol. Physiol.* **290:**L442–L449.

165. **Wagner, R. P., S. M. Levitz, A. Tabuni, and H. Kornfeld.** 1992. HIV-1 envelope protein (gp120) inhibits the activity of human bronchoalveolar macrophages against *Cryptococcus neoformans. Am. Rev. Respir. Dis.* **146:**1434–1438.
166. **Walenkamp, A. M., A. F. Verheul, J. Scharringa, and I. M. Hoepelman.** 1999. Pulmonary surfactant protein A binds to *Cryptococcus neoformans* without promoting phagocytosis. *Eur. J. Clin. Invest.* **29:**83–92.
167. **Weinberg, P. B., S. Becker, D. L. Granger, and H. S. Koren.** 1987. Growth inhibition of *Cryptococcus neoformans* by human alveolar macrophages. *Am. Rev. Respir. Dis.* **136:**1242–1247.
168. **Wormley, F. L., Jr., J. R. Perfect, C. Steele, and G. M. Cox.** 2007. Protection against cryptococcosis by using a murine gamma interferon-producing *Cryptococcus neoformans* strain. *Infect. Immun.* **75:**1453–1462.
169. **Wozniak, K. L., and S. M. Levitz.** 2008. *Cryptococcus neoformans* enters the endolysosomal pathway of dendritic cells and is killed by lysosomal components. *Infect. Immun.* **76:**4764–4771.
170. **Wuthrich, M., H. I. Filutowicz, T. Warner, G. S. Deepe, Jr., and B. S. Klein.** 2003. Vaccine immunity to pathogenic fungi overcomes the requirement for CD4 help in exogenous antigen presentation to CD8+ T cells: implications for vaccine development in immune-deficient hosts. *J. Exp. Med.* **197:**1405–1416.
171. **Yamaguchi, H., Y. Komase, M. Ikehara, T. Yamamoto, and T. Shinagawa.** 2008. Disseminated cryptococcal infection with eosinophilia in a healthy person. *J. Infect. Chemother.* **14:**319–324.
172. **Yauch, L. E., M. K. Mansour, S. Shoham, J. B. Rottman, and S. M. Levitz.** 2004. Involvement of CD14, toll-like receptors 2 and 4, and MyD88 in the host response to the fungal pathogen *Cryptococcus neoformans* in vivo. *Infect. Immun.* **72:**5373–5382.
173. **Zaragoza, O., and A. Casadevall.** 2004. Antibodies produced in response to *Cryptococcus neoformans* pulmonary infection in mice have characteristics of nonprotective antibodies. *Infect. Immun.* **72:**4271–4274.
174. **Zelante, T., A. De Luca, P. Bonifazi, C. Montagnoli, S. Bozza, S. Moretti, M. L. Belladonna, C. Vacca, C. Conte, P. Mosci, F. Bistoni, P. Puccetti, R. A. Kastelein, M. Kopf, and L. Romani.** 2007. IL-23 and the Th17 pathway promote inflammation and impair antifungal immune resistance. *Eur. J. Immunol.* **37:**2695–2706.
175. **Zheng, C. F., L. L. Ma, G. J. Jones, M. J. Gill, A. M. Krensky, P. Kubes, and C. H. Mody.** 2007. Cytotoxic CD4+ T cells use granulysin to kill *Cryptococcus neoformans*, and activation of this pathway is defective in HIV patients. *Blood* **109:**2049–2057.
176. **Zhong, Z., and L. A. Pirofski.** 1998. Antifungal activity of a human antiglucuronoxylomannan antibody. *Clin. Diagn. Lab. Immunol.* **5:**58–64.

Cryptococcus: From Human Pathogen to Model Yeast
Edited by J. Heitman et al.
©2011 ASM Press, Washington, DC

34

Invasion of *Cryptococcus* into the Central Nervous System

FRANÇOISE DROMER AND STUART M. LEVITZ

As detailed elsewhere in this book, meningoencephalitis is the most common clinical manifestation of cryptococcosis. The most characteristic histopathological lesion is the dilation of the Virchow-Robin spaces (Fig. 1), leading to radiological images described as dilated Virchow-Robin spaces, pseudocysts, or masses (5). The basis for the neurotropism of *Cryptococcus neoformans* and the related species, *Cryptococcus gattii*, has been the subject of much research and speculation. The polysaccharide capsule is the major virulence factor of these fungi, and capsule undoubtedly contributes to the ability of *Cryptococcus* to establish residency in the central nervous system (CNS). Capsule interferes with phagocytosis and subsequent killing by host neutrophils and macrophages. Phagocytosis is further impaired in the CNS, where levels of opsonins, particularly complement, are low (51). The major component of capsule, glucuronoxylomannan, shedded during infection around the lesions, also inhibits T-cell function and contributes to the elevated intracranial pressure often observed in cryptococcal meningoencephalitis (25, 54).

Melanin synthesis in *C. neoformans* is catalyzed by phenoloxidase (laccase), an enzyme with a substrate specificity for phenolic compounds, such as L-DOPA and dopamine. As the brain is rich in these substrates, it has been postulated that this could help account for the propensity of phenoloxidase-positive organisms, including *C. neoformans*, to infect the nervous system (21). The contribution of host factors to the neuropathology of cryptococcal meningoencephalitis also should be appreciated. This is perhaps best illustrated by the neurological deterioration observed in some patients with AIDS and cryptococcosis who experience immune reconstitution when they receive antiretroviral therapy (39).

Regardless of the fungal virulence factors and host determinants that contribute to the propensity of *C. neoformans* to cause disease in the CNS, it is clear that the necessary first step is for the fungus to gain entrance into the CNS. The prerequisite for brain invasion is fungemia, which is recorded in almost 50% of HIV-infected patients with cryptococcosis (13). Moreover, fungemia correlated with fungal load in all compartments including the brain 24 h after intravenous inoculation in mice (40). This chapter reviews the current state of knowledge regarding how *C. neoformans* invades into the CNS.

THE BLOOD-BRAIN BARRIER AND MICROBIAL INVASION

An understanding of the unique characteristics of the blood-brain barrier (BBB) is essential to comprehend the potential mechanisms by which blood-borne microbes, including *C. neoformans*, transverse it. It should also be appreciated that because of these properties, microbial entry into the brain is different compared with other organs. Loosely defined, the BBB is a structural and functional barrier that is formed by brain microvascular endothelial cells (BMECs), astrocytes, and pericytes (34). BMECs, which line the capillaries supplying blood to the brain, have unique tight junctions that contribute to the barrier function of the BBB (47). BMECs also have low rates of pinocytosis compared to endothelial cells obtained from other organs. The astrocytes and pericytes are thought to help maintain the barrier properties of the BMECs (34). However, the extent to which astrocytes and pericytes directly affect the translocation of microorganisms across the BBB is unclear. These cell types have received scant attention from researchers studying host defenses against cryptococcosis. In one study, primary cultures of human fetal astrocytes, activated with interleukin-1β plus interferon-γ, inhibited the growth of *C. neoformans* by a nitric oxide-dependent mechanism (37). The astrocytes did not phagocytose the fungi but could be found in close proximity to *C. neoformans* during human infection. Perivascular microglial cells, which are innate immune cells but are not considered part of the BBB, do have anticryptococcal activity under defined conditions (1, 38). Therefore, microglial cells may form a first line of innate defense should the BBB be penetrated.

A major function of the BBB is maintenance of the neural microenvironment by regulating the passage of molecules into and out of the brain. Efficacy and side effects of many

Françoise Dromer, Institut Pasteur, Molecular Mycology Unit, CNRS URA3012, Paris, France. **Stuart M. Levitz,** Department of Medicine, University of Massachusetts Medical School, Worcester, MA 01605.

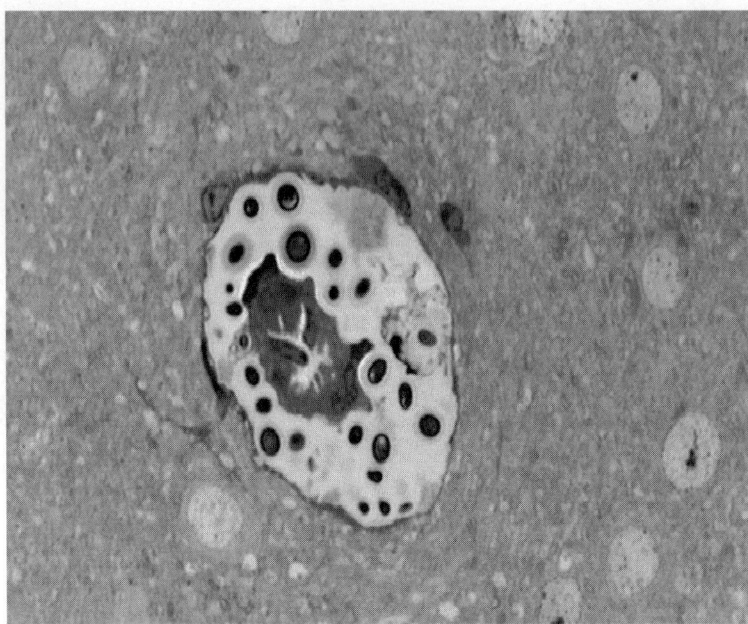

FIGURE 1 Typical aspect of a dilated Virchow-Robin space during cryptococcal meningoencephalitis. Semithin section of the brain cortex of a mouse with severe meningoencephalitis after inoculation with *C. neoformans* showing the pseudocysts centered by a brain capillary and filled with numerous capsulated yeasts either free or inside phagocytes.

medications are largely influenced by their capacity to cross the BBB. Indeed, CNS penetration of antifungal drugs must be considered when choosing regimens for the treatment of cryptococcosis. Drugs that disrupt the BBB are under development for the treatment of neurological diseases. It remains to be seen whether a side effect of these drugs will be an increased incidence of CNS infections due to facilitated passage of microbes.

In vitro models of the BBB have greatly facilitated study of the mechanisms by which microbes traverse the BBB (35, 50). BMECs can be cultured at very high purity, and they retain many of the distinctive properties of their in vivo counterparts, including formation of confluent monolayers with tight junctions. Moreover, BMECs can be seeded on porous tissue culture inserts and grown to confluence (3). This model allows separate access to the upper compartment (blood side) and the lower compartment (brain side), thus mimicking penetration from the vascular system into the brain. Integrity of the monolayer can be assessed by measuring transendothelial electrical resistance and permeability to small molecules (3, 44). Endothelial cells obtained from sources other than the brain (e.g., umbilical vein) should be used with great caution in BBB models, as such cells significantly differ from BMECs. Another caveat to interpretation of in vitro studies is that they generally use static conditions as opposed to dynamic conditions that mimic blood flow through capillaries. Finally, most in vitro models lack astrocytes and pericytes.

The three major mechanisms by which microbes can penetrate the BBB (34) are by "Trojan horse" (infected phagocytes), transcellular penetration, and paracellular penetration, as discussed below. It is important to emphasize that only live cryptococci can cross the BBB (4) and that none of these mechanisms are mutually exclusive. Indeed, there is experimental evidence, reviewed below, that all three contribute to invasion of *C. neoformans* into the CNS (6). Nevertheless, the relative contribution of each of these mechanisms is not clear. Finally, while most of the attention in the literature has been devoted to the barrier function of the BBB, recently it was demonstrated that the BBB may play a role in the immune response to fungal infection. A genomic survey of changes in the levels of gene expression in human BMECs following challenge with *C. neoformans* revealed changes in expression levels of interferon and major histocompatibility complex genes (29).

ENTRY OF *CRYPTOCOCCUS* VIA TROJAN HORSE (INFECTED PHAGOCYTES)

The Trojan horse mechanism refers to microbial penetration of the BBB within infected phagocytes (34). The Trojan horse mechanism of crossing the BBB has been demonstrated for bacterial pathogens, such as *Listeria monocytogenes* (12), and viral pathogens, such as HIV (14). The role of monocytes in CNS invasion by *C. neoformans* relies first on clinical observations. Meningoencephalitis and fungemia are significantly more frequent among HIV-positive than HIV-negative patients (13), and monocytes from HIV-infected patients have diminished anticryptococcal activity (19, 20, 42). Moreover, HIV infection has recently been shown to increase entry of monocytes into the brain (53). In a murine model of cryptococcosis, it was established using Ficoll gradients that yeasts circulate in close contact with macrophages (8). At a stage of severe meningoencephalitis, poorly capsulated yeasts were seen in monocytes circulating in leptomeningeal capillaries, suggesting that the Trojan horse mechanism could be true (Fig. 2).

Experimental studies long focused on tissue macrophages, especially bronchoalveolar macrophages, as vehicles by which *C. neoformans* could travel from the lungs to the blood compartment and subsequently to the brain. Indirect evidence of the role of monocytes was thus found after the use of genetically modified *C. neoformans* with altered survival

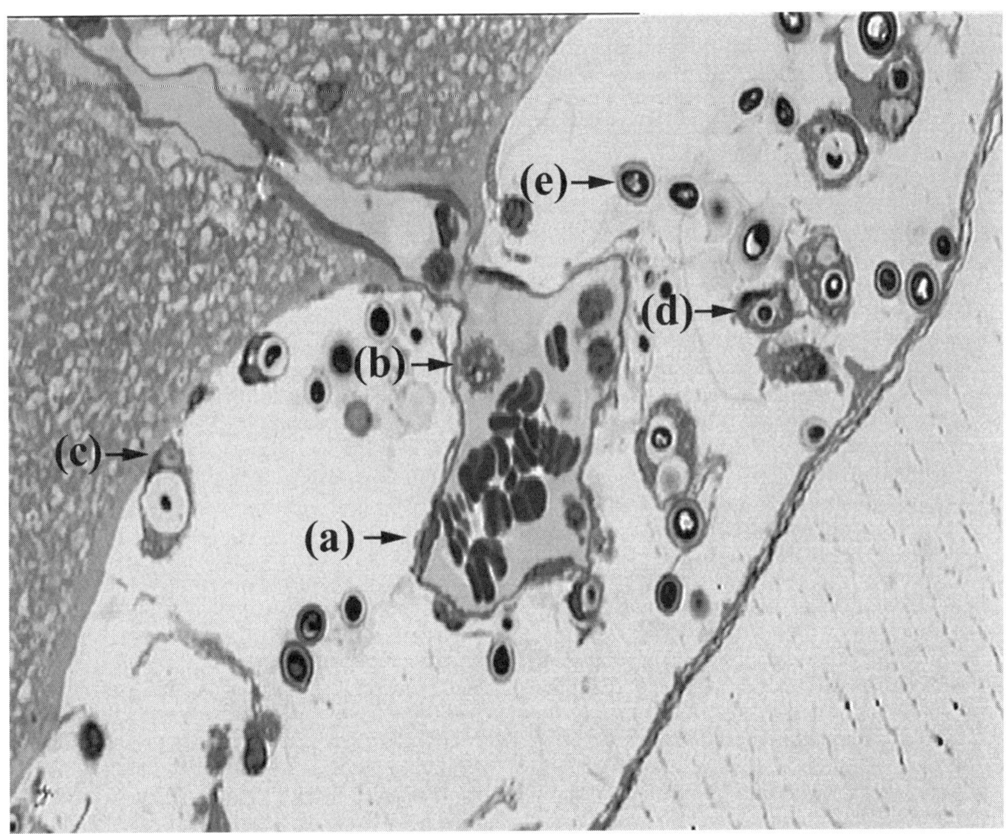

FIGURE 2 Semithin section of leptomeninges from a mouse with severe meningoencephalitis after inoculation with *C. neoformans*. (a) Dilatation of a Virchow-Robin space is visible with the meningeal capillary containing numerous red blood cells. (b) A mononuclear cell probably carrying a yeast cell can be seen inside the capillary. (c) On the left, a small capillary with a yeast inside the cytoplasm of an endothelial cell. Cryptococci are also seen in the leptomeningeal space outside the capillaries, either (d) as poorly encapsulated yeasts inside vacuolated macrophage-like cells apparently touching the outside membrane of the vessel or (e) as free and with thick capsules.

in host phagocytic cells (33, 41, 48). While a phospholipase B-deleted mutant strain (Δ*plb1*) was unable to produce brain infection, this was achieved when bronchoalveolar macrophages from Δ*plb1*-infected mice injected into naive mice were used (48). Decreased fungal load in the brain after injection of anti-CD11b antibody was attributed to decreased fungemia and depletion of CD11b$^+$ cells without demonstration of monocyte depletion (32).

The role of monocytes in BBB crossing by *C. neoformans* was demonstrated using bone-marrow-derived monocytes loaded in vitro with *C. neoformans* and inoculated into naive mice (6). Fungal load in the brains of these animals was significantly higher 24 h after inoculation than in the brains of mice inoculated with free yeasts. It was also established that both free and monocyte-loaded yeasts could invade the brain at the same time and over time.

TRANSCELLULAR PENETRATION OF BMECs BY *CRYPTOCOCCUS*

An in vitro model of the BBB was used to demonstrate that following incubation of *C. neoformans* with a confluent monolayer of human BMECs, yeast cells adhere to the monolayer, following which internalization and transcellular penetration occur (3, 7). The percentage of yeast cells that adhere to BMECs is typically very low (around 1 to 3%), even under the static conditions of the assay. Nevertheless, these studies provide proof that transcellular penetration of *C. neoformans* can occur. Although the integrity of the monolayer was preserved, some evidence of weakening was found in that degradation of the tight junction marker protein, occludin, was demonstrated (7).

The mechanism by which transcellular penetration of *C. neoformans* occurs has been the subject of investigation. Conceptually, transcellular traversal can be broken down into three sequential events: binding (adherence) to the apical (vascular) surface of the BMEC, internalization, and egress from the basolateral (brain) side. An early event following incubation of *C. neoformans* with BMECs is the formation of microvillus-like membrane protrusions on the BMECs (3). Deletion of *CPS1*, a gene that encodes a protein predicted to contain a glycosyltransferase moiety, from *C. neoformans* resulted in an approximately two-thirds decrease in binding of the yeast cells to human BMECs (2). The deleted strain had diminished surface expression of hyaluronic acid (or related molecules). One caveat to the interpretation of these studies is that the knockout strain had a smaller capsule size and grew poorly at 37°C. Subsequent studies demonstrated that binding to BMECs was inhibited by treatment of wild-type yeast cells with hyaluronidase (26). Moreover, the ability of different strains of C. *neoformans* to bind to human BMECs directly correlated with their measured hyaluronic acid content. Interestingly, *CPS1* shares homology with the type 3 polysaccharide synthase encoded by the *cap3B* gene of another neurotropic organism, *Streptococcus pneumoniae* (2).

Several lines of evidence suggest that CD44, a receptor for hyaluronic acid that is present on human BMECs, recognizes C. *neoformans* (28). Blockage of CD44 function by antibodies, inhibitors, or small hairpin RNA abrogated the association of C. *neoformans* with BMECs. Conversely, overexpression of CD44 in human BMECs resulted in increased binding of C. *neoformans*. Finally, CD44 accumulated at the site of interaction of yeast cells with BMECs.

Infection with HIV is the major risk factor for cryptococcal meningitis. Accordingly, the interplay between HIV, BMECs, and C. *neoformans* was studied (30). It was found that the HIV gp41 protein stimulated C. *neoformans* binding to human BMECs in vitro and enhanced brain infection in vivo. Thus, in addition to immunodepletion, HIV infection could directly contribute to the high incidence of cryptococcal meningitis in patients with AIDS. However, it remains unclear whether sufficient quantities of gp41 are typically present during HIV infection to mediate this effect.

Studies examining the interactions of C. *neoformans* with human BMECs in vitro have demonstrated that adherence and internalization are accompanied by remodeling of the actin cytoskeleton (7). Moreover, C. *neoformans* invasion into human BMECs was diminished by inhibition of actin polymerization with cytochalasin D or by blocking protein kinase C-α activation (27). Protein kinase C-α functions as an upstream regulator of actin filament activity. Incubation of human BMECs with C. *neoformans* led to phosphorylation of protein kinase C-α. By immunofluorescence microscopy, phosphorylated protein kinase C-α colocalized with β-actin on the membrane of human BMECs. While inhibition of actin polymerization or protein kinase C-α activation decreased engulfment of C. *neoformans* by BMECs, it did not affect adherence. Thus, the binding and internalization steps are separate events. The mechanism by which C. *neoformans* is extruded from the basolateral (brain) side of the BMECs has not been elucidated.

Histopathological studies provide evidence that transcellular penetration of C. *neoformans* occurs in vivo. Ultrathin brain sections obtained from mice with established cryptococcal meningoencephalitis demonstrated yeast cells inside cells whose shape and structure strongly resembled that of endothelial cells (8). At earlier time points, C. *neoformans* either was associated with endothelial cells or escaped from the brain capillary vessels into the adjacent neuropil by 3 h (3). However, by 22 h, C. *neoformans* could be found in the brain parenchyma away from the blood vessels.

PARACELLULAR TRAVERSAL OF BMECs BY *CRYPTOCOCCUS*

Paracellular traversal is defined as microbial penetration between barrier cells with and/or without evidence of tight-junction disruption (34). As opposed to the Trojan horse and transcellular penetration mechanisms, there are fewer experimental data supporting a paracellular mechanism of entry into the CNS by C. *neoformans*. Indirect evidence comes from a study demonstrating disruption of the BBB in mice challenged intravenously with C. *neoformans* (4). Horseradish peroxidase was used as a marker of leakiness of the BBB. At early (5 minutes and 1 h) time points after challenge, no horseradish peroxidase extravasation was observed. However, by 24 h, areas of horseradish peroxidase leakage and vascular hemorrhage were noted. Furthermore, by immunofluorescent and electron microscopy, intraparenchymal yeasts in the vicinity of damaged basal lamina and endothelial cells were observed (4).

Clinical studies of the integrity of the BBB during cryptococcal meningoencephalitis are scarce. Measurement of serum matrix metalloproteinase-2 levels in the serum of patients with viral, fungal, or bacterial meningitis suggested (albeit with a limited sample size) that cryptococcosis was associated with disruption of the BBB (31). Indirect evidence was also provided by elevated vascular endothelial growth factor (a potential mediator of BBB disruption) levels in cerebrospinal fluid and serum of patients with cryptococcal meningoencephalitis compared with healthy control subjects (9).

Similarly, following intravenous or intratracheal challenge with the highly virulent C. *neoformans* strain H99, wedging of the yeasts in small capillaries, altered structure of microvessel walls, and formation of mucoid cysts in the proximity of damaged microcapillaries were seen (46). As noted above, in an in vitro model of the BBB, evidence of tight-junction weakening was observed (7). Thus, infection with C. *neoformans* can result in damage to the BBB. However, a definitive histological demonstration of C. *neoformans* traversing the BBB between endothelial cells has not been published. Indeed, rather than traversing between BMECs in an intact BBB, paracellular traversal of C. *neoformans* may be more likely to occur in situations where the BBB is severely disrupted, as discussed in more detail below.

OTHER POTENTIAL MECHANISMS

For angioinvasive fungi such as *Aspergillus*, *Candida*, and the agents of zygomycosis, adherence of hyphae to vascular endothelial cells can lead to occlusion of the blood vessel and subsequent hemorrhagic infarction (49). When occlusion occurs in blood vessels supplying the CNS, invasion into damaged tissue can occur. Emboli as a complication of fungal endocarditis can cause a similar pattern of disease. However, this pattern of disease is rarely, if ever, seen with C. *neoformans*.

Nevertheless, mechanical obstruction could contribute to CNS invasion. Capillaries measure 5 to 10 μm in diameter. While red and white blood cells can streamline their shape to "squeeze" through capillaries, fungi are inflexible. The capsule of C. *neoformans* is stimulated by in vivo conditions (17), resulting in organisms whose diameter can exceed that of capillaries. A study of murine brains shortly after inoculation with C. *neoformans* using classical histopathology or two-photon microscopy showed distorted capillary walls around trapped yeasts. Urease was shown to promote yeast sequestration into brain capillaries and subsequent dissemination (46). Recently, using intravital microscopy, it was demonstrated that following intravenous injection of fluorescently labeled C. *neoformans*, yeast cells got stuck in the brain capillaries (M. Shi and C. Mody, unpublished data). Some of the arrested fungi were released to flow freely in the bloodstream a few seconds after initially stopping. However, other yeast cells were observed to cross the capillary wall into the brain. While polystyrene beads of similar size to C. *neoformans* also got stuck in the capillaries, only the live fungi crossed the cerebral vasculature. These data suggest that mechanical trapping of yeast cells in the brain capillaries precedes fungal transmigration into the CNS.

One can speculate that *C. neoformans* with larger-diameter capsules could get irreversibly stuck in capillaries where mechanical pressure from ongoing replication and capsule expansion could disrupt tight junctions. This is supported by histological studies suggesting that *C. neoformans* invades brain parenchyma at sites where microemboli of fungi occlude small capillaries (46).

The choroid plexus, which is found in the ventricles of the brain, is responsible for cerebrospinal fluid production. The choroid plexus contains epithelial cells with tight junctions that surround capillaries and thus form a blood–cerebrospinal fluid barrier. Two studies have examined whether *C. neoformans* gains access to the CNS via the choroid plexus (3, 4). In each, mice were challenged intravenously with *C. neoformans*, and histopathology of the brain was performed at time intervals up to 10 days postinjection. The choroid plexus and meninges remained free of yeast cells at all time points examined. Moreover, there was no evidence of disruption of the epithelial layer or the basal lamina on reticulin-stained choroid plexus sections (4). In contrast, at the early time points, yeast cells were found in close association with brain capillaries (3, 4). These data suggest that following hematogenous dissemination, *C. neoformans* generally reaches the CNS by penetrating the BBB rather than by traversing epithelial cells at the choroid plexus. Additionally, at early time points, invasion of the CNS by *C. neoformans* did not occur from the surface of the brain via the invasion of the blood vessels supplying the leptomeninges (pia mater and arachnoid) (46). However, in a model of severe murine cryptococcosis, yeast cells were found in meningeal capillaries and within host cells touching the outer membrane of the capillaries (8) (Fig. 2).

Cerebrospinal fluid shunts are occasionally placed in patients with cryptococcal meningitis to assist in the management of hydrocephalus or elevated intracranial pressure. Although extremely rare, cerebrospinal fluid shunts placed in patients without cryptococcosis can subsequently get infected with *C. neoformans* (24, 52). In cases such as these, the possibility that the shunt could have served as the source of entry of the fungus into the CNS must be considered. Other unusual routes of entry into the CNS are theoretically possible. Contiguous spread via the optic nerve could occur following intraocular cryptococcosis (11). *C. neoformans* is frequently inhaled. In addition to the lungs, inhaled organisms can lodge in the nose or sinuses. The olfactory and trigeminal nerves innervate the nasal cavity, providing a direct connection to the CNS (18). Similarly, due to their anatomical proximity, contiguous spread of microorganisms from infected sinuses to the brain can transpire. It should be emphasized, however, that the potential mechanisms described in this paragraph are likely to account for a very small fraction of the cases of entry of *C. neoformans* into the CNS. The vast majority of human cryptococcal meningoencephalitis cases are thought to originate following hematogenous dissemination.

VIRULENCE FACTORS THAT FACILITATE CNS INVASION

A study in which live, but not dead, *C. neoformans* crossed the BBB of mouse following intravenous challenge suggests that active fungus-derived processes contribute to CNS penetration (4). As reviewed elsewhere in this book, a myriad of virulence factors have been described in *C. neoformans*. Some virulence factors, such as secreted phospholipase B and laccase, appear to help *C. neoformans* disseminate from the lungs to the bloodstream (45, 48), a necessary prelude to crossing the BBB. Other virulence factors help the organism survive the unique conditions of the CNS, such as low oxygen tension (23). However, for these and most other putative virulence factors, their contribution to the capacity of the fungus to traverse the BBB has not been specifically studied.

Two cryptococcal virulence factors have been implicated in facilitating traversal of the BBB, capsule, and urease. Capsule has been studied using in vitro models of the BBB by comparing acapsular and encapsulated isogenic strains (3, 7). The encapsulated yeast cells were internalized and traversed the human BMEC more efficiently than the acapsular cells. In contrast, a study using human umbilical vein endothelial cells demonstrated internalization of acapsular, but not encapsulated, *C. neoformans* (22). These disparate results may be explained, as noted above, by the significant differences between BMECs and endothelial cells derived from other sources.

The other cryptococcal virulence factor that has been linked to brain invasion is urease, which is produced by nearly all clinical isolates of *C. neoformans* (46). Urease expression contributes to the CNS invasion by enhancing yeast sequestration within cerebral microcapillary beds during hematogenous spread. However, once inside the brain, urease may not be as important for virulence. When *C. neoformans* was directly injected into the cerebrospinal fluid of corticosteroid-pretreated rabbits, urease expression did not affect growth in the CNS (10).

Finally, mating types could influence brain invasion. Mating type α predominates in nature and among clinical isolates, with very few mating type **a** organisms discovered among clinical isolates of var. *grubii*. *C. neoformans* var. *neoformans* mating type **a** is less virulent than mating type α (36). For var. *grubii*, both mating types have the same capacity of producing brain infection when inoculated intravenously or intracranially separately, but mating type **a** loses its ability to penetrate the brain when coinoculated with mating type α (43). Whether coinfection triggers virulence traits (e.g., melanin, urease) in α cells or shuts them down in **a** cells, or whether the interaction results in physical alteration of the **a** cells that prevent their entry into the brain compartment, remains to be determined. This could partially explain the prevalence of α strains in clinical isolates, knowing that the main clinical presentation is meningoencephalitis and that more than one strain can be found in the same environment (15, 16).

CONCLUSIONS

Meningoencephalitis is the most common and serious clinical manifestation of cryptococcosis. Remarkable progress has been made toward elucidating the means by which *C. neoformans* gains access into the CNS. The data strongly suggest that in all but rare circumstances, the fungus enters the CNS by traversing the BBB during hematogenous dissemination. There are three major mechanisms by which organisms can traverse the BBB: Trojan horses, transcellular penetration, and paracellular invasion. Under carefully defined experimental conditions, it has been convincingly demonstrated that *C. neoformans* utilizes Trojan horses and transcellular penetration. The evidence of paracellular invasion is somewhat more circumstantial. Disruption of the BBB and even entire capillaries has been observed, but yeast cells have not been "caught in the act" between BMECs.

Despite evidence of multiple mechanisms by which *C. neoformans* can traverse the BBB, it remains uncertain as to how invasion most commonly occurs during natural infection in humans. Key variables that have not been well studied include fungal strain variation and the influence of host immunity. For example, an isolate with a large capsule may be more likely to get lodged in brain capillaries and mechanically disrupt the BBB. Conversely, the transcellular and Trojan horse mechanisms might be easier for a small capsule isolate. All three mechanisms might be expected to occur repeatedly in an immunocompromised patient with high-grade fungemia. In contrast, for the patient with transient low-grade fungemia, the likelihood that *C. neoformans* will be phagocytosed by circulating neutrophils and monocytes is greater. This would be expected to favor the Trojan horse mechanism. The challenge ahead will be to develop strategies to reduce neurotransmission of *C. neoformans* in high-risk patients. This will be difficult at the level of the BBB if, as the data suggest, the fungus has at its disposal multiple lines of attack.

The authors thank Meiqing Shi and Christopher Mody for generously allowing us to discuss their unpublished data, and Fabrice Chrétien (Neuropathology, Hospital Henri-Mondor, Créteil, France) for sharing his pictures of C. neoformans *invading the CNS.*

REFERENCES

1. **Aguirre, K., and S. Miller.** 2002. MHC class II-positive perivascular microglial cells mediate resistance to *Cryptococcus neoformans* brain infection. *Glia* **39:**184–188.
2. **Chang, Y. C., A. Jong, S. Huang, P. Zerfas, and K. J. Kwon-Chung.** 2006. CPS1, a homolog of the *Streptococcus pneumoniae* type 3 polysaccharide synthase gene, is important for the pathobiology of *Cryptococcus neoformans*. *Infect. Immun.* **74:**3930–3938.
3. **Chang, Y. C., M. F. Stins, M. J. McCaffery, G. F. Miller, D. R. Pare, T. Dam, M. Paul-Satyasee, K. S. Kim, and K. J. Kwon-Chung.** 2004. Cryptococcal yeast cells invade the central nervous system via transcellular penetration of the blood-brain barrier. *Infect. Immun.* **72:**4985–4995.
4. **Charlier, C., F. Chretien, M. Baudrimont, E. Mordelet, O. Lortholary, and F. Dromer.** 2005. Capsule structure changes associated with *Cryptococcus neoformans* crossing of the blood-brain barrier. *Am. J. Pathol.* **166:**421–432.
5. **Charlier, C., F. Dromer, C. Leveque, L. Chartier, Y. S. Cordoliani, A. Fontanet, O. Launay, and O. Lortholary.** 2008. Cryptococcal neuroradiological lesions correlate with severity during cryptococcal meningoencephalitis in HIV-positive patients in the HAART era. *PLoS ONE* **3:**e1950.
6. **Charlier, C., K. Nielsen, S. Daou, M. Brigitte, F. Chretien, and F. Dromer.** 2009. Evidence of a role for monocytes in dissemination and brain invasion by *Cryptococcus neoformans*. *Infect. Immun.* **77:**120–127.
7. **Chen, S. H., M. F. Stins, S. H. Huang, Y. H. Chen, K. J. Kwon-Chung, Y. Chang, K. S. Kim, K. Suzuki, and A. Y. Jong.** 2003. *Cryptococcus neoformans* induces alterations in the cytoskeleton of human brain microvascular endothelial cells. *J. Med. Microbiol.* **52:**961–970.
8. **Chretien, F., O. Lortholary, I. Kansau, S. Neuville, F. Gray, and F. Dromer.** 2002. Pathogenesis of cerebral *Cryptococcus neoformans* infection after fungemia. *J. Infect. Dis.* **186:**522–530.
9. **Coenjaerts, F. E., M. van der Flier, P. N. Mwinzi, A. E. Brouwer, J. Scharringa, W. S. Chaka, M. Aarts, A. Rajanuwong, D. A. van de Vijver, T. S. Harrison, and A. I. Hoepelman.** 2004. Intrathecal production and secretion of vascular endothelial growth factor during cryptococcal meningitis. *J. Infect. Dis.* **190:**1310–1317.
10. **Cox, G. M., J. Mukherjee, G. T. Cole, A. Casadevall, and J. R. Perfect.** 2000. Urease as a virulence factor in experimental cryptococcosis. *Infect. Immun.* **68:**443–448.
11. **Crump, J. R., S. G. Elner, V. M. Elner, and C. A. Kauffman.** 1992. Cryptococcal endophthalmitis: case report and review. *Clin. Infect. Dis.* **14:**1069–1073.
12. **Drevets, D. A., P. J. Leenen, and R. A. Greenfield.** 2004. Invasion of the central nervous system by intracellular bacteria. *Clin. Microbiol. Rev.* **17:**323–347.
13. **Dromer, F., S. Mathoulin-Pelissier, O. Launay, and O. Lortholary.** 2007. Determinants of disease presentation and outcome during cryptococcosis: the CryptoA/D study. *PLoS Med.* **4:**e21.
14. **Fischer-Smith, T., and J. Rappaport.** 2005. Evolving paradigms in the pathogenesis of HIV-1-associated dementia. *Expert Rev. Mol. Med.* **7:**1–26.
15. **Franzot, S. P., J. S. Hamdan, B. P. Currie, and A. Casadevall.** 1997. Molecular epidemiology of *Cryptococcus neoformans* in Brazil and the United States: evidence for both local genetic differences and a global clonal population structure. *J. Clin. Microbiol.* **35:**2243–2251.
16. **Garcia-Hermoso, D., S. Mathoulin-Pelissier, B. Couprie, O. Ronin, B. Dupont, and F. Dromer.** 1997. DNA typing suggests pigeon droppings as a source of pathogenic *Cryptococcus neoformans* serotype D. *J. Clin. Microbiol.* **35:**2683–2685.
17. **Granger, D. L., J. R. Perfect, and D. T. Durack.** 1985. Virulence of *Cryptococcus neoformans*. Regulation of capsule synthesis by carbon dioxide. *J. Clin. Invest.* **76:**508–516.
18. **Hanson, L. R., and W. H. Frey, 2nd.** 2008. Intranasal delivery bypasses the blood-brain barrier to target therapeutic agents to the central nervous system and treat neurodegenerative disease. *BMC Neurosci.* **9**(Suppl 3)**:**S5.
19. **Harrison, T. S., H. Kornfeld, and S. M. Levitz.** 1995. The effect of infection with human immunodeficiency virus on the anticryptococcal activity of lymphocytes and monocytes. *J. Infect. Dis.* **172:**665–671.
20. **Harrison, T. S., and S. M. Levitz.** 1997. Mechanisms of impaired anticryptococcal activity of monocytes from donors infected with human immunodeficiency virus. *J. Infect. Dis.* **176:**537–540.
21. **Hogan, L. H., B. S. Klein, and S. M. Levitz.** 1996. Virulence factors of medically important fungi. *Clin. Microbiol. Rev.* **9:**469–488.
22. **Ibrahim, A. S., S. G. Filler, M. S. Alcouloumre, T. R. Kozel, J. E. Edwards, Jr., and M. A. Ghannoum.** 1995. Adherence to and damage of endothelial cells by *Cryptococcus neoformans* in vitro: role of the capsule. *Infect. Immun.* **63:**4368–4374.
23. **Ingavale, S. S., Y. C. Chang, H. Lee, C. M. McClelland, M. L. Leong, and K. J. Kwon-Chung.** 2008. Importance of mitochondria in survival of *Cryptococcus neoformans* under low oxygen conditions and tolerance to cobalt chloride. *PLoS Pathog.* **4:**e1000155.
24. **Ingram, C. W., H. B. Haywood, 3rd, V. M. Morris, R. L. Allen, and J. R. Perfect.** 1993. Cryptococcal ventricular-peritoneal shunt infection: clinical and epidemiological evaluation of two closely associated cases. *Infect. Control Hosp. Epidemiol.* **14:**719–722.
25. **Jain, N., L. Li, Y.-P. Hsueh, A. Guerrero, J. Heitman, D. L. Goldman, and B. C. Fries.** 2009. Loss of allergen 1 confers a hypervirulent phenotype that resembles mucoid switch variants of *Cryptococcus neoformans*. *Infect. Immun.* **77:**128–140.
26. **Jong, A., C. H. Wu, H. M. Chen, F. Luo, K. J. Kwon-Chung, Y. C. Chang, C. W. Lamunyon, A. Plaas, and S. H. Huang.** 2007. Identification and characterization of CPS1 as a hyaluronic acid synthase contributing to the pathogenesis of *Cryptococcus neoformans* infection. *Eukaryot. Cell* **6:**1486–1496.
27. **Jong, A., C. H. Wu, N. V. Prasadarao, K. J. Kwon-Chung, Y. C. Chang, Y. Ouyang, G. M. Shackleford, and**

S. H. Huang. 2008. Invasion of *Cryptococcus neoformans* into human brain microvascular endothelial cells requires protein kinase C-alpha activation. *Cell. Microbiol.* **10:**1854–1865.

28. **Jong, A., C. H. Wu, G. M. Shackleford, K. J. Kwon-Chung, Y. C. Chang, H. M. Chen, Y. Ouyang, and S. H. Huang.** 2008. Involvement of human CD44 during *Cryptococcus neoformans* infection of brain microvascular endothelial cells. *Cell. Microbiol.* **10:**1313–1326.
29. **Jong, A., C. H. Wu, W. Zhou, H. M. Chen, and S. H. Huang.** 2008. Infectomic analysis of gene expression profiles of human brain microvascular endothelial cells infected with *Cryptococcus neoformans*. *J. Biomed. Biotechnol.* **2008:**375620.
30. **Jong, A. Y., C. H. Wu, S. Jiang, L. Feng, H. M. Chen, and S. H. Huang.** 2007. HIV-1 gp41 ectodomain enhances *Cryptococcus neoformans* binding to HBMEC. *Biochem. Biophys. Res. Commun.* **356:**899–905.
31. **Kanoh, Y., T. Ohara, M. Kanoh, and T. Akahoshi.** 2008. Serum matrix metalloproteinase-2 levels indicate blood-CSF barrier damage in patients with infectious meningitis. *Inflammation* **31:**99–104.
32. **Kawakami, K., Y. Koguchi, M. H. Qureshi, T. Zhang, Y. Kinjo, S. Yara, K. Uezu, K. Shibuya, S. Naoe, and A. Saito.** 2002. Anti-CD11 b monoclonal antibody suppresses brain dissemination of *Cryptococcus neoformans* in mice. *Microbiol. Immunol.* **46:**181–186.
33. **Kechichian, T. B., J. Shea, and M. Del Poeta.** 2007. Depletion of alveolar macrophages decreases the dissemination of a glucosylceramide-deficient mutant of *Cryptococcus neoformans* in immunodeficient mice. *Infect. Immun.* **75:**4792–4798.
34. **Kim, K. S.** 2008. Mechanisms of microbial traversal of the blood-brain barrier. *Nat. Rev. Microbiol.* **6:**625–634.
35. **Kim, K. S.** 2006. Microbial translocation of the blood-brain barrier. *Int. J. Parasitol.* **36:**607–614.
36. **Kwon-Chung, K. J., J. C. Edman, and B. L. Wickes.** 1992. Genetic association of mating types and virulence in *Cryptococcus neoformans*. *Infect. Immun.* **60:**602–605.
37. **Lee, S. C., D. W. Dickson, C. F. Brosnan, and A. Casadevall.** 1994. Human astrocytes inhibit *Cryptococcus neoformans* growth by a nitric oxide-mediated mechanism. *J. Exp. Med.* **180:**365–369.
38. **Lee, S. C., Y. Kress, D. W. Dickson, and A. Casadevall.** 1995. Human microglia mediate anti-*Cryptococcus neoformans* activity in the presence of specific antibody. *J. Neuroimmunol.* **62:**43–52.
39. **Lortholary, O., A. Fontanet, N. Memain, A. Martin, K. Sitbon, and F. Dromer.** 2005. Incidence and risk factors of immune reconstitution inflammatory syndrome complicating HIV-associated cryptococcosis in France. *AIDS* **19:**1043–1049.
40. **Lortholary, O., L. Improvisi, M. Nicolas, F. Provost, B. Dupont, and F. Dromer.** 1999. Fungemia during murine cryptococcosis sheds some light on pathophysiology. *Med. Mycol.* **37:**169–174.
41. **Luberto, C., B. Martinez-Marino, D. Taraskiewicz, B. Bolanos, P. Chitano, D. L. Toffaletti, G. M. Cox, J. R. Perfect, Y. A. Hannun, E. Balish, and M. Del Poeta.** 2003. Identification of App1 as a regulator of phagocytosis and virulence of *Cryptococcus neoformans*. *J. Clin. Invest.* **112:**1080–1094.
42. **Monari, C., F. Baldelli, D. Pietrella, C. Retini, C. Tascini, D. Francisci, F. Bistoni, and A. Vecchiarelli.** 1997. Monocyte dysfunction in patients with acquired immunodeficiency syndrome (AIDS) versus *Cryptococcus neoformans*. *J. Infect.* **35:**257–263.
43. **Nielsen, K., G. M. Cox, A. P. Litvintseva, E. Mylonakis, S. D. Malliaris, D. K. Benjamin, Jr., S. S. Giles, T. G. Mitchell, A. Casadevall, J. R. Perfect, and J. Heitman.** 2005. *Cryptococcus neoformans* {alpha} strains preferentially disseminate to the central nervous system during coinfection. *Infect. Immun.* **73:**4922–4933.
44. **Nizet, V., K. S. Kim, M. Stins, M. Jonas, E. Y. Chi, D. Nguyen, and C. E. Rubens.** 1997. Invasion of brain microvascular endothelial cells by group B streptococci. *Infect. Immun.* **65:**5074–5081.
45. **Noverr, M. C., P. R. Williamson, R. S. Fajardo, and G. B. Huffnagle.** 2004. CNLAC1 is required for extrapulmonary dissemination of *Cryptococcus neoformans* but not pulmonary persistence. *Infect. Immun.* **72:**1693–1699.
46. **Olszewski, M. A., M. C. Noverr, G.-H. Chen, G. B. Toews, G. M. Cox, J. R. Perfect, and G. B. Huffnagle.** 2004. Urease expression by *Cryptococcus neoformans* promotes microvascular sequestration, thereby enhancing central nervous system invasion. *Am. J. Pathol.* **164:**1761–1771.
47. **Rubin, L. L., and J. M. Staddon.** 1999. The cell biology of the blood-brain barrier. *Annu. Rev. Neurosci.* **22:**11–28.
48. **Santangelo, R., H. Zoellner, T. Sorrell, C. Wilson, C. Donald, J. Djordjevic, Y. Shounan, and L. Wright.** 2004. Role of extracellular phospholipases and mononuclear phagocytes in dissemination of cryptococcosis in a murine model. *Infect. Immun.* **72:**2229–2239.
49. **Stergiopoulou, T., J. Meletiadis, E. Roilides, D. E. Kleiner, R. Schaufele, M. Roden, S. Harrington, L. Dad, B. Segal, and T. J. Walsh.** 2007. Host-dependent patterns of tissue injury in invasive pulmonary aspergillosis. *Am. J. Clin. Pathol.* **127:**349–355.
50. **Stins, M. F., N. V. Prasadarao, L. Ibric, C. A. Wass, P. Luckett, and K. S. Kim.** 1994. Binding characteristics of S fimbriated *Escherichia coli* to isolated brain microvascular endothelial cells. *Am. J. Pathol.* **145:**1228–1236.
51. **Truelsen, K., T. Young, and T. R. Kozel.** 1992. In vivo complement activation and binding of C3 to encapsulated *Cryptococcus neoformans*. *Infect. Immun.* **60:**3937–3939.
52. **Walsh, T. J., R. Schlegel, M. M. Moody, J. W. Costerton, and M. Salcman.** 1986. Ventriculoatrial shunt infection due to *Cryptococcus neoformans*: an ultrastructural and quantitative microbiological study. *Neurosurgery* **18:**373–375.
53. **Wang, H., J. Sun, and H. Goldstein.** 2008. Human immunodeficiency virus type 1 infection increases the in vivo capacity of peripheral monocytes to cross the blood-brain barrier into the brain and the in vivo sensitivity of the blood-brain barrier to disruption by lipopolysaccharide. *J. Virol.* **82:**7591–7600.
54. **Yauch, L. E., J. S. Lam, and S. M. Levitz.** 2006. Direct inhibition of T-cell responses by the *Cryptococcus* capsular polysaccharide glucuronoxylomannan. *PLoS Pathog.* **2:**e120.

Cryptococcus: From Human Pathogen to Model Yeast
Edited by J. Heitman et al.
©2011 ASM Press, Washington, DC

35

Cryptococcosis in Experimental Animals: Lessons Learned

KARL V. CLEMONS AND DAVID A. STEVENS

As detailed in the previous edition of this book (18), and in other chapters in the present volume, *Cryptococcus* spp. are a group of organisms that affect humans and animals alike, as an opportunist in the immunosuppressed and also as a primary pathogen in some instances. The advances in molecular biology technologies, the so-called "omics" (genomics, proteomics, metabolomics, etc.), and genetic manipulations have advanced our knowledge of the biology of these organisms greatly and allowed us to perform analyses not possible less than 20 years ago. For example, complete genome sequencing data (108) allow us to "mine" for genes, examine gene expression under various conditions, and do comparative genomics with related organisms in search of potential virulence genes and drug targets. Thus, molecular techniques for profiling the organisms enhance our overall understanding of the biology of these yeasts. Similarly, advances in immunological techniques now allow us to examine cytokine expression by microarray or multiplex PCR assays, as well as determine the types of host cells responding to the organism by the use of fluorescence-activated cell sorters and simultaneous staining of multiple cell surface markers. In spite of these advances, researchers remain limited to a large extent in examining either the organism or the host profiles. New technologies are needed to be able to examine the organism and the host concurrently and will surely be developed in the near future.

Although a great deal of information can be collected from in vitro assays, the contributing role of biological processes or specific genes to the capacity of *Cryptococcus* to cause disease and how the host responds to disease must eventually be examined in the complexity of an in vivo system. Because our ability to study the disease in humans remains limited, the use of surrogates as models of infection is important.

Animal models of fungal infection are an integral part of research into the mechanisms of pathogenesis, host response, disease prevention, and treatment. These models include experimental infection with mycelial or yeast-form organisms, which are primary or opportunistic pathogens in humans and animals (15). Research on cryptococcosis is not an exception, and various studies have been published in each of the areas of interest mentioned above and reviewed in other publications (15, 17, 18, 125). Historically, murine (rodent), rabbit, and guinea pig models of infection have been most widely used, although numerous other mammalian models have been reported (15, 17, 18, 125). More recently, nonmammalian models of infection have been developed and reviewed (61, 133, 186). Each of these models has a role to play in contributing to our understanding of cryptococcosis. Our goal for this chapter is to present salient examples in the use and performance of animal models of cryptococcal infection, particularly murine and rabbit models, and demonstrate how these models in conjunction with the application of emerging technologies have advanced our knowledge about this disease.

WHY STUDY ANIMAL MODELS OF CRYPTOCOCCOSIS?

Although a seemingly simple question, answered in general terms above, it is extremely important to fully define the purpose of studying a model. We, and others, have detailed the philosophy of performing animal models and the approaches to answering specific questions concerning pathogenesis, host response, or therapy for not only *Cryptococcus*, but a variety of other mycoses as well, since the same principles apply regardless of the infection being modeled (11, 15, 17, 24, 31, 32, 34–36, 125, 171, 177–179). In brief, one needs to define the parameters of the model needed to address the aim of the study. The initial question will be what species of animal, and in what numbers, should be used? A second question will be what type of disease is being emulated, and what is the host's immune status? What strain of the organism should be used? And, finally, what end points will be assessed? We caution investigators to answer each of these questions carefully, since the results from a poorly designed model are rarely useful. Thus, the model should mimic clinical disease and be standardized, reproducible, affordable, controllable with respect to severity, and have clear end points.

Karl V. Clemons and David A. Stevens, California Institute for Medical Research, San Jose, CA 95128; Division of Infectious Diseases, Department of Medicine, Santa Clara Valley Medical Center, San Jose, CA 95128; and Division of Infectious Diseases and Geographic Medicine, Department of Medicine, Stanford University, Stanford, CA 94305.

The benefits and weaknesses of animal models for cryptococcal models and fungal models in general have been discussed in detail in previous reviews (15, 18). In brief, animal models have a number of benefits in that we can quickly perform numerous studies using sufficient number of animals to attain statistical data, models are often predictive of clinical results, and the variables of the model including disease severity, infecting strain of fungus, host age and sex, and experimental duration can be closely controlled. Mice have become the animal model of choice for a number of reasons including availability of many syngeneic and immunodeficient strains, affordability, ease of animal husbandry, and availability of immunological reagents. However, one must recognize that no single model can answer all questions and that replicate experiments may not give identical results because of the complexity of the in vivo system. In addition, in some instances models can be too acute in severity or can have target organs of infection that are not the same as in the clinical infection. Lastly, the performance of some models, particularly larger animals such as rabbits, can entail high costs and specialized supportive veterinary care of the animals. The choice of animal species to use then becomes a balance of these factors.

Particular attention should be paid to the end point parameters that will be used to assess the study. In murine models, survival and fungal burden in the target organs are standard objective end points, which are readily tallied and determined. Other parameters used to follow infection are more subjective and include histopathology, body weight, where weight loss is a correlate of disease progression, body temperature, and clinical symptoms. Although it can be somewhat difficult to follow clinical signs of infection in mice, clinical evaluation is readily done on rabbits, where daily weights, body temperatures, and observations of neurological manifestations of disease (e.g., head canting, paresis, motor dysfunction, activity level, etc.) can be routinely evaluated. These parameters can be valuable indicators of disease progression and severity, and our laboratory has found these types of parameters to correlate very closely with disease severity in a rabbit model of coccidioidal meningitis (15, 24, 192). For example, progressive weight loss and increasingly severe signs of neurological dysfunction are indicative of disease progression.

The use of death as an end point deserves particular mention. In the current era of increasingly more stringent regulations on the use of laboratory animals, it is no longer acceptable to most Institutional Animal Care and Use Committees for the investigator to allow the experimental animals to die of infection. Thus, criteria are established, most often on individual protocol bases, to determine when an animal should be euthanatized to alleviate pain and suffering. These criteria may include significant weight loss (e.g., >20%), severe dehydration, paralysis or other immobility problems (e.g., spinning), or coma. These criteria are useful and should be determined and applied in close cooperation with a laboratory animal-certified veterinarian. However, one must be cautious in the application of euthanasia criteria to avoid possible skew of results and potentially drawing the wrong conclusions from the study. Our own experience has shown us that rabbits can have about 20% loss of body weight, yet with supportive therapy of analgesics, subcutaneous administration of fluids, and provision of appetite stimulants such as fresh vegetables these animals can show signs of recovery within a few days and stabilize with respect to progression of disease. In addition, we have observed rabbits that appear moribund, lethargic, and ataxic in the late afternoon, which could be cause for euthanasia, appear alert, responsive, and active the following morning, presumably coinciding with onset of a therapeutic effect. Thus, the decision to euthanatize should be made carefully and with the input of the attending veterinarian.

For mice, the situation can be very similar and the application of euthanasia criteria somewhat more difficult. An excellent example is provided by work on *Cryptococcus neoformans* capsule synthesis and virulence by Liu et al. (107). In those studies capsular synthesis mutants were tested for virulence in a pulmonary model. The mutants induced transient signs of infection lasting 4 to 8 days that included >20% loss of body weight, a criterion for euthanasia in many institutions, yet these same animals went on to recover and to appear normal, with few deaths occurring over the course of the experiment (107). Rigorous application of the euthanasia criteria would have required these same animals to be euthanatized and likely led the authors to draw the conclusion that the capsular mutants were almost equally virulent to the parental strain, rather than the conclusion that the capsular mutants had severely reduced virulence (107). Thus, the investigator should be wary and aware that euthanasia of moribund animals too soon could possibly affect the overall outcome of the study.

MURINE MODELS OF CRYPTOCOCCOSIS

Routes of Infection

Prior to establishing a model of cryptococcosis, the researcher needs to realize that choice of infecting strain, maintenance of the strain, route of infection, genetic background of the host, and immune status all contribute to how the experimental infection will progress. Mice are the most frequently used species for cryptococcal models and can be infected by one of several routes (15, 17, 125). Intraperitoneal infection using 10^5 to 10^7 yeast cells is technically easy to master, and many mice can be infected quickly. However, dissemination from the peritoneal cavity to the organs, particularly the brain, can be inconsistent, making reproducibility an issue. Intravenous inoculation of 10^3 to 10^5 organisms establishes systemic infection immediately and results in consistent multiple organ (e.g., lung, liver, spleen, kidney) and meningeal involvement. Although the inoculation technique is more difficult to master, once adept, a researcher can infect 100 to 200 animals in an hour. Interestingly, the fungal burdens in the visceral organs will decline with time in this model, leaving the brain as the primary target organ showing progressive disease.

Human cryptococcosis is acquired via inhalation, and thus, pulmonary models of infection are better mimics of the natural disease progression. Mice can be infected via inhalation of yeasts (ca. 10^3 to 10^4 in 30 to 50 μl) by placing droplets of inoculum on the nares of an anesthetized mouse and allowing the animal to inhale the droplets. An important aspect of the procedure is the depth of the anesthesia, which needs to be sufficient to suppress the swallow reflex. At the proper plane of anesthesia, a droplet placed on the nose will induce a strong "sniff" response. Animals under too little anesthesia inhale only a small portion of the inoculum, while swallowing the majority of it, or sneeze the inoculum away. Animals under too deep a plane of anesthesia will not readily inhale the inoculum. We have found that the use of an inhalational anesthesia, such as isoflurane, provides the control needed, with the advantage of the mice recovering within minutes of the procedure. Ketamine HCl or pentobarbital are injectable anesthetics that can be used. However, recovery times of up to 1 h can be considered a drawback. Regardless of the anesthesia, it is important to keep the mouse in a vertical position and also ensure the closure of its mouth to force the inspiration of the inoculum. Some investigators use an intratracheal inoculation to infect via the pulmonary route to ensure consistent administration of the inoculum. However, this procedure is technically more difficult, requiring a surgical plane of anesthesia, an incision to expose the trachea for inoculation, and final closure of the wound.

Because cryptococcosis often causes severe meningeal disease, some investigators use a direct intracerebral inoculation technique to establish central nervous system (CNS) infection. In this procedure the animals are anesthetized using an inhalational anesthesia. The site of injection (ca. 10^3 to 10^4 cryptococci in ≤50 μl volume) is 4 to 5 mm posterior to the eyes midline to the cranium to a depth of about 2 mm using a small-gauge needle (e.g., 26 gauge). Once mastered, it is not a difficult technique to perform, although some deaths can occur as a result of the injection, especially if the plane of anesthesia is extremely deep, and areas of hemorrhage on the surface of the brain can also be seen on necropsy. A drawback to this route of inoculation is that it establishes disease in the parenchyma of the brain, as well as in the meninges.

An alternative route of infection, used in our laboratory for studies of meningitis due to *Coccidioides immitis*, is via lumbar injection into the cerebrospinal fluid (CSF) (91). The mouse is anesthetized to a surgical plane, the hair shaved from the back, and an incision made in the skin exposing the L1L2 lumbar spine. A small-gauge needle is used to inject the organisms in ≤50 μl volume. When this is done correctly, the tail will flick slightly. The skin is closed using wound clips (7 mm) or with sutures. This establishes meningeal infection with as few as 10 arthroconidia (91). In preliminary unpublished studies we have used this route to establish meningeal cryptococcosis but have not fully developed this as a model, which would be similar to the well-described rabbit model of cryptococcal meningitis (151). The investigator also should be aware that with either of these methods the organisms will disseminate from the CNS space and can establish infection in other organs.

Mouse Strains

The genetic background of the mouse greatly influences the course of infection, and numerous studies have addressed the susceptibility of the host, as well as the role played by various components of the immune system in resistance to disease. In early studies by Rhodes et al. (158, 159), susceptibility to fatal cryptococcosis by either an intravenous or intraperitoneal route of infection of a single strain of *C. neoformans* was shown to be primarily related to C5 deficiency, where strains deficient in C5, such as DBA/2J or A/J, died sooner than did C5-sufficient strains such as BALB/c, B10, or DBA/1J. It should be noted, however, that there were differences in the susceptibility of the C5-sufficient strains with respect to time to death after intravenous infection (159). Interestingly, Macher et al. (115) noted that complement levels in experimentally infected guinea pigs showed hypocomplementemia, and intracardial injection of heat-killed cryptococci induced the same reduction in complement. Although no association with major histocompatibility complex (MHC) haplotype was noted in those studies, others have found that female B10 MHC congenic mice with a *k/k* haplotype were more susceptible to intraperitoneal infection on the basis of recovery of *Cryptococcus* strain H99 (serotype A) from the liver (122). Interestingly, in those studies the authors reported that infection of the male mice resulted in inconsistent infectious burdens between experiments, which in some instances showed reversed haplotype susceptibility patterns (122). The contrast of the results between the studies (122, 159) is potentially due to the use of different organisms, as well as CFU versus lethality as end points; the possibility of sex-related differences in susceptibility to infection is also raised by the latter study (122). Lortholary et al. (109) have reported differences in cytokine response between male and female mice, as well as the increased susceptibility of young male mice.

Interestingly, a study by Aguirre et al. (3) examined the issue of age-related resistance and found that mice aged 22 to 24 months (laboratory mice have an estimated lifespan of 2 to 3 years) were more susceptible to intravenous or intranasal challenge than were mice 2 to 3 months of age. Furthermore, these authors found that this difference was not related to T-cell function (3). Thus, susceptibility to primary cryptococcal infection in mice appears to show that two extremes of age of mice have increased susceptibility: very young mice (i.e., <6 weeks of age) and very old mice (22 to 24 months of age). It could also be of interest to determine whether altering the route of infection would give the same result with respect to susceptibility pattern. Comparison across various studies is suggestive that by route of infection, the numbers of cryptococci required to cause lethal infection has a rank order of intracerebral ≤ intranasal ≤ intravenous << intraperitoneal << subcutaneous. Studies with other fungal models have shown changes in patterns of susceptibility on the basis of route of infection (15).

In addition to the association of mouse susceptibility with C5 and MHC, various strains of genetically immunodeficient mice have been shown to be exquisitely susceptible. Various studies have demonstrated susceptibility of the nude mouse (T-cell deficient) (68, 88, 127, 128, 141), SCID mouse (T- and B-cell deficient) (16, 25, 65, 66, 68, 74, 75, 88), beige mouse (polymorphonuclear leukocyte degranulation defects and reduced NK cell activity) (120, 166, 168), and beige athymic mouse (168). Models of cryptococcosis studied in B-cell-deficient mice appear to show discordant results between X-linked *xid* mice (B-cell defects and antibody deficiency) (119) and anti-mu-treated mice, which are B-cell deficient and unable to produce antibody (127), as the latter were determined to be no more susceptible to *C. neoformans* than were controls, whereas the *xid* mice were more susceptible than parental (119). Comparative results such as these are suggestive that the production of antibody by B cells is not the sole contributor to resistance to cryptococcosis, but that the B cell contributes in other ways, possibly through cytokine production, to host resistance. In more recent studies by Rivera et al. (161), B-cell-deficient mice were found to be more susceptible to intranasal or intravenous infection and had higher levels of interferon-γ (IFN-γ), MCP-1, and MIP-1α, suggesting that antibody produced by B cells has a regulatory effect on the production of these proinflammatory molecules (161).

The genetically deficient mice mentioned above have naturally occurring mutations that are pleiotrophic in their effects. For example, nude mice are athymic, resulting in the T-cell defect, but are also hairless and show increased NK cell and macrophage activity, and in spite of the T-cell deficiency have shown paradoxical resistance to some experimental infections such as systemic candidosis (41). Similarly, beige mice have degranulation and chemotactic defects, particularly of polymorphonuclear leukocytes, but also show pigmentation dilution (responsible for their lighter coat color as compared to normal littermates) and decreased NK cell activity. Even mouse strains considered as "normal" can show differences in immune response contributing to susceptibility to fungal infection. In studies of coccidioidomycosis it was demonstrated that DBA/2 (resistant) mice showed a strong and early IFN-γ response, whereas BALB/c (susceptible) mice had a strong Th2 response and low IFN-γ production (117). Thus, results obtained with these types of animals must be evaluated carefully. Overall though, the use of these models has clearly shown us the contribution of different cells of the innate immune response, cell-mediated immunity, and the involvement of the humoral arm of acquired resistance against cryptococcal infection.

Mapping the Evolution of the Immune Response to Cryptococcal Infection

An alternative approach to the use of naturally, or antibody-depleted, immunodeficient mice for determination of what parts of the immune response are most important has been the study of murine models of cryptococcal infection for the determination of the temporal development of the immune response to infection. In early studies, murine models of infection were used to study the evolution and specificity of delayed-type hypersensitivity after infection with *Cryptococcus* (72, 157). Subsequent studies have included determining the cell types infiltrating lesions, and T lymphocyte subsets involved in adaptive resistance and examination of cytokine profiles induced by infection. In studies done with a pulmonary model of cryptococcosis, histologic examination demonstrated prominent T-cell-mediated inflammatory response composed of lymphocytes, macrophages, and neutrophils (83, 87). Furthermore, the importance of T cells was demonstrated by adoptive transfer experiments done in SCID mice showing increased pulmonary clearance of the cryptococci in animals given T cells (88).

Additional studies showed a predominant role for $CD4^+$ in regulating the inflammatory response, as well as the involvement of $CD8^+$ cells (85, 87). Similarly, $CD4^+$ cells were found to be important to resistance of immunized mice. Higher numbers of $CD4^+$ cells were present in the brains of intracerebrally infected immune mice than were found in nonimmune mice (12). Furthermore, depletion of the $CD4^+$, but not the $CD8^+$, cells in vivo exacerbated disease, as did depletion of IFN-γ, and impaired cellular infiltrations (12). Depletion of $CD4^+$ cells using antibodies showed decreased survival and increased fungal burden or dissemination after intravenous or intratracheal routes of infection (126). Interestingly, the fungal burden in the intratracheally inoculated mice depleted of $CD4^+$ cells was not exacerbated, although more dissemination from the lungs was noted in these animals (126). We examined the temporal accumulation of T cells, B cells, and monocytes by immunohistochemistry of brain tissue from systemically infected BALB/c mice to follow the development of the immune response in the CNS compartment (116). The initial infiltrating cells were $CD11b^+$ and were determined to be primarily monocytes, which were first detected on day 7 of infection. By day 10, the infiltrates comprised $CD4^+$ and $CD8^+$ T lymphocytes (estimated ratio of 2.5:1), $CD11b^+$ cells, and few $CD45R^+$ B cells. This same pattern of cellular infiltrate was observed through 28 days of infection, but no granuloma formation was evident, nor were giant cells observed (116). These results further corroborated those reported by Dobrick et al. (49), with the influx of $CD11b^+$ cells and the relatively higher ratio of $CD4^+$ to $CD8^+$ cells at site of infection.

The advent of newer technologies, such as PCR, and the availability of enzyme-linked immunosorbent assay kits allowed for the determination of cytokine profiling. In studies of this type, cytokine expression has been measured by mRNA or protein in various tissues and fluids such as bronchoalveolar lavage (13, 77, 84, 86, 94–96, 109, 110, 116, 162, 187, 194). Kawakami et al. (96) examined mRNA cytokine profiles in the lungs of intratracheally infected BALB/c × DBA/2 F1 mice. They found minimal expression of the Th1 cytokines LT or IFN-γ and no expression of interleukin-2 (IL-2) through 14 days of infection, whereas IL-6, IL-4, and IL-10 were detected beginning on day 1 and continuing through day 14 of infection (96). Similar to the Th1 cytokine profile, these authors found little to no expression of tumor necrosis factor-α (TNF-α), IL-12p40, or inducible nitric oxide synthase (iNOS); transforming growth factor-β (TGF-β) was detected only on day 1 of infection (96).

Blasi et al. (10) examined the cytokine response in brain after intracerebral inoculation with heat-killed cryptococci, which induce resistance to intracerebrally inoculated infection, but not intravenously inoculated infection with *Cryptococcus*. Using a PCR methodology, IL-6 and IL-1β mRNA levels were increased in the brains of the heat-killed cryptococci-inoculated mice versus saline-treated animals; interestingly, TNF-α and IL-1α levels did not change (10).

We examined cytokine expression in the brain on a temporal basis by PCR methods (116). Systemically infected BALB/c mice were studied and the progression of meningitis followed in conjunction with the cytokine profile. Interestingly, we found constitutive expression of TGF-β, IL-18, IL-12p40, and IL12p35 in infected and uninfected mice through 28 days of infection. However, IL-12p40 expression increased beginning on day 10 of infection. Induced expression of higher levels of mRNA were found for IL-1α, IL-1β, IL-4, IL-6, IL-10, IFN-γ, TNF-α, and iNOS in the infected mice but not uninfected mice except for barely detectable levels of IL-1α. These cytokines varied in the time of first detection, with IL-1α, TNF-α, and iNOS detectable on day 1 of infection, whereas IL-4, IL-6, and IFN-γ were first detectable on day 5, and IL-1β and IL-10 first on day 10 of infection (116). Our results were suggestive of the lack of a polarized T-cell response in the brain during the 28 days of infection, since both Th1 and Th2 cytokines appeared at about the same times and to the same extent. Interestingly, and similar to the study by Kawakami et al. (96), IL-2 was not detected at any time point, which is suggestive that there is not a localized proliferation of T cells occurring in the brain or in the lung tissues. Uicker et al. (187) compared cytokine profiles in the brain of immunized and nonimmunized mice. They found a Th1 pattern of cytokine expression in intracerebrally infected immunized CBA/J mice and reduced Th2 cytokine expression in comparison with nonimmune infected controls, indicative that the cytokine profile can be driven to a protective Th1 profile.

Overall, studies of the development and mapping of the immune response in mice have demonstrated the protective nature of the Th1 response and more recently the inclusion of Th17 cells (producing IL-23 and IL-17) in this protective response (98), whereas the Th2 response is not protective, and that this profile can be modulated by immunization. Furthermore, the T helper subset studies have shown the importance of $CD4^+$ cells in regulation of the inflammatory infiltrates, particularly in the brain, but also that the profiles of different organs are similar. Future studies using microarray technology to follow the global response of the mouse will be of great interest, particularly to determine whether organ-specific responses occur.

Gene-Knockout Mice

Advances in molecular techniques have allowed the commercial breeding of specific gene-knockout (KO) mice, which are deficient in one or more genes including cytokines, enzymes, cell surface receptors, and lymphocyte subsets (45, 82, 162). Although these mice can be extremely expensive to purchase, they make a number of different types of studies possible. However, one should be aware that the immune system is complex and redundant, which can lead to possible confounding effects and unclear results. In spite of the cost and possible difficulties in interpreting results, the use of various KO mice has been a particularly fruitful area of research in the field of host response to cryptococcal infection. Numerous publications have detailed studies using these animals to begin to dissect the contribution of individual cytokines, chemokines, cell surface receptors, and lymphocyte subsets in the host response to infection.

During the initial stages of infection the organism encounters pattern recognition proteins, such as collectins, Toll-like receptors (TLRs), and Dectin-1, responsible for recognition of the organism and initiation of the immune response, as well as activation of innate mechanisms of response. The collectins, surfactant protein A (SP-A), and SP-D are soluble collectin proteins, which have been shown to be important in the innate immunity to some organisms and regulators of TNF-α (11a, 40). Giles et al. (62) examined the role of SP-A in pulmonary cryptococcosis in SP-A$^{-/-}$ KO mice to determine a possible role. Interestingly, the SP-A$^{-/-}$ mice were found to be no more susceptible to cryptococcal infection nor to have depressed TNF-α production in the lungs (62). These results, although in contrast to other organisms (15, 40, 62), confirmed the poor in vitro binding of SP-A to encapsulated *Cryptococcus* (189), with the animal model demonstrating that SP-A does not appear to play a significant role in the innate response.

Mannose receptor is a C-type lectin receptor that binds mannosylated antigens and participates in antigen presentation by dendritic cells. Dan et al. (42) investigated the importance of the mannose receptor's role in presenting mannosylated antigens (mannoproteins) present on *Cryptococcus*. They found that MR$^{-/-}$ KO mice were more susceptible than control mice to lethal pulmonary infection and had higher CFU in the lungs after 28 days of infection with serotype A organisms and that CD4$^+$ T cells did not proliferate in response to the mannosylated antigen (42). However, the authors point out that the control mice also died, surmising that the mannose receptor contributes, but does not mediate, a completely protective immune response (42). Moreover, the mannose receptor has a role in cellular trafficking, and its absence may lead to immunodeficiency on other bases. Dectin-1 is a type II C-type lectin responsible for binding β-glucan by dendritic cells and is important to the immune response against some fungi. Interestingly, pulmonary or systemic infection of Dectin-1$^{-/-}$ KO mice with serotype A *Cryptococcus* showed these animals to be equivalent in susceptibility to controls and have no difference in IFN-γ or IL-12 production (138). This lack of difference was attributed to the relatively low amount of β-1,3-glucan in the cryptococcal cell wall compared to an organism such as *Candida albicans*. The use of KO mice in the collectin studies cited serves to illustrate that the initial host response to *Cryptococcus* may be somewhat unique among the fungi and that not all collectin pathways are involved.

The TLR family of receptors' recognition of microbial products results in stimulating the production of proinflammatory cytokines through the MyD88-dependent pathway or other NFκB pathways. In vivo studies have demonstrated that MyD88$^{-/-}$ and TLR2$^{-/-}$ KO mice are significantly more susceptible to pulmonary or systemic cryptococcal infection (6, 197) but that TLR4 is not likely to contribute significantly to resistance (6, 197) even though TLR4 does bind glucuronoxylomannan (GXM) (197). Furthermore, the relatively greater susceptibility of MyD88$^{-/-}$ KO mice is indicative of a larger role for it in resistance, as it is also part of other cytokine regulatory pathways, IL-1R and IL-18R, the latter of which ultimately results in IFN-γ production and subsequent activation of macrophages. Thus, these studies demonstrate that the various signaling pathways are not equally important in resistance to cryptococcal infection.

A variety of KO mice deficient for a specific cytokine, chemokine, receptor, or enzyme have also been used in models of cryptococcosis. In our laboratory we found that IFN-γ KO mice were more susceptible to systemic infection with serotype A *Cryptococcus* (181). Others have found similar increased susceptibility with IL-12$^{-/-}$ KO (5, 44, 93) and IL-18 KO (92) mice, both of which are regulators of IFN-γ production, and thus the increased susceptibility to infection is likely the result of the reduced production of IFN-γ rather than a direct effect of the IL-12 or IL-18. More recently, IL-23$^{-/-}$ KO mice have been shown to be more susceptible to intravenous or intraperitoneal infection with serotype D *Cryptococcus*, and these mice also show a reduced production of IL-17, but not an altered Th1/Th2 balance in response or reduction in IFN-γ levels (98). A mechanism by which IFN-γ plays its role in resistance is via the stimulation of iNOS and NO production (111, 112).

The involvement of Th2-associated cytokines has also been investigated using KO mice. Different studies have demonstrated that IL-10$^{-/-}$ mice are less susceptible to cryptococcal infection (5, 7). Studies done with IL-4$^{-/-}$ KO mice are somewhat less clear. It appears that IL-4$^{-/-}$ KO mice may be less susceptible to lethal infection than controls (5, 9, 73) but that the virulence of the infecting strain may influence this result, with less virulent strains cleared more readily from the lung tissues (5, 9, 73). More recently, a similar result has been reported comparing smooth and mucoid colony types, which showed IL-10$^{-/-}$ KO mice were more resistant than controls to the less virulent smooth colony type and equally susceptible to the more virulent mucoid colony type strain (69). Additional evidence for the role played by cytokines regulating the Th2 response comes from the demonstration that IL-13$^{-/-}$ KO mice were more resistant to infection than controls, and the IL-13-overexpressing transgenic mice were more susceptible to pulmonary infection (132).

In addition to the KO mice mentioned in the preceding paragraphs, other cytokine-deficient mice such as IL-6$^{-/-}$ and a variety of cc type chemokine receptor (CCR) KO mice have been examined to determine the relative roles played by these immunomodulatory molecules (5, 17, 35, 36, 81). In some studies, mice have been depleted of a specific cytokine, such as IFN-γ (167, 181), IL-12 (76), or TNF-α (2), by treating immunocompetent animals with a monoclonal antibody directed at the cytokine of interest. Studies such as these have given results very similar to those using the KO mice. However, one must be aware that antibody depletion of a cytokine is somewhat different than the use of a KO mouse, which may have developed compensatory mechanisms to overcome the cytokine deficiency, and that antibody depletion does not always give an equivalent result. One example comes from our work with aspergillosis and the involvement of IL-10 (27). In those studies we found that IL-10 KO mice were more resistant to infection but that the use of an anti-IL-10 known to deplete mice of IL-10 or the use of anti-IL-10R did not make the animals more resistant (27). Thus, it may not be solely the deficient cytokine in the KO mouse that alters its susceptibility, or, alternatively, antibody may incompletely remove a cytokine, and investigators should be cautious in interpretation of results.

One aspect of studies of host response that has been most frequently associated with therapy is that of immunomodulation. IFN-γ and IL-12 have been prime candidates for use as adjunctive therapies for cryptococcosis (15, 26, 28, 32, 36, 97, 99, 114, 180, 181, 198). In our studies, we found that the IL-12 in combination with fluconazole or IFN-γ given alone or in combination with amphotericin B significantly enhanced the antifungal therapy (26, 28, 114). Interestingly, another approach to the administration of pure cytokine has been immunomodulation of the host by the use of the organism. Wormley et al. (195) engineered a strain of *Cryptococcus* to express murine IFN-γ, using this strain in pulmonary infection of mice. They found these animals resolved the primary infection and were resistant to secondary challenge with an altered T-cell profile; animals that had received the IFN-γ-producing strain had a depressed Th2 response (195). We have examined the potential of gene therapy using an adenovirus vector expressing IFN-γ, administering it intracerebrally (30). In those studies we found high concentrations of IFN-γ in the CSF of mice for a period of several days (30). However, preliminary studies showed that these mice did not show greater resistance to systemic infection with serotype A *C. neoformans* (unpublished data). Regardless, the possibility of using an organism, even *Cryptococcus* itself, to alter the immune response, potentially resulting in increased resistance to secondary infection or even treatment of a primary infection, is a subject for additional future studies.

Overall, the studies using various strains and genetically deficient mice have demonstrated the complexity of the immune response to cryptococcal infection. The Th1 response has been predominant in its role, although the humoral response is also a necessary part of resistance. In addition, the immune response has been demonstrated to be tissue specific and also related to the relative virulence of the organism. These lessons further illustrate the utility of well-designed and well-performed models of infection.

STUDIES OF VIRULENCE AND PATHOGENESIS IN MURINE MODELS

The question of what makes *Cryptococcus* virulent has been addressed by numerous investigators for many years and will continue to be studied for many years to come. Included as part of these studies are those of pathogenesis and the evolution of disease, which is intertwined with virulence, making individual discussions of them difficult. Unlike some bacterial infections where a single toxin is responsible for resultant disease (e.g., diphtheria toxin) or cellular constituent (e.g., capsule production for *Streptococcus pneumoniae*) and loss of that single trait eliminates the virulence of the organism, the virulence of eukaryotic fungi is a quantitative trait, with no single gene being responsible. The quantitative nature of fungal virulence is nowhere more evident than with *Cryptococcus*, where many genes and cellular constituents are considered to be virulence factors (8, 15, 17–19, 21, 22, 37–39, 54, 70, 79, 100, 101, 106, 113, 121, 123–125, 142, 145, 153, 164, 165, 169, 174). In general, the virulence of *Cryptococcus* spp. has been related to capsule formation, melanin, calcineurin, phospholipase, urease, and growth at 37°C. How a trait contributes to virulence and to what degree it contributes are difficult questions to answer. Common throughout the years of study of the virulence of *Cryptococcus* is the reliance on animal models of infection, particularly murine models, to help elucidate just what traits make this organism a pathogen and how the disease develops. Because in-depth discussion and review of virulence studies are presented in other chapters in this volume, we will limit our discussion to a few examples of how animal models are used and what they can tell the investigator.

Strain of *Cryptococcus*

The choice of the strain(s) of *Cryptococcus* the investigator makes to use in studies is critical, and organisms that may be tractable on a genetic basis may not be the most virulent, or in some instances may be avirulent. We make this statement based on our experience in the study of the virulence of *Saccharomyces cerevisiae*, where comparison of the relative virulence of clinical, nonclinical, and genetically defined strains showed a range of virulence potentials, with the unfortunate result that the strains most readily used for genetic manipulation of *S. cerevisiae* were the least virulent in CD-1 mice (29). Similarly, the comparative virulence of *Cryptococcus* also varies among the serotypes and within a serotype. An early example of difference in virulence and disease progression was found by Staib and Mishra (172). In these studies using two strains of *C. neoformans*, they noted that after intraperitoneal inoculation of mice, one strain could be recovered primarily from the brain, and only the brain at later time points, but induced no neurological symptoms in the mice and caused only low mortality, whereas a second strain resulted in neurological symptoms, very high mortality, and could be recovered from lung, liver, spleen, kidney, and heart, as well as brain; interestingly, both strains formed capsules in vivo (172). Thus, in the same strain of mouse a substantial difference in the comparative virulence and the classic tissue tropism to the brain exhibited by *Cryptococcus* were demonstrated.

One interesting study has used the different strain serotypes A, B, C, and D of *Cryptococcus* and compared the capacity of these strains to disseminate from a footpad route of infection to the organs and brain in four different strains of mice (89). These authors noted that *C. neoformans* var. *grubii* (serotype A) and *C. neoformans* var. *neoformans* (serotype D) were more invasive, particularly in DBA/2 and BALB/c mice, than was *Cryptococcus gattii* (89). In addition, they found A/J mice and hybrid mice more resistant to dissemination from the footpad. These results clearly demonstrate the influence of strain of *Cryptococcus* and the strain of mouse used on the comparative virulence of the organism and the course of disease. Barchiesi et al. (4) further examined the relationship of serotype to virulence, correlating it with mating type. These authors found serotype A *MAT*α stains to be more highly virulent than Aa or Da types and found that serotype D *MAT*α strains showed a range of virulence in a systemic murine model, as well as in fungal burden in the brain after pulmonary infection (4). Thus, an association of mating type with virulence was established by two different routes of infection.

Further to this point, Clancy et al. (23) found that *C. neoformans* var. *grubii* isolates from AIDS patients range in virulence capacity from highly lethal to hypovirulent in outbred ICR mice, inbred BALB/c mice, and BALB/c nude mice in an intracerebral model. Fries et al. (58) have found differences in virulence in cryptococcal strains serially isolated from AIDS patients. A range of virulence has also been found in environmental isolates of *Cryptococcus* (143). More recently, Lin et al. (106) examined the influence of ploidy in relationship to mating type, demonstrating a minimal effect on virulence with hybrids that were αAAα or αADα or **a**α AD, but a reduced virulence of **a**AD**a** hybrid, further indicating mating type association and also indicating the minimal effect of ploidy.

The differences noted above raise the very important point of whether only one strain of organism and one strain of mouse should be used in virulence studies. Using the example of our own studies of *S. cerevisiae*, we found that changing the host strain of mouse can alter the comparative virulence of strains. In the original study, we used CD-1 mice and found a range of virulence in the strains of *S. cerevisiae* tested based on comparative CFU recovered from the brain after 14 days of infection, but no mortality, whereas the use of DBA/2 mice resulted in some strains of *S. cerevisiae* becoming lethal to the mice, and some strains considered virulent by our definition in the CD-1 mice showed greatly reduced virulence in the DBA/2 mice, indicating the influence of the host and possible multiple strategies for virulence (14, 29).

An excellent example of changes in apparent virulence by *Cryptococcus* comes from the study of Luberto et al. (113). In this work, the authors identified a novel virulence factor, which they named App1 (i.e., antiphagocytic protein 1). An App1 KO strain of *Cryptococcus* was engineered and tested in vivo for virulence in a pulmonary model in two different strains of mice, A/Jcr and Tgε26 (these mice lack functional T and NK cells and mimic AIDS patients [113]). A/Jcr mice infected with the App1 KO strain survived significantly longer than did A/Jcr mice infected with the parental wild-type strain of *Cryptococcus*, (113); mice infected with the App1 KO strain also had lower fungal burdens in the lungs and less dissemination to the brain than those infected with the wild-type *Cryptococcus*. These results are indicative of App1 being a factor associated with virulence in that the KO strain had reduced virulence. However, the investigators received a surprise when infection of the Tgε26 mice with the same strains provided a complete reversal of the comparative virulence. In those animals, the App1 KO strain of *Cryptococcus* was significantly more virulent than the wild-type parental strain, as evidenced by shorter survival of infected animals and higher fungal burdens and dissemination to the brain. Thus, the App1 KO strain of *Cryptococcus* appeared "hypervirulent" in the immunodeficient mice (113).

There are several lessons to be learned from these studies of *Saccharomyces* and *Cryptococcus*. First and foremost is the point that the relative virulence of the organism is highly dependent on the host and the immune status of the host and, thus, the crucial intertwining of fungal virulence with the host response. That a fungal gene or constituent determined to be a virulence factor in a single strain of mouse may be a virulence factor only in that specific strain of mouse should make the investigator cautious in making blanket assumptions concerning the importance of that gene or constituent. Lastly, the use of only comparative survival as a measure of virulence would lead to the conclusion in some studies that the organism is avirulent, since no deaths occurred due to the infecting strains of fungus. Although survival is a very objective parameter, it may be somewhat crude, especially for fungi that are often considered to be of low virulence. The use of comparative CFU recovered from target organs provides a more sensitive measure of the relative capacity of the organisms to proliferate in the host, and if done on a temporal basis gives a profile of infection and relative virulence.

Murine Studies with *C. gattii*

Murine models and virulence studies of *C. gattii* deserve special mention, as the majority of published work has been done using *C. neoformans* var. *grubii* or *C. neoformans* var. *neoformans*. Clinically, *C. gattii* most often infects immunocompetent individuals, whereas *C. neoformans* var. *grubii* and *C. neoformans* var. *neoformans* are most often associated with infection in immunosuppressed/immunodeficient individuals. However, *C. neoformans* var. *grubii* can infect and produce disease in immunocompetent individuals, and it has been suggested that infections with these strains may be attributable to mutations other than *pkr1* (this mutation makes serotype A strains hypervirulent [53]); interestingly, the serotype A strains recovered from the immunocompetent patients were virulent in a pulmonary model of murine infection but were less virulent than the laboratory standard serotype A strain H99, which was originally recovered from a Hodgkin's lymphoma patient (53). As mentioned above, *C. gattii* has been considered to be less virulent than *C. neoformans* var. *grubii* or *C. neoformans* var. *neoformans* in comparative studies (16–18, 89, 101, 125, 185). Similar to the studies of *C. neoformans* var. *grubii*, differences in strain virulence and an association with capsule formation and colony switching has been noted for *C. gattii* (16, 20, 89, 90, 101, 185). Torda et al. (184) established a murine pulmonary model of *C. gattii* infection that mimicked the pathology of that observed in humans, which was composed of a mixed inflammatory response, granulomas, T cells, interstitial pneumonitis, and B cells. What has not been demonstrated by murine studies is the corroboration of the human data that *C. gattii* causes disease in immunocompetent more than in immunodeficient hosts.

In comparative studies, we found that *C. neoformans* var. *grubii* was more virulent than various strains of *C. gattii* in normal BALB/c mice, hydrocortisone-suppressed BALB/c mice, or SCID mice and that *C. gattii* had no predilection for infection in immunosufficient animals (16). We should point out that we were unable to administer the *C. gattii* inocula intravenously at very high numbers of yeast for the inoculum (i.e., ca. 10^7 yeasts) because of deaths within a few hours. These deaths were likely due to physical blockage of the capillary beds by the highly mucoid strains of *C. gattii*. In addition, we noted that one strain of *C. gattii* showed an unusual tissue tropism for mucosal (i.e., gastrointestinal tract) and cutaneous tissue that was observable only after 30 days of infection in BALB/c mice, and interestingly, rectal prolapse was more frequent in immunosuppressed animals (16). Additional studies will be required to clarify this tropism, but it does suggest different mechanisms of virulence among the serotypes of *Cryptococcus*.

In the recent Vancouver Island outbreak, the most prominent *C. gattii* strain (R265) causing disease (95% of cases), a recombinant from a parental strain (R272) that uncommonly causes disease, clearly was more virulent in murine studies compared to the parental strain. This study suggests that *C. gattii* strains in nature can recombine, and this affects their disease-producing fitness in humans and mice (57).

RABBIT MODELS OF CRYPTOCOCCOSIS

The use of rabbits in models of fungal infection has been much more limited than the use of rodents. Several aspects of using rabbits limit their use and include the cost of purchasing and maintaining these animals, the reduced numbers that can be handled in an individual experiment by most investigators and their animal care facilities, the need for more highly skilled technical personnel, the lack of inbred strains (i.e., syngenic), and the paucity of immunological reagents useful for rabbits. In spite of this, the rabbit has been used as a model for a number of fungal infections, including candidosis, aspergillosis, coccidioidomycosis and cryptococcosis (15). The use of the rabbit for models of cryptococcosis was nicely reviewed in the first edition of this book (18), and that review is recommended reading for those wishing to perform a rabbit model of cryptococcosis.

The model is established by direct intracisternal inoculation of the organism into the CSF. Although this may appear to be a daunting route of infection, with proper training and manipulation of the animal, the procedure is readily performed. The animal is anesthetized usually with ketamine or a cocktail of ketamine, xylazine, and acepromazine, or other chemical anesthetics, or with an inhalant anesthesia, usually isoflurane. Once under a deep plane of anesthesia, the rabbit can be placed in a laterally recumbent position, the head manipulated to open the cisterna to access, and the inoculation done using a small-gauge needle (e.g., 25 gauge). Of note, the investigator should be aware that rabbits are sensitive to anesthesia, and the advice of an experienced veterinarian is crucial. In our work on rabbits in a CNS model of coccidioidomycosis, we use the ketamine-xylazine-acepromazine cocktail for the infection process, since this is done in a class II biosafety cabinet, and then use isoflurane when sampling CSF. One should be aware that recovery from anesthesia may take upward of 1 h or more using the ketamine cocktails, whereas recovery from isoflurane is usually a matter of less than 15 minutes until the animal is fully alert and can be returned to the animal facility. It is important to monitor these animals, as their airways can be constricted easily while they are anesthetized, resulting in suffocation. A rabbit has approximately 8 ml of CSF, and samples of up to 1 ml can be taken routinely. The frequency of sampling varies according to experiment, but we have sampled the same animals four times in a single day taking 0.5-ml samples each time. In comparison, CSF sampling from a mouse is usually a terminal procedure, and a clean blood-free sample of about 10 μl of CSF is a good result.

In the initial description of the development of a rabbit model of chronic cryptococcosis, Perfect et al. (151) demonstrated that immunosuppressing the animal with cortisone allowed the organism to flourish in the CNS and be recovered from the CSF in spite of the 39°C core temperature, which is suboptimal for the growth of the organism, and established a meningeal infection that mimics human disease. Although the route of infection was direct intracisternal inoculation of the organism into the CSF, rather than a more natural route of dissemination to the CNS from a pulmonary focus, the model has proven invaluable. For example, numerous studies have been performed to assess the efficacy of different therapeutics (146, 147, 154, 170, 196) and studies of virulence (63, 140, 155, 164, 165, 183, 190, 193).

The rabbit has provided the perfect system for the study of in vivo gene expression by *Cryptococcus* because the organism can be recovered from the CSF, and the CSF of the same animal can be sampled on a temporal basis. Thus, the adaptation of the organism to the environment of the host could be followed. Steen et al. (173) used a serial analysis of gene expression library to characterize in vivo gene expression of *Cryptococcus* serotype A *MATα* strain H99. They used pooled cryptococci recovered from rabbits on three different days of infection and sequenced the tags recovered from these organisms (173). Comparisons of the recovered tags could then be done with serial analysis of gene expression libraries for in vitro growth at different temperatures (i.e., 25 or 37°C), resulting in 1,459 tags shared among the three conditions of in vivo growth and in vitro growth at 25 or 37°C (173). The in vivo expression results showed genes more highly expressed than from in vitro growth for a variety of cellular processes of protein synthesis or catabolism, stress response, respiration, transport, signal transduction, and other metabolic processes (173). This is a seminal study, which lays the groundwork for truly determining what the organism is doing during the infective process, how it acquires the needed nutrients, what metabolic pathways are crucial to in vivo proliferation, and how it responds to the stresses placed on it by the host.

Studies of host response in the rabbit have been more limited (78, 150, 152), and as noted, this is in part due to the lack of immunological reagents. Immunoglobulin G in CSF was found to be an opsonin (78), and CSF macrophages from nonsuppressed animals showed increasing H_2O_2 production and fungal killing, but those from suppressed animals had reduced capacity to kill *Cryptococcus* (150), and vaccination of animals did not result in protective response to CNS infection (152). We are aware of no studies examining the evolution of the immune response of rabbits to cryptococcal infection. Our work on coccidioidal meningitis in rabbits may be useful to future studies on cryptococcal meningitis. We have followed the CSF levels of matrix metalloproteinase-9 (MMP-9), a marker of the inflammatory response, and also developed a set of primers for various immunomodulatory molecules (e.g., interleukins, IFN-γ, MMP-9, etc.), which we used to demonstrate the increased levels of MMP-9 during CNS infection and upregulation of various mediators in coccidioidal infection-induced arteritis of the ventral artery in the brain (191, 199). The application of these tools to the rabbit model of CNS cryptococcosis will be of interest, particularly if the gene expression of the organism is followed in conjunction with the evolution of the immune response. Furthermore, the completion of the rabbit genome sequence project and the annotation of the genome will be valuable for the generation of immunological reagents.

MODELS OF CRYPTOCOCCOSIS FOR THERAPY STUDIES

A primary use of experimental models of cryptococcosis over the past 30 years has been that of testing the efficacy of various antifungal agents. These models have been studied primarily in mice and rabbits, but rats and guinea pigs have also been used. This is an area that has been reviewed in numerous publications, discussing the efficacy of various azoles, polyenes, echinocandins, immunotherapy, and antibody therapy (15, 17, 18, 32, 36, 125, 137, 177, 180, 181). A general protocol for performing an efficacy study, for example in mice, is that of using groups of 8 to 19 animals per treatment arm, which would include untreated or diluent-treated animals, escalating doses of the antifungal to be tested, and one or more drugs used as comparators. Animals are most often infected with an inoculum that will result in >70% mortality, and treatment initiated 1 to 4 days after infection; the route of inoculation has been pulmonary, intravenous, and intracerebral or intracisternal (see references 32 and 180 for more detailed reviews of performing these types of studies). Efficacy of the drug is determined by following survival or determining CFU in the organs at some time after the cessation of therapy. For drugs that may be of modest efficacy, we have found that CFU will result in a more sensitive determination, as a drug may be somewhat inhibitory but not protect sufficiently to enhance survival. The results of animal model trials have directly contributed to the antifungal armamentarium available to the clinician today and will continue to do so in the future.

A primary benefit of doing efficacy studies in animal models of cryptococcosis is being able to do studies not likely to be performed in human clinical trials. For example, conventional deoxycholate-formulated amphotericin B has been used for many years clinically but has substantial drawbacks in its toxicity profile in humans at the needed dosage. Efforts to ablate this toxicity have resulted in the development of three commercially available lipid-carried formulations of amphotericin B (AmBisome, Abelcet, and Amphotec). Separate murine and rabbit studies showed each formulation to have efficacy (1, 67, 71, 80, 148, 149, 154, 182). Although this is useful information, the question asked by many was, Does one formulation have benefit over the others with respect to efficacy? Thus, a question not likely to be answered in a clinical trial was addressed by us in a murine model of systemic cryptococcosis, where a direct comparison of the four amphotericin B formulations could be made using the same dosages (33). The result of the study showed that conventional amphotericin B was the least effective and that the lipid-carried preparations could be administered at higher dosages with a subsequently improved outcome with AmBisome and Amphotec, better than Abelcet (33).

Increasingly in clinical practice the question of the utility of combination therapy and what combination of drugs should be used is asked. Several studies by Larsen have examined different combinations of conventional antifungals (47, 48, 102, 104). These studies have confirmed the utility of amphotericin B and flucytosine, as well as suggested the synergistic action of fluconazole and flucytosine. In addition, Larsen et al. (103) have shown in vitro and in vivo correlates for fluconazole dosage against *Cryptococcus*, which may contribute to prognostic response to cryptococcal infection based on use of the optimal dose of fluconazole. The use of immunotherapy has long been an attractive idea, and adjunctive administration of IL-12, IFN-γ, and granulocyte-macrophage colony-stimulating factor with a conventional antifungal have been tested and have shown enhanced efficacy (28, 64, 114), and the use of IFN-γ as an adjunct has been taken to clinical trial, with some encouraging results (144). Of note has been the demonstration that specific antibodies against the capsular GXM of *Cryptococcus* hold therapeutic promise (5, 50–52, 55, 130, 131, 160, 161, 163). The studies done with animals have directly led to the initiation of clinical trials, which have shown the safety of the antibody (105). Thus, without the animal model data it is unlikely that immunotherapy or antibody therapy would ever come to clinical trial.

Prevention of infection is even more desirable, and vaccine studies in animal models have been done in the field of cryptococcosis. Initial studies done with mice showed protection with acapsular or mutant strains (59, 60, 157). The use of GXM-tetanus toxoid conjugates also showed protection in mice (46, 129), and these types of studies have been further refined by the use of a mimotope of GXM conjugated to protein to induce specific protective antibodies in the mice, which results in prolonged survival upon challenge (43, 56). In addition to the capsular GXM inducing a protective response, mannoproteins have shown protection in murine models (118). As yet, no vaccine preparation for *Cryptococcus* has been used in humans. However, continued studies will hopefully lead to the ability to prevent this disease with a vaccine.

ALTERNATIVE MODELS

Nonmammalian models of infection with *Cryptococcus* are relatively recent in their application to the study of this organism. Several different species have been used in these studies including *Drosophila*, *Galleria*, *Caenorhabditis*, *Dictyostelium*, and *Acanthamoeba* (15, 61, 133, 186). Pathogenesis studies in *Caenorhabditis* have shown the susceptibility of the nematode to *Cryptococcus* and been useful in identifying genes involved in pathogenesis (134, 135, 139, 156), and they have shown that males are more resistant than are hermaphrodites (188). Similarly, *Galleria* has been used as a model system for pathogenesis with the demonstration of genes involved in virulence in mammalian hosts also involved in killing the wax moth larva (136).

Because cryptococci are inhabitants of the environment, they will come into contact with soil amoebas, such as *Acanthamoeba* and the amoebic form of the slime mold, *Dictyostelium*. Studies using these hosts have demonstrated that the amoeba phagocytoses the yeast but that encapsulated cryptococci can replicate in the amoeba, whereas acapsular yeasts are killed by the amoeba (176). Similarly, *Dictyostelium* amoebas can be infected and unable to control encapsulated *Cryptococcus*, whereas mutant forms of the amoeba can be unable to control growth of acapsular mutants; encapsulated *Cryptococcus* can increase its virulence for the amoeba upon passage, similar to what occurs in mammalian systems (175). In addition to showing correlates to mammalian phagocytic cell interactions, these data are also suggestive in explaining how the yeast survives in the environment, i.e., by being resistant to killing by soil amoeba.

The use of these models will continue to add to our knowledge about the virulence and pathogenesis of *Cryptococcus*. They are amenable to rapid screening of mutants of *Cryptococcus* and implication of genes involved in the virulence of the organism for these hosts, and there have been encouraging results showing correlates with genes involved in virulence in mammalian hosts. However, the potential of different mechanisms of virulence by the fungus may not lead to total correlation, and the substitution of these models for the conventional mammalian models cannot routinely be made. In addition, studies of drug efficacy may be used as screens but cannot address the issues of efficacy in specific target organs, or prevention of dissemination, nor will they address whether the drug is more active against the disease at the higher temperatures of the mammalian host and has vastly different pharmacology issues.

CONCLUSIONS

We have endeavored to provide the reader with a brief overview of the many ways animal models of *Cryptococcus* are useful and necessary in our understanding of the disease caused by this organism. If one were to ask what are the most important lessons derived from these studies, the answer may well depend on the area of interest of the responder. For those primarily interested in the host response, the answer would be along the lines of how much we have learned about the immune system and its complexity in the way an immune response to cryptococcal infection evolves, as well as what we currently believe to be most important to the development of protective resistance. Similarly, one most interested in what makes the organism a pathogen will point to the many studies of genes necessary for the organism to cause disease. The clinicians among us may feel that the use of cryptococcal models to determine therapeutic efficacy and guide the development of treatments are most critical. Although each may be right in one way, all of these things are important if one steps back to view the entirety of the picture.

For those readers new to this area, and the young investigators contemplating studies that involve models of *Cryptococcus*, we have tried to pick salient examples of how these models are performed and how the results can be used. Interspersed in the discussion are thoughts and comments arising from our many years of experience in establishing animal models of not only cryptococcal infection, but also many other fungi. These comments were used to draw attention to things we believe are critical and to provide useful advice about how to go about performing a model, such as how the age or sex or genetic background of the host and fungus can significantly change the outcome of the model and, thus, the conclusions drawn from the studies. The number of publications stating that a murine model was used, but providing no details as to the age and sex, and occasionally even the strain, of the mice used in the studies is astounding and unfortunately makes the published results less useful to someone who may wish to replicate the experiments. Careful attention to these details is necessary. Results published about other fungal animal models should also be kept in mind and learned from. We would leave the reader with the thought that animal models of cryptococcal infection are a useful and necessary tool in our studies and that much can be learned from carefully performed and reproducible studies. As technologies advance and improve in how we can manipulate the organism and the host, we will continue to learn more and more about this disease, and it will be through the use of animal models that we will at some point in time be able to control, cure, and prevent this infection.

REFERENCES

1. **Adler-Moore, J., and R. T. Proffitt.** 2002. AmBisome: liposomal formulation, structure, mechanism of action and preclinical experience. *J. Antimicrob. Chemother.* **49**(Suppl 1)**:**21–30.
2. **Aguirre, K., E. A. Havell, G. W. Gibson, and L. L. Johnson.** 1995. Role of tumor necrosis factor and gamma interferon in acquired resistance to *Cryptococcus neoformans* in the central nervous system of mice. *Infect. Immun.* **63:**1725–1731.
3. **Aguirre, K. M., G. W. Gibson, and L. L. Johnson.** 1998. Decreased resistance to primary intravenous *Cryptococcus neoformans* infection in aged mice despite adequate resistance to intravenous rechallenge. *Infect. Immun.* **66:**4018–4024.
4. **Barchiesi, F., M. Cogliati, M. C. Esposto, E. Spreghini, A. M. Schimizzi, B. L. Wickes, G. Scalise, and M. A. Viviani.** 2005. Comparative analysis of pathogenicity of *Cryptococcus neoformans* serotypes A, D and AD in murine cryptococcosis. *J. Infect.* **51:**10–16.
5. **Beenhouwer, D. O., S. Shapiro, M. Feldmesser, A. Casadevall, and M. D. Scharff.** 2001. Both Th1 and Th2 cytokines affect the ability of monoclonal antibodies to protect mice against *Cryptococcus neoformans*. *Infect. Immun.* **69:**6445–6455.
6. **Biondo, C., A. Midiri, L. Messina, F. Tomasello, G. Garufi, M. R. Catania, M. Bombaci, C. Beninati, G. Teti, and G. Mancuso.** 2005. MyD88 and TLR2, but not TLR4, are required for host defense against *Cryptococcus neoformans*. *Eur. J. Immunol.* **35:**870–878.
7. **Blackstock, R., K. L. Buchanan, A. M. Adesina, and J. W. Murphy.** 1999. Differential regulation of immune responses by highly and weakly virulent *Cryptococcus neoformans* isolates. *Infect. Immun.* **67:**3601–3609.
8. **Blackstock, R., K. L. Buchanan, R. Cherniak, T. G. Mitchell, B. Wong, A. Bartiss, L. Jackson, and J. W. Murphy.** 1999. Pathogenesis of *Cryptococcus neoformans* is associated with quantitative differences in multiple virulence factors. *Mycopathologia* **147:**1–11.
9. **Blackstock, R., and J. W. Murphy.** 2004. Role of interleukin-4 in resistance to *Cryptococcus neoformans* infection. *Am. J. Respir. Cell. Mol. Biol.* **30:**109–117.
10. **Blasi, E., R. Barluzzi, R. Mazzolla, L. Pitzurra, M. Puliti, S. Saleppico, and F. Bistoni.** 1995. Biomolecular events involved in anticryptococcal resistance in the brain. *Infect. Immun.* **63:**1218–1222.
11. **Brummer, E., and K. V. Clemons.** 1987. Animal models in systemic mycoses, p. 79–95. *In* M. Miyaji (ed.), *Animal Models in Medical Mycology*. CRC Press, Boca Raton, FL.

11a. **Brummer, E., and D. A. Stevens.** 2010. Collectins and fungal pathogens: roles of surfactant proteins and mannose binding in host resistance. *Med. Mycol.* **48:**16–28.

12. **Buchanan, K. L., and H. A. Doyle.** 2000. Requirement for CD4(+) T lymphocytes in host resistance against *Cryptococcus neoformans* in the central nervous system of immunized mice. *Infect. Immun.* **68:**456–462.
13. **Buchanan, K. L., and J. W. Murphy.** 1994. Regulation of cytokine production during the expression phase of the anticryptococcal delayed-type hypersensitivity response. *Infect. Immun.* **62:**2930–2939.
14. **Byron, J. K., K. V. Clemons, J. H. McCusker, R. W. Davis, and D. A. Stevens.** 1995. Pathogenicity of *Saccharomyces cerevisiae* in complement factor five-deficient mice. *Infect. Immun.* **63:**478–485.
15. **Capilla, J., K. V. Clemons, and D. A. Stevens.** 2007. Animal models: an important tool in mycology. *Med. Mycol.* **45:**657–684.
16. **Capilla, J., C. M. Maffei, K. V. Clemons, R. A. Sobel, and D. A. Stevens.** 2006. Experimental systemic infection with *Cryptococcus neoformans* var. *grubii* and *Cryptococcus gattii* in normal and immunodeficient mice. *Med. Mycol.* **44:**601–610.
17. **Carroll, S. F., L. Guillot, and S. T. Qureshi.** 2007. Mammalian model hosts of cryptococcal infection. *Comp. Med.* **57:**9–17.
18. **Casadevall, A., and J. R. Perfect.** 1998. *Cryptococcus neoformans*. ASM Press, Washington, DC.
19. **Casadevall, A., A. L. Rosas, and J. D. Nosanchuk.** 2000. Melanin and virulence in *Cryptococcus neoformans*. *Curr. Opin. Microbiol.* **3:**354–358.
20. **Chaturvedi, S., P. Ren, S. D. Narasipura, and V. Chaturvedi.** 2005. Selection of optimal host strain for molecular pathogenesis studies on *Cryptococcus gattii*. *Mycopathologia* **160:**207–215.
21. **Chen, S. C., M. Muller, J. Z. Zhou, L. C. Wright, and T. C. Sorrell.** 1997. Phospholipase activity in *Cryptococcus neoformans*: a new virulence factor? *J. Infect. Dis.* **175:**414–420.
22. **Chung, S., P. Mondon, Y. C. Chang, and K. J. Kwon-Chung.** 2003. *Cryptococcus neoformans* with a mutation in the tetratricopeptide repeat-containing gene, CCN1, causes subcutaneous lesions but fails to cause systemic infection. *Infect. Immun.* **71:**1988–1994.
23. **Clancy, C. J., M. H. Nguyen, R. Alandoerffer, S. Cheng, K. Iczkowski, M. Richardson, and J. R. Graybill.** 2006. *Cryptococcus neoformans* var. *grubii* isolates recovered from persons with AIDS demonstrate a wide range of virulence during murine meningoencephalitis that correlates with the expression of certain virulence factors. *Microbiology* **152:**2247–2255.

24. **Clemons, K., J. Capilla, and D. A. Stevens.** 2007. Experimental animal models of coccidioidomycosis. *Ann. NY Acad. Sci.* **1111:**208–224.
25. **Clemons, K. V., R. Azzi, and D. A. Stevens.** 1996. Experimental systemic cryptococcosis in SCID mice. *J. Med. Vet. Mycol.* **34:**331–335.
26. **Clemons, K. V., E. Brummer, and D. A. Stevens.** 1994. Cytokine treatment of central nervous system infection: efficacy of interleukin-12 alone and synergy with conventional antifungal therapy in experimental cryptococcosis. *Antimicrob. Agents Chemother.* **38:**460–464.
27. **Clemons, K. V., G. Grunig, R. A. Sobel, L. F. Mirels, D. M. Rennick, and D. A. Stevens.** 2000. Role of IL-10 in invasive aspergillosis: increased resistance of IL-10 gene knockout mice to lethal systemic aspergillosis. *Clin. Exp. Immunol.* **122:**186–191.
28. **Clemons, K. V., J. E. Lutz, and D. A. Stevens.** 2001. Efficacy of recombinant gamma interferon for treatment of systemic cryptococcosis in SCID mice. *Antimicrob. Agents Chemother.* **45:**686–689.
29. **Clemons, K. V., J. H. McCusker, R. W. Davis, and D. A. Stevens.** 1994. Comparative pathogenesis of clinical and nonclinical isolates of *Saccharomyces cerevisiae*. *J. Infect. Dis.* **169:**859–867.
30. **Clemons, K. V., P. Kamberi, T. M. Chiller, R. A. Sobel, E. Brummer, J. Kolls, and D. A. Stevens.** 2005. Effects of interferon-γ gene therapy in the murine central nervous system and concentrations in cerebrospinal fluid after intrathecal or intracerebral administration. *Biotechnology* **4:**11–18.
31. **Clemons, K. V., and D. A. Stevens.** 2006. Animal models of *Aspergillus* infection in preclinical trials, diagnostics and pharmacodynamics: what can we learn from them? *Med. Mycol.* **44**(Suppl. 1)**:**S119–S126.
32. **Clemons, K. V., and D. A. Stevens.** 2006. Animal models testing monotherapy versus combination antifungal therapy: lessons learned and future directions. *Curr. Opin. Infect. Dis.* **19:**360–364.
33. **Clemons, K. V., and D. A. Stevens.** 1998. Comparison of Fungizone, Amphotec, AmBisome, and Abelcet for treatment of systemic murine cryptococcosis. *Antimicrob. Agents Chemother.* **42:**899–902.
34. **Clemons, K. V., and D. A. Stevens.** 2005. The contribution of animal models of aspergillosis to understanding pathogenesis, therapy and virulence. *Med. Mycol.* **43**(Suppl 1)**:**S101–S110.
35. **Clemons, K. V., and D. A. Stevens.** 2001. Overview of host defense mechanisms in systemic mycoses and the basis for immunotherapy. *Semin. Respir. Infect.* **16:**60–66.
36. **Clemons, K. V., and D. A. Stevens.** 2002. Immunomodulation of fungal infections: do immunomodulators have a role in treating mycoses? *EOS Riv. Immunol. Immunofarmacol.* **22:**29–32.
37. **Cox, G. M., T. S. Harrison, H. C. McDade, C. P. Taborda, G. Heinrich, A. Casadevall, and J. R. Perfect.** 2003. Superoxide dismutase influences the virulence of *Cryptococcus neoformans* by affecting growth within macrophages. *Infect. Immun.* **71:**173–180.
38. **Cox, G. M., H. C. McDade, S. C. Chen, S. C. Tucker, M. Gottfredsson, L. C. Wright, T. C. Sorrell, S. D. Leidich, A. Casadevall, M. A. Ghannoum, and J. R. Perfect.** 2001. Extracellular phospholipase activity is a virulence factor for *Cryptococcus neoformans*. *Mol. Microbiol.* **39:**166–175.
39. **Cox, G. M., J. Mukherjee, G. T. Cole, A. Casadevall, and J. R. Perfect.** 2000. Urease as a virulence factor in experimental cryptococcosis. *Infect. Immun.* **68:**443–448.
40. **Crouch, E., K. Hartshorn, and I. Ofek.** 2000. Collectins and pulmonary innate immunity. *Immunol. Rev.* **173:**52–65.
41. **Cutler, J. E.** 1976. Acute systemic candidiasis in normal and congenitally thymic-deficient (nude) mice. *J. Reticuloendothel. Soc.* **19:**121–124.
42. **Dan, J. M., R. M. Kelly, C. K. Lee, and S. M. Levitz.** 2008. Role of the mannose receptor in a murine model of *Cryptococcus neoformans* infection. *Infect. Immun.* **76:**2362–2367.
43. **Datta, K., A. Lees, and L. A. Pirofski.** 2008. Therapeutic efficacy of a conjugate vaccine containing a peptide mimotope of cryptococcal capsular polysaccharide glucuronoxylomannan. *Clin. Vaccine Immunol.* **15:**1176–1187.
44. **Decken, K., G. Kohler, K. Palmer-Lehmann, A. Wunderlin, F. Mattner, J. Magram, M. K. Gately, and G. Alber.** 1998. Interleukin-12 is essential for a protective Th1 response in mice infected with *Cryptococcus neoformans*. *Infect. Immun.* **66:**4994–5000.
45. **Deepe, G. S., Jr., L. Romani, V. L. Calich, G. Huffnagle, C. Arruda, E. E. Molinari-Madlum, and J. R. Perfect.** 2000. Knockout mice as experimental models of virulence. *Med. Mycol.* **38**(Suppl 1)**:**87–98.
46. **Devi, S. J.** 1996. Preclinical efficacy of a glucuronoxylomannan-tetanus toxoid conjugate vaccine of *Cryptococcus neoformans* in a murine model. *Vaccine* **14:**841–844.
47. **Diamond, D. M., M. Bauer, B. E. Daniel, M. A. Leal, D. Johnson, B. K. Williams, A. M. Thomas, J. C. Ding, L. Najvar, J. R. Graybill, and R. A. Larsen.** 1998. Amphotericin B colloidal dispersion combined with flucytosine with or without fluconazole for treatment of murine cryptococcal meningitis. *Antimicrob. Agents Chemother.* **42:**528–533.
48. **Ding, J. C., M. Bauer, D. M. Diamond, M. A. Leal, D. Johnson, B. K. Williams, A. M. Thomas, L. Najvar, J. R. Graybill, and R. A. Larsen.** 1997. Effect of severity of meningitis on fungicidal activity of flucytosine combined with fluconazole in a murine model of cryptococcal meningitis. *Antimicrob. Agents Chemother.* **41:**1589–1593.
49. **Dobrick, P., K. Miksits, and H. Hahn.** 1995. L3T4(CD4)-, Lyt-2(CD8)- and Mac-1(CD11b)-phenotypic leukocytes in murine cryptococcal meningoencephalitis. *Mycopathologia* **131:**159–166.
50. **Dromer, F., and J. Charreire.** 1991. Improved amphotericin B activity by a monoclonal anti-*Cryptococcus neoformans* antibody: study during murine cryptococcosis and mechanisms of action. *J. Infect. Dis.* **163:**1114–1120.
51. **Dromer, F., J. Charreire, A. Contrepois, C. Carbon, and P. Yeni.** 1987. Protection of mice against experimental cryptococcosis by anti-*Cryptococcus neoformans* monoclonal antibody. *Infect. Immun.* **55:**749–752.
52. **Dromer, F., C. Perronne, J. Barge, J. L. Vilde, and P. Yeni.** 1989. Role of IgG and complement component C5 in the initial course of experimental cryptococcosis. *Clin. Exp. Immunol.* **78:**412–417.
53. **D'Souza, C. A., F. Hagen, T. Boekhout, G. M. Cox, and J. Heitman.** 2004. Investigation of the basis of virulence in serotype A strains of *Cryptococcus neoformans* from apparently immunocompetent individuals. *Curr. Genet.* **46:**92–102.
54. **Erickson, T., L. Liu, A. Gueyikian, X. Zhu, J. Gibbons, and P. R. Williamson.** 2001. Multiple virulence factors of *Cryptococcus neoformans* are dependent on VPH1. *Mol. Microbiol.* **42:**1121–1131.
55. **Feldmesser, M., A. Mednick, and A. Casadevall.** 2002. Antibody-mediated protection in murine *Cryptococcus neoformans* infection is associated with pleotrophic effects on cytokine and leukocyte responses. *Infect. Immun.* **70:**1571–1580.
56. **Fleuridor, R., A. Lees, and L. Pirofski.** 2001. A cryptococcal capsular polysaccharide mimotope prolongs the survival of mice with *Cryptococcus neoformans* infection. *J. Immunol.* **166:**1087–1096.

57. **Fraser, J. A., S. S. Giles, E. C. Wenink, S. G. Geunes-Boyer, J. R. Wright, S. Diezmann, A. Allen, J. E. Stajich, F. S. Dietrich, J. R. Perfect, and J. Heitman.** 2005. Same-sex mating and the origin of the Vancouver Island *Cryptococcus gattii* outbreak. *Nature* **437:**1360–1364.
58. **Fries, B. C., and A. Casadevall.** 1998. Serial isolates of *Cryptococcus neoformans* from patients with AIDS differ in virulence for mice. *J. Infect. Dis.* **178:**1761–1766.
59. **Fromtling, R. A., R. Blackstock, N. K. Hall, and G. S. Bulmer.** 1979. Immunization of mice with an avirulent pseudohyphal form of *Cryptococcus neoformans. Mycopathologia* **68:**179–181.
60. **Fromtling, R. A., A. M. Kaplan, and H. J. Shadomy.** 1983. Immunization of mice with stable, acapsular, yeast-like mutants of *Cryptococcus neoformans. Sabouraudia* **21:**113–119.
61. **Fuchs, B. B., and E. Mylonakis.** 2006. Using non-mammalian hosts to study fungal virulence and host defense. *Curr. Opin. Microbiol.* **9:**346–351.
62. **Giles, S. S., A. K. Zaas, M. F. Reidy, J. R. Perfect, and J. R. Wright.** 2007. *Cryptococcus neoformans* is resistant to surfactant protein A mediated host defense mechanisms. *PLoS ONE* **2:**e1370.
63. **Granger, D. L., J. R. Perfect, and D. T. Durack.** 1985. Virulence of *Cryptococcus neoformans*. Regulation of capsule synthesis by carbon dioxide. *J. Clin. Invest.* **76:**508–516.
64. **Graybill, J. R., R. Bocanegra, C. Lambros, and M. F. Luther.** 1997. Granulocyte colony stimulating factor therapy of experimental cryptococcal meningitis. *J. Med. Vet. Mycol.* **35:**243–247.
65. **Graybill, J. R., P. C. Craven, L. F. Mitchell, and D. J. Drutz.** 1978. Interaction of chemotherapy and immune defenses in experimental murine cryptococcosis. *Antimicrob. Agents Chemother.* **14:**659–667.
66. **Graybill, J. R., and D. J. Drutz.** 1978. Host defense in cryptococcosis. II. Cryptococcosis in the nude mouse. *Cell. Immunol.* **40:**263–274.
67. **Graybill, J. R., and L. Mitchell.** 1980. Treatment of murine cryptococcosis with minocycline and amphotericin B. *Sabouraudia* **18:**137–144.
68. **Graybill, J. R., L. Mitchell, and D. J. Drutz.** 1979. Host defense in cryptococcosis. III. Protection of nude mice by thymus transplantation. *J. Infect. Dis.* **140:**546–552.
69. **Guerrero, A., and B. C. Fries.** 2008. Phenotypic switching in *Cryptococcus neoformans* contributes to virulence by changing the immunological host response. *Infect. Immun.* **76:**4322–4331.
70. **Hamilton, A. J., and J. Goodley.** 1996. Virulence factors of *Cryptococcus neoformans. Curr. Top. Med. Mycol.* **7:**19–42.
71. **Hamilton, J. D., and D. M. Elliott.** 1975. Combined activity of amphotericin B and 5-fluorocytosine against *Cryptococcus neoformans* in vitro and in vivo in mice. *J. Infect. Dis.* **131:**129–137.
72. **Hay, R. J., and E. Reiss.** 1978. Delayed-type hypersensitivity responses in infected mice elicited by cytoplasmic fractions of *Cryptococcus neoformans. Infect. Immun.* **22:**72–79.
73. **Hernandez, Y., S. Arora, J. R. Erb-Downward, R. A. McDonald, G. B. Toews, and G. B. Huffnagle.** 2005. Distinct roles for IL-4 and IL-10 in regulating T2 immunity during allergic bronchopulmonary mycosis. *J. Immunol.* **174:**1027–1036.
74. **Hill, J. O.** 1992. CD4+ T cells cause multinucleated giant cells to form around *Cryptococcus neoformans* and confine the yeast within the primary site of infection in the respiratory tract. *J. Exp. Med.* **175:**1685–1695.
75. **Hill, J. O., and K. M. Aguirre.** 1994. CD4+ T cell-dependent acquired state of immunity that protects the brain against *Cryptococcus neoformans. J. Immunol.* **152:**2344–2350.
76. **Hoag, K. A., M. F. Lipscomb, A. A. Izzo, and N. E. Street.** 1997. IL-12 and IFN-gamma are required for initiating the protective Th1 response to pulmonary cryptococcosis in resistant C.B-17 mice. *Am. J. Respir. Cell. Mol. Biol.* **17:**733–739.
77. **Hoag, K. A., N. E. Street, G. B. Huffnagle, and M. F. Lipscomb.** 1995. Early cytokine production in pulmonary *Cryptococcus neoformans* infections distinguishes susceptible and resistant mice. *Am. J. Respir. Cell. Mol. Biol.* **13:**487–495.
78. **Hobbs, M. M., J. R. Perfect, D. L. Granger, and D. T. Durack.** 1990. Opsonic activity of cerebrospinal fluid in experimental cryptococcal meningitis. *Infect. Immun.* **58:**2115–2119.
79. **Hogan, L. H., B. S. Klein, and S. M. Levitz.** 1996. Virulence factors of medically important fungi. *Clin. Microbiol. Rev.* **9:**469–488.
80. **Hostetler, J. S., K. V. Clemons, L. H. Hanson, and D. A. Stevens.** 1992. Efficacy and safety of amphotericin B colloidal dispersion compared with those of amphotericin B deoxycholate suspension for treatment of disseminated murine cryptococcosis. *Antimicrob. Agents Chemother.* **36:**2656–2660.
81. **Huffnagle, G. B., M. B. Boyd, N. E. Street, and M. F. Lipscomb.** 1998. IL-5 is required for eosinophil recruitment, crystal deposition, and mononuclear cell recruitment during a pulmonary *Cryptococcus neoformans* infection in genetically susceptible mice (C57BL/6). *J. Immunol.* **160:**2393–2400.
82. **Huffnagle, G. B., and G. S. Deepe.** 2003. Innate and adaptive determinants of host susceptibility to medically important fungi. *Curr. Opin. Microbiol.* **6:**344–350.
83. **Huffnagle, G. B., and M. F. Lipscomb.** 1992. Animal model of human disease: pulmonary cryptococcosis. *Am. J. Pathol.* **141:**1517–1520.
84. **Huffnagle, G. B., and M. F. Lipscomb.** 1998. Cells and cytokines in pulmonary cryptococcosis. *Res. Immunol.* **149:**387–396; discussion 512–514.
85. **Huffnagle, G. B., M. F. Lipscomb, J. A. Lovchik, K. A. Hoag, and N. E. Street.** 1994. The role of CD4+ and CD8+ T cells in the protective inflammatory response to a pulmonary cryptococcal infection. *J. Leukoc. Biol.* **55:**35–42.
86. **Huffnagle, G. B., and L. K. McNeil.** 1999. Dissemination of *C. neoformans* to the central nervous system: role of chemokines, Th1 immunity and leukocyte recruitment. *J. Neurovirol.* **5:**76–81.
87. **Huffnagle, G. B., J. L. Yates, and M. F. Lipscomb.** 1991. Immunity to a pulmonary *Cryptococcus neoformans* infection requires both CD4+ and CD8+ T cells. *J. Exp. Med.* **173:**793–800.
88. **Huffnagle, G. B., J. L. Yates, and M. F. Lipscomb.** 1991. T cell-mediated immunity in the lung: a *Cryptococcus neoformans* pulmonary infection model using SCID and athymic nude mice. *Infect. Immun.* **59:**1423–1433.
89. **Irokanulo, E. A., and C. O. Akueshi.** 1995. Virulence of *Cryptococcus neoformans* serotypes A, B, C and D for four mouse strains. *J. Med. Microbiol.* **43:**289–293.
90. **Jain, N., and B. C. Fries.** 2008. Phenotypic switching of *Cryptococcus neoformans* and *Cryptococcus gattii. Mycopathologia* **166:**181–188.
91. **Kamberi, P., R. A. Sobel, K. V. Clemons, D. A. Stevens, D. Pappagianis, and P. L. Williams.** 2003. A murine model of coccidioidal meningitis. *J. Infect. Dis.* **187:**453–460.
92. **Kawakami, K., Y. Koguchi, M. H. Qureshi, Y. Kinjo, S. Yara, A. Miyazato, M. Kurimoto, K. Takeda, S. Akira, and A. Saito.** 2000. Reduced host resistance and Th1 response to *Cryptococcus neoformans* in interleukin-18 deficient mice. *FEMS Microbiol. Lett.* **186:**121–126.
93. **Kawakami, K., Y. Koguchi, M. H. Qureshi, A. Miyazato, S. Yara, Y. Kinjo, Y. Iwakura, K. Takeda, S. Akira,**

M. Kurimoto, and A. Saito. 2000. IL-18 contributes to host resistance against infection with *Cryptococcus neoformans* in mice with defective IL-12 synthesis through induction of IFN-gamma production by NK cells. *J. Immunol.* **165:**941–947.

94. **Kawakami, K., M. H. Qureshi, Y. Koguchi, K. Nakajima, and A. Saito.** 1999. Differential effect of *Cryptococcus neoformans* on the production of IL-12p40 and IL-10 by murine macrophages stimulated with lipopolysaccharide and gamma interferon. *FEMS Microbiol. Lett.* **175:**87–94.
95. **Kawakami, K., K. Shibuya, M. H. Qureshi, T. Zhang, Y. Koguchi, M. Tohyama, Q. Xie, S. Naoe, and A. Saito.** 1999. Chemokine responses and accumulation of inflammatory cells in the lungs of mice infected with highly virulent *Cryptococcus neoformans*: effects of interleukin-12. *FEMS Immunol. Med. Microbiol.* **25:**391–402.
96. **Kawakami, K., M. Tohyama, X. Qifeng, and A. Saito.** 1997. Expression of cytokines and inducible nitric oxide synthase mRNA in the lungs of mice infected with *Cryptococcus neoformans*: effects of interleukin-12. *Infect. Immun.* **65:**1307–1312.
97. **Kawakami, K., M. Tohyama, Q. Xie, and A. Saito.** 1996. IL-12 protects mice against pulmonary and disseminated infection caused by *Cryptococcus neoformans*. *Clin. Exp. Immunol.* **104:**208–214.
98. **Kleinschek, M. A., U. Muller, S. J. Brodie, W. Stenzel, G. Kohler, W. M. Blumenschein, R. K. Straubinger, T. McClanahan, R. A. Kastelein, and G. Alber.** 2006. IL-23 enhances the inflammatory cell response in *Cryptococcus neoformans* infection and induces a cytokine pattern distinct from IL-12. *J. Immunol.* **176:**1098–1106.
99. **Kullberg, B. J.** 1997. Trends in immunotherapy of fungal infections. *Eur. J. Clin. Microbiol. Infect. Dis.* **16:**51–55.
100. **Kwon-Chung, K. J., I. Polacheck, and T. J. Popkin.** 1982. Melanin-lacking mutants of *Cryptococcus neoformans* and their virulence for mice. *J. Bacteriol.* **150:**1414–1421.
101. **Kwon-Chung, K. J., B. L. Wickes, L. Stockman, G. D. Roberts, D. Ellis, and D. H. Howard.** 1992. Virulence, serotype, and molecular characteristics of environmental strains of *Cryptococcus neoformans* var. *gattii*. *Infect. Immun.* **60:**1869–1874.
102. **Larsen, R. A., M. Bauer, A. M. Thomas, and J. R. Graybill.** 2004. Amphotericin B and fluconazole, a potent combination therapy for cryptococcal meningitis. *Antimicrob. Agents Chemother.* **48:**985–991.
103. **Larsen, R. A., M. Bauer, A. M. Thomas, A. Sanchez, D. Citron, M. Rathbun, and T. S. Harrison.** 2005. Correspondence of in vitro and in vivo fluconazole dose-response curves for *Cryptococcus neoformans*. *Antimicrob. Agents Chemother.* **49:**3297–3301.
104. **Larsen, R. A., M. Bauer, J. M. Weiner, D. M. Diamond, M. E. Leal, J. C. Ding, M. G. Rinaldi, and J. R. Graybill.** 1996. Effect of fluconazole on fungicidal activity of flucytosine in murine cryptococcal meningitis. *Antimicrob. Agents Chemother.* **40:**2178–2182.
105. **Larsen, R. A., P. G. Pappas, J. Perfect, J. A. Aberg, A. Casadevall, G. A. Cloud, R. James, S. Filler, and W. E. Dismukes.** 2005. Phase I evaluation of the safety and pharmacokinetics of murine-derived anticryptococcal antibody 18B7 in subjects with treated cryptococcal meningitis. *Antimicrob. Agents Chemother.* **49:**952–958.
106. **Lin, X., K. Nielsen, S. Patel, and J. Heitman.** 2008. Impact of mating type, serotype, and ploidy on the virulence of *Cryptococcus neoformans*. *Infect. Immun.* **76:**2923–2938.
107. **Liu, O. W., M. J. Kelly, E. D. Chow, and H. D. Madhani.** 2007. Parallel beta-helix proteins required for accurate capsule polysaccharide synthesis and virulence in the yeast *Cryptococcus neoformans*. *Eukaryot. Cell* **6:**630–640.
108. **Loftus, B. J., E. Fung, P. Roncaglia, D. Rowley, P. Amedeo, D. Bruno, J. Vamathevan, M. Miranda, I. J. Anderson, J. A. Fraser, J. E. Allen, I. E. Bosdet, M. R. Brent, R. Chiu, T. L. Doering, M. J. Donlin, C. A. D'Souza, D. S. Fox, V. Grinberg, J. Fu, M. Fukushima, B. J. Haas, J. C. Huang, G. Janbon, S. J. Jones, H. L. Koo, M. I. Krzywinski, J. K. Kwon-Chung, K. B. Lengeler, R. Maiti, M. A. Marra, R. E. Marra, C. A. Mathewson, T. G. Mitchell, M. Pertea, F. R. Riggs, S. L. Salzberg, J. E. Schein, A. Shvartsbeyn, H. Shin, M. Shumway, C. A. Specht, B. B. Suh, A. Tenney, T. R. Utterback, B. L. Wickes, J. R. Wortman, N. H. Wye, J. W. Kronstad, J. K. Lodge, J. Heitman, R. W. Davis, C. M. Fraser, and R. W. Hyman.** 2005. The genome of the basidiomycetous yeast and human pathogen *Cryptococcus neoformans*. *Science* **307:**1321–1324.
109. **Lortholary, O., L. Improvisi, C. Fitting, J. M. Cavaillon, and F. Dromer.** 2002. Influence of gender and age on course of infection and cytokine responses in mice with disseminated *Cryptococcus neoformans* infection. *Clin. Microbiol. Infect.* **8:**31–37.
110. **Lortholary, O., L. Improvisi, N. Rayhane, F. Gray, C. Fitting, J. M. Cavaillon, and F. Dromer.** 1999. Cytokine profiles of AIDS patients are similar to those of mice with disseminated *Cryptococcus neoformans* infection. *Infect. Immun.* **67:**6314–6320.
111. **Lovchik, J., M. Lipscomb, and C. R. Lyons.** 1997. Expression of lung inducible nitric oxide synthase protein does not correlate with nitric oxide production in vivo in a pulmonary immune response against *Cryptococcus neoformans*. *J. Immunol.* **158:**1772–1778.
112. **Lovchik, J. A., C. R. Lyons, and M. F. Lipscomb.** 1995. A role for gamma interferon-induced nitric oxide in pulmonary clearance of *Cryptococcus neoformans*. *Am. J. Respir. Cell. Mol. Biol.* **13:**116–124.
113. **Luberto, C., B. Martinez-Marino, D. Taraskiewicz, B. Bolanos, P. Chitano, D. L. Toffaletti, G. M. Cox, J. R. Perfect, Y. A. Hannun, E. Balish, and M. Del Poeta.** 2003. Identification of App1 as a regulator of phagocytosis and virulence of *Cryptococcus neoformans*. *J. Clin. Invest.* **112:**1080–1094.
114. **Lutz, J. E., K. V. Clemons, and D. A. Stevens.** 2000. Enhancement of antifungal chemotherapy by interferon-gamma in experimental systemic cryptococcosis. *J. Antimicrob. Chemother.* **46:**437–442.
115. **Macher, A. M., J. E. Bennett, J. E. Gadek, and M. M. Frank.** 1978. Complement depletion in cryptococcal sepsis. *J. Immunol.* **120:**1686–1690.
116. **Maffei, C. M., L. F. Mirels, R. A. Sobel, K. V. Clemons, and D. A. Stevens.** 2004. Cytokine and inducible nitric oxide synthase mRNA expression during experimental murine cryptococcal meningoencephalitis. *Infect. Immun.* **72:**2338–2349.
117. **Magee, D. M., and R. A. Cox.** 1996. Interleukin-12 regulation of host defenses against *Coccidioides immitis*. *Infect. Immun.* **64:**3609–3613.
118. **Mansour, M. K., L. E. Yauch, J. B. Rottman, and S. M. Levitz.** 2004. Protective efficacy of antigenic fractions in mouse models of cryptococcosis. *Infect. Immun.* **72:**1746–1754.
119. **Marquis, G., S. Montplaisir, M. Pelletier, S. Mousseau, and P. Auger.** 1985. Genetic resistance to murine cryptococcosis: increased susceptibility in the CBA/N XID mutant strain of mice. *Infect. Immun.* **47:**282–287.
120. **Marquis, G., S. Montplaisir, M. Pelletier, S. Mousseau, and P. Auger.** 1985. Genetic resistance to murine cryptococcosis: the beige mutation (Chediak-Higashi syndrome) in mice. *Infect. Immun.* **47:**288–293.

121. **McClelland, E. E., P. Bernhardt, and A. Casadevall.** 2006. Estimating the relative contributions of virulence factors for pathogenic microbes. *Infect. Immun.* **74:**1500–1504.
122. **McClelland, E. E., D. L. Granger, and W. K. Potts.** 2003. Major histocompatibility complex-dependent susceptibility to *Cryptococcus neoformans* in mice. *Infect. Immun.* **71:**4815–4817.
123. **McClelland, E. E., W. T. Perrine, W. K. Potts, and A. Casadevall.** 2005. Relationship of virulence factor expression to evolved virulence in mouse-passaged *Cryptococcus neoformans* lines. *Infect. Immun.* **73:**7047–7050.
124. **Missall, T. A., M. E. Pusateri, and J. K. Lodge.** 2004. Thiol peroxidase is critical for virulence and resistance to nitric oxide and peroxide in the fungal pathogen, *Cryptococcus neoformans*. *Mol. Microbiol.* **51:**1447–1458.
125. **Mitchell, T. G., and J. R. Perfect.** 1995. Cryptococcosis in the era of AIDS: 100 years after the discovery of *Cryptococcus neoformans*. *Clin. Microbiol. Rev.* **8:**515–548.
126. **Mody, C. H., M. F. Lipscomb, N. E. Street, and G. B. Toews.** 1990. Depletion of CD4+ (L3T4+) lymphocytes in vivo impairs murine host defense to *Cryptococcus neoformans*. *J. Immunol.* **144:**1472–1477.
127. **Monga, D. P., R. Kumar, L. N. Mohapatra, and A. N. Malaviya.** 1979. Experimental cryptococcosis in normal and B-cell-deficient mice. *Infect. Immun.* **26:**1–3.
128. **Monga, D. P., R. Kumar, L. N. Mohapatra, and A. N. Malaviya.** 1980. Experimental cryptococcosis in normal and T cell deficient mice. *Indian J. Med. Res.* **72:**641–649.
129. **Mukherjee, J., A. Casadevall, and M. D. Scharff.** 1993. Molecular characterization of the humoral responses to *Cryptococcus neoformans* infection and glucuronoxylomannan-tetanus toxoid conjugate immunization. *J. Exp. Med.* **177:**1105–1116.
130. **Mukherjee, J., M. Feldmesser, M. D. Scharff, and A. Casadevall.** 1995. Monoclonal antibodies to *Cryptococcus neoformans* glucuronoxylomannan enhance fluconazole efficacy. *Antimicrob. Agents Chemother.* **39:**1398–1405.
131. **Mukherjee, J., M. D. Scharff, and A. Casadevall.** 1992. Protective murine monoclonal antibodies to *Cryptococcus neoformans*. *Infect. Immun.* **60:**4534–4541.
132. **Muller, U., W. Stenzel, G. Kohler, C. Werner, T. Polte, G. Hansen, N. Schutze, R. K. Straubinger, M. Blessing, A. N. McKenzie, F. Brombacher, and G. Alber.** 2007. IL-13 induces disease-promoting type 2 cytokines, alternatively activated macrophages and allergic inflammation during pulmonary infection of mice with *Cryptococcus neoformans*. *J. Immunol.* **179:**5367–5377.
133. **Mylonakis, E., and A. Aballay.** 2005. Worms and flies as genetically tractable animal models to study host-pathogen interactions. *Infect. Immun.* **73:**3833–3841.
134. **Mylonakis, E., F. M. Ausubel, J. R. Perfect, J. Heitman, and S. B. Calderwood.** 2002. Killing of *Caenorhabditis elegans* by *Cryptococcus neoformans* as a model of yeast pathogenesis. *Proc. Natl. Acad. Sci. USA* **99:**15675–15680.
135. **Mylonakis, E., A. Idnurm, R. Moreno, J. El Khoury, J. B. Rottman, F. M. Ausubel, J. Heitman, and S. B. Calderwood.** 2004. *Cryptococcus neoformans* Kin1 protein kinase homologue, identified through a *Caenorhabditis elegans* screen, promotes virulence in mammals. *Mol. Microbiol.* **54:**407–419.
136. **Mylonakis, E., R. Moreno, J. B. El Khoury, A. Idnurm, J. Heitman, S. B. Calderwood, F. M. Ausubel, and A. Diener.** 2005. *Galleria mellonella* as a model system to study *Cryptococcus neoformans* pathogenesis. *Infect. Immun.* **73:**3842–3850.
137. **Najvar, L. K., R. Bocanegra, and J. R. Graybill.** 1999. An alternative animal model for comparison of treatments for cryptococcal meningitis. *Antimicrob. Agents Chemother.* **43:**413–414.
138. **Nakamura, K., T. Kinjo, S. Saijo, A. Miyazato, Y. Adachi, N. Ohno, J. Fujita, M. Kaku, Y. Iwakura, and K. Kawakami.** 2007. Dectin-1 is not required for the host defense to *Cryptococcus neoformans*. *Microbiol. Immunol.* **51:**1115–1119.
139. **Nielsen, K., G. M. Cox, A. P. Litvintseva, E. Mylonakis, S. D. Malliaris, D. K. Benjamin, Jr., S. S. Giles, T. G. Mitchell, A. Casadevall, J. R. Perfect, and J. Heitman.** 2005. *Cryptococcus neoformans* {alpha} strains preferentially disseminate to the central nervous system during coinfection. *Infect. Immun.* **73:**4922–4933.
140. **Nielsen, K., G. M. Cox, P. Wang, D. L. Toffaletti, J. R. Perfect, and J. Heitman.** 2003. Sexual cycle of *Cryptococcus neoformans* var. *grubii* and virulence of congenic a and α isolates. *Infect. Immun.* **71:**4831–4841.
141. **Nishimura, K., and M. Miyaji.** 1979. Histopathological studies on experimental cryptococcosis in nude mice. *Mycopathologia* **68:**145–153.
142. **Odom, A., S. Muir, E. Lim, D. L. Toffaletti, J. Perfect, and J. Heitman.** 1997. Calcineurin is required for virulence of *Cryptococcus neoformans*. *EMBO J.* **16:**2576–2589.
143. **Pal, M.** 2005. Pathogenicity of environmental strains of *Cryptococcus neoformans* var *neoformans* in murine model. *Rev. Iberoam. Micol.* **22:**129.
144. **Pappas, P. G., B. Bustamante, E. Ticona, R. J. Hamill, P. C. Johnson, A. Reboli, J. Aberg, R. Hasbun, and H. H. Hsu.** 2004. Recombinant interferon-γ 1b as adjunctive therapy for AIDS-related acute cryptococcal meningitis. *J. Infect. Dis.* **189:**2185–2191.
145. **Perfect, J. R.** 2005. *Cryptococcus neoformans*: a sugar-coated killer with designer genes. *FEMS Immunol. Med. Microbiol.* **45:**395–404.
146. **Perfect, J. R.** 1990. Fluconazole therapy for experimental cryptococcosis and candidiasis in the rabbit. *Rev. Infect. Dis.* **12**(Suppl 3)**:**S299–S302.
147. **Perfect, J. R., G. M. Cox, R. K. Dodge, and W. A. Schell.** 1996. In vitro and in vivo efficacies of the azole SCH56592 against *Cryptococcus neoformans*. *Antimicrob. Agents Chemother.* **40:**1910–1913.
148. **Perfect, J. R., and D. T. Durack.** 1985. Comparison of amphotericin B and N-D-ornithyl amphotericin B methyl ester in experimental cryptococcal meningitis and *Candida albicans* endocarditis with pyelonephritis. *Antimicrob. Agents Chemother.* **28:**751–755.
149. **Perfect, J. R., and D. T. Durack.** 1982. Treatment of experimental cryptococcal meningitis with amphotericin B, 5-fluorocytosine, and ketoconazole. *J. Infect. Dis.* **146:**429–435.
150. **Perfect, J. R., M. M. Hobbs, D. L. Granger, and D. T. Durack.** 1988. Cerebrospinal fluid macrophage response to experimental cryptococcal meningitis: relationship between in vivo and in vitro measurements of cytotoxicity. *Infect. Immun.* **56:**849–854.
151. **Perfect, J. R., S. D. Lang, and D. T. Durack.** 1980. Chronic cryptococcal meningitis: a new experimental model in rabbits. *Am. J. Pathol.* **101:**177–194.
152. **Perfect, J. R., S. D. Lang, and D. T. Durack.** 1981. Influence of agglutinating antibody in experimental cryptococcal meningitis. *Br. J. Exp. Pathol.* **62:**595–599.
153. **Perfect, J. R., B. Wong, Y. C. Chang, K. J. Kwon-Chung, and P. R. Williamson.** 1998. *Cryptococcus neoformans*: virulence and host defences. *Med. Mycol.* **36**(Suppl 1)**:**79–86.
154. **Perfect, J. R., and K. A. Wright.** 1994. Amphotericin B lipid complex in the treatment of experimental cryptococcal meningitis and disseminated candidosis. *J. Antimicrob. Chemother.* **33:**73–81.
155. **Petzold, E. W., U. Himmelreich, E. Mylonakis, T. Rude, D. Toffaletti, G. M. Cox, J. L. Miller, and J. R. Perfect.** 2006. Characterization and regulation of the trehalose

synthesis pathway and its importance in the pathogenicity of *Cryptococcus neoformans*. *Infect. Immun.* **74:**5877–5887.

156. **Reese, A. J., A. Yoneda, J. A. Breger, A. Beauvais, H. Liu, C. L. Griffith, I. Bose, M. J. Kim, C. Skau, S. Yang, J. A. Sefko, M. Osumi, J. P. Latge, E. Mylonakis, and T. L. Doering.** 2007. Loss of cell wall α(1-3) glucan affects *Cryptococcus neoformans* from ultrastructure to virulence. *Mol. Microbiol.* **63:**1385–1398.
157. **Reiss, F., and E. Alture-Werber.** 1976. Immunization of mice with a mutant of *Cryptococcus neoformans*. Characterization of the mutant, actively acquired resistance to experimental cryptococcosis in mice. *Dermatologica* **152:**16–22.
158. **Rhodes, J. C.** 1985. Contribution of complement component C5 to the pathogenesis of experimental murine cryptococcosis. *Sabouraudia* **23:**225–234.
159. **Rhodes, J. C., L. S. Wicker, and W. J. Urba.** 1980. Genetic control of susceptibility to *Cryptococcus neoformans* in mice. *Infect. Immun.* **29:**494–499.
160. **Rivera, J., J. Mukherjee, L. M. Weiss, and A. Casadevall.** 2002. Antibody efficacy in murine pulmonary *Cryptococcus neoformans* infection: a role for nitric oxide. *J. Immunol.* **168:**3419–3427.
161. **Rivera, J., O. Zaragoza, and A. Casadevall.** 2005. Antibody-mediated protection against *Cryptococcus neoformans* pulmonary infection is dependent on B cells. *Infect. Immun.* **73:**1141–1150.
162. **Romani, L.** 2004. Immunity to fungal infections. *Nat. Rev. Immunol.* **4:**1–23.
163. **Rosas, A. L., J. D. Nosanchuk, and A. Casadevall.** 2001. Passive immunization with melanin-binding monoclonal antibodies prolongs survival of mice with lethal *Cryptococcus neoformans* infection. *Infect. Immun.* **69:**3410–3412.
164. **Rude, T. H., D. L. Toffaletti, G. M. Cox, and J. R. Perfect.** 2002. Relationship of the glyoxylate pathway to the pathogenesis of *Cryptococcus neoformans*. *Infect. Immun.* **70:**5684–5694.
165. **Salas, S. D., J. E. Bennett, K. J. Kwon-Chung, J. R. Perfect, and P. R. Williamson.** 1996. Effect of the laccase gene CNLAC1, on virulence of *Cryptococcus neoformans*. *J. Exp. Med.* **184:**377–386.
166. **Salkowski, C. A., and E. Balish.** 1991. Cryptococcosis in beige mice: the effect of congenital defects in innate immunity on susceptibility. *Can. J. Microbiol.* **37:**128–135.
167. **Salkowski, C. A., and E. Balish.** 1991. A monoclonal antibody to gamma interferon blocks augmentation of natural killer cell activity induced during systemic cryptococcosis. *Infect. Immun.* **59:**486–493.
168. **Salkowski, C. A., and E. Balish.** 1990. Pathogenesis of *Cryptococcus neoformans* in congenitally immunodeficient beige athymic mice. *Infect. Immun.* **58:**3300–3306.
169. **San-Blas, G., L. R. Travassos, B. C. Fries, D. L. Goldman, A. Casadevall, A. K. Carmona, T. F. Barros, R. Puccia, M. K. Hostetter, S. G. Shanks, V. M. Copping, Y. Knox, and N. A. Gow.** 2000. Fungal morphogenesis and virulence. *Med. Mycol.* **38**(Suppl 1)**:**79–86.
170. **Schell, W. A., G. M. De Almeida, R. K. Dodge, K. Okonogi, and J. R. Perfect.** 1998. In vitro and in vivo efficacy of the triazole TAK-187 against *Cryptococcus neoformans*. *Antimicrob. Agents Chemother.* **42:**2630–2632.
171. **Sorensen, K. N., K. V. Clemons, and D. A. Stevens.** 1999. Murine models of blastomycosis, coccidioidomycosis, and histoplasmosis. *Mycopathologia* **146:**53–65.
172. **Staib, F., and S. K. Mishra.** 1975. Contributions to the strain-specific virulence of *Cryptococcus neoformans*. Animal experiments with two *C. neoformans*-strains isolated from bird manure. Preliminary report. *Zentralbl. Bakteriol. Orig A* **230:**81–85. (In German.)
173. **Steen, B. R., S. Zuyderduyn, D. L. Toffaletti, M. Marra, S. J. Jones, J. R. Perfect, and J. Kronstad.** 2003. *Cryptococcus neoformans* gene expression during experimental cryptococcal meningitis. *Eukaryot. Cell* **2:**1336–1349.
174. **Steenbergen, J. N., and A. Casadevall.** 2003. The origin and maintenance of virulence for the human pathogenic fungus *Cryptococcus neoformans*. *Microbes Infect.* **5:**667–675.
175. **Steenbergen, J. N., J. D. Nosanchuk, S. D. Malliaris, and A. Casadevall.** 2003. *Cryptococcus neoformans* virulence is enhanced after growth in the genetically malleable host *Dictyostelium discoideum*. *Infect. Immun.* **71:**4862–4872.
176. **Steenbergen, J. N., H. A. Shuman, and A. Casadevall.** 2001. *Cryptococcus neoformans* interactions with amoebae suggest an explanation for its virulence and intracellular pathogenic strategy in macrophages. *Proc. Natl. Acad. Sci. USA* **98:**15245–15250.
177. **Stevens, D. A.** 1996. Animal models in the evaluation of antifungal drugs. *J. Mycol. Med.* **6:**7–10.
178. **Stevens, D. A.** 1997. Animal models of blastomycosis. *Semin. Respir. Infect.* **12:**196–197.
179. **Stevens, D. A.** 1986. Animal models to evaluate antifungal activity: utility of models of chronic deep mycoses, p. 153–157. *In* H. V. B. K. Iwata (ed.), *In Vitro and In Vivo Evaluation of Antifungal Agents*. Elsevier Science.
180. **Stevens, D. A., B. J. Kullberg, E. Brummer, A. Casadevall, M. G. Netea, and A. M. Sugar.** 2000. Combined treatment: antifungal drugs with antibodies, cytokines or drugs. *Med. Mycol.* **38**(Suppl 1)**:**305–315.
181. **Stevens, D. A., T. J. Walsh, F. Bistoni, E. Cenci, K. V. Clemons, G. Del Sero, C. Fe d'Ostiani, B. J. Kullberg, A. Mencacci, E. Roilides, and L. Romani.** 1998. Cytokines and mycoses. *Med. Mycol.* **36**(Suppl 1)**:**174–182.
182. **Takemoto, K., Y. Yamamoto, and Y. Ueda.** 2006. Influence of the progression of cryptococcal meningitis on brain penetration and efficacy of AmBisome in a murine model. *Chemotherapy* **52:**271–278.
183. **Toffaletti, D. L., K. Nielsen, F. Dietrich, J. Heitman, and J. R. Perfect.** 2004. *Cryptococcus neoformans* mitochondrial genomes from serotype A and D strains do not influence virulence. *Curr. Genet.* **46:**193–204.
184. **Torda, A., R. K. Kumar, and P. D. Jones.** 2001. The pathology of human and murine pulmonary infection with *Cryptococcus neoformans* var. *gattii*. *Pathology* **33:**475–478.
185. **Torres-Rodriguez, J. M., Y. Morera, T. Baro, J. M. Corominas, and E. Castaneda.** 2003. Pathogenicity of *Cryptococcus neoformans* var. *gattii* in an immunocompetent mouse model. *Med. Mycol.* **41:**59–63.
186. **Tournu, H., J. Serneels, and P. Van Dijck.** 2005. Fungal pathogens research: novel and improved molecular approaches for the discovery of antifungal drug targets. *Curr. Drug Targets* **6:**909–922.
187. **Uicker, W. C., H. A. Doyle, J. P. McCracken, M. Langlois, and K. L. Buchanan.** 2005. Cytokine and chemokine expression in the central nervous system associated with protective cell-mediated immunity against *Cryptococcus neoformans*. *Med. Mycol.* **43:**27–38.
188. **van den Berg, M. C., J. Z. Woerlee, H. Ma, and R. C. May.** 2006. Sex-dependent resistance to the pathogenic fungus *Cryptococcus neoformans*. *Genetics* **173:**677–683.
189. **Walenkamp, A. M., A. F. Verheul, J. Scharringa, and I. M. Hoepelman.** 1999. Pulmonary surfactant protein A binds to *Cryptococcus neoformans* without promoting phagocytosis. *Eur. J. Clin. Invest.* **29:**83–92.
190. **Wang, P., C. B. Nichols, K. B. Lengeler, M. E. Cardenas, G. M. Cox, J. R. Perfect, and J. Heitman.** 2002. Mating-type-specific and nonspecific PAK kinases play shared and divergent roles in *Cryptococcus neoformans*. *Eukaryot. Cell* **1:**257–272.
191. **Williams, P. L., S. L. Leib, P. Kamberi, D. Leppert, R. A. Sobel, Y. D. Bifrare, K. V. Clemons, and**

D. A. Stevens. 2002. Levels of matrix metalloproteinase-9 within cerebrospinal fluid in a rabbit model of coccidioidal meningitis and vasculitis. *J. Infect. Dis.* **186:**1692–1695.

192. **Williams, P. L., R. A. Sobel, K. N. Sorensen, K. V. Clemons, L. M. Shuer, S. S. Royaltey, Y. Yao, D. Pappagianis, J. E. Lutz, C. Reed, M. E. River, B. C. Lee, S. U. Bhatti, and D. A. Stevens.** 1998. A model of coccidioidal meningoencephalitis and cerebrospinal vasculitis in the rabbit. *J. Infect. Dis.* **178:**1217–1221.

193. **Wills, E. A., I. S. Roberts, M. Del Poeta, J. Rivera, A. Casadevall, G. M. Cox, and J. R. Perfect.** 2001. Identification and characterization of the *Cryptococcus neoformans* phosphomannose isomerase-encoding gene, MAN1, and its impact on pathogenicity. *Mol. Microbiol.* **40:**610–620.

194. **Wormley, F. L., Jr., G. M. Cox, and J. R. Perfect.** 2005. Evaluation of host immune responses to pulmonary cryptococcosis using a temperature-sensitive *C. neoformans* calcineurin A mutant strain. *Microb. Pathog.* **38:**113–123.

195. **Wormley, F. L., Jr., J. R. Perfect, C. Steele, and G. M. Cox.** 2007. Protection against cryptococcosis by using a murine gamma interferon-producing *Cryptococcus neoformans* strain. *Infect. Immun.* **75:**1453–1462.

196. **Wright, K. A., J. R. Perfect, and W. Ritter.** 1990. The pharmacokinetics of BAY R3783 and its efficacy in the treatment of experimental cryptococcal meningitis. *J. Antimicrob. Chemother.* **26:**387–397.

197. **Yauch, L. E., M. K. Mansour, S. Shoham, J. B. Rottman, and S. M. Levitz.** 2004. Involvement of CD14, toll-like receptors 2 and 4, and MyD88 in the host response to the fungal pathogen *Cryptococcus neoformans* in vivo. *Infect. Immun.* **72:**5373–5382.

198. **Zhou, Q., R. A. Gault, T. R. Kozel, and W. J. Murphy.** 2007. Protection from direct cerebral cryptococcus infection by interferon-γ-dependent activation of microglial cells. *J. Immunol.* **178:**5753–5761.

199. **Zucker, K. E., P. Kamberi, R. A. Sobel, G. Cloud, D. N. Meli, K. V. Clemons, D. A. Stevens, P. L. Williams, and S. L. Leib.** 2006. Temporal expression of inflammatory mediators in brain basilar artery vasculitis and cerebrospinal fluid of rabbits with coccidioidal meningitis. *Clin. Exp. Immunol.* **143:**458–466.

Cryptococcus: From Human Pathogen to Model Yeast
Edited by J. Heitman et al.
©2011 ASM Press, Washington, DC

36

Veterinary Insights into Cryptococcosis Caused by *Cryptococcus neoformans* and *Cryptococcus gattii*

RICHARD MALIK, MARK B. KROCKENBERGER, CAROLYN R. O'BRIEN, DEE A. CARTER, WIELAND MEYER, AND PAUL J. CANFIELD

Cryptococcosis can affect a very wide range of animal species—domestic, free-living, and experimental. This chapter will focus on naturally occurring disease in animals. Experimental animal models of cryptococcosis will be covered elsewhere in this monograph. The current list of animal species in which cryptococcosis has been reported by no means excludes the possibility of cryptococcosis occurring in *any* species, given the extraordinary range of animals that have so far been reported to have become infected with this organism.

Susceptible animal species reported to date include:

- Invertebrates: amoebae (96), nematodes (73), insect larvae (caterpillars) (74)
- Amphibians: the toad *Bufo bufo* (91)
- Reptiles: snakes and lizards (40, 42, 75)
- Birds: especially psittacines but also passerines (62, 85)
- Monotremes (echidnas)/marsupials (including koalas) (46, 47, 51, 103)
- Eutherian mammals: mice (30), rats (33), dogs (60, 67), cats (28, 32, 68), ferrets (58), horses (1, 4, 15, 18, 20, 38, 54, 83, 86, 87, 89, 95, 100, 105), goats (16, 36, 94), sheep (52, 53), cattle (31, 82, 84), camels, alpacas (35), llamas (9), cheetahs (11, 72), seals, and toothed whales (97)

This list is not exhaustive but demonstrates the capacity of this organism to cause disease in a plethora of different species, seemingly irrespective of their physiology and lifestyle.

Richard Malik, Centre for Veterinary Education, The University of Sydney, NSW Australia 2006. **Mark B. Krockenberger and Paul J. Canfield,** Faculty of Veterinary Science, The University of Sydney, NSW Australia 2006. **Carolyn R. O'Brien,** Faculty of Veterinary Science, University of Melbourne, Werribee, Victoria, Australia 3030. **Dee A. Carter,** School of Molecular and Microbial Biosciences, The University of Sydney, NSW Australia 2006. **Wieland Meyer,** Molecular Mycology Research Laboratory, The University of Sydney Western Clinical School at Westmead Hospital, Westmead Millennium Institute, NSW Australia 2145.

Rather than providing an exhaustive list of disease features in individual species, this chapter concentrates on an approach based on the concept of comparative pathology and "one medicine/one health." It should be emphasized at the outset that members of the *Cryptococcus neoformans* species complex, comprising the two species *C. neoformans* and *Cryptococcus gattii*, are environmental organisms. The vast bulk of the biomass of these species resides in the environment. Although there is active and ongoing research into the *C. neoformans* species complex in relation to its environmental associations, we still have much to learn about the basic biology of this organism in its ecological niche. However, under the right conditions (whatever these are), cryptococcal species can behave as remarkably effective and cosmopolitan animal pathogens. From a general veterinary perspective, cryptococcosis is a sporadic cause of systemic disease, especially in cats, dogs, ferrets, goats, and horses. It is the most common systemic mycosis of cats worldwide. In certain regions, it is a significant cause of localized disease, specifically mastitis in cattle (31, 82, 84). Veterinarians who treat wildlife occasionally come across cryptococcosis cases in many different species, especially in Australia, South Africa, and most recently in western Canada.

The koala (*Phascolarctos cinereus*), an arboreal marsupial, has probably the highest incidence of cryptococcosis of all species. Necropsy studies reveal that cryptococcosis is the likely cause of death in approximately 3% of koalas. This species is therefore an important naturally occurring model for studying the epidemiology and pathogenesis of this disease (10, 45–51). Information about the pathogenesis of naturally occurring cryptococcosis in animals is critical to understanding the same disease in human hosts, especially the early pathogenesis of the disease after acquisition of the organism from the environment, where equivalent human data are currently lacking.

There is a lot to say about what animals can tell us about the *C. neoformans* species complex and cryptococcosis. But from the outset, let us emphasize one key point: veterinarians are critical players in the provision of clinical isolates to basic researchers interested in population genetics, epidemiology, and disease pathogenesis.

ONE MEDICINE AND CRYPTOCOCCOSIS

It has become fashionable to use the term *one medicine* when contemplating the study of comparative aspects of infectious diseases. The one medicine concept has a rich history extending back to the German pathologist and physician Robert Virchow, who in turn strongly influenced William Osler, James Law, and Ian Beveridge, to name just a few medical and veterinary luminaries. The central tenet was that much could be learned about disease pathogenesis by looking at features of naturally occurring and experimentally induced disease in different species. In other words, one medicine is built on the concepts of comparative pathophysiology and disease pathogenesis.

Observation of disease in a wide range of animal species (including humans), living different lifestyles in different environments, often highlights important aspects of the biology of the fungus. Serendipity may be an important part of this process, insofar as key observations need to be made at the right time and the right place. Once a new hypothesis has been generated through observation, it is vital to test the notion further. Hypotheses developed through observation may be tested by making additional or more detailed observations of the host, environment, or pathogen in naturally occurring disease (e.g., obtaining many isolates for culture and subsequent molecular studies) or by reproducing disease under laboratory conditions. Over the past few decades, there has been a rich history of these types of insights obtained through objective, accurate, and open-minded observations. Without doubt, they have greatly contributed to our understanding of cryptococcosis.

Disease is the result of the interaction between (i) the agent of disease (or pathogen), (ii) the animal host, and (iii) the environment that they share (host/pathogen/environment interactions). In terms of the fungus, all vertebrate animal species (including humans) can be considered as accidental, ectopic, dead-end encounters for *C. neoformans* and *C. gattii*. There are certain key factors, however, that have the potential to reveal much about the biology of the fungus and therefore its capacity to cause disease:

Some animal species are inherently more resistant to cryptococcosis. The reasons for this are still not clear, although this is a well-accepted concept for other primary pathogens, such as *Mycobacterium tuberculosis*. Resistance may occur at the level of phagocytic cell function. For example, rats are much more resistant to cryptococcal infection than mice, as has been shown experimentally on a number of occasions (e.g., reference 33).

Compared to people, most animal species have a relatively fixed geographical range and specific environmental niches. Apart from migratory birds and marine mammals, most animals affected by cryptococcosis have limited ranges. We, and others, have been able to exploit this situation to address various epidemiologic features of cryptococcosis. Specifically, we have often been able to determine where and when individual koalas were infected and the likely period of latency between infection and the development of clinical signs. For companion animals, we can make inferences about the site of infection because many animals (especially cats and ferrets) rarely travel more than a short distance away from their place of domicile.

Animals are exposed to different microenvironments according to their behavior and are therefore useful sentinel species. Surveying companion animals and wildlife in addition to human patients gives insights into disease pathogenesis for all species, especially concerning environmental associations of the pathogen. This might involve detecting nasal colonization, serum antigen levels, or serum antibodies (as markers of previous or current exposure).

The primary site of infection influences whether infective propagules establish asymptomatic colonization, a small limited focus of disease, or an extensive primary focus that rapidly disseminates. The anatomical location of this primary site of infection has a large impact on subsequent disease pathogenesis within the host, and additionally, there are critical anatomic differences between animal species. While human beings tend to have simple nasal passages, the nasal architecture of cats, dogs, horses, and ruminants is generally extensive and convoluted. As a consequence, infectious propagules are much more likely to be filtered by the upper respiratory tract for most animals, compared to humans.

These anatomic differences may explain why mycotic rhinosinusitis is a common manifestation of primary cryptococcosis in many animal species, while it is rare in human patients. Dolphins, on the other hand, breathe through a blow hole; thus, teleologically, their lungs are at increased risk for deposition of infectious propagules directly into the lower respiratory tract (71), even though those propagules may be present in low concentrations. Cryptococcosis is a well-characterized cause of mastitis in dairy cattle and goats, but not in other species. This is likely due to the pendulous nature of the mammary gland in at-risk dairy breeds, and husbandry practices, such as milking machines, which increase exposure of the teat canal to infectious organisms present in soil and other parts of the environment (31, 82, 84).

Different animal species function at different body temperatures, while within a given animal, certain tissues (e.g., skin, testes, and upper respiratory passages) are maintained at a lower temperature than internal organs exposed to core body temperature. It has been argued that *Cryptococcus laurentii* is more likely to involve cooler extremities and rarely cause systemic infections. Koalas have a lower body temperature; this may be relevant to their propensity to get cryptococcosis. In many birds, high body temperature is said to be important in natural resistance to cryptococcosis or restriction of infection to the cooler portions of the upper respiratory tract or skin (see Fig. 1).

Animals come in a huge range of sizes. This affects their physiology and hence the pathophysiology that underlies infection and immunity. For example, gas flow through airways is affected greatly by whether flow is laminar or turbulent, which in turn is inversely related to the radius of the airways. Thus, resistance to airflow and turbulence are much more likely to occur with air passages of small diameter. For this reason, small particles are much more likely to penetrate deep into the respiratory tract of large animals, as opposed to smaller animals. This may in part explain why the sinonasal cavity is the primary site of infection in small animals, while in larger animals the lungs contain the primary focus. Animals of intermediate size can have involvement of either the upper and/or lower respiratory tract. This phenomenon can be studied using scintigraphy and radioactive particles of known size, but to the best of our knowledge, such comparative studies have not been performed. Likewise, there is a dearth of experimental studies on infection of larger mammals using "physiologic doses" of organisms deposited via the respiratory tract.

Ethically, it is much easier to obtain specimens from animals compared to humans, whether they be nasal/nasopharyngeal swabs (to look for asymptomatic colonization), serum specimens (to determine prevalence of subclinical disease), or even necropsy specimens (to detect subclinical

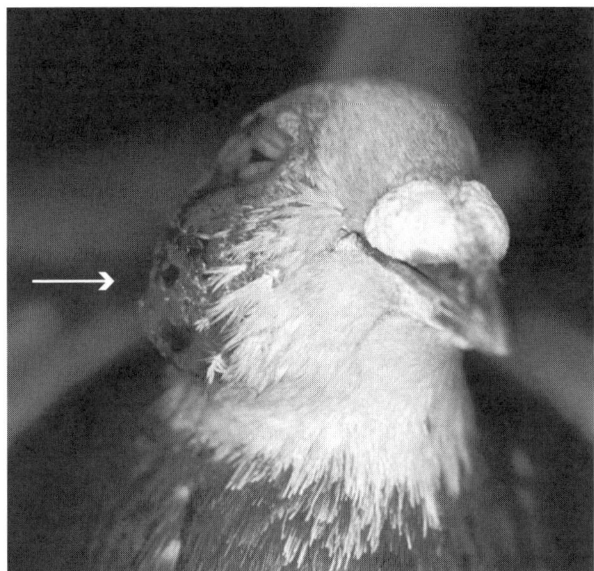

FIGURE 1 Localized subcutaneous *C. neoformans* var. *grubii* infection in a racing pigeon. This lesion (arrow) most likely developed as a sequela to contamination of a peck injury with infective propagules.

internal organ involvement). For example, the authors believe it would be worthwhile to survey indigenous human populations in northern Australia and Papua New Guinea to determine the prevalence of subclinical cryptococcosis using archived serum specimens collected for other reasons. Because of ethical issues, this is not an option, although this could certainly be done prospectively. Such ethical limitations almost never apply to archived sera or tissues from animals.

Unlikely animal specimens, including deaths due to motor vehicles ("road kills"), animals euthanized on humane grounds, and abattoir specimens, may provide very useful insights in relation to subclinical disease. For example, a study of 50 goats in Western Australia concerned principally with the detection of caprine arthritis-encephalitis virus incidentally detected that 6% of individuals necropsied had small focal cryptococcal lesions in their lungs (27).

NEW INSIGHTS INTO CRYPTOCOCCOSIS IN COMPANION ANIMALS SINCE 1998

Most of the salient clinical and pathological features of cryptococcosis in cats, dogs, ferrets, and horses were characterized during the 1970s and 1980s (26, 58, 60, 68, 86, 98). However, the literature up until quite recently was deficient for a number of reasons. First, because many reports were individual case studies or small case series, it was hard to determine the overall prevalence of different clinical entities. Second, little attention had been paid to making a complete mycological diagnosis to the species or molecular type level. It was therefore impossible to determine the serotype, species, variety, or mating type of *C. neoformans* species complex isolates that gave rise to disease in the majority of instances. Third, little or no attempt was made to look at the comparative prevalence of disease in different mammalian hosts in relation to appropriate reference populations (88, 102). It was therefore not possible to say, for example, whether cryptococcosis was statistically more common in one species than another or in one place rather than another.

In Australia (70, 79), and more recently in British Columbia (Canada) (21–25, 39, 44, 55, 56, 97), there has been a concerted effort to collect the necessary epidemiologic information as well as isolates from a meaningful proportion of cryptococcosis cases. These isolates have subsequently been utilized to inform epidemiological studies of cryptococcosis in humans, as well as provide basic biological data about the fungus. Veterinarians, physicians, and scientists in Australia and Canada are thus at the forefront of understanding the link between disease, the fungus, and the environmental niche, especially in relation to *C. gattii*.

CRYPTOCOCCOSIS IN COMPANION ANIMALS IN EASTERN AUSTRALIA

This section is very much driven by the following key questions that relate to host/pathogen/environment interactions:

- Is disease due to *C. gattii* clinically and/or epidemiologically distinguishable from disease caused by *C. neoformans*?
- Why is cryptococcosis more common in one species than in another? Is this due to anatomic differences, physiological differences, different behavior, or a combination of the three? Is gender, lifestyle, or place of domicile important?
- Which part of the respiratory tract is the primary site of infection: upper, lower, or both? Is the alimentary tract or skin ever the primary site of infection?
- Is the pattern of disease unique to Australia or present the world over?
- Immune function testing is almost never carried out in veterinary patients. Disease that remains localized for a substantial period at the primary site of infection is presumed to reflect a likely immune-competent patient; dissemination of disease early in the clinical course and widespread involvement of many tissues, especially skin and multiple peripheral lymph nodes, is suggestive of a possible underlying immune deficit, either genetic (major histocompatibility complex related, Toll-like receptor related, subnormal CD4 count, etc. [104]), drug induced (corticosteroids, cyclosporine, cytotoxic agents) or infectious (feline leukemia virus, feline immunodeficiency virus [FIV], feline infectious peritonitis virus).

Cats and Dogs

The large retrospective study by O'Brien et al. (79) provides a comprehensive picture of cryptococcosis in small companion animals in eastern Australia, with cases mainly drawn from New South Wales and fewer from Victoria and Queensland. This survey encompassed a relatively complete data set concerning 155 cats and 40 dogs diagnosed with cryptococcosis between 1981 and 2001.

Cryptococcosis was the most common systemic fungal disease of cats in eastern Australia. There is a consensus that this is the case the world over. Data from both the east coast of Australia and Western Australia (see below) both suggested that the incidence of cryptococcosis in cats is approximately six times higher than the incidence in dogs. Acute and chronic upper respiratory infections may facilitate either primary colonization and/or invasion of the sinonasal cavity by basiodiospores of *C. neoformans* and *C. gattii* by disrupting normal physical and immunological barriers and other protective mechanisms such as the effectiveness of the sneeze reflex. Admittedly, we have no direct data to support this hypothesis, although it is a compelling explanation as to why an animal as fastidious as the cat, and with such small external nares, commonly develops cryptococcal rhinosinusitis. Stated another way, unlike the dog, feline behaviors rarely extend to inhaling dirt and debris while digging and sniffing the environment. One notable exception, however, is the digging activity of cats when excavating a hole in which to bury their feces.

In the majority of instances, our view is that both *C. neoformans* and *C. gattii* act as primary pathogens in cats. Male cats were slightly overrepresented (considering the reference population of normal cats) (102). Although the age range of affected cats was very broad, the peak incidence was between 2 and 3 years of age, which is consistent with primary exposure as young adults, with emergence of disease either at that time or later (as a result of recrudescence). Siamese, Birman, and ragdoll breeds were significantly overrepresented, suggesting that genetic factors predispose to infection. This may represent a predisposition to infection with cryptococci per se or a predisposition to other conditions (e.g., epitheliotropic viral respiratory infections) which in turn facilitate the development of cryptococcosis.

The role of FIV was hard to determine, as the prevalence of infection (20/96 cats tested; 21%) was similar to that for other nondescript cohorts of "sick cats" but higher than the overall prevalence of FIV in "normal cats" (8%) (77). Longstanding FIV infection may predispose to the development of cryptococcosis, but likely only in a limited proportion of affected cats. In these patients cryptococcosis may become especially severe, progress quickly, or disseminate earlier or more widely than in FIV-negative cats (68). Without doubt, inappropriate administration of corticosteroids causes rapid progression of disease (Color Plate 6A), often precipitating development of central nervous system (CNS) signs.

In eastern Australia, *C. neoformans* var. *grubii* caused 71% of cases, while *C. gattii* accounted for the remaining 29%. Population genetic analysis suggested that *C. neoformans* var. *grubii* isolates from cats in the Sydney region represented part of a recombining population (12). Most *C. gattii* cases (37 cats) were caused by VGI infections, with only five VGII isolates (all from New South Wales) (13). Rural cats were more likely to be infected with *C. gattii* than inner-city or suburban cats. Disease caused by *C. neoformans* and *C. gattii* could not be distinguished on the basis of clinical features.

In accordance with earlier data from Australia and case series from overseas, the nasal cavity was the most common primary site of infection (e.g., Color Plates 6, 7). Nasal signs were subtle in some patients, although positive nasal culture was obtained almost without exception using superficial swabs or washings and culture on Staib's niger seed agar. The rostral part of the nasal cavity was infected more commonly than the caudal nasal cavity, although in a minority of cases prominent nasopharyngeal involvement was documented (63). It would be of some benefit to characterize the extent of tissue involvement further using computed tomography and magnetic resonance imaging (29) to better resolve changes in the sinonasal region (see below) and to look more carefully for concurrent lung involvement.

Critically, infection was generally constrained to the nasal cavity for a considerable time, before extending to nearby contiguous structures (including paranasal sinuses, nasal planum, nasal bridge, retrobulbar space, and brain via the cribriform plate) or disseminating widely via the bloodstream (e.g., to skin, subcutis, and CNS). Dissemination of infection to the CNS was generally a late sequela. Lower respiratory tract involvement was rarely documented (3), although this is a common site for primary infection in free-living and captive cheetahs (Fig. 2 and 3) (11, 72). Very rarely, cats could be seen with localized cutaneous lesions, presumably acquired following a cat scratch injury contaminated by infective propagules (66); these cases clinically resembled other infections caused by bacterial and fungal saprobes following penetrating injury and typically were located on the naso-ocular region, a region commonly subjected to scratch injuries (67).

FIGURE 2 Cheetah with cerebral cryptococcosis. The primary site of infection was the lung (see Fig. 3). The "glassy stare" of cryptococcal optic neuritis is evident. The patient is receiving a subcutaneous infusion of amphotericin B. The patient was treated successfully, regaining its vision. Cheetahs seems to be the only big cat overrepresented in cohorts of wildlife with naturally occurring cryptococcosis. Both captive and free-living cases have been documented.

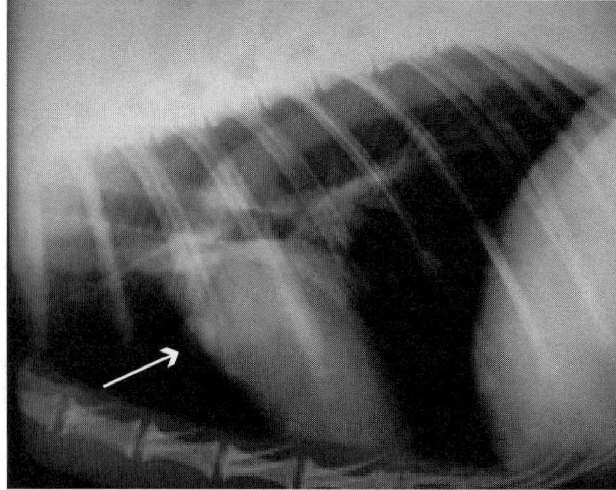

FIGURE 3 Lateral thoracic radiograph of the cheetah with cryptococcal meningoencephalitis in Fig. 2. Note the mass lesion in the lung at the cranial aspect of the cardiac silhouette (arrow).

There was a marked preponderance of large-breed young-adult dogs (less than 5 years) among the cohort of canine cryptococcosis cases. Although the nasal cavity was the most commonly identified primary site of infection, it was not invariably involved, and occasional cases had primary involvement of the lung or gastrointestinal tract (61). CNS involvement and widely disseminated disease were far more common than in the feline cohort; i.e., CNS signs were not always a late development. Certain breeds were statistically overrepresented (e.g., boxers, dalmatians, Doberman pinschers, and German shepherds). The proportion of *C. neoformans* var. *grubii* and *C. gattii* was similar to cats.

Clinical signs in cats and dogs with CNS involvement typically included bilaterally dilated pupils that failed to respond to light, most likely referable to optic neuritis and/or retinitis, additional variable cranial nerve abnormalities, and, in one cat, signs referable to a transverse myelopathy. Vestibular dysfunction was often present in cats and dogs with CNS cryptococcosis (6, 60). Our suspicion was that this group of clinical signs resulted from cryptococcal infection tracking through the cribriform plate to the olfactory bulbs and hence to the optic nerves, which are situated in close proximity. Limited unpublished data from a small number of necropsy examinations is consistent with this speculation. Interestingly, the neuro-ophthalmic findings of indigenous human patients in Papua New Guinea (90) with *C. gattii* infections due to molecular type VGI are remarkably similar to Australian cats and dogs with cryptococcal meningoencephalitis (60, 68). Similar pathomechanisms may be involved.

In general terms, the prognosis for cats and dogs with cryptococcosis was good to guarded, depending on such factors as species (better prognosis in cats), time between presentation and development of first signs, and extent and severity of tissue involvement (especially in relation to the presence or absence of CNS disease) (78). In uncomplicated cases, monotherapy with fluconazole or itraconazole was often effective if given for sufficiently long periods (typically in excess of 6 months and often for more than 1 year). Advanced or severe cases, however, required amphotericin B and flucytosine to "get on top" of the infection, followed by a very long period of consolidation therapy using one of the azoles. In our hands, monotherapy with azoles, although convenient, was inferior to combination therapy with amphotericin B. Indeed, we abandoned azole monotherapy in cases with CNS or disseminated disease after several consecutive cases of this type did not respond until amphotericin B was started. The availability of comparatively inexpensive generic fluconazole (sourced from India, China, or Mexico) through compounding pharmacists (57) has greatly improved the prognosis for cats and dogs with cryptococcosis, because in veterinary medicine a high cost of therapy often results in euthanasia or a course of therapy insufficiently long to effect a permanent cure.

Serum antigen titers proved useful to guide therapy (64), although the fall in titer typically lagged behind clinical improvement. Initial antigen titers correlated quite well with the extent of disease present outside the CNS and were as high as 131,072 in severely affected patients; typical values were 256 to 8,192. As in human patients, infection had the propensity to recur, sometimes following a very considerable period (as long as 13 years in one cat) after apparently successful therapy (45). In such cases, molecular tests confirmed that recurrence involved the inciting cryptococcal isolate, rather than acquisition of a new isolate many years later (45).

Serendipitously, it was discovered that amphotericin B could be administered intermittently as a subcutaneous infusion two or three times a week (59), and this therapy has gained widespread acceptance for treating cryptococcosis and other fungal infections in dogs and cats. In a veterinary setting this is a practical and cost-effective way to administer a potent antifungal systemically, and we have since used this approach in cats, dogs, cheetahs (Fig. 2), and koalas. Perhaps a similar approach using intraperitoneal amphotericin B may prove a useful option in human patients in developing nations, where 2 to 6 weeks of continuous intravenous therapy is not affordable or practical. Focal recalcitrant lesions have been treated using intralesional amphotericin B in concert with oral azoles therapy. Interestingly, dogs (but never cats) invariably developed a severe drug reaction if given flucytosine for a sufficient length of time in concert with amphotericin B (65).

Ferrets

Considering they are a comparatively rare pet in eastern Australia (and Canada), ferrets seem to be at increased risk for developing cryptococcosis compared to dogs and possibly even cats. Perhaps this reflects their domicile and behavior, with close and intimate contact with soil and dirt. Also, like cats, they have small upper airways and a high prevalence of upper respiratory disease, which may be viral in origin.

Ferrets with cryptococcosis commonly have primary involvement of the upper respiratory tract and draining regional lymph nodes (58). Primary infection of the lungs and gastrointestinal tract is also seen, and extension to the CNS can occur via the cribriform plate or hematogenous spread. In Australia, most cases were infections caused by *C. gattii* molecular type VGI, although one ferret with primary alimentary disease was infected by *C. neoformans* var. *grubii*. Ferrets from western Canada were presented for localized disease of the skin and subcutis of their limbs, with infections referable to *C. gattii* isolates responsible for the Vancouver Island epidemic (55, 58).

DNA fingerprinting and microsatellite analysis of two sibling ferrets were of great interest, as clinical disease developed 12 months later in one animal but involved exactly the same strain as that which affected its sibling. Presumably, this was because both animals were simultaneously exposed and infected from a common environmental source, but with different time courses of pathogenesis (58).

Like cats and dogs, antigen testing was found to be reliable, and animals presented sufficiently early in the clinical course could generally be successfully treated with a long course of an azole (58). It may be possible to administer amphotericin B subcutaneously in this species, although we have not had the recourse to do so.

Birds

Very few case series of avian cryptococcosis had been reported until a concerted effort was made to assemble a representative cohort of birds diagnosed by veterinarians in Australia (62). Cases included two Australian king parrots, an *Eclectus* parrot (Fig. 4), two corellas, one Major Mitchell's cockatoo (all *C. gattii* infections), one Gang-gang cockatoo, and a racing pigeon (both *C. neoformans* var. *grubii* infections). These cases were compared and contrasted with cases reported previously. As in cats and dogs, there was a propensity for cryptococcosis to cause invasive disease of the upper respiratory tract (Fig. 4) in parrots domiciled in Australia. Male birds were more commonly affected, and *C. gattii* accounted for most cases, suggesting exposure to eucalyptus material (perhaps in tree branches, bark, and hollows) as a predisposing factor (26). In these cases, *C. gattii* appeared

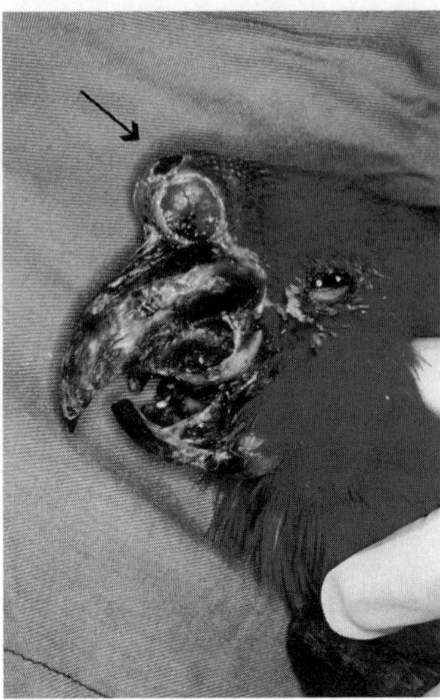

FIGURE 4 Invasive and deforming cryptococcosis affecting the chonae (arrow) of an *Eclectus* parrot. This was a *C. gattii* VGI infection.

to behave as a primary pathogen of immunocompetent hosts. Localized cutaneous disease was observed in one racing pigeon, a *C. neoformans* var. *grubii* infection, likely resulting from a contaminated peck injury (Fig. 1).

In contrast to the Australian avian cohort, parrots, pigeons, and other avian species investigated in Europe and North America had a pattern of disease more suggestive of opportunistic infection of immune-compromised hosts, with lower respiratory involvement and often-widespread dissemination to multiple tissues. Captive kiwis in Australia and New Zealand appeared at risk for developing severe diffuse cryptococcal pneumonia with or without dissemination to other internal organs. Disease in kiwis was a result of *C. gattii* infections, possibly due to the use of eucalyptus material in their artificial enclosures and possibly also as a consequence of their low body temperature.

Avian cryptococcosis has proven difficult to treat. Although long-term fluconazole therapy in concert with surgical debulking of lesions is somewhat successful, many cases are refractory or relapse after apparently successful therapy. Alternative approaches, possibly using intralesional amphotericin B, should be investigated. Studies of pharmacokinetics of azoles in avian patients should be performed to ensure optimal doses are administered, rather than extrapolation via allometric scaling.

Horses

From 1981 to the time of going to press, we have had first-hand experience with only a single horse with cryptococcosis, a pony mare with multifocal cryptococcal mass lesions throughout its pulmonary parenchyma (7). Clinical and radiological findings were attributable to pulmonary infection caused by *C. gattii* molecular type VGI. This patient was cured with intravenous amphotericin B therapy, adapting a protocol used to treat equine pythiosis. Swabs taken from the nasal vestibules of numerous horses situated in the same facility as this pony failed to detect instances of nasal colonization by any species of the *C. neoformans* complex (7).

The literature suggests that horses can develop either upper or lower respiratory involvement, with the respiratory tract or gut as the primary portal for infection. Horses can present for deforming sinonasal or retrobulbar disease, nasopharyngeal involvement, nasal discharge with occasional epistaxis, pneumonia, pleurisy, neurological signs, osteomyelitis, and even uterine infections (1, 4, 15, 18, 20, 38, 83, 86, 87, 89, 95, 100, 105). Despite this plethora of potential presentations, and the likelihood that cases along the east coast of Australia would have come to our attention, equine cryptococcosis in eastern Australia would appear to be very rare. The reason for this is intriguing, as horses have a high prevalence of intercurrent respiratory disease—both viral and bacterial—and often have owners sufficiently motivated to seek veterinary attention and therapy. The situation in Western Australia is quite different and will be presented later.

Koalas

A definitive series of 43 cases of cryptococcosis in captive and free-living koalas from the east coast of New South Wales, Australia, were reviewed (51). All cases were shown to be infections caused by *C. gattii* molecular type VGI, although subsequent to the publication of this study we discovered a pocket of infections caused by *C. gattii* molecular type VGII in koalas residing in a wildlife park on the outskirts of Perth (Western Australia). The relationship between koalas and eucalyptus trees, which probably underpins this strong association, will be explored later in this chapter.

No age or gender predispositions were observed. The respiratory tract was the primary site of infection in 77% of cases. Although the lower respiratory tract was affected most commonly (60% of cases), 30% of koalas had sinonasal involvement (Fig. 5, Color Plate 8), while 14% had both upper and lower respiratory tract involvement. Older anatomical studies and recent investigations using computed tomography demonstrated that koalas have quite a small and simple nasal cavity, with well-developed paranasal sinuses (Fig. 5), suggesting that inhaled infectious propagules may be less likely to be filtered out of inspired air than in animals such as the cat, dog, and ferret.

Local extension to surrounding contiguous tissues, such as nasal planum (Color Plate 8), bridge of nose (Fig. 5), and incisive canal, was common when there was sinonasal involvement. Dissemination to distant tissues, including the CNS (37% of cases), was also common. Two cases involved only localized infection of the skin and subcutis, which were likely secondary to penetrating injuries. Late presentation was probably the reason for the high proportion of cases with disseminated disease. In cases that were followed using antigen testing, the indolent and delayed nature of disease could often be appreciated and will be discussed later in the chapter.

Other Domestic and Wildlife Species

Cryptococcosis has been reported sporadically in other species. Our culture collection includes several cows that developed granulomatous cryptococcal mastitis, presumably as a result of infection tracking up through the teat canal. One alpaca presented with CNS signs referable to cryptococcal meningitis (35). Numerous indigenous mammals including possums, two potoroos, a quokka, a feather-tailed glider, one sticknest rat, one eastern water skink (reptile), and a captive dunnart (Fig. 6) were documented with fatal cryptococcosis (47, 103). The vast majority of these infections were referable to *C. gattii* (almost invariably being caused by molecular type VGI in animals from eastern Australia).

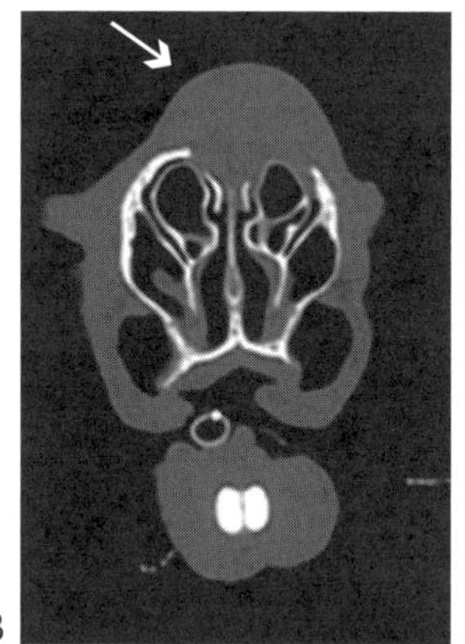

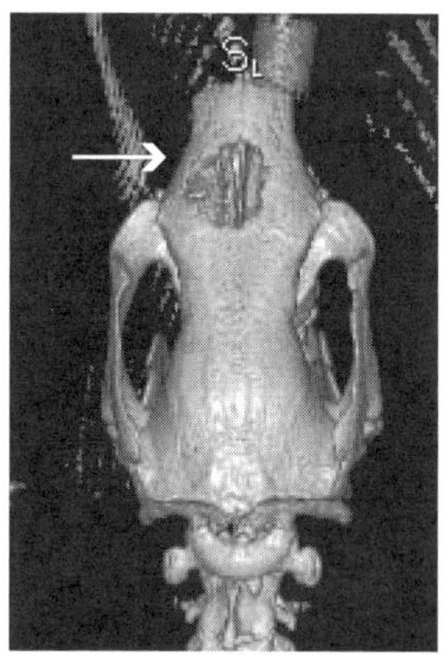

FIGURE 5 Invasive cryptococcal rhinitis in a koala due to *C. gattii* (VGI). (A) Note the prominent swelling (arrow) above the bridge of the nose due to infection penetrating the nasal bones to access the subcutaneous space. We presume this infection started in the dorsal nasal meatus (see thin-slice computed tomography image in panel B), resulting in invasion and destruction of the overlying nasal bone and extension of infection to the subcutaneous tissues of the nasal bridge (arrow, B). (C) A three-dimensional bone-density image of this patient is shown; note the bone loss associated with the infection (arrow). The infected tissues in the nasal bridge were excised surgically, and the patient was subsequently treated using fluconazole (10 mg/kg orally twice daily). Despite this, the defect in the bone persisted, the serum antigen titer remained unchanged at 8 to 16 for about 6 months, and the koala subsequently developed lymphoid leukemia.

Only the sugar glider and one potoroo were infected with *C. neoformans* var. *grubii*.

CRYPTOCOCCOSIS IN WESTERN AUSTRALIA: A DIFFERENT PICTURE

An extensive retrospective study of cryptococcosis cases seen in Western Australia between 1995 and 2006 was recently completed in collaboration with colleagues at Murdoch University and the largest veterinary clinical pathology laboratory in that state. Cryptococcosis was identified as the cause of disease in 155 animals, comprising 72 cats, 57 dogs, 20 horses, 3 alpacas, 2 ferrets, and a sheep. The study was weakened by somewhat incomplete mycological data (70). Despite this limitation, there were some striking differences in the patterns of disease across the range of animal species surveyed. As in eastern Australia, cats were roughly five to six times more likely to develop cryptococcosis than dogs (70, 88, 92). Of great interest was the finding that horses were almost twice as likely as dogs to become infected. Dogs and horses tended to develop disease at an early age (1 to 5 years), while cats were presented over a much wider range of ages. In cats and dogs, the upper respiratory tract was the most common primary site of infection, while horses and alpacas tended to have lower respiratory involvement.

Epidemiologically, the most striking finding was the high frequency with which *C. gattii* was identified by appropriate testing, with infections attributable to this species comprising 5/9 cats, 11/22 dogs, 9/9 horses, and 1/3 alpacas. Preliminary molecular genotyping suggested that most of the *C. gattii* infections in domestic animals (9/9 cases; plus 3 archival isolates from sheep) were of the molecular type VGII. This contrasts to the situation on the eastern seaboard of Australia, where disease attributable to *C. gattii* is much less common and is mainly due to the molecular type VGI. This finding may in part explain why there was an overall more

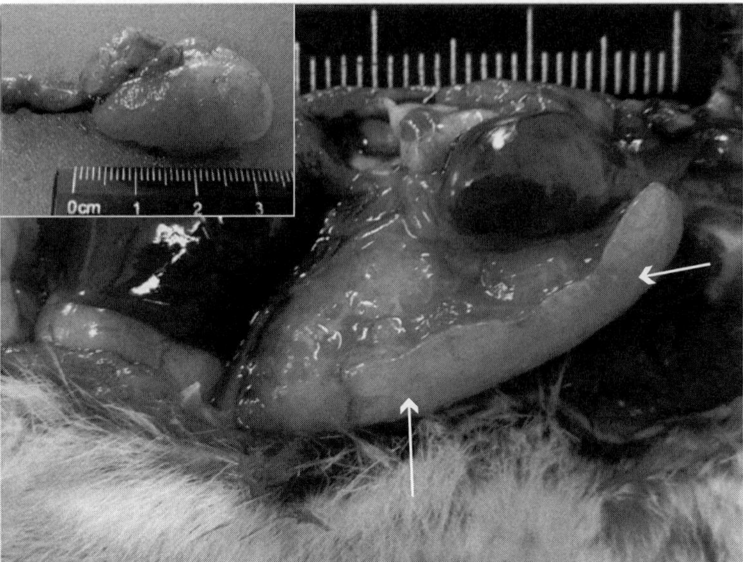

FIGURE 6 Diffuse cryptococcal pneumonia in a captive Julia Creek dunnart. Note the consolidation of an entire lung (arrows)—essentially a confluent cryptococcal granuloma. The liver is evident on the left-hand side, while the heart is situated on top of the consolidated lung. The inset shows the affected lung after it had been removed from the chest cavity.

guarded prognosis for successful therapy in patients from Western Australia (70, 92).

There are intriguing similarities between cryptococcosis in animals in southwestern Western Australia and a somewhat different range of species in British Columbia (Canada), although we suspect in each instance that a variably high prevalence of molecular type VGII infections provides the explanation for the unusually wide host range and infection of both typical and atypical hosts, including horses, small ruminants, and marine mammals. Interestingly, the older literature also identifies Western Australia as a site where cryptococcosis occurs with some frequency in goats—with four cases reported by Chapman et al. in 1990 (16) and a further three by Ellis et al. in 1988 (27). Affected goats had primary respiratory disease, involving the upper (Color Plate 9) and/or lower respiratory tract, including the trachea. It is possible that the high incidence of cryptococcosis in goats in southwestern Western Australia is also related to the high environmental presence there of *C. gattii* molecular type VGII. The second report (27) is especially interesting, as the pulmonary foci of cryptococcosis represented subclinical disease and were detected only because of a study looking for another pathogen (27). Given that only 50 goats were subject to necropsy examination, a prevalence of 6% subclinical cryptococcosis is striking. Early reports of cryptococcosis in this location suggest that strains of molecular type VGII have been in existence there for many years (10, 20, 86, 93, 94).

The reasons why infections caused by *C. gattii* molecular type VGII are more prevalent in southwestern Western Australia and the Northern Territory (17, 43) are unknown. In the vicinity of Perth, companion animals and a captive population of koalas and wombats in a wildlife park functioned as sentinel species to alert us to the presence of *C. gattii* molecular type VGII in this geographical region, whereas in the Northern Territory it was the high prevalence of disease in indigenous human populations that helped focus our attention. Perhaps scrutiny of animals living in close proximity to Aboriginal populations in the top end of Australia (43) will provide additional insights. Routine necropsy of Aboriginal camp dogs that die or are euthanized would provide data and may be a useful surveillance measure. Similarly, camp dogs anesthetized for routine desexing could be surveyed via nasal washing and determination of serum antigen titers. Currently, we have no cogent theory for why these two regions, not to mention Columbia and Vancouver Island (British Columbia), are especially favorable for *C. gattii* molecular type VGII.

CRYPTOCOCCOSIS IN ANIMALS IN NORTH AMERICA, EUROPE, AND JAPAN

There are numerous veterinary schools in North America, Europe, and Japan, and one might expect there to have been large surveys of cryptococcosis in companion animals from such centers. Indeed, there are some excellent retrospective studies from single institutions (32, 41) and also a number of multi-institutional studies (8, 28). Additionally, interesting or unusual case studies continue to be published (e.g., references 76, 81, and 101). Although these case series provide excellent clinical, laboratory, pathologic, and imaging data, almost none makes a systematic attempt to correlate epidemiology and patient characteristics (species, breed, gender, place of domicile) with isolates based on molecular or serological identity. Likewise, little or no attempt is made to correlate the epidemiology of animal disease with human disease in the same vicinity.

Although *C. gattii* is likely to be far less prevalent in the continental United States compared to Australia, examination of isolates and classification as *C. gattii*, *C. neoformans* var. *grubii*, and *C. neoformans* var. *neoformans* might prove informative, as we suspect there are "hot spots" for *C. gattii* (molecular types VGI and/or VGII) along the west coast of North America. Indeed, this may explain in part why disease seems especially common in San Diego and San Francisco and their environs, based on the experience of the

University of California at Davis group and private practitioners such as Keith Richter. There is some suggestion that feline leukemia virus may have once been a cause for immune suppression in the overall cohort of North American cats with cryptococcosis (41), although it is generally agreed that this retrovirus infection is becoming less and less common.

Recent work from the veterinary neurology group at the University of California at Davis has characterized the range of structural changes observed during magnetic resonance imaging in cats and dogs with cryptococcosis (P. Dickinson, K. Vernau, and R. LeCouteur, personal communication). These studies accord with previous pathological studies in suggesting that cryptococcal organisms gain access to the CNS via two alternative routes: (i) via the cribriform plate, with early involvement of the olfactory bulbs and optic nerves, and (ii) hematogenously, resulting in either primary meningeal involvement or cryptococcomas at various levels through the neuraxis (e.g., reference 29).

The situation in Europe is likewise unclear, although there are some very well-documented cases of disease in specific instances. For example, a recent series of outbreaks in goats due to C. *gattii* molecular type VGI in Spain has raised numerous questions about why such case clusters occur. The dearth of large case series of cats and dogs with cryptococcosis from European veterinary institutes suggests that cryptococcosis may be less common there compared to North America and Australia. Specifically, the infection seems quite rare in cats and dogs within the United Kingdom. In Japan, cryptococcosis would appear to be the most important systemic mycoses of cats, with all cases referable to C. *neoformans* α mating type (80).

Veterinary epidemiology and pathology has been at the forefront of investigations of the Vancouver Island outbreak of cryptococcosis over the past decade. This cooperative approach among veterinary pathologists, veterinary clinicians, veterinary epidemiologists, microbiologists, and medical epidemiologists has resulted in data from spontaneous animal disease cases complementing those from human patients and environmental surveys. This has allowed investigators to gain an overarching appreciation of the host/pathogen/environmental interaction in relation to the expansion of C. *gattii* molecular type VGII into a novel environment. This will be discussed in more detail later in the chapter.

THE "KOALA CONNECTION": A MODEL OF THE PATHOGENESIS OF CRYPTOCOCCOSIS IN ALL SPECIES

A wealth of circumstantial evidence suggests that human patients generally encounter infectious cryptococcal propagules as finely dispersed aerosols, or small particles, which are subsequently deposited at the level of respiratory bronchioles following inhalation. A particularly heavy inoculum of infectious propagules presumably favors establishment of a nidus of infection. Intact innate immunity, such as an effective cough reflex and good functionality of the mucociliary escalator, and timely development of acquired cell-mediated and antibody-mediated immunity (14) all favor elimination of the infective agent, either prior to development of tissue invasion or after limited tissue invasion. It is difficult to conceptualize colonization of the lower respiratory tract with cryptococcal organisms, and the development of infection likely requires invasion of the epithelial lining of the smaller airways or organisms reaching the alveolar spaces and hence alveolar macrophages.

The situation is likely different in mammals that have a well-developed upper respiratory tract, which as well as being more effective at olfaction is also more efficient in humidifying and filtering air and therefore trapping inhaled particulate matter prior to it reaching the trachea and lower airways. As a result, it seemed logical to speculate that mucus lining the upper respiratory passages could under certain circumstances become colonized by cryptococcal organisms. These might then function as transient commensals and only occasionally penetrate the epithelium of the nasal passages or paranasal sinuses.

This hypothesis was supported by a small study using randomly sourced dogs and cats, which showed that C. *neoformans* var. *grubii* could be cultured, often in large numbers (>100 colonies per plate), from deep nasal washings in 14% of dogs and 7% of cats in the absence of demonstrable cryptococcal antigen in serum or histological evidence of tissue invasion in biopsies of the nasal passages (69). It was thought that for behavioral reasons, dogs were much more likely to be exposed to dust and dirt as they explored their olfactory environment through sniffing and digging and thus were more likely to become colonized by cryptococcal organisms. These studies were subsequently extended to koalas, an arboreal Australian marsupial adapted to living and foraging in *Eucalyptus* trees (19). It is possible to easily obtain material from koalas living in artificial environments in zoos and wildlife parks and also from free-living animals admitted into veterinary hospitals as a result of disease, trauma, or misadventure. Gentle manual restraint is all that is required to obtain nasal swabs, and likewise it is relatively easy to collect blood from the cephalic vein for serological studies (Fig. 7). Radiographs and cross-sectional imaging studies (Fig. 5), on the other hand, require general anesthesia.

Preliminary work demonstrated that the nasal vestibule and dorsal nasal meatus of koalas was colonized by C. *neoformans* var. *grubii* and/or C. *gattii* depending on the place of domicile of individual animals (19). As a consequence of these similar observations in cats, dogs, and koalas, we developed a concept that we could use animals in general, and koalas in particular, as "biological air samplers" or sentinel species to determine the presence of infectious propagules in their immediate environment (46, 50, 69).

During these investigations, it became clear that husbandry practices somehow permitted koalas to dramatically amplify the amount of C. *gattii* molecular type VGI cells present in many man-made captive koala environments. Subsequent work showed this to be also the case for C. *gattii* molecular type VGII in a captive koala collection near Perth in Western Australia.

As well as virtually 100% nasal colonization by C. *gattii*, the skin was frequently colonized also, and C. *gattii* could be cultured with ease from all areas in the enclosures contacted by koalas, including perches, trees, branches, ropes, feeding apparatus, and soil and eucalyptus material supplied as food (browse) present within the enclosure (50) (Fig. 8). Interestingly, all isolates tested, without exception, were of the α mating type, whereas in natural tree hollows we typically find a variable proportion of isolates to be also of the **a** mating type (37). Our speculation is that koalas, as a result of their movement and scarification of surfaces through climbing, perhaps in concert with favorable actions of urine and feces (maceration, nutrients from digested gum leaves), somehow provide a very suitable environment for clonal proliferation of cryptococcal organisms, thereby resulting in a high environmental presence of organisms and concurrent nasal and cutaneous colonization. Conceptually, this is similar

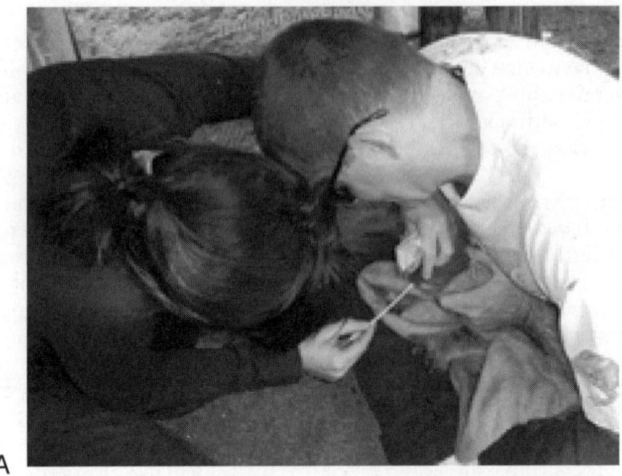

A

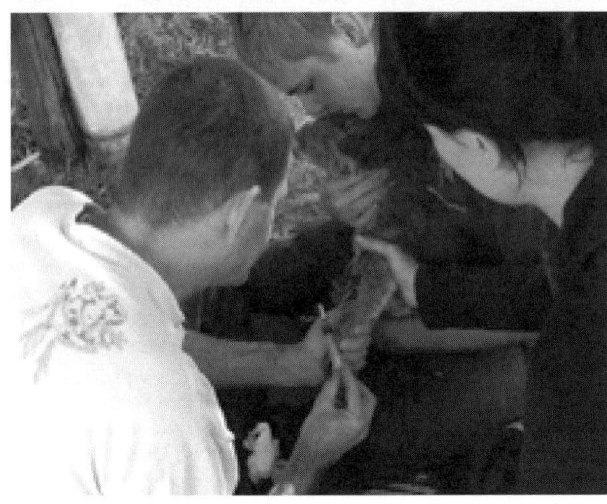

B

FIGURE 7 (A) Collecting swabs from the nasal vestibule/dorsal nasal meatus of a koala. Gentle manual restraint, sometimes using a hessian bag, is all that is required for this innocuous procedure. Preliminary studies by our group demonstrated that in this species, there is good correlation between superficial nasal swabs and deeper nasal washings in terms of detecting colonization by *C. gattii*. (B) Collecting blood from the cephalic vein using a 23-gauge butterfly needle. It is straightforward to collect 1 to 4 ml of blood over 30 s using this technique, sufficient for hematology and serology.

to the way in which pigeons amplify the amount of *C. neoformans* var. *grubii* and *C. neoformans* var. *neoformans* cells in their environments.

It is not our contention to state that *C. gattii* molecular type VGI actually requires koalas for its natural propagation, as there are several parts of Australia that have been sampled that revealed abundant presence of *C. gattii* but no koalas, such as parts of the Murray and Murrumbidgee river systems. Likewise, the molecular type VGI has been found in many parts of the world that do not have koalas (or equivalent animals), such as Spain, western Canada, and South America.

The commonness of nasal colonization and the serological prevalence of subclinical disease in koalas were directly correlated with the extent of cryptococcal presence in the environment in which they lived (48). Subclinical disease was defined as a positive serum cryptococcal antigen titer ($\geq$2) in an animal with a positive nasal culture. Titers could be as high as 64 in these asymptomatic animals. They were seen also in koalas living in the wild, with titers in this group ranging from 4 to 16. Interestingly, koalas with symptomatic cryptococcosis generally had titers of $\geq$128, with a single recent exception of clinical disease in a koala with a titer of only 4 (unpublished data).

Such subclinical disease was observed to progress in three ways: (i) the animal would remain without signs, and the titer would eventually decline to zero; (ii) the animal would remain without signs, but the titer would persist indefinitely at a low level; or (iii) the animal would progress to develop clinical disease, either immediately (within weeks to months) or after a substantial amount of time (years). Opportunistic necropsies of animals without clinical signs of cryptococcosis and low positive serum cryptococcal antigen titers frequently identify limited (microscopic) tissue invasion (see below). Although similar types of scenarios are envisaged to account for seroconversion via subclinical infection in human patients (34), most of the supporting evidence is circumstantial.

FIGURE 8 Koala in a captive enclosure. Note "browse" provided as food and wooden logs provided as perches. This area is richly abundant in *C. gattii* VGIIb. Inset shows a very simple sampling technique: a swab moistened with sterile saline is rubbed vigorously over a log provided as a climbing perch.

Following the natural course of these subclinical infections over a substantial period proved to be a powerful tool in unraveling the complexity of disease progression in this species. For example, one koala with an antigen titer of 8 was transported to a zoo in Japan, where perhaps as a result of stress due to transport or change in husbandry or nutrition, fatal diffuse pulmonary cryptococcosis developed. The isolation of *C. gattii* molecular type VGI at necropsy, which is not endemic to Japan, supported the conjecture of subclinical disease up to 1 year prior to clinical disease, most likely somewhere within the lungs, before the animal left Australia. This has been seen subsequently a number of times.

The nature of occult lesion(s) in animals with subclinical disease could be determined on some occasions when animals died as a result of other causes (e.g., vehicular trauma, intercurrent disease such as chlamydiosis) and where serum (from heart blood) was available for antigen testing (48, 51). For example, a koala with a titer of 16 had a 5-mm-diameter lesion within the ventrocaudal portion of the right maxillary sinus at necropsy, while a titer of 4 was found in a koala with a microscopic pulmonary cryptococcal granuloma. By way of contrast, animals with clinical (symptomatic) cryptococcosis generally have titers in serum ranging from 128 to 16,384 depending on chronicity and the extent of disease. These observations emphasize that koalas are far more likely to develop transient colonization, or self-limiting subclinical disease with timely mobilization of an effective immune response, than clinical disease. Clinical disease, therefore, represents the tip of the iceberg. However, it should be emphasized that, by our estimation, cryptococcosis is the principal cause of death in 3% of koalas that progress to necropsy examination (the second most prevalent infectious disease in this species following chlamydiosis) (47). Cryptococcosis is nonetheless a significant cause of death in koalas, despite the likelihood that koalas have coevolved over millennia with this particular fungus.

As a result of our work, it is now routine to measure the cryptococcal antigen titer during health surveillance checks of captive koalas and to intervene with drug therapy (typically fluconazole 20 mg/kg twice daily) if the titer remains above 32, even in the absence of clinical signs. Therapy is continued until the titer declines to nil. The availability of generic fluconazole through compounding pharmacies has made this approach much more affordable than previously, and indeed we used to use itraconazole for this purpose because historically it was less expensive. Likewise, it has become mandatory to determine the antigen titer in serum before transferring a captive koala to a zoo overseas and to preemptively treat subclinical cryptococcosis prior to export. The evidence from studies of koalas suggests the usefulness of active serological monitoring of persons who are at high risk for developing cryptococcosis due to *C. gattii*, specifically indigenous patients in far northern Australia. Proactive monitoring and early treatment with fluconazole may be prudent and cost-effective, considering the expense of treatment when cases are presented later in the clinical course and the high morbidity and mortality of such presentations.

Preliminary unpublished work suggests that environmental decontamination within koala enclosures using biocides such as F10™ veterinary disinfectant/biocide (benzalkonium chloride and polyhexamethylene biguanide) at standard dilutions can reduce the environmental presence of *C. gattii* substantially. There is a concomitant reduction in the prevalence of both nasal colonization and subclinical disease, as reflected by serum antigen titers, in animals living in treated environments. Time will tell as to whether the prevalence of clinical disease will also be reduced. Unfortunately, systemic treatment with azoles has not proven to be effective at reducing asymptomatic colonization of mucus in the nasal cavity in the absence of strategies directed at the environment. The reasons why this is the case should be further studied, as it may reflect insufficient dosing due to the pharmacokinetic peculiarities of the koala.

Three years ago we were called to assist in the investigation and treatment of a koala with severe invasive cryptococcal rhinosinusitis (Color Plate 8). The animal was domiciled near Newcastle, New South Wales, in eastern Australia, at the time the diagnosis was made. Laboratory investigations of the isolate from this patient proved it to be of a *C. gattii* isolate of the VGII molecular type. Although we had isolated the molecular type VGII from five cats in various parts of New South Wales and shown them to be a potentially recombining population (13), this molecular type had never previously been associated with either colonization or disease in koalas anywhere in Australia. The patient had been born in a wildlife park in Western Australia and had also spent some time in a wildlife park in Sydney, New South Wales. As we were aware of largely unpublished data concerning a cluster of infections caused by the molecular type VGII in a flock of sheep in southwestern Western Australia (93), we considered the wildlife park near Perth as the most likely source of the VGII infection.

Systematic sampling of this wildlife park found it to have a very high environmental presence of VGII, with most animals being colonized and a substantial number of cases with subclinical and even clinical disease. Environmental investigations identified an isolate identical to the disease isolate. The presence of such a hyperendemic and apparently recombining VGII population in a part of Australia not normally inhabited by koalas raised a number of questions. This single patient provided the impetus for us to reach out to veterinary colleagues at Murdoch University and the Vet-Path laboratory and quickly determine a thumbnail sketch of different molecular types of the *C. neoformans* species complex encountered in animals in Western Australia. As stated earlier, the striking finding was that approximately half of canine and feline cryptococcosis cases and all horses and sheep in Western Australia with cryptococcosis were infected by isolates of *C. gattii* molecular type VGII.

The koala work has proven to definitively link environmental "contamination" with clinical disease and elaborate on the early pathogenesis of the disease. Further studies continue to examine the components of the koala host response in different stages of disease, the effect of koala major histocompatibility complex genetic diversity on the susceptibility to cryptococcosis, and investigations into the biology of *C. gattii* in the environment.

CRYPTOCOCCOSIS IS ACQUIRED FROM THE ENVIRONMENT: CLUSTERS OF CASES AND MINOR EPIZOOTICS

While cryptococcosis is considered a sporadic disease, case clustering and minor epizootics have long been recognized in animal species. Yet surprisingly little attention has been paid to a number of small outbreaks of cryptococcosis that have occurred over the years, including the following examples.

The first was in a zoological park in Washington, DC (United States), and involved 18 tree shrews—7 short-eared elephant shrews (*Macroscelides proboscides*), 6 large tree shrews (*Tupaia tana*), and 5 lesser tree shrews (*Tupaia minor*)—over a 30-month period from 1991 to 1993. Clinical signs were absent or included weight loss, shivering, dyspnea, and/or neurologic disease (99). An environmental source of infection was not determined, and the isolates were not characterized sufficiently to know if a single molecular type accounted for all cases. Retrospective molecular biological investigation of the outbreak strains would be of great interest if appropriate specimens have been archived. The second outbreak was a smaller outbreak involving striped grass mice domiciled in the nocturnal house of Antwerp Zoo and referable to environmental colonization by *C. neoformans* (5).

The third outbreak occurred in the early 1990s, when over 100 sheep in a single flock of unknown size in Western Australia developed fatal cryptococcosis. Unfortunately, details of the clinical signs, the chronology of the outbreak, the percentage of animals affected, and other relevant information were never recorded. Culture from three representative cases were, however, available for subsequent molecular testing, and all three were shown to be *C. gattii* molecular type VGII. In each instance, positive cultures were obtained from the lungs and the meninges (93).

Fourth, spatial clustering, if not temporal clustering, has been documented in koalas with cryptococcosis caused by *C. gattii* molecular type VGI (51), and spatial and temporal clustering has been observed in koalas with cryptococcosis caused by *C. gattii* molecular type VGII (N. Saul, personal communication).

Finally, five outbreaks of cryptococcosis in goats in the central western part of the Iberian peninsula of Spain occurred from 1990 to 1994. Affected goats had either lower respiratory and/or neurological signs, and infections were referable to *C. gattii* molecular type VGI isolates in all instances, possibly associated with *Eucalyptus camaldulensis* trees located in this region. The extent of the problem deserves emphasis: in 1990, 12% of 140 goats were affected; in 1991, 2% of 250 goats were affected; in 1994, 10% of 300 goats were affected, while in 1994, 2.5% of 120 goats were affected. All affected goats died, the diagnosis being confirmed, however, in only a small but likely representative number of animals by necropsy (2).

Despite the above examples, it is generally agreed that cryptococcosis in humans, companion animals, and koalas is a sporadic disease. Even in contrived man-made environments, where the number of cryptococcal organisms is amplified by the presence of koalas, it is unusual for more than one animal to have clinical cryptococcosis at any given point in time. In all the instances discussed above, numerous animals were affected at the same time. Presumably, this resulted from exposure of a large number of animals to a heavy inoculum of infectious propagules, resulting in simultaneous development of symptomatic disease in a proportion of exposed animals, and presumably an even higher prevalence of subclinical infection and possibly nasal colonization. It is of great interest that in both instances where isolates had been sufficiently well characterized, the infections were shown to be due to *C. gattii*; the sheep in Australia were infected with molecular type VGII, while the goats in Spain were infected with the molecular type VGI.

The reasons how and why these epizootics occurred require further study, as the phenomenon may have something in common with the situation that developed on a much larger scale on Vancouver Island (outlined below).

From limited molecular typing data and anecdotal reports, it would appear that temporal clustering is more pronounced with disease due to molecular type VGII (koalas, sheep, Vancouver Island outbreak, etc.), although spatial clustering is present more generally in cryptococcosis caused by *C. gattii* and perhaps even to some degree in disease caused by *C. neoformans*.

THE VANCOUVER ISLAND OUTBREAK

There is little doubt that for mycologists around the world, the outbreak or epidemic of cryptococcosis that has been evolving in the region around Vancouver Island (39, 44) has been a phenomenon of great fascination. This is because, historically, incidents such as this often give rise to unique insights that force reevaluation of current concepts and provide cogent insights into microbial evolution and disease pathogenesis. Presumably, such incidents will help us better understand how the *C. neoformans* species complex can suddenly change from a largely environmental organism causing rare sporadic disease to a source of more prevalent human and animal infections (21–25, 39, 44, 55, 56, 97).

The Vancouver Island outbreak has really emphasized the value of one medicine when it comes to unraveling infectious diseases that affect both humans and animals. The first suggestion of a problem was a dramatic increase in the number of veterinary patients diagnosed with cryptococcosis in 1999–2000 (55, 97). Most of these cases were diagnosed in a single veterinary laboratory. Instead of sporadic disease in two or three cats per year, a large number of new cases were being diagnosed consistently, and not just in companion animals. In addition to feline and canine cryptococcosis cases, infections were seen in ferrets, llamas, a horse, birds, and several porpoises. Taking one year as an example, 68 new animal cases of cryptococcosis were diagnosed between January and October in 2003 (55).

Indeed, the number of animal cases has consistently and greatly exceeded the number of human cases. Despite a likely substantial underestimation—due to lack of complete investigation and diagnosis of animal disease—the number of animal cases exceeds human cases by almost 75%. This emphasizes the value of domesticated and free-living animals as sentinels of disease. The story reflects favorably on the Canadian veterinary and scientific communities, and especially on Dr. Sally Lester, a veterinary clinical pathologist in private practice who did much of the initial diagnostic work and helped connect key players in subsequent investigations.

Veterinarians ensured that a representative number of isolates were forwarded to human infectious disease facilities. From there it did not take long to confirm that human and animal cases alike were attributable to *C. gattii*. Subsequent work by independent laboratories confirmed that all cases were due to two subgenotypes of the *C. gattii* molecular type VGII (VGIIa [major type] and VGIIb [minor type]) (30) and that these have further evolved as the infection has spread into the mainland of British Columbia (21–25). Interestingly, the colonial morphology of those *C. gattii* isolates on Sabouraud's plates was generally drier than the Australian isolates of molecular type VGI and had a unique susceptibility pattern (with susceptibility to ketoconazole but often resistance to fluconazole).

Other features of disease in animals infected by the Vancouver strains set them apart from the cases seen in Australia. Dogs were affected more commonly than would be expected from the comparative canine-to-feline prevalence in Australia. In addition, clinical disease in the Vancouver cases tended to be more advanced or disseminated at the time of diagnosis (with more primary neurological presentations), and outcomes were on the whole worse. Indeed, the overall pattern of disease in cats and dogs and involvement of unusual species, such as camelids, ferrets, horses, and marine mammals, was more reminiscent of the pattern of cryptococcosis seen in animals domiciled near Perth, Western Australia, as might be expected from the higher prevalence of infections due to molecular type VGII in both locations.

Veterinary investigators were key players in unraveling the changing epidemiology of cryptococcosis in western Canada. Taking a page from the book of colleagues in Australia, Canadian investigators were proactive in using a range of animals as biological air samplers and sentinel species. Studying animals normally residing on Vancouver Island, 4.3% (4/94) of cats, 1.1% (3/280) of dogs, 2% (2/98) of assorted wildlife species (both gray squirrels) (21), and 1.5% of nasal swabs from horses were shown to have nasal colonization by *C. gattii*. Even more compelling was the finding that 6/84 (7.1%) cats and 2/266 (0.8%) dogs had evidence of subclinical infection (antigen-positive serology) (23). Furthermore, two of these cats subsequently went on to develop overt symptomatic disease over the next few months (24).

The detection of new cases in animals living exclusively on the mainland of British Columbia, and having never visited Vancouver Island, was the first indication that the organism had extended its environmental niche into mainland British Columbia. These animal cases were initially identified in 2003 and became increasingly common from March 2004, 6 months before the first human cases were reported from this area.

The occurrence of infections predominantly due to *C. gattii* molecular type VGII in human and veterinary patients in western Canada forced us to reappraise the selectivity of the immunohistochemical technique we had developed to differentiate *C. neoformans* and *C. gattii* in tissue sections and cytological smear preparations. When Krockenberger et al. (49) developed this technique, the main focus was to determine whether archival preparations from koalas were caused by *C. gattii*, and indeed this proved to be the case. However, in the validation of the technique, no tissues with disease caused by isolates of the molecular type VGII were included because of its low prevalence in the reference population. When VGII-infected tissues were subsequently tested systematically, the labeling pattern using the range of monoclonal antibodies currently available indicated a significant variation from *C. gattii* molecular type VGI isolates, with weak labeling by monoclonal antibodies prepared against *C. neoformans* capsule. Knowledge of this potential and reassessment of the interpretation of test results has reinstated the usefulness of this technique. Most importantly, however, the practical finding of alterations to the structural antigens of the capsule has pointed to yet another important difference between isolates of the two molecular types VGI and VGII.

SUMMARY AND CONCLUDING REMARKS

Since the first edition of this monograph, knowledge about all aspects of cryptococcosis has advanced at a staggering pace. Veterinary clinicians interested in fungal disease are to be commended for having been important players in this quest for new knowledge. There can be no doubt that studying disease in companion animals, production animals, and wildlife can play a critical role in unraveling the environmental associations of the *C. neoformans* species complex. More collaboration between academic veterinarians, practitioners, mycology reference laboratories, and research institutes may provide further information relevant to both human and animal patients. The future of one medicine in medical and veterinary mycology looks secure.

ADDENDUM

Since the submission of this chapter, important new data concerning cryptococcosis in animals from the United States have been obtained by Jane Sykes's group at the University of California-Davis. Evidence of cryptococcal species and strains present in northern California (USA) has come from analysis of organisms isolated from cats and dogs. Interestingly, *C. gattii* molecular type VGIII has predominated as a pathogen of cats, while *C. neoformans* has been isolated from the majority of dogs (S. R. Trivedi, J. E. Sykes, M. S. Cannon, E. R. Wisner, W. Meyer, B. K. Sturges, P. J. Dickinson, and L. R. Johnson, submitted for publication). The occurrence of CNS involvement in both cats and dogs from California appears to be more common than in representative populations from Australia. Heteroresistance to fluconazole has been common among VGIII isolates from cats (J. Sykes, unpublished). *C. gattii* molecular types VGIIa and VGIIb have also been detected occasionally in dogs and cats from California with cryptococcosis. Dogs infected with *C. neoformans* consistently have widely disseminated infection, similar to that reported in HIV-infected humans. Although *C. gattii* molecular type VGII has only been isolated from two dogs, both have had cryptococcomas localized to the caudal nasal cavity with invasion of the cribriform plate, with no evidence of disease elsewhere.

REFERENCES

1. **Barclay, W. P., and A. deLahunta.** 1979. Cryptococcal meningitis in a horse. *J. Am. Vet. Med. Assoc.* **174:**1236–1238.
2. **Baro, T., J. M. Torres-Rodriguez, M. H. De Mendoza, Y. Morera, and C. Alia.** 1998. First identification of autochthonous *Cryptococcus neoformans* var. *gattii* isolated from goats with predominantly severe pulmonary disease in Spain. *J. Clin. Microbiol.* **36:**458–461.
3. **Barrs, V. R., P. Martin, R. G. Nicoll, J. A. Beatty, and R. Malik.** 2000. Pulmonary cryptococcosis and *Capallaria aerophila* infection in an FIV-positive cat. *Aust. Vet. J.* **78:**154–158.
4. **Barton, M. D., and I. Knight.** 1972. Cryptococcal meningitis of a horse. *Aust. Vet. J.* **48:**534.
5. **Bauwens, L., F. Vercammen, C. Wuytack, K. Van Looveren, and D. Swinne.** 2004. Isolation of *Cryptococcus neoformans* in Antwerp Zoo's nocturnal house. *Mycoses* **47:**292–296.
6. **Beatty, J. A., V. R. Barrs, G. R. Swinney, P. A. Martin, and R. Malik.** 2000. Peripheral vestibular disease associated with cryptococcosis in three cats. *J. Feline Med. Surg.* **2:**29–34.
7. **Begg, L. M., K. J. Hughes, A. Kessell, M. B. Krockenberger, D. I. Wigney, and R. Malik.** 2004. Successful treatment of cryptococcal pneumonia in a pony mare. *Aust. Vet. J.* **82:**686–692.
8. **Berthelin, C., C. Bailey, P. Kass, A. Legendre, and A. Wolf.** 1994. Cryptococcosis of the nervous system in dogs. Part 1. Epidemiological, clinical and neuropathological features. *Prog. Vet. Neurol.* **5:**88–97.
9. **Bildfell, R. J., P. Long, and R. Sonn.** 2002. Cryptococcosis in a llama (*Lama glama*). *J. Vet. Diagn. Invest.* **14:**337–339.
10. **Bolliger, A., and E. S. Finckh.** 1962. The prevalence of cryptococcosis in the koala (*Phascolarctos cinereus*). *Med. J. Aust.* **49:**545–547.
11. **Bolton, L. A., R. G. Lobetti, D. N. Evezard, J. A. Picard, J. W. Nesbit, J. van Heerden, and R. E. Burroughs.** 1999. Cryptococcosis in captive cheetah (*Acinonyx jabatus*): two cases. *J. S. Afr. Vet. Assoc.* **70:**35–39.
12. **Bui, T., X. Lin, R. Malik, J. Heitman, and D. Carter.** 2008. Isolates of *Cryptococcus neoformans* from infected animals reveal genetic exchange in unisexual, α mating type populations. *Eukaryot. Cell* **7:**1771–1780.
13. **Campbell, L. T., B. J. Currie, M. B. Krockenberger, R. Malik, D. H. Ellis, and D. Carter.** 2005. Clonality and recombination in genetically differentiated subgroups of *Cryptococcus gattii*. *Eukaryot. Cell* **4:**1403–1409.
14. **Casadevall, A., A. Cassone, F. Bistoni, J. E. Cutler, W. Magliani, J. W. Murphy, L. Polonelli, and L. Romani.** 1998. Antibody and/or cell-mediated immunity, protective mechanisms in fungal disease: an ongoing dilemma or an unnecessary dispute? *Med. Mycol.* **36**(Suppl 1)**:**95–105.
15. **Chandna, V. K., E. Morris, J. M. Gliatto, and M. R. Paradis.** 1993. Localised subcutaneous cryptococcal granuloma in a horse. *Equine Vet. J.* **25:**166–168.
16. **Chapman, H. M., W. F. Robinson, J. R. Bolton, and J. P. Robertson.** 1990. *Cryptococcus neoformans* infection in goats. *Aust. Vet. J.* **67:**263–265.
17. **Chen, S. C., B. J. Currie, H. M. Campbell, D. A. Fisher, T. J. Pfeiffer, D. H. Ellis, and T. C. Sorrell.** 1997. *Cryptococcus neoformans* var. *gattii* infection in northern Australia: existence of an environmental source other than known host eucalypts. *Trans. R. Soc. Trop. Med. Hyg.* **91:** 547–550.
18. **Cho, D. Y., L. W. Pace, and R. E. Beadle.** 1986. Cerebral cryptococcosis in a horse. *Vet. Pathol.* **23:**207–209.
19. **Connolly, J. H., M. B. Krockenberger, R. Malik, P. J. Canfield, D. I. Wigney, and D. B. Muir.** 1999. Asymptomatic carriage of *Cryptococcus neoformans* in the nasal cavity of the koala (*Phascolarctos cinereus*). *Med. Mycol.* **37:**331–338.
20. **Dickson, J., and E. P. Meyer.** 1970. Cryptococcosis in horses in Western Australia. *Aust. Vet. J.* **46:**558.
21. **Duncan, C., H. Schwantje, C. Stephen, J. Campbell, and K. Bartlett.** 2006. *Cryptococcus gattii* in wildlife of Vancouver Island, British Columbia, Canada. *J. Wildlife Dis.* **42:**175–178.
22. **Duncan, C., C. Stephen, and J. Campbell.** 2006. Clinical characteristics and predictors of mortality for *Cryptococcus gattii* infection in dogs and cats of southwestern British Columbia. *Can. Vet. J.* **47:**993–998.
23. **Duncan, C., C. Stephen, S. Lester, and K. H. Bartlett.** 2005. Subclinical infection and asymptomatic carriage of *Cryptococcus gattii* in dogs and cats during an outbreak of cryptococcosis. *Med. Mycol.* **43:**511–516.
24. **Duncan, C., C. Stephen, S. Lester, and K. H. Bartlett.** 2005. Follow-up study of dogs and cats with asymptomatic *Cryptococcus gattii* infection or nasal colonization. *Med. Mycol.* **43:**663–666.
25. **Duncan, C. G., C. Stephen, and J. Campbell.** 2006. Evaluation of risk factors for *Cryptococcus gattii* infection in dogs and cats. *J. Am. Vet. Med. Assoc.* **228:**377–382.
26. **Ellis, D. H., and T. J. Pfeiffer.** 1990. Natural habitat of *Cryptococcus neoformans* var. *gattii*. *J. Clin. Microbiol.* **28:**1642–1644.
27. **Ellis, T. M., W. F. Robinson, and G. E. Wilcox.** 1988. The pathology and aetiology of lung lesions in goats infected with caprine arthritis-encephalitis virus. *Aust. Vet. J.* **65:**69–73.
28. **Flatland, B., R. T. Greene, and M. R. Lappin.** 1996. Clinical and serologic evaluation of cats with cryptococcosis. *J. Am. Vet. Med. Assoc.* **209:**1110–1113.
29. **Foster, S. F., G. Parker, R. M. Churcher, and R. Malik.** 2001. Intracranial cryptococcal granuloma in a cat. *J. Feline Med. Surg.* **3:**39–44.
30. **Fraser, J. A., S. S. Giles, E. C. Wenink, S. G. Geunes-Boyer, J. R. Wright, S. Diezmann, A. Allen, J. E. Sta-**

jich, F. S. Dietrich, J. R. Perfect, and J. Heitman. 2005. Same-sex mating and the origin of the Vancouver Island *Cryptococcus gattii* outbreak. *Nature* **437:**1360–1364.

31. **Galli, G., and A. Socci.** 1969. Further studies on experimental cryptococcal mastitis. *Arch. Vet. Ital.* **20:**3–11.
32. **Gerds-Grogan, S., and B. Dayrell-Hart.** 1997. Feline cryptococcosis: a retrospective evaluation. *J. Am. Anim. Hosp. Assoc.* **33:**118–22.
33. **Goldman, D., S. C. Lee, and A. Casadevall.** 1994. Pathogenesis of pulmonary *Cryptococcus neoformans* infection in the rat. *Infect. Immun.* **62:**4755–4761.
34. **Goldman, D. L., H. Khine, J. Abadi, D. J. Lindenberger, L. Pirofski, R. Niang, and A. Casedevall.** 2001. Serologic evidence for *Cryptococcus neoformans* infection in early childhood. *Pediatrics* **107:**e66.
35. **Goodchild, L. M., A. J. Dart, M. B. Collins, C. M. Dart, J. L. Hodgson, and D. R. Hodgson.** 1996. Cryptococcal meningitis in an alpaca. *Aust. Vet. J.* **74:**428–430.
36. **Gutierrez, M., and J. F. Garcia Marin.** 1999. *Cryptococcus neoformans* and *Mycobacterium bovis* causing granulomatous pneumonia in a goat. *Vet. Pathol.* **36:**445–448.
37. **Halliday, C. L., T. Bui, M. B. Krockenberger, R. Malik, D. H. Ellis, and D. Carter.** 1999. Presence of α and a mating types in environmental and clinical collections of *Cryptococcus neoformans* var. *gattii* from Australia. *J. Clin. Microbiol.* **37:**2920–2926.
38. **Hilbert, B. J., C. R. Huxtable, and S. E. Pawley.** 1980. Cryptococcal pneumonia in a horse. *Aust. Vet. J.* **56:**391–392.
39. **Hoang, L. M. N., J. A. Maguire, P. Doyle, M. Fyfe, and D. L. Roscoe.** 2004. *Cryptococcus neoformans* infections at Vancouver Hospital and Health Sciences Centre (1997–2002): epidemiology, microbiology and histopathology. *J. Med. Microbiol.* **53:**935–940.
40. **Hough, I.** 1994. Cryptococcosis in an eastern water skink. *Aust. Vet. J.* **76:**471–472.
41. **Jacobs, G. J., L. Medleau, C. Calvert, and J. Brown.** 1997. Cryptococcal infection in cats: factors influencing treatment outcome, and results of sequential serum antigen titers in 35 cats. *J. Vet. Intern. Med.* **11:**1–4.
42. **Jacobson, E. R., J. L. Cheatwood, and L. K. Maxwell.** 2000. Mycotic diseases of reptiles. *Semin. Avian Exotic Pet Med.* **9:**94–101.
43. **Jenney, A., K. Pandithage, D. A. Fisher, and B. J. Currie.** 2004. Cryptococcus infection in tropical Australia. *J. Clin. Microbiol.* **42:**3865–3868.
44. **Kidd, S. E., P. J. Bach, A. O. Hingston, S. Mak, Y. Chow, L. MacDougal, J. W. Kronstad, and K. H. Bartlett.** 2007. *Cryptococcus gattii* dispersal mechanisms, British Columbia Canada. *Emerg. Infect. Dis.* **13:**51–57.
45. **Kluger, E. K., H. K. Karaoglu, M. B. Krockenberger, P. K. Della Torre, W. Meyer, and R. Malik.** 2006. Recrudescent cryptococcosis, caused by *Cryptococcus gattii* (molecular type VGII), over a 13 year period in a Birman cat. *Med. Mycol.* **44:**561–566.
46. **Krockenberger, M., P. J. Canfield, and R. Malik.** 2002. What koalas are trying to tell us about cryptococcosis. *Microbiol. Aust.* **23:**29–30.
47. **Krockenberger, M., K. Stalder, R. Malik, and P. J. Canfield** 2005. Cryptococcosis in Australian wildlife. *Microbiol. Aust.* **26:**69–71.
48. **Krockenberger, M. B., P. J. Canfield, J. Barnes, L. Vogelnest, J. Connolly, C. Ley, and R. Malik.** 2002. *Cryptococcus neoformans* in the koala (*Phascolarctos cinereus*): serological evidence for subclinical cryptococcosis. *Med. Mycol.* **40:**273–282.
49. **Krockenberger, M. B., P. J. Canfield, T. R. Kozel, T. Shinoda, R. Ikeda, D. I. Wigney, P. Martin, K. Barnes, and R. Malik.** 2001. An immunohistochemical method that differentiates *Cryptococcus neoformans* varieties and serotypes in formalin-fixed paraffin-embedded tissues. *Med. Mycol.* **39:**523–533.
50. **Krockenberger, M. B., P. J. Canfield, and R. Malik.** 2002. *Cryptococcus neoformans* in the koala (*Phascolarctos cinereus*): colonisation by *C. n.* variety *gattii* and investigation of environmental sources. *Med. Mycol.* **40:**263–272.
51. **Krockenberger, M. B., P. J. Canfield, and R. Malik.** 2003. *Cryptococcus neoformans* var. *gattii* in the koala (*Phascolarctos cinereus*): a review of 43 cases of cryptococcosis. *Med. Mycol.* **41:**225–234.
52. **Laws, L., and G. C. Simmons.** 1966. Cryptococcosis in a sheep. *Aust. Vet. J.* **42:**321–323.
53. **Lemos, L. S., A. S. dos Santos, O. Vieira-da-Motta, G. N. Texeira, and E. C. de Carvalho.** 2007. Pulmonary cryptococcosis in slaughtered sheep: anatomopathology and culture. *Vet. Microbiol.* **125:**350–354.
54. **Lenard, Z. M., N. V. Lester, A. J. O'Hara, B. J. Hopper, and G. D. Lester.** 2007. Disseminated cryptococcosis including osteomyelitis in a horse. *Aust. Vet. J.* **85:**51–55.
55. **Lester, S. J., N. J. Kowalewich, K. H. Bartlett, M. B. Krockenberger, T. Fairfax, and R. Malik.** 2004. Clinicopathologic features of an unusual outbreak of cryptococcosis in dogs, cats, ferrets, and a bird: 38 cases (January to July 2003). *J. Am. Vet. Med. Assoc.* **225:**1716–1722.
56. **MacDougall, L., S. E. Kidd, E. Galanis, S. Mak, M. J. Leslie, P. R. Cieslak, J. W. Kronstad, M. G. Morshed, and K. H. Bartlett.** 2007. Spread of *Cryptococcus gattii* in British Columbia, Canada, and detection in the Pacific northwest, USA. *Emerg. Infect. Dis.* **13:**42–50.
57. **Malik, R.** 2007. Use of fluconazole for treating cryptococcosis in cats. University of Sydney Postgraduate Committee in Veterinary Science. *Control and Therapy Series* **249:**5.
58. **Malik, R., B. Alderton, D. Finlaison, H. Karaoglu, W. Meyer, P. Martin, M. P. France, J. McGill, S. J. Lester, C. R. O'Brien, and D. N. Love.** 2002. Cryptococcosis in ferrets: a diverse spectrum of clinical disease. *Aust. Vet. J.* **80:**49–55.
59. **Malik, R., A. J. Craig, D. I. Wigney, P. Martin, and, D. N. Love.** 1996. Combination chemotherapy of canine and feline cryptococcosis using subcutaneously administered amphotericin B. *Aust. Vet. J.* **73:**124–128.
60. **Malik, R., E. Dill-Macky, P. Martin, D. I. Wigney, D. B. Muir, and D. N. Love.** 1995. Cryptococcosis in dogs: a retrospective study of 20 consecutive cases. *J. Med. Vet. Mycol.* **33:**291–297.
61. **Malik, R., G. B. Hunt, C. R. Bellenger, G. S. Allan, P. Martin, P. J. Canfield, and D. N. Love.** 1999. Intra-abdominal cryptococcosis in two dogs. *J. Small. Anim. Pract.* **40:**387–391.
62. **Malik, R., M. B. Krockenberger, G. Cross, R. Doneley, D. N. Madill, D. Black, P. McWhirter, A. Rosenwax, K. Rose, M. Alley, D. Forshaw, I. Russell-Brown, A. C. Johnstone, P. Martin, C. R. O'Brien, and D. N. Love.** 2003. Avian cryptococcosis. *Med. Mycol.* **41:**115–124.
63. **Malik, R., P. Martin, D. I. Wigney, D. B. Church, W. Bradley, C. R. Bellenger, W. A. Lamb, V. R. Barrs, S. Foster, S. Hemsley, P. J. Canfield, and D. N. Love.** 1997. Nasopharyngeal cryptococcosis. *Aust. Vet. J.* **75:**483–488.
64. **Malik, R., R. McPetrie, D. I. Wigney, and D. N. Love.** 1996. Use of the cryptococcal latex agglutination antigen test for diagnosis and monitoring of therapy in veterinary patients with cryptococcosis. *Aust. Vet. J.* **74:**358–364.
65. **Malik, R., C. Medeiros, D. I. Wigney, and D. N. Love.** 1996. Suspected drug eruption in seven dogs during administration of flucytosine. *Aust. Vet. J.* **74:**285–258.
66. **Malik, R., J. Norris, J. White, and B. Jantulik.** 2006. 'Wound cat.' *J. Feline Med. Surg.* **8:**135–140.

67. **Malik, R., L. Vogelnest, C. R. O'Brien, J. White, C. Hawke, D. I. Wigney, P. Martin, and J. M. Norris.** 2004. Infections and some other conditions affecting the skin and subcutis of the naso-ocular region of cats: clinical experience 1987–2003. *J. Feline Med. Surg.* **6:**383–390.
68. **Malik, R., D. I. Wigney, D. B. Muir, D. J. Gregory, and D. N. Love.** 1992. Cryptococcosis in cats: clinical and mycological assessment of 29 cases and evaluation of treatment using orally administered fluconazole. *J. Med. Vet. Mycol.* **30:**133–144.
69. **Malik, R., D. I. Wigney, D. B. Muir, and D. N. Love.** 1997. Asymptomatic carriage of *Cryptococcus neoformans* in the nasal cavity of dogs and cats. *J. Med. Vet. Mycol.* **35:**27–31.
70. **McGill, S., R. Malik, N. Saul, S. Beetson, C. Secombe, I. Robertson, and P. Irwin.** 2009. Cryptococcosis in domestic animals in Western Australia: a retrospective study from 1995–2006. *Med. Mycol.* **47:**625–639.
71. **Miller, W. G., A. A. Padhye, W. van Bonn, E. Jensen, M. E. Brandt, and S. H. Ridgway.** 2002. Cryptococcosis in a bottlenose dolphin (*Tursiops truncatus*) caused by *Cryptococcus neoformans* var. *gattii*. *J. Clin. Microbiol.* **40:**721–724.
72. **Millward, I. R., and M. C. Williams.** 2005. *Cryptococcus neoformans* granuloma in the lung and spinal cord of a free-living cheetah (*Acinonyx jubatus*). A clinical report and literature review. *J. S. Afr. Vet. Assoc.* **76:**228–232.
73. **Mylonakis, E., F. M. Ausubel, J. R. Perfect, J. Heitman, and S. B. Calderwood.** 2002. Killing of *Caenorhabditis elegans* by *Cryptococcus neoformans* as a model of yeast pathogenesis. *Proc. Natl. Acad. Sci. USA* **99:**15675–15680.
74. **Mylonakis, E., R. Moreno, J. B. El Khoury, A. Idnurm, J. Heitman, S. B. Calderwood, F. M. Ausubel, and A. Diener.** 2005. *Galleria mellonella* as a model system to study *Cryptococcus neoformans* pathogenesis. *Infect. Immun.* **73:**3842–3850.
75. **Namara, T. S., R. A. Cook, J. L. Behler, L. Ajello, and A. A. Padhye.** 1994. Cryptococcosis in a common anaconda (*Eunectes murinus*). *J. Zoo Wild. Med.* **25:**128–132.
76. **Newman, S. J., C. E. Langston, and T. J. Scase.** 2003. Cryptococcal pyelonephritis in a dog. *J. Am. Vet. Med. Assoc.* **222:**174, 180–183.
77. **Norris, J. M., E. T. Bell, L. Hales, J. A. Toribio, J. D. White, D. I. Wigney, R. M. Baral, and R. Malik.** 2007. Prevalence of feline immunodeficiency virus infection in domesticated and feral cats in eastern Australia. *J. Feline Med. Surg.* **9:**300–308.
78. **O'Brien, C. R., M. B. Krockenberger, P. Martin, D. I. Wigney, and R. Malik.** 2006. Long-term outcome of therapy for 59 cats and 11 dogs with cryptococcosis. *Aust. Vet. J.* **84:**384–392.
79. **O'Brien, C. R., M. B. Krockenberger, D. I. Wigney, P. Martin, and R. Malik.** 2004. Retrospective study of feline and canine cryptococcosis in Australia from 1981 to 2001: 195 cases. *Med. Mycol.* **42:**449–460.
80. **Okabayashi, K., R. Kano, T. Watanabe, and A. Hasegawa.** 2006. Serotypes and mating types of clinical isolates from feline cryptococcosis in Japan. *Jpn. J. Med. Sci.* **68:**91–94.
81. **O'Toole, T. E., A. F. Sato, and E. A. Rozanski.** 2003. Cryptococcosis of the central nervous system in a dog. *J. Am. Vet. Med. Assoc.* **222:**1722–1725.
82. **Pal, M., and B. S. Mehrotra.** 1983. Cryptococcal mastitis in dairy animals. *Mykosen* **26:**615–616.
83. **Pearson, E. G., B. J. Watrous, J. A. Schmitz, and R. J. Sonn.** 1983. Cryptococcal pneumonia in a horse. *J. Am. Vet. Med. Assoc.* **183:**577–579.
84. **Pounden, W. D., J. M. Amberson, and R. F. Jaeger.** 1952. A severe mastitis problem associated with *Cryptococcus neoformans* in a large dairy herd. *Am. J. Vet. Res.* **13:**121–128.
85. **Raso, T. F., K. Werther, E. T. Mirandi, and M. J. S. Mendes-Giannine.** 2004. Cryptococcosis outbreak in psittacine birds in Brazil. *Med. Mycol.* **42:**355–362.
86. **Riley, C. B., J. R. Bolton, J. N. Mills, and J. B. Thomas.** 1992. Cryptococcosis in seven horses. *Aust. Vet. J.* **69:**135–139.
87. **Roberts, M. C., R. H. Sutton, and D. K. Lovell.** 1981. A protracted case of cryptococcal nasal granuloma in a stallion. *Aust. Vet. J.* **57:**287–291.
88. **Robertson, I. D., J. R. Edwards, S. E. Shaw, and W. T. Clark.** 1990. A survey of pet ownership in Perth. *Aust. Vet. Pract.* **20:**210–212.
89. **Scott, E. A., J. R. Duncan, and J. E. McCormack.** 1974. Cryptococcosis involving the postorbital area and frontal sinus in a horse. *J. Am. Vet. Med. Assoc.* **165:**626–627.
90. **Seaton, R. A., N. Verma, S. Haraqii, J. P. Wembrii, and D. A. Warrel.** 1997. Visual loss in immunocompetent patients with *Cryptococcus neoformans* var. *gattii* meningitis. *Trans. R. Soc. Trop. Med. Hyg.* **91:**44–49.
91. **Seixas, F., M. D. Martins, M. D. Pinto, P. J. Travassos, M. Miranda, and M. D. Pires.** 2008. A case of pulmonary cryptococcosis in a free-living toad (*Bufo bufo*). *J. Wildlife Dis.* **44:**460–463.
92. **Shaw, S. E.** 1988. Successful treatment of 11 cases of feline cryptococcosis. *Aust. Vet. Pract.* **18:**135–139.
93. **Sorrell, T. C., A. G. Brownlee, P. Ruma, R. Malik, T. J. Pfeiffer, and D. H. Ellis.** 1996. Natural environmental sources of *Cryptococcus neoformans* var. *gattii*. *J. Clin. Microbiol.* **34:**1261–1263.
94. **Sorrell, T. C.** 2001. *Cryptococcus neoformans* variety *gattii*. *Med. Mycol.* **39:**155–68.
95. **Steckel, R. R., S. B. Adams, G. G. Long, and A. H. Rebar.** 1982. Antemortem diagnosis and treatment of cryptococcal meningitis in a horse. *J. Am. Vet. Med. Assoc.* **180:**1085–1089.
96. **Steenbergen, J. N., H. A. Shuman, and A. Casadevall.** 2001. *Cryptococcus neoformans* interactions with amoebae suggest an explanation for its virulence and intracellular pathogenic strategy in macrophages. *Proc. Natl. Acad. Sci. USA* **98:**15245–15250.
97. **Stephen, C., S. Lester, W. Black, M. Fyfe, and S. Raverty.** 2002. Multispecies outbreak of cryptococcosis on southern Vancouver Island, British Columbia. *Can. Vet. J.* **43:**792–794.
98. **Sutton, R. H.** 1981. Cryptococcosis in dogs: a report on 6 cases. *Aust. Vet. J.* **57:**558–564.
99. **Tell, L. A., D. K. Nichols, W. P. Fleming, and M. Bush.** 1997. Cryptococcosis in tree shrews (*Tupaia tana* and *Tupaia minor*) and elephant shrews (*Macroscelides proboscides*). *J. Zoo Wildl. Med.* **28:**175–181.
100. **Teuscher, E., A. Vrins, and T. Lemaire.** 1984. A vestibular syndrome associated with *Cryptococcus neoformans* in a horse. *Zentralbl. Veterinarmed.* A **31:**132–139.
101. **Tiches, D., C. H. Vite, B. Dayrell-Hart, S. A. Steinberg, S. Gross, and F. Lexa.** 1988. A case of canine central nervous system cryptococcosis: management with fluconazole. *J. Am. Anim. Hosp. Assoc.* **34:**145–151.
102. **Toribio, J. A., J. M. Norris, J. D. White, N. K. Dhand, S. A. Hamilton, and R. Malik.** 2009. Demographics and husbandry of pet cats living in Sydney, Australia: results of cross-sectional survey of pet ownership. *J. Feline Med. Surg.* **11:**449–461.
103. **Vaughan, R. J., S. D. Vitali, P. A. Eden, K. L. Payne, K. S. Warren, D. Forshaw, J. A. Friend, A. M. Horwitz, C. Main, M. B. Krockenberger, and R. Malik.** 2007. Cryptococcosis in Gilbert's and long-nosed potoroo. *J. Zoo Wildl. Med.* **38:**567–573.
104. **Walker, C., R. Malik, and P. J. Canfield.** 1995. Analysis of leucocytes and lymphocyte subsets in cats with naturally-occurring cryptococcosis but differing feline immunodeficiency virus status. *Aust. Vet. J.* **72:**93–97.
105. **Watt, D. A.** 1970. A case of cryptococcal granuloma in the nasal cavity of a horse. *Aust. Vet. J.* **46:**493–495.

DIAGNOSIS, TREATMENT, PREVENTION, AND CLINICAL PERSPECTIVES

Cryptococcus: From Human Pathogen to Model Yeast
Edited by J. Heitman et al.

37

Cryptococcosis in Transplant Recipients

NINA SINGH AND BARBARA D. ALEXANDER

Cryptococcosis is one of the most significant opportunistic fungal infections in solid organ transplant (SOT) recipients (37, 52, 72). Disease presentation and principles of management of cryptococcosis in other hosts are relevant in transplant recipients. However, iatrogenic immunosuppressive agents in transplant recipients influence not only the risk but also the extent of disease and outcomes in posttransplant cryptococcosis (33, 66). For example, while calcineurin-inhibitor agents are potently immunosuppressive, they also have intrinsic activity against *Cryptococcus* and may modify disease expression and prognosis in transplant-associated cryptococcosis (66). This chapter summarizes the current knowledge and topical developments in the epidemiologic characteristics, clinical manifestations, diagnosis, and management of cryptococcosis in transplant recipients.

SOT RECIPIENTS

Epidemiology

Cryptococcosis is the third most commonly occurring invasive fungal disease in SOT recipients. The overall incidence of cryptococcosis in SOT recipients is 10/1,000 cases, or ~2.8%, and ranges from 0.3 to 5% (32, 79). Cryptococcosis represented 8% of the invasive fungal diseases in SOT recipients in the Transplant Associated Infection Surveillance Network database (P. G. Pappas, C. Kaufmann, B. Alexander, D. Andes, S. Hadley, T. Patterson, R. Walker, V. Morrison, T. Perl, K. Wannemuehler, and T. Chiller, presented at the 47th Interscience Conference on Anticrobial Agents and Chemotherapy [ICAAC], Chicago, IL, 17 to 20 September 2007). An estimated 20 to 60% of the cases of cryptococcosis in non-HIV-infected patients in the United States (76) and 17.4% in France (18) occurred in SOT recipients.

Cryptococcal infection is acquired by inhalation of the basidiospores from the environment, with subsequent containment of the yeast by granulomatous inflammatory responses. Establishment of latency is proposed to occur in a primary complex consisting of the pulmonary lesion and the hilar lymph nodes (3). A vast majority of cryptococcal disease in transplant recipients is considered to represent reactivation of quiescent infection (22, 28). Pre- and posttransplant assessment of sera for cryptococcal antibodies using an immunoblot assay (Fig. 1) showed that a majority of patients who developed cryptococcosis exhibited serologic evidence of cryptococcal infection before transplantation (62). These patients developed cryptococcosis significantly earlier after transplantation than those without preexistent cryptococcal antibodies, suggesting that a substantial proportion of transplant-associated cryptococcosis resulted from reactivation of latent infection with progression to disease (62). Evidence-based epidemiological investigations have also shown the occurrence of cryptococcal disease following acquisition of primary infection following transplantation (38, 48). Isolates from a pet cockatoo and a renal transplant recipient with cryptococcosis showed identical genotypic profiles, suggesting recent acquisition of the yeast (49). Rare cases of transmission from donor organ and tissue grafts have also been reported (50).

Most cryptococcal disease in SOT recipients is due to *Cryptococcus neoformans* var. *grubii* (serotype A), which has no particular geographic predilection (36). *C. neoformans* var. *neoformans* (serotype D), on the other hand, is prevalent in northern Europe. Until recently *Cryptococcus gattii* was regarded as a tropical and subtropical fungus. Its ecological niche, however, has expanded to temperate regions, and acquisition of cases within the United States has been documented, including in SOT recipients (J. Heitman, presented at 45th Annual Meeting of the Infectious Diseases Society of America [IDSA], San Diego, CA, 4 to 7 October 2007). The incubation period of *C. gattii* disease in Vancouver Island and the Pacific Northwest has been documented to be ~6 months (4, 37, 44).

Calcineurin inhibitors are currently the mainstay of immunosuppression in SOT recipients. These agents do not appear to influence the incidence, but may affect the risk, of dissemination and outcome of cryptococcal disease (32). Patients receiving a calcineurin-inhibitor-based regimen were less likely to have disseminated disease and more likely to have cryptococcosis limited to the lungs (66). These differences were attributed to the anticryptococcal activity of these agents that target the fungal homologs of calcineurin (24, 66). Clinical strains of *C. neoformans* in SOT

Nina Singh, VA Pittsburgh Healthcare System and University of Pittsburgh, Pittsburgh, PA 15260. **Barbara D. Alexander,** Departments of Medicine and Pathology, Duke University Medical Center, Durham, NC 27710.

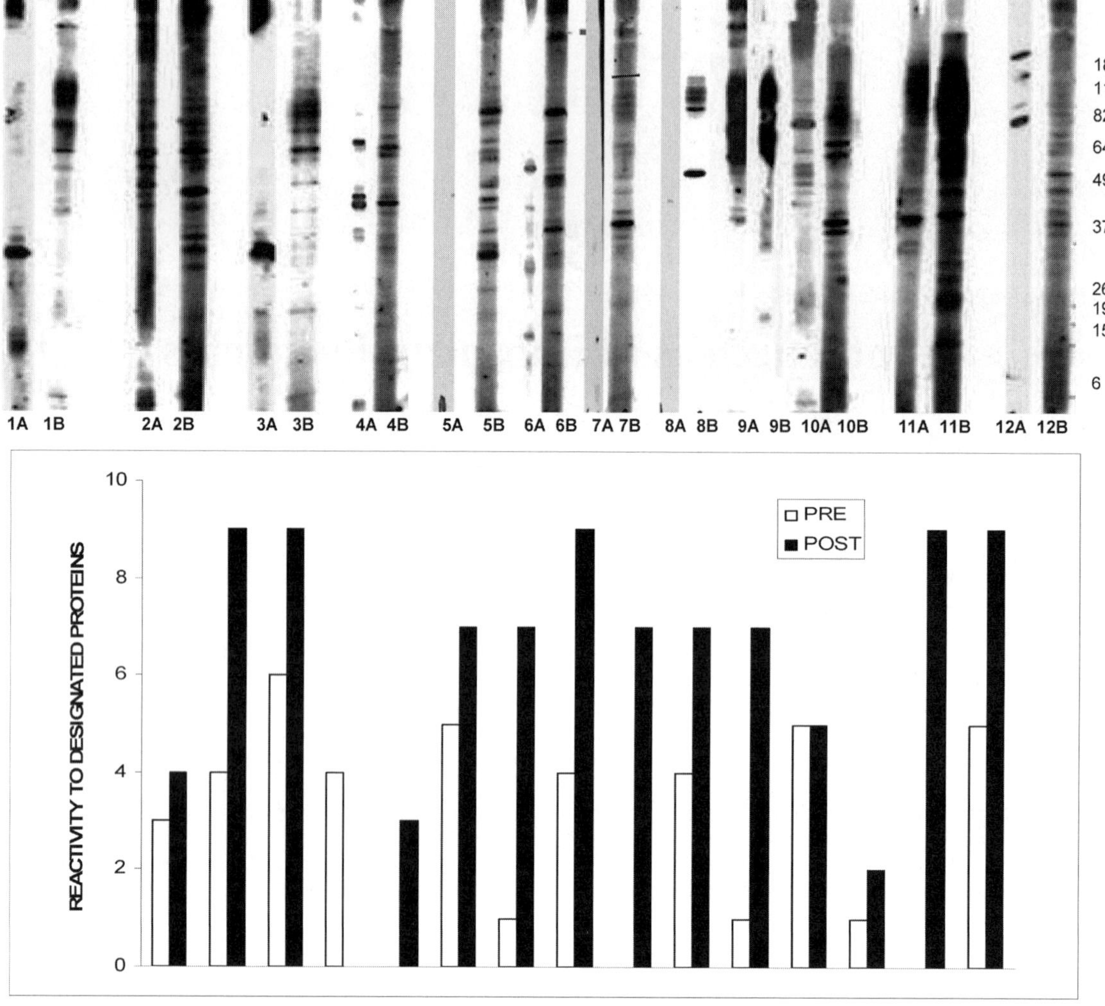

FIGURE 1 (Top) Immunoblots of paired sera from SOT recipients (with blot on the left of each pair made with sera obtained pretransplant and blot on the right made with sera obtained at the diagnosis of posttransplant cryptococcosis) exhibited an increase in reactivity against nine designated proteins in association with cryptococcosis. (Bottom) The corresponding median number of designated proteins recognized by these paired sera is shown.

recipients, however, remain susceptible to calcineurin-inhibitor agents, suggesting that breakthrough infection is primarily due to the immunosuppressive effect on host defense and not due to the selection of drug-resistant strains (6). Corticosteroids (2) are also associated with an increased risk of cryptococcosis in all non-HIV-infected hosts (2, 15, 18, 51, 76). However, the precise daily dose that confers a higher risk in SOT recipients remains unknown. New treatment paradigms in the approach to immunosuppression include increasing reliance on minimally immunosuppressive strategies such as T-cell-depleting antibodies for induction therapy or as treatment of rejection in SOT recipients (64). These agents have also been associated with invasive fungal infections in transplant recipients (53, 75). For example, alemtuzumab, an agent that causes profound and lasting depletion of CD4+ T cells, was associated with a dose-dependent increase in the risk for cryptococcosis (64). The cumulative incidence of cryptococcosis was 0.3% in SOT recipients who did not receive alemtuzumab or antithymocyte globulin, 1.2% in those who received a single dose, and 3.5% in patients who received ≥1 dose of these agents ($P = 0.04$) (64). Invasive fungal infections occurred more frequently in SOT recipients who received alemtuzumab as antirejection as opposed to induction therapy (53).

Cryptococcosis in SOT recipients is typically a late-occurring infection; the median time to onset was 16 to 21 months after transplantation in three studies (32, 66; Pappas et al., ICAAC). The time to onset was earlier for liver and lung compared to kidney transplant recipients and may be due to a higher intensity of immunosuppression in the former subgroups (32). Rarely, cryptococcosis has been observed very early (i.e., within the first month post–liver transplant) and may represent previously unrecognized or asymptomatic disease in cirrhotic patients with end-stage liver disease during transplant candidacy (78, 79).

Clinical Manifestations

Between 53 and 72% of the cryptococcal disease in SOT recipients is disseminated or involves the central nervous system (CNS) (32, 35, 66, 76). Overall, 61% of the SOT recipients in one report had disseminated disease, 54% had pulmonary, and 8.1% had skin, soft-tissue, or osteoarticular cryptococcosis (66). CNS disease has been documented in 25 to 55% of the SOT recipients with cryptococcosis and manifests most frequently as meningoencephalitis with altered mental status (48 to 64%), headache (46 to 62%), fever (46%), meningismus (14%), visual loss (7%), and/or seizures (4%) (32, 79). The duration of symptoms generally ranged from 2 to 30 days in one study (mean, 17 days) (79). Cerebrospinal fluid (CSF) findings typically reveal mild to moderate pleocytosis with mean values of white blood count of 33 to 188 (range, 0 to 1,464) cells/mm^3, glucose 36 to 52 (range, 2 to 181) mg/dl, and protein of 74 to 226 (range, 16 to 1,015) mg/dl (32, 66, 79). Overall, 33 to 39% of the SOT recipients with CNS cryptococcosis have fungemia (32, 66, 79). Patients with CNS disease in one report were more likely to be fungemic than those without CNS involvement (36% versus 5%, $P < 0.001$) (71).

Up to 29 to 33% of the SOT recipients with CNS cryptococcosis may have CNS parenchymal lesions or abnormal neuroimaging findings in association with cryptococcis (35, 71). These comprise patchy or diffuse leptomeningeal enhancement, parenchymal mass lesions or cryptococcomas, hydrocephalus, gelatinous pseudocysts, or dilatation of the perivascular (Virchow-Robin) spaces (71). The frequencies of clinical symptoms such as headache, altered mental status, and visual symptoms were not significantly different for patients with or without CNS lesions in one report (71). However, CSF cryptococcal antigen titers were higher for patients with leptomeningeal compared with parenchymal lesions or hydrocephalus (71). Fungemia was documented in 50% of the patients with meningeal lesions, 17% of those with parenchymal lesions, and none of the patients with hydrocephalus (71).

Approximately 33% of the SOT recipients with cryptococcosis have disease limited to the lungs (4). In others, extrapulmonary disease is present along with pulmonary cryptococcosis (71). Nodules (single or multiple) and infiltrates (alveolar or nodular) are the most common imaging findings in posttransplant cryptococcal disease. Pulmonary cryptococcosis manifesting as acute respiratory failure is associated with a grave prognosis (9). Approximately 38% of the SOT recipients in one report either had no pulmonary symptoms or pulmonary cryptococcosis detected as an incidental finding on imaging studies (47). These patients presented with cryptococcal disease later in the posttransplant period and were receiving significantly lower maintenance dosages of prednisone than symptomatic patients (65). Patients with nodular densities or mass lesions were less likely to be symptomatic than those with pleural effusions or alveolar infiltrates (65).

Cutaneous cryptococcosis can present as papular, nodular, or ulcerative lesions; cellulitis; or necrotizing fasciitis (23, 31). While cutaneous lesions largely represent hematogenous dissemination, skin has also been identified as a portal of entry of *Cryptococcus* and a potential source of subsequent disseminated disease in SOT recipients (48). Such cutaneous infections in France are often due to serotype D (20). Septic arthritis, tenosynovitis, myositis, and osteomyelitis due to cryptococcosis have also been reported in SOT recipients (1, 8, 23, 26). Rare manifestations of cryptococcal disease include choroiditis, pyelonephritis, and prostatitis (5, 29).

Immune Reconstitution Syndrome

It is increasingly apparent that restoration of host immunity, particularly when abrupt, may have adverse sequelae. The host can become gravely ill with symptomatic disease due to immune reconstitution (10). Rapid reduction of immunosuppressive therapy in conjunction with initiation of antifungal therapy in SOT recipients may lead to the development of immune reconstitution syndrome (IRS), the clinical manifestations of which mimic worsening disease due to cryptococcosis (42, 70). Iatrogenic immunosuppressive agents employed in transplant recipients such as calcineurin-inhibitor agents and corticosteroids exert their effect by preferentially inhibiting Th1 (interleukin-2 and gamma interferon) compared to Th2 (interleukin-10) responses (14, 25). Tacrolimus inhibits Th1 to a greater extent than cyclosporine A (16, 21, 30). The biologic basis of IRS in SOT recipients is believed to be the reversal of a Th2 to Th1 proinflammatory response upon withdrawal or reduction of immunosuppression. Potent T-cell-depleting agents such as alemtuzumab have also been recognized as risk factors for IRS (34).

An estimated 5 to 11% of SOT recipients with cryptococcal disease may develop IRS typically 4 to 6 weeks after initiation of antifungal therapy (68, 70). In one study, IRS developers were more likely to have received potent immunosuppression comprising a combination of tacrolimus, mycophenolate mofetil, and prednisone compared with patients without IRS ($P = 0.007$). Additionally, cases with IRS versus those without IRS were more likely to have disseminated cryptococcosis (69). These data are consistent with those in HIV settings where more profound immunosuppression at the onset of infection and greater severity of infection (or disseminated disease) correlated with an increased likelihood of IRS after antiretroviral therapy. IRS may present as lymphadenitis, cellulitis, aseptic meningitis, cerebral mass lesions, hydrocephalus, or pulmonary nodules (42, 70). In kidney transplant patients, the development of IRS was temporally associated with allograft loss (69). The overall probability of allograft survival following cryptococcosis in kidney transplant recipients was significantly lower in patients who developed IRS compared with those who did not ($P = 0.0004$) (69).

IRS needs to be distinguished from worsening cryptococcosis and from other opportunistic infections or drug-related complications. There are no specific markers that can reliably establish the diagnosis of IRS (11). Histopathologic examination often reveals granulomas containing macrophages with or without necrosis, a feature rarely observed at initial diagnosis of cryptococcosis in an immunosuppressed host (70).

Outcome

Mortality rates in SOT recipients with cryptococcosis have typically ranged from 33 to 42% (32) and may be as high as 49% in those with CNS disease (32). Overall mortality in SOT recipients with cryptococcosis in the current era is ~15% (66). In a case series of 28 SOT recipients with cryptococcal meningitis, mortality correlated with altered mental status, absence of headache, and liver failure, and liver failure was an independent predictor for death (79). In a separate study, renal failure at baseline was associated with a higher mortality rate, but receipt of a calcineurin-inhibitor agent was independently associated with a lower mortality rate (66). Improved outcomes with the use of calcineurin-inhibitor agents may be attributable in part to their synergistic interactions with antifungal agents (39). In patients with CNS lesions due to *Cryptococcus*, the mortality rate varied depending on the type of lesion (71). Whereas the mortality rate was 50% in patients with parenchymal lesions, those with leptomeningeal lesions or hydrocephalus had a mortality rate of 12.5% despite a higher cryptococcal antigen titer (71). Antigen titers in the CSF or serum have not been shown to correlate with CSF sterilization at 2 weeks or overall mortality at 90 days in SOT recipients with CNS cryptococcal disease. With cryptococcosis limited to the lungs, the mortality rate is as low as 2.8% (66). However, the mortality rate is higher in patients with symptomatic pulmonary disease than those with asymptomatic or incidentally detected pulmonary cryptococcosis (65).

HEMATOPOIETIC STEM CELL TRANSPLANT RECIPIENTS

Cryptococcosis occurs rarely in hematopoietic stem cell transplant (HSCT) recipients. According to the data from the Transplant Associated Infection Surveillance Network, similar numbers of SOTs and HSLTs were performed in the United States from 2001 to 2005 (17,226 versus 16,390), but cryptococcosis developed in 9% of the SOT recipients versus 0% of the HSCT recipients (40; Pappas et al., ICAAC). In the world's literature, only nine HSCT recipients with cryptococcosis have been identified to date (12, 41, 46, 51, 57, 58, 73). Of seven patients in whom the source of stem cell transplantation was identified, five were autologous and two were allogeneic HSCT recipients (73). Cryptococcal disease appears to occur earlier posttransplant in HSCT compared with SOT recipients; the time to onset of cryptococcosis was 12 days and ranged from 64 days to 4 months posttransplantation (73). Seven patients had disseminated cryptococcosis (meningitis in three and cryptococcemia in four). One patient had cryptococcal pneumonia, and the site was unspecified in another. Notably, all four patients in whom the species of cryptococci were reported had non-*neoformans Cryptococcus* (*C. terreus*, *C. laurentii*, *C. albidus*, and *C. adeliensis*). Outcomes were available in four patients: two were alive and two died at the end of follow-up (73).

The basis for the relative rarity of cryptococcal disease in HSCT recipients is not known. However, it has been proposed that thymic regeneration of transplanted stem cells, a dominant Th1 response in allogeneic stem cell transplant recipients (considered protective against cryptococcosis), and wider employment of fluconazole prophylaxis in the early posttransplant period may account for less frequent occurrence of cryptococcal disease in HSCT compared with SOT recipients (73).

TISSUE TRANSPLANT RECIPIENTS

Approximately 35,000 corneal transplants are performed annually in the United States (27). Despite being a rare complication, corneal transplantation can potentially transmit cryptococcosis (4, 13). To date, five corneal transplant recipients with cryptococcosis have been reported in the literature (4, 13, 55, 56, 59, 73); two were infected with *C. neoformans*, two with *C. albidus*, and one patient had both *C. laurentii* and *Fusarium solani*. Two patients acquired cryptococcal disease through donor-to-host transmission, including one donor with disseminated cryptococcosis (4, 13). Cryptococcosis occurred at a median of 2 months after corneal transplantation (73). In all five patients, the disease was limited to the eye and comprised cryptococcal keratitis in four and endophthalmitis in one. Two patients received both systemic and topical antifungal agents, while two received topical agents only. Two patients required retransplantation, and one required enucleation of the eye despite aggressive medical and surgical treatment. Even though the disease was limited to the eye in all cases, the visual outcomes were uniformly poor. Except for one patient without vision description at the end of follow-up, the remaining four patients had residual visual acuity of 20/40, hand motion, light perception, and blindness, respectively.

DIAGNOSIS

Typical manifestations of infectious processes are often subtle or lacking in transplant recipients, and thus a high index of suspicion for cryptococcosis must be maintained, particularly when transplant patients present with mild respiratory or CNS symptoms or unexplained radiographic findings. As previously mentioned, between 53 and 72% of cryptococcal infections in SOT recipients are disseminated or involve the CNS, and detection of the cryptococcal capsular polysaccharide antigen in serum or CSF represents the most direct means of providing a rapid diagnosis of cryptococcosis. Per Table 1, the majority (88 to 98%) of transplant recipients with CNS involvement will have a positive serum cryptococcal antigen test, and 98 to 100% will have a positive antigen test of the CSF (32, 66, 71, 79). Several well-established, commercial tests are available for detecting and quantitating the circulating cryptococcal antigen (74). These diagnostic tools provide real-time results and excellent sensitivity (93 to 100%) and specificity (93 to 100%) (63). However, the clinician should keep in mind that both false-positive (45) and false-negative (61) results have been described in transplant populations. In fact, the cryptococcal antigen test is not infrequently negative in cases of cryptococcosis limited to the lung (47, 77). In one study, SOT patients with single pulmonary nodules were less likely to have a positive antigen test than were those with all other radiographic presentations (65). Among SOT patients with isolated pulmonary cryptococcosis, lung transplant recipients were less likely to have positive cryptococcal antigen test results than were recipients of other types of SOT (65). Thus, when pulmonary cryptococcosis is suspected but the cryptococcal antigen test is negative, a different diagnostic strategy, such as bronchoscopy or biopsy, may be necessary.

The CSF profile of cryptococcal meningitis (see Table 1) is similar to that of other chronic meningitides and is therefore not particularly discriminatory. Direct microscopic examination of the CSF for the presence of *Cryptococcus* can

TABLE 1 Clinical and laboratory characteristics of SOT recipients with CNS cryptococcosis[a]

CSF findings	Value
Opening pressure (cm H_2O)	27–33 (9–70)
White blood cell count (mm^3)	33–188 (0–1,464)
Glucose (mg/dl)	36–52 (2–181)
Protein (mg/dl)	74–226 (16–1,015)
Positive cryptococcal antigen	98–100%
Positive India ink	50–80%
Positive culture	77–93%
Serum cryptococcal antigen positivity	88–98%
Fungemia	33–39%
Abnormal neuroimaging findings	30%

[a] Data from references 68, 73, and 81.

be quickly performed by adding India ink directly to CSF. The organism appears as a round yeast ranging in size from 2 to 15 μm, typically with a single, narrow-necked bud and with a capsule represented by a halo against a black background (63). When positive, the India ink exam offers initial diagnostic confirmation of cryptococcal disease. However, the India ink exam is negative in up to 50% of cases. In addition, a positive India ink does not discriminate viable from nonviable organisms and therefore cannot be used to judge successful sterilization of the CSF.

Although time-consuming, culturing the organism from body fluids and/or tissue is another important diagnostic tool. Based on available data, 33 to 39% of SOT recipients with cryptococcosis have detectable fungemia (32, 43, 79). For the two most widely used automated, continuous monitoring blood culture systems in the United States, the Bactec (BD Diagnostic Systems, Sparks, MD) and BacT/Alert (bioMérieux, Durham, NC) systems, the mean time to recovery of *C. neoformans* from blood ranges from 2.8 to 3.8 days (L. B. Reller, personal communication). The organism is also readily recoverable from other clinical specimens on solid fungal culture media.

Radiologic imaging is also an important part of the diagnostic workup of cryptococcosis, particularly as findings have implications for outcome and management. For example, identifying evidence of obstruction on CNS imaging would trigger a procedure for decompression. Of 61 SOT patients with CNS involvement, mortality was higher (3/6, 50%) for patients with parenchymal lesions compared with those with leptomeningeal disease (1/8, 12.5%) or hydrocephalus (0/3). In addition, new-onset lesions during the course of therapy were more likely to represent IRS and were associated with better outcomes (71). Based on data from HIV-infected patients, magnetic resonance imaging is more effective than computed tomography scans in identifying CNS cryptococcal lesions (11). Finally, approximately one-third of SOT recipients with pulmonary cryptococcosis may have disease detected as an incidental finding on imaging studies (47, 65). In one series, the most common radiographic presentation of pulmonary cryptococcosis in SOT patients was nodules (18/48, 38%), which were multiple in 10 patients and single in 8 patients (65). As the natural history of such infection is not known, new pulmonary radiographic findings should prompt an aggressive diagnostic evaluation, including biopsy for those patients in whom the serum cryptococcal antigen is negative.

MANAGEMENT

Treatment recommendations for cryptococcal disease in the transplant population are based, in large part, on data extrapolated from clinical trials in other hosts and expert opinion and are outlined in Table 2 (67). Consonant with guidelines from the IDSA (54, 60), a lipid formulation of amphotericin B (liposomal amphotericin B 3 to 6 mg/kg/day or amphotericin B lipid complex 5 mg/kg/day) plus flucytosine (100 mg/kg/day) for at least 2 weeks is highly recommended for CNS disease and moderately severe to severe non-CNS or disseminated disease in transplant recipients. This rapidly fungicidal regimen is strongly encouraged, as transplant recipients can possess a high burden of yeasts on initial presentation, and failure to use flucytosine for 14 days as part of the initial induction regimen has been associated with treatment failure in other immunosuppressed populations (7, 19). Lipid formulations of amphotericin B are recommended in lieu of amphotericin B deoxycholate

TABLE 2 Management of cryptococcal disease in SOT recipients

Disease and phase	Regimen	Duration
Meningoencephalitis		
Induction	Preferred regimen: AmBisome 3–4 mg/kg/day or Abelcet 5 mg/kg/day plus flucytosine 100 mg/kg/day[a]	2 weeks
	Alternative therapy: AmBisome 3–4 mg/kg/day or Abelcet 5 mg/kg/day	4 weeks
Consolidation phase	Fluconazole 400–800 mg/day[a]	8 weeks
Maintenance phase	Fluconazole 200 mg/day	6–12 months
Isolated pulmonary cryptococcosis[b]	Fluconazole 400 mg/day[a]	6–12 months

[a] Dosages should be adjusted for renal insufficiency.

[b] Disseminated disease must be excluded in all patients. Those with disseminated disease, diffuse pulmonary infiltrates, and acute respiratory failure should be treated with the same regimen as cryptococcal meningoencephalitis.

owing to the risk of nephrotoxicity when concurrently administered with a calcineurin inhibitor and owing to the significant number of transplant patients with underlying renal dysfunction at the time of diagnosis. As in other populations, the flucytosine dose should be adjusted for renal dysfunction, and administration should be carefully monitored for toxicity, particularly hematologic side effects. Serum flucytosine levels should be measured after 3 to 5 days of treatment, and levels over 100 μg/ml should be avoided (17).

Patients with a positive CSF culture at baseline should have a repeat lumbar puncture after 14 days of induction therapy, and induction therapy should be continued until the CSF is sterilized. The median time to CSF sterilization in one study of SOT patients with positive CSF cultures at baseline was 10 days (mean, 16 days); CSF cultures were negative at 14 days in 52% (11/21) of patients (69).

Induction therapy should be followed by consolidation therapy with fluconazole dosed at 400 to 800 mg daily for 8 weeks, and subsequently, maintenance therapy with fluconazole at 200 to 400 mg daily to complete 6 to 12 months of total treatment. In all cases, fluconazole doses should be adjusted for renal function. This treatment schedule is based on data suggesting high rates of relapse in SOT patients who did not receive maintenance therapy but a low risk of relapse (1.3%) for those in whom maintenance therapy was administered for at least 6 months (69). SOT patients with asymptomatic or mild disease limited to the lung may be treated with fluconazole monotherapy (equivalent to 400 mg daily). Although data from the SOT population to support this decision are limited, no significant difference in outcome with the use of an amphotericin B formulation compared with fluconazole was found for patients with disease limited to the lung (66).

Reduction in immunosuppressive therapy should be considered in transplant recipients diagnosed with cryptococcosis. However, it is important that such reductions are gradual over time. Some experts advocate slowly tapering corticosteroids before calcineurin inhibitors since the calcineurin inhibitors have anticryptococcal activity (67).

In any case, patients need to be monitored closely during this period for manifestations of IRS, which must be distinguished from worsening cryptococcosis. Unfortunately, and as mentioned previously, there are no specific markers that can reliably establish the diagnosis of IRS. No definitive treatment for minor manifestations of IRS is necessary. However, patients with suspected IRS should be followed closely over the ensuing weeks until symptoms resolve. On the other hand, patients with severe IRS complications such as increased intracranial pressure or other life-threatening symptoms associated with inflammation may be managed with corticosteroids, as adjunctive therapy with an antifungal agent, over a 2- to 6-week period. The duration of such a steroid taper should be tailored based on the individual patient's symptoms and response and should be closely monitored. Insufficient data exist to assess the utility of nonsteroidal anti-inflammatory drugs.

REFERENCES

1. **Anderson, D. J., C. Schmidt, J. Goodman, and C. Pomeroy.** 1992. Cryptococcal disease presenting as cellulitis. *Clin. Infect. Dis.* **14:**666–672.
2. **Baddley, J. W., J. R. Perfect, R. A. Oster, R. A. Larson, G. A. Pankey, H. Henderson, D. W. Haas, C. A. Kauffman, R. Patel, A. K. Zaas, and P. G. Pappas.** 2008. Pulmonary cryptococcosis in patients without HIV infection: factors associated with disseminated disease. *Eur. J. Clin. Microbiol. Infect. Dis.* **27:**937–943.
3. **Baker, R. D.** 1976. The primary pulmonary lymph node complex of cryptococcosis. *Am. J. Clin. Pathol.* **65:**83–92.
4. **Beyt, B. E., Jr., and S. R. Waltman.** 1978. Cryptococcal endophthalmitis after corneal transplantation. *N. Engl. J. Med.* **15:**825–826.
5. **Biswas, J., L. Gopal, T. Sharma, S. Parikh, H. N. Madhavan, and S. S. Badrinath.** 1998. Recurrent cryptococcal choroiditis in a renal transplant patient. *Retina* **18:**273–276.
6. **Blankenship, J. R., N. Singh, B. D. Alexander, and J. Heitman.** 2005. Cryptococcus neoformans isolates from transplant recipients are not selected for resistance to calcineurin inhibitors by current immunosuppressive regimens. *J. Clin. Microbiol.* **43:**464–467.
7. **Brouwer, A. E., A. Rajanuwong, W. Chierakul, G. E. Griffin, R. A. Larsen, N. J. White, and T. S. Harrison.** 2004. Combination antifungal therapies for HIV-associated cryptococcal meningitis: a randomised trial. *Lancet* **363:**1764–1767.
8. **Bruno, K. M., L. Farhoomand, B. S. Libman, C. N. Pappas, and F. J. Landry.** 2002. Cryptococcal arthritis, tendinitis, tenosynovitis, and carpal tunnel syndrome: report of a case and review of the literature. *Arthritis Care Res.* **47:**104–108.
9. **Casadevall, A.** 2004. What is the latest approach to pulmonary cryptococcosis. *J. Respir. Dis.* **25:**188–199.
10. **Casadevall, A., and L. A. Pirofski.** 2003. The damage-response framework of microbial pathogenesis. *Nat. Rev. Microbiol.* **1:**17–24.
11. **Charlier, C., F. Dromer, C. Leveque, L. Chartier, Y. S. Cordoliani, A. Fontanet, O. Launay, O. Lortholary, and French Cryptococcosis Study Group.** 2008. Cryptococcal neuroradiological lesions correlate with severity during cryptococcal meningoencephalitis in HIV-positive patients in the HAART era. *PloS One* **3:**E1950.
12. **Chou, L. S., R. E. Lewis, C. Ippoliti, R. E. Champlin, and D. P. Kontoyiannis.** 2007. Caspofungin as primary antifungal prophylaxis in stem cell transplant recipients. *Pharmacotherapy* **27:**1644–1650.
13. **de Castro, L. E., O. A. Sarraf, J. M. Lally, H. P. Sandoval, K. D. Solomon, and D. T. Vroman.** 2005. Cryptococcus albidus keratitis after corneal transplantation. *Cornea* **24:**882–883.
14. **D'Elios, M. M., R. Josien, M. Manghetti, A. Amedei, M. de Carli, M. C. Cuturi, G. Blancho, F. Buzelin, G. del Prete, and J. P. Soulillou.** 1997. Predominant Th1 cell infiltration in acute rejection episodes of human kidney grafts. *Kidney Int.* **51:**1876–1884.
15. **Diamond, R. D., and J. E. Bennett.** 1974. Prognostic factors in cryptococcal meningitis: a study in 111 cases. *Ann. Intern. Med.* **80:**176–181.
16. **Dilhuydy, M. S., T. Jouary, H. Demeaux, and A. Ravaud.** 2007. Cutaneous cryptococcosis with alemtuzumab in a patient treated for chronic lymphocytic leukaemia. *Br. J. Haematol.* **137:**490.
17. **Drew, R. H., and J. R. Perfect.** 2004. Flucytosine. *In* D. Raoult, R. Weber, and V. L. Yu (ed.), *Antimicrobial Therapy and Vaccines.* Apple Trees Productions, New York, NY.
18. **Dromer, F., S. Mathoulin-Pélissier, A. Fontanet, O. Ronin, B. Dupont, O. Lortholary, and French Cryptococcosis Study Group.** 2004. Epidemiology of HIV-associated cryptococcosis in France (1985–2001): comparison of the pre- and post-HAART eras. *AIDS* **18:**255–262.
19. **Dromer, F., S. Mathoulin-Pélissier, O. Launay, O. Lortholary, and French Cryptococcosis Study Group.** 2007. Determinants of disease presentation and outcome during cryptococcosis: the CryptoA/D study. *PloS Med.* **4:**e21.
20. **Dromer, F., S. Mathoulin, B. Dupont, L. Letenneur, and O. Ronin.** 1996. Individual and environmental factors associated with infection due to *Cryptococcus neoformans* serotype D. *Clin. Infect. Dis.* **23:**91–96.
21. **Ferraris, J. R., M. L. Tambutti, R. L. Cardoni, and N. Prigoshin.** 2004. Conversion from cyclosporine A to tacrolimus in pediatric kidney transplant recipients with chronic rejection: changes in the immune responses. *Transplantation* **77:**532–537.
22. **Garcia-Hermoso, D., G. Janbon, and F. Dromer.** 1999. Epidemiological evidence for dormant *Cryptococcus neoformans* infection. *J. Clin. Microbiol.* **37:**3204–3209.
23. **Gloster, H. M., Jr., R. A. Swerlick, and A. R. Solomon.** 1994. Cryptococcal cellulitis in a diabetic kidney transplant patient. *J. Am. Acad. Dermatol.* **30:**1025–1026.
24. **Gorlach, J., D. S. Fox, N. S. Cutler, G. M. Cox, J. R. Perfect, and J. Heitman.** 2000. Identification and characterization of a highly conserved calcineurin binding protein, CBP1/calcipressin, in *Cryptococcus neoformans*. *EMBO J.* **19:**1–12.
25. **Gras, J., A. Cornet, D. Latinne, and R. Reding.** 2004. Evidence that Th1/Th2 immune deviation impacts on early graft acceptance after pediatric liver transplantation: results of immunological monitoring in 40 children. *Am. J. Transplant* **4:**444.
26. **Hall, J. C., J. H. Brewer, T. T. Crouch, and K. R. Watson.** 1987. Cryptoccal cellulitis with multiple sites of involvement. *J. Am. Acad. Dermatol.* **17:**329–332.
27. [Reference deleted.]
28. **Haugen, R. K., and R. D. Baker.** 1954. The pulmonary lesions in cryptococcosis with special reference to subpleural nodules. *Am. J. Clin. Pathol.* **37:**1381–1390.
29. **Hellman, R. N., J. Hinrichs, G. Sicard, R. Hoover, P. Golden, and P. Hoffsten.** 1981. Cryptococcal pyelonephritis and disseminated cryptococcosis in a renal transplant recipient. *Arch. Intern. Med.* **141:**128–130.
30. **Hodge, S., G. Hodge, R. Flower, and P. Han.** 1999. Methylprednisolone upregulates monocyte IL-10 production in stimulated whole blood. *Scand. J. Immunol.* **49:**548–553.
31. **Horrevorts, A. M., F. T. M. Huysmans, R. J. J. Koopman, F. Jacques, and G. M. Meis.** 1994. Cellulitis as first clinical presentation of disseminated cryptococcosis

in renal transplant recipients. *Scand. J. Infect. Dis.* **26:** 623–626.

32. **Husain, C., M. M. Wagener, and N. Singh.** 2001. *Cryptococcus neoformans* infection in organ transplant recipients: variables influencing clinical characteristics and outcome. *Emerg. Infect. Dis.* **7:**375–381.
33. **Husain, S., M. M. Wagener, and N. Singh.** 2001. *Cryptococcus neoformans* in organ transplant recipients: correlates of variability in clinical characteristics and outcome. *Emerg. Infect. Dis.* **7:**375–381.
34. **Ingram, P. R., R. Howman, M. F. Leahy, and J. R. Dyer.** 2007. Cryptococcal immune reconstitution inflammatory syndrome following alemtuzumab therapy. *Clin. Infect. Dis.* **44:**e115–e117.
35. **Jabbour, N., J. Reyes, S. Kusne, M. Martin, and J. Fung.** 1996. Cryptococcal meningitis after liver transplantation. *Transplantation* **61:**146–167.
36. **Jarvis, J. N., and T. S. Harrison.** 2008. Pulmonary cryptococcosis. *Semin. Respir. Crit. Care Med.* **29:**141–150.
37. **Kanj, S. S., K. Welty-Wolf, J. Madden, V. Tapson, M. A. Baz, D. Davis, and J. R. Perfect.** 1996. Fungal infections in lung and heart-lung transplant recipients: report of 9 cases and review of the literature. *Medicine* **75:**142–156.
38. **Kapoor, A., S. M. Flenchner, K. O'Malley, D. Paolone, T. M. File, Jr., and A. F. Cutrona.** 1999. Cryptococcal meningitis in renal transplant patients associated with environmental exposure. *Transplant. Infect. Dis.* **1:**213–217.
39. **Kontoyiannis, D. P., R. E. Lewis, B. D. Alexander, O. Lortholary, F. Dromer, K. L. Gupta, G. T. John, R. Del Busto, G. B. Klintmalm, J. Somani, G. M. Lyon, K. Pursell, V. Stosor, P. Munoz, A. P. Limaye, A. C. Kalil, T. L. Pruett, J. Garcia-Diaz, A. Humar, S. Houston, A. A. House, D. Wray, S. Orloff, L. A. Dowdy, R. A. Fisher, J. Heitman, N. D. Albert, M. M. Wagener, and N. Singh.** 2008. Calcineurin inhibitor agents interact synergistically with antifungal agents in vitro against *Cryptococcus neoformans* isolates: correlation with outcome in solid organ transplant recipients with cryptococcosis. *Antimicrob. Agents Chemother.* **52:**735–738.
40. **Kontoyiannis, D. P., K. Marr, B. J. Park, B. D. Alexander, E. J. Anaissie, T. J. Walsh, J. Ito, D. R. Andes, J. W. Baddley, J. M. Brown, L. M. Brumble, A. G. Freifeld, S. Hadley, L. A. Herwaldt, C. A. Kauffman, K. Knapp, G. M. Lyon, V. A. Morrison, G. Papanicolaou, T. F. Patterson, T. M. Perl, M. G. Schuster, R. Walker, K. A. Wannemuehler, J. R. Wingard, T. M. Chiller, and P. G. Pappas.** 2007. Prospective surveillance for invasive fungal infections in hematopoietic stem cell transplant recipients, 2001–2006; overview of the TRANSNET database. *Clin. Infect. Dis.* **50:**1091–1100.
41. **Krcmery, V., Jr., A. Kunova, and J. Mardiak.** 1997. Nosocomial *Cryptococcus laurentii* fungemia in a bone marrow transplant patient after prophylaxis with ketoconazole successfully treated with oral fluconazole. *Infection* **25:**130.
42. **Lanternier, F., M. O. Chandesris, S. Poirée, M. E. Bougnoux, F. Mechai, M. F. Mamzer-Bruneed, J. P. Viard, L. L. M. Galmiche-Rolland, and O. Lortholary.** 2007. Cellulitis revealing a cryptococcosis-related immune reconstitution inflammatory syndrome in a renal allograft recipient. *Am. J. Transplant.* **12:**2826–2828.
43. **Lortholary, O., K. Sitbon, F. Dromer, and the French Cryptococcosis Study Group.** 2005. Evidence for HIV and *Cryptococcus neoformans* interactions in the proinflammatory and anti-inflammatory responses in blood during AIDS-associated cryptococcosis. *Clin. Microbiol. Infect.* **11:**296–300.
44. **MacDougall, L., S. E. Kidd, E. Galanis, S. Mak, M. J. Leslie, P. R. Cieslak, J. W. Kronstad, M. G. Morshed, and K. H. Bartlett.** 2007. Spread of *Cryptococcus gattii* in British Columbia, Canada, and detection in the Pacific Northwest, USA. *Emerg. Infect. Dis.* **13:**42–50.
45. **McManus, E. J., and J. M. Jones.** 1985. Detection of a *Trichosporon beigelii* antigen cross-reactive with *Cryptococcus neoformans* capsular polysaccharide in serum from a patient with disseminated *Trichosporon* infection. *J. Clin. Microbiol.* **21:**681–685.
46. **Mendpara, S. D., C. Ustun, A. M. Kallab, F. M. Mazzella, P. A. Bilodeau, and A. P. Jillella.** 2002. Cryptococcal meningitis following autologous stem cell transplantation in a patient with multiple myeloma. *Bone Marrow Transplant* **30:**259–260.
47. **Mueller, N. J., and J. A. Fishman.** 2003. Asymptomatic pulmonary cryptococcosis in solid organ transplantation: report of four cases and review of the literature. *Transplant Infect. Dis.* **5:**140–143.
48. **Neuville, S., F. Dromer, O. Morin, B. Dupont, O. Ronin, and O. Lortholary.** 2003. Primary cutaneous cryptococcosis: a distinct clinical entity. *Clin. Infect. Dis.* **36:**337–347.
49. **Nosanchuk, J. D., S. Shoham, B. C. Fries, D. S. Shapiro, S. M. Levitz, and A. Casadevall.** 2000. Evidence of zoonotic transmission of *Cryptococcus neoformans* from a pet cockatoo to an immunocompromised patient. *Ann. Intern. Med.* **132:**205–208.
50. **Ooi, H. S., B. T. M. Chen, H. L. Cheng, O. T. Khoo, and K. T. Chan.** 1971. Survival of a patient transplanted with a kidney infected with *Cryptococcus neoformans. Transplantation* **11:**428–429.
51. **Pappas, P. G., J. R. Perfect, G. A. Cloud, R. A. Larsen, G. A. Pankey, D. J. Lancaster, H. Henderson, C. A. Kauffman, D. W. Haas, M. Saccente, R. J. Hamill, S. Holloway, R. M. Warren, and W. E. Dismukes.** 2001. Cryptococcosis in human immunodeficiency virus-negative patients in the era of effective azole therapy. *Clin. Infect. Dis.* **33:**690–699.
52. **Paya, C. V.** 1993. Fungal infections in solid-organ transplantation. *Clin. Infect. Dis.* **16:**677–688.
53. **Peleg, A. Y., S. Husain, E. J. Kwak, F. P. Silveira, M. Ndirangu, J. Tran, K. A. Shutt, R. Shapiro, N. Thai, K. Abu-Elmagd, K. R. McCurry, A. Marcos, and D. L. Paterson.** 2007. Opportunistic infections in 547 organ transplant recipients receiving alemtuzumab, a humanized monoclonal DC-52 antibody. *Clin. Infect. Dis.* **44:**204–212.
54. **Perfect, J. R., W. E. Dismukes, F. Dromer, D. L. Goldman, J. R. Graybill, R. J. Hamill, T. S. Harrison, R. A. Larsen, O. Lortholary, M. H. Nguyen, P. G. Pappas, W. G. Powderly, N. Singh, J. D. Sobel, and T. C. Sorrell.** 2010. Clinical practice guidelines for the management of cryptococcal disease: 2010 update by the Infectious Diseases Society of America. *Clin. Infect. Dis.* **50:**291–322.
55. **Perry, H. D., and E. D. Donnenfeld.** 1990. Cryptococcal keratitis after keratoplasty. *Am. J. Ophthalmol.* **110:** 320–321.
56. **Perry, H. D., S. J. Doshi, E. D. Donnenfeld, and G. S. Bai.** 2002. Topical cyclosporin A in the management of therapeutic keratoplasty for mycotic keratitis. *Cornea* **21:**161–163.
57. **Ramchandren, R., and D. E. Gladstone.** 2004. *Cryptococcus albidus* infection in a patient undergoing autologous progenitor cell transplant. *Transplantation* **77:**956.
58. **Rimek, D., G. Haase, A. Luck, J. Casper, and A. Podbielski.** 2004. First report of a case of meningitis caused by *Cryptococcus adeliensis* in a patient with acute myeloid leukemia. *J. Clin. Microbiol.* **42:**481–483.
59. **Ritterband, D. C., J. A. Seedor, M. K. Shah, S. Waheed, and I. Schorr.** 1998. A unique case of *Cryptococcus laurentii* keratitis spread by a rigid gas permeable contact lens in a patient with onychomycosis. *Cornea* **17:**115–118.

60. **Saag, M. S., R. J. Graybill, R. A. Larsen, P. G. Pappas, J. R. Perfect, W. G. Powderly, J. D. Sobel, W. E. Dismukes, for the Mycoses Study Group Cryptococcal Subproject.** 2000. Practice guidelines for the management of cryptococcal disease. *Clin. Infect. Dis.* **30:**710–718.
61. **Safdar, N., C. L. Abad, S. Narayan, D. R. Kaul, and S. Saint.** 2009. Clinical problem-solving. Keeping an open mind. *N. Engl. J. Med.* **360:**72–76.
62. **Saha, D. C., D. L. Goldman, X. Shao, A. Casadevall, S. Husain, A. P. Limaye, M. Lyon, J. Somani, K. Pursell, T. L. Pruett, and N. Singh.** 2007. Serologic evidence for reactivation of cryptococcosis in solid-organ transplant recipients. *Clin. Vaccine Immunol.* **14:**1550–1554.
63. **Shea, Y. R.** 2007. Algorithms for detection and identification of fungi, p. 1745–1761. *In* P. R. Murray, E. J. Baron, J. H. Jorgensen, M. L. Landry, and M. A. Pfaller (ed.), *Manual of Clinical Microbiology*, 9th ed. ASM Press, Washington, DC.
64. **Silveira, E. P., S. Husain, E. J. Kwak, P. K. Linden, A. Marcos, R. Shapiro, P. Fontes, J. W. Marsh, M. de Vera, K. Tom, N. Thai, H. P. Tan, A. Basu, K. Soltys, and D. L. Paterson.** 2007. Cryptococcosis in liver and kidney transplant recipients receiving anti-thymocyte globulin or alemtuzumab. *Transplant Infect. Dis.* **9:**22–27.
65. **Singh, N., B. D. Alexander, O. Lortholary, F. Dromer, K. L. Gupta, G. T. John, R. del Busto, G. B. Klintmalm, J. Somani, G. M. Lyon, K. Pursell, V. Stosor, P. Munoz, A. P. Limaye, A. C. Kalil, T. L. Pruett, J. Garcia-Diaz, A. Humar, S. Houston, A. A. House, D. Wray, S. Orloff, L. A. Dowdy, R. A. Fisher, J. Heitman, M. M. Wagener, and S. Husain.** 2008. Pulmonary cryptococcosis in solid organ transplant recipients: clinical relevance of serum cryptococcal antigen. *Clin. Infect. Dis.* **46:**e12–e18.
66. **Singh, N., B. D. Alexander, O. Lortholary, F. Dromer, K. L. Gupta, G. T. John, R. del Busto, G. B. Klintmalm, J. Somani, G. M. Lyon, K. Pursell, V. Stosor, P. Munoz, A. P. Limaye, A. C. Kalil, T. L. Pruett, J. Garcia-Diaz, A. Humar, S. Houston, A. A. House, D. Wray, S. Orloff, L. A. Dowdy, R. A. Fisher, J. Heitman, M. M. Wagener, S. Husain, and the Cryptococcal Collaborative Transplant Study Group.** 2007. *Cryptococcus neoformans* in organ transplant recipients: impact of calcineurin-inhibitor agents on mortality. *J. Infect. Dis.* **195:**756–764.
67. **Singh, N., F. Dromer, J. R. Perfect, and O. Lortholary.** 2008. Cryptococcosis in solid organ transplant recipients: current state of the science. *Clin. Infect. Dis.* **47:**1321–1327.
68. **Singh, N., G. Forrest, C. Sifri, B. Alexander, L. Johnson, M. Lyon, O. Lortholary, and Cryptococcal Collaborative Transplant Study Group.** 2008. Cryptococcus-associated immune reconstitution syndrome (IRS) in solid organ transplant (SOT) recipients: results from a prospective, multicenter study, abstract #2152, 48th Annual ICAAC/IDSA 46th Annual Meeting Washington, DC 25 to 28 October.
69. **Singh, N., O. Lortholary, B. D. Alexander, K. L. Gupta, G. T. John, K. Pursell, P. Munoz, G. B. Klintmalm, V. Stosor, R. Del Busto, A. P. Limaye, J. Somani, M. Lyon, S. Houston, A. A. House, T. L. Pruett, S. Orloff, A. Humar, L. Dowdy, J. Garcia-Diaz, R. A. Fisher, S. Husain, and the Cryptococcal Collaborative Transplant Study Group.** 2005. Allograft loss in renal transplant recipients with *C. neoformans* associated immune reconstitution syndrome. *Transplantation* **80:**1131–1133.
70. **Singh, N., O. Lortholary, B. D. Alexander, K. L. Gupta, G. T. John, K. Pursell, P. Munoz, G. B. Klintmalm, V. Stosor, R. Del Busto, A. P. Limaye, J. Somani, M. Lyon, S. Houston, A. A. House, T. L. Pruett, S. Orloff, A. Humar, L. Dowdy, J. Garcia-Diaz, A. C. Kalil, R. A. Fisher, S. Husain, and the Cryptococcal Collaborative Transplant Study Group.** 2005. An "immune reconstitution syndrome"-like entity associated with *Cryptococcus neoformans* infections in organ transplant recipients. *Clin. Infect. Dis.* **40:**1756–1761.
71. **Singh, N., O. Lortholary, F. Dromer, B. D. Alexander, K. L. Gupta, G. T. John, R. del Busto, G. B. Klintmalm, J. Somani, G. M. Lyon, K. Pursell, V. Stosor, P. Munoz, A. P. Limaye, A. C. Kalil, T. L. Pruett, J. Garcia-Diaz, A. Humar, S. Houston, A. A. House, D. Wray, S. Orloff, L. A. Dowdy, R. A. Fisher, J. Heitman, M. M. Wagener, and S. Husain.** 2008. Central nervous system cryptococcosis in solid organ transplant recipients: clinical relevance of abnormal neuroimaging findings. *Transplantation* **86:**647–651.
72. **Singh, N., and D. L. Paterson.** 2005. *Aspergillus* infections in transplant recipients. *Clin. Microbiol. Rev.* **18:**44–69.
73. **Sun, H. Y., M. M. Wagener, and N. Singh.** 2009. Cryptococcosis in solid-organ, hematopoietic stem cell, and tissue transplant recipients: evidence-based evolving trends. *Clin. Infect. Dis.* **48:**1566–1576.
74. **Tanner, D. C., M. P. Weinstein, B. Fedorciw, K. L. Joho, J. J. Thorpe, and L. Reller.** 1994. Comparison of commercial kits for detection of cryptococcal antigen. *J. Clin. Microbiol.* **32:**1680–1684.
75. **Uslu, A., A. Nart, I. Coker, S. Kose, A. Aykas, M. C. Kahya, M. F. Yuzbasioglu, and M. Dogan.** 2004. Two-day induction with thymoglobulin in kidney transplantation: risks and benefits. *Transplant Proc.* **36:**76–79.
76. **Vilchez, R. A., J. Fung, and S. Kusne.** 2002. Cryptococcosis in organ transplant recipients: an overview. *Am. J. Transplantation* **2:**575–580.
77. **Vilchez, R. A., W. Irish, J. Lacomis, P. Costello, J. Fung, and S. Kusne.** 2001. The clinical epidemiology of pulmonary cryptococcosis in non-AIDS patients at a tertiary care medical center. *Medicine* **80:**308–312.
78. **Winston, D. J., C. Emmanouilides, and R. W. Busuttil.** 1995. Infections in liver transplant recipients. *Clin. Infect. Dis.* **21:**1077–1091.
79. **Wu, G., R. A. Vilchez, B. Eidelman, J. Fung, R. Kormos, and S. Kusne.** 2002. Cryptococcal meningitis: an analysis among 5521 consecutive organ transplant recipients. *Transplant Infect. Dis.* **4:**183–188.

Cryptococcus: From Human Pathogen to Model Yeast
Edited by J. Heitman et al.

38

Cryptococcosis in AIDS

BETTINA C. FRIES AND GARY M. COX

Infection with HIV is the most common predisposing condition for developing cryptococcosis. In this chapter we will focus on discussing aspects that are specific to the coinfection of HIV and *Cryptococcus neoformans*. These include clinical presentation, treatment regimens, alterations of the host immune response, and concerns about drug toxicities and interactions.

HISTORICAL PERSPECTIVE

HIV is a retrovirus that exists in two forms: HIV-1, which is responsible for the vast majority of the infections worldwide, and HIV-2, which is endemic to certain areas of western Africa. HIV is thought to have evolved from simian immunodeficiency virus, and epidemiologic evidence suggests that there have been repeated transmissions from nonhuman primates to humans occurring in central Africa over the past century. It is likely that HIV had been infecting humans and causing AIDS in the Congo River basin of Africa for several years prior to HIV being recognized as a pathogen in the United States and Europe. One researcher has found evidence of 65 cases of cryptococcosis diagnosed in central Africa from 1947 to 1968 and postulates that these represent the very first cases of cryptococcosis as an AIDS-associated opportunistic infection and a harbinger of what was to occur later on in the 20th century (55). Soon after the HIV pandemic was recognized in the United States and Europe, it became clear that cryptococcosis was an important opportunistic pathogen in patients with AIDS. Together with *Pneumocystis jiroveci* pneumonia, cryptococcosis was, and remains, the most common life-threatening systemic fungal infection in HIV-infected patients (23, 35).

Based on data from sentinel sites in Atlanta, Georgia, and Houston, Texas, the incidence of AIDS-associated cryptococcosis in the United States appears to have peaked some time prior to 1992 (53). Since that time the incidence of cryptococcosis has decreased steadily. It is noteworthy that the decrease in incidence began before the availability of highly active antiretroviral therapy (HAART) in late 1995. The use of HAART has dramatically altered the course of the HIV pandemic in the United States and Western Europe, because patients with AIDS who respond to HAART are able to reconstitute their immune systems so that they are no longer susceptible to opportunistic infections including cryptococcosis. It is usually assumed that the availability of HAART is the main explanation for the decrease in the incidence of AIDS-associated cryptococcosis. Although the introduction of HAART is probably the most significant factor, based on the epidemiologic data mentioned above, other factors likely contributed. One important component of this was the increased use of azole antifungal drugs to treat other fungal infections, such as mucosal candidiasis, that unknowingly also served to prevent the development of cryptococcosis. Not only has HAART contributed to a decrease in incidence in cryptococcosis, but it has also been associated with improved survival in patients diagnosed with AIDS-associated cryptococcosis. A French study that included 240 pre- and 149 post-HAART patients with *C. neoformans*–associated chronic meningoencephalitis (CME) documented that the mortality rate per 100 person years dramatically dropped from 63.8 to 15.3 in their patient cohort once HAART was available (22).

In countries with limited or no access to HAART regimens *C. neoformans* continues to be a major opportunistic fungal pathogen (50). Data from the Centers for Disease Control and Prevention estimate that more than 680,000 deaths worldwide are caused annually by infection with *C. neoformans*, the majority of which are in AIDS patients in Africa (67). Moreover, there are still a substantial number of HIV-infected patients even in developed countries that are diagnosed with cryptococcosis (53). Most often, these patients diagnosed in developed countries are either unaware of their underlying HIV infection or knew of their HIV diagnosis but had not responded to HAART because of issues with medicine adherence or access to medical care (53).

HOST RESPONSE IN AIDS PATIENTS

Our understanding of the host response elicited by HIV infection and the resulting immunodeficiency is evolving. Continual depletion through cytopathic and apoptotic mechanisms

Bettina C. Fries, Departments of Medicine, Microbiology, and Immunology, Albert Einstein College of Medicine, Bronx, NY 10804. **Gary M. Cox,** Department of Medicine Mycology Research Unit, Duke University, Durham, NC 27710.

is thought to drive activation and depletion of the naive T-lymphocyte repertoire. Many other aspects of the humoral and cellular immune response including HLA haplotypes contribute to the pathogenesis of AIDS (36). Some critical events such as T-cell destruction in tissues occur very early and may contribute to imbalanced immunological hyperactivity (64). Interestingly, experimental studies demonstrate bystander activation of T cells that recognize and control other pathogens such as cytomegalovirus, which may contribute to reactivation of these latent infections (44). Except for rare exceptions, the natural immune response against HIV is completely inadequate and results in a relentless progression of the infection.

MOLECULAR CHARACTERIZATION OF *C. NEOFORMANS* STRAINS IN HIV-INFECTED PATIENTS

Worldwide the majority (>90%) of cryptococcosis cases in AIDS patients are caused by C. *neoformans* var. *grubii*, which are serotype A strains (17, 28, 83). A higher prevalence of serotype D strains compared to other collections was reported in France (21). These strains were predominantly isolated from North African immigrants. Extensive molecular typing of C. *neoformans* strains from all over the world determined that the majority of serotype A strains are clonal, although a subpopulation of strains from Botswana also exhibit evidence of recombination as a result of sexual reproduction (47, 48). C. *gattii* infections in AIDS patients are still rare worldwide, and even in endemic areas the majority of AIDS patients present with C. *neoformans* var. *grubii* infections (14, 25, 59).

C. *neoformans* is ubiquitous to many environments and is commonly associated with avian guano and vegetative debris (2, 33). Environmental strains do not differ from clinical strains by molecular typing (18, 47) and are acquired by inhalation of asexually produced spores (87). C. *neoformans* infection in AIDS patients can either be the result of a newly acquired primary infection or alternatively constitute a reactivation of latent C. *neoformans* infection. Several case reports of AIDS and otherwise immunocompromised patients link cryptococcosis to an exposure to a sick pet (8). In one case molecular typing confirmed that the patient and the bird were infected with the same strain (62). Molecular typing of C. *neoformans* strains derived from AIDS patients that immigrated from North Africa to France showed clustering of isolate subtypes from patients originating in Africa compared to those from Europe, indicating that these patients had acquired their infectious C. *neoformans* strains long before their clinical diagnoses were made (32). In support of this conclusion, several studies have shown that both healthy humans and HIV-infected humans exhibit antibodies to cryptococcal polysaccharide (1, 24, 32, 77). In addition, reactivation of C. *neoformans* can be experimentally achieved by treating rats with steroids (34). Hence, although evidence of both scenarios exists, most likely reactivation of dormant infection is more common than primary infection in AIDS patients. Molecular typing of sequential isolates has shown that persistent disease and relapse are usually caused by the same C. *neoformans* strain (42). During chronic cryptococcosis the infecting strain can change phenotypic characteristics that are relevant for virulence (16, 30), a process referred to as microevolution (29). Some strains can even undergo phenotypic switching and generate phenotypic variants that differ in virulence (31).

EFFECT OF HAART ON *C. NEOFORMANS* INFECTION

HAART is proven to be cost-effective in the industrialized world and in South Africa (20, 58). Patients with AIDS who respond to HAART show a greatly reduced incidence of opportunistic infections such as cryptococcosis (53). This effect is not only attributed to the overall improvement of immune parameters, but increasing experimental evidence suggests that protease inhibitors can have a direct effect on microbial pathogens (9, 10, 57, 70). The effect of protease inhibitors on C. *neoformans* was predominantly documented with indinavir. Given its relatively low cost and proven efficacy, indinavir remains a key compound of HAART, especially in resource-limited settings (15). Importantly, more recent studies with tipranavir, a newer-generation protease inhibitor, showed similar effects (10). It is also noteworthy that in vitro studies demonstrated that combinations of antimycotic agents and saquinavir at subinhibitory concentrations can be synergistic (7). Both tipranavir and indinavir inhibited production of C. *neoformans*' polysaccharide capsule as well as urease and protease, but not melanin and phospholipase production.

Indinavir treatment made C. *neoformans* more susceptible to intracellular killing by effector cells and promoted an efficient host response in immunocompromised mice (70). Specifically, pretreated mice exhibited enhanced interleukin-12 (IL-12) production by dendritic cells of the spleen, as well as reduced CD14 and FcγR expression and increased CD86 and CD40 expression. Their splenic T cells also exhibited enhanced production of interferon-gamma (IFN-γ) and IL-2 and increased proliferation in response to fungal antigens (Fig. 1). Furthermore, indinavir treatment induced an expansion of dendritic cells in spleens and increased expression of costimulatory molecules (CD40 and CD80) and IL-12. This promoted the development of an effective T-cell response that lowered fungal burden in the brain (68), and thus mice survive an otherwise lethal challenge (69, 70) (Fig. 1). Taken together, experimental evidence suggests that HAART may benefit C. *neoformans*–infected patients not only through an indirect enhancement of immunity but also through direct effects on the fungi.

EFFECT OF HIV COINFECTION ON THE HOST RESPONSE TO *C. NEOFORMANS*

Coinfection with malaria and HIV promotes the spread of both diseases in sub-Saharan Africa (3), and transient increases in HIV viral load were found to be the result of recurrent coinfection with malaria. For hepatitis B and C, coinfection enhances mortality rate in HIV patients, but the cause is multifold (88). Experimental evidence suggests that C. *neoformans* coinfection can affect HIV replication. HIV infects mainly T cells, macrophages, and occasionally other cells including astrocytes and endothelial cells. In contrast, C. *neoformans* is a eukaryote that does not require a host cell, but as an obligate intracellular pathogen it infects macrophages and other phagocytic cells (27). Cryptococcal polysaccharide significantly increased the infectivity of HIV-1-infected H9 cells and subsequent production of infectious HIV-1 and formation of syncytia in vitro. Similar results were documented in lymphocytes from HIV-1-infected patients (71–73). C. *neoformans* enhanced HIV expression in monocytic cells through a tumor necrosis factor (TNF)-alpha- and NF-κB-dependent mechanism (40), although different studies done with azide-killed C. *neoformans* showed

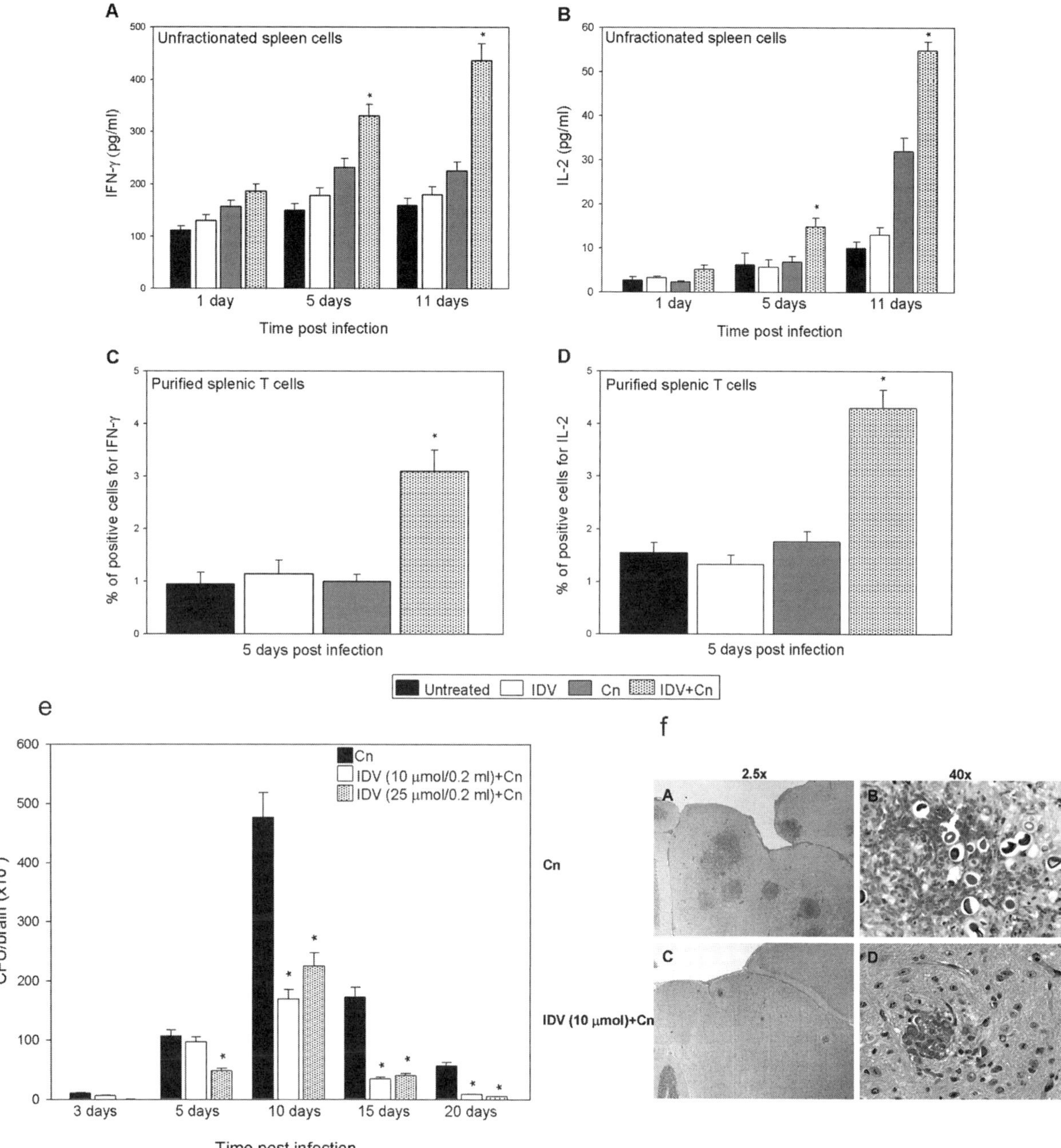

FIGURE 1 Effect of indinavir on *C. neoformans*. Kinetics of IFN-γ and IL-2 production from splenocytes of mice treated with indinavir and challenged with *C. neoformans*. Cells were cultured for 18 h in the presence of heat-inactivated *C. neoformans* (E:T = 1:2), and supernatants were tested for (A) IFN-γ and (B) IL-2 levels by specific enzyme-linked immunosorbent assays. In selected experiments, 5 days after *C. neoformans* infection, unfractionated splenocytes were cultured as above, CD3$^+$ T cells were subsequently purified, and intracellular staining for (C) IFN-γ and (D) IL-2 was evaluated. Data reported are mean ± SE from three separate experiments; *, $P < 0.05$ (indinavir + *C. neoformans* versus *C. neoformans*). (e) CFU recovery from the brain was determined 3, 5, 10, 15, and 20 days after fungal infection; indinavir (25 or 10 μmol/0.2 ml, days 3, 2, 1). Data reported are mean ± SE from five separate experiments; *, $P < 0.05$ (indinavir-treated versus *C. neoformans*). (f) Histological analysis of brains from mice treated with indinavir. Mice were treated with indinavir (10 μmol/0.2 ml, days 3, 2, 1) and subsequently infected with *C. neoformans*. Ten days after infection, animals were killed, brains were excised, and brain sections were stained with periodic acid-Schiff stain. (f-A, f-B) Brain sections from mice infected with *C. neoformans*. (f-C, f-D) Brain sections from mice treated with indinavir and infected with *C. neoformans*. Original magnification, ×2.5 (f-A, f-C); ×40 (f-B, f-D) (68).

that enhancement of HIV replication was not affected by treatment with monoclonal antibody to TNF (63).

CNS Infection

Encephalitis in HIV patients can be caused by opportunistic pathogens such as toxoplasmosis, cytomegalovirus, JC virus, and *C. neoformans* as well as the HIV virus, which can cause encephalitis on its own. In the central nervous system (CNS), HIV replication occurs relatively independently of systemic infection. In one study median HIV-1 RNA levels in cerebrospinal fluid (CSF) were higher in patients with cryptococcal meningitis and HIV, indicating enhanced intrathecal viral replication in patients with cryptococcal meningitis (60). Clinical data indicate that early treatment with HAART does not affect early fungicidal activity (5). Several clinical studies highlight differences in the inflammatory response indices in the CSF of HIV-coinfected patients when compared to HIV-uninfected patients (46, 49). One study measured lower production of pro- and anti-inflammatory mediators, specifically TNF-alpha, IL-6, IL-8, and IL-10, in the CSF of HIV-positive patients with CME compared to HIV-negative patients with CME. Two studies reported higher CSF soluble TNF receptor II levels among HIV-infected patients with CME (49, 84), which may be beneficial by prolonging the effect of proinflammatory cytokines, which are also higher in HIV survivors. In summary, these data suggest that lower cytokine production in CSF was associated with reduced anticryptococcal activity of effector cells. Together with CD4 lymphopenia this may explain the inability of HIV-positive patients to eradicate *C. neoformans*.

Other intriguing questions are whether HIV and *C. neoformans* coinfection (i) increases the risk of HIV- and/or CME-related organic brain syndrome and (ii) facilitates the dissemination of *C. neoformans* to the CNS. Both HIV and *C. neoformans* infection can cause encephalitis in infected humans. Activated but uninfected macrophages and glial cells are the principal mediators of HIV encephalitis and dementia, although these studies pertain to clade B infections, where HIV dementia is more common than in HIV clade C infections, the dominant HIV subtype in Africa and Asia. Double immunolabeling of brain sections of AIDS patients with CME revealed that tissue cryptococcal polysaccharide was mostly localized in macrophages and microglia, and less frequently in reactive astrocytes and endothelial cells (45). Thus, *C. neoformans* infection could promote activation of CNS macrophages and glial cells and potentially enhance cognitive impairment. Furthermore, HIV infection affects the integrity of the blood-brain barrier (81). Murine infection experiments demonstrated that monocytes may facilitate crossing of the blood-brain barrier by *C. neoformans* and thus contribute to fungal dissemination (13). Hence, it is conceivable that HIV infection further loosens the integrity and thus facilitates dissemination to this protected area. Consistent with this hypothesis is the finding that CME is more common in AIDS patients than in solid organ transplant patients with *C. neoformans* infection. Additional studies will be necessary to explore both of these hypotheses.

Lung Infection

Alveolar macrophages (AMs) comprise the initial host defense in the lungs against cryptococcosis and may arrest infection before dissemination occurs. Several studies have demonstrated that HIV-1 infection of AMs impairs their anticryptococcal activity (41). Uninfected AMs from normal donors demonstrated innate fungicidal activity against *C. neoformans*, whereas HIV-1-infected AMs exhibited significant reduction, or even loss, of fungicidal activity (41). This reduced antifungal activity was not due to any cytotoxic effect of HIV-1, impaired binding, or internalization of yeast by AMs but rather involved a defect of intracellular antimicrobial processing. In support of this finding, other studies indicated that gp120 decreases the internalization and fungistasis of *C. neoformans* by human AMs (85). Macrophage function and fungistatic activity declined as the disease progressed. In peripheral blood mononuclear cells, anticryptococcal activity was reduced because both oxidative and nonoxidative effector pathways were affected by coinfection with HIV (39). T cells of HIV-infected patients exhibited defects including defective granulysin production by $CD4^+$ T cells (89). Also, neutrophils from patients with late-stage HIV infection exhibited impaired IL-8 production in response to *C. neoformans* (56), which was partially restored by HAART.

Another factor that may affect HIV-infected patients' ability to mount a successful antifungal host response is the common abuse of methamphetamine. In vitro studies have shown that in macrophages this drug compromises major histocompatibility complex class II antigen processing by the endosomal-lysosomal pathway and antigen presentation. Most importantly, this drug exposure facilitates intracellular replication and inhibits intracellular killing of *C. neoformans* and thus affects pathogenesis in these patients (79) (Fig. 2).

CLINICAL MANIFESTATIONS OF CRYPTOCOCCOSIS IN PATIENTS WITH HIV INFECTION

Cryptococcosis in the setting of HIV infection occurs almost exclusively among patients with very advanced AIDS. The majority of patients will have CD4 lymphocyte counts that are below 100/μl (38, 66), and cryptococcal meningitis has been the AIDS-defining illness for up to 39% of patients (Table 1).

The most common manifestation of cryptococcosis in patients with HIV infection is meningoencephalitis, and CNS involvement is found in the vast majority of patients with AIDS in whom cryptococcal infection is diagnosed. There are some patients with AIDS who appear to have isolated cryptococcal pneumonia, fungemia, or a positive serum antigen titer, but these cases are so unusual that all such patients should be assumed to have simultaneous meningoencephalitis until proven otherwise with a lumbar puncture. The symptoms of meningoencephalitis in patients with AIDS usually evolve in a subacute fashion over a period of 1 to 2 weeks. Table 1 shows cumulative data on 575 patients enrolled in the two largest prospective treatment trials on AIDS-associated cryptococcal meningitis (76, 82). Both of these trials took place in the era before HAART was available, and the data are somewhat skewed based on the entry criteria into these two studies. All patients in these studies were required to have a positive CSF culture for *C. neoformans*, and patients presenting with coma or moderate to severe liver and kidney disease were excluded. The most common symptom in patients in these two trials was headache, which was found in 90% of participants (Table 1). Other symptoms, such as meningismus (42%), visual disturbance (29%), and cough (29%), were seen in a minority of patients. Anecdotal experience has shown that some rare patients with cryptococcal meningoencephalitis can present in a fulminant fashion with obtundation and rapid progression to death.

The initial physical exam for patients with AIDS-associated cryptococcal meningoencephalitis is typically

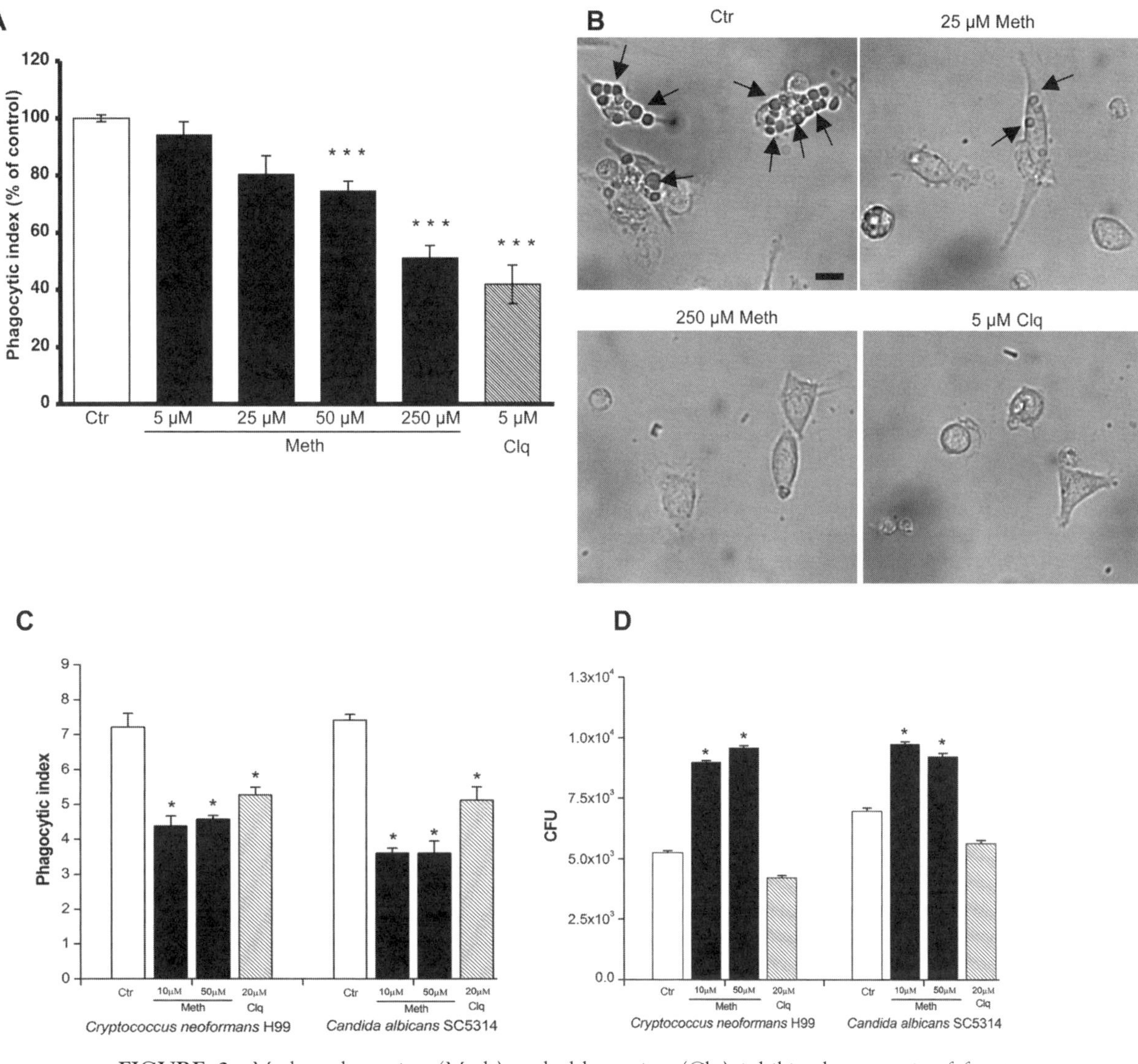

FIGURE 2 Methamphetamine (Meth) and chloroquine (Clq) inhibit phagocytosis of fungi. (A) Cells were incubated in Meth or Clq for 2 h, and then immunoglobulin G (IgG)-coated erythrocytes were added. Extracellular uningested IgG-coated erythrocytes were lysed and removed. Phagocytic index was quantified and inhibition is indicated as percentage of control. Data were collected from four to seven experiments (n = 300 cells; mean ± SEM; ***, $P < 0.0001$, two-tailed ANOVA). (B) Images of macrophage cells impaired in the phagocytosis of opsonized sheep erythrocytes after 2 h of Meth or Clq treatment. Arrows denote phagocytosed erythrocytes. Scale bar, 10 μm. (C, D) J774.16 cells were exposed to phosphate-buffered saline, Clq, or Meth for 2 h followed by incubation with *C. neoformans* or *Candida albicans*. The phagocytic indices were determined after 1 h or 30 min for *C. neoformans* and *C. albicans*, respectively (C). (Statistics: $n = 300$; mean ± SEM; *, $P < 0.05$, two-tailed ANOVA.) CFU after 24 h incubations (D). ($n = 300$ cells; mean ± SEM; *, $P < 0.05$, two-tailed ANOVA.) From reference 79.

significant only for the presence of fever in someone who appears chronically ill. This paucity of physical findings is in contrast to the very high burden of infection seen in patients with AIDS-associated cryptococcosis, and this disconnect is a reflection of the very poor immune response seen in this population. Fever was found in 76% of patients presenting with cryptococcal meningitis (Table 1), and other physical exam findings include diastolic hypertension, which can be a sign of increased intracranial pressure and has been associated with early mortality (26). Papilledema is uncommon, but it is an important finding to recognize since it is also a sign of increased intracranial pressure (37). Most of the patients will have normal mentation at their presentation to medical care (Table 1). Altered mentation, somnolence, and obtundation are poor prognostic signs when present (76, 82). A minority of patients will present with focal neurologic deficits. Some of the focal neurologic deficits are cranial neuropathies presumably related to inflammation of the basilar meninges, and hearing loss and visual changes can also be seen.

The diagnosis of cryptococcal meningitis is established by culturing the organism from CSF. Because patients can

TABLE 1 Characteristics of 575 patients presenting with AIDS-associated cryptococcal meningitis in two large treatment trials from 1988 to 1994

Characteristic	Percentage of patients
Cryptococcosis as AIDS-presenting illness	39%
Blood culture growing *C. neoformans*	7%
CSF with positive India ink exam	79%
Normal sensorium at presentation	84%
Headache at presentation	90%
Fever at presentation	76%
Meningismus at presentation	42%
Visual disturbance at presentation	29%
Cough at presentation	29%

present with subacute, nonspecific symptoms, a high degree of suspicion needs to be maintained in patients with known or suspected AIDS complaining of a headache. A lumbar puncture with measurement of opening pressure should be performed in all patients with suspected cryptococcal meningitis, provided that they do not have some other contraindication for this procedure. In addition to the routine measurements of CSF glucose, protein, and cell counts, the fluid should also be examined with India ink and sent for cryptococcal antigen and fungal cultures. The CSF glucose, protein, and cell counts are usually fairly unremarkable, and up to a third of patients with AIDS-associated cryptococcal meningitis will have a normal CSF profile (19). The CSF glucose levels are usually normal but can be slightly low (37), and the CSF protein levels are also usually normal, but can sometimes be slightly elevated. The CSF white blood cell counts are usually below 20 cells/μl with a lymphocyte predominance. Examination of the CSF with India ink reveals the characteristic encapsulated yeast forms in most patients, and the cumulative data from two treatment trials showed that 79% of patients had a positive India ink exam of CSF on presentation (Table 1). In fact, some patients will have more yeast forms in their India ink preparations than white blood cells. Again, this highlights the very high burden of infection in the face of an extremely weak immune response. The CSF cryptococcal antigen assay is very sensitive and specific for the diagnosis of meningitis. One study comparing various assays found that all of them performed well, with sensitivities ranging from 93 to 100% and specificities of 93 to 98% (4, 80). False-positive tests are usually of very low titer and can result from infection due to fungi from the genus *Trichosporon* or bacteria from the genera *Stomatococcus* and *Capnocytophaga* (11, 51, 86). In patients with AIDS-associated cryptococcal meningitis, the CSF cryptococcal antigen titers are usually quite high, with median values of 1:4,096 (82). The detection of cryptococcal antigen in the CSF can be an important adjunct to the diagnosis, especially in patients with negative India ink exams, since the test can be run soon after the lumbar puncture is performed, and a positive test can suggest the presence of meningitis well before the organism grows in culture.

There are three main prognostic factors in patients presenting with AIDS-associated cryptococcal meningitis. Abnormal mental status, CSF antigen titers >1:1,024, and CSF white blood cell counts <20/μl are all independently associated with increased mortality (76). Patients with increased intracranial pressure also have increased mortality, and this can be reflected by papilledema and/or diastolic hypertension (26).

Patients with AIDS-associated cryptococcal meningitis frequently also have extraneural involvement that reflects the disseminated nature of the infection. The most common site of extraneural infection has been the blood, with up to 57% of patients having fungemia at the time of presentation (Table 1). In some cases, the diagnosis of cryptococcal meningitis is only identified after first isolating *C. neoformans* in blood cultures that were drawn in the workup of fever. Cryptococci can also be cultured in some meningitis patients from sputum, urine, and the skin. When skin lesions due to *C. neoformans* are present, they frequently appear as small papules resembling molluscum contagiosum. However, cryptococcal skin involvement can also appear in many other forms including plaques, abscesses, purpura, and cellulitis (54).

Patients with AIDS-associated cryptococcal meningitis almost always have positive assays for cryptococcal antigen in the serum, and the sensitivity of a positive serum assay for the presence of meningitis is comparable to that obtained with a positive CSF antigen assay. It is important to stress that the serum assays should never be a substitute for lumbar puncture, but these can be helpful screening tests in patients at high risk for cryptococcal meningitis. A study from Cambodia evaluated 327 patients with AIDS who were prospectively screened for serum cryptococcal antigen (52). Assays were positive in 58 patients (18%), and further testing was performed in 52 of these patients, among whom 41 were found to have cryptococcal meningitis, 2 had pulmonary cryptococcosis without apparent CNS involvement, and 10 had isolated positive antigen titers with no evidence of infection in the CNS, blood, or respiratory tract. The patients with isolated positive serum antigen tests tended to have much lower titers as compared to the patients with meningitis (52). Another study from South Africa retrospectively examined blood from 707 patients with advanced HIV infection and found that 7% had a detectable serum cryptococcal antigen (43). Furthermore, the detection of cryptococcal antigenemia was 100% sensitive for predicting the subsequent diagnosis of cryptococcal meningitis within the next year (43).

Radiographic imaging of the brain using computed tomography or magnetic resonance should be performed in patients presenting with focal neurologic signs or symptoms. Mass lesions due to cryptococcosis are fairly unusual in patients with AIDS, and when these masses are detected in patients with cryptococcal meningitis, coincident diagnoses such as toxoplasmosis, lymphoma, and tuberculosis should be considered. A review of various published reports on magnetic resonance imaging in AIDS-associated cryptococcal meningitis reveals that mass lesions were detected in 14 of 54 patients (26%), but it is unclear if all of these can be attributed to cryptococcosis since a significant number of patients in these studies also had toxoplasmosis (12). Hydrocephalus can sometimes be seen, and often magnetic resonance scans are positive only for subtle dilatation of the Virchow-Robin spaces (12).

The treatment of AIDS-associated cryptococcal meningitis is usually divided into three stages: induction, consolidation, and maintenance. There are a variety of treatment guidelines that have been published, including those by the Infectious Diseases Society of America (75) and a consortium of the Centers for Disease Control and Prevention, National Institutes of Health, and the HIV Medicine Association. In general, there is agreement among the various treatment guidelines. The first large prospective study on the treatment of AIDS-associated cryptococcal meningitis compared low-dose intravenous amphotericin B (≥0.3 mg/kg/day) with fluconazole dosed at a relatively low dose of 200 mg/day as the primary therapy (76). Both of these regimens were similar in terms of the primary end point of having two consecutive sterile CSF cultures by the end of the 10-week treatment period, and there was no difference in the overall mortality rates in the two groups. However, there were two important findings from this study: treatment with amphotericin B resulted in faster sterilization of the CSF than fluconazole, and a higher proportion of patients treated with fluconazole died from cryptococcosis within the first 2 weeks.

These findings prompted a second large prospective treatment trial where all patients received a 2-week induction therapy with higher-dose amphotericin B (0.7 to 0.8 mg/kg/day) with or without flucytosine (100 mg/kg/day with renal adjustment) (82). After the induction therapy, the patients were randomized to receive 8 weeks of consolidation therapy with either fluconazole or itraconazole (both dosed at 400 mg/day), followed by indefinite maintenance therapy with the same agent dosed at 200 mg/day. There were no significant differences in outcome between the two treatment groups at the end of the consolidation therapy. However, patients assigned to itraconazole as the maintenance therapy had significantly higher relapse rates, and patients who had received flucytosine during induction therapy not only had significantly lower relapse rates but appeared to sterilize their CSF cultures faster than patients treated with amphotericin B alone. This study was also significant in that the overall mortality rate of 6% was much lower than what had been found in other trials. This improved survival was attributed to the higher doses of amphotericin B used as the induction therapy and to the improved management of increased intracranial pressure.

Current guidelines consider the combination of amphotericin B dosed at 0.7 mg/kg/day plus flucytosine dosed at 100 mg/kg/day in four divided doses as the preferred induction therapy. The induction therapy is given for 2 weeks, at which point most experts would recommend a repeat lumbar puncture. If the CSF cultures are sterile at this 2-week time point, then the patient can be changed to the consolidation therapy with fluconazole (400 to 800 mg/day) for 8 weeks. If the CSF cultures are still positive at the 2-week time point, then most experts recommend another 2 weeks of induction therapy followed by repeat CSF assessment. The lumbar puncture after 2 weeks of induction therapy is used as an important benchmark because clinical experience has shown that failure to sterilize the CSF after 14 days of therapy is associated with increased mortality (74). For the patients on consolidation therapy that have demonstrated steady clinical improvement after 8 weeks, the fluconazole dose can be decreased to maintenance therapy levels of 200 mg/day.

There has also been interest in examining regimens that use fluconazole instead of flucytosine as the induction therapy along with amphotericin B since fluconazole is much more easily obtainable in international settings such as sub-Saharan Africa and Southeast Asia. One randomized trial involving patients with HIV infection in Thailand performed serial quantitative cultures of CSF in patients treated with different induction regimens (6). The combination of amphotericin B (0.7 mg/kg/day) plus flucytosine (100 mg/kg/day) was found to be more rapidly fungicidal than amphotericin B alone, amphotericin B plus fluconazole (400 mg/day), and triple therapy with all three drugs. This trial did not evaluate clinical outcomes, but most people have inferred that sterilizing the CSF more rapidly will lead to increased acute survival in cryptococcal meningitis, and these data have helped strengthen the view that induction therapy with amphotericin B and flucytosine is the optimal combination. However, the limited availability of flucytosine in sub-Saharan Africa and Southeast Asia has prompted trials investigating amphotericin B and fluconazole combination therapies. One such trial compared amphotericin B alone (0.7 mg/kg/day) with amphotericin B combined with fluconazole in one of two dosages (400 mg or 800 mg) as the induction therapy in a randomized, phase II trial of patients with AIDS-associated cryptococcal meningitis in both Thailand and the United States (66). This trial demonstrated that the combination of amphotericin B and high-dose fluconazole was well tolerated and showed promise as an effective treatment strategy for patients living in areas where flucytosine is not readily available.

The most important adjuvant therapy during the induction phase of antifungal therapy is careful control of increased intracranial pressure. More than 75% of patients with cryptococcal meningitis will have opening pressures of >190 mm H_2O measured during initial lumbar puncture (37). Some of these patients are asymptomatic, whereas others will have severe headache, cranial nerve deficits, visual loss, and altered mentation. The mechanism for increased intracranial pressure is not completely understood and is more completely discussed in another chapter. However, it is clear that all patients with symptomatic increased intracranial pressure need to be aggressively managed with daily lumbar punctures to reduce the closing pressure to <200 mm H_2O (37, 75). Some patients may require placement of a lumbar drain to control increased pressures, but this is uncommon. There is no role for any systemic medications, such as acetazolamide, in the treatment of increased intracranial pressure in the setting of cryptococcal meningitis. Another adjuvant therapy that has been tried has been interferon gamma. A placebo-controlled trial randomly assigned patients to receive interferon gamma-1b at either 100 or 200 μg subcutaneously three times per week for 10 weeks or placebo (65). Treatment with interferon gamma resulted in more patients achieving sterile CSF cultures after 2 weeks of therapy, but the study was not sufficiently powered to demonstrate any survival benefit. The role of interferon gamma in cryptococcal meningitis is still unknown, and it cannot be recommended for use at this point.

The time to initiate antiretroviral therapy in patients with cryptococcal meningitis is unclear. Early initiation of therapy can be associated with higher rates of immune reconstitution syndrome, and this will be discussed in detail in another chapter. Many experts recommend that HAART be delayed for 8 weeks to try to avoid precipitating immune reconstitution, and it has not been clearly demonstrated that early HAART is beneficial in patients with cryptococcosis.

In the era before HAART was available, all patients who survived the initial stages of cryptococcal meningitis were kept on maintenance therapy for the rest of their lives out of the fear that infection would relapse without chronic suppression. HAART resulted in dramatic improvements in patients' immune responses, and it became clear that many patients were able to become cured of cryptococcosis after responding to HAART. The safety of withdrawing maintenance therapy was evaluated in a retrospective analysis of 100 patients who had stopped fluconazole suppression once their CD4 lymphocyte counts had increased to greater than 100 cells/μl on HAART (61). There were no cases of recurrent meningitis, and three patients who had negative serum cryptococcal antigen titers at the time of discontinuation had their antigen titers turn back to positive after stopping maintenance therapy (61). Most experts recommend that maintenance therapy can be safely discontinued in patients who have been on at least 6 months of maintenance fluconazole and have also demonstrated an adequate response to HAART as measured by an undetectable HIV serum viral load and an increase of the CD4 lymphocyte count to >100 cells/μl (75).

There are a few drug interactions when using fluconazole and antiretrovirals that are important to consider. The most important of these is the interaction of fluconazole and nevirapine. Coadministration of these results in a greater than 100% increase in nevirapine blood levels, and there are concerns that this could lead to hepatotoxicity (nevirapine package insert; U.S. Department of Health and Human Services and National Institutes of Health; The Living Document: Guidelines for the Use of Antiretroviral Agents in HIV-Infected Adults and Adolescents, retrieved March 18, 2008, from www.aidsinfo.nih.gov). Fluconazole also can increase the area under the curve of zidovudine when these are coadministered (78), but the increase is not thought to be clinically significant (zidovudine package insert). In general, fluconazole can also result in increased levels of protease inhibitors. The only protease inhibitor for which there are cautions in coadministration with fluconazole is tipranavir, and it is recommended that when these are given together, the dose of fluconazole not exceed 200 mg/day (U.S. Department of Health and Human Services and National Institutes of Health; The Living Document: Guidelines for the Use of Antiretroviral Agents in HIV-Infected Adults and Adolescents, retrieved March 18, 2008, from www.aidsinfo.nih.gov).

REFERENCES

1. **Abadi, J., and L. Pirofski.** 1999. Antibodies reactive with the cryptococcal capsular polysaccharide glucuronoxylomannan are present in sera from children with and without human immunodeficiency virus infection. *J. Infect. Dis.* **180:**915–919.
2. **Abou-Gabal, M., and M. Atia.** 1978. Study of the role of pigeons in the dissemination of *Cryptococcus neoformans* in nature. *Sabouraudia* **16:**63–68.
3. **Abu-Raddad, L. J., P. Patnaik, and J. G. Kublin.** 2006. Dual infection with HIV and malaria fuels the spread of both diseases in sub-Saharan Africa. *Science* **314:**1603–1606.
4. **Asawavichienjinda, T., C. Sitthi-Amorn, and V. Tanyanont.** 1999. Serum cryptococcal antigen: diagnostic value in the diagnosis of AIDS-related cryptococcal meningitis. *J. Med. Assoc. Thailand* **82:**65–71.
5. **Bicanic, T., G. Meintjes, R. Wood, M. Hayes, K. Rebe, L. G. Bekker, and T. Harrison.** 2007. Fungal burden, early fungicidal activity, and outcome in cryptococcal meningitis in antiretroviral-naive or antiretroviral-experienced patients treated with amphotericin B or fluconazole. *Clin. Infect. Dis.* **45:**76–80.
6. **Brouwer, A. E., P. Teparrukkul, S. Pinpraphaporn, R. A. Larsen, W. Chierakul, S. Peacock, N. Day, N. J. White, and T. S. Harrison.** 2005. Baseline correlation and comparative kinetics of cerebrospinal fluid colony-forming unit counts and antigen titers in cryptococcal meningitis. *J. Infect. Dis.* **192:**681–684.
7. **Casolari, C., T. Rossi, G. Baggio, A. Coppi, G. Zandomeneghi, A. I. Ruberto, C. Farina, G. Fabio, A. Zanca, and M. Castelli.** 2004. Interaction between saquinavir and antimycotic drugs on *C. albicans* and *C. neoformans* strains. *Pharmacol. Res.* **50:**605–610.
8. **Castella, G., M. L. Abarca, and F. J. Cabanes.** 2008. Cryptococcosis and pets. *Rev. Iberoam. Micol.* **25:**S19–S24 (In Spanish.).
9. **Cauda, R., E. Tacconelli, M. Tumbarello, G. Morace, F. De Bernardis, A. Torosantucci, and A. Cassone.** 1999. Role of protease inhibitors in preventing recurrent oral candidosis in patients with HIV infection: a prospective case-control study. *J. Acquir. Immune Defic. Syndr.* **21:**20–25.
10. **Cenci, E., D. Francisci, B. Belfiori, S. Pierucci, F. Baldelli, F. Bistoni, and A. Vecchiarelli.** 2008. Tipranavir exhibits different effects on opportunistic pathogenic fungi. *J. Infect.* **56:**58–64.
11. **Chanock, S., P. Toltzis, and C. Wilson.** 1993. Cross-reactivity between *Stomatococcus mucilaginosus* and latex agglutination for cryptococcal antigen. *Lancet* **342:**1119–1120.
12. **Charlier, C., F. Dromer, C. Leveque, L. Chartier, Y. S. Cordoliani, A. Fontanet, O. Launay, O. Ortholary, and the French Cryptococcosis Study Group.** 2008. Cryptococcal neuroradiological lesions correlate with severity during cryptococcal meningoencephalitis in HIV-positive patients in the HAART era. *PLoS ONE* **3:**e1950.
13. **Charlier, C., K. Nielsen, S. Daou, M. Brigitte, F. Chretien, and F. Dromer.** 2009. Evidence of a role for monocytes in dissemination and brain invasion by *Cryptococcus neoformans*. *Infect. Immun.* **77:**120–127.
14. **Chen, S., T. Sorrell, G. Nimmo, B. Speed, B. Currie, D. Ellis, D. Marriott, T. Pfeiffer, D. Parr, and K. Byth.** 2000. Epidemiology and host- and variety-dependent characteristics of infection due to *Cryptococcus neoformans* in Australia and New Zealand. Australasian Cryptococcal Study Group. *Clin. Infect. Dis.* **31:**499–508.
15. **Cressey, T. R., N. Plipat, F. Fregonese, and K. Chokephaibulkit.** 2007. Indinavir/ritonavir remains an important component of HAART for the treatment of HIV/AIDS, particularly in resource-limited settings. *Expert Opin. Drug Metab. Toxicol.* **3:**347–361.
16. **Currie, B., H. Sanati, A. S. Ibrahim, J. E. Edwards, Jr., A. Casadevall, and M. A. Ghannoum.** 1995. Sterol compositions and susceptibilities to amphotericin B of environmental *Cryptococcus neoformans* isolates are changed by murine passage. *Antimicrob. Agents Chemother.* **39:**1934–1937.
17. **Currie, B. P., and A. Casadevall.** 1994. Estimation of the prevalence of cryptococcal infection among patients infected with the human immunodeficiency virus in New York City. *Clin. Infect. Dis.* **19:**1029–1033.
18. **Currie, B. P., L. F. Freundlich, and A. Casadevall.** 1994. Restriction fragment length polymorphism analysis of *Cryptococcus neoformans* isolates from environmental (pigeon excreta) and clinical sources in New York City. *J. Clin. Microbiol.* **32:**1188–1192.
19. **Darras-Joly, C., S. Chevret, M. Wolff, S. Matheron, P. Longuet, E. Casalino, V. Joly, C. Chochillon, and J. Bedos.** 1996. *Cryptococcus neoformans* infection in France: epidemiologic features of and early prognostic parameters for 76 patients who were infected with human immunodeficiency virus. *Clin. Infect. Dis.* **23:**369–376.
20. **Dorrington, R., D. Bourne, D. Bradshaw, R. Laubscher, and I. Timaeus.** 2001. The impact of HIV/AIDS on adult mortality in South Africa. Technical report for the South African Medical Research Council.
21. **Dromer, F., S. Mathoulin, B. Dupont, and A. Laporte.** 1996. Epidemiology of cryptococcosis in France: a 9-year survey (1985–1993). French Cryptococcosis Study Group. *Clin. Infect. Dis.* **23:**82–90.
22. **Dromer, F., S. Mathoulin-Pelissier, A. Fontanet, O. Ronin, B. Dupont, and O. Lortholary for the French Cryptococcosis Study Group.** 2004. Epidemiology of HIV-associated cryptococcosis in France (1985–2001): comparison of the pre- and post-HAART eras. *AIDS* **18:**555–562.
23. **Dromer, F., S. Mathoulin-Pelissier, O. Launay, O. Lortholary, and the French Cryptococcosis Study Group.** 2007. Determinants of disease presentation and outcome

during cryptococcosis: the CryptoA/D study. *PLoS Med.* **4:**e21.
24. **Dromer, F., O. Ronin, and B. Dupont.** 1992. Isolation of *Cryptococcus neoformans* var. *gattii* from an Asian patient in France: evidence for dormant infection in healthy subjects. *J. Med. Vet. Mycol.* **30:**395–397.
25. **Ellis, D. H.** 1987. *Cryptococcus neoformans* var. *gattii* in Australia. *J. Clin. Microbiol.* **25:**430–431.
26. **Fan-Havard, P., E. Yamaguchi, S. Smith, and R. Eng.** 1992. Diastolic hypertension in AIDS patients with cryptococcal meningitis. *Am. J. Med.* **93:**347.
27. **Feldmesser, M., Y. Kress, P. Novikoff, and A. Casadevall.** 2000. *Cryptococcus neoformans* is a facultative intracellular pathogen in murine pulmonary infection. *Infect. Immun.* **68:**4225–4237.
28. **Franzot, S. P., J. S. Hamdan, B. P. Currie, and A. Casadevall.** 1997. Molecular epidemiology of *Cryptococcus neoformans* in Brazil and the United States: evidence for both local genetic differences and a global clonal population structure. *J. Clin. Microbiol.* **35:**2243–2251.
29. **Franzot, S. P., J. Mukherjee, R. Cherniak, L. C. Chen, J. S. Hamdan, and A. Casadevall.** 1998. Microevolution of a standard strain of *Cryptococcus neoformans* resulting in differences in virulence and other phenotypes. *Infect. Immun.* **66:**89–97.
30. **Fries, B. C., and A. Casadevall.** 1998. Serial isolates of *Cryptococcus neoformans* from patients with AIDS differ in virulence for mice. *J. Infect. Dis.* **178:**1761–1766.
31. **Fries, B. C., E. Cook, X. Wang, and A. Casadevall.** 2005. Effects of antifungal interventions on the outcome of experimental infections with phenotypic switch variants of *Cryptococcus neoformans*. *Antimicrob. Agents Chemother.* **49:**350–357.
32. **Garcia-Hermoso, D., G. Janbon, and F. Dromer.** 1999. Epidemiological evidence for dormant *Cryptococcus neoformans* infection. *J. Clin. Microbiol.* **37:**3204–3209.
33. **Garcia-Hermoso, D., S. Mathoulin-Pelissier, B. Couprie, O. Ronin, B. Dupont, and F. Dromer.** 1997. DNA typing suggests pigeon droppings as a source of pathogenic *Cryptococcus neoformans* serotype D. *J. Clin. Microbiol.* **35:**2683–2685.
34. **Goldman, D., S. C. Lee, and A. Casadevall.** 1994. Pathogenesis of pulmonary *Cryptococcus neoformans* infection in the rat. *Infect. Immun.* **62:**4755–4761.
35. **Gordon, S. B., A. L. Walsh, M. Chaponda, M. A. Gordon, D. Soko, M. Mbwvinji, M. E. Molyneux, and R. C. Read.** 2000. Bacterial meningitis in Malawian adults: pneumococcal disease is common, severe, and seasonal. *Clin. Infect. Dis.* **31:**53–57.
36. **Goulder, P. J., and D. I. Watkins.** 2008. Impact of MHC class I diversity on immune control of immunodeficiency virus replication. *Nat. Rev. Immunol.* **8:**619–630.
37. **Graybill, J. R., J. Sobel, M. Saag, C. van Der Horst, W. Powderly, G. Cloud, L. Riser, R. Hamill, and W. Dismukes.** 2000. Diagnosis and management of increased intracranial pressure in patients with AIDS and cryptococcal meningitis. The NIAID Mycoses Study Group and AIDS Cooperative Treatment Groups. *Clin. Infect. Dis.* **30:**47–54.
38. **Hajjeh, R. A., L. A. Conn, D. S. Stephens, W. Baughman, R. Hamill, E. Graviss, P. G. Pappas, C. Thomas, A. Reingold, G. Rothrock, L. C. Hutwagner, A. Schuchat, M. E. Brandt, and R. W. Pinner.** 1999. Cryptococcosis: population-based multistate active surveillance and risk factors in human immunodeficiency virus-infected persons. Cryptococcal Active Surveillance Group. *J. Infect. Dis.* **179:**449–454.
39. **Harrison, T. S., and S. M. Levitz.** 1997. Mechanisms of impaired anticryptococcal activity of monocytes from donors infected with human immunodeficiency virus. *J. Infect. Dis.* **176:**537–540.
40. **Harrison, T. S., S. Nong, and S. M. Levitz.** 1997. Induction of human immunodeficiency virus type 1 expression in monocytic cells by *Cryptococcus neoformans* and *Candida albicans*. *J. Infect. Dis.* **176:**485–491.
41. **Ieong, M. H., C. C. Reardon, S. M. Levitz, and H. Kornfeld.** 2000. Human immunodeficiency virus type 1 infection of alveolar macrophages impairs their innate fungicidal activity. *Am. J. Respir. Crit. Care Med.* **162:**966–970.
42. **Jain, N., B. L. Wickes, S. M. Keller, J. Fu, A. Casadevall, P. Jain, M. A. Ragan, U. Banerjee, and B. C. Fries.** 2005. Molecular epidemiology of clinical *Cryptococcus neoformans* strains from India. *J. Clin. Microbiol.* **43:**5733–5742.
43. **Jarvis, J. N., S. D. Lawn, M. Vogt, N. Bangani, R. Wood, and T. S. Harrison.** 2009. Screening for cryptococcal antigenemia in patients accessing an antiretroviral treatment program in South Africa. *Clin. Infect. Dis.* **48:**856–862.
44. **Kaur, A., N. Kassis, C. L. Hale, M. Simon, M. Elliott, A. Gomez-Yafal, J. D. Lifson, R. C. Desrosiers, F. Wang, P. Barry, M. Mach, and R. P. Johnson.** 2003. Direct relationship between suppression of virus-specific immunity and emergence of cytomegalovirus disease in simian AIDS. *J. Virol.* **77:**5749–5758.
45. **Lee, S. C., A. Casadevall, and D. W. Dickson.** 1996. Immunohistochemical localization of capsular polysaccharide antigen in the central nervous system cells in cryptococcal meningoencephalitis. *Am. J. Pathol.* **148:**1267–1274.
46. **Liappis, A. P., V. L. Kan, N. C. Richman, B. Yoon, B. Wong, and G. L. Simon.** 2008. Mannitol and inflammatory markers in the cerebral spinal fluid of HIV-infected patients with cryptococcal meningitis. *Eur. J. Clin. Microbiol. Infect. Dis.* **27:**477–479.
47. **Litvintseva, A. P., L. Kestenbaum, R. Vilgalys, and T. G. Mitchell.** 2005. Comparative analysis of environmental and clinical populations of *Cryptococcus neoformans*. *J. Clin. Microbiol.* **43:**556–564.
48. **Litvintseva, A. P., X. Lin, I. Templeton, J. Heitman, and T. G. Mitchell.** 2007. Many globally isolated AD hybrid strains of *Cryptococcus neoformans* originated in Africa. *PLoS Pathog.* **3:**e114.
49. **Lortholary, O., F. Dromer, S. Mathoulin-Pelissier, C. Fitting, L. Improvisi, J. M. Cavaillon, B. Dupont, and the French Cryptococcosis Study Group.** 2001. Immune mediators in cerebrospinal fluid during cryptococcosis are influenced by meningeal involvement and human immunodeficiency virus serostatus. *J. Infect. Dis.* **183:**294–302.
50. **McCarthy, K., H. H. Crewe-Brown, M. R. B. Maloba, and R. Hajjeh.** 2003. The burden of cryptococcosis in Gauteng: results of population-based active surveillance: 2002–2003. *Comm. Dis. Surv. Bull.* **Nov.:**10–12.
51. **McManus, E., and J. Jones.** 1985. Detection of a *Trichosporon beigelii* antigen cross-reactive with *Cryptococcus neoformans* capsular polysaccharides in serum from a patient with disseminated *Trichosporon* infection. *J. Clin. Microbiol.* **21:**681.
52. **Micol, R., O. Lortholary, B. Sar, D. Laureillard, C. Ngeth, J.-P. Dousset, H. Chanroeun, L. Ferradini, P. J. Guerin, F. Dromer, and A. Fontanet.** 2007. Prevalence, determinants of positivity, and clinical utility of cryptococcal antigenemia in Cambodian HIV-infected patients. *J. Acquir. Immune Defic. Syndr.* **45:**555–559.
53. **Mirza, S. A., M. Phelan, D. Rimland, E. Graviss, R. Hamill, M. E. Brandt, T. Gardner, M. Sattah, G. P. de Leon, W. Baughman, and R. A. Hajjeh.** 2003. The changing epidemiology of cryptococcosis: an update from population-based active surveillance in 2 large metropolitan areas, 1992–2000. *Clin. Infect. Dis.* **36:**789–794.

54. **Mitchell, T., and J. Perfect.** 1995. Cryptococcosis in the era of AIDS: 100 years after the discovery of *Cryptococcus neoformans*. *Clin. Microbiol. Rev.* **8:**515–548.
55. **Molez, J.-F.** 1998. The historical question of acquired immunodeficiency syndrome in the 1960s in the Congo River Basin area in relation to cryptococcal meningitis. *Am. J. Trop. Med. Hyg.* **58:**273–276.
56. **Monari, C., A. Casadevall, D. Pietrella, F. Bistoni, and A. Vecchiarelli.** 1999. Neutrophils from patients with advanced human immunodeficiency virus infection have impaired complement receptor function and preserved Fc-gamma receptor function. *J. Infect. Dis.* **180:**1542–1549.
57. **Monari, C., E. Pericolini, G. Bistoni, E. Cenci, F. Bistoni, and A. Vecchiarelli.** 2005. Influence of indinavir on virulence and growth of *Cryptococcus neoformans*. *J. Infect. Dis.* **191:**307–311.
58. **Moore, R. D.** 2000. Cost effectiveness of combination HIV therapy: 3 years later. *Pharmacoeconomics* **17:**325–330.
59. **Morgan, J., K. M. McCarthy, S. Gould, K. Fan, B. Arthington-Skaggs, N. Iqbal, K. Stamey, R. A. Hajjeh, and M. E. Brandt.** 2006. *Cryptococcus gattii* infection: characteristics and epidemiology of cases identified in a South African province with high HIV seroprevalence, 2002–2004. *Clin. Infect. Dis.* **43:**1077–1080.
60. **Morris, L., E. Silber, P. Sonnenberg, S. Eintracht, S. Nyoka, S. F. Lyons, D. Saffer, H. Koornhof, and D. J. Martin.** 1998. High human immunodeficiency virus type 1 RNA load in the cerebrospinal fluid from patients with lymphocytic meningitis. *J. Infect. Dis.* **177:**473–476.
61. **Mussini, C., P. Pezzotti, and J. Miro.** 2004. Discontinuation of maintenance therapy for cryptococcal meningitis in patients with AIDS treated with highly active antiretroviral therapy: an international observational study. *Clin. Infect. Dis.* **38:**565.
62. **Nosanchuk, J. D., S. Shoham, B. C. Fries, D. S. Shapiro, S. M. Levitz, and A. Casadevall.** 2000. Evidence of zoonotic transmission of *Cryptococcus neoformans* from a pet cockatoo to an immunocompromised patient. *Ann. Intern. Med.* **132:**205–208.
63. **Orendi, J. M., H. S. Nottet, M. R. Visser, A. F. Verheul, H. Snippe, and J. Verhoef.** 1994. Enhancement of HIV-1 replication in peripheral blood mononuclear cells by *Cryptococcus neoformans* is monocyte-dependent but tumour necrosis factor-independent. *AIDS* **8:**423–429.
64. **Paiardini, M., I. Frank, I. Pandrea, C. Apetrei, and G. Silvestri.** 2008. Mucosal immune dysfunction in AIDS pathogenesis. *AIDS Rev.* **10:**36–46.
65. **Pappas, P. G., B. Bustamante, E. Ticona, R. J. Hamill, P. C. Johnson, A. Reboli, J. Aberg, R. Hasbun, and H. H. Hsu.** 2004. Recombinant interferon-gamma 1b as adjunctive therapy for AIDS-related acute cryptococcal meningitis. *J. Infect. Dis.* **189:**2185.
66. **Pappas, P. G., P. Chetchotisakd, R. A. Larsen, W. Manosuthi, M. I. Morris, T. Anekthananon, S. Sungkanuparph, K. Supparatpinyo, T. L. Nolen, L. O. Zimmer, A. S. Kendrick, P. Johnson, J. D. Sobel, and S. G. Filler.** 2009. A phase II randomized trial of amphotericin B alone or combined with fluconazole in the treatment of HIV-associated cryptococcal meningitis. *Clin. Infect. Dis.* **48:**1775–1783.
67. **Park, B. J., K. A. Wannemuehler, B. J. Marston, N. Govender, P. G. Pappas, and T. M. Chiller.** 2009. Estimation of the current global burden of cryptococcal meningitis among persons living with HIV/AIDS. *AIDS* **23:**525–530.
68. **Pericolini, E., E. Cenci, E. Gabrielli, S. Perito, P. Mosci, F. Bistoni, and A. Vecchiarelli.** 2008. Indinavir influences biological function of dendritic cells and stimulates antifungal immunity. *J. Leukoc. Biol.* **83:**1286–1294.
69. **Pericolini, E., E. Cenci, C. Monari, M. De Jesus, F. Bistoni, A. Casadevall, and A. Vecchiarelli.** 2006. *Cryptococcus neoformans* capsular polysaccharide component galactoxylomannan induces apoptosis of human T-cells through activation of caspase-8. *Cell. Microbiol.* **8:**267–275.
70. **Pericolini, E., E. Cenci, C. Monari, S. Perito, P. Mosci, G. Bistoni, and A. Vecchiarelli.** 2006. Indinavir-treated *Cryptococcus neoformans* promotes an efficient antifungal immune response in immunosuppressed hosts. *Med. Mycol.* **44:**119–126.
71. **Pettoello-Mantovani, M., A. Casadevall, and H. Goldstein.** 1993. The presence of cryptococcal capsular polysaccharide increases the sensitivity of HIV-1 coculture in children. *Ann. NY Acad. Sci.* **693:**281–283.
72. **Pettoello-Mantovani, M., A. Casadevall, T. R. Kollmann, A. Rubinstein, and H. Goldstein.** 1992. Enhancement of HIV-1 infection by the capsular polysaccharide of *Cryptococcus neoformans*. *Lancet* **339:**21–23.
73. **Pettoello-Mantovani, M., A. Casadevall, P. Smarnworawong, and H. Goldstein.** 1994. Enhancement of HIV type 1 infectivity in vitro by capsular polysaccharide of *Cryptococcus neoformans* and *Haemophilus influenzae*. *AIDS Res. Hum. Retrovir.* **10:**1079–1087.
74. **Robinson, P. A., M. Bauer, M. A. E. Leal, S. G. Evans, P. D. Holton, D. M. Diamond, J. M. Leedom, and R. A. Larsen.** 1999. Early mycological treatment failure in AIDS-associated cryptococcal meningitis. *Clin. Infect. Dis.* **28:**82–92.
75. **Saag, M. S., R. J. Graybill, R. A. Larsen, P. G. Pappas, J. R. Perfect, W. G. Powderly, J. D. Sobel, and W. E. Dismukes.** 2000. Practice guidelines for the management of cryptococcal disease. Infectious Diseases Society of America. *Clin. Infect. Dis.* **30:**710–718.
76. **Saag, M. S., W. G. Powderly, G. A. Cloud, P. Robinson, M. H. Grieco, P. K. Sharkey, S. E. Thompson, A. M. Sugar, C. U. Tuazon, J. F. Fisher, N. Hyslop, J. M. Jacobson, R. Hafner, and W. E. Dismukes.** 1992. Comparison of amphotericin B with fluconazole in the treatment of acute AIDS-associated cryptococcal meningitis. *N. Engl. J. Med.* **326:**83–89.
77. **Saha, D. C., D. L. Goldman, X. Shao, A. Casadevall, S. Husain, A. P. Limaye, M. Lyon, J. Somani, K. Pursell, T. L. Pruett, and N. Singh.** 2007. Serologic evidence for reactivation of cryptococcosis in solid-organ transplant recipients. *Clin. Vaccine Immunol.* **14:**1550–1554.
78. **Sahai, J., K. Gallicano, A. Pakuts, and D. Cameron.** 1994. Effect of fluconazole on zidovudine pharmacokinetics in patients infected with human immunodeficiency virus. *J. Infect. Dis.* **169:**1103–1107.
79. **Talloczy, Z., J. Martinez, D. Joset, Y. Ray, A. Gacser, S. Toussi, N. Mizushima, J. Nosanchuk, H. Goldstein, J. Loike, D. Sulzer, and L. Santambrogio.** 2008. Methamphetamine inhibits antigen processing, presentation, and phagocytosis. *PLoS Pathog.* **4:**e28.
80. **Tanner, D., M. Weinstein, B. Fedorciw, K. Joho, J. Thorpe, and L. Reller.** 1994. Comparison of commercial kits for detection of cryptococcal antigen. *J. Clin. Microbiol.* **32:**1680–1684.
81. **Toborek, M., Y. W. Lee, G. Flora, H. Pu, I. E. Andras, E. Wylegala, B. Hennig, and A. Nath.** 2005. Mechanisms of the blood-brain barrier disruption in HIV-1 infection. *Cell. Mol. Neurobiol.* **25:**181–199.
82. **van der Horst, C. M., M. S. Saag, G. A. Cloud, R. J. Hamill, J. R. Graybill, J. D. Sobel, P. C. Johnson, C. U. Tuazon, T. Kerkering, B. L. Moskovitz, W. G. Powderly, and W. E. Dismukes.** 1997. Treatment of cryptococcal meningitis associated with the acquired immunodeficiency syndrome. National Institute of Allergy and Infectious

Diseases Mycoses Study Group and AIDS Clinical Trials Group. *N. Engl. J. Med.* **337:**15–21.

83. **Viviani, M. A., M. Cogliati, M. C. Esposto, K. Lemmer, K. Tintelnot, M. F. Colom Valiente, D. Swinne, A. Velegraki, and R. Velho.** 2006. Molecular analysis of 311 *Cryptococcus neoformans* isolates from a 30-month ECMM survey of cryptococcosis in Europe. *FEMS Yeast Res.* **6:**614–619.
84. **Vullo, V., C. M. Mastroianni, M. Lichtner, F. Mengoni, and S. Delia.** 1995. Increased cerebrospinal fluid levels of soluble receptors for tumour necrosis factor in HIV-infected patients with neurological diseases. *AIDS* **9:**1099–1100.
85. **Wagner, R. P., S. M. Levitz, A. Tabuni, and H. Kornfeld.** 1992. HIV-1 envelope protein (gp120) inhibits the activity of human bronchoalveolar macrophages against *Cryptococcus neoformans. Am. Rev. Respir. Dis.* **146:**1434–1438.
86. **Westerink, M., D. Amsterdam, and R. Petell.** 1987. Septicemia due to DF-2. Cause of a false-positive cryptococcal latex agglutination result. *Am. J. Med.* **83:**155.
87. **Wickes, B. L., M. E. Mayorga, U. Edman, and J. C. Edman.** 1996. Dimorphism and haploid fruiting in *Cryptococcus neoformans*: association with the alpha-mating type. *Proc. Natl. Acad. Sci. USA* **93:**7327–7331.
88. **Wyles, D. L.** 2008. Hepatitis virus coinfection in the Strategic Management of Antiretroviral Therapy (SMART) study: a marker for nonliver, non-opportunistic disease mortality. *Clin. Infect. Dis.* **47:**1476–1478.
89. **Zheng, C. F., G. J. Jones, M. Shi, J. C. Wiseman, K. J. Marr, B. M. Berenger, S. M. Huston, M. J. Gill, A. M. Krensky, P. Kubes, and C. H. Mody.** 2008. Late expression of granulysin by microbicidal CD4+ T cells requires PI3K- and STAT5-dependent expression of IL-2Rbeta that is defective in HIV-infected patients. *J. Immunol.* **180:**7221–7229.

Cryptococcus: From Human Pathogen to Model Yeast
Edited by J. Heitman et al.

39

Antifungal Trials: Progress, Approaches, New Targets, and Perspectives in Cryptococcosis

PETER G. PAPPAS

Among the invasive mycoses, cryptococcosis is one of the most studied entities in the realm of clinical mycology. In the decades preceding 1990, published studies of the treatment of cryptococcosis focused on non-HIV-infected patients, including individuals with recognizable underlying disorders associated with immune suppression, as well as otherwise healthy patients who had no evidence of an underlying condition (4, 23). Since 1990, all randomized and prospective trials examining the treatment of cryptococcosis have involved HIV-infected patients exclusively (7, 10, 11, 16, 19, 21, 31, 34, 35, 43, 44, 52, 64, 73, 75, 79, 81, 86, 97, 100, 103). Moreover, every prospective therapeutic trial for the treatment of cryptococcosis published to date has focused on central nervous system (CNS) cryptococcosis as the key criterion for enrollment. As such, there are no randomized studies specifically addressing patients with non-CNS cryptococcosis. The current recommendations for non-CNS cryptococcosis are largely extrapolated from data generated from studies of patients with CNS cryptococcosis, retrospective reviews, and small case series (69, 80).

Thus, there remain significant questions relating to the most effective therapeutic approach to patients with cryptococcosis. These questions are particularly relevant in the current environment, given the therapeutic potential of some of the newer antifungal agents and immunomodulators. The main challenge for clinical investigation is to explore the potential role of these agents in improving long-term outcomes among patients with cryptococcosis through the conduct of well-designed clinical trials. Specifically, these studies should focus on optimizing the use of currently available antifungal agents to determine the therapeutic potential for novel agents with new targets and/or mechanisms of action, to develop a deeper understanding of the influence of immunomodulation on response to antifungal therapy, to explore ways in which to favorably modulate the immune response among patients with clinically apparent infection, to better elucidate the nonpharmacologic management of CNS and pulmonary disease, and to prospectively study the treatment of non-CNS infections and infections in non-HIV-infected patients.

This chapter will briefly describe the history of treatment for CNS cryptococcosis and will discuss novel approaches to therapy with traditional agents, the newer antifungal agents, experimental agents, and the potential role of immunotherapy. Critical clinical needs and the development of strategic approaches to treatment and prevention of cryptococcal disease are also discussed.

THE TREATMENT OF CRYPTOCOCCOSIS: A BRIEF HISTORY

The first attempt to treat cryptococcal meningitis with medical therapy dates back to the mid-1920s, when a young boy was treated unsuccessfully with intrathecal rabbit immune sera (85). A potential role for immune equine sera was also suggested, but the efficacy of this intervention was never clearly determined (96). There remained some interest in administering human immunoglobulin with amphotericin B deoxycholate (AmB-d) for patients with CNS cryptococcosis (28, 29), but once AmB-d became available in the 1950s, the era of systemic antifungal therapy had arrived, and general interest in antibody therapy for cryptococcosis waned.

Despite the early successes with AmB-d, clinical outcomes were imperfect (2). Treatment was difficult to administer and frequently toxic. Following an episode of CNS cryptococcosis, there were frequent therapeutic failures resulting in significant complications including blindness, deafness, cerebral infarctions, significant neurologic impairment, and death (8, 14, 24, 42, 66, 77). The availability of 5-flucytosine (1973) provided the first opportunity to explore combination antifungal therapy with AmB-d. The availability of fluconazole (1990), itraconazole (1992), the lipid formulations of amphotericin B (1995–1996), and the expanded spectrum triazoles (2000s) provided additional options in the antifungal armamentarium (34, 105). Subsequently, there have been prospective clinical trials examining each of these agents including AmB-d (4, 23, 75), lipid formulations of amphotericin B (3, 34), flucytosine (4, 11, 23, 43), and fluconazole (81, 100).

Peter G. Pappas, Division of Infectious Diseases, University of Alabama at Birmingham, Birmingham, AL 35294-0006.

The modern era of comparative therapeutic trials for the treatment of invasive mycoses was ushered in with the two classic studies of Bennett et al. (4) and Dismukes et al. (23). Both of these studies focused on treatment of cryptococcal meningitis in a heterogeneous group of patients, the majority of whom were HIV-negative. The focus of these studies was to determine the effectiveness of combination therapy with AmB-d plus flucytosine compared to AmB-d alone and to determine the appropriate length of combination therapy. The Bennett study demonstrated that AmB-d plus flucytosine administered for 6 weeks was associated with more favorable outcomes than AmB-d alone administered for 10 weeks (4). The drawbacks to this study included relative underdosing of amphotericin B (0.3 to 0.4 mg/kg daily), its open-label design, and the relatively small numbers of patients. However, the study was revolutionary in that it demonstrated that a study of combination antifungal therapy for an uncommon infection could be successfully conducted utilizing a multicenter and randomized design. Moreover, the results of this study became the standard to which other studies of cryptococcosis would be compared.

Utilizing a similar approach in study design, the Dismukes study was published several years later and attempted to define the optimal length of combined therapy with AmB-d plus flucytosine by comparing 4 to 6 weeks of treatment among patients with cryptococcal meningitis (23). The study demonstrated that 4 weeks of combination therapy was adequate for most nonimmunocompromised patients without underlying disease or neurologic complications. The two regimens were associated with relatively low mortality and relapse rates (23). The drawbacks of this study included relatively small numbers of evaluable patients, underdosing of AmB-d (0.3 mg/kg daily plus flucytosine 150 mg/kg daily), and the heterogeneity of the patient population (about two-thirds non-HIV-immunocompromised patients, including diabetics, two HIV-positive patients, and the remainder otherwise healthy patients). The study was instrumental in defining the minimum length of therapy for different categories of patients with CNS cryptococcosis based on underlying disease and clinical/laboratory findings at baseline, and in documenting the propensity of transplant recipients and other immunocompromised patients to develop relapsing disease following short courses (4 weeks) of combination antifungal therapy.

In 1990, the availability of fluconazole, a triazole antifungal with excellent in vitro activity against *Cryptococcus neoformans*, made possible a series of classic studies comparing different strategies for the treatment of HIV-associated cryptococcal meningitis (44, 52, 75, 81). Some early studies demonstrated slower rates of response to fluconazole monotherapy and gave way to studies of short-course "induction" therapy with AmB-d alone or AmB-d plus flucytosine, followed by "consolidation" therapy with fluconazole (100). In aggregate, these studies demonstrated an important role for fluconazole as a step-down therapy following approximately 2 weeks of initial therapy with AmB-d with or without flucytosine. Subsequent observations suggested the safety of discontinuing fluconazole suppression among patients who have experienced immune reconstitution following initiation of antiretroviral therapy (1, 59). Based on much more limited data, the role of itraconazole, another triazole with excellent in vitro activity, versus *C. neoformans* was generally limited to salvage therapy in the management of cryptococcosis (19, 21, 79). Since the completion of these initial trials, subsequent trials have focused on the creative use of existing agents at varying doses and/or combinations in an effort to maximize the therapeutic benefit without causing unacceptable toxicities (5–7, 11, 25, 48, 65, 74, 78, 83) and to most effectively manage complications of immunologic reconstitution (47, 49, 50). Many of these trials have been generated from developing countries where the burden of HIV-associated cryptococcosis and the need for innovative therapies is the greatest.

One of the unintended results of the focus on HIV-associated cryptococcosis over the past two decades has been the relative lack of attention on complicated cryptococcal infections in other groups of patients, leading to the misconception that cryptococcal disease has all but disappeared from the non-HIV-infected population. As demonstrated in several recent multicenter surveys, non-HIV-associated cryptococcosis continues to be an important complication in non-HIV-immunocompromised patients, especially among organ transplant recipients, and warrants much more attention (24, 36, 66, 87, 95, 102, 106).

NEWER ANTIFUNGAL AGENTS

There are several newer antifungal agents that could offer significant advantages over more traditional agents. Specifically, three new triazoles, voriconazole, posaconazole, and isavuconazole, demonstrate excellent in vitro activity against most strains of clinically relevant *Cryptococcus* species (70, 99). There are few experimental or clinical data to suggest significant advantages in vivo or in clinical outcomes. However, each of these agents possesses unique properties that should generate significant interest toward exploring their potential as first-line or salvage agents.

Voriconazole is an expanded-spectrum triazole with excellent in vitro activity versus *C. neoformans* and *Cryptococcus gattii* (70, 101). It also has favorable pharmacokinetics, can be given by parenteral or oral administration, and is dosed twice daily (38). Similar to fluconazole, it achieves excellent absorption (96%) following oral administration and excellent penetration into the cerebrospinal fluid (CSF) (>50% of serum levels) (38). There is limited clinical experience with voriconazole in treatment of cryptococcosis, mostly pertaining to salvage situations among patients who have failed to respond to conventional therapy. In one study, voriconazole demonstrated a favorable response (cured, improved, or stable) in more than 80% of patients (68). There is no prospective randomized study evaluating voriconazole for primary therapy in cryptococcosis, and thus its role in this disorder is largely undetermined. However, based on its in vitro activity and favorable pharmacokinetic profile, it would be a reasonable candidate to be explored among patients with HIV-associated cryptococcosis as a first- or second-line agent in regions where AmB is unavailable or contraindicated. It may also play an important role as a salvage agent for those who have failed prior therapy. The disadvantages of voriconazole include its challenging safety profile (especially photosensitivity rash and elevated liver function tests), unpredictable pharmacokinetics, significant drug-drug interactions, and expense (38). The compound is unavailable and unaffordable in much of the undeveloped world, thus limiting its broader development.

Posaconazole, an expanded-spectrum triazole that is currently only available as an oral suspension, is administered two to four times daily and demonstrates excellent in vitro activity against *Cryptococcus* species (99). The clinical experience with posaconazole in cryptococcosis is limited to a small number of patients with disease refractory to or intolerant of standard antifungal therapy. Among these patients, posaconazole has been associated with a favorable outcome (cured, improved, or stable) in approximately 70% (72, 76). Posaconazole has a more favorable safety profile when compared to voriconazole and probably fewer significant drug-drug interactions (76). However, similar to voriconazole, it has unpredictable pharmacokinetics and is unavailable and/or unaffordable in most of the developing world. Administered orally, posaconazole does not achieve meaningful CSF concentrations despite demonstrated in vivo activity in animals and humans with CNS cryptococcosis. As an oral agent with a very favorable toxicity profile and excellent in vitro activity, this compound could serve as an important primary or salvage therapy for cryptococcosis, but to date no prospective studies to treat cryptococcosis have been performed or are planned.

There are no clinical data concerning isavuconazole, an experimental expanded-spectrum triazole with excellent in vitro activity against *Cryptococcus* species (99). The compound has an antifungal spectrum that is similar to voriconazole and posaconazole, but it has superior pharmacokinetics to both agents: it is given by parenteral or oral administration once daily and has a favorable safety profile and possibly fewer drug-drug interactions than voriconazole. Data on CSF penetration are unavailable at this time. To date, the bulk of clinical experience with isavuconazole has been among patients with invasive aspergillosis and invasive candidiasis. Because of its acceptable safety profile, limited drug interactions, and favorable pharmacokinetics, this compound could be strongly considered for study as a first- or second-line agent for HIV- or transplant-associated cryptococcosis.

IMMUNOTHERAPY

Factors influencing immune function are an important focus of therapy among patients with cryptococcosis. Smaller studies examining attempts to enhance immune function through the administration of cytokines and monoclonal antibodies have produced mixed results (45, 64). An alternative approach has been based on the hypothesis that adjunctive glucocorticosteroids might have an important role in abrogating the inflammatory response among otherwise normal hosts, perhaps leading to fewer neurologic sequelae and better overall outcomes (27). Finally, the influence of the calcineurin inhibitors, cyclosporine and tacrolimus, on fungal cell growth through the fungal calcineurin pathway not only has led to a better understanding of their influence on host response and natural history of cryptococcosis in transplant recipients, but also has provided insights into their potential role as adjuvant antifungal agents (18, 41, 54, 55, 61, 91).

Antibody Therapy

Passive immunotherapy has been advocated as a potentially important adjunct to conventional antifungal therapy in the management of invasive mycoses (13, 26, 96). Interest in specific "serum therapy" for fungal pathogens is longstanding, and the use of "hyperimmune" equine serum was practiced in the 1930s as part of recommended therapy for cryptococcosis (96). More recently, Gordon and Casadevall reported three patients with acute cryptococcal meningitis who were treated with a combination of rabbit antibody to *C. neoformans* and AmB-d (30). The antibody was generally well tolerated, and two patients converted their cryptococcal antigen from positive to negative during therapy. These observations, in part, laid the foundation for the current interest in antibody therapy directed toward *C. neoformans*.

The hypothetical value of cryptococcal antigen removal from the infected host relates to its function as a virulence factor: cryptococcal antigen is known to inhibit leukocyte migration, promote cerebral edema, promote cytokine dysregulation, induce T-lymphocyte secretion of immunosuppressive molecules, and induce the shedding of L-selectin. Thus, its removal by means of monoclonal antibody therapy would be potentially beneficial to the host (46, 57).

In experimental studies, monoclonal anticryptococcal antibodies demonstrate enhancement of opsonization and increased host killing of these organisms (12, 32, 56–58). In a murine model of CNS cryptococcosis, monoclonal antibodies to cryptococcal antigen have been administered with encouraging results. Reduced tissue burden of organisms, enhanced granuloma formation, and prolonged survival have been demonstrated. Moreover, anticapsular antibodies enhance antifungal-mediated killing of *C. neoformans*.

Based on these encouraging data in animal studies, a phase I dose-escalation clinical safety study was performed with 20 HIV-positive patients with a history of successfully treated cryptococcal meningitis (45). In this study, patients received between 0.01 and 2.0 mg/kg of a murine monoclonal anticryptococcal antibody (MAb 18B7) as a single intravenous dose. Most patients tolerated lower doses well; at doses exceeding 0.5 mg/kg, patients more often experienced back pain, myalgias, nausea, and headache. Two patients developed antimurine antibodies without influencing MAb 18B7 serum levels. One patient with preexisting murine antibodies had no measurable levels in serum, and no patients had measurable CSF MAb 18B7 levels. However, there was an important trend toward decreasing serum cryptococcal antigen titers among patients receiving higher doses (1 to 2 mg/kg) of MAb 18B7 (45).

The clinical ramifications of this study are potentially significant, but further development of this compound was stopped due to concerns regarding toxicity and technical difficulty in manufacturing. Despite its theoretic potential, at present there are no other clinical initiatives focused on developing newer, more effective or safer monoclonal antibodies. Thus, despite abundant experimental data suggesting benefit, the future of monoclonal antibodies as adjunctive therapy for serious cryptococcal infections remains uncertain as this time.

Cytokines (IFN-γ and IL-12)

The proinflammatory cytokines interferon gamma (IFN-γ) and interleukin-12 (IL-12) are important mediators of the Th-1 response. IL-12 is a critical proinflammatory cytokine produced by stimulated macrophages and is the key cytokine that promotes IFN-γ production by CD4 lymphocytes, natural killer cells, and CD8 lymphocytes (71). IFN-γ activates NK cells, enhances neutrophil function and survival, stimulates production of other cytokines such as tumor necrosis factor-α, and stimulates granuloma formation. IFN-γ also inhibits the activation of CD4-positive Th-2 cells, and in turn inhibits the production of IL-4, IL-10, and other anti-inflammatory cytokines.

Abundant in vitro data demonstrate the enhancement of antiphagocytic activity by IFN-γ-stimulated macrophages against *C. neoformans* (92, 93). Moreover, in vivo studies have demonstrated improved survival, decreased tissue burden, and enhanced anticryptococcal activity in animals experimentally infected with *C. neoformans* that have been treated with IFN-γ alone or in combination with AmB-d (36, 49, 51).

There are relatively few data for the treatment of human mycoses with IFN-γ and virtually no data for IL-12 among humans with active infections. IFN-γ is approved for use in humans as a prophylactic measure for patients with chronic granulomatous disease for the prevention of infectious complications. IL-12 is currently not approved for use in humans. With the exception of one randomized study, most experience using IFN-γ for the treatment of invasive fungal disease has been limited to anecdotes and includes patients with refractory cryptococcosis, coccidioidomycosis, hepatosplenic candidiasis, aspergillosis, and zygomycosis. In the only randomized trial in the treatment of cryptococcosis, 75 patients with HIV-associated CNS cryptococcosis were randomized to receive 100 μg of IFN-γ, 200 μg of IFN-γ, or placebo subcutaneously three times weekly for 10 weeks in conjunction with conventional antifungal therapy with AmB-d (0.7 mg/kg for 14 days) followed by fluconazole (400 mg daily for 8 weeks) (64). IFN-γ was well tolerated by most patients throughout the study period. The results of this phase II trial demonstrated an important trend toward more rapid clearance of *C. neoformans* from the CSF in IFN-γ recipients compared to placebo recipients ($P = 0.064$) and significantly more rapid decline in CSF cryptococcal antigen. However, there was no clear clinical or survival benefit for IFN-γ recipients, and there were no clinical or mycological trends suggesting that there were better outcomes among those who received higher-dose IFN-γ (64).

In aggregate, these data suggest a potentially important role for IFN-γ in the management of cryptococcosis. Presently, IFN-γ is not approved for use in this condition nor is it part of a routine regimen for uncomplicated CNS cryptococcosis. It is occasionally administered as part of a salvage regimen among patients with refractory disease that is unresponsive to conventional therapy. Unfortunately, despite these data from experimental models, a modest clinical trial, and anecdotal reports of successful outcomes, there are currently no plans to move forward in the further development of this therapeutic approach because of limited interest on the part of the commercial sponsors and federal agencies.

Glucocorticosteroids

There are good data supporting the role of adjuvant corticosteroids in the early management of bacterial meningitis, and this has become a standard approach to patients with proven or suspected disease. However, the role of adjuvant glucocorticosteroids for CNS cryptococcosis has never been elucidated on the basis of clinical trials, and there remains debate as to whether glucocorticosteroids play any role in this setting. Some have suggested a potential role for adjuvant glucocorticosteroids in otherwise healthy patients with CNS cryptococcosis due to the high observed frequency of neurologic complications in this population (27), but to date there are no clinical data supporting this concept. Of course, high doses of corticosteroids are also associated with the appearance of CNS cryptococcosis, so it represents a "two-edged sword."

NEWER THERAPEUTIC TARGETS

Calcineurin Inhibitors

It has been established that the calcineurin signaling pathway is important in the pathogenesis of cryptococcal infections. Current evidence demonstrates the importance of the calcineurin pathway by showing that strains of *Cryptococcus* lacking the calcineurin A gene (CNA1 mutants) are generally avirulent (54, 55). Consistent with these observations, it has been shown that the calcineurin activator calmodulin is essential for *C. neoformans*. The downregulation of calmodulin is associated with impaired growth and reproduction. Thus, blocking this pathway through the use of calcineurin inhibitors, such as cyclosporine A and tacrolimus, could be a potentially useful approach to adjunctive therapy for serious cryptococcal infections. This hypothesis has been explored by several investigators, but there have been conflicting results from animal models of cryptococcal infection. In a rabbit model of cryptococcal meningitis, cyclosporine A exacerbated disease; however, in a murine model with the same cryptococcal strain, cyclosporine A reduced fungal burden in all tissues except for the brain, where drug penetration is poor. Subsequent in vitro studies have also demonstrated that both cyclosporine A and tacrolimus inhibit in vitro growth of *C. neoformans* at 37°C.

The concept of antifungal drug combinations that employ calcineurin inhibition for the treatment of *C. neoformans* infections has been tested in vitro, in animal models, and has been observed in retrospective human series. Tacrolimus demonstrates synergistic activity with several antifungals including bafilomycin A (an experimental agent), caspofungin, and fluconazole (20). The mechanisms of synergy vary with the compounds and are both dependent and independent of calcineurin function.

Clinical experience with calcineurin inhibitors in a retrospective series provides some evidence that patients receiving tacrolimus experience fewer CNS and possibly more skin and subcutaneous infections where a lower temperature might enhance the growth of *Cryptococcus* spp. (87). Moreover, overall mortality among transplant recipients who are also receiving a calcineurin inhibitor has been reported to be lower than for those receiving non-calcineurin inhibitor-based immunosuppression (41, 87). Furthermore, a prolonged exposure to calcineurin inhibitors did not select for drug-resistant *C. neoformans* isolates (9). These observations suggest that calcineurin inhibition through the use of existing or novel agents is a potentially important approach to therapy for cryptococcal infections. However, these infections continue to occur with relatively high frequency, especially among organ transplant recipients. Thus, a balance favoring more antifungal activity and less immunosuppression will be necessary for this class of agents to become useful in the therapy of this infection. The importance of the calcineurin pathway in the virulence of *C. neoformans* is an important step forward in understanding the most effective ways to utilize calcineurin inhibition as a therapeutic tool.

Efungumab

Efungumab is a monoclonal antibody directed at fungal heat shock protein 90 (Hsp90), and in combination with fluconazole and AmB-d, it demonstrates synergistic in vitro activity against *Candida* spp. and *C. neoformans*. Studies in a murine model of invasive candidiasis demonstrated promising synergistic activity between efungumab and AmB-d or fluconazole. The precise mechanism of action of efungumab is unclear. Recent observations from animal models of invasive candidiasis suggest that Hsp90 inhibition enhances the fungicidal effect of caspofungin and AmB-d. Some have also suggested that efungumab exerts its effect through interaction with the fungal calcineurin pathway, but these observations need to be confirmed and further elucidated with efungumab (90).

In humans, efungumab has been studied in a randomized, double-blind, controlled trial among patients with invasive candidiasis (62). In this study, the combination of AmB-d plus efungumab led to significantly better outcomes (84%

versus 48%) when compared to AmB-d alone. A much smaller unpublished study was conducted with efungumab among patients with HIV-associated cryptococcal meningitis. The study was terminated early due to slow patient accrual, and the results have not been published.

At present, it remains unclear whether efungumab will play a role in the treatment of *Cryptococcus* in the future; however, it should be an important consideration as new information concerning the function of Hsp90 and its relationship with the fungal calcineurin pathway are further elucidated.

Other Agents

Miltefosine is an alkyl phosphocholine compound that has in vitro and clinical activity against *Leishmania* spp. and *Trypanosoma cruzi*. It is orally available and is approved for clinical use in patients with these infections (17). Toxicities of the compound include nausea, vomiting, hepatotoxicity, and rash. In vitro testing against several fungal species, including selected *C. neoformans* and *C. gattii* isolates, has demonstrated excellent activity, with MIC_{90}s of 2.0 μg/ml (37). There are also data in a murine model of cryptococcosis suggesting that miltefosine was effective in delaying illness and death in experimentally infected mice (82, 104). The mechanism of antifungal action is unclear, but phosphocholines have structural similarities to the natural substrates of fungal phospholipase B, which is a virulence determinant in *Candida albicans* and *C. neoformans* infections (37). Miltefosine may interfere with cell wall synthesis or cell membrane biochemistry. Experience using miltefosine to treat invasive mycoses in humans is very limited. A recent report suggested efficacy for miltefosine as an adjunctive agent with voriconazole in an immunocompetent child with *Scedosporium prolificans* osteomyelitis who had failed prior antifungal therapy (40). The potential role for miltefosine as an adjunctive agent in cryptococcal disease should be explored, particularly in areas of the world where more traditional antifungal agents are unavailable or unaffordable.

There are several other compounds that are in very early development as potentially useful antifungal agents. Agents that block the beta-oxidation pathway (certain fatty acids), chitinase inhibitors, fatty acid synthase inhibitors, and homoserine transacetylase inhibitors are among the more promising agents.

SPECIFIC AREAS FOR CONSIDERATION

Clinical Trial Design

Cryptococcosis will continue to be an important complication among HIV-infected individuals, organ transplant recipients, other patients with predisposing underlying disorders, and otherwise healthy patients for the foreseeable future. To best address the unique challenges that each of these patient populations present will require a cadre of committed basic and clinical investigators from varied institutions who partner with industry, nonprofit organizations, government agencies, and biostatisticians to achieve clearly defined goals.

An appropriate design is critical to the proper statistical and clinical interpretation of any clinical trial. The randomized, controlled, double-blind design is the gold standard for trials that compare different treatments, and this has been the case for interventional trials in cryptococcosis. Moving forward, it will be important to consider ways to improve the clinical design in order to maximize the efficient conduct of these trials and to make their interpretation more clinically meaningful.

The use of alternative clinical end points represents a novel approach that could influence future trials by requiring fewer evaluable patients while providing insightful clinical data. Appropriately, most studies to date have used mortality and neurologic stability at a defined interval as primary clinical end points. However, few studies have prospectively explored long-term consequences of CNS cryptococcosis, such as visual and hearing abnormalities, learning disorders, and other measures of functional neurologic impairment as a primary outcome. The measurements can be painstaking and require careful follow-up, but these data could provide important insights and a more accurate assessment of the long-term consequences of this disease and the influence of specific therapies on these complications. In addition to overall mortality and "neurologic stability," these measurements could become an important factor in determining "noninferiority" or "superiority" of experimental versus standard treatment.

All future studies will also need to integrate culture data into the definition of success (5, 11, 65). Based on recent studies of CNS cryptococcosis from the United States, Thailand, and South Africa, CSF culture negativity has been used as the mycologic component of a composite measure of success. In addition, several recent studies have measured sequential quantitative CSF cultures and the rate of sterilization in an attempt to correlate with clinical outcome (5, 6, 11). While these studies have demonstrated significant differences among antifungal agents/regimens with regard to rates of CSF sterilization, this has not been shown to be a surrogate marker of clinical success as a primary end point. Thus, one goal of future studies will be to confirm the value of this technique as a predictor of clinical outcome.

HIV-Associated Cryptococcosis

Undoubtedly, the greatest therapeutic need for cryptococcosis is among HIV-infected individuals in the developing world. Globally, approximately one million cases and at least 500,000 deaths occur annually as a result of AIDS-associated cryptococcosis (67). The main obstacles to improving clinical outcomes and controlling disease burden globally relate to poor infrastructure, limited access to care, and limited investment in early diagnosis and intervention.

Due to cost, access to effective antifungal therapy for cryptococcosis in much of the world is generally limited to fluconazole. Access to AmB-d is very limited in many parts of the developing world and requires parenteral administration and hospitalization in most instances. Access to certain older agents such as flucytosine is sporadic outside of the United States, Canada, Australia, and Western Europe. Newer antifungal agents such as voriconazole and posaconazole are extremely limited outside of the developed world.

In many developing countries, particularly in sub-Saharan Africa, a major obstacle to care is the limited ability to adequately manage increased intracranial pressure (60, 94). For many providers, this is a matter of limited supplies for the lumbar puncture procedure, while for others there is a cultural barrier to acceptance for repeat (or a single) lumbar punctures to monitor and treat increased intracranial pressure. Furthermore, neurosurgical procedures for increased intracranial pressure are extremely uncommon in most of the developing world.

To reconcile these remediable problems will require investment from health care organizations, nonprofit groups, governments, and private corporations. For example, there has been very little investment in screening methods that could recognize early disease and encourage early intervention. Data from sub-Saharan Africa suggest that early screening for cryptococcosis using serum cryptococcal antigen assays is effective in identifying those patients who are at greatest risk of developing overt cryptococcal disease (22, 98). More aggressive screening combined with improved access to existing technologies could prove to be an extremely effective strategy for improving outcomes among this highly vulnerable population.

Cryptococcosis in Transplantation

In the developed world, cryptococcosis in transplantation represents a large and inadequately studied complication among transplant recipients. It is the third most common invasive fungal infection among organ transplant recipients (36, 63, 84). To date, there have been no prospective randomized trials examining the treatment of cryptococcosis in a transplant population. The best available data come from retrospective analyses and two prospective observational studies, including the Transplant-Associated Infection Surveillance Network (TRANSNET) (63) and the Collaborative Cryptococcal Study Group (95). The important therapeutic issue confronting those who provide care for transplant recipients pertains to the specifics of antifungal therapy, including the most appropriate agents for induction and step-down therapy and the duration of chronic suppression with antifungals. Important unanswered questions concerning immunosuppression relate to the role (if any) of the calcineurin inhibitors in the natural history and management of cryptococcosis and specific steps to avoid complications relating to immune reconstitution inflammatory syndrome following the reduction of immunosuppression in a patient with cryptococcosis (88, 89).

Cryptococcosis in Nontransplant, Non-HIV-Infected Patients

Nontransplant, non-HIV-infected patients constitute up to 50% of those presenting with cryptococcal disease at some centers (15, 33, 53, 66). Recent data suggest that postinfectious complications, including significant neurologic residua and mortality, occur at a higher rate among these patients when compared to HIV-infected patients and transplant recipients with cryptococcosis (27). The reasons for these disparities in outcomes remain unclear. Some have speculated that these poorer outcomes could relate to an overly responsive and potentially harmful innate immune response. The spectrum of inflammatory responses among these patients is broad but needs to be better defined in the host without obvious predisposing risks for cryptococcosis. The intensity and duration of induction therapy, step-down therapy, and an assessment for the need of chronic suppressive antifungal therapy should be evaluated. Furthermore, it will be important to establish the role, if any, of adjunctive therapy with immunomodulators such as glucocorticosteroids, IFN-γ, and IL-12 in the otherwise healthy host.

REFERENCES

1. **Aberg, J. A., R. W. Price, D. M. Heeren, and B. Bredt.** 2002. A pilot study of the discontinuation of antifungal therapy for disseminated cryptococcal disease in patients with acquired immunodeficiency syndrome, following immunologic response to antiretroviral therapy. *J. Infect. Dis.* **185:**1179–1182.
2. **Appelbaum, E., and S. Shtokalko.** 1957. *Cryptococcus* meningitis arrested with amphotericin B. *Ann. Intern. Med.* **47:**346–351.
3. **Baddour, L. M., J. R. Perfect, and L. Ostrosky-Zeichner.** 2005. Successful use of amphotericin B lipid complex in the treatment of cryptococcosis. *Clin. Infect. Dis.* **40**(Suppl. 6)**:**S409–S413.
4. **Bennett, J. E., W. E. Dismukes, R. J. Duma, G. Medoff, M. A. Sande, H. Gallis, J. Leonard, B. T. Fields, M. Bradshaw, H. Haywood, Z. A. McGee, T. R. Cate, C. G. Cobbs, J. F. Warner, and D. W. Alling.** 1979. A comparison of amphotericin B alone and combined with flucytosine in the treatment of cryptococcal meningitis. *N. Engl. J. Med.* **301:**126–131.
5. **Bicanic, T., G. Meintjes, R. Wood, M. Hayes, K. Rebe, L. G. Bekker, and T. Harrison.** 2007. Fungal burden, early fungicidal activity, and outcome in cryptococcal meningitis in antiretroviral-naive or antiretroviral-experienced patients treated with amphotericin B or fluconazole. *Clin. Infect. Dis.* **45:**76–80.
6. **Bicanic, T., C. Muzoora, A. E. Brouwer, G. Meintjes, N. Longley, K. Taseera, K. Rebe, A. Loyse, J. Jarvis, L. G. Bekker, R. Wood, D. Limmathurotsakul, W. Chierakul, K. Stepniewska, N. J. White, S. Jaffar, and T. S. Harrison.** 2009. Independent association between rate of clearance of infection and clinical outcome of HIV-associated cryptococcal meningitis: analysis of a combined cohort of 262 patients. *Clin. Infect. Dis.* **49:**702–709.
7. **Bicanic, T., R. Wood, R, G. Meintjes, K. Rebe, A. Brouwer, A. Loyse, L. G. Bekker, S. Jaffar, and T. Harrison.** 2008. High-dose amphotericin B with flucytosine for the treatment of cryptococcal meningitis in HIV-infected patients: a randomized trial. *Clin. Infect. Dis.* **47:**123–130.
8. **Blackie, J. D., G. Danta, T. Sorrell, and P. Collignon.** 1985. Ophthalmological complications of cryptococcal meningitis. *Clin. Exp. Neurol.* **21:**263–270.
9. **Blankenship, J. R., N. Singh, B. D. Alexander, and J. Heitman.** 2005. *Cryptococcus neoformans* isolates from transplant recipients are not selected for resistance to calcineurin inhibitors by current immunosuppressive regimens. *J. Clin. Microbiol.* **43:**464–467.
10. **Bozzette, S. A., R. A. Larsen, J. Chiu, M. A. Leal, J. Jacobsen, P. Rothman, P. Robinson, G. Gilbert, J. A. McCutchan, J. Tilles, J. M. Leedom, and D. D. Richman.** 1991. A placebo-controlled trial of maintenance therapy with fluconazole after treatment of cryptococcal meningitis in the acquired immunodeficiency syndrome. The California Collaborative Treatment Group. *N. Engl. J. Med.* **324:**580–584.
11. **Brouwer, A. E, A. Rajanuwong, W. Chierakul, G. E. Griffin, R. A. Larsen, N. J. White, and T. J. Harrison.** 2004. Combination antifungal therapies for HIV-associated cryptococcal meningitis: a randomised trial. *Lancet* **363:** 1764–1767.
12. **Casadevall, A., W. Cleare, M. Feldmesser, A. Glatman-Freedman, D. L. Goldman, T. R. Kozel, N. Lendvai, J. Mukherjee, L. Pirofski, J. Rivera, A. L. Rosas, M. D. Scharff, P. Valadon, K. Westin, and Z. Zhong.** 1998. Characterization of a murine monoclonal antibody to *Cryptococcus neoformans* polysaccharide that is a candidate for human therapeutic studies. *Antimicrob. Agents Chemother.* **42:**1437–1446.
13. **Casadevall, A., A. Cassone, F. Bistoni, J. E. Cutler, W. Magliani, J. W. Murphy, L. Polonelli, and L. Romani.** 1998. Antibody and /or cell-mediated immunity, protective mechanisms in fungal disease: an ongoing dilemma or an unnecessary dispute? *Med. Mycol.* **36**(Suppl 1)**:**95–105.
14. **Chayakulkeeree, M., and J. R. Perfect.** 2006. Cryptococcosis. *Infect. Dis. Clin. N. Am.* **20:**507–544, v–vi.
15. **Chen, S., T. Sorrell, G. Nimmo, B. Speed, B. Currie, D. Ellis, D. Marriott, T. Pfeiffer, D. Parr, and K. Byth.** 2000. Epidemiology and host- and variety-dependent char-

acteristics of infection due to *Cryptococcus neoformans* in Australia and New Zealand. The Australasian Cryptococcal Study Group. *Clin. Infect. Dis.* **31**:499–508.

16. **Coker, R. J., M. Viviani, B. G. Gazzard, B. DuPont, H. D. Pohle, S. M. Murphy, J. Atouguia, J. L. Champalimaud, and J. R. Harris.** 1993. Treatment of cryptococcosis with liposomal amphotericin B (AmBisome) in 23 patients with AIDS. *AIDS*. **7**:829–835.
17. **Croft, S. L., K. Seifert, and M. Duchene.** 2003. Antiprotozoal activities of phospholipid analogues. *Mol. Biochem. Parasitol.* **126**:165–172.
18. **Cruz, M. C., M. DelPoeta, P. Wang, R. Wenger, G. Zenke, V. F. Quesniaux, N. R. Movva, J. R. Perfect, M. E. Cardenas, and J. Heitman.** 2000. Immunosuppressive and nonimmunosuppressive cyclosporine analogs are toxic to the opportunistic fungal pathogen *Cryptococcus neoformans* via cyclophilin-dependent inhibition of calcineurin. *Antimicrob, Agents Chemother*. **44**:143–149.
19. **de Gans, J., P. Portegies, G. Tiessens, M. E. Schattenkerk, J. Karel, C. J. van Boxtel, J. R. van Ketel, and J. Stam.** 1992. Itraconazole compared with amphotericin B plus flucytosine in AIDS patients with cryptococcal meningitis. *AIDS* **6**:185–190.
20. **Del Poeta, M., M. C. Cruz, M. E. Cardenas, J. R. Perfect, and J. Heitman.** 2000. Synergistic antifungal activities of bafilomycin A, fluconazole, and the pneumocandin MK-0991/caspofungin acetate (L-743-873) with calcineurin inhibitors FK506 and L-685, 818 against *Cryptococcus neoformans*. *Antimicrob. Agents Chemother*. **44**:739–746.
21. **Denning, D. W., R. M. Tucker, L. H. Hanson, J. R. Hamilton, and D. A. Stevens.** 1989. Itraconazole therapy for cryptococcal meningitis and cryptococcosis. *Arch. Intern. Med.* **149**:2301–2308.
22. **Desmet, P., K. D. Kayembe, and C. De Vroey.** 1989. The value of cryptococcal serum antigen screening among HIV-positive/AIDS patients in Kinshasa, Zaire. *AIDS* **3**:77–78.
23. **Dismukes, W. E., G. Cloud, H. A. Gallis, T. M. Kerkering, G. Medoff, P. C. Craven, L. G. Kaplowitz, J. F. Fisher, C. R. Gregg, C. A. Bowles, S. Shadomy, A. M. Stamm, R. B. Diasio, L. Kaufman, S. Soong, and W. C. Blackwelder.** 1987. Treatment of cryptococcal meningitis with combination amphotericin B and flucytosine for four as compared with six weeks. NIAID Mycoses Study Group. *N. Engl. J. Med.* **317**:334–341.
24. **Dromer, F., S. Mathoulin, B. Dupont, O. Brugiere, and L. Letenneur.** 1996. Comparison of the efficacy of amphotericin B and fluconazole in the treatment of cryptococcosis in human immunodeficiency virus-negative patients: retrospective analysis of 83 cases. The French Cryptococcosis Study Group. *Clin. Infect. Dis.* **22**(Suppl 2):S154–S160.
25. **Dromer, F., C. Bernede-Bauduin, D. Guillemot, and O. Lortholary.** 2008. Major role for amphotericin B-flucytosine combination in severe cryptococcosis. *PLoS ONE* **3**:e2870.
26. **Dromer, F., J. Salamero, A. Contrepois, C. Carbon, and P. Yeni.** 1987. Production, characterization, and antibody specificity of a mouse monoclonal antibody reactive with *Cryptococcus neoformans* capsular polysaccharide. *Infect. Immun.* **55**:742–748.
27. **Ecevit, I. Z., C. J. Clancy, I. M. Schmalfuss, and M. H. Nguyen.** 2006. The poor prognosis of central nervous system cryptococcosis among nonimmunosuppressed patients: a call for better disease recognition and evaluation of adjuncts to antifungal therapy. *Clin. Infect. Dis.* **42**:1443–1447.
28. **Gordon, M. A.** 1963. Synergistic serum therapy of systemic mycoses. *Mycopathologia* **19**:150.
29. **Gordon, M. A., and E. Lapa.** 1964. Serum protein enhancement of antibiotic therapy in cryptococcosis. *J. Infect. Dis.* **114**:373–378.
30. **Gordon, M. A., and A. Casadevall.** 1995. Serum therapy for cryptococcal meningitis. *Clin. Infect. Dis.* **21**:1477–1479.
31. **Graybill, J. R., J. Sobel, M. Saag, C. van der Horst, W. Powderly, G. Cloud, L. Riser, R. Hamill, and W. Dismukes.** 2000. Diagnosis and management of increased intracranial pressure in patients with AIDS and cryptococcal meningitis. The NIAID Mycoses Study Group and AIDS Cooperative Treatment Groups. *Clin. Infect. Dis.* **30**:47–54.
32. **Grinsell, M., L. C. Weinhold, J. E. Cutler, Y. Han, and T. R. Kozel.** 2001. In vivo clearance of glucuronoxylomannan, the major capsular polysaccharide of *Cryptococcus neoformans*: a critical role for tissue macrophages. *J. Infect. Dis.* **184**:479–487.
33. **Hage, C. A., K. L. Wood, H. T. Winer-Muram, S. J. Wilson, G. Sarosi, and K. S. Knox.** 2003. Pulmonary cryptococcosis after initiation of anti-tumor necrosis factor-alpha therapy. *Chest* **124**:2395–2397.
34. **Hamill, R. J., J. D. Sobel, W. El-Sadr, P. C. Johnson, J. R. Graybill, K. Javaly, and D. E. Barker.** 2010. Comparison of 2 doses of liposomal amphotericin B and conventional amphotericin B deoxycholate for treatment of AIDS-associated acute cryptococcal meningitis: a randomized, double-blind clinical trial of efficacy and safety. *Clin. Infect. Dis.* **51**:225–232.
35. **Haubrich, R. H., D. Haghighat, S. A. Bozzette, J. Tilles, and J. A. McCutchan.** 1994. High-dose fluconazole for treatment of cryptococcal disease in patients with human immunodeficiency virus infection. The California Collaborative Treatment Group. *J. Infect. Dis.* **170**:238–242.
36. **Husain, S., M. M. Wagener, and N. Singh.** 2001. *Cryptococcus neoformans* infection in organ transplant recipients: variables influencing clinical characteristics and outcome. *Emerg. Infect. Dis.* **7**:375–381.
37. **Ivanovska, N.** 2003. Phospholipases as a factor of pathogenicity in microorganisms. *J. Mol. Catalysis B: Enzymatic* **22**:357–361.
38. **Johnson, L. B., and C. A. Kauffman.** 2003. Voriconazole: a new triazole antifungal agent. *Clin. Infect. Dis.* **36**:630–637.
39. **Joly, V., L. Saint-Julien, C. Carbon, and P. Yeni.** 1994. In vivo activity of interferon-γ in combination with amphotericin B in the treatment of experimental cryptococcosis. *J. Infect. Dis.* **170**:1331–1334.
40. **Kesson, A. M., M. C. Bellemore, T. J. O'Mara, D. H. Ellis, and T. C. Sorrell.** 2009. *Scedosporium prolificans* osteomyelitis in an immunocompromised child treated with a novel agent, hexadecylphosphocholine (miltefosine), in combination with terbinafine and voriconazole: a case report. *Clin. Infect. Dis.* **48**:1257–1261.
41. **Kontoyiannis, D. P., R. E. Lewis, B. D. Alexander, O. Lortholary, F. Dromer, K. L. Gupta, G. T. John, R. Del Busto, G. B. Klintmalm, J. Somani, G. M. Lyon, K. Pursell, V. Stosor, P. Munoz, A. P. Limaye, A. C. Kalil, T. L. Pruett, J. Garcia-Diaz, A. Humar, S. Houston, A. A. House, D. Wray, S. Orloff, L. A. Dowdy, R. A. Fisher, J. Heitman, N. D. Albert, M. M. Wagener, and N. Singh.** 2008. Calcineurin inhibitor agents interact synergistically with antifungal agents in vitro against *Cryptococcus neoformans* isolates: correlation with outcome in solid organ transplant recipients with cryptococcosis. *Antimicrob. Agents Chemother*. **52**:735–738.
42. **Kovacs, J. A., A. A. Kovacs, M. Polis, W. C. Wright, V. J. Gill, C. U. Tuazon, E. P. Gelmann, H. C. Lane, R. Longfield, G. Overturf, A. M. Macher, A. S. Fauci, J. E. Parrillo, J. E. Bennett, and H. Masur.** 1985. Cryptococcosis in the acquired immunodeficiency syndrome. *Ann. Int. Med.* **103**:533–538.
43. **Larsen, R. A., S. A. Bozzette, B. E. Jones, D. Haghighat, M. A. Leal, D. Forthal, M. Bauer, J. G. Tilles, J. A. McCutchan, and J. M. Leedom.** 1994. Fluconazole combined

with flucytosine for treatment of cryptococcal meningitis in patients with AIDS. *Clin. Infect. Dis.* **19:**741–745.

44. **Larsen, R. A., M. A. Leal, and L. S. Chan.** 1990. Fluconazole compared with amphotericin B plus flucytosine for cryptococcal meningitis in AIDS. A randomized trial. *Ann. Intern. Med.* **113:**183–187.
45. **Larsen, R. A., P. G. Pappas, J. Perfect, J. A. Aberg, A. Casadevall, G. A. Cloud, R. James, S. Filler, and W. E. Dismukes.** 2005. Phase I evaluation of the safety and pharmacokinetics of murine-derived anti-cryptococcal antibody 18B7 in subjects with treated cryptococcal meningitis. The NIAID Mycoses Study Group. *Antimicrob. Agents Chemother.* **49:**952–958.
46. **Lendvai, N., A. Casadevall, Z. Liang, D. L. Goldman, J. Mukherjee, and L. Zuckier.** 1998. Effect of immune mechanisms on the pharmacokinetics and organ distribution of cryptococcal polysaccharide. *J. Infect. Dis.* **177:**1647–1659.
47. **Lesho, E.** 2006. Evidence base for using corticosteroids to treat HIV-associated immune reconstitution syndrome. *Expert Rev. Anti. Infect. Ther.* **4:**469–478.
48. **Longley, N., C. Muzoora, K. Taseera, J. Mwesigye, J. Rwebembera, A. Chakera, E. Wall, I. Andia, S. Jaffar, and T. S. Harrison.** 2008. Dose response effect of high-dose fluconazole for HIV-associated cryptococcal meningitis in southwestern Uganda. *Clin. Infect. Dis.* **47:**1556–1561.
49. **Lortholary, O., A. Fontanet, N. Memain, A. Martin, K. Sitbon, and F. Dromer.** 2005. Incidence and risk factors of immune reconstitution inflammatory syndrome complicating HIV-associated cryptococcosis in France. *AIDS* **19:**1043–1049.
50. **Lortholary, O., F. Dromer, S. Mathoulin-Pelissier, C. Fitting, L. Improvisi, J. Cavaillon, and B. Dupont.** 2001. Immune mediators in cerebrospinal fluid during cryptococcosis are influenced by meningeal involvement and human immunodeficiency virus serostatus. The French Cryptococcosis Study Group. *J. Infect. Dis.* **183:**294–302.
51. **Lutz, J. E., K. V. Clemons, and D. A. Stevens.** 2000. Enhancement of antifungal chemotherapy by interferon-gamma in experimental systemic cryptococcosis. *J. Antimicrob. Chemother.* **46:**437–442.
52. **Menichetti, F., M. Fiorio, A. Tosti, G. Gatti, M. B. Pasticci, F. Miletich, M. Marroni, D. Bassetti, and S. Pauluzzi.** 1996. High-dose fluconazole therapy for cryptococcal meningitis in patients with AIDS. *Clin. Infect. Dis.* **22:**838–840.
53. **Mitchell, D. H., T. C. Sorrell, A. M. Allworth, C. H. Heath, A. R. McGregor, K. Papanaoum, M. J. Richards, and T. Gotlieb.** 1995. Cryptococcal disease of the CNS in immunocompetent hosts: influence of cryptococcal variety on clinical manifestations and outcome. *Clin. Infect. Dis.* **20:**611–616.
54. **Mody, C. H., G. B. Toews, and M. F. Lipscomb.** 1988. Cyclosporin A inhibits the growth of *Cryptococcus neoformans* in a murine model. *Infect. Immun.* **56:**7–12.
55. **Mody, C. H., G. B. Toews, and M. F. Lipscomb.** 1989. Treatment of murine cryptococcosis with cyclosporin-A in normal and athymic mice. *Am. Rev. Respir. Dis.* **139:**8–13.
56. **Mukherjee, J., M. Feldmesser, M. D. Scharff, and A. Casadevall.** 1995. Monoclonal antibodies to *Cryptococcus neoformans* glucuronoxylomannan enhance fluconazole efficacy. *Antimicrob. Agents Chemother.* **39:**1398–1405.
57. **Mukherjee, J., M. D. Scharff, and A. Casadevall.** 1992. Protective murine monoclonal antibodies to *Cryptococcus neoformans*. *Infect. Immun.* **60:**4534–4541.
58. **Mukherjee, J., L. Zuckier, M. D. Scharff, and A. Casadevall.** 1994. Therapeutic efficacy of monoclonal antibodies to *Cryptococcus neoformans* glucuronoxylomannan alone and in combination with amphotericin B. *Antimicrob. Agents Chemother.* **38:**580–587.
59. **Mussini, C., P. Pezzotti, J. M. Miro, E. Martinez, J. C. de Quiros, P. Cinque, V. Borghi, A. Bedini, P. Domingo, P. Cahn, P. Bossi, A. de Luca, A. A. Monforte, M. Nelson, N. Nwokolo, S. Helou, R. Negroni, G. Jacchetti, S. Antinori, A. Lazzarin, A. Cossarizza, R. Esposito, A. Antinori, and J. A. Aberg.** 2004. Discontinuation of maintenance therapy for cryptococcal meningitis in patients with AIDS treated with highly active antiretroviral therapy: an international observational study. *Clin. Infect. Dis.* **38:**565–571.
60. **Newton, P. N., L. H. Thai, N. Q. Tip, J. M. Short, W. Chierakul, A. Rajanuwong, P. Pitisuttithum, S. Chasombat, B. Phonrat, W. Maek-A-Nantawat, R. Teaunadi, D. G. Lalloo, and N. J. White.** 2002. A randomized, double-blind, placebo-controlled trial of acetazolamide for the treatment of elevated intracranial pressure in cryptococcal meningitis. *Clin. Infect. Dis.* **35:**769–772.
61. **Odom, A., S. Muir, E. Lim, D. L. Toffaletti, J. Perfect, and J. Heitman.** 1997. Calcineurin is required for virulence of *Cryptococcus neoformans*. *EMBO J.* **16:**2576–2589.
62. **Pachl, J., P. Svoboda, F. Jacobs, K. Vandewoude, B. van der Hoven, P. Spronk, G. Masterson, M. Malbrain, M. Aoun, J. Garbino, J. Takala, L. Drgona, J. Burnie, and R. Matthews.** 2006. A randomized, blinded, multicenter trial of lipid-associated amphotericin B alone versus in combination with an antibody-based inhibitor of heat shock protein 90 in patients with invasive candidiasis. The Mycograb Invasive Candidiasis Study Group. *Clin. Infect. Dis.* **42:**1404–1413.
63. **Pappas, P. G., B. D. Alexander, D. R. Andes, S. Hadley, C. A. Kauffman, A. Freifeld, E. J. Anaissie, L. Brumble, L. Herwaldt, J. Ito, D. P. Kontoyiannis, G. M. Lyon, K. A. Marr, V. A. Morrison, B. J. Park, T. Patterson, T. M. Perl, R. A. Oster, M. Schuster, R. Walker, T. J. Walsh, K. A. Wannemuehler, and T. M. Chiller.** 2010. Invasive fungal infections among organ transplant recipients: results of the Transplant-Associated Infection Surveillance Network (TRANSNET). *Clin. Infect. Dis.* **50:**1101–1111.
64. **Pappas, P. G., B. Bustamante, E. Ticano, R. J. Hamill, P. C. Johnson, A. Reboli, J. Aberg, R. Hasbun, and H. Hsu.** 2004. Recombinant interferon-gamma 1b as adjunctive therapy for AIDS-related acute cryptococcal meningitis. *J. Infect. Dis.* **189:**2185–2191.
65. **Pappas, P. G., P. Chetchotisakd, R. A. Larsen, W. Manosuthi, M. I. Morris, T. Anekthananon. S. Sungkanuparph, K. Supparatpinyo, T. L. Nolen, L. O. Zimmer, A. S. Kendrick, P. Johnson, J. D. Sobel, and S. G. Filler.** 2009. A phase II randomized trial of amphotericin B alone or combined with fluconazole in the treatment of HIV-associated cryptococcal meningitis. *Clin. Infect. Dis.* **48:**1775–1783.
66. **Pappas, P. G., J. R. Perfect, G. A. Cloud, R. A. Larsen, G. A. Pankey, D. J. Lancaster, H. Henderson, C. A. Kauffman, D. W. Haas, M. Saccente, R. J. Hamill, M. S. Holloway, R. M. Warren, and W. E. Dismukes.** 2001. Cryptococcosis in human immunodeficiency virus-negative patients in the era of effective azole therapy. *Clin. Infect. Dis.* **33:**690–699.
67. **Park, B. J., K. A. Wannemuehler, B. J. Marston, N. Govender, P. G. Pappas, and T. M. Chiller.** 2009. Estimation of the current global burden of cryptococcal meningitis among persons living with HIV/AIDS. *AIDS* **23:**525–530.
68. **Perfect, J. R., K. A. Marr, T. J. Walsh, R. N. Greenberg, B. DuPont, J. de la Torre-Cisneros, G. Just-Nubling, H. T. Schlamm, I. Lustar, A. Espinell-Ingroff, and E. Johnson.** 2003. Voriconazole treatment for less-common, emerging, or refractory fungal infections. *Clin. Infect. Dis.* **36:**1122–1131.

69. **Perfect, J. R., W. E. Dismukes, F. Dromer, D. L. Goldman, J. R. Graybill, R. J. Hamill, T. S. Harrison, R. A. Larsen, O. Lortholary, M. H. Nguyen, P. G. Pappas, W. G. Powderly, N. Singh, J. D. Sobel, and T. C. Sorrell.** Clinical practice guidelines for the management of cryptococcal disease: 2010 update by the Infectious Diseases Society of America. *Clin. Infect. Dis.* **50:**291–322.
70. **Pfaller, M. A., J. Zhang, S. A. Messer, M. E. Brandt, R. A. Hajjeh, C. J. Jessup, M. Tumberland, E. K. Mbidde, and M. A. Ghannoum.** 1999. In vitro activities of voriconazole, fluconazole, and itraconazole against 566 clinical isolates of *Cryptococcus neoformans* from the United States and Africa. *Antimicrob. Agents Chemother.* **43:**169–171.
71. **Pietrella, D., T. R. Kozel, C. Monari, F. Bistoni, and A. Vecchiarelli.** 2001. Interleukin-12 counterbalances the deleterious effect of human immunodeficiency virus type 1 envelope glycoprotein gp120 on the immune response to *Cryptococcus neoformans*. *J. Infect. Dis.* **183:**51–58.
72. **Pitisuttithum, P., R. Negroni, J. R. Graybill, B. Bustamante, P. Pappas, S. Chapman, B. S. Hare, and C. J. Hardalo.** 2005. Activity of posaconazole in the treatment of central nervous system fungal infections. *J. Antimicrob. Chemother.* **56:**745–755.
73. **Pitisuttithum, P., S. Tansuphasawadikul, A. J. Simpson, P. A. Howe, and N. J. White.** 2001. A prospective study of AIDS-associated cryptococcal meningitis in Thailand treated with high-dose amphotericin B. *J. Infect.* **43:**226–233.
74. **Powderly, W. G., G. A. Cloud, W. E. Dismukes, and M. S. Saag.** 1994. Measurement of cryptococcal antigen in serum and cerebrospinal fluid: value in the management of AIDS-associated cryptococcal meningitis. *Clin. Infect. Dis.* **18:**789–792.
75. **Powderly, W. G., M. S. Saag, G. A. Cloud, P. Robinson, R. D. Meyer, J. M. Jacobson, J. R. Graybill, A. M. Sugar, V. J. McAuliffe, S. E. Follansbee, C. U. Tuazon, J. J. Stern, J. Feinberg, R. Hafner, and W. E. Dismukes.** 1992. A controlled trial of fluconazole or amphotericin B to prevent relapse of cryptococcal meningitis in patients with the acquired immunodeficiency syndrome. The NIAID AIDS Clinical Trials Group and Mycoses Study Group. *N. Engl. J. Med.* **326:**793–798.
76. **Raad, I. I., J. R. Graybill, A. B. Bustamante, J. R. Graybill, R. Hare, G. Corcoran, and D. P. Kontoyiannis.** 2006. Safety of long-term oral posaconazole use in the treatment of refractory invasive fungal infections. *Clin. Infect. Dis.* **42:**1726–1734.
77. **Rex, J. H., R. A. Larsen, W. E. Dismukes, G. A. Cloud, and J. E. Bennett.** 1993. Catastrophic visual loss due to *Cryptococcus neoformans* meningitis. *Medicine* **72:**207–224.
78. **Robinson, P. A., M. Bauer, M. A. Leal, S. G. Evans, P. D. Holtom, D. M. Diamond, J. M. Leedom, and R. A. Larsen.** 1999. Early mycological treatment failure in AIDS-associated cryptococcal meningitis. *Clin. Infect. Dis.* **28:**82–92.
79. **Saag, M. S., G. A. Cloud, J. R. Graybill, J. D. Sobel, C. U. Tuazon, P. C. Johnson, W. J. Fessel, B. L. Moskovitz, B. Wiesinger, D. Cosmatos, L. Riser, C. Thomas, R. Hafner, and W. E. Dismukes.** 1999. A comparison of itraconazole versus fluconazole as maintenance therapy for AIDS-associated cryptococcal meningitis. The NIAID Mycoses Study Group. *Clin. Infect. Dis.* **28:**291–296.
80. **Saag, M. S., R. J. Graybill, R. A. Larsen, P. G. Pappas, J. R. Perfect, W. G. Powderly, J. D. Sobel, and W. E. Dismukes.** 2000. Practice guidelines for the management of cryptococcal disease. *Clin. Infect. Dis.* **30:**710–718.
81. **Saag, M. S., W. G. Powderly, G. A. Cloud, P. Robinison, M. H. Grieco, P. K. Sharkey, S. E. Thompson, A. M. Sugar, C. U. Tuazon, J. F. Fisher, N. Hyslop, J. M. Jacobson, R. Hafner, and W. E. Dismukes.** 1992. Comparison of amphotericin B with fluconazole in the treatment of acute AIDS-associated cryptococcal meningitis. The NIAID Mycoses Study Group and the AIDS Clinical Trials Group. *N. Engl. J. Med.* **326:**83–89.
82. **Santangelo, S., H. Zoellner, T. C. Sorrell, C. Wilson, C. Donald, J. Djordjevic, Y. Shounan, and L. Wright.** 2004. Role of extracellular phospholipases and mononuclear phagocytes in dissemination of cryptococcosis in a murine model. *Infect. Immun.* **72:**2229–2239.
83. **Schaars, C. F., G. A. Meintjes, C. Morroni, F. A. Post, and G. Maartens.** 2006. Outcome of AIDS-associated cryptococcal meningitis initially treated with 200 mg/day or 400 mg/day of fluconazole. *BMC Infect. Dis.* **6:**118.
84. **Shaariah, W., Z. Morad, and A. B. Suleiman.** 1992. Cryptococcosis in renal transplant recipients. *Transplant Proc.* **24:**1898–1899.
85. **Shapiro, L. L., and J. B. Neal.** 1925. Torula meningitis. *Arch. Neurol. Psych.* **13:**174–190.
86. **Sharkey, P. K., J. R. Graybill, E. S. Johnson, S. G. Hausrath, R. B. Pollard, A. Kolokathis, D. Mildvan, P. Fan-Havard, R. H. K. Eng, T. F. Patterson, J. C. Pottage, M. S. Simberkoff, J. Wolf, R. D. Meyer, R. Gupta, L. W. Lee, and D. S. Gordon.** 1996. Amphotericin B lipid complex compared with amphotericin B in the treatment of cryptococcal meningitis in patients with AIDS. *Clin. Infect. Dis.* **22:**315–321.
87. **Singh, N., B. D. Alexander, O. Lortholary, F. Dromer, K. L. Gupta, G. T. John, K. J. Pursell, P. Munoz, G. B. Kintmalm, V. Stosor, R. Del Busto, A. P. Limaye, A. C. Kalil, T. L. Pruett, J. Garcia-Diaz, A. Humar, S. Houston, A. A. House, D. Wray, S. Orloff, L. A. Dowdy, R. A. Fisher, J. Heitman, M. M. Wagener, and S. Husain.** 2007. *Cryptococcus neoformans* in organ transplant recipients: impact of calcineurin-inhibitor agents on mortality. *J. Infect. Dis.* **195:**756–764.
88. **Singh, N., O. Lortholary, B. D. Alexander, K. L. Gupta, G. T. John, K. J. Pursell, P. Munoz, G. B. Kintmalm, V. Stosor, R. Del Busto, A. P. Limaye, J. Somani, M. Lyon, S. Houston, A. A. House, T. L. Pruett, S. Orloff, A. Humar, L. A. Dowdy, J. Garcia-Diaz, R. A. Fisher, A. C. Kalil, J. Heitman, and S. Husain.** 2005. Antifungal management practices and evolution of infection in organ transplant recipients with *Cryptococcus neoformans* infection. *Transplantation* **80:**1033–1039.
89. **Singh, N., O. Lortholary, B. D. Alexander, K. L. Gupta, G. T. John, K. J. Pursell, P. Munoz, G. B. Kintmalm, V. Stosor, R. Del Busto, A. P. Limaye, J. Somani, M. Lyon, S. Houston, A. A. House, T. L. Pruett, S. Orloff, A. Humar, L. A. Dowdy, J. Garcia-Diaz, R. A. Fisher, A. C. Kalil, J. Heitman, and S. Husain.** 2005. An immune reconstitution syndrome-like illness associated with *Cryptococcus neoformans* infection in organ transplant recipients. *Clin. Infect. Dis.* **40:**1756–1761.
90. **Singh, S. D., N. Robbins, A. K. Zaas, W. A. Schell, J. R. Perfect, and L. E. Cowen.** 2009. Hsp90 governs echinocandin resistance in the pathogenic yeast *Candida albicans* via calcineurin. *PLoS Pathog.* **5:**e1000532.
91. **Steinbach, W. J., J. L. Reedy, R. A. Cramer, J. R. Perfect, and J. Heitman.** 2007. Harnessing calcineurin as a novel anti-infective agent against invasive fungal infections. *Nature Rev.* **5:**418–430.
92. **Stevens, D. A.** 1998. Combination immunotherapy and antifungal chemotherapy. *Clin. Infect. Dis.* **26:**1266–1269.
93. **Stevens, D. A., E. Brummer, and K. V. Clemons.** 2006. Interferon-γ as an antifungal. *J. Infect. Dis.* **194:**S33–S37.
94. **Sun, H. Y., C. C. Hung, and S. C. Chang.** 2004. Management of cryptococcal meningitis with extremely high intracranial pressure in HIV-infected patients. *Clin. Infect. Dis.* **38:**1790–1792.

95. **Sun, H. Y., M. M. Wagener, and N. Singh.** 2009. Cryptococcosis in solid-organ, hematopoietic stem cell, and tissue transplant recipients: evidence-based evolving trends. *Clin. Infect. Dis.* **48:**1566–1576.

96. **Taber, K. W.** 1937. Torulosis in man. *JAMA* **108:**1405–1406.

97. **Tansuphaswadikul, S., W. Maek-a-Nantawat, B. Phonrat, L. Boonpokbn, A. G. Mctm, and P. Pitisuttithum.** 2006. Comparison of one week with two week regimens of amphotericin B both followed by fluconazole in the treatment of cryptococcal meningitis among AIDS patients. *J. Med. Assoc. Thai.* **89:**1677–1685.

98. **Tassie, J. M., L. Pepper, C. Fogg, S. Biraro, B. Mayanja, I. Andia, A. Paugam, G. Priotto, and D. Legros.** 2003. Systematic screening of cryptococcal antigenemia in HIV-positive adults in Uganda. *J. Acquir. Immune Defic. Syndr.* **33:**411–412.

99. **Thompson, G. R., N. P. Wiederhold, A. W. Fothergill, A. C. Vallor, B. L. Wickes, and T. F. Patterson.** 2009. Antifungal susceptibilities among different serotypes of *Cryptococcus gattii* and *Cryptococcus neoformans*. *Antimicrob. Agents Chemother.* **53:**309–311.

100. **van der Horst, C. M., M. S. Saag, G. A. Cloud, R. J. Hamill, J. R. Graybill, J. D. Sobel, P. C. Johnson, C. U. Tuazon, T. Kerkering, B. L. Moskovitz, W. G. Powderly, and W. E. Dismukes.** 1997. Treatment of cryptococcal meningitis associated with the acquired immunodeficiency syndrome. The NIAID Mycoses Study Group and AIDS Clinical Trials Group. *N. Engl. J. Med.* **337:**15–21.

101. **van Duin, D., W. Cleare, O. Zaragoza, A. Casadevall, and J. D. Nosanchuk.** 2004. Effects of voriconazole on *Cryptococcus neoformans*. *Antimicrob. Agents Chemother.* **48:**2014–2020.

102. **Vilchez, R. A., J. Fung, and S. Kusne.** 2002. Cryptococcosis in organ transplant recipients: an overview. *Am. J. Transplant.* **2:**575–580.

103. **White, M., C. Cirrincione, A. Blevins, and D. Armstrong.** 1992. Cryptococcal meningitis: outcome in patients with AIDS and patients with neoplastic disease. *J. Infect. Dis.* **165:**960–963.

104. **Widmer, F., L. C. Wright, D. Obando, R. Handke, R. Ganendren, D. H. Ellis, and T. C. Sorrell.** 2006. Hexadecylphosphocholine (miltefosine) has broad-spectrum fungicidal activity and is efficacious in a mouse model of cryptococcosis. *Antimicrob. Agents Chemother.* **50:**414–421.

105. **Wiley, J. M., N. L. Seibel, and T. J. Walsh.** 2005. Efficacy and safety of amphotericin B lipid complex in 548 children and adolescents with invasive fungal infections. *Pediatr. Infect. Dis. J.* **24:**167–174.

106. **Wu, G., R. A. Vilchez, B. Eidelman, J. Fung, R. Kormos, and S. Kusne.** 2002. Cryptococcal meningitis: an analysis among 5521 consecutive organ transplant recipients. *Transpl. Infect. Dis.* **4:**183–188.

Cryptococcus: From Human Pathogen to Model Yeast
Edited by J. Heitman et al.

40

Vaccines and Antibody Therapies from *Cryptococcus neoformans* to Melanoma

ARTURO CASADEVALL, EKATERINA DADACHOVA, AND LIISE-ANNE PIROFSKI

The interaction of *Cryptococcus neoformans* with the human host is characterized by a high incidence of infection and a very low incidence of disease. In urban cities like New York, antibodies to cryptococcal proteins and capsular polysaccharide are prevalent in the majority of healthy adults, providing a marker for prior or concurrent infection in those individuals (1, 47). Since infection is common and disease is rare, one can deduce that normal defense mechanisms are highly efficacious in preventing infection from progressing to disease. This allows us to conclude that normal immunity can readily contain the fungus to the sites of initial infection, which are probably the lungs for the majority of individuals. The ability of the normal immune system to prevent cryptococcal disease indicates that immunity to *C. neoformans* is common, and this in turn suggests that it should be possible to induce protection against disease with vaccines. However, host and microbial factors each signal a need for caution in translating the observation that cryptococcal disease is rare in normal individuals to the possibility of protecting individuals who are at risk with a vaccine. This is underscored by the fact that the majority of individuals who are at risk for cryptococcal disease do not have normal immune systems and that *C. neoformans* pathogenesis is so complex.

HOST DEFENSE AGAINST *C. NEOFORMANS*

Studies of effective natural host immune responses can often provide important cues for vaccine development. For example, correlating the presence of serum antibody with resistance to disease might suggest that a vaccine that elicits similar antibody responses would also confer immunity to the development of disease. However, it is also possible to induce a state of immunity by immunizing with antigens that do not induce an immune response during the course of natural infection (15). For example, antibodies to diphtheria toxin (DT) protect against diphtheria despite the fact that such antibodies do not appear to be associated with immunity during the natural course of disease (15). For *C. neoformans*, disease is extremely rare in hosts without known immunological deficits, but the correlates of immunity remain uncertain.

The widespread consensus in the field that cellular immunity makes a critical contribution to resistance to cryptococcosis is based on the following observations: (i) the overwhelming majority of cases of cryptococcosis occur in individuals with impaired T-cell function such as that found in patients with AIDS (90); (ii) histopathological studies have established that tissue control of *C. neoformans* is associated with granuloma formation, a hallmark of cell-mediated immunity (CMI) (44); (iii) animal models with specific defects in components of CMI are much more susceptible to lethal infection than immunologically intact controls; (iv) adoptive T-cell transfer has been shown to provide partial protection in mice; and (v) *C. neoformans* is a facultative intracellular pathogen that often resides in macrophages, and control of such pathogens often requires intact CMI (37). However, since CMI extends from innate to acquired immune responses and has such a central role in host defense, any impairment in CMI is also likely to affect humoral and innate immune defenses. Consequently, knowing that CMI is important for host defense against cryptococcosis does not necessarily translate into a specific correlate of immunity.

A definitive role for humoral immunity in host defense against natural infection with *C. neoformans* has been more difficult to establish. In contrast to CMI, inborn and acquired defects in humoral immunity are not clearly associated with cryptococcosis, even though there are numerous case reports demonstrating that individuals with X-linked hyper-immunoglobulin M and hypogammaglobulinemia are at increased risk for disease and ample data suggesting that for certain individuals a deficit in antibody-mediated immunity is associated with an increased vulnerability to disease (40, 105, 106). Nevertheless, serological studies have shown that populations at risk for cryptococcosis also have quantitative immunoglobulin defects, suggesting an association between increased vulnerability to disease and

Arturo Casadevall, Departments of Microbiology & Immunology and Medicine, Albert Einstein College of Medicine, Bronx, NY 10461. **Ekaterina Dadachova**, Departments of Nuclear Medicine and Microbiology and Immunology, Albert Einstein College of Medicine, Bronx, NY 10461. **Liise-anne Pirofski**, Departments of Medicine (Infectious Diseases) and Microbiology & Immunology, Albert Einstein College of Medicine, Bronx, NY 10461.

antibody defects. In contrast to nonconclusive human epidemiological and serological data, there is unequivocal evidence from animal models that specific antibody can be protective, whether administered passively or induced by vaccination. Studies from several independent laboratories have shown that passive antibody administration in mice is associated with prolonged survival and reduced fungal burden. In addition, mice with B-cell defects are at increased risk for disease. Hence, there is a current consensus that antibody immunity can contribute to host defense, although its role relative to CMI remains uncertain.

Although vaccine strategies usually aim to protect a host by eliciting adaptive cellular or antibody immunity, innate immunity is likely to be sufficient to prevent most infectious diseases. In light of this, it is noteworthy that innate immune imprinting by immunization with another microbe or antigen can prime innate responses with enhanced activity against *C. neoformans* (115, 117). For example, pulmonary immunization with *Mycobacterium bovis* BCG produces long-term changes in the lung immune response that are associated with protection against subsequent cryptococcal infection (115). In fact, immunization with a single antigen, namely, modified heat-labile toxin from *Escherichia coli*, induced a protective effect that reduced susceptibility to various pathogens, including *C. neoformans* (117). These observations suggest that immunogens that nonspecifically enhance innate immune responses in the lungs might be useful vaccine candidates for *C. neoformans*. One caveat is that such strategies could exacerbate disease if the host damage that occurs in the course of the disease stems from the immune response itself. For cryptococcosis, this is a concern for immune reconstitution inflammatory syndrome–related disease (10, 63, 104).

A particular concern for the design of a vaccine for *C. neoformans* is that certain cellular and antibody responses can be detrimental to the host. For example, overexuberant inflammation has been associated with worse outcomes in mouse models of cryptococcosis and in patients with AIDS and who have received solid organ transplants and develop immune reconstitution inflammatory syndrome after improvement of their immunological function (10, 104). For antibody immunity, nonprotective and disease-enhancing antibodies that differ from protective antibodies in their specificity, isotype, and effect on the inflammatory response are well described (64, 77). A particularly disconcerting finding with one monoclonal antibody (MAb) was that it could be either protective or nonprotective, depending on the mouse genetic background, underscoring that protective antibodies cannot be defined by singular attributes. This raises the possibility that host genetics is among the compendium of factors that could influence antibody efficacy in humans (97) and that some vaccines intended to elicit stronger cell-mediated and/or antibody responses might have the unintended consequence of enhancing the inflammatory response to infection. Furthermore, the possibility that cryptococcosis could reflect an ineffective host response stemming from either a weak or a strong immune response raises another question, namely, whether a single vaccine will be effective or whether the development of more than one vaccine, tailored to the nature of the underlying host response, might be prudent.

PROPHYLACTIC OR THERAPEUTIC VACCINES?

In considering a vaccine for cryptococcosis, the first issue that one must consider is whether the vaccine is intended to function prophylactically or therapeutically. Since cryptococcal infection often occurs in early childhood and latency is probably a common outcome of infection, any vaccine to prevent or interrupt infection or the development of latency would have to be administered very early in life. However, such a strategy may not be acceptable or practical given that cryptococcosis in children and adults is an extremely rare disease in immunologically intact individuals. Moreover, given the proliferation of new information on the importance of microbial communities in mucosal tissues to human health and physiology, future work may reveal that *C. neoformans* contributes to a beneficial homeostasis. On the other hand, prevention of *C. neoformans* infection might protect against its role as a cofactor in asthma (46).

Nonetheless, it would be difficult to develop, justify, and/or use such a vaccine in young children. Therefore, since cryptococcosis is primarily a disease of immunocompromised individuals who acquired *C. neoformans* during childhood infection and developed latency, and disease is often the result of reactivation of latent infection, a vaccine that prevents or reduces the likelihood of disease makes the most sense. By definition, such a vaccine would be a therapeutic vaccine; however, it is conceivable that additional vaccines could be formulated as adjunctive therapy during the course of disease. As evidenced by the foregoing statements, the development of a vaccine for cryptococcosis introduces several entirely novel challenges. For example, given that disease is usually associated with immunosuppressed states, a vaccine to prevent disease would have to confer immunity despite the presence of abnormal or defective immunity and/or provide immunity that endures despite the subsequent development of immune suppression. Encouragement that such a vaccine is possible comes from experience with the varicella-zoster vaccine, which prevents disease in the setting of leukemia and other immune defects in children. In addition, the measles-mumps-rubella vaccine prevents these diseases in HIV-infected children, and though less immunogenic than in immunocompetent children, pneumococcal and *Haemophilus influenzae* conjugate vaccines reduce the incidence of invasive disease in children with HIV infection. An important caveat to these examples is that for each one, vaccine efficacy depends on the production of neutralizing or opsonic antibodies that induce microbial clearance. A vaccine for cryptococcosis will likely have to enhance or reduce the immune response to *C. neoformans* but might not need to induce fungal clearance, a considerably more complex problem.

Experiments in a mouse model support the concept of a therapeutic vaccine for cryptococcosis. Mice chronically infected with *C. neoformans* that were immunized with a vaccine consisting of peptide mimotope of the cryptococcal capsular polysaccharide glucuronoxylomannan (GXM) conjugated to carrier proteins survived considerably longer than control mice (27). The vaccine appeared to enhance survival by eliciting protective antibodies to the polysaccharide and modifying the inflammatory response to de novo infection (27).

EXPERIMENTAL VACCINES

Numerous studies over the past three decades have established that immunizing a naive host with *C. neoformans* antigenic components can elicit immune responses that reduce mortality or significantly prolong survival in mice challenged with lethal inocula. Furthermore, experimental vaccines have been developed that mediate protection by eliciting antibody or cellular immune responses. The ability to elicit useful immune responses from either the antibody or cellular arms of the immune system and experimental evidence that antibody and cellular immunity cooperate against *C. neoformans* raise the tantalizing possibility that a vaccine formulation that elicits combined responses would be significantly more effective than one that elicits responses by only one arm of the immune system. In this chapter we will consider recent developments in the vaccine field and note the publication of two recent comprehensive reviews that include additional information on the problem of vaccination against *C. neoformans* (16, 28). In addition, we note two other reviews on the broader subject of vaccine development to fungi including *C. neoformans* that provide additional insight into the approaches that are being tried in medical mycology (18, 29).

Early attempts to make vaccines from killed cryptococcal cells were generally not successful (103). In fact, killed-cell immunization has sometimes been associated with detrimental immune responses and shorter survival following vaccination and challenge (83). However, immunization with killed cells can induce protective responses in certain hosts, such as rats, whereas immunization with isolated polysaccharide appears to exacerbate infection (4). Older studies demonstrated that live attenuated vaccines provided significant protection in mice, but such vaccines are unlikely to be developed for use in immunocompromised hosts (for an excellent and complete review of those earlier efforts see reference 103). Studies with an avirulent strain of *C. neoformans* deficient in calcineurin revealed that infection did not induce resistance to subsequent challenge with a virulent strain (118). Rapid clearance of this strain suggests that induction of effective immunity necessitates at least the establishment of a temporary infection that is sufficient to elicit Th1-type responses (118). An interesting variation in the live vaccine approach was the development a *C. neoformans* strain that expresses interferon-gamma (119). This strain elicits an effective immune response presumably by expressing interferon-gamma locally that promotes a brisk and strong Th1-type response and also presumably enhances the efficacy of innate immunity.

POTENTIAL VACCINE ANTIGENS

Four types of *C. neoformans* components have been shown to elicit immune responses that modify the outcome of experimental infection to the benefit of the host: polysaccharides, proteins, lipids, and melanin-type pigments. Although only polysaccharides and proteins are likely candidates for vaccine development, all four will be discussed, given that they establish important precedents.

Polysaccharide Antigens

C. neoformans is an encapsulated pathogen, and polysaccharide-based vaccines have an excellent track record in eliciting protective immunity against bacteria with polysaccharide capsules. For example, current vaccines for pneumococcus, *H. influenzae* type B, and meningococcus each use capsular polysaccharide components to elicit antibody responses that mediate protection through a variety of mechanisms, including enhancement of phagocytosis and modification of the host inflammatory response (98). The capsule is also a critical virulence factor for *C. neoformans*, and antibodies to the major capsular polysaccharide GXM of *C. neoformans* protect against experimental infection. However, immunization with polysaccharide antigens is often not effective because polysaccharide antigens are relatively poor immunogens that fail to elicit significant antibody responses (56, 57, 81). Furthermore, there is a large body of data demonstrating that injection of purified cryptococcal capsular polysaccharides into mammalian hosts is detrimental to the immune response and can increase susceptibility to subsequent challenge with *C. neoformans* (112, 113). For example, immunization with capsular polysaccharide increases the susceptibility of mice and rats to cryptococcal infection (4, 8). Consequently, purified cryptococcal polysaccharide antigens do not appear to be suitable vaccine candidates because they are poorly immunogenic and can promote deleterious effects on host immunity.

The poor immunogenicity of capsular polysaccharides can be overcome by linking them to carrier proteins. The first attempt to produce a polysaccharide-protein conjugate vaccine for *C. neoformans* was reported in the mid-1960s and resulted in a compound that was highly immunogenic but not protective against lethal disease, the so-called Goren vaccine (49, 50). The inefficacy of this vaccine remains unexplained, but recent observations suggest plausible explanations. First, the vaccine elicited very high antibody levels and protective antibody loses its efficacy when administered in very high doses due to a prozone-like effect (107, 108). Second, the vaccine used total capsular polysaccharide, which is a complex mixture of GXM and galactoxylomannan. Although GXM is now known to elicit both protective and nonprotective antibodies, galactoxylomannan causes profound immunological derangements that could predispose immunized mice to *C. neoformans*–mediated disease (30). Historically, polysaccharide-protein conjugate vaccines have used the plentiful polysaccharide material (exopolysaccharide) found in culture supernatants as a capsular antigen. This approach is supported by the observation that antibodies to exopolysaccharides bind to intact capsule. However, recent studies using chemical and physical techniques have established significant differences between exopolysaccharides and capsular polysaccharides, raising the question of whether these are in fact interchangeable (vaccine) antigens (42). Consequently, the choice of what polysaccharide preparation to use for vaccine production should be a major consideration in the preparation of polysaccharide-protein conjugates and at this time remains unresolved.

After the failure of the Goren vaccine (49, 50), the next attempt to generate a polysaccharide conjugate vaccine occurred in the early 1990s, catalyzed by the epidemic of cryptococcosis in patients with AIDS (32). The latter vaccine used purified GXM conjugated to carrier proteins such as tetanus toxoid (TT) and produced highly immunogenic GXM-protein conjugates that were protective in immunized mice and elicited antibodies that were protective in naive mice (14, 31). Human MAbs produced from immortalized B cells from GXM-TT-vaccinated individuals were protective in mice (41, 93, 126, 127), and serum antibodies from vaccinated individuals enhanced the antifungal activity of phagocytes against *C. neoformans* (126). Although this promising vaccine was not pursued further, it was a landmark reagent that illustrated the immunogenicity and biological function of capsule-derived vaccines, albeit in immunocompetent individuals. Despite the overall success of this vaccine in immunocompetent mice, it is noteworthy that it also elicited antibodies that were nonprotective in mice (77), and its efficacy in immunologically impaired hosts was never tested. The latter point is important, because conjugate vaccines are often not highly immunogenic in individuals with defects in CMI, which is the type of defect that predisposes patients to cryptococcosis. Importantly, MAbs derived from GXM-TT-vaccinated mice required CD4 T cells to mediate protection in mice (121), raising additional questions as to their potential to be effective in immunocompromised hosts.

Another approach to develop a carbohydrate-based vaccine is to use synthetic oligosaccharides. These compounds have the advantage that they are synthesized chemically and thus constitute structurally defined molecules that are entirely different than heterogeneous, polydisperse polysaccharide preparations. Two groups have synthesized oligosaccharides representing cryptococcal carbohydrate motifs, some of which were found to be immunogenic when conjugated to a protein carrier (2, 3, 43, 89, 114, 124, 125). Immunization of mice with an oligosaccharide conjugate vaccine displaying the predominant motif found in serotype A polysaccharide was immunogenic but elicited nonprotective antibodies (86). One potential limitation of the oligosaccharide approach is that some protective MAbs to GXM appear to bind conformational epitopes (73). However, it may be possible to reconstruct such conformational epitopes by the employment of larger oligosaccharides and/or combinations of GXM structural motifs. Although the first attempt to make an oligosaccharide-based vaccine was disappointing, there is hope that this extremely promising approach will yield a compound that elicits a protective antibody response. Nonetheless, given the complexity of GXM structural motifs, the identification of suitable oligosaccharides is laborious and requires highly sophisticated techniques in organic chemistry.

An alternative approach to eliciting antibodies to capsular polysaccharide is to use small peptides that have the ability to mimic the conformation of carbohydrate antigens. These peptides can be readily identified from phage libraries designed to express enormous combinations of amino acid sequences using MAbs to capsular polysaccharide. Peptides that can bind to the antibody used for selection are known as mimetics, and peptide mimetics that elicit an antibody response that binds to the original antigen defined by the selecting antibody are known as mimotopes. Peptide mimotopes of GXM selected by mouse and human MAbs have been reported (6, 39, 111, 123). An advantage of peptide mimotopes is that they can be made easily and reproducibly and thus provide an attractive option for generating vaccines. Furthermore, the demonstration that a peptide selected by a protective human MAb generated from an individual immunized with the GXM-TT vaccine elicited a protective antibody response in mice (39) provides a proof of principle for the power of this approach. This peptide conjugated to DT also elicited protective antibody responses in mice expressing human immunoglobulin genes (65).

However, there are also limitations to the mimotope approach. In one set of experiments mice immunized with a peptide produced high levels of antibodies reactive with a protective epitope only when they were primed initially with GXM-TT (5). Site-directed mutagenesis of the binding contacts of a protective MAb revealed that different amino acids were involved in binding to polysaccharide and mimotope epitopes (84, 85), raising the possibility that different types of antibodies are elicited by carbohydrate and peptide antigens. In this regard, it was noteworthy that antibody responses elicited by glycosyl determinants in keyhole limpet hemocyanin reacted with GXM but were not protective (72). Hence, like native carbohydrates, antigenic mimicry by peptides or unrelated carbohydrates can elicit both protective and nonprotective antibodies reactive with GXM. This suggests that the phenomenon of antigenic mimicry could be harnessed in vaccine design against *C. neoformans*, with the caveat that the variables and mechanisms that govern the induction of protective antibodies are better understood.

Another group of polysaccharide antigens with potential for vaccine development are beta-1,3-glucans. These polysaccharides are cell wall components for many types of pathogenic fungi. Antibodies to beta-1,3-glucans have been shown to protect against *Aspergillus fumigatus* and *Candida albicans*, raising the intriguing possibility that a vaccine broadly protective against fungal pathogens can be developed (109, 110). Along these lines, MAbs to beta-1,3-glucans have also been shown to inhibit *C. neoformans* growth and to be effective against cryptococcal infection in mice (96). Although the cell wall of *C. neoformans* is covered by the polysaccharide capsule, antibodies can mediate antifungal effects by diffusing to and binding moieties on the cell surface. Consequently, beta-1,3-glucans have emerged as novel polysaccharide vaccine candidates for fungi.

Protein Antigens

In concert with the fact that *C. neoformans* is a facultative intracellular pathogen, the consistent observation that control of cryptococcal infection in tissue involves granuloma formation has catalyzed the search for antigens that elicit cell-mediated (granulomatous) responses. Given that protein antigens have the capacity to induce antigen processing and presentation of peptides, they are considered to be prototypical compounds for eliciting cell-mediated responses. Hence, there has been great interest in identifying protein antigens suitable for vaccine design. Cryptococcal infection elicits strong and rapid antibody responses to fungal proteins (17, 47), such as heat shock proteins (HSPs) (53), but currently there is no evidence that these antibody responses contribute to host protection. On the other hand, protein antigens such as mannoproteins elicit strong cell-mediated responses that have been associated with partial protection against lethal challenge in mice. For example, cryptococcal filtrates produce strong delayed-type hypersensitivity reactions as measured by footpad reactions in mice (12, 80). Although such filtrates are complex, heterogeneous mixtures of fungal products that include many proteins (9, 36), it was established that mannoproteins are the component responsible for delayed-type hypersensitivity reactions (82). Notably, cryptococcal mannoproteins stimulated lymphoproliferation and cytokine release from human peripheral blood mononuclear cells (61, 95), an effect that was also associated with induction of HIV proliferation (88). In vivo, mannoprotein immunization was shown to induce strong cellular immune responses that significantly improved the outcome of experimental cryptococcal infection in mice (58, 68, 91). Subsequent work has established that the so-called mannoprotein fraction in cryptococcal supernatants is a heterogenous collection of proteins that has potent immunological properties (52, 60, 120). The ability of mannoproteins to trigger strong T-cell responses appears to depend on protein glycosylation and the interaction of fungal glycosyl structures with mannose receptors (66, 67).

The potent T-cell activating properties of mannoproteins combined with their proven ability to provide protection against lethal murine cryptococcosis strongly supports their role as potential vaccine candidates. Notably, cryptococcal mannoproteins elicit immune responses that also protect against *C. albicans*, suggesting that it might be possible to exploit shared determinants between fungi to design a vaccine with broad antifungal activity (92). Mannoproteins also elicit strong antibody responses (94). Although the role of mannoprotein-binding antibodies in protection against cryptococcosis is uncertain, patients with HIV and cryptococcosis are less likely to manifest antibody responses to these antigens (94). Whether this deficit reflects a vulnerability in these patients that predisposes them to disease and/or stems from impaired B-cell responses in the setting of immunosuppression is not clear. Given that some mannoproteins are enzymes that presumably have extracellular function, it is conceivable that certain antibody responses to these proteins do contribute to host protection. In this regard, it is noteworthy that an antibody to *C. albicans* mannoprotein has been shown to have direct anticandidal activity (75).

Another type of protein antigen that has emerged as a possible vaccine candidate is HSP. HSPs are major targets of the antibody response to *C. neoformans* (53, 54), although the effectiveness of these antibodies, if any, is uncertain. Antibodies to HSP have now been shown to be protective against *Histoplasma capsulatum* (51), and an antibody fragment recognizing HSP90 is protective against *C. albicans* (70, 71). This antibody is also effective against *C. neoformans* (87), and its efficacy raises the possibility that certain immune responses to HSP do contribute to host defense. Although HSPs are highly immunogenic, their use as vaccine antigens could be impeded by their homology to host proteins, which raises the specter of autoimmunity.

Lipids and Pigments

Antibodies to glucosylceramide (99) and melanin (100) have each been shown to mediate direct antifungal activity against *C. neoformans*.

PASSIVE ANTIBODY THERAPIES

The difficulty in treating human cryptococcosis with antifungal drugs has stimulated interest in the development of adjunctive passive antibody therapies. Historically, antibody therapies were used as adjuncts of amphotericin B early in the antifungal therapy era with promising anecdotal results (48, 62). Efforts to develop adjunctive antibody therapy were discontinued in the 1960s given the increasing efficacy and availability of amphotericin preparations, the difficulties in making immune sera, and the regulatory/economic hurdles of developing therapy for a relatively rare disease. However, in the late 1980s and early 1990s the discovery that certain MAbs (34, 35, 78, 102) could modify the course of experimental cryptococcal infection in mice brought a renaissance of interest in the potential use of passive antibody therapy. The administration of MAbs to the cryptococcal capsule was shown to potentiate the efficacy of amphotericin B, fluconazole, and flucytosine (33, 38, 76, 79). Most intriguingly from a therapeutic standpoint, passive antibody therapy led to the clearance of serum polysaccharide with deposition in hepatic reticuloendothelial cells, suggesting that specific antibody could be used to clear polysaccharide from tissue (45) and possibly reverse the immunological deficits associated with this microbial component. To improve antibody effector function and reduce immunogenicity in humans, mouse-human chimeric antibodies were generated, and several were shown to be protective in mice (7, 74, 122). Fully human antibodies that were protective in mice were made from immortalized B cells from human volunteers who received the GXM-TT conjugate vaccine (41, 93) and human immunoglobulin transgenic mice immunized with GXM-DT (65).

One murine MAb was developed for clinical use and tested in a phase I trial in patients with cryptococcosis (13). This MAb, known as 18B7, was generated from a mouse immunized with the GXM-TT vaccine (78). Despite the availability of mouse-human chimeric antibodies, a decision was made to proceed to clinical trial with the murine MAb because of logistical issues and concern that the exchanging human and murine constant regions would affect the specificity of the immunoglobulin molecule (74). The human trial, which was conducted from 2000 to 2002, demonstrated that MAb administration to HIV-infected patients with measureable serum antigen (GXM) was safe and well tolerated (59). Although this trial was not designed to assess efficacy, a reduction in serum antigen was observed in the patients that received the highest doses of MAb (59).

The clinical trial with MAb 18B7 was historic in the sense that it was the first time MAb therapy was clinically tested against a fungal disease. Despite encouraging results with adjunctive MAb therapy, there has been no further clinical development of MAb 18B7. The lukewarm pharmaceutical interest in passive antibody therapy for cryptococcosis is based on simple economic concerns stemming from various factors that, in combination, have proven to be a formidable obstacle for the clinical development of this therapeutic approach. The advent of highly active antiretroviral therapies reduced the number of patients with cryptococcosis as a complication of AIDS, and this has reduced the potential market size for specific antibody therapy. The availability and introduction of newer antifungal drugs has increased the efficacy of conventional therapy and diminished the need for antibody therapy. Like all biological therapies, antibody therapy is associated with high costs, and this reduces its attractiveness, given that large amounts of immunoglobulin would be needed for each patient. Finally, there is the reality that with existing antifungal therapy, antibody therapy must be developed as an adjunct to conventional therapy, and this means that clinical testing needs to show a benefit of antibody-antifungal therapy over antifungal therapy alone. Since antifungal therapy is effective in the majority of cases, a very large trial would be needed to establish the superiority of combined therapy. Hence, the combination of fewer cases, newer drugs, high costs, and costly clinical trials has conspired to halt further clinical work on MAb 18B7 therapy. However, these developmental hurdles have also catalyzed new approaches to antibody therapy based on improving the fungicidal activity of the immunoglobulin molecule.

Radioimmunotherapy for Cryptococcal Infection

Some of the difficulties encountered in the clinical development of antibody therapy for cryptococcosis could be surmounted if the immunoglobulin could be modified to greatly increase its intrinsic activity against *C. neoformans*. For example, a more efficacious antibody would require lower immunoglobulin doses, and this would make the approach more economically attractive by reducing the cost of therapy. Higher efficacy could also translate into a higher therapeutic margin that would be manifest as more effective combination therapy, which would in turn facilitate clinical trial design to demonstrate the advantage of combined antibody-antifungal therapy. One approach to increase antibody activity is to link the MAb to a radionuclide that emits cytotoxic radiation and then use the antigen-MAb interaction to deliver it to sites of infection. This approach is known as radioimmunotherapy (RIT) and was originally pioneered as a therapy for cancer.

Studies with *C. neoformans* have served to establish the proof of principle for the concept that RIT can be applied to the treatment of infectious diseases (24). Conjugation of MAb 18B7 to either ^{188}Re or ^{213}Bi converted this immunoglobulin into a fungicidal antibody in vitro. The mechanism of action for this phenomenon involved antibody binding to the polysaccharide capsule, placing the attached nuclide in close enough proximity to the fungal cell body so that when radioactive decay occurred, the cryptococcal cell was damaged by the emergent subatomic particles. Treatment of lethally infected mice with MAb 18B7 labeled with either ^{188}Re or ^{213}Bi prolonged survival and reduced fungal burden in both lungs and brain (24). Therapeutic effects of radiolabeled MAb 18B7 were observed even when mice had circulating polysaccharide antigen (GXM), suggesting that the affinity of the antibody for cryptococcal cells and/or tissue polysaccharide in the vicinity of cryptococci was much higher than for soluble antigen. RIT of fungal infection was associated with minimal toxicity consisting solely of a transient reduction in platelet counts. Evaluation of long-term survivors treated with RIT revealed no evidence of late toxicity such as pulmonary fibrosis (20). Subsequent studies have expanded on the initial observations to demonstrate that RIT is effective against high-inoculum infection (11) and cryptococcal biofilms (69) and that it does not select for radiation-resistant mutants among *C. neoformans* cells (11). In addition, it has been established that there are isotype-related differences in the efficacy of RIT (19) and that the predominant fungicidal effect is due to direct killing rather than a cross-fire effect (19).

The success of RIT against *C. neoformans* led to the application of this approach against experimental bacterial infection (21), HIV (26), and virally induced tumors (116). Hence, RIT for cryptococcal disease is a highly promising experimental approach. In contrast to unlabeled MAb, radiolabeled MAbs are potent microbicidal molecules that combine a new activity in the form of emitted cytotoxic radiation with classical antibody functions such as phagocytosis, complement activation, etc. Although there is always the theoretical concern that delivery of radiation to sites of infection will have the double-edged effect of also damaging immune cells, this concern has not been borne out with available evidence. Alpha particles such as those emitted by ^{213}Bi have very short trajectories in tissue, making them more likely to damage fungal cells than more distant immune cells. Even if cryptococci with attached radiolabeled antibody are phagocytosed, there appears to be only minimal damage to macrophages, and it is not clear that such damage is necessarily detrimental, given that live macrophages are major sites for fungal replication. Today there is increasing interest in RIT as a cancer therapy, and it is hoped that as this approach is further developed and its use becomes more commonplace, the logistical setup being put in place can be used for reintroducing antibody therapy for cryptococcosis in the form of RIT.

From Cryptococcal Melanin to Melanoma Therapy

Although not immediately related to vaccines and antibody therapies for *C. neoformans*, we relate an example of how research involving this fungus has led to the development of an experimental therapy against metastatic melanoma that is currently in clinical evaluation. In the late 1990s murine MAbs were made to *C. neoformans* melanin with the goal of using these reagents for the study of melanization in vivo and in vitro (101). Subsequently, it was hypothesized that if these MAbs bound to human melanin, they could be used to deliver tumoricidal radiation to melanoma cells. Since melanin is normally found intracellularly in human pigmented cells, the key insight that catalyzed this approach was to posit that there must be extracellular melanin in melanoma as a result of rapid cell turnover and that such melanin would be accessible to antibody. This hypothesis was validated when it was shown that a radiolabeled antibody to fungal melanin could be used to arrest tumor growth in nude mice harboring human melanoma tumors (25). Validation of this approach led to the development of ^{188}Re-labeled antibody to melanin for the treatment of melanoma, and such an antibody is now in clinical trials in patients with late-stage metastatic melanoma (55). For additional details and reviews of this approach see references 22 and 23.

REFERENCES

1. **Abadi, J., and L. Pirofski.** 1999. Antibodies reactive with the cryptococcal capsular polysaccharide glucuronoxylomannan are present in sera from children with and without HIV infection. *J. Infect. Dis.* **180:**915–919.
2. **Alpe, M., P. Svahnberg, and S. Oscarson.** 2003. Synthesis of *Cryptococcus neoformans* capsular polysaccharide structures. Part IV. Construction of thioglycoside donor blocks and their assembly. *J. Carbohydr. Chem.* **22:**565–577.
3. **Alpe, M., P. Svahnberg, and S. Oscarson.** 2004. Synthesis of *Cryptococcus neoformans* capsular polysaccharide structures. Part V. Construction of uronic acid-containing thioglycoside donor. *J. Carbohydr. Chem.* **23:**411–424.
4. **Baronetti, J. L., L. S. Chiapello, M. P. Aoki, S. Gea, and D. T. Masih.** 2006. Heat killed cells of *Cryptococcus neoformans* var. *grubii* induces protective immunity in rats: immunological and histopathological parameters. *Med. Mycol.* **44:**493–504.
5. **Beenhouwer, D. O., R. J. May, P. Valadon, and M. D. Scharff.** 2002. High affinity mimotope of the polysaccharide capsule of *Cryptococcus neoformans* identified from an evolutionary phage peptide library. *J. Immunol.* **169:**6992–6999.
6. **Beenhouwer, D. O., P. Valadon, R. May, and M. D. Scharff.** 2000. Peptide mimicry of the polysaccharide capsule of *Cryptococcus neoformans*, p. 143–160. *In* M. W. Cunningham and R. S. Fujinami (ed.), *Molecular Mimicry, Microbes and Autoimmunity*. ASM Press, Washington, DC.
7. **Beenhouwer, D. O., E. M. Yoo, C. W. Lai, M. A. Rocha, and S. L. Morrison.** 2007. Human immunoglobulin G2 (IgG2) and IgG4, but not IgG1 or IgG3, protect mice against *Cryptococcus neoformans* infection. *Infect. Immun.* **75:**1424–1435.
8. **Bennett, J. E., and H. F. Hasenclever.** 1965. *Cryptococcus neoformans* polysaccharide: studies of serologic properties and role in infection. *J. Immunol.* **94:**916–920.
9. **Biondo, C., G. Mancuso, A. Midiri, M. Bombaci, L. Messina, C. Beninati, and G. Teti.** 2006. Identification of major proteins secreted by *Cryptococcus neoformans*. *FEMS Yeast Res.* **6:**645–651.
10. **Bonham, S., D. B. Meya, P. R. Bohjanen, and D. R. Boulware.** 2008. Biomarkers of HIV immune reconstitution inflammatory syndrome. *Biomark. Med.* **2:**349–361.
11. **Bryan, R. A., Z. Jiang, X. Huang, A. Morgenstern, F. Bruchertseifer, R. Sellers, A. Casadevall, and E. Dadachova.** 2009. Radioimmunotherapy is effective against high-inoculum *Cryptococcus neoformans* infection in mice and does not select for radiation-resistant cryptococcal cells. *Antimicrob. Agents Chemother.* **53:**1679–1682.
12. **Buchanan, K. L., and J. W. Murphy.** 1993. Characterization of cellular infiltrates and cytokine production during the expression of the anticryptococcal delayed-type hypersensitivity response. *Infect. Immun.* **61:**2854–2865.
13. **Casadevall, A., W. Cleare, M. Feldmesser, A. Glatman-Freedman, D. L. Goldman, T. R. Kozel, N. Lendvai,**

J. Mukherjee, L. Pirofski, J. Rivera, A. L. Rosas, M. D. Scharff, P. Valadon, K. Westin, and Z. Zhong. 1998. Characterization of a murine monoclonal antibody to *Cryptocococcus neoformans* polysaccharide that is a candidate for human therapeutic studies. *Antimicrob. Agents Chemotherap.* **42:**1437–1446.

14. **Casadevall, A., J. Mukherjee, S. J. N. Devi, R. Schneerson, J. B. Robbins, and M. D. Scharff.** 1992. Antibodies elicited by a *Cryptococcus neoformans* glucuronoxylomannan-tetanus toxoid conjugate vaccine have the same specificity as those elicited in infection. *J. Infect. Dis.* **65:**1086–1093.
15. **Casadevall, A., and L. Pirofski.** 2003. Exploiting the redundancy of the immune system: vaccines can mediate protection by eliciting 'unnatural' immunity. *J. Exp. Med.* **197:**1401–1404.
16. **Casadevall, A., and L. A. Pirofski.** 2005. Feasibility and prospects for a vaccine to prevent cryptococcosis. *Med. Mycol.* **43:**667–680.
17. **Chen, L.-C., D. L. Goldman, T. L. Doering, L. Pirofski, and A. Casadevall.** 1999. Antibody response to *Cryptococcus neoformans* proteins in rodents and humans. *Infect. Immun.* **67:**2218–2224.
18. **Cutler, J. E., G. S. Deepe, Jr., and B. S. Klein.** 2007. Advances in combating fungal diseases: vaccines on the threshold. *Nat. Rev. Microbiol.* **5:**13–28.
19. **Dadachova, E., R. A. Bryan, C. Apostolidis, A. Morgenstern, T. Zhang, T. Moadel, M. Torres, X. Huang, E. Revskaya, and A. Casadevall.** 2006. Interaction of radiolabeled antibodies with fungal cells and components of the immune system in vitro and during radioimmunotherapy for experimental fungal infection. *J. Infect. Dis.* **193:**1427–1436.
20. **Dadachova, E., R. A. Bryan, A. Frenkel, T. Zhang, C. Apostolidis, J. S. Nosanchuk, J. D. Nosanchuk, and A. Casadevall.** 2004. Evaluation of acute hematologic and long-term pulmonary toxicities of radioimmunotherapy of *Cryptococcus neoformans* infection in murine models. *Antimicrob. Agents Chemother.* **48:**1004–1006.
21. **Dadachova, E., T. Burns, R. A. Bryan, C. Apostolidis, M. W. Brechbiel, J. D. Nosanchuk, A. Casadevall, and L. Pirofski.** 2004. Feasibility of radioimmunotherapy of experimental pneumococcal infection. *Antimicrob. Agents Chemother.* **48:**1624–1629.
22. **Dadachova, E., and A. Casadevall.** 2005. Melanin as a potential target for radionuclide therapy of metastatic melanoma. *Future Oncol.* **1:**541–549.
23. **Dadachova, E., and A. Casadevall.** 2006. Renaissance of targeting molecules for melanoma. *Cancer Biother. Radiopharm.* **21:**545–552.
24. **Dadachova, E., A. Nakouzi, R. A. Bryan, and A. Casadevall.** 2003. Ionizing radiation delivered by specific antibody is therapeutic against a fungal infection. *Proc. Natl. Acad. Sci. USA* **100:**10942–10947.
25. **Dadachova, E., J. D. Nosanchuk, S. Li, A. D. Schweitzer, A. Frenkel, J. S. Nosanchuk, and A. Casadevall.** 2004. Dead cells in melanoma tumors provide abundant antigen for targeted delivery of ionizing radiation by a monoclonal antibody to melanin. *Proc. Natl. Acad. Sci. USA* **101:**14865–14870.
26. **Dadachova, E., M. C. Patel, S. Toussi, C. Apostolidis, A. Morgenstern, M. W. Brechbiel, M. K. Gorny, S. Zolla-Pazner, A. Casadevall, and H. Goldstein.** 2006. Targeted killing of virally infected cells by radiolabeled antibodies to viral proteins. *PLoS. Med.* **3:**e427.
27. **Datta, K., A. Lees, and L. A. Pirofski.** 2008. Therapeutic efficacy of a conjugate vaccine containing a peptide mimotope of cryptococcal capsular polysaccharide glucuronoxylomannan. *Clin. Vaccine Immunol.* **15:**1176–1187.
28. **Datta, K., and L. A. Pirofski.** 2006. Towards a vaccine for *Cryptococcus neoformans*: principles and caveats. *FEMS Yeast Res.* **6:**525–536.
29. **Deepe, G. S., Jr.** 2004. Preventative and therapeutic vaccines for fungal infections: from concept to implementation. *Expert.Rev. Vaccines* **3:**701–709.
30. **De Jesus, M., A. M. Nicola, S. Frases, I. R. Lee, S. Mieses, and A. Casadevall.** 2009. Galactoxylomannan-mediated immunological paralysis results from specific B cell depletion in the context of widespread immune system damage. *J. Immunol.* **183:**3885–3894.
31. **Devi, S. J. N.** 1996. Preclinical efficacy of a glucuronoxylomannan-tetanus toxoid conjugate vaccine of *Cryptococcus neoformans* in a murine model. *Vaccine* **14:**841–842.
32. **Devi, S. J. N., R. Schneerson, W. Egan, T. J. Ulrich, D. Bryla, J. B. Robbins, and J. E. Bennett.** 1991. *Cryptococcus neoformans* serotype A glucuronoxylomannan-protein conjugate vaccines: synthesis, characterization, and immunogenicity. *Infect. Immun.* **59:**3700–3707.
33. **Dromer, F., and J. Charreire.** 1991. Improved amphotericin B activity by a monoclonal anti-*Cryptococcus neoformans* antibody: study during murine cryptococcosis and mechanisms of action. *J. Infect. Dis.* **163:**1114–1120.
34. **Dromer, F., J. Charreire, A. Contrepois, C. Carbon, and P. Yeni.** 1987. Protection of mice against experimental cryptococcosis by anti-*Cryptococcus neoformans* monoclonal antibody. *Infect. Immun.* **55:**749–752.
35. **Dromer, F., J. Salamero, A. Contrepois, C. Carbon, and P. Yeni.** 1987. Production, characterization, and antibody specificity of a mouse monoclonal antibody reactive with *Cryptococcus neoformans* capsular polysaccharide. *Infect. Immun.* **55:**742–748.
36. **Eigenheer, R. A., L. Y. Jin, E. Blumwald, B. S. Phinney, and A. Gelli.** 2007. Extracellular glycosylphosphatidylinositol-anchored mannoproteins and proteases of *Cryptococcus neoformans*. *FEMS Yeast Res.* **7:**499–510.
37. **Feldmesser, M., Y. Kress, P. Novikoff, and A. Casadevall.** 2000. *Cryptococcus neoformans* is a facultative intracellular pathogen in murine pulmonary infection. *Infect. Immun.* **68:**4225–4237.
38. **Feldmesser, M., J. Mukherjee, and A. Casadevall.** 1996. Combination of 5-flucytosine and capsule binding monoclonal antibody in therapy of murine *Cryptococcus neoformans* infections and in vitro. *J. Antimicrob. Chemother.* **37:**617–622.
39. **Fleuridor, R., A. Lees, and L. Pirofski.** 2001. A cryptococcal capsular polysaccharide mimotope prolongs the survival of mice with *Cryptococcus neoformans* infection. *J. Immunol.* **166:**1087–1096.
40. **Fleuridor, R., R. H. Lyles, and L. Pirofski.** 1999. Quantitative and qualitative differences in the serum antibody profiles of human immunodeficiency virus-infected persons with and without *Cryptococcus neoformans* meningitis. *J. Infect. Dis.* **180:**1526–1535.
41. **Fleuridor, R., Z. Zhong, and L. Pirofski.** 1998. A human IgM monoclonal antibody prolongs survival of mice with lethal cryptococcosis. *J. Infect. Dis.* **178:**1213–1216.
42. **Frases, S., L. Nimrichter, N. B. Viana, A. Nakouzi, and A. Casadevall.** 2008. *Cryptococcus neoformans* capsular polysaccharide and exopolysaccharide fractions manifest physical, chemical, and antigenic differences. *Eukaryot. Cell* **7:**319–327.
43. **Garegg, P. J., L. Olsson, and S. Oscarson.** 1996. Synthesis of oligosaccharides corresponding to structures found in capsular polysaccharides of *Cryptococcus neoformans*—II. *Bioorg. Med. Chem.* **4:**1867–1871.
44. **Goldman, D., S. C. Lee, and A. Casadevall.** 1994. Pathogenesis of pulmonary *Cryptococcus neoformans* infection in the rat. *Infect. Immun.* **62:**4755–4761.

45. **Goldman, D. L., A. Casadevall, and L. S. Zuckier.** 1997. Pharmacokinetics and biodistribution of a monoclonal antibody to *Cryptococcus neoformans* capsular polysaccharide antigen in a rat model of cryptococcal meningitis: implications for passive immunotherapy. *J. Med. Vet. Mycol.* **35:**271–278.
46. **Goldman, D. L., J. Davis, F. Bommartio, X. Shao, and A. Casadevall.** 2006. Enhanced allergic inflammation and airway responsiveness in rats with chronic *Cryptococcus neoformans* infection suggests a potential role for fungal pulmonary infection in the pathogenesis of asthma. *J. Infect. Dis.* **193:**1178–1186.
47. **Goldman, D. L., H. Khine, J. Abadi, D. J. Lindenberg, L. Pirofski, R. Niang, and A. Casadevall.** 2001. Serologic evidence for *Cryptococcus infection* in early childhood. *Pediatrics* **107:**E66.
48. **Gordon, M. A., and A. Casadevall.** 1995. Serum therapy of cryptococcal meningitis. *Clin. Infect. Dis.* **21:**1477–1479.
49. **Goren, M. B.** 1967. Experimental murine cryptococcosis: effect of hyperimmunization to capsular polysaccharide. *J. Immunol.* **98:**914–922.
50. **Goren, M. B., and G. M. Middlebrook.** 1967. Protein conjugates of polysaccharide from *Cryptococcus neoformans. J. Immunol.* **98:**901–913.
51. **Guimaraes, A. J., S. Frases, F. J. Gomez, R. M. Zancope-Oliveira, and J. D. Nosanchuk.** 2009. Monoclonal antibodies to heat shock protein 60 alter the pathogenesis of *Histoplasma capsulatum. Infect. Immun.* **77:**1357–1367.
52. **Huang, C., S. H. Nong, M. K. Mansour, C. A. Specht, and S. M. Levitz.** 2002. Purification and characterization of a second immunoreactive mannoprotein from *Cryptococcus neoformans* that stimulates T-cell responses. *Infect. Immun.* **70:**5485–5493.
53. **Kakeya, H., H. Udono, N. Ikuno, Y. Yamamoto, K. Mitsutake, T. Miyazaki, K. Tomono, H. Koga, T. Tashiro, E. Nakayama, and S. Kohno.** 1997. A 77-kilodalton protein of *Cryptococcus neoformans*, a member of the heat shock protein 70 family, is a major antigen detected in the sera of mice with pulmonary cryptococcosis. *Infect. Immun.* **65:**1653–1658.
54. **Kakeya, H., H. Udono, S. Maesaki, E. Sasaki, S. Kawamura, M. A. Hossain, Y. Yamamoto, T. Sawai, M. Fukuda, K. Mitsutake, Y. Miyazaki, K. Tomono, T. Tashiro, E. Nakayama, and S. Kohno.** 1999. Heat shock protein 70 (hsp70) as a major target of the antibody response in patients with pulmonary cryptococcosis. *Clin. Exp. Immunol.* **115:**485–490.
55. **Klein, M., N. Shibli, N. Friedmann, G. B. Thornton, R. Chisin, and M. Lotem.** 2008. Imaging of metastatic melanoma (MM) with a 188Rhenium(188Re)-labeled melanin binding antibody. *J. Nucl. Med.* **49**(Suppl 1)**:**52P.
56. **Kozel, T. R., and J. Cazin, Jr.** 1972. Immune response to *Cryptococcus neoformans* soluble polysaccharide. *Infect. Immun.* **5:**35–41.
57. **Kozel, T. R., W. F. Gulley, and J. J. Cazin.** 1977. Immune response to *Cryptococcus neoformans* soluble polysaccharide: immunological unresponsiveness. *Infect. Immun.* **18:**701–707.
58. **Lam, J. S., M. K. Mansour, C. A. Specht, and S. M. Levitz.** 2005. A model vaccine exploiting fungal mannosylation to increase antigen immunogenicity. *J. Immunol.* **175:**7496–7503.
59. **Larsen, R. A., P. G. Pappas, J. R. Perfect, J. A. Aberg, A. Casadevall, G. A. Cloud, R. James, S. Filler, and W. E. Dismukes.** 2005. Phase I evaluation of the safety and pharmacokinetics of murine-derived anticryptococcal antibody 18B7 in subjects with treated cryptococcal meningitis. *Antimicrob. Agents Chemother.* **49:** 952–958.
60. **Levitz, S. M., S. Nong, M. K. Mansour, C. Huang, and C. A. Specht.** 2001. Molecular characterization of a mannoprotein with homology to chitin deacetylases that stimulates T cell responses to *Cryptococcus neoformans. Proc. Natl. Acad. Sci. USA* **98:**10422–10427.
61. **Levitz, S. M., and E. A. North.** 1997. Lymphopriliferation and cytokine profiles in human peripheral blood mononuclear cells stimulated by *Cryptococcus neoformans. J. Med. Vet. Mycol.* **35:**229–236.
62. **Littman, M. L.** 1959. Cryptococcosis (torulosis). Current concepts and therapy. *Am. J. Med.* **27:**976–998.
63. **Lortholary, O., A. Fontanet, N. Memain, A. Martin, K. Sitbon, and F. Dromer.** 2005. Incidence and risk factors of immune reconstitution inflammatory syndrome complicating HIV-associated cryptococcosis in France. *AIDS* **19:**1043–1049.
64. **Maitta, R., K. Datta, Q. Chang, R. Luo, K. Subramanian, B. Witover, and L. Pirofski.** 2004. Protective and non-protective human IgM monoclonal antibodies to *Cryptococcus neoformans* glucuronoxylomannan manifest different specificity and gene usage. *Infect. Immun.* **22:**4062–4068.
65. **Maitta, R., K. Datta, A. Lees, S. Belouski, and L. Pirofski.** 2003. Immunogenicity and efficacy of cryptococcal glucuronoxylomannan peptide mimotope-protein conjugates in human immunoglobulin transgenic mice. *Infect. Immun.* **72:**196–208.
66. **Mansour, M. K., E. Latz, and S. M. Levitz.** 2006. *Cryptococcus neoformans* glycoantigens are captured by multiple lectin receptors and presented by dendritic cells. *J. Immunol.* **176:**3053–3061.
67. **Mansour, M. K., L. S. Schlesinger, and S. M. Levitz.** 2002. Optimal T cell responses to *Cryptococcus neoformans* mannoprotein are dependent on recognition of conjugated carbohydrates by mannose receptors. *J. Immunol.* **168:**2872–2879.
68. **Mansour, M. K., L. E. Yauch, J. B. Rottman, and S. M. Levitz.** 2004. Protective efficacy of antigenic fractions in mouse models of cryptococcosis. *Infect. Immun.* **72:**1746–1754.
69. **Martinez, L. R., R. A. Bryan, C. Apostolidis, A. Morgenstern, A. Casadevall, and E. Dadachova.** 2006. Antibody-guided alpha radiation effectively damages fungal biofilms. *Antimicrob. Agents Chemother.* **50:**2132–2136.
70. **Matthews, R., S. Hodgetts, and J. Burnie.** 1995. Preliminary assessment of a human recombinant antibody fragment to hsp90 in murine invasive candidiasis. *J. Infect. Dis.* **171:**1668–1771.
71. **Matthews, R. C., G. Rigg, S. Hodgetts, T. Carter, C. Chapman, C. Gregory, C. Illidge, and J. Burnie.** 2003. Preclinical assessment of the efficacy of mycograb, a human recombinant antibody against fungal HSP90. *Antimicrob. Agents Chemother.* **47:**2208–2216.
72. **May, R. J., D. O. Beenhouwer, and M. D. Scharff.** 2003. Antibodies to keyhole limpet hemocyanin cross-react with an epitope on the polysaccharide capsule of *Cryptococcus neoformans* and other carbohydrates: implications for vaccine development. *J. Immunol.* **171:**4905–4912.
73. **McFadden, D. C., and A. Casadevall.** 2004. Unexpected diversity in the fine specificity of monoclonal antibodies that use the same V region gene to glucuronoxylomannan of *Cryptococcus neoformans. J. Immunol.* **172:**3670–3677.
74. **McLean, G. R., M. Torres, N. Elguezabal, A. Nakouzi, and A. Casadevall.** 2002. Isotype can affect the fine specificity of an antibody for a polysaccharide antigen. *J. Immunol.* **169:**1379–1386.
75. **Moragues, M. D., M. J. Omaetxebarria, N. Elguezabal, M. J. Sevilla, S. Conti, L. Polonelli, and J. Ponton.** 2003. A monoclonal antibody directed against a *Candida albicans*

cell wall mannoprotein exerts three anti-*C. albicans* activities. *Infect. Immun.* **71:**5273–5279.

76. **Mukherjee, J., M. Feldmesser, M. D. Scharff, and A. Casadevall.** 1995. Monoclonal antibodies to *Cryptococcus neoformans* glucuronoxylomannan enhance fluconazole activity. *Antimicrob. Agents Chemother.* **39:**1398–1405.
77. **Mukherjee, J., G. Nussbaum, M. D. Scharff, and A. Casadevall.** 1995. Protective and non-protective monoclonal antibodies to *Cryptococcus neoformans* originating from one B-cell. *J. Exp. Med.* **181:**405–409.
78. **Mukherjee, J., M. D. Scharff, and A. Casadevall.** 1992. Protective murine monoclonal antibodies to *Cryptococcus neoformans. Infect. Immun.* **60:**4534–4541.
79. **Mukherjee, J., L. Zuckier, M. D. Scharff, and A. Casadevall.** 1994. Therapeutic efficacy of monoclonal antibodies to *Cryptococcus neoformans* glucuronoxylomannan alone and in combination with amphotericin B. *Antimicrob. Agents Chemother.* **38:**580–587.
80. **Murphy, J. W.** 1988. Influence of cryptococcal antigens on cell-mediated immunity. *Rev. Infect. Dis.* **10**(Suppl. 2)**:** S432–S435.
81. **Murphy, J. W., and G. C. Cozad.** 1972. Immunological unresponsiveness induced by cryptococcal polysaccharide assayed by the hemolytic plaque technique. *Infect. Immun.* **5:**896–901.
82. **Murphy, J. W., R. L. Mosley, R. Cherniak, G. H. Reyes, T. R. Kozel, and E. Reiss.** 1988. Serological, electrophoretic, and biological properties of *Cryptococcus neoformans* antigens. *Infect. Immun.* **56:**424–431.
83. **Murphy, J. W., F. Shafer, A. Casadevall, and A. Adesina.** 1998. Antigen-induced protective and non-protective cell-mediated immune components against *Cryptococcus neoformans. Infect. Immun.* **66:**2632–2639.
84. **Nakouzi, A., and A. Casadevall.** 2003. The function of conserved amino acids at or near the complementarity determining regions for related antibodies to *Cryptococcus neoformans* glucuronoxylomannan. *Mol. Immunol.* **40:** 351–361.
85. **Nakouzi, A., P. Valadon, J. D. Nosanchuk, N. S. Green, and A. Casadevall.** 2001. Molecular basis for immunoglobulin M specificity to epitopes in *Cryptococcus neoformans* that elicit protective and non-protective antibodies. *Infect. Immun.* **69:**3398–3409.
86. **Nakouzi, A., T. Zhang, S. Oscarson, and A. Casadevall.** 2009. The common *Cryptococcus neoformans* glucuronoxylomannan M2 motif elicits non-protective antibodies. *Vaccine* **27:**3513–3518.
87. **Nooney, L., R. C. Matthews, and J. P. Burnie.** 2005. Evaluation of Mycograb, amphotericin B, caspofungin, and fluconazole in combination against *Cryptococcus neoformans* by checkerboard and time-kill methodologies. *Diagn. Microbiol. Infect. Dis.* **51:**19–29.
88. **Orendi, J. M., A. F. M. Verheul, N. M. de Vos, M. R. Visser, H. Snippe, R. Cherniak, V. V. Vaishnav, G. T. Rijkers, and J. Verhoef.** 1997. Mannoproteins of *Cryptococcus neoformans* induce proliferative response in human peripheral blood mononuclear cells (PBMC) and enhance HIV replication. *Clin. Exp. Immunol.* **107:**293–299.
89. **Oscarson, S., M. Alpe, P. Svahnberg, A. Nakouzi, and A. Casadevall.** 2005. Synthesis and immunological studies of glycoconjugates of *Cryptococcus neoformans* capsular glucuronoxylomannan oligosaccharide structures. *Vaccine* **23:**3961–3972.
90. **Perfect, J. R., and A. Casadevall.** 2002. Cryptococcosis. *Infect. Dis. Clin. North Am.* **16:**837–874, v–vi.
91. **Pietrella, D., R. Cherniak, C. Strappini, S. Perito, P. Mosci, F. Bistoni, and A. Vecchiarelli.** 2001. Role of mannoprotein in induction and regulation of immunity to *Cryptococcus neoformans. Infect. Immun.* **69:**2808–2814.
92. **Pietrella, D., R. Mazzolla, P. Lupo, L. Pitzurra, M. J. Gomez, R. Cherniak, and A. Vecchiarelli.** 2002. Mannoprotein from *Cryptococcus neoformans* promotes T-helper type 1 anticandidal responses in mice. *Infect. Immun.* **70:**6621–6627.
93. **Pirofski, L., R. Lui, M. DeShaw, A. B. Kressel, and Z. Zhong.** 1995. Analysis of human monoclonal antibodies elicited by vaccination with a *Cryptococcus neoformans* glucuronoxylomannan capsular polysaccharide vaccine. *Infect. Immun.* **63:**3005–3014.
94. **Pitzurra, L., S. Perito, F. Baldelli, F. Bistoni, and A. Vecchiarelli.** 2003. Humoral response against *Cryptococcus neoformans* mannoprotein antigens in HIV-infected patients. *Clin. Exp. Immunol.* **133:**91–96.
95. **Pitzurra, L., A. Vecchiarelli, R. Peducci, A. Cardinali, and F. Bistoni.** 1997. Identification of a 105 kilodalton *Cryptococcus neoformans* mannoprotein involved in human cell-mediated immune response. *J. Med. Vet. Mycol.* **35:**299–303.
96. **Rachini, A., D. Pietrella, P. Lupo, A. Torosantucci, P. Chiani, C. Bromuro, C. Proietti, F. Bistoni, A. Cassone, and A. Vecchiarelli.** 2007. An anti-{beta} glucan monoclonal antibody inhibits growth and capsule formation of *Cryptococcus neoformans* in vitro and exerts therapeutic, anti-cryptococcal activity in vivo. *Infect. Immun.* **75:**5085–5094.
97. **Rivera, J., and A. Casadevall.** 2005. Mouse genetic background is a major determinant of isotype-related differences for antibody-mediated protective efficacy against *Cryptococcus neoformans. J. Immunol.* **174:**8017–8026.
98. **Robbins, J. B., R. Schneerson, P. Anderson, and D. H. Smith.** 1996. The 1996 Albert Lasker Medical Research Awards. Prevention of systemic infections, especially meningitis, caused by *Haemophilus influenzae* type b. Impact on public health and implications for other polysaccharide-based vaccines. *JAMA* **276:**1181–1185.
99. **Rodrigues, M. L., L. R. Travassos, K. R. Miranda, A. J. Franzen, S. Rozental, W. De Souza, C. S. Alviano, and E. Barreto-Bergter.** 2000. Human antibodies against a purified glucosylceramide from *Cryptococcus neoformans* inhibit cell budding and fungal growth. *Infect. Immun.* **68:**7049–7060.
100. **Rosas, A. L., J. D. Nosanchuk, and A. Casadevall.** 2001. Passive immunization with melanin-binding monoclonal antibodies prolongs survival in mice with lethal *Cryptococcus neoformans* infection. *Infect. Immun.* **69:** 3410–3412.
101. **Rosas, A. L., J. D. Nosanchuk, M. Feldmesser, G. M. Cox, H. C. McDade, and A. Casadevall.** 2000. Synthesis of polymerized melanin by *Cryptococcus neoformans* in infected rodents. *Infect. Immun.* **68:**2845–2853.
102. **Sanford, J. E., D. M. Lupan, A. M. Schlagetter, and T. R. Kozel.** 1990. Passive immunization against *Cryptococcus neoformans* with an isotype-switch family of monoclonal antibodies reactive with cryptococcal polysaccharide. *Infect. Immun.* **58:**1919–1923.
103. **Segal, E.** 1987. Vaccines against fungal infections. *CRC Crit. Rev. Microbiol.* **14:**229–273.
104. **Singh, N., O. Lortholary, B. D. Alexander, K. L. Gupta, G. T. John, K. Pursell, P. Munoz, G. B. Klintmalm, V. Stosor, R. del Busto, A. P. Limaye, J. Somani, M. Lyon, S. Houston, A. A. House, T. L. Pruett, S. Orloff, A. Humar, L. Dowdy, J. Garcia-Diaz, A. C. Kalil, R. A. Fisher, and S. Husain.** 2005. An immune reconstitution syndrome-like illness associated with *Cryptococcus neoformans* infection in organ transplant recipients. *Clin. Infect. Dis.* **40:**1756–1761.
105. **Subramaniam, K., B. Metzger, L. H. Hanau, A. Guh, L. Rucker, S. Badri, and L. A. Pirofski.** 2009. IgM(+)

memory B cell expression predicts HIV-associated cryptococcosis status. *J. Infect. Dis.* **200:**244–251.
106. **Subramaniam, K. S., N. French, and L. Pirofski.** 2005. *Cryptococcus neoformans*-reactive and total immunoglobulin profiles of human immunodeficiency virus-infected and uninfected Ugandans. *Clin. Diagn. Lab Immunol.* **12:**1168–1176.
107. **Taborda, C. P., and A. Casadevall.** 2001. Immunoglobulin M efficacy against *Cryptococcus neoformans*: mechanism, dose dependence and prozone-like effects in passive protection experiments. *J. Immunol.* **66:**2100–2107.
108. **Taborda, C. P., J. Rivera, O. Zaragoza, and A. Casadevall.** 2003. More is not necessarily better: "Prozone-like" effects in passive immunization with immunoglobulin G. *J. Immunol.* **140:**3621–3630.
109. **Torosantucci, A., C. Bromuro, P. Chiani, F. De Bernardis, F. Berti, C. Galli, F. Norelli, C. Bellucci, L. Polonelli, P. Costantino, R. Rappuoli, and A. Cassone.** 2005. A novel glyco-conjugate vaccine against fungal pathogens. *J. Exp. Med.* **202:**597–606.
110. **Torosantucci, A., P. Chiani, C. Bromuro, F. De Bernardis, A. S. Palma, Y. Liu, G. Mignogna, B. Maras, M. Colone, A. Stringaro, S. Zamboni, T. Feizi, and A. Cassone.** 2009. Protection by anti-beta-glucan antibodies is associated with restricted beta-1,3 glucan binding specificity and inhibition of fungal growth and adherence. *PLoS ONE* **4:**e5392.
111. **Valadon, P., G. Nussbaum, L. F. Boyd, D. H. Margulies, and M. D. Scharff.** 1996. Peptide libraries define the fine specificity of anti-polysaccharide antibodies to *Cryptococcus neoformans*. *J. Mol. Biol.* **261:**11–22.
112. **Vecchiarelli, A.** 2000. Immunoregulation by capsular components of *Cryptococcus neoformans*. *Med. Mycol.* **38:**407–417.
113. **Vecchiarelli, A.** 2007. Fungal capsular polysaccharide and T-cell suppression: the hidden nature of poor immunogenicity. *Crit Rev. Immunol.* **27:**547–557.
114. **Vesely, J., L. Rydner, and S. Oscarson.** 2008. Variant synthetic pathway to glucuronic acid-containing di- and trisaccharide thioglycoside building blocks for continued synthesis of *Cryptococcus neoformans* capsular polysaccharide structures. *Carbohydr. Res.* **343:**2200–2208.
115. **Walzl, G., I. R. Humphreys, B. G. Marshall, L. Edwards, P. J. Openshaw, R. J. Shaw, and T. Hussell.** 2003. Prior exposure to live *Mycobacterium bovis* BCG decreases *Cryptococcus neoformans*-induced lung eosinophilia in a gamma interferon-dependent manner. *Infect. Immun.* **71:**3384–3391.
116. **Wang, X. G., E. Revskaya, R. A. Bryan, H. D. Strickler, R. D. Burk, A. Casadevall, and E. Dadachova.** 2007. Treating cancer as an infectious disease: viral antigens as novel targets for treatment and potential prevention of tumors of viral etiology. *PLoS ONE* **2:**e1114.
117. **Williams, A. E., L. Edwards, I. R. Humphreys, R. Snelgrove, A. Rae, R. Rappuoli, and T. Hussell.** 2004. Innate imprinting by the modified heat-labile toxin of *Escherichia coli* (LTK63) provides generic protection against lung infectious disease. *J. Immunol.* **173:**7435–7443.
118. **Wormley, F. L., Jr., G. M. Cox, and J. R. Perfect.** 2005. Evaluation of host immune responses to pulmonary cryptococcosis using a temperature-sensitive *C. neoformans* calcineurin A mutant strain. *Microb. Pathog.* **38:**113–123.
119. **Wormley, F. L., Jr., J. R. Perfect, C. Steele, and G. M. Cox.** 2007. Protection against cryptococcosis by using a murine gamma interferon-producing *Cryptococcus neoformans* strain. *Infect. Immun.* **75:**1453–1462.
120. **Wozniak, K. L., and S. M. Levitz.** 2009. Isolation and purification of antigenic components of *Cryptococcus*. *Methods Mol. Biol.* **470:**71–83.
121. **Yuan, R., A. Casadevall, J. Oh, and M. D. Scharff.** 1997. T cells cooperate with passive antibody to modify *Cryptococcus neoformans* infection in mice. *Proc. Natl. Acad. Sci. USA* **94:**2483–2488.
122. **Zebedee, S. L., R. K. Koduri, J. Mukherjee, S. Mukherjee, S. Lee, D. F. Sauer, M. D. Scharff, and A. Casadevall.** 1994. Mouse-human immunoglobulin G1 chimeric antibodies with activity against *Cryptococcus neoformans*. *Antimicrob. Agents Chemother.* **38:**1507–1514.
123. **Zhang, H., Z. Zhong, and L. Pirofski.** 1997. Peptide epitopes recognized by a human anti-cryptococcal glucuronoxylomannan antibody. *Infect. Immun.* **65:**1158–1164.
124. **Zhang, J., and F. Kong.** 2003. Facile syntheses of the hexasaccharide repeating unit of the exopolysaccharide from *Cryptococcus neoformans* serovar A. *Bioorg. Med. Chem.* **11:**4027–4037.
125. **Zhang, J., and F. Kong.** 2003. Synthesis of a hexasaccharide, the repeating unit of O-deacetylated GXM of *C. neoformans* serotype A. *Carbohydr. Res.* **338:** 1719–1725.
126. **Zhong, Z., and L. Pirofski.** 1996. Opsonization of *Cryptococcus neoformans* by human anticryptococcal glucuronoxylomannan antibodies. *Infect. Immun.* **64:**3446–3450.
127. **Zhong, Z., and L. Pirofski.** 1998. Antifungal activity of a human anti-glucuronoxylomannan antibody. *Clin. Diagn. Lab. Immunol.* **5:**58–64.

Cryptococcus: From Human Pathogen to Model Yeast
Edited by J. Heitman et al.
©2011 ASM Press, Washington, DC

41

Diagnostic Approach Based on Capsular Antigen, Capsule Detection, β-Glucan, and DNA Analysis

MARA R. DIAZ AND M. HONG NGUYEN

DETECTION OF THE POLYSACCHARIDE CAPSULE

The polysaccharide capsule is a unique feature of *Cryptococcus neoformans* that distinguishes the organism from other medically important yeasts. As such, a variety of techniques that detect the cryptococcal capsule and its components have been developed as diagnostic tests. These techniques include direct visual detection of the capsule through stains, such as India ink and mucin stains, and serological assays.

Visual Detection of Capsule

India Ink

The cryptococcocal polysaccharide capsule can be detected in smear specimens of cerebrospinal fluid (CSF), urine, sputum, tissue, or other infected materials by India ink or colloidal carbon preparations. India ink staining is a rapid test in which equal volumes of clinical sample and India ink are mixed. The capsule excludes the ink particles and thereby appears as a halo around the yeast organisms within a dark background. The yeast cells of *Cryptococcus* are spherical in shape and approximately 5 to 7 μm in diameter; the capsules vary in their thickness and are of a few micrometers on the average. Advantages of India ink staining include its technical simplicity and the rapid turnaround time of a few minutes. Disadvantages include its poor sensitivity. For example, the sensitivity of India ink in diagnosing cryptococcal meningitis in non-HIV-infected patients is approximately 30 to 72% compared with culture (146, 152). The sensitivity in diagnosing disease in HIV-infected patients is higher (approximately 80%) (80). The discrepancy in sensitivities likely stems from greater infectious burdens in HIV-infected patients. Indeed, the lower limit of detection of India ink is approximately 10^3 to 10^4 yeast cells/ml CSF (33). Sensitivity might be improved by testing pellets from centrifuged CSF samples.

India ink preparations can also give false-positive results, particularly with inexperienced readers who might mistake lymphocytes or fat droplets for cryptococci. This problem has been exacerbated over the past few years, as many clinical laboratories have eliminated routine India ink examinations of CSF in favor of exclusive use of the cryptococcal antigen test (see below). If antigen testing is done in real time on CSF samples as they are received, the elimination of India ink is not likely to impact patient care. For clinical laboratories that batch cryptococcal antigen tests or send samples elsewhere for testing, however, India ink remains the major work force, since the timely results might impact management decisions.

Histological Stains

In tissue sections, Gomori methenamine silver and calcofluor white are broad-spectrum fungal histochemical stains that detect fungi regardless of type. Some commonly available histochemical stains, on the other hand, stain certain types of fungal organisms but not others. These more narrow-spectrum stains can be useful for differential diagnosis. Various histochemical stains react with cryptococci, including mucin stains such as mucicarmine, alcian blue, and periodic acid-Schiff that stain the capsule and Fontana-Masson stain, which interacts with the cell wall.

Mucicarmine stains the cryptococcal polysaccharide capsule, revealing organisms as bright carmine red, often with a spiny scalloped appearance (112, 166). Mucicarmine is specific for cryptococci, although it may also weakly stain the cell wall of *Blastomyces*. It does not stain other yeasts such as *Coccidioides*, *Histoplasma*, or *Candida*. Clinicians should recognize that mucicarmine stain may not detect capsule-deficient cryptococci. Alcian blue and periodic acid-Schiff also stain the polysaccharide capsule. Fontana-Masson stain detects melanin and other silver-reducing granules in the cell wall (95) and stains the wall dark brown to black. This stain is particularly useful for detecting capsule-deficient *C. neoformans* (145). Fontana-Masson stain also interacts with the walls of dematiaceous fungi and *Sporothrix schenckii*, and on occasion it will stain spherules of *Coccidioides* dark brown to black. It does not stain *Candida*, *Blastomyces*, *Histoplasma*, *Paracoccidioides*, *Rhizopus*, or *Aspergillus*.

Combinations of histochemical stains have been shown to improve the detection of cryptococci within tissues (101, 102).

Mara R. Diaz, Rosenstiel School of Marine and Atmospheric Science, University of Miami, Miami, FL 33149. **M. Hong Nguyen,** University of Pittsburgh, Pittsburgh, PA 15261.

In particular, combining mucicarmine or alcian blue stains with Fontana-Masson staining has been shown to be superior to any of the stains alone in identifying organisms within tissue samples. This combination stain distinctively highlights both the cell wall (Fontana-Masson) and mucin-positive capsule (mucicarmine or alcian blue stains) (101). An alternative approach is to subsequently use Fontana-Masson stain for mucin-stain-negative samples that might harbor capsule-negative cryptococci. Gomori methenamine silver will also detect capsule-deficient strains.

Cryptococcal Antigen

Detection of capsular polysaccharide antigen is the most valuable rapid test for the laboratory diagnosis of cryptococcosis. During infection, the capsular polysaccharide is solubilized in body fluids such as CSF and serum, and it can be detected and quantified with specific rabbit anti–*C. neoformans* antiserum. The sensitivity of CSF antigen detection is highest for diagnosing cryptococcal meningitis in HIV-infected persons and varies based upon factors such as the fungal burden in the CSF and the presence of well-encapsulated organisms. Almost all AIDS patients with cryptococcal meningitis will have a positive serum cryptococcal antigen. In patients without AIDS, however, the sensitivity of serum cryptococcal antigen in diagnosing cryptococcal meningitis is only approximately 60%.

Urine and fluid recovered from the lungs might also be suitable for testing in selected cases, as markers of disseminated and pulmonary cryptococcosis, respectively. Small studies showed that antigen titers ≥1:8 in bronchoalveolar fluid from AIDS patients and percutaneous pulmonary aspirates were sensitive for cryptococcal pneumonia, although the positive predictive value might be poor, at least in the case of bronchoalveolar fluid (8, 16, 93, 106). There is some evidence that cryptococcal antigen might be positive in the urine of patients with disseminated cryptococcosis, but quantification was lower than seen in serum or CSF (27).

LAT Assays

Antigen detection assays are available in different testing formats, but most common are latex agglutination (LAT) assays. In the LAT assays, latex particles containing high-titered, purified immunoglobulin against cryptococcal polysaccharide antigens agglutinate with antigens within CSF samples, sera, or other samples from infected patients. Purified immunoglobulins are either polyclonal immunoglobulin G (IgG), or monoclonal IgG or IgM (Table 1) (5, 90, 149, 160, 162, 171, 173, 174). LAT assays are easy to perform, have rapid turnaround time, and detect all serotypes. Moreover, antigen detection can be quantitated as the greatest dilution of serum that gives a positive result. FDA-approved commercial kits typically demonstrate 90 to 100% sensitivity and 97 to 100% specificity for CSF compared with culture and clinical diagnosis. Although multivalent binding between IgM antibody and the repeating epitope of the capsular polysaccharide might result in increased stability of the antigen-antibody complex compared with polyclonal or monoclonal IgG detection systems, the ranges of sensitivities and specificities for various kits are similar (90, 162). The level of detection of most LAT kits is at least 10 ng of polysaccharide/ml. Details of commercially available tests are provided in Table 1.

Although uncommon, false-positive test results can be encountered, particularly in testing of sera (12, 70, 74, 123, 173). The most common cause of false-positive tests is the presence of rheumatoid factor (9, 74, 173). In addition, false-positive tests have been associated with malignancy (false positivity rate as high as 0.3%) (76, 91), chronic meningitis, collagen vascular disease, agar syneresis fluid (74), and organisms such as *Trichosporon beigelii* (115, 117), *Stomatococcus mucilaginosus* (26), and *Capnocytophaga canimorsus* (172). To minimize false positives, many kits now include pretreatment of specimens with heat and proteolytic enzymes (pronase) (70) and incorporate a control reagent containing latex beads coated with normal rabbit globulin. In addition, use of IgM monoclonal antibody effectively eliminates reactions with rheumatoid factor. Although these modifications improve the false-positive rates, it is estimated that 2% of HIV-infected patients might have false-positive LAT reactivity that is heat-stable, resistant to pronase, and undetectable by normal rabbit globulin controls (173), the exact cause of which remains unclear. In cases in which clinicians doubt a positive LAT assay result, they should ask the microbiology

TABLE 1 Comparison of selected commercially available LA and EIA assays for detection of cryptococcal antigens

Assay	Heat inactivation (serum, CSF)	Pronase treatment (serum)	Pronase treatment (CSF)	Positive test	Note
Polyclonal					
Crypto-LA (International Biological Laboratories)	Yes	No	No	Distinct clumps	Less sensitive than other LAT assays Inclusion of control LATEX
MYCO-Immune (American Microscan)	Yes	No	No	1–4+	Less sensitive than other LAT assays Inclusion of control LATEX
IMMY (Immuno-mycologics)	Yes	Yes	Yes	1–4+, clumping	Inclusion of control LATEX
CALAS (Meridian Diagnostics)	Yes	Yes	No	2–4, clumping	Inclusion of control LATEX
Monoclonal					
Crypto-LEX (Trinity Laboratories)	Yes	Yes	No	>1+	No need for control LATEX
Pastorex *Cryptococcus* (Sanofi-Diagnostics Pasteur)	Yes	Yes	No	>1+	No need for control LATEX
EIA					
PREMIER EIA (Meridian Diagnostics)	No	No	No	OD > 0.150	More sensitive than LAT assay No need for control LATEX

laboratory to treat the specimen with 2-β-mercaptoethanol, as it eliminates nonspecific reactivity without affecting true-positive results (173).

LAT assays can also yield false-negative reactions. Common causes include low levels of cryptococcal antigen production (as might be seen with low organism burden or has been postulated for capsule-deficient *C. neoformans*) (35), a prozone effect seen with very high organism burdens, or the presence of nonspecific proteins that may mask antigen. For the latter, pronase treatment is helpful. Serial dilutions to detect prozone effects can be considered in cases for which the clinical suspicion of cryptococcosis is very high.

EIA

An FDA-approved enzyme immunoassay (EIA) kit is also available, which detects glucuronoxylomannan, the principal component of polysaccharide. EIA is more expensive than LAT assay kits and requires a plate reader, but it offers potential advantages of not requiring enzymatic pretreatment of samples, not reacting with rheumatoid factor or other unidentified proteins (47), and providing subjective spectrophotometric readings. On balance, it might be slightly more sensitive than LAT assays and, in one study, it detected cryptococcal disease in the serum and CSF at an earlier stage (64). It is approximately 12-fold more sensitive than the LA in detecting serogroups A and B but is equivalent in detecting serogroups C and D. The currently available PREMIER EIA displays a rare prozone effect, usually on samples with cryptococcal antigen titers ≥32,768 (58). The laboratory, therefore, needs to be notified in cases where cryptococcal disease is strongly considered so that sample dilution can be performed appropriately. Similar to the LAT assays, false-positive tests can occur with infection due to *T. beigelii* or contamination with syneresis fluid.

The agreement between PREMIER EIA and commercially available LAT assays is 92 to 98% (58, 64, 149, 160), but it is important to recognize that antigen titers measured by EIA and the different LAT kits are not numerically equivalent. In testing serial samples from individual patients, therefore, it is advisable to use the same kit from a given manufacturer.

Utility of Cryptococcal Antigen

Since the serum cryptococcal antigen is positive in almost all AIDS patients with cryptococcal meningitis, there has been interest in using the test to screen high-risk HIV-infected persons, such as those with fevers, neurological symptoms, and/or low CD4 counts. Reports of the clinical utility of screening strategies have been conflicting, which likely reflects varying prevalence of cryptococcosis in different studies. Along these lines, screening is likely to be of greatest utility in areas of high incidence. In certain parts of the developing world, for example, cryptococcal meningitis remains an important cause of mortality, accounting for up to 20% of all early deaths among AIDS patients initiating antiretroviral therapy (50, 82, 83, 85, 100). In studies from sub-Saharan Africa and Southeast Asia, 7 to 18% of asymptomatic, first-time HIV clinic attendees had positive serum cryptococcal antigen screening tests. Moreover, positive antigen tests were highly predictive of subsequent development of cryptococcal meningitis and mortality (38, 83, 121, 161). In a study from South Africa, a serum cryptococcal antigen ≥1:8 was 100% sensitive and 96% specific for predicting cryptococcal meningitis during the first year of antiretroviral therapy among patients with CD4 counts ≤100 cells/μl. Such results, therefore, suggest that cryptococcal antigen screening before the initiation of antiretroviral therapy might facilitate targeted preemptive antifungal strategies and improve survival. Prospective studies are needed to test this hypothesis, as well as to compare screening-based antifungal interventions with routine prophylaxis strategies based solely on CD4 counts.

Monitoring Cryptococcal Antigen in Patients with AIDS-Related Cryptococcal Diseases

Uncertainty has also centered on the utility of monitoring CSF or serum cryptococcal antigen levels in the management of AIDS-related cryptococcosis (141), an issue that was further confounded by the introduction of antiretroviral therapy. Although it is commonly expected that one should see a decrease in cryptococcal antigen titers after 2 or more weeks of therapy, there is no definitive evidence that titers predict or correlate with clinical and mycological outcomes (1, 4, 20, 141). As such, changes in antigen titers have limited value in the management of cryptococcal meningitis. On balance, cryptococcal antigen is a valuable tool for the initial diagnosis of cryptococcal meningitis, but the utility of serial serum or CSF testing is unclear. Antigen testing should be considered in management decisions only in conjunction with careful clinical assessment, microbiology, and other test results.

BG TESTING

(1,3)-β-D-Glucan (BG) is a polysaccharide component of the cell walls of pathogenic fungi including *Candida*, *Aspergillus*, and *Fusarium*, but not the Zygomycetes. It is present to a limited extent in *C. neoformans*. At the current time, there are four commercially available assays that detect BG in serum. Three are from Japan (Fungitec G-test MK, β-glucan test Wako, β-glucan Star), and one from the United States (Fungitell). These assays use different standard β-glucans and reagents that have variable reactivity to β-glucans (77, 130, 131). The cutoff for test positivity and individual result values cannot be directly compared if different tests are performed. The Fungitell assay (Associates of Cape Cod) is an FDA-approved commercial test that detects BG in blood and other body fluids. The assay can measure picogram amounts of BG in samples from patients with invasive candidiasis and aspergillosis, but its diagnostic capacity for invasive cryptococcosis is marginal (77, 131). In one study, 12 of the 15 patients with cryptococcosis had serum BG below the 60-pg/ml cutoff value (133). In a more recent study of the Fungitec G-test MK assay, BG levels were elevated in five of six patients with cryptococcal fungemia (130). The differing results in these studies might reflect differences in BG assays used and/or the different types of cryptococcosis. The diffusion of BG into the blood may be hindered in cryptococci with large capsule. Given the limited experience to date with BG, there is no role at present for BG detection in the diagnosis of cryptococcosis.

DNA AS A MOLECULAR TOOL FOR THE DIAGNOSIS OF CRYPTOCOCCOSIS

Over the past two decades, a myriad of molecular techniques have been explored for the detection of pathogenic fungi. Many of these techniques, as opposed to classical techniques, have the potential to achieve rapid, sensitive, and specific detection of the *C. neoformans* species complex. Molecular fingerprinting techniques, e.g., randomly amplified polymorphic DNA (RAPD) (15), restriction fragment length polymorphism (RFLP) (29, 89, 119), pulsed-field gel electrophoresis (14), amplified fragment length polymorphism (AFLP) (13, 89), PCR fingerprinting with minisatellite (M13) or microsatellite primers [e.g., $(GACA)_4$ or $(CTG)_5$] (29, 32, 46, 89, 119), karyotypes (15, 49, 65), sequencing (17, 41, 87, 176), multilocus microsatellite typing (17), mating-type locus (31, 49), and multilocus sequence typing (MLST) (17), are among the most popular typing techniques that have been applied to discern and characterize the genetic heterogeneity within the species of the *C. neoformans* species complex. These typing molecular methods have aided investigators to unveil strain relatedness and biodiversity of the species complex. Because of the high discriminatory power, some of the techniques have been quickly adopted as an important epidemiological tool for tracing possible sources, modes, and routes of infection of this medically important etiological agent (48, 89, 163, 164, 167).

PCR Fingerprinting

This technique, which uses a single primer (e.g., minisatellite [M13] or microsatellite [$(GACA_4)$-$(GTG)_5$]) that is specific to a hypervariable repetitive DNA sequence, has been used as the major fingerprinting technique to characterize the genetic heterogeneity and the global molecular epidemiology of *C. neoformans* (32, 46, 119, 120, 144, 169). Based on fingerprinting analysis, the *C. neoformans* species complex consists of eight major molecular types followed by subtypes representing hybrids or strain variants (13, 128). These groupings, which are represented by types (VNI [*C. neoformans* var. *grubii*, serotype A]; VNII [*C. neoformans* var. *grubii*, serotype A]; VNIII [serotype AD]; VNIV [*C. neoformans*, serotype D]; VGI, VGII, VGIII, and VGIV [*Cryptococcus gattii*, serotypes B and C]), have been confirmed in various global epidemiological studies undertaken with clinical, veterinary, and environmental isolates (46, 119, 120). Among the described molecular types, VNI and VNIV were found to be the most common types in patients diagnosed with AIDS, whereas type VNI isolates have been associated with the majority of human cryptococcosis infections worldwide (46).

RFLP

RFLP analysis has been extensively used to study strain relatedness in infectious fungi. The traditional RFLP method delineates interspecies variation based on the generation of species-specific banding patterns through the digestion of whole genomic DNA with endonucleases. The complexity of the RFLP profile pattern depends on the enzyme selection and the number of restriction sites in the genome. The banding pattern, which reflects the genomic distribution of recognition sites, is usually visualized with ethidium bromide or by hybridizing the restriction fragments with a labeling probe that is immobilized in a solid surface (i.e., Southern blotting). Unlike traditional RFLP, which often suffers from poor resolution and complex profiles, the hybridization probe allows better visualization and resolution and a simplified banding pattern. However, with the advent of PCR technology, PCR-RFLP analysis has become a more convenient tool for species identification.

RFLP analysis has characterized *C. neoformans* species complex strains, and the method has been applied on single genes, e.g., orotidine monophosphotase pyrophosphorylase (*OMP*), phospholipase B (*PLB1*), *CAP1*, *TEF1*, *URA5*, and internal transcribed spacer 1 and 2 (ITS1 and ITS2) of rRNA, and the large ribosomal region of the mitochodrial DNA (56, 122, 144, 176, 177). Although these independent studies proved the utility of a single gene as an important tool to establish molecular differences among strains, other studies have questioned the level of strain discrimination that is achieved with single-gene RFLP analysis (36). A study conducted with *C. neoformans* clinical isolates demonstrated that when HindIII-digested DNA was probed with *URA5*, only five molecular patterns were generated. In contrast, a more diverse molecular heterogeneity was documented with SstI-digested DNA and the repetitive CNRE-1 probe (36). The CNRE-1 probe generated 16 different RFLP patterns. The sharp difference in RFLP profiles is not surprising, as CNRE-1 hybridizes with repetitive elements that occur as multiple copies in all chromosomes, whereas the *URA5* gene hybridizes to a single-copy gene (36). Multiple-copy genes (such as rRNA genes) can also show a higher level of complexity because of intragenomic sequence heterogeneity among the copies. Thus, the level of heterogeneity should be taken into consideration when analyzing the data since it can lead to misclassification.

One thing to note is that even though the fingerprinting techniques portray similar molecular type groupings, these fingerprinting methods do not fully correlate genotype with serotype classification data (57, 96, 119). This pattern has also been documented by Diaz et al. (41, 43), who employed sequence analysis of the intergenic spacer region (IGS) of the large-subunit ribosomal DNA (LSU rDNA) to characterize isolates within the *C. neoformans* species complex.

Even though most of the conventional RFLP fingerprinting techniques have been extensively used for genetic diversity analysis, PCR-based methods such as PCR-RFLP and PCR fingerprinting are more attractive methods since they do not always require probing or Southern blotting, require a small amount of DNA, and are relatively faster than conventional RFLP typing methods. Despite these advantages, PCR fingerprinting methods have not been fully adopted in diagnostic laboratories since these methods are not amenable for automation, are difficult to standardize, and need careful validation. Nevertheless, these typing methods have been an essential tool to unveil the complexity of genetic diversity and continue to be instrumental in the epidemiology field.

RAPD

RAPD analysis has also become a popular method for fingerprinting medically important fungi (3, 7, 88). RAPD analysis involves DNA amplification with short oligonucleotide primers that target unknown sequences in the genome. Polymorphic patterns can be distinguished between isolates through the generation of multiple-amplicon-size products. The technique requires testing multiple sets of primers to allow strain differentiation. Those primers that provide the best variability are selected. This DNA fingerprinting method has been extensively used to elucidate the genetic heterogeneity within the *C. neoformans* species complex (14, 15, 19) and has coincided with the subclassification distribution of the *C. neoformans* species complex (87). Using this technique, Crampin et al. (34) described 12 distinct genetic profiles among isolates of *C. neoformans* var. *neoformans*. In combination with pulsed-field gel electrophoresis analyses, this technique achieved individual strain differentiation and geographical relationships between *C. neoformans* var. *gattii* and *C. neoformans* var. *neoformans* (14). Even though the method is relatively easy to perform and costs less than other methods, this technique has certain limitations, including the sensitivity to reaction conditions. Difficulties surrounding the reproducibility of amplification products can result from small differences in the primer-to-template-concentration ratio, the temperatures during the amplification reaction, and the concentration of magnesium in the PCR reaction (67, 148).

AFLP

Other fingerprinting analyses such as AFLP have also proved to be useful to discriminate strains (6, 13, 108). This multilocus genotyping method selectively amplifies restriction fragments in a genomic DNA digest. The method not only combines universal applicability, high discriminative power, and reproducibility, but it also offers the potential to detect large numbers of amplification products, as it does not target specific areas of the genome for marker identification. Thus, a large number of loci are analyzed. When compared to other fingerprinting techniques, such as RAPD and PCR fingerprinting, AFLP has shown higher discriminatory power for the differentiation of genetically related strains (14, 89). As with RAPD analyses, only small amounts of DNA are required. Based on this technique, Boekhout et al. (13) found considerable genetic divergence between the species complex and characterized the presence of six distinct genotypic groups.

This technique, with PCR fingerprinting, was successfully employed in a pilot study of clinical and environmental sources (collected between 1999 and 2002) on Vancouver Island, British Columbia. The purpose was to discern the molecular types and to establish the environmental source responsible for an unprecedented outbreak of cryptococcosis. This outbreak caused illness in hundreds of patients and animals with no apparent immunodeficiency (89, 116, 157). The study revealed that approximately 95% of the isolates belonged to the *C. gattii* VGII/AFLP6 type. Further characterization of the samples revealed two subtypes, namely VGIIa/AFLP6A and VGIIb/AFLP6B (89). The former type accounted for 90% of the encountered VGII isolates. The study also reported that two of the infections were caused by genotype VGI/AFLP4. The subtype VGIIa/AFLP6A has been regarded as very virulent (107, 116) and was recently identified as the primary pathogen on a 51-year-old immunocompetent individual who acquired the disease during a visit to Vancouver Island (107, 116). The presence of *C. gattii* in Vancouver Island is highly unusual, as *C. gattii* is usually restricted to tropical and subtropical climates. This incidence highlights the changing distribution of pathogenic species and stresses the importance of monitoring the potential global dispersal of this medically important pathogen.

Karyotyping

With the advent and invention of pulsed-field gel electrophoresis, chromosome size fragmentation can be induced under an electric field gel electrophoresis. In this system, the cells are pretreated with enzymes and proteases to remove cell walls and proteins. Once the cells are pretreated, the sample undergoes electrophoresis. Based on the migration pattern of chromosome-sized DNA fragments, the karyotypic variability of strains can be determined. The high karyotype variability observed among strains might be linked to mobility of DNA fragments or to the presence of multiple copies of a given gene, i.e., rDNA (65).

This technique has been successfully developed and applied as an epidemiological marker to fingerprint *C. neoformans* (15, 45, 128, 138, 139, 150) and to identify the progeny after mating analysis (49). A remarkable diversity in the karyotyping pattern of *C. neoformans* isolates has been documented (65). This study found 29 different karyotyping patterns out of the 33 isolates of *C. neoformans* var. *neoformans* strains and six different patterns for the 13 analyzed strains of *C. neoformans* var. *gattii*. This study also reported different global genome sizes for the varieties. The var. *neoformans* genome size fluctuated from 14.2 to 20.9 Mb, whereas var. *gattii* size ranged from 7.9 to 16.8 Mb. Since most isolates present significant variation in chromosome sizes, the technique was successful at targeting strains shared between patients. For instance, Pfaller et al. (139) demonstrated that 75% of Ugandan patients with cryptococcosis shared their infecting strain with at least one other patient. The technique is also useful for comparing sequential isolates from patients and monitoring the microevolution of an infected strain (2, 60). In view of the heterogeneity in chromosome sizes between the strains, some researchers have suggested that *C. neoformans* strains have not clonally expanded from one isolate but that the species has undergone evolutionary changes that allowed for chromosomal re-rearrangement (138).

Some studies have stressed the importance of karyotyping as an epidemiological tool because some strains can maintain a stable pattern during successive in vitro and in vivo sequential passages (137, 138). However, others have questioned the utility of karyotyping because some strains undergo rapid karyotypic changes by successive strain passes (62). For instance, a high frequency of karyotypic changes has been reported when a particular strain was passed through a mouse or when a strain was subject to successive repeated passages (62). This chromosome instability and phenotypic variations appeared to be associated with the presence of transposon-rich areas (111). Because of the chromosomal instability in some strains, it is advisable to validate karyotyping analysis with other molecular fingerprinting techniques to corroborate any genetic changes associated with microevolution processes during the course of infection.

Although the described fingerprinting techniques are useful tools for epidemiologists in clinical laboratories, where a high volume of samples are screened, these assays can be laborious and technically demanding and only allow a limited number of samples to be analyzed. In addition, the lack of a standardized fingerprinting method for interlaboratory use and reference laboratories makes even more difficult the adoption of some of the typing methods as a gold standard for routine diagnostic analysis.

DNA SEQUENCING AS A TYPING AND IDENTIFICATION METHOD

More selective analyses have been carried out with DNA gene sequence analyses. This technique, which is considered the gold standard for accurate species identification and utilizes the principle of the dideoxy chain terminator technique, has become an important tool for yeast (55, 94) identification and has prompted the development of numerous molecular techniques that are based on DNA sequence analysis. The technique is based on DNA synthesis with incorporation of deoxynucleotides (dNTPs) as well as dideoxynucleotides (ddNTPs). ddNTPs are modified dNTPs that lack a 3′-hydroxyl group. The lack of this group does not permit subsequent additions of dNTPs, causing a termination of the newly synthesized DNA molecule. The prematurely ended fragments are then electrophoresed and separated based on size. Because different ddNTPs are used in different tubes or are marked with different fluorophores, the DNA sequence composition can be inferred. Overall, this method offers a number of advantages over other typing techniques, e.g., a higher level of reproducibility and less subjectivity in analysis. In addition, the data generated by these analyses are deposited in public central banks, which facilitate accessibility to all users. Even though the method can be highly automated, the investigator has to invest a considerable amount of time proofreading the sequences, which could be an enormous task in high-throughput laboratories.

Nucleic acid sequence diversity within rRNA genes as well as adjacent intergenic sequences has been widely used to infer phylogenies and for species identification. This gene family contains conservative and nonconservative regions that have shed light on the sequence diversity between both the intervariety and the intravariety of the *C. neoformans* species complex. Sequence analysis undertaken by Fan et al. (52) showed that the varieties of *C. neoformans* differed by only two nucleotides in the 16S rRNA and 10 bp in the 26S rDNA. The observed similarities in sequences are not surprising since both regions are regarded as highly conserved. In contrast, significant sequence divergences have been documented in the IGS and the ITS regions of the rRNA gene (41, 43, 53, 87). Sequence divergence in the IGS region allowed the subclassification of the *C. neoformans* species complex into six major genotypes (43), whereas ITS sequences grouped the species complex into seven groups (87). These major subtypes are consistent with the molecular distribution of PCR fingerprinting and AFLP analysis. The comparable molecular types and genotypes are described in Table 2.

In view of the fact that the ITS region has been regarded as one of the most phylogenetically informative regions for species identification, a study was conducted with 373 strains (86 different species), to establish the feasibility of the ITS1 and ITS2 regions to identify yeasts of clinical relevance, including *C. neoformans* (104). This study found that the rates of correct identification based on the ITS1 and ITS2 sequence analysis were 96.8% and 99.7%, respectively. Although the ITS1 region displayed more interspecies variability, it was concluded that the ITS2 region was a better locus to establish species delineation and thus can be an important tool to discern most medically important yeasts species (104). However, analysis of both loci can offer a better comprehensive analysis for genetic differences between strains (30). Some studies reported that differences in length and sequence polymorphisms in the ITS region of fungi can be useful criteria for successful identification of medically important species (30, 63, 165). For instance, length polymorphism analysis of both ITS1 and ITS2 regions permitted identification of 30 clinical pathogenic yeast species with an accuracy rate of 98% (30). Similarly, 30 genus- or species-specific patterns were recognized among the 33 fungal species on the basis of differences in the lengths of amplicons (63). Although both regions provided sufficient information to establish a successful identification of commonly occurring pathogenic species, analysis based solely on ITS2 length polymorphism did not provide enough species delineation for species such as *Candida albicans* and *Candida krusei* (165). These studies confirmed the importance of using both ITS regions, as they provide more length variability than a single-locus analysis.

Other genes such as *URA5* (23, 24, 177); *TOP1* (108); *TOP11* (136); *UB11* and *UB14* (154); *LAC* (108, 175, 177); *mtLRNA* (176); *CAP10, GPD1, TEF1, MSOD1, LAC1, PLB1, URE1, CAP59*, and *MP88* (108); MF**a** (32); and MFα (32) have been sequenced and have led to useful information on strain differentiation, species identification, and phylogeny. The sequences from these genes are available in public databases that continue to expand as advances are made in the molecular field.

As we enter into a new genomic era, whole-genomic sequence data have enriched our understanding of the diversity, phylogeny, pathogenesis, and genetic factors involved in the virulence mechanism of pathogens. Sequence strategies such as shotgun amplification have allowed whole-genome amplifications. The whole-genome shotgun sequences (WGS) approach involves the fragmentation of DNA into millions of fragments, which are sequenced by the chain termination method to obtain "reads." The reads are subsequently reassembled to produce a series of "scaffolds" that are generated by specialized computer programs. Using this approach, physical maps of WGS of various *C. neoformans* strains are being assembled through a collaborative effort of various research groups, e.g., Stanford University (Stanford Genome Technology Center), Duke University (Center for Genome Technology), Genome Canada (Genome Sciences Centre), and The Institute for Genomic Research and the Whitehead Institute (Center for Genome Research).

To date, these consortia have completed the genome sequence of two closely related *C. neoformans* strains, JEC21 and B3501A (111). The sequences, which are publicly available in GenBank, have generated a wealth of information on the genetic structure of *C. neoformans*. For instance, WGS analysis estimated a genome size of 20 Mb and 14 chromosomes for this species (111). Another characteristic of the genome structure is the presence of ~6,500 introns and transposon-rich areas, which represents 5% of the genome size (111). The data also revealed the presence of 6,572 protein-encoding genes and 30 novel genes, which are believed to be linked with the virulence properties of the organism since these genes are probably involved with capsule biosynthesis (111). It can be inferred from these studies that as new information becomes available through WGS, novel genetic markers can be discovered, which can lead to better detection and identification systems.

MLST

Other approaches such as MLST have been explored to study the *C. neoformans* species complex. The MLST technique uses a number of unrelated, phylogenetically informative genes that are amplified and sequenced. With this technique, different strains can be assigned a "barcode" of sequence types based on the divergence of sequences of the studied genes. MLST allows for strain delineation and is used as a routine clinical tool in bacteriology. In fact, a Web-based bacteriological MLST system is available. In contrast, there are scant publicly available MLST data on fungi. A search on these MLST depositories only includes the pathogenic species, *C. gattii, C. neoformans, Aspergillus fumigatus*, and several species within the genus *Candida* (http://www.mlst.net/ and http://pubmlst.org). Other medically important fungi, e.g., *Coccidioides immitis* (92), *Batrachochytrium dendrobatidis* (126), and *Fusarium solani* (178), have been MLST typed, but their data are not publicly available.

MLST studies have been undertaken to study the monophyletic lineages of the *C. neoformans* species complex (17). Analysis on different loci (e.g., IGS1, ITS, *CNLAC1, TEF1, RPB1*, and *RPB2*) revealed similar phylogenetic lineages that are in correlation with AFLP or PCR fingerprinting. Based on the monophyletic lineage distribution of the strains, which was similar for all loci, the authors ruled out the possible occurrence of recombination events between monophyletic lineages and suggested each monophyletic lineage represents a separate taxon. Similar studies undertaken with IGS, ITS, *TOP1*, and *CAP59* also supported the monophyletic distribution of *C. neoformans* var. *neoformans* and var. *grubii*, but according to the molecular phylogenies of the studied loci, *C. gattii* isolates followed a polyphyletic distribution (158).

TABLE 2 Molecular type concordance of *C. neoformans* and *C. gattii*[a]

Species/variety/hybrid	Serotype	PCR fingerprinting molecular type[b]	PCR fingerprinting molecular type[c]	AFLP genotype[d]	AFLP genotype[e]	*URA5* RFLP type[f]	*PLB1* RFLP type[g]	IGS genotype[h]	ITS genotype[i]
C. neoformans var. *grubii*	A	VNI	VN6 (VN5)	AFLP1	VNI	VNI	A1	1A/1B	ITS1
	A	VNII		AFLP1A/AFLP1B	VNB	VNII		1A	ITS1
	A	VNII	VN7	AFLP1A/AFLP1B	VNII	VNII	A2	1C	ITS1
AD hybrid	AD	VNIII	VN3/VN4	AFLP3		VNIII	A3	2C	ITS1/ITS2
C. neoformans var. *neoformans*	D	VNIV	VN1 (VN2)	AFLP2		VNIV	A4	2A/2B/2C	ITS2
C. gattii	B/C	VGI		AFLP4A/AFLP4B		VGI	A5	4	ITS3/ITS7
	B/C	VGII		AFLP6		VGII	A6	3	ITS4
	B/C	VGIII		AFLP5A/AFLP5B/AFLP5C		VGIII	A7	5	ITS5
	B/C	VGIV		AFLP7		VGIV	A8	6	ITS6

[a] Data from reference 118.
[b] References 56, 119, 120.
[c] Reference 169.
[d] Reference 13.
[e] Reference 108.
[f] Reference 119.
[g] Reference 96.
[h] References 41, 43.
[i] Reference 87.

Most recently, MLST analyses have been undertaken to characterize the population structure of environmental and clinical isolates (29, 57, 108). A recent study in China that involved 115 strains from non-HIV patients demonstrated that the majority of the isolates (n = 103) belonged to *C. neoformans* var. *grubii* molecular type VNI (mating type α), followed by VGI (n = 8), VNIII (n = 2), VNIV (n = 1), and VGII (n = 1) (57). This study also found that molecular type VGI is the most causative agent of *C. gattii* infections in China. Similar studies by Chen et al. (29) analyzed the genotype of 129 clinical strains from 16 different provinces in China using MLST analysis (*CAP10*, *GPD*, IGS1, *LAC1*, *PLB1*, *SOD1*, *TEF1*, and *URE1* genes) as well as M13 PCR fingerprinting. The results demonstrated that the majority of the analyzed strains (93%) belonged to *C. neoformans* subtype VNIc (serotype A, MATα), whereas only 7% of the isolates belonged to *C. gattii* and portrayed a typical VGI genotype pattern. Curiously, all *C. neoformans* isolates displayed a remarkable genetic homogeneity, which is in contrast to reports from other countries.

Litvintseva et al. (108) undertook a more comprehensive analysis with a global collection of 1,085 *C. neoformans* var. *grubii* (serotype A) isolates. The analysis, which used 12 unlinked polymorphic loci (i.e., *MPD1*, *TOP1*, *MP88*, *CAP59*, *URE1*, *PLB1*, *CAP10*, *GPD1*, *TEF1*, *SOD1*, *LAC1*, and IGS1 genes), showed a better heterogeneity among the analyzed strains and identified three distinct genetic subpopulations, namely, VNI, VNII, and VNB. Among these subpopulations, VNB was described as a new subpopulation within *C. neoformans* var. *grubii* and was found phylogenetically related to VNI. These isolates, which were described as unique from Botswana, displayed greater diversity than groups VNI or VNII and revealed evidence of recombination. Based on this study, a central Web-based database was created at www.mlst.net (http://cneoformans.mlst.net/).

As part of an initiative to study the genetic diversity of the agents of cryptococcosis and its epidemiological significance, the Cryptococcal Working Group was established in 2007 under the sponsorship of the International Society for Human and Animal Mycology. This working group selected MLST typing of the housekeeping genes *CAP59*, *GPD1*, *LAC1*, *PLB1*, *SOD1*, and *URA5* and the IGS1 region as the method of choice for global strain genotyping. Selection of these loci was based on the degree of polymorphism, allelic diversity, and level of discrimination. This initiative created a central website at www.mlst.net/, where unknown or known sequence types can be deposited or retrieved in a centralized global MLST database. Access is available via the online software NRDB (http://linux.mlst.net/nrdb/nrdb.htm).

DNA Sequencing Kit

In view of the numerous advantages of the sequencing method, commercially available kits such as MicroSeq D2 LSU rDNA (Applied Biosystems, Rotkreuz, Switzerland) have been developed for the identification of medically important fungal species, including *C. neoformans*. The kit can identify species based on unique sequence areas of the LSU rRNA (28S rDNA). According to a study conducted at the Mayo Clinic in Minnesota, the kit was able to identify 93.9% of the clinical isolates at the species level (68). However, one of the pitfalls of the kit is that the MicroSeq fungal library is limited since it excludes some clinically important species (68, 69, 129). In addition, the kit does not have the potential to discriminate the *C. neoformans* species complex based on molecular types. Therefore, the utility as a diagnostic or epidemiological tool is rather limited.

Other alternative sequencing methods, such as sequencing by hybridization or parallel pyrosequencing, have emerged as new technologies for species identification. The pyrosequencing technique uses a cascade of four different types of enzymatic reactions and is based on the detection of released pyrophosphate during DNA synthesis. This sequencing technique holds great promise for single nucleotide polymorphism genotyping and discovery in allelic frequency studies and haplotyping analysis. Advantages of the technology include flexibility, adaptability for different applications, and the ease of automation. But the main drawback is that only 30 to 50 bases of sequence information are provided. The limitation in the read length may not provide sufficient phylogenetic information for species delineation, especially if the analysis is based on a single gene. For instance, in a recent study where the performance of pyrosequencing was tested against 133 isolates of clinically relevant nondematiaceous yeasts, it was demonstrated that although the method was successful at identifying most *Candida* isolates at the species level, the method was unable to identify some *Cryptococcus* and *Trichosporon* species (125). The latter study, which used comparative analysis of the ITS1-5.8S-ITS2 region, showed that conventional cycle sequencing analysis provided higher levels of accuracy than pyrosequencing. For instance, pyrosequencing identified 69.1% of the isolates to the species level, whereas conventional cycle sequence analysis identified 78.9% of isolates (125). Therefore if pyrosequencing is adopted as an identification method, multiple-gene sequencing analysis should be integrated in the assay to achieve accurate species delineation. Other molecular biological methods combine DNA sequence analyses with species-specific hybridization probes (84).

PCR

The PCR method has been extensively used for the diagnosis of opportunistic fungal infections (51, 54, 75, 97). In conventional PCR, a set of primer pairs is selected based on species-specific DNA sequences. The set of primers allows the exponential amplification of the specific target region. The amplified product is detected by comparing the position of the bands with the profile of a positive control that represents the species of interest. Because of its versatility, PCR has been extensively used to perform a wide array of genetic manipulations, diagnostic tests, DNA cloning procedures, DNA fingerprinting (AFLP, RAPD PCR fingerprinting), Southern blotting, DNA sequencing, and recombinant DNA technology. The majority of the developed PCR assays target multicopy genes such as the rRNA genes, LSU rDNA (54, 73), mitochondrial genes (176, 177), the ITS regions (97, 124), IGSs (18, 43), and other genes (96, 159).

Nested PCR

Although conventional PCR techniques have improved diagnosis of cryptococcal meningitis, the technique can present sensitivity problems when applied directly to clinical specimens. To address this problem, nested PCR assays were successfully developed for the detection of *C. neoformans* in CSF (142) and clinical lung specimens, e.g., bronchoalveolar lavage fluid, sputum, bronchial aspirates, and transbronchial lung biopsy (159). The technique uses two sets of amplification primers that target a single locus. The first set of primers amplifies an outer region of the selected locus, whereas the other primer set amplifies an internal region. Since the second amplification reamplifies the product of the first reaction, the sensitivity of the assay can be greatly enhanced. Using this approach, sensitivity levels as low as 10 cells have been reported for CSF samples (142), whereas detection limits of 10 pg or 10^3 CFU were reported for lung clinical specimens (159). Although this detection method has advantages and has been proven to be a good molecular detection system, the method is more prone to carryover contamination since two sequential PCR amplifications are employed.

Multiplex PCR

Multiplex PCR is another type of PCR amplification in which multiple sets of primers are included in the reaction to allow simultaneous detection of species in a single reaction. The method has been used for fungal detection despite the fact that lengthy optimization procedures are required. Using various sets of species-specific ITS primers, Chang et al. (25) developed a multiplex PCR format based on the polymorphic pattern of the ITS species-specific bands of *C. neoformans* and *Candida* spp. The PCR assay, which was validated with 255 blood cultures, showed a test sensitivity of 96.9% and a limit of detection of 20 CFU/ml. Similarly, Luo and Mitchell (114) developed two multiplex formats with various sets of ITS primers for the detection of *Candida* spp., *A. fumigatus*, and *C. neoformans*. The described method provided 100% sensitivity and specificity that allowed the identification of the species directly from intact yeast cells with detection limits that ranged from 1 to 10 yeast cells. As opposed to intact yeast cell amplification, most of the molds tested failed the direct amplification. This was attributable to sturdier cell walls and/or the presence of endogenous nucleases.

To further evaluate the feasibility of multiplex PCR as a detection system, a comparative study was undertaken with other classical methods of identification (e.g., Crypto Check serotyping kit [Iatron] and canavanine-glycine-bromothymol blue agar [CGB]) that are routinely used to identify the species and serotypes of *C. neoformans*. The study demonstrated a good correlation between multiplex PCR and Crypto Check serotyping. However, the CGB method, which is based on the differential growth of CGB, misidentified 6 out of 131 samples (103).

Multiplex PCR was also employed with primers that amplify STE (sterile) gene sequences to discern the serotypes and MAT of *C. neoformans* (22). The results strongly confirmed the utility of the PCR method as a tool for fungal identification and as an alternative approach to discriminate serotypes and MAT types. Analysis of the cost-effectiveness of multiplex PCR versus the Crypto Check serotyping kit method demonstrated that this multiplex PCR approach was more economical and required less reagent. The study also provided further evidence of the role of pigeon droppings as a reservoir for *C. neoformans* and the prevalence of *C. neoformans* var. *grubii* (Aα) among environmental isolates.

Multiplex Tandem PCR

A novel multiplex PCR approach was recently applied in the detection of fungi in culture specimens (98, 99). This approach uses multiplex tandem PCR. The method involves a short cycle amplification with multiplex-specific primers followed by a sequential number of simultaneous PCR amplifications that target several genes. In the second step, each individual gene is identified based on the melting profiles of the target, which are quantified by real-time PCR. The technique resembles nested PCR since the primers for the second PCR are nested inside the primers for the first PCR. However, it differs from nested PCR since multiple primer pairs are employed in the first reaction, and this reaction serves only as a preamplification step because of the limited number of cycles (10 cycles). Using target-specific primers that amplify sequences of ITS1 and ITS2, the elongation factor 1-alpha and the tubulin genes, the authors were able to simultaneously identify numerous fungal pathogens, e.g., the *C. neoformans* species complex, *C. albicans*, *Candida dubliniensis*, *Candida glabrata*, *Candida guilliermondii*, *C. krusei*, the *Candida parapsilosis* complex, *F. solani*, *Fusarium* species, and *Scedosporium prolificans*. The assay proved to be specific and sensitive enough to target 10 cells/ml of blood (98). The observed high levels of specificity and sensitivity of the method can be attributed to the employment of two different sets of specific primers that are used for both rounds of amplification. Some of the method highlights include accurate simultaneous detection of species from blood specimens, identification of the target sequence in polymicrobial samples, and the capability to detect up to 72 targets. The method is amenable for automated operation and uses software that facilitates qualitative and quantitative analysis. In addition, it allows for direct detection of fungal colonies (without DNA isolation) with detection limits of 10^5 CFU/ml (99).

PCR methods have advantages over other diagnostic methods, as low target numbers of fungal DNA can be amplified and detected without the need of viable fungal cells. These assays are more sensitive than conventional blood cultures and are sensitive enough to detect fungal DNA in blood and bronchoalveolar fluid (51, 72, 109, 110, 143). Because the methods are quite sensitive, they can generate false positives caused by contamination from other samples. Also, the presence of DNA or polymerase inhibitors can result in false-negative reactions. To address these problems, which can be common in clinical laboratories, commercial PCR kits have integrated built-in controls to trace any potential inhibitors or contaminants.

Sample preparation from clinical specimens continues to be a key factor for the success of PCR-based assays since it can limit both the sensitivity and specificity of the assays. Improvements in DNA extraction techniques are constantly needed to recover a maximum yield of DNA. In view of this, many extraction techniques have been adapted to clinical specimens, and a myriad of commercially available kits are currently available.

DNA PROBE TECHNOLOGY

DNA probes have been widely adopted as a research and diagnostic tool since this method of identification can be easily adapted and streamlined for clinical applicability. In addition, DNA probe technology provides the required specificity and sensitivity to identify microorganisms in a short time. In view of this, a wide variety of DNA probe techniques have been developed for diagnosis of *C. neoformans* and other fungal etiological agents (18, 40, 134, 135, 147). DNA probe technology is based upon the assumption that complementary strands of the DNA of interest hybridize with a specific probe sequence to achieve a stable double-stranded nucleic acid molecule. Based on this principle, many nucleic acid probes have been developed and tested in different hybridization assay formats (e.g., Southern blot, Northern blot, dot blot, direct hybridization, competitive hybridization) for the detection and identification of etiological agents.

The nucleic acid probes are designed to target DNA or RNA, and their length can range from ~15 to thousands of

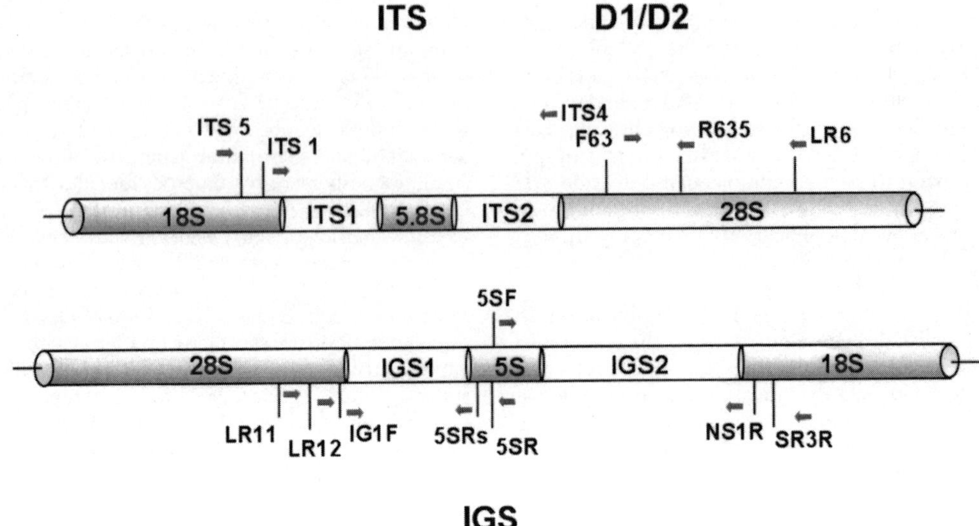

FIGURE 1 A schematic representation of the rRNA gene family organization. Different sets of primers can be used to amplify target regions such as the DI/D2 domain and the ITS and IGS regions. Arrows indicate the direction in which a given primer amplifies.

bases long. In view of the fact that shorter probe sequences hybridize faster with their complementary target sequence and are less prone to secondary structures, many researchers have opted to employ oligonucleotide probes, which are short sequences of chemically synthesized nucleotides (less than 50 bp). Under high stringency conditions for temperature, ionic strength, and pH, many probes can detect DNA sequences that differ by a single base pair (39, 40). To achieve detection, the probes are usually labeled with enzymes (e.g., alkaline phosphatase, horseradish peroxidase), affinity labels (biotin, digoxigenin), antigenic substrates, radioisotopes (^{32}P), or chemiluminescent moieties (acridinium esters). Once hybridization has taken place, a reporter molecule is added to the reaction to detect the labeling moiety attached to the probe.

Probe development has concentrated on several target DNA regions, e.g., a noncoding DNA region, conserved nucleic acid regions, or domains encoding a toxin or a virulence factor. To date, the most commonly used phylogenetic marker is the rRNA gene. This multicopy gene family comprises four nuclear rRNA genes (26/28S-18S-5.8S-5S) arranged as "head to tail" tandem repeats separated by the spacer regions, the ITS and IGS (Fig. 1). As opposed to the 18S and 26S, which contain highly conserved regions that can be used for universal primer selection to amplify a wide range of taxa, the spacer regions (ITS and IGS) and the D1/D2 region of the LSU rRNA are the most variable regions for resolution of species identification.

The sensitivity of the DNA probes can depend upon the chosen assay format. For instance, if the target DNA is not amplified, a large amount of specimen will most likely be required to achieve an appropriate detection. Because a good diagnostic tool should detect the infectious agent at the early stages of the disease, methods of amplification, i.e., PCR, have been coupled to different molecular assay formats. The advantage of PCR is that it can detect DNA or RNA specific to certain microorganisms and does not rely on recovery of the microorganism or the ability of the individual to recognize the pathogen.

Probes: RFLP

Some *C. neoformans* probes have been developed with repetitive elements derived from RFLP generated from cloned genomic cDNA or from specific DNA fragments using the PCR technique. For example, based on middle repetitive DNA sequences, Polacheck et al. (140) developed species- and variety-specific probes (e.g., CND1.4, CND1.7) to identify and trace particular strains of *C. neoformans*. The method, which employed Southern blot analysis and plasmid DNA, distinguished five distinct hybridization patterns.

Another repetitive DNA probe, CNRE-1 has also been developed to target a specific family of repetitive DNA fragments that are found throughout the yeast genome (153, 156). This repetitive sequence is found in all *C. neoformans* chromosomes and is present in 10 to 20 copies (153, 156). CNRE-1 has been used in a variety of studies focused on genetic variation (153), pathogenesis (155), and the relationship of pigeon excreta to clinical isolates (36). Fingerprinting analysis with CNRE-1 provided the first substantial evidence of the genetic division between *C. neoformans* var. *neoformans* and var. *grubii* (153). For instance, when this probe was challenged with *C. neoformans* var. *grubii* SacI-digested DNA, the profile pattern consisted of 11 to 16 fragments, whereas *C. neoformans* var. *neoformans* yielded 5 to 11 bands (59, 153). Similarly, fingerprinting analysis with CNRE-1 and Southern blot of SstI-digested DNA (*C. neoformans* var. *grubii* [ATCC 6352]) documented a distinct banding pattern that consisted of 12 bands with various intensities, whereas *C. neoformans* var. *neoformans* (ATCC 28958) produced fewer bands of much less intensity (59, 153). Various studies with clinical isolates have proven the utility of this probe as an epidemiological tool since CNRE-1 can recognize multiple RFLPs among clinical isolates (28, 153). When used in combination with *URA5* gene sequence analyses, the probe can identify clonal population structures of *C. neoformans* isolates that appeared to be geographically restricted (28). A CNRE-1 probe was also successfully employed to characterize and compare strains from pigeon excreta with those of patients from a limited geographic area in New York (36). This study revealed a remarkable diversity among the isolates.

Through the use of a specific telomere sequence of the plasmid pCnTel-1, Garcia-Hermoso et al. (66) developed the CENTEL probe to establish whether two strains of *C. neoformans* could be considered epidemiologically linked. The authors showed that the CENTEL probe has more discriminatory power at establishing differences between unrelated strains than the CNRE-1 probe. Their observation was based on the total number of fingerprinting profile patterns (CENTEL: 17 profiles; CNRE-1: 15 profiles) generated by 17 different environmental strains. However, the performance of these probes with clinical strains was somehow different, since the fingerprinting profile of strains that were classified as link isolates (strains isolated from the same patient) were more diverse than those obtained from isolates recovered from different patients (unrelated strains). Based on the data, which enabled the calculation of two DICE coefficients to establish the cutoff values of genetic distances between link and unrelated strains, the authors concluded that CENTEL might be an appropriate tool to examine microevolution during infection.

A linear autonomous plasmid DNA probe, UT-4p, has also been developed to fingerprint *C. neoformans* (168). When this probe was hybridized to AccI-digested genomic DNAs, it generated unique DNA fingerprint patterns that correspond to serotype classification. The different fingerprinting profile patterns are believed to be associated to telomeric- or subtelomeric-region length heterogeneity of strains (113). Because of their discriminatory power, probes such as UT-4p and CNRE-1 can be employed as epidemiological markers (28, 36, 66, 167).

Although the above-described probes and assay format have proven highly discriminatory, their use as a routine tool in diagnostic labs is hampered by the complexity and long turnaround times of Southern blot analysis. For instance, the sample requires digestion with restriction endonucleases, followed by separation and transfer of the fragments to a membrane for hybridization. Other drawbacks include the use of radioactive materials and the need for a considerable amount of purified DNA of high molecular weight, which can limit the sensitivity of the method.

Probe-Based Real-Time PCR

Real-time PCR has been shown to be a highly sensitive and specific diagnostic tool for the detection of medically important fungi (10, 11, 78). This PCR amplification method enables both detection and quantification of the target DNA during the early phases of the reaction. Quantification of DNA is undertaken with intercalating DNA fluorescent dyes such as SYBR Green I or sequence-specific probes that are labeled with a quencher dye and a reporter fluorescent dye assay. As the probe binds to the DNA, the DNA elongates and causes the release of the reporter dye. The newly synthesized DNA can be quantified based on the amount of fluorescence emission.

The use of fluorescent reporter probes, e.g., Molecular Beacons, TaqMan, FRET Hybridization, and Scorpion primers, has been found to be a better alternative than SYBR Green I because these probes enhance specificity, allow quantification in the presence of nontarget DNA, and permit a certain degree of multiplexing. Real-time PCR has been successfully used for the detection of fungal pathogens such as *Candida* species, *C. neoformans*, and *Aspergillus* species (10, 11, 78, 81). Many of these studies have documented sensitivities that are better than or comparable to those established by conventional PCR. For instance, a recent study that employed SYBR Green I documented detection limits as low as 1 pg DNA/μl (*C. neoformans* and six *Candida* species). The developed PCR method, which was validated with 58 clinical strains, proved to be specific, and it was concordant with other methods of identification, e.g., germ-tube assay, VITEK yeast biochemical card, and API-32C (78). Another study that used broad-range real-time PCR amplification of fungal 28S rDNA fragments and sequencing documented sensitivity levels ranging from 1 to 15 rDNA copies per PCR (170). However, a reduction in analytical sensitivity was documented when the assay was validated with clinical specimens derived from different biological matrixes, e.g., urine or serum: 10 CFU/400 μl, tracheal secretion: 50 CFU/400 μl, and EDTA-anticoagulated blood: 10 CFU/200 μl. These results illustrated that the nature of the biological matrix can have a profound effect on sensitivity levels and can lead to false-negative results. To increase the level of sensitivity, a noncompetitive internal control probe can be used to monitor amplification in every specimen. The inclusion of an internal control probe can improve the reliability of the assay by decreasing the risk of false-negative results due to PCR inhibition (44, 170).

Despite the sensitivity issues, which can be overcome with an internal control probe, the broad real-time PCR/sequencing approach offered some advantages. For instance, unlike an array of species-specific probes that screens for known targets, this method enables identification of rare or infrequent fungi because it relies on sequencing to establish identification (170). However, the method suffers from inadequate resolution among closely related species. For example, it could not discern species within the genera *Aspergillus* and *Penicillium* because some species share sequence homologies in the 28S rDNA. Therefore, for further identification, the analysis requires a second amplification with a primer/probe set specific for ITS1, which is a more polymorphic region.

Another broad-range PCR assay was recently developed by Bergman et al. (10), who conducted real-time PCR amplifications of the ITS region and employed two-dimensional point analyses based on the melting profiles of amplicons generated on the ITS1 and ITS2 regions. The method, which was tested with 102 clinical isolates and targeted *C. neoformans* and other common pathogenic yeast species, successfully detected and separated 14 different species by their specific melting temperature profiles. Even though the melting point analysis provides a relatively broad-range detection, the method has shortcomings. (i) It is only suitable for culture-based isolates since clinical specimens can promote unspecific SYBR Green I detection. (ii) Detection is limited by the melting temperature T_m of assay conditions. Thus, amplicons that require melting points above the T_m of cycling conditions cannot reach optimal denaturation and cannot be detected (10). (iii) Different species can have similar T_m values, which can hinder detection (151).

Like any other method, real-time PCR has inherent pitfalls and sources of errors, e.g., amplification biases, exponential amplification of errors, mispriming, formation of primer-dimers, and differential cycling efficiency. Additionally, some fluorescent dyes are incompatible with several assay platforms, and the multiplex capability is limited because of the scant number of commercially available fluorescent dyes.

Padlock Probe

A molecular-based method has been developed for the detection of the *C. neoformans* species complex and hybrids (86). This approach used two technologies, padlock probes (PLPs) and hyperbranched rolling circle amplification (RCA). PLPs are approximately 50- to 100-mer probes that contain target complementary sequences separated by a linker. Under the presence of a correct target sequence they can be circularized by enzymatic ligation. To amplify the intensity of the signal generated by PLP, the probe is subsequently amplified with hyperbranched RCA, which is an isothermal target amplification method. As opposed to PCR, RCA produces a single amplified product that remains attached to the target molecule. Based on this approach, the authors designed four PLPs targeting species-specific single-nucleotide polymorphisms at the ITSs of the RNA gene. The assay, which was validated with 99 isolates and used PCR-based amplicons for PLP/ligation reactions, was found to be specific for the detection of the *C. neoformans* species complex. But false-positive signals were obtained when an RCA3-ITS probe was challenged with medically important species such as *Candida albidus*, *C. curvatus*, and *C. luteolus*. Sequence analysis revealed that all three species contained the same single nucleotide polymorphism with *C. neoformans* var. *grubii* at position 223. Therefore, careful design of PLP is necessary to achieve a good specificity. Additionally, the method suffers from dubious amplification of target-independent products that are sometimes associated with excess of PLP or primer. The unligated PLP can act as a template or primer, which can lead to nonspecific DNA synthesis and a subsequent nonspecific fluorescent signal that jeopardizes the accuracy of the assay.

Probe-Based Microarrays

Microarray technology has received enormous attention and is under much review since the technology holds great promise to improve diagnosis and prognosis of disease. Unlike other molecular-based methods of identification, high-density microarray technology allows the simultaneous analysis and identification of thousands of species at a time by immobilizing probes on a solid surface, which are then hybridized with labeled DNA. Because of the high-throughput capability, the arrays have been extensively used for expression of thousands of genes and disease pathology. However, despite the enormous and potential applicability of array technology, a limited number of arrays have focused on fungal detection. The majority of these arrays are limited since they do not cover the whole spectra of pathogenic species and rather were focused on a limited number of fungal species within the genera *Candida* and *Aspergillus*.

Leinberger et al. (105) described an array based on ITS-specific sequences for the detection of the 12 most common pathogenic *Candida* and *Aspergillus* organisms at the species level. Similarly, Huang et al. (79) described a high-throughput microarray to rapidly and simultaneously identify *C. neoformans* and other fungal pathogenic species. The method was based on the reversed solid hybridization of oligonucleotide and used universal ITS primers specific for the conserved regions of the rDNA gene (5.8S and 28S rRNA) and species-specific probes targeting sequence variation of the ITS2 region. The array, which was validated with isolates from clinical specimens and with spiked blood samples, was able to simultaneously detect a total of 20 different pathogenic species with detection limits as low as 15 pg/ml of DNA (5 to 1.2 CFU).

Most recently, Campa et al. (21) recently developed a more comprehensive fungal array system able to identify 24 species belonging to 10 genera. This two-dimensional microarray utilized two technologies, oligonucleotide probes and APEX technology (arrayed primer extension). APEX is an enzymatic genotyping method that uses an extension, performed by a DNA polymerase, along with four different dye terminators. This single-base extension method, which is initiated at the 3′ side of the anchored species-specific probe, was proven to be specific, despite low levels of cross-reactivity within some closely related species, e.g., *Candida metapsilosis*, *Candida orthopsilosis*, *C. parapsilosis*, *A. fumigatus*, and *Aspergillus terreus*.

Microarray technology continues to show promise for the development of powerful diagnostic tools. However, the potential use of microarrays is confounded by its high cost, poor platform flexibility, and slow hybridization kinetics. In addition, there are some other issues that undermine the power of this technology, e.g., inconsistency in sequence fidelity of the spotted probes, low specificity, poor reproducibility, and data noise.

LUMINEX BEAD SUSPENSION ARRAY

Luminex x-map technology (Luminex Corp., Austin, TX) is a detection system that uses a specialized flow cytometer for the identification of DNA targets in a high-throughput format. The technology can detect 500 different analytes in a reaction tube/well and uses multiple color fluorescent microspheres by varied proportions of red and infrared fluorescent dyes within microspheres, which create an array of up to 500 separate bead classifications, each of which can represent a single species. The technology has been applied to the detection of nucleic acid analysis with a variety of formats such as direct DNA hybridization and solution-based enzymatic reaction capture format. The direct hybridization assay uses a liquid suspension hybridization with specific oligonucleotide probes that are modified with a 5′-end amino C-12. The probes are covalently bound via the carbodiimide method to unique sets of fluorescent beads. Upon hybridization, the biotinylated target amplicon is detected by the conjugate, streptavidin R-phycoerythrin. A solution-based enzymatic array uses specific capture sequences that are incorporated into a product by a polymerase or ligase. The incorporation of the capture sequence allows hybridization to a complementary address sequence on the microsphere. Variants of this enzymatic capture format include (i) allele-specific primer extension, (ii) oligonucleotide ligation assay, and (iii) single based chain extension.

Because of the versatility of the system, which can be adapted to a wide variety of nucleic acid and immunoassay analysis, the technology has been employed for the detection of fungal pathogens (18, 37, 39, 40, 42, 132, 134, 135). Using this technology, Diaz and Fell described an eight-plex hybridization array for the detection of the varieties and genotypes within the *C. neoformans* species complex (40). The method, which employed a direct hybridization assay format, proved to be specific, as it allowed discrimination of 1 bp mismatch with no apparent cross-reactivity, permitted the detection of 10^1 to 10^3 genome copies, and allowed simultaneous detection of target sequences. Also, the assay can be carried out directly with yeast cells or from isolated DNA. The described assay format was recently validated with a collection of environmental and clinical isolates (18). The suspension array correctly identified the isolates at the species and genotype level. It also confirmed the identification of hybrid isolates that according to flow cytometric profiles and cloning experiments were classified as diploid or nearly diploid. These hybrid isolates, which were found to contain two *IGS1* alleles, belonged to serotype AD or BD.

Some of the advantages of this technology are accuracy, low cost of operation, high multiplexing capability, faster hybridization reactions, better and more stable surface chemistry than planar DNA microarray platform, amenability to high-throughput screening, and flexibility in platform design. For instance, different clinical testing platforms can be created by combining different sets of microspheres. Any modification to the platform only requires the addition or removal of probes of interest. This is in sharp contrast to density microarrays, where any modification to the testing platform requires the printing of new plates with specialized and expensive equipment. Since a level of redundancy identification is of uppermost importance in clinical diagnosis, the array can also be adapted to include heuristic species-specific probes and broad specific probes that will target multiple species within a taxonomic lineage. This strategy increases the reliability of the detection system and helps to unveil new emergent pathogens.

The technology can easily be adapted for the development of commercial or customized fungal identification kits which, depending on the needs of the user, can be expanded to detect new emergent species. Because of its high-throughput capability and potential for automation, bead-based technologies can bring progress to DNA-based diagnostics. As with any new diagnostic tool, time-consuming validation and adaptation procedures are required.

REFERENCES

1. **Aberg, J. A., J. Watson, M. Segal, and L. W. Chang.** 2000. Clinical utility of monitoring serum cryptococcal antigen (sCRAG) titers in patients with AIDS-related cryptococcal disease. *HIV Clin. Trials* **1:**1–6.
2. **Almeida, A. M., M. T. Matsumoto, L. C. Baeza, R. B. Oliveira e Silva, A. A. Kleiner, M. S. Melhem, M. J. Giannini, and the Laboratory Group on Cryptococcosis.** 2007. Molecular typing and antifungal susceptibility of clinical sequential isolates *of Cryptococcus neoformans* from Sao Paulo State, Brazil. *FEMS Yeast Res.* **7:**152–164.
3. **Anderson, M. J., K. Gull, and D. W. Denning.** 1996. Molecular typing by random amplification of polymorphic DNA and M13 southern hybridization of related paired isolates of *Aspergillus fumigatus*. *J. Clin. Microbiol.* **34:** 87–93.
4. **Antinori, S., A. Radice, L. Galimberti, C. Magni, M. Fasan, and C. Parravicini.** 2005. The role of cryptococcal antigen assay in diagnosis and monitoring of cryptococcal meningitis. *J. Clin. Microbiol.* **43:**5828–5829.
5. **Babady, N. E., J. E. Bestrom, D. J. Jespersen, M. F. Jones, E. M. Beito, M. J. Binnicker, and N. L. Wengenack.** 2009. Evaluation of three commercial latex agglutination kits and a commercial enzyme immunoassay for the detection of cryptococcal antigen. *Med. Mycol,* **47:** 336–338.
6. **Barreto de Oliveira, M. T., T. Boekhout, B. Theelen, F. Hagen, F. A. Baroni, M. S. Lazera, K. B. Lengeler, J. Heitman, I. N. G. Rivera, and C. R. Paula.** 2004. *Cryptococcus neoformans* shows a remarkable genotypic diversity in Brazil. *J. Clin. Microbiol.* **42:**1356–1359.
7. **Bart-Delabesse, E., H. van Deventer, W. Goessens, J. L. Poirot, N. Lioret, A. van Belkum, and F. Dromer.** 1995. Contribution of molecular typing methods and antifungal susceptibility testing to the study of a candidemia cluster in a burn care unit. *J. Clin. Microbiol.* **33:**3278–3283.
8. **Baughman, R. P., J. C. Rhodes, M. N. Dohn, H. Henderson, and P. T. Frame.** 1992. Detection of cryptococcal antigen in bronchoalveolar lavage fluid: a prospective study of diagnostic utility. *Am. Rev. Respir. Dis.* **145:**1226–1229.
9. **Bennett, J. E., and J. W. Bailey.** 1971. Control for rheumatoid factor in the latex test for cryptococcosis. *Am. J. Clin Pathol.* **56:**360–366.
10. **Bergman, A., V. Fernandez, K. O. Holmström, B. E. Claesson, and H. Enroth.** 2007. Rapid identification of pathogenic yeast isolates by real time PCR and two dimensional melting-point analysis. *Eur. J. Clin. Microbiol. Infect. Dis.* **26:**813–818.
11. **Binnicker, M. J., S. P. Buckwalter, J. J. Eisberner, R. A. Stewart, A. E. McCullough, S. L. Wohlfiel, and N. L. Wengenack.** 2007. Detection of *Coccidioides* species in clinical specimens by real-time PCR. *J. Clin. Microbiol.* **45:**173–178.
12. **Blevins, L. B., J. Fenn, H. Segal, P. Newcomb-Gayman, and K. C. Carroll.** 1995. False-positive cryptococcal antigen latex agglutination caused by disinfectants and soaps. *J. Clin. Microbiol.* **33:**1674–1675.
13. **Boekhout, T., B. Theelen, M. Diaz, J. W. Fell, W. C. Hop, C. A. Edwin, E. C. Abeln, F. Dromer, and W. Meyer.** 2001. Hybrid genotypes in the pathogenic yeast *Cryptococcus neoformans*. *Microbiology* **147:**891–907.
14. **Boekhout, T., A. van Belkum, A. C. Lendeers, H. A. Verbrugh, P. Mukamurangwa, D. Swinne, and W. A. Scheffers.** 1997. Molecular typing of *Cryptococcus neoformans*: taxonomy and epidemiological aspects. *Int. J. Syst. Bacteriol.* **47:**432–442.
15. **Boekhout, T., and A. van Belkum.** 1997.Variability of karyotypes and RAPD types in genetically related strains of *Cryptococcus neoformans*. *Curr. Genet.* **32:**203–208.
16. **Bottone, E. J., M. Sindone, and V. Caraballo.** 1998. Value of assessing cryptococcal antigen in bronchoalveolar lavage and sputum specimens from patients with AIDS. *Mt. Sinai J. Med.* **65:**422–425.
17. **Bovers, M, F. Hagen, E. E. Kuramae, and T. Boekhout.** 2008. Six monophyletic lineages identified within *Cryptococcus neoformans* and *Cryptococcus gattii* by multilocus sequence typing. *Fungal Gen. Biol.* **45:**400–421.
18. **Bovers, M., M. R. Diaz, F. Hagen, L. Spanjaard, B. Duim, C. E. Visser, H. L. Hoogveld, J. Scharringa, I. M. Hoepelman, J. W. Fell, and T. Boekhout.** 2007. Identification of genotypically diverse *Cryptococcus neoformans* and *Cryptococcus gattii* isolates using Luminex xMAP technology. *J. Clin. Microbiol.* **45:**1874–1883.
19. **Brandt, M. E., L. C. Hutwagner, L. A. Klug, W. S. Baughman, D. Rimland, E. A. Graviss, R. J. Hamill, C. Thomas, P. G. Pappas, A. L. Reingold, and R. W. Pinner.** 1996. Molecular subtype distribution of *Cryptococcus neoformans* in four areas of the United States. Cryptococcal Disease Active Surveillance Group. *J. Clin. Microbiol.* **34:**912–917.
20. **Brouwer, A. E., P. Teparrukkul, S. Pinpraphaporn, R. A. Larsen, W. Chierakul, S. Peacock, N. Day, N. J. White, and T. S. Harrison.** 2005. Baseline correlation and comparative kinetics of cerebrospinal fluid colony-forming unit counts and antigen titers in cryptococcal meningitis. *J. Infect. Dis,* **192:**681–684.
21. **Campa, D., A. Tavanti, F. Gemignani, C. S. Mogavero, I. Bellini, F. Bottari, R. Barale, S. Landi, and S. Senesi.** 2008. DNA microarray based on arrayed primer extension technique for identification of pathogenic fungi responsible for invasive and superficial mycoses. *J. Clin. Microbiol.* **46:**909–915.
22. **Carvalho, V. G., M. S. Terceti, A. L. T. Dias, C. R. Paula, J. P. Lyon, A. M. Siquiera, and M. C. Franco.** 2007. Serotype and mating type characterization of *Cryptococcus neoformans* by multiplex PCR. *Rev. Inst. Med. Trop. Sao Paulo* **49:**207–210.
23. **Casadevall, A., L. F. Freundlich, L. Marsh, and M. Scharff.** 1992. Extensive allelic variation in *Cryptococcus neoformans J. Clin. Microbiol.* **30:**1080–1084.

24. **Casadevall, A., and M. Fan.** 1992. *URA5* gene of *Cryptococcus neoformans* var. *gattii*: evidence for a close phylogenetic relationship between *C. neoformans* var. *gattii* and *C. neoformans* var. *neoformans*. *J. Gen. Appl. Microbiol.* **38:**491–495.
25. **Chang, H., S. N. Leaw, A. H. Huang, T. L. Wu, and T.C. Chang.** 2002. Rapid identification of yeasts in positive blood cultures by a multiplex PCR method. *J. Clin. Microbiol.* **39:**3466–3471.
26. **Chanock, S. J., P. Toltzis, and C. Wilson.** 1993. Cross-reactivity between *Stomatococcus mucilaginosus* and latex agglutination for cryptococcal antigen. *Lancet* **342:** 1119–1120.
27. **Chapin-Robertson, K., C. Bechtel, S. Waycott, C. Kontnick, and S. C. Edberg.** 1993. Cryptococcal antigen detection from the urine of AIDS patients. *Diagn. Microbiol. Infect. Dis.* **17:**197–201.
28. **Chen, F., B. P. Currie, L. C. Chen, S. G. Spitzer, E. D. Spitzer, and A. Casadevall.** 1995. Genetic relatedness of *Cryptococcus neoformans* clinical isolates grouped with the repetitive DNA probe CNRE-1. *J. Clin. Microbiol.* **33:**2818–2822.
29. **Chen, J., A. Varma, M. R. Diaz, A. P. Litvintseva, K. K. Wollenberg, and K. J. Kwon-Chung.** 2008. *Cryptococcus neoformans* strains and infection in apparently immunocompetent patients, China. *Emerg. Infect. Dis.* **14:**755–762.
30. **Chen, Y. C., J. D. Eisner, M. M. Kattar, S. L. Rassoulian-Barrett, K. Lafe, U. Bui, A. P. Limaye, and B. T. Cookson.** 2001. Polymorphic internal transcribed spacer region 1 DNA sequences identify medically important yeasts. *J. Clin. Microbiol.* **39:**4042–4051.
31. **Cogliati, M., M. C. Esposto, A. M. Tortorano, and M. A. Viviani.** 2006. *Cryptococcus neoformans* population includes hybrid strains homozygous at mating-type locus. *FEMS Yeast Res.* **6:**608–613.
32. **Cogliati, M., M. C. Esposto, D. L. Clarke, B. L. Wickes, and M. A. Viviani.** 2001. Origin of *Cryptococcus neoformans* diploid strains. *J. Clin. Microbiol.* **39:**3889–3894.
33. **Cohen, J.** 1984. Comparison of the sensitivity of three methods for the rapid identification of *Cryptococcus neoformans*. *J. Clin. Pathol.* **37:**332–334.
34. **Crampin, A. C., R. C. Matthews, D. Hall, and E. G. V. Evans.** 1993. PCR fingerprinting by random amplification of polymorphic DNA. *J. Med. Vet. Mycol.* **31:**463–465.
35. **Currie, B., L. Freundlich, M. Soto, and A. Casadevall.** 1993. False-negative cerebrospinal fluid cryptococcal latex agglutination tests for patients with culture-positive cryptococcal meningitis. *J. Clin. Microbiol.* **31:**2519–2522.
36. **Currie, B. P., L. F. Freundlich, and A. Casadevall.** 1994. Restriction fragment length polymorphism analysis of *Cryptococcus neoformans* isolates from environmental (pigeon excreta) and clinical sources in New York City. *J. Clin. Microbiol.* **32:**1188–1192.
37. **Das, S., T. M. Brown, K. L. Kellar, B. P. Holloway, and C. J. Morrison.** 2006. DNA probes for the rapid identification of medically important *Candida* species using a multianalyte profiling system. *FEMS Immunol. Med. Microbiol.* **46:**244–250.
38. **Desmet, P., K. D. Kayembe, and C. De Vroey.** 1989. The value of cryptococcal serum antigen screening among HIV-positive/AIDS patients in Kinshasa, Zaire. *AIDS* **3:**77–78.
39. **Diaz, M. R., and J. W. Fell.** 2004. High-throughput detection of pathogenic yeasts in the genus *Trichosporon*. *J. Clin. Microbiol.* **42:**3696–3706.
40. **Diaz, M. R., and J. W. Fell.** 2005. Use of a suspension array for rapid identification of the varieties and genotypes of the *Cryptococcus neoformans* species complex. *J. Clin. Microbiol.* **43:**3662–3672.
41. **Diaz, M. R., J. W. Fell, T. Boekhout, and B. Theelen.** 2000. Molecular sequence analyses of the intergenic spacer (IGS) associated with rDNA of the two varieties of the pathogenic yeast, *Cryptococcus neoformans*. *Syst. Appl. Microbiol.* **23:**535–545.
42. **Diaz, M. R., T. Boekhout, B. Theelen, M. Bovers, F. J. Cabañes, and J. W. Fell.** 2006. Microcoding and flow cytometry as identification system for *Malassezia* species. *J. Med. Microbiol.* **55:**1197–1209.
43. **Diaz, M. R., T. Boekhout, T. Kiesling, and J. W. Fell.** 2005. Comparative analysis of the intergenic spacer regions and population structure of the species complex of the pathogenic fungus: *Cryptococcus neoformans*. *FEMS Yeast Res.* **5:**1129–1140.
44. **Dreier, J., M. Störmer, and K. Kleesiek.** 2005. Use of bacteriophage MS2 as an internal control in viral reverse transcription-PCR assays. *J. Clin. Microbiol.* **43:**4551–4557.
45. **Dromer, F., A. Varma, O. Ronin, S. Mathoulin, and B. Dupont.** 1994. Molecular typing of *Cryptococcus neoformans* serotype D clinical isolates. *J. Clin. Microbiol.* **32:**2364–2371.
46. **Ellis, D., D. Marriot, R. A. Hajjeh, D. Warnock, W. Meyer, and R. Barton.** 2000. Epidemiology: surveillance of fungal infections. *Med. Mycol.* **38:**173–182.
47. **Engler, H. D., and Y. R. Shea.** 1994. Effect of potential interference factors on performance of enzyme immunoassay and latex agglutination assay for cryptococcal antigen. *J. Microbiol.* **32:**2307–2308.
48. **Escandón, P., A. Sánchez, M. Martínez, W. Meyer, and E. Castañeda.** 2006. Molecular epidemiology of clinical and environmental isolates of the *Cryptococcus neoformans* species complex reveals a high genetic diversity and the presence of the molecular type VGII mating type a in Colombia. *FEMS Yeast Res.* **6:**625–635.
49. **Esposto, M. C., M. Cogliati, A. M. Tortorano, and M. A.Viviani.** 2009. Electrophoretic karyotyping of *Cryptococcus neoformans* AD hybrid strains. *Mycoses* **52:**16–23.
50. **Etard, J. F., I. Ndiaye, and M. Thierry-Mieg.** 2006. Mortality and causes of death in adults receiving highly active antiretroviral therapy in Senegal: a 7-year cohort study. *AIDS.* **20:**1181–1189.
51. **Evertsson, U., H. J. Monstein, and A. G. Johansson.** 2000. Detection and identification of fungi in blood using broad range 28SrDNA PCR amplification and species-specific hybridization. *APMIS* **108:**385–392.
52. **Fan, M., B. P. Currie, R. R. Gutell, M. A. Ragan, and A. Casadevall.** 1994. The 16S-like, 5.8S and 23S-like rRNAs of the two varieties of *Cryptococcus neoformans*: sequence, secondary structure, phylogenetic analysis and restriction fragment polymorphisms. *J. Med. Vet. Mycol.* **32:**163–180.
53. **Fan, M., L. C. Chen, M. A. Ragan, R. R. Gutell, J. R. Warner, B. P. Currie, and A. Casadevall.** 1995. The 5S rRNA and the rRNA intergenic spacer of the two varieties of *Cryptococcus neoformans*. *J. Med. Vet. Mycol.* **33:**215–221.
54. **Fell, J. W.** 1995. rDNA targeted oligonucleotide primers for the identification of pathogenic yeasts in a polymerase chain reaction. *J. Ind. Microbiol.* **14:**475–477.
55. **Fell, J. W., T. Boekhout, A. Fonseca, G. Scorzetti, and A. Statzell-Tallman.** 2000. Biodiversity and systematics of basidiomycetous yeasts as determined by large subunit rD1/D2 domain sequence analysis. *Int. J. Syst. Evol. Microbiol.* **50:**1351–1371.
56. **Feng, X., Z. Yao, D. Ren, and W. Liao.** 2008. Simultaneous identification of molecular and mating types within the *Cryptococcus* species complex by PCR-RFLP analysis. *J. Med. Microbiol.* **57:**1481–1490.
57. **Feng, X., Z. Yao, D. Ren, W. Liao, and J. Wu.** 2008. Genotype and mating type analysis *of Cryptococcus neoformans* and *Cryptococcus gattii* isolates from China that

mainly originated from non-HIV-infected patients. *FEMS Yeast Res.* **8:**930–938.

58. **Frank, U. K., S. L. Nishimura, N. C. Li, K. Sugai, D. M. Yajko, W. K. Hadley, and V. L. Ng.** 1993. Evaluation of an enzyme immunoassay for detection of cryptococcal capsular polysaccharide antigen in serum and cerebrospinal fluid. *J. Clin. Microbiol.* **31:**97–101.
59. **Franzot, S. P., I. F. Salkin, and A. Casadevall.** 1999. *Cryptococcus neoformans* var. *grubii*: separate varietal status for *Cryptococcus neoformans* serotype A isolates. *J. Clin. Microbiol.* **37:**838–840.
60. **Franzot, S. P., J. Mukherjee, R. Cherniak, L. C. Chen, J. S. Hamdan, and A. Casadeval.** 1998. Microevolution of a standard strain of *Cryptococcus neoformans* resulting in differences in virulence and other phenotypes. *Infect. Immun.* 66:89–97.
61. **Franzot, S. P., J. S. Hamdan, B. P. Currie, and A. Casadevall.** 1997. Molecular epidemiology of *Cryptococcus neoformans* in Brazil and the United States: evidence for local genetic differences and a global clonal population structure. *J. Clin. Microbiol.* **35:**2243–2251.
62. **Fries, B. C., F. Chen, B. P. Currie, and A. Casadevall.** 1996. Karyotype instability in *Cryptococcus neoformans* infection. *J. Clin. Microbiol.* **34:**1531–1534.
63. **Fujita, S. I., Y. Senda, S. Nakaguchi, and T. Hashimoto.** 2001. Multiplex PCR using internal transcribed spacer 1 and 2 regions for rapid detection and identification of yeast strains. *J. Clin. Microbiol.* **39:**3617–3622.
64. **Gade, W., S. W. Hinnefeld, L. S. Babcock, P. Gilligan, W. Kelly, K. Wait, D. Greer, M. Pinilla, and R. L. Kaplan.** 1991. Comparison of the PREMIER cryptococcal antigen enzyme immunoassay and latex agglutination assay for detection of cryptococcal antigens. *J. Clin. Microbiol.* **29:**1616–1619.
65. **Garcia-Bermejo, M. J., J. Anton, C. Ferrer, I. Meseguer, J. L. Abad, and M. F. Colom.** 2001. Chromosome length polymorphism in *Cryptococcus neoformans* clinical and environmental isolates. *Rev. Iberoam. Micol.* **18:**174–179.
66. **Garcia-Hermoso, D., F. Dromer, S. Mathoulin-Pelissier, and G. Janbon.** 2001. Are two *Cryptococcus neoformans* strains epidemiologically linked? *J. Clin. Microbiol.* **39:** 1402–1406.
67. **Hadrys, H., M. Balick, and B. Schierwater.** 1992. Applications of random amplified polymorphic DNA (RAPD) in molecular ecology. *Mol. Ecol.* **1:**55–63.
68. **Hall, L., S. Wohlfiel, and G. D. Roberts.** 2003. Experience with the MicroSeq D2 large-subunit ribosomal DNA sequencing kit for the identification of commonly encountered clinically important yeast species. *J. Clin. Microbiol.* **41:**5099–5102.
69. **Hall, L., S. Wohlfiel, and G. D. Roberts.** 2004. Experience with the MicroSeq D2 large-subunit ribosomal DNA sequencing kit for identification of filamentous fungi encountered in a clinical laboratory. *J. Clin. Microbiol.* **42:**622–626.
70. **Hamilton, J. R., A. Noble, D. W. Denning, and D. A. Stevens.** 1991. Performance of cryptococcus antigen latex agglutination kits on serum and cerebrospinal fluid specimens of AIDS patients before and after pronase treatment. *J. Clin. Microbiol.* **29:**333–339.
71. **Hanafy, A., S. Kaocharoen, A. Jover-Botella, M. Katsu, S. Iida, T. Kogure, T. Gonoi, Y. Mikami, and W. Meyer.** 2008. Multi-locus microsatellite typing for *Cryptococcus neoformans* var. *grubii*. *Med. Mycol.* **46:**685–696.
72. **Hayette, M. P., D. Vaira, F. Susin, P. Boland, G. Christiaens, P. Melin, and P. de Mol.** 2001. Detection of *Aspergillus* species DNA by PCR in bronchoalveolar lavage fluid. *J. Clin. Microbiol.* **39:**2338–2340.
73. **Haynes, K. A, T. J. Westerneng, J. W. Fell, and W. Moens.** 1995. Rapid identification of pathogenic fungi by polymerase chain reaction amplification of large subunit ribosomal DNA. *J. Med. Vet. Mycol.* **33:**319–325.
74. **Heelan, J. S., L. Corpus, and N. Kessimian.** 1991. False-positive reactions in the latex agglutination test for *Cryptococcus neoformans* antigen. *J. Clin Microbiol.* **29:**1260–1261.
75. **Hendolin, P. H., L. Paulin, P. Koukila-Kahkola, V. J. Anttitlia, H. Malmberg, M. Richardson, and J. Ylikoski.** 2000. Panfungal PCR and multiplex liquid hybridization for detection of fungi in tissue specimens. *J. Clin. Microbiol.* **38:**4186–4192.
76. **Hopfer, R. L., E. V. Perry, and V. Fainstein.** 1982. Diagnostic value of cryptococcal antigen in the cerebrospinal fluid of patients with malignant disease. *J. Infect. Dis.* **145:**915.
77. **Hossain, M. A., T. Miyazaki, K. Mitsutake, H. Kakeya, Y. Yamamoto, K. Yanagihara, S. Kawamura, T. Otsubo, Y. Hirakata, T. Tashiro, and S. Kohno.** 1997. Comparison between Wako-WB003 and Fungitec G tests for detection of (1–>3)-beta-D-glucan in systemic mycosis, *J. Clin. Lab. Anal.* **11:**73–77.
78. **Hsu, M. C., K. W. Chen, H. J. Lo, Y. C. Chen, M. H. Liao, Y. H. Lin, and S. Y. Li.** 2003. Species identification of medically important fungi by use of real-time LightCycler PCR. *J. Med. Microbiol.* **52:**1071–1076.
79. **Huang, A., J. W. Li, Z. Q. Shen, X. W. Wang, and M. Jin.** 2006. High-throughput identification of clinical pathogenic fungi by hybridization to an oligonucleotide microarray. *J. Clin. Microbiol.* **44:**3299–3305.
80. **Imwidthaya, P., and N. Poungvarin.** 2000. Cryptococcosis in AIDS. *Postgrad. Med. J.* **76:**85–88.
81. **Innings, A., M. Ullberg, A. Johansson, C. J. Rubin, N. Noreus, M. Isaksson, and B. Herrmann.** 2007. Multiplex real-time PCR targeting the RNase P RNA gene for detection and identification of *Candida* species in blood. *J. Clin. Microbiol.* **45:**874–880.
82. **Jarvis, J. N., A. Boulle, A. Loyse, T. Bicanic, K. Rebe, A. Williams, T. S. Harrison, and G. Meintjes.** 2009. High ongoing burden of cryptococcal disease in Africa despite antiretroviral roll out. *AIDS* **23:**1182–1183.
83. **Jarvis, J. N., S. D. Lawn, M. Vogt, N. Bangani, R. Wood, and T. S. Harrison.** 2009. Screening for cryptococcal antigenemia in patients accessing an antiretroviral treatment program in South Africa. *Clin. Infect. Dis.* **48:**856–862.
84. **Jordan, J. A., and M. B. Durso.** 1996. Rapid speciation of the five most medically relevant *Candida* species using PCR amplification and a microtiter plate-based detection system. *Mol. Diagn.* 1:51–58.
85. **Kambugu, A., B. Castelnuovo, B. Wandera, A. Kiragga, and M. R. Kamya.** 2007. Antiretroviral therapy in an urban African cohort does not prevent significant early mortality Abstract WEPEB055. *In* Program and Abstracts of the 4th International AIDS Society, Conference on HIV Pathogenesis, Treatment and Prevention, Sydney, Australia.
86. **Kaocharoen, S., B. Wang, K. M. Tsui, L. Trilles, F. Kong, and W. Meyer.** 2008. Hyperbranched rolling circle amplification as a rapid and sensitive method for species identification within the *Cryptococcus* species complex. *Electrophoresis* **29:**3183–3191.
87. **Katsu, M., S. Kidd, A. Ando, M. L. Moretti-Branchini, Y. Mikami, K. Nishimura, and W. Meyer.** 2004. The internal transcribed spacers and 5.8 S rRNA gene show extensive diversity among isolates of the *Cryptococcus neoformans* species complex. *FEMS Yeast Res.* **4:**377–388.
88. **Kersulyte, D., J. P. Woods, E. J. Keath, W. E. Goldman, and D. E. Berg.** 1992. Diversity among clinical isolates of *Histoplasma capsulatum* detected by polymerase chain reaction with arbitrary primers. *J. Bacteriol.* **174:**7075–7079.
89. **Kidd, S. E., F. Hagen, R. L. Tscharke, M. Huynh, K. H. Bartlett, M. Fyfe, L. MacDougall, T. Boekhout,**

K. J. Kwong-Chung, and W. Meyer. 2004. A rare genotype of *Cryptococcus gattii* caused the cryptococcosis outbreak on Vancouver Island (British Columbia, Canada). *Proc. Natl. Acad. Sci. USA* **101:**17258–17263.

90. **Kiska, D., D. Orkiszewski, D. Howell, and P. Gilligan.** 1994. Evaluation of new monoclonal antibody-based latex agglutination test for detection of cryptococcal polysaccharide antigen in serum and cerebronspinal fluid. *J. Clin. Microbiol.* **32:**2309–2311.
91. **Kontoyiannis, D. P.** 2003. What is the significance of an isolated positive cryptococcal antigen in the cerebrospinal fluid of cancer patients? *Mycoses* **46:**161–163.
92. **Koufopanou, V., A. Burt, and J. W. Taylor.** 1997. Concordance of gene genealogies reveals reproductive isolation in the pathogenic fungus *Coccidioides immitis*. *Proc. Natl. Acad. Sci. USA* **94:**5478–5482.
93. **Kralovic, S. M., and J. C. Rhodes.** 1998. Utility of routine testing of bronchoalveolar lavage fluid for cryptococcal antigen. *J. Clin. Microbiol.* **36:**3088–3089.
94. **Kurtzman, C. P., and C. J. Robnett.** 1998. Identification and phylogeny of ascomycetous yeasts from analysis of nuclear large subunit (26S) ribosomal DNA partial sequences. *Antonie van Leeuwenhoek* **73:**331–371.
95. **Kwon-Chung, K. J., W. B. Hill, and J. E. Bennett.** 1981. New, special stain for histopathological diagnosis of cryptococcosis. *J. Clin Microbiol.* **13:**383–387.
96. **Latouche, G. N., M. Huynh, T. C. Sorrell, and W. Meyer.** 2003. PCR-restriction fragment length polymorphism analysis of the phospholipase B (*PLB1*) gene for subtyping of *Cryptococcus neoformans* isolates. *Appl. Environ. Microbiol.* **69:**2080–2086.
97. **Lau, A., S. Chen, T. Sorrell, D. Carter, R. Malik, P. Martin, and C. Halliday.** 2007. Development and clinical application of a panfungal PCR assay to detect and identify fungal DNA in tissue specimens. *J. Clin. Microbiol.* **45:**380–385.
98. **Lau, A., T. C. Sorrell, S. Chen, K. Stanley, J. Iredell, and C. Halliday.** 2008. Multiplex tandem PCR: a novel platform for rapid detection and identification of fungal pathogens from blood culture specimens. *J. Clin. Microbiol.* **46:**3021–3027.
99. **Lau, A., T. C. Sorrell, O. Lee, K. Stanley, and C. Halliday.** 2008. Colony multiplex-tandem PCR for rapid, accurate identification of fungal cultures. *J. Clin. Microbiol.* **46:**4058–4060.
100. **Lawn, S. D., L. G. Bekker, L. Myer, C. Orrell, and R. Wood.** 2005. Cryptococcocal immune reconstitution disease: a major cause of early mortality in a South African antiretroviral program. *AIDS* **19:**2050–2052.
101. **Lazcano, O., V. O. Speights, J. Bilbao, J. Becker, and J. Diaz.** 1991. Combined Fontana-Masson-mucin staining of *Cryptococcus neoformans*. *Arch. Pathol. Lab. Med.* **115:**1145–1149.
102. **Lazcano, O., V. O. Speights, J. G. Strickler, J. E. Bilbao, J. Becker, and J. Diaz.** 1993. Combined histochemical stains in the differential diagnosis of *Cryptococcus neoformans*. *Mod. Pathol.* **6:**80–84.
103. **Leal, A. L., J. Faganello, M. C. Bassanesi, and M. H. Vainstein.** 2008 *Cryptococcus* species identification by multiplex PCR. *Med. Mycol.* 46:377–383.
104. **Leaw, S. N., H. C. Chang, H. F. Sun, R. Barton, J. P. Bouchara, and T. C. Chang.** 2006. Identification of medically important yeast species by sequence analysis of the internal transcribed spacer regions. *J. Clin. Microbiol.* 44:693–699.
105. **Leinberger, D. M., U. Schumacher, I. B. Autenrieth, and T. T. Bachmann.** 2005. Development of a DNA microarray for detection and identification of fungal pathogens involved in invasive mycoses. *J. Clin. Microbiol.* **43:**4943–4953.
106. **Liaw, Y. S., P. C. Yang, C. J. Yu, D. B. Chang, H. J. Wang, L. N. Lee, S. H. Kuo, and K. T. Luh.** 1995. Direct determination of cryptococcal antigen in transthoracic needle aspirate for diagnosis of pulmonary cryptococcosis. *J. Clin. Microbiol.* **33:**1588–1591.
107. **Lindberg, J., F. Hagen, A. Laursen, J. Stenderup, and T. Boekhout.** 2007. *Cryptococcus gattii* risk for tourists visiting Vancouver Island, Canada. *Emerg. Infect. Dis.* **13:**178–179.
108. **Litvintseva, A. P., R. Thakur, R. Vilgalys, and T. G. Mitchell.** 2006. Multilocus sequence typing reveals three genetic subpopulations of *Cryptococcus neoformans* var. *grubii* (serotype A), including a unique population in Botswana. *Genetics* **172:**2223–2238.
109. **Loeffler, J., H. Hebart, P. Cox, N. Flues, U. Schumacher, and H. Einsele.** 2001. Nucleic acid sequence-based amplification of *Aspergillus* RNA in blood samples. *J. Clin. Microbiol.* **39:**1626–1629.
110. **Loeffler, J., H. Hebart, U. Brauchle, U. Schumacher, and H. Einsele.** 2000. Comparison between plasma and whole blood specimens for detection of *Aspergillus* DNA by PCR. *J. Clin. Microbiol.* **38:**3830–3833.
111. **Loftus, B. J., E. Fung, P. Roncaglia, D. Rowley, P. Amedeo, D. Bruno, J. Vamathevan, M. Miranda, I. J. Anderson, J. A. Fraser, J. E. Allen, I. E. Bosdet, M. R. Brent, R. Chiu, T. L. Doering, M. J. Donlin, C. A. D'Souza, D. S. Fox, V. Grinberg, J. Fu, M. Fukushima, B. J. Haas, J. C. Huang, G. Janbon, S. J. Jones, H. L. Koo, M. I. Krzywinski, J. K. Kwon-Chung, K. B. Lengeler, R. Maiti, M. A. Marra, R. E. Marra, C. A. Mathewson, T. G. Mitchell, M. Pertea, F. R. Riggs, S. L. Salzberg, J. E. Schein, A. Shvartsbeyn, H. Shin, M. Shumway, C. A. Specht, B. B. Suh, A. Tenney, T. R. Utterback, B. L. Wickes, J. R. Wortman, N. H. Wye, J. W. Kronstad, J. K. Lodge, J. Heitman, R. W. Davis, C. M. Fraser, and R. W. Hyman.** 2005. The genome of the basidiomycetous yeast and human pathogen *Cryptococcus neoformans*. *Science* **307:**1321–1324.
112. **López, J. F., and R. F. Lebrón.** 1972. *Cryptococcus neoformans*: their identification in body fluids and cultures by mucicarmine stain (Mayer). *Bol. Asoc Med.* **64:**203–205.
113. **Louis, E. J.** 1995. The chromosome ends of *Saccharomyces cerevisiae*. *Yeast* **11:**1553–1573.
114. **Luo, G., and T. G. Mitchell.** 2002. Rapid identification of pathogenic fungi directly from cultures by using multiplex PCR. *J. Clin. Microbiol.* **40:**2860–2865.
115. **Lyman, C. A., S. J. Devi, J. Nathanson, C. E. Frasch, P. A. Pizzo, and T. J. Walsh.** 1995. Detection and quantitation of the glucuronoxylomannan-like polysaccharide antigen from clinical and nonclinical isolates of *Trichosporon beigelii* and implications for pathogenicity. *J. Clin. Microbiol.* **33:**126–130.
116. **MacDougall, L., and M. Fyfe.** 2006. Emergence of *Cryptococcus gattii* in a novel environment provides clues to its incubation period. *J. Clin. Microbiol.* **44:**1851–1852.
117. **McManus, E. J., and J. M. Jones.** 1985. Detection of a *Trichosporon beigelii* antigen cross-reactive with *Cryptococcus neoformans* capsular polysaccharide in serum from a patient with disseminated *Trichosporon* infection. *J. Clin. Microbiol.* **21:**681–685.
118. **Meyer, W., D. M. Aanensen, T. Boekhout, M. Cogliati, M. R. Diaz, M. C. Esposto, M. Fisher, F. Gilgado, F. Hagen, S. Kaocharoen, A. P Litvintseva, T. G. Mitchell, S. P. Simwami, L. Trilles, M. A. Viviani, and J. Kwon-Chung.** 2009. Consensus multi-locus sequence typing scheme for *Cryptococcus neoformans* and *Cryptococcus gattii*. *Med. Mycol.* **12:**1–14.
119. **Meyer, W., A. Castañeda, S. Jackson, M. Huynh, and E. Castañeda.** 2003. Molecular typing of IberoAmerican

Cryptococcus neoformans isolates. *Emerg. Infect. Dis.* **2:**189–195.

120. **Meyer, W., K. Marszewska, M. Amirmostofina, R. P Igreja, C. Hardtke, K. Methling, M. A. Viviani, A. Chindamporn, S. Sukroongreung, M. A. John, D. H. Ellis, and T. C. Sorrell.** 1999. Molecular typing of global isolates of *Cryptococcus neoformans* var. *neoformans* by PCR-fingerprinting and RAPD. A pilot study to standardize techniques on which to base a detailed epidemiological survey. *Electrophoresis* **20:**1790–1799.
121. **Micol, R., O. Lortholary, and B. Sar.** 2007. Prevalence, determinants of positivity, and clinical utility of cryptococcal antigenemia in Cambodian HIV-infected patients. *J. Acquir. Immune Defic. Syndr.* **45:**555–559.
122. **Mihrendi, S. H., P. Kordbacheh, B. Kazemi, S. Samiei, M. Pezeshki, and M. R. Khorramizadeh.** 2001. A PCR-RFLP method to identification of the important opportunistic fungi: *Candida* species, *Cryptococcus neoformans, Aspergillus fumigatus* and *Fusarium solani. Iranian J. Publ. Health* **30:**103–106.
123. **Millon, L., T. Barale, M.-C. Julliot, J. Martinez, and G. Mantion.** 1995. Interference by hydroxyethyl starch used for vascular filling in latex agglutination test for cryptococcal antigen. *J. Clin. Microbiol.* **33:**1917–1919.
124. **Mitchell, T. G., E. Z. Freedman, T. J. White, and J. W. Taylor.** 1994. Unique oligonucleotide primers in PCR for identification of *Cryptococcus neoformans. J. Clin. Microbiol.* **32:**253–255.
125. **Montero, C. I., Y. R. Shea, P. A. Jones, S. M. Harrington, N. E. Tooke, F. G. Witebsky, and P. R. Murray.** 2008. Evaluation of pyrosequencing (R) technology for the identification of clinically relevant non-dematiaceous yeasts and related species. *Eur. J. Clin. Microbiol.* **27:** 821–830.
126. **Morgan, J. A. T., V. T. Vredenburg, L. J. Rachowicz, R. A. Knapp, M. J. Stice, T. Tunstall, R. E. Bingham, J. M. Parker, J. E. Longcore, C. Moritz, C. J. Briggs, and J. W. Taylor.** 2007. Population genetics of the frog-killing fungus *Batrachochytrium dendrobatidis. Proc. Natl. Acad. Sci. USA* **104:**13845–13850.
127. **Nelson, M. R., M. Bower, C. Smith, D. Shanson, and B. Gazzard.** 1990. The value of serum cryptococcal antigen in the diagnosis of cryptococcal infection in patients infected with human immunodeficiency virus. *J. Infect.* **21:**175–811.
128. **Ngamwongsatit, P., S. Sukroongreung, C. Nilakul, V. Prachayasittikul, and S. Tantimavanich.** 2005. Electrophoretic karyotypes of *C. neoformans* serotype A recovered from Thai patients with AIDS. *Mycopathologia* **159:**189–197.
129. **Ninet, B., I. Jan, O. Bontems, B. Léchenne, O. Jousson, R. Panizzon, D. Lew, and M. Monod.** 2003. Identification of dermatophyte species by 28S ribosomal DNA sequencing with a commercial kit. *J. Clin. Microbiol.* **41:**826–830.
130. **Obayashi, T., K. Negishi, T. Suzuki, and N. Funata.** 2008. Reappraisal of the serum (1->3)-beta-D-glucan assay for the diagnosis of invasive fungal infections: a study based on autopsy cases from 6 years. *Clin. Infect. Dis.* **46:**1864–1870.
131. **Obayashi, T., M. Yoshida, H. Tamura, J. Aketagawa, S. Tanaka, and T. Kawai.** 1992. Determination of plasma (1→3)-β-D-glucan: a new diagnostic aid to deep mycosis. *J. Med. Vet. Mycol.* **30:**275–280.
132. **O'Donnell, K., B. A. Sarver, M. Brandt, D. C. Chang, J. N. Wang, B. J. Park, D. A. Sutton, L. Benjamin, M. Lindsley, A. Padhye, D. M. Geiser, and T. J. Ward.** 2007. Phylogenetic diversity and microsphere array-based genotyping of human pathogenic Fusaria, including isolates from the multistate contact lens-associated U.S. keratitis outbreaks of 2005 and 2006. *J. Clin. Microbiol.* **45:**2235–2248.
133. **Ostrosky-Zeichner, L., B. D. Alexander, D. H. Kett, J. Vazquez, P. G. Pappas, F. Saeki, P. A. Ketchum, J. Wingard, R. Schiff, H. Tamura, M. A. Finkelman, and J. H. Rex.** 2005. Multicenter clinical evaluation of the (1->3) beta-D-glucan assay as an aid to diagnosis of fungal infections in humans. *Clin. Infect. Dis.* **41:**654–659.
134. **Page, B. T., and C. P. Kurtzman.** 2005. Rapid identification of *Candida* and other clinically important yeast species by flow cytometry. *J. Clin. Microbiol.* **43:**4507–4514.
135. **Page, B. T., C. E. Shields, W.G. Merz, and C. P. Kurtzman.** 2006. Rapid identification of ascomycetous yeasts from clinical specimens by a molecular method based on flow cytometry and comparison with identifications from phenotypic assays. *J. Clin. Microbiol.* **44:**3167–3171.
136. **Pazin, J. G., T. H. Rude, C. C. Dykstra, and J. R. Perfect.** 1995. Cloning of the topoisomerase II gene in of *Cryptococcus neoformans*. Abstract 345. 33rd Infectious Disease Society Meeting, San Francisco, CA.
137. **Perfect, J. R., B. B. Magee, and P. T. Magee.** 1989. Separation of chromosomes of *Cryptococcus neoformans* by pulsed field gel electrophoresis. *Infect. Immun.* **57:**2624–2627.
138. **Perfect, J. R., N. Ketabchi, G. M. Cox, C. W. Ingram, and C. Beiser.** 1993. Karyotyping *of Cryptococcus neoformans* as an epidemiological tool. *J. Clin. Microbiol.* **31:**3305–3309.
139. **Pfaller, M. A., J. Zhang, S. Messer, M. Tumberland, E. Mbidde, C. Jessup, and M. Ghannoum.** 1998. Molecular epidemiology and antifungal susceptibility of *Cryptococcus neoformans* isolates from Ugandan AIDS patients. *Diagn. Microb. Infect. Dis.* **32:**191–199.
140. **Polacheck, I., G. Lebens, and J. B. Hicks.** 1992. Development of DNA probes for early diagnosis and epidemiological study of cryptococcosis in AIDS patients. *J. Clin. Microbiol.* **30:**925–930.
141. **Powderly, W.G., G. A. Cloud, W. E. Dismukes, and M. S. Saag.** 1994. Measurement of cryptococcal antigen in serum and cerebrospinal fluid: value in the management of AIDS-associated cryptococcal meningitis. *Clin. Infect. Dis.* **18:**789–792.
142. **Rappelli, P., R. Are, G. Casu, P. L. Fiori, P. Cappuccinelli, and A. Aceti.** 1998. Development of a nested PCR for detection of *Cryptococcus neoformans* in cerebrospinal fluid. *J. Clin. Microbiol.* **36:**3438–3440.
143. **Reiss, E., T. Obayashi, K. Orle, M. Yoshida and R. M. Zancope-Oliveira.** 2000. Non-culture based diagnostic tests for mycotic infections. *Med. Mycol.* **38:** 147–159.
144. **Ribeiro, M. A., and P. Ngamskulrungroj.** 2008. Molecular characterization of environmental *Cryptococcus neoformans* isolated in Vitoria, ES, Brazil. *Rev. Inst. Med. Trop. Sao Paulo.* **50:**315–320.
145. **Ro, J. Y., S. S. Lee, and A. G. Ayala.** 1987. Advantage of Fontana-Masson stain in capsule-deficient cryptococcal infection. *Arch. Pathol. Lab. Med.* **111:**53–57.
146. **Saha, D. C., I. Xess, and N. Jain.** 2008. Evaluation of conventional and serological methods for rapid diagnosis of cryptococcosis. *Indian J. Med. Res.* **127:**483–488.
147. **Sandhu, G. S., B. C. Kline, L. Stockman, and G. D. Robert.** 1995. Molecular probes for diagnosis of fungal infections. *J. Clin. Microbiol.* **33:**2913–2919.
148. **Schierwater, B., and A. Ender.** 1993. Different thermostable DNA polymerase may amplify different RAPD products. *Nucleic Acids Res.* 21:4647–4648.
149. **Sekhon, A. S., A. K. Garg, L. Kaufman, G. S. Kobayashi, Z. Hamir, M. Jalbert, and N. Moledina.** 1993. Evaluation of a commercial enzyme immunoassay for the detection of cryptococcal antigen. *Mycoses* **36:**31–34.

150. **She, X., W. Liu, and N. Guo.** 2006. Karyotype analysis of *Cryptococcus neoformans* with pulse electrophoresis. *J USA China Med. Sci.* **3:**41–45.
151. **Skow, A., K. A. Mangold, M. Tajuddin, A. Huntington, B. Fritz, R. B. Thomson, Jr., and K. L. Kaul.** 2005. Species level identification of staphylococcal isolates by real time PCR and melt curve analysis. *J. Clin. Microbiol.* **43:**2876–2880.
152. **Snow, R. M., and W. E. Dismukes.** 1975. Cryptococcal meningitis: diagnostic value of cryptococcal antigen in cerebrospinal fluid. *Arch. Intern. Med.* **135:**1155–1157.
153. **Spitzer, E. D., and S. G. Spitzer.** 1992. Use of a dispersed repetitive DNA element to distinguish clinical isolates of *Cryptococcus neoformans*. *J. Clin. Microbiol.* **30:**1094–1097.
154. **Spitzer, E. D., and S. G. Spitzer.** 1995. Structure of the ubiquitin-encoding genes of *Cryptococcus neoformans*. *Gene* **161:**113–117.
155. **Spitzer, E. D., S. G. Spitzer, L. F. Freundlich, and A. Casadevall.** 1993. Persistence of initial infection in recurrence of *Cryptococcus neoformans* meningitis. *Lancet* **341:**595–596.
156. **Spitzer, S. G., and E. D., Spitzer.** 1994. Characterization of CNRE-1 family of repetitive DNA elements in *Cryptococcus neoformans*. *Gene* **144:**103–106.
157. **Stephen, C., S. Lester, W. Black, M. Fyfe, and S. Raverty.** 2002. Multispecies outbreak of cryptococcosis on southern Vancouver Island, British Columbia. *Can. Vet. J.* **43:**792–794.
158. **Sugita, T., R. Ikeda, and T. Shinoda.** 2001. Diversity among strains of *Cryptococcus neoformans* var. *gattii* as revealed by sequence analysis of multiple genes and a chemotype analysis of capsular polysaccharide. *Microbiol. Immunol.* **45:**757–768.
159. **Tanaka, K., T. Miyazaki, S. Maesaki, K. Mitsutake, H. Kakeya, Y. Yamamoto, K. Yanagihara, M. A. Hossain, T. Tashiro, and S. Kohno.** 1996. Detection of *Cryptococcus neoformans* gene with pulmonary cryptococcosis. *J. Clin. Microbiol.* **32:**2826–2828.
160. **Tanner, D. C., M. P. Weinstein, B. Fedorciw, K. L. Joho, J. J. Thorpe, and L. B. Reller.** 1994. Comparison of commercial kits for detection of cryptococcal antigen. *J. Clin. Microbiol.* **31:**1680–1684.
161. **Tassie, J. M., L. Pepper, and C. Fogg.** 2003. Systematic screening of cryptococcal antigenemia in HIV-positive adults in Uganda. *J. Acquir. Immune Defic. Syndr.* **33:** 411–412.
162. **Temstet, A., P. Roux, J. L. Poirot, O. Ronin, and F. Dromer.** 1992. Evaluation of a monoclonal antibody-based latex agglutination test for diagnosis of cryptococcosis: comparison with two tests using polyclonal antibodies. *J. Clin. Microbiol.* **30:**2544–2550.
163. **Tintelnot, K., K. Lemmer, H. Losert, G. Shar, and A. Polak.** 2004. Follow up of epidemiological data of cryptococcosis in Austria, Germany and Switzerland with special focus on the characterization of clinical isolates. *Mycoses* **47:**455–464.
164. **Trilles, L., M. S. Lazéra Mdos, B. Wanke, R. V. Oliveira, G. G. Barbosa, M. M. Nishikawa, B. P. Morales, and W. Meyer.** 2008. Regional pattern of the molecular types of *Cryptococcus neoformans* and *Cryptococcus gattii* in Brazil. *Mem. Inst. Oswaldo Cruz* **103:**455–462.
165. **Turenne, C. Y., S. E. Sanche, D. J. Hoban, J. A. Karlowsky, and A. M. Kabani.** 1999. Rapid identification of fungi by using the internal transcribed spacer 2 genetic region and an automated fluorescent capillary electrophoresis system. *J. Clin. Microbiol.* **37:**1846–1851.
166. **Vance, A. M.** 1961. The use of the mucicarmine stain for a rapid presumptive identification of *Cryptococcus* from culture. *Am. J. Med. Technol.* **27:**125–128.
167. **Varma, A., D. Swinne, F. Staib, J. E. Bennett, and K. J. Kwon-Chung.** 1995. Diversity of DNA fingerprints in *Cryptococcus neoformans*. *J. Clin. Microbiol.* **33:**1807–1814.
168. **Varma, A., and K. J. Kwon-Chung.** 1992. DNA probe for strain typing of *Cryptococcus neoformans*. *J. Clin. Microbiol.* **30:**2960–2967.
169. **Viviani, M. A., H. Wen, A. Roverselli, R. Caldarelli-Stefano, M. Cogliati, P. Ferrante, and A. M. Tortorano.** 1997. Identification by polymerase chain reaction fingerprinting of *Cryptococcus neoformans* serotype AD. *J. Med. Vet. Mycol.* **35:**355–360.
170. **Vollmer, T., M. Störmer, K. Kleesiek, and J. Dreier.** 2008. Evaluation of novel broad-range real-time PCR assay for rapid detection of human pathogenic fungi in various clinical specimens *J. Clin. Microbiol.* **46:**1919–1926.
171. **Warren, R. J., A. Perceval, and B. W. Dwyer.** 1993. Comparative evaluation of cryptococcal latex tests. *Pathology* **25:**76–80.
172. **Westerink, M. A., D. Amsterdam, R. J. Petell, M. N. Stram, and M. A. Apicella.** 1987. Septicemia due to DF-2. Cause of a false-positive cryptococcal latex agglutination result. *Am. J. Med.* **83:**155–158.
173. **Whittier, S., R. L. Hopfer, and P. Gilligan.** 1994. Elimination of false-positive serum reactivity in latex agglutination test for cryptococcal antigen in human immunodeficiency virus-infected population. *J. Clin. Microbiol.* **32:**2158–2161.
174. **Wu, T. C., and S. Y. Koo.** 1993. Comparison of three commercial cryptococcal latex kits for detection of cryptococcal antigen. *J. Clin. Microbiol.* **18:**1127–1130.
175. **Xu, J., G. Luo, R. J. Vilgalys, M. E. Brandt, and T. G. Mitchell.** 2002. Multiple origins of hybrid strains of *Cryptococcus neoformans* with serotype AD. *Microbiology* **148:**203–212.
176. **Xu, J., R. Ali, D. A. Gregory, D. Amick, S. E. Lambert, H. J. Yoell, R. J. Vigalys, and T. G. Mitchell.** 2000. Uniparental mitochondrial transmission in sexual crosses in fungus *Cryptococcus neoformans*. *Curr. Microbiol.* **4:**269–273.
177. **Xu, J., R. J. Vilgalys, and T. G. Mitchell.** 2000. Multiple genealogies reveal recent dispersion and hybridization in the human pathogenic fungus *Cryptococcus neoformans*. *Mol. Ecol.* **9:**1471–1481.
178. **Zhang, N., K. O'Donnell, D. A. Sutton, A. Nalim, R. C. Summerbell, A. A. Padhye, and D. M. Geiser.** 2006. Members of the *Fusarium solani* species complex that cause infections in both humans and plants are common in the environment. *J. Clin. Microbiol.* **44:**2186–2190.

Cryptococcus: From Human Pathogen to Model Yeast
Edited by J. Heitman et al.

42

Management of Cryptococcal Meningoencephalitis in Both Developed and Developing Countries

JOSEPH N. JARVIS, TIHANA BICANIC, AND THOMAS S. HARRISON

This chapter will discuss current management of cryptococcal meningoencephalitis in terms of antifungal drug therapy, the management of complications, notably raised cerebrospinal fluid (CSF) pressure, the issues of the timing and choice of antiretroviral therapy (ART) in HIV-infected patients, and the management of symptomatic relapse, including immune reconstitution syndromes. The outcomes associated with current management are reviewed, and opportunities for improved intervention, including for earlier diagnosis and treatment, are highlighted. The emphasis is on HIV-associated infection in resource-limited settings, which constitutes the majority of the global burden of cryptococcal infection (113).

OUTCOMES OF CURRENT MANAGEMENT

A wide range in clinical outcomes has been reported for treatment of cryptococcal meningitis from different settings and studies, making it important to bear in mind the many factors that can affect morbidity and mortality rates. These include (i) the underlying diagnosis; (ii) the severity of infection at the start of treatment, perhaps best assessed by the baseline organism burden and mental status, which may be affected by access to medical services, complex cultural factors that affect the time at which patients present to available services, and any delays in appropriate investigation; (iii) the availability of antifungal drugs and facilities for monitoring therapy and measuring and managing raised CSF pressure; (iv) facilities for supportive care and treatment of concomitant illness, especially common in the HIV-infected population; and (v) the degree to which the population studied is selected and the completeness of follow-up data.

In developed-country settings, the combination of relatively easy access to care, facilities for prompt diagnosis, close monitoring, and optimal supportive care, as well as a full range of antifungal drugs, results in reported 10-week mortality in the range of 10 to 26% (Table 1). The lower end of this range represents the results of randomized trials, notably the landmark study of the Mycoses Study Group (150), which by their nature are selective to some degree, tending to exclude the sickest patients. Higher mortality has been reported in some less-selective series (87, 132).

In resource-limited settings mortality rates are generally higher (Table 2): in Southeast Asia and Latin America, between 19 and 43% at 10 weeks (28, 40, 62, 127). In Africa, where resources are most restricted, 10-week mortality has ranged from 24 to 95% (18, 21, 71, 85, 104, 136). Examination of Table 2 illustrates that a high proportion of patients with abnormal mental status at presentation is associated with high mortality irrespective of the antifungal treatment available.

Of note, before the availability of ART, cryptococcal meningoencephalitis was usually a late-stage event in the course of untreatable HIV infection. As such, management was essentially palliative rather than curative in the developed world (108), as was the case until very recently in many resource-limited settings. Fortunately, access to ART has increased dramatically in recent years, and ART programs in resource-limited areas have yielded results comparable to those seen in developed-country settings (13). Access to ART, which to date has usually been started 2 to 10 weeks into antifungal therapy, has not yet been shown to have reduced the acute mortality of ART-naive patients presenting with cryptococcal meningitis (87). However, ART has transformed the long-term prognosis of HIV patients with cryptococcal meningitis (18, 87). The result is that if patients anywhere in the world with access to ART survive the initial critical months of cryptococcal meningoencephalitis, they have an excellent long-term prognosis (Fig. 1). Thus, the challenge is to reduce the unacceptably high early mortality associated with cryptococcal meningitis, so as to raise the proportion of patients who survive to benefit from and become established on ART.

PROGNOSTIC FACTORS, DETERMINANTS OF CLINICAL OUTCOME, AND STRATEGIES TO REDUCE MORTALITY

Prior to the HIV epidemic, factors identified as predicting a poor outcome in cryptococcal meningitis included underlying

Joseph N. Jarvis, Tihana Bicanic, and Thomas S. Harrison, Centre for Infection, St George's University of London, London SW17 ORE, United Kingdom.

TABLE 1 Outcomes of therapy for HIV-associated cryptococcal meningitis in developed-country settings

Country	Reference	Year	Induction treatment	Abnormal mental status	ART available	2-week mortality	10-week mortality	Comments
USA	135	1988–1989	Amphotericin B 0.4–0.5 mg/kg/day +/– flucytosine 150 mg/kg/day or fluconazole 200–400 mg/day	27%*	No	12%	17%	*Lethargy or obtundation Comatose patients and those unlikely to survive 2 weeks excluded.
USA	150	1991–1994	Amphotericin B 0.7 mg/kg/day +/– flucytosine 100 mg/kg/day	11%	No	5.5%	10–23%*	Comatose patients excluded. *Exact 10-week mortality unknown as only subset of patients re-randomized at 2 weeks.
USA	132	1986–1993	Amphotericin B 0.3–0.7 mg/kg/day + flucytosine 150 mg/kg/day	—	No	12%	26%	
France	87	1990–1996	Amphotericin-based therapy 75%, fluconazole monotherapy 25%	—	Only in 1996	—	19%	No difference between pre-ART and ART era (21% vs. 18%).
France	49	1997–2001	Amphotericin B + flucytosine 52%, amphotericin B monotherapy or fluconazole monotherapy in the remainder	33%	Yes	6.5%*	15%**	*Just HIV-positive patients. **12-week data

disease (hematologic malignancy or corticosteroid use), absence of headache, abnormal mental status, high organism burden (India ink positivity or high cryptococcal antigen titer), poor host inflammatory response (CSF white cell count <20 per ml), and raised CSF opening pressure (45, 47). In non-HIV-associated cryptococcal meningitis, after the introduction of fluconazole, mortality was associated with chronic renal or liver failure or hematologic malignancy as predisposing factors, absence of headache, male gender, and altered mental status (112). In series of HIV-associated cryptococcal meningitis, abnormal mental status and high organism load, measured by quantitative CSF culture or CSF antigen titer, are the most consistent and important baseline factors associated with death (19, 28, 135), while abnormal neurology and brain imaging, raised CSF opening pressure, and low CSF white cell count have also been associated with poor outcome (49, 57, 135).

In addition to the baseline organism load, the initial mycological response to treatment has been shown to be important in subsequent outcome. In a multivariate analysis, 2-week CSF culture status was the most important factor in determining response (survival with negative culture) at week 10 (132). In addition, in a combined cohort of over 250 patients whose response to treatment was assessed by serial quantitative CSF cultures, the rate of clearance of infection over the first 2 weeks was shown to be a third factor, independent of altered mental status and baseline organism load, that was associated with survival at 2 and 10 weeks (19). A model of how the factors affecting clinical outcome may interrelate is shown in Fig. 2, based on this analysis.

Figure 3 summarizes some of the reasons underlying the high mortality of patients with cryptococcal meningitis, including the constraints on optimal management in developing-country settings. While most deaths within the first 2 weeks are directly related to cryptococcal infection, between then and patients becoming established on ART, other HIV-related opportunistic infections and complications increasingly contribute. Also listed in the figure are some strategies for addressing these issues more effectively in the future. In the following sections, as well as discussing current management, we have included such strategies and research needs.

ANTIFUNGAL DRUG THERAPY

Amphotericin B was the first effective therapy for cryptococcal meningoencephalitis (8) and remains the mainstay of initial therapy (134). Earlier pivotal trials, prior to the HIV epidemic, examined dose, duration, and combination therapy with flucytosine. Of note, in the absence of an effective, safe, oral agent for maintenance therapy, the aim was radical cure without later recurrence. The results of these studies suggested that 6 weeks of the combination of amphotericin B (0.3 mg/kg/day) and flucytosine (150 mg/kg/day) was more effective and as well tolerated as 10 weeks of amphotericin B (0.4 mg/kg/day) alone (14) and was associated with fewer

TABLE 2 Outcomes of therapy for HIV-associated cryptococcal meningitis in resource-limited settings

Country	Reference	Year	Induction treatment	Abnormal mental status	ART available	2-week mortality	10-week mortality	Comments
Thailand	62	1996–1997	Amphotericin B 0.5–0.8 mg/kg/day	—	No	—	43%	
Thailand	127	1997–1999	Amphotericin B 0.7 mg/kg/day	7%	No	16%	~40%*	*Exact 10-week figure not reported
Thailand	28	2002	Amphotericin B +/– flucytosine 100 mg/kg/day and/or fluconazole 400 mg/day	19%	No	14%	22%	
Cambodia	99	2004	Amphotericin B 0.7 mg/kg/day	—	Yes	10%*	37%**	*3- and **12-week figures
Peru	40	1998–2001	Amphotericin B 0.7 mg/kg/day	28%	—	13%	19%	
Brazil	109	1995–1997	Amphotericin B–based therapy 80% Fluconazole plus flucytosine 20%*	11.4% confusion 8.6% fits	Yes	31%	63%**	*Exact doses of treatment not specified **Deaths during "second phase" of treatment Timing not specified
Zambia	104	1998–1999	Fluconazole 200 mg/day*	13% confusion 9% fits 3% paralysis	No	39%	96%**	*Data only reported for those who received treatment 400 mg stat dose initially **12-week figure
Uganda	96	1994	Fluconazole 200 mg/day	48%	No	40%	64%*	*8-week figure
			Fluconazole 200 mg/day + flucytosine			16%	44%*	
South Africa	136	1999–2002	Fluconazole 200 mg or 400 mg/day	40%*	No	25%**	~50%***	*Includes focal neurology **In hospital *** Median follow-up for discharged patients 36 days, lost = censored
South Africa	18	2005	Amphotericin B 1 mg/kg/day*	24%	Yes	17%	37%	*Amphotericin only given for 1 week 5 of 54 patients received fluconazole as initial therapy
Uganda	71	2001–2002	Amphotericin B 0.7 mg/kg/day	38%	No	46%	—	Comatose patients excluded
		2006–2007			Yes	20%	43%	

(Continued on next page)

TABLE 2 *(Continued)*

Country	Reference	Year	Induction treatment	Abnormal mental status	ART available	2-week mortality	10-week mortality	Comments
South Africa	21	2005–2006	Amphotericin B 0.7–1 mg/kg/day + flucytosine 100 mg/kg/day	13%	Yes	6%	24%	
Botswana	23	2005–2006	Amphotericin B 1 mg/kg/day	12%	Yes	17%*	—	*In-hospital figure
Uganda	85	2005–2007	Fluconazole 800 mg/day	47%	Yes	37%	60%	
			Fluconazole 1,200 mg/day			22%	48%	
Malawi	63	2008	Fluconazole 1,200 mg/day	39%	Yes	37%	58%	
			Fluconazole 1,200 mg/day + flucytosine 100 mg/kg/day			10%	43%	

relapses than the same regimen given for 4 weeks (47). The development of fluconazole in the 1980s led to a direct comparison of fluconazole with amphotericin B–based treatment in patients with HIV-associated infection. The trial showed no mortality difference between the arms of the trial at 2 and 10 weeks, although trends were in favor of amphotericin B, and the study was only powered to detect a 30% lower response rate in the fluconazole arm (135). The dosage of both drugs was low (mean amphotericin B dose 0.4 to 0.5 mg/kg/day, with 14% of patients also receiving flucytosine, versus fluconazole 200 mg/day, with 34% of patients having the dose increased to 400 mg/day after a median of 22 days) compared to current recommendations (amphotericin B 0.7 to 1 mg/kg/day and fluconazole 800 to 1,200 mg/day), and the results were disappointing in both arms of the trial. In particular, the median time to sterilization of CSF was 42 days with amphotericin B and 64 days with fluconazole. A small single-center study did find amphotericin B (initially 0.7 mg/kg/day) plus flucytosine to be clinically superior to fluconazole 400 mg/day (79).

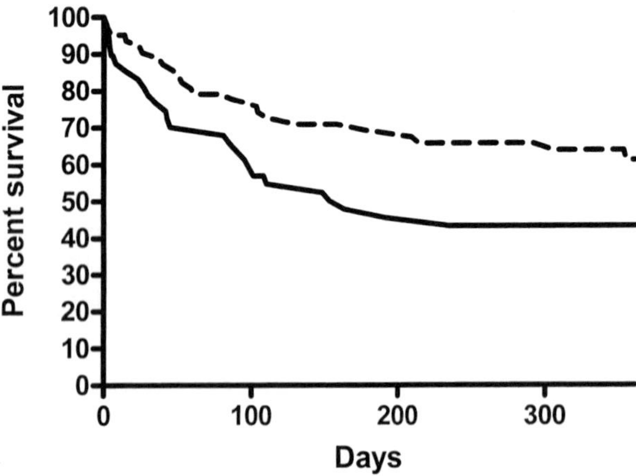

FIGURE 1 Survival to 1 year for patients with HIV-associated cryptococcal meningitis in Cape Town, South Africa. Solid line: nonselected cohort, induction treatment with amphotericin B 1 mg/kg/day alone for 1 week. Dashed line: randomized trial, induction treatment with amphotericin B (0.7 or 1 mg/kg/day) plus flucytosine. Data from references 21 and 24, respectively.

This was the background of the landmark HIV-associated cryptococcal meningitis study of van der Horst and colleagues, the rationale of which was to try to get early control of infection with rapidly fungicidal higher-dose amphotericin B (0.7 mg/kg/day), with or without flucytosine (100 mg/kg/day), but then to avoid the toxicities associated with these drugs by switching to consolidation and then maintenance treatment with the better-tolerated azoles, either fluconazole or itraconazole (150). The results, published over a decade ago, remain the best to date from a phase III trial and form the basis of current guidelines (Table 3). Over half of the patients had sterile CSF confirmed at 2 weeks, and overall 2-week mortality was 5.5%. Addition of flucytosine was associated with a trend ($P = 0.06$) toward a higher proportion of patients with sterile CSF at 2 weeks and with a lower rate of later relapse. The significance of this last observation has been questioned, however, since most relapses occurred in patients subsequently randomized to itraconazole rather than fluconazole (J. E. Bennett, presented at the 7th International Conference on Cryptococcus and Cryptococcosis, Nagasaki, Japan, 2008), whereas, as a result of the trial, fluconazole is established as the azole of choice for consolidation and maintenance (133).

Thus, debate continues as to the benefit of flucytosine. The more rapid fungicidal activity of the combination of amphotericin B and flucytosine was confirmed in a trial that directly assessed the rate of clearance of infection from CSF, based on serial quantitative CSF cultures (28). Despite the relatively small numbers of patients studied, the rate of clearance of infection, or early fungicidal activity, was significantly more rapid for amphotericin B plus flucytosine compared with amphotericin B alone or in combination with

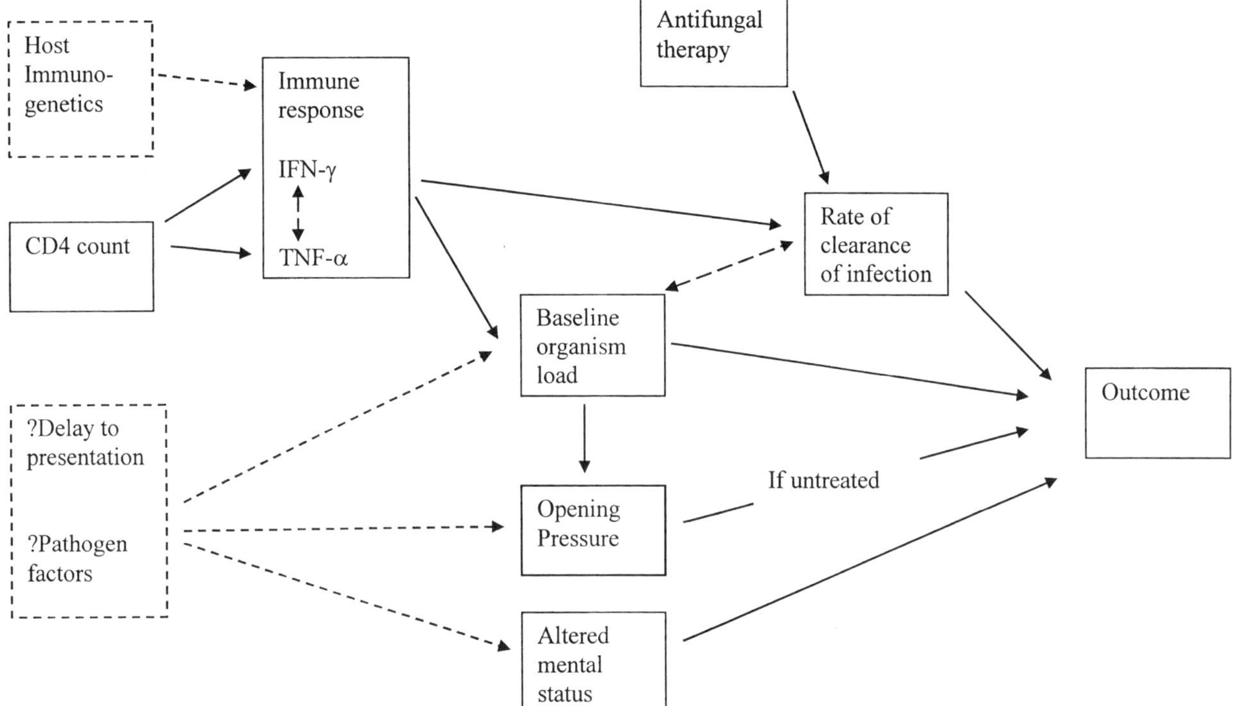

FIGURE 2 A model illustrating possible relationships between factors associated with rate of clearance of infection and survival. Proposed causal links are shown with solid arrows, noncausal associations with long-dashed arrows, and speculative associations with short-dashed arrows. TNF, tumor necrosis factor. From reference 22 with permission.

- The inadequacy of current antifungal drugs and regimens
    - Phase III studies of amphotericin B plus fluconazole
    - Studies to optimize oral regimens, and compare optimal oral regimens with standard and short-course (1 week) induction with amphotericin B
  - Inadequate access to antifungal drugs and the required facilities and resources to monitor patients optimally
    - Increased access to amphotericin B, flucytosine, and resources for monitoring

- Unrecognized and untreated complications
  - Raised cerebrospinal fluid pressure
    - Increased access to manometers
  - Cryptococcal IRIS
    - Further work to address immunological basis, risk factors, and outcome of cryptococcal IRIS

- Other HIV-related illness: absence of data on optimal timing of ART, delays in accessing ART
    - Trials of the timing of ART in resource-limited settings
    - Continued expansion of ART programs

- Late presentation
    - Point-of-care diagnostic test
    - Pre-emptive treatment for antigen-positive patients, through screening prior to ART if CD4 cell count <100

FIGURE 3 Reasons for high mortality and strategies to reduce mortality in resource-limited settings.

TABLE 3 Treatment recommendations for cryptococcal meningitis in immunocompromised patients

Regimen	Treatment[a]	Duration
Initial antifungal regimen: D-AmB induction	D-AmB 0.7–1.0 mg/kg/day plus 5-FC 100 mg/kg/day (75 mg/kg/day if intravenous formulation used)	2 weeks
	Alternatives:	
	In renal dysfunction: L-AmB 3 mg/kg/day; preferred formulation in transplant patients	2 weeks
	If 5-FC is not available: D-AmB 1 mg/kg/day intravenously or D-AmB 0.7–1 mg/kg/day intravenously plus fluconazole 800 mg/day[b]	2 weeks (1 week better than no D-AmB[c])
Consolidation with D-AmB-based regimens	Fluconazole 400 mg/day[b]	8 weeks
If D-AmB is not available	Fluconazole 1,200 mg/day[b] plus 5-FC 100 mg/kg/day orally (if available)	2–10 weeks[d] 2 weeks
Maintenance therapy	Fluconazole 200 mg/day[b], start ART at 2–6 weeks. Consider discontinuing maintenance after a minimum of 1 year if CD4+ cell count >100 /ml³ and HIV viral load suppressed	≤ 1 year
	Alternative: Itraconazole 200–400 mg/day	

[a] D-AmB, amphotericin B deoxycholate; 5-FC, flucytosine; L-AmB, liposomal amphotericin B.

[b] Increase fluconazole dose by 50% if patient is on rifampin.

[c] Consider fluconazole 800 mg/day as consolidation if only 1 week of treatment with D-AmB.

[d] Fluconazole increases nevirapine levels, and the safety of high-dose fluconazole with nevirapine is unknown. Efavirenz is preferred. Consider reducing fluconazole to 400 mg/day with full-dose nevirapine.

fluconazole at 400 mg/day or all three drugs given together. In addition, in a large, prospective, unselected but uncontrolled study, patients treated with amphotericin B plus flucytosine had a much lower rate of treatment failure (death or positive CSF culture) at 2 weeks than patients treated with amphotericin B alone (48).

While a reduction in mortality has not been demonstrated in a randomized study, most experts and current guidelines recommend addition of flucytosine to amphotericin B, given the evidence above and the fact that in this patient population, and with this dose and duration, flucytosine is reasonably well tolerated. In the Mycoses Study Group trial there was only a 3% drug discontinuation rate within the first 2 weeks, and this was mostly related to amphotericin B (150). In the Thai study, there was no significant difference in blood count parameters between patients who did or did not receive flucytosine and no drug discontinuations within 2 weeks (28, 31). A substudy of that trial may help explain the lack of toxicity observed (31). Comparison of flucytosine drug levels between patients treated with intravenous versus oral formulation showed that the bioavailability of oral flucytosine in these late-stage, HIV-infected Thai patients was only around 50%, lower than in previous studies in normal subjects and non-HIV patients in North America (24, 38). This resulted in serum levels (20 to 30 mg/liter) that were lower than those usually associated with toxicity but were nevertheless sufficient to stay above the MIC for the whole dosing interval. Consistent with the proposed concentration-independent pharmacodynamics of flucytosine (7, 152), oral formulation resulted in the same boost to fungicidal activity as was seen with the higher flucytosine levels achieved with the intravenous formulation. If intravenous flucytosine is used, as is common in Europe, 100 mg/kg/day may be in excess of that required for maximal fungicidal activity, and lower doses, for example, 75 mg/kg/day, are probably sufficient. Further studies of the bioavailability of oral flucytosine are needed in late-stage HIV-infected patients.

Although the numbers are very small, despite the lower serum flucytosine levels, 5-fluorouracil was detected more often in patients on the oral compared with the intravenous formulation (31). 5-Fluorouracil is produced from flucytosine by cytosine deaminase, present in fungi and bacteria but not mammalian cells, and is thought to be involved in the bone marrow toxicity of flucytosine (46). This finding is consistent with the fact that gut bacteria may play a role in the generation of 5-fluorouracil in patients given oral flucytosine (59, 153).

Of note, monitoring of flucytosine in these trials was done with blood counts rather than real-time flucytosine serum levels. Nevertheless, if available rapidly, serum drug concentrations can add another level of safety to the use of flucytosine and are recommended in current Infectious Diseases Society of America (IDSA) guidelines (134). Target levels are 30 to 80 μg/ml, although, as discussed above, trough levels of 20 μg/ml may be adequate.

Liposomal amphotericin B has been the most widely used of the lipid formulations for cryptococcal meningitis. In a study in a rabbit model of CNS *Candida* infection, comparing all the lipid formulations (amphotericin B colloidal dispersion, Amphotec; amphotericin B lipid complex, Abelcet; liposomal amphotericin B, AmBisome; and conventional amphotericin B deoxycholate), liposomal amphotericin B achieved the highest brain tissue levels and resulted in the lowest brain tissue CFU (57). Although less nephrotoxic, it appears to be no more effective than the conventional formulation. In a large randomized trial, comparing 3 mg/kg/day and 6 mg/kg/day with 0.7 mg/kg/day of conventional formulation, there were no differences in efficacy (2 week culture conversion, 2 and 10 week survival) between the three arms of the trial (58). Liposomal amphotericin B was less nephrotoxic than conventional drug, and 3 mg/kg/day liposomal amphotericin B was less nephrotoxic than 6 mg/kg/day. Thus, liposomal amphotericin B is an alternative for those patients with preexisting renal impairment, on other nephrotoxic agents, or who develop renal impairment on the conventional formulation. It is preferred in transplant patients, most of whom are on nephrotoxic calcineurin inhibitors (144).

Some experts and guidelines suggest a routine lumbar puncture at 2 weeks, with prolongation of induction phase therapy if the CSF culture is not yet sterile (134). This is a very reasonable strategy, given evidence of the importance of sterilization for later, 10-week, outcome. However, it has not been studied and necessarily implies that all patients will continue induction until CSF culture results are available. Given the dose- and duration-dependent nature of the most important amphotericin B and flucytosine toxicities, even a few extra days of induction treatment will carry increasing risks of clinically significant side effects. Alternative strategies, also requiring but worthy of study, could be to tailor the duration of induction based on the initial organism load, as determined by quantitative cultures, hemocytometer counts, or less precisely, CSF antigen titers (30) or to increase the efficacy of consolidation-phase therapy by increasing the fluconazole dose.

Unfortunately, flucytosine is not widely available at present in Africa and Asia. The choice in these areas, in centers that can give amphotericin B, is between using amphotericin B alone or in combination with fluconazole at 800 mg/day, as discussed below.

In contrast to flucytosine, evidence suggests that amphotericin B has concentration-dependent activity (6). As discussed above, sterilization of CSF was much more rapid in recent trials with 0.7 mg/kg/day (150) than the prior studies with lower doses (135). Further, a dose of 1 mg/kg/day, as used for some other systemic fungal infections, has been used in noncomparative cohort studies and appeared reasonably well tolerated (41, 127). Dosages of 0.7 and 1.0 mg/kg/day, both given with flucytosine, were directly compared in a recent early fungicidal activity study in Cape Town (21). The higher dose was associated with significantly more rapid clearance of infection. The study was not powered for clinical or toxicity end points, but trends in laboratory side effects suggested slightly more anemia and possibly renal impairment at the higher dose. These side effects were reversible, and the authors considered that any potential increase in toxicity remained manageable (22). Interestingly, anemia on amphotericin B treatment was more common in women.

The study provided evidence to retain the 0.7- to 1-mg/kg/day dosage range in updated IDSA guidelines (117), and in settings where flucytosine is not available to boost fungicidal activity, it provided data to support use of the 1-mg/kg dosage. Use of this dose, if amphotericin B is used alone, has been endorsed by the Southern African HIV Clinicians Society (142). Further data on the safety of the 1-mg/kg/day dose in larger numbers of patients will be available from a large study in Vietnam in which all patients will receive the higher dose (ISRCTN95123928, www.controlled-trials.com).

A 1-mg/kg/day dose of amphotericin B may also be preferable if resources preclude a full 2-week induction course. In this case, a shortened course of amphotericin B may be better than no amphotericin B. A very significant reduction in the burden of infection in terms of the CSF CFU counts of between 3 and 4 logs can be achieved with 7 days of amphotericin B at 1 mg/kg/day (18). In addition, renal impairment and anemia, which are dose-related and usually manifest during the second week of induction (21), may be reduced. Although small and underpowered, a randomized study from Thailand found induction with 1 week of amphotericin B not to be substantially inferior to 2 weeks (146).

Despite the antagonism between amphotericin B and azoles that has been observed in some systems (11), most animal model data suggest that amphotericin B plus fluconazole is a very effective combination against *Cryptococcus neoformans* (44, 77). In Thailand, amphotericin B and fluconazole at 400 mg/day was more rapidly fungicidal than amphotericin B alone, although the difference did not reach statistical significance, but was less rapidly fungicidal than amphotericin B plus flucytosine (28). Amphotericin B plus fluconazole at 800 mg/day is more promising. In a recent phase II study by Pappas and colleagues, amphotericin B plus fluconazole at 800 mg/day was associated with the highest proportion of successful outcomes (combined end point of sterile CSF, survival, and neurological stability or improvement) at 2, 6, and 10 weeks, in comparison with amphotericin B alone and amphotericin B plus fluconazole 400 mg/day (111). Phase III studies of this combination are underway (ISRCTN95123928, www.controlled-trials.com) and planned, but in the meantime, given the widespread availability, affordability, and tolerability of fluconazole, amphotericin B plus fluconazole at 800 mg/day is an alternative to high-dose amphotericin B alone where flucytosine is not available. Amphotericin B plus fluconazole 800 mg/day is to be used as an induction therapy in an NIH-funded trial of the timing of ART after cryptococcal meningitis to be conducted at African sites (D. Boulware, personal communication). Interestingly, in the phase II study, there was a trend toward inferiority for the amphotericin B plus fluconazole 400 mg/day arm compared to the other two arms at early time points. Although it is a preliminary and unexplained finding, amphotericin B plus fluconazole 400 mg/day need not, and probably should not, be used.

Whether used alone or with a second drug, and at whichever dose, it is vital that amphotericin B therapy is monitored closely and potassium replaced, as required, and that saline fluid supplementation is given, provided there is no contraindication, to reduce nephrotoxicity (27, 119). To reduce the need for oral potassium supplements that can exacerbate nausea, it has been our practice to add a small amount (20 mmol) of potassium to a daily 1 liter of saline prehydration fluid. The importance of monitoring and fluids may be illustrated by an important study from Kampala, Uganda, where 2 weeks of induction with amphotericin B was used throughout (71). Despite use of amphotericin B, 2-week mortality was high (51%) in a cohort of patients treated in 2001, but significantly reduced (20%) in a cohort from 2006. Measures of severity did not appear to have changed between the two cohorts, and introduction of ART, started after the 2-week time point, could not have contributed. Instead, a possible reason for the improvement was efforts by the investigators to make fluid loading routine and a switch from once-weekly to thrice-weekly blood monitoring in the later cohort.

Unfortunately, in many centers in Africa, 2 weeks of induction therapy with amphotericin B is not yet feasible as a safe and sustainable treatment. To date, amphotericin B has not usually been provided as part of the HIV program of major donors. In addition to the costs of the drug, which may be substantial in local terms (20), are the requirements for hospitalization; intravenous drug administration, including nursing time and expertise in siting and maintaining intravenous access for a drug that causes considerable phlebitis; saline fluids and electrolyte replacement; and regular, rapid, and reliable laboratory monitoring for renal function, electrolytes, and hemoglobin.

In the absence of routine fluid and saline loading and close monitoring, amphotericin B cannot be given safely for 2 weeks. Whether a short, 5- to 7-day course of amphotericin B could be given safely without reliable monitoring is an important question under investigation. In the meantime, many centers in Africa continue to rely on fluconazole, available free through the Pfizer donation program (156) and also in inexpensive generic form. The question for such centers is what dose should be used. The earlier and largest studies used 200 to 400 mg/day. As discussed above, although there was no significant difference in mortality comparing this fluconazole dose with low-dose amphotericin B, the results in the fluconazole arm were poor (135). Results with fluconazole at 200 to 400 mg/day have also been poor in Africa. In a series from Zambia the median survival with fluconazole 200 mg/day monotherapy was 19 days compared to 10 days in untreated patients (104). Mortality with fluconazole 200 mg/day in a Ugandan trial was 40% in the first 2 weeks and 64% at 2 months (96). And in Cape Town, a retrospective study with incomplete follow-up, with patients lost to follow-up censored, nevertheless reported around 50% 10-week mortality with 200 or 400 mg/day (136). This represents a minimum estimate, and the 10-week mortality in unselected patients treated with fluconazole at up to 400 mg/day is probably significantly more than 50% in most African settings. In light of these data, investigators in a recent global burden study estimated the 3-month mortality associated with cryptococcal disease in sub-Saharan Africa to be 70% (113).

In Cape Town, in a small group of patients, with serial quantitative cultures, fluconazole at 400 mg/day was shown to be essentially fungistatic, at least over the first 2 weeks of treatment (18). The resulting prolonged period with a high viable organism load may predispose to the development of secondary fluconazole resistance. Such resistance is a significant problem when initial therapy is with fluconazole at 400 mg/day (16). A further concern is that prolonged active infection could also increase the risk of immune reconstitution reactions following introduction of ART. Although data on this point are lacking, of relevance, the risk of developing culture-negative symptomatic relapse after ART was shown to be associated with the burden of infection, as assessed by serum cryptococcal antigen titer, prior to starting ART (67).

Thus, fluconazole at 400 mg/day appears inadequate as an induction treatment. However, animal model data suggest a dose-response effect with fluconazole (74), and plasma concentrations of fluconazole in patients with fungal infection are known to increase linearly with doses up to 2 g/day (5). In terms of safety, 800 mg/day has been shown to be safe in randomized trials (111, 131), 1,200 mg/day is given routinely by many experts in coccidioidal meningitis (70), and fluconazole doses up to 2 g have been given to small numbers of patients with cryptococcal meningitis and other fungal infections (5, 100). The numbers treated at the highest doses are very small, but serious side effects seem not to be very frequent, at least up to 1,600 mg.

In terms of efficacy, accepting the caveats of comparing across trials and the small numbers of patients in the higher-dose cohorts, there is some suggestion from earlier studies that time to sterilization of CSF becomes shorter as the dose of fluconazole is increased from 200 to 800 mg/day. Median time to CSF sterilization was 64 days with 200 to 400 mg/day (135), mean time to sterilization was 41 days with 400 mg/day (79), and median times were 21 and 33 days with 800 mg/day (61, 98). In a small study examining fluconazole dose escalation from 800 mg to 2 g/day, the percentage of patients alive with a negative CSF culture at 10 weeks increased up to 1,600 mg/day (100). In South West Uganda, 1,200 mg/day was significantly more rapidly fungicidal than 800 mg/day (85). There was no suggestion of increased toxicity, in particular, no liver function disturbance, at the higher dose, although the numbers involved (30 patients per trial arm) means continued vigilance is needed. However, the rate of clearance with fluconazole 1,200 mg/day was still considerably less rapid than with amphotericin B–based regimens (Table 4).

An additional strategy to try to increase the efficacy of an oral antifungal regimen is addition of flucytosine. Based on encouraging murine data (3), a cohort of patients was treated with fluconazole 400 mg/day plus flucytosine 150 mg/kg/day for 10 weeks (78). The median time to CSF sterilization was relatively short at 23 days. By 10 weeks at this dose, 28% of patients had side effects requiring discontinuation of flucytosine. However, over 95% of participants tolerated 2 weeks of therapy. In Uganda, flucytosine for 2 weeks was well tolerated and additive with low-dose fluconazole (200 mg/day) (96). In a study by Larsen and colleagues, flucytosine 100 mg/kg/day was given for 4 weeks with increasing doses of fluconazole and again had a benefit (higher percent of patients alive with a negative CSF culture at 10 weeks) that was most pronounced with fluconazole at 800 to 1,200 mg/day (100). In that study, grade 4 neutropenia ($<500 \times 10^9$/liter) occurred in 18% of patients given flucytosine for 4 weeks, without evidence of increased infection. In comparison, in ongoing studies in South Africa and Malawi using 2 weeks of flucytosine with amphotericin B and/or fluconazole, 5% of 123 patients developed grade 4 neutropenia in the first 2 weeks (authors' unpublished data).

In Malawi, subsequent to the fluconazole dosage study in Uganda, the fungicidal activity of fluconazole 1,200 mg/day was compared with and without the addition of oral flucytosine 100 mg/kg/day. Flucytosine led to a very significant further increase in early fungicidal activity, to –0.28 log CFU/day, the nearest any oral regimen has come to the fungicidal activities measured for amphotericin B–based therapy (Table 4) (63). Although not powered for clinical end points, there were fewer deaths in the combination arm at 2 weeks that reached borderline significance ($P = 0.05$). There were more episodes of neutropenia in the combination arm, although no increase in infection-related serious adverse events. Analysis of flucytosine and fluorouracil levels from the study will explore whether bioavailability or gut conversion of flucytosine to fluorouracil is different in patients in Africa compared with Southeast Asia.

Given the poor outcomes with lower doses, some guidelines and countries have adopted 800 mg/day as the standard fluconazole dose, in light of the large amount of safety data for this dosage (105, 142). However, as discussed above, current evidence suggests that the optimal oral regimen is fluconazole 1,200 mg/day plus flucytosine, and in the absence of flucytosine, fluconazole at 1,200 mg/day (117). A dose-escalation study, exploring doses above 1,200 mg/day, based on Larsen and colleagues' earlier work, has been proposed. Given the results with fluconazole and flucytosine, a clinical end point trial in resource-limited settings of this combination against 1- and 2-week amphotericin B induction is also needed.

TREATMENT OF TRANSPLANT PATIENTS AND NON-HIV, NONTRANSPLANT PATIENTS

Treatment of transplant patients with cryptococcal meningoencephalitis usually follows the same induction, consolidation, and maintenance phases as for HIV-infected patients (see chapter 37). Given that most patients are on nephrotoxic calcineurin inhibitors, liposomal amphotericin B is preferred to the conventional formulation (117). Maintenance treatment is usually stopped after 1 year. Whether the same induction, consolidation, and maintenance strategy should be used for non-HIV, nontransplant patients, including those with no apparent immune defect, is less clear. Prior to the HIV epidemic, these patients were treated with 4 to 6 weeks of amphotericin B and flucytosine with the aim of radical cure, based on the results of two landmark randomized trials (14, 47). However, since the time of those

TABLE 4 Studies determining the early fungicidal activity of antifungal regimens in treatment of HIV-associated cryptococcal meningitis

Reference	Country	Treatment	n	Early fungicidal activity[a]	Comments
28	Thailand	Amphotericin B 0.7 mg/kg	14	–0.31	Randomized controlled trial
		Amphotericin B 0.7 mg/kg + fluconazole 400 mg/day	11	–0.39	
		Amphotericin B 0.7 mg/kg + fluconazole 400 mg/day + flucytosine 100 mg/kg/day	15	–0.38	
		Amphotericin B 0.7 mg/kg + flucytosine 100 mg/kg/day	12	–0.54	
18	South Africa	Fluconazole 400 mg/day	5	–0.02	Observational cohort
		Amphotericin B 1 mg/kg	49	–0.48	Amphotericin B only given for 1 week, followed by fluconazole 400 mg
21	South Africa	Amphotericin B 0.7 mg/kg + flucytosine 100 mg/kg/day	30	–0.45	Randomized controlled trial
		Amphotericin B 1 mg/kg + flucytosine 100 mg/kg/day	34	–0.56	
85	Uganda	Fluconazole 800 mg/day	30	–0.07	Dose-escalation cohort
		Fluconazole 1,200 mg/day	30	–0.18	
63	Malawi	Fluconazole 1,200 mg/day	20	–0.11	Randomized controlled trial
		Fluconazole 1,200 mg/day + flucytosine 100 mg/kg/day	21	–0.28	

[a] Early fungicidal activity is a measure of cryptococcal clearance rate, calculated using a summary statistic for each patient, defined as the decrease in log CFU per milliliter of CSF per day, by use of the slope of the linear regression of log CFU against time for each patient. Early fungicidal activity is the mean rate of decrease in CSF cryptococcal CFU per day (28).

trials, the doses of amphotericin B used have increased, those of flucytosine decreased, and safe oral consolidation/maintenance therapy with fluconazole has become available. Thus, patients may be treated with 4 to 6 weeks of induction based on the earlier trials, but using modern dosing, or a 2-week induction as used for HIV-infected patients. There are no data to compare these alternatives in this patient group. Current IDSA guidelines favor the longer induction, still followed by fluconazole consolidation and maintenance for 1 year (117). Given the duration of relatively high-dose amphotericin B, a significant proportion of patients will require a switch to liposomal formulation, or alternate-day amphotericin B dosing, or to an earlier than planned switch to fluconazole, in order to avoid significant renal impairment.

In apparently immunocompetent patients, especially with *Cryptococcus gattii* infection, cerebral cryptococcomas are more common (143). Large lesions may respond poorly to medical therapy, prompting long induction and consideration of surgery in some cases. Adjunctive steroids may be used if there is significant surrounding cerebral edema (125) (see also chapter 44).

Non-HIV, nontransplant patients without an apparent immune defect or risk factor need investigation of immune function including HIV testing and CD4 cell count as a minimum. There is some evidence that some non-HIV, nontransplant patients with cryptococcal meningitis have a specific defect in proinflammatory cytokines, with low interferon gamma (IFN-γ) and high interleukin-10 production in response to the infection (29, 106). Since this group of patients is likely to be highly heterogeneous, measurement of CSF cytokines should be done before considering adjunctive therapy with IFN-γ. However, adjunctive IFN-γ given for 10 weeks appeared to be safe and was associated with a trend toward an increase in the proportion of patients with sterile CSF at 2 weeks in a phase II study in HIV-associated cryptococcal meningitis (110). An ongoing study, also in HIV-associated infection, is looking at much shorter courses of IFN-γ using a rate of clearance end point (ISRCTN72024361, www.controlled-trials.com).

MAINTENANCE THERAPY

In HIV-infected patients, prior to recovery of immune function with combination ART, maintenance therapy is required to prevent relapse since relatively short amphotericin B–based induction treatment is unlikely to completely sterilize the infection. Fluconazole 200 mg/day was shown to be superior to either itraconazole 200 mg/day, or weekly amphotericin B (129, 133). Given the poor bioavailability of itraconazole, 400 mg/day may be more effective than 200 mg/day.

Two newer azoles, voriconazole and posaconazole, provide additional alternatives for the rare patient who either cannot tolerate fluconazole or develops fluconazole resistance. Secondary fluconazole resistance was a significant problem in Cape Town when low-dose fluconazole (400 mg/day) monotherapy remained the standard for initial treatment (16). However, since amphotericin B–based induction was introduced at the site, such resistance has very rarely been documented (64). Both voriconazole and posaconazole have low MICs for *C. neoformans*, including against

TABLE 5 Data on activity of posaconazole and voriconazole in cryptococcal infection[a]

Determination	Posaconazole (148)	Voriconazole
In vitro activity[b] (10, 116, 121–124)	MIC_{90}: 0.25–0.5 μg/ml Also active against fluconazole-resistant strains (10, 50, 116, 121, 123)	MIC_{90}: 0.12–0.25 μg/ml Also active against fluconazole-resistant strains (75, 124, 151)
Animal models	Equivalent efficacy to fluconazole in rabbit model (116) Active in murine model; at doses tested, appeared inferior to amphotericin B (12)	Good activity in murine model (149), in high doses equivalent efficacy to amphotericin B (95, 138)
CNS penetration[c]	CSF: Poor. Undetectable CSF levels in animal models, despite therapeutic plasma levels and in vivo efficacy (116) Brain tissue: No data	CSF: Good. High CSF levels demonstrated in animal models and immunosuppressed patients (89) Brain tissue: High levels demonstrated in guinea pig models and immunosuppressed patients (151)
Clinical data	Case series of activity in fungal CNS infections (2, 126); clinical success in 14 of 29 patients (48%)[d] with refractory cryptococcal meningitis (126, 130)	Case series of activity in fungal CNS infections (9, 137); successful response in 7 of 18 (39%) patients[e] with refractory cryptococcal meningitis (120)

[a] Adapted from reference 66, with permission.

[b] For comparison, MIC_{90} for itraconazole 0.5 μg/ml, fluconazole 8–16 μg/ml.

[c] Relevance of CSF penetration is unclear given that the concentration of amphotericin B, a very active drug, is very low in CSF (83, 118).

[d] Success required resolution of (seen in four patients) or improvement in (10 patients) attributable symptoms, signs, and radiographic abnormalities. All successes also had negative CSF culture. Of the other 15 patients, 6 classified as stable.

[e] Success required demonstration of reduction in antigen titer; of the other 11 patients, 10 classified as stable; >90% of patients were alive at 90 days.

fluconazole-resistant strains (Table 5) (121, 124). Animal model data show good activity, although without any clear advantage over the established drugs (12, 95, 116, 138). Voriconazole develops good CSF and brain tissue levels (89). Both drugs have been used successfully in patients refractory to or intolerant of primary therapies (120, 126). However, to date, no comparative clinical studies have been published comparing these drugs with each other or with fluconazole. Both have extensive drug-drug interactions (32). In common with all azoles, both inhibit the CYP3A4 isoenzyme. Voriconazole also inhibits CYP2C9 and CYP2C19 and is itself a substrate for CYP2C19, CYP2C9, and CYP3A4. Posaconazole is metabolized by phase II conjugation reactions. Interactions with antiretrovirals and rifamycins complicate and constrain use of either drug in patients with HIV and/or tuberculosis (up-to-date information available at www.hiv-druginteractions.org). For example, concomitant voriconazole and efavirenz leads to increased efavirenz levels and a reduction of >50% in voriconazole: doses of 300 mg/day for efavirenz and 400 mg twice a day for voriconazole are advised, with drug-level monitoring (39, 84). Posaconazole levels are similarly reduced by approximately 50% by efavirenz, although there is no effect on efavirenz (76). Neither drug can be used with rifampin.

Prior to the availability of combination ART, maintenance treatment was continued indefinitely in HIV-infected patients. Since the availability of ART, a number of studies have suggested that relapse is very unlikely if fluconazole is discontinued after a minimum 1 year of antifungal therapy if the CD4 cell count is above 100 cells/μl and the viral load is suppressed (1, 87, 103, 154). In a large retrospective study of patients discontinuing fluconazole with a CD4 cell count >100 cells/μl, just four episodes of recurrent cryptococcal infection were reported during a median 28 months of follow-up (1.5 events per 100 person-years) (103). In addition, only one of four episodes was culture confirmed, making exclusion of late immune reconstitution difficult in at least two cases. At the time of relapse, in one patient the viral load was again detectable, and in another the CD4 cell count had fallen to <50 cells/μl. Two patients had reverted from serum cryptococcal antigen negative to positive. Thus, continued monitoring of the response to ART is important, with reinstitution of maintenance fluconazole if ART fails. Remembering that the median time to serum cryptococcal antigen negativity is approximately 3 years (87), a recurrently positive serum cryptococcal antigen, or marked rise in titer, may be useful to detect early relapse in those who have discontinued maintenance (117).

MANAGEMENT OF RAISED CSF PRESSURE

Raised CSF pressure is very common in patients with cryptococcal meningitis and is associated with increased symptoms and signs and increased short-term mortality (56). Clinically, it is not possible to predict which patients have severely raised pressure, emphasizing the need for routine measurement of opening pressure in all patients at presentation. In an analysis of the last large Mycoses Study Group trial, CSF pressure >20 cm occurred in approximately 75%, and >35 cm in approximately 25%, of cases (56). Very similar percentages, 63% and 19%, respectively, have been reported in a combined cohort from Thailand and South Africa (15).

Where resources and facilities allow, a computed tomography or magnetic resonance imaging head scan should be done in all patients with suspected or proven cryptococcal meningitis prior to any diagnostic or therapeutic lumbar puncture. Where access to such scans is limited, patients with reduced consciousness level or focal neurological signs should be prioritized. Of note, however, there is consensus that the balance of risk favors both diagnostic and therapeutic lumbar punctures in this clinical setting in resource-limited areas without a facility for computed tomography or magnetic resonance imaging head scans (117).

The pathophysiological basis of raised pressure most likely is obstruction to CSF outflow at the level of the arachnoid granulations by fungal cells and possibly shed capsular polysaccharide (42, 92). This would explain why the ventricles in patients with raised pressure are usually not dilated, since there is no pressure gradient between CSF in the ventricles and CSF over the convexities of the brain. In animal models and human in vitro systems, while passage of particles, including *Saccharomyces* yeasts (3 to 6 μm) and a proportion of red blood cells, across arachnoid granulations can be demonstrated, larger particles are increasingly excluded, and high concentrations of particles have been shown to impede CSF flow (55, 72, 155, 161). In a small postmortem series from Cape Town of patients dying with cryptococcal meningitis, large numbers of fungal cells were concentrated within the arachnoid granulations compared to the rest of the brain parenchyma, and, although the series was very small, the number of organisms tended to positively correlate with the premortem CSF pressure (88).

Possible additional or alternative factors that have been proposed include cerebral edema secondary to cytokine-mediated inflammation and increased vascular permeability and an osmotic effect of fungal metabolites (37, 42, 97). It is possible that cerebral edema is important in some cases, especially when inflammatory responses are more marked, for example, in non-HIV-associated disease and at presentation as opposed to later time points (141). However, at least in HIV-related cryptococcal meningitis, brain scans do not commonly suggest marked cerebral edema (35). In addition, CSF vascular endothelial growth factor, a mediator of vascular permeability, and CSF osmolality have not been correlated with CSF pressure (15, 37, 92). A high organism load is necessary but not sufficient for development of high CSF pressure (15), suggesting that other factors, perhaps relating to the phenotype of the *C. neoformans* isolate, may be important (53). It is also interesting that baseline and day 14 pressure, although both related to organism load, are not themselves closely correlated (15), suggesting that different additional factors may play a role at baseline compared with on treatment.

If CSF outflow obstruction is the basis of raised CSF pressure, lumbar puncture with careful mechanical drainage is a rational therapeutic approach. Therapeutic lumbar puncture is often immediately effective in relieving headache and other symptoms and signs of raised pressure and indeed may be the only, or the most effective, way to relieve headache. Although controlled studies have not been done, given the association of raised CSF pressure with adverse outcome, expert opinion recommends serial lumbar punctures for patients with pressures over 25 cm (56, 134). Compliance with these guidelines has been shown to be associated with improved outcomes in a small study in the United States (140). Further evidence of the efficacy of serial lumbar punctures comes from a recent analysis of data combined from studies in Thailand and Cape Town using serial quantitative CSF cultures to assess response to treatment. In these studies repeated lumbar punctures (a median of four per patient) were done on days 3, 7, and 14 as part of the study protocol, and in addition, further daily lumbar punctures were done in patients with high pressure. In contrast to the earlier Mycoses Study Group trial, in this data set there was no association of raised pressure with mortality (15), suggesting that serial lumbar punctures do abrogate the negative effect of raised pressure on outcome.

IDSA guidelines recommend drainage of sufficient CSF to reduce the pressure by 50% (134). The safe limit for CSF removal at a single lumbar puncture is not defined, but many would not drain more than 30 ml. The relationship between volume of CSF drained and fall in CSF pressure is approximately linear, but there are patients in whom removal of modest volumes is associated with significant falls in pressure (15). Therefore, it is safest to remeasure pressure after every 10 ml of CSF drained and stop drainage once the pressure is 20 cm or below. Some experts recommend using a large-bore lumbar puncture needle on the basis that some continued drainage of CSF via a dural leak will more readily occur after the needle is removed.

Median CSF pressures show a tendency to settle in the first week and then rise again slightly by day 14 (15), and severely raised CSF pressure not infrequently develops on treatment and despite successful sterilization of CSF. Thus, clinicians need to have a low threshold for repeating a lumbar puncture to recheck pressure in any patients with persistent symptoms attributable to raised pressure. Indeed, given that raised pressure may be manifest by headache only, this is an additional argument for doing a routine lumbar puncture at day 14 in all symptomatic patients.

While repeat lumbar puncture is effective in many patients to relieve headache and control pressure, at the severe end of the spectrum, there are patients in whom daily lumbar punctures are insufficient. In this case, a temporary lumbar drain can be used to safely remove an order of magnitude more CSF per day (51, 90, 94). Such drains are routinely used in neurosurgical practice. Nursing and medical staff need to be appropriately trained, and patients requiring a lumbar drain should ideally be managed on a neurosurgery ward. In addition, the patient needs careful counseling. Strict aseptic technique is required in their placement and care. Nevertheless, such drains have been used successfully in resource-limited settings (94). In the context of cryptococcal meningitis, the drain can be set to a pressure level (for example, 20 cm above the insertion of the drain with the patient supine) by adjusting the height of the collecting manometer. Drainage of 10 ml of CSF per h and over 200 ml per day is not unusual. It is unclear how long such drains are usually needed in the context of cryptococcal infection and how duration of drainage relates to infection risk. In our experience, removal after around 7 days of drainage has not been followed by recurrence of severely raised CSF pressure. In the largest reported series, from Thailand, 7 days was also the median duration, with an interquartile range of 4 to 10 days (94). On the other hand, symptoms of raised pressure returned in most patients who had an external lumbar drain clamped after only a day or so (51). In the Thai study, 3 of 54 patients (6%) developed an infected drain, all successfully treated following removal, and one patient with low platelets developed bilateral subdural hematomas. At 3 months after lumbar drainage, 82% of patients were known to be alive, which represents a favorable outcome given that these patients would be expected to have a relatively high mortality.

If a temporary drain is not possible, then a permanent ventricular or lumbar shunt will also be effective, although it leaves the patient with a permanent device (51, 160). Other medical interventions for high CSF pressure, including steroids, mannitol, and acetazolamide, are not recommended. Steroids appeared to be associated with worse outcome, although the data are uncontrolled (56), and a trial of acetazolamide was stopped early because of metabolic side effects and lack of efficacy (107).

Ventricular shunts are required for the rarer patients who develop true hydrocephalus due to an obstruction earlier in the pathway of CSF circulation. This is more common in patients with non-HIV-associated disease, in whom there is generally a more vigorous inflammatory response. Such shunts do not spread or prevent cure of infection (114).

ART

Randomized studies are needed to determine the optimal timing of ART after a diagnosis of cryptococcal meningitis. While the evidence suggests that cryptococcal immune reconstitution inflammatory syndrome (IRIS) is potentially fatal (81) and more common if ART is started earlier (86), this needs to be weighed against the fact that patients will continue to develop and die from the complications of HIV infection while not on ART. In developing-country settings, in particular, this ongoing mortality before, and soon after the start of, ART is very high (26, 80). Thus, the risk-benefit of earlier ART may depend on the setting, and studies need to be done in both developed and resource-limited settings. Risk-benefit may also be affected by increased awareness of and experience managing IRIS reactions. There is increasing evidence that if recognized early and managed appropriately, IRIS reactions related to various opportunistic infections can usually be managed successfully (102).

In a mainly U.S.-based study, early (median 2 weeks) versus later (median 6 weeks) initiation of ART after opportunistic infections, excluding tuberculosis, were compared. The most frequent opportunistic infection was *Pneumocystis*, but 35 patients (12% of the total) with cryptococcal infection were included. The results favored earlier initiation, with fewer deaths plus progressions to AIDS and no apparent increase in IRIS episodes (162). On the other hand, a recent small randomized study from Zimbabwe, presented in abstract, found that ART started within 3 days of initial antifungal therapy with fluconazole 800 mg/day was associated with significantly increased mortality compared to starting ART after 10 weeks of antifungal treatment (91). It is possible that fluconazole, being much less rapidly active than amphotericin B, may have increased the risk of IRIS reactions in this study. This result cautions against introduction of ART concomitant with antifungal therapy and very early use of ART in the context of fluconazole induction.

On the basis of current evidence, the IDSA panel has left a fairly wide window of 2 to 10 weeks for initiation of ART after cryptococcal meningitis (117). The Southern African HIV Clinicians Society suggest 2 to 4 weeks (142). A study comparing initiation of ART as an inpatient (7 to 14 days) versus early outpatient (4 weeks) treatment in Africa in the context of amphotericin B treatment is planned (D. Boulware, personal communication).

In terms of the choice of antiretroviral regimen, clinicians should be aware of an interaction between fluconazole and nevirapine. In a pharmacokinetic study in Cape Town with HIV-infected patients who were stable on ART with a suppressed viral load and CD4 count >100 cells/μl (54) and in patients with cryptococcal infection in Thailand (93), fluconazole at 200 to 400 mg/day has been shown to cause about a doubling of nevirapine levels. While increased nevirapine toxicity was seen in the Cape Town study, this has not been seen to date in patients treated with fluconazole for cryptococcal infection (93). Of note, the hepatotoxicity of nevirapine is not closely associated with blood levels (73). There are no data on the effect of higher doses of fluconazole, as may be considered for treatment beyond the initial 2 weeks when antifungal therapy is initiated with an oral regimen, on nevirapine. In this case, efavirenz, with which there is no significant interaction at low fluconazole dosage, would be the non-nucleoside reverse transcriptase inhibitor (NNRTI) of choice, so as to enable optimal antifungal treatment with no delay to ART. If nevirapine is the only NNRTI available, it may be safest, pending further data, to use a maximum of 400 mg/day fluconazole with full-dose nevirapine.

PATIENTS ALREADY ON ART AT THE TIME OF PRESENTATION

In centers in Africa with increasing access to ART, up to one-third of patients now presenting with cryptococcal meningitis are already on ART, most having started ART within the previous 1 to 3 months (18, 23). Such cases have been termed "unmasking IRIS" (159), and although these patients have active infection with positive CSF cultures, IRIS may be playing some role in the pathophysiology and presentation. In Cape Town, these patients were found to have substantially lower fungal burdens in CSF (median 34,000 versus 235,000 CFU/ml CSF in ART-naive patients) (18). Despite this good prognosis indicator, their acute, 10-week mortality was similar to that of patients naive to ART at the time of cryptococcal presentation. However, 1-year survival was substantially higher than in ART-naive patients. The data suggest that the benefit of already being on ART was related to treatment of other HIV-related illness rather than any benefit on the acute cryptococcal infection. On the other hand, in Botswana, Bisson and colleagues did find that acute, in-hospital mortality appeared reduced in this group compared to ART-naive patients (23).

At present, management of these patients does not differ from ART-naive patients. We do not yet understand any difference in the pathophysiology of the infection in this group on which to base a tailored, rational approach to therapy. In Cape Town, the baseline CSF cytokine profile was not different for patients on and not on ART at the time of initial cryptococcal diagnosis (18).

MANAGING SYMPTOMATIC RELAPSE AND CRYPTOCOCCAL IRIS

Cryptococcal IRIS occurs in around 10 to 20% of patients with HIV-associated cryptococcal meningitis who subsequently start ART (17, 86, 139). The median onset is 1 to 2 months after ART (17, 139, 145), but much later presentations can occur (86). Manifestations include recurrent fever, headache, meningismus, and high CSF pressure. CSF white cell counts may be raised. Although CSF cultures may be negative, active infection can coexist with IRIS, especially if initial antifungal therapy has been with slowly fungicidal fluconazole. Outside the CNS, pulmonary infiltrates and lymphadenopathy have been described. Risk factors include fungemia, low initial CD4 cell count, a rapid rise in CD4 cell count on ART, and starting ART within 2 months of cryptococcal diagnosis (17, 86, 139).

Since IRIS is a diagnosis of exclusion, it needs to be differentiated from other causes of symptomatic relapse, namely nonadherence with, or failure to prescribe, secondary fluconazole prophylaxis (64), development of secondary resistance to fluconazole (16), and an alternative noncryptococcal cause for the symptoms, in particular concomitant tuberculous meningitis (69).

Thus, patients need reassessment with lumbar puncture and CSF pressure measurement and CSF cryptococcal antigen, fungal and mycobacterial cultures, as well as routine cultures and analyses. Although to date only used in a research setting, this is one instance in which repeat quantitative CSF cultures may help in immediate management: a sharp rise in CFU counts since the last on-treatment culture would suggest nonadherence or fluconazole resistance, whereas even if the culture is not sterile, a further fall in counts is consistent with IRIS as the predominant problem (85).

Antiretroviral medication should be continued. Amphotericin B may be restarted if available, especially if the patient is ill enough to warrant readmission pending the CSF culture results. Any raised CSF opening pressure should be managed with repeat lumbar punctures with controlled CSF drainage. If the CSF culture is negative, fluconazole maintenance treatment can be continued or resumed. While IRIS is often self-limiting, if patients deteriorate despite these measures, no alternative diagnosis is apparent, and the CSF culture result (negative, low, or falling organism burden) supports IRIS, then a short, 2- to 6-week, course of reducing corticosteroids has been used successfully (17, 86) and should be considered. The extent to which cryptococcal IRIS affects the overall prognosis of HIV-associated cryptococcal disease is not clear (17), but any detrimental effect will be reduced by increased awareness of the issue, prompt recognition, and appropriate management.

If a patient with symptomatic relapse has a positive CSF culture, if available, relapse and original isolates should be tested in parallel for fluconazole sensitivity. Reinduction therapy with amphotericin B should be given until cultures are sterile followed by maintenance therapy with standard or higher doses of fluconazole or alternative azoles, depending on the sensitivity results, the availability of alternative agents, and any requirement for concomitant, interacting drugs. Interpretation of fluconazole MIC for *C. neoformans* is not well established, but some evidence suggests an MIC of >16 mg/ml is more likely to be associated with therapeutic failure (4, 158), and a more than fourfold rise in MIC of the relapse compared with the initial isolate would be consistent with secondary resistance (142).

PROPHYLAXIS, SCREENING AND PREEMPTIVE TREATMENT, ISOLATED ANTIGENEMIA, AND FACILITATING EARLIER DIAGNOSIS

HIV-associated cryptococcal disease could be prevented if all HIV-infected patients were diagnosed and given antiretroviral treatment prior to critical falls in CD4 cell count. Widespread availability of ART and earlier testing and treatment have reduced the incidence of cryptococcal disease in developed-country settings (101). However, although ART is increasing in availability across the developing world, HIV programs in many areas, particularly sub-Saharan Africa, are, as yet, struggling to keep pace with the epidemic (65). For example, the Western Cape, which has one of the most rapidly expanding ART programs in South Africa, still has more individuals entering stage 4 disease than are being started on ART, and although the proportion of patients with a CD4 cell count <100 cells/μl at the time of initiation of ART is decreasing, the absolute number of such patients is still increasing (25).

Thus, the issue of HIV-associated cryptococcal disease needs addressing specifically (60, 113). Work has focused on strategies to prevent clinical disease or preemptively treat early subclinical infection. In the absence of ART, primary prophylaxis with fluconazole or itraconazole in patients with a low CD4 cell count has been shown to prevent fungal infections including cryptococcal meningitis but not to be associated with a clear mortality benefit (34, 36, 128). Prophylaxis was not adopted in developed-country settings because of this lack of effect on survival, the relatively low incidence of disease, and concerns over drug interactions and drug resistance, as well as cost. However, in high-incidence areas, the WHO recommended "consideration" of azole prophylaxis (157), and Thailand adopted such a policy (fluconazole 400 mg, once per week, in those with a CD4 cell count <100 cells/μl) prior to the widespread availability of ART in that country. In a more recent primary prophylaxis study in East Africa, reported in abstract, fluconazole had to be given to a large number of patients to prevent one case and again had no effect on mortality (115). The study was conducted partly before ART became available, and when ART was available, there was a median of 11 weeks between enrolment and initiation of ART, which may explain why a small number of patients in the trial who were initially cryptococcal antigen negative in serum developed cryptococcal meningitis.

An alternative strategy to across-the-board prophylaxis is screening for cryptococcal antigen prior to initiation of ART in patients with a CD4 cell count <100 and targeted preemptive treatment only for those testing positive (66, 67). In Africa and Southeast Asia, cryptococcal antigenemia has been found in 7 to 18% of first-time HIV clinic attendees (43, 82, 99, 147). It is a specific and sensitive test and, importantly, is positive a median of 22 days prior to the development of clinical disease (52). A recent study from South Africa provides strong support for antigen screening (67). Blood samples were available to test retrospectively from a well-characterized cohort of over 700 patients entering an ART program. Of those with a CD4 count <100 cells/μl, 13% tested positive for cryptococcal antigen, and 28% of antigen-positive patients with no prior history of cryptococcal disease developed cryptococcal meningitis in the year after starting ART. Importantly, of 661 patients testing antigen negative, none subsequently developed cryptococcal disease, giving 100% sensitivity for the test in predicting cryptococcal meningitis in this cohort, with prompt initiation of ART (median 2 weeks after screening).

Thus, in a setting where ART is available and is started promptly, a screen and preemptive treatment strategy could be an effective and highly attractive alternative to primary prophylaxis with fluconazole, allowing more intensive treatment to be targeted to the 10% or so of patients at substantial risk of clinical disease. Prospective studies are needed. In addition, how best to manage asymptomatic antigen-positive patients needs to be defined. A conservative approach, and one that most experts would recommend in a developed-country setting, outside of a screening program, would be to lumbar puncture all such patients and treat those with evidence of meningitis with amphotericin B and those without with high-dose fluconazole. However, in many developing countries, as part of a screening program, this would impose a further considerable burden on already overstretched HIV services and might not be acceptable to all asymptomatic patients. It may be that asymptomatic antigen-positive patients, or those with an antigen titer below a certain cutoff, could be safely treated with high-dose fluconazole without lumbar puncture. Of note, in a primary prophylaxis strategy without screening, such patients would be treated with low-dose fluconazole. Furthermore, in the retrospective study, two-thirds of antigen-positive patients did not develop clinical cryptococcal disease, appearing to control their infection with ART alone (67). A successful screen and preemptive treatment strategy could prevent the third of cryptococcal cases that now present in some centers after patients have started ART.

Development of a point-of-care immunodiagnostic test that could be used for antigen screening would greatly facilitate implementation of an antigen screening program. Ideally such a test would use a noninvasive sample such as urine, which has been shown to contain cryptococcal antigen in infected patients, although at lower concentration than serum (33). A point-of-care antigen-based diagnostic test could also greatly facilitate earlier diagnosis and treatment of all patients. Diagnosis of HIV-associated cryptococcal disease is not a problem if a lumbar puncture is performed. However, the usual presenting symptoms of headache and fever are very nonspecific, and often lumbar puncture is deferred until the disease is advanced and the prognosis poor. In a valuable study from Uganda, probably reflective of the situation in much of sub-Saharan Africa, of 100 consecutive patients presenting to a hospital with presumed CNS infection, half had sought care prior to coming to the hospital, and the mean duration of symptoms was 13 days (149). Further delays occurred after presenting to the hospital: only 57 of 88 lumbar punctures ordered on the first day were done on the first day, and the average time to diagnose and initiate treatment for confirmed cryptococcal meningitis was 3.5 days. Efforts are urgently needed to try to address these prehospital and postpresentation delays in diagnosis and therapy.

REFERENCES

1. **Aberg, J. A., R. W. Price, D. M. Heeren, and B. Bredt.** 2002. A pilot study of the discontinuation of antifungal therapy for disseminated cryptococcal disease in patients with acquired immunodeficiency syndrome, following immunologic response to antiretroviral therapy. *J. Infect. Dis.* **185:**1179–1182.
2. **Al-Abdely, H. M., A. M. Alkhunaizi, J. A. Al-Tawfiq, M. Hassounah, M. G. Rinaldi, and D. A. Sutton.** 2005. Successful therapy of cerebral phaeohyphomycosis due to *Ramichloridium mackenziei* with the new triazole posaconazole. *Med. Mycol.* **43:**91–95.
3. **Allendoerfer, R., A. J. Marquis, M. G. Rinaldi, and J. R. Graybill.** 1991. Combined therapy with fluconazole and flucytosine in murine cryptococcal meningitis. *Antimicrob. Agents Chemother.* **35:**726–729.
4. **Aller, A. I., E. Martin-Mazuelos, F. Lozano, J. Gomez-Mateos, L. Steele-Moore, W. J. Holloway, M. J. Gutierrez, F. J. Recio, and A. Espinel-Ingroff.** 2000. Correlation of fluconazole MICs with clinical outcome in cryptococcal infection. *Antimicrob. Agents Chemother.* **44:**1544–1548.
5. **Anaissie, E. J., D. P. Kontoyiannis, C. Huls, S. E. Vartivarian, C. Karl, R. A. Prince, J. Bosso, and G. P. Bodey.** 1995. Safety, plasma concentrations, and efficacy of high-dose fluconazole in invasive mold infections. *J. Infect. Dis.* **172:**599–602.
6. **Andes, D., N. Safdar, K. Marchillo, and R. Conklin.** 2006. Pharmacokinetic-pharmacodynamic comparison of amphotericin B (AMB) and two lipid-associated AMB preparations, liposomal AMB and AMB lipid complex, in murine candidiasis models. *Antimicrob. Agents Chemother.* **50:**674–684.
7. **Andes, D., and M. van Ogtrop.** 2000. In vivo characterization of the pharmacodynamics of flucytosine in a neutropenic murine disseminated candidiasis model. *Antimicrob. Agents Chemother.* **44:**938–942.
8. **Appelbaum, E., and S. Shtokalko.** 1957. Cryptococcus meningitis arrested with amphotericin B. *Ann. Intern. Med.* **47:**346–351.
9. **Bakleh, M., A. J. Aksamit, I. M. Tleyjeh, and W. F. Marshall.** 2005. Successful treatment of cerebral blastomycosis with voriconazole. *Clin. Infect. Dis.* **40:**e69–e71.
10. **Barchiesi, F., D. Arzeni, A. W. Fothergill, L. F. Di Francesco, F. Caselli, M. G. Rinaldi, and G. Scalise.** 2000. In vitro activities of the new antifungal triazole SCH 56592 against common and emerging yeast pathogens. *Antimicrob. Agents Chemother.* **44:**226–229.
11. **Barchiesi, F., A. M. Schimizzi, F. Caselli, A. Novelli, S. Fallani, D. Giannini, D. Arzeni, S. Di Cesare, L. F. Di Francesco, M. Fortuna, A. Giacometti, F. Carle, T. Mazzei, and G. Scalise.** 2000. Interactions between triazoles and amphotericin B against *Cryptococcus neoformans*. *Antimicrob. Agents Chemother.* **44:**2435–2441.
12. **Barchiesi, F., E. Spreghini, A. M. Schimizzi, M. Maracci, D. Giannini, F. Carle, and G. Scalise.** 2004. Posaconazole and amphotericin B combination therapy against *Cryptococcus neoformans* infection. *Antimicrob. Agents Chemother.* **48:**3312–3316.
13. **Bekker, L. G., L. Myer, C. Orrell, S. Lawn, and R. Wood.** 2006. Rapid scale-up of a community-based HIV treatment service: programme performance over 3 consecutive years in Guguletu, South Africa. *S. Afr. Med. J.* **96:**315–320.
14. **Bennett, J. E., W. E. Dismukes, R. J. Duma, G. Medoff, M. A. Sande, H. Gallis, J. Leonard, B. T. Fields, M. Bradshaw, H. Haywood, Z. A. McGee, T. R. Cate, C. G. Cobbs, J. F. Warner, and D. W. Alling.** 1979. A comparison of amphotericin B alone and combined with flucytosine in the treatment of cryptococcal meningitis. *N. Engl. J. Med.* **301:**126–131.
15. **Bicanic, T., A. E. Brouwer, G. Meintjes, K. Rebe, D. Limmathurotsakul, W. Chierakul, P. Teparrakkul, A. Loyse, N. J. White, R. Wood, S. Jaffar, and T. Harrison.** 2009. Relationship of cerebrospinal fluid pressure, fungal burden and outcome in patients with cryptococcal meningitis undergoing serial lumbar punctures. *AIDS* **23:**701–706.
16. **Bicanic, T., T. Harrison, A. Niepieklo, N. Dyakopu, and G. Meintjes.** 2006. Symptomatic relapse of HIV-associated cryptococcal meningitis after initial fluconazole monotherapy: the role of fluconazole resistance and immune reconstitution. *Clin. Infect. Dis.* **43:**1069–1073.
17. **Bicanic, T., G. Meintjes, K. Rebe, A. Williams, A. Loyse, R. Wood, M. Hayes, S. Jaffar, and T. Harrison.** 2009. Immune reconstitution inflammatory syndrome in HIV-associated cryptococcal meningitis: a prospective study. *J. Acquir. Immune. Defic. Syndr.* **51:**130–134.
18. **Bicanic, T., G. Meintjes, R. Wood, M. Hayes, K. Rebe, L. G. Bekker, and T. Harrison.** 2007. Fungal burden, early fungicidal activity, and outcome in cryptococcal meningitis in antiretroviral-naive or antiretroviral-experienced patients treated with amphotericin B or fluconazole. *Clin. Infect. Dis.* **45:**76–80.
19. **Bicanic, T., C. Muzoora, A. E. Brouwer, G. Meintjes, N. Longley, K. Taseera, K. Rebe, A. Loyse, J. Jarvis, L. G. Bekker, R. Wood, D. Limmathurotsakul, W. Chierakul, K. Stepniewska, N. J. White, S. Jaffar, and T. S. Harrison.** 2009. Independent association between rate of clearance of infection and clinical outcome of HIV-associated cryptococcal meningitis: analysis of a combined cohort of 262 patients. *Clin. Infect. Dis.* **49:**702–709.
20. **Bicanic, T., R. Wood, L. G. Bekker, M. Darder, G. Meintjes, and T. S. Harrison.** 2005. Antiretroviral roll-out, antifungal roll-back: access to treatment for cryptococcal meningitis. *Lancet Infect. Dis.* **5:**530–531.
21. **Bicanic, T., R. Wood, G. Meintjes, K. Rebe, A. Brouwer, A. Loyse, L. G. Bekker, S. Jaffar, and T. Harrison.** 2008. High-dose amphotericin B with flucytosine for the

treatment of cryptococcal meningitis in HIV-infected patients: a randomized trial. *Clin. Infect. Dis.* **47:**123–130.

22. **Bicanic, T., R. Wood, G. Meintjes, K. Rebe, A. Brouwer, A. Loyse, L. G. Bekker, S. Jaffar, and T. Harrison.** 2008. Reply to Pasqualotto. *Clin. Infect. Dis.* **47:**1110–1111.
23. **Bisson, G. P., R. Nthobatsong, R. Thakur, G. Lesetedi, K. Vinekar, P. Tebas, J. E. Bennett, S. Gluckman, T. Gaolathe, and R. R. Macgregor.** 2008. The use of HAART is associated with decreased risk of death during initial treatment of cryptococcal meningitis in adults in Botswana. *J. Acquir. Immune Defic. Syndr.* **49:**227–229.
24. **Block, E. R., and J. E. Bennett.** 1972. Pharmacological studies with 5-fluorocytosine. *Antimicrob. Agents Chemother.* **1:**476–482.
25. **Boulle, A., P. Bock, M. Osler, K. Cohen, L. Channing, K. Hilderbrand, E. Mothibi, V. Zweigenthal, N. Slingers, K. Cloete, and F. Abdullah.** 2008. Antiretroviral therapy and early mortality in South Africa. *Bull. W. H. O.* **86:**678–687.
26. **Braitstein, P., M. W. Brinkhof, F. Dabis, M. Schechter, A. Boulle, P. Miotti, R. Wood, C. Laurent, E. Sprinz, C. Seyler, D. R. Bangsberg, E. Balestre, J. A. Sterne, M. May, and M. Egger.** 2006. Mortality of HIV-1-infected patients in the first year of antiretroviral therapy: comparison between low-income and high-income countries. *Lancet* **367:**817–824.
27. **Branch, R. A.** 1988. Prevention of amphotericin B-induced renal impairment. A review on the use of sodium supplementation. *Arch. Intern. Med.* **148:**2389–2394.
28. **Brouwer, A. E., A. Rajanuwong, W. Chierakul, G. E. Griffin, R. A. Larsen, N. J. White, and T. S. Harrison.** 2004. Combination antifungal therapies for HIV-associated cryptococcal meningitis: a randomised trial. *Lancet* **363:**1764–1767.
29. **Brouwer, A. E., A. A. Siddiqui, M. I. Kester, K. C. Sigaloff, A. Rajanuwong, S. Wannapasni, W. Chierakul, and T. S. Harrison.** 2007. Immune dysfunction in HIV-seronegative, *Cryptococcus gattii* meningitis. *J. Infect.* **54:**e165–e168.
30. **Brouwer, A. E., P. Teparrukkul, S. Pinpraphaporn, R. A. Larsen, W. Chierakul, S. Peacock, N. Day, N. J. White, and T. S. Harrison.** 2005. Baseline correlation and comparative kinetics of cerebrospinal fluid colony-forming unit counts and antigen titers in cryptococcal meningitis. *J. Infect. Dis.* **192:**681–684.
31. **Brouwer, A. E., H. J. van Kan, E. Johnson, A. Rajanuwong, P. Teparrukkul, V. Wuthiekanun, W. Chierakul, N. Day, and T. S. Harrison.** 2007. Oral versus intravenous flucytosine in patients with human immunodeficiency virus-associated cryptococcal meningitis. *Antimicrob. Agents Chemother.* **51:**1038–1042.
32. **Bruggemann, R. J., J. W. Alffenaar, N. M. Blijlevens, E. M. Billaud, J. G. Kosterink, P. E. Verweij, and D. M. Burger.** 2009. Clinical relevance of the pharmacokinetic interactions of azole antifungal drugs with other coadministered agents. *Clin. Infect. Dis.* **48:**1441–1458.
33. **Chapin-Robertson, K., C. Bechtel, S. Waycott, C. Kontnick, and S. C. Edberg.** 1993. Cryptococcal antigen detection from the urine of AIDS patients. *Diagn. Microbiol. Infect. Dis.* **17:**197–201.
34. **Chariyalertsak, S., K. Supparatpinyo, T. Sirisanthana, and K. E. Nelson.** 2002. A controlled trial of itraconazole as primary prophylaxis for systemic fungal infections in patients with advanced human immunodeficiency virus infection in Thailand. *Clin. Infect. Dis.* **34:**277–284.
35. **Charlier, C., F. Dromer, C. Leveque, L. Chartier, Y. S. Cordoliani, A. Fontanet, O. Launay, and O. Lortholary.** 2008. Cryptococcal neuroradiological lesions correlate with severity during cryptococcal meningoencephalitis in HIV-positive patients in the HAART era. *PLoS ONE* **3:**e1950.
36. **Chetchotisakd, P., S. Sungkanuparph, B. Thinkhamrop, P. Mootsikapun, and P. Boonyaprawit.** 2004. A multicentre, randomized, double-blind, placebo-controlled trial of primary cryptococcal meningitis prophylaxis in HIV-infected patients with severe immune deficiency. *HIV Med.* **5:**140–143.
37. **Coenjaerts, F. E., M. van der Flier, P. N. Mwinzi, A. E. Brouwer, J. Scharringa, W. S. Chaka, M. Aarts, A. Rajanuwong, D. A. van de Vijver, T. S. Harrison, and A. I. Hoepelman.** 2004. Intrathecal production and secretion of vascular endothelial growth factor during cryptococcal meningitis. *J. Infect. Dis.* **190:**1310–1317.
38. **Cutler, R. E., A. D. Blair, and M. R. Kelly.** 1978. Flucytosine kinetics in subjects with normal and impaired renal function. *Clin. Pharmacol. Ther.* **24:**333–342.
39. **Damle, B., R. LaBadie, P. Crownover, and P. Glue.** 2008. Pharmacokinetic interactions of efavirenz and voriconazole in healthy volunteers. *Br. J. Clin. Pharmacol.* **65:**523–530.
40. **Dammert, P., B. Bustamante, E. Ticona, A. Llanos-Cuentas, L. Huaroto, V. M. Chavez, and P. E. Campos.** 2008. Treatment of cryptococcal meningitis in Peruvian AIDS patients using amphotericin B and fluconazole. *J. Infect.* **57:**260–265.
41. **de Lalla, F., G. Pellizzer, A. Vaglia, V. Manfrin, M. Franzetti, P. Fabris, and C. Stecca.** 1995. Amphotericin B as primary therapy for cryptococcosis in patients with AIDS: reliability of relatively high doses administered over a relatively short period. *Clin. Infect. Dis.* **20:**263–266.
42. **Denning, D. W., R. W. Armstrong, B. H. Lewis, and D. A. Stevens.** 1991. Elevated cerebrospinal fluid pressures in patients with cryptococcal meningitis and acquired immunodeficiency syndrome. *Am. J. Med.* **91:**267–272.
43. **Desmet, P., K. D. Kayembe, and C. De Vroey.** 1988. The value of cryptococcal serum antigen screening among HIV-positive/AIDS patients in Kinshasa, Zaire. *AIDS* **3:**77–78.
44. **Diamond, D. M., M. Bauer, B. E. Daniel, M. A. Leal, D. Johnson, B. K. Williams, A. M. Thomas, J. C. Ding, L. Najvar, J. R. Graybill, and R. A. Larsen.** 1998. Amphotericin B colloidal dispersion combined with flucytosine with or without fluconazole for treatment of murine cryptococcal meningitis. *Antimicrob. Agents Chemother.* **42:**528–533.
45. **Diamond, R. D., and J. E. Bennett.** 1974. Prognostic factors in cryptococcal meningitis. A study in 111 cases. *Ann. Intern. Med.* **80:**176–181.
46. **Diasio, R. B., D. E. Lakings, and J. E. Bennett.** 1978. Evidence for conversion of 5-fluorocytosine to 5-fluorouracil in humans: possible factor in 5-fluorocytosine clinical toxicity. *Antimicrob. Agents Chemother.* **14:**903–908.
47. **Dismukes, W. E., G. Cloud, H. A. Gallis, T. M. Kerkering, G. Medoff, P. C. Craven, L. G. Kaplowitz, J. F. Fisher, C. R. Gregg, C. A. Bowles, et al.** 1987. Treatment of cryptococcal meningitis with combination amphotericin B and flucytosine for four as compared with six weeks. *N. Engl. J. Med.* **317:**334–341.
48. **Dromer, F., C. Bernede-Bauduin, D. Guillemot, and O. Lortholary.** 2008. Major role for amphotericin B-flucytosine combination in severe cryptococcosis. *PLoS ONE* **3:**e2870.
49. **Dromer, F., S. Mathoulin-Pelissier, O. Launay, and O. Lortholary.** 2007. Determinants of disease presentation and outcome during cryptococcosis: the CryptoA/D study. *PLoS Med.* **4:**e21.
50. **Espinel-Ingroff, A.** 2003. In vitro antifungal activities of anidulafungin and micafungin, licensed agents and the investigational triazole posaconazole as determined by NCCLS methods for 12,052 fungal isolates: review of the literature. *Rev. Iberoam. Micol.* **20:**121–136.

51. **Fessler, R. D., J. Sobel, L. Guyot, L. Crane, J. Vazquez, M. J. Szuba, and F. G. Diaz.** 1998. Management of elevated intracranial pressure in patients with cryptococcal meningitis. *J. Acquir. Immune Defic. Syndr. Hum. Retrovirol.* **17:**137–142.
52. **French, N., K. Gray, C. Watera, J. Nakiyingi, E. Lugada, M. Moore, D. Lalloo, J. A. Whitworth, and C. F. Gilks.** 2002. Cryptococcal infection in a cohort of HIV-1-infected Ugandan adults. *AIDS* **16:**1031–1038.
53. **Fries, B. C., S. C. Lee, R. Kennan, W. Zhao, A. Casadevall, and D. L. Goldman.** 2005. Phenotypic switching of *Cryptococcus neoformans* can produce variants that elicit increased intracranial pressure in a rat model of cryptococcal meningoencephalitis. *Infect. Immun.* **73:**1779–1787.
54. **Geel, J., J. Pitt, C. Orrell, M. Van Dyk, and R. Wood.** 2004. The effect of fluconazole on nevirapine pharmacokinetics. XV International AIDS Conference. Bangkok, Thailand, 11 to 16 July. **1:**369.
55. **Glimcher, S. A., D. W. Holman, M. Lubow, and D. M. Grzybowski.** 2008. Ex vivo model of cerebrospinal fluid outflow across human arachnoid granulations. *Invest. Ophthalmol. Vis. Sci.* **49:**4721–4728.
56. **Graybill, J. R., J. Sobel, M. Saag, C. van Der Horst, W. Powderly, G. Cloud, L. Riser, R. Hamill, and W. Dismukes.** 2000. Diagnosis and management of increased intracranial pressure in patients with AIDS and cryptococcal meningitis. The NIAID Mycoses Study Group and AIDS Cooperative Treatment Groups. *Clin. Infect. Dis.* **30:**47–54.
57. **Groll, A. H., N. Giri, V. Petraitis, R. Petraitiene, M. Candelario, J. S. Bacher, S. C. Piscitelli, and T. J. Walsh.** 2000. Comparative efficacy and distribution of lipid formulations of amphotericin B in experimental *Candida albicans* infection of the central nervous system. *J. Infect. Dis.* **182:**274–282.
58. **Hamill, R., J. Sobel, W. el-Sadr, P. Johnson, J. R. Graybill, K. Javaly, and D. Bardker.** 1999. Presented at the Program and abstracts of the 39th ICAAC, 26 to 29 September, San Francisco, CA. Abstract 1161.
59. **Harris, B. E., B. W. Manning, T. W. Federle, and R. B. Diasio.** 1986. Conversion of 5-fluorocytosine to 5-fluorouracil by human intestinal microflora. *Antimicrob. Agents Chemother.* **29:**44–48.
60. **Harrison, T. S.** 2009. The burden of HIV-associated cryptococcal disease. *AIDS* **23:**531–532.
61. **Haubrich, R. H., D. Haghighat, S. A. Bozzette, J. Tilles, and J. A. McCutchan.** 1994. High-dose fluconazole for treatment of cryptococcal disease in patients with human immunodeficiency virus infection. The California Collaborative Treatment Group. *J. Infect. Dis.* **170:**238–242.
62. **Imwidthaya, P., and N. Poungvarin.** 2000. Cryptococcosis in AIDS. *Postgrad. Med. J.* **76:**85–88.
63. **Jackson, A., J. Nussbaum, D. Namarika, J. Phulusa, J. Kenala, C. Kenyemba, J. Jarvis, S. Jaffar, M. Hosseinipour, C. Van der Horst, and T. Harrison.** 2009. Flucytosine plus high dose fluconazole is superior to high dose fluconazole alone: results of a randomized trial comparing cryptococcal meningitis treatments in Malawi, abstr. LBPEB02. 5th IAS Conference on HIV Pathogenesis, Treatment and Prevention, Cape Town, South Africa, 19 to 22 July.
64. **Jarvis, J., G. Meintjes, Z. Williams, K. Rebe, and T. Harrison.** 2009. Symptomatic relapse of HIV-associated cryptococcal meningitis in South Africa: the role of inadequate secondary prophylaxis. *S. Afr. Med. J.* **100:**378–382.
65. **Jarvis, J. N., A. Boulle, A. Loyse, T. Bicanic, K. Rebe, A. Williams, T. S. Harrison, and G. Meintjes.** 2009. High ongoing burden of cryptococcal disease in Africa despite antiretroviral roll out. *AIDS* **23:**1181–1185.
66. **Jarvis, J. N., and T. S. Harrison.** 2007. HIV-associated cryptococcal meningitis. *AIDS* **21:**2119–2129.
67. **Jarvis, J. N., S. D. Lawn, M. Vogt, N. Bangani, R. Wood, and T. S. Harrison.** 2009. Screening for cryptococcal antigenemia in patients accessing an antiretroviral treatment program in South Africa. *Clin. Infect. Dis.* **48:**856–862.
68. **Jarvis, J. N., S. D. Lawn, R. Wood, and T. S. Harrison.** 2009. Correspondence. Reducing mortality associated with opportunistic infections among patients with advanced HIV infection in sub-Saharan Africa: reply to DiNubile. *Clin. Infect. Dis.* **49:**812–813.
69. **Jarvis, J. N., A. Williams, T. Crede, T. S. Harrison, and G. Meintjes.** 2009. Adult meningitis in a setting of high HIV and TB prevalence: findings from 4961 cases, abstr. 2915. 5th IAS Conference on HIV Pathogenesis, Treatment and Prevention, Cape Town, South Africa, 19 to 22 July.
70. **Johnson, R. H., and H. E. Einstein.** 2006. Coccidioidal meningitis. *Clin. Infect. Dis.* **42:**103–107.
71. **Kambugu, A., D. B. Meya, J. Rhein, M. O'Brien, E. N. Janoff, A. R. Ronald, M. R. Kamya, H. Mayanja-Kizza, M. A. Sande, P. R. Bohjanen, and D. R. Boulware.** 2008. Outcomes of cryptococcal meningitis in Uganda before and after the availability of highly active antiretroviral therapy. *Clin. Infect. Dis.* **46:**1694–1701.
72. **Kapoor, K. G., S. E. Katz, D. M. Grzybowski, and M. Lubow.** 2008. Cerebrospinal fluid outflow: an evolving perspective. *Brain Res. Bull.* **77:**327–334.
73. **Kappelhoff, B. S., F. van Leth, P. A. Robinson, T. R. MacGregor, E. Baraldi, F. Montella, D. E. Uip, M. A. Thompson, D. B. Russell, J. M. Lange, J. H. Beijnen, and A. D. Huitema.** 2005. Are adverse events of nevirapine and efavirenz related to plasma concentrations? *Antivir. Ther.* **10:**489–498.
74. **Kartalija, M., K. Kaye, J. H. Tureen, Q. Liu, M. G. Tauber, B. R. Elliott, and M. A. Sande.** 1996. Treatment of experimental cryptococcal meningitis with fluconazole: impact of dose and addition of flucytosine on mycologic and pathophysiologic outcome. *J. Infect. Dis.* **173:**1216–1221.
75. **Klepser, M. E., D. Malone, R. E. Lewis, E. J. Ernst, and M. A. Pfaller.** 2000. Evaluation of voriconazole pharmacodynamics using time-kill methodology. *Antimicrob. Agents Chemother.* **44:**1917–1920.
76. **Krishna, G., A. Moton, L. Ma, M. Martinho, M. Seiberling, and J. McLeod.** 2009. Effects of oral posaconazole on the pharmacokinetics of atazanavir alone and with ritonavir or with efavirenz in healthy adult volunteers. *J. Acquir. Immune Defic. Syndr.* **51:**437–444.
77. **Larsen, R. A., M. Bauer, A. M. Thomas, and J. R. Graybill.** 2004. Amphotericin B and fluconazole, a potent combination therapy for cryptococcal meningitis. *Antimicrob. Agents Chemother.* **48:**985–991.
78. **Larsen, R. A., S. A. Bozzette, B. E. Jones, D. Haghighat, M. A. Leal, D. Forthal, M. Bauer, J. G. Tilles, J. A. McCutchan, and J. M. Leedom.** 1994. Fluconazole combined with flucytosine for treatment of cryptococcal meningitis in patients with AIDS. *Clin. Infect. Dis.* **19:**741–745.
79. **Larsen, R. A., M. A. Leal, and L. S. Chan.** 1990. Fluconazole compared with amphotericin B plus flucytosine for cryptococcal meningitis in AIDS. A randomized trial. *Ann. Intern. Med.* **113:**183–187.
80. **Lawn, S., A. Harries, X. Anglaret, L. Myer, and R. Wood.** 2008. Early mortality among adults accessing antiretroviral treatment programmes in sub-Saharan Africa. *AIDS* **22:**1897–1908.
81. **Lawn, S. D., L. G. Bekker, L. Myer, C. Orrell, and R. Wood.** 2005. Cryptococcocal immune reconstitution disease: a major cause of early mortality in a South African antiretroviral programme. *AIDS* **19:**2050–2052.

82. **Liechty, C. A., P. Solberg, W. Were, J. P. Ekwaru, R. L. Ransom, P. J. Weidle, R. Downing, A. Coutinho, and J. Mermin.** 2007. Asymptomatic serum cryptococcal antigenemia and early mortality during antiretroviral therapy in rural Uganda. *Trop. Med. Int. Health* **12:**929–935.
83. **Liu, H., H. Davoudi, and T. Last.** 1995. Determination of amphotericin B in cerebrospinal fluid by solid-phase extraction and liquid chromatography. *J. Pharm. Biomed. Anal.* **13:**1395–1400.
84. **Liu, P., G. Foster, R. R. LaBadie, M. J. Gutierrez, and A. Sharma.** 2008. Pharmacokinetic interaction between voriconazole and efavirenz at steady state in healthy male subjects. *J. Clin. Pharmacol.* **48:**73–84.
85. **Longley, N., C. Muzoora, K. Taseera, J. Mwesigye, J. Rwebembera, A. Chakera, E. Wall, I. Andia, S. Jaffar, and T. S. Harrison.** 2008. Dose response effect of high-dose fluconazole for HIV-associated cryptococcal meningitis in southwestern Uganda. *Clin. Infect. Dis.* **47:**1556–1561.
86. **Lortholary, O., A. Fontanet, N. Memain, A. Martin, K. Sitbon, and F. Dromer.** 2005. Incidence and risk factors of immune reconstitution inflammatory syndrome complicating HIV-associated cryptococcosis in France. *AIDS* **19:**1043–1049.
87. **Lortholary, O., G. Poizat, V. Zeller, S. Neuville, A. Boibieux, M. Alvarez, P. Dellamonica, F. Botterel, F. Dromer, and G. Chene.** 2006. Long-term outcome of AIDS-associated cryptococcosis in the era of combination antiretroviral therapy. *AIDS* **20:**2183–2191.
88. **Loyse, A., H. Wainwright, J. Jarvis, T. Bicanic, K. Rebe, G. Meintjes, and T. Harrison.** 2009. Histopathology of the arachnoid granulations and brain in HIV-associated cryptococcal meningitis: correlation with cerebrospinal fluid pressure. *AIDS* **24:**405–410.
89. **Lutsar, I., S. Roffey, and P. Troke.** 2003. Voriconazole concentrations in the cerebrospinal fluid and brain tissue of guinea pigs and immunocompromised patients. *Clin. Infect. Dis.* **37:**728–732.
90. **Macsween, K. F., T. Bicanic, A. E. Brouwer, H. Marsh, D. C. Macallan, and T. S. Harrison.** 2005. Lumbar drainage for control of raised cerebrospinal fluid pressure in cryptococcal meningitis: case report and review. *J. Infect.* **51:**e221–e224.
91. **Makadzange, A., C. Ndhlovu, K. Takarinda, M. Reid, M. Kurangwa, V. Chikwasha, and J. Hakim.** 2009. Early vs delayed ART in the treatment of cryptococcal meningitis in Africa, abstr 36cLB. 16th Conference on Retroviruses and Opportunistic Infections, Montreal, Canada.
92. **Malessa, R., M. Krams, U. Hengge, C. Weiller, V. Reinhardt, L. Volbracht, F. Rauhut, and N. H. Brockmeyer.** 1994. Elevation of intracranial pressure in acute AIDS-related cryptococcal meningitis. *Clin. Investig.* **72:**1020–1026.
93. **Manosuthi, W., C. Athichathanabadi, S. Uttayamakul, T. Phoorisri, and S. Sungkanuparph.** 2007. Plasma nevirapine levels, adverse events and efficacy of antiretroviral therapy among HIV-infected patients concurrently receiving nevirapine-based antiretroviral therapy and fluconazole. *BMC Infect. Dis.* **7:**14.
94. **Manosuthi, W., S. Sungkanuparph, S. Chottanapund, S. Tansuphaswadikul, S. Chimsuntorn, P. Limpanadusadee, and P. G. Pappas.** 2008. Temporary external lumbar drainage for reducing elevated intracranial pressure in HIV-infected patients with cryptococcal meningitis. *Int. J. STD AIDS* **19:**268–271.
95. **Mavrogiorgos, N., O. Zaragoza, A. Casadevall, and J. D. Nosanchuk.** 2006. Efficacy of voriconazole in experimental *Cryptococcus neoformans* infection. *Mycopathologia* **162:**111–114.
96. **Mayanja-Kizza, H., K. Oishi, S. Mitarai, H. Yamashita, K. Nalongo, K. Watanabe, T. Izumi, J. Ococi, K. Augustine, R. Mugerwa, T. Nagatake, and K. Matsumoto.** 1998. Combination therapy with fluconazole and flucytosine for cryptococcal meningitis in Ugandan patients with AIDS. *Clin. Infect. Dis.* **26:**1362–1366.
97. **Megson, G. M., D. A. Stevens, J. R. Hamilton, and D. W. Denning.** 1996. D-mannitol in cerebrospinal fluid of patients with AIDS and cryptococcal meningitis. *J. Clin. Microbiol.* **34:**218–221.
98. **Menichetti, F., M. Fiorio, A. Tosti, G. Gatti, M. Bruna Pasticci, F. Miletich, M. Marroni, D. Bassetti, and S. Pauluzzi.** 1996. High-dose fluconazole therapy for cryptococcal meningitis in patients with AIDS. *Clin. Infect. Dis.* **22:**838–840.
99. **Micol, R., O. Lortholary, B. Sar, D. Laureillard, C. Ngeth, J. P. Dousset, H. Chanroeun, L. Ferradini, P. J. Guerin, F. Dromer, and A. Fontanet.** 2007. Prevalence, determinants of positivity, and clinical utility of cryptococcal antigenemia in Cambodian HIV-infected patients. *J. Acquir. Immune Defic. Syndr.* **45:**555–559.
100. **Milefchik, E., M. A. Leal, R. Haubrich, S. A. Bozzette, J. G. Tilles, J. M. Leedom, J. A. McCutchan, and R. A. Larsen.** 2008. Fluconazole alone or combined with flucytosine for the treatment of AIDS-associated cryptococcal meningitis. *Med. Mycol.* **46:**393–395.
101. **Mirza, S. A., M. Phelan, D. Rimland, E. Graviss, R. Hamill, M. E. Brandt, T. Gardner, M. Sattah, G. P. de Leon, W. Baughman, and R. A. Hajjeh.** 2003. The changing epidemiology of cryptococcosis: an update from population-based active surveillance in 2 large metropolitan areas, 1992–2000. *Clin. Infect. Dis.* **36:**789–794.
102. **Murdoch, D. M., W. D. Venter, C. Feldman, and A. Van Rie.** 2008. HIV immune reconstitution syndrome in sub-Saharan Africa. *AIDS* **22:**1689–1690.
103. **Mussini, C., P. Pezzotti, J. M. Miro, E. Martinez, J. C. de Quiros, P. Cinque, V. Borghi, A. Bedini, P. Domingo, P. Cahn, P. Bossi, A. de Luca, A. d'Arminio Monforte, M. Nelson, N. Nwokolo, S. Helou, R. Negroni, G. Jacchetti, S. Antinori, A. Lazzarin, A. Cossarizza, R. Esposito, A. Antinori, and J. A. Aberg.** 2004. Discontinuation of maintenance therapy for cryptococcal meningitis in patients with AIDS treated with highly active antiretroviral therapy: an international observational study. *Clin. Infect. Dis.* **38:**565–571.
104. **Mwaba, P., J. Mwansa, C. Chintu, J. Pobee, M. Scarborough, S. Portsmouth, and A. Zumla.** 2001. Clinical presentation, natural history, and cumulative death rates of 230 adults with primary cryptococcal meningitis in Zambian AIDS patients treated under local conditions. *Postgrad. Med. J.* **77:**769–773.
105. **National AIDS Commission (NAC) and Malawi Ministry of Health and Population.** 2008. *Treatment of AIDS: Guidelines for the use of Antiretroviral Therapy in Malawi*, 2nd ed. Lilongwe, Malawi: Ministry of Health. http://www.hivunitmohmw.org/Main/AntiretroviralTherapy. [Online.]
106. **Netea, M. G., A. E. Brouwer, E. H. Hoogendoorn, J. W. Van der Meer, M. Koolen, P. E. Verweij, and B. J. Kullberg.** 2004. Two patients with cryptococcal meningitis and idiopathic CD4 lymphopenia: defective cytokine production and reversal by recombinant interferon-gamma therapy. *Clin. Infect. Dis.* **39:**e83–e87.
107. **Newton, P. N., L. H. Thai, N. Q. Tip, J. M. Short, W. Chierakul, A. Rajanuwong, P. Pitisuttithum, S. Chasombat, B. Phonrat, W. Maek-A-Nantawat, R. Teaunadi, D. G. Lalloo, and N. J. White.** 2002. A randomized, double-blind, placebo-controlled trial of acetazolamide

for the treatment of elevated intracranial pressure in cryptococcal meningitis. *Clin. Infect. Dis.* **35:**769–772.

108. **Panther, L. A., and M. A. Sande.** 1990. Cryptococcal meningitis in the acquired immunodeficiency syndrome. *Semin. Respir. Infect.* **5:**138–145.
109. **Pappalardo, M. C., R. C. Paschoal, and M. S. Melhem.** 2007. AIDS-associated central nervous system cryptococcosis: a Brazilian case study. *AIDS* **21:**1971–1972.
110. **Pappas, P. G., B. Bustamante, E. Ticona, R. J. Hamill, P. C. Johnson, A. Reboli, J. Aberg, R. Hasbun, and H. H. Hsu.** 2004. Recombinant interferon- gamma 1b as adjunctive therapy for AIDS-related acute cryptococcal meningitis. *J. Infect. Dis.* **189:**2185–2191.
111. **Pappas, P. G., P. Chetchotisakd, R. A. Larsen, W. Manosuthi, M. I. Morris, T. Anekthananon, S. Sungkanuparph, K. Supparatpinyo, T. L. Nolen, L. O. Zimmer, A. S. Kendrick, P. Johnson, J. D. Sobel, and S. G. Filler.** 2009. A phase II randomized trial of amphotericin B alone or combined with fluconazole in the treatment of HIV-associated cryptococcal meningitis. *Clin. Infect. Dis.* **48:**1775–1783.
112. **Pappas, P. G., J. R. Perfect, G. A. Cloud, R. A. Larsen, G. A. Pankey, D. J. Lancaster, H. Henderson, C. A. Kauffman, D. W. Haas, M. Saccente, R. J. Hamill, M. S. Holloway, R. M. Warren, and W. E. Dismukes.** 2001. Cryptococcosis in human immunodeficiency virus-negative patients in the era of effective azole therapy. *Clin. Infect. Dis.* **33:**690–699.
113. **Park, B. J., K. A. Wannemuehler, B. J. Marston, N. Govender, P. G. Pappas, and T. M. Chiller.** 2009. Estimation of the current global burden of cryptococcal meningitis among persons living with HIV/AIDS. *AIDS* **23:**525–530.
114. **Park, M. K., D. R. Hospenthal, and J. E. Bennett.** 1999. Treatment of hydrocephalus secondary to cryptococcal meningitis by use of shunting. *Clin. Infect. Dis.* **28:**629–633.
115. **Parkes-Ratanshi, R., A. Kamali, K. Wakeham, J. Levin, C. Nabiryo Lwanga, N. Kenya-Mugisha, A. Coutinho, J. Whitworth, H. Grosskurth, and D. Lalloo.** 2009. Successful primary prevention of cryptococcal disease using fluconazole prophylaxis in HIV-infected Ugandan adults, abstr. 32. 16th Conference on Retroviruses and Opportunistic Infections Montreal, Canada.
116. **Perfect, J. R., G. M. Cox, R. K. Dodge, and W. A. Schell.** 1996. In vitro and in vivo efficacies of the azole SCH56592 against *Cryptococcus neoformans*. *Antimicrob. Agents Chemother.* **40:**1910–1913.
117. **Perfect, J. R., W. E. Dismukes, F. Dromer, D. L. Goldman, J. R. Graybill, R. J. Hamill, T. S. Harrison, R. A. Larsen, O. Lortholary, M. H. Nguyen, P. G. Pappas, W. G. Powderly, N. Singh, J. D. Sobel, and T. C. Sorrell.** 2010. Clinical practice guidelines for the management of cryptococcal disease: 2010 update by the infectious diseases society of America. *Clin. Infect. Dis.* **50:**291–322.
118. **Perfect, J. R., and D. T. Durack.** 1985. Comparison of amphotericin B and N-D-ornithyl amphotericin B methyl ester in experimental cryptococcal meningitis and *Candida albicans* endocarditis with pyelonephritis. *Antimicrob. Agents Chemother.* **28:**751–755.
119. **Perfect, J. R., M. H. Lindsay, and R. H. Drew.** 1992. Adverse drug reactions to systemic antifungals. Prevention and management. *Drug Saf.* **7:**323–363.
120. **Perfect, J. R., K. A. Marr, T. J. Walsh, R. N. Greenberg, B. DuPont, J. de la Torre-Cisneros, G. Just-Nubling, H. T. Schlamm, I. Lutsar, A. Espinel-Ingroff, and E. Johnson.** 2003. Voriconazole treatment for less-common, emerging, or refractory fungal infections. *Clin. Infect. Dis.* **36:**1122–1131.
121. **Pfaller, M. A., S. A. Messer, L. Boyken, R. J. Hollis, C. Rice, S. Tendolkar, and D. J. Diekema.** 2004. In vitro activities of voriconazole, posaconazole, and fluconazole against 4,169 clinical isolates of *Candida* spp. and *Cryptococcus neoformans* collected during 2001 and 2002 in the ARTEMIS global antifungal surveillance program. *Diagn. Microbiol. Infect. Dis.* **48:**201–205.
122. **Pfaller, M. A., S. A. Messer, L. Boyken, C. Rice, S. Tendolkar, R. J. Hollis, G. V. Doern, and D. J. Diekema.** 2005. Global trends in the antifungal susceptibility of *Cryptococcus neoformans* (1990 to 2004). *J. Clin. Microbiol.* **43:**2163–2167.
123. **Pfaller, M. A., S. A. Messer, R. J. Hollis, and R. N. Jones.** 2001. In vitro activities of posaconazole (Sch 56592) compared with those of itraconazole and fluconazole against 3,685 clinical isolates of *Candida* spp. and *Cryptococcus neoformans*. *Antimicrob. Agents Chemother.* **45:**2862–2864.
124. **Pfaller, M. A., J. Zhang, S. A. Messer, M. E. Brandt, R. A. Hajjeh, C. J. Jessup, M. Tumberland, E. K. Mbidde, and M. A. Ghannoum.** 1999. In vitro activities of voriconazole, fluconazole, and itraconazole against 566 clinical isolates of *Cryptococcus neoformans* from the United States and Africa. *Antimicrob. Agents Chemother.* **43:**169–171.
125. **Phillips, P., K. Chapman, M. Sharp, P. Harrison, J. Vortel, T. Steiner, and W. Bowie.** 2009. Dexamethasone in *Cryptococcus gattii* central nervous system infection. *Clin. Infect. Dis.* **49:**591–595.
126. **Pitisuttithum, P., R. Negroni, J. R. Graybill, B. Bustamante, P. Pappas, S. Chapman, R. S. Hare, and C. J. Hardalo.** 2005. Activity of posaconazole in the treatment of central nervous system fungal infections. *J. Antimicrob. Chemother.* **56:**745–755.
127. **Pitisuttithum, P., S. Tansuphasawadikul, A. J. Simpson, P. A. Howe, and N. J. White.** 2001. A prospective study of AIDS-associated cryptococcal meningitis in Thailand treated with high-dose amphotericin B. *J. Infect.* **43:**226–233.
128. **Powderly, W. G., D. Finkelstein, J. Feinberg, P. Frame, W. He, C. van der Horst, S. L. Koletar, M. E. Eyster, J. Carey, H. Waskin, T. M. Hooton, N. Hyslop, S. A. Spector, S. A. Bozzette, for The NIAID AIDS Clinical Trials Group.** 1995. A randomized trial comparing fluconazole with clotrimazole troches for the prevention of fungal infections in patients with advanced human immunodeficiency virus infection. NIAID AIDS Clinical Trials Group. *N. Engl. J. Med.* **332:**700–705.
129. **Powderly, W. G., M. S. Saag, G. A. Cloud, P. Robinson, R. D. Meyer, J. M. Jacobson, J. R. Graybill, A. M. Sugar, V. J. McAuliffe, S. E. Follansbee, et al.** 1992. A controlled trial of fluconazole or amphotericin B to prevent relapse of cryptococcal meningitis in patients with the acquired immunodeficiency syndrome. The NIAID AIDS Clinical Trials Group and Mycoses Study Group. *N. Engl. J. Med.* **326:**793–798.
130. **Raad, I., S. Chapman, R. Bradsher, V. Morrison, M. Goldman, and J. Graybill.** 2004. Posaconazole salvage therapy for invasive fungal infections, abstr. M-669. 44th Interscience Conference on Antimicrobial Agents and Chemotherapy, 30 October to 2 November. American Society for Microbiology, Washington, DC.
131. **Rex, J. H., P. G. Pappas, A. W. Karchmer, J. Sobel, J. E. Edwards, S. Hadley, C. Brass, J. A. Vazquez, S. W. Chapman, H. W. Horowitz, M. Zervos, D. McKinsey, J. Lee, T. Babinchak, R. W. Bradsher, J. D. Cleary, D. M. Cohen, L. Danziger, M. Goldman, J. Goodman, E. Hilton, N. E. Hyslop, D. H. Kett, J. Lutz, R. H. Rubin, W. M. Scheld, M. Schuster, B. Simmons, D. K. Stein,**

R. G. Washburn, L. Mautner, T. C. Chu, H. Panzer, R. B. Rosenstein, and J. Booth. 2003. A randomized and blinded multicenter trial of high-dose fluconazole plus placebo versus fluconazole plus amphotericin B as therapy for candidemia and its consequences in nonneutropenic subjects. *Clin. Infect. Dis.* **36:**1221–1228.

132. **Robinson, P. A., M. Bauer, M. A. Leal, S. G. Evans, P. D. Holtom, D. A. Diamond, J. M. Leedom, and R. A. Larsen.** 1999. Early mycological treatment failure in AIDS-associated cryptococcal meningitis. *Clin. Infect. Dis.* **28:**82–92.

133. **Saag, M. S., G. A. Cloud, J. R. Graybill, J. D. Sobel, C. U. Tuazon, P. C. Johnson, W. J. Fessel, B. L. Moskovitz, B. Wiesinger, D. Cosmatos, L. Riser, C. Thomas, R. Hafner, and W. E. Dismukes.** 1999. A comparison of itraconazole versus fluconazole as maintenance therapy for AIDS-associated cryptococcal meningitis. National Institute of Allergy and Infectious Diseases Mycoses Study Group. *Clin. Infect. Dis.* **28:**291–296.

134. **Saag, M. S., R. J. Graybill, R. A. Larsen, P. G. Pappas, J. R. Perfect, W. G. Powderly, J. D. Sobel, and W. E. Dismukes.** 2000. Practice guidelines for the management of cryptococcal disease. Infectious Diseases Society of America. *Clin. Infect. Dis.* **30:**710–718.

135. **Saag, M. S., W. G. Powderly, G. A. Cloud, P. Robinson, M. H. Grieco, P. K. Sharkey, S. E. Thompson, A. M. Sugar, C. U. Tuazon, J. F. Fisher, et al.** 1992. Comparison of amphotericin B with fluconazole in the treatment of acute AIDS-associated cryptococcal meningitis. The NIAID Mycoses Study Group and the AIDS Clinical Trials Group. *N. Engl. J. Med.* **326:**83–89.

136. **Schaars, C. F., G. A. Meintjes, C. Morroni, F. A. Post, and G. Maartens.** 2006. Outcome of AIDS-associated cryptococcal meningitis initially treated with 200 mg/day or 400 mg/day of fluconazole. *BMC Infect. Dis.* **6:**118.

137. **Schwartz, S., M. Ruhnke, P. Ribaud, L. Corey, T. Driscoll, O. A. Cornely, U. Schuler, I. Lutsar, P. Troke, and E. Thiel.** 2005. Improved outcome in central nervous system aspergillosis, using voriconazole treatment. *Blood* **106:**2641–2645.

138. **Serena, C., F. J. Pastor, M. Marine, M. M. Rodriguez, and J. Guarro.** 2007. Efficacy of voriconazole in a murine model of cryptococcal central nervous system infection. *J. Antimicrob. Chemother.* **60:**162–165.

139. **Shelburne, S. A., 3rd, J. Darcourt, A. C. White, Jr., S. B. Greenberg, R. J. Hamill, R. L. Atmar, and F. Visnegarwala.** 2005. The role of immune reconstitution inflammatory syndrome in AIDS-related *Cryptococcus neoformans* disease in the era of highly active antiretroviral therapy. *Clin. Infect. Dis.* **40:**1049–1052.

140. **Shoham, S., C. Cover, N. Donegan, E. Fulnecky, and P. Kumar.** 2005. *Cryptococcus neoformans* meningitis at 2 hospitals in Washington, D.C.: adherence of health care providers to published practice guidelines for the management of cryptococcal disease. *Clin. Infect. Dis.* **40:**477–479.

141. **Siddiqui, A. A., A. E. Brouwer, V. Wuthiekanun, S. Jaffar, R. Shattock, D. Irving, J. Sheldon, W. Chierakul, S. Peacock, N. Day, N. J. White, and T. S. Harrison.** 2005. IFN-gamma at the site of infection determines rate of clearance of infection in cryptococcal meningitis. *J. Immunol.* **174:**1746–1750.

142. **Southern African HIV Clinicians Society.** 2007. Guidelines for the prevention, diagnosis and management of cryptococcal meningitis and disseminated cryptococcosis in HIV-infected patients. *S. Afr. J. HIV Med.* **28:**25–35.

143. **Speed, B., and D. Dunt.** 1995. Clinical and host differences between infections with the two varieties of *Cryptococcus neoformans. Clin. Infect. Dis.* **21:**28–34; discussion 35–36.

144. **Sun, H. Y., M. M. Wagener, and N. Singh.** 2009. Cryptococcosis in solid-organ, hematopoietic stem cell, and tissue transplant recipients: evidence-based evolving trends. *Clin. Infect. Dis.* **48:**1566–1576.

145. **Sungkanuparph, S., U. Jongwutiwes, and S. Kiertiburanakul.** 2007. Timing of cryptococcal immune reconstitution inflammatory syndrome after antiretroviral therapy in patients with AIDS and cryptococcal meningitis. *J. Acquir. Immune Defic. Syndr.* **45:**595–596.

146. **Tansuphaswadikul, S., W. Maek-a-Nantawat, B. Phonrat, L. Boonpokbn, A. G. Mctm, and P. Pitisuttithum.** 2006. Comparison of one week with two week regimens of amphotericin B both followed by fluconazole in the treatment of cryptococcal meningitis among AIDS patients. *J. Med. Assoc. Thai.* **89:**1677–1685.

147. **Tassie, J. M., L. Pepper, C. Fogg, S. Biraro, B. Mayanja, I. Andia, A. Paugam, G. Priotto, and D. Legros.** 2003. Systematic screening of cryptococcal antigenemia in HIV-positive adults in Uganda. *J. Acquir. Immune Defic. Syndr.* **33:**411–412.

148. **Torres, H. A., R. Y. Hachem, R. F. Chemaly, D. P. Kontoyiannis, and I. I. Raad.** 2005. Posaconazole: a broad-spectrum triazole antifungal. *Lancet Infect. Dis.* **5:**775–785.

149. **Trachtenberg, J. D., A. D. Kambugu, M. McKellar, F. Semitala, H. Mayanja-Kizza, M. H. Samore, A. Ronald, and M. A. Sande.** 2007. The medical management of central nervous system infections in Uganda and the potential impact of an algorithm-based approach to improve outcomes. *Int. J. Infect. Dis.* **11:**524–530.

150. **van der Horst, C. M., M. S. Saag, G. A. Cloud, R. J. Hamill, J. R. Graybill, J. D. Sobel, P. C. Johnson, C. U. Tuazon, T. Kerkering, B. L. Moskovitz, W. G. Powderly, and W. E. Dismukes.** 1997. Treatment of cryptococcal meningitis associated with the acquired immunodeficiency syndrome. National Institute of Allergy and Infectious Diseases Mycoses Study Group and AIDS Clinical Trials Group. *N. Engl. J. Med.* **337:**15–21.

151. **van Duin, D., W. Cleare, O. Zaragoza, A. Casadevall, and J. D. Nosanchuk.** 2004. Effects of voriconazole on *Cryptococcus neoformans. Antimicrob. Agents Chemother.* **48:**2014–2020.

152. **Vermes, A., H. J. Guchelaar, and J. Dankert.** 2000. Flucytosine: a review of its pharmacology, clinical indications, pharmacokinetics, toxicity and drug interactions. *J. Antimicrob. Chemother.* **46:**171–179.

153. **Vermes, A., E. J. Kuijper, H. J. Guchelaar, and J. Dankert.** 2003. An in vitro study on the active conversion of flucytosine to fluorouracil by microorganisms in the human intestinal microflora. *Chemotherapy* **49:**17–23.

154. **Vibhagool, A., S. Sungkanuparph, P. Mootsikapun, P. Chetchotisakd, S. Tansuphaswaswadikul, C. Bowonwatanuwong, and A. Ingsathit.** 2003. Discontinuation of secondary prophylaxis for cryptococcal meningitis in human immunodeficiency virus-infected patients treated with highly active antiretroviral therapy: a prospective, multicenter, randomized study. *Clin. Infect. Dis.* **36:**1329–1331.

155. **Welch, K., and M. Pollay.** 1961. Perfusion of particles through arachnoid villi of the monkey. *Am. J. Physiol.* **201:**651–654.

156. **Wertheimer, A. I., T. M. Santella, and H. J. Lauver.** 2004. Successful public/private donation programs: a review of the Diflucan Partnership Program in South Africa. *J. Int. Assoc. Physicians AIDS Care* **3:**74–79, 84–85.

157. **WHO.** 2008. Essential prevention and care interventions for adults and adolescents living with HIV in resource-limited settings. ISBN 978 92 4 159670 1, http://www.who.int/hiv/pub/guidelines/EP/en/index.html.

158. **Witt, M. D., R. J. Lewis, R. A. Larsen, E. N. Milefchik, M. A. Leal, R. H. Haubrich, J. A. Richie, J. E. Edwards, Jr., and M. A. Ghannoum.** 1996. Identification of patients with acute AIDS-associated cryptococcal meningitis who can be effectively treated with fluconazole: the role of antifungal susceptibility testing. *Clin. Infect. Dis.* **22:**322–328.

159. **Woods, M. L., 2nd, R. MacGinley, D. P. Eisen, and A. M. Allworth.** 1998. HIV combination therapy: partial immune restitution unmasking latent cryptococcal infection. *AIDS* **12:**1491–1494.

160. **Woodworth, G. F., M. J. McGirt, M. A. Williams, and D. Rigamonti.** 2005. The use of ventriculoperitoneal shunts for uncontrollable intracranial hypertension without ventriculomegally secondary to HIV-associated cryptococcal meningitis. *Surg. Neurol.* **63:**529–531; discussion 531–532.

161. **Yamashima, T.** 1986. Ultrastructural study of the final cerebrospinal fluid pathway in human arachnoid villi. *Brain Res.* **384:**68–76.

162. **Zolopa, A., J. Andersen, W. Powderly, A. Sanchez, I. Sanne, C. Suckow, E. Hogg, and L. Komarow.** 2009. Early antiretroviral therapy reduces AIDS progression/death in individuals with acute opportunistic infections: a multicenter randomized strategy trial. *PLoS ONE* **4:**e5575.

Cryptococcus: From Human Pathogen to Model Yeast
Edited by J. Heitman et al.

43

Public Health Importance of Cryptococcal Disease: Epidemiology, Burden, and Control

BENJAMIN J. PARK, SHAWN R. LOCKHART, MARY E. BRANDT, AND TOM M. CHILLER

Over the past 30 years, the epidemiology of cryptococcosis has changed dramatically. Once primarily seen as a rare infection in immunocompetent hosts, the incidence of cryptococcosis exploded globally with the AIDS epidemic. However, in the past decade, countries with widespread access to antiretroviral (ARV) medications have seen a substantial decrease in the incidence and burden of cryptococcosis. This is not the case in resource-limited settings, especially in sub-Saharan Africa, where access to ARVs is still very limited. These countries continue to have a high incidence of cryptococcosis with many fatalities, and in some areas the disease is a more frequent cause of death than tuberculosis. Today, cryptococcosis remains one of the most important opportunistic infections among persons living with HIV/AIDS, particularly in less-developed countries. Public health agencies, academicians, and clinicians should focus resources and efforts on reducing the morbidity and mortality of this infection.

Another substantial shift in the epidemiology of cryptococcosis was the emergence of *Cryptococcus gattii* in North America as a primary pathogen. Although this chapter will focus primarily on HIV-associated cryptococcosis, due to its much greater public health burden, the challenges to public health agencies in addressing the threat of *C. gattii* will also be discussed.

GLOBAL EPIDEMIOLOGY OF HIV-ASSOCIATED CRYPTOCOCCOSIS

The vast majority of disease caused by *Cryptococcus* is found in persons who are infected with HIV, especially those who develop AIDS. The epidemiology of HIV-associated cryptococcosis varies substantially in different parts of the world, with the highest numbers of cases occurring in areas where there are also the highest numbers of persons living with AIDS. Despite the public health successes of dramatically decreasing the incidence of cryptococcal meningitis in some regions, the incidence and mortality related to this disease remain substantial in resource-limited settings.

Benjamin J. Park, Shawn R. Lockhart, Mary E. Brandt, and Tom M. Chiller, Mycotic Diseases Branch, Centers for Disease Control and Prevention, Atlanta, GA 30333.

Sub-Saharan Africa

The greatest burden of HIV globally is in sub-Saharan Africa, with over 25,000,000 persons infected with HIV. There have been few studies done to evaluate the epidemiology of cryptococcosis in this region, although all have generally shown a very high burden. Reported incidence has ranged from 1.4% among persons with AIDS (47) to 12.2% among persons routinely screened using a serum cryptococcal antigen (CrAg) test (17). The lowest incidence was reported in South Africa, where population-based, laboratory-based surveillance was conducted among an ARV-naive population in metropolitan Johannesburg (47). This study reported an incidence of 1.4% among persons with AIDS and 0.1% among persons with HIV. However, it is not known how complete the case finding was, as poor access to health care may have hindered presentation to hospitals, and disease may have gone unrecognized. Reports from sub-Saharan Africa have documented cryptococcal disease as a major under-recognized cause of death among persons with AIDS (38).

Other studies have demonstrated a much higher burden. One study of 1,372 HIV-infected adults, conducted in Uganda from 1995 to 1999, found an incidence of 4.04% per year (21). Cryptococcal meningitis has also been reported as the most common cause of meningitis in many areas of sub-Saharan Africa, more common than *Streptococcus pneumoniae* or *Neisseria meningitidis*, even in areas in the "meningitis belt" (3, 4, 22, 28, 29, 62).

There are even fewer reports examining trends in cryptococcal disease over time. One surveillance network, conducted by the South African National Institute for Communicable Diseases, has performed long-term surveillance. These data suggest an increase in incidence of cryptococcal meningitis, despite increasing ARV medication availability and coverage (23). Reasons for this increase are not clear but may be related to improvement in surveillance techniques, or, alternatively, due to a high or increasing prevalence of AIDS. Measuring trends will be extremely important now that concerted efforts are under way to increase the population receiving ARV medication.

The high mortality associated with the disease in sub-Saharan Africa has also been demonstrated. Given the resource-intensive nature of treatment, which requires frequent lumbar punctures to manage increased intracranial pressure, intravenous antifungals, and long-term secondary prophylaxis, it is not surprising that cryptococcal meningitis is one of the most frequent causes of death among persons with AIDS in sub-Saharan Africa (20, 39). Case fatality of cryptococcal meningitis has been shown to approach 100% (3) in some areas. Other cohort studies, including a cohort of South African miners with high autopsy rates (11), have shown that 13 to 44% of deaths among persons with HIV/AIDS in this region are due to cryptococcal disease (11, 21, 38, 54). Considering that over 2 million persons with HIV/AIDS die each year in sub-Saharan Africa (70), annual *Cryptococcus*-related deaths may range from 270,000 to 920,000.

South and Southeast Asia

This region, with the contribution of a large and evolving HIV epidemic in India, has the second-largest burden of cryptococcal disease. India alone is home to an estimated 3.1 million to 9.4 million persons living with HIV (67). Reports from India have documented high incidence rates for cryptococcosis among persons with HIV/AIDS. Incidence ranged from 1.7% in northern India (68) to 4.7% (36) in the south. Among a group of hospitalized persons with HIV, approximately 2% had cryptococcal meningitis; of note, one-third of these persons were also coinfected with tuberculosis (37). Coinfection with tuberculosis may be a serious problem, because often, once the diagnosis of tuberculosis is made, there is no further effort to diagnose other concurrent infections. Patients may therefore go for prolonged periods without further diagnostic workup and may not be diagnosed at all until death.

In Thailand, a survey of over 100,000 persons with AIDS showed that nearly 20% of those infected had cryptococcal meningitis (7). Studies have also described up to a 36% prevalence among hospitalized persons with AIDS (2) and have reported *Cryptococcus* as the most common opportunistic infection (2, 30). Others have found it to be the most common cause of chronic meningitis (27). Studies from Vietnam and Cambodia have shown similar high rates (34, 35, 43, 63).

North and South America

The latest estimates of cryptococcal meningitis incidence in North America are low. Population-based surveillance from Atlanta and Houston, conducted from 1992 to 2000, found a mean incidence of 4 to 5 cases per 100,000 population in 1992 (50). In addition, cohort studies of HIV-infected persons have demonstrated similar rates. The CDC's Adult and Adolescent Spectrum of Disease project, a cohort study of persons with HIV, reported a rate from 1992 to 1998 of about 0.6% (31), and the Multicenter AIDS Cohort Study, conducted from 1990 to 1998, reported an incidence of 1.5% (59). Epidemiological studies conducted in Latin America have reported a higher incidence. One study, the Chilean AIDS Cohort, showed a 3.4% incidence among ARV-naive persons. A single-center study from Brazil showed a higher prevalence of 5.2% among persons with AIDS admitted to a tertiary care hospital.

One of the most impressive aspects of the epidemiology of cryptococcosis in North America is the marked decline beginning in the late 1990s. The reduction in the incidence of cryptococcosis is clearly a public health success story. In population-based surveillance conducted in Atlanta and Houston, the incidence of cryptococcosis among persons with AIDS declined from 66/1,000 persons with AIDS in 1992 to 7/1,000 in 2000 (50). Other cohort studies have corroborated this finding. Similarly, the Adult and Adolescent Spectrum of Disease project documented a decrease in cryptococcosis among persons living with HIV/AIDS, from 2% per year in 1992 to 0.1% in 1998 (31).

Reasons for this decline are multifactorial but are probably related to the introduction of ARVs in 1996, as the risk for cryptococcal infection is dramatically lower among persons whose CD4 count is >100. Additionally, improved care and treatment of persons with HIV in general probably contributed to the reduction.

Europe

Studies in Europe also showed a remarkable decrease in the burden of cryptococcosis and other opportunistic infections, largely coinciding with the introduction of ARVs. In Spain, an 8-year study conducted from 1989 to 1997, largely prior to the era of ARVs, found a low incidence of 0.2 to 0.7% per year among persons with AIDS (60). This study found the overall incidence of opportunistic infections to be decreasing, largely due to improved patient care, although the incidence of cryptococcosis remained constant. The EuroSIDA Cohort followed 9,803 persons with HIV/AIDS in 70 centers across Europe from 1994 to 2002 and documented a dramatic decrease in cryptococcal disease pre- and post-ARVs (15). Incidence of cryptococcosis declined from 1% in 1995 to 0.1% in 2001. Other studies have confirmed this decrease. The pre- and post-ARV eras were compared in a study conducted in France, where a 46% decrease in cryptococcosis cases was detected in the post-ARV era (1997 to 2001), compared to the pre-ARV era (1985 to 1996) (19).

East Asia

Few data exist from East Asia. Two studies from South Korea demonstrate a 1.7% incidence rate among HIV-infected persons from single centers (33, 53). The epidemiology in Japan and China is largely unknown. China, with its expanding HIV/AIDS epidemic, may have a large number of cases. Further surveillance should be conducted in these areas to understand trends in the incidence of cryptococcosis.

Oceania

Studies in Oceania, which includes Australia and New Zealand, have shown a low incidence of cryptococcosis. One national surveillance project has reported an incidence of 6.6 per 100,000 population in Australia and 2.2 per 100,000 in New Zealand. Among persons with AIDS, rates ranged from 1.4 to 3.9% in Australia and 3.8 to 4.9% in New Zealand (9). Declines in opportunistic infections, including cryptococcosis, have also been described beginning in 1996, probably corresponding with the introduction of ARVs (18).

ESTIMATION OF THE GLOBAL BURDEN OF CRYPTOCOCCAL MENINGITIS

Using available epidemiological data, we recently estimated the global burden of cryptococcal meningitis in persons with HIV in terms of yearly cases and deaths (55). Global burden of disease studies are frequently performed by public health agencies to allow for comparison with other diseases, in order to help set public health priorities. The CDC and the World Health Organization have conducted numerous studies on a host of diseases to measure disease-specific burden throughout the world (12, 13, 70).

TABLE 1 Estimated cryptococcal meningitis cases and deaths among 10 UNAIDS global regions by using published incidence rates from studies conducted in those regions[a]

Region	HIV prevalence in 1,000s	Estimated yearly cryptococcal meningitis cases (range), in 1,000s	Assumed 90-day case-fatality rate (%)	Estimated deaths (range) in 1,000s
Sub-Saharan Africa	22,500	720 (144.0–1,296.0)	70	504.0 (100.8–907.2)
East Asia	800	13.6 (2.7–24.5)	9	1.2 (0.2–2.2)
Oceania	75	0.1 (0.0–0.1)	9	0.009 (0.0–0.009)
South and Southeast Asia	4,000	120 (24.0–216.0)	55	66.0 (13.2–118.8)
Eastern Europe and Central Asia	1,600	27.2 (5.4–49.0)	55	15.0 (3.0–27.0)
Western and Central Europe	760	0.5 (0.1–1.0)	9	0.045 (0.009–0.09)
North Africa and Middle East	380	6.5 (1.3–11.6)	55	3.6 (0.7–6.4)
North America	1,300	7.8 (1.6–14.0)	9	0.7 (0.1–1.3)
Caribbean	230	7.8 (1.6–14.1)	55	4.3 (0.9–7.8)
Latin America	1,600	54.4 (10.9–97.9)	55	29.9 (6.0–53.8)
Global	33,200	957.9 (371.7–1,544.0)		624.7 (125.0–1,124.9)

[a] Adapted from Park et al. (55).

In order to estimate the disease burden for cryptococcal meningitis among persons with HIV, we identified appropriate studies on disease incidence published since 1996 and were able to extrapolate the rates to the general population, using the total numbers of people with HIV as the denominators as determined by the 2006 UNAIDS report (67). The ranges were estimated by calculating +/− one standard deviation.

We estimated that approximately 958,000 cases of cryptococcal meningitis occur each year (range, 371,700 to 1,544,000) (Table 1) (55). The region with the greatest number of cases was sub-Saharan Africa, with 720,000 cases per year, followed by South and Southeast Asia, with 120,000 cases per year. Western and Central Europe (500 cases) and Oceania (100 cases) had the fewest estimated cases (55).

We were also able to estimate the number of deaths associated with *Cryptococcus*. To do this, we assumed case-fatality rates based on published clinical trials (45, 69) and by reviewing case series, surveillance reports, and studies of outcomes of cryptococcal meningitis (3, 21, 24, 28, 32, 38, 42, 45, 47, 50, 52, 54, 65). We also consulted with clinical experts directly involved in the care and treatment of patients with cryptococcal meningitis. Based on this research, we were able to estimate a case-fatality rate of 9% in developed countries, 55% in less-developed countries, and 70% in sub-Saharan Africa.

Using these rates, we estimated that a total of 625,000 deaths occur globally each year due to cryptococcal meningitis (Table 1) (55). The region with the most deaths again was sub-Saharan Africa, with over 500,000 deaths per year. The fewest deaths were estimated to occur in Western and Central Europe and Oceania, at <50 per year (55).

As mentioned previously, the utility of global burden of disease estimates is that they allow public health agencies to place particular diseases in the context of other diseases. Using data from the WHO Global Burden of Disease project, we compared the burden due to *Cryptococcus* with other infectious diseases (not including HIV) (70), to demonstrate how *Cryptococcus* infection compares to other important infectious diseases that are considered to have a high burden of disease globally. The nearly 1 million cases of cryptococcal meningitis that are projected to occur annually are much less frequent than diarrheal diseases (4.5 billion cases), malaria (408 million cases), or tuberculosis (7 million cases), but more than hepatitis B (914,000 cases) and bacterial meningitis (448,000 cases) (70), illustrating that globally, *Cryptococcus* appears to be the most common pathogen causing meningitis.

Given the high case-fatality rate associated with cryptococcal meningitis, we wanted to also compare the total number of deaths due to the disease with other common diseases. When comparing deaths in sub-Saharan Africa (70), which is the region with the greatest burden of cryptococcosis (55), it is clear that it is one of the leading causes of infection-related mortality, possibly even causing more deaths than more common infections such as tuberculosis (Fig. 1).

IMPROVING PUBLIC HEALTH CAPACITY TO DETECT AND MANAGE HIV-RELATED CRYPTOCOCCOSIS

The large global burden of cryptococcal disease presents a number of challenges to public health, particularly in the resource-poor regions of high HIV prevalence in sub-Saharan Africa and South/Southeast Asia. Multiple strategies are required to combat this public health challenge.

Surveillance

Public health surveillance is an essential element in public health practice and allows factual insight into the epidemiology of a disease in order to inform decisions and take action. Data from effective surveillance systems are useful, cost-effective tools for targeting resources and evaluating programs. A key objective of surveillance is to provide information to guide interventions.

The methods used for surveillance of infectious diseases often progress along with economic development of the country. It is useful to divide surveillance into a spectrum of activities consisting of three distinct levels. Each level is more complex, and has greater capacity for controlling and detecting disease, but also depends on more resources and infrastructure (Fig. 2).

Population-based surveillance, the ability to look for disease occurrence within a defined population, is widely used

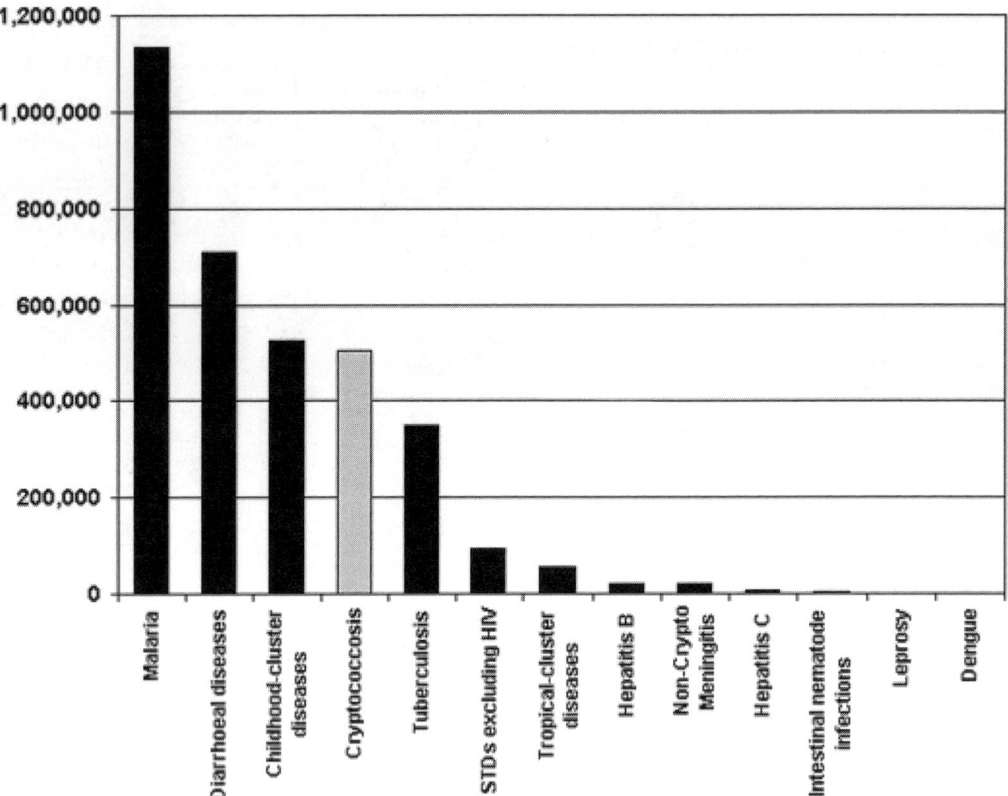

FIGURE 1 Comparison of deaths in sub-Saharan Africa due to HIV-related cryptococcosis and common infectious diseases excluding HIV (adapted from Park et al. [55]).

in the United States for reliable detection and tracking of diseases and has been used for cryptococcal meningitis (26, 50). However, population-based surveillance is dependent upon adequate case-finding and having accurate estimates of the total population from which those cases arise, especially for estimation of disease burden. In regions with poor access to health care, where persons presenting with subacute meningitis may not always seek medical attention and where appropriate diagnostic procedures (e.g., lumbar puncture) are not always performed, the number of detected cases may be lower than the actual case burden.

Another method to estimate burden is through cohort studies. Cohort studies involve following a group of individuals over time in order to detect who develops disease. Case ascertainment and follow-up is likely to be more complete in these studies. For example, in South Africa (47), population-based surveillance estimated rates of cryptococcal disease that were much lower than rates from cohort studies in nearby countries (0.1% versus 3.2 to 4.0%), probably as a result of the incomplete case-finding that may have resulted from health care access issues (21, 46, 47). However, because these studies are resource-intensive, they are difficult to sustain.

Regardless of the method, a clear benefit of ongoing surveillance is the ability to track trends. Additionally, ongoing prospective surveillance can be useful to monitor trends in susceptibility to drugs in the population under surveillance. Although a few reports have demonstrated that the prevalence of cryptococcal isolates with elevated MICs to fluconazole remains low (5, 16, 57), there are some reports from areas with high fluconazole usage that suggest that MIC values for fluconazole are increasing among cryptococcal isolates (56, 61). Currently, there is no clear consensus on whether MIC values can predict response to therapy in cryptococcal infections (1, 14, 71), and MIC breakpoint guidelines are not yet available for *Cryptococcus*.

Ongoing prospective surveillance can be useful to monitor trends in susceptibility in the population under surveillance. Reports have demonstrated that prevalence of cryptococcal isolates with elevated MICs to fluconazole is low and not increasing (5, 16, 57), although a report from Cambodia has suggested an increase (61). Nevertheless, future surveillance programs are needed to understand antifungal susceptibility patterns, particularly in geographic areas where use of fluconazole has increased. In particular, it will be important to understand patterns of susceptibility in Thailand, where primary prophylaxis using an azole medication has been recommended. It is important to monitor for the emergence of any resistance, as azole medications are, and will likely remain, a mainstay of treatment and maintenance in many areas.

Laboratory Capacity

Another fundamental component in improving public health surveillance is the ability of laboratories to identify the etiological agent of infection. This is essential for appropriate public health action, because there are many causes of meningitis, especially in areas hardest hit by *Cryptococcus*.

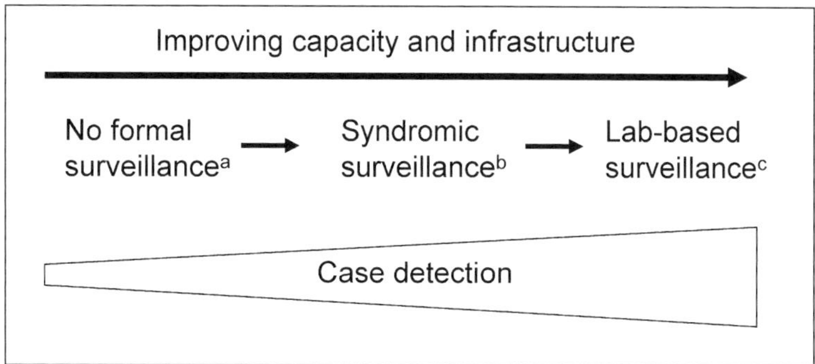

FIGURE 2 Spectrum of case-based disease surveillance. (a) Without a formal surveillance, only large outbreaks or unusual disease can be detected. (b) Syndromic surveillance is based on groups of signs or symptoms indicative of a common diagnosis, such as acute meningitis. (c) Laboratory-based surveillance relies on laboratory-confirmed pathogens.

Currently, many of the countries with the highest burden of infection do not have the laboratory capacity to reliably detect *Cryptococcus*. Many of these laboratories, which may be small, one-room facilities in rural areas of sub-Saharan Africa, are poorly equipped and staffed by persons with minimal training. Public health agencies such as the CDC, in partnership with nongovernmental organizations such as the American Society for Microbiology, have been expanding laboratory capacity as a component of PEPFAR, the President's Emergency Plan for AIDS Relief. This partnership has resulted in the improvement of the physical and human capacity in many laboratories that receive PEPFAR support, but more help in many more countries is needed.

Efforts are mainly focused on improving capacity for India ink staining and diagnosis by CrAg latex agglutination testing at central and provincial hospitals. However, rural and district hospitals in many countries in sub-Saharan Africa are likely able to perform CrAg latex agglutination testing if test kits are available, despite the lack of sophisticated equipment. Since many sites are familiar with syphilis rapid plasma reagin testing, which is also a latex agglutination test, they may be able to reliably perform CrAg testing. Expansion of the availability of the CrAg latex agglutination test may help to identify more cases in remote areas, which may lead to improved recognition and survival. Still, these efforts to improve diagnosis of *Cryptococcus* face many challenges; persistence, patience, and resources will continually be needed.

Management: Realities in the Field

Although Infectious Diseases Society of America guidelines recommend treatment of cryptococcal meningitis with amphotericin B and 5-fluorocytosine, the reality in sub-Saharan Africa is that in many countries, these medications are simply not available or are cost-prohibitive to use. Additionally, the medical infrastructure required for the complex management of these patients, with frequent lumbar punctures to manage intracranial pressure and monitoring of renal function among those on amphotericin B, is not realistic in many areas. Therefore, creative treatment and management solutions will need to be developed.

One promising strategy is the use of fluconazole at high doses for treatment. Fluconazole is available at little cost or through the Pfizer Diflucan Partnership program, which provides fluconazole to countries at no cost. Recent data have shown that high doses of fluconazole, at dosages of up to 1,200 mg/day, are safe and effective (41). In areas where amphotericin B or 5-fluorocytosine are not available, these treatment strategies may be considered. Public health agencies should work with ministries of health to determine the most cost-effective strategies given the numerous competing resources for HIV/AIDS care and treatment.

PUBLIC HEALTH STRATEGIES TO REDUCE MORBIDITY AND MORTALITY OF HIV-RELATED CRYPTOCOCCOSIS

Because cryptococcal meningitis carries such a high burden globally, and treatment options are so limited in the most affected countries, public health should also focus efforts on prevention of disease or prevention of severe sequelae. A few potential strategies exist that are worthy of further study.

Primary Prophylaxis

The first method that deserves further investigation is primary prophylaxis. Primary prophylaxis trials performed in the 1990s in economically developed countries (United States, Europe, and Australia) using either fluconazole or itraconazole showed a reduction or elimination of risk for development of cryptococcal infection but never showed an overall survival benefit (48, 58, 64). As a result, primary prophylaxis was never recommended as a prevention strategy. However, these studies were all performed in developed countries in optimized clinical conditions where the incidence and attributable mortality from cryptococcal disease among the cohorts was likely fairly low, relative to the current reality in areas like sub-Saharan Africa. In these resource-limited countries, with a higher incidence and higher case fatality, the survival benefit of primary prophylaxis may be greater and therefore more likely to achieve significance.

Further studies are therefore needed in developing nations, particularly where the burden of cryptococcal infection is high. Two studies have been performed recently in Thailand, with differing results. The first study, involving 129 patients with CD4 counts <300 cells/μl receiving either itraconazole prophylaxis or placebo, did not show a survival benefit of receiving antifungal medication, although no patients in the itraconazole arm developed cryptococcosis (8). Another small study, which enrolled patients with CD4 counts of <100 cells/μl, did suggest a survival benefit (10).

In this study, 90 patients were randomized to receive either fluconazole 400 mg/week or placebo. Interestingly, there was no benefit in preventing cryptococcosis, and some persons in the treatment arm developed cryptococcal disease (3/44, 6.8%) (10). To date, no randomized trials to evaluate prophylaxis have been performed in sub-Saharan Africa. Given the high mortality of persons with cryptococcal disease in this region, further investigation of this strategy is warranted.

Preemptive Treatment

Another strategy worth exploring is one to identify and preemptively treat persons before development of cryptococcal meningitis. Most patients with cryptococcal meningitis in sub-Saharan Africa present very late, often with very low CD4 counts and with signs of advanced disease, and survival is extremely poor (21, 29, 47). If treatment can be initiated prior to developing advanced disease, outcomes are likely to be better. One method of identifying such persons is to perform serum CrAg testing among asymptomatic or mildly symptomatic individuals.

Cryptococcal antigenemia has been shown to be prevalent in persons without symptomatic disease. Studies using prospective serum CrAg screening or retrospective review of available serum have reported a prevalence of 6 to 18% (17, 38, 40, 49, 66). These studies have been conducted among persons with WHO stage 3 or 4 disease (66), with a CD4 count <200 cells/μl (49) or ≤100 cells/μl (40), or even among all HIV-infected individuals (17). Many of these antigenemic persons have few to no symptoms of meningitis. In Uganda, one study demonstrated 5.8% prevalence of antigenemia among all asymptomatic individuals with CD4 counts of ≤100 cells/μl. Another study from Uganda reported that 38% of antigenemic persons were believed to be asymptomatic (66). A study from Cambodia reported that 10.8% of persons with CD4 counts of <100 cells/μl and no symptoms of neuromeningitis had a positive CrAg (49).

Antigenemia has also been shown to precede clinical disease and independently predict poor outcomes. In one study from Uganda, CrAg positivity preceded clinical symptoms by a median of 22 days (range, 5 to 234), with 11% being positive for greater than 100 days (21). Another study found that after controlling for numerous confounders including CD4 cell count, HIV viral load, anemia, presence of active tuberculosis, and body mass index, asymptomatic cryptococcal antigenemia was associated with a higher risk of death (RR 6.6, 95% CI 1.9 to 23.6) and had a population-attributable risk similar to that of active tuberculosis (40).

Identifying antigenemic persons with few or no symptoms would allow for early treatment using an oral agent, such as fluconazole, which as mentioned previously is widely available and inexpensive. The study from Cambodia treated 10 persons with asymptomatic antigenemia with fluconazole 200 mg/day for 12 weeks (49). By 12 weeks after initiation of therapy, none of these had developed cryptococcal meningitis. One of the benefits of preemptive therapy may be the prevention of immune reconstitution inflammatory syndrome, which may contribute substantially to early mortality among persons initiating ARV medication.

Further study of this "screen and treat" strategy is urgently needed. Future trials should determine the benefit of screening for and preemptively treating cryptococcal antigenemia, the dose and type of antifungal agent that produces the best outcomes and tolerability, and the optimum follow-up for such patients.

PUBLIC HEALTH CHALLENGES RELATED TO NON-HIV-ASSOCIATED *CRYPTOCOCCUS* INFECTION

Immunocompromised Persons

Cryptococcal disease is a well-known complication of patients on immunosuppressive medications such as corticosteroids. Cryptococcosis is also a known risk among persons undergoing solid organ transplantation. Furthermore, reports of cryptococcal infection are increasing among persons receiving other forms of immunosuppression, such as immunomodulatory medications (25, 51). In the United States, one study among all non-HIV-infected persons in the Atlanta, Georgia, area estimated an incidence of 0.4 to 5/million population/year (50). Among solid organ transplant recipients, the CDC's TRANSNET surveillance system estimated the cumulative incidence to be approximately 0.2% at 1 year posttransplant in the United States (54a). Future surveillance for cryptococcosis among non-HIV-infected persons should focus on high-risk groups such as transplant recipients or persons undergoing therapy with immunomodulatory medications.

C. gattii

From a public health standpoint, one of the more challenging aspects of the epidemiology of *Cryptococcus* is the emergence of the primary pathogen *C. gattii* in North America. Efforts are under way to understand the scope and magnitude of the problem and to raise awareness of the pathogen in the clinical setting in areas where it may not have been previously seen. Since its initial identification on Vancouver Island in 1999 (44), estimates of incidence have been calculated for Vancouver Island and mainland British Columbia in Canada, although no broader incidence estimates or trends for the spread of this organism have been determined. Nonetheless, cases of *C. gattii* are now occurring in the Pacific Northwest of the United States, with cases being reported from the states of Washington, Oregon, and Idaho in both humans and animals (6, 44). Case identification may be particularly challenging, because while cases in immunocompetent individuals are more easily recognized, cases of *C. gattii* in immunosuppressed patients may be mistaken for *Cryptococcus neoformans*, a pathogen not uncommon in this patient group.

In order to fully understand the epidemiology of this emerging infection in the United States, both human and animal public health agencies at the federal, state, and local levels are working together to form a surveillance network. *C. gattii* has already been made a reportable disease in Washington state. Coordinated surveillance for this infection will be essential to monitor and predict the epidemiology of this emergent disease.

The findings and conclusions in this presentation/report are those of the author(s) and do not necessarily represent the official position of the Centers for Disease Control and Prevention.

REFERENCES

1. **Aller, A. I., E. Martin-Mazuelos, F. Lozano, J. Gomez-Mateos, L. Steele-Moore, W. J. Holloway, M. J. Gutierrez, F. J. Recio, and A. Espinel-Ingroff.** 2000. Correlation of fluconazole MICs with clinical outcome in cryptococcal infection. *Antimicrob. Agents Chemother.* **44:**1544–1548.
2. **Amornkul, P. N., D. J. Hu, S. Tansuphasawadikul, S. Lee, B. Eampokalap, S. Likanonsakul, R. Nelson, N. L. Young, R. A. Hajjeh, K. Limpakarnjanarat, and**

T. D. Mastro. 2003. Human immunodeficiency virus type 1 subtype and other factors associated with extrapulmonary cryptococcosis among patients in Thailand with AIDS. *AIDS Res. Hum. Retroviruses* **19:**85–90.

3. **Bekondi, C., C. Bernede, N. Passone, P. Minssart, C. Kamalo, D. Mbolidi, and Y. Germani.** 2006. Primary and opportunistic pathogens associated with meningitis in adults in Bangui, Central African Republic, in relation to human immunodeficiency virus serostatus. *Int. J. Infect. Dis.* **10:**387–395.
4. **Bogaerts, J., D. Rouvroy, H. Taelman, A. Kagame, M. A. Aziz, D. Swinne, and J. Verhaegen.** 1999. AIDS-associated cryptococcal meningitis in Rwanda (1983–1992): epidemiologic and diagnostic features. *J. Infect.* **39:**32–37.
5. **Brandt, M. E., M. A. Pfaller, R. A. Hajjeh, R. J. Hamill, P. G. Pappas, A. L. Reingold, D. Rimland, D. W. Warnock, and Cryptococcal Disease Active Surveillance Group.** 2001. Trends in antifungal drug susceptibility of *Cryptococcus neoformans* isolates in the United States: 1992 to 1994 and 1996 to 1998. *Antimicrob. Agents Chemother.* **45:**3065–3069.
6. **Byrnes, E. J., III, R. Bildfell, S. A. Frank, T. G. Mitchell, K. Marr, and J. Heitman.** 2009. Molecular evidence that the Vancouver Island *Cryptococcus gattii* outbreak has expanded into the United States Pacific Northwest. *J. Infect. Dis.* **199:**1080–1086.
7. **Chariyalertsak, S., T. Sirisanthana, O. Saengwonloey, and K. E. Nelson.** 2001. Clinical presentation and risk behaviors of patients with acquired immunodeficiency syndrome in Thailand, 1994–1998: regional variation and temporal trends. *Clin. Infect. Dis.* **32:**955–962.
8. **Chariyalertsak, S., K. Supparatpinyo, T. Sirisanthana, and K. E. Nelson.** 2002. A controlled trial of itraconazole as primary prophylaxis for systemic fungal infections in patients with advanced human immunodeficiency virus infection in Thailand. *Clin. Infect. Dis.* **34:**277–284.
9. **Chen, S., T. Sorrell, G. Nimmo, B. Speed, B. Currie, D. Ellis, D. Marriott, T. Pfeiffer, D. Parr, and K. Byth.** 2000. Epidemiology and host- and variety-dependent characteristics of infection due to *Cryptococcus neoformans* in Australia and New Zealand. Australasian Cryptococcal Study Group. *Clin. Infect. Dis.* **31:**499–508.
10. **Chetchotisakd, P., S. Sungkanuparph, B. Thinkhamrop, P. Mootsikapun, and P. Boonyaprawit.** 2004. A multicentre, randomized, double-blind, placebo-controlled trial of primary cryptococcal meningitis prophylaxis in HIV-infected patients with severe immune deficiency. *HIV Med.* **5:**140–143.
11. **Corbett, E. L., G. J. Churchyard, S. Charalambos, B. Samb, V. Moloi, T. C. Clayton, A. D. Grant, J. Murray, R. J. Hayes, and K. M. De Cock.** 2002. Morbidity and mortality in South African gold miners: impact of untreated disease due to human immunodeficiency virus. *Clin. Infect. Dis.* **34:**1251–1258.
12. **Crump, J. A., S. P. Luby, and E. D. Mintz.** 2004. The global burden of typhoid fever. *Bull. WHO* **82:**346–353.
13. **Crump, J. A., F. G. Youssef, S. P. Luby, M. O. Wasfy, J. M. Rangel, M. Taalat, S. A. Oun, and F. J. Mahoney.** 2003. Estimating the incidence of typhoid fever and other febrile illnesses in developing countries. *Emerg. Infect. Dis.* **9:**539–544.
14. **Dannaoui, E., M. Abdul, M. Arpin, A. Michel-Nguyen, M. A. Piens, A. Favel, O. Lortholary, F. Dromer, and F. C. S. Group.** 2006. Results obtained with various antifungal susceptibility testing methods do not predict early clinical outcome in patients with cryptococcosis. *Antimicrob. Agents Chemother.* **50:**2464–2470.
15. **d'Arminio Monforte, A., P. Cinque, A. Mocroft, F. D. Goebel, F. Antunes, C. Katlama, U. S. Justesen, S. Vella, O. Kirk, J. Lundgren, and EuroSIDA Study Group.** 2004. Changing incidence of central nervous system diseases in the EuroSIDA cohort. *Ann. Neurol.* **55:**320–328.
16. **Datta, K., N. Jain, S. Sethi, A. Rattan, A. Casadevall, and U. Banerjee.** 2003. Fluconazole and itraconazole susceptibility of clinical isolates of *Cryptococcus neoformans* at a tertiary care centre in India: a need for care. *J. Antimicrob. Chemother.* **52:**683–686.
17. **Desmet, P., K. D. Kayembe, and C. De Vroey.** 1989. The value of cryptococcal serum antigen screening among HIV-positive/AIDS patients in Kinshasa, Zaire. *AIDS* **3:**77–78.
18. **Dore, G. J., Y. Li, A. McDonald, H. Ree, J. M. Kaldor, and National HIV Surveillance Committee.** 2002. Impact of highly active antiretroviral therapy on individual AIDS-defining illness incidence and survival in Australia. *J. Acquir. Immune Defic. Syndr.* **29:**388–395. (Erratum, **30:**368.)
19. **Dromer, F., S. Mathoulin-Pelissier, A. Fontanet, O. Ronin, B. Dupont, O. Lortholary, and French Cryptococcosis Study Group.** 2004. Epidemiology of HIV-associated cryptococcosis in France (1985–2001): comparison of the pre- and post-HAART eras. *AIDS* **18:**555–562.
20. **Etard, J. F., I. Ndiaye, M. Thierry-Mieg, N. F. Gueye, P. M. Gueye, I. Laniece, A. B. Dieng, A. Diouf, C. Laurent, S. Mboup, P. S. Sow, and E. Delaporte.** 2006. Mortality and causes of death in adults receiving highly active antiretroviral therapy in Senegal: a 7-year cohort study. *AIDS* **20:**1181–1189.
21. **French, N., K. Gray, C. Watera, J. Nakiyingi, E. Lugada, M. Moore, D. Lalloo, J. A. Whitworth, and C. F. Gilks.** 2002. Cryptococcal infection in a cohort of HIV-1-infected Ugandan adults. *AIDS* **16:**1031–1038.
22. **Gordon, S. B., A. L. Walsh, M. Chaponda, M. A. Gordon, D. Soko, M. Mbwvinji, M. E. Molyneux, and R. C. Read.** 2000. Bacterial meningitis in Malawian adults: pneumococcal disease is common, severe, and seasonal. *Clin. Infect. Dis.* **31:**53–57.
23. **Group for Enteric, Respiratory and Meningeal Disease Surveillance in South Africa.** 2007. GERMS-SA Annual Report. http://www.nicd.ac.za/units/germs/germs.htm. [Online.]
24. **Gumbo, T., G. Kadzirange, J. Mielke, I. T. Gangaidzo, and J. G. Hakim.** 2002. *Cryptococcus neoformans* meningoencephalitis in African children with acquired immunodeficiency syndrome. *Pediatr. Infect. Dis. J.* **21:**54–56.
25. **Hage, C. A., K. L. Wood, H. T. Winer-Muram, S. J. Wilson, G. Sarosi, and K. S. Knox.** 2003. Pulmonary cryptococcosis after initiation of anti-tumor necrosis factor-alpha therapy. *Chest* **124:**2395–2397.
26. **Hajjeh, R. A., L. A. Conn, D. S. Stephens, W. Baughman, R. Hamill, E. Graviss, P. G. Pappas, C. Thomas, A. Reingold, G. Rothrock, L. C. Hutwagner, A. Schuchat, M. E. Brandt, and R. W. Pinner.** 1999. Cryptococcosis: population-based multistate active surveillance and risk factors in human immunodeficiency virus-infected persons. Cryptococcal Active Surveillance Group. *J. Infect. Dis.* **179:**449–454. (See comment.)
27. **Helbok, R., S. Pongpakdee, S. Yenjun, W. Dent, R. Beer, P. Lackner, P. Bunyaratvej, B. Prasert, A. Vejjajiva, and E. Schmutzhard.** 2006. Chronic meningitis in Thailand. Clinical characteristics, laboratory data and outcome in patients with specific reference to tuberculosis and cryptococcosis. *Neuroepidemiology* **26:**37–44.
28. **Heyderman, R. S., I. T. Gangaidzo, J. G. Hakim, J. Mielke, A. Taziwa, P. Musvaire, V. J. Robertson, and P. R. Mason.** 1998. Cryptococcal meningitis in human immunodeficiency virus-infected patients in Harare, Zimbabwe. *Clin. Infect. Dis.* **26:**284–289.
29. **Hovette, P., T. O. Soko, G. Raphenon, P. Camara, P. R. Burgel, and O. Garraud.** 1999. Cryptococcal meningitis

in AIDS patients: an emerging opportunistic infection in Senegal. *Trans. R. Soc. Trop. Med. Hyg.* **93:**368.

30. **Inverarity, D., Q. Bradshaw, P. Wright, and A. Grant.** 2002. The spectrum of HIV-related disease in rural Central Thailand. *Southeast Asian J. Trop. Med. Public Health* **33:**822–831.
31. **Kaplan, J. E., D. Hanson, M. S. Dworkin, T. Frederick, J. Bertolli, M. L. Lindegren, S. Holmberg, and J. L. Jones.** 2000. Epidemiology of human immunodeficiency virus-associated opportunistic infections in the United States in the era of highly active antiretroviral therapy. *Clin. Infect. Dis.* **30**(Suppl 1)**:**S5–S14.
32. **Khanna, N., A. Chandramuki, A. Desai, V. Ravi, V. Santosh, S. K. Shankar, and P. Satishchandra.** 2000. Cryptococcosis in the immunocompromised host with special reference to AIDS. *Indian J. Chest Dis. Allied Sci.* **42:**311–315.
33. **Kim, J. M., G. J. Cho, S. K. Hong, K. H. Chang, J. S. Chung, Y. H. Choi, Y. G. Song, A. Huh, J. S. Yeom, K. S. Lee, and J. Y. Choi.** 2003. Epidemiology and clinical features of HIV infection/AIDS in Korea. *Yonsei Med. J.* **44:**363–370.
34. **Klotz, S. A., H. C. Nguyen, T. Van Pham, L. T. Nguyen, D. T. Ngo, and S. N. Vu.** 2007. Clinical features of HIV/AIDS patients presenting to an inner city clinic in Ho Chi Minh City, Vietnam. *Int. J. STD AIDS* **18:**482–485.
35. **Kong, B. N., J. I. Harwell, P. Suos, L. Lynen, S. Mohiuddin, S. Reinert, and D. Pugatch.** 2007. Opportunistic infections and HIV clinical disease stage among patients presenting for care in Phnom Penh, Cambodia. *Southeast Asian J. Trop. Med. Public Health* **38:**62–68.
36. **Kumarasamy, N., S. Solomon, T. P. Flanigan, R. Hemalatha, S. P. Thyagarajan, and K. H. Mayer.** 2003. Natural history of human immunodeficiency virus disease in southern India. *Clin. Infect. Dis.* **36:**79–85.
37. **Lakshmi, V., T. Sudha, V. D. Teja, and P. Umabala.** 2007. Prevalence of central nervous system cryptococcosis in human immunodeficiency virus reactive hospitalized patients. *Indian J. Med. Microbiol.* **25:**146–149.
38. **Lara-Peredo, O., L. E. Cuevas, N. French, J. W. Bailey, and D. H. Smith.** 2000. Cryptococcal infection in an HIV-positive Ugandan population. *J. Infect.* **41:**195.
39. **Lawn, S. D., L. Myer, C. Orrell, L. G. Bekker, and R. Wood.** 2005. Early mortality among adults accessing a community-based antiretroviral service in South Africa: implications for programme design. *AIDS* **19:**2141–2148.
40. **Liechty, C. A., P. Solberg, W. Were, J. P. Ekwaru, R. L. Ransom, P. J. Weidle, R. Downing, A. Coutinho, and J. Mermin.** 2007. Asymptomatic serum cryptococcal antigenemia and early mortality during antiretroviral therapy in rural Uganda. *Trop. Med. Int. Health* **12:**929–935.
41. **Longley, N., C. Muzoora, K. Taseera, J. Mwesigye, J. Rwebembera, A. Chakera, E. Wall, I. Andia, S. Jaffar, and T. S. Harrison.** 2008. Dose response effect of high-dose fluconazole for HIV-associated cryptococcal meningitis in Southwestern Uganda. *Clin. Infect, Dis.* **47:**1556–1561.
42. **Lortholary, O., G. Poizat, V. Zeller, S. Neuville, A. Boibieux, M. Alvarez, P. Dellamonica, F. Botterel, F. Dromer, and G. Chene.** 2006. Long-term outcome of AIDS-associated cryptococcosis in the era of combination antiretroviral therapy. *AIDS* **20:**2183–2191.
43. **Louie, J. K., N. H. Chi, L. T. Thao, V. M. Quang, J. Campbell, N. V. Chau, G. W. Rutherford, J. J. Farrar, and C. M. Parry.** 2004. Opportunistic infections in hospitalized HIV-infected adults in Ho Chi Minh City, Vietnam: a cross-sectional study. *Int. J. STD AIDS* **15:**758–761.
44. **MacDougall, L., S. E. Kidd, E. Galanis, S. Mak, M. J. Leslie, P. R. Cieslak, J. W. Kronstad, M. G. Morshed, and K. H. Bartlett.** 2007. Spread of *Cryptococcus gattii* in British Columbia, Canada, and detection in the Pacific Northwest, USA. *Emerg. Infect. Dis.* **13:**42–50.
45. **Mayanja-Kizza, H., K. Oishi, S. Mitarai, H. Yamashita, K. Nalongo, K. Watanabe, T. Izumi, J. Ococi, K. Augustine, R. Mugerwa, T. Nagatake, and K. Matsumoto.** 1998. Combination therapy with fluconazole and flucytosine for cryptococcal meningitis in Ugandan patients with AIDS. *Clin. Infect. Dis.* **26:**1362–1366.
46. **Mbanya, D. N., R. Zebaze, E. M. Minkoulou, F. Binam, S. Koulla, and A. Obounou.** 2002. Clinical and epidemiologic trends in HIV/AIDS patients in a hospital setting of Yaounde, Cameroon: a 6-year perspective. *Int. J. Infect. Dis.* **6:**134–138.
47. **McCarthy, K. M., J. Morgan, K. A. Wannemuehler, S. A. Mirza, S. M. Gould, N. Mhlongo, P. Moeng, B. R. Maloba, H. H. Crewe-Brown, M. E. Brandt, and R. A. Hajjeh.** 2006. Population-based surveillance for cryptococcosis in an antiretroviral-naive South African province with a high HIV seroprevalence. *AIDS* **20:**2199–2206.
48. **McKinsey, D. S., L. J. Wheat, G. A. Cloud, M. Pierce, J. R. Black, D. M. Bamberger, M. Goldman, C. J. Thomas, H. M. Gutsch, B. Moskovitz, W. E. Dismukes, and C. A. Kauffman.** 1999. Itraconazole prophylaxis for fungal infections in patients with advanced human immunodeficiency virus infection: randomized, placebo-controlled, double-blind study. National Institute of Allergy and Infectious Diseases Mycoses Study Group. *Clin. Infect. Dis.* **28:**1049–1056.
49. **Micol, R., O. Lortholary, B. Sar, D. Laureillard, C. Ngeth, J. P. Dousset, H. Chanroeun, L. Ferradini, P. J. Guerin, F. Dromer, and A. Fontanet.** 2007. Prevalence, determinants of positivity, and clinical utility of cryptococcal antigenemia in Cambodian HIV-infected patients. *J. Acquir. Immune Defic. Syndr.* **45:**555–559.
50. **Mirza, S. A., M. Phelan, D. Rimland, E. Graviss, R. Hamill, M. E. Brandt, T. Gardner, M. Sattah, G. P. de Leon, W. Baughman, and R. A. Hajjeh.** 2003. The changing epidemiology of cryptococcosis: an update from population-based active surveillance in 2 large metropolitan areas, 1992–2000. *Clin. Infect. Dis.* **36:**789–794.
51. **Muñoz, P., M. Giannella, M. Valerio, T. Soria, F. Díaz, J. L. Longo, and E. Bouza.** 2007. Cryptococcal meningitis in a patient treated with infliximab. *Diagn. Microbiol. Infect. Dis.* **57:**443–446.
52. **Mwaba, P., J. Mwansa, C. Chintu, J. Pobee, M. Scarborough, S. Portsmouth, and A. Zumla.** 2001. Clinical presentation, natural history, and cumulative death rates of 230 adults with primary cryptococcal meningitis in Zambian AIDS patients treated under local conditions. *Postgrad. Med. J.* **77:**769–773.
53. **Oh, M. D., S. W. Park, H. B. Kim, U. S. Kim, N. J. Kim, H. J. Choi, D. H. Shin, J. S. Lee, and K. Choe.** 1999. Spectrum of opportunistic infections and malignancies in patients with human immunodeficiency virus infection in South Korea. *Clin. Infect. Dis.* **29:**1524–1528.
54. **Okongo, M., D. Morgan, B. Mayanja, A. Ross, and J. Whitworth.** 1998. Causes of death in a rural, population-based human immunodeficiency virus type 1 (HIV-1) natural history cohort in Uganda. *Int. J. Epidemiol.* **27:**698–702.

54a. **Pappas, P. G., B. D. Alexander, D. R. Andes, S. Hadley, C. A. Kauffman, A. Freifeld, E. J. Anaissie, L. M. Brumble, L. Herwaldt, J. Ito, D. P. Kontoyiannis, G. M. Lyon, K. A. Marr, V. A. Morrison, B. J. Park, T. F. Patterson, T. M. Perl, R. A. Oster, M. G. Schuster, R. Walker, T. J. Walsh, K. A. Wannemuehler, and T. M. Chiller.** 2010. Invasive fungal infections among organ transplant recipients: results of the Transplant-Associated Infection Surveillance Network (TRANSNET). *Clin. Infect. Dis.* **50:**1101–1111.

55. **Park, B. J., K. A. Wannemuehler, B. J. Marston, N. Govender, P. G. Pappas, and T. M. Chiller.** 2009. Estimation of the current global burden of cryptococcal

meningitis among persons living with HIV/AIDS. *AIDS* **23:**525–530.

56. **Pfaller, M. A., D. J. Diekema, D. L. Gibbs, V. A. Newell, H. Bijie, D. Dzierzanowska, N. N. Klimko, V. Letscher-Bru, M. Lisalova, K. Muehlethaler, C. Rennison, M. Zaidi, and the Global Antifungal Surveillance Group.** 2009. Results from the ARTEMIS DISK global antifungal surveillance study, 1997–2007: 10.5-year analysis of susceptibilities of noncandidal yeast species to fluconazole and voriconazole determined by CLSI standardized disk diffusion testing. *J. Clin. Microbiol.* **47:**117–123.
57. **Pfaller, M. A., S. A. Messer, L. Boyken, C. Rice, S. Tendolkar, R. J. Hollis, G. V. Doern, and D. J. Diekema.** 2005. Global trends in the antifungal susceptibility of *Cryptococcus neoformans* (1990 to 2004). *J. Clin. Microbiol.* **43:**2163–2167.
58. **Powderly, W. G., D. Finkelstein, J. Feinberg, P. Frame, W. He, C. van der Horst, S. L. Koletar, M. E. Eyster, J. Carey, and H. Waskin.** 1995. A randomized trial comparing fluconazole with clotrimazole troches for the prevention of fungal infections in patients with advanced human immunodeficiency virus infection. NIAID AIDS Clinical Trials Group. *N. Engl. J. Med.* **332:**700–705.
59. **Sacktor, N., R. H. Lyles, R. Skolasky, C. Kleeberger, O. A. Selnes, E. N. Miller, J. T. Becker, B. Cohen, J. C. McArthur, and Multicenter AIDS Cohort Study.** 2001. HIV-associated neurologic disease incidence changes: Multicenter AIDS Cohort Study, 1990–1998. *Neurology* **56:**257–260.
60. **San-Andres, F. J., R. Rubio, J. Castilla, F. Pulido, G. Palao, I. de Pedro, J. R. Costa, and A. del Palacio.** 2003. Incidence of acquired immunodeficiency syndrome-associated opportunistic diseases and the effect of treatment on a cohort of 1115 patients infected with human immunodeficiency virus, 1989–1997. *Clin. Infect. Dis.* **36:**1177–1185.
61. **Sar, B., D. Monchy, M. Vann, C. Keo, J. L. Sarthou, and Y. Buisson.** 2004. Increasing in vitro resistance to fluconazole in *Cryptococcus neoformans* Cambodian isolates: April 2000 to March 2002. *J. Antimicrob. Chemother.* **54:**563–565.
62. **Schutte, C. M., C. H. Van der Meyden, and D. S. Magazi.** 2000. The impact of HIV on meningitis as seen at a South African Academic Hospital (1994 to 1998). *Infection* **28:**3–7.
63. **Senya, C., A. Mehta, J. I. Harwell, D. Pugatch, T. Flanigan, and K. H. Mayer.** 2003. Spectrum of opportunistic infections in hospitalized HIV-infected patients in Phnom Penh, Cambodia. *Int. J. STD AIDS* **14:**411–416.
64. **Smith, D. E., J. Bell, M. Johnson, M. Youle, B. Gazzard, S. Tchamouroff, G. Frechette, W. Schlech, S. Miller, D. Spencer, W. Seifert, M. Peeters, K. De Beule, and Itraconazole Prophylaxis Study Group.** 2001. A randomized, double-blind, placebo-controlled study of itraconazole capsules for the prevention of deep fungal infections in immunodeficient patients with HIV infection. *HIV Med.* **2:**78–83.
65. **Sun, H. Y., M. Y. Chen, C. F. Hsiao, S. M. Hsieh, C. C. Hung, and S. C. Chang.** 2006. Endemic fungal infections caused by *Cryptococcus neoformans* and *Penicillium marneffei* in patients infected with human immunodeficiency virus and treated with highly active anti-retroviral therapy. *Clin. Microbiol. Infect.* **12:**381–388.
66. **Tassie, J. M., L. Pepper, C. Fogg, S. Biraro, B. Mayanja, I. Andia, A. Paugam, G. Priotto, and D. Legros.** 2003. Systematic screening of cryptococcal antigenemia in HIV-positive adults in Uganda. *J. Acquir. Immune Defic. Syndr.* **33:**411–412.
67. **UNAIDS.** 2006. 2006 Report on the Global AIDS Epidemic. http://www.unaids.org/en/KnowledgeCentre/Resources/Publications/ [Online.]
68. **Vajpayee, M., S. Kanswal, P. Seth, and N. Wig.** 2003. Spectrum of opportunistic infections and profile of CD4+ counts among AIDS patients in North India. *Infection* **31:**336–340.
69. **van der Horst, C. M., M. S. Saag, G. A. Cloud, R. J. Hamill, J. R. Graybill, J. D. Sobel, P. C. Johnson, C. U. Tuazon, T. Kerkering, B. L. Moskovitz, W. G. Powderly, and W. E. Dismukes.** 1997. Treatment of cryptococcal meningitis associated with the acquired immunodeficiency syndrome. National Institute of Allergy and Infectious Diseases Mycoses Study Group and AIDS Clinical Trials Group. *N. Engl. J. Med.* **337:**15–21.
70. **WHO.** 2002. Revised Global Burden of Disease (GBD) 2002 Estimates. http://www.who.int/healthinfo/global_burden_disease/estimates_regional_2002_revised/en/. [Online.]
71. **Witt, M. D., R. J. Lewis, R. A. Larsen, E. N. Milefchik, M. A. Leal, R. H. Haubrich, J. A. Richie, J. E. J. Edwards, and M. A. Ghannoum.** 1996. Identification of patients with acute AIDS-associated cryptococcal meningitis who can be effectively treated with fluconazole: the role of antifungal susceptibility testing. *Clin. Infect. Dis.* **22:**322–328.

Cryptococcus: From Human Pathogen to Model Yeast
Edited by J. Heitman et al.

44

Clinical Perspectives on *Cryptococcus neoformans* and *Cryptococcus gattii*: Implications for Diagnosis and Management

TANIA C. SORRELL, SHARON C.-A. CHEN, PETER PHILLIPS, AND KIEREN A. MARR

Human cryptococcosis is most commonly due to infection with *Cryptococcus neoformans* and *Cryptococcus gattii*, with *C. neoformans* var. *grubii* (serotype A) accounting for the great majority of infections worldwide (9, 32, 58). Clinical differences between the two species are influenced by differences in ecology, epidemiology, biology, and host association (9, 23, 52, 73, 100).

DISTINGUISHING *C. NEOFORMANS* FROM *C. GATTII* INFECTION

The epidemiology and burden of cryptococcosis are detailed in chapter 43 of this volume. Nevertheless, this chapter is prefaced by a brief discussion of the topic, especially with regard to *C. gattii*, since this is relevant to understanding clinical issues and intervention strategies. Unlike cryptococcosis due to *C. neoformans*, which occurs worldwide, *C. gattii* infection was initially thought to be restricted to subtropical or tropical regions including northern Australia, Papua New Guinea, Central Africa, and Malaysia and then to include the temperate climatic zones of Australia, southern California, and South America (15, 31, 56, 60, 79, 89). Occasional apparently autochthonous cases have been reported in a number of other countries in southern Europe and Southeast Asia (19, 98, 109). However, the recent emergence of *C. gattii* infection on Vancouver Island, Canada, in 1999 and spread to the Pacific Northwest of the United States underscores the ability of this species to exploit new and different environments (21, 51, 70, 101).

Over the past 10 years, host-associated risk factors and species-specific features of cryptococcosis have been delineated, diagnostic tests have been improved, novel imaging techniques have been developed for diagnosis, and management strategies have been refined. These developments are summarized in this chapter and have informed the global therapeutic guidelines for the diagnosis and management of cryptococcosis (85a, 91).

Tania C. Sorrell and Sharon C.-A. Chen, Centre for Infectious Diseases and Microbiology, Westmead Hospital and the University of Sydney, Westmead, NSW 2145, Australia. **Peter Phillips,** Infectious Diseases Unit, St. Paul's Hospital, Vancouver, BC, Canada V6Z 1YC. **Kieren A. Marr,** Johns Hopkins University School of Medicine, Baltimore, MD 21205.

PATIENT RISK GROUPS AND PREDISPOSING FACTORS

Species-dependent differences in the clinical presentation of human cryptococcosis are well defined and are largely but not solely due to differences in host immunological status (15, 100). Although it generally holds true that *C. neoformans* is an opportunistic pathogen of immune-compromised hosts and that *C. gattii* is a primary pathogen of patients without immune compromise, *C. neoformans* causes disease in a significant number of apparently healthy hosts, and cases of cryptococcal meningoencephalitis caused by *C. gattii* occur in patients immunosuppressed by AIDS, organ transplantation, or other causes of immune deficiency (12, 14, 15, 42, 78, 100, 110). The major patient comorbidities and risk factors for cryptococcosis and ranking of their clinical importance are summarized in Table 1. An overview of the relationship between host predisposing factors and causative cryptococcal species is provided below.

HIV Infection

Globally, HIV/AIDS remains the dominant risk factor for cryptococcosis despite widespread use of combined antiretroviral therapy (83; chapter 38, this volume). In this population, disease is overwhelmingly caused by *C. neoformans* var. *grubii*. Approximately 2% of cases of HIV-associated meningoencephalitis have been due to *C. gattii* as determined in population-based studies in Australia and the Gauteng province of South Africa (15, 66, 78). More recently, a relatively high prevalence of *C. gattii* (12.3%) was noted following a retrospective analysis of strains from AIDS patients in southern California (12), and during the Vancouver Island outbreak, patients with *C. gattii* infection were more likely

TABLE 1 Patient risk groups and predisposing risk factors for development of cryptococcosis[a]

Risk group condition	Frequency (importance)[b]
HIV infection	++++
Corticosteroids (≥20 mg prednisone equivalent per day)	++++
Solid organ transplantation	+++
Solid tumor	++
Hematologic malignancy[c,d]	+++
Connective tissue disease[d]	++
Chronic obstructive lung disease[d]	++
Diabetes mellitus	+/–
Sarcoidosis	+
Cirrhosis	?
Pregnancy	+/–

[a] Modified and updated from Casadevall and Perfect (9).

[b] ++++, must consider in differential diagnosis of illness; +++, a complicating infectious disease factor; ++, major risk group with or without severe immunosuppression; +, more than a dozen cases reported.

[c] Includes lymphoma, chronic leukemias.

[d] Steroids add to the risk.

than the general population to be infected with HIV (chapter 23, this volume). It should be noted that only subsets of isolates were cultured and identified to species level in most of these studies, and hence our understanding of the impact of C. *gattii* in HIV infection is not complete, and contemporary prospective studies are warranted.

Organ Transplantation

Organ transplantation remains a major risk factor for cryptococcosis, accounting for 10 to 20% of cases of disseminated infection in HIV-negative patients (for a detailed discussion see chapter 37). Renal transplants have constituted the largest subset, partly due to the high number performed and partly because of the relatively high use of corticosteroids (see below) in this group (9). Cryptococcosis is uncommon in hematopoietic stem cell transplantation; in a large multicenter study in the United States, only 2 of 306 cases were associated with hematopoietic stem cell transplantation, and 54 (18%) with solid organ transplantation (SOT) (82). As with HIV infection, the great majority of cases have been caused by C. *neoformans* var. *grubii* (15, 82), although 18% of cryptococcal infections from SOT recipients in France were due to C. *neoformans* var. *neoformans* (94). Notably, a small number of SOT recipients who presumably had exposure within the Pacific Northwest of the United States were infected with C. *gattii* (110).

Malignancy

While cryptococcosis is generally uncommon in the setting of malignancy, patients with lymphomas and chronic leukemias are at particular risk (see Table 1) (44, 74, 111). In a contemporary review of 31 patients from a single cancer center, most had lymphopenia (61%) and immune dysfunction due to prior steroid (51%) or fludarabine (36%) use (52). As in the older series, hematologic malignancies, especially lymphoma, were associated with the highest risk (reviewed in reference 74). Breast and lung cancer have also been associated with cryptococcosis (9, 15). Where species identification was performed, nearly all cases were caused by C. *neoformans* var. *grubii*. Larger prospective surveys are required to elucidate species-specific differences in this group. It is notable that in the Vancouver Island outbreak, patients with C. *gattii* infection were more likely than the general population to have a history of invasive cancer, though many had received corticosteroids (see below and chapter 23, this volume).

Corticosteroid Therapy

After HIV infection, corticosteroid use is the single biggest risk factor for cryptococcosis in humans (9). Any condition associated with prolonged or high-dose corticosteroid exposure predisposes to cryptococcosis. Other than malignancy and organ transplantation, such conditions include connective tissue disorders (e.g., systemic lupus erythematosus and vasculitides), chronic obstructive pulmonary disease, and sarcoidosis (36, 50, 85). In one study, steroids and other immunosuppressive therapies accounted for almost 31% of cases in HIV-negative immunosuppressed individuals. All were infected with C. *neoformans*, with the exception of four with underlying sarcoidosis (15).

Sarcoidosis

Sarcoidosis has been considered an independent risk factor for cryptococcosis due to the associated defect in cell-mediated immunity, and patients occasionally present with cryptococcosis (9). However, the small risk associated with sarcoidosis may be due mainly to corticosteroid use (4). Early reports linked sarcoidosis with cryptococcal bone disease (4). In most cases cryptococcal isolates were not identified to species level, although it can be assumed that cases from the northeast of the United States, where autochthonous cases of C. *gattii* infection have not been reported, were caused by C. *neoformans* (56). Only 2.5% of HIV-negative, immunosuppressed patients with cryptococcosis had underlying sarcoidosis in a more recent survey; interestingly, four of the five were infected with C. *gattii* (15).

Idiopathic CD4+ Lymphopenia

In the absence of HIV infection, idiopathic CD4+ lymphopenia was proposed as a risk factor after it was observed that CD4+ lymphocyte counts are low in some patients with cryptococcosis (9, 54). In such cases, infections are usually caused by C. *neoformans*, but at least one case of C. *gattii* has been described (96). Concurrent or sequential opportunistic infections are frequent. The numbers of these CD4+-deficient patients are small but may be increasing (114).

Novel Immunosuppressive Therapies

Cases of cryptococcosis, all due to C. *neoformans*, have been reported following treatment with infliximab; the anti-CD11a agent efalizumab; the new anti–tumor necrosis factor-α monoclonal antibody adalimumab; and alemtuzumab, an anti-CD52 lymphocyte ablative agent (41).

Other Potential Risk Diseases

Cryptococcosis in pregnancy is uncommon, with only 25 to 35 cases described in the English literature (summarized in references 9 and 30). Two occurred in patients with a defined immunodeficiency, one with HIV infection and one receiving corticosteroid and azathiaprine therapy (30). Species identification was not performed. In two studies from Australia and Papua New Guinea, 2 of 26 and 1 of 49 immunocompetent patients with C. *gattii* infection were pregnant (15, 74).

Diabetes mellitus has been associated with cryptococcosis (9), although it was not an independent risk factor in a case-control study of AIDS-associated cases in the United States

(35). Results from an Australian population-based study also failed to show a significant association with diabetes per se (15). However, in the same study, *C. gattii* infections were overrepresented in the Aboriginal population, in whom diabetes is common. Study of more patients is required to confirm an association between diabetes and cryptococcosis.

Patients with chronic lung disease also appear to be at risk of cryptococcosis when receiving corticosteroids. Changes in lung function associated with smoking may contribute to this risk, since in the Vancouver Island *C. gattii* outbreak, more patients had underlying lung disease and were current smokers than was the case in the general population (chapter 23).

End-stage liver disease may also be an independent risk factor for cryptococcosis, though this is not yet proven. Expert opinion suggests that cryptococcosis is more difficult to treat in this group and that outcomes are poor (9).

DEMOGRAPHY AND CRYPTOCOCCOSIS: APPARENTLY HEALTHY HOSTS

C. gattii typically affects hosts without immune deficiency and is considered to be a primary pathogen (15, 90, 98, 100). Nevertheless, despite a clear association of *C. neoformans* with immunosuppression, in Oceania the incidence in healthy hosts is higher than that of *C. gattii* except in certain regions such as the Northern Territory of Australia and in Papua New Guinea (15, 31, 60). Many studies have confirmed the propensity of *C. gattii* to infect healthy hosts (15, 31, 42, 59, 100). Furthermore, a minority of cases acquired on Vancouver Island were overtly immunosuppressed (chapter 23, this volume). In contrast, approximately one-half of the small number of cases identified to date in the American states of Washington and Oregon have occurred in individuals with a measurable immunodeficiency (e.g., SOT or receipt of steroids for other reasons) (K. Marr, personal communication).

Gender

Cryptococcosis is reported more frequently in men than in women, independent of the disproportionate representation of males with AIDS in the data sets from Western countries (9, 61). Among immunocompetent Australian patients, male sex was a significant risk factor for cryptococcosis (male:female ratio of 1.9:1 overall), but this was significant only in patients with *C. gattii* infection (male:female ratio, 3.3:1). A similar observation in Papua New Guinea was attributed to increased exposure of adult males to environmental sources of *C. gattii* (92). Gender differences were not apparent in HIV-negative immunosuppressed patients (15). In recent cohort studies of cancer patients and unselected HIV-negative patients (79% had underlying immunocompromise) in the United States, the ratio of males to females was 1.4:1 (52, 82). On Vancouver Island, just over half the cases (55.5%) of *C. gattii* infection affected males (chapter 23).

Ethnicity and Genetic Factors

Australian Aborigines have been consistently overrepresented in reports of cryptococcosis from centers in northern Australia (15, 16, 31, 42). The increased risk of infection due to *C. gattii* in Aboriginal people occurs among those living in rural or semirural areas, consistent with increased environmental exposure (see below) (15). However, the incidence of infection with *C. neoformans* var. *grubii* is also higher in Aboriginal people, suggesting that variables such as genetic susceptibility, socioeconomic status, or other factors may play a role (15). Neither human leukocyte antigen class I nor class II phenotypes were associated with cryptococcosis due to *C. gattii* in Papua New Guinea, though a trend to increased susceptibility was associated with human leukocyte antigen B*5601 (108). In Papua New Guinea, as in Australia, environmental exposure appears to be the dominant risk factor for *C. gattii* infection (92).

Environmental Exposure

The ecological associations of *C. neoformans* and *C. gattii* have been studied extensively (see chapter 18). *C. neoformans* has been isolated from diverse environmental sources including weathered pigeon droppings. In contrast, *C. gattii* infection has been associated with exposure to host trees. The first hint of an environmental link with Australian eucalypts was reported in the early 1990s (27, 29) when Ellis and Pfeiffer noted that the distribution of human disease corresponded to the distribution of two species of redgum, *Eucalyptus camaldulensis* and *Eucalyptus tereticornis*. The concentration of *E. camaldulensis* along water courses and the association of rural-dwelling Australian Aborigines with these trees was thought to explain the high prevalence of *C. gattii* infection in this group (see below). This epidemiological link was subsequently confirmed by clinical and molecular epidemiological studies (15, 97). The association of *C. gattii* infection with residence in or working in a rural/semi-rural domicile (15, 42) has been maintained (S. C. A. Chen, and T. C. Sorrell, personal communication). *C. gattii* has also been isolated from eucalypts in countries with small numbers of human cases including the United States, Brazil, and Italy (27, 75, 77). The association with a rural domicile was not evident in cases from the Pacific Northwest of the United States, suggesting that the epidemiology is different (K. Marr, personal communication).

Cases of *C. gattii* also occur in regions where eucalypts are not endemic, most evidently on Vancouver Island, where coastal Douglas fir and coastal western hemlock trees have been implicated as an environmental niche and in areas such as in Malaysia and the "top end" of the Northern Territory of Australia and South America (15, 47; see chapter 23).

Summary

The major clinical differences between infections due to *C. neoformans* versus *C. gattii* are the respective associations with immunosuppression and healthy hosts. Despite the emergence of *C. gattii* in new climatic zones, its propensity to cause disease in healthy hosts remains strong, as does the association between *C. neoformans* and immunocompromise. However, *C. neoformans* also affects healthy hosts, causing 56% (52 of 93) of cases in one Australian study (15).

CLINICAL MANIFESTATIONS OF DISEASE

Incubation Period

Studies of travelers to Vancouver Island revealed a median time to presentation with cryptococcosis of 6 to 7 months (range 2 to 11 months) (69). This is much shorter than the estimated 110 months for *C. neoformans* var. *grubii* (33). The longer latent period may reflect a higher incidence of reactivation of dormant infection associated with immunosuppression in the case of *C. neoformans* (chapter 31). In a cohort of SOT recipients, a proportion of cases were seropositive prior to transplantation, suggesting that cryptococcal disease resulted from reactivation following the use of immunosuppressive therapy (94).

Sites of Disease

Clinically apparent disease due to *C. gattii* and *C. neoformans* most often involves the lung or central nervous system (CNS). In the largest comparative case series of 312 episodes in Australia and New Zealand, CNS involvement occurred in 64% (30/47) of *C. gattii* infections and 65% (171/265) of those due to *C. neoformans* (15). In other series, figures as high as 97% or as low as 16% have been reported (Table 2). Species-dependent differences in the clinical and imaging manifestations of cryptococcal disease (see below and Table 2) are mainly, but not solely (see below), determined by the immune status of the host (15, 74, 100).

C. gattii presents more often with lung involvement (65 to 84% of cases versus 4 to 44%; $P < 0.001$) (3, 15, 100) (Table 2) as well as with concurrent lung and brain infection (32% versus 10%, $P < 0.001$) (15). Compared with *C. neoformans*, it is significantly more likely to cause large mass lesions (cryptococcomas) of the lung ($P < 0.001$) and/or brain ($P = 0.001$) (Table 2). Examples of such mass lesions, which mimic neoplastic disease, are shown in Fig. 1 and Fig. 2. To date, these have been described only with *C. gattii*. Single cerebral cryptococcomas have been mistaken for pyogenic abscesses or brain tumors and biopsies (73) have been undertaken without appropriate microbiologic studies being performed.

After lung or brain, *C. neoformans* is most often cultured from blood, urine, and/or skin lesions (Table 2). Various other specimens have occasionally yielded *C. neoformans* including bone marrow, joint fluid, ascitic fluid, feces, larynx, lymph node (15), prostate (7), and bone (4). At autopsy, disease has been described in kidneys, liver, spleen, adrenals, heart, ovary, and muscle (62). Sites outside the

TABLE 2 Comparison of *C. neoformans* and *C. gattii* infection: sites of disease, symptoms, signs, spinal fluid, and head CT findings

Feature	*C. neoformans*				*C. gattii*	
	HIV positive	References	HIV negative	References	All subjects	References
Site of disease[a]						
CNS[b]	65–84%	20, 35, 53, 115	38–72%	23, 24, 48, 62, 82	16–≥97%[c]	3, 16, 78, 99
Brain			9%	16	33%[d]	16
Lung	4–18%	20, 35, 53, 115	7–44%	23, 48, 62, 82	65–84%[e]	3, 16, 99
Blood	12–74%	53, 115	9–46%	23, 24, 48, 62	0–60%[f]	16, 78, 99
Urine	0–57%	53, 115	3–44%	23, 24, 28, 62	0%	16, 99
Skin	0–7%	53, 115	3–15%	23, 48, 62	1–6%	3, 16
Features (CNS infection)						
Fever	56–95%	20, 34, 111, 115	67–73%	23, 62, 82	54–61%	3, 73, 78, 99
Headache	67–100%	20, 34, 53, 111	50–85%	23, 62, 82	59–88%	3, 73, 78, 99
Altered mentation	10–23%	20, 34, 115	42–52%	23, 62, 82	31%	73
Seizure	4–9%	20, 34, 115	5%	82	38%[e]	73
Symptoms (lung infection)						
Fever	66%	8	29–63%	48, 82	31%	3
Cough	58–66%	8,13	17–61%	48, 82	68%	3
Dyspnea	33–50%	8,13	27–48%	48, 82	66%	3
Chest pain	25%	13	19–44%	48, 82	50%	3
CSF in CNS cases						
WBC median (cells/mm^3)	4–11	34	73–89	24, 82	NA	
India ink positive	66–88%	20, 53, 111, 115	51–52%	24, 82	61–≥96%	73, 78
CSF CRAG positive	91–100%	20, 53, 13	97–100%	24, 82	≥87%	73, 78
OP (mm H_2O)	≥250 in 60%	34	230–270[g]	24, 82	NA	
Serum CRAG positive	94–100%	20, 53, 115	86–87%	24, 82	NA	
CT head in CNS cases						
Mass or abscess	3–9%	20, 115	NA		33–58%[e]	16, 73, 99
Hydrocephalus	0%	20, 34, 115	NA		17–36%[e]	16, 73, 99

[a] Multiple sites may be involved.
[b] Abbreviations: CNS, central nervous system; CRAG, cryptococcal antigen titer; CSF, cerebrospinal fluid; NA, not available; OP, opening pressure; WBC, white blood cell.
[c] %CNS cases not specified for reference 78 but was at least 97%.
[d] Comparison of frequency of brain cryptococcomas in *C. neoformans* versus *C. gattii* infection; $P = 0.001$.
[e] $P < 0.001$.
[f] Only a subset of patients had blood cultures performed.
[g] Median value.

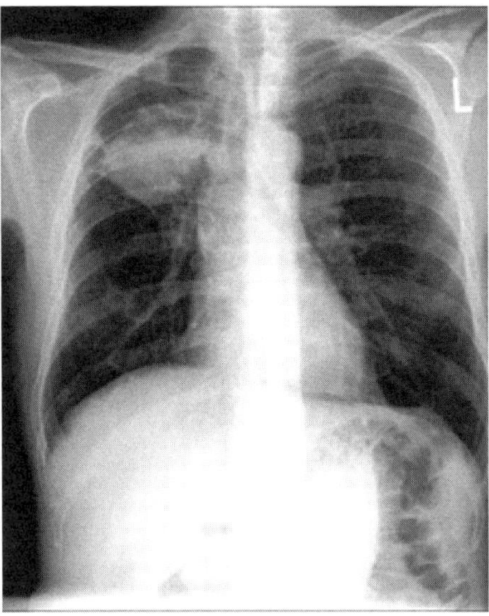

FIGURE 1 Chest radiograph of an immunocompetent 47-year-old male showing a right-upper lung zone mass lesion from which *C. gattii* was isolated on percutaneous needle biopsy.

lung or CNS are more often culture-positive in immunocompromised hosts and hence only occasionally involved in *C. gattii* disease (15, 100). When *C. gattii* affects patients with AIDS, the presentation resembles that due to *C. neoformans* (78).

Clinical Findings

Fever, pulmonary, or neurological symptoms per se are not useful in distinguishing between the two species of *Cryptococcus* (Table 2). The clinical features of meningitis and lung disease associated with AIDS are often mild, at least in developed countries, presumably reflecting the diminished contribution of the host response to symptomatology in this group. This assertion is supported by their predisposition to develop an immune response inflammatory syndrome with improvement in immunological function. Seizures, however, are more commonly encountered in *C. gattii*, as opposed to *C. neoformans*, infection (38% versus 4 to 9%, $P < 0.001$) (Table 2). The most frequent symptoms of lung infection are cough, dyspnea, and chest pain. There are multiple reports of asymptomatic lung infection (which may present with a mass lesion[s] on chest X ray) associated with both cryptococcal species (3, 48, 65, 71, 82). Very large lesions have been described only in disease due to *C. gattii*, which in one instance involved the upper lobe of the right lung and presented with Pancoast's syndrome (72).

Neurological Complications

In two retrospective Australian studies, papilledema, focal neurological deficits including cranial nerve palsies, ataxia, and seizures were all more common in *C. gattii* infections, as were neurological sequelae that resulted in significant disability (73, 100) (Table 3). Whether these complications truly represent a species-dependent effect or differences in host immune status and/or delays in presentation is uncertain. Certainly, numerous ocular complications have been described in cryptococcosis caused by *C. neoformans* and *C. gattii*, especially in patients with raised intracranial pressure (ICP) and/or those who present late, as has often been the case in early studies from developing countries. Papilledema is the most common manifestation (6, 59, 67, 73, 80, 87, 93), but blurred vision, visual field defects, extraocular muscle paresis, impaired pupillary and accommodative function, multifocal choroiditis, endophthalmitis, and optic atrophy have all been described. Visual loss is one of the most serious, debilitating sequelae (if not the most serious) of CNS cryptococcosis. It is thought to result primarily from invasion of the optic nerve by cryptococci (18, 55, 80) and/or the effects of raised ICP (see below) (87, 102) or possibly from local arachnoiditis leading to nerve infarction and cryptococcal infiltration of the choroid (2, 34, 64). Earlier reports estimated visual loss to occur in 1.1 to 3.3% of patients (43, 76, 87, 105), but a study from Rwanda documented rates of 9% (49), and particularly high rates (>50%) were noted in *C. gattii* infection in Papua New Guinea prior (93) to the AIDS epidemic. These high rates have not been observed in other areas endemic for *C. gattii*, including Australia (31, 73, 100).

Increased ICP and Hydrocephalus

Elevated ICP is observed in neurocryptococcosis caused by both *C. neoformans* and *C. gattii*. It is a common complication of cryptococcal meningitis in both HIV-positive and HIV-negative patients, being reported in 30 to 75% (34, 91). When it occurs early, raised ICP is a poor prognostic indicator (34, 59, 91, 93, 102, 113). Manifestations include severe headache, papilledema, loss of vision, hearing loss, other neurologic abnormalities, and death, presumably secondary to brain ischemia (43, 59, 87, 91). Rapid diagnosis and treatment is the key to achieving good outcomes. Late presentations of cryptococcal meningitis with symptomatic hydrocephalus, ataxia, deteriorating mental function, dilated cerebral ventricles, and meningitis also occur and generally respond well to relief of pressure by insertion of a cerebrospinal fluid (CSF) shunt (10, 73, 88, 100). Delays in

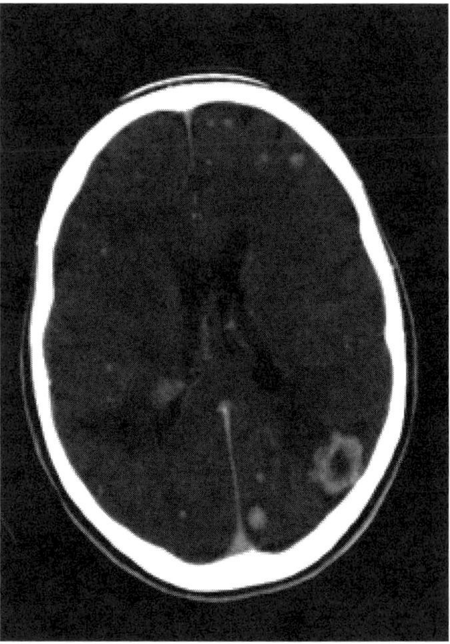

FIGURE 2 Contrast CT head scan of an immunocompetent patient showing several ring-enhancing lesions in the cerebral hemispheres and in the regions of the basal ganglia due to *C. gattii*.

TABLE 3 Clinical features and outcome of neurocryptococcosis in healthy hosts, based on cerebral CT scan[a]

Group	Clinical presentation	CSF findings				Outcome
		Leukocytes	Glucose	CRAG	Culture	
1	Meningitis	Raised	Low	+	+	Good
2	Single brain "abscess/tumor"	Normal/raised	Normal	Low or –	–	Good (with surgical removal)
3	Multiple brain cryptococcomas	Raised	Low	+	+	Poor
4	Hydrocephalus	Raised	Low	+	+	Good (with shunting)

[a] Based on Mitchell et al. (73).

diagnosis and treatment may lead to serious and potentially permanent loss of cognitive function (22, 63). Cryptococcomas due to *C. gattii* have caused or contributed to raised ICP or obstructive hydrocephalus via a mass effect, resulting in progressive neurological deficit (73, 84). Mitchell et al. found that a quarter of the patients with *C. gattii* infection and hydrocephalus had "silent" cryptococcal meningitis, presenting with features of hydrocephalus, dilated ventricles on computed tomography (CT) scan, and an unexpected finding of cryptococcal meningitis on CSF examination (Table 3) (73). The association between *C. gattii* and hydrocephalus was confirmed in a subsequent nationwide study in the same country (15) (Table 2). Factors that predispose patients with cryptococcal meningitis to develop hydrocephalus remain unclear, but most patients with this complication have indolent meningitis (84).

Within-Species Comparison of *C. neoformans* (Serotype A versus D)

Although less pronounced than the differences between *C. neoformans* and *C. gattii*, Dromer et al. observed some differences in disease due to *C. neoformans* serotype A (var. *grubii*) and serotype D (var. *neoformans*) (23, 24). Among HIV-positive individuals in France, serotype A disease was more common in those born in Africa, and was less often associated with hyponatremia, than serotype D disease (51% versus 71%, $P = 0.031$). Serotype D infection was more common in those >60 years old, injecting drug users, those with skin lesions, and those with hyponatremia (23, 24). In HIV-negative individuals serotype D was associated with higher CSF antigen titers and mortality (four of five versus none of five, $P = 0.048$) (24).

IMAGING IN CRYPTOCOCCOSIS

As with clinical manifestations, imaging findings in *C. neoformans* and *C. gattii* infection are generally different because of the strong association between *C. neoformans* and immunocompromise and between *C. gattii* and healthy hosts.

Lung

The chest X ray is often normal in patients with AIDS or resembles that in HIV-negative immunocompromised patients. In this group, alveolar and interstitial pulmonary infiltrates account for approximately 70% of lung lesions, compared with only 17% in *C. gattii* infections in healthy hosts (15). The predominant pulmonary manifestations in immunocompetent patients are circumscribed cryptococcomas, which are also more commonly due to *C. gattii* than *C. neoformans* both in patients with lung infection (OR 12.2, 95% CI 3.3 to 46, $P < 0.001$) and those with combined lung and CNS disease (Table 4) (15).

CNS

Computerized Tomography (CT)

Overall, cerebral involvement (defined as CT scan findings of reduced attenuation, contrast enhancement, or edema), cryptococcomas, hydrocephalus, and focal neurologic findings are more often associated with *C. gattii* infection (Tables 2 and 4) (15, 73, 100). Mass lesions or "abscesses" were observed in cranial CT scans in 33 to 58% of cases of *C. gattii* disease, most commonly in the basal ganglia, thalamus, and cerebellum. However, the association of cerebral cryptococcomas with *C. gattii* infection disappeared when the analysis was confined to immunocompetent hosts (15).

TABLE 4 Comparison of occurrence of lung and brain cryptococcomas due to *C. neoformans* and *C. gattii* in immunocompetent persons

Clinical findings	% of patients with finding: *C. neoformans*	% of patients with finding: *C. gattii*	P value[a]	References
Lung involvement				
Lung cryptococcomas among pulmonary infection	29	83	<0.001	16
Lung cryptococcomas among CNS infection	22	77	0.05	73
CNS involvement				
Brain cryptococcomas among CNS cases	24	36	0.4[b]	16
Brain cryptococcomas or obstructive hydrocephalus	33	69	0.05	73

[a] By chi square analysis with *C. neoformans* as the reference category.
[b] Odds ratio 2.0 (95% confidence interval 0.5–7.3).

In contrast, concurrent pulmonary and cerebral cryptococcomas, and hydrocephalus, were more frequently associated with *C. gattii*, independent of host status (15, 73); obstructive hydrocephalus was present in 17 to 36% of *C. gattii* neurological infections (15, 73, 100). Cerebral CT scans frequently show no features of cryptococcosis in patients with AIDS. In one series, obstructive hydrocephalus or mass lesions were rare despite ICPs of ≥250 mm water in 54% of HIV-positive patients with *C. neoformans* meningitis (34).

MRI

Cryptococcal infection spreads from the basilar cisterns, with the accumulation of cryptococci and gelatinous capsular material along the perivascular (Virchow-Robin) spaces, resulting in dilatation of these spaces. This finding is readily seen on magnetic resonance imaging (MRI), most commonly in the region of the basal ganglia. It is present in almost 50% of cases of HIV-associated cryptococcosis (11) and is a useful clue in the diagnosis in such patients (11, 81, 95). MRI scans are more sensitive than CT scans in detecting small mass lesions and may show basilar meningeal enhancement after injection of gadolinium (11).

New Imaging Techniques

Nuclear magnetic resonance (NMR) spectroscopy is a technology that generates complex data based on chemical composition and metabolite profiles of cells, including human and microbial cells. In clinical practice it can be linked to MRI to identify the spatial location of brain lesions and their metabolic "fingerprint" in a single examination. This form of neurospectroscopy is already in use for noninvasive diagnosis of brain tumors and pyogenic abscesses (99). NMR spectra of cryptococci cultured in vitro are characterized by an abundance of the disaccharide α,α-trehalose, which contributes to their identification in vivo (25, 37, 38). However, NMR spectra from living cells contain hundreds of metabolites that are not visible to the naked eye but that can be differentiated reliably by the use of computerized pattern recognition methods and that when applied will increase diagnostic accuracy. In vivo it has been demonstrated that identification of cerebral cryptococcomas is feasible, though acquisition of spectra from more cases is required to achieve an unambiguous diagnosis (99).

MICROBIOLOGICAL DIAGNOSIS

Almost all cases of cryptococcosis are readily diagnosed by culture and/or a positive serum or CSF cryptococccal antigen (CRAG) titer (57). False-negative CRAG titers may result from a prozone effect (where there is an excess of antibody in the blood that can be overcome by appropriate sample dilution) or occasionally with low titers in *C. gattii* infection; this is due to a lower affinity for the antibodies used in the latex CRAG test, which were developed against *C. neoformans*. In general, standard microbiological tests have similar sensitivities in AIDS and other immunocompromised patients and in healthy hosts. Serological tests are discussed in detail in chapter 41.

Molecular Diagnostics

Molecular tests can distinguish between *C. neoformans* and *C. gattii* but are seldom required for diagnosis of cryptococcosis due to either species. Simple tests such as growth on concanavanine bromothymol blue agar or serotyping are sufficient for speciation in the routine clinical laboratory. In the uncommon event that the isolate does not grow or is seen only in histological sections, molecular methods may be used (chapter 24, chapter 41). These include PCR-based assays to detect and identify the organism directly from clinical specimens. They typically exploit nucleotide polymorphisms in the internal transcribed spacer region of the fungal rDNA complex and require sequencing analysis (46). A novel approach employing hyperbranched rolling circle amplification targeting single nucleotide polymorphisms (SNPs) in the internal transcribed spacer region has been used successfully to distinguish *C. neoformans* from *C. gattii* and between the varieties of *C. neoformans* (45).

ANTIFUNGAL SUSCEPTIBILITIES

Currently, there are no published guidelines with interpretative breakpoints for antifungal susceptibility testing of *C. neoformans* or *C. gattii*. Only *C. neoformans* has been included in the guidelines of the CLSI for testing of yeasts. However, MICs, most commonly determined by broth microdilution, have been reported for both *C. neoformans* and *C. gattii*.

A decrease in susceptibility to azole drugs, especially fluconazole, was noted among isolates of *C. neoformans* var. *grubii* from patients with AIDS, in parallel with widespread use of fluconazole prior to the advent of highly active antiretroviral therapy (HAART). However, following the introduction of HAART in Africa and the use of amphotericin B and 5-flucytosine as induction therapy, virtually all isolates have been susceptible to currently available antifungal agents including fluconazole (the echinocandins have no clinically useful in vitro activity) (1, 5, 15, 28, 86).

Many studies have shown that *C. neoformans* and *C. gattii* have similarly low MICs to antifungal drugs (1, 15, 104, 112), which have not increased in the past decade (1, 28, 86). In contrast, a small number of studies (from Brazil, Malaysia, Spain, and Taiwan) have reported that *C. gattii* is less susceptible to at least some antifungal agents (17, 103, 106, 107). Trilles et al. found significantly higher geometric mean MICs for fluconazole (9.54 mg/liter versus 3.89 mg/liter) and voriconazole (0.15 mg/liter versus 0.06 mg/liter) against *C. gattii* compared with *C. neoformans* as well as for amphotericin B and 5-flucytosine (107). In Taiwan *C. gattii* was less susceptible to 5-flucytosine and amphotericin B (17), and in a contemporary Spanish survey, not only were MICs of the three azoles tested (fluconazole, voriconazole, and posaconazole) significantly higher for *C. gattii*, but the minimal fungicidal concentrations of fluconazole and voriconazole MICs for treatment were higher for *C. gattii* serotype B (106). Increased fluconazole MICs have been reported from the Pacific Northwest of the United States (110) but not from across the border in British Columbia, Canada. The implications of the higher fluconazole MICs for therapeutic dosing and clinical outcomes are yet to be determined.

MANAGEMENT

The overall management of cryptococcosis is discussed in detail in published guidelines, for example, those published by the Infectious Diseases Society of America (91). These now include a global perspective and discussion of special situations such as management of cryptococcomas and *C. gattii* infection (85a). There is evidence that when *C. gattii* affects patients with AIDS, it presents with features similar to those of *C. neoformans* var. *grubii* (78), suggesting that management should be similar. This presumption is supported by data showing that the course and outcomes of cryptococcal meningitis (without cerebral mass lesions or hydrocephalus) are independent of the causative cryptococcal species (73).

Cryptococcosis in AIDS and organ transplantation is discussed in chapters 37 and 38 of this book.

Neurocryptococcosis in Hosts without Classic Immunocompromise

This group of patients is heterogeneous, and there are no randomized controlled trial data regarding therapy. Intracranial infection with *C. gattii* has been associated with more neurological sequelae (see above), a delayed response to therapy, and more frequent neurosurgical intervention than that due to *C. neoformans* (68, 73, 100). However, species-dependent differences in presentation and hence approach to management are largely determined by associations with host immune status and the duration of illness before presentation. Thus, the slow response of *C. gattii* brain infection therapy is primarily due to the presence of cerebral cryptococcomas (100). Late presentations are associated with higher morbidity and mortality independent of the causative species of *Cryptococcus* or host status (15).

Management of CNS cryptococcosis is best guided by the appearance of the cerebral CT scan, rather than cryptococcal species, and CT appearances at presentation have been correlated with outcome (73). In addition, since one of the most critical determinants of the outcome of cryptococcal meningoencephalitis is control of CSF pressure, CSF pressures should always be measured as part of the lumbar puncture procedure, and adequate control of raised ICP achieved by repeated lumbar puncture and/or by early shunting procedures (34, 91).

For patients with meningitis due to *C. neoformans*, induction therapy with amphotericin B and flucytosine for 2 (if favorable prognostic factors) (91) or 4 to 6 weeks is recommended provided that the patient has no neurological complications and the CSF cultures are negative after 2 weeks (85a). Experts treating *C. gattii* meningitis favor a 4- to 6-week course of induction therapy based on clinical experience. Eradication therapy with fluconazole 400 mg/day for 8 weeks and then maintenance therapy for 6 to 12 months is recommended to prevent relapse (85a).

Symptomatic hydrocephalus visible on imaging requires antifungal therapy plus early reduction of ICP by shunting. In the absence of cryptococcomas, the antifungal regimen should be as for meningitis.

Large, surgically accessible cryptococcomas presenting as pyogenic abscesses or tumors with substantial surrounding edema, with or without mass effect, should be treated with combination antifungal therapy and early surgical removal, as the response to antifungal therapy is poor and slow, and residual morbidity is high (73). Multiple mass lesions require prolonged induction and eradication therapy. Adjunctive corticosteroids may be required if there is substantial surrounding edema, especially in the presence of neurological deficits.

The duration of induction and eradication antifungal therapy depends primarily on clinical and mycological responses. Reliance on imaging can be misleading, as intracerebral cryptococcomas persist for prolonged periods (39). Reimaging is indicated in apparent cases of relapse and may reveal new lesions, enlargement of lesions, and/or increased perilesional edema despite effective antifungal therapy. This appears to result from an immune response inflammatory-like syndrome, as has been reported in a case of *C. gattii* neurocryptococcosis (26).

Recombinant interferon gamma has been tried as an additional modality in patients with *C. gattii* infection unresponsive to repeated or prolonged courses of multiple antifungal drugs. Its contribution to subsequent outcomes is uncertain.

In summary, there is evidence that neurocryptococcosis due to *C. gattii* requires more intensive and prolonged therapy than that due to *C. neoformans*, but this is mainly due to differences in underlying host status and hence clinical presentation. An effect on outcome of the reduced susceptibility to fluconazole of some isolates of *C. gattii* has not yet been demonstrated.

Pulmonary Disease

Most healthy patients with isolated pulmonary disease will respond to fluconazole, whereas the immunocompromised groups should be treated as for neurocryptococcosis (91). Large mass lesions in the lung cause substantial general debility or local complications and may require surgical resection resulting in significant in-hospital and subsequent morbidity related to patient rehabilitation. In one retrospective study that examined the need for thoracic surgery in patients with cryptococcosis, 69% of patients with *C. gattii* infection underwent surgical resection of the lesion compared with none with *C. neoformans*. Despite few evidence-based data to guide surgical management of pulmonary cryptococcomas, the development of complications such as respiratory obstruction or uncontrolled hemorrhage is clearly an indication for surgical intervention.

SUMMARY

Clinical differences between *C. neoformans* and *C. gattii* depend mainly on differences in ecology, epidemiology, and host immune status (9, 15, 23, 73, 100). Neurological disease carries the highest morbidity and mortality in all cases. The major determinants of outcome include neurological status, ICP, the presence of cerebral mass lesions, and cryptococcal load at presentation. There is evidence that induction therapy with amphotericin B and flucytosine is superior to other regimens for neurological disease. Longer courses of induction and eradication therapy are used in *C. gattii* infection, based on clinical experience. Early relief of high ICP is a major determinant of improved outcome. Corticosteroids and/or surgical shunting may be required to prevent coning or relieve symptomatic hydrocephalus. Surgical removal of accessible cryptococcomas, usually due to *C. gattii*, may be required. The clinical relevance of higher MICs of fluconazole observed in some small series of *C. gattii* infection is yet to be determined.

REFERENCES

1. **Aller, A. I., R. Claro, C. Castro, C. Serrano, M. F. Colom, E. Martín-Mazuelos, and Grupo de Epidemiología de la Cryptococcosis.** 2007. Susceptibility of *Cryptococcus neoformans* isolates in HIV-infected patients to fluconazole, itraconazole and voriconazole in Spain: 1994–1996 and 1997–2005. *Chemotherapy* **53:**300–305.
2. **Andreola, C., M. P. Ribeiro, C. R. de Carli, A. L. Gouvea, and A. L. Curi.** 2006. Multifocal choroiditis in disseminated *Cryptococcus neoformans* infection. *Am. J. Ophthalmol.* **142:**346–348.
3. **BC Centre for Disease Control.** 2007. BC *Cryptococcus gattii* surveillance summary, 1999–2006. http://www.bccdc.org/topic.php?item=109. [Online.] Accessed 28 April 2009.
4. **Behrmann, R. E., J. R. Masci, and P. Nicholas.** 1990. Cryptococcal skeletal infections: case report and review. *Rev. Infect. Dis.* **12:**181–190.
5. **Bicanic, T., T. Harrison, A. Niepieko, N. Dyakopu, and G. Meintjes.** 2006. Symptomatic relapse of HIV-associated

cryptococcal meningitis after initial fluconazole monotherapy: the role of fluconazole resistance and immune reconstitution. *Clin. Infect. Dis.* **43:**1069–1073.

6. **Blackie, J. D., G. Danta, T. Sorrell, and P. Collignon.** 1985. Ophthalmological complications of cryptococcal meningitis. *Clin. Exp. Neurol.* **21:**263–270.
7. **Bozzette, S. A., R. A. Larsen, J. Chiu, M. A. Leal, J. G. Tilles, D. D. Richman, J. M. Leedom, and J. A. McCutchan.** 1991. Fluconazole treatment of persistent *Cryptococcus neoformans* prostatic infection in AIDS. *Ann. Intern. Med.* **115:**285–286.
8. **Cameron, M. L., J. A. Bartlett, H. A. Gallis, and H. A. Waskin.** 1991. Manifestations of pulmonary cryptococcosis in patients with acquired immunodeficiency syndrome. *Rev. Infect. Dis.* **13:**64–67.
9. **Casadevall, A., and J. R. Perfect.** 1998. *Cryptococcus neoformans*. ASM Press, Washington, DC..
10. **Chan, K. H., K. S. Mann, and C. P. Yue.** 1989. Neurosurgical aspects of cerebral cryptococcosis. *Neurosurgery* **25:**44–48.
11. **Charlier, C., F. Dromer, C. Leveque, L. Chartier, Y.-S. Cordoliani, A. Fontanet, O. Launay, and O. Lortholary.** 2008. Cryptococcal neuroradiological lesions correlate with severity during cryptococcal meningoencephalitis in HIV-positive patients in the HAART era. *PLoS ONE* **3:**e1950.
12. **Chaturvedi, S., M. Dyavaiah, R. A. Larsen, and V. Chaturvedi.** 2005. *Cryptococcus gattii* in AIDS patients, Southern California. *Emerg. Infect. Dis.* **11:**1686–1692.
13. **Chechani, V., and S. L. Kamholz.** 1990. Pulmonary manifestations of disseminated cryptococcosis in patients with AIDS. *Chest* **98:**1060–1066.
14. **Chen J., A. Varma, M. R. Diaz, A, P. Litvintseva, K. K. Wollenberg, and K. J. Kwon-Chung.** 2008. *Cryptococcus neoformans* strains and infection in apparently immunocompetent patients, China. *Emerg. Infect. Dis.* **14:**755–762.
15. **Chen, S., T. Sorrell, G. Nimmo, B. Speed, B. Currie, D. Ellis, D. Marriott, T. Pfeiffer, D. Parr, K. Byth, and the Australasian Cryptococcal Study Group.** 2000. Epidemiology and host- and variety-dependent characteristics of infection due to *Cryptococcus neoformans* in Australia and New Zealand. *Clin. Infect. Dis.* **31:**499–508.
16. **Chen, S. C., B. J. Currie, H. M. Campbell, D. A. Fisher, T. J. Pfeiffer, D. H. Ellis, and T. C. Sorrell.** 1997. *Cryptococcus neoformans* var. *gattii* infection in Northern Australia: existence of environmental source other than known host eucalypts. *Trans. R. Soc. Trop. Med. Hyg.* **91:**547–550.
17. **Chen, Y. C., S. C. Chang, C. C. Shih, C. C. Hung, K. T. Luhd, Y. S. Pan, and W. C. Hsieh.** 2000. Clinical features and in vitro susceptibilities of two varieties of *Cryptococcus neoformans* in Taiwan. *Diagn. Microbiol. Infect. Dis.* **36:**175–183.
18. **Cohen, D. B., and B. J. Glasgow.** 1993. Bilateral optic nerve cryptococcosis in sudden blindness in patients with acquired immune deficiency syndrome. *Ophthalmology* **100:**1689–1694.
19. **Colom, M. F., S. Frases, C. Ferrer, A. Jover, M. Andreu, S. Reus, M. Sanchez, and J. M. Torres-Rodriguez.** 2005. First case of human cryptococcosis due to *Cryptococcus neoformans* var. *gattii* in Spain. *J. Clin. Microbiol.* **43:**3548–3550.
20. **Chuck, S. L., and M. A. Sande.** 1989. Infections with *Cryptococcus neoformans* in the acquired immunodeficiency syndrome. *N. Engl. J. Med.* **321:**794–799.
21. **Datta, K., K. H. Bartlett, and K. A. Marr.** 2009. *Cryptococcus gattii*: emergence in western North America: exploitation of a novel ecological niche. *Interdiscip. Perspect. Infect. Dis.* **2009:**176532, p. 1–8.
22. **Diamond, R. D., and J. E. Bennett.** 1974. Prognostic factors in cryptococcal meningitis. *Ann. Intern. Med.* **80:**176–181.
23. **Dromer, F., S. Mathoulin, B. Dupont, L. Letenneur, O. Ronin, and the French Cryptococcosis Study Group.** 1996. Individual and environmental factors associated with infection due to *Cryptococcus neoformans* serotype D. *Clin. Infect. Dis.* **23:**91–96.
24. **Dromer, F., S. Mathoulin-Pelissier, O. Launay, O. Lortholary, and the French Cryptococcosis Study Group.** 2007. Determinants of disease presentation and outcome during cryptococcosis: the Crypto A/D/ study. *PLoS. Med.* **4:**e21.
25. **Dzendrowskyj, T. E., B. Dolenko, T. C. Sorrell, R. L. Somorjai, R. Malik, C. E. Mountford, and U. Himmelreich.** 2005. Identification of cerebral cryptococcoma using a computerized analysis of ^{1}H NMR spectra in an animal model. *Diagn. Microbiol. Infect. Dis.* **52:**101–105.
26. **Einsiedel, L., D. L. Gordon, and J. R. Dyer.** 2004. Paradoxical inflammatory reaction during treatment of *Cryptococcus noeformans* var. *gattii* meningitis in an HIV-seronegative woman. *Clin. Infect. Dis.* **39:**e78–82.
27. **Ellis, D., and T. Pfeiffer.** 1994. Cryptococcosis and the ecology of *Cryptococcus neoformans*. *Jpn. J. Med. Mycol.* **35:**111–122.
28. **Ellis, D., T. Sorrell, and S. Chen.** 2007. Impact of antifungal resistance in Australia. *Microbiol. Aust.* **28:**171–175.
29. **Ellis, D. H., and T. J. Pfeiffer.** 1990. Natural habitat of *Cryptococcus neoformans* var. *gattii*. *J. Clin. Microbiol.* **28:**1642–1644.
30. **Ely, E. W., J. E. Peacock, Jr., E. F. Haponik, and R. G. Washburn.** 1998. Cryptococcal pneumonia complicating pregnancy. *Medicine* **77:**153–167.
31. **Fisher, D., J. Burrow, D. Lo, and B. Currie. 1993.** *Cryptococcus neoformans* in tropical northern Australia: predominantly variant *gattii* with good outcomes. *Aust. NZ. J. Med.* **23:**678–682.
32. **Franzot, S. P., I. F. Salkin, and A. Casadevall.** 1999. *Cryptococcus neoformans* var. *grubii*: separate varietal status for *Cryptococcus neoformans* serotype A isolates. *J. Clin. Microbiol.* **37:**838–840.
33. **Garcia-Hermoso, D., G. Janbon, and F. Dromer.** 1999. Epidemiological evidence for dormant *Cryptococcus neoformans* infection. *J. Clin. Microbiol.* **37:**3204–3209.
34. **Graybill, J. R., J. Sobel, M. Saag, C. van der Horst, W. Powderly, G. Cloud, L. Riser, R. Hamill, W. Dismukes, and the NIAID Mycoses Study Group and AIDS Cooperative Treatment Groups.** 2000. Diagnosis and management of increased intracranial pressure in patients with AIDS and cryptococcal meningitis. *Clin. Infect Dis.* **30:**47–54.
35. **Hajjeh, R. A., L. A. Conn, D. S. Stephens, W. Baughman, R. Hamill, E. Graviss, P. G. Pappas, C. Thomas, A. Reingold, G. Rothrock, L. C. Hutwagner, A. Schuchat, M. E. Brandt, R. W. Pinner, and the Cryptococcal Active Surveillance Group.** 1999. Cryptococcosis: population-based multistate active surveillance and risk factors in human immunodeficiency virus-infected persons. *Clin. Infect. Dis.* **179:**449–454.
36. **Hedderwick., S. A., H. F. Bonilla, S. F. Bradley, and C. A. Kauffman.** 1997. Opportunistic infections in patients with temporal arteritis treated with corticosteroids. *J. Am. Geriatr. Soc.* **45:**334–337.
37. **Himmelreich, U., T. E. Dzendrowskyj, C. Allen, S. Dowd, R. Malik, B. P. Shehan, P. Russell, C. E. Mountford, and T. C. Sorrell.** 2001. Cryptococcomas distinguished from gliomas with MR spectroscopy: an experimental rat and cell culture study. *Radiology.* **220:**122–128.
38. **Himmelreich, U., C. Allen, S. Dowd, R. Malik, B. P. Shehan, C. M. Mountford and T. C. Sorrell.** 2003. Identification of metabolites of importance in the pathogenesis of pulmonary cryptococcomas in rat lung using magnetic resonance spectroscopy. *Microbes Infect.* **5:**285–290.

39. **Hospenthal, D. R., and J. E. Bennett**. 2000. Persistence of cryptococcomas on neuroimaging. *Clin. Infect. Dis.* **31:**1303–1306.
40. [Reference deleted.]
41. **Jarvis, J. N., F. Dromer, T. S. Harrison, and P. Lortholary.** 2008. Managing cryptococcosis in the immunocompromised host. *Curr. Opin. Infect. Dis.* **21:**596–603.
42. **Jenny, A., K. Pandithage, D. A. Fisher, and B. J. Currie**. 2004. *Cryptococcus* infection in tropical Australia. *J. Clin. Microbiol.* **42:**3865–3868.
43. **Johnston, S. R. D., E. L. Corbett, O. Foster, S. Ash, and J. Cohen.** 1992. Raised intracranial pressure and visual complications in AIDS patients with cryptococcal meningitis. *J. Infect.* **24:**185–189.
44. **Kaplan, M. H., P. P. Rosen, and D. Armstrong.** 1977. Cryptococcus in a cancer hospital. *Cancer* **39:**2265–2274.
45. **Kaocharoen, S., B. Wang, K. M. Tsui, L. Trilles, F. Kong, and W. Meyer.** 2008. Hyperbranched rolling circle amplification as a rapid and sensitive method for species identification within the *Cryptococcus* species complex. *Electrophoresis* **29:**3183–3191.
46. **Katsu, M, S. Kidd, A. Ando, M. L. Moretti-Branchini, Y. Mikami, K. Nishimura, and W. Meyer.** 2004. The internal transcribed spacers and 5.8S rRNA gene show extensive diversity among isolates of the *Cryptococcus neoformans* species complex. *FEMS Yeast Res.* **4:**377–388.
47. **Keah, K. C., S. Parameswari, and Y. M. Cheong.** 1994. Serotypes of clinical isolates of *Cryptococcus neoformans* in Malaysia. *Trop. Biomed.* **11:**205–207.
48. **Kerkering, T. M., R. J. Duma, and S. Shadomy.** 1981. The evolution of pulmonary cryptococcosis: clinical implications from a study of 41 patients with and without compromising host factors. *Ann. Intern. Med.* **94:**611–616.
49. **Kestelyn, P., H. Taelman, J. Bogaerts, A. Kagame, M. Abdel Aziz, J. Batungwanayo, A. M. Stevens, and P. Van de Perre.** 1993. Optical manifestations of infections with *Cryptococcus neoformans* in patients with the acquired immunodeficiency syndrome. *Am. J. Ophthalmol.* **116:**721–727.
50. **Khan, M. A., and S. Sbar.** 1975. Cryptococcal meningitis in steroid-treated systemic lupus erythematosus. *Postgrad. Med. J.* **5:**660–662.
51. **Kidd, S. E, F. Hagen, R. L. Tscharke, M. Huynh, K. H. Bartlett, M. Fyfe, L. MacDougall, T. Boekhout, K. J. Kwon-Chung, and W. Meyer.** 2004. A rare genotype of *Cryptococcus gattii* caused the cryptococcosis outbreak on Vancouver Island (British Columbia, Canada). *Proc. Natl. Acad. Sci. USA* **101:**17258–17263.
52. **Kontoyiannis, D. P, W. K. Peitsch, B. T. Reddy, E. E. Whimbey, X. Y. Han, G. P. Bodey, and K. L. Rolston**. 2004. Cryptococcosis in patients with cancer. *Clin. Infect. Dis.* **32:**e145–150.
53. **Kovacs, J. A., A. A. Kovacs, M. Polis, W. C. Wright, V. J. Gill, C. U. Tuazon, E. P. Gelmann, H. C. Lane, R. Longfield, G. Overturf, A. M. Macher, A. S. Fauci, J. E. Parrillo, J. E. Bennett, and H. Masur.** 1985. Cryptococcosis in the acquired immunodeficiency syndrome. *Ann. Intern. Med.* **103:**533–538.
54. **Kumlin, U., L. G. Elmquist, M. Granlund, B. Olsen, and A. Tarnvik.** 1997. CD4 lymphopenia in a patient with cryptococcal osteomyelitis. *Scand. J. Infect. Dis.* **29:**205–206.
55. **Kupfer, C., and E. McCrane.** 1974. A possible cause of decreased vision in cryptococcal meningitis. *Invest. Ophthal.* **13:**801–804.
56. **Kwon-Chung, K. J., and J. E. Bennett.** 1984. Epidemiologic differences between the two varieties of *Cryptococcus neoformans. Am. J. Epidemiol.* **120:**123–130.
57. **Kwon-Chung, K. J., and J. E. Bennett.** 1992. *Medical Mycology*. Lea & Febiger, Philadelphia, PA.
58. **Kwon-Chung, K. J., T. Boekhout, J. W. Fell, and M. Diaz.** 2002. Proposal to conserve the name *Cryptococcus gattii* against *C. hondurianus* and *C. bacillisporus* (*Basidiomycota, Hymenomycetes, Tremellomycetidae*). *Taxon* **51:**804–806.
59. **Lalloo, D., D. Fisher, S. Naraqi, I. Laurenson, P. Temu, A. Sinha, A Sawerie, and B. Mavo.** 1994. Cryptococcal meningitis (*C. neoformans* var. *gattii*) leading to blindness in previously healthy Melanesian adults in Papua New Guinea. *Q. J. Med.* **87:**343–349.
60. **Laurenson, I. F., A. J. Trevett, D. G. Lalloo, N. Nwokolo, S. Naraqi, J. Black, N. Tefurani, A. Saweri, B. Mavo, J. Igo, and D. A. Warrell.** 1996. Meningitis caused by *Cryptococcus neoformans* var. *gattii* and var. *neoformans* in Papua New Guinea. *Trans. R. Soc. Trop. Med. Hyg.* **90:**57–60.
61. **Levitz, S. M.** 1991. The ecology of *Cryptococcus neoformans* and the epidemiology of cryptococcosis. *Rev. Infect. Dis.* **13:**1163–1169.
62. **Lewis, J. L., and S. Rabinovich.** 1972. The wide spectrum of cryptococcal infections. *Am. J. Med.* **53:**315–322.
63. **Liliang, P.-C., C.-L. Liang, W.-N. Chang, H.-J. Chen, T.-M. Su, K. Lu, and C.-H. Lu.** 2003. Shunt surgery for hydrocephalus complicating cryptococcal meningitis in human immunodeficiency virus-negative patients. *Clin. Infect. Dis.* **37:**673–678.
64. **Lipson, B. K., W. R. Freeman, J. Beniz, M. H. Goldbaum, J. R. Hesslelink, R. N. Weinreb, and A. A. Sadun.** 1989. Optic neuropathy associated cryptococcal arachnoiditis in AIDS patients. *Am. J. Ophthalmol.* **107:**523–527.
65. **Liss, H.P., and D. Rimland.** 1981. Asymptomatic cryptococcal meningitis. *Am. Rev. Respir. Dis.* **124:**88–89.
66. **Litvintseva, A. P., R. Thakur, L. B. Reller, and T. G. Mitchell.** 2005. Prevalence of clinical isolates of *Cryptococcus gattii* serotype C among patients with AIDS in sub-Saharan Africa. *Eukaryot. Cell* **192:**888–892.
67. **Lo, D.** 1976. Cryptococcosis in the Northern Territory. *Med. J. Aust.* **2:**825–828.
68. **Lopez-Martinez, R., J. L. Soto-Hernandez, L. Ostrosky-Zeichner, R. Castanon-Olivares, V. Angeles-Morales, and J. Sotelo.** 1996. *Cryptococcus neoformans* var. *gattii* among patients with cryptococcal meningitis in Mexico. First observations. *Mycopathologia* **134:**61–64.
69. **MacDougall, L., and M. Fyfe.** 2006. Emergence of *Cryptococcus gattii* in a novel environment provides clues to its incubation period. *J. Clin. Microbiol.* **44:**1851–1852.
70. **MacDougall, L., S. Kidd, E. Galanis, S. Mak, M. J. Leslie, P. R. Cieslak, J. W. Kronstad, M. G. Morshed, and K. H. Bartlett.** 2007. Spread of *Cryptococcus gattii* in British Columbia, Canada and its detection in the Pacific Northwest, USA. *Emerg. Infect. Dis.* **13:**42–50.
71. **Miller, K. D., J. M. Mican, and R. T. Davey.** 1996. Asymptomatic solitary pulmonary nodules due to *Cryptococcus neoformans* in patients infected with human immunodeficiency virus. *Clin. Infect. Dis.* **23:**810–812.
72. **Mitchell, D. H., and T. C. Sorrell.** 1991. Pancoast's syndrome due to pulmonary infection with *Cryptococcus neoformans* variety *gattii*. *Clin. Infect. Dis.* **14:**1142–1144.
73. **Mitchell, D. H., T. C. Sorrell, A. M. Allworth, C. H. Heath, A. R. McGregor, K. Papanoum, M. J. Richards, and T. Gottlieb.** 1995. Cryptococcal disease of the CNS in immunocompetent hosts: influence of cryptococcal variety on clinical manifestations and outcome. *Clin. Infect. Dis.* **20:**611–616.
74. **Mitchell, T. G, and J. R. Perfect.** 1995. Cryptococcosis in the era of AIDS: 100 years after the discovery of *Cryptococcus neoformans*. *Clin. Microbiol. Rev.* **8:**515–548.
75. **Montagna, M. T., M. A. Viviani, and A. Pulito.** 1997. *Cryptococcus neoformans* var. *gattii* in Italy. Environmental investigation related to an autochthonous clinical case in Apulia. *J. Mycol. Med.* **7:**93–96.

76. **Montejo, M., K. Aguirrebengoa, R. Preita, J. C. Ibañez de Maeztu, J. Oñate, and P. Gonzalez de Zarate.** 1995. Blindness in patients with cryptococcal meningitis and AIDS. *Eur. Neurol.* **35:**239–240.
77. **Montenegro, H., and C. R. Paula.** 2000. Environmental isolation of *Cryptococcus neoformans* var. *gattii* and *Cryptococcus neoformans* var. *neoformans* in the city of Sao Paulo, Brazil. *Med. Mycol.* **38:**385–390.
78. **Morgan, J., K. M. McCarthy, S. Gould, K. Fan, B. Athington-Skaggs, N. Iqhal, K. Stamey, R. A. Hajjeh, and M. E. Brandt, for the Gauteng Cryptococcal Surveillance Initiative Group.** 2006. *Cryptococcus gattii* infection: characteristics and epidemiology in a South African province with high HIV seroprevalence, 2002–2004. *Clin. Infect. Dis.* **43:**1077–1080.
79. **Nishikawa, M., M. Lazera, G. G. Barbosa, L. Trilles, B. R. Balassiano, R. Macedo, C. Bezerra, M. A. Perez, P. Cardarelli, and B. Wanke.** 2003. Serotyping of 467 *Cryptococcus neoformans* isolates from clinical and environmental sources in Brazil: analysis of host and regional patterns. *J. Clin. Microbiol.* **41:**73–77.
80. **Okun, E., and W. T. Butler.** 1964.Ophthalmologic complications of cryptococcal meningitis. *Arch. Ophthalm.* **71:**52–57.
81. **Ostrow, T. D., and P. A. Hudgins.** 1994. Magnetic resonance imaging of intracranial fungal infections. *Top. Magn. Reson. Imaging.* **6:**22–31.
82. **Pappas, P. G., J. R. Perfect, G. A. Cloud, R. A. Larsen, G. A. Pankey, D. J. Lancaster, H. Henderson, C. A. Kauffman, D. W. Haas, M. Saccente, R. J. Hamill, M. S. Holloway, R. M. Warren, and W. E. Dismukes.** 2001. Cryptococcosis in human immunodeficiency virus-negative patients in the era of effective azole therapy. *Clin. Infect. Dis.* **33:**690–699.
83. **Park, B. J., K. A. Wannemuehler, B. J. Marston, N. Govender, P. G Pappas, and T. M. Chiller.** 2009. Estimation of the current global burden of cryptococcal meningitis among persons living with HIV/AIDS. *AIDS* **23:**525–530.
84. **Park, M. K., D. R. Hospenthal, and J. E. Bennett.** 1999. Treatment of hydrocephalus secondary to cryptococcal meningitis by use of shunting. *Clin. Infect. Dis.* **28:**629–633.
85. **Perfect, J. R.** 1989. Cryptococcosis. *Infect. Dis. Clin. N. Am.* **3:**77–102.

85a. **Perfect, J. R., W. E. Dismukes, F. Dromer, D. L. Goldman, J. R. Graybill, R. J. Hamill, T. S. Harrison, R. A. Larsen, O. Lortholary, M. H. Nguyen, P. G. Pappas, W. G. Powderly, N. Singh, J. D. Sobel, and T. C. Sorrell.** 2010. Clinical practice guidelines for the management of cryptococcal disease: 2010 update by the Infectious Diseases Society of America. *Clin. Infect. Dis.* **50:**291–322.

86. **Pfaller, M. A., S. A. Messer, L. Boyken, C. Rice, S. Tendolkar, R. J. Hollis, G. V. Doern, and D. J. Diekma.** 2005. Global trends in the antifungal susceptibility of *Cryptococcus neoformans* (1990 to 2004). *J. Clin. Microbiol.* **43:**2163–2167.
87. **Rex, J. H., R. A. Larsen, W. E. Dismukes, G. A. Cloud, and J. E. Bennett.** 1993. Catastrophic visual loss due to *Cryptococcus neoformans* meningitis. *Medicine* **72:**209–224.
88. **Richardson, P. M., A. Mohandas, and N. Arumugasamy.** 1976. Cerebral cryptococcosis in Malaysia. *J. Neurol. Neurosurg. Psychiatry* **39:**330–337.
89. **Rozenbaum, R., A. R. Goncalves, B. Wanke, M. J. Caiuby, H. Clemente, M. D. Lazera, P. C. Monteiro, and A. T. Lonfero.** 1992. *Cryptococcus neoformans* varieties as agents of cryptococcosis in Brazil. *Mycopathologia* **119:**133–136.
90. **Rozenbaum, R., and A. R. Goncalves.** 1994. Clinical epidemiological study of 171 cases of cryptococcosis. *Clin. Infect. Dis.* **18:**369–380.
91. **Saag, M., R. J. Graybill, R. A. Larsen, P. G. Pappas, J. R. Perfect, W. G. Powderly, J. D. Sobel, W. E. Dismukes, for the Mycoses Study Group Cryptococcal Subproject.** 2000. Practice guidelines for the management of cryptococcal disease. *Clin. Infect. Dis.* **30:**710–718.
92. **Seaton, R. A., A. J. Hamilton, R. J. Hay, and D. A. Warrell.** 1996. Exposure to *Cryptococcus neoformans* var. *gattii*: a seroepidemiological study. *Trans. R. Soc. Trop. Med. Hyg.* **90:**508–512.
93. **Seaton, R. A., N. Verma, S. Naraqi, J. P. Wembri, and D. A. Warrell.** 1997. Visual loss in immunocompetent patients with *Cryptococcus neoformans* var. *gattii* meningitis. *Trans. R. Soc. Trop. Med. Hyg.* **91:**41–49.
94. **Singh, N., F. Dromer, J. R. Perfect, and O. Lortholary.** 2008. Cryptococcosis in solid organ transplant recipients: current state of the science. *Clin. Infect. Dis.* **47:**1321–1327.
95. **Smith, A. B., J. G. Smirniotopoulos, and E. J. Rushing.** 2008. Central nervous system infections associated with human immunodeficiency virus infection: radiologic-pathologic correlation. *Radiographics* **28:**2033–2058.
96. **Sorrell, T. C.** 2008. East meets west. Cryptococcosis in Australia, abstr. SY-01-08, p. 42. Program and Abstracts of the 7th International *Cryptococcus* and Cryptococcosis Conference, 11 to 14 September, Nagasaki, Japan.
97. **Sorrell, T. C., S. C. A. Chen, P. Ruma, W. Meyer, T. J. Pfeiffer, D. H. Ellis, and A. G. Brownlee.** 1996. Concordance of clinical and environmental isolates of *Cryptococcus neoformans* var. *gattii* by random amplification of polymorphic DNA analysis and PCR fingerprinting. *J. Clin. Microbiol.* **34:**1252–1260.
98. **Sorrell, T. C.** 2001. *Cryptococcus neoformans* variety *gattii*. *Med. Mycol.* **39:**155–168.
99. **Sorrell, T. C., and U. Himmelreich.** 2008. NMR in Mycology. *Curr. Fungal Inf. Rep.* **2:**149–156.
100. **Speed, B., and D. Dunt.** 1995. Clinical and host differences between infections with the two varieties of *Cryptococcus neoformans*. *Clin. Infect. Dis.* **21:**28–34.
101. **Stephen, C., S. Lester, W. Black, M. Fyfe, and S. Raverty.** 2002. Multispecies outbreak of cryptococcosis on southern Vancouver Island, British Columbia. *Can. Vet. J.* **43:**792–794.
102. **Tan, C. T.** 1988. Intracranial hypertension causing visual failure in cryptococcal meningitis. *J. Neurol. Neurosurg. Psychiatry* **15:**944–946.
103. **Tay, S. T., H. T. Tanty, K. P. Ng, M. Y. Rohani, and H. Hamimah.** 2006. In vitro susceptibilities of Malaysian clinical isolates of *Cryptococcus neoformans* var. *grubii* and *Cryptococcus gattii* to five antifungal drugs. *Mycoses* **49:**324–330.
104. **Thompson, G. R., III, N. P. Wiederhold, A. W. Fothergill, A. C. Vallor, B. L. Wickes, and T. F. Patterson.** 2009. Antifungal susceptibilities among different serotypes of *Cryptococcus gattii* and *Cryptococcus neoformans*. *Antimicrob. Agents Chemother.* **53:**309–311.
105. **Torres, O. H., E. Negredo, J. Ris, P. Domingo, and A. M. Catafau.** 1999. Visual loss due to cryptococcal meningitis in AIDS patients. *AIDS* **13:**530–534.
106. **Torres-Rodriguez, J. M., E. Alvarado-Ramirez, F. Murciano, and M. Sellart.** 2008. MICs and minimum fungicidal concentrations of posaconazole, voriconazole and fluconazole for *Cryptococcus neoformans* and *Cryptococcus gattii*. *J. Antimicrob. Chemother.* **62:**205–210.
107. **Trilles, L., B. Fernandez-Torres, M. dos Santos Lazera, B. Wanke, and J. Guarro.** 2004. In vitro antifungal susceptibility of *Cryptococcus gattii*. *J. Clin. Microbiol.* **42:**4815–4817.
108. **Van Dam, M. G., R. A. Seaton, and A. J. Hamilton.** 1998. Analysis of HLA association in susceptibility to

infection with *Cryptococcus neoformans* var. *gattii* in a Papua New Guinea population. *Med. Mycol.* **36:**185–188.

109. **Velagraki, A., V. G. Kiosses, H. Pitsouni, D. Youkas, V. D. Danilidis, and N. J. Legakis.** 2001. First report of *Cryptococcus neoformans* var. *gattii* serotype B from Greece. *Med. Mycol.* **39:**419–422.
110. **West, S., E. Brynes, S. Mostad, R. Thompson, R. Barnes, C. Beiser, J. Heitman, and K. Marr.** 2008. Emergence of *Cryptococcus gattii* in the Pacific Northwest United States, abstr. M1849, p. 658. 48th Annual Interscience Conference on Antimicrobial Agents and Chemotherapy, Washington, DC.
111. **White, M., C. Cirricione, A. Blevins, and D. Armstrong.** 1992. Cryptococcal meningitis with AIDS and patients with neoplastic disease. *J. Infect. Dis.* **165:** 960–966.
112. **Widmer, F., L. C. Wright, D. Obando, R. Handke, R. Ganendren, D. H. Ellis, and T. C. Sorrell.** 2006. Hexadecylphosphocholine (miltefosine) has broad-spectrum fungicidal activity and is efficacious in a mouse model of cryptococcosis. *Antimicrob. Agents Chemother.* **50:**414–421.
113. **Yu, Y. L., Y. N. Lau, E. Woo, K. L. Wong, and B. Tse.** 1988. Cryptococcal infection of the nervous system. *Q. J. Med.* **66:**87–96.
114. **Zonios, D. I., J. Falloon, C. Y. Huang, D. Chaitt, and J. E. Bennett.** 2007. Cryptococcosis and idiopathic CD4 lymphocytopenia. *Medicine* **86:**78–92.
115. **Zuger, A., E. Louie, R. S. Holzman, M. S. Simberkoff, and J. J. Rahal.** 1986. Cryptococcal disease in patients with acquired immunodeficiency syndrome. Diagnostic features and outcome of treatment. *Ann. Intern. Med.* **104:**234–240.

Index

G